HANDBOOK OF INDUSTRIAL ENGINEERING

ADVISORY COMMITTEE

HANDBOOK OF INDUSTRIAL ENGINEERING

Edited by
GAVRIEL SALVENDY
Purdue University

A Wiley-Interscience Publication
JOHN WILEY & SONS
New York • Chichester • Brisbane • Toronto • Singapore

Library of Congress Cataloging in Publication Data:

Main entry under title:

Handbook of industrial engineering.

"A Wiley-Interscience publication."
Includes index.
1. Industrial engineering—Handbooks, manuals, etc.
I. Salvendy, Gavriel, 1938–
T56.23.H36 658.5 81-23059
ISBN 0-471-05841-6 AACR2

Printed in the United States of America

10 9 8 7 6 5 4

CONTRIBUTORS

Virgil L. Anderson
Professor
Department of Statistics
Purdue University
West Lafayette, Indiana

Thomas J. Armstrong
Assistant Professor of Industrial Hygiene, Center for Ergonomics
The University of Michigan
Ann Arbor, Michigan

Jeffrey L. Arthur
Assistant Professor
Department of Statistics
Oregon State University
Corvallis, Oregon

Guy J. Bacci
Manager
Industrial Engineering
International Harvester
Chicago, Illinois

Andrew D. Bailey, Jr.
Professor and Chairman of Department of Accounting
University of Minnesota
Minneapolis, Minnesota

Moshe M. Barash
Professor
School of Industrial Engineering
Purdue University
West Lafayette, Indiana

Franklin H. Bayha
Private Practice
Ann Arbor, Michigan

Richard L. Behling
Consultant
Booz, Allen & Hamilton, Inc.
New York, New York

Corwin A. Bennett
Professor
Department of Industrial Engineering
Kansas State University
Manhattan, Kansas

William E. Biles
Professor and Chairman
Department of Industrial Engineering
Louisiana State University
Baton Rouge, Louisiana

Robert A. Boehmer
Manufacturing Manager
Speedway Plants
Detroit Diesel Allison
Indianapolis Operations
Division of General Motors Corporation
Indianapolis, Indiana

Harold N. Bogart
Retired Director of Manufacturing Engineering Systems—Manufacturing Staff
Ford Motor Company
Dearborn, Michigan

Walter C. Borman
Executive Vice President
Personnel Decisions Research Institute
Minneapolis, Minnesota

Evan F. Bornholtz
Associate Professor
Department of Industrial Administration
General Motors Institute
Flint, Michigan

Chester L. Brisley
Professor and Associate Chairman
Department of Engineering and Applied Science, University Extension
The University of Wisconsin
Milwaukee, Wisconsin

Philip S. Brumbaugh
President
Quality Assurance, Inc.
St. Louis, Missouri

James R. Buck
Professor and Chairman
Systems Division
The University of Iowa
Iowa City, Iowa

William J. Burgess
Manager
Industrial Engineering Department
Tennessee Eastman Company
Kingsport, Tennessee

Donald C. Burnham
Retired Chairman of the Board
Westinghouse Electric Corporation
Pittsburgh, Pennsylvania

Robert W. Burns
Director
Personnel Relations and Community Affairs
American Motors Corporation
Southfield, Michigan

Robert E. Busby
Senior Industrial Engineer
Tennessee Eastman Company
Kingsport, Tennessee

James P. Caie, Jr.
Senior Project Engineer
Manufacturing Development
Technical Center
General Motors Corporation
Warren, Michigan

Don B. Chaffin
Chairman and Professor
Department of Industrial Engineering and Operations Engineering
Director of the Center for Ergonomics
The University of Michigan
Ann Arbor, Michigan

Daniel O. Clark
Western Regional Director
MTM Association for Standards and Research
West Coast Office
Newport Beach, California

Guy C. Close, Jr.
Staff Senior Industrial Engineer
Hughes Aircraft Company
Radar Systems Group
Manufacturing Division
Los Angeles, California

Joel D. Cohen
Senior Project Engineer
Manufacturing Development
Technical Center
General Motors Corporation
Warren, Michigan

E. Nigel Corlett
Professor and Head
Department of Production Engineering and Production Management
University of Nottingham
Nottingham, England

Merton D. Corwin
Manager
Robot Research and Development
Cincinnati Milacron, Inc.
Cincinnati, Ohio

Douglas C. Crocker
Technical Advisor
Management Services Division
Eastman Kodak Company
Rochester, New York

Malcolm J. Crocker
Assistant Director
Acoustics and Noise Control
R. W. Herrick Laboratories
Professor of Mechanical Engineering
Purdue University
West Lafayette, Indiana

Ralph E. Cross, Sr.
Chairman of the Board
Cross & Trecker Corporation
Fraser, Michigan

Adam W. Cywar
Manager
Information Systems Planning and Administration
IBM Corporation—Austin Plant
Austin, Texas

Louis E. Davis
Professor and Chairman
Center for Quality of Working Life
Institute of Industrial Relations
University of California
Los Angeles, California

David J. DeMarle
Supervisor
Management Services Division
Eastman Kodak Company
Rochester, New York

Gene J. D'Ovidio
Associate
Booz, Allen & Hamilton, Inc.
New York, New York

Colin G. Drury
Professor
Department of Industrial Engineering
State University of New York at Buffalo
Buffalo, New York

Karl Eady
Director Midwest Office
MTM Association for Standards and Research
Park Ridge, Illinois

Samuel Eilon
Professor and Head
Department of Management Science
Imperial College of Science and Technology
London, England

Hamed K. Eldin
Professor
Department of Industrial Engineering and Management
Oklahoma State University
Stillwater, Oklahoma

Mitchell Fein
President
Mitchell Fein, Inc.
Hillsdale, New Jersey

Orlando J. Feorene
Director
Management Services Division
Eastman Kodak Company
Rochester, New York

Charles E. Geisel
Manager
Corporate Industrial Engineering Department
Container Corporation of America
Carol Stream, Illinois

James Gerlach
Doctoral Student
Krannert School of Management
Purdue University
West Lafayette, Indiana

Kenneth M. Gettelman
Editor
Modern Machine Shop
Cincinnati, Ohio

William Gomberg
Professor of Labor Relations
Wharton School of Management
University of Pennsylvania
Philadelphia, Pennsylvania

Christian H. Gudnason
Professor of Production Management
Head of Institute of Production Engineering and Mechanical Technology
Technical University of Denmark
Lyngby, Denmark

Inyong Ham
Professor
Department of Industrial and Management Systems Engineering
The Pennsylvania State University
University Park, Pennsylvania

Walton M. Hancock
Professor
Department of Industrial and Operations Engineering
The University of Michigan
Ann Arbor, Michigan

Robert H. Harder
Staff Development Engineer
Manufacturing Development
Technical Center
General Motors Corporation
Warren, Michigan

Clyde Holsapple
Assistant Professor
School of Management
University of Illinois
Urbana, Illinois

Knut Holt
Professor
Division of Industrial Management
University of Trondheim
The Norwegian Institute of Technology
Trondheim, Norway

Michael R. Hottinger
Director
Product Scheduling Task Force
Information Systems
Chevrolet Motor Division
General Motors Corporation
Warren, Michigan

John J. Jarvis
Professor
School of Industrial and Systems Engineering
Georgia Institute of Technology
Atlanta, Georgia

Takeji Kadota
Director
Japan Management Association
Tokyo, Japan

Eliezer E. Kamon
Professor
Noll Laboratory for Human Performance Research
The Pennsylvania State University
University Park, Pennsylvania

Kailash C. Kapur
Professor
Department of Industrial Engineering and Operations Research
Wayne State University
Detroit, Michigan

William J. Kennedy, Jr.
Associate Professor
Department of Mechanical and Industrial Engineering
University of Utah
Salt Lake City, Utah

Hugh D. Kinney
Executive Vice President
SysteCon, Inc.
Norcross, Georgia

James L. Knight
Research Psychologist
Bell Telephone Laboratories
Holmdel, New Jersey

Dev. S. Kochhar
Associate Professor
Department of Industrial and Operations Engineering
Center for Ergonomics
The University of Michigan
Ann Arbor, Michigan

Edward J. Kompass
Editor
Control Engineering
Barrington, Illinois

Stephan A. Konz
Professor
Department of Industrial Engineering
Kansas State University
Manhattan, Kansas

Ilkka Kuorinka
Assistant Director
Department of Physiology
Institute of Occupational Health
Helsinki, Finland

Tarald O. Kvalseth
Associate Professor
Department of Mechanical Engineering
University of Minnesota
Minneapolis, Minnesota

Ferdinand F. Leimkuhler
Professor
School of Industrial Engineering
Purdue University
West Lafayette, Indiana

Baruch Lev
Professor and Dean
Faculty of Management
Tel-Aviv University
Ramat-Aviv
Tel-Aviv, Israel

Kari Lindström
Departmental Director
Department of Psychology
Institute of Occupational Health
Helsinki, Finland

Roy F. Lomicka
Technical Advisor
Management Services Division
Eastman Kodak Company
Rochester, New York

Henry C. Lucas, Jr.
Professor and Chairman
Computer Applications and Information Systems Area
Graduate School of Business Administration
New York University
New York, New York

Raymond P. Lutz
Executive Dean of Graduate Studies and Research
University of Texas at Dallas
Richardson, Texas

R. Preston McAfee
Assistant Professor of Economics
Krannert School of Management
Purdue University
West Lafayette, Indiana

Ernest J. McCormick
Professor Emeritus of Psychological Sciences
Purdue Univesity
West Lafayette, Indiana

Robert A. McLean
Professor
Department of Statistics
University of Tennessee
Knoxville, Tennessee

John J. Mariotti
Professor
Department of Industrial Engineering
General Motors Institute
Flint, Michigan

Richard A. Mathias
President
Macotech Corporation
Seattle, Washington

Wayne E. Mechlin
Manager of Application Engineering
Industrial Robotic Division
Cincinnati Milacron, Inc.
Cincinnati, Ohio

Deborah J. Medeiros
Assistant Professor
Department of Industrial and Management Systems Engineering
The Pennsylvania State University
University Park, Pennsylvania

David Meister
Associate Director
Design of Manned Systems
U.S. Navy Personnel Research and Development Center
San Diego, California

James M. Miller
Associate Professor
Department of Industrial and Operations Engineering
The University of Michigan
Ann Arbor, Michigan

Davendra Mishra
Manager of Player Technology Center Operations
RCA "Selecta Division" Video Disks Operations
Indianapolis, Indiana

Colin L. Moodie
Professor
School of Industrial Engineering
Purdue University
West Lafayette, Indiana

Robert G. Morris
Professor and Head
Department of Industrial Engineering
General Motors Institute
Flint, Michigan

Marvin E. Mundel
President
M. E. Mundel & Associates, Inc.
Silver Spring, Maryland

Rintaro Muramatsu
Professor
Department of Industrial Engineering
School of Science and Engineering
Waseda University
Tokyo, Japan

Katta G. Murty
Professor
Department of Industrial and Operations Engineering
The University of Michigan
Ann Arbor, Michigan

Gerald Nadler
Professor
Department of Industrial Engineering
University of Wisconsin
Madison, Wisconsin

Seiichi Nakajima
Executive Director and General Secretary
Japan Institute of Plant Engineers
Tokyo, Japan

Benjamin W. Niebel
Professor Emeritus
Department of Industrial and Management Systems Engineering
The Pennsylvania State University
University Park, Pennsylvania

Shimon Y. Nof
Assistant Professor
School of Industrial Engineering
Purdue University
West Lafayette, Indiana

Stanley D. Nollen
Associate Professor
School of Business Administration
Georgetown University
Washington, D.C.

Phillip F. Ostwald
Professor
Department of Mechanical Engineering
University of Colorado
Boulder, Colorado

Irvin Otis
Manager
Industrial Engineering
Corporate Manufacturing Staff
American Motors Corporation
Detroit, Michigan

Joseph A. Panico
President
American Productivity Improvement Systems, Inc.
Columbus, Ohio

Eleanor S. Pape
Associate Professor
Department of Industrial Engineering
University of Texas at Arlington
Arlington, Texas

George M. Parks
Professor and Dean
School of Business Administration
Emory University
Atlanta, Georgia

Richard G. Pearson
Professor
Department of Industrial Engineering
North Carolina State University
Raleigh, North Carolina

Norman G. Peterson
Vice President
Personnel Decisions Research Institute
Minneapolis, Minnesota

Don T. Phillips
Professor
Department of Industrial Engineering
Texas A & M University
College Station, Texas

John V. Pilitsis
Division Manager
Inventory Management Systems Development and Implementation
American Telephone and Telegraph Company
Parsippany, New Jersey

A. Alan B. Pritsker
President
Pritsker & Associates
West Lafayette, Indiana

M. Raghavachari
Professor and Dean (Planning)
Indian Institute of Management
Vastrapur, Ahmedabad, India

H. Donald Ratliff
Professor
School of Industrial and
Systems Engineering
Georgia Institute of Technology
Atlanta, Georgia

A. Ravindran
Professor and Director
School of Industrial Engineering
University of Oklahoma
Norman, Oklahoma

Gintaras V. Reklaitis
Professor
School of Chemical Engineering
Purdue University
West Lafayette, Indiana

James A. Richardson
Assistant Director
Management Services Division
Eastman Kodak Company
Rochester, New York

W. J. Richardson
Professor
Department of Industrial Engineering
Lehigh University
Bethlehem, Pennsylvania

Randall P. Sadowski
Associate Professor
School of Industrial Engineering
Purdue University
West Lafayette, Indiana

Gavriel Salvendy
Professor
School of Industrial Engineering and
Chairman, Human Factor Program
Purdue University
West Lafayette, Indiana

Byron W. Saunders
Professor Emeritus
School of Operations Research
and Industrial Engineering
Cornell University
Ithaca, New York

Herbert D. Schwetman
Associate Professor
Department of Computer Science
Purdue University
West Lafayette, Indiana

Joseph Sharit
Doctoral Student
School of Industrial Engineering
Purdue University
West Lafayette, Indiana

Sheldon Shen
Assistant Professor
Faculty of Accounting
College of Administrative Sciences
Ohio State University
Columbus, Ohio

M. Larry Shillito
Technical Advisor
Management Services Division
Eastman Kodak Company
Rochester, New York

E. Ralph Sims, Jr.
President
E. Ralph Sims, Jr. & Associates
Lancaster, Ohio

Alfred H. Smith
Executive Vice President
Delphi Corporation
Plymouth, Michigan

Karl F. Speitel
Technical Associate
Management Services Division
Eastman Kodak Company
Rochester, New York

Kathryn E. Stecke
Assistant Professor of Policy and Control
Graduate School of Business Administration
The University of Michigan
Ann Arbor, Michigan

Arnold L. Sweet
Professor
School of Industrial Engineering
Purdue University
West Lafayette, Indiana

Yoshihiko Tanaka
Research Fellow
Department of Industrial Engineering
Waseda University
Tokyo, Japan

Jose M. A. Tanchoco
Associate Professor
Department of Industrial Engineering
Virginia Polytechnic Institute and State University
Blacksburg, Virginia

Ronald L. Tarvin
Research Associate
Robot Research and Development
Cincinnati Milacron, Inc.
Cincinnati, Ohio

Daniel Teichroew
Professor
Department of Industrial and Operations Engineering
The University of Michigan
Ann Arbor, Michigan

Gerald J. Thuesen
Professor
School of Industrial and Systems Engineering
Georgia Institute of Technology
Atlanta, Georgia

James A. Tompkins
President
Tompkins Associates, Inc.
Raleigh, North Carolina

Wayne C. Turner
Professor
Department of Industrial Engineering and Management
Oklahoma State University
Stillwater, Oklahoma

Gerald J. Wacker
Personnel Research Coordinator
The Aerospace Corporation
Los Angeles, California

Urban Wemmerlöv
Assistant Professor
School of Business
University of Wisconsin
Madison, Wisconsin

Andrew B. Whinston
Professor of Economics
Computer Science and Management
Krannert School of Management
Purdue University
West Lafayette, Indiana

John A. White
Professor
School of Industrial and Systems Engineering
Georgia Institute of Technology
Atlanta, Georgia

Gary E. Whitehouse
Professor and Chairman
Department of Industrial Engineering and Management Systems
University of Central Florida
Orlando, Florida

Theodore J. Williams
Professor of Engineering and Director
Laboratory for Applied Industrial Control
Purdue University
West Lafayette, Indiana

James R. Wilson
Assistant Professor
Department Mechanical Engineering
University of Texas
Austin, Texas

Leroy H. Wulfmeier
Retired Director of Purchasing
Chevrolet Division
General Motors Corporation
Warren, Michigan

FOREWORD

This *Handbook of Industrial Engineering* is being published at a very opportune time because the mission of industrial engineering is undergoing a change in magnitude and emphasis. For the past century, industrial engineering has been responsible for much of the economic progress in manufacturing. The industrial engineers have studied the way people work in our factories and the relationship of those workers to their machinery and tools. The focus has been on individuals and how to improve the effectiveness of their work.

In the future, industrial engineers will continue to study how the individual works, but much greater emphasis will be placed upon studying the systems within which the work is performed in order to optimize the operation of the total system. New technologies brought about by the computer, such as robots and automated systems, and the computerization of much of their own work, will require industrial engineers to continually restudy the application of their knowledge.

Industrial engineers in the future must greatly expand their activities beyond studying the workers in manufacturing. The percentage of the work force engaged in the production of goods passed its peak more than a decade ago, and the people working in service industries, which include hospitals, banking, insurance, post office, hotel, restaurant, armed services, government, universities, distribution, marketing, and so forth, now number twice as many as those working in manufacturing. These service workers and the systems they work in require industrial engineering techniques to improve their productivity, just as manufacturing still requires it. Industrial engineers will have to use the word "industry" in its broadest sense.

It is now recognized that productivity is a key factor in determining our standard of living, in maintaining our competitiveness in world markets, and in alleviating inflation. Industrial engineers must play a major role in improving productivity since they deal with the factors that affect it most: how people work and how they use technology. This *Handbook* deals not only with the "human use of human beings" and with the interaction between humans and machines, but with the entire system that productively makes a product or performs a service.

The chapters of the *Handbook* were written by scores of experts who, through example and theory, provide a source of ideas for improvement. The *Handbook* should be valuable to industrial engineers and other engineers, as well as to all levels of managers. The industrial engineering principles that are outlined are timeless and basic and should prove useful to corporations, both large and small, to continuous process as well as discrete part manufacturers, and especially to those working in the service industries where most of the jobs are today.

Industrial engineering serves people and can lead all workers to improved productivity and to a higher standard of living.

D. C. BURNHAM, *Retired Chairman of the Board*
Westinghouse Electric Corporation

PREFACE

Claude S. George, Jr. in *The History of Management Thought* notes the application of motion study, layout, and materials handling by Cyrus in 400 B.C., the principle of work specialization by Plato, and in Caesar's time the use of job specifications by Cato and Varro. James Lee in his book *The Gold and Garbage in Management Theories and Prescriptions* makes reference to assembly lines at the arsenal of Venice in 1440.

To people such as Frederick Taylor, Henry L. Gantt, and Frank and Lillian Gilbreth in the United States, Henri Fayol in France, and others in Sweden, France, and Germany go the credit for describing and structuring a disciplined process that eventually spawned what is now contemporary industrial engineering practice. Their search for a rigorous methodology was embraced by innovative engineers and managers throughout the industrialized world. The summation of this effort over the past 100 years has generated a body of knowledge in the following broad areas of specialization:

Organization and job design.
Methods engineering.
Performance measurement and control of operations.
Evaluation, appraisal, and management of human resources.
Ergonomics/human factors.
Manufacturing engineering.
Quality assurance.
Engineering economy.
Facilities design.
Planning and control.
Computers and information systems.
Quantitative methods.
Optimization.

The foregoing shows how broad the field has become. It shows, too, that industrial engineering has broadened its concern to include many highly relevant management issues and questions such as the measurement and improvement of productivity and quality of working life and optimal utilization of the available resources. As such the *Handbook* should be of value to all industrial engineers and managers—whether they are in profit motivated operations or in other nonprofit fields of activity.

Such a breadth of subject matter presents a serious challenge to successfully represent the entire field of industrial engineering with a single handbook. I did not believe in 1978, when this all began, that any one person could properly select the subjects to be included in the *Handbook* without serious distortions to fit his or her own particular areas of knowledge and bias. Accordingly, an Advisory Committee composed of experts in the more important areas of industrial engineering was invited to assist the Editor in planning the contents of the *Handbook*. The Advisory Committee members are listed on page ii. I sincerely appreciate their excellent counsel and advice during the preparation of the *Handbook*. Nonetheless, any sampling deficiencies that remain are, of course, my own responsibility.

As should be apparent from a perusal of the names of the 133 authors of the 107 chapters of this *Handbook*, we made every effort to obtain authors with diverse training and professional affiliations, from the United States and other countries throughout the world. Each author contributing to this *Handbook* was guided by a set of objectives developed by the Advisory Committee. The objectives that each author received included the following:

1. The *Handbook* should serve the following people:
   a. Practicing industrial engineers.
   b. Practitioners with no formal training in industrial engineering.
   c. Nonindustrial engineering type personnel.
   d. Educators and extension course leaders.
2. The *Handbook* should serve the following types of organizations:
   a. The very small, medium, and large corporations.
   b. Continuous process and discrete part manufacturing industries.
   c. Service industries including hospitals; banking; insurance; post office; hotel, motel, and restaurants; armed services; local, state, and federal government; universities; distribution; marketing; and labor relations.
3. It is planned that the following six notions will run, as a theme, through the *Handbook:*
   a. Techniques and methods have evolved that should assist supervision and management in the design and operation of a work environment requiring increased interaction between humans and "intelligent" machines.
   b. Experience indicates that "human use of human beings" can lead to greater productivity and is in fact compatible with the profit motive.
   c. The last half century has seen the effective application of techniques that subdivide activities to improve operations. The next decade will see a sharp emphasis on the study of "total systems" in order to optimize operations through the integration of subsystems or parallel systems. The *Handbook* will describe this expansion of analytical capability. The need for maintaining both approaches will be indicated, since they are complementary and not mutually exclusive concepts.
   d. Since the purpose of the *Handbook* is the application of knowledge to solving real-world problems, it should present all available useful tables, graphs, nomographs, and formulas pertaining to the application and use of industrial engineering methodologies. The scope and limitations of each methodology should be reviewed and a step-by-step description of its use should be illustrated.

e. Examples should be utilized to demonstrate the use of various methodologies. From these examples the reader may draw inferences on the use of the presented methodology to specific work situations.
f. Since industrial engineering and the methodologies of industrial engineering are practiced and utilized in both manufacturing and service industries, it is therefore imperative to recognize both of these areas during the writing of the various chapters.

In addition to the *Handbook* objectives outlined above, chapters authors were made cognizant of the Industrial Engineering Terminology (ANSI Z94.1–294.12).

The many contributing authors came through magnificently. I thank them all most sincerely for agreeing so willingly to create this *Handbook* with me.

Each submitted chapter was studied by two independent reviewers and myself. Much of the reviewing was done by the Advisory Committee. In addition, the following individuals have kindly contributed to the review process:

William P. Adams
E. Nigel Corlett
Joseph El-Gomayel
I. Ham
Eliezer E. Kamon
W. Karasek
Stephan A. Konz
Ernest J. McCormick
Colin L. Moodie
James M. Moore
W. A. Pesch
John V. Pilitsis
A. Ravindran
Gerald J. Wacker
Howard Weiss

The index to this *Handbook* was prepared by the most able Jack Posey, of Purdue University, in cooperation with all of the chapter authors.

I am most pleased that prior to my undertaking the position as Editor of this *Handbook* I had the opportunity of visiting with and gaining valuable inputs from Walton M. Hancock, John Mariotti, and Salash El-Magraby. I am most indebted to Orlando J. Feorene for his effective written communications, parts of which are included in this preface. During the various phases of the preparation of this *Handbook*, I was most fortunate to receive a number of excellent suggestions from David Beldin, Executive Director, and James L. Wolbrink, Director of Publications, The American Institute of Industrial Engineers.

I have had the privilege of working with Thurman R. Poston, our John Wiley Editor, a truly conscientious first-rate editor. I was fortunate to have, during the preparation of this *Handbook*, the able assistance of Joy Taylor in carrying out a variety of diversified secretarial tasks. Jean Blackburn of Hamptons Editorial Services has done an outstanding job in copy editing the *Handbook*.

Finally, I would like to express special thanks to my wife Catherine for her kind assistance in the preparation of the *Handbook;* to my parents, Paul and Katarina; and to my children Laura and Kevin who make it all possible.

GAVRIEL SALVENDY
West Lafayette, Indiana
January 1982

CONTENTS

HANDBOOK OF INDUSTRIAL ENGINEERING

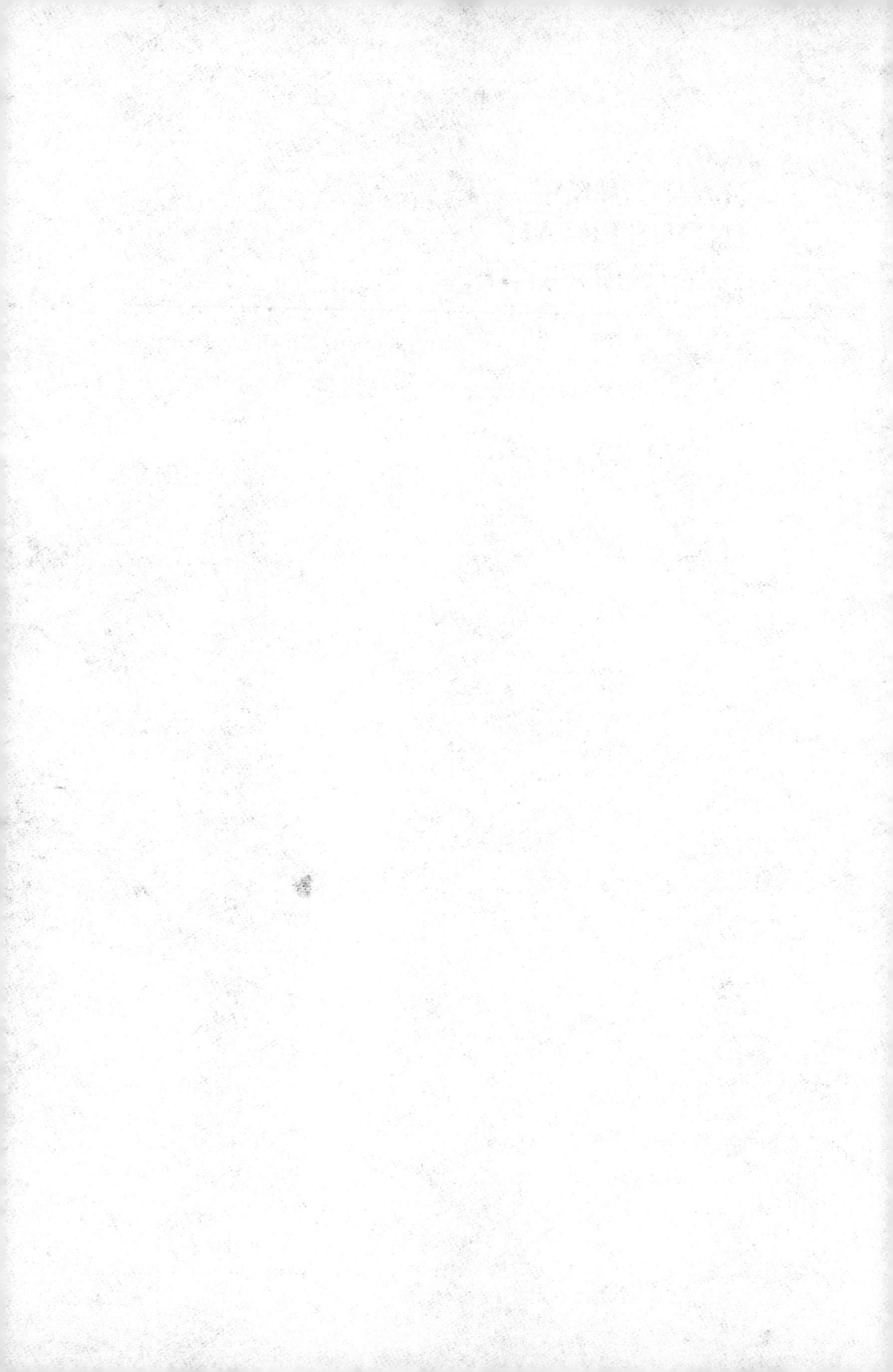

SECTION 1

INDUSTRIAL ENGINEERING FUNCTION

CHAPTER 1.1

The Industrial Engineering Profession

BYRON W. SAUNDERS
Cornell University

1.1.1 WHAT IS INDUSTRIAL ENGINEERING?

Engineering is defined by the Accreditation Board for Engineering and Technology [ABET, formerly Engineers Council for Professional Development (ECPD)] as

> *the profession in which a knowledge of the mathematical and natural sciences gained by study, experience and practice is applied with judgment to develop ways to utilize economically, the materials and forces of nature for the benefit of mankind.*

The key words seem to be "mathematical and natural sciences," "applied with judgment," and "economically." The American Institute of Industrial Engineers (AIIE) in turn has defined the special field of industrial engineering as

> *concerned with the design, improvement and installation of integrated systems of people, materials, equipment and energy. It draws upon specialized knowledge and skill in the mathematical, physical and social sciences together with the principles and methods of engineering analysis and design to specify, predict and evaluate the results to be obtained from such systems.*

This includes all elements of the general definition of engineering, elaborates in several ways on what is meant by "applied with judgment" and in most respects is comparable to definitions of other fields of engineering. Broadly defined, the function of industrial engineers is to bring together people, machines, materials, and information to facilitate an effective operation. Essentially, the industrial engineer is engaged in the design of a system and his or her function is primarily that of management. However, the element that is unique to industrial engineering as defined above is the explicit reference to people and to the social sciences in addition to the natural sciences. This expands the field of knowledge required and the types of systems with which industrial engineers are concerned. Industrial engineering is concerned, therefore, not merely with the design, installation, evaluation, and redesign of things or systems of things but also with people interacting in and with the system, so that people are part and parcel of the operating elements. The interfacing of people with machines, where the design of the total system must include not only the physical elements of the machines but also the behavioral characteristics, stress-strain relationships, load carrying, energy characteristics, and motivational responses of the people who are vital links in the system (just as much as cams and gears are links in a mechanical system), is what differentiates industrial engineering from the other engineering disciplines. In addition to people as components of an operating system, there is also the effect of people on the operations of the system as external forces. Goods from the manufacturing industries and/or services from the service industries all depend ultimately on some demand function created by people (customers or clients). The response of these external people to various stimuli (inflation, advertising, bonuses, price breaks, quality, etc.) creates fluctuations in the demand for goods and services, which can be amplified many times as the demand works its way back to the beginning at the opposite end of the distribution, manufacturing, or service system.

The industrial engineering task is to design and redesign, through study, analysis, and evaluation, the components that make up man-machine systems. The components are then brought together to design the total system by a proper integration of the individual components. However, the mechanical components–the machines–are designed by the mechanical engineer and the machine designer, and the industrial engineer must coordinate and cooperate with these specialists. The instrumentation and driving forces are generally provided by the electrical or electronics engineer, chemical processes are provided by the chemical engineer, and other engineering specialities provide for the design of components that fall within their special areas of expertise. Nowhere, however, had the people components been appropriately studied in order to design their tasks and to integrate them properly with the mechanical/electrical/chemical components so that each held a proper place in the system. It was this lack of people orientation to the design of tasks that stimulated the earliest work by the pioneers in the field that later evolved into industrial engineering. It is because of this necessity to deal with the people aspect of any engineering design that the handbook devotes Sections 2, 3, 4, 5, and 6 to various aspects of human problems and interaction needed to make any system operative. In addition to the explicit functions and concerns of the industrial engineering profession, the human side must be considered paramount in both its uniqueness among engineering disciplines and its importance.

One other aspect must also be included to answer the question "What is industrial engineering?" From its earliest days it has been concerned with the improvement of whatever was being designed and/or evaluated. If it was an individual human task, industrial engineers would try to make it more efficient, less tiring, and its motions easier, more productive, and with less waste, energy, and effort. If it was a series of tasks, it would try then to make them more uniform and integrated with better flow and a minimum of stop and go action. If it was a handling task, it would try to eliminate or reduce the amount of movement involved through changing the size of load, by rearranging the layout of the shop, office, or service area, or by the use of different parts or components that would allow for better timing, flow, or similarity in processing required. If it was a manufacturing task, it would try either to redesign or to use different materials in order to use better or newer production methods and provide better integration and flow between processing steps. The examples could go on and on, but underlying them all would be the fundamental notion of cost reduction and of more efficient use of resources, be they human, material, physical, or financial. Indeed, some of the early industrial engineers were referred to as "efficiency engineers" and while this term was used generally in a derogatory way for reasons that need not be discussed here, there was a certain truth to the title because there were many charlatans attempting to break into the field who had neither proper training, orientation, nor interest except for quick financial benefit. The term is rarely used today. Instead we speak of *productivity and productivity improvement*. If industrial engineers had to focus on only one concept to describe their field of interest, objectives, and proper frame of reference, it would have to be productivity improvement. This has become both a national and international concern and one that is crucial to the health of the world's economies. Industrial engineers accept as their primary mission that of productivity improvement, which, broadly defined, implies a more efficient use of resources, less waste per unit of input supplied, higher levels of output for fixed levels of input supplied, and so on. The inputs may be human effort, energy in any of its myriad forms, materials, invested dollars, or others. Succinctly stated, its mission would be to try to produce more or to serve better without increasing the resources being consumed.

1.1.2 HOW HAS INDUSTRIAL ENGINEERING CHANGED AND DEVELOPED?

Industrial engineering has undergone several significant changes since the middle of the twentieth century. Both the scope of the field and the methodology used by its practitioners has changed to a degree that is unusual even in an era when rapid change has been commonplace. New and expanded needs of public and private organizations and the availability of new tools and skills mean new requirements and opportunities for industrial engineers. In the period since 1950, the field is probably the most rapidly growing one in engineering.

Prior to the mid-1950s, industrial engineers were primarily concerned with human interactions in the design of manufacturing plants, construction facilities, methods of cost and quality control, production control, materials processing and handling, and related manufacturing operations to produce goods or services. Historically, the design of certain subsystems such as individual work centers and production centers were also a significant part of the industrial engineer's task. The design procedures then in use were more qualitative than quantitative in nature, and a heavy reliance was placed on empirical evidence of what would work to accomplish a given result.

The decade of the 1950s saw a complete reorientation of the field of industrial engineering. As new mathematical and statistical techniques and equipment became available, the emphasis shifted from qualitative methods to a greater reliance, emphasis, and research effort directed toward more quantitative methods of problem solving. The high speed, stored program digital computer became an indispensable tool for the industrial engineer. The increase in emphasis on basic science and the shift from empirical toward more quantitative analytical methods was not, of course, peculiar to industrial engineering, but it is likely that the changes were more drastic in this field, since industrial engineering emerged from the World War II period as probably the least quantitatively oriented of all the engineering fields.

A concurrent change of equal importance has taken place in the kind of projects on which industrial engineers work. Before 1950 nearly all industrial engineering was practiced in the manufacturing phase of the mechanical goods industries. The expansion of the field is suggested by the many new titles used in place of the simple designation of industrial engineering. Today there are areas such as operations research; manufacturing, production, or automation engineering; and even human engineering. There are systems engineers, management or administrative engineers, and operations engineers, and they are found in the fields of transportation, distribution, military logistics, weapons systems analysis, finance, public health, and the service industries, as well as in manufacturing.

It is also apparent that the scope of the field is changing. Traditionally, the concern was with relatively small systems, but current research is more concerned with macrosystems, and people entering the profession are inclined in this direction. Technological developments, together with the previously successful work of industrial engineers, have reduced the need for direct labor, and the emphasis has shifted away from the former concentration on the design and measurement of individual workplace activities. Most significantly, methods are available that make the rational design and analysis of larger systems feasible. There has emerged the whole new area of operations research or operations analysis, which is based on many new mathematical techniques and on providing systematic ways of handling the complex operations found in modern industry, government, or service organizations.

For a long time industrial engineering suffered, in contrast to other fields of engineering, from the relatively undeveloped state of the sciences upon which it was based. In effect, there simply did not exist any dominant science upon which industrial engineering could rest.

The sciences that are relevant to industrial engineering are diverse and tend to be less quantitative than other scientific areas. The social sciences, toward which industrial engineering looks for information about the behavior of the human elements of its systems, are still not sufficiently well developed to support an engineering discipline in a scientific sense. The relevant mathematical areas—discrete mathematics and the mathematics of uncertainty—are still not well developed, although significant advances have been made in the last 30 years. Perhaps the most promising of all the recent developments is the emergence of what this writer's colleagues have called a "science of operations", which is quite distinct from both mathematics and the social sciences. It appears likely that a number of different operating system types can be identified, defined, and described. These system types exist completely independently of the area of application. They can be studied in general terms to determine their fundamental properties, which can then be used as the basis for initial analysis and ultimate design of new and more complex systems.

It is this last notion, the concept of a science of operations and the ability to use this science in analyzing and synthesizing any type of operating system, that has resulted in the ability of the field of industrial engineering to expand its horizons and the opportunities for its practitioners to the point that they have become almost limitless. The operating systems to which the science is presently being applied and to which the practitioners have digressed from industrial systems alone has been impressive. One area of service that has seen wide application of industrial engineering analysis is the hospital and general health care system. A tremendous amount of human activity is involved, costs are escalating, and facilities are expensive and technically complex. Productivity improvement in the health care system is a natural for industrial engineers and is desperately needed by society. The industrial engineers have moved rapidly into this field during the last two or three decades. The banking system is another area of service in which no manufacturing operations in the usual sense of metal removal or deformation, parts assembly, or finishing and coating take place. However, the banking system is involved in thousands of manual and machine operations daily, where information storage, retrieval, and transmission must be immediate and precise, where quality control is crucial, where inventory and scheduling functions peculiar to banking abound, and, most importantly, where people are essential to the smooth operation of the system. Here, too, industrial engineers are finding a natural place for their analytical talents and methodology and the opportunity to synthesize and design new operating modes and systems to provide for the more efficient flow of money and information.

A similar rationale can be made for any type of operating system. The individual components of each system may vary, but with the developing science of operations, which is universal and describes in abstract terms any type of system, the industrial engineer need only develop a specialized model from the general abstraction, substitute the explicit conditions for the more abstract parameters, and then proceed with the analysis and design of a new or modified system to improve the productivity of whatever is involved. It does not matter whether inventory systems are imbedded in a manufacturing entity, a hospital, a bank, a department store, an educational enterprise, or an airline—to suggest some of the possible locations that inventory problems arise. Similarly, quality monitoring systems, scheduling procedures and systems, personnel evaluation and manning systems, production systems, or the myriad types of operating systems and entities with which life deals can be and are all described, modeled, evaluated, and synthesized by the modern day industrial engineer.

1.1.3 MEASUREMENT—A BASIC NEED OF INDUSTRIAL ENGINEERS

The art and science of the problems of measurement have been practiced by engineers and artisans from earliest times. In the early days all measurement methods were crude and imprecise. As time went on the methods improved and the equipment used to make measurements likewise improved. By the nineteenth century such routine problems of measurement as linear distance, volumes, rates of flow from fixed vessels with standard orifices, conventional height and weight problems, and similar problems related to fixed sources with nonvarying elements and parameters were pretty well established and standardized. Where such measurements were necessary for the industrial engineer, they presented no more of a problem than they did for any other type of engineer. However, it should also be noted that the problems of measurement were also quite crucial and errors of any amount could lead to serious problems for the engineer/designer. Incorrect measurement of the spacing of rails for a railroad could lead to serious problems to cite only one example.

The industrial engineering measurement problem arose because human beings became vital cogs in the systems with which the industrial engineers were concerned, and their systems responded to people and their motivations and responses to various stimuli. As a result, the measurement of people, their activities, and their responses to the varying stimuli encountered with and by the systems were of crucial importance. The time that it would take a quart of fluid of known viscosity at a known temperature to flow from a given size orifice could be quite easily determined, and for the same conditions it would be constant. On the other hand, the time that it might take a single worker to assemble two pieces of metal using some type of fastener to secure them together would be a highly variable thing, even if the part sizes were absolutely standard, the temperature, air quality, and other environmental conditions were all constant, and even with the same worker performing the task. The individual's concentration on his or her task could vary significantly for literally dozens of reasons, some physiological, some mental, or some external. Whatever the reason, the measurement of the amount of concentration, the reasons for its change, and the changes that occur between cycles of performance of the task present some very significant measurement problems. If different people were used to perform the same task there would be additional elements of variability with which the industrial engineer must contend. Even the time of day that the task is being performed can contribute another variable. As a result, a serious concern on the part of industrial engineers has been a methodology for attempting to establish consistent standards of measurement of human activity. (For a more detailed discussion of performance measurement see Section 4.) While it may become a task for a technician rather than the industrial engineer, the fact is that without good, consistent, defensible standards for the activities performed by people, the entire system design that would be predicated on such measures is apt to break down and be useless. What, then, are some of the critical areas of measurement?

One area that has been traditionally associated with industrial engineering, but of itself is not and should not be considered industrial engineering, is the field of time study and the measurement and performance of human work. The measurement of the time it would take to turn a piece of steel from one diameter to a lower diameter at some known feed and speed is easily calculated for the machine portion of the task on automatic equipment. To establish the amount of time a worker would take to perform the task on a manual lathe with manual loading and unloading, where the worker controls his or her own movement, is a far different calculation. The fact that one worker would perform the same task in varying times does not add to the simplicity. Then when one tries to establish a time for design purposes that involves several workers, all of whose time will vary, the complexity rapidly becomes apparent. Much data must be used in the process, and this is one area where the field of statistics becomes of critical importance to the industrial engineer.

Another important point should be made at this time, however, which is that the time a worker might take or be allowed to perform a given task is useful for several activities. The usual thought is that a time must be established for payroll purposes. This is true, but it is only one of many purposes and in many ways it is the least important of them. The time it takes to perform a task is crucial for the proper scheduling of activities. Without proper time values, planning becomes a hopeless mess for manpower, inventories, and transport of raw materials or finished goods into or out of an operation. In a clinic or in hospital facilities, the scheduling of limited space and equipment cannot be established in any reasonable way if satisfactory and defensible time values are not available. In financial institutions the time it takes to perform certain operations and to clear information and paperwork through the system can result in a profit or loss of several hundred thousand dollars if attention to these matters is not given. One nonprofit institution with which the author is familiar saved over $500,000 one year based solely on the timing that was involved and the better control of its financial operations related to cash flow items. It all happened after a study of the system disclosed where the manual operations were being performed, the time it took to perform them, and how some controls could be instituted to insure that they were performed according to a preplanned schedule. The time necessary to perform human tasks is, therefore, a crucial piece of information the industrial engineer must have. Where it is not available it must be obtained by whatever means necessary to get it. The engineer may have to establish procedures to make sure that the figures obtained are the proper ones for the uses that he or she might have in mind. However, it should be repeated that the establishment of time values is not the responsibility of the industrial engineer, however much some may believe it to be. In the early days of industrial engineering there was much effort devoted to the problems of time study and time standards and much work is still going on today in that field by specialists in the art and science of time study. However, time study practitioners are not industrial engineers, if that is the limit of their activity and expertise.

There is another area that must be identified with the problems of measurement, the measurement of future events or activities. In the case of existing or past activities, actual events have taken place that can be observed, and the conditions that exist can be recorded and replicated when needed. However, for future events, there may be additional variables if a certain event does or does not happen as expected. This, then, introduces the concept of probability, or the likelihood of an event happening in accordance with our expectations. If the event does not occur at a time and with conditions as planned, then alternative results might occur. The engineer must know about these possibilities and, where necessary, include their likelihood in his or her planning. One of the early uses of the theory of probability and its relationship to the timing problem of people and their activities is frequently found in queuing problems and applications of queuing theory. Job shop scheduling, which depends so much on the establishment of proper time values to perform the shop operations and recognizes the variability that will exist, is in reality one big queuing problem. Probability theory is another of the mathematical disciplines that it is crucially important for the industrial engineer to have in his or her "tool kit."

There are a great many other problems of measurement for the industrial engineer and they all require in some form or other an ability to deal with the probabilistic nature of future events or to deal with masses of data concerning present or past events using proper statistical tools. As examples, there are problems associated with machines that are completely independent of the human operators such as questions of quality produced by a machine and its ability to maintain consistent quality, or the rate of decomposition in quality. The rate of throughput and its variability with time (time of day), and the maintenance costs experienced by individual machines or processes over time and their decomposition are measures that are vital for proper maintenance planning and replacement (of equipment) purposes.

In one form or another, all of the measurement problems suggested and those that are dealt with in more detail later in this volume are different ways of measuring productivity. In the classical economic sense, productivity is generally expressed in the man-hours required to achieve a certain level of output. More simply and generally stated, productivity would have to include any one of a number of possible measures whereby a certain input is used to create a certain output. If this input level is decreased so that less is required to maintain a fixed level of output, then we conclude that productivity has improved. Conversely, if a fixed level of input is used and a higher level of output is achieved, we can still conclude that productivity has improved. The real problem for the industrial engineer is the choice of parameters to be used as the proper input and output measures in a particular situation. If the input parameters change because of a change in technology, it may not be readily apparent whether or not productivity has in fact improved. As a general rule we want to measure those things that would identify waste and lead to waste reduction. As stated earlier, we want to make all of our efforts more efficient, more productive, and less wasteful. We not only want to make our output greater in terms of the man-hours expended but we also want higher levels of output or improved quality of output in terms of per dollar of invested capital, per square foot of space occupied, per kilowatt of energy consumed, per mile of transport used, per machine-hour expended, or per any combination of hundreds of other possible measures that could be used to evaluate whether we are in fact making an improvement in our mode and methods of operation. There may be no single proper measure for productivity evaluation. The individual case may call for something new in the concept of measurement because of new or different conditions encountered. Frequently a composite index of several parameters is necessary, but one fact must always be paramount and that is that the measures used must be relevant to the situation at hand. The mere fact of an improvement (reduction) in man-hours required to perform a given task does not tell the whole story, if in the process of achieving the result, an excessively high cost and investment might be involved. The gain in productivity resulting from such high-cost investment possibilities may reduce the man-hours required, but do it at a prohibitive investment cost. It is part of the industrial engineer's job, therefore, to assess the total impact of any design changes or improvements in such a way that all elements of the system are evaluated in any productivity evaluation and that all relevant items are included in whatever type of composite index is used.

In keeping with the basic definition of engineering, the economic factors and impact cannot be ignored but, on the contrary, are crucial to any evaluative process.

Inasmuch as *design* has been identified as the unique engineering activity that distinguishes it from similar fields (operations research and management science, for example), the art and science of measurement becomes even more crucial for the engineer. If any designs are based on irrelevant data or on some parameters that are irrelevant to the situation, then no matter how creative or innovative a design might be it could still be inadequate for the task it was intended to perform. It may perform the task or serve the function but it would rarely be considered to be efficient. For these reasons, the importance of measurement cannot be overemphasized. Many of the standard measures used by other engineers are still used by industrial engineers such as those in horsepower calculations, energy requirements, length, height, and weight, and the like. The one area that is different and in which there is still much research needed has to do with better ways and means to evaluate and measure human effort and productivity. When better and more precise measures are available, they can then be useful in describing and modeling the operations involved and in optimizing, by any of several possible optimization methodologies, possible alternative solutions to a problem. A final caution must be restated, however. In spite of conceptions that some individuals might have to the contrary, the technique and act of measuring human effort is not by itself industrial engineering. In the early days of industrial engineering, as we subsequently point out, there was much effort and energy given to this phase of industrial engineering—so much so that in the eyes of many the subject of time study did in fact become synonymous with industrial engineering. It has taken a long time to dissuade the general public, industrial management, and even many industrial engineers from this notion, but it is true and must be recognized. The time study technician is a vital link in the information chain, without whose time information, supplied by the proper use of one of the available techniques to produce it, the industrial engineer's job is much more difficult and open to challenge by those who have a vested interest in the time value assigned to the operations.

1.1.4 A BRIEF HISTORY OF INDUSTRIAL ENGINEERING

To this point we have tried to define the field of industrial engineering and have emphasized the importance of measurement problems. How, then, did the field evolve and why, as just suggested, is the topic of time study incorrectly perceived by some to be synonymous with industrial engineering? A point to start with would be an acknowledgment of the fact that all engineering disciplines in their early days arose because of the necessity of solving some particular problem. The solutions to such problems were created generally by practical people who did things because they knew from experience that they would work, not because of any underlying science from which they designed their systems. They worked from empirical evidence—either of their own or of some one who might have preceded them. For example, the men who built the viaducts that brought water to Rome some 2000 years ago did not design them knowing the frictional forces and corresponding losses in flow that would be experienced by using different materials to make and/or line the pipes and open channels. They just knew that if they built them using a certain size, pitch, and materials, they could get enough water to meet the needs of the Romans. In much more recent times, electrical engineers did not know in the beginning how to plan for or to protect power lines from lightning strikes or other power surges. They just knew that if they put in protection of a certain capacity, it would take care of the normal surges and overloads. Today, however, the understanding of transients in an electrical system is a well-understood phenomenon that can be designed for using appropriate electrical theory. The fact is that all engineering disciplines were developed from empirical evidence and as a result of research and understanding that evolved, a more scientific base was gradually established.

Industrial engineering is certainly no different. It, too, started with an empirical base and only since about 1950 has industrial engineering been able to gradually develop its own scientific base. In this sense, industrial engineering is no different from any of the other engineering disciplines. It is only that the scientific base was much later in arriving for industrial engineering than it was for other engineering disciplines. Industrial engineering had to wait for developments in the underlying sciences that were necessary before the science of operations could take shape and provide the insight necessary to establish the scientific bases for much of industrial engineering design. As this science of operations began to take shape and as computer technology became more sophisticated and able to meet the needs of the engineer, the empiricism with which the industrial engineer had been confined gradually began to disappear. What, therefore, were the principal events and who were the participants who first developed the empirical base and then participated in the evolution of the field into the more substantive science we know it as today?

Historians of science and technology might argue as to the beginnings of industrial engineering. The generally accepted beginnings relate to the work done by Frederick W. Taylor, who was concerned primarily with concepts of productivity, even though he did not refer to it in those terms.

Prior to his work, however, there were others whose writings referred to concepts that ultimately became associated with industrial engineering, whose impact on Taylor is difficult to assess. One of the earliest of these is Adam Smith's treatise *The Wealth of Nations*, published in 1776. The concepts he expressed concerning the proper division of labor, while not original, nevertheless became an important factor in the unfolding of the impending Industrial Revolution. How much influence Adam Smith had on the creation of the factory system is debated by economic historians, but it is clear that his writings and those of both his students and contemporaries were important milestones in the development of the factory system and of the Industrial Revolution which it created. He was an economist not an engineer, and as a result, his writings came from this perspective. There were other authors, largely economists, who wrote during the Industrial Revolution whose ideas probably had some influence on those we normally consider the pioneers in the field of industrial engineering. The classic writers of the time who dealt with the topic of "economic science," as it was referred to in England and which, by inference, is being suggested could have influenced the thinking of Taylor and others, would include Malthus and his *Essay on Population*, published in 1798, Ricardo's *Principles of Political Economy and Taxation* of 1817, and John Stuart Mill's *Principles of Political Economy* of 1848.

A more direct line to the pioneering group in industrial engineering might be provided by Charles W. Babbage. Babbage was a professor of mathematics at Cambridge University who had wide-ranging interests that went well beyond the limits of pure mathematics. As a result of research he had undertaken for a different project, he wrote *On the Economy of Machinery and Manufactures* in 1832. This volume has a tremendous range of insights resulting from his observations of manufacturing plants, and surely must have impressed any of those early workers in the field of industrial engineering who had read it. For instance, he discusses such subjects as the time required for learning of a particular task, the effects of subdividing tasks into smaller and less detailed elements on learning time, and the effects of learning on the generation of waste. Other insights he provides have to do with the time saving realized in changing from one task to another, the effect of requiring workers to change tools, and of the advantages to be realized by repetitive tasks. These notions were quite revolutionary in the early nineteenth century, although some of them, based on current knowledge, have proven to be less than desirable. However, considering the time and place that they were first enunciated and used, they were radical departures from the conventional wisdom. In other chapters Babbage discusses such notions as wage payment and the effects of different approaches to wage payment by considering profit sharing plans. Another discussion suggests some of the relations and conflicts that exist between labor and management in connection with the introduction of machinery (or, as we call it, automation) into the system of manufacturing.

Perhaps one of the most important contributions to industrial engineering that Babbage made, although it was not so recognized at the time, was his attempt to build a computer—or as he referred to it, an "analytical calculating machine." He was not successful in this attempt except in the most rudimentary way, yet his basic notion that a machine could be designed and built that would perform many mathematical operations was a vision that was far in advance of the technology needed to deliver it. It took over 100 years before a useful computer would be available. Babbage's machine was completely mechanical, whereas the first significant computers that became operational were dependent on some necessary technological developments in electrical and electronic theory and devices, rather than on the conceptualization of what was to be accomplished with such a machine. Clearly, once a computing machine became available, many more applications were found for it than were originally foreseen. This should not detract, however, from the powerful insights that were necessary to provide Babbage's original vision of what could be done if such equipment were available.

During the latter half of the nineteenth century, there were others, primarily in the United States, who clearly provided the impetus and thinking that led to interest in the start of formal education in the field of industrial engineering. One such person was Henry R. Towne, who was associated with the Yale and Towne Manufacturing Company and with the American Society of Mechanical Engineers (ASME). Towne emphasized the economic aspects and responsibilities of the engineer's job in a paper that he presented to the ASME.

It is important to note that he used ASME as the professional society to which he presented his views and expressed the conviction that a need existed for a professional group with interest in the problems of manufacturing and management. This suggestion ultimately led to the creation of the Management Division of ASME, one of the groups active today in promoting and disseminating information about the art and science of management, including many of the topics and activities industrial engineers find themselves engaged in. It also points up the fact that ASME was the breeding ground for industrial engineering. Many of the early papers on topics that were later associated with industrial engineering were given before ASME, and several of the early figures in the field of industrial engineering later served as presidents of ASME.

In addition to his interest in the management of the industrial enterprise, Towne also was concerned specifically with wage payment plans and the remuneration of workers. Another active worker and writer on this last topic, who also presented his views before the ASME, was Frederick A. Halsey, the father of the Halsey premium plan of wage payment. His motivation in proposing this plan was to increase productivity measured in terms of the labor cost. His plan also included the notion that some of the gains realized from such productivity increases be shared with the workers creating them, according to a formula he proposed. A third individual who laid much of the groundwork for the developing activity that has come to be known as industrial engineering was Henry L. Gantt. He, too, used the meetings of ASME as the vehicle to present his ideas, which covered a wider range than some of his predecessors. He was interested not only in costs but also in the proper selection and training of workers and in the development of proper incentive plans to reward them. He also was interested in scheduling problems and was the originator of the Gantt chart, which in its modern form uses probabilistic information and procedure. The evolution of Performance Evaluation Review Technique (PERT) and Critical Path Method (CPM) as scheduling devices is a development that goes well beyond Gantt's original idea, but was only possible because of developments in the field of probability and the availability of appropriate computing technology. (For further discussion of these topics see Sections 11, 13, and 14.)

Probably the most often quoted and generally acknowledged instigator of studies that have led to the discipline of industrial engineering was Frederick W. Taylor. While he did not use the term "industrial engineering" in his work and was himself a mechanical engineering graduate of the Stevens Institute of Technology, his writings and talks under the aegis of ASME are generally credited as being the beginning of the discipline. Taylor, however, preferred the term "scientific management" to describe the work in which he was engaged and to which he devoted so much effort. One cannot presume to be well versed in the origins of industrial engineering without reading Taylor's books *Shop Management* and *The Principles of Scientific Management*. Essentially what Taylor was proposing was a more rational and planned approach to the problems of production and shop management. He did not confine his activities to management problems alone but was active in research on metal cutting and the technical problems of production as well. While much of his work in both fields was crude by modern standards, considering the state of the art and knowledge in the relevant fields at the time, he was very advanced and in many instances misunderstood. It is difficult to summarize his work (and hence the beginnings of industrial engineering) in a single or even a few sentences, but primarily he was vitally interested in better and much more complete planning by management, better selection and training of workers, more mutual understanding and respect between workers and management, and a proper incentive for workers when they had performed according to the plans set forth. His interest in both time study and motion study was not for the studies themselves, but for the part they played and the information they supplied for the planning of activities. He was really trying to develop a "science of planning" or, as was expressed earlier, a science of operations, but without the necessary relevant basic sciences on which to build. As a result, his "science" was totally empirical in nature but one that he was able to demonstrate would produce very significant results in productivity improvement. This is, after all, what he was after and what the profession is still after today, but with a markedly different and better set of scientific tools with which to deal. Taylor by general acknowledgment, if not by formal designation, is considered to be one of the two giants in the field of industrial engineering who started it on its long road to where it is today.

The other giant of the early days was Frank Bunker Gilbreth. He, too, was an engineer and obviously had been impressed by the work and writings of Taylor. However, his interest in improving the efficiency by which work was done had a different emphasis from Taylor's. Taylor, as indicated, focused on planning and organization of work, and while this entailed both the study of the methods whereby work was done and the time in which to do it, these were not his primary interests. Taylor in fact has been referred to by some as the father of time study. While this may or may not be true, it is a much more accepted condition that the Gilbreths (husband and wife) were the instigators of motion study and the scientific study of work and workers. In addition to motion study, the Gilbreths were also noted for their work in skill study, fatigue study, as well as in time study. These last three, however, all seemed to be incidental to the primary concern of motion study and the finding of "the one best way" of performing work by individuals and/or groups. In his work Gilbreth was enthusiastically joined by his wife, Dr. Lillian M. Gilbreth. Together they formed a very effective and remarkable team. The unique feature of industrial engineering that makes it different from other engineering disciplines—the attention paid to human values, human interaction, and the human response to environmental and physiological limitations of work and the workplace—was achieved naturally by them because Lillian Gilbreth had obtained a Ph.D. in psychology and was able to contribute very effectively and cooperatively to the human problems associated with her husband's studies. Frank Gilbreth's work had a profound effect on many people in the field and stimulated much research and activity in the field of motion study that continues to this day. One of their significant contributions—small as it may seem today—was their definition of the elements of motion, which enabled individual motions to be studied and dealt with more effectively rather than by trying to deal with work simply by looking at aggregate motions only. Thus subdivision of motions into "therbligs" (Gilbreth spelled backward) was a distinct step forward in the scientific analysis of human work.

The first Ph.D. granted in the United States in the field of industrial engineering was the result of research done in the area of motion study. It was awarded to Ralph M. Barnes by Cornell University as recently as 1933. Barnes' thesis was then rewritten in book form as the well-known textbook, *Motion and Time Study*, which has gone through many revisions and editions and is still looked on as the "bible" on motion study. His thesis title was "Practical and Theoretical Aspects of Micro-Motion Study" and was supervised by Dexter Kimball, who was his major professor.

When one looks at the writings of the Gilbreths and then adds to that the stimulus given to Barnes and the many students who followed in his footsteps in the field of motion study, one can only conclude that the Gilbreths have indeed probably had a greater impact on the field than any other individual or group during the first 50 years of the development of industrial engineering. The work they did was the precursor of much that is reported in this handbook in Sections 2, 3, and 4.

There were many others who should be recorded in any detailed history of the field of industrial engineering, however, space limitations preclude more than a mention of their names so that any interested individuals will at least have a clue whom they should look for in any library searches they might wish to undertake. They are Hugo Diemer, Charles B. Going, Harrington Emerson, Robert Hoxie, Dexter S. Kimball, George H. Shepard, Arthur G. Anderson, L. P. Alford, and, in a somewhat later period but still prior to World War II, Alan G. Mogenson, Ralph M. Barnes, Marvin G. Mundel, and Harold B. Maynard. While this list is not meant to be exhaustive, it will lead the reader to others of the same or earlier periods. Those enumerated have all had a significant impact on the field of industrial engineering—some through academic pursuits and others through their consulting and work with industry.

From those listed, however, specific mention should be made of Mogenson because of his activities in teaching and trying to bring the concepts of motion study to the workers in the factories of America and of the world. His approach is what he chose to call "work simplification." His thesis was very simple. The people who know any job the best are the workers doing that job. Therefore, if the workers are trained in the simple steps necessary to analyze and challenge the work they are doing, then they are also the ones most likely to be able to make improvements in it. The approach was, therefore, to train key people in manufacturing plants, which he did through his Work Simplification Conferences in Lake Placid, New York. These trained people would then return to their home plants and, in turn, conduct training programs in plant for top managers as well as workers. By giving the analytical tools to those on the job, he felt (and history seems to confirm) that the most simple manual operations, requiring little beyond simple jigs and fixtures, could be improved significantly by the workers themselves, and did not need the skills of an industrial engineer until much greater degrees of complexity were involved. This concept of taking motion study training directly to the workers through the work simplification programs was a tremendous boon to the war production effort during World War II, and its value in terms of productivity was of inestimable value in the prosecution of that conflict.

Most of the leaders of the early work in industrial engineering focused their activities on motion study and related areas of work at the individual workplace to make it more productive. However, there was another area that deserves mention even in such a brief review as this must be. Statistics as a subject could not be applied in a major way to the engineering problems of industry even though it was a discipline started one hundred years ago. In the first 20 years of the twentieth century work was done on the theory of sampling and in 1931 Dr. Walter Shewhart of Bell Telephone Laboratories published *Economic Control of the Quality of Manufactured Product* based on sampling theory. It was assembled from numerous papers that he had issued as internal memos or journal articles that had been published during the 1920s to describe his approach to controlling quality by sampling at various spots in the production process. Depending on the particular sampling plan, the size, and the resulting calculations, considerable insight could be gained into the quality of an entire batch without requiring 100% inspection. While the ideas he expressed were general knowledge, they were not taken seriously or widely applied until World War II. Since then, however, a great many other texts have been written and much research done to extend and enlarge on the concepts that Shewhart proposed. Hence, modern statistical quality control was definitely a pre-World War II development, which spawned many practitioners as well as a separate professional society—the American Society for Quality Control. The requirements and plans for the proper control of quality are a necessary component and consideration in the analysis and design of manufacturing systems. Some of the concepts originally conceived for the control of quality have now been extended into other areas, and the control chart has found applications in inventory planning and control, marketing analysis and control, and in financial control and accounting, to suggest just a few of the areas of expansion.

During and following World War II, the developments in motion study, time study, work simplification and in quality control, along with some topics dealing with the personnel functions of wage and salary administration, job evaluation, merit rating, plant layout and materials handling, and the production control activities of routing and scheduling constituted the essence of industrial engineering activity. In some manufacturing organizations, perhaps only one or two of the above functions was recognized, while in others there might have been a rather complete coverage of all these topics. From an organizational standpoint, the activities identified as industrial engineering might have been located in anyone of several possible locations. In some firms, the industrial engineering function was located within the engineering organization, in others it was part of the manufacturing organization and had relatively little contact with engineering. In some cases, the group was located in the personnel organization when the functions served were primarily those of a personnel nature. The net effect of all this was a dispersed discipline that had too little focus, was built largely on empiricism, had no national organization or group to bring it together and provide a centralized focus, and generally was perceived to be at best a subprofessional activity.

Without attempting to chronicle all the events in their proper order, this situation started to change shortly after World War II. In 1948 the American Institute of Industrial Engineers (AIIE) was founded in Columbus, Ohio. The requirements for membership were such that engineers were eligible primarily by completing a proper college level program *or* by equivalent experience to give the breadth and understanding that comes from engineering experience. Prior to the establishment of AIIE, there had existed other groups. The most important of these was probably The Society for the Advancement of Management, the successor to the original Taylor Society, although it did not require engineering credentials of its members. It was more management than engineering oriented. The American Society for Quality Control already mentioned was founded at the close of World War II. The establishment of these two societies requiring professional credentials for membership began to give the focus that had been lacking and had resulted in the dispersion of effort to advance the profession up until that time. The only other group that had attempted to meet the needs of engineers was ASME as mentioned previously. This organization had divisions of interest but apparently never really satisfied the needs of its industrial engineering members—hence other professional societies were spawned.

Of greater importance, however, was the release of previously classified material dealing with some of the analyses that had been made during the course of the war itself. The field of operations research had its start during the war when certain scientists from a broad range of disciplines were asked to use scientific analysis on some of the operational problems of prosecuting the war. As a result, both physical and social scientists delved into the problems presented using methods known to them. Where known methods were not readily available, research had to be conducted to produce viable ones. As a result of these efforts, significant advances were made in the understanding of the operational problems and alternative courses of action available to those responsible for decision making. Hence the field of operations research emerged. The analysis of the operations involved provided the decision makers—the admirals, generals, and politicians—with the various options that might be available to them in certain operational situations and the tradeoffs and likely outcomes if certain options were followed. As these documents describing the operational problems and studies during the war were declassified, it became apparent to some of the practitioners in the field of industrial engineering that there were some striking similarities between the operational problems in a war and the operational problems of producing and distributing goods. A slightly different twist to the Gilbreths' "one best way" was simply the finding of the "optimal" strategy to follow in a number of different production and marketing situations. Some of the operations researchers from war time then extended their area of activity to include industrial problems. They were frequently unsuccessful, however, because of the greater number of variables in industry on the one hand, and the lack of military discipline among the workers on the other. Although this is an oversimplification, nonetheless, there were significant differences and many adaptations had to be made both in the methodologies used and in the implementation of the results.

The decade of the 1950s, therefore, was a most active one in the transition from the prewar era of empiricism to the more quantified methods available after World War II. It was also during this period that two more organizations were established—the Operations Research Society of America and The Institute of Management Sciences. Both of these organizations tended to have a more academic and theoretical orientation—more so for the former than the latter—and both had less emphasis on applied activities and reporting of results in the manner in which the engineering societies were normally operating. As a result, there was a further splintering of effort in the extension of research into the field of application and getting application information into the hands of people "on the firing line" in industry who were in the best position to put the concepts into immediate use. Although this may be another slight oversimplification, it is true that the number of profit seeking organizations that were supporting applied operations research activities and enjoying the benefits from the new methodologies was far less than it might have been or should have been. At the same time, the engineering organizations were slow to pick up the newer, advanced approaches. However, the gap between theoretical research being done in the universities, the government installations, and in some of the larger industries and the actual applications on a large scale was quite great.

By the 1960s, however, much of this apathy and reluctance to delve into the new had been dissipated and some of the methodologies originally associated with operations research came to be much more standard procedure for industrial engineers. More mathematics was found in the curricula of most industrial engineering schools and the approach to the analysis and design of industrial (and nonindustrial) systems began to change. The concepts of designing, analyzing, describing, and synthesizing the operations by building and manipulating a proper mathematical model of the system was generally accepted. With advances in the several fields of mathematics, in mathematical programming for studying optimization problems, in probability for studying problems where uncertainty exists, and in statistics for analysis and prediction based on data analysis, a whole new era was emerging and many of the classical approaches to industrial engineering problems were being superseded by the newer methods. The old empiricisms were being replaced by what has been described as the science of operations built largely on developments in the field of mathematics.

Concurrently with and crucial to these developments, however, was another very important milestone. This was the availability of the high-speed, stored program digital computer. Prior to the availability of the computer, even if the developments in mathematical techniques for handling large problems had been available, they would not have been of much use to the industrial engineer because of an inability to process and handle the data and experiment with the models that were being designed to describe operational systems. However, the developments in computer technology changed all that and the benefits were obvious. First, as a high-speed calculating device, the computer was able to handle calculations in a few minutes that would otherwise have taken weeks or months to do, if at all. Frequently, even if answers could have been obtained manually, the time required was such that the situation calling for a decision might well have passed. The tremendous increase in calculating speed was of significance to all engineers and to industrial engineers in particular.

A second benefit, the computer's ability to store data and then recall it at any time, allowed for procedures that heretofore had been unavailable. The ability to make comparisons with previously stored data to answer "what if" questions introduced a whole series of opportunities that the engineer had not had access to prior to the availability of the computer. This storage capability, the ability to make calculations and store the answers for later use or comparison, and the opportunity to have whole programs available for standard calculations (a least-squares subroutine, for instance) meant that, through the use of subroutines, much more powerful techniques and procedures were available to the engineer. In most cases, once the problem had been defined and properly modeled, the calculations could be carried out by technicians leaving the engineer free to carry on with the more creative elements of his or her task.

A third and in many ways one of the most significant benefits of the computer to the industrial systems engineer was the ability to experiment with large systems, which he or she had been unable to do prior to the computer age, when mechanical engineers had not been limited in their ability to experiment to anywhere near the degree that industrial engineers had been. If the industrial engineers in the Gilbreths' or in Taylor's time wished to experiment with an individual worker or even with several workers doing manual or semiautomatic tasks, this was possible. However to experiment with a particular plant layout or a special material handling system, tying together several machine tools, trying several alternate production processes and methods, and tying up the production capabilities of a manufacturing facility for experimentation purposes was impossible. It could not be done in a scaled-down version because of the human problem. The human being just cannot be reduced to one half or one fourth scale size.

The net effect of this was that the industrial engineer did not have the freedom to experiment with possible system configurations or to try pilot plant operations to the degree that the mechanical engineer, the electrical engineer, or the chemical engineer did with his or her type of system. Chemical engineers were able to build small scale models or pilot plants and then extrapolate the results to a larger system. Mechanical engineers and electrical engineers were able to set their equipment either in scaled-down or full-size models in the laboratory to study and understand the physical properties and relationships of what was going on within the system. Until the availability of the stored program digital computer, the industrial engineer did not have the luxury of this kind of experimentation. However, with the large storage capacities that became available and with sufficient insight and creative imagination, it did become possible for the industrial engineer to describe, in logical and mathematical terms when needed, the behavior and relationships of various elements within their systems. They could then change the parameters of the systems, as described in the system logic, and simulate a day's, week's, month's, or year's operation of the systems, measure the results, and compare the results with alternative system designs. By this process, the industrial engineers obtained the ability to experiment with the large and even not too large systems that had been denied them until the introduction of the high-speed, stored program digital computer.

It was primarily these two developments—the mathematical advances and their applications in the field of operations research and the development of the high-speed, stored program digital computer—that literally changed industrial engineering from a nonquantitative empirical science to one of considerable mathematical sophistication and caused it to be considered a hard science. As stated earlier, the industrial engineering profession is now founded on its basic and engineering sciences to as great a degree as its companion engineering disciplines. The relevant engineering sciences for the industrial engineer would be a different set from that most relevant to the mechanical engineer, although, of course, there would be many commonalities. The relevant techniques and sciences and their current state of sophistication are discussed in detail in this handbook in Sections 12, 13, and 14, and the ability to use and understand this material is crucial for the modern industrial engineer. The application of some of these concepts is presented in some of the functional sections as well, especially in Sections 7, 8, 9, 10, and 11.

One additional element should be considered in reviewing the development of industrial engineering and the acceleration of efforts coming out of the war years. Mention has already been made of the presence of the human element in industrial engineering that differentiates it from other engineering disciplines. While this factor was present from the very beginning as developed by Taylor and the Gilbreths, the problems that developed during World War II caused a considerable expansion of effort in this area. The speed of warplanes had gotten so high that reaction times had to be reduced, and in any situation in which a human being was confronted with a control panel where decisions had to made, for example, the layout and arrangement of the controls and recorders being used became increasingly important. As a result, research was accelerated in this sphere first by the United States Air Force and then by others both in and out of government. From these needs, as well as from the recognition that the human being is a very complex system that must be considered when incorporating a human into another system, the whole topic of "human engineering" or "human factors" or "ergonomics," as it is better known in Britain and Europe, was born. Today, this rapidly expanding speciality within industrial engineering is the focus of efforts by many engineers. Because of the types of problems involved and the stresses on the human system, the engineers are joined by psychologists, physiologists, biomechanical specialists, and others. It is an important area. An all too familiar example of a situation on which more attention should have been given to this aspect of system design is the control panel design associated with the nuclear generating plant at Three Mile Island where the accident occurred in the spring of 1979 that has caused so much debate about nuclear issues. Another example is the midair collision of two planes over the wide open spaces above the Grand Canyon in June 1956. Section 6 discusses this topic in much more detail and should be carefully referred to in all situations where human beings are part of the operating components of a system.

Another important area of specialization having to do with the human side of the engineered system is what has been identified as *job design*. This is treated in much more detail in Section 2, Chapters 2.1 and 2.5. This concept has been advocated by Professor L. E. Davis of the University of California at Los Angeles based on research he has undertaken to improve the systems being designed by paying more attention to the job to be done by the people doing it than has been true heretofore. A slightly different and somewhat enlarged concept that is focused on the design of the total work system—the overall concept of what has to be accomplished—is what Professor G. Nadler of the University of Wisconsin (Madison) has called *work design*. His "ideals concept," while not addressed specifically in this volume, is also worthy of study.

Finally, attention is also directed to the problems of motivation of people in Section 2, Chapters 2.2 and 2.3. Human potential is great indeed when sufficiently challenged to cause people to want to do something. The problem of motivating them sufficiently to want to do it, however, is an extremely complex issue. It is one that has been studied by psychologists, sociologists, engineers, and the entire range of management people. The basis for a motivational appeal to the worker—the carrot, if you will—can vary markedly between people performing the same task and for the same person performing different tasks. Research efforts in these two areas, the design of work and the motivation of people to perform, is a constant search and inquiry into the behavior patterns of people in order to make the task more satisfying and comfortable for the people involved and by so doing to realize further gains in productivity improvement.

It was probably the decade of the 1950s that resulted in the greatest expansion of interest in industrial engineering and the greatest leap forward in the building of a more complete scientific base upon which the discipline could rest. The decades of the 1960s and 1970s expanded the knowledge base so that today (1980) the field of industrial engineering has a firm base in mathematics, which should allow for improved and better understanding of mathematical modeling.

As a result of these developments, the industrial engineer of the 1980s has many more sophisticated tools with which to analyze his or her problems and to design new and improved systems. In the process, however, the industrial engineer has had to specialize to a greater degree than ever before, and industrial engineering is now breaking down into subspecialities just as mechanical engineering did during the first half of the twentieth century, when industrial engineering itself was a spinoff from mechanical engineering. Within the industrial engineering family of specialists are the quality control people. Also operating from the statistical and probabilistic base are the reliability specialists. However, it should be noted that reliability notions are equally applicable to other engineering disciplines and industrial engineers do not and should not claim this area as the sole province of industrial engineering. Another subspecialty is value analysis. Value analysis concepts were developed in order to provide a basis for more attention to the proper and efficient use of materials. In one sense this is a mechanical designer's or an electrical designer's problem, but it has rapidly become an industrial engineering problem, when the production engineer analyzed the materials used in production and the various production processes that could be used to make a given part or assembly. This is an important concept and is discussed in greater detail in Section 7.

One area that was always somewhat of a subspecialty in industrial engineering but is much more so today and has even spawned another professional society (the American Production and Inventory Control Society, or APICS) is the subject of production and inventory control. The use of inventories as an integral part of the production sequence is a most important concept. If production could be created instantaneously, then no inventories would be required. However, as the time to produce is lengthened, the use of inventories becomes more important. Where alternative processes and methods are available, which have differing times associated with each process, the question concerns tradeoffs between slower, generally less expensive processes that require more investment in inventories versus less investment in inventories but higher-speed production, generally requiring more expensive machinery. The proper use and placement of inventories as part of the production strategy is crucial to the success of an enterprise. The proper identification of where in the production process inventory storage should be is part of this problem. Should inventory be only at the end of the process—at two or three stages back so the finishing steps must be undertaken only when orders are in hand? This has many economic implications and the proper resolution of questions such as these can have a significant bearing on the profitability and productivity of an enterprise. Part of this consideration, therefore, is also the question of material handling methods and the physical layout of the production facility.

Mention has also been made of the human engineering/human factors/ergonomics factor as an important subspecialty in industrial engineering, and this should be reemphasized. While there is no specific computer subgroup or orientation for it that can be identified, this is not because it is not an important area but rather because it is important to *all* subareas of industrial engineering. Anyone professing to be an industrial engineer in the environment of the 1980s has to be conversant not only with many computer software concepts in order to handle the calculations and simulations that are necessary for a broad range of problems, but also with some forms of hardware because of the necessity of designing systems to accommodate certain portions of the information chain and to have those systems respond to information that is generated by the processes of production, by the processes of sales and orders from external sources, and by the needs or demands for "services" that would be presented to a service organization.

There are many other reference points in the history of industrial engineering, but because of the necessity that this review be brief, we cannot discuss them here, believing that at least some of the major influences have been mentioned. For more information, consult the selected bibliography at the end of the chapter to expand on this brief coverage.

1.1.5 THE EXPANDING ROLE OF THE INDUSTRIAL ENGINEER

As has been suggested, the modern industrial engineer is no longer confined to industry. While the genesis of the activity was from industry and much of the early work was done in industry, this is no longer a limitation, so students or practitioners who are no longer motivated to an industrial career in the strict sense of the term need not look with dismay on industrial engineering as a limited field to be practiced only in the industrial and manufacturing industries. Because modern industrial engineering is based on a science of operations, then anywhere that "operations" are found that require systems of people, machines, and processes of some type, the industrial engineer has a natural outlet for his or her talents. As a result, the field of banking is one area that has been receiving the attention of industrial engineers during the last few years. Between the manual operations that must be performed, which in several areas are now being replaced by machines (computers and microprocessors), and the information flow and accuracy that are crucial to the banking business, this forms a natural outlet for the industrial engineer. The list of government agencies and activities into which industrial engineers have migrated is great indeed for the reason that the industrial engineer has found a home in the banking field. The objectives of government agencies may be different from those of the profit seeking sector of our economy because they tend to be less concerned with cost (rightly or wrongly) and more concerned with services to people and the political motivations behind their actions. Engineers designing systems for government agencies generally would have different parameters and objectives from those for industry, but they still have problems of productivity, inventories of services, information flow of human activity, so that the generalized methodologies of the industrial engineer can all be applied; only the coefficients of the generalized and abstract models need to be modified to accommodate the new conditions.

It would be both dangerous and short-sighted to attempt to enumerate all the possible outlets and applications of design methodologies for industrial engineering work because, on the one hand, an attempt at complete enumeration would inevitably leave out some that should be included and, on the other, would tend to foreclose future developments that would expand the field even further. Suffice it to say that the modern industrial engineering methodology will find use and application wherever the concepts found in the original definitions at the beginning of this chapter are found. This, then, would include those situations with a human parameter, but it would also include situations wherever various materials are used or where alternatives exist. Additionally, it would include problems in which systems of alternative energy sources or uses are found and where various options exist for the selection of equipment, processes, or other technical choices for accomplishing a given objective or set of objectives. As stated originally, the basic objective at all times for the industrial engineer is the design or redesign of a system regardless of its size that will improve the productivity of that system so that the consumption of the inputs or components that go into the system are used as efficiently, minimally, and expeditiously as possible. No facet of activity is, therefore, excluded from the application of industrial engineering analysis and design. Active industrial engineering programs are found in utilities, airlines, bus and freight transportation companies, hospitals and other health care facilities, banks, food and agricultural areas, and, of course, in the "hospitality" industry of hotels, motels, restaurants, and the like, to merely suggest the range of areas in which industrial engineers are currently active. Perhaps one of the areas in which industrial engineering activity has not been as prominent as it should be is in education where productivity is not a primary concern.

1.1.6 EDUCATIONAL OPPORTUNITIES IN INDUSTRIAL ENGINEERING

To conclude this brief introduction to industrial engineering and to this handbook, it seems appropriate to suggest what educational opportunities are available for the reader interested in further study. There are several modes of study that can be pursued and the proper path for any individual clearly would depend on such factors as age, previous education, current level of mathematical ability, experience, present location, and other related personal factors.

The most complete education would be that provided by the 4-year engineering college or university leading to a bachelor's degree in industrial engineering, or one of its closely related degree titles. There are at the present time (as listed in the 47th Annual Report of ECPD) a total of 78 industrial engineering or IE-related accredited undergraduate degree programs. In addition, there are other closely related programs such as computer and systems engineering that are not counted in the above totals. These programs are offered by schools in every section of the country, making them easily accessible to all, and include private schools as well as state supported institutions.

Outside of the United States, there are several schools teaching industrial engineering in Canada, which are accredited by the Canadian equivalent of the American accreditation process for engineers, in Mexico, and in other Latin and South American countries. Abroad, identifying available schools might be a little more difficult, although there are several appropriate degree programs in Western Europe. In Great Britain there are a great many comparable programs, and while the system of higher education and the types of certification (i.e., honors degrees, first and second degrees, diplomas) are somewhat different from those in the United States, there is a great diversity of educational opportunity that leads to the same type of overall training and follows roughly the same classification of schools. For any individual interested in study in Great Britain, the best source of information would be The Institution of Production Engineers.

In the United States there are 13 programs in IE technology accredited by ABET as well as a few others with closely related titles. These programs are generally of 4 years' duration, lead to a technology degree rather than an engineering degree, and do not require quite the depth of knowledge in the basic sciences. For those seeking careers as technicians or engineering assistants these programs should do an excellent job of preparation. For the 4-year schools of engineering leading to the accredited baccalaureate degree there are many that give advanced training and degrees that terminate with the Doctor of Engineering or the Ph.D. The former is the most advanced academic degree for the practitioner while the latter is the most advanced degree for those embarking on a career in research, generally with an academic career in mind. Inquiry to any of the undergraduate schools or to ABET would provide additional information about the educational opportunities and the types of programs in which the several schools tend to specialize in their graduate and advanced degree programs.

Differing substantially from the formal degree programs are a great many that can generally be called "short courses" or "continuing education programs," which are sponsored by different groups or organizations. Probably the best overall source of such programs would be the list of professional societies at the end of this chapter. Professional society mailing lists are frequently used to advertise such programs, so the societies know what programs are planned, where they are given, and for what type of audience they are intended. In addition, several of the professional societies themselves sponsor programs and publish monographs dealing with specialized topics. Active participation and membership in one or more of the professional societies would be an important step, therefore, for one who is interested in pursuing or enlarging a career in the field of industrial engineering. Another source of the short course type of program is the private consulting firm. Such firms have built up mailing lists over the years, and their courses are known to most of the major industrial firms as well as to the professional societies. Such programs are frequently given in hotels in or near the major metropolitan centers, where there is a reasonably large industrial population. These short courses vary considerably in their content, coverage, and level of sophistication with which they discuss the subjects covered. Caution should be exercised, therefore, before enrolling in such programs. The duration of these programs can also vary considerably, most frequently running 1 or sometimes 2 days. Some, however, will run longer, up to a week. In the case of some of the better and more well-known courses, there is a concentrated 4- to 6-week in-depth exposure. All of these programs fit varying needs, are distinct contributions to the profession, and are especially helpful to the practitioner who is interested in keeping up with the latest developments in the field in order to maintain his or her own skills at a maximum level.

Another but not the least method of enhancing one's education in the field is by attending the annual (or more frequent and many times specialized) conferences of the professional societies. Generally speaking, annual meetings contain a broad range of technical discussions that emphasize recent applications and installations of complex methodologies or unique solutions. In addition to annual meetings, which give rather broad topical coverage, there are also specialized conferences of 2 or 3 days, where the focus is more apt to be directed toward one or two related topics. These society programs are advertised regularly to members and the fees to those with society membership are reduced from those fees charged to nonmembers.

An alphabetical listing of professional societies that specialize in industrial engineering activities, or in some distinct branch of IE, follows.

Accreditation Board for Engineering and Technology (formerly ECPD)
345 East 47th Street
New York, NY 10017

American Institute of Industrial Engineers
25 Technology Park/Atlanta
Norcross, GA 30092

American Production and Inventory Control Society
Watergate Building, Suite 504
2600 Virginia Ave., NW
Washington, DC 20037

American Society for Engineering Education
One Dupont Circle, Suite 400
Washington, DC 20036

American Society of Mechanical Engineers
345 East 47th Street
New York, NY 10017

American Society for Quality Control
161 W. Wisconsin Ave.
Milwaukee, WI 53203

Council for National Academic Awards (Great Britain)
344–354 Gray's Inn Road
London, England WC 1X 8BP

Human Factors Society
P.O. Box 1369
Santa Monica, CA 90406

The Institute of Management Sciences
146 Westminster Street
Providence, RI 02903

Institution of Production Engineers (Great Britain)
Rochester House
66 Little Ealing Land
London, England W5 4XX

International Ergonomics Association (Great Britain)
Five Lyncroft Gardens
Houslow, Middlesex, England TW3 2QT

International Material Management Society
3310 Bardaville Drive
Lansing, MI 48906

Method Time Measurement Association for Standards and Research
9–10 Saddle River Road
Fair Lawn, NJ 07410

Operations Research Society of America
428 E. Preston Street
Baltimore, MD 21202

Society of Manufacturing Engineers
P.O. Box 930
Dearborn, MI 48128

BIBLIOGRAPHY

The bibliographic references that follow are not exhaustive but rather are a few of the principal sources that will lead interested readers to other relevant material dealing with the history and development of industrial engineering and its transition from an empirical to a science-based discipline.

GILBRETH, FRANK B., *Motion Study*, D. Van Nostrand Co., New York, 1911.

HICKS, PHILIP E., *Introduction to Industrial Engineering and Management Science*, McGraw-Hill, New York, 1977.

HOXIE, ROBERT FRANKLIN, *Scientific Management and Labor*, D. Appleton & Co., New York, 1915.

PIKE, E. ROYSTON, *Human Documents of the Industrial Revolution*, George Allen & Unwin Ltd., London, 1966.

RITCHEY, JOHN A., *Classics in Industrial Engineering*, Prairie Publishing Co., Delphi, Indiana, 1964.

SCHULTZ, ANDREW Jr., "The Quiet Revolution," *Engineering, The Cornell Quarterly*, Vol. 4, No. 4 (1970), pp. 2–10.

TAYLOR, FREDERICK W., *Shop Management*, Harper & Row, New York, 1947.

CHAPTER 1.2

Organization and Administration of Industrial Engineering

O. J. FEORENE

Eastman Kodak Company

1.2.1 INTRODUCTION

An effective industrial engineering organization must be responsive to the specific needs of the organization it serves. It must demonstrate the capability of delivering a competent level of professional and technical support, often uniquely developed to suit these needs. As the needs of the client organization change so must there be a corresponding change in the way the industrial engineering function is organized and in the services offered.

It should be evident then that line management needs specialized help and it should also be evident that the nature of this help differs between enterprises. Functions or specialized services effective in one organization may be dysfunctional or unsuitable in another. As a result, a variety of industrial engineering functions and a diversity of organizations have taken root in business and industry. The names used to identify these activities add up to a lengthy list of descriptive and sometimes confusing titles.

Yet beneath all this apparent nonuniformity there is a common preoccupation, a single-minded attentiveness to methods of improving the effectiveness of management in utilizing the resources at their disposal, in the development of alternatives, and in the optimization of the decision process.

All leaders, all managers have limited resources to meet their goals. The continuous search for the most effective use of people, materials, equipment, facilities, time, money, and information has evolved a rationale for a disciplined process and the emergence of a unique body of knowledge. It is the implementation of this discipline and the utilization of this knowledge that characterize the role of industrial engineering in an organization.

What all managers of industrial engineering have in common are the problems associated with planning, organizing, directing, and controlling the technical resources assigned to them. This common concern transcends the differences in organizational structures and responsibilities and allows an examination of the administrative aspects of managing the industrial engineering effort.

This chapter will examine those activities that all industrial engineering managers must recognize in carrying out their responsibilities, regardless of the number of industrial engineers employed or the nature of the products or services offered by the parent organization.

1.2.2 ORGANIZING THE INDUSTRIAL ENGINEERING DEPARTMENT

A clear sense of purpose is an essential factor in a review or restructuring of the industrial engineering organization. The mission of the organization should be clearly described in terms as free of ambiguity as possible. The functions should be understood and accepted by general management as well as by the client organizations. If an industrial engineering staff is to be effective, it should continuously reexamine and reassess the services it offers to its clients. In the absence of a self-imposed critique, even an older organization with a long and successful history of work measurement and methods improvement can experience an erosion in its support role.

There are a number of ways of establishing an industrial engineering staff where none currently exists. One procedure, certainly not uncommon, is an announcement by top management that an industrial engineering staff is to be formed and an outline of the functions and responsibilities to be assigned to the new organization. The announcement may be preceded by a study conducted by an outside consultant or based on visits to other companies. It may well be simply the result of management concerns regarding the need for a concerted staff capability to assist in the management of improvement throughout the company.

Another frequently used procedure is the appointment of a senior executive to a newly created position of manager of an industrial engineering function, leaving to the new manager the responsibility of developing and proposing the specific activities and the role of the new organization. In the absence of a qualified executive, some firms have recruited experienced managers of industrial engineering from outside the company. On occasion these new managers surround themselves with former associates as a nucleus for the new industrial engineering organization.

In either event, it is not unlikely that the new manager will draw heavily on a hand picked cadre of competent engineers and analysts as assistants in defining and specifying the new role that has been assigned. Some quite successful industrial engineering organizations have come into being through such efforts.

It would seem, however, that the existing staff organization should not be the sole determinant of the role expected of the staff in an enterprise. Poor line/staff relations can often be traced to the unilateral behavior of either partner-to-be in a new joint venture. The involvement and commitment of the client in the development of the industrial engineering role establishes and augments the interdependence of line and staff. A shared responsibility in developing and establishing the new function should minimize the possibility of future misunderstandings or ambiguities regarding the role of the industrial engineering organization.

It is, therefore, highly desirable to include in the study team representatives from general management, the client organizations, as well as senior personnel from the new industrial engineering organization. It is also desirable that the study team investigate as many other industrial engineering organizations as possible to avoid a narrow definition and to take full advantage of the experiences of other firms regardless of where they are in the world. Peter Drucker, in the Preface to his text on management, states:

> *I have particularly stressed the Japanese experience—not only because far too few managers in the West understand Japanese management and organization, but also because an understanding of the often very different ways in which Japan, the only non-Western developed country, tackles a common task (e.g. the determination of profitability, the organization of work and workers, or the making of decisions) may help the Western manager to understand better what he himself is trying to do. The basic conviction of this book throughout is that each country's managers can and need to learn from the best others have to offer.*[1]

We need only to look at the remarkable improvements in productivity introduced by many companies in Japan and West Germany to test the efficacy of this method. Study teams were organized in these countries comprised of a cross section of management somewhat along the lines just discussed. These teams spent years visiting industrial engineering organizations in many of the leading American companies during the 1950s and 1960s. They were good students, keen observers, and learned their lessons well. Many of these companies are today worldwide leaders in their industries and have much to teach American companies in the effective application of industrial engineering concepts and methodologies.

1.2.3 TYPICAL FUNCTIONS

Regardless of the size of the parent organization or even the type of business or industry in which the firm is primarily engaged, the industrial engineering staffs in the United States and in a number of other countries tend to have a number of functions in common. Almost all manufacturing organizations have staff activities or services dealing with work measurement, methods engineering, and facilities planning.

An unpublished survey conducted in 1979, involving 27 companies represented on the Council of Industrial Engineering of the American Institute of Industrial Engineers, indicated that most of the industrial engineering managers were responsible for the following functions:

Facilities planning and design.

Methods engineering.

Work systems design.

Production engineering.

Management information and control systems.

Organization analysis and design.

Economic analysis.

Operations research.

Work measurement.
Wage administration.
Quality assurance.

The second most frequently mentioned group of functions appeared to more uniquely satisfy the needs or goals of specific firms:

Project management and support.
Cost controls and standards.
Inventory controls.
Energy conservation.
Computerized process controls.
Product packaging, handling, and testing.
Tool and equipment selection.
Production control.
Product improvement studies.
Preventive maintenance programs.

Some of these activities are extensions of techniques or services emerging from the more traditional industrial engineering specialties. In total they reflect the dynamic aspects of contemporary industrial engineering practice.

A third less frequently reported group of activities by these same companies further supports the trend toward a widening horizon of management needs:

Profit planning.
Capital program analysis.
Distribution systems.
Consulting services to suppliers.
Evaluation of prospective suppliers.
Management audits and operations.
Reviews.
Safety programs.
Training programs.

The companies participating in this survey ranged in size from a few thousand employees to some of the largest employers in the industrial world. Some manufactured and assembled consumer goods, some provided services, while others produced basic commodities.

It is particularly significant to note that industrial engineering practice in most of these companies had its start in the shop floors and factories. At the present time, most provide services to other organizations within the firm such as marketing, distribution, finance, research, legal and patent departments, industrial relations–in short, most if not all of the component units making up a company or institution.

The extension of industrial engineering into other segments of the economy is reflected in the industrial engineering representatives from banks, hospitals, military organizations, and state and federal agencies who each year attend professional conferences and seminars in ever increasing numbers. The growth in the service sector is a contributing factor to the changing role of industrial engineers and to the variety of functions carried out by them. The exchange of experiences between industrial engineers in both business and service sectors has hastened the introduction of improved analytical techniques. As a result, management in general has benefited by the expansion of industrial engineering into new fields of endeavor.

A comprehensive survey conducted by Neville Harris of industrial engineering practice in 667 firms in the United Kingdom reveals a strongly similar pattern of functional responsibilities.[2] The Harris survey examines the popularity of the techniques that were used frequently and regularly. Exhibit 1.2.1 is a popularity table prepared by Harris of the specialized activities reported by 401 of the 667 companies. Of these 401 companies, 127 had additional specialized functions as shown in Exhibit 1.2.2. All of the 667 respondents indicated the degree to which each used 32 techniques provided in the survey. Exhibit 1.2.3 is a ranking of these management techniques. It indicates, for example, that 87% or 580 companies made use of methods studies. The wide variety of specialties and management techniques reported by Harris appears to reflect the pattern seen in the United States.

Exhibit 1.2.1 Popularity Table of Specialisms

Specialism	No.	%
Work study	336	84
Organization and methods	287	72
Remuneration systems	181	38
Computers	143	35
Systems analysis	129	32
Statistics	115	29
Project network techniques	109	27
Production engineering	98	24
Operational research	85	21
Value analysis	85	21
Ergonomics	37	9
Method study	23	6
Work measurement	26	6
	401	

Exhibit 1.2.2 Nonstandard Divisions, Units, Sections

Major Groupings	Examples	Total No.
Personnel/industrial relations	Communication, personnel, industrial relations, training	59
Office/administrative services	Typing, telecommunications, printing, reprography, office services	39
Production/manufacture	Production control, quality, stock control, stores, technical support	35
Costing/finance	Cost estimating, internal audit	18
Computers/systems	Manual systems, business methods, programming, data processing, operating manuals, management information science	16
Layout/materials handling	Factory and office layout materials handling, shipping services	14
Long-range planning Corporate planning	Corporate planning, long-range capacity planning, new product planning	11
Research/intelligence	Research and development, information	6
Sales/marketing	Market research, quotations	5
Purchasing	Office equipment purchasing, central purchasing	3
Miscellaneous	Organization and business consultancy, water usage, special projects, management by objectives	12

In 1977, W. A. Reynolds and M. K. Cheung reported the results of a pilot survey of 20 plants of the use of industrial engineering techniques in Hong Kong.[3] Their analysis indicates that layout and materials handling techniques, work study, plant and manufacturing engineering, and production and quality control were the most widely used techniques. Some of the other techniques reported in use are incentive pay, systems analysis, forecasting, value engineering, critical path analysis, and linear programming. The wide range of activities that characterize industrial engineering in the West is again evident in the Reynolds and Cheung sample.

1.2.4 CHANGES IN INDUSTRIAL ENGINEERING FUNCTIONS

There have been significant and important changes during the past 30 years in the functions normally assigned to the industrial engineering staff. These changes have been molded by the growing body of knowledge about the behavioral aspects of work design, an intensification in the application of quantitative analytical methods, and the maturation of computer technology. The emergence of an increasingly sophisticated managerial clientele seeking more powerful tools to deal with

Exhibit 1.2.3 Popularity Table of Management Techniques, Indicating Percentage of Respondents using the Techniques Regularly

Technique	Rank Order	(%) Respondents
Method study	1	87
Work measurement (direct)	2	79
Incentive application	3	71
Layout studies	4	66
Design of forms	5	66
Materials handling problems	6	58
Development of information systems	7	58
Cost-benefit analysis	8	56
Work measurement (indirect)	9	51
Choice of materials handling equipment	10	46
Organization studies	11	43
Job evaulation	12	42
Choice of office equipment	13	41
Management development	14	38
Systems analysis	15	33
Inventory/stock control analysis	16	31
Computer programming	17	26
Use of networks for project control	18	26
Use of networks for planning	19	25
Work measurement on the office	20	23
Motion economy	21	21
Management by objectives	22	21
Value analysis	23	19
Use of networks for resource allocation	24	15
Ergonomics	25	12
Group technology	26	12
Hazard and operability studies	27	12
Simulation	28	12
Photographic/filming	29	7
Linear programming	30	7
Queuing	31	6
Risk analysis	32	6

complex alternatives has in itself been a prime factor in the restructuring of the industrial engineering staff.

Another powerful force in shaping the changing role of the industrial engineer is the widening scope of activities that brought the industrial engineer out of the factory into such nonmanufacturing organizations as marketing, distribution, finance, and product development. Within industry the practice of industrial engineering has become a pervasive influence in the search for improvements in productivity.

The forces of change have brought about a reassessment of the efficacy and value of the traditional methods and techniques used by industrial engineers in the analysis and resolution of management problems. Some work measurement techniques have changed or disappeared completely, while simulations and systems analyses have become potent weapons in the armory of productivity. The subdivision of work into smaller, more manageable units as an optimization technique has been challenged by systems engineering concepts that pursue the computerized integration of operations as a more optimal strategy.

As the American economy passes through the portals separating the Industrial Age from a new world dominated by service organizations and knowledge workers, the industrial engineer has kept pace and is today found assisting managers in merchandising, social, health care, banking, and government institutions.

Whatever form the industrial engineering organization takes, it must be one that lends itself to continuous change in its functions, its client relations, and in its primary mission. The staff support and capability must reflect the needs and interests of the client organization. It has been suggested that the ability to adapt to change is a gross measure of intelligence in most life forms. A parallel may well be found in the life cycle of a staff organization.

There are a number of steps that can be taken to insure that the industrial engineering organization is sensitive to the need for adapting to changing needs and technologies. A simple and yet effective move is to assign to one or more persons the responsibility for staying abreast of new techniques or methodologies in the functions charged to the industrial engineering organization. Larger industrial engineering staffs can establish a technology center or a corps of specialists with this same responsibility. What is basic is not the size of the company or the number of industrial engineers involved but recognition of the need for establishing a life-renewing flow of technical knowledge if the staff is to avoid atrophy. If left to chance or to the initiative of individual engineers, there is a very real possibility that the pressures of daily responsibilities will take precedence. The process of holding someone accountable for new knowledge, even on a part-time basis, provides a foundation for insuring the devotion of a minimum amount of time in maintaining technical competence.

Scheduling seminars and conferences for the industrial engineering staff and for general management on relevant techniques is another way of keeping abreast of new knowledge. These conferences can be low-budget activities utilizing internal resources and can be held on site. There are a number of qualified consultants who specialize in the preparation of such sessions and almost all engineering colleges offer a variety of such programs through their extension programs. In many instances there is some value in having these programs conducted on site, although "technical retreats" held off site appear to be gaining in popularity. Once again, what is important to the industrial engineering manager is not the format or locale of such programs but the realization that technical proficiency must be assiduously pursued and that there is a planned effort to insure that the organization is keeping up with new concepts and techniques.

Attendance at local and national meetings of technical and professional societies should be planned and budgeted as carefully as other training requirements. There has been an impressive increase in the number of meetings conducted by such professional societies as the American Institute of Industrial Engineers and The Institute of Management Sciences. These meetings are generally open to the public and held at a number of different places. The level of quality of the presentations tends to be quite high in keeping with the high professional standards of the societies. These programs facilitate the acquisition of new knowledge plus fostering an exchange of experiences among the attendees. Needless to say, active participation in the programs of the professional societies is also a strong stimulus in developing professional competency. These are but some of the steps that should be considered to insure that the functional responsibilities of the industrial engineering organization continues to be sustained by concepts and techniques that reflect the best of current experience and research.

1.2.5 ORGANIZING THE INDUSTRIAL ENGINEERING DEPARTMENT

There are a number of organizational formats that have been adopted by industrial engineering organizations both in the United States and abroad. The variety seems to be as extensive as the diversity in functions. Nor does there appear to be a distinct pattern as to the placement of the function within the firm or enterprise. In manufacturing organizations, where there has been a longer history of industrial engineering work and perhaps more commonality in the way workshops are organized, the industrial engineering manager usually reports to the plant manager or whoever is responsible for operations. This relationship reflects the traditional early employment of industrial engineers in improving the productivity of factory operations.

Knut Holt undertook a study of industrial engineering practice in seven European countries for the purpose of comparing European and American organizations.[4] His report suggests that administrative and organizational functions are essentially similar. Most of the companies participating in his survey had centralized industrial engineering departments, although work was organized in different ways; some companies had organized their industrial engineering departments by specialized functional groups, others by the operating departments served.

In many medium-sized companies, and perhaps more frequently in larger manufacturing firms, there usually is a small headquarters group with an advisory or overseeing role, while the majority of the industrial engineers are on the staffs of the managers of the operating units. The headquarters group not infrequently provides a professional recruiting and training service for the branch organizations as well as undertaking or directing company-wide studies affecting a number of the company units. A headquarters office can play an important role in managing the career development plans of all the industrial engineers and provide a brokerage function in expediting the transfer and reassignment of technical personnel.

As a member of the top management team the headquarters director of industrial engineering is in a good position to develop and disseminate company policies affecting industrial engineering practice throughout the company and on occasion to undertake special studies for company officers. Quite often the establishment of work measurement programs, premium pay plans, conditions of work, and other management policies of concern to the company as a whole emanate from the headquarters office for implementation throughout all the company units. Exhibit 1.2.4 schematically depicts, in a somewhat simplified manner, the relationship between the headquarters group and the other industrial engineering organizations within the company.

Exhibit 1.2.4 Relationship between Headquarters Group and Centralized Industrial Engineering Organization

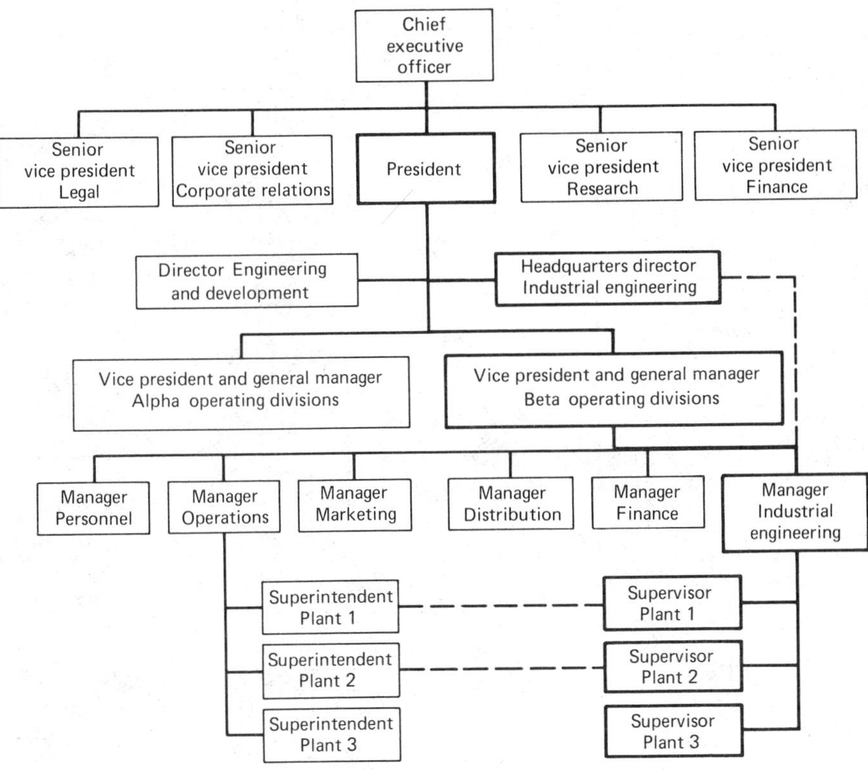

In this arrangement the industrial engineers are members of the centralized divisional staff. However, most of the engineers are assigned and perhaps even physically located within the client area. The fluctuating needs of the smaller plant managers are absorbed more readily by the centralized divisional industrial engineering staff. An annual estimate of technical assistance is negotiated by the plant manager with the manager of industrial engineering. The engineers assigned serve as industrial engineering generalists to the plant manager. This format encourages a close identity with the goals and objectives of the client and facilitates communication between the line and staff organizations. The industrial engineers are usually treated as members of the client organization and enjoy the benefits of a friendly environment, while retaining the objectivity of a third party in the resolution of problems.

Some companies employ a decentralized structure in which the industrial engineers are permanently assigned to the lowest level of managerial responsibility within the manufacturing organization. These engineers make up the staff under the direction of the local manager. On occasion, a dotted line relationship is retained with a second-level manager of industrial engineering who in turn may hold a similar relationship with the headquarters or corporate director of industrial engineering. Such an arrangement insures that there is a communication channel for the uniform interpretation of company policies and practice—at the same time permitting local control over the technical staff. Exhibit 1.2.5 outlines schematically the communication and authority channels of such a format.

There is another organizational option open to industrial engineering departments consisting of more than 10 or 12 engineers. This alternative is based on the formation of specialty groups within the industrial engineering department. A technical specialty is assigned to one or more engineers who, by dint of skill, experience, or personal interest, have elected to become experts in a specific technology. One group might have a materials handling responsibility, for instance, another group might consist of work measurement specialists, a third may specialize in the behavioral sciences, and so on. These specialists are then assigned to study teams on the basis of the specific skills required by the problem at hand. A parallel exists in the construction industry where a building contractor schedules excavators, masons, plumbers, or electricians in the order required to fabricate the building. (See Exhibit 1.2.6.)

Exhibit 1.2.5 Decentralized Industrial Engineering Staff

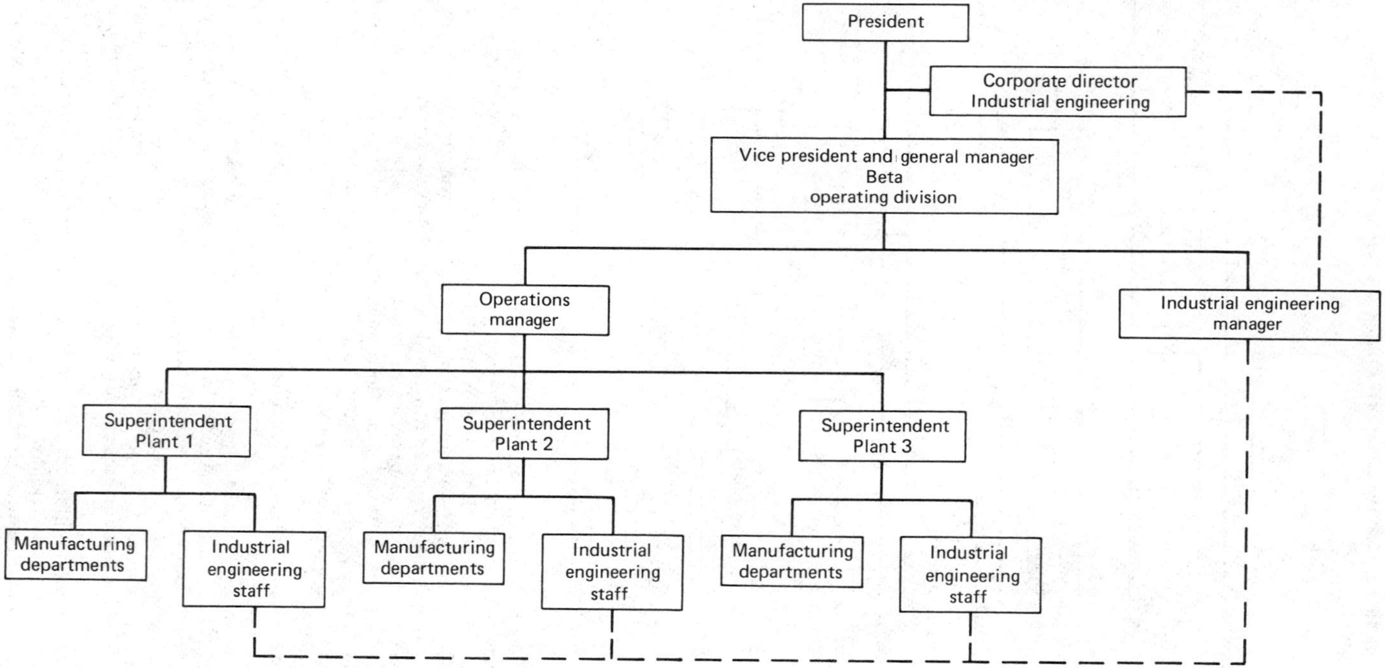

Exhibit 1.2.6 Industrial Engineering Department Organized by Specialized Functions

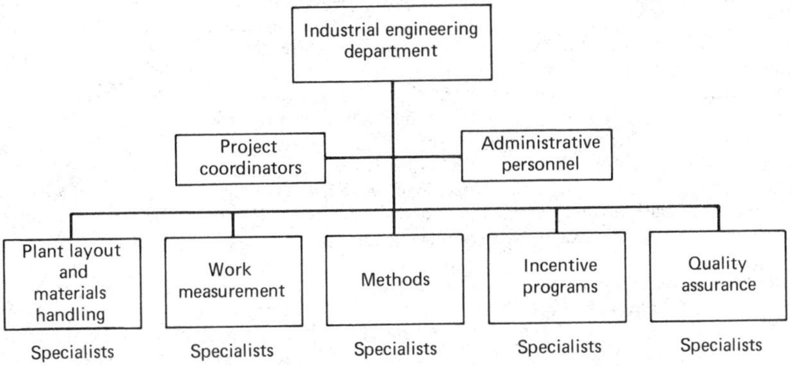

Industrial engineering department

Project coordinators

Administrative personnel

Plant layout and materials handling

Work measurement

Methods

Incentive programs

Quality assurance

Specialists

Specialists

Specialists

Specialists

Specialists

This type of organization relies on a project coordinator (the general contractor) to effectively program and utilize the necessary technical skills. Such a coordinator needs to be well versed in the technologies available. If project coordinators are not available from the industrial engineering department, the client will have to orchestrate the industrial engineering skills. This may not be a reasonable alternative in light of the overall technical knowledge required of the coordinator. In addition, the coordination of a major project can put heavy demands on the time of the project leader. Client personnel capable of managing a project must give up their line responsibilities if they are to undertake this assignment. Part-time attention to the project and to normal line duties could mean less than satisfactory performance in either situation.

Companies with fairly large staff departments have tried a number of organizational patterns that best suit their unique requirements. Considerable attention has been given to a matrix (or team) type of industrial engineering organization that tries to capture the best features of these variations (Exhibit 1.2.7). In this type of structure, one or more engineers are assigned as generalists to a specific department, plant, or division. Usually the engineers are members of a central industrial engineering staff, although they might be physically located in the operations area. At any given moment in time, the studies underway in a specific client area could have needs for a certain mix-

Exhibit 1.2.7 Matrix Type of Industrial Engineering Organization

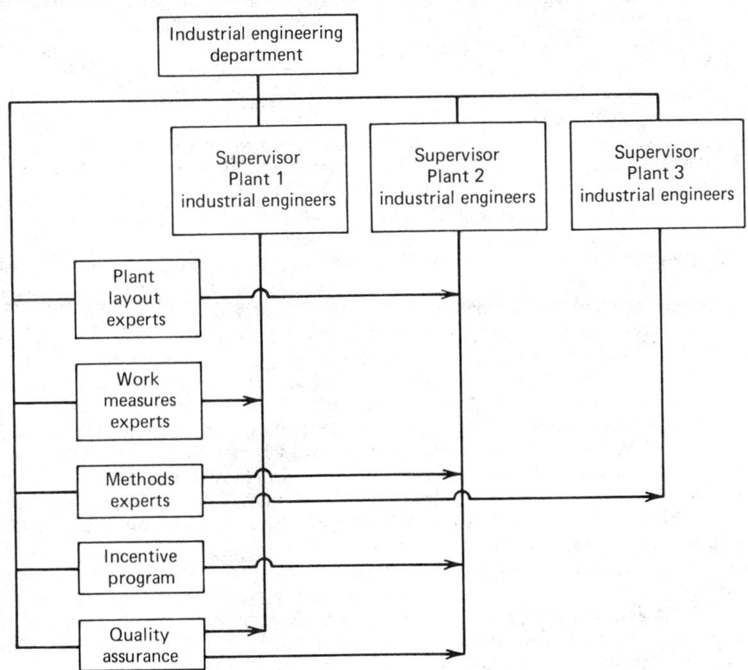

ture of specialized assistance. Exhibit 1.2.7 illustrates one such moment. In the example shown, the industrial engineering studies in Plant 1 require the services of an expert in work measurement and assistance from the quality assurance group. These specialists are in addition to the regular work force of industrial engineers. At the same moment, the supervisor at Plant 2 needs plant layout, methods, incentive, and quality assurance support. Plant 3 requires only some additional methods assistance at that particular moment.

The supervisors act as general contractors and subcontract the necessary support they need from the group of experts. The supervisor limits his contracts to the expert service and engineering time needed to get the job done. This keeps the supervisor's own group at a low staffing level and yet allows the size to fluctuate with the work load. Admittedly, a matrix structure places unusual demands on both the individual engineers and on supervision. However, the increasing number of organizations relying on matrix structures in managing projects attests to the effectiveness of this kind of organization.

The matrix approach has an additional feature in its favor: It appears to support more easily a sustained effort in keeping abreast of new concepts and technologies. The experts tend to read and publish in their own specialized fields of interest. They tend to stay familiar not only with trends but with other experts in and out of their own companies. Other engineers and clients gravitate to these experts on the strength of their knowledge and experience. They are natural leaders for seminars, workshops, and conferences. They are excellent consultants or advisors and make good training instructors. All of these positive features exist in some degree in virtually any organization. However, a matrix structure seems to foster a collaborative, mutually supportive environment that can be highly productive where a number of disciplines need to be focused on a problem.

It should be noted that a growing number of companies are combining industrial engineering, management information systems, and operations research departments into one centrally managed technical corps. The services of these consolidated groups are usually available on a fee-for-service basis to all levels of management. These staff consolidations appear to be of particular value in smaller companies where it is more difficult to acquire a critical mass of technical talent. The widespread acceptance of total systems analysis concepts that draw upon the combined knowledge of a number of disciplines has contributed to the desirability of pooling technical talent wherever possible.

Although in many companies operations managers still direct the industrial engineering function, the growth of centralized and consolidated staff groups will probably bring about changes in the more traditional line/staff relations. The 1979 survey made of the members of the Council on Industrial Engineering of the AIIE noted earlier reflects a growing interest on the part of company management in establishing internal consulting organizations offering a broad array of professional and technical services. Words other than "industrial engineering" are being chosen to better describe these much broader roles. Yet in almost every case in this relatively small sample, the industrial engineering activities remain at the core of the work undertaken by these departments. The department titles selected by these companies give some indication of the intent behind the words: management services, managament systems, corporate consulting services, manufacturing consulting division, corporate services, general manufacturing services, productivity services, manufacturing consultants, and corporate consulting services.

The Harris study of management services in the United Kingdom[2] also reveals a variety of departmental titles used by 104 out of the 667 companies and, like the smaller American survey, they appear to describe a greater scope of responsibility than the traditional industrial engineering functions. A few of the titles are management and computer systems, productivity and systems services, and organization and productivity services.

The increased responsibilities being assigned to the industrial engineering organization is further disclosed by the management level direction of the functions. In at least one medium-sized North American company the industrial engineering manager reports directly to the chief executive officer. This assignment was made because of the high level of concern and attention that this chief executive officer believed should be directed at the improvement of productivity throughout the entire corporation. Other members of top management in the United States and Canada having direct responsibilities for the industrial engineering function are president, vice president of operations, vice president and general manager, vice president of corporate services, group vice president of corporate technology and venture management, and vice president of manufacturing. Some other members of management at a lower level within the firm having responsibility for the industrial engineering function are director of planning, general manufacturing services manager, plant operations staff, director of administration, manager of manufacturing consulting division, and manager of manufacturing engineering.

It is evident that industrial engineering organizational structures are uniquely designed to suit the needs of the parent organization. Certainly one starting point is a definition of the purpose and mission of the industrial engineering organization. Picking someone else's model may not prove to be the best fit in all cases; yet an awareness and understanding of organizational practices in other firms may stimulate and promote beneficial modifications that might not otherwise be considered. Industrial engineering organizations should be designed with flexibility in mind in order to adapt to the changes taking place in businesss and industry.

1.2.6 GOAL SETTING PROCESSES

The reasons for establishing a new industrial engineering organization or perpetuating an existing function should be developed through a group process involving line and staff participants. Setting goals for a staff organization without client and management participation is destined to be a sterile process that will not bear fruit. Active client involvement tends to assure a high level of cooperation and commitment in both the development and effective utilization of the staff service.

The explicit directives and the implicit behavior of top management and senior officers of the enterprise describe the goals of the organization as a whole. In order to effectively respond to these goals, all levels of management down to and including the first-level supervisor must develop a set of objectives consistent with these overall goals. In the effort to attain these objectives, managers must have access to specific technical skills and resources. The functional activities available from the industrial engineering group should match these requirements.

In a completely new environment, a series of workshops may be held to develop a mutual understanding of objectives, problems, and anticipated technical support. As suggested, the principals at these workshops should be representatives from all levels of concerned management as well as the senior industrial engineering representative. These discussions should produce a set of requirements that the industrial engineering manager can use as a guide in organizing the functions of his department.

In established organizations, a continuous dialogue involving the principals who make up the organizational hierarchy of the enterprise is clearly indicated if goals, objectives, and resources are to be consonant. Regularly scheduled operations planning or strategy meetings, for example, materially assist in focusing the efforts of line and staff as a team upon meeting the goals of the enterprise. The exclusion of staff from such councils in the belief that the staff role is secondary and that staff should respond only when called augers for poor or even ineffectual support from staff groups.

The communications that most effectively weld line and staff into a strong team are those that take place at every interface level: the engineer's relations with the client, the industrial engineering supervisor's daily contacts with middle management, and the industrial engineering manager's attendance at and participation in management councils and committee meetings. The needs of the clients are continuously changed by shifting priorities, new projects, unexpected problems, and the pressures of the marketplace. The satisfaction of these changing needs must dictate the internal goals and objectives of the industrial engineering organization. In the absence of client participation, a unilateral posture on the part of the industrial engineering manager could easily generate a mismatch between line needs and staff support services.

There is a tendency on the part of some technical staffs to withdraw into an isolated sphere of behavior in the belief that their specialized knowledge can best be managed and employed only by technical experts. This aloofness acts as a wall between effective line/staff relationships and unless destroyed may prove to be insurmountable.

The group process establishes the commonality of concerns supporting the creation of a new industrial engineering staff or the modifications required to improve the effectiveness of an existing staff. A firm foundation is formed thereby for assessing existing skills and techniques. Plans can be formulated for the identification and implementation of training and professional development programs. The acquisition of new techniques or methodologies generating from this joint effort are directed at improving the effectiveness of the client served rather than in satisfying technical competence as an end in itself. In addition, the professional profile of the industrial engineering group will not be a carbon copy of some other organization but rather one that uniquely serves the parent company.

1.2.7 SELECTING INDUSTRIAL ENGINEERING SUPERVISORS AND MANAGERS

The highest possible qualifications should be sought in the selection and appointment of the manager of the industrial engineering function. This person should meet the same high standards as those set for the peer group of managers with whom the industrial engineering manager is expected to relate. At stake is not only the future level of technical competency of the industrial engineering staff but, just as important, the vitality of the line/staff relationship.

A manager holding a degree in industrial engineering from any of the colleges or universities with a well-recognized academic program should be well-qualified in terms of the technical knowledge required to carry out his or her responsibilities. A degree in any of the engineering disciplines, along with some training or experience in business administration, has proven to be an excellent background for a number of industrial engineering managers in industrial organizations. People holding graduate level degrees in industrial engineering or in the management sciences should be ideal candidates for promotion to positions of responsibility within industrial engineering organizations in almost any kind of firm or institution.

In the selection of a manager, consideration should be given to a person with training or experience in the computer sciences in addition to an industrial engineering background. The pervasiveness of computerized operations, whether in the office, the laboratory, or the factory, has introduced new and higher levels of complexity to the design of work.

The search for leaders should not overlook production managers with good technical or engineering backgrounds. The combination of practical experience and technical training is highly desirable and proficiency in both the line and staff environments should prove to be of real value for a future member of top management.

As stated earlier, the manager of the staff function should be as technically and professionally groomed as the managers in the line organization. The same special training or management development opportunities should be available to all managers regardless of responsibility. There can be no mutual respect or credible acceptance if the staff manager is intellectually or professionally incapable of keeping up with his or her counterparts in the line organization.

Depending on the experience and education of the person selected to be manager of industrial engineering, managerial effectiveness can be considerably enhanced through attendance at educational conferences and seminars conducted by professional societies and colleges of engineering. Even the smallest of organizations can take advantage of the many short courses directed at improving the manager's ability to manage his or her function. In addition, there is available today a veritable wealth of information in the industrial engineering journals literature, texts are easily available, and a good technical reference library is within everyone's means. However, training courses and technical libraries are, at best, supplements that may make a good leader better but by themselves will not produce managers. The identification and selection of the kind of industrial engineering manager best suited for a particular firm has to follow the same dictates that govern the selection of any other manager in that organization. The staff should not be a refuge for second-class managers.

The industrial engineering manager must have, as assistants, supervisors who are also chosen from the best technical talent available. Usually supervisors are selected from the professional and technical ranks of the industrial engineering organization. This would be the natural promotional path for the outstanding engineers in the department. However, in order to get the best possible leaders, it may be necessary to recruit from other technical organizations within the company or elsewhere.

Good supervisors may be born but every industrial engineering manager should establish a well-planned training program for developing leaders. Potential leadership skills should be identified and strengthened as early as possible. Smaller companies may wish to utilize programs developed by consultants or other outside agencies to prepare their outstanding engineers for increased responsibilities. Larger firms can draw upon their more extensive training resources. The primary concern of all industrial engineering managers should be the creation of an in-depth progression ladder to insure that leaders are being identified and prepared.

1.2.8 RECRUITMENT AND PLACEMENT OF PROFESSIONAL AND TECHNICAL PERSONNEL

Policies regarding the recruitment and placement of professional and technical personnel are usually well established in most firms. Some organizations have a central office responsible for the recruitment of all technical personnel while others leave this responsibility with the local manager. Regardless of the practice followed, the mechanics of hiring industrial engineering personnel should be occasionally reviewed to insure that the process fulfills the need for qualified personnel.

Recruitment at colleges and universities for most of the newly hired engineers is practiced by many companies. Future misplacement problems develop at the outset of such programs if the recruiters do not have a reasonably good understanding of the practice of industrial engineering. Too often, in the interest of economy, the recruiter interviews a number of college graduates in a variety of technical disciplines while on campus. It is unrealistic to expect a recruiter to have a sound understanding of every technical position within any given company. As pointed out earlier, the practice of industrial engineering varies widely in the United States and an applicant, after interviewing with a number of companies on campus, may be confused by the range of activities and opportunities that are available. Experienced recruiters can usually help the applicant in determining whether the initial interview should be followed by an invitation for a more intensive interview in the industrial engineering department. To assist in this process, efforts are usually made to educate the full-time recruiter as thoroughly as possible in current industrial engineering practice. Some personnel managers borrow industrial engineers and train them to be part-time recruiters for their departments. Still others send out teams made up of engineers and recruiters in order to improve the initial screening process on campus. In general, the campus interviews are, at best, considered to be just the first of a series of steps in the process of evaluating the qualifications of an applicant. The really critical evaluations take place when the applicant visits the industrial engineering department. For this reason these visits must be well planned in advance so that the agenda for each applicant is consistent with his or her academic background and work experience. The following composite of a format followed by a number of companies in interviewing college graduates describes this process in a little more detail.

A host is usually selected to meet and shepherd the applicant during the company visit—preferably someone having the most in common with the visitor, for example, schooling, degrees, marital status, and the like. An introduction over a meal and a tour of the community may well be in order. If possible, the first person to interview the applicant should be the one assigned the personnel activities within the industrial engineering department. This initial discussion should include a review of the proposed agenda to ensure that the interests of the applicant as well as those of the company are being met. Special interest in a particular function or activity should be acknowledged and plans rearranged to allow the applicant to explore that interest in depth with one of the engineers directly associated with the field. Within one day, three or four such tailored experiences can be scheduled with as many different practitioners. In total, including the host, five or six people will have been exposed to the applicant.

Bearing in mind that two sets of interests must be satisfied—the applicant's and the company's—the introduction of a number of people into the process enriches the validity of the assessment process. The applicant must be given every encouragement to learn as much possible not only about the technical demands that must be faced but also about the work environment—how projects are undertaken, communication channels, personal appraisals and performance reviews, promotional opportunities—especially as seen by the practitioners who will be members of his or her peer group if the applicant joins the organization.

At the end of the day each of the company participants completes an evaluation form about the applicant. These are consolidated by the personnel assistant and serve as a basis for a decision as to whether an offer should be made to the applicant or whether further interviews are in order.

If the applicant has had practical experience in addition to academic work, the process is modified to match the applicant's level of technical seniority. Otherwise, the procedure is basically the same.

Another source of potential candidates for the industrial engineering staff are those technical people displaced by changes in operational plans and those who are looking for a change in their career paths within the company. The interview process may have to be altered for this type of applicant. However, the opportunity for a mutual evaluation of a possible match should still prevail. In some instances, a trial transfer for a period of 6 months or 1 year can be arranged between the departments involved to better evaluate the desirability of a permanent transfer. There are many variations to this practice of relocating capable and productive personnel within a company. Every new applicant should clearly be the best possible person who can be recruited to fit the needs of the industrial engineering department. This is particularly true in accepting transferees from within the company. The industrial engineering department should not be a place for misfits.

A properly staffed professional organization, regardless of size, should reduce much of the administrative load on engineering personnel by providing adequate clerical and stenographic assistance. The careful selection and training of these people can materially improve the effective utilization of scarce and talented engineering resources.

There is a specific role that can be filled by technicians to further improve the productivity of the industrial engineering organization. Almost every community has a 2-year junior college or technical institute within commuting distance. Most of these schools award associate degrees in the applied sciences and have proven to be excellent sources for technicians and analysts. Here, too, the selection and continued training of these people is an important factor in rounding out the total capability of the organization. Through on-the-job experience and part-time schooling, many of these technicians graduate into the professional ranks within the company.

The employment of summer students and students enrolled in cooperative educational programs offers many advantages to firms and to the students. The work blocks allow a much better assessment to be made of the student as a candidate for future employment. The student, in turn, gains a realistic view of what the world of work is really like first hand. Not only can career plans be directly affected but the learning process can be greatly enhanced when the work experience is tailored to suit the academic progress of the student. The danger of "make-work" situations, where students are either left to shift for themselves or are asked to do menial tasks having little or no bearing on their technical training, needs no further amplification as to its effect on the student. These student experiences require the same level of management interest and concern as do those of the permanent professional employees.

1.2.9 THE EMPLOYMENT OF TECHNICAL CONSULTANTS

The employment of outside specialists can help fill the gaps in technological capability that industrial engineering departments face from time to time. The outside consultant can also be of great value in the development of a conscious effort to build and strengthen the internal industrial engineering organization. These needs will surface regardless of the size of the industrial engineering organization or the level of technological sophistication. As the scope of industrial engineering activities expands, so will the variety of specialized techniques that can be employed grow in number and complexity. Outside specialists can help reduce learning time in the acquisition and application of new knowledge.

There are occasions when additional personnel are required for a limited amount of time or to meet a critical target date. The availability of experienced personnel on a temporary basis may be the difference between success and failure.

Complete reliance on outside support can be counterproductive. Professional competency grows with experience; the more exposure the internal staff has to complex problems the stronger will be their ability to respond to such problems in the future. Outside consultants take the experience they acquire on the job with them when they leave. This experience contributes to their technical strength and, therefore, adds to their value to their clients. However, this gain to the outside consultant is a lost opportunity in becoming technically more self-reliant internally. It is unrealistic, however, to expect any internal staff at a given moment to be expert in all matters related to its function. Such an end may not be economically attractive. To meet this need the outside specialist does provide a very real service that is otherwise difficult to acquire.

It is possible to undertake a mutually satisfactory working arrangement between the internal staff and the outside specialists. A team made up of consultants and staff can be quite effective in the resolution of problems. The internal staff can not only learn and gain experience by working with the consultants but, with their knowledge of the company, can also immeasurably reduce the time that an outside consultant must spend just learning the operations and sources of information. Technical experience gained from such joint ventures not only accrues to the benefit of the consultant but is also acquired by the internal industrial engineering staff. This experience will prove to be useful after the consultants leave if there are changes in the study at hand or in troubleshooting problems that may develop at some later date.

1.2.10 COLLEGE AND UNIVERSITY RELATIONSHIPS

Relationships with colleges and universities can take many forms and tend to be beneficial to both parties. The teaching institutions have a very real need to stay in close touch with the changing trends and needs in the world of business and industry. Relevancy between courses of study and actual practice is a requirement of the first order of concern. Practitioners, for their part, not only look to the schools for their recruits but also for their continuing educational needs. They must have access to refresher programs and technical seminars in order to stay abreast of new theoretical concepts and techniques. Attempts to satisfy these two needs account for a large majority of the different liaisons that have sprung up throughout the country.

A relatively expedient way for teachers and professors to keep up with industrial engineering practice without leaving their faculty posts is to make themselves available as part-time consultants or as supplementary technical resources on study teams in local government, business, or industrial organizations. In so doing, they introduce an outside objectivity and at the same time enrich their own capability as teachers, writers, and lecturers. Such an arrangement can also help satisty a company's short-term needs for selected technical skills and avoid unnecessary overstaffing.

A number of graduate schools look for opportunities that would enable their students to undertake actual problems from business and industry rather than to solve textbook exercises. Such programs allow a number of graduate students to work together as a team under the guidance of both the faculty advisor and a supervisor from the cooperating company. Such programs are of substantive value to small as well as large companies, since by so doing they keep in touch with the latest technological and theoretical developments.

A wide variety of short courses and seminars are offered throughout the year by most colleges and universities. If there is a specific interest in a field of study for which there appears to be no public offering, a little searching can reveal a number of schools willing to develop a special program just for that purpose. Many of the training courses offered to the general public are specifically directed at helping small organizations by providing a convenient medium for the active interchange of ideas. Some long-lasting affiliations have been fostered through such programs. The popularity of these short courses and seminars suggests that many organizations–regardless of size–derive benefit in participating.

The strong interest in part-time undergraduate and graduate study has resulted in the formation of branches of state universities throughout their respective states. These extension facilities have considerably expanded the ability to participate in college level classroom work even for those at some distance from the university centers.

Industrial engineers are often invited to speak as guest lecturers at university sponsored seminars open to the general public or to students in the classrooms. This is another instance in which all parties profit. The practitioner can bring the pragmatics of actual experiences to the lecture hall more effectively than an instructor with little or no cognitive exposure to management problems. The student with the help of his or her instructor should be able to question and understand better the interface between theoretical concepts and actual practice. The practitioner, in turn, may well be put into a learning mode, since he or she will be challenged to rationalize practical applications in terms of contemporary theory.

Some companies and colleges have exchanged faculty and senior engineers. These exchanges are difficult to arrange; family situations must be considered, a successful engineer may not necessarily prove to be a successful teacher or vice versa, the costs of making the moves may be prohibitive, and the probability of finding a match of abilities to suit the needs of the two organizations is marginal. Nevertheless, efforts continue to be made to find feasible solutions to these problems.

Occasions arise when a detached observer can be of value in assessing the technical capability of the industrial engineering staff. An analysis of the client environment by visiting professors and their judgment of the ability of the local industrial engineering staff to function effectively in that environment provides the industrial engineering manager with a perspective otherwise difficult to attain.

There is a growing interest within the academic world in establishing advisory groups made up of senior engineers and managers representing a cross section of business, industry, and service sectors of the economy. The responsibilities of these advisory groups vary from school to school. In general, they are free to examine and advise on curriculum, such topical areas of interest within courses as methodologies or techniques, the extent of the research and development effort in the field of industrial engineering, faculty publications, faculty competence, faculty/student relations, and other activities such as the administration and planning of the industrial engineering program at that school.

University relations can have a positive influence on the level of competence of an industrial engineering organization. An effectively planned program insures that these relations will continue to be actively supported and managed.

1.2.11 DEVELOPING NEW CLIENTS

The practice of industrial engineering had its early success in the shops and factories of the late 1800s. The passage of time witnessed the testing and adoption of a number of work measurement and improvement techniques, some of which eventually formed the basis for much of the technology still in use today. Along the way it also became evident that the application of industrial engineering principles outside the manufacturing environment could lead to equally impressive improvements in productivity.

World War II stimulated the tempo of research in decision making processes and in the simulation and modeling of alternatives. During the 1950s and 1960s improved understanding of the behavioral aspects of work design broadened the scope and complexity of industrial engineering studies still further.

While these events were taking shape and having their impact in extending the practice of industrial engineering, the computer pushed open the doors to the realities of on-line interactive management information systems. At the same time new notions surfaced leading to the formulation of concepts that could deal with the analysis and design of total systems–concepts that recognized no boundaries in the search for optimization. Most recently, the congruence of sensor technology and microprocessor technology has brought within the reach of management that promise of a computerized world of work.

The industrial engineer's potential field of study has, by virtue of these new concepts and technologies, expanded far beyond the factory floor. The horizontal extension of the analytical process to include all the related operations in a manufacturing process introduces the engineer to potential clients not otherwise reached under normal study circumstances. So there is a natural extension of technical services encouraged by the very process of undertaking analyses of the system as a whole.

The principal responsibility of industrial engineering managers over the years has been to assist management in the process of improving performance. In the past, managers' efforts were addressed to individual operations. Today, to be successful in carrying out their responsibilities, managers' efforts must be virtually boundless. Since the managers have the specialized skills at their disposal, it is their responsibility to bring this knowledge and capability to the attention of their peers within the management hierarchy of the enterprise. This responsibility is an active rather than a passive one. The industrial engineering manager is often in the best position to translate the possible application of the resources and techniques within the staff organization to the problems faced by the line manager. It is unreasonable to assume that all managers at all levels are sufficiently conversant with all technical disciplines to always employ the most appropriate tools or resources in carrying out their responsibilities. This reason alone makes it incumbent on the manager of industrial engineering to not only identify potential areas in which professional and technical expertise may be brought to bear but also to educate all managers on the state of the art so that they can intelligently utilize these resources in their operations. Needless to say, the manner in which this obligation is addressed depends not only on the personal and professional attributes of the managers of industrial engineering but also on the managerial styles of their peers. If the environment within the company is conducive to a team effort, then there usually is reasonable assurance that all of the resources available will be applied in a concerted effort to meet the goals of the company.

There are a number of ways to develop this mutually supportive environment. An interchange of executives between the industrial engineering and operating units has been effective in dispelling the "we/they" feelings that often polarize line/staff relationships to the detriment of the organization. The careful selection and placement of these leaders contributes immeasurably to instilling a feeling of mutual respect at all levels.

Satisfied clients are one of the best sources of support in extending the practice of industrial engineering into new areas. Knowledgeable clients can often easily related with similar problems experienced by their colleagues elsewhere in the organization. Through their normal channels of communication they can reach potential clients outside the usual flow of industrial engineering activities. Clients' own lay versions of problems and their possible resolutions may on some occasions carry more weight than expert opinion.

The experience of clients can also be put to effective use in technical seminars or executive briefings by having them participate in the presentations made to other supervision and management. The review of actual studies and open discussions on the pragmatics of a problem by the client introduces a level of credibility often difficult to achieve by the practitioner in a new or unfamiliar organizational environment.

Some companies circulate technical abstracts to all operating and staff personnel. Where this is an accepted practice, the industrial engineering manager includes brief statements describing the successful application of a new technique or the resolution of a problem having some commonality throughout the company. The preparation of an industrial engineering newsletter is also a fairly popular practice. The more effective letters are directed at helping managers understand the nature of the skills and expertise available to them.

Many industrial engineering departments prepare periodic listings of all new projects. These listings are used primarily internally for informing all industrial engineers and supervisors of projects of possible interest in carrying out their assignments. Since many of these listings are brief but descriptive statements of proposed studies, they have been useful in familiarizing potential clients with the possible applicability of a similar study in their own area of responsibility. This is an example of what might be termed the "soft sell" in informing potential clients about the services available from the industrial engineering organization. Such material has the further advantage of avoiding the cost of specially prepared brochures or manuals.

Short informational assignments are another tactic quite effective in educating managers and key personnel to line and staff operations throughout the entire company. A series of informational meetings are relatively easy to schedule and have the advantage of mixing potential clients with representatives from organizations already effectively using the resources of the industrial engineering organization. The personnel invited to attend these sessions are in some instances those selected for management grooming or executive development programs. The seminars may be held off plant, away from the distractions of the company environment. There also may be advantages in scheduling these informational workshops at different company sites on a rotational basis. Here again, local circumstances dictate the format of these meetings, and the details regarding timing or location will tend to be custom tailored.

A serious detriment to the effective extension of industrial engineering support can arise through restrictive position profiles or job descriptions prepared by personnel departments. In the interest of standardization or wage and salary evaluation procedures, the industrial engineering functions have in some firms been limited exclusively to time study or work simplification studies. Well-prepared position profiles or job descriptions are very useful not only for managing salary equity within a company but also as a guide in the recruitment and placement of professional and technical staff. If written too narrowly in terms of job responsibilities and necessary skills, however, competent applicants may not be interested in offers for employment. Technical and professional personnel already employed may even consider leaving the organization in frustration at their inability to utilize their skills and knowledge. The training of industrial engineers may tend to be limited in scope in order to conform with the position profiles, and the ability to adapt to new industrial engineering concepts and technologies may be placed in jeopardy.

The industrial engineering manager who is limited to only a few functions in the factory has little to offer either manufacturing or nonmanufacturing managers beyond a few basic skills. This limitation of technical skills should be a matter of concern to all levels of management within the firm. Effective industrial engineering practice must recognize the need to identify not only the benefits that accrue from improving the performance of a single worker but also the impact of such change on the larger related entity. It is not enough to improve the labor content in a single station if that action generates increased quality costs or additional capital expenditures elsewhere in the system.

It is against this conflicting background of goals and objectives that the industrial engineering organization must prioritize its own activities. Yet the final priorities must be consonant with the principal mission of the enterprise. For some projects, the prioritization is predetermined. For example, corporate guidelines may be issued automatically, eliminating further effort on all projects failing to satisfy a minimum rate of return on the investment required. Other projects may be aborted by management for any one of a number of reasons–product defects, manufacturing costs, late market entry, technological changes, and so on. On the other hand, company directives may mandate that certain activities be undertaken and take precedence over all other projects.

Normally, the work that must be done and that which must cease poses little difficulty for the industrial engineering manager. The real difficulties are those arising from the process of allocating the remaining technical staff to the balance of work requiring attention, since few staff organizations are permitted the luxury of having all of the necessary technical skills to satisfy all of the clients all of the time.

Management steering committees are formed in some instances to determine which of the remaining projects are staffed and to establish a precedence schedule or ranking of the projects. In other cases, the industrial engineering staff is expected to establish the ranking using top management guidelines as a basis for the allocation of technical personnel to specific projects. A still different process involves a series of meetings between the clients and the industrial engineering staff at the lowest organizational level within the company. The preliminary prioritized listing of projects at this level includes estimates of the engineering skills required and the number of technical personnel to be assigned to each project. Since the clients are part of a larger organization, the prioritization of all the clients at the same organizational level has to be repeated and, with limited industrial engineering technical resources, some projects will have to be dropped in order to accommodate those with greater potential or having a higher order of urgency. This process is repeated until all levels within the company hierarchy have had the opportunity to participate in the prioritization of the demand for staff support. Depending on how the company is organized and whether the industrial engineering staffs are centralized or decentralized, this process reiteration may stop at the departmental, plant, or divisional level of the company.

Regardless of the process for prioritizing the projects, the industrial engineering organization may not have the proper skill mix to respond to all the technical demands that the final list of projects requires. As indicated earlier, the ability of the industrial engineering manager to anticipate the type of support necessary to meet client needs may be the crucial factor in establishing a firm basis for effective line/staff relationships.

The assessment of the technological skills required must perforce be unending especially in light of the lead time that is needed to properly recruit and train the engineers and analysts. Even a simple logging of the work undertaken by the industrial engineering organization can reveal those techniques or methodologies most likely to be required in future studies. One method in use requires engineers to use a classification code when preparing their daily work log. The code facilitates analyses by clients served, by the nature of the study, or by the specific service rendered, as well as providing information regarding time spent on various projects. This procedure can be done either manually or on computer. It is particularly simple to adopt if the staff organization is on a fee-for-service basis. The mechanics of billing the clients on a periodic basis can produce, at the same time, data for managing and directing the efforts of the staff organization. Trend charts can be prepared and are useful for predicting staffing and training needs for the near future.

Section 1.2.12, dealing with methods of developing new techniques and services, touched on the importance of analyzing management reports and records in anticipating future demands on staff support. It was also pointed out that a skills inventory of the industrial engineering staff would highlight potential deficiencies in specific techniques or methods that would be useful input for recruiting and training plans.

Training programs based on long-range needs should encourage those personnel seeking a career in staff work and should tend to insure the existence of a competent and stable technical resource. Larger organizations can develop internal training programs using experienced personnel or company training facilities. Other firms may have to engage consultants, faculty at local colleges, or perhaps send their professional and technical personnel to more distant academic institutions to acquire the necessary training. The budget plans of a well-managed industrial engineering organization should include funds for developing the capability of the staff. The half-life of an industrial engineer has been estimated by some managers to be as low as 3 years. Even if it were 6 years, the rapidity of social, cultural, and technological change places training high on the list of management concerns.

1.2.14 SUPERVISORY AND MANAGERIAL ROLES

Industrial engineers, technicians, analysts, and stenographers are usually grouped into a logical work unit under the direction of a supervisor or group leader. The supervisor assigns the work, advises the engineer on applicable techniques, reviews the proposals, and, on occasion, presents the final recommendations to the client. The supervisor is also responsible for rating the performance of the personnel in the unit and arranging for their training. The engineering supervisor has in the past been the primary force in guiding the technical work from analysis through design to implementation. Many successful industrial engineering organizations still rely on the supervisor as the technical expert in the group of engineers.

In many organizations, the industrial engineering supervisor is more apt to be in a transitional phase—in the role of a resource or skills manager as well as that of a technical leader. The supervisor's role is being reshaped by the increasing complexity of industrial operations and the need for a multiplicity of techniques and skills to deal with difficult and intricate problems. This environment has contributed to the growth of multidisciplinary project teams made up of professional and technical resources from a number of staff organizations. The industrial engineering supervisor may be asked to assign some engineers to one or more of these project teams where their work will be supervised by a project leader. From such assignments a mixture or matrix of managerial and supervisory relationships have evolved to cope with the problems of organizing the work of the technical resources on the team.

Another factor influencing change in the supervisor's role is the change in the behavior of the senior industrial engineers. This mature group of professionals not only has less need for direct technical supervision but in addition, many of them have become experts in a specialized field of industrial engineering and may be more knowledgeable than the supervisor about that particular technology. Moreover, these professionals tend to be guided by the client in organizing and planning their work. In a decentralized organization, the engineers are conveniently part of the same plant or facility. However, even highly centralized industrial engineering organizations tend to locate the industrial engineer physically in the client's area of operations. This proximity allows the industrial engineer to become intimately familiar with the operating areas.

The more closely the engineer relates to the client served, the more often will the engineer have opportunities to develop and demonstrate the capability of undertaking assignments of increasing professional responsibility. By the very nature of the engineer's work, he or she is often in the best position to assess potential problems in the client area and to initiate the studies that could minimize the impact of the problem or perhaps even rectify the situation. The engineer is, in effect, working directly for the client organization—not the industrial engineering organization. Yet, the engineer looks to his or her supervisor for the necessary support to carry out the assignments. The supervisor, as a result, must adapt a more versatile role. The supervisor must remain knowledgeable regarding the progress and results of the work performed by senior engineers. Newly hired engineers also need the supervisor to guide them in carrying out their initial work assignments.

Supervisors must plan for the training and professional development needs of all their personnel. They tend to move away from technical problems knowing that their experts can deal with the techniques. Their supervisory behavior tends toward a supportive role rather than one that is directive. They find themselves spending more time with management to stay abreast of operational problems and prepare plans in anticipation of demands for technical support. They are seen more in the role of resource managers rather than of technical experts, since they are expected to organize the required talent into effective study teams and assist in the prioritization of projects. The increasing amount of time spent by the supervisors in planning sessions with management forces them to delegate more responsibilities to the senior industrial engineers in their groups. This satisfies both the need of the engineers for more responsibility and the need of the managers to raise the ability of the staff to anticipate and respond to management's goals and objectives.

The managers of industrial engineering departments should be even further removed from the technical problem solving details. Their responsibilities are to the enterprise as a whole and their sphere of functional influence should extend to all levels of management. They are to create and support line and staff interactions along vertical lines within the company as well as those that take place along horizontal channels of communication. They must themselves be physically present at many of these managerial crossroads. They can not marshal their resources effectively if they exist in a vacuum of management information. This could easily happen if they enmesh themselves in the technical details of the work of their staff. Their presence is needed both in their organization and among their peers.

There are any number of management councils, advisory bodies, or operations committees that deal with long-range planning, manufacturing options, marketing strategies, or operating problems. The industrial engineering managers' participation in these sessions brings to full circle the staff relationships that began with engineers' assignments at the operating level and supervisors' interactions at the middle management interface.

There are differences between companies in the manner in which the industrial engineering managers relate to their superiors regarding the work of their organization. In many firms, the industrial engineering manager is held responsible for the design and maintenance of the work standards and the premium pay plan. The industrial engineers conduct the time studies and frequently initiate methods changes so that the work standards are based on the most productive way of performing the work. This can and does lead to changes in the layout of work areas to take advantage of the methods improvements. The work standards are audited on some regular basis by the industrial engineers to avoid deterioration due to loose standards or poor methods. The industrial engineering manager authorizes any and all changes in the work standards. As a result, the standards are perceived as belonging to the industrial engineering organization. There is little or no felt ownership of the work measurement or incentive pay plans by the line supervision or management. Furthermore, the manager of industrial engineering is expected to prepare reports on the effectiveness of these plans in meeting performance standards for the plant manager or the general management of the company. The manager also reviews cost reductions or increases with management.

Many companies report their satisfaction in assigning the responsibility for labor performance to the industrial engineering organization. Others admit to some discomfort in placing the control and improvement of the costs of labor under the direction of a staff manager. On occasion some concern is voiced over the apparent division in accountability, especially when there are problems in meeting the labor standards or a perceived impact of the work measurement plans in other areas of concern such as quality or safety. The direct reporting of cost reductions to the top management of the company by the industrial engineering manager appears to some as though the manager might be seeking credit for improvements achieved by the line organizations. Some of these concerns have led many companies to reexamine the reporting relationships regarding line/staff activities.

A somewhat different reporting relationship is found in those companies that delegate the responsibility for managing improvement to the lowest possible level of line authority, whether it be through work incentives, methods changes, technological innovations, or the employment of motivational techniques. In these companies, the industrial engineering staff continues to be used extensively in setting standards or other aspects of the improvement process. The industrial engineering work is commissioned by the operating manager, who is the sole authorizer of any change subject to the approval of the line superiors. The goals are set by the operating manager, who holds total ownership.

The performance of the operating unit is reported through normal line channels. This often takes the form of regularly scheduled reviews of the labor, material, and machine costs of each product compared to previously established standards, the comparison of actual and budgeted inventory levels, quality standards, or service goals. These accomplishments are not reported by the staff but by the line management, although considerable staff assistance is required in preparing the information or in bringing about the improvements. This support is acknowledged by the line managers and recognized by general management as evidence of effective line/staff teamwork.

The two totally different reporting and authorizing procedures just described by no means cover the variety of ways in which the industrial engineering manager relates with clients or communicates with superiors. They do exemplify two different management styles and may be useful as guides in analyzing the way in which line and staff organizations share a reporting responsibility.

A brief reference was made earlier to the responsibilities of the industrial engineering manager to the enterprise as a whole. In discharging this responsibility the manager must assess the level of understanding held by management of the technical capabilities of the industrial engineering staff. The manager has to take the initiative first, in recognizing the applicability of a specific technique to a problem situation, and second, in explaining or demonstrating the uniqueness or power of this approach to those unfamiliar with the technology involved. This might come about quite naturally in the process of discussing operating problems at meetings where the industrial engineering manager is expected to share knowledge as an active participant. On other occasions, he or she may stage a review of particular techniques useful to a given situation. The manager is, by virtue of his or her association with the general management of the enterprise, in a position to initiate activities not otherwise possible by other members of the industrial engineering organization. Typically, such industrial engineering activities are those that would generally affect operations as a whole. To assist in this process, the industrial engineering manager may have discretionary funds budgeted for his use as "seed" money in exploring potential opportunities for improving the effectiveness of operations. These limited funds permit the accumulation or preparation of sufficient information to support the feasibility of a more complete study.

1.2.15 ACCOUNTING FOR EFFORT

There are five principal groups who have an interest in the plans and status of the work undertaken by the industrial engineering staff. They are the industrial engineers, the clients, industrial engineering supervisors and managers, the financial analysts, and the general management of the company. The engineers who must plan and carry out the studies need information that will allow them to exercise some control over their effort. The clients are interested in the progress of the work as it relates to their objectives and goals. The industrial engineering supervisor and manager must have access to timely information in order to direct and control the resources within the industrial engineering organization. The financial analysts need to distribute actual staff costs either directly to a product bill or to proper burden accounts. Finally, general management must have information that quantifies the utilization of staff support in terms of people, costs, and effectiveness. The size of the industrial engineering department makes some difference in the methods used to plan and control its effort. However, regardless of size, someone at some time must devote some attention to this responsibility. The mechanics may differ, but the notion of control is generally accepted.

The preparation of operating budgets is usually preceded by a review of work already underway that will carry over to the next budget year. Estimates are prepared of people required and the time span necessary to do this work. New projects are defined and the estimates of people and time for these projects are added to the information already assembled. Estimates are made for training time and professional development activities as well as the time required for vacations, holidays, illnesses, and other absences. The net available engineering time required to do the work is calculated and used as a basis for determining personnel additions or reassignments. Historical attrition records provide data to further modify the personnel requirements.

This process is considerably simplified if plans for the coming budget year begin with a zero base of requirements, in which each year every project, old and new, is challenged as a candidate for staffing and funding. This is an oversimplification of the methods in use in zero base accounting procedures. However, it does typify the effort to identify all industrial engineering work, the department requesting the service, and the value to the company of the study of the projects undertaken. In a cost conscious environment, those projects for which a good business case can be made are approved, while other projects are deferred, aborted, or not undertaken at all.

One way to control indirect staff support is to count the number of people associated with every function in the company. A functional analysis should reveal the extent of redundant or overlapping staff activities. Economies can result by eliminating the duplication or by consolidating the functions into groups that can be more efficiently organized and managed. The number of people identified with each function becomes the base number, subject to further assessment in terms of overall personnel needs within the company. In preparing budgets for future operations, deviations from the base number should be substantiated by increases or decreases in the need for the staff function. This simple and direct approach to controlling the size of indirect operations is practiced in some of the largest corporations in the United States as well as in much smaller ones. As in other methods of control, it is subject to abuse. The convenience of issuing directives specifying a fixed percent in the decrease of all personnel will have varying degrees of impact on the staff organizations. In times of economic constraints, such staffs as industrial engineering that contribute to improvements in productivity might best be exempt from such decreases or perhaps even increased in size in an effort to improve profits. However, it is difficult to argue the fact that any function can be limited by simply limiting the personnel assigned to it. When policy mandates that the number of personnel must be decreased, a simple head count is effective in determining the degree of compliance and in pinpointing those organizations having difficulty in responding to the directive within the time period desired by general management.

The procedures for allocating the cost of industrial engineering services can be grouped into two general categories:

1. The use of overhead or burden accounts to accumulate all the industrial engineering costs. These indirect costs are then distributed over all the direct operating organizations using direct labor hours, product bill, or some other accounting basis for the distribution.
2. Direct billing or charge-for-service accounts that accumulate industrial engineering costs by specific using organization.

The pros and cons of accounting procedures for professional staff services runs the gamut from emotional reactions to rational logic. The proponents for charging all the costs of a staff service, regardless of who is using it, to a general burden account, which is later distributed in some fashion as an addition to the cost of doing business, point to the simplicity of the accounting process as an advantage. The argument is also advanced that if a line manager using the industrial engineering staff were to be charged for the full cost of the staff service he or she would choose not to use it. Indeed, the line manager might even reason that he could hire his own industrial engineers at less cost, since he would thereby not be paying for the administrative costs of a central industrial engineering department. As a result, the early history of most industrial engineering departments reflects these concerns and the cost of the staff support was absorbed into general burden accounts. Since there was no direct billing to the user of the service and since the line manager was not required to submit a budget for controlling the cost, the industrial engineers were seen as a "free" service accountable only to general management. In the beginning, this free service was undoubtedly a great aid in encouraging line managers to utilize a relatively new technical staff technique in improving the effectiveness of their operations. Without this support it may not have been possible in some instances to demonstrate the value of the industrial engineering staff.

However, absorbing the costs of staff into a general burden account makes it difficult to determine accurately which products or organizations are responsible for generating the staff costs. A staff service for which there is no local accounting tends to degenerate into requests for support for nonessential or low-priority projects because it is perceived either as a "free" or, at best, "fixed" cost that is accrued whether it is used or not. There is little control of a general burden account at the lower organizational levels. The justification for the staff activity gravitates to the highest management level responsible for the total operating budget, since the total burden accounts are distributed at that level. This forces the staff director to review these activities at a fairly high level within the organization. This may not always be consistent with efforts to delegate the management of operating costs to the lowest possible level within an organization.

Technical and professional staff organizations that charge for their services point out that a client billed directly for technical support is inclined to have a higher level of interest in the prioritization and progress of their projects. Charging for services, in their opinion, stimulates the identification and selection of projects having high value to the firm, the reasoning apparently being that a costly staff service should be used on those projects having the highest possible level of return.

The advantages of a full bill-out system for industrial engineering services are most apparent when operating budgets are subject to particularly critical reviews. The preparation of budgets forces the industrial engineering manager and the clients to be as quantitative as possible in the selection of projects and studies to improve productivity. The industrial engineering procedures for accumulating the necessary information for billing clients lend themselves admirably to historical performance analyses useful for predicting and anticipating projected staff support.

The mechanics of a bill-out system are relatively straightforward. The time spent on specific studies is posted daily or weekly on a log sheet by the engineer. The log sheet identifies the engineer, the department served, the date, the time spent, and a brief descriptive title of the project. The engineer also accounts for the balance of his or her time spent on other activities such as training, supervision, absences, and vacations. This data base can be compiled and arranged manually or electronically into a variety of displays useful for managing the operations of the industrial engineering department. Trends in client needs, training requirements, allocation of professional and technical personnel by product lines or other groupings are but a sampling of the control data that can be assembled. Such a system simplifies periodic reviews with clients of actual costs of industrial engineering services as compared to budgeted plans. The data is also useful input in assessing project priorities and the possible re-allocation of staff personnel.

A billing system has direct value to the staff engineers who are consultants to the clients and need information to better manage their efforts. Clients are willing to pay for good performance and their willingness to continue to fund the work of the engineering consultant is a positive reinforcement of their satisfaction with the support they are receiving.

1.2.16 INFORMATION AND COMMUNICATION NEEDS

An effective and responsive industrial engineering organization must have visible and easily accessible channels for the transmission of information vital to its staff members. The personnel within the industrial engineering organization have need for information that affects them as individuals as well as for information that guides and shapes the business of the industrial engineering department. They have an interest in personnel recruitment and placement plans, training opportunities, organizational achievements on major projects, allocation of resources, and the prioritization of work. In short, most of the information required to manage the industrial engineering organization needs to be shared with all personnel to gain their understanding and commitment to department goals. Since the industrial engineering department is made up of new recruits, senior engineers, clerks, analysts, and technicians, a variety of formats has to be considered in satisfying the needs of these different groups for information.

Regularly scheduled staff meetings for supervisors and senior engineers at which administrative matters can be discussed, seminars on new techniques, annual presentations by the industrial engineering manager to all personnel highlighting achievements, describing goals and objectives, the circulation of information regarding budgetary activities, and notices of organizational changes and promotions suggest the wide range of informational opportunities that can be effectively explored. There must also be an environment that is conducive to sharing matters of mutual interest and concern. The supervision and management levels within the industrial engineering department should make themselves accessible to any individual or group desirous of exploring organizational or administrative matters that are of concern to them. An open environment is as important as planned meetings in developing and maintaining good communication channels. To this end, some departments schedule a series of meetings with new employees usually held during the first 2 or 3 years of employment. At these meetings career interests are explored and matters related to job satisfaction discussed. In addition, questions regarding the administration of the industrial engineering department are encouraged and freely answered. These meetings are quite helpful in alleviating some of the anxieties and concerns often held by new employees. At the same time, this experience makes it easier for employees at some future time to come in on their own initiative if they should feel the need to discuss a matter of concern.

Clients also have information needs particularly in regard to the services they have contracted from the industrial engineering staff. Most clients are quite specific about the information they need and the format that best suits their requirements. The administrative procedures within the industrial engineering department should be as flexible as possible to satisfy these needs as well as those necessary for the operation of the department. Generally speaking, projects or studies undertaken for a client are usually prepared in some documentary form that includes a description of the requested study along with time estimates and critical dates. Such documentation is a matter-of-course procedure in those companies where there is a charge for the use of staff services. At the same time, the written job requests can be reviewed by those having an interest or a role in the assignment to insure agreement on the scope of the study.

On a regular basis, progress reviews are scheduled by the engineer with the client. Depending on the circumstances, these progress reviews may be verbal briefings or formal, written progress reports. This information allows the client to continuously reassess options or introduce changes in the scope of the assignment as a result of the data currently available. At the completion of the study, a closure process usually takes place. A final report may be requested to accompany proposals for capital expenditures or to serve some other interest of the client. These reports also have value as reference material to other engineers working on similar studies. It may even be worthwhile to have a cross-indexed reference file to facilitate the search and retrieval of major projects conducted by the industrial engineering department.

There is an obligation on the part of technical staffs to provide overviews or state of the art briefings to clientele and to management. The importance of this educational responsibility can not be over emphasized. The effective employment of knowledge workers is directly related to the level of understanding of the technology held by the employer—in this case, the client. It is not enough that the industrial engineering manager and staff are well versed in the latest techniques and methodologies in the management sciences. It is essential for all levels of management to have some knowledge and understanding of the power and potential of the tools that they can command through the industrial engineering staff. The doctor/patient relationship—if it ever was a viable course of behavior—is simply not suited for contemporary line/staff relationships. The notion of a technical staff prescribing a method of operation that the line manager does not understand, is to accept without question, and yet be held accountable for the final results of is unreal. There must be mutual understanding of the distinct roles for which each organization is responsible. A team approach in which a participative and collaborative environment is fostered appears to be effective in developing good line/staff relations.

As described in Section 1.2.11, some industrial engineering departments regularly circulate technical abstracts or newsletters within their firms. Such efforts at communications need to be well edited and couched in a relevant format to avoid the fate of "junk mail." Not all company environments are receptive to the mass circulation of staff bulletins. The same concerns apply to annual staff reports usually prepared for higher levels of management. These can be very useful as informational documents in communicating with managers not normally exposed to the day-to-day activities of the technical staff. Well-written annual reports can help educate management to technological trends and accomplishments, or set the scene for new staff ventures. There is a danger, however, that staff reports may be redundant in reporting events that management may have received information about through other channels or that the reports may be perceived to be promotional and self-seeking in nature. If such reports leave the impression that the staff organization is taking sole credit for accomplishments in the line organization, cooperation may well deteriorate to the point where line/staff relations may be adversely affected.

Perhaps a final point should be made regarding communications between technical staffs and management: Clarity in speech and in writing is an essential ingredient in promoting understanding. If there ever was a time and place for lucidity, for simplicity, for plain unadorned speech, it surely would seem to be at the moment when the technical expert is explaining his or her work to a client. Among themselves, professionals and technicians relax with a native jargon born out of convenience and quite useful in sharing ideas. To the uninitiated it may sound like gibberish and, at best, somewhat pretentious. To the confused client it may develop a sense of helpless frustration that could extinguish any hope for further collaboration.

Effective communications are essential in helping management understand the capability and competence that is available from the industrial engineering staff. The single biggest challenge facing industrial engineering managers is the structuring of a process that leads to the intelligent and effective use of the staff by all levels of management. It is to this end that the resources are marshaled and the operations organized. A new tool is fashioned for management, designed in a way that is uniquely suited for the work at hand.

PUBLICATIONS

ACROSS THE BOARD
The Conference Board, Inc.
845 Third Avenue
New York, NY 10022

ADMINISTRATIVE MANAGEMENT
Geyer-McAllister Publications
51 Madison Avenue
New York, NY 10010

ADMINISTRATIVE SCIENCE QUARTERLY
Graduate School of Business & Public Administration
Cornell University
Ithaca, NY 14853

AIIE TRANSACTIONS
American Institute of Industrial Engineers
25 Technology Park/Atlanta
Norcross, GA 30092

AMERICAN DEMOGRAPHICS
American Demographics, Inc.
P.O. Box 68
Ithaca, NY 14850

BEHAVIORAL SCIENCE
Behavioral Science Systems Science Publications
University of Louisville
Louisville, KY 40208

BEHAVIORAL SCIENCES NEWSLETTER
Roy W. Walters & Associates, Inc.
60 Glenn Avenue
Glenn Rock, NJ 07452

BIOMEDICAL COMMUNICATIONS
United Business Publications, Inc.
750 Third Avenue
New York, NY 10017

BIOMETRIKA
Biometrika Office
University College of London
Gower Street
London WCLE 6BT, England

CALIFORNIA MANAGEMENT REVIEW
Graduate School of Business Administration
University of California
Berkeley, CA 94720

CHEMICAL ENGINEERING
McGraw-Hill, Inc.
1221 Avenue of the Americas
New York, NY 10020

CHEMICAL AND ENGINEERING NEWS
American Chemical Society
1155 16th Street, NW
Washington, DC 20036

COMPUTER DIGEST (Formerly *Data Channels*)
Phillips Publishing, Inc.
7315 Wisconsin Avenue
Washington, DC 20014

COMPUTERS CONTROL AND INFORMATION THEORY
National Technical Information Service
U.S. Department of Commerce
5825 Fort Royal Road
Springfield, VA 22161

COMPUTERS AND PEOPLE
Berkely Enterprises, Inc.
815 Washington Street
Newtonville, MA 02160

DATACOMM ADVISOR
Communications Field, Inc.
214 Third Avenue
Walton, MA 02154

DATA PROCESSING DIGEST
Data Processing Digest, Inc.
6820 La Tijera Boulevard
Los Angeles, CA 90045

DYNAMICA
University of Bradford, Management Centre
Systems Dynamics Research Group
Emm Lane, Bradford 9
West Yorkshire, England

ECONOMETRICA
Econometric Society
Department of Economics
Northwestern University
Evanston, IL 60201

ELECTRONIC NEWS
Fairchild Publications, Inc.
7 East 12th Street
New York, NY 10003

ELECTRONICS
McGraw-Hill, Inc.
1221 Avenue of the Americas
New York, NY 10020

ENGINEERING ECONOMIST (THE)
The American Society for Engineering Education
Engineering Division
300 West Chestnut Street
Ephrata, PA 17522

ENGINEERING EDUCATION
American Society for Engineering Education
One Dupont Circle, Suite 400
Washington, DC 20036

ENGINEERING NEWS RECORD
McGraw-Hill, Inc.
1221 Avenue of the Americas
New York, NY 10020

FOOD ENGINEER FOR MANAGEMENT
The Chilton Company
Chilton Way
Radnor, PA 19089

FOOTNOTES TO THE FUTURE
Futuremics, Inc.
1629 K Street, NW, Suite 5129
Washington, DC 20006

FUTURES
300 East 42nd Street
New York, NY 10017

FUTURIST (THE)
The World Future Society
P.O. Box 30369, Bethesda Branch
Washington, DC 20014

GRAPHIC ARTS MONTHLY
Technical Publications
Dunn & Bradstreet Division
666 Fifth Avenue
New York, NY 10019

GROUP ORGANIZATION STUDIES
University Associates, Inc.
7596 Eads Avenue
La Jolla, CA 92037

HANDLING AND SHIPPING
Penton/IPC, Inc.
1111 Chester Avenue
Cleveland, OH 44114

HARVARD BUSINESS REVIEW
Harvard University
Graduate School of Business Administration
Boston, MA 02163

HOUSEHOLD & PERSONAL PRODUCTS
Rodman Publishing Corp.
Box 555
26 Lake Street
Ramsey, NJ 07446

IBM JOURNAL OF RESEARCH AND DEVELOPMENT
International Business Machines Corp.
Armonk, NY 10504

IBM SYSTEMS JOURNAL
International Business Machines Corp.
Armonk, NY 10504

IEEE TRANSACTIONS OF ENGINEERING MANAGEMENT
The Institute of Electrical and Electronics Engineers, Inc.
445 Hoes Lane
Piscataway, NJ 08854

IEEE TRANSACTIONS OF SONICS AND ULTRASONICS
Institute of Electrical and Electronics Engineers, Inc.
445 Hoes Lane
Piscataway, NJ 08854

INDUSTRIAL AND LABOR RELATIONS REVIEW
The New York State School of Industrial and Labor Relations
Cornell University
Ithaca, NY 14853

INDUSTRIAL RESEARCH AND DEVELOPMENT
Technical Publishing
1301 South Grove Avenue
Barrington, IL 60010

INDUSTRIAL ROBOT
IFS Publications, Ltd.
35–39 High Street
Kempston, Bedford MK42 7BT, England

INTERFACES
The Institute of Management Sciences
146 Westminster Street
Providence, RI 02903

INTERNATIONAL STATISTICAL REVIEW
Longman Group, Ltd.
43–45 Annandale Street
Edinburgh EH7 4AT, Scotland

JOURNAL OF ACADEMY MANAGEMENT
Academy Management
P.O. Drawer KZ
Mississippi State University
Mississippi State, MS 39762

JOURNAL OF ADVERTISING RESEARCH
Advertising Research Foundation
3 E. 54th Street
New York, NY 10022

JOURNAL OF THE AMERICAN HOSPITAL ASSOCIATION
American Hospital Publishing, Inc.
P.O. Box 10483
Chicago, IL 60610

JOURNAL OF THE AMERICAN STATISTICAL ASSOCIATION (JASA)
American Statistical Association
Business and Technical Department–Editorial Office
806 15th Street, NW
Washington, DC 20005

JOURNAL OF APPLIED BEHAVIORAL SCIENCE
NTL Institute for Applied Behavioral Science
P.O. Box 9155, Rosslyn Station
Arlington, VA 22209

JOURNAL OF APPLIED PSYCHOLOGY
American Psychological Association, Inc.
1200 Seventeenth Street, NW
Washington, DC 20036

JOURNAL OF INDUSTRIAL ENGINEERING
American Institute of Industrial Engineers
25 Technology Park/Atlanta
Norcross, GA 30092

JOURNAL OF MARKETING RESEARCH
American Marketing Association
Edwards Brothers, Inc.
Ann Arbor, MI 48104

JOURNAL OF ORGANIZATIONAL BEHAVIOR MANAGEMENT
Behavioral Systems, Inc.
3300 Northeast Expressway, Suite 1 P
Atlanta, GA 30341

JOURNAL OF QUALITY TECHNOLOGY
American Society for Quality Control
Plankinton Building
161 W. Wisconsin Avenue
Milwaukee, WI 53203

JOURNAL OF SYSTEMS MANAGEMENT
Association for Systems Management
24587 Bagley Road
Cleveland, OH 44138

LABORATORY MANAGEMENT
United Business Publications, Inc.
750 Third Avenue
New York, NY 10017

LONG RANGE PLANNING
Pergamon Press, Ltd.
Maxwell House, Fairview Park
Elmsford, NY 10523

MANAGEMENT SCIENCE
The Institute of Management Sciences
146 Westminster Street
Providence, RI 02903

MATHEMATICS OF OPERATIONS RESEARCH
The Institute of Management Sciences
146 Westminster Street
Providence, RI 02903

MICROGRAPHICS
National Micrographics Association
8728 Colesville Road
Silver Spring, MD 20910

MINICOMPUTER NEWS
Benwill Publishing Corp.
1050 Commonwealth Avenue
Boston, MA 02215

MODERN MATERIALS HANDLING
Cahners Publication Company
Division of Reed Holdings, Inc.
221 Columbus Avenue
Boston, MA 02116

MODERN OFFICE PROCEDURES
Modern Office Procedures
P.O. Box 95759
Cleveland, OH 44101

MODERN PACKAGING
Cahners Publication Company
Chicago Division
5 S. Wabash Avenue
Chicago, IL 60603

MODERN PLASTICS
McGraw-Hill, Inc.
1221 Avenue of the Americas
New York, NY 10020

NEXT
Andrew Reinbach
49 W. 11th Street
New York, NY 10011

OFFICE (THE)
The Economics Press, Inc.
12 Daniel Road
Fairfield, NJ 07006

OMEGA
Pergamon Press, Ltd.
Hennock Road
Marsh Barton
Exeter, Devon EX 2 8RP, England

OPERATIONS RESEARCH
Operations Research Society of America
428 East Preston Street
Baltimore, MD 21202

OPERATIONS RESEARCH/MANAGEMENT SCIENCE ABSTRACTS
Executive Sciences Institute, Inc.
P.O. Drawer M
Whippany, NJ 07981

ORGANIZATIONAL DYNAMICS
American Management Association
Box 319
Saranac Lake, NY 12983

PERFORMANCE IMPROVEMENT
Performance Improvement Publishing Co.
Box 128
500 Main Street
Ridgefield, CT 06877

PERSONNEL ADMINISTRATOR (THE)
American Society for Personnel Administration
30 Park Drive
Berea, OH 44017

PERSONNEL JOURNAL
Personnel Journal
866 W. 18th Street
Costa Mesa, CA 92627

PERSONNEL MANAGEMENT ABSTRACTS
Graduate School of Business Administration
Office of Publication
University of Michigan
Ann Arbor, MI 48109

PROJECT MANAGEMENT QUARTERLY
Project Management Quarterly Institute
P.O. Box 43
Drexel Hill, PA 19026

PRINT
R.C. Publications, Inc.
355 Lexington Avenue
New York, NY 10017

PRINTING IMPRESSIONS
North American Publishing Co.
401 N. Broad Street
Philadelphia, PA 19108

PSYCHOLOGICAL BULLETIN
American Psychological Association, Inc.
1200 17th Street, NW
Washington, DC 20036

QUALITY CONTROL AND APPLIED STATISTICS
Executive Sciences Institute, Inc,
8 Ford Hill Road
Whippany, NJ 07981

RADIO ELECTRONICS
Gernsbach Publications, Inc.
200 Park Avenue South
New York, NY 10003

RESEARCH MANAGEMENT
Interscience Publishers
Division of John Wiley and Sons, Inc.
605 Third Avenue
New York, NY 10016

SALES AND MARKETING MANAGEMENT
Sales and Marketing Management
633 Third Avenue
New York, NY 10017

SCIENTIFIC AMERICAN
Scientific American, Inc.
415 Madison Avenue
New York, NY 10017

SCIENTIFIC NEWS
Science Service, Inc.
1719 North Street, NW
Washington, DC 20036

SCRAP AGE
Three Sons Publishing Co.
6311 Gross Point Road
Niles, IL 60648

SECRETARY (THE)
The National Secretaries Association
2440 Pershing Road, Suite G10
Kansas City, MO 64108

SIMULATION
Simulation Councils, Inc.
Box 2228
La Jolla, CA 92038

SLOAN MANAGEMENT REVIEW
Sloan Management Review Association
Alfred P. Sloan School of Management
Massachusetts Institute of Technology
50 Memorial Drive
Cambridge, MA 02139

SOCIETY
Evans Press
P.O. Box 2033
Fort Worth, TX 76113

SOLID WASTES MANAGEMENT
Cook College
Rutgers–The State University
Box 231
New Brunswick, NJ 08903

TECHNOLOGICAL FORECAST AND SOCIAL CHANGE
Elsevier-North Holland, Inc.
52 Vanderbilt Avenue
New York, NY 10017

TECHNOLOGY REVIEW
Massachusetts Institute of Technology
50 Memorial Drive
Cambridge, MA 02139

TECHNOMETRICS
Technometrics Management
Committee of the American Society for Quality Control
806 15th Street, NW
Washington, DC 20005

TELEVISION/RADIO AGE
Television Radio Editorial Corp.
20th and Northampton
Easton, PA 18042

TRAINING AND DEVELOPMENT JOURNAL
American Society for Training and Development, Inc.
P.O. Box 5307
Madison, WI 53705

TRANSPORTATION AND DISTRIBUTION MANAGEMENT
Traffic Service Corporation
Washington Building
Washington, DC 20005

WAREHOUSING AND PHYSICAL DISTRIBUTION PRODUCTIVITY REPORT
Marketing Publications, Inc.
529 14th Street, SW
217 National Press Building
Washington, DC 20045

WHARTON MAGAZINE
C. Lynn Coy Associates
220 East 54th Street
New York, NY 10022

WIRELESS WORLD
Electrical-Electronic Press, Ltd.
Dorest House
Stamford Street
London SE1 9LU, England

WORD PROCESSING WORLD
Geyer-McAllister Publications
51 Madison Avenue
New York, NY 10010

WORLD OF WORK REPORT
Work in America Institute, Inc.
700 White Plains Road
Scarsdale, NY 10583

REFERENCES

1. PETER F. DRUCKER, *Management: Tasks, Responsibilities, Practices*, Harper and Row, New York, 1973, p. xiii.
2. NEVILLE HARRIS, *Management Services in the United Kingdom*, The Institute of Management Services, Enfield, Middlesex, England, 1979.
3. W. A. REYNOLDS and M. K. CHEUNG, "Some Aspects of the Organization and Use of Industrial Engineering Techniques in Hong Kong," *International Journal of Production Research*, Vol. 15, Nos. 5 & 6 (1977).
4. KNUT HOLT, "Industrial Engineering: A Dynamic Response to Change?", *OMEGA, The International Journal of Management Science*, Vol. 3, No. 5 (1975), pp. 523–540. (This paper is an abridged version of a report by K. Holt entitled "Industrial Engineering–A Dynamic Response to Change? Organization and Practice of Industrial Engineering in Selected American and European Companies," published in the report series of The Division of Industrial Management, University of Trondheim, The Norwegian Institute of Technology.)

BIBLIOGRAPHY

In addition to the references listed above, the following is a useful source:

MORRIS, WILLIAM T., "Implementation Strategies for Industrial Engineers," Grid Publishing Co., Columbus, OH, 1979.

CHAPTER 1.3

Improving the Effectiveness of Industrial Engineering Practice

GERALD NADLER

University of Wisconsin

1.3.1 INTRODUCTION

If you had thorough knowledge of all the techniques and principles in this handbook, would you be assured of practicing industrial engineering (IE) effectively?

Based on even the simplest level of understanding about practice, your likely answer is "no." When differences among characteristics are considered from one organization to another–history, mythology, management style, the personalities of the industrial engineers (IEs), the knowledge of IE techniques and principles possessed by others in the organization–"no" becomes firmer. When the many externalities are considered–actions of competitors, conditions of the economy, supplies of materials–a "no" answer is reinforced.

Then *how* is an IE to practice the profession effectively? If knowledge of techniques and principles on the one hand provides no assurance, and the surroundings (externalities and company characteristics) on the other hand provide "interference," a lose-lose situation seems to emerge. Fortunately, we know from the many successful IEs and IE programs in organizations that win-win conditions can and do exist.

There seems to be a conversion-to-practice linkage not often described or taught that bridges the gap between the overwhelming number of techniques and tools of the IE trade and the awesomeness of the surroundings. Take the linchpin away, and we have the sterility and negativism of the "technological fix" of the technique specialists, or the immobility and helplessness of the "we can't do anything about it anyway" of the status quo protectors.

The key link is a "total approach" to finding effective and innovative responses and solutions that are likely to be accepted or implemented. A total approach shows *how* to decide when the many IE techniques, measures, and models are or are not useful, and *how* to interrelate appropriately with the real-world organization and environment.

1.3.2 PRACTICING IE EFFECTIVELY

What constitutes effective IE practice is addressed often but insufficiently in almost every chapter of this handbook. All techniques and procedures, for example, are presented with criteria for gauging how well they are *applied*. Statements are offered regarding "acceptable levels" of accuracy and precision for almost all of the *techniques and procedures*. Definitions of IE and IE departments are provided, in Section 1 especially, to set out objectives and goals that the *IE field* and its efforts should forward. Well-stated principles of what a "good" solution ought to "look" like (e.g., an organization structure, a manufacturing layout, a user-based information system, a quality assurance program) are proposed for *adoption and installation* in a real-world situation.

Yet all of us know about many failures even when the "best" techniques and models were used, "accurate and precise" measurements were obtained, "high-level" objectives and goals were guides for the IEs, and the "most modern" principles were used in developing the solution.[1] At the same time, all of us know about the successes that have occurred when far from "pure" ideas were involved. As a matter of fact, no references are available to show that any significant real-world re-

---

Much of this chapter is adapted from G. Nadler, *The Planning and Design Approach*, Wiley, New York, 1981.

sults ever match the pure ideas.[2] As necessary as pure ideas are for stimulation and a knowledge base, they are far from sufficient in defining effective practice of IE. Industrial engineers should stop talking about improving decision making with pure techniques and models and talk instead about helping clients with their problems.

A three-pronged concept of effectiveness in IE practice emerges when considering this background:

1. To maximize the quality of recommended solutions for or responses to the real-world organization.
2. To maximize the likelihood of real-world implementation of the solutions or managerial acceptance of the responses.
3. To maximize the effectiveness of all real-world and IE resources used in the efforts.

The first prong of the effectiveness concept incorporates many pure ideas within the word *quality*. Most of us define quality in terms of incorporating the best techniques and highest levels of accuracy and precision in a response, or incorporating the most modern principles into a solution. Other descriptors are frequently encountered: reliable, adaptable, innovative (a creative solution that is somewhat inefficient is often much better than a standard or unnecessary solution that is very efficient), pluralistic, satisfactory to client, high benefit/cost ratio, efficient use of resources needed in a solution (machines, money, material, energy, information, personnel, and facilities), and simplicity. A quality solution or response also contains the seeds of ideas for its own continuing improvement. Each organization and manager has individually developed criteria for defining the quality of solutions or responses. (Because so many criteria are involved, optimization of tradeoffs among them are inevitable. Optimization at the operational level of selecting particular specifications does not conflict with the conceptual maximization involved with this prong. Maximization would seek first to favorably assess whether or not the right problem is being worked on, the appropriate people are being involved, and necessary interfaces are specified. Then optimization can take place within the maximized framework.)

In the second prong, failure to implement effective solutions or to get a client to accept a response (e.g., evaluation report, description of the cause of a machine failure, synthesis of knowledge in a particular topic) is *prima facie* evidence that something went awry in the IE process. Effective IE practice is often portrayed as the product of multiplying the quality of the solution or response by the implementation or acceptance. Even if the solution or response quality is 100, nonimplementation or nonacceptance (zero) produces no effectiveness of practice. Implementation and acceptance involve changing the behavior of real-world people from their state at the time of their initially perceived need through selection, approval, installation, and use of the solution or response to a state of continuing search for its improvement.

For the third prong of the effectiveness concept, it is conceivable that with unlimited amounts of time and money many problems could be solved or responses could be obtained. But it is highly unlikely that any organization would commit itself to such vast expenditures. Limited time and money demand maximizing the effectiveness and efficiency of IE professionals and the real-world people brought into the IE efforts. This constitutes the praxiology of IE,[3] an essential *quid pro quo* offered by IE professionals to the organization in return for the problem solving responsibility delegated to them. Exhibit 1.3.1 illustrates only one way of depicting the benefits of this third prong. It describes the result of a survey of 48 companies,[4] and shows that over twice the economic savings per staff member can be obtained with the conversion-to-practice ideas developed in this chapter. Conversely, a given level of required economic savings can be obtained with roughly half the number of IE and other staff personnel.

Attempting three concurrent maximizations immediately identifies the need for the conversion-to-practice total approach linkage in IE. There is just no rigorous theory incorporating techniques for triple maximization. Quantitative methods for identifying tradeoffs within and among the many factors do not exist, and scientific measurements of the factors themselves are very often not available. Clearly, IE does not even begin to have such theories and concepts. Interrelationships in an organization and environmental conditions are so complex and human aspects so difficult, that the probability of ever having such theories and measures is infinitesimally small.

What remains, then, is that the effectiveness of IE practice is just as much a *perception* on the part of humans (managers, workers, customers, suppliers, client, users, and the like) as it is measureable (e.g., cost reductions, productivity improvements, time-delay elimination). It can be proven that 23% of the costs can be saved *now*, but if the workers do not perceive the truth of the measurements or solution, a costly strike can occur. Effective IE practice in this case would require that the costs and long-term negativism of a strike be included in the original calculations, most likely leading to the adoption of a phased-in installation of changes.[5] Similarly, a favorable human perception of IE's contribution to the organization's effectiveness and viability leads to numerous problems being referred to IE where no cost or time savings or objective measures of effectiveness are readily available, for example, design of new facilities, site location, acquisition evaluations, new product feasibility studies, and so on.

Exhibit 1.3.1 Survey of 48 Companies Illustrating How Direct Program Staff is Effectively Utilized

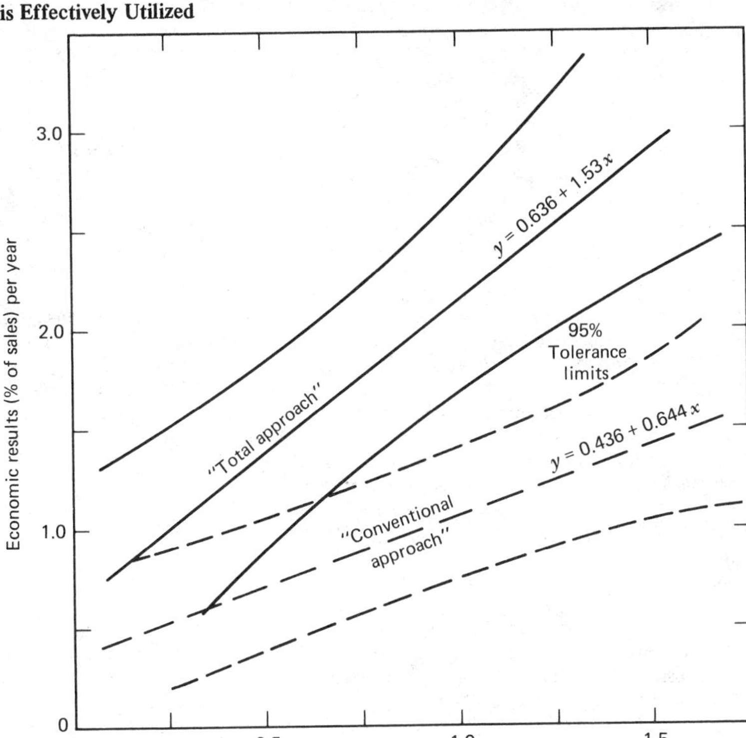

1.3.3 A TIMELINE VIEW OF PRACTICE

Building a conversion-to-practice link between the techniques and the surroundings requires *a timeline perspective of what an IE ought to be doing.*[6] Time is irreversible. It cannot be slowed or speeded up. It illustrates perfectly negative and positive infinity. The grand concept of the passage of chronological time is the only basis for understanding how IE should be practiced. "Time has no divisions to mark its passage. . . . It is only we mortals who ring bells and fire off pistols."[7]

A continuous (rather than discrete) timeline is the fundamental basis for understanding the past, present, or future of any phenomenon. Assume that a straight line or vector in space with an arrow at one end represents a timeline. Any arbitrary point can locate the present (second, minute, hour, day, week, month, or whatever unit), which automatically defines the past and the future.

Many techniques (such as formulas, pictures, drawings, or graphs) are symbolic forms to abstract, model, or describe the past, present, or hoped-for conditions of the phenomenon of interest at arbitrarily selected previous, present, or future points of time.

Each phenomenon description thus far is static. Most contain only a limited perspective about the phenomenon, even omitting knowledge already available about its past and present conditions, but are deemed to focus elsewhere. Descriptions of a future are typically predictions of what the static conditions may be or what someone or some group wants the conditions to be in the future. Names and labels are often assigned to such snapshots of the future such as "post-industrial society," "automation," or "autonomous work groups." All of the visions of good solutions projected by the principles in this handbook illustrate types of static future snapshots.

Such static snapshots of future outcomes are most often inappropriate because each snapshot for a particular phenomenon assumes that it is "right." But rightness is usually based on the techniques, models, or principles which are *arranged and synthesized on the basis of the past*! In addition, an arrangement or synthesis aggregates data from many past cases, each of which was "successful" in varying degrees and with different groups of people. *Industrial engineering practice must look forward to what the specific human in the specific institution will perceive as right or successful solutions or responses in solving their problems.* This must surely be the objective an IE seeks, certainly *not* a commitment to apply techniques or increase measurement accuracy per se. No assurance can be provided about the efficacy of any technique or model in this forward-looking perspective, nor can a priori criteria assure selection of which technique would be most appropriate. Many techniques, models, and measurements will still be useful, but many others ought not be used at all. This background provides a substantive explanation of what is meant by the exhortations to be sure one is "asking the right questions" and "working on the right problem." It also sets the stage for utilizing the timeline view of IE practice.

In effect, the world of IE practice works along a timeline *in parallel* to the client, departmental, or real-world situation in which the problem or need emerges. The real world continues to operate and produce products or services even as it and the IE decide to work on a problem. In addition, many other staff or indirect entities or worlds (e.g., long-range planning, finance, organizational development, product engineering, data processing) are working in parallel to both the operating and producing real world and the IE world. The problem may occur in one of the staff worlds. At the very least, the IE world must interface with several worlds day by day, week by week, and month by month as the problem is being worked on.

Throughout the efforts along a timeline, the operating real world insists on obtaining its results, that is, the outcome that it was set up to produce in seeking to achieve its purposes. The IE must help solve problems the organization feels are important to continue its outputs or improve some aspects of them. To get the organization to agree to utilize the quality solutions or responses effectively developed through the IE's involvement means that much more than a snapshot of the solution or response is needed. Just as critical is the day by day, week by week, month by month transition plan to assure eventual operation or utility.

Further explanation of how the timeline view of practice can help an IE now depends on some definitions of the word *problem*. What is clear is that almost everyone agrees there are different types of problems,[8] but what types there are is not clearly settled.

1.3.4 PROBLEMS

The word *problem* has been given different definitions. Several illustrative ones include a "felt difficulty,"[9] "a problem is a stimulus situation for which an organism does not have a steady response,"[10] stresses and strains in thought structures,[11] "gap,"[12] "dissatisfaction" with a purposeful state,[13] and an obstacle to be circumvented.[14]

Another typology calls problems well structured or ill structured. Other classifications are sometimes suggested: programmed or nonprogrammed, constrained or nonconstrained, open or closed, routine or nonroutine, technological or human, or deterministic or stochastic. Almost all of these are inappropriate if not outright misleading. They concern classifications based mainly on the techniques or models available, not on the problem or need of the real world.

Engineering often defines four types of problems its professionals face:

1. To improve an existing system (e.g., reducing size, weight, or cost, or improving performance or appearance).
2. To diagnose and remedy some trouble or find the cause of an accident in an operating product or working system.
3. To develop a new system or combination of objects, information, humans, and energy to achieve a purposeful, situation-specific solution.
4. To develop a new use for existing devices, information, or systems.

These categories are useful to IEs and are incorporated later in a more encompassing view.

The dictionary is probably the best source from which to draw a definition that brings together all of the ideas. It defines a problem as a *substantive matter about which there is concern*. A substantive matter may be a question, situation, phenomenon, person, or issue. A concern may be an uncertainty, obstacle, desire, difficulty, or doubt. Whatever the formulation, a simple definition seems to remain, *something* about which there is concern.

From here on, "something" is designated the substantive aspect, and "concern" the values aspect. The substantive aspect includes both the type of problem it is and its specific locus, the locus being the "what, where, when, and who" unique to each problem situation. The values aspect encompasses the desires, aspirations, and needs that have made the substantive aspect a matter of concern.

The substantive aspect classifies[15] problems in the context of human purposeful activities. This enables IEs to identify that kind of problem the real world is asking them to solve and the appropriate methodology for solving it. The values aspect explores the human motivations that make something a problem. There are important benefits in this definitional format:

1. Defining the substantive aspect of a particular problem gives greater assurance that an appropriate methodolgy will be used. For example, a particular situation may pose a planning and design problem. If it is approached with a research methodology, the end product is likely to be

a series of studies, not an operational solution. Strict attention to a problem type and locus significantly reduces the probability of an error of the third kind,[16] in other words, working on or finding the perfect solution for the wrong problem.

2. A clear idea of a problem's locus focuses the problem solving effort on the specifics of each unique situation. Rather than transferring one solution to another situation, each one is custom tailored to specific needs, values, and resources.

3. Designating a problem's value aspects places problem solving squarely in the context of human aspirations and needs. This forestalls the unhappy tendency of IEs to become specialists without a structure of ethical responsibility. Such specialists assume that problems will be defined simply to fit into available techniques.

These benefits lead to the following criteria for determining how substantive matters may be defined:

1. Categories should minimally overlap but still encompass all the problems associated with all activities in which humans engage.

2. While identifying problems, categories should focus on the purposes with which humans confront a situation rather than on problems. To do otherwise unnecessarily restricts the solution finding space.

3. The categories should produce a prescriptive understanding of what to do about the problem, that is, one that suggests a methodology.

4. Categories should enhance the probability of working on the right problem and developing creativity in solution finding.

As a result of the imperatives of these criteria, the following classification organizes substantive matters on the basis of purposeful activities, that is, on the repertoire of behavior we employ day in and day out in the process of living.

1.3.5 PURPOSEFUL ACTIVITIES AND VALUES

Purpose implies aim or intention. Purposeful means "having the qualities of" and "characterized by" purpose (mission, aim, direction, primary concern). In addition, purpose has the characteristics of constructive, legitimate, organizationally needed, socially needed, and socially acceptable. Activities are the behaviors associated with aim or intention. The following categories emerge when considering purposes of humans, thus identifying the following fundamental human purposeful activities:

1. Assure self-preservation and survival; *self-preservation*.
2. Operating and supervising an existing good solution or system; *operating and supervising*.
3. Create or restructure a situation-specific solution or system; *planning and design*.
4. Search for generalizations or find causes; *research*.
5. Evaluate performance of previous solutions or other purposeful activities; *evaluation*.
6. Gain skills or acquire knowledge about existing information and generalizations; *learning*.
7. Experience leisure; *leisure*.

Purposeful activities (1) and (7) are not discussed further. Activity (1) is a basic, instinctive one that leads to one or more of the others, once a level of physical and social well-being is attained. Except for doing nothing, (7) also gets converted to one of the remaining five, once a person decides what leisure really means personally. Exhibit 1.3.2 lists some problems that illustrate each of the five remaining purposeful activities.

The five major purposeful activities could concern any specific object or locus. A shipping department, as an example, may be associated with all five. The purposeful activities enable an IE to identify whether the shipping department poses a problem of operating and supervising, planning and design, learning, evaluation, or research. The purposeful activity may be planning a layout, operating and supervising it, or evaluating it. The five are not mutually exclusive. Each may be involved with, and depend on, the other. Successful planning and design frequently requires, at various points in a project, research, evaluation, and learning.

Industrial engineering incorporates more types of problems than most engineering branches (see Section 1.3.4 above). It still helps to improve systems (with the operating and supervising and the planning and design purposeful activities), diagnose or find causes (research), develop new systems (planning and design). and develop new uses (planning and design). An IE also does evaluation, research, and learning.

The primary purposeful activities encompass a number of secondary ones, which are not exclusive to any single primary activity but occur frequently in all. They include:

Exhibit 1.3.2 Problems Illustrating the Different Purposeful Activities

Operate and Supervise

Determine if the quality control limits should be relaxed to accept more parts.
Establish two "downside" budgets at 5% increments of sales levels.
Write up the specifications for the fourth-quarter working capital loan.
Update standard cost files to reflect new materials and labor rates.

Planning and Design

Develop a national gypsy moth pest management system for the United States.
Determine if six factory locations should be consolidated into one new manufacturing facility.
Recommend if a $40,000 cost to automate each loading dock at each of 24 warehouses (with an 8-month payback) should be approved.
Set up an information system for admitting patients and concurrently setting up all record files.

Research

Find the relationship between sales of original equipment throughout the country and parts sales.
Determine the cause of the accident with the crane operator.
Identify why the entry level skills of new employees are declining.
Establish a profile of customer attitudes toward products.

Evaluation

Determine whether or not the 1300 service centers throughout the country are effective.
Assess the performance (costs, output levels, and so forth) of the 1-year-old factory.
Inspect nursing homes at random to determine if each is in compliance with state quality standards.
Evaluate the effectiveness of the cost reduction program set up 2 years ago.

Learning

Provide to the vice president a summary of automation methods currently available.
Identify the range of desk-top computer terminals available today.
Determine the state of the art regarding regional planning methods.
Arrange for teachers to practice team-building skills.

Make a decision.
Maintain a standard of achievement (control).
Resolve a conflict.
Model or abstract a phenomenon.
Develop creative ideas.
Establish priorities.
Practice and exercise individual efforts.
Focus and motivate individual efforts.

The fundamental purposeful activity sets the context for the secondary ones. Make a decision about what—operating and supervising, or planning and designing a solution? Resolve a conflict about what—evaluation or learning? Develop creative ideas about what? Model a situation for what purpose? Establish priorities for what reason? Practice a skill for what reason? Focus and motivate individual efforts to achieve what primary purposeful activity?

Why does something become a problem? Why do humans seek to better the world and themselves? What does "better" mean, anyway? What is the "concern" aspect of a problem? Even if we cannot know the causes of human motivations and values, we can assess, albeit incompletely and inaccurately, their expression.

Contemporary theorists argue that values, ethics, and motivation stem from human instincts, wants, and desires for certainty and security. The nature of these needs has been explored by various authors. The needs hierarchy is one of the best known.[17] The hierarchy starts with (1) physiological existence and (2) security needs, and is then followed by the need levels of (3) social and affiliation, (4) esteem and reputation, (5) autonomy and independence, and (6) self-actualization. Two other needs are suggested but not emphasized: (7) cognitive need to learn and understand, and (8) esthetic. As people's needs are met at one level, they seek the next.

In discussing human motivations it is essential to define the word *better*. Humans are almost always seeking something better, which explains how problems get started. Someone becomes concerned because something better is thought possible.

Better is multifaceted. It never reflects a single desire or need. For example, improved productivity is a "better" that means concern for costs, material usage, machine utilization, impacts on supplier, and so forth. When the desire for better is directed toward some thing or situation (e.g., the shipping department), it can be expressed on three levels: as values, as objectives, and as goals. In reference to the productivity example, better also encompasses the *values* of "learning for its own sake" and the desire to apply knowledge about productivity improvement to "meet human needs." Values are beliefs, desired endstates, societal and individual aspirations, and desiderata. These values are then expressed in specific objectives and goals (measures). *Objectives* are the criteria for determining how well a particular value is achieved (for the productivity example, some objectives may be to reduce material cost per unit, increase machine utilization, improve gross margin per unit of sales). *Goals* are the performance levels or amounts of an objective to be attained within specified time or cost limits (e.g., decrease material cost per unit by 10% within 8 months).

In defining a problem, IEs need to establish the values, objectives, and goals of the particular purposeful activity at a specific locus. It is the desire for better as expressed in the values and measures of human purposeful activities that makes a problem out of a particular phenomenon. An effective solution will reflect and deliver more of the values than were present at the rise of the problem. Objectives and goals operationalize values for a specific locus.

The values, objectives, and goals associated with each of the five purposeful activities are different. A manager who asks an IE to determine the relationship between the capital cost of various pieces of equipment and the percent of downtime for them may be motivated by curiosity or by previous maintenance reports. The manager may not have any idea how the information will be used, but thinks it may be useful in the future. Thus, a *research* purposeful activity tends to have different values, objectives, and goals than would an operating and supervising one, and so on. If the manager asks an IE to look into keeping the maintenance system in good order, the values, objectives, and goals associated with that *operating and supervising* purposeful activity are different.

Acknowledging that there is a values aspect to all problems means that the IE cannot adopt an objective stance. Finding a solution to a problem always incorporates subjectivity and human concerns. This view should remove IE efforts from the realm of narrow disciplines and techniques. It forces the solution measures to transcend the merely quantifiable and to incorporate critical subjective factors.

1.3.6 ACHIEVING PURPOSEFUL ACTIVITIES (SOLVING PROBLEMS)

Critical now to improving IE practice is recognizing that *a different approach should be followed in trying to achieve each of the purposeful activities*. In generic terms, this means that a different problem solving mental set is needed for the different kinds of problems (including the ever-present one of deciding *not* to do anything at all for the problem). In operational terms, this means that the types of questions asked and the nature of the information obtained will vary greatly.

These differences in approach can be described in terms of a timeline scenario. The real world (or an IE sorting through data about the real world) identifies a problem (substantive matter and concerns). A problem seldom appears suddenly. Dissatisfactions, desires, poor performances, or uncertainties usually build up to the point where the real world (organization, department, key organizational committees on which the IE hopefully serves, and the like) decides (or the IE suggests) that something needs to be done. Doing something almost always involves some time limitations. The IE immediately tries to sort out with people in the real-world situation just what type of purposeful activity is really being considered. Values and motivations for proclaiming that a difficulty represents a problem can be helpful in identifying the purposeful activity involved because a different *scenario could be tracked along the timeline for each purposeful activity*.

The problem is then "transferred" to the research world, or planning and design world, or evaluation world, and so on. The purpose to be achieved is the critical feature. The IE today, using the same approach for all problems, usually proceeds by collecting large amounts of data and then withdrawing to an office to solve the problem. Contact with the real-world client or department occurs almost always on a discrete, disjointed, and occasional data source basis: get the sales statistics, fit the situation into a taxonomy or classification scheme, determine the number of distribution points, and so on. People in the real world are very seldom told the context for which the information is needed, so they start almost immediately to develop defenses and skepticism about the efforts. This often leads them to providing biased data that they feel puts them in a favorable light (whatever that may be).

As the IEs in their usual problem solving world go on, commitment becomes firmer and firmer regarding what they believe are the perceptions and assumptions structuring the client's decision to seek help in solving the problem. The IE perceptions at this later point are usually different from those the IEs had at the start of the project. The IEs begin to develop innovative responses or solutions as their perspective, assumptions, and knowledge base change in a way different from those of the client.

Over the same time span, the real world changes, and people's perceptions, priorities, and understanding of the problem there will thus be different. These modifications start to occur almost immediately after the IE problem solving effort begins. "Disturbances," "normal operating changes," and "new technology and knowledge" from all parts of the environment influence and change the client's world. The problem itself may change, or its importance may diminish or increase. Most of these changes are imperceptible, but they still alter the subconsciousness of the individuals in the organization or department. As more of them occur, the real world itself becomes different, and the perceptions of the scope and context of the project can be dramatically modified.

The IEs operating in this conventional mold eventually present the recommended solution or the asked-for response to the real world. It is at this point that the failure of so many IE efforts finally shows up. The IE world and the real world have each greatly modified their perceptions of the problem but almost inevitably in different directions. The client rejects what the IE submits or proposes because the premises underlying how each views the proposal at that point in time are just too divergent. The solution finally adopted, if any, is far too often a compromise that neither side particularly likes. Lack of creativity, unachieved purposes, defensiveness, unmet or barely met objectives, hostility, and unneeded procedures often characterize the results.

Although this usual scenario appears to apply to large projects, similarly changing perspectives occur with small projects. The reasons for the differences in perspective also explain why an attempt to transfer a solution from one situation to another is almost never successful. Industrial engineering efforts to sell or adapt a solution developed for another situation are inappropriate because no two situations are really alike.

Continuing interactions are needed between the client and IE worlds at every step and phase of all the approaches. Such interactions are needed even if people from the real world serve on the project team. Real-world people on teams become IE-world people as the project proceeds, because their perceptions are very likely to be IE oriented soon after the project's start. All of the other people in the real world need to be kept involved with joint interactions at many points in the process. The major responsibility for the interaction falls to the IE, who must adapt to and continually seek this joint interchange of perspectives.

This frequent sharing of perceptions and understanding is greatly aided by considering the particulars of a different approach for achieving each purposeful activity. The first concept involves recognizing that many conceptual schemes are represented in problem solving literature—rational, affective, reductionist, chance, intuitive, incremental, and so on. The second is to recognize that each one or each combination may be better for different purposeful activities. The third is that all of the conceptual schemes need to be incorporated in varying ways into a different total approach for each purposeful activity.

The main reason that all approaches (rational, affective, reductionist, and so on) need to be incorporated into a total one is that humans know and use them all on occasion as methods of solving problems. To claim one or the other as "it" belies human perceptions. The need to incorporate all approaches into a total one is buttressed by the growing realization, even among scientists, that the rationalist tendency of the West "to divide the perceived world into individual things . . . is not a fundamental feature of reality. . . ."[18]

The next question concerns *how* a total approach is to be detailed. Human perceptions are again the most likely source of answers, especially from those humans who have previously exhibited quite successful solution finding behavior and approaches. The evidence shows that determining those factors, within the holism of a particular purposeful activity, that describe a total approach will never be complete. A review of the strands of human processes for change and the operational concerns of practitioners in moving over time toward response generation or solution finding leads to the proposition that five factors might suffice in describing a total approach. That is, a total approach is an integrated whole of simultaneous attention to five intertwined factors at each point in time:

1. Pursuing a strategy and protocol of action phases and steps. (What set of steps or thinking phases should one follow as the approach proceeds in time? What is the decision making flow along the timeline, to put it in terms of human perceptions?)
2. Specifying and presenting the solution or response in a framework of properties and attributes suitable for the purposes. (A "system" is often used as the desired framework, but a research solution, a planning and design solution, an operating and supervising solution, an evaluation solution,

and a learning solution are so obviously different, that a different "systems" framework for each may be expected.)

3. Involving people from the client or organizational world while pursuing the strategy so that acceptance or implementation is likely. (Varying amounts of involvement will occur depending on the type of problem and the number of affected people. People should continuously be given an *opportunity* to participate and interact.)

4. Identifying and using knowledge and information most effectively for the situation. (Studies, experience, legends/myths, and the like have a wide range of potential utility in each of the purposeful activities. For example, research or describing what happened uses different information and knowledge than does operating and supervising or predicting the pertinence and consequences of deliberate actions or changes.)

5. Arranging for continuing search for change and improvement of even the forthcoming solution or response. (Searching for changes and improvements in all areas of the real world and implementing those that are feasible and worthwhile avoids future shock, which only occurs when change is imminent for people not expecting it. Each solution or response should contain its own projected changes and improvements. Continuing *search* for change in all solutions and areas of an institution is stability.)

The different total approach to each purposeful activity can be explored initially by identifying a theme that describes the holism of each approach. Each one appears to need different reasoning processes and techniques if the five factors in each are to become operational. Detailing what each factor in the total approach might entail for each purposeful activity builds on a variety of doubting/believing perspectives and philosophical reasoning from the various individual approaches. Exhibit 1.3.3 presents this structure, briefly summarizing ideas about each factor for the respective purposeful activities. Most of the information is familiar to the reader, but its display in a different perspective should provide relevance for problem solving. No claim is intended for completeness or accuracy, since much more could be written about each of the total approaches. Yet the exhibit formalizes the basic concept that *the approach or response to a problem* varies according to *the purposeful result that is sought*. Purpose is always the governing factor.

The five factors for each purposeful activity are broadly arranged in Exhibit 1.3.3 along a timeline. That is, the first step under *pursuing the strategy* holistically roughly interweaves the first item in each of the other four factors. Then the second step interrelates approximately the second item of each factor. Looping and jumping among the steps are to be expected, thus signifying that parts of all factors jump and loop within several steps of the strategy.

Each factor varies in its importance at different times in an approach and from one purposeful activity to another. Involving people in planning and design is more important in all phases than it is in, say, research. Or the standards of specifying a solution are more important in research than in operating. Or using current information and knowledge is more important at the beginning of evaluation than later.

A much more complete timeline scenario can portray the total approach for each purposeful activity. Exhibit 1.3.4 shows the planning and design timeline scenario.[19] In addition to illustrating the five factors, Exhibit 1.3.4 shows the continuing and jointly needed interactions between the planning and design world (in this case) and the real world. Point 2, design a solution finding structure, is particularly beneficial for an IE seeking to improve professional performance. It says, in effect, that solving each problem should be initiated by defining how the solution response will be sought—who should be involved, what purposes should the solution finding really achieve, what inputs are to be used, when the solution finding structure or system will work, how much money is needed, and so on. Then this structure can be put into operation by the IE with good assurance of support for the effort because, in almost all cases, the solution finding structure is developed by the appropriate managers and decision makers in conjunction with the IE.

Determining which purposeful activity and thus approach are to be considered when "a problem walks in the door" or data indicates to an IE that a problem may exist can be aided by asking several questions: What is the nature of the uneasiness? What needs are not being satisfied? What result is really being sought? What are the expectations of the manager? What are the values that best express the benefits sought? What does the group or manager consider important? What purpose is to be achieved? What purpose would that achieve?

Most of these responses are expressions of purposes, values, and objectives/goals. By arranging them in a hierarchy from smaller to larger purposes or values/objectives as perceived by the group or client, the IE or group has an excellent opportunity to review the contextual scope of the problem and thus be able to select the right purposeful activity and related approach. Each person will have a chance to use his or her own criteria in selection of "the" problem within the whole context, even though the criteria include such political ones as "whose ox is gored," frequency of occurrence of difficulty, and personal factors. Of course, identifying one approach is only the beginning, since several purposeful activities may be involved *over time* with the same problem.

Points 1, 1a, and 2 in Exhibit 1.3.4 are three that would appear on the timeline scenario for each purposeful activity.

Exhibit 1.3.3 Summaries of the Intertwined Factors of a Total Approach to Each of the Purposeful Activities

Purposeful Activities	Intertwined Factors of a Total Approach				
	Pursuing the Strategy	Specifying a Solution	Involving People	Using Information	Arranging for Change
Operating and supervising	Learn objective and norms of the system; Obtain resources and measures of systems performance; Lead, motivate, control, and correct operational processes.	Performance objectives; Specifications for inputs, equipment, personnel, and information systems; Operating policies; Personnel policies; Specifications for the sequence of operations.	Active manager direction together with clients, employees, and others; Experts to analyze resource tradeoffs, plan sequence, design communication, coordination, and measurement systems.	Measurement and information systems; Organization theory; Engineering and marketing knowledge; Financial analysis; Timely and well-focused information about current operations within overall plan of of operations.	Establish long-range planning to improve the system; Obtain resources for continuing improvement of operations (consultants, in-house seminars, personnel development and management improvement programs, courses, conferences, etc.); Develop change attitudes.
Planning and design	Select purpose(s) to be achieved from hierarchy(ies); Develop feasible ideal solution target for regularity conditions; Detail solutions to achieve purpose(s) for all conditions.	Purpose hierarchy and dimensions of selected purpose(s); List of ideal and possible solutions; Specify solution with 8 elements and 6 dimensions of system matrix.	Decision makers, users, implementers, and others affected by the solution in defining the purpose, generating ideas, and selecting the solution; Professionals used mainly as process coordinators, with "expert" roles used in response to specific needs.	Planning and design theory and techniques guide process; Other information is gathered only if relevant to achieving the planning and design purpose in specific setting.	Build adaptation into solutions; Periodic replanning; Develop organizational support for ongoing improvement of satisfactory systems rather than crisis correction.
Evaluation	Identify evaluation users' purposes and perspective; Establish or reconfirm values and objectives; Measure performance; Interpret the results.	Definition of users; Descriptions and operational scales for their values and objectives; Specifications for measurement procedures; Measurement data and interpretation.	Users to define perspective; Clients to clarify and operationalize values and measures; Professionals to measure; Users, clients, and influentials to analyze and interpret with professionals.	Techniques for clarifying and operationalizing objectives, gathering measurement data, and analyzing measurement results.	Develop ongoing evaluation arrangements; Build evaluation into new systems; Provide adaptability to changing values, objectives, and reference groups; Upgrade participants' evaluation skills; Periodically evaluate the evaluation system against *its* objectives and make necessary changes.

Research	Review background about area of research; Formulate general hypothesis; Determine operational implications of the hypothesis; Devise tests; Verify or refute the generalization.	Definition of the subject area; Investigator perspectives; Causal or statistical hypothesis; Statements of assumptions and implications to test; Empirical results.	Investigators and resource people (sources of background information); Experts for design of experiments; A larger professional community to shape the research agenda and interpret results.	Background resources ("the literature"); Techniques for developing models and designing experiments; Existing theories; Information about the subject area to be used comprehensively, because the sequence of discovery is guided by its empirical results, in unpredictable directions.	Develop a long-range research agenda; Seek unifying higher-order generalizations; Use the results of the present study as the starting point for further investigations.
Learning	Decide on intended information or skill; Determine level to attain; Implement learning activities; Assess information or skill acquisition.	Forms of information, concepts, or new applications of ideas; Improved ability to form new concepts; Choices of cognitive, affective, and psychomotor objectives; Stated sequence of learning; Internalized results within personal "maps."	Learner as primary participant with resource people (teachers, consultants, librarians, etc.) to assist.	Learning theory; Information about particular areas of study; Existing techniques and strategies for different types of study; Recommended curricula; Varieties of previous experiences and learning; Accumulated experience of resource people; Information recorded and indexed in various forms.	Long-range focus on "learning to learn"; Periodically review short-range objectives in relation to larger purposes; Identify major affective goals such as overcoming learning aversions; Revise learning plans to adapt to changing needs.

Exhibit 1.3.4 Illustration of One of Five Total Approaches, Shown in Relation to the Real World with which an Industrial Engineer Works

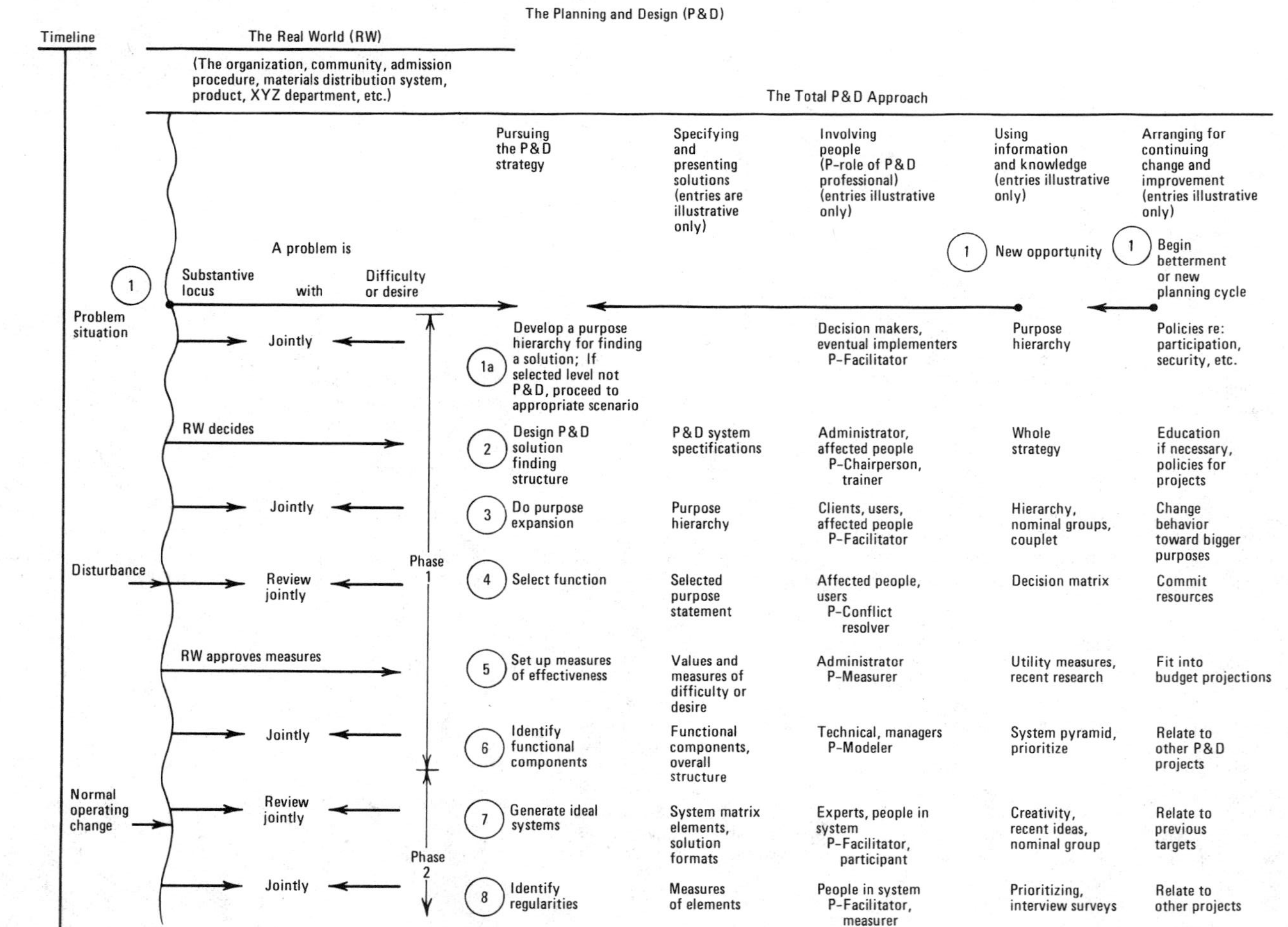

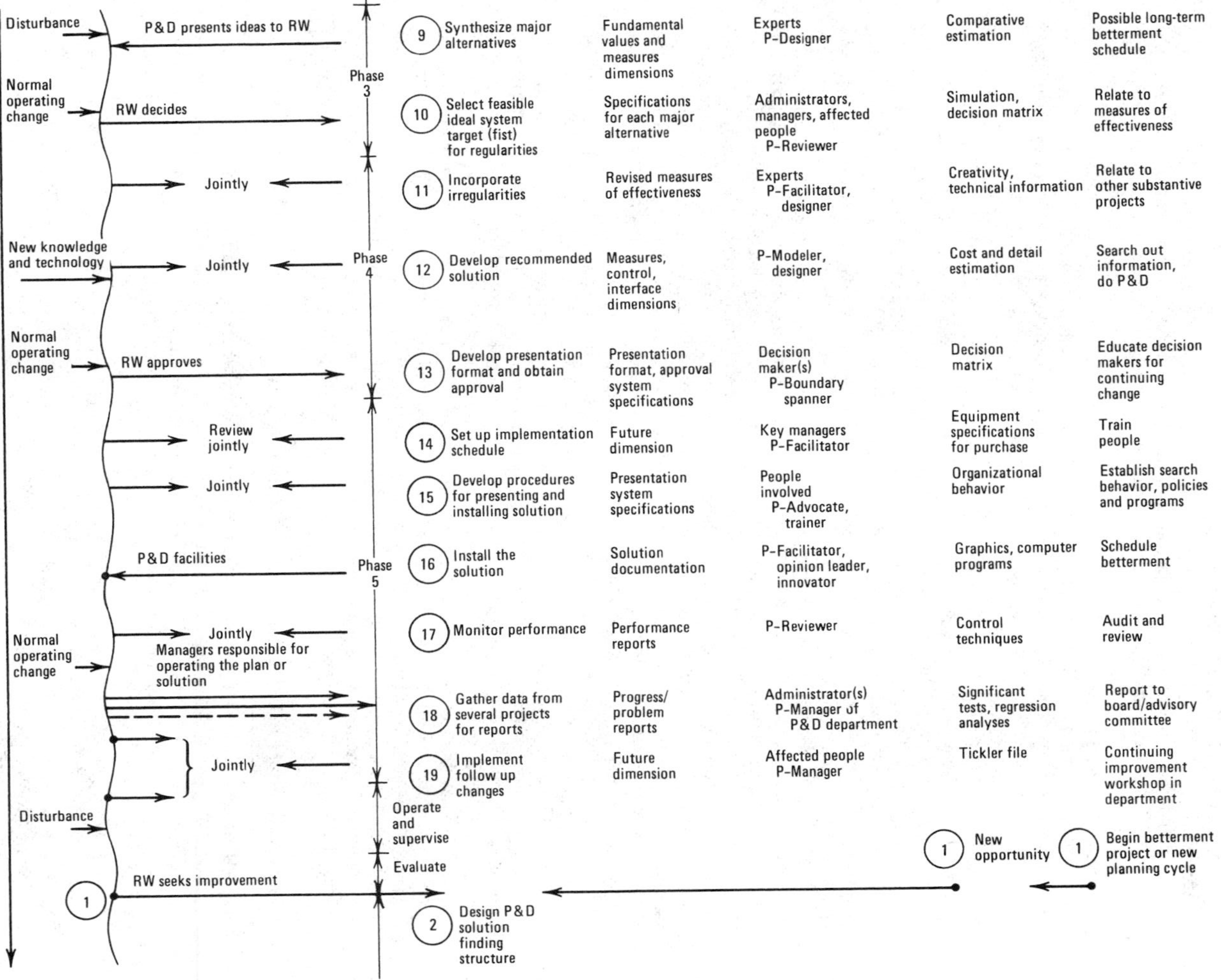

Disturbance
P&D presents ideas to RW
Normal operating change
RW decides
Jointly
New knowledge and technology
Jointly
Normal operating change
RW approves
Review jointly
Jointly
P&D facilities
Jointly
Managers responsible for operating the plan or solution
Normal operating change
Jointly
Disturbance
RW seeks improvement
1
Phase 3
Phase 4
Phase 5
Operate and supervise
Evaluate
9 Synthesize major alternatives
Fundamental values and measures dimensions
Experts P-Designer
Comparative estimation
Possible long-term betterment schedule
10 Select feasible ideal system target (fist) for regularities
Specifications for each major alternative
Administrators, managers, affected people P-Reviewer
Simulation, decision matrix
Relate to measures of effectiveness
11 Incorporate irregularities
Revised measures of effectiveness
Experts P-Facilitator, designer
Creativity, technical information
Relate to other substantive projects
12 Develop recommended solution
Measures, control, interface dimensions
P-Modeler, designer
Cost and detail estimation
Search out information, do P&D
13 Develop presentation format and obtain approval
Presentation format, approval system specifications
Decision maker(s) P-Boundary spanner
Decision matrix
Educate decision makers for continuing change
14 Set up implementation schedule
Future dimension
Key managers P-Facilitator
Equipment specifications for purchase
Train people
15 Develop procedures for presenting and installing solution
Presentation system specifications
People involved P-Advocate, trainer
Organizational behavior
Establish search behavior, policies and programs
16 Install the solution
Solution documentation
P-Facilitator, opinion leader, innovator
Graphics, computer programs
Schedule betterment
17 Monitor performance
Performance reports
P-Reviewer
Control techniques
Audit and review
18 Gather data from several projects for reports
Progress/ problem reports
Administrator(s) P-Manager of P&D department
Significant tests, regression analyses
Report to board/advisory committee
19 Implement follow up changes
Future dimension
Affected people P-Manager
Tickler file
Continuing improvement workshop in department
2 Design P&D solution finding structure
1 New opportunity
1 Begin betterment project or new planning cycle

1.3.7 ROLES OF THE IE

One aspect of Exhibit 1.3.4 needs additional explanation. The column representing the factor *involving people* includes in each entry an illustrative role and function of the IE professional. Many different roles are indicated there, and the exhibit only represents the planning and design purposeful activity. As one might expect, other roles may be necessary to encompass all the purposeful activities with which an IE is involved.

What is critical, though, is that effective practice of IE *requires* the individual to portray different roles and functions at various points in time during each scenario for each problem or purposeful activity. No longer can an IE adopt only one mantle of objectivity, modeler, facilitator, or design expert and hope to be successful. All of the roles and functions are needed at one time or another, as illustrated in Exhibit 1.3.4.

Exhibit 1.3.5 lists most of the different types of roles and functions an IE may need to take on. More than one may occur at one time, such as measurer/participant/advocate in *operating and supervising* a productivity improvement program in a company; boundary spanner/facilitator/trainer in *planning and design* of an improved patient admission and charging system as part of a hospital cost reduction program; or advisor/analyst/reviewer in *evaluating* the effectiveness of a new factory built a year ago. Most important, to repeat, is the *change* from one role to another as the project proceeds along the timeline (even though the IE may be more comfortable with just certain roles). Each step in a purposeful activity almost assuredly needs a different IE role. Industrial engineers need to be aware of all roles as the basis for knowing when to change roles, when to call in someone else who may have a particular role capability, or when to gain the skills to add to their repertoire.

The experience of an IE does not change any roles to be portrayed but does influence the scope and level of the problems or purposeful activities handled. A new IE is likely to have small and first-line projects (report on learning progress, develop methods on a machine). An IE with 3 to 7 years experience gets tactical-level assignments (automated quality assurance reporting, small factory layout, company cost reduction objectives for the year), one with 8 to 12 years experience gets strategic assignments (consolidating four plants into one, marketing survey design, budget development), and over 12 years, IEs get policy assignments (corporate planning, new product and acquisition assessment, positions as assistants to presidents or members of Boards of Directors). Improving effectiveness of practice at all levels remains essentially the same.

Successful IE practice obviously involves more than just different roles and functions within an appropriate purposeful activity. The organization or client has roles they must portray if effective results are to be obtained. First and foremost for the IE to ascertain, is that there is a client, *a person in the real world who is committed to doing something* about the problem if a solution or response is found. Be extremely wary of starting any project that you alone, as IE, feel is important unless there is someone in the real world who wants to and can take action at *the end of the effort*. You may be absolutely correct in assessing the importance of achieving a needed purpose in the hierarchical context of purposes for the company, department, building. But a company, department, or building can do absolutely nothing about working on a problem and implementing change. People in the real world are the only instruments who can do this, and this commitment at the beginning is essential for effective interaction along the timeline and for utilization of ideas to get results. You can try to generate this commitment, but drop the project if it is not forthcoming.

Exhibit 1.3.5 Roles and Functions of an Industrial Engineer

1. ADVISOR/CONSULTANT–be available as needed by others for clarification of other roles, interpretation of data, review, and the like.
2. ADVOCATE/ADVERSARY/ACTIVIST/LOBBYIST–promote actively (a) a particular party or position, (b) a solution or outcome, or (c) a process or approach (e.g., planning and design for finding a solution).
3. ANALYST–separate a whole into parts and interactions, and examine them to explore for insight and characteristics.
4. BOUNDARY SPANNER–bridge the information/style/interests gaps between IE and user/client/adopter. Serve as integrator or "linchpin" to explain ideas to, say, physicians about computers and computer specialists about health care and physician needs.
5. CATALYST/MOTIVATOR–provide stimulus and skill availability to a group or individual.
6. CHAIRPERSON–be responsible for an IE project and manage and facilitate the team/group/task force.
7. DECISION MAKER–select a preference from among alternative possibilities for topic of concern.

Exhibit 1.3.5 (*Continued*)

8. *THE* DESIGNER/PLANNER–produce the solution specifications and serve as advocate of the solution through the implementation phase.
9. EXPERT–provide a high level of knowledge, skill, and experience in a specific content/topic area of application with or without comparable capabilities in the IE approaches.
10. FACILITATOR/COORDINATOR–provide appropriate purposeful activity approach (all five factors) guidance, and structure to a group.
11. HELPING PROFESSIONAL–combine empathy for a person (or small group) with knowledge, skill, and experience in guiding the individual (or group) to a resolution of personal/group difficulties.
12. INFORMATION RESOURCE PERSON–be familiar with the categorization and availability status of data related to field of the IE effort.
13. INNOVATOR/INVENTOR–seek to produce a creative/unique/advanced technology solution and advocate its use all the way through implementation.
14. MANAGER OF IMPROVEMENT SEARCH PROGRAM–operate and supervise the program for continuing search for change and improvement in the organization.
15. MEASURER–obtain data and facts about existing conditions (usually a "quantitative determinist").
16. MEDIATOR–serve to conciliate different perspectives of two or more parties in an IE effort, especially when conflicts arise.
17. MODELER–produce abstraction of existing or desired phenomenon.
18. OPINION LEADER–seek to influence others regarding their contribution to IE or to the efficacy of a solution or response, however it was developed.
19. ORGANIZER/PROMOTER–develop a need, a plan or design to meet it, and a program to get the solution adopted/sold/used.
20. PARTICIPANT/COLLABORATOR–provide input occasionally as a group member in the project based on normal citizen or organizational knowledge.
21. PROJECT MANAGER–operate, supervise, and continuously evaluate the usually large-scale and complex project structure or system.
22. RESEARCHER–seek to develop a generalization about a particular phenomenon of concern in the project or overall efforts.
23. REVIEWER/EVALUATOR/CRITIC–assess the phenomenon (plan, previously implemented solution, process being followed, and the like) in terms of its adherence to desired values, objectives, and goals.
24. SURROGATE–present views of others not attending a meeting.
25. TRAINER/EDUCATOR–have people involved learn the skills and knowledge of IE and the various approaches to purposeful activities.
26. ARBITRATOR–see roles and functions 7 (Decision Maker), 9 (Expert), 14 (Manager of Improvement Search Program), 16 (Mediator), and 21 (Project Manager).
27. BARGAINER–see roles and functions 4 (Boundary Spanner), 5 (Catalyst/Motivator), and 16 (Mediator).
28. CHALLENGER–see roles and functions 3 (Analyst), 12 (Information Resource Person), 16 (Mediator), and 23 (Reviewer/Evaluator/Critic).
29. CONFLICT RESOLVER–see roles and functions 1 (Advisor/Consultant), 4 (Boundary Spanner), 11 (Helping Professional), and 16 (Mediator).
30. CONSULTANT–see role and function 1 (Advisor/Consultant).
31. COORDINATOR–see role and function 10 (Facilitator/Coordinator).
32. DATA GATHERER–see roles and functions 3 (Analyst), 12 (Information Resource Person), 15 (Measurer), 17 (Modeler), and 22 (Researcher).
33. INTEGRATOR–see roles and functions 4 (Boundary Spanner), 6 (Chairperson), 10 (Facilitator/Coordinator), 14 (Manager of Improvement Search Program), 21 (Project Manager), and 25 (Trainer/Educator).
34. LOBBYIST–see role and function 2 (Advocate/Adversary/Activist/ Lobbyist).
35. MOTIVATOR–see role and function 5 (Catalyst/Motivator).
36. NEGOTIATOR–see roles and functions 4 (Boundary Spanner), 5 (Catalyst/Motivator), 10 (Facilitator/Coordinator), and 16 (Mediator).

The world will not beat a path to your door for the greatest mousetrap design unless they *want and will use* an improved result. This idea keeps the IE focused on projects with high probabilities of success as important ingredients in improving the effectiveness of IE practice.

Given this commitment, other roles are to be portrayed by clients: They must be open in expressing difficulties and expectations, provide situation-specific information expertise, assist in data collection, make timely decisions, designate key people to involve in the project, allocate resources, and so on.

The timeline perspective with a total approach for each problem is a constant reminder of a role IEs must perform well—interrelationships with other staff resources similarly seeking to improve or better the organization (see Section 1.3.3). These people are critically needed on project teams, or they must be involved in frequent joint interaction with the IE comparable to that in which the IE engages with the real world.

Professionalism for the IE is a great deal more than having expert skills in the techniques and knowledge of the principles well presented in the rest of this handbook. As necessary as they are, this chapter has not needed to mention any techniques and tools with which IEs are usually associated (e.g., information systems, work measurement, materials handling, methods engineering, simulation, economic evaluation). All of these need to be oriented for possible use with the total approach for obtaining real world results. Chapter 3.1 on methods design is a good illustration of how an old friend, methods engineering, can be transformed using the total approach into a dynamic force for achieving modern results. As important as techniques, principles, and models are in the *education* of an IE, they are relatively unimportant in the *practice* of IE (as every survey of people who graduated 3 to 5 years back also shows).

1.3.8 SUMMARY

Effectively practicing IE requires a conversion-to-practice linkage between the huge number of techniques, principles, and models, on the one hand, and very complex surroundings, on the other hand. A total approach is the key link because it deals with the most critical condition of effectiveness—the *perception* on the part of people in the real world that IE is *helping* to achieve their purposes and values. A total approach in conjunction with each purposeful activity gives the IE an opportunity to select the most effective approach rather than merely assuming, as too often occurs, that collecting data and modeling are necessary. The majority of problems you are asked to work on *may* be solved within the context of the original statement of need. That is, you could use the operating and supervising approach when the manager *says* it is an operating problem, or the planning and design approach when the manufacturing director *says* the layout should be designed, and so on.

But the timeline perspective shows there is no way to know *in advance* where a breakthrough may occur. Because a change in viewpoint may lead to a significant result, the IE needs to review *each* problem with purposes and values from the start so that he or she can clarify what the right problem or purposeful activity is.

Exhibit 1.3.6 summarizes the concepts and principles of a total approach which an IE can in-

Exhibit 1.3.6 Concepts and Principles of a Total Approach for Improving the Effectiveness of Industrial Engineering Practice

1. Humans perform purposeful activities that influence and are influenced by the time-variant objectives and goals they seek to attain.
2. Five primary purposeful activities are the major types IEs face—operating and supervising, planning and design, research, evaluation, and learning.
3. The IE world works in parallel along a timeline with the real-world entity in which the problem or need (purposeful activity) emerges.
4. IE includes incorporation of proposed solutions into operating activities or acceptance of responses by managers. Solutions and responses, by themselves, are insufficient as IE outcomes.
5. IE is part of at least one hierarchy of purposes. IEs must deal with human perceptions about how well IE contributes to the larger organizational and human purposes in the hierarchy.
6. IE efforts parallel other organizational activities and purposes.
7. IE, as part of its larger system (e.g., a company), at any point in time can be in one of three conditions of existence—future, satisfactory, or unsatisfactory. IE will tend toward unsatisfactory existence if deliberate efforts are not introduced to improve the effectiveness of its practice.
8. The structure of an IE effort operating satisfactorily in one setting should not be transported to another setting. Each IE program and problem solving effort needs a unique development.
9. A total approach in seeking solutions or responses is a holistic view that, for now, can be described with five interwoven factors operating along a timeline: (1) pursuing a strategy,

Exhibit 1.3.6 ***(Continued)***

(2) specifying and presenting solutions, (3) involving people, (4) using information and knowledge, and (5) arranging for continuing change and improvement.

10. A different set of concepts for the five interwoven factors in a total approach to each purposeful activity along a timeline significantly increases the probability of maximizing the quality of a recommended solution or response, the likelihood of its implementation or acceptance, and the effectiveness of resources used in the IE effort.
11. Each problem or purposeful activity should be approached by initially developing a solution-finding or response-generating structure that can be put into operation for the unique conditions. This helps to ascertain the commitment of people to *do* something about the results. Otherwise, the project is most likely not worth doing.
12. An IE needs to portray many different roles as a project proceeds. No single one is sufficient, nor will the roles remain the same from one project to another.

corporate in all professional efforts to significantly improve IE practice. A three-pronged concept of effectiveness in IE practice is:

1. To maximize the quality of recommended solutions for or responses to the real world organization.
2. To maximize the likelihood of real world implementation of the solutions or managerial acceptance of the responses.
3. To maximize the effectiveness of all real world and IE resources used in the efforts.

REFERENCES

1. For just two illustrations, see H. N. SHYCON, "All Around the Model," *Interfaces*, Vol. 8, No. 3 (May 1978), pp. 45–47; and G. NADLER, "Is More Measurement Better?", *Industrial Engineering*, Vol. 10, No. 3 (March 1978), pp. 20–25.
2. G. J. WACKER and G. NADLER, "Myths About Implementing Quality of Working Life Programs," *California Management Review*, Vol. 22, No. 3 (Spring 1980), pp. 15–23.
3. T. KOTARBINSKI, *Praxiology: An Introduction to the Sciences of Efficient Action*, Pergamon Press, New York, 1965.
4. O. FRIEDMAN, "The Economic Effect of Cost Control Programs in the Mid-West Industry," unpublished master's thesis, University of Wisconsin-Madison, 1973.
5. NADLER, "Is More Measurement Better?", pp. 20–25.
6. G. NADLER, "A Systems Engineering Approach to Securing Real-World Changes: A Timeline Perspective," *Journal of Applied Systems Analysis*, Vol. 6, No. 2 (April 1979), pp. 89–100.
7. T. MANN, *The Magic Mountain*, translated by H. T. Lowe-Porter, Knopf, New York, 1962.
8. H. MINTZBERG, D. RAISINGHANI, and A. THEORET, "The Structure of 'Unstructured' Decision Processes," *Administrative Science Quarterly*, Vol. 21, No. 2 (June 1976), pp. 246–275.
9. J. DEWEY, *How We Think*, D.C. Health, Boston, 1910.
10. G. DAVIS, *Psychology of Problem Solving: Theory and Practice*, Basic Books, New York, 1973.
11. M. WERTHEIMER, *Productive Thinking*, enl. ed., Harper & Brothers, New York, 1959.
12. K. KOFFKA, *Principles of Gestalt Psychology*, Harcourt, Brace, New York, 1935.
13. R. L. ACKOFF and F. EMERY, *On Purposeful Systems*, Aldine-Atherton, Chicago, 1972.
14. N. R. F. MAIER, Problem Solving and Creativity, in *Individuals and Groups*, Brooks/Cole, Belmont, California, 1970.
15. Criteria and explanations for this assertion are the subject of NADLER, *The Planning and Design Approach*, chap. 2.
16. A. W. KIMBALL, "Errors of the Third Kind in Statistical Consulting," *Journal of the American Statistical Association*, Vol. 57, 1957, p. 133.
17. A. MASLOW, *Motivation and Personality*, Harper & Brothers, New York, 1954.
18. A. ROSENFELD, "When Man Becomes As God," *Saturday Review*, 10 December, 1977, p. 15.
19. G. NADLER, "A Timeline Theory *of* Planning and Design," *Design Studies*, Vol. 1, No. 5 (July 1980), pp. 299–307.

CHAPTER 1.4

Productivity: An Overview

D. C. BURNHAM

Retired Chairman of Westinghouse Electric Corporation

1.4.1 WHAT IS PRODUCTIVITY?

Webster defines productivity as "the physical output per unit of productive effort; the degree of effectiveness of industrial management in utilizing facilities for production; the effectiveness of utilizing labor and equipment." John Kendrick in his book *Understanding Productivity* defines it as "the relationship between output of goods and services and the inputs of resources, human and non-human, used in the production process.[1] Jackson Grayson, head of the American Productivity Center, defines it simply as "what you get out of an activity for what you put in." Perhaps the simplest definition is "output divided by input."

It is possible to calculate the productivity of labor, the productivity of capital, the productivity of energy, and the productivity of materials, since all of these are involved in most production of articles or services. There are systems whereby the productivity of an operation can be measured in total, weighting each of these factors and combining them into one overall productivity measurement. (The measurement of productivity is discussed in some detail by Marvin E. Mundel in Chapter 1.5.)

The definition and measurement of productivity can become rather complex and most managers usually like to look at it in a relatively simple manner as "goods or services produced by an individual in a given time." Capital and energy are regarded as aids to individuals to make them more productive, and the use of materials is usually measured separately.

1.4.2 THE OBJECTIVE IS PRODUCTIVITY IMPROVEMENT

Although it is important to know the specific productivity of an operation or of a country so that it can be compared to other operations or countries, the major objective in productivity is improvement. Most industrial engineers are concerned with raising the productivity in the organization in which they work–raising it relative to other comparable organizations and raising it measured against its own performance in a previous period. As is pointed out in the discussion on productivity measurement (Section 1.4.11) an absolute measurement, although desirable, is not essential. Improvement over a period of time can be measured if a suitable base period is chosen and the same factors measured in subsequent periods. The improvement in productivity is usually expressed by a percentage, which is determined by dividing the current productivity by the productivity in the base period.

1.4.3 THE IMPORTANCE OF PRODUCTIVITY IMPROVEMENT

Why is productivity improvement important? Because we can have only what we produce. Some people think that just redistributing, spreading what we have, will somehow give everybody more. That's a myth. Unless we produce more goods and services next year than we did this year, we will not have more, no matter what happens to prices or wages. Productivity improvement comes about when the individual produces more goods or services in the same amount of time.

Improving productivity has a tremendous cumulative effect. Improvements made next year are added to the improvements of this year and last year. So by boosting the productivity rate a few percentage points above our rate of population growth, remarkable results are possible. For example, if we improve our output of goods and services per person by just 2% per year, the cumulative effect of that annual 2% increase would produce a 724% increase in productivity in 100 years, a relatively short span in the course of history. Over such periods of time it is the rate of productivity improvement that determines a nation's progress.

The difference between the standard of living in the United States and that of the developing countries is explained by the fact that we in the United States have improved our productivity and they have not. And that difference began with productivity improvement in agriculture. United States agriculture offers the greatest example of continuous productivity improvement in history.

About 100 years ago, the United States government began setting up land grant colleges and initiating a long-range program to improve farming. Better seeds, better farm equipment, soil enrichment, crop rotation all added up to a basic productivity improvement program. During the last 100 years, that program was so successful that now it requires less than 4% of the work force in the United States to raise the food needed to feed the rest of the population. In 1880, nearly 50% of the work force was needed to do that job. The great productivity advance on our farms made possible America's industrial growth by freeing people to move into cities and go into industry. This provided the manpower needed for our industrial development–development that has made the United States the greatest industrial nation in the world.

It is true that improvement in our standard of living in terms of material goods or services depends directly upon our improvement of productivity in producing those goods and services. Productivity improvement is vital to lessening inflationary pressures on prices of goods and services to consumers for the long term. A high rate of growth in output per man-hours allows wages and salaries to be increased without raising unit labor costs and prices of goods and services. More efficient use of energy, materials, and capital make it possible to offset the rising prices of these inputs. By lessening long-term inflationary pressures through productivity improvements, we can help to reverse erosion of real incomes and living standards.[2] Improvement in productivity helps industries to be competitive in world markets. In this way productivity improvement is a significant factor in maintaining a suitable balance of trade.

1.4.4 PRODUCTIVITY VERSUS INFLATION[3]

In the arsenal of weapons against inflation, gains in productivity are more potent than is generally assumed. The primary role that gains in productivity play is well understood: An increase in productivity brings an increase in aggregate supply, which holds down unit labor costs. This, in turn, brings downward pressure on the average price of goods. But to a degree not widely recognized, increases in productivity can have "multiplier" effects on moderating inflation. A one-unit increase in the longer-run growth of productivity can produce a more than one-unit decrease in the rate of inflation over a period of time.

This magnified effect of productivity gains is due to the workings of the so-called "wage-price spiral." An increase in wage rates can push up prices; and the increase in prices can, in turn, operate to push up costs and, as a result, prices. The increase in prices once again operates to push up wage rates, which act to push up prices again–and so the spiral continues. But whenever an increase in productivity growth occurs, it will act as a more than one-time brake on the spiral.

Let us assume that the increase in productivity growth occurs at a point where wages are acting to push up prices. On the first round, the increase in productivity growth will moderate the wage-induced rise in prices. On the second round, by holding down the initial price increase, the earlier productivity gain can moderate subsequent wage increases; likewise, the resulting price increase would be more moderate–and so on through each round of the spiral.

Productivity gains are a direct offset to wage increases. If we assume, in one case, no productivity gain and, in the second, a 3% productivity increase, the difference in unit labor costs is 3%.

	Case 1	Case 2
Wages rise	8%	8%
Productivity increases	0%	3%
Unit labor costs will rise	8%	5%

Since labor costs are a dominant component of total business costs, it follows that in Case 1, inflation will be roughly 8%, and in Case 2, about 5%.

Exhibit 1.4.1 shows that the relevant statistics bear out these illustrations:

As discussed earlier, productivity improvements played a vital role in containing unit labor costs between 1955 and 1977. Thereafter, as productivity gains slackened, unit labor costs began spiraling upward.

The rise in labor costs has gone hand in hand with inflation. While inflation has not always coincided exactly with rising unit labor costs, the two statistics have tracked each other quite closely.

Exhibit 1.4.1 Annual Percent Changes in Wages, Productivity, and Unit Labor Costs in Nonfarm Business and Consumer Prices, 1955–1977[a]

	Compensation per Hour Worked	minus	Output per Hour Worked (Productivity)	equals	Unit Labor Costs	Consumer Price Index (Inflation)
1955	3.7		4.1		−0.4	−0.4
1957	5.9		2.2		3.7	3.6
1959	4.4		3.7		0.7	0.8
1961	3.5		2.8		0.7	1.0
1963	3.7		3.5		0.2	1.2
1965	3.4		3.3		0.2	1.7
1967	5.8		1.9		3.9	2.9
1969	6.5		−0.2		6.7	5.4
1971	6.6		2.9		3.5	4.3
1973	7.5		1.5		6.0	6.2
1975	9.9		1.9		7.9	9.1
1977	8.3		1.9		6.7	6.5

[a]Source: U.S. Department of Labor, Bureau of Labor Statistics.

1.4.5 UNITED STATES PRODUCTIVITY VERSUS PRODUCTIVITY IN OTHER COUNTRIES

Productivity in the United States for many decades improved at a rate of 3.2% per year for the private business economy. Exhibit 1.4.2 shows this rate of improvement for 1947 to 1967. Since 1967, the rate of improvement has dropped off to nearer $1\frac{1}{2}$% per year and in some periods, productivity has actually dropped. This lower productivity growth rate in recent years has been a real threat to the United States economy. It has aggravated the inflation problem, it has reduced profits so that less money is available for research and development and capital investment, and it has made our industries less competitive in world markets.

A few years ago, the National Center for Productivity and Quality of Working Life held a conference at which the future trends in productivity were discussed by the experts most knowledge-

Exhibit 1.4.2 Output per Man-Hour in the Private Business Economy, 1950–1978[a]

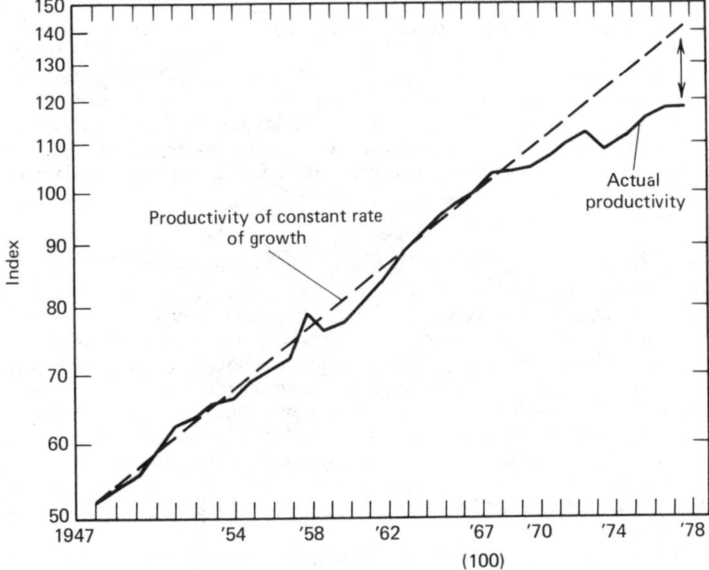

[a]Basic data: U.S. Bureau of Labor Statistics.

Exhibit 1.4.3 Projections of Domestic Output per Employee Hour in the United States and Four Other Industrial Nations for Selected Years[a]

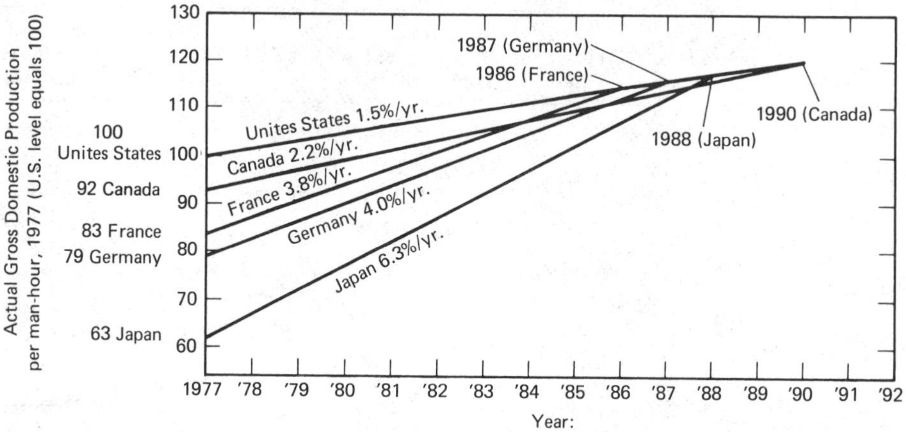

[a]Based on actual output levels in 1977 and projected productivity growth rates.

able in this field. A conclusion of this conference was that productivity in the United States through the 1980s would improve at a rate of about $1\frac{1}{2}$ or 2% per year. It was forecasted that productivity in Germany and France and some of the other western European countries would improve at about 4% per year through the 1980s, and that productivity in Japan would improve at a rate of about 6% per year. These forecasts appear to be coming true, so the status of the United States in the world economy is deteriorating relative to many other countries. At currently projected productivity growth rates, four international competitors will overtake the United States in production per employee by or before 1990 as shown in Exhibit 1.4.3.

The ability of American industry to maintain its competitive position in the expanding world markets is being clouded by rapid increases in manufacturing productivity in Japan, West Germany, and many other nations with whom the United States trades. Although the average level of productivity in the United States is still higher than that in other industrial countries, as shown in Exhibit 1.4.3, the productivity growth in manufacturing has been lagging. Our productivity advantage is narrowing. In some key industries such as radio, television, and small automobiles, it has been eliminated with respect to Japan.

1.4.6 FACTORS AFFECTING PRODUCTIVITY

Factors Affecting National Productivity

A statement of national policy on productivity was issued in 1975 by the National Commission on Productivity and Work Quality. It identified three major issues affecting productivity: (1) human resources; (2) technology and capital investment; and (3) government regulation.

Human Resources

The general level of education is an important factor in national productivity. The use of computers and other sophisticated equipment and systems requires better educated employees.

Employees need to be motivated to be productive. Pay is not enough. They need to have good, safe, working conditions and to be recognized as the most vital part of the enterprise. It is becoming recognized that all employees want to participate in planning how the work is done and that they can make positive contributions to productivity improvement.

Labor unions and management may be adversaries in negotiating pay and benefits but they can cooperate in seeking productivity improvements, to the benefit of all. Government, too, can help by sponsoring more education, especially in fields that directly affect productivity.

Technology and Capital Investment

The major factor in long-range continuing productivity improvement is technology, and new technology depends on research and development (R & D).

The federal government has sponsored much of the R & D effort through military, space, and agriculture programs. It could also help the private business efforts in R & D through tax relief for these activities. More direct support of research at universities would help in developing new technology. For industry or services to put new technology into use they must invest in new machinery and equipment and other facilities. The government can do much to aid this investment:

Encourage personal savings so the capital is available to invest.

Reduce the taxes on profits so there will be an incentive as well as available dollars to invest in new facilities.

Allow depreciation rates that will provide cash flow for new investment.

Directly encourage new investment through increased investment tax credits.

Government Regulation

An excessive amount of government regulation has had a detrimental effect on productivity through the diversion of talent and investment to activities that do not improve productivity. Government could do much to eliminate unneeded regulations and to make cost-benefit analyses to determine the necessary regulations such as those on health and safety.

Factors Affecting Productivity in Manufacturing and Services

Productivity improvement has made the United States the greatest industrial nation in the world. The organized approach to productivity improvement in American industry began with Frederick W. Taylor, who has been called the father of industrial engineering. He showed how the proper organization of work effort, based on detailed knowledge of how a job is done and breaking it down into its basic elements, can bring improved efficiency.

Another member of that group of pioneers in scientific management who appeared during the quarter century following the Civil War was Henry Laurence Gantt, whose special talent was the development of knowledge that could be used to motivate and control human effort. His task and bonus incentive pay plan became well known throughout the world, and the Gantt chart, a simple device for precisely anticipating and recording performance, speeded the building of cargo ships during World War I. This was the forerunner of the Performance Evaluation Review Technique (PERT) charts, which helped put American astronauts on the moon in our day.

Then came the Gilbreths–Lillian and Frank–who carefully studied people's movements as they worked, examining movies frame by frame. They broke jobs down into very small elements, even to finger movements, so that each movement could be questioned and perhaps simplified or eliminated.

The aim of this early scientific management effort–and the aim of the continuing productivity effort today–is not to get people to work harder, but to work smarter. People do not work any harder today than they did 50 years ago–in fact the work is generally easier. There is a limited amount of improvement to be obtained from human effort. Once a person is giving full normal effort, no more can be expected. The only way to get continuous, cumulative productivity improvement is through changes in method, and in industrial or services productivity this involves four factors: (1) product or system design; (2) machinery and equipment; (3) the skill and effectiveness of the worker; and (4) production volume. A word about each of these elements will clarify the part they play in productivity improvement.

Product or System Design

If, through better product design, a product can be simplified by eliminating some of its parts or pieces, it is obvious that the material these pieces are made of will no longer be needed. Nor will the equipment, tooling, and labor to make them be required. Value analysis can bring out many product design changes that improve productivity.

Research and development is a vital contributor to improved product design. Studies of R & D and its effect on productivity show an important relationship. Research can turn up entirely new principles that permit a function to be performed in a new way at a much lower cost. Think how today's office copy machines make it so much easier for secretaries to make copies of documents without the use of carbon paper or the need for retyping.

Standardization of the product and the use of group technology are still other design factors that make possible greater productivity in the factory. When many products can be made the same, the engineering cost can be spread over many more units. More refined tooling can be applied and the worker can do his or her job better. and maintenance in the field becomes easier.

All of these factors affecting product design have a cumulative effect over the years. Forty years ago a large electric transformer had a capacity of 72,000 kilovolt-amperes (kVA). Through research and development of new materials and design improvements that same-size transformer today has a capacity of 500,000 kVA. Today's model transforms seven times more electric power than did the old transformer of the same size and weight.

The design of the system can effect productivity in the services just as the product design affects manufacturing productivity. The layout of supermarkets, the planning of airports, and the system of scheduling and handling patients in a doctor's office are examples of services that vary greatly in productivity from one systems design to another.

Machinery and Equipment

Once the product is designed, then how it is made offers the next opportunity for productivity improvement in industry. The equipment used–machines, tools, conveyors, robots, the way the factory is laid out–all are important.

The importance of machinery and equipment on productivity in services can be demonstrated by looking at a kitchen or laundry room. A homemaker a generation ago had to devote a full day to the family washing, using tubs and a wringer. Now, the clothes can be put into an automatic washing machine and other activities can be pursued. Productivity improvement has washed the term "wash day" right out of our vocabulary. This day of service work has become an hour.

The computer is an essential tool in modern manufacturing today. It helps design the products, it helps operate the complicated machine tools, and it controls the inventory of materials and parts. It has become an essential ingredient in productivity improvement.

By each year adding new equipment and improving old equipment, we achieve the cumulative effect that makes possible the continuous rise in productivity.

Skill and Effectiveness of the Worker

Basic to industrial or services productivity is the skill and effectiveness of the people on the job. Getting people to work harder is not the objective. People do need to be trained in the best way of doing a job; the best way is not done through instinct. Varying degrees of effectiveness can be observed in the work of waitresses who serve in restaurants. The trained and experienced waitress can do the same job in a much shorter time and with far greater effectiveness than a new one who fills in on weekends.

Even the well-trained employee, however, must be motivated, must want to do his or her best on the job. This is an area of great concern in recent years and one that deserves considerable attention in the future. Employee participation in such programs as "quality circles" not only motivates employees but also develops effective productivity ideas and can have a significant positive effect on motivation and productivity.

Production Volume

In many companies today, there are as many salaried people as there are hourly workers. The number of people who actually work directly on building the products are in the minority in most industrial operations.

Assume that the volume of output is to be doubled. The number of direct workers would have to be doubled and a few indirect workers might also be needed. But there would probably not be a need for more engineers, research scientists, headquarters staff people, or other support personnel. So if the output is doubled, the productivity of these support people is in effect doubled! The volume effect can be tremendous.

This is one of the things Henry Ford and the early automobile pioneers discovered. There used to be several thousand automobile manufacturers back in the early days of that industry. As soon as some of them began getting high volume, however, the volume effect on their productivity was so great that the many low-volume producers could not compete. Now we have only three large companies in this industry in the United States, and they are producing cars at a far smaller cost per unit than a thousand small companies possibly could.

An example of the volume effect in the services is the fast food chain. Here the systems planning, detailed training procedures, facilities design, and procurement are spread over thousands of retail stores that are identical.

One manager of an electric motor plant who had experienced a sharp increase in volume explained his higher profits this way. "Our volume went up so fast," he said, "that we didn't have time to build up our expenses."

1.4.7 PRODUCTIVITY AND QUALITY

There is little doubt that anyone would not consider the quality of a new Rolls-Royce to be high, and doubtful that anyone would question the statement that many, many more man-hours are spent producing one of these cars than the more mass-produced car that most people drive. The price of $100,000, which is 10 or 20 times as much as a typical car, would indicate that the productivity of manufacturing a Rolls-Royce is relatively low, while the quality is high.

Quality and productivity have often been considered to be in conflict. Marketing people sometimes ask, "If you put in that change, yes, it will lower the cost, but won't it also lower the quality of the product?" Productivity and quality usually go hand in hand. It is not quality *versus* productivity, but quality *with* productivity.

French statesman Alexis de Tocqueville, after visiting the United States in the 1840s, observed that workers in a democratic society concentrated on inventing methods to enable them to work not only better but also more quickly and cheaply, with the objective of producing goods of simple quality on a large scale at a price the average person could afford. He contrasted the American way with that of European crafters, who concentrated on a few high-quality, handmade objects at a high cost for a handful of rich customers. Over the past 100 years both productivity and quality in the United States have improved substantially. Industry, using technology, has succeeded in mass producing quality goods. Refrigerators, automatic washing machines, automobiles, computers, and cameras are examples of products in which the quality has improved year after year, and in which the productivity of the people making these products has also continually improved.

Technological changes that have contributed to gains in productivity have directly enhanced product quality. Mechanization, automation, and computer-aided design and manufacturing have involved transfer of human dexterity to more reliable and more accurate equipment with higher productivity. Standardization, specialization, and simplification of products have facilitated mass production and product uniformity. Productivity and quality have improved together.

If one looks at the productivity of human effort and the use of our material resources over the lifetime of a product rather than at just the initial cost, the conclusion will be reached that quality is a very important factor in attaining overall improved productivity for society. We tend to measure productivity in industry by considering the total number of man-hours, both direct and indirect labor, required to produce a product. If we can reduce the required man-hours, we consider that we have improved the productivity. This may be a fair measurement as far as the individual or the factory is concerned.

Assume that the product is an automatic washing machine and that the normal life of that product in use by the customer is 6 years. If that washing machine could be engineered and built with a quality that would give it a normal lifetime of 12 years, it probably would not change the man-hours required to produce the product very much, nor would the amount of material required to make the product be changed very much. However, if the product actually lasted twice as long, performed its function for the customer for twice as many years, from a society standpoint we would have practically doubled the productivity of the people who produced that machine. It is quite possible that if productivity were considered on the broader basis of getting maximum usage from the energy, materials, capital, and human effort that go into producing a product, we would strive for ever higher quality in order to have ever higher productivity as far as society is concerned. Customers may eventually adopt this overall viewpoint and not use just the narrow consideration of initial cost only.

A quality defect can negate thousands of hours of highly productive work. Productivity must really be measured on what the product does when performing its service for the ultimate user. Productivity measured up to the door of the shipping floor is not enough.

Large generators for electric utilities cost millions of dollars. They require scores of thousands of man-hours by skilled engineers and workers and are so large that the final assembly and inspection must be done where the generator is installed. Several years ago, after one company had shipped a large unit that was hydrogen cooled, one of the final field inspections was designed to see that the passages through which the hydrogen circulated were not blocked in any way. This was done by blowing air through the passages. Some of the passages had a Y in them, so the practice was to close off one arm of the Y and check the flow through the other arm; then to close off the already checked arm of the Y to check the air passage through the first arm. All the passages checked satisfactorily, but after the tests, the inspector neglected to remove one $0.20 piece of copper that had been used to block off the second passage to check the flow in the first passage. Control of this inspection operation was not satisfactory. The machine was placed into operation, and within a day it had overheated, burned out the coils, and caused a million dollars worth of repair work. In this case, inadequate quality control negated thousands of hours of productive work.

The Japanese have provided several good examples of quality and productivity going hand in hand. The production of higher-quality television sets and small automobiles attracted customers and gave Japanese manufacturers a significant share of the market. The higher volume helped the productivity of the capital and of the indirect workers. The volume justified the investment in high technology facilities and this further increased the productivity of the workers in the factories.

1.4.8 PRODUCTIVITY AND EMPLOYMENT[4]

Workers today have a greater potential than ever before for making large contributions to productivity. They are considerably better educated and more widely traveled than their predecessors, and television has expanded their range of experience and information. These factors have also raised and changed the nature of workers' job expectations. There is a growing belief all over the world that the traditional organization of work and the workplace itself is changing to satisfy the physical, economic, social, and psychological needs of the modern work force. It is only by satisfying these needs that progress can be made in realizing the full potential of modern technology.

The recent emphasis on meeting the expressed and unexpressed desire of workers for a more satisfying and safer work environment is a new episode in the long history of workplace reforms. Overtime, the efforts of unions, progressive employers, and social legislators have brought about better working conditions: child labor and sweatshops are relics of the past, and shorter hours, vacations, safer workplaces, and many other improvements are the rule. Seniority, grievance procedures, and the right to negotiate collectively the rates and rules for work have helped counter the alienation and powerlessness that infect people in a highly organized, technological society. However, changing circumstances give rise to new expectations and new interest in alternative ways of working.

The United States Department of Labor's Quality of Employment surveys. taken in 1969 and 1972–1973, ranked pay and job security high on the list of job expectations. But it also found that workers want many other opportunities: to receive training; to use their talents more fully; to have greater flexibility in work hours, education, leisure, and retirement; to have greater protection against health and safety hazards on the job; and to exercise greater control over the way their work is performed. Only a small minority of those surveyed—not more than 20%—expressed dissatisfaction with their jobs. This minority view is noteworthy, however, for it was expressed by young, educated workers, whose views may dominate in the future.

Today both labor and management are searching for new ways to accommodate the aspirations of the work force and, at the same time, meet the needs of economic survival. Group incentive systems, flexible work schedules, autonomous work teams, job redesign, goal setting, and other new techniques are being tried, with varying degrees of success. According to a Work in America Institute study of 103 such experiments, the programs showing the most promise for improving both productivity and job satisfaction seem to be comprehensive ones aimed at the social, psychological, physical, and technical aspects of the work environment: performance recognition, skill training, participation in planning the work, safety and health protection, stress reduction, and appropriate equipment.

Employees today have a large stake in the productivity and survival of their firms. A large share of employee compensation is in the form of pensions, health and welfare benefits, and other wage supplements that depend on continuous employment with a particular firm. Improving a firm's competitiveness and the quality of life on the job is in the mutual interest of workers, unions, and industry management.

To a great degree, the prospects that labor and management will cooperate to improve productivity depend on assuring workers that their jobs are secure. Many employees see higher productivity as a job threat, and they are not likely to give their full support to improvement efforts unless they have some confidence that they will still be employed. Worker displacement is not the inevitable consequence of higher productivity. If output is increased, or if work hours are reduced, employment need not suffer, and it might even expand. Historically, industries in which productivity has risen faster than the national average often are the same industries in which employment has risen by a larger-than-average percentage; conversely, many industries with lagging productivity have had to cut employment.

The record of Japan and many western European countries is one of low unemployment and high rates of productivity improvement. This suggests that an expanding economy and a positive labor market policy can provide sufficient jobs for all. Rising national income tends to raise demand for all goods and services. This helps sustain, or even increase, employment in companies or industries with rapidly rising productivity.

Even though an industry's total employment level is not adversely affected, technological, market, and other economic changes can bring hardships for particular groups of workers. When major changes are taking place, the personal hardships they cause often can be alleviated by advance planning. If the impact of the change is small enough, normal attrition by retirements, deaths, or voluntary turnover often obviates the need to lay off workers.

There is some concern about not having a work force that will be capable of operating with the new automated and highly technical equipment that is going into modern plants and offices. The available work force should be able to do the job, as indicated by this quotation from *The New Science of Management Decision* by Herbert A. Simon, who received the Nobel Prize in economics in 1979. Dr. Simon has studied the impact of automation on the work force and writes:

> *The evidence indicates that automation does not in fact substantially change the distribution of skill levels in the factory or office. A large fraction of the jobs in our present economy, perhaps as many as 70%, require no skills more complex than those involved in driving a car*

but the latter skills are acquired by almost all of the adults in our society. Most people, appropriately motivated, are able to acquire impressive new skills in a great hurry and with modest amounts of formal training.[5]

1.4.9 OPPORTUNITIES FOR PRODUCTIVITY IMPROVEMENT

No matter what kind of work is being studied, whether it be in the factory or in a service operation, there are great opportunities for making productivity improvement. Exhibit 1.4.4 pictures the opportunities that exist in nearly any activity for improving productivity from 1980 to 1985. Assuming that productivity is 100% in 1980, if nothing is done differently from then, it will stay at that 100% level. Without any capital investment and without any new technology there is usually an opportunity to move productivity up 10% in the next 2 years by just working with the people and inspiring them to want to do a better job. This does not mean that they must work harder or faster, just more productively. It is important to get everyone really interested in doing a good, productive job. Specific improvement is accomplished by cutting down on absenteeism, starting to work on time, improving the quality of the work, reducing the amount of inspection required, reducing the amount of maintenance needed, making sure that delays do not occur because materials are not available to work on, and so on. All of these steps, plus minor methods changes made by the employee, can really make a significant difference and raise the whole level of productivity.

This improvement potential for people applies to the white collar work force just as much as it applies in the shop. If the office people become enthusiastic about doing a productive job, they will eliminate many of the existing diversions and focus on doing the important things. In most cases, they can easily be 10% more productive without any major change in the facilities, the computers, or the systems that are provided for them to do their work.

Getting everyone on the team pulling together can often improve productivity much more than 10%. In emergencies, people often perform at productivity rates that are three or four times above what they normally do. When a machine or a piece of vital equipment breaks down in a plant, maintenance crews often do as much in an hour as they might normally do in a day. When the General Motors hydramatic transmission plant at Livonia, Michigan burned down many years ago, the whole operation of the Oldsmobile division was in jeopardy. Oldsmobile engineers redesigned the housings and controls, the model shop got castings made and machined and cars were being tested with a different transmission, the Dynaflow, in less than a week. This would have normally been a 6-month project, but red tape was cut, decisions were made quickly, and everyone formed a team to get the job done. The productivity of these white collar employees was many times higher than normal during that week.

The target of a 10% improvement in productivity for all people in the organization in the next 2 years is a practical one and should be the base upon which to build a continuing productivity improvement from better technology.

Adding technology to the base of productive people to get a continuing improvement rate of 5% per year should be possible in most organizations (see Exhibit 1.4.4). The application of robots to an assembly operation or the use of a computer on new activities in the office may bring about high percentages of improvement in productivity in specific cases. The overall improvement should

Exhibit 1.4.4 Projected Productivity of People and Technology

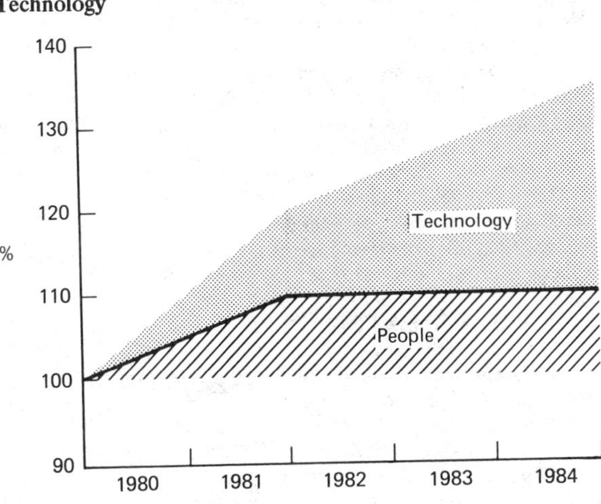

come out to 5% each year. But the benefits of this wonderful technology may not be realized unless the people are cooperative.

It is important that everyone in the organization is informed about new technology that will be used. People do not like to be surprised, especially unpleasantly. Workers and managers as well fear things that they do not understand. The people who will be directly affected should be in on the planning. They should be asked for their suggestions and ideas before the changes are made. If a union is involved, the union officers should be informed about what is planned in the way of technological improvements before those improvements are put into operation.

The introduction of new technology should be planned so that it will not seem a threat to employees. The new technology could be introduced when the volume of product required is increasing. The changes could be made on a gradual, incremental basis so that attrition can take care of the displacement of personnel. An active program may be needed to retrain people for new jobs that are still needed after the new technology is in operation. This people planning may be vital to the success of the new technology.

It is hardly ever the case that significant improvement is not possible this year because a good job of improvement was done last year. Usually the department that makes a 5% improvement in productivity in one year will probably make a 5% improvement in productivity the following year, and one that made only a 1% improvement in one year will probably make only a 1% improvement the following year. This is because there is always opportunity for improvement and the accomplishment depends upon the effort that is put into finding improvements and putting them into effect.

1.4.10 PRODUCTIVITY IN THE SERVICES

Just as productivity improvement in agriculture reduced the percentage of the working population in agriculture from 50% 100 years ago to 3.4% in 1980, the work on productivity improvement in industry is reducing the percentage of the work force required to make the things needed. Before the end of this century, the number of people required in industry will have been reduced to 20% of the work force. This means that over 75% of the work force will be working in the services, including health care, government, education, retailing, maintenance, recreation, and the like. This is where the people are going to be working and this is where there is an urgent need to improve productivity. Over 2 million people work in eating and drinking places now, and this is four times as many people as are employed in the entire steel industry. The job of productivity improvement has not been tackled in the services the way it has been in agriculture and industry, but some of the same techniques can be used, and it should be possible to get the same kind of continuous improvement.

Consider the people working in a hospital. They are doing the same kinds of things that the people are doing in the factories and in the offices of industry. They are keeping records, moving things around, running machines, performing work at benches, and so on. The same techniques can certainly be tried in the services that have worked so well in industry. Although it is more difficult to measure, performance in the services can be measured, at least against the preceding year if absolute standards cannot be used. Method studies can be applied to the services area and certainly more capital can be invested in equipment to help do the jobs.

The white collar workers in industry are in many ways "service" workers. The IBM Corporation has an excellent system for measuring and improving productivity in the white collar area. They have divided all of the service or white collar work into about 130 different logical elements. For each of these elements they measure head count against a logical base. For example, they measure the number of secretaries against the base of the number of salaried exempt employees. They measure the number of janitors against a base of the square footage of office area that is serviced. These ratios are plotted each year on a chart for each one of the 130 factors. Each of their more than 30 plants would have a dot on this chart. When these charts are circulated through the corporation, each manager can see who is doing a better job than he or she is on each factor and is motivated to emulate the best. Each manager can also see from his or her own measurement of the preceding year compared to the current year, the kind of progress that is being made in improving each factor of his or her white collar or service work. This system has stimulated significant annual improvements in white collar productivity at IBM.

1.4.11 MEASURING PRODUCTIVITY

The measurement of productivity is discussed in detail in Chapter 1.5 by Dr. Mundel, but this section emphasizes the importance of measurement and covers the use of relatively crude measurements if a precise measurement is not available.

The continual improvement in productivity of agriculture in this country has meant that only 3.4% of the nation's working population now grows more than enough food for all of us. Measurement of input and output of farming has, during this time, been quite specific and understandable.

Bushels of wheat per acre and the cost to grow a pound of beef have been precise enough to allow farmers to select ever more efficient ways of growing their products.

In industry, the blue collar worker has been measured very accurately since the time of Taylor and Gilbreth. This precise measurement has enabled industry to continually improve the productivity of the blue collar worker. The pieces per hour coming from a machine or the seconds and fractions of a second required to assemble a part have formed a basis for measuring improvement so that better methods can be selected and put into effect. The productivity improvement in industry has been great enough so that the percentage of workers in the United States in industry passed its peak by the period 1965 to 1970 and it will not be long before fewer than 25% of our workers will be making all the things that we need.

The services (government, education, health care, transportation, etc.) comprised nearly 70% of our work force in 1980. Not as much progress has been made in improving productivity in this area, and one of the major reasons for this slower progress is the difficulty and lack of measurement of performance. Most services are harder to measure than farms or factories. The input can be measured fairly well, but it is usually not possible to make a simple count of the output because the output is not usually just a quantity measurement. The quality or usefulness of the output is of vital importance, may vary greatly, and is difficult to measure. Performance measurement of a police officer or a nurse or a teacher cannot be as precise as 63.1 bushels per acre or 0.042 minutes per piece.

Logical, believable measurement inspires improvement. Roger Bannister or anyone else probably would never have run a mile in 4 minutes if he were not being timed precisely. Measurement and improvement go hand in hand. To achieve improvement, any measurement, even if it is relatively crude, is better than no measurement at all. Where accurate standards cannot be set, comparisons have been useful. Relatively crude standards used for comparison have helped get the productivity improvements that have been attained in some services. If a job can be measured accurately enough so that the performance can be compared to the performance of a similar job in another organization, the organization turning out the best performance can be used as a standard. Those who do not measure up to it can be asked to emulate the best performer. At the very least, performance one year can be measured against performance of the preceding year. Present cost or head count can be related to the comparable cost or to the number of people required to do the same job the preceding year. Budgeting systems often give this type of measurement and comparison.

There is not much incentive to do better if you do not know whether or not you are doing better. To optimize performance, believable, simple, accurate measurements and equally qualified standards are needed. In a service such as computing services a simple count of the number of printed sheets turned out is not meaningful by itself. The service is not to provide sheets of data but to provide information the customer desires. Customer satisfaction must enter into the productivity formula to measure the true worth of the service.

To be useful in promoting productivity improvement, measurements must be understood and their validity believed by the people using them. There is a great difference between saying, "this project will improve the productivity of the operation a great amount" and saying, "this project will improve the productivity of the operation by 22%."

1.4.12 THE INDUSTRIAL ENGINEER'S RESPONSIBILITY

The industrial engineer is best equipped from the standpoint of education, interest, and attitude toward work to take the leadership in improving productivity in any organization. To do this, the industrial engineer will have to move out of the traditional role of setting standards for the direct laborers, and will have to study every person in the organization. He or she will also have to see that productivity studies are made on energy and materials. He or she will have to work with the manufacturing engineers on new technology and with the product design engineers on designing products for improved productivity.

This broadened responsibility means that many industrial engineers, who would normally go to work in industry, must go to work in the services, where there are great opportunities for productivity improvement and where there are more than twice as many employees as there are in manufacturing. It is important for society that productivity improvement be accelerated, and the industrial engineer should take up the responsibility for seeing that this acceleration takes place.

1.4.13 THE EFFECT OF PRODUCTIVITY ON SOCIETY[6]

Exhibit 1.4.5 shows the change in the distribution of United States employment since 1870. In 1870, nearly half of the workers were in agriculture, about one fourth in industry, and the other fourth in the services. The great improvement in productivity in agriculture has brought the number of people working on the farm in 1980 down to less than 4%. The number of people producing goods, that is, in industry, construction, and mining, increased until about 1950; but due to

Exhibit 1.4.5 Distribution of U.S. Employment

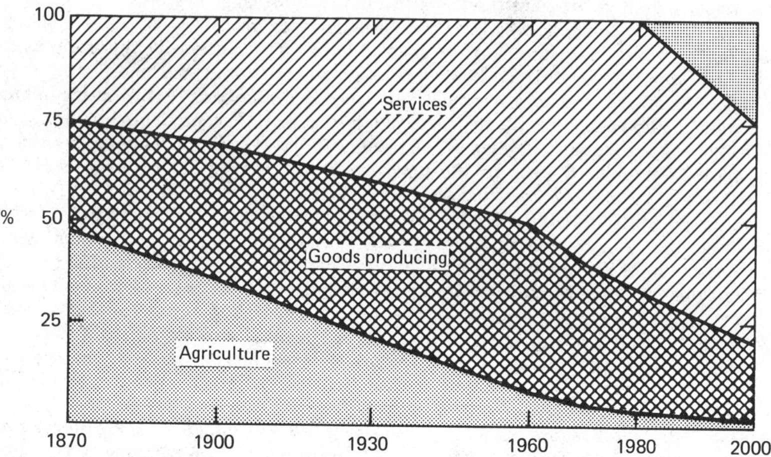

productivity improvements in the goods producing industries, that category is down to about 28% of the total work force in 1980. This leaves 68% of the work force producing services.

The exhibit extends to the year 2000. Agriculture at that point requires only 2 or 3% of the work force, and all of the goods are produced by less than 20% of the work force. This leaves over 75% of the working population of our country to perform services. If even modest progress is made between 1980 and 2000 in improving productivity in the services, by the year 2000 we can have all the services we have now and still have the equivalent of 25% of the work force available for something else, as shown by the triangle in the upper right-hand corner of Exhibit 1.4.5.

This available time could be used by us to further our education. It could be used to support added people occupied full- or part-time in the arts or cultural fields, for more leisure activity, or to help people in developing areas to improve their standards of living. A similar exhibit was shown to Dr. Charles Malik of Lebanon, former President of the United Nations General Assembly. He suggested that the following be added to the use of this available time: time for contemplation, conversation, community life, time with family and time for fellowship, and to be more human to each other.

There is no foreseeable limit to productivity improvement, and it can have a significantly beneficial effect on the quality of our lives in the future.

REFERENCES

1. JOHN W. KENDRICK, *Understanding Productivity*, The Johns Hopkins University Press, Baltimore, MD, 1977.
2. This paragraph is from the July 1, 1974 report of the National Commission on Productivity and Work Quality.
3. This section is adapted from *Reaching a Higher Standard of Living*, The New York Stock Exchange, Office of Economic Research, 1979.
4. Most of this section is extracted from "Productivity in the Changing World of the 1980's," the final report of the National Center for Productivity and Quality of Working Life.
5. HERBERT A. SIMON, The New Science of Management Decision, Prentice-Hall, Englewood Cliffs, NJ, 1977.
6. Much of this section is from D. Burnham, "Productivity Improvement," the Benjamin F. Fairless Memorial Lecture at Carnegie-Mellon University, Pittsburgh, PA, 1972.

BIBLIOGRAPHY

Manufacturing Productivity Solutions, Proceedings of the Society of Manufacturing Engineers, October 2–3, 1979. Available from the Society of Manufacturing Engineers, One SME Drive, P.O. Box 930, Dearborn, MI 48128.

Productivity Perspectives, American Productivity Center, Inc., Houston, TX, 1979.

Reaching a Higher Standard of Living, Office of Economic Research, The New York Stock Exchange, New York, 1979.

CHAPTER 1.5

Productivity Measurement and Improvement

MARVIN E. MUNDEL

M. E. Mundel & Associates, Inc.

1.5.1 INTRODUCTION

As a preliminary to improving productivity one must measure the current productivity status so as to have a datum from which to measure change. Further, the nature of the desired change must be carefully defined so that change represents a desired form of improvement. Productivity, therefore, must be measured in a way that reflects changes in the measured situation that are considered desirable.

Hence Section 1.5.2 contains definitions of the special terms employed in productivity measurement, followed by various definitions of productivity measures as a preliminary to a discussion of the optimization of productivity (Section 1.5.3). This is followed by a Section 1.5.4, which discusses technologies basic to productivity measurement. Examples of productivity measurement are described next (Section 1.5.5), and last, the problem of improving productivity is addressed (Section 1.5.6) by indicating a general procedure that also indicates the significance of the other materials in this handbook.

1.5.2 BASIC DEFINITIONS

Productivity

Productivity is the ratio of the outputs produced for use outside of an organization, with due allowances for the different kinds of products, divided by the resources used, all divided by a similar ratio from a base period. Hence it is an index; it has no dimension. Mathematically, the productivity index is

$$\frac{AOMP/RIMP \quad (1)}{AOBP/RIBP \quad (2)} \times 100 \quad (A) \qquad \text{or} \qquad \frac{AOMP/AOBP \quad (3)}{RIMP/RIBP \quad (4)} \times 100 \quad (B)$$

where $AOMP$ = Aggregated outputs, measured period
$RIMP$ = Resource inputs, measured period
$AOBP$ = Aggregated outputs, base period
$RIBP$ = Resource inputs, base period

Either formulation produces an identical value, although the subordinate ratios have different meanings. Subordinate ratio (1) is called the *current performance index*, ratio (2) is called the *base performance index*, ratio (3) is referred to as the *outputs index*, and ratio (4) the *inputs index*.

In formulation (A), the ratios of (1) and (2) may be computed from periods of different lengths without disturbing the meaning of the productivity computation or of the subordinate ratios. In formulation (B), if the subordinate ratios are to have meaning, all data must be from equal periods of time.

Kinds of Productivity Measures

1. Labor Productivity. In this formulation, the resource inputs are aggregated in terms of labor hours. Hence the index is relatively free of changes caused by wage rates and labor mix.

2. Direct Labor Cost Productivity. In this formulation the resource inputs are aggregated in terms of direct labor costs. This index will reflect the effect of both wage rates and changes in the labor mix. However, "constant dollars" may be used to eliminate this distortion.

3. Capital Productivity. Several formulations are possible. In one, the resource inputs may be the charges during the period to depreciation; in another, the inputs may be the book value of capital equipment.

4. Direct Cost Productivity. In this formulation all items of direct cost associated with resources used are aggregated on a monetary value basis. Constant dollars may be used.

5. Total Cost Productivity. In this formulation all resource costs, including depreciation, are aggregated on a monetary basis. Constant dollars may be used.

6. Foreign Exchange Productivity. In this formulation the only resource cost considered is the amount of foreign exchange required.

7. Energy Productivity. In this formulation the only resource considered is the amount of energy consumed in btu or kW, as may be most convenient.

8. Raw Materials Productivity. In this formulation the numerators are usually weight of product; the denominators are the weight or value of raw materials consumed.

Hence it may be seen that many different productivity indexes may be constructed. The list given above is not exhaustive.

Effectiveness

The term *effectiveness* is used to describe how well the outputs achieve the desired goals; how much *results* are obtained because of the outputs. Measures of effectiveness should be delineated prior to the identification of outputs in that until such measures are established, one can hardly identify what to count as outputs or how to count them.

For instance, to measure a purchasing office solely in terms of the number of purchase orders issued, cost per purchase order, or first cost of material purchased would misdirect the subsequent productivity measure. The purchasing department should be charged with obtaining a lowest total cost as determined by what is purchased. Timeliness to avoid work delays, the avoidance of inferior quality, which might increase processing costs or reduce raw material productivity, should be given primacy.

Misidentification of measures of effectiveness frequently leads to ludicrous results. For instance, one purchasing department was measured in terms of "amount of discount negotiated." This led them to buy books from one dealer who quoted $15 per book but who offered to sell them for $11, if 100 books were ordered, rather than buying the books from another source who offered the books at $9 each with no room to negotiate a discount.

Worth

Worth refers to the desirability of achieving the specified results. There is no absolute measure of this. Social and personal value systems are used to provide an answer.

Outputs

Outputs are the goods and services produced for use outside of the organization, that are for delivery to the marketplace or the served sector of the society, geography, or economy, and that are intended to directly achieve the purpose of the organization. When the productivity of a part of an organization is measured, care must be taken to avoid suboptimization, a topic that is examined in depth in Section 1.5.3.

Aggregating Outputs, General

In any formulation of a productivity measure or index, the outputs must be aggregated for both the base and measured periods. The method must be the same in both cases. The various options include:

1. By labor hour standards, in hours, using base year values, weighting the individual outputs before adding.

2. By profit margin, weighting each kind of output by its current profit margin for the measured period and base period outputs by their base year profit margins.
3. By the market value of the outputs, weighting current outputs with current values and base period outputs with base year values.
4. By the weight of the outputs.
5. By a simple count, if only one kind of output is involved (but this is very seldom the case).

Aggregating Output, Specific

The outputs must be aggregated in a way that reflects their contribution to the effectiveness of the organization. In many cases, methods other than common methods need to be employed. The old adage, "You cannot add apples and oranges," may not hold. If the fruit were intended as missiles, four oranges and five apples make nine. One could even add in tomatoes.

A typical error is to identify and count functions or activities rather than outputs, as defined previously. Typical of such lists are such items as:

1. Meetings attended.
2. Letters responded to.
3. Phone calls responded to.
4. Planning actions taken.
5. Policy recommendations made.

It is not that these do not take time. Further, they may well be (except for items 4 and 5) for use outside the organization. However, they seldom are designed to directly achieve the purpose of the organization; the number of each could be vastly changed without necessarily affecting the effectiveness of the organization. It would be like counting the productivity of a machine shop in pounds of metal removed, a foundry in terms of metal melted, the use of a bus in terms of number of stops and starts, or a lawyer in terms of cases litigated. (*Note:* I would think a good lawyer, a great deal of the time, helps people avoid creating situations that lead to litigation.)

Aggregating Substantive Outputs

Where the organization being measured produces substantive outputs of intrinsic value, the delineation of outputs almost always precedes their actual production and is documented by drawings, parts lists, and so forth.

Aggregating Service Outputs

With service and government organizations the outputs must be identified prior to any attempt to measure productivity. One relatively new technique for assisting in this effort for service activities is called *work-unit analysis* and is described in Section 1.5.5.

1.5.3 OPTIMIZING PRODUCTIVITY

Definition

Optimization means achieving a maximum or minimum with respect to a given criterion or criteria. If criteria are used, some way of weighting the relative *worth* of each criterion must also be established. Further, it should be recognized that choosing a particular productivity index may predicate the basis of optimization.

For instance, to compare the results of using different measures of productivity let us assume the following facts (simple numbers have been chosen to simplify the computations):

Item	Year 1	Year 2
Number of outputs (all of one kind)	10	15
Direct labor hours	2,000	4,000
Direct labor cost	20,000	23,000
Capital depreciation	6,000	7,000
Capital book value	18,000	36,000
Total direct cost	30,000	38,000
Total costs	40,000	53,000
Foreign exchange used	$4,000	$10
Energy used	1,000 kW	1,400 kW
Raw material used	10,000 lbs	15,000 lbs

Item	Year 1	Year 2
Labor productivity index	$= \dfrac{15/4000}{10/2000} \times 100 = 75\%$	
Direct labor cost productivity index	$= \dfrac{15/23{,}000}{10/20{,}000} \times 100 = 130\%$	
Capital depreciation productivity index	$= \dfrac{15/7000}{10/6000} \times 100 = 129\%$	
Capital book value productivity index	$= \dfrac{15/36{,}000}{10/18{,}000} \times 100 = 75\%$	
Total direct cost productivity index	$= \dfrac{15/38{,}000}{10/30{,}000} \times 100 = 118\%$	
Total cost productivity index	$= \dfrac{15/53{,}000}{10/40{,}000} \times 100 = 113\%$	
Foreign exchange productivity index	$= \dfrac{15/10}{10/4000} \times 100 = 60{,}000\%$	
Energy used productivity index	$= \dfrac{15/1400}{10/1000} \times 100 = 107\%$	
Raw material productivity index	$= \dfrac{15/15{,}000}{10/10{,}000} \times 100 = 100\%$	

Basis of Optimization

In the private manufacturing sector, the normal basis of optimization is current profits (while protecting future profits). Hence the most common formulation is *total cost productivity*. However, even in this setting, *labor productivity* as defined is a useful index evaluating the effect of motivation and diligence on operations. In the private sector, there are, however, many staff activities that more closely resemble the public sector.

In government and service operations there are many activities that resemble the private sector. The work of a government printing office or highway department are examples. The outputs are tangible, easily identifiable, and readily quantified. It is in the staff activities in the private sector and in the service activities of government, which predominate, that much work must be done to usefully aggregate the outputs in order to develop and use a *labor productivity index*, the most commonly used index. Optimization is measured, therefore, in terms of the maximum aggregated output per unit of labor resource used, within various other constraints.

Constraints on Optimization

Most frequently, constraints on optimization appear in the service sector. For example, a store might increase the productivity of its sales clerks by having fewer clerks. However, service may become so poor that a more than an offsetting amount of profit is lost.

When a part of an organization is studied, suboptimization is frequently an important danger to avoid. For instance, if productivity improvement focuses on clerk-typists, service to the professional staff may be so poor that they do their own clerical work or waste time waiting for service. A hospital laboratory, in an effort to increase its productivity may at times backlog work, increasing the average length of stay of patients, a cost all out of proportion to the savings.

Timeliness, percent of services that must be provided, and quality of services are frequently features that must be considered as constraints on the optimization of part of an organization so as to avoid suboptimizing the total organization.

Motivating the Necessary Behavior

For top management in the private sector, in most cases the pressures of the profit and loss statement and the balance sheet supply the motivating forces.

For top management in government, there is actually positive demotivation to optimize productivity in many cases. There may be pressure for goal attainment for which a surplus of resources is an advantage. Also, a past history of across-the-board staff reductions may motivate the carrying of excess staff. It is axiomatic that a hard, tough budget line is needed to motivate such organizations.

For staff services, private and government, the remarks made with respect to top management in government are equally applicable.

By and large, workers have frequently resisted efforts to improve productivity. There is insufficient understanding that if wages rise faster than productivity, then only inflation can ensue. It is axiomatic that most people will not exert effort toward goals that do not directly concern their own well being. It is to this end that wage incentive schemes, profit sharing schemes. and productivity improvement sharing schemes are all directed. Note, however, that productivity improvement sharing schemes require that productivity be measured.

The Senior Executive Service, part of the United States Civil Service Reform Act of 1978, includes a scheme designed to provide financial incentive for good management. Its effect on productivity remains to be seen.

1.5.4 BASIC SOURCES OF INFORMATION FOR DEVELOPING MEASURES OF PRODUCTIVITY

An examination of the various measures of productivity described in Section 1.5.2 indicates three major sources of information for constructing such indexes:

1. Product identification information.
2. Accounting information.
3. Work measurement information.

Product Identification Information

For products that are tangible in nature and of intrinsic value, the product catalogs and drawings serve to provide a framework for identifying the different kinds of products prior to weighting each kind of output in the mix. Only after proper weighting can the outputs be aggregated. For services, however, output identification information is seldom available, and for these some considerable analysis may be needed as a first step in developing a productivity index. Indeed, in developed nations such as the United States, where approximately 80% or more of employment is in the indirect or service sector, the necessary analyses may be a major requirement. A suitable technique for this is included in Section 1.5.5 of this chapter.

Accounting Information

Depending on the sophistication of the accounting system in use, the weighting of each kind of output may or may not be feasible from accounting records alone. With a detailed cost accounting system, all the requisite information may be available. (See also Section 9 of this handbook, "Engineering Economy.")

If, however, detailed cost accounting data allocating labor, material, and overhead costs to each kind of product are not available, the development of the necessary information may be a laborious task. Because of the pressure of profits or cost, such information is most usually available for tangible products of intrinsic value, however.

In the case of service or indirect activities, for which output identification is seldom available, even sophisticated cost accounting systems tend to lump together functional costs regardless of the product mix or its change from period to period. Hence subsequent to the use of some technique for identifying the different kinds of service outputs, some type of work measurement is usually necessary to assist in allocating the different costs (weightings) to the different kinds of products prior to aggregation.

Work Measurement Information

Work measurement is used here to refer to the use of any technique to determine the amount (and kind, if desired) of labor required to produce each kind of output in a base period. Such information is needed to complete almost all productivity computations other than those for raw materials described in Section 1.5.2 of this chapter. (See Section 4 of this handbook, "Performance Measurement and Control of Operation.")

1.5.5 IDENTIFYING SERVICE OUTPUTS: WORK-UNIT ANALYSIS[1]

Definitions

In the concept *work-unit analysis*, the hyphenated term *work-unit* has a special meaning. A variety of other special terms are also required to facilitate a meaningful discussion.

Work-Unit. A *work-unit* is an amount of work, or the result of an amount of work, that is conveniently treated as an integer (an each) when examining work from a quantitative point of view.

In this definition, the word *convenient* means that it provides a useful basis for

1. Applying other industrial engineering techniques.
2. Substantively supporting staff resource aspects of budgets.
3. Developing unit labor and other costs.
4. Planning and assigning work.
5. Continually reviewing work load forecasts and ongoing staff resource utilization (as well as the utilization of any associated adjuncts).
6. Continually comparing performance with plans.
7. Measuring the productivity of an organization.

It should be obvious that within the definition given a wide variety of sizes of work-units could exist, covering a range from "one motion of the arm" to "the total annual work of an organization."

Orders of Work-Units. To facilitate an orderly discussion of work in work-unit terms, a series of definitions identifying work-units of various sizes must be established. Each different size is referred to as a different *order of work-unit*.

Work-Units of Different Orders. Exhibit 1.5.1 gives a series of definitions for work-units of various sizes suitable for discussing, in a consistent fashion, the quantitative aspects of any type of work. This list starts from the eighth-order and goes down to the first-order work-unit to conform to the sequence of analysis. Also, it seems appropriate to use smaller numbers for smaller work-units. It should be noted that not all of these orders of work-units will necessarily be involved in a work-unit analysis. On the other hand, in complex situations, there may be a need for orders of

Exhibit 1.5.1 Definitions of Basic Orders of Work-Units

Numerical Designation	Name	Definition
Eighth-order work-unit	Results	What is achieved because of the outputs of the activity.
Seventh-order work-unit	Gross output	A large total of end products or completed services of the working group.
Sixth-order work-unit	Program output	A group of like outputs or completed services that represent part of a seventh-order work-unit but are a more homogeneous subgroup.
Fifth-order work-unit	End product	A unit of final output; the units in which a program is quantified; the smallest output, produced for use outside of the organization, which contributes to the achievement of the objective without further work being done on that output.
Fourth-order work-unit	Intermediate product or component	A part of a unit of final output; the intermediate product may become part of the final output or merely be required to make it feasible to achieve the final output.
Third-order work-unit	Task	Any and all of the activity and things associated with the performance of a unit of assignment by either an individual or a crew, depending on the method of assigning.
Second-order work-unit	Element	The activity associated with the performance of part of a task, which it is convenient to separate to facilitate the designing of the method of performing the task or the time study of the task.
First-order work-unit	Motion	The performance of a human motion. This is the smallest work-unit usually encountered in the study of work. It is used to facilitate job design or time study and never appears in control systems above this level of use.

work-units between the ones given here. In such cases, decimals would be added (e.g., an order of work-unit between the seventh-order and the sixth-order would be identified as the 6.5th-order).

Exhibit 1.5.1 is offered not as a strict exercise in the taxonomy of work, but as a guide in facilitating nomenclature. It is designed to provide a language for use when tracing the path from objectives to outputs and required resources when there is no obvious natural method. One must remember that a work-unit is defined as an amount of work *convenient* with respect to providing a basis for applying motion and time study techniques and assisting in managerial control. The concept of convenience suggests that different numbers of orders of work-units may, at times, be desirable. Productivity is not measured below the fifth-order work-unit.

Work-Unit Analysis. *Work-unit analysis* is the delineation of the outputs and subparts of outputs of an organization in work-unit terms. The analysis starts with the objective of the organization, the statement of the measures of effectiveness, and proceeds through the delineation of smaller and smaller units of output (orders of work-units) until the following criteria are met:

1. There is a clearly visible relationship between objectives and outputs.
2. A suitable level of detail is reached such that meaningful forecasts can be made of outputs required for future periods.
3. A level of detail is reached such that other industrial engineering techniques may be employed.
4. At each level of detail the list of outputs is all-inclusive.
5. At each level of detail the items in the list are mutually exclusive.

Work-Unit Structure. A *work-unit structure* is the hierarchical list of work-units resulting from a work-unit analysis. It is also called a *work-unit hierarchy*.

Procedure for Applying Work-Unit Analysis

Step 1. Stating the Objective (Eighth-Order Work-Units)

To begin the delineation of the work-unit structure, a statement of the objective of the organization being studied should be set forth. This later forms the basis for the measurement of effectiveness. In that higher levels of management will be confronted with many such statements of objectives, a specific format should be adhered to so as to assist in clear and rapid communication. The suggested format has seven headings, which are subsequently explained:

1. Type of service.
2. Mission area.
3. Purpose: Intent.
4. Purpose: Dimension
5. Goals.
6. Limitations.
7. Freedoms.

Type of Service. In industrial service organizations there are seven types of service:

1. Product output, directly related.
2. Output facilitative.
3. Labor control.
4. Finance control.
5. Research and development.
6. Customer service.
7. Material acquisition.

These categories are self-defining. Work-unit analysis is usually applied to categories 2, 3, 4, 5, 6, and 7. Category 1 is the classical area of application of industrial engineering.

In government there are four basic types (or modes) of service, although an organization may use a combination of these. If a combination of these is used, the situation should be so identified. The four types and their definitions are:

1. Constructive service–do for the public that which it would otherwise have to do for itself.
2. Social constraint–restrain people, firms, or groups from undesirable or illegal activities.
3. Grants, awards, and financial assistance–achieving any effect by funding people or organizations outside the government to undertake a desired activity.
4. Internally consumed service–service that has no program relationship with the public; its

efforts are totally consumed in a larger part of the total organization and are not a direct part of a final output (e.g., the internal budget section).

Mission Area. This refers to the sectors of the enterprise, its market, its customers, or suppliers, where the results of the organization being studied are achieved, and where the desired effects or "impacts" are achieved. This is not to be confused with the actual effects, which are the subject of the next three subsections. With government organizations this term refers to the sectors of the society, the economy, or the geography "impacted."

Purpose: Intent. Under this heading the type of effect to be achieved in each of the impact areas listed under Mission Area should be described. Several intents may be listed, if appropriate. However, there should be at least one intent given for each mission area. If not, one can hardly think of it as an "impacted" area.

Purpose: Dimension. Under this heading describe the quantifiable attributes or characteristics of each intent described under the subhead Intent that are of value in quantifying the results achieved in the impact areas. At least one or more dimensions should be given for each intent. It is important to note that these dimensions need not be (and are most usually not), related to the staff resources required by the organization. Many useless budget arguments are generated by a failure to note this usual lack of relationship.

Goals. Under this heading indicate the amount to be achieved in terms of each of the previously listed dimension or dimensions of each intent. An organization usually has two types of goals: one long-range and one for each budget year. True managerial control requires the explicit stating of the goal for each forecast year as well as for the current year. To assist in managerial control it is desirable, however, to have a statement of the long-range goal with which each year's goal must conform.

Limitations. Under this heading list the unique (not typical of all organizations) restrictions on the operation of the organization being studied and its outputs. This statement should not include staff and budget limitations, in that this is hardly unique. More properly one may find, "Can only recommend change."

Freedoms. Under this heading list the areas of action in which the organization being studied has some unique freedom of choice, such as if a contracts group is "Not limited to selecting the lowest bidder."

Step 2. Delineating Gross Outputs (Seventh-Order Work-Units)

In order to achieve the purposes stated previously with respect to a convenient work-unit structure, it is desirable first to divide the totality of outputs into a number of large categories. The separated categories should clarify the significance of each outputs group with respect to

1. Different mission areas served.
2. Different (radically) methods of producing the outputs.
3. Totally independent or different cost systems (e.g., in industry, volume of product related versus fixed funding base; in government, federally funded versus reimbursed work).
4. Different fund accounts separately appropriated.
5. Different benefits obtained.
6. Different mixes of intents.
7. Different types (or modes) of action.

To avoid confusing these outputs with tasks, particularly when the outputs are services, it is most convenient to describe them by means of a past-tense verb together with appropriate adjectives and nouns, for example, "Product designs made," "Economic evaluations reported," "Factories inspected and reported," "Audit services provided." Further, a simple phrase such as "Product design made" must be thought of in the broadest sense. It is not merely the final making of a design. Rather, it is all the activities and services produced that finally culminates in "Product design made." It includes the sick and annual leave of all employees associated with the output, their payroll services, and so forth. The small list of seventh-order work-unit categories must be thought of as representing the net result of all the resources expended by the organization. The list must be devised to meet this concept.

Step 3. Delineating Program Outputs (Sixth-Order Work-Units)

Each seventh-order work-unit should be broken into two or more sixth-order work-units (of which the seventh is the sum) unless such a separation would serve neither the main criteria nor the additional criteria given here. The separation of additional categories at this point, while continuing to serve the main criteria of a work-unit structure, should make more clear the significance of the outputs with respect to

1. Independent, separate cause systems that generate work.
2. Useful subaggregations for decisions with respect to a balance of benefits.
3. Subclasses of the separations initiated at the seventh order, for the same reasons as for seventh-order work-units.
4. Outputs that appear alike but require different amounts of resource support.
5. Subgroups within a seventh-order work-unit that are more alike within the subgroup with respect to mission area served, purpose, and so forth, than the remainder of the outputs in the seventh-order work-unit of which they are a part.

Step 4. Delineating Units of Output (Fifth-Order Work-Units)

In all cases, sixth-order work-units are aggregations of outputs. That which it is convenient to identify as an "each" of the "all" contained in any sixth-order work-unit is called a fifth-order work-unit. Separation at the fifth-order should, in addition to continuing to serve the main criteria, assist in

1. Identifying a normal "each" of actual output with respect to utility.
2. Identifying an "each" in a manner related to the management decision making process with reference to "do/not do" types of decisions.
3. Provide a useful, meaningful basis for "pricing out" the outputs (unit costs).
4. Provide a basis for a work count for that which is contained in each sixth-order work-unit.
5. Provide a work-unit convenient for some type of workload forecasting, method improvement, work measurement, and so forth.

Additional Statements Related to Main Criteria

1. The complete list of any order of work-unit must represent the totality of work that results from the activity of the organization being studied, with the detail increasing as the lower orders are reached. As was stated, the categories in each order must, therefore, be all-inclusive. In addition, they must not overlap; they must be mutually exclusive.
2. The work-units that are forecast for budgeting or counted for performance measurement must be either
   a. End-product outputs (a unit of service or product on which no further work will be done within the organization being studied and which is assumed to contribute to the achievement of the objective).
   b. Outputs having a known and fixed relationship with end products.
   c. Related to resource usage.
3. The work-unit structure must be acceptable to the program personnel and must be related to how they do or should think of outputs when making program decisions. If the program personnel have participated in developing the work-unit structure, these criteria should be met readily.

Case I—Work-Unit Analysis Applied to a National Productivity Board

Introduction

Work-unit analysis can seldom be accomplished without a considerable involvement of program personnel in that they are usually the only ones with the necessary detailed knowledge of the programs. The concept of "participation," in most cases, must be used fully to accomplish this analysis successfully.

Further, the organizational segment to be analyzed must be defined. The analysis may be applied to a section, a branch, an office, or a department. The most useful approach would entail application to the totality of the organization that has responsibility for programs that interface with the remainder of the organization or with the public. (In applying the analysis to government organizations, the interface is usually with a sector of the economy, society, or geography, where the effects achieved are evident.)

In the case examined here, a National Productivity Board applied the analysis to itself in order to have a basis for applying more detailed techniques for self-improvement. They also wished to improve their own internal managerial control.

The Work-Unit Structure

Following the procedure given above, the industrial engineering staff of the board described its work-unit structure as shown in Exhibits 1.5.2 through 1.5.11.

Exhibit 1.5.2 National Productivity Board (NPB)

MODE: Internally consumed.

Mission Area (Impact Area)

1. Productivity of the private sector, industry, in Singapore.
2. Productivity of the private sector, wholesale and retail establishments, in Singapore.
3. Productivity of the service sector, other than above, in Singapore.
4. Productivity of the government sector, in Singapore.

Purpose (Results Wanted)

Intent (Nature of Results):

1. To increase the productivity of industry.
2. To increase the productivity of the wholesale and retail sector.
3. To increase the productivity of the service sector, other than above.
4. To increase the productivity of government organizations.

Dimensions (How Results are Quantified):

Industry

1.1 Percent of establishment requests responded to (consultancies).

1.2 Percent of total population of industrial establishments served (consultancies).

1.3 Percent of requests responded to in not more than 2 months.

1.4 Percent of services resulting in gains exceeding cost, within 1 year (consultancies).

1.5 Percent of services resulting in gains exceeding cost, within more than 1 year but less than 5 years (consultancies).

1.6 Percent of clientele who place a repeat call for services, within 5 years, for new consultancy projects.

1.7 Percent of requests for consultancy services that result in at least a 5 to 10% increase in productivity in the project area.

1.8 Percent of requests for consultancy services that result in a 10 to 25% increase in productivity in the project area.

1.9 Percent of requests for consultancy services that result in an increase in productivity in excess of 25% in the project area.

1.10 Value from consultancy projects.

1.11 Percent of trainees who represent "repeat" companies.

1.12 Number of enrollees in NPB courses and seminars.

1.13 Number of valid client complaints.

Wholesale and Retail

2.1 Percent of establishment requests responded to (consultancies).

2.2 Percent of total population of establishments served (consultancies).

2.3 Percent of requests responded to in not more than 2 months.

2.4 Percent of services resulting in gains exceeding cost, within 1 year (consultancies).

2.5 Percent of services resulting in gains exceeding cost, within more than 1 year but less than 5 years (consultancies).

2.6 Percent of clientele who place a repeat call for services, within 5 years, for new consultancy projects.

Exhibit 1.5.2 (*Continued*)

2.7 Percent of requests for consultancy services that result in at least a 5 to 10% increase in productivity in the project area.

2.8 Percent of requests for consultancy services that result in a 10 to 25% increase in productivity in the project area.

2.9 Percent of requests for consultancy services that result in an increase in productivity in excess of 25% in the project area.

2.10 Value from consultancy projects.

2.11 Percent of trainees who represent "repeat" companies.

2.12 Number of enrollees in NPB courses and seminars.

2.13 Number of valid client complaints.

Service Sector (Other Than Preceding, Private)

3.1 Percent of establishment requests responded to (consultancies).

3.2 Percent of total population of establishments served (consultancies).

3.3 Percent of requests responded to in not more than 2 months.

3.4 Percent of services resulting in gains exceeding cost, within 1 year (consultancies).

3.5 Percent of services resulting in gains exceeding cost, within more than 1 year but less than 5 years (consultancies).

3.6 Percent of clientele who place a repeat call for services, within 5 years, for new consultancy projects.

3.7 Percent of requests for consultancy services that result in at least a 5 to 10% increase in productivity in the project area.

3.8 Percent of requests for consultancy services that result in a 10 to 25% increase in productivity in the project area.

3.9 Percent of requests for consultancy services that result in an increase in productivity in excess of 25% in the project area.

3.10 Value from consultancy projects.

3.11 Percent of trainees who represent "repeat" companies.

3.12 Number of enrollees in NPB courses and seminars.

3.13 Number of valid client complaints.

Government Organizations

4.1 Percent of establishment requests responded to (consultancies).

4.2 Percent of total population of establishments served (consultancies).

4.3 Percent of requests responded to in not more than 2 months.

4.4 Percent of services resulting in gains exceeding cost, within 1 year (consultancies).

4.5 Percent of services resulting in gains exceeding cost, within more than 1 year but less than 5 years (consultancies).

4.6 Percent of clientele who place a repeat call for services, within 5 years, for new consultancy projects.

4.7 Percent of requests for consultancy services that result in at least a 5 to 10% increase in productivity in the project area.

4.8 Percent of requests for consultancy services that result in a 10 to 25% increase in productivity in the project area.

4.9 Percent of requests for consultancy services that result in an increase in productivity in excess of 25% in the project area.

4.10 Value from consultancy projects.

4.11 Percent of trainees who represent "repeat" organizations.

4.12 Number of enrollees in NPB courses and seminars.

4.13 Number of valid client complaints.

Goals (Amount of Results Sought)

Industry

1.1 100%.

1.2.1 20% of those under 49 employees.

1.2.2 10% of those from 50 to 199 employees.
1.2.3 5% of those over 199 employees.
1.3 100%.
1.4 85%.
1.5 100%.
1.6 20%.
1.7 100%.
1.8 25%.
1.9 5%.
1.10 S$2,500,000[a].
1.11 25%.
1.12 10% increase.
1.13 0.

Wholesale and Retail

2.1 100%.
2.2.1 5% of those under 49 employees.
2.2.2 10% of those over 50 employees.
2.3 100%.
2.4 80%.
2.5 100%.
2.6 10%.
2.7 100%.
2.8 10%.
2.9 5%.
2.10 S$200,000.
2.11 25%.
2.12 10% increase.
2.13 0.

Service Sector (Other Than Preceding, Private)

3.1 100%.
3.2.1 5% of those under 49 employees.
3.2.2 10% of those from 50 to 199 employees.
3.2.3 5% of those over 199 employees.
3.3 100%.
3.4 90%.
3.5 100%.
3.6 10%.
3.7 100%.
3.8 25%.
3.9 10%.
3.10 S$100,000.
3.11 25%.
3.12 10% increase.
3.13 0.

Government Organizations

4.1 100%.
4.2.1 90% of those under 49 employees.
4.2.2 10% of those with 50 to 199 employees.
4.2.3 5% of those over 199 employees.

Exhibit 1.5.2 (*Continued*)

4.3 100%.
4.4 90%.
4.5 100%.
4.6 10%.
4.7 50%.
4.8 10%.
4.9 5%.
4.10 S$800,000.
4.11 25%.
4.12 10%.
4.13 0.

Limitations (Numbers not Keyed to Preceding Lists)

1. Has no authority to initiate consultancies.
2. Has no authority to enforce installation of recommendations.
3. Cannot force enrollment at seminars and courses.
4. Must suffer "political" consequences of rejecting an applicant for seminar or course.
5. Internal politics in client organization external to NPB.
6. Must use Public Service Commission pay scale.

Freedoms (Numbers not Keyed to Preceding Lists)

1. Can initiate and advertise courses and seminars.
2. Can "tailor" seminar to clients' needs or requests.
3. Can control assignments of staff.
4. Can reject applications to seminars if enrollee does not meet requirements.
5. Can subcontract-out for courses and seminars.
6. Can make use of technical expert (TE) services via Asian productivity organization (APO).
7. Can establish industry advisory groups.
8. Can publish and disseminate.
9. Does not have to acquire employees through the Public Service Commission.

[a]Total of items 1.10, 2.10, 3.10, and 4.10 equals 2 times operating deficit including overhead.

Exhibit 1.5.3 Seventh-Order Work-Units (Gross Product Groupings)

01. Industry consultancies completed.[a]
02. Wholesale and retail consultancies completed.
03. Service sector, private, other than above, consultancies completed.
04. Government organization consultancies completed.
05. Training, formal, provided.
06. Publications provided.
07. Government sponsored information projects completed.
08. General productivity advocacy provided.

[a]"Completed" means brought to a stage where the evaluation of the "results" is feasible.

Exhibit 1.5.4 Sixth-Order Work-Units (Groups of "Like" Products)

0101. Industry, industrial engineering, production improvement consultancies completed.[a]

0102. Industry, industrial engineering, new process, new product layout, and so on consultancies completed.

0103. Industry, industrial engineering, product improvement consultancies completed.

0104. Industry, mechanization and automation application, consultancies completed.

0105. Industry, management information system consultancies completed.

0106. Industry, management system, for example, job evaluation scheme, and the like consultancies completed.

0201. Wholesale and retail, industrial engineering, production improvement consultancies completed.

0202. Wholesale and retail, industrial engineering, new service procedure or layout, and so on consultancies completed.

0203. Wholesale and retail, mechanization and automation consultancies completed.

0204. Wholesale and retail, low-cost automation consultancies completed.

0205. Wholesale and retail, management information system consultancies completed.

0206. Wholesale and retail, managerial systems, for example, job evaluation schemes, and the like consultancies completed.

0301. Service sector, industrial engineering, production improvement (cost of services) consultancies completed.

0302. Service sector, industrial engineering, new service procedure or layout, and so on consultancies completed.

0303. Service sector, service improvement consultancies completed.

0304. Service sector, mechanization and automation consultancies completed.

0305. Service sector, management information system consultancies completed.

0306. Service sector, management systems, for example, job evaluation schemes, and the like consultancies completed.

0401. Government organization, industrial engineering, production improvement (cost of services) consultancies completed.

0402. Government organization, industrial engineering, new service procedure or layout, and so on consultancies completed.

0403. Government organization, service improvement consultancies completed.

0404. Government organization, mechanization and automation consultancies completed.

0405. Government organization, management information system consultancies completed.

0406. Government organization, management systems, other than above, consultancies completed.

0501. In plant,[b] group of organizations, special multidisciplinary training courses provided.[c]

0502. In plant, single organization, multidisciplinary training courses provided.

0503. In plant, marketing training programs provided.

0504. In plant, sales training programs provided.

0505. In plant, management and executive development training programs provided.

0506. In plant, financial management training programs provided.

0507. Thirty-hour modules, certificate courses provided.

0508. Twenty-hour modules, certificate courses provided.

0509. "General admissions" training courses provided.

0510. Training courses provided via contract services.

0511. Training courses completed via APO TE services.

0701. Periodical publications published and disseminated.[d]

0702. Special publications, bulletins (1 to 10 pages), published and disseminated.

0703. Special publications, monographs (10 to 99 pages), published and disseminated.

0704. Special publications, books (100 pages or more), published and disseminated.

0705. Productivity advocacy publications prepared and disseminated.

Exhibit 1.5.4 (*Continued*)

0801. Government sponsored statistical collection and analysis researches completed and reports presented.

0802. Government sponsored other research undertakings completed and reports presented.

0901. Productivity advocacy verbal presentations made.

0902. Productivity advocacy soundtape (and/or slide-tape) presentations prepared.

0903. Productivity advocacy soundtape (and/or slide-tape) presentations arranged for and made.[e]

0904. Productivity advocacy videotape presentations prepared.

0905. Productivity advocacy videotape presentations arranged for and made.[e]

[a]See footnote *a* to Exhibit 1.5.3 for a definition of "completed."
[b]"In plant" means "in the organization, whatever kind."
[c]"Provided" means "with NPB staff."
[d]Does not include periodic financial or performance reports of the NPB.
[e]Regardless of who does the actual presentation.

Exhibit 1.5.5 Fifth-Order Work-Units (Unit outputs: the smallest outputs of value, with respect to achieving the objective, without further work being done on the output)

010101. An industry industrial engineering, production improvement consultancy completed.

010201. An industry industrial engineering, new process, new product layout, and so on, consultancy completed.

010301. An industry industrial engineering, product improvement consultancy completed.

010401. An industry mechanization and automation application consultancy completed.

010501. An industry management information system consultancy completed.

010601. An industry management system, for example job evaluation, and the like, consultancy completed.

010602. A productivity committee established in an organization.

020101. A wholesale and retail industrial engineering, production improvement consultancy completed.

020201. A wholesale and retail industrial engineering, new service procedure or layout, and so on, consultancy completed.

020301. A wholesale and retail mechanization and automation consultancy completed.

020401. A wholesale and retail low-cost automation consultancy completed.

020501. A wholesale and retail management information system consultancy completed.

020601. A wholesale and retail management system, for example, job evaluation, and the like, consultancy completed.

020602. A productivity committee established in an organization.

030101. A service sector industrial engineering, production improvement (cost of services) consultancy completed.

030201. A service sector industrial engineering, new service procedure or layout, and so on, consultancy completed.

030301. A service sector service improvement consultancy completed.

030401. A service sector mechanization and automation consultancy completed.

030501. A service sector management information system consultancy completed.

030601. A service sector management system, other than above, consultancy completed.

030602. A productivity committee established in an organization.

040101. A government organization industrial engineering, production improvement (cost of services) consultancy completed.

Exhibit 1.5.5 (*Continued*)

040201. A government organization industrial engineering, new service procedure or layout, and so on, consultancy completed.

040301. A government organization service improvement consultancy completed.

040401. A government organization mechanization and automation consultancy completed.

040501. A government organization management information system consultancy completed.

040601. A government organization management system, other than above, consultancy completed.

040602. A productivity committee established in an organization.

050101. An in plant[a] group of organizations, special multidisciplinary training course provided.[b]

050201. An in plant single organization, multidisciplinary training course provided.

050301. An in plant marketing training program provided.

050401. An in plant sales training program provided.

050501. An in plant management and executive development training program provided.

050601. An in plant financial management training program provided.

050701. Thirty-hour module, certificate course provided.

050801. Twenty-hour module, certificate course provided.

050901. A "general admissions" training course provided.

051001. A training course provided via contract services.

051101. A training course provided via APO TE services.

070101. An issue of a periodical publication published and disseminated.[c]

070201. A special publication, bulletin (1 to 10 pages), published and disseminated.

070301. A special publication, monograph (10 to 99 pages), published and disseminated.

070401. A special publication, book (100 pages or more), published and disseminated.

070501. A productivity advocacy publication prepared and disseminated.

080101. A government sponsored statistical collection and analysis research completed and report presented.

080201. A government sponsored other research undertaking completed and report presented.

090101. A productivity advocacy verbal presentation made.

090201. A productivity advocacy soundtape (and/or slide-tape) presentation prepared (count in units of 10 minutes).

090301. A productivity advocacy sound tape (and/or slide-tape) presentation arranged for and made.[d]

090401. A productivity advocacy videotape presentation prepared (count in units of minutes).

090501. A productivity advocacy videotape presentation arranged for and made.[d]

---

[a]"In plant" means "in the organization, whatever kind."
[b]"Provided" means "with NPB staff."
[c]Does not include periodic financial or performance reports of the NPB.
[d]Regardless of who does the actual presentation.

Exhibit 1.5.6 System Implementation

A. Effectiveness Reporting

1. This may start immediately.
2. Use preprinted form such as that shown in Exhibit 1.5.7.
3. Each manager to use one blank copy of Exhibit 1.5.7 as a tally sheet during month, entering either tallies (as for item .1) or have population figures (as for item .2).
4. The tally sheets are to be converted to a form with numbers.
5. These forms of item 4 may be converted to quarterly and annual reports, but monthly updating will spread out the workload and permit more timely verification of questionable items.
6. *Note:* Many of the entries, for example, .7 and .8, will refer only to projects completed during the reporting period. These data will come from the consultant and professional reports, Exhibit 1.5.9.

B. Work Measurement Alternatives for NPB

1. Use of Past Data
   - 1.1 Use assignment record, billings, and reports to allocate professional staff time for past year (either from end of past month, calendar year, or fiscal year, as feasible) to the work-units 010101 through 080201 inclusive. Do not include managerial or clerical time. (Account for all professional staff time except for leave.)
   - 1.2 For the same period as 1.1 above, determine the work count of work-units 010101 through 080201 with the following special counting methods:

     010101 through 040601: One each is a count of 1.
     050101 through 050505: One day of course is count of 1.
     050601 and 050701: One module is a count of 1.
     050801: One day of course is count of 1.
     050901 and 051001: A course is a count of 1.
     070101 through 080201: One each is a count of 1.
   - 1.3 Divide the work time allocation for each work-unit (from 1.1 above) by the work count for the work-unit (from 1.2 above) to get the base period "standard time."
   - 1.4 Obtain the following data for the same period (include all time except leave):

     a. Total professional staff time = Tp
     b. Total clerical and support time = Tc
     c. Total manual support time = Tm
     d. Total managerial time = Td

     Determine the following as decimal values:

     a. Tc/Tp = ______
     b. Tm/Tp = ______
     c. Tc/Tp = ______
2. Use of Logging, if Past Data is Not Accessible
   - 2.1 Have all professional staff keep work log as shown in Exhibit 1.5.8 for a base period of at least 6 months.
   - 2.2 Using the logs of 2.1 above, determine the data for 1.1 and 1.2.
   - 2.3 Follow computations of 1.3 and 1.4.
3. In either case the work report form of Exhibit 1.5.9 is to be initiated immediately and put into continuous use.

Exhibit 1.5.7 Effectiveness Evaluation Sheet

From ________ To ________

Item No.	Dimension	Goal (%)	Base No.	Score No.
.1	Percent of establishment requests responded to (consultancies).	100		
.2	Percent of total population of industrial establishments served (consultancies).	20 (under 49)		
		10 (50 to 199)		
		5 (200 or more)		
.3	Percent of requests responded to (in a satisfactory way) in not more than 2 months (+ client is satisfied).	100		
.4	Percent of services (consultancies) resulting in gains exceeding cost within 1 year.	85		
.5	Percent of services resulting in gains exceeding cost in more than 1 but not less than 5 years (consultancies).	100		
.6	Percent of clientele who place a repeat call for services within 5 years for new consultancy project.	20		
.7	Percent of requests for consultancy services that result in a 10 to 25% increase in productivity in the project area.	100		
.8	Percent of requests for consultancy service that result in a 10 to 25% increase in productivity in the project area.	25		

Exhibit 1.5.8 Work Log

Name ________________ Week of ________________

Work-Unit	Day	Hours	Day Total	Work-Unit	Day	Hours	Day Total

See Instruction Sheet

Exhibit 1.5.8 (*Continued*)

Work Log Instruction Sheet

1. Use standard work-unit codes (see Section 1.5.5).
2. Do not record events of less than 0.5 hours separately.
3. Do not record in units other than multiples of 0.5 hours.
4. Account for all work time per day.
5. Use the following as a work-unit number for the indicated activity:

   01 Personal.
   02 Sick leave.
   03 Annual leave.
   04 Receive training or study non–work-unit related.
   05 Give training, internal.
   06 Administrative.
   07
   ( )
   08 (May be assigned later.)
   ( )
   09
6. Stop use of log after base period and only continue the form shown in Exhibit 1.5.9. This form will be in continuous use both during and after the study and used without concern as to the work measurement method used.

Exhibit 1.5.9 Project Completion Report

Date of Report:
Work-Unit Code:
Date Completed:
Client Name:
Summary of Type of Study:
Special Work Count for 050101 to 051001:

See objectives dimension sheet (Exhibit 1.5.2) and enter appropriate number next to dimension or check, as may be appropriate.

.4
.5
.6
.7
.8
.9
.10 S$ =
.11
.12

Name of consultant(s) ______________________

____

____

Exhibit 1.5.10 Productivity Reporting

1. Use form such as Exhibit 1.5.11.
2. Enter work counts from project completion sheets for period.
3. Compute earned hours.
4. Add ratio hours for clerical, manual, and managerial time.
5. Divide by hours actually worked to obtain productivity indexes.

Exhibit 1.5.11 Productivity Report

From ______________ To ______________ Page ____ of ____ pps.
(Date) (Date)

Work-Unit No.	Work-Unit Description	Unit of Count	Standard Time	Work Count	Earned Hours	Worked Hours	Productivity
010101	Industry, industrial engineering, production improvement consultancies completed.						
010201	etc.						
080201	A government sponsored, other research undertaking completed and report presented.						
	Total for Professional Group						
	Total for Clerical Group[a]			b	c		
	Ratio for Manual Group[a]			b	c		
	Ratio for Management Group[a]			b	c		
	Grand Total						

[a] Enter appropriate ratio.
[b] Enter professional earned hours.
[c] Enter appropriate [a] × [b].

Each of the seventh-order work-units was assigned a 2-digit number. Two additional digits were affixed at each successive order of work-unit so as to give each work-unit at each order a unique identifier. If these numbers are prefixed by another two digits to indicate the suborganization, a hierarchical chart of accounts result.

Results

The eighth-order work-unit clearly states the basis for measuring the organization's success. This cannot do other than help manage; it shows the way to go. The seventh-order work-unit list serves as a basis for examining the allocation of resources (aggregated to this level) and for evaluating different allocation strategies. The sixth-order work-unit list serves as a basis for developing forecasts. The fifth-order list serves as a basis for detailed managerial control. Further, the fifth-order list delineates the service outputs in a manner suitable for the application of other, more detailed industrial engineering techniques (see Sections 1 through 14, this handbook).

1.5.6 Managing with the Aid of a Productivity Measurement System

Introduction

The information in this section is designed to assist managers at all levels in making effective use of a formal productivity measurement system, and in their efforts to improve the productivity and effectiveness of their organizations. (As was noted in Section 1.5.2, *effectiveness* is included, in that outputs cannot be identified until the criterion or criteria of effectiveness have been stated. The term *productivity measurement system* is used to describe a system that monitors both productivity and effectiveness; to monitor one without the other would be myopic.)

This section describes a general pattern for using the system data and for monitoring the effects of applying the information and technologies contained in the other chapters of this handbook. The material that follows is divided into seven parts, each part relating to one feature of a productivity measurement system.

These seven features do not improve productivity; it is the managers who make decisions who improve the organization's performance–decisions using data that the systems can produce at desired intervals, together with judicious use of the other materials in this handbook.

A Summary of the Seven Features of a Productivity Measurement System

A productivity measurement system that is properly implemented has at least the seven basic components below. These components are phrased in a manner that enables them to be applied to any kind of organization, for example, manufacturing, mining, service, government–profit seeking or non-profit seeking. The features of the system are listed here and are subsequently examined in detail. The detailed examination, like the summary, is phrased so as to apply to all kinds of organizations.

1. A statement of the objectives of the organization.
2. A list of the units of output of the organization, which serve to achieve the organization's objectives.
3. A set of base year standards, one for each kind of output, that includes standard time, standard cost, standard raw material use, equipment use, tool use, and so forth, depending on the requirements of the appropriate productivity index.
4. A method of building a zero base budget using forecasts of outputs, standard times, and forecasts of the productivity in various appropriate terms.
5. A means of computing the productivity indexes at selected intervals.
6. A means of comparing output forecasts with actual outputs at selected intervals.
7. A means of adding resource usage data and associated productivity indexes in a meaningful fashion related to outputs to reduce the details in reports as data go to higher- and higher-level managers.

A Statement of the Objectives of the Organization

Nature of the Statement

The statement of the objectives is constructed for each organization to show the kinds of results the organization is charged with achieving, how these results are to be quantified, and how much is desired during a program year. In profit seeking organizations these will be couched in monetary terms. In nonprofit organizations these need to be couched in substantive, quantitative terms appropriate to the nature of the results. Hence the statements of objectives of nonprofit organizations will show more uniqueness than those of profit seeking organizations.

Basic Purposes of the Statement

A comparison of what is achieved because of the outputs, with respect to the statement of desired objectives, will assist in evaluating:

1. The relative achievements that could be obtained with alternative raw inputs, output design or mix, processes, facilities, facilities utilization, staff resources, or staff motivation. *Note:* To generate these alternatives the technologies described in the remaining chapters of this handbook are applicable. Alternative "sales" efforts may be an appropriate avenue of change. This subject is not treated in this handbook.

2. Underachievement (if this occurs) may be caused by too few outputs resulting from defects in raw inputs, output design or mix, processes, facilities, facilities utilization, staff resources or staff motivation, sales efforts, or economic changes. Means of determining the problem areas and remedying the defects are contained in all of the remaining chapters of this handbook.

3. Possible achievement of more than anticipated may be caused by good management or some unpredicted change in one or more of the factors listed under item 1 above. The cause should be identified and "institutionalized." The value of the overachievement should be ascertained. In profit seeking organizations, overachievement may indicate some larger fault creating the greater need for the service; each case should be examined individually. The material in Section 9 on engineering economy and Section 14 on optimization of this handbook may be particularly useful.

Effectiveness Evaluation

Management Actions Required. The manager should assign to a specific individual responsibility for evaluating results (effectiveness) and for reporting the findings to him or her. The manager should periodically review this report in view of these basic purposes. (*Note:* In profit seeking organizations this is the assignment of the responsibility for preparing the financial statements and is almost always properly performed. In nonprofit organizations, the financial reports are also usually adequately prepared but these are not measures of effectiveness; rather they are an evaluation of the ability of the organization to exist. The statements of true effectiveness are additionally required.)

Possible Decisions. In conducting the evaluation the manager needs to decide between the desirability of current outputs or mixes of outputs and alternative outputs or mixes of outputs.

Records That Are Desirable. A record of all decisions made with respect to the areas of alternatives should be maintained in order to provide a basis for evaluating future proposals for change.

A List of Units of Output of the Organization, Which Serves to Achieve the Organization's Objectives

Nature of the List

This is a list of outputs. It starts at the basic planning level. Hierarchical grouping methods are indicated. The list is called a work-unit structure. It provides ways of counting outputs from a basic unit level as well as in various degrees of meaningful aggregation. (It is developed from the top down; it is used from the bottom up.) It is subject to change when the outputs of the organization change.

Note that the language used above includes the concepts of the manufacturing company's catalogs and parts lists. However, the terms used are also applicable to service organizations and governmental units.

Purposes of the List

1. It provides a basis for the evaluation of

a. The desirability of the current outputs or current mix of outputs.

b. The desirability of alternative outputs or alternative mixes of outputs.

2. It provides a set of categories of outputs for

a. Forecasting future workloads.

b. Comparing events with forecasts.

c. Developing time standards (using any one of a rather extensive list of work measurement techniques) or other standards such as standard costs, standard material use, and so forth.

d. Developing a system of management by objectives.
e. The creation of a chart of accounts for a cost accounting system for computing unit costs.
f. Examining various strategies of staff, equipment, facility, financial resource allocations. (See Section 14 on optimization in this handbook.)
g. Directing, in output related terms, the allocation of resources.

Management Actions Required

1. Authority and responsibility delegations. The manager should assign responsibility to specific individuals to

a. Maintain a current, hiearchical list of outputs—the work-unit structure. This list, if used for managing, must be up to date; if used for budgeting, it must reflect the anticipated outputs.
b. Describe and evaluate alternatives of all factors for achieving the results described in the statement of objectives.
c. Forecast, at a suitable level of detail, the amount of each output to be produced during future periods.

2. Decision making

a. The authority delegations of the preceding item all refer to situations wherein alternatives exist. The manager must undertake the responsibility for those decisions with respect to the selection of alternatives that maintain or increase the effectiveness of the organization while at the same time either maintaining or improving the productivity of the organization (measured appropriately) or providing desirable unit-cost-productivity tradeoffs.
b. Problems of the quality of working life (see Sections 5 and 6 of this handbook) may interact with these and should be considered. However, it would appear erroneous to consider improved effectiveness, productivity, and unit costs as counterproductive per se to morale and quality of working life.

Quality of Outputs

None of the preceding or following procedures should be thought of as changing, in any way or manner, the inherent responsibility of a manager for maintaining the necessary and desirable level of quality of any and all outputs. The systems in no way interfere with, usurp, or abrogate any previously imposed or assigned authority or responsibility for quality (see Section 8 of this handbook). However, this should not preclude the consideration of quality-quantity tradeoffs.

Records That are Desirable

A record of all decisions made with respect to the preceding areas of choice should be maintained in order to provide a basis for establishing the quality of management.

A Set of Base Year Standards, One for Each Kind of Output, That Includes Standard Time and Standard Cost, Depending on the Requirements of the Appropriate Productivity Index

Definitions

A *base year* is any year previous to the current year. The specific year to use as a base year is usually designated by a higher authority. A *standard time* is the actual working time (staff hours) required to produce each kind of unit of output. This is not to be confused with the time from the initiation to completion of a unit of output; this latter time is called *duration*. Duration commonly exceeds standard time. If a productivity index other than labor productivity is employed, a different base year value is required.

Purpose of Base Year Standard Times or Other Base Year Standard

Base year standard times (or other base year values) provide a datum, a baseline from which progress may be measured. A comparison of actual performance with base year standards provides a total reflection (with respect to quantity of performance) of all factors impinging on the outputs of the organization.

Management Actions Required

Responsibility for the development of base year standards must be assigned to a specific individual.

Records That Are Desirable

A record of all decisions made with respect to the preceding areas of choice should be maintained in order to provide a basis for establishing the quality of management.

Zero Base Budgets for Staff Resources

Definition

A *zero base budget* is one developed by forecasting the amount of all outputs required for a budget year, multiplying each output by an appropriate base year standard time, dividing the result by the budget year anticipated productivity, and, finally, dividing the last figure by the staff hours of time available for productive work, per staff member, to find the number of staff required. In addition, alternative outputs must be considered.

Requirements for Zero Base Budgets

As defined above, zero base budgets require

1. A list of units of output to forecast, including alternative outputs.
2. Forecasts of how many of each kind of output will be required in the budget year.
3. Base year standard times (see Section 4 of this handbook).
4. Productivity forecasts for the budget year (see Section 13 of this handbook).
5. Selection of the most appropriate outputs for budget formulation.

Management Responsibilities

1. Managers should assign specific individuals responsibility for items 1 through 5 above.
2. Managers should examine the narrative justifications accompanying the forecasts and satisfy themselves as to the acceptability of these narratives.
3. Managers should assign responsibility for a continuing examination of the methods of producing the outputs forecast as well as alternative outputs, in view of the ever present obligations of a manager to continually seek more economical ways of reaching assigned objectives. (Sections 2 through 14 of this handbook are pertinent to this seeking of alternatives.)

Records That Are Desirable

A record of all decisions made with respect to the preceding areas of choice should be maintained in order to provide a basis for establishing the quality of management.

Computing Productivity Indexes at Selected Intervals

Definition and Timing

The productivity index is defined as the ratio of objective-oriented outputs produced to staff hours, cost, or other inputs, compared to a similar ratio for a base year. In order to react to changes in productivity, the index should be computed either weekly, every 4 weeks, or quarterly as may be appropriate to obtaining a timely and meaningful figure free of random variations.

Management Responsibilities

Management should assign the responsibility for

1. Generating the necessary reports of work output accomplished.
2. Collecting output data and computing productivity (see Section 12 of this handbook).
3. Examining all deviations from base year or forecast productivity and the reasons for them, and taking corrective actions, as feasible.
4. Reporting on item 3 above to higher authorities.
5. Continuously striving to improve productivity while maintaining both quality, as required, (see Section 8 of this handbook) and service levels, including control of duration required to provide service.

Records That Are Desirable

A record of all decisions made with respect to the preceding areas of choice should be maintained in order to provide a basis for establishing the quality of management.

Comparing Forecasts with Actual Outputs

Purpose of the Comparison

Budgets are most properly based on a forecast of work load. Changes in productivity reflect changes in the ratio of work load to staff or other resource. Hence one important set of data to monitor during a program year is the relationship between forecast work load and actual work load performed. Backlog should also be monitored. (See Section 12 of this handbook.)

Managerial Responsibilities

Management should assign to specific individuals responsibility for reporting at suitable intervals with respect to:

1. A comparison of the quantities of outputs forecast, produced, and needed for selected periods of operation and a recommendation of actions, as may be needed, to cause plans and events to conform to the maximum extent practicable.
2. Changes in backlog.
3. Volume of backlog versus plan.
4. Changes in duration.
5. Maintaining both base year standard values and new standard values for new outputs (to evaluate productivity) and current year standard inputs or productivity indexes (to support future realistic zero base staff resource budgets, staff allocations, and suitable staff workloads).
6. Developing and maintaining a work assignment schedule.
7. Collecting and examining unit costs and their comparison with previous and planned costs.

Management Decisions Required

1. The need for decisions may be caused by
   a. Discrepancies between planned and actual workload.
   b. Changes in backlog.
   c. Volume of backlog.
   d. Changes in duration.
2. The necessary decisions may require
   a. Reallocation or borrowing of staff, facilities, or equipment. (See Section II on planning and control in this handbook.)
   b. Allowing staff attrition to take place.
   c. Requesting (with substantive data support) more staff, equipment, facilities, and so forth, as may be appropriate.
   d. Changing the service level (shorter or longer duration).

Records That Are Desirable

A record of all decisions made with respect to the preceding areas of choice should be maintained in order to provide a basis for establishing the quality of management.

Reports

Special Features

The list of outputs, referred to previously, have been described as having hierarchical groupings. The list of outputs, which serves as a report for a supervisory manager, has predefined ways of being aggregated for reporting to higher-level managers. Several such "detail collapsing levels" are defined in the output lists (work-unit structures). (See Section 12 of this handbook.)

Basis of Aggregation of Outputs

It is well understood that adding a work count of one kind of output to a work count of another kind of output may produce a meaningless number. In a productivity measurement system, however, each unit of output has a base year standard value. Adding the base year standard equivalent of a unit work count of one kind of output to the base year standard equivalent of one or more other kinds of output produces a logical total—a total that may be at all levels of aggregation

1. Compared to the resources used.
2. Compared to the resources budgeted.

3. Used to compute a productivity index, which may be compared to base year performance or budgeted productivity forecast.

Management Responsibilities

1. Management must assign to specific individuals the design of the hierarchical reporting system.
2. Management must assign to specific individuals the preparation of appropriate reports as required to support the total reporting system.

Management Decisions

Management must make decisions with respect to workload, method, and staff in order either to

1. Correct discrepancies between plans and events at each level of aggregation within the reporting systems.
2. Minimize the damage caused by differences between planned and actual workload, backlog, and productivity at all levels of reporting.

Records That Are Desirable

A record of all decisions made with respect to the preceding areas of choice should be maintained in order to provide a basis for establishing the quality of management.

REFERENCE

1. M. E. MUNDEL, Adapted from *Motion and Time Study–Improving Productivity*, Prentice-Hall, Englewood Cliffs, NJ, 1978, chap. 9.

BIBLIOGRAPHY

A list of books relevant to productivity measurement was compiled by Frederick L. Haynes and appeared in the March 1976 issue of *IE*, the monthly publication of the American Institute of Industrial Engineers.

Periodical literature is also available from the following organizations, alphabetized by country.

AUSTRALIA. Productivity Promotion Council, National Committee, GPO Box 475D, Melbourne, Victoria, 3001.

Department of Productivity, Anzac Park, West Building, Constitution Avenue, Prakes, A.C.T., 2600.

AUSTRIA. Osterreichisches Zentrum fur Wirtschaftlichkeit und Produkitivitat, Hohenstaufengasse 3, 1014 Vienna.

BELGIUM. Office Belge Pour l'Accroissement de la Productivite, Rue de la Concorde, 60, 1050 Bruxelles.

BULGARIA. National Centre for the Social Productivity of Labour, Bld G. Dimitrov, 52, Sofia.

CANADA. Department of Industry, Trade and Commerce, Productivity Branch, Ottawa KIA OH5.

DENMARK. Danish Productivity Council on Industry, Handicrafts, and Commerce, Danmarks Erhvervsfond, Codanhus–G1., Kongevej 60, 1850 Copenhagen V.

ENGLAND. International Council for the Quality of Working Life, London Graduate School of Business Studies, Sussex Place, Regent's Park, London NW1 4SA.

GERMANY, FEDERAL REPUBLIC OF. Rationalisierungs-Kuratorium der Deutschen Wirtschaft, Gutleutstrasse 163-167, 6000 Frankfurt (Main).

GREECE. Greek Productivity Centre, Kapodistiou 28, Athens 147.

HONG KONG. Hong Kong Productivity Council, Rooms 512–516, Gloucester Building, Des Voeux Road, C, P.O. Box 16-132.

INDIA. National Productivity Council, Productivity House, Lodi Road, New Delhi 110 003.

IRELAND. Irish Productivity Centre (I.P.C.), 35–39 Shelbourne Road, Dublin 4.

ISRAEL. Israel Institute of Productivity (IIP), P.O.B. 33010, Tel-Aviv.

ITALY. Instituto Nazionale per l'Incremento delia Produttivita (INIP), Piazza Indipendenza, 11B, 00185 Roma.

JAPAN. Asian Productivity Organization, Aoyama Dai-Ichi Mansions, Minato-ku, Tokyo, 107.

Chubu Productivity Center, No. 10-2-Chome, Sakae, Nagoya.

Japan Productivity Centre, 1, 3-chome, Shibuya, Shubuya-ku, Tokyo.

KOREA. Korean Productivity Center, 10, 2-GA, Pil-dong, Hung-gu, Seoul.

LUXEMBOURG. Office Luxembourgeois pour l'Accroissement de la productivite (OLAP), Rue A. Lumiere, 18.

MEXICO. Centro Nacional de Productividad de Mexico, A.C., Anillo Perferico 2143, Mexico 20, D.F.

NETHERLANDS. Commissie Opvoering Produktiviteit (COP), Bezuidenhoutseweg, 60, The Hague.

NEW ZEALAND. Productivity Centre, c/o Department of Trade and Industry, Private Bag, Wellington.

NORWAY. Norsk Produktivitetsinstitutt (NPI), Boks 8401, Hammersborg, (Akersgatan, 64), Oslo 1.

Work Research Institutes, Gydas vei 8, Oslo 1.

SOUTH AFRICA. National Productivity Institute (NPI), Private Bag 191, Pretoria.

SPAIN. Direccion General de Promocion Industrial y Technologia, Ministerio de Industria, Calle Ayala 3, Madrid 1.

SWEDEN. Arbetslivscentrum, Box 5606, 114 86 Stockholm.

Psykologiska Institutionen, Universitet Stockholm, Box 6706, 113 85 Stockholm.

TURKEY. Milli Produktivite Merkezi (MPM), Mithatpasa Cadessi 46, Yenisehir, Ankara.

UNITED STATES of AMERICA. American Center for the Quality of Work Life, 3301 New Mexico Avenue, NW, Suite 202, Washington, DC 20016.

American Productivity Center, 1700 West Loop South, Houston, TX 77027.

Center for Government and Public Affairs, Auburn University at Montgomery, Montgomery, AL 36117.

Center for Productive Public Management, John Jay College of Criminal Justice, City University of New York, 445 West 59th Street, New York, NY 10019.

Center for Quality of Working Life, Institute of Industrial Relations, University of California, 405 Hilgard Avenue, Los Angeles, CA 90024.

Committee on Productivity, American Institute of Industrial Engineers, 25 Technology Park, Norcross, GA 30092.

Georgia Productivity Center, Engineering Experiment Station, Georgia Institute of Technology, Atlanta, GA 30332.

Harvard Project on Technology, Work, and Character, 1710 Connecticut Avenue, NW, Washington, DC 20009.

Institute for Productivity, 592 De Hostos Avenue, Baldrich, Hato Ray, Puerto Rico 00918.

Management and Behavioral Science Center, Wharton School, University of Pennsylvania, Vance Hall, 3788 Spruce Street, Philadelphia, PA 19104.

Manufacturing Productivity Center, IIT Center, 10 West 35th Street, Chicago, IL 60616.

Maryland Center for Productivity and Quality of Working Life, College of Business and Management, University of Maryland, College Park, MD 20742.

Massachusetts Quality of Working Life Center, 14 Beacon Street, Suite 712, Boston, MA 02108.

National Academy of Sciences, Office of Publications, 2101 Constitution Avenue, NW, Washington, DC 20418.

Oklahoma Productivity Institute, School of Industrial Engineering and Management, Oklahoma State University, Stillwater, OK 74074.

The Productivity Council of the Southwest, STF 124, 5151 State University Drive, Los Angeles, CA 90032.

Productivity Institute, College of Business Administration, Arizona State University, Tempe, AZ 85281.

Productivity Research and Extension Program, North Carolina State University, P.O. Box 5511, Raleigh, NC 27607.

Purdue Productivity Center, School of Industrial Engineering, Purdue University, Grisson Hall, West Lafayette, IN 49707.

Quality of Working Life Program, Center for Human Resource Research, The Ohio State University, 1375 Perry Street, Suite 585, Columbus, OH 43201.

Quality of Working Life Program, Institute of Labor and Industrial Relations, University of Illinois at Urbana-Champaign, 540 East Armory Avenue, Champaign, IL 61820.

South Florida Productivity Center, New World Center Campus, 300 NE Second Avenue, Room 1402, Miami, FL 33101.

Utah State University Center for Productivity and Quality of Working Life, UMC 35, Utah State University, Logan, UT 84321.

Work in America Institute, Inc., 700 White Plains Road, Scarsdale, NY 10583.

YUGOSLAVIA. Jugoslovenski Zavoda Za Produktivnost Rada, (Yugoslave Productivity Institute), Uzun Mirkova, 1, Belgrade.

CHAPTER 1.6

Creative Problem Solving

KNUT HOLT

University of Trondheim

1.6.1 INTRODUCTION

The design of a new product, or a new system, whether it is of a technological or social nature, can be viewed as a problem solving process. A rather common approach consists of the following steps: definition of the problem, collection of data, analysis of data, development of alternatives, and selection and implementation of the solution. The process is of an iterative character; one has to move back and forth, collect more information, analyze anew, redefine the problem, modify previous conclusions, and so on.

The first step of the process is defining the problem. Usually this is the most important part, since it determines scope and direction of the following steps. However, in actual practice this part is often neglected. The reason for this may be that this stage in itself is a highly complex process, consisting of several iterative steps where needs of those involved, available resources and existing constraints have to be balanced properly.

By making certain assumptions and simplifications, the industrial engineer is able to solve many problems by means of mathematical models (see Sections 13 and 14). These models, which represent an idealization of the real structure, vary from simple functional relationships to complex equation systems. Some of them make it possible to find optimal solutions to problems with many variables. When it is not possible to use analytical models, one may be able to develop simulation models. By changing the values of important parameters, it is then possible to study various alternatives and their consequences and thereby create a foundation for selection of a satisfying solution.

The majority of problems facing the industrial engineer involve technical, economic, and social factors. Under such circumstances, it is often not possible to find a satisfactory solution by means of mathematical models. The industrial engineer then has to rely on his or her own knowledge and on general methods for systematic problem solving, for example, as presented by Kepner and Tregoe.[1] Logical thinking will play an important role, but it should be supplemented by creative techniques. Such techniques, often called "divergent" or "lateral thinking," stimulate the problem solver to look at things in new ways. Conventional thinking has, according to de Bono,[2] patterns of expectations that fix the way of looking at things. The purpose of lateral thinking is to escape from these fixed ideas and generate new ones. The process is fundamentally different from that of logical thinking, where one must be right at each stage. In lateral thinking one is allowed to be wrong in order to get far enough away from fixed ideas so that one can find new ones.

Logical analytical thinking is effective in many cases, but there are also cases where it should be replaced or supplemented by creative thinking processes. These processes can be used both by individuals and by groups, and they have a wide area of application, ranging from minor improvements to genuine innovations.

1.6.2 BASIC CONCEPTS AND MODELS

Creativity is closely linked to innovation and generation of ideas. Since there is much confusion regarding these terms, they must be defined before a meaningful discussion can take place.

Creativity

Although millions of words have been spoken and written about creativity, knowledge about the topic is still limited. One lacks even a commonly accepted definition. Among the many suggested, the one shown in Exhibit 1.6.1, first introduced by Taylor,[3] appears to be best for practical purposes. One weakness is that it depends on a subjective assessment of what is worthwile. On the

Exhibit 1.6.1 Definition of Creativity

Creativity is that thinking which results in the production of ideas that are both novel and worthwhile.

other hand, it is dynamic and action oriented. It stresses that the ideas generated must be such that they can be used for practical purposes.

Innovation

Creativity is closely linked to innovation, which is a dynamic process that can be illustrated by a simple model, as shown in Exhibit 1.6.2. The iterative character of a trial-and-error process with many feedback loops is indicated by arrows. The impact of creative behavior is usually strongest in the first stage, the generation of the basic idea. This is a crucial point, since it gives the direction of the whole process. However, creativity is needed at all stages, as stressed by Morton[4]:

> *Innovation is not just one simple act. It is not just a new understanding or the discovery of a new phenomenon, not just a flash of creative invention, not just the development of a new product or manufacturing process; nor is it simply the creation of new capital and markets. Rather innovation involves related creative activity in all these areas. It is a connected process in which many and sufficient creative acts, from research through service, are coupled together in an integrated way for a common goal.*

Generation of Ideas

The generation of an innovative idea, that is, an idea that has the potential of becoming an innovation, is in itself a complicated process. A simple model of this process is given in Exhibit 1.6.3. Engineers often focus their attention on the problem solving aspect, on creating the right answer by finding an appropriate concept for the solution. However, as indicated, an innovative idea is a fusion of a need and a possibility for fulfillment of the need. Therefore it is necessary for the engineer to pay attention to the needs of those involved. This part of the idea generation activity is concerned with defining the problem by asking the right question.

1.6.3 NEED FOR CREATIVITY

The need for creativity can be assessed from the point of view of those involved with the problem. To illustrate, in the development of a new product or service, which represents a rather complex

Exhibit 1.6.2 Model of the Innovation Process

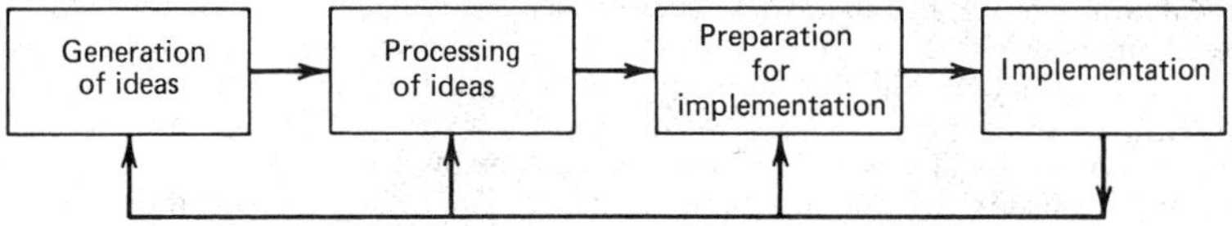

Exhibit 1.6.3 The Fusion Model of Idea Generation

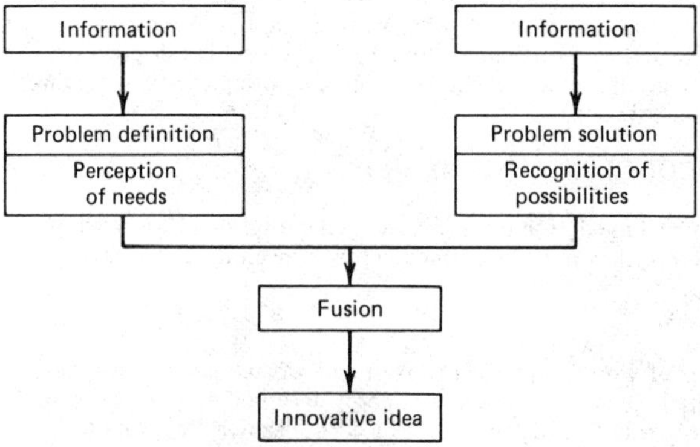

situation, one can distinguish between the following interests that will be directly or indirectly influenced by the solution: the firm, the users, the employees, and society.

It is difficult to assess the needs of those involved. One has to simplify a complex reality. In real life a person or a group is usually motivated by the interaction of several needs, some of which may have opposing effects. The needs also vary over time, from organization to organization, from group to group, and from person to person.[5]

The Organization

According to Holt,[6] organizational needs can be classified in a hierarchical order, as indicated in Exhibit 1.6.4. At the bottom is the need for survival. It provides strong motivation to find solutions that will keep the organization alive. Under normal conditions behavior is dominated by the need for maintaining current operations. However, if economic or social objectives are not fulfilled, the need for improvements will be strongly felt. As the law of diminishing returns makes its influence, or as changes in the environment require new approaches, there will be a need for new ideas and solutions. The first reaction usually will be to adopt or adapt approaches developed outside the firm. If this does not solve the problem, the need for developing new solutions from within will be strongly felt.

Whereas short-term needs related to current operations are easily perceived, long-term needs

Exhibit 1.6.4 Hierarchical Model of Organizational Needs

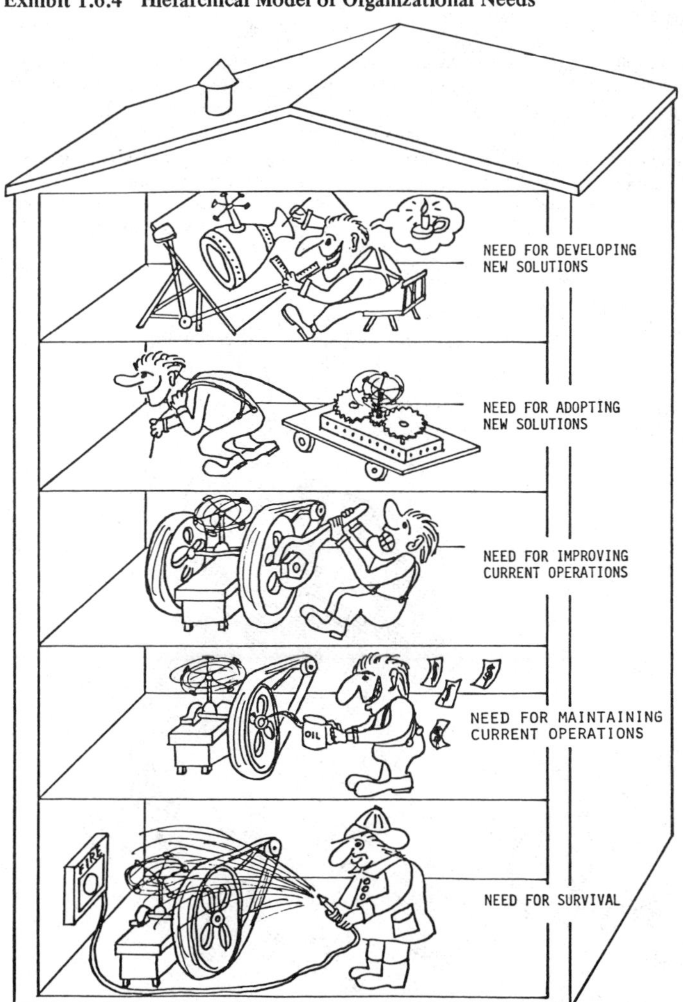

requiring new solutions are often vague in the beginning. However, if action is delayed too long, it may be very difficult, if possible at all, to find a satisfactory solution. The company may then find itself in a critical situation where the need for survival will dominate its behavior.

The Users

The needs of potential users are often neglected. This fact was demonstrated by Robertson[7] in a study of 34 firms that failed. Four of the firms had made no inquiries to potential users, six had made too few inquiries, two had ignored the results, two had misinterpreted the answers, six had been committed to preconceived designs, and three had failed to understand the environment to which their products were to be subjected.

Because of the negligence of user needs, many products, as indicated in Exhibit 1.6.5, suffer from a "quality gap," that is, a gap between desirable and actual product characteristics such as function, appearance, safety, maintainability, reliability, and durability. According to Freeman,[8] the gap appears to be larger for consumer products than for industrial products. This may be because industrial users meet a supplier on more equal terms, whereas most producers are in a superior position in relation to the consumer because of their know-how, status, advertising, and so on. In both cases new and better solutions are required in order to satisfy important user needs.

The Employees

The needs of employees can, according to Maslow,[9] be arranged in a hierarchical order, as indicated in Exhibit 1.6.6. At the bottom are physiological needs. When satisfied, they no longer motivate behavior. New, higher-level needs then emerge with greater strength. The first of these is the need for safety, that is, freedom from physical, psychological, and economic threats. Next on the ladder are social needs, which encompass a sense of belonging, friendship, and love. On the next highest level comes the need for esteem, including self-respect through mastery and achievement and respect of others through recognition and approval. At the top comes the need for self-actualization, which is concerned with the realization of one's full potential through self-expression.

Exhibit 1.6.5 Need of the User for Better Quality

Exhibit 1.6.6 Maslow's Need Hierarchy

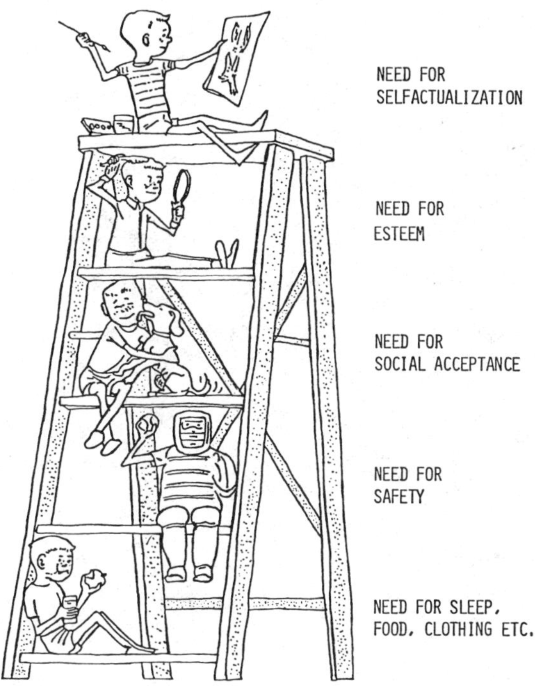

Society

Societal needs include a vast number of collective needs at the local, national, and global levels, such as those related to energy, resources, environment, housing, transportation, safety, medical care, education, and quality of life. In many places these needs have not been satisfactorily fulfilled. As indicated in Exhibit 1.6.7, the situation for some of them even appears to be getting worse. This is particularly true for the limited amount of natural resources that will have to support a rapidly growing world population. The threat of "more mouths and less food" requires new and radical solutions. There is a strong need for gradually changing the present society into a "recycling society" by replacing current manufacturing methods with new processes that make recycling possible with a minimum of environmental pollution and with a minimum use of energy.

Need Fulfillment

In their work industrial engineers should aim at solutions that allow for optimal satisfaction of needs, considering the many, often conflicting interests of the organization, the users, the employees, and society. They must face complex need patterns in which the influences of several needs are acting at the same time and with varying strengths. Further, the needs vary, not only among organizations, groups, and persons, but also over time. In addition, several situational factors of a technical, marketing, economic, and societal nature influence the solution of the problem. Therefore it is hard to give general recommendations. However, most problems have one thing in common: They require new ideas and creative solutions in order to satisfy the needs of those involved. Creativity thus appears to be a key concept in the fulfillment of societal, organizational, and individual needs.

1.6.4 THE CREATIVE INDIVIDUAL

Several theories have been developed to explain creative behavior.[10] Currently prevalent is the psychological theory, which assumes that creative ideas have their origin in previous experiences. The formation of new patterns or combinations is the result of nonrational, uncontrolled thinking

Exhibit 1.6.7 The Need for a Better Society ("And This They Once Called Flowers")

processes of an intuitive nature in the subconscious mind. All people have a certain creative ability. This ability is distributed among the population along a Gaussian curve.

Highly Creative Engineers

A particularly important group is highly creative engineers, who combine sensitivity to needs with technical ingenuity. These engineers are often hampered in their creative behavior by the bureaucratic organization of the place where they work. However, they can make important contributions of an innovative nature if they are given freedom and are stimulated to use their talents.

One approach is to apply the concept of "smuggled research" ("moonlight projects"). Rabinov[11] supports this solution and claims that everything that is done illegally is done efficiently: "I have always had very good bosses. They let me steal money as long as I did not put it into my pocket. Out of these thefts eventually came big projects."

A legal approach for supporting gifted engineers is used by Bucy,[12] who stresses that new ideas must be carefully nurtured. He therefore has allocated money to several "idea programme representatives" throughout the company. When an employee comes up with a new idea, he or she has only to convince the representative about its value. The employee will then get money to carry out a feasibility study within an agreed-upon time. Usually an arrangement is made with the employee's boss so that he or she can continue the regular job while working on the idea. The employee may also be temporarily transferred to another unit, for example, the central engineering center, in order to work with his or her idea.

A third approach is to give the engineer the opportunity to spend a certain amount of time working on projects selected by himself or herself. Some firms have obtained good results in this way. Others have had negative experiences, perhaps because the engineer often gets time for his or her own project only after all other tasks are finished.

Other Employees

Although great effort should be made to support highly creative individuals, one should not neglect the creative talents that are latent in the great majority of employees. Most firms have in these employees an untapped, but powerful, resource. However, in order to utilize it, a systematic approach is required. It can vary from a simple suggestion system to an elaborate action program. In some cases such a program may lead to new solutions or radical improvements. However, most ideas are likely to be of a more "bread-and-butter" type. There are plentiful opportunities for

Exhibit 1.6.8 Model of a Systematic Idea Conception

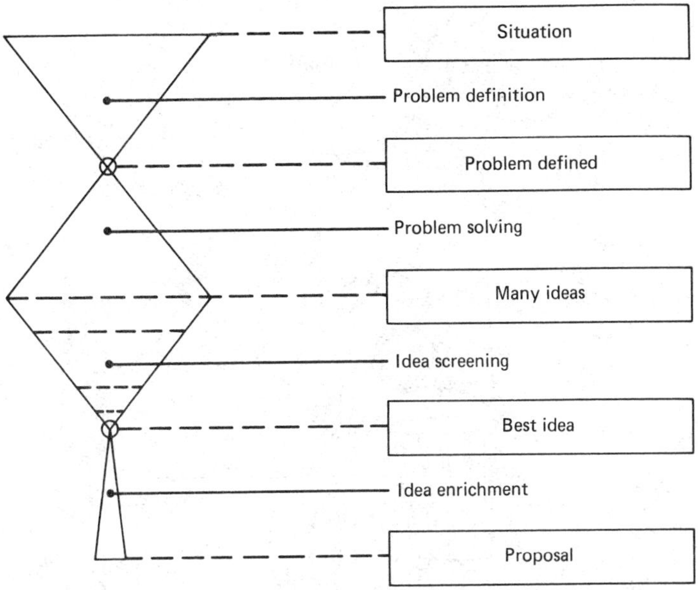

employees to demonstrate creative thinking by improving existing products and current activities. Even a routine design or a simple operation can be improved by combining known elements in many different ways.

1.6.5 CREATIVE THINKING

Several techniques are available to firms that want to stimulate employees' creative thinking. As shown in Exhibit 1.6.8, these techniques can be referred to in terms of four groups: problem definition, problem solving, idea screening, and idea enrichment.

Problem Definition

A problem is characterized by a gap between a wanted and an existing situation that is of such magnitude that it requires corrective action. This discrepancy depends on the strengths of the needs that are perceived and on the degree to which they are fulfilled. Therefore the first step in the problem solving process is to define the problem by assessing the needs of those directly and indirectly influenced by the problem. Next, one has to analyze the factors involved in the situation, by considering external and internal opportunities and constraints and by evaluating the nature of the problem. If necessary, the total problem should be divided into subproblems, with priorities assigned to them.

Problem Solving

Having defined the problem properly, the next step is the creation of a large amount of ideas. The fantasy and imagination of those participating are stimulated by the use of creative techniques such as brainstorming, brainwriting, morphological analysis, and many others.[13] it has been demonstrated by Osborne[14] that creative behavior by individuals and groups can be greatly increased by such techniques.

The most widely used technique is brainstorming. It is based on "free association," but certain rules have to be followed, as indicated in Exhibit 1.6.9. If there are many participants, a type of competitive brainstorming may be used where the participants are divided into small groups with five to six members in each. In charge of the groups are leaders who have been briefed about the problem situation beforehand.

Another technique is brainwriting, which is a kind of written brainstorming. Each member of the group writes during a short period–for example, 4 minutes–three ideas on a paper form. The form is then circulated to the person sitting to the right or left, who writes three new ideas, and so on. A variation of this is the brainwriting pool.[15] Here each participant, as indicated in Exhibit 1.6.10, takes a form from a pool in the middle of the table, notes his or her ideas on the form, puts it back

Exhibit 1.6.9 Brainstorming Session

into the pool, takes another form, adds new ideas, returns it to the pool, and so on. In recent years brainwriting has found increasing application. This may be because it is often difficult in actual practice to get good results from a brainstorming session, which requires a very good leader. This is particularly true when there are one or several persons who dominate the group or who are not able to cooperate in a proper manner.

Morphological analysis is a method of structuring a problem. Problems with two major parameters are depicted in a two-dimensional diagram or matrix, often called a "morphological box." Relevant values of the parameters are listed in the front column and at the top field of the matrix. If the problem has three major parameters, a three-dimensional model is required. If more than

Exhibit 1.6.10 Brainwriting Pool

Exhibit 1.6.11 Morphological Analysis

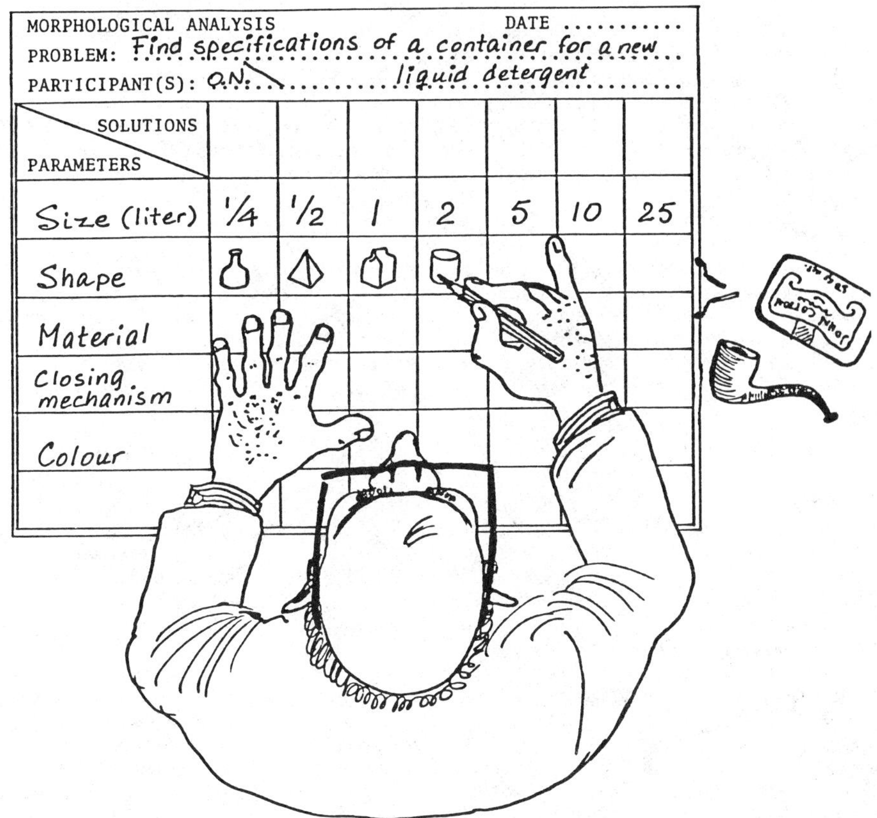

three parameters are involved, a table has to be used, as shown in Exhibit 1.6.11. Morphological analysis is well suited for individual use, but may also be applied by groups with up to eight persons.

MORPHOLOGICAL ANALYSIS
DATE
PROBLEM: Find specifications of a container for a new liquid detergent
PARTICIPANT(S): O.N.
SOLUTIONS
PARAMETERS
Size (liter)
1/4
1/2
1
2
5
10
25
Shape
Material
Closing mechanism
Colour

Techniques for creative thinking are increasingly being used by progressive organizations for problem solving purposes. Several of them can also be used for defining problems where one of the most important tasks is to assess the needs of those involved.

Idea Screening

By systematically applying techniques for creative thinking, a large number of ideas are created. The next step is to find the best one among them. This is a difficult task, which requires a large effort; as a rough guide, the screening and evaluation of ideas takes about three times as much time as their conception.

The search for the best idea is done by progressive elimination through a series of evaluations. Basically, these evaluations are made by selecting appropriate criteria related to the technical, marketing, economic, business, and societal aspects of the problem.[16]

During the early evaluation stages, ideas that do not fit company policies are screened out. Used first are simple criteria, for example, criteria that can be answered by "yes" or "no." As the number of ideas decreases, more effort is put into the evaluation. Costly information is restricted to the final evaluation stages, in which only a few alternatives are left.

Idea Enrichment

The result of the screening process is one idea, which theoretically should be the one that best fulfills the needs of those concerned. However, before the idea is presented as a proposal, an attempt should be made to enrich it, for example, by means of value analysis or negative brainstorming. The latter is a form of brainstorming in which the idea is critically examined by studying all thinkable ways in which it can fail.

1.6.6 ORGANIZATIONAL CLIMATE

Training employees in techniques for creative thinking can be a great stimulus to creativity. However, the result will be bad if the organizational climate is such that it does not foster creative behavior. According to Taguiri,[17] the organizational climate is a relatively enduring quality of the internal environment that is experienced by the members of the organization and that influences their behavior. The climate varies among and within organizations. Some have a warm, friendly, and supporting climate. Others have a cold and bureaucratic climate that hampers creative behavior. In the latter case it is possible to improve the existing climate by measuring and diagnosing it.

Climate Measurements

The organizational climate can be measured by various methods. One of them, presented by Holt[18] and based largely on the practical experience of engineers and engineering managers, is indicated in Exhibit 1.6.12. The climate here is characterized by 13 factors, which are measured by means of a questionnaire in two versions, one for managers and one for employees. For each climate factor the respondent indicates perceived and desired climate. By analysis of the data thus provided, one gets a quantitative basis for discussions regarding action to be taken in order to improve the climate.

Key Factors

Measurements of the organizational climate in several organizations, both large and small, indicate that one of the most important factors is the attitude of the organization toward new ideas and proposals. Top management comes strongly into the picture, but even more important appears to be the attitude of the immediate supervisor. His or her ability to give encouragement and support, to help in the generation of ideas, and to represent his or her group in a proper way is of great importance. Here is probably one of the most challenging and difficult tasks facing engineering managers who want to promote creative behavior.

A second important factor is time for creative activity. Good ideas do not necessarily come during working hours. Behind the generation of an idea is a complex psychological process, partly subconscious, where old patterns are formed into new relationships. It cannot be controlled and requires a certain amount of information, which can be provided and evaluated through discussions, reading, thinking, calculating, experimenting, and so forth. Such activities require time.

A third factor that many find important is recognition of creative behavior. It can be given in the form of financial and nonfinancial rewards. For the creative engineer it is often the intrinsic aspects of the job that appeal most. Possible approaches to providing recognition are through participation in innovation groups; increased freedom with regard to working hours and contact with other people; more influence on choice of projects and determination of work schedules and methods; support for development of ideas; and opportunities for professional growth, for example, by allowing "free time" or fellowship for pursuing special studies. Whatever approach is used, they all share the belief that the good idea should be rewarded, even if the final result proves to be bad.

Application

Proper application of climate measurements can be a powerful means of organizational development in engineering and other departments. Great attention should be given to the implementation of the measurements and their utilization. The major responsibility should be clearly fixed. Further, provision should be made for the active participation of those involved in planning and implementing the study and in utilizing the results. An example of such an approach from a large shipyard is given by Holt.[19] Here a program was planned and organized as broadly indicated by the list on the following page.

Exhibit 1.6.12 Basic Factors Characterizing the Organizational Climate

Time for creative thinking	Recognition of creativity
Freedom from restrictions	Physical environment
Freedom of choice	Interaction with others
Reception of new ideas	Composition of staff
Attitude of supervisor	Method of problem solving
Attitude of the organization	Contact with the project
	Type of project

Introductory Phase. Management discussions concerning objectives of the program; organization (project manager, steering committee, advisory board); information for employees.

Measurement Phase. Measurement of organizational climate and job attitude; feedback to employees of results of measurements.

Problem Definition Phase. Analysis of measurement data within the various units; listing of needs and problem areas; formulation of projects.

Problem Solution Phase. Selection of projects; organization of project groups; training in creative problem solving; development of alternatives; selection of solutions and formulation of proposals.

Implementation Phase. Decision on proposals; responsibility for implementation; organization; resource allocation; motivation; information and instruction.

Follow-up. New measurements of organizational climate; interviews; modifications.

By measuring the existing climate, one was able to identify weak points, define problems, develop solutions, and create a better climate for creative behavior, thereby improving idea generating activities.

1.6.7 ACTION PROGRAMS

Creativity can be promoted in many ways. A rather simple approach is to send an employee, for example, an industrial engineer, to a seminar on creative techniques and to leave the rest to him or her. Success may be obtained from such an approach, but it is not likely. To get good results, great attention should be given to selecting proper activities and to organizing them in an action program.

Activities

There is no patent solution for the design of a program for promoting creativity, but some activities to be considered are shown in Exhibit 1.6.13, which should be useful as a starting point. As one gains experience, new ideas and possibilities will emerge. Whatever design is chosen, those concerned must face the fact that all innovations–including action programs for promoting creativity–are characterized by risks. A basic requirement is, therefore, that one is willing to take risks and to experiment with various solutions.

In developing an action program, one should keep in mind that not only each company, but also each individual, is different and has unique problems. Some people are highly achievement oriented–they are willing to take risks and get satisfaction by proposing radical solutions. Others are more safety minded and limit their creative behavior to minor improvements. Because of these differences, all those concerned should participate in analysis and decision making about possible approaches. Flexible solutions allowing for individual adjustments should be given high priority.

Organization

An action program for promoting creativity should, according to Geschka,[20] be supported by a power promoter (*Machtpromotor*) and a professional expert (*Fachpromotor*). The power promoter is a high-level manager who has enough power and motivation to provide the necessary resources, but who does not have to engage himself or herself in the implementation of the program. The professional expert is a staff member who has been trained in the use of creative techniques and who is motivated to apply them in his or her own work and to help others to use them. He or she must have enough competence to act as a moderator for the groups and to take the responsibility for the program, including the promotion and coordination of creative activities in the firm.

The program may start in a small way, for example, by introducing one technique in one depart-

Exhibit 1.6.13 Activities in an Action Program for Promotion of Innovation and Creativity

Development of a positive attitude toward innovation and creativity by diagnosis and improvement of the organizational climate

Training of managers and employee representatives in the management of innovation and creativity

Training of managers and employees in idea generating techniques

Development of an efficient suggestion system

Development of a project organization for evaluation, development, and implementation of innovative ideas ment, and then gradually expand as one gains experience. Another approach is to plan a full and integrated action program for the whole organization. In that case it should be organized as a project, with a budget, a time schedule, and a full- or part-time manager. Industrial engineers should have many of the qualifications required for such a task. A broad advisory committee, with several levels and functions represented, and a small steering committee composed of motivated, knowledgeable persons may also be useful.

In organizations that are highly influenced by the change processes, one may organize a separate unit for idea generation activities. One example is a large firm that employs a full-time coordinator assisted by a small secretariat. The major activities are to organize, provide training, and assist several "innovation groups" in their creative efforts. Each group, consisting of six to seven persons, spends about one day per month on idea generating activities for clients within the firm. Such an approach is best suited for large firms, but there are small firms of about 400 employees that have found it worthwhile to use a full-time coordinator for such purposes.

1.6.8 CONCLUSION

Most industrial engineering departments have, up to now, focused their attention upon improvement of manufacturing operations by use of traditional tools. The philosophy has focused on costs and efficiency. Although this has worked in the past, it does not meet the requirements of the future.

Both social and political developments require a new philosophy characterized by an innovative attitude and a human-oriented approach. One of the greatest minds, Albert Einstein, had already envisaged this in 1934, when he said: "Concern for man himself and his faith must always be the chief interest of all technical endeavours. Never forget this in the midst of diagrammes and equations." If industrial engineers are able to acquire this attitude, they will have a good base for solving the problems of today and tomorrow. They will be able to demonstrate that they are equipped to handle one of the greatest challenges facing man today—sensitivity to human needs in a time that is more and more influenced by automation and data control. They will be motivated to develop strategies and tactics that stimulate creative behavior. The company that will succeed in a rapidly changing world is the one that is able to utilize the creative talents of employees most effectively for the benefit of all concerned.

REFERENCES

1. C. H. KEPNER and B. B. TREGOE, *The Rational Manager*, McGraw-Hill, New York, 1965.
2. E. DE BONO, "Creativity and the Role of Lateral Thinking," *Personnel*, May–June 1971, pp. 8–18.
3. D. W. TAYLOR, "Thinking and Creativity," *Annals of the New York Academy of Sciences*, 1960, pp. 108–127.
4. J. A. MORTON, *Organizing for Innovation*, McGraw-Hill, New York, 1971.
5. J. P. CAMPEL and R. D. PRITCHARD, *Motivation Theory*, *Handbook of Industrial and Organization Psychology*, Rand McNally, Chicago, 1976.
6. K. HOLT, "Creativity—A New Challenge to the Industrial Engineer," *International Journal of Production Research*, September 1977, pp. 411–421.
7. D. ROBERTSON, "The Marketing Factor in Successful Industrial Innovation," *Industrial Marketing Management*, 1973, No. 4, pp. 369–374.
8. C. FREEMAN, *Innovation in a Changing World*, Proceedings of the International TNO-Conference, Rotterdam, 1973.
9. A. H. MASLOW, *Motivation and Personality*, Harper & Row, New York, 1954.
10. J. ROSSMAN, *Industrial Creativity. The Psychology of the Inventor*, University Books, New York, 1964.
11. J. RABINOV, in M. F. WOLFF, "Managing the Creative Engineer," *IEEE Spectrum*, August 1977, pp. 52–57.
12. F. BUCY, "Managing Innovation," *Electronic Design*. September 1979, pp. 108–110.
13. T. RICKARDS, *Problem Solving Through Creative Analysis*, Gower Press, Epping, England, 1974.
14. A. F. OSBORNE, *Applied Imagination*, Scribner's, New York, 1963.
15. H. GESCHKA, G. R. SCHAUDE, and H. SCHLICKSUPP, "Modern Techniques for Solving Problems," *International Studies of Management & Organization*, Vol. 6, No. 4 (Winter 1976/77), pp. 45–63.
16. N. H. GIRAGOSIAN, *Successful Product and Business Development*, Dekker, New York, 1978.

17. R. TAGUIRI and G. LITWIN, Eds., *Organizational Climate*, Harvard Business School, Boston, 1968.

18. K. HOLT, "Creativity and Organizational Climate," *Work Study & Management Services*, September 1971, pp. 576–583.

19. K. HOLT, *The Scanship Case. A Programme for Promotion of Innovation*, International Institute for the Management of Technology, Milan, 1972.

20. H. GESCHKA, "Implementierungsproblemen bei der Anwendung von Ideenfindungsmethoden in der Praxis der Unternehmen," in H. C. PFOHL and B. RÜRUP, Eds., *Anwendungsprobleme moderner Planungs- und Entscheidungstechniken*, Hanstein, Königstein, 1979.

SECTION 2

ORGANIZATION AND JOB DESIGN

CHAPTER 2.1
Organization Design

LOUIS E. DAVIS
University of California, Los Angeles

2.1.1 INTRODUCTION

Most activities of an advanced society and many aspects of the lives of its people are funneled through organizations. The structure of organizations is central to the well-being of both society and individuals. Most Western societies are suffering from an individual-organization crisis—a crisis surrounding the societies' arrangements for getting their work done. The crisis centers on the mismatch between the individual's needs and expectations and the organization's goals and practices.[1] The growth of this crisis was stimulated largely by the rapid changes in society that have an impact on both individuals and organizations, while organizations have continued to be designed with too little regard to the changes in society and in individual values. Therefore, organizational planning in the form of organization design and redesign (renewal) is becoming a high-priority management response to the crisis situation in society today.

Organizations are social inventions, that is, created units of society where people are brought together and provided a technology and structure for doing work. Organizations and parts of organizations are invented and reinvented every day. In this respect, organization design as a process of invention is mundane activity. From the viewpoint of the organization's members, however, organization design is the invention not only of a place to work, but also of a small society in which they will live and through which they will achieve their expectations and goals and acquire status and rewards.

The consequences of the performance and problems of an organization are absorbed by the society and its members. Well-performing organizations provide society with benefits and opportunities and treat organizations as ecosystems, whereas problem-laden, poorly performing organizations burden society, affecting all members, sometimes disastrously. Therefore, to society as a whole, the quality of organization designs is crucial. As societal entities, organizations have to be designed to respond to the conditions of the 1980s.

The focus of this chapter is on the design process, that is, on the process of inventing or creating or forming an organization within which the efforts of many people are combined and coordinated to achieve the goals crucial to the organizations's survival and success. Rational organization design requires a process that creates valid alternatives, evaluates them in relation to constraints and consequences, and provides the means for selecting the preferred alternatives. The design process is necessarily iterative, proceeding from the general to the specific elements of the organization. Each decision must be integrated with prior decisions to achieve a congruently evolving system.

The present common practice, which is no alternative to rational organization design, is to copy bits and pieces of other organizations and to attempt to fit them together without considering their applicability, congruence, or fit to the organization as a system. Copying also forecloses the opportunity, really the necessity, to address the unique aspects of each organization and its environment. Effective organization design is a learning process that must be pursued systematically. The process of organization design becomes a focal point for bringing together a variety of demands, requirements, concepts, regulations, and so on, each of which contributes to or constrains the process (see Exhibit 2.1.1).

The fundamental propositions for organization design that are of consequence to managers, staff, and engineers as designers are:

Assisted by Eli Berniker.

Exhibit 2.1.1 Requirements Influencing Design of Organizations

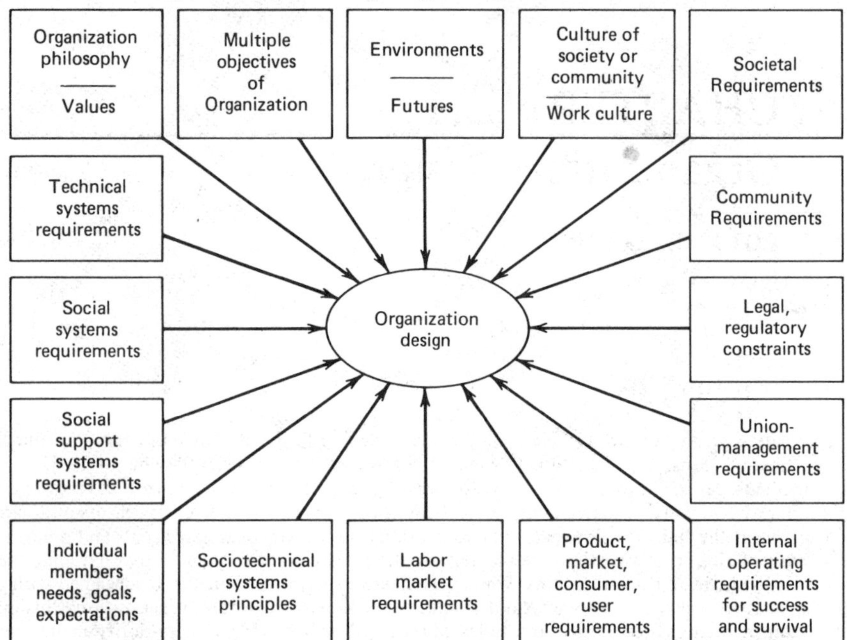

1. There is a structured process of organization design for (re)inventing, (re)creating, or (re)designing organizations that is suitable to emerging external and internal demands (see Section 2.1.9).

2. The phases of the design process proceed from the general to the detailed (see Section 2.1.10).

3. There is an indicative model of comprehensive, integrative organization design (see Section 2.1.5).

4. Organizations operate as four simultaneously existing entities, and organization design must address the requirements of each (see Section 2.1.2).

5. Organization structures reflect the theories and values extant at the time they were invented. Current organizations are dominated by past theories and outdated values (see Section 2.1.3).

6. Effective organizations operate as open sociotechnical systems (see Section 2.1.4).

7. Organizations are both part of and influenced by complex, unstable environments[2] (see Section 2.1.6).

8. For survival and success, organization structures should be designed to cope with instability and uncertainty in their environments as well as with internal and external stability and predictability[3] (see Section 2.1.6).

9. To cope with instability and uncertainty, the organization has to be capable of responding to environmental demands and multiple objectives generated by its own goals and by the goals of its members and other stakeholders (see Section 2.1.6).

10. For the design of organizations, the unit of analysis is the organization-environment set[2] (see Section 2.1.6).

11. Organizations are social entities, as well as work (production/service) entities, which are parts of the larger society. Organization design is therefore influenced by societal and local values and beliefs and by perceived futures (see Section 2.1.7).

12. An organization philosophy is needed as a guide for design of an organization and for its operation (see Section 2.1.7).

13. A comprehensive, integrated design process is required for effective design (see Section 2.1.5).

14. Sociotechnical systems principles of organization design are available to aid the designer (see Section 2.1.8).

15. There are additional considerations and requirements for redesigning existing organizations (see Section 2.1.10).

2.1.2 DEFINITION OF AN ORGANIZATION

In Western societies the structure of organizations is central to the well-being of society and its individual members. Organization and job structures are increasingly seen as crucial to the survival and success of individual firms, agencies, and other components of any society and thus to the future form of that society. Effective and responsive forms of organization have to be designed and continually redesigned to meet rapidly changing conditions in their environments, which cause both external instability and new internal demands.

What is an organization? There are many different answers, depending upon the organizational context and purpose of the definition. Each definition contains some aspects of what constitutes an organization. When these various definitions are taken together, they give a full picture of what is meant by an organization. Different definitions of an organization include the following:

1. *Organizations are social inventions*, that is, they are created by people for specific purposes at a particular time in history in order to provide some needed goods or services, frequently through the application of some particular technology. As social inventions, organizations reflect the culture and beliefs of the period and the locality in which they were first established. The structure of an organization reflects the objectives it sought to accomplish. Over time, the organization's structure is modified by growth, by the particular problems and difficulties imposed on it, and by changes in its environment to which it must respond. From this point of view, every organization design is out of date, since its environment is constantly changing.

2. *Organizations are production or transformation agencies* where materials, information, or people are imported, transformed into desired outcomes, and exported. In this context, organizations exist to do work (the work of transforming). However, since organizations seek to survive, they must do work without injuring their capacity to do future work. In fact, the test of organizational effectiveness lies in the extent to which the capacity to do work in the future is enhanced. Therefore organizations must perform other functions and activities besides those of transformation.

3. *Organizations in the private sector are economic entities* that exist only to satisfy economic goals. They use and are required to account for all resources provided to them to achieve the desired economic goals. The test of achieving economic goals is whether the organization can make a profit. Profitability is seen as the test of organizational effectiveness. Present changes in the environment of organizations require that they account for more and more of the resources provided to them, including people, atmosphere, water, money, and so on. As such, to survive and succeed, even economic organizations have to satisfy more than purely economic goals.

4. *Organizations are social entities, or "minisocieties."* They are frameworks within which people are brought together to work on achieving the goals, economic or otherwise, of the organization. Members of organizations develop ways of working together, of solving problems, and of dealing with conflicts that arise and develop the means of sustaining the organization as a social entity. These patterns of interaction become the roles and role relationships that give structure to the organization. It is the structure that distinguishes an organization from an assemblage of individuals and gives it particular systems characteristics. However, different people carry out these roles in different ways, modifying the roles, relationships, and the structure of the organization as a social entity. The organization functions as a minisociety for the organization's members, providing an arena for achieving personal goals and for distributing rewards such as recognition and status. These rewards are valuable inside the minisociety as well as in the larger society outside the organization.

5. *Organizations are collections of individuals*, each with different needs, expectations, and goals, placed in a cooperative situation to achieve the goals of the organization. However, in doing so, the individuals expect to be able to achieve some of their personal goals. There is no basis for expecting individuals' prior commitment to the goals of the organization unless their roles and the organization's structure, design of jobs, and management style support a reasonable expectation of achieving their individual goals. An organization needs to be designed to satisfy not only its own objectives, but also the multiple objectives of the many people who have a stake in its future. This has led to the inclusion of quality of working life criteria as a critical part of organization design.

6. *Organizations are open sociotechnical systems.* In many ways this definition encompasses all of the preceding definitions. Organizations are taken to be systems open to the demands, forces, and requirements of the surrounding environment. As such, organizations are making constant adjustments and place a high value on the adaptability of their members in this social system. At best, organizations are in a quasi-stable state of internal adjustment, seeking to achieve desired outcomes in an uncertain and ever-changing environment. Organizations are seen as living systems and, like plants, animals, and people, display remarkable adaptability and variety of effective responses to different situations. To achieve their goals, they perform their functions or do their work through the joint action of their technical and social systems. Their technical systems are

engineered and operate in a deterministic way. Their interacting social systems are required to make the necessary adjustments in the face of changing conditions in order to achieve desired outcomes. Thus both the technical and the social systems are equally needed and must be designed jointly to avoid suboptimization. This definition is expanded later as the basis for organization design.

In contrast, over the past 50 to 100 years engineers have tended to perceive organizations too narrowly as production systems of a particular kind. These production systems are made up of machine elements and human elements whose activities have been designed as if they were mechanical entities. The primary role of people has been to execute those functions not yet mechanized, automated, or programmed. Narrow, highly specialized human tasks are as completely specified as those of machines, and the organization is expected to operate as an elaborate "clockworks." People have been seen to be its least reliable elements, and therefore their jobs must be completely specified, if the organization is to be productive and efficient. This classic machine theory of organization, which is still dominant in Western societies, will be further explored. Its application results in relatively inflexible organizations, which poorly utilize the human resources put at their disposal.

The reality of the growing turbulent—that is, unpredictable, interactive, and in some ways unmanageable—state of the environment now confronting organizations in Western societies indicates that prudent organization design treat organizations as ecosystems, referred to in the introduction to this chapter. As such, any narrow view that considers only some of the aspects of organizations may well produce organization designs inadequate to existing conditions. Therefore such organizations are likely to be at risk in achieving effectiveness and in surviving. None of the definitions of an organization explored in this section exclusively reveals what an organization is, but each contains some aspects. In effect, an organization is all of the preceding definitions and yet not any one exclusively. Taken together, the combined definitions set the task that organization design must accomplish, namely, how to include all the aspects identified as constituting an organization.

2.1.3 THEORIES OF ORGANIZATION

It is commonplace among managers and engineers to view the design of organizations as a practical endeavor carried on without theoretical considerations. Although this position is called "realistic," the reality is that it is impossible to design an organization, or any of its parts, without a theory or model of organization, even if it be only a private or an ad hoc one of "it has worked for me." Of course every manager and engineer has just such an implicit theory or model without calling it such. By denying that theoretical considerations are necessary to the design of organizations, the implicit underlying assumptions of the "practical" designer need never be exposed to examination.

However, there is no way of appreciating the opportunities and choices available in organization design without making explicit the underlying models in each approach. The design of any organization reflects some model of human behavior and motivation, some model of the interactions between people and technical systems, and some model of the interactions between the organization and its environment. The outcomes of the process of organization design depend on the theories of organization employed, on the assessments of the present and future states of the environment, and on the culture of the locality, country, or society of which the organization is to be a part, as well as on the constraints, laws, and regulations of the locality.

The theories of organization themselves generally reflect the culture of the place and of the historical period in which they evolved. Theories keep changing rapidly now in response to changes in social values and increased knowledge. The oldest theories are still in use in many settings. Organizations based on them are seen to be rickety and quaint, needing constant shoring up with a variety of props in order to continue to operate. The present situation reveals the particularly poor fit between the most popular (and oldest) theories and the very different needs and expectations of younger (18 to 35 years) workers. The gap is so large between what organizations offer and what younger employees expect from them that a substantial individual-organization crisis exists.[1]

Organizations as social inventions have been around since the dawn of human history. The pyramids, the many temples that are the remains of the ancient civilizations, and the large Polynesian sailing craft of those primitive societies were all products of the many people with a large variety of skills and competences whose efforts were integrated and coordinated—in short, organized. However, what characterizes all of these examples of "organizations" is that they were not deliberately designed, as is the case in modern societies. The "organizations" evolved as tradition over the years in stable or slowly changing cultures, based on societal allocation of roles and thus of tasks and skills. Much of what is taken for granted in modern organizations owes its origin to the late eighteenth-century industrial culture that broke with tradition and introduced man-made organizations.

Division of Labor

The direct forebear of classical organization theory is Adam Smith, who in his *Wealth of Nations*[4] first codified the notion of "division of labor," under which the skills in a given craft are fragmented into small elements or activities so that each may be performed with minimum skill and experience at lower immediate cost, if not total cost. The division of labor is the first example of the explicit design of organizations in the modern context.

Smith's concepts were extended by Charles Babbage, as reported in *On the Economy of Machinery and Manufactures*,[5] which must be considered the first manufacturing management text. Babbage developed a complete program for the design and management of manufacturing organizations. It reflects the ideologies, principles, and culture of his period, 1800–1835, including the employment of small children in factories. A theme that pervaded most organization design in the first Industrial Revolution, which began about 1780, was the need to establish control over workers' behavior and to reduce the dependence of manufacturers on the skills of employees. This was the major driving force behind fragmentation of crafts and the mechanization of production.

Classical Organization

The currently dominant, or most popular, theory of organization has been in use for so long that it is referred to as the "classical theory of organization." It has two theoretical roots, each about 100 years old. One of these is the theory of scientific management, based largely on the pragmatic innovations of Frederick W. Taylor,[6] and the other is the theory of bureaucracy, based on the scholarly research of Max Weber.[7] Classical theory considers the organization to be closed to the outside environment and composed of isolated segments held together by a hierarchy. Therefore it emphasizes the span of control (rule of seven) for determining levels of organization. The major function of supervisors and managers is to provide the bonding or glue needed to hold the different parts together. The fragility of classical organizations is just that they depend upon numerous supervisors to hold the various pieces of the organization together instead of the cohesiveness of the parts of the organization built into the design of the organization itself.

There are three fundamental aspects of classical organization theory that are interrelated and that have come to be accepted as standard for organizations:

1. The individual and his or her task are the basic building blocks of the organization. They are bound together by supervisors.
2. Individual task units are grouped by function, location, or by time segment, such as a shift. This upsets the integration requirement.
3. Each jobholder is under the individual control of a supervisor (unity of command), thus reserving authority to higher levels in the organization.

Scientific Management

This form of production organization was the contribution of Frederick W. Taylor[6] between 1890 and 1914. With Frank Gilbreth,[8] who followed, Taylor carried the division of labor (fragmentation of skills) one step further, introducing quantitative rules for making a single task into a job. To the division of labor were added the complete specification of task content, down to individual motions, and the strict separation of planning of work from its execution. Contributed to classical organization theory were superspecialization and complete job specification, suggesting that organizations should be designed as if they were elaborate production clockworks.

Bureaucracy

Bureaucracy[7] was one of the most notable social inventions of its time. It provided four major contributions to the structuring and operation of organizations. The first was the substitution of rules for arbitrary decisions, permitting organizations to record useful practices and further develop these practices as a means of accomplishing the goals of the organization. Second, bureaucracy formally limited and defined authority, so that it became less capricious and less personalized. Third, bureaucracy provided for the treatment of individuals according to universalistic, rather than particularistic, criteria, provided a uniform treatment for individuals and eliminating personal privilege and position in the placement of people and in the evaluation of their performance. Fourth, bureaucracy called for a hierarchical pyramid in the organization, consisting of managers of hundreds, managers of fifties, managers of tens, and so on. The hierarchical pyramid was based on a rationalization of the work to be done and provided simultaneously a system for delegating authority and a career ladder to be climbed by members of the organization.

Classical organization theory has contributed to making organizations rigid and incapable of change as needed. Such organizations are inefficient because they absorb large amounts of manpower and energy in order to control and regulate those who have the work to do, underutilize and misuse members of the organization, and fail to meet the needs and expectations of members. Finally, the costs of rigidity within organizations are increasing, as is the threat to their survival.

Human Relations

The first large-scale challenge to classical organization theory came from the Hawthorne experiments (1927–1932), which led to the formulation of the human relations approach to management.[9] Discovered in the Hawthorne experiments was the existence of an internal environment, or informal organization, which dealt with the host of purposes, objectives, values, motivations, and expectations that people bring with them to the workplace. Additionally, the informal organization, which parallels the formal or official organization, operates through sets of social controls that workers exercise in order to govern the amount of output, the degree of absenteeism, and so on, that they consider to be legitimate. Finally, the experiments discovered that improvements in productivity were related to the motivational effects on workers of having attention given to their needs. Workers appeared also to respond favorably to a more humane, more participative or supportive style of supervision.

Actually, the human relations approach had a relatively small effect on the design of organization structures, but did influence reward schemes, supervisory levels, personnel control approaches, communication approaches, management style, and supervisor training. The overall result of the human relations development was a classical form of organization with more participation, more information sharing, somewhat better trained supervisors, and more sensitive management styles.

Matrix Organizations

The special requirements of particular kinds of industries, such as aerospace, which developed after World War II, brought on some new organization designs.[10,11] These kinds of industry are based on advanced research and development, which essentially produces designs rather than products. (Very small amounts of any product are made.) Such organizations require the contributions of a variety of highly trained and specialized professionals, who work under relatively unstructured conditions to develop innovative products (designs) based on the application of the latest scientific developments. The matrix organization was evolved as an appropriate organization structure to suit this set of external (products) and internal (profession/specialist) requirements. If the work is done through projects, the specialized human resources can be kept available and rearranged in different mixes as required. The matrix, or project, form of organization has spread to other settings where different combinations of many skill inputs are needed for each product made or service rendered.

Open Systems Theories of Organizations

Two developments contributed to open systems theory as a successor to classical organization theory. The developments were (1) sociotechnical systems,[12] which originated in the early 1950s, and (2) the concept of organic structures that Burns and Stalker [13] proposed in 1961. Section 2.1.4 reviews the concepts of sociotechnical systems theory.

2.1.4 OPEN SOCIOTECHNICAL SYSTEMS THEORY

Organizations as Open Systems

Open sociotechnical systems take the organization to be an open system engaged in transforming inputs to achieve desired outcomes, that is, as doing work to achieve objectives. As such, these organizations have permeable boundaries exposed to the environment in which they exist, so that developments in the environment permeate, or enter into, the organization along with the inputs that the organization acquires. Most frequently, the changes in the environment enter an organization through the people who work in it, as well as through the marketing or sales functions of the organization or through the materials or other input functions.

As open systems, organizations can be conceived to be in constant interaction or constant commerce with their environment across their boundaries. Organizations import, or receive, inputs from their environment, transform these into the desired outputs, and export the outputs to their environment. From the environment the organization receives feedback mediated by forces in the environment, which may or may not be immediately useful to the organization. The same concept may be applied to each unit of the organization. In this instance, however, the environment of the unit is that of the larger organization of which it is a part. Organizations bring two critical factors to bear in the transformation process, namely, technology in the form of a technical system and people in the form of a social system.

Technical Systems

The technical system is one of two parts of the transformation system essential to achieving the desired outcomes of an organization. A technical system consists of a set of artifacts, tools, machines, facilities, methods, programs, and procedures, which are the means by which the people in the organization transform the inputs. As such, the technical system and its functioning represent a bridge that limits and channels many of the interactions between an organization and its markets in the supporting environment. The effective operation of the technical system is the means by which the organization achieves success and continued access to resources (inputs). Once designed, the technical system is stable, and any disturbances that occur either in the technical system itself or in the environment have to be dealt with by the people or social system.

Each transformation process has an underlying technology that allows a variety of technical systems designs. The choice of particular designs both influences the design organization structure and is influenced by social assumptions. It is a sociotechnical choice, reflecting cultural, economic, and social values.[14,15] For example, a critical assumption in industrial engineering task design is, "It must be possible for each individual worker to be held responsible for his or her individual performance by a supervisor." This is a social assumption implying poor performance by workers, which necessitates supervision and control to achieve desired outcomes. It leads to building up an organization from prescribed individual, controllable jobs and then adding a supervisory hierarchy to control performance. Technical system design based on this assumption creates individual work stations, routes information feedback to management, and separates quality control functions from production functions. Such organizations perform poorly, reinforcing the original assumption, that is, the self-fulfilling prophecy.

Social Systems

People individually, or in small or large groups, are necessary to operate technical systems at the level of the machine, the machine group, or the overall transformation system. People have interrelationships not only with the technical system, but with each other, forming systemic relationships, that is, social systems.

The social system consists of both people and structure. It consists of the set of members of the organization acting in their roles. It also consists of the set of roles and their role relationships, which make up the structure. In carrying out the requirements of their roles, the role occupants (people) generate additional role relationships beyond those required to do the work of the organization. Social systems contain not only the roles or jobs held by individuals and the role relationships, but also the authority structure, communication structure, adaptation mechanisms, learning mechanisms, social system maintenance mechanisms, career structure, and so on, associated with a unit of the organization and its technical system.

In addition to operating and maintaining the technical system, social systems solve problems that arise and adapt to disturbances or to changes in the environment by adjusting the technical system or finding different ways of accomplishing the goals. In open systems terms, the social system is the mediator between the limits and capacities of the technical system and the requirements of the environment. Both the social and the technical systems are essential to the functioning of an organization. Some have defined an organization as the outcome of embedding the technical system in the social system, or a sociotechnical system.

Sociotechnical Systems

Sociotechnical systems view organizations as transformation agencies. Transformation requires both a technical and a social system whose existence and interaction are essential to accomplishing the goals or achieving the desired outcomes of the organization. The technical system defines the tasks to be performed, and the social system prescribes the way in which they are performed. Each interacts with the other at every person-machine interface or group-machine-function interface. Both the technical and the social systems operate under "joint causation," meaning that they both are affected by causal events in the environment. The technical system, once designed, is stable; it is the social system that has the adaptive capabilities.

Organizations are seen as seeking to maintain a state of quasi-stable equilibrium in order to achieve the goals that have been accepted. Therefore the issue of how goals will be achieved in the face of disturbances or of environmental changes strongly influences design choices that are made. This is particularly so since the adaptive element is the social system; thus the means for adaptation have to be particularly designed.

The concept of joint causation leads to the related sociotechnical system concept of "joint optimization." It states that, when two independent systems—the technical and the social—are directively correlated, that is, they respond jointly to causal events or interact under joint causation, then optimizing one system and fitting the second to it will result in suboptimization of the joint system. Therefore maximizing the effectiveness of the organization requires joint optimization of the technical and social systems rather than maximizing the technical system and fitting the social system to it. This is one of the fundamental departures from scientific management concepts. Joint optimization carries with it the prospect of degrading either the technical system or the social system in order to achieve a better joint outcome.

The need for joint optimization has led to the joint design of the technical and social systems of organizations to make it possible to consider the best possible fit between the technical and the social system, given the objectives and the requirements of each. On the part of the technical system, joint optimization means making available to the social system sources of variances or disturbances, access to the sources of variances or disturbances, the means for overcoming internal and external variances or disturbances at the source, a cognitive map of how the technical system functions, and relevant information and feedback mechanisms to permit self-regulation and problem solving.

Further, technical system design in itself depends upon and requires the resolution of social system questions. These questions center on how the particular machine, tool, and so on, will be used; what information the item must provide; and how it may be adjusted, and so on, by those who are going to use the equipment or the equipment system. Here again, joint design is called for. Further, unless joint design is undertaken, the acquisition or importation of a technical system designed elsewhere or independently will inevitably mean that to a greater or smaller degree a social system will be imported into the organization together with that technical system. The issue, of course, is whether the social system so imported will be congruent with the goals and philosophy of the organization being designed.

Characteristics of Organizations as Open Sociotechnical Systems

Organizations operating as open sociotechnical systems constantly seek to achieve a condition of steady state operation. They seek to maintain desired throughput by continuously adapting to a considerable range of external changes that permeate their boundaries. In seeking to achieve steady state operation in the face of changing demands from their environments, organizations rely on the following open sociotechnical systems properties:

1. **Equifinality.** The ability to follow different suitable paths to achieve accepted outcomes when confronted by variable or unstable inputs, disturbances, and unanticipated demands.
2. **Requisite Response Capability.**[16] The organization, its units, and its members possess the repertoire of knowledge, skills, and authorities to match the variety of demands or environmental conditions faced by the organization.
3. **Self-Regulation.** The organizational unit decides how and when it will apply its response capability so that it can maintain its present and future capacity to function effectively and meet the goals for which it is responsible.
4. **Relevant Boundaries.**[14] Each of the preceding properties depends upon how each organizational unit is defined or established. This requires that the boundaries around each organizational unit be located so as to include all the means necessary to achieve the desired organizational outcomes. The present common practice in organizations is to fragment their adaptive capabilities by separating functions of organizational subunits in order to maintain control by management. The unchallenged assumption is that more is to be gained from the ability to control than from the capability of adapting. The separation of quality control activities from production is a typical case.[17]

2.1.5 COMPREHENSIVE, INTEGRATED DESIGN OF ORGANIZATIONS

Consideration of (1) the many functional roles of organizations within society, (2) the various constraints, demands, and requirements coming from their environments, and (3) the systemic properties they require as operating sociotechnical systems suggests two important criteria for an effective organization design process. The process must be comprehensive in its analysis of all aspects of the organization in its interactions with the many relevant elements of its environment. The process must also be integrative, so that the needed systems characteristics and response capabilities are designed as a coherent system relevant to each organizational entity to enable it to meet the various existing and anticipated demands of the environment. The comprehensive, integrated design model achieves these requirements through a series of phases, as indicated in Exhibit 2.1.2.

Comprehensive, integrated design requires designers to develop the organization design requirements and criteria by (1) analyzing the future organization as part of its organization-environment set and (2) developing an organization philosophy as a statement of a shared set of values and purposes of the organization. This phase is followed by the organization design phase, which involves the joint design of the technical and social systems and the design of the social support systems of the organization. Finally, the design is implemented. Although the process is described

Exhibit 2.1.2 Comprehensive, Integrated Design Model

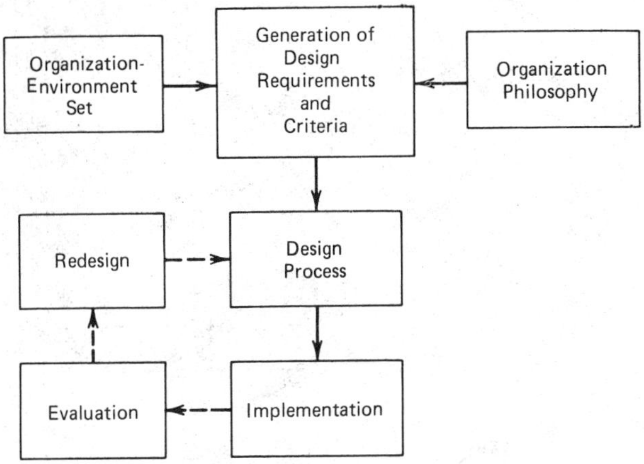

Organization-Environment Set
Generation of Design Requirements and Criteria
Organization Philosophy
Redesign
Design Process
Evaluation
Implementation

as linear, it is interactive and iterative. At any point, earlier analyses and decisions are subject to reevaluation and modification.

The first and second steps of the comprehensive, integrative design process are interlocked and interactive, so that in individual circumstances one or the other may be undertaken first. Values are the basis for appreciating[18] and understanding the complexities of the organization-environment set. They focus analysis on what is important and on relevant specific domains and issues. The analysis of the organization-environment set poses issues for designers that require them to make explicit the values and assumptions underlying their own decisions. Without shared values it is difficult to carry on the process of analysis in an open and coherent manner. The process of analyzing the organization-environment set is discussed in Section 2.1.6. The process of developing shared values and of creating a statement of organization philosophy is reviewed in Section 2.1.7.

2.1.6 ORGANIZATION-ENVIRONMENT SET

Modern organization design must consider all the components of the organization-environment set, that is, the organization embedded in its environmental context. This is particularly the case when organizations are conceived to function as open sociotechnical systems.

A model of the organization-environment set (Exhibit 2.1.3) indicates three major sources of demands, requirements, and constraints confronting an organization. First, developments in relevant environments create issues to which the organization will have to respond. If those environments are unpredictable and unstable, as opposed to predictable and placid, effective design must emphasize organizational adaptability and response variety. An aim of the model is to indicate the requisite variety of capabilities needed within the organization to respond to environmental contingencies. Second, the stakeholders in the organization's environment include stockholders, lenders, members of the organization, customers, users, government agencies, and others. Each of these has interests in the organization and a potential for action that can significantly affect the organization's future. Therefore the design must anticipate that the organization will have to satisfy multiple objectives of its many stakeholders. Finally, as an operating system, the organization has to perform some internal functions necessary to ensure success and continuity. These are the many elements to be designed in the process of inventing an organization.

Another way of looking at multiple objectives is to view the organization as occupying several roles in its environmental context.[18] It has to respond to different demands of particular aspects of the environment and satisfy needs of different stakeholders. For example, as a transformation agency, the organization's technical system has to respond to technological, economic, market, and other aspects of the environment; has to meet economic and other needs of various stakeholders; and has to satisfy needs of workers for interesting work, careers, and so on.

Organizational Environments

Comprehensive, integrative organization design requires that environmental states, present and future, be analyzed carefully and that assumptions made in the analyses be made explicit and be tested. Designers require answers to these questions: What issues for design derive from the future impacts of different aspects of environments on the organization? What will be the impact of the

Exhibit 2.1.3 Organization-Environment Set

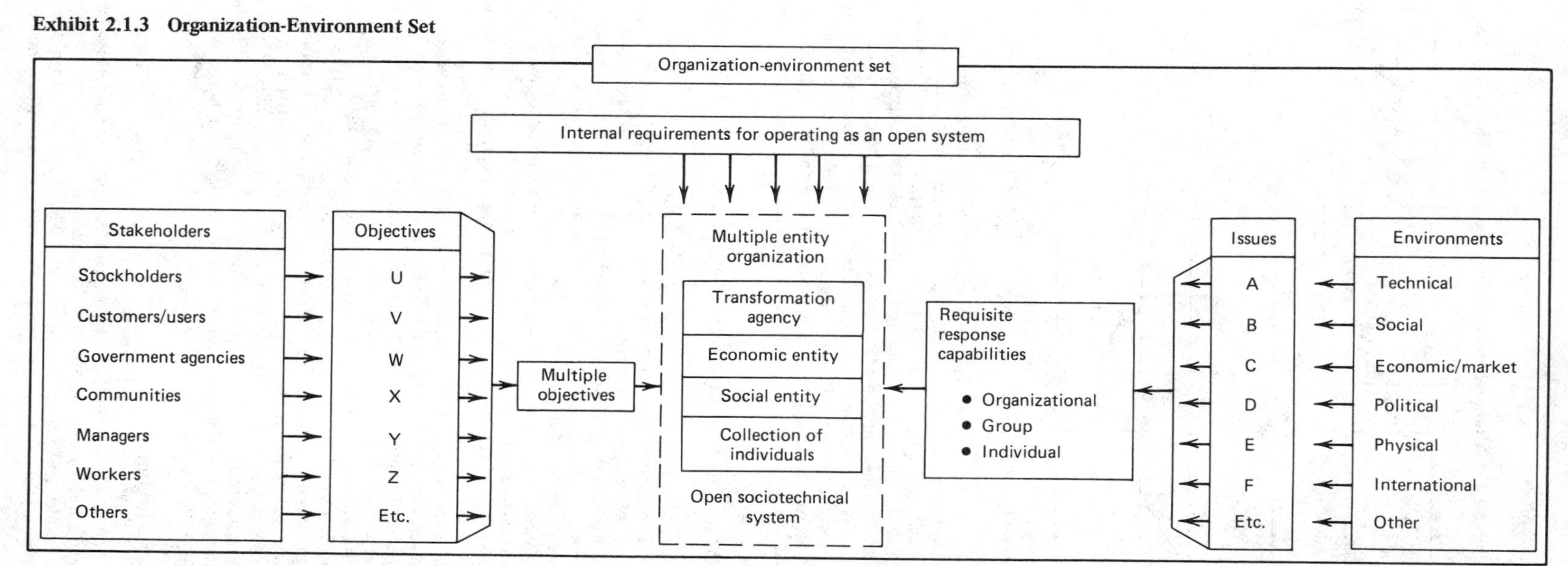

organization on its environments? What are the contingent developments to which the organization may be required to respond? The ability of an organization to survive will depend on its ability to adapt to future environmental states. Although the goal of an organization may be to provide a product or service needed by the society, the structure of the organization will be strongly influenced by the state of the environment and the specific demands that the environment is generating. (For an in-depth discussion of the changing states of organizational environments, see Emery and Trist.[2])

Modern organization design therefore places great emphasis on identifying and assessing the effects of changes in environments and the implications of future developments, becoming involved with the methods for studying the future and the data derived from such studies. Examinations of environmental changes can best be undertaken by conceiving of the organizational environment as being divided into a number of environments, such as the political, the societal, the technological, the economic, and the physical environments. Each of these environments is changing, interacting, and generating demands for organizations. Understanding these environments and assessing the implications in depth is crucial for effective organization design.

There is a great attraction to the use of generalizations about environments, thereby avoiding the confrontation of concrete issues and the testing of the assumptions underlying the generalizations. The very turbulence and complexity of present environments invalidate generalizations. Comprehensive, integrative design requires that each organization analyze its unique environments and assess the issues so raised.

Technological Environment

This environment stimulates the largest amount of environmental change and instability, particularly in products and markets. It is characterized by a still-increasing rate of development, by increasing capitalization in order to apply the more sophisticated developments, and by growing constraints on its use. Its high material-providing capabilities are such that plenty, and how to consume plenty, has displaced scarcity as an expectation.

Additionally, technological changes are introducing severe changes in the kinds and meaning of work, in the meaning of efficiency or productivity, and in the lack of differentiation among white collar, blue collar, and professional work. Advanced technology is changing not only what is meant by work, but also what is meant by skills, replacing manipulative skills with conceptual skills in problem solving, monitoring, and information analyses. Increased capitalization and reduced numbers of workers have increased the significance of each worker's contribution, commitment, and identification with the organization's goals. The higher the level of technology employed by an organization, the greater its dependence on its workers. In the design or redesign of organizations, the comprehensive evaluation of developments in the technological environment must address interactive effects on other environments and relevant stakeholders.

Social Environment

This environment is, most broadly, the society and culture within which the organization will function. A critical segment of the social environment is the work force from which comes the organization's members. Changes in values, expectations, educational levels, consumption, leisure activities, and economic well-being are all generating new requirements for organization design. Each organization design or redesign calls for the collection of data regarding changes in social environment, since these are changing so rapidly. Although much can be learned from national surveys and statistical reports,[19] designers of an organization must also evaluate the local community from which it actually will draw its members. Finally, in multiple-plant companies valuable insights may be gained by analyzing the impacts of social issues on the organizations of existing plants.

Economic Environment

In the market economy almost every aspect of an organization's operations are affected by changes in the economy. A major investment program will usually be justified by an economic analysis of future developments in the economic environment. If such an analysis assumes stable conditions for the organization during the payback period, the program may be designed too rigidly to achieve the planned return. Evidence is accumulating that organizations, particularly larger ones, can no longer depend on environmental stability in assessing the future impact of the economic environment.

Political Environment

The political institutions of society are increasingly active in regulating organizational activities. In the face of some overregulation that is stifling some developments, citizens continue to seek redress of grievances through political institutions. Furthermore, internal to organizations, the U.S. government and its various agencies have by law and regulation chosen the workplace as the arena for changing society. The problems of racial inequality, sexual discrimination, the integration of minorities, and the utilization of the unskilled have all been made requirements to which the place of employment must respond. American society has conferred upon government agencies the status of stakeholders in economic organizations.

Environmental Turbulence

Organizational environments have been changing rapidly and are continuing to change at an increasing rate. Not only are the interactions with the environment changing, but the interactions between various external elements in the environments are changing the very rules under which organizations must operate. In fact, there is high uncertainty as to what are appropriate rules for relationships outside the organization. This is what is meant by environmental turbulence.[3]

Turbulence increases the relative instability and uncertainty within which organizations must survive and succeed. This poses a new requirement for the design of organizations. Historically, organizations based on classical theory presupposed stability in their environments and could be reassured of success by optimizing efficiency as measured internally. Turbulent environments pose survival problems for organizations in addition to efficiency problems, since uncertainty cannot be effectively reduced or stability assured.

In the face of these conditions, the most effective survival strategy may be that of organizational adaptability. Organizational adaptability requires a wide response capability (requisite variety of responses)[20] to meet a wide array of contingencies. Additionally, the organization must be capable of self-renewal or self-redesign. This requires the organization to be sufficiently flexible to learn from its experiences and to disseminate learnings widely in order to permit rapid self-reorganization to meet new contingencies.

Stakeholders

The second major source of organization design requirements is the relevant stakeholders. Stakeholders, though part of the environment of an organization, are specific actors who confront an organization with particular objectives. In support of their objectives, stakeholders may react or initiate actions that may be supportive of or detrimental to the organization. The organization's continued viability depends critically on the participation and support of many significant stakeholders.

The analysis of this part of the organization-environment set begins with identifying significant relevant stakeholders and the demands and requirements they make on the organization. Members of the organization, including managers, are shown as stakeholders in Exhibit 2.1.3. They, together with other groups, have objectives they associate with their participation in the organization. The analysis of stakeholders' demands and requirements should be sufficiently detailed to reveal differences that might be masked by broad categories.

Multiple Objectives

A summary of the stakeholders' implied objectives, derived from demands and requirements, makes it clear that the organization must be designed to satisfy multiple objectives. The comprehensive, integrated design process is directed toward generating outcomes that will be satisfactory for most stakeholders and for the organization. For members of the organization at all levels, one of the important goals of the design process is to create the organization structure and job contents so that a high quality of working life is available concurrently as the organization strives to meet economic and other effectiveness and survival goals.

Functional Requirements of Organizations

There are requirements that must be met if organizations are to function effectively in order to produce desired outcomes as well as to maintain their capacity for future operation. Roles have to be designed, relationships must be defined, information must be communicated, and many support systems must operate to achieve organizational functioning. These are the aspects of an organization that define its configuration and its capacity to interact with both stakeholders and environments.

The requirements may be grouped into four general categories:

1. Goal Attainment. The core function of any organization is to produce value by transforming inputs into outputs of greater value to its environment. This requires the design of a technical system and of a social system through which its members can effectively operate the technical system.

2. Adaptation. Mechanisms are necessary so that the organization can adapt itself to changing requirements and to emerging conditions. These mechanisms include learning processes, feedback, planning, and ongoing means to change its structures, roles, and activities.

3. Integration. Given the complexity of an organization's internal activities and relationships with its environment, mechanisms are necessary to enable integration among all these activities and events. These include communication, coordination, conflict resolution, reward systems, and control procedures.

4. Organizational Maintenance. The continuity of organizational functioning depends on the organization's ability to people its roles with members adequately prepared to execute its many activities. Organizational maintenance functions include recruitment, selection, training, reward systems, career advancement systems, and disciplinary procedures.

The analysis of the three major parts of the organization-environment set–organization environments, stakeholders, and organizational functional requirements–creates a listing of the required response capabilities, the multiple objectives, and the operational requirements that must be designed for the organization.

2.1.7 ORGANIZATIONAL VALUES AND ORGANIZATION PHILOSOPHY

Organizational Values

Organizations are social inventions reflecting the culture and the values of the society at the time of their invention. Given the same products or services to be provided, and the same technologies, organizations will be different, based on different values influencing the design of their technical and social systems. To be successful the process of inventing organizations must be responsive to the issues and multiple objectives posed by the organization-environment set and must proceed from an explicit set of shared values and an agreed-upon view of the future (see Exhibit 2.1.3).

Changes in Societal Values

Advanced Western societies are in a state of rapid change from their more traditional form of the last 200 years into new, but yet undefined, forms. The values and the culture of the earlier period are under unrelenting challenge. The culture of a society, as distinct from its social goals, specifies how things should be done rather than what should be done, that is, the means rather than the ends. It is the choice of means rather than the choice of ends that gives a society or an organization its flavor or its climate.

Various significant cultural changes are taking place, and it appears that they will continue to take place, stimulated by changes in values in the influential population group, the 25- to 35-year-olds. The emerging values, the postindustrial values, appear to add up to a shift in the basis of accommodation between industry and society and between organization and community. The emerging postindustrial values that are crucial to the design of organizations are shown in Exhibit 2.1.4. Such values are giving rise to the very different needs and expectations being brought to the workplace by the coming majority of the U.S. work force. So large are the differences that this group is being viewed as the new breed of worker.[21]

Values in Technical System Design

The technical system influences the design of the structure of the organization but does not determine it. The choice of structure is a sociotechnical choice.[15] The design of technical systems is not value free. It is affected and limited by four major sets of determinants, only one of which derives from the available technologies. The technological determinants derive from the state of development of the technology on which the technical system is based. The major influence here is the status of development of the science or sciences underlying the technology. If the technology is underdeveloped because the underlying science cannot provide information about cause and effect relationships, stability, predictability, and so on, then the technical systems derived from such an underdeveloped technology will require particular relationships with the social system[22] that are not required with highly developed technology. Contrast such organizational features as control, autonomy, discretion, personal power, and so on, between, for example, cooking and automobile assembly. The other three value determinants are:

1. Various physical and economic constraints as well as legal and regulatory requirements set by governments at different levels.
2. Societal values that express and legitimate what may be required of people in doing work and which of their own goals members of an organization can expect to satisfy in the workplace.
3. The designer's own values and assumptions about how the organization ought to function

Exhibit 2.1.4 Agrarian, Industrial, and Postindustrial Values

Traditional Agrarian	Modern Industrial	Future Postindustrial
1. Affectivity Immediate gratification Expressiveness in actions, relationships	Affective neutrality Deferred gratification Instrumental approach to work, relationships	Self-expression Do your own thing Interpersonal instrumentalism *or* authenticity?
2. Collectivism Priority of collective goals	Individualism Priority of individual goals	New communitarianism
	Transitional–self actualization	Changed concept of destiny of of man
3. Particularism Duty owed on basis of kinship, membership of collectivity	Universalism Duty owed to all (Bureaucratically) equal treatment	Compensatory particularism; "fairness" of outcome preferred to equality of treatment; correcting past injustices to members of "disadvantaged" categories
4. Ascription Status given by "who you are"	Achievement Status given by "what you have done"	Situation–limited status and authority Equal value of all kinds of "achievement"
5. Diffuseness of role	Role specificity; separation of work from leisure, community roles	Denial of role, insistence on identity

Source. ALBERT CHERNS, "Social Change and Social Values," *Journal of Social and Economic Studies*, Vol. 5, No. 1 (1977), p. 85.

(the designer's own theory of organization) and what the roles are of people in relation to machines and machine systems, that is, how the technical system will be operated.[23]

The determinants are shown in Exhibit 2.1.5.

The Role of Values in Social System Design

Social system design defines the roles of the organization's members and how their activities will be integrated. It further determines how people will be managed, controlled, and rewarded. All of these design decisions are influenced by the social values of the designers as well as by the assumptions they make about the organization's members.[24]

The changes in values of society and the response of organizations to these changes are most notably visible in the design of organizations' social systems. A social system provides the opportunities for satisfying to a substantial degree the needs and expectations of its members. A crucial issue for advanced Western societies today is that of individual commitment to the goals of organizations. Commitment may be expected to be given and to be strongly positive when members of an organization perceive that they will have opportunities to achieve, through their membership in the

Exhibit 2.1.5 Determinants of Technical System Design

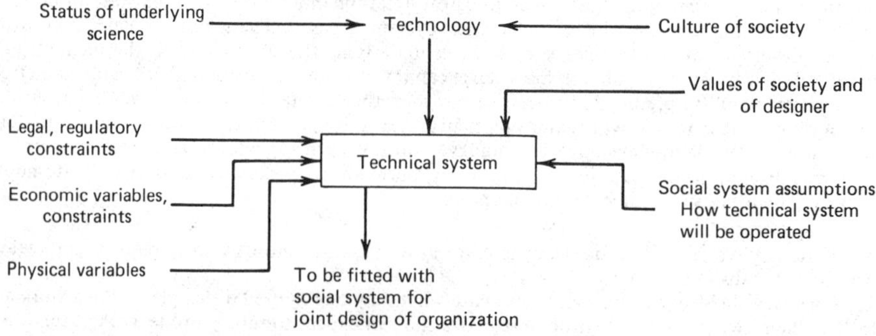

organization, goals and objectives important to them. That is, commitment can be expected to grow out of the circumstance of having one's own needs and objectives satisfied while satisfying the needs and objectives of the organization. It is this dual expectation that has given rise to the notion of designing organizations and their social systems to achieve or satisfy multiple objectives, most important of which are the objectives and goals of the organization, but others of which are the goals and expectations of the organization's members.

Congruence between design values and current values in society is fundamental to effective organization design. It may be achieved by developing an organization philosophy.

Organization Philosophy

An organization philosophy is a statement of shared values that will guide the design of the organization.[24] It is an enabling document and serves as a constitutional document, defining "what kind of unique minisociety is to be created." The document guides decisions about the technical system, social system, structure of the organization, roles of its members, social support systems, and relationships with the larger society. Without such a statement or guide, no design effort can proceed to explore all the possibilities, to create new approaches, and to assess these using the guiding statements of the philosophy as criteria.

Copying the organizational philosophy statements of successful organizations is of little value and may be harmful for two reasons. First, each organization faces different environments and markets and uses different technologies. Second, the understandings implied by these statements are broader and more complex than can be encompassed in simple statements. Such understandings are created in the preparatory process of developing an organization philosophy. Here different executives of the organization and members of design teams, including technical specialists, personnel specialists, marketing specialists, and union officials as appropriate, examine the environments that the organization will have to face; explore their assumptions about the role and functions of organizations and their members in such environments; and develop a shared set of agreements about values, purposes, and futures that will have to be faced. The process of exploring and agreeing on underlying values provides the basis for developing commitment to the goals and structure of the organization to be designed. Documenting the shared values in the form of an organization philosophy statement permits passing these on to those who will join the organization later.

An illustration of an organization philosophy statement is provided by Paul Hill in his book *Towards a New Philosophy of Management*.[24] Hill provides the most extensive treatment available of organization philosophy and its impact on the process of redesign.

Developing an Organization Philosophy Statement

There are a number of useful techniques, some more structured and others less so, for analyzing the organization-environment set and developing shared values. The dialectic method[25] and open systems planning[26] address both the characteristics of the organization-environment set and the values and assumptions of the designers. The Delphi method[27] plus scenario building, or policy Delphi, is suited for the assessment of future environmental states and stakeholders' objectives, without considering values, assumptions, or the internal functional requirements of the organization. A less-structured approach, illustrated here, focuses on discrepancies between the organization and its environment. Which of the methods will be most useful depends on the organization's managers, designers, and consultants.

The illustration is an actual case of developing an organization philosophy statement antecedent to the designing of a new manufacturing plant by a design team (see Section 2.1.9 for description of design team). Generally, a design team includes representatives of all branches of a firm whose functions have a bearing on the future operation of the new plant. The design team is sanctioned by a higher-level policymaking body and brings to the design process the following:

1. Organization and management experience.
2. Expectations about future developments.
3. Understandings about the realities and problems of organizational functioning.
4. A sense of environmental requirements.
5. Private theories of organization, that is, what makes an organization work.
6. Dissatisfaction with how organizations operate and how they underutilize the talents of their people.

The design team has the sanction to explore freely the organization-environment set without considering existing policies and practices as constraints. The first step is to review this set so that team members can deepen their appreciation of the issues arising from the environment (see Exhibit 2.1.3). This is followed by a review of the principles and theories underlying more traditional organization design. The purpose is to expose and review the private theories held by the members of the team and to relate these to members' experiences of organizational functioning. The purpose of these two extended reviews is to make explicit the mismatch between existing organization designs and the demands coming from the environment. These reviews proceed iteratively as different aspects of the organization-environment set are considered and compared to espoused organization theories.

The process produces two important outcomes: an increased understanding of the organization-environment set and value statements provided by members of the group. Such statements of good intentions usually are seen to be too idealistic and are thus ignored; in this process, however, these statements are essential. They are recorded and critically tested with concrete instances from the past experiences of members of the design team. This generates an appreciation of the impacts of the statements by associating a valuation with factual experience. It also ensures that statements will not be overlooked or lost. Sets of tested value statements are collected. The exploration of discrepancies continues until the potential for generating value statements is exhausted.

This step is followed by an exploration of the goals of the organization in order to help generate a set of explicit purposes. These are tested with concrete examples anchored in the experience of the firm's management. All of the preceding steps are reiterated and reviewed in relation to examinations of forecasts of near-future changes in societal environment, markets, technology, and so on. Value statements are likely to be modified on the basis of this review of likely future needs and changes.

The final step is to codify the set of tested value statements and purposes into an organization philosophy statement, which, after agreement by the sanctioning body, becomes the charter under which the design process will continue. Although it serves as a charter for design decisions, it is continuously subject to modification and review based on additional learnings of the design team.

2.1.8 SOCIOTECHNICAL SYSTEMS PRINCIPLES OF ORGANIZATION DESIGN

The general principles or guidelines for the design of organizations presented here are derived from the practice of designing new and redesigning existing organizations, using open sociotechnical systems theory. Some of the principles have been described by Cherns[28] and others by Davis.[14] The principles are as follows:

Systems

1. Organization Philosophy. The process of organization design requires the guidance of an agreed-to set of values and purposes explicitly stated as an organization philosophy.

2. Compatibility. The process of design or redesign must be compatible with its objectives. If the objective of design is an organization capable of self-modification, of adapting to change, and of making the greatest use of the creative capacities of its members, then a constructively participative organization is needed. A necessary condition for this type of organization is that people be given the opportunity to participate in the design of the jobs they are to perform. This principle is more applicable in the redesign of organizations rather than in the design of new ones. However, even in new design, it is possible to design in outline and to allow those who will be affected by the design to participate in the completion of it at a later date.

3. Open System. As open systems, organizations continuously adapt to requirements flowing from their environments. This requires the design of flexible structural features and of mechanisms for capturing and using organizational learning. The properties of the organization should reflect the salient properties of its internal and external environments.

4. Systemic Integrity. The structure of the organization and its roles reflect the recognition that all aspects of organizational functioning are interrelated; the organization design process must ensure the integrity of interrelated roles and structures.

5. Human Values–Quality of Work Life. An objective of organization design is to provide a high quality of work life. The objective implies that not everyone wants the same values from a work situation and that therefore options should be provided for accommodating individual's preferences, needs, and expectations to the extent possible.

Quality of work life means the quality of the relationship between worker and total working environment, with human dimensions added to the usual technical and economic dimensions. How to include criteria for quality of work life in design or redesign is reviewed in Chapter 2.5. The important, general criteria for quality of work life are:

Security.
Equitable pay and rewards.
Justice in the workplace.
Relief from bureaucratic coercion.

Meaningful and interesting work.

Variety.

Challenge.

Control over self, work, and workplace.

Own area of decision making.

Learning, growth.

Feedback, knowledge of results.

Work authority: authority to do that for which one is held responsible.

Recognition for contributions: financial, social, and psychological rewards; status; advancement.

Social support: can rely on others when needed and can expect sympathy and understanding when needed.

Futures that are viable (no dead-end jobs).

Ability to relate what one does at work to social life outside the workplace.

6. **Participation in Design and Operation.** Successful implementation of an organization design or change (redesign) depends substantially upon ownership of the design. Participation in the design process is essential for those who have the responsibility for achieving successful operation. As participants, managers, supervisors, and workers not only bring individual and organizational learning to the design process, but also convert organization philosophy into concrete reality. Thus they are inventing working lives of the organization's members. As stated by managers who, participating in a recent organization design, chose to leave to incoming workers the design of the details of their jobs, "We are not in the business of inventing other people's lives."

7. **Organizational Uniqueness.** The structure of the organization or of its component units and roles should suit the specific organization's situation and requires individualized design rather than imported or copied solutions.

Organizational Structuring

8. **Self-Maintaining Organizational Units.** Adaptive organizations require self-maintaining units as their basic building blocks. A self-maintaining organizational unit is one that has the capability to perform all the activities required to achieve its specific objectives under a wide variety of contingencies. It can maintain its internal structure and adapt itself to the changing demands impinging upon it from its environment. Such units may exist as groups with supervisors, semiautonomous teams, or autonomous teams.

9. **Boundary Location.** Locating internal boundaries determines the composition of self-maintaining organizational units. Boundary location is a crucial early activity in organizational design. The choice of boundaries can facilitate or impede the achievement of many organizational objectives. Internal organizational boundaries should be located so that:

a. Within an organizational unit, those responsible for achieving the outcomes can have access to and exercise control over the disturbances or variances that occur in performing work.

b. Members of a unit can develop some autonomy or a substantial degree of control over their own activities in achieving the desired goals of the unit.

c. Members of a unit can have access to all the information they need to solve the unit's problems and to assess its performance (feedback).

d. The boundaries are between the main transformation processes rather than cutting through a process.

e. They are at the completion of a process, product, or subdivision of the product.

f. The outcomes of work activities can be measured at their boundaries in order to provide feedback needed by the unit to regulate itself.

g. Members of a unit can develop an identity with the product, process, or outcomes.

h. Coordination between activities and people can be accomplished within the organizational unit, leaving integration to boundary managers.

i. Members of a unit can develop face-to-face relationships in carrying out the work of the unit.

j. The requisite skills and activities needed to perform assigned work and to maintain the technical and social systems of the organizational unit are within the boundary.

k. The need for external control and external coercion is minimized. Increased opportunities are available for self-regulation in achieving desired outcomes.

10. **Boundary Management.** This is a managerial and supervisory function essential to the successful performance and survival of a self-maintaining organizational unit. The function involves maintaining and protecting the unit's boundary by buffering the unit from external disturbances or changed demands beyond its response capability (or preparing it to cope), while providing its members with resources, training, information, and so on, that are required to accomplish its goals.

11. Joint Optimization. This recognizes that organizations function as sociotechnical systems. The design process requires joint evaluation of the impact of the technical system design on the social system and the impact of the requirements of the social system on the operation of the technical (work) system. Optimizing one of the systems and fitting the other to it will suboptimize the outcomes for performance of the total organization. Optimizing the outcomes of the total organization requires the joint design of the technical and the social systems.

12. Make Large Small. Organizational and physical structures should provide smaller and more intimate units and environments for individuals or groups.

Organizational Functioning

13. Minimal Critical Specification. This principle has two aspects, positive and negative. The positive aspect requires that what is essential be identified; specifying more than what is needed closes options that should be kept open for effective design. The negative aspect states simply that no more should be specified than is absolutely essential. The principle of minimal critical specification is essential to the design of adaptive organizations with flexible people. It is obviously not applicable to the design of machines or of organizations taken to be machines. When designing machines, every last gear, cog, and screw must be designed or else the machine will not function. By contrast, in designing a social entity such as an organization, designing every last element is overdesign, for it denies the learning and the innovation that the individual develops in the organization.

14. Variance Control for System Stability. Variances are unprogrammed events or disturbances that are likely to arise in any organization while it is doing its work, that is, achieving its goals. The deviations may be associated with the quality of the incoming raw material, with missing information, with machine failure, and so on. Identifying variances or disturbances and determining which must be controlled for the success of the process requires analysis that provides crucial information for the design of instrumentation, control systems, information systems, and so on.

Disturbances or variances that cannot be eliminated must be controlled or regulated as near to their point of origin as possible; otherwise they will be exported by the organizational unit. The application of the sociotechnical criterion requires that the organization's members have access to the sources of variances, the means of controlling or regulating the variances, and the necessary authorities and competencies for doing so.

15. Multifunctionalism (Organism Versus Mechanism). The application of this principle also is essential to the design of adaptive organizations. Traditional or classical forms of organization rely heavily on the concept of people as redundant parts, requiring that they perform highly specialized, fragmented tasks. Of course such people become easily replaceable. However, when response to unpredictable events (as in the case of high technology) or to new situations (as in the case of adapting to change) is required, then a large repertoire of performances becomes necessary. In these circumstances it is more effective for each element or organizational unit to possess more than one function. The same function can be performed in different ways by using different combinations of elements. There are several routes to the same goal, the principle sometimes being described as equifinality. Complex natural organisms have all followed this route of development. Complex systems, such as computers, have also been designed on the principle of multifunctional parts.

16. Information Flow. The objective of information flow is to provide information (data, feedback) to those who have to take action rather than to those who control the actions of others. Properly directed, sophisticated information systems can provide organizational units or work teams with the right type and amount of feedback to enable them to learn to control variances that occur within the scope of their spheres of responsibility and competence and to anticipate events that are likely to have a bearing on their performances.

17. Complementarity. The design of work and equipment should be based on the complementarity of people and machines and not on the competition between them.[29] Recognition and utilization of the unique capabilities of people to act as the adaptive elements in systems of machines and people are essential to the design and effective functioning of organizations.

Support Systems

18. Support Congruence. Systems of social support, such as rewards, advancement, and so on, should be designed so as to reinforce the behaviors that the organization design required for success. If cooperation between individuals is required for success, then the system should not be designed to reward competition. If group or team responsibility is required for success, then a payment system rewarding individual performance would be incongruent. Payment systems as well as selection, training, conflict resolution, work measurement, performance assessment, promotion, and so on, can either reinforce or contradict the behaviors that are called for by the organization design.

19. Minimal Status Differentials. Differences in privileges and status that are unnecessary to role performance and organizational effectiveness should be minimized, if not eliminated.

Organizational Continuity

20. Transitional Organization. Depending upon the amount of experience and skills available at the time of implementation of a design or redesign, transitional or start-up organizations need to be designed. Such temporary organizations must be congruent with, and support the attainment of, the desired organization structure and roles.

21. Incompleteness. Design is an iterative process. The closure of options opens new ones. As soon as a design is implemented, its consequences indicate the need for redesign. The means for redesign or renewal need to be built into the structure of an organization at all levels so that it can deal with changes it must make on an ongoing basis.

2.1.9 CREATING THE TEMPORARY DESIGN ORGANIZATION

The design of an organization, when following comprehensive, integrated design, is likely to take substantial time and to require the efforts of a large number of functional staff members, who need be brought together in a temporary design organization. Various agreements with senior management are required to create this temporary structure and to enable it to pursue its mission effectively. The process of creating the temporary design organization is illustrated in Exhibit 2.1.6. Although the process is pictured as relatively linear, many of the steps take place concurrently, with considerable interaction among them.

Steps in the Creation Process

Agreement on Mission

This step is undertaken to define the scope of the project to be designed and the conduct and expected outcomes of the design process. Agreements are necessary to sanction the joint design of the technical and social systems of the organization, the necessary participation by staff, and the temporary organization needed to design and implement the new or revised organization.

Exhibit 2.1.6 Creating the Temporary Design Organization

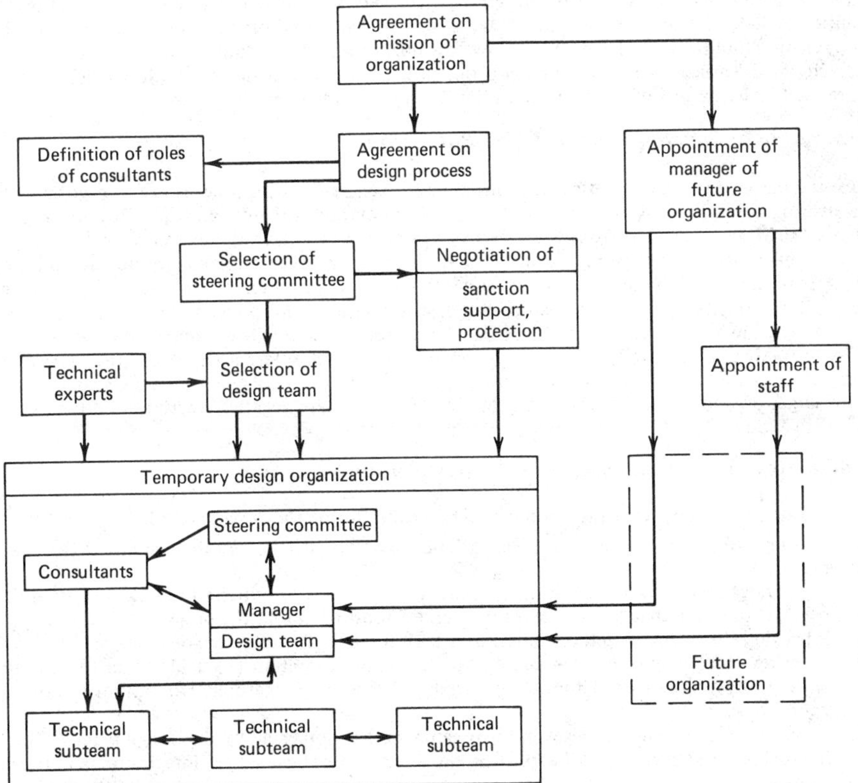

Agreement on Design Process

This agreement is a charter for all that follows. Agreements are generated among the project manager, the internal or external consultant, and senior management as to how the design process will be structured, guided, and managed. These agreements provide the basis for the next steps in creating the temporary design organization.

Definition of the Roles of the Consultants

Experience has indicated that comprehensive, integrated design is most effecitvely carried out with the aid of consultants as guides. The consultant may be internal or external. Agreement is required concerning the relationship of this role to the roles of other participants in the process.

Creation of a Temporary Design Organization

To support the design process, a temporary design organization is created, which exists until the start-up of the new organization. The temporary organization consists of a steering committee, a design team, consultants as needed, and various ad hoc subteams of functional experts appointed by the design team.

The steering committee is usually composed of senior executives or policymakers who represent each of the major functions of the organization that will be effected by the new design.

The design team members initially come from each function that has a bearing on design and implementation, including design engineering, process engineering, manufacturing, finance, quality control, personnel, and management. In unionized settings, under certain conditions, union representatives have been members of design teams. At the beginning, except for the manager of the future organization and the union representatives, the design team consists of functional experts. As the different parts of the organization are designed, functional and general managers replace them. By the time the design is completed, the design team consists primarily of managers and supervisors of the future organization.

Appointment of Manager of Future Organization

The lengthy design process must be effectively managed by someone whose overview and responsibilities include the implementation of the design in the future organization. Unless impossible, this person should be the head of the new organization, which requires that the individual be selected much earlier than is common practice in industry. The manager should participate in all phases of the design process, providing continuity in the implementation phase.

Negotiation of Sanction, Support, and Protection

Once the design team and steering committee have been selected, it is necessary to negotiate with the steering committee the sanction, support, and protection of the design team. Of critical importance is sufficient sanction from the steering committee to enable the design team to creatively explore many alternatives, some of which may differ from accepted company policies and practices. Typically, such sanction is conditional on the steering committee's receiving sufficient feedback and on its eventual approval of the design decisions and policies. The temporary design organization must have support in the form of resources–time for its participants to devote to the design efforts and accessibility to experts, both internal and external, as needs arise. The design team requires the protection of the steering committee as proposals develop and become known in the larger organization. Creative new solutions often generate negative reactions in other parts of the organization.

Functioning of the Temporary Design Organization

The steering committee performs a number of functions during the design period:

1. It brings together the various organizational interests, needs, and demands that the design must address.
2. It provides resources in the form of funding, allocation of experts to work on the project when needed, and time that the design team members must devote to the project.
3. It provides feedback to and communicates with the board of directors of the firm.
4. It buffers the design team, providing sanction and protection to enable them to explore a large variety of alternatives without the need to answer prematurely to authorities in the organization.
5. It acts to initiate policy changes in the organization as requested by the design team.
6. It provides guidance to the design team on matters of policy, values, integration, and the use of resources.
7. It provides sanction for the organization philosophy and subsequent design decisions.

The functions of the consultants in the design organization and of the ad hoc teams of experts are defined by the design team. One aspect of their roles is to advise the design team. No engineer, architect, consultant, manager, or other individual is permitted, on the basis of his or her functional expertise, to decide unilaterally on any aspect of the design. Each technical or organizational proposal is examined by the entire team and is evaluated in terms of its expected and unintended effects on the sociotechnical system.

The design team functions as a committee of the whole sociotechnical design of the organization, chaired by the manager of the future organization. It reserves to itself both the assignment of analysis and design to ad hoc subteams and the final decision making on all aspects of the organization. In effect, the design team asks experts and ad hoc design subteams to develop design alternatives, which form the bases for assessment and decision regarding acceptance of various designs. The design team may begin its work on a part-time basis, which becomes full-time as design demands increase. For 80 to 90% of its existence, the design team meets periodically at increasingly frequent intervals to develop its own designs, particularly in determining the organization philosophy, the principal technology, and the main boundaries of the organization. Later it will meet with the experts on the various ad hoc specialized subteams who develop the alternatives and data that the design team needs for making its decisions.

Continuity of the Design Process

Values regarding the role of individuals and learning acquired during the start-up experience require the structuring of a participative, continuing design process and a redesign process.

Very frequently, the design team, in dealing with the question of who "owns" the design and its various elements, and therefore who should participate in the design process, will choose to proceed on the basis of evolutionary design. Evolutionary design begins with accepting that the design process will be based on the principle of minimal critical specification (see Principle 13 under Section 2.1.8), that is, specifying the minimum critical functions and aspects of the organization needed to tie the various elements, functions, and roles together. Second, the design team accepts that completely specifying the details may, in fact, cause harm if done prematurely. Therefore the position is adopted that details will be added in the growing-in or start-up period of the organization. Third, the design team proceeds with the recognition that various aspects of the design can be decided only on the basis of the learning that takes place during the actual operation of the new organization. This means that the new organization will be going through planned trial stages at the start-up and that an apparatus for capturing the learning will be put into place for use in completing the details and for finally choosing the best alternatives. Therefore the design process may continue for as long as a year after start-up in order to collect information or receive feedback from experiences in the functioning of the organization.

Additions and modifications that derive from the learning that takes place during the start-up of the new organization usually require that a redesign occur at 6 to 12 months after the start-up begins. A planned process of redesign or modification of design needs to be made visible in order to permit a wide degree of participation on the part of those who have the responsibility for the organization and who have to live with it on an intimate basis.

2.1.10 PHASES OF THE ORGANIZATION DESIGN PROCESS

There are several major phases in the process of designing organizations, each of which includes several steps, which are shown in Exhibit 2.1.7. The phases include the development of preliminary data, the generation of design criteria, the designing of the organization, and the implementation of the design.

Development of Preliminary Data

The process begins with the gathering of preliminary data about the environment within which the future organization will be situated. This information is relevant to the exploration of the organization-environment set discussed in Section 2.1.6. As part of the feasibility studies associated with a new project, preliminary studies will often be undertaken before the creation of the temporary design organization.

Assay of the Internal Environment

In new organizations that will be part of a larger firm, a scan (survey or examination) of the larger firm is necessary since it will be, to a great extent, the major environment of the organization. Of particular concern is the examination of the policies and values of the larger organization in order to identify potential areas of conflict or incongruence. In such cases boundary protecting mechanisms will be required to ensure that the new organization will have sufficient opportunity to operate according to its values and structure while reporting to and being directed by the larger organization.

Exhibit 2.1.7 Phases of the Comprehensive, Integrated Organization Design Process

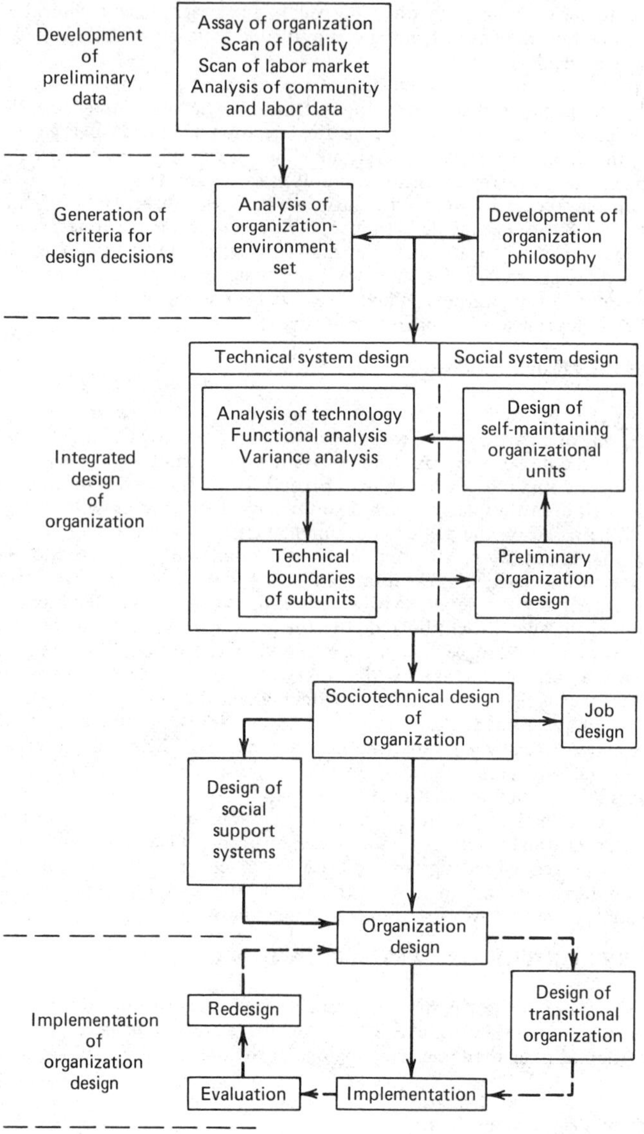

Scan of Locality or Community

An examination of the immediate environment of the new organization provides essential data needed for many design decisions. Needs, expectations, and requirements of the local community need to be identified.

Scan of Labor Market

This may be undertaken in conjunction with the survey of the community and will provide information about available manpower resources, travel distances, educational facilities, and wage patterns of the locality.

Analysis of Community and Labor Data

This analysis identifies the needs, expectations, and constraints that the designers will face at the outset. It will permit evaluation of an important issue: the extent to which the organization will commit itself to the training and development of local labor resources as opposed to importing workers from other areas.

Generation of Criteria for Design Decisions

In this stage the criteria for evaluating the design decisions are created. Essentially, this requires the interactive exploration of the organization-environment set and the development of an organization philosophy, as discussed in Sections 2.1.6 and 2.1.7. The agreed-upon philosophy statement, together with the shared understanding developed in analyzing the organization-environment set, becomes the design charter for the design team. The charter (organization philosophy) is sanctioned by the steering committee after it has developed understandings at its level and with the design team.

Integrated Design of the Organization

Design consists of three activities, one of which is creative and inventive, and the other two of which are analytic:

1. The creation of alternatives.
2. The evaluation of alternatives.
3. The selection of preferred alternatives.

A critical issue in design is the management of constraints. The early imposition of constraints inhibits the evolution of design alternatives that might more ideally achieve the goals stated in the organization philosophy. The most effective approach is that of constraint-free design, followed by cost-benefit analyses of the consequences and costs of either keeping or eliminating particular constraints. The development of alternatives proceeds until most or all of the design criteria have been met. However, at least two viable alternatives must be developed for each aspect and unit of the organization for evaluation. With two alternatives, comparison and evaluation will expose hidden assumptions, that is, constraints, and provide opportunities for further learning.

Design is an iterative process that proceeds from the general to the specific. The design charter may be thought of as a blank canvas, limited only by a general specification of what its content is to be. The design process proceeds to outline the structure of the organization, entering each decision onto the canvas. Each past decision becomes part of the input for subsequent decisions. At critical junctions, it may be found that contradictions exist between decisions and between decisions and the charter. In such cases it becomes necessary to undertake another iterative cycle.

The four phases of the organization design process occur interactively. For emphasis, these are shown serially in Exhibit 2.1.7. It is useful to start the design process with the work the organization is expected to do, the required transformation processes, and the knowledge and tools available to accomplish those transformations.

Design of the Technical System

The physical design of a plant and its technical system are generated by design subteams composed of appropriate experts under the direction of the design team. The work of these subteams can best serve the purposes of the design team if the design team itself includes individuals of sufficient technical expertise to monitor the work of other contributors whose roles are to provide alternative proposals and information for critical evaluation. Three analyses are undertaken by the design team.

The first is analysis of the technology. Technology is the body of knowledge upon which are built the transformation process and the methods used to execute it. Of critical importance are the limitations of this knowledge, the variability of inputs into the system, and the types of problems that dominate production processes based on the technology. Examples are poorly understood conversion processes, random equipment breakdowns, and complex pollution or energy control requirements. The essential issue is: Where must the energies of the organization be focused in order to exploit the technology in such a way as to achieve the organizations's goals?

The second analysis is of the functional requirements of the technical system. The operation of the proposed technical system will require that many activities be performed by people. At this stage these are a listing of functions that require human intervention. They are important in comparing alternative designs.

The third analysis is variance analysis. Variances are disturbances in the production process that must be brought under control.[30] They are the exceptional events associated with the functional activities listed in the second analysis. They differ from the technological characteristics of the first analysis in their specificity and association with each step in the transformation process. Information about variance control is crucial for making decisions about instrumentation, feedback, measurement, and the location of controls and control loops. Furthermore, variance control requirements are likely indicators of the skill requirements of operators and the boundary location between organizational units.

The information gathered in these analyses is used to guide the preliminary bounding of organizational subunits. The analysis of the technology identifies the critical response capabilities that the organization must be able to mobilize to meet major contingencies. The other two analyses identify logical process boundaries that can be the basis for viable organizational subunits.

The Bounding of Subunits. The bounding of the organizational subunits with respect to the technical system focuses on critical change points in the transformation process where clear state changes or outputs can be identified. The technical systems analysis divides the transformation process into a series of unit operations, each of which identifies a change of state of the throughput. Boundaries should be located between unit operations at significant change points in the system. Examples are points where changes occur in physical location, where processing technology information is required, or where shifts occur from batch to continuous production. Ideally, boundaries should be located where a definable output of controllable quality can be identified. (See Section 2.1.8 for other guides to boundary location.)

Design of the Social System

The design of the social system is concerned with the creation of structure and roles and with the joining of the roles and structure with the technical system. First, a preliminary social system is designed, which, together with the preliminary design of the technical system, is reiterated to achieve the goal of joint optimization. The most crucial choice to be made by organization designers at this level is the choice of boundaries that will determine what are to be the basic units of the organization.

Preliminary Organization Design. The outcome of the technical system design and the preliminary location of boundaries are input for the initial social system design. The overall structure of the organization can be outlined based on these inputs. The number of subunits, their relative size, the requirements for integrating their activities, and some indication of the management structure of the organization may all be developed at this point. Uusually the major subdivisions of the organization and the organizational units at the bottom of the organization can be identified. Fitting the middle levels of organization occurs later.

Design of Self-Maintaining Organizational Units. At this point it becomes possible to evaluate the proposed organizational subunits as self-maintaining units. In the organization design approach developed here, the basic unit of the organization is the self-maintaining organizational unit, as contrasted with organization designs build up from individual jobs. The requirements for effective functioning of self-maintaining units are indicated in Section 2.1.8. These units are viewed as mini-organizations so that the requirements of systems, organizational structuring, and organizational functioning can all be fruitfully applied to developing proposed organizational units.

Iteration of the Design Process. The output of the first iteration of the design process will generally lead to the discovery of many constraints, difficulties, and discrepancies between the technical system as proposed and the social system design that would be required to operate it. The design process, as it proceeds from general design decisions to more specific aspects, may require the generation of additional alternatives to meet the requirements uncovered by the analyses at each step. Successive iterations of the process, generally shorter than the first, are required to resolve these difficulties and to arrive at organization designs that jointly optimize the functioning of the technical and social systems.

As was indicated in the prior section, the membership of the design team is likely to change as the work of the temporary design organization progresses. As members of the future organization are selected, they will tend to replace specialist members of the team. This is particularly appropriate as the design decisions become more specific. This is one expression of the principle of minimum critical specification, which suggests that, by leaving the design as open as possible, the adaptive potential of the future members of the organization will be maximized. The new members of the design team become responsible for the very design decisions that they will have to implement and work within. Having participated in the design process, they will have learned to address many of the issues that they will face when developments necessitate redesign.

Sociotechnical Systems Organization Design

After preliminary choice of boundaries, social system design creates the structure and roles of the self-maintaining units, the basic units of the organization. Following design of the basic units, and the roles of their members, comes the design of the boundary maintenance and control rules of the organization and of its managers and supervisors. At this point, the number of levels and the functions of the different levels are resolved.

Self-maintaining units of organization, particularly if designed as minisocieties, are the basic building blocks of the organization, valuable for a number of reasons related to effectiveness and individual satisfaction. Such organizational units can continue to work to achieve their agreed-upon goals even with substantial disturbances by reorganizing themselves as they deem appropriate. Such units, frequently called teams, provide participation for their members in all activities that the unit must perform or control. For the individual, a unit provides membership in a small organization on a face-to-face basis and places a wide variety of tasks, functions, and responsibilities in the hands of each member. Further, the characteristics of the units reflect multiple options, minimum bureaucratic apparatus and perquisites, high participation, futures, open and multiple careers for achieving these futures, social support systems, and opportunities to relate life inside the organization to life outside.

To develop self-maintaining units as the basic building blocks of organizations, boundaries around such units should be located as indicated in Section 2.1.8.

Locating boundaries, testing the consequences, relocating them, and reiterating this process will ultimately permit the design of teams, groups, and roles.

Job Design

At this stage, it is possible to begin the design of jobs through the structuring of roles and the determination of role contents and relationships and by interlocking these with the tasks to be performed as identified in technical system analysis. This is followed by assigning to roles or groups the production tasks, equipment maintenance tasks, inspection tasks, and organizational maintenance tasks such as communication, coordination, and training. (See Chapter 2.5 for a complete discussion of job design.)

A critical issue that must be decided is whether it is necessary to specify individual jobs as part of the organization design. The decision has important implications for the quality of working life of the members of the future organization. The design should be sufficiently specific to ensure that all necessary tasks can be executed. It is not required that these tasks be allocated to individuals as jobs. One alternative among several is to let the future members of the organization design their own allocations of tasks as members of the teams that are to operate as self-maintaining organizational units. This alternative provides advantages in flexibility and adaptability and also meets the expectations of organization members.

The roles of managers and supervisors in the organization can be specified, and the unique skills required of them in this type of organization can be identified. The team must consider the requirements of the manager role as a buffer between the organizational unit and parallel units and other parts of the organization, which requires the manager to mediate the impacts of the environment on the organizational unit. Additional requirements are considered for such manager roles as resource providers, technical experts, trainers, agents for securing goal agreement between the unit and the larger organization, evaluators, auditors, and integrators of the activities between units.

Design of the Social Support Systems

Given the sociotechnical design of the organization, it is necessary to address the design of the many support functions required to sustain the organization. These include the reward systems; career path systems; systems of advancement, training, discipline, justice, and constitutionality; and functions such as communication, planning, and scheduling, as well as group and individual norms.

Each of these elements must be designed so as to support the purposes and values of the organization structure. They, too, must be tested by the criteria developed in the organization philosophy. For example, if it is intended that people allocate their tasks among themselves in a cooperative team, a reward system based on individual incentives would operate at cross-purposes with that design. If multiple skilling of team members is an important means of achieving flexible and effective operation, then a seniority system coupled with specific job assignments may restrict most opportunities for on-the-job training.

Organization Design

In this step the sociotechnical organization design and the social support systems are formed. The available preliminary design of the organization will now permit the development of scenarios as to the functioning of the organization under different contingencies. The shortcomings and difficulties that the scenarios indicate become the basis for reiterating various theses of the design. During this step the final design of the technical system can be settled, including tentative design of the middle of the organization to fit the bottom to the top. Again, scenarios may have to be developed to permit the choice of a final design for the top of the organization, the bottom of the organization, and finally the middle of the organization.

Implementation of Organization Design

Most frequently, new organization designs as well as redesigns encounter great difficulties in implementation. Many are abandoned or are severely modified so that whatever is innovative about the designs is removed. The difficulties and failures can be attributed largely to the lack of designing the necessary components for implementation. The components to be designed by the design team are (1) a transitional organization, if needed; (2) recruitment; (3) selection; and (4) pre-start-up training.

Design of Transitional Organization

It may be necessary to design a transitional, or start-up, organization to implement the organization design. In new organizations this need arises when the people available do not possess all of the skills required to carry out the design. In redesigns of existing organizations existing structures may not permit an immediate changeover to the new design. In either case a transitional organization is required to implement the organization design.

The critical issue during implementation, and therefore for the design of the transitional organization, is to ensure congruence between the implementation and the ultimate design. A common mistake is to establish temporary practices for the transition or start-up periods that conflict with the organization design. These practices often become permanent features of its operations. The procedures and practices established for start-up should facilitate achievement of the intended design.

During the transitional period, the emphasis is on training and development of the organizational units and their members. It is a period that may be characterized by substantially larger numbers of trainers, advisors, temporary skilled staff, and modified job designs that permit people to apply the pre-start-up training they received. It is very important that people have available objective indicators that signal the end of the transitional period, when they may expect to receive the benefits of steady state operation. Such information serves the purpose of setting goals to be reached and allows people to monitor their progress in implementing the design.

Recruitment. A recruitment process that is congruent with the values, structures, and operation of the organization is required. Organizations built around self-maintaining organizational units or teams usually require a recruitment scheme based on staged or phased information giving, to permit prospective members of the organization to "buy into" the organization. Such designs have been very successful in recruiting people willing to commit themselves to membership in such organizations.

Selection. The criteria for selection also have to be congruent with the structure, values, goals, and operation of the organization. The criteria should reflect the main attributes of the organization as reflected in the roles of its members. In organizations having self-maintaining units or teams, members of the teams frequently participate in the selection process. Individual commitment to a team is built in part on participating in the membership granting or removal process.

Training. Two kinds of training need to be designed: technical training and social system training. The kind of technical training will depend on the skills of the incoming work force progression schemes, and so on. Continuing training schemes, not necessarily of the classroom variety, are also required to support progression. The social system training, in addition to indoctrination in the organization philosophy, consists of problem solving, conflict resolution, planning, counseling, instruction, and other skills necessary to operate and maintain the social system.

Implementation. It is very important that the organization capture the learning that is generated during implementation. A fruitful outcome of the participation of organization members both on the design teams and in the implementation process is a very enhanced capacity to appreciate the significance of start-up events, to learn from the adaptations required, and to utilize this learning in modifying the organization design. This is the basis for the continued evolution of the organization design by the members to meet the changing demands of the environment.

Evaluation

Measures of success and measures of effectiveness for purposes of providing feedback to the organization and its sanctioners need to be designed. The evaluation measurements again have to be congruent with the values, structure, and roles within the organization. For example, managers whose role may include the development of subordinates should be evaluated on this objective just as they are evaluated on other measures reflecting the bottom line, namely, costs and profits of the organization.

Redesign

No organization design can be complete. There is the expectation that changes will need to be made, both to overcome deficiencies in the design and to capture and use the learning that has taken place during the start-up and early operating phases of operation. Particularly, redesign or completion of design may be necessary if the organization has been structured on the basis of specifying the minimal critical structural, informational, and job requirements. Such designs permit a high degree of participation and learning to take place, resulting in valuable improvements. Members of participative organizations expect that improvements will take place and that they will be able to make contributions to such improvements. The design team therefore is required to design a process and to indicate when the process will be used for either concluding the design or improving or redesigning the organization at various periods in its ongoing life. Designs are always incomplete from an evolutionary point of view. Therefore modification and redesign are ongoing processes in an organization for which provision is made in self-maintaining organizational units.

Differences. The concepts and principles used in the design of new organizations are applicable to the redesign of existing organizations. However, there are specific differences that must be considered.

1. Members of the organization exist and they have acquired privileges that they value.
2. The social system exists with a shared history, existing relationships, and implied futures.
3. There are existing organizational rules, procedures, and practices that reflect design philosophies of the past.
4. There are formal and informal contractual relationships with unions and individuals.

All of these arrangements will be disturbed in the redesign process. The very initiation of the redesign process constitutes a step in eventual implementation.

Major Requirements and Concerns. This poses two major additional requirements in a redesign effort. First, in addition to exploring the external environment of the organization as a source of demands and requirements, it is necessary to explore the existing organization and to consider its past history. The constraints found should not be allowed to inhibit the process of generating new alternatives, but they must be considered when decisions are made. Second, even though full support, sanction, and protection may be obtained from the upper levels of an organization, the redesign process itself as an intervention in the organization can generate effects that will preclude many innovations or even completely undermine the whole effort. Therefore it is necessary to consider the future implications for the organization redesign of the practices pursued in redesigning the organization. A useful approach is to evaluate the proposed design process and its steps in terms of the organization philosophy. This is more likely to ensure congruence between the redesign process and its redesign goals.

On the positive side, the existence of on-site knowledge, experience, and skills can be invaluable in developing and evaluating design alternatives. A process is necessary that can tap into the wealth of knowledge within the organization at all levels. This can be achieved if all levels of the organization are well represented on the design team.

In many instances union participation in the temporary redesign organization has proved useful and effective in redesign efforts. Managers may resist this, but it must be recognized that it is very difficult to impose cooperation and teamwork on a group of people. Where unions are resistant to participation in redesign efforts, a useful approach may be to ask the union representatives to consult with other unions who have successfully participated in quality-of-work-life programs.

Another useful strategy available in redesign efforts is to test alternatives that might otherwise be rejected because of uncertainty over consequences. The design team seeks agreement for trials that can expose costs and benefits of organizational innovations. Under these circumstances many innovations can be generated, tried out, and modified to make them both acceptable and useful. Again, such experiments should be pursued only with full recognition of their possible consequences, since it is frequently impossible to undo some changes, no matter how temporary they were intended to be.

Redesigns often undertaken piecemeal have unintended effects on other parts of the organization. This raises the issue of the orderly diffusion of innovations in an organization, which requires that diffusion issues be part of the prechange strategic planning for organization redesign. Diffusion problems arise only when there are successful redesigns. Failures are more readily managed by disposing of or isolating them.

A major concern in redesign is the diffusion through the rest of the organization of the innovation developed. A particular issue requiring careful planning is how a redesign that significantly changes a part of an organization can be implemented and allowed to evolve even though it is quite different from the organization structure, job structure, and practices in surrounding organizational units. Some form of shelter needs to be provided to allow the proposed change to be introduced and become stabilized, in order to determine its value. Additionally, strategy needs to be developed as to how surrounding organizational units may begin to use the organizational concepts and change approaches so that they, too, can develop redesigns that will be congruent with the organizational unit that has been changed. Major differences in organization structure, jobs, and management style do not survive for very long in isolated organizational units. Unless diffusion strategies are developed to spread innovations into other units, the redesign may be encapsulated and eventually throttled, with the changed unit reverting back to the structure and style of the rest of the organization. Viewing the organization as a system indicates that a successful redesign will have an impact on all other units of the organization. Therefore redesign strategies must be the issue of what to do with success. Successful redesigns generate a need for futher redesigns to move the whole organization into congruence with new values and futures.

Emerging Theories

There are a growing number of books on organization design of which Galbraith[31] and Connor[32] are examples. However, none of them deals with the process of organization design. Their authors use the term "organization design" in the same sense that "organization structure" is used in this chapter. The books do present different specific theories upon which to structure organizations. Since this chapter focuses on the designing of organizations and not on a review of the various theories of organization, these books were not emphasized. However, readers interested in emerging theories will wish to refer to the bibliography at the end of the chapter.

REFERENCES

1. L. E. DAVIS, "Individuals and the Organization," *California Management Review*, Spring 1980, p. 5.
2. F. E. EMERY and E. L. TRIST, *Toward a Social Ecology*, Plenum, New York, 1973.
3. F. E. EMERY and E. L. TRIST, "The Causal Texture of Organization Environments," *Human Relations*, Vol. 18, 1965.
4. ADAM SMITH, *The Wealth of Nations*, Penguin, London, 1970 (originally published 1776).
5. CHARLES BABBAGE, *On the Economy of Machinery and Manufactures*, Augustus M. Kelly, New York, 1965 (4th edition originally published 1835).
6. F. W. TAYLOR, *The Principles of Scientific Management*, Harper & Row, New York, 1911.
7. M. WEBER, *The Theory of Social and Economic Organization*, translated by A. M. Henderson and T. Parsons, Free Press, Glencoe, NY, 1964.
8. F. GILBRETH, *Motion Study, A Method for Increasing the Efficiency of the Worker*, Van Nostrand, New York, 1911.
9. F. J. ROETHLISBERGER and W. J. DICKSON, *Management and the Worker*, Harvard University Press, Cambridge, MA, 1947.
10. D. R. KINGDON, *Matrix Organizations: Managing Information Technologies,* Tavistock Publications, London, 1973.
11. J. R. GALBRAITH, Ed., *Matrix Organizations: Organization Design for High Technology*, MIT Press, Cambridge, MA, 1971.
12. F. E. EMERY and E. L. TRIST, "Sociotechnical Systems," in C. W. Churchman and M. Verhulst, Eds., *Management Science: Models and Techniques*, Vol. 2, Pergamon, Oxford, 1960.
13. T. BURNS and G. M. STALKER, *The Management of Innovation*, Tavistock Publications, London, 1961.
14. L. E. DAVIS, "Evolving Alternative Organization Designs: Their Sociotechnical Bases," *Human Relations*, Vol. 30, No. 3 (1977).

15. E. L. TRIST, G. W. HIGGIN, H. MURRAY, and A. B. POLLOCK, *Organizational Choice*, Tavistock Publications, London, 1963.
16. W. R. ASHBY, *Design for a Brain*, Wiley, New York, 1960.
17. F. E. EMERY, "The Assembly Line–Its Logic and Our Future," in L. E. Davis and J. C. Taylor, Eds., *Design of Jobs*, 2nd ed., Goodyear, Santa Monica, CA, 1979.
18. G. VICKERS, *The Art of Judgement*, Basic Books, New York, 1965.
19. C. KERR and J. M. ROSOW, Eds., *Work in America: The Next Decade*, Van Nostrand Reinhold, New York, 1979.
20. W. R. ASHBY, *An Introduction to Cybernetics*, Chapman and Hall, London, 1956.
21. D. YANKELOVICH, "Work, Values and the New Breed," in C. Kerr and J. M. Rosow, Eds., *Work in America: The Next Decade*, Van Nostrand Reinhold, New York, 1979.
22. L. E. DAVIS and J. C. TAYLOR, "Technology, Organization and Job Structure," in R. Dubin, Ed., *Handbook of Work, Organization and Society*, Rand McNally, Chicago, 1976.
23. L. E. DAVIS and J. C. TAYLOR, "Technology Effects of Jobs, Work and Organizational Structure: A Contingency View," in L. E. Davis, and A. B. Cherns, Eds., *Quality of Working Life*, Vol. 1, Free Press, Glencoe, NY, 1975.
24. P. HILL, *Towards a New Philosophy of Management*, Gower, Essex, 1971.
25. R. O. MASON, "A Dialectical Approach to Strategic Planning," *Management Science*, Vol. 15, No. 8 (1969).
26. G. K. JAYARAM, "Open Systems Planning," in W. A. Pasmore, and J. J. Sherwood, Eds., *Sociotechnical Systems*, University Associates, La Jolla, CA, 1978.
27. M. TUROFF, "The Policy Delphi," in H. A. Linstone and M. Turoff, Eds., *The Delphi Method*, Addison-Wesley, Reading, MA, 1975.
28. A. B. CHERNS, "The Principles of Sociotechnical Design," *Human Relations*, Vol. 29, 1976.
29. N. JORDAN, "Allocation of Functions Between Man and Machines in Automated Systems," in L. E. Davis and J. C. Taylor, Eds., *Design of Jobs*, 2nd ed., Goodyear, Santa Monica, CA, 1979.
30. P. ENGELSTAD, "Sociotechnical Approach to Problems of Process Control," in L. E. Davis and J. C. Taylor, Eds., *Design of Jobs*, 2nd ed., Goodyear, Santa Monica, CA, 1979.
31. J. R. GALBRAITH, *Organization Design*, Addison-Wesley, Reading, MA, 1977.
32. P. E. CONNOR, *Organizations: Theory and Design*, Science Research Associates, Chicago, 1980.

BIBLIOGRAPHY

GALBRAITH, J. R., *Designing Complex Organizations*, Addison-Wesley, Reading, MA, 1973.

KHANDWALLA, P. N., *The Design of Organizations*, Harcourt Brace Jovanovich, New York, 1977.

LAWRENCE, P. R. and J. W. LORSCH, *Developing Organizations: Diagnoses and Action*, Addison-Wesley, Reading, MA, 1969.

PFEFFER, J., *Organization Design*, AHM Publishing Co., Arlington Heights, IL, 1978.

THOMPSON, JAMES D., Ed., *Approaches to Organization Design*, University of Pittsburgh Press, Pittsburgh, PA, 1966.

TUSHMAN, M. L. and D. A. NADLER, "Information Processing as an Integrating Framework in Organization Design," *Academy of Management Review*, Vol. 3, 1978, pp. 613–621.

CHAPTER 2.2

Nonfinancial Motivation: Creating a Work Environment for Effective Human Performance

JAMES A. RICHARDSON
ROY F. LOMICKA
Eastman Kodak Company

2.2.1 INTRODUCTION

This handbook contains two chapters pertaining to the subject of motivation. The other chapter, 2.3, is entitled "Financial Motivation." A few introductory remarks are in order to explain this dual treatment.

Industrial engineering has had a long history of association with motivation in the form of wage incentive plans. For its first 50 years, up to and through World War II, probably in excess of 50% of all industrial engineering activity was concerned with the installation and maintenance of wage incentive plans. A substantial part of industrial engineering is to this day still so engaged.

Since the mid-1950s, there has grown a body of theory and practice that broadens the substance of work motivation beyond financial incentives. Stimulated by the results of academic research and prodded by the negative side effects of overreliance on financial incentives, many organizations sought and have adopted motivational alternatives. This chapter embraces many aspects of these alternatives and presents what the authors hope is a comprehensive and useful view.

We wish to be clear that this chapter is not in "opposition" to financial incentives. In fact, the "conditions" put forth herein for high productivity require the existence of a perceived equitable relationship between compensation and performance. What the authors take issue with is the oversimple notion that money is the sole source of human motivation to work. This assumption apparently supported many traditional financial incentive plans.

An alternative format for this chapter could have been a summary of the past 25 years' research from behavioral science origins. This body of research has contributed many powerful ideas that have had an impact on the world of work. The ideas range from the broadly relevant "hierarchy of human needs" of Abraham Maslow to the "hygiene/motivation" work theories of Frederick Herzberg and include the more recent work-design-oriented ideas of Richard Hackman and Greg Oldham. A difficulty in presenting a summary of such theory is that much remains "unproved"—unproved except as synthesized into more broadly comprehensive "working" models, which can only be tested empirically. It is only as many theories are combined and reduced to working practice that the basic theoretical ideas become useful.

As writers of a "handbook," we therefore address our contribution to those who wish to make *use* of new motivational ideas in the work environment.

2.2.2 ORIGINS OF THIS CHAPTER

This chapter represents a compressed and distilled version of the 20 year experience of a group that has attempted to understand and apply the many relevant and sometimes conflicting theories that have come from academic research. It also represents considerable cross-checking with others in the industrial community who have tried to do the same. The distillate is the product of success and failure, with continuous revisions made accordingly, converging on the model presented here. Were this to be written a year from now it would differ certainly in detail and possibly in structure, but not in general context or concept.

Exhibit 2.2.1 A Model for Performance Improvement Through People

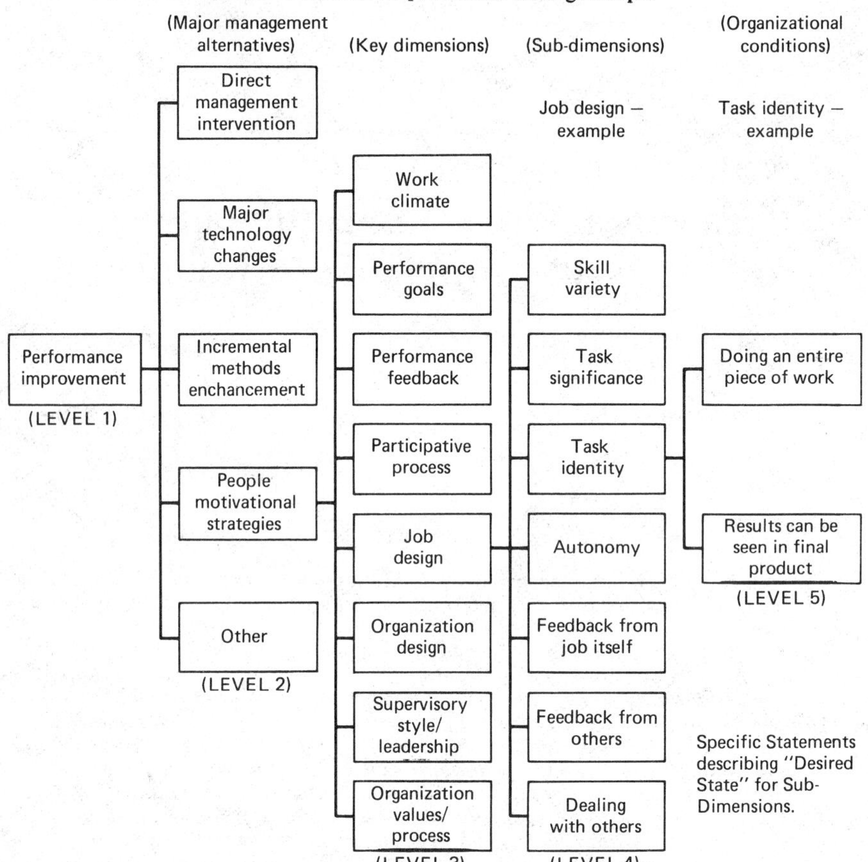

2.2.3 MODEL PRESENTATION

The centerpiece of this chapter is a *model* (Exhibit 2.2.1) explicating the *conditions* under which people are most likely to achieve organizational objectives and satisfy their own needs at the same time. The model therefore describes an *environment*.

The presentation is in two formats:

1. A diagrammatic overview, which maps the dimensions and subdimensions of the environment (Exhibit 2.2.1).
2. A word description of the dimensions, which explains the environment in more detail (Exhibit 2.2.2).

Exhibit 2.2.1 is an attempt to show, on one piece of paper, a performance improvement model with five levels of detail. The levels are as follows:

Level 1, performance improvement, is put forth as the ultimate objective from management's point of view and assumes that performance improvement will be defined in terms of whatever management states its legitimate goals to be. There is no further development of this notion in this chapter.

Level 2, major management alternatives, is meant to display those major means available to management by which it acts to achieve the ultimate objective. Note that the thrust of this chapter is the development of only one of these alternatives—that of people motivational strategies. The other alternatives are not developed here.

Levels 3 and 4, key dimensions and subdimensions, respectively, list dimensions of the "people" alternative in two levels of detail. Level 4 further details job design only, as an example. The other key dimensions are detailed in the word description that follows.

Level 5, organizational conditions, contains the detailed descriptions of the conditions under which it is asserted that people will be most likely to achieve organizational and personal objectives. There are 115 of these condition statements, which are the primary focus of this chapter. They are put forth as a desired state, or in a sense an ideal, toward which the organization should be moving.

The model (word description), described in 8 key dimensions, 39 subdimensions, and 115 desired conditions, is shown in Exhibit 2.2.2.

Exhibit 2.2.2 A Motivational Environment for Performance Improvement

I. Work Climate

A. *Competence of Supervisor*

1. Supervisors are technically competent, that is, they demonstrate general expertise and job knowledge relevant to the function being supervised.[1]
2. Supervisors do a good job of telling others about the capabilities and availabilities of their people.[1]
3. People receive equitable treatment and demonstrated respect from their supervisors.

B. *Relationship With Supervisor*

1. Supervisors are friendly and easy to talk with.
2. Supervisors are genuinely interested in listening to what their people have to say.[1]
3. Supervisors help their people solve problems that occur on the job, by providing information, offering ideas, and so on.

C. *Relationship With Peers*

1. Work group members are friendly and easy to work with.[1]
2. Work group members are interested in listening to what each has to say.[1]
3. Work group members help each other find ways to do their jobs better, by providing information, offering ideas, and so on.
4. Work group members encourage each other to give their best effort.[1]

D. *Working Conditions*

1. The work area is free from annoying distractions.
2. The temperature of the work area is comfortable, and the ventilation is good.
3. The layout of the work area is convenient, and the housekeeping is good.
4. The work area is safe.
5. The work area is pleasant and conducive to task accomplishment.

E. *Supplies and Equipment*

1. People can obtain tools and supplies without difficulty as they are needed on the job.
2. The equipment people have to work with is adequate, efficient, and well maintained.

F. *Pay and Benefits*

1. People feel they are fairly paid for what they contribute to their organization.
2. People view their employee benefit programs as adequate.

G. *Personal Life*

1. The demands of the job do not unduly keep people from doing things they would like to do in their personal lives.
2. The kind of work people do is viewed as worthwhile by their friends and neighbors.

H. *Security*

1. People in the organization have job security.

I. *Policies and Practices*

1. People view the organization's policies as fair and not detrimental to job accomplishment.

J. *Status*

1. The organization's use of "status symbols" is functionally related to effective job performance rather than a means of highlighting differences in job levels.

II. Performance Goals

A. *Organization (Work Group) Goals*

1. The organization operates with goals, that is, goals for the organization (work group) and goals for individuals.

Exhibit 2.2.2 (*Continued*)

2. The organization (work group) has clearly communicated and reasonable goals for
   a. Quantity (production).
   b. Quality/waste.
   c. Service.
   d. Cost.
   e. Improvement.
   f. Personnel development.
   g. Absence.
   h. Safety.
3. The organization's (work group's) goals are established jointly by work group members and supervisors through group participation and discussion.
4. People receive all the information they need to participate in setting challenging, but realistic, goals, for example, production schedules, quality requirements, customer (user) reactions, and constraints (cost, legal, policy, time).

B. *Control Systems ("Score Keeping")*
1. A control system exists to "track and trend" performance data against the organization's (work group's) goals.
2. People know how well their organization (work group) is meeting its goals.
3. Control data are viewed by the organization (work group) as helpful information for self-direction and control rather than as a source of punishment.
4. The organization (work group) actively engages in problem solving activities in order to "close the gap" between actual and desired performance.

C. *Individual Goals*
1. Individuals have clearly communicated and reasonable goals.
2. Individuals understand how their goals contribute to the organization's (work group's) goals.
3. Goal accomplishments of individuals are used as performance criteria by supervisors.

III. Performance Feedback (to the Performer)

A. *Expectations*
1. People clearly understand what their supervisors and others in their work groups expect of them (tasks and behaviors).[1]
2. The people are functionally able to fulfill most of the expectations of their supervisors and of others in their work group.
3. People know what they are being measured against in their jobs.

B. *Feedback Data*
1. People are able to track and trend their own performance on the job.
2. People receive performance information, that is, deviation from goals or standards, before their supervisors receive this information.

C. *Feedback Schedule*
1. Supervisors or information systems give feedback to people as soon as possible after availability of performance results.
2. Supervisors schedule continuous feedback to their people at frequencies that take into account individual differences.

D. *Reinforcement*
1. The organization positively reinforces good performance of its members.
2. When people do their jobs well, this positively affects how much money they will make.[1]
3. When people do their jobs well, they get a feeling of personal satisfaction in job accomplishment.[1]
4. When people do their jobs well, they receive positive feedback and demonstrated respect from their supervisors and from others in their work groups.[1]
5. There are good opportunities for advancement in the organization for people who do a good job.
6. The organization analyzes its systems and procedures so as to minimize or eliminate negative reinforcement for good performance.

Exhibit 2.2.2 (*Continued*)

IV. Participative Process
  A. *Goal Setting*
    1. Individuals help set performance goals important to their work, that is, quantity, quality, service, cost, improvement, and so on.
  B. *Planning and Work Methods*
    1. Individuals help plan how their work is to be done.
    2. Individuals help select and evaluate the work methods used on their jobs.
  C. *Problem Solving*
    1. Individuals are encouraged to think up better ways to carry out their work.
    2. Individuals are provided structured opportunities (time and techniques) to work on improvements in their jobs and to solve problems.
  D. *Information Flow*
    1. Individuals receive the information necessary to set goals and to plan their own work.
    2. Supervisors let their people know, in advance, of any changes that will affect them in their jobs.
    3. Work groups receive information about what is going on in other departments or shifts with whom they must work collaboratively.
    4. Work groups receive information about their organization's performance, customer or user reactions, future plans, and so on, that affects the "success" of their organization.

V. Job Design[2]
  A. *Skill Variety*
    1. Individuals' jobs require them to do a variety of tasks using a number of talents and skills.
    2. Individuals' jobs require them to use complex or higher-level thinking skills; that is, the job is not simple and manually repetitive.
  B. *Task Identity*
    1. Individuals' jobs are arranged so that they have a chance to do a segment of work from beginning to end.
    2. The results of individuals' activities can be seen in the final product or service.
  C. *Task Significance*
    1. The outcomes of individuals' work affect other people in important ways.
    2. The work of individuals can be seen as significant in the broader scheme of things.
  D. *Autonomy*
    1. Individuals' jobs give them some discretion for deciding how and when the work is to be done.
  E. *Feedback From the Job Itself*
    1. The work itself provides direct and clear information to individuals about how well they are doing (aside from any feedback from co-workers or their supervisors).
  F. *Feedback From Others*
    1. Supervisors and co-workers provide individuals with timely and continuous feedback about how well they are doing.
  G. *Dealing With Others*
    1. Individuals' jobs require them to work closely with other people (internal and external) in carrying out their work activities.

VI. Organization Design
  A. *Mission/Goals*
    1. The organization's mission and goals are clear to its members.[3]
    2. The time and energy of organization members are spent doing those things that are consistent with the organization's mission and goals.
  B. *Roles/Responsibilities*
    1. People are sure about what responsibilities they have and about what others in their work group are supposed to be doing.[3]

Exhibit 2.2.2 (*Continued*)

2. Group members discuss their role expectations with each other so that they may work together effectively.
3. Group members ensure that responsibilities important to goal accomplishment are assigned to individuals, that is, that responsibilities are not "falling through the cracks."

C. *Skills and Abilities*

1. People receive adequate training for their jobs.
2. People have continuing opportunities to learn and grow on their jobs.
3. The organization (work group) has in-depth competence.
4. The organization (work group) has, or has access to, the knowledge and skills it needs to analyze and solve most of its own problems.

D. *Promotional/Career Path*

1. Good opportunities exist for people to get ahead in their organization (work group) in terms of
   a. Technical structure.
   b. Hierarchical structure.
2. "Getting ahead" in the organization (work group) is based primarily on ability and competence.
3. The organization actively facilitates and/or removes barriers to individual growth.

VII. Supervisory Style/Leadership

A. *Assumptions About People (Held by Supervisors)*

1. People are social beings motivated by a desire to achieve and behave responsibly. They value their independence and their capacity to control their own fate.[4]
2. Most people have drives toward personal growth and development if the environment is supportive and challenging. Most people want to become more of what they are capable of becoming.[4]
3. People have a need for recognition and respect for what it is they are as persons. They gain satisfaction from growth and application of new skills.[5]
4. People want the right to question what is going on and to participate in changing things if things need changing in order to develop a caring, effective, healthy system.[5]
5. Most people desire to make and are capable of making a higher level of contribution to the accomplishment of organization goals than most organization environments will allow.[4]
6. People want to belong to a stable group of friends who share a common productive purpose in order to be helpful and supportive toward one another.[5]

B. *Behaviors of Supervisor*

1. Supervisors communicate information about production schedules, quality requirements, customer (user) reactions, and constraints (cost, legal, policy, time) as a basis for goal setting and problem solving.[6]
2. Supervisors participate with their people in goal setting, problem solving, and establishing work methods.[6]
3. Supervisors allow people to manage their own work once mutually agreed upon goals are established.[6]
4. Supervisors involve people in defining criteria for effective performance and feedback methods for self-evaluation of performance.[6]
5. Supervisors provide performance feedback to people in a climate of candor that permits natural acknowledgment of success and a discussion of improvement opportunities.[6]
6. Supervisors explain rules and consequences of violations. They try to understand violations and make them learning experiences for themselves and the violators. They make sure, to the best of their ability, that disciplinary actions are fully understood.[6]
7. Supervisors listen to what their people have to say and share information and offer ideas (joint problem solving) rather than give advice.
8. Supervisors encourage people to give their best efforts and, where people need each other for goal accomplishment, to work as a team.

Exhibit 2.2.2 ***(Continued)***

9. Supervisors create learning experiences for their people, but also encourage people to be proactive in taking charge of their own careers.[6]

VIII. Organization Values/Processes

A. Values/Assumptions

1. The organization has a real interest in the quality of work life of those who work there.
2. The organization values collaborative efforts of its members (as distinct from individual competition).
3. The organization assumes that most of the energy and ingenuity for solving problems resides in the people of the organization and needs only to be unleashed.

B. Improvement Processes

1. The improvement process of the organization includes examining the *way things are* and the *way things ought to be* (by organization members) in such areas as organization, communication, decision making, relationships, and leadership and identifying improvement areas for problem solving.
2. The improvement process of the organization includes problem solving activities by task groups of organization members to improve organization performance.

C. Change Process

1. Adequate sources of feedback exist to tell the organization about the satisfaction or dissatisfaction of its "customers" (users of the organization's products or services).
2. The organization is flexible; that is, it is capable of making fast changes in response to new or changing customer needs.

D. Decision Making Process[1]

1. Information is widely shared in the organization so that those who make decisions have access to the best available information.
2. Decisions are made at those levels where the best information exists.
3. Work group members have an opportunity to influence decisions related to their work.
4. Organizations (work groups) who share goals plan together and coordinate their efforts.

E. Group Process[7]

1. Real teamwork exists within work groups and/or between organizations who share common goals.
2. People feel free to express their opinions, say how they feel about an issue, ask questions that may display ignorance, or disagree with any position without concern for retaliation, ridicule, or negative consequences.
3. People have genuine concern for each other's job welfare, growth, and personal success; that is, a person need not waste time or energy protecting himself or herself from others in accomplishing the organization's (work group's) goals.
4. Individuals feel appreciated by the other members of the organization (work group) when doing a good job; when things are not going well, people make an effort to help each other.
5. A climate exists such that people need not be guarded or cautious about what is communicated to other members of the organization (work group); people do not "play games" with each other.
6. People view conflict as normal and natural and as an asset because they believe most growth and innovation is derived from conflicts.
7. When disagreements arise, the disagreeing individuals or groups get together, talk through their points of view until each can see some logic in the other's ideas, and then come to an agreement that makes sense to everyone.
8. When problems are being worked on by group members, they understand what the issue is, what they have decided to do about it (after discussion), and what each member's responsibilities are.
9. Group members effectively carry out the group's decisions.
10. Group members respect individual differences; that is, they do not force unnecessary conformity on group members.

2.2.4 THE PURPOSE AND OBJECTIVES OF THE MODEL

The purpose of this model is to initiate, *with a supervisor group*, a planned and long-term effort to improve performance through people. Specifically, the model is helpful in accomplishing the following objectives or action steps:

1. Communicate organizational conditions that have an impact on the performance of people (based on the authors' views and experiences).
2. Gain understanding (not necessarily agreement) of the organizational conditions.
3. Assess a target area, for example, a department or work center, against the organizational conditions.
4. Identify improvement opportunities where supervisors have a readiness or a commitment to change.
5. Solve problems and implement changes (managed by the supervisor group).

The performance-improvement-through-people strategy is designed as an organizationwide effort; however, the initiating target areas selected should be those where there exist (1) reasonable agreement with the concepts and (2) a relative absence of other major change programs.

However, before any actions are taken with the supervision of a target area, steps should be taken to obtain higher-level management's endorsement and support of the desired states shown in the model. The model reflects a value system and beliefs about people and work that may differ from those held by management. These steps provide a rich opportunity to discuss values and beliefs, to clarify statements for understanding, and to modify statement wording to fit the target organization, for example, manufacturing, maintenance, or research.

There is a need to emphasize to the management group that performance improvement through people is a long-term effort and that problems inevitably will arise to create obstacles to long-term management support. For example, there may be a short-term need to lower costs of operations because of an economic situation or for other competitive reasons. How will this need affect management's priorities for continued improvement efforts through people, which may not provide short-term results? Potential problems regarding long-term management support need to be dealt with openly at this time.

When higher-level management endorsement and long-term support are reasonably assured, a second preliminary step is taken to convert the model statements (desired states) into a question format, with a scaling or scoring provision enabling a supervisor group to evaluate or diagnose a target area for improvement opportunities. An example of the question format is shown in Exhibit 2.2.3 (taken from the Performance Feedback dimension of Exhibit 2.2.2).

2.2.5 USING THE MODEL IN THE DIAGNOSTIC FORMAT

The model in the diagnostic format closely resembles a typical organizational survey or questionnaire given to employees for assessment of their work situations. The authors support getting information from employees where their work situations are affected, but this document is *not* designed for that purpose; consequently, it should *not* be given to professional or hourly persons for their assessments.

The document *is* designed for supervisors and/or designated staff persons for their assessments of the target area. The intention of the process is that the supervisor group be viewed by its people as knowledgeable and proactive, not reactive, in making social systems changes, concurrent with technical systems changes.

A recommended process for using the diagnostic model is as follows:

1. Communicate the objectives and process to the first- and second-line supervisor group (normally five to eight people) of the target area. The communications should also reflect higher-level management endorsement and support for the concepts and process.
2. Engage the supervisor group in a "first-cut" assessment of the target area. *Note*: The authors make the following points for "scoring" the document:

   a. Assume *all* the conditions are valued and important, that is, that they show the desired state.
   b. It is "OK" to disagree; in fact, disagreements are sought out in planned group discussions.
   c. The scoring, 1 (needs improvement) to 5 (doing well), should show how much a condition would need to be strengthened or improved to achieve the desired state, assuming the condition is valued and important.
   d. Score the condition even though the change cannot be managed at the first- or second-line supervisor level.
   e. The scoring is not the critical step; the group discussions that follow are. Consequently, leave blanks where one does not have information or is "uncomfortable."

Exhibit 2.2.3 An Example of the Question Format Used to Diagnose a Target Area for Improvement

III. Performance feedback (to the performer)	Needs Improvement (−)				Doing Well (+)
A. Expectations					
1. Do people clearly understand what their supervisors and others in their work groups expect of them (tasks and behaviors)?	1	2	3	4	5
2. Are the people able to fulfill most of the expectations of their supervisors and of others in their work groups?	1	2	3	4	5
3. Do people know what they are being measured against in their jobs?	1	2	3	4	5
B. Feedback Data					
1. Are people able to "track and trend" their own performance on the job?	1	2	3	4	5
2. Do people receive performance information, that is, deviation from goals or standards, before their supervisors receive this information?	1	2	3	4	5

f. The purpose of the group discussions is to locate improvement opportunities where the supervisor group has a readiness and a commitment to change.

3. Summarize and feed back responses to the supervisor group. An illustration of how to do this is shown in Exhibit 2.2.4.

4. Plan and facilitate a series of discussion meetings with the supervisor group to identify potential improvement opportunities.

Exhibit 2.2.4 An Illustration of How to Summarize and Feed Back Supervisors' Responses[a]

	Needs improvement →				Doing well
III. Performance Feedback (to the Performer)	1	2	3	4	5
A. Expectations					
1. Understanding supervisors'/peers' expectations.			/	~~////~~	/
2. Ability to fulfill expectations.				~~////~~/	/
3. Knowledge of performance measures.		/	////	//	
B. Feedback Data					
1. "Track and trend" performance by individual.	//	///	//		
2. Aware of performance information before supervisor.	/	///	//	/	

[a]Specific statements describing "desired state" for subdimensions.

Exhibit 2.2.5 An Example Worksheet to Assist Supervisors in Data Collection on Specific Situations

Job dimension ____________________

Describe a specific situation that needs improvement. (What? Where? When? Extent?)	What is the effect of this situation on the attitudes, feelings, and behaviors of people?

a. Experience indicates that it takes approximately 8 hr of supervisors' discussion to complete a diagnosis of a target area. This is usually accomplished in four 2 hr meetings held 1 week apart.

b. The supervisors are asked to work in pairs in preparing for the discussion meetings by describing *specific situations* in the target area that need improvement (influenced by the summary of responses to the diagnostic questions) and the effect of the current situation on the attitudes, feelings, and behaviors of people. An example worksheet is shown in Exhibit 2.2.5.

c. The outcome of the discussion meetings is a list of 15 to 25 potential improvement opportunities and the effects of the current situation on people as expressed by the supervisor group. These effects generally have a negative impact on organization performance. One example showing supervisors' comments in the Performance Feedback dimensions is given in Exhibit 2.2.6.

5. Plan and facilitate a meeting with the supervisor group to select from the set of potential improvement opportunities two to five items for change in the current year. Some suggested criteria for selection are the supervisory group having:

a. Knowledge, skills (or access to knowledge and skills), and authority for problem solving and implementing change.

b. A personal willingness to work on the selected items in problem solving task groups.

c. A belief that working to improve these items would make a difference in the performance of people.

6. Designate a supervisor to *manage* the problem solving and implementation process. This step does not require the supervisor to be the problem solver, but he or she is responsible for getting the problem solved and the change implemented using the best available resources.

a. Using the best available resources often means including the workers in the problem solving process. This should always occur if the workers' jobs are likely to be affected by the solution.

b. The industrial engineer assists the responsible supervisor in two areas: (1) provides a plan for problem solving and guides or facilitates the group problem solving meetings and (2) adds to content alternatives where he or she has expertise.

Exhibit 2.2.6 An Example Showing Supervisors' Comments in the Performance Feedback Dimension

Job dimension Performance feedback (to the performer)

Describe a specific situation that needs improvement. (What? Where? When? Extent?)	What is the effect of this situation on the attitudes, feelings, and behaviors of people?
Feedback Data	
1. There is a lack of feedback data from engineers to technicians as to job results, particularly when an extra effort is put in to save an experiment.	1. Technicians lose interest in the experiment, not knowing if efforts are worthwhile. Leads to not caring.
2. Information about errors made in customers' work are passed on to supervisors, not the technicians.	2. Technicians have feelings of inferiority, and lose respect for engineers.
3. No consistent department policy exists for frequent, periodic feedback to technicians.	3. No opportunities exist for technicians to compare performance to standards.

7. Regularly repeat the diagnostic and problem solving process for continuous improvement. Based on the authors' experiences, the work never ends.

2.2.6 THE ROLE OF THE INDUSTRIAL ENGINEER

On several occasions the authors have used the words "plan and facilitate" in reference to, for example, a discussion meeting. This refers to having and utilizing skills to help groups become effective in working together on tasks, decision making, or problem solving. The industrial engineer's role in these situations is that of a "process consultant," not an "expert consultant." It is strongly recommended that the industrial engineer who uses this model be one who is broadly familiar with the underlying behavioral science concepts and principles, whose values and beliefs are congruent with these concepts and principles, and who has process consulting skills for helping groups to become effective.

2.2.7 PLANNING INTERVENTION

The authors wish that the state of the art were at a stage such that it were simply a matter of following a diagnosis with an appropriate intervention, much as a physician would prescribe following a medical diagnosis. This is not so, and the best we can do here is to provide a suggested array of interventions that situationally may be appropriate for the selected item(s) for problem solving and implementation. The suggested interventions are:

1. Supervisory training.
2. Team building/development.
3. Role analysis technique.
4. Action-research model.
5. Group problem solving approaches, for example, Kepner Trego Models.
6. Structured participative processes, for example QC circles, work simplification.
7. Job enrichment/design.
8. Goal setting processes, for example, management by objectives.
9. Alternative work schedules, for example, flexitime.
10. Open systems analysis.
11. Consequence and feedback analysis.
12. Information systems analysis.

2.2.8 SUMMARY AND CONCLUSION

We have attempted here to present a nonfinancial motivational strategy that stresses the following:

Focusing on the "whole" organizational environment as the basis of motivation.

Emphasizing the creation of an environment made up of a set of desired (and defined) conditions under which people are most likely to work toward achieving organizational and personal goals.

Suggesting a diagnostic and educational process that highlights those aspects of the environment that most need remedial action.

Suggesting some intervention alternatives to improving those environmental aspects.

The authors' experience has indicated certain strengths and obstacles to using this "environmental" approach to motivation. The *strengths of the approach* are as follows:

1. Is comprehensive, that is, includes much of what has been learned and experienced in applying behavioral science concepts.
2. Describes the organizational conditions in terms of desired state.
3. Requires supervisory assessment of organizations before taking action.
4. Communicates a common "model" to all organizations and to all levels of supervision.
5. Provides an opportunity to examine supervisors' values and assumptions about people and work.
6. Allows actions to be tailored to the needs of the "target" organization based on supervisors' readiness and commitment.
7. Keeps action choices visible, that is, shows what *is* and *is not* being worked on.
8. Facilitates a systematic and long-range plan for performance improvement through people.

The *obstacles to the approach* are as follows:

1. Requires a long-term effort; it is hard to play a long-term game.
2. Requires a high level of first-line supervisory involvement and commitment.

3. Requires continuing support from higher levels of management. This is hard to get because other improvment activities often take priority.
4. Require attention to many variables, *not* a few, this can be "mind-blowing."
5. Requires patience and care by many people. It is difficult to evaluate and (often) does not provide short-term results.
6. Requires some risk and (perhaps) few rewards to initiators.

REFERENCES

1. *Survey of Organizations*, Institute of Social Research, University of Michigan, Ann Arbor, 1974.
2. R. HACKMAN and G. OLDHAM, "A New Strategy for Job Enrichment," *California Management Review*, Summer 1975, pp. 57–71.
3. I. RUBIN, M. PLOVNIK, and R. FRY, *Task Oriented Team Development*, Situation Management Systems, Inc., Boston, MA, 1973, 1975, 1977.
4. W. L. FRENCH and C. H. BELL, *Organization Development: Behavioral Science Interventions for Organization Improvement*, Prentice-Hall, Englewood Cliffs, NJ, 1973, pp. 65–66.
5. J. V. CLARK and C. G. KRONE, *Open Systems Redesign* (unpublished).
6. F. S. MYERS, *Managing Without Unions*, Addison-Wesley, Reading, MA, 1976, p. 111.
7. J. P. JONES, *The Ties That Bind*, National Association of Manufacturers, New York, 1967.

BIBLIOGRAPHY

BECKHARD, R., *Organization Development Strategies and Models*, Addison-Wesley, Reading, MA, 1969.

CLARK, J. V., and C. G. KRONE, "Towards an Overall View of O.D. in the Early Seventies," in J. Thomas and M. Beenis (eds.), *Management of Change and Conflict*, Penguin, New York, 1973.

COOPER, R., *Job Motivation and Job Design*, Institute of Personnel Management, London, England, 1974.

FORD, R. N., *Motivation Through Work Itself*, AMA, New York, 1969.

GELLERMAN, S., *Motivation and Productivity*, AMA, New York, 1963.

GLASER, E. M., *Productivity Gains Through Worklife Improvement*, Harcourt Brace Jovanovich, New York, 1976.

HERTZBERG, F., "The Motivation To Work–One More Time, How Do You Motivate Employees?," *Harvard Business Review*, January–February 1968, pp. 53–62.

HUGHES, C. L., *Goal Setting*, AMA, New York, 1965.

MASLOW, A. H., *Motivation and Personality*, Harper & Row, New York, 1970.

MCADAM, J., "Behavior Modeling, A Human Resource Management Technique," *Management Review*, October 1975.

MCGREGOR, D., *The Human Side of Enterprise*, McGraw-Hill, New York, 1960.

MCGREGOR, D., *The Professional Manager*, McGraw-Hill, New York, 1967.

MILLER, L. M., *Behavior Management*, Wiley, New York, 1978.

MYERS, F. S., "Every Employee a Manager," *California Management Review*, Spring 1968, pp. 9–20.

MYERS, F. S., "Who Are Your Motivated Workers?," *Harvard Business Review*, 1964, pp. 73–88.

ODIORNE, G., *MBO II*, Faran Pitman, Belmont, CA, 1979.

RUMMLER, G. A., "Troubleshooting Performance Problems," *Bureaucrat*, July 1974, pp. 182–195.

SCHEIN, E., *Organizational Psychology*, Prentice-Hall, Englewood Cliffs, NJ, 1970.

SKINNER, B. F., *Contingencies of Reinforcement*, Appleton-Century-Crofts, New York, 1969.

VANDURA, A., *Principles of Behavior Modification*, Holt, Rinehart & Winston, New York, 1969.

WALTON, R., "Quality of Working Life, What Is It?," *Sloan Management Review*, Fall 1973, pp. 11-21.

CHAPTER 2.3
Financial Motivation

MITCHELL FEIN
Mitchell Fein, Inc.

2.3.1 INTRODUCTION

Managers are incentive-minded and apparently believe that pay by performance motivates employees. Yet the use of incentives is restricted. Managers have different opinions on how to improve employee effectiveness.

> A study of more than 400 plants in the United States found that, when these plants instituted work measurement, productivity rose an average of 14.6%. When plants instituted wage incentives where there previously was work measurement, productivity rose an added 42.9%. The average increase from no measurement to incentives was 63.8%.[1] Yet many companies do not even measure worker effectiveness.
>
> Though financial incentives increase productivity, only 26% of factory employees work on incentives; 74% are paid by the hour or on salary; practically no white collar employees are on incentive.[2–5]
>
> About 75% of outside salespeople are on some form of incentive.[6]
>
> Approximately 82% of manufacturing companies have executive bonus plans; the median bonus for the three top manufacturing executives averages 45% of their base pay.[7]
>
> In 1978 the two top executives at General Motors (GM) and Ford were paid annual salaries of about $350,000 and additional bonuses and incentives of about $650,000, equal to 185% of their base pay; yet years ago GM eliminated incentives for factory employees.
>
> The marketing employees of International Business Machines (IBM) who sell products and services work on quota bonus plans; all other IBM employees are on straight salary.
>
> A study of executive compensation of 1100 companies listed on the New York Stock Exchange found that companies that had formal incentive plans for their executives earned on the average 43.6% more pretax profit than did the nonincentive companies.[8]

The use pattern shows that management favors incentives for salespeople and for top management.

Financial motivation is usually thought of in mechanical terms, like piecework: produce more pieces and earn more. Sales incentives are like piecework: sell more units and your earnings increase. The concept and mechanics of establishing these forms of incentives are simple.

Financial motivation is much more than just different forms of piecework. It is an extension of a company's compensation policies, which are determined by its managing policies. A comprehensive discussion of financial motivation must start with the managing policies of companies to ensure that incentive practices are in agreement with and supported by these policies.

For a full understanding of the role of financial incentives in managing a company and of how these motivate employees to higher productivity, the main factors relevant to motivation acting on and affecting the employees must be examined so that each is seen in relation to the others and to the whole.

Raising Employee Productivity

The main ways to raise productivity are through management's efforts and through employees' efforts. Management's efforts can be categorized as either fully reliant on its staff of engineers, technicians, and other specialists, without employee cooperation, or fully reliant on this staff, but also encouraging employees to become involved in raising productivity. Employees can be motivated to higher productivity by:

Setting time standards to achieve a fair day's work for a fair day's pay and enforcing the standards.

Traditional wage incentives for individual employees and small groups.

Plantwide productivity sharing plans.

Work improvement efforts such as quality-of-work-life projects, quality control circles, labor management committees, and job enrichment.

All activities and practices, especially those involving employees, are affected by or based on managing policies.

2.3.2 MANAGING PRACTICES

Programs to motivate employees to raised productivity must be consistent with the managing policies of the company. Management's beliefs and policies must form the basis for motivation programs. The extent to which management either unilaterally controls the company or encourages the employees to become involved in the operations depends on the approach management believes will create more effective operations.

Companies in this country are managed on the cardinal principle that management makes all decisions and bears responsibility for results; employees not in management are not involved in or responsible for decisions or results. Full control of the operations, commonly referred to as "management's rights," is vested solely in management. Under this principle a range of alternatives is available, from the unilateral action and responsibility of management to the involvement of hourly and nonexempt salaried employees in improving productivity and reducing costs. Regardless of the degree of employee involvement, control of the operations is not shared. Whether decisions and responsibility for results are narrowly centered at the top or are extended to broader and lower levels of management does not alter the basic tenet that management controls all operations and decisions.

Worker participation in Europe is different from the practice in this country. Codetermination in Germany is the legal right given to workers to exercise a voice in making decisions on how the company is operated. The legal status of worker participation differs among European countries. American trade unions are opposed to worker participation in company operating and policy decisions and prefer to maintain their traditional adversary role.

Clarifying Meanings

Vaguely defined terms used in the management and behavioral science literature in this country create misconceptions concerning employee involvement efforts. The term "participation" is applied to a wide range of activities in which employees voluntarily become involved with management to solve problems, and "participative management" is used to describe the attitudes of managers who favor the involvement process.

Participation does not adequately describe the true relationship between nonmanagement employees and management in these voluntary efforts. Participation implies that those involved have a share of responsibility for results—which they do not have. Involvement without accountability is not participation, but consultation. That is what really occurs when employees are encouraged to suggest improvements; management may accept or reject employees' ideas, because only management has authority to make decisions.

"Involved in decisions" is another term used vaguely in the literature, and this, too, is misleading. Employees can make suggestions and are consulted; they may influence decisions, but they are not involved in decision making in the sense in which the term is used in managerial practice. Using these terms loosely can harm cooperation efforts. The term "participative managing" should be replaced by the more appropriate and descriptive "consultative managing."

Traditional and Consultative Managing Practices

From the wide range of views presented in the management and behavioral science literature describing the differences between traditional and consultative managing practices, it appears there are numerous and sharp differences between them. However, these can be reduced to one basic difference: whether, and the extent to which, management encourages employees to become involved and to cooperate with management in improving operations.

Traditional and consultative are the two main approaches to managing in this country. Few companies use consultative practices, but the trend is increasing. In an overview of managerial practices covering relations between management and employees, and ways of involving employees in cooperative activities with management, Strauss presents a range of experiences, opinions, and practices that the reader will find useful.[9]

Under traditional managing, management has full authority and responsibility for the operations.

Managements's rights are not diluted.

Management unilaterally makes all operating and policy decisions.

Management has sole responsibility for operations and for productivity improvement.

Employees' views are not sought, nor are employees involved in decisions.

Employees do not share directly in productivity gains.

Conventional financial incentives may be offered to employees on production operations.

With a union, management's actions and responsibilities to the employees are constrained by a labor contract. Without a union, management must balance the company's needs with the employees' needs to satisfy the employees.

Consultative management practices include all of these characteristics, with changes to provide nonmanagement employees with opportunities to become involved in developing and implementing improvements; employees may also share in productivity gains.

Consultative managing is not traditional managing with diluted management rights; it is based on the belief that management's goals will be most effectively met when employees are voluntarily involved in helping to raise productivity.

Design of Incentive Programs

The design of an incentive program depends on the organization's compensation and management policies. Traditional and consultative managing can be employed with or without incentives or sharing. The method of wage payment used does not affect managing policies; rather, incentive practices depend on policies.

Whether or not to share productivity gains with employees has staunch proponents on both sides. Traditional managers often take the position that a responsible company provides good wages and working conditions, that productivity gains are shared in annual wage and working conditions improvements, and that an incentive program therefore is not needed.

Consultative managers, in addition to providing good wages and working conditions, may be willing to share with employees the productivity gains created by them. Productivity sharing, as with Scanlon, Rucker®,* or Improshare®† plans, is generally a plant or company plan covering all employees. Such plans encourage employee involvement in productivity improvement and are used only in consultative managing.

2.3.3 TRADITIONAL MANAGING PRACTICES

To improve managing effectiveness, managers should understand the underlying precepts, shortcomings, and strengths of the practices they use and of those that are available. This discussion highlights traditional managing shortcomings and problems.

Trends in Traditional Managing

Managing arts and lore are moving toward becoming a management science. The trend in managing is to closer and tighter control over operations. The paradox of this trend is that, just as sociologists and behavioral scientists are counseling that employees should have more involvement and freedom of expression in their jobs, management science is enabling management to exercise greater control over the workplace and workers.

The move toward more sophisticated and tighter controls is supported by managers who see no viable alternative approach to productivity improvement other than the close controls now utilized, which are designed to operate in the adversary relations environment of the workplace, to compel higher productivity from workers who are not motivated to raise their output and help achieve management's goals.

Traditional managing theory places full responsibility for productivity improvement on management. Workers have no share or voice in raising productivity, nor are their opinions sought. Overall productivity is raised by concentrating primarily on the low areas and by bringing these into line.

Problems With Traditional Approaches

A study of 300 research studies on the relationship between job satisfaction and motivation by a National Science Foundation supported New York University (NSF/NYU) team pinpointed major shortcomings of approaches to raise worker productivity and found that increased productivity depends on two propositions:

*As used here, Rucker® is the registered service mark of the Eddy-Rucker-Nickels Company.
†As used here, Improshare® is the registered service mark of Mitchell Fein.

The key to having workers who are both satisfied and productive is motivation, *that is, arousing and maintaining the will to work effectively–having workers who are productive not because they are coerced but because they are committed.*

Of all the factors which help to create highly motivated/highly satisfied workers, the principal one appears to be that effective performance be recognized and rewarded–in whatever terms are meaningful to the individual, be it financial or psychological or both. [10]

This reads well, makes complete sense, and is irrefutable. Yet across the country most managers do just the opposite. Managing policies are based on coercion in most plants; workers are seldom rewarded financially for more effective performance. The realities in the workplace are diametrically opposite to what is needed to raise job satisfaction and motivation. Though unintentionally, it works out that most workers are generally penalized by loss of overtime and layoffs for doing a better job, so they oppose management's objectives. Management senses the antagonism, and its managing and control systems are designed to operate in a hostile environment, to apply pressure and coercion to workers in order to get them to do more.

Workers readily see that, if they assist in raising productivity, some of them will be penalized; if they improve productivity, reduce delays and waiting time, and reduce crew sizes, some will be displaced, and the plant will require fewer employees. They will receive no financial gains for their efforts, nor are they persuaded that increased company profits will benefit them in the future. What employee will assist in raising productivity, only to be penalized for his or her diligence?

The "exempt" employees–executives, administrators, professionals, and salespeople–are treated differently. A manager does not work himself or herself out of a job by superior performance, nor is a salesperson's job security threatened because he or she sells too much. An engineer does not cause the layoff of other engineers by being too creative. These employees expect rewards for their creativity and effectiveness.

When workers excel and raise productivity, the company benefits and management is pleased, but the workers do not benefit. On the contrary, in the short term their economic interests are threatened, and some suffer loss of income. When exempt employees are more effective, they are covered with glory; their economic security is enhanced, not threatened. Ironically, the relationship between management actually provides workers with the incentive not to cooperate in productivity improvement. Without realizing it, all that most companies offer their employees for greater dedication and for raising productivity is the opportunity to reduce their earnings and job security. No wonder workers oppose productivity improvement. The system operates perfectly to demotivate workers. A more effective system could not have been designed to cause workers to oppose management's goals.

Effectiveness of Traditional Managing

Despite the shortcomings and problems of traditional managing, companies continue to improve their operations. The increasing difficulties in operating companies are caused by the changing work force and the increasing complexity of production technologies and equipment. Greater personal freedom in society generates pressures for more democracy in the workplace. As work becomes more complex, more cooperation is needed from workers.

Increased support from employees for management's goals can be gained by establishing labor relations practices that employees see meet their needs. The NSF/NYU study provided data obtained from a nationwide study of manager and employee attitudes toward various factors that affect job satisfaction and motivation. On a question involving the importance placed by employees on 20 workplace factors, employees rated the following as their top priorities, in descending order: better communications from management, giving employees greater job security, higher pay, better provisions to protect employees from arbitrary and unfair treatment, better treatment by supervisors, improved working conditions, and more opportunities for advancement.[10]

The tremendous advances made in this country since early in this century were all accomplished using traditional managing practices and largely under rigid and tough management's rights positions. Without watering down essential management rights to manage, old-time, abrasive, and arrogant managing practices are being replaced with commonsense practices more in tune with the times.

"Humanized" labor relations policies are not to be confused with consultative managing practices. Creating effective two-way communication channels with employees, listening to their complaints and needs, training supervisors to be effective leaders and employees to upgrade their skills, and otner such items will ımprove job satisfaction, reduce job tensions, and make work more satisfying. Adversary relations will diminish. Though the effect of such changes on productivity is not clearly established in studies, common sense says productivity should improve.

2.3.4 CONSULTATIVE MANAGING PRACTICES

Consultative managing encourages employee involvement in day-to-day operations, with employees questioning how work is performed and suggesting changes. This significant change from traditional managing affects many other business policies. For consultative managing to be successful, it must be accepted by top management and believed in and followed down the line to supervisors.

Consultative managing does not require management to give up any of its rights to manage; management still has the last word in all decisions. Managers give up only the right not to listen and take on an obligation to consider employees' proposals. Most important, managers' actions and attitudes must show that they mean what they say.

Better Communications From Management

The need for improved communication between employees and management was highlighted in the NSF/NYU study which found, in a nationwide study of decision maker beliefs, including those of management and union leaders, that "better communications from management" ranked on a level with higher pay and job security as very useful in improving employee attitudes and motivation.[10]

Better communications from management should not be confused with offering employees a consultative voice in productivity improvement. The purpose of each is quite different, as are the methods for establishing the process. Better communications are useful and effective in both traditional and consultative managing.

Improvement in the Quality of Work Life

Improving the quality of work life is increasingly used to describe all sorts of desirable outcomes. A successful quality-of-work-life program requires that management encourage these programs' activities.

Articles on such programs in management and behavioral science journals report improved employee relations; some discuss improved operations and productivity, but the data are not clear. None that the author has read involved sharing productivity gains with the employees.

Quality Control Circles

Quality control circles were refined and popularized in Japan to improve product quality, with raising productivity a high-order goal. Worker involvement in resolving quality problems and dedication to solving production difficulties are continually stressed. A commonly accepted definition of a quality control circle is a small group of employees who perform similar or related work and who meet voluntarily to solve workplace-related problems. The total involvement of employees is usually small.

Though similar in some respects to quality-of-work-life programs, quality control circles are highly oriented to production and operations, with sharp attention to bottom-line results. Working together in small groups, these employees usually concentrate on operations with which they are familiar. Quality control circles use practices developed many years ago by industrial engineers, such as motion study, work simplification, and other refined, analytical methods improvement techniques.

Quality control circles are usually highly structured within a company framework, with committees, leaders, facilitators, and specialists. As with quality-of-work-life programs, companies operating quality control circles do not offer to share savings with employees.

Labor-Management Committees

Labor-management committees are groups of employees and managers established to accomplish a specific task. These committees have been used for years, but relatively few operate on a formal and continuing basis; they are often established to deal with a crisis.

Labor-management committees are excellent for communicating between employees and management. Dealing with committees is a way of life in union plants; committees are integral to union-company relations. Such committees are incongruous in nonunion companies, which constitute about 80% of companies in this country. Management in these companies is usually reluctant to establish such a committee, if only because the committee can readily become the form for representation of the employees.

Efforts by management to communicate directly with all employees, even in small groups, are effective when limited to several a year, but when the meeting frequency is increased, employee interest drops off sharply. Small committees of active and concerned members will sustain high-level interest, which can readily be communicated to a larger group.

Management's Rights

Literature describing worker involvement and cooperation with management in improvement projects emphasizes that it is vital for workers to have a say in how things are done; the language often used is that "they should be involved in decision making." The proponents of democratizing the workplace miss the distinction that workers can only make suggestions; they are not involved in decision making in the full sense of the term.

The preservation of management's rights is extremely important to managers of all organizations in the private and public sectors. Diluting some rights or sharing decision making is strongly resisted by management. Loose use of language undoubtedly has killed off many projects where managers were led to believe workers would become partners in decisions.

2.3.5 MOTIVATION AND COMPENSATION PRACTICES

The general notion in industry regarding motivation is that managers must motivate their employees. Actually, it works the other way around: Motivation results from the urge within a person to do something, to accomplish a task, to reach a goal. At most, managers can provide a climate that fosters motivation in employees. A critical element is the perception that employees have of the working environment and their readiness to view management's goals as in their best interests.

Motivation and compensation are closely related in employees' minds. Compensation methods are based on managing policies. The two methods of compensation are payment by time (no incentives) and payment by performance (incentives).

In payment by time, where there is a union, employees' wages are established periodically by agreement. In the absence of a union, wage scales are set by management, in keeping with industry and community practices. Productivity is affected by management efforts and is not reflected in wages. When employee productivity is measured, payment by time is called "measured day work" (MDW).

When employees are paid by performance, they receive higher pay when their performance improves. The work measurement practices for MDW and incentives are similar; the main difference is the method of wage payment.

The main approaches to encouraging workers to improve their performance and the overall productivity of the organization are:

Fair day's work.
Financial incentives.
Job enrichment/satisfaction.

The fair day's work concept is used mainly for management's benefit, to attain desired productivity levels. It is sometimes used by employees to justify wage increases.

Financial incentives date back thousands of years as pay by the piece. This concept is easily understood and readily communicated. Many companies use incentives; some have difficulties, which are discussed later.

Job enrichment came into prominence about 20 years ago. Many claims were made for it, but what came through loudest was that it would improve job satisfaction, would solve many workplace and labor relations problems, and might even substitute for pay increases.

Fair Day's Work

The concept of fair day's work has two aspects: To the employee it means receiving a fair day's pay for a fair day's work. To management it means the employee should produce a fair day's work for a fair day's pay. A fair day's pay in a union plant is defined in labor relations terms as the pay and working conditions bargained by the employees and management. A fair day's pay in a nonunion plant is generally interpreted in the same way.

The concept is used by management to establish benchmark productivity levels to be maintained by employees, which management is entitled to for the employees' pay level. Work measurement time standards are then set so that normal equals 100%, or acceptable productivity level (APL), to correspond to the fair day's work concept.[11]

These definitions and concepts are established for management's benefit. Management bears the burden and responsibility to operate a measured day work system to monitor individual employee productivity and to maintain productivity at desired levels.

Wage Incentives

Managers wishing to improve productivity with money as a motivator will find support in the findings of behavioral science and industrial engineering studies, which agree that money can be a prime way to increase employee motivation. The NSF/NYU study team developed six critical ingredients of effective systems to raise job satisfaction and worker motivation, headed by the following: "financial compensation of workers must be linked to their performance and to productivity gains."[12] The team found that, when workers' pay is linked to their performance, the motivation to work is raised, productivity is higher, and they are likely to be more satisfied with their work.

By any measure, pay tied to productivity is the most powerful motivator of improved work performance. Yet only 26% of U.S. workers work under financial incentives. In some industries such as basic steel and sewn products, incentives cover more than 80% of the work force; in many industries no incentives are employed. Few nonmanufacturing operations are on incentive plans.[2–5]

The work itself serves as a motivator and should not be overlooked. Though money is a powerful motivator, many people work for more reasons than money alone.

Job Satisfaction/Enrichment Concepts

The dominant theme in the behavioral science literature of the past 20 years was that work had been so simplified that workers' desires to work were turned off. The remedy proposed was that skills and judgments must be added to the denuded jobs. "Make work more interesting and demanding, remove the monotony and drudgery, and the will to work will be raised" became the theme for numerous studies and research papers. Job enrichment became a slogan. Management and behavioral science journals reported glowingly about a number of companies where jobs were enriched, citing improvements in productivity, work quality, absenteeism, and labor relations.

Contrary Job Enrichment Concepts

Studies were published during this period with contrary data and positions, but these did not receive the prominence given the job enrichment proposals. The NSF/NYU study on the relationship between job satisfaction and motivation found that:

> *If there is any one fact that stands out clearly from the massive accumulation of data–the hundreds of studies encompassed in this report–it is that worker job satisfaction and productivity do not necessarily follow parallel paths. This does not mean that the two objectives are incompatible, for there is evidence that it may be possible to achieve them together. Nor does it mean that the two goals are totally independent of one another. Under certain conditions, improving productivity* will *enhance worker satisfaction and improvements in job satisfaction* will *contribute to productivity. What it does mean is that there is no automatic and invariant relationship between the two. Indeed, the two objectives are so loosely coupled, there are so many intervening links between them, and the relationship is so indirect, that efforts which aim primarily at improving worker satisfaction on the assumption that productivity will thereby automatically increase are more likely than not to leave productivity unchanged, or at best to improve it marginally, and may even cause it to decline.*[13]

Four combinations of satisfaction/motivation are possible: a satisfied worker may be highly motivated or weakly motivated; a dissatisfied worker may be highly motivated or weakly motivated. The data do not show a negative or a positive relationship, but rather, no relationship. The NSF/NYU study found that, though common sense and logic suggest that improved job satisfaction leads to raised motivation, workers do not necessarily react that way. A satisfied worker may just be a satisfied worker, not any more motivated to produce than before.

A study by Fein of job enrichment and job change studies and papers found that many were flawed and contained unsubstantiated statements.[14]

Though job enrichment supporters proposed that workers would prefer enriched jobs to less demanding ones, experience did not bear this out. Simonds and Orife reported in a study that:

> *No statistically significant preference for less routine (more enlarged) jobs was shown either with or without increased pay. Required physical effort proved not to be a significant factor. The assumption that job enrichment and/or enlargement is sought and desired by employees has become widespread in recent years. This is despite the fact that empirical studies of what workers want in their jobs have been limited, have often been based on questionable methodology, and have at times given conflicting results.*[15]

The 1970 *Survey of Working Conditions* prepared by the Survey Research Center (SRC) of the University of Michigan[16] is widely cited as the authority that interesting work is preferred by workers over pay. An analysis by Fein of the SRC data found that the SRC researchers had averaged all the data and called the average a "composite worker." When the data were analyzed by occupation, good pay and job security moved up, and interesting work dropped, for factory workers; professionals and managers may place a higher priority on their work.[17]

A study by Edwin A. Locke et al. found that money is a more powerful motivator than is generally believed:

> *Our findings may surprise or even shock many social scientists. For the last several decades ideological bias has led many of them to deny the efficacy of money as a motivator and to*

emphasize the potency of participation. The results of research to date indicate that the opposite viewpoint would have been more accurate.[18]

Failure of Job Enrichment

The proof of a social theory ultimately rests on how it is accepted by those who are affected by it. The concepts of the Maslow, McGregor, Herzberg school did not hold up when measured against whether the many experiments applying the concepts are still in effect. The job change experiments were quietly discontinued; very few articles were published describing the final results and reasons for termination.

Workers have not supported behavioral scientists' views on satisfaction/dissatisfaction and motivation. The simple proof is that no union in this country has ever demanded that management restructure jobs or the way the work is performed or has ever raised other issues included under the broad umbrella of job enrichment.

The realities of the workplace were minimized by the behavioral scientists who proposed that enriched work and greater job autonomy would encourage the increased interest of workers in their jobs and in managements' goals. The poor results obtained by the job change experiments provide ample proof that more than job design changes are needed to raise workers' motivation to work.

The failure of job enrichment to attract entire plants does not mean that more interesting work and involvement in work improvement is not important to many workers. On the contrary, as is discussed later, under changed conditions in which workers have a stake in productivity improvement, and where management encourages worker involvement, larger numbers of workers prefer job changes to their present jobs. The important difference is that workers must be rewarded for doing a better job; they must receive a piece of the action. That was not done in the job enrichment programs.

2.3.6 A PROGRAM FOR COMPANYWIDE IMPROVEMENT

A program to achieve companywide productivity improvement with employee involvement can be developed based on experiences and data available from numerous companies.

Creating Congruent Goals

An examination of who gains and who loses as productivity improves forces the conclusion that, for workers to support productivity improvement, a program must reward workers for improved performance and eliminate practices that penalize them as productivity rises.

Productivity improvement is a goal of management, not of workers. Management strives for raised output with reduced labor input. Workers have no such goal; they mainly seek increased pay and improved job security.

Workers and management strive in opposite directions. The traditional relationship between workers and management is a zero sum game; it is win or lose. When workers gain wage increases, these eat into management's profits. Holding wage increases down benefits the company and the stockholders.

Workers and unions understand that in the long run increased productivity also benefits workers. Reduced costs ensure new orders, jobs, and the company's ability to pay increased wages and benefits. But workers' attitudes at the workplace toward productivity improvement are usually hostile to management's goals.

Management's productivity goal can become a worker goal when workers share in productivity improvement. Sharing productivity gains is simple in concept and practice. A labor cost base per unit to make a product is established, and when costs are reduced, the gains are shared. To involve all workers, consider the entire plant as one group, comparing only the value of the labor input to the value of the output. This approach to sharing creates significant changes in workers' vested interests. Under the traditional relationship of workers to production, they are motivated to enhance their narrow interests; they want more of what they have and are little concerned with how production fares around them.

With plantwide sharing, however, their interest focuses on how many units were completed in how many total input hours. They are rewarded for reducing overall costs. Before sharing, when there were losses through work delays, equipment breakdowns, or spoilage, only management suffered the loss. Under the sharing plan, workers share gains and losses and are motivated to minimize production impediments. Most significant, their concerns shift to the whole group, to all areas that affect input and output. In this sense their interests parallel management's. Both gain and lose together.[19]

The Case for Sharing Productivity Gains

Quality-of-work-life programs, quality control circles, and various behavioral science supported worker involvement programs do not provide financial rewards to workers when they improve the operations. Their only recompense is a feeling of satisfaction from having done a good job and helped the company.

When the company gains financially through workers' efforts beyond what their jobs require of them, the workers should be entitled to a fair share. Management violates universally accepted principles for establishing pay rates by encouraging employees to work at higher skill levels without commensurate extra compensation. All pay scales in the private and public sectors are set to reflect the skill and effort needs of the job. Many companies determine job rates through formal job evaluation plans. The procedure always followed is to prepare a job description delineating the principal requirements of the job and then to evaluate the job against a point evaluation plan. Production jobs do not require employees to use judgment and initiative to improve work performance and to reduce costs; these tasks are the responsibility of technicians and engineers, who are properly paid to do such work. When additional higher-level tasks are added to a job, it is reevaluated, and a new job rate is set.

When employees are encouraged to use higher-level skills to innovate work improvements, it makes sense that they should share in extra compensation for their efforts. If workers feel good because they did a good job, they will feel better with extra pay in their pockets. The extra pay is an excellent way for management to thank employees who did more than was expected of them.

If improvements reduce operating costs and workers do not receive a share of the gains, management's credibility and good faith are questioned. It is one thing for everyone to cooperate and work hard to prevent a plant from closing; everyone gains as the plant and jobs are saved. It is quite another thing for a company not in dire straits to pocket the gains from cooperative efforts.

Differences Between Day Work and Productivity Sharing

When working under day work, whether hourly or salaried, employees' pay does not change. Under productivity sharing, all employees have an opportunity to share in productivity gains.

Considerable data on rewards and pay by performance show that people improve their performance when they can share in the gains. Numerous studies show that employees' work attitudes and satisfaction also improve. Reports from companies that have established productivity sharing plans, such as Scanlon, Rucker, and Improshare show that labor relations always improve; complaints and grievances decrease.

Those interested in data and reports on employee attitudes to productivity and goals should read Edward E. Lawler's material; he is the most prolific of the researchers.[20]

Differences Between Traditional Incentives and Productivity Sharing

Under productivity sharing, gains over the productivity level of a base period are shared between employees and the company. Productivity sharing is designed to create conditions under which workers and management benefit by moving on parallel paths to a common goal: more units of product made in fewer man-hours of work; only finished good units are counted.

The rationale for traditional wage incentive plans stems directly from traditional managing practices; it is rooted in the adversary relations environment of the workplace. The setting of time standards under traditional managing, whether used for measured day work or for incentives, is probably the greatest single cause of worker grievances. Everything about traditional incentives is involved in the we-they relations on the plant floor. Tremendous sums of money depend on how time standards are established.

Productivity sharing is not an incentive plan; it is a philosophy of managing that encourages employees to become involved in productivity improvement. Productivity sharing creates a work environment in which employees see improved productivity as beneficial to them. Under the philosophy of productivity sharing, worker productivity goals and management goals become congruent. Measurement standards are established as the average of a past base period for an entire product; using the past base eliminates the disputes on the plant floor to loosen the standards.

A major difference between traditional incentives and productivity sharing is that under incentives individual workers are motivated only to make more money; they have no interest in raising productivity or in management's productivity goals. Under productivity sharing, workers may still be primarily motivated to make more money, but since productivity sharing is based on achieving overall improvement in productivity goals, more money can be earned only by raising productivity.

Traditional incentives and work measurement practices require that changes in work conditions be reflected in changed time standards. Workers readily circumvent management's rules as they earn more for less work. It is very difficult to prevent traditional time standards from deteriorating over a long period. With productivity sharing, workers are encouraged to use their ingenuity and to change methods, exactly what management attempts to prevent under incentives, and they share in the gains, measured from standards established in a base period.

In most plants using traditional incentives, only the direct labor workers are covered. Those not on an incentive plan are alienated because they do not have an opportunity to increase their earnings. Traditional incentives are usually designed for individual employees or small groups; this encourages workers to build protective walls around their operations to maximize their earnings. They are very creative in innovating ways of "making money with a pencil," which are practically impossible to eliminate; these substantially increase costs. Incentives fractionate the work force and create conditions that accentuate workers' narrow interests. Since employees are measured only on their output, they have little interest in overall quality, spoilage, and plantwide problems.

The sharing plan creates an opposite spirit. Employee interests spread to the plant when the entire group is rewarded for its gains. Counting output only as finished units packed in cartons ready for shipment and rewarding all workers for productivity gains over a base period focus employees' attention on the need to reduce labor input and increase product output.

As workers' interests shift toward management's, the rationale for traditional managing starts to fade. When workers become concerned with final outcomes, they will also be more interested in how operations proceed throughout the plant and in the many details and production impediments that occur around them, which, in the past, they ignored or even encouraged. Since productivity gains are shared whether innovated by workers or management, conceivably industrial engineers, who are disparaged by workers, will instead be welcomed, because workers will gain from engineers' improvement efforts. That would indeed signal a significant change in the attitude of workers and managers toward each other's interests and needs.

In a casual reading of the preceding comparisons, it may seem that everything is wrong with traditional incentives and right with productivity sharing. That is true only where traditional incentives and time standards have deteriorated badly with long-established past practices. When traditional incentives and time standards are maintained soundly, the incentives will yield excellent results for both employees and management, and productivity may exceed the overall productivity attained under a productivity sharing plan. Maintaining a sound incentive plan and time standards is difficult, but it can be done, as attested to by many thousands of companies that operate with such incentives.

Productivity Sharing as a Way of Life

Companies that rely on the incentive aspects of productivity sharing to motivate employees to actively help in raising productivity will probably get mediocre results. Although employees value increased earnings, they must see that management believes in and is dedicated to the principles of productivity sharing; involving employees in these efforts is very important.

At the start of a productivity sharing plan. the interest of managers and of employees is usually at a high level. With time, employee interest wanes, until a small number of employees, perhaps 15%, remain active. Among managers, and particularly among supervisors, the decline in interest is sometimes just as noticeable, even when they share in the gains. It is difficult to pinpoint the reasons; knowing the drop-off will occur, managers must encourage employee-company cooperative activities to ensure that the plan's principles and goals are emphasized and practiced.

Where plans operate well for many years, it is because employees and managers want the plan to succeed. Where plans operated at a low level or eventually failed, the cause can always be traced to an indifferent attitude among managers, which was reflected by the employees. Though traditional incentives can be successful with only the prospects of financial gains, a productivity sharing plan cannot succeed solely on the basis of the potential financial benefits. Productivity sharing is not an incentive plan, but a way of life at the workplace; it requires that managers always consider that success of the plan is important.

Productivity Plans in Use

Three productivity sharing plans are used in this country: the Scanlon plan, since about 1936; the Rucker plan, since World War II; and the Improshare plan, since 1974. All are broad group-sharing plans, with significantly different productivity measurement systems.

Scanlon Plan

The Scanlon plan, developed in the depths of the Depression in the mid-1930s by Joe Scanlon, then research director of the Steelworkers Union, is the most widely known plan in this country. Many articles, studies, and texts have been prepared on the plan. The term "Scanlon plan" is often used generically, referring to productivity sharing in general.

McGregor's essay, "The Scanlon Plan Through a Psychologist's Eyes,"[21] captures the philosophy of the plan and its potential benefits for workers, management, and society. Scanlon joined the staff of the Massachusetts Institute of Technology (MIT) in 1946 and for 10 years worked to refine and extend the plan to industrial plants. After his untimely death in 1956, several associates carried on his work; Frederick Lesieur and Carl Frost are the most prominent.

The Scanlon plan measures productivity improvement by a change in the computed ratio of total payroll dollars divided by the total dollar sales value of production. Since shipment dollars may be different from production dollars, each month the net sales figure is adjusted by the change that month in work-in-process and finished goods inventories, to obtain the sales value of production.

The single measurement ratio is affected by factors such as:

Changes in product mix, especially with differing labor content products.

Changes in selling prices caused by market and competition as well as those caused by changes in materials.

Wage increases.

Changes in production methods, tooling, and capital equipment.

Changes in functions and staffing.

In actual applications the ratio measure is varied to suit the circumstances. Plants with a few simple products may not have difficulties with this measure. When product mix and manufacturing processes change, as in most plants, a single ratio is not a valid measurement of productivity. Revising the ratio to correct for changes in labor costs or selling prices not in the control of employees requires employee approval; this is sometimes not easy to obtain. Another drawback is the need to open the company's accounting books to the employees.

Most Scanlon plans distribute 75% of the gains to the employees and 25% to the company, though the percentage can be changed. Of the monthly gains, 25% is placed into a pool to absorb loss months; at the end of the year the entire pool is distributed. Other plan details relating to how and when sharing gains are paid and the highly structured suggestion plans and labor management committees do not significantly differentiate the plan from others.

Here is a simplified example of calculations. Assume \$5,000,000 net sales and \$1,000,000 total payroll in the base year.

$$\text{Base ratio} = \frac{\$5{,}000{,}000}{\$1{,}000{,}000} = 20\%$$

In a typical month when the plan operates, suppose the following occurs:

Net sales	\$500,000
Increase in finished goods	25,000
Value of production	\$525,000
Allowed payroll costs (20%)	105,000
Actual payroll	80,000
Bonus pool	\$ 25,000
Company share (25%)	6,250
Subtotal	\$ 18,750
Reserve for deficit months (25%)	4,688
Employee share for distribution	\$ 14,062
Bonus percentage (\$14,062/\$80,000)	17.6%

Moore and Ross present considerable operating details and different ways to measure productivity changes to overcome the single ratio measure.[22] Frost, Wakely, and Ruh describe Scanlon plans in operation.[23] Both books are excellent as source materials and contain exhaustive bibliographies of published materials on the subject. White presents a concise analysis of the writings of others in the field.[24]

Rucker Plan

The Rucker plan was developed by Allan W. Rucker of the Eddy-Rucker-Nickels Company in Cambridge, Massachusetts, in the late 1940s. The Rucker measure of productivity is called

> *. . . economic productivity—the output of value added by manufacture for each dollar of input of payroll costs. Value added by manufacture is the difference between sales income from goods produced and the costs of the materials, supplies, and outside services consumed*

in the production and delivery of that output. Payroll costs are all employment costs paid to, because of, or on behalf of the employee group measured. Thus, economic productivity may measure the financial effectiveness of a plant's hourly-rated employees, its total employment, or some blend of hourly and salaried people. Flexible extra pay programs may be designed, using the principles described here for plant people only, for a mixture of plant and office people, for office people only, or for managers only.[25]

Measuring productivity change as the change in the dollar value added per payroll dollar provides a more reliable measure than the Scanlon measure of payroll dollars per production value dollar because under the value-added approach all purchased materials are excluded. The calculations are similar to those of the Scanlon measure, except that instead of using sales dollars, the figures used are sales less all purchased materials. Under Rucker, productivity measurement with a single ratio presents the same problems as under Scanlon. Several papers describe details of the Rucker measurement base.[26,27]

Improshare Plan

The Improshare plan was developed by Mitchell Fein and was first used in 1974; parts of the plan were used for more than 20 years. Improshare is derived from improved productivity through sharing. Improshare productivity measurements use traditional work measurement practices modified to a selected base period. Productivity measurement with Improshare is considerably different from that with Scanlon or Rucker. The Improshare plan was developed to eliminate the measurement difficulties encountered in complex operations with changing products, technology, and capital equipment. Details of Improshare are presented in Section 2.3.7.

Profit Sharing

Sharing profits with employees is one way of providing a share of the gains that they helped to produce. More than 300,000 companies in this country have profit sharing plans that provide benefits as cash payments, as payments deferred for retirement, or as a combination of both.

Bert Metzger, president of the Profit Sharing Research Foundation, writes:

Profit sharing is an organization incentive program uniquely designed to increase productivity and share the gains with all those who contribute to corporate success. Profit sharing plans take their place alongside Improshare, Scanlon, and Rucker plans as alternate or complementary ways to bring about a more productive operation and a mutually beneficial relationship between management and employees. All of these can be described as "system incentive" programs because they tend to unite all those in the enterprise in pursuit of common goals.[28]

Profit sharing creates a working climate that employees see as beneficial to them; they benefit as the company does. Profit sharing by itself may not provide sufficient attraction in the short term to motivate workers to higher productivity levels, although some companies report excellent productivity improvement with profit sharing. When used together with productivity sharing or traditional incentives that measure and reward in the short term, profit sharing can round out a complete package.

Metzger makes a strong point that the most effective results will be obtained from combining direct incentives and profit sharing as a systems incentive program for the entire organization: individual and group incentives for the narrow interests in the workplace and profit sharing to create organizationwide interests and teamwork. Using both types of incentive strengthens each and helps to overcome the inherent shortcomings of each. The Profit Sharing Research Foundation has published a number of excellent research and information papers containing numerous cases, typical plans, and legal aspects of profit sharing.*

2.3.7 THE IMPROSHARE PLAN

The Improshare plan is significantly different from the Scanlon and Rucker plans; it is a plan based on work measurement, which permits close productivity measurement under changing conditions.

Plantwide Productivity Measurement

The most important element of a productivity sharing plan is the measurement of work performed. Money values should not be used in productivity measurement because many factors that affect

*Profit Sharing Research Foundation, 1718 Sherman Avenue, Evanston, Illinois 60201.

costs do not affect productivity. A plan to share productivity gains must measure the contributions and inputs of the employees and processes being measured and exclude factors outside their control.

Traditional work measurement establishes the time it "should take" to perform a given task under prescribed conditions, not how long it took to perform the work in the past. Such normal, or fair day's work, standards are established through performance rating, with a stopwatch time study or predetermined standards, against a defined measurement base. This leveling or normalizing of observed data is the keystone of traditional work measurement; it must be employed.

The arguments that arise in setting traditional time standards are avoided by measuring productivity against the average level of an agreed-upon period. Using a method called "measurement by parameters,"[29] standards are set at the average of the past, using historical data within a place of work, with no need to performance rate the work performance data. The rationale for this approach is that "yesterday's" performance is established as the APL. Measurements in the future will be made against this APL base.[30]

Traditional measurement and measurement by parameters are two different measurement systems; one has no bearing or influence on the other. Not having to performance rate the data in this approach does not obviate the need to performance rate the data in traditional measurement. It is as valid to measure against the past average productivity as it is to establish any other level that may be determined by altering the past average or by performance rating the observed data obtained from time studies, provided that the measurement base used is defined. The past average is a valid base if that is the base management decides to use; it is the APL.

Measuring productivity by comparing the labor hour value of completed production to the total labor hour input is indisputable and valid. Only acceptable product, packed and ready for shipment, is counted. Everyone in the labor force is included. In large plants a department can be the group, and the measure is the time value of labor added to the product in the department compared to the total labor input in that department. This overall approach to measuring productivity avoids the arguments and rationalizations that occur with conventional accounting practice, which separates workers into those who work directly on the product and those who do support and service work. Engineers traditionally follow accounting practice and mainly measure productive operations. Labor that goes into services, product repairs, maintenance, and other such work is usually not measured. Traditional incentives usually cover only the production operations.

Productivity Measurement for Improshare

When measuring groups or a plant under the Improshare plan, a reliable measurement base is the average productivity over a past period. Considering the total output of the group against the total hours worked by the group permits the establishment of valid measures that include all employees.

This principle of measuring and of productivity sharing is shown by a simplified example. A single-product plant of 100 employees produced 50,000 units over a 50 week period in which employees worked a total of 20,000 hrs. The average time per unit is 200,000/50,000 = 4.0 hrs. Suppose an Improshare plan is introduced under which employees and management share productivity gains fifty-fifty below the past cost of 4.0 hrs/unit. In a given week, if 102 employees worked a total of 4080 hrs and produced 1300 units, the value of the output would be 1300 × 4.0 hrs/units = 5200 hrs. The gain would be 5200 – 4080 hrs worked = 1120 hrs with one half, or 560 hrs going to the employees. Translated into pay, this would be 560/4080 = 13.7% additional pay to each employee, based on each employee's weekly pay. Management also would gain 560 hrs. Where originally the unit cost of the product was 4.0 hrs, the new unit cost, including productivity sharing payments, is (4080 + 560)/1300 = 3.57 hrs. Thus costs, including productivity sharing payments to the employees, have been reduced.

Similar results could have been obtained by using labor and production data in dollars, but as changes are made in wage rates or selling prices, the data would have to be adjusted or else the employees will have gained or lost because of factors beyond their control. This occurs with the Scanlon and Rucker plans.

In plants with multiple products, a measurement base must be established that will reflect the past average productivity for all products and for the entire plant. This was done for a company with 350 factory hourly employees, not including salaried, which produced 475 different products made of machined and sheet metal components. The plant operated under MDW; that is, no incentives were used, but conventional engineered time standards measured individual employee's productivity. Since these standards included only the work of productive employees and omitted about one third who did all sorts of so-called nonproductive work, it was necessary to compute the composite productivity of the entire plant.

The engineered time standards for all the operations to produce each product, established using traditional work measurement methods, were totaled to obtain the overall engineered standard time by product. Working from records of finished products transferred from production to the warehouse, the total for the year was obtained for each product made, which was then multiplied by the respective total standard time for each product, to obtain the total standard hours produced for all output. The payroll ledgers were used to obtain the total hours worked by all employees in the plant for the year, including all the nonproductive workers.

During the year the workers produced 367,500 standard man hr based on product time standards and worked a total of 700,000 man hr. The hours worked are much higher than the produced hours because the time standards did not include any nonproductive work, such as receiving and shipping, maintenance, materials handling, machine setups, waiting for work, and scrap and salvage; also, employees were below management's 100% measurement standards. To convert the engineered standards to reflect the previous year's productivity, and to factor in all nonproductive time, a base productivity factor (BPF) was computed:

$$\text{BPF} = \frac{\text{total actual hours worked}}{\text{total standard hours produced}}$$

The BPF represents the relationship in the base period between the actual hours worked by all employees in the group and the value of the work in man-hours produced by these employees, as determined by the measurement standards used in the base period. In effect, the BPF is a means to "use up" all hours worked and to factor into the original standards all occurrences that were not included in the standards. This approach is equitable to employees and to managers when managers agree that they are willing to use the past average productivity as the measurement base from which to measure productivity improvements.

Members of national accounting firms agree that this approach produces valid product costs, which include all labor costs, and reflects the productivity of the base period. In this example, the BPF = 700,000/367,500 = 1.904. Multiplying all engineered product time standards by 1.904 creates base standards to be used for the Improshare plan, which groups the entire plant. All 350 hourly workers, down to the floor sweeper, are included in the product costs and the productivity measurements.

The same result is obtained by using the accounting department's product standard costs, expressed in man-hours per unit of product; these standard costs are used to transfer products from work-in-process to finished goods inventory on the accounting records. The BPF is calculated in the same way.

Relationship of Base Productivity Factor to Time Standards

The standard costs multiplied by the BPF will fully reflect the average operating conditions and productivity that prevailed in the base period. These modified standard costs, called "Improshare product standards," are then used to measure productivity in any other period. The BPF includes all plant labor hours not included in the standard costs, excluding holidays, vacations, and nonworked time, all at the average productivity level of the base period.

Productivity measurement must be made against a defined base or else the measurements will have no meaning. This requires measurement standards to be frozen, or at least clearly identified as to the base. For example, suppose overall productivity of a plant is raised by 10%. If the measurement standards are updated at the end of the year and multiplied by 0.909 (1.0/1.1), the plant productivity will show no improvement.

Productivity measurements for Improshare must be made against frozen measurement standards at the beginning of the Improshare program. The accounting department will continue to update its standard costs following accounting practice, but the standards used as the basis for the Improshare product standards must be maintained.

In the previous example the BPF was calculated as 1.904, based on 700,000 total hr worked and 367,500 hr transferred to finished goods, representing standard cost. Suppose the standard cost standards in this plan were cut in half; the calculations would be

$$\text{BPF} = \frac{700{,}000}{367{,}500/2} = 3.810$$

The BPF is doubled. This is exactly what is intended by the BPF—to reflect the average operating conditions and productivity that prevailed in the base period. Productivity is always measured against specific time standards. Once the BPF is established, it is carried on into the future with no change. The assumption is that the relationship between standard costs and indirect cost is fairly constant.

New operation and product time standards in the future must be set to the same work performance base used to set standards in the base period. Reducing either time standards or the BPF will tighten the measurement system; such tightening is not proper under Improshare.

The BPF's in several plants cannot be compared without knowing exactly how the time standards in the plants were established and what the measurement bases were to which the standards were set. As shown previously, the BPF in one plant can be double that in another plant by changing how the time standards are set. The BPF is not a unit of measurement by itself.

Essentials of the Improshare Plan

An Improshare plan can be developed for any operations. The plan can be applied to one person or to a thousand, to small groups or to an entire plant. It can be used to supplement conventional incentive plans; several plans can operate in a single plant. The versatility of the Improshare plan comes from the way productivity is measured: hours input against hours output.

A full Improshare plan contains complete details of how to establish measurement standards, how to calculate productivity changes, and how to make calculations under changing conditions. The main features of the plan are:

Increased productivity is shared by employees in the group, usually the entire plant or company.

The input is the total man-hours worked by the group.

The value of the output of the group is the total good units produced multiplied by the past average man-hours standard. With multiple products, the total output is the sum of all the products completed multiplied by their respective standards.

Productivity improvement is shared fifty–fifty between employees and the company.

Gains are calculated weekly, with a moving average to span several weeks to create a stable output level. Productivity is shared and paid weekly. Losses are absorbed into the moving average.

The past average productivity level is used as the measurement base. The average man-hours required during a base period to produce a unit of product is established as standard. This includes all so-called nonproductive time, such as the work done by material handlers, setup men, inspectors, and others involved in the group.

Man-hour standards are frozen at the average of the base period. Standards will not be changed when operations are changed by either management or employees, except for capital equipment and technology changes, which are specifically defined. Increased productivity will be shared, with no attempt to pinpoint whether employees or management created the savings.

An agreed-upon ceiling is established on productivity sharing earnings. The excess over the ceiling will be carried forward to future weeks; eventually the standards may be "bought back" from the workers by cash payments.

The main constraints on the plan are:

Total unit man-hour costs under the plan cannot exceed unit costs in the past. Costs must decrease as productivity is raised.

Management rights are not changed. All changes in methods and quality must be approved by management. Production levels, schedules, assignment of employees, and so on, are vested in management, as they were before.

Union contractual agreements are not altered.

These summary details expand to about 25 typewritten pages to describe a full plan, tailored to a particular company.

The plan obligates management to a set of rules, but puts no limitations on the workers. It is not an agreement in a technical sense, because the employees are not held to any conditions. The Improshare plan does not require the signature of a union representative or that workers follow any new rules. The plan's ground rules specify how productivity will be measured and shared, who will be included, how various types of production changes will be handled, and other such details. All of these are binding only on management.

Placing the onus on management does not make the arrangements more favorable to the employees. The proposition to share does not obligate management to any payments unless productivity actually increases as measured by management's yardsticks and records. The plan clearly provides that management's rights are not changed. Management does not enter into a blind arrangement, nor are its traditional prerogatives and rights diminished.

The Improshare plan measures only final results, usually as finished products ready for shipment. The system encourages employees to move into areas that are not now open to them. Workers will not take on management's responsibilities and start running the plant. To make more good products in fewer man-hours, they will start using skills and abilities that today may be wasted. They will voluntarily do things they would refuse to do if ordered to by management. When workers become bottom-line oriented, all sorts of improvements are possible.

Control Over Improshare Plan

The greatest problem in controlling the operation of a traditional incentive plan is that the time standards deteriorate over time, causing loosened standards. To obscure the looseness, workers hold production back and often work less than a full day in order to maintain an even level of incentive earnings so that management will not be alerted to the loosened standards. In some plants this process seriously retards overall productivity.

The Improshare plan creates opposite conditions, since employees are encouraged to use their ingenuity, change how the operations are performed, and keep raising their output. As productivity is measured overall for the plant, what occurs on a few operations is not important; it is the overall result that is measured.

All Improshare plans require measurement standards to be frozen at the average of the base period and not to be changed except for capital equipment and technology changes or the buy-back of standards. Control over the Improshare plan is maintained by:

A ceiling on productivity improvement at 160%, which is 30% in earnings.

Cash buy-back of measurement standards.

An 80/20 share on improvements created by capital equipment.

All Improshare plans have a 30% ceiling on earnings, or 160% productivity. When productivity exceeds the ceiling in any measurement period, the excess hours produced are banked and moved ahead to the next period. Banking the excess is an inducement for the employees to produce as much as they can and to create a cushion for subsequent periods when productivity may be lower than the ceiling. Should productivity rise above the ceiling and continue at a high level, the Improshare plan provides a simple formula for a one-time cash buy-back of all the measurement standards, for one year's payment of the savings created by the changed standards. The process is voluntary, and employees and managers must agree to the buy-back.

Suppose productivity averages 180% and the ceiling is 160%; the excess hours produced, represented by the 20 percentage points, is banked for future periods. If productivity remains over the ceiling, the standards can be bought back with the employee's agreement, when all the time standards involved are reduced by a factor, so that, in this case, 180% becomes 160%. The employees receive a cash payment of 50% of the 20%, projected for a year, at their regular pay. A $5-an-hour employee would receive a cash payment of $5 × 2000 hrs × 50% × 20% = $1000. Simultaneous with the buy-back, all time standards are reduced by a multiplier of 1.6/1.8 = 0.8889.

Capital equipment and technology gains are also shared. Expenditures for equipment of $10,000 or more are identified as capital change. A total of 80% of the cost savings attributed to the equipment is removed from the measurement standards and 20% is left in. Since productivity gains are shared fifty-fifty, management is returned another 10%, so that management receives 90% of the gains created by the equipment. Workers share 50% of all gains from equipment that costs less than $10,000 and 10% of gains from equipment costing more than $10,000. Technology changes are treated in the same way. Over several years the shared gains from capital equipment can be substantial.

Why Fifty-Fifty Sharing?

Workers who ask why gains should be shared fifty-fifty when, they claim, they do all the work do not fully understand the sources of productivity improvement. In introducing an Improshare plan, management revises a cardinal rule regarding how time standards are established. Under traditional work measurement practices, when changes are made in methods, procedures, tools and dies, or other factors that affect how operations are performed, new operation time standards are established; the company gets all the gains.

Under Improshare, operation standards are frozen at the base period and are not changed when operations are changed by either managers or employees, except for capital equipment and technology changes or buy-backs of standards. Increased productivity is shared, with no attempt to pinpoint whether employees or managers created the savings.

Since management personnel continually make changes in operations, when the Improshare plan goes into effect, even if the employees do not contribute, they still receive 50% of the gains that the company always retained in the past.

There are many good uses for the company's share of the gains. Some can be used to develop improved products, new and better tooling, and better services; to provide price reductions to customers; and to strengthen the company's position in the market in other ways. A portion can be used to create better jobs and job security as the company's finances improve. Scanlon plans, which pass on 75% to the employees and 25% to the company, do not provide sufficient return to the company to warrant introducing a productivity sharing plan.

Measuring Job Shops

Job shops and single contracts are the most difficult to measure using traditional work measurement practices. With Improshare, these are the easiest. Job shops estimate and bid every job. Improshare uses the labor content of the estimate as the standard. All sorts of contracted and one-time jobs can be handled this way, including such diversified work as writing computer programs, fabricating structural steel, and carrying out construction projects; there is no limit.

The goal is to beat the hours estimated in the quotation and to share the gains. Since the estimate is used as the basis for the quotation, the company is protected against inflated estimates; the customer will keep the estimators honest.

Improshare for Nonmanufacturing Work

Improshare can be applied to any work for which unit work counts can be developed. For example:

Bank Operations. Check encoding: The total man-hours required to process checks over a base period is the standard. Include hours of service people, trainers, supervisors, batch totalers, and others.

Surface Mine. Two standards are required: one for removing and replacing overburden and the other for removing the mineral or coal. Use man-hours per cubic yard of overburden and man-hours per ton of mined material. Include all employees in one or the other. Pool total input and output.

Warehouse. Man-hours per case shipped is the standard; include hours of all employees. If receipts are considerably out of phase with shipments, with peaks for each, set man-hours for receipts separate from shipments.

Other examples are provided in "Establishing Time Standards by Parameters."[29]

Using Improshare With Traditional Incentives

An Improshare plan can be designed to operate in conjunction with traditional incentives, or the incentives can be replaced by Improshare. The following alternatives are possible:

1. The incentive plan can be discontinued and replaced by Improshare.
2. All or some of the incentives can continue, and employees who do not participate in the incentives can be included in Improshare.
3. The incentives can continue, and an Improshare plan can cover all employees, including those on incentives; several variations are possible.

Concept of Charged Hours

When a plant operates without incentives, with or without measurement, the average productivity of all the employees and operations during the base year is reflected in the BPF. Dividing the total produced hours at standard into the total hours worked by all employees included in the plan provides a BPF that reflects the base period productivity.

When an incentive plan operates before the introduction of an Improshare plan, in calculating the BPF it is necessary to include, in addition to the total hours worked by all employees, the total incentive premium hours earned by all employees on incentive in order to obtain the average productivity of the plant during the base period.

The total hours worked plus incentive hours earned as incentive premium are called the "charged hours." The standards, or the standard costs used to generate the standard hours produced for the year, would show the productivity level at standard as equal to the standard productivity level at which incentive earnings start. Since the employees earned incentive pay, the past productivity level was higher than standard. The average level for the base period by which productivity exceeds the standard base is calculated mathematically from the average incentive earnings for the year for all employees on an incentive plan expressed in hours. If incentives are paid in dollars, the equivalent hours earned can be calculated by dividing the total dollar incentive pay earned by the average hourly rate of all employees on the incentive plan.

This can be illustrated in another way by an example. Assume 200 employees, 52 on incentive and 48 nonincentive; incentive employees are at 125% average productivity.

Incentive direct labor hours	(52 × 40)	2080
Nonincentive hours	(48 × 40)	1920
Total hours worked		4000
Incentive premium hours (25% × 2080)		520
Charged hours (total hours worked + premium hours)		4520

The proof of the charged hours concept is that, if the incentive employees were all at 100%, to maintain the same output, instead of 52 employees at 125% there would have to be (52 × 125%)

65 employees. The total hours worked would then be (48 + 65) × 40 = 4520 hr, which is the same as the preceding calculations in which the incentive premium hours were added.

In calculating the BPF for a plant in which incentives will be retained or replaced by Improshare, the numerator of the fraction must be charged hours, equal to the total hours worked by all employees in the plan, plus the premium incentive hours earned during the base period by all incentive employees. In calculating productivity after the plan is established, the charged hours concept must also be followed.

Improshare in Place of Traditional Incentives

A traditional incentive plan can be discontinued and replaced by Improshare. To protect the employees who have been on incentive against reduced earnings, they are guaranteed their former average hourly earnings as an additional hourly rate, called a personal red circle add-on. All employees, those formerly on incentive and those not on it, will then receive the same percentage share from productivity improvement.

The company is protected against loss by the charged hours concept; all red circles paid each period are added to hours worked. Output hours must exceed the charged hours for productivity sharing to pay off.

Improshare for Nonincentive Employees Only

When incentives are retained, and Improshare will involve only nonincentive employees, the productivity calculations are the same as for the entire plant, except that only the nonincentive employees share in the gains.

Productivity of the entire plant is calculated by total product made. The input hours are the charged hours of the entire plant, which include all hours worked plus incentive premium hours. In calculating the productivity share to all employees eligible to share, divide the hours gained by the hours worked by the eligible employees.

Improved Job Security

A program for productivity improvement must provide for workers' job security. Although the idea is not new that economic insecurity is a restraint on the will to work, its effect often is minimized by managers, behavioral scientists, and industrial engineers involved in productivity improvement.

Managers must view job security not only in the social sense of how it affects workers' lives, but also as important to high levels of productivity. In plants without job security, workers stretch out the work if they do not see sufficient work ahead of them. They will not work themselves out of their jobs. When workers stretch out their jobs, though it is hidden from view, it is reflected in costs.

Managers historically have considered job security as a union demand to be bargained, as are other issues. This has been a tragic error, because whenever job security is lacking, labor productivity is restrained. Paradoxically, job security should be established as a demand of management.

Productivity gains from the Improshare plan can be used to fund a job security plan on the principles of supplemental unemployment benefit (SUB) plans that have operated successfully in the major industries since 1955. By setting aside two percentage points of the productivity gains from Improshare, one percentage point each from the employees and from the company, a sufficient SUB fund can be established.

An important twist is reverse seniority on layoffs; highest seniority employees are laid off first. Instead of a layoff being a penalty, it becomes a reward to the old-timers. Information on SUB plans can be obtained from the U.S. Department of Labor[31]; other good sources are the personnel departments of local plants in the automotive, farm equipment, steel, rubber, and glass industries.

Productivity sharing provides employees with the opportunity to help ensure their job security by assisting in raising productivity and reducing costs. In a competitive market even small cost reductions can make a big difference in being able to secure new orders. Ultimately, the best assurance of job security is working for a profitable company that stays in business.

Tantalizing Prospects

Consultative managing and productivity sharing are alternatives to traditional managing. Productivity sharing develops conditions and practices that are diametrically opposite to the tight cost control methods conventionally employed to raise productivity.

Traditional managing requires exact time standards, individual employee measurements, job descriptions, and detailed job instructions to ensure that workers perform effectively. Increased productivity does not directly benefit workers. Productivity sharing establishes a loosely structured environment in which workers perform effectively because they want to and because they benefit from their efforts.

Traditional managing does not rely on worker cooperation; managers marshall their resources and act unilaterally to achieve management's goals. Productivity sharing rewards employees for helping to achieve a primary management goal: more units of products or service output with fewer units of labor input. Congruent goals can be developed through productivity sharing that will enable workers and management to work together for mutual gain.

Achieving higher productivity through productivity sharing is so simple that it appears incredible. Increase workers' will to work by rewarding them for more effective performance; simultaneously eliminate practices that penalize them as they progress. This is all that need be done. Productivity sharing is the essence of simplicity. Share gains starting with today's labor costs. The only magic in the plan is that workers and management put into it: the will to gain together.

2.3.8 TRADITIONAL WAGE INCENTIVE PLANS

Traditional incentives follow traditional managing practices. Time standards are set by traditional work measurement techniques designed to work in an adversary relations work environment; management unilaterally sets time standards, and when employees exceed the standards, they earn extra pay.

There are no good or bad incentive plans; what is effective in one company may work poorly in another, under different circumstances and management policies. Some managers swear by incentives, and some swear at them.

Employees' perception of the incentive plan and how it affects them is a critical factor in the results obtained. Incentives must be discussed in the context of actual conditions in the workplace. The work environment affects the success or failure of the incentive plan. The same set of incentive rules can create different results under different conditions. When employees have confidence in management's integrity, all kinds of changes can be made with little questioning by the employees. If relations are poor and trust is lacking, employees resist changes and hold back their output.

Traditional Incentive Practices

Traditional incentive practices may vary, but commonly accepted principles are used by most companies. Unions and employees are accustomed to working with incentives, and labor relations policies have been established to permit the incentives to operate.

The approaches used to establish and operate traditional incentives are almost the opposite of those used for productivity sharing. The most important aspect of incentives is the standards used to measure employee productivity and from which to calculate incentive earnings; these are established through highly structured methods. Under productivity sharing, measurement standards are established as the average of "yesterday," without attempting to ascertain whether yesterday was high or low. Time standards set by traditional measurement are "how long it should take" standards developed through well-established and accepted work measurement principles and practices, using the stopwatch and performance rating, predetermined standards systems [such as Methods Time Measurement (MTM), and work factor], standard data established by the company, and other such means. The essential principle is that management unilaterally establishes a fair day's work standard above which the employees will earn increased pay at an agreed-upon rate.

Traditional incentives are most often established for individual employees on the basis that each person is entitled to the benefits of his or her production. There are no published definitive data to show that the total output of a group working on individual incentives is higher than the output of the same group pooling its output; it just makes sense that individual output should be higher.

In working on individual incentives, employees are primarily interested in their own output and are not concerned with production in other areas of the company, if only because they are not affected. The point is made in discussing group production under productivity sharing that there are advantages to pooling output for a group or for the entire company.

A serious deficiency of traditional incentives is that the rules for establishing time standards give the employees the incentive to innovate methods and work improvements and to put these into effect surreptitiously. This enables them to earn more with less effort, or to maintain their earnings level with less effort. If they turn their ideas in to the company, the industrial engineers revise the time standards so that the company receives the entire advantage. Slowly introduced methods changes by the employees cause creeping changes, the most serious cause of the loosening of time standards; these and other incentive problems are discussed in later sections.

Survey of Wage Incentive Practices

A study conducted jointly by Patton Consultants, Inc., and the AIIE, published in 1977,[32] provided useful data on current work measurement and wage incentive practices. Robert S. Rice, who authored the study, compared the data obtained to those of a similar study he had conducted for *Factory* magazine in 1959,[33] to indicate trends.

The AIIE/Patton study included 1500 usable responses from companies of AIIE members and a list provided by Patton; these are not a cross section of American industry, but include a large proportion of companies that use work measurement and wage incentives. The data therefore do not represent companies in general, but they are useful as a guide. Use care in comparing data from the 1977 study with those of the 1959 study—the list of companies is different, and there is no way to ascertain how the sampled companies are skewed.

On the use of incentives, the data of the 1977 study showed 7 percentage points' decrease in wage incentive companies during the 17-year period, from 51% to 44%. This study, however, showed an increase in incentive coverage—the percent of time that employees are on incentive—which compensates for the decrease in the number of companies using incentives.

In manufacturing, 95% of the companies in the 1977 study used work measurement, and 59% used wage incentives. In nonmanufacturing, 69% used work measurement, and only one company used wage incentives. Bureau of Labor Statistics (BLS) studies show that, on the average, 26% of U.S. workers work on incentive.[2–5]

Two thirds of the respondents started work measurement 10 or more years ago; 37% of the installations were 25 years or older. During the past 5 years, 20% started a work measurement program. Of all the companies, 78% used their own personnel to establish work measurement programs.

A total of 89% of the companies used work measurement for estimating and costing; 59% used it for wage incentives; 55% used it for production scheduling; and 41% used it for performance measurement only. Other uses were spread out thinly.

The prevalence of wage incentives by type was as follows: standard hour plans, 61.1%; straight piecework, 35.9%; sharing plans, 18.6%; plantwide bonus plans, 5.4%; MDW (as incentive), 3%; profit sharing, 8.4%; all others, 6.0%.

Time standards were established by the following means: time study, 89.5%; standard data, 61.4%; estimate based on historical experience, 44.2%; predetermined time standards system, 32.2%; work sampling, 21.3%; others, 3.0%. Many companies used a combination of methods, accounting for the overlapping of data.

On measurement of indirect work, the most useful methods were work sampling, 43%; ratio to direct labor, 43%; direct standards, 34%; historical data, 31%; none, 13%; short interval scheduling, 1%; predetermined time standards, 1%; others, 2%.

Auditing of standards increased over the 17-year period from 54% in 1959 to 82% in 1977. The frequency of auditing was random and/or continuous basis, 74%; longer than annually, 30%; interval of 6 months to a year, 19%; 6-month interval or less, 4%; daily, 1%.

Auditing was performed by time study, 48%; performance reports, 24%; work sampling, 5%; predetermined standards, 3%; review standards, 2%; review methods, 2%; cost comparisons, 2%; others, 14%. Audit studies were conducted by industrial engineering, 77%; management, 4%; accounting, 4%; department supervisors, 3%; consultants, 3%; production analysts, 3%; auditors, 3%; others, 3%.

Conditions that triggered a revision of the standards were changes in methods, materials, and so on, 75%; low performance due to tight standards, 65%; high performance due to loose standards, 54%. Most companies revised the standards when the effect of a change in operating conditions was about 5%. A significantly large number did not review standards unless the effect was from 10 to 15%. Standards changes were initiated by management, 79%; management and union, 24%; union, 23%; employees, 3%.

The targeted and actual work performances for work measurement only were in ranges from 80 to 110%, with an average performance of 101%. Target for wage incentive performance ranged from 90 to 140%, with an average performance of 119%. Actual performance on MDW ranged from 70 to 110%, with an average of 93%. Actual performance on wage incentives ranged from 90 to over 150%, with an average of 123%.

The range from low to high performance, expressed as a percentage of average performance, was 49% for work measurement and 53% for wage incentives. Note that this contradicts the often-cited spread of 2:1.

The expected earnings percentage on incentive for manual and machine-controlled operations was shown as a range, with the average for manual controlled, 36%, and for machine controlled, 33%.

The data on allowances ranged widely. Personal allowance was generally in the range of 4 to 6%, with some plants providing allowances of up to 20%. Fatigue allowances varied widely with the type of work, but most reported from 2 to 20%.

Incentive earnings were calculated and guaranteed as follows: daily calculated, 65.5%, daily guaranteed, 56.9%; weekly calculated, 25.1%, weekly guaranteed, 33.9%.

The effects of wage incentives on productivity costs and so on are shown in Exhibit 2.3.1. Wage incentives were highly endorsed as a means of raising productivity and reducing costs for direct labor; most believed direct labor incentives improved morale and supervisor effectiveness. The support for indirect labor was less than for direct, but nevertheless appreciable.

Exhibit 2.3.1 What Effects Are Your Wage Incentives Getting?

	Direct Labor		Indirect Labor	
	Yes	No	Yes	No
Improved productivity	95.1%	1.9%	43.2%	5.6%
Reduced costs	94.4	1.2	42.6	5.6
Improved employee morale	60.5	21.0	28.4	13.6
Better supervisory effectiveness	58.6	27.8	26.5	18.5
Improved quality of output	25.3	50.6	2.5	
None of the above (ineffective)	3.7		4.3	

Payment practices for employees on incentive for work not covered by standards are shown in Exhibit 2.3.2. In the 17 years between the studies there was an approximate 30% increase in the practice of paying average earnings for work not covered by standards; there was a corresponding reduction in companies paying base rate for such work.

On grievances, in about half the union plants the union participates in some manner in establishing standards; union representatives in about 29% of the union plants take independent time studies. Nearly half of the companies report that wage incentives grievances are less than 10% of all grievances; 80% of companies report that they are less than 30%. Most employee complaints are settled before they become formal grievances.

Approximately 25% of the companies stated that in the past they had either revised or dropped an incentive plan; 15% had canceled an incentive plan; 85% said they would make the change again. Major reasons cited for failures include poor maintenance of standards and loose standards, mechanization or major methods changes, and lax administration. Where changes were made in an existing incentive plan, 57% instituted a new incentive plan, 25% established MDW, 8% established day work at average earnings, 8% established day work at base rate, 1% used day work with an added increment, and 1% used a bonus plan.

Attitudes toward incentives are shown in Exhibit 2.3.3; 75% of top management favored incentives. Support for incentives increases downward through the organization. The 1977 study shows that the attitudes are about 15% less favorable than they were in the study done 17 years before.

The data showed that the coverage of workers by both MDW and wage incentives is increasing, with predicted further increase in time. Estimates of coverage for indirect work are lower.

Exhibit 2.3.2 How Do You Pay Employees on Incentive for Work Not Covered by Standards?

	Base Rate	Average Earnings	Other
Nonstandard conditions (material, machine, etc.)	45%	33%	22%
R & D work on new product, tooling, etc.	40	43	17
Trainees and beginners	68	9	23
Work that has no standard	57	28	15
Machine breakdown	73	14	13
Submarginal performance (handicapped, etc.)	79	7	14
Holidays and vacations	29	63	8
Transfer for convenience of the company	27	60	13
Transfer for convenience of the individual	70	7	23

Exhibit 2.3.3 What Are the Attitudes Toward Incentives in This Company Unit?

	For		Against	
	Strongly	Mildly	Mildly	Strongly
Top management	50%	25%	11%	14%
Middle management	52	27	13	8
First-line supervisors	45	36	13	6
Nonmanagement employees	36	48	10	6
Union leaders	44	26	17	13

2.3.9 CONTROLLING WORK MEASUREMENT AND INCENTIVE SYSTEMS*

Control of time standards and the work measurement system depends on management's ability to control changes in operation methods and procedures. When a competent analyst sets a time standard, in all probability the standard accurately represents both the work being performed at the time and the work measurement criteria used in the plant. As the analyst leaves the work site, things start happening: The employees use their ingenuity to loosen the standard. Methods changes are made by supervisors and design and tool engineers; material specifications are changed; and so on. Often, some of the changes are not fed back, and what are commonly called "creeping changes" set in.

Creeping changes are a most insidious influence in eroding time standards and the work measurement system. Although management welcomes operation changes that raise productivity, when changes are made that are not reflected in the time standards, the standards are loosened, and in the long term other standards will also be affected.

Managing Time Standard and Incentive Problems

Problems of controlling time standards in either MDW or incentive plants arise from the same basic causes. Under incentives workers press to earn as much as they can for as little work as they can get away with. Workers in the MDW plants also want lightened work loads, though their earnings are not directly involved.

Excessively high earnings and runaway standards, withholding of production, excessive grievances, and numerous other difficulties are all symptoms of deteriorating standards and incentives. Attempts to eliminate these symptoms are useless unless the root causes are pinpointed and dealt with.

Causes of Deterioration in Standards and Incentives

The most serious problem in administering standards and incentives is their tendency to deteriorate with time. When this occurs, managers believe it is caused by the pressure of workers to obtain concessions from management, as occurs in other employee-employer relations.

To ensure the sound administration and maintenance of time standards and incentive plans, the following principles should be utilized:

1. A time standard represents the work requirements, consisting of materials, equipment, methods, and work conditions.
2. When a change is made in work requirements, a corresponding change must be made in the time standard.
3. In MDW, the principle followed is a fair day's work for a fair day's pay.
4. Incentive pay is earned for increased productivity measured by time standards.
5. The incentive potential should be specified for manual work when working at an incentive pace and for machine- or process-controlled operations.
6. The plan must be fair to employees and to management.

When these principles are maintained, the time standards and incentive plan will not deteriorate. There may be labor relations problems and bargaining, but the standards and incentives will con-

*Much of the material presented in the remainder of this chapter has been adapted from M. Fein, "Wage Incentive Plans," in H. B. Maynard, ed., *Industrial Engineering Handbook*, McGraw-Hill, New York, 1971.

tinue to benefit employees and management. It is management's responsibility to uphold these principles and the integrity of the plan.

Proscribed practices that negate these principles and are outside the standards and incentive agreements occur when (1) employees find there are other ways to increase their earnings than through increased productivity, the only method provided by the plan, or (2) management does not fulfill its responsibility to administer a sound plan. Practices that most frequently contribute to deterioration of standards and incentives involve:

Changes in work requirements that are not reflected in the time standards.

Establishment of time standards that vary from established practice or from the incentive earnings potential in the agreement.

Cheating by employees.

Deterioration rarely arises from negotiated changes in the standards or in the incentive plan. More frequently, it is caused by shortsighted managers who permit degenerative practices to creep in. Others are introduced by workers surreptitiously, but usually with some knowledge of plant supervision. Probably more standards and incentive plans are wrecked by erosion of the relationship between the time standard and the work requirements through creeping changes than by all other causes combined.

Serious degeneration occurs when management knowingly sets time standards that exceed standard practice or the agreed-upon potential incentive earnings. Many plans provide that machine- or process-controlled operations are to permit a given level of incentive earnings, say 30%. Invariably, actual earnings are from 40 to 75%. When standards are set for employees to maintain their previous average, the incentive agreement is violated, and the standards degenerate.

No one condones cheating, and anyone caught can be fired; the union will rarely intercede. But "making money with the pencil" is widely practiced in various forms.

Preventing Deterioration

Taking the following precautions will prevent the deterioration of standards and incentive plans:

1. Do not permit dilution, through prohibited practices, of the six essential principles for administering time standards and incentive plans.
2. Develop a fair plan; clearly delineate all aspects.
3. Administer the plan equitably; do not attempt to gain unfair advantage.
4. Attend to grievances and complaints promptly; keep workers informed.
5. Administer the plan with a capable staff; employ sound work measurement techniques.
6. Establish an audit program to monitor time standards and incentive operations continuously.

The most important precaution to take to prevent deterioration is the first one listed. Do not permit dilution of the essential principles through prohibited practices. This requires managers to comprehend the difference between prohibited practices and legitimate bargaining.

Some aspects of time standards and incentives are proper subjects of bargaining between workers and management. Concessions in such areas no more dilute management's control over standards and incentives than do wage increases For example, the incentive earnings opportunity when working at an incentive pace, or for machine-paced operations, is bargainable. Unless time standards are specifically excluded from bargaining in the labor-management contract, management cannot take the position that its standards are "correct" or "engineered" and therefore not subject to bargaining.

Bargaining and making concessions in individual standards need not erode the plan. The setting of loose standards by compromise loosens the involved standards and will permit excessive earnings on these jobs, but this should not affect other standards, which are still governed by the agreement or by standard practice.

Some aspects of standards and incentives must not be bargained; otherwise the plan will be seriously impaired. The principle of not permitting changes in work content without changing the standard is not a bargainable issue. This principle is a main source of protection for workers against abuses of increased work loads. It is also management's main protection to ensuring that improvements in methods and process are recovered.

Grievances and Arbitration

Procedures for handling grievances and arbitration of time standards and the operation of wage incentive plans are generally detailed in union contracts. Nonunion companies handle these matters using a range of methods from informal procedures to formal rules; some include compulsory arbitration for differences not settled in grievance steps.

Gottlieb and Werner[34] discuss the statutory obligations of an employer under federal labor laws. The Elkouri and Elkouri text[35] on arbitration is a classic, with much useful information. Wiggens[36] provides material on handling industrial engineering disputes. Berenbeim[37] provides case material from nonunion companies.

Preventing Creeping Changes

When changes are made by management personnel that are not reflected in the standards, the workers quietly adjust to the changes and do not "kill the rates," which would alert the industrial engineers. Changes innovated by workers are put into affect slowly in order to avoid detection by supervisors. Workers oppose management on this issue because they have every incentive to withhold production and hide their innovations. An extra benefit to workers is that loose standards become leverage to use in loosening other standards. Management must protect its position by providing in the standards and incentive agreement that a change is a change, even if innovated by employees.

Matching Practices to Agreement

Success of an incentive plan depends on the employees' wanting to produce more; they must see the plan as fair to them and must have confidence in management's integrity. Where management attempts to obtain unfair advantages, employees will lose confidence in the plan. They will react by increasing proscribed practices, which become sanctified in their eyes.

Many time standards and incentive problems arise through carelessness and lack of attention. When such problems go to arbitration, it is usually found that, though arbitrators are not bound by hard-and-fast rules, there is fairly close agreement on basic principles. Arbitrators will not rescue managers who had remedies available to correct standards loosened by creeping changes but who allowed several years to pass without taking action. Arbitrators often give greater weight to standards and incentive past practices than to the written agreement.

Practices that are clearly against management's interests and that would never have been agreed to in negotiations sometimes creep in slowly over time, mainly through lack of attention by management. Examples include unreasonable restrictions on how company engineers are to take time studies and calculate time standards.

The subject of equitable earnings opportunity with incentives often finds management on the short end of the bargain after a number of years. An incentive plan starts out with the proposition that extra pay is earned for extra effort. Employees complain, with justification, that when a machine breaks down they are denied, through no fault of their own, the opportunity to earn extra pay. When management agrees to pay some form of average hourly earnings, it is usually difficult to restrict the practice thereafter. Many incentive plans suffer when average hourly earnings payments go to extremes. The remedy is to reduce downtime to a minimum and to maintain high incentive coverage, which reduces nonincentive work and minimizes the effect of average hourly earnings.

Where management agrees to increased incentive earnings potentials, this should be incorporated into the incentive agreement and adhered to in practice. Even if management is saddled with high incentive earnings through past practice, this should be recorded in writing so that it can be maintained. If this is not done, the earnings level will surely rise still further. It is the clinging to the fantasy of one set of standards for the agreement and another for plant practices that destroys the sanctity of the agreement.

Management Responsibilities

Many managers have an oversimplified concept of work measurement and believe that there are simple, universally accepted techniques to establish valid time standards. They often believe problems with time standards come primarily from employees' general opposition to all things that favor management.

When management undertakes the establishment of a comprehensive MDW or incentive plan, it must recognize the magnitude of the undertaking and the skill and quality of the personnel that will be needed for its administration. Many incentive difficulties arise because the program is not supported by adequate budgets and competent industrial engineers.

Supervisors must be trained so that they fully understand the plan and are able to cope with the problems that arise in their departments. Too often, supervisors are not involved in the standards and incentive plan, placing a tremendous load on the industrial engineers.

Administration of the incentive plan is a high-order management responsibility. Standards and incentives cannot be operated as a part-time activity. Because labor relations are always affected, critical decisions must be made at a high level.

There are often significant differences in management attitudes under incentives from those that exist in MDW plants. In the incentive plant, management sometimes takes advantage of the workers' desire to earn extra pay, counting on this to keep the workers going in the face of difficulties. In an MDW plant, workers will probably stop the operation if they have difficulties. Under MDW, management relies on supervision to obtain higher productivity. Greater priority is given to the training and upgrading of supervisors. In the incentive plant, management expects the motivation of workers to obviate the need for close supervision.

Nothing can take the place of good supervision, especially not incentives. Where managers establish incentives to make up for deficiencies in supervision, the incentives will probably deteriorate and add to management's difficulties. Sound operation of an incentive plan over a long period requires competent supervision. The same plant on MDW may require greater supervisory attention to maintain high productivity. Although some managers of incentive plants cite the pressures of employees to earn more for less effort as the major cause of problems, the primary causes of incentive problems are more often found in deficiencies in managing.

Exercising Control Over Time Standards and Incentives

It is difficult to control time standards and incentives primarily because the stability of the plan is based on the maintenance of given conditions and practices, whereas the elements involved in the incentive plan are constantly changing; operations change, tooling is revised, quality requirements are altered, pressures are exerted by the employees, and standards concessions are made by management. Changes that conform with the plan do not cause problems, but when changes violate the plan, erosion sets in.

Control of the time standards and incentive plan depends largely on the control of change. Because many of the changes are not readily detected when they occur, or because the long-term effect may not be discernible, appropriate procedures must be established to detect and evaluate changes. The most effective way to detect changes involving the time standards and incentive plan is through audit and control procedures.

Purpose of Audit

Audit of time standards and incentives is a continuous review and appraisal of work measurement and incentive practices to determine if these are carried out in conformance with management policy and objectives. The primary purposes of audit are:

Detection of change so that it can be examined.

Evaluation of change.

Provision of meaningful data upon which to formulate decisions for acting on change.

The information provided by audit will enable management to formulate decisions on remedial steps that should be taken. Remedial action is a management responsibility and not a purpose of audit. An effective audit program will:

Ensure that time standards and incentives operate as designed.

Determine how changes arose and the effect these have on the time standards and incentive plan, on costs, and on employee earnings.

Provide data to evaluate the effectiveness of the work measurement system and wage incentive plan.

By preventing erosion of the time standards and of the incentive plan, audit will assist in keeping the productivity level and the incentive earnings in line with the time standards and incentive agreement and with management's objectives. It is vital that, with incentives, audit is not used to limit or depress employee earnings. Audit will, in the long run, hold earnings in line, the line being the incentive agreement between the employees and management.

Management's Right to Audit

Audit is a basic management right that is indispensable to monitoring operations and gathering information in order to evaluate activities and make decisions. However, when incentives are involved, workers will not accept management's prerogative without assurances that it will not be misused.

Management should clearly explain the purpose of the program to the employees, how it will work, and the actions that management expects to take. Employees' rights must be clearly established so that they are fully protected. Because audit can become a sensitive issue that can easily be misinterpreted, management should be completely candid with the employees so that they and management personnel fully understand the audit purpose and procedures. The reception given to management's audit program will depend on the relationship between employees and management and the trust employees have in management.

Criteria for Audit

The essential criteria for audit are:

The time standards or wage incentive agreement, including the relevant sections of the labor-management contract, if there is one.

Definitions of how to measure, particularly the benchmarks for "normal."

The work requirements and conditions of the operation when the time standard was originally established.

The criteria must be clearly delineated so that they are not lost sight of or changed with the passage of time. Most important of the criteria is the incentive agreement that establishes the basis for the incentive program.

Practices are sometimes introduced that are not included in the agreement. Where these violate the written agreement, management should clarify and remove discrepancies in order to avoid administration problems and to delineate the criteria for auditing. For example, with incentives the agreement may specify that, when working at an incentive pace, the employees have the opportunity to earn 30% extra pay. But when the plant averages 50%, and the employees on the average are not "superskilled," management must examine its definitions. The engineers who establish such time standards violate the agreement. If management has, in effect, agreed to a higher incentive opportunity through past practice, this should be incorporated into the agreement so that the engineers always follow it.

Management benefits when it updates the agreement to reflect current practice. Where most of the employees earn 50% incentive pay, and the agreement provides for 30%, management has as much chance of reducing the incentive earnings as it has of convincing the employees to take a 20% cut in their hourly pay. When incentive earnings later inch upward above 50%, management will have no agreed-upon basis for setting standards. Where the incentive agreement does not quantitatively specify the relationship between incentive earnings and incentive pace, management must expect a continual rise in earnings year after year.

Many difficulties in work measurement and audit arise from the subjective basis of the measurement process, which cannot be avoided in measuring human work performance. One way of minimizing such problems is by definitions of normal, or APL, related quantitatively to incentive pace. Sound practice is to establish benchmark operations, preferably on films, which will create permanent visual records of APL. Predetermined time standards such as MTM, work factor, and Basic Motion Times (BMT) provide such yardsticks, when the employees and management agree to these measures.

An important concern of audit is whether any changes have been made in the operation methods, tooling, and conditions since the original time standard was established. The audit procedure seeks to determine the extent of changes, since these affect the time standards.

Conducting the Audit

Audit techniques should be organized to achieve the main purpose: detection of change and deviations from standard practice. Audit studies are primarily concerned with operator performance on specific operations—to ascertain whether changes have occurred in the work requirements of the operation as it was established when the time standard was set. In auditing a time standard, the criteria are the original work methods and conditions of the operation. The level of earnings of the worker at the time of the audit is not germane to the audit, although the earnings may be symptomatic of something that has occurred. Sound audit requires that the data supporting the time standards be clearly set forth and available.

Effective audit requires the following steps:

1. **Select the Operation to Audit.** Studies should be made on randomly selected operations. When the analyst is in a department, if he or she will randomly select any operation that is then being performed on which to conduct a study, a random selection will have been made. The analyst merely walks over to an operation in process and proceeds to conduct an audit study. On some days the analyst will have no time for audit; on others he or she may have ample time. With this approach, audit tasks are flexible and can be performed without detracting from the analyst's ability to set standards and handle his or her other assigned responsibilities.
2. **Give No Prior Notice of Audit.** Neither shop management nor employees should be notified

in advance that an audit study will be made, so that changes are not made in how the operations are performed. The purpose of the audit is to sample the operation as it is currently being performed.

3. Prevent Auditor Bias. The auditor should not consult any files or background material concerning the operation before the study is made. The less he or she knows about the operation and how the standard was established, the better for purposes of audit.

4. Study the Operation. The auditor samples the operation to determine how it is currently performed; all pertinent information is obtained. A time study is made of the operation in the conventional manner, and a time standard is calculated reflecting the existing conditions. The audit study need not be as detailed or as long as the original study; several cycles may be sufficient for the audit. Record all element descriptions, and identify all the work performed for the operation, including infrequent elements.

5. Compare Audit Data with Original Data. After the study has been completed, the original data should be obtained from the files and the audit study compared to the original study, element by element. Where differences are found in the element comparisons, they should be flagged for discussion and investigation. When the auditor finds the audited standard to be proper, all study papers should be clipped to a cover sheet that describes the findings, and the batch should be placed into the standards file.

6. Investigating Differences in Practices. Differences in work conditions and work requirements should be investigated in order to determine how and when these arose. This investigation is an important part of audit, because it will turn up practices and looseness that undoubtedly affect other standards. Management inadequacies will also be uncovered.

7. Take Action. The audit study is a fact finding function that does not require the auditor to take corrective action. Action is a management responsibility. All the facts and recommendations of the auditor should be submitted to management. Work measurement analysts and industrial engineers must separate the purpose of the audit procedure from taking corrective actions. With the conclusions of the study before them, management will make decisions concerning the advisability of taking corrective actions and the steps that should be initiated. Analysts should not assume these management prerogatives. Actions that may be deemed advisable and rational by work measurement analysts may be seen differently by management. A record of conclusions and actions taken by management should be filed with the audit data, becoming part of the permanent time standards file. This information may be important at a future date.

8. Maintain an Audit Log. A log should be kept by auditors and by departments listing all audit studies taken, showing date, auditor's name, and department and operation, to provide a record of the frequency of audit studies. This will permit management to audit the auditors to ensure that sufficient audit studies are being taken.

Relationship of Time Standards to Base Rates

Time standards in MDW plants usually are not permitted to become loose to offset low base rates, since these employees have no way of earning compensating pay. Low incentive base rates may be acceptable to employees if the time standards are sufficiently loose to permit adequate take-home pay.

The relationship between base rates and time standards is rarely expressed formally by management and employee representatives, yet it is a pervading influence in the setting of standards. The practice of gauging time standards' propriety by take-home pay adequacy seems to negate the admonition that the time required to perform an operation should not be related to the pay of the employee.

Detecting Changes Through Productivity Report Analysis

Many companies attempt to monitor time standards by evaluating operator productivity overall by time periods and by operations. Clerks performing the task are instructed to signal low and high variations beyond established productivity levels. Exception reports are then prepared so that analysts can check the high-productivity operations in order to ascertain whether the performance by particular operators truly reflects their productivity or whether excessive productivity is caused by loose standards. Low productivity is also investigated.

Productivity report analysis is useful in revealing tight standards, where operation difficulties are encountered, and permits management to investigate and take corrective actions. Low performances may show improperly trained employees, excessive downtime, nonmeasured work, occasions when standards cannot be used because of nonstandard conditions, and other such factors. This information is necessary to raising productivity.

As far as auditing time standards is concerned, analyzing daily employee productivity will hardly reveal creeping changes and loose standards. Smart operators know exactly what the reports show, and they either will manipulate their reported times so that their calculated productivity is within limits or will withhold output. In either event, loose standards are obscured.

2.3.10 REMEDIES FOR DETERIORATED TIME STANDARDS

Time standards are generally considered deteriorated when employees can perform at an APL with substantially less effort than should be required, based on the criteria for the plant. There are no productivity figures or limits to define "loose standards."

Evaluating the Effect of Loose Time Standards

When management believes its time standards are broadly loose, the first step in planning a course of action is to make a complete study of the work measurement system from the origin, to show where and how various practices arose to erode the plan. Prohibited practices should be analyzed to show the effect each has on manufacturing costs, productivity, and operation of the work measurement system.

A detailed analysis should be made of the extent of the time standards' looseness by setting new standards for a cross section of the operations and determining the weighted percentage by which all time standards are loose. The study should include standards that may be tight, which are absorbed by employees and compensated for by loose standards. Management will then have an evaluation of the potential for increased productivity and reduced costs when production meets properly established time standards.

The study should pinpoint management concessions in setting time standards that inflate and loosen the standards, including the reasons for the concessions. It is especially necessary in incentive plants to consider the relationship between base rates and time standards, discussed previously, to avoid mislabeling time standards as loose when they may not be loose in relation to an equitable take-home pay base.

Outside consultants and auditors who conduct studies should be sure they have the full history of the work measurement system and incentive plan before recommending remedies. When long-standing time standards are checked against sharply defined work measurement criteria, the original standards may appear far out of line. In checking with management old-timers, it may be found that, for valid reasons at the onset of the measurements, various concessions were made. This often occurs in MDW plants, when operation productivity is found to be 60 to 65%. Apprehensive that undue pressures may create labor relations problems, management introduces extra allowances and practices to make 60% equal to 80%. Thirty years later conditions are vastly different, but the old practices still continue.

Loose time standards, for whatever cause, increase costs and penalize management. Under MDW, loose standards do not benefit employees except by easing their work load. Under incentives, loose standards permit increased take-home pay up to a level deemed safe by the employees; above that, employees hold back and gain reduced work loads. Correcting loose incentive standards invariably reduces take-home pay and is strongly resisted by employees. When management tightens MDW standards, workers do not directly lose pay, though some may be displaced through higher productivity and may lose overtime pay; there are ways to minimize these losses. Employees on MDW usually resist standards changes less than do incentive employees.

Correcting Loose MDW Standards

A one-time revision across the board of loose time standards is a major undertaking, even with good labor relations. It is considerably different from routine efforts to update standards. The approach adopted by management depends on relations with the employees and on the extent of mutual trust and good will. In some plants, when a standards change program is fully disclosed to the employees, they accept the need for changes and cooperate. In others violent eruptions occur when management attempts even to investigate loose standards, let alone make changes. Correcting standards in such plants is extremely difficult. The following are general approaches that are used:

1. **Eliminate Ineffective Operations and Equipment.** Reduce effect of loose standards by eliminating the operations that are loose through introducing radically different methods, new tooling, mechanization, new products, and so on.
2. **Gain Employees' Agreement.** Obtain an understanding with employees that management may adjust loose standards. Different ground rules can be established, including a joint employee-management study group to oversee the program and handle grievances expeditiously.
3. **Raise Hourly Rates.** Increase base hourly rates by a specified amount in exchange for an agreement to revise standards across the board or for specific operations. These agreements are sometimes made during contract negotiations. Since contract negotiations are hectic, better progress is usually made between contracts.
4. **Buy Back Loose Standards.** Some companies have "bought" the loose standards from the employees with cash bonuses; the mechanics are described in detail later.
5. **When MDW is Used, Establish Wage Incentives.** Many companies have changed from MDW to wage incentives and in the process have eliminated all the MDW time standards. Managers

are often amazed when employees who vigorously defended the propriety of MDW standards will accept substantially tighter time standards for incentives with little complaint. Before an incentive program is undertaken, managers must be sure they have a clear understanding with the employees on how the incentive standards will be established. Accepting several standards as a trial may not reflect the employees' attitude when the changeover is in full swing. Experience has shown that, when companies change from MDW to incentives, new studies must be taken and new time standards established. The new standards should preferably not be related to or based on prior MDW standards.

6. **Apply Pressure.** Managers sometimes resort to various forms of pressure to obtain assent by employees to standards revisions, by threatening subcontracting of work or the closing down of operations, departments, or an entire plant.

Buying Back Loose MDW Standards

Revising standards would present fewer difficulties if managers fully understood why workers resist changes. Omitting plants where worker opposition is deep seated, workers in most plants want a piece of the action as gains are made. Frequently, loose standards are created by worker ingenuity and resourcefulness. When management tightens such standards, workers feel cheated of their gains; they must work harder, and they get nothing for their efforts.

A simple and effective approach to standards changes is to pay employees a one-time cash bonus related to the amount by which the standards are reduced. This buy-back, as it is called, works as follows: Suppose a study shows that, if new time standards were set in a department, on the average these would be 89% of present standards. If the employees produced at 100% to the new standards (assuming these are APL standards), the increase in production would be 12.36% (1.0/.89 = 1.1236). Under the buyback, management would pay each involved employee 12.36% of his or her annual base pay as a bonus. A \$5-an-hour employee would receive 0.1236 × \$5 × 2000 hrs = \$1,236. Providing that the employees kept to their bargain and maintained 100% productivity on the new standards, the buy-back would cost management nothing, since the increase in productivity of the employees would have paid for it.

Managers should note that the buy-back is a one-time payment for continuous savings. The payment is covered by reduced time standards; the cost is recovered by one year's increased production to the new standard. Some managers may see these payments as a bribe. Battle-weary managers who have spent years haggling over loose standards will readily agree that buying them back can be the fastest and most practical way to correct standards.

A fair way to pay the buy-back is to offer the employees one third on introduction of the new standards, one third 3 months later, and the final third after 6 months. Extending the payments permits management and the employees to show good faith. Since management initiates the plan, the company's good will is indicated by the initial one-third payment, even before the employees increase productivity.

The buy-back is a simple and equitable method of permitting employees to share in increased productivity. The payments cost management nothing, since the increased productivity balances the cash payments. Extending the payments over 6 months is assurance to employees and managers that neither is buying a pig in a poke and that both have ample opportunity to express themselves on various standards and to reach final agreements. The buy-back need not be plantwide.

A one-time cash payment is a small price to pay for an across-the-board tightening of standards accomplished with employee approval, permitting management to move quickly to make the changes. The payment is recovered from the reduction in time standards and so costs nothing. Thereafter, the reductions in standards are all cost savings. An increase in base rates, described previously, is a cost that can never be recovered.

Correcting Loose Incentive Time Standards

Loose incentive time standards are far more difficult to correct than MDW standards because tightened standards may reduce employees' take-home pay as well as the potential for increased pay should the employees decide to take advantage of the loose standards. The general approaches to correcting incentive time standards are similar to those used for MDW standards.

The methods that follow are used in various ways by managers to correct loose incentive time standards. No method is superior to the others. Approaches that work well in one plant may cause serious labor problems in another; methods turned down by employees in some plants may succeed in others. Past relations between employees and management greatly affect their attitude toward correcting loose time standards. The general approaches are:

1. **Eliminate Ineffective Operations and Equipment.** The approach is similar to that used for MDW, except that the new standards may have to maintain the earnings potential of the replaced standards.
2. **Gain Employees' Agreement.** With good relations between managers and employees, many

companies have openly discussed the problems regarding competition from domestic and foreign companies and have demonstrated to their employees that increased output is needed to protect jobs. Where communication between employees and management is good, employees understand management's problems and needs for cost reduction. Various approaches are used to demonstrate that revised time standards will not cause undue burdens on the employees. Labor-management study teams are often useful in facilitating two-way communications between employees and management and in overcoming obstacles. Management may have to make some concessions to employees.

3. **Raise Hourly Rates.** Standards may be revised in exchange for specified base rate increases. This method will increase or reduce costs, depending on whether the reduction in time standards exceeds the base rate increases. Increases in hourly rates are a useful way of giving employees a share of the gains that are obtained when time standards are revised.
4. **Consider Base Rate Adjustment Factor Method.** Time standards are tightened in the same ratio that base rates are increased. Managers and industrial engineers should carefully study the relationship between their base rates and standards to determine whether this type of adjustment would clear up what appear to be loose standards.
5. **Buy Back Loose Standards.** The buy-back described for MDW companies can also be applied to incentive operation, but there is a significant difference. With MDW, employees' pay does not vary with output. With incentives, when standards are tightened and employees produce more, they lose pay they would have earned under the old standards; there may be resistance to the buy-back. This is discussed in greater detail later.
6. **Replace Incentives With MDW.** Some companies that had difficulty in controlling their incentive plans discontinued the plans and instituted MDW. The problems of this approach are described later.
7. **Apply Pressure.** The method most relied upon by managers to enforce acceptance of time standard revisions is pressure applied to the employees, usually coupled with statements that work will be subcontracted, departments will be closed, product lines will be eliminated, and so on. These are usually not idle threats, but reflect difficulties management encounters in meeting competition. Employees see these moves as coercion; management usually states that it has no recourse but to curtail losses and unprofitable operations. Incentive employees are usually more militant than MDW employees, and greater difficulties are encountered in resolving loose standards with them.

Buying Back Loose Incentive Standards

The approach for buying back loose standards described under MDW applies also to incentives, with one major difference. Under incentives, when standards are loose, employees can hold their productivity at a level that they believe is safe and will not be questioned by management. The potential for increased earnings is always there, even though they may not convert it into earnings. When such time standards are revised, even though employees' take-home pay is reduced, and that often causes employee objections.

Control Through a Ceiling on Incentive Earnings

Practically all incentive plans incorporate the principle of no ceiling on incentive earnings; a worker's incentive pay is limited only by his or her skill and diligence. Management guarantees that time standards will not be reduced because incentive earnings are high. The no ceiling principle is widely accepted as sacred and inviolate.[16]

When the operation of incentive plans is examined closely over time, it is found that (1) the no ceiling principle is a myth–real ceilings are established by the employees in most incentive plants, and (2) the no ceiling principle is a major cause of deterioration in the incentive time standards.

The no ceiling guarantee is interpreted differently by employees and management.

Employees' Position

They take the guarantee literally: There is no limit to what they can earn.

They are entitled to increased pay for increased output.

Management's Position

The guarantee is made in good faith.

Standards are guaranteed against change, provided that the basis for the standard does not change.

The definition of a time standard is critically important to the guarantee. It is one of the essential principles for the sound administration of time standards and incentive plans, discussed previously, provided that the time standard for an operation does not represent pieces per hour, but the work required to produce the pieces. There is an important difference between the two measures.

Workers understand this difference. When management adds to work requirements, workers expect extra time to be allowed for the operation. This is an inviolate requirement for the maintenance of equitable time standards, which workers never relinquish. When management improves the work methods of an operation, workers accept a reduction of the standard. When workers make improvements, however, they change their position and adopt the stance that standards represent pieces per hour. They claim that, since they made the improvements, they are entitled to the gains. Powerful social forces operate on the plant floor to influence employees' motivation. If there is withholding of production because of employee differences with management, an individual will probably go along with the group.

No ceiling on earnings is really a myth. Real and effective ceilings operate in most incentive plants. The concept that an individual worker's ability to produce is limited only by his or her own desire and skill is not what it appears to be. When time standards are soundly set and conform to work measurement definitions and criteria, effective limits are established because the range of human work capacity has physiological limits. Relatively few highly skilled workers will reach much higher levels than others.

Management's pronouncement that there is no ceiling on incentive earnings justifies in employees' eyes all actions to increase earnings except cheating.

When employees use their ingenuity to increase their output beyond that possible with management's prescribed methods, they firmly believe the extra output belongs to them. Since they know management would appropriate their ingenuity if it were discovered, they have no compunction about hiding their improvements. They see withheld production as necessary to protecting their productivity gains and to preventing management from learning that they have innovated changes.

Employee ingenuity, even when it is the attribute of a single employee, is considered the collective property of all employees. When management uncovers employee methods changes and attempts to revise the official operations sheet and the time standard, all employees indignantly protest the change. They believe management is appropriating their creativity. No argument will convince employees that management has a right to their extra productivity.

Increasing the Effectiveness of Incentives

Withheld production as an obstacle to increased productivity can be removed by making two major changes in current incentive practice: (1) establish a formal ceiling on incentive earnings and (2) buy back the increased productivity that employees innovate above the ceiling.

These simple changes will establish an entirely different set of rules under which employees will have no motivation to violate the various conditions of the incentive plan to increase their earnings. Once a ceiling is set, increased output beyond the ceiling will only bring idle time during the day; it will not add to employee take-home pay. Since the ceiling is expressed in productivity per hour, employees cannot produce 8 hours of work in 4 hours and then sit around till the end of the day.

The major advantages of a ceiling are:

The incentive to beat the game is removed.

The deterioration process that results from creeping changes is halted.

Uneven earnings in departments and between various groups are eliminated. This is a major source of difficulty when incentive earnings run away on relatively simple work, which then pays higher earnings than more skilled work.

The rat race of workers trying for higher and higher earnings is eliminated. Older workers, who cannot produce at the rate they could when they were younger, are not pressed by younger workers with greater stamina.

A change in standards is less hotly contested by employees at the ceiling because it will not reduce their earnings, although they must increase their output.

Establishing Ceilings in Incentive Plants

When installing an incentive plan in a plant that has never had one, a ceiling can readily be established, since the incentive plan provides a substantial increase in earnings over base earnings.

Where incentives already exist, problems may be encountered in adopting a ceiling, depending on the level of earnings. To determine where to establish a ceiling, management should prepare a spread sheet showing the average incentive hourly earnings of each employee for about a 12 month period and record the incentive earnings by 5% intervals. This will show how many earned from 0 to 4.9%, from 5 to 9.9%, from 10 to 14.9%, and so on. The ceiling should be established to encompass about 75 to 85% of all employees. When this is done, the employees below the ceiling will not suffer any loss of earnings.

Employees whose earnings exceed the ceiling can be compensated in two ways: by red circle "add-on" or by cash buy-back. The red circle add-on is commonly used to protect high-incentive-earning employees. Those whose past earnings exceed the ceiling can receive the difference between their past earnings and the ceiling as an hourly addition to their incentive earnings, for all hours worked. These employees' earnings will thereby be ensured.

2.3.11 MEASURED DAY WORK AS AN INCENTIVE PLAN

Measured day work is generally thought of as an arrangement under which an employee is paid a fixed hourly rate or salary independent of the employee's productivity. Before World War II, MDW had another meaning:[38] It was a bona fide incentive plan under which an employee was paid a fixed wage rate for a given period. The employee's average productivity for the period then determined the employee's wage rate for the next period, which could be a month or even 3 months. Suppose the employee had a base rate of $5.00/hr. If he or she worked at an average of 110% for a period, the base rate for the next period would become $5.50/hr. Meanwhile, the employee would be establishing his or her base rate for the next period. The minimum would be guaranteed, permitting varying earnings above the $5.00/hr base pay.

This type of MDW was an incentive plan with productivity calculations over a long period. It was widely practiced in this country up to World War II, when rigid regulations by the War Labor Board were imposed on permissible wage increases. The board only approved the introduction of bona fide incentive plans based on work measurement time standards. The incentive MDW concept was discouraged because it was an easy way for a company to get around regulations. Measured day work then came to mean that an employee worked against time standards on an hourly basis with no incentive opportunity.

The two definitions of MDW are still continued, but mainly for the record. The definitions established by the American National Standards Institute (ANSI) are:

Work performed for a set hourly nonincentive wage on which production standards have been established (most frequently used).

An incentive plan wherein the hourly wage is adjusted up or down and is guaranteed for a fixed future period (usually a quarter) according to the average performance in the prior period (infrequently used).[39]

Incentive Features of MDW

A combination of the best features of nonincentive MDW and conventional incentives may provide conditions that employees will prefer to either MDW or incentives and that will prove beneficial to management. Pre-World War II MDW can be adapted to plants on incentives that wish to discontinue their incentives and to plants that employ no incentives.

Consider a plant on incentives: Management proposes to eliminate the incentives and to permit each employee to select the hourly rate he or she wishes to receive, provided that each maintains a level of productivity to justify his or her selected wage level. To get the new wage payment system operating, each employee is permitted to establish his or her base rate based on the average of the three highest productivity weeks of the past 13 weeks. If the average was 120% and the employee's base rate is $5/hr, his or her new hourly rate would be $1.2 \times \$5 = \$6$. The employee would be paid $6/hr for the next 13 weeks, regardless of his or her productivity during these weeks. Productivity in the meantime is measured week by week, and a new average for the current 13 weeks is calculated, which then establishes a new base rate to be paid to the employee for the next 13 weeks.

A similar approach can be followed at the start when the employees are on nonincentive MDW. Instead of using the highest 3 weeks to establish the average productivity, each employee can select a base rate he or she believes would be warranted by the productivity he or she expects to maintain during the next 13 weeks.

Permitting employees to select in advance the pay levels they believe they can justify by their future productivity levels is proposed as a gesture of good faith by management, to encourage employees to view productivity improvement as beneficial to them. Asking workers to take the first step and work at a higher productivity level in advance of receiving a higher base rate may defeat the proposition.

When this incentive MDW approach is examined, it is seen that the main objections raised by employees against conventional incentives have been removed:

Each employee knows exactly the extra incentive pay he or she will receive for the next 13 weeks.

The constant pressure to produce is removed, and an employee can work at different productivity levels, depending on how he or she feels during the day, or from day to day. If an employee falls behind for several days, the work can be made up on subsequent days. The day-to-day pressures of conventional incentives are removed.

Adopting longer measurement periods will facilitate the establishment of productivity measures for service and support operations, which cannot readily be measured by micromeasurement techniques.

Employees can review their productivity by day and by week to be sure they will attain the 13-week average they have selected in advance to justify their new hourly rate. If measurement is by group, members of the group can meet to discuss the changes needed to bring the group average in on target.

Reactions to Incentive and Nonincentive MDW

Employees usually accept that management requires work measurement and labor reporting in order to manage effectively. If given a choice, they would prefer to work without measurement. They view measurement as coercion to raise their productivity.

Given the choice of working under nonincentive MDW or conventional incentives such as piecework, and earning the same take-home pay, there is no question that employees will select the nonincentive conditions. When the choice is between the two pay systems, and increased pay opportunity is offered, most employees select the incentive method. Although no case data are available on employees' reasons, it is safe to assume that the increased pay is the main reason.

When employees work under nonincentive MDW, they usually do not express a preference for being measured by individuals or by a group. If there is a preference, it will more often be for the group measurement. In changing over from nonincentive MDW to incentive MDW, group measurement will usually be preferred by most in the group, especially if the service and support employees are included.

When traditional individual incentives are replaced by group incentive MDW, the high-earning employees will invariably complain if the group average is below the productivity level they previously achieved. The most frequently used approach to protect these employees who are above the group average is to guarantee their past average earnings by red circles. These employees will receive a personal add-on per hour, decided in advance, to ensure against a loss in personal earnings. These red circles may increase operation costs, and management may object to this approach. When going from traditional individual to individual incentive MDW, these problems do not arise.

There are insufficient case study data to predict definitively how employees who were previously on individual incentive will react when placed on group incentives. Studies of companies that replaced conventional incentive plans with nonincentive MDW clearly show that productivity on the whole dropped, often by large amounts.[40] Based on the author's experience, when traditional incentives are replaced by incentive MDW, productivity will be held at higher levels than when replaced by nonincentive MDW.

Some industrial engineers will contend that, as the measurement period is increased from daily to weekly, and more so for a 13-week period, productivity will decline, especially for groups. However, with a clear understanding among the work force of how all will benefit by the incentive MDW plan, and with good faith on all sides, productivity should, in fact, rise, especially if the employees have the opportunity to discuss their progress and take steps to raise their productivity.

Workers generally pace themselves at a level of productivity they find comfortable and can maintain. Extending the measurement period to 4 weeks, or even to 13 weeks, gives workers the advantage that they can vary their productivity levels from day to day, yet maintain their overall average at the level they wish to attain. The longer measurement period is also advantageous to management. In most plants productivity is calculated daily. Under a standard hour incentive plan, if an employee produces less than 8 hours' work, he or she is still paid for 8 hours. The loss is absorbed by management as make-up pay. If productivity is averaged over a long period, make-up pay will be averaged out. Some agreement between employees and management may be necessary on this point.

2.3.12 DESIGN OF TRADITIONAL INCENTIVE PLANS

Most wage incentive plans in this country are one-for-one plans, meaning that a 1% increase in productivity increases earnings by 1%. Piecework is the simplest incentive in which the standard is expressed as dollars per piece. A standard hour plan is the same as piecework, except that standards are expressed as minutes or hours per piece. If an employee works 8 hours and produces 10 hours' worth of work, he or she is paid for 10 hours at an agreed-upon base rate.

Different types of incentive plans can be designed with the earnings ratio different from the one-for-one arrangement. The mathematics and details of such plans are provided by Fein in another text.[41] Niebel, Barnes, and Nundel describe incentive plans in their texts.[42–44]

Most traditional incentive plans reward individual employees. Incentives can be designed for group work or to reward an entire department or even a plant.

Most incentive plans in this country operate successfully and benefit employees with higher earnings and companies with higher productivity and lower costs. Eliminating these incentives would reduce pay levels and raise costs. Where incentive plans have deteriorated, as described previously in this chapter, changes can be made to correct the problems and permit the incentives to operate. There is no question that pay by performance motivates employees to higher performance; this reduces costs and benefits employees, companies, and the public that buys their products.

Most industrial engineers use 100% normal as a self-supporting base in work measurement. This approach is valid within a given plant, but it may not always be used to compare time standards or incentive plans in different plants. In such comparisons the measurement systems of the plants must be set to the same measurement base.

Major Factors in Design of Incentive Plans

The major factors in plan design are (1) criteria to establish time standards and to evaluate performance, based on the relationship of normal to incentive pace, and (2) the ratio of labor participation in the increased productivity above normal.

Measurement and Incentive Plan Criteria

Measurement Reference Base. The most frequently used work measurement base is "normal," which may be defined by benchmarks. Incentive pace is less frequently used, but more definitive.

Incentive Pace. This is the work pace of a worker motivated by incentive pay to produce at a high productivity level, which can be maintained day after day. This pace is a subjective judgment, which may be supported by benchmarks or previously agreed-upon time standards for specifically defined operations or fundamental work elements.

Motivated Productivity Level (MPL). This is the work pace of a motivated worker who possesses sufficient skill to do the job properly; who is physically fit for the job, after adjustment to it; and who works at an incentive pace that can be maintained day after day without harmful effect.[45]

Normal or APL. This is the work pace established by management, or jointly by management and employees, at a level that is considered satisfactory; it is established at a given motivated productivity level. The APL is usually the productivity level at which incentive pay starts.

Standard. Standard is an established measurement base, usually stated as time per production unit, from which productivity measurements are made. Standard is usually set at either normal (APL) = 100% or incentive pace (MPL) = 100%. Technically, it may be any base that can be defined. Standard need not coincide with either normal (APL) or incentive pace (MPL).

Labor Participation Ratio. This is the percentage of increased productivity above standard that employees receive as incentive pay for their share of increased productivity. The most frequently used ratio is one-for-one, in which employees receive 100% above standard.

Note in Exhibit 2.3.4 that whether normal is expressed as 100% (column 2) or incentive pace as 100% (column 8), the pieces/hr at incentive pace (columns 5 or 9) remains at 100 pieces/hr. Also note that for each level of incentive expectancy percent, the pieces/hr for normal (columns 3 or 7) is the same whether normal is 100% (column 2) or the equivalent lower percent (column 6) when incentive pace is 100%. For example, at 25% incentive expectancy, normal is 80 pieces/hr (columns 3 or 7) whether normal is called 100% (column 2) or 80% (column 6). At 30% expectancy, normal is 77 pieces/hr (columns 3 or 7) whether normal is called 100% (column 2) or 77% (column 6).

Suppose each horizontal line in the table represented a different plant and that management in each plant had established as the incentive expectancy percent for its plant the figure shown on that line. Note that if normal is expressed as 100% (column 2), then each plant designates its nor-

Exhibit 2.3.4 Relationship Between Acceptable Productivity Level (APL) and Motivated Productivity Level (MPL) for Wage Incentive Plans of Various Incentive Expectancies

	Normal as 100%				Incentive Pace as 100%			
	Normal (APL)		Incentive Pace (MPL)		Normal (APL)		Incentive Pace (MPL)	
Incentive Expectancy Percent (1)	% (2)	Pieces/hr (3)	% (4)	Pieces/hr (5)	% (6)	Pieces/hr (7)	% (8)	Pieces/hr (9)
20	100	83.3	120	100	83.3	83.3	100	100
25	100	80.0	125	100	80.0	80.0	100	100
30	100	77.0	130	100	77.0	77.0	100	100
40	100	71.4	140	100	71.4	71.4	100	100
50	100	66.7	150	100	66.7	66.7	100	100

mal as 100%. Yet the pieces/hr (column 3) for each plant's normal is different. Such normals as measurement bases cannot be compared among companies. But when normal is expressed in relation to incentive pace as 100% (column 6), the pieces/hr at incentive pace (column 9) for each plant does not change. Now these normals for each plant fully represent the relative productivity level of each plant with respect to the others–assuming that the incentive pace concepts are the same, which they properly must be.

Analysis of Wage Incentive Plans

The basic incentive plans used are analyzed to show the relationship among productivity, earnings, and costs under the various plans. The following notations are used in the incentive plan formulas:

x = ratio of any given productivity level to the measurement base of the incentive plan, which can be normal (APL) = 100%, incentive pace = 100%, or any other definable base. In comparing plans, only incentive pace = 100% should be used.

y_w = ratio of wages at any productivity level to the base rate. The base rate will be assumed to be the wage paid at the productivity at which incentive begins. At this point, $y_w = 1.00$.

y_c = ratio of labor cost at any point to the labor cost at normal (APL).

p = ratio of labor participation in incentives. It is calculated as (% incentive pay)/(% increase in productivity above normal at point measured).

s = ratio of normal (APL) to incentive pace (MPL) for the plan.

General Comments About Incentive Plans

The cases that follow are all based on normal (APL) = 100% as the start of incentive payment. Therefore only two cases, 100% participation and other than 100% participation, are needed to describe all plans, except plans that combine both. When several plans are compared, the base is changed to incentive pace = 100%. Each incentive plan described guarantees the employee's base rate should productivity fall below normal (APL).

Case 1. "Day work" is a method of wage payment, not an incentive plan. It is discussed here because aspects of day work apply to incentive operations when productivity is below normal and base rates are guaranteed. Under day work the employee is paid the same hourly rate, regardless of productivity. The relationships are therefore as follows (see Exhibit 2.3.5):

Earnings $$y_w = 1 \tag{1}$$

Costs $$y_c = \frac{1}{x} \tag{2}$$

Exhibit 2.3.5 Relationship Between Costs, Earnings, and Productivity in the Daywork Wage Payment Plan

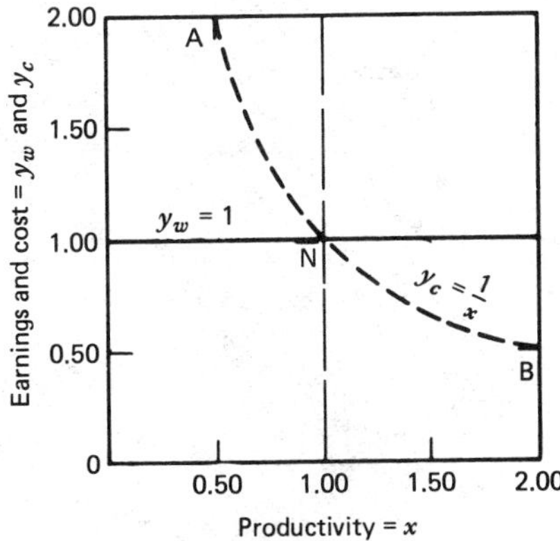

Cost – – –

Earnings ——

Case 2. *Incentive starting at normal (APL) = 100%, with 100% participation*, increases pay 1% for each 1% in productivity above normal, as shown by line NE in Exhibit 2.3.6.

	Below Normal		Above Normal	
Earnings	$y_w = 1$	(1)	$y_w = x$	(3)
Costs	$y_c = \frac{1}{x}$	(2)	$y_c = 1$	(4)

Exhibit 2.3.6 Relationship Between Costs, Earnings, and Productivity in the Incentive Plan with Incentive Beginning at Normal (APL) = 100%, with 100% Participation

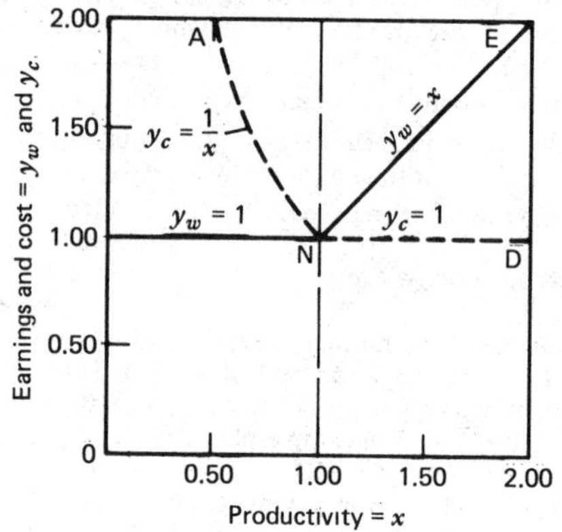

Case 3. *Incentive starting at normal (APL) = 100%, with other than 100% participation*. These plans are practically always less than 100% participation and are called "sharing plans."

Earnings are shown on line NF in Exhibit 2.3.7. Line NE shows earnings for Case 2, as a comparison with Case 3. When normal (APL) = 100%, the following applies:

	Below Normal		Above Normal	
Earnings	$y_w = 1$	(1)	$y_w = 1 + p(x - 1)$	(5)
Costs	$y_c = \frac{1}{x}$	(2)	$y_c = \frac{y_w}{x}$	
			$y_c = \frac{1 + p(x - 1)}{x}$	(6)

When incentive pace = 100%, with normal expressed in relation to incentive pace, equations 1, 2, 5, and 6 change to:

	Below normal		Above Normal	
Earnings	$y_w = 1$	(1)	$y_w = 1 + p\left(\frac{x}{s} - 1\right)$	(7)
Costs	$y_c = \frac{s}{x}$	(9)	$y_c = \frac{s[1 + p(x/s - 1)]}{x}$	(8)

Exhibit 2.3.7 Relationship Between Costs, Earnings, and Productivity in the Incentive Plan with Incentive Beginning at Normal (APL) = 100%, with Less Than 100% Participation.

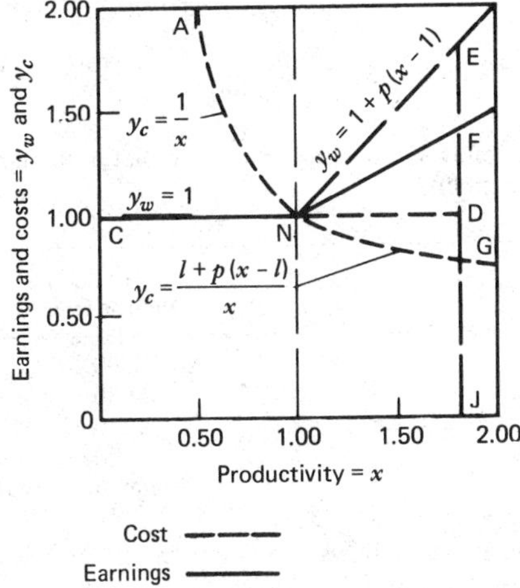

The participation ratio can be increased to over 100% for special cases where automatic machines are involved and the objective of the plan is to secure maximum operation of the machines. For example, a plan could offer a 3% increase in pay for a 1% increase in productivity, or some other such ratio. If the APL were set a 85%, with maximum output of the machine at 100%, $s = 0.85$. The participation ratio = 3. At 100% productivity, the earnings potential of the plan, following equation 7, would be

$$y_w = 1 + p\left(\frac{x}{s} - 1\right)$$
$$= 1 + 3(1/0.85 - 1)$$
$$= 1.529 = 52.9\%$$

Case 4. *Incentive beginning at lower than 100% productivity, but with 100% participation.* This plan in reality is the plan described in Case 2 with a reduced normal (APL). Such situations arise when management decides that the normal (APL) established at a given incentive expectancy is too high with respect to preincentive productivity. The various levels shown in Exhibit 2.3.7 fit this case. Suppose management decided initially that 25% incentive expectancy was appropriate for its plant. Normal (APL) = 100% would then be set, using Exhibit 2.3.7 as an example, at 80 pieces/hr. With preincentive production at 62.5, the employees would have to increase their output by 28% to reach normal.

If the productivity gap is too wide to be bridged under the circumstances, management can reduce the gap by reducing the normal (APL) to any other lower level shown in Exhibit 2.3.7. This increases the incentive expectancy of the plan and preserves meaningful guidelines. Some engineers suggest that the time standards not be changed, but that the incentive plan start paying off at 90%, 80%, or the like, rather than at 100%. If this is done, standards set at 100% are misleading with respect to incentive potential when incentive payments start at 80%.

Comparison of Different Incentive Plans

To compare incentive plans that have different incentive expectancies, use incentive pace (MPL) = 100% as the measurement base. The earnings above normal for such plans are obtained from equation 7:

$$y_w = 1 + p\left(\frac{x}{s} - 1\right)$$

To determine the productivity level at which two plans, A and B, have equal earnings, calculate equation 7 for each plan:

$$1 + p_a\left(\frac{x}{s_a} - 1\right) = 1 + p_b\left(\frac{x}{s_b} - 1\right)$$

This reduces to

$$x = \frac{p_a - p_b}{p_a/s_a - p_b/s_b} \tag{10}$$

Compare two plans with the following specifications; note that both plans are set to incentive pace (MPL) = 100% as the common point.

	Symbol	Plan A, One-for-One	Plan B, Sharing
Incentive pace		1.00	1.00
Normal (APL)	s	0.80	0.667
Incentive expectancy		25%	25%
Participation ratio	p	1.00	0.50
Base rates		same under both plans	

Comparison of Plans A and B is as follows: To determine the incentive earnings level at which Plans A and B are the same, use equation 10.

$$x = \frac{1.00 - .050}{1.00/0.80 - 0.50/0.667} = \frac{0.50}{0.50} = 1.00 = 100\%$$

Plans A and B will yield the same incentive earnings at 1.00, which is incentive pace = 100%.

Plans A and B are very different. Plan A is a relatively tight one-for-one plan, starting incentive payment at 80% of incentive pace with 25% expectancy. Plan B pays the same incentive pay at incentive pace as Plan A, but it starts to pay at 66.7% of incentive pace and at 50% participation.

Suppose management wants to design a Plan C to fit in between A and B, with the following features: incentive expectancy = 30%; participation ratio = 75%; s = 0.714, which is calculated as follows, using equation 7:

$$y_w = 1 + p\left(\frac{x}{s} - 1\right)$$

$$s = \frac{px}{y_w - 1 + p}$$

At what productivity level will Plans A and C break even in earnings? Using equation 10,

$$x = \frac{1.00 - 0.75}{1.00/0.80 - 0.75/0.714} = \frac{0.25}{0.20} = 1.25 = 125\%$$

The basic characteristics of the three plans compare as follows:

	A	B	C
1. With MPL = 100%, plans start payment at s	80%	66.7%	71.4%
2. Incentive expectancy	25%	25%	30%
3. Participation ratio (p)	100%	50%	75%
4. Break-even in earnings			
A with B		At 100% MPL	
A with C		At 125% MPL	

Inspection of the wage lines on Exhibit 2.3.8 shows that below 100% MPL Plan B is more liberal to the employees than Plan A; above 100%, the reverse occurs. Plan C is more liberal than Plan A for all levels below 125% MPL. To compare costs at any level, use the appropriate cost equation.

Exhibit 2.3.8 Relationship of Earnings to Productivity for Different Incentive Plans

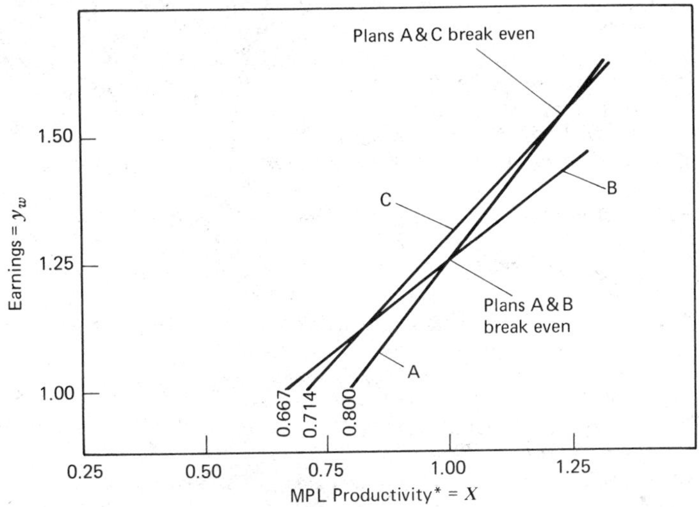

*Note: Because plans are compared at MPL = 100%, x in this graph is not at the same base as in Exhibits 2.3.5, 2.3.6, and 2.3.7, which are at APL = 100%.

REFERENCES

1. MITCHELL, FEIN, "Work Measurement and Wage Incentives," *Industrial Engineering*, September, 1973.
2. EARL L. LEWIS, "Extent of Incentive Pay in Manufacturing," *Monthly Labor Review*, May 1960.
3. U.S. DEPARTMENT OF LABOR, Bureau of Labor Statistics, *Wages and Related Benefits*, Part 2, Bulletin 1345-83, Washington, D.C., June 1964.
4. GEORGE L. STELLUTO, "Report on Incentive Pay in Manufacturing Industries," *Monthly Labor Review*, July 1969.
5. MITCHELL FEIN, personal correspondence with Bureau of Labor Statistics, 1969–1973.
6. DAVID A. WEEKS, *Compensating Salesmen and Sales Executives*, Report No. 579, The Conference Board, New York, 1972, p. 6.
7. H. FOX, *Top Executive Compensation*, Report No. 753, The Conference Board, New York, 1978, p. 4.
8. L. J. BRINDISI, "Survey of Executive Compensation," *World*, Spring 1971, p. 52.
9. GEORGE STRAUSS, "Managerial Practices," in Richard J. Hackman and Lloyd J. Suttle, Eds., *Improving Life at Work*, Goodyear, Santa Monica, CA, 1977.
10. R. A. KATZELL, D. YANKELOVICH, M. FEIN, O. A. ORNATI, and A. NASH, *Work, Productivity, and Job Satisfaction*, The Psychological Corporation, New York, 1975, p. 12.
11. MITCHELL FEIN, "Wage Incentive Plans," in H. B. Maynard, Ed., *Industrial Engineering Handbook*, 3rd ed., McGraw-Hill, New York, 1971.
12. KATZELL, *Work, Productivity, and Job Satisfaction*, p. 36.
13. KATZELL, *Work, Productivity, and Job Satisfaction*, p. 12.
14. MITCHELL FEIN, "Job Enrichment, a Reevaluation," *Sloan Management Review*, Vol. 15, No. 2 (1974).
15. ROLLIN H. SIMONDS and JOHN N. ORIFE, *Administrative Science Quarterly*, Vol. 20, December 1975.
16. SURVEY RESEARCH CENTER, University of Michigan, *Survey of Working Conditions*, U.S. Government Printing Office, Washington, D.C., November 1970.

17. MITCHELL FEIN, "The Real Needs and Goals of Blue Collar Workers," *Record*, The Conference Board, February 1973.
18. E. A. LOCKE, DENA B. FEREN, VICKIE M. McCALEB, KARYLL N. SHAW, and ANNE T. DENNY, "The Relative Effectiveness of Four Methods of Motivating Employee Performance," in K. D. Duncan, M. M. Bruneberg, and D. Wallis, Eds., *Proceedings of the NATO Industrial Conference*, August 1979, Wiley, London, 1980.
19. MITCHELL FEIN, *Rational Approaches to Raising Productivity*, Monograph No. 5, American Institute of Industrial Engineers, Norcross, GA, 1974.
20. EDWARD E. LAWLER, III, "Reward Systems," in *Improving Life at Work*, Goodyear, Santa Monica, CA, 1977.
21. DOUGLAS MCGREGOR, "The Scanlon Plan Through a Psychologist's Eyes," in Frederick G. Lesieur, Ed., *The Scanlon Plan*, The MIT Press, Cambridge, MA, 1958.
22. BRIAN E. MOORE and TIMOTHY L. ROSS, *The Scanlon Way to Improved Productivity*, Wiley, New York, 1978.
23. C. F. FROST, J. H. WAKELY, and R. A. RUH, *The Scanlon Plan for Organization Development*, Michigan State University Press, Ann Arbor, MI, 1974.
24. KENNETH J. WHITE, "The Scanlon Plan: Causes and Correlates of Success," *Academy of Management Journal*, Vol. 22, No. 2 (1979), pp. 292–312.
25. CARL HEYEL, Ed., *The Encyclopedia of Management*, 2nd ed., Van Nostrand Reinhold Company, New York, 1973, p. 895.
26. A. W. RUCKER, *Progress in Productivity and Pay*, The Eddy-Rucker-Nickels Company, Cambridge, MA, 1952.
27. A. W. RUCKER, *Gearing Wages to Productivity*, The Eddy-Rucker-Nickels Company, Cambridge, MA, 1962.
28. BERT L. METZGER, *The Future of Profit Sharing*, Profit Sharing Research Foundation, Evanston, IL, 1979, p. 89.
29. MITCHELL FEIN, "Establishing Time Standards by Parameters," in *Proceedings*, American Institute of Industrial Engineers 1978 Annual Conference, Norcross, GA.
30. MITCHELL FEIN, "Work Measurement Today," *Industrial Engineering*, August 1972 and September 1972.
31. U.S. DEPARTMENT OF LABOR, *Supplemental Unemployment Benefit Plans and Wage-Employment Guarantees*, Bulletin No. 1425-3, Washington, D.C., June 1965.
32. ROBERT S. RICE, "Survey of Work Measurement and Wage Incentives," *Industrial Engineering*, July 1977.
33. ROBERT S. RICE, "Survey of Work Measurement and Wage Incentives," *Factory*, April 1959.
34. BERTRAM GOTTLIEB and CHARLES WERNER, *Statutory Obligation of an Employer to Furnish Information to a Union*, American Institute of Industrial Engineers, Norcross, GA, 1975.
35. FRANK ELKOURI and EDNA A. ELKOURI, *How Arbitration Works*, 3rd ed., The Bureau of National Affairs, Washington, D.C., 1973.
36. RONALD L. WIGGENS, *The Arbitration of Industrial Engineering Disputes*, The Bureau of National Affairs, Washington, D.C., 1970.
37. D. BERENBEIM, *Nonunion Complaint Systems: A Corporate Appraisal*, Report No. 770, The Conference Board, New York, 1980.
38. MITCHELL FEIN, "Measured Day Work as an Incentive Plan," *Industrial Engineering*, January 1979.
39. AMERICAN NATIONAL STANDARDS INSTITUTE, *ANSI Industrial Engineering Terminology, Work Measurement and Methods*, ANSI Z94.12; and revised draft, American Society of Mechanical Engineers, New York, 1972.
40. RALPH M. BARNES, "Industrial Engineering Survey 1967," *The Journal of Industrial Engineering*, December 1967.
41. FEIN, "Wage Incentive Plans," *Industrial Engineering Handbook*.
42. BENJAMIN W. NIEBEL, *Motion and Time Study*, 6th ed., Irwin, Homewood, IL, 1976.
43. RALPH M. BARNES, *Motion and Time Study*, 7th ed., Wiley, New York, 1980.
44. MARVIN E. MUNDEL, *Motion and Time Study*, 5th ed., Prentice-Hall, Englewood Cliffs, NJ, 1978.
45. MUNDEL, *Motion and Time Study*, p. 74.

CHAPTER 2.4

Job and Task Analysis

ERNEST J. McCORMICK

Purdue University

2.4.1 INTRODUCTION

Job and task analysis (generally referred to here as job analysis) deals with the study of human work. This field has had a rather checkered history, characterized by periods of increasing and decreasing interest and by the occasional development of significant new methods and procedures. The recent years have been noteworthy for two reasons, namely, because of the development and use of certain new methods and because of increased interest in the use of job-related data. In part the increased interest in job analysis has been the result of certain provisions of the Civil Rights Act of 1964, especially the provisions relating to employment and compensation.

This chapter deals primarily with methods of analyzing human work, with particular emphasis on methods that provide for the quantification of job-related data. Although the chapter touches on certain applications of job-related data, the primary discussions of such data are in Chapters 2.5, 3.1, 3.2, 3.3, 5.2, 5.3, and 5.4.

In general terms, job analysis is concerned with the gathering, evaluating, and recording of job-related data. Although job analysis is the study of the work activities of workers, it is not the study of the workers themselves. In the analysis of human work it is also important to maintain a clear distinction between what the worker does and what gets done as the consequence of his or her work activities. Further reference to this point is made later.

Terminology

Terminology in the field of job study is far from being precise and definitive, since certain terms have been used with various shades of meaning. In this regard Melching and Borcher,[1] referring to the use of job analysis in curriculum development, express this state of confusion as follows:

> *While job analysis experts employ concepts such as task, function, responsibility, duty, etc. as though the distinctions among them were both obvious and fixed, this is simply not true. The curriculum designer should be warned that any attempt to place these terms into a reliable hierarchy may not turn out to be very rewarding.*

Realizing such risks, it can be said that in at least some circumstances a duty sometimes is considered to be broader in content than a task, a task is considered broader than an element (or job element), and an element is considered broader than an elemental motion. But one should keep in mind that the distinctions are very thin and may even depend on the context in which the terms are used, including the degree of specialization of a job. The *Handbook for Analyzing Jobs* (1979 draft),[2] for example, suggests that the activity "slices cold meats and cheese" could be an element in the job of short order cook, a task in the job of sandwich maker, and the total job of a deli cutter-slicer.

Portions of this chapter are based on, or reprinted by permission from, the following sources: ERNEST J. MCCORMICK. *Job Analysis: Methods and Applications.* AMACOM, a division of American Management Association, New York, 1979) pp. 15–149; and ERNEST J. MCCORMICK, "Job Information: Its Development and Applications," in *ASPA Handbook of Personnel and Industrial Relations*, copyright © 1979 by the Bureau of National Affairs, Inc., Washington, D.C.

2.4.2 USES OF JOB-RELATED INFORMATION

Any program dealing with the collection and analysis of job-related information should be undertaken only if it would serve some relevant purposes. In this regard it is helpful to view the uses from the point of view of the user, such as employment organizations with uses for personnel selection and placement, job design, and job evaluation; unions that are interested in uses for contract negotiations, jurisdictional matters, and so on; government agencies with uses in the operation of public employment offices, establishing standards for licensing, and equal employment opportunity; and individuals whose interests are in such matters as vocational counseling and preparation.

2.4.3 ASPECTS OF JOB ANALYSIS PROCESSES

The development of job-related information typically involves a two-stage process. The first stage consists of eliciting the information from a "source," such as by observation and interview of a job incumbent or by the reporting of such information by the incumbent through use of a questionnaire. The second stage consists of organizing and presenting the information in the desired format. In the case of conventional job analysis procedures, this format usually is the common job description. In the case of certain methods the format may consist of a computer output, illustrations of which are presented later.

The purposes of any given job analysis program should serve as the basis for making decisions about the specific aspects of the program. In making such decisions one should start by considering the type of information and format desired as an "end product" and then plan the initial analysis procedures to ensure the fulfillment of this objective. When planning for the initial data collection, the following four questions need answering.

1. What *type* of information is to be obtained?
2. In what *format* is the information to be obtained and presented?
3. What *method* of analysis will be used?
4. What *agent* will be used? Usually the "agent" is an individual, such as a job analyst, a supervisor, or the job incumbent. In special circumstances it may be a device such as a camera.

Methods of Job Analysis

There are various methods of collecting job-related information. The most common are the interview and observation, which are frequently used in combination. Sometimes group interviews are used, in which two or more job incumbents are interviewed at the same time by the job analyst. In recent years various types of structured job analysis questionnaires have been developed. These usually consist of a listing of job-related items, such as tasks, from which incumbents, supervisors, or analysts are asked to indicate something about the involvement of incumbents with each item. Occasionally, open-ended questionnaires are used, with incumbents being asked to describe their jobs. In other circumstances work diaries are used, with incumbents keeping a record of how they spend their time.

Discussion of Methods

Although the observation-interview method, with a job analyst serving as the "agent," is the most common method used for manual type jobs, it is fairly common practice for people in professional and managerial jobs to write their own job descriptions. It should be added, however, that in recent years there has been an increased use of structured job analysis questionnaires that are completed by job incumbents, analysts, or supervisors. This chapter deals particularly with the conventional job analysis procedures that typically involve observation and interview methods and with structured job analysis questionnaires.

2.4.4 PREPARATION FOR A JOB ANALYSIS PROGRAM

As with most things, success of a job analysis program depends largely on the care that goes into the planning and preparation stage. A few aspects of this phase are discussed briefly.

Crystallizing the Program Objectives

A job analysis program typically should be considered by an organization only if there is some recognized need for it. Although this need might relate to one of the several possible objectives or uses of job-related information, the organization should carefully consider other possible legitimate uses of such data and then set forth the specific objectives that it wishes to achieve with the program. In this regard, however, it should be recognized that, although the information obtained with a given program might well serve several different purposes, there are limits to the purposes that can reasonably be fulfilled with any given program.

Development of Job Analysis Materials

Once the objectives of a job analysis program have been crystallized, the next stage is to develop the various materials that will be required. Paramount to this stage is the development of the job analysis format that is to be used and of the instructions that are to be followed in the actual analysis of jobs and in the preparation of final job descriptions, computer printouts, or other end products. The job analysis format and instructions should provide for the collection of the *type* of information that would be required to fulfill the stated objectives, the *format* in which such information is to be obtained (and later presented), the *method* of analysis, and the *agents* (the individuals–or devices–to be used in the data collection phase).

In the case of a conventional job analysis program, the format would be a simple form that provides space for the desired information. This form can be used for the recording of notes by the analyst and also for preparing a final job description. However, in the case of a structured job analysis procedure, appropriate questionnaires need to be selected or developed. Some examples are given later.

Instructions need to be prepared for all persons who will be involved, such as job analysts and job incumbents, if the latter are to complete any questionnaires. The instructions should be as simple and straightforward as possible and preferably should be pretested with a sample of people before they are prepared in final form.

Selection and Training of Analysts

When analysts are to be used, care should be taken in their selection. They should be people with analytical ability, writing skill, and personal qualities that would be useful in interviewing and dealing with others. Whenever possible, the analysts should be persons who have some familiarity with the jobs that are to be analyzed. If the analysts are not already familiar with the jobs, one aspect of the training program should consist of reading relevant background material relating to the industry in question, the processes with which the jobs are involved, and the organizational structure in which the jobs exist.

The analysts should be trained in the job analysis procedures that are to be followed and should try a few "practice" job analyses if they have never done any before.

Advance Information About the Program

When a job analysis program is to be inaugurated, there should be appropriate communications to all persons who will be involved, such as department heads, supervisors, and job incumbents. In this regard, it is usually good practice for management to prepare a letter or some other form of announcement that assures all interested persons of management support of the program.

When a job analyst is to analyze jobs in a particular unit of the organization, arrangement should be made through the normal chain of command. Whoever makes such arrangements should make clear to the various officials the reasons for the analysis and should indicate what will be involved, including the time required on the part of the job incumbents. In turn, the supervisor of the job incumbents should advise the incumbents themselves of the analyses and should schedule the analyses at such times as are mutually convenient with the incumbents, the job analyst, and the work activities of the unit. Thus the way should be completely paved for the study by allaying any qualms on the part of the incumbents.

If the analyst uses the combined observation and interview method, it is frequently the practice to observe the worker first and conduct the interview afterwards. However, the two may be accomplished at the same time, or the analyst may prefer to interview the worker first in order to identify in advance the key aspects of tasks that warrant close attention. The sequencing of the two processes will depend on the nature of the job and on the judgment of the analyst as to what approach would be most useful. The interview can be carried out at the work site if that location provides adequate privacy and if it is reasonably quiet and safe. Otherwise it should be carried out at some other location that satisfies these requirements.

To be safe, the analyst should again make sure that the individual whose job is to be analyzed has been advised in advance by his or her supervisor of the purpose of the analysis and that the individual is willing to have his or her job analyzed.

2.4.5 OBSERVATION AND INTERVIEWING IN JOB ANALYSIS

Various job analysis methods involve the observation of incumbents performing their jobs and/or the conducting of interviews with incumbents, with their supervisors, or with others who are knowledgeable about the job in question.

Interviewing in Job Analysis Processes

Since interviewing is an important part of various methods of job analysis, the individuals who serve as analysts need to develop interviewing skills that will enable them to get the most out of each interview.* Although interviews normally involve an interviewer and an interviewee, in some instances more persons may be involved in either or both roles.

Interviews can vary in the degree of "structure" from unstructured to highly structured. In job analysis processes, semistructured interviews usually are most appropriate, especially if the interviewer uses a job schedule that provides for eliciting information on each of several aspects of a job. Such a schedule typically consists of a prepared list of questions or items about which information is to be obtained. Although the job schedule provides a basic structure around which the interview can be conducted, the interviewer should be prepared to adapt his or her interview approach to the personality of the interviewee and to the nature of the job being discussed.

Principles of Good Interviewing

There are three basic principles of good interviewing. First, the initiative should always be with the interviewer, but not to the point of overpowering the interviewee. Second, the interviewer's manner and attitude should reflect sincere and genuine interest in the interviewee. Third, the interviewer should guide the interview toward the purpose of obtaining the desired information.

Developing Interviewing Skills

Skill in interviewing is based largely on asking the right questions at the right time in the right words. But, in addition to asking the right questions, the interviewer must develop the ability to listen actively to the responses from the interviewee. This requires sufficient sensitivity to understand what is being said and the ability to recall the important points to be recorded. Kuriloff et al.[3] suggest that questions used by interviewers be checked against the following criteria:

The question should be related to the purpose of the analysis. The wording should be clear and unambiguous.

The question should not "lead" the respondent; that is, it should not imply that a specific answer is desired.

The question should not be "loaded" in the sense that one form of response might be considered to be more socially desirable than another.

The question should not ask for knowledge or information the interviewee cannot be expected to have.

There should be no personal or intimate material that the interviewee might resent.

The questions asked should encourage the interviewee to do most of the talking. Any lack of clarity may be based on a lack of understanding of the question, on the use of unfamiliar language, or on the lack of ability to express oneself. In any event, the interviewer should probe further, with simple, diplomatic questions, until the point is clarified.

Kuriloff et al.[4] suggest that, while "listening" to the respondent, the interviewer can engage in any of four mental activities, as follows:

Thinking ahead of the speaker, that is, trying to anticipate what the discourse is leading to and what conclusions will be drawn from the words spoken at the moment.

Weighing the evidence given by the interviewee that supports the points being made, by asking mentally, "Is this point valid? Is the evidence complete?"

Listening between the lines for meaning that is unspoken.

Paying close attention to nonverbal signals such as facial expressions, gestures, tone of voice, and emphasis, to see if the spoken meaning has been altered in any way.

General Guidelines in Interviewing

Realizing that there are no simple cookbook rules for conducting good interviews, the following few guidelines or reminders are given for whatever value they may be to persons who will be serving as job analysts.

*This discussion of the interview is based in part on material from A. H. KURILOFF, and C. H. STONE, *Training Guide for Observing and Interviewing in Marine Corps Task Analysis. Evaluation of the Marine Corps Task Analysis Program*, Technical Report No. 2., California State University, Los Angeles, August 1975; and U.S. DEPARTMENT OF LABOR, Manpower Administration, *Handbook for Analyzing Jobs*, Washington, D.C., 1972.

Preparing for the Interview

1. Build the interviewee's interest in advance through well-prepared announcements, and be sure that each interviewee is advised in advance by his or her supervisor of the arrangements for the interview.
2. Select proper accommodations that ensure privacy for the interview.
3. Avoid or minimize the use of status symbols that earmark the interviewer as having a higher "status" than the interviewee.

Opening the Interview

1. Put the worker at ease by learning his or her name in advance, introducing yourself and discussing general and pleasant topics long enough to establish rapport. Be at ease.
2. Make the purpose of the interview clear by explaining why the interview was scheduled, what is expected to be accomplished, and how the worker's cooperation will help in the production of occupational analysis tools used for placement and counseling.
3. Encourage the worker to talk by always being courteous and showing a sincere interest in what he or she says.
4. Relate the interview to goals that the interviewee holds important.

Steering the Interview

1. Help the worker to think and talk according to the logical sequence of the duties performed. If duties are not performed in a regular order, ask the worker to describe the duties in a functional manner by taking the most important activity first, the second most important next, and so on. Request the worker to describe the infrequent duties of the job, ones that are not part of his or her regular activities, such as the occasional setup of a machine, occasional repairs, or infrequent reports. Infrequently performed duties, however, do not include periodic or emergency activities such as an annual inventory or the emergency unloading of a freight car.
2. Ask questions to encourage the respondent to talk.
3. In unscheduled portions of the interview, use probing techniques to keep the conversation alive: an expectant pause, brief assenting comments, unobtrusive neutral questions, summarizing what the respondent just said, or repeating a question.
4. Allow the worker sufficient time to answer each question and to formulate an answer. He or she should be asked only one question at a time.
5. Phrase questions carefully, so that the answers will be more than "yes" or "no."
6. Leading questions should be avoided.
7. Conduct the interview in plain, easily understood language.
8. Show honest personal interest in the interviewee.
9. Do not be aloof, condescending, or authoritative.
10. Keep a steady, consistent pace.
11. Be sure to obtain complete and specific information of all the types that are required for the analysis.
12. Include in the analysis reference to other related jobs, if such information is relevant to the analysis of the job in question.
13. Control the interview with regard to the subject matter (the information required for the analysis) and with regard to time. If the interviewee strays too far from the subject, the interviewer can bring him or her back by summarizing the information already obtained.
14. Conduct the interview patiently and with regard to the reactions of the interviewee as reflected, for example, by his or her nervousness or lack thereof. The analyst should attempt to establish a friendly, but businesslike, relationship with the worker. There is no prescribed method of accomplishing this; the analyst must develop techniques for dealing with workers in a way that establishes rapport and encourages cooperation. The analyst who conducts the interview in a courteous and interested manner, who listens carefully to what the worker has to say, and who is neither patronizing nor overbearing usually is able to gain the worker's trust and confidence and obtain the desired information.

The analyst must be alert to the worker's reaction to being observed or interviewed. Although most workers can be expected to be slightly nervous—at least initially—while being studied, those who become upset, distracted, annoyed, belligerent, or uncooperative should not be subjected to continued study. The analyst should select, or ask the supervisor to select, other workers in the same job.

Matters unrelated to job analysis, such as grievances, labor-management conflicts, safety and health violations, and wage classification problems should not be discussed by the analyst. If a worker brings up such topics, the analyst should tactfully steer the interview back to the analysis of the job. The analyst should avoid making comments or suggestions about improving work flow, plant layout, work methods, and the design of the jobs.

Closing the Interview

1. Indicate the approach to the end of the interview by the kinds of questions you ask and by the inflection of your voice.
2. If relevant, summarize the information obtained from the worker, indicating the major duties performed and the details concerning each of the duties.
3. Conclude by pointing out the value of the information the respondent has given.
4. Close the interview on a friendly note.

Miscellaneous Do's and Don't's for Interviewers

1. Do not take issue with the worker's statements.
2. Do not show any partiality to grievances or conflicts concerning the employer-employee relations.
3. Do not show any interest in the wage classification of the job.
4. Show politeness and courtesy throughout the interview.
5. Do not "talk down" to the worker.
6. Do not permit yourself to be influenced by your personal likes and dislikes.
7. Be impersonal. Do not be critical or attempt to suggest any changes or improvements in organization or methods of work.
8. Talk to the worker only with permission of his or her supervisor.
9. Verify job data, especially technical or trade terminology, with supervisor or department head.
10. Verify completed analysis with proper official.

The analyst should make notes as he or she observes and interviews the worker, but should do so in as unobtrusive a manner as possible, combining the note taking with the conversational aspect of the interview. Some specific suggestions for effective note taking are as follows.

1. Notes should be complete and legible and should contain data necessary for the preparation of the job analysis schedule.
2. Notes should be organized logically, according to job tasks and the categories of information required for a complete analysis.
3. Notes should include only the facts about the job, with emphasis on the work performed and on the worker traits involved. Use only words, phrases, and sentences that impart necessary information.

Following the observation and interview with the worker, the analyst usually would interview the supervisor to obtain certain types of additional information (such as that dealing with experience, training, relationships to other jobs, etc.) and to clarify any points on which he or she is in doubt.

The first-line supervisor, rather than the worker, should be interviewed when factors such as safety, high noise levels, and language barriers hinder oral exchange of information between the analyst and the worker; when ground rules prohibit interviews with the worker; and when the supervisor is better able to describe and explain the job observed.

Ensuring Completeness of Collected Data

Sometimes, observation shows only a portion of the worker's activities because the total range of activities occurs over a period of several hours, days, weeks, or longer. Questioning often reveals that the worker performs many additional activities that are not seen during the observation phase. Some activities may be unobservable because of the work location, time of day (e.g., at the start and the end of the work shift), or infrequency of performance.

Keeping Records of Observations or Interviews

If an analyst is observing a job or interviewing a worker, he or she should take notes or obtain other records that can be used later in the preparation of the final job description. Although a good memory is a definite asset in job analysis studies, it should not be relied upon too heavily. When interviewing a worker, however, the analyst should make the note taking as unobtrusive as possible. In addition to notes that might be taken, the analyst might consider using a tape recorder and in the case of some jobs even a camera. Such equipment, however, should be used only with the knowledge and consent of the worker and his or her supervisor and only when justified by the nature and purpose of the job analysis. Further, in the case of some jobs the analyst might draw rough sketches of machines and equipment or obtain already printed material relating to them.

2.4.6 JOB ANALYSIS WRITING

The end product of certain job analysis processes consists of some form of written material. Needless to say, the objective of the analyst in writing such material should be that of conveying the intended meaning in as reliable and valid a fashion as possible. Some of the differences in the nature of job analysis material that are characteristic of different methods are discussed and illustrated in later chapters. For the purposes here, however, I will touch on three aspects of job analysis writing that apply in various ways to different methods of job analysis. These aspects are organization, sentence content and structure, and selection of words.

Organization

The organization of job description material depends upon the job analysis method being used and upon the nature of the job in question. In the case of most job description material, the information is organized into related major job segments or in the sequence in which the job activities are carried out, if there is such sequence. Lacking any other logical basis for organization, the descriptive information may be arranged in terms of judged importance of the various activities or the time devoted to them. In the case of certain structured job analysis procedures, such as task inventories that consist of listings of tasks, the tasks may even be arranged alphabetically, in terms of the "duties" with which they are involved or in terms of functional relationships.

Sentence Content and Structure*

The content of sentences used in job description varies greatly with the method being used. In the case of some task and methods analysis procedures, for example, the sentences are very simple, consisting of only a verb (in the third person, present tense) and an object (possibly with an adjective modifying the object), such as "installs antennas" or "operates power saw." In the case of conventional job descriptions, however, the sentence structures may be complex, compound, or compound-complex, as in the following examples:

Complex sentence (containing one main clause and one or more subordinate clauses): Replaces tube when test indicates that present tube is not good.

Compound sentence (containing two or more main clauses and no subordinate clauses): Removes tire from rim and inspects for defects.

Compound-complex sentence (containing two or more main clauses and at least one subordinate clause): At end of month, or when all accounts receivable and accounts payable records are received, prepares a listing of each and computes totals.

In connection with job descriptive material, it should be noted that, in some instances, as in some task inventories that are illustrated later, the verb is in the first person, present tense form, as "Solder minor leaks in radiator." In such instances a job incumbent can assume "I" as the subject. In some instances statements are used that do not constitute complete sentences. This is especially the case with certain structured job analysis procedures in which items of equipment, such as "blow torch," or descriptions of activities, such as "use of keyboard devices," are used.

Selection of Words

In preparing job descriptive material, the analyst should be careful to select words that minimize the possibility of ambiguity while achieving the desirable objective of brevity. A few general guidelines regarding the use of words are as follows:†

Prefer the simple word to the farfetched.

Prefer the concrete word to the abstract.

Prefer the single word to circumlocution.

Prefer the short word to the long.

Use technical words of special significance only if the intended audience would understand them. Otherwise they should be avoided or explained.

Use adjectives sparingly, since they may tend to reflect opinions. When their use is supported by reasonably objective evidence, however, they can be used if they add significant meaning.

*Adapted in part from A. H. KURILOFF and D. YODER, *Communications in Task Analysis. Evaluation of the Marine Corps Task Analysis Program*, Technical Report No. 9, California State University, Los Angeles, 1975, p. 59.

†Adapted in part from KURILOFF and YODER, *Communications in Task Analysis*, pp. 62–70.

Minimize the use of gerunds and participles. These are words derived from verbs that usually end in "-tion," "-ion," "-ing," and "-ment." A gerund is a verb used as a noun, such as in the sentence "The investigator's *findings* indicated that . . .". A participle is a verb used as an adjective, such as in the sentence "While observing the bread *baking* . . .". There are times, however, when gerunds and participles provide the most effective and efficient method of description.

Limit the use of imprecise words such as "condition," "situation," "facilitate," and "inadequacy."

2.4.7 PREPARATION AND REVIEW OF JOB ANALYSIS MATERIAL

In the case of many job analysis programs, the persons who serve as the job analysts, that is, the "agents" referred to earlier, prepare descriptive material such as conventional job descriptions, listings of tasks performed by job incumbents, and various types of "forms" listing the operations that are carried out. In such instances it is usually the practice to present drafts of such materials to supervisors, experts, or management personnel for review, possible modification, and, in some cases, formal approval. In the use of structured job analysis questionnaires that are completed by job incumbents, the questionnaires sometimes are reviewed by supervisors. As mentioned previously, for certain structured job analysis procedures, the final format of information may consist of computer printouts.

2.4.8 CONVENTIONAL JOB ANALYSIS PROCEDURES

The conventional job analysis programs that are used in many organizations typically involve the collection of job-related information by observing and/or interviewing job incumbents and the preparation of job descriptions, which usually are written in essay form.

The organization that has had the most extensive experience in conventional job analysis is the U.S. Employment Service (USES) of the Employment and Training Administration (ETA) of the U.S. Department of Labor. Although certain of the procedures it uses are only relevant for its objective, some of its practices and guidelines may have general applicability for other organizations that are planning conventional job analysis programs. Thus the primary focus of the discussion of conventional job analysis here is based on USES practices.

Content of Conventional Job Analyses

Generally, most conventional job analysis programs provide for obtaining two major types of job information, namely, what the USES calls "work performed" and "work characteristics," there being more specific classes of both of these, which the USES refers to as "job components."[5] In the case of each of these components, the USES provides a systematic procedure for characterizing, coding, or rating any given job. Although these procedures are not included here, the concepts implied by these components would be relevant to many conventional job analysis programs.

Components of Work Performed

The components of work performed include those listed here. (In certain instances the current terminology as used by the USES is different from that used before; in such instances the former terminology, as was used in the 1972 edition of the *Handbook for Analyzing Jobs*, is also given.)

Worker Functions. These are the ways in which a job requires the worker to function in relation to data, people, and things, as expressed by mental, interpersonal, and physical worker actions. The USES provides a "structure" of these worker functions, there being a sequence for each of the three classes that forms something of a hierarchy, as presented in Exhibit 2.4.1. These and other similar types of verbs are used in job descriptions to indicate *what* the worker does to accomplish some phase of his or her job.

Work Fields. These are different types of technologies and socioeconomic objectives that reflect how work gets done and what gets done as the result of the activities of a job, or, in other words, the purpose of the job. Examples based on specific technologies are "electroplating" and "abrading"; on overall social objectives are "accommodating" and "health caring"; on the types of objects dealt with are "animal propagating" and "plant cultivating"; and on combinations of specific related technologies are "machining" and "structural fabricating-installing-repairing."

Work Devices (formerly machines, tools, equipment, and work aids–MTEWA). These include the machines, equipment, tools, and work aids used by the worker to carry out the specific activities of the job.

Materials, Products, Subject Matter, and Services (MPSMS). This category includes (1) basic materials being processed, such as fabric, metal, or wood; (2) final products being made, such as automobiles or baskets; (3) data, when being dealt with or applied, such as in economics or physics; and (4) services, such as barbering or dentistry.

Exhibit 2.4.1 Structure of Worker Functions as Used by the U.S. Employment Service

Data	People	Things
0 Synthesizing	0 Consulting	0 Precision working[b]
1 Planning	1 Directing	1 Setting up[c]
2 Analyzing	2 Negotiating-persuading[a]	2 Operating-controlling[a,c]
3 Differentiating	3 Instructing	3 Driving-operating[a,c]
4 Computing	4 Supervising	4 Manipulating[b]
5 Compiling	5 Treating	5 Tending[c]
6 Copying	6 Performing	6 Feeding-offbearing[a,c]
7 Sensing	7 Selling	7 Handling[b]
	8 Serving	
	9 Communicating	

Source. U.S. DEPARTMENT OF LABOR, Employment and Training Administration, *Handbook for Analyzing Jobs* (1979 draft), Washington, D.C., September 1979, p. 118.

[a] This function is a single function.

[b] This function is a non-machine-oriented function.

[c] This function has a machine-oriented relationship.

Worker Characteristics. This includes job analysis components that reflect characteristics required of the worker for successful job performance and characteristics of the work activities themselves and of the environment in which they are performed. Those provided for in the USES procedures include the following: general educational development, job training time, aptitudes, interests, temperaments, and physical demands of the job.

Environmental Conditions. These are the specific environmental conditions to which the worker is exposed while performing assigned duties, such as cold, heat, noise, and vibration.

Discussion

The nature of the specific content of conventional job descriptions naturally varies with the purposes in mind, but normally the descriptions of the work performed include material relevant to all of the components of work performed mentioned previously, usually at least certain worker characteristics (such as important ones required for successful job performance) relevant environmental conditions. The work-performed portions of job descriptions frequently include a job summary and more detailed descriptions of the tasks included in the job.

In job analysis jargon one frequently hears reference to the "what, how, and why" of job analysis processes. This refers to the desirability, in describing human work, of being sure that the description covers what the worker does, how he or she does it, and why he or she does it. Although in some circumstances the "how" and "why" may be implicitly obvious, if there is any question at all about these aspects, they should be explicitly brought out in the description.

What the worker does is characterized by statements regarding the physical and mental activities that are performed on the job. As Butler points out,[6] physically the worker may transport materials, cut, grind, set up, regulate, finish, or otherwise change the position, shape, or condition of the work by the expenditure of physical effort; mentally he or she may engage in such activities as planning, computing, judging, or directing, including in some instances the governing of the expenditure of his or her own or others' physical effort.

The *how* of the work activities that are performed deals with the methods or procedures used to carry out the job tasks. In the case of physical activities, this may involve the use of machinery and tools or other equipment, the following of certain procedures or routines, or the execution of certain physical responses such as hand movements. In the case of mental activities, this may involve the use of calculations or formulas, the exercise of judgment, or the selection and transmittal of thought. In considering the *how* of the job, Butler[7] suggests that the analyst should try to cover the following questions:

What tools, materials, and equipment are used to accomplish all of the tasks of the job?

Are there other tools, materials, and equipment that have not been observed? If so, how do they work?

What methods or processes are used to accomplish the tasks of the job?

Are there other methods or processes by which the same work can be done?

The *why* of the job analysis process, the basic purpose(s) of the job, should be one of the first things the analyst seeks to determine and should be brought out in any job summary. Besides describing in the job summary why the job exists, the description of the specific tasks should also include some indication as to why the individual tasks are performed in case this is not clearly implied in the description of what and how. The *why* of the individual tasks generally would characterize the purpose of the task as related to the fulfillment of the overall objectives of the job as incorporated in the job summary.

The following portions of a job description, dealing with the job summary and a couple of tasks, are given to illustrate the manner in which the *what*, *how*, and *why* are brought out.[8]

Job: Engine lathe operator–first class

Job Summary: Sets up and operates an engine lathe to turn small airplane fittings from brass or steel barstock or from unfinished aluminum or magnesium alloy castings (*why*), finishing fitting down to specified close tolerances (*what, how*).

Work Performed (descriptions of two tasks):

1. Sets up lathe (*what*); carefully examines blueprints (*what*) to determine the dimensions of the part to be machined (*why*), using shop mechanics (*how*) to calculate any dimensions (*what*) not given directly on the print (*why*) or to calculate machine settings (*why*).
2. Sets up lathe to turn stock held in chuck (*what*); attaches to lathe the accessories, such as chuck and tool holder (*what*), necessary to perform the machining (*why*), threading and locking the chuck and the head stock spindle (*how*) and setting.

Writing Style in Conventional Job Analysis

A rather definitive writing style has become somewhat common in job analysis circles, especially in writing conventional job descriptions. This is described in the *Handbook for Analyzing Jobs* (1979 draft)[9] as it would apply to the description of tasks:

> *The style to be followed in recording the descriptions of tasks should conform to the following basic rules:*
>
> 1. *The style is terse and direct, omitting all unnecessary words.*
> 2. *The present tense is used throughout.*
> 3. *"The worker," the subject of each sentence, is implied (unstated), and no pronoun is used in place of "the worker."*
> 4. *Each sentence begins with an action verb, as specific as possible to indicate what the worker does. (Exceptions to this convention are certain adverbs such as "occasionally" and "manually," which may precede an action verb at the beginning of a sentence.)*
> 5. *Words are chosen for exactness and for having but one interpretation.*
> 6. *Articles ("the," "a") are omitted.*
> 7. *Superlatives ("most," "best"), certain types of adverbs ("very," "perfectly"), and attributes ("complex," "large," "heavy," "small") are not used.*
> 8. *Jargon, verbose phrasing, and little-known words (in place of simpler, better known synonyms) are avoided.*
> 9. *Technical and little-known terms and special or unusual machines, equipment, tools, work aids, materials, and products are defined, either parenthetically or by underlining the term the first time it appears and defining it in a supplemental section. The background of the intended users of a job description determines which terms need defining.*
> 10. *In detailed job descriptions, flag (lead) statements introduce and summarize each task. (They may not be needed for brief job descriptions.)*

There are, however, some circumstances in which some qualifying word or phrase should precede the verb, such as in specifying the circumstances under which a particular activity is performed, such as "At the end of each week, compiles reports to show . . . ," or when an adverb is needed to modify the verb, such as "Verbally assigns . . . ".

Basic Sentence Structure

In the case of most job description material, there is also a basic sentence structure that has become somewhat standardized in job description practice. This basic sentence structure, as set forth in the *Handbook of Analyzing Jobs* (1972), is given in Exhibit 2.4.2, with an example of a "job worker" situation, the operation of a telephone switchboard. The job worker situation is defined briefly, and the "analysis" consists of the following components:

Verb (the "worker function")

Immediate object (typically materials, tools, equipment or work aids; data; or people)

Infinitive phrase
  Infinitive (a "work field")
  Object of the infinitive (some material, product, subject matter, or service)

Some other examples are as follows:

Verb	Immediate Object	Infinitive	Object of Infinitive
Analyzes	examination papers	to evaluate	knowledge of law candidates.
Compiles	credit information	to determine	credit rating.
Operates	saw	to cut to size	metal materials.
Feeds	blending machine	to blend	flour.
Describes	features of interest	to inform	visitors to factory.
Computes	hours, pay scale, and so on	to calculate and post	wages.

In describing most work activities, the analyst should keep in mind the writing style and the type of sentence structure discussed previously, characteristically using a "worker function" verb, indicating the immediate object of the verb, followed by an infinitive phrase to reflect the "purpose," or "why," of the activity (this phrase consisting of an infinitive and its object). The analysis given in Exhibit 2.4.2 and the other examples provided are admittedly rather stilted. Such a sentence structure, when used in actual job descriptions, usually would be worded in such a fashion as to fit in with other descriptive material. In some instances the infinitive phrase can be omitted if the purpose of the activity is obvious.

The Job Summary

The sentence analysis technique as applied to a job summary is illustrated as follows:

1. Begin with a worker action verb, such as those referred to in connection with basic activities. Remember that "the worker" is always the implied subject of the verb. Example: "*Tends* . . . ".

2. Follow the action verb with the machine or equipment used, the immediate object of the verb. Example: "Tends *injection molding machine* . . . ".

3. Next, indicate the purpose of the worker action by an infinitive phrase, beginning with the word "to." The purpose, thus stated, should reflect the basic method used in the task. Example: "Tends injection molding machine *to mold* . . . ".

4. Next, indicate the materials and/or products as the objects of the infinitive phrase. Example: "Tends injection molding machine to mold *resin pellets* into *plastic bottles*."

Exhibit 2.4.2 Example of the Analysis of the Basic Sentence Structure of a Description of Work Activity, in Particular, the Activities of a Switchboard Operator

Job Worker Situation: Operates cord or cordless switchboard to relay incoming, outgoing, and interoffice calls. On cordless switchboard, pushes switch keys to make connections and relay calls. On cord-type equipment, plugs cord in jacks mounted on switchboard. Supplies information to callers and records messages.

Analysis

Verb (Worker Function)	Immediate Object	Infinitive Phrase	
		Infinitive (Work Field)	Object of Infinitive
Compares	switchboard operation with standards	to relay	calls.
Converses with	callers	to convey, to receive	information.
Operates	cord or cordless switchboard	to relay	incoming, outgoing, and interoffice calls.

Source. U.S. DEPARTMENT OF LABOR, Manpower Administration, *Handbook for Analyzing Jobs*, Washington, D.C., 1972, p. 201.

Descriptions of Tasks

Task descriptions usually start off with a flag statement that consists of a verb and an object. It serves to orient the reader to the scope and content of the task by showing in general terms what the worker does. The task description that follows elaborates on the flag statement through specific action verbs, as well as with other categories of information that tend to reflect the what, how, and why of job analysis processes. An example of a task description is as follows:

> *Fells and limbs trees: Pulls cord to start gasoline-powered chain saw. Squeezes trigger to bring chain saw up to cutting speed. Cuts notch in standing tree in direction of desired fall. Cuts halfway through tree in center of notch. Walks around tree and completes cut from other side (backcut), stepping back to avoid possible jump or twist from stump. Walks alongside felled tree and limbs tree (saws off all branches), squeezing and releasing chain saw trigger to increase or decrease cutting speed, depending on diameter of branch and resistance of wood. Holds chain saw firmly against tree throughout cutting process to avoid kickback from knots and tough bark.*

A flag statement, "Fells trees," would cover the activities described in the first five sentences of the task description. The sixth sentence, beginning with "Walks alongside felled tree and limbs tree . . . ," describes an activity that is performed after the tree is felled, namely, the limbing of trees. (The last sentence of the description pertains to both felling and limbing.) The flag statement therefore should read, "Fells and limbs trees."

Flag statements may be written in basic-activity terms, such as "Compiles data," "Instructs student," and "Tends machine"; or they may have other verbs, such as "Maintains files," "Prepares report," and "Responds to customer inquiries."

Task descriptions that are extremely brief or for which little detail is given may not require flag statements.

Example of Job Description

The purpose of a job description is to convey to the reader as accurate an impression of the job as possible. The analyst therefore should organize and write the description in whatever manner would best serve this purpose; he or she should not be a slave to any given "model" job description. Recognizing this, the following example of parts of one job description, that of a tool designer, is given.[10]

JOB SUMMARY (Tool Designer)

Designs special tools, dies, jigs, and fixtures for use on all types of production machines.

Body

1. Studies tooling problem to determine basic part and machine specifications governing design of the tool: Reads tool order, examines blueprint of finished part, and analyzes sequence of operations sheet to determine machining operations required of new tool. Studies part and part blank and computes dimensions of part or parts before and after machine operation for which tool design is required. Sketches part in relationship to tool as guide to tool design. Examines machine for which tool is to be made and confers with chief tool designer to gather information to make decisions relative to designing the tool needed to perform necessary machining operations. Draws sketches of machine and part incorporating basic decisions made, including dimensions, clearances, and tolerances (see comments).

2. Designs tools: Develops form and shape of tool by studying tool design drawings of other similar tools, by comparing own ideas of tool's design with accumulated part and machine specifications, and by drawing rough and semidetailed sketches. Calculates final detail dimensions, clearances, and tolerances of tool, using design reference book, machinists' handbook, mechanical engineers' handbook, trigonometry, slide rule, and standard formulas. Draws general assembly drawings of complete tool, showing top, front, and side views; tool, machine, and part in actual use; and all dimensions, tolerances, and clearances.

3. Writes tool specifications covering materials and processes: Selects commercial tool items for purchase from vendors, considering part machined, tolerances, allowances of tool and part, speed of machine, tool coolants used, estimated tool life, and cost in relation to specifications desired. Selects type of material, such as tool steel, to use in fabricating tool, using same criteria in selecting material as purchased part. Determines fabrication or construction specifications.

4. Assigns general assembly drawing to design drafter for preparation of detail drawings of tool parts, advises design drafter concerning drafting techniques and procedures, and reviews completed detail drawings.

5. Advises toolmaker on problems, such as sizes, tolerances, clearances, material selection, and other problems encountered in fabricating and assembling tool.
6. Redesigns tools that fail to meet machining requirements.

Job Requirements

Experience: None

Acceptable: Tool design drafter

Training data: Minimum training time

a. Inexperienced workers. 6 months
b. Experienced workers. 6 months

Training:

In-plant (on job) training: Tool designer apprentice

Vocational training: Technical school or vocational school. Course must include algebra, geometry, trigonometry, mechanical drawing, machine shop practice, and shop mathematics.

Specific job skills acquired through training: Fundamental principles of tool designing. Experience in designing tools. Calculating dimensions and reading blueprints and specifications. Knowledge of algebra and other mathematics that are helpful in computations.

Apprenticeship:

Formal: X
Informal:
Length required: 5 years
Design drafter

Responsibility. Responsible for proper and efficient design of tools; drill jigs; turning fixtures; grinding fixtures for surface grinders and cylindrical grinders; and blanking, piercing, and forming dies and gages. Responsible for the correct dimensions and calculation of tools, jigs, fixtures, dies, and gages. Responsible for assigning assembly drawings to design drafters. Responsible for verifying that completed drawings comply with specifications and blueprints.

Job Knowledge. Must have knowledge of shop mathematics and drafting techniques. Must be able to read and understand blueprints. Must be able to use micrometers, verniers, height and depth gages, and calipers. Must have knowledge of tool design, toolmaking methods, and machining and other properties of metals.

Mental Application. Must be able to develop new ideas in tool design and to adapt existing designs. Must be alert and able to concentrate on fine detail. Must be able to exercise independent judgment and confer with others to resolve problems. Must exercise initiative in solving design problems.

Dexterity and Accuracy. Must be accurate in making calculations and in designing tools that meet very fine specifications, often to within 0.0001 of an inch. Must be able to read instructions, specifications, and various measuring instruments with absolute accuracy.

General Comments. In general, the tool designers develop all types of tools for various production machines used in the factory. The tool designers are not recognized as specialists in such areas as tool or cutter design, jig or fixture design, or die design, although some have more talent, interest, and experience in some of these special areas, and the supervisor routes such jobs to them. At times two or more tool designers will work on one tool order, each designing one particular tool of the many making up the tool order.

Vocational school graduates and technical high school graduates who have training in algebra, geometry, trigonometry, mechanical drawing, machine shop practice, and shop mathematics can be trained on the job to become tool designers so that mechanical engineers can utilize the fuller scope of their training on machine design or in a supervisory or advisory capacity on design activities.

2.4.9 STRUCTURED JOB ANALYSIS METHODS

The widespread use of conventional job descriptions is a reflection of the fact that they do serve useful purposes in personnel management and vocational guidance, despite some limitations. The limitations of such descriptions arise primarily from their dependence upon verbal material that is largely in essay form. Even with the most skillful use of language in essay form, there can be some "slippage" in conveying to the reader the meaning that is intended.

Because of such shortcomings, there have been efforts over the years to develop job analysis methods that are more "systematic" and that tend to be more quantitative than qualitative. Generally, these efforts have been aimed at the development of job analysis procedures that provide for the identification and/or measurement of "units" of job-related information, such as tasks and worker attributes. This would make it possible to compare jobs in terms of similarities and differences, to group jobs in terms of their similarities, and to otherwise "manipulate" job-related information conceptually and statistically. Such approaches are becoming known as structured job analysis procedures.

These efforts have been carried out to develop methods that can serve certain purposes better than conventional job descriptions and other purposes that cannot be served at all by conventional descriptions. It is expected that conventional job descriptions will continue to be useful for personnel management and other purposes, especially for characterizing the "role" or objective of jobs and for reflecting an "integrated" impression of the job activities. Actually, some job analysis programs consist of combinations of conventional and structured methods. In fact, the procedures used by the USES, although basically of the conventional job analysis approach, include some structured methods.

Structured job analysis procedures provide for analyzing jobs in terms of specific "units" of job-related data. The analysis results in either (1) a determination as to whether any given "unit" of data does or does not apply to a job or (2) a numerical rating of the degree to which it does apply. Two types of structured job analysis procedures are discussed here, namely, task inventories and the position analysis questionnaire (PAQ).

Task Inventories

A task inventory is a form of structured job analysis questionnaire that consists of listing the tasks within some occupational field. These questionnaires typically provide for job incumbents, supervisors, or analysts to report some job-related information about the incumbent's involvement with each task. They also lend themselves to other uses, as is discussed later.

Task inventories of one type or another have been developed and used over many years, but their primary development and use has been by the United States Air Force during and since the 1960s, under the direction of Raymond E. Christal, Air Force Human Resources Laboratory, Brooks Air Force Base, Texas.[11–13] The methodology has been adopted by other military services in the United States and other countries, other government agencies, universities, some private companies, and certain trade and professional organizations.*

The Nature of Task Inventories

A task inventory is characterized by two features, namely, a list of tasks for the occupational field in question and provision for some type of response to each task. The list of tasks usually consists of all or most of the tasks that can be performed by incumbents within the occupational field. Typically, the individual task descriptions consist only of a statement of *what* is done, in job-oriented terms, and do not include indications of *how* and *why*.

The tasks usually, but not always, are grouped into broader "duties" such as "maintaining and repairing braking systems." An example of part of a task inventory as presented by Melching and Borcher[14] is shown in Exhibit 2.4.3. In their final form, task inventories might include anywhere from a few dozen to several hundred tasks for an occupational area.

Response Scales Used With Task Inventories

There are essentially two types of response scales that can be provided for with task inventories. The first provides for some indication about the involvement of the job incumbent with each task. The second type provides for some response about the task in terms of a judgment or attitude about it.

Response scales of the first type are sometimes referred to as "primary rating factors." Various types of such scales can be used, such as the following:

Importance (importance of the task to the job).

Part-of-job scale, as follows:

0 Definitely not a part of the position.
1 Under unusual circumstances may be a minor part of the position.
2
3

*Directories of task inventories have been developed by the Center for Vocational Education, The Ohio State University, 1960 Kenny Road, Columbus, Ohio 43210. In addition, *Task Analysis Inventories*, published by the Manpower Administration (now the Employment and Training Administration) of the U.S. Department of Labor in 1973, consists of a compilation of task inventories for several occupational areas.

4 A substantial part of the position.
5
6
7 A most significant part of the position.

Performance (whether the incumbent does or does not perform the task).

Frequency of performance (how often the task is performed per unit of time, such as per day, per week, or per month).

Time spent (time spent on the task when it is performed, as in minutes).

Relative time spent (estimated time spent on each task relative to time spent on other tasks).

The United States Air Force uses the following scale:

1. Very small amount.
2. Much below average.
3. Below average.
4. Slightly below average.
5. About average.
6. Slightly above average.
7. Above average.
8. Much above average.
9. Very much above average.

Variations of these can be developed, such as the total time spent on a task during the individual's total work career, as contrasted with the time spent on it in the present job.

In the reporting of time spent on various tasks, it has been the experience of the Air Force that a *relative* time-spent scale, such as previously shown, usually is better than an *absolute*, or percentage, time-spent scale. When a relative time-spent scale is used, the responses can be converted into an estimate of percentage of work time. The procedure for estimating these percentages is given by Archer.[15] The values so derived are regarded as estimates of the percentage of work time spent by each incumbent on the tasks in question.

The second type of response scale, sometimes referred to as "secondary taskrating factors," provides for judgmental or subjective responses about the tasks themselves, as contrasted with reports about the involvement of incumbents with the tasks. Some of the secondary rating factors that have been used are:

Complexity of the task.

Exhibit 2.4.3 Examples of a Few Tasks From a Task Inventory for Automobile Mechanics[a]

Automotive Mechanics Task Inventory	Page 19 of 23 Pages	
Listed below are a duty and the tasks which it includes. Check all tasks which you perform. Add any tasks you do which are not listed. Then rate the tasks you have checked.	Check	Time Spent
M. Maintaining and repairing braking systems		1. Very much below average 2. Below average 3. Slightly below average 4. About average 5. Slightly above average 6. Above average 7. Very much above average
1. Repair master cylinder		
2. Repair wheel cylinder	✓	4
3. Replace brake hoses and lines	✓	1
4. Replace brake shoes	✓	6
5. Resurface brake drums		
6. Adjust brakes	✓	7

[a]From Melching and Borcher, reference 14.

Criticality of the task.

Difficulty of learning the task.

Where the incumbent learned the task (training course, on the job, etc.).

Where the respondent believes the task should be learned.

Special training required (amount) to do the task.

Time considered necessary to learn the task (ranging from "can do now" to "can learn in a few hours" to "would take more than a year to learn").

Difficulty of performing the task.

Technical assistance required in task performance.

Supervision required in task performance.

Satisfaction in performing task.

Usually, responses to the secondary rating factors are made by job incumbents. Thus an incumbent might be asked to use one primary rating factor, such as the relative time spent on individual tasks, and also to use one or more secondary rating factors, such as difficulty of learning the task or satisfaction in performing the task. However, such secondary rating scales can be completed by supervisors, "experts," or others to elicit their judgments about, or subjective reactions to, the individual tasks in the abstract. In either case the responses to the secondary rating scales can be used to derive an index of the factor as related to the individual tasks. An example of such an index comes from an occupational survey of the television equipment repair career ladder in the Air Force.[16] Experienced personnel were asked to rate the tasks in an inventory on task difficulty. The average difficulty indices for several tasks are given below. The original indices were converted to values with an arbitrary mean of five and a standard deviation (SD) of one.

Task	Average Difficulty Index
Align receiver tuners	7.07
Install microwave relay systems	6.81
Perform operational checks of helical VTRs	5.01
Adjust video processing amplifiers	4.98
Remove or replace loudspeakers	3.13
Operate turntables	2.91
Clean film splicers	2.29

Development of Task Inventories

As indicated previously, a task inventory usually is developed for a specific occupational area. The procedures involved are time consuming, frequently taking several weeks, and usually involve the collaboration of a job analyst and a technical expert. Usually, experimental forms of inventories are administered to samples of job incumbents in order to elicit their comments and suggestions and to obtain data that might be useful in the preparation of a final form. (See Melching and Borcher[14] for general instructions regarding the development of inventories.)

Administration of Task Inventories

The administration may be arranged for by bringing groups of job incumbents together under the guidance of a job analyst or by distributing the inventories to the incumbents who are asked to complete and return them to a central office. In most instances in which task inventories are used, the respondents are asked to record their responses on forms that can facilitate any subsequent analysis. Thus mark-sensing forms or optical scanning forms are used frequently, in order to facilitate the later computer processing of the resulting data.

In planning the administration of task inventories for large organizations, it is often desirable to select a sample of the incumbents to respond, rather than the entire work force.

Analysis of Task Inventory Data

Data returned from task inventories can be subjected to various types of analysis by the use of appropriate computer programs. One of the primary types of analysis results is the preparation of a group job description that summarizes the responses to the tasks for all individuals within whatever job "group" has completed the inventory. An example is shown in Exhibit 2.4.4. Another type of

Exhibit 2.4.4 Portion of a Group Job Description–U.S. Air Force Television Equipment Repair Specialist

Representative Task	Percent Performing[a]
Solder or desolder connectors or hardwire circuits	85
Interpret schematic diagrams	85
Remove or replace plug-in or screw-in electronic components, such as transistors, tubes, or indicator lights	84
Remove or replace soldered electronic components on etched circuit boards	83
Adjust receiver or monitor operating controls, such as focus or centering	83
Interpret block diagrams or component location diagrams	79
Isolate malfunctions of receiver or monitor circuits	79
Isolate malfunctions of receiver or monitor subassemblies	71
Adjust camera operating controls, such as target beam or setup	70
Isolate malfunctions to receiver or monitor subassemblies	70
Remove or replace receiver or monitor subassemblies	67

Source. OCCUPATIONAL SURVEY BRANCH, USAF Occupational Measurement Center, *Occupational Survey Report: Television Equipment Repair Career Ladder*, Randolph Air Force Base, Texas, AFPT 90-304-376, October 1979.
[a] Representative tasks performed by 258 personnel.

analysis results in job-type description. This analysis is based on statistical procedures, such as cluster or factor analysis, that result in the identification of "job types," that is, the groups of job incumbents who perform somewhat similar combinations of tasks. An example is given in Exhibit 2.4.5.

There are many other types of analyses that can be carried out with task inventory data, these depending partly on the nature of the primary and secondary rating scales that are used and on the objectives of the analysis.

Position Analysis Questionnaire*

The PAQ is a structured job analysis questionnaire that provides for analyzing jobs in terms of 187 job elements. The elements are of a "worker-oriented" nature, which tends to characterize, or to imply, the human behaviors that are involved in jobs.[17] As such, the PAQ lends itself to use in the analysis of a wide variety of jobs.

Organization of the PAQ

The job elements in the PAQ are organized in six divisions as follows (examples of two job elements from each division are included):

1. **Information Input.** Where and how does the worker get the information used in performing his or her job? Examples: Use of written materials; near-visual differentiation.
2. **Mental Processes.** What reasoning, decision making, planning, and information processing activities are involved in performing the job? Examples: Level of reasoning in problem solving; coding/decoding.
3. **Work Output.** What physical activities does the worker perform, and what tools or devices does he or she use? Examples: Use of keyboard devices; assembling/disassembling.
4. **Relationships With Other Persons.** What relationships with other people are required in performing the job? Examples: Instructing; contacts with public customers.

*The PAQ was developed under the provisions of a series of contracts between the Office of Naval Research, Personnel and Training Research Programs Branch, and the Purdue Research Foundation and is copyrighted by the Purdue Research Foundation. The PAQ and related materials are available through the University Book Store, 360 West State Street, West Lafayette, Indiana 47906. Further information regarding the PAQ is available through PAQ Services, Inc., 1625 North, 1000 East, Logan, Utah 84321. Computer processing of PAQ data is available through the PAQ data processing division at the Logan, Utah, address.

Exhibit 2.4.5 Parts of Job-Type Descriptions for Three Job Types in U.S. Air Force Television Equipment Repair Career Ladder

Representative Task	Percent Performing
Job Type: Quality Control Inspectors	
Draft correspondence or reports	91
Perform technical inspection of equipment	91
Evaluate proposed equipment modifications	91
Evaluate completed equipment modifications	91
Interpret policies, directives, or procedures for subordinates	82
Evaluate inspection reports or procedures	82
Perform quality control inspections of equipment	82
Job Type: Instructors	
Conduct resident course classroom training	100
Evaluate progress of resident course students	100
Write test questions	83
Counsel trainees on training progress	83
Develop course curricula or plans of instruction (POI)	83
Maintain training records, charts, or graphs	83
Demonstrate how to locate technical information	67
Administer or score tests	50
Job Type: Camera Maintenance	
Interpret schematic diagrams	100
Adjust camera linearity controls	100
Adjust camera operating controls, such as target beam or setup	100
Adjust camera sync processing circuits	100
Isolate malfunctions of camera sync processing circuits	100
Isolate malfunctions to camera interconnecting cables	100
Adjust receiver or monitor operating controls, such as focus or centering	100

Source. OCCUPATIONAL SURVEY BRANCH, USAF Occupational Measurement Center, *Occupational Survey Report: Television Equipment Repair Career Ladder*. Randolph Air Force Base, Texas, AFPT 90-304-376, October 1979.

5. Job Context. In what physical or social contexts is the work performed? Examples: High temperature; interpersonal conflict situations.

6. Other Job Characteristics. What activities, conditions, or characteristics other than those already described are relevant to the job? Examples: Specified work pace; amount of job structure.

Rating Scales Used With the PAQ

Provision is made for rating each job on each job element. Six types of rating scales are used, as follows:

Letter Identification	Type of Rating Scale
U	Extent of Use
I	Importance to the job
T	Amount of Time
P	Possibility of occurrence
A	Applicability
S	Special code (used in the case of a few specific job elements)

A specific rating scale is designated to be used with each job element, in particular, the scale that is considered to be most appropriate to the content of the element. All but the "A" (Applicability) Scale are 6-point scales, with "0" being used for "Does not apply," illustrated as follows:

Importance to the Job	
N	Does not apply
1	Very minor (importance)
2	Low
3	Average
4	High
5	Extreme

The "A" Scale (Applicability) provides for responses of "Applies" or "Does not apply" and is used only in the case of certain job context elements, such as "Regular work hours."

Analysis of Jobs With the PAQ

The analysis of jobs with the PAQ is typically carried out by job analysts, methods analysts, personnel officers, or supervisors, and in some instances job incumbents are asked to analyze their own jobs, especially in the case of managerial, professional, and other white collar workers.

Exhibit 2.4.6 Job Dimensions Based on Principal Components Analyses of PAQ Data for 2200 Jobs

Division Dimensions

Division 1: Information Input

1. Interpreting what is sensed.
2. Using various sources of information.
3. Watching devices/materials for information.
4. Evaluating/judging what is sensed.
5. Being aware of environmental conditions.
6. Using various senses.

Division 2: Mental Processes

7. Making decisions.
8. Processing information.

Division 3: Work Output

9. Using machines/tools/equipment.
10. Performing activities requiring general body movements.
11. Controlling machines/processes.
12. Performing skilled/technical activities.
13. Performing controlled manual/related activities.
14. Using miscellaneous equipment/devices.
15. Performing handling/related manual activities.
16. General physical coordination.

Division 4: Relationships With Other Persons

17. Communicating judgments/related information.
18. Engaging in general personal contacts.
19. Performing supervisory/coordination/related activities.
20. Exchanging job-related information.
21. Public/related personal contacts.

Division 5: Job Context

22. Being in a stressful/unpleasant environment.
23. Engaging in personally demanding situations.
24. Being in hazardous job situations.

Division 6: Other Job Characteristics

25. Working nontypical versus day schedule.
26. Working in businesslike situations.
27. Wearing optional versus specified apparel.
28. Being paid on a variable versus salary basis.
29. Working on a regular versus irregular schedule.
30. Working under job-demanding circumstances.
31. Performing structured versus unstructured work.
32. Being alert to changing conditions.

Overall Dimensions

33. Having decision, communicating, and general responsibilities.
34. Operating machines/equipment.
35. Performing clerical/related activities.
36. Performing technical/related activities.
37. Performing service/related activities.
38. Working regular day versus other work schedules.
39. Performing routine/repetitive activities.
40. Being aware of work environment.
41. Engaging in physical activities.
42. Supervising/coordinating other personnel.
43. Public/customer/related contacts.
44. Working in an unpleasant hazardous/demanding/environment.
45. Unnamed.

Job Dimensions Based on the PAQ

Analyses using PAQ of a sample of 2200 jobs were used by Mecham[18] as the basis for identifying the underlying "dimensions" of jobs, these having been identified by statistical factor analysis. Any given dimension can be thought of as constituting a combination of certain job elements that tend to "go together" in jobs. In statistical terms, they are correlated with each other. It is not feasible here to define each dimension or to list the job elements that tend to characterize each dimension, but their titles are listed in Exhibit 2.4.6. Some impression of their nature can be inferred from their titles.

For any given job, it is possible, with a computer program, to derive a score on each dimension. Thus any job can be "described" quantitatively in terms of job dimension scores. Since these are expressed quantitatively, they can be used for any of several different purposes. One such purpose is that of identifying job families, jobs that have similar profiles of job dimension scores. The job dimension scores can also be used for deriving estimates of pay rates for jobs (thus eliminating the need for conventional job evaluation processes), as discussed in Chapter 5.3. In addition, the PAQ can be used for deriving estimates of the aptitude requirements for jobs, thus eliminating the need for conventional test validation procedures as discussed in Chapter 5.3. The statistical basis for such predictions is discussed by McCormick et al.[19] and will not be described here. However, some illustrative data from this study are given. Predicted mean test scores on four aptitudes were derived statistically from PAQ data for incumbents on about a hundred jobs. These predicted scores were then correlated with the actual mean test scores of incumbents on tests of these aptitudes, the resulting correlations being as follows:

Aptitude	Correlation	No. of Jobs
General intelligence	.75	34
Verbal aptitude	.71	50
Numerical aptitude	.67	65
Spatial aptitude	.70	30

The results from various studies and from experience have demonstrated that structured job analysis procedures such as the PAQ can have substantial practical utility for various personnel management functions.

REFERENCES

1. M. H. MELCHING, and S. D. BORCHER, *Procedures for Constructing and Using Task Inventories*, Research and Development Series No. 91, The Ohio State University, Center for Vocational and Technical Education, Columbus, March 1973.
2. U.S. DEPARTMENT OF LABOR, Employment and Training Administration, *Handbook for Analyzing Jobs* (1979 draft), September 1979.
3. A. H. KURILOFF, D. YODER, and C. H. STONE, *Training Guide for Observing and Interviewing in Marine Corps Task Analysis. Evaluation of the Marine Corps Task Analysis Program*, Technical Report No. 2, California State University, Los Angeles, August 1975.
4. KURILOFF et al., *Training Guide*.
5. *Handbook for Analyzing Jobs* (1979 draft), p. 7.
6. J. L. BUTLER, *Job Analysis: (What + How + Why = Skills Involved)*, private manuscript, Stamford, CT, 1975, pp. 37–38.
7. BUTLER, *Job Analysis*, p. 39.
8. BUTLER, *Job Analysis*, p. 43.
9. *Handbook for Analyzing Jobs* (1979 draft), p. 47.
10. *Handbook for Analyzing Jobs* (1979 draft), pp. 374–376.
11. R. E. CHRISTAL, *New Directions in the Air Force Occupational Research Program*. USAF, AFHRL, Personnel Research Division, Lackland Air Force Base, Texas, 1972.
12. J. E. MORSH, *Computer Analysis of Occupational Survey Data*. USAF, AFHRL, Personnel Research Division, Lackland Air Force Base, Texas, 1969.
13. J. E. MORSH and W. B. ARCHER, *Procedural Guide for Conducting Occupational Surveys in the United States Air Force*. ASAF, AMD, Personnel Research Laboratory, Lackland Air Force Base, Texas, PRL-TR-67-11, 1967.
14. MELCHING and BORCHER, *Procedures for Constructing and Using Task Inventories*, p. 35.

15. W. B. ARCHER, *Computation of Group Job Descriptions from Occupational Survey Data*. USAF, AMD, Personnel Research Laboratory, Lackland Air Force Base, Texas, PRL-TR-66-12, 1966.

16. OCCUPATIONAL SURVEY BRANCH, USAF, Occupational Measurement Center, *Occupational Survey Report: Television Equipment Repair Career Ladder*, Randolph Air Force Base, Texas, AFPT 90-304-376, October 1979.

17. E. J. MCCORMICK, P. R. JEANNERET, and R. C. MECHAM, "A Study of Job Characteristics and Job Dimensions as Based on the Position Analysis Questionnaire (PAQ)," *Journal of Applied Psychology*, Vol. 56, No. 4 (1972), pp. 347–368.

18. R. C. MECHAM, Unpublished report, February 1977.

19. E. J. MCCORMICK, A. S. DENISI, and J. B. SHAW, "Use of the Position Analysis Questionnaire for Establishing the Job Component Validity of Tests," *Journal of Applied Psychology*, Vol. 64, No. 1 (1979), pp. 51–56.

CHAPTER 2.5

Job Design

LOUIS E. DAVIS

University of California, Los Angeles

GERALD J. WACKER

The Aerospace Corporation

2.5.1 INTRODUCTION

At the heart of any organization are its jobs. The division of the work to be done by the organization into jobs sets forth relationships among people and between people and technology. This chapter examines the factors and decision processes that lead to the design of jobs. Emphasized are approaches to the design of effective and satisfying jobs, which meet both the organization's needs for effectively achieving its goals through the use of its human resources and the individual's needs, expectations, and goals.

2.5.2 SYSTEMS APPROACH TO JOB DESIGN

Research into the factors underlying job design,[1,2] and the effects of job designs on productivity and employee satisfaction,[3–7] suggests that jobs can be understood only in relation to a whole enterprise or agency as a system. This section reviews the systems perspective for the design of jobs.

Multiple Dimensions of Jobs

An organization, whether a manufacturing plant, a service firm, or a government agency, exists in several dimensions. First, it is a *production entity*, consisting of buildings, equipment, and technology (technical systems) that have been brought together for the purpose of transforming certain inputs into desired outputs. Of course an organization is more than just a production entity. It is also a *social or institutional entity*, or "miniature society," consisting of roles, traditions, conflicts, long-term strategies, and relationships with other institutions. Finally, an organization is a *collection of particular individuals.* Each employee has goals, commitments, and lifestyles that only partially intersect with the organization. To design jobs that are effective, account must be taken of the production dimension, the organizational dimension, and the individual dimension as they pertain to the various kinds of decisions that go into job design.

Definition of Organization and Its Jobs

An organization is a system of roles or jobs and a structure of role relationships deliberately brought together at a particular time to accomplish desired outcomes. To achieve its desired outcomes, an organization does work. To do work, an organization brings together people (social system) and the means (technical system) through which they can jointly accomplish the desired outcomes. To perform effectively, these subsystems are required to act as a joint system, frequently referred to as a sociotechnical system.

Definition of Job

A job is a cluster of tasks assigned to a role in an organization. It is a part of the organization with which its members are identified. Over the last 100 years, jobs were defined as clusters of work tasks performed by individuals, whereas roles were defined as clusters of work tasks *plus* other kinds of tasks, which needed to be done because work was done not simply by an individual, but by an individual in an organization. In organizations designed on the basis of contemporary organization theory, the difference between job and role has disappeared. An organization is functioning when its roles are occupied by people carrying out the requirements of each role or job. An organization remains virtually unchanged even though its roles or jobs are occupied by different people at different times.

Definition of Job Design

Job design is the process of making the decisions that determine (1) what tasks are to be performed by the work force, (2) what tasks are to be clustered into what jobs, and (3) how the jobs are to be linked together.

Technical and Social Support Systems

Job design decisions are interdependent with decisions regarding technical systems and with personnel policy. A number of technical systems alternatives may be derived from a conversion technology. A technical system is a set of procedures, techniques or methods, instructions, equipment, tools, and layout chosen for carrying out a particular process of converting a material, information, or human input into a desired output. Choices made regarding the content and configuration of a technical system are in part job design decisions. How technical system choices create tasks and influence the clustering of tasks into jobs is reviewed in the remainder of this section.

The design of jobs involves not only technological choices but also social choices. The technical system used to convert material, information, or human inputs into desired outputs generates tasks that must be performed. Similarly, the social system within which the technical system is embedded also generates tasks that must be performed. Both kinds of tasks, to different extents, come to be included in jobs. However, what the tasks are and how they are configured originate with the choices made regarding the design of the social system, also referred to as the social support system.

For machines to function properly, they must be accommodated to the organization. Technical support systems include data processing devices, dust collectors, dehumidifiers, spare parts inventory systems, technical manuals, and even access to technical consultants. Analogously, a social support system helps people to accommodate to the organization so that they can function properly in their jobs. Social support systems include aids for recruitment, training, safety, and security; arrangements for counseling, skill maintenance, discipline, and justice; facilities for rest, recreation, eating, and hygiene; reward systems for compensation and advancement; employee benefits; provisions for connecting employees' work and nonwork roles (such as the availability of telephones for personal calls, car pools, etc.); and provisions for managing the organization as a miniature society.

Because of the interdependence of technical system design and social support system design with job design, these decision making areas are included within the realm of job design (see Exhibit 2.5.1). Decision making areas that lead to the establishment of work roles include those that determine:

The materials, equipment, information, and resources to be used by jobholders.

The latitudes of decision making authority awarded to jobholders.

The means for adapting to difficulties or unexpected events.

The means for scheduling, supervising, evaluating, and controlling work.

Paths of advancement and promotion.

The provisions for maintaining and elaborating the organizational structure.

Job Design Under Conditions of Human and Technological Uncertainty

Underlying job design are assumptions about human and technological certainty and predictability. People differ from one another in size, skill, experience, viewpoint, and taste. In addition, the availability and alertness of any one person can vary greatly from day to day, because of health, mood, outside commitments, and so on. Designers can generally predict the actions and capabilities of a machine with much greater certainty than those of a person. However, this fact can be quite misleading in job design, as several authors have noted.

Jordan[8] pointed to the futility of trying to categorize routine tasks into those best suited to machines versus those best suited to humans. Tasks that are so routine as to be reduced to a formula are almost always more efficiently done by machine. The primary advantage of humans is their ability to deal with *technological* uncertainty, such as mechanical breakdowns, unusual pro-

Exhibit 2.5.1 Social and Technological Decisions Underlying Job Design

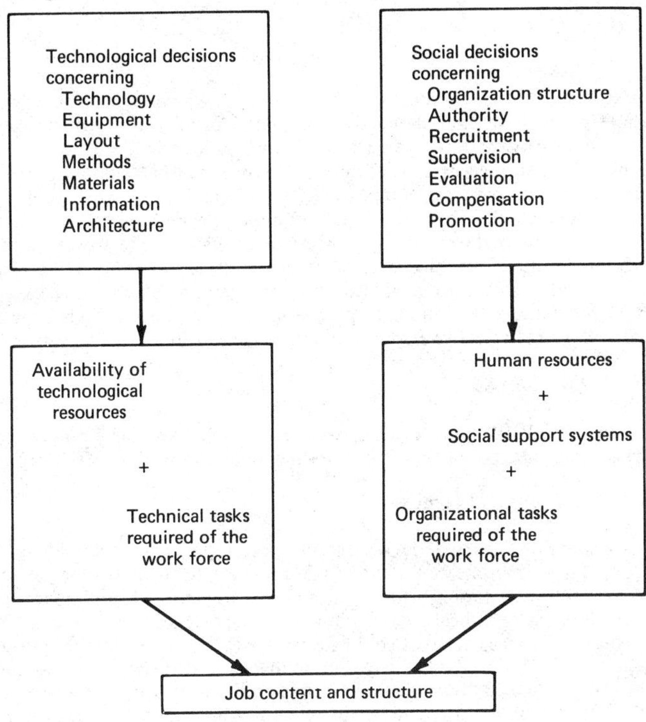

duction requirements, disruptive events, and processes that involve value judgments. Davis and Taylor[2,9,10] discussed the job design implications of automated technology. Human interactions with automated machinery are less routine and less predictable than interactions with older technologies. When situations arise that require responses beyond the capability of the technical system itself, then the presence of humans, with their versatile response capability, is essential to providing responses that the organization needs if it is to succeed. The same applies when the social system or organization has been upset by absence of people or by other events. "The human system is far overdesigned, overengineered for simple jobs," wrote Waddel in the 1956 edition of the *Industrial Engineering Handbook.*[11]

Additionally, people have a perennial tendency to form social bonds and to help each other. If jobs are designed in recognition of this tendency, then some of the human uncertainty due to individual differences, health, and mood can be counteracted by mutually supportive relationships among employees. It is important that the job designer give maximal consideration to the social and creative characteristics of humans, especially when technical systems do not behave with complete certainty.

An anecdote may serve as an object lesson. A computerized information system for housekeeping and accounting was installed in a hospital. Soon afterward, the chief of surgery requested of the night receptionist that a particular patient be assigned a particular room for the next day. Unfortunately, the computer system had already reserved that room for someone else, and no provision had been made for the receptionist to reexamine the decision and exercise her own judgment.

Two scenarios can be envisioned. First, the receptionist could shrug her shoulders and tell the surgeon, "I only work here; if you want something, then *you* can try talking to the computer." The surgeon might have to wait until morning and then spend time searching for the proper forms and approvals. The second scenario is what actually occurred. The receptionist had previously discovered that, after five successive invalid entries into the computer, it would display the message "revert to manual." The old manual procedure involved calling a floor nursing supervisor and arranging room assignments with her. So the receptionist "subverted" the computer system in order to respond to an unpredicted event. Had the receptionist not discovered how to control her technical system accidentally, had she not been familiar with the manual procedure and acquainted with the floor supervisor, and had she not been motivated, then an ineffective outcome for the organization would have occurred. In this case the technical system for room assignment was unable by itself to respond to an unusual and unpredicted event and thus required an adaptation by the organization's social system through the coordinated actions of the surgeon, the receptionist, and the floor supervisor.

Summary

Jobs contain three dimensions: a production dimension, an organizational dimension, and an individual dimension. Each dimension should be considered in all job design decisions. These decisions determine what tasks are to be performed by the work force, how tasks are clustered into jobs, and how jobs are linked together. This framework is outlined in Exhibit 2.5.2.

An organization is a joint system, composed of people and equipment, tools, and so on. A technical support system enables the equipment to function properly in the organizational setting, and a social support system enables employees to accommodate to the the organization. Because the behavior of humans and of technical systems is not totally certain and predictable, jobs should be designed to enable the organization to deal with uncertainty. Whereas machines can perform routine tasks more predictably than humans, people have the potential to act more creatively and to coordinate their actions in adaptive response to nonroutine events.

2.5.3 SOURCES OF TASKS

Jobs are composed of tasks and the conditions under which they can be accomplished. These tasks derive from several sources, which reflect the three dimensions of an organization.

Production and Technological Needs

The most obvious source of tasks derives from the production and technological needs of the organization. From this source one can identify tasks relating to the performance, inspection, and evaluation of work; the preparation, operation, and maintenance of equipment; the recording, processing, and retrieval of technical information; the transportation and care of materials; and the control of technological variables that could adversely affect the production process.

Several mediating factors can affect production and technical system task requirements. Technical system uncertainty or instability places demands on the role of people in the production system. Technological uncertainty is affected by the reliability of equipment, the variability of inputs and outputs, and the unpredictability of events. In addition, the rate of technological innovation introduces long-term uncertainty about the particular tasks that will be required in the future of the work force. The complexity of the technology, the degree of automation, and the scale of production (from custom to mass production) are other mediating factors.

Organizational Needs

In addition to tasks that contribute directly to the production of goods and services, another source of tasks derives from the needs of the organization. These needs include the coordination of activities, the maintenance of the social order, the management of the organization as a miniature society, and the short- and long-term adaptations of the organization to changing conditions in its environment. Organizational needs call for tasks relevant to the hiring and training of a work force, communication, the discovery and exchange of knowledge, and the constructive resolution of interpersonal and organizational conflicts.

Organizational task requirements are affected by the following factors: labor turnover rate, absentee rate, union-management requirements, short- and long-term goals of the organization, the stability of these goals, financial requirements, and legal/regulatory requirements. The first two variables have both mediating and outcome effects. On the one hand, poorly designed jobs are likely to exacerbate labor turnover, absenteeism, and grievances. On the other hand, frequent turnover and absenteeism–regardless of cause–give rise to greater need for hiring, training, coordination, and evaluation.

Individual Needs

A third source of tasks derives from the needs of individuals. People have physical, psychological, social, and economic needs. The particular directions and strengths of each of these needs differ greatly from individual to individual. Needs are a function of personality and of the commitments, requirements, and lifestyle preferences that come from outside the organization. For example, needs for elbow room or self-control expressed by younger workers influence demands for providing areas of decision making that a worker can call his or her own. These needs are influenced by past experiences,[3] expectations,[12] social reference groups,[13] fads, fashions, values, and the culture. The organization is not alone in providing for individual needs; other institutions in society help people to satisfy their various needs.

Exhibit 2.5.2 Systemic Factors in Job Design

Sources of Tasks	Task Clusters	Job Linkages
	As a Production Entity[a]	
Production and Technological Needs	*Imposed by the Technological Design*	*Technological Resources*
Operation Maintenance Quality assurance Control of variance Information processing Reliability of equipment Variability of inputs and outputs Predictability of events Complexity of the technology Level of automation Scale of production Rate of technological innovation	Availability of data Timeliness of data Location of readouts Location of controls Layout Manual overrides Ability of equipment to be adjusted by its operators	Examples–conveyor belt, telephone
	As an Organizational Entity[a]	
Organizational Needs	*Imposed by the Organization Design*	*Organizational Resources*
Hiring Training Coordination Social maintenance Adaptation to change Organizational learning Exchange of information Conflict resolution Flexibility Labor turnover rate Absence rate Union-management requirements Stability of goals Long-term goals Financial requirements Legal/regulatory requirements	Reward opportunities Career paths Who reports to whom Latitudes of authority and responsibility	Reporting relationships Promotion paths Meetings Newsletters Records
	As a Collection of Individuals[a]	
Needs of Individuals	*Carried into the Enterprise by Individuals*	*Personal and Interpersonal Resources*
Physical needs Psychological needs Social needs Economic needs Individual differences Experiences Expectations Values Culture Other need providing institutions	Existing clusters of skill and experience Expectations about jobs and organizations Work habits Lifestyles	Acquaintanceships Acceptance of goals Social skills Understandings of others' tasks

[a]Dimension of the enterprise.

One of the most obvious needs of individuals that needs to be addressed in job design is that of financial compensation (see Chapter 2.3). In addition, organizations must concern themselves with employees' needs for safety, health, comfort, promotion and career advancement, socialization with co-workers, and job satisfaction. These needs have been discussed in the job design literature under the rubric of "quality of working life."[14] A comprehensive checklist of quality-of-working-life criteria, to serve as an aid to designers, is provided in Exhibit 2.5.3. Job designs directed toward providing a high quality of working life incorporate various attempts to provide better job security, equity, and rewards and to satisfy the increasingly articulated psychological needs of all workers. These needs have been stated before by Englestad,[15] and the list provided here was enlarged by Davis[16] to include satisfaction of individual differences.

1. The need for the content of the work to be *reasonably demanding* in terms other than of sheer endurance, yet to provide a minimum of *variety* (not merely novelty).
2. The need to be able to learn on the job and to go on learning. At a minimum, this requires outcome specifications, standards of performance, and knowledge of results (feedback).
3. The need for some area of decision making that an individual can call his or her own. Within

Exhibit 2.5.3 Quality-of-Working-Life Criteria Checklist

Physical Environment

Safety
Health
Attractiveness
Comfort

Compensation

Pay
Benefits

Institutional Rights and Privileges

Employment security
Justice and due process
Fair and respectful treatment
Participation in decision making

Job Content

Variety of tasks
Feedback
Challenge
Task identity
Individual autonomy and self-regulation
Opportunity to use skills and capacities
Perceived contribution to product or service

Internal Social Relations

Opportunity for social contact
Recognition for achievements
Provision of interlocking and mutually supportive roles
Opportunity to lead or help others
Team morale and spirit
Small-group autonomy and self-regulation

External Social Relations

Job-related status in the community
Few work restrictions on outside lifestyle
Multiple options for engaging in work (e.g., flexible work hours, part-time options, shared jobs, and subcontracting)

Career Path

Learning and personal development
Opportunities for advancement
Multiple career path possibilities

this area one can exercise one's own discretion and be evaluated on the basis of objective outcomes.

4. The need for social support in the workplace: the need for an individual to know that he or she can rely on others for help when needed in performing the job and that sympathy and understanding will be available when needed.

5. The need for recognition, within the organization, of one's performance and contributions.

6. The need to be able to relate what one does and what one produces to one's social life outside the organization.

7. The need to feel that one's job leads to some sort of desirable future, that is, no dead-end jobs, but available career paths.

8. The need to know that choices are available in the organization by which one can satisfy one's needs and achieve one's objectives—that is, options as to kinds of jobs, localized structuring of work, many career paths, options for progressing at different rates, and so on.

Some tasks, such as oiling a machine or filling out a form for vacation, can be clearly identified with a single type of need. Other tasks serve several needs. For example, the task of planning a weekly work schedule has to take into account the satisfaction of production needs, organizational needs, and individuals' needs. It is important that the design of jobs enhance the ability of the jobholder to take all relevant needs into account when performing his or her assigned tasks.

2.5.4 SOCIAL AND TECHNICAL SYSTEMS DECISIONS THAT CREATE TASKS

Exhibit 2.5.1 shows various social and technological decisions that underlie job design. How these decisions create tasks are reviewed here; how they influence the clustering of tasks into jobs and how they link jobs together are discussed later.

Technological Decisions

A technology consists of the physical and informational resources by which people can systematically bring about some desired result. Throughout human history there have been incidents of both the accidental discovery of technology and the purposeful, painstaking development of technology. In the modern era more than ever, technological development is influenced by hidden assumptions about people and jobs.

For example, Davis and Taylor[2] reported a technical system design, comprised of machine and "human machine" elements, in which there were a number of parallel, partially automated machines for filling and capping aerosol spray cans. Each machine, with one operator in attendance, was placed so that there was no communication among the operators. Most of the operator's time was spent inserting a small plastic tube into the large hole in the top of each upright can, brought to the operator on a circular conveyor belt. The second, less frequent intervention—and the basic reason for the presence of the operator at all—was to press, when needed, a stop switch on a post directly in front of him or her. In the event of perceptible trouble anywhere in the machine, the operator was expected to shut off the machine and to seek help to resolve the problem.

The job design called for workers to be human machine elements in this production system. They performed in isolation the technologically unnecessary and tedious task of inserting tubes, which could have been easily done by the machine. The human task of inserting tubes into cans was developed simply because the primary task of sensing and diagnosing a problem required human eyes and ears, and by hiring those, the company acquired a set of hands, which were not to be left idle. It is difficult to know whether this "truncated" technological development was based on the state of the art or on assumptions held by the designers about people. Failure to develop servomechanical devices and/or remote sensing and control mechanisms for automatic shutoff was a major factor leading to an extremely poor job design. Perhaps if the designers had taken into account the needs of the social being to whom the eyes, ears, and hands were attached, a different technological design would likely have emerged.

In contrast to the preceding case is one involving the design of a food processing plant.[17] With the existing technology, there was a task that involved the manual unloading of 100 lb sacks of raw material from boxcars and the stacking of the sacks on pallets, to be stored until the dumping of their contents into hoppers. From that point, automated equipment sorted, cooked, processed, and packaged the product. In existing plants the unloading and dumping tasks were performed by men, whereas the operation of the automatic equipment was done primarily by women. This was unacceptable for a number of reasons. One reason was the desire of all individuals to have a chance to learn all phases of the operation. With the backing of top management, the designers initiated an industrywide conversion to bulk packaging of the raw material so that mechanical equipment could handle it. Additional automated machinery was developed so that all the routine human tasks in the plant could be performed by either men or women.

Social System Decisions

Perhaps the most dramatic instance of a set of social decisions creating tasks was the framing of the U.S. Constitution. The then-radical criterion that government representatives were to be popularly elected created the multitude of tasks involved with the conducting of elections. On a smaller scale, a food processing plant[17] was designed so that its social support system included peer performance appraisal. This created organizational tasks pertaining to the administration and facilitation of the performance evaluation process.

Decisions about methods of recruitment not only make human resources available to fill jobs, but also stimulate employee expectations regarding job security, career opportunities, and interpersonal relationships. Choices of the methods for supervision, compensation, promotion, and conflict resolution set forth their own tasks. At the food processing plant, a set of social decisions led to the identification of the following tasks to be performed by workers, which became part of their jobs: problem solving, counseling, task assignment scheduling, record keeping, communicating between work teams and between shifts, convening and leading meetings, safety coordination, security coordination, and training.

2.5.5 SOCIAL AND TECHNOLOGICAL DECISIONS THAT CONSTRAIN THE CLUSTERING OF TASKS

Technological Decisions

Technological decisions can add or eliminate options in designing jobs. Even such an apparently simple decision as the location of a switch or meter, or the mode of input and output for a computer, allocates technological resources to particular work stations and thus constrains the clustering of tasks into jobs through the physical locations of the means for controlling the production process. It is typically thought to be effective, for example, to construct a railroad spur and loading dock of a plant so that shipping and receiving tasks are performed in one physical location. In one recently designed paper mill, however, this arrangement was deemed inappropriate to the job design, since it would not have permitted the group of employees responsible for converting the raw material to control its inputs as to quality and time in order to achieve the outputs desired. From a purely technical viewpoint, it would have been desirable to build a single railroad spur and loading dock for both receiving and shipping. The broader systemic viewpoint prevailed, and the extra cost was more than repaid by building separate spurs and docks.

All too often, engineers have designed technical systems with hardly a glance at job design criteria. Underlying this neglect is an ideology of "technological determinism," meaning that technology does—and should—develop in isolation of and unconstrained by social considerations and that social structure is—and should be—fitted to the technical system whether or not the consequences may be dysfunctional. The folly of technological determinism is often illustrated by citing the case of the Chevrolet Vega plant at Lordstown, Ohio, in 1972. Although hailed in 1970 as "the world's most technologically advanced assembly line," the plant was soon gripped by labor discontent, sabotage, absenteeism, and carelessness. This culminated in a strike in 1972, which attracted national attention because it centered on the consequences of the job designs for a young, rural work force whose members said money alone was not enough and whose local union leaders publicly decried the dehumanized jobs and dizzying speeds. Since then, several GM divisions have sought ways to develop technical systems designs that meet job design requirements as well as more conventional criteria. In many industries, assembly technologies once characterized by long conveyors have been succeeded by U-shaped configurations, buffer stages, and self-propelled assembly carriers.[18] These technical innovations were stimulated in part by the need to make jobs more flexible and appealing.

Reviewing the empirical literature regarding the relationship between technology and job structure, Davis and Taylor[2] concluded that a self-fulfilling prophecy appears to be operating. Production systems that are designed in accordance with assumptions that people are reliable and intelligent and that they desire variety and challenge yield jobs that reflect those assumptions. Similarly, when technical systems are designed on the basis of assumptions that people are unreliable, that they should be isolated, and that they have few tasks to do, these assumptions lead to technological choices that ensure their own fulfillment, thus reinforcing the original unfounded assumptions.

Social Decisions

Social decisions can bias the allocation of tasks to jobs. The designers of a new food processing plant[17] incorporated both training and operational tasks into production jobs, so that workers could cross-train each other. In other plants this would have been constrained by a social support system that did not encourage workers to share their skills and knowledge other than informally. In the new design, pay levels were graded according to the number of skills and tasks an individual had mastered, regardless of the individual's particular work assignment at a particular time. Under such a reward structure, workers are meeting their own needs for pay and advancement, while simultaneously meeting the organization's needs for flexibility, by cross-training each other.

2.5.6 CLUSTERING CONSTRAINTS CARRIED INTO THE ORGANIZATION BY INDIVIDUALS

Job design takes account of the existing clusters of skills and experience in the labor market. This in itself can be a social decision, since designers often choose a particular segment of the labor market for which to design jobs.

More subtle are requirements or constraints coming from individuals' expectations, values, and past experience. After World War II many of the operations in the coal mines of Durham, England, were mechanized, replacing manual methods of coal mining.[19] Accompanying the mechanization was job redesign. For generations miners in that region had worked in small, relatively autonomous crews. The dangers and production uncertainties inherent in coal mining required high levels of mutual dependence, cooperation, and trust among co-workers. The new job design called for a one-man/one-job mode of working that did not fit the mining culture. Each miner was accustomed to performing all the tasks required to work a section of the mine. The miners were unwilling to accept a narrowing of task assignment or to give up control over the selection of their work mates. Subsequently, a crew type of job design was reinstated, still retaining the new technology, and productivity vastly improved.

A lively controversy has developed around the question of employees' desires for variety, challenge, and autonomy in their jobs. Some research suggests that deep-seated personality traits may underlie employee preferences for one or another type of job.[20] Other experiments and field studies indicate that employee desires and expectations are simply reflections of culture, familiar experiences, and reference group values.[5,7,12,13,21] Many job design endeavors have proceeded on the negative principle of the self-fulfilling prophecy or on the positive principle of rising expectations, namely, that employees' expectations tend to grow as society makes more opportunities available inside and outside the workplace.

2.5.7 RESOURCES FOR LINKING JOBS TOGETHER

Technological resources help transfer materials and information from person to person. The conveyor belt, for example, is one device that links jobs by virtue of transporting materials between work locations. Information is transferred by such technical devices as telephone, teletype, and computer.

Organizational resources help transmit decisions, instructions, feedback, and rewards among jobholders. Communication, promotion and career paths, meetings, and newsletters all serve these functions.

Personal and interpersonal resources are frequently overlooked linkages between jobs. Acquaintanceships, personal commitments to organization goals, social skills, and familiarity with tasks outside one's own job all help to integrate the many jobs into a single system.

Technological and social decisions create job boundaries by creating tasks and constraining clustering of tasks. These decisions also provide linkages across job boundaries. Some boundary predispositions are brought into the enterprise from outside. It is the task of the job designer to ensure that the technological, organizational, and individual boundaries complement and reinforce each other. After reviewing the main types of job designs, we present some methods of analysis to aid in the alignment of job boundaries.

2.5.8 TYPES OF JOB DESIGNS

Undesigned Jobs

Certainly one way to design jobs is to let them "grow like topsy." This laissez-faire approach to job design characterized production systems before the Industrial Revolution (1780). Jobs, or rather "trades," or sets of skills, evolved slowly, changing little from generation to generation. The particular allocation of tasks to jobs, the tools and techniques for task performance, and the quality standards were based on traditions and rules of thumb put forth by those who had mastered the skills. They were enforced and regulated by associations or guilds of master craftsmen. Even today many job designs derive from tradition. As an illustration, the roles of physician and nurse are *not* designed simply as an extension of the technology of medicine. These job designs also rest on social traditions that are now protected by legislation and by guilds of doctors and nurses.

Probably the best-known critic of undesigned jobs was Frederick W. Taylor.[22] He urged that traditional rules of thumb give way to more deliberate methods for making job design decisions. Traditions, argued Taylor, are rarely tested by controlled experimentation, and they are improved only by a very slow and haphazard process of trial and error. Taylor attributed this to the tradesman's lack of scientific education as well as to social support systems that did not encourage improved work methods. From 1890 to 1910 Taylor went on to develop another approach to job design, as a part of his system of scientific management.

Machine Model of Jobs

The machines that emerged during the Industrial Revolution had two major impacts on job design. The first impact was technological. The proliferation of new machines meant wholly new tasks and thus new job designs. The second impact was social. Engineers and inventors were the heroes of the day, and optimism ran high that the perspectives of engineering and physical science could be fruitfully applied to the design of social systems. Organizations came to be visualized as precise, clockworklike mechanisms, which contained human cogs as well as mechanical cogs. The most descriptive metaphor is that of the organization as an elaborate clockworks.

Functional Specialization

The machine metaphor of organization led to the job design principle of functional specialization. This principle originated with Adam Smith[23] and Charles Babbage in 1790[24] and was reified by Taylor 100 years later.[22] It derived from the then-radical proposition, breaking with past tradition, that only the economic criterion should be used in assigning work to people. This meant, as with other single, variable answers, that work was to be done only on the basis of the cheapest way (in the short run), which was to break up or fragment skills.

Smith, Babbage, and later Taylor urged owners and managers to design each job as narrowly and precisely as possible. Precise prescription was necessary so that workers could be controlled by managers and supervisors. By logical extension of this principle, all the tasks deriving from the needs of individuals become vested in specialized "personnel" jobs, the tasks deriving from organizational needs are given to specialized clerical and staff jobs, and all tasks derived only from production needs are given to production jobs. Specialization is taken even further—all maintenance needs are assigned only to maintenance jobs, all inspection tasks are clustered into the job of inspector, and so on. A principle related to functional specialization, which has guided clustering of tasks, is that all tasks within a job should be at the same skill level.[1]

The alleged advantage of the functional specialization principle is that scarce knowledge and skill can be put to its widest use. Its disadvantages are a highly fragmented work force, which is costly to establish and coordinate; the possible loss of information between job boundaries when many specialties are needed in a single production process; and jobholders who are alienated by the narrowness and dead-endedness of their jobs, which reduces their willingness to contribute maximally to the organization.

These disadvantages can lead to what is sometimes called "job myopia," or the "It's not my job" syndrome. Some needs go unmet because those who are aware of a problem shrug it off as not part of their narrow jobs, which is likely the case. Those whose jobs call for them to deal with the needs do not learn of them until a problem of sufficient magnitude has developed. Additionally, specialists ignore and/or create problems according to their own work loads rather than according to the real needs of the organization. Another symptom of job myopia is that the human energy that would ideally be devoted to cross-training and spontaneous problem solving is instead diverted to protecting and/or aggrandizing individual specialties.

Deterministic Job Specification

Another principle to come out of the machine model of organizations and jobs is that of deterministic job specification. In the design of a machine nothing can be left to chance; engineers specify every detail. By analogy, in the design of jobs a designer may see the need to specify every task, movement, and interaction, thus prescribing everything in the job, leaving no discretion to the employee. The analogy is fallacious. Unanticipated events often require employees purposely to deviate from their job specifications, that is, to use their discretion. Incidents of "working to rule" or "malicious obedience" show that discretion may be the most important ingredient that people bring to their jobs, even to routine jobs.

In Britain, railroad workers, instead of striking a few years ago, critically impeded train service by doing exactly what their job descriptions and standard operating procedures called for, no more and no less. They stopped the railroad from operating by leaving discretion at home when they came to work. Similarly, in New Jersey, police enforced all traffic laws without any exercise of discretion and thus managed to cripple traffic flow on key arteries.

Of course no model of job design can prevent employees from expressing their grievances in one way or another; however, that is no reason for limiting discretion. A set of jobs will result in the achievement of an organization's goals not only because of the specified or prescribed aspects of jobs, but also because of the ability and willingness of employees to use discretion.

Taylor's Approach

The machine model of organizations and jobs was developed to its most elaborate form by Frederick W. Taylor. He based his approach on the vision that a value-free science of human work existed and that it was his mission to develop that science. Taylor confidently foretold of a time when questions about the amount of work expected of a particular jobholder would be determined no differently than the calculation of the rising and setting of the sun, that is, scientifically.

The mainstay of Taylor's "science" was the fragmentation of tasks or skills into their simplest elements. These were then measured in units of time. Using Adam Smith's and Charles Babbage's economic criterion for assigning work to people, the sole criterion for creating and clustering tasks became minimum time per task or movement. This practice enabled the principles of functional specialization and deterministic job specification to be carried to their ultimate.

Taylor further insisted that the organizational tasks required to administer his scientific system be allocated to a new cadre of specialist-supervisors, thus adding the further organizational principle of separation of planning (thinking) from doing. Taylor's job designs thus separated the planning and conceptualization of work from its execution. To Taylor's critics that separation constituted a fundamental dehumanization of work. Nevertheless, after 70 years of widespread application of Taylor's principles of scientific management,[22] the discretionary element of jobs has not been replaced, and recognition of its need is increasing. Taylor's misconceived concept of a value-free science of human work became the basis for dehumanized work, which is so roundly rejected now.

Social and Technological Requirements

In order to design jobs according to the machine model, a number of social and technological decisions are required. Emery[25] listed four. The first is the development and installation of transfer devices, the classic example being Henry Ford's automobile assembly conveyors. Transfer devices (1) move each object requiring work to each worker, making it feasible for each worker to perform but a single task, and (2) enable managers to control directly the behavior of each worker, in this instance the pace of performing tasks.[26]

The second requirement is the standardization of both the product (or service) and the means to be used to make it. Since a production process is fragmented into many tasks, and since each task becomes a job assigned to a worker, workers thus become the interchangeable parts of the organization. Standardization requires an elaborate planning process so that the established specifications or norms will be applicable to all the cases that workers, customers, and clients will encounter. If customized products and services are needed, or if conditions of technological uncertainty prevail, then standardization will more than likely not be cost-effective.

Third are decisions inherent in "balancing the line." A set of single-task jobs is designed so that an exact number of human components will produce an exact number of product or service outputs per unit of time. These numbers cannot be expanded or contracted without a major system redesign, constraining the flexibility of the organization. Any breakdown, human or technical, that "unbalances" the system leads to downtime for the whole process. With downtime comes enforced idleness of production workers, and because it is caused by the design itself, such idleness is excused as "unavoidable delay" in the short run. In the long run come layoffs, with costly unemployment insurance claims and social disruptions. A tightly balanced production line is also vulnerable to unanticipated absences, thus requiring a surplus of workers to be available.

The fourth requirement is to develop elaborate supervisory jobs. Coercive supervision is needed to make people adhere to the machine model of organizations and jobs. Since only supervisors may undertake organizational tasks while workers perform single production task jobs, many supervisors and layers of supervisors are needed to coordinate the work and to deal with the human and technological uncertainty that inevitably arises.

Job Enlargement and Job Enrichment

During World War II the massive amounts of military material to be produced, including complex equipment, required rapid, large-scale expansion of the labor force, bringing an influx of people who were unskilled and inexperienced in industrial work. Widespread reliance upon the machine model of job design in war-related industry led to discoveries of the limits of that model. Researchers developed new approaches to job design, called "job enrichment" and "job enlargement."

In job enlargement, the principles of functional specialization and deterministic job specification are somewhat moderated. Instead of striving to have each job consist of a single task or as small a fragment of work as possible, job enlargement clusters larger groups of tasks and allows jobholders certain degrees of discretion. Conant and Kilbridge[27] reported that job enlargement was beneficial since it reduced nonproductive work and assembly line balance delays, increased the quality of the product and worker satisfaction, and provided savings in labor costs and greater production

Exhibit 2.5.4 Staffing of Bag Making Shift With Initial Fragmented Jobs

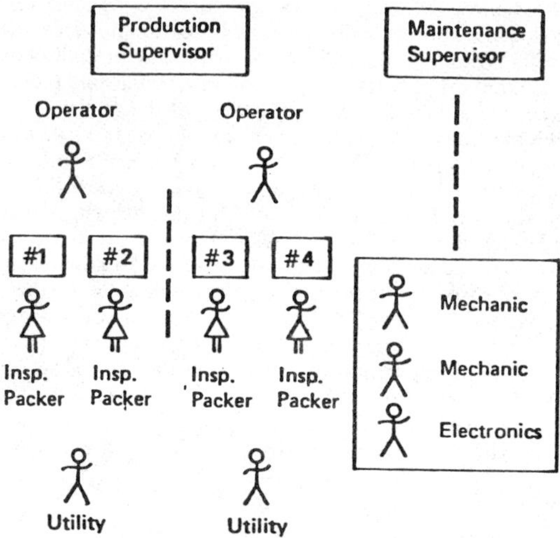

flexibility. Given the needs and expectations of workers today, it is doubtful that job enlargement can be adequate to the challenges of the 1980s as it was to those of the 1950s.

From psychological studies of worker motivation, Frederick Herzberg[28] developed job enrichment as an approach to worker satisfaction. According to this concept, a single job should encompass not only production tasks, but also many of the setup, scheduling, maintenance, and control tasks related to the operation. In addition, each jobholder should be given the resources and responsibility for results relevant to his or her operation. Herzberg emphasized that job enrichment entails a "vertical" as well as a "horizontal" clustering of tasks, so that a job contains not just a variety of tasks, but planning and control tasks and a range of skill levels. The central criteria for job enrichment are that the job offer a challenge to the jobholder and that the jobholder be provided with feedback to gauge his or her own personal accomplishment.

A plastic bag plant[29] tripled its number of automatic bag making machines while enriching the jobs in its bag making department. The unenriched jobs (Exhibit 2.5.4) clustered the tasks associated with the front end of the machines into one set of jobs, and back-end tasks into another set of jobs. Since the front-end jobs required more skill, the positions, entitled "operators," were filled by men at higher pay. The back-end jobs, entitled "inspector-packers," were filled by women at lower pay. Much of the work on the machines involved dealing with breakdowns and errors. It was felt that these could be greatly reduced if each operator were assigned to perform all the tasks associated with one particular machine. In addition to this redesign of the jobs (Exhibit 2.5.5), a new compensation plan was introduced in which all workers were paid a monthly salary rather than an hourly wage. Both men and women held the newly enriched operator jobs, and the company reported an increase in both productivity and job satisfaction.

Although job enlargement and job enrichment represented major steps toward more efficient uses of human resources as well as increased humanization of work, each has significant shortcomings. These approaches do not take explicit consideration of the relationships between jobs. The one-person/one-job unit of analysis is retained from the machine model of organization. Neither do they deal with the needs and expectations that people have about their jobs (as distinct from work tasks), about coercive working environments, and about their futures.

Self-Maintaining Work Teams

Beginning with studies of English coal mines in the 1950s, researchers at the Tavistock Institute, London, developed a body of literature[30] aimed at incorporating groups as well as individuals into job design. That literature draws on parallel work in the United States[3,6] and has become known as sociotechnical systems theory. It has led not only to the modern job design option of self-supervising teams, but more importantly to the underlying principle of *self-maintaining work units.*

The organization is divided into segments that can operate as fairly autonomous service or

Exhibit 2.5.5 Staffing of Bag Making Shift With Current Enriched Jobs

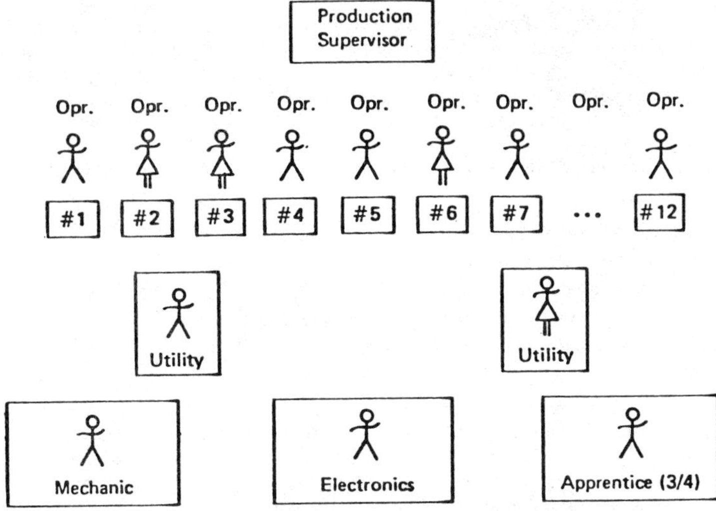

Total: 17 (plus 1 supervisor)

production units, or "minibusinesses." Each of these organizational segments or units is designed to contain all the tasks and resources necessary to meet its service or production and organization needs as well as many of its members' needs. In some cases all the tasks of a self-regulating work unit can be incorporated into one job, which may be the ultimate in modern job design. However, designing a work unit around a single person leaves it vulnerable to human variability in terms of the health, personality, availability, and so on, of the person. In many, if not most, cases it is preferable to design self-regulating work units around small groups of people, thus making them into self-maintaining organizational units. This enables the natural forces of group cooperation and cohesiveness to counteract both the human and the technological uncertainty that arises within the work unit.

A second principle of sociotechnical systems theory is *minimal critical specification*, the exact antithesis of the machine model's principle of deterministic complete job specification.[31] The self-maintaining work unit is designed so as to allow it as much flexibility and discretion as feasible. Job designers specify only the minimum that is absolutely critical for the work unit to function as a part of the whole organization. Other job design decisions are left as options for the work team to exercise, based on the experiences of its members. In this way, production, organizational, and individual needs can be integrated into local decision making processes. In some of the latest organization job designs, self-maintaining work teams take responsibility for organizational learning, cross-training, peer counseling, internal coordination, and adaptation to uncertainty. The job design decisions that are typically specified and those that are typically left as options for work teams are listed in Exhibit 2.5.6.

Exhibit 2.5.6 Job Design Decisions for Self-Maintaining Work Teams

Typically Specified by Job Designers, With Participation by Workers When Available	Typically Left for Team Decision Making, With Manager Consultation When Needed
Measurable input specifications	Task assignment scheduling
Measurable output specifications—quantity and quality	Work methods
Equipment and resources	Work pace
Work station layout options	Work hours
Compensation and aavancement plan	Counseling and discipline
External information flows	Internal information flows
	Internal leadership
	Team membership

Illustration of Self-Regulating Team

The use of self-regulating work teams is illustrated by a field experiment conducted in a textile plant in India following installation of automatic looms. Rice[32] reports that the engineers conducted an intensive time study of the automatic looms in order to lay out equipment and assign work loads to people on the basis of narrow tasks. These looms did not produce the quantity and quality levels already attained by the nonautomatic looms, let alone attain the improvements expected.

The automatic weaving sheds contained 240 looms, and work to be done was divided into 10 one-task jobs:

A weaver tended approximately 30 looms.

A battery filler served about 50 looms.

A smash hand tended about 70 looms.

A gater, cloth carrier, jobber, and assistant jobber were assigned to 112 looms.

A bobbin carrier, feeler-motion fitter, oiler, sweeper, and humidification fitter were each assigned to 224 looms.

These tasks were highly interdependent, and the utmost coordination was required to maintain continuity of production. However, the worker-machine assignments created organizational confusion. Each weaver had to relate to five eighths of a battery filler, three eighths of a smash hand, one fourth of a gater, one eighth of a bobbin carrier, and so on. The jobbers who carried out on-line maintenance reported to shed management through a separate supervisory channel from weavers, and there were no criteria to establish whose looms should have priority when breakdowns and other trouble occurred.

A redesign was undertaken to provide a self-regulating group responsible for the operation and maintenance of a specific bank of looms. Geographic, rather than functional, division of the weaving shed produced interaction patterns that enabled regularity of relationships among those with interrelated jobs. Individuals could now be held responsible for the production of their teams. The redesign was suggested by the workers themselves based on discussions with a consultant, supervisors, and managers. The consolidated teams reported to a single shift supervisor, who reported to an overall shed manager.

As a result of these changes, efficiency rose from an average of 80% to 95%, and damage dropped from a mean of 32% to 20% after 60 working days. In the adjacent part of the weaving shed, where job design changes were not made, efficiency dropped for a while to 70% and never rose above 80%, while damage continued at an average of 31%. The whole shed was then converted, and the improvements were permanently maintained. When it became clear that there was improvement, a way was found to introduce consolidated loom groups throughout the large number of nonautomatic sheds. A third shift, which had been previously resisted by the union, could then be introduced. Within loom groups, status differences were reduced; the less skilled were given opportunities to learn the roles of the more skilled, so that a promotion path was created. Wages were increased as substantially as costs were decreased.

Compatible Social and Technological Decisions

Self-maintaining work teams require compatible social and technological decisions. At Volvo's Kalmar plant in Sweden, a new technology for automobile assembly was developed to support the team form of job design.[18] Computer-monitored carriers enable each work group to lay out its own work and to control the pace of work. Volvo estimated that the new technology would cost 10% more than a conventional assembly line, but expected to recover the costs in increased flexibility and higher worker motivation. At Philips in the Netherlands, televisions are assembled in "production islands," which are U-shaped assembly lines that enable assembly workers to interact and cooperate.[33] In a new polypropylene plant,[34] a process control computer displays decision aiding information rather than making decisions. The responsibility for final decisions rests with work teams. This design was chosen because of the technological uncertainty contained in the large number of uncontrollable variables flowing in the system. If these could come to be controlled, large economic advantages would accrue. In this instance economic success was seen to be correlated with learning—the more variables that workers could learn to control, the higher the efficiency.

Compensation Plans

Three types of compensation plans have been used to support team job designs. The first, used primarily in Europe, is group piecework payment. The whole team is paid an amount to be disbursed among individual members according to internal decisions.

The second type of compensation plan is called the "group bonus," and if applied to an entire plant, it is the Scanlon plan.[35,36] Here individuals are paid on a conventional basis, but they receive bonuses when their respective teams exceed a certian production or cost target or invent improved methods that result in cost savings. Under the Scanlon plan the sharing of cost savings is plantwide.

The third type of compensation plan is called "pay by skill" or "pay by knowledge." Tasks and skills–not whole jobs–are graded, and then standards for training and testing are developed. Individuals' pay levels are determined according to their demonstrated skills and knowledge. This system serves to qualify an individual to perform certain tasks for his or her team and to train other team members in those tasks. The pay differentials do not, however, bestow authority over other team members, since the team operates as a group of peers and makes decisions by consensus.

Advantages and Disadvantages of Work Teams

One of the major advantages of self-maintaining work teams is the ability of a small group to accommodate variations in individual and organizational needs. Some members may prefer complex, multiskilled tasks; others, simple tasks; and still others, a variety of tasks. Team members, if they have adequate knowledge, skills, resources, flexibility, and social support, can usually develop mutually acceptable solutions, which often include job rotation and flexible working arrangements. This localized, participatory decision making process increases the likelihood that decisions will be accepted by employees without the need for coercion.

For self-maintaining work teams to function at their potential, their members must possess considerable social skills and have attained a fairly high degree of emotional maturity. The present-day labor force, both white and blue collar, does largely possess the social and emotional sophistication needed for team job design. In addition, a team job design should provide in its social support system for social as well as technical training. It is not unusual for a new work force at a new plant to undergo 40 hours or more of training in social and organizational skills and team development prior to start-up. Refresher sessions and problem solving workshops are frequently provided.

One of the pitfalls of team job design is inadequate attention to team makeup. For team members to deal with intrateam and interteam problems, action should be taken to minimize the entry of conflicts present in the larger society into the organization or team. In one new plant in the southeastern United States, subteams were self-selected along clique lines, which also happened to be racial lines. Within days after start-up, before the whole team could develop either cohesiveness or complete understanding of its tasks, problems within the team began to take on racial overtones. The team recognized that its makeup enabled old, deep-seated external societal conflicts to impede its internal problem solving processes. It reassigned its members to subteams so that its internal organization cut across racial and clique lines, thus creating work-centered communication links between its parts.

It is important that a self-maintaining work team develop a practical understanding of consensus decision making. One of the most critical and difficult areas of team decision making is that of dealing internally with members who do not meet team membership norms or standards (excessive absence, unwillingness to help others, etc.). For teams to develop a style of consensus decision making, managers must deal with teams in a manner that is neither neglectful nor interfering. A facilitative management style is one of the social system requirements in support of team job design.

Matrix Structure

The concept of matrix organization originated in large aerospace enterprises.[37,38] The complexity of interrelationships among the different parts of an aerospace project, coupled with the great technological uncertainty faced by aerospace firms, may make it infeasible to segment the organization into autonomous minibusinesses. In the matrix structure, human resources are managed by two superimposed organizational structures. One of these is based on functional specialty or technology, and the other on particular projects or interdisciplinary problems. This is illustrated in Exhibit 2.5.7.

Kingdon[39] reported a matrix job design in which the "workers" were engineering analysts with doctoral degrees and computer programmers with master of science degrees. Although a functional supervisor was nominally responsible for homogeneous groups of specialists, project assignments were determined by project managers who "hired" specialists from the functional departments for as long as they were needed. This practice was called the "job shop." Heterogeneous groups of specialists often worked together for several weeks in a place known as the "bat cave," from which they would emerge only when a critical problem had been solved.

Several recently designed manufacturing plants have instituted matrix structures within their production work forces. In one instance clusters of technical tasks and production responsibilities

Exhibit 2.5.7 Matrix Job Structure

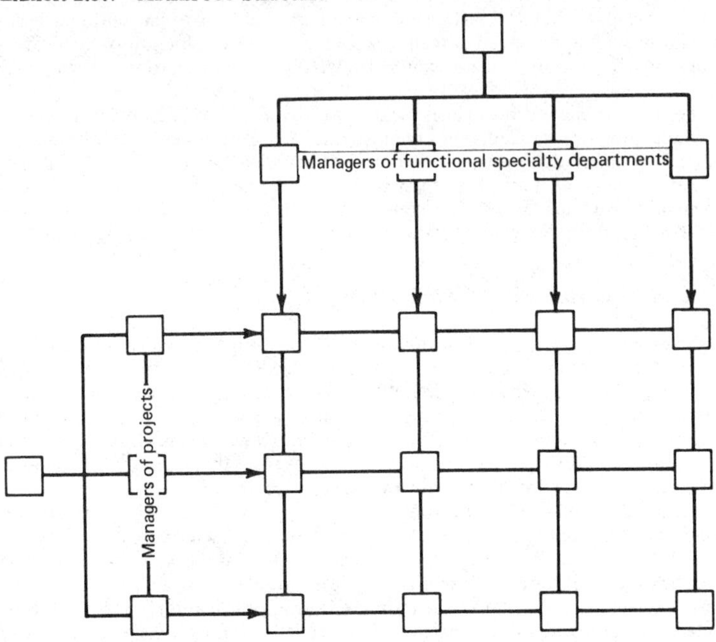

were allocated to five self-maintaining work teams, consisting of 4 to 28 members per team. The matrix concept was used to bring representatives of each team together to deal with the organizational tasks that required interteam coordination. On each team one member was elected to fill each of several "coordinator" roles, such as quality-control coordinator, communication coordinator, and safety coordinator. These were not full-time duties; it was each team's collective responsibility to allocate members' time so that both organizational and production needs would be met. From time to time all teams' quality control coordinators would meet with the plant's quality control manager, safety coordinators with the personnel manager, and so on. Communicators from each team jointly met with the operations manager every day. Special task forces composed of ad hoc representatives from all teams (including management) were created to solve a variety of plantwide problems.

Matrix structure is the most flexible form of job design, but it makes very great demands on both workers and managers. These demands stem from the ambiguity of intersecting channels of authority. Whereas many people thrive on the freedom and responsibility accorded to jobholders in a matrix structure, others prefer more rigidly structured and less responsible jobs. Managers sometimes lack the social skills necessary to manage a matrix job structure or are uncomfortable in a system in which they do not have continuing authority over a particular fixed group of subordinates. Despite the great demands of matrix structure, careful recruitment methods have turned up more than enough qualified applicants for those organizations in which jobs were so designed.

2.5.9 ANALYTIC METHODS IN SUPPORT OF JOB DESIGN

The aim of this section is to present some methods for identifying tasks and task clusters. The overriding objective of these methods of analysis is to help job designers to align the technological and organizational boundaries between jobs.

Transformation Flowchart

To conceptualize the organization as a production entity, job designers may wish to draft a flowchart showing how the product or service progresses from its input state to its output state. Industrial engineering flowcharts tend to break down actions too finely, often presuming a prior selection and layout of equipment, and sometimes even a prior design of jobs. The purpose of a transformation flowchart is to enable the designers to conceptualize production requirements without being locked into specific technological or social decisions.

The transformation flowchart, an example of which is given in Exhibit 2.5.8, shows (1) the progressive states of the product or service and (2) the "unit operations" by which the product

Exhibit 2.5.8 Transformation Flowchart for a Dairy

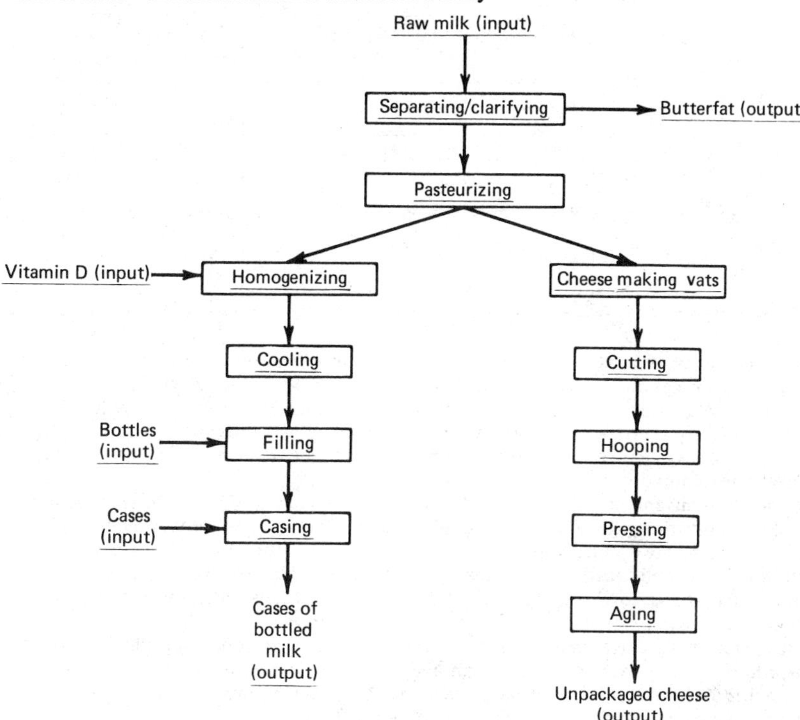

or service is transformed from one state into another. Product or service states are of three types: inputs, throughputs, and outputs. An input is a raw material or entry state that is to be transformed by the organization. An output is a finished product, final state, or waste material that leaves the process. A throughput is an intermediate state of work within the process.

Unit operations are self-contained segments of processes describing the transformation of an object, in the form of material, information, or person, from an input to an output state. They describe state changes of the object being transformed. To carry out the state change, tasks need to be performed. Thus the unit operation indicates the tasks to be performed, which the job designer then clusters or assigns to a job.

Unit operations are expressed in terms of the transformation or service being rendered. If particular machines connote particular changes in product state, then it may be useful to express the unit operations in terms of these machines. Care must be taken not to exclude technical system alternatives. Inspection is usually not shown as a separate unit operation if it serves merely to verify that a transformation occurred. On the other hand, creating or updating permanent records, such as a medical history, that are not used merely to verify transformation or to control the production process may be considered separate unit operations. Storage is usually not considered a unit operation unless some change of product state—whether desired or undesired—could occur during the storage. Decision making, calculation, setup, positioning, and so on, are not expressed as unit operations by themselves, but rather are considered tasks belonging to a unit operation.

The selection and layout of equipment and the design of jobs should enable employees to visualize the transformation flowchart as they work. Job or team boundaries should encompass one or more unit operations, so that each job or team is allocated a distinct piece of the transformation process, one having identifiable and measurable outcomes, and is responsible for an actual transformation of the product or process from one state to another.

Variance Analysis

With complete technological certainty, there would be no deviation or disturbance in the throughput states. Most organizations, however, are faced with pockets of technological uncertainty where human intervention is critical for controlling variability. The purpose of variance analysis is to identify the areas of technological uncertainty in the transformation or production process and ultimately to design jobs so that jobholders, individually or in a team or group, can effectively control variability where it arises.

Exhibit 2.5.9 Variance List Format

Unit Operation	Product State Variances	System Variances

A "variance" is an unwanted discrepancy between a specification or desired state and an actual state. Often specifications appear as ranges of tolerance within which deviation is permissible. A variance then occurs only when a specified range of tolerance is exceeded. There are two types of variances: product state (or service state) variances, which pertain to input, throughput, and output specifications, and system variances, which pertain to equipment and the ambient physical environment (heat, dust, humidity, etc.).

Although the concept of variance could easily be extended to discrepancies in human performance, skill, motivation, and so on, this would confound too many variables together. The concept of variance is therefore applied only to the technical system. The process of variance analysis first identifies the kinds and locations of human interventions needed to control technological variability at its source and, second, identifies the information, skills, and decisions needed for successful control or regulation of variances (discrepancies or disturbances).

The first step in variance analysis is to list all variances that could impede the production or service process. Trivial variances can be ignored, but when in doubt, include the variance on the list. Often it is useful to organize the variance list in the format shown in Exhibit 2.5.9.

Here are some guidelines for identifying variances:

A variance should be listed with reference to the unit operation in which it would occur, regardless of where the variance is ultimately detected.

It is sometimes desirable to express variances as deviations rather than merely as variables, for example, "water too cold" rather than "water temperature." Some analysts, however, prefer simply to list variable names.

Variances are better expressed in state terms rather than in process terms, for example, "dirty dishes" rather than "unwashed dishes."

It is useful to express variances so as to reflect the degree of precision and objectivity of the specifications. For example, temperature can be a subjective variance, stated as "water too cold," or an objective one, stated as "water less than 92°C."

The second step in variance analysis is to identify the dependency or causal relationship among the variances. An aid for this is a variance interrelatedness matrix, an example of which is shown in Exhibit 2.5.10, used in the redesign of jobs in a paper mill. The format of the matrix is analogous to that of an intercity mileage chart on a road map. Each cell of the matrix shows the degree of relatedness between two variances. A blank cell indicates no relatedness, while a "3" indicates relatedness of great importance, such as between the variances numbered 22 and 42 in Exhibit 2.5.10. The matrix helps the job designers to understand how the parts of the transformation or production system are dependent on each other. In designing jobs, tasks should be clustered and jobs linked, so as to reflect the dependency relationships among variances.

The third step is the identification of "key variances," those whose control is most critical to the successful outcome of the production system. Job design should focus on control of key variances. Paraphrasing the Pareto principle, a large percentage of the problems in a production system are caused by a small percentage of the variances. A key variance has the following attributes: (1) its actual, not theoretical, occurrence can seriously impair the quantity, quality, or cost of production, human resources, or technological resources; (2) it interacts with, or causes disturbances in, many other variables; (3) it occurs stochastically—the time, place, frequency, or intensity of occurrence cannot be predicted with certainty; and (4) it can be detected, prevented, corrected, or otherwise controlled by timely appropriate human action.

The fourth step in variance analysis is to construct a table of key variance control, the format of which is shown in Exhibit 2.5.11. This should contain brief descriptions of how, where, and

Exhibit 2.5.10 Variance Interrelatedness Matrix[a]

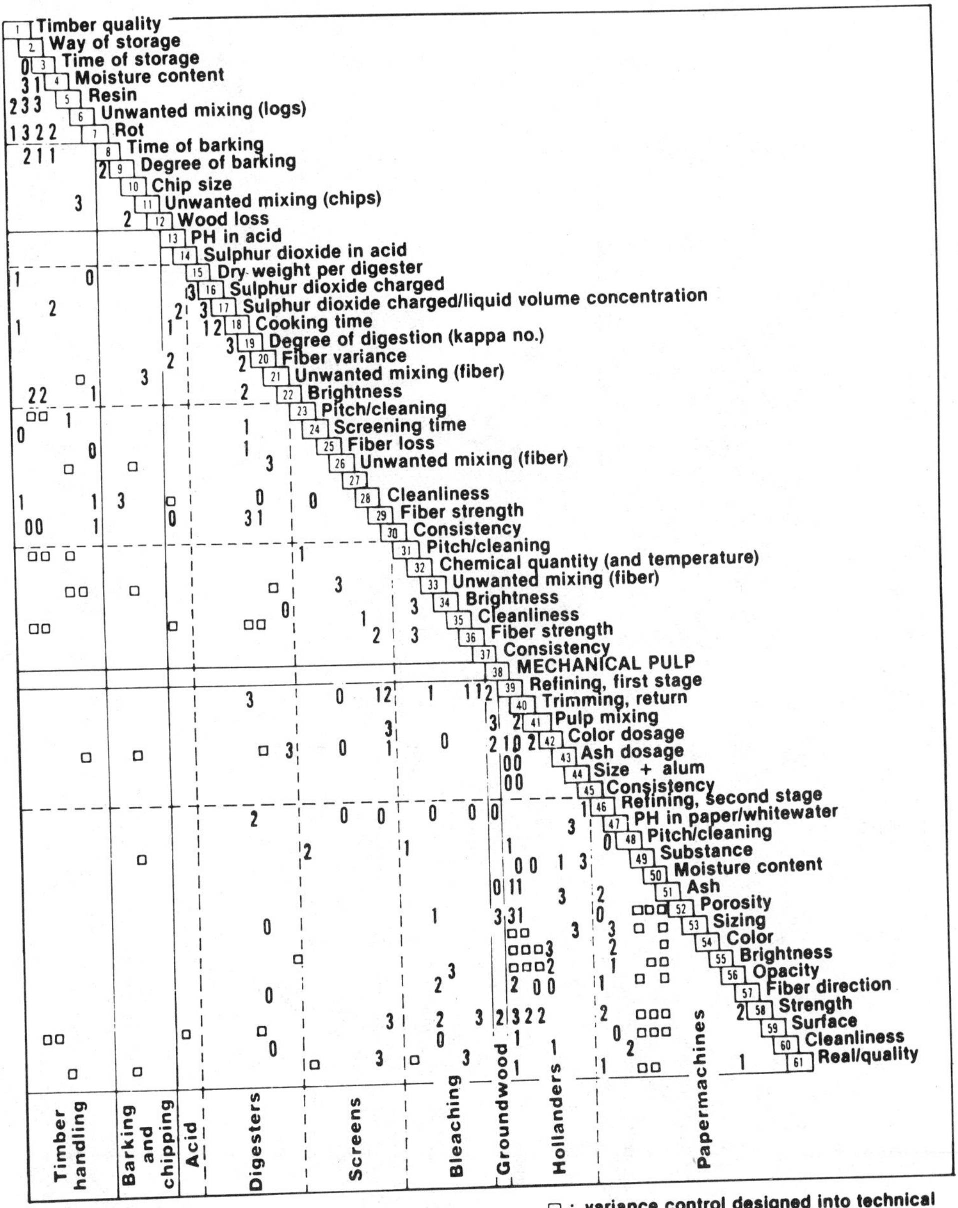

1 Timber quality
2 Way of storage
3 Time of storage
4 Moisture content
5 Resin
6 Unwanted mixing (logs)
7 Rot
8 Time of barking
9 Degree of barking
10 Chip size
11 Unwanted mixing (chips)
12 Wood loss
13 PH in acid
14 Sulphur dioxide in acid
15 Dry weight per digester
16 Sulphur dioxide charged
17 Sulphur dioxide charged/liquid volume concentration
18 Cooking time
19 Degree of digestion (kappa no.)
20 Fiber variance
21 Unwanted mixing (fiber)
22 Brightness
23 Pitch/cleaning
24 Screening time
25 Fiber loss
26 Unwanted mixing (fiber)
27
28 Cleanliness
29 Fiber strength
30 Consistency
31 Pitch/cleaning
32 Chemical quantity (and temperature)
33 Unwanted mixing (fiber)
34 Brightness
35 Cleanliness
36 Fiber strength
37 Consistency
38 MECHANICAL PULP
39 Refining, first stage
40 Trimming, return
41 Pulp mixing
42 Color dosage
43 Ash dosage
44 Size + alum
45 Consistency
46 Refining, second stage
47 PH in paper/whitewater
48 Pitch/cleaning
49 Substance
50 Moisture content
51 Ash
52 Porosity
53 Sizing
54 Color
55 Brightness
56 Opacity
57 Fiber direction
58 Strength
59 Surface
60 Cleanliness
61 Real/quality

Timber handling | Barking and chipping | Acid | Digesters | Screens | Bleaching | Groundwood | Hollanders | Papermachines

Meaning of variances

0- of theoretical interest only
1- of little practical importance
2- of medium practical importance
3- of great practical importance

□ : variance control designed into technical system, human intervention excluded
or
□ : indirect dependence, intervening variable controls variance transmission

[a]*Source.* P. H. ENGLESTAD, "Sociotechnical Approach to Problems of Process Control," in L. E. Davis and J. C. Taylor, *Design of Jobs*, Goodyear, Santa Monica, Calif., 1979, p. 205.

Exhibit 2.5.11 Format of Key Variance Control Table

Key Variance	Occurrence	Detection	Feedforward	Correction	Feedback	Prevention	Suggestions for Better Control	
							Technical	Social
Unit operation: Variance:	Where: How: By whom:	Where: How: By whom:	Channels: Lead time:	Where: How: By whom:	Channels: Lag time:	Where: How: By whom:		
Unit operation: Variance:	Where: How: By whom:	Where: How: By whom:	Channels: Lead time:	Where: How: By whom:	Channels: Lag time:	Where: How: By whom:		

by whom each key variance can occur and can be detected, corrected, and prevented. It should also describe how information about each key variance can be transmitted. "Feedforward" is the information that is transmitted from the point where a variance is detected to the point where it can be corrected or kept under control. "Feedback" is the information that is transmitted from the point of detection to the point where further occurrence can be prevented.

The fifth step is to construct a table of the skills, knowledge, information, and authority required for people to control key variances. The format for such a table is shown in Exhibit 2.5.12. Jobs should be designed so that jobholders have the skills, knowledge, information, and authority needed to control key variances.

Technological Assessment

"Technology" has been defined as the physical and informational resources by which people can systematically bring about some desired result. It has been noted that one of the primary functions of people in an organization is to deal with technological uncertainty. Whereas variance analysis can help to identify specific areas of technological uncertainty, technological assessment seeks a broader description of the general characteristics of the technology and their implications for job design. A number of dimensions of technology relevant to job design are described here. A suggested format for this analysis is shown in Exhibit 2.5.13.

Automation

Automation is technology that does not require human assistance. There are three phases of automation. The first is the operational phase, in which human physical movement is replaced or extended by machine. Second is the sensory phase, in which human senses are replaced by equipment sensors. The third phase of automation is the logic phase, in which information processing tasks are performed by equipment. Those tasks that are not automated, including variance control tasks that exceed the response capability internal to the technology itself, constitute the production-related tasks that are clustered into jobs. A typical effect of automation is a decrease in employee time spent on operational tasks and a relative increase in the proportion of employee time spent on maintenance, regulation, planning, and control.[2,40] Some unit operations may require primarily operational tasks of people, and others primarily regulatory tasks. This may suggest ways in which team job designs can make flexible use of human resources.

Programmability

Some technologies are based on an exact science, so there is complete information about what needs to be done in order to achieve a specified result—there is a recipe to be followed. In other technologies the underlying science is inexact or incomplete, and the necessary actions entail some intuition or trial and error. The job designer should assess the areas of high and low programmability of action so that jobs contain elements of each. Human discretion is required when tasks are nonprogrammable.

Subjectivity

With some technologies, output quality can be determined objectively, but with others, specifications are vague and variance detection is subjective. A food processing engineer remarked casually, "We could almost train monkeys to operate the equipment; what really takes skill is knowing how to recognize the color and taste of a good product and what to do to get it consistently good." The jobs needed to be designed to provide feedback between equipment operation and product results so that operators could develop a continuous sensitivity to product quality.

Stability

Although most variances occur during the performance of some unit operation, some can occur "spontaneously" during storage, transport, waiting, and so on. Variance control is more critical when input material, product states, or equipment are unstable.

Equivalency

Engelstad[41] reported on a paper mill in which four pulp digesters were treated as equivalent or interchangeable pieces of equipment. In fact, however, one of the digesters was particularly efficient with certain kinds of wood fiber. A few operators discovered this, but their jobs were designed so that they had no incentive for sharing this information with others. On the other hand, jobs in an aluminum smelter[42] were redesigned to take account of the fact that each smelting

Exhibit 2.5.12 Format of Table of Skills, Knowledge, Information, and Authority Needed to Control Key Variances

Key Variance to be Controlled	Skills Needed	Knowledge Needed	Information Needed	Local Authority Needed

Exhibit 2.5.13 Technological Assessment Format

Unit Operation	Automation	Programmability	Subjectivity	Stability	Equivalency	Scale	Degradation	Implications for Job Design

furnace tended to develop its own "personality." Workers performed a wide range of tasks for a small group of furnaces in order to learn the peculiarities of each one.

Scale

In general, technologies designed for producing small lots or batches call for different job designs than those designed for continuous mass production.[43] Indeed, one of the ways to generate alternative technical system choices is to design two hypothetical production systems, the first for producing the product in a lot of one, and the second for automated mass production.

Degradation

Some technologies function in either an "up" or "down" state. A computer, for example, rarely makes random errors; either it is not functioning at all, or if it makes errors when functioning, they are made consistently, given the same inputs and programs. Other technologies function in intermediate states, such as an automobile in need of a tune-up. When many diverse technical components are united into a technical system, a breakdown in one of the components can degrade the whole system. Under some conditions the whole system will go down. Under other conditions the degradation is less severe—the system can continue to function, albeit in a slower or less powerful mode.

People, of course, usually degrade quite gracefully—rarely do they abruptly stop functioning.[8] The job designer should place people in such a way as to enable the system as a whole to degrade less abruptly in the case of a breakdown in one part of the system. The conditions of abrupt degradation should be identified and social system backup procedures designed. Jobs should be designed to take advantage of people's flexibility. One example is the integration of both operation and maintenance activities into jobs; thus when operations cannot be performed, jobholders can engage in maintenance activities.

Task Ratings

The principal tasks to be performed by the work force can be subjectively rated according to the quality-of-working-life criteria shown in Exhibit 2.5.3 (also see Hackman and Oldham[44]). These ratings can be used to guide the clustering of tasks into jobs so that each job or team contains a reasonable balance of both undesirable and desirable tasks, isolated work and teamwork, simple and complex tasks, and so on.

Task ratings can be misleading in job design. A job is not simply the sum of its tasks, but also the interrelationship among the tasks so as to form a meaningful whole job. Attempting to design jobs solely according to the summation of task ratings, ignoring task interrelationships, is contrary to the systems perspective of job design.

Mobility Analysis

An important source of job design information resides in the actual capabilities of employees. Mobility analysis addresses a simple question: Who can move to what other jobs should the need arise? The format for mobility analysis is shown in Exhibit 2.5.14. The example shown suggests that jobs C, D, E, and F might possibly be linked into a team job design.

Responsibility Analysis

Concomitant with the clustering of tasks into jobs is the allocation to each role of certain responsibilities for making decisions, for use of resources, and for achieving results. The allocation of these responsibilities should be explicitly considered in job design. Exhibit 2.5.15 illustrates a format for compiling such information.

In a study of the redesign of supervisory jobs,[45] two modifications in supervisors' responsibilities were introduced separately into a number of aircraft instrument repair shops. One of the modifications was to allocate to the supervisor's unit responsibility for all tasks required to complete the instruments processed in his or her shop. This represented a change from a function-based to a product-based division of the organization into smaller unit-shops. The other modification involved not only the shift to a product-based organizational unit, but the addition of quality responsibility. Inspection tasks and authority for final acceptance of the instruments were added to the supervisor's unit. The supervisors under the second condition initially performed the inspection themselves and later delegated it to subordinates. Compared with control shops in which no changes took place, attitudes, productivity, and quality improved in the modified shops. As supervisors concerned themselves with planning and controlling their increased responsibility, their subordinates took on some of the characteristics of self-regulating work teams.

Exhibit 2.5.14 Mobility Analysis Format

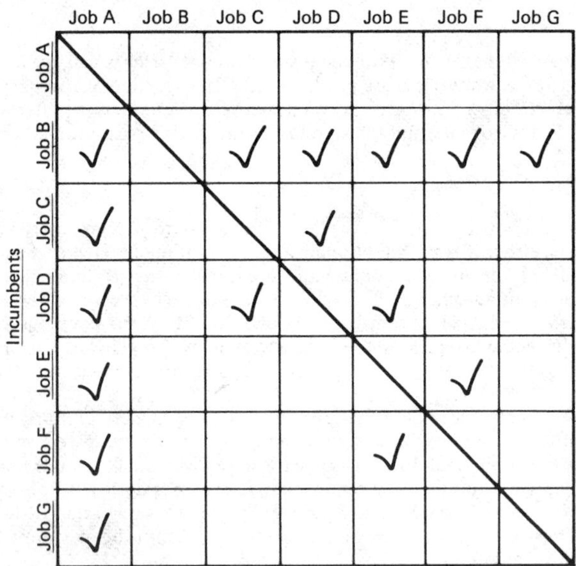

Cross-Boundary Interaction Analysis

In an existing job structure, the goodness of fit between actual job boundaries and organization needs can be investigated by analyzing the interactions across job and group boundaries. A format for such an analysis is shown in Exhibit 2.5.16. Jobholders are asked to describe the frequencies and reasons for work-related interactions with others in the organization. These responses are summarized and recorded in the appropriate cell in the table. This analysis may provide information on task interdependence and variance control as well as on social relationships that have evolved among jobholders.

Loci of Organizational Stability and Instability

Social system stability indicators, such as labor turnover rate, absenteeism rate, grievance rate, incidents of antisocial behavior, and attitude measures, can point to special needs for job design. The increasing heterogeneity of the work force (age, sex, ethnicity, and education dimensions), together with the faster rates of social change and shorter lead times of market shifts and technological innovations, has focused attention on internal stability. There is both the desire to preserve employee skills and commitments within the organization and the desire to adapt to changing conditions. Social system instability presents a dual challenge for job design: first, to design jobs that abate the causes of instability and, second, to design jobs that enable the system to cope with instability.

The analysis of turnover is both important and problematic. On the one hand, turnover has a negative connotation in its disruption of continuity of role incumbents and interpersonal relations. On the other hand, it has a positive connotation if individuals move to better jobs along a career path. A format for the analysis of labor turnover is presented in Exhibit 2.5.17. Exit interviews provide a source of data. Turnover analysis can suggest needs for more challenging jobs and for better paths of advancement.

2.5.10 CRITERIA FOR CLUSTERING TASKS AND ALIGNING BOUNDARIES

Following the analysis of the technological and organizational characteristics of an organization, tasks are clustered into jobs, and provisions are made for linkages between jobs. This is in essence a partitioning of the organization, which requires an alignment of technological and organizational boundaries, as discussed previously in this chapter.

A number of criteria should be considered in this partitioning process. It is unlikely that all

Exhibit 2.5.15 Responsibility Analysis Format

Job or Team	Tasks	Discretionary Aspects of Task Performance	Latitude for Decisions	Available Resources	Accountability for Results	Sources of Feedback

Exhibit 2.5.16 Format of Cross-Boundary Interaction Analysis

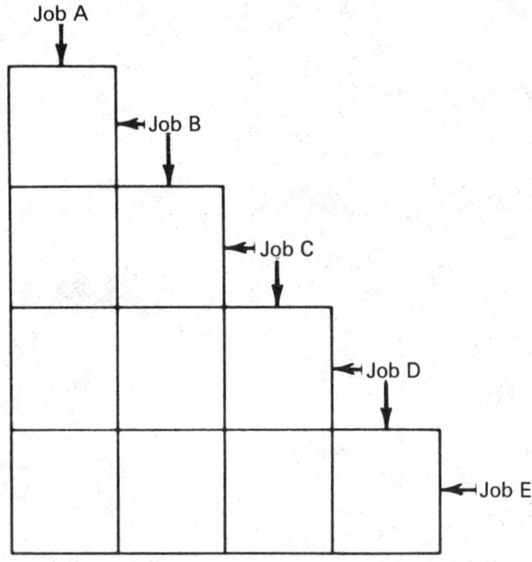

criteria can be optimized; rather, job designers must deal with tradeoffs. The criteria for clustering tasks into jobs and teams and for aligning boundaries are as follows:

Each task cluster forms a meaningful unit of the organization—a minibusiness or an identifiable craft.

Jobs or teams are separated by stable buffer areas.

Each job or team has definite, identifiable, measurable inputs and outputs.

Each job or team has appropriate units of performance evaluation—such as production, quality, maintenance, cost, waste, errors, absenteeism, turnover, cross-training, skill acquisition, customer complaints, machine utilization, and downtime.

Timely feedback about output states and feedforward about input states are available.

Each job or team is provided resouces to measure and control throughput states.

Short feedback and feedforward loops localize variance control within the boundaries of jobs or teams.

There is more interdependence among tasks clustered together within a boundary than across the boundary. The same holds for jobs linked together.

Tasks are clustered around mutual cause-effect relationships.

Tasks are clustered around common skills, knowledge, or data bases.

Jobs are linked together to facilitate cross-training.

Jobs incorporate opportunities for skill acquisition relevant to career advancement.

Exhibit 2.5.17 Labor Turnover Analysis Format

Job	Jobholder	Dates Job Held	Prior Job	Successive Job	Reasons for Leaving

Jobs or teams each contain a balance of group and individual tasks, desirable and undesirable tasks, and simple and complex tasks.

Jobs or teams each contain a balance of stressful and comfortable work environments.

Jobs or teams are capable of self-regulation and of adaptation to variations in their environments.

The work stations of a team are geographically and/or temporally proximate to each other.

A team can maintain social relationships with face-to-face interactions.

2.5.11 PROCESSES AND STRATEGIES FOR DESIGNING NEW JOBS

When a new organization is designed, or when an existing organization is markedly expanded, the design of its jobs should be systematically integrated into other design decisions (see Chapter 2.1). As a general rule, these design projects pass through four phases.

Phase 1: Formation of the Steering Committee

The design of an organization cannot be undertaken solely by design specialists. All parties who may be affected by the design should be involved in design decisions.

Previously we reviewed the case of a food processing plant whose designers initiated an industry-wide conversion to bulk packaging of their input material in order to accommodate job design criteria. This decision was beyond the latitude initially given to the project engineer and required the involvement of several corporate executives. Other decisions for that plant's design required the involvement of quality control executives, employee relations executives, and others who were members of a steering committee.

To set the stage for wide involvement in the design process, a steering committee is formed, consisting of key individuals with a direct stake in the new organization. The functions of the steering committee are (1) to provide the necessary resources, support, and protection to the design project team and the resources for implementation; (2) to engender cooperation between the design task force and the parties who are affected by its design decisions; (3) to bring design objectives into focus and to assist in future diffusion; and (4) to oversee and guide the design process.

Phase 2: Design Task Force

Data gathering, analysis, and generation of design alternatives is done by a design task force. The task force often includes outside architects, engineers, and job design consultants. The design progresses by identifying the technological and environmental characteristics of the enterprise, identifying some "sketches" of various job design alternatives, and then considering more specific technical and social design decisions. Sometimes certain technical and social design decisions confront existing institutional precedents, constraints, or restrictions. It is then that the philosophy statement is given its test as a viable document.

Phase 3: Philosophy Statement

The initial task of the design task force is to draft a statement of organizational philosophy that acts as a charter for the design project once it is approved by the steering committee. The philosophy statement addresses the following issues:

Stakeholders in the project and their interests in it.

Purposes of the design project.

Production and technological needs and requirements.

Organizational needs and requirements.

Explicit recognition of the needs and requirements of individuals who, as members of the organization, will fill jobs.

Values and assumptions about the organization's environment and future.

Values and assumptions about people and work.

Values and assumptions about management and managers.

Relationship between the organization to be designed and existing institutions.

Processes for decision making during the design.

Processes for decision making during and following startup.

Evolutionary aspects of the project—what is not to be designed until after startup.

Resources for implementation.

Risk bearing and responsibility for results.

Expectations of participation, cooperation, and support.

There are various methods for drafting the philosophy statement, depending on the degree to which values are already shared. Chapter 2.1 reviews methods of developing organization philosophy.

Phase 4: Evolution

In job redesign in an existing organization, and after managers and supervisors have been appointed in the design of a new organization, the process of job design shifts to a more participatory mode. The underlying strategy for this shift was discussed previously as the principle of minimal critical specification. One member of a steering committee expressed it more colloquially: "We want to avoid dictating other people's work lives. If we were designing this plant for ourselves to work in, we would want to leave a lot of options open." As the organization evolves, its jobholders use their experiences to help perfect its design. Continual redesign of jobs becomes one of the organizational tasks of the work force.

2.5.12 PROCESSES AND STRATEGIES FOR REDESIGNING EXISTING JOBS

Although existing organizations might do well to begin job redesign with a reexamination of managerial strategy, technology, and organizational structure, there are usually a number of constraints that cannot be removed. Job redesign is often just part of a continuous stream of changes the organization makes to meet requirements coming from its environment. One of the objectives of job redesign is to provide an experience that sets a precedent for future redesign needs.

The designer should identify all relevant change forces, both internal and external to the organization and both past and future. Usually there is a complex system of forces that underlies the job redesign endeavor. In an aluminum smelter,[42] rising labor turnover rates, increasing difficulty in recruiting workers, technological obsolescence, and more aggressive government regulation of occupational safety and health conditions and equal employment opportunities for minorities and women led to efforts to automate some of the most physically demanding and undesirable tasks. The new automated technology required less manpower and new task clusters. But the new technology alone did not solve the organization's problems of turnover, recruitment, and efficiency. Thus the job redesign was directed not only at suiting the new technology, but also at creating more attractive and challenging jobs, at attaining work force flexibility so that the organization could function with a broad range of turnover rates, and at training and advancement geared to individual readiness, regardless of vacancies created by turnover.

Employee suggestions for, and their reactions to, redesigned jobs are influenced by their past experiences and their expectations for the future. In one plant employees saw redesign as being accomplished by petitioning for a second time clock so that they would not have to lose time in clocking out for lunch. That there were more fundamental aspects to an improved quality of working life was beyond their experiences. In the aluminum smelter cited previously, the job redesign resulted in fewer status differences between smelter jobs. Although most of the employees welcomed the new opportunities and new technology, some complained about the loss of their former perquisites. Attitudes improved among both the employees whose jobs were changed and those whose jobs had not been changed, because the latter group looked forward to an imminent changeover–their jobs now had a future.

The phases of redesign are analogous to those of new design, with the following additions:

Representatives of existing jobholders, or of their union, are usually included in the redesign project task force. Sometimes the redesign team's membership represents a "diagonal slice" of the organization, with various levels of the hierarchy represented such that no one is on the committee along with his or her direct superior.

If several separable redesign projects are considered, each has its own design task force; a single steering committee coordinates the projects.

Implementation planning is more critical in redesign, and implementation issues are considered concomitantly with design decisions. Probably the most important aspect of implementation is *ownership of change*. Those responsible for carrying out and living with proposed redesigns should feel that the changes have come out of their own interests and efforts. A widespread commitment to a proposal is a vital factor to its ultimate success, for two reasons: (1) emotional identification with the redesign and the desire to put forth the effort to make it work, and (2) understanding the details of the redesign sufficiently to estimate and minimize the effects, that is, to reduce the uncertainty over the consequences for the individual.

A number of methods can enable jobholders to contribute to the redesign. These include "nominal group technique"[46]; sociotechnical analysis approaches, including variance analysis; redesign principles; and quality-of-working-life characteristics.

Role of Consultants

In both new design and redesign, the job design consultant has a very different role from that of a conventional engineering consultant. Because the design process is multiphasic and includes many parties, the consultant performs five basic functions:

Training and Guidance. The principles of job design and the methods of analysis are taught to members of design task forces and steering committees, who, in turn, make design decisions. The consultant does not design the jobs, but rather guides the principals through the job design endeavor.

Process Facilitation. The consultant helps plan and facilitate the processes in all phases of the design. For example, steering committee and task force members may be unacquainted with their roles, and the consultant helps clarify these issues as the need arises and aids them in working together.

Mediating. When several distinct institutions are party to a redesign project, for example, company and union, the consultant may act as a neutral broker and go-between in order to ensure consensual agreement for design and implementation.

Research. The assessment of job design results may require research methods for which the consultant has special training and experience. In such cases he or she may collect and analyze data and present findings useful for the teams.

Change Agent. As innovative job designs are developed, the consultant acts as a carrier of these innovations to other parts of the institution of which the organization is a part. The role here is not one of "expert" so much as "idea broker."

2.5.13 CONCLUSION

Job design is a process establishing the content of roles or jobs, which are then assigned to individuals to perform. In the design of jobs for new organizations, the process is inseparable from organization design. In the redesign of jobs, organizational boundary issues as well as technical system design issues must be examined, since they may place unnecessary constraints on the design. The designer must remember that the organization is a system and that the jobs are component parts of this system. Further, the system embraces both the organization and its environment, bringing into focus the fact that job design has to concern itself not only with the technical and economic needs of the organization, but with the needs and expectations of the jobholders on whose task performances the organization depends for the achievement of its goals.

REFERENCES

1. L. E. DAVIS, R. CANTER, and J. HOFFMAN, "Current Job Design Criteria," *Journal of Industrial Engineering*, Vol. 6, No. 2 (1955), pp. 5–11.
2. L. E. DAVIS and J. C. Taylor, "Technology and Job Design," in L. E. DAVIS and J. C. TAYLOR, Eds., *Design of Jobs*, 2nd ed., Goodyear, Santa Monica, CA, 1979.
3. L. E. DAVIS and R. R. CANTER, "Job Design Research," *Journal of Industrial Engineering*, Vol. 7. 1956, pp. 275–282.
4. L. E. DAVIS and R. WERLING, "Job Design Factors," *Occupational Psychology*, Vol. 34, 1960, pp. 109–120.
5. G. I. SUSMAN, "Job Enlargement: Effects of Culture on Worker Responses," *Industrial Relations*, Vol. 12, 1973, pp. 1–15.
6. L. E. DAVIS and E. L. TRIST, "Improving the Quality of Working Life: Sociotechnical Case Studies," in J. O'Toole, Ed., *Work and the Quality of Life*, MIT Press, Cambridge, MA, 1974.
7. R. DUNHAM, "Reactions to Job Characteristics: Moderating Effects of the Organization," *Academy of Management Journal*, Vol. 20, 1977, pp. 42–65.
8. N. JORDAN, "Allocation of Functions Between Man and Machine in Automated Systems," *Journal of Applied Psychology*, Vol. 47, 1963, pp. 161–165.
9. L. E. DAVIS, "The Coming Crisis for Production Management: Technology and Organization," *International Journal of Production Research*, Vol. 9, 1971, pp. 65–82.
10. J. C. TAYLOR, "Some Effects of Technology in Organizational Change," *Human Relations*, Vol. 24, 1971, pp. 105–123.
11. H. L. WADDEL, "The Fundamentals of Automation," in H. B. Maynard, Ed., *Industrial Engineering Handbook*, McGraw-Hill, New York, 1956, pp. 325–331.
12. C. ORPEN, "The Effects of Job Enrichment on Employee Satisfaction, Motivation, Involvement, and Performance: A Field Experiment," *Human Relations*, Vol. 32, 1979, pp. 189–217.

13. G. R. OLDHAM and H. E. MILLER, "The Effect of Significant Other's Job Complexity on Employee Reactions to Work," *Human Relations*, Vol. 32, 1979, pp. 247–260.
14. L. E. DAVIS and A. B. CHERNS, Eds., *The Quality of Working Life*, Vol. 1 and 2, Free Press, New York, 1975.
15. P. H. ENGELSTAD, "Sociotechnical Approach to Problems of Process Control," in L. E. DAVIS and J. C. TAYLOR, Eds., *Design of Jobs*, 2nd ed., Goodyear, Santa Monica, CA, 1979.
16. L. E. DAVIS, "Evolving Alternative Organization Designs," *Human Relations*, Vol. 30, 1977, pp. 261–273.
17. L. E. DAVIS and G. J. WACKER, *Comprehensive Socio-Technical System Design: A Case Study*, in preparation.
18. P. G. GYLLENHAMMAR, *People at Work*, Addison-Wesley, Reading, MA, 1977.
19. E. L. TRIST, G. W. HIGGIN, H. MURRAY, and A. B. POLLOCK, *Organizational Choice*, Tavistock, London, 1963.
20. J. W. LORSCH and J. J. MORSE, *Organizations and Their Members*, Harper & Row, New York, 1974.
21. R. M. KANTER, *Men and Women of the Corporation*, Basic Books, New York, 1977.
22. F. W. TAYLOR, *The Principles of Scientific Management*, Harper & Row, New York, 1911.
23. A. SMITH, *The Wealth of Nations*, Penguin, London, 1970 (originally published 1776).
24. C. BABBAGE, *On the Economy of Machinery and Manufacturers*, Augustus M. Kelly, New York, 1965 (4th ed. enlarged; originally published 1835).
25. F. E. EMERY, "The Assembly Line–Its Logic and Our Future," in L. E. DAVIS and J. C. TAYLOR, Eds., *Design of Jobs*, 2nd ed., Goodyear, Santa Monica, CA, 1979.
26. L. E. DAVIS, "Pacing Effects on Manned Assembly Lines," *International Journal of Industrial Engineering*, Vol. 4, 1966, pp. 171–180.
27. E. H. CONANT and M. D. KILBRIDGE, "An Interdisciplinary Analysis of Job Enlargement: Technology, Costs, and Behavioral Implications," *Industrial and Labor Relations Review*, Vol. 18, 1965, pp. 377–390.
28. F. HERZBERG, *Work and the Nature of Man*, World, Cleveland, 1966.
29. L. E. DAVIS and A. B. CHERNS, "Transition to More Meaningful Work–A Job Design Case," in L. E. DAVIS and A. B. CHERNS, Eds., *The Quality of Working Life*, Vol. 2, Free Press, New York, 1975.
30. F. E. EMERY, Ed., *Systems Thinking*, Penguin, London, 1969.
31. P. G. HERBST, *Socio-Technical Design*, Tavistock, London, 1974.
32. A. K. RICE, *Productivity and Social Organization: The Ahmedabad Experiment*, Tavistock, London, 1958.
33. J. F. DEN HERTOG, "The Search for New Leads in Job Design," *Journal of Contemporary Business*, Vol. 6, No. 2 (1977), pp. 49–66.
34. L. E. DAVIS and C. S. SULLIVAN, "A Labour-Management Contract and Quality of Working Life," *Journal of Occupational Behavior*, 1980, Vol. 1, pp. 29–41.
35. J. N. SCANLON, "Profit Sharing Under Collective Bargaining: Three Case Studies," *Industrial and Labor Relations Review*, 1948, Vol. 2, p. 58 ff.
36. F. G. LESIEUR and E. S. PUCKETT, "The Scanlon Plan has Proved Itself," *Harvard Business Review*, Vol. 47, No. 5 (1969), pp. 109–118.
37. D. I. CLELAND and W. R. KING, *Systems Analysis and Project Management*, McGraw-Hill, New York, 1968.
38. J. GALBRAITH, *Designing Complex Organizations*, Addison-Wesley, Reading, MA, 1973.
39. D. R. KINGDON, *Matrix Organization*, Tavistock, London, 1973.
40. R. J. HAZLEHURST, R. J. BRADBURY, and E. N. CORLETT, "A Comparison of the Skills of Machinists on Numerically Controlled and Conventional Machines," *Occupational Psychology*, Vol. 43, No. 3 (1969), pp. 169–182.
41. ENGELSTAD, "Sociotechnical Approach to Problems of Process Control," 1979.
42. G. J. WACKER, "Evolutionary Job Design: A Case Study," working paper, Department of Industrial Engineering, University of Wisconsin, Madison, 1979.
43. J. WOODWARD, *Industrial Organization: Theory and Practice*, Oxford University Press, New York, 1965.
44. J. R. HACKMAN and G. R. OLDHAM, "Development of the Job Diagnostic Survey," *Journal of Applied Psychology*, Vol. 60, 1975, pp. 159–170.

45. L. E. DAVIS and E. S. VALFER. "Supervisorv Job Design," *Ergonomics*, Vol. 8, 1965, p. 1.

46. A. L. DELBEQ, A. H. VAN DE VEN, and D. H. GUSTAFSON, *Group Techniques for Program Planning*, Scott, Foresman, Glenview, IL, 1976.

BIBLIOGRAPHY

CSIKSZENTMIHALYI, M., *Beyond Boredom and Anxiety*, Jossey-Bass, San Francisco, 1976.

FORD, R. N., *Why Jobs Die and What to Do About It*, AMACOM, New York, 1979.

HERZBERG, F., *The Managerial Choice: To Be Efficient and Human*, Dow Jones, New York, 1976.

HILL, P., *Towards a New Philosophy of Management*, Gower, London, 1976.

MUMFORD, E., and H. SACKMAN, Eds., *Human Choice and Computers*, North Holland, Amsterdam, 1975.

NADLER, D. A., J. R. HACKMAN, and E. E. LAWLER, *Managing Organizational Behavior*, Little, Brown, Boston, 1979.

SCHON. D. A. *Beyond the Stable State*, Random House, New York, 1971.

SHEPPARD, H. L. and N. Q. HERRICK, *Where Have All the Robots Gone?* Free Press, New York, 1972.

SECTION 3

METHODS ENGINEERING

CHAPTER 3.1

Methods Design

CHARLES E. GEISEL

Container Corporation of America

3.1.1 THE ROLE OF THE METHODS DESIGN

A primary concern of industrial engineers is accomplishing necessary functions, purposes, or goals with a minimum use of resources. *How* this is done is called the "method," which is a description of how we use resources to accomplish our purposes.

Methods are an integral part of our lives since we use them to accomplish anything we want done at home, work, or play. Much of what we get out of life individually and collectively is determined by (1) how well our methods utilize our limited resources, such as time, energy, materials, and money; (2) how our methods physically and psychologically affect us; and (3) the quality of the results or outputs of our methods by the service or product. The primary resource that methods use is time.

Time is the most critical of all resources, since each person has exactly the same amount of time in each day. The methods used in our daily activities determine how much we will get out of life, since time is the essence of life. In our personal life the less time required to do that which may be unpleasant, the more time left to do what we enjoy. Time is also the most frequently used basis of compensation; therefore, methods that require less time reduce the cost of a unit of product or service. A television set requiring only 20 labor hours to produce can cost less than one requiring 30 labor hours.

Methods can also determine the amount of materials, energy, and capital used (utilization of equipment). The methods design that can reduce the consumption or waste of these resources also reduces the cost of a unit of product or service. If, through better cutting methods, the cloth wasted in making a suit is reduced by 5%, its cost is lowered.

Increases in productivity through better methods design can yield an improved standard of living. In any society, regardless of political system, it is possible to have a low standard of living with high productivity if the output of the productivity is poorly distributed, but it is impossible to have a high standard of living with low productivity. We cannot provide any of the social benefits many people desire, such as good food, clean air and water, and adequate shelter, without maximizing the output from our limited resources through improved methods design.

The design of the method determines the effort expended by the people using the method, their probability of strain and injury, and how they feel about what they are doing. A method with long reaches or requiring the body to be in uncomfortable positions will tire the people using it. It can make them less productive; introduce the probability of strains, tenosynovitis, or other injuries; and possibly cause the people to develop a negative attitude about the task and the people related to it.

The design of the method can also determine the quality of the output produced by the method, be it a physical product or a service. For example, a poor assembly method for a radio may cause a part to be occasionally misplaced, which may cause the radio to fail. A bank teller using a poor method to serve a customer may cause customers to think the bank is careless with their money. If the output meets the expected quality specification, it can be a source of pride and job satisfaction for the person using the method.

3.1.2 DESCRIPTION OF HOW SYSTEMS OPERATE

A method describes how resources are used to process inputs into outputs in order to accomplish purposes or functions through a network of activities called a "system." A system is the operational utilization of human, physical, and information resources, which converts or works on inputs to arrive at outputs that achieve a purpose or purposes in a sociological and physical environment. Systems can be any size, from the small system or method of filing a folder in a file

Exhibit 3.1.1 Example of a System Hierarchy: Small to Large Systems[a]

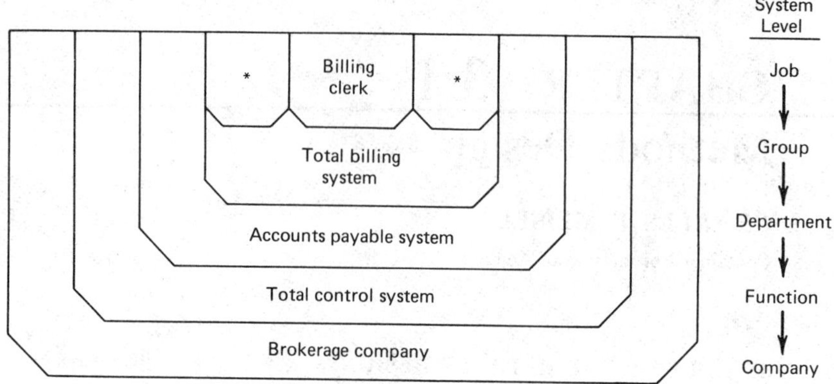

[a]Asterisked portions signify horizontal systems.

drawer to the large system or methods of running a large corporation or agency. Every system is part of a larger system where it interfaces with other subsystems (see Exhibit 3.1.1). This means that if a method is designed for a large system, it will encompass the methods of the smaller subsystems. If a large system is too complicated to design, it can be broken down into subsystems, and methods can be designed for each of these.

To design a method totally, each of the eight elements of any system, regardless of size, must be considered.[1] Each element has five dimensions, which form a systems matrix. This matrix and two guide columns–constraints and regularity–form the methods design framework (Exhibit 3.1.2). This framework provides a guide and checklist for methods design and encourages complete specifications.

The eight system elements are as follows:

1. **Purpose.** The function, mission, aim, or need for the system. What the system is to accomplish, not how.
2. **Input.** The physical items, people, and/or information that enter the system, to be processed into the output, such as ore, toaster parts, paper, customers, or patients.
3. **Output.** That which the system produces to accomplish its purpose; products of the system, such as finished steel, assembled toasters, boxes, and served customers or treated patients; and by-products such as slag or scrap.
4. **Sequence.** The steps required to convert, transform, or process the input to the output. (This is the heart of the methods design.) In small systems the sequence is the description of how the human agent, or worker, does his or her job.
5. **Environment.** Condition under which the system operates, including physical, attitudinal, organizational, contractual, cultural, political, and legal. Examples are temperatures, lighting, items of the union contract that would effect the design, or Occupational Safety and Health Administration (OSHA) regulations.
6. **Human Agents (workers).** The people who aid in the steps of the sequence without becoming part of the output. This includes what they do, their skills, responsibilities, and rewards.
7. **Physical Catalysts (equipment).** The equipment and physical resources that aid in the steps of the sequence without becoming part of the output; the equipment and layout used to make the system operate.
8. **Information Aids.** Knowledge and information resources that aid in the sequence without becoming part of the output. Examples are instructions, specification sheets, and checklists.

The five dimensions of each system element are as follows:

1. **Fundamental.** The basic description of the element, its physical characteristics and specifications–the what, how, where, or who.
2. **Rate.** The time-based measures of the element during operation of the system, such as parts per minute or customers per hour.
3. **Control.** The evaluation of the element, and procedures for modifying activities to maintain specifications, such as minimum weight requirements per unit.
4. **Interface.** The reaction of the element to other system dimensions, for example, how complicated equipment affects hiring practices.
5. **Future.** Planned changes and research needs–what changes should be provided for.

Exhibit 3.1.2 The System Matrix

	Dimensions					Guides	
Elements	Fundamental	Rate	Control	Interface	Future	Constraints	Regularity
Purpose							
Input							
Output							
Sequence							
Environment							
Human agents (workers)							
Physical catalysts (equipment)							
Information aids							

The two guides used in methods design for each element are:

Constraints. The minimum restrictions or limitations that must be considered in the design of each element; those conditions or specifications that must be part of the final method specifications.

Regularity. The most important, regular, or usual conditions and specifications of each element; those that the method will normally deal with.

3.1.3 A STRATEGY FOR METHODS DESIGN

The following is a strategy for designing new methods or for improving existing methods.[2]

Determining the Purpose of the Method

The most important element of any existing or proposed system or method is the function or purpose. It will indicate whether the method is necessary and will guide the design. Methods design should begin with the question "What purpose or function should be achieved with the appropriate method?" If the function of the method cannot be defined, then the method is not needed and therefore does not have to be designed. Starting a methods design by determining the purpose or function avoids the implied criticism and accompanying resistance or defensive behavior usually caused by studying and questioning a present method. It also saves time, since only data needed for design are collected. Analysis of the present system is done only to gather information for the design or for the justification of an improved method. Fact finding is no substitute for thought. When you start by defining the function, you are not looking for trouble in the present system, an approach that can send you down the wrong path.

People who are or will be concerned with the system or affected by it should be involved in this first phase, particularly those who will use the method. They can bring their years of experience to bear in finding the true function. The functional approach also gives them an insight into their jobs. When people are involved in these systems, leadership must be provided to balance the time invested in design with the results anticipated from the design.

The first step in determining the function is to give a name to the system to be studied or designed, such as "patient medication system."

The second step is to determine the most immediate or direct function of that system—its reason, mission, purpose, or objective. It refers to *what* is to be accomplished, not *how*. The function of a system is not the output. For example, the function of a hamburger stand is to relieve hunger; the output is hamburgers. The function of a corrugated die cut system is to create openings in a sheet of corrugated board; the output is a die cut board.

How the function statement is worded is important, for it serves as a guide to design. It should start with an action verb and be specific, but nonlimiting. For example, the function statement "drill a hole" will give a mental picture of a method that usually includes a drill. The statement "create a hole" can imply different methods, but usually a round hole remains. The statement "create a void" usually brings another, different set of methods to mind. Do not include goal-like statements such as "process 40 customers an hour." This should be handled as a constraint or requirement of the rate of output of the method. Do not restrict the opportunity for design or design space by restrictive words in the function statement. For example, do not say "stamp a can end," but say "form it"; do not "file information," but "store it"; do not "make a drawing," but "communicate dimensions and form."

A maintenance group started thinking about methods of preventive maintenance by realizing that their function was not to repair equipment, but to keep it running. When machine crews were overexposed to noise, a massive study was proposed to quiet down the machinery by the method of putting expensive soundproof "enclosers" around them. When asked what the function was, the workers answered that it was to "provide a quiet workplace." When asked why there should be a quiet workplace, they answered that it was to "reduce workers exposure to noise to under noise limits." Starting with this function, the job methods and machine control locations were redesigned to enable the worker to perform essential functions outside noisy areas, which were identified by noise contour charts.

One of the problems encountered when trying to identify the function is that there may be more than one; for example, the functions of a packaging system may be to provide a vapor barrier and to give a pleasing appearance. There are two functions, so it must be decided which is the more important and which is used as a constraint.

After the most immediate or direct function has been determined, the third step is to ask of that function, "What is the function of this function?" or "Why is this done?" The answer to this question will usually define the function of the next larger system. This is repeated several times until functions are determined for systems too large to work in, such as the distribution system of the

Exhibit 3.1.3 Function Expansion–Die Cut System

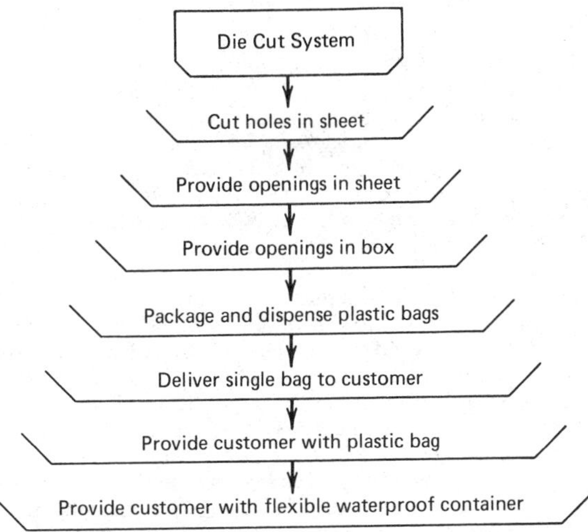

United States. This procedure is called "function expansion". Always expand functions far beyond levels that might be remotely considered for the selected function. This gives an insight into how the function you select fits into the system hierarchy and demonstrates an ability to look beyond the present problem.

After the functions are ranked in order of system size, the level at which the method can be designed is selected. There are usually six criteria to use in this selection: (1) potential savings or improved services, (2) management desires, (3) time limitations, (4) organizational factors, (5) control factors, and (6) capital availability. The selection of the function statement and its level is critical to the methods design and to its impact in the organization in which it operates. It also focuses the design effort on the purpose rather than on the initial problem.

After the function is selected, check to be sure (1) that everyone understands the statement without having to refer to the starting project; (2) that the function statement does not specify other system characteristics; and (3) that it is possible to design a method that can achieve the statement without reference to some other need or purpose.

Exhibit 3.1.3 shows the function expansion of a die cut system for cutting holes in sheets of corrugated board, which later becomes a box for dispensing plastic bags. By selecting "package and dispense plastic bags" as a function statement, designers came up with a method that produced a box (output), which did not need to go through the die cut operation at all, but which used the flaps cut in the first printing and slotting operation to create the opening.

Exhibit 3.1.4 Function Expansion–Paper Usage System

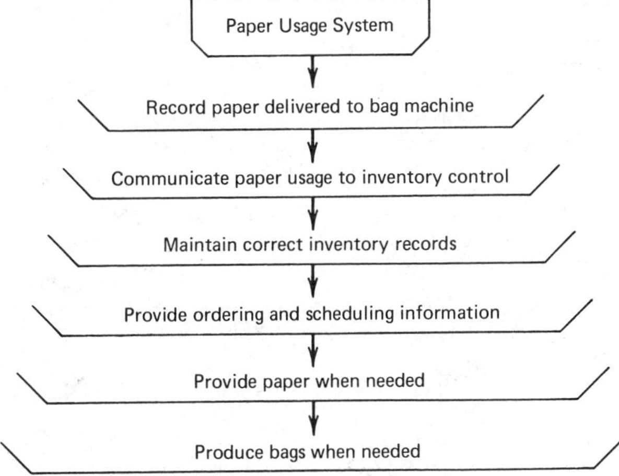

Exhibit 3.1.4 shows the function expansion for a system of recording paper delivered to a bag making machine.

Conceptualizing Ideal Methods

Once the function level has been determined, and the purpose of the method has been clearly defined and judged absolutely necessary, the design phase can begin. During the design phase, it is valuable to include workers concerned with and affected by the method, for it enables designers to use workers' experience with the system and to approach solutions from many different aspects. Consulting checklists will stimulate the flow of ideas during the design phase. The most complete collection of checklists broken down by system elements can be found in *Work Design–A Systems Concept*, by Gerald Nadler.[3]

Individual and group creativity can be stimulated by directed brainstorming, generating as many broad statements of systems as possible to accomplish the function or purpose, without criticism of the ideas at this point. This can yield unique methods (see Chapter 1.5). As many ideas as possible should be generated on how to design a method to accomplish the purpose or function of the system, even extreme ideas, which may ignore the constraints and which may not, at this time, be technologically workable. The focus should be on what should be rather than on what is. All ideas should be recorded and divided into ultimate ideas, both those that depend on technology not currently available and those that are technologically workable. For example, for the paper recording system shown in Exhibit 3.1.4, an ultimate idea was a device that could automatically record all the physical characteristics of a roll of paper by looking at the roll. The mental exercise of this phase (1) stimulates creativity, (2) reduces the inclination to defend the existing system or past practices, and (3) prepares people for future changes.

Identifying Constraints and Regularity

The constraints for each system element should be listed. A constraint or restriction is an assumed or actual condition that must be part of the final method, such as government regulations, union restrictions, company policies, or physical limitations. The necessity of each of these limitations is challenged. The fewer the restrictions, the greater the design space. Some of the constraints may be considered tentative and be temporarily eliminated in order to test the success of a method designed without them. If handicapped people are to work in the system, their limitations, for example, could be included as a constraint.

This is also the time to define regularity, that is, the conditions of each system element that would represent a large proportion or frequency of occurrences for which we can design methods. Many a good method has been abandoned because of an odd or infrequent situation that it would not accommodate. Irregular conditions may be accommodated by a separate system. The regularities should be handled before irregularities. The constraint and regularity columns form the guides (Exhibit 3.1.5) for developing the methods outline.

Exhibit 3.1.5 An Example of the Guides That Would Be Used to Design a Method to "Communicate Paper Usage To Inventory Control"

	Constraints	Regularity
Input	Must identify paper, order, and machine.	Kraft paper 50 in. diameter rolls.
Output	Capable of being recorded on cards by noon of the day following use.	100 transactions per day.
Sequence	Paper not to be reported until it is placed on machine.	All paper completely consumed.
Environment	None.	Local plant.
Equipment	A computer cannot be considered at this time.	All paper delivered by truck.
Human factor	Inventory record must be kept by a salaried employee only.	All workers fully trained.
Information aids	None.	Current standard tables.

Outlining Practical Methods

The ideas can be further developed by considering the cells in the system's matrix and applying some of the following principles:

1. Purpose. Try to eliminate the need for the purpose.
2. Input. Design the fewest and lowest-cost items or information.
3. Output. Design the fewest and lowest-cost output that satisfies the purpose or function.
4. Sequence. Minimize the number of times information, material, products, or people are handled or altered and the number and cost of operations required to process output from input. Once control of an object is gained, as much as possible should be done in processing the object, since it costs time and effort to release and regain control.
5. Environment. Minimize the number of changes required in, or constraints imposed by, the environment.
6. Human Agents. Maximize the utilization of worker skills. Minimize all movements, operations, holds, verifications, and delays. Minimize the number of labor hours and special skills required within the specifications of quality of product or service. Simplify all necessary operations, movements, measurements, verifications, and manipulations. Minimize steps, reaches, lifts, weight lifted, and distance moved.
7. Physical Catalysts. Utilize equipment at maximum capacity. Run it at optimum speed, with minimum downtime. Employ automated equipment if justified (see Chapter 7.5). Minimize the amount and cost of equipment in keeping with costs and quality of output.
8. Information Aids. Use automatic data processing if justified, and never record information more than once (Chapter 12.2).

Exhibit 3.1.6 may be used as a guide in this phase.

Selecting the Best Method Outline

After several alternative ideas have been outlined, they can be evaluated using five criteria: (1) hazard (Section 6), (2) economic (Section 9), (3) control (Section 8), (4) psychological (Section 2), or (5) organizational (Section 2).

Formulating Details of the Selected Method Outline

After a method has been selected, it should be further detailed by the use of specifications, drawings, layouts, charts, and descriptions and should be improved by use of work simplification techniques, principles, and checklists.[4] The system matrix should be used as a guide to ensuring that all facets of the method are considered in the design. However, only information that is necessary for answering specific questions should be gathered.

Inputs and Outputs

Inputs and outputs, whether they be materials, products, information, or people, should be described and specifications established in order to ensure that the purpose of the system is accomplished (see Chapter 7.2). The rate at which the inputs enter the system and the outputs are processed must be established. Control of the quantity and quality of inputs and outputs should be considered, as should the inputs' and outputs' effects on or interface with adjoining systems. Value

Exhibit 3.1.6 Principles for Methods Design

1. Design only to accomplish necessary purposes in the most ideal way.
2. Consider all systems elements and dimensions.
3. Design for regularity before considering exceptions.
4. Focus on what should be rather, than on what is.
5. Consider layout and equipment design.
6. Eliminate or minimize all body movements.
7. Keep people's backs straight and their hands close to their navels.
8. Handle objects and record information only once.
9. Gather only information necessary for design or justification.
10. Minimize use of all resources.

engineering can be used in the design of products, which may be inputs and/or outputs (see Chapter 7.3).

Sequence

The sequence of events that processes the input into the output can be described by symbolic and systematic representations or models, called "charts" (Chapter 3.3). Material handling techniques can be applied to methods of handling materials or products throughout the sequence (Chapters 10.3 and 10.4). Performance evaluation and review technique (PERT) and critical path method (CPM) can be used to chart large sequences.[5]

Product process charts or person process charts[6] should be used to describe and plot the steps necessary for modifying the product or processing a person through the system. These charts have a symbol of each operation, movement, storage, delay, and inspection of the product or person. The symbols are linked together to form the sequence.

Form process charts[7] describe and plot the processing of information usually contained in forms. These charts have a symbol for the origin of each form or datum and for the operations, movements, storage, and disposal of forms and the transmission of information. The rate at which products, people, or information move through the sequence, the controls in the form of inspections and verifications, and the interfacing with other systems are specified.

Human Agents

Human agents or human activities can also be represented or described in charts (see Chapter 3.3). Process charts portray the step-by-step procedure a person follows in doing a job when moving from place to place. There is a symbol for each operation, movement, delay, hold, or verification involved. Operation charts[8] describe and plot the activity of a worker in one place. There is a symbol for each operation, movement, hold, and delay. The activities for each hand are plotted side by side, usually on a time scale. When work with a machine is involved, or if the machine is controlling, a man and machine chart is used. It contains one symbol for when the machine is running and one for when it is idle. The machine activity is plotted alongside the worker's activity. For several workers on one machine, a multiman and machine chart is used, and for several machines, multimachine charts are used (see Chapter 3.5).

All the movements of a worker should be designed using the principles of motion economy developed by R. M. Barnes[9] and the prerequisites of biomechanical work tolerance proposed by E. R. Tichauer[10] (see Section 4). To translate some of the biomechanical concepts into terms that may be more easily understood and remembered by nontechnical people, use the following body biomechanic[11] rules:

1. **The Straight Back Rule.** Design the task to keep the back and neck straight, even if the worker must bend over. Keeping the back straight also applies to twisting or sidestep movements, which should be avoided.

2. **The Belly Button Rule.** When lifting or handling items or controls, keep the hands close to the navel.

a. This keeps the weight close to the lumbosacral joint, which is in line horizontally with the navel. Therefore the closer the hands to the navel, the smaller the lifting movement (weight × distance to the spine) and thus the less the back strain.

b. When the hands are active near the navel, the elbows are down, reducing muscle strain. Also, as the hands move forward, away from the navel, the bicep is stretched out, causing it to lose its mechanical advantage and to become fatigued.

3. **The Swinging Arm Rule.** Movements of the arm should follow the normal swing of the arm, since it takes four times as much time and effort to move an object on a straight line. The motion should then be stopped by a barrier rather than the muscular action.

4. **The Straight Wrist Rule.** Grasping, holding, or rotating the hand with a flexed wrist, or doing fine manipulations with a straight wrist, should be avoided. When the wrist is flexed, the tendons are bent and become subject to stress and friction while opening and closing the hand.

5. **The Skin Rule.** Pressure concentration on small skin areas should be avoided. Prolonged pressure will restrict circulation, which can injure small blood vessels and restrict circulation, causing numbness and tingling.

6. **The Lazy Foot Rule.** Methods involving removing and replacing guards or locking out switches should be designed so that the guards are easy to replace and the switches convenient and easy to lock out. Otherwise, the worker will have a great tendency not to make the effort to move his or her lazy feet and put back the guard or lock out the switch.

7. **The No Brain Rule.** Ask of each element in the task, "If the worker didn't think when he or she performed the element, could he or she get hurt?" Then ask, "How can we design the method, machine, or workplace to keep the worker from getting hurt?"

8. **The Body versus Machine Rule.** Consider whether the machine can injure the worker while the worker is performing each element. Look for stored energy in moving parts, automatic controls that can activate movement, nip point, protrusions, hot spots, sharp edges and points, and locations of safe-stop buttons.

The interface with other systems that provides the worker with services and supplies also must be considered in the design.

Physical Catalysts

Physical catalysts (equipment) of a system have significant effect on methods design. The design of tools and machinery often dictates the method of use, and the configuration of the workplace influences the methods used by the people working there. The design of the handtools, desks, counters, workbenches and chairs affects the methods used by the worker and should be part of the methods design in order to minimize movements and cumulative pathogenic stress vectors (see Chapter 6.9).

Machinery should be designed in consideration of the method the operator will use at the interface between worker and machine (see Chapter 6.8). Controls, levers, pedals, and handwheels should be placed within easy reach, so that the operator does not have to take steps, bend over, or make long reaches. Instruments should be capable of being read easily, without the operator's making head or body movements. Feed hoppers, discharges, conveyors, and any equipment surface that a worker must use should be at an appropriate height so that the worker does not have to bend over. The feed and discharge rates, machine control and maintenance, and planned equipment changes must be taken into consideration in the methods design.

Finally, one must also take into account the interface with other systems that receive the outputs or that provide the inputs, maintenance service, supplies, waste, and by-product removal.

Information Aids

Information aids required by the system must be provided in the design along with the rate and control of the information. Only the minimum number of critical details and controls should be specified. The interface dimension outlines how the information comes in from other systems.

Analyzing the Proposed Method for Improvement

After each cell of the system matrix has been considered, and the necessary specifications, layouts, charts, drawings, descriptions, crewing, and so on, are developed, then each should be checked to ensure that its design or specification is the optimum for accomplishing the purpose of the method. Each detail and element of the method, proposed or existing, should be analyzed according to the following questions of work simplification (see Chapter 3.3): *What* is the purpose? *Why* is it necessary? *Where* should it be done? *When* should it be done? *Who* should do it? *How* should the method be best accomplished? Each element should also be examined to see if the materials, products, equipment, information, steps in the sequence, or human activity in the system can be eliminated, combined, simplified, or rearranged to improve the method.

The safety of the worker and of the equipment should always be uppermost when designing methods. However, once the elements of a method have been designed, each should be analyzed by asking the question "How could the person performing the method get hurt?" Then, if hazards exist, either they should be designed out of the method or cautions should be put into the methods' descriptions. The procedure for this analysis is called "job safety analysis" or "task hazard analysis" and can be found in the National Safety Council's *Accident Prevention Manual* and other texts.[12]

3.1.4 METHODS DESCRIPTIONS

Methods descriptions are written for records, further improvements, time study, and training. They should have the following features:

1. A simple title that identifies the method, such as "medication administration system."
2. The purpose of the system, such as "Get the right medicine to the right person at the right time."
3. The date and the name of the person who wrote it.
4. The inputs, outputs, by-products (including waste), tools and equipment specifications (including packaging), and maintenance need not be included in the methods description, but references should be made to drawings, manuals, and other documents that will contain the necessary information.
5. A workplace layout, or reference to where one can be found.

6. The environmental conditions under which the method will operate.
7. The rates of inputs, outputs, and machine speeds.
8. A list of needs from other systems, such as pallets, services, information, forms, and maintenance.
9. An elemental breakdown of each person's activities.

The regular elements should be listed in the order of their occurrence. Irregular elements–those that occur less than once per cycle–and their frequency of occurrence should be listed after the regular elements. The description of each element should be brief, but complete. The body of the description should be preceded with an action verb, such as "grasp," "open," or "lift." Also to be recorded are the tools and equipment used, the distances traveled by the person or hands, what is handled, its characteristics, and its weight.

3.1.5 LAYOUT OF THE WORKPLACE

As geography makes history, the layout of the workplace influences the design of the methods, the utilization of space, and the time required to perform the method.

The size of the workplace depends on the size of the system. For a city delivery system, the workplace could be the whole city, including warehouse locations and delivery sites along with individual dock layouts. For a plant system, it would be the whole plant, including the warehouse. For a single work station in the plant, it would be the location of materials, tools, feeds, discharges, and controls. For an office or service counter, it would include the location of files, equipment, baskets, phones, and shelves. A desk is a workplace and should accommodate the methods used there.

The layout of the workplace determines:

1. The distances that materials, products, waste, people, or equipment will move.
2. The amount of storage space (in process or otherwise), which, in turn, can determine how often certain elements such as replenish supplies, or setups, will occur.
3. The body movements, such as reaches and steps.
4. The delays created by interference with people or equipment caused by their locations.
5. To some extent, how people feel about their jobs.
6. In service industries where the customer interfaces with the method, how the customer reacts to the service rendered.

The design of the method should be followed by the design of the layout. Failure to do this can force the method to accommodate the layout, possibly requiring more labor and delays, poor utilization of space, and even requiring the worker to make unsafe movements.

Changing existing layouts to accommodate new or improved methods must always be considered as part of the methods design and evaluation process (see Chapters 10.2 and 10.6).

3.1.6 INVOLVEMENT OF PEOPLE IN METHODS DESIGN

People and their abilities and talents should be an integral part of any methods design. It should be recognized that most people at all levels can understand many of the most involved techniques and situations and can contribute to the design. People involved with, affected by, or concerned with the planned or existing method should be asked for ideas and should provide information about designing or improving it. When possible, they should take an active part in the design of a better method, help install it, and follow up on its progress (see Chapter 7.8).

Some of the management policies that must be established to involve people effectively in the methods design process are:

1. Management must be willing to share information with employees.
2. Management must realize that employees are capable and must permit them to participate in workshop groups that will design systems for the organization.
3. Employees displaced by new methods designed by the groups will not be immediately laid off, but the work force will be adjusted through attrition or volume fluctuations.

3.1.7 THE USE OF STANDARDS TO BRING ABOUT METHODS IMPROVEMENT

The principal reason for measuring performance is to motivate people to make improvements.[13] Standards or other measures of performance are needed to identify where improvement is needed, to set goals, and to measure progress (see Chapter 4.1). They can (1) evaluate the effectiveness of the design efforts, (2) predict the method's performance by providing measures and standards for the rate and control dimensions for the elements of the system, and (3) encourage methods improvements by telling people what is expected of them. However, caution must be exercised not to expend more time and effort in developing or administering the measures or in building more accuracy than is required in order to accomplish the purposes of the measures.[14]

3.1.8 CONTINUING METHODS IMPROVEMENT

To encourage methods changes on an ongoing basis, a continuing productivity improvement program should be established. The program normally has a person appointed as an improvement coordinator (usually an industrial engineer), who is the leader and resource person to encourage and assist people in making changes. The emphasis is on making continual and deliberate changes to reduce the waste of labor, machine time, materials, and product and to improve services to the customer. The input to the program can come from a suggestion system or a program that expects suggested changes from everyone on a periodic basis. Recognition should be given to those making suggestions, particularly for those suggestions that lead to successful installation of changes.

Management must support a continuing methods improvement program if it is going to work. The greater the management support, the more successful any methods design effort will be.

REFERENCES

1. C. E. GEISEL and G. NADLER, "The Best Method is Not Good Enough," *Proceedings of the Fall Industrial Engineering Conference*, AIIE, Atlanta, 1978, p. 3.
2. G. NADLER, *The Planning & Design Professions: An Operational Theory*, Wiley, New York, 1981.
3. G. NADLER, *Work Design–A Systems Concept*, Irwin, Homewood, IL, 1970, pp. 662–667.
4. A. H. MOGENSEN and H. F. GOODWIN, "Work Simplification," *Factory*, July 1958.
5. R. D. ARCHIBALD and R. L. VILLORIA, *Network Based Management Systems*, Wiley, New York, 1967.
6. NADLER, *Work Design*, p. 331.
7. NADLER, *Work Design*, p. 254.
8. M. E. MUNDEL, *Systematic Motion and Time Study*, Prentice-Hall, Englewood Cliffs, NJ, 1947, p. 48.
9. R. M. BARNES, *Motion and Time Study*, Wiley, New York, 1958, p. 214.
10. E. R. TICHAUER, *The Biomechanical Basis of Ergonomics*, Wiley, New York, 1978, p. 33.
11. C. E. GEISEL, "Ergonomics," Paper presented at the meeting of the Safety Bulletin Composite Can and Tube Institute, Washington, D.C., June–July 1978, p. 1.
12. F. E. BIRD, JR., *Management Guide to Loss Control*, Institute Press, Atlanta, 1974, pp. 60–76.
13. C. E. GEISEL, "Productivity Measurement, A Prelude To Improvement," *TAPPI*, Vol. 61, No. 8 (September 1978), p. 33.
14. C. E. GEISEL, "Is Work Measurement Effective Today?," *IMS Clinic Proceedings*, Industrial Management Society, Des Plaines, IL, 1973, p. 65.

CHAPTER 3.2

Motion Study

DANIEL O. CLARK

MTM Association for Standards and Research

GUY C. CLOSE, Jr.

Hughes Aircraft Company

3.2.1 INTRODUCTION

The performance of work by man or machine is usually accomplished by movement. The effectiveness of this movement in terms of both accuracy and time is determined by the distance moved, the control exerted, and the conditions under which the movement is made.

Motion study is the application of various techniques to examine thoroughly the movements involved in work. The study may apply to the movement of people, processes, parts, or paperwork. Reaching, bending, walking, and moving may involve excess motions, which slow down manual operations. If cycles of these motions are repeated with high frequency, a significant reduction in potential output will result. Similarly, faulty operation of a mechanical device can decrease yield and increase waste. It could also be a hazard to equipment and personnel. The objective of motion study is to discover and understand motion deficiencies related to both human effort and system or mechanical function in order to increase the effectiveness of each facet of the action. The study should result in fewer labor hours per unit produced, reduced human effort, fewer rejects, and optimal costs.

Selection of the motion study technique to be used in any particular investigation depends on the cost-effectiveness of anticipated results. For example, a micromotion analysis of a motion picture film taken at 1000 frames/min will frequently require as much as 12 hours of an analyst's time for each minute of actual film-observed time. This involves a frame-by-frame analysis of 25 ft of film. Such a study may be warranted only for extremely high production, short-cycle manual operations. At the other extreme, a process diagram of paperwork flow throughout the entire organizational system may satisfy the needs of an investigation of this total function. The technique selected must be appropriate to the problem encountered.

It must be recognized at the outset that, as with most industrial engineering effort, motion study will result in actions involving people. The effect of the possible changes should be anticipated prior to the study, and steps should be taken to alleviate negative reactions. Since the impact of most studies will be a reduction of labor time and cost per unit produced, the well-being of people must be considered in every case. Frequently these changes will allow for higher productivity with the same work force. If the work to be performed is limited, some people will be removed from the particular task. The engineer should be in a position to assure the involved people that the management has other equivalent work for them to perform.

In those cases where replacement work is not available, it may be more acceptable to time the reductions in the labor force to coincide with natural attrition and turnover. In any event the methods changes brought about by the motion study should make the work assignment easier as well as more expeditious. Increased productivity should result, without requiring an expenditure of increased effort by involved personnel.

Another consideration of the study, and part of any analysis, is the cost of implementing the change. Changes can be categorized loosely into three types:

1. Those that involve a minimal outlay of funds, such as minor moves of existing equipment and additions of small, expendable handtools.

2. Those requiring the building of fixtures and the addition of ancillary equipment that can be written off as an expense in the given year.

3. Those requiring investment in capital equipment. Payoff is expected over a period of years.

This chapter provides information that will allow the engineer to select an appropriate methodology based on the facts of the situation.

3.2.2 BACKGROUND OF MOTION STUDY

The forces of competition have continually required industry and business to seek out better and easier ways to perform work. Investigation of the utilization of human effort indicates that about 40% of the manual work performed in the home, shop, and office results in wasted effort that adds no value to the product.[1]

The pioneers in the field of motion study were Frank B. and Lillian M. Gilbreth. They initiated the first recorded research in the field of motion study, the cornerstone of methods engineering. Frank Gilbreth developed the technique of micromotion study in which motion picture film of an operation was analyzed frame by frame and assigned a coding representing elemental hand motions. These elemental motions were called "therbligs," an anagram using the letters in his name. They represented causative actions or events (Exhibit 3.2.1). H. B. Maynard, G. J. Stegemerten, and J. L. Schwab utilized these basic elements in developing the methods-time standards for the Methods Time Measurement (MTM) system.[2]

A second camera study method is "memo motion" or "time-lapse photography." This technique was developed at Purdue University by Marvin E. Mundel.[3] Henry W. Parker, civil engineering professor at Stanford University, introduced the technique to the building trade.

In effect, use of time-lapse photography in its various forms is a method of recording a series of nonrandom events for purposes of analysis. This technique results in nondisputable evidence of what is actually occurring during the period of the study. It is a method of sampling the population of events.

More recently, the use of closed-circuit television in conjunction with a videotape recorder has been developed by Akiyuki Sakima at Keio University.[4]

Exhibit 3.2.1 Therbligs

Operation	Abbreviation	Operation	Abbreviation
Gilbreths		*American Society of Mechanical Engineers*	
		Physical Basic Elements	
Transport empty	TE	Reach	R
Transport loaded	TL	Move	M
		Change direction[a]	CD
Grasp	G	Grasp	G
Hold	H	Hold[a]	H
Release load	RL	Release load	RL
Pre-position	PP	Pre-position[a]	PP
Assemble	A		
Disassemble	DA	Disengage	D
		Semimental Basic Elements	
Position	P	Position[a]	P
Search	Sh	Search[a]	S
Select	St	Select[a]	SE
		Mental Basic Elements	
Plan	Pn	Plan[a]	PL
Inspect	I	Examine	E
		Objective Basic Element	
Use	U	Do	DO
		Delay Basic Elements	
Avoidable delay	AD	Avoidable delay[a]	AD
Unavoidable delay	UD	Unavoidable delay[a]	UD
		Balancing delay[a]	BD
Rest for overcoming fatigue	R	Rest for overcoming fatigue	F

Source. Courtesy of International Labour Office, *Introduction to Work Study*, Atar, Geneva, 1960.

[a]These are ineffective elements of movement; wherever they occur, an attempt should be made to eliminate them completely if at all possible.

3.2.3 IDENTIFYING THE PROBLEM

Economic Considerations

A motion study project can lead to excessive technical labor costs. To avoid this problem, the analyst may apply the Pareto principle, which states, for example, that 20% of any group of items contain 80% of the value. A preliminary survey to ascertain how and where the 20/80 principle exists allows the analyst to minimize the effort for maximum benefit. The selection of appropriate techniques with their assorted equipment requirements becomes a most important cost consideration for the chosen study.

Preliminary Survey

To select the proper motion study technique that will provide the detail needed for problem solution, a preliminary survey is mandatory. Such a survey outlines the scope of potential improvement relative to products, people, machines, and the assorted workplace. It will assist in establishing priorities of items for investigation. It will establish parameters required to identify the problems. A meaningful preliminary survey can avoid unnecessary time expenditures on low-priority problem areas.

The preliminary survey of a given work activity will provide data to justify the expected study. Further investigation of these data will help determine approaches to solving the problems and will lead to prudent selection of motion study techniques.

The overall approach to problem identification will be designed from the findings of the preliminary survey. The techniques and the degree of study depth relative to people interacting with equipment, materials, and the work environment can be determined from preliminary survey results. It is very important to understand that the time spent during an organized preliminary survey will prove highly beneficial during the actual motion study.

Establishing a Base Point of Reference

Having too much data at the beginning will not be a problem, but insufficient or inadequate data can establish a baseline that will give inaccurate comparisons. Methods and processes, number of people involved, equipment settings, quality requirements, and input and output data are examples of good baseline data. This initial documentation is essential for comparing proposed improvements resulting from the completed study.

3.2.4 TECHNIQUES FOR GATHERING AND ANALYZING INFORMATION

Charting

Instructions for completing several types of analytical charts are discussed in Chapter 3.3. However, there are certain features that will help you utilize these charts to their best advantage. Charting organizes the analyst's thoughts and frequently points the way toward improving the function under study. Charting lists sequentially, one detail at a time, allowing the analyst to concentrate on each of the details—one at a time.

Process Charts

Process charts serve to document existing conditions with a minimum of writing. They can summarize a mass of information graphically and provide a baseline for the operation as it exists. Process diagrams relating to the charts pictorialize the operation being studied. Recommendations for modifying the process or method can be annotated on the chart in proper sequence. Charts can be used to compare quantitatively the proposed method with the present method.

The process chart of a proposed change will allow the engineer to explain a complex situation. It is a tool for convincing others that an idea is worthwhile and well thought out; it makes it easy for management to say "yes" to a proposal. The chart indicates where the obvious improvements can take place and which tasks or operations warrant a more detailed type of study, such as film analysis.

In analyzing the process chart, look for the following situations:

Unnecessary operations.

Long moves between operations.

Two or more moves between operations.

Two or more inspections in sequence.

Changes in direction of flow; backtracking.

Large-volume items on long route, with small-volume items on most direct line of flow.

Possible combination of operations.

Location of storage areas relative to work areas.

Delays caused by delivery schedules for parts and materials.

Sequence changes that would improve productivity.

Process charting is a rather gross technique, but one that is very effective in reducing a complicated process or a series of operations into manageable segments. This logical approach gives more and better results than intuitive solutions to work problems.

Right- and Left-Hand Charts

Sometimes known as a "workplace chart," the right- and left-hand chart provides a visual method for analyzing the relationship and balance of the work performed by the hands. Despite automation, there are many work stations in the office and shop that require the application of physical human effort. The right- and left-hand chart breaks the task down into combinations of motions called "elements." Each element must have an easily identified starting and ending point. It is a convenient, recognizable, short subdivision of a work cycle.

The following ideas are suggested in order to improve your analysis:

Make a summary statement of what the operation accomplishes.

How long does it take to perform the operation with the present method? This is a benchmark from which to measure improvement.

What quantity is made or processed through this operation per day, per week, per month?

Check the preceding and succeeding operations to ensure that the work is arriving and leaving in the proper condition.

Record all tools, machines, equipment, and materials used, and sketch their locations in the work area.

Multiple-Activity Charts

People work collectively on assembly lines, in construction crews, in typing pools, in machine centers, in dental offices, and in other groups. They combine their efforts with one another and with their equipment to perform useful work. Multiple-activity and man-machine charting provides a means with which to analyze these jobs for improvement.

To make use of this technique, it is necessary to add a time parameter to the data. The multiple-activity chart stresses time and the combined activity of a group of people and/or machines. The time values may be obtained from existing standard data; from a high-level, predetermined time system such as MTM-2 or MTM-3; or by photographic techniques (see the following section). These charts are very similar to the "simo chart" developed by the Gilbreths. Current practice is to use decimal minutes as the time unit rather than the "wink" value utilized by Gilbreth. (The wink equaled 1/2000 min.) More detailed analysis can be performed using the film frame count and converting totals to decimal minutes or other units of time. Objectives of multiple-activity analysis are to:

Establish an optimum work team size.

Provide a fair allocation of job duties or work assignments.

Reduce idle or delay time.

The objective of work is to add salable value to the product. The overall requirement of multiple-activity analysis is to increase the ratio of value adding time to the time spent on nonproductive activities such as handling, setup, and waiting.

Data Acquisition Using Photographic Pratices

Overview

Perhaps the most useful practices in motion study are those utilizing the motion picture camera. Motion pictures taken in a technical manner provide the engineer with a means with which to record action as it happens. These films can be observed over and over again at various speeds of projection as the engineer studies methods to improve the use of time.

Motion pictures allow us to compress or expand time. The camera records action and data in relation to a time base established by the speed at which the film moves through the camera. Various means are available for accurately recording the time interval; for example, pulse generators can be connected to drive timing lamps, which, in turn, register timing blips on the edge of the film.

Pictures projected at the same rate of speed at which they were taken appear to be normal. To compress time, the action is photographed at a rate slower than the planned projection speed. This time compressing method is called time-lapse, pulse, or interval photography. For example, an engineer on a construction site may photograph a work crew at one frame every 4 sec. At this rate, 4 hours of activity can be recorded on a super-8 mm film cartridge. Projection time at 18 frames/sec will be 3.3 min, a compression of 72 to 1. The resulting film allows the analyst to observe undesirable movements—or *lack* of movement that would go unnoticed when casually observed. The discovery of nonmotion or delays frequently proves most important. If a cement pouring and finishing crew is observed to regularly wait 20 min for a concrete mixing truck, quick action will be taken by the contractor to correct the situation.

To expand or slow down time, the pictures are taken at a faster rate than the projection rate. This allows the study of actions that occur too quickly for the human eye to perceive. This method is known as time magnification.

$$\text{time magnification} = \frac{\text{camera picture frequency (frames/sec)}}{\text{projection frequency (frames/sec)}}$$

For example, a mechanical device that takes 2 sec to complete its cycle of loading a component into a carrier is misplacing a component every third or fourth cycle. We film the action at 200 frames/sec and project the developed film at 16 frames/sec.

$$\text{time magnification} = \frac{200}{16} = 12.5$$

The time for a cycle of action on the viewing screen will be 12.5×2 sec = 25 sec. Such slowing down allows the analyst to observe in slow motion the relationship of movement among individual parts of the mechanism—in this case, the components being handled. Time magnification frequently results in the discovery of the cause of failure or defect and allows corrective action.

Exhibit 3.2.2 depicts a motion picture spectrum showing frames per second and ranges of shutter speeds from time-lapse, or pulse, photography to very high speed photography. Typical applications are listed for the various phases of technical photography.

A word of caution is in order. There is a certain basic skill level required to put data on film and to analyze it effectively. If the engineer has the inclination and the time to learn this skill and become proficient, the results will be very rewarding. If there is a time constraint requiring that one succeed in trapping the data on the first try, it is best to obtain the help of an experienced professional.

Successful technical filming depends on the right combination of film speed, filters, shutter speed, shutter angle, aperture, distance to subject, framing, focusing, and lighting. In the pages that follow, suggestions will be given that will assist the engineer in understanding technical motion picture filming and analysis. The advantages gained in using motion pictures should make the aforementioned effort worthwhile. Advantages of technical motion picture use are that it:

Records facts unobtainable by other means.

Can provide quick study results.

Allows individual analysis of simultaneous actions.

Permits analysis of detail in quiet surrounding.

Provides a permanent record of the method observed.

Can provide material for training, grievance data, and skill transfer.

Equipment Selection

The gathering of photographic data requires a complete photographic system—considerably more than just a camera. For some studies the system requirements may be very modest; other studies may require very elaborate systems. The system depends on the task to be performed. Basic filming equipment includes the camera, lens, lens filters, light meter, tripod, analytical projector, and editing equipment. In high-speed and ultra high speed photography, special lighting will be required. However, for time-lapse, normal, or fast filming, it is recommended that existing light be utilized, combined with fast film and film forcing if necessary.

Exhibit 3.2.2 Motion Picture Spectrum

	← Projection looks faster		Projection looks slower (projected at normal speed) →		
Frames/sec →	$\frac{1}{10}$–10	16–24	32–48	200–500	1000–11,000
	Pulse/time lapse	Normal	Fast	High speed	Very high speed
Shutter speeds (sec)[a] →	$\frac{1}{25}-\frac{1}{50}$	$\frac{1}{25}-\frac{1}{100}$	$\frac{1}{100}-\frac{1}{250}$	$\frac{1}{500}-\frac{1}{18,000}$	$\frac{1}{3000}-\frac{1}{150,000}$
	Pin-registered film, Intermittent advance (spans the four columns from Pulse/time lapse to High speed)				Rotating prism Continuous advance
Applications →	Long-cycle study Crew coordination Gross time data Work load analysis	Short-cycle study, work study Equipment evaluation Skill transfer Time data	Motion research High skill transfer	Mechanical device analysis Automotive crash study Saw chip dispersion	Very high speed mechanism analysis Explosive devices Projectile velocity study

[a]Varies with degree of shutter opening.

The engineer's organization may wish to acquire some of the general-purpose camera and ancillary equipment. On the other hand, it is usually practical to rent or lease the more specialized equipment. It may be advantageous to hire a photographer who specializes in the type of photographic analysis being planned. In this case he or she would normally provide the necessary equipment.

If the engineer plans to acquire company-owned equipment, the first decision will be to purchase either a super-8 mm or a 16 mm camera and accessories. Perhaps a closed-circuit videocassette television system should be considered. Exhibit 3.2.3 lists some evaluations that should help make fair comparisons.

Keep in mind that consumer equipment is built to a price. A home movie camera may be used to take 50 or fewer cartridges of film during its useful life. It is conceivable that the analyst will take that many cartridges in a year as part of regular work assignments. Professional equipment is built to perform accurately for many feet of film. It is also adjustable and repairable. Of course the professional camera may cost from five to ten times what a home movie outfit would cost.

In any camera, 8 mm, 16 mm, or video, there are certain features to be recommended. The capability of through-the-lens viewing is a necessity. The video camera has an electronic viewfinder for monitoring your shots. The viewfinder image should be bright, clear, and easy to use. Automatic apertures are useful for unattended camera photography where available light will increase or decrease during the filming. A zoom lens has some advantages, but is not too useful in technical photography. The single focal length lens generally accepts more light because of the simpler lens system. The shutter may be fixed or variable. The fixed shutter, for example, may have an angle of opening of 135°; the variable shutter may be adjustable from 2 to 160° of opening.

The conventional camera should have an accurate intermittent, pin-registered mechanism. For time-lapse/pulse photography the camera must have an accessory or built-in intervalometer.

Built-in light meters are usually adequate for outdoor photography. A direct reading incident light meter is more accurate when taking indoor motion pictures.

Steady pictures require a strong tripod. Since most technical film will be shot without tilts and pans, features for this practice can be foregone in favor of rigidity. Of course facilities for properly positioning the camera are essential.

One super 8 mm cinecamera is available that will perform at speeds of 10 to 250 frames/sec. It can also be switched to time-lapse, or pulse, mode of 1 to 10 pulses/sec with built-in intervalometer. The shutter speed is constant for all pulse rates. For immediate playback, a similar camera is available that uses the Polaroid film system.

An analytical projector compatible with the camera equipment will be required to study the film. Such projectors will operate from 1 frame/sec to 24 frames/sec with flickerless projection. They will project a single frame without film damage and can be stopped or reversed with the push of a button. Digital frame counters assist in analysis.

Editing equipment such as rewind reels, frame counter, viewer, and splicer may be added as required.

Basic equipment for a videocassette system is a video camera, tripod, videocassette recorder-player, and a monitor. A $\frac{3}{4}$ in. tape time-lapse recorder-player is available that not only will record at normal speeds, but will record from 1 to 4 days—24 to 96 hr—on a 72 minute tape. This provides a range in compression ratio from 1440/72 = 20 to 5760/72 = 80.

Exhibit 3.2.3 Comparisons of Various Photographic Equipment

Features	Super-8 mm Camera	16 mm Camera	Video Camera
Equipment cost[a]	100%	200%	500%
Approximate weight (portable equipment)	10 lb	25 lb	100 lb
Picture quality (resolution)	Good	Excellent	Fair
Film/tape cost (includes processing)[a]	100%	225%	15% ($\frac{3}{4}$ in. tape)
Frames/reel or cartridge	3600/50 ft	4000/100 ft	108,000/hr
Film/tape reuseable	No	No	Yes
Processing delay	Overnight	Overnight	Instant viewing
Sound recording	Available	Available	Available
Slow motion projection	Good	Good	Need special equipment
Technical skill required	Some	High	High

[a]As a percentage of super-8 mm.

In other words, it will record from 1 frame every 0.67 seconds to 1 frame every 2.67 seconds in the time-lapse mode. This unit is fully flexible for viewing with still frame projection, forward, and reverse at a range of speeds.

A video color camera is available that will take satisfactory pictures with as little as 10 ft-c of light on the subject (100 lux), using an aperture of 1.4/f. This camera weighs 13 lb without batteries, which makes it fully portable for shop work.

Making the Study

Preplan what you want to discover with the film—what data you wish to collect. A good job of "pre-editing" to determine the filming sequence will save time and film. Have a clear understanding with the management of the involved work area as to what is to be filmed—and frequently, what is not to be filmed. Clarification and explanation as to the reasons for filming should be made to anyone who will appear in the film. It is advisable to get written clearance from individuals who are identifiable in the film.

Recruit other needed personnel to help in the filming. An area expert should be available to clarify questionable points regarding the operation or process. Electricians may be necessary if lighting additions are required.

A schedule should be established, including data, time, filming sequence, and personnel to be on hand.

Film type and the quantity needed for the project must be decided. Black-and-white film gives the sharpest detail and is available in extremely fast speeds. It can be force developed to a greater extent than color film. On the other hand, color film helps in identifying particular features during analysis and is more acceptable to viewers. The faster the film, the greater the tendency toward graininess. Super 8 mm cartridges are available in color with a film speed of 160 ASA. A very satisfactory 16 mm color film has a 400 ASA tungsten rating. Fluorescent lights cause changes in color that can be corrected by filtering. The aforementioned 400 ASA film can be color corrected with an FLB filter; however, its effective speed is reduced to 200 ASA. For analytical purposes, the higher speed is usually more desirable than accurate color rendition. Some correction can be made during the film duplicating process.

Exhibit 3.2.4 is useful in determining the amount of film that will be necessary for the study. Since the film price is the least of the cost of analysis, it is preferable to have an excess of film rather than to run short during the study.

It is suggested that every effort be made to conduct the study with available light. This is known in the trade as cinema verité. Photographic lights distract personnel in the vicinity of the study and require a larger film crew as well as cumbersome equipment and special electrical hookups for high current draw. Recommended lighting for most inside work areas ranges from 50 to 200 ft-c. Thus most work can be filmed without additional lighting.

Exhibit 3.2.4 Film Running Times at Various Shutter Speeds

	Seconds		Minutes		Hours	
Frames/ Sec	Super-8 mm 50 ft Cartridge[a]	16 mm 100 ft Roll[b]	Super-8 mm 50 ft Cartridge	16 mm 100 ft Roll	Super- 8 mm 50 ft Cartridge	16 mm 100 ft Roll
$\frac{1}{10}$	36,000	40,000	600.0	667.0	10.0	11.1
$\frac{1}{2}$	7,200	8,000	120.0	133.0	2.0	2.2
1	3,600	4,000	60.0	67.0	1.0	1.1
2	1,800	2,000	30.0	33.0	0.5	0.6
4	900	1,000	15.0	17.0	0.25	0.3
6	600	667	10.0	11.0	–	–
12	300	333	5.0	6.0	–	–
18[c]	200	222	3.3	3.7	–	–
24[d]	150	167	2.5	2.8	–	–
48	75	83	1.3	1.4	–	–
100	36	40	0.6	0.7	–	–
200	18	20	0.3	0.3	–	–

[a] 50 ft super-8 mm cartridge has 3600 frames, 72 frames/ft.
[b] 100 ft 16 mm roll has 4000 frames, 40 frames/ft.
[c] "Silent" speed.
[d] "Sound" speed.

It is highly recommended that a test film be taken and developed prior to the final filming. Such a test shows the analyst whether everything is right, including the following:

Are the operator's hands covering some detail you wish to see?
Is the lighting intense enough or too intense?
What about depth of field?
Are features you need to analyze in focus?
Have you excluded a detail you wish to capture on film?
Have you included a detail that you wish to exclude?

Variations in camera position should be made during the test and recorded. If your method is time-lapse, you may wish to alter the filming interval. The test film gets the subject used to the camera and reduces the temptation to exaggerate motions during final filming. The test also verifies the listed film speed. Slow chemical changes in film emulsion can alter the realized speed. Adjustments in aperture can be made to alleviate this problem on your final study film.

You are ready for your final filming. Follow camera instructions exactly in film loading. If the camera mechanism allows, carefully clean aperture and film track before loading each roll or cartridge. Precision equipment is fragile. Close tolerances preclude forcing the gates and latches. Your first few feet of film should be a take of a "slate"—an identity card showing date, department, subject name(s), photographer, and necessary camera and film information. You may wish to develop a checklist.

Make your focusing on the "long side." For example, if your actual distance to subject 20 ft, a 30 ft setting will give greater depth of field—area in good focus—than a setting of 15 ft.

Start with a "long shot"—an overview that orients the viewer and helps identify the area being studied. Next, move in with medium or close-up shots. If your lighting allows a lens opening of f/5.6 or smaller (f/8, f/11, etc.), your lens will have all items in sharp focus at 8 ft and beyond. Minimize the use of zoom, pan, and tilt. It is usually desirable to move the camera between shots. This gives the appearance of continuity to your film. Try to avoid such distractions as reflections from polished surfaces, crooked framing, and extraneous material. Gum chewing and smoking on the part of the subject should be discouraged. Watches and rings can distract during close-ups. Always use a tripod or other solid camera mount to eliminate unwanted movement.

In high-speed filming, adequate lighting and related depth of field can be a problem. Each time you double your shutter speed, you must double your light. You obtain this by opening the aperture one stop. Each decrease in aperture stop number doubles the light exposed to the film; conversely, each increase in aperture number reduces the light by 50%. A proper balance of filming variables must be achieved by the engineer-cameraman to produce a satisfactory film for analysis.

Once the successful film is made, it is suggested that a work copy be made from the original. This allows additional copies to be made—if needed—from an unscratched original.

Time-Lapse Photography

"A motion picture actually consists of a series of still pictures called frames, which are taken at *regular* intervals and which are subsequently viewed at the same, or at different regular intervals."[5] By extending the time interval between frames, by allowing more than normal *time* to lapse between photographs, we compress the real time of the subject photographed, as mentioned previously. The interval frequency you use depends on the speed of the action you wish to study. It also depends on the detail you need to determine appropriate actions for improvement. This may vary from 1 frame taken every 15 sec to 10 frames taken every second. You may ask yourself, "Do the actions I wish to investigate take longer than 15 sec to complete?" If some of the needed actions take place in less than 15 sec, you may well miss their occurrence in your photography. In this case you must shorten the interval to more nearly match the time span of actions to be recorded. When in doubt, decrease your interval. The most commonly utilized interval for construction studies is 4 sec. At this speed, two rolls of film will run for 8 hr.

Time-lapse photography is frequently used for gross studies taken over a period of several hours or days. Color film is recommended, because, it shows more detail than black-and-white film. It is recommended that a slight overexposure of the film be employed; otherwise, important details occurring in the shadows may be underexposed. This can be accomplished by setting the aperture one half stop wider than indicated by the light meter. The camera should be mounted to view the work area from an elevated position that gives an unobstructed view. Such location also offers a perspective of the operation that makes interrelationships easier to follow. If your lens does not cover the total area needed for the study, a wide-angle lens may correct the problem.

Not only would the filming cost be prohibitive if these studies were made at normal speeds, but the time for film analysis would be excessive. An example of this use of time-lapse photography was that performed by industrial engineers for the city of Dallas. Hydrant maintenance crews and concrete patching crews were filmed to determine the work habits and effectiveness of the various crew members. Multiple-activity charts based on the film data resulted in redesign of the job functions, which allowed for a reduction in crew sizes in both cases.

An industrial engineer who specializes in methods improvement of construction jobs gets the crew together before filming to explain the objectives of the study. After filming, the crew views the picture and is encouraged to suggest ideas for improvement. The film is used as an irrefutable reference as to how the work was performed.

Such studies can be used to analyze work flow, crew balance, operational delays, handling of materials, warehousing operations, and the utilization of equipment and work force. The camera will frequently point out those activities that contribute negatively to the objectives of the business. In cases where multiman crews and/or multimachines work in parallel synchronization, the time-lapse technique can provide accurate and complete data both for method and gross time values.

Time-lapse films are taken in speed ranges of 4 to 10 frames/sec when the study must perceive starting and ending points of quick manual motions. A consortium study of test station operations for testing electronic devices was filmed at 6 frames/per sec. Excessive reaches, moves, and trunk and head movements became obvious at this time-lapse speed.

Advantages of the time-lapse technique listed by one study group are as follows:

A permanent record is provided. The film can be reviewed over and over again to double-check data.

The film records all activity within its range for the full period of the study.

After an initial period, the personnel being studied are not affected by the camera and go about their work in the habitual manner.

The analyst can perceive exactly what each member of the group did or did not do.

Time-lapse photography is a relatively inexpensive method of gathering an accurate mass of data.

High-Speed Photography

The industrial engineer may wish to use high-speed photography in solving problems incurred with mechanical devices, as compared with the use of slower filming techniques to study and analyze problems involving people.

To slow down a high-speed operation, we must take the film at a rate faster than the planned projection speed. "The human eye cannot resolve motion that occurs in less than $\frac{1}{4}$ second. Actions or movements that occur too quickly for the eye of a human observer to follow can be slowed down by high speed photography so that all aspects of the action can be studied."[6] High-speed photography can save much trial and error by exposing problems invisible to the human eye.

The selection of frame speed depends on the velocity of the motion to be studied. For example, the peripheral speed of a circular saw may have a tip velocity of 150 ft/sec. To slow down action so that the motion of individual teeth and saw chip patterns could be perceived, it would be necessary to film at 600 to 1000 frames/sec. Researchers in ballistics and explosives may require film speeds of 1.4×10^4, requiring highly specialized equipment. However, the type of high-speed studies made for motion study purposes will normally be in the 50 to 500 frames/sec range. These speeds can be provided by intermittent, pin-registered cameras utilizing essentially the same mechanism used by conventional movie cameras.

To select the proper equipment for a particular job, the photographer must understand the objectives of the study. What do you expect to learn from the film? Careful evaluation of information needed will frequently help avoid the delay and cost of reshooting the operation. It is suggested that normal-speed motion pictures of the study be included in order to clarify the relationship to real time.

In high-speed photography it is desirable to position the camera at right angles to the subject's line of motion. The camera-to-subject distance should be close enough to ensure adequate image size, and yet far enough back to encompass the total part of the motion being studied.

The time frame within which the motion occurs must be recorded. This is usually accomplished by a timing lamp that is built into the camera, which flashes a light streak along the sprocket margin at a constant known rate. On some equipment, fiber optics or light emitting diodes producing a digital display have simplified data reduction problems.

High-speed photography requires high-intensity lighting. Power requirements and circuit capacities must be ascertained prior to filming. Accessory transformers may be required to provide extra circuits and to avoid shutting down the power in the department where the study is taking place.

It is highly recommended that the industrial engineer secure professional assistance or consultation before attempting to conduct a high-speed photography motion study.

Video Tape Recording

Many industrial engineers are making use of video tape recording. Black-and-white filming has been used predominately in these cases because of lighting requirements and system costs of color equipment. However, color cameras that will satisfactorily operate under most existing light conditions are now available at nominal cost.

A video tape recording has the advantages of credibility. Being much harder to edit than film and being immediately available for playback, it is far less suspect on the part of the employee and his or her supervisor. It is sometimes advantageous to use videotape to study an overall situation and to use filming techniques to detail specific areas pointed out as critical by the videotape. For shop operations that occur at infrequent intervals, the videotape has been used to record methods and to establish time values for a single cycle.

The use of videotape is a real people-involvement tool. People are accustomed to television and accept it as a management tool.

The playback feature can be used as a beginning point for a motion study brainstorming session with the people involved in operations being studied. The engineer can take advantage of the addition of sound. Audio can be used by the operator to explain the operation being performed.

Operation of the video equipment parallels the operation of filming equipment. Of course it is necessary to have a power source for cameras and monitors.

Present-day equipment provides steady stills and slow-motion playback. Tapes can be reused or filed as a permanent record of the operation under study.

As with photographic analysis, the engineer will find it necessary to study and practice in order to become proficient with video tape recording and analysis.

Film Analysis

Standard Analysis

The gathering of photographic data has been completed, and motion study data are now ready to be compiled and analyzed. The exact technique for this analysis is the next decision to be made. But certain questions should first be explored in this decision making process, including, the following:

What are the objectives of your analysis?
What do you propose to do with your findings?
What assistance can be obtained from the preliminary survey data?
What are the time and budget constraints for the project?
How many feet of film have been processed?
Which actions can be covered with a cursory observation?
Which operations or tasks need detailed analysis?
Which of the charting techniques could be helpful in your breakdown?
To what degree must you record qualitative and quantitative measurements?
What are the appropriate forms for recording the information?

Answering these questions will be helpful in establishing a logical approach and in determining a film analysis technique needed to complete the scope of the assignment.

Standard film analysis is associated with rolls of film of various lengths that are to be viewed for data acquisition. It must be recognized that the minimum required film speed for the detail needed is necessary to reduce analysis time. Detailed data are acceptable at the start of the study filming and until it is determined what degree of exactness is necessary.

As previously described, film analysis projectors for the frame-by-frame method of analysis are available that provide 1 to 24 frames/sec in the forward and reverse modes. Speed and direction can be quickly changed to observe a given problem area. Analysis in the reverse mode can uncover motion problems not observed in the conventional forward mode. These projectors also provide digital frame count displays for ease of counting. Elapsed frames multiplied by the time per frame, based on camera speed, gives the elapsed time for the motion element being analyzed.

The method of observation and analysis can sometimes be easily determined following several viewings of the subject film. A concept of the percentage breakdown by function can be obtained. Problem areas become readily identified. Various speeds can be explored, corresponding to the detail needed.

Frame-by-Frame Analysis

As previously mentioned, the analyst should view all frames of a filmed operation or activity sequence. However, the frame-by-frame technique refers to the examination of a large group of frames as a unit or the analysis of single succeeding frames for the proper determination of element content.

By first viewing the film, the beginning and ending points of the concerned activities can be established. Each critical section should be approached as follows:

Project the film section forward and in reverse several times in order to understand the content and procedure.

Select or develop the proper form to collect your data.

Analyze one person or event with a step-by-step approach, recording the frame counts of each step for the completed cycle under study.

As you analyze the subject film, note motions or elements that can be eliminated, improved, or changed in the sequence or reassigned.

Follow the same procedure for analyzing the additional people or events for the same complete cycle and frame count period.

Plot simultaneous right- and left-hand motion patterns if required, following one hand at a time.

Convert frame count units to time units.

Review composite findings in order to detect potentially effective changes.

The techniques used for film analysis are similar to the charting techniques explained previously in this section and detailed in Chapter 3.3. Typical frame analysis forms are shown in Exhibits 3.2.5, 3.2.6, and 3.2.7.

Film Loop Analysis

The filming of a short task or operation can be cut at the beginning and end of the working cycle. By splicing this section of film together with a section of leader film stock, you form a loop. The blank leader indicates the beginning and ending points of the sequence. The film loop is threaded through a projector's guides and sprockets. The long length of the loop is placed around a remote sheave-type device such as a film reel or makeshift bearing. The film loop is then projected on the screen, producing a repetitive viewing of the same operation. One must be careful to ensure that the film is supported away from the projector's light source to prevent damage to the film. The analytical-type projector is the only equipment that can project total light, single-frame images without burning the film.

The analysis of a film loop allows for exact documentation of the motions. The film can be run in the reverse mode to assist in the recognition of motion breakpoints. Reverse projection may identify unusual and awkward motions that can be eliminated or corrected. The analyst should also study the total sequence for possible improvements in motion.

The general approach for analyzing film loops is to note the first event, element, or motion that is observed at the beginning. On the next loop cycle, check that your first observation was correct and observe the next segment for orientation. As the next passing of this segment appears, it is so noted, and this procedure of analysis goes forth until all of the detail breakdown has been captured on the selected form.

The continued study of a loop will provide skill transfer for workers expected to follow an explicit motion pattern when performing a complicated task. The same concentrated study can provide times for elements of the operation by applying a predetermined time system that has estimated distance ranges. Formal training for MTM-2 and MTM-3 practitioners includes the film loop analysis method for motion recognition and system application. Using such time systems allows the analyst to compare current and proposed methods relative to time and methods improvement.

Film loops can provide supervisors an opportunity to view their operations with engineers and employees. This promotes thorough understanding of the motion pattern designed for the operation. Films and projectors involved in loop analysis are produced in both 8 mm and 16 mm format. Cassette loop devices are available for continuous running of longer film loops. This overcomes the limitation of loop length support is previously mentioned. Based on the detail of analysis selected, the engineer can adapt an analysis form (see Exhibits 3.2.5, 3.2.6, and 3.2.7).

Time-Lapse Analysis

Film analysis of compressed real time that has been documented by means of time-lapse photography appears quite different to the analyst. The long interval frequency of picture frames produces

Exhibit 3.2.5 Process Analysis Form

Operation	Adjust potentiometer for precise voltage	Film number	ET 14
Segment/cycle	Adjusting voltage	Camera speed	24 frames/sec
Start frame	– 0 – Operator RCR	Scale Division	1 = 5 frames
End frame	634 Analyst B. Edsel	Date	3 - 6

Frames	Elapsed Frames	Element Description	Code[a]	Time Conversion	Notes/comments
20	20	Get probe from test panel	R	23	
56	36	Probe to point		42	
96	40	Tuning tool to screw	L	46	Held in idle hand
124	28	Identify meter reading		32	ET @ HD M U T (lng)
222	98	Fine adjust		114	(2 decimal)
262	40	Probe to test panel	R	46	
274	12	Check reading		14	
286	12	Get panel knob		14	
330	44	Course adjust calibration		51	
346	16	Get probe		19	
382	36	Probe to test point		42	
402	20	Identify reading		23	ET @ HD (med)
440	38	Tuning tool to screw	L	44	
464	24	Identify meter reading		28	
576	112	Fine adjust		130	(2 decimal)

an unrealistic speed of action. Once this viewing adjustment is made, analysis of gross studies taken over a period of hours or days can be accomplished. These films provide a record of long processes and the people involved. It is, in effect, a continuous sampling of the work being studied. It can reveal many problem areas involving groups of people, task assignments, avoidable delays, handling of materials, and utilization of equipment. When a higher rate of frames per second for this technique is used, the body motions of people at work can be discerned for a more detailed analysis. Exhibit 3.2.2 displays the camera speed range of $\frac{1}{10}$ to 10 frames/sec in which projection of the resultant film looks faster than normal. The general applications listed are long-cycle study, crew coordination, gross time data, and work load analysis. The analysis of body motion would require 6 to 10 frames/sec, whereas a building construction or earth moving process could use a speed range as low as 1 frame every 10 seconds.

Film analysis approaches previously mentioned are applicable to the time-lapse technique. In a

Exhibit 3.2.6 Multiple-Column Analysis

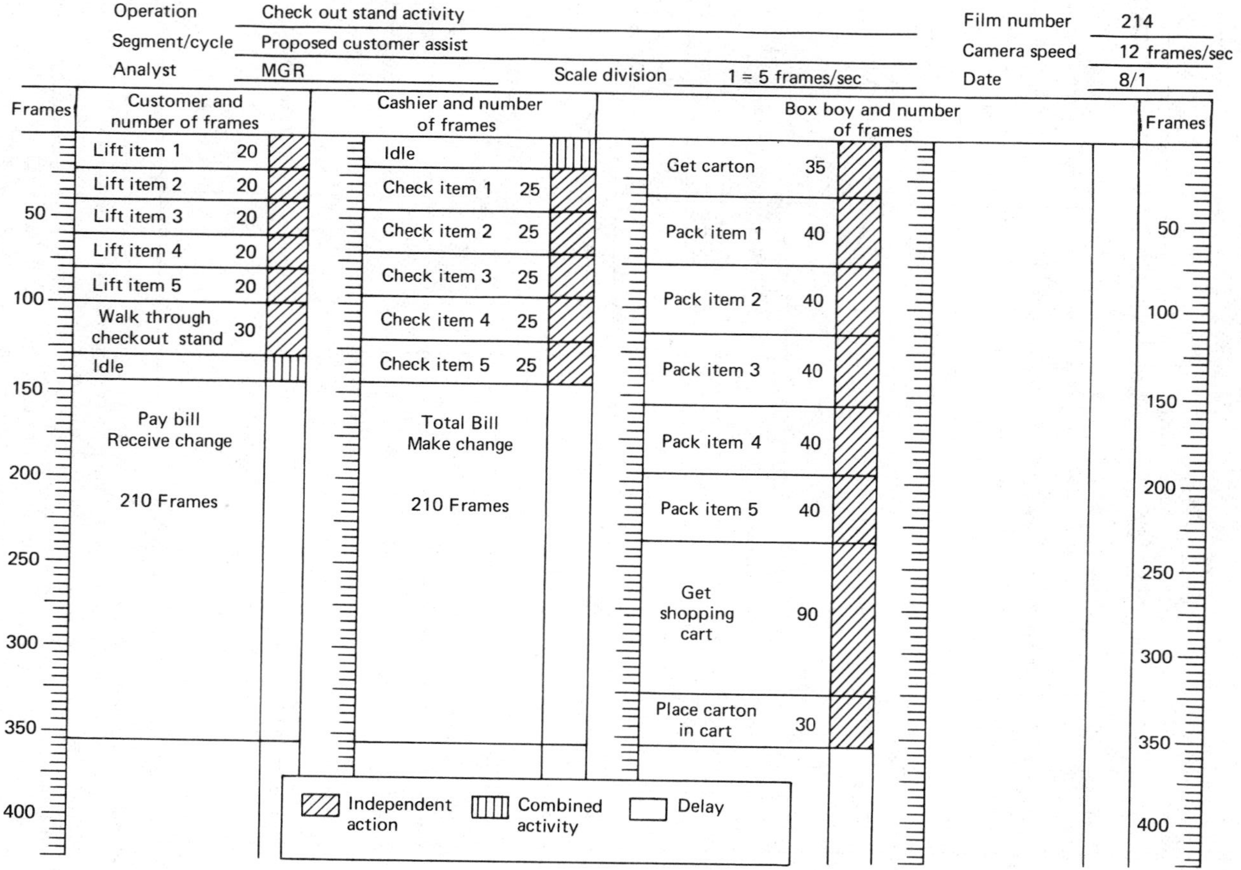

Operation: Check out stand activity
Segment/cycle: Proposed customer assist
Analyst: MGR
Scale division: 1 = 5 frames/sec
Film number: 214
Camera speed: 12 frames/sec
Date: 8/1

Customer and number of frames		Cashier and number of frames		Box boy and number of frames	
Lift item 1	20	Idle		Get carton	35
Lift item 2	20	Check item 1	25	Pack item 1	40
Lift item 3	20	Check item 2	25	Pack item 2	40
Lift item 4	20	Check item 3	25	Pack item 3	40
Lift item 5	20	Check item 4	25	Pack item 4	40
Walk through checkout stand	30	Check item 5	25	Pack item 5	40
Idle		Total Bill Make change	210 Frames	Get shopping cart	90
Pay bill Receive change	210 Frames			Place carton in cart	30

Frames scale: 50, 100, 150, 200, 250, 300, 350, 400

Independent action | Combined activity | Delay

Exhibit 3.2.7 Simo Chart[a]

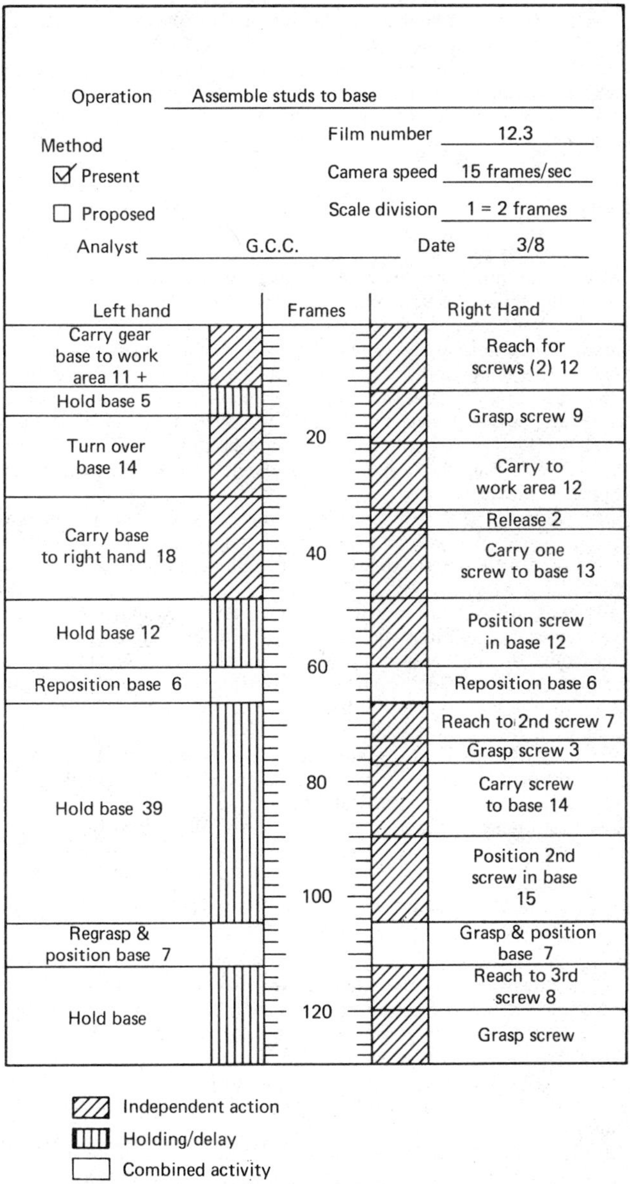

Operation: Assemble studs to base

Method

☑ Present

☐ Proposed

Film number: 12.3

Camera speed: 15 frames/sec

Scale division: 1 = 2 frames

Analyst: G.C.C.

Date: 3/8

Frames: 20, 40, 60, 80, 100, 120

Left hand
Carry gear base to work area 11 +
Hold base 5
Turn over base 14
Carry base to right hand 18
Hold base 12
Reposition base 6
Hold base 39
Regrasp & position base 7
Hold base

Right Hand
Reach for screws (2) 12
Grasp screw 9
Carry to work area 12
Release 2
Carry one screw to base 13
Position screw in base 12
Reposition base 6
Reach to 2nd screw 7
Grasp screw 3
Carry screw to base 14
Position 2nd screw in base 15
Grasp & position base 7
Reach to 3rd screw 8
Grasp screw

Independent action

Holding/delay

Combined activity

[a]In this analysis one scale division equals two frames. The elapsed frames for each hand's activity are listed under each description.

chosen work activity, the film frames can be analyzed at various speeds in order to understand the present method of operation. Obvious changes and improvements are immediately recognizable during the initial viewing of the film. The analyst utilizes the basic principles of work simplification and motion economy to obtain improved productivity.

Analysis of time-lapse films also provides the opportunity to evaluate the utilization and performance of the crew members relative to their assigned task. Utilizing the multicolumn analysis format, and following each individual or process under study, the analyst can determine the percentage of time for such activities, which will reveal the amount of time spent on each activity being evaluated. These data are usually expressed as a percentile. Such information can assist in improved manpower allocations and future training. The analyst should be reminded that the film can be run in both forward and reverse modes at variable speeds in order to determine and assess problem areas. Other uses of the time-lapse technique may involve materials flow, paperwork flow, records filing and retrieval, material handling and staging, and storage operations.

An advantage of the time-lapse technique is that it produces a mass of data at low cost; also, the corresponding analysis requirement is less time consuming. Another advantage is that managers and employees can become involved in the improvement process, which will generate deeper understanding and acceptance.

High-Speed Analysis

An advantage of high-speed photography is that you can obtain two methods of analysis in one filming. First, there is the important subjective analysis that is obtained when the analyst simply views the action on the screen. A basic understanding of the action or problem is now aided by viewing the action in a time frame that is slowed down. In Exhibit 3.2.2, the photographic speed in frames per second above normal shows resultant slow viewing when projected at normal speeds. This allows the viewer to see individual segments of the action in their relationship to each other. It is then possible to correct any errors in the action or even form a new concept of it.

Second, the film can provide a means of quantitative analysis of the action if the proper steps were taken beforehand to ensure that such data could be obtained. Proper setup of the subject, proper recording of the filming data, and the placement of other recording devices can yield a wide range of important measurements.

Since normal projection of 16 or 24 frames/sec produces an extremely slow rate of analysis, the time expenditure for high-speed and very high speed analysis can be excessive and costly. It has been found that high-speed filming for motion studies will fall into the 32 to 48 frames/sec filming speed.

A research study that investigated manual motions performed under stereoscopic microscopes was filmed at 40 frames/sec. Filming was conducted with two cameras synchronized to provide simultaneous data from both insides and outside the microscope.[7] The motions of the hands and tools were in the view of the "inside" camera. The outside motions and electronic data collecting equipment were within the view of the "outside" camera. Exhibit 3.2.8 shows a complete display of a typical film analysis procedure for this research involving the microscope. In high-speed filming, analysis of such motions as finger tremor and eye movement can be captured and recorded. Note the column captioned "change of direction numbers" that recorded observed tremor. Eye motions were recorded in the "C.E. (critical event) code" and the "critical event description" columns. The eye motion information was derived from an oscilloscope, using the electrooculography technique.[8]

Videotape Analysis

Videotape analysis is becoming more acceptable in motion study because of the state of the art in equipment advancements. Videocassette recorder and playback equipment is available that will play the tapes in various forward slow-motion modes as well as in reverse. This enables the engineer to study the tape recording in a manner similar to that used in film analysis.

Videotape analysis has a great advantage over film analysis in that the tape is immediately available for continuous playback after filming. This allows involvement of those being photographed. Possibilities for immediate improvement can be discussed and in some cases implemented or simulated for subsequent filming during the same setup.

Frame count display is an integral part of the equipment, which enables the analyst to return the operation to the zero point for a repeat viewing. This can be done over and over, giving the effect of a film loop. Timing of the videotaped activities can be conducted externally with standard time study techniques or with electronic clocks or digital time reading that is automatically recorded on the tape during the filming process. The charts displayed in Exhibits 3.2.5, 3.2.6, and 3.2.7 can be adapted to the videotape analysis of motion study.

A particular advantage of videotape recording is that the engineer can incorporate sound recording with video recording. This gives a second dimension to the analytical capability. Operational characteristic noises can indicate beginning and ending points. Sounds of parts being placed in and out of the process can signal breakpoints that assist the overall analysis. By use of electronic timers with multiple readout capability, a series of functions can also be recorded to establish time or percentage data for evaluation.

The state of the art of video techniques is improving and will continue to give the engineer additional motion study options.

Exhibit 3.2.8 Data Sheet for Magnification Project Film Analysis

FILM NUMBER (1–8)	CYCLE NUMBER (9)	SHEET NO. (10–11)	DEXTERITY (12–13)	PERFORMANCE (14–15)	TREMOR RATE (16–17)	ILLUMINATION (18–20)	DECIBEL (21–22)	CAMERA SPEED: OUT (23–25)	CAMERA SPEED: IN (26–28)
AØ32Ø15D	3	12	6Ø	2Ø	Ø7	1ØØ	4Ø	398	4ØØ

OPERATOR R. Mathews KEYPUNCHED BY ______

FILM ANALYST RAM VERIFIED BY ______

DATE 2/10 SHEET 8 OF 8

(29–30)	FILM SECTION	FRAME NUMBER START (31–34)	FRAME NUMBER STOP (35–38)	(39)	CHG. DIR. NO. (40–41)	EYES H (42–43)	EYES V (44–45)	C.E. CODE (46–50)	CRITICAL EVENT DESCRIPTION (51–65)	TARGET SIZE (66–67)	HAND L-R-B (68)	DISTANCE INCHES (69–73)	TOOL CODE (74)	TOOL HOLD DISTANCE (75–76)	PART SIZE INCHES (77–80)	
0.1	NA-2	17Ø3	219Ø	B				HA57	APPLY CEMENT			.		.	.	1 Total Major Work Pattern
0.2		1846	1896	I	6			MTL3	APP TO M-L	3	R	.154	M	.7	.ØØ5	2
0.3		18ØØ	1839	O	1			HTL6	TO PAPER	3	R	12.ØØØ		.7	.	3 M = Plastic Probe
0.4		18Ø2	18Ø8	O		R5	U3	EM	EYES-UP-RT			.		.	.	4
0.5		18Ø8	1815	O				EM	OFF SCREEN			.		.	.	5
0.6		1815	1847	O		L2	U3	EM	EYES-UP-L			.		.	.	6
0.7		1889	1892	O		R1		EM	EYES-RT			.		.	.	7
0.8		18ØØ	1818	O				HL2Ø	HD-MOT-2ØDEG			.		.	.	8
0.9		18ØØ	1851	O				HD1Ø	HD-MOT-1ØDEG			.		.	.	9 (End of Cycle)
1.0												.		.	.	10
1.1												.		.	.	11
1.2												.		.	.	12
1.3												.		.	.	13
1.4												.		.	.	14
1.5												.		.	.	15
1.6												.		.	.	16
1.7												.		.	.	17
1.8												.		.	.	18
1.9												.		.	.	19
2.0												.		.	.	20
2.1												.		.	.	21
2.2												.		.	.	22
2.3												.		.	.	23
2.4												.		.	.	24
2.5												.		.	.	25

SYSTEMS 3876

3.2.5 OPTIMIZING THE WORK METHOD

To reiterate the objectives of motion study, the analyst is attempting to develop work methods that will result in (1) fewer hours per unit of work produced; (2) reduced effort on the part of the employee; and (3) a minimal amount of waste, scrap and rework. These objectives should be carried out in such a manner as to increase the effectiveness of human activity by removing those features of the job and workplace that promote inefficiency and physical stress.

A critical examination of the details brought forth by the charting and filming activities should raise questions that suggest desirable changes. The development of optimal methods requires imagination and ingenuity as well as a practical understanding of the work under examination.

It is assumed that recommendations to correct the obviously wasteful activities will be proposed as soon as they are substantiated. Interest by the people involved may result in suggestions for changing the more apparent defects in the work cycle.

Engineers are usually well informed on the characteristics of the manufacturing process and the functioning of the machines or equipment. The functional characteristics of the human being in the man-machine relationship are not as well comprehended. Troubles relating to the lack of understanding and consideration of the capabilities and limitations of the employee within the work system occur with unwarranted frequency. This is the basic reason for the development of the discipline of ergonomics. (Section 6 should be consulted for details regarding this field of knowledge.) The principles that follow incorporate some of these considerations.

Principles for Motion Improvement*

In 1923 Frank Gilbreth published a list of principles of motion economy based on his research and perceptive intuition. Gilbreth's list has been expanded and restructured by Ralph Barnes and others. These principles are summarized in this section and fall into three categories:

Principles relating to the human being.
Principles relating to the work environment.
Principles concerning time conservation.

Motion principles relating to the human being are as follows:

1. *The hands and arms should follow smooth, continuous, curved motion patterns.* Such curvilinear motions require low control and are easier and faster than controlled motions. Tossing a small part in a container is less time consuming than threading a needle–a highly controlled motion. Sudden stops and sharp changes of direction break work rhythm.
2. *The hands should preferably move simultaneously, rhythmically, and symmetrically in opposite directions, beginning and ending their motions together.* Simultaneous motions in rhythmic motion paths encourage prompt establishment of effective habit patterns. They provide for natural motions with respect to the physiological characteristics and balance of the body. This suggests that all people performing the same work should be encouraged to follow a standard moti n procedure.
3. *Within practical limits, movements should be confined to the shortest distance possible.* The farther the hands move from the body, the less accurate and the more time consuming the move. Reach and move motions maintained at less than 16 in. reduce the need for shoulder and trunk motions. Bending, stooping, carrying, lifting, and walking should be avoided or reduced in the work cycle.
4. *Both hands should be used for productive work.* This principle is seldom questioned in such activities as rowing a boat, using a typewriter, or playing a musical instrument. In other forms of work, the hand is frequently used as a vice or holding fixture. Hands can frequently be relieved of holding by providing a fixture or clamp operated by a foot pedal. For work requiring a standing position, the foot pedal should be replaced by a lever actuated by another body member.
5. *Work requiring the use of the eyes should be maintained within the field of normal vision.* The eyes direct much of the work of the hands. It is preferable to locate such work directly in front of the operator. Storage of parts should also be in the line of vision for easy pickup. Parts disposal seldom requires eye direction and can therefore be located at the right and left of the work center. If work can be located in a restricted area, a minimum of eye fixations will be required, and the distance between fixations will be shortened. Movements requiring visual control for termination take more time than the same motion that is terminated with physical stops, or within the field of normal vision.

*Adapted from GUY C. CLOSE, Jr., "Principles for Motion Improvement," *Work Improvement*, Wiley, New York, 1960, pp. 235–269.

6. *Actions should be distributed among the muscles of the body in accordance with the inherent capabilities of body members.* Upper limbs have speed and accuracy; lower limbs have strength and stability. Providing for intermittent use of different muscle groups allows the opportunity to rest the unused muscles.

7. *Muscular force required for motions should be minimized.* Body momentum should be utilized to the best mechanical advantage. However, momentum should be reduced to a minimum when it must be overcome by muscular effort. Parts can be slid instead of lifted. Weight reducing holes can be placed in hand-moved fixtures and equipment to reduce the effective net weight handled. Avoid covert lifting tasks. "In a modern industrial environment, the heaviest article normally handled by man at work is his own body or segments thereof. . . . In most instances the mass of an object moved is quite insignificant when compared with the weight of the body segment involved in the operation. . . ."[9] A 1 lb. soldering iron is held in position by an 11 lb. arm.

Principles relating to the work environment are as follows:

1. *The workplace should be designed with motion economy considerations.* Fixed stations should be established for the location of tools, parts, and materials. This permits habit formation, which is also a safety feature. Work follows a habitual safe motion pattern. Tools pre-positioned in toolholders can be retrieved and replaced after use without visual assistance. Materials and parts positioned in the sequence in which they will be used will reduce decision making requirements. Workplace height should be designed to allow for a standing or sitting position.

2. *Tools and equipment should be designed and/or selected with ergonomic constraints in mind.* Tools and machine handles should provide maximum hand contact surface. Tool handle designs are available that allow torquing tools, such as screwdrivers, to be turned with a straight-line force: forearm to wrist to tool. Pliers, diagonal cutters, and similar tools are designed with handles altered to allow the user to perform the operation without bending the wrist. Levers, handwheels, and other machine controls should allow activation with the least body relocation and with the greatest mechanical advantage. Large, mushroom-shaped control buttons operated with the butt of the thumb are preferred over those requiring finger manipulation. The analyst should be cautioned that installation of ergonomically sound tools and equipment may not be readily accepted by the employee. Proper use of a particular tool and the advantages of its use must be carefully explained to the operator.

3. *Materials handling methods should be selected in accordance with time-weight factors.* Mechanical devices should be used for heavy lifting. Slides, guides, stops, quick-action clamps, hoists, ejectors, and other mechanical devices can be incorporated within the work station to reduce both time and effort. Handling devices, such as bins and conveyors, should deliver material close to the point of use. Drop delivery facilities allow quick disposal of finished work, scrap material, and rejects. Materials and parts pre-positioned for succeeding operations simplify pickup and safeguard parts.

Principles concerning time conservation are as follows:

1. *Any hesitations or temporary cessations of motion should be questioned.* Unavoidable delays occurring with regularity may provide time to perform an additional operation. The metal set time in permanent molding causes a delay that frequently allows the assignment of gate removal to the operation. Work should be performed during the machine or process cycle time. Perhaps the operator can tend two or more machines or process units.

2. *The number of motions should be minimized.* The motion pattern that required the fewest steps or elements will usually result in the shortest performance time. Elimination of elements or steps and combination of these steps are two recognized methods to reduce the labor content of the job. If several operators are using a variety of motion patterns to complete the same work assignment, it should be assumed that one method is better than the others. A study of the various methods may result in a composite method that will encompass the best features of each.

3. *Processing more than one part at a time should be considered.* Many small bench assemblies can be processed in dual "left-and-right-hand" fixtures. Dual working gives better rhythm and body balance.

Motion Reduction and Time Balancing

Multiple-activity charting applications and film analysis, particularly time-lapse photography, have provided techniques for gathering and analyzing the motion activities of multiple-man operations and man-machine operations. The purpose of this section is to suggest ways of developing improved motion/methods for these types of work assignments. As previously stipulated, the overall objective is to increase value adding time and to reduce idle and delay time by coordination of each individual's activities. Value adding time is that part of the work sequence that alters the product in such a way as to increase its value.

The film and charting analyses show the time spent by each individual and each machine as idle, handling, setup, and value adding work time. The charting analysis also indicates those motion cycles that must be accomplished by two or more individuals working as a team, or by an individual working with a machine; for example, two men load a sack of heavy coins into a security truck, and an operator replaces a broken needle in a machine. Such activities are most difficult to relocate within the work sequence. On the other hand, work being done by an individual independent of a machine, or work being done by other members of the group, can frequently be changed in sequence, or the work can be reassigned to a member who has excessive waiting time.

The first technique in film analysis is for the analyst to follow and chart each individual for the full work cycle. This exercise shows which team member has the most time consuming work load and which member the least. It suggests other areas to which parts of the work can be assigned. The redistribution of assignments to give each member a balanced load will reduce the job cycle time and the total man minutes per unit produced.

A second method worth considering is the assignment of additional personnel to the crew. In a study of man-machine activity on the shearing of magazine pages on a Lawson flat bed cutter, it was determined that the use of *two* operators would almost double press production. This removed a bottleneck and eliminated the need to purchase an additional cutter. People prefer work activity to enforced idleness. They will usually be willing to cooperate in establishing equal work loads for each member.

Next, consider the man-machine activity. By reducing the setup time, a larger part of the day can be utilized in producing units of work. Setup time can be minimized by:

Standardizing tooling to simplify setup.

Installing quick-action clamps and locating pins to lock fixtures into position.

Providing tools, fixtures, and cutters at the work station *prior* to need.

Supplying setup instructions.

The machine can be devised to perform useful work for a larger portion of the cycle by:

Utilizing air-operated clamps for holding parts in the fixture.

Reducing "cutting air" by starting and stopping the table feed at the beginning and end of the cutting pass.

Installing dual fixtures, such as a lazy Susan or shuttle, to allow loading/unloading during the machine cycle.

Rearranging the sequence of operations to reduce cycle time.

Providing quick-change tool holders when more than one tool is required.

A milling operation called for face milling a part on two different horizontal planes. This required the table to be raised and lowered once during each cycle. By altering the sequence for two successive parts, only one table movement per cycle was required; for example, part 1–mill surface A, then B, part 2–mill surface B, then A. The operator can keep the machine active by preparing the next part during the machining cycle. For example, he or she may deburr a casting so that it

Exhibit 3.2.9 Various Ways of Improving the Work Through Motion Study

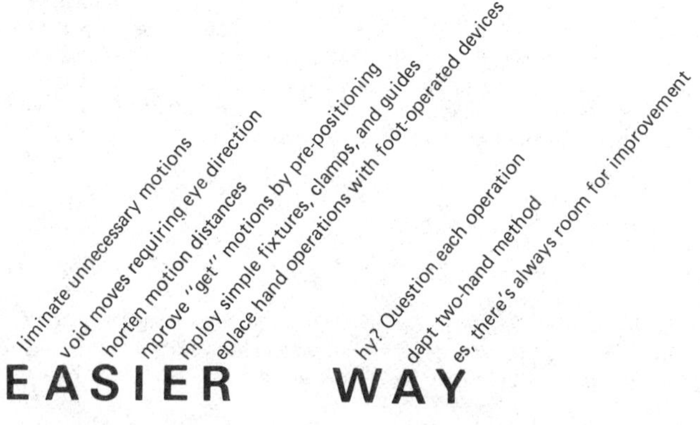

will be accommodated by the fixture. Disposal of the finished unit can be delayed until the next part is loaded and the machine set in operation.

The work flow through the machine is always subject to question. Speeds and feeds, depths of cuts, number of process passes, finish tolerances, and so on, should be checked with design specifications. The engineer should ascertain whether increased fixture rigidity or the placing of vibration dampers would allow increased speeds and feeds. Some of these comments may appear beyond the realm of the motion analyst, but as an industrial engineer, he or she must carry the project to the most cost-effective conclusion.

Exhibit 3.2.9 may serve as a reminder of the various ways of improving the work through motion study.

3.2.6 QUANTIFYING PROPOSED METHODS

Visualization with Predetermination

Visualization is the conception of a mental picture showing the motions that will be required to perform a given task. A complete understanding of the procedure captured on film or in charts with the aid of predetermined time data and visualization techniques leads to proposals for improvements. Therefore the film represents the "old method," and the utilization of tools and approaches previously discussed form the basis of a visualized "new method."

Recording or charting of the film analysis data provides documentation of the elements of the activity. Time for these elements can be applied by means of predetermined time systems or standard data that directly correspond to elemental methods. With the aid of these data, a proposed method can be established. The cycle time for the old and the proposed method can then be compared for feasibility and implementation.

The use of computers for retrieval of standard data assists the visualization of proposed methods. Storage files of common standard data with known content can be called upon quickly to establish the new or revised method time. These times can also aid in establishing the priority of items that should be corrected in the proposed visualization process.

In simultaneous motion analysis, computers can be programmed to the correct logic pertaining to a given situation. There is a methods assignment index (MAI) in the 4M computer program that applies MTM-1 system time values and rules. This index shows how completely both hands have been assigned work during the work cycle. An index of 50% would represent effective use of only one hand at a time, whereas 100% would indicate perfect utilization of both hands.[10] The detail from the MAI calculation will aid the analyst in visualizing and comparing various alternate methods.

Test Run Verification

Prior to implementation, a proposed method requires testing to prove its validity. It is recommended that a laboratory mock-up of the new operation be set up for evaluation prior to actual production or use. Additional improvements may be developed, with a resultant time reduction. This methods improvement approach will be enhanced by utilizing generic predetermined time systems such as MTM-1, MTM-2, or MTM-3. Exact times can be developed with these techniques, as well as the final method appraisal.

There are other ways of testing the proposed method. The assistance of management can be enlisted to participate in a test run during actual production. This creates an atmosphere of involvement for acceptance of the proposal. Additional improvements can be made during this test run session. The charting techniques can then be explained to the participants and appropriately and benefically utilized.

Of course, time checks of the new method can be accomplished by time study techniques. The proper number of cycles should be selected in order to satisfy desired accuracy. Greater confidence will be established in the validity of the data when many cycles have been time studied.

The use of published standard data elements for the specific application to the method can be used in the test run. In such cases complete backup of the standard data is recommended for traceability to verify the method.

Test run verification of proposed, new, or revised methods adds validity to the application of motion study.

Selection for Cost-Effectiveness

Previous sections have discussed techniques for gathering and analyzing data for a work problem. Methods for optimizing solutions have been proposed. Procedures for quantifying and verifying time values have been suggested. The final test is to establish the cost-effectiveness of the proposal. Will it reduce costs and increase productivity?

At this point, final questions should be considered in the decision making process:

Have several solutions to the problem been considered?
How will production volume be affected?
Have employees been involved?
Are all labor-related expenses known?
Will the proposal forestall the need for additional equipment?
Have alternate sources for purchased tooling been investigated?

For present and proposed methods, comparative time values have been established. Comparison time for each unit of production, without cost-additive equipment considerations, provides basic calculations for cost-effectiveness. An example calculation is as follows:

$$\text{annual savings} = (C - P) \times U \times L$$

where

C = current time per unit (decimal hours)
P = proposed time per unit (decimal hours)
U = units produced per year
L = total labor expense per hour

When methods improvement requires purchased equipment or tooling as part of the proposal, the annual amortized cost must be subtracted from the gross annual savings.

With this information, management can make bottom-line decisions on the cost-effectiveness of alternate proposals. The selected proposal should not only result in increased productivity, but also maintain product quality and employee satisfaction. These are important ingredients in convincing management of the proposal's value.

3.2.7 IMPLEMENTATION FOR OBTAINABLE RESULTS

An idea–a penetrating ingenious analysis of the problem–remains only an idea unless it is satisfactorily implemented. Value is measured in terms of attainment. Implementation of a new idea relating to a work method presents two problems: the technical problem and the human problem.

Involvement of People

If the engineer has carefully carried out the study and analysis, the technical problem should be solved. The idea will work, technically. On the other hand, the people involved in helping to make the idea work must have a degree of "ownership" in the idea to ensure its success.

> *Under really successful management, it is realized that the employee has an inquiring mind, and that unless this inquiring tendency . . . is recognized and his curiosity is satisfied he can never do his best work. Unless the man knows why he is doing the thing, his judgment will never reinforce his work. He may conform to the method absolutely but his work will not enlist his zeal unless he knows exactly why he is asked to work in the particular manner prescribed.*[11]

This contemporary thought was offered by Lillian Gilbreth in 1914. Allan Mogensen, creator of work simplification, realized early in his career that positive motivation can be achieved by allowing the employee to build his or her self esteem through participation: "Let's take advantage of man's desire to be important. Institute a program that will result in complete participation and one that is based on recognition of accomplishment."[12] A more modern recognition of the advantages of involvement and participation is taking place in sections of the needle trades industry. Industrial engineers are taking videotapes of existing motion patterns and methods being used by a group of operators. An immediate playback is shown to the individual after filming. Later the group is brought together to observe the recordings of all the operations under study. The videotape becomes a catalyst for a brainstorming session. This chance to participate is appreciated by the employees and has resulted in many work improvements.

A second example is that of a crew balance problem on a metal shearing operation. This operation had been running for years at a constant production rate. A time-lapse study was taken to search out any possible production improvements. The processed film was shown to the crew during lunch break. Many amusing remarks were made regarding how hard the lead man worked and how "soft" the helpers had it. A multiple-activity chart, prepared from the film, was made, showing, chronologically, the activities of each person and each machine. Idle and delay segments were indicated by red shading. The crew was allowed to reorganize its work under the direction of its foreman. Better balance of work assignments resulted in a 20% improvement. Instead of resenting the change, the crew considered it a result of its own accomplishment.

Training in the New Method

Implementation includes the training of the people who are to perform in accordance with the new method and the development of written standard practices and instructions.

Frequently missed by the engineer is the fact that the employee may have to replace a frequently repeated habitual motion pattern with a different pattern requiring mental decision and direction. The new pattern, more effective on the analyst's chart, may cause hesitations and errors. The temporary loss of work rhythm may discourage the employee to the point that he or she returns to the previous method. An explanation to the employee as to what to expect–that a certain amount of difficulty may be experienced at first–may make the difference between rejection or acceptance of the new method. The breaking and replacing of habit patterns must be carefully handled.

New motion patterns can be demonstrated on a one-to-one basis, "live," or by use of video or motion pictures. Repeated projection in slower-than-normal speed will allow the operator to follow the work pattern.

Instruction sheets and visuals, such as slide projection or picture book planning, will help maintain the method sequence.

The engineer who has the opportunity to instruct the new employee in the proper work method has the advantage of not having to break old habits. One organization with severe training problems resulting from turnover has established a pilot unit. Each new employee spends several days learning the exact method, sequence, and motion pattern of the operation to be performed on the regular job assignment. Similar training centers are utilized by aerospace companies to ensure proper techniques in soldering, welding, and so on. The addition of motion pattern learning should be considered under these circumstances.

Followup With Continuous Improvement

After a new work pattern is established, some employees are immediately tempted to revert to past practices. There is also the long-term erosion of good motion/methods practices caused by such conditions as changes in personnel, lack of proper maintenance of the equipment, and changes in purchased material.

Immediate alterations of the new method can be discouraged by monitoring the work area. Additional operator training may be required.

To decrease the impact of long-term tendencies toward motion practice erosion, an audit procedure is recommended. "The audit should be performed frequently enough to remind supervisors of their responsibility to monitor methods."[13] Methods discrepancies found during the audit must be resolved. No purpose is served if the audit results are ignored.

Operation instruction sheets or visuals should be current and available. The motion pattern content and sequence in use should be checked against the original analysis proposal. Equipment and tooling should be inspected for modification and wear. A methods program to increase and maintain productivity can remain effective only if the operations being performed continue to reflect the optimum method. An auditing procedure can help.

Engineers cannot be satisfied with any existing work situation. Competition at home and abroad, inflation, and product improvement require a continuing search for better methods. They must maintain a constant awareness of new equipment and processes and must search out new ways to solve work problems. The techniques and practices described in this chapter will help the engineer realize that there is always room for improvement.

REFERENCES

1. H. SKERRY HALL, "Putting Work Simplification to Work," *University of Illinois Bulletin*, Vol. 53, November 30, 1956, p. 7.
2. H. B. MAYNARD, G. J. STEGEMERTEN, and J. L. SCHWAB, *Methods-Time Measurement*, McGraw-Hill, New York, 1948.
3. M. E. MUNDEL, *Motion & Time Study*, 5th ed., Prentice-Hall, Englewood Cliffs, NJ, 1978.
4. AKIYUKI SAKIMA, "A New Industrial Engineering Tool: The Use of the Video Tape Recorder," *The Journal of Industrial Engineering*, April 1966, pp. 209–215.
5. KODAK, *High Speed Photography*, Kodak Publication No. G-44, Rochester, NY, 1975, p. 4.
6. KODAK, *High Speed Photography*, p. 4.

7. MTM ASSOCIATION FOR STANDARDS AND RESEARCH, *Interim Report, MTM Magnification Research Project*, Fairlawn, NJ, October 1970.
8. DANIEL O. CLARK, "Industry Launching MTM Microscope Research," *The MTM Journal*, Vol. 14, No. 4, p. 34.
9. E. R. TICHAUER, *Biomechanical Basis of Ergonomics*, Wiley, New York, 1978, p. 48.
10. MTM ASSOCIATION FOR STANDARDS AND RESEARCH, *4M Mod II Users Manual*, Fairlawn, NJ, February 1980.
11. LILLIAN M. GILBRETH, "Psychology of Management," in Spriegel and Myers, *Writings of the Gilbreth's*, Irwin, Homewood, IL, 1953, p. 431.
12. ALLAN H. MOGENSEN, "What Incentive?," *Aircraft Production*, Vol. 1, No. 1 (August 1943).
13. JOHN R. ANTONIEWICZ, "Auditing Approach to Methods and Standards," *MTM 1976 Fall Conference Proceedings*, Fairlawn, NJ, p. 2.

BIBLIOGRAPHY

BENSINGER, CHARLES, *The Video Guide*, Video-Info Publishing Company, Santa Barbara, CA, 1980.

CLOSE, GUY C., Jr., *Work Improvement*, Wiley, New York, 1960.

DOXIE, FLOYD T., "Biomechanics Used to Avoid Labor Relations Problems," *AIIE Journal*, Vol. 28, No. 11.

INTERNATIONAL LABOR OFFICE, "Introduction to Work Study," Atar, Geneva, 1960.

MTM ASSOCIATION FOR STANDARDS AND RESEARCH, *MTM Magnification Research Project Final Report*, Fairlawn, NJ, 1972.

RICE, I. M., "Management Improvements Cut Cost," *Water and Wastes Engineering*, May 1975.

WALSH, PHILIP A., "Take Another Look at Memo Motion," *Industrial Engineering*, May 1978.

CHAPTER 3.3
Charting Techniques

TAKEJI KADOTA
Japan Management Association

3.3.1 INTRODUCTION

Purpose of Charts

Charts are the graphical presentation of work that has been broken down into basic components or units. They are one of the most important tools in methods engineering.

Charts aid in analyzing and improving the present method. The basic methods study procedure is as follows:

1. *Select* the work to be studied.
2. *Record* all the relevant facts.
3. *Examine* these facts critically.
4. *Develop* the most practical, economical, and effective method.
5. *Install and maintain* that method.

The charts are useful in recording, examining, and developing stages. The method of recording is explained first, and the rest discussed later.

Charts are also descriptive and communicative aids for understanding the process and activities. Clear and concise visualization by standard symbols and conventions facilitates comprehension and understanding of the process and activities. For example, charts may be used to present improved methods to management, as handy sources of process information in plant layout work, to train employees in standard methods, or to simplify the overall perspective of complicated office procedures.

Charting Methods

Analysis of Chronological Sequence

This method involves breaking down a process under study into events or activities chronologically. There are two types of analysis according to the subject, that is, a product (or material) or a person. Typical charts are as follows:

Flow process charts (product-type).
Flow process charts (person-type).
Operation process charts.
Form process charts.

Movement and Flow of Activities

Diagrams are used to indicate the path of movement. The subject of movement may be a product, material, and/or person. These diagrams are usually confined to the process sequence information.

Time Interrelationships for Multiple Activities

Time interrelationships among multiple activities involving different subjects are graphically shown in the same time scale, so that the interactions of related events of different subjects are clearly indicated. Subjects may be persons, limbs of a person, and/or machines. Typical charts are as follows:

Multiple-activity process charts.
Right- and left-hand charts.
Network diagrams.

Data Recording Devices

Film and electronic devices such as video tape recorders (VTRs) have become widely used in methods study work. They are particularly useful for micromotion study, the study of long work cycle activities, or multiple activities. Since the information recorded is usually transcribed into one of the aforementioned charts, this equipment is a useful aid rather than a substitute for charting techniques.

3.3.2 FLOW PROCESS CHARTS

A flow process chart is a graphic, symbolic representation of the work performed or to be performed on a product as it passes through some or all of the stages of a process. Typically, the information included in the chart is quantity, distance moved, type of work done (by symbol with explanation), and equipment used. Work times may also be included.[1]

Synonyms for flow process charts are process charts, flow charts, and process (or person) analysis.

The flow process chart is one of the techniques used to record a process sequence, that is, a series of events or activities in the order in which they occur. It is the most general application and is typical of similar sequence charts.

There are three types of charts that depend on the subject of the flow charted.

1. **Product-type (or Material-type) Chart.** The process of events involving a product or material (Exhibit 3.3.1).[2]
2. **Person-type Chart.** The process involving a person's activities (Exhibit 3.3.2).[3]
3. **Equipment-type Chart.** The process of events that occur on equipment.

For the analysis of transportation equipment, such as forklift trucks, this type of charting is useful. The combined flow process chart for triple resources involving a person, material, and equipment is shown in Exhibit 3.3.3.[4]

Because it is similar to a product-type chart in convention and practice, the equipment-type chart is not treated separately in this chapter.

Charting Conventions and Practices

Symbols

Events or actions are classified into the following five groups[5]:

○ **Operation.** An operation occurs when an object is intentionally changed in any of its physical or chemical characteristics; is assembled or disassembled from another object; or is arranged or prepared for another operation, transportation, inspection, or storage. An operation also occurs when information is given or received or when planning or calculating takes place.

⇨ **Transportation.** Transportation occurs when an object is moved or a person moves from one location to another, except when such movement is part of the operation or is caused by the operator at the work station.

□ **Inspection.** An inspection occurs when an object is examined for identification or is verified for quality or quantity in any of its characteristics.

D **Delay.** A delay occurs when an object or person waits for the next planned action.

▽ **Storage.** Storage occurs when an object is kept and protected against unauthorized removal.

Exhibit 3.3.1 Material-type Flow Process Chart: Engine Stripping, Cleaning, and Degreasing (Original Method)[a]

FLOW PROCESS CHART	~~MAN~~/MATERIAL/~~EQUIPMENT~~ TYPE			
CHART No. *1* SHEET No. *1* OF *1*	SUMMARY			
	ACTIVITY	PRESENT	PROPOSED	SAVING
Subject charted: *Used bus engines*	OPERATION ○	*4*		
ACTIVITY: *Stripping, cleaning and degreasing prior to inspection*	TRANSPORT ⇨	*21*		
	DELAY D	*3*		
	INSPECTION □	*1*		
	STORAGE ▽	*1*		
METHOD: PRESENT/~~PROPOSED~~	DISTANCE (m)	*237.5*		
LOCATION: *Degreasing Shop*	TIME *(man-min)*	—	—	—
OPERATIVE(S): CLOCK Nos. *1234* *571*	COST	—		
	LABOUR	—		
CHARTED BY:	MATERIAL	—		
APPROVED BY: DATE:	TOTAL	—	—	—

DESCRIPTION	QTY.	DISTANCE (m)	TIME (min)	○	⇨	D	□	▽	REMARKS
Stored in old-engine store								•	
Engine picked up					•				*Electric crane*
Transported to next crane		*24*			•				" "
Unloaded to floor					•				
Picked up					•				" "
Transported to stripping bay		*30*			•				" "
Unloaded to floor					•				
Engine stripped				•					
Main components cleaned and laid out				•					
Components inspected for wear; inspection report written							•		
Parts carried to degreasing basket		*3*			•				
Loaded for degreasing					•				
Transported to degreaser		*1.5*			•				*Hand crane*
Unloaded into degreaser					•				
Degreased				•					
Lifted out of degreaser					•				" "
Transported away from degreaser		*6*			•				" "
Unloaded to ground					•				
To cool						•			
Transported to cleaning benches		*12*			•				*By hand*
All parts cleaned completely				•					
All cleaned parts placed in one box		*9*			•				*By hand*
Awaiting transport						•			
All parts except cylinder block and heads loaded on trolley					•				
Transported to engine inspection section		*76*			•				*Trolley*
Parts unloaded and arranged on inspection table					•				
Cylinder block and head loaded on trolley					•				
Transported to engine inspection section		*76*			•				*Trolley*
Unloaded to ground					•				
Stored temporarily awaiting inspection						•			
TOTAL		*237.5*		*4*	*21*	*3*	*1*	*1*	

[a]Reference 2.

Exhibit 3.3.2 Person-type Flow Process Chart: Serving Dinners in a Hospital Ward[a]

FLOW PROCESS CHART		MAN/~~MATERIAL/EQUIPMENT~~ TYPE			
CHART No. *7* SHEET No. *1* OF *1*		SUMMARY			
		ACTIVITY	PRESENT	PROPOSED	SAVING
Subject charted: *Hospital nurse*		OPERATION ◯	*34*	*18*	*16*
		TRANSPORT ⇨	*60*	*72*	*(—12)*
ACTIVITY: *Serve dinners to 17 patients*		DELAY D	—	—	—
		INSPECTION ☐	—	—	—
		STORAGE ▽	—	—	—
METHOD: PRESENT/PROPOSED		DISTANCE (m)	*436*	*197*	*239*
LOCATION: *Ward L*		TIME (man-h)	*39*	*28*	*11*
OPERATIVE(S):	CLOCK No.	COST:	—	—	—
		LABOUR	—	—	—
CHARTED BY:	DATE:	MATERIAL *(Trolley)*	—	*$24*	—
APPROVED BY:	DATE: –	TOTAL *(Capital)*		*$24*	

DESCRIPTION	QTY. *(plates)*	DIST-ANCE (m)	TIME (min)	◯	⇨	D	☐	▽	REMARKS
ORIGINAL METHOD									
Transports first course and plates – kitchen to serving table on tray	*17*	*16*	*.50*						*Awkward load*
Places dishes and plates on table	*17*	—	*.30*						
Serves from three dishes to plate	—	—	*.25*						
Carries plate to bed 1 and return	*1*	*7.3*	*.25*						
Serves	—	—	*.25*						
Carries plate to bed 2 and return	*1*	*6*	*.23*						
Serves	—	—	*.25*						
(Continues until all 17 beds are served. See figure 32 for distances)									
Service completed, places dishes on tray and returns to kitchen	—	*16*	*.50*						
Total distance and time, first cycle		*192*	*10.71*	*17*	*20*	—	—	—	
REPEATS CYCLE FOR SECOND COURSE		*192*	*10.71*	*17*	*20*	—	—	—	
Collects empty second course plates		*52*	*2.0*	—	*20*	—	—	—	
TOTAL		*436*	*23.42*	*34*	*60*				
IMPROVED METHOD									
Transports first course and plates – kitchen to position A – trolley	*17*	*16*	*.50*						*Serving trolley*
Serves two plates	—	—	*.40*						
Carries two plates to bed 1; leaves one; carries one plate from bed 1 to bed 2; returns to position A	*2*	*1.5* *0.6* *1.5*	*.25*						
Pushes trolley to position B	—	*3.0*	*.12*						
Serves two plates	—	—	*.40*						
Carries two plates to bed 3; leaves one; carries one plate from bed 3 to bed 4; returns to position B	*2*	*1.5* *0.6* *1.5*	*.25*						
(Continues until all 17 beds are served. See figure 32 and note variation at bed 11)									
Returns to kitchen with trolley	—	*16*	*.50*						
Total distance and time, first cycle	—	*72.5*	*7.49*	*9*	*26*				
REPEATS CYCLE FOR SECOND COURSE	—	*72.5*	*7.49*	*9*	*26*				
Collects empty second course plates	—	*52*	*2.00*	—	*20*				
TOTAL	—	*197*	*16.98*	*18*	*72*				

[a]Reference 3.

Exhibit 3.3.3 Use of a Preprinted Form for a Triple Resource Chart[a]

CHART NO: *XY/17* PRESENT METHOD

PROCEDURE: *Sawing of material to length*

PLACE: *Factory 2*

DATE: *12th February, 1971*

OPERATOR: *AB*

CHARTED BY: *CD*

CHART BEGINS: *Fork-lift truck to store*

MATERIALS USED:

CHART ENDS: *Material to next process*

TOOLS, EQUIPMENT: *Fork-lift truck, sawing machine*

ACTIVITY	MAN	MATERIAL	EQUIPMENT
Fork-lift truck goes to store			
Material collected			
Take material to machine			
Unload material			
Operator cuts material			
Measure material			
Load on truck			
Material to next process			

[a]Reference 4.

Combination. Symbols have been combined to show activities performed concurrently.

Inspection is performed within an operation.

An operation is performed while a product is in motion.

These symbols are used for either product-type or person-type charts.

When unusual situations outside the range of the definitions are encountered, the intent of the definitions summarized in the following tabulation will enable the analyst to make the proper classifications.

Classification	Predominant Result
Operation	Produces or accomplishes
Transportation	Moves
Inspection	Verifies
Delay	Interferes
Storage	Keeps

Forms

Charts are usually drawn on blank paper or cross-ruled paper. The latter is more convenient when we express events involving more than one item of material, the activities of more than one person, or alternative routes or procedures (Exhibit 3.3.4).[5]

Preprinted forms may be used for charting single items for convenience and time saving in recording (Exhibits 3.3.1, 3.3.2, and 3.3.3).

Identification and information are recorded in the heading, such as the chart type (product or person), the present method or the proposed one, the name of the subject charted, and the location or the department name of the study.

Chart Construction

In the person-type flow process chart, there are no horizontal lines representing the entrance of material into a process, as was shown in Exhibit 3.3.4, and no storage symbol is generally used.

Although the same symbols are used for both objects and persons, the active voice of verbs is used for each brief description in person-type chart entries, and the passive voice in product- or material-type charts (cf. Exhibits 3.3.1 and 3.3.2).

Symbols for events or activities may be numbered in the sequence in which they are charted for identification and reference purpose (Exhibit 3.3.4).

Disassembly operations may be similarly portrayed to that utilized in Exhibit 3.3.4 for product processing. Charts of this type are often used in repair shops, chemical plants, or food processing and meat industries. Materials disassembled or extracted are represented as flowing from the process by a horizontal material line drawn to the right from the vertical flow line and a little below the symbol for the operation.

Recording

First, the scope of the study, that is, the starting and ending points of the process, has to be determined. For the purpose of analyzing the present methods, the chart must be made from direct observation, not by guesswork. While recording each step in the process, apply a questioning attitude. Although critical examination is conducted later, any ideas or inspirations for improvement should be written down as they occur.

3.3.3 OTHER TYPES OF FLOW PROCESS CHARTS

Operation Process Charts

An operation process chart is a graphic, symbolic representation of the act of producing a product or providing a service, showing operations and inspections performed or to be performed, with their sequential relationships and materials used. Operation and inspection time required and location may be included.[5] A synonym for this chart is outline process chart.

This chart is considered to be an abbreviated form of the flow process chart (product-type) previously described, since only the major events such as principal operations and inspections are recorded. The conventions for charting are the same as those for flow process charts (Exhibit 3.3.5).[5] It provides an overall view of the entire process from beginning to end at a glance.

Form Process Charts

A form process chart is a graphic, symbolic representation of the process flow of paperwork forms. It is similar to a flow process chart, except that the item of interest is one or more forms. A form process chart may show organizations, operations, movement, temporary and controlled storage, inspection or verification, disposal of all forms charted, and the source and type of information transmitted between forms. Flow process chart symbols may be adapted to reflect the form processing activity.[5]

Synonyms for form process charts are information process analysis, functional form analysis, form analysis charts, paperwork flow charts, and procedure flow charts.

As compared with the regular product-type flow process chart, the "product" in this case is paperwork and/or information, and the manufacturing process is the clerical procedure or system.

There is no set standard for symbols, and some differences exist between authorities. The symbols, however, are quite similar to those used in the ordinary flow process chart, except for the "relationship" symbol. This is a kind of information transfer and the key operation of form process charts, because the true subjects of office procedures are not forms, but information. Since forms are a means of conveying information, what we are pursuing is the flow of information, not merely clerical activities.

Exhibit 3.3.4 Material-type Flow Process Chart Showing Manner in Which Several Components Are Processed and Brought Together[a]

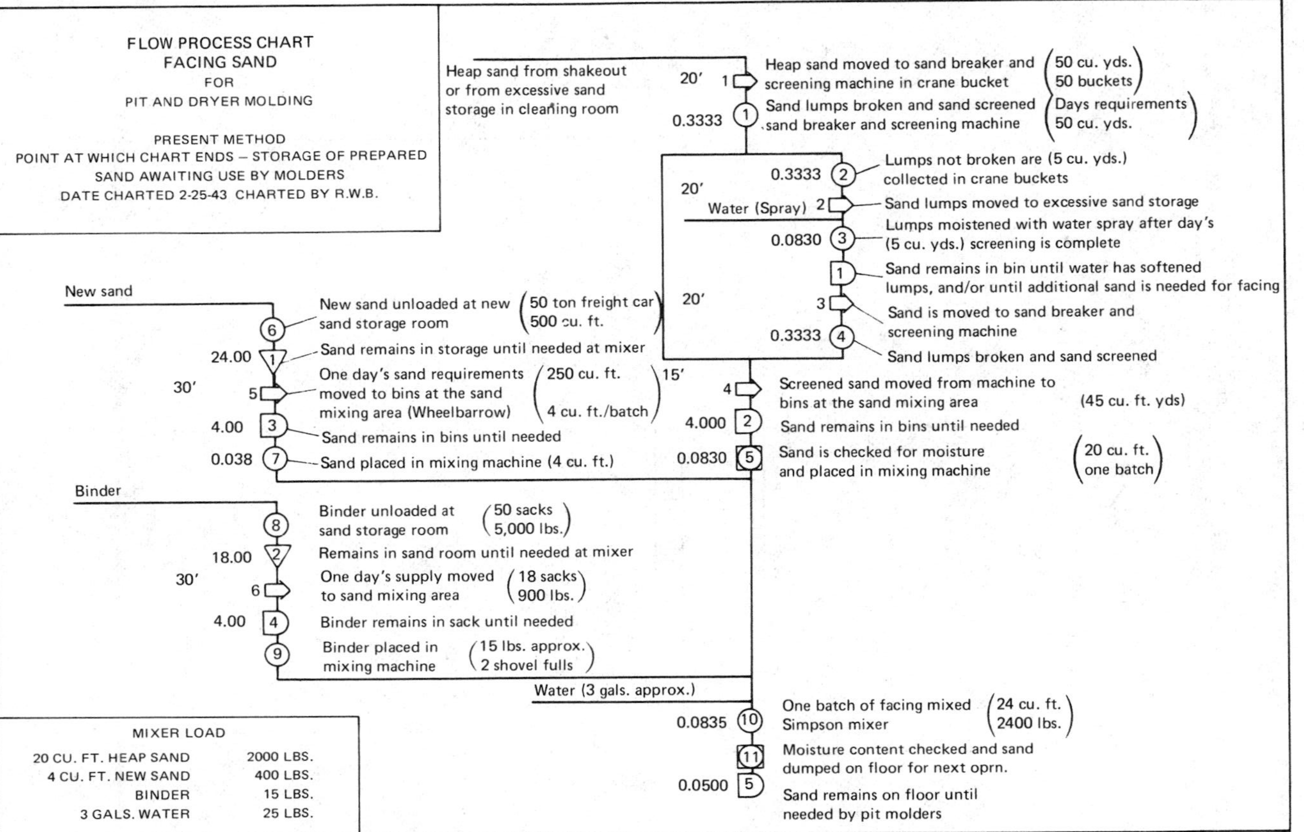

[a]Reference 5.

Exhibit 3.3.5 Typical Operation Process Chart[a]

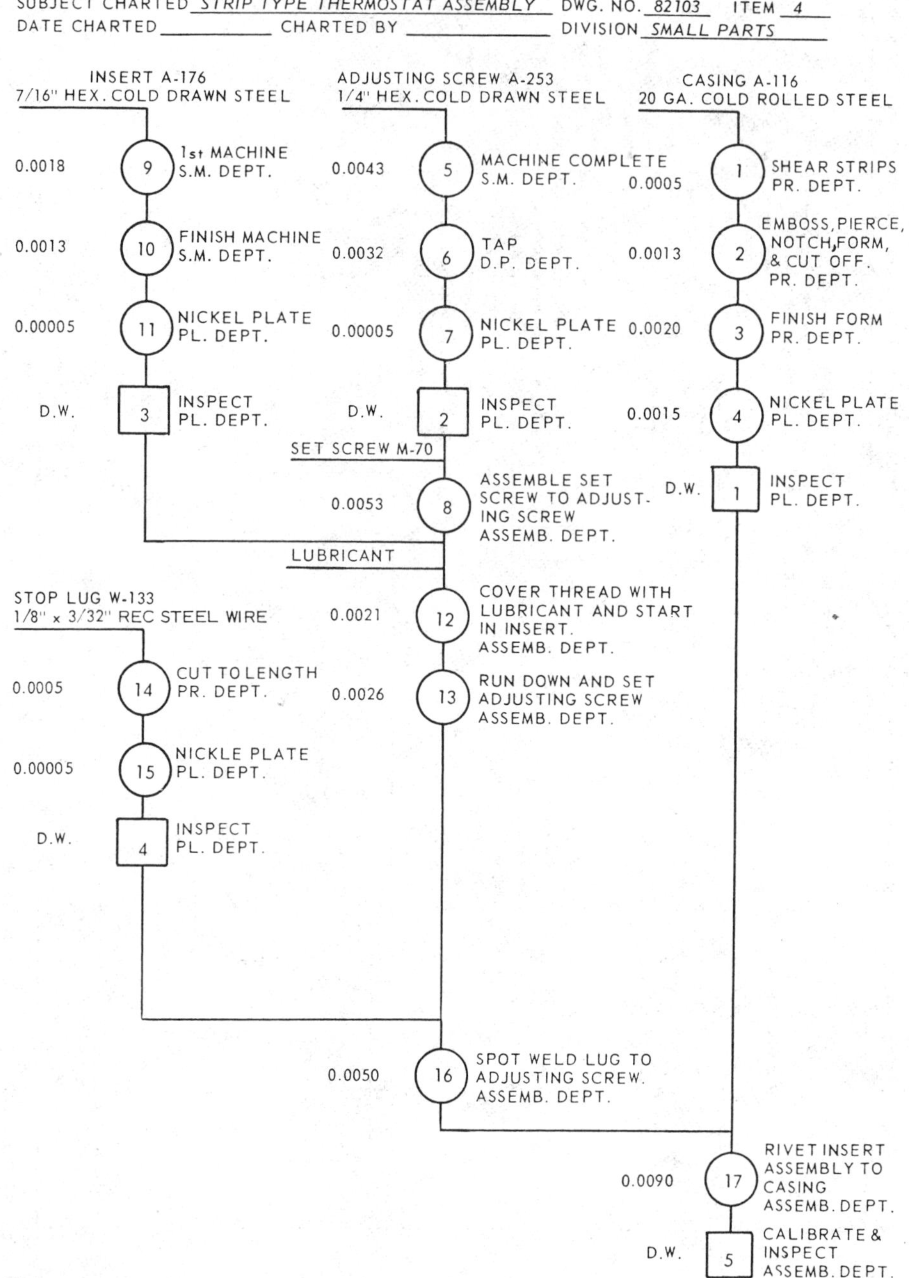

[a]Reference 5.

Charting Conventions and Practices

Symbols

G. C. Close, Jr., suggests the following symbols, which were developed by the Standard Register Company.[6]

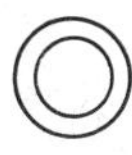

Origin. This is similar to the operation symbol. The small circle is added to indicate the origin of a record. The *first* time information of any kind is placed on a form, a single piece of paper, or a multiple-part record, whether it be handwritten, typed, punched or stamped, this symbol is used.

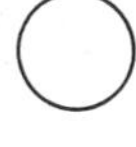

Adding to the Record. When subsequent writing or other means of placing *additional* information on the form occurs, it is designated by this crosshatched operation symbol.

Handling. The plain operation symbol is used for such nonproductive operations as sorting, folding, stapling, separating copies, and matching.

Move. The small circle was the original Gilbreth move, or transportation symbol. It is used (in place of the arrow) for movement between departments or work centers.

Inspection. The square is used when the record itself is checked for accuracy *and* when errors found will be corrected. If the record is to be corrected, accepted, rejected, or reprocessed in some way because of this particular step, the inspection symbol is used. It should be noted that this is for the inspection of paperwork. If material is being inspected, the accompanying record is usually *delayed.*

Delay. The triangle, which on the flow process chart stands for storage, is used on the procedure flow chart to represent delay. This identifies a point or time wherein the record is in an inactive status. When a record is filed, held in a desk, awaiting mail pickup, or destroyed, the step is identified with the delay symbol.

Relationship. In addition to the six other symbols, a "V" is required in paperwork charting to denote *relationship.* This symbol shows that one form *causes* something to happen to another. The form being charted has an effect on, or a relationship to, the other documents.

If necessary, some modification of these symbols may be used in order to differentiate activities. For instance, a "C" is added to the crosshatched operation symbol for "adding to the record" to indicate computation, a "T" is combined with the relationship symbol to identify the telephone as the source of information, and so forth.

Chart Construction and Recording

In gathering facts, the flow of each copy of each form under study has to be recorded on a separate sheet. The chart for each sheet is similar to the ordinary flow process chart. The final form process chart will be constructed out of these individual charts, as is shown in Exhibit 3.3.6.[7]

Sometimes chart forms are divided into vertical columns representing departments of the organization through which the forms travel. In some cases the departmental columns are further divided into specific persons.

Computerized Information Processing

Manual clerical work has been rapidly replaced by electronic data processing systems in offices. The form process chart, however, is still highly useful in these situations for analyzing and designing office procedure, although additional symbols may be necessary to take care of different activities.

For instance, the IBM system flow chart manual provides the symbols shown in Exhibit 3.3.7,[8] which comply with the requirements of the International Organization for Standardization (ISO) for computerized information processing. An example of charting is shown in Exhibit 3.3.8.[9]

Exhibit 3.3.6 Procedure Flow Chart for Stores Requisition[a]

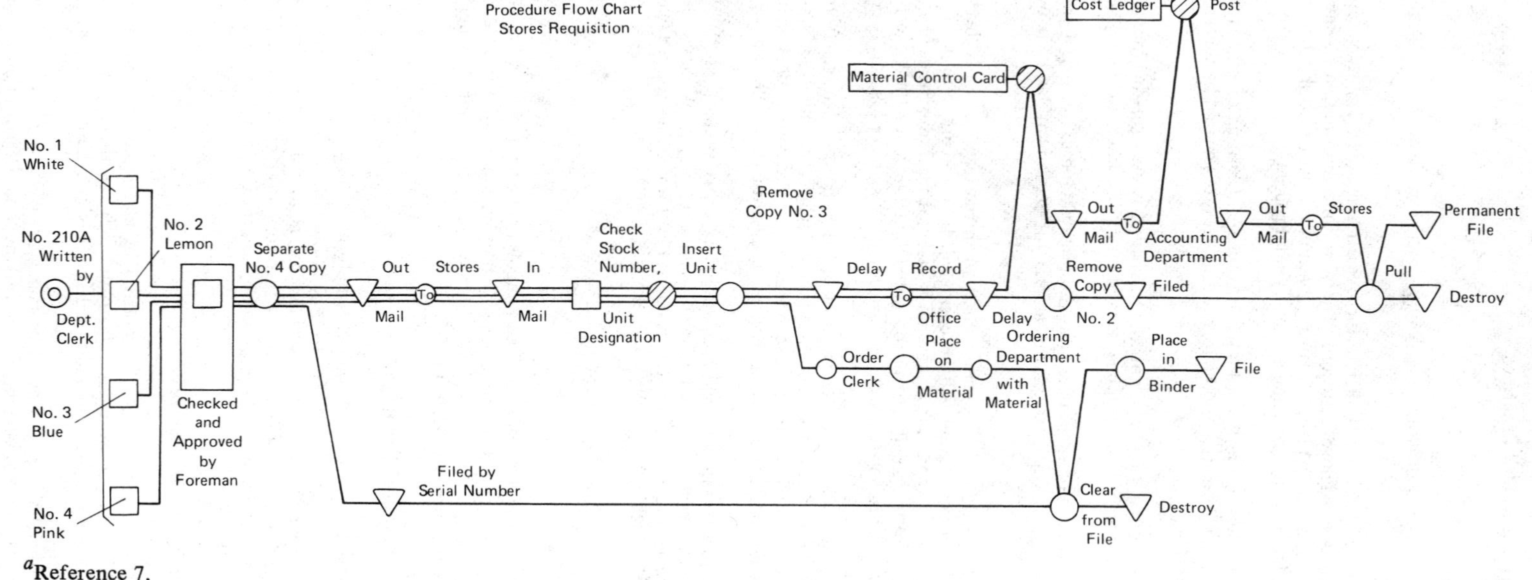

[a]Reference 7.

Exhibit 3.3.7 Flowcharting Template[a]

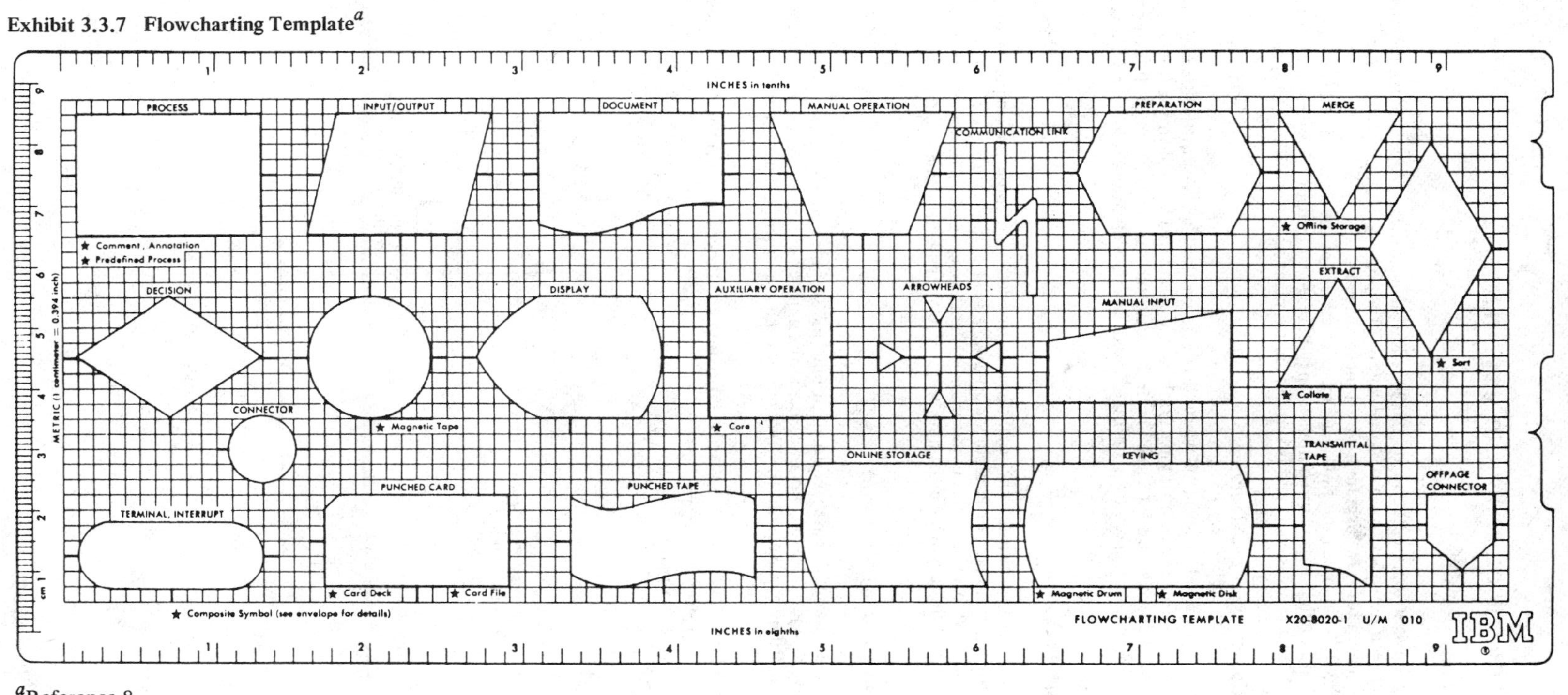

[a]Reference 8.

3.3.4 DIAGRAMS OF FLOW AND MOVEMENT

Flow Diagrams

A flow diagram is a representation of the location of activities or operations, and the flow of materials between activities, on a pictorial layout of a process. Usually used with a flow process chart.[5] A synonym for flow diagram is flow process diagram.

Exhibit 3.3.9[10] illustrates the flow diagram. The subject of the flow charted is usually a product or material, although a person or equipment may be selected instead. This diagram is often used to supplement a flow process chart and is a useful aid for plant layout work (see Chapter 10.2).

Charting Conventions and Practices

Each activity is specified in terms of where it occurs on the plan view or the drawing of the work area involved, sometimes accompanying symbols and numbers identical to the flow process chart. Small arrows are inserted on the flow line to indicate the direction of travel. In cases where the flows of several subjects are shown on the same diagram, a different color can be used for each subject.

String Diagrams

This diagram is used to measure the total distance traveled or to count the frequency of the movements of workers, materials, or equipment on a scale plan or model using a thread and pins. It is a special form of flow diagram and is used for analyzing the aggregate of flow patterns or movements of various subjects in a quick and simple way. A synonym for string diagram is trip frequency diagram.

Conventions and Practices

A thread is tied around pins inserted to all turning and stopping points on the path in the sequence of movement. By measuring the length used and the number of strings connected between adjacent pins, the total distance moved and the frequency of the movement are determined.

Exhibit 3.3.8 Off-line Clerical Procedure at a Processing Checkpoint[a]

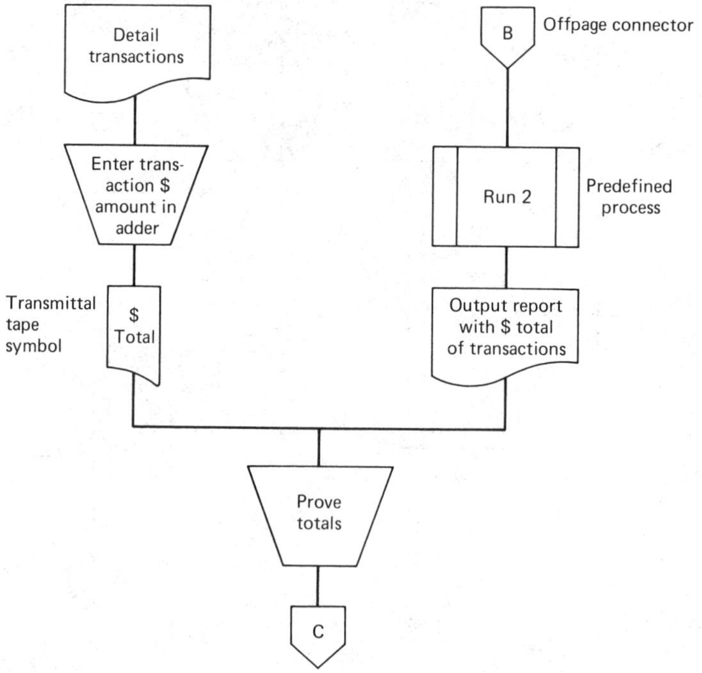

Detail transactions
B
Offpage connector
Enter transaction $ amount in adder
Run 2
Predefined process
Transmittal tape symbol
$ Total
Output report with $ total of transactions
Prove totals
C

[a]Reference 9.

Exhibit 3.3.9 Flow Diagram: Engine Stripping, Cleaning, and Degreasing[a]

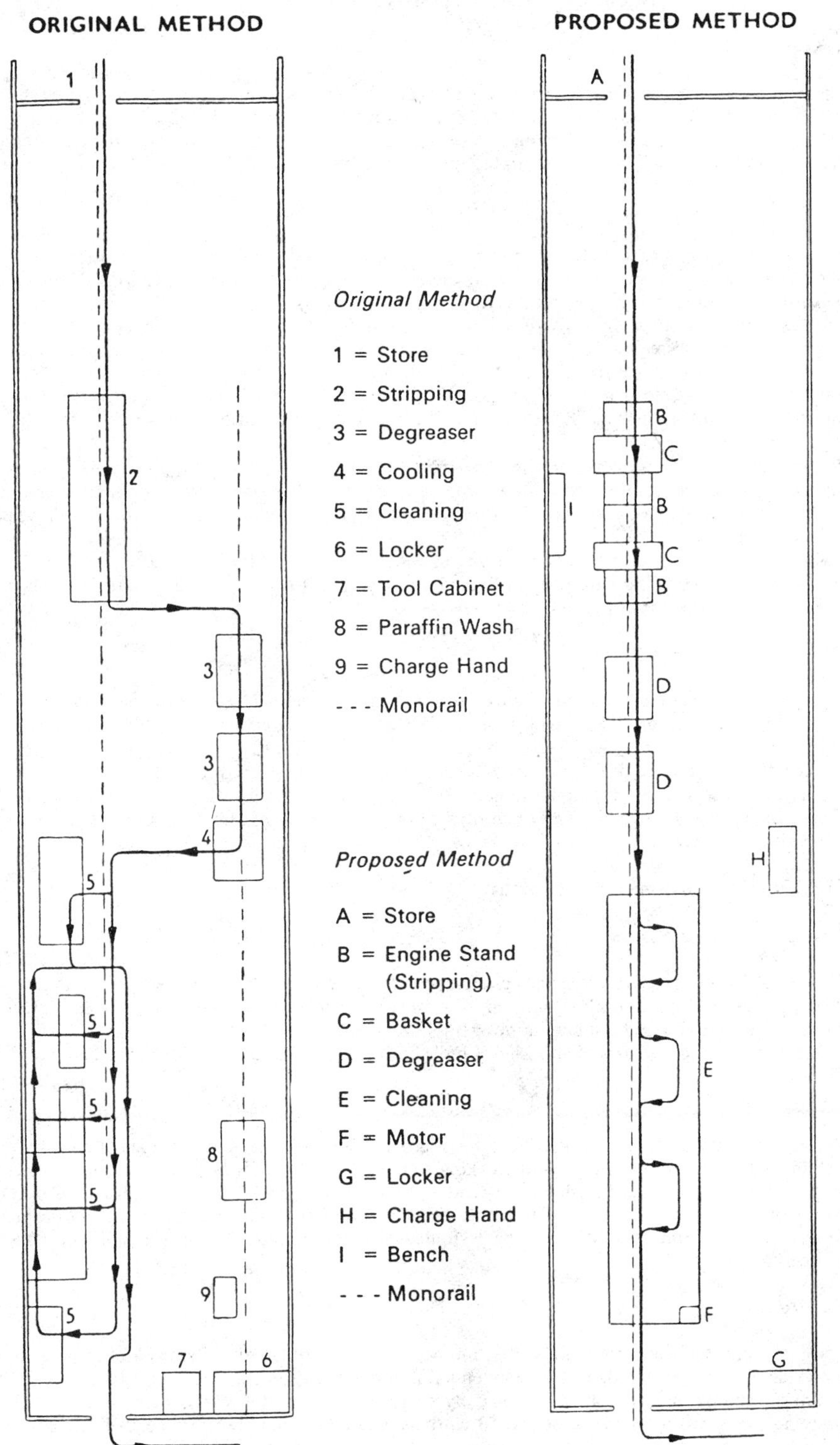

[a]Reference 10.

String diagrams are most useful for nonstandard work situations such as job shops, repair shops, warehouses, offices, and numerous activities in service industries.

Travel Charts

The chart presents, in a matrix form, quantitative data on movements taking place between any two work stations. The units are usually weight or quantity transported and the trip frequency.

The travel chart is a sort of tabular form of the string diagram. It is often used for material handling and layout work. A synonym for this chart is trip frequency chart.

3.3.5 MULTIPLE-ACTIVITY PROCESS CHARTS

A multiple-activity process chart is a chart of the coordinated synchronous or simultaneous activities of a work system of one or more machines and/or one or more persons. Each machine (and/or person) is shown in a separate, parallel column, indicating its activities as related to the rest of the work system. Examples are a multiman process chart, a Gantt chart, a multiman-machine process chart, a man-machine process chart, and a man-multimachine process chart.[5] Synonyms for this chart are multiple-activity charts and multiple-activity operation charts.

When a work system is made up of more than one subject, that is, persons and/or machines, the interrelationships of their activities have much effect on the utilization of manpower and machine capacities.

A multiple activity process chart shows graphically the simultaneity of these activities on a common time scale. In contrast to a flow process chart, a multiple-activity process chart is usually applied only to one work station at a time, except when jobs require a person to move from place to place cyclically.

When a person or persons work with one or more machines, this chart is often referred to as a "man-machine chart." In case of coordinated work crews, it is called a "gang chart." An assembly line may be regarded as a special case of crew work.

Advantages of the Charts

Although every charting technique is useful in selling and demonstrating new methods to people, the multiple-activity process chart is more so. With the interrelated activities of several persons and machines, it is difficult to demonstrate new methods to people without this chart.

Since the multiple-activity process chart clarifies unavoidable delay time for both workers and machines in the present methods, it helps us to improve the utilization of these resources. This chart can specify the overall cycle time of a work system with multiple activities, which becomes the basis of the standard time.

Charting Conventions and Practices

Forms

The blank form has separate vertical columns, one for each person or machine involved in the work system. Provided within each column are a vertical bar representing the types of steps (or breakdowns) of an activity and a subcolumn for brief description.

One of the simplest charts of this kind is in Exhibit 3.3.10,[11] a man-machine chart.

Codes

Various shadings or colors are used in a vertical bar to identify the types of steps of an activity. In Exhibit 3.3.10 these codes represent working time and idle time for both a person and a machine. M. E. Mundel suggests using the set of codes shown in Exhibit 3.3.11,[12] as advocated by D. B. Porter of New York University. The length of each code bar shows the time spent on the step of an activity, and simultaneity of steps is indicated by horizontal alignment with the common vertical time scale.

Recording

For analyzing the present methods, the time value may be obtained by stopwatch. One complete job cycle is to be recorded. It is recommended that one start the cycle at the beginning of the first operation among all activities performed by every member of the working crew under study. In studying the coordinated work of several workers, a number of analysts are needed to record their individual times.

Exhibit 3.3.10 Multiple-Activity Chart–Man and Machine: Finish Mill Casting (Original Method)[a]

MULTIPLE ACTIVITY CHART

			SUMMARY			
CHART No. *8*	SHEET No. *1*	OF *1*		PRESENT	PROPOSED	SAVING
PRODUCT: *B. 239 Casting*		DRAWING No. *B. 239/1*	CYCLE TIME	(min)		
			Man	*2.0*		
PROCESS: *Finish mill second face*			Machine	*2.0*		
			WORKING			
			Man	*1.2*		
			Machine	*0.8*		
MACHINE(S): *Cincinnati No. 4 vertical miller*	SPEED *80 r.p.m.*	FEED *15 in./min*	IDLE			
			Man	*0.8*		
			Machine	*1.2*		
			UTILISATION			
OPERATIVE:		CLOCK No. *1234*	Man	*60%*		
CHARTED BY:		DATE:	Machine	*40%*		

TIME (min)	MAN	MACHINE	TIME (min)
0.2	*Removes finished casting* *Cleans with compressed air*	*Idle*	0.2
0.4	*Gauges depth on surface plate*		0.4
0.6	*Breaks sharp edge with file* *Cleans with compressed air*		0.6
0.8	*Places in box* *Obtains new casting*		0.8
1.0	*Cleans machine with compressed air*		1.0
1.2	*Locates casting in fixture:* *starts machine and auto feed*		1.2
1.4	*Idle*	*Working* *Finish mill second face*	1.4
1.6			1.6
1.8			1.8
2.0			2.0
2.2			2.2
2.4			2.4
2.6			2.6
2.8			2.8
3.0			3.0
3.2			3.2
3.4			3.4
3.6			3.6
3.8			3.8

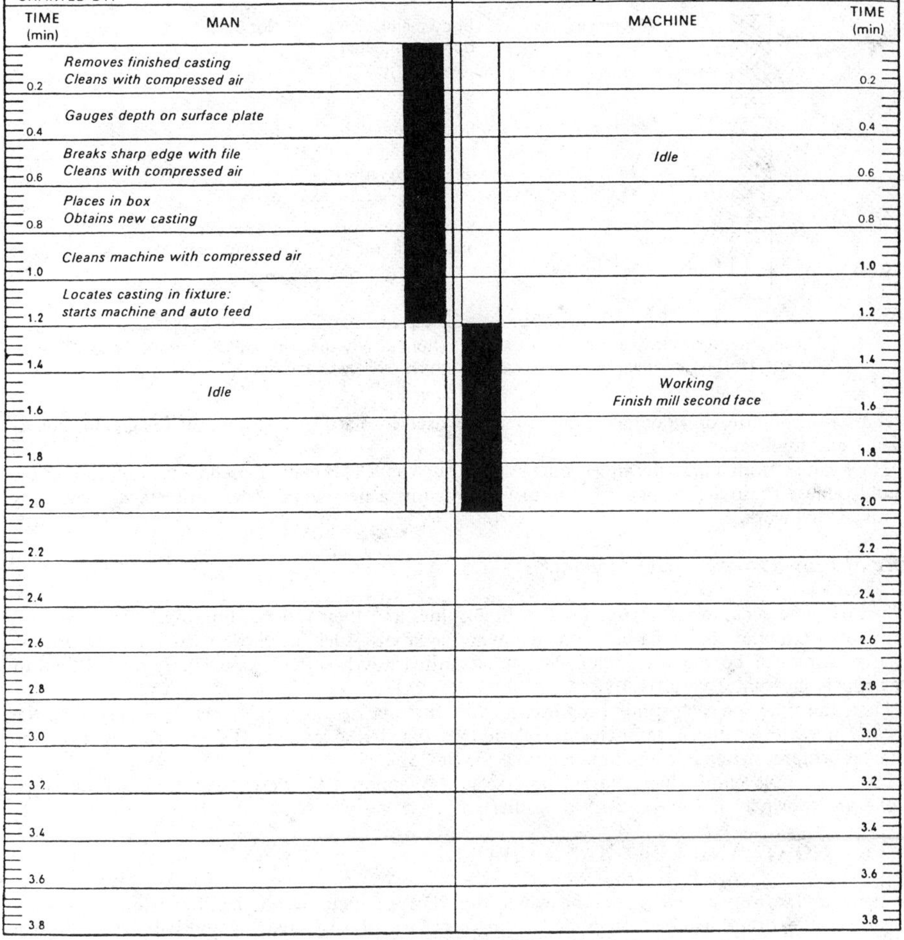

[a]Reference 11.

Exhibit 3-3.11 Codes for Multiple Activity Process Chart

Symbol[a]	*Name*	*With Man Activities Is Used to Represent*	*With Machine Activities Is Used to Represent*
(solid black)	Suboperation	Body member or operator doing something at one place.	Machine working ("on" time), machine paced.
(half black)	Suboperation	Not used.	Machine working ("on" time), operator paced.
(cross-hatched)	Movement	Body member or operator moving toward or with an object.	Not used.
(diagonal hatched)	Hold	Body member maintaining an object in a fixed position.	Not used.
(blank)	Delay	Body member or operator is idle.	Machine is idle (downtime).

[a] The amount of shading is chosen to suggest automatically the general usefulness of the step. The less shading, the probable greater undesirability of the step.

A motion picture or videotape camera may be used in place of a stopwatch for greater convenience and timing accuracy.

Time values from a predetermined time system, such as MTM, may be used instead of direct time study values. These are particularly useful for designing a new work system prior to the beginning of the actual work.

Use of Flow Process Chart Symbols

When extreme accuracy is not required in time values and their interrelationships, their graphical representation may be replaced with geometric symbols, such as those used for flow process charts. Time can be roughly obtained from an ordinary wristwatch. An example of this kind of charting is shown in Exhibit 3.3.12.[13]

First, the time for each symbol representing the first operator is roughly measured, and then the second operator's time is estimated in relation to the first operator. The vertical length of the chart is not proportional to the time required for the job.

Previous charts employing accurate time scales, as opposed to those using symbol approximations, are known as multiple-activity time charts.

3.3.6 RIGHT- AND LEFT-HAND CHARTS

A right- and left-hand chart is one on which the motions made by one hand in relation to those made by the other hand are recorded, using standard process chart symbols or basic therblig abbreviations or symbols.[5] Synonyms for this chart are two-handed process charts, operation charts, and operator process charts.

The chart records the repetitive manual work of an individual worker at a work station. It shows all activities of both hands and their relations with one another. Whenever necessary, the feet may also be charted.

The chart reveals any unnecessary idleness of the hands resulting from the unbalanced assignment of work to each hand. It is therefore a special version of the multiple-activity chart previously described.

Although any manual activity controlled by a person can be studied, this chart is recommended only for highly repetitive tasks.

Exhibit 3.3.12 Multiman and Machine Process Chart With Time Columns for Making Large Paper Bag Containers[a]

ORIGINAL — MULTI-MAN AND MACHINE PROCESS CHART — __OF__

OF MAKING LARGE PAPER BAG CONTAINERS

Date 11/20/50 Part P-3 PAPER Operator MM-EG Mach. ____

By S.F.C. No. 4521

Dist.	Sec.		FOLDER-1ST OPERATOR	Dist.	Sec.		GLUER-2ND OPERATOR	GLUE MACHINE	
3'	4	o	To Right Side of Paper		4	○	Remove Bag from Form	44	▽
	5	○	PU Edge	7'	5	⊖	To Press		
3'	2	⊖	To Center of Table		3	○	Raise Press with Foot		
	3	○	Fold Side Over Form		8	○	Place Bag Under Press		
3'	1	o	To Left Side						
	2	○	PU Edge						
3'	1	⊖	To Center of Table						
	5	○	Fold Side Over Form		2	○	Smooth Stack		
	7	○	Get Tape		2	○	Smooth While Lower Press		
					2	○	Smooth Side of Stack		
				6'	4	o	To Base Supply		
	3	○	Position Tape		9	○	PU Base		
	10	○	Glue on Tape	3'	5	⊖	To Glue Machine		
	9	○	Fold Up Bottom		5	▽	Other Side of Base-Glue 1st side	5	○
					3	○	Turn Base (Rotate)	3	▽
	6	○	Fold End		5	▽	2nd Side	5	○
	4	○	Get Base		3	○	Turn Base (Rotate)	3	▽
	6	○	Insert Base		5	▽	3rd Side	5	○
					3	○	Turn Base (Rotate)	3	▽
	6	○	Smooth Bottom Stop		5	▽	4th Side	5	○
	8	○	Smooth Sides		2	○	Turn Base Over	9	▽
				4'	2	⊖	To Table		
					3	○	Move Base on Table Over, Place New		
				3'	2	o	To 1st Operator		

	1ST	2ND	TOTAL	MACH.
○	20	15	35	4
o	2	2	4	–
⊖	2	3	5	–
▽	–	0	0	20
▽ (marked)	–	4	4	–
	24	24		

[a]Reference 13.

Charting Conventions and Practices

Symbols

The activity to be studied is sequentially divided into work elements performed by each body member. The symbols used to represent these steps are similar to those used for flow process charts.

○ **Operation.** An operation occurs when the hand is used for activities such as picking up, positioning, use, assembly, and so on.

⇨ **Transportation.** Transportation occurs when the hand is moved to or from the object.

D **Delay.** Delay occurs when the hand is idle or waiting for the other hand.

▽ **Hold.** Hold occurs when the hand holds the object in a fixed position in order to help the other hand's work.

Although the symbols are similar to those used for flow process charts, the meaning of each symbol is slightly different. For instance, the inspection done by a worker is classified as an "operation" in the right- and left-hand chart.

Occasionally, therblig symbols are used in place of process chart symbols in which case it is called a "therblig chart." In the same way, the predetermined time system, such as MTM, use similar analysis charts with their basic motion codes.

Forms and Chart Construction

Although charts may be drawn on blank paper, preprinted forms such as those shown in Exhibit 3.3.13[14] are more convenient. A symbol in every horizontal row for either hand is connected downward in sequence of occurrence. Although each horizontal row does not indicate the time taken by the step, the relationship of two hands is indicated by the horizontal level of the rows. The concurrent action of each hand is recorded on the same row.

Recording

Right- and left-hand charts are usually constructed from direct observation. Time values are taken with a stopwatch, but only for determining the precedence and concurrency of each step taken by both hands.

The beginning of a work cycle is usually the moment immediately after the disposal of the finished work. Charting starts with the hand that does the most work. The other hand is then charted on the appropriate rows in relation to the steps it takes. Only one hand is charted at a time.

3.3.7 OTHER CHARTS FOR MANUAL OPERATIONS

Operator Process Charts

This chart is regarded as a simplified version of the right- and left-hand chart, without the differentiation of activities between each hand. It is used to study the overall activity of repetitive manual work performed by an individual worker. A synonym for this chart is process chart.

The breakdown of operations is expressed sequentially using the same symbols as those used for right- and left-hand charts. Time values are usually not shown on the chart.

In general, the operator process chart is used for activities involving longer work cycle times.

Since the method of charting is similar to that used for flow process charts (person-type), with the exception of scope and magnitude of steps, the same form as that shown in Exhibit 3.3.2 is often used for operator process charts.

Simo Charts

This chart is similar to a right- and left-hand chart, with finer motion symbols plotted vertically against a time scale. Synonyms for simo charts are micromation analysis (or study) and micromotion data.

Therbligs are used in place of the process chart symbols for each element, and not only the two hands, but also the other body part details may be analyzed. These data may be transcribed from the micromotion analysis supported by movie film or videotape.

Only highly repetitive jobs with short work cycles economically warrant this kind of charting. An analysis chart of a predetermined time system such as MTM may be used for the same purpose instead of a simo chart (see Chapter 3.2).

3.3.8 NETWORK DIAGRAM

A network diagram is a graphic representation of the relationships of interdependent activities in a process or project. It is also called a critical path diagram, a PERT chart, an activity network, or a link diagram.

The process or project under scrutiny usually consists of a large number of activities. In general, the objective is to optimize the entire time span subject to the limited availability of resources, such as manpower or money.

The most popular diagram of this type, PERT, involves complex procedures with electronic data processing. This network is usually used for large projects involving hundreds or thousands of components.

Exhibit 3.3.13 Two-Handed Process Chart: Cutting Glass Tubes (Original Method)[a]

TWO-HANDED PROCESS CHART	
CHART No. *1* SHEET No. 1 OF *1,*	WORKPLACE LAYOUT
DRAWING AND PART: *Glass tube 3 mm dia., 1 metre original length*	ORIGINAL METHOD
OPERATION: *Cut to lengths of 1.5 cm*	GLASS TUBE — POSITION FOR MARK — JIG
LOCATION: *General shop*	
OPERATIVE:	
CHARTED BY: DATE:	

LEFT-HAND DESCRIPTION	RIGHT-HAND DESCRIPTION
Holds tube	*Picks up file*
To jig	*Holds file*
Inserts tube to jig	*File to tube*
Presses to end	*Holds file*
Holds tube	*Notches tube with file*
Withdraws tube slightly	*Holds file*
Rotates tube 120°/180°	*Holds file*
Pushes to end jig	*Moves file to tube*
Holds tube	*Notches tube*
Withdraws tube	*Places file on table*
Moves tube to R.H.	*Moves to tube*
Bends tube to break	*Bends tube*
Holds tube	*Releases cut piece*
Changes grasp on tube	*To file*

METHOD	PRESENT		PROPOSED	
	L. H.	R. H.	L. H.	R. H.
Operations	*8*	*5*		
Transports	*2*	*5*		
Delays	–	–		
Holds	*4*	*4*		
Inspections	–	–		
Totals	*14*	*14*		

[a]Reference 14.

Applications to Methods Study

Typical applications of network diagrams are scheduling problems and progress control of complex projects, such as new plant construction or marketing of a new product. Even the simple manual-type network diagram described here, however, is also useful and economical, particularly for methods study. Applications include maintenance overhaul of large facilities, setup operation of rolling mill stands, and heart surgery.

Although the interdependency of activities is much more complicated, and the duration longer, the network diagram may be regarded as a special type of regular multiple-activity chart.

Charting Conventions and Practices

Chart Construction

A simple network diagram is shown in Exhibit 3.3.14. It expresses *activities* (represented by lines) and *events* (represented by circles, and occasionally by squares).

Events are positions or statuses in the course of completion, including the origin and the termination of the project. A square may be used in place of a circle to differentiate important events, as shown in the exhibit. Each event is numbered according to its precedence.

Activities, identified by lines, are operations or groups of operations in a department that have to be performed in order to obtain the status of an event from the preceding event. The direction of the arrow indicates the precedence relationship between two events. The expected time (minutes, hours, days, weeks) for performing each activity is entered under the activity identification. In Exhibit 3.3.14 activities A, B, and C are 5, 5, and 10 days, respectively.

In the PERT system three kinds of time estimates–that is, optimistic, the most likely, and pessimistic–are given for each activity in order to calculate a probability distribution of the time values.

To represent a correct sequence, an activity with zero time or no cost may be charted with a dashed line. This is referred to as a "dummy activity" (activity D in Exhibit 3.3.14).

Critical Path and Slack Time

The longest path of the network is called the critical path," which determines the minimum project time. In Exhibit 3.3.14 the critical path from event 1 to event 8 is indicated by heavy lines (that is, C-G-I-K), and its value is 27 days. There can be more than one critical path.

Activities that do not lie on the critical path have a certain flexibility in time, they can be delayed without affecting completion of the total project. This leeway is referred to as "slack time." It does not appear under each activity, but is indicated in parentheses only under the final activity of the noncritical path, as shown in Exhibit 3.3.14.

For example, the critical path from event 1 to event 5 is the pathway C-G. The slack time of the adjoining noncritical pathway B-F is 8 weeks (16 – 8).

Exhibit 3.3.14 Simple Network Diagram

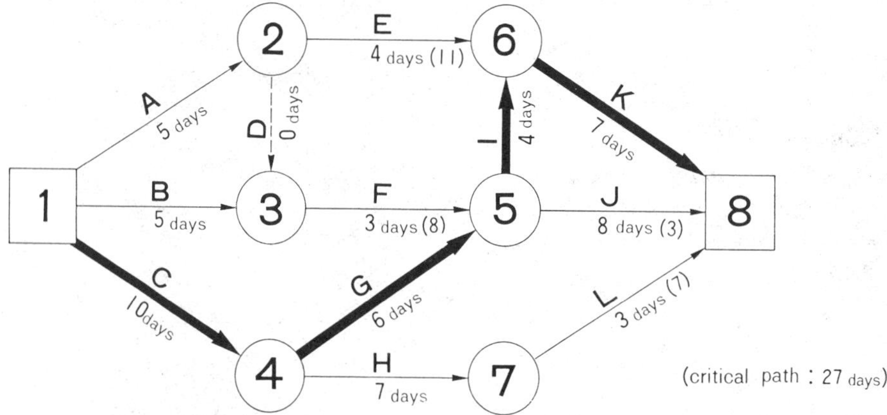

Use of Critical Path

When the objective is to reduce the completion time of the project, it is better to concentrate efforts on activities that lie on the critical path. Expending effort on other pathways is useless. "Excessive" reduction of time for the critical path over the duration of noncritical paths is also wasteful.

In Exhibit 3.3.14 the critical path from event 1 to event 5 (16 days) may be reduced by as much as 8 days, that is, the slack time of noncritical pathway B-F, but by no more, as long as the improvement is confined to pathway C-G.

A noncritical path is likely to result in the excessive allocation of resources such as manpower or money. Since some delays are permitted without extending the whole project time, saving the excess resources may be realized by trading off this slack time.

The questionnaire technique for improvement discussed later is also applicable to network diagrams.

3.3.9 APPLICATION OF CHARTING TECHNIQUES

Selection of Techniques

What kind of chart to use for descriptive and communicative aid does not present a problem because the need itself is direct and specific. However, it can be difficult to determine what chart to use as an aid in analyzing methods because the problem to be solved or method of approach is obscure at the beginning of the study. Here are some hints on selecting the appropriate charting for particular studies.

As described previously, charting techniques are classified into three categories; these are represented by flow process charts, multiple-activity process charts, and flow diagrams.

Each process or activity consists of a series of constituents, or elements. The present method is merely one series of elements among many other possibilities. To disclose fictitious restrictions in the chronological sequence of elements in the present process or operations, flow process charts are convenient and useful. Combining and rearranging the constituents (i.e., activities) of the present method are easily contemplated with these charts, to say nothing of the improvement of the constituents themselves. This provides opportunities to develop new ideas. The same applies to the other techniques in this category, such as form process charts or even operator process charts.

Multiple-activity process charts, which constitute another charting category, help to disclose the time difference between simultaneous activities. Whenever a work station or an operation consists of more than two subjects, whether person or machine, and other body members, multiple-activity charts are always useful for realizing a better balancing between subjects (i.e., person, machine, and hand).

The third category of charting, using the flow diagrams, records information supplementary to flow process charts or the analysis of materials handling and physical flow in the work area.

Level of Breakdown

For purposes of analysis, a process is sequentially broken down into subdivisions or steps expressed by five symbols: operation, transportation, inspection, delay, and storage. This becomes a flow process chart. One of these subdivisions or steps may be further divided into operations, making it an operator process chart. Each manual operation is again divided into the activities of each hand in a right- and left-hand chart. Further breakdown takes place until the individual therbligs (therblig charts) or other basic motion codes with finer time values (simo charts) are realized.

Similar breakdown occurs within the category of multiple-activity charts; that is, multiple-activity charts to right- and left-hand charts to therblig charts to simo charts.

Determining the level of breakdown needs some consideration. In general, the coarser the breakdown, the more significant the savings that can be realized in the case of improvement. For instance, to "eliminate" a whole operation in the flow process chart is much more profitable than to "eliminate" a therblig in the simo chart. The finer breakdown of activities also takes too much time in analysis, making it difficult for analysts to focus attention on the specific elements, because the flood of detailed data has to be dealt with. By the same token, long work cycle jobs have to be broken down into coarse steps. For this purpose, operator process charts are more suitable than right- and left-hand charts, and in some cases grosser techniques, such as person-type flow process charts, may be even more useful.

The production volume or the life span of the activity to be studied should be related to the level of the breakdown of the job. Detailed analysis, such as a simo chart, is justified only for large-quantity production or lasting activities.

The grosser techniques, however, may not always be preferable. Improvement on the grosser steps, such as an activity of a symbol in flow process charts, sometimes requires a large investment or technological innovation that is not always possible or permissible. On the other hand, any improvement of the elements in a right- and left-hand chart usually requires little investment, and it is also rather easy to develop new ideas for elements in the activity.

In short, grosser techniques, such as flow process charts or man-machine charts, are applied first, before conducting more detailed analyses, such as right- and left-hand charts, and not vice versa.

Defining Specific Objectives

Before analyzing the present methods, the objectives of the study must be specified. Improvement alone is too broad and vague. Improving what? Labor productivity? Output per hour? Project duration? Throughput time, or what else? To try to improve everything is likely to result in improving nothing or, at best, in the ineffective use of analysts' time with minimal gain. The technique used, the method of recording, how to analyze, the evaluation of alternatives, and the like are closely related to the objectives of the study.

The objectives in the flow process analysis may be manpower saving, the reduction of throughput time, or others. Even the use of network charts is not always for reducing project duration, but also for saving manpower in a noncritical pathway.

Multiple-activity charts may be used for improving machine utilization or increasing the rate of production instead of balancing work or saving labor. The approach and solutions may differ somewhat in each case.

Importance of Using "Normalized" Facts

Most textbooks on methods engineering emphasize the importance of getting the real facts from actual observation when charting the present method. That is, chart what *is actually happening* from direct observations, and not what the analyst thinks *should be happening*.

It is quite correct to caution the analyst not to draw imaginary charts based on memory without any on-the-spot investigation. This does not mean, however, that charting of the present method should be likened to taking candid photos, that is, recording indiscriminately whatever the analyst happens to witness. The facts used for charting are not the "raw" facts on the shop floor.

The analyst should be concerned with the methods themselves, not with discipline or the training of specific employees. Methods distorted by personal caprice, carelessness, or unskillfulness of the workers being observed have to be correct in charting according to the analyst's judgment.

Smooth production can be hindered by many interruptions and tardiness beyond the control of the workers that occur in the manufacturing process during the course of a working day. These include machine breakdowns, defective materials, and shortage of work. These irregular and ancillary problems have to be deleted in charting the present method if they are contingent and unusual incidents not occurring regularly.

Since the proposed method will be illustrated in the same manner, comparison between the present and the proposed method becomes logical and meaningful in the charts.

In conclusion, the present method has to be charted from actual observation, not from memory or the imagination. The observed method, however, has to be "normalized" by adjustment and correction made at the analyst's discretion in order to represent a true picture of the present *method.*

3.3.10 BASIC QUESTIONNAIRE TECHNIQUE

Questioning Attitude

The charting techniques previously mentioned are merely means of recording. The use of charting in relation to "examine" and "develop" stages in methods study described at the beginning of the chapter is discussed here.

Habitual experience and familiarity with the work are likely to make individuals blind to any possibilities for improvement. One of the best ways to avoid this pitfall is to inspire the ability to inquire. The questionnaire technique is the systematic application of different questions concerning the job in order to search for better ideas. This questioning attitude, a kind of organized creative thinking, is tied to the four general approaches, that is, eliminate, combine, rearrange, and simplify, as shown in Exhibit 3.3.15. This technique applies to each step of any of the charting methods described so far.

Exhibit 3-3.15 Six Questions Tied to Four Approaches

Primary Questions	Secondary Questions	Approaches	
What is the purpose?	*Why* is it necessary?	Eliminate	
Where is it done?	*Why* is it done there?	Combine and Rearrange	The places
When is it done?	*Why* is it done then?	Combine and Rearrange	The sequences
Who does it?	*Why* does this person do it?	Combine and Rearrange	The persons
How is it done?	*Why* is it done this way?	Simplify	

Procedures

Every step of the process or activity is challenged by asking five questions: what, where, when, who, and how. The "why" question is incorporated into all five. The analyst must concentrate fully on each question in the following sequence:

(Purpose)	*What* is the purpose? *Why*? Why should it be done? What would happen if it weren't done? What other methods would accomplish the same result? What should be finally done?
(Place)	*Where* is it done? *Why*? Are there other places where it might be performed better? Where should it be finally done?
(Sequence, Time)	*When* is it done? *Why*? Could it be done in a different sequence or at another time more economically? When should it be finally done?
(Person)	*Who* does it? *Why*? Who else could do it more efficiently? Who should finally do it?
(Means)	*How* is it done? *Why*? Are there any other safer and more profitable ways of doing it? How should it be finally done?

As shown in Exhibit 3.3.15, such systematic questioning leads to the following general approaches, and every step of the process or activity in any type of chart is scrutinized for these improvement possibilities:

1. *Eliminate* all unnecessary operations or elements: It is not unusual to find unnecessary operations being performed because of lack of communication or sheer habit. In general, no preparation or investment in this type of improvement is needed. It is the best improvement that can be made among these four actions, and the most important.
2. *Combine* operations or elements: Different operations performed at different locations by more than two persons may be assigned to one person or to the same workbench.
3. *Rearrange* operations or elements: The most probable improvement in this approach is to change the sequence of operations or elements. Change of work area or person should also be examined.
4. *Simplify* the necessary operations or elements: After examining, without success, the improvement possibilities previously described, methods of simplifying and improving the individual operations or elements themselves are looked into. This approach seeks not only to make manual operations more effective, but also to improve the effective use of tools and equipment.

Since these inquiring approaches are arranged to follow a descending order of importance in terms of the magnitude of probable improvement, they should be used in the same order.

Questionnaire Practices

The more alternatives, the better. Since the first new idea thought of is not necessarily the best, one better idea is not enough. Analysts have to create as many alternative ideas as possible.

Operations or elements may be divided into three types: do, make ready, and put away. "Do" operations are those that involve any change in the shape or chemical and physical condition of the product. Conduct the "do" operation first.

"Make ready" operations are preparing for "do," such as set up or load machine, and "put-away" operations are those during which the work is moved from the machine or workplace, such as cleaning up, tossing aside, or replacing supplies.

All the "do" operations or elements have to be examined first, since all others depend on "do" operations.

3.3.11 SIMPLE EXAMPLES

Case 1: Engine Stripping, Cleaning, and Degreasing[15]

The process is illustrated in the flow process chart (Exhibit 3.3.1) and the flow diagram (Exhibit 3.3.9), previously described. It was done in a bus engine workshop, where an engine was stripped, degreased, and cleaned for inspection.

Examination of the flow process chart shows a very high proportion of "nonproductive" activities. There are, in fact, only 4 operations and 1 inspection, whereas there were 21 transportations and 3 delays. Out of 29 activities, excluding the original storage, only 5 can be considered "productive."

Let us apply the questionnaire technique to the first transportations:

Q. What is the purpose?

A. The engine is carried part of the way through the store by one electric crane, is placed on the ground, and is then picked up by another, which transports it to the stripping bay.

Q. Why should it be done?

A. Because the engines are stored in such a way that they cannot be directly picked up by the monorail crane that runs through the stores and degreasing shop.

Q. What other methods would accomplish the same result?

A. The engines could be stored so that they are immediately accessible to the monorail crane, which could then pick them up and run directly to the stripping bay.

Q. What should be done?

A. The preceding suggestion should be adopted.

The flow process chart and the flow diagram of the improved method are shown in Exhibit 3.3.16[16] and Exhibit 3.3.9, respectively.

Case 2: Finish Mill Casting on a Vertical Miller[17]

The operation is the finish milling of one face of a cast-iron casting. The opposite face is used for locating it in the fixture. The original method is shown in Exhibit 3.3.10. When examined critically, this chart (original method) shows that the machine remains idle for nearly three quarters of the operating cycle. This is because the operator carries out all activities with the machine stopped, but remains idle while the machine is running on an automatic feed.

Exhibit 3.3.17[18] shows the improved method of operation. It will be seen that gauging, deburring the edges of the machined face, placing the casting in the box of finished work, picking up an unmachined casting, and placing it on a worktable so that it is ready to locate in the fixture are now all done while the machine is running.

A slight gain in time has been made by placing the boxes with the finished work and with the work to be done next to one another, so that one casting can be put away at the same time at which the new one is lifted from its box. The cleaning of the machined casting with compressed air has been deferred until after the sharp edges have been broken down, thus saving an extra operation.

The result is a saving of 0.64 min on 2 min, or a gain of 32% in the productivity of the milling machine and operator without any capital expenditure.

3.3.12 CHECKLISTS

Application of the questionnaire techniques is rather complex in right- and left-hand charts and man-machine charts. M. E. Mundel suggests the following checklists for these cases.

Checklist for Right- and Left-handed Chart*

Basic Principles

A. Reduce total steps to a minimum.
B. Arrange in best order.
C. Combine steps where feasible.

*Operation chart" here is a synonym of right- and left-hand chart.

Exhibit 3.3.16 Material-type Flow Process Chart: Engine Stripping, Cleaning, and Degreasing (Improved Method)[a]

FLOW PROCESS CHART			MATERIAL TYPE			
CHART No. *2*	SHEET No. *1*	OF *1*	SUMMARY			
Subject charted: *Used bus engines*			ACTIVITY	PRESENT	PROPOSED	SAVING
ACTIVITY: *Stripping, degreasing and cleaning prior to inspection*			OPERATION ○	*4*	*3*	*1*
			TRANSPORT ⇨	*21*	*15*	*6*
			DELAY D	*3*	*2*	*1*
			INSPECTION □	*1*	–	*1*
			STORAGE ▽	*1*	*1*	*1*
METHOD: PROPOSED			DISTANCE (m)	*237.5*	*150.0*	*87.5*
LOCATION: *Degreasing shop*			TIME *(man-min)*	—	—	—
OPERATIVE(S):	CLOCK Nos. *1234* *571*		COST LABOUR MATERIAL			
CHARTED BY: APPROVED BY:	DATE:		TOTAL	—	—	—

DESCRIPTION	QTY.	DISTANCE (m)	TIME (min)	○	⇨	D	□	▽	REMARKS
Stored in old-engine store		—	—					•	
Engine picked up					•				*Electric hoist on mono-rail*
Transported to stripping bay		*55*			•				
Unloaded on to engine stand					•				
Engine stripped				•					
Transported to degreaser basket		*1*			•				*By hand*
Loaded into basket					•				*Hoist*
Transported to degreaser		*1.5*			•				,,
Unloaded into degreaser					•				,,
Degreased				•					
Unloaded from degreaser					•				,,
Transported from degreaser		*4.5*			•				,,
Unloaded to ground					•				
Allowed to cool						•			
Transported to cleaning benches		*6*			•				,,
All parts cleaned				•					
All parts collected in special trays		*6*			•				
Awaiting transport						•			
Trays and cylinder block loaded on trolley					•				
Transported to engine inspection section		*76*			•				*Trolley*
Trays slid on to inspection benches and blocks on to platform					•				
TOTAL		*150*		*3*	*15*	*2*	—	*1*	

[a]Reference 16.

Exhibit 3.3.17 Multiple Activity Chart–Man and Machine: Finish Mill Casting (Improved Method)[a]

MULTIPLE ACTIVITY CHART

CHART No. *9* SHEET No. *1* OF *1*

PRODUCT *B 239 Casting* DRAWING No. *B.239/1*

PROCESS *Finish mill second face*

MACHINE(S) *Cincinnati No. 4 vertical miller* SPEED *80 r.p.m.* FEED *15 in./min*

OPERATIVE: CLOCK No. *1234*

CHARTED BY: DATE:

SUMMARY	PRESENT	PROPOSED	SAVING
CYCLE TIME	(min)		
Man	2.0	1.36	0.64
Machine	2.0	1.36	0.64
WORKING			
Man	1.2	1.12	0.08
Machine	0.8	0.8	—
IDLE			
Man	0.8	0.24	0.56
Machine	1.2	0.56	0.64
UTILISATION			Gain
Man	60%	83%	23%
Machine	40%	59%	19%

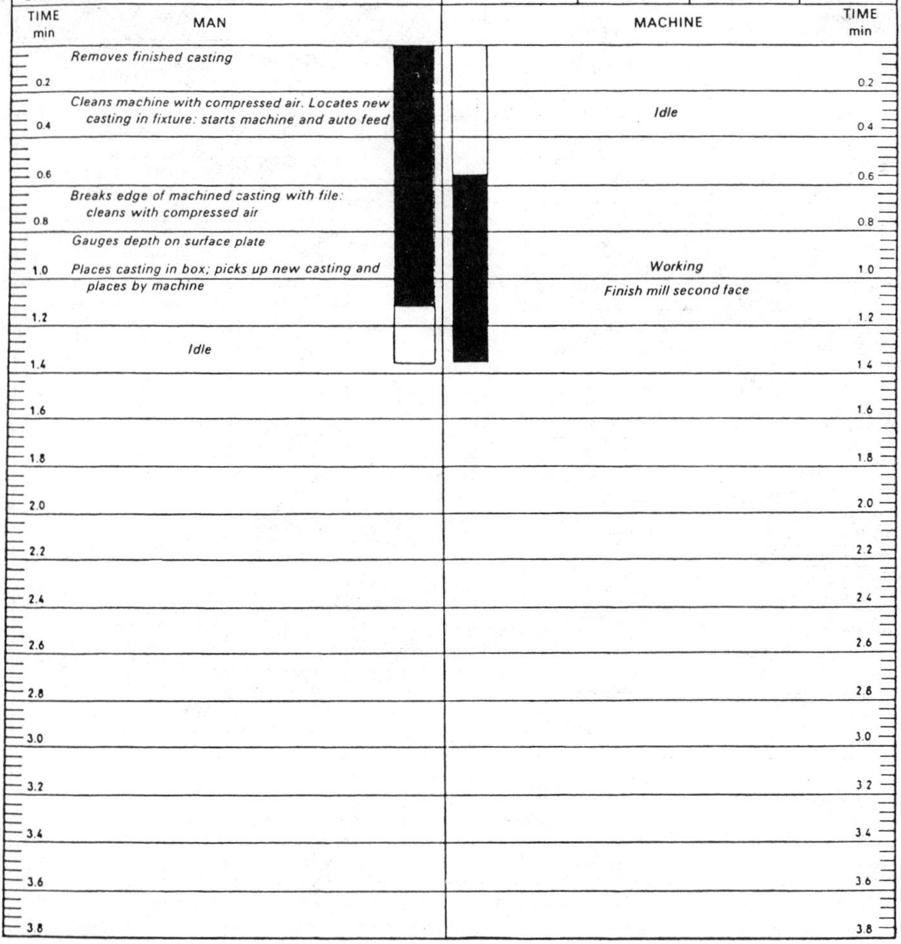

TIME min	MAN	MACHINE	TIME min
0–0.2	*Removes finished casting*	*Idle* (0–0.56)	
0.2–0.56	*Cleans machine with compressed air. Locates new casting in fixture: starts machine and auto feed*		
0.56–0.8	*Breaks edge of machined casting with file: cleans with compressed air*	*Working* *Finish mill second face* (0.56–1.36)	
0.8–0.96	*Gauges depth on surface plate*		
0.96–1.12	*Places casting in box; picks up new casting and places by machine*		
1.12–1.36	*Idle*		

[a]Reference 18.

D. Make each step as easy as possible.
E. Balance the work of the hands.
F. Avoid the use of the hands for holding.
G. The workplace should fit human dimensions.

1. Can a suboperation be eliminated?
   a. As unnecessary?
   b. By a change in the order of work?
   c. By a change of tools or equipment?
   d. By a change of layout of the workplace?
   e. By combining tools?
   f. By a slight change of material?
   g. By a slight change in product?
   h. By a quick-acting clamp on jig, if jigs are used?

2. Can a movement be eliminated?
   a. As unnecessary?
   b. By a change in the order of work?
   c. By combining tools?
   d. By a change of tools or equipment?
   e. By a drop disposal of finished material? (The less exact the release requirements, the faster the release.)

3. Can a hold be eliminated? (Holding is extremely fatiguing.)
   a. As unnecessary?
   b. By a simple holding device or fixture?

4. Can a delay be eliminated or shortened?
   a. As unnecessary?
   b. By a change in the work that each body member does?
   c. By balancing the work between the body members?
   d. By working simultaneously on two items? (Slightly less than double production is possible with the typical person.)
   e. By alternating the work, each hand doing the same job, but out of phase?

5. Can a suboperation be made easier?
   a. By better tools? (Handles should allow maximum flesh contact without sharp corners for power; easy spin, small diameter for speed on light work.)
   b. By changing leverages?
   c. By changing positions of controls or tools? (Put into normal work area.)
   d. By better material containers? (Bins that permit a slide grasp of small parts are preferable to bins that must be dipped into.)
   e. By using inertia where possible?
   f. By lessening visual requirements?
   g. By better workplace heights? (Keep workplace height below the elbow.)

6. Can a movement be made easier?
   a. By a change of layout, shortening distances? (Place tools and equipment as near place of use and as nearly in position of use as possible.)
   b. By changing direction of movements? (Optimum angle of workplace for light knobs, key switches, and handwheels is probably 30° and certainly between 0° and 45° to plane perpendicular to plane of front of operator's body.)
   c. By using different muscles? (Use the first muscle group in this list that is strong enough for the task.)
      (1) Finger? (Not desirable for steady load or highly repetitive motions.)
      (2) Wrist?
      (3) Forearm?
      (4) Upper arm?
      (5) Trunk? (For heavy loads, shift to large leg muscles.)
   d. By making movements continuous rather than jerky?

7. Can a hold be made easier?
   a. By shortening its duration?
   b. By using stronger muscle groups, such as the legs, with foot-operated vises?

Checklist for Man–Machine Charts[20]

Basic Principles

A. Eliminate steps.
B. Combine steps.
C. Rearrange in best fashion.
D. Make each step as easy as possible.
E. Raise percentage of cycle of machine running time to maximum.
F. Reduce machine loading and unloading to minimum.
G. Raise machine speed to economic limit.

(The first seven questions that follow are similar to those used with right- and left-handed charts, where more detail was given; hence, the reader is also referred to them. The bare questions are given here so as to provide, at one place, all the checklist items to be used.)

1. Can a suboperation be eliminated?
   a. As unnecessary?
   b. By a change in the order of work?
   c. By a change of tools or equipment?
   d. By a change in layout of the workplace?
   e. By combining tools?
   f. By a slight change of material?
   g. By a slight change in product?
   h. By a quick-acting clamp on the jigs or fixtures?

2. Can a movement be eliminated?
   a. As unnecessary?
   b. By a change in the order of work?
   c. By combining tools?
   d. By a change of tools or equipment?
   e. By a drop disposal of finished material?

3. Can a hold be eliminated? (Holding is extremely fatiguing.)
   a. As unnecessary?
   b. By a simple holding device or fixture?

4. Can a delay be eliminated or shortened?
   a. As unnecessary?
   b. By a change in the work each body member does?
   c. By balancing the work between the body members?
   d. By working simultaneously on two items?
   e. By alternating the work, each hand doing the same job, but out of phase?

5. Can a suboperation be made easier?
   a. By better tools?
   b. By changing leverages?
   c. By changing positions of controls or tools?
   d. By better material containers?
   e. By using inertia where possible?
   f. By lessening visual requirements?
   g. By better workplace heights?

6. Can a movement be made easier?
   a. By a change of layout, shortening distances?
   b. By changing the direction of movements?
   c. By using different muscles? Use the first muscle group in this list that is strong enough for the task:
      (1) Finger.
      (2) Wrist.
      (3) Forearm.
      (4) Upper arm.
      (5) Trunk.
   d. By making movements continuous rather than jerky?

7. Can a hold be made easier?
   a. By shortening its duration?
   b. By using stronger muscle groups, such as the legs, with foot-operated vises?

8. Can the cycle be rearranged so that more of the handwork can be done during running time?
   a. By automatic feed?
   b. By automatic supply of material?
   c. By change of man and machine phase relationship?
   d. By automatic power cutoff at completion of cut or in case of tool or material failure?

9. Can the machine time be shortened?
   a. By better tools?
   b. By combined tools?
   c. By higher feeds or speeds?

3.3.13 GROUP CREATIVITY AND ITS APPLICATIONS TO INDUSTRY

Brainstorming

The group approach is gaining popularity in generating creative ideas, and the most popular method is brainstorming. It is a conference technique by which a group attempts to create a solution for a given problem. The volume and variety of ideas generated by the members attending are greater than would be gathered from the same number of persons working by themselves. This technique is often used together with the questionnaire technique in analyzing and improving a present method (see Chapter 1.5).

Deliberate Change

The Procter and Gamble Company conducts a companywide methods change program. The savings per member of management was $400 in 1946, the first year of the program, and $43,000 in 1977–1978. The rate of return has been about 1000 times.

One of the unique approaches this company developed was the principle of "deliberate change."[21] Deliberate change is quite different from improvement. Improvement means performing a method more effectively; change means developing and using a new method. Even though an operation is carried out perfectly, there is still a potential for savings by making a deliberate change.

The following principles form the basis of the deliberate change approach to profit improvement:

1. Perfection is no barrier to change.
2. Every dollar of cost must contribute its fair share of the profits.
3. The savings potential is the full existing cost.
4. Never consider any item of cost necessary.

Elimination Approach

The Procter and Gamble Company also has a formal procedure that it calls the "elimination approach."[22] Although the company is constantly improving methods and simplyfing work, it believes that the ideal solution is to eliminate the cost. Its approach to cost elimination is as follows:

1. Select the cost for questioning.
2. Identify the basic cause.
3. Question the basic cause for elimination.
   a. Discard the basic cause.
   b. Apply "why?" questioning.

Improvement in Design Approach

A methods design procedure for improving multiple activities or production lines, called *organized design for line and crew systems* (ORDLIX),[23] is well known in Japan because of its successful applications. It is not only a method of balancing time, but also a method that helps to develop new ideas by stimulating creative thinking of analysts.

Devices to remove psychological strain and obstacles to creativity are incorporated into a step-by-step procedure, such as the following:

1. *Inequitable division of study effort according to priority.* It is not effective to make an all-around study or to dig into trifling details.

2. *The simpler the problem, the higher the creativity.* Only major elements are used to develop the skeleton of the new method. This avoids troublesome details and makes the analyst focus attention on only the essential parts of the method's framework.

3. *Better ideas can be squeezed.* Setting a target for the analysts makes them exert more effort to achieve the target. Adding minor secondary elements to the tentatively balanced method framework forces analysts to improve bottleneck activity.

3.3.14 LIMITATIONS OF CHARTING TECHNIQUES

The charting techniques described in this chapter are a means of analyzing the present method, which is merely one of many alternative solutions to a problem (or an objective). The current solution, that is, the present method, sometimes prevents the analyst from generating entirely different or superior solutions to the problems (or objectives), because the analyst is too familiar with the existing solution to be free from it in order to create new ideas.

The design process is replacing the traditional analytical approach in methods improvement work. It seeks a new method of transforming the original state of affairs (i.e., input or starting point) into another one (i.e., output, objective, or result), which is irrelevant to the existing method. In this approach, analysis is attempted, and the problem (i.e., objective) itself is analyzed rather than a solution (i.e., present method), as is the case in traditional analysis.

Even in this design process, however, these charts are a useful and indispensable means of analyzing the problem, especially in gathering facts.[24] In the idea generation stage of the design process, the creative thinking or questionnaire technique previously discussed are used in almost the same way as in analytical methods improvement.

REFERENCES

1. AMERICAN NATIONAL STANDARD, Industrial Engineering Terminology, *Work Measurement and Methods* (Z94.12-1972), American Society of Mechanical Engineers, New York, 1973.
2. INTERNATIONAL LABOUR OFFICE (ILO), *Introduction to Work Study*, 3rd ed., ILO Publications, Geneva, Switzerland, 1978, p. 98.
3. ILO, *Work Study*, p. 135.
4. S. JOHNSON and G. OGILVIE, *Work Analysis*, The Butterworth Group, London, 1972, p. 34.
5. ASME, ASME Standard, *Operation and Flow Process Charts* (ANSI Y15.3-1974), New York, 1972.
6. G. C. CLOSE, JR., *Work Improvement*, Wiley, New York, 1960, pp. 140–141.
7. CLOSE, *Work Improvement*, pp. 144–145.
8. IBM, *Flowcharting Techniques* (C20-8152-1), IBM Technical Publications Department, White Plains, NY, 1969, p. 7.
9. IBM, *Flowcharting Techniques*, p. 15.
10. ILO, *Work Study*, p. 103.
11. ILO, *Work Study*, p. 140.
12. M. E. MUNDEL, *Motion and Time Study*, 5th ed., Prentice-Hall, Englewood Cliffs, NJ, 1978, p. 249.
13. G. NADLER, *Work Simplification*, McGraw-Hill, New York, 1947, p. 122.
14. ILO, *Work Study*, p. 164.
15. ILO, *Work Study*, pp. 102–105.
16. ILO, *Work Study*, p. 105.
17. ILO, *Work Study*, pp. 139–142.
18. ILO, *Work Study*, p. 141.
19. M. E. MUNDEL, *Motion and Time Study*, pp. 251–252.
20. M. E. MUNDEL, *Motion and Time Study*, pp. 230–231.
21. R. M. BARNES, *Motion and Time Study*, 7th ed., Wiley, New York, 1980, pp. 523–528.
22. R. M. BARNES, *Motion and Time Study*, pp. 53–54.
23. T. KADOTA, "ORDLIX and its Application to Japanese Industry," *Proceedings–of the Spring Annual Conference*, American Institute of Industrial Engineers, Atlanta, GA, 1979, pp. 591–597.
24. E. V. KRICK, *Methods Engineering*, Wiley, New York, 1962, p. 43.

BIBLIOGRAPHY

CARLSON, G. B., H. A. BOLZ, and H. H. YOUNG Eds., *Production Handbook*, 3rd ed., Ronald, New York, 1972.

CURIE, R. M., and J. E. FARADAY, *Work Study*, 4th ed., Pitman Publishing Ltd., London, England, 1977.

MAYNARD, H. B., Ed., *Industrial Engineering Handbook*, 3rd ed., McGraw-Hill, New York, 1971.

NADLER, G., *Work Design: A Systems Concept*, Rev. ed., Irwin, Homewood, IL, 1970.

NIEBEL, B. W., *Motion and Time Study*, 6th ed., Irwin, Homewood, IL, 1976.

SHAW, A. G., *The Purpose and Practice of Motion Study*, Columbine Press, Buxton, England, 1960.

CHAPTER 3.4
Assembly Line Balancing

COLIN L. MOODIE
Purdue University

3.4.1 THE CONCEPT OF PRODUCT ASSEMBLY

A dictionary definition of "assembly" that is suitable to a manufacturing environment is the following: to fit or put together the parts of ____. Although most firms sell products that consist of assemblages of several or more parts, the typical person usually thinks of the industrial assembly function in terms of automobiles, home appliances, or other high sales volume products that contain many individual parts. The assembly function would seem to be the moment of truth for a manufacturing firm: All components that have been fabricated and/or purchased are combined into a working unit or subunit. Defective components, incorrect components, or missing components can halt the assembly function.

If assembly is performed progressively by manual labor, then an even distribution of the assembly work elements along the manual assembly stations is important. Assembly balancing, more often termed "assembly line balancing," is an important, preassembly, microproduction planning function, which facilitates the smooth flow of assemblies along a progressive assembly system.

3.4.2 HISTORICAL PERSPECTIVE OF PROGRESSIVE ASSEMBLY

The contemporary assembly line has an interesting history. Probably the two most important manufacturing developments of the past that led to progressive assembly are the concept of interchangeable parts and the concept of the division of labor. The latter development preceded the former by several centuries, but it is the two of them combined that allows the design of assembly stations along a moving conveyor, where operators perform specific assembly tasks, often selecting component parts from hundreds, maybe thousands, of identical parts in containers adjacent to the work station.

To detail the complete (and often fascinating) history of progressive assembly would be inappropriate in this chapter; however, Exhibit 3.4.1, from Wild,[1] can provide at a glance a worthwhile overview of how the past has helped shape the present.

3.4.3 BASIC CONCEPTS OF ASSEMBLY LINE BALANCING

The basic idea of an assembly line is that a product is progressively assembled as it is transported, past relatively fixed assembly stations, by a material handling device such as a conveyor. The work elements, which have been established through the division of labor principle, are assigned to the work stations so that all stations have nearly an equal amount of work to do. Each worker, at his or her station, is assigned certain of the work elements. The worker performs them repeatedly on each production unit as it passes the station. Exhibit 3.4.2, from Tuggle,[2] shows a layout template of a typical assembly line.

The generally accepted definition of the assembly line balancing problem is one attributed to Salveson[3]: "to minimize the total amount of idle time or equivalently to minimize the number of operators to do a given amount of work for a given assembly line speed." This is known as "minimizing the balance delay." "Balance delay" is defined as the amount of idle time for the entire assembly line resulting from unequal total task times assigned to the various work stations. In those rare situations where it is possible to achieve a perfect balance, no idle time would be created by the balance.

Kilbridge and Wester[4] studied the variation in idle times at stations caused by different assembly line balances. They showed that high balance delay was associated with a wide range of work element times and a high degree of line mechanization. They tentatively concluded that the three predominant contributors to high balance delay for an assembly line system for a specific product

Exhibit 3.4.1 Stages in the Development of Flow Production

1260	Division of labor in Venice commented on by Dante and Marco Polo.		
1438	Flow line at Venice arsenal described.		
1496	Mass production of needles by Leonardo da Vinci.		
1617	Use of automatic straight-line process in Spanish mint.		
1717	Unsuccessful attempt to manufacture guns using interchangeable parts in France.		
1731	Manufacture of buttons and pins on flow line in Moscow.		
1746	Description of flow-line production of pins in England.		
1785	Oliver Evans designs "automatic" flour mill.		
1785	Production in France of interchangeable parts for muskets.	1793 and 1796	Outbreak of war between France and Britain, and Spain and Britain, respectively.
1798	Eli Whitney's first contract for 10,000 guns. (Later used interchangeable parts.)		
1799	Government contract for gun manufacture to Simeon North. (Later used interchangeable parts.)		
1804	Manufacture of ships' biscuits on flow line in England.	1803 to 1815	Napoleonic Wars.
1809	Mass production of ship's blocks in England.		
1830	Manufacture of brass clocks with interchangeable parts–Chauncey Jerome, United States.	1830	"Penny press" launched.
1837	Assembly line layout principle used at Bridgewater foundry, England.		
1839	Use of flow-line principle at Chorlton Mills, England.		
1846	Use of interchangeable parts in sewing machines.	1845 to 1848	America at war with Mexico.
1847	Use of interchangeable parts in farm machinery in United States.		
1848	Use of interchangeable parts in manufacture of watches in United States.		
1851	Crystal Palace Exhibition–interchangeable parts demonstrated.		
1855	Enfield and British South Africa arms factories modeled on Colt systems.	1854 to 1856	Crimean War.
1861	Flow-line production in meat processing in Chicago.	1861 to 1865	American Civil War.
1891	Manufacture of freight cars on flow line.		
1899	Design of "low-cost" Oldsmobile.		
1906	Olds and Cadillac cars made in large quantities.		
1908	First "Model T" Ford made.		
1913	Use of first assembly line at Ford plant.		
1922	Use of transfer line at A. O. Smith Corporation, United States.		
1923	Use of "hand" transfer line at Morris Engines, England.		
1924	Use of automatic transfer line at Morris Engines.		

Source. Reference 1.

Exhibit 3.4.2 Typical Configuration of an Assembly Line[a]

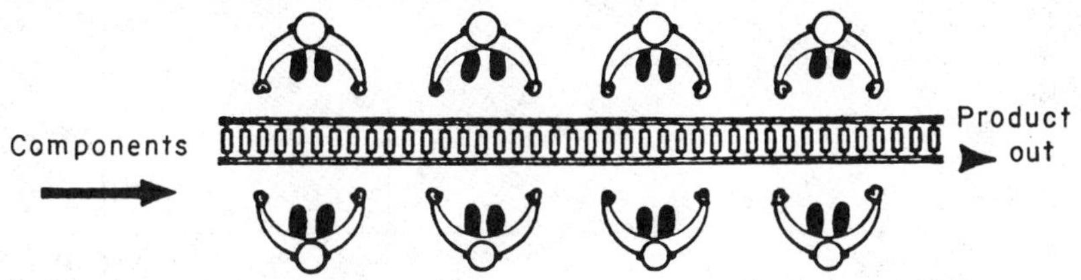

[a]Reference 2.

were a wide range of work element times, a large amount of inflexible line mechanization, and indiscriminate choice of cycle times. However, as shall be seen, the cycle time is often dictated by a specific, desired production rate, which, in turn, may not lead to a low balance delay.

3.4.4 PARAMETERS USED FOR MODELING THE ASSEMBLY LINE SYSTEM

To better understand the assembly line balancing problem, and the computer-oriented line balancing procedures, it is essential to define the problem symbolically. The following symbols are introduced to serve this purpose:

c = cycle time
k = work station number $1 \leqslant k \leqslant K$
i = work element identification number $1 \leqslant i \leqslant N$
T_i = time value for work element i
S_k = amount of time assigned to station k
d_k = delay (idle time) at station k
D = balance delay for entire assembly line

The cycle time defines the rate at which assembled products emerge from the end of the assembly line. It is also the maximum amount of time that a product being progressively assembled is available to an assembly station as it passes by. Given a product to be assembled on a conveyor line, the cycle time can be determined as follows:

$$C = \frac{H}{P}$$

where H = hours per planning horizon (day, work, etc.) and P = production volume desired in H hours, including rework and scrap.

Using this value of C, we have the following minimum possible number of stations for an assembly line:

$$K_{\min} = \frac{\sum_{i=1}^{N} T_i}{C} + r\,(0 \leqslant r \leqslant 1) = \text{an integer}$$

Thus if the division of assembly labor, $\Sigma_{i=1}^{N} T_i/C$, for a given cycle time has a remainder r, a perfect balance, that is, one in which station time equals cycle time for all stations, is not possible. C for perfect balance, if possible, is:

$$C = \frac{\sum_{i=1}^{N} T_i}{K_{\min}}$$

The balance delay for the total line (spread over all stations) is defined symbolically in the following form:

$$D = \sum_{k=1}^{K} dk = \sum_{k=1}^{K} (C - S_k)$$

Exhibit 3.4.3 shows the relationship among d, c, and S for a single station.

Exhibit 3.4.3 Relationship Among d, c, and S

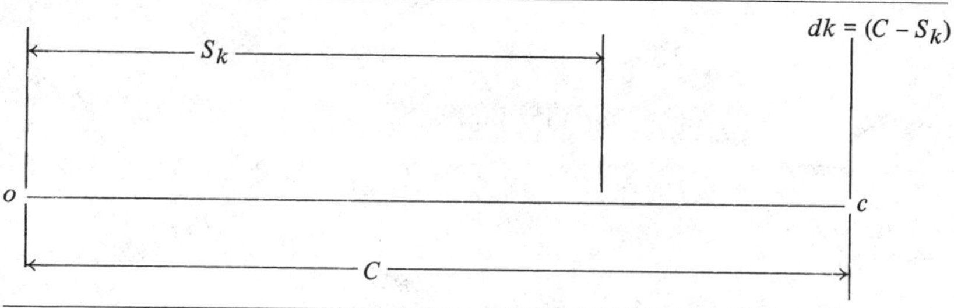

It is easy to show that, if the balance delay is minimized, the number of stations will also be minimized.

$$\text{Min} \sum_{k=1}^{K} (C - S_k) = KC - \sum_{k=1}^{K} S_k$$

$$= KC - \sum_{k=1}^{K} \sum_{i=k} T_i$$

$$= KC - \sum_{i=1}^{N} T_i$$

$$= KC - \text{a constant}$$

3.4.5 PRECEDENCE RESTRICTIONS TO THE ASSIGNMENT OF ELEMENTS TO STATIONS

The assignment of work elements to stations for a specific cycle time (determined by the desired production rate) will be complicated by the element ordering restrictions. It should be reemphasized here that in most cases an assembly work element is the result of a rational division of the total assembly work required to complete the product. This division can create work elements that have been reduced to a minimum number of basic components, such as reach, grasp, travel, and insert. Some work elements, because of local conditions, may not be subdivided to this extent.

A precedence diagram can graphically define the precedence restrictions among work elements for visual observation. The diagram shown in Exhibit 3.4.4 for a simple, nine-element assembly, appeared in Hoffman.[5] This diagram shows that elements 2 and 3 cannot be performed prior to the completion of element 1, but that they can be completed in any order with respect to themselves after element 1 has been completed. The other work elements display similar precedence restrictions.

With these restrictions regarding when certain elements can be performed, it should be noted that there will be more than a single sequence in which they may be performed in order to complete the assembly. In the example shown in Exhibit 3.4.4, there are 24 different sequences.

Exhibit 3.4.4 Precedence Diagram[a]

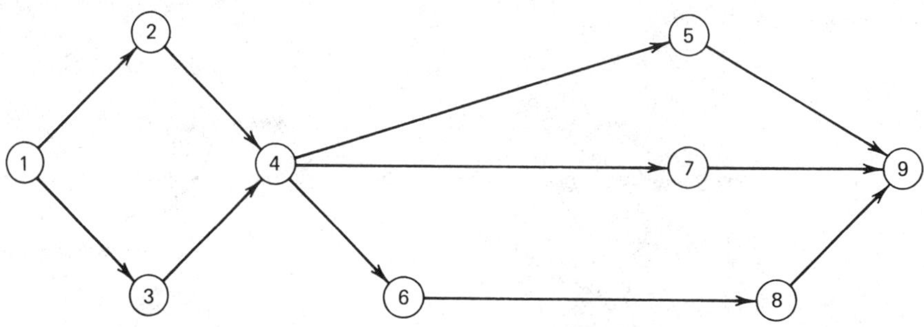

[a]Reference 5.

The precedence diagram information can be more compactly contained (without the visual properties, however) in a precedence matrix. The matrix in Exhibit 3.4.5 shows the precedence relationships among the nine elements shown in Exhibit 3.4.4.

Exhibit 3.4.5 Precedence Matrix

i \ j	1	2	3	4	5	6	7	8	9
1	0	+1	+1	+1	+1	+1	+1	+1	+1
2	−1	0	0	+1	+1	+1	+1	+1	+1
3	−1	0	0	+1	+1	+1	+1	+1	+1
4	−1	−1	−1	0	+1	+1	+1	+1	+1
5	−1	−1	−1	−1	0	0	0	0	+1
6	−1	−1	−1	−1	0	0	0	+1	+1
7	−1	−1	−1	−1	0	0	0	0	+1
8	−1	−1	−1	−1	0	0	0	0	+1
9	−1	−1	−1	−1	−1	−1	−1	−1	0

In this matrix a "+1" indicates a must-precede relationship for the ith element relative to the jth element. For example, element 3 must precede element 4. A "0" indicates no relationship. This is an obvious entry for the matrix cells that make up the diagonal; however, it also indicates no-relationship element pairs such as 5 and 6, 7 and 8, and so on. The "−1" entries in the matrix indicate a must-follow relationship of the i element relative to the j element. This information is really redundant, since it is implied by the +1 relationships. Other forms of the precedence matrix do not carry −1 relationships.

3.4.6 AN ASSEMBLY LINE BALANCING METHOD

The precedence matrix and the functional relationships among the assembly line parameters (previously discussed) are the basic inputs to an assembly line balancing algorithm. The basic algorithms, which are all that are discussed in this chapter, usually operate on the precedence matrix using either a heuristic method or some form of an optimization procedure, that is, dynamic programming, branch and bound, and so on. Many computer-oriented procedures have been proposed in the literature; some are of academic interest only, but others have found significant use in industrial applications. The ranked positional weight method, a heuristic procedure, is discussed here.

The ranked positional weight procedure reported by Helgeson and Birnie[6] in 1961 will serve the purpose of showing how a basic assembly line balancing algorithm works. It must be realized, however, that this is a simple algorithm as compared to those procedures utilized in industrial applications, which must be more complex because of large numbers of elements and precedence restrictions, as well as physical assembly line and tooling constraints, and so on.

Using the previously defined nine-element problem, the precedence matrix can be augmented with two additional columns to give the element time information required by the Helgeson-Birnie assembly line balancing method (Exhibit 3.4.6). The first column of this matrix contains the operation time (in hours) for the element represented by that row of the matrix. The last column of the matrix (column 11) contains the positional weights of the elements represented by the respective rows of the matrix. The positional weight of an element is the sum of the element's time value and the time values of all elements with which it has a +1 relationship. For example the positional weight of element 4 is calculated as 0.05 + 0.01 + 0.04 + 0.05 + 0.04 + 0.06 = 0.25.

Exhibit 3.4.6 Positional Weight Matrix

	T_i	1	2	3	4	5	6	7	8	9	PW[a]
1	0.05	0	1	1	1	1	1	1	1	1	0.37
2	0.03	0	0	0	1	1	1	1	1	1	0.28
3	0.04	0	0	0	1	1	1	1	1	1	0.29
4	0.05	0	0	0	0	1	1	1	1	1	0.25
5	0.01	0	0	0	0	0	0	0	0	0.1	0.07
6	0.04	0	0	0	0	0	0	0	0	.1	0.10
7	0.05	0	0	0	0	0	0	0	0	1	0.11
8	0.04	0	0	0	0	0	0	0	0	1	0.10
9	0.06	0	0	0	0	0	0	0	0	0	0.06

[a]Positional weight.

The basic logic of the ranked positional weight method of assembly line balancing is to assign elements to a station, until the cycle time of that station is about to be exceeded, by assigning elements in order of decreasing positional weight, as allowed by precedence constraints.

If the product defined by the nine-element precedence diagram had a scheduled production rate of 285 units for each 40 hr period, how would the ranked position weight assembly line balancing method set up the assembly line? The cycle time would be calculated as follows:

$$C = \frac{H}{P} = \frac{40 \text{ hr}}{285 \text{ units}} = 0.14 \text{ hr/unit}$$

For that particular cycle time, the optimal number of stations is

$$K = \frac{\sum T_i}{C} = \frac{0.37}{0.14} = 2.64 \Rightarrow 3 \text{ stations}$$

The noninteger result indicates that a perfect (no idle time) balance is not possible and that the minimum number of stations is three.

As previously stated, the thrust of the Helgeson-Birnie method is to assign elements to stations in order of decreasing value of positional weight when precedence restrictions and remaining unassigned idle time in the station permit. The flow diagram shown in Exhibit 3.4.7, from Wild,[1] defines the element-to-station assignment steps. A computer program could be written from this diagram. Following the logic defined by the flow diagram, the three-station balance shown in Exhibit 3.4.8 will be the result.

3.4.7 BALANCING ASSEMBLY LINES FOR SIMULTANEOUS PRODUCTION OF MORE THAN ONE MODEL

Since many industrially produced products can have several or many style variations that have simultaneous demand, it may, at times, be more practical to produce more than one model style simultaneously on an assembly line. A good example of this is in the automobile industry, where several models of the same basic car are moving on the assembly line at the same time. Generally, since each model can have a different precedence diagram, the amount of work required for each of the models will be different; this can create an uneven flow of work along the line, which will

Exhibit 3.4.7 Positional Weight Flow Diagram[a]

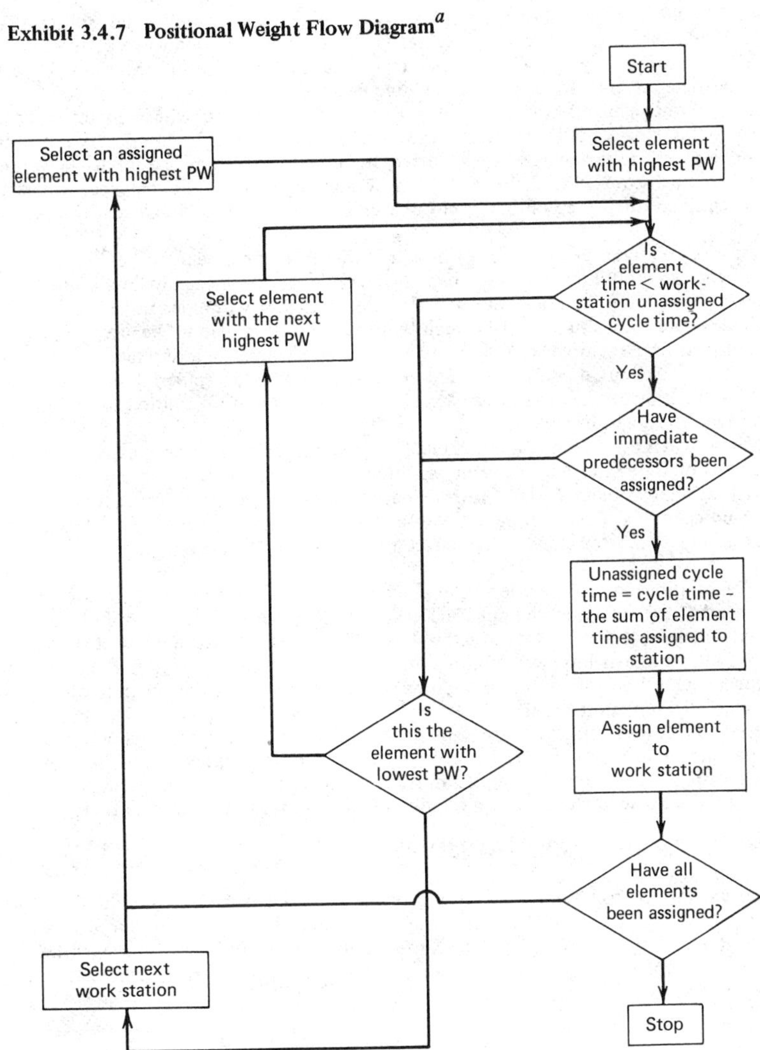

[a]Reference 1.

Exhibit 3.4.8 Nine-Element Solution by Positional Weight

Work Station (k)	Element (i)	Positional Weight	Immediate Predecessors	Element Time T_i	Station Time ΣT_i	Balance Relay $C - S_k$
1	1	0.37	—	0.05	0.05	0.09
1	3	0.29	1	0.04	0.09	0.05
1	2	0.28	1	0.03	0.12	0.02
2	4	0.25	2.3	0.05	0.05	0.09
2	7	0.11	4	0.05	0.10	0.04
2	6	0.10	4	0.04	0.14	0
3	8	0.10	6	0.04	0.04	0.09
3	5	0.07	6	0.01	0.05	0.09
3	9	0.06	5, 7, 8	0.06	0.11	0.03

be reflected in the unevenness of the work element assignments at individual stations. There are two major problems in model mix balancing: (1) determining the sequence of the units flowing down the line and (2) assigning all the elements for all models to specific stations. The latter will be seen to be different than for a single-model line balance.

In situations where mixed-model assembly line balancing is feasible, it can be an economical alternative to the single-model batch mode, where inventories will build up between runs of the various models, and setup charges are incurred for changeover from one model to another. The question that must be answered, however, is whether the mixed-model balance, which will have more balance delay than the combination of several single-model assembly line balances, is more cost-effective.

Determination of the order (sequence) in which the different models to be assembled will progress down the assembly line is a major part of the mixed-model assembly line balancing problem. The different work stations along the line contain operators with specific skills, fixed tools, or both, so the sequence of units must be such that a given station will not be alternately overloaded and underloaded. An example of this could be a station where soldering takes place. A possible sequence of six different models could require this station to do soldering on three consecutive units and no soldering on the next three. A sequence that required this station to perform soldering on every other unit might be superior.

A procedure whereby the units are launched onto the line to begin progressive assembly at unequal intervals of time (and hence causing different spacing of the units on the line) could reduce the severity of the problem defined in the preceding paragraph. However, a specific known sequence of units on the line is desired because it will facilitate the coordination with feeder assembly lines and purchased and fabricated components, which must be available at the proper time.

A number of procedures have been proposed for determining good sequences of units for progressive assembly in a mixed-model line. The goal of these methods is to determine the sequence of models that will balance the work load of the stations over the different models. If the work station is such that the worker is mobile (walks a short distance while working on the unit as it moves), a good sequencing method will minimize the number of extreme movements required by an operator between successive units on the line. Methods proposed by Thomopoulos,[7] Dar-El and Clother,[8] and others work toward improving on this criterion. Exhibit 3.4.9, from Thomopoulos,[9] shows possible operator movement caused by a model sequence on a mixed-model line for four types of stations with different physical constraints. Note that if the operator is unable to complete assembly by the time he or she reaches a fixed boundary, a utility worker must do it.

3.4.8 MIXED-MODEL RELATIONSHIPS

Let the various models to be simultaneously produced on the progressive assembly line be denoted by $A, B, C, \ldots,$* and the total number of units of each model to be produced in a given time period T be denoted $N_a, N_b, N_c, \ldots$. If t_{ij} is the duration of the ith work element in the jth model ($j = A, B, C, \ldots$), then the minimum number of operators, n, required to produce the given number of units in time period T is

$$n = \frac{\sum_j \left(N_j \sum_i t_{ij}\right)}{T}$$

If n is a fraction, it is raised to the next higher integer.

If the N_js are not given, but it is known that the various models are to be produced in the proportions $f_a{:}f_b{:}f_c \ldots$, the maximum production of each model in time period T with n operators is

$$\text{Max } N_j = \frac{nT}{\sum_j \left(f_j \sum_i t_{ij}\right)} f_j \qquad j = A, B, C, \ldots$$

It is assumed that the balancing process will evenly divide the assembly work among the n operators so that the amount of time each operator works on a given model is approximately the same. This amount of time is defined as the "model cycle time." It is designated by C_j and is

*Nomenclature used here is that of Webster and Kilbridge, reference 10.

Exhibit 3.4.9 Four Types of Stations[a]

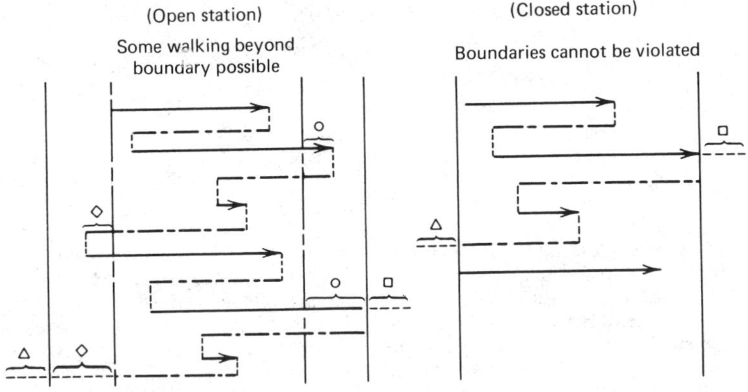

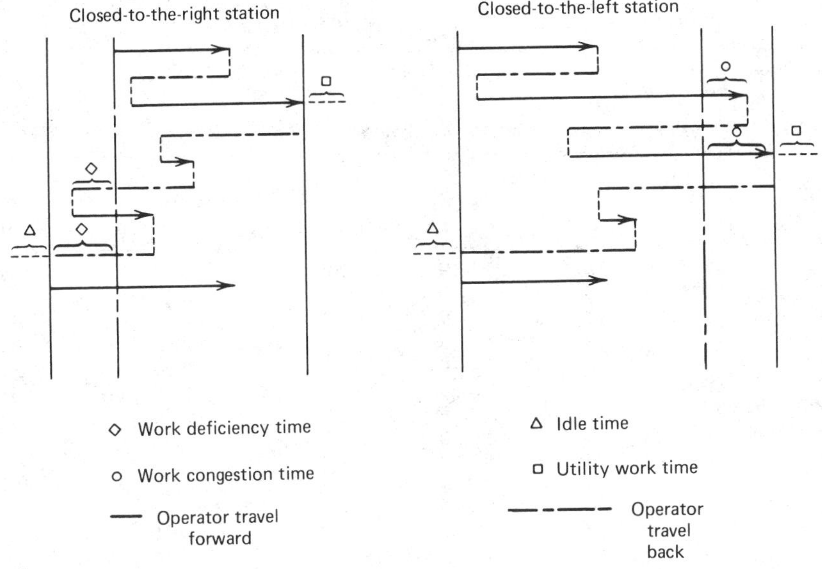

[a]Reference 9.

determined for each model by

$$C_j = \frac{\sum_i t_{ij}}{n} \qquad j = A, B, C, \ldots$$

For a given number of stations, the maximum model cycle time will be for the model that has the maximum total amount of assembly work. For the fixed-rate launching system, the maximum model cycle time will be the time interval between the launching of consecutive units.

3.4.9 AN EXAMPLE

The following example from Thomopolous[7] illustrates certain aspects of the mixed-model balancing problem. A day's production schedule for a specific assembly line is as follows:

Type of Model	Required No. of Units
A	42
B	28
C	14
D	2
E	5
F	9
	100

If we consider *all* of the work elements associated with the six models, that will be a large number. However, even though the different models may be similar, successive units moving down the assembly line will not necessarily require the same work elements to be performed on them. This will tend to invalidate the assembly line balancing methods that work well for single-model line balancing. Assigning elements to stations on a total (for all units) time, rather than a cycle time (one unit), basis is one way of accomplishing element assignment to stations working on mixed models.

In the six-model, 100-unit example, the assembly line has been divided into four sections, primarily by labor groups. The example will consider only the first labor group. If a station has 450 min of assembly time available during a shift week (the duration of this 100-unit schedule), then the station would be well balanced if its assigned elements added up to 450 min for the week. Using this element assignment philosophy, the relationship of the various work elements assigned to one station for the different models is given in Exhibit 3.4.10. With the proper sequence of different models moving down the assembly line, it should be possible for an operator, working at this station, to complete all elements for all models.

Exhibit 3.4.10 Work Elements Assigned to Mixed-Model Station

Element No.	No. of Units of Model						Total Number of Units	Element Time Duration (Min)	Total Time (Min)
	A	B	C	D	E	F			
1	42	28	14	2	5	9	100	0.32	32.00
2	0	0	0	0	0	9	9	0.11	0.99
3	42	28	14	2	5	9	100	0.44	44.00
4	42	28	14	2	5	9	100	0.62	62.00
5	0	0	0	0	5	0	5	0.26	1.30
6	42	28	14	2	5	9	100	0.45	45.00
7	42	28	14	2	5	9	100	0.26	26.00
8	0	28	0	0	0	0	28	0.07	1.96
9	0	0	0	0	5	0	5	0.31	1.55
10	0	28	0	0	0	0	28	0.05	1.40
11	42	28	14	2	5	9	100	0.20	20.00
12	0	0	0	0	0	9	9	0.04	0.36
13	42	28	14	2	5	9	100	0.89	89.00
14	0	0	0	0	5	0	5	0.16	0.80
15	42	28	14	2	5	9	100	0.24	24.00
16	0	28	0	0	0	0	28	0.02	0.56
17	0	0	0	0	5	0	5	0.12	0.60
18	42	28	0	0	0	0	70	0.05	3.50
19	0	0	0	0	0	9	9	0.04	0.36
20	0	0	0	2	0	0	2	0.17	0.34
21	42	28	0	0	0	0	70	0.04	2.80
22	42	28	14	2	5	9	100	0.48	48.00
23	42	28	14	2	5	9	100	0.08	8.00
24	42	0	0	0	5	0	47	0.30	14.10
25	42	28	14	2	5	9	100	0.00	0.00
26	0	0	0	0	5	0	5	0.15	0.75
27	0	0	14	2	0	9	25	0.20	5.00
28	0	0	0	2	5	0	7	0.46	3.22
31	0	0	0	2	5	9	16	0.40	6.40
									443.99

Exhibit 3.4.11 Five Conveyor Configurations[a]

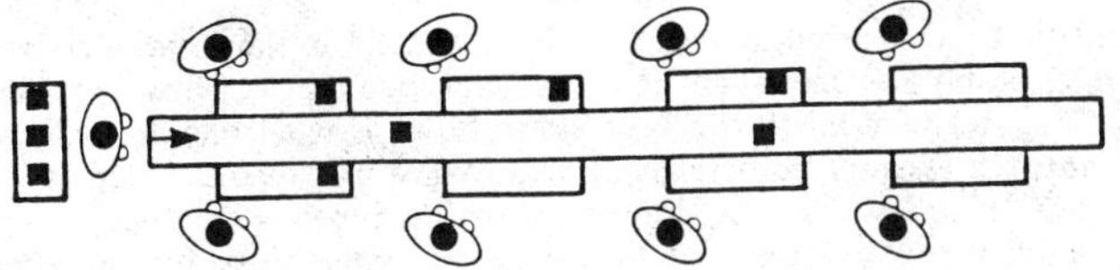

Belt conveyor, with diverters at each station, sends batches to assemblers from control station. Good way to assemble components and subassemblies.

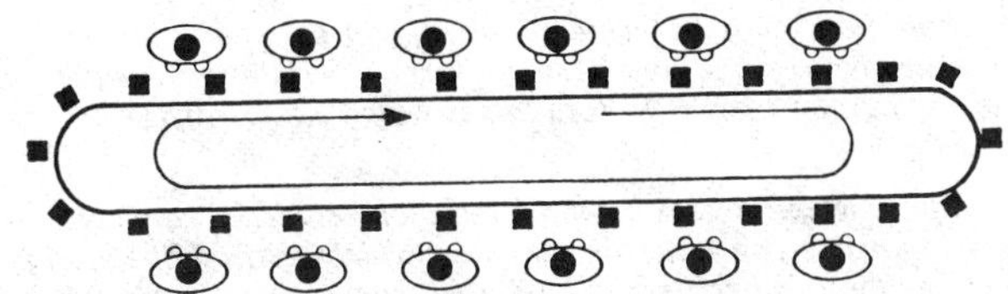

Carousel circulates assembly material to workers. Very flexible in terms of variations in numbers of assembly steps and people needed.

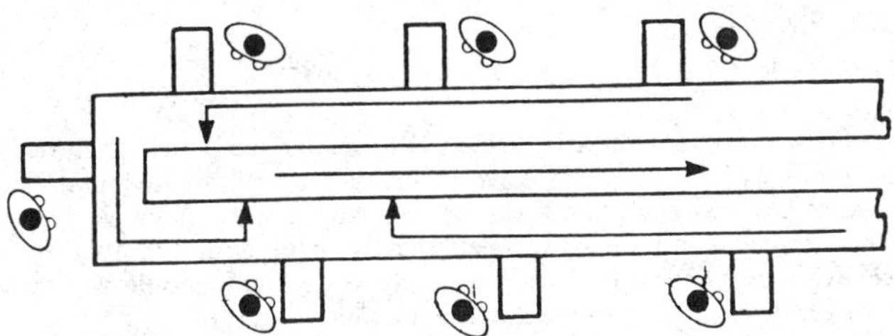

Multiple-path conveyors enable several products to be built at once, with variable numbers of stations and paths of flow between them.

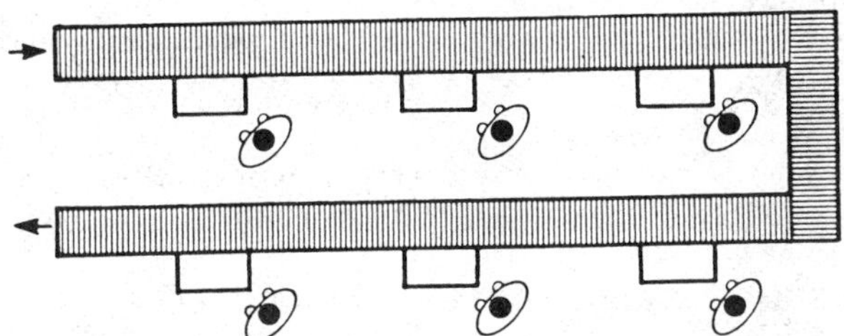

Gravity roller conveyors, for sequential assembly, are linked with powered cross conveyors. Very flexible in terms of layout.

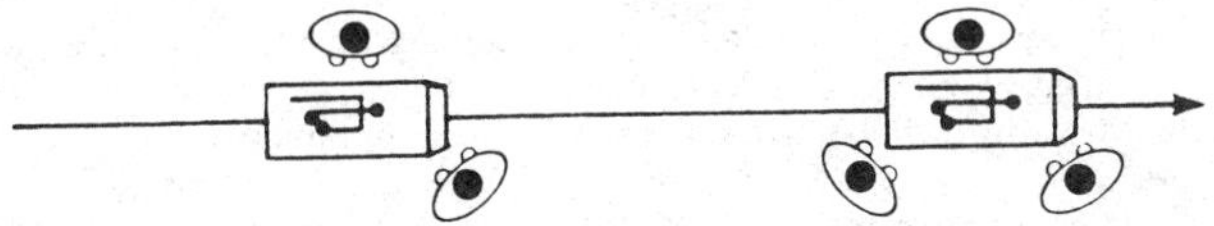

Towline conveyor approach moves assemblies from one group of assemblers to another. Assemblers may work on moving assemblies, or carriers may be taken off-line. In-floor chain, overhead towline, and air-film supported carriers are typical equipment used.

[a]Reference 11.

3.4.10 DIFFERENT CONFIGURATIONS OF CONVEYORS FOR ASSEMBLY

It is apparent that mechanized material handling, usually in the form of conveyors, is essential for a progressive assembly system. There are, in general, two types of flow patterns for assembly lines: progressive flow and random flow. "Progressive flow" is what one normally expects to see in an industrial plant; the assembly progressively takes place in sequential steps as the unit moves along the conveyor. The earliest assembly lines were of this type, as are many modern ones. "Random flow" is a relatively new development that allows materials to be moved between stations in any sequence. This might be desirable where some assemblies do not require certain operations.

Modern Materials Handling[11] notes some additional procedures for moving units between stations:

> *Between-station flow may be achieved using conventional conveyors equipped with diverters. A roller conveyor "loop" can be formed within an assembly area, for example. Any materials may be carried in tote boxes which are identified by movable retroflective tabs.*
>
> *Scanners or photoelectric sensors mounted along side the conveyor read the tables and signal a programmable controller or computer which controls the storing. Because the conveyor is a closed loop, materials may be routed from any one station to another.*

Although there are a number of ways to configure a conveyorized, progressive assembly system, *Modern Materials Handling*[11] diagrammed five that are used in industrial applications.

3.4.11 COMPUTER-ASSISTED ASSEMBLY LINE

Early assembly lines were all tediously balanced by paper-and-pencil analysis by the industrial engineer. Since about 1960, computers have in many cases eliminated that chore so that many balancing configurations and cycle times can be analyzed rapidly. Now we are seeing miniprocess and microprocess control computers used beneficially in modern assembly line systems. The industrial engineer who is to be involved with assembly line systems should be aware of the potential of computer control. A few examples are described here.

A 1979 example of these innovative assembly lines is that which was designed for the assembly of Ford transaxles. The Bendix-built system is the largest nonsynchronous, automated assembly system for automatic transaxels in the world. Its 500 ft of conveyor will connect 156 assembly, inspector, checking, and testing stations, arranged to form three oval-shaped sections. Although many of the stations will be automatic, a total of 67 operators will be required to perform various part placement, visual, and noise inspections to assemblies.[12]

Fiat, in Italy, has an automated assembly line for producing automobile bodies. Called the "robogate line," this system employs a computer-directed, self-propelled pallet system for producing engines.[13] This new flexible assembly system will be able to produce 1500 engines daily in any mix of about 100 types and variations. A total of 37 self-propelled pallets will shuttle between 10 assembly islands, carrying parts, work in progress, and completed engines. The travel of the pallets is controlled by the system's control computer.

3.4.12 JOB ENLARGEMENT ON THE ASSEMBLY LINE

Early in this chapter it was noted that the idea of the division of labor, a concept that dates back to Adam Smith, had much to do with establishing the feasibility of the early (late nineteenth and early twentieth century) assembly lines. Within the past two decades, however, there have been some successful applications of assembly lines that attempt to enlarge the amount of work provided each work station. In such applications a moving conveyor may bring work to a station and remove it when assigned operations have been completed; however, in such situations a buffer inventory of units waiting to be worked on will be maintained by the work station to absorb variation in work time.

These "new" types of assembly lines, made possible by innovative conveyor configurations, are called "modular" assembly systems by Tuggle[2] and will usually take different physical configurations for different applications. Some advocates of this form of assembly line development indicate that quality can be improved by emphasizing craftsmanship and creating worker identification with the product. This type of line, though, may require a larger number of duplicate tools and fixtures because some workers may be doing similar tasks. Two examples of a modular line, from Tuggle,[2] are given in Exhibits 3.4.12 and 3.4.13.

Exhibit 3.4.12 Modular Stations[a]

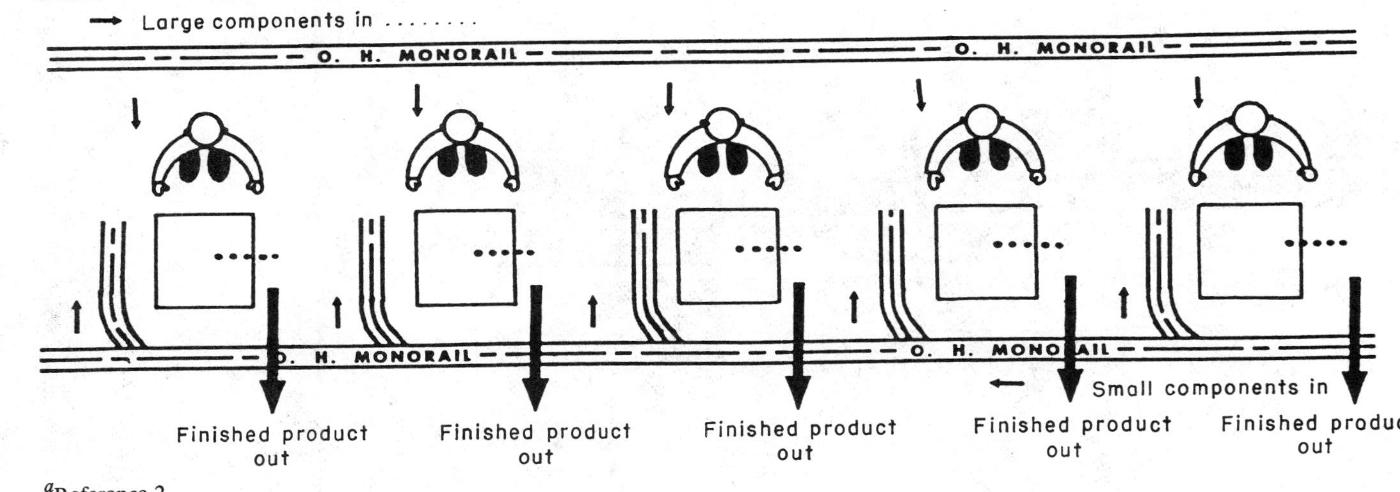

[a]Reference 2.

Exhibit 3.4.13 Assembly Teams Configuration[a]

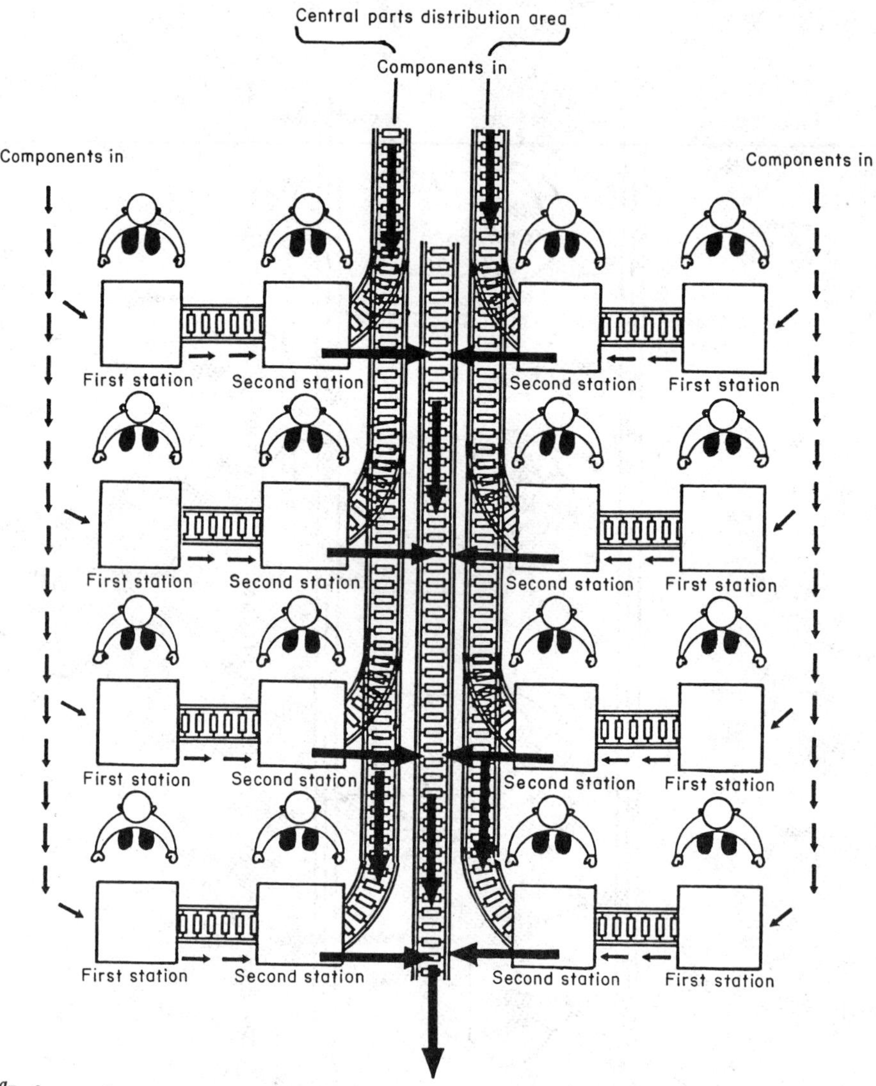

[a]Reference 2.

REFERENCES

1. R. WILD, *Mass-Production Management*, Wiley, New York, 1972.
2. G. TUGGLE, "Job Enlargement Cuts Assembly Line Inefficiencies," *Industrial Engineering*, February 1969.
3. M. L. SALVESON, "The Assembly Line Balancing Problem," *Transaction of A.S.M.E.*, Vol. 77, August 1955.
4. M. KILBRIDGE and L. WESTER, "The Balance Delay Problem," *Management Sciences*, Vol. 8, No. 1, 1962.
5. T. R. HOFFMAN, "Permutations and Precedence Matrices with Automatic Computer Applications to Industrial Processes," unpublished doctoral thesis, University of Wisconsin, Madison, 1959.
6. W. B. HELGESON and D. P. BIRNIE, "Assembly Line Balancing Using the Ranked Positional Weight Technique," *Journal of Industrial Engineering*, Vol. 12, No. 6, 1961.

7. N. THOMOPOULOS, "Line Balancing-Sequencing For Mixed-Model Assembly," *Management Science*, Vol. 14, No. 2, 1967.
8. E. DAR-EL and R. CLOTHER, "Assembly Line Sequencing for Model Mix," *International Journal of Production Research*, Vol. 13, No. 5, 1975.
9. N. THOMOPOULOS, "Mixed Model Line Balancing with Smoothed Station Assignments," *Management Science*, Vol. 16, No. 9, 1970.
10. L. WESTER and M. KILBRIDGE, "The Assembly Line Model-Mix Sequencing Problem," *Proceedings of 3rd International Conference on Operations Research*, Paris, 1964.
11. MODERN MATERIALS HANDLING, "How Conveyors Organize Assembly Operations," November 1979, p. 114.
12. A. WRIGLEY, "Automated Line Engineered for Ford Transaxles," *American Metalworking News*, October 1, 1979.
13. E. MASSAI, "Fiat Continuing Beyond Conveyor Assembly," *American Metalworking News*, October 15, 1979.

BIBLIOGRAPHY

ARCUS, A. L., "COMSOAL: A Computer Method of Sequencing Operations for Assembly Lines," *International Journal of Production Research*, Vol. 4, No. 4 (1966).

DAR-EL, E. M., "Mixed Model Assembly Line Sequencing Problems," *Omega*, Vol. 6, No. 4 (1978).

DAVIS, L. E., "Pacing Effects on Manned Assembly Lines," *International Journal of Production Research*, Vol. 4, No. 3 (1966).

HELD, M., R. M. KARP, and R. SHARESHIAN, "Assembly Line Balancing–Dynamic Programming with Procedence Constraints," *Operations Research*, Vol. 2, No. 3 (1963).

MACASKILL, J. L. C., "Production Line Balances for Mixed-Model Lines," *Management Science*, Vol. 19, No. 4 (1972).

MANSOOR, E. M., "Assembly Line Balancing–An Improvement Over the Ranked Positional Weight Technique," *Journal of Industrial Engineering*, Vol. 15, No. 2 (1964).

MASSAI, E., "Fiat Robogate Welding System Hits Output Target," *American Metalworking News*, September 12, 1979.

MISHRA, R., K. C. SAHU, and S. SAHU, "Multimodel Assembly Line Balancing–A Case Study," *Industrial Engineering and Management*, October–December 1972.

NEVINS, A. J., "Assembly Line Balancing Using Best Bod Search," *Management Science*, Vol. 18, No. 9 (1972).

PRENTING, T. O., and N. T. THOMOPOULOS, *Humanism and Technology in Assembly Line Systems*, Hayden, Rochelle Park, NJ, 1974.

CHAPTER 3.5

Machine Interference: Assignment of Machines to Operators

KATHRYN E. STECKE

The University of Michigan

3.5.1 INTRODUCTION

Within some types of manufacturing systems, a problem of significant practical importance is due to a phenomenon known as machine interference. In a simple form of the problem, suppose that one operator or repairperson is in charge of several similar machines. Periodically a machine stops and will not resume production until it is fixed (attended) by the repairperson (operator). If two or more machines happen to be stopped (down) at the same time, only one of them can be serviced at a time. Because the remaining down machines have to wait to be repaired, these idle machines are nonproductive. This waiting time is what is referred to as "interference time" (i.e., the down machines are "interferring" with one another).

Production is lost because of both repair (service) time and interference time. Production rate (of the machines and the operator) is a function of the following:

1. The distribution of time between failures.
2. The distribution of time of repairs.
3. The distribution of time for the repairperson to move between machines.
4. The number of machines assigned to the operator.

The basic machine interference problem is to decide on the best number of machines to assign to an operator. A related problem is to decide on the number of repairpersons for a particular number of machines so as to optimize some measure or combination of objectives. Some such objectives are to maximize production, minimize loss due to idle machines and idle labor, and minimize costs, subject to constraints such as meeting demand or staying within a budget.

There are other questions that might be important, such as:

1. How is the interference time of a machine calculated or measured?
2. How are the cycle time (or total average time to produce a workpiece) and hence the expected production rate calculated?
3. How should an operator decide which task or repair to perform next?

This chapter reviews many methods and models to help the industrial engineer answer these and related questions. A cross section of the previous research in the area is presented. Many of the methods are described in detail along with the required assumptions, the range of applicability, and the input needed and the output obtained. All of this information is summarized later in Exhibits 3.5.13 and 3.5.14. Some methods can be used directly to get an estimate of machine interference time; other techniques are presented for which some information is given to aid a reader in using the existing literature for additional details.

3.5.2 IMPORTANCE OF THE PROBLEM

The problem of assigning the appropriate number of machines to a person is an important one. Some of the effects of assigning too many machines are:

1. The operator is overworked and will sometimes have a serious backlog of down machines. She or he becomes tired and the work pace slows down, thus aggravating the problem.

2. The output expected by management is not produced.
3. If there is an incentive plan to make more money for extra work, the bonus could rarely be achieved.
4. As machines and productivity go down, so does the operator.

Likewise, there are problems resulting from underassigning machines to workers. The amount of effort required to keep the machines running at the expected efficiency level is then minimal. With not enough work with which to keep busy, the operator can be bored (possibly causing daydreaming and inattention). A lighter work assignment could become the norm. In this case even the incentive to achieve a higher production rate than required (that is, more money) is sometimes resisted by experienced operators because they realize that continuous high output might cause their work standard to be increased.

Also, the development of reasonable work standards and an equitable incentive plan requires first an estimate of the degree of machine interference idleness that is expected to occur for a given assignment of machines. This information will aid in making correct decisions concerning the management of the people and the machines.

Finally, an assessment of the amount of machine interference is required to determine machine efficiencies, the amounts of production, and hence profits, which are obtained by varying different factors, such as changing the queue discipline (see Section 3.5.6), using a different quality of the required raw material (which could change the production or repair times), and changing the number of machines assigned to an operator, as well as other such system changes.

3.5.3 MACHINE INTERFERENCE PROBLEMS

The textile industries were a major source for the original procedures developed to handle machine interference problems. There are many situations in various industries where problems of "machine" interference occur, as illustrated by the following examples:

1. The people from various departments of a company go to a common storeroom to obtain their required tools or materials. The people are the "machines", and the storeroom is analogous to the operator or repairperson. In this case the time for an operator to move between machines is usually zero.
2. The ships that pull into a port each require one of the limited number of berths for a certain period of time.
3. Weaving cloth on a loom requires periodic checking by an operator because yarn breaks or the supply cone holding the yarn requires replacing when empty. Such jobs require little time, so several machines are assigned to each operator. Yet occasionally, many looms may be stopped simultaneously because an operator can attend to only one at a time.
4. Consider a computer system consisting of N processors (CPUs) and O memory modules such that each CPU can access each memory module. The CPUs are the machines, and the memory modules are the operators. In such a system, different computer programs may simultaneously request access to the same memory module, which can cause interference.
5. Deciding on the number of tables to assign to a waitress, the proper staffing of nurses at an emergency room, and assigning the best number of clerks to handle baggage at an airport are similar in nature to machine interference problems.

In all of these situations, there is a random arrival of customers, each with a variable service time (which depends on the type of service required). When a customer does not get immediate attention, idle time (waiting time, interference time) occurs. As more customers demand service, a queue is formed. As we will see, many of the solution methods for such interference problems are based on queuing theory (see Chapter 13.7).

3.5.4 PERFORMANCE MEASURES

Production, cost, and idle time are some usual performance measures. In terms of money, there is a loss due to both idle labor and down machines. Labor costs are inversely related to production costs (or costs from less production caused by idle machines), that is, losses because of machine interference. As more machines are assigned to an operator, the labor costs decrease (the operator is busier), but machine costs increase (more machines must wait longer for required service). Therefore one version of the machine interference problem is deciding on the "economical" number of machines to assign to an operator. The aim is to balance the labor cost with the net productivity of the machines.

Although cost is usually the major concern, a current high demand may require a higher level of production. In such a situation, labor costs would increase as a result of assigning fewer machines to an operator. Alternatively, when market demand is low, a lower production rate may be desirable (if inventory cost is a factor). Then a larger number of machines may be assigned to an operator. In these cases a reasonable objective would be to assign machines so that the overall production is close to that desired, depending on the current demand and inventories.

3.5.5 MACHINE INTERFERENCE MODELS

Many techniques to handle various types of machine interference problems are presented here. Assumptions under which each is applicable differ for the different methods. The usual assumptions are discussed in terms of their applicability and relevance in Section 3.5.13. Many of the analytic methods are based on probability (see Chapter 13.7), queuing theory, or simulation methodology (see Chapter 13.11).

Models of machine interference problems can be classified with respect to the pattern of machine breakdown. When the time between breakdowns and service times is constant, the situation is known as a "regular system." In other words, the time at which a machine should stop is predictable. Automatic screw machines and plastic molding presses are typical types of machines with regular patterns. Machine assignments may be made based on the cost of an idle machine (loss of production) versus the labor cost by analyzing the resulting workload of an operator. On the other hand, if the time until breakdown is not known, or the service time is unknown, the situation is called "random." Mathematically speaking, if the probability that a running machine will stop in the next instant is independent of the length of time it has been running, the time between breakdowns is an observation from an exponential distribution (see Chapter 13.7). This is a common situation, and thus most of the methods that follow assume exponentiality. Suppose that the time between breakdowns depends on the quality of the operator's repair when the machine was previously fixed. If done right, the machine will not break down for a long time; if not, it might require adjustment sooner than expected. The distribution of such breakdown times might then be hyperexponential.[1] Another common assumption is that the times are constant.

Some of the procedures are concerned with deciding how many machines to assign to one operator. Such problems are referred to as "multiple machine–single operator problems." Other methods are concerned with the number of operators to assign to a given number of machines. This can be termed a "multiple machine–multiple operator problem."

3.5.6 OPERATOR QUEUE DISCIPLINES

If several machines are down, how does an operator choose which machine to service next? This decision may affect the amount of interference. The rule by which the operator decides is known as the "queue discipline." Some of the possibilities follow.

Random Tending

Suppose that the operator has just repaired one of several down machines. If each of the remaining stopped machines has the same probability of being the next one to be fixed, the decision process will be called "random tending" or "random servicing." In other words, the down machines are randomly serviced.

Cyclic Patrol

Another queuing discipline, called "cyclic patrol," requires that the operator stop to check each machine in his or her assignment by way of a path, called a cycle, even if the machine is not down or stopped. Some industries require tasks to be inspected or checked periodically, particularly in processes where faulty supervision can result in the production of a significant amount of a defective product. In the textile industry, the cloth has to be of a certain quality, for example, woven evenly. There are several types of cyclic patrol (sometimes referred to as "cycle tending" or "regular patrolling").

Closed Cyclic Patrol

A closed cyclic patrol requires that the operator first move from an initial machine in her or his assignment, to a second, and consecutively to the last, visiting each machine once. This path becomes a cycle when she or he walks from the last machine back to the first (see the solid line in Exhibit 3.5.1). Any machines that are found down during the patrol are repaired and restarted.

Alternating Cyclic Patrol

A similar discipline is an alternating cyclic patrol. In this case the operator walks from a first machine along a path to a last machine and then reverses direction, patrolling the same set of machines

Exhibit 3.5.1 Closed Cyclic Patrol

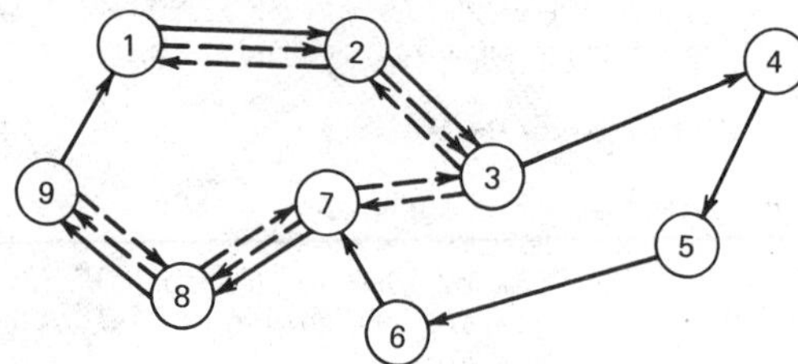

in reverse order. Suppose that, in Exhibit 3.5.1, the operator was assigned machines 1, 2, 3, 7, 8, and 9. Then, following the dashed lines, a possible cycle of an alternating cyclic patrol could be the following machines:

$$1 \to 2 \to 3 \to 7 \to 8 \to 9 \to 8 \to 7 \to 3 \to 2 \to 1$$

The operator would start over again, servicing downed machines encountered during the patrol in both phases of the cycle.

Distance Priority

A different kind of queue discipline allows priorities to be assigned to some or all machines. One such discipline, the distance priority, has an operator service the nearest down machine. In addition to manufacturing systems (where this rule is commonly practiced), such a decision rule is meaningful in law enforcement and guerrilla warfare (where a random service discipline could be harmful if applied by the law or by a guerrilla). Such a priority makes more sense for (groups of) repairpersons maintaining computer equipment at different locations. The walking distance will be shorter than that from any cyclic or random tending assignment.

Nonpreemptive Priority

Another type of priority discipline, nonpreemptive priority, can be described as follows: Suppose that one operator is assigned to tend N different groups of machines, with N_i machines in the ith group, $i = 1, \ldots, N$. Assume that the groups are ordered such that the ith group has a higher priority than the $(i + 1)$st group. The priority discipline is nonpreemptive in the sense that, if some machine is currently being serviced, the breakdown of a machine of higher priority does not interrupt service on the current machine (the current machine is not preempted). In effect, a machine's priority is considered only when an operator has finished servicing one machine and needs to know which machine he or she is to service next. In this case the operator will choose (randomly—that is, random tending) a machine from the highest priority group that contains a down machine. If all machines are up, then the operator is idle until the next machine breaks down. A special case of such a system occurs when the number of machines in each group is one, that is, $N_i = 1$ for $i = 1, 2, \ldots, N$. Then there are N machines assigned to the operator.

Preemptive Priority

A preemptive priority discipline is similar to a nonpreemptive priority; the difference is that, if a machine breaks down that is of a higher priority than the machine currently being serviced, then the operator interrupts repair on the current machine to fix the one of higher priority. (We say that the current machine has been preempted.) When the operator resumes service to the preempted machine, service can either start where it left off or begin again, depending on the service discipline.

Shortest Service Time Priority

Sometimes it is known which kinds of machine breakdowns take a long or short time to service. Using this discipline the next machine to be serviced is the one that can be repaired fastest.

Comparison of Disciplines

There are many other possible queue disciplines. For additional information and mathematical results, see Jaiswal.[2]

In general, the shortest service time and shortest distance priorities produce better results than, say, random tending. Which of these two might be better depends on the problem-specific data concerning lengths of service time and walking time, but both of these are good for (1) keeping machines up and (2) keeping interference down.

Likewise, preemptive priorities applied in an appropriate way can similarly be better than non-preemptive priorities or random tending. Finally, in practice, even randomly tending machines is usually better than any of the cyclic patrols to achieve keeping machines up and interference down. Yet, for reasons already mentioned, some types of machines require periodic checking and supervision, which makes cyclic patrol disciplines applicable.

3.5.7 DEFINITION OF TERMINOLOGY

The notation used in the following methods of handling machine interference problems is defined and described. Any consistent scaling is valid. The time quantities are sometimes defined as ratios with respect to service time.

The length of time to repair (service, attend, remove, fix, adjust) a machine includes anything required of the operator after arriving at a down machine. In some methods it includes the time it takes an operator to get to the machine after it has stopped running. The time to service a machine is sometimes assumed to be a constant; at other times, it is random. Let

$$S = \text{average service time for repair (in some time units, say minutes)}$$

Then

$$\mu = \frac{1}{S} = \text{average service rate per machine (in units per minute)}$$

The length of time from when a machine is restarted until it stops running is called the "production time" (or sometimes the machine running time or the time between breakdowns). It is usually assumed that the machines go down independently of each other, that is, that there is no common cause of stopping, such as heat, electrical power failures or surges, or a common supply stream. A machine is said to break down regardless of whether the failure is such that the machine actually stops or continues to run and produces an unsatisfactory product. Let

$$P = \text{average production time per machine (in time units)}$$

Then

$$\lambda = \frac{1}{P} = \text{average breakdown rate (in pieces per unit running time)}$$

$$= \text{average number of breakdowns per minute of machine running time}$$

A servicing factor can be defined by letting

$$\rho = \frac{\lambda}{\mu} = \frac{S}{P} = \text{average number of time units (minutes) of service required of an operator to keep one machine running for a time unit (minutes)}$$

We also have

$$\frac{1}{\rho} = \frac{\mu}{\lambda} = \frac{P}{S} = \text{average amount of production (or running) time per unit of service time}$$

When a machine breaks down, if the operator is servicing another down machine, a period of interference (or waiting or queuing) is begun, which lasts until the operator is able to begin service on this down machine. Let

$$I = \text{average interference time per machine (in time units)}$$

The walking (or patrolling) time is the time it takes an operator to proceed from one machine to the next, assuming that no repair action is required. Walking time can include inspection time. Sometimes, as an approximation, the average walking time is incorporated into the average service time. Let

$$W = \text{average walking time from machine to machine (in time units)}$$

At times an operator might have some duties to perform other than servicing the machines. Some of the jobs could be done while the machine is running. Such duties do not affect the production times of the machines; they affect only the calculations of operator busy or idle time. On the other hand, there are some jobs that must be done while the machines are running. Some such tasks can be termed "ancillary duties" and can include retooling or adjusting. We will call other types of tasks "maintenance jobs." Let

A = average ancillary time per machine (in time units)

M = scheduled average maintenance time per machine (in time units)

The total average amount of cycle time is the sum of the production, service, interference, patrolling, ancillary, and maintenance times,

$$C = P + S + I + W + A + M$$

assuming that no events other than P, production, occur while a machine is running.

The ratio of the production time to the cycle time is the efficiency of the machine (Me), sometimes called "machine availability." Machine efficiency is calculated as a fraction of the cycle time.

$$Me = \frac{P}{C} = \frac{P}{P + S + I + W + A + M}$$

Often the patrolling time is averaged in with the service times, and ancillary and scheduled maintenance are ignored; then

$$Me = \frac{P}{C} = \frac{P}{P + S + I}$$

Perfect production (no idle or service time required) occurs when

$$Me = \frac{P}{P + S + I}$$

$$= \frac{P}{P + 0 + 0}$$

$$= 1$$

Me is a normalized measure of the actual production obtained and is often smaller than perfect production because of queuing or service or interference time.

The average number of machines up is

$$N(Me) = N \times \frac{P}{C}$$

If all of the machines are similar, the operator spends an average of one time unit servicing each machine during the cycle time C. The operator efficiency is the proportion of clock time during which the operator services the machines, or

$$Oe = \frac{S + W + A + M}{C} = \frac{S + W + A + M}{P + S + I + W + A + M}$$

Efficiency loss is the amount of production that is lost because of service and interference time. Let

$$E = \frac{S + I}{P + S + I + W + A + M}$$

or if

$$E = \frac{S + I}{P + S + I}$$

then

$$E = 1 - Me$$

Note that all of the time quantities were expressed not in minutes or hours, but in some arbitrary "time units." A rescaling of parameters is valid, and some relations would change proportionately.

3.5.8 COLLECTION AND USE OF DATA

How are all of the values for the various time quantities (i.e., service, interference, walking, production, etc., times) known? For some kinds of systems, a computer can be used to observe, measure, collect, record, and quantify the required data. For some of the industries and problems with which we are concerned, people with stopwatches are on the shop floor to measure and record the relevant data. For a given machine, they can record the lengths of time it is (1) running, (2) down and being serviced, and (3) down and waiting for service (periods of interference time). Likewise, the amount of time it takes an operator to walk from machine to machine can be recorded along with any other desired measures. Another data collection method, ratio-delay study, requires a person to sample machines randomly and gather data (see Chapter 3.2). The data are used to estimate the quantities needed. For example, an average service time is calculated by first summing the values of all of the observed service times and then dividing this quantity by the number of observations of service time recorded. Likewise, the frequency of breakdowns, average length of breakdowns, average time between breakdowns, average downtime (which can be broken up into the average service and average interference times) can be calculated.

A technique such as ratio-delay requires first recording the data mentioned and then either calculating S and I as a percentage of downtime, D, or calculating S, I, and P as a percentage of the average clock time, C. Since $S + I + P = C$ (when patrolling and other times are not taken into consideration), it can be seen that S/C, I/C, and P/C are all less than 1. Moreover, $S/C + I/C + P/C = 1$ (or, when multiplied by 100 to become percentages, they add up to the whole or 100%).

For example, suppose that during an 8 hr shift a machine works for 6 hr, is serviced for $1\frac{1}{2}$ hr, and is idle for $\frac{1}{2}$ hr. Then (P/C) 100% = 75%, (S/C) 100% = 18.75%, and (I/C) 100% = 6.25%.

It has been noted that many of the quantities of interest are derived from actual shop data. It was early recognized that the machine interference time was inconvenient to obtain this way because of the following:

1. It can be difficult to time simultaneous events, such as service times, interference times, and patrolling times, for several machines.
2. It would be necessary to develop interference tables and curves covering a whole range of possible assignments of different sizes (*many* stopwatch studies and perturbations of the system).
3. Even if a time study only took measurements of the (usually upredictable and variable) breakdowns of N machines, the period of the timing might not have been long enough to be representative of the actual conditions of the plant.
4. Ratio-delay studies are not exact; times are randomly sampled.

Therefore many methods attempt to define the amount of machine interference by use of a mathematical relationship involving the number of machines assigned and the service, production, and other times. Some techniques treat the problem deterministically and derive empirical relationships between variables of interest. Other techniques take the inherent variability into account, utilizing queuing theory (see Chapter 13.7) and binomial expansions (see Chapter 13.1).

3.5.9 CLASSICAL TECHNIQUES TO CALCULATE MACHINE INTERFERENCE

There have been many methods, graphs, equations, charts, and tables developed to handle various aspects of the machine interference problem. Some calculate an average amount of machine interference, and others relate machine efficiency (actual average production time as a percentage of total average clock time) to average service time, average interference time, number of machines assigned to (one or more) operator(s), and other relationships. Some are generalized to take patrolling time, personal time, maintenance time, ancillary time, and other quantities into consideration.

Some of these techniques and the associated assumptions are presented in the next sections, followed by a discussion of the assumptions under which each technique is valid. Exhibits 3.5.13 and 3.5.14, presented later, summarize many of the methods and models and indicate when each might be applicable.

Deterministic Situation

Let us start with a simple example* to demonstrate the basic concepts. The following analysis assumes that the machines are similar. Suppose that it takes an operator 0.15 min to load and/or

*John Mariotti, personal communication, 1981.

unload a workpiece to/from a machine and 0.05 min to inspect the machine and walk to the next. Once loaded, the machine is running for 0.35 min. Then we have

$$P = 0.35$$
$$S = 0.15$$
$$W = 0.05$$

Note that the service is *regular* and *constant*. The operator's walking and inspection can be done while a machine is running, but this is not always possible. In general, when the cycle time is

$$C = S + P$$

(there is no interference time), the number of machines that should be assigned to an operator is

$$N' = \frac{S + P}{S + W}$$
$$= \text{perfect assignment}$$

since for N' machines there is neither operator nor machine idle time. For our problem

$$N' = \frac{0.15 + 0.35}{0.15 + 0.05} = 2.5$$

When N' is not an integer, the number of machines assigned is either N or $N + 1$ machines, where

$$N < N' < N + 1$$

For our problem

$$2 < 2.5 < 3$$

The actual machine assignment (of N or $N + 1$ machines) will depend on the results of an economic analysis. Let us chart the operator's and the machine's time utilization for $N = 2$ (Exhibit 3.5.2) and $N + 1 = 3$ (Exhibit 3.5.3).

If the operator is taking care of two machines, the cycle time of each machine is

$$C = P + S + I$$
$$= 0.35 + 0.15 + 0$$
$$= 0.50 \text{ min}$$

All walking can be done while the machines are running. The percentage of time during which the operator is idle is

$$1 - Oe = \left(\frac{0.5 - 0.4}{0.5}\right) 100\%$$
$$= 20\%$$

Notice that there is no idle time on either machine.

Let us find the number of pieces produced per hour per machine. The production rate per machine is

$$Pr' = \frac{\text{time}}{C}$$
$$= \frac{60 \text{ min/hr}}{0.5 \text{ min/piece/machine}}$$
$$= 120 \text{ pieces/hr/machine}$$

The actual amount of production achieved from two machines is then

$$Pr = (120 \text{ pieces/hr/machine}) \times (2 \text{ machines})$$
$$= 240 \text{ pieces/hr}$$

Exhibit 3.5.2 Operator/Machine Chart for Two Machines

Time	Operator Duties	Operator	Machine No. 1	Machine No. 2
0.1	Unload and load machine 1 (0.15)		0.15	IDLE
0.2	Walk to machine 2 (0.05)			
0.3	Unload and load machine 2 (0.15)			0.15
0.4	Inspect and walk to machine 1 (0.05)			
0.5	Operator idle, both machines processing (0.1)	IDLE	0.35	
0.6	Unload and load machine 1 (0.15)		0.15	
0.7	Inspect and walk to machine 1 (0.05)			0.35
0.8	Unload and load Machine 2 (0.15)			0.15
0.9	Inspect and walk to machine 1 (0.05)			
1.0	Operator idle both machines processing (0.1)	IDLE	0.35	

Legend

Operator service

Operator walking

Machine idle

Operator service

Machine idle

Suppose that

$$c_{\ell} = \text{labor rate of pay} = \$14.00/\text{hr}$$
$$c_N = \text{cost of running the machine} = \$15.00/\text{hr/machine}$$

Then the operator cost per workpiece is (\$14.00/hr)/(240 pieces/hr) = \$0.058/piece, and the machine cost per piece is [(\$15.00 + \$15.00)/hr] /(240 pieces/hr) = \$0.125/piece. The total cost per piece is \$0.058 + \$0.125 = \$0.183.

In general, the cost per piece if we choose N machines is

$$\text{cost}_N = \frac{c_{\ell} + Nc_N}{60 \times N/C}$$

Finally, personal time off (for lunch, breaks, etc.) for the operator can be allowed. Assume that the operator has 15% of the time off. Then the adjusted cycle time becomes (0.5) (1.15) = 0.575 min/piece/machine, and the quantities calculated can be adjusted accordingly.

Another analysis can be performed to find similar information when three machines are assigned to the operator. In this case the operator is always working, but the machines are periodically idle. As before, the operator is patrolling while the machine is running, but notice that the walking time now causes idle time or interference. In this case

$$\begin{aligned} C &= P + S + I \\ &= 0.35 + 0.15 + 0.1 \\ &= 0.60 \end{aligned}$$

Exhibit 3.5.3 Operator/Machine Chart for Three Machines

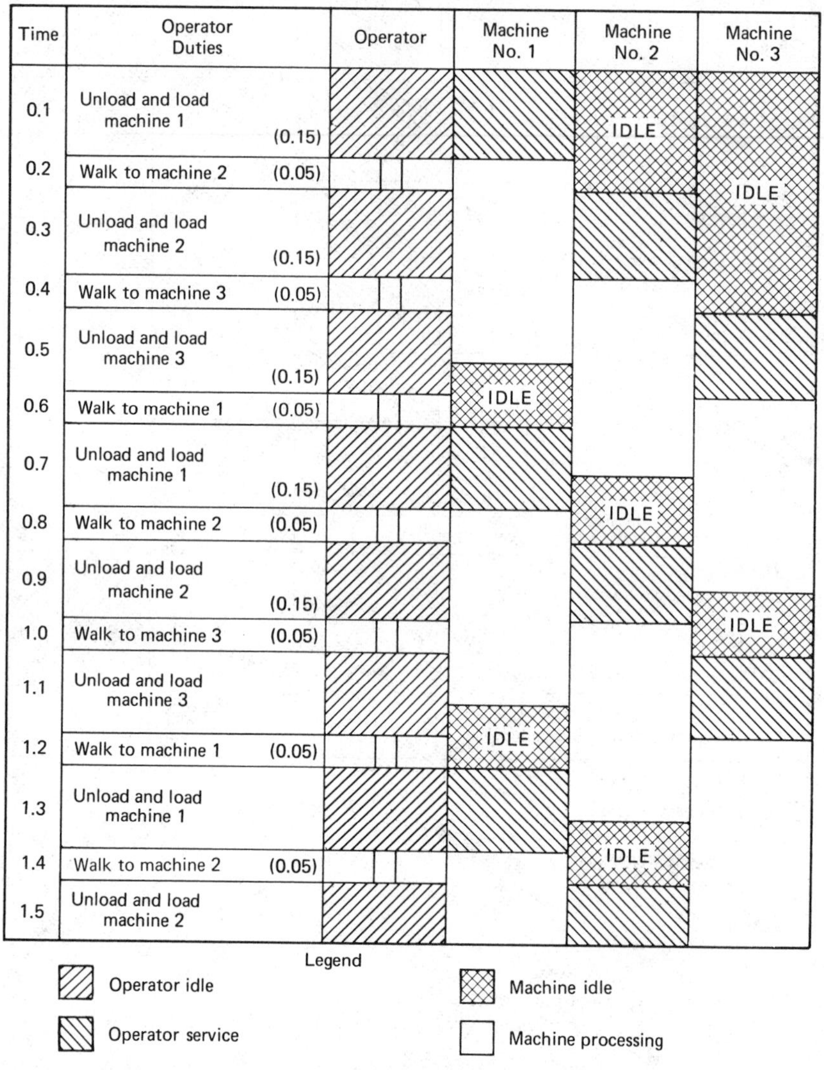

Time	Operator Duties		Operator	Machine No. 1	Machine No. 2	Machine No. 3
0.1	Unload and load machine 1	(0.15)			IDLE	
0.2	Walk to machine 2	(0.05)				IDLE
0.3	Unload and load machine 2	(0.15)				
0.4	Walk to machine 3	(0.05)				
0.5	Unload and load machine 3	(0.15)				
0.6	Walk to machine 1	(0.05)		IDLE		
0.7	Unload and load machine 1	(0.15)				
0.8	Walk to machine 2	(0.05)			IDLE	
0.9	Unload and load machine 2	(0.15)				
1.0	Walk to machine 3	(0.05)				IDLE
1.1	Unload and load machine 3					
1.2	Walk to machine 1	(0.05)		IDLE		
1.3	Unload and load machine 1					
1.4	Walk to machine 2	(0.05)			IDLE	
1.5	Unload and load machine 2					

Legend

Operator idle

Operator service

Machine idle

Machine processing

The production rate is

$$Pr = \left(\frac{60 \text{ min/hr}}{0.6 \text{ min/piece/machine}} \right) \times (3 \text{ machines})$$
$$= 300 \text{ pieces/hr}$$

The operator cost per piece is (\$14.00/hr)/(300 pieces/hr) = \$0.047/piece. The machine cost per piece is (3 × \$15.00/hr)/(300 pieces/hr) = \$0.15/piece. The total cost per piece is \$0.197. In general, the cost per piece if we choose $N + 1$ (e.g., 3) machines is

$$\text{cost}_{N+1} = \frac{c_\ell + (N+1)\, c_N}{[60(N+1)]/[(N+1)\, S]}$$

since the cycle time is now $C = (N + 1)\, S$ because the operator is not idle, but the machines now are. When $\text{cost}_N/\text{cost}_{N+1} > 1$, an assignment of $N + 1$ machines is best; when $\text{cost}_N/\text{cost}_{N+1} < 1$, an assignment of N machines is best.

If 15% of the time is allowed for personal time off, the adjusted cycle time for three machines is (0.6) (1.15) = 0.69 min/piece/machine.

It can be seen that the operator cost per piece is more for two machines (because fewer pieces are being produced). The machine cost per piece is more expensive for three machines (because more machines are required, although more pieces are being produced). The total cost per piece is less for two machines.

Such an analysis is useful to answer many types of questions. One situation could be the following. Suppose that two machines are owned already and that it would cost \$40,000 to buy another machine. Considering only the labor cost, what is the minimum number of pieces that have to be produced to recover this initial cost? The answer is \$40,000/(\$0.058 – \$0.047) per piece = 3,636,364 pieces.

A next logical question is the following: If the third machine is bought, how many years would it take to make up the money spent for the new machine? The answer is 3,636,364 pieces/[(300 pieces/hr) × (8 hr/shift) (2 shifts/day) × (240 days/year)] = 3.16 years. It may not be desirable to purchase another machine.

This system is completely deterministic. All time quantities are constant, or regular. It is assumed that machines are similar and do not break down. This example is used later to demonstrate how to use other methods. Of course the assumptions will sometimes differ. The production and/or service times may be random rather than constant. The average production time is the average time between services. If breakdowns occur randomly, then the production time, P, will be the average time between breakdowns.

Expected Workload Analysis

Another procedure is based on an analysis of the expected workload of an operator, given the number of machines assigned. In this method the average time to produce a part (the production plus service plus interference times) weighted by the machine efficiency (which requires a knowledge of the interference time) is divided by an expected work load of the operator (which could include service and walking and ancillary times), resulting in the number of machines to assign to an operator.[3,4]

At first glance this method seems ideal, but it has drawbacks in that interference time must be known in order to choose N; likewise, N must be known before I can be measured or calculated. This problem spurred much research into various ways (formulas, graphs, tables) to calculate I given N.

Many of the methods described shortly give I as some function of N and λ under various assumptions.

Machine Interference Formula

One of the earlier techniques was a formula developed by Wright,[5] which determines the amount of machine interference (as a percentage of the average service time) as a function of:

1. $X = P/S = 1/\rho$ = ratio of average production time to the average service time and
2. N = number of machines assigned to *one* operator.

His equation is adapted from Fry's[6] solution of a problem of telephone line congestion. The required assumptions are that:

1. service times are *constant* and
2. machines are *randomly serviced* or *tended* by an operator. Wright's formula is

$$\frac{I}{S} \times 100 = 50\left\{[(1 + X - N)^2 + 2N]^{1/2} - (1 + X - N)\right\}$$

The formula was empirically checked with an analysis of more than 1100 hr of actual collected shop data. The study included eight different kinds of machines in four different industries and so is considered to be a general study. It concluded that the formula was accurate for assignments of six or more machines to one operator, but not for assignments of fewer than six machines. Wright then developed the empirical curves shown in Exhibit 3.5.4 to use when the number of machines is less than or equal to six. With respect to his formula, the curves for two to five machines give smaller values for I; for six machines, the results from the graph and the formula are identical except at the limiting points.

As an example, suppose that an operator has been assigned 70 machines. The average production time per unit of output has been determined by a stopwatch study to be 150 min, and the mean service time per unit of output, 5 min. Then the average interference time (as a percentage of the

Exhibit 3.5.4 Interference When the Number of Machines Assigned to One Operator is Six or Fewer[a]

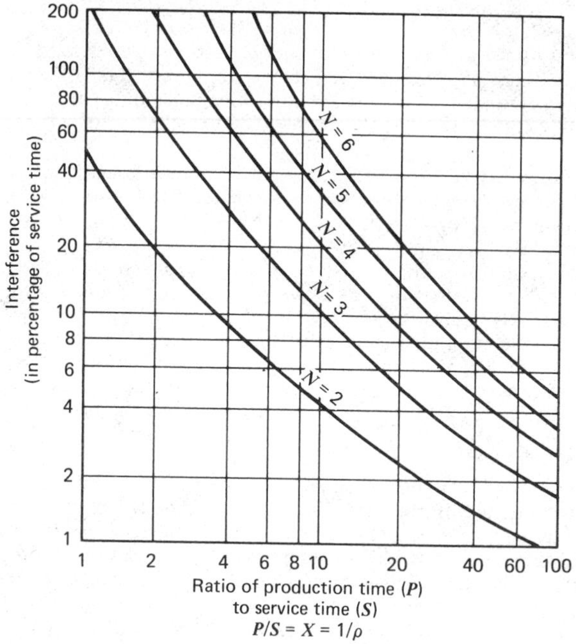

[a]Reproduced by permission from reference 5.

average service time) is

$$\frac{I}{S} \times 100 = 50\left\{[(1 + X - N)^2 + 2N]^{1/2} - (1 + X - N)\right\}$$

$$= 50\left\{\left[\left(1 + \frac{150}{5.0} - 70\right)^2 + 140\right]^{1/2} - \left(1 + \frac{150}{5.0} - 70\right)\right\}$$

$$= 3987.7\%$$

Then we have the following (*average* values, in minutes):

Production time (per unit)	150.00
Service time (including personal time, walking time, etc.)	5.00
Interference time (39.88 × 5.0 = 199.4)	199.40
Total time (for 70 machines)	354.40
Time per machine (354.4/70 = 5.06)	5.06

An example that finds the economical number of machines to assign to an operator based on assigning costs to both the operator (labor cost) and the machines (production cost) is given in Carson et al.[4] Given the cost information, the economical number of machines to assign to an operator can be read from a graph (see Exhibit 3.5.5)[5] where the input required is the ratio of production (machine) cost to labor cost and the ratio of production time to service time ($P/S = 1/\rho$).

In 1965 Smith[7] modified Wright's formula to get

$$\frac{I}{S} \times 100 = 50\left\{[(N - 1.5 - X)^2 + 2(N - 1)]^{1/2} + (N - 1.5 - X)\right\}$$

A comparison of the solutions obtained by this formula with those obtained by Palm,[8] Ashcroft,[9] Wright,[5] and others indicated that Smith's formula was most similar to Palm's and Ashcroft's.

We now apply Smith's formula to the example of Section 3.5.9 for both $N = 2$ and $N = 3$. For

Exhibit 3.5.5 Diagram for Determining the Number of Machines to Assign to an Operator[a]

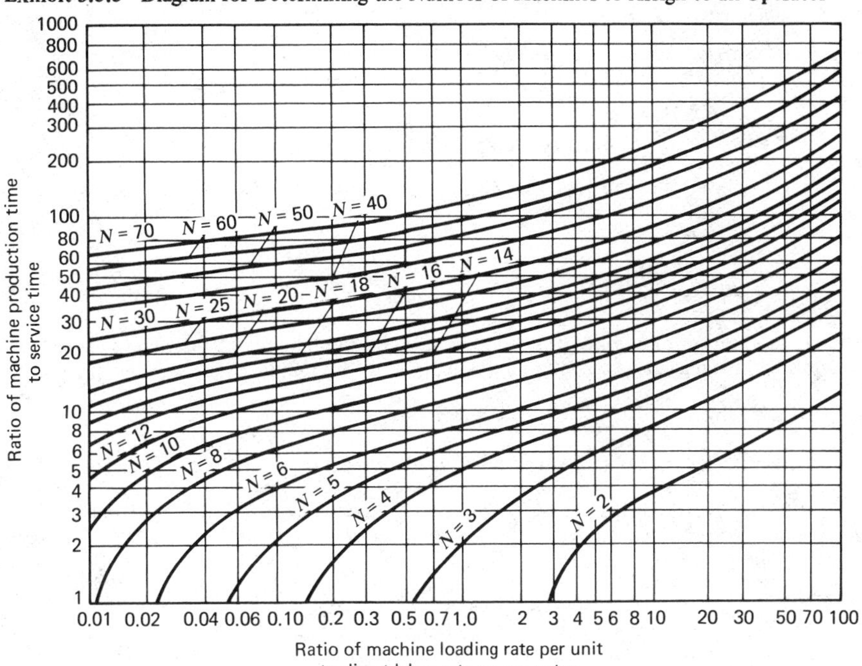

[a]Reproduced by permission from reference 5.

this problem, the walking time, $W = 0.05$, is combined with the service time, $S = 0.15$, to get

$$S = 0.15 + 0.05 = 0.2$$

$$P = 0.35$$

$$X = \frac{P}{S} = \frac{0.35}{0.2}$$

$$N = 2$$

The interference as a percentage of average service time is

$$\frac{I \times 100}{S} = 50 \left\{[(N - 1.5 - X)^2 + 2(N - 1)]^{1/2} + (N - 1.5 - X)\right\}$$

$$= 50 \left\{\left[\left(2 - 1.5 - \frac{0.35}{0.2}\right)^2 + 2(1)\right]^{1/2} + (2 - 1.5 - 1.75)\right\}$$

$$= 50 \left\{[1.5625 + 2]^{1/2} - 1.25\right\}$$

$$= 50 \{1.887 - 1.25\}$$

$$= 50 \{0.637\}$$

$$= 31.873$$

Then

$$I = (0.3187)(0.2)$$

$$= 0.064 \text{ min}$$

Notice how the random nature of breakdowns introduced some interference time into the situation where none had occurred in the deterministic system.

If the operator were to be in charge of three machines, the average interference time could be similarly calculated.

$$\frac{I \times 100}{S} = 50 \{[(3 - 1.5 - 1.75)^2 + 2(2)]^{1/2} + (3 - 1.5 - 1.75)\}$$

$$= 50 \{2.016 - 0.25\}$$

$$= 88.278$$

Then

$$I = (0.883)(0.2)$$
$$= 0.177 \text{ min}$$

The random nature of breakdowns tends to increase the amount of interference time. In the deterministic three-machine situation of Section 3.5.9, the interference time was only 0.1 min.

Ashcroft's Method

In 1950 Ashcroft[9] developed tables that determined machine efficiency as a function of

1. $Y = S/P = \rho$ = ratio of the average service time to the average production time and
2. N = the number of machines assigned to *one* operator.

The assumptions are:

1. Service times are *constant* (hence the machines are assumed to be similar).
2. The machines are *randomly tended.*
3. The time between breakdowns of each of the machines is exponentially distributed.

For his formulas and tables, Ashcroft measured the machine output to be the average number of production hours per hour when N machines are assigned to one operator. If we call this quantity A_N, then the machine efficiency is

$$Me = \frac{A_N}{N}$$

Two of Ashcroft's tables appear as Exhibits 3.5.A1 and 3.5.A2 in the appendix to this chapter. Additional tables and graphs are found in O'Connor[10]; a good account of the derivation of Ashcroft's equation can be found in Eilon.[11]

Applying the data from the example in Section 3.5.9,

$$Y = \frac{S + W}{P} = 0.2/0.35 = 0.57$$

Interpolating from Exhibit 3.5.A1 in the appendix,

$$A_2 = 1.17$$
$$A_3 = 1.53$$

These values are the average number of machine hours per hour when two and three machines, respectively, are assigned to the operator. From these

$$Me \text{ (for } N = 2) = 1.17/2$$
$$= 0.585$$

$$Me \text{ (for } N = 3) = 1.53/3$$
$$= 0.51$$

Of course, in general, an assignment of two machines has to be more "machine efficient" than that of three machines, since there would be less interference time. There is more output from three machines, but more operator idle time when only two machines are assigned. Also, three machines cost more than two.

Probability-Based Models

The next technique takes into consideration the variability of the interference times by using the binomial distribution (see Chapter 13.2) as follows: Let

$$D = \text{Prob[one of the } N \text{ machines is down at any given moment]}$$
$$1 - D = \text{Prob[one of the } N \text{ machines is working at any given moment]}$$

The probability that any n of the N machines are down at any moment is

$$\binom{N}{n} D^n (1 - D)^{N-n}$$

where

$$\binom{N}{n} = \frac{N!}{n!\,(N-n)!}, \quad N! = N(N-1)(N-2)\ldots 3 \times 2 \times 1$$

= the number of ways N machines can be partitioned into 2 groups

where

group 1 contains n machines (the down machines)
group 2 contains the remaining $N - n$ machines (those working). Let

$$D^n = \text{Prob}[n \text{ of the } N \text{ machines are down}]$$
$$(1 - D)^n = \text{Prob}[N - n \text{ of the machines are up}],$$

if the machines are independent.

The average interference time is calculated as follows: If n of the N machines are down, then $n - 1$ machines have to wait (because of interference). Then

$$I_n = \text{Prob}[n \text{ of the } N \text{ machines are down}] \times \text{number of interference waits}$$

and

$$I = \frac{1}{N} \sum_{n=0}^{N} I_n = \frac{1}{N} \sum_{n=0}^{N} \binom{N}{n} D^n (1 - D)^{N-n} (n - 1)$$

Note that this is the expected value of $n - 1$ interference waits where n has a binomial distribution. That is, the average interference is found by adding the products of the probabilities and the corresponding number of waits.

An example will demonstrate how machine interference is calculated. Suppose that one operator is in charge of six machines. From the production obtained at the end of a week, it was seen that the average downtime per machine was 20% of the time that the machines were operating. In this case we have

$$N = 6$$
$$D = \text{Prob[one of the } N \text{ machines is down]} = \tfrac{1}{5} = .2$$
$$1 - D = \text{Prob[one of the } N \text{ machines is up]} = \tfrac{4}{5} = .8$$

Let us look at one case. Suppose that three out of the six machines are down ($n = 3$). While one of the machines is being repaired, the other two must wait. Then

$$\begin{aligned} \text{Prob[3 of the 6 machines are down]} &= \binom{N}{n} D^n (1 - D)^{N-n} \\ &= \binom{6}{3} (.2)^3 (1 - .2)^{6-3} \\ &= 20(.008)(.8)^3 \\ &= .08192 \end{aligned}$$

The average interference time from three down machines is calculated as follows: Since three of the six machines are down, two machines have to wait (because of interference); then average interference for three machines down is

$$\begin{aligned} I_3 &= \text{Prob}[3 \text{ of the 6 machines are down}] \times \text{number of interference waits} \\ &= .08192\ (2) \\ &= .16384 \end{aligned}$$

A working table is formed to calculate interference (see Exhibit 3.5.6). Since there are six machines, the average percentage of interference per machine, $\tilde{I}$, would be

$$\tilde{I} = .462/6 = .077 = 7.7\%$$

and the average fraction of time in service, $\tilde{S}$, is

$$\tilde{S} = D - \tilde{I} = .2 - .077 = .123 = 12.3\%$$

The assumptions of this model are:

1. The machines are *randomly tended.*
2. Service times are *randomly distributed*, with a mean service time S for all machines.
3. Machines fail independently.

Jones[12] derived formulas for calculating interference (in average percentage of total time) per machine and operator idle time (in percentage of total time) as functions of

1. the operator's average service time per machine as a percentage of total time, S' (i.e., $S' = S/C$) and
2. N = the number of machines assigned to one operator.

Jones' formula for I is derived as follows:

1. Recall that

$$\begin{aligned} (1 - D)^N &= \text{Prob}[\text{all } N \text{ machines are up}] \\ &= \text{Prob}[\text{the operator is idle}] \end{aligned}$$

2. Therefore

$$\begin{aligned} 1 - (1 - D)^N &= \text{Prob}[\text{one or more of the machines are down}] \\ &= \text{Prob}[\text{the operator is working}] \end{aligned}$$

3. $[1 - (1 - D)^N]/N$ = the proportion of total time that the operator will service each machine, which is also $S'(1 - I)$.
4. The probability that any machine is down is

$$D = S'\ (1 - I) + I\,*$$

Exhibit 3.5.6 Interference Time

n No. of Machines Down Together	$\binom{N}{n}$		D^n		$(1 - D)N_{-n}$		Interference Waits	Interference
6	1	×	$.2^6$			×	5	.00032
5	6	×	$.2^5$	×	.8	×	4	.006144
4	15	×	$.2^4$	×	$.8^2$	×	3	.04608
3	20	×	$.2^3$	×	$.8^3$	×	2	.16384
2	15	×	$.2^2$	×	$.8^4$	×	1	.24576
1	6	×	$.2^1$	×	$.8^5$	×	0	
0	1			×	$.8^6$	×	0	
							Total Interference =	.462144

*If one operator serviced only one machine, the average percentage service time would be S', and the machine would never have interference idleness. Since an operator actually services several (N) machines, interference becomes a factor; in this case the average percentage service time is $S'(1 - I)$, and so the average downtime $D = S + I = S'(1 - I) + I$.

5. From steps 3 and 4, we have

$$S'(1 - I) = D - I = \frac{1 - (1 - D)^N}{N}$$

Therefore, average percentage interference is

$$I = D - \frac{1 - (1 - D)^N}{N}$$

A table of values from this formula can be found in Exhibit 3.5.A3 in the appendix at the end of this chapter. This work assumes that each machine in any possible assignment has the same service time requirements, that is, that machines are similar. Jones relaxed this assumption in a later paper.[13] He developed a simulation of assignments on 2 to 10 machines randomly tended by one operator. The output consisted of the resultant machine interference and operator idle time. Results from 100 runs, whose machine assignments involved variable machine average service times, implied that, for any given number of machines, the average interference per machine decreases as the degree of difference in the average service times of the assigned machines increases. Adjustment factors were derived, which could be applied to his previous table (Exhibit 3.5.A3), to arrive at corrected interference values. The adjustment factors were graphed for easier use. The equations and tables can be found in Maynard.[14]

3.5.10 QUEUING THEORY APPROACHES

Among the first mathematical discussions of machine interference is one (1933) due to Khintchine.[15] Patrolling times were ignored or incorporated into the average service times. Other early papers are those of Palm[8] and Kronig.[16]

The Classical Model

This section presents the classical queuing theory approach to solving a machine interference problem. Suppose that there are N machines and O operators. (For the methods reviewed so far, $O = 1$). As an example of such a model, consider a computer system consisting of N independent processors (CPUs) and O independent memory modules such that each CPU can access each memory module. In such a system, independent programs may simultaneously request access to the same memory module and interference will occur. The assumptions of the model are:

1. The time between breakdowns (or production time) of any one of the machines is a sample from a *negative exponential* probability distribution with mean $1/\lambda$ (or mean rate λ). A breakdown is random and is independent of the operating behavior of the other machines. Then, when there are n machines not working at time t,

$$\text{Prob[one of the } M - n \text{ machines goes down in the interval } (t, t + \Delta t)] = (N - n)\,\lambda\,\Delta t + O(\Delta t),$$

where Δt is a small increment of time.

2. Any one of the n down machines requires only one of the O operators to fix it. The service time distribution is *negative exponential* with mean $1/\mu$ for each machine and each operator. The service times are mutually independent and also independent of the number of down machines. Then

$$\text{Prob[one of the } n \text{ down machines is fixed in an interval } \Delta t]$$

$$= \begin{cases} n\mu\,\Delta t + O(\Delta t) & \text{for } 1 \leqslant n \leqslant O \\ O\mu\,\Delta t + O(\Delta t) & \text{for } 0 < n \leqslant N \end{cases}$$

3. The machines are *randomly tended.*

The system can be pictured as shown in Exhibit 3.5.7. Note that the model is that of a finite source, closed queuing model in which the arrival (breakdown) rate decreases as the number in the system (number of down machines, n) increases.

The queue contains $n - O$ down machines not being serviced. Let the average queuing (waiting, interference) time be denoted by I. Let

$$N_q = \text{expected (or average) number of down machines in the queue}$$

Exhibit 3.5.7 ***M*****-Machine Service Queuing Systems With** ***O*** **Operators**

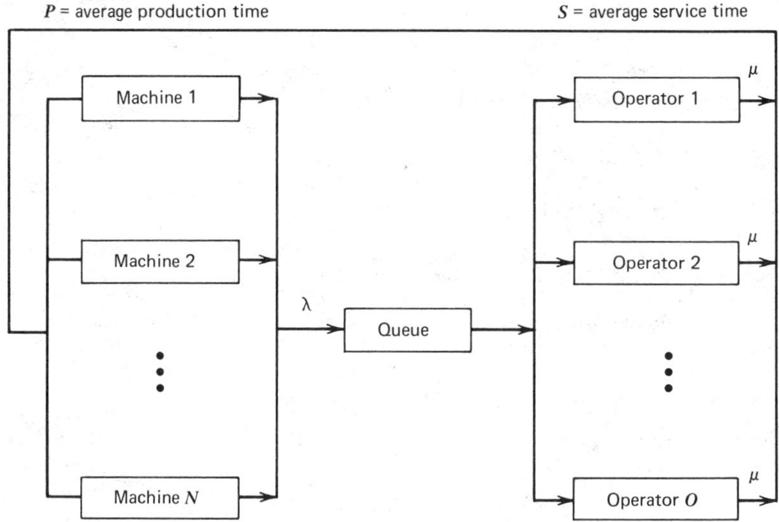

$E[n]$ = expected (or average) number of down machines

$M(t)$ = the number of down machines at time t

Define

$$P_n(t) = \text{Prob}[M(t) = n \mid M(0) = i]$$

A transition from state $P_n(t)$ to $P_{n+1}(t + \Delta t)$ is caused by a breakdown of one of the $N - n$ working machines; a transition from state $P_n(t)$ to $P_{n-1}(t + \Delta t)$ means that one of the down machines has been fixed. The state $P_0(t)$ occurs when all of the machines are up. Then the stochastic process, $M(t)$, can be modeled as a birth and death process, with rates

$$\lambda_n = \begin{cases} (N - n)\lambda, & n = 0, 1, \ldots, N \\ 0, & n > N \end{cases}$$

$$\mu_n = \begin{cases} n\mu, & n = 1, 2, \ldots, 0 \\ O\mu, & n = 0 + 1, \ldots, N \end{cases}$$

The transition diagram is as shown in Exhibit 3.5.8.

The forward Kolmogorov equations of the birth-death process are

$$P'_0(t) = N\lambda P_0(t) + \mu P_1(t)$$

$$P'_n(t) = -\{(N - n)\lambda + n\mu\} P_n(t) + (N - n + 1)\lambda P_{n-1}(t) + (n + 1)\mu P_{n+1}(t), \qquad 1 \leqslant n < 0$$

$$P'_n(t) = -\{(N - n)\lambda + O\mu\} P_n(t) + (N - n + 1)\lambda P_{n-1}(t) + O\mu P_{n+1}(t), \qquad 0 \leqslant n < N$$

$$P'_N(t) = -O\mu P_N(t) + \lambda P_{N-1}(t)$$

This finite system of ordinary differential equations can be solved (for the equilibrium values of P_n) by setting these first derivatives equal to zero while noting that the equilibrium (or stationary

Exhibit 3.5.8 State Transition Diagram for Machine Breakdown and Repair

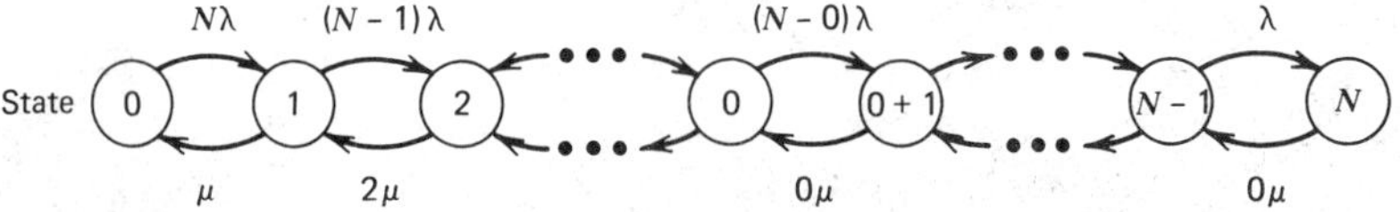

or steady state) values are

$$P_n = \lim_{t \to \infty} P_n(t)$$

The flow balance equations become

$$N\lambda P_0 = \mu P_1$$

$$\{(N-n)\lambda + n\mu\} P_0 = (N-n+1)\lambda P_{n-1} + (n+1)\mu P_{n+1}, \qquad 1 < n < 0$$

$$\{(N-n)\lambda + O\mu\} P_0 = (N-n+1)\lambda P_{n-1} + O\mu P_{n+1}, \qquad 0 \leqslant n < N$$

$$O\mu P_N = \lambda P_{N-1}$$

These equations are solved recursively using the relationship

$$(N-n)\lambda P_n = \begin{cases} (n+1)\mu P_{n+1}, & n < 0 \\ O\mu P_{n+1}, & n \geqslant 0 \end{cases}$$

Letting $\rho = \lambda/\mu$ (the servicing factor), the steady state probabilities are

$$P_n = \begin{cases} \binom{N}{n} \rho^n P_0, & n = 0, 1, \ldots, O \\ \binom{N}{n} \dfrac{n!}{O!O^{n-O}} \rho^n P_0, & n = O+1, \ldots, N, \end{cases} \qquad (1)$$

where P_0 is obtained by solving $\Sigma_{n=0}^{N} P_n = 1$ to get

$$P_0 = \left[\sum_{n=0}^{O} \binom{N}{n} \rho^n + \sum_{n=O+1}^{N} \binom{N}{n} \frac{n!}{O!O^{n-O}} \rho^n \right]^{-1}$$

This solution is a special case of the birth and death process as given in Bhat[17] and Feller,[18] for example.

The expected (average) number of down machines is

$$E[n] = \sum_{n=0}^{N} nP_n = \sum_{n=1}^{N} (n-1) P_{n-1} = \sum_{n=1}^{N} nP_n - (1 - P_0)$$

It can be seen that there is no closed-form expression for $E[n]$ in general, but for a particular problem (system), $E[n]$ is easily computed. There is a closed-form expression for single-server (only one operator) systems. In this case

$$E[n] = N + \frac{\lambda + \mu}{\lambda}(1 - P_0)$$

Some system performance measures are:

1. Machine efficiency is

$$Me = \frac{E[n]}{N}$$

or percentage of average production obtained (or the fraction of total production time on all machines).

2. Average operator utilization is

$$Oe = \sum_{n=0}^{O} \frac{nP_n}{O} + \sum_{n=O+1}^{N} P_n$$

or fraction of time an operator would be working.

3. Average number of idle operators is

$$\sum_{n=0}^{O} (O - n) P_n$$

4. Average number of machines waiting is

$$\sum_{n=O+1}^{N} (n - O) P_n$$

By dividing measure 3 by the number of operators, O, and measure 4 by the number of machines, N, some related measures are

3.′ Coefficient of loss for operator is

$$\frac{\sum_{n=0}^{O} (O - n) P_n}{O}$$

or percentage of idle operators.

4.′ Coefficient of loss for machines is

$$\frac{\sum_{n=O+1}^{N} (n - O) P_n}{N}$$

or percentage of interference time.

The purpose of the following example is to show the advantages obtained in system performances and productivity from the pooling of operators. In this case several operators have the same assignment of machines.

Exhibit 3.5.9 from Bhat[17] has values for operator utilization for pairs of (N, O) parameters that have the same machine per operator ratio ($N/O = 4$ and then 15).

Notice that the operator utilization is increasing for a given ρ even though the ratio of the number of machines per operator stays the same. This is an indication that it is better, when feasible, to pool operators rather than to assign a particular number of machines to each operator individually. A good example that shows similar advantages from pooling can be found in Feller.[18] The example considers two cases: (1) 6 machines serviced by one operator and (2) 20 machines serviced by three operators. The results show that, even though the workload per operator increased from system 1 (6 machines/operator) to system 2 ($6\frac{2}{3}$ machines/operator), the machines were serviced more efficiently in system 2. The advantages of pooling are well known. For a discussion of flexibilities and advantages from pooling machines, see Stecke and Solberg.[19]

Exhibit 3.5.9 Operator Utilizations for Proportional Parameters

ρ	N	O	Operator Utilization
0.45	4	1	0.881
	8	2	0.934
	16	4	0.994
0.05	15	1	0.656
	30	2	0.682
	60	4	0.705

Reproduced by permission from reference 17.

Notice in Exhibit 3.5.9 that, for a given number of machines per operator (assuming N/O is integer), as O (and therefore N) increases, the operator utilization slowly increases. Likewise, under the same conditions, the machine efficiency will slowly increase.

Some of the theory and results regarding machine interference from queuing theory are presented in some of the standard operations research textbooks, such as that of Wagner.[20]

Parts of the preceding model as well as several generalizations can be found in Allen.[21] In particular, a closed queuing network problem is re-solved under different assumptions:

1. the service time per machine by one operator, S, is *constant*, and
2. the production time per machine, P, is *constant*. Formulas similar to equation 1 are derived.

The limiting equations (equation 1) have been solved recursively[22–26] to prepare tables useful in deciding the economically optimal number of operators for a given number (N) machines. A table of D.C. Palm[22] is given as Exhibit 3.5.A4 in the appendix of this chapter. The assumptions (repeated from the beginning of this section in briefer form) are:

1. The time between breakdowns of any one of the machines is a sample from a *negative exponential distribution* with mean P.
2. The service times are *exponentially distributed* with mean S.
3. The machines are *randomly tended*.

Peck and Hazelwood[26] offer mostly machine interference tables. Their tables were generated on a computer (UNIVAC I®) using formulas developed from the previous closed queuing model. In reviewing some old notation and introducing some new notation, recall that

I = average interference time per machine
S = average service time per machine
P = average production time per machine
Me = machine efficiency = fraction of the machines running = $(S + P)/(S + P + I)$ (slightly different from the definition in Section 3.5.7; usually the service time is not included)
$X = S/(S + P)$
D = Prob[delay] = Prob[a down machine incurs interference time]
N_S = average number of machines being serviced
N_P = average number of up machines (producing)
N_I = average number of machines waiting for service

Then

$$N_S = MeNX = N\left(\frac{S+P}{S+P+S}\right)\left(\frac{I}{S+P}\right) = N\left(\frac{I}{S+P+S}\right)$$

$$N_P = (1 - X)\,NMe$$

$$N_I = N(1 - Me)$$

Before demonstrating the use of the tables with an example, it should be pointed out that they are very extensive. The number of machines tabulated ranges from 4 to 250, and $X \in [0.001, 0.950]$ in varying increments (see Exhibit 3.5.A5 in the appendix).

Suppose that the Diedinthewool textile firm has 14 looms. From time studies it was found that on the average the machines are running for 64 min and require 36 min of service. Then

$$N = 14$$

$$X = \frac{S}{S+P} = \frac{36}{36+64} = 0.36$$

$$N_S = MeNX = Me(14)\,0.36 = 5.04\,Me$$

$$N_I = N(1 - Me) = 14(1 - Me)$$

$$N_P = (1 - X)\,NMe = 0.64(14)\,Me = 8.96\,Me$$

Each machine causes a profit of \$10.00 for each hour producing (longer than an hour of real time). The cost to service the machines is \$5.00/hr/operator. The operators get paid whether or not they are working. Using Exhibit 3.5.A5 we will construct our own table, as shown in Exhibit 3.5.10.

Exhibit 3.5.10 Fourteen Machines, Three to Nine Operators

O	9	8	7	6	5	4	3
N_S	5.035	5.02	4.975	4.83	4.51	3.886	2.99
N_I	0.014	0.056	0.182	0.574	1.47	3.206	5.698
N_P	8.95	8.924	8.844	8.593	8.02	6.91	5.313
Labor cost: \$5 × O	45.00	40.00	35.00	30.00	25.00	20.00	15.00
Profit: \$10 × N_P	89.50	89.24	88.44	85.93	80.20	69.10	53.13
Net profit (\$/hr)	44.50	49.24	53.44	55.93	57.20	49.10	38.13

To see how the entries from this table were calculated, look at the N_P row and the column for seven operators.

$$N_P = (1 - X)\,NMe = 0.64\,(14)\,Me = 8.96\,Me$$

From Exhibit 3.5.A5 we see that, for $X = S/(S + P) = 0.36$ and $O = 7$, $Me = 0.987$. Then $N_P = 8.96$, $Me = 8.96(0.987) = 8.844$. The other entries are similarly calculated. The maximum profit per hour is \$57.20, which is attained when five operators are tending the 14 machines.

Suppose that a policy of preventive maintenance for the 14 machines is introduced, costing \$10.00/hour. Because of this extra care, the service time required decreases from 40 to 20 hr, and the production time increases from 60 to 80 hr. Now

$$X = \frac{S}{S+P} = \frac{20}{100} = 0.2$$

$$N_S = MeNX = 14\,(0.2)\,Me = 2.8\,Me$$

$$N_I = N(1 - Me) = 14\,(1 - Me)$$

$$N_P = (1 - X)\,NMe = 0.8\,(14)\,Me = 11.2\,Me$$

The profit per hour producing remains the same (\$10.00). The service cost is now

$$\text{labor cost} = \$5.00 \times O + \$10.00$$

We will construct a new table (Exhibit 3.5.11) utilizing the same Exhibit 3.5.A5 in the appendix.

As before, the maximum profit per hour is attained when the 14 machines are tended by five operators. Note that the cost for three operators and maintenance is the same as the cost for five (the optimum) in the previous example, yet the net profit is more than 30% higher because the machines are utilized more.

These tables can be used for many types of problems. One such application is in determining the number of operators required in order to meet some demand or production requirement. As in our examples, the cost of additional labor can be calculated and compared with the cost of lost production. Additional costs might be added if the required production is not met.

Trade-offs between assigning N machines to one operator versus, say, an assignment of $2N$ machines to two operators can be examined. All things being equal, the second assignment must be better. But all things are not equal. For example, the patrolling time or the average time to walk from machine to machine could be larger for a larger group of machines. This increase can be absorbed into the average service time.

Exhibit 3.5.11 Fourteen Machines, Two to Six Operators

O	6	5	4	3	2
N_S	2.797	2.78	2.724	2.526	1.952
N_I	0.014	0.098	0.378	1.372	4.242
N_P	11.19	13.902	10.898	10.102	7.806
Labor cost: \$5 + \$10	40.00	35.00	30.00	25.00	20.00
Profit: \$10 × N_P	111.90	139.02	108.98	101.02	78.06
Net profit (\$/hr)	71.90	104.02	78.98	76.02	58.06

Morse[1] contains a graph to aid in optimizing operator crew size for a set of N machines based on balancing the cost of the crew against the production costs, assuming exponential breakdown and service times.

Extensions of the Basic Model

The queuing theory results presented so far assume that if walking or patrolling time had been measured, it had been incorporated in with the average service time. Some studies are now mentioned that relax or change some of the usual assumptions.

In 1957 Mack et al.[27] studied the machine interference problem. Their assumptions were:

1. Service time is a *constant* and machines are similar.
2. Queue discipline is *cyclic patrol* and *closed*.
3. Walking (patrolling) time is a *constant*.
4. There is one operator.

Howie and Shenton[28] examined machine efficiency under a closed cyclic patrol queue discipline with constant walking time. Lawing[29] determined the optimum number of machines (quill winders) to assign to a single operator also under a closed cyclic patrol. Time between breakdowns and service times is assumed to be *exponentially distributed*. Ben-Israel and Naor[30] also considered regular patrolling. Grant[31] compared random tending to a cyclic patrol queuing discipline for several operators.

The model of Jaiswal and Thiruvengadam[32] (basically that proposed in Benson and Cox[23]) investigated systems of machines that were subject to two kinds of breakdowns (e.g., older versus newer machines) to decide which type should have a higher priority. The assumptions were:

1. Service times are *general* (need not be exponentially distributed).
2. Queue discipline is a *nonpreemptive priority*.
3. Time between breakdowns is *exponentially distributed*.

This model could be adopted to handle limited operator availability. If the operator is required to perform other duties in addition to servicing the machines, the operator is not always available. This was demonstrated in the example from Section 3.5.9.

The model of Hodgson and Hebble[33] assigns one operator to service N groups of machines with N_i machines in each group i, $i = 1, \ldots, N$. The assumptions are:

1. Service time distribution is *general*.
2. Queue discipline is a *nonpreemptive priority*.
3. The time between breakdowns is *exponentially distributed*.

Reynolds[34,35] decided that in several problems the walking distance to perform a service might be an important consideration. The assumptions of his model are:

1. Service times are *exponentially distributed*.
2. Queue discipline is a *distance priority* (service the nearest down machine).
3. There are several operators.
4. The time between breakdowns is *exponentially distributed*.

Reynolds compared the performance of his model to that of several others:

1. Palm's[36] model that *ignores* patrolling time.
2. Palm's[36] *random tending* model.
3. Lawing's[29] *patrolling* model.
4. Reynolds'[35] *distance priority* model.

All four models assume *exponentially distributed* service times and breakdown times. The performance measure is the effective number of operating machines, given the number of machines (N) and the number of operators (O) (which, when divided by N, is the machine efficiency).

The result of the comparisons was that the distance priority produced the largest effective number of working machines.

3.5.11 SIMULATION/REGRESSION APPROACH

Deakin[37] has used simulation to create a regression model for deciding on the number of machines to assign to an operator. His model is concerned with the cyclic patrol disciplines, both closed and alternating, for a single operator. In addition, under both queuing disciplines, he allows for a "relief operator" to relieve the operator during two 10 min breaks and for a 20 min lunch period, and thereby down machines continue to be serviced. The assumptions are:

1. Service times are *constant.*
2. Queue discipline is *cyclic patrol:*
   a. Closed cycle:
      (1) With a relief operator provided.
      (2) Without a relief operator provided.
   b. Alternating cycle:
      (1) With a relief operator.
      (2) Without a relief operator.
3. Patrolling time is *constant.*
4. Time between breakdowns is *exponentially distributed.*

Assumption 4 implies that the number of times N machines (assigned to one operator) break down each running hour is a sample from a Poisson distribution with mean $N\lambda$.

The objective is to assign an appropriate number of machines to an operator such that the production obtained is close to that desired (which is dependent on demand or inventory). This objective assumes that a particular machine efficiency is the target for which management is aiming. Machine efficiency indirectly specifies the desired production and so makes the characterization of demand less problem-specific.

Since the model has not been documented elsewhere, it is presented here in detail. A set of inputs and corresponding outputs (describing a 2^4 fractional factorial design with a center point—see Chapter 13.4) to a simulation of a cycle tending model are given in Exhibit 3.5.12.

For each of the four types of systems, the procedure consists of making nine simulation runs with two sets of parameters of the input, with that of the ninth run being the mean values of each of the two sets. The output from each run is the machine efficiency, which becomes input into a multiple linear regression model. An equation is fitted to each set of nine points. The natural log of the input is used in the regression to account for nonlinear relationships among the variables. The results are equations relating

N = the number of machines assigned to one operator,

linearly to natural logs of the:

S = service time (constant)
E = expected operator efficiency loss = $(1 - Me)\,100$
W = patrolling time (constant)

Exhibit 3.5.12 Simulation of Cycle Tending Assignments

Simulation Inputs[a]				Output Average Machine Efficiency		
S (Constant)	λ (Poisson)	W (Constant)	N	Closed Cycle	Alternating Cycle	Relief Provided[b]
0.4	1	0.08	30	95.34	94.38	96.87
0.6	1	0.03	30	96.33	95.69	97.87
0.4	3	0.03	30	85.10	85.59	91.74
0.6	3	0.08	30	71.33	71.07	74.85
0.4	1	0.03	24	97.64	97.48	98.61
0.6	1	0.08	24	95.49	94.83	97.02
0.4	3	0.08	24	84.25	82.55	88.97
0.6	3	0.03	24	83.36	83.95	89.68
0.5	2	0.055	27	89.66	89.10	93.39

Reproduced by permission from reference 37.
[a]S = constant service time in minutes; λ = breakdown rate (average machine stops per run hour); W = constant patrolling time in minutes; and N = number of machines assigned.
[b]Closed or alternating cycle.

The resultant equations are:

1. Closed cycle model:

$$N = \frac{\text{EXP}[1.166 - 0.525 \ln(S) + 0.702 \ln(E) - 0.268 \ln(W)]}{\lambda} \tag{2}$$

2. Alternating cycle:

$$N = \frac{\text{EXP}[0.759 - 0.538 \ln(S) + 0.753 \ln(E) - 0.348 \ln(W)]}{\lambda} \tag{3}$$

3. Relief-operator-provided, closed and alternating cycles:

$$N = \frac{\text{EXP}[1.005 - 0.609 \ln(S) + 0.677 \ln(E) - 0.416 \ln(W)]}{\lambda} \tag{4}$$

These equations provide good estimates for $N \in [24, 30]$, $W \in [0.03, 0.08]$, $\lambda \in [1, 3]$, and $S \in [4, 6]$. Although in this particular case the range of applicability is extremely limited, in similar or very different environments, analogous equations can easily be developed.

As an example, to illustrate the use of the equations, suppose that a group of machines breaks down twice per running hour on the average. The operator requires exactly 30 sec (0.5 min) to repair and start a down machine. The operator takes 3 sec (0.05 min) to walk from machine to machine in a closed cycle.

Suppose that production requirements indicate a need for machine efficiency to be 90%. We need to determine the number of machines to be assigned to the operator in order to achieve the required production levels. The assignment parameters are

$$\begin{aligned} S &= 0.5 \text{ min} \\ W &= 0.05 \text{ min} \\ E &= 100\% - 90\% = 10\% \text{ (loss is allowable)} \\ \lambda &= 2 \text{ breakdowns/run hour} \end{aligned}$$

Solving equation 2, the number of machines to assign to the operator is

$$\begin{aligned} N &= \frac{\text{EXP}[1.166 - 0.525 \ln(S) + 0.702 \ln(E) - 0.268 \ln(W)]}{\lambda} \\ &= \frac{\text{EXP}[1.166 - 0.525 \ln(0.5) + 0.702 \ln(10) - 0.268 \ln(0.05)]}{2} \\ &= \frac{51.9}{2} \\ &= 25.95 \text{ machines} \end{aligned}$$

Since $N = 26$ machines is within the range for which the equation is valid (24 to 30), we may assume that such an assignment will result in an average loss of efficiency (due to interference time and service time) of about 10% of the total machine run time available for the given input parameters.

Let us see what the effect is of providing a relief operator to the original to take over when she or he goes on breaks. With the same parameters and equation 4,

$$\begin{aligned} N &= \frac{\text{EXP}[1.005 - 0.609 \ln(S) + 0.677 \ln(E) - 0.416 \ln(W)]}{\lambda} \\ &= \frac{\text{EXP}[1.005 - 0.609 \ln(0.5) + 0.677 \ln(10) - 0.416 \ln(0.05)]}{2} \\ &= 34.4 \text{ machines} \end{aligned}$$

More machines (30% more) can be assigned to the operator (when relief is provided) with no loss of machine efficiency (the desired production is still attained) because down machines do not now pile up on the operator while she or he is on a break.

Since $N = 34$ is outside the region of applicability for the equation, we expect the error associated with the solution to be greater than when N was 26.

Similarly, the range of values used for S, W, λ, and N should not be exceeded because of likely errors from such extrapolation. However, Deakin has had good experiences using the equations despite the obvious extrapolations. Of course simulation may be used directly when the parameters are significantly different.

3.5.12 SIMULATIONS

The three studies described in this section use very detailed simulation programs in their approach to solving machine interference problems. The first study, by Haagensen,[38] used simulation to determine interference time and its effect on machine efficiency. The input to the simulation includes:

1. The number and mix of machines in a proposed assignment.
2. Speed of the machines.
3. The number of operators (restricted to one or two) per assignment.
4. Operator efficiencies.
5. Breakdown times of the machines, which can be

a. Regular—the time between breakdowns is a constant.
b. Not regular—the time between breakdowns is a sample from a
   (1) Uniform distribution, or
   (2) Normal distribution.

For each machine there can be up to seven types of breakdowns. The output is:

1. The amount of the product produced.
2. The time required to produce this amount.
3. Service times.
4. Machine production, idle, and interference times.

The output is not a solution, merely information that is useful in making decisions. The program has been used by the Phoenix Cable Company to aid in:

1. Determining the assignment of machines to one or two operators.
2. "Maximizing" production or "minimizing" costs.
3. Establishing wage incentive payments.
4. Investigating operator learning curves.

The simulation by Freeman et al.[39] was more general than Haagensen's[38] since it allowed for different types of operators (such as mechanics, regular operators, relief operators) of various skill levels. The problem being investigated essentially was this: Given a set of machines and information about required duties and skill levels of operators, what is the optimum number of operators of each skill level required to tend the machines? The computer program is flexible; some of the relaxations are:

1. There may be tasks that

a. Can be done only while a machine is running.
b. May be done while a machine is running.
c. Can or must be done by several operators.

2. The time between breakdowns (or some other task)

a. May depend on running time.
b. May be a function of total (clock) time.
c. May be a sample from a more appropriate distribution, such as normal, constant, or Erlang, depending on the actual data.

3. Some tasks may have slack time (in which case, if the task is done during the slack time, there would be no interference time associated with it).

4. Machines may be of different types and may have different average breakdown rates.

5. The queue discipline can be different for different machines or tasks. Some may be preemptible; others may have a nonpreemptive priority discipline.

6. Personal time and breaks are allowed in the model.

7. Partial operator availability can be modeled.

8. During operator breaks
   a. Machines may be shut down or
   b. A relief operator may be provided or
   c. No relief operator is provided and machines are then allowed to break down as usual.

This simulation has been used for as many as six machines and 25 different tasks. The storage requirements for such a problem are 25,000 integer words and 5 to 10 min/run on a CDC 3300®.

Another general simulation was done at Western Electric.[40] As before, the objective is to attain the goals set by management, which may relate to cost or production criteria, machine efficiency, or interference. The simulation was written in GASP IV[41] (a discrete/continuous FORTRAN-based simulation language) and required about 14,000 words of memory to run on a Xerox Sigma-9® computer. Each 12,000 simulated min required, on the average, 0.63 min of CPU time (see Chapter 13.11).

Simulation is a very useful tool for analyzing a particular situation and the various options in detail. Simulation makes it possible to study the effects of different changes on the operation of a manufacturing system by making changes in the simulation model rather than directly experimenting on the system itself. If a manufacturing system is sensitive to the values of the parameters, simulation is a good method to use in investigating different configurations because it is flexible and can contain many details of the actual system. Approximate queuing and probability methods can be used to provide initial input values when using simulation to find a good assignment. A summary of many of the available techniques is given in Exhibits 3.5.13 and 3.5.14.

3.5.13 DISCUSSION OF THE ASSUMPTIONS

To decide which of the available techniques (many listed in Exhibits 3.5.13 and 3.5.14) might be used to handle a particular machine interference problem, the assumptions compatible with the problem should be determined. For example, are the machine breakdowns random or regular? If they are random, then the queuing theory results[26,36,42–44] could be applicable. This section examines some of the assumptions made in developing the various methods with respect to their justification in a real situation.

Suppose, for example, that a down machine has just been repaired. Then one might expect this machine to run for some time before stopping again. The assumption of exponential breakdown times implies that the time remaining until breakdown is independent of however long the machine has been running.

Because service events and times can then be planned, congestion for regular breakdowns is usually less than that for random breakdowns. Comparing the example solved in Section 3.5.9 with the solution obtained using Smith's formula, such differences in interference times were observed. With several machines under the supervision of one operator, the superposition of several sequences of regular events (breakdowns) will result in a random sequence of breakdowns (for all practical purposes) unless the separate sequences keep in phase.

If there is only one type of breakdown, then the service time could be nearly constant. On the other hand, if there are several breakdown types such that those with shorter service times occur much more often than those with longer service times, the assumption that the pooled distribution is similar to the exponential is good. In such a case queuing theory approaches can be applicable, unless some priority queue discipline is used.

Another usual assumption is that the operator is always available to service the down machines. The simulations were able to take into consideration other operator duties, personal time off (lunch and breaks), or other ancillary duties. Sometimes other duties can be done while the operator is idle.

Some work does tie up the operator's time. Such work can be independent of the total amount of production (such as going for tools) or proportional to the total production (such as personal time off). If

$$A = \text{proportion of time spent on ancillary work}$$

then the real servicing factor would be

$$\rho' = \frac{\rho}{1 - A} = \frac{S}{P(1 - A)}$$

Other sources offer additional discussions on the effects and classifications of various types of ancillary work.[7,23,42]

Exhibit 3.5.13 Summary of Many Methods, Assumptions, Ranges of Applicability, and Input/Output

Author (Reference Number)	Year	Service Time/Breakdown Time Distributions	Queue Discipline	Model/Formula	Range of Applicability (Number of Machines)	Input/Output
Wright (5)	1936	Constant service time; one operator	Random tending	$I = 50\{[(1 + X - N)^2 + 2 \times N]^{1/2} - (1 + X - N)\}$	6 to 200	I (% of S) as a function of $P/S = 1/\rho$ and N
				Exhibit 3.5.4	1 to 6	
Wright (5)	1936	Constant service time; one operator	Random tending	Exhibit 3.5.5	2 to 70	N as a function of $P/S = 1/\rho$ and \$(mac)/\$(labor)
Jones (12)	1946	Negative exponential service time; one operator; similar machines	Random tending	Exhibit 3.5.A3 in the appendix	2 to 100 (formulas otherwise)	I and $1 - Oe$ as functions of N and S'
Ashcroft (9)	1950	Constant service time; negative exponential breakdown distribution; one operator	Random tending	Exhibits 3.5.A1 and 3.5.A2 in the appendix	1 to 20	$N \times (Me)$ as a function of $S/P = \rho$ and N
Palm (22, 36)	1947; 1958	Negative exponential service and breakdown times; one operator	Random tending	Exhibit 3.5.A4 in the appendix	1 to 144	I, P, and Oe as a function of $S/P = \rho$ and N
Benson and Cox (23)	1951; 1952	Negative exponential service and breakdown times; one operator	Random tending	Tables (not included here)	1 to 15	Oe as a function of $S/P = \rho$ and N
Peck and Hazelwood (26)	1958	Negative exponential service and breakdown times; several operators	Random tending	Book of tables; three page sample is Exhibit 3.5.A5 in the appendix	4 to 250	Me and I as a function of $S/(S + P)$, O, and N
Smith (7)	1965	General service distribution; negative exponential breakdown distribution; one operator	Cyclic patrol or distance	$I = 50[\{(N - 3/2 - X)^2 + 2(N - 1)\}^{1/2} + (N - 3/2 - X)]$	Any number of machines	I as a function of $P/S = 1/\rho$ and N
Deakin (37)	1980	Constant service times; negative exponential breakdown times; constant walking times; one operator	Cyclic patrol (both closed, alternating, with and without relief)	Equations 2, 3, and 4	24 to 30	N as a function of S, E, and W

Exhibit 3.5.14 Queuing Disciplines and References[a]

Random Tending	Cyclic Patrol	Nonpreemptive Resume	Preemptive Resume	Distance Priority	General (Any)
Khintchine; 1933 (15)	Mack, Murphy, and Webb; 1957 (27)	Jaiswal and Thiruvengadam; 1963 (32)	Jaiswal; 1968 (2)	Smith; 1965 (7)	Haagensen; 1970 (38)
Wright, Duvall, and Freeman; 1936 (5)	Howie and Shenton; 1959 (28)	Hodgson and Hebble; 1967 (33)		Reynolds; 1969, 1975 (34, 35)	Freeman, Hoover, and Satia; 1973 (39)
Kronig and Mondria; 1943 (16)					
Palm; 1943, 1947, 1958 (8, 22, 36)	Lawing; 1959 (29)				Bredenbeck, Ogden, and Tyler; 1975 (40)
Weir; 1944 (42)	Ben-Israel and Naor; 1960 (30)				
Jones; 1946, 1949 (12, 13)					
Ashcroft; 1950 (9)	Grant; 1960 (31)				
Benson and Cox; 1951, 1952 (23)	Smith; 1965 (7)				
Fetter; 1955 (25)	Deakin; 1980 (37)				
Peck and Hazel-wood; 1958 (26)					
Grant; 1960 (31)					
Allen; 1978 (21)					

[a]Reference numbers in parentheses.

The breakdown rate could be a factor of concern; even if the breakdowns are random, the breakdown *rate* could vary. For example, as a machine gets older, it might break down more often. Different operators work at different paces, which implies that the average service time could change over time. If an operator gets tired during the day, her or his average service time might increase. Perhaps several interference problems could be solved, each using the present breakdown *rate*.

Sometimes the average walking time is absorbed in the average service time. The results may not be very accurate if the distances between the machines are very different; then more detail should be incorporated in the model.

3.5.14 CONCLUSIONS

A detailed analysis of the particular manufacturing system is recommended to aid in a correct choice of a method that could be applicable and appropriate given the values for the input parameters. For many systems, the approximate techniques are sufficient (and cheaper) to aid in deciding on a correct machine assignment.

Many procedures have been presented to help in deciding on a correct number of machines to assign to an operator. The choice of a relevant method depends on the particular industrial system. Exhibits 3.5.13 and 3.5.14 summarize many of the techniques and the assumptions under which each is applicable. The choice of an appropriate method is system dependent. There can be no universally correct method or formula.

Published works on machine interference not explicitly discussed here are cited in references 45–66.

3.5.15 ACKNOWLEDGMENTS

The author would like to acknowledge the helpful comments of John Mariotti.

APPENDIX 3.5

Exhibit 3.5.A1 Ashcroft Numbers: $N = 1$ to 10; $Y = 0.01$ to 1; Average Number of Machine Hours per Hour, A_N, for an Operator Having the Care of N Similar Machines

$Y = S/P$	$N = 1$	$N = 2$	$N = 3$	$N = 4$	$N = 5$	$N = 6$	$N = 7$	$N = 8$	$N = 9$	$N = 10$
0.00	1.00	2.00	3.00	4.00	5.00	6.00	7.00	8.00	9.00	10.00
0.01	0.99	1.98	2.97	3.96	4.95	5.94	6.93	7.92	8.91	9.90
0.02	0.98	1.96	2.94	3.92	4.90	5.88	6.85	7.83	8.81	9.78
0.03	0.97	1.94	2.91	3.88	4.84	5.81	6.77	7.74	8.70	9.66
0.04	0.96	1.92	2.88	3.84	4.79	5.74	6.69	7.64	8.58	9.52
0.05	0.95	1.90	2.85	3.79	4.74	5.67	6.61	7.53	8.45	9.37
0.06	0.94	1.88	2.82	3.75	4.68	5.60	6.51	7.42	8.31	9.19
0.07	0.93	1.86	2.79	3.71	4.62	5.52	6.42	7.29	8.15	8.99
0.08	0.93	1.85	2.76	3.67	4.56	5.44	6.31	7.16	7.98	8.76
0.09	0.92	1.83	2.73	3.62	4.50	5.36	6.20	7.01	7.78	8.50
0.10	0.91	1.81	2.70	3.58	4.44	5.28	6.08	6.85	7.57	8.21
0.11	0.90	1.79	2.67	3.53	4.38	5.19	5.96	6.68	7.33	7.89
0.12	0.89	1.77	2.64	3.49	4.31	5.10	5.83	6.50	7.08	7.55
0.13	0.88	1.76	2.61	3.44	4.24	5.00	5.69	6.31	6.81	7.19
0.14	0.88	1.74	2.58	3.40	4.18	4.90	5.55	6.10	6.53	6.83
0.15	0.87	1.72	2.55	3.35	4.11	4.80	5.40	5.90	6.25	6.48
0.16	0.86	1.71	2.52	3.31	4.04	4.70	5.25	5.68	5.97	6.14
0.17	0.85	1.69	2.50	3.26	3.97	4.59	5.10	5.47	5.70	5.82
0.18	0.85	1.67	2.48	3.22	3.90	4.48	4.94	5.26	5.44	5.52
0.19	0.84	1.66	2.44	3.17	3.83	4.37	4.79	5.05	5.19	5.24
0.20	0.83	1.64	2.41	3.12	3.75	4.26	4.63	4.85	4.95	4.99
0.21	0.83	1.62	2.38	3.08	3.68	4.15	4.48	4.66	4.73	4.75
0.22	0.82	1.61	2.35	3.03	3.61	4.04	4.33	4.47	4.53	4.54
0.23	0.81	1.59	2.33	2.98	3.53	3.94	4.18	4.30	4.34	4.34
0.24	0.81	1.58	2.30	2.94	3.46	3.83	4.04	4.13	4.16	4.16
0.25	0.80	1.56	2.27	2.89	3.39	3.72	3.90	3.98	4.00	4.00
0.26	0.79	1.55	2.24	2.85	3.31	3.62	3.77	3.83	3.84	3.84
0.27	0.79	1.53	2.22	2.80	3.24	3.52	3.65	3.69	3.70	3.70
0.28	0.78	1.52	2.19	2.75	3.17	3.42	3.53	3.56	3.57	3.57
0.29	0.77	1.51	2.16	2.71	3.10	3.33	3.42	3.44	3.45	3.45
0.30	0.77	1.49	2.14	2.67	3.03	3.23	3.31	3.33	3.33	3.33
0.31	0.76	1.48	2.11	2.62	2.97	3.14	3.21	3.22	3.22	3.22
0.32	0.76	1.46	2.09	2.58	2.90	3.06	3.11	3.12	3.12	3.12
0.33	0.75	1.45	2.06	2.53	2.84	2.98	3.02	3.03	3.03	3.03
0.34	0.75	1.44	2.03	2.49	2.77	2.90	2.93	2.94	2.94	2.94
0.35	0.74	1.42	2.01	2.45	2.71	2.82	2.85	2.86	2.86	2.86
0.40	0.71	1.36	1.89	2.25	2.43	2.49	2.50	2.50	2.50	2.50
0.45	0.69	1.30	1.78	2.07	2.19	2.22	2.22	2.22	2.22	2.22
0.50	0.67	1.24	1.67	1.90	1.98	2.00	2.00	2.00	2.00	2.00
0.55	0.64	1.19	1.57	1.76	1.81	1.82				
0.60	0.62	1.14	1.48	1.63	1.66	1.67				
0.65	0.61	1.10	1.40	1.51	1.54	1.54				
0.70	0.59	1.05	1.32	1.41	1.43	1.43				
0.75	0.57	1.01	1.25	1.32	1.33	1.33				
0.80	0.55	0.97	1.19	1.24	1.25	1.25				
0.85	0.54	0.94	1.13	1.17	1.17	1.18				
0.90	0.53	0.91	1.07	1.11	1.11	1.11				
0.95	0.51	0.87	1.02	1.05	1.05	1.05				
1.00	0.50	0.84	0.98	1.00	1.00	1.00				

Reproduced by permission from reference 9.

Exhibit 3.5.A2 Ashcroft Numbers: N = 11 to 20; Y = 0.005 to 0.27; for N = 17 to 20 and Y Between 0.145 and 0.27, Use N = 16

Y	N = 11	N = 12	N = 13	N = 14	N = 15	N = 16	N = 17	N = 18	N = 19	N = 20
0.000	11.00	12.00	13.00	14.00	15.00	16.00	17.00	18.00	19.00	20.00
0.005	10.94	11.94	12.93	13.93	14.92	15.92	16.91	17.91	18.90	19.89
0.010	10.88	11.87	12.86	13.85	14.84	15.83	16.82	17.80	18.79	19.78
0.015	10.82	11.80	12.79	13.77	14.75	15.73	16.71	17.69	18.69	19.65
0.020	10.76	11.73	12.71	13.68	14.65	15.62	16.59	17.56	18.53	19.50
0.025	10.69	11.66	12.62	13.58	14.54	15.50	16.46	17.41	18.37	19.32
0.030	10.62	11.57	12.53	13.48	14.42	15.37	16.31	17.24	18.17	19.10
0.035	10.54	11.48	12.42	13.36	14.29	15.21	16.13	17.04	17.94	18.82
0.040	10.46	11.39	12.31	13.23	14.13	15.03	15.92	16.79	17.64	18.48
0.045	10.37	11.28	12.18	13.08	13.95	14.82	15.66	16.48	17.27	18.03
0.050	10.27	11.16	12.04	12.91	13.75	14.57	15.35	16.10	16.81	17.45
0.055	10.17	11.04	11.89	12.71	13.51	14.27	14.98	15.64	16.23	16.75
0.060	10.05	10.90	11.71	12.49	13.23	13.92	14.54	15.09	15.56	15.93
0.065	9.93	10.74	11.51	12.24	12.91	13.52	14.04	14.47	14.80	15.04
0.070	9.80	10.57	11.29	11.96	12.55	13.06	13.47	13.78	14.00	14.14
0.075	9.65	10.38	11.05	11.65	12.15	12.56	12.87	13.08	13.20	13.28
0.080	9.50	10.18	10.79	11.30	11.72	12.03	12.25	12.38	12.45	12.48
0.085	9.33	9.96	10.50	10.94	11.27	11.49	11.63	11.71	11.74	11.76
0.090	9.15	9.72	10.19	10.55	10.80	10.96	11.05	11.09	11.10	11.11
0.095	8.96	9.47	9.87	10.16	10.34	10.45	10.49	10.52	10.52	10.52

0.100	8.76	9.21	9.54	9.76	9.89	9.96	9.98	9.99	10.00	10.00
0.105	8.55	8.94	9.21	9.38	9.46	9.50	9.52	9.52	9.52	9.52
0.110	8.34	8.67	8.88	9.00	9.06	9.08	9.09	9.09	9.09	9.09
0.115	8.12	8.39	8.56	8.64	8.68	8.69	8.69	8.69	8.69	8.69
0.120	7.89	8.12	8.24	8.30	8.32	8.33	8.33	8.33	8.33	8.33
0.125	7.67	7.85	7.94	7.98	7.99	8.00	8.00	8.00	8.00	8.00
0.130	7.44	7.59	7.65	7.68	7.69	7.69	7.69	7.69	7.69	7.69
0.135	7.22	7.34	7.38	7.40	7.41	7.41	7.41	7.41	7.41	7.41
0.140	7.01	7.09	7.13	7.14	7.14	7.14	7.14	7.14	7.14	7.14
0.145	6.80	6.86	6.89	6.89	6.90	6.90	6.90	6.90	6.90	6.90
0.150	6.59	6.64	6.66	6.66	6.67	6.67				
0.160	6.21	6.24	6.25	6.25	6.25	6.25				
0.170	5.86	5.88	5.88	5.88	5.88	5.88				
0.180	5.55	5.55	5.55	5.55	5.55	5.55				
0.190	5.26	5.26	5.26	5.26	5.26	5.26				
0.200	5.00	5.00	5.00	5.00	5.00	5.00				
0.210	4.76	4.76	4.76	4.74	4.76	4.76				
0.220	4.54	4.54	4.54	4.54	4.54	4.54				
0.230	4.35	4.35	4.35	4.35	4.35	4.35				
0.240	4.17	4.17	4.17	4.17	4.17	4.17				
0.250	4.00	4.00	4.00	4.00	4.00	4.00				
0.260	3.85	3.85	3.85	3.85	3.85	3.85				
0.270	3.70	3.70	3.70	3.70	3.70	3.70				

Reproduced by permission from reference 9.

Exhibit 3.5.A3 Random Machine Interference Table[a]

Number of Machines	Operator's Total Percentage Interference Causing Work Load[b] on Individual Attention Basis														
	50	55	60	65	70	75	80	85	90	95	100	105	110	115	120
2	*3.9*	*4.9*	*5.9*	*7.0*	*8.2*	*9.5*	*10.7*	*12.4*	*13.9*	*15.5*	*17.3*	*18.8*	*20.8*	*22.6*	*24.4*
	52	48	44	40	36	32	29	25	23	18	17	15	13	11	9
3	*3.7*	*4.7*	*5.7*	*6.7*	*7.8*	*9.2*	*10.5*	*12.2*	*13.8*	*15.4*	*17.2*	*18.8*	*20.8*	*22.6*	*24.4*
	52	48	43	39	35	32	28	25	22	18	17	15	13	11	9
4	*3.2*	*3.9*	*4.9*	*5.8*	*6.9*	*8.2*	*9.4*	*10.9*	*12.6*	*14.2*	*15.9*	*17.7*	*19.6*	*21.6*	*23.4*
	52	47	43	39	35	31	28	24	21	18	16	14	12	10	8
5	*2.7*	*3.5*	*4.3*	*5.2*	*6.2*	*7.3*	*8.6*	*9.9*	*11.4*	*13.0*	*14.7*	*16.5*	*18.5*	*20.6*	*22.6*
	52	47	43	39	34	31	27	23	20	17	15	12	11	8	7
6	*2.3*	*3.1*	*3.7*	*4.5*	*5.5*	*6.6*	*7.8*	*9.1*	*10.6*	*12.2*	*13.8*	*15.7*	*17.6*	*19.6*	*21.7*
	51	47	42	38	34	30	26	23	20	17	14	11	9	7	6
7	*2.1*	*2.7*	*3.3*	*4.0*	*5.0*	*6.0*	*7.1*	*8.4*	*9.8*	*11.3*	*13.0*	*14.9*	*16.8*	*18.8*	*20.9*
	51	46	42	38	34	30	26	22	19	16	13	11	8	7	5
8	*1.9*	*2.4*	*3.0*	*3.7*	*4.6*	*5.4*	*6.6*	*7.7*	*9.0*	*10.6*	*12.3*	*14.1*	*16.0*	*18.2*	*20.2*
	51	46	42	37	33	29	25	22	18	15	12	10	8	6	4
9	*1.7*	*2.2*	*2.8*	*3.3*	*4.3*	*5.0*	*6.0*	*7.1*	*8.5*	*9.9*	*11.6*	*13.4*	*15.4*	*17.5*	*19.6*
	51	46	42	37	33	29	25	21	17	15	12	9	7	5	4
10	*1.5*	*2.0*	*2.6*	*3.1*	*4.0*	*4.6*	*5.6*	*6.6*	*8.0*	*9.4*	*11.0*	*12.8*	*14.8*	*17.0*	*19.0*
	51	46	42	37	33	28	25	21	17	14	11	8	6	4	3
11	*1.4*	*1.9*	*2.4*	*2.9*	*3.7*	*4.3*	*5.3*	*6.2*	*7.5*	*8.9*	*10.5*	*12.4*	*14.4*	*16.5*	*18.5*
	51	46	41	37	33	28	24	20	17	13	11	8	6	4	3
12	*1.3*	*1.8*	*2.2*	*2.7*	*3.4*	*4.1*	*5.0*	*5.9*	*7.1*	*8.5*	*10.1*	*12.0*	*14.0*	*16.1*	*18.2*
	51	46	41	37	32	28	24	20	16	13	10	7	5	4	2
13	*1.3*	*1.7*	*2.1*	*2.5*	*3.2*	*3.9*	*4.7*	*5.6*	*6.8*	*8.1*	*9.7*	*11.6*	*13.6*	*15.7*	*18.0*
	51	46	41	37	32	28	24	20	16	13	9	7	5	3	2
14	*1.2*	*1.6*	*2.0*	*2.4*	*3.0*	*3.7*	*4.4*	*5.3*	*6.4*	*7.7*	*9.3*	*11.2*	*13.3*	*15.4*	*17.8*
	51	46	41	37	32	28	24	20	16	12	9	7	5	3	2

15	*1.1*	*1.5*	*1.9*	*2.2*	*2.8*	*3.5*	*4.2*	*5.1*	*6.1*	*7.4*	*8.9*	*10.8*	*13.0*	*15.1*	*17.6*
	51	46	41	36	32	28	23	19	15	12	9	6	4	3	1
16	*1.1*	*1.4*	*1.8*	*2.1*	*2.6*	*3.2*	*4.0*	*4.8*	*5.9*	*7.0*	*8.5*	*10.5*	*12.6*	*14.8*	*17.4*
	51	46	41	36	32	27	23	19	15	12	9	6	4	2	1
17	*1.1*	*1.3*	*1.7*	*2.0*	*2.5*	*3.1*	*3.8*	*4.6*	*5.6*	*6.7*	*8.2*	*10.2*	*12.3*	*14.5*	*17.2*
	50	46	41	36	32	27	23	19	15	11	8	6	3	2	1
18	*1.0*	*1.2*	*1.6*	*1.9*	*2.4*	*3.0*	*3.6*	*4.4*	*5.4*	*6.5*	*7.9*	*9.9*	*12.0*	*14.4*	*17.1*
	50	46	41	36	32	27	23	19	15	11	8	5	3	2	1
19	*1.0*	*1.1*	*1.5*	*1.9*	*2.3*	*2.9*	*3.5*	*4.2*	*5.2*	*6.2*	*7.7*	*9.6*	*11.8*	*14.3*	*17.0*
	50	46	41	36	32	27	23	19	15	11	8	5	3	1	—
20	*0.9*	*1.1*	*1.4*	*1.8*	*2.2*	*2.8*	*3.4*	*4.1*	*5.0*	*6.0*	*7.5*	*9.3*	*11.5*	*14.1*	*16.9*
	50	46	41	36	32	27	23	18	15	11	8	5	2	1	—
25	*0.8*	*0.9*	*1.2*	*1.5*	*1.8*	*2.3*	*2.8*	*3.4*	*4.2*	*5.2*	*6.6*	*8.4*	*10.8*	*13.7*	*16.3*
	50	45	41	36	31	27	22	18	14	10	7	4	2	—	—
30	*0.7*	*0.8*	*1.0*	*1.2*	*1.6*	*2.0*	*2.4*	*2.9*	*3.7*	*4.6*	*5.9*	*7.7*	*10.2*	*13.4*	*16.7*
	50	45	41	36	31	26	22	17	13	9	6	3	1	—	—
40	*0.5*	*0.6*	*0.8*	*1.0*	*1.2*	*1.5*	*1.8*	*2.3*	*2.9*	*3.7*	*5.0*	*6.8*	*9.4*	*13.2*	*16.6*
	50	45	40	36	31	26	21	17	13	9	5	2	—	—	—
50	*0.3*	*0.5*	*0.6*	*0.8*	*0.9*	*1.2*	*1.5*	*2.0*	*2.5*	*3.2*	*4.3*	*6.1*	*9.3*	*13.1*	*16.6*
	50	45	40	36	31	26	21	17	12	8	4	1	—	—	—
75	*0.2*	*0.3*	*0.4*	*0.5*	*0.6*	*0.8*	*1.0*	*1.3*	*1.7*	*2.1*	*3.3*	*5.2*	*9.1*	*13.0*	*16.5*
	50	45	40	35	30	26	21	16	12	7	3	—	—	—	—
100	*0.2*	*0.2*	*0.3*	*0.4*	*0.5*	*0.7*	*0.8*	*1.0*	*1.2*	*1.6*	*2.8*	*5.1*	*9.0*	*13.0*	*16.5*
	50	45	40	35	30	26	21	16	11	7	3	—	—	—	—

Reproduced by permission from reference 14.

[a]This table shows average percentage (of total time) interference time per machine (in italic type) and percentage (of total time) operator idle time (in nonitalic type) in randomly serviced multimachine assignments tended by one operator when servicing demands are random and approximately the same for each assigned machine.

[b]Figure the work load at expected operator servicing time and exclude deferrable "internal" duties (those that can be performed when all machines are producing): This gives the correct estimate of machine interference when there is little variation of the work loads of assigned products. Then, to estimate percentage operator idle time, deduct from the respective estimated percentage operator idle time the estimated percentage of time to be spent on the ancillary duties.

Exhibit 3.5.A4 D. C. Palm's Table, Giving I, P, and Oe as a Function of N and ρ

$\rho = 0.01$

N	I	P	Oe
1	0.0	99.0	1.0
2	0.0	99.0	2.0
3	0.0	99.0	3.0
4	0.0	99.0	4.0
5	0.0	99.0	5.0
6	0.1	99.0	5.9
7	0.1	99.0	6.9
8	0.1	98.9	7.9
9	0.1	98.9	8.9
10	0.1	98.9	9.9
11	0.1	98.9	10.9
12	0.1	98.9	11.9
13	0.1	98.9	12.9
14	0.1	98.9	13.8
15	0.2	98.9	14.8
16	0.2	98.8	15.8
17	0.2	98.8	16.8
18	0.2	98.8	17.8
19	0.2	98.8	18.8
20	0.2	98.8	19.8
21	0.2	98.8	20.7
22	0.3	98.8	21.7
23	0.3	98.7	22.7
24	0.3	98.7	23.7
25	0.3	98.7	24.7
26	0.3	98.7	25.7
27	0.3	98.7	26.6
28	0.4	98.7	27.6
29	0.4	98.6	28.6
30	0.4	98.6	29.6
31	0.4	98.6	30.6
32	0.4	98.6	31.6
33	0.4	98.6	32.5
34	0.5	98.6	33.5
35	0.5	98.5	34.5
36	0.5	98.5	35.5
37	0.5	98.5	36.4
38	0.6	98.5	37.4
39	0.6	98.4	38.4
40	0.6	98.4	39.4

86	3.5	95.5	82.2
87	3.7	95.4	83.0
88	3.8	95.2	83.8
89	4.0	95.1	84.6
90	4.2	94.9	85.4
91	4.4	94.7	86.2
92	4.6	94.5	86.9
93	4.8	94.3	87.7
94	5.0	94.1	88.4
95	5.2	93.8	89.1
96	5.5	93.6	89.8
97	5.8	93.3	90.5
98	6.0	93.0	91.2
99	6.3	92.7	91.8
100	6.7	92.4	92.4
101	7.0	92.1	93.0
102	7.3	91.8	93.6
103	7.7	91.4	94.2
104	8.1	91.0	94.7
105	8.5	90.6	95.2
106	8.9	90.2	95.7
107	9.3	89.8	96.1
108	9.8	89.4	96.5
109	10.2	88.9	96.9
110	10.7	88.4	97.3
111	11.2	87.9	97.6
112	11.7	87.4	97.9
113	12.3	86.9	98.2
114	12.8	86.3	98.4
115	13.4	85.8	98.6
116	13.9	85.2	98.8
117	14.5	84.6	99.0
118	15.1	84.1	99.2
119	15.7	83.5	99.3
120	16.3	82.9	99.4
121	16.9	82.3	99.5
122	17.5	81.7	99.6
123	18.1	81.1	99.7
124	18.8	80.4	99.8
125	19.4	79.8	99.8
126	20.0	79.2	99.8
127	20.6	78.6	99.9

27	1.8	96.3	52.0
28	1.9	96.1	53.8
29	2.1	96.0	55.7
30	2.2	95.9	57.5
31	2.4	95.7	59.3
32	2.5	95.6	61.2
33	2.7	95.4	63.0
34	2.9	95.2	64.7
35	3.1	95.0	66.5
36	3.3	94.8	68.2
37	3.6	94.6	70.0
38	3.8	94.3	71.7
39	4.1	94.1	73.4
40	4.3	93.8	75.0
41	4.7	93.5	76.7
42	5.0	93.2	78.3
43	5.3	92.8	79.8
44	5.7	92.4	81.3
45	6.1	92.0	82.8
46	6.6	91.6	84.3
47	7.1	91.1	85.7
48	7.6	90.6	87.0
49	8.1	90.1	88.3
50	8.7	89.5	89.5
51	9.3	88.9	90.7
52	10.0	88.3	91.8
53	10.7	87.6	92.8
54	11.5	86.8	93.8
55	12.3	86.0	94.6
56	13.1	85.2	95.4
57	14.0	84.3	95.1
58	14.9	83.4	96.8
59	15.9	82.5	97.8
60	16.8	81.5	97.8
61	17.9	80.5	98.3
62	18.9	79.5	98.6
63	19.9	78.5	98.9
64	21.0	77.5	99.2
65	22.0	76.4	99.4
66	23.1	75.4	99.5
67	24.2	74.4	99.6
68	25.2	73.3	99.7

40	17.4	80.2	96.2
41	18.8	78.9	97.0
42	20.1	77.5	97.7
43	21.6	76.2	98.2
44	23.0	74.8	98.7
45	24.4	73.4	99.0
46	25.9	72.0	99.3
47	27.3	70.6	99.5
48	28.7	69.2	99.7

$\rho = 0.04$

N	I	P	Oe
1	0.0	96.2	3.8
2	0.2	96.0	7.7
3	0.3	95.9	11.5
4	0.5	95.7	15.3
5	0.7	95.5	19.1
6	0.9	95.3	22.9
7	1.1	95.1	26.6
8	1.3	94.9	30.4
9	1.5	94.7	34.1
10	1.8	94.4	37.8
11	2.1	94.1	41.4
12	2.4	93.8	45.0
13	2.8	93.5	48.6
14	3.2	93.1	52.1
15	3.6	92.7	55.6
16	4.0	92.3	59.1
17	4.5	91.8	62.4
18	5.1	91.3	65.7
19	5.7	90.7	68.9
20	6.4	90.0	72.0
21	7.1	89.3	75.0
22	8.0	88.5	77.9
23	8.9	87.6	80.6
24	9.9	86.7	83.2
25	11.0	85.6	85.6
26	12.2	84.5	87.9
27	13.4	83.2	89.9
28	14.8	81.9	91.7
29	16.3	80.5	93.3
30	17.9	79.0	94.7

6	2.0	92.5	33.3
7	2.5	92.0	38.6
8	3.1	91.4	43.9
9	3.7	90.8	49.0
10	4.5	90.1	54.1
11	5.3	89.4	59.0
12	6.2	88.5	63.7
13	7.3	87.5	68.2
14	8.4	86.4	72.6
15	9.7	85.2	76.6
16	11.2	83.8	80.4
17	12.8	82.3	83.9
18	14.6	80.6	87.0
19	16.5	78.8	89.8
20	18.6	76.8	92.2
21	20.8	74.7	94.1
22	23.1	72.5	95.7
23	25.5	70.3	97.0
24	27.9	68.0	98.0
25	30.3	65.8	98.7

$\rho = 0.07$

N	I	P	Oe
1	0.0	93.5	6.5
2	0.4	93.1	13.0
3	0.9	92.6	19.5
4	1.4	92.1	25.8
5	2.0	91.6	32.1
6	2.7	91.0	38.2
7	3.4	90.3	44.2
8	4.3	89.5	50.1
9	5.2	88.6	55.8
10	6.3	87.6	61.3
11	7.5	86.4	66.6
12	8.9	85.1	71.5
13	10.4	83.7	76.2
14	12.2	82.1	80.4
15	14.1	80.3	84.3
16	16.2	78.3	87.7
17	18.5	76.2	90.6
18	21.0	73.9	93.1
19	23.5	71.5	95.1

N	I	P	Oe
41	0.6	98.4	40.3
42	0.7	98.4	41.3
43	0.7	98.3	42.3
44	0.7	98.3	43.3
45	0.7	98.3	44.2
46	0.8	98.3	45.2
47	0.8	98.2	46.2
48	0.8	98.2	47.1
49	0.9	98.2	48.1
50	0.9	98.1	49.1
51	0.9	98.1	50.0
52	1.0	98.1	51.0
53	1.0	98.0	52.0
54	1.0	98.0	52.9
55	1.1	98.0	53.9
56	1.1	97.9	54.8
57	1.1	97.9	55.8
58	1.2	97.8	56.8
59	1.2	97.8	57.7
60	1.3	97.8	58.7
61	1.3	97.7	59.6
62	1.4	97.7	60.6
63	1.4	97.6	61.5
64	1.5	97.6	62.4
65	1.5	97.5	63.4
66	1.6	97.4	64.3
67	1.7	97.4	65.3
68	1.7	97.3	66.2
69	1.8	97.3	67.1
70	1.8	97.2	68.0
71	1.9	97.1	69.0
72	2.0	97.0	69.9
73	2.1	97.0	70.8
74	2.1	96.9	71.7
75	2.2	96.8	72.6
76	2.3	96.7	73.5
77	2.4	96.6	74.4
78	2.5	96.5	75.3
79	2.6	96.4	76.2
80	2.7	96.3	77.1
81	2.8	96.2	77.9
82	3.0	96.1	78.8
83	3.1	96.0	79.6
84	3.2	95.8	80.5
85	3.4	95.7	81.3
128	21.2	78.1	99.9
129	21.8	77.5	99.9
130	22.4	76.9	99.9
131	22.9	76.3	100.0
132	23.5	75.7	100.0
133	24.1	75.2	100.0
134	24.6	74.6	100.0
135	25.2	74.1	100.0
136	25.7	73.5	100.0
137	26.3	73.0	100.0
138	26.8	72.5	100.0
139	27.3	71.9	100.0
140	27.9	71.4	100.0
141	28.4	70.9	100.0
142	28.9	70.4	100.0
143	29.4	69.9	100.0
144	29.9	69.4	100.0

$\rho = 0.02$

N	I	P	Oe
1	0.0	98.0	2.0
2	0.0	98.0	3.9
3	0.1	98.0	5.9
4	0.1	97.9	7.8
5	0.2	97.9	9.8
6	0.2	97.8	11.7
7	0.3	97.8	13.7
8	0.3	97.7	15.6
9	0.4	97.7	17.6
10	0.4	97.6	19.5
11	0.5	97.6	21.5
12	0.5	97.5	23.4
13	0.6	97.5	25.3
14	0.6	97.4	27.3
15	0.7	97.4	29.2
16	0.8	97.3	31.1
17	0.9	97.2	33.1
18	0.9	97.1	35.0
19	1.0	97.1	36.9
20	1.1	97.0	38.8
21	1.2	96.9	40.7
22	1.3	96.8	42.6
23	1.4	96.7	44.5
24	1.5	96.6	46.4
25	1.6	96.5	48.3
26	1.7	96.4	50.1
69	26.2	72.3	99.8
70	27.2	71.3	99.9
71	28.2	70.4	99.9
72	29.2	69.4	99.9

$\rho = 0.03$

N	I	P	Oe
1	0.0	97.1	2.9
2	0.1	97.0	5.8
3	0.2	96.9	8.7
4	0.3	96.8	11.6
5	0.4	96.7	14.5
6	0.5	96.6	17.4
7	0.6	96.5	20.3
8	0.7	96.4	23.1
9	0.8	96.3	26.0
10	1.0	96.2	28.8
11	1.1	96.0	31.7
12	1.3	95.9	34.5
13	1.4	95.7	37.3
14	1.6	95.5	40.1
15	1.8	95.4	42.9
16	2.0	95.2	45.7
17	2.2	94.9	48.4
18	2.4	94.7	51.2
19	2.7	94.5	53.9
20	3.0	94.2	56.5
21	3.3	93.9	59.2
22	3.6	93.6	61.8
23	3.9	93.3	64.4
24	4.3	92.9	66.9
25	4.7	92.5	69.4
26	5.2	92.1	71.8
27	5.7	91.6	74.2
28	6.2	91.1	76.5
29	6.8	90.5	78.7
30	7.4	89.9	80.9
31	8.1	89.2	83.0
32	8.9	88.5	84.9
33	9.7	87.7	86.8
34	10.6	86.8	88.5
35	11.6	85.9	90.2
36	12.6	84.9	91.6
37	13.7	83.8	93.0
38	14.9	82.6	94.2
39	16.1	81.4	95.3
31	19.6	77.4	95.9
32	21.3	75.7	96.9
33	23.0	74.0	97.7
34	24.8	72.3	98.4
35	26.6	70.6	98.8
36	28.4	68.9	99.2
37	30.1	67.2	99.5

$\rho = 0.05$

N	I	P	Oe
1	0.0	95.2	4.8
2	0.2	95.0	9.5
3	0.5	91.8	14.2
4	0.7	94.5	18.9
5	1.0	94.3	23.6
6	1.4	94.0	28.2
7	1.7	93.6	32.8
8	2.1	93.3	37.3
9	2.5	92.9	41.8
10	3.0	92.4	46.2
11	3.5	91.9	50.6
12	4.1	91.4	54.8
13	4.7	90.8	59.0
14	5.4	90.1	63.1
15	6.2	89.3	67.0
16	7.1	88.5	70.8
17	8.1	87.6	74.4
18	9.1	86.5	77.9
19	10.4	85.4	81.1
20	11.7	84.1	84.1
21	13.1	82.7	86.9
22	14.7	81.2	89.3
23	16.5	79.6	91.5
24	18.3	77.8	93.4
25	20.2	76.0	95.0
26	22.2	74.1	96.3
27	24.3	72.1	97.3
28	26.4	70.1	98.1
29	28.5	68.1	98.7

$\rho = 0.06$

N	I	P	Oe
1	0.0	94.3	5.7
2	0.3	94.0	11.3
3	0.7	93.7	16.9
4	1.1	93.3	22.4
5	1.5	92.9	27.9
20	26.2	69.0	96.6
21	28.9	66.5	97.7

$\rho = 0.08$

N	I	P	Oe
1	0.0	92.6	7.4
2	0.5	92.1	14.7
3	1.2	91.5	22.0
4	1.9	90.9	29.1
5	2.7	90.1	36.1
6	3.5	89.3	42.9
7	4.5	88.4	49.5
8	5.7	87.3	55.9
9	7.0	86.1	62.0
10	8.5	84.8	67.8
11	10.1	83.2	73.2
12	12.0	81.4	78.2
13	14.2	79.5	82.7
14	16.5	77.3	86.6
15	19.0	75.0	90.0
16	21.8	72.4	92.7
17	24.6	69.8	94.9
18	27.6	67.1	96.6
19	30.5	64.4	97.8

$\rho = 0.09$

N	I	P	Oe
1	0.0	91.7	8.3
2	0.7	91.1	16.4
3	1.4	90.4	24.4
4	2.3	89.6	32.3
5	3.3	88.7	39.9
6	4.5	87.7	47.3
7	5.8	86.5	54.5
8	7.3	85.1	61.3
9	9.0	83.5	67.6
10	10.9	81.7	73.6
11	13.1	79.7	78.9
12	15.6	77.5	83.7
13	18.3	75.0	87.8
14	21.2	72.3	91.1
15	24.2	69.5	93.8
16	27.4	66.6	95.9
17	30.6	63.7	97.4

Exhibit 3.5.A4 (*Continued*)

$\rho = 0.10$

N	I	P	Oe
1	0.0	90.9	9.1
2	0.8	90.2	18.0
3	1.8	89.3	26.8
4	2.8	88.3	35.3
5	4.1	87.2	43.6
6	5.5	85.9	51.5
7	7.1	84.4	59.1
8	9.0	82.7	66.2
9	11.2	80.8	72.7
10	13.6	78.5	78.5
11	16.3	76.1	83.7
12	19.3	73.4	88.0
13	22.5	70.4	91.6
14	25.9	67.4	94.3
15	29.4	64.2	96.4

$\rho = 0.15$

N	I	P	Oe
1	0.0	87.0	13.0
2	1.7	85.5	25.7
3	3.6	83.8	37.7
4	6.0	81.8	49.1
5	8.7	79.4	59.6
6	11.8	76.7	69.0
7	15.4	73.5	77.2
8	19.5	70.0	84.0
9	23.8	66.2	89.4
10	28.4	62.3	93.4

$\rho = 0.20$

N	I	P	Oe
1	0.0	83.3	16.7
2	2.7	81.1	32.4
3	5.9	78.4	47.0
4	9.8	75.2	60.2
5	14.2	71.5	71.5
6	19.2	67.4	80.8
7	24.6	62.8	88.0
8	30.3	58.1	93.0

$\rho = 0.30$

N	I	P	Oe
1	0.0	76.9	23.1
2	5.1	73.0	43.8
3	11.1	68.4	61.6
4	18.0	63.1	75.7
5	25.4	57.4	86.1
6	33.0	51.6	92.8

$\rho = 0.40$

N	I	P	Oe
1	0.0	77.4	28.6
2	7.5	66.0	52.8
3	16.3	59.8	71.8
4	25.6	53.1	85.0
5	34.9	46.5	93.0

Reproduced by permission from reference 36.

Exhibit 3.5.A5 Illustration of Peck and Hazelwood's Tables for $N = 14$, Giving Me and I as a Function of ρ, N, and O

$s/(s+p)$	O	I	Me
.060	1	.714	.902
.062	3	.046	.999
	2	.219	.990
	1	.732	.894
.064	3	.050	.999
	2	.231	.989
	1	.750	.885
.066	3	.054	.998
	2	.244	.988
	1	.766	.876
.068	3	.058	.998
	2	.256	.987
	1	.782	.866
.070	3	.062	.998
	2	.269	.985
	1	.798	.856
.075	3	.074	.998
	2	.301	.982
	1	.833	.830
.080	3	.088	.997
	2	.333	.977
	1	.863	.803
.085	4	.021	.999
	3	.102	.996
	2	.367	.973
	1	.890	.775
.090	4	.026	.999
	3	.117	.995
	2	.401	.967
	1	.912	.746
.095	4	.031	.999
	3	.133	.994
	2	.435	.961
	1	.931	.718
.100	4	.036	.999
	3	.151	.992
	2	.469	.954
	1	.946	.690
.105	4	.043	.999
	3	.169	.991
	2	.502	.947
	1	.958	.663
.110	4	.050	.998
	3	.189	.989
	2	.536	.938
	1	.967	.637
.115	4	.058	.998
	3	.209	.987
	2	.569	.929
	1	.975	.613
.120	4	.066	.997
	3	.230	.985
	2	.601	.919
	1	.981	.589
.125	5	.017	.999
	4	.075	.997
	3	.252	.982
	2	.632	.909
	1	.986	.567
.130	5	.020	.999
	4	.085	.996
	3	.274	.980
	2	.662	.897
	1	.989	.547
.135	5	.024	.999
	4	.096	.995
	3	.297	.976
	2	.691	.885
	1	.992	.527
.140	5	.028	.999
	4	.107	.994
	3	.321	.973
	2	.719	.873
	1	.994	.509
.145	5	.032	.999
	4	.119	.994
	3	.345	.969
	2	.745	.859
	1	.995	.492
.150	5	.036	.999
	4	.132	.992
	3	.370	.965
	2	.770	.846
	1	.997	.476
.155	5	.041	.998
	4	.145	.991
	3	.395	.961
	2	.793	.831
	1	.997	.460
.160	5	.047	.998
	4	.159	.990
	3	.420	.956
	2	.815	.817
	1	.998	.446
.165	5	.053	.998
	4	.174	.988
	3	.445	.950
	2	.835	.802
	1	.999	.433
.170	6	.015	.999
	5	.059	.997
	4	.189	.987
	3	.470	.945
	2	.854	.787
	1	.999	.420
.180	6	.019	.999
	5	.074	.996
	4	.222	.983
	3	.521	.932
	2	.886	.757
	1	.999	.397
.190	6	.025	.999
	5	.090	.995
	4	.257	.978
	3	.570	.918
	2	.913	.727
.200	6	.032	.999
	5	.109	.993
	4	.295	.973
	3	.619	.902
	2	.934	.697
.210	6	.040	.998
	5	.130	.991
	4	.333	.967
	3	.665	.885
	2	.951	.669
.220	7	.013	.999
	6	.049	.997
	5	.153	.969

Exhibit 3.5.A5 *(Continued)*

s/(s + p)	O	I	Me
	4	.374	.960
	3	.708	.867
	2	.963	.642
.230	7	.016	.999
	6	.060	.997
	5	.178	.986
	4	.415	.952
	3	.748	.848
	2	.973	.616
.240	7	.020	.999
	6	.073	.996
	5	.205	.983
	4	.457	.943
	3	.785	.828
	2	.981	.592
.250	7	.025	.999
	6	.087	.995
	5	.234	.980
	4	.499	.932
	3	.818	.807
	2	.986	.569
.260	7	.032	.998
	6	.103	.993
	5	.265	.975
	4	.541	.921
	3	.848	.786
	2	.990	.548
.270	7	.039	.998
	6	.120	.992
	5	.297	.970

s/(s + p)	O	I	Me
	4	.582	.909
	3	.874	.764
	2	.993	.528
.280	8	.012	.999
	7	.047	.997
	6	.139	.990
	5	.331	.965
	4	.622	.896
	3	.896	.743
	2	.995	.510
.290	8	.015	.999
	7	.056	.997
	6	.160	.987
	5	.366	.958
	4	.661	.882
	3	.916	.723
	2	.996	.492
.300	8	.019	.999
	7	.067	.996
	6	.183	.985
	5	.402	.951
	4	.698	.868
	3	.932	.702
	2	.998	.476
.310	8	.023	.999
	7	.078	.995
	6	.208	.982
	5	.439	.944
	4	.732	.852
	3	.945	.682
	2	.998	.461

s/(s + p)	O	I	Me
.320	8	.028	.998
	7	.092	.994
	6	.234	.978
	5	.476	.935
	4	.765	.837
	3	.957	.663
	2	.999	.446
.330	8	.034	.998
	7	.106	.992
	6	.262	.974
	5	.513	.926
	4	.795	.820
	3	.966	.645
	2	.999	.433
.340	8	.041	.998
	7	.123	.991
	6	.291	.970
	5	.550	.916
	4	.822	.804
	3	.973	.627
	2	.999	.420
.360	9	.016	.999
	8	.057	.996
	7	.160	.987
	6	.353	.959
	5	.622	.895
	4	.670	.771
	3	.984	.593
.380	9	.023	.999
	8	.078	.995
	7	.203	.981

s/(s + p)	O	I	Me
	6	.419	.947
	5	.690	.871
	4	.907	.738
	3	.991	.563
.400	9	.033	.998
	8	.104	.992
	7	.252	.975
	6	.486	.932
	5	.751	.845
	4	.936	.706
	3	.995	.535
.420	10	.012	.999
	9	.046	.997
	8	.135	.989
	7	.307	.967
	6	.555	.915
	5	.806	.818
	4	.957	.675
	3	.997	.510
.440	10	.017	.999
	9	.063	.996
	8	.172	.985
	7	.366	.956
	6	.622	.896
	5	.852	.790
	4	.971	.647
	3	.998	.487
.460	10	.025	.999
	9	.084	.994
	8	.215	.980
	7	.429	.945

Reproduced by permission from reference 26.

REFERENCES

1. P. M. MORSE, *Queues, Inventories and Maintenance*, Wiley, New York, 1962, pp. 167–174.
2. N. K. JAISWAL, *Priority Queues*, Academic Press, New York, 1968.
3. B. W. NEIBEL, *Motion and Time Study*, 6th ed., Irwin, Homewood, IL, 1976, pp. 382–385.
4. G. B. CARSON, H. A. BOLZ, and H. H. YOUNG, Eds., *Production Handbook*, 3rd ed., Ronald, New York, 1972, pp. 1277–1285.
5. W. R. WRIGHT, W. G. DUVALL, and H. A. FREEMAN, "Machine Interference," *Mechanical Engineering*, Vol. 58, No. 8 (August 1936), pp. 510–514.
6. T. C. FRY, *Probability and Its Engineering Uses*, Van Nostrand, Princeton, NJ, 1928.
7. J. T. SMITH, "Machine Interference and Related Problems," *Work Study*, Vol. 14, Nos. 6–10 (June–October 1965), pp. 9–16, 21–28, 28–36, 25–34, 25–32.
8. D. C. PALM, "Intensitatschwankungen in Fernsprechverkehr," or "Analysis of the Erlang Traffic Formula for Busy-Signal Arrangements," *Ericsson Technics*, Vol. 6, 1943, pp. 39.
9. H. ASHCROFT, "The Productivity of Several Machines Under the Care of One Operator," *Journal of the Royal Statistical Society B*, Vol. 12, No. 1 (1950), pp. 145–151.
10. T. F. O'CONNOR, *Productivity and Probability*, Emmott, Manchester, England, 1952.
11. S. EILON, *Elements of Production Planning and Control*, Macmillan, New York, 1970, pp. 291–302.
12. D. W. JONES, "A Simple Way to Figure Machine Downtime," *Factory Management and Maintenance*, October 1946.
13. D. W. JONES, "Mathematical and Experimental Calculation of Machine Interference Time," *The Research Engineer*, Georgia Institute of Technology, Atlanta, GA, January 1949.
14. H. B. MAYNARD, Ed., *Industrial Engineering Handbook*, McGraw-Hill, New York, 1971, pp. 3-92–3-112.
15. A. JA. KHINTCHINE, "Uber die mittiere Dauer des Stillstandes von Maschinen," *Matematiceski Sbornic*, Vol. 40, No. 2 (1933), pp. 119–123.
16. R. KRONIG, "On Time Losses in Machinery Undergoing Interruptions," Part 1, Part 2 (with H. Mondria), *Physica*, Vol. 10, 1943, pp. 215–224, 331–336.
17. U. N. BHAT, *Elements of Applied Stochastic Processes*, Wiley, New York, 1972, pp. 238–241.
18. W. FELLER, *An Introduction to Probability Theory and Its Applications*, Vol. 1, 3rd ed., Wiley, New York, 1968, pp. 462–468.
19. K. E. STECKE and J. J. SOLBERG, *Scheduling of Operations in a Computerized Manufacturing System*, Report No. 10, NSF Grant No. APR74 15256, School of Industrial Engineering, Purdue University, West Lafayette, IN, December 1977.
20. H. M. WAGNER, *Principles of Operations Research*, 2nd ed., Prentice-Hall, Englewood Cliffs, NJ, 1975, pp. 888–889.
21. A. O. ALLEN, *Probability, Statistics and Queueing Theory with Computer Science Applications*, Academic Press, New York, 1978.
22. D. C. PALM, "Arbetskraftens Fordelning Vid Betjaning av Automatmaskiner," or "The Distribution of Repairmen in Servicing Automatic Machines," *Industritidningen Norden*, Vol. 75, 1947, pp. 75–80, 90–94, 119–125.
23. F. BENSON and D. R. COX, "The Productivity of Machines Requiring Attention at Random Intervals," *Journal of the Royal Statistic Society B*, Vol. 13, 1951, pp. 65–82; Vol. 14, 1952, pp. 200–219.
24. F. BENSON, J. G. MILLER, and M. W. H. TOWNSEND, "Machine Interference," *Journal of the Textile Institute*, Vol. 44, 1953, pp. 619–644.
25. R. B. FETTER, "The Assignment of Operators to Service Automatic Machines," *Journal of Industrial Engineering*, Vol. 6, No. 5 (September–October, 1955), pp. 22–30.
26. L. G. PECK and R. N. HAZELWOOD, *Finite Queuing Tables*, Wiley, New York, 1958.
27. C. MACK, T. MURPHY, and N. L. WEBB, "The Efficiency of N Machines Uni-directionally Patrolled by One Operator When Walking Times and Repair Times Are Constants," *Journal of the Royal Statistical Society B*, Vol. 19, 1957, pp. 166–172.
28. A. J. HOWIE and L. R. SHENTON, "The Efficiency of Automatic Winding Machines With Constant Patrolling Time," *Journal of the Royal Statistical Society B*, Vol. 21, 1959, pp. 381–395.
29. W. D. LAWING, "A Mathematical Method for Determining the Optimum Assignment of Quill Winders to a Patrolling Tender," master's thesis, North Carolina State University, Raleigh, 1959.

30. A. BEN-ISRAEL and P. NAOR, "A Problem of Delayed Service, I, II," *Journal of the Royal Statistical Society B*, Vol. 22, 1960, pp. 245–276.

31. R. O. GRANT, JR., "A Comparison Between the Random and the Patrolled Walk in the Assignment of Operators to Automatic Machines," master's thesis, North Carolina State University, Raleigh, 1960.

32. N. K. JAISWAL and K. THIRUVENGADAM, "Simple Machine Interference With Two Types of Failure," *Operations Research*, Vol. 11, 1963, pp. 624–636.

33. V. HODGSON and T. L. HEBBLE, "Nonpreemptive Priorities in Machine Interference," *Operations Research*, Vol. 15, No. 2 (March–April 1967), pp. 245–254.

34. G. H. REYNOLDS, *An M/M/c Queue for the Distance Priority Machine Interference Problem*, Operations Research Center Report 69-35, University of California, Berkeley, November 1969.

35. G. H. REYNOLDS, "An M/M/m/n Queue for the Shortest Distance Priority Machine Interference Problem," *Operations Research*, Vol. 23, No. 2 (March–April 1975), pp. 325–341.

36. D. C. PALM, "Assignment of Workers in Servicing Automatic Machines," *Journal of Industrial Engineering*, Vol. 9, No. 1 (January–February 1958), pp. 28–42.

37. G. R. DEAKIN, "Simulation/Regression Approach to Machine Interference," unpublished report, Chemical and Industrial Engineering Management Services, Chester, VA, 1980.

38. G. E. HAAGENSEN, "The Determination of Machine Interference Time Through Simulation," *The Western Electric Engineer*, Vol. 14, No. 2 (April 1970), pp. 35–40.

39. D. R. FREEMAN, S. V. HOOVER, and J. SATIA, "Solving Machine Interference by Simulation," *Industrial Engineering*, Vol. 5, No. 7 (July 1973), pp. 32–38.

40. J. E. BREDENBECK, M. G. OGDEN III, and H. W. TYLER, "'Optimum' Systems Allocation: Applications of Simulation in an Industrial Environment," *Proceedings of the Midwest AIDS Conference*, 1975, pp. 28–32.

41. A. A. B. PRITSKER, *The GASP IV Simulation Language*, Wiley, New York, 1974.

42. W. F. WEIR, "Figuring Most Economical Machine Assignment," *Factory Management and Maintenance*, Vol. 102, No. 12 (December 1944), pp. 100–102.

43. G. BLOM, "Some Contributions to the Theory of Machine Interference," *Biometrika*, Vol. 50, 1963, pp. 135–143.

44. F. BENSON, "Machine Interference–A Mathematical Study of Some Congestion Problems in Industry," unpublished doctoral dissertation, University of Birmingham, England, 1957.

45. I. ADIRI and B. AVI-ITZHAK, "A Time-Sharing Queue With a Finite Number of Customers," *Journal of the Association for Computing Machinery*, Vol. 16, No. 2 (April 1969), pp. 315–323.

46. J. E. ARONSON, "Heuristics for the Deterministic, Single Operator, Multiple Machine, Multiple Run Scheduling Problems," Technical Report OREM 80019, Department of Operations Research and Engineering Management, Southern Methodist University, Dallas, 1980.

47. R. W. CONWAY, W. L. MAXWELL, and H. W. SAMPSON, "On the Cyclic Service of Semi-Automatic Machines," *Journal of Industrial Engineering*, Vol. 13 (March–April 1962), pp. 105–107.

48. D. R. COX and W. L. SMITH, *Queues*, Wiley, New York, 1961.

49. R. DUBE and E. A. ELSAYED, "A Multi-Machine Labor Assignment for Variable Operator Service Times," *Computers and Operations Research*, Vol. 6, 1979, pp. 147–154.

50. I. L. HAINES and C. F. ROSE, "Use of Queueing Theory in Setting Production Standards," *Journal of Industrial Engineering*, Vol. 13, No. 6 (November–December 1962), pp. 456–459.

51. N. K. JAISWAL, "Preemptive Resume Priority Queue," *Operations Research*, Vol. 9, 1961, pp. 732–742.

52. J. KILLINGBACK, "Cyclic Interference Between Two Machines on Different Work," *International Journal of Production Research*, Vol. 3, 1964, pp. 115–120.

53. J. R. KING, "On the Optimal Size of Workforce Engaged in the Servicing of Automatic Machines," *International Journal of Production Research*, Vol. 8, No. 3, 1970, pp. 207–220.

54. J. G. MILLER and W. L. BERRY, "The Assignment of Men to Machines: An Application of Branch and Bound," *Decision Sciences*, Vol. 8, No. 1, 1977, pp. 56–72.

55. P. NAOR, "On Machine Interference," *Journal of the Royal Statistical Society B*, Vol. 18, 1956, pp. 280–287.

56. P. NAOR, "Normal Approximation to Machine Interference With Many Repair Men," *Journal of the Royal Statistical Society B*, Vol. 19, 1957, pp. 334–341.

57. P. NAOR, "Some Problems of Machine Interference," *Proceedings First International Conference, Operations Research*, English University Press, Oxford, 1957.

58. L. TAKÄCS, "Probabilistic Treatment of the Simultaneous Stoppage of Machines With Consideration of the Waiting Times," *Magyar Tud. Akad. Mat. Fiz. Oszt. Kozl.*, Vol. 1, 1951, pp. 228–234.

59. K. THIRUVENGADAM, "Queueing With Breakdowns," *Operations Research*, Vol. 11, 1963, pp. 62–71.

60. K. THIRUVENGADAM, "A Generalization of Queueing With Breakdowns," *Defence Science Journal* (India), Vol. 14, 1964, pp. 1-16.

61. K. THIRUVENGADAM, "A Priority Assignment in Machine Interference Problems," *OPSEARCH* (India), Vol. 1, 1964, pp. 197–216.

62. K. THIRUVENGADAM, "Machine Interference Problem With Limited Server's Availability," OPSEARCH (India), Vol. 2, 1965, pp. 65–84.

63. K. THIRUVENGADAM, "Studies in Waiting Line Problems," doctoral thesis, University of Delhi, Delhi, India, 1965.

64. K. THIRUVENGADAM and N. K. JAISWAL,"The Stochastic Law of Busy Periods for the Simple Machine Interference Problems," *Defence Science Journal* (India), Vol. 13, 1963, pp. 263–270.

65. K. THIRUVENGADAM and N. K. JAISWAL, "Application of Discrete Transforms to a Queueing Process of Servicing Machines," *OPSEARCH* (India), Vol. 1, 1964, pp. 87–105.

66. L. J. WATTERS, ''Queueing Theory Applied to an Idle-Time Utilization Problem, *Journal of Industrial Engineering*, Vol. 17, No. 7 (July 1966), pp. 394–388.

SECTION 4

PERFORMANCE MEASUREMENT AND CONTROL OF OPERATION

CHAPTER 4.1

Work Standards: Establishment, Documentation, Usage, and Maintenance

JOSEPH A. PANICO
American Productivity Improvement Systems

4.1.1 THE WORK MEASUREMENT PROCEDURE

Developing a work measurement system involves four distinct chronological steps. A work standard, or production quota, must be established. Once established, a method must be developed to document that the quota is met. This process of documentation allows management to use work standards to improve productivity by eliminating problems that inhibit output. When these problems appear with some regularity, they trigger a process of maintenance, which in essence keeps the system viable and truly predictive of what management can expect from its production processes. Although each of the four steps may be analyzed separately, all four must dovetail, working in concert to build what is called a "total work measurement system."

A procedure for measuring work must be chosen. The measurement technique may be quite arbitrary, using only past experience to set production quotas. Conversely, it may be extremely sophisticated, using the sum of motion patterns detailed to the 0.0006 min. The degree of sophistication is determined by the work environment. A method of verification will also range in complexity from the honor system to simple time cards to elaborately detailed computer analysis. The thrust of the whole process is to improve and control costs through better information and subsequent accountability. Once the system is functioning, it must be continually audited to make certain that productivity is complying with the targeted demands. In essence, a vigil must be maintained over the whole process to ensure the desired or expected productivity levels.

A study of each part of this total process of establishing, documenting, using, and maintaining work standards will provide insight into how each functions separately and how they function jointly to form a total procedure for enhancing productivity.

4.1.2 ESTABLISHING A WORK MEASUREMENT SYSTEM

Introduction

Hours, days, weeks, or other time segments are used to measure output as a mark of employee achievement. These measurements may be subjective or objective. An informal way to distinguish objectivity from subjectivity is found in the way a sentence is prefaced. "I believe . . ." usually prefaces a statement couched in subjectivity. "The facts are . . ." usually prefaces an objective statement. However, arguments will always arise regarding belief versus fact; thus it is established that objectivity is quantified and subjectivity is qualified.

A further elaboration is necessary. Subjective measurements of employee performance embody opinion, prejudice, likes and dislikes. Objective measurements are supposedly limited to discernible output. To imply that formal work measurement eliminates all subjectivity is incorrect. Yet, the opposite must be considered too. Informal, qualified measurements contribute greatly to employee dissatisfaction and business risk. Many events may cause employee dissatisfaction. It is generally found that employees respond better in an environment with goals that are readily known and achievable. When employees respond favorably to these goals, output becomes more predictable. Thus raw materials, goods-in-process, and finished goods inventories become better stabilized.

In total, management may be able to operate within the framework of greater predictability and less risk when employees equal or exceed the stipulated standard.

Scientific work measurement examines the employee's work in finer detail. Techniques that use average or other course measurements are replaced by methods that divide the task into measurable parts. This allows the job to be synthesized and thus made easier. A better method is found. Each employee is given access to the information and is trained in the procedure. In this proper setting the employee achieves the goals with the help of management, and a rapport develops. Scientific work measurement is used throughout the world. Its scope and benefits are so broad that it is used to establish precise goals for business, clerical, manufacturing, retailing, service, and other operations. In essence, measurement of work pervades all fields and disciplines.

Proceeding to Engineered Work Standards

Standards Established Subjectively

Many organizations in their inception phase do not operate with any formalized measurement of work. In this informal setting management's relationships with the production worker are very close. Thus social, rather than formal, methods control worker output. As the organization grows, these intimate structures tend to break down, and the worker may feel somewhat divorced from the whole process. Few workers knowingly curtail work. The sociological structure has changed. Interrelationships have changed, and what was once an informal, sharing workplace may now be replaced by formality. Obviously, the humanistic needs of the worker in the workplace require considerable attention. Management must understand and help to meet these needs. However, what was once a highly productive, informal structure may now be formal and somewhat less productive. Systems, products, and work assignments have changed. Supervisors have changed. Top management may not be seen for months. Thus in this setting willing workers may not produce to capacity because goals are not understood and workers have not shared in establishing them.

Expanding organizations try to continue operations in this informal setting because it worked before, but diminished productivity soon dictates a need for change. The evolutionary process begins. Subjective criteria are the basis for setting the goals at first. These criteria are tolerated by the workers. For every subjective standard there seems to be a subjective list that reasons why the subjective standard goals cannot be met. Operating in this way, management must resort to pressure in the form of rules. Rules seem to be made to be bent, as do subjective work standards, so management seeks out individual standards in terms of averages.

Using Averages to Determine Productivity

This second step in the evolutionary process may be the most demeaning. An average discourages, through peer pressure, those who perform above the norm. It usually does not stimulate to higher levels those performing below the norm, who are doing so not because of indifference, but more so because of inability. Averages do not analyze work. They seldom bring about training of the worker in a prescribed method for improved work performance. Managers who set goals in this manner leave workers alone. Operating in this way, workers are further isolated from management, and resentment seems to mount. As managers feel worker pressure, they may lower requirements, accept excuses, or turn their heads, generally not seeing a truly bad performance.

Averages can work, but they require many of the techniques usually associated with formalized work measurement. Often an organization may believe its work situation is so unique that formalized work measurement procedures will fail. This is unusual. If averages must be used, they, too, should be preceded by methods study. Management's responsibility is not only to provide goals, but to train, develop, and encourage workers to reach the quotas set for them. A further sophistication of averages is statistical analysis of historical information on a time-related basis. This technique tends to squeeze standards higher, based on learning theory and capital investments. Here again, the full thrust to achieve could be erroneously placed solely on the worker.

There are many plans in use today based on averages. Those that succeed incorporate formalized analysis of work on a continuing basis.

Analysis of Formalized Procedures

The final major step in establishing work standards may involve formalized procedures incorporating time study, predetermined time systems, standard data, or computerized standard data. All of these systems employ a precise method for evaluating work. Although these techniques are magnificent tools to aid productivity, they carry the stigma of "efficiency expert." This expression, coined many years ago, still haunts the field today.

Time Study. Time study uses a stopwatch or electronic watch to aid in the evaluation of work. A task is broken down into basic divisions of accomplishments, called elements. The sum of these elements constitutes the total time in which the worker performs the job. Added to this total are personal, unavoidable, and fatigue times. These increase the time on the task because they are a deterrent to sustained daily productivity. Stopwatch time plus personal, unavoidable, and fatigue time are coupled with a performance rate, which is a measure of worker intensity and accomplishment. The job analyst combines these to provide the worker and management with a standard that stipulates how long a task should take or how many pieces should be produced in a unit of time.

The job is broken into subtasks in order to facilitate method study. In a job consisting of 16 elements, one worker may perform an element or sequence in less time than another worker, yet the second worker may outperform the first in a different element or sequence. The analyst who is evaluating the job's elements may also see further refinement that could aid productivity. Combining these individual and better ways to work will usually result in improved procedures. Training others in these better methods will result in increased worker productivity. Such gains would have gone unnoticed had the job not been separated into the smaller groupings and timed in order to see which segment of work was performed better.

A stopwatch, especially to the uninitiated, is sometimes an imposing device. Analysis of work involving "being clocked" for 2 hrs. or more is often misunderstood by the worker and summarily rejected. Managers who may not favor accountability for themselves may sympathize with the worker resulting in the work measurement program being conducted in a climate of hostility. Measurement is paramount to successful operations. The need for measurement ultimately suppresses personal feelings; consequently, management proceeds with formalized measurement in order to increase productivity. The gains are manyfold. Accomplishment levels are objectively defined, better job methods are established, and a concordance is established between the worker and management.

Predetermined Time Systems. A more detailed analysis of work is obtained when predetermined time systems are used to measure and improve worker performance. In these systems the predominant motions of workers are defined in motion patterns. A time study element may consist of 10 motion patterns. Defining these motions and assigning times to them will magnify the job into sequences so that many unnecessary deterrents to productivity may be isolated and subsequently corrected. The finer the detail, obviously the more time spent in job analysis.

Predetermined time systems minimize formal timing procedures.[1] The use of a stopwatch is limited to the study of process times or special tasks. Predetermined time systems are excellent for improving motions and motion sequences, called methods. The time to carry out a formal predetermined analysis is extremely long. Volume, savings, and worker acceptability may justify this type of analysis. The nature of the job dictates which procedure—time study or predetermined time systems—should be used. In a manufacturing setting, predetermined systems may be rejected by the union because of long-term, usually unwarranted disapproval.[2] Predetermined systems continue to be employed, however, and are gaining wider acceptance.

Standard Data and Computerized Systems. Use of standard data is an attempt to find family relationships for various jobs performed in similar ways on identical equipment or for a particular product. Through this technique it is possible to establish work standards without personally investigating the job as it is performed on the floor. The system is superb when used correctly. Interpolations are generally accurate enough to establish times, but they fall short when operator training is considered. The problem with standard data is the temptation to extrapolate, wherein the analyst conjectures from the known data in order to establish worker goals and anticipated prices.

Computerized standard data are somewhat more refined because they delve into the relationships of motion sequences rather than elemental sequences. Linear regression, multiple regression, correlation analysis, and F tests assist in analyzing the relationships of data, but statistics do not make right what was wrong in the inception. To use standard data correctly, the analyst must be exposed to the floor or the job. Remember that the principal reason for work standards lies in method/motion analysis. Unfortunately, this cannot be conducted in absentia; the analyst must observe the job firsthand.

Choosing the Measurement System

The Transition From Subjective to Objective Work Standards

Most organizations begin with subjective standards. These are often intuited by the supervisor. Frequently, the employees being measured do not understand the criteria for measurement or the points assigned to each criterion. An attempt to objectify these criteria by developing tabular rating forms is still laden with subjectivity because the supervisor's feelings remain at the point of the pencil.

Any organization wants good employee relationships. Subjective standards tend to destroy these bonds. An organization must use measurement, however, in order to minimize risk through predictability.* It also must use measurement to improve productivity through goal setting. The thrust of objective standards is to develop the methods and procedures necessary for the employee to attain work quotas. Thus, in situations where subjective standards are failing, an organization may seek out techniques for establishing a more reliable basis for prediction. When making this transition, suspicion may lead to rumor, and a feeling of uneasiness may arise. Thus it is imperative that the change be made openly. Employees should be formally informed about the new system. Management should not trust someone down the line to inform another, who, in turn, informs yet another. The message will disintegrate, with possible unfavorable repercussions.

Fitting the Plan to the Work Environment

Service Environment. When making the transition from subjective to objective standards, one must seek out the work standard system that best fits the operation. Some business environments may be antagonistic toward stopwatch studies. Generally, service institutions fall into this category. In this environment the employee may have traded real for psychic income. The stopwatch tends to destroy this trade-off.†

Thus many organizations choose to use predetermined systems or standard data for establishing the work quotas. The sophistication of a predetermined system seems to be more acceptable to employees working in a service-type environment. A tremendous amount of data is available in these fields to assist those who are implementing a new system.‡ Numerous regional or national meetings, where papers are given pertaining to these specific service-type work measurement areas, are available to those wishing to gain further understanding.

Clerical and Retail Environment. In the clerical or retail setting, the resistance to stopwatch studies diminishes. This is not to imply that work standard procedures are openly embraced; it serves only to indicate that, once a determination is made to proceed to objective standards, and once this fact is communicated to the employees, one system versus another does not generally become an issue. Time study has good reliability and can be performed quickly. With this technique, many jobs can be analyzed for individual standards, the information can be developed into standard data, and the process can be explained somewhat easier to those who are being analyzed. It must be emphasized again that the work measurement procedure not only provides the time spent on the task, but the method patterns for meeting the prescribed goal. In a service operation the work may not be as cyclical as in a clerical or retail operation. Within the sociological frame of reference, predetermined systems seem to be better accepted in a service setting. Also, the work tends away from complete repetition. A predetermined system therefore lends itself well to situations with these restrictions.

Clerical and retail operations may use either system successfully. The employees in these fields of work may be formally organized and bound contractually. This contract may stipulate that only time study may be used. Many contracts make this stipulation. Another contractual restriction commonly found stipulates that, in the event of a work standards dispute, predetermined systems must be verified through stopwatch studies. Although this contractual requirement is found more in manufacturing agreements, it is a consideration for the clerical and retailing operations, too. The union representing these groups is a very large organization. They have a formal industrial engineering staff specializing in work measurement. Thus, when proceeding to formalized standards for clerical or retail operations, it is imperative to fit the system to the contract—if one exists.

Manufacturing Environment. Manufacturing is in itself unique. Work measurement is used extensively in this setting. From a historical perspective, work standards, subjective or objective, have strong roots in this environment. At one time in the history of work, labor was in great abundance and was available for long working hours. Whenever more productivity was required, the

*A work standard is used to forecast production levels. Inventories, prices, and manpower are often based on this forecast of what should be expected in terms of productive output from each employee.

†Many service institutions and some manufacturing establishments have refused to use a stopwatch for developing work standards. They reason that an employee "being clocked" would develop feelings much like those of a blue collar worker. Furthermore, it could destroy the employees' feeling that they are an integral part of the managerial process.

‡Training courses are offered in MTM-C-A two-level clerical standard data system by the MTM Association for Standards and Research, Fairlawn, NJ.

situation was met simply by placing more labor onto the job. This inundation method for reaching production quotas prevailed until the Industrial Revolution. "Have countries" were faced with labor shortages and brought in many immigrants. Division of labor techniques were employed to accelerate production. There was a trend toward more humanistic working conditions. Labor costs increased as a portion of the total product cost. Restrictions on work were imposed by law, government, and unions. With this series of historical events came a greater need for work efficiencies, and thus time and motion study took on greater importance as a technique for measuring and improving work performance.

Originally, the techniques required for measurement were coarse. Most jobs were manual. When the machine assumed a greater portion of the work cycle, measurement became more sophisticated because of increasing cost relationships of man to machine. As industry changed from mechanical to electrotechnics, once again another refinement in measurement occurred. Within this sequence, traditional measurement was refined, new techniques for measurement were established, and controls were maintained closer to the job time sequence.

With this long history of work standards, manufacturing operations tend to be less disturbed with the principle of measurement. While workers' attitudes toward measurement are caustic, they are somewhat more receptive to the system when compared with other work environments. Small manufacturing operations usually commence with subjective standards. As the operations become more repetitious, more objectivity is sought. Many manufacturing establishments are bound contractually when work standards are considered. The specific contractual language must be assessed to determine which work standards may be employed. Also, the contract may stipulate break times, methods of payment, disciplinary procedures, allowances, and a host of other terms that restrict the company's work standards practices. In the absence of a contract or a union, a manufacturing establishment should proceed to work standards with complete formality. The first person a system answers to is the worker—with or without a union.

Conclusion

Predetermined systems have been used in practically every type of manufacturing operation. The predominant system in use, however, is time study. Whichever system is employed, it is always best to have a plan for implementation. The system would be better served if implementation were to proceed department by department, rather than by trying to establish standards plantwide. Rather than waiting to have all standards set for every job and every operation, the firm could proceed by establishing the procedure for each machine or workplace in a specific department, continuing in this manner until the whole operation is covered.

Obviously, the product type and product mix will determine the feasibility of the plan for implementation. The worst procedure is randomization, whereby standards are placed on the job on a first come, first serve basis. Just as there must be logic in how the job is broken into parts, there must be logic in how to proceed within the department and within the facility. Sometimes it is wise to choose those jobs or products that constitute the preponderance of sales. Depending on the operation, 20% of products may provide 80% of sales. Thus, if standards are set on these operations first, the impact of a work standards program on profits will be realized more quickly.

From a historical base, the process of standard implementation has been from subjective systems to average systems to time study. Once an objective system has been established, predetermined systems, standard data, or computerized standard data may be further refined. The process may entirely bypass time study with great success. Again, it must be emphasized that the predominant system in use today is time study. It is not obsolete. When used correctly, it is a very reliable work standards system. Some of the inequities found in time study are minimized with predetermined systems. Conversely, the inequities of predetermined systems cannot be overlooked. Each system has its rightful place. There is much literature in the field that provides the advantages and disadvantages of one system versus another. The system chosen will not eliminate the need for exercising common sense about people, their relationship to work, their needs and aspirations, and their sincere wish for a job well done. Within a properly applied and communicated work standard system lies greater productivity and a happier employee. If the system does not work, the cause is external. The system usually is not the problem.

Selecting a System for Motivation and Control

Increasing Productivity

A work standards system will not function well without precise controls. Once goals are set, the worker will usually try to achieve them. Controls in the form of production reports will indicate whether a given employee or department is working up to the prescribed production levels. The report will usually present data arranged in columns, indicating the actual time spent by the employee or department in fulfilling the task, the number of jobs or work assignments undertaken, and the obstacles to sustained productive effort. A production report may have 15 columns indicating a variety of information about the worker, the department, or the job. The importance of this report does not rest solely with worker accountability, although management certainly wishes to see that work is being performed at profitable levels. The main reason for the report is to make obvious the deterrents to daily production.

Knowledgeable managers are aware that their responsibility is to eliminate delays and thus to provide more time for work. A workday may be comprised of 60% productive time and 40% nonproductive time. If a plan to stimulate production results in the employees producing at 15% above anticipated levels, this will not usually diminish nonproductive time. It may mean that they have increased their output within the available 60% productive time. If the report is designed to segment the 40% nonproductive time, then management may isolate those factors causing productive delays and thus change the ratio from 60:40 to 80:20. It is hoped that this additional 20% will be used by the employee for producing at the same 15% above anticipated levels. Work standards are an integral portion of this report. They provide the targeted production levels. Business organizations determine production or service costs based on the employees' achieving 100% of the anticipated levels. Lower productive levels must be absorbed by the organization or the market they serve.

Comparison of Plans

In an attempt to ensure that employees work to prescribed production levels, management may provide inducement in many forms. The two most commonly used forms are MDW and financial incentives. Both are incentive plans. The first involves enticement with normal or average productivity demands. The second uses money coupled to highly intensified productivity.

Measured Day Work Plans. Straight day work involves paying the employee's hourly rate usually in the absence of objective work standards. Measured day work also pays the employee's hourly rate, but output is measured and controlled objectively. In a pure sense, work standards are established using time study, predetermined systems, or other measurement methods. These standards are made available to employees and to those who supervise them. Controls are developed to measure output and for listing those factors that restrict productivity. The process continues to follow this chronological sequence. Each employee, group, or department is kept informed, either privately or by posted charts, of progress made. Since enticement rests on a desire to attain prescribed goals, employees, supervisors, and other managerial departments work together in this endeavor. This joint effort produces better communication and understanding between the employees and management. Once goals are reached and maintained, the whole process is scientifically squeezed in order to attain even higher goals. The pros and cons of MDW are offered with great persuasion. Success depends probably more on managerial attitude than on the purity of one system versus another.

There are specific applications where MDW is preferred and works better than its counterpart—financial incentives. There is no set rule, however, since some MDW plans work exceptionally well in what was originally the domain of the financial. The converse also holds. Measured daywork is a widely used incentive program. Its proponents are quick to point out that the costs of maintaining the system are substantially lower than those for a program using financial incentives. Other suppositions, such as quality improvements, better labor harmony, and worker honesty, are offered as reasons for adopting MDW plans. Some work standards specialists believe that work intensity—the concept of what constitutes a fair day's work—should be somewhat less for measured daywork plans when comparing these two systems. Using this philosophy, 100% as an objective measure of normal work for financial plans could, as a hypothesis, be called 110% or more for MDW plans. In essence, MDW would require less productivity when objectively determining what a "normal" operator should produce than would comparable norms objectively generated for financial plans.

This is not a universally agreed-upon concept. Many practitioners advocate the same norm. Often MDW precedes financial incentive plans. Conversely, failing or aged financial plans may be replaced by MDW. In the case where MDW precedes financial plans, it would be difficult to tell employees that the normal work requirement will increase 10 to 30% in the transition to financial plans. All along, the employees had been told that the standards in use were fair and attainable. Furthermore, there may have been grievances and arbitration surrounding these standards. Thus the claim that standards once considered fair are too easy now that financial rewards are involved is somewhat difficult for workers to accept.

Measured day work requires discipline in the event that workers are not reaching the specified goals. In contractual situations there is a step-by-step procedure to follow. In noncontractual situations the procedures are less formal, but they still exist. When using MDW, the expectation for the attainment level of employees may be lower than that for financial plans. Assuming a norm of 100%, the MDW expectation could be 80 to 90% of this value. The expectation with financial plans will be 100% or better.

Methods Time Measurement is a predetermined system. It is used more extensively than any other work standards plan, with the exception of time study. All motion patterns normally associated with work are cataloged, and the time is listed on a card for use by certified practitioners. There is not one card for MDW and another for financial plans. No modifier exists with which to change the values of the card when using MDW. Thus 100%, or normal, is similar for both incentive systems. The same is true for time study. A performance rate of 100% should be the same when using MDW or financial plans. A few multiplant corporations have chosen to differentiate what constitutes normal performance, or 100%, depending on which incentive plan is in use. This is their exclusive choice and should not be misconstrued as the operating philosophy of the entire profession.

Many companies set MDW standards at 100% and discipline according to this level. Much arbitration has authenticated this approach to be a correct operating procedure. The concept of accepting lower attainment levels when employing MDW schemes is sociological. In any task plan involving lower attainment levels, 100% is easily attainable and may be reached or exceeded "day in and day out without undue stress or fatigue." Thus accepting lower performance levels when using MDW is by choice, not by design. Accepting 80% to 90% is erroneously justified because there is no financial reward for levels exceeding 100%. Unfortunately, this belief has caused MDW norms to slide lower. When comparing the output of MDW plans versus that of financial plans there is a pronounced difference. This is a general statement; obviously, there are a few exceptions. Setting MDW norms lower, essentially viewing, for example, 85% as 100%, is convenient for standard costing, pricing, production control, and other managerial functions. But who is to say that, over the years when workers are attaining only 85% of these already lowered goals, the same sociological reasoning would not prevail to thus further erode MDW norms?

Justifying MDW on the basis of lower maintenance costs, better labor harmony, worker honesty, improved quality, and other factors may not be universally true. If an 85% norm is accepted, then these factors for justification would have to result in producing a tremendous amount of money when compared with financial plans that result in returns at 125% levels. Also, use of MDW requires a larger and somewhat more sophisticated supervisory staff. In many cases labor harmony and worker honesty are reportedly better when using MDW. But if discipline is invoked when 90 to 100% levels are not maintained, these advantages tend to diminish very quickly. Also, consider that the workers in a slowdown situation are hurting only the company or the business, not themselves. Under MDW, they get paid the same wages. Contrast this situation to that in which an incentive plan is used; here part of the payment for the slowdown comes out of the employees' pockets. Another factor for consideration involves methods changes or an error in establishing the work standard. Assume that employees are working at the 100% level under MDW plan and that either a methods change or an error in the work standard occurs so that employees now could produce to the 115% level. Given these conditions, the employees will maintain output at the 100% level; thus productivity in a true sense diminishes.

Financial Plans. Rewarding an employee with additional monies for achievement above the prescribed task level is broadly called a "financial work standards system." Money has been used throughout history as an inducement. Attempts to remove money as a stimulant have been discussed in economic, philosophic, and theologic writings. Experiments will continue, especially in business and particularly with respect to productivity. Nonfinancial rewards seemingly work well only in those cases in which there is esprit de corps. Thus taking one such plan that is working well in one setting and transferring it to another may cause a productivity decrease. Money is a universally understood medium of exchange, and this may be the reason for its widespread use as an inducement for higher productivity.

The popularity of work standards, methods engineering, and motion analysis led to the development of many financial compensation plans. The majority of these plans were successful in a particular and closely monitored environment. Generally, plans could be classified as one-for-one, group, and gains sharing plans. An early set of specialized systems included the Halsey, Rowan, and Taylor Differential Piece Rate plans. As the work environment changed, modifications to this set occurred. A few of the newer plans were the Gantt Task and Bonus System and the Emerson Efficiency Bonus System. The procedure of seeking a fair compensation system was then further refined, again to meet a changing work environment, into plans for equalizing compensation throughout the plant or business organization. The Bedaux Point, Haynes-Manet, and Parkhurst Differential Bonus plans are a few of these equalizing systems.[3] All of these plans worked exceptionally well when closely monitored and in select situations. Group plans, and combinations of group and individual bonuses, also constituted a popular specialized compensation system. Historically, group plans abounded, seemingly for every type of business and every changing decade. Again, it must be emphasized that all of these plans were successful only in particular situations. As universal systems of compensation, the probability of their success was extremely low.

Today, more than 90% of wage payment systems are one-for-one plans. They are easily understood by the employee and generally are more competitive than other types of plans. In the one-for-one plan, a task level is determined, and the individual is rewarded in direct proportion for attainment levels above the norm. For example, an individual employee may perform at a level 25% above standard. A direct and proportional payment for this attainment level would be $1.25 for every $1.00 of base rate. In many cases corporations have an incentive base and a day work base. The incentive base is lower. An employee may have to perform at levels 10% above the norm to achieve the day work base.

Other systems may have an incentive base, a day work base, and an afterpayment. Using $1.00 for comparative purposes, this type of plan may pay $1.00 to the employee for those times he or she cannot produce because of delays beyond his or her control. For producing times, the employee will be compensated at ($1.25)($0.40) = $0.50 plus an afterpayment of $0.60, for a total of $1.10 for the hour. In such situations the worker is producing at 25% above task level, but is being compensated only 10% above the base. This type of plan often curtails productivity, encourages cheating, and results in supervisory complicity, particularly with respect to downtime.

These three one-for-one schemes are just a few of a multitude of such systems. They, too, are troublesome and require considerable attention in order to keep them effective as stimulants for greater productivity.

In group plans individual payment is predicated on the total group's performance above a prescribed standard. Gain sharing plans make partial payment. These plans may be group or individual payment schemes. For example, the plan may call for 50% of the incentive bonus to be given to the company and 50% to the employee. One-for-one plans begin payment at a defined task level. Incentive is paid in direct proportion to the achievement above the defined level. A 25% increase in productivity, for example, will result in a 25% increase in payment from a prescribed monetary point.

The prevalent system is one-for-one, with incentive payment beginning at base rate. Using this program of payment, 100% equals base rate. A performance level of 25% above the standard will result in a payment of $1.25 (base rate). A performance level of 80%, a rate lower than expected, will be paid at base rate. The employee's pay is guaranteed at 100%. Most one-for-one systems stipulate that any long-term performance below 100% will result in disciplinary action, with subsequent discharge if continued. Organizations with one-for-one plans that vary from direct base rate payment usually seek change to the prescribed formal and proper format, which is a one-for-one over base rate type of plan. Business organizations, particularly manufacturing firms, employing other variants to a direct one-for-one system are the exception. The use of specialized plans is diminishing. Historically, they require considerable maintenance and over the years may not keep pace with expected investment to productivity ratios. One-for-one plans retain their popularity and seemingly remain the greater stimulant to productivity.

Summary of Payment Plans. Financial plans may be troublesome. Conversely, MDW programs that truly employ disciplinary procedures are equally troublesome. Data showing a trend away from use of one system in favor of another may not be statistically correct. Financial plans are used extensively in the industrial setting. Manufacturing establishments have a long history with the principles and practices of formalized work standards procedures. With this ability, they can essentially "make book" on productivity levels, tying work standards to a monetary plan. Once other segments of business also become more familiar and comfortable with work standards, they too may employ financial plans. At the turn of the century, who would have been bold enough to pronounce that work standards would be used in hospitals, insurance companies, banks, scientific research establishments, and many other seemingly nonmeasurable areas? Also, who would have ventured the prediction that automobile repairs, stocking of grocery operations, meat cutting, and general maintenance would be tied to a financial incentive plan? It is premature to hypothesize about which system will predominate. What can be said is that the procedure of formalized work standards is spreading into many new fields and expanding in scope.

Coupling Work Standards with Job Evaluation

Problems Associated with Afterpayment Incentive Plans

The validity of most work standards diminishes as a function of time. What was a good, crisp, reliable standard 3 years ago may be incorrect today. The workplace may have changed. Methods could have changed too. Numerous events could cause a previously good standard to age.

Consider a company that determines that its work standards do not truly reflect productivity. Furthermore, a large afterpayment is being added to daily earnings. Consider, too, that production workers are falsifying reports, with the supervisor essentially looking the other way. Here, then, is a company with a work standards system that is out of control.

To correct this problem, the company bolsters its work measurement staff to restudy all prevailing standards. These changes will seriously curtail employee earnings because of the huge afterpayment. In this hypothetical situation the union will agree, provided that a new job evaluation plan is implemented. Through this technique all jobs are quantitatively analyzed, which leads ultimately to a new base rate for each job class. Demands for productivity increase tremendously when the new work standards plan is implemented. Originally the demand was 20 pieces/hr. Earnings for this were ($0.40)(200%) = $0.080 plus an afterpayment of $0.60. For each dollar of base rate, the employee was earning $1.40, or 40% additional earnings. By using downtime advantageously, these earnings could be increased another $0.20; thus for each dollar of base rate, the employee would earn $1.60, or 60% additional pay.

Assume that, using a new evaluation plan, the same dollar would increase to $1.10. Originally the dollar comprised $0.40 incentive plus $0.60 afterpayment. Now for 100% performance the plan will pay $1.10 instead of $1.00. There will be no afterpayment, and the company is determined to control downtime reporting and any falsification of records. A properly installed incentive system has a potential of 25% to 35% above base rate earnings. An employee who originally could earn $1.60, or 160%, may now have the potential to earn only ($1.10)(135%) = $1.42 for each dollar of original base. Furthermore, the employee will have to work a full day because downtime has been minimized. Here, then, is a case where both base rate and demands for productivity have been increased.

Job Evaluation and Buy-out Problems

It is always a temptation in these types of situations to balance out the system by maintaining under the new work standards and job evaluation system previous amounts of take-home pay. It is felt that anything less will be too harsh for the workers. This same argument is used throughout industry when there is an attempt to curtail overtime—particularly in cases where it has been abused. In such situations some companies choose to adopt an MDW plan whenever a new job evaluation is implemented. A company reasons that in time it will look again at financial incentives. In actuality, this transition to incentives rarely takes place, and a company may be faced with diminished productivity at higher base rates. "Sweetening" base rates to solicit workers' consent for changing standards is another technique that is sometimes advocated. This strategy calls for increasing base rates coupled with a change in work standards to effect an increase in productivity. All of these situations have worked at some time in a given facility. Success is usually tied to the particular moment and specific environment. There are cases where work requirements have been intensified and base rates decreased. All of these are the exceptions. Ultimately, the company must face reality.

The company must establish new standards and couple them with a base rate that will enable the firm to remain competitive in the marketplace. When a union is involved, these changes must be negotiated into a new contract. If the changes are scheduled within a contractual period, then the union must concur. Management has the right to restudy jobs. Within the framework of the contract are procedures for changing work standards. In the event of disputes, the union may present a formal grievance, which could ultimately lead to arbitration. For operations where employees are not represented by a union, the situation surrounding a change of work standards is usually more critical. The same holds for new job evaluation plans. Generally, in nonunion environments the organization has subjective rules that are often more restrictive than those established under a union contract.

This is not an easy problem to solve. A company may falsely believe that, if no work standard program existed, then most difficulties would disappear. One group of approximately 200 companies has found that financial incentives generate 48% more productivity than day work and 29% more productivity than MDW. This group is heavily machine controlled. In situations where jobs are less dependent on machines, the ratios are even higher. Work standards have a pronounced impact on productivity. The cost of having quotas established this meticulously may be well worth some of the supposed inconveniences. A job evaluation program establishes a relationship between various categories of work. The wage payment line determines the hourly base for each job class. Work standards provide anticipated productivity levels. The proper combination of these three factors will keep the employees' pay and the company's market position competitive.

Drafting a Plan, Obtaining Approval, Selecting and Training a Staff

Any new plan or change in an existing system must be communicated to both management and employees. These original thoughts should be drafted into a preliminary proposal for presentation. The secret to selling is involvement. Management has a broader perspective on finance, sales, and other factors that may affect the plan. It may also have concrete recommendations for improving the plan itself. With management's involvement, the first hurdle is not so imposing. Once these changes are made, management and the supervisory group should be given a final look at the written draft. This report should contain the techniques and time for implementation. Employees or their representatives are next in this communication scheme. They may object, but frequently this group will offer new ideas or recommendations for change. Again, the procedure of redrafting the report may follow. Ultimately everyone is involved in or knows about the proposal. Other formal procedures for notification should follow. Although this process may seem imposing, and to some, condescending, it usually results in a smoother installation of the standards.

Additional staff are required when starting, expanding, or restructuring a work standards system. The composition of the staff is very important. Usually organizations choose to mix the staff's experience levels with respect to both work standards and the product line. Some members may be chosen from the employee group because of their product knowledge. Others may be recruited from outside the organization for their experience in the field.

Formal training is essential for those chosen from within. Many organizations have their own training facilities and staff. Some choose to train new staff members on a one-to-one basis, whereas others seek outside professional help either to train in-house or at a formal training center. Whichever training method is employed, it is imperative that it be completed quickly and formally. Sending an inadequately trained technician onto the floor or into an area scheduled for measurement could damage the analyst's reputation and also the credibility of the standards program. Many costs may result from an incorrectly established work standard. A loose work standard could be one cost. Another could occur with work standard arbitration. Spending adequate time in developing staff in the beginning will yield greater dividends over the long term.

4.1.3 DOCUMENTING A WORK STANDARDS SYSTEM

Introduction

Control is essential to any goal setting system. Management sets goals and bases its overall forecast on the workers' attaining a prescribed production level. This forecast may include manpower budgeting, plans for expansion, and market penetration. One of the key variables in this forecast is the cost of labor. Costs and time are related. Thus the ability of labor to reach or exceed formulated production expectations will have a direct realtionship to cash flow, scheduling, inventory, and other critical control functions.

Management goes to the marketplace betting on labor's being able to achieve these standards. If a specific standard is set at 20 pieces/hr, and the actual productivity is 16 pieces/hr, the cost of labor increases by 25%. The decrease in productivity causes the facility and the machine to be underutilized. Inventories rise, therefore acting as a buffer for unpredictability. Many basic inventory equations use alpha (α) levels, where α equals the probability of a stock-out. Incorporating this uncertainty into Poisson predicting equations would result in increased inventory demands because standards are set at one level while the actual production is lower, with greater variability. The problem compounds if one department meets the quota and another does not. Inconsistency causes goods-in-process inventory to climb. This is an inventory of partially completed parts. Goods-in-process inventory, like any inventory, may keep money tied up because it limits the turns per year. Another extremely important application of work standards is in the development of manpower projections based on forecasts. Without precise work standards, the ability to predict becomes too probabilistic, and management must resort to overkill, which results in the employment of excessive amounts of labor. The attainment of goals must be monitored. This is why companies resort to many types of controls for documenting that a system is working according to the prescribed plan.

Developing Workplace Controls

Labor may be classified as direct and indirect. In a manufacturing setting direct labor changes the part. The cost for this change may be directly assigned to the specific product. Indirect labor is ancillary work which cannot be assigned or amortized to each part. In a service environment, instead of cutting off metal, a part is altered with each additional entry onto a form. This is change too. The labor hours to complete an insurance claim are the result of many combined efforts. At each stage in the process, labor is added to the form. As the form reaches various stages, higher-level labor reviews what has been accomplished thus far and then adds its labor onto it. In another example, the cost to complete a sales order may be higher than the markup on the product when small quantities are involved. For this reason, many corporations place what is in essence a nightclub coverchange, in terms of a minimum order cost, onto the product sold. The methods used to account for direct labor will vary, depending on the business or organization. The ultimate goal is to gain greater control.

Manufacturing and Service Workplace Controls

Example From a Service Environment

Insurance companies may have 200 employees involved in the claims process. Incoming mail is addressed in the field to specific box numbers to assist in presorting. After this initial process, the mail is opened; hand sorted by claim type and state; hand stamped into batches of 50, with coding by year, day, batch number, and item number; microfilmed by batch; sent to pending files with batches entered onto cathode-ray tubes (CRTs); checked by examiners; keyed into the computer; and then sent to storage or paid directly to the claimant. All of the intricate substeps are not listed, but this procedure requires both work standards and controls. Each person keying in various information must use his or her identification number, or else the material will not be accepted. From here, a computer program will measure the individual effectiveness against a prescribed standard for each of the functions on the one report that is being completed. Each form does not require the same number of steps, but the computerized program will recognize the activity and credit the individual with a given number of production minutes.

These predetermined standards are developed by analysts for every function and operation in the process. Standards are established for key stroking, calculating, ink stamping, deciding and reacting, and eye focusing, to name but a few. Being able to automatically identify the operator with the work performed allows the organization to measure the system's effectiveness accurately. With this information, the company can generate up to 25 control-type reports. Every activity is coded, such as prepare duplicate copies, copy from microfilm, verify insured's address, and work duplicates. From the coding and other information, control reports may be developed pertaining to weekly production efficiency against standards by employee or group, categories and type of work, ranking analyses, manpower projections against forecast, and modification of standards to learning theory. This is a partial listing of the controls.

Examples From Manufacturing Environments

The key to manufacturing controls is also identification. All departments, including product design, manufacturing, and sales, should use similar identifying numbers. This will bring about better communication among various departments and will enhance the reporting procedure. Work standards will be identified with these numbers, and so will employees and departmental performance.

In a manufacturing setting a company may develop an extensive coding system to identify direct or indirect labor activities. It is not unusual to find hundreds of codings. The engineering or product number identifies the part. Each part may have numerous direct labor operations throughout various departments. Good coding will allow management to know exactly how much of the product is completed, which manufacturing stage is next, what and how much material is used, actual versus proposed labor, and other information vital to a profitable operation. A company wishes to know costs. More important, it wishes to know the reasons for costs. With this knowledge, management may remedy problem areas. However, such a control system must provide quick feedback, or else the remedies may be too late. Consider the process of taking a shower. If the hot and cold knobs controlled valves that were a mile away from the shower head, it would be highly unlikely that the desired temperature could be reached. When the valves are close in, they can be fine-tuned for exacting temperatures.

Every supervisor, and to a lesser extent the production employee, will be given instructions on how to use the coded system. The code book may contain identifying numbers for absences, direct labor, downtime, emergencies, indirect labor, inspection, maintenance, material handling, setup, union activities, and other categories. All of these are redefined into subcodes for further clarification. The routing sheet triggers the whole process. This sheet records the sequence of jobs necessary to complete the part. With each step the routing sheet will also show and identify the department, drawing number, machines, material, operation code, operation description, part name, quantities, setup, weight, work standard, and other details. By the time the part is released for production, almost every aspect necessary for its manufacturing and control is known.

Knowing and achieving are two separate things, which is the reason most systems have a detailed production card for the workers to fill out daily. A computer card may be divided into two parts. On one side the production worker fills out columns for account number, base rate, clock number, date, department, hours worked, job class, order number, name, pieces produced, setup, standard hours, and supervisory approval. At day's end and after formal approval, this detailed card is sent to the data processing area, where all of this information is entered into a master program. On some systems the second half of the card is keypunched. Other systems have a direct entry procedure. In small organizations this whole procedure is tabulated by a production clerk without the aid of computers. At this point, therefore, the procedure is complete. All data have been coded. Production information about individual work is recorded, and a detailed analysis has been compiled.

Analyzing the Production Report

Contents of a Production Report

The information in the production report is used to monitor worker, departmental, and plant efficiencies. Columns define the activities, and rows identifying the operator's performance with respect to the columns. A report may have columns that detail the department, the employee's name and number, day work hours, direct labor, earned hours, gross average, gross earnings, hours worked, hourly rate, indirect labor, labor grades, normal average, normal earnings, overtime hours, part number, percent efficiency, pieces produced, shift differential, standard hours, the week's efficiency, and the week's utilization.

Every type of company seemingly has a favorite way of listing and placing these data on the report. Most programs usually begin a row with the department, the empolyee's name, the clock number, and the part numbers worked on for the day. Using these data, management may ascertain how well the organization is performing according to the prescribed goals. Such a report may be extremely beautiful, with all its columns tucked neatly onto the approximately 11 in. × 15 in. page. But all these data may be meaningless if the input information has been incorrectly entered or falsified.

Problems Regarding Falsification of Data

Problems relating to the falsification of data are so severe and universal that most union contracts or employees' manuals have a statement that stipulates the following: "Employees shall be responsible for reporting accurate amounts of work performed and shall be required to record the type, quantity, and other pertinent facts on the time cards furnished by the company. No recording of downtime will be allowed, unless approved by the supervisor as it happens." If the work standards system has aged, the employees may be using downtime either to increase earnings or to peg production. Conversely, if the work standards are too tight, or if the wage system includes a huge afterpayment, then downtime may be used to pad earnings. Supervisors, who may have previously been production workers, soon look the other way in either situation. They sometimes relate to production employees more than does management. They are on the firing line. A system may be out of control while the production report indicates everything is fine.

Downtime Reporting

Downtime is usually a sign of trouble. It may be abused by the production worker. As a result, management may make sweeping edicts limiting downtime reporting. Supervisors, fearing what may happen to their own jobs, could become overzealous in applying management's rules. Given this sequence of events, the workers could become irate, particularly in those cases where downtime is justified. Not being allowed to account for realistic downtime could drastically affect earnings. In extreme cases the workers may feel so helpless that they fail to report what has been happening.

Management expects a quantity of daily downtime. Managers provide for it in allowances and will also tolerate a limited amount above that which has been formally listed in a work standards system. But systems go awry. The aim of management is not to police the workers, but to find ways of eliminating those events that strip away productive time. A correct downtime reporting procedure will allow the data to be tabulated for remedial actions.

In the foundry the molders may work at extremely high levels to get ahead of the pourers. Since they cannot fill the lines with more molds, they seek to ring out on downtime. This helps to increase earnings. In a ground beef operation the chopper-mixer-blenders also work at extremely high levels and also get ahead of the plastic-casing operation. This work is not on incentive, so the operator takes a break. Milling machine standards are too tight, so the operator works 6 hr of production and turns in 3 hr of assorted downtime or additional setups. Workers may get off to a bad start during the day, so they do not report all production the day it is produced. On the bad day they accept day work. On the next day they turn in, on paper, the pieces saved from the day before. These are downtime abuses.

Alternatively a worker may honestly list downtime because of lack of materials, tool breakage, parts for assembly not to specification, safety harness malfunction, and other causes. Here are problems that may be corrected when known. Again, it must be emphasized that downtime listings may be used advantageously by management to limit production problems. If abuses are flagrant, it may signify that a particular work standard or the total system is faltering. In situations where certain types of downtime persist, these, too, may be investigated for corrective action. Too frequently management correlates downtime with deviousness rather than with managerial or engineering error. Management's responsibility is to correct, not to criticize.

4.1.4 USING WORK MEASUREMENT SYSTEMS

Introduction

The field of work measurement is expanding in scope and advancing technologically with each passing year. Yet there are many companies that still do not use this valuable technique. Instead, they rely on history. What has been done in the past governs their thinking. Companies may continue this way, using profit as their single yardstick. It is entirely possible, however, that a company will remain profitable while losing its share of the total market. Eventually this trend will cause the company to die, essentially having been gobbled up by its competitors.

Look at the newspapers of 100, 50, or even 25 years ago. The majority of major or minor companies advertised do not exist today. The number of companies currently listed on the New York Stock Exchange that were listed 50 years ago is relatively small. One principal thesis of Darwin was that an organism must adapt to its environment or die. This is seemingly true for business organizations. They, too, must adapt. But the worst time to adapt is during crisis. As an analogy, assume that a heavily laden wagon is on an incline, and that 20 men with ropes are restricting any movement. Suppose that, in a momentary lapse, the wagon inches forward. For each second of relaxation, the force required to restrain it once again may have quadrupled. As another analogy, a high-speed fan may be easily held prior to spinning, but reaching in after it has reached full revolutions per minute will result in the fingers being severed.

Darwin's thesis and the laws of physics for which examples were given also apply to companies. They, too, must be keen, adaptive, and flexible, all within the framework of precise control. Work standards measure output. They also provide input to the control function. Properly applied work standards will assist the company in making quick and correct analyses for both broad and specific control areas. From the broad perspective, work standards contribute information to assist in the functions of budgeting, costing, estimating, forecasting, product expansion, and developing quotations. Specifically, work standards are used for improving efficiency through downtime analyses, equipment utilization, flow of work, labor utilization, plant layout, line balancing, material handling, measuring supervisory efficiency, methods, minimizing production interruptions, motion patterns, productivity audits, tooling, worker disqualifications, and workplace layout. This whole area pertaining to specific controls deals exclusively with accounting for and improving worker productivity. These are the areas where work standards will have the greatest impact. With precisely engineered standards, a company may develop ways to instantaneously verify its competitive position and to control a potential slide before it gains momentum.

Analyzing Worker Performance

Reasons for Irregularities in Performance

The production report measures worker performance according to the engineered work standard. When one analyzes this report, some inequities may soon become apparent. If one worker's earnings undulate, this may indicate indifference, stockpiling, or irregularities in work standards. This worker's percent performance may, for example, be 95, 115, 90, 117, 87, 128, 92, 118, 93, and 114 for a 2 week period. Looking at the run of work subjectively, it would appear that this series is a controlled productivity. Statistical tests made on the run should the process continue for $n \geqslant 20$ days could also verify that this series is not a random process.

Runs of this type are not usually this obvious. Careful scrutiny of a report may show apparent control over the long run, which could indicate that workers are making a product one day and combining it with the next day's production. This may indicate supervisory complicity with the production worker. It could also mean that the standards are incorrect or that the production workers are abusing downtime by making false entries on their daily time cards. The objectives of this analysis are to isolate error and to make corrections, regardless of who is at fault–management, labor, the work standard, or the system. Assigning fault is not remedial. Corrective action is the key to increasing productivity.

An analysis of work using a production report may show that one worker is consistently below a groups' performance. When supervisory management is quizzed, the retort may be that this particular worker's performance is really not of high quality and that the worker cannot be discharged because of performance slightly above 100%. The rest of the personnel working on similar jobs may be consistently earning at the 125 to 135% level. Many times, the extent of time supervisors spend for training a new worker is so minimal that it is almost like a propositional calculus statement–an argument solved through methods employing symbolic logic–which may read: If most humans can swim, and if these are humans, then throw them into the water.

In one typical case a worker was pointed out as being totally inept, but steady. Productivity reports indicated a work level at 40% below that of others. Yet this worker remained at the work station all day and seemed to produce at slightly above normal levels. Why, then, was there such a discrepancy? This particular worker's training was as follows: Watch the other workers for a while and then do it on your own, but do a good job. So this worker was hammering into a product four times the number of nails necessary for seven different operations. In this case the particular worker finished the day dripping with perspiration, only to earn 40% less than others who finished pine-scented fresh.

To think this does not occur in any type of operation is the height of naiveté. In the foundry one operator grinds the gate with a pressure bar, and another hooks the piece. A brake operator may make one bend and index the piece to make a second bend, while another operator may first make the bend on the back gauge and then pull the piece forward to the front gauge for the final bend. This does not require indexing the piece. When seeing that one operator's performance is so apparently low in comparison with that of others, the analyst or management should make a concerted effort to make firsthand observations.

Looking at Day Work, Overtime, and Downtime as Measures of Efficiency

The production report should compare the time on standards versus time worked. Often this will provide indexes of downtime usage in terms of ratios. Other systems actually tabulate downtime. It is the objective of most work standards systems to minimize day work hours. At times, inefficiencies on day work may be isolated by machine, product, work group, supervisor, or manager. This report may single out day work in comparison with overtime pay. Day work generally implies that units were not produced. The production report may show large amounts of day work when overtime is provided on a demand basis. This could indicate featherbedding to overtime. If day work is in the ratio of 2:5 for normal hours, it could be reasonable to assume the same ratio for overtime. If investigations of overtime hours consistently indicate day work in the ratio of 1:10, then closer attention may be warranted with regard to regular work hour practices.

The supervisor and other floor-oriented managerial personnel are basically accountable for these problems and abuses. A production report based on reliable work standards will quickly point out those areas requiring immediate attention. A supervisor requires training more than does a production employee. Managers are quick to criticize the supervisor, much as the supervisors are quick to defend errant workers. Both are wrong. A trained, knowledgeable supervisor will act with confidence. This will be obvious to all who investigate the production report that measures both worker and supervisory efficiency.

Comparisons of Day Work, MDW, and Incentive Performance Levels

Descriptive statistical analysis of productive work indicates that day work is normally distributed, with the mean substantially below 100%. Furthermore, the best day workers produce twice that of the worst day workers. An MDW plan's mean is also below 100%, but the distribution is slightly skewed right-positive. The distribution for incentive workers has much greater right-positive skewness, with means usually in the 120% category. These data, shown in Exhibit 4.1.1, clearly reveal that within the set of all production workers there are select subsets that constitute MDW and incentive workers.

These subsets are not random. It takes years for a company to recruit incentive workers. A newly established air conditioning plant, which used predetermined systems as a basis for establishing work standards, required 7 years to stabilize its production force. After many initial screenings, a newly hired employee may be released to the production floor for training. Companies with a union contract have training and disqualification procedures that are specifically defined. Without a formal contract a company still needs to disqualify those newly hired employees who apparently will never quite be able to make good production employees. Disqualification is easier in the beginning because a union is not involved. Later on, the procedure becomes more difficult because of union affiliation.

Subjective disqualification is disastrous. Objective disqualification, though somewhat more harsh, provides a better production force. The procedure is fair when viewed over many years. The production report will isolate those workers who fail to meet work standards during a fair trial period. It will also isolate those employees who do not reach standards after moving into new jobs. Sometimes even the previously good workers start to falter. In this case the cause may be outside the work environment. Maybe an individual's health is failing or there are problems at home. In this framework the production report may isolate a problem that the company may help remedy through compassion.

Exhibit 4.1.1 Performance Distribution as a Function of (*a*) Day Work, (*b*) MDW, and (*c*) Incentive Work

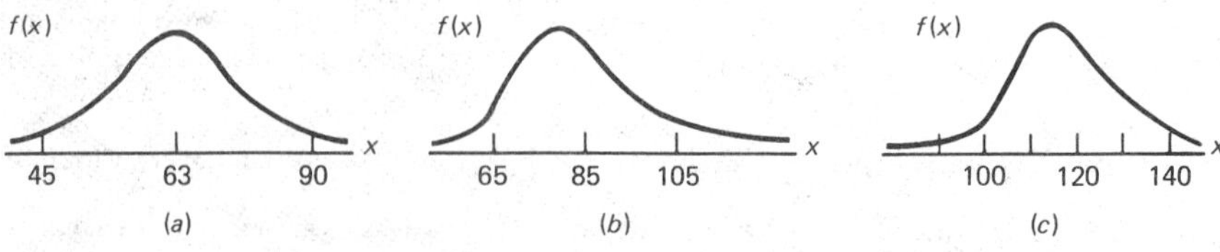

Problems Associated with Coverage

When developing work standards, the recommended and classical approach requires the analyst to study how the material arrives at the workplace, who is responsible for its delivery, how it is taken away, and who is responsible for removing it. This whole procedure of analyzing equipment, material flow, tooling, and workplace layout is embodied in the topic of methods engineering. Every work standards analyst should include this type of analysis as a precondition prior to the formal establishment of a work standard. Many times work standards are set without regard to methods. Some of this haste is caused by management's requirement for coverage. This coverage concept is often worn like a badge of courage by those engaged in establishing work standards. How much coverage seems to generate more concern than the quality of the standard.

The era of the computer brought on many new managerial controls. Inventory control analysis, forecasting, manpower loading, bill-of-material processors, and other software required greater knowledge of anticipated productivity for each stage of product growth. This placed tremendous pressure on the work standards department. With this surge the department soon became buried in work, particularly with each change in product. A philosophy of "work standards at any cost" then prevailed. This burden caused individual standards to fail because they were studied in an "as is" condition. Methods were discarded for coverage in many cases.

The Impact of Incorrect Work Standards Practices on Cost and Market Position

Precision is essential. The work standards must be reliable and descriptive of the correct method. Prediction is "bouyed up" on authenticity. With correctly engineered work standards and methods analyses, a company may go to the marketplace with more authority.*

To determine pricing, an analysis of labor, burden, material, and general and administrative costs is required. An incorrect standard will destroy this whole pricing procedure. "Burden" is usually defined as a function of direct labor; thus one error magnifies another. Work standards are a measure of productive efficiency. Using this information, a company will be in a better position to determine how to budget labor and costs for the next quarter. Looking ahead like this requires forecasting. For example, a determination to expand requires that an analysis of labor and equipment be synthesized to improve predictability. The heart of any equation designed to formulate productivity levels is contained within the term used for predicting labor's output. Work standards are an individual expression of anticipated productivity. Their reliability is paramount to the whole process.

4.1.5 MAINTAINING A WORK STANDARDS SYSTEM

Introduction

Change is functionally related to the dimension of time. What is believed to be true today may be disproved tomorrow. It seems as though change is a driving force of humanity. Build a wall to restrict a path, and ultimately it will be scaled. Establish a goal, and methods will be devised to attain even higher levels.

Placed in a proper setting, this desire for change can be used advantageously, particularly in business. Quote a time or method to complete a task, and these limits, like the wall, are challenging too. During the inception phase, management will describe what is to be done, how it should be done, what machine and tools to use, and the time allowed for completion. In a work environment the challenge begins once the employee is trained, and a sufficient time is spent in learning the specifics of the task. Very few jobs exist today exactly as they were yesterday. The production worker, designer, and process engineer constantly seek change.

Given these conditions, a work standard may not reflect current procedures. Work standards are affected by change. Thus a total work standards system may deteriorate without precise methods to audit or communicate that a change has in fact taken place. Management issues a challenge in terms of work quotas. The changes that occur may be subtle, but taken together, they may amount to a significant improvement over what was quoted originally. Employees who make changes are often reluctant to disclose the new procedures. They reason that management will take these ideas, exploit them, and establish new and more restrictive work quotas; consequently, they conceal these new techniques and procedures. Union contracts often state that the "agreement is that process changes devised by the operator shall not be used as a method for reducing legitimate earning." The fear exists, however, so that a cautious employee may stretch out the work day or produce for only 6 out of 8 hr. This causes lower facility and machine utilization.

*"Authority" is defined here as the ability to bid or price new jobs based on work standards practices that yield realistic labor costs.

A manager having worked previously with an uncontrolled standards system may opt for day work or some other plan, believing that deterioration, falsification, and other ills are the ultimate plight of incentive-type work standards programs. In reality, any environment—measured or unmeasured, incentive or nonincentive—is affected this way. Any work is challenging. Work standards may intensify the challenge. A game-type atmosphere is created when work is measured, goals are established, and procedures are implemented for control purposes. A game implies a bet, and knowledgeable managers are aware that, in order to capitalize on this particular bet, they must first be willing to make the wager. Business managers should not be afraid of what employees may do in order to beat this game. This desire and drive may be used most advantageously. The majority of precisely measured work standards systems are working quite well. Obviously, like the finest of automobiles, a work standards system requires planned and periodic maintenance to ensure that everything is functioning correctly. It seems that notoriety prevails. One bad work standards system may offset 25 good ones. Those that are good have been looked after, cared for, and maintained.

General Reasons for Deterioration in Work Standards

Change in Methods

As methods, design, and processes change, the time required for a given unit of work may change too. A change in motion sequence, workplace layout, delivery of material, or other aspects of operations may be classified as methods changes. This type of change implies operator control. In most environments the employee's job content and procedures are controlled by management. A new employee is trained to complete the task using the techniques as prescribed. In sophisticated systems the computer may print out left- and right-hand motion sequences indicating how to proceed.

Many times, an experienced employee will work closely with a supervisor in establishing the methods for completing a new job. Ultimately, the employee is working alone. In many instances employees start to experiment with different work procedures and usually are quite keen in developing ways to beat the system. Many times they guard these improvements and are reluctant to share them with other workers or management. Performance reports, line supervisors, or formal audits by members of the work standards department may detect these changes. If the job is restudied, the operator may return to the previously established methods. Any attempt to force "their own" techniques back onto them may be met with hostility.

This situation could lead to a grievance and subsequently to arbitration. It is an extremely explosive issue. Many union contracts specify that a job cannot be restudied unless there is a 5% change in the method. The restriction goes further by stipulating only times for job elements with a 5% change may be reevaluated. No sequential relationship is recognized. Thus a 5% change in elements 1, 3, and 5 will not allow the restudy of elements 2 and 4. The interpretation of these contractual statements may vary from company to company and from arbitrator to arbitrator. That such restrictions are included in many union contracts is testimony to the magnitude of the problem. Employees fear rate cutting or the raising of job quotas, regardless of the environment or the system in use. Management must change, however, to improve profits and market position.

These two philosophies collide. Most assuredly, the majority of employees wish to play a part in the success of their business. Management also wishes its employees well. To infer that this is not a competitive state would be incorrect. Yet, within this climate, events have consistently improved for both groups.

A change in methods affects both management and employees. As previously explained, spontaneous change is readily apparent, but is harder to correct. Over the years small subtle changes occur in the job. These "creeping methods changes" usually are not the domain of one person. Employees change to new jobs to seek out higher pay or a position to which they are better suited. Thus many individuals or a whole group may have previously worked on one or a series of jobs. Each change they had introduced may have resulted in new ideas and improved methods. As a consequence, there is a better opportunity for each new employee working this job to increase productivity.

Performance reports may indicate that a change has occurred. Supervisors, who should periodically check the methods being used, could also report that work procedures are not the same. Also, a formal standards auditing plan would isolate those jobs with a changed working procedure. As a result, jobs can be restudied to establish new work standards, since these cumulative changes do not belong solely to one individual or group. It is imperative that exacting procedures are formulated to discover or communicate that a change has occurred. Audits by the work standards department should be scheduled. The supervisor should also be required to subjectively audit jobs for any change. Working together, they will help increase productivity to everyone's advantage.

Change in Design

Work standards age and deteriorate as methods change.* Management may have been responsible for some change, but usually it results from worker initiative in either short-term or long-term situations. Design or process changes are instigated principally by management. From a broad perspective, work standards may deteriorate because of changes brought on by the employee or by management. In the manufacturing environment, changes in parts or materials are design changes. The changing of a form—for example, reducing the number of entries or their placement on the form—may qualify as material or design change in the nonindustrial setting. Engineering drawings usually precede a change in material and parts. In an industrial operation, product engineering should notify the work standards department about a forthcoming design change. In many cases these changes are simply instituted without formal notification.

Work standards are seriously affected when tools, equipment, or other manufacturing processes change. In retailing or service operations, new pricing machines may be purchased, the cashier may use a touch versus a touch-and-look register, and older microfilm machines may be replaced by newer models which eliminate many manual operations. In the foundry a change in grinding wheels, increasing surface feet per minute, and adopting pressure bars could combine to drastically affect the time to grind gates.

Poor Communication

Again, all of these changes could occur without a change in work standards. Usually a company has formal methods that guard against changes going unnoticed. Routing sheets show the sequence of operations and machines used. Supervisors are formally trained to make known any variation that could affect standards. Performance reports also show variances that may indicate a change has occurred. Audits by the work standards department will isolate change too. Furthermore, most companies have regularly scheduled meetings where topics affecting production are discussed.

All of these checks on change going unnoticed can be embodied in one word—communication. Formal methods for communication are essential. The initialing and the authorization of a work standards study by the manager of industrial engineering, the departmental supervisor, and the superintendent are forced methods of communication. So are grievances. Change is occurring and will continue to occur. Management must take advantage of change in order to remain competitive.

Some Prevalent Factors in System Deterioration

Problems with Allowance Computation

Individual standards are part of the total work measurement system. When a system falters, all individual standards are affected. One major contributor to a system's demise is the misapplication of allowances. A work day comprises productive and nonproductive times. Some nonproductive times are contracted to include fixed break times for morning and afternoon plus makeready and cleanup times. In the case of three shift operations, nonproductive times usually include a paid lunch.

A typical work day has 480 min. Assume the contract stipulates that these nonproductive times constitute 32 min for a single-shift operation. This means that the worker has contracted to work only 480 - 32 = 448 min/day. This averages to a 56 min hr when considered with respect to equal time segments. Some segments may contain almost 60 min of work, whereas others, particularly those where the fixed break time occurs, could be substantially shorter. As an average, however, each hour starts with 56 min available for work—by contract.

Other nonproductive times may be contractual too. A work standards system that includes personal, unavoidable, and fatigue allowances is also negotiated into the contract. Personal allowance is primarily time for relieving normal physical functions. Another deterrent to sustained productivity is worker fatigue. Considerable studies have been conducted in this area. From a broad perspective, fatigue varies with the length of cycle, job, and working conditions. Unavoidable delays are interruptions to work that are outside the control of the operator. The type, nature, and class of work usually affect this total amount of delay.

These three factors—personal, fatigue, and delay allowances—consume time and restrict productivity. The minutes used for personal and unavoidable delays may be found objectively through sampling procedures or all-day work studies. Subjectivity governs the time allowed for fatigue. In practice it has been found that 10% to 17% of total allowance will cover the majority of work situations. Obviously, these are generalities. In assembly line work, the operator may be "spelled

*"Aging" is defined as the process wherein workplace procedures, methods, and motion patterns have changed significantly since the time of the original study.

off," that is, relieved by another at specified intervals or as required. Extremely hot work at the top of a glass tank may have a routine of 15 min on and 15 min off. Allowances must be studied and applied as scientifically as possible. Documentation is essential.

By contract the worker is only made available to the company for an average of 56 min/hr. Also by contract the time for personal, unavoidable, and fatigue allowances may be stipulated, as, for example, 15% of the production time. Assume that the time to complete one job is 389.5652 normalized min. This time does not include allowances. Given that the allowances are extensions of the production time, the 389.5652 min* would be increased by 15% to 448.0000 allowed min. This is the total time allowed to work by contract. Dividing into the 56 min of available work would yield 56/448 – 0.1250 pieces/min. Total daily production would be (8)(0.1250) = 1.0000 pieces/day. Thus 15% allowances would equal 58.4348 daily min. If the time for another part were 25 normalized min, the total time provided by allowances would remain at 58.4348 min, provided that computations were carried out in this prescribed manner.

Allowances compiled in this way are considered to extend the time required to produce one piece. In other applications 15% is considered to be part of the total daily minutes available for work. Computations using this concept would be (0.15)(448) = 67.20 min. The work hour is reduced to 47.6 min, or a 17.647% extension to productive minutes. Consider a piece requiring 25 normalized min to produce. Using the concept that allowances are extensions of productive minutes, the number of units required per hour would be (56)/(25)(1.15) = 1.9478 pieces. Allowances as a percentage of the total day would result in a demand for 56/(25)(1.17647) = 1.9040 pieces. This is a significant difference when considering all employees and all operations. Instead of using percentages, some systems use actual minutes. If the contract stipulates, or if statistical analysis indicates, that 24.00 min be allowed daily for personal needs, 20.22 min for fatigue, and 14.13 min for unavoidable delays, the total minutes available for work would be reduced by 58.45 min. Here the intent is clear, and the actual minutes may have been objectively determined. Again using the concept that 58.45 min are not available for production, the total daily minutes would be reduced by 58.45/(448.00 – 58.45) = 0.15 = 15%.

Allowance computations should be the same with or without a contract. They are not gifts to the employees and should be truly representative of times not available for work. A company unquestionably wishes to know, with authenticity, the cost and projected output for each product. Correct computations of allowances will provide greater clarity. Many companies consider that allowance computations should be made on the basis of 480 min/8 hr day. In essence, this philosophy takes back in the allowance those times that were essentially subtracted from the workday in formal negotiations. Consider again that 32 min are not available for work, but that allowance computations are based on 480 min. Productivity requirements, with allowances considered as extensions to normal minutes, would be for the previously cited 25 min. Thus 480/(25)(1.15) = 16.5716 pieces/day, or 2.0715 pieces/hr. Using allowances as part of the total day the requirement would be 2.0400 pieces/hr. Obviously, the company is taking back minutes in the allowances that were essentially negotiated out as time not available for work. Although formal break times reduce fatigue and enable one to meet personal needs, they cannot be taken back subjectively. Using direct minutes will eliminate many computational problems and errors.

Incentive systems with properly evaluated work standards may increase earnings for the better workers at 125% to 135% levels. Assume a runaway incentive system paying 271% to these same workers. Allowances pyramid in these cases. When working at these levels, employees will receive twice the allowances of a normal worker. Using 25 min as a normal time, this could become 28.75 min with 15% allowances. Producing at 271%, the task would be completed in 10.63 min. For each unit produced the operator earns 3.75 min of allowances. By completing one piece in 10.63 min, a unit of allowances equaling 3.75 min is also earned. A second piece would also equal 10.63 + 3.75 = 14.38 min. Together they constitute 28.75 min, but two units of allowances were earned. The operator would use this time for production, and then pyramiding would increase even further. If, as a hypothesis, allowances were paid as a separate check, this problem would be minimized. Even at normal incentive levels they are distributed unequally. This analysis serves only to show the importance of proper allowance interpretation to a total work standards system. In one example they are too restrictive, whereas in another, too loose. Either misapplication will weaken a total system.

Process Time Payment Errors

On jobs with long machine cycles, the work sequence may be load = 2 min, machine time = 20 min, and unload = 3 min. With machine time fixed, the operator would not be able to earn incentive on a greater portion of the total time. If load time and unload time were performed at 25% above normal, the 5 min of external time would reduce to 5 min/1.25 = 4.00 min. The machine time

*A watch may read to two-decimal accuracy and a TMU to four places, but most practicing analysts advocate four decimal places in the development of standards. Demands of 1000 to 1500 pieces/hr could alter weekly pay significantly when considering only two-decimal accuracy.

would remain constant. The operator would earn 25/24 = 1.04, or 4% incentive. In many applications this situation represents a lost incentive opportunity. By contract or policy, the machine time is increased, for example, by 125%, to provide an opportunity to earn incentive throughout the cycle. The pay cycle is 5 min external plus (20)(1.25) = 25 min machine time for 30 min of the total cycle. At normal the cycle would be completed in 25 min. At 25% incentive, the cycle would be 24 min. The concept here obviously is to provide equal incentive opportunity and stimuli to keep expensive equipment occupied for a greater portion of the day.

The internal portion of the cycle, machine time, may be utilized for additional work. During this 20 min of cycle, the machine operator may be required to carry out functions necessary to the efficient operation of the machine. "Clear chips" is one such function. Another is attention time. At critical portions of the cycle, the operator may have to exercise considerable attention to see that all is going well for part, machine, and tools. During other, less critical portions of the cycle, the operator could be required to pre-position the next piece, gauge a previous part, or run a second or third machine.

When more than one similar machine is assigned, the model may be explored binomially for machine interference. Other modeling for similar or dissimilar machine assignments can be developed using Poisson, exponential, Erlang, or simulation techniques.[4] Again, this mathematical modeling is to determine machine interference. When assigning two or more machines to one operator, there are times when the second, third, or both the second and third machines require service or attention from the operator. With an operator working on the first machine, the second or third must wait. This interference queue extends the cycle time of the waiting machines, thus reducing the daily productive time available. Machine interference may range from 2% to 25% per machine. The factors governing the extent of acceptable interference when modeling a process are based primarily on labor to cost ratios.

The internal portion, or machine time, of the total cycle can be used to occupy the worker in other activities. The prevailing thought is to utilize worker time since the company is paying throughout the cycle. This is particularly true when incentives are paid on machine time. Many arbitrators have ruled that the internal portion can only be "loaded" at normal rather than incentive levels. Thus, when making calculations using arbitrators' philosophies to ascertain how to utilize internal times, they must be determined at normal, or average, levels of performance. The company in the case cited previously may load the internal 20 min at normal. Usually, this total portion of the cycle cannot be fully utilized, and waiting time occurs.

The question of how to pay the operator during these inherent delays is one that plagues work measurement analysts. It can cause workers to be overpaid or underpaid. A work standards system may degenerate dramatically if this particular concept is treated incorrectly. At times a system is built with so much mathematical sophistication that it in fact pays labor 140% to 160% for 100% performance.

Consider a variable, bit-type, screwdriver which has additional tools in the hollowed-out handle. Assume that in the assembly process the operator turns this screwdriver over, and the tools dump out onto the table for subsequent selection. The dump portion of this cycle is outside the control of the operator. The time from the moment the bits slide down and out of the inverted screwdriver until they hit the workbench is called "process time." With sophisticated measuring, this portion of the process may be timed. Using a predetermined system, both the right and the left hand are isolated as being idle for this extremely short period of the cycle. This is process time. The operator could not hurry this portion of the sequence, and there would be no incentive opportunity on this portion of the cycle.

According to some work standards systems, this portion of the cycle should be expanded by, for example, 25%, because no incentive opportunity existed because of operator waiting time. This description is simplified and leaves out many items, such as reaction time, eye travel, and eye focus. But these, too, are somewhat subjective in this type of analysis. Should this time be expanded to provide incentive? Obviously not! This is not the intent of process allowance. Yet, this is exactly what some companies do in their work standards applications. Assume that an extremely high-speed photographic process was employed to analyze work. Using this technique, all times when the right hand and left hand were stationary concurrently could be factored out. This could represent a substantial portion of the total cycle when totaled. In very short cycles, it would represent 30% of the total time; in longer cycles, possibly 20%. Should these cumulative times be paid process allowance? Again the answer is no. Although this latter example violates the precise definition of process time, it serves to illustrate how detrimental this concept may be in establishing accurate work cycles.

In the foundry, process allowances of 133% have been paid on jolt cycle, on the time it takes sand to drop into a mold from a gravity feed chute, on squeeze time, and on many other facets of the operation. In the jolt sequence, time is set by the operators, and they spread the sand during its total cycle; thus the job is not restricted. Similarly, the operator is active in the sand filling process, pulling a lever, spreading the sand, and vibrating the chute.

In sheet metal operations, the running of a brake press is restricted somewhat by safety devices, which may stop the operator's rhythm of motions. Also, during the portion of time that the press goes up and a portion of the time it goes down, the operator is turning to pick up a new piece and is pre-positioning it for insertion. Since the operator time is less than the machine time, the cycle is inflated by process allowance because the operator cannot work faster because of machine restrictions. Yet, the operator's work time at 100% performance used 95% of the machine time.

Consider a powered assembly line where each station is set at 1.00 normalized min. Should the workers be paid 125% because this is machine controlled? Remember that arbitration has ruled that 100% is all that can be asked for both external and internal work. This assembly line does not, however, provide incentive pay at 1 min. In applications of this type, the line is usually sped up or more parts are introduced per unit of time. The worker will truly earn incentive. But what if other portions of the shop are paid process allowance on the machine-controlled portion of the cycle? Should not the assembly line workers also be paid incentive for 100%?

These are perplexing questions. Care must be exercised in developing formulas for paying process time. In some applications, process time not only is inflated, but also is given allowances. Again, computational errors of this type must be guarded against lest they become locked up contractually. The original intent of process allowance was based on the fact that a machine operator could not increase the feeds or speeds without destroying both tools and products. It also implied absolute attention time during a greater portion of the cycle. In this example, machine time could not be increased because of the physical constraints. Essentially, the cycle was locked up. For other applications these restraints do not exist because cycle times can be altered. In many misapplications of process allowances, the workers in a facility may be performing at 110%, but their pay is at the 160% level.

Misinterpretation of Contractual Incentive Opportunity Clause

A union contract may contain a sentence or section stipulating that an average incentive worker, while on incentive work, will have the opportunity to earn 25% above the occupational hourly rate. There are many variations to this contractual statement. In most cases the intent is quite clear. It means the system will be set so that a worker performing with average skill and average effort will earn 100%. Here 100% is defined as normal. A 100% system, so defined, will provide the 25% opportunity as required. The contract is essentially dictating a low task plan. The statement as given is very difficult to interpret—particularly for those not acquainted with work standards systems. During negotiations or during a grievance, the company may unknowingly consent to another interpretation, which could result in payment of 125% for 100%. A large number of companies are currently in this position. Therefore, if 125% is used to multiply the allowed time, this results in $(1.15)(1.25) = 1.44$, or 44% allowances when 15% personal, fatigue, and delay allowances are used. This pyramiding is even more severe in other cases employing higher values. The error magnifies many times when coupled with a misapplication of process allowances.

The "opportunity" statement in the union contract was originally one of concern. It was intended as a protective device to insure that companies did not performance rate at lower values when observing average performance. It also was used in high risk plans to indicate that 100% would earn 25% or some other stipulated incentive. It was never intended to mean that normal performance would earn an automatic 25% incentive pyramided over the allowances. These problems could be avoided in the next contract by defining the section with greater clarity. Some companies choose to fight the misinterpretation based upon sections within the contract that nullify standards with computational error. Other companies have chosen arbitration by deliberately targeting a few jobs for testing the legitimacy of the statement. Another method is to lock all current standards to the employees, essentially red circling them, until another group of workers takes the job.

But in the meantime, all new standards will be developed in the absence of the opportunity factor. The problem is severe. It curtails production. It pays for more than what was originally intended. Some statements are so definitively written that arbitration would not provide recourse. It is impractical in a true sense to believe that currently established standards could be reduced by 25%. Future standards or aged standards may be studied without using this boosting value, but first, test it in arbitration or restructure the language during negotiations or possibly reach an accord with the union. Openness is important. It is not in the best interest of the company or the union to market their product with labor costs that are 25% above those of their competitors.

Organizational Weakness of Work Standards Department

In many businesses the work standards department is organizationally weak. It is the "rate department," the "timing section," or the "clocking group." Production workers often refer to the work standard procedure as "being clocked." The structure is so distinct in some operations that the methods department and work standards department are separated. Each reports to a different department head and is at a different organizational level. In one organization the office may be found in some obscure corner, with rotary calculators on a desk top carved with local graffiti. In another there may be computer terminals, carpeted floors, and an up-to-date library. Pay rates vary considerably. Some organizations require college degrees. Many other disparities could be noted, but these few serve to illustrate that a substantial number of businesses still consider the department to be basically unimportant.

When a work standard has been established everyone becomes expert. The worker knows the standard is incorrect and threatens a grievance. The supervisor believes the standard requires more allowances. The operations manager asks, "What's the take-home pay?" The industrial relations department requests a middle-of-the-road position. So essentially this individual standard is negotiated.

Although these practices have diminished over the years, they are characteristic of what plagues the field. Management erroneously believes that it sets the labor cost for each piece produced. In reality it is the work standards analyst who makes this contractually binding determination. Work standards analysts or their department heads are not generally given input during negotiations. This may be the reason why incentive opportunity statements or process allowance procedures mysteriously appear in a new contract. Many times new language is required to correct some past problems. Yet no one seems to communicate with the responsible department for its input.

The problems causing system deterioration have been classified as technical, applicational, and organizational. Of them all, the one creating the greatest havoc with productivity is that associated with organizational weakness. In one business the production workers may be completing their workday in 5 hr. Another operation may have flagrant overall cheating on the time cards. This list could be expanded. Why do problems like these remain when they are known? Management listens to the tax accountant because the rules must be followed or else they will be punished. They also listen to the engineer, particularly when discussing topics pertaining to product liability. Power is vested in these and other functions from external sources. They must take heed. In the case of bad work standard practices, management may or may not provide sufficient attention. Given a strong department organizationally, management would be forced internally to correct these problems at the time they occur. It is often too late when the bottom line is red.

Techniques for Isolating the Specific Problem Area

Auditing Procedures–Formal

Any work measurement system or specific work standard will age. What was a good, crisp, reliable work standard yesterday may be quite loose today. Ingenuity fosters change, and most of it is advantageous. Random or stratified auditing procedures may be used for determining what has changed and the magnitude of the change. A method for establishing which jobs to audit may be devised through the use of the variance report and a random number generator. This technique for job selection may be done plantwide or by department. Precaution must be exercised in this procedure in order to eliminate bias. A completely random audit would choose operations without respect to sales or frequency. Stratified methods could incorporate a Lorenz analysis* coupled with random selection of jobs for audits with respect to sales of frequency.

Demand auditing is not random. This technique calls for audits of jobs that indicate excessive earnings. These operations are selected because of pure earnings or improper ratios of earnings to downtime. The problem in this job selection procedure comes from the fact that workers soon learn what performance levels cause "spies to descend on them." Thus they cap their earnings or use other methods to disguise their good fortune. Another method for determining which jobs to audit comes from a modified Z formulation. Using hypothesis testing and Z distribution statistics, a computer printout will indicate which standards are essentially out of line–which are too loose or too tight–based on variability.† This formula is modified with $Z = 0.84$ for MDW when

*Graphing the cumulative percentage of orders versus the cumulative percentage of sales will isolate those products making the greatest contribution to revenue. See W. Allen Wallis and Harry V. Roberts, *Statistics–A New Approach*, Free Press, New York, 1956, pp. 257–264.

†See Section 13 of this text for more information pertaining to Z and t distributions. The use of the formula for determining out-of-line work standards is proprietary to one computerized predetermined system, but follows the general form of

$$Z_0 = \frac{[\bar{X} - 0.84\,\hat{S})] - \mu}{(\hat{S}/\sqrt{n})}$$

where n = number of pieces produced against this one standard for over one week; $n > 100$
$\hat{S}$ = sample SD
$\bar{X}$ = mean operator time for this one standard
$\bar{\mu}$ = actual engineered standard time
$Z_\alpha = \pm 1.96$

Test the hypothesis:
$H_0: \mu =$ the MDW standard
$H_1: \mu \neq$ the MDW standard

80% is assumed to be normal. Some analysts prefer *t* statistics for these computations. This whole concept of determining which jobs to audit in order to make a fair judgment about the validity of current standards is extremely difficult. Greater variability in performance may indicate loose or tight jobs based on incorrectly established work standards. It also could indicate pegged production or worker dissatisfaction. A computer would be flagging these jobs for auditing, when in fact they may be absolutely correct. Nothing will surpass a good communications link with the floor when selective auditing is to be employed.

Of all these usable systems, random auditing seems to be the best. Random auditing will pick up many secondary jobs. The thrust of auditing is *manyfold*. Certainly costs on major jobs are important. But when considering workers' attitudes, they may become disillusioned with minor as well as major producing jobs.

When using time study methods, a formal audit should be of sufficient duration to include many of the irregular elements, delays, and other factors that could restrict productivity. Statistical formulas that dictate the length of a time study based on element variability are quite often misleading. These shorter types of studies assume that delays are correct. In essence, they attempt only to validate elemental times.

Assume that a 2 hr time study was taken in the formal auditing process. Within this time study could be such element as wait for materials, tool breakage, interpret production orders, and irregular elements. If, for example, 50 such studies were taken, they would represent 100 hr of analysis. But more important, when a composite picture of these 50 studies is made, a very good representation of what constitutes the complete nonproduction time would be forthcoming. From 50 studies it would be possible to make a pie chart showing the producing times and most segments of the nonproducing times. One 2 hr study therefore would have substantially good predictability. If 50 studies of 15 min duration were taken, it would represent 12.5 hr. A composite picture from these may be extremely biased when considering predictions with respect to nonproducing times. Each individual study, 15 min in length, would not be statistically reliable of the true composite. Yet, the 15 min study could show little element variability and by formulation be quite significant.

This is not to demean the statistical process. It serves as a caution to those engaged in formal audits. In many jobs change has occurred outside of the cyclical elemental procedure. A short study may not find these changes and may erroneously be considered fair. For every hour of time study, it usually requires 1 to $1\frac{1}{2}$ hr of in-office computational work. Management consulting firms usually estimate two to three studies per day in making quotations. It is easily recognized, therefore, that many audits may be done in haste because of these time constraints. An audit must check that all aspects of the job are performed as defined in the previous study. If this were a grievance, or if the study were going into arbitration, then considerable time would be spent in preparing a study. Why not be this diligent when auditing for management?

Auditing Procedures–Statistical

Performance auditing is another technique that may be used in a cursory way to ascertain if production standards are out of line. This is a quick and reliable method so long as the technique is limited in scope. A company may generate statistical data about the performance level of employees, the department, or the total production facility. These computer analyses provide a wide range of information.

Assume that company documents indicate that production workers are producing at 110%. They are disappointed. The work standard system supposedly provides for 25% opportunity. The workers are also upset because of company pressure. They file a grievance because the system is not providing the 25% opportunity as stipulated in the contract. Here the company blames the worker, and the workers blame the company. In cases like this, a performance rating audit could determine if something were wrong. A fever thermometer may indicate that something is in fact wrong with the body. It substantiates the statement "I don't feel well." It does not isolate the cause. Performance auditing is also a substantiating technique.

Estimation and hypothesis testing are the techniques employed in this auditing procedure. Assume company records indicate that the production employees as a group are attaining an average of 110%. A random sample of 30 production jobs is performance rated, with the results of $\overline{X} = 122\%$ and $\hat{S} = 8\%$. Using *t* statistics, a confidence interval of $P(119 \leqslant \mu \leqslant 125) = .95$ was developed. When the SD is not known, $\hat{S}$ may be used in its place. The *t* distribution demands normalcy and homogeneity.[5] The idea here is not to delve into statistical theory. The concept is that 30 jobs were performance rated, which resulted in a prediction that the true performance level would fall somewhere in the interval 119 through 125%. Obviously, argument could be raised regarding the validity of the performance rater or falsification by the production workers. But these exist in full-fledged auditing procedures. In reality, it exists when standards are established. For this particular case, the question to answer is why are production workers earning 110% when they should be earning at an average of 119% to 125%?

A look at the work standards system being used may indicate that 5 min for picking up materials were provided in the delay allowances based on three changeovers per day. With short interval scheduling, the workers may change over between seven to nine times per day. Also, the company has 30 min of contractual break time. In this particular time study application, the normal time is expanded by 15% and then divided into 60 min, which in essence takes back some of the break time and condenses the time necessary for delays.*

The idea behind performance auditing is to establish with some objectivity that the system is functioning correctly or incorrectly according to the prescribed standards. In actual applications, the system has been admitted in arbitration, has been used to prove that work standards are too loose or too tight, and also has been remedial in that it has caused the system to be scrutinized more intensely.

Auditing Predetermined Standards

The selection of predetermined jobs for auditing follows the same philosophy and format. Auditing by time study is not recommended, but some union contracts stipulate the procedure. In the cases requiring time study verification, follow the same practices recommended for time study auditing. Predetermined systems falter too. The external causes are basically the same as those causing time study failure. Some analysts fail to account for the irregularities that occur from one cycle to another. For example, when positioning one part to another, it may be found that only in 7 of 10 times do they mate perfectly; yet in the analysis, the motion sequence required to correct the alignment three out of ten times is not proportioned into the total time. This will make the work standard too restrictive. One-handed conventions that indicate that all motion sequences of one hand are limited out by another are frequently too restrictive. Consider an MTM analysis requiring 40 TMUs = 0.024 min. If the analyst errantly includes a regrasp, G2 = 5.6 TMU, in the motion sequence, the mistake will be $\frac{5.6}{40}$ = 14%. Some MTM analyses that contain many body motions are subject to greater variances. These and others are technical failings that must be covered in the auditing process. Predetermined systems or time study usually fail when standards are set on the worker's method rather than on one engineered by the analyst. Predetermined systems are excellent work measurement and methods procedures. When applied correctly, they are superb. But they too are vulnerable to change and analyst error. Thus auditing is essential.

Auditing to Determine if Workers are Limiting Their Production

At times, production workers peg their productivity, which means they deliberately control output. Examples may be found in various production environments. On an assembly line, for example, workers curtail production for 6 months to prove their case prior to negotiations. In some automobile manufacturing operations, the production workers will work to 150% levels and then quit for the day. Filing insurance claims is measured against a prescribed standard. Here, too, workers may meet the standard and then quit.

Although all of these conditions are well known to managers, they seem at times quite helpless in controlling the situation. The contract may restrict any attempt to redevelop standards. Labor relations may choose to placate the worker, essentially looking the other way. Managers may feel this is not the right timing because they are caught up in the philosophy of "getting it out the back door." Thus the workers may go their merry way. In some situations it is so bad that one production worker will ring out the other department members who left the facility earlier. Another situation found the production workers leaving almost the minute they arrived because there was not enough work at the machine; consequently, they could make $2\frac{1}{2}$ days in 1 day of work. This pegging of work occurs in all fields. A telephone repair person may sit in his or her truck for an hour because the $1\frac{1}{2}$ hr job was completed in one half hour. Controlled production is damaging. The facility or machines are not totally utilized. Workers are not always this flagrant, but they stretch out the workday to protect their good deal. This is particularly true when management is using techniques to investigate productivity in order to regain control of the manufacturing function.

Determining whether workers are subtly controlling productivity is not an easy task. It may be known, but cannot be proved. Conversely, it may not be known. In these situations a nonparametric statistical test, runs up and down, may help to determine that daily levels of production are not random. This test, contained in the theory of runs, is time related in this particular application. Assume that productivity from a worker-paced assembly line given in daily units per hour, was 10.91, $\overset{+}{11.03}$, $\overset{+}{12.02}$, $\overset{-}{8.97}$, $\overset{+}{11.86}$, $\overset{+}{12.44}$, $\overset{-}{11.58}$, $\overset{-}{11.50}$, $\overset{-}{10.42}$, $\overset{+}{10.74}$, $\overset{+}{11.37}$, $\overset{+}{11.70}$, $\overset{+}{12.46}$, $\overset{+}{13.10}$, $\overset{-}{12.33}$, $\overset{-}{10.09}$, $\overset{+}{10.95}$, $\overset{+}{12.15}$, $\overset{-}{9.68}$, $\overset{+}{10.87}$, $\overset{-}{10.69}$, $\overset{-}{10.52}$, $\overset{-}{10.32}$, $\overset{+}{11.54}$, $\overset{-}{11.13}$, $\overset{-}{10.89}$, $\overset{-}{10.56}$, $\overset{+}{11.33}$, $\overset{-}{9.72}$, $\overset{+}{11.01}$, $\overset{+}{11.35}$, $\overset{+}{11.46}$, $\overset{+}{12.08}$, $\overset{-}{11.46}$, $\overset{+}{11.94}$, $\overset{-}{9.79}$, $\overset{-}{9.31}$, $\overset{+}{10.35}$, $\overset{+}{10.83}$, $\overset{-}{10.00}$, $\overset{-}{9.97}$, $\overset{-}{9.79}$, $\overset{+}{10.35}$, $\overset{+}{11.35}$, $\overset{+}{11.41}$, $\overset{-}{10.63}$, $\overset{-}{8.98}$, $\overset{+}{9.83}$, $\overset{-}{9.65}$, $\overset{+}{12.78}$, $\overset{+}{13.00}$, $\overset{+}{17.04}$, $\overset{-}{9.17}$, $\overset{+}{9.44}$, $\overset{+}{9.45}$, $\overset{+}{12.35}$, $\overset{-}{11.56}$,

*Problems associated with this type of computation were discussed in the section on some prevalent factors in system deterioration.

12.63, 9.93, 8.66, 11.68, 11.92, 12.43, 9.70, 11.83, 11.30, 11.86, 11.20, 10.21, 11.67, 10.65, 11.71, 10.11, 11.30, 11.23, 10.32, 13.64, 7.60, 8.28, 12.06, 12.74, 9.58, 9.48, 9.46, 10.86, 12.59, 13.17, 14.34, 11.44, 11.43, 10.02, 12.30, and 12.54 for $n = 93$ days of production. The plus sign indicates an increase from the previous day, and a minus sign a decrease.[6] Thus

$$R = \text{the number of changes in sign} = 49$$
$$E(R) = \tfrac{1}{3}(2n - 1) = \tfrac{1}{3}[2(93) - 1] = 61.67$$
$$E(V) = \tfrac{1}{90}(16n - 29) = \tfrac{1}{90}[16(93) - 29] = 16.21$$
$$Z = \frac{R - E(R)}{E(V)} = \frac{49 - 61.67}{4.03} = -3.14$$

where R = actual runs
$E(R)$ = expected runs
$E(V)$ = expected variance
n = number of observations or sample size

To calculate the number of changes in sign, look at the series, which begins

1 2 3 4 5 6 7
\+ + − + + − − − + + + + + − − + +

Here there are seven changes in signs.

The solution to this problem lies in hypothesis testing. Assuming $\alpha = 0.01$, $Z_\alpha = -2.33$; thus $-3.14 < -2.33$, and the concept of randomness is rejected for this particular test.

Based on the results of this test, a work standards analyst may suspect that workers are limiting their daily output. These standards may have been questioned in a grievance, and production workers may be engaging in a controlled, nonrandom productive activity prior to arbitration. It could be that they are also protecting loose standards. Obviously this is a good situation, because their productive levels and high pay may be achieved with little effort. The workers usually know what excesses will cause management to study their jobs. Thus they curtail production. This test could also indicate that the standards are too tight. An employee will control production at lower levels at a point that curtails any further inquiry.

That this test does not indicate randomness may justify a conclusion that workers are deliberately controlling production. A union contract usually stipulates the conditions under which a new standard may be developed. Changes (in methods, materials, or design), clerical errors, and other factors may contractually justify management's restudying and resetting the work standard. Limiting production, or pegging, is usually a sign that one of the factors affecting change is present. This statistical test is one way to investigate possible controlled worker output. Auditing jobs on the basis of this test may prove that the work standard does not truly reflect performance.

REFERENCES

1. DELMAR W. KARGER and FRANKLIN H. BAYHA, *Engineered Work Measurement*, (2nd Edition), Industrial Press, New York, 1966, p. 268.
2. UNITED AUTO WORKERS, Time Study–Engineering and Education Departments, *Is Time Study Scientific?*, Publication No. 325, Solidarity House, Detroit, 1972.
3. NATIONAL INDUSTRIAL CONFERENCE BOARD, INC., *Systems of Wage Payment: Studies in Industrial Relations*, Author, New York, 1930, pp. 102–115.
4. JOSEPH A. PANICO, *Queuing Theory–A Study of Waiting Lines for Business, Economics, and Science*, Prentice-Hall, Englewood Cliffs, NJ, 1969, pp. 18–91.
5. PAUL G. HOEL, *Introduction to Mathematical Statistics*, Wiley, New York, 1962, pp. 262-296.
6. YA-LUN CHOW, *Statistical Analysis*, Holt, Rinehart & Winston, New York, 1975, pp. 536–577.

BIBLIOGRAPHY

BROWN, ROBERT G., *Management Decisions for Production Operations*, Dryden Press, Hinsdale, IL, 1971.

GILBRETH, FRANK B., *Motion Study*, Van Nostrand, New York, 1911.

LAUFER, ARTHUR C., *Operations Management*, South-Western Publishing, Cincinnati, 1975.

MAYNARD, H. B., *Industrial Engineering Handbook*, 3rd ed., McGraw-Hill, New York, 1971.

MOGENSEN, ALLAN H., *Common Sense Applied to Motion Time Study*, McGraw-Hill, New York, 1932.

SMITH, GEORGE L., *Work Measurement–A Systems Approach*, GRID Publishing, Columbus, OH, 1978.

WIGGINS, RONALD L., *The Arbitration of Industrial Engineering Disputes*, The Bureau of National Affairs, Washington, DC, 1970.

CHAPTER 4.2

Measuring and Controlling Machine Performance

ALFRED H. SMITH
Delphi Corporation

IRVIN OTIS
American Motors Corporation

4.2.1 PURPOSE OF MEASURING AND CONTROLLING MACHINE PERFORMANCE

The primary purpose of management is to effectively utilize machines, manpower, materials, and money to produce a product or provide a service. One area of concern is the effective utilization of machines or equipment.

Chapter 11.7 discusses the impact of maintenance upon capacity. Industry must ensure that optimum utilization of machines has been achieved before spending additional monies for additional capacity. The costs of a new facility must be carefully evaluated against the optimum utilization of existing equipment.

This chapter is devoted to the maintenance techniques used across a broad spectrum of industries to measure and control machine performance.

4.2.2 TYPES OF MAINTENANCE MANAGEMENT

Breakdown Management

Some industries utilize "breakdown" maintenance management technique as a programmed managerial decision. The basic premise of this technique is to spend minimum maintenance manpower and money to keep equipment running. Most repairs are done on a "bailing-wire-and-gum" or "quick-fix" basis until the equipment's condition requires complete rebuilding, overhaul, or replacement. These companies maintain a minimum staff of maintenance personnel who specialize in breakdown repairs and generally contract outside services for rebuilding or overhauls.

This technique minimizes the maintenance labor and material costs that must come out of current operating profits. This approach is used in situations such as the following:

1. Small manufacturing facilities that have general-purpose machine tools that can be repaired quickly with minimum loss of machine capacity.
2. Manufacturing facilities that have obsolete or excess capacity because of changes in the market.
3. Manufacturing facilities that have multiple or on-line spares. Some companies install three pumps (with valving) in an application that requires only one. This permits continued and spared operation with one pump broken down.
4. Manufacturing facilities that are scheduled for phaseout in the near future.
5. Operations that are auxiliary to production and that do not control the ability to produce parts. Examples of this situation are parts washers or fork trucks where the parts could be processed manually until the repairs are completed.

The use of breakdown management techniques is not necessarily bad management. It might be the correct management approach in the situations cited.

Preventive Maintenance Management

"Preventive maintenance" is a term that has received considerable attention, but that is not clearly understood. As used in this chapter, preventive maintenance consists of the following essential features:

1. **Lubrication.** The paramount feature of all preventive maintenance programs is a lubrication program. This program must be developed and administered by people trained in the lubrication of equipment. Skilled lubrication personnel are essential.
2. **Inspections.** The second essential feature is the development and use of checklists by trained personnel in performing regularly scheduled inspections and in providing feedback on conditions of equipment. This feedback provides maintenance management with criteria regarding physical changes in equipment, such as vibration, heat, and wear. Management can evaluate the significance of these changes and establish overhauls or shutdowns for anticipated repairs or replacements.
3. **Overhauls or Scheduled Shutdowns.** The third essential feature is the short- and long-range planning of projected overhauls or shutdowns. Short-range planning usually encompasses primarily production equipment, such as transfer lines, heat-treat furnaces, presses, and reactors. Long-range planning usually involves plant utility equipment, such as air compressors, transformers, and pollution or effluent systems.

Current use of the preventive maintenance management technique (within the framework of this chapter's definition) is very low. The steel, aircraft, and chemical industries are leaders in its use. Manufacturing facilities usually are not faced with the necessity of using this type of maintenance management until the equipment has aged 7 to 10 years. However, steps should be taken to prepare for this eventual necessity by training maintenance supervisors in the technique during these years.

Predictive Maintenance Management

Predictive maintenance management is a technique of forecasting failure and replacement or repair *just prior* to failure. The aircraft, chemical, steel and iron, and similar continuous-process industries utilize this technique extensively, as follows:

1. **Aircraft Industry.** Overhauls and routine visual and minor preventive maintenance inspections are performed based upon running hours and checklists. Redundant systems function as on-line spares and are sometimes used in lieu of detailed preventive maintenance inspections.
2. **Chemical Industry and Refineries.** On-line monitoring of vibration, heat, flows, pressures, and so on, are used in lieu of preventive maintenance inspections. Changes in condition are the basis for planning and scheduling shutdowns and overhauls. On-line spares are used to extend equipment running time to permit repairs without shutting down entire units.
3. **Steel and Iron Industries.** Running hours and on-line as well as off-line manual monitoring are used in lieu of preventive maintenance inspections.

In general, an industry begins to reduce preventive maintenance inspection manhours by implementing equipment monitoring with established parameters and limitations such as amount of vibration, heat, and pressures. There are specific applications of predictive maintenance, such as offshore oil drilling platforms in isolated areas, that require the use of monitoring devices rather than preventive maintenance inspections.

Planned and Scheduled Maintenance Management

Planned and scheduled maintenance is the performance of needed repairs or shutdowns, as identified by preventive and predictive maintenance monitoring systems, in order to minimize machine downtime and maximize maintenance manpower performance. Foundries, steel mills, and most basic chemical facilities utilize this technique to schedule and thus ensure the availability of the extensive maintenance manpower required for shutdowns or turnarounds of equipment. Refineries also utilize this technique to ensure the availability of the parts and of the specialized manpower and equipment required to overhaul cracking towers, evaporators, reactors, and so on.

Composite Type of Maintenance Management

Most large maintenance organizations must simultaneously utilize *all* of the previously discussed management techniques. Illustrations of this are as follows:

1. **Breakdown.** Many manufacturing facilities do not have maintenance coverage for production operations for 21 shifts (3 shifts × 7 days/week). In general, the second shift (3:30 p.m. to

11:30 p.m.) has some maintenance personnel to handle preventive, predictive, and scheduled maintenance carrying over from the first shift. For the third shift (midnight to 7:30 a.m.), maintenance coverage for production operations usually are breakdown fixers, which function as such by management decision or plan.

2. **Preventive Maintenance.** Those production operations of a special-purpose configuration that do not have on-line spare equipment have aggressive and effective lubrication, inspection, and overhaul programs. Personnel must be trained to perform these functions on a scheduled basis.

3. **Predictive Maintenance.** Plant utilities, such as air compressors, transformers, and gas, water, oil, and effluent systems, must have maintenance programs that *schedule* replacement or repairs to occur while production is not occurring. Unexpected failures of this type of equipment results in massive losses and schedule disruptions. This equipment may have built-in devices to predict failures.

4. **Planned and Scheduled Maintenance.** Planning and scheduling of the major overhauls and repairs are required to ensure effective utilization of both available manpower and machine-hours.

A well-organized and well-managed maintenance operation will normally utilize *all* of the management techniques cited.

4.2.3 ESTABLISHMENT OF PARAMETERS FOR MEASUREMENT AND CONTROL

Breakdown Maintenance

The primary parameters for measuring and controlling machinery maintained on a breakdown basis are:

1. The cost of the lost machine-hours in terms of lost production units and the effect upon profits.
2. The differentials between the cost of a preventive maintenance program and continued breakdown repairs. This comparison of differentials is sometimes referred to as the "break-even point." When the combined costs of breakdown repairs and rebuilding costs exceed the costs of a preventive maintenance program, the break-even point has been reached.
3. The impact upon direct labor productivity in situations where the operation is manually controlled or paced. Loss of direct labor productivity must be minimized.

In general, the primary goal of breakdown maintenance management is to maximize return on direct labor dollars with minimum capital cost outlays.

Preventive Maintenance

The key objectives of preventive maintenance management are:

1. To extend the economic life of capital equipment through effective utilization of lubrication systems, preventive maintenance inspection programs, and overhaul/shutdown schedules.
2. To minimize the impact of unexpected shutdowns resulting from equipment failures. Breakdowns must be reduced to a minimum.
3. To maintain equipment history records and machine performance data that are essential to identifying and acting upon changes in equipment conditions that indicate required repairs or overhauls.

The primary goal of preventive maintenance management is to maximize return on invested capital dollars in process- or equipment-controlled industries.

Predictive Maintenance

The primary goal of predictive maintenance is to provide continuous, uninterrupted service through the use of monitoring and scheduling backup units to take over the operation. The costs of monitoring both on-line and off-line operation are not the primary concern. Unanticipated equipment failures must be prevented. The penalty costs of such failures are excessive and can lead to the economic collapse and/or failure of the business. The chemical, petroleum, and public utility companies will sometimes schedule a major operating unit down because of the uncertainties involved in continuing operations. They will bring a standby or spare unit into operation, switch all production to the standby or spare unit, and phase down the primary unit. The utility industry must do this, even though the primary unit may still be operational, when they can no longer predict continued reliable operation.

Parameters that ensure predictability are the primary goal, and the aircraft industries are the leaders in this field. Examples of these parameters are discussed later in this chapter.

Planned and Scheduled Maintenance

The most significant, and perhaps the most unique, characteristic of planned and scheduled maintenance is the existance of documented backlogs of projected maintenance work. The chemical, petroleum, and steel and iron industries usually have planned and scheduled backlogs showing projections of maintenance manpower, materials, and special equipment requirements for 12 to 18 months into the future. These projections are mandated because of the necessity to place orders for spare parts and specialized outside contract labor months in advance. This, in turn, requires the banking or storing of raw materials and finished products to minimize disruptions to customers and suppliers.

The projections regarding the amounts of these banks or stores are based upon two critical factors: (1) the market demands anticipated during the scheduled downtime and (2) the length of time that the unit is scheduled to be down. It is very important that a unit's downtime schedule be met and that all needed repairs be completed during the downtime. Millions of dollars are invested in these types of shutdowns or turnarounds. The unit must come back on stream and run for an extended period. If it does not meet schedules and run as planned, markets can be lost because of the loss of products.

The primary goal of planned and scheduled maintenance management is to meet schedules and keep equipment on stream for the needed economic run period.

4.2.4 MEASUREMENT TECHNIQUES

Breakdown Maintenance

The primary objective of breakdown-oriented maintenance is to keep equipment running. The techniques for measuring the effectiveness of this type of maintenance are as follows:

1. **Equipment Utilization Recording.** All breakdowns are recorded either through a log of machine breakdowns or a work order system. These logs or work orders are analyzed by equipment number to determine if the break-even point (repair versus rebuild) has been reached. The primary concern is not maintenance costs, but rather the effect of continued use upon direct labor performance.
2. **Direct Labor Variance Analyses.** Each time an operation has to be shut down for maintenance, the direct labor operators incur lost time. Detailed reports of labor variances are analyzed to determine the effect of breakdowns upon direct labor costs.
3. **Work Sampling.** Some industries use work sampling of both direct and indirect labor operations rather than direct labor variance reporting. Work sampling is also used to determine the causal factors of downtime.
4. **Manufactured Information Systems.** Some industries use electronic reporting systems for analysis of performance of direct labor operations. Examples of these types of systems are discussed later in this chapter.

Preventive Maintenance

Managements that use preventive maintenance have the most difficult task because the measurements include

1. Economic life of equipment.
2. Direct labor variance.
3. Maintenance repair/spare parts inventories.
4. Maintenance labor productivity.

The scope and degree of evaluation techniques used in this type of maintenance management vary from industry to industry. The following items for evaluation are examples, but they are not all-inclusive:

1. Return on invested capital dollar.
2. Percentage of maintenance labor and materials costs per sales dollar.
3. Percentage of maintenance costs per capital invested dollar.
4. Percentage of maintenance costs (labor only) per direct labor dollar.
5. Work sampling of maintenance operations.
6. Universal maintenance standards for standard data measurement of maintenance productivity.

In general, chemical, petroleum, glass, steel, and iron (continuous-process) industries concentrate on items 1, 2, and 3. Basic industries and manufacturing concentrate on items 4, 5, and 6.

Managements that use preventive maintenance are also faced with extensive training programs and large repair and/or spare parts inventories. Measurements of the effectiveness of these factors are also of primary concern.

Predictive Maintenance

The primary goal of predictive maintenance is to predict performance of equipment in order to prevent unscheduled failures. This type of maintenance management has seen spectacular growth in the last decade. The development and application of the hardware for such monitoring systems has been extensive. Some of these techniques are:

1. **Vibration Monitoring.** This is perhaps the earliest form of predictive maintenance monitoring. For decades maintenance management used the "listen, look, and touch" method to predict mechanical failures. The problem with this method was that damage had already occurred. The lead time needed to obtain required parts, plan the repairs, and schedule the shutdown *prior* to failure had been lost.

   The use of extremely sensitive on-line vibration monitoring devices to warn about significant changes in vibration has increased dramatically in the last decade. Power plants, refineries, and chemical plants have used this technique for years on large units such as turbines and compressors. Coupled with the significant development of monitoring devices has been the development of vibration analyzers. When the monitor indicates a significant change in vibration patterns, the vibration analyzer can perform a series of tests to identify the *source* of the vibration. Plans can then be initiated to take corrective action prior to the occurrence of damage, and needed lead time for parts, plans, schedules, and so on, is achieved.
2. **Heat Monitoring.** Rotating and reciprocating mechanical equipment generates heat. Lubricants are used to reduce friction *and* control heat. Significant changes in mechanical equipment temperatures are indications of potential mechanical problems. On-line monitoring of temperature changes in mechanical equipment is an old and proven technique.

   Recent developments in infrared monitoring of electrical switch gear have proven very useful in predicting switch gear failure. Infrared photography of glass furnace walls can be used to predict refractory life.
3. **Electrical Monitoring.** Monitoring demand and resistance of electrical components is another old and proven technique. Significant changes in electric motor-driven drive trains are indications of potential mechanical and/or electrical problems. Changes in amperes for large fan motors indicate potential restrictions in the duct or pipe systems. There are many such electrical monitoring systems, all of which are designed to monitor changes in order to predict and plan for potential failures.
4. **Monitoring of Flows or Pressures.** Hydraulic and pneumatic systems are designed to operate within a range of pressures and flows. Significant changes in the components' pressures or flows are indicators of potential problems. It is possible to predict failure by plotting and analyzing such changes. This monitoring technique is used extensively in chemical, petroleum, steel, glass, and manufacturing industries.
5. **Monitoring of Corrosion or Wear.** Monitoring of the corrosion or wear of pipes, reactors, refractory, vessels, rolls, and gears has been used for many years to predict potential failures. Measurements are taken on a continuing basis, and projections of future failure are plotted. These projections provide needed lead times to plan and schedule overhauls.

Planned and Scheduled Maintenance

The four major factors involved in measuring the effectiveness of planned and scheduled maintenance are:

1. **Backlog.** The size and type of documented backlog must be evaluated continuously. Significant changes in size or type can indicate a need to adjust manpower or equipment projections.
2. **Schedule Compliance.** Continuous measurement of schedule compliance is another device used extensively. The ability of maintenance operations to complete scheduled work within scheduled time frames is paramount.
3. **Productivity.** Planned and scheduled maintenance with documented backlog provides the environment that permits measurement of the maintenance personnel's productivity. The use of standards for repetitive maintenance jobs has been accepted in industries that have documented backlog. Some industries, such as steel, have perfected incentive systems for maintenance jobs.
4. **Staffing Levels.** Some industries measure maintenance performance as a ratio to direct labor hours or assets employed. Other industries have developed highly specialized criteria that have

proven useful for establishing staffing levels by crafts. Chemical, petroleum, glass, and steel industries have developed systems that utilize outside contractors for peak loads, such as major unit shutdowns or turnarounds. They maintain captive maintenance staffing levels to accomplish normal repairs and routine maintenance at prescribed levels of production. American corporations incur *billions* of dollars per year for maintenance. Staffing levels must be constantly measured to ensure adequate productivity as previously mentioned.

4.2.5 MEASUREMENT DEVICES—EQUIPMENT AND TECHNIQUES

Vibration Monitoring

Concepts

The planet earth rotates and vibrates as it makes its journey around the sun. The rotation and vibration present no problem as long as they remain within the present parameters of frequency, velocity, and displacement. Changes in any one of these parameters could result in damage.

For years maintenance management used the "listen, look, and touch" method to detect changes in vibration that might indicate potential problems. However, this method was always too late! If a change is perceptible to the eyes, ears, or fingers, damage has already occurred.

The objective of developing the equipment for and the technique of vibration analysis is really threefold:

1. To predict equipment failure prior to damage.
2. To determine the cause of the failure.
3. To determine potential damaged parts and plan and schedule shutdown.

The vibration analysis equipment manufacturers have developed excellent brochures and training programs that explain this technique in detail.

Analysis of Modes

Most vibration equipment manufacturers provide the ability to measure vibration in:

1. **Mils or Inches Displacement.** Peak-to-peak motion or amount of vibration.
2. **Velocity Inches per Second.** How fast it is vibrating with respect to a fixed reference. This mode is a function of both displacement and frequency.

In general, vibration *monitoring* uses velocity inches per second to detect changes, whereas vibration *analysis* uses mils displacement to identify the causes of such changes. The velocity mode is especially effective in detecting small, high-frequency vibration such as would be produced by rotating gears, bearings, and so on. The technique of monitoring involves repeated analysis of horizontal, vertical, and axial vibrations in the velocity inches per second mode at prescribed pickup points. Significant changes in the readings could indicate the following problems:

1. Bearing failure is imminent.
2. Coupling is misaligned.
3. Shaft is misaligned.
4. Fan blades are out of balance.
5. Belts are slipping.
6. Driver (motor) is source of vibration.
7. Driven part (gear, fan, etc.) is source of vibration.

Industry Applications

The petroleum, refining, and power plant industries utilize steam-driven turbines as the drivers for huge pumps or electrical generators. These steam-driven turbines have built-in vibration monitoring devices that automatically shut a unit down when programmed vibration limits are reached. The glass industry utilizes huge air handling fans to maintain glass furnace melting and refractory wall temperatures. These applications have built-in devices that activate alternate similar units to take over the service prior to automatic shutdown. These units are termed "on-line devices."

Some industries provide multipoint analyzers that allow an operator to take vibration readings across an entire power train. These readings are plotted to project the effect of significant changes in vibration. These multipoint analyzers and portable single-point recorders are termed "off-line devices." These units require overt action on the part of an operator to take down a specific unit and bring into service a backup spare. These units may have "self-destruct" protections, but they do not automatically activate a backup unit.

Electrical Monitoring

Concepts

Monitoring of electrical equipment generally includes vibration, as discussed previously, and additional factors such as load and temperature changes. The most commonly used electrical monitoring technique (after vibration) is temperature monitoring.

Dirt is a principal cause of electrical failure. In almost every instance dirt will affect resistance. Because of higher resistance, heat occurs, which is detrimental to electrical insulation. Monitoring this increase in heat is one of the most commonly used techniques for electrical apparatuses, switch gear, and so on. For example, failures of main breakers, power and distribution transformers, and substation apparatuses can usually be traced to insulation breakdown. Failures of motors (other than mechanically caused) usually are the result of insulation breakdowns. Some electrical switch gear have built-in devices to measure both load and temperature. A review of all the various techniques used would require a complete textbook. For simplicity, a review of electric motor "meggering" might illustrate one technique.

The megger insulation tester is a small, portable instrument used to measure insulation resistance in ohms or megohms. The value of the insulation is reduced as the ground insulation becomes saturated or contaminated with moisture, dust, dirt, and oil. Insulation that has deteriorated because of age *or* because of the pressure of temperature in excess of normal reduces the value of insulation. Monitoring motor megohm readings on a scheduled basis will provide warnings of potential motor damage and/or failure. Any persistent downward trend is usually a warning of trouble ahead.

Techniques

One technique used for monitoring primary and secondary switch gear and distribution systems involves the use of infrared photography. The entire system is photographed with infrared film to locate "hot" spots, which appear as bright white areas on prints and could indicate potential failures resulting from heat. This excess heat is resulting from some changes in load or resistance.

The design and application of electrical equipment must take account of ambient temperatures and dirt inherent in the operation. Foundries, steel mills, glass factories, heat treaters, and so on, should utilize the electrical equipment designed specifically for such environments.

To summarize, the techniques of vibration and electrical monitoring involve the performance of tests on a scheduled basis. The results of these tests are evaluated to identify significant *changes* in the equipment. These tests can be performed on line by use of recorders or off line by use of testing devices. The importance of measuring mechanical and electrical machine performance cannot be overemphasized.

Industry Applications

The steel and aluminum industries are consumers of enormous amounts of electrical energy. As discussed previously, infrared photographs of inaccessible switch gear are used to highlight hot spots in primary distribution systems.

The chlorine and caustic chemical industry is also a user of huge amounts of electrical energy. Significant variations in voltage and/or amperes can cause irreparable damage to multimillion dollar installations. Variations in loads placed upon large electrically driven conveyors are monitored on a continuing basis. Significant changes in amperes could indicate potential mechanical failure.

There are unlimited illustrations of how industry monitors electrical energy. The cost of such energy has now become a major factor, and monitoring these costs to reduce peak load demands and consumption is critical.

Pyrometers

There are many forms of pyrometers used to monitor equipment performance. Some are direct contact types that are concerned primarily with surface temperatures. Contact pyrometer readings are taken on large critical equipment, such as motors and drive trains, in chemical, glass, steel, and petroleum product handlers or transporters. Elevated bearing surface temperatures (above 180°F) are clear indications of potential equipment problems.

Noncontact pyrometers, such as optical pyrometers, are used in glass and steel industries for monitoring the process of making a product. These same devices are also used to monitor equipment performance.

Obviously, power plants monitor their boilers or reactors with on-line devices. The use of pyrometers throughout industry for measuring and controlling equipment performance is varied.

Pressure and Flows

Pressure regulators are used to control the flow of materials, gases, liquids, and so on. Pressure gauges are used to measure the effect of this flow upon a regulated system. There are many devices available to measure flows, pressures, and so on. There are also many devices that automatically correct or adjust equipment components experiencing abnormal pressures for flows. Some illustrations of these devices are:

1. Hydraulic filter bypasses (pressure and flow).
2. Tank or vessel relief valves (gas and air flows).
3. Maxitrol valves (gas and air flows).
4. Hydraulic or pneumatic directional valves (flows).
5. Anemometers (air flows).
6. Air line check valves (loss of flow).
7. Pressure- or flow-actuated solenoids (air, hydraulic).
8. Weight-controlled valves (air cylinders, scales, etc.).

There are hundreds of such devices available. All of these devices are used to measure and control equipment performance.

Noise

Recent improvement in noise detection has prompted development of measuring and controlling devices actuated by sound waves. Some illustrations of these are:

1. Sonic-level controllers that open or close feeding valves or chutes based upon sound used in the textile and raw materials bulk handling industries.
2. Sonic testing of equipment component parts to detect flaws. Used in aircraft, automotive, and glass industries.
3. Sonic testing to measure thickness of pipes, plates, and vessels from one side. Used in chemical, petroleum, and steel industries.
4. Ultrasonic testing (nonresonated) of equipment components such as high-speed spinners or winders. Used in the aircraft and textile industries.

The use of noise-actuated devices has reached the stage where computers can now respond directly to voice inputs.

Metal Erosion or Growth

The normal techniques for measuring wear or growth of rotating or reciprocating equipment basically involve measuring physical changes over prescribed time spans. The wear of gears, pinions, sheaves, cylinders, valves, and so on, can be predicted by maintaining physical data records and projecting expected life. This technique is as old as the pyramids, and devices for such measurements range from simple micrometers to electron microscopes. The measurement and projection of the wear of pipelines, vessels, reactors, and so on, has also been in use since the pyramids. Recent developments in nuclear meters and similar devices now permit continuous on-line readings of pipe and vessel wear.

Distilling of cracking towers contain metal trays to achieve desired product separations. These trays wear out and must be replaced. Projected life can be determined as a direct function of the type and amount of product processed.

Large conveying systems experience wear in chain links, rollers, pins, and so on. This wear results in travel distance increases (growth), which must be adjusted by automatic idlers or take-ups. The amount of take-up required can be measured in linear distance, electrical resistance, or hydraulic or pneumatic pressure variations. These variations can be plotted to predict when repair or rebuilding is necessary.

Metal tubes exposed to atmospheric-controlled heat over extended periods of process time tend to grow or change in configuration. Visual optics or monitoring devices are used to measure such changes in order to predict life.

Foundry and glass furnaces contain refractories that deteriorate as a result of the combustion systems, products handled, rates of flows, and so on. The measurement of these refractories by means of optical pyrometers, infrared photography, platinum sensors, and so on, has been an accepted industry practice for many years. The ability to predict the life of such furnaces is critical to economic survival. Coal crushers and pulverizers are crucial to the operation of coal-fired power plants. The deterioration of blades, hammers, rods, balls, and so on, in these mills can be predicted almost as a direct function of the type and amount of product processed. The same conditions exist in mining operations for such ores as bauxite, copper, and iron ore.

These types of measurements and techniques are only a small sample from the maintenance viewpoint, illustrating, the need for and benefits of measuring and controlling machine performance.

Maintenance Staffing Levels

As discussed previously in this section, the ability to control wide fluctuations in demands on the maintenance manpower is paramount to effective maintenance productivity. The most inefficient maintenance technique is to work on unplanned and unscheduled breakdowns of major product units. Unforeseen or uncontrolled breakdowns result in

1. Direct labor inefficiencies.
2. Product and/or market losses.
3. Inefficient maintenance productivity.
4. Loss of productive capacity.

Such breakdowns are usually met by working maintenance crews overtime. If this does not satisfy these peak demands, outside contractors are used to augment existing manpower. Eventually pressures build up within the affected organization to increase maintenance manpower productivity and, if necessary, staffing levels. These pressures serve only to aggravate the peak and valley conditions. When the equipment has been brought back into reliable condition, deep "valleys" occur in maintenance demands. This results in a surplus of maintenance personnel, who become engaged in realtively unimportant busywork. If the equipment continues to operate without major breakdowns, management usually reduces manning levels. The vicious cycle starts again when the next major failure occurs, and maintenance management is again forced to resort to unplanned, unscheduled breakdown maintenance.

From a maintenance management viewpoint, the measuring and controlling of machine performance is crucial to the effective utilization of maintenance manpower. The ability to develop and operate within limited resources (budgets) depends primarily upon the success or failure of measuring and controlling machine performance.

4.2.6 PRODUCTION COUNTING DEVICES

Mechanical Counters

There are many devices available for recording pieces produced. These counting devices can be actuated by mechanical, electrical, light, sound wave, heat, pressure and flow impulses. These impulses, in turn, can be transmitted to digital or analog counters, recorders, cathode tubes, discs, tapes, computers, and so on.

To simplify the description of these devices, let us assume that *any* impulse, once picked up, can be processed into *any* digital or analog counter, recorder, and so on. The primary difference is in the impulse of the pickup device. Mechanically actuated pickups for production counting have been used since the dawn of man. The variations on mechanical pickups are as widespread and as old as man. These pickups are usually the result of physical contact between the pickup device and

1. Raw material feed-in.
2. In-process material movement.
3. Finished product out.
4. Rotation of driver.
5. Rotation of driven part.

The primary characteristic of mechanically actuated production counter's is *physical contact* based upon movement, strokes, manual work, and so on.

Electrical Counters

Electrical counters can be impulsed by contact or noncontact with any of the five movements, strokes, and so on, cited for mechanical counters. Electrical relay contact can be made with limit switches (contact) or proximity switches (noncontact). The sizes, types, and applications of electrical counters far exceed those of mechanical counters.

The primary characteristic of electrically actuated production counters is that *physical contact is not required* to detect movements, strokes, or manual work. These devices can be impulsed by changes in electrical load, demand, volts, amperes, and so on, for the driver (motor). They can be impulsed by the interruption of a magnetic field or by the dropout of current.

Electronic Counters

Electronic counting devices are basically those devices impulsed by light or sound waves. The use of photoelectric eyes has been well established for years. The recent innovation in laser beams has opened up a completely new field. Lasers can be used to perform work and simultaneously check the workpiece produced for proper dimensions, tolerances, and so on. The combining of photoelectric eyes, laser beams, and industrial robots with self-monitoring of equipment performance by electrical and mechanical devices has laid the foundation for a new industrial revolution.

Computers

The application of computers to on-line control and measurement of equipment has been an accepted industrial practice since the early 1960s. Steel mills, foundries, glass furnaces, and so on, have used on-line computers to control additions of raw materials, alloys, oxygen lancing, inoculation, fluxing, and so on. Drug firms have used on-line computers to control bacterial processes, culture growth, and so on.

The use of computers for assimilating, collating, and evaluating manufacturing planning effectiveness, operations performance, and equipment utilization has created a new discipline, sometimes termed "manufacturing information systems." The explosion of computer-assisted design and computer-assisted manufacturing techniques in the 1980s will make the 1970s look like the Stone Age.

BIBLIOGRAPHY

BLANCHARD, B. S., Jr., and E. E. LOWERY, *Maintainability–Principles and Practices*, McGraw-Hill, New York, 1969.

BOGLE, H. A., "Application of Time Study and Methods to Maintenance," in *Proceedings of Seventh Annual Time and Methods Conference*, Society of Advanced Management, New York, 1962.

CARSON, GORDON B. Ed., *Production Handbook*, 3rd ed., Ronald, New York, 1972.

CUNNINGHAM, C. E., and WILBERT COY, *Applied Maintainability Engineering*, Wiley, New York, 1972.

"How Maintenance Managers Feel About Using Work Standards," *Factory*, December 1967.

LEWIS, BERNARD T., "Developing Maintenance Time Standards," *Industrial Education Institute*, Boston, 1967.

LEWIS, B. T., and W. W. PEARSON, *Maintenance Management*, Riders, 1963.

MAYNARD, H. B., Ed., *Handbook of Modern Manufacturing Management*, McGraw-Hill, New York, 1970.

MAYNARD, H. B., Ed., *Industrial Engineering Handbook*, 3rd ed., McGraw-Hill, New York, 1971.

MILLER, ELMER J., and J. W. BLOOD, Eds., *Modern Maintenance Management*, American Management Association, New York, 1963.

MOORE, F. G., *Manufacturing Management*, 4th ed., Irwin, Homewood, IL, 1965.

MORROW, L. G., Ed., *Maintenance Engineering Handbook*, 2nd ed., McGraw-Hill, New York, 1960.

NEWBROUGH, E. T., *Effective Maintenance Management*, McGraw-Hill, New York, 1967.

SMITH, ALFRED H., "Boost Maintenance Efficiency With Simplified Scheduling," *Foundry*, July 1972.

STANIAR, WILLIAM, Ed., *Plant Engineering Handbook*, 2nd ed., McGraw-Hill, New York, 1959.

"Universal Maintenance Standards," *Factory Management and Maintenance*, November 1955.

WILKINSON, JOHN J., "Measuring and Controlling Maintenance Operations," *Journal of Methods-Time Measurement*, March–April 1966.

WILKINSON, JOHN J., "Maintenance Management," *Plant Engineering*, 4 parts, January 9, April 17, May 15, and June 12, 1969.

CHAPTER 4.3

The Learning Curve

WALTON M. HANCOCK

The University of Michigan

FRANKLIN H. BAYHA

Private Practice, Ann Arbor, Michigan

4.3.1 INTRODUCTION

A learning curve is the phenomenon whereby, as the number of cycles increases, the time per cycle or the cost per cycle decreases for a large number of cycles. Learning can be separated into two major areas: that which occurs while a person is doing a task repetitively and that which occurs while an organization produces many units of a particular product. The former area is called "human learning," and the latter the "production progress function," which is the term generally used in the literature. Fortunately, the mathematics is the same for both areas. This chapter presents the appropriate equations and then discusses human learning and production progress functions. Examples of the use of the equations also are given.

4.3.2 THE MATHEMATICS OF LEARNING

Learning curve equations generally take the form of

$$Y = KX^{-A} \tag{1}$$

where Y = the time per cycle
K = the time for the first cycle
X = the number of cycles
A = a constant for any given situation. The value is determined by the learning rate.

If we take the log of both sides of equation 1, we get a straight line.

$$\log Y = \log K - A \log X \tag{2}$$

Thus, if we plot the equation on log-log paper, A will be the slope and K will be the intercept. One of the useful properties of this equation is that every time X (the number of cycles) is doubled, Y (the time per cycle) decreases by a fixed percentage. This is the source of the commonly used term "percent learning curve." For example, every time the units double, the value of Y for a 90% curve is 90% of the previous value. Suppose the first unit had a cycle time of 10 min; then we would get the results shown in Exhibit 4.3.1 by successively doubling the number of units.

Exhibit 4.3.2 is a plot of a 90% curve using linear coordinates, and Exhibit 4.3.3 is the same equation plotted on log-log paper. Examination of Exhibits 4.3.1, 4.3.2, and 4.3.3 reveals that the rate of learning decreases as the number of cycles increases. Learning, however, continues in many situations for a very long period. This fact has a significant impact on the setting of production standards and on the potential for cost reduction of high-volume production. In the past the equations were usually solved by graphical means, but with the widespread use of hand calculators that can compute logarithms, graphs are no longer necessary. The following are examples of the most typical calculations:

1. What is the average cycle time for units N_1 to N_2? Assume cycle times are in minutes. Let AV = the average time. Then

Exhibit 4.3.1 An Example of the Cycle Times for a 90% Curve

Number of Units (X)	Cycle Time (Y)
1	10.0
2	9.0
4	8.1
8	7.3
16	6.6
32	5.9
64	5.3
128	4.8
256	4.3
512	3.9

$$AV = \frac{\int_{N_1-1/2}^{N_2+1/2} KX^{-A}}{(N_2 + 1/2) - (N_1 - 1/2)} = \frac{[K(N_2 + 1/2)^{1-A}/1 - A] - [K(N_1 - 1/2)^{1-A}/1 - A]}{N_2 - N_1 + 1}$$

$$= \frac{K[(N_2 + 1/2)^{1-A} - (N_1 - 1/2)^{1-A}]/1 - A}{N_2 - N_1 + 1}$$

$$= \frac{K[(N_2 + 1/2)^{1-A} - (N_1 - 1/2)^{1-A}]}{(1 - A)(N_2 - N_1 + 1)} \tag{3}$$

where N_1 is the first unit of the run and N_2 is the last unit.

Using the conditions of Exhibit 4.3.1, where $K = 10$ min, and a 90% curve, what would be the average cycle time for the first 50 cycles? We would proceed as follows:

a. Determine the value of A using equation 1 and any value of X except 1, where the solution for A is indeterminate.

$$Y = KX^{-A}$$

Exhibit 4.3.2 Plot of 90% Curve (Arithmetic)

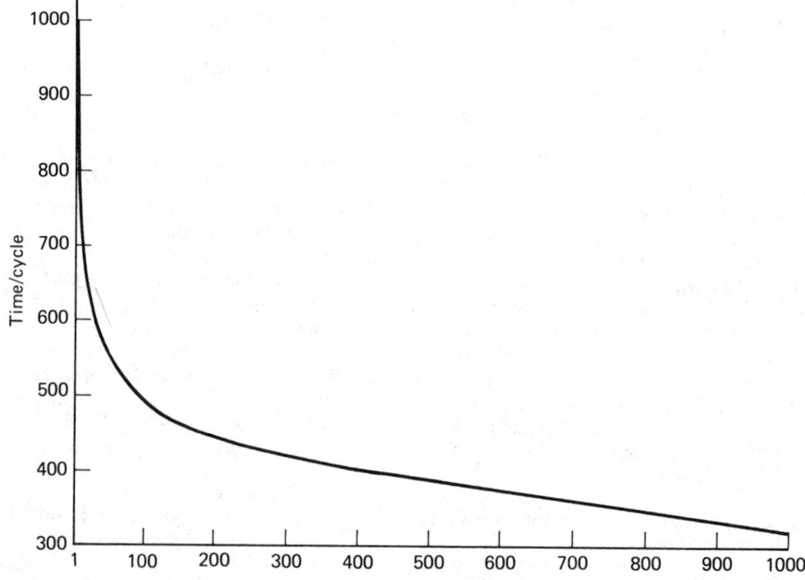

Exhibit 4.3.3 Plot of 90% Curve (Logarithmic)

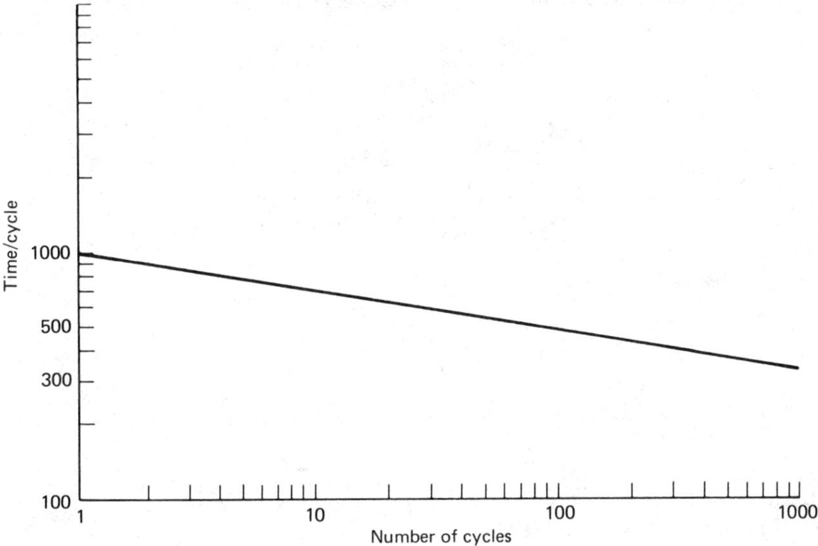

Let $X = 2$; then $Y = 9.0$.

$$9 = (10)(2)^{-A}$$

$$\log 9 = \log 10 - A \log 2$$

$$A = \frac{\log 10 - \log 9}{\log 2} = 0.1520$$

b. Substituting in equation 3,

$$AV = 10 \frac{[(50 1/2)^{0.848} - (1/2)^{0.848}]}{(0.848)50}$$

$$= 10 \frac{(27.82 - 0.556)}{42.40} = 6.43 \text{ min}$$

2. What would be the cycle time for the fiftieth cycle, using equation 1?

$$Y = (10)(50)^{-0.1520} = \frac{10}{(50)^{0.1520}} = 5.52 \text{ min}$$

3. If the standard time is 3.5 min, and if K is 2.5 times the standard, how many cycles would it take to attain the standard time, assuming a 90% curve? Let S = the standard time. Then

$$S = 2.5SX^{-0.1520}, \quad X^{0.1520} = \frac{2.5S}{S} = 2.5$$

or

$$\log X = \frac{\log 2.5}{0.1520}$$

or

$$X = \text{antilog} \frac{\log 2.5}{0.1520} = 415 \text{ cycles}$$

4. How long should it take an operator to attain the standard in question 3? We would use the numerator of equation 3 for the answer. Let C = the cumulative time.

$$C = \frac{K[(N_2 + 1/2)^{1-A} - (N_1 - 1/2)^{1-A}]}{1 - A}$$

$$= 10\left[\frac{(415 + 1/2)^{0.848} - (1 - 1/2)^{0.848}}{0.848}\right]$$

$$= 10\left[\frac{166.17 - 0.56}{0.848}\right] = 1952.98 \text{ min} \qquad (4)$$

If there were 7.5 working hours in 1 day, then

$$\frac{1952.98}{7.5 \times 60} = 4.34 \text{ days}$$

5. Of the 4.34 days (1952.98 min), how much of the time could be considered training costs? Exhibit 4.3.4 might be helpful.

Since the total time under the learning curve would be 1952.98 min, and the time to do the task if the operator were performing at standard would be 415 × 3.5 = 1452.5 min, the training time would be 1952.98 − 1452.5 = 500.48 min.

6. Suppose an operator would do 50 cycles, and then 2 weeks from now another 150 cycles. What would be the cumulative and the average times for the 150 cycles? Exhibit 4.3.5 depicts the situation.

The problem is somewhat complex because the operator will forget a certain amount of what he or she has learned as a result of having a break. This is called "remission." The amount of remission has been found to be a function of where the operator is on the learning curve when the break occurs. An approximation to the point of remission where the operator will start the second run can be made by drawing a straight line between the time for the first cycle and the standard

Exhibit 4.3.4 The Training Portion of the Learning Curve

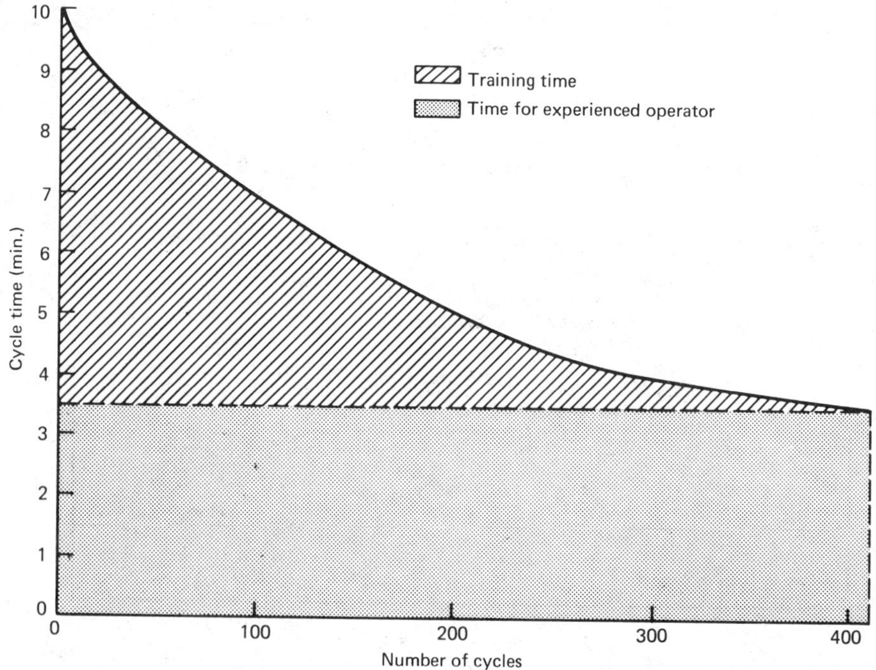

Exhibit 4.3.5 The Effect of Breaks on Operator Learning

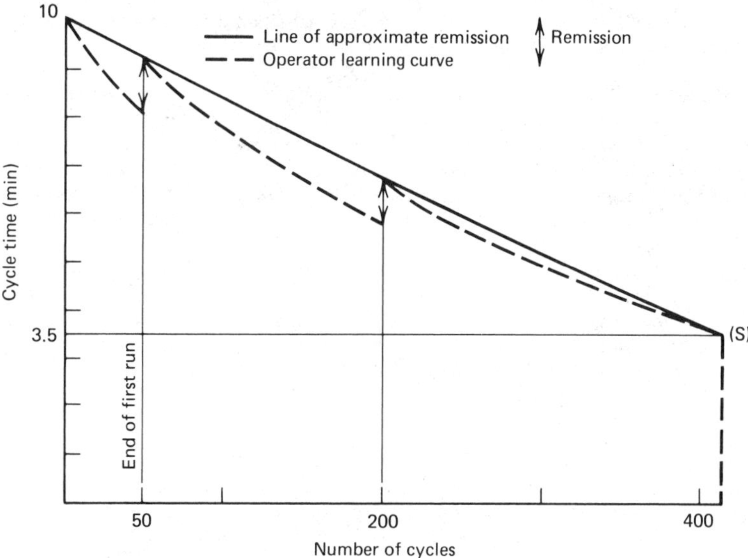

time (S). The equation for this line is

$$R = K - \frac{K - S}{CS} X_i \tag{5}$$

where R = the time of the first cycle after the break
CS = the number of cycles to standard computed for the first run of 50 cycles
X_i = the cycle number of the first cycle after the break.

Since the second run started at the fifty-first cycle, then $R = 10 - 0.0157(51) = 9.20$ min. For the 150 cycles, 9.20 then becomes the new value for K.

The value of A also changes because the rate of learning increases[1]. The value of A is estimated by assuming that S will be attained in 365 cycles. Thus

$$3.5 = 9.20(365)^{-A}$$

$$A = 0.1638$$

The times for the one hundred fiftieth cycle would be

$$AV = 9.20 \left[\frac{(150 + 1/2)^{0.836} - (1 - 1/2)^{0.836}}{0.836(150)} \right]$$

$$= 4.81 \text{ min}$$

$$C = AV \times CY \tag{6}$$

where CY is the number of cycles involved and C is the cumulative time. $C = 4.81 \times 150 = 722.16$ min.

4.3.3 HUMAN LEARNING

There are a number of factors that affect the rate at which people learn to do repetitive jobs. The complexity or chance effect inherent in task performance of the job affects the rate of learning. The capabilities of the worker also have an impact. Much has been written about both topics, and the results are only summarized here. (See Chapter 6.1 for the basic concepts of psychomotor performance.)

Complexity of the Job

Job complexity can be examined as a three-dimensional situation, because there are three major variables that affect complexity from a learning viewpoint. These are:

1. **Cycle Length.** Normally, longer jobs are considered to be more complex, because the worker will forget more between repetitive acts. Cycle length is at least partially accounted for in the learning curve equations, because the cumulative time increases as a function of cycle length.

2. **Amount of Uncertainty in the Motions Involved.** Uncertainty is generally measured by the number of higher-skilled motions involved, such as the more difficult positions, simultaneous motions and grasps. The more uncertainty in the task, the longer it will take the operator to learn to do the job. This aspect has been investigated using the MTM-1 system.[2]

3. **Amount of Prior Training.** In many operations the worker may have developed high skills at certain subtasks. Once a worker has sufficient practice opportunity to become skilled, the remission rate is very low. An example is the ability to hit a ball with a bat. Typical examples in working situations are the ability to use a calculator, a soldering gun, or a micrometer. The MTM-1 learning curve methodology also takes this aspect into account.[1]

Capabilities of People

For the vast majority of working situations, the human capabilities of primary interest are the psychomotor capabilities. These capabilities, which constitute the ability of people to use their hands and feet in conjunction with their sensory functions, vary among people. The ability to learn to use one's psychomotor capabilities also varies. Some of the factors that affect learning are:

1. **Person's Age.** Many, although certainly not all, older people learn psychomotor skills slower than younger people. The rate of learning seems to be relatively constant from approximately 18 to 35 years, when it then starts to decline for many people.[3,4]

2. **Amount That People Have Been Required to Learn in the Past.** There is some evidence to indicate that, if a person stops learning to do new jobs as he or she grows older, then his or her ability to learn decreases. "Decrease" means that it will take longer to learn to do a new task, but that it can still be learned.[4]

3. **Person's Nervous System and Physical Capabilities.** The quality of a person's nervous system declines with age. People who have good nervous systems when they are young tend to show less decline with age than people who have relatively poor nervous systems when they are young. A person's physical capabilities decline with age also, but they start to decline at a later age, and the rate of decline is not nearly as critical to job performance. This is why many people, as they grow older, tend to choose physical jobs over jobs with high information content.[4]

How People Learn

With all of the preceding complexities, how can the learning rate of a person or a group of persons be predicted? The answer is that, although the various factors affect learning, the range of learning rates, as measured by the percent learning function, seems to lie, generally, between 88 and 92% learning curves for those parts of a job that have to be learned. If one has older people in the work force, then it would probably be desirable to choose the higher number.[1]

Computation of the number of days to learn a job may provide estimates that are considerably shorter than those that are currently used in many situations. The reason for the difference will vary, but one important aspect is the method used to teach the people "threshold learning." This is the learning that occurs prior to the worker's barely knowing how to do the job without external assistance. Threshold learning is not included in the learning curve equations because the methods used are so variable that they cannot be predicted with any accuracy. (See Chapter 5.2 for the effect of training methods.) Thus the threshold learning time has to be added to the time predicted by the equations. The learning curve equations contain the "conditioned learning time," which is the time that it takes the worker to learn after he or she barely knows how to do the work.

Observation and measurement of threshold times have provided considerable evidence that trial-and-error learning, where the worker is told to figure out how to do the job himself or herself without any formal instruction, results in the longest threshold learning times. Unfortunately, this is the method most commonly used in industry. Considerable cost savings are possible by using formal instruction given by people who are thoroughly familiar with the operations to be performed and who have the ability to teach the method sequences to the worker properly.

Although the learning curve equations suggest that people learn at a smooth rate, this is not usually the case. Exhibit 4.3.6 gives an example of a person learning a task. The actual curve appears jagged, primarily because the number of eye movements are decreasing, but at an uneven rate

Exhibit 4.3.6 A Comparison of Actual Learning With the Predicted Learning Rate

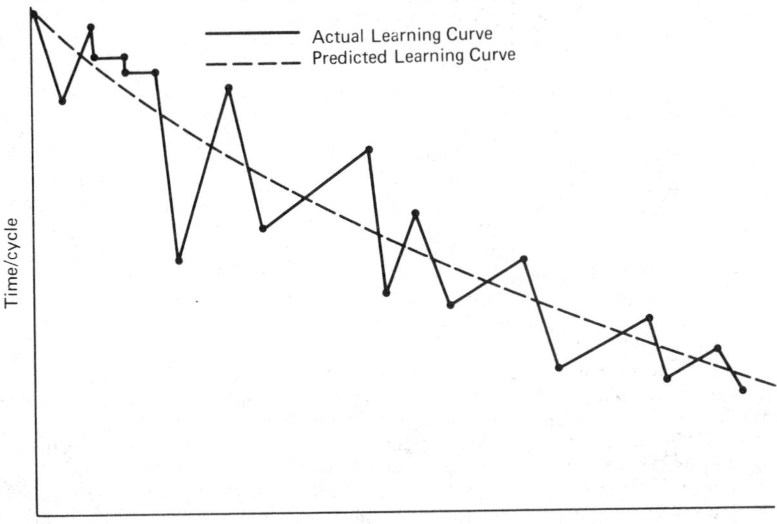

during the learning process. When humans begin to perform a motor task, they need to use their eyes very frequently to get information; as they continue to repeat the task, they "chunk" information. Chunking is a process whereby people get more information per eye fixation and thus need to use their eyes less and less with increasing numbers of cycles. The jagged curve occurs because in the process of attempting to reduce eye fixations, people frequently reduce the number too much and then have to slow down, and when this happens, they will use more eye fixations in the next cycle.

People not only attempt to reduce cycle times by chunking information, but also attempt to get the information they need by using their lower-order senses, especially their kinesthetic (sense of position) and tactile sense (sense of touch). The motivation for this effort is that the use of the eyes consumes a large amount of time as compared to the use of the lower-order senses. Observations of experienced workers performing a task reveal that they use their eyes a minimum amount. Therefore the industrial engineer can tell the experience level of a worker by watching his or her eyes. The worker who can look all around and still work without interruption is a skilled worker. This situation bothers many managers because they feel that the workers are not paying attention to what they are doing and sometimes discipline them for it. Of course, if the workers respond to the discipline and use their eyes more often, their productivity will probably decrease.

Effect of Human Learning on Development of Time Standards

An understanding of human learning is very important for the industrial engineer responsible for setting production standards. For the engineer using time study, it is especially important because of the consequence of conducting a time study on an inexperienced worker. Referring to example 1 in Section 4.3.2, the average time for the first 50 cycles was 6.43 min, whereas the average time for the next 150 cycles was 4.81 min (example 6, Section 4.3.2). Thus a time study conducted on a worker with limited practice opportunity will result in a loose standard, unless that worker or any other worker will not have any more practice opportunity than when the time study was conducted.

The situation is further complicated by the fact that, while a person is learning, he or she appears, to the industrial engineer who is not knowledgeable about learning curves, to be working at a fast pace. The results are much faster pace rating than should be given, which inflates the standard further. Using example 2 in Section 4.3.2, where the fiftieth cycle is predicted to be 5.52 min, a person at the point on the curve would usually appear to be exerting much effort. If the industrial engineer were to rate the person at 130, then the standard would be 5.52 × 1.30 = 7.18 min! The best procedure is for the time study engineer to estimate the pace, which will usually be fast, and the skill level, which will usually be low (too many eye fixations, hesitations, and possible fumbles), and to average the two estimates.

If we assume, as many time study engineers do, that the standard applies to an experienced person, then how many cycles must be performed before a person becomes experienced? The cycle

Exhibit 4.3.7 Reduction in Cycle Time as a Function of the Number of Cycles Performed Where $K = 10$ and for a 90% Learning Curve

Number of Cycles	Cycle Time
1	10.0
100	5.0
1,000	3.5
10,000	2.5
100,000	1.7
1,000,000	1.2
2,000,000	1.1
3,000,000	1.0

time, of course, decreases and does not reach a constant time. Studies of people doing repetitive-cycle work for several million cycles find that the cycle time continues to decrease, as shown in Exhibit 4.3.7. Thus a time standard presumes certain run lengths. If the run lengths are longer, the standard will be loose, and if the run lengths are shorter, the standard will be tight. If run length stays constant, but turnover increases, the standards will also be harder to achieve.

What, then, is an appropriate procedure to follow? Perhaps the most practical method for many types of operations where time study is used is the following[5,6]:

1. Determine the acceptable period in which a worker should have to be able to attain the standard. (The example to follow shows the importance to the resultant standard time of establishing this period).
2. Subtract from step 1 the time the worker should take to start the conditioned learning phase.
3. Take two or more time studies at different numbers of cycles for several workers who are in the conditioned learning phase. Performance rate the studies, being very careful to compensate for the higher effort and lower skill that occurs during learning. Also be careful to record the cycle number where the time studies were taken.
4. Obtain the values of A and K for the learning curve by using the following equations, where Y_1 and Y_2 are the normal times at X_1 and X_2 cycles, respectively.

$$Y_1 = KX_1^{-A}$$
$$Y_2 = KX_2^{-A}$$

5. Solve equation 4 for N_2, where C = the conditional learning time found in step 2. Assume $N_1 = 1$.

$$N_2 = \left[C\frac{(1-A)}{K} + 0.5^{1-A}\right]^{1/(1-A)} - \frac{1}{2} \tag{7}$$

6. The standard time is the cycle time at N_2. We obtain the standard time using equation 1, where $N_2 = X$.

As an example of the preceding methodology, suppose we have the following data:

normal time at 50 cycles = 6.23 min
normal time at 250 cycles = 5.01 min
conditioned learning time = 2400 min (40 hr)

Then $6.23\,(50^A) = K$, and $5.01\,(250^A) = K$.

$$\frac{6.23}{5.01} = \frac{250^A}{50^A}$$

$$\log\frac{6.23}{5.01} = A \log 250 - A \log 50$$

$$A = 0.1354$$

Substituting,

$$K = 6.23\,(50^{0.1354})$$

$$K = 10.5810$$

solve equation 7 for N_2.

$$N_2 = \left[2400\ \frac{(1-0.1354)}{10.5810} + 0.5^{1-0.1354}\right]^{1/(1-0.1354)} - \frac{1}{2}$$

$$= 449.20 \text{ cycles}$$

The time standard will then be $Y = 10.5810(449.20)^{-0.1354} = 4.63$ min.

Prediction of Learning Rates Using Predetermined Time Systems

Of the predetermined time systems, the MTM-1 system has a methodology to determine how long conditioned learning should take to attain the standard. The methodology is explained in detail elsewhere,[1,7] and there are computer programs available from the MTM Association to aid in application. Linear prediction equations are used for most of the MTM-1 elements. They can easily be added to attain linear approximations to equation 1. The rate of learning is a function of the relative frequency of the various MTM motions in the standard. Research on the learning rates of the various MTM motions has found that the higher-order motions, such as grasp and position, take much longer to learn than the lower-order motions, such as reach and move. The methodology provides for submotion sequences that are repeated within a cycle, as well as those subsequences where practice opportunity exists from other work situations. A methodology for handling repeated work opportunities while the worker is learning is also included.

Once the prediction equations are developed for a particular application, the number of cycles to standard, the cycle time for a given cycle, and the average time for a number of cycles can be computed. In the MTM-1 methodology, the value of K is usually assumed to be 2.5, whereas in the time study methodology, K is usually computed. The value of K is assumed to be 2.5 because in the industrial studies of the application, the learning curves have indicated that the first cycle is usually performed at a cycle time that is approximately 2.5 times the MTM-1 standard. The methodology can also be used if another value of K is found to be more appropriate.

Determining the Standard and the Learning Times for a Work Group

One of the situations that may be encountered in the application of learning curves by time study is that the normal standard time, or the MTM-1 standard time, is what is known as "low task" time. That is, they are standard times that can be attained and sustained for long periods by a large majority of the healthy, experienced working population.

Exhibit 4.3.8 is a histogram of the relationship between the standard time and the performance of the working population while working at the maximum sustainable pace. Different authors have

Exhibit 4.3.8 The Relationship Between the Motivated Performance of the Work Force and Low Task Standards

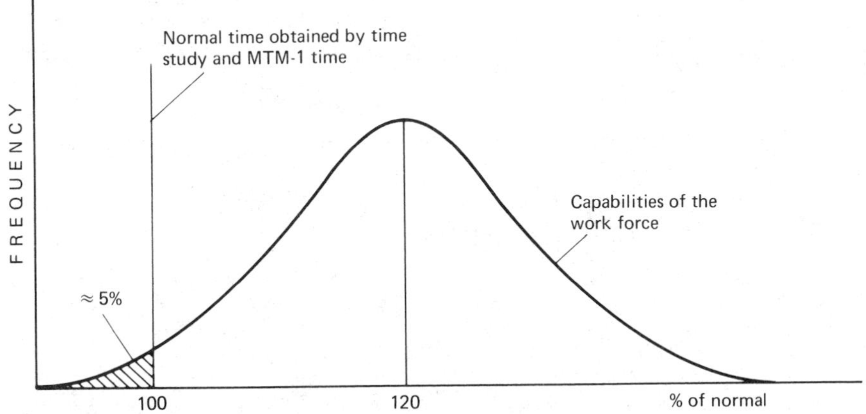

slightly different histograms of the average output of the work force under high-motivation conditions. Exhibit 4.3.8 is derived from research data and industrial experiences, where the allowances are less than or equal to 5%.[8] Other authors [9,10] estimate the average motivated performance of the work force to be from 125 to 130%. Unfortunately, the allowances that are used are not stated, but they are probably in the 10 to 15% range.

Since a large percentage of the population can equal or exceed the normal times, most of the people who have just learned sufficiently to attain the normal times will not be fully learned. Thus the number of cycles found in the time study example of 449.20 cycles is the number of cycles that the average person should take to attain the low task standard. Since in a statistical sense 50% will learn faster and 50% slower, the following questions might be asked:

1. How long will it take to learn the job so that the large majority (95% or more) can attain the standard of 4.63 min?
2. What should be the normal time standard if 4.63 min is the average accomplishment after 4 cycles, and if everyone should be able to accomplish the standard at this point?

The answer to question 1 is as follows:

$$\frac{4.63}{1.20} = \text{average population capability} = 3.86 \text{ min}$$

$$3.86 = 10.5810X^{-0.1354}$$

$$X = 1716 \text{ cycles}$$

Thus 1716 cycles is the estimate of the number of cycles that it would take for approximately 95% of the work force to attain at least 4.63 min.

We then use equation 4 to get the cumulative time:

$$C = 10.5810\left[\frac{(1716 + 1/2)^{1-0.1354} - (1 - 1/2)^{1-0.1354}}{1 - 0.1354}\right]$$

$$= 7656 \text{ min}$$

Thus 7656 min, or 127.60 hr, will be required for 95% of the people to do the task at the standard time of 4.63 min.

Question 2 is answered by assuming that, when the average of the group is 4.63 min, the poorest learner will be 20% slower, or 4.63 × 1.20 = 5.56 min. Thus, if a large percentage of the group should attain the standard in 2400 min, the standard should be 5.56 min.

4.3.4 PRODUCTION PROGRESS FUNCTIONS

Production progress functions are a method of measuring and estimating the rate at which an active organization learns to produce a product. This type of learning has been found to follow the same negative power functions that are used for human learning. However, the learning rate is considerably faster, the most common rate being an 80% curve. Also, the dependent variable is usually cost per unit of time per unit.

A number of studies have been reported in the literature concerning the percent production process found in various industries. De Jong offers the most comprehensive list,[6] which is summarized in Exhibit 4.3.9. The factors that caused the large reductions in unit times in the industries listed varied from situation to situation. The following were the major reasons:

1. Organizational improvements.
2. Improvement in the dimensions of the pieces to be assembled.
3. Improvement in work methods.
4. Improvements in the means of production (new machines).
5. Increase in the skill of the employees (human learning).

Production progress thus occurs in many industries at a rapid rate. It continues for large numbers of units. Although the examples given in Exhibit 4.3.9 are in terms of time per unit, many organizations experience the same percentage production progress rates where costs per unit are plotted versus number of units. In a real sense, production progress is the antithesis of standard costs, because with production progress causing costs to drop 20% or more every time the units are doubled, standard costs do not stay standard for very long. Knowledge of production progress

Exhibit 4.3.9 A Sample of the Percent Production Progress Found in Various Industries[a]

Source	Percent Production Progress
Volkswagen, 1945 to 1949	40
Volkswagen, 1950 to 1954	20
Twenty light alloy products	20
Repair of goods wagons	20
Home construction	14–27
Welding of thin steel	30
Airplane production	25–30
Shipbuilding	10–26
Vehicle bodies	20–30
German armament industry	18–35
Railway carriages	7–25

[a]The percent learning curve would be 100 minus percent production progress.

functions is very important to those who are involved in bidding, cost analysis, and the pricing of products. For example, suppose we have two competing companies, both of which decide to produce a very similar product. Let us further assume that both companies start production at the same time; however, company A experiences a 75% progress curve, and company B an 80% curve, as depicted in Exhibit 4.3.10. It is obvious, without doing any computations, that company A will have lower costs per unit, provided that company A produces at the same rate as company B.

The equations presented previously can be used to compute a number of factors as follows:

1. What is the unit cost of the two-hundredth unit of company B if $K = 100$? Using equation 1, with $A = 0.3219$ for an 80% curve,

$$Y = 100(200)^{-0.3219} = \$18.17$$

Exhibit 4.3.10 A Plot of 75% to 80% Production Progress Curve

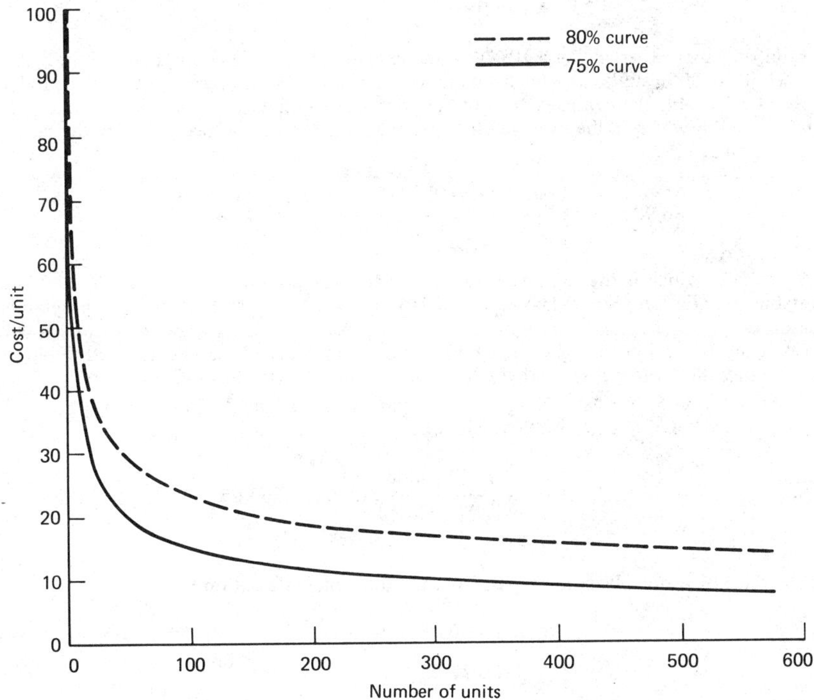

2. What would be the total cost of the first 200 units of company B if $K = 100$? Using equation 4,

$$C = 100 \left[\frac{(200 + 1/2)^{1-0.3219} - (1 - 1/2)^{1-0.3219}}{1 - 0.3219} \right]$$

$$= \$5275.31 = \text{total cost of the first 200 units}$$

3. What would be the average cost of the first 200 units of company B? Using the result in step 2 and dividing by 200,

$$\frac{5275.31}{200} = \$26.38$$

4. Production is planned for 20,000 units next year. At the beginning of the year, we will have already produced 3000 units. What should be the budget for company B for this product? Using equation 4,

$$C = 100 \left[\frac{(23{,}000 + 1/2)^{1-0.3219} - (3001 - 1/2)^{1-0.3219}}{1 - 0.3219} \right]$$

$$= \$100{,}163.31$$

5. What should be our unit cost in step 4?

$$\frac{100{,}163.31}{20{,}000} = \$5.01 = \text{unit cost}$$

6. There is a change in planning for company B. Instead of 20,000 units, 40,000 units are planned for the next year. What will be the unit cost? Using equation 3,

$$\text{average} = 100 \left[\frac{(43{,}000 + 1/2)^{1-0.3219} - (3001 - 1/2)^{1-0.3219}}{(1 - 0.3219)(43{,}000 - 3000)} \right]$$

$$= \$4.27 = \text{the unit cost}$$

7. Company B is experiencing a 10%/year increase in the cost of materials and labor. If the unit cost is \$7.60 at 3000 units, how many units would have to be produced so that the production progress would enable the company to continue to sell the product at \$7.60? Solution: $X \times 1.10 = \$7.60$, $X = \$6.90$, which is the average unit cost that will have to be realized over the next year. Using equation 3,

$$\$6.90 = 100 \left[\frac{(N_2 + 1/2)^{1-0.3219} - (3001 - 1/2)^{1-0.3219}}{(1 - 0.3219)(N_2 - 3000)} \right]$$

$N_2 - N_1 = 2240$, which is the number of units needed to be produced to counter the cost increase attributable to inflation. *Note:* The equation is best solved by iteration, using a programmable calculator.

8. Company B is a job shop producing 3000 units. Three months later it is asked to bid on the production of 2500 more units. What would be the estimated costs for the 2500 units?

a. Using equation 5, but using dollars for the units of S, R, and K, and assuming S is \$6.08 at $X_1 = 6000$, the value of R at $X_2 = 3001$ would be

$$R_1 = 100 - \left(\frac{100 - 6.08}{6000} \right)(3001) = \$53.02$$

where $CS = X_1$.

b. The value of A must be recomputed from the following calculation:

$$6.08 = 53.02\,(6000 - 3000)^{-A}$$

$$A = 0.2705$$

c. Using equation 4, where $R = K, N_2 = 2500$, and $N_1 = 3001$,

$$C = \frac{53.02}{0.7295} [(2500 + 1/2)^{0.7295} - (1 - 0.5)^{0.7295}]$$

$$= \$21{,}847.64 \text{ for 2500 units.}$$

REFERENCES

1. WALTON M. HANCOCK and PRAKASH SATHE, *Learning Curve Research on Manual Operations*, Research Report 113A, MTM Association for Standards and Research, Fair Lawn, NJ, 1969.
2. DON B. CHAFFIN and WALTON M. HANCOCK, *Factors in Manual Skill Training*, Research Report 114, MTM Association for Standards and Research, Fair Lawn, NJ, 1966.
3. ROBERT B. CLIFFORD and WALTON M. HANCOCK, "An Industrial Study of Learning," *Journal of Methods Time Measurement*, Vol. 9, No. 3, pp. 12–27.
4. A. T. WELFORD, *Aging and Human Skill*, Oxford University Press, New York, 1958.
5. J. R. DE JONG, "Increasing Skill and Reduction of Work Time," *Time and Motion Study*, September 1964, pp. 28–41.
6. J. R. DE JONG, "Increasing Skill and Reduction of Work Time–Concluded," *Time and Motion Study*, October 1964, pp. 20–33.
7. WALTON M. HANCOCK, "The Learning Curve," in H. B. Maynard, Ed., *Industrial Engineering Handbook*, 3rd ed., McGraw-Hill, New York, 1971, pp. 7-102–7-114.
8. WALTON M. HANCOCK and ULF ÅBERG, *Design Criteria of Predetermined Time Systems with Special Reference to the MTM System*, International MTM Directorate, Stockholm, Sweden, 1968.
9. RALPH M. BARNES, *Motion and Time Study, Design and Measurement of Work*, 5th ed., Wiley, New York, 1964, p. 324.
10. MARVIN E. MUNDEL, *Motion and Time Study Principles and Practices*, 4th ed., Prentice-Hall, Englewood Cliffs, NJ, 1970, p. 303.

BIBLIOGRAPHY

ABRAMOWITZ, J. B., and G. A. SHATTUCK, JR., *The Learning Curve, A Technique for Planning, Measurement and Control*, Report No. 31.101, IBM, 4th ed., Harrison, NJ.

BARNES, RALPH M., and HAROLD T. AMRINE, "The Effect on Practice of Various Elements Used in Screwdriver Work," *Journal of Applied Psychology*, April 1942, pp. 197–209.

KARGER, D. W., and F. H. BAYHA, *Engineered Work Measurement*, 3rd ed., Industrial Press, 1977.

NANDA, R., and G. L. ALDER, *Learning Curves, Theory and Application*, American Institute of Industrial Engineers, Norcross, GA, 1977.

CHAPTER 4.4
Time Study

BENJAMIN W. NIEBEL
The Pennsylvania State University

4.4.1 DEFINITION AND OBJECTIVES

Time study is a technique of establishing an allowed time standard for performing a given task, based upon measurement of the work content of the prescribed method, with due allowance for fatigue and for personal and unavoidable delays. The objective of time study is to determine reliable standards for all work, both direct and indirect, being undertaken by the enterprise for the efficient and effective management of the operation.

With reliable time standards, work can be scheduled so as to maximize output over time, thus obtaining high utilization of both labor and equipment. A system of variance reporting can be introduced, thus simplifying good management. Management can investigate the differences between actual and standard times and take appropriate action where necessary.

Standard times facilitate methods engineering. Since time is a common measure for all jobs, time standards are a basis for comparing various methods of doing the same piece of work.

Standard times serve as a basis for wage incentive plans. Certainly it would be impractical to introduce any incentive system where the worker is rewarded in proportion to output, without having standard times. Furthermore, standard times serve as a means of securing an efficient layout of the available space. Since time is the basis for determining how much of each kind of equipment is needed, accurate time standards provide a means of determining plant capacity and of balancing the work force with the available work. They provide a basis for purchasing new equipment and for improving production control.

A further objective of the development of reliable time standards is to initiate the procedure of accurate cost determination in advance of production.

Other good management practices that may be enhanced with the application of measured time standards include budgetary control, the development of supervisory bonuses, and the ensuring of the maintenance of quality requirements.

4.4.2 TIME STUDY EQUIPMENT

The equipment needed to develop reliable standards is minimal. All that is needed is an accurate, reliable stopwatch; a well-designed work study form; and an electronic calculator to work up the study.

Several types of stopwatches are in use today. The majority of those being used fall into one of the following four classifications:

1. Electronic stopwatch.
2. Decimal minute stopwatch (mechanical–0.01 min).
3. Decimal hour stopwatch (mechanical–0.0001 hr).
4. Decimal minute stopwatch (mechanical–0.001 min).

Each of these stopwatches has advantages and disadvantages, depending upon the nature of the operation being studied. Instead of a firm's using just one type of watch to establish standards, it is usually desirable to have at least two of the types on hand.

In addition to these types of stopwatches, there are several time recording devices that offer some advantages over the stopwatch. These include time recording machines, motion picture cameras, and videotape equipment.

Totally electronic stopwatches provide resolution to one hundredth of a second and accuracy to 0.003%. They weigh about 0.25 kg and are about 13 cm long by 5 cm wide and 5 cm deep. They permit the timing of any number of individual elements, while also measuring the total elapsed time. Electronic stopwatches operate on rechargeable batteries. Typically the batteries must be recharged after about 14 hr of continuous service. The only disadvantage, other than cost, is that some difficulty may be encountered in reading the display of the electronic stopwatch in studies made in the bright sunlight. The limited battery life may result in an untimely interruption of a study.

The decimal minute stopwatch (mechanical–0.01 min) tends to be a favorite with the majority of time study analysts. It is economical, portable, and dependable over long periods. It has 100 divisions on its face, and each division is equal to 0.01 min. Although the analyst must read a moving hand when conducting continuous-type studies, it is relatively easy to read accurately in view of the size of the dial and the speed of the sweep hand.

There is a special adaptation of the decimal minute watch that many time study analysts find convenient to use when taking continuous-type studies since it allows reading a stopped hand. This watch has two sweep hands that move simultaneously from zero when the watch is started. At the termination of the first element a side pin is depressed, which will stop the lower sweep hand only. The analyst can now read the elapsed time of the element being measured while the upper hand is continuing to measure the cycle time. The analyst will then depress the side pin, and the lower hand will rejoin the upper hand, which has been moving uninterruptedly.

The decimal hour stopwatch (mechanical–0.0001 hr) is similar to the decimal minute stopwatch, except for the unit of time. If the industry or business prefers to issue standards in terms of decimal hours per piece, then this watch would be advantageous over the decimal minute (0.01) type.

The decimal minute stopwatch (mechanical–0.001 min) is a specialty watch used for timing only one element of a cycle or a portion of a cycle. In this watch each division is equal to 0.001 min. Since the sweep hand moves rapidly (6 sec to circle the dial), the analyst always stops the watch immediately at the end of the element being measured so that the watch can be read. This watch is useful in developing standard data.

Time recording machines are helpful devices that may be used in the absence of a time study analyst to measure the time that a facility is productive. The machines include chart paper on which a stylus continuously records the state of the machine. Sensors will close only when the machine or activity is productive. Time recording machines are available with multiple channels so that the state of several machines can be recorded continuously over the work day.

Videotape and motion picture camera equipment are useful in time study work. They are especially helpful in recording operator methods and elapsed time. However, the cost of film and the delay necessitated by having to send the film out to be developed prohibit the use of the motion picture camera in many instances. The high initial cost of quality videotape equipment may preclude its use.

Both of these picture taking methods are especially useful in establishing standards by one of the fundamental motion techniques, such as MTM or Work-Factor. By taking pictures of the operator at his or her work station, and then studying them in detail a frame at a time, the analyst can record the exact details of the method used and assign basic motion time values.

It is important to use a well-designed form for recording elapsed time and working up the study. All details of the study should be recorded on the form. This can be done by setting up an operator process chart (right- and left-hand chart) on the form, as shown in Exhibit 4.4.1. In addition to providing a permanent record of the tools and materials in the work area, the form should accommodate such methods data as feeds, depths of cuts, speeds, tool type and geometry, and inspection specifications. The form should also include space for the operator's name and number, operation description, machine name and number, special tools used and their respective numbers, department where the operation is performed, and prevailing working conditions.

The form should be designed so that the analyst can conveniently record watch readings, foreign elements, and rating factors, and still use the sheet to calculate the allowed time. Exhibit 4.4.2 shows the front of a time study form developed to accommodate the study of any type of operation.

4.4.3 REQUIREMENTS FOR EFFECTIVE TIME STUDY

Several fundamental requirements must be met before a time study is undertaken. First, the operator should be thoroughly acquainted with the method to be followed before the study is conducted. This method should be endorsed by the industrial engineering department and be standardized at all points at which it is to be used before the study begins. Furthermore, the union steward, the departmental foreman, and the operator should be advised that the job is to be studied.

The foreman should check the method in advance of the study to ensure that the correct tooling is being used; that the tools have the correct geometry; that the correct feeds, speeds, and depths

Exhibit 4.4.1 Back of Time Study Form

SKETCH

STUDY NO. ________ DATE ________

OPERATION ________

DEPT. ________ OPERATOR ________ NO. ________

EQUIPMENT ________

________ MCH. NO. ________

SPECIAL TOOLS, JIGS, FIXTURES, GAGES ________

CONDITIONS ________

MATERIAL ________

PART NO. ________ DWG. NO. ________

PART DESCRIPTION ________

ACT BREAKDOWN	
LEFT HAND	RIGHT HAND

ELEM. NO.	SMALL TOOL NUMBERS, FEEDS, SPEEDS, DEPTH OF CUT, ETC.	ELEMENTAL TIME	OCC. PER CYCLE	TOTAL TIME ALLOWED

EACH PIECE ________ TOTAL

SET-UP ________ HRS. PER C

FOREMAN ________ INSPECTOR

OBSERVER ________ APPROVED BY

Exhibit 4.4.2 Front of Time Study Form

DATE / /
STUDY NO.____
SHEET NO.____
OF ____
SHEETS

ELEMENTS

NUMBER																																								
NOTES	CY. NO.	T	R	F	T	R	F	T	R	F	T	R	F	T	R	F	T	R	F	T	R	F	T	R	F	T	R	F	T	R	F	T	R	F	T	R	F	T	R	F

SUMMARY

TOTALS													
OBSER.													
AVE. TIME													
LEV. FACT													
L.F. X AVE.T.													
% ALL.													
TIME ALLOWED													

REMARKS

FOREIGN ELEMENTS

SYM	R	T	DESCRIPTION
A			
B			
C			
D			
E			
F			
G			
H			
I			
J			
K			

RATING CHECK

SYN. VAL. = ____
OBS. VAL. = ____ = ____ %

ALLOWANCE SUMMARY

PERSONAL	
UNAVOIDABLE	
FATIGUE	
TOTAL ALLOWANCE %	

STUDY STARTED	STUDY FINISHED	OVERALL TIME

of cut are being used; that lubrication is being applied in accordance with specifications; and that material of the proper quality is on hand.

If several operators are available for study, the foreman should determine to the best of his or her ability which one will provide the most satisfactory study.

The union steward should ensure that only trained, competent operators are selected for study. He or she should accept the responsibility for explaining to the operator why the study is being undertaken and should answer questions posed by the operator regarding the study.

The time study procedure is a sampling technique whereby a random sample of data is collected and analyzed to determine a value that has a significant impact on the operator, on his or her foreman, on the success of the product, and on the success of the company. Because of the importance of the work measurement procedure, the following personal characteristics can be considered essential for the successful work measurement analyst: good judgment, analytic ability, honesty, resourcefulness, self-confidence, tact, patience, optimism, pleasing personality, enthusiasm, and neat appearance. The analyst should also regularly meet the following requirements:

1. To study the existing method carefully before conducting the time study to ensure that the method is correct.
2. To review with the foreman all aspects of the operation in order to obtain his or her endorsement of the tooling, materials, and procedure being utilized by the operator.
3. To respond to any questions posed by the operator, the union steward, or the foreman regarding the time study procedure.
4. To record on the time study form all details of the method under study.
5. To record accurately an adequate sample of element times in order to allow the development of an equitable standard.
6. To evaluate the performance of the operator honestly and fairly.
7. To apply an appropriate allowance to all normal times where applicable.
8. To calculate accurately the standard times for each element of the time study.
9. To conduct himself or herself in such a manner that he or she will obtain and hold the respect and confidence of the representatives of both labor and management.

4.4.4 SELECTING THE OPERATOR TO BE STUDIED

In beginning a time study the initial approach is made through the departmental foreman or line supervisor. The foreman will then advise the union steward that a time study is to be conducted on his or her operation or operations. The job should be reviewed by the time study analyst and the foreman, who must then agree that the operation is ready to be studied from a methods engineering viewpoint.

Often, more than one operator will be performing the operation that is to be studied. When this is the case, the operator selected should be one who represents above-average or somewhat above average performance within the group. This operator should be well trained and experienced in the proper method. He or she should have demonstrated during the prestudy method observation by the time study analyst and the foreman that he or she performs the work systematically and consistently. The operator selected should be familiar with time study procedures and practice and should have confidence in time study methods as well as in the time study analyst. The operator should have a cooperative spirit and accept positive suggestions given by the foreman and the analyst.

Sometimes the time study analyst will not be able to select an operator for study since only one worker will be performing the job to be studied. In these cases the analyst will need to be especially careful in rating the operator's performance since he or she may be performing at either end of the rating scale, and there will be no other operator to study in order to help validate normal performance.

Once the operator to be studied has been determined, he or she should be approached in a friendly manner and be informed that the operation is to be studied. He or she should be given the opportunity to ask any questions as to the timing procedure, the method of performance rating, and the application of allowances. In cases where the operator has not been studied before, it is sound practice to explain the procedure patiently. It is important that a good rapport exist between the worker and the time study analyst. The analyst should strive to obtain both the confidence and the respect of the worker.

4.4.5 ANALYSIS OF METHODS AND MATERIALS

Complete details of the method being utilized and the specifications and conditions of all materials must be carefully recorded on the time study form. This is extremely important since any change in method is usually justification for restudying the job. If the method under study is not positively identified, any future method improvement may not be studied, and this will lead to loose rates and all the problems related to loose time standards.

The time study should include a sketch of the work area drawn to scale. This sketch will help identify the method under study since it will show the position of materials, tools, fixtures, and so on, in relation to the operator. An operator process chart (right- and left-hand chart) of the method being studied should be recorded below the sketch. It is desirable to complete this chart before one begins to record the data.

Complete information should be recorded concerning the machines, tooling, handtools, jigs or fixtures, gauges, materials used, and working conditions, since each of these factors has an impact on method. It is also desirable to describe carefully the operation being performed and to record the name and clock number of the operator, the department in which the study is conducted, the date of the study, and the name of the time study analyst. The completed time study becomes a source of much valuable information for developing standard data, for designing time formulas, and for the method improvement of other work. It has this complete utility only if all the important details are identified and recorded.

4.4.6 DIVIDING THE OPERATION INTO ELEMENTS

The operation being studied should be divided into groups of therbligs known as "elements." An element is a division of work that can be measured with stopwatch equipment and that has readily identifiable terminal points. To divide the operation into its individual elements, the time study analyst should carefully observe the operator for several cycles. If the cycle time is relatively long (over 30 min), the analyst will need to observe only one or two cycles in order to break the job down into its elements. In extremely long cycles the analyst will be able to write the description of the elements while conducting the study. However, it is desirable to determine, before the start of the study, the elements into which the operation is to be divided.

It is a good idea to break the operation down into elements that are as fine as possible. This increases the value of the study because of the potential of using the allowed elemental values for standard data. Of course the elements should not be so fine that accuracy of reading the watch will be affected. Elemental divisions of about 0.04 min are about as fine as can be read consistently by the experienced time study analyst. However, if the preceding and succeeding elements are relatively long, an element as short as 0.02 min can be readily timed.

To identify elemental end points, and to develop consistency in reading the watch from one cycle to the next, sound as well as sight should be considered in elemental breakdown. In most cases terminal points can be identified by both sound and sight. The sound of laying a tool down and the sound that is heard when a tool begins a cut or finishes a cut are examples of terminal points that are more easily identified by sound than by sight.

Several basic rules should be followed by the time study analyst in connection with elemental breakdown. These are:

1. Ensure that all the elements being performed are really necessary. If it appears that one or more are unnecessary, the time study should be discontinued and a methods study undertaken in order to develop the proper method.
2. Elements should not combine machine time with manual or nonmachine time.
3. Constant time should not be combined with variable time in the same element. A "constant element" is an element whose performance time does not vary significantly when changes in the process or in the dimensions of the product occur. A "variable element" is one whose performance time is affected by one or more characteristics, such as size, shape, hardness, or tolerances, so that as these conditions change, the time required to perform the element changes.
4. Elements should be selected so that terminal points can be identifed by a characteristic sound.
5. Select elements that are as fine as possible, yet of adequate duration so that they can be accurately measured.

4.4.7 DETERMINING THE NUMBER OF CYCLES TO STUDY

Time study is a sampling procedure. As such, it is important that an adequate-sized sample of data be collected so that the resulting standard is reasonably accurate.

From an economic standpoint, both the duration of the cycle and the activity of the work need to be considred in determining the number of cycles to be observed. Exhibit 4.4.3 can serve as a guide to determining the number of cycles that should be studied in the development of a standard.

The number of cycles that should be studied in order to ensure an adequate sample may be determined statistically. This value, considered along with the guidelines in Exhibit 4.4.3, can help the analyst to decide the length of the observation.

Exhibit 4.4.3 Number of Cycles Required for a Time Study

Cycle Time	Minimum Number of Cycles to Study for Given Activity: More Than 10,000/year	5000 to 10,000/year	1000 to 5000/year	Less Than 1000/year
More than 60 min	6	5	4	3
40 to 60 min	8	7	6	5
20 to 40 min	10	9	8	7
10 to 20 min	12	11	10	9
5 to 10 min	20	18	16	15
2 to 5 min	25	22	20	18
1 to 2 min	40	35	30	25
Less than 1 min	60	50	45	40

It is known that averages of sample $\bar{x}$ drawn from a normal distribution of observations are distributed normally about the population μ. The variance of $\bar{x}$ about the population mean μ equals σ^2/n where n equals the sample size and σ^2 equals the population variance. Normal curve theory leads to the following confidence interval equation:

$$\bar{x} \pm z \frac{\sigma}{\sqrt{n}} \tag{1}$$

Equation 1 assumes that the SD of the population is known. This, of course, is not the case when conducting a time study. However, the SD of the population can be estimated by computing the SD s of a sample taken from the population.

$$s = \sqrt{\frac{\Sigma x_i^2}{n-1} - \frac{(\Sigma x_i)^2}{n(n-1)}}$$

When estimating σ from a sample SD, we are dealing with the quantity $(\bar{x} - \mu)/(s/n^{1/2})$,[1] which is not normally distributed except in cases where the sample size is relatively large ($n > 30$). Its distribution is the student's t distribution. The confidence interval equation then is

$$\bar{x} \pm t \frac{s}{\sqrt{n}}$$

The required number of observations for a given degree of accuracy can be calculated by solving for n as a percentage of $\bar{x}$. Thus $k\bar{x} = ts/n^{1/2}$ where k = an acceptable percentage of $\bar{x}$. So if we let N = the number of observations to study and n = the number of observations in the sample that permitted the calculation of $\bar{x}$ and s, we can compute for N from the following equation:

$$N = \left(\frac{st}{k\bar{x}}\right)^2$$

The value of t is, of course, determined from percentage points of the t distribution and is based on the sample size n and the probability (p). For example, an analyst may conduct a study of a job whose cycle time has been estimated at 4 min. Based on Exhibit 4.4.3, the analyst decides to study 25 cycles, since the activity of the work has been estimated to be 10,000/year. The analyst would like to know if this sample size is adequate if he wants assurance that $\bar{x}$ is within ± 10% of μ. Let us assume that the mean of the element with the greatest variation is 0.25 min and that its SD is .05. Then we could calculate as follows:

$$N = \left[\frac{(.05)(2.06)}{(.10)(0.25)}\right]^2 = 16.97 \text{ or } 17$$

Thus the sample of 25 was more than adequate. Note that the value of t in the sample was based on an .05 probability (see Exhibit 4.4.4). For further explanation of the calculation of N, see Niebel, reference.[2]

Exhibit 4.4.4 Percentage Points of the *t* Distribution

	Probability (P)[a]												
n	·9	·8	·7	·6	·5	·4	·3	·2	·1	·05	·02	·01	·001
1	·158	·325	·510	·727	1·000	1·376	1·963	3·078	6·314	12·706	31·821	63·657	636·619
2	·142	·289	·445	·617	·816	1·061	1·386	1·886	2·920	4·303	6·965	9·925	31·598
3	·137	·277	·424	·584	·765	·978	1·250	1·638	2·353	3·182	4·541	5·841	12·941
4	·134	·271	·414	·569	·741	·941	1·190	1·533	2·132	2·776	3·747	4·604	8·610
5	·132	·267	·408	·559	·727	·920	1·156	1·476	2·015	2·571	3·365	4·032	6·859
6	·131	·265	·404	·553	·718	·906	1·134	1·440	1·943	2·447	3·143	3·707	5·959
7	·130	·263	·402	·549	·711	·896	1·119	1·415	1·895	2·365	2·998	3·449	5·405
8	·130	·262	·399	·546	·706	·889	1·108	1·397	1·860	2·306	3·896	3·355	5·041
9	·129	·261	·398	·543	·703	·883	1·100	1·383	1·833	2·262	2·821	3·250	4·781
10	·129	·260	·397	·542	·700	·879	1·093	1·372	1·812	2·228	2·764	3·169	4·587
11	·129	·260	·396	·540	·697	·876	1·088	1·363	1·796	2·201	2·718	3·106	4·437
12	·128	·259	·395	·539	·695	·873	1·083	1·356	1·782	2·179	2·681	3·055	4·318
13	·128	·259	·394	·538	·694	·870	1·079	1·350	1·771	2·160	2·650	3·012	4·221
14	·128	·258	·393	·537	·692	·868	1·076	1·345	1·761	2·145	2·624	2·977	4·140
15	·128	·258	·393	·536	·691	·866	1·074	1·341	1·753	2·131	2·602	2·947	4·073
16	·128	·258	·392	·535	·690	·865	1·071	1·337	1·746	2·120	2·583	2·921	4·015
17	·128	·257	·392	·534	·689	·863	1·069	1·333	1·740	2·110	2·567	2·898	3·965
18	·127	·257	·392	·534	·688	·862	1·067	1·330	1·734	2·101	2·552	2·878	3·922
19	·127	·257	·391	·533	·688	·861	1·066	1·328	1·729	2·093	2·539	2·861	3·883
20	·127	·257	·391	·533	·687	·860	1·064	1·325	1·725	2·086	2·528	2·845	3·850
21	·127	·257	·391	·532	·686	·859	1·063	1·323	1·721	2·080	2·518	2·831	3·819
22	·127	·256	·390	·532	·686	·858	1·061	1·321	1·717	2·074	2·508	2·819	3·792
23	·127	·256	·390	·532	·685	·858	1·060	1·319	1·714	2·069	2·500	2·807	3·767
24	·127	·256	·390	·531	·685	·857	1·059	1·318	1·711	2·064	2·492	2·797	3·745
25	·127	·256	·390	·531	·684	·856	1·058	1·316	1·708	2·060	2·485	2·787	3·725
26	·127	·256	·390	·531	·684	·856	1·058	1·315	1·706	2·056	2·479	2·779	3·707
27	·127	·256	·389	·531	·684	·855	1·057	1·314	1·703	2·052	2·473	2·771	3·690
28	·127	·256	·389	·530	·683	·855	1·056	1·313	1·701	2·048	2·467	2·763	3·674
29	·127	·256	·389	·530	·683	·854	1·055	1·311	1·699	2·045	2·462	2·756	3·659
30	·127	·256	·389	·530	·683	·854	1·055	1·310	1·697	2·042	2·457	2·750	3·646
40	·126	·255	·388	·529	·681	·851	1·050	1·303	1·684	2·021	2·423	2·704	3·551
60	·126	·254	·387	·527	·679	·848	1·046	1·296	1·671	2·000	2·390	2·660	3·460
120	·126	·254	·386	·526	·677	·845	1·041	1·289	1·658	1·980	2·358	2·617	3·373
∞	·126	·253	·385	·524	·674	·842	1·036	1·282	1·645	1·960	2·326	2·576	3·291

Source. Reprinted from Table 3 of R. A. FISHER and F. YATES, *Statistical Tables for Biological, Agricultural, and Medical Research*, Oliver & Boyd, Ltd., Edinburgh, by permission of the authors and publishers.
[a]Probabilities refer to the sum of the two tail areas; for a single tail, divide the probability by 2.

4.4.8 CONDUCTING THE STUDY

Recording Elemental Time Values

There are two methods of recording elemental time values while conducting the time study: the "continuous method" and the "snap-back method." Each has certain advantages and disadvantages.

When using the continuous method, the stopwatch is allowed to run for the entire duration of the study. The watch is read at the breaking point of each element while the hands of the watch are moving. In the snap-back method the watch is read at the termination point of each element, and then the hands are snapped back to zero. As the next element takes place, the hands move from zero. Since elapsed elemental values are read directly from the watch at the end of each element, no clerical time is required for making successive subtractions, as is necessary with the continuous method. Also, with the snap-back method it is simpler to record times of elements that are performed out of order by the operator. A third advantage of this method is that elapsed elemental times during repeated cycles can be readily compared for consistency. Thus the analyst is better able to quickly estimate when he or she has taken an adequate sample of readings.

There are two disadvantages to the snap-back method. First, there is no overall recording of the time taken to perform the study. This information is desirable since it can be used as an overall check on the accuracy of the recording of times and since it represents a value that gives the operator a feeling of the fairness of the study when he or she compares this value with the product of the standard time developed and the number of pieces produced during the duration of the study.

Second, in the snap-back method some time is lost while the hand is being snapped back to zero. It has been estimated that the hand of the watch remains stationary for approximately 0.004 sec. This would be a 5% error in elements of 0.08 min. The new electronic watches do not result in this loss of time since they permit the continual running of the watch, thus resulting in the display of exact elapsed times element by element as each occurs.

The principal advantage of the continuous over the snap-back method is that it gives a complete record of the entire observation period. Consequently, this type of study is more appealing to the union and the operator since it is apparent that no time has been left out of the study and that all delays and foreign elements have been recorded. The continuous method is also more suitable for measuring short elements (0.06 min or less). Since no time is lost in snapping the hand back to zero, accurate values can be obtained on successive elements as short as 0.04 min and on elements of 0.02 min when followed by a relatively long element.

More clerical work is involved in calculating the study when the continuous method is used. Since the watch is read at the breaking point of each element while the hands of the clock continue their movements, it is necessary to make successive subtractions of the consecutive readings in order to determine elapsed elemental times.

Variations in the Established Sequence of Elements

During the course of the study, the analyst will occasionally encounter four types of variations in the sequence of elements that he or she originally established. First, the analyst may miss an element. When this happens, the procedure is to record immediately an "M" in the R column of the time study form (Exhibit 4.4.2). Second, the operator will sometimes omit an element. This should happen very infrequently; it indicates that the operator is inexperienced. When it does happen, the procedure is to record a short, horizontal line through the space in the R column.

A third variation that may occur is the performance of an element or elements in a different sequence from what had originally been recorded. When this happens, the observer should record immediately in the R column of the element being performed out of order the time the element was begun and the time it was completed. This procedure is repeated for each element performed out of order and for the first element that is performed back in the normal sequence.

The fourth variation that is encountered is the introduction, during the course of the study, of elements that had not been anticipated and that do not constitute part of the work cycle. They are referred to as "foreign elements." Foreign elements can occur either during the course of a planned element or between elements. Typical foreign elements include such unavoidable delays as a tool breakage, an inadvertent dropping of a part or a tool, and an interruption by the foreman to query the operator. When a foreign element occurs during a planned element, the practice is to signify the event by an alphabetical designation in the T column of this element. When a foreign element occurs between two elements, the alphabetical designation is recorded in the T column of the work element following the interruption. The letter "A" can be used to identify the first foreign element, the letter B the second, and so on. Most time studies of 30 min or morewill contain several foreign elements.

After the foreign element has been identified with an appropriate symbol in the T column at the point of occurrence, a short description of the element and the time of its completion are recorded in the R column of the foreign element section on the time study form. The duration of each foreign element is recorded when working up the study. The time required for foreign elements, especially those that are unavoidable delays, represents important information in connection with the development of equitable standards. The time consumed by foreign elements is provided for by adding appropriate allowances to the normal time. As will be shown later, these allowances typically range from 15 to 20%. If a study showed an unusual number of foreign elements occurring, the analyst would want to take into consideration their total time in order to develop an allowance that would be fair.

4.4.9 RATING THE OPERATOR'S PERFORMANCE

Perhaps the most important step of the entire time study procedure is rating the operator's performance. The reliability of the standard developed cannot be assured unless the time study analyst has done a precise job of performance rating during the entire course of the study. In the majority of the time studies made, the performance of the operator will deviate from that which has been defined as normal. Consequently, it is necessary that an adjustment be made to the mean observed time in order to determine the time for the normal operator to do the job when working at an average pace. The time taken by the above-average operator must be increased to that required by the normal operator. Similarly, the time taken by the below-average operator needs to be reduced to that required by the normal operator.

Today there is not a universally accepted method of performance rating. The majority of rating systems being used are not completely objective, since they depend largely on the judgment of the time study analyst. It is primarily for this reason that the time study man reflects the high personal qualifications previously cited.

The Concept of Normal Performance

Unfortunately, just as there is no one accepted method of performance rating, there is no universal concept of normal performance. Those industries utilizing highly repetitive motion patterns to produce a highly competitive product, such as the garment industry, will undoubtedly have a different concept of normal operator performance than those job shop industries producing products covered by patents.

A typical description of normal performance developed by the author[3] is

> *A workman who is adapted to the work, and has attained sufficient experience to enable him to perform his job in a workmanlike manner with little or no supervision. He possesses coordinated mental and physical qualities which enable him to proceed from one element to another without hesitation or delay, in accordance with the principles of motion economy. He maintains a good level of efficiency by his knowledge and proper use of all tools and equipment related to his job. He cooperates and performs at a pace best suited for continuous performance.*

It is desirable for a company to identify benchmark operations in terms of normal performance so that consistency in performance rating can be developed by the various time study analysts in the company. For example, an operator carrying a 20 lb load a distance of 25 ft may be expected to take 0.095 min when working at a normal pace. A normal time to index a hex turret of a particular size and style of turret lathe may be established at 0.04 min. A half minute to deal a four-hand bridge deck (52 cards) is considered by many to be typical of normal performance. Having a number of familiar operations identified completely as to method to be employed and to established values of what is considered normal performance does much to sell the validity of the performance rating system.

Application of the Learning Curve

Industrial engineers have long recognized that learning takes time (see chapter 4.3). Consequently, it is much better to conduct a time study on an experienced employee than on a new one who has not become proficient and who would necessarily have to be leveled negatively. In rather simple work, such as light assembly involving relatively few parts, an operator can become very proficient in just a few days. In other situations where work is complex, it may take several weeks before the operator can achieve coordinated mental and physical qualities enabling him or her to proceed from one element to the next without hesitation or delay.

Exhibit 4.4.5 illustrates a typical learning curve. Note that in this case the cumulative average time diminishes slowly after 160 units have been produced. It would be easier to develop a fair performance rating factor if the operator were studied after at least 160 units had been produced. Experience has proved that, when an operator is performing at either a normal or close-to-normal pace, performance rating errors are minimized. Of course there will be many times when it is not convenient to wait for the development of a time standard. Thus it may be necessary for the analyst to study an operation being performed by a new operator who is near the steepest slope of his or her learning curve. At this point the operator's performance may be only 70 or 80% of normal. It is on such occasions that the analyst will need to exercise acute powers of observation and utilize his or her best judgment, based upon thorough training and experience, so as to assign a performance rating factor that results in the computation of an equitable normal time.

It is desirable for a company to plot learning curve data for the various classes of work being performed. This information can be quite helpful in determing when it would be desirable to conduct a study and can also indicate levels of productivity that can be anticipated as additional learning takes place. Data for the learning curve, which tends to be hyperbolic, can be plotted on logarithmic paper, thus linearizing the plotting. Exhibit 4.4.6 illustrates the data from Exhibit 4.4.5 plotted on logarithmic paper.

The analyst should recognize that experience in similar work can provide learning on subsequent and different work. A new turret lathe operator may need a 3 week experience on a job before his or her learning curve beings to "flatten out." However, an experienced turret lathe operator may need only 3 days for his or her learning curve to flatten out on the same job.

Learning curve theory proposes that, when the total quantity of units produced doubles, time per unit declines at some constant percentage. If, for example, it is anticipated that an 85% rate of improvement will be experienced, then, as production doubles, the average time per unit will decline 15%. Exhibit 4.4.7 shows the cumulative average hours per unit of output at various cumulative production levels under an 85% rate of improvement.

Exhibit 4.4.5 Typical Learning Curve

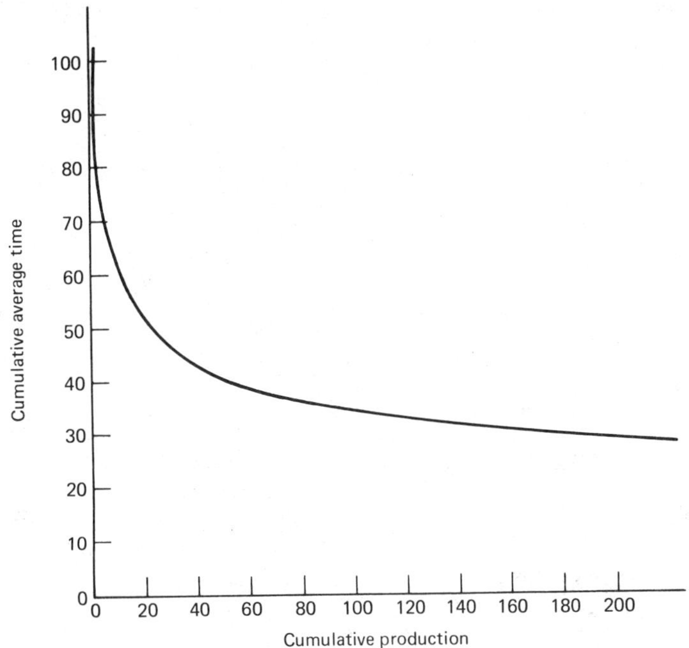

Exhibit 4.4.6 Typical Learning Curve

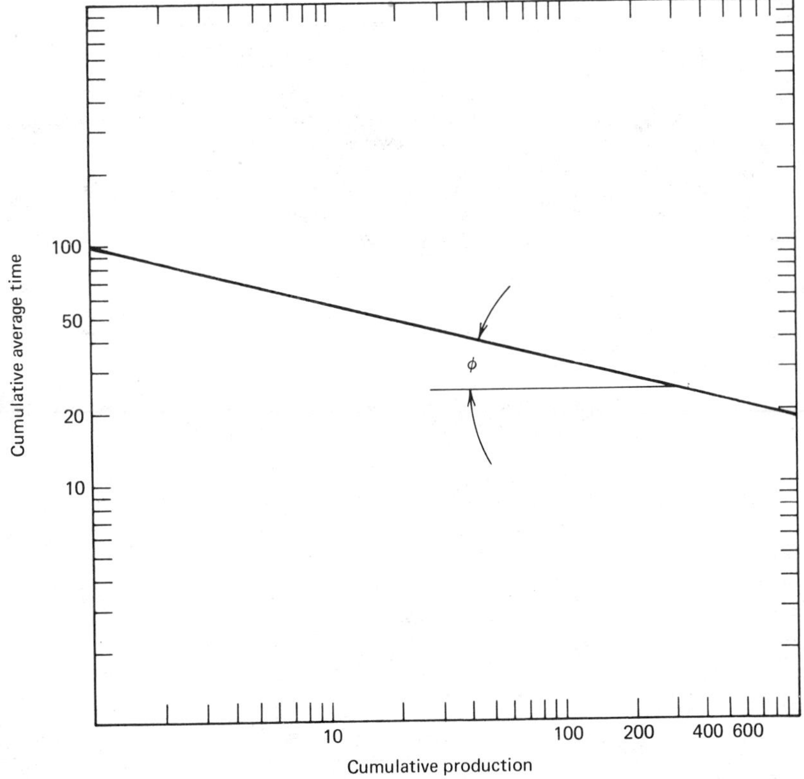

Exhibit 4.4.7 Effects of Learning on Production Line When an 85% Rate of Improvement Occurs

Cumulative Production	Cumulative Average Hours per Unit	Ratio to Previous Cumulative Average
1	100.00	–
2	85.00	85
4	72.25	85
8	61.41	85
16	52.20	85
32	44.37	85
64	37.72	85

Typically the learning curve is a hyperbola of the form $y_x = cx^n$. On log-log paper the learning curve can be presented as $\log y_x = \log c + n \log x$, where

y_x = cumulative average value of x units
c = value in time of the first unit
x = number of units produced
n = exponent representing the slope (tan ϕ in Exhibit 4.4.6)

By definition, the percent learning is y_{2x}/y_x, or $c(2x)^n/c(x)^n = 2^n$. Taking the log of both sides of the equation,

$$n = \frac{\text{log of percent learning}}{\log 2}$$

Thus for 85% learning: n = log of 0.85/log of 2 = –0.070581/0.301029 = –0.2345, and arc tan 0.2345 = 13.1974°.

The slopes of learning curve percentages frequently encountered with conversion factors to calculate unit times are shown in Exhibit 4.4.8.

It has been pointed out that we are working with the cumulative average time per piece as learning takes place. There are times when we want to know the estimated time to produce a specific unit. Thus we would be interested in plotting the "unit learning curve." This curve is asymptotically parallel to the log plot of the cumulative curve once the cumulative number of parts produced reaches some quantity, such as 15. Thus the individual curve on log-log paper curves downward from unit 1 until it becomes parallel to the cumulative average line, which usually takes place somewhere between the tenth and twentieth unit.

The conversion factor to calculate the unit time from the cumulative time can be obtained by adding 1 to the value of n. Thus the conversion factor for a 90% curve would be equal to 1.0000 + (–0.1520) = 0.8480. Exhibit 4.4.8 provides the conversion factors for some of the common learning curves.

An application of the preceding theory may be helpful. Let us assume that a 95% learning curve is anticipated on an assembly operation. An operator takes 8.45 hr to assemble the first unit. We want to know the average time to produce 100 units and the assembly time for the one hundredth unit.

$y_x = (8.45)(100)^{-0.074} = 6.0100$ average time per unit to assemble 100 units
$(6.0100)(1 + (-0.074)) = 5.5652$ hr assembly time for one hundredth unit

Exhibit 4.4.8 Slopes of Common Learning Curves and Conversion Factors (Its Slope) to Calculate Unit Times.

Learning Percentages	Slope	Conversion Factor
95	–0.0740	0.9260
90	–0.1520	0.8480
85	–0.2345	0.7655
80	–0.3219	0.6781
75	–0.4150	0.5850
70	–0.5146	0.4854

Characteristics of a Dependable Rating System

To be successful, a rating system must be reasonably accurate. This means it should not deviate more than ±10% of the correct rating factor (which is never known). Second, the rating system should promote consistency. If an analyst rates consistently high or consistently low, it is relatively easy to train him or her to become more accurate. However, if the system lends to an analyst's rating too high (10% or more) some of the time, it will probably not succeed. Third, the rating system should be simple, easily explained, easily understood, and easily learned.

When to Rate

The rating of the operator's work performance should be done during observation of the elemental times. Since many studies take some time to complete, the frequency of the rating of the operator's performance is of concern. Certainly it is important to rate as often as necessary in order to obtain a good evaluation of the performance being demonstrated.

On short-cycle, repetitive operations where the entire study is completed in 30 min or less, the entire study should be performance rated and the rating factor recorded for each element in the space provided. Power feed or machine-controlled elements are, of course, not rated. Their performance rating factor is always 1.00 since this time will not be adjusted.

When studies are relatively long, more than 30 min, yet many of the elements are of short duration, then it is best to performance rate each cycle of the study.

Sometimes the study will be quite long (more than 30 min), and the majority of the elements relatively long (more than 0.20 min). In such cases, it is probably desirable to performance rate each element of each cycle.

Rating Systems

Westinghouse Performance Rating

A rating system that has had wide application, especially on short-cycle, repetitive operations where performance rating of the entire study takes place, is the Westinghouse performance rating plan. The characteristics and attributes that the plan considers are classified under (1) dexterity, (2) effectiveness, and (3) physical application. These three major classifications do not in themselves carry any numerical weight, but have in turn been assigned attributes that do carry numerical weight. Nine attributes are evaluated under this system: three are related to dexterity, four to effectiveness, and two to physical application. Exhibit 4.4.9 provides values for each of these attributes at various levels of performance.

Dexterity. The first attribute under the classification of dexterity is displayed ability in use of equipment and tools and in assembly of parts. When the analyst evaluates this attribute, he or she considers primarily the effectiveness of the operator after obtaining and having under control those tools and parts utilized in performing the operation. The analyst is careful not to evaluate the "get" (reach, grasp, and move) elements.

The second attribute under the dexterity classification is certainty of movement. Here the analyst gives consideration for positive rating to the freedom from hesitation, pauses, or unecessary long moves.

The third attribute related to dexterity is coordination and rhythm. Here the analyst looks for smoothness of motions as evidenced by freedom from movements of acceleration and deceleration during the course of the operation.

Effectiveness. "Effectiveness," the second major classification, is defined as efficient, orderly procedure. Four attributes are considered under this classification. The first of these is displayed ability to continually replace and retrieve tools and parts with automaticity and accuracy. In evaluating this attribute the analyst notes the degree of freedom from the basic divisions of accomplishment of "search" and "select." For a positive rating he or she expects the worker to repeatedly place tools and materials in specified locations and positions from cycle to cycle. The analyst also looks for the operator to retrieve parts and tools with no hesitation.

The second of the effectiveness attributes is displayed ability to facilitate, eliminate, combine, or shorten motions. In evaluating this attribute the analyst observes the proficiency of the following basic divisions: position, pre-position, release, and inspect. The skilled operator, because of his or her manipulative ability, will be able to perform position in noticeably short times.

The third attribute that is evaluated under the effectiveness classification is displayed ability to use both hands with equal ease. The analyst here gives special attention to the left hand in right-handed operators and vice versa.

The fourth attribute in the effectiveness classification is displayed ability to confine efforts to necessary work. This attribute carries only a negative weight, for no percentage is added when the

Exhibit 4.4.9 Values of Attributes of a Performance Rating System

	Performance Level Values				
Attribute	+ Above		0 Expected	– Below	
Dexterity					
Displayed ability in use of equipment and tools and in assembly of parts	6	3	0	2	4
Certainty of movement	6	3	0	2	4
Coordination and rhythm		2	0	2	
Effectiveness					
Displayed ability to continually replace and retrieve tools and parts with automaticity and accuracy	6	3	0	2	4
Displayed ability to facilitate, eliminate, combine, or shorten motions	6	3	0	4	8
Displayed ability to use both hands with equal ease	6	3	0	4	8
Displayed ability to confine efforts to necessary work			0	4	8
Physical Application					
Work pace	6	3	0	4	8
Attentiveness			0	2	4

work is confined to the necessary work because this condition is expected. Unnecessary work may occur during the course of the study because of inexperience on the part of the operator. Unnecessary work involves unnecessary cleaning of workpiece or work station, unnecessary rechecking of a dimension, unnecessary filing of a part preparatory to placing it in a chuck, and so on. The duration of the unnecessary work is too short to remove from the study as a foreign element and consequently is compensated for in a performance rating factor.

Physical Application. The third major classification, "physical application," has been defined as the demonstrated rate of performance. Two attributes are included under this classification: work pace, and attentiveness. Work pace is rated entirely on the speed of movement, whereas attentiveness is rated as the degree of displayed concentration.

Training. It should be apparent that the Westinghouse rating system requires considerable training. A course involving 30 hr of intensive effort is the minimum that is required of time study analysts who have a background equivalent to at least 2 years of engineering technology. This 30 hr course should involve 25 hr of rating films and of discussing the attributes and the degree to which each is displayed.

The recommended procedure is as follows:

1. A film is shown and the operation is explained.
2. The film is reshown and rated.
3. The individual ratings of those taking the course are compared and discussed.
4. The film is reshown, and the attributes are pointed out and explained.
5. Step 4 is repeated as often as is necessary to reach understanding and agreeement.

Speed Rating

Today speed rating is probably the most widely used rating system. Under this method the time study analyst considers only the rate of accomplishment per unit of time. He or she compares the performance being demonstrated with his or her concept of normal performance for the operation being studied.

Under speed rating, 100% represents normal. If an operator were performing at a pace 25% faster than normal, the rating would be 1.25. Similarly, if the performance were 75% of normal, the operator would be rated 0.75.

To be consistently accurate in using speed rating, the time study analyst must be familiar with the work being studied. He or she should have a variety of benchmarks that are used for comparing with the performance being observed.

Under speed rating a comprehensive training program is recommended before the analyst is assigned to make independent studies. This training program would involve about 25 hr of rating films and live operators and then discussing the results with professional time study engineers who conduct the course.

The practice followed by time study engineers when speed rating is to make two judgments. The analyst first appraises the performance in order to determine whether it is above or below his or her concept of normal. The analyst then determines in what precise position on the rating scale the performance should be placed.

Objective Rating

This rating system was developed by Marvin E. Mundel.[4] It was proposed in order to eliminate the obvious difficulty of establishing a normal speed criterion for every type of work. Under this rating system a work assignment with an agreed performance is established. Then all other jobs are compared to this "base" job as to pace. After an assignment of pace performance, a secondary factor is assigned to the job to allow for its relative difficulty. Those factors influencing the difficulty adjustment that have been assigned are (1) amount of body used, (2) foot pedals, (3) bimanualness, (4) eye-hand coordination, (5) handling or sensory requirements, and (6) weight handled or resistance encountered.

Numerical values have been assigned for a range of degrees of each factor. The sum of the numerical values for each of the six factors constitutes the secondary adjustment. Thus the normal time would be

$$N = (P)\,(D)\,(T)$$

where N = computed normal time for element
P = pace rating factor
D = difficulty adjustment factor
T = mean observed elemental time

Objective rating usually will provide consistency in the rating procedure since comparing the pace of a job to the pace of an operation that one is completely familiar with is more easily accomplished than judging all the attributes of an operation, which the analyst may not be completely familiar with. The secondary factor will not affect consistency since this factor merely adjusts the rated time by a percentage, and the percentage is based on established job difficulties.

Synthetic Rating

With the development of predetermined motion time systems (see Chapter 4.5), some concerns have successfully used the synthetic leveling procedure. Under this procedure a performance factor for several effort elements of the work cycle are established by comparing actual elemental observed times to times developed by fundamental motion data. The mean of these established rating factors is then used as the performance factor of all effort elements composing the study. For example, let us assume an analyst is making a 15-element time study. He or she may choose three of the short elements—such as "pick up 2 lb casting, place in air chuck, and close"; "back off cross slide and index turret"; and "open chuck, remove piece, and place in tray"—as those that will serve as the basis for the established leveling factor. Using fundamental motion data times, the analyst will establish normal times for these three elements. For example, the following could be obtained:

Element Number	Description	Fundamental Motion Data Time	Measured Time	Performance Factor
2	Pick up 2 lb casting, place in air chuck, and close.	0.12	0.11	1.09
7	Back off cross slide and index turret.	0.08	0.07	1.14
15	Open chuck, remove piece, and place in tray.	0.10	0.09	1.11

The mean of the three elemental performance factors would be used as the rating factor to be applied to the entire study:

$$\frac{1.09 + 1.14 + 1.11}{3} = 1.11 \text{ synthetic performance factor}$$

An objection to this rating method is the time required to make a right- and left-hand chart and to assign and summarize the fundamental motion times. One might suggest that it would be more expeditious to compute the normal time for the entire study with one of the fundamental motion data systems. This certainly would be appropriate when the cycle time is relatively short–for example, less than 1 min. However, considerable time would be saved using the synthetic performance rating system described when the cycle times of the effort elements are 5 min or longer.

Guide to Selecting a Rating System

From a practical standpoint, only the speed rating and the objective rating techniques are applicable if one wishes to performance rate each element of each cycle. All the techniques can be readily used where performance rating of each cycle or periodic rating takes place. Certainly the majority of time studies of 30 min or less are taken by applying only one performance factor to all effort elements of the study. (Machine-controlled elements are always rated 1.00.)

In general, the performance rating plan that is easiest to explain, understand, and apply is speed rating when augmented by standard data benchmarks. The standard data benchmarks may be determined from previous standards developed by the stopwatch procedure or from fundamental motion data.

Under speed rating, normal should always be considered to be 100. The range of performance of the typical speed rating scale is 0.60 to 1.40. Seldom would a cooperative operator perform below 0.60, and certainly one would very seldom see performances greater than 140 demonstrated.

Five criteria should be used in order to ensure the success of the speed rating procedure:

1. **Experience by the time study analyst in the class of work being performed.** This does not mean that the analyst need have had experience as an operator in the work being studied, although ideally this would be excellent. It does mean that he or she has experience as an observer and is familiar with all details of the method being utilized. For example, if it is a metal cutting operation, the analyst should be familiar with the tooling, including the tool geometry, feeds, speeds, and depth of cuts, that is characteristic of the operation being performed. He or she should understand the capability of the coolant being used and of the surface finish expected.
2. **Benchmark standard data on two or more of the elements being studied.** Through the development and orderly cataloging of standard data elements, the analyst should know what normal performance is on several of the elements being studied. This information will be a helpful guide to establishing the performance factor of the entire study.
3. **Regular training speed rating by observing representative films or live operators.** It is important that all practicing time study analysts receive regular training in performance rating. This training should involve observation of films or videotape illustrating a variety of operations characteristic of the company. The films or videotape should illustrate the full range of the rating scale–usually from 0.60 to 1.40. The various analysts should make their ratings independently and then discuss their results with other analysts to determine reasons for deviation from the correct rating.
4. **Selection of an operator who has had adequate experience.** If possible, the operator to be studied should have had sufficient experience to have reached the "flat" portion of his or her learning curve–for example, beginning at 60 units in Exhibit 4.4.5. He or she should be identified as a cooperative employee, who in the past has performed regularly at normal or above normal.
5. **Use of the mean value of three independent studies.** It is good practice when establishing standards on high-production jobs to conduct more than one study before arriving at the standard. The total error due to both the performance rating and the determination of an average elemental elapsed time that deviates from the population mean is reduced when the averages of several independent studies are used in computing the standards. Of course economics will not always permit this procedure.

Training for Rating

The key to good time standards is regular training in performance rating by all practicing time study engineers. The time study analyst is expected to regularly establish standards within ±5% of the correct standard (which is really never known). However, if the analyst is able to regularly performance rate films or videotape depicting operations characteristic of his or her plant within ±5% of what has been agreed upon as being a certain performance, then it is highly probable that he or she will maintain a record of setting standards that will be accepted by both labor and management.

The company should require regular training of all of its time study analysts. This will help ensure not only the validity of developed standards, but also the consistency in rating between the various analysts as well as the improvement in consistency of each analyst with reference to his or her own work.

The most extensively used training technique in performance rating is the observation of videotape or motion picture films illustrating a variety of operations characteristic of those being performed in the company. A complete range of the performance scale should be demonstrated by the videotape or films. Each viewing will have a known performance level, and immediately after it is shown on the screen, the various trainees will rate the performance independently. Their respective ratings will then be compared with the known ratings. For those analysts who deviate substantially from the known rating, the operation will be reshown and discussed as to the rationale for the correct rating.

Analysts should plot their ratings on a chart so as to keep a record of their performance and to note their improvement after training sessions. Exhibit 4.4.10 illustrates the type of chart recommended. Note that the area identified by ±5% from the correct area broadens as the performance of the operator exceeds 120% or falls below 80%. We can expect a time study analyst who is thoroughly trained to performance rate within ±5% of the correct rating as long as the performance being demonstrated is somewhere between 0.85 and 1.15 of normal. If the performance of the operator is outside this range, it becomes increasingly difficult to performance rate accurately.

The company should have an extensive library of rating videotapes or films. Greater flexibility in use of the company visual aids can be obtained by showing them at different speeds at the various training sessions. Thus the analyst will not be able to depend on his or her memory as to the rating of a particular film, since the rating that it demonstrated on its last viewing will be different from the rating of the present viewing.

Exhibit 4.4.10 Chart Showing a Record of Seven Studies, With the Analyst Tending to Rate High on Studies 1, 2, 4, and 6[a]

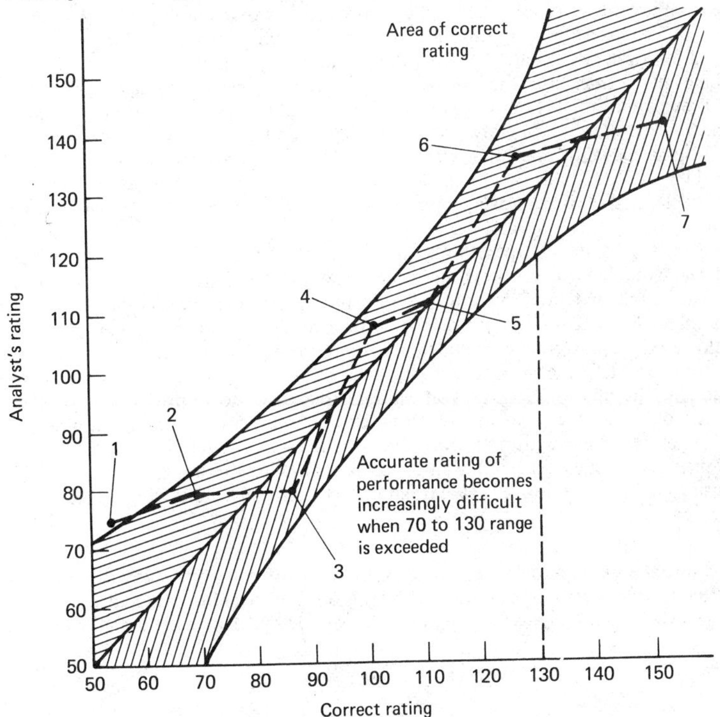

[a]Courtesy of NIEBEL, *Motion and Time Study*, p. 349.

4.4.10 ALLOWANCES

Next to performance rating, the magnitude of the applicable allowance to be assigned to the normal time is the most controversial aspect of the time study procedure. The application of unrealistic allowances can destroy all the accuracy and care taken in developing equitable normal times. Fair allowances must be added to normal times in order to develop allowed times that are realistic for all-day performance by the operator. Since the stopwatch study is conducted over a relatively short period, and abnormal readings, unavoidable delays, and time for personal needs are removed from the study in determining the average or selected time, it is necessary to make some additions to provide for unavoidable delays and other legitimate lost time.

Allowances are provided to cover time taken for personal needs, unavoidable delays, and a general slowdown of performance because of fatigue. It is important that fair allowances be determined as accurately as possible. Allowances should not be a bargaining issue in labor-management negotiations.

In general, the applied allowance will not be the same percentage for all elements of the time study. Certain allowance factors will apply to all elements, other factors will apply to the effort elements only, and others will apply to the machine-controlled elements only.

Two methods are used for determining the allowance structure for personal and unavoidable delays. These are the all-day production study and the work sampling methods.

Under the production study method, the observer spends the full day observing a small group of workers (usually three or four) doing the same class of work. The observer records the duration of and reason for each idle interval. After taking a representative sample of data (this may involve several days), he or she computes the proportion of time utilized for each interruption. For example, the observer would compute the percentage time lost for interruptions to the worker, tool repair, taking care of personal needs, machine downtime not under the worker's control, and so on. These interruption times will need to be performance rated also, since the data obtained in this fashion, like those for any time study, must be adjusted to the level of normal performance.

The production study method is exceptionally tiresome to both the observer and the operators being studied. Being closely watched for several days, from the time one gets to his or her work station in the morning until punching out on the job in the late afternoon, can be quite disconcerting. Another disadvantage of the production study method is that data are often gathered for a period of only 2 or 3 days. This may be an inadequate sample or even a sample that deviates considerably from the population.

A better method of developing equitable allowances is work sampling (see chapter 4.6). This is a technique used to investigate the proportions of total time devoted to the various activities that constitute a job or work situation.[5] With the work sampling method, a large number of observations (usually more than 2000) taken over an extended period (usually 2 weeks or more) are taken at random times and of different operators. If the total number of legitimate occurrences, other than work, that involve the operators is divided by the total number of working observations, the result will equal the allowance that the group of operators are currently taking as a percentage of the working time. Exhibit 4.4.11 illustrates a summary of a work sampling taken for determining unavoidable and personal delay allowances in a heavy machine shop involving 14 cutting facilities.

Note that in this study the operators were taking only 2.5 % of the work day for personal allowances. The unavoidable delay, "wait for crane," was involving 4.8% of the work day. The reader should be cautioned that the 2.5% and 4.8% do not represent the allowance added to the normal time since these allowances are based on the entire 8 hr day. Allowances to be added to the normal time should be based on a percentage of that normal time.

In the work sampling method no stopwatch is used, since the observer merely walks through that portion of the plant under study at random times and notes precisely what the operator or operators are doing at the time of the observation. Thus work sampling requires only the part-time services of the analyst during the course of the study, differing from the production study when the analyst is busily engaged during the entire course of the study.

When using work sampling to gather allowance data, the following precautions should be exercised:

1. Make sure the union and operators are aware the work sampling study is being conducted.
2. Confine the work sampling study to similar groups of machines or facilities.
3. Use as large a sample size as is practical. The sample size can be estimated from the equation

$$n = \frac{\hat{p}(1 - \hat{p})}{\sigma_p^2}$$

Exhibit 4.4.11 Performance Measurement and Control of Operation

Date________________

Observer________________

Machine	Cutting	Setup	Machine Idle	Crane Wait	Wait—Inspection	Wait—Tools Not Available	Wait—Tool Trouble	Confer With Other Shift	Tool Handling	Get or Grind Tools	Confer With Foreman	Wait for Job	Remove Chips, Clean Table	Personal	Avoidable	
20 ft VBM	101	7	14	2	3	1		2	37	5	3			6	35	216
16 ft VBM	102	34	14	15	3	2		1	28	5	1	7	4			216
28 ft BVM	119	34	10	5	5				20	2	1	2			18	216
12 ft VBM	109	24	12	13	6	1		3	26	6	2	3	3	2	6	216
16 ft planer	127	17	6	9	2				22		2	15		4	12	216
8 ft IMM	64	18	17	16	3			2	30	7	3			28	28	216
16 ft VBM	147	19	10	14	3	1			15	2			1	1	3	216
14 ft planer	140	8	5	7	2			2	17	3		3		11	18	216
72 in. e. lathe	99	13	12	7	3			1	32	8	2			3	36	216
96 in. e. lathe	89	9	29	18	11	1		2	29	8	3	4		3	10	216
96 in. e. lathe	109	14	12	8	10	3	3		32	9	8	2		1	5	216
160 in. e. lathe	72	34	13	14	6	1	2	4	21	3	3	1	1	4	37	216
11½ in. planer	106	35	11	10	4			1	11	4	5	3	2	8	16	216
32 ft VBM	151	23	8	7	1			1	10	2	1	5	2	5		216
	1535	289	173	145	62	10	5	19	330	64	34	45	13	76	224	3024
Percentage	50.8	9.6	5.7	4.8	2.1	0.3	0.2	0.6	10.9	2.1	1.1	1.5	0.4	2.5	7.4	100%

where n = total number of random observations upon which p is based
$\hat{p}$ = estimated percentage occurrences of the element being sought, expressed as a decimal
σ_p = SD of a percentage.

For development of this equation, see Niebel.[2]

4. When making the random observations, use care so as not to anticipate what is happening, but to record only what is happening at the exact time of the random observation.

5. Be sure that the individual observations are made at random times for all the working hours of the shift under study. (A table of random numbers can be used for this purpose.)

6. Take the daily observations over a reasonably long period. (Two weeks or longer is suggested.)

Personal Delays

Every allowed time should contain time for the employee to maintain his or her general well-being. These times are used for trips to the rest room and drinking fountain, and they typically accumulate to about 20 min during the working day. The amount of time needed by the operator for personal reasons will depend upon the working conditions; the class of work being performed; and the age, physical condition, and habits of the operator. Usually, an allowance of 5% of the working day (24 min) will prove to be adequate. Thus if the normal working time (exclusive of allowances) were 420 min, and we wanted to provide the employee an allowance of 5% of the working day, we would add 5.7% allowance to the working time (24/420 = 5.7%).

Companies should not bargain for the size of the personal daily allowance. Its magnitude should be based on either work sampling or production studies. Companies that have provided a 10 or 15 min coffee break should regard these benefits as a shortening of the workday, not as an allowance to be reckoned with in the development of a standard.

Today a 5% allowance for personal delays is appropriate in the majority of the work environments in the United States. This allowance is customarily applied to all elements of the time study.

Fatigue

It is also considered good practice to provide an allowance for fatigue on the effort elements of the time study. It is not applied to machine-controlled elements, since during these periods the operator can be resting and overcoming fatigue that has set in. Fatigue, which may be defined as a lessening in the capacity for work, is most difficult to measure. The amount of fatigue experienced varies significantly not only from one person to the next, but also in the same person from one day to the next. It is not homogeneous in any respect; it ranges from strictly physical to purely psychological and includes combinations of both the physical and the psychological. It will have a marked influence on some and very little on others.

Three major factors cause fatigue. The first is the general working environment, including the amount of light, the temperature level, the relative humidity, and the air freshness. Other conditions related to the working environment include the noise level and its duration, the color scheme of the work area, and the immediate environment.

The second factor that causes fatigue is the nature of the work being performed. The degree of monotony of the body movements has a bearing on the amount of fatigue, as do the actual physical effort and the muscle tiredness resulting from the stressing of the muscles.

Third, the general health of the worker has an important bearing on the amount of fatigue that is experienced. Both physical health and mental health are related to fatigue. Thus fatigue can be caused by the home conditions of the employees as well as by their emotional stability, the amount of rest they have had, their diet, their age, and their general physical condition.

The amount of physical fatigue that takes place in industry as a whole is diminishing because of improvements that have been made in both the designs of jobs and the working conditions. In particular, much of the heavy work has been taken from jobs through increased mechanization and automation. Unfortunately, however, most fatigue is not physical, but psychological. By careful job placement through the efforts of industrial relations departments, the amount of psychological fatigue can be minimized.

Proper allowance for fatigue should be made for working conditions and job design, which both directly influence the amount of fatigue that is experienced. No fatigue allowance should be provided for the general health factors that influence the degree of fatigue. Thus such conditions as emotional stability, rest, diet, age, physical stature, and strength should be considered in employee selection.

Typically there is a drop in productivity near the end of the working day. This falloff in productivity is largely attributable to fatigue. Invariably the rate of production tends to increase during the early part of the day but then begins to decline after the third hour. Usually we can anticipate a short period of increased productivity after the lunch break. However, this soon begins to drop off, and then output will continue to decline for the balance of the working day.

The International Labor Office has tabulated the effect of working conditions in order to arrive at an allowance factor for both personal delays and fatigue. These values are shown in Exhibit 4.4.12. A study of this exhibit will point out that the appropriate allowance should be computed for each element of the study. For example, in the study of the machining of a 30 lb casting, the 5% "use of force" allowance would apply only to the elements "pick up casting and place in chuck" and "remove casting from chuck and lay aside."

Exhibit 4.4.12 Effects of Working Conditions on Determining Allowances

Allowance	Value (%)
Constant Allowances	
Personal allowance	5
Basic fatigue allowance	4
Variable Allowances	
Standing allowance	2
Abnormal position allowance	
Slightly awkward	0
Awkward (bending)	2
Very awkward (lying, stretching)	7
Use of force or muscular energy (lifting, pulling, or pushing)—weight lifted, in pounds:	
5	0
10	1
15	2
20	3
25	4
30	5
35	7
40	9
45	11
50	13
60	17
70	22
Bad light	
Slightly below recommended	0
Well below	2
Quite inadequate	5
Atmospheric conditions (heat and humidity)—variable	0–10
Close attention	
Fairly fine work	0
Fine or exacting	2
Very fine or very exacting	5
Noise level	
Continuous	0
Intermittent—loud	2
Intermittent—very loud	5
High-pitched—loud	5
Mental strain	
Fairly complex process	1
Complex or wide span of attention	4
Very complex	8
Monotony	
Low	0
Medium	1
High	4
Tediousness	
Rather tedious	0
Tedious	2
Very tedious	5

Courtesy of International Labor Office, Geneva, Switzerland.

The methods engineer, when designing the work station, should be cognizant of the potential savings through ideal work conditions and the consequent reduction of fatigue allowances in the standard (see Chapter 2.5).

Designing the work station so that the operator can work in a seated position, utilizing little force application, with adequate light and temperature control, can result in as much as a 45% increase in productivity for certain of the work elements.

Unavoidable Delays

Allowance for unavoidable delays applies only to the effort elements of the time study. Typical causes of unavoidable delays include the several interruptions made by the foreman, dispatcher, inspector, material handler, and so on, during the working day. Then, too, on occasion there will be material irregularities where the material may be somewhat larger or harder than, or in a different location from, what is considered standard and what was being used when the time study was being made.

*Machine Interference**

Another important reason for adding an allowance for unavoidable delays when the operator is assigned more than one machine to operate is to allow for "machine interference." Interference allowance provides for time when one facility or more must wait until the operator completes the work on another facility that he or she is servicing. The more facilities assigned to an operator, the greater should be the "interference" delay allowance. Of course the amount of interference that occurs is directly related to the performance of the operator. When the operator is performing at a level above normal, there will be less interference taking place than when he or she is performing poorly by taking more time than normal to attend the stopped machine. It is the time study analyst's responsibility to determine the normal interference time that would be added to the machine running time required to produce one unit of output and the normal time required by the operator to service the stopped machine, in order to compute the cycle time. Thus

$$C = T_1 + T_2 + T_3$$

where C = cycle time to produce one unit of output
T_1 = running time to produce one unit of output
T_2 = normal time to service a stopped machine
T_3 = time lost by normal operator working because of machine interference

The cycle time divided into the running time of each machine multiplied by the number of machines assigned to the operator provides the average machine running hours per hour. Thus

$$M = \frac{nT_1}{C}$$

where M = machine running hours per hour, and
n = number of machines assigned to the operator.

By use of queuing theory techniques (see Chapter 13.7), tables have been developed where the interval between service time is exponential and where the service time is either exponential or constant. Exhibit 4.4.13 provides these values for various ratios of T_2/T_1, which has been designated k. For example, if $n = 25$, the running time were 120 min, and the service time (determined by work measurement) proved to be 3.60 min, then k would be

$$\frac{3.60}{120} = 0.03$$

By referring to Exhibit 4.4.13 and by assuming the service time was exponential, we find T_3 = 4.7% of the cycle time, and T_1 = 92.5% of the cycle time. Thus

$$\begin{aligned} C &= T_1 + T_2 + T_3 \\ &= 120 + 3.60 + 0.047C \\ 0.953C &= 123.60 \\ C &= 129.70 \text{ min} \end{aligned}$$

*For a more detailed discussion of this topic see Chapter 3.5, "Machine Interference: Assignment of Machines to Operator."

Exhibit 4.4.13 Tables of Waiting Time and Machine Availability for Selected Servicing Constants[a, b]

n	A T_3	A T_1	B T_3	B T_1
		$k = 0.01$		
1	0.0	99.0	0.0	99.0
10	0.1	99.0	0.1	98.9
20	0.1	98.9	0.2	98.8
30	0.2	96.8	0.4	98.6
40			0.6	98.4
50			0.9	98.1
60			1.3	97.8
70			1.8	97.2
80			2.7	96.3
85			3.4	95.7
90			4.2	94.9
95			5.2	93.8
100			6.7	92.4
105			8.5	90.6
110			10.7	88.4
115			13.4	85.8
120			16.3	82.9
121			16.9	82.3
122			17.5	81.7
123			18.1	81.1
124			18.8	80.4
125			19.4	79.8
126			20.0	79.2
127			20.6	78.6
128			21.2	78.1
129			21.8	77.5
130			22.4	76.9
131			22.9	76.3
132			23.5	75.7
133			24.1	75.3
134			24.6	74.6
135			25.2	74.1
136			25.7	73.5
137			26.3	73.0
138			26.8	72.5
139			27.3	71.9
140			27.9	71.4
141			28.4	70.9
142			28.9	70.4
143			29.4	69.9
144			29.9	69.4
		$k = 0.02$		
1	0.0	98.0	0.0	98.0
5	0.1	98.0	0.2	97.0
10	0.2	97.8	0.4	97.6
15	0.4	97.7	0.7	97.4
20	0.6	97.5	1.1	97.0
25	0.8	97.2	1.6	96.5
30	1.2	96.9	2.2	95.9
35			3.1	95.0
40			4.3	93.8
45			6.1	92.0
50			8.7	80.5
51			9.3	88.9
52			10.0	88.3

n	A T_3	A T_1	B T_3	B T_1
		$k = 0.02$ (*cont.*)		
53			10.7	87.6
54			11.5	86.3
55			12.3	86.0
56			13.1	85.2
57			14.0	84.3
58			14.9	83.4
59			15.9	82.5
60			16.8	81.5
61			17.9	80.5
62			18.9	79.5
63			19.9	78.5
64			21.0	77.5
65			22.0	76.4
66			23.1	75.4
67			24.2	74.4
68			25.2	73.3
69			26.2	72.3
70			27.2	71.3
71			28.2	70.4
72			29.2	69.4
		$k = 0.03$		
1	0.0	97.1	0.0	97.1
5	0.2	96.9	0.4	96.7
10	0.5	96.6	1.0	96.2
15	1.0	96.2	1.8	95.4
20	1.6	95.5	3.0	94.2
25	2.8	94.4	4.7	92.5
26	3.1	94.1	5.2	92.1
27	3.4	93.7	5.7	91.6
28	3.8	93.4	6.2	91.1
29	4.3	92.9	6.8	90.5
30	4.8	92.4	7.4	89.9
31			8.1	89.2
32			8.9	88.5
33			9.7	87.7
34			10.6	86.8
35			11.6	85.9
36			12.6	84.9
37			13.7	83.8
38			14.9	86.8
39			16.1	81.4
40			17.4	80.2
41			18.8	78.9
42			20.1	77.5
43			21.6	76.2
44			23.0	74.8
45			24.4	73.4
46			25.9	72.0
47			27.3	70.6
48			28.7	69.2
		$k = 0.04$		
1	0.0	96.2	0.0	96.2
2	0.1	96.1	0.2	96.0

Exhibit 4.4.13 (*Continued*)

n	A T_3	A T_1	B T_3	B T_1
		k = 0.04 (*cont.*)		
3	0.2	96.0	0.3	95.9
4	0.2	95.9	0.5	95.7
5	0.3	95.8	0.7	95.5
6	0.5	95.7	0.9	95.3
7	0.6	95.6	1.1	95.1
8	0.7	95.5	1.3	94.9
9	0.8	95.4	1.5	94.7
10	1.0	95.2	1.8	94.4
11	1.1	95.1	2.1	94.1
12	1.3	94.9	2.4	93.8
13	1.5	94.7	2.8	93.5
14	1.8	94.5	3.2	93.1
15	2.0	94.2	3.6	92.7
16	2.3	94.0	4.0	92.3
17	2.6	93.6	4.5	91.8
18	3.0	93.3	5.1	91.3
19	3.4	92.9	5.7	90.7
20	3.9	92.4	6.4	90.0
21	4.5	91.8	7.1	89.3
22	5.2	91.2	8.0	88.5
23	6.0	90.4	8.9	87.6
24	6.8	89.6	9.9	86.7
25	7.9	88.6	11.0	85.6
26	9.0	87.5	12.2	84.5
27	10.4	86.2	13.4	83.2
28	11.9	84.7	14.8	81.9
29	13.6	83.0	16.3	80.5
30	15.5	81.3	17.9	79.0
31			19.6	77.4
32			21.3	75.7
33			23.0	74.0
34			24.8	72.3
35			26.6	70.6
36			28.4	68.9
37			30.1	67.2
		k = 0.05		
1	0.0	95.2	0.0	95.2
2	0.1	95.1	0.2	95.0
3	0.2	95.0	0.5	94.8
4	0.4	94.9	0.7	94.5
5	0.5	94.7	1.0	94.3
6	0.7	94.6	1.4	94.0
7	0.9	94.4	1.7	93.6
8	1.1	94.2	2.1	93.3
9	1.4	93.9	2.5	92.9
10	1.6	93.7	3.0	92.4
11	2.0	93.4	3.5	91.9
12	2.3	93.0	4.1	91.4
13	2.7	92.6	4.7	90.8
14	3.2	92.2	5.4	90.1
15	3.8	91.7	6.2	89.3
16	4.4	91.0	7.1	88.5

n	A T_3	A T_1	B T_3	B T_1
		k = 0.05 (*cont.*)		
17	5.2	90.3	8.1	87.6
18	6.1	89.5	9.1	86.5
19	7.1	88.5	10.4	85.4
20	8.4	87.3	11.7	84.1
21	9.8	85.9	13.1	82.7
22	11.5	84.3	14.7	81.8
23	13.4	82.5	16.5	79.6
24	15.3	80.5	18.3	77.8
25	17.8	78.2	20.2	76.0
26	20.3	75.9	22.2	74.1
27	22.8	73.6	24.3	72.1
28	25.3	71.2	26.5	70.1
29	27.9	68.8	28.5	68.1
		k = 0.06		
1	0.0	94.3	0.0	94.3
2	0.2	94.2	0.3	94.0
3	0.4	94.0	0.7	93.7
4	0.6	93.8	1.1	93.3
5	0.8	93.6	1.5	92.9
6	1.1	93.3	2.0	92.5
7	1.4	93.1	2.5	92.0
8	1.7	92.7	3.1	91.4
9	2.1	92.4	3.7	90.8
10	2.6	91.9	4.5	90.1
11	3.1	91.4	5.3	89.4
12	3.8	90.8	6.2	88.5
13	4.5	90.1	7.3	87.5
14	5.4	89.2	8.4	86.4
15	6.5	88.2	9.7	85.2
16	7.8	87.0	11.2	83.8
17	9.3	85.6	12.8	82.3
18	11.1	83.9	14.6	80.6
19	13.2	81.9	16.5	78.3
20	15.6	79.7	18.6	76.5
21			20.8	74.7
22			23.1	72.5
23			25.5	70.3
24			27.9	68.0
25			30.3	65.8
		k = 0.07		
1	0.0	93.5	0.0	93.5
2	0.2	93.2	0.4	93.1
3	0.5	93.0	0.9	92.6
4	0.8	92.7	1.4	92.1
5	1.1	92.4	2.0	91.6
6	1.5	92.1	2.7	91.0
7	1.9	91.7	3.4	90.3
8	2.4	91.2	4.3	89.5
9	3.1	90.6	5.2	88.6
10	3.8	89.9	6.3	87.6

Courtesy of NIEBEL, *Motion and Time Study*, p. 702.

[a] All tables assume random calls for service. Column A is for constant servicing time and column B for an exponential distribution of servicing times.

[b] Values are expressed as percentages of total time, where $T_1 + T_2 + T_3 = 100\%$.

and

$$T_3 = 0.047C$$
$$= 6.10 \text{ min}$$

Thus 6.10 min of interference delay time should be added to the running time and service time in order to determine the allowed cycle time.

Avoidable Delays

No allowance is added to the normal time for avoidable delays. Typical avoidable delays include visiting with other employees for social reasons; idleness other than rest to overcome fatigue; and time taken for personal reasons, such as smoking and eating a sandwich, that are beyond the personal delay allowance. It should be understood that avoidable delays are permitted, but when taken, they are detrimental to the operator's productivity. In general, the operator is entitled to take as much avoidable delay time as he or she chooses, so long as his or her output for the day is equal to or higher than standard. However, under favorable wage payment and good supervision, seldom would a worker restrict his or her output by taking an undue proportion of avoidable delay time.

Extra Allowances

There are two classes of extra allowances. The first is an allowance added to take care of an unusual situation and applied only to a particular lot or run. For example, a shipment of castings that deviate from standard may be received, but because of the pressing need for the product, they are released to the production floor. These castings may require some additional filing time in order to have them fit the tooling. Or, because of the unavailability of a pneumatic stamp on a particular run of parts, the operator may be obliged to hand stamp this order. Any deviation, resulting in a small amount of extra work, from what the time study method identified can be handled by adding an "extra allowance." If the deviation in method is substantial, then a new time standard should be established, with clear identification of the current methods. Extra allowances of this type should be clearly identified so that they are applied only to the particular order, lot, or run for which the extra work is necessitated.

The second class of extra allowance is the provision of an allowance during the "attention time" of the cycle. The purpose here is to reward the operator who maintains full utilization of the machine-controlled facility he or she is running. With this attention time allowance, the operator on a machined-controlled element can earn an incentive pay or be given an efficiency rating comparable to a worker who is working predominantly on an effort-paced assignment. Without this extra allowance, it would not be possible for the operator to earn a performance beyond his or her normal personal fatigue or unavoidable delay allowance, even though his or her facility maintained full operation during the entire work shift. The amount of this attention time allowance should be based on the proportion of the work cycle requiring operation attention. With this extra allowance the operator should be able to achieve approximately the same performance as the average performance of those direct labor employees who have not been assigned to machine-paced work.

Exhibit 4.4.14 provides typical allowances that have been used by a variety of companies. These allowances have been applied in the development of acceptable standards in these concerns. They may or may not be adequate in a particular plant environment. They will, however, provide a guide to establishing equitable allowances to be added to normal times in order to develop reliable standard times.

4.4.11 CALCULATING THE STANDARD TIME

The standard time for the operation under study is equal to the sum of the elemental standard times. This time can be defined as the time required for an average operator, who has been fully trained to handle the work assignment and who is working at a normal pace, to perform the operation.

Elemental allowed times are computed from the average element times. These mean values are multiplied by a conversion factor, which is equal to the product of the performance factor and one plus the appropriate percentage allowance. For example, a given element of study may have an average time of 0.22 min. The assigned performance factor may have been 110%, and the allowance to be added may be 16% of the normal time. The allowed time for this element would be

$$\text{allowed time} = (0.22)(1.10)(1.16) = 0.281 \text{ min}$$

This allowed time would be added to the allowed time of the other elements composing the study in order to arrive at a standard time to perform the operation.

Exhibit 4.4.14 Representative Percentage Allowances for Typical Industrial Operations

Operation	Method or Facility	Total Applied to Effort Time	Total Applied to Machine Time	Personal	Clean Work Station	Oil Machine	Shutdown	Tool Maintenance	Unavoidable Delays and Fatigue
Anneal	Oven (gas and oil fired)	13	–	5	$\frac{1}{2}$	–	–	–	$7\frac{1}{2}$
Assembly	Bench	15	–	5	$\frac{1}{2}$	–	–	–	$9\frac{1}{2}$
Assembly	Floor	16	–	5	$\frac{1}{2}$	–	–	–	$10\frac{1}{2}$
Blacksmith	Drop forge	21	–	7	1	–	–	–	13
Brake, press	Power	15	–	5	$\frac{1}{2}$	–	–	–	$9\frac{1}{2}$
Drill	Hand feed	15	–	5	$\frac{1}{2}$	$\frac{1}{2}$	$\frac{1}{2}$	2	$6\frac{1}{2}$
Drill	Power feed	15	12	5	$\frac{1}{2}$	$\frac{1}{2}$	m-2	m-4	e-9
Engrave	Pantograph	15	–	6	$\frac{1}{2}$	$\frac{1}{2}$	e-$\frac{1}{2}$	e-2	e-$6\frac{1}{2}$
Lathe	Engine	15	15	5	2	1	m-2	m-5	e-7
Lathe	Turret	17	15	5	2	1	m-2	m-5	e-9
Milling	Horizontal and vertical	16	15	5	2	1	m-2	m-5	e-8
Grinding	Blanchard	15	15	5	2	1	m-2	m-5	e-7
Grinding	Thread	17	15	5	2	1	m-2	m-5	e-9
Grinding	External and internal	16	15	5	2	1	m-2	m-5	m-8
Punch press	Up to 100 tons	14	–	5	$\frac{1}{2}$	1	$\frac{1}{2}$	–	7
Saw	Circular	14	–	5	$\frac{1}{2}$	$\frac{1}{2}$	$\frac{1}{2}$	1	$6\frac{1}{2}$
Saw	Do-all	15	–	5	$\frac{1}{2}$	$\frac{1}{2}$	$1\frac{1}{2}$	2	$5\frac{1}{2}$
Shear	Square	15	–	5	$\frac{1}{2}$	$\frac{1}{2}$	1	–	8
Welder	Spot	17	–	5	$\frac{1}{2}$	–	2	3	$6\frac{1}{2}$
Paint	Spray	17	–	5	2	–	1	1	8

Courtesy of NIEBEL, *Motion and Time Study*, p. 390.
m: applies to machine time only.
e: applies to effort time only.

Using the Electronic, Hand-Held Calculator

Today, with the electronic, hand-held calculator, work standards may be developed with both accuracy and speed. This tool is a must for the time study analyst (see chapter 4.7).

The advanced professional calculator permits easy calculations with a constant such as a performance rating factor, allowance, or conversion factor. A key is available that permits storing a number and an operation for use in repetitive calculations. Typically the calculations of +, −, ×, ÷, Y^x, $\sqrt[x]{Y}$, and Δ% are made.

The procedure is to first enter the operation, which would be "multiply" in the case of converting average observed elemental times to allowed elemental times. Then the repetitive number would be entered into storage. This would be the conversion factor as previously explained.

After the constant (conversion factor) is stored, additional calculations for determining the allowed elemental times are completed by entering the variable (mean elemental time) and pressing the "equal" key. For example, a given time study's effort elements may have been performance rated 1.10, and an allowance of 15% is to be added to the normal time. The analyst wishes to compute the allowed time for the effort elements whose mean times we will assume are as follows: 0.161, 0.052, 0.314, 0.081, 0.128, and 0.097.

The procedure for most calculations, except those using reverse polish notation, would be as follows:

Number	Enter	Display
1	Clear calculator	0
2	Performance rating factor	1.100
3	Multiply	1.100
4	Allowance factor	1.150
5	Equal	1.265
6	Multiply	1.265
7	Constant key	1.265[a]
8	0.161	0.161
9	Equal	0.204
10	0.052	0.052
11	Equal	0.066
12	0.314	0.314
13	Equal	0.397
14	0.081	0.081
15	Equal	0.103
16	0.128	0.128
17	Equal	0.162
18	0.087	0.087
19	Equal	0.123

[a]Stores 1.265.

The use of the hand-held calculator has reduced the arithmetic of calculating a time standard by more than 50% of the longhand and slide rule methodology that made the establishment of good standards so costly. Today it is quite feasible for a time study analyst to establish four time studies (each representing about 30 min of observation) during an 8 hr working day.

Expressing the Standard Time

Work standards of relatively short duration are expressed in hours to produce 100 units. Expressed in this form, the standard is more compatible with the various reporting systems of the company or business. For example, a work standard of 3.27 min/piece would be expressed as 5.45 hr/100 pieces. In this form it is very easy to compute operators' efficiency and daily earnings if an incentive plan is in operation. If an operator produced in 1 day 164 pieces based on the preceding standard, and the operator's base rate of pay were \$8.50/hr, his or her efficiency and daily earnings would be computed as follows:

$$\text{efficiency} = \frac{5.45 \times 1.64}{8} = 111.7\%$$

$$\text{earnings} = \$8.50 \times 8 \times 1.117 = \$75.96$$

It is a good idea to express the work standard in both minutes per piece and hours per 100 pieces since the operator is able to relate to the standard more easily if he or she knows the number of minutes allotted to produce one unit of output.

4.4.12 TEMPORARY STANDARDS

Some times it will be necessary to establish a standard on an operation that the operator is not completely familiar with. Furthermore, the nature of the work will be such that the time study analyst will not have standard data or formulas on file for use in establishing the work standard. Since the operator will be at the "steep" portion of the learning curve, the analyst will be reluctant to use standard data, even if they do exist, to establish the standard because he or she will realize that the operator will not be able to achieve standard performance until after a longer break-in period. The desirability of having a standard will be obvious, but the analyst will know that the operator requires more experience before a permanent standard can be established.

A solution in such cases is the establishment of a "temporary" standard. This standard would apply only to the existing order or perhaps only to a finite number of pieces on the existing order. This standard would be more liberal than the permanent standard since it would take into consideration in the performance rating procedure that the operator was in the early stages of his or her learning curve and would not reach the flat portion for some time.

Temporary standards, when released to the production floor, should be clearly identified as being applicable only to a fixed quantity. It is a good idea to issue temporary standards on vouchers of a different color from that of permanent standard vouchers so as to clearly indicate that the rate is temporary. Upon the expiration of the temporary standards, they should be immediately replaced by permanent standards.

4.4.13 SETUP STANDARDS

It is important that the time study analyst use the same care and precision in studying the setup, teardown, and put-away elements as he or she does in making the time study of the production work elements. Those work elements included in the setup usually involve all or many of the following: punch in on the job, get tools from crib, obtain operation card and drawing from dispatcher, setup the machine or facility, punch out on the job, remove tools from machine, and return tools to crib. The analyst needs to be especially careful in studying the setup elements since he or she will be observing but one cycle and will need to record them as they occur. Setup elements, on the average, will be considerably longer than elements performed during the production study, so that the alert analyst will be able to identify, record, measure, and performance rate them as they occur.

Setup standards should always be identified as separate allowed times and should not be combined with each-piece times. It is usual practice to record the setup time in hours on the operation card, since the hour is the unit of time referred to in connection with money base rates. It may be desirable to show the setup time in both decimal hours and decimal minutes for the convenience of the operator.

4.4.14 MAINTENANCE OF TIME STANDARDS

Once a work measurement system is introduced, it needs to be well maintained. Time standards are always based upon a specific method, and as time passes, minor method improvements will be introduced. The source of these method changes will vary. Some may be introduced by the foreman, the inspector, the product engineer, the methods engineer, or the operator. No matter who introduces the method change, that a change has been made should signal that the affected portion of the operation should be restudied. As soon as the time study analyst has been advised of a method change, he or she should study that portion of the operation that is affected by the change and recalculate the standard. There is no question that the principal reason for a standards program's becoming obsolete is because of looseness in the standards brought about by creeping methods changes that have been introduced with no corresponding restudy of the operator by the analyst.

To ensure that the methods being used when the time study was conducted are still being employed, and that the established rate is equitable, a regular audit schedule should be observed. The more active the standard is, the more frequent the audit should be. The following schedule can be used as a guide in establishing the frequency of the time standard audit:

Hours of Application of Standard in One Year (Standard Time × Number of Pieces Produced)	Frequency of Audit
More than 700 hr	Every 6 months
More than 100 hr, less than 700 hr	Every year
More than 50 hr, less than 100 hr	Every 2 years
Less than 50 hr	Every 3 years

The auditing procedure is to obtain a copy of the original time study in order to obtain details of the method studied. Then the analyst should observe the method as it is currently being carried out. If there are any changes in the way the operation is being performed, then that portion of the operation that is affected should be restudied and the new standard be immediately introduced. In the event that the method has not changed, the observer should check the overall time of two or three cycles to verify that the normal time being required is in accord with the developed time standard. The time study audit does not necessitate the conducting of a new time study. It is a sampling procedure to verify that the developed rate is in line and that the prescribed method is being followed or has been improved. If improved, or changed, then a new detailed study must be made.

4.4.15 STANDARD DATA

To gain the maximum use of developed time standards that have proved to be satisfactory, the analyst should develop, classify, code, and file standard data (see Chapter 4.8). Standard data are elemental time values taken from time studies or calculated from proven previous studies. These elemental values are classified and coded so that they can be readily retrieved and accumulated in order to determine an equitable standard on an operation without having to measure the time required to perform that operation. The term "standard data" refers not only to tabulated data, but also to algebraic expressions, curves, alignment charts, and tables that allow the rapid determination of an elemental time value.[6]

Standards that are developed using standard data will be consistent since the performance rating procedure is not required. Furthermore, the standards so developed will be acceptable to both labor and management since they are derived from time studies that have proved to be satisfactory. Since the values are tabulated, and it is only necessary to accumulate the required elements in establishing a standard, the various time study analysts within a given company will arrive at identical standards of performance for a given method.

In general, standards on new work can be established much more rapidly by standard data than by measurement using the time study stopwatch procedure. Consequently, after an adequate inventory of standard data elements is accumulated, it is feasible to establish standards on indirect labor operations such as shipping, receiving, and maintenance.

Classification of Standard Data

Standard data should be classified and coded for rapid retrieval by the engineer. It usually is advantageous to classify standard data first by facility operation, such as drill press, milling, turret lathe, or bench assembly; second by facility style, such as 17 in. single spindle, 21 in. three spindle, or 21 in. radial drill; and third by facility size, such as single spindle 27 in., single spindle 21 in., single spindle 17 in., single spindle 12 in., and so on; and fourth by manufacturer, such as Leland-Gifford, Cincinnati, or Delta.

The standard data that are recorded for a specific machine, such as a Warner and Swasey number 5 turret lathe, should be broken down into setup elements and each-piece elements. Exhibit 4.4.15 provides a listing of standard data elements applicable to a specific plant for an Allen 17 in. vertical single-spindle drill press.

It is possible to combine standard data elements to allow more rapid computation of a standard. For example, Exhibit 4.4.16 illustrates combined setup data for a Warner and Swasey number 5 turret lathe that are applicable in a specific plant. If a certain job needed a facing, turning, and grooving tool in the square turret, and a drill, boring tool, and collapsible die in the hex turret, the allowed setup time would be 81.6 min + (2 × 8.63) = 98.86 min. The 81.6 min (line 10) allows time to set up the facing, turning, and grooving tools in the square turret and the collapsible die in the hex turret. A total of 8.63 min for each tool beyond the first tool in the box-tool is allowed. Since a drilling and boring operation from the hex turret was required, the time 8.63 min (line 13) is allowed.

Variable standard data can be recorded in either tabulated or equation form. Variable data are usually computed from feed and speed information that has been specified. When using such data, it is important that the analyst always consider the lead and overrun of the cutting tool as a portion of the length of cut. For example, the lead of a 1 in drill is 0.30 in; this distance must be added to the length of the drilled hole when computing the total cutting time.

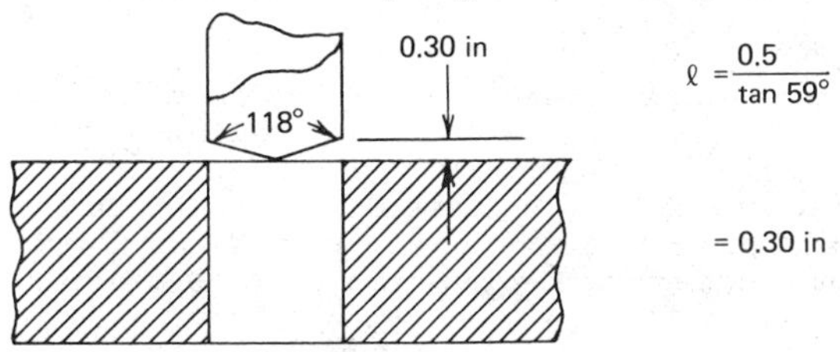

Exhibit 4.4.15 Standard Data Elements Applicable to a Specific Plant for an Allen 17 in Vertical Single-Spindle Drill[a]

Elements	Minutes
Setup Elements	
Study drawing.	1.250
Get material and tools and return and place ready for work.	3.750
Adjust height of table.	1.310
Start and stop machine.	0.090
First-piece inspection (includes normal wait time for inspector).	5.250
Tally production and post on voucher.	1.500
Clean off table and jig.	1.750
Insert drill in spindle.	0.160
Remove drill from spindle.	0.140
Each-Piece Elements	
Grind drill (prorate).	0.780
Insert drill in spindle.	0.160
Insert drill in spindle (quick-change chuck).	0.050
Set spindle.	0.420
Change spindle speed.	0.720
Remove tool from spindle.	0.140
Remove tool from spindle (quick-change chuck).	0.035
Pick up part and place in jig.	
Quick-acting clamp.	0.070
Thumbscrew.	0.080
Remove part from jig.	
Quick-acting clamp.	0.050
Thumbscrew.	0.060
Position part and advance drill.	0.042
Advance drill.	0.035
Clear drill.	0.023
Clear drill, reposition part, and advance drill (same spindle).	0.048
Clear drill, reposition part, and advance drill (adjacent spindle).	0.090
Insert drill bushing.	0.046
Remove drill bushing.	0.035
Lay part aside.	0.022
Blow out jig and part and lay part aside.	0.081
Plug gage part.	0.120 per hole

Courtesy of NIEBEL, *Motion and Time Study*, p. 426.
[a]Work size: Small work—up to 4 lb in weight and such that two or more parts can be handled in each hand.

A similar consideration must be made in computing cutting time in connection with milling machine work. For example, Exhibit 4.4.17 shows the lead in the work being machined to be CB in. This distance must be added to the 10 in cut made by the machine in order to calculate the entire cutting time.

If the cutter were 4 in in diameter, the distance BC would be

$$\sqrt{(AC)^2 - (AB)^2} = \sqrt{(2)^2 - (1.75)^2} = 0.97 \text{ in}$$

A nomogram or system of curves can be useful in recording variable time. Such a nomogram is illustrated for turning and facing on a lathe in Exhibit 4.4.18.

Development of Standard Data

In developing standard data, the analyst reviews all time studies taken in the past that have proved to be satisfactory. Like elements from the various studies are analyzed statistically, and mean values are coded, classified, and tabulated for future use.

Sometimes desired elemental values do not exist, or if they do, the sample size or dispersion of the data is such that the analyst may question its validity. In such instances, the analyst will resort to work measurement of the particular element in question. Since the analyst will be studying only

Exhibit 4.4.16 Standard Data for Setup Elements for Warner and Swasey Number 5 Turret Lathe

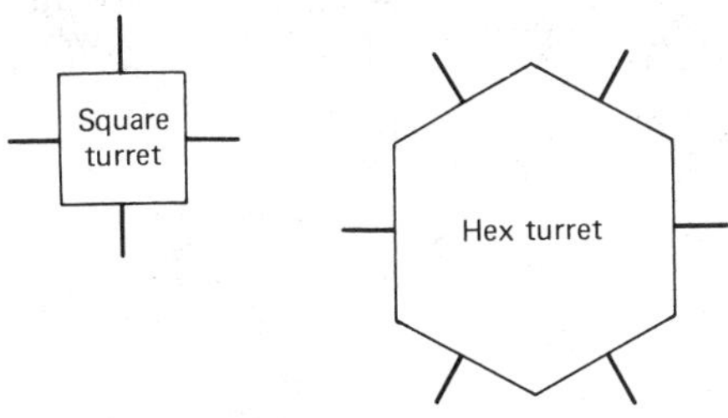

Basic Tooling

No.	Square Turret	Partial	Chamfer	Bore or Turn	Drill	S. Tap or Ream	C. Tap	C. Die
		Hex Turret						
1.	Partial.	31.5	39.6	44.5	48.0	47.6	50.5	58.5
2.	Chamfer.	38.2	39.6	46.8	49.5	50.5	53.0	61.2
3.	Face or cut off.	36.0	44.2	48.6	51.3	52.2	55.0	63.0
4.	Tn bo grv rad.	40.5	49.5	50.5	53.0	54.0	55.8	63.9
5.	Face and chf.	37.8	45.9	51.3	54.0	54.5	56.6	64.8
6.	Fa and cut off.	39.6	48.6	53.0	55.0	56.0	58.5	66.6
7.	Fa and tn or tn and cut off.	45.0	53.1	55.0	56.7	57.6	60.5	68.4
8.	Fa, tn, and chf.	47.7	55.7	57.6	59.5	60.5	69.7	78.4
9.	Fa, tn, and cut off.	48.6	57.6	57.5	60.0	62.2	71.5	80.1
10.	Fa, tn, and grv.	49.5	58.0	59.5	61.5	64.0	73.5	81.6

11. Circled basic tooling from above. ______
12. Each additional tool in square. $4.20x$ ______ = ______
13. Each additional tool in hex. $8.63x$ ______ = ______
14. Remove and setup three jaws. 5.9 ______
15. Set up subassembly or fixture. 18.7 ______
16. Set up between centers. 11.0 ______
17. Change lead screw. 6.6 ______

Total setup ______ min

Courtesy of NIEBEL, *Motion and Time Study*, p. 410.

Exhibit 4.4.17 The Lead in the Work Being Machined to be CB In.

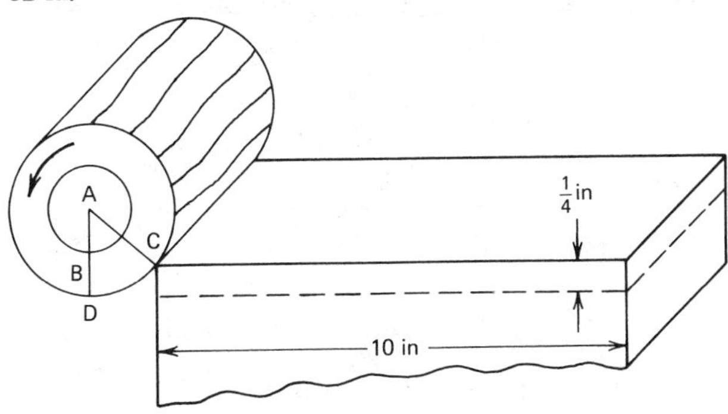

Exhibit 4.4.18 Nomogram for Determining Facing and Turning Time[a]

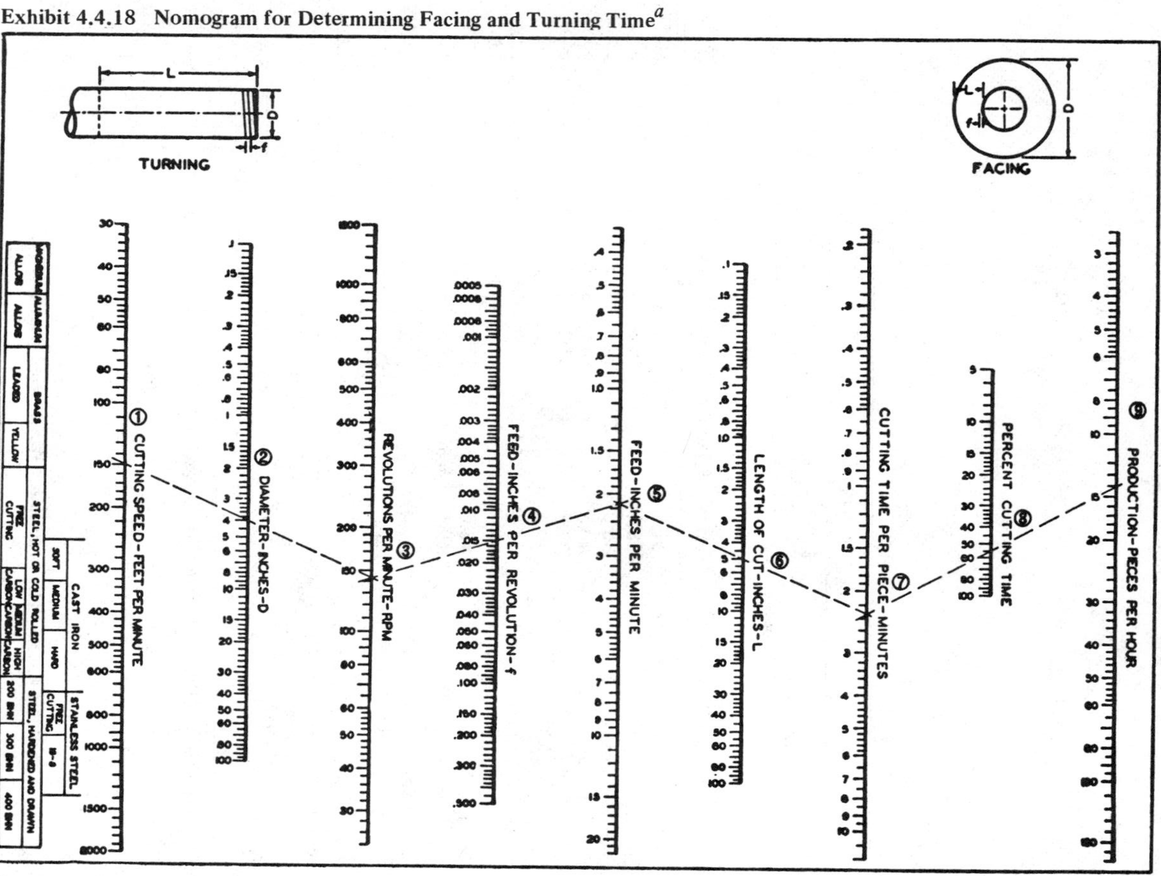

[a]Courtesy of Crobalt, Inc.

one or perhaps two elements of the cycle, he or she frequently uses a watch that measures to the closest 0.001 min. In such cases the snap-back method is employed. Upon completion of the observations (an adequate sample should be taken), the elemental values are summarized and the mean is determined. A performance rating factor is applied in order to arrive at equitable normal time values. Frequently allowances are not added, because the standard data will be more flexible if allowance values are applied at the time the standard data are used. For example, the standard data element "pick up small casting (up to 4 lb) and place in two-jaw 14 in air chuck" may have a different value in an environment where the temperature and humidity are such that a 10% personal delay allowance is required than in an environment characteristic of a typical machine shop, where a 5% personal delay allowance is appropriate.

The finer the standard data, the more flexibility they can provide in establishing standards on new work. However, the finer the element is, the more difficult it is to measure. By measuring groups of fine standard data and then computing their individual values by using simultaneous equations, standard data can be developed for very brief portions of elements or elements of short duration.

For example, we might want to determine standard data values for the five fine elements performed in succession on a given facility. These five elements might be (1) reach 20 in and bring a 4 lb casting to work station, (2) place casting in jig over two locating pins, (3) close drill jig cover, (4) start drill, and (5) advance spindle. The time for these five elements may be so short that they can be timed accurately only in groups. For example, we could easily measure and performance rate elements 1, 2, and 3 when timed collectively, but would have much difficulty in measuring their duration individually. Let us assume elements 1, 2, and 3 together had a mean time of 0.078 min, which we could assign as group element A. Elements 2, 3, and 4 might equal 0.064 min and be assigned B; elements 3, 4, and 5 might equal 0.060 min and be assigned C; elements 4, 5, and 6 might equal 0.076 min and be assigned D; and finally, elements 5, 1, and 2 might in combination equal 0.082 min, which we will designate E.

Then

$$A + B + C + D + E = (3)(\text{element 1}) + (3)(\text{element 2}) + (3)(\text{element 3}) + (3)(\text{element 4}) + (3)(\text{element 5}) = 0.360$$

$$\text{element 1} + \text{element 2} + \text{element 3} + \text{element 4} + \text{element 5} = 0.120$$

Therefore

$$A + \text{element 4} + \text{element 5} = 0.120$$

$$\text{Element 4} + 5 = 0.120 - A = 0.120 - 0.078 = 0.042 \text{ min}$$

Since element 3 + element 4 + element 5 = 0.060, element 3 = 0.060 − 0.042 = 0.018 min. Similarly, element 4 + element 5 + element 1 = 0.076, and element 1 = 0.076 − 0.042 = 0.034 min. By substituting in equation for A; .078 = 0.034 + element 2 + 0.018; therefore, element 2 = 0.026.

4.4.16 FORMULA CONSTRUCTION

To shorten the rather laborious procedure of adding up a large number of standard data elements, a usable formula may usually be designed to establish standards over the range of work characterized by the data. A formula as applied to time study involves the development of an algebraic expression or system of curves that can be used to establish a standard prior to the beginning of production.

A technician is able to establish consistent time standards quite rapidly using formulas. Furthermore, standards established by formulas are less susceptible to error since less arithmetic is involved in their solution than in conducting a stopwatch time study or in summarizing standard data elements.

To develop a reliable formula, the analyst must have an adequate sample of data taken from the complete range of work to which the formula is to apply. Usually 10 to 15 time studies that have proved to be satisfactory represent a sample size that will give good results. It is very important that like elements in the time studies being used for developing the formula be consistent in their end points. Those studies in which the elements' end points are inconsistent should not be used.

Once the analyst has selected an adequate sample of reliable data consistent in their end points, he or she should post the data on a work sheet for analysis of the constants and variables. The constants are combined, and a mean is established for incorporating into the formula. The variables are then studied in order to determine which one or ones are affecting the time to perform the work. In many cases only one variable is responsible for the time taken. The data should be plotted with time being the dependent random variable whose distribution depends on the independent variable x. It should be understood that, in most relationships that we are estimating, x is not random; it is fixed for all practicality, and the analyst is concerned with the mean of the corresponding distribution of time y (the successive elements observed through stopwatch analysis) for the given x.

The plotted data may take several forms—straight line, parabola, hyperbola, ellipse, or exponential forms—or no geometric form. If the data do not plot in any geometric form, it is highly probable that more than one independent variable is having an impact on the dependent random variable time. Frequently, it is helpful to test the data to see if they are characterized by the power functions $y = bm^x$. This can be done by plotting them on semilogarithmic paper to see if the points will approximate a straight line on a transformed scale. If the paired data give a straight line when plotted on semilog paper, then the curve of y on x is exponential.

At times, plotting the data on logarithmic paper will result in a linear relationship. Here $y = ax^m$ where $\log y$ is linear with $\log x$.

Solving When There is More Than One Independent Variable

When time data result in a scatter plotting against the supposed independent variable, it is quite likely a second independent variable is affecting the plotting. If two independent variables are involved in a linear relationship, multiple regression may be used. Here a plane is fitted to a set of n points so as to minimize the sum of squares of the vertical distances from the points to the plane. Here we are minimizing

$$\sum_{i=1}^{n} [y_i - (b_0 + b_1 x_i + c_1 z_i)]^2$$

where y = dependent variable time
x = first independent variable
z = second independent variable
b_1 = coefficient of x
c_1 = coefficient of z
b_0 = constant

The normal equations are:

$$\Sigma y = nb_0 + b_1 \Sigma x + c_1 \Sigma z$$

$$\Sigma xy = b_0 \Sigma x + b_1 \Sigma x^2 + c_1 \Sigma xz$$

$$\Sigma zy = b_0 \Sigma z + b_1 \Sigma xz + c_1 \Sigma z^2$$

Graphic Methods

When using multiple regression, the form of the equation selected and the inclusion of cross products of the variables in the equation when interaction effects are present may complicate the analysis. At times a graphic solution may be simpler and just as reliable. The procedure is to identify the independent variables that have an impact on the dependent variable time. A graphic plotting is made for each independent variable. One independent variable is first selected, and the data are reviewed to see if a number of studies can be identified where the second independent variable is relatively constant. Then time is plotted against the first independent variable for only those points where the second independent variable has shown little, if any variation.

Once this plotting has been completed, a new time scale should be constructed on the y axis. This scale can be thought of as a "time correcting scale." By extending the bottom point of the plotting to this time correcting scale, unity for this scale will be established. Corresponding distances are marked off on this scale.

The next step is to determine a time correcting value for all data points by using the plotting. This value is then divided into the time value of each data point in order to determine a corrected time. The resulting corrected time is now plotted against the second independent variable.

To use the two curves to predict the variable elemental time on new work, the analyst uses the first curve to select a time correcting value and the second plotting for determining a corrected time value. The product of the corrected time value and the time correcting value equals the normal time of the variable element being sought.

Exhibit 4.4.19 illustrates this method of graphic solution where the time to wind copper coils on collapsible mandrels was found to depend on both wire gage and lineal length of wire wound

Exhibit 4.4.19 Method of Graphic Solution Showing (*a*) Relationship Between Time and Wire Gage for Lengths of Wire Approximately 1500 Ft and (*b*) Relationship Between Corrected Time and Wire Length

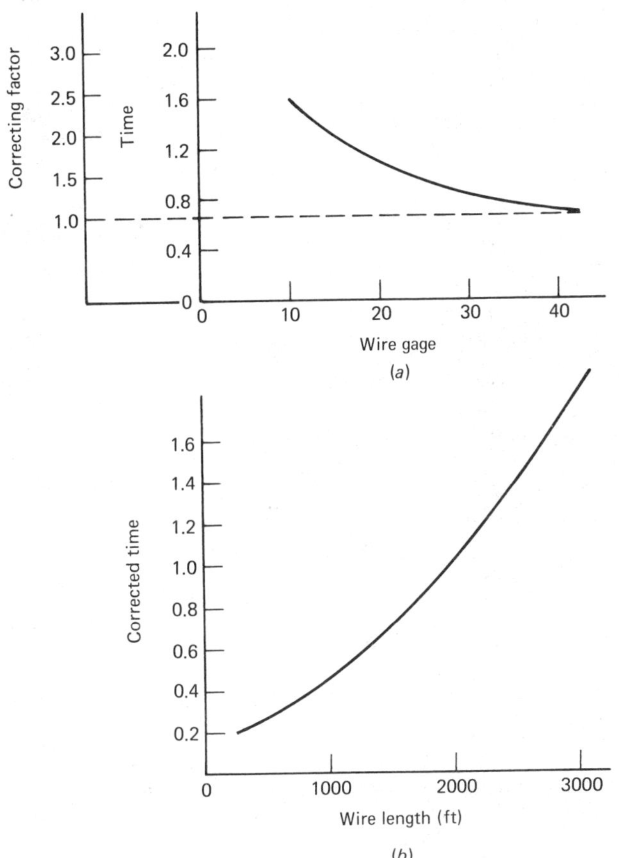

on the mandrel. Exhibit 4.4.19*a* shows the relationship of wire gage and time for constant lengths (approximately 1500 ft). Exhibit 4.4.19*b* shows the relationship between the corrected time and the length of wire wound. To use this system of curves, one first obtains a correcting factor from Exhiit 4.4.19*a* for the wire gage being used and then obtains a corrected time from Exhibit 4.4.19*b* for the length of wire being used. The product of the correcting factor and the corrected time will equal the normal time for the job under study.

4.4.17 STOPWATCH STUDIES FOR INDIRECT STANDARDS

To establish standards for indirect work such as maintenance, shipping and receiving, or toolroom operations, the cost of utilizing stopwatch studies, standard data, and formulas may exceed the benefits resulting from the standards if conventional methods are utilized. A technique known as "universal indirect standards" has been developed that allows the establishment of adequately reliable standards in advance of performing the work (see Chapter 4.9).

The principle behind universal standards is assigning the majority of indirect work that is performed to an appropriate slot.[7] Each slot will have its own standard, which will have been determined by studying similar work in the past using time study and other conventional methods such as standard data and formulas. When the range of indirect work is approximately 20 hr, a total of 20 slots or standards usually will prove adequate for the establishment of relatively accurate standards.

The procedure used in developing a universal standard system is to study a reasonably large sample of the indirect work for which the system is being developed. Typically, 200 or 300 standards are established using the stopwatch procedure. These standards should cover the full time range experienced by the indirect labor department. These measured standards are referred to as "benchmark standards." The benchmark standards are then arranged in numerical order. If 20 universal standards are to be developed, the analyst distributes the benchmark standards over a normal or gamma distribution. Practice has proved that the normal distribution will give satisfactory results, although the gamma distribution is more analogous to the true distribution of the benchmark jobs.

If the normal distribution is used, and 20 slots are to be used, it is necessary to divide the distribution into 20 segments. Each segment then determines which of the benchmark jobs will be assigned to a given slot. The mean of the assigned benchmark jobs to a given slot will be the value of the indirect standard characterized by that slot. For example, if 300 benchmark jobs were available, and if 20 slots were desired, the value of each slot based on the normal distribution would be computed as follows:

Z (Standard Values)	Area	Benchmark Jobs for Slot	Slot Value Based on Mean of These Benchmark Jobs (BMJ)
−3.0 to −2.7	0.0022	(0.0022)(300) = 0.66	BMJ 1
−2.7 to −2.4	0.0047	(0.0047)(300) = 1.41	BMJ 2–3
−2.4 to −2.1	0.0097	(0.0097)(300) = 2.91	BMJ 4–6
−2.1 to −1.8	0.0180	(0.0180)(300) = 5.40	BMJ 7–11
−1.8 to −1.5	0.0309	(0.0309)(300) = 9.27	BMJ 12–20
−1.5 to −1.2	0.0483	(0.0483)(300) = 14.49	BMJ 21–35
−1.2 to −0.9	0.0690	(0.0690)(300) = 20.70	BMJ 36–56
−0.9 to −0.6	0.0902	(0.0902)(300) = 27.06	BMJ 57–83
−0.6 to −0.3	0.1078	(0.1078)(300) = 32.34	BMJ 84–115
−0.3 to 0	0.1179	(0.1179)(300) = 35.37	BMJ 116–150
0 to 0.3	0.1179	(0.1179)(300) = 35.37	BMJ 151–185
0.3 to 0.6	0.1078	(0.1078)(300) = 32.34	BMJ 186–217
0.6 to 0.9	0.0902	(0.0902)(300) = 27.06	BMJ 218–244
0.9 to 1.2	0.0690	(0.0690)(300) = 20.70	BMJ 245–265
1.2 to 1.5	0.0483	(0.0483)(300) = 14.49	BMJ 266–280
1.5 to 1.8	0.0309	(0.0309)(300) = 9.27	BMJ 281–289
1.8 to 2.1	0.0180	(0.0180)(300) = 5.40	BMJ 290–294
2.1 to 2.4	0.0097	(0.0097)(300) = 2.91	BMJ 295–297
2.4 to 2.7	0.0047	(0.0047)(300) = 1.41	BMJ 298–299
2.7 to 3.0	0.0022	(0.0022)(300) = 0.66	BMJ 300

These 20 slot values (universal indirect labor standards) will be the basis of establishing standards on all new indirect labor work. When studying a new work assignment, the analyst will fit the job to a category (slot) where similar jobs (benchmark jobs) have been studied and standards established.

4.4.18 USES OF STANDARDS

Good standards for both direct and indirect labor are essential for the continuous effective operation of a business or enterprise. They are the principal means for the following:

1. Determining plant capacity.
2. Balancing the work force with available work.
3. Controlling production.
4. Determining costs.
5. Introducing a standard cost system.
6. Introducing and maintaining an incentive wage payment system.
7. Providing budgetary control.
8. Introducing a supervisory bonus system.
9. Purchasing the most productive equipment.
10. Securing an efficient plant layout.
11. Comparing alternative methods.
12. Measuring the effectiveness of management.

REFERENCES

1. RONALD E. WALPOLE and RAYMOND H. MYERS, *Probability and Statistics for Engineers and Scientists*, Macmillan, New York, 1972.
2. BENJAMIN W. NIEBEL, *Motion and Time Study*, 6th ed., Homewood, IL, Irwin, 1976.

3. NIEBEL, *Motion and Time Study*, p. 333.
4. MARVIN E. MUNDEL, *Motion and Time Study: Principles and Practices*, 4th ed., Prentice-Hall, Englewood Cliffs, NJ, 1960.
5. R. E. HOLLAND and W. J. RICHARDSON, *Work Sampling*, McGraw-Hill, New York, 1957.
6. PHILIP F. OSTWALD, *Cost Estimating for Engineering and Management*, Prentice-Hall, Englewood Cliffs, NJ, 1974.
7. RICHARD M. CROSSMAN and HAROLD W. NANCE, *Master Standard Data: The Economic Approach to Work Measurement*, rev. ed., McGraw-Hill, New York, 1972.

BIBLIOGRAPHY

BARNES, RALPH, *Motion and Time Study*, 7th ed., Wiley, New York, 1980.

KRICK, EDWARD V., *Methods Engineering*, Wiley, New York, 1962.

MUNDEL, MARVIN E., *Motion and Time Study: Principles and Practices*, 5th ed., Prentice-Hall, Englewood Cliffs, NJ, 1978.

NADLER, GERALD, *Work Design: A Systems Concept*, rev. ed., Irwin, Homewood, IL, 1970.

CHAPTER 4.5

Predetermined Motion Time Systems

CHESTER L. BRISLEY

University of Wisconsin–Extension

KARL EADY

MTM Association for Standards and Research

4.5.1 INTRODUCTION

The measurement of work has been of major interest to industrial engineers and their predecessors since Frederick W. Taylor introduced the stopwatch in 1883. Prior to that time, techniques for measuring work were limited to the use of historical data and estimating by those more or less familiar with the work to be measured. Time study became the first "scientific" means of measuring work. Since Taylor's time, additional scientific tools have been added to the work measurement field, notably work sampling and the use of predefined motions and their associated time values. This latter technique has come to be known as predetermined motion time systems (PMTSs).

Although all of the above-mentioned work measurement techniques are still being used in varying degrees in just about every industry, be it manufacturing- or service-oriented, PMTSs are becoming increasingly popular both as additional tools and as replacements for the other techniques for the following reasons:

1. The systems generally provide a practical vehicle for analyzing and improving methods for performing work.
2. The work methods may be designed prior to the initiation of production.
3. The predetermined time values associated with precisely described motions provide for greater consistency in the establishment of standards than is possible with other work measurement techniques.
4. When used for developing standard multiuse elements, PMTSs are generally faster.

4.5.2 EARLY DEVELOPMENT OF PMTSs

The Gilbreths

Current PMTSs have been developed from the original research and work of Frank B. and Lillian M. Gilbreth. In 1912 Frank Gilbreth presented a paper to the American Management Association wherein he explained the principles of motion economy, the systematic elimination of inefficiencies, and the concept of determining performance times by analyzing the motions required to perform work.[1]

Gilbreth classified human accomplishment according to 18 fundamental motions which he called "therbligs" ("Gilbreth" in reverse). While most of these define motions of the hands and arms, some denote mental reactions and others indicate periods of inactivity, such as rest and delays.

Gilbreth made the first attempt at measuring times by motions required to perform an operation. He placed ruled paper in the motion path and took motion pictures of the motions performed. Thus it was possible to determine the distance through which each motion was made. This technique constituted a great refinement in motion analysis and went a great deal further than anything that had been developed previously.

Gilbreth also used a cyclograph, a mechanism consisting of small electric bulbs that were fastened to the fingers of the operator. These bulbs flashed at regular intervals, and by taking pictures with a stereoscopic camera, the movements of the hands and fingers were recorded on the plate of the camera in three dimensions. Since the flashes occurred at known time intervals, it was possible to determine the time involved and the distance covered by the movement between each dot appearing in the photograph.

A. B. Segur

One of the pioneers in the development of PMTSs was Asa Bertrand Segur.[2] In 1922 Segur began development of his PMTS, which he called "Motion-Time-Analysis" (MTA), by analyzing micromotion films taken of expert operators during World War I. These films were originally taken with the view of discovering a means of training blind and other handicapped workers to perform useful industrial tasks after the war. The films were made of workers who were the best available in their industry. At the time these analyses were made, Gilbreth's motion classification was available as an aid. In fact, Segur worked with Gilbreth on this project of training the blind and handicapped soldiers.

Joseph H. Quick

Between 1934 and 1938 Joseph H. Quick developed the first of a family of systems that he called Work-Factor.® This system was developed in Philadelphia, Pennsylvania, from original motion time studies using stopwatches, photo timers, films, and fast-film snapshots. Seventeen thousand motion times were recorded of approximately 1100 workers in a wide variety of operations, such as machine shops, punch press shops, assembly plants, wood mills, plastic works, plating shops, and offices.

Although Segur and Holmes[3] had some interest in the time for mental processes, it was Joseph Quick who engaged in considerable research to this area.[4] Data for determining mental process time were obtained in laboratories and in visual inspection, printing, and engineering departments.

Harold B. Maynard

In 1946 Harold B. Maynard began the development of MTM at the Westinghouse Electric Corporation in collaboration with Gustave J. Stegemerten and John L. Schwab. Maynard had been commissioned by Westinghouse to develop a system for describing and evaluating operation methods. He began by taking motion pictures of sensitive drill press operations and analyzing them in terms of Gilbreth's therbligs. In so doing he discarded those therbligs that were not associated with manual motions and renamed many of those remaining.

The pictures were leveled as to performance by a number of engineers chosen for their experience in pace rating. The performance was leveled by a rating system designed by Lowry, Maynard, and Stegemerten.[5] The system involves the development of a composite rating as a percentage of "normal," based on the factors of skill, effort, consistency, and conditions. Method efficiency was not included in the evaluation since this was to be the main purpose of MTM. Plotting of the leveled film times as a function of distance for the various cases of "transport empty" and "transport loaded" revealed a remarkably high coefficient of correlation. Thus MTM became a PMTS admirably suited not only for method evaluation, but also for the establishment of a consistent time standard for any given method.

Other Developers

In the 1950s Gerald B. Bailey and Ralph Presgrave developed the PMTS known as Basic Motion Timestudy (BMT). This system also follows the concepts of the Gilbreths and is similar in many respects to Maynard's work. Bailey and Presgrave were awarded the Society for the Advancement of Management's Gilbreth Medal as a result of their work in this field.

Many additional pages could be devoted to all the others who have contributed to the improvement of existing PMTSs and the development of new ones, but space restrictions do not permit more than to mention Ulf Åberg and Walton M. Hancock, who proposed, in the 1960s, a method of determining the statistical accuracy of PMTSs.[6]

4.5.3 DEFINITION OF PMTSs

Before specific PMTSs available to the industrial engineer today can be presented, commonly used terms must be defined and general system characteristics explained. In this regard the following items will be brought to the reader's attention:

1. The language of PMTSs.
2. Leveling.
3. Classification of PMTSs.

4. Common characteristics.
   a. Accuracy.
   b. Speed of application.
   c. Method description.
   d. Instruction requirements.

The Language of PMTSs

The language of PMTSs is the language of "action" words—words that are used as concise descriptions of specific elementary manual movements (therbligs). Each PMTS has its own set of action words which must be defined in detail if a common understanding of any PMTS is to be achieved.

The action words used to describe specific basic motions in some of the more prominent PMTSs are shown in Exhibit 4.5.1.

Establishment of Work Pace

When using PMTSs, performance rating is no longer necessary, since the data of PMTSs currently in use were leveled to a common base at the time of development. Motion times established for use in the Work-Factor tables represent times required by an average of experienced workers as leveled to an incentive pace by an average of experienced engineers. Final data were resolved into curves, and formulas were derived for various body members.[7] Methods Time Measurement data were leveled at their development to represent a 100% (normal) performance level.

Classification of PMTSs

Classification by Work Type

Predetermined motion time systems may be classified as generic, functional, or specific. A generic system is one that is intended to be understood by all users of work measurement and that is not restricted in application. Alternative terms for "generic" are "general" and "universal." The action words of a generic system are in themselves generic; that is, they give no indication as to what type of work is being measured. Examples of generic action words are "reach," "transport," "grasp," and "select." Among the generic systems in use today are Maynard Operation Sequence Technique (MOST), MTM, Work-Factor Systems, and Master Standard Data (MSD).

A functional system is one that is adapted to a particular type of activity, such as clerical work, tool usage, or microassembly. Element names of functional systems very often reveal the function for which the system is meant. For example, "measure" is an element of a functional system intended for machine shop measurement. "Sweep" is an action word in a custodial work measurement system. "File" is a common element name in clerical work measurement systems.

Specific systems are those that are largely proprietary and that may have been developed for a particular industry or organization. Specific systems also include proprietary standard data systems that may have been developed from other generic and/or functional systems.

Classification by Element Level

From a technical viewpoint, PMTSs may also be classified by the level of element complexity. Under practical conditions no single system will have all of its elements defined at the same level of comprehensiveness, but most of them should have. Classification by element complexity is of great help in determining situations in which each level can be used most efficiently.

Basic-Level Systems. Basic-level systems are those whose elements consist mostly of single motions that cannot be further subdivided. One can deduce from this fact that analyzing an operation with a basic-level system can be quite time consuming, especially if the operation is relatively long. This is because many elements must be used to describe the operation completely. Since the elements of most PMTSs currently in use have several variables, such as distance, object weight, or degree of precision required at the end point of the motion, the decision making process can be quite complex, thus adding further to the time required to make an analysis.

Higher-Level Systems. By combining two or more of the single elements of a basic-level system into a multimotion element, a second-level system is developed. Second-level systems are faster and easier to use because the number of variables has been reduced in the combining process and not so many need be used to analyze an operation.

Third- and fourth-level systems may also be generated by continuing the combining process. As the system level increases, the number of elements decreases, and the size of each element in terms of work content increases.

Exhibit 4.5.1 Comparison of Terms for PMTSs

Definition	Gilbreth's Therblig	A. B. Segur's Motion Time Analysis	Detailed Work-Factor	MTM-1	Basic Motion Timestudy
The act of:					
Moving a transportation means without a load	Transport empty	Transport empty	Transport	Reach	Reach
Moving a transportation means with a load or against resistance	Transport loaded	Transport loaded	Transport	Move	Move
Gaining complete managing control	Grasp	Grasp	Grasp	Grasp	Included in reach
Rearranging the part being transported to have it in readiness for continuing the main operation	Pre-position	Pre-position	Pre-position	Included in position	Included in move
Guiding actions with sensory movements		Direct			
Bringing two parts to an exact and predetermined relationship with each other	Position	Position	Assemble	Position (primary engagement)	Precision factor
Placing a positioned object in or on another object from which it is later removed, such as a part in a fixture or a wrench on a nut	Assemble			Position (secondary engagement)	
Performing a mechanical or chemical operation	Use	Use	Use		
Removing one object from another object with which it had been "assembled"	Disassemble		Disassemble	Disengage (if recoil is present)	

Determining the location of anything	Search find	Search		Reach case C	
Making a choice between two or more pieces that are in a known location	Select	Select	Mental process	Reach case C	Visual direction
Examining the characteristics of anything	Inspect	Inspect		Eye focus Eye travel	Eye time
Determining a method for accomplishing anything	Plan	Plan			
Letting go of the object by laying it down, dropping it, or throwing it	Release load	Release load	Release	Release	Included in move
To overcome the effect of weight, friction, and the like, or to exert precise control				Apply pressure and static component	Force
The delay in:					
Retaining an object in a fixed position without moving it	Hold	Hold			
The operation that permits elimination of fatigue	Rest	Rest			
The operation that is beyond the control of the operator	Unavoidable delay	Unavoidable delay			
The operation that is under the control of the operator	Avoidable delay	Avoidable delay			
The operation caused by the nervous limitations of the human body		Balance delay			

Generic systems exist on the basic level as well as on higher levels. Functional systems that have been derived from basic-level generic systems exist only as higher-level data. Functional systems derived from basic research of human motions, however, may be classified as basic-level systems.

Interactive Characteristics

All PMTSs have certain interactive characteristics or attributes. Among the more important of these, three merit some discussion because of their importance in aiding the industrial engineer in selecting a PMTS for a given work measurement project that will meet the demands the project imposes. These important attributes are accuracy, speed of application, and degree of method description.

Accuracy

Predicting the accuracy of a PMTS is not an easy task. There are many ways in which this accuracy can be expressed. One such way is to state the plus-or-minus percentage deviation in time from a "true" value. As was previously noted, Walton M. Hancock devised a statistical means for determining the accuracy of a PMTS. The results of his work were published in the MTM journal.[6] When the accuracy of a PMTS is expressed in this way, it is necessary to state the confidence level and the length of the operation being analyzed. (The deviation from the "true" value usually decreases as the length of the operation increases, and vice versa.)

The accuracy of basic-level systems is generally greater than that of higher-level systems for a given nonrepetitive cycle length. This is true because the shorter time values of the single-motion elements have much smaller SDs than those of the longer elements of the higher-level systems. Exhibit 4.5.2 is an accuracy chart used by the MTM Associative to specify the relative accuracy of MTM systems to MTM-1.

Speed of Application

Application speed is directly related to the size of the individual motions or elements making up a PMTS. The shorter the average element time of a system, the longer it takes to make an analysis. For this reason basic-level systems take the longest time, whereas higher-level systems are proportionately faster. With manually applied systems, the speed of application is inversely proportional to the accuracy. Exhibit 4.5.3 lists the application speeds of the MTM systems.

Exhibit 4.5.2 Relative Accuracies of the MTM Systems Compared to MTM-1

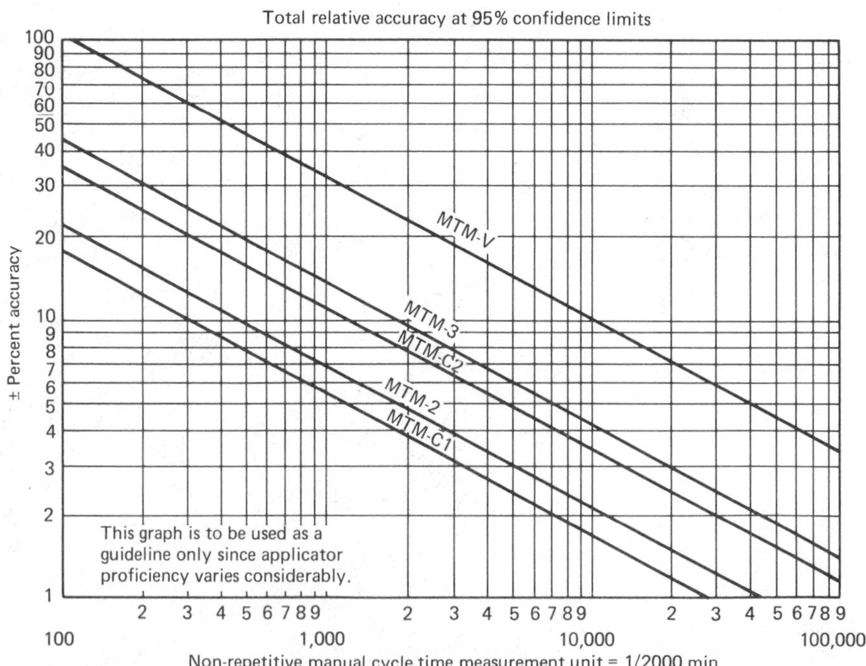

Exhibit 4.5.3 Relative and Absolute Speeds of Application of the MTM Systems

System	Speed of Application			
	Absolute		Relative (Approximate)	
MTM-1	250	times the cycle time	—	
MTM-2	100	" " " "	2	times as fast as MTM-1
MTM-3	35	" " " "	7	" " " " "
MTM-V	10	" " " "	23	" " " " "
MTM-C1	125	" " " "	2	" " " " "
MTM-C2	75	" " " "	4	" " " " "
4 M Data	15 to 60	" " " "	4 to 15	" " " " "

Level of Method Description

The third decision criterion that is inherently a characteristic of the measurement system is the level of methods description. This criterion, like accuracy and speed of application, is directly related to element or motion length. Because of this fact, basic-level PMTSs have the highest degree of method description, whereas higher-level systems have the least.

4.5.4 COMPUTERIZATION

Computerized PMTSs have been in existence since the 1970s. Most, but not all, of these are computerized versions of manual systems. The first computerized PMTS required a mainframe computer. In recent years computerized PMTSs are available also for minicomputers and the increasingly popular desktop microprocessing units.

Mainframe Systems

Several PMTSs have been developed for mainframe computers. These are the computers that service entire plants or corporations. Predominant in the field are systems available from Management Sciences Inc., Appleton, Wisconsin; Science Management Corporation, Moorestown, New Jersey; MTM Association, Fair Lawn, New Jersey; A. T. Kearney Inc., Chicago, Illinois; Rath & Strong, Lexington, Massachusetts; and H. B. Maynard Company, Pittsburgh, Pennsylvania.

Minicomputers

Most of the systems available for mainframe computers can also be utilized on minicomputers. Minicomputers are generally used by small companies and departments within larger corporations. The main advantage of the minicomputer is the opportunity for greater industrial engineering deparmental control.

Desktop Microprocessors

Predetermined motion time systems have been developed for desktop microprocessing units, consisting of a display unit, a keyboard, disc drives, and high-speed printers. Systems are also available for use on programmable hand-held calculators. Among the first to develop PMTSs for desktop devices were General Analysis Corporation, Inc., Los Angeles, California; Science Management Corporation; and the MTM Association.

4.5.5 PREDETERMINED MOTION TIME SYSTEMS

The remainder of this chapter is devoted to those PMTSs in common use today.

Manual Systems

Basic Motion Timestudy

Basic Motion Timestudy is a generic, basic-level system that was developed in the 1950s by Gerald B. Bailey and Ralph Presgrave, consulting partners of Woods, Gordon & Company, a Canadian management consulting firm with its main office in Toronto, Canada. The system and its application rules are described in detail in a book, *Basic Motion Timestudy*, written by Bailey and Presgrave.[8]

In developing the system, the authors defined the term "basic motion" as a movement of any body member starting and ending at rest. This term was chosen as the basic unit of motion because it applies to all body members and actually describes the way they move. It can also be readily recognized.

The basic motion "reach" includes the Gilbreth therbligs "transport empty" and "grasp" whenever the grasp occurs at the end of "transport empty" without stopping. If the hand stops before making the grasp, "reach" includes only the therblig "trasport empty," and the grasp is analyzed separately.

The basic motion "move" includes the placing of an object if there is uninterrupted movement. "Move," then, is equivalent to the therbligs "transport loaded," "pre-position and position," or "transport loaded" only, if the hand stops before the actual placing of the object occurs.

If the hand stops before making the grasp, there will be additional motions, which are treated as separate moves or reaches.

As an example, we may reach to a tote box full of identical objects. The initial contact (terminating the reach) may be made against the supply of objects with the fingers open. A second motion is then used to draw the fingers together until they grasp one or more of the objects. This ends the second motion. If the action has resulted in securing more than one object, additional motions may be required to reject the surplus objects.

Actually, no distinction (other than for purposes of description) is made between Reach and Move. The concept is of a hand moving through space without reference to purpose. The time taken to move a given distance varies according to several influences. The essential variable is the length of the movement.

Muscular Control. The degree of muscular control is dealt with by dividing moves into three fundamental types.

Class A motions are the simplest in type and the lowest in time values. They are the motions that are stopped without muscular effort. They are stopped by impact and thus do not contain an element of deceleration. Thus all muscular effort is directed toward carrying the arm forward. None is directed toward slowing or stopping. Typical of such motions are hammering, slamming a door while keeping the hand in contact, and punching with the fist.

Class B motions, on the other hand, are stopped entirely by muscular control, usually in space. Common examples include the upstroke in hammering, tossing an object aside, opening a door or drawer, or any other situation in which the motion is ended without the hand (or object being moved) coming into contact with another object. The element of deceleration is introduced, with the result that class B motions take somewhat longer to perform than do class A motions.

Class C motions take the longest time to perform because they end in touching an object or surface at the terminal point. This contact is usually a grasping or placing action. Greater control is required than in class B motions. Class C motions are more frequently encountered than any other types of motion in manual activities. Simple examples are reaching for a desk pen or telephone or putting down a paperweight.

Classes B and C have subclasses, BV and CV, respectively.

Visual Direction. The function of the eye in certain motions is the second variable.

The difference is entirely a question of whether or not the eyes move while the motion that needs eye attention is being performed. If the eyes do not have to move during the motion, but can be focused on the end point of the motion in advance, the time to complete the motion is not affected by the eye attention. If the eyes cannot be focused on the end point of the motion before the motion begins, the motion itself is delayed, and the time to perform it is longer. By definition, it then becomes a visually directed motion.

"Precision" and "Application of Force" Allowances. Additional time is allowed if extra care or more exact muscular control is required at the end of a motion in the placing or grasping action. This is covered by a "precision" allowance, which increases with the motion length and tolerances or limits within which the fingertips must be located in the ending action. Time data are provided for five degrees of precision, ranging from $\frac{1}{2}$ to $\frac{1}{32}$ in.

Additional time values are also given for situations where the application of force is required, such as in moving heavy objects or in tightening or loosening actions.

Simultaneous Motions. The system also recognizes that under certain circumstances simultaneous actions can be performed in the same time that is required for one of the arms to perform its motion. That is, it takes no longer for both arms to move than for one arm to move when the motions are identical and when neither (or only one) requires visual direction.

However, when both arms and hands require visual direction to complete their motions, additional time will be required if the end points of the motions are separated from one another. This

Exhibit 4.5.4 Range of Detailed Work-Factor Times

Elements	Time Ranges (Minutes)
Transport (reach and move)	0.0016 to 0.0236
Grasp (Gr)	0 to 0.0189
Pre-position (PP)	0 to 0.0120
Assemble (Asy)	0.0018 to 0.0130
Use (Use)	(Varies with process)
Disassemble (Dsy)	0 to 0.0088
Release (Rl)	0 to 0.0033
Mental process (MP)	0.0020 to 0.0030

is because the eyes have to shift from one end point to the other before one arm and hand can complete its motion.

The time allowances for simultaneous arm motions therefore take into account separation distance or the distance between the end points of the motion's. Also, they take into account the precision required in the ending action.

Whenever body motions occur with arm motions, they are usually complementary to the arm motions and thus do not require separate identification. A common example is bending the trunk in order to complete a long arm motion. Side steps and body turns may also be used in this manner. However, the system provides application rules to deal with overlapping motions when required.

Time Values. In all BMT data tables the time values are expressed in ten thousandths of a minute (0.0001). The tables do not include personal, fatigue, and delay allowance factors.

Short System. Basic Motion Timestudy has adopted a short second-level system, which is less accurate but faster to apply. Also, special clerical data have been developed with very comprehensive reading times and data on the use of keytype office equipment.

Standard Data. Specialized BMT standard data have been developed for specific applications, notably for securing and orienting small parts in high-volume bench assembly operations.

Detailed Work-Factor System

This generic, basic-level system was developed from original motion time studies using stopwatches, photo timers, films, and fast-film snapshots. The detailed time unit is 0.0001 min (see Exhibit 4.5.4). Information on this and other Work-Factor systems is available from Science Management Corporation.

The system was developed under the direction of Joseph H. Quick. Rating was based on the evaluation of skill and effort of operators while being studied. Exhibit 4.5.5 is the analysis of an operation utilizing the Detailed Work-Factor System.

Mento-Factor® System

The Mento-Factor System is a basic-level system that was developed to determine the time required for measuring mental processes such as those involved in decision making.

Ready Work-Factor® System

This generic, second-level system was developed by simplifying Detailed Work-Factor time values. The Ready Work-Factor time unit is 0.001 min. The elements and time range in minutes are shown in Exhibit 4.5.6.

The system is particularly suitable for measuring medium- to long-run operations with cycles of 0.15 min and greater. It can be taught to supervisors and employees relatively quickly.

Exhibit 4.5.7 is an illustration of the use of the Ready Work-Factor System.

Brief Work-Factor® System

A generic, third-level system, the Brief Work-Factor system was developed for measuring nonrepetitive work. The Work-Factor foundation lists these features:

1. It uses time values applied to segments of work rather than individual work motions.
2. There are six time values and 27 classifications.

Exhibit 4.5.5 Detailed Work-Factor Analysis of a Drawing Operation

Part	Sheet No.	Company	Section No.	Part No.	Sub.	Oper. No.
CASE	1 OF 2	JOHN DOE AND SON	37	48-719	0	7

Operation Name & Description: 1ST DRAW– 2 PIN DIE — MACH. #1031 BLISS DOUBLE ACTION DRAW PRESS 240 TONS- 72.5 RPM. BLANK - CR STEEL THK .050 ± .003," DIA. 19.25," WT. 4.04 LBS

No.	LEFT HAND Elemental Description	Motion Analysis	Elem. Time	Cumulative Time		Elem. Time	Motion Analysis	RIGHT HAND ELEMENTAL DESCRIPTION	No.
1	R FOR BLANK	A20D	80	80	80	80	A20D	R FOR BLANK	1
2	GR BLANK	F1W	23	103	103	23	F1W	GR BLANK	2
3	M BLANK TO DIE	A40WSD	159	262	262	159	A40WSD	M BLANK TO DIE	3
4	RL AND CLEAR FINGERS	F3W	28	290	290	28	F3W	RL AND CLEAR FINGERS	4
5	PLACE FINGERS ON BLANK	F3D	28	318	399	109	A40D	R FOR TRIP LEVER	5
6	PUSH BLANK AGAINST PINS	A2P	29	347	415	16	F1	GR LEVER	6
7	WITHDRAW HAND (A10) + WAIT	BD	146	493	493	78	A10WW	PULL LEVER TO START PRESS (7 LBS)	7
8	M HAND TO HOLD BLANK (TURN SIMO)	A30D	96	589	589	96	A30D	R FOR OIL RAG	8
9	PRESS DOWN TO HOLD BLANK	A1W	26	615	606	17	F2	GR RAG	9
10	HOLD BLANK (AT CENTER)	-	-		691	85	A12UD	M RAG + DIP IN OIL PAN	10
11	" " " "	-	-		723	32	A6	M RAG FROM OIL PAN	11
12	" " " "	-	-		749	26	A4	SHAKE RAG (SQUEEZE SIMO)	12
13	" " " "	-	-		804	55	A18	M RAG TO STACK of BLANKS	13
14	" " " "	BD	298	913	913	109	A40U	APPLY OIL (CIRCULAR MOTION)	14
15	RL BLANK	A1W	26	939	955	42	A10	RAISE RAG FROM BLANK	15
16	M HAND FROM BLANK	A16	52	991	997	42	A10	STRIKE TO DISLODGE BLANK	16
17	AP BLANK	A3D	32	1023	1048	51	A15	M RAG TO SIDE	17
18	GR BLANK	F1W	23	1046		—	—	HOLD RAG	18
19	TURN BLANK OVER (RL SIMO)	2A14W	138	1184		—	—	" "	19
20	M HAND TO BLANK CENTER	A13D	67	1251	1222	174	BD	" "	20
21	PRESS DOWN TO HOLD BLANK	A1W	26	1277	1277	55	A18	M RAG TO BLANK	21
22	HOLD BLANK AT CENTER	BD	109	1386	1386	109	A40U	APPLY OIL (CIRCULAR MOTION)	22
23	RL BLANK	A1W	26	1412	1437	51	A15	M RAG ASIDE NEAR PAN (TOSS)	23
24	WAIT	—	-		1517	80	A20D	R FOR TRIP LEVER	24
25	"	—	-		1533	16	F1	GR LEVER	25
26	"	BD	381	1793	1781	248	BD	WAIT FOR END OF MACHINE CYCLE	26
27	M HAND TO PIECE ON PUNCH	A20D	80	1873	1873	92	A15WW	PUSH LEVER TO STOP PRESS (10 LBS)	27
28	CATCH PIECE ON PALM (GRAVITY)	REACT	20	1893	1896	23	F1W	RL LEVER	28
29	M PIECE TO CHUTE (BALANCE)	A40WPD	159	2052		—	—	WAIT	29
30	TOSS TO CHUTE	A5W	43	2095	2095	199	BD	"	30
31	TURN TO WORK TABLE	T14D	100	2195	2195	100	T14D	TURN TO WORK TABLE	31
32									32

MACHINE CYCLE = .1380 MIN. (RH elements 8 to 27)

The Work-Factor® COMPANY	H.B. AMSTER, ENGINEER	TOTAL 2195; DATE 1-20-	Select Time MINUTES .2195	Multiplier 2.45	STD. HOURS PER 100 .538	STD. PCS PER HR.

® TRADE MARK — FORM 101

Part	Sheet No.	Company	Section No.	Part No.	Sub.	Oper. No.
CASE	2 of 2	JOHN DOE AND SON	37	48- 719	0	7

Operation Name & Description: 1ST DRAW (CONTINUED)

LEFT HAND — Elemental Description — Motion Analysis — Elem. Time — Cumulative Time — Elem. Time — Motion Analysis — RIGHT HAND ELEMENTAL DESCRIPTION

NOTES

1. MACHINE TIME IS .1380 MINUTES (OCCURS DURING RH ELEMENTS 8 to 27 INCLUSIVE)

2. TIME FOR ELEMENT #26 IN RH (248) DETERMINED AS FOLLOWS:

(493 + 1380 - 1533 - 92)

WHERE: 493 IS CUMULATIVE TIME AT END OF ELEMENT #7 IN RH
1380 IS MACHINE TIME
1533 IS CUMULATIVE TIME AT END OF ELEMENT #25 IN RH
92 IS TIME FOR ELEMENT #27 IN RH (PUSH LEVER TO STOP PRESS)

The Work-Factor® COMPANY	ENGINEER	TOTAL; DATE	Select Time	Multiplier		

® TRADE MARK — FORM 101

Exhibit 4.5.6 Range of Times for Ready Work-Factor System

Elements	Time Ranges (Minutes)
Transport (reach and move)	0.002 to 0.017[a]
Grasp (Gr)	0 to 0.008[a]
Pre-position (PP)	0 to 0.009[a]
Assemble (Asy)	0 to 0.013
Use (Use)	(Varies with process)
Disassemble (Dsy)	0 to 0.010
Release (Rl)	0 to 0.002
Mental process (MP)	0.002 to 0.003

[a]The differences between element values in the various systems are not significant since they are offset by the statistical characteristics of the system.

3. The system includes an even simpler format, using only four time values.
4. Comparative terms are not used to classify work difficulty. All classifications are numerical, eliminating judgments and providing high consistency in application.
5. It is applied to nonrepetitive work.
6. The system was compiled from Detailed Work-Factor.
7. It is a third-level system compatible with other Work-Factor systems.
8. Training in the Brief Work-Factor system requires from 5 to 15 classroom hr, depending on time study experience.

Methods Time Measurement Systems

MTM-1. The MTM-1 system is a generic, basic-level system. Information on MTM-1 and other MTM systems is available from the MTM Association.

In 1940 a large group of time study analysts completed a methods improvement program conducted by the Methods Engineering Council. The analysts who were trained in the MTM system achieved substantial cost reductions in applying it to production systems. However, the inventors of MTM, in making an analysis of these results, became convinced that the cost reductions were actually the result of methods correction rather than an outgrowth of true methods engineering.

Maynard, Stegemerten, and Schwab therefore searched for a means by which good methods might be established in advance of production. They deduced that, if operators learned the best method as they began a new task, the need for marked improvements later would be lessened. Training costs would also be lower. This would be a boon to managers plagued by production problems, labor difficulties, lack of training guides, and little usable knowledge for correct methods establishment prior to starting production.

They decided to study common industrial operations and endeavored to develop "methods formulas." Their initial choice was sensitive drill press operations. They intended to extend the same approach into other areas if they were successful in building suitable formulas in their first effort, so that there would eventually result a body of information desired by work analysts. Research was then conducted to expand the results into one of the first predetermined time systems to gain general public acceptance and usage.

The MTM originators took a practical approach to the stumbling block of standard time measurement through the use of performance rating. First, while actual shop runs of drilling operators were being photographed, several seasoned raters used a system developed by Lowry, Maynard, and Stegemerten to rate independently the parts of the operation and the operation as a whole. Second, the film was analyzed for the motion content of the operation in question. Third, the consensus of ratings was applied to the frame counts to yield normal motion times. Essentially, then, everyone who applies the MTM data times to motions equivalent to the well-defined categories set up by the MTM system tacitly agrees that the ratings of the original observers constitute a standard of normal. This means that all persons who properly apply MTM are using the same yardstick.

The time represented by one motion picture frame naturally depended on the speed at which the film was taken and projected. Constant speed equipment was used to ensure uniform time increments for each frame. By applying the performance rating in percentage terms, it was then possible to find the average time consumed by an average operator in the frame in question.

However, it was of practical necessity to assign time values in units that could be easily used and that yielded numerical results that could readily be used to supply the input to cost systems. Most industries rely on decimal minutes or decimal hours for their measurement and subsequent costing of labor. Time units such as these, however, would have been difficult to use because many of the

Exhibit 4.5.7 Ready Work-Factor Analysis of Assembling Pegs to Board

WORK-FACTOR TWO HAND ANALYSIS FORM

Part Name	Sheet No.	Company	Department	Part No.	Sub.	Oper. No.
Pegboard - 30 Pegs	1 of 1					

Operation Name & Description: Asy pegs to board - 1 hand

No.	LEFT HAND Elemental Description	LEFT HAND Analysis	LEFT HAND Time Units	Cumulative Time		RIGHT HAND Time Units	RIGHT HAND Analysis	RIGHT HAND Elemental Description	No.
1								R to 1st peg from edge of	1
2					7	7	20-1	table	2
3					10	3	2-	Gr 1st peg	3
4					12	2	0-50%	P P peg	4
5					18	6	10-2	M peg to hole	5
6								Asy beveled end of	6
7					23	5	CT-.4-3/8	peg to hole	7
8					24	1	0-	Rl peg	8
9					29	5	10-1	R to 2nd peg	9
10					46	17	El 3-8	Gr and Asy 2nd peg	10
11					442	396	18 x EL 9-10	PU and Asy 18 more pegs	11
12					447	5	10-1	R to 21st peg	12
13					448	1	0-	Gr peg (isolated)	13
14								M peg to board -	14
15					454	6	10-2	PP internal	15
16					459	5	CT-.4-3/8	Asy peg to hole	16
17					460	1	0-	Rl peg	17
18	Hold Board	BD	622	622	622	162	9 x EL 12 - 17	PU and Asy 9 more pegs	18
19									19
20									20
21									21
22									22
23									23
24									24

wofac	Date: 1 July 76	Analyst: E. Boepple	Total: 622	Time in Minutes: .622	Multiplier		13.5/2/ (69/1)

basic motions were very short in elapsed time of performance and many zeros would have been needed between the decimal point and the first significant digit. This is clearly illustrated by the film speed times expressed in commonly used time units. Since the film speed used in the original research was 16 frames/sec, each frame covered an unrated elapsed time of 0.0625 sec, 0.0010417 min, or 0.00001737 hr.

The obvious way to avoid such unwieldy time units was to recognize that units are arbitrary by nature, with the necessity for conversion to other desired units being the only real limitation. Maynard, Stegemerten, and Schwab therefore invented a new time unit, known as the time measurement unit (TMU), and assigned 0.00001 hr as the value of one TMU. Since most wages are in dollars per hour, the TMU can be multiplied by the hourly rate and the decimal point then shifted five places to the left to find the cost of labor directly. Also, the hours required to produce 100 motions (or pieces) can be found by shifting the decimal two places to the left.

As a result of the unit chosen, the following time conversions are valid:

1 TMU = 0.00001 hr   1 hr = 100,000 TMU
= 0.0006 min   1 min = 1667 TMU
= 0.036 sec   1 sec = 27.8 TMU

The research was verified by Maynard, Stegemerten, and Schwab. However, validation was possible only by an independent, unbiased source. Such validation followed quickly on publication of the MTM textbook by the inventors.[5] Cornell University conducted an independent investigation and reported on it for the Management Division of the American Society of Mechanical Engineers at its annual meeting held in New York City November 26–December 1, 1950.[9] A copy of this report (paper number 50-A-88) can be obtained from the ASME.

Exhibit 4.5.8 illustrates the use of MTM-1 in analyzing a simple operation.

MTM-2. The generic MTM-2 system was developed by the International MTM Directorate, an organization composed of 12 national MTM associations. It is based on MTM-1 and constitutes the second level of the MTM family, with 39 time values. The system has a speed of analysis twice that of MTM-1, but a somewhat lower precision in time prediction.

In developing MTM-2, frequency distributions were made of more than 22,000 MTM-1 motions collected from companies using MTM in the United States, Sweden, and Great Britain. The distri-

Exhibit 4.5.8 An MTM-1 Analysis of Sharpening Pencil With Hand-Held Sharpener

MTM ASSOCIATION
FOR STANDARDS
AND RESEARCH

SHARPEN PENCIL

Sheet 1 of 1
SYSTEM: MTM-1
STUDY NO.
DATE: 10-20-80
ANALYST: [illegible]

LEFT HAND DESCRIPTION	F	LH MOTION	TMU	RH MOTION	F	RIGHT HAND DESCRIPTION
Reach to sharpener		R6B	8.6	R5B		Reach to pencil
Grasp		G1A	2.0	G1A		Grasp
Toward Pencil		(M4B)	10.3	M6C		To sharpener
			11.2	P1 SD		In sharpener
			1.7	mMfA		Extra insertion
			5.6	G2		Secure hold
			34.0	T120S	5	Turn to sharpen
			6.0	RL1	3	Release pencil
			20.4	T120	3	Turn hand back
			6.0	G1B	3	Grasp
			7.5	D2E		Remove pencil
Sharpener Aside		M6B	8.9	(M4B		Pencil aside
		I45S	–	F45S)		
			122.2	= 4.4		seconds

Exhibit 4.5.9 An MTM-2 Analysis of Sharpening Pencil With Hand-Held Sharpener

MTM ASSOCIATION FOR STANDARDS AND RESEARCH

SHARPEN PENCIL

Sheet 1 of 1
SYSTEM: MTM-2
STUDY NO.
DATE: 10-20-80
ANALYST: Kgk

LEFT HAND DESCRIPTION	F	LH MOTION	TMU	RH MOTION	F	RIGHT HAND DESCRIPTION
Get sharpener		GB6	10	GB6		Get pencil
			26	PC6		Place in sharpener
			3	PA2		Extra insertion
			30	PA6	5	Sharpen
			30	GB6	3	Get pencil
			6	PA6		Remove pencil
Sharpener aside		PA6	6	PA6		Pencil aside
			110	= 4.0 seconds		

bution was found to be essentially the same in the three countries. The developers then used this information in the development of MTM-2.

Exhibit 4.5.9 illustrates the MTM-2 analysis of the same operation analyzed with MTM-1 in Exhibit 4.5.8.

MTM-3. The generic MTM-3 system was also developed by the International MTM Directorate from the same 22,000 MTM-1 motions used in developing MTM-2. The system has 10 time values and is the third level of the MTM family of systems. It has a speed of analysis that is seven times faster than MTM-1. It can be used in situations where a less detailed methods description is required and where reduced precision can be tolerated.

Exhibit 4.5.10 shows the MTM-3 analysis of the operation previously analyzed with MTM-1 and MTM-2.

MTM-GPD. The first higher-level system to be developed under the auspices of the MTM Association is called "MTM General Purpose Data" (MTM-GPD). This system is both generic and functional in character and has data on two levels.

The generic data of the second level were derived from specific motion patterns in MTM-1 using average distance ranges having midpoints at 1, 6, 12, 18, and 24 in. All motions of MTM-1 are included in the elements of these data.

Exhibit 4.5.10 An MTM-3 Analysis of Sharpening Pencil With Hand-Held Sharpener

MTM ASSOCIATION FOR STANDARDS AND RESEARCH

SHARPEN PENCIL

Sheet 1 of 1
SYSTEM: MTM-3
STUDY NO.
DATE: 10-20-80
ANALYST: Kgk.

LEFT HAND DESCRIPTION	F	LH MOTION	TMU	RH MOTION	F	RIGHT HAND DESCRIPTION
Sharpener to work area		(H-)	34	HB6		Pencil to sharpener
			7	TA6		Extra insertion
			14	TA6	2	Reverse & first turn
			54	HA6	3	Sharpen pencil
			7	TA6		Pencil out
Sharpener aside		TA6	7	TA6		Pencil aside
			123	= 4.4 seconds		

Exhibit 4.5.11 An MTM-GPD Analysis of Assembling Bolt To Fixture With Wrench

MTM ASSOCIATION FOR STANDARDS AND RESEARCH

ASSEMBLE BOLT IN FIXTURE WITH HAND TOOL

Sheet 1 of 1
SYSTEM: MTM-GPD
STUDY NO.
DATE: 10-20-80
ANALYST: KKK.

LEFT HAND DESCRIPTION	F	LH MOTION	TMU	RH MOTION	F	RIGHT HAND DESCRIPTION
Get bolt from tray		BGT-JO-12	25			Get wrench from bench
Place bolt in fixture		BPL-CS-12	31			
			35	BPL-CN-12		Place wrench on bolt
			37	BTL-WB-15		First turn - 120°
			96	BTL-WB-16	2	Two additional turns - 120°
			13	BPL-AL-12		Wrench aside to bench top
			13	BPL-AL-12		Fixture aside to bench top
			250 = 9 seconds			

The functional data of the second level are contained on a second data card and pertain essentially to the use of hand tools. The elements were also developed from specific motion patterns of MTM-1.

The third level is called multipurpose data and contains both generic and functional data. The generic elements combined the "get" and "place" elements of the second level. The functional elements cover the activities of clamping and vising.

Exhibit 4.5.11 illustrates the use of MTM-GPD.

MTM-C. This functional clerical data system is a full clerical work measurement system at two levels of job description, precision, and speed of analysis. The system was developed by a consortium of banking and service industries.

The level 1 (second-level) data are comprehensive in scope, covering activities in nine areas: getting and placing, opening and closing, fastening and unfastening, filing, reading and writing, typing, miscellaneous handling, body motions, and machine operation. The element codes are six-digit numeric, each of which is documented with a specific MTM-1 motion pattern.

Level 1 also serves as documentation for level 2, which covers the same activities at a higher level (third level). Distance ranges are reduced to one, and codes are simplified alphanumeric and mnemonic.

Examples of level 1 and level 2 operations are shown in Exhibits 4.5.12 and 4.5.13.

MTM-V. This functional work measurement standard data system based on MTM-1 was developed for machine tool users. The MTM-V system is of the fourth level and contains time values for handling and adjusting work pieces of any weight and size, including machine tool setup, attaching crane hooks, and other mechanical handling equipment. Process-controlled activities are not included. The system's 12 elements are of two types: those that can be accomplished by the hand and finger alone, and those that require the use of a hand tool to accomplish the objective. The technique is stated to be more than 20 times faster to apply than MTM-1. Exhibit 4.5.14 is a typical example of an MTM-V operation.

MTM-M. The functional, basic-level MTM-M system is specifically designed for use where assembly is performed under stereoscopic microscopes (microassembly). It has proved highly advantageous not only for time determination, but also for method improvement in this type of work. This system was developed by a consortium of industrial members in cooperation with the University of Michigan.

Data were developed from original research and are unique to the use of stereoscopic microscopes in assembly work. The data are contained in four tables, which specify motion direction as they pertain to entering and leaving the microscopic field. A section of the data card is shown in Exhibit 4.5.15.

Exhibit 4.5.12 An MTM-C1 Analysis of Replacing Page in Three-Ring Binder

MTM-C OPERATION ANALYSIS

MTM ASSOCIATION FOR STANDARDS AND RESEARCH

VALIDATION
Sheet of

MTM-C Level 1

Replace page in 3-ring binder

DEPARTMENT: Clerical | ANALYST: CNR | DATE: 11/77

No.	Description	Reference	Element TMU	Occurrence per Cycle	TMU per Cycle
1	OPEN BINDER				
	Get binder from shelf	113 520	21	1	21
	Aside to desk	123 002	22	1	22
	Get cover	112 520	14	1	14
	Open cover	212 100	15	1	15
2	LOCATE CORRECT PAGE				
	Read on first page	510 000	7	2	14
	Locate approximate	451 120	16	3	48
	Identify page number	440 630	22	3	66
	Locate correct page	450 130	18	4	72
	Identify pages	440 630	22	3	66
3	REPLACE PAGES				
	Get binder rings	112 520	14	1	14
	Open rings	210 400	21	1	21
	Get old sheet	111 100	10	1	10
	Aside sheet to basket	123 002	22	1	22
	Get new sheet	111 100	10	1	10
	Insert sheet in binder	462 104	64	1	64
	Get rings	112 520	14	1	14
	Close rings	222 400	21	1	21
4	CLOSE COVER AND ASIDE BINDER				
	Get cover	112 520	14	1	14
	Close cover	222 100	13	1	13
	Get binder	112 520	14	1	14
	Aside binder to shelf	123 002	22	1	22
				TOTAL TMU PER CYCLE	577
				ALLOWANCES ___ %	
				STANDARD HOURS PER ___ UNIT	
				UNITS PER HOUR	

577 TMU = 20.8 seconds

Exhibit 4.5.13 An MTM-C2 Analysis of Replacing Page in Three-Ring Binder

MTM ASSOCIATION FOR STANDARDS AND RESEARCH

MTM-C OPERATION ANALYSIS

VALIDATION
Sheet of

MTM-C Level 2

Replace page in 3-ring binder

DEPARTMENT: Clerical | ANALYST: CNR | DATE: 2/77

No.	Description	Reference	Element TMU	Occurrence per Cycle	TMU per Cycle
	Get and aside binder	G5A2	29	1	29
	Open cover	O1	29	1	29
	Read first page	RN2	14	1	14
	Locate pages	LI2	129	1	129
	Identify pages	I3	22	6	132
	Open rings	O4	35	1	35
	Remove sheet	G1A2	32	1	32
	New sheet on rings	HI14	84	1	84
	Close rings	C4	34	1	35
	Close cover	C1	27	1	27
	Aside binder	G5A2	29	1	29
				TOTAL TMU PER CYCLE	575
				ALLOWANCES ___%	
				STANDARD HOURS PER ___ UNIT	
				UNITS PER HOUR	

575 TMU = 20.7 seconds

Modular Arrangement of Predetermined Times

The MTM system has produced other functional and specific systems developed by various consultants and associations.

In 1964 the Australian Association for Predetermined Time Standards and Research (AAPTSR) was formed to develop higher-level data sets based upon the two major PMTSs employed in Australia at the time: MSD and MTM. Chris Heyde acted as general director of the AAPTSR.

In 1965 the International MTM Directorate issued MTM-2, and the AAPTSR conducted a field

Exhibit 4.5.14 An MTM-V Analysis of Assembling Bolt to Fixture With Wrench

MTM ASSOCIATION FOR STANDARDS AND RESEARCH

ASSEMBLE BOLT IN FIXTURE WITH HAND TOOL

Sheet 1 of 1
SYSTEM: MTM-V
STUDY NO. ______
DATE: 10-20-80
ANALYST: KRh

LEFT HAND DESCRIPTION	F	LH MOTION	TMU	RH MOTION	F	RIGHT HAND DESCRIPTION
Get bolt from bench and place in fixture			200	FLE22		Get wrench from bench and tighten bolt with 3 turns. Aside wrench and fixture to bench top.
			200 =	7.2 seconds		

Exhibit 4.5.15 A Section of the MTM-M Data Card

11–INSIDE FIELD to INSIDE FIELD

CODE RANGE (D/T)				1	2	3	4	5	6	7	8	9	10	11	
RANGE DISTANCE ÷ TOLERANCE				TO 0.75	0.75 TO 1.5	1.5 TO 3.0	3.0 TO 6.0	6.0 TO 12	12 TO 25	25 TO 50	50 TO 100	100 TO 200	200 TO 350	350 TO 725	SIMO ADDITIVE
TOOL AND COND		CODE	CHARACTERISTIC												
GRASPING	ET	C*	GTC (TOOL CONTACT GR)	4.3	5.7	7.5	9.2	11.0	12.8	14.6	16.3	18.1	19.6	21.3	2.1
		G	GT (TOOL GRASP)	3.7	7.6	12.2	16.7	21.3	26.0	30.7	35.2	39.8	43.8	48.2	12.8
	LT	N*	NO RELEASE–SCRUB	3.7	4.2	4.7	5.2	5.7	6.2	6.8	7.3	7.8	8.2	8.7	+
		S	RLTC–2 DIM MOVE	3.0	3.5	4.2	4.8	5.4	6.1	6.7	7.3	8.0	8.5	9.1	3.9
		V	RLTC–3 DIM MOVE	5.6	6.1	6.8	7.4	8.0	8.7	9.3	9.9	10.6	11.1	11.7	3.9
		O	RLT (TOOL RELEASE)	4.5	6.4	8.6	10.7	12.9	15.2	17.4	19.5	21.7	23.6	25.7	9.6
	EF	A	ALL CONDITIONS	10.6	12.6	15.0	17.5	19.9	22.4	24.9	27.3	29.7	31.8	34.2	0.0
	LF	N	NO RELEASE	3.3	5.5	8.0	10.5	13.0	15.7	18.3	20.8	23.3	25.5	28.0	15.9
		M*	RLM OR RLMC	4.1	7.9	12.5	17.0	21.5	26.3	30.9	35.4	40.0	43.9	48.3	5.6
	ED	A	ALL CONDITIONS	5.0	8.6	12.7	16.9	21.1	25.4	29.7	33.9	38.1	41.7	45.8	6.0
PROBING	ES	A	ALL CONDITIONS	8.9	11.8	15.3	18.9	22.4	26.0	29.6	33.1	36.6	39.7	43.1	0.0
	LS	A	ALL CONDITIONS	2.0	3.1	6.7	10.3	13.9	17.7	21.4	25.0	28.6	31.8	35.3	21.5
	EP	A	ALL CONDITIONS	4.6	6.4	8.6	10.8	12.9	15.2	17.4	19.6	21.7	23.6	25.7	0.0
	LP	A	ALL CONDITIONS	2.8	6.5	10.8	15.2	19.5	24.0	28.4	32.8	37.1	40.9	45.1	2.2
CUTTING	EC	A	ALL CONDITIONS	–	–	–	30.0	30.0	30.0	36.8	53.4	69.9	84.3	100.3	+
	LC	A	ALL CONDITIONS	67.3	69.9	73.0	76.1	79.2	82.4	85.5	88.6	91.7	94.4	97.4	+
STRIPPING	EZ	A	THERMAL–ALL CONDITIONS	2.0	4.7	10.0	15.3	20.5	26.0	31.3	36.6	41.9	46.4	51.5	+

+NO SIMULTANEOUS MOTIONS INCLUDED IN DATA BASE

*POWER FACTOR–MAGNIFICATION POWER HAS AN EFFECT ON TMU VALUES ON THOSE DATA LINES LISTED BELOW. EACH CHART VALUE IS ADJUSTED AS FOLLOWS:

ETC–WHEN MAG POWER EXCEEDS 20× ADD 0.68 TMU FOR EACH POWER GREATER THAN 20×, e.g., 30×, ADD 0.68(30 – 20) = 6.8 TMU

LTN–WHEN MAG POWER EXCEEDS 5×, SUBTRACT 0.06 TMU FOR EACH POWER GREATER THAN 5×, e.g., 20×, SUBTRACT 0.06(20 – 5) = 0.9 TMU

LFM–WHEN MAG POWER EXCEEDS 5×, ADD 0.94 TMU FOR EACH POWER GREATER THAN 5×, e.g., 20×, ADD 0.94(20 – 5) = 14.1 TMU

10–INSIDE FIELD to OUTSIDE FIELD

CODE RANGE (D/T)				1	2	3	4	5	6	7	8	9	10	11	12	
RANGE DISTANCE ÷ TOLERANCE				TO 0.75	0.75 TO 1.5	1.5 TO 3.0	3.0 TO 6.0	6.0 TO 12.0	12.0 TO 25.0	25.0 TO 50.0	50.0 TO 100	100 TO 200	200 TO 350	350 TO 725	725 TO 3000	SIMO ADDITIVE
TOOL AND COND		CODE	CHARACTERISTIC													
ALL	RL	M*	ASIDE AND RLM OF TOOL OR OBJECT	10.5	14.0	18.1	22.2	26.3	30.6	34.7	38.8	42.9	46.5	50.5	57.8	5.3
GRASPING	ET	C	GTC AND NONE	10.9	13.0	15.4	17.9	20.4	23.0	25.5	28.0	30.5	32.6	35.0	39.6	0
		G	GT (TOOL GRASP)	43.2	44.9	46.9	49.0	51.0	53.2	55.3	57.3	59.4	61.2	63.1	66.9	+
	LT	A	ALL EXCEPT RLM	15.3	17.4	19.8	22.3	24.7	27.3	29.8	32.2	34.7	36.8	39.2	43.6	0
	EF	A	ALL EXCEPT RLM	14.8	17.4	20.5	23.7	26.7	30.0	33.1	36.2	39.4	42.1	45.1	50.7	0
	LF	A	ALL EXCEPT RLM	12.5	14.7	17.4	20.0	22.6	25.3	28.0	30.6	33.2	35.5	38.1	42.9	0
	ED	A	ALL EXCEPT RLM	7.4	12.5	18.6	24.7	30.7	37.0	43.2	49.3	55.4	60.6	66.5	77.4	0
PROBING	ES	A	ALL EXCEPT RLM	14.5	18.1	22.3	26.5	30.8	35.2	39.5	43.7	48.0	51.7	55.8	63.5	0
	EP/LP	A	ALL EXCEPT RLM	14.2	17.7	21.8	25.9	30.0	34.3	38.5	42.6	46.7	50.3	54.3	61.8	0
CUTTING	EC	A	ALL EXCEPT RLM	7.1	10.6	14.8	18.9	23.0	27.3	31.5	35.7	39.8	43.4	47.4	54.9	0
STRIPPING	EZ	A	ALL EXCEPT RLM	7.0	15.8	26.1	36.5	46.8	57.6	68.1	78.4	88.8	97.8	107.8	126.3	+

+NO SIMULTANEOUS MOTIONS INCLUDED IN DATA BASE
*POWER FACTOR: WHEN MAG POWER EXCEEDS 5× ADD 0.61 TMU FOR EACH POWER GREATER THAN 5×

test of MTM-2 compared to MSD in order to determine which might be the better system upon which to develop higher-level data sets. No firm conclusions were reached because MTM-2 emerged slightly better in some tests, whereas MSD did better in others. Thus the AAPTSR decided to develop a new system.

The basic research in developing distance-time matrices and spatial arrangements led to the development of the Modular Arrangement of Predetermined Times (MODAPTS), a generic and functional second-level system. The basic MODAPTS research data, in their original unedited form, are on file in Sydney in the offices of the AAPTSR. The system was tested by AAPTSR members in their shop environments and compared favorably with MTM-1, MSD, and MTM-2. It was released in 1966.

The basic unit in MODAPTS is a simple finger movement. All other activities are expressed in terms of this finger movement or module. There are only eight different values—0, 1, 2, 3, 4, 5, 17 and 30 mods. The basic mod value is 0.129 sec. The eight mod values are applied to 21 types of activities derived from movements of fingers, limbs, body, and eyes.

In MODAPTS the first identification is the class of movement, and the second tag or label is that which happens at the end of the movement, or the "terminal activity." There are two categories of terminal activities—"obtaining control" and "things to destination." Each category has three different mod values, which are selected based upon the type of terminal activity.

Exhibit 4.5.16 demonstrates the use of MODAPTS in setting a standard for a simple U-bolt assembly operation. The MODAPTS analysis sheet describes the operational steps and the time values in mods allowed for each movement or activity.

Office MODAPTS and transit MODAPTS are the two MODAPTS subsystems. The latter is a recent development in the area of warehousing and materials handling standard data.

Master Standard Data

Serge A. Birn Company developed the generic, second-level MSD system in the late 1950s. It was developed to set standard MTM-based data on manually controlled operations where production was less than 100,000 units/year or a few thousand units per week. Between production runs, the operator would lose most of the skill he or she would develop. Statistically, a very high percentage of work in industry falls within this limited practice category. Master Standard Data was developed

Exhibit 4.5.16 MODAPTS Analysis Sheet

Operation: Assembly of $\frac{1}{4}$ in. diameter steel rod U-bolt, with flat bearing plate, two plain washers, and two $\frac{1}{4}$ in. hexagon nuts

Step	Code	Frequency[a]	Mod Units
1. Pick up U-bolt (left hand) and plate (right hand), bring together, put legs of U-bolt through the two holes in plate.	4, 3, 4, 5	1	16
2. Pick up first washer (right hand), put onto one leg of U-bolt (held in left hand).	4, 3, 4, 2	1	13
3. Ditto other washer on other leg.	4, 3, 4, 2	1	13
4. Pick up first nut, put on threaded leg.	4, 1, 4, 5	1	14
5. Screw down nut until thread shows.	1, 0, 1, 0	5	10
6. Pick up and put on second nut.	4, 1, 4, 5	1	14
7. Screw down second nut as above.	1, 0, 1, 0	5	10
8. Put aside completed U-bolt assembly.	4, 0	1	4
Total mod units			94
Multiply by 0.129 for normal seconds			12.1

[a]The sum of the code numbers is multiplied by the frequency to obtain mod units.

by statistically studying all motions; consequently, because many motions studied occur rarely, they can be included as random variables.

Master Standard Data comprises the most common MTM-1 motions, the B, C, and D reaches; all the grasps, with the exception of the G1C and the nonsymmetrical positioning; A, B, and C moves; P1 and P2 positions; TS turns; and both releases and "apply pressure." Likewise, simultaneous motions are unlikely with the lack of practice; therefore a simplified simultaneous motion chart was constructed which assumes "no practice opportunity." Also, six simplified tables combined with decision charts established the tables:

Obtain–O, place-P, rotate, use, finger shift, and body motions.

Master Standard Data became one of the first higher-level horizontal data systems. Much of the work done by Serge A. Birn Company contributed to the development of MTM-GPD; MSD is also similar to MTM-2. Exhibit 4.5.17 illustrates an MSD analysis.

Exhibit 4.5.17 An MSD Analysis of Blanking and Forming Operation

Code 7565-10

M. S. D. ANALYSIS SHEET

Department or Activity: Dept. #5 - Fabricating

Operation: Blank and Form Bushing from Strip

Conditions: Minster #4 Press, 20 bushings from 2" x 42" x 18 ga. strip, stock on table at right of operator, bushings blown out of die, skeleton asided to tote box at left of press.

Prepared By: RMC Approved By: HWN Date: 9/1/61 Sheet 1 of 1

Seq.	Description	Data Code	Time	Freq.	Total
1	Skeleton	O12H2	38	1	38
2	From die	EF	11	1	11
3	Get blank (R.H.)	O12H1	25	1	25
	Aside skeleton	P12G	-	-	-
4	Blank to L.H.	P6O	11	1	11
5	Blank to die	P12L1	21	1	21
6	Hands to trip	O12S2	17	1	17
7	Press cycle	MT	30	1	30

Total:	153
% Allowance	23
Total Allowed	176
Std. Time	.0018
Prod. Per Hour	555

Exhibit 4.5.18 An MCD Analysis of Recording and Posting Operation

Sheet 1 of 1 | CLERICAL METHODS ANALYSIS SHEET | Operation No. 4

Division Comptroller | Department Sales
Section Customer Service | Job Order Indexing
Operation Record and Post Branch Order
Date 2/17/60 Analyst Moll | Supervisor H. Payne

MCD CODE	DESCRIPTION	VAR.	Work Units	Freq.	Total Units	Seq. No.
G-MT	Pencil	Env.	33	1	33	
W-D	Total Number (2 digits)		18	2	36	
G-MT	Order Book		33	1	33	
G-VF	Aside Order Book		46	1	46	1
O-BC	Open and Close Book		48	1	48	
R-D	Read Order Number (10 digits)		2	10	20	
W-D	Record Number (10 digits)		18	10	180	396
		BO	396	1/55	7	

CODE	Identification	Frequencies
Env.	Envelope	55 BO/Envelope
BO	Branch Order	

TOTAL WORK UNITS 7

A set of standard data developed from MTM is called Master Clerical Data (MCD), illustrated in Exhibit 4.5.18. It is functional in nature, dealing exclusively with the clerical function.

Maynard Operation Sequence Technique

This technique was developed by the Swedish Divison of H. B. Maynard and Company, Inc., in 1967–1972 and was introduced in the United States in 1974.* The MOST work measurement system is applicable for any cycle length and repetitiveness, as long as there are variations in the motion pattern from one cycle to another. Based on the structure and theory of MTM-1 and MTM-2, MOST systems can be applied to direct productive work, such as machining, fabrication, and assembly, as well as to material handling, distribution, maintenance, and clerical activities.

All necessary procedures and principles for development of suboperation data in any work area are available as MOST application systems. The major features are:

1. Compilation of work conditions in work management manuals.
2. Filing of suboperation data in a central data bank, based on a uniform coding system.
3. Construction of data application forms, that is, work sheets or spread sheets based on statistical principles.

The system employs a small number of predetermined models of fixed activity sequences which cover practically all aspects of manual activity.

*Information regarding this proprietary technique was contributed by Kjell Zandin, senior vice president, H. B. Maynard and Company, Inc. The technique, based on MTM-1 and MTM-2, was developed in Sweden under Zandin's direction.

The differences between the levels are the multipliers. Identical index numbers are applied on all levels. The multipliers are as follows:

1. Basic sequence models (basic MOST) = multiplier 10.
2. Bridge cranes and wheeled trucks = multiplier 100.
3. Job preparation, and so on = multiplier 1000.

General Move. The first sequence model—"general move"—is defined as moving an object from one location to another freely through the air. This can account for as much as 35% of the work of a machine operator and even more for an assembly worker. This activity is represented by the following sequence of letters or subactivities:

A B G A B P A

A = action distance (mainly horizontal hand or body motions)
B = bend (mainly vertical body motions)
G = grasp
P = position

The variations for each subactivity are indicated by an index figure, for instance,

$$A_6 \quad B_6 \quad G_1 \quad A_1 \quad B_0 \quad P_3 \quad A_0$$

A_6 = walk 3 to 4 steps
B_6 = bend and rise
G_1 = grasp one light object with one hand
A_1 = move within reaching distance
B_0 = no bend
P_3 = place objects with adjustments
A_0 = no return move

The index values representing various elements are memorized from an index table. The developers state that all values on this and other index tables are derived from MTM-2 or MTM-1.

The time value for the sequence is obtained simply by adding together the index numbers and multiplying the sum by 10. For instance, the standard time for the preceding sequence is

$$6 + 6 + 1 + 1 + 0 + 3 + 0 = 17 \qquad 17 \times 10 = 170 \text{ TMU}$$

Controlled Move. "Controlled move" is a movement of the object that is restricted in at least one direction by contact with or an attachment to another object.

Tool Use. "Tool use" covers not only conventional hand tools such as wrenches, screwdrivers, gauges, and writing tools, but also fingers and mental processes.

Consistency. Brinckloe[10] emphasized that MTM-1 precision and its years of trial by use make it a body of knowledge in which industrial engineers can have high confidence. The MTM-based systems (including MOST) that draw on this body of knowledge owe their legitimacy to this heritage.

The MOST system is an MTM-based system designed above all for consistency and with its table subdivisions selected so that they will span typical small variations in method and layout. The design meshes well with the instinctive "rounding" tendency of experienced applicators and avoids the time consuming "fine-tuning" analyses in areas where they may be too detailed for variations encountered in practice, while retaining good accuracy in areas where such accuracy pays off.

It appears that there is a large category of operations for which the consistent behavior of MOST will produce standards with accuracy approximately equivalent to that of higher-level systems, but with far less effort and in far shorter time.

Exhibit 4.5.19 is an example of a MOST-developed operation.

Computerized Systems

WOCOM System

The generic, basic-level WOCOM system provides automated application of the two most recognized systems of predetermined times: Work-Factor and MTM. It is designed for simple operation

Exhibit 4.5.19 A MOST Analysis of Changing Punch and Die in Strippit Press

MOST-calculation

Code: 1 1 1 0 4 4 0 3 0 1
Date: 3/19/75
Sign.: KZ
Page: 1 / 1

Area: Fabrication Department

Activity: Exchange Punch and Die in Strippit Press

No	Method
1	Loosen Tools in Holder
2	Remove and Aside Tools to Drawer (A); Close Drawer (B)
3	Open Drawer (A), Get New Tools and Position in Holder (B)
4	Secure Tools in Holder

No	Sequence model	Fr	Time
2A	$A_1\ B_0\ G_3\ A_3\ B_3\ P_1\ A_0$		110
3B	$A_1\ B_3\ G_3\ A_3\ B_0\ (P_3)\ A_0$	(2)	160
1	$A_1\ B_0\ G_1\ M_3\ X_0\ I_0\ A_0$		50
2B	$A_1\ B_0\ G_1\ M_1\ X_0\ I_0\ A_0$		30
3A	$A_1\ B_0\ G_1\ M_3\ X_0\ I_0\ A_0$		50
4	$A_1\ B_0\ G_1\ M_3\ X_0\ I_3\ A_0$		80

TIME = .29 ~~TMU~~/MINUTES | 480

and flexibility in meeting varying individual company needs and does not require computer equipment or previous experience in work measurement methods. Programs are available through the General Electric time-sharing network or for operation on in-house computers. Information on the system is available from the Work-Factor Foundation.

The system consists of eight modular computer programs that can be used to analyze human and machine work, alternate work methods, and assembly line operations and to maintain labor standards. A test installation in a division of the RCA Corporation produced the following results:

1. The system proved to be accurate and faster than the existing method of establishing new Detailed Work-Factor standards.
2. Training time for personnel on the new system was less than half of that required for manual application of Detailed Work-Factor.
3. The system provided the standards group with more time to devote to industrial engineering projects, since most of the effort devoted to the establishment and maintenance of standards could be accomplished by personnel other than industrial engineers.
4. A 50% reduction in the clerical phase of establishing Detailed Work-Factor standards was achieved.

UnivEl®—Data Base for UnivAtion® Systems

In 1964 Willard L. Kern organized Management Science, Inc., of Appleton, Wisconsin, for the purpose of developing and marketing computer software for industrial engineering applications utilizing new concepts.

The systems utilize universal formulas to generate precise elemental standard times. The system allows the formulas to be resident in the computer core to eliminate the need to develop, store, and retrieve standard data. Because values were developed through the use of mathematical formulas, only numerical input data are needed in the system. This eliminates the task of developing standard data stored and coded through alphanumerically organized reference tables and of maintaining the data as well as the significant alphanumeric coding structures as future changes are required.

The totally integrated systems are named "UnivAtion Systems" and are made up of modules for various manufacturing controls in which PMTSs are employed using UnivEl as a data base.

Output information automatically generated or mass changed through the use of the UnivEl system consists of elemental methods instructions and standards and a complete "network data base" of all manufactured products and operations.

Micro-Matic Methods and Measurement Data System (4M Data)

Westinghouse Electric Corporation, Pittsburgh, Pennsylvania, developed the logic and procedures that are the basis for the Micro-Matic Methods and Measurement (4M) Computerized Data System.[11] The 4M Data System acquired its name as follows:

"Micro": The system retains all the variables incorporated in the MTM procedures of reach, move, grasp, and position.

"Matic": The methods analysts need no longer refer to a data card or table of values since the simple codes required to use the system are based on easily remembered categories. In addition, the system determines automatically a near-optimum method, subject only to restrictions imposed by the analyst in describing the general method.

"Methods": The ability to develop and define sound manual methods is inherent in MTM-1.

"Measurement": While "get" and "place" input codes are utilized to minimize the analysis time, the codes are designed so that the computer can interpret them, recognize their basic-level components, and apply them correctly for the two hands.

Optimal criteria for the selection and application of MTM procedures have been incorporated in the computer program. As a consequence, the program logic develops from the analyst's get/place inputs the more detailed motion components that are required to apply MTM-1 precisely. The program then applies these microsegments of data in close compliance with established MTM-1 rules for two-hand motion patterns.

A number of method improvement indexes are generated by the system to assist the industrial engineer in recognizing places where methods could be improved. These indexes include:

1. MAI—a percentage that indicates to what degree both hands are utilized.
2. RMB—a percentage of the total element or operation cycle time directly influenced by "reach" and "move" distances and the performance of body motions.

3. GRA–cycle time indicating the complexity involved in performing grasps.
4. POS–cycle time identifying the precision required in object placement in jigs and fixtures.
5. PROC–a ratio of waiting-for-process time to total cycle time.

The system reduces the application time of MTM-1 by a factor of 1 to 4. The precision of the MTM-1 system is maintained. However, a number of second-level elements are also included, which, if used exclusively, reduce application time still further, possibly by a ratio of 1 to 15. Precision with the second-level elements is somewhat less than with MTM-1, but greater than with any of the manual second-level systems.[11] The system is specifically designed for the development and application of standard data elements, but individual operation studies can be made with equal ease.

Output information includes several versions of operation standard reports, operator method instructions, and listings of element and part-operation data.

Maynard Operation Sequence Technique

Through the use of a minicomputer quite often located directly in the industrial engineering department, or a mainframe IBM in the data processing department, with on-line interactive access for the industrial engineer, MOST time standards can be calculated from workplace data and a method description. The computer will thus do all the work measurement, relieving the industrial engineer of practically all paperwork.

The MOST computer systems, which are generic and second- and third-level, include all necessary steps and procedures to develop a complete time standard for an operation, such as:

1. Development of suboperation data based on MOST work measurement systems.
2. Calculation of time standards, including manual times, process times, and allowances.
3. Printing of method instructions and route sheets.

The core of the program consists of comprehensive filing and editing procedures for retrieval, revision, and mass updating of time standards. The output formats can be tailor-made to customer conditions and requirements.

The H. B. Maynard Company asserts that, by using the computer, 1 hr of work can be measured with 1 to 5 hr of analysis, and a time standard can be calculated in an average of 5 min.

Process Time. The computer can generate a series of optional modules to calculate process time for such operations as machining, welding, line balancing, and labor reporting. These times can be integrated at the proper point to give a complete standards calculation.

Simulation. Finally, the computer gives the user simulation capabilities. Proposed work standards changes can be displayed to determine whether the new method or equipment will add or subtract time from the standard. Hypothetical conditions can be created for the method, workplace, and process.

CUE–Modified MTM-1 Data

The generic, basic-level CUE system consists of modified MTM-1 data programmed into Texas Instruments' hand-held calculator, TI-59. Information on the system can be obtained from General Analysis, Inc.

The CUE-I system is set up using the predominant hand (or the hand that performs the longer motion); CUE-II is a two-hand methods analysis version. The later program applies the modified data on a two-hand basis, performing detailed, simultaneous motion analyses and computing extent of hand idle time, average motion complexity, average motion length, and external process time.

The TI-59 has available five buttons with which to handle 10 user labels. The buttons are marked "A" to "E"; by pressing a second key, these same buttons will handle programs A′ to E′. Therefore higher-level data are directly inputted into the calculator either in observing an operation or in simulating an operation while estimating the time.

A 12 in. "get" would be inputted by pressing buttons 1 and 2, then A, or 12A; a 6 in. "move" would be 6B; and an "apply pressure" (AP 1) 7E.

Besides the accuracy evaluations of General Analysis, Inc., three national companies have made independent comparisons of CUE to their MTM-1 studies. A summary of the results (combined) is shown in Exhibit 4.5.20.

The average difference shown in Exhibit 4.5.21 is the average of the absolute percent deviations of CUE from MTM-1. This measure does not allow plus and minus differences to cancel.

The operation description is handwritten, but the output is printed on the TI 100A® printer to which the TI-59 calculator is attached.

Exhibit 4.5.20 Turn Casting in Lathe Annotated CUE Output

Operation Description	CUE Code	CUE	Output
Obtain and Pull Down Lever	12A	12.	GET
	6B	6.	MOVE
Get Next Casting to Machine	20A	20.	GET
(10 Lbs.)	20.10B	20.1	MOVE
Wait for Machine Cycle	.05C′	0.0550	MIN.
Loosen Holder	4A	4.	GET
	7E		AP1
	90A′	90.	TURN
Remove Finished Casting	4A	4.	GET
	A	0.	GET
To Bin	15.10B	15.1	MOVE
Place New Casting in Fixture	15A	15.	GET
	4B	4.	MOVE
	9C	9.	POSN
Turn and Tighten Holder	6A	6.	GET
	90A′	90.	TURN
	7E		AP1
		329.	TMU
		0.1973	MIN.
		15	%PFD
		0.2269	TOT

Exhibit 4.5.21 CUE Application and System Error

Cycle Time (min)	Absolute Average Difference, Percentage MTM-1 System
⩽0.02	7.7
0.021 to 0.04	4.0
0.041 to 0.06	2.6
0.061 to 0.08	2.5
0.081 to 0.10	2.3
>0.10	2.3

Automatic Data Application and Maintenance

The Automatic Data Application and Maintenance (ADAM) system of PMTS, a generic and functional second-, third-, and fourth-level system, was developed for use on desktop microprocessing units. It represents the state of the art in computerized work measurement. The core program is used with modules developed for specific PMTSs. The first module to be developed was the computerized version of MTM-C, a clerical work measurement version. The ADAM version is called ADAM-C. Other modules that have been developed are ADAM-2, computerized MTM-2, and ADAM-V, computerized MTM-V. The modules may be used separately or in any combination. Information about them is available from the MTM Association.

Inputs to and outputs from the system are not rigidly structured. Adaptability to specific needs is a prominent feature of the system. Data entry may also be accomplished in several ways—directly by keyboard, through a data collection device, or by a "wand" passed over bar codes.

Exhibit 4.5.22 is a flowchart of the system, and Exhibit 4.5.23 illustrates a typical output.

4.5.6 A FORWARD LOOK AT WORK MEASUREMENT TECHNOLOGY

At the beginning of each decade, there is always an interest in trying to look ahead and to predict the possibilities of future endeavors. All predictions are based upon past and current situations from which we extrapolate trends for the future. The formulation of a good forecast therefore depends upon a thorough analysis of trends. The only way we can predict the future is to have power to shape that future.

Exhibit 4.5.22 Flowchart of ADAM System

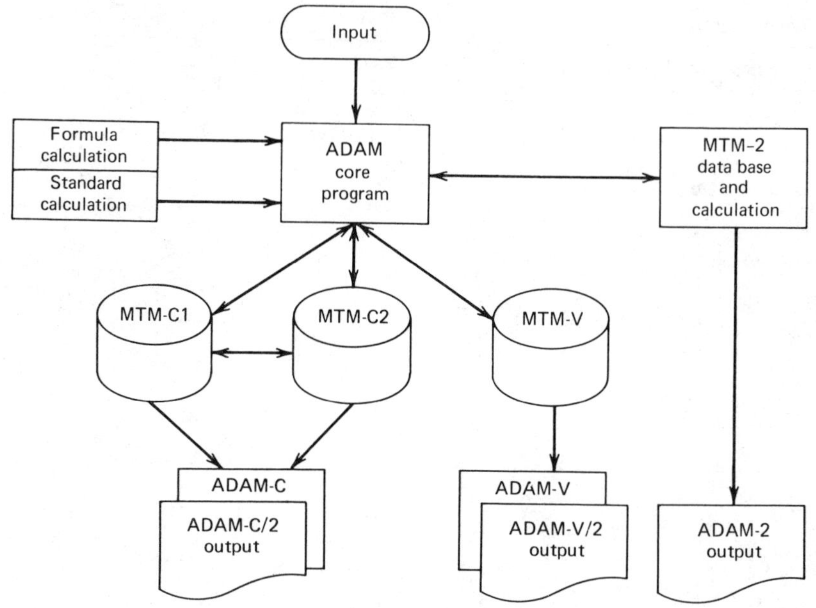

Total ADAM system ADAM—C/V/2

Exhibit 4.5.23 ADAM-C Analysis of Replacing a Sheet in a Three-Ring Binder

MTM ASSOCIATION FOR STANDARDS & RESEARCH

DEPARTMENT:	CLERICAL
FUNCTIONAL AREA:	FILING
TASKNAME:	REPLACE PAGE IN 3-RING BINDER
TASKCODE:	123-XXX-321
DATE: 02-25-80	REVISION DATE: –
UNIT OF MEASURE:	SHEETS REPLACED
BY: DOUGLAS M TOWNE	APPROVED: R. JONES
REMARKS:	MTM-C LEVEL 2 ANALYSIS

TOTAL TMU	575.0000
PF&D ALLOWANCE	0.1500
STND. TMU	661.2500
STND. HRS./UNIT	0.0066
UNITS/STND. HR.	151.2287

LINE	DESCRIPTION	ELE.	TIME	FREQ	TOTAL
1	GET AND ASIDE BINDER	G5A2	29	1/1	29
2	OPEN COVER	O1	29	1/1	29
3	READ FIRST PAGE	RN	7	2	14
4	LOCATE PAGES	LC12	129	1/1	129
5	IDENTIFY PAGES	I30	22	6	132
6	OPEN RINGS	O4	35	1/1	35
7	REMOVE SHEET	G1A2	32	1/1	32
8	NEW SHEET ON RINGS	HI14	84	1/1	84
9	CLOSE RINGS	C4	35	1/1	35
10	CLOSE COVER	C1	27	1/1	27
11	ASIDE BINDER	G5A2	29	1/1	29

We are seeking correct answers about the future. Although we would like a foolproof method of doing this, an alternative is to base our forecast on the views of many authorities in comparison with our own.

At the first Industrial Engineering Managers' Seminar conducted by the AIIE in Detroit, Michigan, on March 12, 1979, one of these authorities, the keynote speaker, F. James McDonald, executive vice president of GM, pointed out what a top executive thinks is happening in the field of industrial engineering, including work measurement.

He pointed out that the image of the industrial engineer as a time standards guy with a stopwatch and a clipboard is obsolete. The word "guy" does not even fit anymore, since many women are entering the field. Likewise, the stopwatch will be replaced by electronic data collectors that have great memory capacity and that can be patchcorded to a large frame computer for further processing of data.

He emphasized that industrial engineers are interested in quality and productivity improvement and therefore have a fascination for robots. The use of robots and programmable machines to perform some of the repetitive and unattractive manual operations will increase and profoundly influence the way a work measurement analyst functions.

McDonald accentuated the fact that energy considerations will be more and more imperative. He stressed that the matter of energy will have profound implications for products, processes, and production facilities. On the other hand, he underscored the fact that industrial engineers' greatest challenge is to build maximum flexibility into their programs in order to facilitate the shifts in production that will be required.

The Effect of MIL-STD 1567

On June 30, 1975, MIL-STD 1567 (USAF) Work Measurement was published. This military standard, initiated by the Air Force, has since been adopted by the U.S. Department of Defense to be applied to all major system production contracts of $100 million annually. Because of this thrust, more industries and service organizations will be considering more seriously the statistical accuracy of their work measurement systems. Therefore PMTS, time study, work sampling, and standard data will be required to be at a stipulated degree of statistical accuracy.

Clerical Work Measurement

Banks, hospitals, public utilities, financial and insurance companies, and government institutions are growing rapidly. The size of the clerical work force is requiring a large portion of our nation's resources. The word-processing units are creating an ever-increasing avalanche of clerical paperwork.

According to Sylvia Huth,[12] future businesses must tackle this deluge of paperwork by storing, condensing and reducing the time it takes to shuffle it.

Current PMTSs relating to clerical work measurement will see greatly increased usage.

Videotape Equipment

A number of companies are making use of videotape equipment for the collection of work measurement data and the documentation of methods. Direct computer data collection is being explored by Notre Dame University in connection with video tape recording. Built-in digital timers can supply time in hours or in hundredths of a minute. Therefore it is possible to review repeatedly just how much time was required for any element of work.

Photographic Data Collection

Another method of collecting data is by means of time-lapse video tape recording and photographic methods. Time-lapse is a technique that uses a camera to expose film at intervals. Using time-lapse, 8 hr of work can be reviewed in approximately 15 min. These time-lapse films can be used to record or monitor any type of activity for methods improvement.

Gregory A. Howell, President of Timelapse, Inc., Mountain View, California, is making use of time-lapse films on construction operations throughout the United States. He states that by means of time-lapse photography he is able to ascertain the amount of nonproductive and productive time on various construction jobs using fixed interval work sampling.

By making use of the time-lapse technique in the construction of a gigantic maintenance hangar, problems were identified that enabled the contractor to reduce man-hours from triple the estimate to 30% under the estimate, according to Howell.

Technology is already available that would permit the direct transmission of videotape and photographic data to a computer for further documentation and analysis. The future of work measurement systems looks challenging indeed!

REFERENCES

1. A. G. SHAW, "Motion Study," in H. B. Maynard, Ed., *Industrial Engineering Handbook*, 3rd ed., McGraw-Hill, New York, 1971.
2. A. B. SEGUR, "The Use of Predetermined Times," *Industrial Management Society Clinic Proceedings*, Chicago, 1964.
3. W. G. HOLMES, *Applied Time and Motion Study*, Ronald, New York, 1938, pp. 251–256.
4. J. H. QUICK, J. H. DUNCAN, and J. A. MALCOLM, *Work-Factor Time Standards*, McGraw-Hill, New York, 1962, pp. 157–221.
5. H. B. MAYNARD, G. J. STEGEMERTEN, and J. L. SCHWAB, *Methods Time Measurement*, McGraw-Hill, New York, 1948.
6. W. M. HANCOCK, "The System Precision of MTM-1," *The Journal of Methods-Time Measurement*, Vol. 15, No. 3, pp. 4–10.
7. QUICK et al., *Work-Factor Time Standards*, p. 13.
8. G. B. BAILEY and R. PRESGRAVE, *Basic Motion Timestudy*, McGraw-Hill, New York, 1958.
9. K. C. WHITE, *Predetermined Elemental Motion Times*, ASME, Detroit, 1950.
10. W. D. BRINCKLOE, "The Impact of Variation in Method or Workplace on the System Precision of MTM-Based Standards," H. B. Maynard and Company, Inc., Pittsburgh, unpublished, July 1979.
11. J. C. MARTIN, " The 4M Data System," *Industrial Engineering*, Vol. 16, No. 3 (March 1974), pp. 32–38.
12. S. HUTH, "MTM-C Management Tool/Measurement Tool," *The Journal of Methods-Time Measurement*, Vol. 6, No. 3 (1979), pp. 12–18.

BIBLIOGRAPHY

ARNWINE, W. C., and W. F. FIELDER, JR., "Determining Requirements and Measuring Standards Accuracy," *MTM Association Spring Conference Proceedings*, Los Angeles, 1978.

AULANKO, V., J. HOTANEN, and A. SALONEN, (English version by K. EADY), *Standard Data Systems and Their Construction*, originally published by Finnish MTM Association, rights acquired by the MTM Association for Standards and Research, 1977, pp. 2-11 to 2-13.

BAYHA, F. H., W. M. HANCOCK, and G. D. LANGOLF, "More Evaluation Parameters for MTM Systems," *The Journal of Methods-Time Measurement*, Vol. 2, No. 1 (1975), pp. 18–30.

BIRN, S. A., R. M. CROSSAN, and R. W. EASTWOOD, *Measurement and Control of Office Costs*, McGraw-Hill, New York, 1961, p. 171.

CROSSAN, R. M., and H. W. NANCE, *Master Standard Data*, McGraw-Hill, New York, 1962, p. 123.

DASCHBACH, J. M., and E. W. HENRY, "Computerized Video Work Measurement," *Computers and Industrial Engineering*, Vol. 4, No. 1 (1980), pp. 13–18.

EADY, K., "What is the MTM Family Really Like?," *The Journal of Methods-Time Measurement*, Vol. 1, No. 2 (1974), pp. 42–55.

EADY, K., "MTM System Accuracy and Speed of Application," *The Journal of Methods-Time Measurement*, Vol. 6, No. 4, pp. 7–10.

HANCOCK, W. M., "New Research Techniques in Work Measurement," *Journal of Methods-Time Measurement*, Vol. 8, No. 1 and 2 (1961), pp. 1–5.

HANCOCK, W. M., "An Assessment of the Research Activities of the U.S./Canada MTM Association 1948–1971," *The MTM Journal of Methods-Time Measurement*, Vol. 16, No. 5 (1971), pp. 4–19.

MUNDEL, M. E., "Motion and Time Study, Synthesized Standards from Basic Motion Times," in W. G. Ireson and E. L. Grant, Eds., *Handbook of Industrial Engineering and Management*, Prentice-Hall, Englewood Cliffs, NJ, 1955, pp. 373–378.

"Predetermined Time Systems," Army Ordnance Corps., Section VIII, Washington, DC, 1954.

PRESGRAVE, R., and G. B. BAILEY, "Basic Motion Timestudy," in H. B. Maynard, Ed., *Industrial Engineering Handbook*, McGraw-Hill, New York, 1963, pp. 5-97 to 5-100.

QUICK, J. H., *The Work-Factor Systems, A Response to the British Employment Department*, questionnaire on predetermined time systems, Science Management Corporation, New York, 1973.

SEGUR, A. B., "Synthetic Times and Methods Design," *Industrial Management Society Clinic Proceedings*, Chicago, 1973.

SHAW, R. J., "MODAPTS–A New Work Measurement System," Peat, Marwick, Mitchell, New York, April 1971.

WEAVER, R. F., J. J. KOLLMAN, and E. A. BOEPPLE, JR., "Developing Standards by Computer," *Industrial Engineering*, January 1978, pp. 26–31.

CHAPTER 4.6

Work Sampling

W. J. RICHARDSON
Lehigh University

ELEANOR S. PAPE
University of Texas at Arlington

4.6.1 INTRODUCTION

In its most basic form, work sampling is one of the simplest work measurement techniques available to the industrial engineer, and yet it can be adapted to analyze sophisticated models of work pattern variation. The supervisor who makes 10 trips through his or her shop, finds a particular machine idle six times, and hence estimates that the machine is idle 60% of the time is engaging in a simple work sampling analysis. The supervisor knows the machine is not idle exactly 60% of the time, but this is the best estimate available, and if quick action is necessary, it will probably be taken based on this estimate.

This chapter deals with methods of making the snap observations to find such an estimate and gives procedures for determining how close that estimate is to the true percentage (or proportion) being estimated. Obviously the estimate will improve as more observations are taken. Thus the same procedures can be used to calculate the number of observations necessary to produce an estimate as close (in probability) to the true percentage as is desired.

Work sampling is particularly useful in the analysis of nonrepetitive or irregularly occurring activities where no complete methods or frequency descriptions are available. Because a work sampling study normally extends over a long (2 to 4 week) period, occasional irregularities do not overly affect results.

The origin of work sampling is generally attributed to L. H. C. Tippett, who worked with the English textile industry during the early 1930s.[1] The technique originated by Tippett was introduced in the United States in 1941 by R. L. Morrow,[2] who called the method "ratio delay." At the outset the technique did not gain wide acceptance. The name "work sampling" was coined in articles by C. L. Brisley[3] and H. L. Waddell[4] in 1952. Not only did the more descriptive name help the technique gain attention, but the article was timely because increased attention was then being paid to indirect labor, an area in which work sampling is particularly useful. Also, increasing numbers of able practitioners in the field of industrial engineering meant many people were able to make discriminating use of the technique presented in Brisley's article.

4.6.2 DEFINITIONS FOR BASIC WORK SAMPLING

"Work" is defined as any activity being studied. "Categories of work" are completely defined, mutually exclusive and exhaustive descriptions of activity such that an observation can be classified as belonging to one and only one category. Selection of categories is discussed in Section 4.6.9 of this chapter. Other definitions are as follows:

K = total number of categories into which one activity is classified

p' = true proportion, or fraction, of time that a worker (or machine or process) is engaged in a specific category of work (This is the quantity to be estimated in a study.)

$p'_1, + p'_2, \ldots, p'_K$ = true proportions of time occupied by the 1st, 2nd, . . . , Kth work categories to be estimated in a study ($P'_1 + p'_2 + \ldots + p'_K = 1$)

N = total number of observations made in order to estimate p' (Sometimes two or more studies are conducted at the same time; e.g., if 100 observations of machine A find it idle 32 times, and if 100 observations of worker Q find him or her walking 17 times, N is 100 for each distinct study, not 200. However, if two similar machines are being observed in order to estimate the amount of idle machine time, and if machine A is idle 32 times while machine B is idle 17 times, $N = 200$.)

X = total number of observations that find a worker (or machine or process) in a specific category

$X_1, X_2, \ldots, X_K$ = total number of observations of category 1, category 2, . . . , category K $(X_1 + X_2 + \ldots + X_K = N)$

p = an estimate of p' found by X/N

$p_1, p_2, \ldots, p_K$ = estimates of $p'_1, p'_2, \ldots, p'_K$ found by $X_1/N, X_2/N, \ldots, X_K/N$

4.6.3 GOALS OF WORK SAMPLING STUDIES

The immediate goal of a work sampling study is to produce "good" estimates of a proportion p', or the set of $p'_1, p'_2, \ldots, p'_K$ proportions corresponding to the K different categories into which a job's activities have been subdivided. Although statisticians have developed a great deal of theory to determine properties that make an estimator "good," we will concentrate on only two qualities: lack of bias and low variance.

A biased estimator is one that is apt to be too high (or too low) because of the procedures used to produce it. For example, if one work category is "clean up," and if the work sampling observer always leaves before the end of the work shift, the estimate p of p' for this category will probably be too low, and thus p is called a biased estimator of p'. Likewise, if workers are concerned about results of a study and can anticipate the times of observations, p can be biased by workers.

Good work sampling studies are conducted in such a way as to produce unbiased estimators and to get those estimators to be close to the desired quantities we wish to estimate. That is, if 20 different studies were conducted by the same procedure to estimate the same p', the objective would be to have the 20 p estimates resulting from the 20 studies fall very close together. If the hypothetical p estimates would be very close, p would be a low-variance estimator of p'.

If p is as apt to be higher than p' as it is to be lower than p' (unbiased) and if it is close to p' (low variance), then p will be considered to be a good estimate of p'. Efforts are taken in designing a work sampling study to minimize both bias and variance of p, although, as is explained in Section 4.6.6, some methods that decrease variance sometimes increase bias.

Although the immediate goal of a work sampling study is to estimate some p' or set of $p'_1, p'_2, \ldots, p'_K$ so that the set of estimates $p_1, p_2, \ldots, p_K$ may be thought of as goals, it is important that the practitioner not lose sight of the ultimate goals of a study. Work sampling studies are undertaken for a number of purposes, among which are obtaining general information, justifying proposed changes, and setting standards.

Obtaining General Information on Process

Information is not, of course, a goal in and of itself. The implied goals are cost reduction through improved utilization of facilities, either equipment or employees, and identification of areas in need of increased facilities.

A general information study might be undertaken when an industrial engineer is new to a firm or when management raises overall questions about effective utilizations. Good design of a general information study is difficult because the designer does not know, when identifying categories, what the study may reveal. For example, at the end of a study, the category "idle waiting for material" may be found inadequate, and hindsight may reveal a need for knowing what type of material was missing and how it was supposed to be delivered to the work station, for example, by conveyor or fork truck.

Studies of this sort are more useful if they are carefully introduced to management in orientation sessions involving open discussion, explanation, and examples that can be simplified with prepared films.*

A general information study should begin with a short (2 or 3 day) test study, after which categories may need to be redefined.

Some general information studies are used to establish or monitor standards for indirect labor, in which case pace rating is done at the time of observation, and some unit (or units) of output must be collected, as described in Section 4.6.4.

*_How To Conduct a Work Sampling Study_, Industrial Education Films, P.O. Box 398, Harwich Port, MA 02646 (black and white, 20 min).

Justifying Proposed Changes

A frequent use of work sampling by industrial engineers is the gathering of data to substantiate a subjective opinion. The engineer may wish to install a new material handling system, but must justify the expenditure with a reputable estimate of the amount of time (and hence money) being spent on material handling with the current system.

Design of such a special-purpose study is easier than a general information study, but problems may arise with the objectivity of observers who are attempting to prove a point and thus induce bias into estimates.

Setting Standards

Predetermined time standard systems are generally superior to work sampling for setting standards on any well-defined job. Likewise, stopwatch studies are considered preferable for short-cycle, repetitive jobs. However, work sampling can be used if other systems are not appropriate, and work sampling is a very useful tool for determining the unavoidable delay allowances for use in setting standards by any method. (Unavoidable delay allowances are discussed in Chapter 4.4.) Because a work sampling study extends over many days, many different causes of delay can be detected, and hence the information gained in a work sampling study can be useful in establishing allowances.

Work sampling is frequently used to set job standards for jobs having irregular components that vary in the amount of time devoted to any one unit of output. For example, time for reworking defective units can vary, depending on severity of defects. By extending the work sampling study over a long period, a large population of jobs is sampled, and the established standard will fit the mean of that population. The standard will be valid only if the population sampled is representative of the population of future jobs. Standards established by work sampling may be more appropriate for formulating budgets for indirect labor departments than for incentive pay systems.

If standards are to be set, it is generally considered necessary to have observations made by a technician trained in performance rating (see Chapter 4.4) and to establish a unit of output that can be readily counted. These procedures are illustrated in Section 4.6.8.

4.6.4 METHODS OF COLLECTING DATA

The methods of collecting work sampling data vary with the size and purpose of the study and with the availability of special equipment, such as computers, for data analysis.

Self-Observation

If the purposes of a work sampling study can be achieved simply by having the workers who are being studied become aware of the results, self-observation can be effective. The time for an observation is signaled by a buzzer inside a work area or by a portable beeper for workers away from a central work area. The worker then notes on a simple ledger what he (or she) was doing opposite the number of the observation.

Although the worker should know exactly what category of work was being performed at the time of the buzzer (even to the point of constructive or nonconstructive thinking), estimates produced by this method may be subject to a great deal of bias. This technique is most effective with semiprofessional personnel who may be unaware of their own poor utilization of time and who have an interest in improving their own efficiency.

Self-observation has been used successfully with clerical workers when introduced by management consultants seeking justification for new word processing or photocopying equipment.

Observation by Trained Industrial Engineer or Technician

If a trained industrial engineer has been working for some time in an area, he or she is usually the best observer. The engineer has been trained in measurement and is accustomed to rapid appraisal. Further, the engineer has an appreciation of the need for objectivity and may better understand the purpose of work sampling. In production shops, particularly, it is common to use the industrial engineer as the observer. If an engineer must be introduced into an area, it will first be necessary for him or her to learn the types of work and to become acquainted with the people.

If the work sampling study has a wide scope, covering several areas, a trained observer will be necessary, and where any performance rating or leveling is being done as part of the study, the trained observer is essential. Assigning the trained observer to work sampling full time is often more efficient because of practical difficulties in effectively utilizing time between observations.

When observations are to be collected on a number of workers or machines, attention must be

Exhibit 4.6.1 Data Collection Form for Trained Observer

Work Sampling Study Data Sheet

Categories:
1- Avoidable delay
2- Unavoid "A"
3- Unavoid "B"
4- Run
5- Set-up

Observer Hammond Date 6/13 Sheet 1 of 1
Study no. 012 Study title ______

Obs. No.	Time of Observation	1 (#204 table 6) Cat.	Rtg.	2 (#128 table 7) Cat.	Rtg.	3 (#542 table 7) Cat.	Rtg.	4 (#137 table 9) Cat.	Rtg.	5 (#600 table 9) Cat.	Rtg.	6 (#627 table 12) Cat.	Rtg.	7 (#106 table 13) Cat.	Rtg.	8 (#221 table 13) Cat.	Rtg.	9 (#403 table 15) Cat.	Rtg.	10 Cat.	Rtg.
1	7:07	2	–	2	–	2	–	1	–	2	–	2	–	2	–	2	–				
2	7:28	1	–	4	80	3	–	4	60	1	–	4	90	1	–	1	–	5	80		
3	7:41	4	80	3	–	5	90	4	60	4	85	3	–	1	–	4	80	4	90		
4	7:36	4	90	4	75	4	90	3	–	4	90	3	–	4	80	4	80	4	90		
5	8:16	3	–	4	65	4	95	2	–	4	90	4	90	4	80	4	90	3	–		
6	8:23			4	60	4	90	4	80	4	90	4	90	2	–	4	85	4	95		
7	8:–				' –	4	90	4	90	4	95	4	85	1				'	–		
8							–	4	85	4	9'										
9																					

given to the system of data collection, considering (1) the ease of tabulation, (2) the likelihood of incurring errors, and (3) the method to be used to reduce and analyze data.

Currently a variety of systems are in use, including three described here.

Pencil and Paper Form

Any type of watch is used to identify the times of observation (or the start of observation rounds), which have been predetermined. Observations are recorded on a grid-type form similar to that shown in Exhibit 4.6.1, where two adjacent columns, one for category and one for rating, are used for each worker, and each row is used for an observation round.

Data from this form must be reduced by totaling the number of times each category occurs and by averaging the observed ratings on every rated category. Category totals may be desired for each worker, and if the alternate variance estimators (Section 4.6.6) are used, it will also be necessary to total the number of times each category occurs on each observation round.

Computers can be utilized to reduce the data from these forms, but if data are collected in this manner, it will be necessary for someone to input the information from the form to the computer, incurring an additional source of error.

*Punched or Mark-Sense Cards**

Using a watch to identify the predetermined times, the observer records observations on cards that can be read directly by a computer. If a computer program is to be used to reduce data, this will obviously save time. It is wise to have a form similar to that in Exhibit 4.6.1 printed by the computer that can be scanned for obvious errors, such as nonexistent category numbers, by the observer who collected the data.

Electronic Clipboards[5]

Clipboards with built-in watches to signal computer-selected observation times are used to store data electronically until the board is connected to the computer to update the data file for the study (or for several simultaneously taken studies).

*"How to Conduct A Work Sampling Study," Industrial Education Films.

Observation by Supervisor or Other Associate of the Work Group

If rating is not to be undertaken, and if the desired study centers on a single area, some associated member of the work group, generally the supervisor, can be enlisted to collect the sampling data.

The form used for data collection might look more like that in Exhibit 4.6.2, although punched cards or mark-sense cards can be used by supervisors as well as engineers. The choice of data collection method depends more on the amount of data that is to be collected by any one observer than on the background of the selected observer.

The supervisor, of course, has the requisite knowledge of the work and the people in the area. If the supervisor acts as the observer, he or she is doing, in a systematic fashion, that which he or she already does in an informal way; therefore no new personnel will be introduced into the area, and no time will be required to train the observer in the work habits of those being observed. Having taken the data, the supervisor should readily accept the results of the study. Whether or not the supervisor can retain objectivity in data collection is a point for consideration by the engineer initiating the study.

4.6.5 METHODS OF DETERMINING OBSERVATION TIMES

Principles of Sampling

Two principles are important in the selection of observation times: randomization and stratification. An observer who decides to make 24 observations a day for 30 days and who selects 24 random times on each of the 30 days is randomizing within the days and stratifying by days.

When random times are desired, one can use tables of random numbers published in many handbooks and textbooks or the simple random number generators programmed into many hand-held

Exhibit 4.6.2 Data Collection Form for Supervisor

Work Sampling of Payroll Office

Observer Mays Date 4/3

Observation of Random Times

	Employee	8:30	9:02	10:30	11:42	1:50	3:15	4:35
1.	Hard	8	6	10	5	4	7	1
2.	List	4	3	9	7	2	3	6
3.	Walker	9	5	3	4	3	2	4
4.	Smith			Absent				
5.	Hays	3	2	3	7	5	3	1
6.	Barnes	3	3	5	2	4	8	7
7.	Schmidt	9	8	4	2	10	7	6
8.	Jackson	5	4	8	6	5	5	4
9.	Murphy	5	8	4	1	2	10	3
10.	Gaines	2	9	6	5	3	7	1
11.	Harris			Absent				
12.	Bell	2	5	2	8	6	5	9

Categories:

1. Photocopying
2. Filing
3. Typing
4. Filling in ledger

engineering calculators. The last three or four digits of telephone numbers can also be used with reasonable assurance of randomness. If 24 random times, to the closest minute, are required, a simple procedure is to examine three-digit random numbers until 24 are found that correspond to times during the work shift. With a few special rules, the simple correspondence that makes 429 into 4:29; 637 into 6:37, and 933 into 9:33 can be used (see Exhibit 4.6.3). The numbers 783 and 987 would be discarded because 7:83 and 9:87 are not times. The first digit 0 can be used for 10, so 048 becomes 10:48, but two "unused" digits must be coded for use as 11 and 12 if the work shift extends across 11 and/or 12 o'clock. In Exhibit 4.6.3 the 7:45 to 4:15 shift uses the digit 5 for 11 and the digit 6 for 12, so that 637 becomes 12:37, and 531 becomes 11:31. The 9:00 to 5:15 shift uses 6 for 11 and 7 for 12, so that 637 becomes 11:37. The 11:00 to 7:15 shift uses 8 for 11 and 9 for 12, so that 813 becomes 11:13. The times that fall outside the work shift are discarded, as would any time that fell during a lunch period.

The 24 times must then be ordered, since it is impossible to observe the 4:29 reading before the 2:52 reading. It is also possible to generate the random times in order by the procedure suggested by Moder and Kahn.[6]

The randomization is undertaken to reduce the bias introduced by worker anticipation of observation times, and the stratification is undertaken to reduce the variance of the estimates. Variance reduction by stratification is a basic sampling principle and is explained by Moder[7] with specific regard to work sampling. Variance reduction by stratification is understood by most practitioners who realize intuitively that spreading the observations out, not allowing them all to occur in one time period, will give better estimates in the sense of avoiding extremely high or extremely low estimates. Additional variance reduction could be gained by the observer's selecting three random times in each hour of each 8 hr workday, thus engaging in stratification by hours. To carry the point further, one random time could be selected in each 20 min period. To ensure even more spread in the observations, one observation could be made every 20 min on the 20 min, starting at some randomly selected time during the first 20 min period of the day, in which case systematic sampling is being practiced.

The variance reducing properties of systematic sampling are very good,[7] provided that the sampling cycle does not coincide with some natural cycle in the work process. If the procedure in the preceding example were to observe every 20 min on the 20 min for thirty 8 hr days without rerandomizing the starting time, we would expect the same activities to be detected at the beginning and end of each day. That is, if a worker fills out an initial day's job card 2 min after the start of the shift every day (say at 7:02), and if the observer makes an observation at 2 min after the start of every 20 min period during the shift (7:02, 7:22, 7:42, 8:02, etc.), then in 30 days this rare activity would be observed 30 times. But if the observations were made at 6 min after the start of each period (7:06, 7:26, 7:46, etc.), the activity would never be observed. Because the 20 min cycle coincides with the natural 8 hr cycle (24 × 20 min = 8 hr), variance is not reduced but increased; extremely large or extremely small values of the estimate can easily occur. Randomizing the starting time each day gives some protection against this variance increase.

A second problem with systematic sampling is the ease with which observation times can be anticipated. However, when making work sampling studies that are not directly related to workers, such as machine utilization, systematic sampling can produce unbiased, low-variance estimates with

Exhibit 4.6.3 Procedure for Finding Times From Random Numbers

Random Numbers	Example A 7:45 to 4:15 Shift (5 → 11; 6 → 12)[a]	Example B 9:00 to 5:15 Shift (6 → 11; 7 → 12)	Example C 11:00 to 7:15 Shift (8 → 11; 9 → 12)
783	~~7:83~~	~~7:83~~	~~7:83~~
429	~~4:29~~	4:29	4:29
637	12:37	11:37	6:37
048	10:48	10:48	~~10:48~~
933	9:33	9:33	9:33
077	~~10:77~~	~~10:77~~	~~10:77~~
531	11:31	~~5:31~~	5:31
252	2:52	2:52	2:52
504	11:04	5:04	5:04
987	~~9:87~~	~~9:87~~	~~9:87~~
813	8:13	~~8:13~~	11:13

[a]Coded digits.

decreased effort on the part of the observer, who can better organize the observation task when times are regular.

Process for Selecting a Sampling Procedure

J. J. Moder has proposed a decision process[7] for selecting a sampling procedure for each day of a work sampling study (see Exhibit 4.6.4). Moder assumes that observations will be stratified by days under all plans.

With the following definitions, Exhibit 4.6.4 may be of assistance to the inexperienced practitioner:

SyRS. Systematic random sampling is practiced if, when r observations rounds (r observations per worker) are to be taken during a day of t min duration, a time is selected at random during the first t/r min of the day, and subsequent observations are made at intervals of exactly t/r min. In the example given in the first part of Section 4.6.5, $t = 8 \times 60 = 480$ min, $r = 24$, and $t/r = 20$ min.

StCRS. In stratified, continuous random sampling the use of the word "continuous" implies that the observer is assigned 100% to work sampling and thus works "continuously" at work sampling. The observations are still snap observations, and no continuous time study is implied. Randomization is achieved by (1) random selection of the starting point of each observation round (numbers can be assigned to each station and a number selected at random), (2) random selection of several different routes (assign numbers to various feasible routes), and (3) coin flip

Exhibit 4.6.4 Decision Tree Diagram for the Selection of a Sampling Procedure for Work Sampling Studies

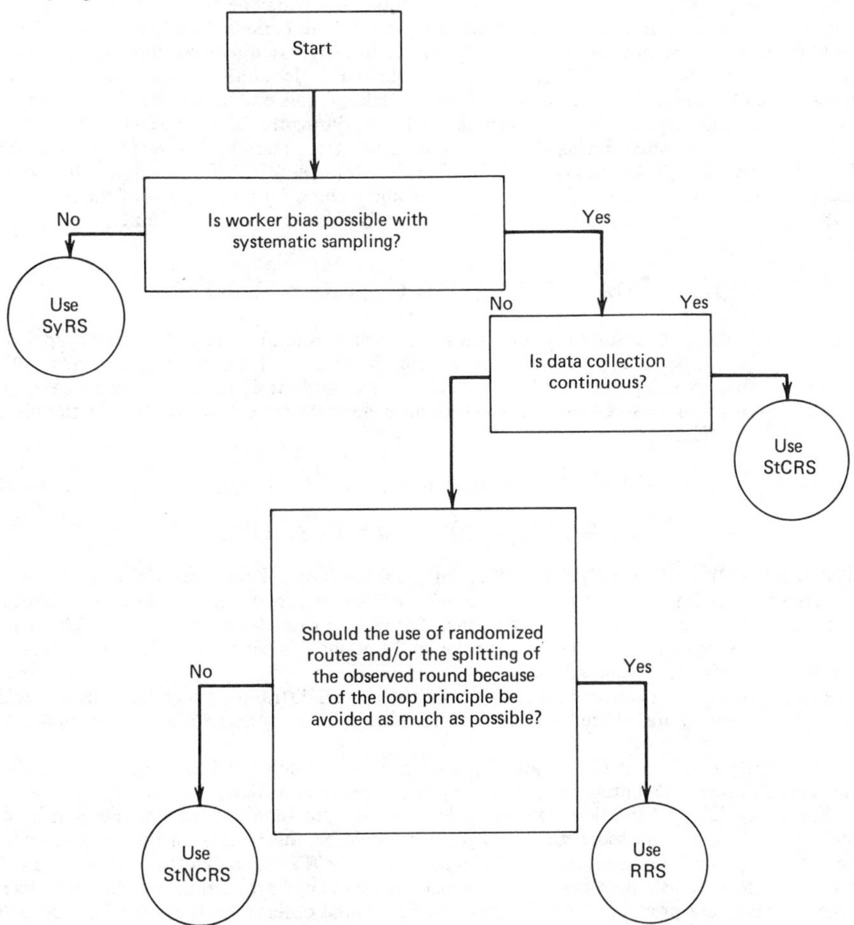

determination of the direction a route will be taken (clockwise or counterclockwise). How much randomization is needed to deter worker bias is a decision that can be made only by those who are familiar with the job process being studied.

StNCRS. Although Moder used the notation "stratified, noncontinuous random sampling" for the procedure when a single random time is selected during each t/r min interval during the day, it could also apply to the practice of selecting three random times in each hour when 24 times are desired during an 8 hr day. If observation rounds are being made, routes can be randomized as with StCRS.

RRS. In restricted random sampling, "random sampling" implies selection of r times at random from the total time period of duration t. When an observer is making a full observation round and two times are selected too close together to allow completion of the earlier round, the restriction implies that the second time *selected* is rejected and replaced with another random time. Since this procedure is adopted at the beginning, no bias is induced by the method, and variance may be slightly decreased.

Loop Principle. When observation rounds of approximately 5 min or more are being made, some bias will be incurred if the loop principle is not recognized. A time selected by any of the preceding procedures is customarily taken to be the starting point of an observation round to be made during a specific period (a day for RRS and SyRS, t/r minutes or "hours" for StNCRS). However, to keep the round within the specific period, that round cannot begin near the end of the period, and similarly the latter part of an observation round will never be observed during the first few minutes of the period. Thus biases can be induced when end and beginning activity differs from midperiod activity.

To remove this source of bias, consider the last observation round of the period as extending into the first few minutes of the same day. That is, if a round takes 10 min, and if the last observation time of the day is 3 min before closing, the last 70% of the round is conducted during the first 7 min of the workday and before the "first" round of the day begins.

If StNCRS is practiced randomizing within shorter time periods, such as hours or 20 min periods, the loop principle should be considered within each period to avoid these biases and may become too confusing to be of benefit. For example, if a single round is to be conducted at some random time during each 20 min period, if a round takes 6 min, and if the random start time "17 min" is selected, only half the round can be completed during the 20 min period, and the second half should be conducted during the *first* 3 min of that same period. That is, if a period starts at 1:20, the time from 1:20 through 1:23 would be used for the last half of the determined round, and the time from 1:37 (20 + 17 = 37) to 1:40 would be used for the first half of the round.

If a work sampling study is being conducted of only a few workers or machines, the loop principle can be ignored.

4.6.6 PROCEDURE FOR CALCULATING CONFIDENCE INTERVALS

At the end of a study, or at any time during a study, estimates can be made of any time proportion desired; that is, $p_1, p_2, \ldots, p_k$ can be found for each worker or for a group of workers together. Recognizing that these estimates are not exactly the proportions desired, it is customary to estimate the precision of these estimators by assuming a normal distribution for each of the p's and claiming 95% confidence that

$$p - [1.96 \times (\mathrm{SD})] < p' < p + [1.96 \times (\mathrm{SD})] \tag{1}$$

or

$$p - [Z_{\alpha/2} \times (\mathrm{SD})] < p' < p + [Z_{\alpha/2} \times (\mathrm{SD})]$$

The quantity SD is the standard deviation (or square root of the variance) of the variable p. The true variance of p depends on the manner in which observations are made and on whether or not day-to-day differences are considered to be included in p' or excluded from p', that is, whether the p' desired is the proportion of time during the study period subject to daily fluctuation or some mean value of daily fluctuating values.

The quantity $Z_{\alpha/2}$ is a constant associated with the normal distribution such that the probability of any one normally distributed quantity being more than $Z_{\alpha/2}$ SDs from its true value is less than α.

Several different values of $(1 - \alpha)$ and $Z_{\alpha/2}$ are given in Exhibit 4.6.5, but the 95% confidence obtainable by using 1.96, approximately 2, is most frequent in practice.

The quantity $[Z_{\alpha/2} \times (\mathrm{SD})]$ is referred to here as the "precision" of the estimate p as an estimator of p'. If p is an unbiased estimator of p', as it should be, this same quantity could be referred to as the "accuracy" of the estimate p. Consider a study that is conducted in such a way that worker or observer bias can be expected to increase the measured proportion of productive time. If workers are actually productive 70% of the time, but around observation times are productive 80%

Exhibit 4.6.5 Confidence Levels and Constants

Confidence $(1 - \alpha)$	Constant (Z_α)
0.9000	1.645
0.9500	1.960
0.9546	2.000
0.9900	2.576
0.9974	3.000

of the time, then p would be estimating 0.8, not 0.7, and the quantity $[Z_{\alpha/2} \times (\text{SD})]$ would tell how close p is to 0.8, not 0.7. In a well-conducted work sampling study, precision is accuracy.

The question remains as to how the SD of each p value should be estimated. This question is answered three ways, one of which is presented here, and two of which are given in the appendix to this chapter.

Binomial Assumption

If one is content with estimating a p' that represents the proportion of time during the study period occupied by a specific category, and if observations can be assumed to be independent, the variance of a binomial random variable is a reasonable model, and $\text{SD} = \sqrt{p(1-p)/N}$, or, more properly, for the kth category, $\text{SD}_A = \sqrt{V_A}$ where $V_A = p_k(1 - p_k)/(N - 1)$. The subscript A is used to identify this first estimate of the SD and the variance, V, of p. The subscripts B and C are used in the appendix.

Then if 350 random observations of a machine find it operating 210 times, being set up 68 times, and idle 72 times, it can be estimated that the machine was idle $(72/350) \times 100 = 20.57\%$ of the time during the total time period of the study. That is, $p = 0.2057$ estimates some unknown p'.

Then with 95% confidence it can be stated that

$$0.2057 - 1.960\sqrt{\frac{(0.2057)\,(0.7943)}{349}} < p' < 0.2057 + 1.960\sqrt{\frac{(0.2057)\,(0.7943)}{349}}$$

or $0.163 < p' < 0.248$.

The quantity 0.7943 in the preceding inequality is the quantity $(1 - p)$, and the quantity 1.960 is the $Z_{\alpha/2}$ value for 95% confidence.

This is the estimator (SD_A) of standard deviation that has been used with good results by industrial engineers for more than 40 years. The good results come not so much because the binomial model fits the sampling procedures used as because the high and low biases tend to balance out.

This estimate of the SD of p_k was easy to calculate during the slide rule era of engineering and is even easier to calculate in a day when inexpensive hand-held calculators have square root keys.

The variance estimate, V_A, can be constructed by totaling only X_k values from the data sheets, eliminating the need for much data manipulation.

Without computer data analysis, the calculation of the binomial method presented here more than offsets theoretical advantages of the methods presented in the appendix when it is realized that the immediate goals are the p_k values, not their precision measurements. However, when computer analysis is being made of data, better variance estimates, referred to as V_B and V_C in the appendix to this chapter, can be calculated without using the binomial assumption.

4.6.7 METHODS OF DETERMINING NUMBER OF OBSERVATIONS NEEDED

By far the most important factors in determining the total number of observations needed in a study are the practical considerations. What is the cost of data collection, and what is the value of the information? All too often someone with very good intentions decides that a ±0.01 precision is desired, without realizing the cost of 0.01 precision over 0.05 or even 0.10, which might be quite sufficient for purposes of a study. In general, the precision estimate is inversely proportional to the square root of the number of observations taken, so that four times as many data would be required to decrease a precision measurement to half its length.

Work sampling practitioners differ markedly in their approach to determining sample size. Some specify precision and solve for sample size, whereas others determine a feasible sample size based on economic constraints and then check to see if the resulting precision will justify performing the study.

Precision is calculated as $Z_{\alpha/2} \times$ (SD). Hence, letting D = precision,

$$N_A = 1 + \frac{p'(1-p')(Z_{\alpha/2})^2}{D^2} \tag{2}$$

The addition of 1 in this formula is neglected as being insignificant in all subsequent formulations. For example, if $p' = .1$ and a precision of $D = .01$ is desired with 95% confidence,

$$N_A = \frac{(0.1)(0.9)(1.96)^2}{(0.01)^2} = 3458$$

total observations will be needed. The theoretically correct addition of 1 to N_A has no appreciable effect.

As an alternate, some practioners find it easier to specify the *relative* precision of estimate desired in a study. Instead of wanting $p' \pm .01$, they might want to find $p' \pm .1p'$, that is, within 10% of p'. The quantity D, then, is termed "absolute precision" (or "accuracy" if p is unbiased), and $R = D/p'$ is termed the "relative precision" (or "accuracy").

$$N_A = \frac{(1-p')(Z_{\alpha/2})^2}{(p')R^2} \tag{3}$$

The desire for a precision within 10% of p' makes $R = 0.10$, and the required number of observations for $p' = 0.10$ is again

$$N_A = \frac{0.9\,(1.96)^2}{0.1\,(0.10)^2} = 3458$$

However, if $D = 0.01$ and $R = 0.10$, as in the preceding examples, but p' is 0.90 instead of 0.10, the number of observations necessary to produce (with 95% confidence) the absolute precision $D = 0.01$ will still be

$$\frac{(0.9)(0.1)(1.96)^2}{(0.01)^2} = 3458$$

but $R = 0.10$ requires a precision of 10% of 0.90, which is 0.09, not 0.01, and hence the required sample size is only

$$N_A = \frac{0.1\,(1.96)^2}{0.9\,(0.1)^2} = 43$$

The quantity N_A from equations 2 and 3 is the traditional quantity used at the beginning of a study to calculate the total number of observations needed with 1.96 (or 2) used for $Z_{\alpha/2}$ to give 95% confidence. Since p' is not only unknown at the beginning of the study, but never really known for sure, a rough estimate of p' is used. Example values of D and R, absolute and relative precision values at 95% confidence, are given in Exhibit 4.6.6 for N_A = 100, 500, 1000, and 5000, and for 11 different values of p'. Using the exhibit as well as the equations requires an estimate of the unknown quantity p'.

The equations

$$D = Z_{\alpha/2}\sqrt{\frac{p'(1-p')}{N_A}}$$

and

$$R = Z_{\alpha/2}\sqrt{\frac{(1-p')}{p'N_A}}$$

can be used to find values not in Exhibit 4.6.6 for other confidence levels (see Exhibit 4.6.5 for $Z_{\alpha/2}$ values) and for other N_A values.

For example, $p' = 0.3$, $Z_{\alpha/2} = 1.96$, and $N_A = 1000$ produce the exhibit entries

$$D = 1.96\sqrt{\frac{(0.3)(0.7)}{1000}} = 0.0284$$

Exhibit 4.6.6 Absolute and Relative Precision (95% Confidence)

	N_A							
	100		500		1000		5000	
p'	D	R	D	R	D	R	D	R
0.05	0.043	0.86	0.019	0.38	0.014	0.27	0.006	0.12
0.10	0.059	0.59	0.026	0.26	0.019	0.19	0.008	0.08
0.20	0.079	0.39	0.035	0.17	0.025	0.12	0.011	0.06
0.30	0.090	0.30	0.040	0.13	0.028	0.09	0.013	0.04
0.40	0.096	0.24	0.043	0.11	0.030	0.08	0.014	0.03
0.50	0.098	0.20	0.044	0.09	0.031	0.06	0.014	0.03
0.60	0.096	0.16	0.043	0.07	0.030	0.05	0.014	0.02
0.70	0.090	0.13	0.040	0.06	0.028	0.04	0.013	0.02
0.80	0.079	0.10	0.035	0.04	0.025	0.03	0.011	0.01
0.90	0.059	0.06	0.026	0.03	0.019	0.02	0.008	0.009
0.95	0.043	0.04	0.019	0.02	0.014	0.01	0.006	0.006

and

$$R = 1.96\sqrt{\frac{(0.7)}{(0.3)\,(1000)}} = 0.0947$$

But $p' = 0.3$, $Z_{\alpha/2} = 3$ (indicating 99.74% confidence), and $N_A = 1000$ give the values

$$D = 3\sqrt{\frac{(0.3)\,(0.7)}{1000}} = 0.043$$

and

$$R = 3\sqrt{\frac{(0.7)}{(0.3)\,(1000)}} = 0.145$$

which do not appear in Exhibit 4.6.6.

A nomograph, courtesy of A. D. Moskowitz, appears in Exhibit 4.6.7. To use the nomograph, find the element percentage, $p' \times 100\%$, on the left-most column (e.g., if $p = 0.2$, find 20%) and find the required precision interval percentage, $D \times 100\%$, on the next column to the right (e.g., if $D = 0.04$, find 4%). Connect these two points and extend that line to the vertical line in the center of the exhibit. This intersection identifies the pivot point. Connecting that pivot point by a new line with the desired confidence level on the short scale, for example, 95%, and extending this new line to the right-most scale yields the necessary sample size, which for this example is $384 = N_A$. By a similar procedure, the nomograph can be used to solve for precision or for confidence level, given the other three quantities.

Both Exhibits 4.6.6 and 4.6.7 are based on the binomial assumption, so that only the proportion p' need be estimated to find a variance estimate and hence a precision value. If the binomial assumption does not hold, it is necessary to gather some data from which to estimate the variance before the sample size or precision can be determined.

It should be apparent that the projected sample size, N_A, will depend on which category (which p'_k) is selected for attention. Some practitioners prefer to look at the smallest p_k of interest (the conservative procedure when using relative precision), some prefer to look at the p_k closest to 0.5 (the conservative procedure when using absolute precision), and some simply select the key category that is of greatest interest in the study.

The determination of N_A can be improved as data are gathered, and p' can be estimated, not just with a guess, but with p, the best estimate at that point in the study.

After a study has been taken for I days

$$N_A(I) = \frac{p(1-p)\,(Z_{\alpha/2})^2}{D^2}$$

but N observations will already have been taken. Hence the additional number of observations necessary to produce the absolute precision, D, will be

$$N_A(I) - N = N\left\{\left(\frac{p(1-p)}{N}\right)\frac{Z^2}{D^2} - 1\right\} = N\left(V_A\frac{Z^2}{D^2} - 1\right)$$

Exhibit 4.6.7 Nomograph for Determining the Number of Observations Needed for a Given Absolute Precision and Confidence Level[a]

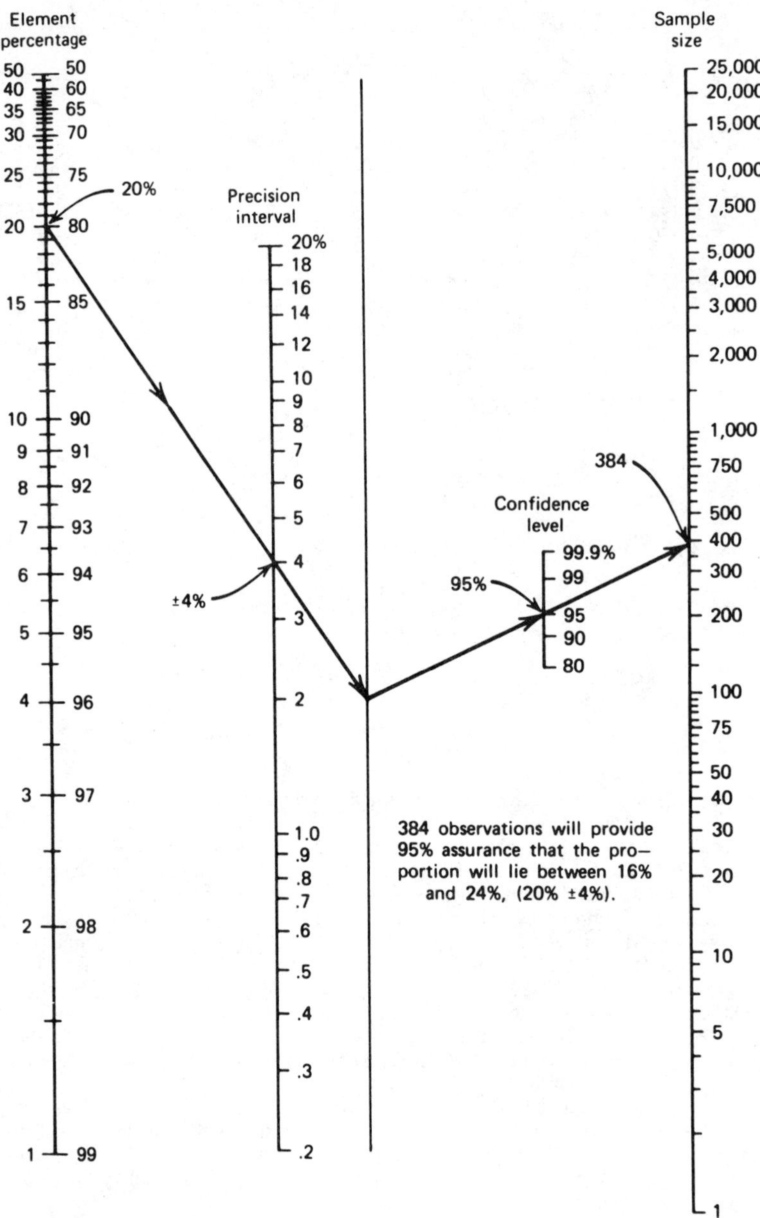

[a]Reproduced by permission from reference 8.

And if observations continue to be taken at the average rate of N/I observations per day, producing the absolute precision D will require an additional number of days equal to

$$(\text{additional days})_A = I\left(V_A\frac{(Z_{\alpha/2})^2}{D^2} - 1\right) \tag{4}$$

For example, if a precision of 0.02 with 95% confidence is desired, and if 5 days of data have produced 125 observations of a particular category during 500 total worker observations, so that $p = 125/500 = 0.25$, then a total of $(0.25)(0.75)(1.96)^2/(0.02)^2 = 1801$. Therefore 1801 observations will be needed. But 500 observations have already been taken, so that an additional $1801 - 500 = 1301$ observations will be needed. Since it has taken 5 days to get the first 500 observations (100 observations/day), it is reasonable to expect the study to take an additional $1301/100 = 13.01$ days.

This figure results immediately from equation 4 as

$$5\left(\frac{(0.25)(0.75)(1.96)^2}{500(0.02)^2} - 1\right) = 13.01$$

After a study is under way, it is also reasonable to calculate, if a computer program is being utilized, the additional days of study necessary to achieve the absolute precision D calculated by V_B and V_C, the alternate variance estimators introduced in the appendix. Procedures for calculating the necessary additional days of study to produce a specified precision are illustrated in Section 4.6.3A of the appendix.

4.6.8 PROCEDURE FOR OBTAINING STANDARD TIMES

Although sampling is not recommended for the purpose of setting work standards for direct labor, it can be very helpful in determining departmental budgets for indirect labor activities that can be related to some unit of output.

If standards are to be set, it is necessary to have observations made by an engineer or technician trained in performance rating (see Chapter 4.4). Rating in a work sampling study is more difficult than rating in a stopwatch study because work sampling observations are theoretically instantaneous. It is typical of observed work sampling studies that many ratings are missed by even the best technicians. That is, the technician feels competent to determine the category of the work being performed, but not competent to rate the effort or skill being used by a particular worker on a particular observation round.

Some work categories, for example, "idle," and machine-controlled categories are never rated. The remaining categories are rated so that the amount of time devoted to a particular job during the study period can be adjusted to an amount that would have been spent if work had been performed at 100%. That is, if $p_3 = 520/2000 = 0.26$, and if $T = 1600$ actual man-hours logged during the study period, then it is estimated that $p_3T = 416$ actual hours were spent on category 3. But if the average of the recorded ratings for category 3, R_3, is 83.5, then the actual man-hour total is adjusted as follows:

$$\frac{R_3 p_3 T}{100} = \frac{(83.5\%)(416)}{100}$$

$$= 348.16 \text{ adjusted hours}$$

It may be necessary to combine several categories of the study into one composite category for setting standards. The composite category should include all productive elements leading to the production of some countable unit of output, for example, boxes shipped, motors repaired, reports filed.

For example, if categories 2, 3, and 5 are combined into one composite category, $p_C = (x_2 + x_3 + x_5)/N$ (see Exhibit 4.6.8). The average rating for the composite category can be found by summing all the ratings observed for all three categories and dividing by the total number of ratings recorded. The number of observations rated for each category in Exhibit 4.6.8 is less than X because some ratings are missed by observers. If one or more of the categories cannot be rated, ratings of 100% should be imposed for all observations of that category. The resulting average ratings can be termed R_C. Note that R_C is *not* a simple average of R_2, R_3, and R_5 found by adding the three ratings and dividing by 3, which would be $(90.7\% + 83.5\% + 76.0\%)/3 = 83.4\%$.

Then the standard time is

$$\frac{AF \times R \times p \times T}{100 \times OP} \text{ hr/unit}$$

or

$$\frac{OP \times 100}{AF \times R \times P \times T} \text{ units/hr,}$$

Exhibit 4.6.8 Example Calculations of Standard Time

Category	X	Total of Ratings	Number of Observations Rated[a]	Average Ratings
#2 Time stamping of order	460	39,905	440	90.7
#3 Recording the order	520	42,685	510	83.5
#5 Filing the order	620	45,630	600	76.0
#6 Unrelated to orders	300	23,600	295	80.0
#7 Idle	100	–	–	–
Total number of observations (N)	2000			
Total number of rated observations			1845	

Categories 2, 3, and 5 are combined into c.

$$P_c = \frac{460 + 520 + 620}{2000} = 0.80$$

$$R_c = \frac{39{,}905 + 42{,}685 + 45{,}630}{440 + 510 + 600} = \frac{128{,}220}{1550} = 82.72\%$$

T = total man-hours during study = 461
AF = allowance factor = 1.16
OP = 1000 orders processed

$$\text{Standard} = \frac{AF \times R_c \times P_c \times T}{100 \times OP} = \frac{1.16 \times 82.72 \times 0.80 \times 461}{100 \times 1000}$$

$$= 0.354 \text{ hr/order}$$

[a]Number of observations rated is less than X because of missed ratings.

where AF = allowance factor in decimal form, 1 + allowance (e.g., 1.10 or 1.15 for 10% or 15% allowance)
R = average rating (percent)
$p = X/N$ = time proportion estimate for the study (in decimal form)
T = total man-hours (or machine-hours) for the study
OP = total units of output for the study related to the time proportion p

When a large number of observations are made to find the estimate p, the precision of that estimate is good. If there is a question about the precision of p because of the work sampling procedure, some practitioners modify p to include, say, two SDs of p before it is used to find the standard. That is,

$$p^* = p + 2\sqrt{\frac{p(1-p)}{N}}$$

Using the example of Exhibit 4.6.8, the effect of this procedure would be to make

$$p_c^* = 0.8 + 2\sqrt{\frac{0.8\,(0.2)}{2000}} = 0.8 + 0.018 = 0.818$$

Substituting 0.818 for 0.800 in the standard time calculation of Exhibit 4.6.8 yields the modified standard 0.362 hr/unit.

This adjustment procedure, from p to p^*, is quite conservative. It should be noted that p might just as reasonably be modified downward. Likewise, it should be noted that this adjustment compensates for uncertainty about p as an estimate of p', but does not consider any uncertainty in R, the rating.

4.6.9 CONSIDERATIONS IN SELECTING WORK CATEGORIES

Categories should be consistent with the end use of the study, initially recognizable by sight, and carefully defined so as to be mutually exclusive (nonoverlapping). If the study is being undertaken to set standards, efforts must be made to ensure that the categories correspond to some readily available measure of output.

Fundamentally, the results of a work sampling study take the form of a series of categories, with the proportion of the total observations recorded in each. The director of the study should work with management to decide what facts are to be sought before creating the categories. If comparisons are to be made among the results of different work sampling studies, the categories should be consistent among studies. But the definition of a specific activity and the precision with which observers classify an activity are of the upmost importance. For example, if a study had two categories, "work" and "idle," it would be very difficult to classify much nonproductive activity (such as cleanup or walking) because the person observed is not "idle." At the same time, walking may not be an integral part of productive work. So great care in selecting and defining categories is recommended. Categories are recorded in writing not only to preserve consistency among observers, but also because studies made in the future for comparison and measurement of change should have similar categories.

A good rule of thumb is that in a fact finding study about half of the categories should be of "productive" activity and the other half "nonproductive." Sometimes a third of the categories is "productive," a third "necessary but not desirable," and a third "nonproductive." The important point is that the study should not be solely to find out where employees are wasting their time, but should also focus on all the categories of the work activity and on the necessary, but not desirable, categories. The working time can be used as the basis for setting standards and for comparison with standard data.

For example, in one study only one category was given for "work," while 15 were listed for various kinds of "wait" and "idle." As a result, all materials handling, setup of equipment, and inspection observations were thrown into the "work" category, and the usefulness of the study was lessened considerably.

An example list of categories for a study of machinists is:

1. Plan–study prints and so on.
2. Setup and cleanup at workplace.
3. Work–use tools and so on.
4. Wait–for other material, for equipment, and so on.
5. Travel–walk, ride.
6. Personal–idle, conversation, and so on.
7. No contact–out of area.

An example list of categories for a clerical study, with primary categories determined by observation and supplementary categories that result from questioning the clerk, is as follows:

1. Use computer terminal.
   a. Customer.
   b. Other external.
   c. Internal (interoffice, reprint, etc.).
2. Type or write.
   a. Customer.
   b. Other external.
   c. Internal (interoffice, reports, etc.).
3. Handle papers, files.
   a. Customer.
   b. Other external.
   c. Internal (interoffice, reports, etc.).
4. Walking.
5. Telephone.
6. Conversation.
7. Personal idle.
8. No contact, out of area.

For example, the observer would make an initial observation of "typing" and then ask the clerk what form is being used (if it is not immediately apparent) in order to determine that the typing was for "customer" work. An honest answer can be expected, and no particular resentment should be aroused, because the clerk knows that he or she has been observed doing productive work.

Categories for special studies should take advantage of what is already known about the work situation. For example, machine recorders may have been installed in an area, providing the means for a partial check in machine utilization studies. If an objective of the study is to check job content against a job evaluation plan, or to check actual division of service time against cost accounting or cost center classifications, these should be considered when creating the categories. To summarize, the end results of a work sampling consist primarily of the category proportions; hence the categories should be selected with care.

4.6.10 EXAMPLE STEP-BY-STEP PROCEDURE

Because work sampling is a flexible tool adaptable to both very simple and very complex studies, it is impossible to create a single step-by-step procedure for designing and conducting a study that will fit all uses of work sampling. If steps are included that are beneficial in a large-scale study, the procedure may appear too complex and cumbersome for the small, quick study. Hence the following list of steps is presented as an example procedure rather than as a comprehensive guide. Many of the steps refer the reader to other sections of this chapter for more detailed explanation.

1. *Define objectives.* Forcing an engineer to write a definition of objectives structures thinking and reduces ambiguities (Section 4.6.4). It is necessary at the beginning to have some rough idea of the scope in terms of days of study and area to be studied. When step 9 is reached, the number of days selected may need to be changed.
2. *Establish measure of output (if indicated).* If the objectives of the study include establishing or monitoring any standard related to output, the method of measuring output must be determined (Section 4.6.).
3. *Determine who will collect data.* Although it is not necessary at this point to know the exact individual who will collect data, before proceeding to step 4, it is helpful to know whether an outside technician will be utilized or if the supervisor will be asked to make observations (Section 4.6.4).
4. *Obtain approval of supervisor and announce study.* Effort expended at this step can aid in acceptance of study recommendations and ease tensions that might bias study results. Great care must be taken in determining how a study is announced to keep from arousing fear and suspicion. Most experienced practitioners advocate open, honest announcement to all who might possibly be concerned. Secret studies can cause serious personnel problems, and generally only novice work samplers are tempted to try to collect data without the general knowledge of all employees involved.
5. *Determine the general method to be used to record data.* In general, this will be easily determined by the available equipment. For example, the purchase of a new computer system for a specific study would cause unreasonable delay.
6. *Train observer(s).* Make sure observers understand the necessity for objectivity, the general principles of sampling, and the importance of adhering to time schedules. If performance ratings are to be made, untrained observers must receive special training, and experienced observers should receive review training in performance rating.
7. *Classify activity into categories.* Selection of categories is best done with the aid of individual(s) who will actually be making the observations. This selection of categories becomes part of the training procedure (Section 4.6.9).
8. *Prepare for the physical aspect of data collection.* Design the form(s) to be used for "pencil and paper" recording or program the computer for the number of workers (machines) and the number of categories to be observed (Section 4.6.4).
9. *Decide on the number of observations needed.* Using a preliminary estimate of p' for some key category resulting from step 7 and some desired precision that is implied by the objectives of step 1, determine the approximate number of observations that will be required. Dividing this number by the number of days of the study (or, better, by the number of study days less 3, to allow for a test study) and by the number of observations per round will give the number of observation rounds per day that are needed. If this figure is unreasonably large, either the number of days or the desired precision must be reevaluated (Section 4.6.7).
10. *Decide on sampling method to be used.* Based on the number of observation rounds per day to be made and on knowledge of the area and the workers to be studied (with regard to possible bias), a sampling method—that is, systematic, stratified, random—must be selected (Section 4.6.5).
11. *Select necessary random times.* Using tables, calculators, or computers, establish the necessary random times for the method selected in step 10. Although it is not necessary to generate random times for the entire study at the outset, the times for the first few days should be established to make the sampling method clear to the observer(s) (Exhibit 4.6.6).
12. *Set up the necessary clerical (or programming) procedures to process results.* It is important that data handling procedures be readily accessible and as free of error sources as possible.
13. *Take a short test study.* The first day's data, and possibly those of the second and third if they are recognized to contain bias, will need to be discarded. Individual circumstances such as familiarity of the work force with the work sampling and the expense of data collection will determine the length of the test study.
14. *Set up control chart(s).* As the study progresses, control charts, as discussed in Chapter 8.3, can be set up to plot the daily estimates of p' (p_i in the terminology of Chapter 8.3) for the purpose of monitoring one or more of the p values for key categories. As the study continues, $\bar{p}$ or p'

(terminology of Chapter 8.3) would be modified, perhaps weekly. If many p values plot close to the control limits, a significant random-day effect is indicated, and the use of V_c for confidence intervals is desirable.

15. *Reevaluate precision of estimates.* As the study progresses, confidence intervals can be calculated periodically. If alternate variance estimators are used, results may allow terminating the study early or may suggest that it be extended beyond the original planned length if these variance estimates are appreciably smaller or larger than the one based on a binomial assumption (Section 4.6.8 and appendix to this chapter).

16. *Write up and file results.* The form of the report will, of course, depend on the objectives of the study, but it will include estimates of the proportion of time allocated to various categories, the precision of these estimates, and in some cases standard times (Section 4.6.8). Reports should be retained for future reference in similar or follow-up studies.

APPENDIX

4.6.1A CORRELATED WORK CREWS[9]

When several workers are observed at (or near) the same time and their observations are totaled into the same x values (and hence p values), their individual reading cannot be considered to be independent. Even if the workers appear to act separately, their activity is correlated (not independent) within an observation round simply because the observations occur at the same time of day. For example, an observation round at the end of the day is likely to find all workers in a cleanup mode.

This correlation can be compensated for by calculating for the kth category the alternate standard deviation $SD_B = \sqrt{V_B}$ where

$$V_B = \frac{\sum_{j=1}^{J} [Y(k,j)]^2/m(j) - Np_k^2}{N(J-1)}$$

The following definitions describe the quantities in this equation:

J = total number of observation rounds made during a study

$m(j)$ = total number of workers (or machines) observed on the jth observation round of the study (For example, $m(5)$ is the number of workers observed on the fifth observation round of the study. Ideally this quantity would be constant throughout a study, but in practice it rarely is.)

$Y(k, j)$ = total number of workers (or machines) found in the kth work category on the jth observation round (For example, $Y(3, 15)$ is the number of workers found in category 3 on the fifteenth observation round of a study. If only one worker is being observed, Y will be only 0 or 1 for each k and j.

The variance estimator V_B of this section is essentially a scaled sample variance of the $Y(k, j)$ values. Its calculation is greatly simplified with a computer, but can be illustrated with the data of Exhibit 4.6.1A, where the category "idle" is taken to be category 1. Using the SD_B value from Exhibit 4.6.7 on equation 1, it can be concluded with 95% confidence that

$$0.168 - (1.96 \times 0.0326) < p' < 0.168 + (1.96 \times 0.0326)$$

or

$$0.104 < p' < 0.232$$

For purposes of comparison, note that $V_A = (0.168)\,(1 - 0.168)/189 = 0.0007396$ and that $\sqrt{V_A} = SD_A = 0.0272$, so if worker correlations are ignored, implying use of SD_A in equation 1, it can be concluded with 95% confidence that

$$0.168 + 1.96 \times 0.0272 < p' < 0.168 + 1.96 \times 0.0272$$

or

$$0.115 < p' < 0.221$$

When, as in Exhibit 4.6.1A, only 15 observation rounds are made in a total study, it is appropriate to compensate for the small number by using a percentage point from a t distribution instead of a

Exhibit 4.6.1A Example Calculation of SD_B

Round	Number of Workers Observed $m(j)$	Number of Workers Observed "Idle" $Y(1,j)$	$\frac{Y(1,j)^2}{m(j)}$
1	12	6	3.000
2	12	2	0.333
3	12	3	0.750
4	12	1	0.083
5	12	2	0.333
6	13	1	0.077
7	13	3	0.692
8	13	2	0.308
9	13	2	0.308
10	13	0	0.000
11	13	1	0.077
12	13	0	0.000
13	13	4	1.231
14	13	2	0.308
15	13	3	0.692
Totals	190	32	8.192

$$P_1 = \frac{32}{190} = 0.168$$

$$N = 190$$

$$J = 15$$

$$V_B = \frac{8.192 - 190\,(0.168)^2}{190\,(15 - 1)} = 0.001064$$

$$SD_B = \sqrt{V_B} = 0.0326$$

Z distribution. In practice, however, more than 30 rounds of a study will be made, and the difference between a t value and a Z value will be trivial.

4.6.2A RANDOM-DAY EFFECTS[9]

When the p' value for which an estimate is desired is not simply the proportion of time occupied by a category during the study period, but the mean value of proportions from which individual daily proportions would be expected to fluctuate, it is necessary to estimate the precision of p as an estimator of that mean value p' on the basis of daily totals.

Random-day effects can be included in the precision estimate by calculating for the kth category

$$SD_C = \sqrt{V_C}$$

where

$$V_C = \frac{\sum_{i=1}^{I} [Z(k,i)]^2/n(i) - Np_k^2}{N(I-1)}$$

The following definitions describe the quantities in the preceding equations:

I = total number of days over which a study extends

$n(i)$ = total number of worker observations made on the ith day (For example, $n(6)$ is the number of worker observations on the sixth day. Ideally this number would remain constant throughout a study, but in practice it rarely does.)

$Z(k, i)$ = total number of worker (or machine) observations found in the kth category on the ith day of the study (For example, $Z(2, 6)$ is the number of workers found in category 2 on the sixth day of the study.)

The variance estimator, V_C, is essentially a scaled sample variance of the $Z(k, i)$ values. Its calculation is greatly simplified with a computer, but can be illustrated with the data of Exhibit 4.6.2A

Exhibit 4.6.2A Example Calculations of SD_c

Day	Number of Worker Observations Made $n(i)$	Number of Observations Made of "Idle" $Z(1, i)$	$\frac{(Z(1,i))^2}{n(i)}$
1	190	32	5.389
2	202	34	5.723
3	200	31	4.805
4	180	27	4.050
5	193	33	5.642
6	175	26	3.863
7	191	28	4.105
8	200	34	5.780
9	192	30	4.688
10	201	29	4.184
11	200	33	5.445
12	200	30	4.500
13	183	25	3.415
14	190	28	4.126
15	185	35	6.622
16	195	34	5.982
17	192	31	5.005
18	195	38	7.405
19	187	28	4.193
20	185	29	4.546
Totals	3836	615	99.414

$$P_1 = \frac{615}{3836} = 0.1603$$

$$N = 3836$$

$$I = 20$$

$$V_c = \frac{99.414 - 3836(0.1603)^2}{3836\,(20 - 1)} = 0.00001158$$

$$SD_c = \sqrt{V_c} = 0.0034$$

where the category "idle" is taken to be category 1. Using SD_C in equation 1, it can be concluded with 95% confidence that

$$0.160 - (1.96 \times 0.0034) < p' < 0.160 + (1.96 \times 0.0034)$$

or

$$0.153 < p' < 0.167$$

For purposes of comparison, note that $V_A = (0.16)\,(1 - 0.16)/3836 = 0.00003504$ and that $SD_A = 0.0059$, so if daily fluctuations are ignored, implying use of SD_A in equation 1, it can be concluded with 95% confidence that

$$0.160 - (1.96 \times 0.0059) < p' < 0.160 + (1.96 \times 0.0059)$$

or

$$0.148 < p' < 0.172$$

That V_C in this case is less than V_A implies that daily differences were very small and that some sort of stratified sampling was practiced so that the binomial assumption gives a conservatively large estimate of the variance of p. When simple random sampling is practiced, V_C is usually larger than V_A.

A disadvantage of this procedure is the necessity of having about 20 days of observations before a reliable estimate of precision can be obtained. When formulating confidence intervals based on

V_C when I is less than 30, better results can be achieved by substituting a $(1 - \alpha/2)$ 100 percentile point of a t distribution with $I - 1$ degrees of freedom for $Z_{\alpha/2}$, which will enlarge the confidence intervals slightly (see Chapter 13.5 for examples).

An advantage of this procedure is that reductions in the variance of p achieved by stratified or systematic sampling will be expected to result in smaller values of V_C and hence a smaller estimate of precision.

The V_c estimate can be shown[7] to be conservative, a bit large, if p is subject to recurring fixed-day effects, for example, Monday effects and Friday effects.

4.6.3A ADDITIONAL DAYS OF STUDY

The procedure of Section 4.6.7 can be readily adapted for use with V_B and V_c, the alternate variance estimators.

When enough data have been collected to get reasonable estimates of V_B and V_c, the additional days of study necessary to achieve the absolute precision D are given by

$$(\text{additional days})_B = I\left(V_B \frac{(Z_{\alpha/2})^2}{D^2} - 1\right) \tag{5}$$

$$(\text{additional days})_C = I\left(V_C \frac{(Z_{\alpha/2})^2}{D^2} - 1\right) \tag{6}$$

Referring to Exhibit 4.6.7, where V_B is calculated to be 0.001064 based on 1 day of study, if a precision of 0.02 is desired with 95% confidence, equation 5 suggests that an additional

$$1\left[(0.001064)\frac{(1.96)^2}{(0.02)^2} - 1\right] = 9.2 \text{ days}$$

will be needed.

Referring to Exhibit 4.6.2A, where V_C is calculated to be 0.00001158 based on 20 days of data, equation 4 suggests that a precision of 0.005 with 95% confidence will be achieved in an additional.

$$20\left[0.00001158\frac{(1.96)^2}{(0.005)^2} - 1\right] = 15.3 \text{ days}$$

If the precision 0.02 is used in this equation instead of 0.005, the result will be negative; that is, a negative number of additional days are needed, implying that the precision of 0.02 has already been reached.

Equivalent formulas for additional days using V_A, V_B, and V_C can be constructed based on relative precision R by substituting $(P \times R)$ for D in the formulas given in equations 4, 5, and 6.

REFERENCES

1. L. H. C. TIPPETT, "A Snap-Reading Method of Making Time Studies of Machines and Operatives in Factory Surveys," *Journal of the Textile Institute Transactions*, Vol. 26, February 1935, pp. 51-55.
2. R. L. MORROW, "Ratio Delay Study," *Mechanical Engineering*, Vol. 63, No. 4 (April 1941), pp. 302–303.
3. C. L. BRISLEY, "How You Can Put Work Sampling to Work," *Factory Management and Maintenance*, Vol. 110, No. 7 (July 1952), pp. 84–89.
4. H. L. WADDELL, "Work Sampling–A New Tool to Help Cut Costs, Boost Productivity, Make Decisions" (editorial), *Factory Management and Maintenance*, Vol. 110, No. 7 (July 1952), p. 83.
5. C. L. BRISLEY and R. DOSSETT, "Computer Use and Nondirect Labor Measurement Will Transform Profession in the Next Decade," *Industrial Engineering*, Vol. 12, No. 8 (August 1980), pp. 34–43.
6. J. J. MODER and H. D. KAHN, "Selection of Work Sampling Observation Times: Part II–Restricted Random Sampling," *AIIE Transactions*, Vol. 12, No. 1 (March 1980), pp. 32–37.
7. J. J. MODER, "Selection of Work Sampling Observation Times: Part I–Stratified Sampling," *AIIE Transactions*, Vol. 12, No. 1 (March 1980), pp. 23-31.

8. A. D. MOSKOWITZ, *A Monograph for Work Sampling*, Work Study and Management Services, Vol. 9, 1965, pp. 349–350.

9. E. S. PAPE, "Work/Activity Sampling–Contemporary Design Analysis Methodology and Applications Part II–Work Sampling Calculations Revisited," AIIE, *1979 Fall Industrial Engineering Conference Proceedings*, Norcross, GA, 1979.

BIBLIOGRAPHY

BARNES, R., *Work Sampling*, 2nd ed., Wiley, New York, 1957.

BARNES, R., *Motion and Time Study*, 7th ed., Wiley, New York, 1980.

HEILAND, R. E., and W. J. RICHARDSON, *Work Sampling*, McGraw-Hill, New York, 1957.

NIEBEL, B. W., *Motion and Time Study*, 6th ed., Irwin, Homewood, IL, 1976.

RICHARDSON, W. J., *Cost Improvement, Work Sampling, and Short Interval Scheduling*, Reston Publishing Company, Reston, VA, 1976.

CHAPTER 4.7

Computerized Work Measurement

DAVENDRA MISHRA

RCA Corporation

4.7.1 WORK MEASUREMENT—A PERSPECTIVE

Since the advent of Frederick Taylor's scientific management, a basic tool for productivity improvement has been work measurement. Work measurement is the application of systematic techniques to determine the work content of a defined task and the corresponding time required for its completion by a qualified worker. Various techniques have evolved to satisfy management objectives of labor quantification, such as historical data, estimates, stopwatch time study, predetermined times, standard data, work sampling, and mathematical techniques. The fundamental tasks of work measurement are:

1. Observation and/or analysis of the job to identify the necessary physical and mental work.
2. Actual measurement of the work using a stopwatch or by applying predetermined elemental times to the tasks enumerated.
3. Determination of operation time.
4. Application of allowances to yield a work standard.
5. Documentation of method of operations and relevant information.
6. Maintenance and updating of the information.

Traditionally, the industrial engineering task of work measurement requires considerable manual effort in data gathering, mathematical and statistical analysis, and generation of operation details to be used by employees. Very often, these time consuming, routine tasks do not leave enough time for analytical creativity. In addition, the person responsible for developing work standards is often burdened by product and process changes, which have an adverse impact on the accuracy, timeliness, and thoroughness of the work measurement system. Also, the inevitable responsibility for maintaining a system of work standards requires dedicated effort which is seldom recognized. Finally, the manual system of standards has remained outside the management information system structure of a business enterprise, preventing greater utilization of labor standards.

4.7.2 COMPUTERS IN WORK MEASUREMENT

The efforts of industrial engineers to improve their own productivity are beginning to see the light of day with the application of computers in work measurement. The computer makes it possible to relieve the industrial engineering analyst of the routine and repetitive manual activities associated with work measurement and to provide a greater opportunity for more comprehensive and creative analysis. Finally, it has become economically feasible to incorporate the work measurement system into the overall input information system of a company and to benefit from its integration with accounting and production control.

The growing applications of automation in work measurement can be broadly classified into the following areas:

Automation of data gathering in time study, work sampling, and so on.

Mathematical and statistical analysis of work measurement data, including application of allowances, performance rating factors, and so on.

Development and organization of standard data.

Development of standard data using mathematical and statistical techniques.

Documentation of methods of operations and relevant information.

Audit control of work measurement inputs, standard data, and related information.

Storage and retrieval of standard data.

Maintenance and updating of a work measurement system.

Establishment of labor standards for assembly lines.

Computation of indexes for optimization of labor standards.

Utilization of a work measurement system for effective management control of an operation.

4.7.3 WORK MEASUREMENT SUBSYSTEM FOR MANAGEMENT CONTROL

The incorporation of the work measurement subsystem into the overall management information system of a business enterprise is growing in popularity. The data base created by computerized work measurement has gone beyond the traditional objectives of generating standard labor data and labor efficiency measurement. In conjunction with production control and accounting systems, it has provided the following tools for effective management control: (1) manpower planning and scheduling; (2) standard cost of fabricated component parts, completed subassemblies, and finished products; (3) labor cost variance reporting; (4) pricing of products; (5) work-in-process control; and (6) utilization of machine work centers.

These computerized features in an integrated information system enhance the profitability of an operation by enabling management to respond systematically to rapid product and process changes, short product life cycles, a variety of products and processes, extreme load fluctuations, and cost changes. For instance, instead of product standard costs being updated once or twice a year, they could be continually updated in the modern, dynamic environment. Computerization of work measurement has certainly made it economical to update labor-oriented information.

4.7.4 AUTOMATION OF DATA GATHERING

Establishment of labor standards using time study or work sampling entails the basic steps of designing the study, observing activities, recording observations, analyzing and validating data, and reporting results. Some improvements have been made in the recording, analysis, and reporting of work measurement data by integrating electronic data gathering devices with computers. One such device is the Datamyte 900®, developed by the Electro General Corporation.* Another solution, known as the Mechanized Activity Sampling Technique (MAST),[1] was developed by A. J. Taylor at the Chase Manhattan Bank.

Datamyte 900

The Datamyte 900 is a general-purpose, hand-held data collection device with a solid-state memory capable of storing up to 32,000 characters in a computer-readable format. A rechargeable battery allows it to operate in a portable mode for 8 hr. At the conclusion of data gathering, the instrument transfers its data to a computer with the help of an interface cable. Subsequently, the computer can be programmed to convert the data to desired analytical information. The Datamyte has been widely used for time studies, work sampling, downtime recording, inventory accounting, production reporting, statistical audits, and so on. Exhibit 4.7.1 displays the components of the instrument.

The keyboard of the Datamyte 900 consists of 14 keys, namely, numbers 0 to 9, plus C, F, underline, and space. Combinations of these characteristics are entered according to a predetermined code. The entry is displayed in up to 12 characters and then entered into memory upon receiving the command "enter." Data can be entered in four basic modes. The simplest mode is where data are accepted as entered, to be subsequently recalled and edited if desired. In the second mode, while data are captured, time is automatically recorded in units of 0.01 min. This information cannot be recalled or edited. The third mode of data entry is where the analyst making the observation is prompted by the display panel of the Datamyte. This feature is achieved by having the computer preload the Datamyte memory with prompt commands and sequence. The prompt feature minimizes the amount of information to be gathered and ensures its completeness. With the exception of prompt display, all the data gathered can be retrieved and edited. Finally, data collection can be accomplished with the features of prompt instructions and time recording. In this case data captured can be recalled and edited.

Use of the Datamyte is quite simple and efficient. In case of a time study, a code number for the element under observation is entered at the beginning. Upon completion of the element, the enter key is depressed. Instantaneously, the element code and elapsed time are recorded. Codes for work elements, noncyclic elemements, foreign elements, and delay elements can be preassigned

*Datamyte 900, *The Computer-Age Answer to Easier Data Collection and Instant Processing/Reporting*, Product Specification, Electro General Corporation, 14960 Industrial Road, Minnetonka, MN 55343.

Exhibit 4.7.1 Components of Datamyte 900

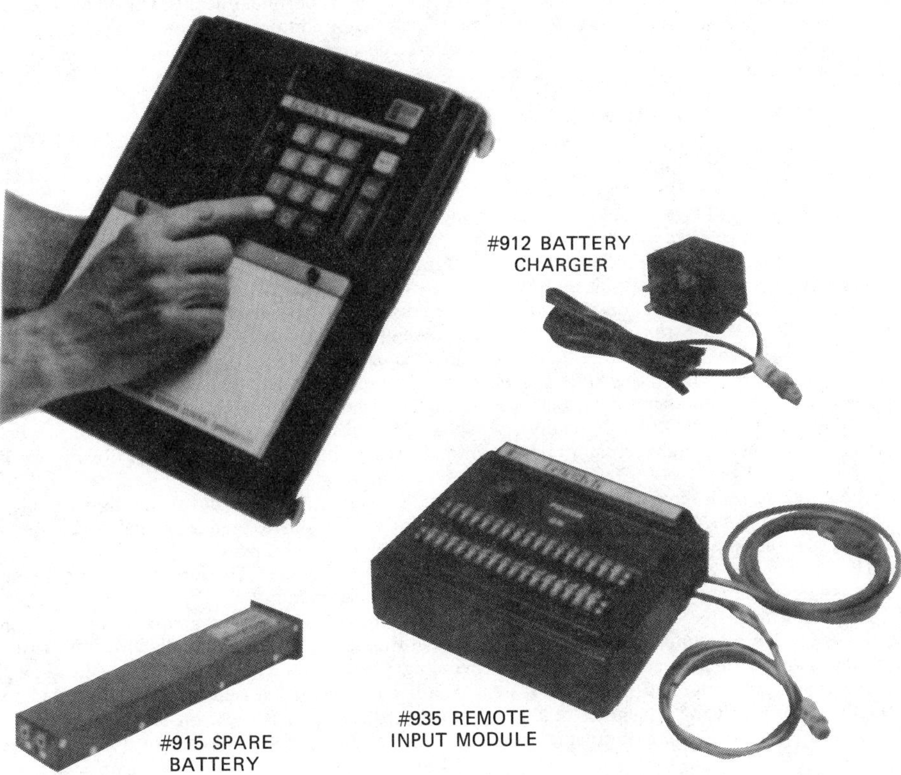

or established during the study. Ratings and piece counts are entered as required, along with allowances and the header card containing date, commencement time, observer identification, and so on. Element descriptions are entered through the computer terminal, and the data gathered are summarized for the analyst.

Work sampling studies are taken without recording time. A random-time-of-day list can be generated by a computer to indicate random sampling periods. In case of fixed interval sampling, the Datamyte 900 interval timer can be used. During the observation cycle, codes are entered for employee, machine, location, activity, cost center, rating, and so on. Then statistical analysis of the work sampling data is performed and reported by the computer.

Mechanized Activity Sampling Technique

Work sampling observations and associated data are directly captured on 40-column IBM cards, using the IBM 3000® portable punch. A mylar overlay is used to assist data recording in the desired format. Interface with a computer provides analysis of the data and generation of control charts.

Considerations for Electronic Data Gathering

An automated data gathering system must have some of the following characteristics to be widely useful:

Ability to accept inputs from stopwatch time study and work sampling techniques.

High speed, accuracy, and reliability of data acquisition.

Portability.

Ability to provide interface with a computer for data transfer, analysis, and storage.

Ability to capture analyst's observations relevant to the study.

Capability to prompt the observer.

Sufficient data storage capacity.

Ability to audit the data being gathered.

The most beneficial outcome of the use of automated devices is that the industrial engineer is able to devote more time to observing an operation for all its pertinent details, with the computer interface ensuring accuracy and virtually instant analysis. In addition, the considerable reduction in time spent on observations and analysis permits the engineer to expand or rerun his or her data collection when necessary.

4.7.5 COMPUTERIZED WORK MEASUREMENT SYSTEMS

State of the Technology

Today, several well-known sources are offering computerized work measurement systems with growing acceptance from users.

The headquarters Industrial Engineering Department of the Westinghouse Electric Corporation created the Micro-Matic Methods and Measurement (4M) System,[2] which is now available through the nonprofit MTM Association. This system, based on MTM-1, is widely used in industry.

In 1970 Rath and Strong, Inc., reported the development of a computer package called Computerized Standard Data,[3] which aids in the analysis, computation, documentation, and maintenance of industrial engineering time standards. It serves as a system for data entry and file maintenance for creating and maintaining a manufacturing information base.

In the early 1970s, WOFAC Company, a division of Science Management Corporation, developed WOCOM®, which permits computerized work measurement using Detailed and Ready Work-Factor® and MTM.[4]

The computerized Maynard Operation Sequence Technique (MOST),[5] developed by H. B. Maynard and Company, provides another alternative for utilization of predetermined data using a unique sequence model aimed at simplifying computation of time standards.

The Automated Advanced Office Controls (Auto-AOC) System* from the Nolan Company, Inc., is a computerized work measurement system for clerical operations.

UniVation Systems® of Management Science, Inc.,† represents an integrated management control system built around a comprehensive manufacturing data base. Mathematically derived time standards are utilized to achieve manpower scheduling, pricing, labor variance analysis, and so on.

The inputs, outputs, and operating system details of the work measurement systems cited are briefly described in the remainder of this section. Bear in mind that these are some of the better-known proprietary systems available today and that they do not represent an exhaustive list.

Micro-Matic Methods and Measurement

The 4M Data system,‡ a computerized work system now available from the MTM Association, adheres to MTM procedures so as to achieve the same precision level of the MTM-1 system, even for the most complex motion patterns.

The MTM motions that constitute 90% of the manual work performed in the common industrial situation are reach, grasp, move, position, and release. These motions are combined into the motion aggregates of GET and PLACE. The computer program accepts simple GET or PLACE notations that describe the general method and then converts these motions to equivalent MTM-1 notations. An extensive electronic data processing (EDP) logic program then utilizes the increments, as a competent analyst should, to develop near-optimum methods and standards within the confines of the general method outlined by the analyst. The times are applied at a preselected task level.

Data Notations. The notations transmitted to the computer for the 4M system provide a precise description of the aggregates GET and PLACE. This is made possible by two features of the system. First, there are precise entries for distance and weight. The 4M distances may be listed to the nearest inch or centimeter, and weight is recorded to the nearest pound or kilogram. Second, there is an indication of terminating components within GET/PLACE. In PLACE notations where POSITION occurs, a maximum of three digits are used to define this ending motion within the PLACE combination without reference to a data card.

Release values have been included within the GET aggregate on a pro rata basis by determining the percentage of release increments not overlapped by other notions in typical studies. Simplifica-

**Introducing Auto-A.O.C.*, General Specifications, Robert E. Nolan Company, Inc., 90 Hopmeadow Street, Simsbury, CT 06070.

†UniVation Systems, General Specifications, Management Science, Inc., 4321 West College Avenue, Appleton, WI 54911.

‡*4M DATA: A Computerized Work Measurement System*, General Specifications, 3233-80, MTM Association for Standards and Research, 9–10 Saddle River Road, Fairlawn, New Jersey 07410.

Exhibit 4.7.2 4M DATA Notations

MMMM DATA

1 MU=.000001 hour	1 hour = 1,000,000 MU
=.00006 minute	1 minute = 16,667 MU
=.0036 second	1 second= 278 MU
=.1 TMU	1 TMU = 10 MU

GET

GXXX xxx

xxx — Distance of Reach, inches or centimeters

- **0–1** A Reach, Contact
- **2** B Reach, Contact
- **3** E Reach, No Contact
- **4** D Reach, Contact
- **1–1** A Reach, Pickup Grasp, Easy
- **2** B Reach, Pickup Grasp, Easy
- **3** D Reach, Pickup Grasp, Flat/Very Small Object/Careful
- **C–1** B Reach, Pickup Grasp w/ Interference <1>.5 inch dia.
- **2** B Reach, Pickup Grasp w/ Interference ≤.5 ≥ .25 inch dia.
- **3** B Reach, Pickup Grasp w/ Interference <.25 inch dia.
- **2** Regrasp, No distance
- **3** A Reach, Transfer Hand-to-Hand
- **4–1** C Reach, Select Grasp, Object > 1 cu. in.
- **2** C Reach, Select Grasp, Object < 1 ≥ .01 cu. in.
- **3** C Reach, Select Grasp, Object < .01 cu. in.

PLACE

PXXX xxx w

w → ENW if over 2.5 lb or 1kg

xxx → Distance of C Move, inches or cm

- **0–1** A Move, No Position
- **2** B Move, No Position
- **2 T** Toss, No Position
- **3** C Move, No Position
- **1 _ _** Clearance ≤ .700 ≥ .300
- **2 _ _** Clearance < .300 ≥ .050
- **3 _ _** Clearance < .050 ≥ .010
- **_ 1 _** Symmetrical(> 10 ways to engage)
- **_ 2 _** Semi-Symmetrical (2-10 ways to engage)
- **_ 3 _** Non-Symmetrical (1 way to engage)
- **_ _ A** Surface Alignment
- **_ _ 0** Easy Insertion to .125 inches inclusive
- **_ _ 1** Easy Insertion to .75 inches inclusive
- **_ _ 2** Easy Insertion to 1.25 inches inclusive
- **_ _ 3** Easy Insertion to 1.75 inches inclusive
- **_ _ 5** Difficult Insertion to .125 inches inclusive
- **_ _ 6** Difficult Insertion to .75 inches inclusive
- **_ _ 7** Difficult Insertion to 1.25 inches inclusive
- **_ _ 8** Difficult Insertion to 1.75 inches inclusive

 FORM 2404-80

tion of the release analysis is the only compromise in the 4M system. The matrix for establishing all the notations of GET and PLACE is illustrated in Exhibit 4.7.2.

The buildup of a GET notation is obtained by the use first of a "G" to identify the aggregate. The second digit specifies the type of grasp. The third digit identifies the size of the part when the search and select type of reach is used. The fourth, fifth, and sixth digits are used to specify the distance of reach in inches or centimeters.

The buildup of a PLACE notation is similar to that described for GET. After the use of a "P" to denote the aggregate, the second digit will be 1, 2, or 3, used to represent the radial clearance of the two objects being positioned. In the absence of POSITION, the second digit is zero. The third digit denotes a (1) symmetrical, (2) semisymmetrical, or (3) nonsymmetrical fit. If a zero was used in the second digit position, the third digit indicates the case of MOVE. The fourth digit denotes the ease or difficulty with which the positioning takes place and, as utilized in the supplementary position table of MTM-1, the depth of insertion up to $1\frac{1}{2}$ in. The fifth and sixth digits are used to indicate the distance of MOVE. The seventh digit incorporates the net weight per hand when the object being transported weighs 2.5 lb or more.

Analysis of Motions and System Inputs. The total input requirement for 4M is much simpler than that recorded for MTM-1 analysis. A set of motions to be analyzed can vary from a single motion by one hand, as shown in Exhibit 4.7.3, to several lines of complex motion aggregates. In the latter case an abbreviated input form is faster to use and especially suited to long-cycle measurements, which require a minimum of line-by-line instruction.

In addition to applying the MTM-1 simultaneous motion rules, the system is programmed with the other necessary functions that an analyst must carry out. These programmed features are:

1. Application of factors to modify the normal MTM standard to satisfy wage incentive requirements or recognize the learning curve.
2. Application of allowances.
3. Handling of metric distances or weights.
4. Application of line and element frequencies.
5. Printing of words describing the 4M notations.
6. Computation of unabsorbed process time printed out as a percentage of total cycle time.
7. Development of method improvement indexes.

Insertion and Retrieval of Information. Elements may be pulled out from studies on system file individually or in blocks. Standard data elements may be developed and inserted into the system for future retrieval by using a 10-character alpha-mnemonic code. Total studies or elements of studies may also be retrieved.

System Outputs. Six reports are generated by the latest version of 4M DATA, called MOD II. The 4M operation analysis report, presented in Exhibit 4.7.4, is the most comprehensive. In addition to reproducing the input information for each line of analysis, the report contains the time for each hand and the total allowed time. It also includes the total line without allowance and the standard time, including allowances and production rate. Finally, it includes the improvement indexes that are described shortly.

Exhibit 4.7.3 4M System Input Format

STANDARD 4M ELEMENT DATA ENTRY

DATE 1 8 80 BY TNG PAGE 1 OF 1

ELEMENT HEADING DATA:

F	T	ELEMENT CODE	M	4M ELEMENT DESCRIPTION	MTR	PO.	LNG
1	2	AGP-04	M	ASSEMBLE RIVETS, NUT, BRACKET		Y	100
1			M				
1			M				
1			M				
1			M				

ANALYSIS DETAIL DATA:

L — NOTES OR OTHER COMMENTS
T — MISCELLANEOUS TIME ENTRY — MISCELLANEOUS DESCRIPTION — MISC. TIME
R — RETRIEVAL OF SUB-ELEMENT — ELEMENT CODE
P — PROCESS TIME ENTRY — PROCESS DESCRIPTION — PROCESS TIME
M 2 — LH — DESCRIPTION CONTINUATION — RH

FILE	TRANS	ELEMENT CODE	LINE SEQUENCE NUMBER	M	1	X	LH DESCRIPTION	LH NOTATION	DIST.	WT.	LH	NV	DC	RH NOTATION	DIST.	WT.	X	RH DESCRIPTION	FREQUENCY	P	B/C
2	2	AGP-04	010	M			RIVET FROM TRAY	G43	07					G43	07			RIVET FROM TRAY			
2			020					SM				N		P316	06			RIVET IN RING			
2			030					SM						G01	03			CONTROL BUTTON			
2			040					SM						P01	01		X	PUSH BUTTON			
2			050				BRKT. FROM TRAY	G42	06					G42	03			NUT FROM TRAY			
2			060		1		BRKT. TO RH FOR	P01	12					P221	09			NUT IN BRACKET			
2			060		2		POS'N.														
2			070											G2				ASSEMBLY			
2																					

Exhibit 4.7.5 illustrates the operator instruction report. The operation standard report in Exhibit 4.7.6 prints the element headings with corresponding left-hand, right-hand, and total time values. The final optional output, the MTM analysis report, displayed in Exhibit 4.7.7, duplicates the 4M operation analysis report with the 4M aggregates shown.

Method Improvement Indexes. A major feature of the 4M system is the generation of five indexes or characteristics of the method analyzed, which highlight areas for possible methods improvement. The Motion Assignment Index (MAI) is a number indicating how effectively both

Exhibit 4.7.4 4M Operation Analysis Report

```
                                                  MICROMATIC ANALYSIS                              DATE 06/02/72

STUDY NO.    105     OPERATION NO.   01    ASSEMBLE COMPENSATORS AND MAGNETS
DEPT. NO.            IDENTIFICATION                       METER FRAME
MACH. NO.            TOOLING ASSY. FIXTURE                                             ANALYST  TRAINING FILM
LNG. LEVEL MTM       PRACTICE OPPORTUNITY                 ALLOWANCE .08

                     LH MOTIONS                                         RH OR BODY MOTIONS              FREQ.    LH    RH TOTAL
                                                                        PROCESS TIME OR DATA INSERT

01

    01   MOVE    ASSEMBLY TO TABLE         P03 -20    G12 -27    GET     FRAME FROM CONTAINER                              *
    02   GET     LARGE COMPENSATOR         G42 -15    P132-22    PLACE   FRAME IN FIXTURE                                  *
    03   PLACE   COMPENSATOR IN CAVITY     P131- 7    G42 - 4    GET     SMALL COMPENSATOR                                 *
    04   GET     LARGE MAGNET              G42 - 7    P131- 7    PLACE   COMPENSATOR IN CAVITY                             *
    05   PLACE   MAGNET IN CAVITY          P131- 8    G42 - 8    GET     SMALL MAGNET                                      *
    06   GET     FRAME                     G11 - 4    P131- 8    PLACE   MAGNET IN CAVITY                 1400  1689  2164*
                                                                                                ST                     2164

TOTAL     2164  MU                                          RMB 56 PERCENT
STANDARD TIME  .00233 HOURS              MAI 71 PERCENT     GRA 15 PERCENT       PROC  0 PERCENT
CYCLES PER HOUR    429                                      POS 30 PERCENT
```

Exhibit 4.7.5 Operator Instruction Report

```
                                                 OPERATORS INSTRUCTIONS                    DATE 06/02/72

                                                                                                      12/10/73
STUDY NO.    101     OPERATION NO.    13     ASSEMBLE MAGNET AND CORE
DEPT. NO.            IDENTIFICATION                           F-TRIP ASSEMBLY
MACH. NO.            TOOLING RATCHET FIXTURE                                             ANALYST TRAINING FILM

                 LH MOTIONS                          RH OR BODY MOTIONS          FREQ.

01  ASIDE FINISHED PART AND ASSEMBLY 2 PLATES IN RH

    01                                        REACH     HANDLE
    02   GET       PART IN FIXTURE            MOVE      HANDLE TO STOP
    03   TOSS      PART ASIDE                 MOVE      HANDLE TO RETURN
    04                                        MOVE FT   PEDAL DOWN
    05                                                  PRESS ACTION
    06   GET       PLATE 1                    GET       PLATE 2                *
    07                                        MOVE FT   PEDAL UP
    08   MOVE      PLATE TOWARD RH            PLACE     PLATES TOGETHER        *
    09   ASSIST    RIGHT HAND                 GET       PLATES ALIGNED         *

02  ASSEMBLE PIN TO PLATES AND PLACE IN FIXTURE
                                                                             *
    01   GET       PIN
    02   PLACE     PIN IN PLATES
    03                                        PLACE     ASSY IN FIXTURE

                                                                        APPROVED BY ---------------------
STANDARD TIME   .00178 HOURS/   .10680 MIN.
                                                                                    ---------------------
CYCLES PER HOURS   561.8
```

Exhibit 4.7.6 Operator Standard Report

```
                                           SUMMARY OPERATOR INSTRUCTIONS            DATE 06/02/72

                                                                                              02/10/77
STUDY NO.    112    OPERATION NO.    09     ASSEMBLE BRACKETS TO RING
DEPT. NO.    30     IDENTIFICATION     20W1301                   WATT-HOUR METER
MACH. NO.           TOOLING PRESS DIE                                             ANALYST TRAINING FILM

                                                                   FREQ.     LH     RH    TOTAL

                                                                             669    849     905
01  PLACE RING ON FIXTURE
                                                                  2.0000   1904   2394    3348
02  ASSEMBLE RIVETS, NUT, BRACKET
                                                                             779    921    1062
03  REMOVE RING. PLACE BRACKET, NUT IN FIXTURE. REPLACE RING
                                                                             349    301     467
04  ASIDE COMPLETED ASSEMBLY. PLACE BRACKET, NUT IN FIXTURE

                                                          RMB 44 PERCENT
TOTAL      5782  MU                                       GRA 22 PERCENT          PROC  0 PERCENT
STANDARD TIME  .00624 HOURS/   .37440 MIN.   MAI 70 PERCENT   POS 34 PERCENT
CYCLES PER HOUR    160.3
```

Exhibit 4.7.7 MTM Analysis Report

```
                                                    MTM ANALYSIS                                            DATE 06/02/72

STUDY NO.    112    OPERATION NO.    09    ASSEMBLE BRACKETS TO RING                                          06/06/77
DEPT. NO.    30     IDENTIFICATION      20W1301         WATT-HOUR METER
MACH. NO.           TOOLING PRESS DIE                  METRIC                              ANALYST TRAINING FILM
LNG. LEVEL MTM      PRACTICE OPPORTUNITY              ALLOWANCE .08

                LH MCTIONS                                                 RH OR BODY MOTIONS                FREC.   LH    RH TOTAL
                                                                           PROCESS TIME OR DATA INSERT

01  PLACE RING ON FIXTURE

    01                                          G12 -12   GET       RING ON BAR                                                8
                                                                      R12B G1A

    02   TRANSFER SIDE OF RING      G3  -24     P01 -10   MOVE      RING TO LH                                  141   164   2208
                  R24A G3                                             M10A

    03                                          G3  - 3   TRANSFER RING FOR BETTER GRASP                              83     83
                                                                      R3A  G3

    04   REACH    TO CONTACT RING   G02 - 4     P02 - 4   MOVE      RING TOWARD FIXTURE                          34    40     40
                  R4B                                                 M4B

    05   ASSIST   RIGHT HAND        SP          P232- 1   PLACE     RING ON FIXTURE                             280   280    280
                  M1C  P22NS4                                         M1C  P22NS4

    06   ASSIST   RIGHT HAND        SP          APA       PRESS     RING ON FIXTURE                             106   106    106
                  APA                                                 APA

                                                                                            ST                             729
```

hands are utilized during the operation analyzed. An operation performed by only one hand has a 50% MAI, whereas perfect use of both hands yields a 100% MAI. The GRA (GRASP, RELEASE, and APPLY PRESSURE) Index denotes the percentage of cycle time that deals with grasping, releasing, and applying pressure. A high value of the index suggests that potential improvements can be made by simplifying grasp requirements. The POS (POSITIONING) Index yields the percentage of total cycle time involved with positioning. A high value of the index suggests the potential for use of fixtures or the need for liberalizing tolerances.

The RMB (REACH, MOVE, and BODY MOTION) Index expresses the labor content of reaches, moves, and body motions as a percentage of the total cycle time. The impact of reduced distances and body motions can therefore be assessed using the index. The final index computed by the system is the Waiting-On-Process Intervals Index, which permits an analyst to seek methods for reducing long waiting intervals.

Standard Data Maintenance. Four basic reports from the system effectively enable the maintenance of standard data. The where-used report, displayed in Exhibit 4.7.8, is obtained for elements and standard data codes containing study, element, frequency, part, operation, and department where used. Revision of specified standard data elements or codes is possible at one time, along with a printout of the revised information.

Two index reports are available. One lists all elements or ranges of the elements on file, and the

Exhibit 4.7.8 Where-Used Report

4M STANDARD DATA CODES WHERE-USED

STD CODE	DESCRIPTION	WHERE-USED	FREQ.	PART NO.	OP.NO.	DEPT.	MU
SAGJM11	GET(2) JUMBL. 0-4 IN	10001/10/01	2.0000	50P0501	PUNCH		1234
		10001/10/02	3.0000	50P0501	PUNCH		1851
		10002/11/01	3.0000	50P0502	PUNCH		1851
		10002/11/02	2.0000	50P0502	PUNCH		1234
		10002/11/03		50P0502	PUNCH		617
SAGJM12	GET(2) JUMBL. 4-8 IN	10001/10/03		50P0501	PUNCH		216
		10001/10/04	2.0000	50P0501	PUNCH		432
		10001/10/05	2.0000	50P0501	PUNCH		432
		10003/12/01		50P0503	PUNCH		216
		10004/13/01	.5000	50P0504	PUNCH		108
		10004/13/02		50P0504	PUNCH		216
SAGJM14	GET(2) JUMBL. 12-16IN	10001/10/07		50P0501	PUNCH		271
		10001/10/08		50P0501	PUNCH		271
		10001/10/09		50P0501	PUNCH		271
		10001/10/10		50P0501	PUNCH		271
		10003/12/02	4.0000	50P0503	PUNCH		1084
		10004/11/02		50P0504	PUNCH		271
SFP0111	STACK PART ASIDE	10001/10/11		50P0501	PUNCH		208
		10006/15/01		50P0506	PUNCH		208

Exhibit 4.7.9 Element Index File Report

4M PART NO. OPERATION/STUDY CROSS REFERENCE 01/07/76

PART NO.	OP.NO.	STUDY	DEPT.	MACH.	DESCRIPTION	ANALYST
		DC 1			SHARPEN PENCIL	TRAINING
		DC 2			ASSEMBLE BRACE AND BIT	TRAINING
		DC 3			ASSEMBLE EXPOSURE METER	TRAINING
		DC 14			REPLACE ROLL IN ADDING MACHINE	TRAINING
		DEMO			CHANGE TIRE ON RIGHT REAR OF VW SEDAN	SCHMIDT
		JAGJM			GET PART & MOVE TO ASSEMBLY LOCATION	T O'NEILL
		JFP01			SIMO:ASIDE,REACH,POSITION AGAINST SIDE AND BACK STOP	S RAHMES
05R1501	01	RETT1	22		ASSEMBLE BRACKETS TO RING WATT HOUR METER	STEVENSON
05R1502	15	RETT2	02		ASSEMBLE AUTOMATIC CONTROLS	JONES
05R1503	25	TREFP	03		ASSY HYDRO GENERATOR	
05R1504	25	TS1AA	35		ASSEMBLE CONNECTOR TO HOUSING	S LYONS
10S2504	12	104	25		ASSEMBLE 5 PARTS AND RIVET	TRAINING FILM
10S2504	13	101	30		ASSEMBLE MAGNET AND CORE	TRAINING FILM
10S2504	14	113	25		ASSEMBLE MAGNET, FIBRE WASHER, FIBRE INSERT	TRAINING FILM
15A1501	17	102	30		ASSEMBLE COVER TO SHROUD	TRAINING FILM
20F1521	01	105	21		ASSEMBLE COMPENSATORS AND MAGNETS	TRAINING FILM
20F1521	02	116	21		ASSEMBLE 5 PARTS AND ADJUST	TRAINING FILM
20W1301	06	110	91		ASSEMBLE VOLT AND CURRENT COILS	TRAINING FILM
20W1301	07	106	30		ASSEMBLE DISC TO SHAFT	TRAINING FILM
20W1301	09	112	30		ASSEMBLE BRACKETS TO RING	TRAINING FILM
20W1301	10	111	91		GROUND TEST, CLOSE TEST LINK	TRAINING FILM
20W1301	11	109	91		PACK METERS IN CARTON	TRAINING FILM
25B1018	08	107	22		ASSEMBLE ELECTRODES, SCREEN, SPRINGS	TRAINING FILM
25P1017	03	103	30		ASSEMBLE & STAKE 4 RIVETS IN VOLT COIL LAMINATIONS	TRAINING FILM
25P1017	05	108	25		ASSEMBLE 3 PARTS AND RIVET	TRAINING FILM
35M1705	19	115	30		ASSEMBLE MOTOR END BELL	TRAINING FILM
35M1705	20	114	35		FIRST OPERATION, MOTOR ASSEMBLY (ALTERNATE CYCLES)	TRAINING FILM
40W1902	22	117	40		WELD CONTACT BUTTON TO SPRING	TRAINING FILM
50P0501	01	10001		PUNCH PRESS	JUMBL. PTS ONE HAND, (1) JUMBL.PT IN OTHER, 0-4 IN	
50P0502	02	10002		PUNCH PRESS	JUMBL. PTS ONE HAND, (1) JUMBL.PT IN OTHER, 6-8 IN	
50P0503	03	10003		PUNCH PRESS	JUMBL. PTS ONE HAND, (1) JUMBL.PT IN OTHER, 8-12 IN	
50P0504	04	10004		PUNCH PRESS	JUMBL. PTS ONE HAND, (1) JUMBL.PT IN OTHER, 12-16 IN	
50P0505	05	10005		PUNCH PRESS	JUMBL. PTS ONE HAND, (1) JUMBL.PT IN OTHER, 16-20 IN	
50P0506	06	10006		PUNCH PRESS	PART ASIDE,REACH TO STACKED PART(12-18 INS)	

other lists all or a range of the part-operation standards on file. The latter may be used as the basis for a work center routing development. Exhibit 4.7.9 illustrates an element index file report.

Finally, Exhibit 4.7.10 is an operation master file report for all the part-operation standards on file.

System Specifications. The 4M DATA system software is written in COBOL for various types of computers having 50*K* core storage for the application programs. The software consists of more than 20 source programs and is available for instantaneous on-line processing or for deferred batch processing.

The WOCOM System

The WOCOM system consists of nine modules to be used individually or in any combination to analyze man-machine activities, develop different work methods, configure assembly lines, and maintain labor standards. Descriptions of the modules are given here.

Work Measurement by Work-Factor (Modules 1 and 2). The manual or machine operation is identified by the user in a simple input format, and the program determines the motions required to perform the operation, along with the times for each. The user receives the detailed motion analysis in Work-Factor and a manufacturing instruction sheet containing each major step in the operation. This module is available in both Detailed and Ready Work-Factor.

Work Measurement by MTM. This module is identical to module one, except that the application rules and time data of MTM are employed.

Application of Learning Allowances. The learning curve allowance is established for the operation variables of the cycle time.

Analysis of Mental and Perpetual Work. Repetitive mental operations, such as inspection or clerical work, are analyzed, and time values are determined.

Measurement of Highly Variable Operations. This module uses the tool of multivariate analysis for complex operations that are not amenable to analysis by the other work measurement techniques. In this case the user has to provide times required to perform the operations under varying conditions which are defined.

Interactive Line Balancing. By varying the allocation of work elements to the work stations of the assembly line, this module generates alternative configurations for it.

Exhibit 4.7.10 Operation Master File Report

STUDY	EL.	PART NO.	OP.NO.	DEPT.	MACH.	4M ELEMENT INDEX ELEMENT DESCRIPTION	12/31/75 FREQ.
1			12			*ASSEMBLE 5 PARTS AND RIVET	
	01					ASIDE COMPLETED PART. PLACE FIRST SPRING IN FIXT. GET 3RD SPG	
101		10S2504	13	30		*ASSEMBLE MAGNET AND CORE	
	01					ASIDE FINISHED PART AND ASSEMBLE 2 PLATES IN PH	
	02					ASSEMBLE PIN TO PLATES AND PLACE IN FIXTURE	
102		15A1501	17	30		*ASSEMBLE COVER TO SHROUD	
	01					ASIDE ASSEMBLY & PLACE COVER ON SHROUD	
103		25P1017	03	30		*ASSEMBLE & STAKE 4 RIVETS IN VOLT COIL LAMINATIONS	
	01					ALIGN IRON AND PLACE 4 RIVETS	
	02					STAKE RIVETS. PLACE NEXT COIL AND IRON IN FIXTURE	
104		10S2504	12	25		*ASSEMBLE 5 PARTS AND RIVET	
	01					ASIDE COMPLETED PART. PLACE FIRST SPRING IN FIXT. GET 3RD SPG	
	02					PLACE 2ND SPRING, TOP PLATE, 3RD SPRING ON FIXTURE	
	03					RIVET TWICE	
105		20F1521	01	21		*ASSEMBLE COMPENSATORS AND MAGNETS	
	01						
106		20W1301	07	30		*ASSEMBLE DISC TO SHAFT	
	01					ASIDE ASSEMBLY, GET DISC, ASSEMBLE TO SHAFT	
	02					LOAD ASSEMBLY IN FIXTURE AND DIE CAST	
107		25B1018	08	22		*ASSEMBLE ELECTRODES, SCREEN, SPRINGS	
	01					PLACE 2 ELECTRODES, SCREEN, BASE IN FIXTURE	
	02					PLACE 2 SPRINGS, RIVET. ASIDE ASSEMBLY	
108		25P1017	05	25		*ASSEMBLE 3 PARTS AND RIVET	
	01					ASIDE ASSEMBLY. PLACE 2 PARTS ON CURRENT IRON	
	02					ASSEMBLE QUADRANT, 2 RIVETS. TRIP PRESS	
109		20W1301	11	91		*PACK METERS IN CARTON	
	01					GET CARTON & OPEN	
	02					CLOSE CARTON END	
	03					INVERT BOX, FOLD 2 SIDES	
	04					FOLD FRONT AND BACK CARTON FLAPS	
	05					GET LINER, BEND 3 TIMES, INSERT INTO CARTON	
110		20W1301	06	91		*ASSEMBLE VOLT AND CURRENT COILS	
	01					PLACE RIVETS ON FIXTURE PINS AND PLACE BRACKET ON RIVETS	
	02					SET VOLT COIL AND PLACE IN FIXTURE	
	03					GET CURRENT COIL AND PLACE IN FIXTURE	
111		20W1301	10	91		*GROUND TEST, CLOSE TEST LINK	
	01					PLACE METER IN TEST BOX	
	02					CLOSE LINK & DRIVE 2 SCREWS	
	03					ASIDE METER TO TEST FIXTURE	
112		20W1301	09	30		*ASSEMBLE BRACKETS TO RING	
	01					PLACE RING ON FIXTURE	
	02					ASSEMBLE RIVETS, NUT, BRACKET	2.0000

Batch Assembly Line Balancing. Based on user supplied inputs, consisting of operation precedence relationships, operation grouping parameters, and desired work station cycle time, the module provides an assembly line design.

Revision and Maintenance of Labor Standards. This ninth module permits selective or across-the-board change to the system of standards as a result of a revision to an element. In addition, an analyst can examine in a test mode the effects of contemplated changes and can update commands identifying all standards that have changed by more than a specified percentage from their previous base times.

Use of the System. The user has to establish the degree of accuracy required and the reports to be generated. The WOCOM system is available on a time-sharing service. Exhibits 4.7.11 and 4.7.12 display typical WOCOM input and output of the system for the task of assembling a toy set where the pieces, called "little people," consist of three parts–hair, face, and body. Exhibit 4.7.13 denotes the manufacturing instruction sheet generated by the system.

Maynard Operation Sequence Technique

In 1974 H. B. Maynard and Company introduced the MOST, a predetermined motion time system that was faster than previous techniques, including the MTM series on which it was based. The reason for the system's speed is the basic premise under which it operates. It assumes that all manual work is performed with a standard sequence of activities. Motion times for each of the activities are predetermined and assigned to a "sequence model." For example, the first part of many manual jobs is transporting a workpiece to the workplace, which may consist of walking a few steps, bending, grasping the part, carrying it back, and placing it on the work surface. The MOST defines each of these actions by a letter, followed by a number denoting its complexity. Adding the numbers and multiplying by 10 gives the time standard in TMU for that operation.

Three basic types of sequence models have been established, namely, general move, controlled

Exhibit 4.7.11 Typical WOCOM Input

```
C PICKUP PART
GET -1 -2 -3 -4
PP -1 -5
M -1 TO -6 -7 -8 -9
SAVE T 1
ER
C GET MORE PARTS
SAVE T 236 1530
ER
C FILL OUT PRODUCTION REPORT
SAVE T 237 648
ER
OBS
 1 30  HAIR
 2 29  FACE
 3 84  BODY
 4  1  FIXTURE
00
C ASSEMBLE LITTLE PEOPLE TO FIXTURE ON CONVEYOR
T 1 1 8 R 0 75 4 8 1 0 / SIMO
ASY 1 TO 4 C .936 .950 DB 3 INDEX / SIMO
RL 1 / SIMO
T 1 2 8 R 0 75 4 8 1 0 / SIMO
ASY 2 TO 1 AT 4 C .312 .343 DB 3 / SIMO
M 2 TO SEAT 1 VF X 2 / SIMO
RL 2 / SIMO
T 1 3 8 R 0 50 4 8 1 0 / SIMO
ASY 3 TO 1 AT 4 C .312 .39 DB 3 / SIMO
M 3 TO SEAT 1 VF X 2 / SIMO
RL 3 / SIMO
C GET MORE PARTS AS REQUIRED
T 236 X .012
T 237 X .004
DETL
PCS 2
ALLOW 19
MIS
```

Exhibit 4.7.12 Typical WOCOM Analysis

```
                         MOTION  ANALYSIS

      LEFT HAND            TIME      LH       RH   TIME      RIGHT HAND

  1 ◆ ASSEMBLE LITTLE PEOPLE TO FIXTURE ON CONVEYOR
  2 PICKUP PART
    GET HAIR                                       GET HAIR
     A8D                     54       54       54    54     A8D
     GR-R           S        48      102      102    48     GR-R            S
    PP  HAIR                                       PP  HAIR
     PP-D-75%       S        54      156      156    54     PP-D-75%        S
    M   HAIR TO FIXTURE                            M   HAIR TO FIXTURE
     A8SD                    70      226      226    70     A8SD
  3 ASY HAIR TO FIXTURE                            SIMO
     CT0.950R0.985  S         5      231      231     5     CT0.950R0.985  S
     ALN V0.25A1S             7      238      238     7     ALN V0.25A1S
     DB 3.00                  2      240      240     2     DB 3.00
     UP A1S                  26      266      266    26     UP A1S
     IND F1S                 23      289      289    23     IND F1S
     INS A1P                 26      315      315    26     INS A1P
  4 RL  HAIR                                       SIMO
     0.50F1                   8      323      323     8     0.50F1
  5 PICKUP PART
    GET FACE                                       GET FACE
     A8D                     54      377      377    54     A8D
     GR-R           S        52      429      429    52     GR-R            S
    PP  FACE                                       PP  FACE
     PP-D-75%       S        54      483      483    54     PP-D-75%        S
    M   FACE TO FIXTURE                            M   FACE TO FIXTURE
     A8SD                    70      553      553    70     A8SD
  6 ASY FACE TO HAIR AT FIXTURE                    SIMO
     CT0.343R0.910  S        26      579      579    26     CT0.343R0.910  S
     ALN V1.50A1S            39      618      618    39     ALN V1.50A1S
     DB 3.00                 12      630      630    12     DB 3.00
     UP A1S                  26      656      656    26     UP A1S
     INS A1                  18      674      674    18     INS A1
  7 M   FACE TO SEAT TIMES  2.00                   SIMO
     A1-X2.00                36      710      710    36     A1-X2.00
  8 RL  FACE                                       SIMO
     0.50F1                   8      718      718     8     0.50F1
  9 PICKUP PART
    GET BODY                                       GET BODY
     A8D                     54      772      772    54     A8D
     GR-R           S        52      824      824    52     GR-R            S
    PP  BODY                                       PP  BODY
     PP-D-50%       S        36      860      860    36     PP-D-50%        S
    M   BODY TO FIXTURE                            M   BODY TO FIXTURE
     A8SD                    70      930      930    70     A8SD
 10 ASY BODY TO HAIR AT FIXTURE                    SIMO
     CT0.390R0.800  S         9      939      939     9     CT0.390R0.800  S
     ALN V0.50A1S            13      952      952    13     ALN V0.50A1S
     DB 3.00                  4      956      956     4     DB 3.00
     INS A1                  18      974      974    18     INS A1
 11 M   BODY TO SEAT TIMES  2.00                   SIMO
     A1-X2.00                36     1010     1010    36     A1-X2.00
```

move, and tool use. "General move" is defined as moving all objects from one location to another freely through the air. This can account for as much as 35% of the work of a machine operator and for even more for an assembly operator. "Controlled move" is a sequence that is applicable when the object retains contact with another object during the move, that is, a lever, crank, or push button. "Tool use" covers not only conventional handtools, such as wrenches, screwdrivers, and gauges, but also fingers and mental processes. Other higher-level, special sequence models have also been developed for the use of material handling equipment.

In 1977 the basic MOST was computerized. The system is the integration of six self-contained modules. The main module is MOST analysis, which carries out the basic work measurement, including the workplace layout, method description, and determination of standard time. The second module is the suboperations data base, where, for each job analyzed, the computer asks a series of questions which form an identifying sentence or title. An example of a title would be "exchange workpiece in three-jaw clutch with T-wrench." This job might contain a number of

Exhibit 4.7.13 WOCOM Manufacturing Instruction Sheet

```
ELEM.
NO.            LEFT HAND                              RIGHT HAND

  1 * ASSEMBLE LITTLE PEOPLE TO FIXTURE ON CONVEYOR
  2 PICKUP PART
    GET HAIR                               GET HAIR
    PP  HAIR                               PP  HAIR
    M   HAIR TO FIXTURE                    M   HAIR TO FIXTURE
  3 ASY HAIR TO FIXTURE                    SIMO
  4 RL  HAIR                               SIMO
  5 PICKUP PART
    GET FACE                               GET FACE
    PP  FACE                               PP  FACE
    M   FACE TO FIXTURE                    M   FACE TO FIXTURE
  6 ASY FACE TO HAIR AT FIXTURE            SIMO
  7 M   FACE TO SEAT TIMES  2.00           SIMO
  8 RL  FACE                               SIMO
  9 PICKUP PART
    GET BODY                               GET BODY
    PP  BODY                               PP  BODY
    M   BODY TO FIXTURE                    M   BODY TO FIXTURE
 10 ASY BODY TO HAIR AT FIXTURE            SIMO
 11 M   BODY TO SEAT TIMES  2.00           SIMO
 12 RL  BODY                               SIMO
 13 * GET MORE PARTS AS REQUIRED
 14 GET MORE PARTS     TIMES  0.01
 15 FILL OUT PRODUCTION REPORT
      TIMES  0.00

ALLOW    S/U HRS     MIN/PC        HRS/PC       PCS/HR TOT. TIME
  19.0                0.062    0.0010303       970.55       1039
```

suboperation method steps, but the entire analysis would be filed in computer memory under this title. It could be retrieved by any one of the key elements in the title, such as "exchange" or "T-wrench." The storage and the retrieval of data are performed through use of a plain language format.

Next, the third module, the operation time calculation, comes into play. The computer locates and organizes all suboperations for a particular workplace and displays them either on the video tube or on printed copy. This serves as the shopping list for assembling a time standard. The analyst selects the appropriate titles from the sheet in their sequence, arranges them in order of occurrence, and assigns a frequency. Subsequently the computer determines the standard and generates the method instruction sheet. With the fourth module the standards are stored in computer memory by part number, part noun, cost center, machine number, operation name, and so on. The fifth module provides mass updating and maintenance of standards. The final module interrelates the five basic ones.

Six optional modules are available that make the industrial engineering job easier. The machining module permits the selection of optimum machines and operating characteristics, yielding a standard time for the operation. There is a similar program for welding. Assembly line balancing is covered by another module. Daily, weekly, or monthly labor reports can be generated and man-machine analysis can be performed. Finally, the work processing module aids in the preparation and maintenance of the work management manuals.

Automated Advanced Office Controls System

The widely used Advanced Office Controls (AOC) System, based on predetermined time standards for clerical operations, has been automated by the Nolan Company, Inc. The new version, Auto-AOC, is a complete package of hardware and software. As a simple microcomputer-based system, it utilizes two programs to maintain the master file and establish standards.

The AOC master file contains, for each element, a phrase describing each cycle and the corresponding time value. Standard AOC element codes are used to access the system. The file maintenance activities include (1) the reviewing on the video display of selected elements, descriptions, and TMU values, (2) the deletion and addition of elements, (3) the revision of element descriptions and/or TMU values, and (4) the listing of the file. The second program takes the basic inputs of task identification and details average daily volume of activity, average batch size, item of count, allowances, and so on. Subsequently the system produces three documents. The task outline is a specific procedure which explains how each task should be performed in the defined work environment. The task analysis presents how the standard was established. The final output, the task summary, combines all of the steps into a single standard, utilizing their relative importance in the task.

UniVation Systems

UniVation Systems, a work-measurement-based management control system developed by Management Science, Inc., is an integrated system consisting of a manufacturing planning data base, bills of materials, routings, operation method instructions, costs, assembly precedence relationships, and facilities loading and scheduling information. Instead of using standard data, the system generates time standards using mathematical formulas. The UniVation Systems are made up of nine modules, which are briefly described here.

The UnivEl® System. This module generates precise elemental time standards using mathematical relationships and method instructions. In addition, it creates an integrated data base for a manufacturing planning and control network.

The Uni-CAMⓉ System. This module provides for the interactive development of process planning information utilizing advanced group technology and parts classification concepts. A stand-alone minicomputer system is utilized. The system creates mathematically generated elemental methods and standards, routings, and an entire manufacturing data base. Mass updating of the entire manufacturing data base is also a feature of the Uni-CAM System. Automated process planning without the use of computer graphics is accomplished by Uni-CAM. Precosting of routings and methods are a by-product of the Uni-CAM System.

The UniComp® System. Using an algebraic language, this module feeds and maintains the mathematical relationships and allowances characterizing processes in the UnivEl System.

The VariCompⓉ System. This module is the software in the intelligent computer terminal for interactive development of input data to the UnivEl time, methods, and data base generator system. It edits the inputs and provides an efficient terminal operation.

The MultiCompⓉ System. This module is the workhorse responsible for the maintenance and mass updating of the overall manufacturing planning and control data base. A simplified data base language allows a product- or operation-oriented change to be reflected in the relevant data base with simple input instruction. This ensures a single source, or responsibility, for the maintenance of the manufacturing data base.

The UniPlanⓉ System. This module provides computerized assembly line balancing. Utilizing the time standards generated by UnivEl, this program develops the optimal design of the assembly line. Its outputs include detailed method instructions for the assembly line operators; tool, drawing, and parts lists for each work station; and revised assembly line configurations when labor crew or production requirements change.

The Routing Data Base® (RTG). This module, generated automatically from data transferred by the UnivEl System, includes network and scheduling information, standard cost data, materials list and usage points, engineering change control information, tooling and process routing data, inventory information, and manufacturing operation measurement standards.

The UniCostⓉ System. This module computes standard and current costs at the operational level in terms of material, labor, and overhead. It offers the capability for costing work-in-process, scrap, and proposed changes.

The Performance Audit and Review (PAR)Ⓣ System. This module generates a labor performance report based on actual operating data from the shop. This report can be by department, shift, and employee. Audit reports are also generated to identify discrepancies in data entry and calculation.

Inputs and Outputs. The inputs and outputs of the UniVation Systems can be vividly seen by examining an example of a milling machine. Exhibit 4.7.14 shows a UnivEl coding sheet consisting of constant data, such as weight and distance. The industrial analyst codes the variable data pertaining to part configurations that make one standard different from another. Exhibit 4.7.15 depicts all the data sent to the computer for generating the standard, the method instruction

Exhibit 4.7.14 UnivEl Coding Sheet

UNIVEL® CODING SHEET — PAGE ___ OF ___

Card	Field	Entry
AL00	SEQ. NO.	
A	FUNC. NO.	00109
A	FUNC. CTR	ASY06
A	DEPT. NO.	460
A	REV. NO. PART / OP	B6
A	PART NUMBER	107120-3
A	OPER. NO. PRIME	40
A	RATE TYPE	RUC
B	PART NAME	RAM LOGIC BOARD
B	UNIT MEAS.	EACH
B	SCE CODE	MM
C	PRIMARY OPERATION DESCRIPTION	HAND INSERT COMPONENTS

Other fields on the form (left blank): LOOP PART, NEW PART, I.D. NO., SET NO., MO DA YR, ENGINEER, SET-UP STD., SET-UP JOB CL., SET-UP MACH. MEN, RUN STD, RUN JOB CL., RUN MACH. MEN, OP. UNIT, TOOL NUMBER, TOOL DESCRIPTION, NUMBER TOOLS, COMPONENT PART, QTY-PER-UNIT, ALTERNATE MATERIAL, AM ST, AM-CODE, SCE CODE, COMPONENT DESCRIPTION, UNIT MEAS, CURRENT COST, DRAWING NUMBER, FAMILY NUMBER.

Seq.	Col. 7	Col. 8	Conditions	Location	Distance 1 (in.)	Distance 2 (in.)	Effective Net Weight	Control Dim	Frequency	•	UnivEl Literals
20	1	2	22	34	130	130			/	,	FIXTURE, SUPPLY, LT, BENCH
30	1	2	11	46	112	112	2	3	/	,	BOARD, SUPPLY ON LEFT, LT, FIXTURE
40	1	2	11	25	3	2	2	1	1002	,	WING NUT, LOC, LT, TIGHTEN
50	1	2	11	23	2	1	3	1	1002	•	BOARD, LOC, LT, STRAIGHTEN
60				5	112				/	:	DIODE, BIN #1 WHILE/INSERTING TRSFMER
70				8	112				/	?	DIODE, INSERT IN BOARD W/LH
80				5	112				/	,	RESISTER, BIN #2 W/RH
90				7	112				/	?	RESISTER, INSERT W/RH
100				4	112				/	,	JUMPER WIRE, BIN #3/W/LH
110				7	112				/	,	JUMPER WIRE, INSERT/W/LH
120				5	112				/	:	DIODE, BIN #4 & 5 /
130				8	112				/	?	DIODE, INSERT /
140				4	112				/	,	RESISTER, BIN #6 & 7
150				7	112				/	,	RESISTER, INSERT /
160				5	116				/	:	CAPACITOR, BIN #8 & 9
170				7	116				/	,	CAPACITOR, INSERT /
180				5	116				/	:	DIODE, BIN #10 & 12/W/RH
190				8	116				/	,	DIODE, INSERT W/RH /
200				5	116				/	:	DIODE, BIN #11 & 13/W/LH
210				8	116				/	,	DIODE, INSERT W/LH /
220				6	116				/	:	1/2 WATT RESISTOR, BIN #14 & 15
230				7	116				/	?	RESISTOR, INSERT /
240				6	116				/	,	WW RESISTOR, BIN #16 & 17
250				7	116				/	?	WW RESISTOR, INSERT/
260				6	124				/	,	CAPACITOR, BIN #18 & 19

Form columns also include: VC, DEGREES OR TURNS, COMMENT FREQUENCY, COMMENT DECIMAL HOURS, COMMENT DECIMAL MINUTES; FORMULA NUMBER 1–17.

• CODE: , COMPUTE, PRINT SUBTOTAL & CLEAR · COMPUTE AND ACCUMULATE TIME ? SIMO – COMPARE TO FOLLOWING ELEMENT : COMPUTE & PRINT TIME SEPARATE FROM ACCUMULATION ; COMPUTE & PRINT CURRENT TOTAL

(UnivAtion) MANAGEMENT SCIENCE INC APPLETON WISCONSIN

FORM NO. U115

Exhibit 4.7.15 Typical Input for the UnivAtion System

```
AC       06803 MIL14     96      002 A4   040276                   100-13019.000    20   RUC
BC  SHAFT                                        EACH MM                                ESS
CC  MILL KEYWAY
DC                        ML-UN2631 UNIVERSAL MILLING FIXT.      1
  20   1211   34    18   20       20 12     1002,TWO SHAFTS,MACH TABLE,SAME,FIXTURE
  30   1211   22    10    6       10  1          ,CLAMPS,FIXTURE,SAME,POSITION
  40   1211   35    10   12       10 12     1002,WRENCH,MACH,SAME,NUT
  50   111     5    10            10 12     1002,WRENCH,NEXT NUT
  60   1211   25    10   10       10 12     3001,NUT W/WRENCH,FIXT,SAME,TIGHTEN
  70   111     2    18            10 12     1002,WRENCH,ASIDE
  80   121 303      24            50 12 20  1002,CRANK,MOVE TABLE TO DEPTH
  90   1211   21    18    1        5  1     1002,BUTTON,MACH,SAME,START MACH
 100   111     5     1            10 12     1002,HANDLE,POSITION TABLE FINAL POSITION
 110   1211   21    14    4       10  1     1002,LEVER,MACH,SAME,ENGAGE FEED
 120   F,11150200 10001501450   1    10040   12    1    10200 125    1    17500    1 100
 130   F11115    10500
 140   121 303      24            50 12 20  1002,CRANK,MOVE TABLE DOWN TO CLEAR
 150   1211   21    18    1        5  1     1002,BUTTON,MACH,SAME,STOP MACH
 160   1211   35    10   12       10 12     1002,WRENCH,MACH,SAME,NUT
 170   111     5    10            10 12     1002,WRENCH,NEXT NUT
 180   1211   25    10   10       10 12     3001,NUT W/WRENCH,FIXT,SAME,LOOSEN
 190   111     2    18            10 12     1002,WRENCH,ASIDE
 200   1213   22    14   40       20 12     1002,TWO SHAFTS,FIXTURE,SAME,BENCH
 210   1231   22    40   14        3  4     1002,BRUSH,MACH,SAME,CLEAN FIXTURE
 220   111     2    10             3  4     3001,BRUSH,CLEAN FIXTURE
 230   111     2    14             3  4     1002,BRUSH,ASIDE
 240   C TIME TO DEBURR AND ASIDE IS INT TO PROCESS TIME
 250   F,9030
9999   E
```

EXAMPLE 2

VARIABLES ENTERED IN FORMULA 1115 ARE:

A.) DIAMETER OF CUTTER 2.00 INCHES

B.) SURFACE FEET/MINUTE 150 SFPM

C.) LENGTH OF CUT 2.00 INCHES

WHEN USING THE VARICOMP SYSTEM, THE ENGINEER NEEDS ONLY TO FILL IN THE VARIABLE FIELDS UNDERLINED IN THE METHODS SET. THE TERMINAL AUTOMATICALLY INSERTS THE REMAINDER.

NOTE: THE CODING SHEET IS ON THE PRIOR PAGE

sheet, and the routing. These data are subsequently transferred to the MultiComp file for mass updating. Exhibit 4.7.16 presents the standard and the method instruction generated by the computer. The file routing, consisting of tooling and material-used information, is displayed in Exhibit 4.7.17. Finally, Exhibit 4.7.18 shows a printout of the current cost of the product.

4.7.6 BENEFITS

The currently available computer-aided work measurement systems are beginning to enhance the productivity of industrial engineers substantially by providing efficient data gathering devices and speedy tools for analyzing the mass of data and by considerably minimizing the manual effort required for reporting, maintaining, and updating time standards. Some of the benefits being realized are as follows:

Reduction in the effort required to specify motions that define a desired method.

Audit control of thoroughness and completeness of data gathering.

Reduced time required for development of standards, yielding faster and greater coverage with standards.

Increased accuracy, consistency, and repeatability of work standards developed by different analysts.

Considerably reduced training time for work measurement personnel.

Greater effectiveness of industrial engineers by relegating clerical tasks to other personnel.

Identification of potential methods improvement through the automatic computation of methods improvement indexes.

Generation of uniform methods of operations for supervisors in order to train operators in the best method.

A common data base of standards for a high degree of accessibility and ease of updating and maintenance.

Accurate and up-to-date base for cost engineering data.

Mathematical techniques providing reliable relationships between production output and work input where other techniques of work measurement fail.

Incorporation of work measurement in the overall management information system.

Enhancing of the motivation of industrial engineers because the system provides job enrichment. Mundane tasks of data manipulation are replaced by opportunities for creativity.

4.7.7 THE FUTURE

Electronic data gathering devices with intelligence and data storage capability will rapidly cause the stopwatch to become obsolete. Two potential developments are envisioned for the field of automated data gathering in the not-too-distant future. First, the combination of video tape recorders and computers will aid time and motion analysis. It is envisioned that the video information will be digitized for computer analysis. Second, the successful advent of voice or speech recognition will minimize the need for data entry during a work measurement exercise.

The author does not envision the proliferation of dedicated minicomputers for work measurement, but rather the utilization of time-sharing services, either by the different plants of a corporation or by the small companies. This will bring about the establishment of centralized work measurement systems with computerized storage, retrieval, and updating of standard data, as well as the preparation of operational documents performed through terminals that interface with remotely situated computers. In addition, assembly line work measurement and alternative method evaluations will be conducted through computer terminals.

In the machine-cycle-dictated operations, such as machine shop and welding, microcomputer-based systems will provide optimization of processes and development of time standards. In the area of indirect labor, such as material handling, warehousing, transportation, and office work, labor standardization will be brought about by application of mathematical and statistical techniques. These will serve as a basis for manpower planning and control.

The management information systems of the future will increasingly incorporate labor standards, in a dynamic mode, into the data base for scientific planning and control of manpower, production scheduling, product costing, and labor variance analysis.

In summary, application of computers in work measurement will significantly improve the productivity of the industrial engineer, who will devote more time to developing a better method rather than to quantifying the existing one.

Exhibit 4.7.16 Standard and Method Instruction

```
                    M A N A G E M E N T   S C I E N C E,   I N C.

PART NAME  SHAFT                                  SOURCE CODE  MM    REVISION NO.   A4/      PART-NO.        100-13019.000
OPER DESC. MILL KEYWAY                            SET      002   FUNCTION  06803    DEPT.     96    CTR. MIL14     OPER    20
TOOL NO.                 ML-UN2631    TOOL DESC.  UNIVERSAL MILLING FIXT.     NUMBER OF TOOLS   1    JOB CL.         CREW
DWG. NO.                          FAMILY NO.                RATE TYPE R U C     ENGINEER  RSS               DATE  04/02/76

NO.         ELEMENT DESCRIPTION                                                                FREQ.    HRS/PC     MIN/PC
  1 OBT.  TWO SHAFTS FROM MACH TABLE 18 IN DIST, MOVE SAME TO FIXTURE 20 IN DIST.              1/  2  .0003120    .01872
  2 OBT.  CLAMPS FROM FIXTURE 10 IN DIST, MOVE SAME TO POSITION 6 IN DIST.                     1/  1  .0002250    .01350
  3 OBT.  WRENCH FROM MACH 10 IN DIST, MOVE SAME TO NUT 12 IN DIST.                            1/  2  .0002777    .01666
  4 MOVE WRENCH TO NEXT NUT 10 IN DIST.                                                        1/  2  .0001878    .01127
  5 OBT.  NUT W/WRENCH FROM FIXT 10 IN DIST, MOVE SAME TO TIGHTEN 10 IN DIST.                  3/  1  .0014658    .08795
  6 MOVE WRENCH TO ASIDE 18 IN DIST.                                                           1/  2  .0001379    .00827
  7 OBT.  CRANK AND TURN TO MOVE TABLE TO DEPTH 24 IN DIST.                                    1/  2  .0003550    .02130
  8 OBT.  BUTTON FROM MACH 18 IN DIST, MOVE SAME TO START MACH 1 IN DIST.                      1/  2  .0000946    .00568
  9 MOVE HANDLE TO POSITION TABLE FINAL POSITION  1 IN DIST.                                   1/  2  .0001388    .00833
 10 OBT.  LEVER FROM MACH 14 IN DIST, MOVE SAME TO ENGAGE FEED 4 IN DIST.                      1/  2  .0000986    .00592
 11 F NO.  1115      OPEN END KEYWAY          2.00 DIAMETER OF END MILL      1.00 MATERIAL FACTOR
               150 SURFACE FEET/MIN          286 RPM OR NEXT CLOSEST      .0040 CHIP LOAD/TOOTH
                12 TOOTH CUTTER          13.7514 FEED RATE IPM              2.00 LENGTH OF CUT
              .125 APPROACH @ FEED RATE         1 NUMBER OF PASSES          7500 TRAVERSE TIME-IPM
             .0100 TOOL ALLOWANCE %                                                          0.500  .0013030    .07818
```

NOTE PRINTOUT OF FORMULA FIXED AND VARIABLE DATA

12 OBT. CRANK AND TURN TO MOVE TABLE DOWN TO CLEAR 24 IN DIST.	1/	2	.0003550	.02130
13 OBT. BUTTON FROM MACH 18 IN DIST, MOVE SAME TO STOP MACH 1 IN DIST.	1/	2	.0000946	.00568
14 OBT. WRENCH FROM MACH 10 IN DIST, MOVE SAME TO NUT 12 IN DIST.	1/	2	.0002777	.01666
15 MOVE WRENCH TO NEXT NUT 10 IN DIST.	1/	2	.0001878	.01127
16 OBT. NUT W/WRENCH FROM FIXT 10 IN DIST, MOVE SAME TO LOOSEN 10 IN DIST.	3/	1	.0014658	.08795
17 OBT. WRENCH FROM ASIDE 18 IN DIST.	1/	2	.0000845	.00507
18 OBT. TWO SHAFTS FROM FIXTURE 14 IN DIST, MOVE SAME TO REAR TO BENCH 40 IN DIST.	1/	2	.0003007	.01804
19 OBT. BRUSH AT REAR FROM MACH 40 IN DIST, MOVE SAME TO CLEAN FIXTURE 14 IN DIST.	1/	2	.0002829	.01697
20 MOVE BRUSH TO CLEAN FIXTURE 10 IN DIST.	3/	1	.0004839	.02903
21 MOVE BRUSH TO ASIDE 14 IN DIST.	1/	2	.0001023	.00614
TIME TO DEBURR AND ASIDE IS INT TO PROCESS TIME				
	TOTAL		.0082314	.49388
F NO. 9030 15.0 PER CENT ALLOW.			.0012347	.07408
	TOTAL		.00947	.568

.94661 HRS/100 PCS — STANDARD

105.6 PCS/HOUR — OUTPUT AT 100%

PG. 2

THIS IS THE METHOD INSTRUCTION AND STANDARD WHICH IS AUTOMATICALLY GENERATED BY THE UNIVEL SYSTEM FROM THE VARICOMP INPUT.

Exhibit 4.7.17 File Routing and Assembly Parts List

```
PGM-A425                         MANAGEMENT SCIENCE INCORPORATED
                             FILE ROUTING AND ASSEMBLY PARTS LIST
REVISION-NUMBER  A4-                                          PART-NUMBER        100-13019.000 MM

REVISION-DATE  04/02/76                   PART-DESCRIPTION  SHAFT
                                            FAMILY--NUMBER
SIMILAR-PART                                DRAWING-NUMBER                          PAGE NUMBER--01
*************************************************************************************************

 OPER    FUN     WC  DEPT OP             OPERATION                        SETUP      RUN   ASSGN   REV
  NUM    NUM    NUM   NUM CD            DESCRIPTION                       TIME       TIME  MH-MN CB NO

  10    74801  TRN74   54   CHAMFER-THREAD-CUTOFF                        .31400    1.92000 01-01 1C
                                                                                   52.08   PCS/HR

                              - - - - -TOOLING- - - - -
                               TN-00042693  001  CUTOFF TOOL
                               TN-04232749  001  THREADING TOOL
                               TN-27986201  001  CHF TOOL

                              - - - - -MATERIAL- - - - -
                         ST-86790CR PR      8.25000 LB.  3/4 INCH 1020 CR BAR STOCK

  10-A  74804  TRN74   54   CHAMFER-THREAD-CUTOFF                        .35800    2.35000 01-01 1C
                                                                                   42.55   PCS/HR

                              - - - - -TOOLING- - - - -
                               TN-00042693  001  CUTOFF TOOL
                               TN-04232749  001  THREADING TOOL
                               TN-27936202  001  CHMF TOOL                                  NEW STANDARD

  20    06803  MIL14   96   MILL KEYWAY                                  .76300     .94661 01-01 UC
                                                                                  105.64   PCS/HR

                              - - - - -TOOLING- - - - -
                                 ML-UN2631  001  UNIVERSAL MILLING FIXT.

  30    27401  DRL07   69   DRILL 1/8 INCH HOLE                          .15000    1.37000 01-01 1C
                                                                                   72.99   PCS/HR

                              - - - - -TOOLING- - - - -
                                DR-721643P  001  DRILL JIG

                                                                    SETUP-TIME   RUN-TIME
                                                            TOTAL-    1.22700     4.23661


                                  COPY OF THE FILE ROUTING WHICH IS AUTOMATICALLY
                                  GENERATED BY THE UNIVEL SYSTEM
```

Exhibit 4.7.18 Standard Cost Output

04/06/76 | MANAGEMENT SCIENCE INCORPORATED | CONFIDENTIAL | PAGE - 1

STANDARD COSTING SYSTEM

PGM-C608 | C U R R E N T   J O B   C O S T | INTERNAL USE ONLY

DESCRIPTION
SHAFT

U/M EACH | CB C | PART NO. 100-13019.000 MM

OPER	COMP-NO	SC	DESCRIPTION/OP	QTY/STD	MATERIAL	PACKAGING	LABOR	VAR-OHD	SUB-TOTAL	FIXED-OHD	TOTAL
	ST-86790CR	FR	3/4 INCH 1020 CR BAR STO	8.25000	225.22500				225.22500		225.22500
	74801	54	CHAMFER-THREAD-CUTOFF	1.92000			12.07680	11.00160	23.07840	11.96160	35.04000
10					225.22500		12.07680	11.00160	248.30340	11.96160	260.26500*
	74804	54	CHAMFER-THREAD-CUTOFF	2.35000			14.78150	13.46550	28.24700	14.64050	42.88750
10-A					225.22500		12.07680	11.00160	248.30340	11.96160	260.26500*
	06303	96	MILL KEYWAY	.94661			5.32941	7.68647	13.01588	11.29306	24.30894
20					225.22500		17.40621	18.68807	261.31928	23.25466	284.57394*
	27401	69	DRILL 1/8 INCH HOLE	1.37000			6.78150	2.64410	9.42560	7.85010	17.27570
30					225.22500		24.18771	21.33217	270.74488	31.10476	301.84964*

NOTE: OPERATION 10-1 (AN ALTERNATE) IS COSTED AS AN INDIVIDUAL OPERATION BUT NOT ADDED TO TOTAL.

AFTER EACH CHANGE TO THE DATA BASE, THE UNICOST SYSTEM AUTOMATICALLY UPDATES THE PART COST.

4.7.8 CONCLUSION

Computerized work measurement is evolving; when it comes of age, it will be possible to reduce drastically the need for human participation in routine clerical work. This chapter has briefly outlined how computerized systems and electronic aids for input, analysis, reporting, and maintenance of work measurement data are proving to be extremely beneficial. It is believed that the industrial engineer is beginning to improve his or her own productivity by utilizing the computer.

REFERENCES

1. A. J. TAYLOR, "Computer-Aided Work Sampling," Work Sampling Workshop, University of Wisconsin-Extension, Milwaukee, 1977.
2. J. C. MARTIN, "The 4M Data System," *Industrial Engineering*, March 1974, pp. 32–38.
3. P. MURPHY, "A Computerized Standard Data System," *Industrial Engineering*, October 1970, pp. 10, 17.
4. R. F. WEAVER and E. A. BOEPPLE, "WOCOM and Quick Work-Factor: The State-of-the-Art in Predetermined Systems," A.I.I.E. Fall Conference, 1979.
5. K. P. ZANDIN, "Relieving the Productivity Shortage," paper presented at Industrial Management Society meeting, Arlington Heights, November 1979.

CHAPTER 4.8

Development and Use of Standard Data

ADAM W. CYWAR

IBM Corporation

4.8.1 OVERVIEW

The preceding chapters in this section have explored the three fundamental techniques for setting time standards: time study, predetermined motion time systems, and work sampling.

This chapter focuses on a fourth technique for establishing standards to measure performance of work, that is, the use of standard data. The essential differences between standard data and the three fundamental techniques are as follows:

1. Standard data are a set of synthesized time values or a mathematical model, either of which may use time values established by one or more of the fundamental techniques.
2. Standard data in the form of a mathematical model use other parametric data and may also use time values that have not been established by the fundamental techniques.
3. Standard data are used to establish standards without the further use of any fundamental technique or timing device.

In this chapter, standard data that are a set of synthesized time values are referred to as "synthetic standard data." Standard data in the form of a mathematical model are called "analytical standard data."

Following this chapter overview, case studies will be used to demonstrate the techniques of developing standard data for various categories of work. The techniques for developing synthetic standard data are well known and have been described fully in many textbooks and other publications[1-6]; therefore minimal treatment is accorded here. The techniques for developing analytical standard data are less familiar, and relatively little discussion of them can be found in the literature. For that reason most of the case studies are devoted to demonstrating the analytical technique.

This chapter concentrates on techniques for developing and using standard data. Information on the mechanization of these techniques through the use of computer systems can be found in Chapters 4.7 and 12.5.

Purpose

Development and use of synthetic and analytical standard data to set standards have been parallel activities throughout the evolutionary development of the fundamental work measurement techniques. Since the earliest development of synthetic "standard elements" for use in taking a stopwatch time study, and the development of early analytical models in the 1950s, work measurement practitioners have developed many shapes and forms of synthetic and analytical standard data.

The primary purpose of using either type of standard data is to minimize the expense of setting time standards and thereby lower the overhead cost of producing a product or rendering a service. When properly developed and maintained, standard data can significantly reduce the expense of establishing standards to measure performance. When poorly developed and/or maintained, the improved productivity of the workers based on the standard data can be more than offset by the effects of faulty performance standards (see section on advantages and limitations of standard data).

Exhibit 4.8.1 An Example Used To Illustrate the Calculation of Synthetic and Analytical Approaches

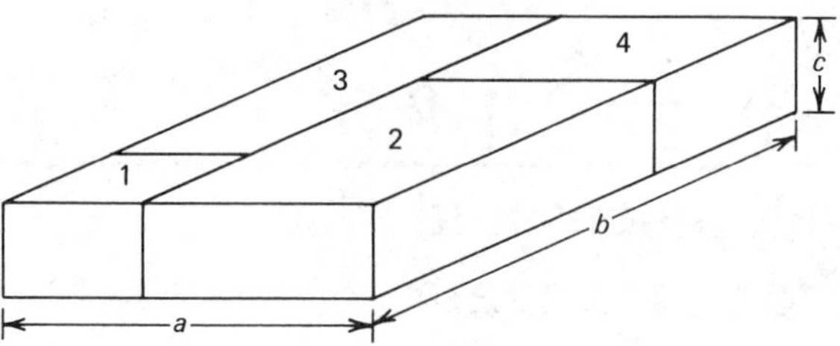

The Two Types of Standard Data

Synthetic and analytical standard data have been developed for many classes of direct and indirect work. To help distinguish between the methods for developing each type of standard data, consider the following simplistic analogy.

The volume of the block depicted in Exhibit 4.8.1 can be calculated in two ways. The synthetic approach would involve summing up the volumes of the four individual subblocks, or

$$V = v_1 + v_2 + v_3 + v_4$$

An analytical approach to calculating the volume would employ the use of the a, b, and c dimensions, or

$$V = (a)(b)(c)$$

In a similar fashion, synthetic standard data are constructed by taking smaller pieces of time data and adding them together to achieve larger blocks of time. Using Mundel's concept of the basic orders of work units,[7] the subblocks in Exhibit 4.8.1 could be element-level (second-order work unit) standard data that were synthesized by combining basic motions (first-order work units). Exhibit 4.8.2 depicts a hierarchy of standard data using the orders of work units concept.

The subblocks described could be at a higher level. Specifically, consider the work of a receiving clerk in a warehouse operation where the objective is to establish a standard time for receiving shipments. The subblocks could then represent task-level (third-order) standard data that were synthesized from lower orders of work units (motions and/or elements). Very possibly, task-level standard data could already exist, in which case it would not be necessary to synthesize data from lower levels. The standard time, in any event, would then be established by adding together the time values contained in the third-level synthetic standard data (see Exhibit 4.8.3).

Exhibit 4.8.2 Hierarchy of Standard Data

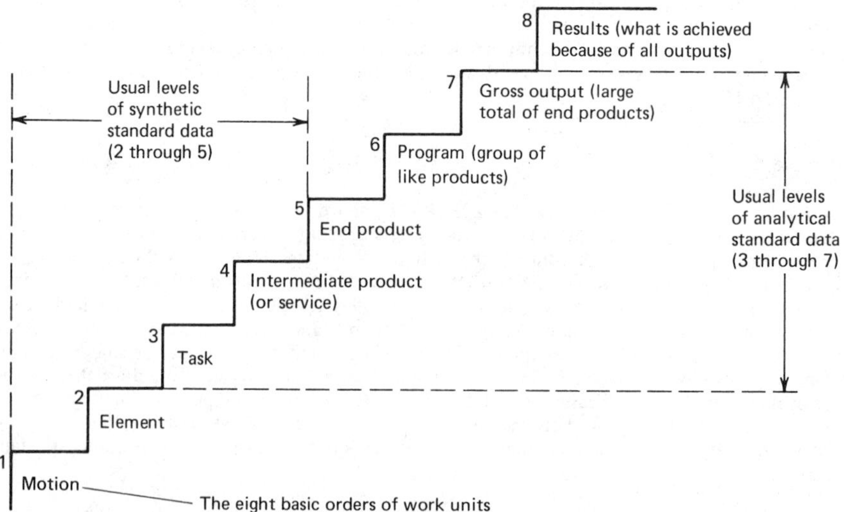

Exhibit 4.8.3 Pictorial Representation of Synthetic Standard Data Construction

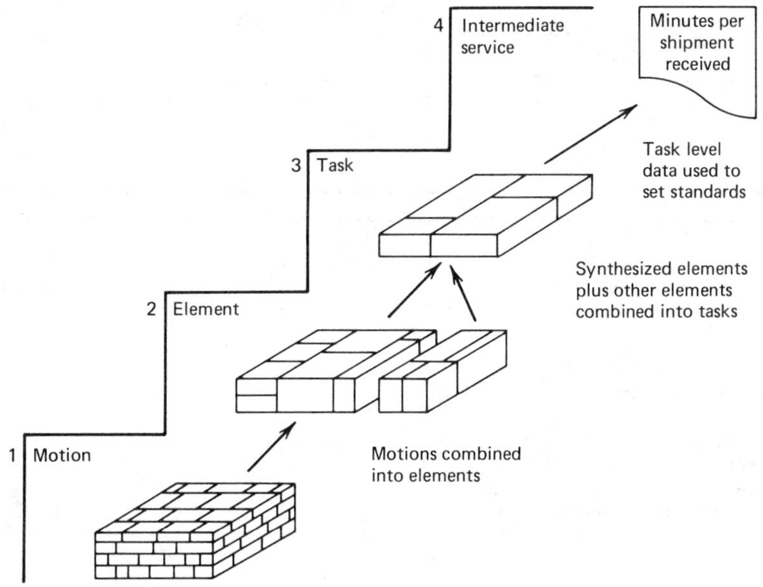

The alternative approach to establishing the standard would be to develop and use an analytical standard data model. This, in fact, was actually done. Continuous time studies were conducted, and data were collected that were analagous to the *a*, *b*, *c* dimensions in Exhibit 4.8.1 This *a*, *b*, *c* type of data (see Exhibit 4.8.4) described the receiving clerk's job in terms of parameters or variables that relate to the variations in the time it takes to receive a shipment. The analytical model was then developed by using multiple regression analysis (see Chapter 13.6), which established the relationship between the parameters and the time to perform the operation.

To calculate the standard minutes for any shipment, the parameter counts for that shipment are multiplied by their respective coefficients and added to the constant value in the model. Appropriate allowances would also be added. As shown in Exhibit 4.8.4, there is the possibility of being in error by $\pm 20\%$ when using the model to calculate the standard time for receiving any one shipment. However, calculating the standard minutes over a longer measurement period (1 day, 1 week, etc.) reduces the error to a tolerable level. A detailed explanation of the error calculation and a further discussion of this model appear in Section 4.8.3.

Exhibit 4.8.5 illustrates the approaches to developing standard data and the relationship to the eight basic orders of work units.

Advantages and Limitations

The use of standard data is an advantage when it lowers both the labor and the overhead cost of producing a product or rendering a service.

The overhead cost of setting standards will be lowered by using standard data. However, if the standards that are set from the standard data are not equitable, the negative impact on employee morale can lead to an increase in labor cost well beyond the savings achieved in the work measurement function.

Standards established from standard data tend to be more consistent than standards established by means of the fundamental techniques. This consistency is fundamental to employees' acceptance of the standards as being fair and equitable. When the standards established from standard data are excessively loose or tight, the obvious inequities are damaging, in spite of consistency. For this reason, standard data must be thoroughly tested prior to use to ensure that standards being set from the data will be within close proximity (usually $\pm 5\%$) to the standards that would result from direct measurement. Testing higher-level standard data by means of one of the fundamental techniques can be extremely time consuming and nearly impossible in some cases. Depending on contractual agreements and other environmental considerations, this inability to test can severely limit the use of standard data above the fourth level.

There is no single technique for developing standard data that is universally applicable to all work situations. Prior to developing and using standard data of any type, the nature of the work and the

Exhibit 4.8.4 Receiving Clerk–Analytical Standard Data Model With Seven Parameters

Study No.	Normal Time (min.)	Parameters						
		P_1	P_2	P_3	P_4	P_5	P_6	P_7
1	12.1	2	6	32,000	0	1	2500	3
2	21.6	3	10	7,997	1	0	0	0
3	154.3	28	54	83,377	16	4	3577	4
4	8.5	1	8	10,763	0	0	0	0
23	41.6	6	19	31,546	2	2	1463	2
24	6.4	1	2	6,576	0	1	6576	0
25	13.1	2	17	81,643	0	2	7430	7

Parameters

P_1 = number of lots in the shipment
P_2 = number of cartons
P_3 = number of pieces
P_4 = number of urgent lots
P_5 = number of lots that are 100% inspected
P_6 = number of pieces actually counted
P_7 = number of discrepancies found

Model

Standard minutes per shipment received = $1.086 + 5.954(P_1) + 0.0358(P_2) + 0.00024(P_3) + 1.192(P_4) - 5.617(P_5) + 0.00041(P_6) - 1.3027(P_7)$

Model Reliability (95% Confidence Level)

$S = 2.450$ = standard error of estimate

$r = 0.911$ = coefficient of correlation

$T = 24.702$ = average time per shipment

L = measurement period in minutes (same units as T)

Measurement Period		Error
1 shipment	$L = 24.702$	±20%
1 day	$L = 480$	±4.5%
1 week	$L = 2400$	±2.0%

Exhibit 4.8.5 Synthetic Versus Analytical Approaches to Developing Standard Data and the Relationship to the Eight Basic Orders of Work Units

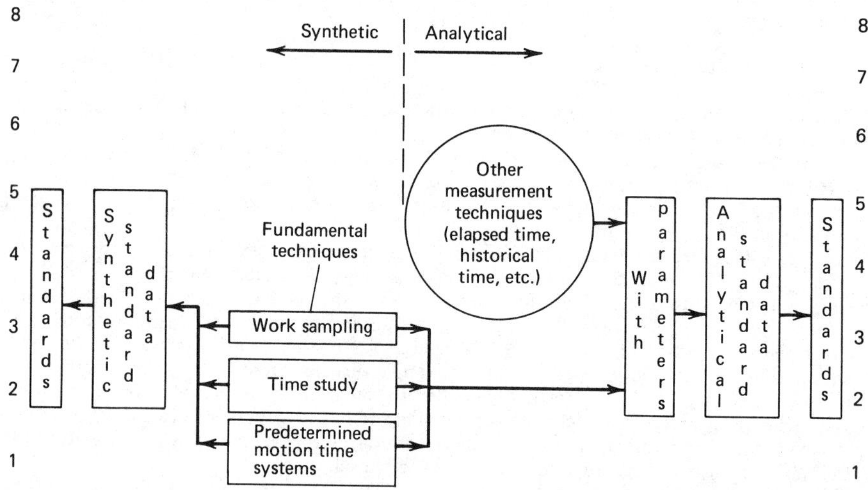

Exhibit 4.8.6 Standard Data Selection Guidelines

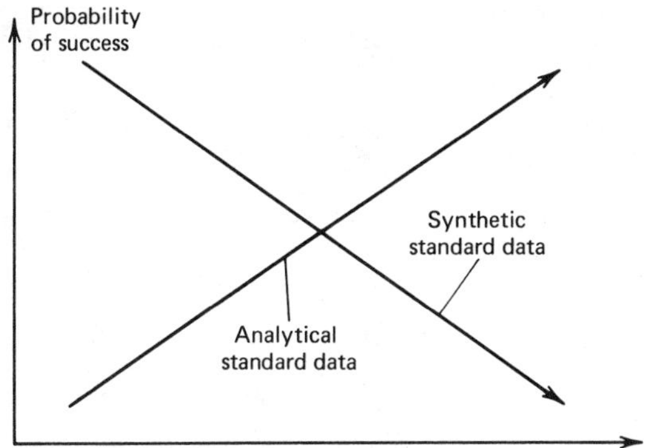

Highly repetitive	Type of operation	Very nonrepetitive
Very short	Length of cycle	Very long
Very frequent	Methods/process changes	Very infrequent
High	Cost of labor (% of total cost)	Low
Many	Number of people doing identical work	Few
Incentives	Method of payment	Day work

operating environment must be examined. This examination will then help to determine what type of standard data is suitable–if any at all.

An operating environment conducive to the successful use of standard data is one where the existing standards–if any–are in good condition; that is, the productivity level resulting from the standards is consistent with management expectations and considered fair by employees. New standards established from standard data will be judged in relationship to the existing standards. If the existing standards are in poor condition, it is extremely difficult to fairly evaluate and gain acceptance of the new standards.

The procedures for documenting major changes in work methods and processes are also important. When these procedures are weak, unknown changes can occur and render a set of standard data invalid. Unless these procedures are firmly in place and strictly adhered to, an excessive amount of time will have to be spent on auditing the standard data. The expense of auditing in this situation may exceed the cost to continue using direct measurement techniques.

Once a decision is reached to proceed with the development of standard data, the choice of whether to use synthetic or analytical data must be made. The characteristics displayed in Exhibit 4.8.6 are the major items to be considered. Although there may be exceptions, if the nature of the work is such that all the characteristics at the low end of the x axis are present, the synthetic approach should probably be followed. Conversely, if the characteristics are all at the high end of the x axis, analytical techniques may be appropriate. These are not hard-and-fast rules; they are general guidelines to assist in the selection process.

4.8.2 DEVELOPING SYNTHETIC STANDARD DATA

Synthetic standard data can be developed or constructed by using the following: elemental data from time studies; basic motions or groups of motions from predetermined motion time systems; and work sampling studies. Very often a combination of these sources is used. The case study used in this section utilized MTM-1 basic motions and time study elements. The building block concept[8] employed here combined first-order work units (motions) to achieve second-order work units (elements), and then additional second-order time study elements were added. This set of second-order standard data was then used to set standards at the fourth-order work unit level. The operations in this illustration were the incentive-paced, manual assembly of electrical components. Although not demonstrated here, it would have been possible to structure larger building blocks at the third-order level (task) and then to have used that data to establish the standards. This was not done since an inordinate amount of third-level data would have had to have been constructed to cover the many different varieties of assembled components.

Gathering Inputs

There are several possible approaches to developing synthetic standard data that are going to be used for establishing standards at the fourth level. In this case the options were narrowed to the following:

1. Using basic motions of predetermined motion time systems (MTM-1), construct element-level data.
2. Use time study elements.
3. Use third-level predetermined motion time systems (universal standard data, etc.—see Chapter 4.5).
4. Construct third-level data using first- and second-order data.

As noted previously, a combination of options 1 and 2 was used to develop the elemental data shown in Exhibit 4.8.7. Time study data were used for the various soldering process times and for certain allowances, such as element 8.30. All other elements were synthesized from basic MTM-1 motions.

It is important to recognize the trade-offs that were made in the preceding selection process. Items relating to the accuracy of the ultimate standards, the cost to build the standard data, the maintainability and auditability, employee and management acceptance, and so on, had to be considered. Given a different plant and operating environment, a choice of other options could have been equally likely. The choice that was made represented a compromise between employees and management. Management would have preferred the use of option 3, whereby existing third-level predetermined motion time systems data could have been purchased, modified or supplemented where necessary, and applied at a lower cost than that for applying lower-level data. The employees preferred option 2; time study had been the principal method of measurement up to this point, and they felt more comfortable with it as a basis for standard data. After considerable discussion, the combination of options 1 and 2 was agreed upon. Unless precluded by contractual obligation or other factors, gaining this agreement is essential to the successful use of standard data.

Structuring the Data

The result of building elemental data from basic motions is illustrated in Exhibit 4.8.8. The first step requires that basic motions for each element be tabulated covering the range of distances and other variables that can be encountered at the operator workplaces. In the case study, individual MTM analyses of various operations were examined to ascertain the patterns and frequency of motions that were being used. Elements, such as 8.30, that utilized time study data were likewise tabulated, and the average times from time study spread sheets were inserted.

Using the tabulated motions, the next step is to synthesize the motions for each element and to calculate the time values for each range of distance. For instance, using the "Tabulation of Motions" for element 1.10 (see Exhibit 4.8.8), the synthesized time value for the 3 in. to 4 in. range was calculated by adding the times for an R4C, a G1A, a G2, an M4C, and an RL1; 0.0040 + 0.0020 + 0.0027 + 0.0038 + 0.0010 = 0.0125 min.

The number of ranges to establish for each element is determined by the magnitude and frequency of occurrence of that element during the operation with the shortest cycle time. For instance, if the shortest cycle time is 0.20 min, and total distortion should not exceed 5%, then an element that accounts for 20% of the cycle time should not contribute more than 0.002 min of distortion (0.20 min × 0.05 × 0.20 = 0.002 min). If the element can occur up to four times, then the ranges should not be incremented greater than every 0.0005 min (0.002/4 = 0.0005 min).

The final step in the process is construction of a specification sheet (see Exhibit 4.8.9) that will be used to establish individual standards. The specification sheet—in this case a single form printed on both sides—contains all the synthesized data shown in the "Synthesis of Motions" column of Exhibit 4.8.8. The form allows the analyst to circle the appropriate time values and then transpose the data to the column on the right for tabulation.

Documentation of all conditions and of the environment surrounding the work area is extremely important. This documentation would include workplace sketches or photographs, operational procedures, material handling requirements, work rules, and so on. This information is vital to the analyst in order to ensure that conditions are still the same prior to setting a standard from the standard data. This information will also be needed during future audits of the data.

Using the Data to Set Standards

The procedures for setting standards with standard data will vary among companies, plants, or other business units. In general, the following steps are basic to the process:

Exhibit 4.8.7 Index of Synthesized Elemental Data for Manual Assembly of Electrical Components

Element No.	Description
1.00	Get part
1.10	Not simo
1.20	Partly simo
1.30	Add–search and select
2.00	Move part
3.00	Dispose part or tool
3.10	Dispose–from hand
3.12	Add–when not in hand
3.13	Add–stack dispose (no positioning)
3.14	Add–stack dispose (positioned)
4.00	Position part
4.10	Class 1
4.20	Class 2
4.30	Class 3
4.40	Add–difficult to handle
4.50	Add–requires visual orientation
4.60	Add–additional location points or stops
4.70	Add–apply pressure required
5.00	Wiring–without tools
5.10	Get wire into hand and/or fingers
5.11	Add–difficult to grasp
5.20	Straighten wire
5.30	Move wire–approximate location
5.40	Connect wire
5.41	Add–difficult to handle
5.42	Add–when wire is wrapped
5.43	Add–tight fit
5.44	Add–bend wire after threading
5.45	Add–crimp with tool
6.00	Wiring–with tools
6.10	Get tool
6.20	Align tool on wire
6.30	Move/pull/align wire
6.31	Add–sequence reposition
6.40	Cut wire
6.41	Add–difficult to cut
6.42	Add–sequence reposition
6.50	Hook end of wire
6.51	Add–sequence reposition
6.60	Connect wire
6.61	Add–crimp after position
6.62	Add–crimp after wrap or thread
6.63	Add–bend with tool after thread
6.64	Add–sequence reposition
7.00	Soldering–with iron
7.10	Get iron
7.20	Get spool solder
7.30	Solder connection
7.40	Add–iron holds part
7.50	Add–additional crystallization
7.60	Add–extreme accuracy
8.00	Dip soldering
8.10	Apply flux–dip
8.20	Apply solder–dip
8.30	Allowance to clean scum/add solder
9.00	Trip production counter
10.00	Open and close–quick-clamp

Exhibit 4.8.8 Example of Synthesized Data Using Basic MTM Motions and Time Study Elements

		Tabulation of Motions				Synthesis of Motions		
Element No.	Element Description	Description	Range	Motion	Time (min.)	Description	Range	Time (min.)
1.00	Get part							
1.10	Get part not simo—get part from pile, tray, container, bench, or conveyor, and move part to a work area.	Left or right hand				R_C + G1A + G2 + M_C + RL1		
		Reach to part	Up to 2″	R2C	0.0028	Get part, not simo	Up to 2″	0.0100
		Reach to part	3″ to 4″	R4C	0.0040	Get part, not simo	3″ to 4″	0.0125
		Reach to part	5″ to 6″	R6C	0.0048	Get part, not simo	5″ to 6″	0.0144
		Reach to part	7″ to 8″	R8C	0.0055	Get part, not simo	7″ to 8″	0.0159
		Reach to part	9″ to 10″	R10C	0.0062	Get part, not simo	9″ to 10″	0.0174
		Reach to part	11″ to 12″	R12C	0.0068	Get part, not simo	11″ to 12″	0.0188
6.30	Move, pull, or align wire—move wire with tool to an approximate location.	Left or right hand				M_B + G2		
		Move wire with tool	Up to 4″	M2B	0.0022	Move-pull-align wire	Up to 4″	0.0049
		Move wire with tool	5″ to 8″	M6B	0.0043	Move-pull-align wire	5″ to 8″	0.0070
		Move wire with tool	9″ to 12″	M10B	0.0059	Move-pull-align wire	9″ to 12″	0.0086
		Open tool to release wire		G2	0.0027			
8.30	Allowances to clean scum off solder pot and to add solder to solder pot.	Clean scum off pot—0.08/1000 $\frac{1}{2}$″ lengths of wire dipped				Allowances—per wire dipped		
		Length of wire	Up to $\frac{1}{2}$″	T.S.	0.00008	Length of wire	Up to $\frac{1}{2}$″	0.0002
		Length of wire	Up to 1″	T.S.	0.00016	Length of wire	Up to 1″	0.0004
		Length of wire	Up to $1\frac{1}{2}$″	T.S.	0.00024	Length of wire	Up to $1\frac{1}{2}$″	0.0005
		Length of wire	Up to 2″	T.S.	0.00032	Length of wire	Up to 2″	0.0007
		Length of wire	Up to $2\frac{1}{2}$″	T.S.	0.00040	Length of wire	Up to $2\frac{1}{2}$″	0.0009
		Length of wire	Up to 3″	T.S.	0.00048	Length of wire	Up to 3″	0.0011
		Add solder—0.10/1000 $\frac{1}{2}$″ lengths of wire dipped						
		Length of wire	Up to $\frac{1}{2}$″	T.S.	0.00010			

Exhibit 4.8.9 Example of Specification Sheet Showing (*a*) Top of Front Side and (*b*) Bottom of Reverse Side

Analyst ____________ Date ____________ Ref. No. ____________

Manual Assembly–Electrical

Part/Assembly No. ____________ Operator No. ____________ Dept. No. ____________

Element No.	To 2"	3"–4"	5"–6"	7"–8"	9"–10"	11"–12"	13"–14"	15"–16"	17"–18"	19"–20"	21"–24"	25"–28"	Time
1.10	0.0100	0.0125	0.0144	0.0159	0.0174	0.0188	0.0203	0.0219	0.0233	0.0248	0.0263	0.0293	0.0188
1.20	0.0050	0.0063	0.0072	0.0080	0.0087	0.0094	0.0102	0.0110	0.0117	0.0124	0.0132	0.0147	0.0065
2.00	0.0025	0.0038	0.0049	0.0057	0.0065	0.0073	0.0081	0.0090	0.0098	4.10–S	4.10–SS	4.10–NS	0.0007
3.13	0.0013	3.14	0.0040	1.30–L	–	1.30–M	0.0007	1.30–S	0.0025	0.0027	0.0044	0.0050	0.0061
3.10	0.0032	0.0043	0.0053	0.0061	0.0069	0.0074	0.0080	0.0086	0.0092	0.0097	0.0103	0.0115	0.0048
3.12	0.0029	0.0039	0.0044	0.0048	0.0052	0.0056	0.0060	0.0065	0.0068	0.0073	0.0077	0.0086	0.0044

(*a*)

7.10	To 4"	5"–8"	9"–12"	15"–18"	19"–24"	6.62	0.0150	6.63	0.0113	6.64	0.0132	0.0079	
	0.0074	0.0094	0.0124	0.0155	0.0193	C	1"–2"	3"–4"	5"–6"	7"–8"	9"–10"	11"–12"	0.0076
7.20	0.0041	0.0051	0.0065	0.0082	7.3	0.0263	0.0015	0.0028	0.0039	0.0047	0.0055	0.0063	0.0260
8.10	To 4"	5"–8"	9"–12"	13"–16"	17"–20"	8.30	½"	1"	1½"	2"	2½"	3"	0.0008
	0.0065	0.0076	0.0092	0.0108	0.0125	One	0.0002	0.0004	0.0005	0.0007	0.0009	0.0011	0.0055
8.20	To 6"	7"–8"	9"–10"	11"–12"	13"–14"	Two	0.0004	0.0008	0.0010	0.0014	0.0018	0.0022	0.0268
	0.0236	0.0250	0.0260	0.0269	0.0278	Three	0.0006	0.0012	0.0015	0.0021	0.0027	0.0033	
	15"–16"	17"–18"	19"–20"	21"–22"	9.00	To 4"	5"–10"	11"–16"	17"–24"	10.0	Total Min.		0.4612
	0.0286	0.0295	0.0303	0.0312		0.0055	0.0062	0.0076	0.0089	0.0268			

(*b*)

Using a checklist, verify the workplace conditions against the data contained in the standard data documentation.

Prepare a workplace sketch, list of tools used, and detailed description of the operation and parts. Cross-references to bills of material or other process documents should be made.

Using the standard data backup detail (see Exhibit 4.8.8), verify that distances and operator methods are correct.

Circle the appropriate values on the specification sheet and, if possible, check several cycles with a stopwatch.

Tabulate the time values on the specification sheet to determine the standard time and add any allowances not covered by the standard data. Use the stopwatch check cycles as a sanity test prior to issuing the standard.

It is possible to use standard data to estimate or predict time standards prior to actual operation. This is especially helpful when setting up operator stations on an assembly line or when developing cost estimates for future products.

Maintaining the Standard Data

Maintenance of synthetic standard data falls into two categories: modification to the data because of specific changes to the operations and periodic audits.

All changes that significantly affect the workplace layout, methods, tooling, procedures, and so on, should be formally communicated to the work measurement function. When these changes affect cycle times by a certain percentage—usually greater than ±5% or as contractually stipulated—it is necessary to modify those portions of the standard data that are no longer valid. Many—if not most—of the changes that reduce cycle times are instituted by the people doing the work. Allowing the worker to share in the benefits of these improvement ideas will normally result in a more timely modification of standard data and therefore more valid data on an ongoing basis. The use of suggestion programs that pay employees a percentage of the savings resulting from their improvement ideas is strongly recommended.

Periodic audits of standards set from the standard data are necessary to verify that workplace conditions are still the same as when the standard data were developed. Creeping changes (small changes over time) that have occurred since the standard data were developed or that have occurred between audits will normally be detected. When employees do not share in the benefits of their own methods improvements, an audit is probably the only time when these improvements will be detected, and at that, only a small percentage will be visible.

The magnitude and frequency of audits should depend on the potential cost exposure that can result from the standard data's becoming invalid. Where standards for a large percentage of the work force are being set from the standard data, it may be necessary to review many standards on at least a quarterly basis. Where coverage with the data is on a smaller scale, semiannual audits of fewer standards may be appropriate. Unless there is very limited use of the standard data, a good rule of thumb is that audits be done at least once a year.

4.8.3 DEVELOPING ANALYTICAL STANDARD DATA

The use of mathematical techniques is not new in the development of standard data. Multiple regression analysis has been used to develop elemental data, such as "get part," where parameters such as the location of the part or weight of the part have been mathematically related to the time to get one part.[9] This elemental data would then be used with other data to develop synthetic standard data.

The difference here is in the use of a mathematical technique to develop a standard data model at the task or higher level. This concept—also not new—was being applied more than 20 years ago.[10] The use of analytical standard data models for measurement is becoming more common because although the nature or composition of work is changing, the need for measurement remains constant. As machines gradually replace people doing direct work, the percentage of the work force doing indirect work steadily mounts. The fundamental need for measurement is not changing; however, the basis for measurement and the techniques of measuring are expanding to meet this need.[11,12]

Development of an analytical standard data model requires two inputs: (1) a set of time values associated with an amount of completed work and (2) the parameters or characteristics describing that completed work.

Using these inputs, mathematical techniques are employed to develop the standard data model. The most popular techniques that have been used to date are multiple linear regression analysis and linear programming. Curvilinear regression analysis, nonlinear programming, and queuing models

Exhibit 4.8.10 Classification of Analytical Standard Data Models

		Order of Work Unit (Level)				
Time[a]	Size of Work Group	3	4	5	6	7
Less than 1 hr	One person (A)	X	X	X		
1 hr to 8 hr	One or more persons (B)		X	X	X	
8 hr or greater	More than one person (C)			X	X	X

[a]Time = cycle time for time per unit of output models or time period for time per time period models.

have also been successfully used, but to a lesser extent. Section 13 contains a detailed explanation of these mathematical techniques.

The case studies that follow display the use of multiple linear regression. It should not be assumed that other mathematical techniques would not have been equally or more appropriate. The criteria for selecting a technique are becoming clearer, but it is still a flexible process; Mundel's discussion of choosing between multiple regression and linear programming,[13] for instance, clearly points out the nonmathematical, judgmental considerations that are part of the process.

The case studies presented here demonstrate that the development and use of analytical standard data is practical; development of practical criteria for selecting a mathematical technique requires additional research.

Analytical standard data models normally range from the third up through the seventh order of work unit. Also, depending on the measurement technique that is used, the models may have one of the following formats: (1) standard time per unit of output or service (minutes per piece, etc.) or (2) standard time per time period (hours per day, etc.).

Exhibit 4.8.10 portrays a classification of analytical models according to time, size of work group, and the orders of work units. Typically, for models where the cycle time or time period is less than 1 hr and the work is performed by one individual, the model can be at the third, fourth, or fifth level and is classified 3A, 4A, or 5A. Each of the models in the case studies that follow is classified according to this scheme.

Third-Level Model

The most common development of analytical standard data using multiple regression has been at the task level. Prior to the widespread availability of computers, the tedious mathematical calculations caused the number of parameters in the model to be severely limited. Usually, no more than three parameters could be considered, although a five-parameter model could be developed manually within an 8 hr period. Anything greater than five parameters was out of the realm of manual computation.

The model illustrated in Exhibit 4.8.11 contained five parameters. The standard error of estimate (S) of the model indicated that applying the model to any one order picked could result in a 21% error at the 95% confidence level.

The amount of error encountered (at the 95% confidence level) in applying the model to determine standard minutes earned is calculated using the following formula:

$$\%\ \text{error} = \frac{(2)(S)(100)}{\sqrt{(L)(T)}}$$

where S = standard error of estimate

L = length of measurement period (same units as T)

T = average time to pick one order

Applying the standard on a daily basis (L = 1 day at 480 min/day = 480) would yield the following:

$$\text{error} = \frac{(2)(1.41)(100)}{\sqrt{(480)(13.359)}} = \pm 3.5\%$$

On a weekly basis (L = 5 days at 480 min/day = 2400) the error would be:

$$\text{error} = \frac{(2)(1.41)(100)}{\sqrt{(2400)(13.359)}} = \pm 1.6\%$$

Exhibit 4.8.11 Order Picking–Third-Level Analytical Standard Data Model (3A)[a]

Study No.	Normal Time (min.)	Parameters				
		P_1	P_2	P_3	P_4	P_5
1	11.51	23	39	104	1	3
2	19.22	31	42	308	5	1
3	4.15	4	13	38	1	0
4	32.41	58	98	442	12	1
28	17.76	28	53	296	5	1
29	20.31	42	75	259	4	0
30	3.33	6	10	13	1	0

Parameters

P_1 = number of different items on the order
P_2 = total number of units on the order
P_3 = total weight of the order
P_4 = number of items out of stock
P_5 = number of items in secured area

Model

$$\text{Standard minutes per order picked} = 0.9115 + 0.0906(P_1) + 0.0182(P_2) + 0.0288(P_3) + 0.8928(P_4) + 1.2785(P_5)$$

Model Reliability (95% Confidence Level)

$S = 1.410$ = standard error of estimate

$r = 0.985$ = coefficient of correlation

$T = 13.359$ = average time per order

L = measurement period in minutes (same units as T)

Measurement Period		Error
1 order	$L = 13.359$	±21%
1 day	$L = 480$	±3.5%
1 week	$L = 2400$	±1.6%

[a]Third level–time per unit of output.

This task-level model was calculated without the use of a computer. Had the computer been utilized, the calculations would have been performed in a matter of a few seconds. More important, it would have been possible to calculate all of the 31 possible models (see Exhibit 4.8.12). Very possibly, a model with fewer parameters that is more accurate could have been found. The next case study will expand on this.

The application of the preceding model to calculate standard hours under a measured day work system was done on a weekly basis. The total number of orders picked and the associated parame-

Exhibit 4.8.12 Number of Possible Models for a Given Number of Parameters

Number of Parameters	Number of Possible Models
1	1
2	3
3	7
4	15
5	31
6	63
7	127
8	255
9	511
10	1023

ter volumes for a week were multiplied by the model constant and parameter coefficients (adjusted for allowances) to produce the weekly earned hours.

Fourth-Level Model

The receiving clerk case study cited earlier in this chapter is revisited here and is shown in Exhibit 4.8.13. Note that the model displayed now contains but two parameters. In this case the computer calculated all of the 127 possible models using all combinations of the seven parameters and ranked the models according to standard error (S).

The results indicated that a model with the specific parameters P_1 and P_3 had the lowest standard error. Although there was no significant reduction in standard error between the two-parameter model and the seven-parameter model, the use of this two-parameter model required less record keeping to determine earned hours.

The seven-parameter model also contained negative coefficients for two of the parameters (P_5 and P_7). Mathematically there is no problem; however, it is usually wise from a psychological standpoint to avoid models with negative coefficients. Related to this is the tendency to infer that individual parameter coefficients represent some type of "standard" for that parameter and can be used alone without the rest of the model. This is not true. The values of the constant and the coefficients in the model are only meaningful as a group or composite. (See Chapter 13.6 for a detailed explanation of the mathematics.)

Twenty-five studies were made in this case to gather data for development of the model. Normally, 20 to 30 pieces of data are required for developing a model. As a minimum, the number of pieces of data should never be less than the number of parameters contained in the data.

The parameter and cycle time data that are collected should cover the entire range of values that

Exhibit 4.8.13 Receiving Clerk–Fourth-Level Analytical Standard Data Model (4B)[a]

Study No.	Normal Time (min.)	Parameters						
		P_1	P_2	P_3	P_4	P_5	P_6	P_7
1	12.1	2	6	32,000	0	1	2500	3
2	21.6	3	10	7,997	1	0	0	0
3	154.3	28	54	83,377	16	4	3577	4
4	8.5	1	8	10,763	0	0	0	0
23	41.6	6	19	31,546	2	2	1463	2
24	6.4	1	2	6,576	0	1	6576	0
25	13.1	2	17	81,643	0	2	7430	7

Parameters

P_1 = number of lots in the shipment
P_2 = number of cartons
P_3 = number of pieces
P_4 = number of urgent lots
P_5 = number of lots that are 100% inspected
P_6 = number of pieces actually counted
P_7 = number of discrepancies found

Model

Standard minutes per shipment received = $1.262 + 5.625(P_1) + 0.00003(P_3)$

Model Reliability (95% Confidence Level)	Measurement Period		Error
$S = 2.193$ = standard error of estimate	1 shipment	$L = 24.702$	±18%
$r = 0.915$ = coefficient of correlation	1 day	$L = 480$	±4.0%
$T = 24.702$ = average time per shipment	1 week	$L = 2400$	±1.8%

L = measurement period in minutes (same units as T)

[a]Fourth level–time per unit of service.

can be encountered in the performance of the operation. Using the model when cycle times and/or parameter values are outside the ranges contained in the original data is not valid.

The reader may wish to review published accounts of models developed for brush painting various-sized parts[14] and for the packaging of assorted-sized finished goods,[15] which are similar to the case study described here.

Fifth-Level Model

The delivery service model shown in Exhibit 4.8.14 is an example of using the analytical approach to develop a single model to cover an entire class of work under a measured day work system. The elapsed times and parameter data were taken from existing records. The elapsed times represented 100% performance against existing standards which were established from a voluminous set of synthetic standard data.

A review of the 127 possible models that were calculated showed that a model with five parameters (P_1, P_2, P_3, P_6 and P_7) could be used with no significant decrease in model reliability. The model with all seven parameters had a standard error of estimate of 0.237 compared to 0.239 for the model shown here.

The measurement period is stated in terms of number of routes since the model applied only to the time spent while out on the route. Other standards were applied to work performed while not on a route.

A comparison was made between the standard hours calculated from the existing standards and the standard hours calculated from the model for each of the 35 routes. The greatest deviation

Exhibit 4.8.14 Delivery Service–Fifth-Level Analytical Standard Data Model (5B)[a]

Route No.	Elapsed Time (hr)	Parameters						
		P_1	P_2	P_3	P_4	P_5	P_6	P_7
1	5.89	52	10.4	2	518	0	8.9	8
2	6.17	47	12.2	2	187	19	7.7	7
3	4.90	67	5.7	6	284	0	2.9	4
4	5.61	52	8.9	2	400	4	6.9	6
33	4.52	42	3.9	2	103	1	7.8	5
34	6.18	51	8.7	1	81	21	12.0	9
35	6.04	46	12.0	1	159	6	6.2	8

Parameters

P_1 = total number of deliveries made
P_2 = route area–square miles
P_3 = number of deliveries brought back (nondeliverable)
P_4 = total weight of all deliveries–pounds
P_5 = number of commercial deliveries
P_6 = vehicle travel distance–miles
P_7 = number of CODs

Model

$$\text{Standard hours per route} = 0.9481 + 0.0329(P_1) + 0.2112(P_2) + 0.0405(P_3) + 0.0501(P_6) + 0.0982(P_7)$$

Model Reliability (95% Confidence Level)

$S = 0.239$ = standard error of estimate

$r = 0.940$ = coefficient of correlation

$T = 5.624$ = average time per route

L = measurement period in routes (same units as T)

Measurement Period		Error
1 route	$L = 5.624$	±8.5%
5 routes	$L = 28.120$	±3.8%
20 routes	$L = 112.48$	±1.9%

[a]Fifth level–time per unit of service.

between standard hours based on an existing standard and standard hours calculated from the model was 8.9%; the smallest deviation was 0.3%; and the average deviation was 2.0%.

Sixth-Level Model

The model displayed in Exhibit 4.8.15 was developed for a four-person group engaged in photographic work in support of a printing operation. The standard data model was used to measure the relative change in productivity of the group from month to month. Payment was on a day work basis, and there was no intention on the part of management to change this.

A labor claiming system whereby employees recorded the time spent on the various tasks was instituted to gather the time values for the model. Only net productive time was used in the model. Time claimed for meetings, personnel matters, and miscellaneous clerical tasks was excluded. The parameter data were already being collected for other record keeping purposes.

Data were collected for 40 workdays, and then the model was calculated. The level of performance during that 40 day period constituted the base for gauging productivity increases or decreases in the future.

The model was used on a monthly basis (approximately 22 workdays) to calculate standard hours for the period. These hours were then compared to actual hours claimed in order to determine the percentage of standard achievement. The percentages were then plotted monthly to determine the overall direction of the group's productivity.

Here again, although data for eight parameters were originally collected, review of the 255 possi-

Exhibit 4.8.15 Photography–Sixth-Level Analytical Standard Data Model (6C)[a]

Day No.	Claimed Time (hr)	Parameters							
		P_1	P_2	P_3	P_4	P_5	P_6	P_7	P_8
1	22.4	36	52	16	0	42	4	37	79
2	25.7	40	93	11	1	74	3	45	68
3	24.6	18	63	9	0	51	1	32	51
4	18.1	12	34	14	2	27	5	38	44
38	20.3	63	82	10	0	9	4	77	30
39	33.6	57	76	8	1	68	2	63	49
40	18.2	24	41	13	2	47	2	29	33

Parameters

P_1 = number of copy checked and delivered
P_2 = number of proofs run
P_3 = number of 2X copy processed
P_4 = number of bath changes made
P_5 = number of blank type ordered
P_6 = number of photo orders made up
P_7 = number of copy opaqued
P_8 = number of prints made

Model

$$\text{Standard hours per day} = 2.552 + 0.111(P_1) + 0.0171(P_2) + 0.180(P_5) + 0.330(P_6) + 0.0922(P_7) + 0.0273(P_8)$$

Model Reliability (95% Confidence Level)

$S = 2.255$ = standard error of estimate

$r = 0.934$ = coefficient of correlation

$T = 21.045$ = average time per day

L = measurement period in hours (same units as T)

Measurement Period		Error
1 day	$L = 21.045$	±21%
5 days	$L = 105.22$	±9.6%
22 days	$L = 462.99$	±4.6%

[a]Sixth level–time per time period (group standard).

ble models indicated that a model with the six particular parameters shown had a standard error only slightly greater than the model with eight parameters.

Seventh-Level Model

At this level model, many types of products and/or services are considered under one umbrella. The case study exhibited in Exhibit 4.8.16 is an example of using a single model to cover a group performing a variety of manual and clerical jobs; the output of an entire packing and shipping operation working three shifts a day was covered.

The hours worked and the parameter data were collected for 98 shifts over a 3 month time span. This large amount of data was used since the variety of parameter and time ranges was very extensive. There was no discernible pattern to the data that would allow easy determination of a narrower subset to cover the extremes of the data. Also, the difference in computer processing time to handle this amount of data versus a smaller subset was trivial. The parameter data all came from existing production records. Using the 10 parameters, 1023 possible models were calculated, and ultimately the one shown with eight parameters was selected.

The constant in the model is expressed in number of hours worked in order to facilitate computation of earned hours. Converting the original constant of hours per shift to a per-hour-worked basis was accomplished by dividing the hours per shift (38.184) by the average hours worked per shift (59.19).

Earned hours were calculated weekly using a 4 week (approximately 60 shifts) moving average.

Exhibit 4.8.16 Packing and Shipping–Seventh-Level Analytical Standard Data Model (7C)[a]

Shift No.	Hours Worked	Parameters									
		P_1	P_2	P_3	P_4	P_5	P_6	P_7	P_8	P_9	P_{10}
1	72.0	0	0	15,201	10,913	120	1767	1275	0	156	9720
2	56.0	0	42,468	10,876	8,004	104	2406	0	3805	340	0
3	96.5	130	0	31,328	9,902	0	1300	364	0	1051	1356
4	3.8	120	0	0	2,000	0	0	0	106	0	0
96	88.0	0	34,036	25,323	16,691	0	0	2074	439	2315	1769
97	43.0	959	0	0	20,525	0	2250	0	0	1780	0
98	64.2	0	0	0	8,068	66	1762	1364	0	0	7914

Parameters

P_1 = units of water tube bundled
P_2 = pounds of raw stock received
P_3 = pounds of finished goods shipped
P_4 = pounds shipped internally
P_5 = pieces of refrigerator coil annealed
P_6 = pounds packed in cartons
P_7 = pounds of water tube annealed
P_8 = pounds of refrigerator coil annealed
P_9 = pounds packed in crates
P_{10} = total pieces annealed

Model

$$\text{Standard hours} = 0.645\,(\text{no. of hours worked}) + 0.00545\,(P_1) + 0.00006\,(P_2) + 0.00044\,(P_3) + 0.00064(P_4) + 0.06488(P_5) + 0.00221(P_6) + 0.00440(P_9) + 0.00024(P_{10})$$

Model Reliability (95% Confidence Level)

S = 10.198 = standard error of estimate

r = 0.733 = coefficient of correlation

T = 59.19 = average hours per shift

L = measurement period in shifts (same units as T)

Measurement Period		Error
1 shift	L = 59.19	±34%
15 shifts	L = 888	±8.9%
60 shifts	L = 3552	±4.4%

[a]Seventh level–time per time period (group standard).

The error that would be encountered with that length of measurement period was ±4.4% at the 95% confidence level.

In this example—as in the previous one—management elected to use past performance as a basis for measurement. The use of past average productivity as a base to establish broad measurements—or the use of any other base—is a management or management and employee decision. The case study demonstrates that the decision to measure is not constrained by technique.

Selection of Parameters

The selection of parameters for use in building analytical models is a process of identifying those items that either *cause* the variation or *reflect* the variation in the time to perform work.

Consider, for example, the parameter "number of vendor visits made" for a purchasing function. This parameter would cause work load to vary. The parameter "purchasing travel expense," however, would reflect the variation in work load.

Similarly, in a food service operation, the parameter "number of meals served" would cause variation, whereas "cases of napkins used" would reflect variation.

Either type of parameter may be used, and all data that are being recorded—whether causal or reflective in nature—can be considered. Often parameters are rejected because there seems to be very little direct relationship between the single parameter and the work being done. However, combinations of individual parameters that seem unrelated when viewed separately result in excellent models.

Generally, parameters used in lower-level models tend to be causal in nature. At this third and fourth level it is fairly easy to identify the relatively few causes of variation. As the level of the model increases, the volume of causal parameters will likewise increase. It then becomes simpler to use reflective parameters.

Consider the case of a high-level model covering an entire maintenance function. The number of parameters that cause variation is great. Here, the use of reflective parameters, such as "dollar value of spare parts inventory" or "total production volume," becomes important.

Occasionally it is useful to create data from other parameters. An example of this occurred in a rolling mill process where the weight, gauge, and width of the finished product were known, but the length of material produced was not. Since it was thought to be significant, a "length unit" parameter was created by dividing the total weight of the order by the cross-sectional area.

Qualitative information may also be converted to parameter data. When developing a task-level model for the operation of pulling samples from incoming goods, a great deal of variation in the cycles is related to the packaging of the goods. Here, a packaging parameter was created with a ranking of 0, 1, and 2 to differentiate among stand-alone cartons, cartons taped to a pallet, and cartons steel-banded to a pallet.

Maintaining Analytical Standard Data

Maintenance of analytical models is required for all of the same reasons previously identified with maintaining synthetic standard data. Normally, changes to a set of synthetic standard data are done on an incremental basis. That is, when changes occur in the workplace environment, it is necessary to modify only that portion of the synthetic data that is directly affected. A change in the workplace that results in bins of parts being rearranged, for instance, may require a change in the "get part" portion of the data only.

With analytical standard data, however, changing any segment of the operating environment means that a new model must be calculated. Also, the model must be recalculated when the value of the parameters goes beyond the range of values in the original data. For example, in the case of the photography model (Exhibit 4.8.15), the range of values for parameter P_8 was 30 to 79. If something occurs that causes the actual values to be less than 30 or more than 79, the model must be recalculated to include the broader range. Likewise, if the size of the group changes, or if other factors affect the amount of "claimed time," the model must be recalculated.

Maintenance of analytical models is very inexpensive, since in most cases all that is required is recalculation of the model with a different range of parameter and/or time values. Additional effort is required if the recalculated model now has unacceptable reliability. In this event different parameters may be required or—depending on the type of change—it may be simpler to split the operation and construct two different analytical models.

A good procedure for tracking the validity of an analytical model is to maintain, in the same format as the original data used to calculate the model, a breakdown of the actual numbers that are used to calculate standard time. These data can then be used subsequently to automatically recalculate the model on a fixed schedule. When the recalculated model differs significantly from the original model, investigative action is required.

In the case of the photography model, a program was written to calculate the standard hours

Exhibit 4.8.17 Example of Resource Planning Data Developed From Analytical Standard Data

No. of Samples Pulled (P_1)	Millions of Units in the Lots Sampled (P_2)															
	10	15	18	20	22	24	26	35	36	37	38	40	45	50	55	60
	Hours per 1000 Lots Sampled (Page 1)															
300,000	439	443	446	447	449	451	452	460	461	462	462	464	468	472	477	481
315,000	455	460	462	464	465	467	469	476	477	478	479	480	485	489	493	497
330,000	472	476	478	480	482	483	485	492	493	494	495	497	501	505	509	513
. . .	. . .	. . .	. . .	. . .	. . .	. . .	. . .	. . .	. . .	. . .	. . .	. . .	. . .	. . .	. . .	. . .
1,200,000	1412	1416	1419	1421	1422	1424	1426	1433	1434	1435	1436	1437	1441	1446	1450	1454
	Hours per 1100 Lots Sampled (Page 2)															
300,000	450	454	456	458	460	461	463	471	471	472	473	475	479	483	487	491
315,000	466	470	473	474	476	478	479	487	488	489	489	491	495	499	504	508
330,000	482	486	489	491	492	494	496	503	504	505	506	507	511	516	520	524
. . .	. . .	. . .	. . .	. . .	. . .	. . .	. . .	. . .	. . .	. . .	. . .	. . .	. . .	. . .	. . .	. . .
1,200,000	1423	1427	1430	1431	1433	1435	1436	1444	1445	1445	1446	1448	1452	1456	1460	1465
	Hours per 1200 Lots Sampled (Page 3)															
300,000	460	465	467	469	470	472	474	481	482	483	484	485	490	494	498	502
315,000	477	481	483	485	487	488	490	497	498	499	500	502	506	510	514	518
330,000	493	497	500	501	503	505	506	514	515	515	516	518	522	526	530	535
. . .	. . .	. . .	. . .	. . .	. . .	. . .	. . .	. . .	. . .	. . .	. . .	. . .	. . .	. . .	. . .	. . .
1,200,000	1433	1438	1440	1442	1443	1445	1447	1454	1455	1456	1457	1458	1463	1467	1471	1475

at the end of each month. The data base for the program was structured to keep the parameter values and claimed time information on a daily basis. At the end of each month, that month's data were automatically fed into the regression analysis program, and a new model was calculated. The new model was then compared to the original model at the end of each month to see if any action was required.

4.8.4 PLANNING WITH ANALYTICAL STANDARD DATA

The structure of an analytical model makes it very convenient for use in simulation models to predict the amount of time required for varying combinations of parameter volumes. Consider the data shown in Exhibit 4.8.17, which was developed by using an analytical model with two parameters. Here a simple program was written that exploded the model into a set of tables which could be quickly referenced for resource planning purposes, cost estimating, and so on. The first three pages of the planning data are shown. To find, for example, the hours required to pull 330,000 samples from 1100 lots where the total quantity is 26 million, one would refer to page 2 of the tables. The hours required would be 496.

The analytical model used in this case covered the operation of pulling samples for inspection from incoming shipments (lots) of small parts. The model was:

$$\text{minutes per lot} = 6.3875 + 0.06487\ (P_1) + 0.00005\ (P_2)$$

where P_1 = number of samples pulled and P_2 = total quantity in the lot.

REFERENCES

1. R. M. BARNES, *Motion and Time Study*, 5th ed., Wiley, New York, 1968.
2. P. CARROLL, *How to Chart Data*, McGraw-Hill, New York, 1960.
3. E. V. KRICK, *Methods Engineering–Design and Measurement of Work Methods*, Wiley, New York, 1966.
4. H. B. MAYNARD, *Industrial Engineering Handbook*, McGraw-Hill, New York, 1970.
5. M. E. MUNDEL, *Motion and Time Study*, 5th ed., Prentice-Hall, Englewood Cliffs, NJ, 1978.
6. B. W. NIEBEL, *Motion and Time Study*, 6th ed., Irwin, Homewood, IL, 1976.
7. MUNDEL, *Motion and Time Study*, p. 116.
8. MAYNARD, *Industrial Engineering Handbook*, pp. 3–122.
9. MAYNARD, *Industrial Engineering Handbook*, pp. 3–134.
10. "A Long Time Coming, But Now Those Unmeasureable Jobs Are Measureable," *Factory*, February 1959, pp. 28–30.
11. C. L. BRISLEY and R. J. DOSSETT, "Computer Use and Non-direct Labor Measurement Will Transform Profession in The Next Decade," *Industrial Engineering*, August 1980, pp. 34–43.
12. MITCHELL FEIN, "Establishing Time Standards By Parameters," *Proceedings–Spring Conference of The American Institute of Industrial Engineers*, American Institute of Industrial Engineers, Atlanta, GA, May 1978.
13. MUNDEL, *Motion and Time Study*, pp. 705–706.
14. E. V. KRICK, *Methods Engineering–Design and Measurement of Work Methods*, pp. 338–342.
15. M. D. SALEM, "Multiple Linear Regression Analysis for Work Measurement of Indirect Labor," *The Journal of Industrial Engineering*, May 1967, pp. 314–319.

BIBLIOGRAPHY

CYWAR, A. W., "A Computerized Statistical System for Indirect Labor Performance Measurement," master's thesis, New Jersey Institute of Technology, Newark, 1965.

RICHARDSON, W. J., *Cost Improvement, Work Sampling and Short Interval Scheduling*, Reston Publishing, Reston, VA, 1976.

THELWELL, R. R., "An Evaluation of Linear Programming and Multiple Regression for Estimating Manpower Requirements," *The Journal of Industrial Engineering*, Vol. 18, March 1967, p. 227.

CHAPTER 4.9

Indirect Operations: Measurement and Control

GUY J. BACCI

International Harvester

4.9.1 INTRODUCTION—THE NEED

Indirect operations are an integral part of the total operations of all enterprises. They represent a sizable investment of company funds and a tremendous potential for ensuring successful return on sales. A number of factors have contributed to the growing importance of these operations:

1. The technology explosion leading to automation of direct operations, which contributes to a lessening need for direct workers and creates a growing need for service or support-type workers.
2. The need to supplement this technology expansion by filling the demand for scientists, engineers, and technicians.
3. The expansion of industries supplying services to consumers.
4. The mounting of clerical requirements to satisfy the legislated regulations of federal and local governments, whose roles are ever increasing.
5. The sizable investment in plant, equipment, facilities, and other assets that demand continuing improvement by management.

These factors have contributed to the huge rate of increase in indirect workers since the turn of the century, a rate more than double that for direct workers.[1]

4.9.2 THE PROBLEM—AWARENESS

Improving the control of indirect operations is important and could contribute significantly to easing the economic problems that confront all business managers. With the industrial and emerging nations facing escalating energy costs and inflation, the necessity to continuously improve productivity is mandatory.

> *The real definition of productivity, however, is simply output over input; the relationship between total physical output of a factory, an industry, or a nation and one or more of the factors of input—labor, invested capital, materials or the effort and ingenuity of management.*[2]

This demand gives foundation to the development of a strategy for improving asset utilization for indirect operations.

4.9.3 THE OPERATIONAL ENVIRONMENT—ITS MANAGEMENT AND EMPLOYEE IMPLICATIONS

Understanding the operating environment of a particular organization is essential. This environment is the result of the working relationship and understanding between management and employees. This interrelationship was discussed by David L. Conway, manager of indirect manpower measurements, IBM Eurocoordination, in a 1979 address on measuring indirect functions.

> *I believe that our most difficult challenge is to create a proper understanding at all levels of the company, and to do this in a credible way. Employees always respond to challenges they understand and to causes they believe in.*[3]

This understanding must be coupled with the skillful management of indirect operations in order to lead to improved asset utilization and results in meeting planned profit objectives. The achieving of these profit or cost objectives is accomplished through maintaining proper balance between the attendant costs of providing indirect services and the resultant value-added output.

4.9.4 THE CRITICAL AREAS OF INDIRECT COSTS

Indirect operations are necessary to support the primary purpose of any organization. The problem for management is to ascertain which services are productive and contribute to overall profitable operations. The productive services should be improved whereas the unproductive or wasteful ones should be eliminated. The strategy for study and improvement presented in this chapter addresses all aspects of indirect services and costs. These services can be categorized in the following areas:

Administration. Management and clerical functions.

Energy. Purchase, distribution, and utilization of utilities.

Housekeeping. Janitorial services, sanitation, and waste disposal.

Human Resources. Services and facilities associated with personnel.

Maintenance. Upkeep and repair of facilities and equipment.

Material Control and Handling. Purchase, handling, storage, and disbursement of material.

Paperwork. Information dissemination and storage.

Reliability and Quality Assurance. Design, testing, and control aspects of product quality.

Security. Asset protection.

Supplies. Perishable tools and support materials.

4.9.5 INDEXES OF EFFECTIVENESS

The areas cited represent the majority of indirect services and certainly an opportunity for improvement. This improvement must be measured in all-encompassing terms that take into consideration the pertinent factors contributing to overall organizational effectiveness. A number of indexes have application throughout many of the areas and should be an integral part of the control and evaluation of the service provided.

Control and Evaluation Reports

Management must maintain cognizance of the performance and costs of indirect services, and certain basic reports will prove beneficial.

1. The actual cost versus budgeted cost on a monthly basis compared to the average monthly cost during the previous year is displayed in Exhibit 4.9.1. This form can also be used to show actual hours versus budgeted hours.
2. The percent performance measured against job standards and reported results can be tracked monthly and compared to the average of the past year, as shown in Exhibit 4.9.2.
3. The current backlog in hours for the services to be performed can be plotted monthly and compared to the average of the past year, as shown in Exhibit 4.9.3.

4.9.6 MOVITATIONAL ASPECTS

In addition to the quantitative measures of effectiveness, there are a number of intangible aspects to the services rendered by indirect operations that must be considered when attempting to evaluate the overall contribution to the organization. A breakdown of budgeted and actual costs into the cost categories of material, labor, and attributable overhead gives further insight into the value of services. Although subjective, a performance evaluation of the services each indirect area performs undertaken by the direct areas that require the service and coupled with a feedback mechanism leads to an understanding of overall organization objectives and to team cooperation. This evaluation can be conducted by industrial engineers or other staff members, but must consider the response time of the service, the quality of the service performed, and the timeliness in meeting estimates and promises. In team building efforts the importance of utilizing the contributing disciplines cannot be overemphasized. What better task force could there be for studying the results of indirect services than representatives of the service function cooperating with representatives of the function to be serviced!

Exhibit 4.9.1 Typical Chart Displaying Actual Cost (A) Versus Budgeted Cost (B)

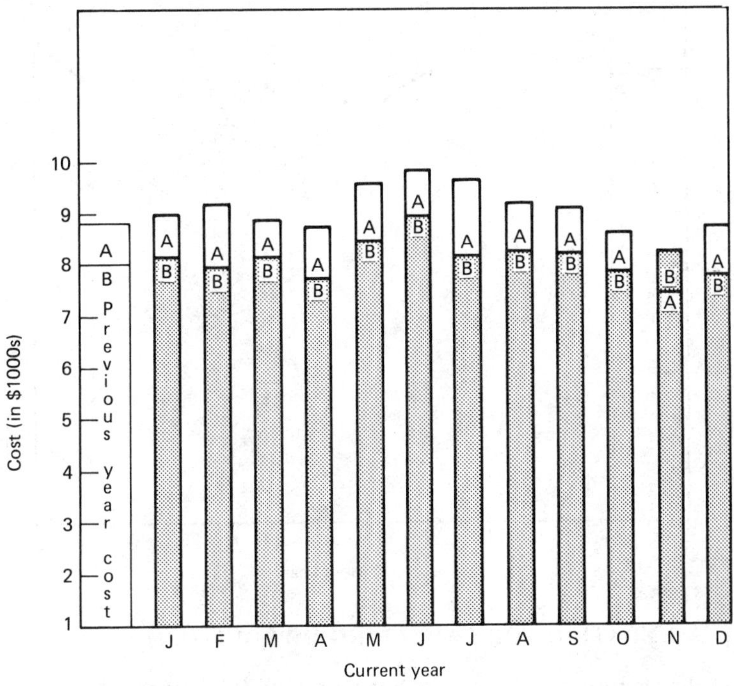

Exhibit 4.9.2 Graph Showing Actual Performance Compared to Standards of Performance

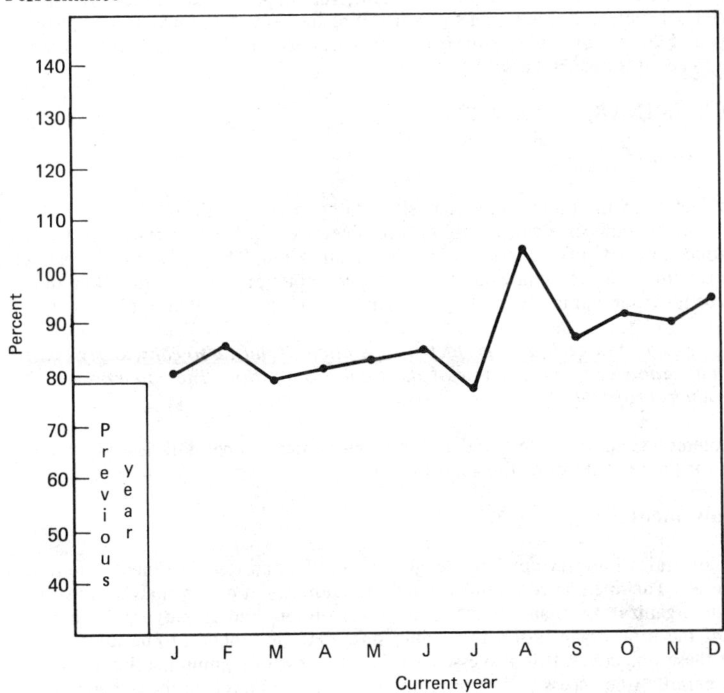

Exhibit 4.9.3 Pictorial Representation of Work Backlog

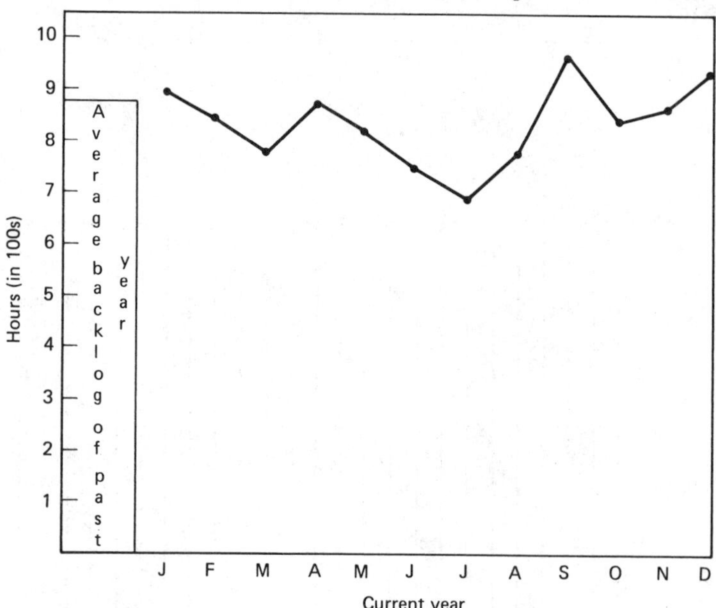

4.9.7 THE FOUNDATION FOR LAUNCHING IMPROVEMENT

Knowledge of the need for indirect operational services, of their impact on the overall organizational results, and of the environment in which these services are performed becomes a foundation for launching improvement. Throughout this handbook many chapters point out various aspects that can be utilized in the quest for improvement, and all should be understood and utilized. All sections offer concepts and techniques for controlling indirect operations. In this chapter, however, the strategy and tactics for applying these concepts are integrated into an implementation plan that leads to improved resource and asset utilization.

4.9.8 PRELIMINARY ANALYSIS

Communication

The first objective of the preliminary analysis is to open communication between the function to be studied and the analysis team. Attaining this objective requires that the analyst be cognizant of the communication network functioning in the organization. The importance of establishing common goals cannot be overemphasized, since the viewpoints of both the function and the analysis team must be incorporated into the study. This importance was pointed out by John C. Werner.

> *It is not enough that each unit perfect its operation, it must also consciously contribute to the simplification and effectiveness of the total organization. The common goal is to produce products profitably.*[4]

Once the common goals are understood and the communication network is evident, the analyst can better plan for the next stages of the analysis.

Total Involvement

This stage integrates the previous knowledge with the human resource involvement that is necessary for success. The opening communication or announcement of the analysis must involve all segments of the organization–management, supervisors, union, and employees. It should include the objectives of the total organization and should state that the function to be studied can contribute to meeting these objectives. It is also essential that the flow of communication be open in all directions. This permits free access to information available at all levels in the organization and the use of this information to ensure the success of the project.

4.9.9 COMPOSITION OF THE ANALYSIS TEAM

Ideally, the project analysis team will be composed of representatives of the four segments of the organization's human resources. Each brings a view to the project that must be considered for total involvement and eventual success. However, there may be situations in which the union and employees do not want to participate in the actual analysis. Then management, supervisors, and staff personnel should carry on the assignment. This situation should not alter the overall communication objectives of the project. If the analysis team requires training in any of the technical aspects, this should be handled well in advance of the conducting of the study. However, if special expertise is required for a short duration, then specific staff expertise can be utilized in an ad hoc situation.

4.9.10 INFORMATION GATHERING

It is of fundamental importance that the analysis team have an overall understanding of the important parameters surrounding the function to be studied. Information from the viewpoint of the total organization can be obtained from many of the staff services supporting the function. This information must cover the important aspects of finances, industrial engineering, compensation, employee relations, materials management, and so on. Once the general information is collected and analyzed, specific functional data will supplement it and enable the analysis team to view the involvement of various operations and the relationship of one operation to another. These specific functional data must answer the important operational questions, as follows:

1. What work is done?
2. Who does it?
3. How much time is spent?
4. How is each job done, and what steps are followed?
5. What is the volume of work?

Personnel Interviews

The specific data can be gathered from the manager, supervisor, and employee by means of interviews and completion of questionnaires designed for each level. Each of these interviews and questionnaires is designed to provide an understanding by responses to specific pertinent questions. From the manager's viewpoint, the general parameters of mission, objectives, planned actions, areas of concern, and performance measurement criteria are key to this understanding. An example of a management questionnaire is given in Exhibit 4.9.4.

The supervisor's questionnaire is oriented more to the operational aspects of the functions. It requires information on major tasks, departmental relationships, potential improvements, and performance evaluation criteria. An example of a supervisor's questionnaire is given in Exhibit 4.9.5.

The employee questionnaire requests more specific information concerning assigned duties and tasks performed. It requests the employee's perception of operating information required to complete the job and how this job might be improved. Information on the method of performance evaluation and on how it could be improved also is required. An example of an employee job analysis questionnaire is given in Exhibit 4.9.6.

Activity Lists

The general information coupled with the specific data gathered from the personnel interviews is supplemented by the activity list. This list is prepared by the analysis team working in conjunction with both supervisors and employees. It outlines the major functions performed and gives the analysis team insight into what the supervisor and employees perceive to be their primary activities. Sample activity lists for janitorial and clerical functions are given in Exhibits 4.9.7 and 4.9.8, respectively.

Task Lists

Since the major emphasis of the analysis is to continue to foster employee involvement, the task list is completed by the employee. The form is designed to list each task performed, the frequency with which the task is performed in a specific period, and the time required to perform the task. The general information required includes the employee's name, occupational or job title, classification, department, section, immediate supervisor, and date. The specific information required includes the following:

Exhibit 4.9.4 Management Questionnaire

This form is used to gain an understanding of the department manager's perception of the department's role and how he or she directs efforts toward accomplishment of that role. It is completed by the individual manager.

1. Name	The manager's name
2. Date	The date the questionnaire was completed
3. Title	The manager's official title
4. Location	The geographic location of the department
5. Mission of the department	The reason for the department's existence and the roles and responsibilities it has been given or has assumed
6. What are the management objectives for the current fiscal year?	The improvements and/or changes in organization, personnel development, work content processes, and/or staffing planned for the current fiscal year
7. Discuss planned actions for the current fiscal year.	A listing of the tasks planned to accomplish the management objectives stated earlier
8. Describe areas of concern.	A description of the significant factors that currently or will have an impact on the effectiveness of the department in accomplishing its mission
9. What operational management reporting do you currently have, and how might it be improved?	A discussion of the tools (i.e., reports, meetings) used to manage the department and the improvements (content, scope timeliness, accuracy) that might be appropriate to increase management control
10. What performance measurement criteria (both qualitative and quantitative) would be most useful to you?	A description of existing and/or desirable criteria that are and/or should be used to measure performance (These would relate to major volume indicators.)

Courtesy of Kearney Management Consultants, Chicago, Illinois.

Exhibit 4.9.5 Supervisor's Questionnaire

This form is used to gain an understanding of how the supervisor perceives his or her responsibilities and to identify the content of the work. It is completed by the individual supervisor.

1. Name	The supervisor's name
2. Department	The supervisor's department
3. Job title	The formal organizational title of the position held
4. Section	The section of the department in which the supervisor works
5. Superior	The person to whom the supervisor reports organizationally
6. Today's date	The date the form is completed
7. Briefly describe the general nature of your job.	A description of what the supervisor's responsibilities are
8. What is your normal work week?[a]	The number of hours per week the supervisor typically works

Exhibit 4.9.5 ***(Continued)***

9. What days of the week?[a]	A listing of the days typically worked
10. Do you have secretarial support?[a]	What type of secretarial support, if any, is available to the supervisor
11. List the major tasks you perform and the estimated hours spent on each.[a]	A list of the major tasks the supervisor performs and the average hours per day that are spent (A personal diary kept for 1 week may be helpful in filling out this section.)
12. Regular weekly activities[a]	A list of the tasks that the supervisor routinely performs on a weekly basis and the hours spent on those tasks
13. Other activities[a]	A list of the tasks that the supervisor routinely performs in a period (quarterly, semiannually, annually, etc.) and the time spent each time the task is performed (The frequency of the task should be identified along with the average hours spent per occurrence.)
14. Departmental relationship	A description of the positions (by job title), and number of people in the supervisor's department and of the nature of the relationship that the supervisor has with other personnel in the department (The direct relationship refers to personnel who are directly supervised by the individual completing the questionnaire.)
15. Briefly describe the potential improvement ideas and/or current problem areas.	A discussion of those areas that would benefit from a thorough examination by the project team (The areas may relate to organization structure, policies, procedures, methods, and/or processes.)
16. What training programs exist today, and how might they be improved?	A list of the programs available to departmental personnel and suggestions for improvement (content, presentation, availability, duration, etc.)
17. What specific operating information do you require, and how might the flow or format of that information be improved?	A description of the information used by the supervisor to plan and control those activities for which he or she is responsible, including recommendations to improve the adequacy of the information
18. By what criteria is your position currently evaluated?	A description of the supervisor's perception of how his or her performance is evaluated
19. How could the performance evaluation criteria for your position be improved?	Suggestions for improving the method for assessing performance (This might include additional or revised criteria or might relate to the movement and/or feedback technique.)
20. By what criteria do you evaluate your subordinates?	A description of how the supervisor assesses subordinates' performance
21. How could the evaluation criteria for your subordinates be improved?	Suggestions to improve the adequacy and quality of the performance evaluation process
22. Contacts outside immediate department	A list of the individuals who have substantive contacts with the supervisor in the course of his or her performing assigned tasks

Courtesy of Kearney Management Consultants, Chicago, Illinois.

[a]Optional questions, to be asked only if the supervisor's position is being evaluated as part of the project.

Exhibit 4.9.6 Job Analysis Questionnaire

This form is used to record the activities performed by an individual worker. It typically is used as the basis for individual interviews.

1. Employee's name	The employee's name
2. Date	The date the form is completed
3. Job title	The formal organizational title of the position held
4. Department	The employee's department
5. Section	The section of the department in which the employee works
6. Immediate supervisor	The name and job title of the employee's immediate supervisor
7. Do you supervise others?	Whether the employee formally supervises other employees
8. Employees supervised	Information about the employees supervised by the individual completing the form
9. Phone calls[a]	The number of telephone calls made and received on a daily basis
10. Tasks performed	The tasks (by name or by the activity number performed during the time of day indicated—not less than 15 min)
11. Assigned duties not performed this week	A description of the tasks and time by time category that were not covered during the sample week. The time should be expressed in hours or quarter hours.
12. Total phone calls[a]	The total number of telephone calls made or received during the week
13. Summary description of daily tasks	A summary of the hours spent by day for each of the family of tasks (Group related tasks into families. The family may contain more than one task.)
14. How can your job be improved?	A description by the employee of ways to improve the job (i.e., changes in procedures, policies, responsibilities, form, schedules)
15. Overtime: Does your job require any overtime?	Mark yes or no
16. If yes, is overtime due largely to holidays?	Mark yes or no
17. Duty or task requiring overtime	List tasks routinely requiring overtime (indicating how much) not caused by holiday work buildup
18. What operating information does your job require?	Identify information required to plan and control your work, along with any improvements that might be helpful to you
19. How does your supervisor evaluate your work?	A description of the basis upon which your performance is evaluated
20. How could this be improved?	A description of improvements to the evaluation process. (They could relate to the basis for evaluation, the standards of performance, and/or feedback with you.)

Courtesy of Kearney Management Consultants, Chicago, Illinois.
[a]Clerical positions.

Exhibit 4.9.7 Activity List Showing Work Elements Typically Included in a Janitorial Function

Department name ________________ Date ______

Activity No.	Activity Description
1	Sweep floors
2	Mop floors
3	Clean washrooms
4	Clean locker rooms
5	Empty waste containers
6	Dust furniture, files, and so on
7	Requisition supplies
8	Wax floors
9	Buff floors
10	Perform outside maintenance
11	Miscellaneous

Exhibit 4.9.8 Activity List Showing Work Elements Typically Included in a Clerical Function

Department name ________________ Date ______

Activity No.	Activity Description
1	Issue certificates class 127A
2	Issue certificates class 127B
3	General public inquiry service
4	Furnish case data to compliance division
5	Administration
6	Employee welfare
7	Miscellaneous

1. Task Number. The tasks should be numbered so as to avoid confusion. They should also be listed in order of their frequency, with each daily task first, followed by those occurring regularly on a weekly, monthly, quarterly, and so on, basis.

2. Task Description. The task should be specifically described. It should be a complete task and not subdivided into its individual work elements. Any other pertinent information, such as machine or tool description, should also be included.

3. Frequency or Quantity per Week. This should indicate the number of times this task is performed per week or per other period.

4. Hours per Week. This should be an estimate of the time required to perform the task.

Examples of completed task lists for janitorial and clerical functions are given in Exhibits 4.9.9 and 4.9.10, respectively.

The task list prepared by the employee presents an excellent opportunity to maintain involvement and keep communications open. The list should be reviewed by the analysis team for completeness and correctness and should be discussed with the employee. This strategy will provide the employee with an opportunity to make suggestions for work improvement, which should be strongly encouraged. These suggestions can be used as a springboard for improving overall functional effectiveness.

Work Distribution Chart

After the activity lists and task lists have been prepared, the information can be consolidated into a work distribution chart. It represents all activities of the function and all personnel responsible for performing these activities. Charting the activities in this manner can identify a number of problem

Exhibit 4.9.9 Task List of Work Elements for Janitorial Function, Showing Frequency and Related Times

Name William Wright	Occupation or title Utility worker	Classification	
Department	Section	Supervisor	Date

Task No.	Description	Quantity per Week	Hours per Week
1	Sweep executive offices (24 in. push broom)	5	5.0
2	Sweep accounting section (24 in. push broom)	5	2.5
3	Sweep standards and methods section (24 in. push broom)	5	5.0
4	Mop executive offices	3	4.5
5	Mop accounting section	2	2.0
6	Mop standards and methods section	2	3.0
7	Clean accounting men's washroom	5	2.5
8	Clean accounting women's washroom	5	2.5
9	Clean engineering men's washroom	5	2.5
10	Requisition clean towels	3	1.5
11	Bring supplies to stationery supply room	1	1.5
12	Empty waste containers–second-floor offices	5	5.0
13	Dust executive offices	5	2.5
	Total	–	40.0

areas. Analysis of the information on the chart enables one to answer the following important questions:

1. What activities take the most time?
2. Is there misdirected effort?
3. Are skills used properly?
4. Are the employees performing unrelated tasks?
5. Are tasks spread too thinly?
6. Is the work distributed evenly?

The analysis first concentrates on the activities that take the most time. The elimination of unnecessary work is the second step, which is followed by a determination of whether employees are working at the proper level or whether they are performing many incidental duties. Next, one determines whether there is an out-of-balance work distribution, and if so, whether the work can be reallocated to even the work load. Examples of work distribution charts are given in Exhibits 4.9.11 and 4.9.12.

Exhibit 4.9.10 Task List of Work Elements for Clerical Function, Showing Frequency and Related Times

Name: Mary Moody	Occupation or title: Correspondence clerk	Classification: CAF 3	
Department	Section	Supervisor	Date

Task No.	Description	Quantity per Week	Hours per Week
1	Write acknowledgments to 127A applications	32	16
2	Get out weekly report for field division director	1	2
3	Add up figures for special reports	36	3
4	Distribute reasons on all 127B cases	44	11
5	Complete cumulative cash report	1	1
6	Answer questions for the public when supervisor is absent	24	2
7	Leave records (filing) (checking)	12	3
8	Sign the mail	80	4
9	Make up scratch pads	1	1
10	Carry special attention cases from desk to desk so they can be		
	gotten out in a hurry	10	5
	Total	–	48

4.9.11 SYNTHESIS AND APPLICATION

Within the framework of total employee involvement, the synthesis of collected data leads to a series of recommendations for improvement. From the employee's viewpoint, these improvements are in the area of working conditions, the employee-supervisor relationship, individual and organizational goal compatibility, and so on. The supervisor's approach to improvements is oriented more to departmental goals and the control system to achieve these goals. Management takes a more macro view of individual and departmental needs.

The results of the analysis can quickly demonstrate the importance of keeping people involved and interested. A corollary objective to the motivational aspects of the analysis is the establishment of a benchmark from which to measure improvements. Also, this analysis will result in a number of improvements that can be implemented immediately. This aspect was discussed by R. Keith Martin in an article on clerical productivity.

> *During the survey it is important to establish a benchmark of present efficiency against which to measure future improvements. This need be no more than an agreement with the*

Exhibit 4.9.11 An Example of the Total Work Distributed to a Maintenance Work Crew

Activity	Total Man-Hours	William Wright Utility Worker	Man-Hours per Week	George Stone Utility Worker	Man-Hours per Week	James Green Utility Worker	Man-Hours per Week	John Richmond Utility Worker	Man-Hours per Week
Sweep floors	39.5	Executive offices Accounting section Standards and methods	5.0 2.5 5.0	Front lobby and stairs Mechanical engineering section Plant engineering section Personnel section	1.5 5.0 5.0 2.5	Dispensary Cafeteria	3.0 5.0	Cafeteria	5.0
Mop floors	32.5	Executive offices Accounting section Standards and methods	4.5 2.0 3.0	Front lobby and stairs Mechanical engineering section Plant engineering section Personnel section	2.5 3.0 3.0 1.5	Dispensary Cafeteria	3.0 5.0	Cafeteria	5.0
Clean washrooms	27.5	Accounting men's Accounting women's Standards and methods men's	2.5 2.5 2.5	Personnel women's Mechanical engineering men's	2.5 2.5	Cafeteria men's Cafeteria women's	2.5 2.5	Machine shop men's Metallurgy men's	5.0 5.0
Clean locker rooms	6.0							Machine shop Metallurgy	3.0 3.0
Empty waste containers	12.5	Second-floor offices	5.0	First-floor offices	7.5				
Dust furniture, files, and so on	4.0	Executive offices	2.5	Front lobby	1.5				
Requisition supplies	4.5	Clean towels	1.5	Clean towels	1.0	Clean towels	1.0	Clean towels	1.0

Wax floors	4.0					First- and second-floor offices	4.0		
Buff floors	8.0					First- and second-floor offices	8.0		
Perform outside maintenance	18.0					Water lawn	5.0	Cut grass Sweep walks	8.0 5.0
Miscellaneous	3.5	Carry supplies to stationery supply room	1.5	Clean glass–doors front lobby	1.0	Clean grill–cafeteria	1.0		
Total	160.0		40.0		40.0		40.0		40.0

Exhibit 4.9.12 An Example of the Total Work Distributed to a Clerical Unit

Activity	Total Man-Hours	Frank Stapleton Section Chief CAF-9	Man-Hours per Week	Thomas Freeman Analyst CAF-8	Man-Hours per Week	William Sullivan Case Director CAF-7	Man-Hours per Week	Mary Moody Correspondence Clerk CAF-5	Man-Hours per Week	Grace Hoffman Head–Steno Pool CAF-5	Man-Hours per Week	Mary O'Rourke Steno (in Pool) CAF-2	Man-Hours per Week
Issue certificates class 127A	121	Policy review recommendations Final review and signing of recommendations	17 6	Preparing recommendations for action	19	Checking drafts form Checking final statements for form	6 10	Dictating acknowledgments Signing acknowledgments	16 4	Proofreading acknowledgments	6	Checking addresses Typing answers, drafts, statements Taking dictation	6 20 11
Issue certificates class 127B	54	Reviewing and signing	7	Checking for form Rechecking approvals	7 2	Preparing approval notices	13	Tabulating	11	Proofreading Selecting field applications Typing "51" forms	2 3	Checking applications for address changes Numbering applications	4 2
General public inquiry service	19	Interviewing callers	2	Preparing daily report Interviewing callers Dictating replies to special inquiries	6 2 1	Interviewing callers	1	Interviewing callers Preparing cumulative report	2 1	Assembling printed materials	3	Interviewing callers	1
Furnish case data to compliance div.	24	Reviewing	2	Checking for form Gathering data	1 1	Gathering data Dictating Revising	5 2 3	Tabulating and checking figures	3	Checking field representatives' names Laying out reports Coding information requests	2 2 2	Tabulating data	1
Administration	34	Conferring with personnel office Conferences Preparing budget request	6 4 3	Staff conferences	2			Posting attendance records Preparing administrative reports	3 2	Hearing grievances Training new employees Proofreading steno work	1 3 10		
Employee welfare	17	Making speeches	1	Making car pool arrangements Keeping credit unit records	2 3	Arranging blood bank Writing newsletter	1 1			Bond selling records Collecting hospital payments Collecting health payments	3 2 1	Handling arrangements for girl's baseball team	3
Miscellaneous				Analyzing operating reports of other sections	2	Making security inspections Controlling rented materials	4 2	Special messenger service Cutting up old forms to use as scratch paper	5 1	Sorting old files Keeping phone directory up to date	3 2		
Total	288		48		48		48		48		48		48

supervisors on the average month's volume of work and the man-hours required to produce it. The analyst should maintain a list of suggestions for improving operations, whether self-originated or received from others, because it is desirable to be prepared to make some recommendations for improvements, as occasions may offer, as quickly as is practical.[5]

4.9.12 PRESENTATION OF DATA

The foundation for improved asset utilization has been set by the analysis study. A number of improvements resulting from the study have been implemented. The study must now be presented to all levels of employees to permit understanding of the next important steps. This presentation should include face-to-face communication between the analysis team and the members of the function studied. The presentation must be structured to demonstrate cognizance of the function and logic in developing improvements. The structure for discussion should include the following:

1. Introduction–purpose of the analysis.
2. Objectives of the analysis.
3. Organization of the study.
4. Methodology of the study.
5. Findings and conclusions.
6. Recommendations–general and specific.
7. Costs and benefits–tangible and intangible.

In a short time all human resource levels will become cognizant of the function and can then have an impact on the results to be achieved. As part of this discussion, again with employees at all levels, the overall development of an implementation plan and strategy should evolve.

4.9.13 IMPLEMENTATION PLAN AND STRATEGY

Mission

The strategy framework that creates the operating environment must begin with all levels of the organization understanding the mission of the installation and the implementation plan.

Organization

The concept of multidisciplinary task forces has recently met with great success because of the constant involvement of management, staff, line supervisors, and functional employees. The use of such task forces not only promotes involvement, but also:

Reinforces the sense of urgency.
Permits more rapid program completion.
Ensures a uniform approach.
Minimizes functional difficulties.
Provides cross-discipline training.

Installation Methods, Procedures, and Techniques

An improved system of planning activities for the function, measuring these activities, and monitoring them must be installed. This system will utilize the most modern and ever-changing concepts, which must include the appropriate work measurement technique and operational control system.

Evaluation

The necessary ingredient to successful measurement and control of indirect operations is the feedback, report, and follow-up process. This is the mechanism that triggers action at all levels and that ensures attainment of objectives.

4.9.14 MEASUREMENT TECHNIQUES AND STANDARDS

Factors

In the selection of a measurement technique or a combination of techniques, a number of factors must be considered in the decision process. A clear understanding of the activity to be measured is paramount, and this is normally provided by the employee closest to the activity, such as the operator, group leader, or supervisor. Typical factors include:

Complexity of the operations.
Coverage required to control the operations.
Consistency of measurement between operations.
Accuracy of measurement.
Costs associated with measurement technique.

Selection of Technique

Throughout this handbook, entire chapters are devoted to techniques that are an integral part of the measurement and control of indirect operations. Section 4 covers work standards, time study, predetermined motion time systems, work sampling, and the use of standard data, all of which are particularly important basic techniques. The principles and concepts of these measurement techniques are not developed here, but application in an indirect operational function is demonstrated.

In recent years the methodologies of queuing regression and correlation have been utilized in studying the factors and problems inherent in indirect operations. Once the factors have been isolated, a computer simulation demonstrates which elements need refinement and correction for optimum productivity. These methodologies are found in Section 13 of this handbook.

Use of Historical Data

This technique employs average time values for particular work packages based on past experience. The records of previously completed work packages are assembled and categorized by complexity and time required to complete. Then the groupings are assigned a degree of difficulty and an average time estimate based on past performance. Simple tasks are ranked in the first category and include standards of 0 to 1.5 hr, or an average of 0.75 hr. Other groupings are similarly categorized. New tasks are initially placed in categories by degree of difficulty, and the task standard for this category is utilized. Advantages of this method include consistency, low administration costs, simple training in their usage, and total coverage of tasks. Among the disadvantages are that past cost inadequacies are built into the system, alternate tasks are difficult to compare, and new work is difficult to assess. An example of standards determined by historical data is given in Exhibit 4.9.13.

Estimating

This technique is widely used in indirect operational measurement, involves estimating the time required to perform a task, based on the best judgment of the person making the estimate. Obvi-

Exhibit 4.9.13 An Example of the Use of Historical Data to Establish Time Standards

Year	Employees	Valves Inspected
1974	5	317,479
1975	6	384,821
1976	7	502,979
1977	8	558,519
	26	1,763,380
Yearly average	6.5	440,950

Therefore in 1 year, one inspector inspects 440,950/6.5 = 67,838 valves.

If a unit time standard is required, use the following calculation:

$$1 \text{ employee represents } (50 \text{ weeks}) \times (40 \text{ hr}) = 2000 \text{ hr}$$

Therefore the time required to inspect one valve is

$$\frac{2000 \text{ hr}}{67{,}838 \text{ valves}} = 0.029 \text{ hr/valve}$$

$$0.029 \text{ hr/valve} \times 50 \text{ min} = 1.74 \text{ min/valve}$$

Courtesy of Kearney Management Consultants, Chicago, Illinois.

Exhibit 4.9.14 An Example of a Self-Logging Form Prepared by a Clerical Employee

Department: Credit | Prepared by: Clerk | Date: 10/21

Task	Unit Measured	Number of Units	Actual Time (hr)
1. Collection correspondence	Letters	10	2.4
2. Credit investigation	Form 2179CI	2	3.4
3. Deduction and remittance advice	Form 2180DR	–	–
4. Dictation and transcription for credit manager	Letters	2	1.7
5. Cash receipt	Checks	–	–
6. Open mail	Pieces	–	–
Total time			7.5

Courtesy of Kearney Management Consultants, Chicago, Illinois.

ously, the personal experience, knowledge, and ability of the estimator will determine the quality of the estimate. Advantages of this technique include the relative ease of development, low administrative cost, and complete coverage of operations. Among the disadvantages are that time values may be inaccurate and inconsistent, methods comparisons are impractical, training is difficult, verification is almost impossible, and the quality of estimates depends completely on the capabilities of the estimators.

Self-Logging

This technique involves the employee in the development of the time value, which offers positive motivational aspects. It has the advantage of being easy to install and understand, so that with simple training it can be utilized. Among the disadvantages are the inclusion of present inefficiencies, the possibility of errors, and the resultant loose time standards. A self-logging form is illustrated in Exhibit 4.9.14.

Short Interval Scheduling or Batching

This technique involves the scheduling of work in small, timed batches. The employees are informed of the completion times for each batch and are requested to keep a record of performance. This is also commonly referred to as "batching."

The first step in developing time estimates for the tasks involves a procedure similar to self-logging. The employees involved in a function are requested to keep track of tasks performed. These performance times are analyzed, and task standards are developed. Then tasks are assigned in time increments so that progress and performance can be checked by verifying the batches completed. As the time standards are more accurately determined by actual measurement, the short interval scheduling system becomes more practical.

The advantages of this system are that the employee becomes involved through self-measurement and evaluation and that the system itself becomes a basis for scheduling. Among the disadvantages are that the standards utilized when only logging is used are inaccurate and that the standard is based on a current method that may require improvement. An example of a work sheet used to collect data for short interval scheduling standards is given in Exhibit 4.9.15.

Wristwatch Study

This technique is commonly utilized in the measurement of indirect operations because it requires less skill and is less disruptive than time study. It is most applicable to functions that include many varied tasks and that are not highly repetitive. A high degree of variation makes it difficult to methodize the task and permits the wide range of elemental time values obtained from this form of work measurement. The advantages are that it is relatively accurate, is detailed, and requires minimum training. The disadvantages are the disruption it causes in operations and its lack of precision. An illustration of a wristwatch study is given in Exhibit 4.9.16.

Stopwatch Time Study

This technique, as described in Chapters 4.1 and 4.4, permits the obtaining of an acceptable level of control. It is one of the best techniques available when high repetition is evident and a degree of standardization is prevalent. Its advantages are that it is highly accurate, it is relatively fast for short durations of task times, and it permits the accurate maintenance of method records. Among

Exhibit 4.9.15 A Typical Work Sheet Utilized in Short Interval Scheduling[a]

Short Interval Scheduling Work Sheet — Activity: Stockroom — Period from 8/9 to 8/13

Date and Hours	Number of Regular Orders Filled	Number of Immediate Orders Filled	Number of Priority Orders Filled	Number of Bins Restocked	Number of Sku's Received	Number of Purchase Orders Completed				
Date 8/9 Hours 72	32 34 27 21 26 54 194	16 9 37 62	62 58 75 65 54 314	64 45 109	3 60 63	52 17 69				
Date 8/10 Hours 72	42 19 45 109	10 40 2 52	101 28 42 62 233	45 8 76 129	21 25 23 74 143	56 56				
Date 8/11 Hours 71	76 68 38 27 209	7 86 7 28 128	15 111 31 21 178	2 36 1 39	9 7 89 105	8 13 2 18 41				
Date 8/12 Hours 55½	50 39 25 27 27 168	76 76	85 95 69 249	9 98 107	32 32	42 30 72				
Date 8/13 Hours 55½	39 63 57 159	33 19 99 71	82 52 78 222	37 37	14 4 18	43 15 58				
Date Hours										

[a]Courtesy of Kearney Management Consultants, Chicago, Illinois.

Exhibit 4.9.16 A Typical Observation and Calculation Form for a Wristwatch Study

Title: Process invoice
Area: Accounts payable
Equipment: Calculator and ledger

Sheet 1 of 1	Unit of measure
No. 3 activity	Invoice

No.	Elemental Task Description
1.	Receive invoice from supervisor and return to desk.
2.	Compare invoice number to ledger number.
3.	Transcribe dollar quantity.
4.	Apply date stamp to invoice.
5.	Write "ok," "hold," or "defer" on invoice.
	A. Ok = all items compare.
	B. Hold = question concerning information or clarity of data.
	C. Defer = does not match.
6.	Return invoice to supervisor, discuss status of "hold" or "defer" invoices, and receive instructions.

No.	Observations				Avoidable Delay Time	Applied Time	Units Completed	Minutes per Unit	Use
	Date	Start	Stop	Elapse					
1.	15 Feb.	815	819	4	0	4	15	0.26	–
1.	16 Feb.	950	956	6	0	6	20	0.30	0.25
1.	17 Feb.	1115	1118	3	0	3	14	0.21	–
2.	15 Feb.	819	822	3	To answer personal question 1	2	15	0.13	–
2.	16 Feb.	956	1000	4	0	4	20	0.20	0.20
3.	15 Feb.	822	830	8	0	8	15	0.53	–
3.	16 Feb.	1000	1040	40	Coffee break 20	20	20	1.00	0.75
4.	16 Feb.	1040	1041	1	0	1	20	0.05	0.10
5.	15 Feb.	832	857	25	0	25	15	1.66	–
5.	16 Feb.	1041	1108	27	0	27	20	1.35	1.50
6.	16 Feb.	1108	1110	2	0	2	20	0.10	0.10
				Totals	21	102	49	Goal 2.90 min.	

Notes: Recommend activity for automatic data processing.

the disadvantages are that it disrupts operations and possibly antagonizes the employee being studied, it gives unsatisfactory results on operations of long cycles, and it is expensive on low-volume activities.

Work Sampling

Work sampling, described in Chapter 4.6, is a widely used technique in analyzing and measuring indirect operations. It highlights the inefficiencies that are apparent in the function. It defines activities and quickly categorizes them to permit analysis and improvements. The advantages are that it is relatively inexpensive, it is quick and nondisruptive, it requires little training, and its findings are highly reliable. Its disadvantages include the lack of methods improvement inherent in its pure usage and the shortcoming in providing a measure of an activity, but not a measure of performance.

Use of Standard Data and Predetermined Time Systems

These techniques should be utilized on highly repetitive tasks of both short and long cycles. The basics and applications of standard data and predetermined time systems are covered in Chapters 4.8 and 4.5, respectively. They both permit the classification of work elements and the analysis of motions. These factors, coupled with the frequency of occurrence, are developed into a time standard. The advantages are that they give an accurate method description, they offer a high degree of precision and low maintenance cost, and they are easy to apply. The disadvantages include the training required to utilize the techniques and the expense of development.

4.9.15 SYSTEM IMPLEMENTATION

The analysis has established the environment for change. The involvement of supervisors and employees has created an atmosphere of cooperation. The analysis team is cognizant of the unique parameters of the function to be studied. The measurement technique or techniques have been researched and selected. The study and planning aspects of the implementation plan for systems change is completed and in accordance with the master plan. What remains is to complete the detail work plan and to concentrate on the major system deficiencies as indicated by the analysis. A concept that permits mass data assimilation is the 20/80 rule, where 80% of the costs or contributing factors are a direct result of 20% of the activities. The analysis has pointed out the important 20% of system deficiencies. The process used to conduct the analysis to determine causes is illustrated in Exhibit 4.9.17.

Idea Generation

The action plan takes shape after a number of early "idea generators" are completed. These generators include a review of important factors concerning the function or department. This checklist of questions is essential for function knowledge and action plan development. An example of such a checklist is given in Exhibit 4.9.18.

To complete the idea generation phase, the work flow must be studied in more detail than it was during the analysis. This is accomplished through the use of a flow process chart, as discussed in Chapter 3.3. An example of a typical flow process chart is shown in Exhibit 4.9.19.

4.9.16 OPERATIONAL SYSTEM

The analysis has highlighted areas of concern. The various forms of work measurement have quantified these concerns. The final review and flow chart have reemphasized the needs and weaknesses.

Exhibit 4.9.17 An Illustration of the Logic Process to Conduct an Analysis[a]

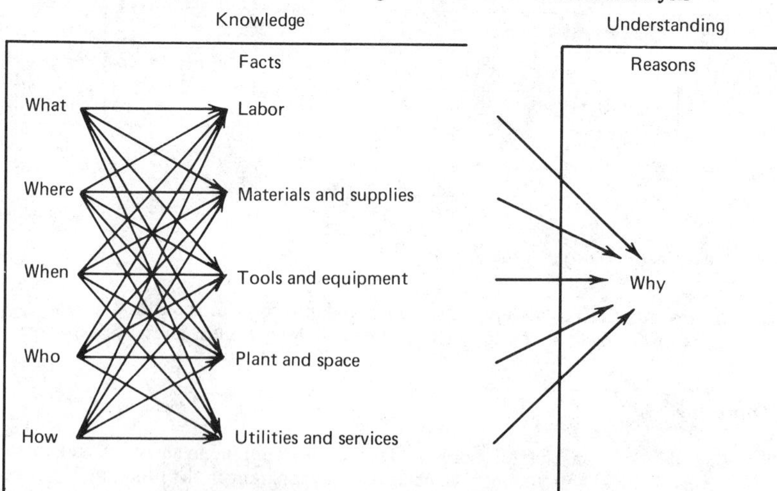

[a]Courtesy of Kearney Management Consultants, Chicago, Illinois.

Exhibit 4.9.18 Checklist of Questions

Things to Ask About a Department—Work Force

1. How many shifts are necessary?
2. How many supervisors are necessary?
3. Is there a need for training programs?
4. Are the workers physically qualified?
5. What determines the size of the work force, both fixed and variable?
6. Are schedules prepared in advance?
7. Is overtime a constant condition?
8. Is the work force distributed for overhaul?
9. Is there a variation in the working hours of the various trades?
10. What is the percentage of absenteeism?
11. Are pools or area assignments effective?
12. Are jobs assigned long enough in advance to give workers time to get proper tools and instructions, to make their sets up, and so on?
13. Is there an assigned routine of fixed and repetitive duties?
14. How long does it take for a worker to get to the job after checking in?
15. How do tools and equipment reach the job?
16. What about lunchtime, coffee breaks, and end of shift—too much time?
17. Are workers working within their classification?
18. Should workers be reclassified higher or lower?
19. How much interference occurs in completing a job?
20. Could work be done cheaper by outside contractor?
21. Could work done by outside contractor be done cheaper by us?
22. Can a once-a-day routine be done less often—once a week perhaps?
23. Is crew size standard and inflexible, and if so, how is it determined?
24. Are repairs done as complete jobs or as patch jobs?
25. Are central force work crews doing jobs that substation or area personnel should do, or vice versa?
26. Who schedules central skills?
27. Do central and area personnel all work the same hours and have the same lunchtimes?
28. Who inspects and approves a central service job?
29. Who reviews cost?
30. Does location of shop and zone area permit the most efficient use of manpower?
31. Is the work area well laid out?
32. Are there sufficient machine tools in the area?
33. Are the right kinds of tools available?
34. Is production equipment regularly scheduled for maintenance downtime?
35. Do you estimate time for a maintenance job?
36. Do you compare the actual performance of the job and the estimated time?
37. Who expedites the spare parts before repair day?
38. What records are kept of repetitive processes?
39. Do you regularly supplement area crews for downtime repair work?
40. Who makes inspections and analysis as a followup to your preventive maintenance program?
41. What personnel control spare parts in the maintenance unit?
42. What is the total value of spare parts inventory?
43. What is the spare parts turnover?
44. Are spare parts stored in easily accessible locations?
45. Are records accurate?
46. Are minimums and maximums established for spare parts inventory?
47. Who orders and purchases new spare parts?
48. To what extent is standardization of spare parts possible?
49. Is it possible to standardize through minor redesign of equipment?
50. How much time is lost by waiting at a crib for supplies?
51. Are the practices for ordering tools and supplies clearly defined?
52. What is the extent of emergency "not-in-stock" items?
53. What delivery schedule is established for zone or area cribs?
54. Who is responsible for receiving tools and equipment into either a zone or a central maintenance unit?
55. Are storerooms and cribs convenient?
56. What tools are issued as personal tools?
57. What records and reports are prepared by the maintenance department?
58. What is the purpose of the report? Who gets it? Is it of value? Does it provide enough information?

Exhibit 4.9.18 *(Continued)*

Things to Ask About a Department–Work Force

59. Do we have a maintenance budget?
60. What control records are available to people responsible for maintenance cost?
61. How much time do supervisors spend on analysis of cost?
62. Do your supervisors understand elementary cost accounting?

Things to Ask About a Department–Supervision

1. Do you have an organization chart?
2. Does everyone know to whom to report?
3. Does the unit work like the chart says?
4. Do supervisors have too large a span of control?
5. Is there overlapping responsibility?
6. Are other persons doing supervisors' work?
7. Are supervisors' duties spelled out?
8. Are supervisors losing effectiveness because of having too much ground to cover?
9. Do supervisors participate in planning their budgets?
10. Do supervisors receive budget data?
11. How much nonmanagerial (clerical) work does a supervisor do?
12. Do work schedules favor a certain supervisor at certain work loads?
13. Do supervisors start shifts at the same time as the workers?
14. Is supervision present on each shift?
15. Are jobs analyzed for most efficient crew size?
16. Is the general foreman performing foremen's duties?
17. How much advance planning is done?
18. How are several groups of workers coordinated on a job?
19. Does the supervisor know the contract?
20. Who gives labor relations advice?

What Makes a Job Hard?

1. Restricted space in which to move around. Also, working alone in a confined place.
2. Poor or improper tools and equipment.
3. Condition of material.
   a. Rusty or dirty.
   b. Oily.
   c. Sharp.
   d. Not to tolerance (machining).
4. Lack of worker skill.

Motion Economy

1. Movements should be simultaneous.
2. Movements should be symmetrical.
3. Movements should be natural.
4. Movements should be rhythmical.
5. Movements should be habitual.

The next step is to improve the operational system by involving all aspects of human resources. This is accomplished by improvements in the planning, scheduling, control, and reporting aspects of the systems.

Work Order Control

Work orders usually fall into one of three categories: (1) prescheduled work, which is typical of preventive maintenance or backlogged work in other indirect areas; (2) regular work, which is scheduled and assigned upon notification; and (3) emergency work, which requires immediate attention. Whatever category is involved, initiating a work order request for services from the indirect function is paramount to control. Even in the case of emergency work, the work order request should be completed even after the work has been completed.

The work order must be specific and describe clearly the work to be performed. It is forwarded to a control analyst, who logs the information and enters the request on a schedule. A typical work order and the flow of the order are displayed in Exhibits 4.9.20 and 4.9.21, respectively.

Exhibit 4.9.19 Typical Flow Process Chart[a]

NO.______
PAGE___OF___

FLOW PROCESS CHART

JOB ____________________

□ MAN OR □ MATERIAL ____________________

CHART BEGINS ____________________

CHART ENDS ____________________

CHARTED BY ____________________ DATE __________

SUMMARY

	PRESENT		PROPOSED		DIFFERENCE	
	NO.	TIME	NO.	TIME	NO.	TIME
○ OPERATIONS						
⇨ TRANSPORTATIONS						
□ INSPECTIONS						
D DELAYS						
▽ STORAGES						
DISTANCE TRAVELLED	FT.		FT.		FT.	

	DETAILS OF (PRESENT / PROPOSED) METHOD	OPERATION	TRANSPORT	INSPECTION	DELAY	STORAGE	ANALYSIS: DISTANCE IN FEET	ANALYSIS: QUANTITY	ANALYSIS: TIME	NOTES	ACTION			
1		○	⇨	□	D	▽								
2		○	⇨	□	D	▽								
3		○	⇨	□	D	▽								
4		○	⇨	□	D	▽								
5		○	⇨	□	D	▽								
6		○	⇨	□	D	▽								
7		○	⇨	□	D	▽								
8		○	⇨	□	D	▽								
9		○	⇨	□	D	▽								
10		○	⇨	□	D	▽								
11		○	⇨	□	D	▽								
12		○	⇨	□	D	▽								
13		○	⇨	□	D	▽								
14		○	⇨	□	D	▽								
15		○	⇨	□	D	▽								
16		○	⇨	□	D	▽								
17		○	⇨	□	D	▽								
18		○	⇨	□	D	▽								
19		○	⇨	□	D	▽								
20		○	⇨	□	D	▽								
21		○	⇨	□	D	▽								
22		○	⇨	□	D	▽								
23		○	⇨	□	D	▽								
24		○	⇨	□	D	▽								
25		○	⇨	□	D	▽								

[a]Courtesy of International Harvester.

Job Planning and Scheduling

The initial step in the planning stage is the analysis of the work order. With good cooperation from those initiating work orders, sufficient information can be obtained to permit effective planning. Proper training of the job planner, which results in a questioning attitude and a checklist of factors to be considered on all work orders, ensures the adequate provision of information even on emergency-type orders. The planner then estimates or calculates the required standard time and schedules the work to an area services function or directly to a craft. This scheduling requires the maintaining of up-to-date information on demands, a good knowledge of material availability, and

Exhibit 4.9.20 Example of a Work Order

SHIFT

WORK ORDER

NAME

ORIGINATOR	CERT. COMPLETE	W.O./A.F.E.	DATE REQUIRED	COMP. DATE	PRIORITY 1 2 3 4 5

ITEM	DEPT.	CRAFT	CREW SIZE	CLOCK No.				UNMEASURED CODE	UNMEASURED HRS.	COST CENTRE	STD. HRS.	REVISED STD. HRS.
1	51	M	02							6103-53	1.6	
2	51	P	01							6103-53	1.6	
3												
4												
INDUSTRIAL ENGINEERING SIGNATURE								TOTAL UNMEASURED			TOTAL STD.	TOTAL NON STD.

TITLE: M0353 ROTARY FURNACE CHANGE BOTTOM DOORSILL

ITEM	DESCRIPTION	CHECK
1	OBTAIN NEW SILL. STORES STOCK ITEM 2205-2240	
	REMOVE BOLTS FROM SILL.	
	LIFT OUT SILL AND REPLACE WITH NEW SILL.	
	INSTALL BOLTS	
	PLACE OLD SILL IN WELDING SHOP AREA OR SCRAP BOX.	
2	DISCONNECT PIPING. HOOK UP PIPING AFTER SILL IS INSTALLED.	
	RECORD LOCATION -	

Exhibit 4.9.21 Flow Chart of Typical Work Order

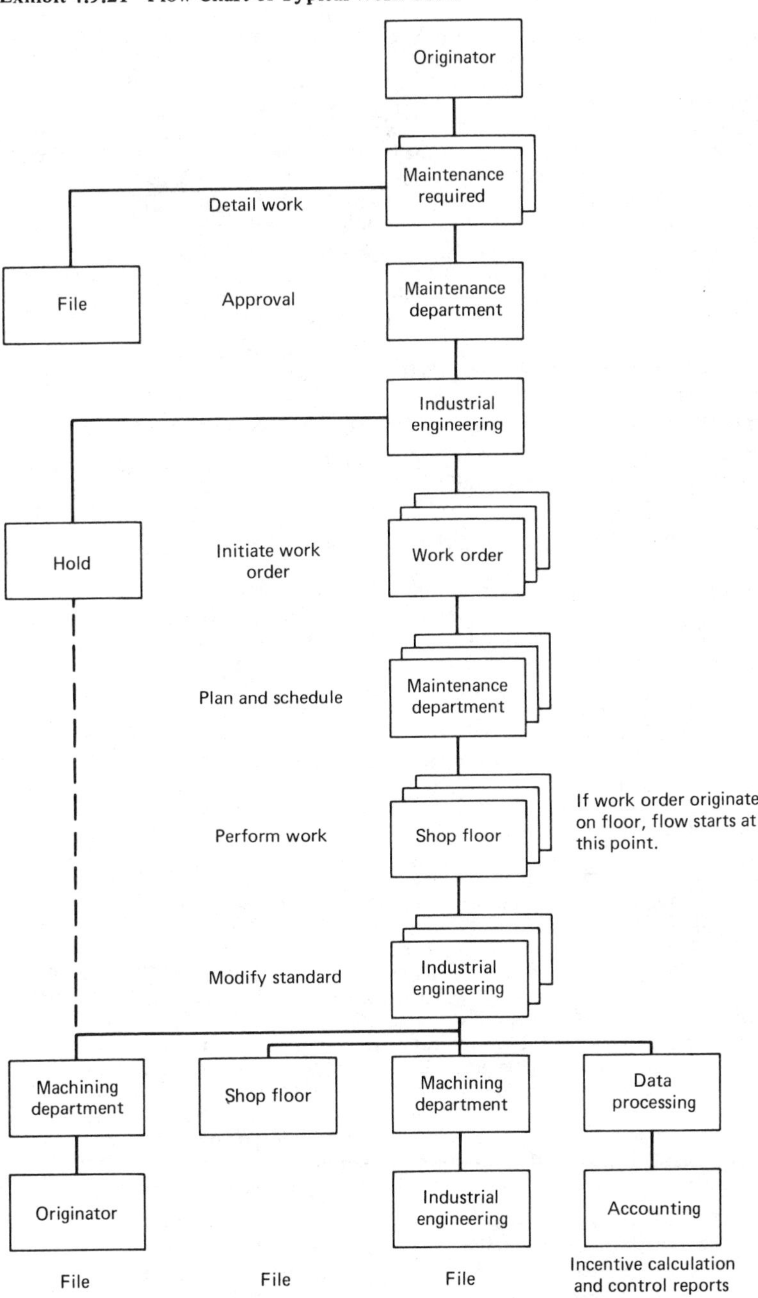

current feedback on work-in-process status. These are accomplished by maintaining effective, yet simple, inventory records to materials. The work orders are then placed in queue for the supervisor to review and assign. These assignments are made by priority sequence, which reflects the analysis of the job planner. The craftsman or craftswoman takes the order to the job, and when finished, fills in the time taken to complete the assignment. The order is returned to the analyst for summarizing and reporting. This system is also used to maintain records of materials requisitioned and utilized. The indirect function supervisor can monitor the results of the analyst's report and counsel with employees as necessary.

Area Versus Craft Supervisor

The main purpose of indirect operations is to support and service the production operations. To fulfill this purpose, the indirect function should be located as close as possible to the functions requiring the service. However, in some cases these servicing functions require considerable floor space of their own for machinery, material storage, and so on. Therefore they are located in areas remote from the production facilities. The planning, scheduling, and control of the remotely located indirect functions are performed in the office area connected with the function; however, much of the work is performed at the location of the production machines and facilities. Supervision of these activities is difficult since they are remote and spread out. This arrangement of indirect services is categorized as "centrally located and dispatched."

In many organizations the service function is located right in the production facility, with sufficient minimum equipment to perform general services. This arrangement has been growing in practice during recent years. The responsibilities of area supervisors under this arrangement include the following:

Inspection of all service requests in their areas.
Diagnosis of service requests.
Estimating the magnitude of the job and the types of crafts required.
Authorizing the procurement of materials, tooling, and parts.
Assignment of personnel to service required.
Supervision of personnel.
Reporting feedback of results, events, and performance.

The advantages of this arrangement for providing indirect services are as follows:

1. Multicraft control of the work will broaden the supervisor's abilities.
2. Services can be provided quickly in areas where a steady demand justifies stationed crews.
3. Travel time between jobs for the crafts is saved.
4. An effective control and reporting system is viable with real-time information.
5. A flexible pool of workers can be used to supplement any minimum area crews when needed.
6. Supervision is improved by eliminating the duty of separate craft supervisors of surveying or partrolling job sites.

Major Volume Indicators

The operational system begins with work order control, is more formalized through the job planning and scheduling, and is based on the most appropriate measurement technique available. The selection criteria for the measurement technique to be utilized are accuracy, consistency, and administrative cost. These criteria must be balanced in the selection process. The selected technique is then utilized to measure "key tasks," or major volume indicators (MVIs). These indicators have been identified from the preliminary analysis, the work measurement activity, possibly a time study or work sampling, or the analysis of work flow. The indicators have a high degree of correlation with predictable work load. Each indicator has two highly significant attributes: (1) it represents that factor that controls the majority of the work load and that varies proportionately with the work load, and (2) it is clearly defined and economically measurable. The MVIs exist in every function. They are predictable factors from which the frequency of occurrence of tasks can be determined as a result of a known degree of correlation. The information upon which the indicators are based is available largely in the form of business trends, plans, forecasts, schedules, and many reporting statistics. Typical examples of MVIs for various indirect functions are:

Functions	MVI
Maintenance	Work order
Service parts disbursement	Parts requisition
Office services	Pieces of mail handled
Accounts payable	Invoices processed
Purchasing	Purchase orders
Expediting	Parts resolved
Specifications	Pages processed
Tool design	Design blueprints
Industrial engineering	Performance standards
Material scheduling	Documents processed
Internal audit	Procedures checked

In all of these functions, other activities occur that take time and resources, but these indicators are most closely correlated to function performance. An example summary form that points out various volume indicators is given in Exhibit 4.9.22. The form displays the activity, the unit of measure (or volume indicator), and the resultant time. By analyzing the form, one can select the MVI, namely, activity 9, "transactions."

The next step is to develop a staffing table based on the MVI of transactions. To accomplish this requires the separation of constant and variable activities. The constant activities occur once per time period, but all other activities are a function of the transactions. Therefore the transactions are directly proportional to the staffing level, as illustrated in Exhibit 4.9.23.

Exhibit 4.9.22 Summary Form Depicting Major Volume Indicator

Department: Accounting | Sheet: 1 of 1
Area: Accounts Payable Clerk | Period: 172 hr month

No.	Activity Description	Unit of Measure	Rep. Minutes	Volume	Monthly Hours
1.	Process voucher	Voucher	6.00	220	22.00
2.	Prepare backlog record	Entry	5.00	22	1.83
3.	Process invoice	Invoice	4.75	290	22.96
4.	File voucher charge-back	Charge	1.40	15	0.35
5.	Make phone inquiry–call	Call	3.00	97	4.85
6.	Write daily summary sheet	Daily	15.00	22	5.50
7.	Write monthly report	Report	150.00	1	2.50
8.	Attend monthly training	Session	100.00	1	1.67
9.	Write account completion	Transaction	6.00	1008	100.80
10.	Verify payment date	Monthly	1.25	110	2.29
			Page total:		164.75
			Grand total:		164.75

Exhibit 4.9.23 Example of a Staffing Table

Department Accounting

Job Description Accounts payable clerk

Indicator Transactions (activity #9)

Constant hours Monthly = 16.81 activities 5, 6, 7, 8, and 10

Variable hours = 147.94

147.94 hr/1008 transactions = 0.146 hr/transaction
172 hr = 1178 transactions

Indicator Quantity	Hours	Staff
0	16.81 (constant)	0.10
1013	172.00	1.0
2191	344.00	2.0
3369	516.00	3.0
4547	688.00	4.0
5725	860.00	5.0

Note: First employee does the constant activities and variable activities. All additional employees do the variable activities only.

Work Count System

It is equally important to establish a work count system of the MVIs. Existing work load reporting systems must be carefully reviewed to determine whether required items are being reported in the defined form and for compatible time periods. To ensure an accurate and usable work count, the following considerations are essential:

The instructions must clearly specify what constitutes a unit of count.

The source of count, or a point in a process at which a unit of count results, must be established.

The count or reporting frequency must be compatible with, or adjustable to, the anticipated length of the measurement period. This is especially important if work sampling is prescribed.

Safeguards that will minimize the possibility of a duplicate or missed count must be established. An example would be a random external audit of the MVI count.

The self-logging form that was shown in Exhibit 4.9.14 is indicative of a work count system.

Since the MVI and the work count system are the core of the performance measurement and reporting system, it is mandatory that the task force periodically review them prior to their formal establishment in the operational system. During this review, the task force should check the following:

Work content of MVIs.

Frequencies of occurrence.

Allowed times to perform tasks associated with MVIs.

Correlation of MVIs to predictable work loads.

4.9.17 PRESENTATION TO OPERATIONAL MANAGEMENT AND EMPLOYEES

The type of presentation, whether handled informally with simple handouts or formally with 16 mm slides and a formal report, depends on the complexity of the indirect function studied and the operational system recommended. It is imperative that this presentation start at the basic operational level, with employees, supervisors, and the task force members an integral part of the presentation. These sessions must be oriented to open discussion and questioning at any time, permitting feedback in order to establish the environment for improvement. The presentation must be well thought out and planned; with clear identification of all the significant consequences of each proposed course of action. These recommended courses of action should be tested in cooperation with the function's management and employees. This involvement should include the opportunity for participative decision making in the working affairs of the function. Clear and detailed discussion of the project's parameters and of the reasoning behind the recommendations should be included. Operational staff should participate in this discussion and should reexamine the analysis and the evidence. This can be accomplished by incorporating the following points in the presentation and discussion:

Project a decision into detailed plans.
Reconsider assumptions.
Review alternatives previously discarded.
Listen to the "devil's advocate."

4.9.18 FINAL REPORT AND PRESENTATION

The format for the final report and presentation should be properly organized and should cover the reason for the study, the condition of environment at the time of the study, the proposed improvements, and the action necessary to incorporate them. The following outline generally covers the important elements.

1. Introduction
2. Objectives and scope of the study
   a. Definition
   b. Areas of investigation
   c. Considerations excluded
   d. Methods of study
   e. Special data collected
3. Summary of findings
   a. Present organizational structure
   b. Function demand and work load
   c. Scheduling and assignment of work
   d. Supervisory control, information, and reporting system
4. Conclusions
   a. Organizational structure
   b. Working relationships and procedures
   c. Performance measurement
   d. Systems installations
   e. Organized approach
   f. Supervisor and employee involvement
   g. Training requirements
5. Recommendations
   a. Basic concepts
   b. Guidelines
   c. Productivity potential
   d. Human resource implications
   e. Improved asset utilization
6. Costs and benefits
   a. Tangible
   b. Intangible
7. Action plan
8. Feedback and evaluation

Action Plan

The action plan separates the overall system implementation into its important activities and phases. It assigns specific responsibilities to the multidisciplinary task force, the functional or operational personnel, or the support-type functions. It details the resource requirements for personnel, capital, and time. The activities are scheduled, and milestone events are established as checkpoints through the action plan. An example of an action plan is given in Exhibit 4.9.24.

Organization Development Implications

As is evidenced by the final report and action plan, implementation of the system will have a tremendous impact on the operating environment of the function. Any change in procedure will contribute to that impact. Maintaining cognizance of the changing working environment is extremely important. The change in operating system is self-evident, but the organizational change must be monitored and controlled. A number of training sessions will bring together the people who contribute to the function's success. It is at these sessions that motivational training is necessary, both through classroom work and through actual demonstration of day-to-day interrelationships among all levels in the functional group. The supervisors must be trained not only in the technical aspects of the changing operational system, but also in the motivational aspects that support the system in day-to-day activities.

4.9.19 FOLLOW-UP, SYSTEM READJUSTMENT, AND SUPPORT

The follow-up aspect of measuring and controlling indirect operations commences with an evaluation of the technical aspects of the system. This evaluation in the early stages after implementation should be repeated often, and results should be fed back to the operations people quickly. It should cover the entire spectrum of the system, from quantifiable results to the important subjective aspects. A typical outline for use as a guide to evaluating an indirect operation system would be as follows:

1. Statistics–base period, this period, next period
   a. MVI
   b. Volume
   c. Hours
   d. Performance
2. Reasons for variance
3. Plans for Improvement
4. Human resource environment
5. Technical system compliance
6. Major accomplishments–this period
7. Key events in progress
8. Next period's key events

Accountability Training and Compliance

Throughout the implementation phase, the supervisors and employees have been trained in systems aspects and behavioral concepts. These are now combined and put to the test in the operational mode. The evaluation of system statistics can measure the quantitative results of the total effort, but the working environment may be deteriorating at the same time. So that management may remain cognizant of the actual day-to-day interrelationships, compliance workshops are conducted with employees and supervisors to ensure they have accurate knwoledge and to adjust the program as necessary. Supervisor and employee attitudes are surveyed anonymously through an attitude study. The program is changed and reinforced to reflect these attitudes. As a result, the operational system is fine-tuned and is eventually accepted, because there has been total involvement throughout the analysis, development, evaluation, training, and readjustment phases.

Procedural Documentation

One of the last remaining steps in implementing the operational system is the documentation of system changes. Before the multidisciplinary task force disbands, this step must be carried out. It is the link to ensuring understanding between the task force and the operations people. Many times the analysis is carried out capably, the development is practical, and the implementation is successful, but it is found that a year later the system cannot be reconstructed and improved. The documentation of system procedures and the study environment is the necessary bond to the final step of incorporating the system into the management operating system.

Management Operating System

The operational system has been tested, evaluated, and documented. The last remaining effort for the multidisciplinary task force is to ensure incorporation of the new system into the indirect function's real-time operating system. The system must be periodically evaluated; as a part of the management operating system, this will become routine along with other systems. The task force assignments have been met, which has ensured attainment of the overall objective of improved asset utilization.

Exhibit 4.9.24 Example of a Well-Organized Action Plan

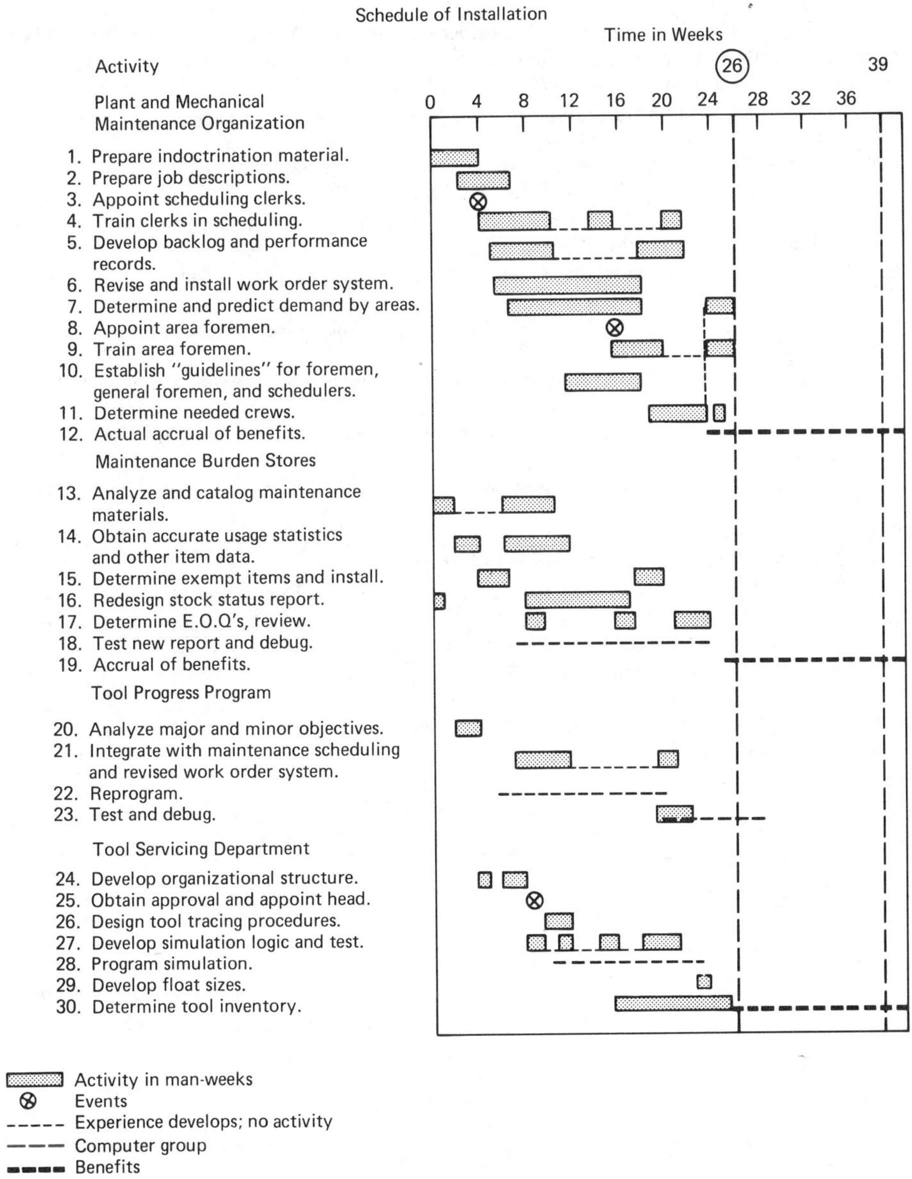

Schedule of Installation

Time in Weeks

Activity (26) 39

0 4 8 12 16 20 24 28 32 36

Plant and Mechanical Maintenance Organization

1. Prepare indoctrination material.
2. Prepare job descriptions.
3. Appoint scheduling clerks.
4. Train clerks in scheduling.
5. Develop backlog and performance records.
6. Revise and install work order system.
7. Determine and predict demand by areas.
8. Appoint area foremen.
9. Train area foremen.
10. Establish "guidelines" for foremen, general foremen, and schedulers.
11. Determine needed crews.
12. Actual accrual of benefits.

Maintenance Burden Stores

13. Analyze and catalog maintenance materials.
14. Obtain accurate usage statistics and other item data.
15. Determine exempt items and install.
16. Redesign stock status report.
17. Determine E.O.Q's, review.
18. Test new report and debug.
19. Accrual of benefits.

Tool Progress Program

20. Analyze major and minor objectives.
21. Integrate with maintenance scheduling and revised work order system.
22. Reprogram.
23. Test and debug.

Tool Servicing Department

24. Develop organizational structure.
25. Obtain approval and appoint head.
26. Design tool tracing procedures.
27. Develop simulation logic and test.
28. Program simulation.
29. Develop float sizes.
30. Determine tool inventory.

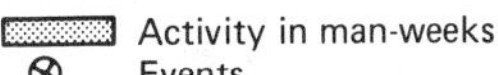 Activity in man-weeks
⊗ Events
- - - - - Experience develops; no activity
— — — Computer group
▬ ▬ ▬ Benefits

REFERENCES

1. BENJAMIN W. NIEBEL, *Motion and Time Study*, Irwin, Homewood, IL, 1972, p. 542.
2. WILLARD C. BUTCHER, "Closing Our Productivity Gap Key to U.S. Economic Health," *Industrial Engineering*, Vol. 11, No. 12 (December 1979), p. 30.
3. DAVID L. CONWAY, "Measuring Indirect Functions," address to the Second Annual Productivity Conference, American Productivity Center, Paris, France, September 19–20, 1979.
4. JOHN C. WERNER, "Operations Control–The Organized Way to Improve Productivity," *Industrial Engineering*, Vol. 10, No. 7 (July 1978), p. 26.
5. R. KEITH MARTIN, "Don't Overlook Clerical Productivity," *Industrial Engineering*, Vol. 9, No. 2 (February 1977), p. 28.

BIBLIOGRAPHY

BITTEL, LESTER R., *Encyclopedia of Professional Management*, McGraw-Hill, New York, 1978.

CANNAN, BERNARD W., "New Frontier in Productivity Improvement: White Collar Workers," *Industrial Engineering*, December 1979, pp. 34–37.

CLAIRE, FRANK V., *Control of Shop Indirect Labor*, American Management Association, New York, 1975.

CONNER, DENIS A., "Measurement and Control in the Office," *Methods-Time-Measurement Conference Proceedings*, MTM Association, Fairlawn, NJ, 1977.

GRAYSON, C. JACKSON, Jr., "Productivity: More Action, Less Talk," American Productivity Center Productivity Conference, Chicago, IL, 1979.

HAMLIN, JERRY L., "Productivity Appraisal for Maintenance Center," *Industrial Engineering*, September 1979, p. 41–45.

HICKS, HERBERT G., and C. RAY GULLETT, *The Management of Organizations*, 3rd ed., McGraw-Hill, New York, 1976.

KOOP, JOHN E., "Indirect Labor Incentives Pay Off," *Industrial Engineering*, February 1977, pp. 28–30.

MAYNARD, HAROLD B., *Handbook of Modern Manufacturing Management*, McGraw-Hill, New York, 1970.

MAYNARD, HAROLD B., *Industrial Engineering Handbook*, 3rd ed., McGraw-Hill, New York, 1971.

NEWMAN, WILLIAM H., CHARLES E. SUMMER, and E. KIRBY WARREN, *The Process of Management*, 3rd ed., Prentice-Hall, Englewood Cliffs, NJ, 1970.

NIEBEL, BENJAMIN W., *Motion and Time Study*, 5th ed., Irwin, Homewood, IL, 1972.

RATHE, ALEX W., and FRANK A. GRYNA, "Applying Industrial Engineering to Management Problems," American Management Association Research Report No. 97, AMA, New York.

STAIRS, D. LEONARD, "Opportunity—The Maintenance Department," address to Productivity Seminar, Charlotte, NC, March 4, 1977.

WALTON, RICHARD E., "Work Innovations in the United States," *Harvard Business Review*, July–August 1979, pp. 88–94.

SECTION 5

EVALUATION, APPRAISAL, AND MANAGEMENT OF HUMAN RESOURCES

CHAPTER 5.1

Subjective Aspects of Performance

RICHARD G. PEARSON
North Carolina State University

5.1 INTRODUCTION

The objective of this chapter is to review the variety of subjective responses made by workers, supervisors, and managers as they relate to work performance and, ultimately, to the effectiveness (productivity) of an enterprise. Both formal and informal techniques for obtaining subjective response data are discussed, as is the utility of such data for improving work performance. Specific topics include fatigue, stress, effort, comfort, annoyance, job satisfaction, and methodological pitfalls to be avoided.

Subjective Responses—Their Importance and Place

In the context of this chapter, the term "subjective" can be defined as existing in the mind. Thus subjective responses include such verbal and written expressions as:

Feelings and moods.
Opinions.
Attitudes.
Judgments.
Complaints.
Appraisals.

Subjective responses can provide insight into the effectiveness of individual task situations, group performance, or the total organization or industrial enterprise. As we shall see from examples in this chapter, some subjective responses (e.g., appraisals and feelings) may be used routinely in assessing effectiveness or in guiding day-to-day work operations. Other responses (judgments and complaints) can be useful in the design or redesign of individual work spaces or the larger work environment. Still others (e.g., opinions) can be used to appraise the overall effectiveness—or, perhaps more commonly, the ineffectiveness—of an organization's performance. In this latter case the situation may be analogous to that of the physician's taking the pulse of an ill patient. Something is judged (by management) to be wrong with the organization, and the finger is pointed at its human components. One way to find out what is wrong, then, is to conduct a survey of the "people part" of the organization, assessing feelings, attitudes, complaints, and dissatisfactions. In short, subjective responses can be regarded as "output measures" or indexes of the efficiency of an enterprise.

At this point it may be useful to interject a comment on *objective* measures, as distinct from *subjective*. In the sense that the industrial engineer commonly is most concerned with human *productivity*, as opposed to *individual* human welfare and performances per se, he or she is more likely to focus attention on objective indexes of systems performance. For example, in the industrial plant the engineer will be concerned with units produced, items rejected by quality control, defective products returned by vendors and consumers, downtime, scrap rate, tool breakage, and the like. Such examples fall within the definition of "objective": that which can be known, that which is unbiased, that which is free from personal feelings. The contrast here with subjective responses, insofar as these do involve personal feelings, presents a challenge for the typical industrial engineer in that he or she is not typically educated in terms of those techniques used to acquire subjective response data. In dealing with humans, one must bear in mind that they are (at times) emotional, irrational, and unpredictable beings. Such concerns fall more within the province of the industrial psychologist than within that of the industrial engineer.

Human subjective responses have long been known to be at the core of such "hidden" costs to production as absenteeism, labor turnover, accidents, grievances, and industrial sabotage. During the 1970s concern for the "well-being" of the industrial worker was elevated in importance by the passage of OSHA and by the emerging interest in what became known as the "quality of working life." More than ever, industrial engineers must be concerned with worker well-being. This view has two implications: (1) the industrial engineer should work closely with industrial psychologists and ergonomists toward the goal of performance improvement, and, in order to do this, (2) he or she will have to become familiar with the use of subjective response data.

The Problem of Individual Differences

One of the basic points made in the study of human psychology is that people are different. We are different in numerous ways–genetic background, personality, abilities, interests, values, motivation, and emotional responsiveness. Superimposed upon this matrix of differences are other variables such as age, sex, body size and shape, and physical condition. When it comes down to predicting the performance of an individual employee, no wonder, then, that the task is so difficult!

Consider the case of a simple model or equation–that Performance = Ability × Motivation. Although we may be able to measure an individual's ability level at some point with reasonable success, our ability to measure motivation often must be qualified for specific instances for the same individual; that is, motivation is a much more elusive, variable, and unstable entity–one that is commonly reflected by the kinds of subjective responses discussed in this chapter. The following two examples are presented to illustrate the role of individual differences (and of subjective responses) as related to human performance.

First, take the case of an autonomous work group. The goal of this approach to productivity is skill development among individuals within the group to the point that tasks can be interchanged among members on some periodic basis (e.g., daily or within a day)–ostensibly to deal with the problem of industrial boredom. But, at the start of each new "shift," decisions have to be made as to who will do what. This requires discussion. It would be nice if we could assume that all workers enjoy and/or are motivated to participate in such group discussions and decisions, but this is not the case. Very simply, there are those workers who could not care less. Furthermore, there is slim hope that one can motivate them to change. So this is one reason why autonomous work groups do not always operate effectively–or at least why some individuals simply do not belong in them. It also constitutes a rationale for assessing attitudes toward the formation of, or participation in, such work groups.

Now let us consider the design of communication channels (networks) within an organization.[1] Some theorists might argue for egalitarian, multichannel networks among individuals composing a group (division, branch) within an organization. True, some individuals (especially those in positions of power within the hierarchy) thrive on being the focus, or center, of attention, that is, of communication. But remember that there also may be some who do not share such joy; they would just as soon be left alone to do their work because interpersonal communication is something they choose to avoid. Again, one cannot effectively motivate such people to communicate. Equally obvious is the importance of identifying such individuals for proper placement within the organization. To the extent that such people are "misplaced" and may contribute to the inefficient operation of communication channels, subjective response data (e.g., opinions and complaints) may be of help in identifying such individual differences in human motivation as they relate to work performance.

5.2 SUBJECTIVE CORRELATES OF WORK PERFORMANCE

Industrial engineers, psychologists, and ergonomists share a common interest in attempts to understand human performance. Clearly there are both physiological and psychological by-products of work that must have some bearing on the future course of work. Numerous questions arise. Why is the performance of worker A not up to par today? Why does worker B show a decrement in his work efficiency over time? Why is it that worker C can sustain her effort, while worker D cannot, and indeed while worker E even improves his efficiency over the workday? Are rest periods necessary? When should they be scheduled? How is it that Joe can complain about being tired on the job and then go bowling after work? Why is it that, after feeling so sluggish at work, one evening I doze off while reading the paper, whereas the next evening I feel quite refreshed after mowing the lawn?

There are answers to these questions. We do have an understanding of the determinants and underlying processes of human performance. The problem is that we do not have a simple, direct measure that enables us to gauge an individual's efficiency at any point. It would be nice to have a probe that we could attach to a worker, and a meter that would read his or her efficiency on some scale from 0 to 10. But no such technique exists. Approaches involving the use of physiological measures have been the subject of much research, but results are disappointing with regard to correlates of work.* The physiological processes that underlie performance are not readily accessible for measurement, but even if they were, recording them in an individual worker on a routine basis remains a practical problem. Thus subjective (psychological) measures continue to be a common choice for evaluating human performance. The challenge then is to "objectify" the use of subjective, or self-report, techniques as discussed in this section.

Occupational Stress and Strain

The word "stress" has enjoyed popular use for many years. Such terms as environmental stress, work stress, emotional stress, mental stress, psychological stress, speed stress, load stress, physiological stress, physical stress, life stress, occupational stress, aircrew stress, and combat stress—to name a few—are known to all of us. But do we really know what stress is? Unfortunately, scientific concern with the subject has been accompanied by a variety of definitions and conceptions of the term.

Nevertheless, it is probably safe to say that most of us have a reasonable appreciation for the term, as when we admit we are "under stress." We recognize that we are in some undesirable or uncomfortable situation that disturbs or interferes with our normal functioning. Typically we are aware of pressure, threats, conflicts, worries, or other noxious elements of the environment that make us "uptight." We feel anxious and complain of being tense. In this context we recognize "stress" as being the *cause* of our problems, and our complaints, both the physically and the psychologically based, as the *effects*.

Recent Studies on Stress

One of the prominent researchers in the field, Hans Selye, has defined stress in terms of its effects. He further defines the cause of stress to be "stressors." Selye's work,[2] which focuses on a syndrome of physiological responses to a stressor, has had a profound effect on research in this field. Selye asserts that all stress is not necessarily bad; this recognizes that some people appear to operate more effectively under certain levels of stress. In recognizing individual differences in response to stress, Selye admonishes each of us to seek our optimum "stress level." A shortcoming of this view, however, has been the problem of defining when stress is "good" or "bad."

A different conception that has emerged in recent years defines stress in terms of an interaction between the person and a situation; stress here is a cognitive perception—something *perceived* by the individual in a specific situation. The *consequences* of stress then are defined as "strain." Such definitions are clearly more palatable to engineers, who are used to defining stress as a load or system of forces that act upon an object to produce a resulting strain in that object (see also Singleton[3]).

It is normal for individuals to respond to stressful situations with some activation of the autonomic nervous system, which ultimately is reflected in endocrine gland activity (e.g., increased adrenalin output), increases in heart and respiratory rates, and decreased digestive activity. If the situation persists, it also is normal to experience some muscular tension and anxiety. When the situation persists, and *coping* with it is inadequate, the symptoms of strain begin to emerge—increased tension, upset stomach, headaches, backaches, depression, insomnia, and possibly ulcers and hypertension (high blood pressure). Thus it is the relative effectiveness of coping responses from one person to another that accounts, in part, for individual differences in reacting to stress. But bear in mind the necessary precondition: The individual must *perceive* the situation as stressful in the first place; if the situation is not perceived as such, then stress does not exist for that person.

Changes in one's lifestyle, whether pleasant or not, require adjustment. Major changes have been related in research studies to higher incidences of emotional disorders, physical symptoms, and illness. As measured by a Life Change Scale, relationships have been found between illness and the prior occurrence within 1 year of stressful events such as death of a spouse, divorce, death of family members, pregnancy, loss of job, change of job, retirement, and marriage—to name a few.[4]

Exhibit 5.1.1 portrays the relationships among a number of measures commonly involved in studies of occupational stress, strain, and illness. The figure is based on the definitive study, *Job Demands and Worker Health*, conducted by Caplan and his associates at the University of Michigan. It should be noted that some of the measures can be determined objectively, whereas others must be derived from subjective response data. Consult their publication[5] for specific details on construction of the several indexes developed for this research. Of special interest to the reader of this chapter should be those specific scales that were developed to assess the subjective strain responses, for example, Job Dissatisfaction, Boredom, Somatic Complaints, Anxiety, Depression, and

*Exceptions are to be noted in the case of heavy work involving energy expenditure, that is, high levels of muscular activity.

Exhibit 5.1.1 Relationships Among Occupational Stress, Strain, and Illness[a]

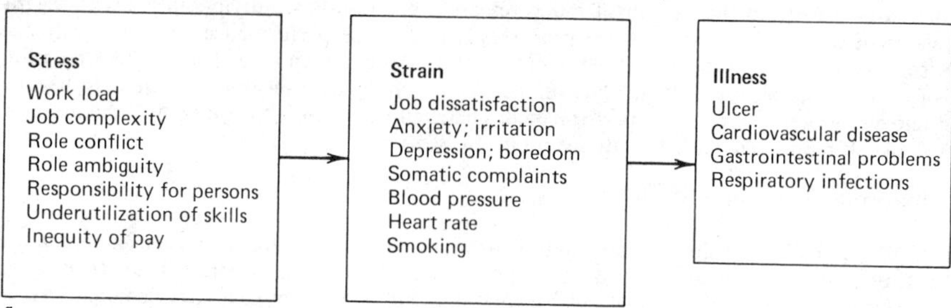

[a]After Caplan et al., reference 5.

Irritation. Although details on the techniques used to develop such scales are beyond the scope of this handbook, it is important to know that such scales exist. (For scale development methodology, see Nunnally.[6])

Another recent study, conducted by some British researchers,[7] was concerned with the stress associated with repetitive work. The subjects were more than 500 shop floor workers engaged in repetitive work in engineering firms and a pharmaceutical company. The nature of the work was mainly unskilled assembly, machine operation, inspection, or packing. A large majority of the sample were women. A job description checklist was developed to assess workers' perceptions of their work. It contained a list of 55 adjectives, each accompanied by the following scale:

Always	4
Often	3
Sometimes	2
Rarely	1
Never	0
Cannot Decide	

Using the checklist, researchers asked workers to describe how they viewed their jobs. Data from the survey were analyzed by factor analysis techniques, which identified four major factors. The factors as labeled by the investigators, and the typical variables associated with each factor, are as follows:

1. **Pleasantness.** Exciting, satisfying, great, enjoyable, fun.
2. **Tedium.** Pointless, dreary, dull, boring, worthwhile, varied.
3. **Pressure.** Fast, tiring, demanding, pressured, slow.
4. **Difficulty.** Difficult, complicated, worrying, easy, simple, easy to cope with.

Some degree of bipolarity is to be noted with regard to factor composition. The four factors identified in this study provide important insight to the types of stress associated with repetitive work as perceived by workers.

The types of strain measures associated with stress research are representative of one type of approach to the subject of this section. The challenges inherent in working with measures of stress and strain should be apparent to the reader and should serve as a caution that rigorous assessment of subjective responses is not all that simple. An extended discussion of the subject of stress appears in Chapter 6.6.

Fatigue

The term "fatigue" is much maligned and misused. Indeed, this author debated even whether to use it as a heading here. There are several objections to the use of this ambiguous term. Some writers equate fatigue with task duration, so that a "long" task necessarily becomes "fatiguing." If the average performance of a group of subjects is observed to show a decrement over time, then the subjects must have become "fatigued." Obviously, if we want a fatiguing task for research purposes, we make it a long one. Note both the circularity and the absurdity of reasoning in these statements. In some cases fatigue is implied; it is sometimes seen to be the cause (of poor performance), whereas at other times it is considered the effect (of a long task). Few researchers have taken the trouble to examine both (1) the *performance* of individual subjects at a particular time and (2) how those individuals *feel* at that time. Briefly, the relationship between how an individual says he or she feels at a particular time (call this "subjective fatigue" if you will) and his or her level of task performance may not reveal a high positive correlation. Further, one's subjective state is not necessarily predictive of future performance. In short, measures of subjective fatigue cannot be taken as either correlates or predictors of performance.

As discussed here, fatigue is a psychological term—a *feeling* reflecting one's disposition toward the task at hand.* In essence, it is a reflection of task aversion, as exemplified by the statement "I'd rather be doing something else." How else can one explain the fact that a "fatigued" worker can run to his or her car when the whistle blows, ending the workday?

It thus becomes necessary when dealing with this topic to make the following specific distinctions:

1. **Fatigue.** Subjective feelings of tiredness.
2. **Work Decrement.** Decline in task performance or efficiency.
3. **Impairment.** Compromises of physical ability (i.e., an incapacity) to perform the task—for example, finger cramps, changes in vision or hearing.

In this context the industrial engineer's prime concern should be with problems of work decrement and their mitigation. Second, the industrial engineer should be aware that impairments of visual, auditory, and motor (muscular) abilities may be causative factors in inefficiency, or in accidents. Finally, complaints of fatigue should not be ignored, because these *may be* symptoms of job dissatisfaction or a lack of motivation (e.g., the task lacks challenge).

In summary, fatigue is a subjective thing—a feeling—and a symptom. It is part of a dimension that extends from feelings of exhaustion and weariness on one end to those of freshness and peppiness on the other. It is not to be confused with other affective dimensions, such as drowsiness (sleepiness), bordeom, and drive (e.g., feeling motivated). To elaborate further: (1) A tired person may have difficulty going to sleep, yet a person may doze off (e.g., during a boring speech) even though he or she does not feel tired; (2) a "bored" person may not necessarily feel "tired"—and vice versa; and (3) a "tired" person may be motivated to work longer, especially if the rewards are high. Finally, it is interesting to note that an effective antidote to feelings of fatigue in many instances is work! That is, the "tired" executive may actually be refreshed by a game of tennis at the end of the workday.

An example of a Feeling Tone Checklist used to assess an individual's level of subjective fatigue is given in Exhibit 5.1.2. The scale was developed utilizing psychophysical and scaling methodologies.[8] A score can be obtained by weighting responses as follows: "better than"– 2; "same as"–1; "worse than"–0. In effect, then, scores can range from 0 to 20. In research conducted by the author and his students, the checklist has been shown to be sensitive to such things as task difficulty, work load and work rest schedules, stimulant and depressant drugs, alcohol, prolonged task performance, and circadian rhythms. An effective critique of this scale, other "fatigue" scales, and the topic of subjective symptomatology is presented by Kinsman and Weiser.[9] Although somewhat dated, the book by Bartley and Chute is perhaps still the best critique and review of the subject of fatigue and related mood states.[10] For another viewpoint, see the work of Grandjean.[11]

To close this section on a positive note, insofar as subjective measures are suggestive of task dissatisfaction or industrial boredom, we do know that much can be done to change the picture for the better. Increasing the variety of sensory inputs (stimulation) through job rotation, enlargement, or enrichment may help. Involving people in decision making through participative management is another approach. Finally, the ergonomist can do much through task design to make the workplace

Exhibit 5.1.2 Feeling Tone Checklist

No.	Better Than	Same As	Worse Than	Statement
1	( )	( )	( )	Very lively
2	( )	( )	( )	Extremely tired
3	( )	( )	( )	Quite fresh
4	( )	( )	( )	Slightly pooped
5	( )	( )	( )	Extremely peppy
6	( )	( )	( )	Somewhat fresh
7	( )	( )	( )	Petered out
8	( )	( )	( )	Very refreshed
9	( )	( )	( )	Fairly well pooped
10	( )	( )	( )	Ready to drop

*Subjective fatigue here is distinguished from muscular (physiological) fatigue, which can be measured objectively (see Chapter 6.4).

a more exciting place in which to perform. Such changes, which should be evaluated in terms of costs and trade-offs relative to expected gains in performance, are not discussed in detail here; the reader should refer to their treatment in Section 2 of this handbook.

Other Mood States

Inherent in the preceding discussion is the fact that there is a multiplicity of mood states that can be subjectively referenced by an individual at some particular time. In short, mood is a multidimensional concept, so that, when one is asked how he or she "feels," the responses may tap any number of dimensions, such as bored, tense, anxious, depressed, tired, and irritable.

Numerous efforts have been devoted to the development of mood or affective checklists involving many items (and presumed dimensions). One major use of such instruments is to determine the side effects of drugs; another is to assess human subjective responses to environmental extremes. The Japanese, among others, have developed multidimensional instruments for use in assessing industrial tasks; other instruments (e.g., the Physical Activity Questionnaire) have been developed for laboratory studies of physically demanding tasks.[9,12] Psychometric techniques (factor analysis and cluster analysis) have been applied to data involving some of these instruments in order to delimit the underlying dimensions or unique mood states that they assess. One hopes that this type of research may one day provide us with a clearer understanding of the complex dimensions of subjective response as related to work, enabling researchers then to explore relationships between such dimensions and variations in human performance.

Effort

For any physically demanding task—one that requires considerable energy expenditure—it is desirable to determine the amount of effort or exertion involved, that is, the energy cost to the individual. Such data are useful for a number of purposes: (1) for preemployment screening to select those who can perform the task and those who cannot, or who should not, for reasons of potential self-injury; (2) for task design involving the location of work surfaces and the selection of proper aids for more efficient manual material handling; and (3) for proper scheduling of work-rest cycles. Commonly, physiological measures of cardiovascular, respiratory, and muscular activity are involved in such determinations—for example, the electrocardiogram (EKG) or heart rate, oxygen consumption, or the electromyogram.

Considerable data exist on energy expenditure (expressed in terms of either kilocalories per minute or oxygen consumption in liters per minute) involving different work activities, working postures, work pace, and task demands (e.g., load handled). Formulas exist for relating energy cost and total working time to requirements for rest periods. Such data can provide guidance to the industrial engineer concerned with physically demanding work activity. (See Chapter 6.4 for expanded treatment of this subject.) If they are inadequate for particular purposes, however, the industrial engineer should recognize that the collection of data appropriate for his or her own needs requires instrumentation of workers (subjects), relatively expensive equipment, and professional assistance in data interpretation. Because of these requirements, it is often not feasible or practical to use this approach in the industrial situation.

An alternative approach involves the use of a rating scale of perceived effort. One such scale, often referred to as the Rating of Perceived Effort (RPE) Scale, or Borg Scale (after its developer), is shown in Exhibit 5.1.3. This 15-point scale has been developed so that the heart rate of a normal, healthy middle-aged man can be predicted if the RPE value is multiplied by 10. Physical working capacity can be defined in terms of the load level yielding a heart rate of 170 beats/min; the corresponding RPE score for this load level has been empirically determined to be 16.5. In practice, each of the scale points is deemed to be equivalent to 10 heart beats/min. Thus a rating of 9 on the RPE Scale (very light) would correspond roughly to a heart rate of 90 beats/min.

A wide range of data has been accumulated using the RPE Scale. (For details and critique, see Kinsman and Weiser[9] and Borg.[13]) Studies have involved such tasks as bicycling with various loads, repetitive lifting, pushing a wheelbarrow with various loads, walking, and running. While the results of research with the RPE Scale support its use as a valid and reliable tool for assessing effort and also confirm its high correlation with heart rate as an index of strain, Ulmer and his associates present data that show perceived exertion to depend considerably more upon stress (the load) than on strain.[14] In assessing the relationships among stress, strain, and perceived exertion shown in Exhibit 5.1.3, they conclude that the influence of stress on RPE scores predominates over the influence of strain (as measured by heart rate). Indeed, there is evidence that under some conditions the relationship between heart rate and RPE does not hold up. Others also have raised questions regarding the nature of the physical symptoms and sensory cues that are experienced by the person and subjectively reflected in the RPE Scale. Notwithstanding such concerns, the utility of the scale for assessing effort in physically demanding work remains apparent.

Exhibit 5.1.3 Relationships Among Stress, Strain, and Borg's RPE Scale.

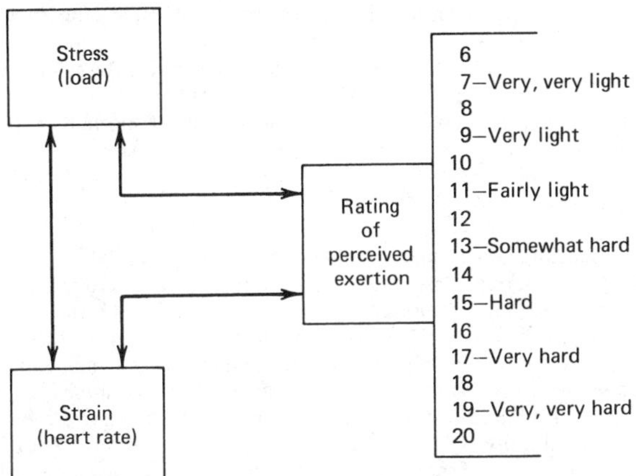

Comfort

Comfort is most commonly an issue in environments with adverse conditions, such as heat, acceleration, and vibration. Concerns typically range from comfort at one extreme, through varying degrees of discomfort, to pain at the other extreme. Various point rating scales have been used by researchers to measure subjective responses. Commonly, these scales are labeled with such terms as "noticeable but not objectionable," "barely acceptable," "tolerable," "intolerable," and "unbearable" to guide the rater. For further discussion of comfort as related to environmental stress, the reader should refer to the chapters on noise and vibration (6.10), illumination (6.11), and climate (6.12). Kinsman and Weiser also review research on pain and on thermal discomfort.[9]

The assessment of comfort also is important in cases where task conditions and workplace design lead to complaints of postural discomfort. An important aspect of workplace comfort—sitting posture—has been the subject of a symposium. The proceedings of this symposium contain a chapter by Shackel and his colleagues that describes various procedures used to assess chair comfort.[15] In their work, which evaluated 10 chair designs, they used the following four subjective measures.

1. **General Comfort Rating.** An 11-statement rating scale was used, including such phrases as "completely relaxed," "quite comfortable," "cramped," and "sore and tender," each preceded by the words "I feel."
2. **Body Area Comfort Rating.** An answer sheet included a diagram of a mannequin divided into 15 body areas. The subject was instructed to rate (rank) the 15 areas in terms of five groups of ratings, ranging from the three most comfortable areas down to the three least comfortable areas.
3. **Chair Feature Checklist.** Specific chair features each were rated on one of three possible answers; for example, seat height could be rated "too high," "correct," or "too low," and seat shape could be rated "poor," "adequate," or "good."
4. **Direct Ranking.** Here the subject sat in each chair, then successively sitting, comparing, grouping, and ordering all until a final rank order was deemed satisfactory to him or her.

Included in Shackel et al.'s chapter is a review and critique of subjective methods used in their own research as well as those used by others.

Turning now to the workplace itself, the industrial engineer should be concerned with tasks that require unusual (sometimes abnormal) postures and with workers who unconsciously assume undesirable postures, for example, flexing the upper torso over the work surface. Some form of discomfort rating technique, such as the one developed by Corlett, may be used here. In Corlett's approach[16] a diagram of the human body marked off into 12 parts (neck, shoulders, upper back, buttocks, thighs, and legs) is used by workers to indicate periodically (e.g., every half hour), during normal work activity, those body areas that are "most painful," then those "next most painful," and so on, until no further areas of discomfort are rated. According to Corlett, the "number of different groups of body parts which were identified before 'no discomfort' was reported represented the number of intensity levels of the pain experienced."[17] Thus each separately reported group of complaints is seen as being identified with a distinct difference in discomfort. This weights the most painful discomforting areas in proportion to the number of levels reported. For evaluation purposes, data may be grouped by body areas, averaged across the number of workers, and then plotted on a time scale for the work period. When changes in equipment and workplace are introduced, the rating process then can be repeated in order to compare the "new" design with the "old."

Corlett's approach represents only one of several techniques useful for assessing postural discomfort at the workplace. Refer to Chapter 6.5 for further discussion of other techniques and of the general subject of discomfort, of the effects of work load, and of methods for reducing and preventing complaints of discomfort.

Annoyance

Another subjective dimension, annoyance, is characterized by complaints that something in the workplace is disturbing, distracting, bothersome, objectionable, and so on. Annoyance complaints predominate in situations that involve noise. Research on the subject by the author[18] involved the development of an Annoyance Rating Scale, as shown in Exhibit 5.1.4. In this research individuals in groups were exposed to noise and then were asked to rate (check) their annoyance on the 25-point scale. (Values shown in the exhibit are for scoring and did not appear on the rating scale; a response scored 15 is shown.) Significant differences were found among average ratings given to six different noises, although all noises were equated in terms of peak sound pressure level. Noise from a pneumatic chipping hammer, for example, had the highest mean annoyance rating, 19.16; jet aircraft flyover noise was rated 12.48; factory noise, 10.76; a truck climbing a hill and shifting gears, 8.00. Thus it should be observed that all "loud" noises are not necessarily equally annoying. The characteristics (periodicity, intermittency) of the noise and its frequency spectrum are major factors in determining the degree of annoyance.

Another major finding of the study was the great range of individual differences among subjects in their annoyance ratings. Individual mean ratings across the six stimuli varied from a low of 1.65 to a high of 22.20. Translated, this means that some subjects were little, if at all, annoyed by the various noises, whereas others were extremely annoyed by the same group of stimuli.

In terms of its effect on people, noise clearly is the most idiosyncratic of the environmental stresses. It must be remembered that noise is a psychological attribute. Furthermore, noise should not be arbitrarily regarded as necessarily bad, that is, as "unwanted." Some people, especially those who live alone, prefer some noise over complete quiet. Then too, some noise provides the worker with necessary feedback concerning equipment operation, for example, the "noise" of a drill bit boring through metal stock or the noise from a faulty bearing. Finally, there is little in the way of

Exhibit 5.1.4 Annoyance Rating Scale

Value		Response
25	–	Unbearable and intolerable
	–	
	–	
22	–	Extremely annoying
	–	
	–	
19	–	Very annoying
	–	
	–	
16	–	Quite annoying
	–	
	–	
13	–	Annoying
	–	
	–	
10	–	Moderately annoying
	–	
	–	
7	–	Somewhat annoying
	–	
	–	
4	–	Slightly annoying
	–	
	–	
1	–	Noticeable but not objectionable

research evidence showing an adverse effect of noise on performance. Indeed, performance often is better under noise–perhaps it keeps us awake! In this latter context one might question the practice of giving "points" to noise (more points are bad) in the work environment for purposes of job evaluation. Industrial engineers take note!

Other Subjective Phenomena

Space does not permit coverage of all the scales and subjective dimensions one could cite, but let us briefly consider two more. With regard to the preceding examples, we have been concerned largely with negative issues–prevention of discomfort, fatigue, exertion, and so on. A positive goal of design is pleasantness. To the extent that a pleasant work environment, whether an office or an industrial workplace, is desired, some consideration needs to be given to the aesthetics of the situation. Although aesthetic pleasantness is an elusive, multidimensional concept, recent research holds out promise that evaluation dimensions can be developed. For relevant commentary here, the reader is referred to a discussion by Bennett.[19]

Finally, it should be noted that much research has been done on the subject of "handling" in manual control situations. Here we are concerned with the "feedback," the cues, or the impression one has from manual control operation of some equipment or vehicle (automobile, ship, or airplane). Handling, insofar as it reflects an operator's acceptance of the effectiveness (e.g., smoothness, stability, responsiveness, precision) of a control system, can be an important element to evaluate in the system design process. Although perhaps this subjective dimension is beyond the scope of this handbook, it should be noted that scales and checklists do exist for rating it.[20]

5.3 ATTITUDES AND JOB SATISFACTION

The preceding section focused on specific subjective dimensions. We now turn to a more general subject–attitudes.

Attitudes and Opinions

Both attitudes and opinions reflect subjective states of the individual. The terms can and should be differentiated.

1. Opinions are views or thoughts held by individuals on specific matters or subjects. They involve one's beliefs or judgments about objects, concepts, events, or relationships. In principle, opinions do not have an affective component; that is, they do not convey positive or negative implications.
2. Attitudes involve feelings toward situations, objects, or persons and, as such, contain an affective component; that is, they reflect pleasantness or unpleasantness.

Perhaps some examples will help clarify the distinctions. Opinions can be exemplified by response to such questions as the following: Should the department organize a bowling team? Should the company change its slogan? Should the plant have its own in-plant cafeteria? Attitudes, on the other hand, take a different form, exemplified by response to these questions: What do you think of the food in the company cafeteria? How good are the company's recreational opportunities?

Insofar as attitudes reflect the behavior, or the potential behavior, of an individual, it can be important to assess them in the industrial situation. Attitudes held by employees toward their work, co-workers, supervisors, and management may be related to their work efficiency and, ultimately, to productivity. In short, negative attitudes toward the job or company may be reflected in a lack of motivation to perform effectively. To obtain such information, various approaches may be used, for example, interviews, questionnaires, and telephone surveys; more is said about these details later.

Systematic attempts to collect opinion and attitude data usually involve some type of questionnaire survey and direct contact with individuals from the target population. Some survey questionnaires contain both attitude and opinion items. Generally, the results of attitude surveys are averaged across individuals in order to develop an index for a specific group, plant, or division of a corporation; on the other hand, the focus of opinion surveys is on the response distribution for individual items.

The Importance of Attitudes

The Hawthorne studies in the 1930s have long been considered a landmark in recognizing the importance of attitudes toward working conditions. These studies confirmed the view that workers' attitudes toward their work situation may be related to how effectively they perform their work. Literally several thousand studies involving assessment of the attitudes of industrial employees have since been reported in the literature. The vast majority of these are referred to as studies of "job satisfaction," a term used to embrace that group, or subset, of attitudes held by members of an organizational entity toward their jobs.

Thus attitude surveys, or studies of job satisfaction, commonly are undertaken when there is some concern for the effectiveness of an organization's human component. It is one way of "taking the pulse." If "sickness" is discovered, then some "treatments" are to be recommended. Major symptoms of job dissatisfaction, which may dictate the need for a study, include high rates of grievances, absenteeism, and labor turnover; accidents; widespread worker complaints and unrest; industrial sabotage; work slowdowns; and unexplainable, significant declines in productivity.

The rationale for conducting surveys of job satisfaction is that *attitudes can be changed.* Thus it is important to evaluate survey findings carefully in order to delimit those changes that can be implemented to affect attitudes in a positive direction. Sometimes management's predictions regarding problem areas are not validated by survey results; indeed, problem areas may be obscure, and findings may contain a few surprises. The author recalls the case of a company that purchased adjacent land on which it had a parking lot constructed for the convenience of company employees. Yet, employee satisfaction with this new convenience did not appear to be overwhelming. What had management done wrong? A survey found that company employees were incensed over the fact that people from other, nearby plants were using "their" lot also. The lot had not become available for their exclusive use, and because management was oblivious to this attitude, no steps had been taken to keep out the "foreigners."

Apart from the costs involved in conducting them, job satisfaction surveys generally can be expected to do no harm. There is some value in giving employees a chance to ventilate their feelings. Additionally, the show of interest by the company in the welfare of its employees can enhance its image in their minds. There also may be a "placebo" effect in that the attention given to the employees exerts a positive effect on their motivation. There are hazards, however, and these involve the expectations that employees have relative to presumed findings of the survey. Many employees will look for improvements that will better their state, whereas others anticipate some form of change. The word "change" may carry with it some apprehension and a connotation of threat to security (stability); indeed, change often is regarded as a conflict since it can imply modifications in working conditions, a new supervisor, or different co-workers.

When should a job satisfaction study be undertaken? This is a difficult question to answer apart from issues involving the major symptoms noted here. There is no question that people in management should "keep their ears to the ground" in order to detect growing job dissatisfaction. Yet, these people should try to follow some middle ground between being overly concerned and unconcerned. One should not be indifferent to one's employees, nor should one be overly sensitive to their minor grumblings either. For example, although different leadership styles may entail different levels of complaints, the ultimate "costs" to the company may not be all that different. The autocratic supervisor may have a good production record, but at the cost of "unhappy" workers, absenteeism, and turnover; on the other hand, the "democratic," human-relations-oriented supervisor, while liked by his or her workers, may have a poorer production record.

It should be noted that the degree of job satisfaction found in companies does not correlate very highly with performance. How "happy" should workers be while they are earning their wages? Perhaps job satisfaction has been overemphasized. Some good food for thought here is contained in a paper by Singleton, who observes that a job may be unpleasant, yet satisfactory to the employee because of extrinsic factors such as pay.[21] Then, too, there are factors that compound the picture and that should be considered with regard to the need to assess job satisfaction. Young workers invariably complain more than older workers. Men may complain about women performing similar jobs, whereas women may be afraid to complain to a male supervisor for fear of losing their jobs.

The Ingredients of Job Satisfaction

Although it may be obvious what kinds of attitudes are tapped in a survey of job satisfaction, it may be desirable to at least list them here. The following dimensions are among those most commonly assessed:

Work.

Pay.

Promotions.

Recognition.

Working conditions.

Benefits.

Supervision.

Co-workers.

Company and management.

In one respect this list mixes apples and oranges. As noted by Locke,[22] two different levels of analysis are involved: events or conditions (the first six elements in the list) and agents (the last three elements). The distinction has theoretical implications which are not discussed here. But it does give us reason to pause and recognize that there are many theoretical views and models of job satisfaction, including Maslow's need hierarchy theory and Herzberg's motivator-hygiene (or two-factor) theory, for example. Locke's chapter contains a good review and critique of these models.

Measurement Techniques

Attitudes and opinions can be assessed in a number of ways: through face-to-face or telephone sessions using a questionnaire format, through mail surveys, and through in-plant group administration of the questionnaire. Additionally, one has the choice of using a commercially available, or "canned," questionnaire or of developing a custom-made instrument. Whatever choices are made here, it is clear that a good knowledge of questionnaire development and survey technique is essential for the effective conducting of any attitude, opinion, or job satisfaction study. Warwick and Lininger[23] provide effective guidance on both questionnaire and survey design as well as on data collection and analysis technique. A chapter by Bouchard includes a discussion and critique of various methodological issues and a commentary on difficulties encountered and on pitfalls to be avoided.[24]

Most surveys use some form of direct verbal self-report. Formats commonly involve a list of adjectives requiring a "yes," "no," or "?" response; an odd-numbered interval scale (e.g., 1–2–3–4–5–6–7), and Likert or Thurstone-type scales. Locke[22] also notes two other approaches: the open-ended, in-depth interview (which does not involve use of a questionnaire format per se) and use of the critical incident technique, exemplified by Herzberg's research.

The Thurstone Scale involves use of a psychophysical method which "arranges" items along a continuum (usually of 7, 9, or 11 steps) from least favorable attitudes or viewpoints on the low end to most favorable on the high end. "Expert" judges are used in the process. Items in the final scale are selected on the basis of two considerations: (1) consistency of rater agreement in placing an item within the same general position on the scale; and (2) adequate coverage of each point on the scale. The average (mean or median) rating given to a selected item by the judges is used to define its "scale value." When the survey is used, each respondent checks those items with which he or she agrees. The score, then, is the average value of all items checked.

In the Likert approach each attitude statement involved in the questionnaire is followed by five choices, as follows:

5–Strongly agree
4–Agree
3–Undecided
2–Disagree
1–Strongly disagree

The score for an individual is computed by summing the numerical values of item responses selected.

Some test publishers, consulting firms, and individuals market for commercial sale standardized survey instruments. Among them are:

Job Diagnostic Survey
The Job Description Index
The Brayfield-Rothe Scale of Job Satisfaction
The Science Research Associates (SRA) Attitude Survey

As an example, the SRA Attitude Survey contains a standardized core of 78 items. These may be supplemented by additional items, such as 31 items exclusively for supervisors or others that can be custom-selected from a pool of 21 items to meet unique organizational concerns. Another component of the survey, "anonymous comments," permits employees to express in writing opinions on certain topics.

Some surveys may use an "unstructured," or open-ended, approach in which the respondent replies orally to the interviewer's questions. Comments can be noted by the interviewer or tape-recorded for later analysis. In such cases there is a need to use content analysis procedures to categorize responses for all respondents. Briefly, this involves generating classifications of key phrases, such as "inadequate pay," "poor communication among management," "too much job pressure," and "confusing policy," and then tallying respondents' comments that fit them.

Exhibit 5.1.5 offers the reader some examples of a variety of such item formats. Although the items here solicit *opinion* responses taken from a safety survey conducted by the author, they

Exhibit 5.1.5 Examples of Opinion Survey Items Selected From a Safety Survey

Item	Question	Response Choices
3d.	How easy is the safety manual to understand?	Extremely easy–quite easy–somewhat easy–quite hard–extremely hard
4a.	How do you rate training program coverage of safety topics?	Excellent–good–fair–poor
4b.	Should the training program devote more time to safety topics?	Yes–no
4e.	How much do you learn about safety while on the job?	A lot–some–a little–none
8i.	How often do you provide feedback on safety conditions to your manager?	Always–usually–sometimes–rarely–never
10c.	How much time is devoted to the topic of safety at safety meetings?	Too much–just right–too little
10d.	How effective are these meetings?	Extremely–very–somewhat–not effective
14e.	On a scale from 1 to 7 (where 1 = never and 7 = always), how often do you have concern for your own personal safety on the job?	1–2–3–4–5–6–7

should suffice for illustrative purposes. In this example the questions and response options were read by an interviewer, who then recorded the worker's responses.

5.4 CONCLUDING REMARKS

Performance Evaluation

Informal judgment of the performance of one person by another is an everyday experience. In industry it is common practice to "objectify" such subjective responses periodically through use of some type of formal, systematic procedure. Such procedures carry a number of different names, including performance evaluation, merit rating, personnel evaluation, and performance appraisal. McCormick and Ilgen[25] describe a number of techniques, including rating scales, rank order and paired-comparison systems, the critical incident technique, and behavioral checklists and scales. Such techniques are used in a number of ways: in assessing personnel development, in determining training needs, in determining wages and salaries, in determining promotions, and as criteria for research evaluations. Since these techniques involve the use of subjective judgments, many of the same approaches to scale development and use that have been noted in this chapter also apply to performance evaluation. Indeed, in the larger context of organizational performance, performance evaluations and attitude and opinion surveys related to company practice may be viewed as complementary processes insofar as they both assess dimensions of the system's effectiveness.

Since the subject of performance evaluation is treated in detail in Chapter 5.4, it is not elaborated on here.

Some Cautions

Not addressed in this chapter are a number of issues that relate to the development and administration of scales and questionnaires. These include such considerations as rater bias, the halo effect, appraisal fear, weighting of responses, validity, and reliability. With regard to assessing subjective responses, there are many "good" practices that need to be observed if one is to obtain quality data (see McCormick and Ilgen for discussion[25]). If you are not familiar with these issues and practices, seek out someone who is.

When Help is Needed

In the event that the industrial engineer or members of the industrial engineering department lack expertise in the area of subjective response measurement (and this is likely to be the case), help will be needed. In large corporations the personnel department should be able to help, especially if its staff includes a qualified industrial or organizational psychologist. In any case, liaison with this department should be pursued; indeed, the department may want to assume (or may insist on assuming) responsibility for the study.

This does not mean that either the personnel or the industrial engineering department will necessarily *conduct* the study. Some studies of the subjective aspects of work as discussed in this chapter (e.g., stress, fatigue, effort, comfort, and annoyance) can be effectively conducted by in-house personnel. It must be recognized, however, that in the area of work attitudes and job satisfaction employees are reluctant to speak freely about such topics as supervisory relations, management effectiveness, and company policy. They fear that their remarks may become available to their supervisors. In this case employees simply are not going to provide valid data. Invariably, then, studies can be conducted only by personnel outside the company when such issues as gaining the trust of employees and ensuring the confidentiality of data are at stake. Sources of help that can be considered include:

1. Consulting firms with competent professionals in the industrial/organizational, ergonomics, and/or personnel evaluation areas.
2. Test publishers–a few of the larger corporations have their own staffs which can provide these services.
3. University industrial extension services–in some cases.
4. Individual university professors.

Since it is not the author's role to recommend specific sources, the industrial engineer is well advised to check out the credentials of any consultant or firm he or she considers. In the case of individual psychologists, it should be noted that most states today require some type of certification or licensing for professional practice. Finally, if the industrial engineer has no experience in working with consultants, he or she should refer to the guide edited by Kubr.[26]

With regard to the commercial availability of standardized scales and job satisfaction surveys, such as the SRA Attitude Survey mentioned previously, the reader is referred to Appendix D of McCormick and Ilgen's book,[27] which lists addresses for a number of sources.

REFERENCES

1. H. J. LEAVITT, *Managerial Psychology*, 4th ed., University of Chicago Press, Chicago, 1978.
2. H. SELYE, *The Stress of Life*, rev. ed., McGraw-Hill, New York, 1978.
3. W. T. SINGLETON, "The Measurement of Man at Work With Particular Reference to Arousal," in W. T. Singleton, J. G. Fox, and D. Whitfield, Eds., *Measurement of Man at Work*, Taylor & Francis, London, 1971.
4. T. H. HOLMES and R. H. RAHE, The social readjustment rating scale. *Journal of Psychosomatic Research*, Vol. 11, 1967, pp. 213–218.
5. R. D. CAPLAN, S. COBB, J. R. P. FRENCH, Jr., R. VAN HARRISON, and S. R. PINNEAU, Jr., *Job Demands and Worker Health*, NIOSH Publication No. 75-160, National Institute for Occupational Safety and Health, DHEW, Washington, DC, 1975.
6. J. C. NUNNALLY, *Psychometric Theory*, 2nd ed., McGraw-Hill, New York, 1978.
7. T. COX and C. J. MACKAY, "The Impact of Repetitive Work," in R. G. Sell and P. Shipley, Eds., *Satisfactions in Work Design: Ergonomics and Other Approaches*, Taylor & Francis, London, 1979.
8. R. G. PEARSON, "Scale Analysis of a Fatigue Checklist," *Journal of Applied Psychology*, Vol. 41, 1957, pp. 186–191.
9. R. A. KINSMAN and P. C. WEISER, "Subjective Symptomatology During Work and Fatigue," in E. Simonsen and P. C. Weiser, Eds., *Psychological Aspects and Physiological Correlates of Work and Fatigue*, Charles C Thomas, Springfield, IL, 1976.
10. S. H. BARTLEY and E. CHUTE, *Fatigue and Impairment in Man*, McGraw-Hill, New York, 1947.
11. E. GRANDJEAN, *Fitting the Task to the Man*, 3rd ed., Taylor & Francis, London, 1980.
12. K. HASHIMOTO, K. KOGI, and E. GRANDJEAN, Eds., *Methodology in Human Fatigue Assessment*, Taylor & Francis, London, 1971.
13. G. BORG, Ed., *Physical Work and Effort*, Pergamon Press, Oxford, England, 1977.
14. H. V. ULMER, U. JANZ, and H. LOLLGEN, "Aspects of the Validity of Borg's Scale. Is It Measuring Stress or Strain?," in G. Borg, Ed., *Physical Work and Effort*, Pergamon Press, Oxford, England, 1977.
15. B. SHACKEL, K. D. CHIDSEY, and P. SHIPLEY, "The Assessment of Chair Comfort," in E. Grandjean, Ed., *Sitting Posture*, Taylor & Francis, London, 1976.
16. E. N. CORLETT and R. P. BISHOP, "A Technique for Assessing Postural Discomfort," *Ergonomics*, Vol. 19, 1976, pp. 175–182.
17. CORLETT and BISHOP, "Assessing Postural Discomfort," p. 179.

18. R. G. PEARSON, F. D. HART, and J. F. O'BRIEN, *Individual Differences in Human Annoyance Response to Noise*, CR-14491, National Aeronautics and Space Administration (NASA) Langley Research Center, Hampton, VA, July 1974.
19. C. BENNETT, *Spaces for People: Human Factors in Design*, Prentice-Hall, Englewood Cliffs, NJ, 1977, Chap. 2.
20. L. R. YOUNG, "Human Control Capabilities," in J. F. Parker, Jr., and V. R. West, Eds., *Bioastronautics Data Book*, 2nd ed., NASA SP-3006, Washington, DC, 1973.
21. W. T. SINGLETON, "Some Conceptual and Operational Doubts About Job Satisfaction," in R. G. Sell and P. Shipley, Eds., *Satisfactions in Work Design: Ergonomics and Other Approaches*, Taylor & Francis, London, 1979.
22. E. A. LOCKE, "The Nature and Causes of Job Satisfaction," in M. D. Dunnette, Ed., *Handbook of Industrial and Organizational Psychology*, Rand McNally, Chicago, 1976.
23. D. P. WARWICK and C. A. LININGER, *The Sample Survey: Theory and Practice*, McGraw-Hill, New York, 1975.
24. T. J. BOUCHARD, JR., "Field Research Methods: Interviewing, Questionnaires, Participant Observation, Systematic Observation, Unobtrusive Measures," in M. D. Dunnette, Ed., *Handbook of Industrial and Organizational Psychology*, Rand McNally, Chicago, 1976.
25. E. J. McCORMICK and D. ILGEN, *Industrial Psychology*, 7th ed., Prentice-Hall, Englewood Cliffs, NJ, 1980, pp. 63–99.
26. M. KUBR, Ed., *Management Consulting: A Guide to the Profession*, International Labour Office, Geneva, Switzerland, 1976.
27. McCORMICK and ILGEN, *Industrial Psychology*, pp. 63–99.

CHAPTER 5.2

Selection and Training of Personnel

WALTER C. BORMAN and **NORMAN G. PETERSON**

Personnel Decisions Research Institute

5.2.1 INTRODUCTION

This chapter describes the proper role of personnel selection and training in establishing and maintaining an effective organization. It also provides practical guidelines for successfully utilizing selection and training methods in organizations. Several different types of personnel tests and other employee selection devices are reviewed along with their "track records" in helping to identify persons who prove to be effective in jobs. Also reviewed are strategies for evaluating the validity and practical utility of selection procedures. Finally, the chapter discusses training approaches and methods for evaluating their effectiveness.

5.2.2 PERSONNEL SELECTION STRATEGIES: MATCHING PERSONS AND JOBS

In organizations people get placed in jobs one way or another. The proper role of personnel selection is to identify persons most likely to succeed in those jobs, with the overall purpose of enhancing organizational effectiveness.

A question that might be asked at the outset is this: If good selection decisions are made over time in an organization, will it make a meaningful difference? That is, can we expect worthwhile increments in worker productivity with improved matching of persons and jobs? Recent analyses suggest that the answer to these questions is an unequivocal yes. Schmidt and his colleagues[1,2] have estimated dollar increases in productivity to be expected with increased accuracy in estimating individuals' potential for success on jobs. For the U.S. economy, these estimates range from approximately $25 billion to well over $100 billion annually, even when relatively modest increases in accuracy are assumed! Schmidt et al.[2] have also made such dollar productivity gain estimates for a single job: computer programmer in the U.S. government. Assuming, again, modest improvement in selection accuracy, they estimate productivity gains of $2.6 million to $5 million/year, or $4175 to $8350 per employee/year. Schmidt et al. conclude

> *It* does *make a difference–an important practical difference–how people are selected. We conclude that the implications of valid selection procedures for work force productivity are much greater than most of us (personnel psychologists) have realized in the past.*[3]

Later in the chapter we discuss in more detail the practical economic utility of personnel selection procedures and the impact of various factors on utility; the point made here, however, is that substantial gains in productivity can be realized with increased accuracy in identifying persons who are well qualified for jobs. Thus improved personnel selection is in general a highly worthwhile goal.

Overview of the Personnel Selection Process

Let us now consider what is actually involved in matching persons and jobs. The proper first step is to analyze the target job, using one of the several effective job analysis methods. (See Chapter 2.4 or Salvendy and Seymour[4] for detailed discussions of job analysis.) A thorough job or task analysis identifies the critical performance requirements of the job and suggests the kinds of knowledge, skills, abilities, and other personal attributes (KSAOs) that might be important to effective performance on the job–that is, what we should be looking for in job candidates. These important

KSAOs in turn suggest the tests or other kinds of selection devices that are likely to be good measures of the target personal characteristics. The tests or other measures are administered to job candidates, and the scores obtained are used to aid in selecting persons to fill that job.

The next critical step in proper personnel selection practice is to evaluate the validity of these tests or other selection devices, that is, the accuracy with which they are *in fact* identifying persons who prove to be effective performers on the job. Test validation results can then be used to modify the selection procedures as necessary.

The process described briefly here must always be placed in the context of certain practical constraints, including economic utility considerations. For example, the availability of candidates for the target job, the job's level of difficulty, the importance of performing competently on the job (i.e., costs of failure and benefits of successful job performance), and the organization's stance regarding equal employment and affirmative action are all factors that should be considered carefully in developing and implementing personnel selection programs. The next sections describe personnel selection and test validation technology and then introduce utility concepts that are extremely important in evaluating the economic gain and/or increases in productivity related to the use of personnel selection procedures.

Use of Tests in Selection

The choosing and/or development of predictor measures (i.e., tests or other devices to measure differences among people) in personnel selection should be done systematically and on the basis of job analysis information. Recent equal employment guidelines[5] and court decisions (e.g., *Griggs* v. *Duke Power Company*[6]) point to the need for tests to be *job related;* basing predictor choice or development on job analysis data is the way to accomplish this.

How might we go about selecting predictor measures based on job analysis information? The goal is to link together job performance requirements and person requirements (i.e., KSAOs important for effective job performance). Once these links are established, and once certain personal characteristics have been identified as important for success on the target job, we can select or develop predictors to measure each important KSAO. Methods exist to discover these links, as shown in the examples that follow.

Bownas and Heckman[7] conducted a nationwide job analysis of the position of entry level fire fighter. Based upon interviews with 124 fire fighters and their supervisors, Bownas and Heckman identified 204 tasks that reflect all job activities engaged in by fire fighters. Ninety-three fire fighters then completed a checklist developed to gather information about the importance of each of these tasks to carrying out fire fighter job duties. Results showed that 120 of the original 204 tasks were sufficiently important to warrant further consideration.

Next, the 120 tasks were grouped into homogeneous clusters according to fire fighters' ratings of the similarities in job content between tasks. (For example, two tasks related to emergency care of fire casualties would likely be rated as similar.) Specifically, mean similarity ratings for all pairs of the 120 tasks were submitted to a Ward-Hook hierarchical cluster analysis,[8] resulting in 17 interpretable task clusters (e.g., salvage and overhaul; apparatus operation; emergency care). These represent, in summary form, the important task requirements of the fire fighter job. Exhibit 5.2.1 presents one of the clusters—rescue—along with several of the tasks that were grouped under it.

The next step in this work was to identify KSAOs likely to be important to success as a fire fighter. Twenty such attributes were identified and carefully defined. Examples are physical strength, coordination, responsibility, and manual dexterity. Bownas and Heckman then established the critical link between the task and personal attribute domains by asking persons knowledgeable about the fire fighter job to rate the importance of each of the 20 attributes to performing effectively on each task dimension. Interjudge agreement was very high for these importance ratings, and therefore the mean ratings were used to select target KSAOs (those with high importance ratings) to measure with personnel tests or other selection devices. This is a good example of the systematic identification of KSAOs based upon job analysis information. Tests could then be selected or developed to measure the target KSAOs, and once the validity (i.e., accuracy of predic-

Exhibit 5.2.1 Example Tasks of One Task Cluster—Rescue—for the Job of Fire Fighter

Move heavy objects or materials to gain access to or to free trapped victims.

Locate and dig to free victims trapped or unconscious in tunnels, pipes, sewers, and so on.

Carry conscious, unconscious, or deceased victims down ladder or stairs using drags, slings, cots, scoops, chairs, stretchers, or improvised equipment.

Remove victims using life gun, lines, and belts.

Assist emergency medical personnel by carrying victims to ambulances or other emergency vehicles.

Rescue drowning persons using poles, ropes, buoys, boats.

tion) of these tests was confirmed, test scores could be used to identify persons with high potential for success as a fire fighter.

Another example is provided by Bosshardt and Lammlein's recent study.[9] They analyzed the machinist job in a metal manufacturing company by interviewing and observing at work a number of incumbent machinists. This analysis resulted in a list of tasks that these machinists performed in the course of carrying out their job duties. The job incumbents subsequently rated the importance of each task, and those tasks judged to be the most important to getting the job done were considered the "performance requirements" for the job.

Results of this step showed that several important tasks appeared to require mechanical ability. For these important tasks, Bosshardt and Lammlein helped the incumbent machinists write items to tap the abilities identified. The machinists were directed to write items that focused on the tasks themselves in an effort to make them as job related as possible. Also, machinist supervisors reviewed the items to ensure their job relatedness. The supervisors documented each of those items that they judged to be job related by providing specific examples of machinists performing tasks highly related to the item content. Thus the development of items for the selection test was clearly designed to be job related. (See Exhibit 5.2.2 for samples of these items.)

Notice that this example is different from the first. The linkup between performance require-

Exhibit 5.2.2 Sample Mechanical Ability Items for Selection Test for Machinists

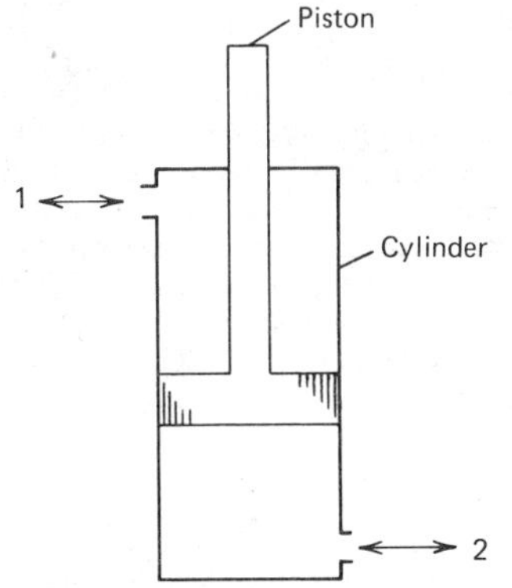

To raise the rod in the cylinder, one would

a. Input pressurized air at 1 and output air at 2.
b. Input pressurized air at 2 and output air at 1.
c. Input equally pressurized air at 1 and 2.
d. Output air equally at 1 and 2.

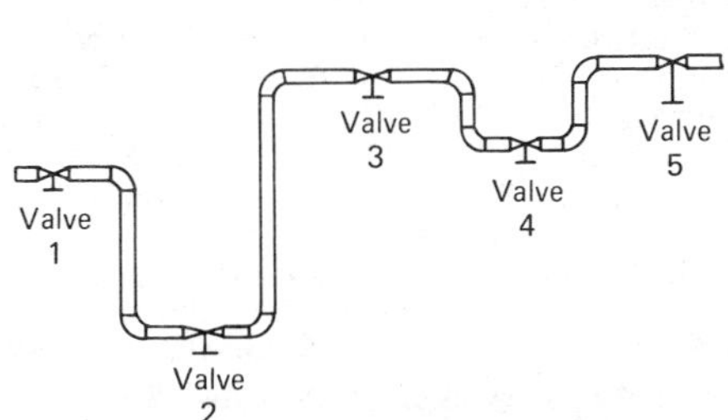

Which valves must be open to completely drain the pipe?

a. Valves 2 and 4.
b. Valves 2, 3, and 4.
c. Valves 2, 3, 4, and 5.
d. Valves 1, 2, 3, 4, and 5.

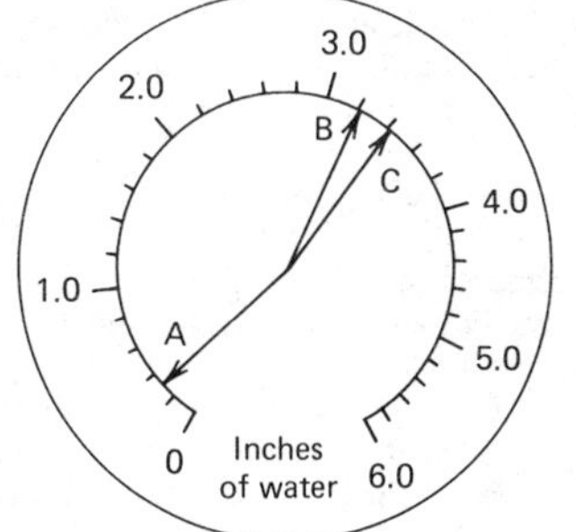

Which needle shows 3.2 inches of water?

a. Needle A.
b. Needle B.
c. Needle C.
d. None of these.

ments and KSAOs has been largely bypassed in that test items were written to reflect directly the abilities to do specific tasks. The important point here, however, is that a job analysis again led directly to the development of test items, and, accordingly, these items are highly job related and likely to yield a valid indication of potential for job success.

Still another example of a method that links job performance requirements with KSAOs to form a sound foundation for choosing or developing selection measures is provided by the Position Analysis Questionnaire. McCormick and his colleagues have linked personal attributes to task dimensions according to their importance for performing competently on those dimensions. (See Chapter 2.4 for more detail.) Thus, when job analysis by means of the PAQ reveals that certain task dimensions are important for that job, inferences can be made about the personal attributes required, based upon the links between task dimensions and attributes discovered in the considerable PAQ research. In this manner the PAQ takes us directly from results of a job analysis to the target personal characteristics to measure with selection tests.

Other systems exist for making the important links between performance requirements and person requirements (e.g., Desmond and Weiss[10]). Also, Dunnette[11] and Peterson and Bownas[12] discuss in more general terms linkages between job performance and worker attributes, and Dunnette and Borman[13] present an idealized job-person match system that might be developed to discover the kinds of links referred to here.

Once the KSAOs likely to be important to job success have been identified on the basis of a job analysis, tests must be selected or developed to measure these target KSAOs.

Purpose of Personnel Testing

The assumptions underlying testing for personnel selection are that persons' abilities, personalities, vocational interests, and so on, differ in measurable ways and that these differences have a bearing on the potential for successful performance on many jobs. Thus the purpose of personnel testing is to measure these individual differences in order to make predictions about future job performance effectiveness. As was discussed previously, a job analysis should form the basis for identifying target KSAOs to measure for personnel selection purposes, and then the selection decision maker should choose or develop the best possible measure(s) of those target KSAOs. This section discusses the various kinds of tests and reviews their track records for accurately identifying persons with high potential for success on jobs.

Types of Tests

Ability Tests. Perhaps the most common type of ability test is the intelligence quotient, or IQ, test. These tests purport to measure a high-level cognitive ability, but what is in fact being measured is in some dispute (e.g., Jensen[14]). Actually, the content of intelligence tests depends largely on the "theory of intelligence" embraced by the test developer and on the purpose for which the test is to be utilized. Refer to Dunnette[11] for a discussion of these matters.

In practice, one thing that *does* seem clear is that scores on intelligence tests consistently predict training performance. That is, persons with high measured intelligence are typically more likely than those with lower measured intelligence to do well in training.[15] This is not too surprising, since an early goal of group-administered paper-and-pencil IQ testing was to predict the ability to learn in school. Thus, when success in training is important to subsequent effectiveness on the job, intelligence tests may provide useful information for selection decisions.

In general, however, IQ tests have been overused in personnel selection. Historically they have been employed many times to help select people for jobs that clearly require no high-level cognitive activity or learning ability, and this has unfairly excluded persons from jobs for which they are qualified. Referring again to the theme of our approach to personnel selection, intelligence tests may be properly used when a job analysis suggests that intelligence, learning ability, or the like are important to successful job performance.

Tests have been developed to measure many other kinds of specific abilities or aptitudes. A partial list appears in Exhibit 5.2.3. As in the case of intelligence, measures of these abilities may be very useful in selection when a job analysis indicates their importance to job performance.

Personality Tests. Paper-and-pencil personality tests have often been criticized for their application in personnel selection. The general line of reasoning has been that they are better suited to the clinical setting and not appropriate for predicting job performance. However, a review of personality testing in selection[20] concludes not that personality measures are useless in the selection context, but that very little evidence has been provided regarding the accuracy with which they predict job performance (i.e., their validity). Instead, personality test publishers often point to the "obvious validity" of their measures without bothering to check on that validity. This is a serious omission; thus later in this chapter we provide a number of strategies for evaluating test validity.

Exhibit 5.2.3 Examples of Aptitudes Measured by Four Prominent Aptitude Batteries

Name of Test	Aptitudes Measured
General Aptitude Test Battery (GATB)[16]	Numerical aptitude Verbal aptitude Spatial aptitude Intelligence Form perception Clerical perception Motor coordination Finger dexterity Manual dexterity
Differential Aptitude Test (DAT)[17]	Verbal reasoning Numerical ability Abstract reasoning Clerical speed and accuracy Mechanical reasoning Space relations Spelling Grammar
Primary Mental Abilities (PMA)[18]	Verbal meaning Number facility Reasoning Perceptual speed Spatial relations
Employee Aptitude Survey (EAS)[19]	Verbal comprehension Numerical ability Visual pursuit Visual speed and accuracy Space visualization Numerical reasoning Verbal reasoning Word fluency Manual speed and accuracy Symbolic reasoning

Regarding limiting their application to clinical settings, however, a number of personality tests have been developed to help predict behavior in "normal" settings (e.g., schools, jobs). Exhibit 5.2.4 provides a brief list of some personality variables along with a sampling of the tests that measure them.

Scales measuring these variables have typically been developed using an empirical keying method. This means that the test developer who wants to measure dominance, for example, chooses for the

Exhibit 5.2.4 Examples of Personality Variables and Tests That Measure Them

Personality Variable	Tests Measuring Variable
Dominance	California Psychological Inventory[21] Personality Research Form[22]
Extraversion	Comrey Personality Scales[23] Eysenck Personality Inventory[24]
Need for achievement	Edwards Personal Preference Schedule[25] Adjective Check List[26]
Emotional stability	Gordon Personal Profile[27] Guilford-Zimmerman Temperament Indicator[28] Sixteen Personality Factor Questionnaire[29]
Flexibility	Personality Research Form California Psychological Inventory

scale those items to which persons thought to have high dominance (through peer ratings or other means) respond differently than persons thought to be less dominant.

On the face of it, the variables in Exhibit 5.2.4 would appear to be important to success on interpersonally oriented jobs especially. The authors' experience suggests just that—that scores on personality tests can be used to provide reasonably accurate predictions of job performance on some jobs. For example, on an autonomous sales job that the authors studied, that of military recruiter,[30] scores on the following personality scales successfully identified effective recruiters: (1) Exhibition and Order from the Personality Research Form; (2) Socialization and Achievement via Conformance from the California Psychological Inventory; and (3) Impulsiveness from the Differential Personality Questionnaire.[31] This study and several other successful applications of personality tests in government and industrial jobs (see Gough[32] for references) suggest that personality tests might be considered when choosing selection measures, especially if the target job has a strong interpersonal component. Again, the key is to use measures of variables that tap personal attributes a job analysis suggests are important.

Psychomotor Tests. It seems reasonable to suppose that performance on jobs requiring psychomotor skills could be predicted using tests that assess general psychomotor abilities. However, as we shall see, the accuracy of predicting job performance using psychomotor tests has been disappointingly low.

First, what *are* psychomotor skills? Fleishman conducted extensive research of these kinds of skills[33] and concluded that there are 11 reasonably independent groupings of psychomotor skills. These appear in Chapter 6.1. Tests of these various skills have been developed and used in attempts to predict task performance. A central result emerging from Fleishman's and others' research[34–36] is that the acquisition of psychomotor abilities in task performance requires different skills at different stages of practice. This makes it difficult to predict performance on tasks because a variety of skills is needed in gaining proficiency on a task. Another important finding from this research is that the skills required to perform tasks successfully tend to be very task-specific. What this means for personnel selection is that tests measuring the broad psychomotor skills (e.g., control precision, reaction time) are not very useful in predicting task performance. Instead, work samples reflecting very closely the task characteristics of the target job are necessary to obtaining good accuracy in predicting job performance. Work samples as tests in personnel selection are discussed later in the chapter.

Other Types of "Tests"

Although not typically thought of as tests, several other kinds of procedures and data are used to select people for jobs, including biographical information, the employment interview, and work samples.

Biographical Information. The philosophy associated with the use of biographical data is that a view of a person's past "life path" provides a good prediction of what he or she is likely to do in the future. Owens[37] is the main proponent of this view of biographical data. He invokes the behavior consistency notion—that is, that past performance is a good indicator of future performance—to argue that relevant information about a person's previous life experiences provides meaningful data that may be used to predict job performance.

Exhibit 5.2.5 provides examples of the kinds of biographical items used in personnel selection. Notice that there are two different types of items—factual biographical data (e.g., sex, number of

Exhibit 5.2.5 Examples of Biographical Items Used in Personnel Selection

Number of jobs you have held in the past 5 years:

________ 0	________ 2–3	________ more than 5
________ 1	________ 4–5	

Years of military service:

________ 0	________ 3–6	________ 10–20
________ 1–3	________ 6–10	________ more than 20

In high school you had

________ no close friends.
________ a few close friends.
________ many close friends.

In a group I have

________ preferred to follow rather than to lead.
________ provided leadership at times and preferred to follow at other times.
________ often provided leadership

brothers) and information about previous experiences, activities, and so on (e.g., leadership experience in high school). Both types may be useful in personnel selection, but the latter is more likely to be directly job related. Pace and Schoenfeldt[38] have, in fact, made suggestions about how to create biographical items on the basis of a job analysis.

Although we recommend using job analysis information to select or develop biographical items, another procedure, empirical keying, is often employed and has certainly been successful in many selection settings. With the empirical keying approach, items are selected according to differences in the ways effective and ineffective job incumbents respond to them. For example, if most effective performers on a sales job respond that they had previous sales experience, whereas less effective performers respond that they did not, this item may prove to identify top sales people for this job. Likewise, other items are selected according to their ability to differentiate between effective and ineffective job incumbents. Applicants responding most closely to the "key"–that is, to the way effective performers respond–receive the highest "scores," and these scores are used subsequently to aid in selection for the job.

Ghiselli's review of employment tests[15] indicates that biographical data are the most successful of all the predictor types in forecasting effectiveness on industrial jobs. Thus biographical items should be seriously considered for personnel selection.

The Employment Interview. The employment interview may be considered a test in that decisions on whether or not to hire are often made on the basis of this method. Unfortunately, the interview has been notoriously poor as a selection device. Research has shown that in making selection decisions interviewers often attend to factors that are unrelated to the job (e.g., sex, race, and similarity to the interviewer in background, attitudes, etc.). Recent reviews of research on the interview[13, 39] discuss the many distortions that are likely to occur when an interviewer attempts to make an accurate estimate of a person's potential for job success. Some very basic suggestions for employment interviewers made by Schmitt[39] indicate the generally primitive state of the art in interviewing. These include: (1) decide what the purpose of the interview is to be; (2) know the requirements of the job to be filled; and (3) allow the applicant time to talk. That these kinds of suggestions have to be made at all reflects badly on employment interview practices, but the fact is that selection interviews are often conducted without regard to even very simple principles.

Yet, increased accuracy in predicting success on jobs may be possible if some of these principles are adhered to. Landy[40] found a reasonable degree of accuracy in predicting the job success of police officers when interviewers were very familiar with the performance requirements from a job analysis and used a structured interview format with questions focused on those performance requirements. (See Guion[41] for a more general discussion of this kind of approach to employment interviewing.) Exhibit 5.2.6 provides some guidelines to aid in conducting selection interviews.

Thus the employment interview has often been severely criticized as a selection device. Because of the way it is typically conducted, these criticisms seem warranted; thus we warn against the casual use of the interview in making selection decisions. However, this selection "test" may be made more useful when some of the suggestions in Exhibit 5.2.6 are followed within the general framework outlined by Guion.[41]

Work Samples. The work sample is a simulation of a job or some part of that job. Work samples require that persons being tested perform the simulated job in order to evaluate potential for performing effectively on the real job. One common example of a work sample is a typing test used to predict performance as a typist. Many other kinds of work samples have been developed, and examples appear in Exhibit 5.2.7.

The simple rationale for using work samples as selection tests is that the KSAOs required for effective performance on the target job are also required for successful performance on the work

Exhibit 5.2.6 Some Guidelines for Selection Interviewing

Use a structured, rather than an unstructured, interview format; that is, know what you are going to ask and elicit the same kinds of information from each candidate.

Know the requirements of the target job.

Relevant past experience, interpersonal skill, and motivation of applicant are generally what interviews are best at assessing; other kinds of KSAOs should probably be tested in other ways.

Take care not to let a first impression of an interviewee overinfluence the selection decision.

Beware of a contrast effect when interviewing several candidates; that is, an average applicant may look *good* if interviewed after an unqualified person or *bad* if interviewed after a highly qualified person.

Be alert to possible bias in selection because of race, sex, attractiveness, or attitudes of the interviewee that are either very similar to or widely divergent from those of the interviewer.

Exhibit 5.2.7 Examples of Work Samples

A lathe test, a drill press test, and a tool dexterity test for machine operators.[42–44]

A sewing machine test for sewing machine operators.[45,46]

Tests for mechanics, such as installing belts and pulleys, disassembling and repairing a gear box, installing and aligning a motor, and pressing a bushing into a sprocket and reaming it to fit a shaft.[47]

A clothes making test.[48]

A screw board test for machine operators in a register manufacturing company.[49]

A test for tracing trouble in a complex circuit.[50]

A packaging test for production machine operators.[51]

A programming test for computer operators.[52]

A role playing test that simulates telephone contacts with customers.[53]

The in-basket test for managers.[54]

sample. Asher and Sciarrino[55] review work samples as selection devices and provide guidance for developing samples based on a "point-to-point" model, which emphasizes matching each work sample element with a corresponding element from the job. Likewise, Schwartz[56] has discussed how job analysis and the identification of KSAOs can lead directly to the development of work samples. This is in keeping with the approach we are advocating, that is, selecting tests based upon careful job analysis and the identification of important KSAOs for performing successfully on the job.

The Asher and Sciarrino review suggests that work samples have provided generally accurate estimates of performance effectiveness of the job—estimates whose accuracy is, in fact, second only to that of biographical data. One should keep in mind, of course, that it may be difficult or inappropriate to develop work samples for some jobs and selection situations. Among the reasons for this is that, for jobs whose performance requires considerable training, work samples are typically not justifiably applied. Also, the "art" of developing work samples is such that some may be more successful than others in providing a faithful test of the target KSAOs.

In passing, we should mention the assessment center concept for selecting managers.[57,58] The assessment center contains a series of work samples designed to tap important KSAOs associated with managerial effectiveness. An exercise typically included is the in-basket test, which consists of letters, memos, phone messages, and other items that the job candidate must react to as he or she plays a manager. Also often included is the leaderless group discussion—a problem solving session with a number of job candidates attempting to contribute to a group discussion of possible solutions to a hypothetical problem.

To conclude this section, we emphasize that no one type of test is superior in all selection situations. Application of particular selection procedures depends upon results of a job analysis and the nature of KSAOs required for job success, as well as upon several other practical considerations discussed later. A reference that is most helpful to learning about the quality of individual tests is Buros' *Mental Measurement Yearbook.*[59]

In sum, developing a personnel selection program is a complicated business, and we recommend seeking the counsel of an experienced professional who is well qualified in job analysis and personnel selection. We recommend *against* using a "pet" test to select people, with no attention paid to the factors mentioned here.

Validation

Purpose of Validation

Previously we referred to tests as being "accurate in predicting job performance" and the like. This section describes how that accuracy, or validity, of tests can be properly evaluated to provide a good picture of how well our selection procedures are doing what they were designed to do.

We strongly advocate the validation of selection procedures. Checking on the accuracy of their predictions makes obvious good sense, enabling adjustments in the procedures if they fail to result in high-quality selection decisions. Validation of selection procedures is often required under Equal Employment Opportunity (EEO) legislation and guidelines, providing another critical reason for validation.

Before reviewing specific test validation strategies, let us discuss briefly the measurement of job performance, the classic "criterion" in personnel selection research. (This is discussed more completely in Chapter 4.4). The reason for being concerned about job performance measurement is that most validation strategies require us to correlate test scores made by individuals with job performance "scores." A test is judged "good" (or valid) if persons who score high on the test tend to perform well on the job, that is, if there is a positive correlation between test scores and performance scores. In this case the test has succeeded in identifying persons likely to perform successfully on the job. A test is considered invalid if a very low or zero correlation is found between test and performance scores. In such a case the test is obviously not useful for identifying persons likely to perform well.

Job Performance Measurement

The goal of performance measurement is to obtain accurate measures of individuals' effectiveness on the job. Accurate performance measurement makes possible a fair estimate of a test's validity when test scores are correlated with performance scores. Inaccurate or biased indicators of performance effectiveness necessarily lead to poor estimates of test validity. Accordingly, obtaining good measures of job performance is critical.

Two Types of Performance Measures. Two types of performance measures should be mentioned–*objective indexes* (e.g., for some assembly jobs, number of units a worker builds per 8 hr day and number of his or her units per day rejected by quality control checks) and *performance ratings.* Objective indexes of a worker's performance are preferred to the subjective impressions provided by performance ratings, but, as we shall see, good objective measures are hard to come by.

The difficulties with the vast majority of objective performance measures are that they are almost invariably *deficient* and/or *contaminated.* By deficient, we mean that the measure provides only a partial picture of the worker's effectiveness on the job; that is, there are important aspects of the job left untapped by the objective measure. Regarding the assembly worker example just cited, the numbers of units built and rejected may be important indexes of effectiveness, but if helping inexperienced workers perform the job correctly and willingness to work overtime during heavy production schedules are also important to job success, then the former two measures, individually or together, do not adequately measure effectiveness on the job. They are deficient.

Contamination of objective measures occurs when factors affecting how well persons do with respect to the measure are beyond the control of that person. Referring again to the assembly worker example, suppose that the number of units built depends somewhat on the timely delivery of certain parts that go into the unit, and that the worker has no control over those deliveries. The number of units built measure therefore provides an "impure" index of effectiveness–it is contaminated. Unfortunately, these are very common problems with objective performance measures, even for relatively low-level production-type jobs where objective indicators are likely to be available. Identifying or developing good, objective indexes for higher-level jobs is even more difficult. (See Gilmer and Deci[60] for a more extended discussion of this topic.) Thus, although objective measures are generally preferrable to subjective ratings of performance, ratings may often be the only means available to obtain performance effectiveness scores for job incumbents.

Of course ratings have their own set of problems, which are discussed in Chapter 4.4. Briefly, factors that lead to inaccuracies in ratings include the following:

1. Ratings are obtained from persons in a poor position to make judgments about incumbent performance.
2. Some raters simply lack the observational and/or judgment skills necessary to make accurate evaluations.
3. Raters provide biased ratings, based not so much on performance, as on race, sex, background, similarity to the rater in attitudes, and so on.
4. Raters commit rating errors, such as evaluating everyone as very effective when, in fact, some workers are performing poorly.
5. Raters fail to use the definitions of the performance dimensions, instead employing their own idiosyncratic beliefs about what it takes to perform effectively and then rating persons accordingly.

Nonetheless, in our judgment, if a few simple principles are followed in gathering performance ratings in selection research, the accuracy of these ratings can at least be maximized within the limitations of the "state of the art." These principles include:

1. Developing the rating scales with great care taken to reflect all important performance requirements of the job.
2. Creating dimensions that are clearly job related and that represent performance factors raters can readily observe in those being rated.
3. Providing clear, simple directions for using the rating scales.
4. Gathering the ratings for research purposes only (rather than for any administrative purpose) and making this clear to the raters.
5. Selecting raters who have good opportunity to observe the performance of those being rated, which may mean considering peers or even subordinates of the person to be rated in addition to their supervisors.

6. Collecting, if possible, ratings of each person being rated from more than one rater so that interrater agreement may be assessed in order to provide at least a rough estimate of the accuracy of the ratings.

Multiple Dimensions of Performance. Recall our previous statement that accurate performance measurement enables us to correlate selection test scores with performance scores in order to obtain proper estimates of the test's validity. Before discussing in more detail strategies for evaluating validity, one more important concept about performance measurement should be introduced, the notion of multiple dimensions of performance.

Like Guion[61] and Dunnette,[62,63] we advocate attempting to predict performance on individual dimensions of the job rather than trying to predict overall job performance. An example should help to illustrate this idea. As mentioned, Borman et al.[30] performed an extensive analysis of the military recruiter job. Several analyses of performance ratings revealed the same three-dimension system for describing three distinctly different performance requirements of the job: selling, relating to others, and organizing. Because of considerable differences in the kinds of tests likely to predict performance in the three different parts of the job, tests were chosen to focus on individual performance dimensions. Thus the following personality scales were selected for these dimensions (and subsequently found to be valid predictors of performance on these dimensions):

Selling–Exhibition

Relating to others–Social Closeness

Organizing–Order, Socialization, Achievement via Conformance

Once different predictor tests are confirmed as valid, as in the preceding example, it is possible to estimate performance for individual tested applicants on each dimension, and this information may be used to help make selection decisions. It is important to note that this approach provides more information to selection decision makers than does a single predicted performance score for overall performance. One can assess in an applicant potential strengths and weaknesses in different aspects of the job that may be useful in planning training for him or her or in matching persons to positions according to the special requirements of those positions (e.g., matching a person with especially good potential in the organizing area to a position that has heavy requirements in that area).

Other "Success" Criteria. One final comment regarding "performance measurement"–in some selection situations we may be interested in using tests to predict behaviors related not to job performance, but to other "success" criteria, such as attendance at work, tenure with the organization, or avoidance of accidents on the job. For example, if costs to train persons to perform a particular job were very high, tenure might prove very important as a criterion of success because of the expense to the organization related to turnover of trained job incumbents. Problems exist with these kinds of criteria as well, and thus, as is the case with objective performance indicators and performance ratings, great care should be taken to develop indexes of these success criteria.

Classic "Types" of Validity

In general, we concur with Dunnette's contention[11] that distinctions between "types" of validity should be de-emphasized, and instead, attention directed toward validation as "a process of learning as much as possible about the behavioral inferences that may properly be made from sources on any selection measure. . . ."[64] Information about what our selection tests are measuring is always useful, no matter what particular validation strategy is employed.

For convenience, however, we discuss validation methods according to a frequently used category system: predictive, concurrent, content, and construct validity. Remember that the overall purpose of evaluating the validity of our selection procedures is to determine the accuracy with which we are predicting job performance (or other success criteria) and to learn how we might improve that accuracy of prediction.

Predictive Validity. Information on predictive validity is probably the best kind to have in making judgments about a test's validity in predicting job performance. Exhibit 5.2.8 shows how this strategy works. Applicants are hired at random alfter taking the test(s), and some time later their performance is assessed. Correlations between the test scores and the performance then provide excellent information about the validity of the test(s). High correlations indicate that test scores are good indicators of future job success, whereas low or zero correlations demonstrate that scores tell us little about how well an applicant is likely to perform later on the job.

An obvious problem with the predictive validation strategy is that organizations are typically unwilling to hire people on a random basis. They want to use *some* kinds of criteria for selection. Also, organization decision makers usually do not want to wait for validation results, and this

Exhibit 5.2.8 Predictive Validation Paradigm

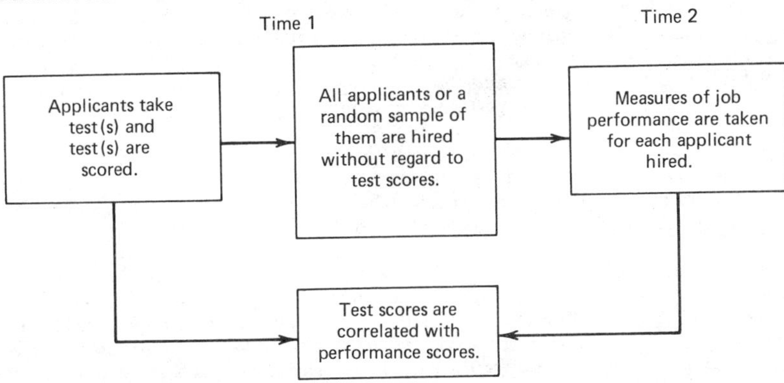

strategy demands a waiting period between Time 1 and Time 2 (Exhibit 5.2.8) during which the performance of those hired can be reasonably measured, perhaps 6 months to 2 years later or even longer after hire. For these reasons, predictive validity is a strategy that, sadly, is seldom used in personnel selection. As a substitute, concurrent validation is often employed.

Concurrent Validity. Exhibit 5.2.9 depicts this strategy. Job incumbents, rather than applicants, are used in concurrent validity research, eliminating the necessity of waiting to gather job performance information. This is the main advantage of the method–as soon as test(s) can be administered and scored, and measures of job performance taken, we can move to assess the validity of the test(s). On the other hand, the comparability of concurrent and predictive validity results has been widely questioned (e.g., Guion[65]), and this is a serious problem because, as mentioned, we are really interested in the *predictive* validity of tests.

Recently, empirical evidence has been presented suggesting good correspondence between concurrent and predictive validity results.[66] If this proves to be a common finding, difficulties with the use of the concurrent validity strategy will be less serious, and we will be able to utilize this method with considerably more confidence.

Content Validity. The content validation strategy has been employed most often, explicitly or often implicitly, in developing achievement tests in educational settings. A teacher wants to ensure that his or her test is tapping the intended course content, that is, those concepts that were to be learned in the course. Thus items are developed to test knowledge of those concepts. Content validity is demonstrated to the degree that the content of the test samples the target knowledge, skill, and so on. Although there are quantitative methods for evaluating content validity (e.g., obtaining expert ratings of the relevance of each test item to a target body of knowledge), typically it is assessed on a subjective basis.

In personnel selection, content validity has traditionally been accorded less status than predictive or concurrent validity. This is because content validity measurement of a particular domain (e.g., achievement in some aspect of mathematics) says nothing about the domain's importance to job

Exhibit 5.2.9 Concurrent Validation Paradigm

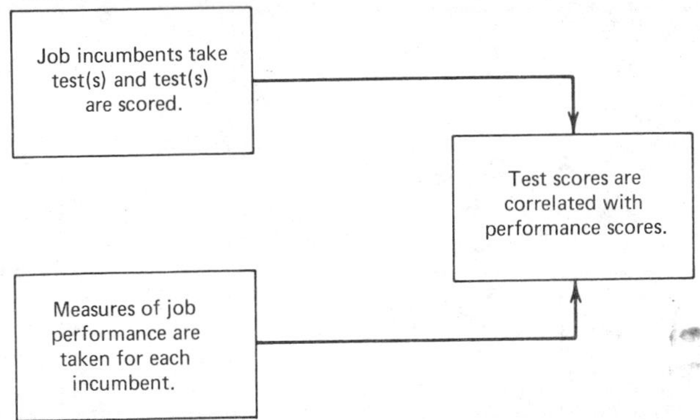

performance. Nonetheless, the broad notion of content validity fits nicely into our treatment of personnel selection in this chapter. We have emphasized the use of a thorough job analysis to pinpoint performance requirements, in turn leading to identification of KSAOs likely to be important to job success. The development of predictors to measure these KSAOs can be viewed as an effort to produce content validity predictors, although with this use of content validation, the predictor items are tied to the job performance requirements through their link with the target KSAOs.

Thus content validity can be a useful concept in personnel selection. However, we advocate empirical confirmation of relationships between test items developed within a framework of content validation and job performance criteria whenever the test sample is large enough to provide a reasonably accurate estimate of the predictive or concurrent validity.

Construct Validity. Assessing the construct validity of a predictor test involves developing hypotheses about what it measures and then evaluating these hypotheses. In the personnel selection context, construct validity is concerned with understanding what selection tests are measuring and with the scientific study of relationships between tests scores and job behavior. Predictive, concurrent, and content validation efforts can each be viewed as providing *some* evidence regarding a test's construct validity, but the concept of construct validity is broader in that it includes many ways of learning about what selection tests are measuring.

Accordingly, construct validity is most important as a strategy for gaining scientific understanding of scores obtained on selection measures and of the links between these scores and job behavior. In our view this strategy is extremely useful, potentially leading to significant increases in the accuracy with which we predict job performance. It is important to note, however, that in personnel selection practice the most critical consideration is accurate prediction of job performance or other job behaviors. Consequently, it is not sufficient to demonstrate good construct validity for a predictor measure; a direct link must be established between a predictor and job-related behavior.

Statistics of Validation

When evaluating the predictive, concurrent, and, in some cases, construct validity of a test, the Pearsonian correlation coefficient (r) is often employed to describe the relationship between test and performance scores. The value of r can vary between +1.0 and −1.0, with +1.0 indicating a perfect correspondence between the two sets of scores, and −1.0 a perfect correspondence also, but with high scores on one variable going with low scores on the other variable. A zero correlation means that the two sets of scores are unrelated, that is, that test scores are useless for predicting performance scores. Two examples and the computational formula for this coefficient appear in Exhibit 5.2.10.

For a moment, let us assume that both of these examples provide results of a predictive validity study. Ten persons were tested and hired, and then some time later their performance was assessed. In example 1 (Exhibit 5.2.10*a*) there is good correspondence between tests and performance

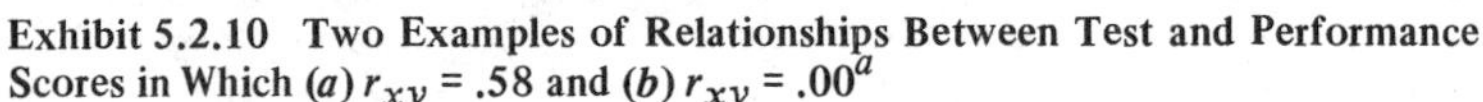

Exhibit 5.2.10 Two Examples of Relationships Between Test and Performance Scores in Which (*a*) r_{xy} = .58 and (*b*) r_{xy} = .00[a]

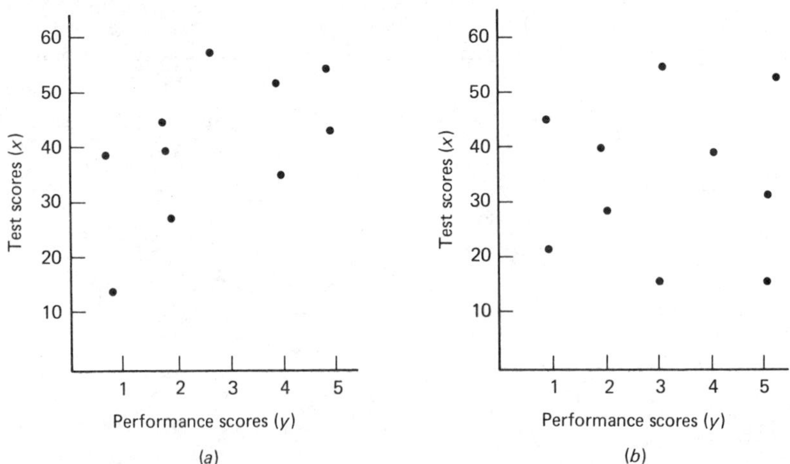

[a]Possible range of scores for tests is 0 to 60 and for performance ratings is 1 to 5.

scores. Individuals who made high scores on the test were likely also to be rated as effective on the job. Thus we could evaluate this particular test as valid for identifying persons likely to perform well on the job. In example 2 (Exhibit 5.2.10*b*) correspondence between the two sets of scores is poor; knowing a person's test score tells us very little about his or her likely performance level. (See Ghiselli[15] for a summary of the actual levels of validities that have been obtained in practice.)

The exhibit examples are "scatterplots" depicting in graphic form the relationships between test scores and performance. Each dot represents a person who has both a test score and a performance score. For instance, in Exhibit 5.2.10*a* the bottom left dot means that this person had a test score of 13 and was rated at the level 1, indicating poor performance on the job. Correlations can be computed using the following formula:

$$r_{xy} = \frac{(\Sigma x_i y_i / N) - m_x m_y}{s_x s_y}$$

where r_{xy} = the correlation between the x and y variables
$m_x m_y$ = the mean of the x and y variables, respectively
$s_x s_y$ = the SD of the x and y variables, respectively

An important point must be made here. The r obtained in each of the examples in Exhibit 5.2.10 is derived from a group of only 10 individuals. Accordingly, this value may well be a poor estimate of the correlation we would obtain if, for example, 500 persons were tested and their performance was evaluated. Without getting into the technical details (see Hays[67] for a technical treatment), large samples of individuals provide better estimates of validity to be expected over the long run (other factors being equal–i.e., the job remains the same, a similar applicant pool is maintained, etc.); therefore, whenever feasible, validity estimates should be based on large numbers of persons. A rule-of-thumb minimum number, 30, has been offered, but recent thinking[68] suggests that much larger samples should be used in order to provide an accurate picture of a test's "true validity" (e.g., 200 to 300 persons). Schmidt's point is that validities based on 30, 50, or even 100 persons yield correlations that may well prove to be poor estimates of validity. We have more to say about this later.

Another important point is that the correlation coefficient often provides a reasonable summary of the relationship between test scores and performance, but under certain conditions this may not be the case. It is beyond the scope of this chapter to discuss these conditions in detail. See Dunnette[63] for a treatment of other ways to characterize relationships between predictors and job performance scores.

So far we have been talking about the case in which we have a single test and a single performance criterion. Often we want to employ several predictors in attempts to predict effectiveness on a performance criterion. An example should help explain why this is so.

Exhibit 5.2.11 is a Venn diagram. It shows that performance in the organizing part of the job is related to scores on *each* of three "tests." The shaded area represents the correlation between the predictors *taken together* and performance in organizing. Notice that this correlation (in graphic form) is greater than the correlation between any one of the "tests" and performance. This phenomenon often occurs in personnel testing practice, and thus it makes good sense, in many cases, to use more than a single test to predict performance. However, problems may arise in estimating the correlation between tests taken together and performance, that is, in estimating the tests' validity in combination.

In general, a problem arises any time that weights are assigned to tests on the basis of data from the sample (e.g., in a concurrent validity study). For example, we might be able to raise the correlation between our three tests taken together and performance by weighting double the scores on one of the three tests. Wherry[69] provides a recent detailed discussion of this general problem; we simply observe that, when we use more than one test and wish to combine the test scores in some way to gain increased accuracy in predicting performance scores, any weights we select based on data from the sample will take advantage of sample-specific configurations in the data. This will lead to overestimates of the validity to be expected in the long run.

Two approaches can be used to "correct" these validity estimates. One involves deriving the desired weighting scheme from one half of the sample, applying the same weights to the test scores in the second half of the sample, and then obtaining the correlation between the weighted test scores and performance in this second sample. This "shrunken" correlation–it is always smaller–can then be used as the validity estimate, usually called the "cross-validity estimate." The other approach is to use shrinkage formulas that are available to correct these validity coefficients. (See Schmitt et al.[70] for examples.)

As can be seen, the issues in attempting precise estimates of validity are technically complex when more than one variable is used as a predictor. See Dunnette and Borman[13] for a recent summary of these issues.

Exhibit 5.2.11 Venn Diagram Showing Relationship Between Scores of Three Predictors and Job Performance in Organizing Work

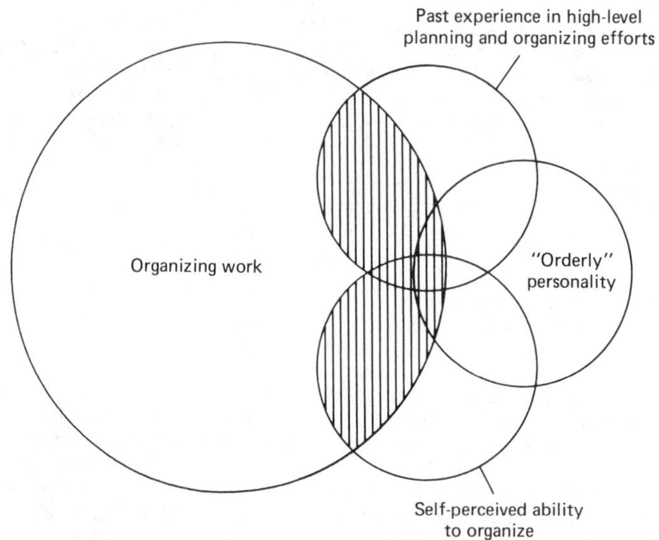

So much for the case of the multiple predictor, single performance criterion. What do we do when multiple performance criteria have been identified (e.g., selling, relating to others, and organizing, as given in a previous example)? Some have advocated weighting the performance criteria according to their importance to overall effectiveness and combining these weighted performance scores to obtain a single performance score for each person. Then the validity of the test(s) may be estimated using the approaches just discussed. However, as mentioned previously, we suggest attempting to predict performance on each dimension separately and therefore obtaining validity estimates for each dimension individually. Using the military recruiter example again, those three tests in the Venn diagram might be used to predict the criterion organizing, and their validity might be estimated using multiple correlation and the cross-validation strategy. Likewise, two or three *other* tests might be used to predict selling performance and still another set of tests to predict performance in relating to others. In each case the validity of these tests would be determined in order to assess the accuracy of prediction.

The technology is available[71] for then scoring all of the tests of an applicant and computing *predicted performance scores* on each performance dimension. This can be accomplished by first scoring the tests for each applicant according to the scoring key, which results in a distribution of scores for those tested (i.e., some applicants with relatively high scores, and others with lower scores). These scores (for each performance dimension) may then be transformed directly into predicted performance scores. We will not get bogged down in the statistical manipulations associated with this transformation, but the principle involved is simple–high test scores result in high predicted performance scores, and lower test scores result in lower predicted performance scores. As mentioned, selection decision makers can now use these predicted performance levels to aid in making decisions on whether to hire or reject.

This is a good time to emphasize that there are many other things besides predicted performance scores for such a decision maker to consider in making a selection decision. Also, we have, up to now, largely ignored other considerations that should be attended to in deciding whether to use selection technology at all, and if it is to be utilized, how best to do so. The next part of the chapter discusses these considerations.

Conditions Affecting Selection Strategies

So far we have discussed the *technology of personnel selection*. In a sense the procedures described are reasonably simple: perform a job analysis; infer from the job's performance requirements the important KSAOs required for successful performance; select and/or develop trial predictor measures to tap these important KSAOs; use an appropriate validity strategy to evaluate the predictor measures; and utilize for selection those predictors that prove to be valid. Of course many of the steps require a highly technical understanding of certain selection and validation principles, but the steps we have described are basically straightforward and easy to understand.

In this section we must, unfortunately, muddy the waters somewhat by introducing the several factors that affect the development of personnel selection systems and their application to real-world selection problems in organizations.

Economic Utility Considerations. The accuracy with which selection procedures are identifying applicants who do well on the job is a very important consideration in assessing the usefulness of those procedures. High accuracy in this context (i.e., good validity for our tests or other kinds of measures) makes likely the selection of highly qualified persons with the appropriate skills for performing successfully on the job. However, for profit making organizations, we, along with many others before us,[1,72–75] argue that economic utility considerations supersede considerations of accuracy or validity. Also, for many organizations in which productivity of individuals is important, though perhaps economic gain is not, similar kinds of utility thinking should enter into judgments about selection procedures.

In particular, the following factors are important in deciding on the relative merits of selection procedures:

1. The validity of the procedures.
2. The selection ratio (i.e., the number of applicants per opening).
3. The base rate of performance on the job (i.e., how adequately incumbents selected under the present procedure can do the job).
4. The impact that good performance versus mediocre or poor performance has on organization earnings or productivity.
5. The costs of selection, both for developing the tests or other procedures and for administering the selection program.

Let us touch on these factors one by one. The validity of our selection procedures is obviously important, as this chapter has stressed, and no more need be said about this factor. The selection ratio is also very important; for example, if it is 1.0—that is, the number of applicants equals the number of openings—selection is not necessary. As will be seen in a succeeding example, as the number of applicants increases in relation to the number of openings, valid selection procedures become more useful. Regarding factor 3, if *everyone* is capable of performing adequately on the job, as is sometimes the case with very low level jobs, special selection efforts are unnecessary. *Any* strategies for selecting persons will result in the same levels of worker performance.

Factor 4 is also important. Valid selection is especially useful from an economic point of view if good or superior performance on the target job makes a big difference (in comparison to mediocre or poor performance) in determining organization profit, dollar savings, or productivity. For example, in certain highly technical quality control work, a competent worker, in contrast to a poor performer, may save the company many dollars by identifying minor defects in electronic circuitry that could have caused serious and expensive component breakdown later. The greater the cost savings, or the increases in profit or productivity, the more useful are accurate, valid selection procedures. Finally, the costs of developing and administering selection procedures obviously affect their usefulness. In general, highly expensive procedures (e.g., a simulator work sample for a nuclear power plant operator job) need to be more accurate in identifying effective performers than procedures that are less expensive (e.g., paper-and-pencil tests).

How can these factors be systematically considered in exploring the usefulness of a selection device? One relatively simple method, the Taylor-Russell tables,[76] takes into account validity, selection ratio, and base rate of competent performers on the job. Suppose that our company's present hiring practices result in only 40% of the workers on a particular job being competent. If the selection ratio is .60—that is, we hire 6 out of every 10 who apply—and the validity of a new selection procedure is .40, then we can expect an increase in the proportion of competent employees to 50%, according to Exhibit 5.2.12. Notice that if the selection ratio were .10, with the same level of validity, we could expect 69% to be competent. One should be careful in interpreting too literally data from these tables,[65] but they do show the considerable effect of selection ratio and base rate, as well as validity, on the usefulness of personnel selection procedures.

Brogden[72] and Cronbach and Gleser[74,75] have presented approaches to evaluating the economic utility of tests, taking into account the implications of good versus poor performance on a job and the costs of testing, as well as the other factors previously mentioned. Unfortunately, these strategies, which are much more satisfactory than the Taylor-Russell approach, have been used only rarely. The main problem with applying them seems to be ignorance about how to apply the complicated equations and difficulty in costing some of the elements of those equations, most notably the SD of job performance in dollars (related to factor 4). Recently, Hunter and Schmidt[1] have offered at least partial solutions to these problems, and we look forward to further developments in this critically important area.

The main point to be made here is that, no matter what the exact approach, *utility thinking* should be brought to bear in considering personnel selection strategies for jobs. The glaring eco-

Exhibit 5.2.12 Taylor-Russell Table Indicating Proportion of Employees Considered Satisfactory with Varying Selection Ratios and Tests of Varying Validities.[a]

r	.05	.10	.20	.30	.40	.50	.60	.70	.80	.90	.95
.00	.40	.40	.40	.40	.40	.40	.40	.40	.40	.40	.40
.05	.44	.43	.43	.42	.42	.42	.41	.41	.41	.40	.40
.10	.48	.47	.46	.45	.44	.43	.42	.42	.41	.41	.40
.15	.52	.50	.48	.47	.46	.45	.44	.43	.42	.41	.41
.20	.57	.54	.51	.49	.48	.46	.45	.44	.43	.41	.41
.25	.61	.58	.54	.51	.49	.48	.46	.45	.43	.42	.41
.30	.65	.61	.57	.54	.51	.49	.47	.46	.44	.42	.41
.35	.69	.65	.60	.56	.53	.51	.49	.47	.45	.42	.41
.40	.73	.69	.63	.59	.56	.53	.50	.48	.45	.43	.41
.45	.77	.72	.66	.61	.58	.54	.51	.49	.46	.43	.42
.50	.81	.76	.69	.64	.60	.56	.53	.49	.46	.43	.42
.55	.85	.79	.72	.67	.62	.58	.54	.50	.47	.44	.42
.60	.89	.83	.75	.69	.64	.60	.55	.51	.48	.44	.42
.65	.92	.87	.79	.72	.67	.62	.57	.52	.48	.44	.42
.70	.95	.90	.82	.76	.69	.64	.58	.53	.49	.44	.42
.75	.97	.93	.86	.79	.72	.66	.60	.54	.49	.44	.42
.80	.99	.96	.89	.82	.75	.68	.61	.55	.49	.44	.42
.85	1.00	.98	.93	.86	.79	.71	.63	.56	.50	.44	.42
.90	1.00	1.00	.97	.91	.82	.74	.65	.57	.50	.44	.42
.95	1.00	1.00	.99	.96	.87	.77	.66	.57	.50	.44	.42
1.00	1.00	1.00	1.00	1.00	1.00	.80	.67	.57	.50	.44	.42

Source. Reference 76.
[a]Proportion of employees considered satisfactory *without* the selection test = .40.

nomic errors possible when cost-effective selection practices are not employed, or when cost *ineffective* selection practices *are* employed, argue strongly for considering the utility of such practices.

"Effects" of Previous Selection Study Results. A widely held belief in personnel psychology has been that test validities vary considerably from job to job because even subtle differences in jobs and/or applicant groups affect relationships between tests and job performance. A test can be valid for predicting success on job A, but when the same test is employed for predicting performance on job B, a job and selection context very similar to job A's, the correlation between test scores and performance may be near zero.

Recently, Schmidt and his colleagues[68,77] have questioned this "situation specificity" belief. They argue convincingly that the range of validity coefficients found in test validation studies would not be nearly so great if the numbers of persons in such studies were larger to allow stable estimates of validity. With this and related arguments, Schmidt et al. are saying in effect that we should not rely so much on results of individual validity studies, especially when the sample size for the study is small (e.g., fewer than 200 to 300 persons). Instead, *true validities* for tests and particular job performance measures should be estimated from past studies using a Bayesian approach. (See Hunter and Schmidt[1] for an explanation of Bayesian statistics applied to this problem.) Within such a framework, if evidence for past validity is strong (i.e., many studies with high validities), a validity study may not even be necessary. If a study is necessary, the Bayesian approach allows the researcher to use both past validity information and the validity obtained in the study to estimate the true validity of the test for use in similar circumstances.

This somewhat heretical position has not received complete acceptance among personnel selection scientists and practitioners, but considerable attention has been and will continue to be directed toward these notions. What *is* clear from this work is that the sample sizes necessary for providing reasonably stable estimates of validity are larger than has been assumed, and we should place less credence on validities obtained in individual little studies with small samples.

Another approach to the problem of evaluating the validity of tests in jobs with few incumbents is to group together jobs with similar performance requirements and perform validation research on the *cluster* of jobs. For example, jobs can be clustered on the basis of similarities between importance of tasks. In this manner jobs with similar patterns of task importance are treated as a single job, and the whole sequence of test development and test validation is conducted for the cluster of jobs.

Still another strategy aimed at increasing sample size for validation research is synthetic validity.[65] With this method, jobs or positions within jobs are grouped together when they share a particular job performance requirement, and the research is conducted for those jobs as a cluster only for that performance requirement. An example of this method is provided by the Peterson et al. study.[78] A large-scale job analysis of 8000 entry level positions in the insurance industry resulted in the identification of 50 performance dimensions reflecting all important performance requirements for these many positions. For example, the dimension "Looks up codes, information in tables, charts, and so on," is relevant for a large number of the positions; therefore tests were developed to predict performance on this dimension, and the validity of these tests will be evaluated using all of the persons in these jobs. The same procedure will be followed with each of the other performance dimensions.

EEO Considerations. A brief background statement may be useful here. During the 1940s and 1950s the test publishing industry grew by leaps and bounds. Tests measuring abilities, aptitudes, and personality were developed at a fast pace. Unfortunately, tests were frequently used to discriminate against black persons and other minorities in selecting people for jobs. At times this was done purposely; at other times the "testing craze" blinded employers to what was happening regarding discrimination in hiring practices.

The Civil Rights Act of 1964 began to change all this. This act, along with updates to it, results of court cases related to discrimination in hiring and promoting, and guidelines issued by the Equal Employment Opportunity Commission (EEOC) and the Office of Federal Compliance Commission, among others, have all had a great impact on personnel selection practices in the United States. Essentially, federal, state, and local governments have stepped in and announced that tests should not be used to discriminate against blacks, women, or other minorities. Instead, everyone should have an equal opportunity to employment in our country's jobs. This very worthwhile principle has enlarged the scope of factors an organization must consider in selecting people for jobs.

The considerations, taken in detail, are very complex. Attending to constantly changing government guidelines, precedents from new court decisions, and revisions in professional standards[5] is a very difficult and complex business. A central concern, however, is simply this. Tests or other selection procedures used for selection are legal if they do not show "adverse impact" on any minority group protected under Title VII of the 1964 Civil Rights Act. "No adverse impact" means that the proportion of minorities hired is not substantially less than the proportion of majority group members hired. (The rule of thumb is that the minority proportion should be no less than four fifths of the majority proportion.) However, when minority group members are hired at a significantly lower proportion than are majority group members (e.g., white males), adverse impact has been demonstrated, and the test is considered illegal, *unless* the organization shows the test to be valid in predicting important job criteria (e.g., job performance). Even if the test is found to be valid, the employer must demonstrate that he or she has made a reasonable attempt to find other tests that are also valid, but that show less adverse impact.

So what does all this mean to the employer engaged in the selection of personnel? Our main suggestion is simply that employers ensure the establishment of valid selection practices. Paying attention to the sequence outlined in this chapter would be most helpful: careful job analysis, systematic identification of important KSAOs, development or choosing of tests or other measures to tap these KSAOs, and assessment of the validity of these measures. We might add that complying with the spirit of the federal guidelines is certainly a worthwhile goal and *may* help a company avoid expensive lawsuits. Such compliance might include the following actions: (1) making an honest effort to recruit qualified minorities, (2) attempting to use selection procedures that are valid *and* that do not show adverse impact, and (3) establishing affirmative action goals for hiring and promoting minority persons.

Beyond the scope of this chapter are more detailed discussions of several issues in the area of EEO and personnel selection; the reader is referred to other sources for more in-depth information about the following topics: (1) test fairness models–statistical methods for ensuring fairness in selection decisions[13,79]; (2) value judgments and/or utility concerns that must be addressed before settling on a test fairness model[79,80]; and (3) differential validity problems–possible difficulties of a test's being valid for one group (e.g., the majority group) and not valid for another group (e.g., a minority group).[81,82]

5.2.3 PERSONNEL TRAINING

In the first part of this chapter we presented personnel selection as a strategy for increasing organizational productivity and effectiveness. We showed that accurate selection of persons for jobs can lead to dramatic increases in productivity and effective job performance. Such dramatic improvements can also be effected through training. Raising the skill levels of workers by training them produces the same effects on productivity as raising the overall skill levels of the work force through selection. Of course training can also have an effect on other things, such as the quality of work and safety in the workplace. For example, Komaki et al.[83] showed impressive improvements in the safe performance of job tasks by vehicle maintenance workers. Before training, about 50% of all job tasks were performed safely, compared to about 75% after training. Such improvement has obvious desirable consequences, such as reduced worker job time lost and reduced hazard to individual workers.

Although training and selection both lead to desirable outcomes, they can be thought of as diametrically opposed strategies to achieving the same ends. Personnel selection seeks to capitalize on individual differences by selecting individuals with higher levels of skills and abilities, whereas training seeks to "smooth out," or level, individual differences by improving each individual's skills. Indeed, there are organizations that almost always prefer to emphasize selection over training, or vice versa.

It does not require great wisdom to observe that both selection and training efforts are essential to organizational success and that both are enhanced by closer coordination and cooperation. Selection serves to identify not only the individuals most likely to benefit from training and to succeed on the job, but also the strengths and weaknesses of individuals entering training. Training increases the probability of job success and productivity for each individual.

When confronted with a "problem," for example, lower-than-desirable productivity or high turnover, clearly an organization always faces a choice among the strategies of selection, training, or others (e.g., job redesign, job elimination). As we have implied previously, the choice is seldom either-or; usually it is one of emphasis. The point is this: Organizations should not automatically assume that training or selection (or other strategies) is *the* answer to a particular problem, but should recognize that most likely a combination of approaches may be optimal.

Overview of the Personnel Training Process

Just as the proper first step in selection is a job analysis aimed at identifying the KSAOs important to effective job performance, the proper first step in the training process is analysis aimed at determining training needs. This needs analysis aids in choosing or developing the appropriate training application, comparable to the choosing or developing of appropriate selection tests. The training program is then applied, and its effects on individual and organizational productivity, efficiency, or other criteria of success are measured, comparable to the validation techniques applied to personnel selection measures.

We discuss, in turn, three components of the training process: needs assessment, training techniques, and training evaluation. First, however, we should note that space is insufficient to present an exhaustive review of the voluminous literature relevant to training. Goldstein[84] has completed a recent literature review in this area, and Campbell[85] has reviewed the pertinent literature from the 1960s. McGehee and Thayer's[86] work is a classic in this field and provides an excellent picture of the state of the art in the early 1960s. In addition, McGehee[87] has recently updated the 1961 work. Craig[88] presents a wealth of "nuts-and-bolts" training information, and Salvendy and Seymour[4] present a human factors orientation to training that is particularly appropriate for industrial blue collar jobs.

Unfortunately, most reviews of training research and application have been pessimistic about the rate of knowledge acquisition in this area and have bemoaned the faddish nature of training in the organization setting,[86,89] though the most recent review by Goldstein[84] is slightly more optimistic. Other criticisms of industrial training efforts focus on (1) the relatively small amount of attention devoted to needs assessment, that is, to determining *what* should be trained; (2) the even smaller degree of effort directed toward the evaluation of training programs; and (3) the undeserved reliance on a single canned technique or a few such techniques as the answer to all or most training problems. Regarding criticism 3, these canned techniques usually come and go, making training in organizations a somewhat faddish enterprise.

The lessons from the reviews for a person contemplating the development of a training course and program for an organization are relatively clear:

1. Training needs must be clearly detailed through appropriate analysis techniques.
2. Some kind of evaluation of training success beyond "happiness sheets" (questions about the satisfaction trainees derived from the training course) should be accomplished.
3. The use of one or a few training techniques as "the" answer(s) to training problems should be avoided.

Identifying Training Needs

Most training texts recommend three kinds of analyses to identify training needs. These analyses are variously labeled, but are usually called (1) organizational analysis; (2) job, task, or operations analysis; and (3) person analysis.

Organizational Analysis

Generally, the intent of an organizational analysis is the identification of problems within the organization that may be susceptible to solution through training. As have others,[87,90] we urge a broader viewpoint here—organizational analysis should identify problems, period. As stated previously, some problems may be most amenable to training solutions, some to selection, some to a combination of these approaches, and still others to additional organizational interventions. Chapter 2.1 of this handbook is devoted to organizational analysis and design, so we refer the reader there for a complete treatment of this topic. However, we do want to mention a particular form of organizational analysis called the "personnel audit." The personnel audit, as defined by Mahler,[91] is a "comprehensive analysis of all aspects of personnel work in an organization," including training and selection. The thrust of the audit is to gather information to guide improvements in the personnel system. Mahler provides a description of this technique, as well as samples of interview questions and the kinds of conclusions reached from an audit.

The point of organizational analysis, then, is to become aware of an organization's personnel-related problems by formally or informally monitoring "symptoms," such as turnover, profit, employee attitudes, and productivity, and/or actively seeking out the problems by means of job satisfaction surveys, personnel audits, projections of future personnel needs, and so on. Once these problems have been identified, it is then appropriate to consider a training approach as a possible solution among other possible solutions.

Job, Task, or Operations Analysis

We have already discussed the importance of job analysis to personnel selection. Job analysis, or task analysis, is also an important prerequisite to the design of training programs. Previously we discussed the study by Bownas and Heckman[7] to illustrate the use of job analysis in selection. We showed that job tasks can be grouped by similarity of content so that homogeneous clusters of tasks might be formed. The clusters of tasks, one of which was illustrated in Exhibit 5.2.1, can also serve as definitions of terminal behaviors for a training program. For example, Exhibit 5.2.1 shows a cluster of fire fighter tasks labeled "rescue," which includes such tasks as "Move heavy objects or materials to gain access to or to free trapped victims" and "Carry conscious, unconscious, or deceased victims down ladder or stairs using drags, slings, cots. . . ." Each of these task statements is a succinct description of a desired terminal behavior for a person undergoing fire fighting training.

There are many methods of job or task analysis that might be used in this context (see, especially, Chapter 2.4, as well as Chapters 3.2, 4.1, 4.5, and 4.6). We believe, however, that the method used should provide task, knowledge, or skill descriptions in sufficient detail that terminal training behaviors can be clearly specified. The analysis should also provide some method of determining the training priorities of the tasks, knowledge, or skills, such as ratings by incumbents or supervisors of importance to overall job success or of amount of time spent on the task or operation. Finally, as McGehee[87] points out, the analysis should specify the level of behavior or performance that is required, as well as providing a description of what is done on the job. Returning to our illustration of the fire fighter job, the task statement "Move heavy objects or materials to gain access to or to free trapped victims" is not complete. We must also specify the weights and shapes of the objects to be moved and the speed with which they should be moved.

An important aspect of the job analysis is establishing the content validity of the training objectives. As mentioned previously, EEO guidelines apply to selection tests and require employers to demonstrate the job relatedness of these tests. The same guidelines apply to training whenever success in training is a prerequisite to obtaining a position.[84] Perhaps even more important, however, is that it simply makes good sense, from an organizational effectiveness standpoint, to ensure that the training objectives or terminal behaviors are actually job related because of the obvious waste of training behaviors or skills that have little or no effect on job performance.

An example of a highly job-related needs analysis is provided by Peterson et al.[92] They performed an analysis of the correctional officer (prison guard) job in a Midwestern adult penal institution. Using a job analysis questionnaire developed from interviews with correctional officers and supervisors, they identified 79 critical job tasks. They also identified 18 worker characteristics (or KSAOs) in a similar manner. Exhibit 5.2.13 shows examples of these 79 tasks and 18 KSAOs.

Then, to determine the importance of the 18 KSAOs to performing the job, 25 experts were asked to rate the importance of each KSAO to successful performance on each of the 79 critical tasks. For example, the experts rated the importance of "communicating, counseling, and advising" for "reporting suspicious activities," for "conducting strip searches," and so on for each of the 79 tasks. Importance was rated on a 5-point scale where 1 = no importance and 5 = very high importance.

The overall importance of each KSAO was then determined by computing the mean of all ratings for the KSAOs. In this manner the most important KSAOs were identified, and training objectives

Exhibit 5.2.13 Examples of Critical Job Tasks and Relevant Worker Characteristics for the Correctional Officer Job

Critical Job Tasks

Report suspicious activities inside and outside the fence (e.g., persons or vehicles loitering near the fence).

Conduct strip search of inmates after visits.

Break up fights between inmates.

Relevant Worker Characteristics

Knowledge and skills relevant to dealing with violent inmates–firearms use, crowd control techniques, first aid principles and procedures, self-defense techniques.

Communicating, counseling, and advising–able to recognize needs of and differences between individuals; able to counsel and advise inmates and to handle personality conflicts among inmates; able to simply talk with and listen to inmates about such things as rule and regulation enforcement and inmate explanations for actions.

---

were set according to these ratings–that is, the most important KSAOs received the highest priority for future training efforts.

Thus job, task, or operations analysis serves to identify clearly what it is that should be trained. The end product should be a clear description of terminal behaviors, skills, or knowledge, ranked with regard to overall criticality for success on the job.

Person Analysis

The objectives of person analysis, variously labeled "personnel" or "manpower" analysis, are to evaluate the effectiveness with which individual employees are performing their jobs and, if a low level of effectiveness is revealed, to determine factors amenable to correction by training.[87]

We have already discussed the problems of accurately measuring the effectiveness of employee job performance. We emphasize that it is just as important to accurately measure job performance for purposes of assessing training needs as it is for purposes of evaluating the validity of selection tests.

If deficiencies in individuals' job performance are found, it becomes necessary in person analysis to diagnose the reasons for these deficiencies. The "objective" methods of performance measurement (production figures, wastage counts, etc.) provide little information in themselves about the reasons for low quality or quantity of work. However, if carefully developed and administered supervisory ratings are used, the reasons for deficiencies should be easier to infer. Recall our previous discussion of the military recruiter job. Three basic performance dimensions were discovered for this job: selling, human relations, and organizing skills. Ratings by supervisors using these job performance dimensions make it possible to isolate deficiencies in a particular recruiter's performance.

Another method of determining reasons for ineffective job performance is to test employees. This can be accomplished using work samples, achievement tests, or even aptitude or ability tests. Work samples will most readily identify those tasks or operations of the job that an employee is performing poorly. Achievement tests will be most useful for identifying deficiencies in knowledge or skills (mathematical knowledge, reading skills, etc.) that are necessary for effective job performance. Aptitude tests are probably best used to identify deficiencies in employees that may cause difficulties in learning a particular task or job. (See Maslow[93] for a review of the role of testing in training and development.)

Training Techniques

Catalog of Techniques

Once training needs have been identified and put into the form of training or instructional objectives, one must design a training program to fulfill those objectives. It would be extremely useful to have a taxonomy of training techniques such that the appropriate technique could be readily identified for each training problem or objective. Sadly, no such taxonomy exists. We quote Hinrichs:

> *It is in this area of selecting training techniques and developing programs that training administration today is more art than science. It is here that very little is known, as research*

evaluation has not revealed which techniques or combinations of techniques are most effective for achieving specific objectives.[94]

There are a number of ways to classify available training techniques. For our purposes, we have chosen the approach taken by Hinrichs, who groups training techniques into three categories: (1) content-oriented–techniques designed to impart substantive knowledge on a cognitive level; (2) process-oriented–techniques intended to change attitudes, develop awareness of oneself and others, and enhance the trainee's interpersonal skills; and (3) mixed–techniques that attempt both to transmit information and to alter attitudes.

Exhibit 5.2.14 presents brief descriptions of 10 common training techniques classified under these three categories, along with the advantages and disadvantages of each. As Hinrichs has said, the utilization of these techniques for particular applied training problems is largely an art, but we believe that the information in this exhibit provides a good starting place for identifying a proper training strategy.

The Process of Training

Together with choosing the particular techniques to be utilized in a training course, the trainer must be concerned with the *process* of training. Several authors have advocated the use of principles of learning (derived primarily from academic laboratory research) as guides for structuring the learning process in a training course.[85,87,95] Hinrichs[90] points out, however, that this advice has been largely ignored by those responsible for applied training. We, too, favor more consistent application of what is known about human learning to development of training programs in organizations. Accordingly, we present seven principles of learning often discussed by training authorities.

Principles of Learning. We draw primarily from descriptions of these principles provided by Campbell et al.[96] (The reader is referred to Chapter 4.3 for a discussion of topics related to learning curves.)

1. Distributed or Spaced Training Periods. How should one schedule the training sessions–8 hr/day for 5 days in a row or 2 hr/day for 20 days? Available research favors distributing shorter sessions over a longer period, but this evidence is based on research with motor skills or rote memorization. Little research focusing on this principle has been performed on more complex, cognitive types of learning.

2. Whole Versus Part Learning. Research indicates that there is an optimal amount of training that should be presented at one time. Unfortunately, the research offers little help in deciding what that amount might be for a particular training program.

Gagné's approach[97–99] can be thought of as a sophisticated form of the "part" learning approach. His strategy places primary emphasis first on content, or what is to be learned, and second on the order of its presentation. He emphasizes (a) clearly identifying the component tasks that make up global job duties or "final performance"; (b) ensuring that trainees achieve full competence on each of these component tasks; and (c) arranging the training program such that component tasks learned early in the program mediate or enhance the learning of component tasks worked on later (i.e., learning task A, which is taught first, makes the learning of task B, taught second, easier and faster).

Clearly, Gagné's approach places considerable importance on job or task analysis. It is also clear that this approach is best suited for jobs that can be easily divided into major tasks consisting of component subtasks and perhaps less well suited for jobs in which the duties are not so easily delineated.

3. Reinforcement. Those trainee behaviors that are rewarded will be comparatively well learned and utilized in other situations. Punishment is an ineffective reinforcer, and its use should generally be avoided in training programs. The problems here are: (a) What constitutes an effective reward to the trainee? and (b) Who or what should dispense the reward? Possibilities range from simple verbal praise from the trainer to promotions given by superiors for excelling in training. Campbell et al.[96] conclude that this principle may be the most valuable, but that it is difficult to apply optimally.

4. Feedback. Also called "knowledge of results," this principle enhances learning by making the training task more interesting and by allowing the trainee to correct his or her own mistakes. The trainer should provide feedback as quickly as possible after the trainee's response *and* should give the trainee reasons for mistakes and ways to avoid those kinds of mistakes in the future.

5. Motivation. This principle can be summarized by saying that a person must want to learn, but not too much. Little motivation or excessively high motivation will usually impair learning. Generally, motivation can be enhanced by making the training itself interesting, exciting, and so on, or by providing an external motivator, such as monetary rewards for high levels of performance.

Exhibit 5.2.14 Outline of Common Training Techniques

Technique	Advantages	Disadvantages
	Content-Oriented Techniques	
Lecture. As typified in usual formal classroom settings in high school and college.	Economical. Good preliminary training method to provide cognitive awareness. Good information giving device, especially when the lecturer is skilled and possesses knowledge not otherwise available to trainees.	No provisions for individual differences. Learning is not self-controlled. Probably limited transfer from lecture to actual skills, except for cognitive tasks. High verbal nature may threaten some trainees, especially those with low verbal ability. No systematic reinforcement provisions. Little opportunity for trainees to participate.
Audiovisual instruction. For example, films, television, records, slides.	Often presents material that cannot otherwise be seen. Tapes or cassettes can be useful in feedback. Can cover a large number of trainees at relatively low cost per trainee. Uniformity of presentation of training content. Provides variety in training content.	Trainees treated passively. Not really a technique in itself; usually a supplement or aid to other techniques. Professional guidance must be sought to prepare films.
Autoinstruction. Programmed instruction (PI), computer-assisted instruction (CAI). These employ two major approaches: a linear, stepwise approach and a branching technique.	Actively involves trainee. Individualized, self-paced. Gives immediate, private feedback. Can update at will, based on experience with the program. Writing the program requires careful organization. Saves training time.	Costly to prepare. Limited to situations where content is clear and objectives are readily identifiable. Low social involvement makes it of limited use for training social interaction skills.
	Mixed Techniques	
Conference or discussion. Discussion of topics, problems, and so on, by a group (i.e., a conference).	Can lead to considerable trainee involvement. Provides opportunities for clarification. Feedback is possible from conference leader and other participants.	Limited to small groups. In practice, often poorly organized. Requires considerable skill and finesse from the conference leader.

Case study or incident process. Study, analysis, and discussion of some area of activity, that is, finance, management, research, production, and so on.	Can be a dynamic, involving experience for trainee. Requires trainee to use logic, analysis, judgment. Trainee receives feedback through discussion. Can be useful for imparting knowledge and for teaching decision making and problem solving approaches.	Difficult to know how much information to include in a case. Role of the discussion leader is very important; must attempt to focus case study on understanding, not just a solution.
Simulations. Ranges from complex machine and computer simulations of space environment, flight training, and so on, involving motor skills to simpler, noncomputerized business games not involving motor skills.	Especially useful where there are interdependencies among people and hardware or where teamwork is essential. Provide transfer of training to job situation. Dynamic–permit viewing consequences of actions–yet are nonpunitive because mistakes are harmless rather than costly or injurious. Intrinsically motivating.	Often do not allow normal approaches to problems. Trainees become overinvolved and do not critique the effectiveness of their behavior. Tendency to "lock in to" a certain strategy to "win" the game. Degree of realism may be costly. Motor skill simulations are very expensive.
On-the-job training methods. There are several of these techniques: Job instruction training–trainer explains task, observes trainee's performance, and offers feedback. Orientation training–systematic effort to give new employee all basic job information. Apprentice training–work under a supervisor for specified period to reach "journeyman" status. Performance appraisal–provide job performance and behavior feedback by means of a formal appraisal system. Coaching–training and learning in day-to-day employee-supervisor relationship. Job rotation–move from one job to another, spending a relatively small amount of time on each job. All of these are based on the philosophy that people learn a job best by doing it.	Excellent exposure to actual job tasks. More or less immediate feedback from the job situation and from the trainer.	May be inefficient; low productivity and high waste possible. May be less involvement of trainee (especially in situations like job rotation). Instruction may not be as competent as that received in the training department. Training may be secondary to getting the job done.

Exhibit 5.2.14 (*Continued*)

Technique	Advantages	Disadvantages
	Process-Oriented Techniques	
Role playing. Trainees assume parts in a realistic situation and act as they believe the person in their role would act.	Considerable feedback received. Highly involving for trainee. Provides practice in interpersonal interaction and problem solving under "real-life" conditions.	Trainees may feel the exercise is childish. Trainees may "overact," neglecting problem solving aspects. Trainer has no control over reinforcements (which are in the hands of the other trainees in the situation). Time consuming. Relatively expensive. Only a limited number can participate.
Sensitivity (T-group) training. There is no single method. In the classic model, participants meet without an agenda to discuss questions about the "here-and-now" of the group process (why participants behave as they do, how they perceive each other, why they are experiencing emotions and feelings).	Highly motivating for some persons. Trainees may change job behavior.	Motivation for what? Reinforcement not controlled at all. Only small number can participate. Considerable psychological stress is generated.
Modeling. Behavior to be learned is shown by models through films or videotapes. Past focus of this technique has been on supervisory behaviors in interpersonal situations. Sequence: Clear statement of the behaviors to be applied. A filmed model or demonstration situation of the skills being applied. Practice through role playing for each trainee. Social reinforcement of correct behavior in the practice situation. Planning by each trainee in how to transfer skills back to his or her job situation.	Combines advantages of role playing and sensitivity training with more traditional methods (lecture, audiovisual, discussion). As practiced, emphasizes direct job relatedness of new skills and transfer to job situation.	Demands much of trainer; may require special training for him or her. May be expensive to construct adequate, appropriate film or demonstration models.

Source: Based on reference 90.

6. Transfer of Training. This principle states that the behaviors or skills learned in the training will be more readily utilized on the job to the extent that the elements of the skill or behavior as presented in training are similar to those required on the job. This means that, as much as possible, the training context should simulate conditions on the job. Care should be taken to ensure that performance requirements on the job are being closely attended to in developing training course content. Incidentally, this is reminiscent of our discussion of work samples versus psychomotor skill tests for personnel selection. Work samples have been more accurate indicators of subsequent job performance, presumably because the skills required to perform effectively on them are more similar to the actual skill demands on the job.

7. Practice. The new behaviors learned in training must be used repeatedly so they will not be forgotten. The lesson here for trainers is simply to allow ample time in the training program for practicing the new skills or behaviors.

It should be apparent by now that Hinrichs' opinion about training course design being more art than science is justified. We do believe, however, that thorough needs analyses, particularly job analysis and person analysis, will go a long way toward clarifying the choices that must be made regarding training techniques and the arrangement of the learning experience.

Evaluation of Training

The evaluation of a particular training effort or program is no different in theory than the evaluation of any other intervention, for example, a new drug intended to cure or alleviate the effects of disease, a treatment program for alcoholism, or a new fertilizer to improve plant growth. The objective of any evaluation is to determine whether the intervention had an effect and to estimate the strength of that effect. Generally, the desired effect of industrial training efforts is to improve the job performance of those exposed to the training. McGehee[87] points out that there are two central reasons for properly evaluating training programs: (1) training costs money, and it is likely that organization members responsible for training will be called upon to justify such expenditures; and (2) EEO laws and guidelines apply to training, though their focus has heretofore been primarily on employment testing. Organizations are likely to be increasingly required to show evidence that their training programs are nondiscriminatory and that the knowledge, skills, and behaviors taught in training programs are job related.

We mentioned previously that most scholars reviewing the training field have directed sharp criticisms toward evaluation of training efforts. Many training programs are simply never evaluated. Many others are evaluated by inadequate methods, usually no more than a brief questionnaire completed by trainees at the end of the program. These questionnaires often ask a few questions about overall reactions to the program, the competence of instructors, perceived benefits of participation, and so on. Although this information is of some value in determining whether trainees enjoyed the training experience, such evaluations offer no information about knowledge or skill acquisition by the trainees or about improved subsequent performance by trainees on the job.

Kirkpatrick[100] labels this brief, end-of-course evaluation an "evaluation of reactions." It is only the first of the following four kinds of evaluations that he believes should be accomplished:

1. *Reaction.* *How well did the [trainees] like the program?*
2. *Learning.* *What principles, facts, and techniques were learned?*
3. *Behavior.* *What changes in job behavior resulted from the program?*
4. *Results.* *What were the tangible results of the program in terms of reduced cost, improved quality, improved quantity, etc.?*[101]

According to Kirkpatrick, then, there are four layers of evaluation. The first two, reactions and learning, are much easier to obtain than the latter two, behavior and results. We can easily devise end-of-course questionnaires to tap trainee reactions. It is a bit more difficult to develop written or work sample tests to measure trainee progress in learning what we set out to teach them, but both of these types of measures can be applied before trainees leave the training program and the trainer's control. The real difficulty comes in attempting to measure changes in on-the-job behaviors or work results and in designing evaluation studies that permit the conclusion that such changes, if found, *are a result of the training program*, and not a result of other events that occurred coincidental to the training. Unfortunately, the evaluation criteria of most interest are these latter two. The real payoff of training occurs when a person's performance effectiveness back on the job is enhanced.

One reason for the greater difficulty in evaluating on-the-job performance and work results is the problem of adequately measuring individuals' job performance. We have already dealt with this problem. Recall that the objective performance indexes are almost invariably deficient or contaminated, and supervisory ratings of job performance may suffer from such problems as halo effect and leniency errors. However, these difficult problems can at least be alleviated, as indicated previously in the chapter.

Assuming we can obtain reasonably accurate measures of individuals' job performance, then it is possible to conduct studies that lead to excellent evaluations of training program effectiveness, provided that a proper experimental design is employed. We do not discuss experimental design in depth (see Campbell and Stanley[102] and Cook and Campbell[103] for full explications), but should briefly mention three important topics in this area: threats to validity, control groups, and randomization.

Threats to validity are factors that, when present in an experiment, such as the evaluation of a training program, may provide alternative explanations for any changes or "effects" that are found. Thus these factors are "threats" in that they, in addition to the training program itself, are plausible reasons for changes in trainee behavior. What are some of these threats? Campbell and Stanley[102] list 12 of them, among which are these examples: (1) "reactive effects of testing"—the effects of taking a pretest early in training on test scores at the end of training; and (2) "history"—events occurring during the training period other than the training program itself.

A major feature of a good experimental design is the use of control groups or comparison groups. Regarding training evaluation, a control group is a group of untrained workers performing jobs similar to those of workers who are undergoing training. (The workers undergoing training are called the "experimental group.") The job performance of persons in the control group and in the experimental group (after training) can then be measured and compared, and an estimate of the effect of training on job performance can then be obtained.

A very important feature of all experimental designs is randomization, or the random assignment of persons to experimental (training) and control (no training) groups. If it is not possible to make such random assignments, the usefulness of most experimental designs is greatly reduced. The main difficulty is an inability to separate effects resulting from the training program from effects resulting from the manner in which persons were selected for training.

In industrial settings appropriate control groups often cannot be formed, or it is impossible to make random assignments of employees to training programs. When this is the case, quasi-experimental designs should be utilized to evaluate training.[103] These designs do not eliminate as many threats to validity as do "true" experimental designs, but employing such designs is much better than conducting no evaluation at all.

To conclude this section, remember that the purpose of evaluating a training program is to answer the deceptively easy questions "Did the training make any difference?" and "If so, how much?" We again urge practitioners to evaluate training programs as routine practice. It may not be possible to design and conduct elegant studies to answer conclusively all research questions regarding a program's true worth, but some attempt should be made to determine whether or not the time and money spent on such programs is really justified.

Uses of External Training Resources

Large companies often have the resources to conduct training themselves, but smaller companies and even larger, well-staffed organizations may occasionally need to seek outside training support. On this topic, Whitlock[104] discusses the proper role of universities, colleges, and other educational institutions; Parry and Ribbing[105] review the use of outside training consultants; and Cantwell et al.[106] discuss external training programs and packages. These sources are helpful, but there is really no systematic evidence to provide clear-cut guidance on using such external resources; rather, the same common sense and healthy skepticism employed in making any large purchase of goods or services are in order.

The first question the organization must answer when considering the use of an external training resource is, "What kind of training do we want?" We have discussed the three major parts of the training process—needs analysis, training techniques, and evaluation—and external consultants are available to perform all three of these activities. We believe, however, that organization decision makers should play the major role in defining the training needs. Indeed, if the organization cannot supply a clear definition of its training needs, deciding upon a consultant or product supplier to approach will certainly be difficult.

Once an organization has a clear definition of its training needs, evaluating the appropriateness of various external resources becomes easier. Choice of a training strategy might range from a short film on a narrowly defined topic to a long-term, live-in program for managers at a university. Parry and Ribbing[105] point out that a consultant is usually called in for one or more of three reasons: (1) there is an urgent need ("time is short and stakes are high") that cannot be met with internal staff; (2) there is a requirement for specialized expertise or facilities not available internally and/or too costly to develop internally; (3) a political need exists for the neutrality or credibility that is more readily attributed to outside consultants than to internal training staff.

Certainly, consultants are not always necessary. For example, college and university courses or programs are frequently useful sources of "broadening" courses for managers and others, and vocational schools or junior colleges offer training in specific skills, such as welding, mechanical drawing, bookkeeping, computer operations, and so on.

Finally, the organization should not omit the evaluation phase simply because the training was not done in-house. The same critical stance should be taken toward external programs as toward internal programs.

5.2.4 CONCLUDING COMMENTS

Selection and training have been discussed as strategies for enhancing the effectiveness of organizations. We have suggested that efforts devoted to selection and training can effect dramatic increases in work force productivity. We repeat our previous statement that selection and training activities should be closely coordinated for maximum benefit to the organization.

Throughout this chapter, we have emphasized the importance of job analysis prior to implementation of selection procedures or training programs. Too often, organizations will reach for solutions (a test or a training package) before they have adequately defined the problem. Careful job analysis will go a long way toward ensuring the eventual success of selection and training efforts.

Good business practices and relatively new government laws and guidelines both call for careful evaluation of the success of selection procedures and training programs. We have provided guidance for carrying out these evaluations and strongly urge organizations to assess the validity of their selection procedures and to evaluate their training programs.

Finally, we note that the selection and training functions are parts of a larger system, that is, the total organization. Many other approaches are available to increase individuals' productivity and organizational effectiveness, as demonstrated in the other chapters of this handbook. The wise organization decision maker employs a variety of human resource utilization strategies, as appropriate, to enhance the overall effectiveness of the organization.

REFERENCES

1. J. E. HUNTER and F. L. SCHMIDT, "Implications of Job Assignment Strategies for National Productivity," in E. A. Fleishman, Ed., *Human Performance and Productivity*, in press.
2. F. L. SCHMIDT, J. E. HUNTER, R. C. MCKENZIE, and T. W. MULDROW, "Impact of Valid Selection Procedures on Work-Force Productivity," *Journal of Applied Psychology*, Vol. 64, 6 (1979), pp. 609–626.
3. SCHMIDT et al., "Impact of Valid Selection Procedures," p. 624.
4. G. SALVENDY and W. D. SEYMOUR, *Prediction and Development of Industrial Work Performance*, Wiley, New York, 1973.
5. "Uniform Guidelines on Employee Selection Procedures," *Federal Register*, Vol. 43, 1978, p. 38290.
6. *Griggs* v. *Duke Power Company*, 401 U.S. 424, 1971.
7. D. A. BOWNAS and R. HECKMAN, *Job Analysis of the Entry-Level Firefighter Position*, Personnel Decisions Research Institute, Minneapolis, 1976.
8. J. H. WARD and M. E. HOOK, "Applications of an Hierarchical Grouping Procedure to a Problem of Group Profiles," *Educational and Psychological Measurements*, Vol. 23, 1963, pp. 69–81.
9. M. J. BOSSHARDT and S. E. LAMMLEIN, *Development of a Selection Test Battery for Millwrights*, Personnel Decisions Research Institute, Minneapolis, 1979.
10. R. D. DESMOND and D. J. WEISS, "Supervisor Estimation of Abilities in Jobs," *Journal of Vocational Behavior*, Vol. 3, 1973, pp. 181–194.
11. M. D. DUNNETTE, "Aptitudes, Abilities, and Skills," in M. D. Dunnette, Ed., *Handbook of Industrial and Organizational Psychology*, Rand McNally, Chicago, 1976.
12. N. G. PETERSON and D. A. BOWNAS, "Human Characteristics, Task Structure, and Performance Acquisition," in E. A. Fleishman, Ed., *Human Performance and Productivity*, Lawrence Erlbaum Associates, Hillsdale, NJ, in press.
13. M. D. DUNNETTE and W. C. BORMAN, "Personnel Selection and Classification Systems," *Annual Review of Psychology*, Vol. 30, 1979, pp. 477–525.
14. A. R. JENSEN, *Bias in Mental Testing*, Free Press, New York, 1980.
15. E. E. GHISELLI, *The Validity of Occupational Aptitude Tests*, Wiley, New York, 1966.
16. B. J. DVORAK, "The General Aptitude Test Battery," *Personnel and Guidance Journal*, Vol. 35, 1956, pp. 145–154.
17. G. K. BENNETT, H. G. SEASHORE, and A. G. WESMAN, *Counseling From Profiles: A Casebook for the Differential Aptitude Tests*, Psychological Corporation, New York, 1951.

18. L. L. THURSTONE, "Primary Mental Abilities," *Psychometric Monographs*, No. 4 (1938).
19. F. L. RUCH and W. W. RUCH, *Employee Aptitude Survey: Technical Report*, Psychological Services, Los Angeles, 1963.
20. R. M. GUION and R. F. GOTTIER, "Validity of Personality Measures in Personnel Selection," *Personnel Psychology*, Vol. 18, 1966, pp. 135–164.
21. H. G. GOUGH, *Manual for the California Psychological Inventory*, Consulting Psychologists Press, Palo Alto, CA, 1957.
22. D. N. JACKSON, *Personality Research Form Manual*, Research Psychologists Press, Goshen, NY, 1974.
23. A. L. COMREY, *Comrey Personality Scales*, Educational and Industrial Testing Service, San Diego, CA, 1970.
24. H. J. EYSENCK and S. B. G. EYSENCK, *Eysenck Personality Inventory*, University of London Press, London, 1963.
25. A. L. EDWARDS, *Manual for the Edwards Personal Preference Schedule*, Psychological Corporation, New York, 1953.
26. H. G. GOUGH and A. B. HEILBRUN, JR., *The Adjective Check List Manual*, Consulting Psychologists Press, Palo Alto, 1965.
27. L. V. GORDON, *Gordon Personal Profile–Inventory*, The Psychological Corporation, New York, 1978.
28. J. P. GUILFORD and W. S. ZIMMERMAN, *The Guilford-Zimmerman Temperament Survey*, Sheridan Supply Company, Beverly Hills, CA, 1949.
29. R. B. CATTELL, H. W. EBER, and M. M. TATSOUKA, *Handbook for the Sixteen Personality Factor Questionnaire*, Institute for Personality and Ability Testing, Champaign, IL, 1970.
30. W. C. BORMAN, J. L. TOQUAM, and R. L. ROSSE, *Development and Validation of an Inventory Battery to Predict Navy and Marine Corps Recruiter Performance*, Report No. 22, final report submitted to Navy Personnel Research and Development Center, Personnel Decisions Research Institute, Minneapolis, 1978.
31. A. TELLEGEN, *The Differential Personality Questionnaire: A Preliminary Manual*, unpublished manuscript, University of Minnesota, Minneapolis, 1976.
32. H. GOUGH, "Personality and Personality Assessment," in M. D. Dunnette, Ed., *Handbook of Industrial and Organizational Psychology*, Rand McNally, Chicago, 1976.
33. E. A. FLEISHMAN, "Dimensional Analysis of Psychomotor Abilities," *Journal of Experimental Psychology*, Vol. 48, 1954, pp. 437–454.
34. E. E. GHISELLI and M. HAIRE, "The Validation of Selection Tests in the Light of the Dynamic Character of Criteria," *Personnel Psychology*, Vol. 13, 1960, pp. 225–231.
35. C. H. FREDERICKSON, "Abilities, Transfer, and Information Retrieval in Verbal Learning," *Multivariate Behavioral Research Monographs*, Vol. 69, 1969. p. 2.
36. K. M. ALVARES and C. L. HULIN, "Two Explanations of Temporal Changes in Ability-Skill Relationships: A Literature Review and Theoretical Analysis," *Human Factors*, Vol. 14, 1972, pp. 295–308.
37. W. A. OWENS, "Background Data," in M. D. Dunnette, Ed., *Handbook of Industrial and Organizational Psychology*, Rand McNally, Chicago, 1976.
38. L. A. PACE and L. F. SCHOENFELDT, "Legal Concerns in the Use of Weighted Applications," *Personnel Psychology*, Vol. 30, 1977, pp. 159–166.
39. N. SCHMITT, "Social and Situational Determinants of Interview Decisions: Implications for the Employment Interview," *Personnel Psychology*, Vol. 29, 1976, pp. 79–101.
40. F. J. LANDY, "The Validity of the Interview in Police Officer Selection," *Journal of Applied Psychology*, Vol. 61, 1976, pp. 193–198.
41. R. M. GUION, "Recruiting, Selection and Job Placement," in M. D. Dunnette, Ed., *Handbook of Industrial and Organizational Psychology*, Rand McNally, Chicago, 1976.
42. G. K. BENNETT and R. A. FEAR, "Mechanical Comprehension and Dexterity," *Personnel Journal*, Vol. 22, 1943, pp. 12–17.
43. W. F. LONG and C. H. LAWSHE, "The Effective Use of Manipulative Tests in Industry," *Psychological Bulletin*, Vol. 7, 1941, pp. 385–397.
44. J. TIFFIN and R. J. GREENLY, "Experiments in the Operation of a Punch Press," *Journal of Applied Psychology*, Vol. 23, 1939, pp. 450–460.
45. M. L. BLUM, "Selection of Sewing Machine Operators," *Journal of Applied Psychology*, Vol. 27, 1943, pp. 35–40.

46. G. C. INSKEEP, "The Use of Psychomotor Tests to Select Sewing Machine Operators–Some Negative Findings," *Personnel Psychology*, Vol. 24, 1971, pp. 707–714.

47. J. E. CAMPION, "Work Sampling for Personnel Selection," *Journal of Applied Psychology*, Vol. 56, 1972, pp. 40–44.

48. E. J. CROFT, "Prediction of Clothing Construction Achievement of High School Girls," *Educational and Psychological Measurement*, Vol. 19, 1959, pp. 653–655.

49. D. L. EKBERG, "A Study in Tool Usage," *Educational and Psychological Measurement*, Vol. 7, 1947, pp. 421–427.

50. D. L. GRANT and D. W. BRAY, "Validation of Employment Tests for Telephone Company Installation and Repair Occupations," *Journal of Applied Psychology*, Vol. 54, 1970, pp. 7–14.

51. F. W. UHLMANN, "A Selection Test for Production Machine Operators," *Personnel Psychology*, Vol. 15, 1962, pp. 287–293.

52. G. P. HOLLENBECK and W. J. MCNAMARA, "CUCPAT and Programming Aptitude," *Personnel Psychology*, Vol. 18, 1965, pp. 101–106.

53. S. GAEL and D. L. GRANT, "Employment Test Validation for Minority and Non-Minority Telephone Company Service Representatives," *Journal of Applied Psychology*, Vol. 56, 1972, pp. 135–139.

54. D. W. BRAY and D. L. GRANT, "The Assessment Center in the Measurement of Potential for Business Management," *Psychological Monographs*, 1966, 80(17 Whole No. 625).

55. J. J. ASHER and J. A. SCIARRINO, "Realistic Work Sample Tests: A Review," *Personnel Psychology*, Vol. 27, 1974, pp. 519–533.

56. D. J. SCHWARTZ, "A Job Sampling Approach to Merit System Examining," *Personnel Psychology*, Vol. 30, 1977, pp. 175–185.

57. D. W. BRAY, R. J. CAMPBELL, and D. L. GRANT, *Formative Years in Business: A Long-Term Study of Managerial Lives*, Wiley, New York, 1974.

58. R. B. FINKLE, "Managerial Assessment Centers," in M. D. Dunnette, Ed., *Handbook of Industrial and Organizational Psychology*, Rand McNally, Chicago, 1976.

59. O. K. BUROS, Ed., *The Eighth Mental Measurement Yearbook*, The Gryphon Press, Highland Park, 1978.

60. B. V. GILMER and E. L. DECI, *Industrial and Organizational Psychology*, McGraw-Hill, New York, 1977.

61. R. M. GUION, "Criterion Measurement and Personnel Judgments," *Personnel Psychology*, Vol. 14, 1961, pp. 141–149.

62. M. D. DUNNETTE, "A Note on *the* Criterion," *Journal of Applied Psychology*, Vol. 47, 1963, pp. 251–254.

63. M. D. DUNNETTE, *Personnel Selection and Placement*, Wadsworth, Belmont, CA, 1966.

64. M. D. DUNNETTE and W. C. BORMAN, "Personnel Selection and Classification Systems," p. 486.

65. R. M. GUION, *Personnel Testing*, McGraw-Hill, New York, 1965.

66. K. PEARLMAN, F. L. SCHMIDT, and J. E. HUNTER, "Validity Generalization Results for Tests Used to Predict Job Proficiency and Training Success in Clerical Occupations," *Journal of Applied Psychology*, Vol. 65, 1980, pp. 373–406.

67. W. L. HAYS, *Statistics for Psychologists*, Holt, Rinehart & Winston, New York, 1963.

68. F. L. SCHMIDT, J. E. HUNTER, and V. W. URRY, "Statistical Power in Criterion-Related Validity Studies," *Journal of Applied Psychology*, Vol. 61, 1976, pp. 473–485.

69. R. J. WHERRY, SR., "Underprediction From Overfitting: 45 Years of Shrinkage," *Personnel Psychology*, Vol. 28, 1975, pp. 1–18.

70. N. SCHMITT, B. W. COYLE, and J. RAUSCHENBERGER, "A Monte Carlo Evaluation of Three Formula Estimates of Cross-Validated Multiple Correlation," *Psychological Bulletin*, Vol. 84, 1977, pp. 751–758.

71. M. D. DUNNETTE and S. J. MOTOWIDLO, *Development of a Personnel Selection and Career Assessment System for Police Officers for Patrol, Investigative, Supervisory, and Command Positions*, prepared for Law Enforcement Assistance Administration, Personnel Decisions, Inc., Minneapolis, 1975.

72. H. E. BROGDEN, "When Testing Pays Off," *Personnel Psychology*, Vol. 2, 1949, pp. 171–183.

73. H. E. BROGDEN and E. K. TAYLOR, "The Dollar Criterion: Applying the Cost Accounting Concept to Criterion Construction," *Personnel Psychology*, Vol. 3, 1950, pp. 133–154.

74. L. J. CRONBACH and G. C. GLESER, *Psychological Tests and Personnel Decisions*, University of Illinois Press, Urbana, 1957.

75. L. J. CRONBACH and G. C. GLESER, *Psychological Tests and Personnel Decisions*, 2nd ed., University of Illinois Press, Urbana, 1965.

76. R. C. TAYLOR and J. T. RUSSELL, "The Relationship of Validity Coefficients to the Practical Effectiveness of Tests in Selection," *Journal of Applied Psychology*, Vol. 23, 1939, pp. 565–578.

77. F. L. SCHMIDT and J. E. HUNTER, "Development of a General Solution to the Problem of Validity Generalization," *Journal of Applied Psychology*, Vol. 62, 1977, pp. 529–540.

78. N. G. PETERSON, J. S. HOUSTON, and M. D. DUNNETTE, *Job Analysis of Entry-Level Insurance Industry Positions: An Interim Report*, Personnel Decisions Research Institute, Minneapolis, 1979.

79. J. E. HUNTER and F. L. SCHMIDT, "Critical Analysis of the Statistical and Ethical Implications of Various Definitions of Test Bias," *Psychological Bulletin*, Vol. 83, 1976, pp. 1053–1071.

80. N. S. PETERSEN and M. R. NOVICK, "An Evaluation of Some Models for Culture-Fair Selection," *Journal of Educational Measurement*, Vol. 13, 1976, pp. 3–29.

81. J. E. HUNTER and F. L. SCHMIDT, "Differential and Single-Group Validity of Employment Tests by Race: A Critical Analysis of Three Recent Studies," *Journal of Applied Psychology*, Vol. 63, 1978, pp. 1–11.

82. R. A. KATZELL and F. J. DYER, "On Differential Validity and Bias," *Journal of Applied Psychology*, Vol. 63, 1978, pp. 19–21.

83. J. KOMAKI, A. T. HEINZMANN, and L. LAWSON, "Effect of Training and Feedback: Component Analysis of a Behavioral Safety Program," *Journal of Applied Psychology*, Vol. 65, 1980, pp. 261–270.

84. I. L. GOLDSTEIN, "Training in Work Organizations," *Annual Reviews of Psychology*, 1980.

85. J. P. CAMPBELL, "Personnel Training and Development," *Annual Review of Psychology*, Vol. 22, 1971, pp. 565–602.

86. W. MCGEHEE and P. W. THAYER, *Training in Business and Industry*, Wiley, New York, 1961.

87. W. MCGEHEE, "Training and Development Theory, Policies, and Practices," in D. Yoder and H. G. Heneman, Jr., Eds., *ASPA Handbook of Personnel and Industrial Relations*, Bureau of National Affairs, Washington, DC, 1979.

88. R. L. CRAIG, Ed., *Training and Development Handbook*, McGraw-Hill, New York, 1976.

89. W. R. MAHLER, "Executive Development," in R. L. Craig, Ed., *Training and Development Handbook*, McGraw-Hill, New York, 1976.

90. J. R. HINRICHS, "Personnel Training," in M. D. Dunnette, Ed., *Handbook of Industrial and Organizational Psychology*, Rand McNally, Chicago, 1976.

91. W. R. MAHLER, "Auditing Pair," *ASPA Handbook of Personnel and Industrial Relations*, Bureau of National Affairs, Washington, D.C., 1976, pp. 2–91.

92. N. G. PETERSON, J. S. HOLTZMAN, M. J. BOSSHARDT, and M. D. DUNNETTE, *A Study of the Correctional Officer Job at Marion Correctional Institution, Ohio: Development of Selection Procedures, Training Recommendations and an Exit Information Program*, Personnel Decisions Research Institute, Minneapolis, 1977.

93. A. P. MASLOW, "The Role of Testing in Training and Development," in R. L. Craig, Ed., *Training and Development Handbook*, McGraw-Hill, New York, 1976.

94. J. R. HINRICHS, "Personnel Training," p. 842.

95. M. L. BLUM and J. C. NAYLOR, *Industrial Psychology*, Harper & Row, New York, 1968.

96. J. P. CAMPBELL, M. D. DUNNETTE, E. E. LAWLER, and K. E. WEICK, *Managerial Behavior, Performance, and Effectiveness*, McGraw-Hill, New York, 1970.

97. R. M. GAGNÉ, "Military Training and Principles of Learning," *American Psychologist*, Vol. 17, 1962, pp. 83–91.

98. R. M. GAGNÉ, *The Conditions of Learning*, Holt, Rinehart & Winston, New York, 1963.

99. R. M. GAGNÉ, *Essentials of Learning for Instruction*, Dryden Press, Hinsdale, IL, 1974.

100. D. L. KIRKPATRICK, "Evaluation of Training," in R. L. Craig, Ed., *Training and Development Handbook*, McGraw-Hill, New York, 1976.

101. D. L. KIRKPATRICK, "Evaluation of Training," pp. 18–22.

102. D. T. CAMPBELL and J. C. STANLEY, "Experimental and Quasi-Experimental Designs for Research on Teaching," in N. L. Gage, Ed., *Handbook of Research on Teaching*, Rand McNally, Chicago, 1963. (Also published as *Experimental and Quasi-Experimental Designs for Research*, Rand McNally, Chicago, 1966.)

103. T. D. COOK and D. T. CAMPBELL, "The Design and Conduct of Quasi-Experiments and True Experiments in Field Settings," in M. D. Dunnette, Ed., *Handbook of Industrial and Organizational Psychology*, Rand McNally, Chicago, 1976.

104. G. H. WHITLOCK, "The Role of Universities, Colleges, and Other Educational Institutions in Training and Development," in R. L. Craig, Ed., *Training and Development Handbook*, McGraw-Hill, New York, 1976.

105. S. B. PARRY and J. R. RIBBING, "Using Outside Training Consultants," in R. L. Craig, Ed., *Training and Development Hardbook*, McGraw-Hill, New York, 1976.

106. J. A. CANTWELL, J. D. HOSTERMAN, and H. R. SHELTON, "Using External Programs and Training Packages," in R. L. Craig, Ed., *Training and Development Handbook*, McGraw-Hill, New York, 1976.

CHAPTER 5.3

Job Evaluation

ERNEST J. MCCORMICK

Purdue University

5.3.1 INTRODUCTION

The problem of establishing satisfactory wage and salary scales is one that haunts the management of many organizations. Wage and salary policies of course have an impact upon every employee of an organization and therefore are of concern to employees on an across-the-board basis. The reader interested in further discussions of job evaluation systems is referred to any of several relevant texts, such as Dunn and Rachel,[1] Livy,[2] Sibson,[3] Rock,[4] and Zollitsch and Langsner.[5]

5.3.2 THE CONCEPT OF EQUITY IN COMPENSATION

In discussions and theories relating to compensation policies, the notion of "equity" is perhaps the most common thorn that keeps bobbing up. Without getting too deeply into the theoretical aspects, let us consider briefly formulations relative to equity in pay.

Adams' Equity Theory

One of the formulations of equity theory is that set forth by Adams.[6] He makes the point that individuals' perceptions of equity in pay are based primarily on a comparison of their own situation with that of others. His equity theory is predicted on the relationship between peoples' perceived *outcomes* from their work involvement and their perceived *inputs* into their jobs, as compared to the outcomes and inputs of other "comparison" persons. The outcomes include any aspects of the job situation that are viewed as being of value, such as pay, fringe benefits, status, and intrinsic interest in the job. The inputs, on the other hand, include any "costs" to the individuals, such as education, skill, personal qualifications, and effort. Adams hypothesizes that this comparison can be expressed as a pair of ratios, as follows:

The Individual		The Comparison Person(s)
$\frac{\text{outcome}}{\text{input}}$	versus	$\frac{\text{outcome}}{\text{input}}$

If we express the outcomes and inputs as simply high (H) or low (L), the following ratios illustrate conditions of equity or inequity:

Equity: $\frac{H}{H}$ versus $\frac{H}{H}$ $\frac{L}{H}$ versus $\frac{L}{H}$ $\frac{H}{H}$ versus $\frac{L}{L}$

Inequality (overreward): $\frac{H}{L}$ versus $\frac{L}{L}$ $\frac{L}{L}$ versus $\frac{L}{H}$ $\frac{H}{H}$ versus $\frac{L}{H}$

Inequality (underreward): $\frac{L}{L}$ versus $\frac{H}{L}$ $\frac{L}{H}$ versus $\frac{L}{L}$ $\frac{H}{H}$ versus $\frac{H}{L}$

This chapter is based in large part on material from the following sources: ERNEST J. MCCORMICK, *Job Analysis: Methods and Applications* (New York: AMACOM, a division of American Management Associations, 1979), pp. 306–329; and from ERNEST J. MCCORMICK and DANIEL R. ILGEN, *Industrial Psychology*, 7th ed., Prentice-Hall, Inc., Englewood Cliffs, NJ, 1980, Chapter 12. Appreciation is expressed to the publishers of these books, AMACOM and Prentice-Hall, Inc., for permission to use these sources.

Adams' equity theory is based on a broad concept called "cognitive dissonance." In the realm of pay, a person presumably would experience some dissonance, or inequity, if the ratios were those characterized as conditions of inequity. Such dissonance can, of course, be in either direction, reflecting "overreward" or "underreward," although it is probably only human to be more aware of "underreward" conditions than the reverse!

Jacques' Equitable Payment Concept

In Great Britain Jaques has postulated that equitable pay for jobs is based primarily on what he calls the "time span of discretion" (TSD). In his scheme he differentiates between the *prescribed* content of jobs (those elements of the work about which the person has no authorized choice) and the *discretionary* content (those elements about which a choice as to how to accomplish them is left to the person). The TSD is defined by Jaques as follows:

> *Time-span of Discretion (TSD): the longest period which can elapse in a role before the manager can be sure that his subordinate has not been exercising marginally sub-standard discretion continuously in balancing the pace and the quality of this work.*[7]

In effect, this is the maximum period during which the manager relies on the discretion of his or her subordinate and the subordinate works on his or her own account. This concept, in turn, depends very much upon the "review of discretion," which Jaques defines as follows:

> *Review of Discretion: review by the immediate manager, or by someone acting on his behalf who is accountable for reporting to him, of the discretion exercised by a subordinate in carrying out a task, as shown in the completion time and the quality of the result.*[8]

The review of a person's work may be either direct (as by the manager) or indirect (as by an inspector, by someone else in the organization, or by the complaints of a customer, for example).

In the vernacular it might be said that the TSD can be thought of as the time before peoples' sins catch up with them. In the case of a production job, the inspector can detect unsatisfactory work within a matter of hours, whereas in the case of a president of a large corporation, it might be several years before performance can be adequately reviewed (by the board of directors). The procedures for measuring the TSD are set forth by Jaques[9–11] and will not be repeated here, but generally, the TSD is expressed in terms of minutes, hours, days, weeks, months, and years.[12]

Given the TSD of jobs, Jaques proceeds to make the point that such measures are highly correlated with the opinions of people regarding the rates of pay that are considered to be fair or equitable for particular jobs.

Although Jaques' concept of the TSD has been subject to certain criticisms and has not been used extensively as the basis for establishing compensation rates, its relevance to this discussion is that it focuses on the theme of equity in pay. Although Jaques' formulation is different from Adams' equity theory, both of them bring to our attention that "equity" in pay is very much based on the perceptions of people regarding the "worth" of jobs, including their perceptions of the worth of their own jobs in relation to that of other jobs. The implication of this is that the going rates of jobs are substantially influenced by such perceptions through the employment process, in which people accept or reject jobs at rates that are offered. If the offered rates are not considered to be reasonably equitable, the job offers usually would be rejected.

5.3.3 OBJECTIVES OF COMPENSATION POLICY

Having briefly discussed Adams' equity theory and Jaques' TSD, let us now return to the matter of job evaluation and to crystallizing the appropriate objectives of compensation policy. To be successful, the compensation policy should provide rates of pay for jobs that are perceived as being reasonably "equitable." In terms of practical considerations, there are two aspects of equity related to the rates of pay that are established on the basis of a job evaluation system. One of these aspects deals with relative differentials in earnings among the jobs *within* an organization. The other deals with differentials in earnings between jobs within the organization and corresponding jobs *outside* the organization. We can think of these as *internal* and *external* considerations. This distinction may help to crystallize what can be thought of as the two objectives of compensation policy.

Internal Compensation Objectives

For a compensation policy to be "acceptable" to those whose earnings are to be determined by it, there must be some scheme whereby the *relative* differences in pay for various jobs are generally recognized by most employees as being reasonably equitable. The relative pay of several jobs might

Exhibit 5.3.1

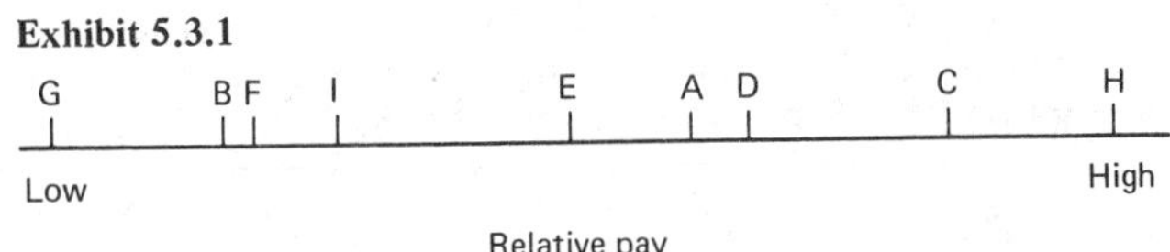

be depicted as falling along a line, such as that shown in Exhibit 5.3.1, in which the letters represent different jobs.

If, by some means, one can order jobs along a scale in such a fashion that the employees generally recognize and accept pay differentials such as those shown in the exhibit, then the compensation policy can be said to have fulfilled one of the objectives. Note that the acceptance must include not only the *order* of jobs along the scale, but also the relative *magnitude of the differences between* the jobs.

External Compensation Objectives

It is not enough that the employees recognize and accept the internal pay differentials of the job as being equitable. Usually the pay scales for the various jobs also need to have some reasonable relationship to those of "corresponding" jobs in the outside world, in particular, to those in the labor market from which the employees come. Let us see what would happen if there were no such "reasonable" relationship.

If the pay scales for jobs within an organization were appreciably *lower than* those of the corresponding jobs elsewhere, one would expect the organization to have difficulty in hiring and keeping people for the jobs in question, since the people could generally earn more someplace else. On the other hand, if the pay scales within an organization were *much higher than* the prevailing pay scales for corresponding jobs elsewhere, the organization could well experience financial problems and might not be able to compete in the market for the sale of its goods and services because of excessively high wage and salary costs. Although the pay for any given job within the organization need not be exactly the same as for the same job elsewhere in the labor market, it needs to be within some "reasonable" range. This reasonable range is discussed later.

Discussion

If we accept these two objectives of compensation policy as being valid, we must, by implication, accept the notion that compensation rates for various types of jobs are primarily influenced by supply and demand factors in the labor market. There are some factors that at least partly upset the supply and demand influence, such as union influences and changes in cultural values. But generally it seems reasonable to believe that the dominant factor influencing pay for various jobs is that of supply and demand.

Given this assumption, the typical problem for wage and salary administrators is one of developing or adopting a procedure that would result in the establishment of pay scales for jobs that fulfill both of the objectives cited here. If an organization has only a few different jobs, and if those jobs exist in other organizations in the labor market, the organization might not need any system for setting its pay scales. It could just pay the going rate for the jobs in question. But if it has many different jobs, some systematic job evaluation system usually is desirable. This is particularly the case with organizations that have some "unique" jobs, that is, jobs that do not exist in their labor market. In such instances an appropriate job evaluation system can provide pay scales for those unique jobs on the basis of the same job characteristics as those of jobs that have identical counterparts in the labor market. Thus the pay for such jobs presumably also would be perceived as being reasonably equitable.

If an organization elects to use a job evaluation program, then it should be one that provides for the evaluation of jobs in terms of job variables that collectively reproduce or predict the going rates for jobs existing in both the organization and the labor market.

This point is illustrated in Exhibit 5.3.2, which shows, for two hypothetical job evaluation systems (A and B), differences in the degree to which the job evaluation point values for a sample of 15 jobs are related to going rates for those jobs. In the case of job evaluation system B (Exhibit 5.3.2*b*), jobs are much farther from a "line of best fit" than in the case of system A (Exhibit 5.3.2*a*). The vertical distance of any given job from the diagonal line illustrates how much the evaluation point value is "off" in predicting going rates, being either too high if below the line or too low if above the line. The predicted rate for any given job as based on either system (A or B) would be on the diagonal line directly above or below the dot representing the job. The amounts of overprediction and underprediction are illustrated using two jobs in Exhibit 5.3.2*b*.

In the case of these two hypothetical systems, it is clear that system A comes much closer to the prediction of going rates than does system B. The accuracy of predicting going rates is essentially a

Exhibit 5.3.2 Illustration, for Two Hypothetical Job Systems (A and B), of Differences in the Predictability of Going Rates of Pay From Job Evaluation Point Values Based on Those Systems. System A (*a*) Is Clearly Superior to System B (*b*).

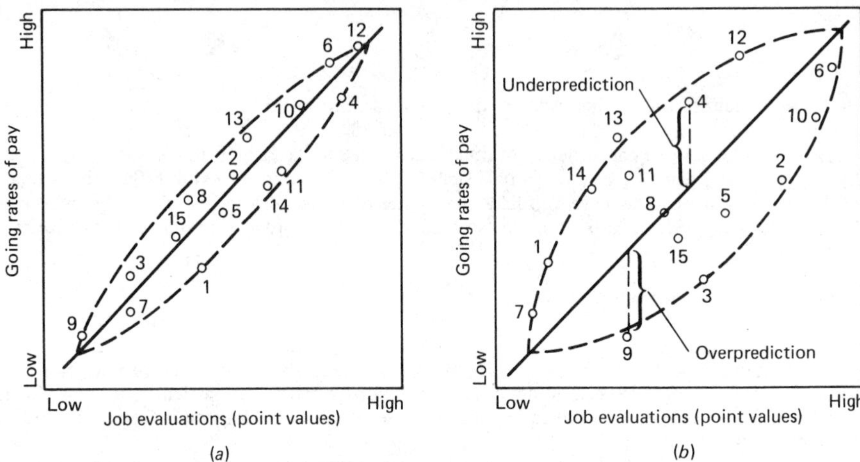

function of the extent to which the system "measures" the job variables that have contributed to the establishment of the going rates in the labor market, or at least measures job characteristics that are correlated with those variables.

Thus the going rates of pay for jobs in the labor market normally would serve as a standard or criterion against which to judge the adequacy (i.e., the validity) of a job evaluation system. Although such rates would usually be the most appropriate criterion, it should be added that in certain circumstances the rates paid within an organization can be used, since such rates usually have "settled down" at values that are reasonably in line with rates in the outside labor market. The factor comparison systems described later involve the use of such a criterion. The use of such a criterion of total job values as the basis for judging the adequacy of a job evaluation system is essentially a "policy capturing" approach. A job evaluation system developed to reproduce such rates can be thought of as capturing the prevailing pay policy of the organization.

In certain special circumstances, ratings by experts of overall job values have been used as the criterion against which job evaluation systems have been compared or judged.

5.3.4 TRADITIONAL JOB EVALUATION SYSTEMS

Job evaluation systems generally provide for the derivation of indices of relative job values within an organization, these indices usually being based on judgments of individuals about the jobs or on certain characteristics of the jobs. The judgments usually are made by members of a job evaluation committee. In turn, these indices are used as the basis for determining wage rates for the jobs covered by the system. Usually this conversion to wage rates is made on the basis of a wage survey to determine going rates of a sample of jobs. In effect, then, a job evaluation program provides a systematic basis for an organization to establish compensation rates for its jobs so that they are in reasonable alignment with going rates for corresponding jobs in the labor market.

The four traditional job evaluation methods are:

1. Ranking method.
2. Classification method.
3. Point method.
4. Factor comparison method.

Ranking Method

In the ranking method, jobs are compared with each other, usually on the basis of judged overall worth. Most typically, these judgments are obtained by a simple ranking of jobs—hence the name "ranking method." However, because jobs can be judged relative to others by the use of other procedures, such as the paired comparison procedure, this method could more appropriately be called the "job comparison method." The reliability of the evaluations usually is enhanced by having several individuals—preferably people who are already familiar with the jobs in question—serve as evaluators. When there are many jobs to be evaluated, however, it usually is difficult to find individuals who are familiar with all of them, and then it is necessary to rely more heavily on key jobs that are known by all raters.

Classification Method

The classification method consists of establishing several categories of jobs along a hypothetical scale. Each such category usually is defined and sometimes is illustrated. In using this method, each job is classified in a special category on the basis of its judged overall worth and its relation to the descriptions of the several categories. The classification method is a rather simple one to develop and use.

Point Method

The point method is probably the most commonly used procedure. It is characterized by the following features: (1) the use of several job evaluation factors; (2) the assignment of "points" to varying "degrees," or levels, of each factor; (3) the evaluation of individual jobs in terms of their

Exhibit 5.3.3 Scoring System for the MIMA Job Evaluation Plan for Shop Jobs

Factors and Degrees

Job Factors	Degrees				
	1st	2nd	3rd	4th	5th
Skill					
Knowledge	14	28	42	56	70
Experience	22	44	66	88	110
Initiative and ingenuity	14	28	42	56	70
Effort					
Physical demand	10	20	30	40	50
Mental and/or visual demand	5	10	15	20	25
Responsibility					
Equipment or process	5	10	15	20	25
Material or product	5	10	15	20	25
Safety of others	5	10	15	20	25
Work of others	5	10	15	20	25
Job Conditions					
Working conditions	10	20	30	40	50
Hazards	5	10	15	20	25

Grade Ranges

Score Range	Grades
139	12
140–161	11
162–183	10
184–205	9
206–227	8
228–249	7
250–271	6
272–293	5
294–315	4
316–337	3
338–359	2
360–381	1
Maximum points	500

Source. Job Evaluation Plan for Production and Related Jobs, Report No. 100, Midwest Industrial Management Association, Chicago, n.d.

degree or level of each factor, and the assignment to each job of the number of points designated for the degree or level on the factor; and (4) the addition of the point values for the individual factors to derive the total point value for each job. This total point value then serves as the basis for conversion to the corresponding wage or salary rate.

Examples of two point systems, those of the Midwest Industrial Management Association (MIMA) for shop jobs and office jobs, are shown in Exhibits 5.3.3 and 5.3.4, respectively. For each system there is a listing of the job factors that make up the system, the point values for different degrees of the various factors, and the score ranges (point values) for possible pay grades. Each factor and its degrees are defined as shown in the example in Exhibit 5.3.5. The definitions here are for the factor "complexity of duties" and its various degrees from the MIMA job evaluation system for office jobs. The members of a job evaluation committee typically evaluate each job on each factor, designating the degree that is considered most appropriate. (Usually the final degree assigned is based on the ratings of the several committee members.) The total point value for a job consists of the sum of the point values of the degrees assigned the job for the several factors. Exhibits 5.3.3 and 5.3.4 show the grades for jobs that correspond to various total score ranges.

Exhibit 5.3.6 shows, for the MIMA system for office jobs, an example of the job evaluation specifications for the job of executive secretary. This includes a job description and a list of fac-

Exhibit 5.3.4 Scoring System for the MIMA Job Evaluation Plan for Office Jobs

Factors and Degrees

	Degrees						
Job Factors	1st	2nd	3rd	4th	5th	6th	7th
Knowledge	15	30	45	60	75	100	
Experience	20	40	60	80	100	125	150
Complexity of duties	15	30	45	60	75	100	
Supervision necessary	5	10	20	40	60		
Effect of errors	5	10	20	40	60	80	
Contact with others	5	10	20	40	60	80	
Confidential data	5	10	15	20	25		
Mental and/or visual demand	5	10	15	20	25		
Working conditions	5	10	15	20	25		
Add for Supervisory Jobs Only							
Type of supervision	5	10	20	40	60	80	
Extent of supervision	5	10	20	40	60	80	100

Grade Ranges

Score Range	Grades
100 and under	1
100–130	2
131-160	3
161-190	4
191-220	5
221-250	6
251-280	7
281-310	8
311-340	9
341-370	10
371-400	11
401-430	12
431-460	13
461-490	14
491-520	15
521-550	16
Maximum points	825

Source. Job Evaluation Plan (Office) for Clerical, Technical, and Supervisory Positions, Report No. 200, Midwest Industrial Management Association, Chicago, n.d.

Exhibit 5.3.5 Example of the Definitions of a Job Evaluation Factor and Its Degrees

Complexity of Duties

This factor evaluates the complexity of the duties in terms of the degree of independent action; the extent to which the duties are standardized; the exercise of judgment; the type of decisions the job requires; and the exercise of discretion, resourcefulness, or creative effort in devising methods, procedures, products, scientific applications, and so on.

1st Degree–Little Judgment

Understand and follow simple instructions and use simple equipment involving few decisions.

2nd Degree–Some Judgment

Perform repetitive or routine duties working from detailed instructions and under standard procedures. Requires the making of minor decisions.

3rd Degree–Simple Analytical Judgment

Plan and perform diversified duties requiring an extensive knowledge of a particular field and the use of a wide range of procedures. Involves the exercise of judgment in the analysis of facts or conditions regarding individual problems or transactions to determine what action should be taken within the specifications of standard practice.

4th Degree–Complex Analytical Judgment

Plan and perform a wide variety of duties requiring general knowledge of company policies and procedures applicable within area of responsibilities and including their application to cases not previously covered. Requires considerable judgment to work independently toward general results, devising methods, modifying or adapting standard procedures to meet different conditions, and making decisions based on precedent and company policies.

5th Degree–Advanced Analytical Judgment

Plan and perform difficult work where only general methods are available. Involves highly technical or involved projects presenting new or constantly changing problems. Requires outstanding judgment and initiative in dealing with complex factors not easily evaluated and also the making of decisions for which there is little precedent.

6th Degree–Advanced Judgment and Ingenuity

Plan and perform complex work involving new or constantly changing problems where there is little accepted method of procedure. Involves participation in the formulation and carrying out of company policies, objectives, and programs for major divisions or function. Considerable ingenuity and exceptional judgment required to deal with factors not easily evaluated, interpret results, and make decisions carrying a great deal of responsibility. Direct and coordinate the work of subordinate supervision in order to attain objectives.

Source. Job Evaluation Plan (Office) for Clerical, Technical, and Supervisory Positions, Report No. 200, Midwest Industrial Management Association, Chicago, n.d.

tors with substantiating information that supports the assignment of degree ratings to the job (and the point values that correspond to those ratings).

Factor Comparison Method

The factor comparison method was originally developed and described by Benge, Burk, and Hay.[13] This discussion of it is based on its original formulation, with particular reference to the procedures involved in its development. The development process is time consuming and complex, but once the system is developed, its implementation is relatively straightforward.

The first step consists of the selection of perhaps 15 or 20 "key" jobs. These jobs should be relatively representative of the types of jobs to be covered by the system and should be those whose rates of pay are considered "satisfactory" and not subject to any dispute. These jobs are then judged, by two processes, by members of the job evaluation committee in terms of a set of certain factors, the five original factors being the following:

Mental requirements.
Skill requirements.
Physical requirements.

Exhibit 5.3.6 Example of the Job Evaluation Specifications for the Job of Executive Secretary, Including the Job Description and the Data Substantiating the Evaluations on Various Factors

Grade 7

Job title: Secretary, Executive

Total points 275

As secretary to company executives, take dictation and transcribe letters, memos, notices, and announcements from outline notes or verbal instructions, or compose letters independently from knowledge of circumstances and policy. Maintain personal files and records, perform personal services for executives, such as expenses, checks on dues, contributions, and so on. Open, stamp, and deliver mail for executives.

Maintain personnel records for office and plant employees, including attendance records, references, assignments; reconcile plant and office payrolls with bank account, prepare checks, make bank deposits. Take minutes of board meetings, transcribe for superior and board members. Keep appointment record for superiors, furnish reminders for business, social, and civic meetings.

Arrange for travel of company executives and other company personnel to cover itinerary to include plane, train, and hotel reservations. Perform other miscellaneous clerical duties of diversified nature as required.

Factors	Substantiating Data	Deg.	Pts.
Education	Knowledge of stenography, typing, transcription from written copy or other sources. Sufficient knowledge of English composition to avoid or detect errors. Familiar with ordinary filing, posting, clerical routines. Equivalent to high school, plus additional business training.	2	30
Experience	More than 3 and up to 4 years.	4	80
Complexity of duties	Diversified semiroutine duties. As a secretary to company executives, take dictation; transcribe letters, memos, notices; maintain personnel and personal files for company employees; make travel arrangements; take board minutes. Judgment required in the analysis of facts in situations to determine proper action and procedure within the limits of standard practice.	3	45
Supervision necessary	Under direction, plan and arrange own work within objective setup, referring only unusual cases to superior.	3	20
Effect of errors	Probable errors in writing letters, keeping records and files, making reservations, may result in loss of time, delays in trips, occasionally affect outside relationships.	3	20
Contacts with others	Regular and frequent contact with company executives, supervisors, employees, outsiders, and all types of positions requiring considerable tact and judgment.	4	40
Confidential data	Regular access to confidential data of all kinds, executive decisions, personnel information, other information of major importance, disclosure of which would be detrimental to company's interests.	4	20
Mental and/or visual demand	Stenography and typing require coordination of manual dexterity with normal mental and visual attention.	3	15
Working conditions	Usual office working conditions.	1	5

Source. Job Evaluation Plan (Office) for Clerical, Technical, and Professional Positions, Report No. 200, Midwest Industrial Management Association, Chicago, n.d.

Responsibility.
Working conditions.

The key jobs are first *ranked* on each of the factors mentioned, with all of the jobs appearing on each of the factor lists. The rankings usually are made independently by several people, with differences in rankings being resolved by consensus. Next, the jobs are subjected to a *rating* process in which the going rate (hourly or salary) is "allocated" to the individual factors on the basis of the judgments of the job evaluation committee members about how much of the going rate is being "paid" for each of the factors. This is done independently by the various raters, and the averages of these values are then used as the final "rate" for each job for each factor. In turn, these money rates for the key jobs are then assigned rank orders.

From these two procedures, then, there are two rank orders of the key jobs on each of the factors, the first being based on the direct ranking of the key jobs on each factor, and the second being the rank order of the money values resulting from the rating process. A hypothetical example of the results of these processes is given in Exhibit 5.3.7 for six jobs. (Usually 15 or 20 or more key jobs are used.) This exhibit shows the going rate for each job, the money value for each job "allocated" to the five factors, the rank order for each factor of these money values for the six jobs, and the rank order for each factor as assigned by the direct ranking procedure.

A major objective in deriving these two sets of rank orders is to identify any inconsistencies between them. In the examples given, such inconsistencies are shown for the jobs of poleman and rammer on the factor of physical requirements. When such inconsistencies are identified, they must be reconciled, or one or both of the jobs in question must be eliminated from the key jobs that are to represent the system. Once such inconsistencies are resolved, the remaining key jobs are formed into five scales, one for each factor, in which the money values of those jobs represent points along the scale. For example, the following scale is for the factor of mental requirements for the four key jobs that would be retained following the elimination of the two for which inconsistencies were found:

Job	Dollar Value on Mental Requirements
Pattern maker	1.85
Substation operator	1.35
Machinist	1.15
Laborer	0.30

In the actual application of the system, other jobs are evaluated on each individual factor by comparing them with the jobs that represent the scale for that factor, each job being assigned a value of each factor. The sum of these values for the five factors for any given job is the total value for the job. These are actually money values, relative to the rates of pay for the key jobs, but because of the now rather common inflationary tendency that would cause such values to become obsolete, there are procedures for making adjustments to account for inflation effects.

Since the original development of the factor comparison system, Edward N. Hay and Associates have developed what is called the "Hay Guide Chart–Profile Method" for use in the evaluation of managerial and professional jobs. Actually, the system has its origins in both the factor comparison and point systems, but in its present form it is rather distinct from both of its origins. The system provides for the comparison of jobs in terms of three factors, defined by Van Horn[14] as follows:

1. *Know-how* is the sum total of all knowledge and skills, however acquired, needed for acceptable performance.
2. *Problem solving* is the amount of original, self-starting thinking required by the job for analyzing, evaluating, creating, reasoning, and arriving at conclusions.
3. *Accountability* is the answerability for actions and for consequences of those actions.

In addition to the three main factors there are modifiers that apply to the ordinate of a double-entry chart for each main factor. The body of each chart contains point values at each intercept. Once a job is slotted into an agreed-upon position on each of the three charts, the point values from these three "positions" are added to produce total points.

5.3.5 JOB COMPONENT METHOD OF JOB EVALUATION

As indicated previously, the traditional methods of job evaluation involve making judgments about jobs or job "factors." Such judgments usually are based on information contained in job descrip-

Exhibit 5.3.7 Illustration of Data for a Hypothetical Sample of Key Jobs as Used in the Factor Method of Job Evaluation

		Mental Requirements			Skill Requirements			Physical Requirements			Responsibility			Working Conditions		
			Rank Order			Rank Order			Rank Order			Rank Order			Rank Order	
Key Job Title	Going Rate of Pay[a]	$[b]	$[c]	D[d]	$	$	D	$	$	D	$	$	D	$	$	D
Pattern maker	$6.30	1.85	1	1	2.30	1	1	0.70	5	5	1.05	2	2	0.40	5	5
Machinist	5.35	1.15	3	3	1.70	2	2	0.95	4	4	1.00	3	3	0.55	4	4
Poleman	5.20	0.55	4	4	0.90	4	4	1.85	②	①	0.70	4	4	1.20	2	2
Substation operator	4.50	1.35	2	2	1.60	3	3	0.15	6	6	1.25	1	1	0.15	6	6
Rammer	4.20	0.25	6	6	0.40	5	5	1.90	①	②	0.40	5	5	1.25	1	1
Laborer	3.50	0.30	5	5	0.25	6	6	1.60	3	3	0.30	6	6	1.05	3	3

Source. Adapted from H. G. ZOLLITSCH and A. LANGSNER, *Wage and Salary Administration*, 2nd ed., South-Western Publishing, Cincinnati, 1970. Table 7.5, p. 183.

Note. Titles and rates are illustrative and for comparison purposes only.

[a] Going rate of pay: prevailing hourly rate for the job.

[b] $: amount of pay judged to be paid for the factor in question as based on rating process.

[c] Rank order–$: rank order of these amounts for the jobs.

[d] Rank order–D: rank order of jobs as based on the "direct" ranking process.

tions and on the knowledge the evaluators may already have about the jobs in question. In recent years, however, certain procedures have been used either experimentally or operationally that provide for deriving "job evaluations" directly from structured job analysis questionnaire data, thereby completely bypassing the need for evaluations based on judgments. Such procedures have been referred to as the "job component method of job evaluation." This is a slight misnomer since no "evaluation" is required, but because the process of establishing wage and salary rates is typically called job evaluation, the term "job component method of job evaluation" is used here.

The basic scheme involves the following processes:

1. The development or selection of an appropriate structured job analysis questionnaire that provides for the analysis of jobs in terms of various "units" (i.e., components) of job-related information. Such questionnaires can consist of job-oriented work elements, such as task inventories, or of worker-oriented elements, such as those in the PAQ that was described in Chapter 2.4.
2. The derivation of numerical "weights" for the various components that reflect the values of the individual components as "contributors" to a criterion of total job values. In some instances such weights have been derived from the ratings of the individual job components, and in other instances by a statistical analysis of their importance as predictors of the criterion of total job values.
3. The analysis of the jobs in question using the structured job analysis questionnaire in question. Such an analysis typically results in a quantitative "score" for each job on each component, or at least an indication of its presence or absence in the job.
4. The use of an appropriate statistical procedure for building up an index of total job values from the combination of the weights of the individual components and of the scores for the individual components as related to the individual jobs.

Although there are variations on this central theme, such procedures result in *statistically derived indices* of total job values from quantitative job data based on structured job analysis questionnaires. Thus the traditional process of "evaluating" jobs is entirely eliminated. Certain examples of this method will illustrate the potential application.

Clerical Task Inventory

An early example of this approach, reported by Miles,[15] was the use of a checklist of office operations, which is now called the Clerical Task Inventory (CTI).* The CTI is a listing of 139 tasks that are performed in clerical and other office-type jobs. In using it the analyst rates the importance of each of the tasks to the job in question. These tasks had been rated by psychologists in terms of relative monetary worth. The mean of these ratings for each task is used as an index of its "worth."

In the case of the CTI for job evaluation purposes, each job in question is analyzed in terms of the importance of each CTI task. The importance rating of each task to each job is multiplied by the index of worth for the task. The products of these multiplications for the various tasks are then added together to derive a total "weighted" value for each job. In practice, it has been found that these total values for the five most important tasks result in the optimum correlation with a criterion of going rates for jobs.

Position Analysis Questionnaire

In a more generalized application of the job component method, McCormick, Jeanneret, and Mecham[16] used the PAQ† with a sample of 340 jobs of various kinds in various industries in various parts of the country. The PAQ is a structured job analysis questionnaire that provides for analyzing jobs in terms of 187 worker-oriented job elements. In this particular study, job dimension scores were derived statistically for the 32 factors that resulted from a previous factor analysis of the PAQ. A statistically weighted combination of scores on nine of these job dimensions produced correlations in the upper .80s with actual rates of pay for two subsamples consisting of 165 and 175 of the jobs as well as for the total sample of 340. With a larger sample of more than 800 varied jobs, the correlation with actual rates of pay was .85, as reported by Mecham.[17] For a sample of 79 jobs in an insurance company, Taylor[18] reported a correlation of .93 between a weighted combination of PAQ job dimension scores and actual rates of pay for the jobs in question.

In using the PAQ for job evaluation purposes, the scores of jobs on the various job dimensions are used in deriving total "job evaluation points" for individual jobs, along with statistically deter-

*The Clerical Task Inventory (originally called the "Job Analysis Check List of Office Operations") is copyrighted by C. H. Lawshe, Jr., and is available through the Village Book Cellar, 308 West State Street, West Lafayette, IN 47906.

†The PAQ is copyrighted by the Purdue Research Foundation and is available through the University Book Store, 360 West State Street, West Lafayette, IN 47906. Further information about the PAQ may be obtained through PAQ Services, 1625 North 1000 East, Logan, UT 84321.

mined "weights" for the individual job dimensions. There are two alternate approaches that can be used in deriving the weights for the individual job dimensions. The first approach is based on data for a large number of varied jobs as described previously; the result of the statistical analysis (technically, regression analysis) is an equation incorporating weights for the dimensions that collectively best predict a criterion of going rates for the jobs in the large and varied sample.

The second approach is one that uses as a criterion the rates of a sample of jobs within the organization rather than the going rates of a broad and varied sample. In this instance the statistical analysis results in the derivation of an equation incorporating weights for the dimensions that collectively best predict the rates for the sample of jobs within the organization.

A slight variation of this approach is represented by a study by Robinson et al.[19] This study dealt with a sample of 19 jobs in a medium-size city. The variation consisted of obtaining data on the going rates for those 19 jobs from 21 cities of similar size and of using the median rates for the individual jobs as the rates to be "captured" in the statistical analysis. The correlation coefficient between the predicted point values based on the PAQ job dimensions and the median rates for the jobs (from the 21 other cities) was .945. In addition, the 19 sample jobs were evaluated by four other methods of deriving compensation rates. The intercorrelation coefficients between the various methods ranged from .82 to .95. The job component method of job evaluation in this instance resulted in values for the various jobs that were as highly correlated with the going rates of jobs as the values derived by any of the other traditional methods of job evaluation and that were more highly correlated than some of the methods.

The PAQ has been used operationally for job evaluation by various private organizations, including those in public utilities, finance, insurance, service industries, manufacturing, and transportation. It has also been used by organizations in the public sector, especially in state and local government units.

Discussion

The job component method of establishing pay rates for jobs basically provides for deriving some index of values for various individual components of jobs, such as tasks and job dimensions, thus making it possible to derive total job evaluation values by combining the values of the individual components of the job. In other words, it is based on the direct statistical use of quantitative job analysis data from a structured job analysis procedure. The bridge between job analysis data and job values is then strictly statistical, thereby eliminating the need for making judgments about jobs as provided for by the traditional job evaluation methods.

5.3.6 CONVERTING JOB EVALUATION RESULTS INTO PAY SCALES

Most job evaluation systems result in jobs being placed along some hypothetical scale of job values. In the case of the point and job component methods, the values are expressed in terms of points. The rank order and classification methods result in rankings or job categories. These values, in turn, need to be converted to actual money values in order to establish the rates of pay for the individual jobs. The conventional use of the factor comparison method is the only exception to the need for such conversion, since it results in values that are actual money values.

The conversion of job evaluation results into money values typically involves establishing a criterion of total job values, which usually is based on going rates in the labor market, and developing a going rate curve and an organization rate curve.

Developing a Going Rate Curve

The going rate curve typically is based on data for a sample of jobs in the organization that have their counterparts in the labor market. It shows the basic relationship between the evaluations for those jobs, as based on the evaluation system used by the organization, and some representative index of their rates of pay in the labor market. The index of the going rate for any given job usually is based on the rates for that job in various organizations and may be the mean, the median, or a weighted index of those rates. Sometimes data on the going rates for jobs are already available in the organization, from local management organizations, from published government sources, or from other sources. If such data are not available for a representative sample of jobs, the organization may carry out a survey of its own.

In general, the going rate data, from whatever source, are represented in a form such as that shown in Exhibit 5.3.8, as presented by Dunn and Rachel.[1] This shows the relationship between the point values of a sample of jobs as based on their evaluations and the weighted average wage rates in the labor market. Dunn and Rachel argue for the use of a weighted average of the going rate for each job because they believe it is the most representative value, but the mean and median values are used in many circumstances. The line in the figure is the line of best fit and may be

Exhibit 5.3.8 Illustration of the Relationship Between Job Evaluation Points and Weighted Average Wage Rates in the Labor Market for a Sample of Jobs[a]

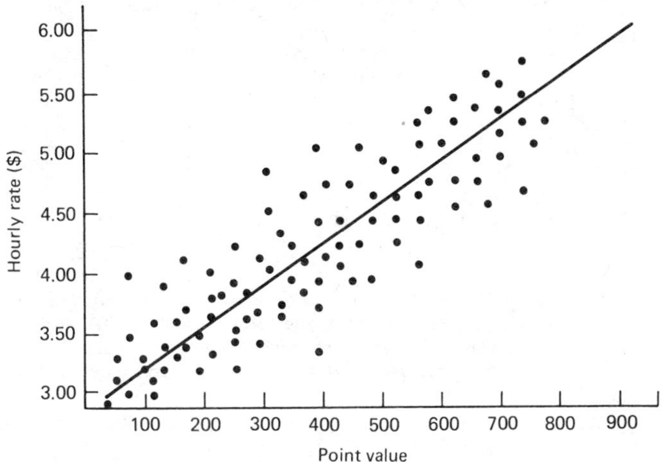

[a]Adapted from J. D. DUNN and F. M. RACHEL, *Wage and Salary Administration: Total Compensation Systems*, McGraw-Hill, New York, 1971, Figure 13-1, p. 223.

drawn by inspection or derived statistically. Although the relationship shown in Exhibit 5.3.8 is linear, in some instances it may be curvilinear.

Setting an Organization Rate Curve

The next stage is that of establishing a wage or salary curve for the specific organization. (In the case of private companies, this is called the company wage or salary curve.) This curve, derived on the basis of the going rate curve, sets the general pattern of rates for the jobs covered by the job evaluation system. Although this curve is based on data for a sample of jobs that exist in other organizations in the labor market, it is of course used in establishing rates of pay for all of the jobs covered, including those that are unique to the organization. This ensures that all the jobs covered will have their rates of pay established on the same basis. Where this curve is actually set with respect to the going rate curve in any given case, however, is a function of various considerations, including economic conditions, contract negotiations, and fringe benefits. Thus it can be at the level of the going rate curve as such or at some level above or below.

There is, of course, the question of how close the organization rate curve should be to the going rate curve. In this regard, Jaques[20] indicates that individuals whose actual payment bracket remains within 3% of "equity" tend to feel that they are being reasonably paid relative to others. In turn, those whose actual payment bracket falls 5% below equity tend to feel that they are being treated somewhat unfairly, and those whose earnings are 10% below equity definitely feel that they are being treated unfairly. Persons whose earnings fall as far as 15% below equity would be expected to seek a change in employment if opportunities exist. The implication of Jaques' interpretations is that pay scales generally should not fall much below 5% of equity, equity being defined by Jaques in terms of TSD. If we equate his concept of equity to the going rate, it would then seem that an organization wage curve should not fall below the going rate curve by more than about 5%.

In converting job evaluation points to actual rates of pay, different practices may be followed. It would be possible to take the evaluated points for a given job and derive the corresponding exact rate that would be applicable. Thus every slight difference in points would result in some difference in hourly rate. In practice, most organizations, including unions, feel that the inherent lack of perfect accuracy in the judgments that underlie a set of job evaluations makes it desirable to bracket together jobs of approximately the same point value and to consider these jobs as equal in setting up the wage structure. This bracketing results in so-called labor grades. The number of labor grades found in specific wage structures varies from around 8 or 10 to 20 or 25. The tendency of most current union demands in wage contract negotiations is to favor a relatively small number of labor grades.

In converting evaluations to pay grades, however, there are many variations that can be considered. Certain variations of these are discussed by Dunn and Rachel[1] and are illustrated in Exhibit 5.3.9. There are pros and cons to these and many other possible variations that should be considered in developing the particular scheme that would be optimum. In the case of unionized jobs, the pattern of conversion may well be a matter for negotiation.

The range in values within individual pay grades reflects the variability that usually is relevant in establishing rates for individuals whose jobs fall within the pay grade. There are two primary bases for such variations, one being performance evaluations of individuals, and the other, job or organization seniority. Some organizations use an automatic progression schedule under which increases automatically become effective after a specified period of time on the job. This principle is employed most frequently in the lower labor grades and with new employees, but it is sometimes used with jobs in higher levels as well. Some organizations use a combination of time and performance evaluations for designating rates for individuals.

Exhibit 5.3.9 Illustrations of Various Patterns for Converting Job Evaluations Into Pay Grades: (*a*) Wage Structure, No Overlap; Constant Range and Width. (*b*) Wage Structure, 50% Overlap; Constant Range and Width. (*c*) Wage Structure, 50% Overlap; Constant-Percentage Range; Constant Width. (*d*) Wage Structure, 50% Overlap; Constant-Percentage Range; Decreasing Width[a]

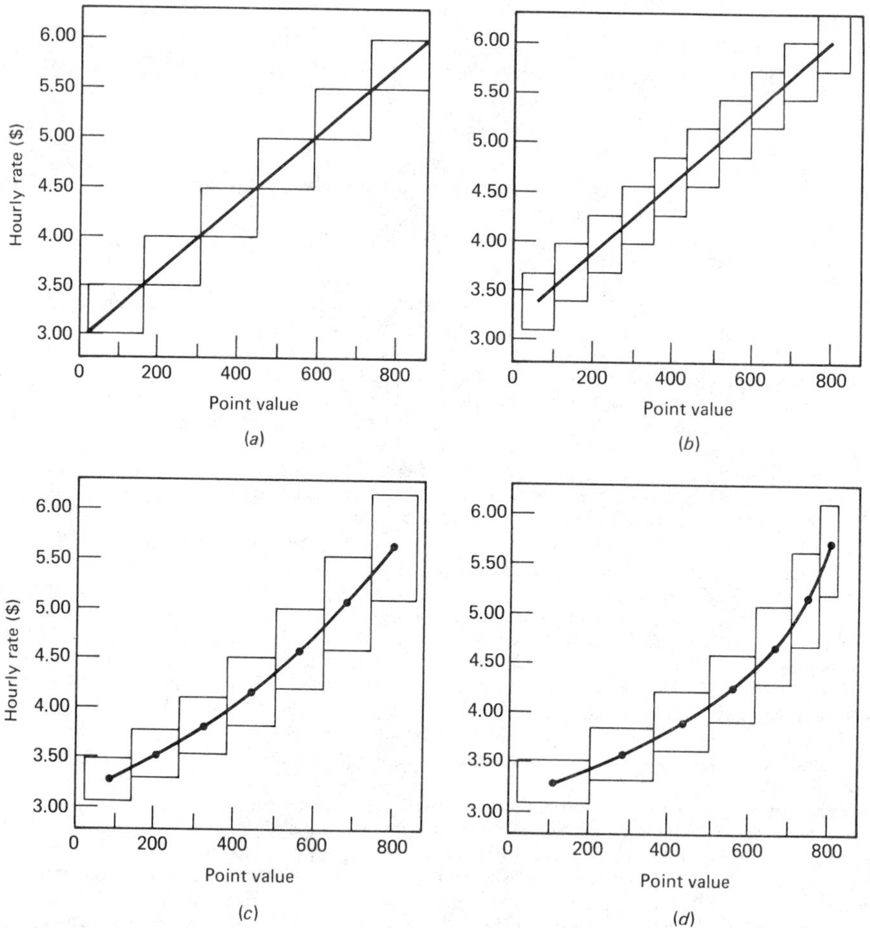

[a]Adapted from J. D. DUNN and F. M. RACHEL, *Wage and Salary Administration: Total Compensation Systems*, McGraw-Hill, New York, 1971, Figures 13-2, 13-3, 13-4, and 13-5, pp. 224, 225, 226, and 227, respectively.

Implementation of Job Evaluation System

Once a job evaluation system has been selected or developed, all of the jobs to be covered are analyzed and then evaluated. Usually the evaluation is done by members of a job evaluation committee. Once the evaluations are completed, the rate for any given job is determined on the basis of the conversion schedule that has been developed. Usually this means that a job is assigned to a given pay grade and assigned the rate that is appropriate for that grade.

For organizations that are implementing a job evaluation system where one does not already exist, or for those adopting a new system, it usually turns out that some jobs are being paid different rates than the (new) system provides for. If the job is being paid less than the schedule indicates, it is usually the practice to raise its rate to the minimum of the pay grade in which it falls. If a job is being paid more than the schedule indicates, it is usually the practice to retain the employees at these "red circle" rates for at least some guaranteed period, such as a year, and to grant them no raises until the general pay level catches up to their present rates. In some instances efforts are made to retrain such employees for jobs whose rates are in line with their current pay levels.

5.3.7 FEATURES OF A DESIRABLE JOB EVALUATION SYSTEM

Let us now crystallize the features of a desirable or satisfactory job evaluation system.

Job Characteristics

Clearly, a critical feature of a job evaluation system is that of the job characteristics on which the evaluations are based. This brings us back to the notion of equity, which has served as a common denominator in the previous discussion. As indicated previously, equity as related to pay is essentially a subjective perception on the part of people regarding the fairness of rates of pay for jobs. Equity is roughly reflected in the going rates of pay for various jobs in the labor market, since such rates are largely the consequence of the interaction between the demand and supply of qualified people who are willing to continue working at the going rates.

By implication, in most circumstances a desirable job evaluation system would be one providing for the evaluation of jobs in terms of the job characteristics that collectively add up to total values that are reasonably correlated with a criterion of going rates in the labor market. In both the point and factor comparison methods, the characteristics used are job factors. In the job component method, they are components such as tasks and job dimensions. For our discussion, let us think in terms of factors such as those used in the point method, although much of what is said about factors would also be relevant to other types of job characteristics.

First, most job evaluation systems tend to result in evaluations of jobs that are reasonably comparable. This generalization is supported by the study by Robinson et al.[19] in which the correlations between the results of four different methods of establishing job rates ranged from .82 to .95. Even if this supposition were true, one should probably select or develop a system that would be most valid in terms of predicting the total job values represented by the criterion of going rates in the labor market. Naturally, ease of understanding, amount of analyst training required, and cost of maintaining the evaluation system also should influence the selection of the system.

It is possible to determine statistically how effective any given system would be in achieving this objective of validity or even to develop a system that would do so. This really involves the identification, by statistical procedures, of the factors and of their appropriate statistical weights that collectively result in the highest correlation with whatever criterion of total job worth is used, such as going rates in the labor market. The scheme, as applied to either an existing system or an "experimental" system that has been developed, involves the following steps:

1. The selection of a representative sample of jobs in the organization.
2. The evaluation of the jobs using the factors of the system that has been selected or developed.
3. The derivation of criterion values of total job worth for the jobs in the sample, based on going rates in the labor market as determined by a wage or salary survey or some other criterion, such as rates within the organization or ratings by experts.
4. The application of appropriate statistical procedures, usually some form of regression analysis. The factors so identified and their statistically determined weights could then be used in the organization with reasonable assurance that the rates of pay based on the system would have a reasonable relationship with rates in the labor market.

A dominant feature of the job component method is that the optimum combination of job components and their weights is determined by statistical analysis, specifically, the relationship, for a sample of jobs, between values on such components and the criterion of total job worth, such as going rates. The results of this analysis are then converted into an equation for deriving indices of job values for all jobs to be evaluated from data based on whatever structured job analysis procedures are used.

Reliability of Job Evaluations

Still another aspect of job evaluation systems that can influence their utility is that of the reliability of the evaluations. In this context "reliability" refers to the degree of relationship between or among the evaluations of two or more independent evaluators, or between separate evaluations made at different times by the same evaluator. Reliability usually is measured by correlating pairs of independent evaluations for a sample of jobs or by correlating separate evaluations made by the same evaluator at different times. It is usually good practice to determine the reliability of evaluations in order to see if they are reasonably adequate. The correlations between and among different evaluators preferably should be reasonably high, such as around .85 or .90 or above. It should be added, however, that the pooling of the evaluations of several *good* evaluators made independently can produce composite evaluations that are more reliable than those of any one given evaluator. Such pooled reliability tends to increase, particularly with up to three or four evaluators, and then to increase more gradually with the addition of more evaluators, up to 10 or so.

Discussion

Wage and salary administrators must feel that they are continuously walking a tightrope because of the conflicting pressures that impinge upon them. A wage and salary program simultaneously must provide positive work incentives for the employees, must be generally acceptable to employees, must be reasonably competitive with conditions in the labor market, and must keep the organization solvent. Obviously, there are no pat and simplistic resolutions to meet these various objectives. However, insight and knowledge relevant to the problem can aid the process of developing a satisfactory program.

REFERENCES

1. J. D. DUNN and F. M. RACHEL, *Wage and Salary Administration: Total Compensation Systems*, McGraw-Hill, New York, 1971.
2. B. LIVY, *Job Evaluation: A Critical Review*, Halsted, New York, 1973.
3. R. E. SIBSON, *Compensation: A Complete Revision of Wages and Salaries*, AMACOM, American Management Associations, 1974.
4. M. L. ROCK, Ed., *Handbook of Wage and Salary Administration*, McGraw-Hill, New York, 1972.
5. H. G. ZOLLITSCH and A. LANGSNER, *Wage and Salary Administration*, 2nd ed., South-Western Publishing, Cincinnati, 1970.
6. J. S. ADAMS, "Wage Inequities, Productivity and Work Quality," *Industrial Relations*, Vol. 3, 1 (October 1963), pp. 9–16.
7. E. JAQUES, *Time-Span Handbook*, Heineman Educational Books, London, 1964, p. 11.
8. JAQUES, *Time-Span Handbook*, p. 11.
9. JAQUES, *Time-Span Handbook.*
10. E. JAQUES, *Equitable Payment*, 2nd ed., Carbondale, IL, Southern University Press, 1970.
11. E. JAQUES, *Measurement of Responsibility*, Wiley, New York, 1972.
12. JAQUES, *Time-Span Handbook*, pp. 107–108.
13. E. J. BENGE, S. L. BURK, and E. N. HAY, *Manual of Job Evaluation*, 4th ed., Harper & Row, New York, 1941.
14. C. W. G. VAN HORN, "The Hay Guide Chart–Profile Method," in M. R. Rock, Ed., *Handbook of Wage and Salary Administration*, McGraw-Hill, New York, 1972, pp. 2(86)–2(97).
15. M. C. MILES, "Studies in Job Evaluation: Validity of a Check List for Evaluating Office Jobs," *Journal of Applied Psychology*, Vol. 36, 1953, pp. 97–101.
16. E. J. MCCORMICK, P. R. JEANNERET, and R. C. MECHAM, "A Study of Job Characteristics and Job Dimensions as Based on the Position Analysis Questionnaire (PAQ)," *Journal of Applied Psychology*, Vol. 56, 1972, pp. 347–368.

17. R. C. MECHAM, personal communication, 1972.
18. L. R. TAYLOR, personal communication, 1972.
19. D. D. ROBINSON, O. W. WAHLSTROM, and R. C. MECHAM, "Comparison of Job Evaluation Methods: A Policy-Capturing Approach Using the Position Analysis Questionnaire (PAQ)," *Journal of Applied Psychology*, Vol. 59, 4 (1974), pp. 633–637.
20. JAQUES, *Equitable Payments*, pp. 154–155.

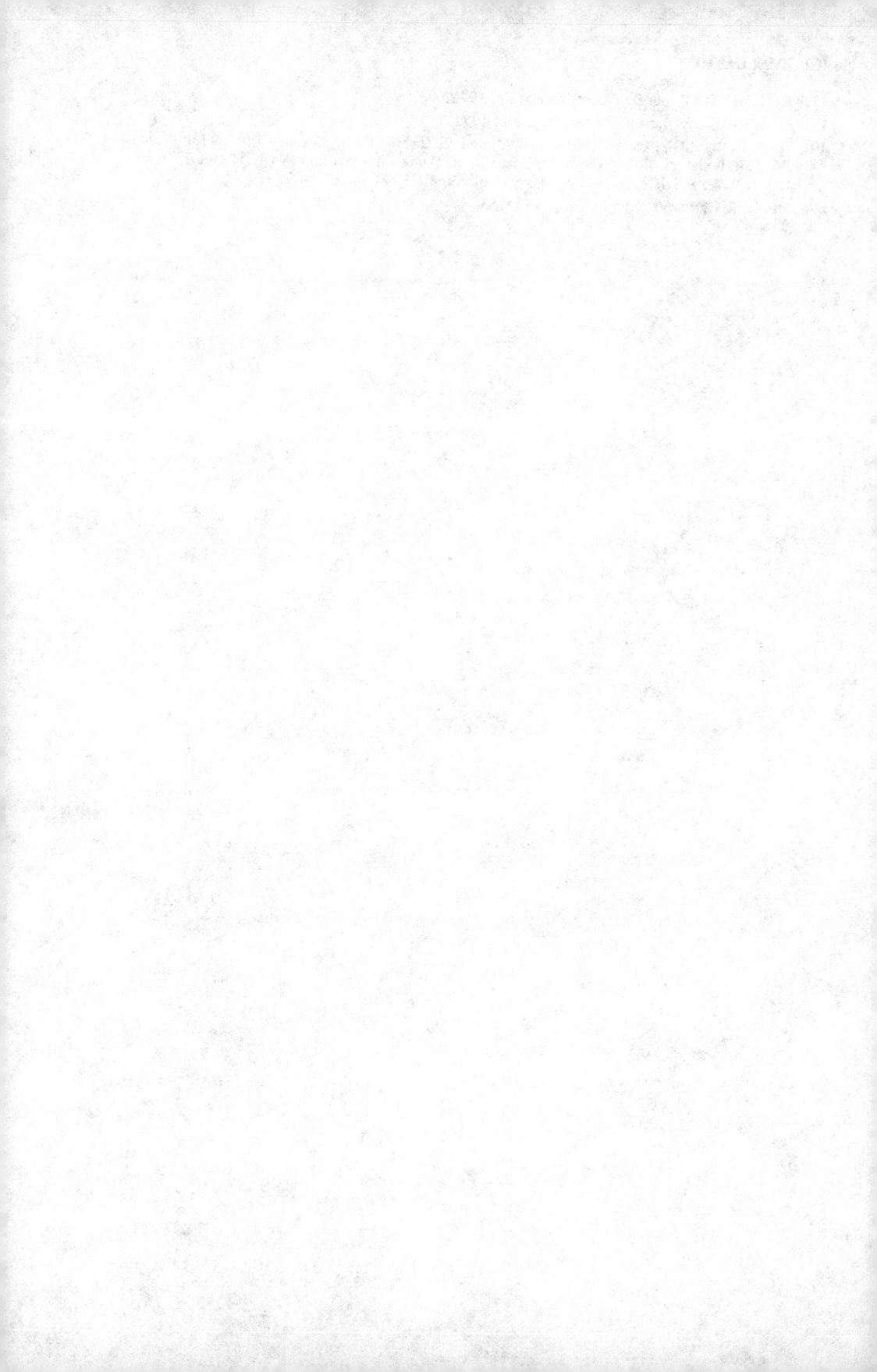

CHAPTER 5.4

Performance Appraisal

IRVIN OTIS
ROBERT W. BURNS

American Motors Corporation

5.4.1 INTRODUCTION

This chapter covers the subject of performance appraisal in a very broad manner to enable the reader to become familiar with the various formalized programs, methods, and techniques used to evaluate the performance of people at work.

American industry must adjust to an increasingly better educated, more talented work force. It is faced with increasing payroll costs, further government involvement, decreasing productivity, and other social and economic pressures. Therefore the techniques involved in measuring and improving employee performance become more critical to an organization's success or failure.

5.4.2 DEFINITION AND PURPOSE

Performance appraisal is a systematic process of evaluating an employee's performance on the job. The process has a multiplicity of titles that have the same basic meaning, that is, performance evaluation, personnel rating, performance review, employee evaluation, efficiency rating, and performance appraisal, among others.

The first step in developing a performance appraisal program is to establish its purposes, uses, and objectives. These may be:

To measure or judge an individual's job contributions and accomplishments.

To encourage performance improvement by identifying both strengths and weaknesses in current performance and by providing the motivational reinforcement to encourage improvement.

To identify employee development and training needs.

To aid in manpower planning by evaluating the supply of talent available within the organization to meet future replacement and other organizational needs.

To identify candidates for promotion.

To serve as a basis for judgments concerning salary increases or bonus incentive awards.

To enhance communication between supervisor and subordinate on individual performance expectations and to provide the framework for a dialogue about personal goals and concerns as well as organizational goals and needs.

To identify employees needing corrective performance consultation or disciplinary action.

To identify employees for reduction in force (layoff) and recall.

To document situations and decisions that relate to Equal Employment Opportunity activity.

5.4.3 EVALUATION METHODS

The evaluation method selected should relate to the objectives that have been established for the performance appraisal program. This section discusses the various methods that are available.

Narrative or Essay

The rater develops in narrative form one or more paragraphs describing such factors as individual strengths, weaknesses, promotional potential, accomplishments, and developmental needs.

The primary advantage of this method is that the rater must devote a considerable amount of thought to his or her response, and as a result, the response is often more valid and better documented than responses obtained through other techniques.

If used exclusively, this technique has some distinct drawbacks:

It is time consuming.

It is sometimes too lengthy.

The evaluator's writing skill, or lack thereof, can influence how impressive the actual performance appears to the reader.

Comparisons with other employees is difficult.

A typical essay question might read: Assess this employee's strengths and weaknesses with regard to administrative ability—planning, organizing, delegating, time effectiveness, reports, follow-up, and so on—allowing 8 to 10 lines for narrative comment.

Critical Incident Method

Under this method the supervisor is required to maintain a log documenting actual incidents that occur throughout the year indicating favorable or unfavorable behavior and results. The log generally specifies categories for classifying and recording supervisory observations. Specific categories may be quality of work, initiative, job knowledge, ability, and so on.

Advantages of this method are:

The appraisal concentrates on evidence and facts relating to behavior and results.

Performance, not traits or characteristics, is emphasized.

It provides a sound basis for improvement and development discussions.

The method's shortcomings are that recording incidents becomes a chore and that the method may lead to oversupervision.

Forced Choice Ratings

Although there are many variations of this method, the usual practice is to ask the rater to choose from a group of four or five statements those which best or least fit the individual being rated. The statements used may have weighted values known only by a third party, who determines the final overall rating.

The advantages of this method are that it establishes standards of comparison between individuals being rated and that supervisory bias is reduced because of the rater's lack of knowledge about the weighting of each statement.

Disadvantages are:

Raters may be irritated by the secrecy about weightings.

Development of forms is complicated and costly.

Its use in providing employee feedback is questionable without more specific information.

A typical statement that might be used under this method is:

The most descriptive statement of performance is indicated by the number 1, and the least descriptive by the number 4.

___Displays good job knowledge.

___Does not work to limit of ability.

___Adapts to changing conditions.

___Is physically unable to meet the job requirements.

Ranking Comparison Technique

There are four methods that may be used when a comparison of individuals is desired either within the same group or with other groups. The basis for judgment of ranking is purely subjective in that it is determined by overall impressions. These methods are often used for choosing individuals for promotions and for determining the size of salary increases. Their major weaknesses are that they do not provide information on why one individual is the best or on what the worst performers are doing. This technique provides no information that can be used in coaching, developing, or providing employee feedback.

Straight Ranking

The names of individuals in the group are listed. The supervisor, after considering the entire list of employees, places them in rank order.

An example of how the ranked list of employees might appear is as follows:

Rank	Employee's Name
Best	1 ___
Next best	2 ___
Next best	3 ___
Next best	4 ___

Alternation Ranking

This ranking is a variation of the straight ranking method. The rater first selects the best or most valuable employee and then the worst or least valuable employee, continuing with this alternate procedure until everyone in the group is ranked. This variation may seem easier to use because raters are asked to do first the thing they do best, which is to identify extremes in performance.

An example of this method would appear as follows:

Employee Name	Number	Rank	
		Best performer	1 ___
		Next best performer	2 ___
		Next best performer	3 ___
		Next best performer	4 ___
		Next best performer	5 ___
		Next best performer	6 ___
		Next worst performer	6 ___
		Next worst performer	5 ___
		Next worst performer	4 ___
		Next worst performer	3 ___
		Next worst performer	2 ___
		Worst performer	1 ___

Paired Comparison Ranking

This technique is more complex in that each individual is compared, one at a time, with every other individual. The individual who receives the most tallies is the best employee, and the individual with no tallies is the worst employee. The main drawback of this method is the amount of time and statistical exercise involved. For example, 45 comparisons are required to rank 10 employees, but 105 comparisons are required to rank 15 employees. For example, if four employees, A, B, C, and D, were compared, the following comparisons would be required:

A with B	B with C	C with D
A with C	B with D	
A with D		

Forced Distribution Method

Forced distribution requires the rater to distribute ratings based upon a bell-shaped distribution curve. The rater must allocate 10% of the employees being rated to the highest performance category, 20% to the next highest, 40% to the middle, 20% next, and 10% to the lowest. As described, this is a rigid system and may be considered unrealistic in that raters feel that the actual performance of their employees should not distribute according to the normal curve. However, this method can be customized by skewing the distribution to match organizational perceptions. When small groups of employees are involved, this method becomes meaningless.

Rating Scale Methods

Conventional Rating System

This method rates the individual on a number of characteristics, that is, quality of work, quantity of work, job knowledge, communications, initiative, and other job factors, according to some graphic scale. The scale consists of either descriptive phrases (outstanding, good, satisfactory, marginal, unsatisfactory) or points (1 to 10). To substantiate a rating, some formats require that the rater explain the reason for the rating.

This method is one of the most popular forms because of its consistency and acceptance by raters. However, weaknesses of this method are:

Raters may interpret definitions differently.

Raters may tend to let the rating on one performance factor influence the rating on other factors (halo effect).

The rating of characteristics or factors does not provide effective employee feedback unless the rater explains the reason for the rating.

An example of this method is shown in Exhibit 5.4.1.

Behavior-Anchored Rating Scales

This technique incorporates the critical incident approach, with raters participating in the development by analyzing the job as to what is considered very successful performance. Job behavior "anchors" for specific performance activities or dimensions of a job are established. The anchors for each dimension of a job are established. The anchors for each dimension are then scaled from highly effective performance to highly ineffective performance.

The advantages of behavior-anchored rating scales are:

Commitment of raters is greater because of their participation in development.

Reliability of ratings is greater because of the direct correlation with the job.

Provides employee feedback relating to specific aspects of performance.

The disadvantages are:

The time requirements for development are considerable.

Raters must be committed to their development.

Each job requires a separate format developed specifically for that job.

Group Judgment Method

The group judgment method utilizes an appraisal group consisting of the supervisor of the employees being judged, two or three other supervisors who have knowledge of the job performance of these employees, and a member of the administrative or personnel staff. The group discusses job duties, performance standards, actual performance of the incumbent, causes of performance, and ideas for improvement and defines a specific action plan for performance improvement or employee development. The ratings given then are based on the consensus of the group.

One advantage of this method is that by using several raters the validity of the appraisal is improved because it reduces the effect of bias by any one rater. Another advantage is that the multiple raters tend to provide better ideas for improvement of performance than would a single supervisor. On the negative side, the process is very time consuming.

Assessment Method

The objective of the assessment method is to predict employees' potential for promotion and the future performance of candidates for hire. A number of candidates are brought together for several days in work environment situations, which may include tests, games, or exercises simulating actual situations that would be encountered on the new job. Trained assessors observe behavior, make judgments, and prepare a final evaluation of each candidate.

The assessment method has good validity in predicting future performance. The obvious drawbacks are the costs associated with developing a professionally designed program and the heavy commitment of time on the part of both assessors and participants.

Exhibit 5.4.1 Example of Conventional Rating System[a]

Performance-Related Factors	O	G	S	M	U	NA	Note below any significant examples for review with the employee. Marginal or unsatisfactory ratings require explanation.
Interpersonal relationships							
Job knowledge							
Quantity of work against objectives							
Quality of work against objectives							
Communications both within and outside department							
Application of analytical ability							
Initiative/resourcefulness							
Adaptability to changing conditions							
Personal development							
Decision making ability							
Developing employees							
Planning and organizing work							
Directing and delegating to employees							
Effectiveness in meeting EEO/AAP responsibilities							

Check (✓) Appropriate Box

[a]Performance level code: O–outstanding; G–good; S–satisfactory; M–marginal; U–unsatisfactory; NA–not applicable.

Management by Objectives

Management by objectives is a participative, results-oriented performance program which involves both the supervisor and the subordinate in establishing mutually acceptable, realistic goals or objectives. This process includes establishing qualitative and quantitative criteria that can be used to measure the degree of goal attainment and time schedules within which the goals will be accomplished. At the conclusion of the scheduled time period, the subordinate makes an appraisal of accomplishments compared to the established goals. The superior and the subordinate then have the appraisal interview.

This technique provides experience for supervisors as to what can reasonably be expected from people, which is useful in manpower planning. It also provides better control of the operation and emphasizes results rather than traits.

Most of the concerns about the technique relate to managements' manipulative use of the program in imposing its standards or objectives. Also, a considerable amount of follow-up is required by supervisors in order to adjust the time schedule as unforeseen events alter the original plan.

An example of how this method might be recorded is as follows:

Performance Objectives	Results
List the individual's most significant assignments, projects, and objectives during this period and the appropriate time and cost factors.	Indicate results achieved in terms of time, costs, and any factors that prevented achievement.

Work Standards Approach

Work standards can be developed for all elements of any job and any level of the organization. Although work standards and objectives are closely related, work standards cover those areas where there is little economic striving for improvement. A work standard approach is a statement of conditions that will exist when a job is properly completed, and it is aimed at maintaining a specific level of, or improving, productivity.

Work standards must be made clear, visible, and fair. Their use is less threatening than the use of other techniques during the performance appraisal interview.

Their weakness lies in the element of comparability, which is directly related to who establishes and measures the standards. Some ranking method is necessary when it comes to decisions on salary increases or promotions.

5.4.4 IMPLEMENTATION

Procedures

Procedures are necessary to provide raters with uniform instructions on how to prepare for the appraisal process, complete the required forms properly, conduct the appraisal interview meeting, resolve any conflict, and dispose of the completed forms.

Communications

The success of the performance appraisal system is closely related to the amount of time given to communicating the purposes, needs, and procedural elements of the program.

Announcing the Program

The announcement of the program should go to all supervisors who will utilize the program and all employees who will be covered by it. The announcement should indicate that both rater training sessions and employee orientation meetings will be held. It should also state that the program is supported by the organization's top management.

Training of Supervisors

The objectives of training supervisors for the performance appraisal are to inform them of the rationale for such a program, to describe the role and responsibility of the rater in the process, to explain the use of evaluation information, and to teach the raters how to conduct a successful performance appraisal interview. An appraisor's handbook or manual should be developed for use not only as part of the training process, but also as a reference.

The significance of thorough rater training cannot be overemphasized, because the interest, abilities, and attitude displayed by the supervisor is far more important than the objectives, methods, mechanics, or forms of any appraisal program.

Orientation of Subordinates

It is suggested that employee orientation meetings be held, regardless of the size of the organization, to explain the purpose and the procedural elements of the appraisal program and to define the employees' role in the process. Ample time should be scheduled to permit questions and answers.

Problems

Any performance appraisal system will have weaknesses and problems that may result in failure to accomplish some of the objectives for which it was established.

Judgment

Even in a well-defined appraisal program, performance judgments may be subjective and biased because of how raters feel about each of their subordinates. Evaluations have also been found to be influenced by raters' perceptions of their own performance. Another conflict occurs as a result of each rater's dual role of judge and coach.

Raters

Varying perceptions of how work is performed have a significant influence on the raters' evaluations. Incomplete information or observations may prevent supervisors from rendering objective evaluations.

Criteria

Performance criteria are not always specific enough or consistent enough to ensure that the employee being rated understands fully the accepted standards of performance or the objectives.

Policy

Multipurpose objectives lead to rater confusion. If decisions relating to promotion, salary increases, layoffs, and so on, are actually based on the outcome of the performance appraisal, then the appraisal has significant importance. Program credibility is destroyed when the employees and raters do not clearly understand the format, the criteria, and the importance of the evaluation as it relates to these decisions.

Legal and EEO Requirements

Performance appraisal systems are subject to close scrutiny by the various branches of government that have investigative authority over alleged discrimination charges. As a result, employers may be called upon to prove that their performance appraisal methods are reasonable and that they do not discriminate against affected class members in any manner.

The significance of objective, rather than subjective, performance criteria becomes apparent when the results of the performance appraisal are used in determining promotions, transfers, layoffs, discipline, merit increases, and training selections.

5.4.5 POST-APPRAISAL INTERVIEW

The environment in which the postappraisal interview is held, the time involved in planning what will be said, and the follow-up on performance issues are extremely significant because it is at this meeting that the concepts and work involved in developing a good, objective program can be destroyed if the meeting is handled poorly. It is a critical step because it has the potential to be the most productive element of the entire appraisal process and thus can be most rewarding to both the rater and the subordinate if conducted properly.

Planning the Interview

Planning for the interview should include a review of the written appraisal and the transforming of that information into language that can be used during the interview. The rater needs to anticipate how the employee will react to the appraisal and to develop a plan for responding to these reactions. The rater should also:

Notify the employee of the appraisal interview at least a day or two in advance. This provides the employee with the opportunity to prepare for the meeting.

Schedule the meeting so that an adequate amount of time is available.

Ensure that the meeting is held in privacy, without any interruptions.

Conducting the Interview

During the interview the rater should point out the employee's strengths, weaknesses, and goal attainments or lack thereof. Specific instances of strong and weak performance should be discussed. Suggestions should also be made on how weak areas can be overcome or improved upon. Remember that the purposes of the interview are to obtain an agreement on the evaluation, to develop an action plan for improved performance, and to establish mutually acceptable objectives for the future.

The interview should conclude on a positive and encouraging note, with the expressed willingness on the part of the rater to talk further if there are any questions or suggestions that arise at a later date.

Follow-up

Performance appraisal should not be a once-a-year process, but rather, an ongoing series of discussions with the employee when either positive or negative performance is observed.

5.4.6 MONITORING AUDITING, AND UTILIZING RESULTS

Performance appraisal programs are usually monitored and audited by the personnel department to ensure that they are completed on time, that the forms are properly completed with the appropriate signatures, and that the employee has acknowledged, in some manner, that the appraisal has been reviewed.

The performance appraisal system and the ratings that result must be used for the purposes that were communicated to all employees. In particular, if deviations are permitted in application of appraisal results that act to the detriment of the rated employee, the credibility of the program is destroyed.

BIBLIOGRAPHY

ALLAN, P. and S. ROSENBERG, "Formulating Usable Objectives for Manager Performance Appraisal," *Personnel Journal*, November 1978, pp. 626–629, 640, 642.

BALL, R. R., "What's the Answer to Performance Appraisal?," *The Personnel Administrator*, July 1978, pp. 43–46.

BAYLIE, T. N., C. KUJAWSKI, and D. M. YOUNG, "Appraisal of People Resources," *ASPA Handbook*, Vol. 1, Bureau of National Affairs, Washington, DC, 1974.

BIRNBRAUER, H., "Taking the Sting Out of Performance Reviews," *Machine Design*, September 1977, pp. 106–109.

BUZZOTTA, V. R., and R. E. LEFTON, "How Healthy is Your Performance Appraisal System?," *The Personnel Administrator*, August 1978, pp. 48–51.

CALHOON, R. P., "Components of an Effective Executive Appraisal System," *Personnel Journal*, August 1969, pp. 617–622, 648.

COLBY, J. D., and R. L. WALLACE, "Performance Appraisal: Help or Hindrance to Employee Productivity?," *The Personnel Administrator*, October 1975, pp. 37–39.

COWAN, J., "A Human-Factored Approach to Appraisals," *Personnel*, November–December 1975, pp. 49–56.

DANZIG, S. M., "What We Need to Know About Performance Appraisals," *Management Review*, February 1980, pp. 20–24.

GAVTSCHI, T. F., "Performance Appraisal," *Design News*, July 19, 1976, pp. 113–114.

GAVTSCHI, T. F., "Performance Appraisal Techniques," *Design News*, August 9, 1976, pp. 95–96.

HAYDEN, R. J., "Performance Appraisal: A Better Way," *Personnel Journal*, February 1978, pp. 606–613.

HAYNES, M. E., "Developing an Appraisal Program," *Personnel Journal*, February 1978, pp. 66–67, 104, 107.

KINDALL, A. F. and J. GATZA, "Positive Program for Performance Appraisal," *Harvard Business Review*, November–December 1963, pp. 73–80.

LAZER, R. I. and W. S. WIKSTROM, *Appraising Managerial Performance: Current Practices and Future Directions*, Conference Board Report No. 723, The Conference Board, New York, 1977.

LEVINSON, H., "Management by Whose Objectives," *Harvard Business Review*, July–August 1970, pp. 97–106.

MAYFIELD, H., "In Defense of Performance Appraisals," *Harvard Business Review*, March–April 1960, pp. 26–32.

MCAFEE, B., "Selecting a Performance Appraisal Method," *The Personnel Administrator*, June 1977, pp. 61–64.

MCGREGOR, D., "An Uneasy Look at Performance Appraisals," *Harvard Business Review*, September–October 1972, pp. 133–138.

OBERG, W., "Make Performance Appraisals Relevant," *Harvard Business Review*, January–February 1972, pp. 61–67.

SCHNEIER, C. E. and R. W. BEATTY, "Integrating Behaviorally-Based and Effectiveness-Based Methods," *The Personnel Administrator*, July 1979, pp. 65–76.

SCHNEIER, C. E. and R. W. BEATTY, "Developing Behaviorally-Anchored Rating Scales (BARS)," *The Personnel Administrator*, August 1979, pp. 59–68.

SCHNEIER, D. B., "The Impact of EEO Legislation on Performance Appraisals," *Personnel*, July–August 1978, pp. 24–34.

CHAPTER 5.5

Human Resources Accounting: Measurement and Utilization

BARUCH LEV
Tel-Aviv University

The most valuable of all capital is that invested in human beings.
Alfred Marshall, *Principles of Economics*

5.5.1 INTRODUCTION

Human resources accounting (HRA) involves the application of economic and accounting concepts to the area of personnel management. It is an information system aimed at providing data on the cost and value of employees to the organization. The system may serve both internal and external users, providing management (internal users) with relevant data on which to base recruiting, training, and other personnel development decisions, and supplying investors, lenders, and other external users of financial statements with information concerning the investment in and utilization of human resources in the organization. Given the specific interest of industrial engineers, the following discussion is mainly concerned with the internal, or managerial, uses of HRA.

The chapter opens with a discussion of the desirability to recognize the investment in human beings as a capital investment. It then proceeds to discuss the estimation of this investment on a conceptual and a practical level, followed by summary descriptions of organizational applications and tests of HRA models.

5.5.2 THE INVESTMENT IN HUMAN BEINGS: A CAPITAL INVESTMENT

The dichotomy in conventional accounting between human and nonhuman capital is fundamental; the latter is recognized as an asset and is therefore recorded in the books and reported in the financial statements, whereas the former is totally ignored by accountants. Economists, on the other hand, express a different view on this issue. Milton Friedman,[1] for example, states:

> *From the broadest and most general point of view, total wealth includes all sources of "income" or consumable services. One such source is the productive capacity of human beings, and accordingly this is one form in which wealth can be held.*

The definition of wealth as a source of income inevitably leads to the recognition of human capital as one of several forms of holding wealth, such as money, securities, and physical (nonhuman) capital. This attitude toward human capital has a broad range of applications in economics. For example, human capital is recognized as an important factor in explaining and predicting economic growth.[2] Also, human capital estimates are used to calculate the private and social returns on the investment in education.[3] Thus in modern economic theory human capital is treated on a par with other forms of earning assets.

Why, then, is the investment in the firm's human resources treated by management differently from the investments in capital assets? Specifically, why are rates of return on the investment in recruiting, training, and other personnel development programs for employees so rarely calculated by managers to determine the desirability and efficiency of such investments? Why are all the costs associated with personnel management expensed, rather than capitalized, during the year in which they are incurred? Why is the firm's investment in human resources missing from the balance sheet?

Two reasons are generally given for treating human resources differently from nonhuman assets. First, very serious measurement problems are encountered in estimating the costs, and particularly the future benefits, of the investment in human resources. Uncertainty as to the future benefits of such investment and the fact that some of the investment attributes, such as employee morale and job satisfaction, are not readily quantifiable are the major reasons for these measurement difficulties.

Second, in perfect labor markets (i.e., where complete information about all job opportunities is freely available to all employees, and no costs are involved in moving from one job to another) every employee will obtain, in the form of wages and other benefits (e.g., training, subsidized housing), the exact value of his or her marginal product to the firm. In this case the total benefits from the firm's human resources will equal its costs, and the investment in human resources will, by definition, be zero. Accordingly, all costs associated with employees are, in this case, justifiably treated as expenses (but not necessarily during the period they occur).

On closer examination, however, it turns out that these two objections to treating the investment in human resources equally with other investments can be overcome. The desirability of overcoming these objections has been pointed out frequently. For example:

> *Explicit recognition of the asset value of human resources would enable managers to consider their employees within the framework of asset management. . . . Managers could then make meaningful choices between various types of human investment and investments in other assets, and in addition give due consideration to an adequate return to these assets. This approach is preferable to the current system which regards human beings as a wage or salary expense, to be used efficiently, but to be written off as an expense in the current period.*[4]

The next section discusses these two issues: measurement of the investment in human resources and the conditions under which such investment exists in organizations.

5.5.3 THE ECONOMIC APPROACH TO THE ESTIMATION OF HUMAN CAPITAL

Definition and Measurement of Human Capital

"Capital" is generally defined as a source of future income (or net cash flow) stream, and its worth is the present, or disconnected, value of this income stream. Thus the value of human capital embodied in a person of age τ is the present value of his or her remaining future earnings from employment. This value for a discrete income stream is

$$V_\tau = \sum_{t=\tau}^{T} \frac{I(t)}{(1+r)^{t-\tau}} \qquad (1)$$

where V_τ = the human capital value of a person τ years old
$I(t)$ = the person's annual earnings up to retirement (This series is known as the "earnings profile.")
r = a discount rate (cost of capital) appropriate to the person or the firm
T = retirement age

The key variable in equation 1 is, of course, the earnings profile, $I(t)$. How can such an earnings profile be estimated? The best source of information for earnings profiles is current data on earnings distribution classified by relevant characteristics, such as age, education, skill, and geographical area. Consider, for example, the problem of estimating the future earnings series of an industrial engineer who is 25 years old. We have current data (from the census and other sources) on average earnings of industrial engineers 25 years of age, 26 years of age, and so on, up to retirement. (See Levand and Schwartz[5] on the estimation and availability of earnings profiles for various professions, trades, skilled employees, etc.) We can therefore estimate next year's earnings of our 25-year-old engineer on the basis of current earnings of an equivalent engineer 26 years of age. The estimate of earnings 2 years hence will be based on current average earnings of a 27-year-old engineer, and so on.* Thus the observed *across persons* earnings profile can be transformed into an *overtime* earnings profile. Such earnings profiles for fine classifications of employees (by skill, trade, education, geographic location, etc.) are commonly constructed by economists dealing with human capital issues, such as returns to education. (See, for example, Lazear[6] and the list of references

*Proper adjustments should, of course, be made for expected events, such as the increase in nominal earnings resulting from inflation.

therein.) The basic information for the earnings profiles is generally derived from U.S. census data. Given the practical availability of earnings profiles for a large number of professions and trades, equation 1 for the value of an individual's human capital can be computed.

Equation 1 estimates the value of human capital embodied in an individual employee. Switching to the *firm's* human resources, dynamic elements have to be incorporated into equation 1, reflecting the probability of the death or exit of employees from the firm prior to retirement and, more important, the movement of employees over time across organizational roles and positions. These dynamic elements have been modeled by use of a Markov chain stochastic process reflecting the movements of employees among organizational roles (see Flamholtz,[7] Auerbach and Sadan,[8] and Friedman and Lev[9]).

Summarizing, various models are available that reflect the total value of employees associated with the firm. This total value will be affected by such factors as the number and age of employees; their degree of skill, education, and training; and the probability that each employee will occupy each organizational position (including the "exit" position) at specified future times. Accordingly, changes in the total value of employees, or in the values of subgroups of employees, will indicate important organizational changes in the firm, such as aging of the firm's labor force, changes in the level of skill and education of employees, and changes in promotion policies. The managerial uses of such information is discussed later.

Human Capital as an Organizational Asset

Recall the two basic objections mentioned previously for treating the investment in the firm's human resources as an asset:

1. It is difficult to estimate in monetary terms the value of this investment.
2. Even if estimated, in a nonslave society this investment belongs to the employees rather than to the firm. Stated differently, in perfect labor markets firms will pay employees in the form of wages and other benefits the total value of their marginal product, and hence no incremental benefits from employee education and training will accrue to firms. Consequently, the investment in human resources does not result in an asset to the firm.

Regarding the first argument, it has been shown that practical and widely used approaches are available for the estimation of human capital values. We turn now to the second argument, concerning who *owns* the investment in human resources.

An investment in human resources yields an asset when the firm is able to pay its employees amounts that are lower than the marginal revenue of their product (MRP). The value of this investment will be the discounted value of the future stream of *differences* (gains) between MRP and employee cost. There are two general situations under which the direct costs associated with the labor force (wages, housing, etc.) will be lower than employee contributions to the firm (MRP): when a firm engages in firm-specific training and when there are restrictions on employee mobility.

Firm-Specific Training

Firms often engage in the training of their employees since perfectly trained persons are rarely readily available in the labor markets. Such on-the-job training is usually conceptualized on a scale that runs from "general" to "specific," depending on the marketability of the resultant skills (see Becker[2]). "General training" is defined as training that increases the worker's productivity to *many firms* by the amount equal to the increased productivity realized by the firm providing the training. On the other hand, "specific training" increases productivity more in the firm providing it than in other firms, and "completely specific training" is training that has no effect on an employee's productivity anywhere except in the firm providing the training.

Obviously, when firms invest in general training, they will find it hard to recoup the benefits of this training (in the form of paying future wages that are lower than employee productivity), since employees can switch to other firms that will pay them the full value of their increased (by training) marginal productivity. However, when investment is made in specific training, the firm can reap the benefits of such investment in the form of relatively low wages, since the higher employee productivity resulting from training is restricted to the firm that has provided the training. Hence investment in specific training can create an asset to the firm, and the value of this asset will be determined by the future stream of wage savings (relative to market wages).

Restrictions on Employee Mobility

Perfect labor markets, where employees are completely mobile and fully informed on all job opportunities, are obviously nonexistent. Information on all job opportunities is not readily and costlessly available to all employees. Furthermore, there are all kinds of restrictions on employee mobility, resulting from labor union arrangements, unwillingness of employees to harm seniority and job security,* costs of transition, and so on. Imperfect information and restrictions on mobility will allow firms to pay wages that are to some extent lower than employee productivity, thereby enabling them to recoup the value of their investment in human resources.

Summary

The existence of specific training and the real-life restrictions on employee mobility create opportunities for firms to recoup the benefits of their investment in human resources. Thus the human capital value of employees, measured by the nationwide average earnings profiles, minus the present value of the firm's commitment to future wages can be regarded as an organizational asset.† The value of this asset will be affected by the amount and kind (specific versus general) of training provided by the firm, recruitment and promotion policies of the firm, fringe benefits, and so on.

Applications of the Economic Measurement of Human Resources

Estimation of the value of a firm's human resources along the lines described will provide management with relevant information for the optimal allocation of human as well as nonhuman resources. Following are some interesting applications of such information, which are further discussed in Lev and Schwartz[5] and Friedman and Lev.[9]

Human resource values will provide information about changes in the structure of the labor force. For example, differences over time in the values of a firm's human resources may result from changes in the age distribution of employees (i.e., "vintage") and/or in the employees' level of education. The firm's time series of human capital values thus contains information about changes in the structure and quality of the labor force.

Differences between a firm's actual wage structure and the average wages prevailing in the relevant labor markets are caused mainly by the firm's personnel policies, extent of training, and indirect compensation systems. These actions represent the firm's investment in human resources. The future series of differences between the firm's actual wages and the average market wages can thus be viewed as the returns on the firm's investment in human resources. Stated differently, had the firm not engaged in specific training and compensation programs *different* from those of other firms in its group, its wage structure would have been very close to the average market wages. Accordingly, measurements of the value of human resources based on market data and those based on the internal wage scale of the firm will allow computation of the firm's investment in human resources, assessment of the efficiency of this investment (returns on the investment), and computation of the depreciation rate of the investment. (See Friedman and Lev[9] for elaboration on such computations and data sources.) All these measurements are unavailable in conventional accounting and cost accounting systems.

5.5.4 ADDITIONAL APPROACHES TO HRA

The economic approach to HRA outlined here is based on two premises: (1) that the "value" of an employee may be derived from his or her future earnings stream, and (2) that the firm's investment in human resources (training, etc.) is reflected by the aggregate difference between the earnings streams employees could obtain in the market (e.g., average U.S. earnings for employees of specific skills, age, seniority, geographic location, etc.) and the actual wage streams paid by the firm.‡ Additional approaches to HRA are available and are briefly discussed here. (For elaboration on these approaches, see Flamholtz.[10])

In addition to estimating the present value of the firm's investment in human resources, management needs data on the cost and value of the labor force to facilitate personnel planning and decision making and to evaluate the effectiveness with which human resources have been devel-

*For example, risk-averse employees would likely discount prospective wages in alternative firms using some factor that reflects the job uncertainties pertaining to those wages. Consequently, the wage necessary to induce mobility must be greater than the existing wage by a sufficiently large risk premium.

†The magnitude of this investment is, of course, an empirical question. More on this issue is given in D. A. DITTMAN, H. A. JURIS, and L. REVSINE, "On the Existence of Unrecorded Human Assets: An Economic Perspective," *Journal of Accounting Research*, Spring 1976, pp. 49–65.

‡This investment can be negative, suggesting, among other things, negative returns to the firm's employee development programs. It should, of course, be noted that the economic measures discussed here are based on the relative costs of employees; they do not reflect directly changes in employee productivity. For a discussion of this issue, see Friedman and Lev, reference 9.

oped, conserved, and utilized throughout the organization. Data are needed for the following phases of personnel management:

1. *Acquisition of human resources*, involving recruiting, selecting, and hiring people to meet present and expected future manpower needs. The first step in the development of relevant data is to forecast manpower requirements* and then translate these needs into a "manpower acquisition budget." Data should also be accumulated on the standard cost of recruiting, selecting, and hiring employees.
2. *Establishment of development policy*, informing management of the cost of recruitment from outside as opposed to development from within (on-the-job training).
3. *Allocation of human resources*, namely, the process of assigning people to various organizational roles and tasks. Various trade-offs are involved in the allocation decision. In general, the most qualified employee should be allocated to a given task. Sometimes, however, management may wish to provide people with the opportunity to develop their skills through on-the-job training, not necessarily assigning them to the task for which they are currently most qualified. Management may also wish to allocate employees to jobs that satisfy their needs. Human resource account- should quantify the variables involved in the allocation decision in order to assist management in understanding the trade-offs involved.
4. *Conservation of human resources*, or the process of maintaining the capabilities of employees and the effectiveness of the human system developed by an organization. In the short run, for example, a divisional manager may put pressure on employees to temporarily increase their productivity or reduce costs, with the effects upon employee motivation, attitudes, and labor relations going unmeasured. Currently, conservation of human resources is measured in terms of turnover rates, but such measures are not totally satisfactory. Human resource accounting can measure certain (social-psychological) indicators of the condition of the human organization, and management can anticipate trends in these variables prior to the actual occurrence of turnover or decreased productivity.

5.5.5 APPLICATIONS AND TESTS OF HRA MODELS

The number of reported practical applications of HRA models to organizations and of tests of the empirical validity of these models has so far been rather limited. This is understandable, given the complexity of the issue and the fact that interest in human resource measurements is a relatively recent phenomenon. Following are summaries of the major publicly reported applications and tests of HRA models.

Applications

A system reflecting the cost of human resources (recruiting, hiring, training, etc.), the capitalization of these costs, and their amortization was developed at the R. G. Barry Corporation (a large shoe manufacturer). In discussing the system's applications, an executive of the corporation stated that the system has affected managerial behavior by (1) heightening management's interest in the economic importance of people, (2) dramatizing the human asset depletion resulting from employee turnover, (3) improving development of people by viewing them as capital items rather than as expenses, and (4) facilitating the accuracy of income measurement by including the effects of changes in human assets on the operating statement.[12] The HRA application at the Barry Corporation is probably the most comprehensive one reported to the general public.

Alexander[13] has developed an HRA model that integrates the original cost of hiring and training employees with the opportunity cost (cost of time lost in engaging employees in nonproductive tasks) and that is designed primarily to monitor the effects of employee turnover. The model has been applied in an office of a large certified public accountant (CPA) firm and is designed primarily for people-intensive service organizations. Acccording to Alexander, the system has provided the firm with facts that have led to some reassessment of its traditional approach to staff mix and resource allocation.

Pyle[14] reported the development of an HRA system by a publicly held soft goods manufacturer. The total costs of the firm were first classified into two components: human resource costs and other costs. The human resource costs were then separated into their expense and asset components. For a cost to be treated as an asset, it must be expected to provide benefits to the company beyond the current accounting period. The human assets were then separated into one of the following seven accounts: (1) recruiting outlay costs, (2) acquisition costs, (3) formal training and familiarization costs, (4) informal training costs, (5) familiarization costs, (6) investment building experience costs, and (7) development costs. Rules and procedures have been developed to de-

*This may be done, for example, by a Markov chains model (see Vroom and MacCrimmon, reference 11).

preciate these investments over their expected future service life. This system makes possible, among other things, the calculation of the rate of return on investment in human resources.

Four applications of HRA systems—two in "big eight" CPA firms, one in an insurance company, and one in a bank—are described in Flamholtz[15] and in Flamholtz and Wollman.[16] These models reflected, in addition to the original costs of human resources, the "replacement cost," namely, the cost to replace existing employees. Another application to a CPA firm is reported in Flamholtz and Lundy.[17]

Tests

In addition to the actual applications reported here, various empirical tests have been conducted on the effects of HRA on managerial decisions. Zaunbrecher[18] performed an experiment to determine the impact of HRA cost information on the personnel selection decision. Results indicate that HRA information, when provided, was used by managers in personnel decision making. Tomassini[19] studied the effects of presenting HRA cost information in a managerial decision involving whether or not to implement a layoff. It was found that managers with access to both conventional and HRA data had different decision preferences than those with access to conventional data only. Flamholtz[20] reports the results of an empirical study designed to determine the impact of human resource value numbers on management decisions. To date, it is the only published study of the effect of human resource *value* (rather than cost) measurement. The results showed statistically significant differences in decisions made using traditional personnel data and either nonmonetary (psychological) human resource value numbers or monetary human resource value numbers.

Various empirical studies have examined the effects of HRA data on investor (rather than managerial) decisions. Elias[21] examined the impact of HRA information on stock investment decisions, reporting somewhat inconclusive results. Hendricks,[22] conducting a similar study, found that stock investment decisions were affected by the addition of HRA information to conventional accounting data. Schwan[23] studied differences in financial decisions made using information in which human resource costs were allocated over several periods compared to decisions made using conventional financial information (all costs expensed during the period of outlay). He concluded that (1) inclusion of HRA information probably does affect the decision maker's ratings of the preparedness of the firm's management to meet the challenges and opportunities of the future and that (2) inclusion of HRA information also probably affects the predictions of future net income. Acland[24] reported that HRA information affected the decisions of financial analysts. It can be concluded, therefore, that HRA information is potentially useful to managers and investors.

5.5.6 SUMMARY

As stated in the introduction to this chapter, human resource accounting involves the application of economic and accounting concepts to personnel management. There are two levels of such application. The elementary level involves accumulating and analyzing relevant cost data on the various phases of personnel management, such as recruiting, placing, and training. The objective here is to make decision makers aware of such costs and to enable them to make the proper choices (e.g., having your own recruiters or using an outside recruiting agency).

The more advanced level of application is aimed at assessing the cost and benefits of the firm's investment in its labor force. This application goes far beyond the intelligent accumulation of available data described in the preceding paragraph. Estimation of the investment in human resources is aimed at providing management (and investors) with tools with which to evaluate, in monetary terms, the changes in the value of employees to the organization induced by various personnel policies, such as hiring, choice of on-the-job training (e.g., specific or general), and promotion. Such an evaluation, involving both the cost and the benefits of personnel policies, will obviously contribute to improved decisions by focusing on the size and the changes in the investment in human resources, the rate of return on this investment (and on alternative personnel policies), and the depreciation rate of the human resource investment.

This chapter has dealt mainly with the second, more advanced level of HRA application, which is, of course, the more challenging and potentially rewarding one. Various models for valuing human resources have been discussed, particularly the economic model of human capital, because of its relatively advanced state of application. Real-world applications of HRA models and various empirical tests of such models have been described. Despite the relatively small number of these applications and tests, it can be concluded that HRA information affect both managers' and investors' decisions, and hence such information is potentially useful. It seems worthwhile, therefore, to pursue the study, development, and construction of HRA models in business organization.

REFERENCES

1. M. FRIEDMAN, "The Quantity Theory of Money–A Restatement," in *Studies in the Quantity Theory of Money*, The University of Chicago Press, Chicago, 1956, p. 4.
2. G. S. BECKER, *Human Capital*, National Bureau of Economic Research, No. 80, Columbia Univeristy Press, New York, 1964.
3. T. JOHNSON, "Time in School: The Case of the Prudent Patron," *The American Economic Review*, December 1978, pp. 862–872.
4. P. MCCOWEN, "Human Asset Accounting," *Management Decision*, Summer 1968, pp. 86–89.
5. B. LEV and A. SCHWARTZ, "On the Use of the Economic Concept of Human Capital in Financial Statements," *The Accounting Review*, January 1971, pp. 103–112.
6. E. LAZEAR, "The Narrowing of Black-White Wage Differentials is Illusory," *American Economic Review*, September 1979, pp. 553–564.
7. E. FLAMHOLTZ, "A Model for Human Resource Valuation: A Stochastic Process With Service Rewards," *The Accounting Review*, April 1971, pp. 253–267.
8. L. R. AUERBACH and S. SADAN, "A Stochastic Model for Human Resources," *California Management Review*, Summer 1974, pp. 24–31.
9. A. FRIEDMAN and B. LEV, "A Surrogate Measure for the Firm's Investment in Human Resources," *Journal of Accounting Research*, Autumn 1974, pp. 235–250.
10. E. FLAMHOLTZ, "Human Resource Accounting: State-of-the-Art and Future Prospects," *Annual Accounting Review*, Vol. 1, 1979, pp. 211–261.
11. V. H. VROOM and K. R. MacCRIMMON, "Toward a Stochastic Model of Managerial Careers," *Administrative Science Quarterly*, Vol. 13, 1968, pp. 26–46.
12. R. L. WOODRUFF, JR., "Human Resource Accounting, "*Canadian Chartered Accountant*, September 1970, pp. 2–7.
13. M. O. ALEXANDER, "An Accountant's View of the Human Resource," *The Personnel Administrator*, November–December 1971, pp. 9–13.
14. W. C. PYLE, "Monitoring Human Resources–On Line," *Michigan Business Review*, July 1970, pp. 19–32.
15. E. FLAMHOLTZ, "Human Resource Accounting," pp. 243–246.
16. E. FLAMHOLTZ and J. B. WOLLMAN, "The Development and Implementation of the Stochastic Rewards Model for Human Resource Valuation in a Human Capital Intensive Firm," *Personnel Review*, Summer 1978, pp. 20–34.
17. E. FLAMHOLTZ and T. S. LUNDY, "Human Resource Accounting for CPA Firms, *CPA Journal*, Vol. 45, October 1975, pp. 45–51.
18. H. C. ZAUNBRECHER, "The Impact of Human Resource Accounting on the Personnel Selection Process," unpublished doctoral dissertation, Louisiana State University, Baton Rouge, LA, 1974.
19. L. A. TOMASSINI, "Assessing the Impact of Human Resource Accounting: An Experimental Study of Managerial Decision Preferences," *The Accounting Review*, October 1977, pp. 904–914.
20. E. FLAMHOLTZ, "The Impact of Human Resource Valuation in Management Decisions: A Laboratory Experiment," *Accounting, Organizations and Society*, Vol. 1, 1976, pp. 153–165.
21. NABIL ELIAS, "The Effects of Human Asset Statements on the Investment Decision: An Experiment," *Empirical Research in Accounting: Selected Studies (1972)*, Supplement to the *Journal of Accounting Research*, 1972, pp. 215–233.
22. J. HENDRICKS, "The Impact of Human Resource Accounting Information on Stock Investment Decisions: An Empirical Study," *The Accounting Review*, Vol. 51, April 1976, pp. 292–305.
23. E. S. SCHWAN, "The Effect of Human Resource Accounting Data on Financial Decisions: An Empirical Test," *Accounting, Organizations and Society*, Vol. 1, 1976, pp. 219–237.
24. D. ACLAND, "The Effects of Behavioral Indicators on Investor Decisions: An Exploratory Study," *Accounting, Organizations and Society*, Vol. 1, 1976, pp. 133–142.

CHAPTER 5.6

Labor Relations: The Special Problems of the Industrial Engineer

WILLIAM GOMBERG

University of Pennsylvania

5.6.1 LABOR RELATIONS, PERSONNEL MANAGEMENT, AND HUMAN RESOURCES MANAGEMENT— THEIR RELATIVE SCOPE

Three terms are at times used interchangeably: labor relations, personnel relations, and human resources management. The term "personnel relations" is the oldest of the three and lent its name to a movement following World War I. The persons associated with this movement, Ordway Tead, Frank Metcalf, and Mary Parker Follett, espoused a philosophy of welfare capitalism and humane treatment of workers. Though it was not the intent of these founders, the movement became associated with measures to persuade labor that unions were not necessary and that workers could rely on the benevolence of employers, equipped with such scientific techniques as job classification, job training, insurance programs, and other welfare measures, to administer unilateral justice. Personnel departments embraced the entire working force of the organization, including junior executives, white collar workers, and blue collar workers.

When unions became revitalized during the 1930s, separate groups of management experts were designated as "labor relations" representatives. Their attentions were generally confined to the blue collar group. They represented management in collective bargaining, and the relationship between themselves and the personnel department remained ill defined. At times they were part of the personnel department; at other times they constituted a separate group.

The Post-World War II Redefinition of Functions

During World War II the inability of the work force to negotiate cash wage increases led to a demand for deferred cash benefits, such as pension funds, supplemental unemployment benefits, and hospitalization, medical, dental, and life insurance. Negotiation of such complex payments did not respond to the relatively simple technique of old-fashioned crisis bargaining. In addition, behavioral scientists became associated with many companies and, along with attorneys, merged to create a new department devoted to the management of human resources in a "union-free" environment. In some cases labor relations continued as a subsidiary unit of this department. In others labor relations remained a small, independent group, as much as to say, "This remnant reminds us of past failures of which some unionization remains a reminder. Our new Department of Human Resources will safeguard us against such future mistakes and, we hope, eliminate the remainder of past errors by persuading unionized workers that their membership has become obsolete."

The labor relations function is dominated by trade unions in both the unionized plant and the nonunion plant. The function's influence in the unionized plant is apparent and requires little amplification. Its influence in the nonunion plant stems from the nonunion personnel director's attempt to anticipate what conditions will govern the unionized plant in order to avoid the direct unionization of the plant in his or her charge.

5.6.2 THE SELF-IMAGE OF THE UNION AND COLLECTIVE BARGAINING LABOR (TRADE UNIONS)*

The institution of unionism is rooted in the national concept of democracy. The principal distinctions that separate American and British democracies from autocracies may be subsumed under two basic concepts: (1) the participation of the governed in the formulation of laws and regulations that accommodate individual freedom to group needs, and (2) the concept of *due process*, or a set of procedures instituted to resolve differences between the governed and the governors, with the added protection of having a culminating decision vested in a judiciary that is completely independent of any pressure exercised by either group.

These concepts have been basic to political life in the United States since 1776, when the 13 colonies declared their independence from the British crown, and were later institutionalized in the form of a national constitution.

Political Citizens and Economic Subjects

Paradoxically, political citizens in the 1790s who were beginning to experience, in their political life, due process and participative decision making, albeit incomplete and rudimentary, were subjected to complete autocracy in their occupational roles as employees. If their employers were benevolent, they enjoyed benevolent despotism. If their employers were harsh, they suffered harsh despotism.

From the very beginning of the Republic, some men revolted against this schizoid duality in their roles as political citizen and economic subject if their labor was for hire. Their status as employee derived from the ancient law of master and servant. The contradiction between their political and economic roles led to the ultimate rejection of this status as servant. The vehicles that these workers created and that embodied their efforts at participation in deciding their fate became known as "unions." They were industrial analogues of the legislature in the political domain.

Wages and Conditions

Because the economic demands that unions voiced were so much easier for the public to understand than the work rule demands, workers' revolts were generally reported as a demand for a raise in wages. The demand for changes in rule making, so that the workers could participate in the formulation of the rules governing the workplace, received attention largely from the specialized scholars; nevertheless, this second demand was present from the start as one of the twin pillars of the labor movement.

The first record of an American strike goes back to 1786, when the Philadelphia printers struck for a minimum income of $6 a week and, indirectly, for future participation in the determination of conditions of employment. They lost on both issues.

Following this inconspicuous beginning, the labor movement waxed and waned throughout the years, assuming its modern form in 1881 with the formation of the Organized Trades and Labor Unions, renamed the American Federation of Labor (AFL) in 1886.

Emergence of Collective Bargaining

In 1890 the United Mine Workers (UMW) of America pioneered the concept of the collective agreement, an instrument achieved through collective bargaining with the employer and representing a focusing of what were hitherto diffuse union tactics into an effective means of addressing union problems. Surprisingly, many of the notions of a collective agreement, however natural they may seem to contemporary Americans, represented a new departure in labor-management relations.

Definition of Collective Bargaining

The process of collective bargaining has been defined by the leading scholar in the field, Sumner Slichter, late of Harvard. He divides collective bargaining into two separate basic functions.

First, it is a method whereby labor as union and capital as management define the price of labor. Second, it is a method of introducing civil rights for labor into industry, that is, of requiring that management be conducted by rule rather than by arbitrary decision. In a sense it is a method of introducing a system of jurisprudence into industry in very much the same manner in which the Glorious Revolution of 1688 substituted parliamentary supremacy for the divine right of kings in

*From W. Gomberg, "Labor (Trade Unions)," in E. Krende and B. Samoft, Eds., *Encyclopedia of Professional Management*, McGraw-Hill, New York, 1979, pp. 603–606, based upon author's chapter in *Military Unions and the U.S. Armed Forces*, University of Pennsylvania Press, Philadelphia, 1977.

English political life. The labor movement extends this constitutional concept to industrial life in which positive law is defined and administrative procedures for carrying it out are outlined, and thus complement both statute law and administrative rule with a system of judicial review whose agent is *free* of any tie or obligation to either side.[1]

The rise of the labor movement antedates the emergence of the institution of collective bargaining, which evolved gradually.

Evolution of Collective Bargaining

The unions first arose as a blind protest imbued with a purely adversary attitude. The progression from being independent craft workers, owning and working with their own tools, to being hired workers of faceless capital was stormy and had few precedents. It was only after a long period of unions' experimentation with other devices and tactics that the institution of collective bargaining evolved. Earlier programs were predicated upon an anti-private-enterprise program.

The Producer Cooperative Movement

The unions opposed the wage system as an institution and promoted producer cooperatives in order to escape the thralldom of this system. They rapidly became disenchanted with these cooperatives, however, when they discovered that most of them failed because of lack of managerial competence. Those that succeeded presented an even more troublesome problem. Those who had pioneered the venture were loath to take on additional cooperators and to give them the same ownership privileges enjoyed by the pioneers. Within a short time there was little to distinguish the enterprise that started as a producer's cooperative from any other private enterprise. For all purposes, the producer's cooperative had become a private partnership of the pioneers, who then employed latecomers on the same basis as any other private enterprise. This historical progression has been repeated in some of the manufacturing-oriented kibbutzim in Israel in the last decade. The modern participative management movement is largely a revival of this institution.

Disappointment With Political Activity

Similarly, attempts to solve problems by way of the political route through alliance with farmers and small businesses led working people to discover that, when the coalition succeeded, small business resented organization even more bitterly than did big business because the former operated at such a close margin to survival.

Management Under Law

The adoption of collective bargaining was finally predicated on the frank acceptance of capitalism as an institution and on the desire to improve workers' status under that economic system. This meant abandoning any class-war doctrine and adopting a creed that acknowledged a simultaneous interest with the employer in the prosperity of the enterprise and conflict of interest in who would take how much from the fruits of the enterprise. In addition, it raised fundamental problems about the governance of the enterprise and the extent to which the employers participated in that governance. Management was acknowledged to be the industrial ruler, but this ruler was obliged to rule under law, or working rules, as they were called in industry. Conflicts over the interpretation of these laws, or rules, has to be resolved in accordance with the principle of due process.

This industrial law now pervades every facet of industrial management, including (1) entrance into the trade, (2) the method of production, and (3) the terms of introducing technological change. Each industry constitutes a local culture of its own and reflects the wide diversity of practices that fall under the rubric of industrial rights.

5.6.3 REVITALIZATION OF AMERICAN LABOR UNIONS AFTER 1933

The American labor movement as we now know it can be dated from 1933, when President Roosevelt's election led to the revival of the labor organizations, many of which were left moribund by the Great Depression of the early 1930s.

The Norris-LaGuardia Act of 1932 had set the stage for this revival by sharply restricting the terms under which federal courts could issue labor injunctions, which had previously been capricious devices used by management to hobble labor.

Section 7a of the National Recovery Act of 1934, followed by the National Labor Relations Act of 1935, set the legislative climate in which unions could revive and flourish. The movement confined itself largely to the private sector.

Arbitrable and Negotiable Issues

From the very beginning, management generally has stood for a containing strategy designed to restrict the subjects about which it is willing to talk in collective bargaining. It has sought to restrict the area of collective bargaining to wages and hours, arguing that all remaining areas constitute management prerogative. The labor movement, on the other hand, has argued that management is obliged to negotiate in any area that has an impact on worker welfare.

A Classic Confrontation

Some years ago these points of view received a formal expression in a controversy between Mr. Phelps, a Bethlehem Steel executive, and Mr. Goldberg, at the time the attorney for the United Steelworkers of America, later the secretary of labor under President Kennedy, and still later Justice Goldberg of the U.S. Supreme Court. Phelps claimed that in the beginning all rights belonged to management, and that management was therefore obligated to discuss only those areas that labor had managed to tear away and insert in the collective agreement. Goldberg dissented, stating that in the beginning management was able to impose an absolute dictatorship on its workers, that collective bargaining broke this usurpation, and that equity had become the criterion for determining the area of collective bargaining in which management was compelled to engage in joint decision making with the union.

Landmark Legislation

The conflict over the permissible areas of collective bargaining has resulted in a stormy history of U.S. industrial relations. In 1919, following World War I, President Wilson called an industrial relations conference to head off a threatened outbreak of nationwide strikes. Management acknowledged the right of any individual worker to join a union but insisted that it remained management's right to deal with or to refuse to deal with the union.

After a series of bloody recognition strikes over the issue of whether management should or should not deal with the union, the U.S. government settled the controversy with the passage of the National Labor Relations Act of 1935. The Act imposed upon management the obligation to bargain collectively with representatives of the workers who had been certified in an election procedure.

Arbitrable Issues

Now the controversy was transferred to the areas in which management was obligated to bargain. Almost every arbitration dispute over working conditions was bedeviled by management's claim that the dispute was nonarbitrable because the areas were not specifically treated in the collective agreement.

A conference similar to the Wilson industrial relations conference of 1919 was called by President Truman in 1945 to avert anticipated post-World War II strikes. The conference foundered on the issue of management's insistence that labor carefully restrict the subject area of collective bargaining and recognize all other areas as management prerogatives. Labor demurred, insisting that such a move was impractical in a time when a rapidly changing economy disclosed that areas that had hitherto been management preserves now were having a powerful impact on working conditions.

In 1947 the arbitrator Harvey Shulman, late dean of Yale Law School, was called upon to resolve a strike called by the United Automobile Workers (UAW) against the Ford Motor Company over an alleged speedup of the assembly line. Shulman resolved this problem by drawing a distinction between absolute management rights and conditional management rights. He defined the setting of production standards as a conditional management right, and this right was restricted to initially proposing a standard over which the union could present grievances. As an example of an absolute management right, he listed the location of a plant.

Narrowing Management Prerogatives

Yet, even this apparently compelling management right was subject to question. The garment workers' union had a restriction in its agreement with the garment employers that was of pre-1947 vintage. It imposed upon management the obligation not to move its plant outside the "5 cent fare" zone in New York City. This provision had been prompted by the predilection of garment employers to sign an agreement on Friday and then escape the union by moving their plant out of town over the weekend.

The automobile workers, who thought nothing of plant location in 1947, were concerned about it by 1958 and obsessed with it in 1980. Thus the nature and definition of what constitutes a management prerogative are clearly changeable. This has been accentuated in 1980 with the inroads that foreign manufacturers have made into the domestic markets of all manufactured products.

Implicit and Explicit Matters

The matter of what areas are relevant to collective bargaining is generally linked to what is arbitrable under a collective agreement. The U.S. Supreme Court, in a trilogy of decisions, struck down the strict construction that arbitrators are confined only to those issues specified in the contract. It defined the industrial relationship as a form of constitutional government in which the scope of arbitration is expanded to include contract implications as well as specific areas of agreement.

5.6.4 U.S. LEGISLATIVE RECORD IN COLLECTIVE BARGAINING

The National Labor Relations Act of 1935, the Wagner Act, was amended by the Taft-Hartley Act of 1947 and the Laudrum-Griffin Act of 1959.

The Taft-Hartley Act protected the rights of workers who did not want to organize and extended the concept of unfair labor practices from managements exclusively to trade unions as well. Title VII of the Laudrum-Griffin Act further amended the National Labor Relations Act, notably strengthening the prohibition of secondary boycotts by unions. The remainder of the Laudrum-Griffin Act provided criteria for the internal government of trade unions.

Most important, all three acts continue the U.S. Government's policy of encouraging the practice of collective bargaining, a practice that is widely resisted by both legal and illegal means by many American managements. The Stevens Cotton Mills have been able to resist successfully all efforts at unionization despite denunciaton of their practices by successive decisions of the Federal Circuit Court of Appeals. A proposed labor reform act, designed to remedy some loopholes in the law, was defeated in 1979 by a Senatorial filibuster sustained by two votes short of the three fifths required to end it. This has embittered relations between management and labor and has led Lane Kirkland, president of the American Federation of Labor and Congress of Industrial Organizations (AFL-CIO), to state that American managers seem to be the only Marxists who believe in the class struggle.

5.6.5 THE SPECIAL PROBLEM OF THE INDUSTRIAL ENGINEER IN ADAPTING TECHNIQUES TO COLLECTIVE BARGAINING*

The industrial engineer works at the bridgehead where technological problems merge with social questions. The civil engineer, the electrical engineer, and the mechanical engineer have always been applied economists. Their jobs have been to build a dam, design a dynamo, and fabricate metal shapes to yield a maximum output at a minimum cost.

The industrial engineer's techniques go beyond the mechanical cost factor. He or she is a designer of organizational structures and administrative techniques to achieve specific industrial purposes and is therefore saddled with all the problems attending human relationships. This means that the effective industrial engineer must become knowledgeable in the fields of anthropology, sociology, and psychology, among others. Above all, in a democratic society he or she must understand the relationship between efficiency and consent.

Many of the techniques that the engineer wants to use cannot be used because of his or her inability to secure consent for their use from the working group. The labor movement today is the principal organized force through which the engineer speaks to and negotiates with workers. For example, the industrial engineer wants to raise productivity. Who can be against increased productivity? It's like being against the Ten Commandments. He asks the trade unionist, "You're in favor of increased production, aren't you?" Then the industrial engineer goes to work and devises a job evaluation system. He proposes the system and is laughed at. He demonstrates with movies, if you please, that it is easier to move a piece of metal 10 in than 10 ft, and still the worker insists upon using the old 10 ft method.

Yes, the other engineers may have their problems, but they are nothing like this. If the civil engineer wants to understand what it is like to be an industrial engineer, let him visualize a dark, eerie Halloween night on which spirits, animate and inanimate, are abroad. Along comes the bridge, which is his pride and joy, spanning a majestic river. It addresses him in these words, "Hey, do you know that I could have remained standing and have carried just as big a load if you had used one quarter the tonnage of steel that my poor piles must hold up?"

This is just an everyday experience for the industrial engineer. The appearance of the trade

*From W. GOMBERG, "Trade Unions and Industrial Engineering," in W. G. Irenson and E. L. Grant, Eds., *Industrial Engineering Handbook,* Prentice-Hall, Englewood Cliffs, NJ, 1955, pp. 1121–1122.

union in the picture merely gives open, organized expression to this feeling. It at least means that the engineer can deal with this problem through an organized group.

The principal areas in which the industrial engineer is in contact with the trade union are as follows:

1. The development of job analysis and job evaluation systems.
2. The development and administration of production standard setting techniques.
3. The development and administration of wage incentive plans.
4. The evolution of the adversary model of collective bargaining into a cooperative mutual problem solving relationship.

5.6.6 JOB EVALUATION PLANS

The problems that engineers installing job evaluation plans will encounter in dealing with unions is recounted in a *Collective Bargaining Report* of the AFL-CIO, excerpts from which follow.*

The rapid increase of formal job evaluation plans in the last two decades has brought a variety of special problems to unions. A major danger is that these plans, which use technical and often inflexible methods to decide how much each job is "worth" as compared to other jobs, may supplant or unduly cramp or limit collective bargaining on wage rates for individual jobs.

The difficulties in working with job evaluation plans are compounded by the fact that there are no generally accepted uniform practices. There are wide variations in types of plans and in their administration.

Also, since the plans tend to be plantwide, rather than industrywide, or even companywide, the problems normally have to be dealt with on a local basis. Bargaining on job evaluation is carried on primarily at the local union level, although there is one major exception, the steel industry, in which the United Steelworkers have negotiated a plan extending through the basic steel industry.

Union Attitudes

Union experience with job evaluation plans has varied widely. As a result, there is no single or overall union attitude or policy toward job evaluation. Some unions have resisted its introduction, whereas others have joined in or have even requested the installation of job evaluation plans. For unions that already have such plans in effect, the immediate question is not whether to accept or reject job evaluation, but rather how the union can represent its members most effectively under the plan now in effect. Should the union participate in it, and if so, how?

The degree to which unions desire to participate varies considerably. In some cases unions have decided on a job evaluation plan jointly with management; they have analyzed jobs, written job descriptions, evaluated jobs, and determined labor grades and rate ranges all on a joint basis. Other unions have allowed management to perform these functions, but have reserved the right to question the results of each step through the grievance procedure. In still other cases unions have simply stated that they are only interested in the wages paid to their members and that management may use whatever means it desires to arrive at what it thinks wages should be. The union simply reserves the right to bargain on the results of any management wage determination, regardless of how the determination is made.

Whether unions are for or against job evaluation on principle, all agree that, where a plan exists, it is necessary to provide protection against any arbitrary and abusive application by management. This protection must be gained primarily through collective bargaining on any questionable results of application of the plan and on the makeup of the plan itself.

Job Evaluation and Collective Bargaining

No job evaluation plan should be considered sacred; no job evaluation finding should be considered as "scientific fact" not subject to challenge. The fact is that job evaluation plans involve considerable human judgment, with room for wide margins of variation. Also, the very nature of such plans ordinarily does not enable them to take into account adequately the special human and economic factors that may arise from time to time.

If the plan is regarded as a rigid formula from which the parties cannot depart, it becomes the master rather than the servant, a hindrance rather than a tool of sound labor relations. A plan should be considered a guide for collective bargaining, not the final authority. If findings are unsatisfactory, the plan or the findings in question should be adjusted.

Job evaluation must be subordinate to collective bargaining or it should not be used at all. Where wage adjustments for particular jobs are arranged for special reasons, a job evaluation plan should not be permitted to stand in the way of such adjustments.

*From AFL-CIO, *Collective Bargaining Report,* Vol. 2, No. 6 (June 1957).

Unions accepting job evaluation plans not only must retain the right to question through the grievance procedure and to gain adjustment of the results of job evaluation, but also must be able to bargain on and change the plan itself. Job evaluation must be flexible, to enable it to meet the practical demands of any specific situation.

Accuracy of Job Evaluation

Many companies have taken the position that job evaluation is a "scientific" process yielding "factual" results. Since facts are not bargainable, these companies would say the results of job evaluation are not proper matters for discussion at the bargaining table.

Union pressure has resulted in some companies' modifying this position. Many companies, however, still try to develop the attitude among union officers and members that job evaluation yields results that are superior to those gained by collective bargaining.

Job evaluation plans are not scientific in any sense of the word. They represent the subjective judgments of fallible human beings. These judgments, whether made by an industrial engineer or a supervisor, should be subject to collective bargaining between union and management.

Right to Information on Job Evaluation Plan

Wherever a job evaluation plan is used, unions must have information on all aspects of the plan. This includes all information on job descriptions, the nature of the plan, labor grades and corresponding wages, and the number of workers in each job and labor grade. Such information is a necessary tool for the union, to enable it to bargain on a reasonably informed and intelligent basis and to represent its members adequately.

The right of unions to such information is apparent and has been upheld by the National Labor Relations Board (NLRB), by various court actions, and in arbitration cases. To help avoid wrangling in individual cases where information would be necessary, some unions have negotiated contract provisions spelling out their right to complete information.

The right to information should not be crippled by unreasonable limitations. Thus unions should not be limited merely to seeing the information, but rather should be entitled to receive copies for use as their own time and resources require. A union should be able to study, check, and analyze at its own conceivable data submitted by the company so that it may fully understand all phases of the plan and any problems.

Red Circle Rates

Since companies are interested in gaining employees' acceptance of a new job evaluation plan, most will guarantee that no wage reductions will take place upon installation of job evaluation. Where an employee's wage rate is higher than the rate determined by job evaluation for the job he or she is working on, the company will agree to retain the higher rate as a personal rate. Personal rates that are higher than evaluated rates have become known as "red circle" rates.

Selling the plan to the union and its members may also indicate the promise of raises to those below the evaluated rates and may even include an across-the-board wage increase. These promises, coupled with the red circle guarantee, may lead unions to accept the adoption of a job evaluation plan as a way of getting wage increases for most workers while protecting those members carrying "high" personal rates.

Actually, when red circle rates are present at completion of a job evaluation, they should be inspected closely for many reasons. Account must be taken in advance of problems they will likely present in the future. For one thing, when a job evaluation rate is significantly lower than a personal rate, it may be an indication that, instead of the personal rate's being too high, the job was evaluated incorrectly and should carry a higher rate. The job may have been evaluated "properly" according to the job evaluation plan, but the error may be in the plan itself. The plan may not take into consideration certain aspects peculiar to this job, or it may not consider them properly in their relation to other job factors.

If there are many red circle situations, then the plan itself may not fit the plant at all. After all, it is foolish to say that a person is being paid more than his or her job is worth if the company would have to pay as much to replace the person. In other words, if results of job evaluation are completely unrealistic in terms of wages for similar jobs in other companies, then it is obvious that the job evaluation is wrong, not the present rate.

Even when a union believes the job has been properly evaluated, a red circle rate raises special questions. How long can the individual retain the higher red circle rate? "As long as the individual remains on that particular job," the company says. But will the company allow the individual to stay there, or will it find some means of displacing him or her? How will the red circle employee fare with future wage increases?

What about red circle rates that will come up in the future as jobs change or new jobs are created? How will they affect future wage levels? And how will they affect the value of future negotiated wage increases?

Every red circle rate presents a challenge to management. Unions have often found that red circlers become marked men or women and that management will find some excuse for eliminating them or their wage "advantage." Some managements have transferred such persons to higher-rated jobs that are in the same wage bracket as their red circle rate. Often, because of lack of training, the persons may not be able to perform the higher-skill job and later on may have to accept a lower-paying job in order to remain employed.

Some managements have been willing to give a red circler an opportunity to learn the skills required on the higher-ranked job, and some have even provided this training. These job transfers are fine until they clash with established seniority practices. Is a less senior employee to be trained and transferred to a higher-rated job simply because he or she carries a red circle rate?

Then there is the problem of future wage increases. Management will want to eliminate the differences between evaluated job rates and red circle rates as quickly as possible. It often proposes that, when wage increases are negotiated, they should apply to everyone except those whose personal rates exceed the evaluated rates.

That unions cannot accept this procedure is obvious. Confronted with union refusal, management may offer to compromise by seeking union approval of a plan whereby some part of future wage increases would not be applied to these higher personal rates. This type of procedure means that some union members receive substantially less in wage increases over a period, which may result in the development of dissatisfied groups within the union, with possible serious frictions leading to a decrease in its future collective bargaining strength.

Changes in Job Content

Red circle rates occasioned by the introduction of a job evaluation plan will in time disappear as a result of normal turnover, transfers, promotions, or job eliminations. However, this does not eliminate red circle problems.

As management improves production methods, many job changes are reflected in lower job requirements. The resulting job evaluations are reflected in job rates lower than are currently being paid. This creates new red circle rates.

Although many unions have been able to negotiate red circle protection when a job evaluation plan is installed, they have found it more difficult to do so for red circle rates resulting from reevaluation after a job's content has been diminished. Many unions have discovered that much, if not all, of negotiated wage increases may be eaten away by the downgrading of jobs and workers' wage rates as job content is changed. The factors in plans such as those of the National Metal Trade Association are weighted in such a way that the usual changes in jobs, brought about by methods improvement and new equipment, will normally reduce the total number of points. This is a strong reason why it is necessary for the job evaluation plan itself, as well as the application of the plan, to be subject to negotiation.

Changing of job content by management after the installation of a job evaluation plan and after agreement has been reached on the wage structure poses other problems. Does the changed job description portray actual changes in the job itself? If the job has changed, does the change require a reevaluation?

Unions should check any changed descriptions or evaluations carefully to make certain that they reflect actual changes and are not used as a handy excuse for watering down the job and its rating. They should check, too, to see if any duties were added to the job that did not appear in the changed description and that may balance any deletions.

A change in a job often may result in a particular task being performed less frequently, but still being required as part of the job. The position of some companies has been that this is grounds for decreasing point allocations for almost all factors. Is this a legitimate position?

There are complex considerations in weighing education and experience factors. The frequency with which a job requires certain specific knowledge, skill, or ability should not be the criterion. If the job requirements are such that a worker must have the knowledge, skill, and ability to cut close tolerance threads on an engine lathe, then the company must be willing to pay for these attributes, regardless of how often they must be exhibited. If the company wants someone on the job who can do this work when needed, it should pay for the knowledge, skill, or ability required.

Job evaluation results have been challenged by equal rights advocates who believe that they are "loaded" against women, confining women's jobs to low classifications. This controversy–very much like the chicken-or-the-egg argument–remains unresolved. The National Academy of Sciences appointed a special panel to evaluate the validity of job evaluation plans. Their conclusions were less than definitive.

5.6.7 WORK MEASUREMENT AND PRODUCTION STANDARDS

The union generally will express indifference about the method the employer uses to propose a standard. He can use a Ouija board, astrology, or MTM. The union will reserve the right to evaluate the standard by its own methods and criteria. It would prefer the acceptance of the worker's estimate. However, if pressed, it will settle for the stopwatch study in which it reserves the right to challenge the rating factor, the delay allowances, and all other variables. This method has the virtue of openness and lends itself to participative bargaining.

The problems that industrial engineers will encounter in the setting of production standards is outlined in a *Collective Bargaining Report*, *Time Study and Union Safeguards*, excerpts from which follow.*

Time Study

Time Study and Union Safeguards

Time study is supposed to be a method of determining the time that should be allowed for a worker to perform a defined job according to a specific method and under prescribed conditions. It is widely used for determining workloads and wage incentive standards, but is an imprecise tool that can be easily abused. Unions confronted with it must be continually on guard against the use of arbitrary, unreasonable, and unrealistic time study results.

Of the whole field of so-called scientific management, time study is the area on which most of labor's distrust and suspicions center. Ever since the introduction of stopwatch time study in the 1880s, most unions have opposed its use. Labor's distrust stems from its practical experiences with time study. Problems arise because of the inherent shortcomings of the time study process itself as well as the method of application of the technique in industry. Time study is usually represented to unions as "scientific," but it produces results that are simply judgments. They are not, and cannot be, scientific or accurate; at best, they represent approximations, and at worst, they are not better than wild guesses.

Although international unions can and do provide expert assistance and information to their locals, the investigation and processing of time study disputes remain primarily local union problems. Based on experience, local unions have devised approaches to time study, which fall into these categories:

1. Some locals prohibit time study altogether.
2. Some locals allow management to use any desired method of setting job standards, but the locals reserve the right to bargain on the results.
3. Some locals participate directly with management in making time studies and in setting standards.
4. A majority of locals allow management to make time studies, but insist on bargaining on both the methods used and their application.

Access to Time Study Data

Unions faced with time study must be certain they have complete information on how it is used by their companies. To provide essential protection to members, this information must not be limited in any respect. It must include not only the results of the time study, that is, the individual job standards, but also the plan in use by the company and the exact procedures followed.

That this information is essential to collective bargaining is apparent. Without it, a union would be unable to process grievances sensibly or to discuss pertinent contract clauses.

The union's legal right to such information has been clearly established by arbitrators, the NLRB, and the federal courts. The NLRB, with court approval, has found that:

1. The employer's obligation to grant a union's request for original time studies and job evaluation data applicable to particular jobs is well established.
2. The union's right to such relevant wage information is not dependent upon processing a particular grievance.
3. The employer's obligation to furnish the data is not limited to the period of contract negotiations, but continues after a collective bargaining agreement has been signed.
4. The union is under no obligation to show a specific immediate or prospective need for this information so long as the company uses time study and job evaluation in a manner affecting production standards and wage rates.

*From AFL-CIO American Federationist, *Collective Bargaining Report*, November 1965.

5. The employer's obligation to furnish this data exists by virtue of the statute (National Labor Relations Act). While the union may waive its statutory right to such information, the waiver must be clear and unmistakable. The union's right is not waived by the union's inability to negotiate a clause in the agreement spelling out its right to the data.[2]

In spite of this, some unions still find it difficult to secure all the necessary data. Some managements still claim that such information is confidential. When union pressure forces compliance, some managements attempt to restrict the information they provide or the means of providing it.

To ensure immediate availability of time study data, without question or limitation, most unions insist on a contract clause such as the following:

The company shall furnish to the union a copy of the time study plan currently being used. It shall make available for inspection by the union any and all records pertaining to time study and the setting of production standards, including original time study observation sheets. Upon request of any shop steward or union officer, the company will furnish copies of any of the preceding information, including copies of time studies.

Some companies have been required by contract to furnish the union with photostatic copies of the time study observation sheet each time a study is made.

In contrast to the many cases involving the union's right to data, there have been only a few involving refusal by a company to permit a union to make its own time study of a disputed job. One recent case involved the head of the union's time study department, who was assisting a local union in investigating some time study grievances. The union engineer was shown the company's time study sheets and all other data requested. Company engineers answered any questions raised by the union engineer, but the union engineer was not allowed to observe and study the disputed jobs on the shop floor. The union's position, which was sustained by the NLRB, was that the right to study the disputed jobs exists on the same statutory basis as the right to data.

The NLRB said:

We find, as did the trial examiner, that the information which the union sought to obtain by means of time studies was not only relevant but also necessary to enable the union to make an intelligent decision whether to proceed to arbitration.

It is well settled that Section 8(a) (5) of the Labor Management Relations Act imposes an obligation upon an employer to furnish upon request all information relevant to the bargaining representative's intelligent performance of its function. This obligation extends to information which the union may require in order "to police and administer existing agreements." The time studies requested by the union herein were in the nature of requests for such information. It is clear that the information requested was both relevant and necessary to enable the union to fulfill its function as the bargaining representative and that it was within the power of the respondent to make such information available to the union. We are of the opinion that compliance with the good faith bargaining prescribed by the Act required the respondent to cooperate with the union by making plant facilities available to the union for the conduct by the latter of its own time studies.

Here the record shows that the time studies were relevant and necessary to the union's administration of the grievance machinery of the contract and that the needed information was not available to the union through alternative channels. Moreover, where, as here, there are no adequate alternative sources of information to which the union may refer, it is clear that respondent's refusal to permit time studies of the disputed operations constituted an unreasonable impediment upon the union's performance of its statutory function.[3]

Accuracy of Time Studies

"We do not bargain standards!" is a frequent assertion by management when faced with union time study grievances.

Management claims that time studies produce facts, and "of course facts are not subject to bargaining or compromise." If time study did result in "facts," then this management position might be sound, and the union might be able to bargain only on whether or not time study should be used in the plant.

But such is not the case. At best, standards developed from time study are only approximations. They involved the use of considerable judgment by the time study analyst at every step of the time study procedure.

Unions should be informed on the following variables in order to have some idea of how accurate the time studies are:

1. The selection of the worker to be studied.
2. The conditions under which the work is performed during the study.
3. The manner in which the operation is broken down into parts or elements.
4. The method of reading the stopwatch.
5. The duration of the study.
6. The rating of the worker's performance.
7. The amount of allowances for personal needs, fatigue, and delays.
8. The method of applying the allowances.
9. The method of computing the job standard from the timing data.

Weaknesses in the Rating Process

After completing the stopwatch timing, the time study analyst has obtained a figure that represents the time, on the average, that it took the particular worker observed to perform the job. This time could be used to set standards only if the worker who has been time studied could be considered a "qualified," "average" employee, working at a "normal" pace and exhibiting a "normal" amount of skill. If the worker does not meet these specifications, then the observed time must be adjusted to make it conform to the "normal" time. This adjusting procedure most frequently is called "rating," but is sometimes called "leveling" or "normalizing."

The very nature of rating opens it to abuse. By manipulating this rating factor, it is easy for a time study analyst to end up with practically any result he or she chooses. In fact, as many unionists know, the rating factor is often used to enable an analyst to end up with a standard that was determined before the time study was taken. In other words, time study is often used to "prove" to the workers that a work load or standard set by the company is "fair." It is practically impossible for a union to prove this kind of deliberate deceit since the "normal" operator is only a figment of the analyst's imagination.

Thus it is easy to see how time study analysts can readily manipulate time studies. Unfortunately, the situation is not much better even when management and its analysts are trying to be completely honest and objective.

The Fallibility of Time Study Analysts

Most time study analysts claim they can judge worker pace within 5%, and some even have claimed that, with experience, this could be reduced to an average error of 2%. But these claims of precision have been proved false by university researchers and management groups who studied the ability of time study analysts to rate workers' performance. These studies show that analysts will, in more than half their ratings, misjudge by more than 10% the variations in work pace. "Errors" of as much as 40% are not at all uncommon among "qualified" and experienced analysts. Some managements have agreed that wide variations and inconsistencies can be expected, but not from full-time, experienced analysts.

A study conducted by the Society for the Advancement of Management (SAM) answered this question. It proved that length of experience in conducting time studies had no relation to the time studies' accuracy. In fact, the average error of those with more than 15 years' experience was greater than for those with less than 6 months' experience. And analysts who spent less than half their time taking time studies rated slightly more accurately than those spending all their time in time study work.

Workers cannot leave the determination of a "normal" pace to management. Progressive engineers such as Mitchell Fein have concluded that an objective concept of normal is a fallacy. They propose that a definition of normal be established by collective bargaining. This normal then becomes the criterion against which all studies are made.

5.6.8 THE BURDEN ON GRIEVANCE PROCEDURES

Time study has become an important issue in union-management relations. Unions have found that grievance handling problems are greatly increased in plants where time study exists. Not only are there more grievances, but a greater amount of time must be spent in investigating and processing time study grievances. And, in spite of a general decline in the number of plants using work measurement, there has been no corresponding decline in the percentage of grievances going to arbitration. This figure has remained at about 20% for the last 10 years, according to published reports of both the American Arbitration Association and the Federal Mediation and Conciliation Service.

Although local unions may seek expert help in special cases, the majority of time study grievances can and should be successfully handled by shop stewards. Union judgment is on a par with management's. The worker on the job is as good an intuitive judge of the propriety of any job standard as is management.

Because management does not like to bargain on standards, and because so little about time study is factual, a large proportion of time study grievances are taken to arbitration. Unfortunately, although many arbitrators may not be biased in favor of the company or union, they all too often accept time study as being scientific and mathematically precise. They do not recognize the shortcomings of time study and therefore are more likely to accept its results, particularly if, as professional industrial engineers, they have an unconscious vested interest in promoting some special technique that is questionable.

Unions and Time Study Grievances

Despite these difficulties and shortcomings, local unions can do a good job of protecting workers against unfair management time studies. To do an effective job, however, locals need informed and alert officers and stewards. Above all, union representatives should not be snowed under by the so-called scientific procedures and arguments of time study analysts.

Time study grievances should be handled in the same manner in which other grievances are handled. The most important step is that of getting the facts. Although some knowledge of time study is helpful, it is not necessary for the shop steward or other union representative to be a time study analyst in order to process a time study grievance. The union representative should do the following:

1. Secure a copy of the company's record of the operation in dispute.
2. Make certain that the record of job conditions and the job description are complete. If either is incomplete, it will be impossible to reproduce the job as it was when the time study was made, and therefore the company's time study cannot be checked. This alone is grounds for rejecting the study.
3. If the time study sheet does contain sufficient information as to how and under what conditions the job was being performed when the time study was made, then it is necessary to determine whether the job is still being performed in exactly the same way.
4. Usually the total operation of the job cycle is broken down for timing purposes into parts called elements. Check the descriptions of each element to see if they describe what the operator is currently required to do. Any change affecting time invalidates the original study.
5. Make sure that everything the operator is required to do as part of the job has been timed and reported on the time study sheet. Watch for tasks that are not part of every cycle.
6. Check for "strike-outs." A time study finding is based on a number of different timings of the same job. The time study analyst may discard some of the timings as "abnormal." If the analyst has discarded any of the recorded times, he or she must record the reason for doing so. This enables the union to determine intelligently if the strike-out is valid. That a particular time is longer or shorter than other times for the same element is not sufficient reason for discarding it.
7. Determine if the time study was long enough to reflect accurately all of the variations and conditions that the operator can be expected to face. Was it a proper sample of the whole job? If not, the time study should be rejected, since its results are meaningless.
8. See that only a simple average, and not the median, mode, or other arithmetic device, was used in calculating the elemental times. The simple average is the only proper method for time study purposes.
9. Check the rating factor on the time study sheet. Try to find out if the analyst recorded the rating factor before leaving the job or after computing the observed times. Ask the operator who was time studied if he or she feels the rating factor is a proper one. Watch this operator work at the pace he or she considers proper and then at the pace required to produce the company's work load. The judgment of the worker and the steward are as valid as that of the time study analyst.
10. Make sure that allowances for personal time, rest, and delays have been provided for in proper amounts.
11. Finally, check all the arithmetic.

Most union representatives have found it unwise to take additional time studies themselves as a check, except as a last resort. It is more effective to show the errors in management's study than to try to prove that a new union time study is a proper one. A union time study is still only the result of judgment. Even when proper methods are used, they tend merely to reduce the inconsistencies of time study, not to eliminate them.

Historically, "time study" has referred to the process described in this chapter, which involves the use of a stopwatch or other timing device. Some companies have introduced other methods of setting production standards, such as standard data and predetermined motion time system. When unions object, companies have claimed that these systems are simply newer forms of time study and may rightfully be used even if a contract has a clause such as "Standards shall be set by time study."

To guard against this type of subterfuge where stopwatch time study is currently used, unions should make sure that future contracts contain this type of clause: "All production standards are to be set by stopwatch time study, or the union is not a party to any predetermined time system."

Resolution on Industrial Engineering Adopted by Fourth Constitutional Convention of AFL-CIO

The AFL-CIO was plagued for years by complaints about predetermined time standards and other work measurement techniques. In response, it passed a resolution on December 12, 1961, from which the following excerpts are extracted.

The last decade has seen tremendous changes in work and in the work environment. Many of these changes are associated with accelerated mechanization and rapidly changing technology and have had profound effects on workers and their unions, on working conditions, on employment and unemployment, on the general character of the work force, and on collective bargaining.

Concurrently with these "automation" changes, other techniques have been introduced into the work place and the collective bargaining area, which, while not generally accompanied by as much fanfare and publicity, have had equally disturbing effects. These are the innovations generally "credited" to the industrial engineer and include predetermined motion time systems, standard data, work sampling, and other "statistical" techniques for setting production standards, as well as electronic timing and control devices. Labor, after several years of trial, is forced to conclude that these newer techniques are based on the same or similar false assumptions that have historically characterized other techniques of the management engineer. When applied to the workplace, these techniques yield results that are not accurate, reliable, or valid.

Unions have historically opposed the use of stopwatch time study, piecework and wage incentive systems, and job evaluation. This opposition was originally based on the arbitrary and abusive use of these methods by management. But unions learned that, even when attempts were made to use the systems in an objective manner, the results were inequitable. The systems themselves, as well as their applications, were found lacking.

Labor has found that the new techniques are nothing more than subtle forms of the old. "Improvements" represent little more than techniques of confusion, making it more and more difficult for workers to understand and cope with problems raised by the "new methods." All too often workers find that the "scientific" method merely is a device to circumvent established collective bargaining arrangements.

In spite of proven shortcomings, work measurement techniques, new and old, are still being used to set unfair production standards and work loads. Incentive systems of questionable validity are used to induce excessive work pace, and job evaluation techniques are being used to downgrade workers and reduce legitimate wage gains.

Research into the basic problems of human work, the measurement and effects of physical and mental fatigue, and the effects of the automated workplace, of job design and job enlargement, and of fitting the job to the worker are all largely ignored by the industrial engineer.

When these management schemes exist, unions must be prepared to negotiate maximum contractual protection in order to minimize harmful effects. But contract language is not enough. Unionists must be as fully informed as possible in order to cope with the "traditional" as well as the "modern" industrial engineering techniques.

5.6.9 WAGE INCENTIVE PAYMENT PLANS

The designing and acceptance of wage incentive plans by unions is detailed in an AFL-CIO *Collective Bargaining Report*,* excerpts from which follow.

Union Attitudes

With few exceptions, unions are opposed to wage incentive plans, both because of the damaging past experience with abuses under such plans and because of the difficulties and ill effects inherent in the plans. Unions that have accepted them or permitted them to continue have usually done so only with reluctance and misgivings. It simply has not always been practical or expedient to oppose or eliminate such plans. A few unions, primarily in the rubber and needle trade industries, where wage incentives are most firmly entrenched, have at least temporarily accepted incentives as part of their collective bargaining programs.

Many industrial engineers agree that wage incentives have been abused in the past, but maintain

*From AFL-CIO, *Collective Bargaining Report*, Vol. 2, No. 12 (December 1957).

that such abuse no longer occurs. In fact, however, although some of the worst abuses have been toned down, primarily through union action, a variety of ill effects and strains on workers is still very much the rule.

The presence of an incentive system invariably means special problems of education, representation, and protection of workers. It puts a strain on the entire collective bargaining process, making it more difficult, more complex, and more costly. The value of any such plan is also highly questionable because, even though it may initially yield increased earnings, it inevitably requires a speedup of work efforts, creates friction between workers, and produces continual wrangling over production standards.

Nature of Incentive Plans

Employers install incentive plans because they expect them to lead to higher profits through reduced production costs. Essentially, these plans all attempt to induce workers to produce more than a fair day's work by promising a monetary reward. They are also based on the notion that workers will not perform an "honest" day's work unless they are "bribed" by the promise of extra money. Management attempts to sell these plans to workers by claiming and emphasizing that an incentive system will result in earnings higher than straight hourly wages.

While it is true that, in theory, management and labor can both gain by way of the incentive route, this seldom happens in practice over an extended period. Labor has found that the disadvantages will normally outweigh the advantages to workers.

Value to Management

Actually, it is questionable, when all things are considered, whether even management really gains enough from incentive plans to make them worthwhile.

Management does have some costs that remain relatively constant over a fairly large range of production. These costs are usually called "factory overhead" or "factory burden." They include such items as executive and supervisory salaries, certain taxes, and machinery and building costs. Naturally, as workers increase production, these "fixed" costs are allocated to a larger number of units. The resulting decrease in fixed cost per unit produced can result in substantial savings to management. It is these savings that appeal to managements and that cause them to install incentive plans.

Once incentives are installed, however, management finds that, in practice, there are many offsetting increases in cost. It is common knowledge that in many cases the costs of setting up time study and wage incentive departments, coupled with such typical by-products of incentive plans as inferior quality, customer discontent, lower employee morale, more grievances, and increased accident insurance rates, may more than offset any savings.

That some managements recognize these incentive shortcomings is evidenced by studies of incentive plans at 416 companies. Of 100 companies surveyed in one study, 40% acknowledged that they felt their incentive plans were not satisfactory. Another 17% said they were only partially satisfied; their comment was that incentive benefits fell short of justifying the bonuses paid to workers.[4]

A study of the experience of 316 companies, covering a 15 year period, found that 78% of the wage incentive plans in these companies had either failed or had developed such major weaknesses that the companies were completely dissatisfied.[5]

Some companies have expressly stated that efficient supervision and reasonable labor-management relations can accomplish more efficient and less expensive production than reliance on incentive systems.

Shortcomings of Incentive Plans

Although there are many types of plans, varying in detail and with different descriptions, they are fundamentally similar. It must be kept in mind that the reasonableness or practical nature of any plan cannot be determined merely by examining a written description of the way the plan is supposed to work. Some of the basic shortcomings shared by virtually all types of plans in practice are discussed here.

Incentive plans require the setting of production "standards." Usually the standards are set on the basis of time study, with all its inconsistencies, inaccuracies, and unreliability.

The union may negotiate certain basic wage guarantees, but actual earnings under an incentive plan will depend on time standards or rates set by time study analysts. The inability of analysts to set standards accurately or consistently means that actual earnings will bear little, if any, relation to workers' efforts.

Some workers will be assigned to jobs having loose rates, where exceptionally high earnings can be made. On other jobs rates will be tight, so that equal or greater effort will not yield incentive earnings as high as on jobs with looser rates. Frequently, workers on jobs with tight, unrealistic production standards must work at a killing pace in order to just "make standard" and to earn base rates.

The supervisor who allocates work has a tremendous weapon in his or her hands when workers are on incentives. It is easy to give friends and informers the gravy jobs and to give the real trade unionists the tight ones. On the other hand, some managements may hope to influence shop stewards or union officers in favor of incentives by alloting them jobs with loose rates.

In any case wage incentives lend themselves to abuse. Workers with similar jobs and pay grades frequently experience large differences in earnings. The opportunities for discrimination are obvious.

Rate Cutting and Speedup

There is also considerable room for jockeying with standards as jobs gradually change. Piece rates or job standards that may be "fair" when they are set often tend to become loose as time goes on, as workers gain experience, improve their skills, and develop new shortcut work methods, and as their earnings rise accordingly.

Although managements usually assure workers that standards will not change simply because earnings go up, some seek an excuse to tighten the rates on jobs where earnings increase by synthesizing trivial methods changes, justifying a retiming of the job. Frequently, the very improvement developed by a worker to increase his or her own earnings becomes management's excuse for cutting the rate!

Speedup is easily accomplished throughout the plant in the same manner. A periodic tightening up of rates, under the guise of methods improvements, forces workers to work harder and faster in order to maintain previous earnings.

Another shortcoming is a different type of steady pressure on workers to work harder. According to the theory of incentives, workers determine their own work pace. They are supposed to be free to work at a faster-than-normal pace or not, as they see fit. But again, theory and practice are different.

Workers who produce less than some predetermined amount, usually some "average," are accused of loafing and are goaded into greater efforts. Management tries continually to raise production by eliminating "below-average" workers. In such ways workers are forced to work faster in order to keep up with the always rising "average" worker. Older workers, particularly, find such competition difficult. All workers are also faced with the fact that the risk of accidents is increased when the work pace is stepped up.

Group Plans' Pressure on Workers

Some incentive plans tie wages to the production not of the individual worker, but of a group of workers. Incentive earnings are paid to all members of the group on the basis of the group's total output. The group may be large or small, involving as few as two or as many as several hundred, or even thousands, of workers. It is sound practice to keep such groups as small as possible.

Group incentive plans can be especially troublesome, since our worker's earnings depend, at least partially, on another's work. No matter how hard a worker may work, factors beyond his or her control or beyond the group's control may limit total output and his or her incentive earnings. These plans easily degenerate, causing workers to assume management's functions.

Management plays workers against each other and tries to get workers to "police" each other. Management may expect one union member to pressure another who may have caused group earnings to decrease because of absenteeism, lateness, and so on. Young workers become impatient with older workers who cannot "keep up."

Incentive Versus Nonincentive Workers

Incentive plans may work to provide higher earnings to a semiskilled group than to skilled workers or to create disproportionate differences between incentive and nonincentive workers. Such situations can contribute to a serious breakdown in worker morale, with the wage incentive systems pitting one group of union members against another.

The problem can be somewhat alleviated by providing for additional payments to service workers. Often such payments are determined by the level of productivity of the production workers.

Weakening Effect on Union

Wage incentives de-emphasize the union's role in securing higher wages. Combined with the creation of friction among workers and the development of conflicting factions, this may threaten the union's entire existence. Its ability to secure decent contractual protection in other areas, such as grievance procedures, seniority, and vacations, may be placed in jeopardy.

There are also the serious difficulties of properly and adequately protecting and representing workers under these plans. The complicated nature of many wage incentive plans, as well as their dependence on time study and other work measurement techniques, makes it difficult for local union officers to handle day-to-day problems. Union representatives usually have to be called in to spend a disproportionate amount of time on incentive shops as compared to day work shops, with a resulting drain on the time needed for other grievance problems, preparation for negotiations, organizing, and other union activities.

Elimination of Incentive Plans

The dangers and inequities of wage incentive plans and their possible antiunion effects have led many unions to attempt to eliminate them at every opportunity. Many unions have been successful in bargaining incentive plans out of contracts where they previously existed, and others have succeeded in persuading management against installing new plans.

There is one bright spot on the incentive horizon. Most "experts" agree that incentive plans may well be on the way out because of technological improvements and automation. As production becomes more and more automatic, workers have less and less control over output. The supposed gains to management of incentive plans will decrease, and presumably many of these plans will be eliminated.

Some companies lacking newer automatic equipment may try to cut costs through incentive plans in order to compete with their better-mechanized and better-automated competitors. Workers in such companies will be forced to work faster and harder, but they will not be able to win a race with automatic equipment. Workers cannot produce enough to save incompetent or inefficient management.

Unions desiring to eliminate incentives must be careful that they do not trade one set of inequities for another. Some companies have agreed to do away with incentives, provided that they could substitute MDW. In other words, management sets the standards of production (usually by time study) that workers must meet. The workers receive straight hourly earnings with no incentive pay. Frequently the production standards are set at a level requiring workers to work at an incentive pace for nonincentive pay. Management can thus secure the equivalent of incentive production without paying for it.

When incentives are eliminated, unions should insist that, if there are to be any production standards required on the nonincentive operations, such standards be developed through direct bargaining with the union. The union must also concentrate on the level of the all-important new hourly pay scales. With or without incentives, guaranteed hourly wages must be negotiated high enough to yield a fair day's pay to workers.

Protection of Workers Under Incentive Plans

It is impossible to build an incentive plan that will be fair to all workers, or one that does not create serious problems for unions. If a union must accept a wage incentive plan, it should insist as strongly as it can on certain principles and practices necessary to minimize arbitrary and abusive applications of such a plan. The refusal of a company to accept and observe these principles must be recognized as a sign that it is seeking to manipulate the plan unfairly to its own advantage at the expense of its workers.

The principles and practices are as follows:

1. There must be adequate, guaranteed mainimum hourly payments to all incentive workers. Workers should be assured of certain reasonable minimum earnings, regardless of the amount they produce in any particular period. Such an hourly guarantee is normally referred to as the "guaranteed base rate" in an incentive program.
2. The base rates on which incentives are built must be realistic. Frequently workers and unions are lulled into a false sense of security by periods of "high" incentive earnings. They neglect to insist that the guaranteed base rates be kept up to date. Management may deliberately allow a loosening up of incentive plans in order to keep from raising base rates. Workers become more and more dependent on incentive earnings as the guaranteed base rate lags farther and farther behind.
3. Incentive base rates should be the same as nonincentive hourly rates. A realistic incentive base rate is one that is at least equal to the hourly rate for nonincentive workers on similar jobs in the industry or area.
4. To apply general wage increases properly to incentive workers, they should be added to the base rates upon which incentive earnings are figured. As a worker increases production, his or her pay per piece decreases, not only for the pieces above the standard, or for the increased production, but for every piece produced.
5. Incentive plans must provide a realistic opportunity for earnings above hourly rates. Any incentive plan that does not provide all workers with an opportunity to earn reasonable extra

earnings is unfair. There must be no artificial barriers to incentive earnings and no ceiling on earnings.

6. Incentive payments should increase at least in direct proportion to production. In fact, since it is always more difficult to produce extra units, and since cost reductions are greater as production increases, payments to workers should really be increased more than output.

7. Incentive earnings should not be "averaged" for the purpose of reducing total earnings. Incentive earnings should never be calculated for periods longer than one day. Days of high earnings should not be lowered by poor days. In no case should below-standard production be used to equalize above-standard production.

When workers normally work on short-run jobs, their earnings should be calculated by the job. A job with a fair rate should not be used to cover up or equalize a job with a tight rate. This can only result in reduced earnings for the worker.

8. Changed standards should provide the same earning opportunity as original standards. One of the ways of limiting unreasonable management tampering with production standards is to provide that, when a legitimate change in methods or equipment occurs, only the time for that portion of the job which is actually changed should be adjusted and only in the amount of the actual change. Management should not be permitted to use a minor change in a job as an excuse for revising the production standard for the entire job.

Most important, the new standard must allow the operator to earn at least as much as before. Changes in earning opportunities should be made, if at all, by mutual agreement through collective bargaining. They should never be determined by time study.

9. Fair payments must be made to incentive operators when assigned to nonincentive work. In most cases the base rate is not enough. Actually, workers on an incentive plan should be assured that their usual incentive earnings will not be cut if, for reasons beyond their control, their production efforts are hindered or they are temporarily shifted to nonincentive work.

The most equitable arrangement to use when an incentive worker is placed on a nonincentive job is to pay his or her prior average hourly earnings, including incentive pay, or the hourly rate on the nonincentive job, whichever is higher.

10. The whole incentive plan must be easily understood by workers and their representatives. No plan should be applied if it can be understood only by engineers and mathematicians. Workers must be able to understand how the standards of production set by the company were arrived at and to calculate their own pay easily and quickly.

Many managements have attempted to prove the "fairness" of an incentive plan by the absence of incentive grievances. Unfortunately, the absence of incentive grievances is often a result of the complicated nature of the incentive plan. Workers cannot intelligently process a grievance over something they cannot understand.

11. The final point is most important. Management must permit every phase of its incentive plan to be reviewed by the union through collective bargaining procedures. Unions of course have the legal right to bargain with management on anything that affects the earnings of union members. Unions must be furnished with all the information that the company has on the wage incentives, as well as on the methods, such as time study, upon which the incentives are based.

Refusal by management to give a union requested wage incentive or time study data, to discuss individual and general problems associated with incentives, or to resolve incentive complaints through the grievance procedure (including arbitration, if necessary) indicates that management has something to hide, that it is not dealing in good faith with the union, and that the plan involved is hardly likely to be beneficial to the union's members.

5.6.10 LABOR'S RIGHT TO TECHNICAL DATA*

The reluctance of management to furnish unions with technical data resulted in an AFL-CIO document advising unions on several points, which are discussed here.

Unions are, with increasing frequency, being faced with management's use of a variety of industrial engineering and other techniques of a technical nature. These techniques are used in estimating production standards and work loads, crew sizes, manpower requirements, line speeds, wage incentives, job evaluation, and merit rating. There is no question but that management's use of these techniques affects wages and working conditions. Therefore every aspect of these techniques is legally subject to collective bargaining.

Unions have long recognized the need for adequate contract language and grievance procedures to protect workers against arbitrary and abusive applications of these management tools. Unfortunately, many union representatives have experienced difficulty in writing and negotiating contract language and in handling grievance and arbitration cases because of the refusal or reluctance of some managements to furnish unions with information and data on these company activities.

Union appeals through the NLRB and the courts have established the union's right to relevant

*From AFL-CIO American Federationist, *Collective Bargaining Report*, October 1963, pp. 19–22.

data where management uses time study and other production standard setting techniques or where an incentive plan or job evaluation plan exists.

There is no question that the employer has a clear obligation to furnish to the union relevant data that are not unduly burdensome. However, many problems are posed by attempts of employers to define "furnish," "relevant," and "burdensome" in such a way as to prevent the union from getting the data it needs to perform its legal obligation of representing workers in a bargaining unit.

Statutory Law

Section 8(a)(5) of the National Labor Relations Act makes it an unfair labor practice for an employer to refuse to bargain with a union representing his or her employees. Section 8(d) defines this obligation as follows:

> *For the purpose of this section, to bargain collectively is the performance of the mutual obligation of the employer and the representative of the employees to meet at reasonable times and confer in good faith with respect to wages, hours, and other terms and conditions of employment, or the negotiation of an agreement, or any question arising thereunder.*

These provisions constitute the statutory basis for the employer's duty to provide a union with information needed in the negotiation and administration of a collective bargaining contract. If an employer violates the statutory obligations, the NLRB is empowered to order him or her to cease and desist from refusing to bargain with the union and to take the affirmative action of furnishing the union with the job information or related data.

General Principles

Every case involving an employer's duty to furnish a union with job information and industrial engineering data has to be decided on its own individual facts. Different facts may mean different results; therefore no single set of rules can hope to cover all situations. Following are some of the general principles developed by the NLRB in dealing with an employer's obligation to supply job and wage data to unions. Knowledge of these general principles should provide a union representative with some notion of the union's rights and the employer's duties under the law. But the principles should be considered as guidelines rather than as final answers.

Scope of Employer's Duty

An employer has the duty to furnish a union, upon request, with all the relevant information pertaining to jobs that is needed by the union in negotiating and administering a collective contract. An employer cannot refuse to supply information on the ground that it is confidential and solely for internal management use or that furnishing it would be burdensome, unless he or she can show compelling need to keep it confidential or that furnishing it would be unduly burdensome.

Certain data, such as current wage information, are presumed to be relevant to the negotiating and administering of a labor contract. Thus a union asking for this type of data does not have to make an initial showing of how the information fits into its bargaining needs, unless the particular data sought clearly appear to be irrelevant. But a union may place itself in a more sympathetic light by disclosing why certain information is desired.

Duration of Employer's Duty

A union's right to job data is not limited to pending contract negotiations. A union is entitled to information needed for the intelligent discharge of three distinct aspects of its obligation to represent employees in a bargaining unit: (1) the actual conduct of bargaining over the terms and conditions of a new contract; (2) the administration of a current contract, including the handling of disputes through the grievance machinery and the resolution of new problems not covered by the existing agreement; and (3) the preparation for coming negotiations.

Manner of Furnishing Information

An employer must comply with all reasonable union requests for relevant job information in such a manner as to make the furnished data meaningful and understandable without undue difficulty. However, this does not necessarily mean that the employer has to supply information in the exact form specified by the union.

This is one of those areas in which the facts of the individual case are far more important than the general rules. In effect, the NLRB will try to balance how burdensome, costly, and time consuming it will be for the employer to supply the information in the particular form requested against how necessary and appropriate it is for the union to have the information in that form. For example, sometimes it is sufficient for an employer to convey information orally, whereas other times, when complicated data are involved, a written presentation of the information is needed.

As a practical matter, a union should try to avoid filing an unfair labor practice charge just because an employer refuses to make data available in the exact form requested the first time. The union should suggest alternative methods of presenting the information or a meeting with the employer to let him or her propose alternative methods of presenting the information. This will demonstrate the union's good faith and, if charges are eventually necessary, will increase the likelihood of success before the NLRB.

Waiver of Union's Right to Information

A union can waive its right to information, for example, through an express clause in the labor contract, spelling out the union's right to receive only certain limited data. But any such waiver must be very clear.

The inclusion in a contract of a provision requiring the employer to supply certain specified data does not mean the union has waived its legal right to other information. Similarly, the inclusion of a general grievance procedure does not relieve the employer of his or her legal obligation to provide necessary information.

Access to Particular Data

Job Descriptions

A union is entitled to access to the existing job descriptions in the bargaining unit it represents. This is true even though the descriptions contain mistakes that the employer is correcting. Having access to unrevised, faulty job descriptions might become important in the processing of a grievance or in contract negotiations where the union is claiming that inequities exist by reason of the errors in question. Where changes are made in job descriptions, the old as well as the new or changed description also may be needed to determine whether changes in the description conform to actual job changes.

In exceptional cases an employer may prove that the job descriptions contain information on secret processes or that the company's competitive position might be endangered if a union removes copies of job descriptions from the employer's premises for examination. In such situations a company may then be able to defeat a charge that it is refusing to bargain, at least where the union has adamantly refused to listen to any of the employer's proposals regarding alternative ways of communicating the needed information.

Job Classifications and Wage Rates

The NLRB and the courts have broadly sustained a union's right to be supplied data on job classifications, wage rates, and rate ranges, including the number of employees in each rate range. A union also has been upheld in its demand for a breakdown of the point ratings used in evaluating all the jobs in the plant.

The NLRB frequently has required employers to provide employee's names and wage rates in such a way that a union could determine exactly which employee received which rate. Occasionally, for some special reason, the NLRB will allow an employer to provide the number of employees receiving each wage rate without linking specifically named employees with particular rates.

The company may not always be required to present data in the precise breakdowns wanted by the union. For example, where a company kept figures on the average earnings of all its employees on a quarterly basis, a union was not upheld in its demand that the employer provide monthly average earnings of employees in each of some two dozen separate bargaining units going back 2 years. The NLRB said that the information was not readily available in the form requested and that it was unduly burdensome to assemble.

Wage Data on Other Employees

An employer was ordered to furnish a union with wage rate data collected by the employer about other employees in the industry. A union also was held entitled to payroll information on the employer's employees at another plant not within the bargaining unit when the employer stated during negotiations that he wanted to keep the two plants in step on wage increases.

Incentive Wages, Merit Ratings, and Piece Rates

The duty of an employer to provide wage data may vary, depending on the type of wage plan in effect or under consideration. Thus an employer has to supply a union with all wage information used as a basis for establishing and maintaining an incentive wage plan and needed by the union in order to understand and deal with the plan.

A union also has a right to information on bonus payments. Where a merit system is in operation, the employer must make available data on the standards used in determining the merit ratings, as well as information on the performance ratings of each of the employees covered by the system.

In the case of piece rates, the employer has to provide breakdowns of the itemized rates on all piece rate operations, depending on the particular product being made. This also would cover methods and rates of time work payments for lost time resulting from machine breakdowns and other reasons.

Time Study and Job Evaluation Data

The NLRB has ruled that a union is entitled to be furnished with the original time study data developed by the employer's experts in determining job standards and that it is entitled to have its own time study expert examine the job in dispute, to check out the employer's figures and procedures. The facts of this particular case, however, resulted in a reviewing court's refusing to allow the union's expert to go on the employer's premises to conduct an independent study, although it did agree that the union had a right to the data.

In one situation in connection with handling pending grievances, the NLRB upheld the union's right to:

1. Original time study sheets and other documents relative to both the prior rates and the new rates.
2. All documents, studies, and other information used to determine the rate of pay for each job.
3. All documents, studies, and other information used to evaluate such jobs, both prior to the change and thereafter.

Furthermore, in connection with both the handling of the grievances and the general administering of the labor contract, the NLRB upheld the union's right to:

1. Time study manuals, instructions, and procedures used in conducting time studies of jobs in the employer's plants, including full information as to the weights given to each factor used to arrive at a final decision on the established rate and as to the factors considered in making such decisions.
2. Manuals, instructions, and procedures used in the development of standard data and in their application in the development of job rates in the plant.

In the same case, the NLRB expressly held that the mere fact that the contract contained a grievance procedure providing for the adjustment of "any complaint" did not prevent a union from going to the NLRB in a dispute over the extent of the union's statutory right to bargaining information.

However, a union may not be entitled to time study information in connection with a particular grievance if the arbitrator has not yet resolved whether the union even has the right under the terms of a given labor contract to make the issue in question the subject of a grievance. This is simply another aspect of the principle that states that whatever data a union seeks must be relevant to the union's function of representing the workers in a particular unit. If it has no right to bargain or grieve over a particular matter, it may have no right to information connected with that matter.

Independent Union Time Study

In contrast to the many cases involving the union's right to data, there have been very few involving a request by a union to make its own time study of a disputed job. Unions consider that the same legal basis exists for the conducting of time studies as exists for the right of unions to time study data. Many unions have negotiated this right into labor-management agreements. A sample of such a clause is as follows:

> *At any step in the grievance procedure, whether on disputes over the standards or over changes in the standards, the union shall have the right to call in a representative of the international union, any outside time study analyst, or both. Such representative, outside time study analyst, or both shall be allowed to time study the job or changes in the job or*

help the union determine whether the union's position is valid. Such representative of the union, outside time study analyst, or both shall then have the right to be present at all steps of the grievance procedure, including arbitration.

In most instances, before requesting a union time study, it would be natural for the union to request that the company supply all necessary data. A refusal will result in a charge to the NLRB, based on the refusal to furnish the requested data. When such a charge is finally resolved in the union's favor, unions generally have experienced little difficulty in making studies or other checks necessary to evaluate the company's data.

How, then, should a union approach the problem of time study data? It should secure as much information as possible from the employer's sources and, if it wishes, should request the right to have its own time study expert make his or her own examination. Particularly in connection with the latter request, and in spite of the fact that it considers the latter right a legal right, the union should strive to show the genuine necessity for an independent survey to confirm or challenge the employer's conclusions.

The union should be as reasonable as possible in its request. If met at first with an employer's refusal to permit the union's time study analyst on the premises, the union might, before it turns to the NLRB, express a willingness to bargain with the employer over the time, manner, and extent of the proposed study in order to minimize any possible inconvenience to the employer.

This discussion on time study also applies to the need of unions to observe and study actual jobs in order to handle job description and job evaluation cases intelligently.

Employer's Ability to Pay and Changes in Productivity

An employer who claims that he or she is financially unable to meet the union's bargaining demands may, depending on the circumstances, have to furnish costs and sales data or other relevant information to back up this claim. Where an employer asserts inability to pay, a union's request for a copy of the employer's financial statement will ordinarily be upheld. An employer has no obligation to supply a union with sales and production figures, however, if he or she does not claim financial inability to meet the demands.

The NLRB has ordered an employer to provide data on changes in the company's rate of productivity so that a union could gauge its wage requests accordingly. This is likely, however, to be the type of information that an employer will have to furnish in some cases and not in others, depending on the way the bargaining negotiations have proceeded.

Whenever a union seeks to have the NLRB order an employer to provide job data, the formal charge against the employer will be that he or she has "refused to bargain in good faith." Unless the information is of the kind to which a union is unquestionably entitled (such as current wage data in a form readily available to the employer), it may be difficult to pin a refusal-to-bargain charge on the employer as long as he or she is willing to discuss the matter with the union and continues to make counterproposals as to what information should be supplied and how to supply it.

The union's general approach, therefore, should ordinarily be to outdo the employer in acting reasonably and in bargaining. This may sometimes even have the quick, happy result of persuading the employer to turn over the desired data. If it does not, at least the union will appear in the most favorable light when it comes time to file the refusal-to-bargain charge with the NLRB.

In seeking information from the employer, a union should consider the following rules (like most "rules," these are made to be broken now and then):

1. If difficulty is expected or encountered in obtaining needed information, submit a formal written, dated request so that there will be no later doubt about just what was requested and when the request was made.

2. Make a request in two parts:

a. Enumerate specifically every item of information and every report, record, study, survey, manual, directive, or other document that is wanted.

b. Ask generally for all other relevant and available information, reports, records, studies, surveys, manuals, directives, or documents pertaining to the subject matter of the union's inquiry.

3. Sometimes an employer will at first refuse to provide requested information and may give a fairly plausible reason for the refusal or indicate that he or she cannot comply in exactly the way proposed. In such cases the union should express its willingness, preferably in writing, to discuss alternative methods of securing the desired data when this can be done without depriving the union of access to the necessary information.

4. If the union desires to have an outside expert enter the employer's premises to conduct an independent study, inform the employer of the union's readiness to arrange the time, extent, and manner of the outsider's examination so as to cause no more interference with the employer's operations than is absolutely necessary.

5. Keep copies of all correspondence with the employer and complete notes on all meetings, setting forth the date, time, and place of the meeting; the names of persons present; and what was said or agreed to by the various parties.

Finally, a word of caution. The obligation of the employer to furnish the type of data discussed here is a legal obligation, as defined by the National Labor Relations Act and by interpretations of the Act by the NLRB and the courts. The conclusions drawn and the principles stated here are general. Before taking legal action by filing charges with the NLRB, the facts of a particular situation should be reviewed by competent legal advisers.

The arbitration of disputes between management and labor over problems involving production standards and wave incentive payment plans is developed by the author in a paper[6] delivered before the National Academy of Arbitrators, of which the author is a member.

5.6.11 THE PRESENT STATUS OF ARBITRATION UNDER WAGE INCENTIVE PAYMENT PLANS

Although more and more jobs in newer, sophisticated technological industries make the worker a monitor of an automatic process whose effort is unrelated to the level of production, there remain a sufficient number of jobs in the older industries where worker effort and production are related. It is in these older industries that the great majority of wage incentive payment plans continue to be found.

Ronald Higgins attempted to bring arbitration of this entire industrial engineering area up to date in his Bureau of National Affairs (BNA) publication, *The Arbitration of Industrial Engineering Disputes* (1970). Information beyond 1970 may be gleaned from the BNA *Labor Arbitration Reports* following Volume 51 of the series under the appropriate index entry.

Labor attorney Owen Fairweather's definition of an incentive payment plan as extra pay for extra work may be compared with that of Robert Roy at Johns Hopkins University, who pointed out that what was extra pay to management was regularly expected pay for workers operating under an incentive plan. Trade unions suggest that these concepts could be accommodated if the two systems of wage payment were defined as time work and production work systems of payment omitting the loaded term "incentive." This approach avoids a clash over so-called scientifically objective concepts of normal effort as against an equity concept of a fair day's work. Progressive industrial engineers have upgraded their understanding of the word "scientific." Mitchell Fein, AIIE vice president and research chairman of its Work Measurement and Methods Engineering Division, denies any scientific concept of normal. Instead, he states:

> *Managers and industrial engineers throughout the country who are experienced in collective bargaining recognize that a bargain about how hard employees should work is implicit in the vast majority of collective bargaining agreements. . . . The pace-effort bargain is generally a corollary of the wage bargain. . . . This is the essence of the principle of a fair day's work.*

Arbitrators will continue to be called upon to resolve cases based upon conflicts over conventional rating and leveling systems implying a scientifically set measurement of normal effort and an implied reward for extra effort. The same may be said for conflicts over the application of micromotion systems of standard data. Micromotion standard data systems do not eliminate rating; they simply conceal the rating system for all parties except the "priest experts."

Only one arbitrator has imposed MTM-developed standards on a union, arguing that it was an accepted, tried and true method that was approved by the engineering profession and that has proved itself by its survival these many years. This may be a tribute to the marketing effectiveness of the MTM Association, but as an indicator of scientific validity, it leaves much to be desired.

This arbitrator's argument makes as much sense as the curative claims of the peddlers of the late Lydia E. Pinkham's vegetable compound being based on its long market survival. To be sure, if the parties have included the use of the MTM system in their contract, then the application of its rules is justified. It is quite another matter to saddle the parties with this "pseudoscience" in the name of equity when neither of the parties has included it in the contract.

This was exactly the issue between the parties in what must be considered the most important arbitration over production standards since the arbitration in 1947 between the UAW and the Ford Motor Company over the speed of the assembly line. It was this arbitration that made the setting of production standards a matter for arbitration even where the subject was barred from the contract. Harry Shulman defined management's right to set a production standard as a right to propose a standard that is subject to protest by grievance.

During the 1970s arbitration between the National Association of Letter Carriers (NALC) and the U.S. Postal Service arose under Article XXXIV of the collective agreement that governs the relationship between parties, giving the Postal Service the right to set production standards. The postal authorities imposed the MTM system of measuring letter carrier work over the opposition of the union. The difference between the parties became the subject for a national arbitration before Sylvester Garrett, who issued his award in Case NB-NAT-6462 on August 6, 1976.

Mr. Garrett ruled that "the arbitrator could not review the studies on which the MTM values originally were based and surely could not accept a concept of 'normal' pace without any knowledge of, or study of, letter carrier work." He went on to state that the Postal Service cannot in the absence of an agreement with the NALC, base work or time standards for letter carriers upon MTM values instead of the results of adequate time or work standards. He continued,

> *As far as the impartial chairman is concerned, the words fair, reasonable, and equitable have no practical meaning for purposes of developing time or work standards except as they are applied to specific employees or groups of employees performing specific tasks under defined conditions.*

Another substantial contribution to the resolution of conflicts over the application of wage incentives to an entire industry was made when 11 steel companies and the United Steelworkers created a three-man panel of arbitrators, consisting of William Simkin, Ralph Seward, and Sylvester Garrett, to set up guidelines to be used by the parties to extend incentive wage coverage to steel production and maintenance workers at the 11 companies who were not then covered by wage incentives and to revise incentives that might be termed too low or too high.

When it was determined that a job, hitherto based on timework, would be covered by incentive, the affected employees were given a 10 cent/h increase retroactive to August 1, 1968, to continue until the jobs in question had been placed on appropriately designed incentive opportunities. Their award, released on August 1, 1969, was a model of a pragmatic program escaping ideological constraints of pseudoscientific engineering techniques.

The steel agreement of July 30, 1968, had set joint incentive study groups for each of the 11 companies, made up of three union and three management representatives. These committees were to determine, among other matters, jobs properly subject to incentive coverage and those not suitable for coverage, the definition of equitable incentive earning opportunities, the adjustment of incentive standards from time to time to maintain equity, and, finally, a set of procedures to implement the preceding sets of principles. All of these joint committees had reached an impasse. The arbitration panel was charged with resolving this impasse.

There then followed a set of pragmatic instructions in the award, unburdened by an opinion that could be misused to legitimize anybody's favorite industrial engineering ideology. In addition to the usual classification of direct and indirect incentive workers, the panel added a third classification: secondary indirect incentive jobs. These, though not qualified for the usual direct or indirect incentives, were defined as those jobs where there normally was an opportunity to make an appreciable and demonstrable contribution to efficiency beyond nonincentive performance.

The importance of this case does not lie in the specific procedures evolved by the panel that are applicable to the steel industry, but in the nonideological problem solving philosophy triumphing over industrial engineering rigidity. The panel's approach is an ideal example of what Professor John R. Commons of Wisconsin had in mind when he enjoined the parties to break the tyranny of the experts.

It should be noted that, since the award of the steel panel, its principles have been fleshed out in a number of cases in which Sylvester Garrett has handed down decisions in specific cases. Garrett's former associate and successor, Alfred Dybeck, has described the significant principles that have been developed therein.

Review of Principles Governing Arbitration of Wage Incentive Plans and Production Standards

Some of the principles governing the arbitration of wage incentive payment plans and production standards since Harvey Shulman handed down his historic decision in the Ford Motor case are reviewed here.

Much of the detail work in agreements spelling out the determination of production standards uses so many undefined and nonoperational words that they are all but useless, particularly when elastic words such as "equitable," "normal," and "fair" are used, so that for all intents and purposes, the arbitrator makes decisions *de novo* with what guidance he or she can get from past practice.

Where the parties have spelled out the use of some predetermined microscopic motion time system in the agreement, the arbitrator is obligated to follow the dictates of the system despite his or her personal disbelief in its efficacy. It should be remembered that the actual application of these systems leaves plenty of room for judgment in the listing of applicable elemental motions and their assigned times.

The arbitrator can be an engineer, if he or she has proper regard for the limitation of his or her own measuring tool, or a lay person who can distinguish equity from rigorous authoritarian pseudoscience.

The arbitrator should be aware that facts so emphatically emphasized by management as the basis of its standards are seldom hard and fast. A fact is but a selective description of a total experience.

The arbitrator can also be reminded that a trade unionist's proclamation of the demand for equity in the setting of standards conceals a technique to gain an incremental wage increase barred by ordinary methods. This approach once did yeoman work for trade unionists, enabling them to escape the constraints of wage control during World War II, and there is little doubt that its use as a tool has increased as is evident by later practices of the 1980s.

Substitution of the Cooperative Model for the Adversary Model of Collective Bargaining*

The problems between American management and labor in which industrial engineers find themselves are derived from the widely accepted adversary model defined in 1941 by Professor Sumner Slichter. Prior to that some industrial engineers pioneered the thinking for a more cooperative model.

In 1916 Robert G. Valentine, a member of the scientific management establishment, published a classic paper, "The Progressive Relationship Between Efficiency and Consent," in which he stated that, under constitutional industrial relations, the unions will contest the share in the management and the share of the product between themselves and the consumer. Separating the problems of scientific management into two categories, he called on the Taylor Society, which functioned until World War I and was concerned with the advancement of scientific management, to provide a planning department to treat two classes of problems: (1) those relating to the determination of the best way of performing an operation under a given set of conditions and (2) those relating to the social, industrial, and moral effects of putting into operation an organization or methods that scientific investigation has determined to be technically the best. Finally, he defined the relationship between efficiency and consent—the doctrine that the worker individually and in organized groups has a right to share in the determination of the conditions under which new technical methods and apparatuses shall be put to use.

Shortly after publication of his paper, Valentine received an opportunity to apply his techniques experimentally amid a good deal of hope and optimism. In January 1916, a board of arbitration, of which Louis D. Brandeis was a member, handed down an award revising "a protocol of peace governing the relationship between the Dress and Waistmakers Association of New York City and the International Ladies' Garment Workers' Union (ILGWU). The award provided for a board of protocol standards charged with, among other duties, supervising the making and testing of piece rates and the assignment of standard times to different operations in the garment industry. Robert G. Valentine was chosen as the first chairman of this board. The board was set up in March 1916, but by September disagreements and rejection of its finding by the employers' association had led to complete ineffectiveness. By 1917 we were in World War I; Valenting had died, and a new era in the importance of organized labor in federal affairs had begun.

World War I opened up new opportunities to leaders of the labor movement. President Wilson instructed the secretary of labor to set up a federal board of arbitration. The secretary of labor set up a war labor conference board on which there was an equal number of representatives from the National Industrial Conference Board and the AFL. The conference suggested the organization of the National War Labor Board.

These measures lent labor a degree of security. Finally, when AFL founder Samuel Gompers himself was chosen a member of the advisory commission of the Council of National Defense, it meant that the scope of his activities reached outside the immediate employer-employee relationship. Other labor leaders were called upon to serve on various policy setting boards. In addition, a friendship developed between Morris L. Cooke and Gompers. Cooke was one of Taylor's original associates and was prominent in the affairs of the Tayor Society and the ASME. It was this social contact that led to the reconciliation between the leaders of the scientific management movement and the organized labor movement. The experience that organized labor had developed in its participation in joint labor-management committees led it to look forward to a constructive relationship with management following World War I.

The orthodox stream of management, however, took its leadership from Elbert Gary of United States Steel, who stood for union busting. The group around the Taylor Society became a progressive management group which continued to encourage an experimental approach to the theory

*From W. GOMBERG, in W. G. Irenson and E. L. Grant, Eds., *Industrial Engineering Handbook*, Prentice-Hall, Englewood Cliffs, NJ, 1955, pp. 1129–1132.

of organization and administration of industrial enterprises and to labor participation in these functions.

In 1919 Gompers agreed to edit jointly with Cooke and Fred Miller, president of the ASME, a set of papers expressing the points of view of the industrial scientists and the representatives of organized labor. This volume was published by the American Academy of Political and Social Science.

The overwhelming majority of management representatives, however, took their cue from the Elbert Gary school of union destruction. A great union busting campaign was undertaken, which reduced the importance of the unions to a shadow of their World War I strength.

The publication of *Waste in Industry* in 1921 marked another milestone in bridging the gap between organized labor and the leadership of the scientific management movement. This volume was the outcome of a proposal, made by Herbert Hoover to the Council of the Federated American Engineering Societies, that a group of engineers conduct an organized survey of a number of chosen industries and make recommendations for the elimination of wasteful practices. The committee assayed the wastes in industry and made recommendations for the participation of labor in administrative decisions that would eliminate waste. In other words, instead of merely philosophizing about labor's participation in industry, it pioneered a blueprint for such participation.

The leaders of organized labor protested in vain their newfound interest in production. As late as 1925, labor still proclaimed:

> *There is still a more important service than the union can render, that of participating in finding better methods of production and greater production economies. A group of workers cannot enter into this type of cooperation unless they know the results of their work will not be used to their disadvantage. We recommend that the Federation keep in touch with such engineers and industrial experts as may be helpful in developing the information and procedure necessary to union-management cooperation.*

Gompers was invited to make speeches to the ASME, and Green, his successor, to the Taylor Society. Although a number of interesting experiments emerged out of this cooperation between management's left wing and the American labor movement, the 1920s were an "ice age" for labor, with an occasional warm island known as an experiment in union-management cooperation.

The First Joint Labor-Management Cooperative Experiments

Under the influence of men such as Morris L. Cooke, Otto Beyer, and Geoffrey Brown, several joint experiments were taken up in the 1920s. The agreement between the ILGWU and the Cleveland Manufacturers Association, the Baltimore and Ohio experiment, and the Naunkeag experiment were the most heavily publicized of these ventures.

The agreement between the Cleveland Manufacturers Association and the ILGWU called for the setting of piece rates by time study techniques.

Under the Baltimore and Ohio experiment, committees of workers were set up independent of grievance committees to make operating suggestions in the maintenance shops of the railroads.

The agreement between the United Textile Workers Union of America and the Naunkeag Textile Mills in New England likewise called for the use of time study techniques to increase work loads.

The experiences that these unions accumulated with the techniques were later to give rise to the positive development of a trade union philosophy of industrial engineering. The common denominator of these experiments was that the unions frankly acknowledged an interest in increasing production, and administrative machinery was set up in each case to help the parties carry through their objectives.

Each of the experiments had terminated by 1931, when the Great Depression made any concern with productivity by either management or labor appear completely obsolete. The course of the experiments during their successful phase left very little impression on American industry, although it did cause some commotion in the intellectual world. At the time of the demise of the experiments, the fundamental industrial relations policies of large American mass production corporations remained antiunion, and it was they, rather than the intellectual, who dictated the business environment.

It is difficult to draw a fine line between the area where ordinary collective bargaining ceases and union-management cooperation begins. Collective bargaining itself starts with a pure conflict psychology, and then, in a developing relationship, the conflict is attenuated and a new relationship develops out of the old. For example, John Mitchell, president of the UMW, in 1890 defined trade agreements as "the terms of a truce determining the conditions under which the workers will permit the owners to operate their properties." Note that the whole emphasis is on conflict, either active or suspended.

Ordinary collective bargaining begins to evolve into union-management cooperation when the union is ready to discuss subjects such as production and sales outside the immediate employer-employee relationship. All through the 1920s, the AFL emphasized its devotion to the concept of union-management cooperation. It frankly was using cooperation in good faith as an organization technique. The southern organizing drive of the 1920s made its appeal to employers by offering the services of the AFL to reduce wastes and increase work loads by setting up committees to give workers a democratic voice in decisions affecting them.

The campaign won almost no buyers, and the AFL, along with the rest of the labor movement, turned toward more classical organization techniques when the passage of Section 7a of the National Industrial Recovery Act in 1934 made the organization of workers once more feasible. These experiences with union-management cooperation influenced very deeply the attitudes of the growing trade unions toward productivity as a concept and toward industrial engineering techniques as tools of collective bargaining.

It is already obvious from what has been said that the concept of union-management cooperation did not carry enough appeal to sell unionism to employers except in some very exceptional cases. Both the Naumkeag experiment and the Cleveland garment workers' experiment broke down because the Depression made it impossible for workers threatened with unemployment to see any particular advantage in participating in high-productivity schemes. The railroad experiments continued on a much reduced plane and with a declining enthusiasm, for the same reasons.

The new unions, which were growing rapidly, and the old unions, which were increasing their membership rapidly, were formulating, by their activities, their own philosophy of production.

Trade Union Ambivalence Toward Technology

The union institution must resolve two contradictory forces that the members press upon the leadership. Membership wants at the same time both a defensive safeguard against technological innovation and some of the gains that come from that innovation. Obviously, in an expanding economy the union can emphasize the gains available. It is not afraid of unemployment.

Cooperative Experimentation and the Organization Drives of the 1930s

Joe Scanlon and Clinton Golden, formerly research director and assistant to the president of the Steel Workers Union and later members of the faculties of MIT and Harvard University, respectively, had dramatic success with some union-management cooperation plans for which both had been responsible.

A careful analysis of the application of the Scanlon plan discloses nothing very unique except the development of a new human relationship between the management and the workers under the leadership of the very forceful personality of Joe Scanlon. Mr. Scanlon's persuasiveness led the management to permit the union to participate in a whole host of activities that would be listed elsewhere as management prerogatives. The particular group incentive plan in which workers augment their wages by attempting to raise productivity by reducing actual costs below a standard cost has been used elsewhere with mixed results. It would therefore be a mistake to confuse the mechanics of the Scanlon plan of union-management cooperation with the effective ingredient, which is Mr. Scanlon's personality. The idea of a Scanlon plan operating in a nonunion plant would make as much sense to Joe Scanlon as a rabbi issuing a decree koshering pork. Cooperative thinking was further developed in the years immediately before World War II in two books: *Organized Labor and Production*, by Morris L. Cooke and Philip Murray, and *The Dynamics of Industrial Democracy*, by Clinton Golden and Philip Murray.

Labor-Management Committees During World War II

Most of the so-called war production committees—or Nelson committees, as they were called, after Donald R. Nelson, chairman of the War Production Board—never really functioned despite the plethora of publicity that they were accorded.

Post-World War II Experiments in Union-Management Cooperation

Three outstanding examples of parties who abandoned the adversary mode of operation for a mutual problem solving approach after World War II are as follows:

1. The negotiation and administration of the agreement of October 18, 1960, on mechanization and modernization between the Pacific Coast Longshoremen's Union and the Pacific Maritime Association.
2. The Kaiser Steel Sharing Plan between the Kaiser Steel Corporation and the United Steelworkers of America in 1959.
3. The Automation Agreement of 1959 between the United Packinghouse Food and Allied Workers, the Amalgamated Meat Cutters and Butcher Workmen of North America, and the Armour Corporation.

The Cooperative Experiment Between the Pacific Longshoremen and the Pacific Maritime Association

The Pacific Longshoremen's Union, headquartered in San Francisco and headed by the avowed Marxist Harry Bridges, between its organization in the 1930s and 1960 maintained an adversary arm's length relationship with the Pacific Maritime Association. In the course of negotiations over the years, the union had instituted a great many work-restrictive practices, including unrealistic limitations on the maximum weight of single sling loads, in the name of health and safety.

In 1959 loading and unloading ships had become so expensive in the port of San Francisco, that the waterfront workers were losing all their work to rival ports. Paul St. Sure, the new head of the Pacific Maritime Association, approached Bridges with the question, "Harry, what would it cost us to get back all the work rules that we should not have yielded to you in the first place?" Bridges, priding himself that his Marxism lent him a realistic picture of what unions could and could not do, agreed to enter into an agreement with the employers, virtually eliminating all past restriction in return for an automation fund; eased early retirement privileges; and a guaranteed annual wage for the central core of his union membership.

The purpose of the automation fund was to provide a cushion for the containerization revolution in loading and unloading ships that would lead to substantial manpower displacement. The plan operated successfully; the port of San Francisco was revitalized, and the impact of technology on the central core of Bridge's membership much eased.

The Human Relations Committees of the Steel Employers and the United Steelworkers Union

The long steel strike of 1959 seemed to be proceeding endlessly, with no termination in sight, when Henry Kaiser and his son Edgar, under the leadership of three prominent labor relations experts, David L. Cole, John Dunlop, and George W. Taylor, agreed to abandon the employers' hard line and to enter into a novel kind of agreement with the United Steelworkers. The Kaiser Steel Sharing Plan, as it became known, provided the following for employees:

1. Greater protection than ever against loss of jobs or income because of technological change.
2. A share in any cost reduction brought about through increased efficiency, split 67.5% for the company and 32.5% for the employees.
3. Wage and benefit increases equal to or better than might be granted by the rest of the steel industry.
4. A lump sum payment to workers who would be ready to surrender their rights in an absolute unbalanced runaway incentive payment plan.

The plan provided the following for management:

1. A curtailment of, and an opportunity to eliminate, the incentive plan.
2. A chance to change work practice with less resistance from employees.
3. Four years of freedom from detailed bargaining.

This thinking spread to the rest of the steel industry. The majors, led by U.S. Steel, organized human relations committees with the union. These committees rationalized the bargaining over the technical details of pension plans, supplemental unemployment benefits, and so on, subsections that did not lend themselves to crisis bargaining.

This approach culminated in Bethlehem Steel's announcement in 1980 that it would engage in an extended program of participative, or more properly, consultative management in an effort to break out of the economic steel crisis that threatened both management and labor's future in the United States.

The Armour Corporation and Packinghouse Workers Union Experiment

A third example of this collaborative approach to collective bargaining is represented by the Packinghouse Workers, the Allied Butcher Workmen, and the Armour Company. The two unions have since combined, along with others, into the United Food Workers Union. The parties organized a tripartite committee with representatives from each organization and chaired by Clark Keer, president of the University of California, and Robbin Fleming, later president of the University of Michigan, who were later replaced by George Schultz, professor at the University of Chicago and the secretary of labor in the Nixon administration, and Arnold Webber, the assistant secretary of labor in the Nixon administration.

The purpose of this committee was to administer the installation of technological changes in such a way as to optimize the benefits to both parties and to cushion the economic impact on those who were displaced.

The organization of the committee was prompted by an announcement in the early summer of 1959 of Armour's closing down six plants, terminating 5000 employees. One year later Armour closed its Oklahoma City plant, displacing an additional 420 employees.

The committee contracted with three university research groups to develop training programs for these displaced persons. Unfortunately, the program was not too successful in placing any substantial number of them during a period of relative economic stagnation. However, the work of this committee did signal the need for a governmental manpower program—that the problems created by mass displacement were beyond the means of private parties.

This problem solving approach spared management the bitterness of irrational strikes and made workers aware that the problems that they faced could not be assuaged, but at least much bitter, useless strife was avoided.

5.6.12 THE "NEW" PARTICIPATIVE MANAGEMENT MOVEMENT

The participative management movement is a new name for the revival of new examples of some of the old labor management experiments. A more technically minded, but ahistorical, generation has invented a new term, "sociotechnical systems," to lend a spirit of eureka to an old idea under the illusion that what they propose is something new.

In the 1970s the Rushton Coal Mine experiment foundered upon the differential rewards that were available to different groups of workers. It ran into much the same kinds of problems as the Kaiser experiment, which was bedeviled by workers who were pressured to abandon the old incentive systems under which they earned much higher premiums than was possible under the new cooperative scheme.

Similarly, the Bolivar Company, confined to the manufacture of automobile mirrors, has been severely hit by the domestic automobile manufacturing depression of the 1980s. Will its participative management program be able to survive this contraction, or will it suffer the same fate as the Baltimore and Ohio and Naumkeag experiments, casualities of the Great Depression of the 1930s?

The GM Corporation has built upon a tradition pioneered by its president of the 1940s, Charles Wilson. Wilson developed the concepts built into the UAW-GM contracts of automatic cost-of-living and productivity share allowances, bought by Walter Reuther, then president of the UAW.

More recently, under the joint leadership of Stephen Fuller, vice president of GM, and Irving Bluestone, vice president of the UAW and director of the GM UAW department, the management of GM has set up a corporationwide joint union-management committee to encourage local experimentation. The Tarrytown plant, formerly a grievance-plagued facility, was completely turned around, and both sides boast proudly of the increase in productivity and improvement of quality that has flowed from this new approach. The Tarrytown facility assembles the Citation®, GM's revolutionary answer to the fuel-efficient, small, front-wheel drive cars with which the Europeans and Japanese have captured such a large share of the U.S. domestic market. Tarrytown is insulated from the unemployment plague that has assailed GM's other facilities because of the competitive demand for this car.

The Ford Motor Company has granted shop stewards the right to stop assembly lines when badly manufactured cars have been permitted to pass by supervisors who have been pressured by financial managers to meet quotas that reflect the slogan, "Get out the iron and all else will care for itself." Such a revolutionary surrender of a fundamental managerial prerogative is mind boggling.

The Chrysler Corporation, recently threatened with bankruptcy, solicited the influence of the UAW in floating a rescue loan from the U.S. Treasury and has elected its president, Douglas Fraser, to the corporate board of directors. Mr. Fraser, in justifying this abandonment of the last vestige of ideological class conflict, asked the rhetorical question, "Who suffers more than our workers when board decisions are made to order plant closures?"

5.6.13 FINAL CONCLUSION ON COLLECTIVE BARGAINING IN A TRANSITIONAL STATE

Collective bargaining is still in the course of experimental development. What started as a private procedure between management and labor has affected the public to such an extent that the government has begun to insist on a third chair at the bargaining table, presumably representing the public interest. President Nixon's invocation of a wage freeze for the first time in peacetime has left a precedent that will continue to reshape collective bargaining for years to come. President Carter attempted similar approaches without any final or conclusive results. Meanwhile, economic pressure is imposing a mutual problem solving approach to collective bargaining on the parties, attenuating its conflict component where the parties react responsibly and rationally.

REFERENCES

1. S. SLICHTER, *Union Policies and Industrial Management*, Brookings Institute, Washington, DC, 1941.
2. *J. I. Case Co.* v. *NLRB*, 253 F 2d 149 (7th Cir. 1958).
3. *Fafnir Bearing Co.*, 146 NLRB 179, 56 LRRM 1108 (1964).
4. "Is Your Incentive Plan Headed for Success or Failure?" *Factory Management and Maintenance*, May 1955, pp. 128–130.
5. B. PAYNE, "Incentives That Work," in *Proceedings of the Annual Fall Conference*, New York, Society for the Advancement of Management, 1951, pp. 23–38.
6. W. GOMBERG, "Wage Incentive Payment Plan," in *Proceedings, National Academy of Arbitrators*, Vol. 32, Bureau of National Affairs, Washington, DC, 1979, pp. 116–124.

SECTION 6

ERGONOMICS/HUMAN FACTORS

CHAPTER 6.1

Psychomotor Work Capabilities

GAVRIEL SALVENDY
Purdue University

JAMES L. KNIGHT
Bell Telephone Laboratories

6.6.1 INTRODUCTION

The purpose of this chapter is to acquaint practitioners with the nature and characteristics of psychomotor performance. A good understanding of psychomotor performance contributes to the effective establishment of job designs, work standards, financial incentive systems, methods improvements, and personnel selection and training.

Definition of Psychomotor Performance and Skill

Psychomotor tasks are all those that require the operator to use controlled movements of his or her body for their accomplishment. Two terms should be defined initially. "Psychomotor performance" refers to the level of achievement attained by the operator in completing a task. Thus one may refer to a high or low level of psychomotor performance. "Psychomotor skill" refers to the potential level of performance (i.e., achievement) of which an operator is capable. High levels of skill are apparent when the operator shows smoothly coordinated, fine, and rapid movements. As skill continues to increase, measurable changes in observable behavior become smaller, but performance becomes increasingly easy, requiring less and less attention of the operator.

Examples of Tasks Involving Psychomotor Performance

Psychomotor performance is widespread, but does not occur in all work activities. Many activities predominantly or exclusively require mental and decision making skills. Most managerial and many inspection tasks are of this variety.

However, psychomotor performance is a key ingredient in the following broad classes of tasks[1]:

Handwork (e.g., wrapping).

Handwork with tools (e.g., using a screwdriver, handwriting).

Single-purpose machine work (e.g., operating a coil winding machine; computer data entry, or keypunching; typing; driving).

Multipurpose machine work (e.g., industrial sewing machine operation).

Group machine work (e.g., controlling weaving and spinning machine systems).

Nonrepetitive work (e.g., equipment repair).

6.1.2 RANGE OF INDIVIDUAL PERFORMANCE DIFFERENCES

Variation in psychomotor performance levels occurs both among different operators and within a single operator over a period of time. This variation arises from three general classes of operator characteristics:

Experience and training.

This chapter was prepared while the second author was employed at Purdue University and does not necessarily reflect the opinions of Bell Telephone Laboratories.

Enduring mental and physical characteristics.
"Transitory" mental and physical characteristics prevailing at the time of task performance.

Transitory characteristics are influenced by many specific factors, including the following:

Motivation.
Temporary illness.
Fatigue.
Stress.
Alcohol and other drugs.
Hours of work (e.g., overtime, shift work).
Physical, social, and psychological work environments.
Food intake.

Psychomotor performance also is influenced by task characteristics such as equipment variability, defects, and malfunctions, and, especially among different operators, the methods employed by operators to perform their tasks.

The combined impact of these various factors on the performance variability of an individual operator (i.e., within-operator variability) has been documented among blue collar workers in manufacturing industries. These studies[2] indicate that reliability* of production output varies from .7 to .9, with a mean of .8. This implies that about 64% (i.e., $.8^2 \times 100$) of an operator's performance in 1 week can be predicted by his or her performance observed during a prior week. Conversely, 36% of the operator's performance cannot be explained in this manner, but is apparently explained by such factors as those previously listed.

It should be noted that individual variability within a working day is markedly smaller than between working days. Furthermore, performance variability within a workday is smallest from mid-morning to early afternoon (Exhibit 6.1.1). During this period, performance fluctuation around a mean level is only about 5% (of the mean), but this variability increases markedly on either side of the mid-morning to early afternoon period. These patterns of within-operator variability, as well as warm-up and slowdown at the beginning and end of the workday, must be accounted for in the establishment of sound time standards (see Chapter 4.1).

Based on many studies, it is well known that psychomotor performance variability *among* operators is much larger than that observed *within* the same operator over successive observations. Generally a performance range of 2 to 1 encompasses 95% of the working population.[3] However, in practical work situations the range encountered is likely to be much smaller than this because of preemployment selection, attrition of some low-performance operators, and peer pressures that may limit the output of high-ability operators.

Thus, when these limiting factors are not operating, in a group of 200 workers, if the highest-performing 5 and the lowest-performing 5 are not considered, then in the remaining 190 operators, the highest-performing will not perform more than twice as well as the lowest-performing (and, conversely, the lowest-performing operator will do at least half as well as the highest-performing operator). The recognition of this range of performance levels is critical to the maintenance of effective incentive systems (Chapter 2.3) and to the development of effective production planning and control techniques (Section 11). The impact of this range of performance levels on financial incentive systems in the absence of effective personnel selection (Chapter 5.2) is illustrated in Exhibit 6.1.2. This exhibit shows, for example, that, in a company with mean incentive earnings of 25%, 10% of the labor force would not earn any incentive pay, whereas 2% would earn more than 63% incentive pay.

6.1.3 CHARACTERISTICS OF PSYCHOMOTOR PERFORMANCE

Components of Psychomotor Performance

Based on the statistical tool of factor analysis,[4] 11 distinct elemental psychomotor abilities have been identified,[5] as portrayed in Exhibit 6.1.3.

These 11 factors are independent of each other: having high or low performance in any specific factor has no bearing on the expected performance on any other of the 10 remaining factors. The existence of this factor structure underlying complex psychomotor performance has broad implications for the selection, placement, job rotation, and training of personnel.

*The reliability coefficient is a measure of consistency determined by the extent to which two successive samples of same-task performance provide similar results. Thus, for example, reliability of performance may be obtained by correlating one week's performance with another's.

Exhibit 6.1.1 Comparison of Output Curve and Ratings Made on One Operator During a Repetitive Manual Operation (Thread-Roll Bulb Holder) During a Working Day, Utilizing Continuous Time Studies. Similar Results Were Obtained for Other Manual Repetitive Tasks and for Other Operators.

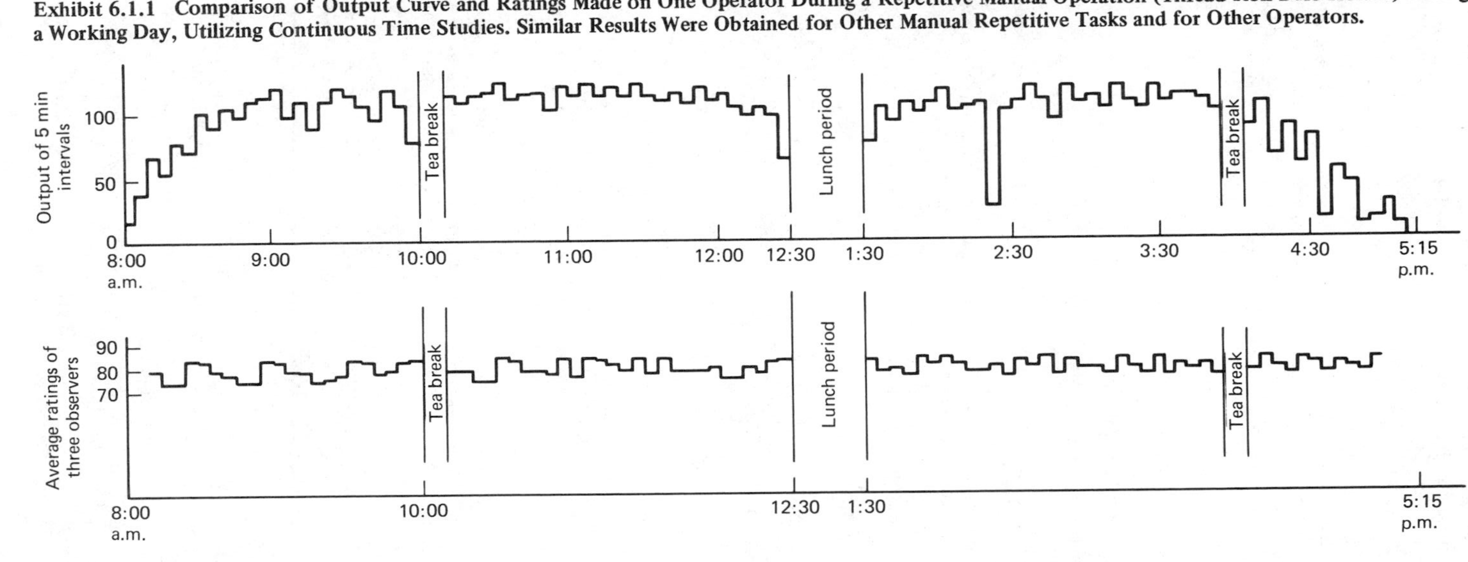

Source. Reference 6.

Exhibit 6.1.2 Psychomotor Performance Ability Distribution and Its Effects on Financial Incentive Earnings in the Absence of Preemployment Selection, Attrition of Some Low-Performance Operators, and Peer Pressures That May Limit the Output of High-Ability Operators.

Company's Percentage Mean Financial Incentive Level Above the 100 Time Standard Rate	Percentage of Workers Who Cannot Meet Time Standard	Two Percent of the Labor Force Would Earn Above the Percentage Incentive Level Indicated Below
15	20	50
20	14	57
25	10	63
30	7	70
35	5	77

Specifically, this factor structure implies that:

Different personnel selection tests and procedures are required for tasks requiring proficiency in different basic psychomotor ability factors.

In rotating personnel from job to job, greatest transfer of training, and hence greatest productivity, will be achieved when operators are transferred to new tasks requiring many of the same basic ability factors as their old tasks.

Training procedures must be tailored to the specific ability factors required by particular jobs.

Shift in Factor Structure With Practice

Initially, performance depends greatly on mental factors, including the ability to understand the task directions, remember instructions, concentrate one's attention on task demands, and perceive important task details. As training progresses, the elemental abilities listed in Exhibit 6.1.3 increas-

Exhibit 6.1.3 Factor Structure of Psychomotor Abilities

Factor	Description
Control precision	Common to tasks that require highly controlled and precise muscular adjustments of controls where larger muscle groups are involved, extending to arm-hand as well as to leg movements.
Multilimb coordination	Ability to coordinate the movements of a number of limbs simultaneously in operating controls. It is general to tasks that require coordination of the two feet, two hands, or hands and feet.
Rate control	Involves the precise timing of continuous responses relative to changes in speed and direction of a continuously moving target or object.
Arm-hand steadiness	Ability to make precise arm-hand positioning movements where strength and speed are minimized. It is general to tasks requiring steady limb position or moving the limb steadily in a lateral or to-and-from plane.
Finger dexterity	Ability to make skillful, controlled manipulations of tiny objects involving primarily the fingers.
Manual dexterity	Skillful, well-directed arm-hand movements are involved in manipulating fairly large objects under speed conditions.
Reaction time	Speed with which the individual is able to respond to a stimulus when it appears.
Response orientation	General to tasks requiring rapid selection of controls and of the direction in which they are to be moved.
Speed of arm movement	Represents simply the speed with which an individual can make a gross, discrete arm movement; accuracy is not required.
Wrist-finger speed	Requires rapid tapping of the pencil in relatively large areas.
Aiming	Best measured by highly speeded printed tests requiring dotting in series of small circles.

ingly account for performance. Furthermore, as the operator becomes more skillful, the specific elemental abilities most important for task performance will change.[7] This is accounted for partly by changes in the methods used by inexperienced and experienced operators to perform a given task.

This shift in the elemental abilities or factor structure has important implications for personnel selection and training. An operator who scores high early in training may not be a high performer later in training: The abilities that produced high initial test scores may not be the abilities needed to perform well after extensive task experience. For example, initial performance of a power press operator depends heavily on eye-hand coordination, but higher performance levels depend upon development of kinesthetic sensitivities. It is therefore crucial that selection tests and training procedures be derived from the study of expert task performers. Selection tests should focus on abilities that will be needed to achieve high-level performance, rather than on those needed at early stages of training. An example that illustrates the benefits derived from adopting such a procedure for the selection of personnel is discussed elsewhere.[8]

Elemental Time Analysis

Psychomotor performance often is analyzed in terms of the elemental motions (e.g., reach, grasp, move, and position), or elements, of which it is, at least conceptually, constituted. This commonly is done in the establishment of time standards using predetermined time systems or MTM (Chapter 4.5). The following information related to elemental psychomotor motions is critical to the industrial engineer's approach to elemental time analysis:

1. All elemental motions should be clearly defined, with specific beginning and end points. No overlaps or omissions between elements should occur.

2. Using elemental motions in the analysis of psychomotor performance is advantageous in that the motions are operationally definable and visually detectable. However, they have a major disadvantage in that the beginning and end points of an element do not necessarily coincide with the beginning and end points of the physiological and mental work associated with the performance of an element.

3. As the operator's psychomotor skill increases, not all elemental motions in a task improve equally. The highest potential for improvement occurs in those task elements requiring the greatest cognitive activity (i.e., imposing the greatest mental load) and therefore showing the greatest performance variability and in those least influenced by chance effects and equipment characteristics.

Improvement in elemental times shows a characteristic pattern: It does not come about because of a general increase in elemental speed, but rather because excessively slow instances of a particular element are progressively eliminated. This effect can be seen in the shifting of histogram patterns presented in Exhibit 6.1.4. These histograms depict elemental times accumulated through the performance of 50,000 cycles of the One-Hole Test,[8] which requires grasping a small, cylindrical object, moving it, and positioning it into a hole with close tolerance.

4. Studies of the additivity of elemental times[9] indicate that errors in work standards due to additivity are measurable, but quite small. These errors decrease significantly as the number of elemental times constituting a task cycle increases.[9]

5. Even though complex psychomotor performance is composed of many relatively minute elemental motions, cumulative changes in the way motions are performed can have very large effects on overall performance. The detailed analysis of the high-speed filming of the performance of a task (the One-Hole Test[9]) shows how changes in elemental motions resulted in a 33% performance improvement–from 100 to 133.[10]

a. Decrease in the knowledge of results sought by the operator. Operators required much less feedback on their performance than they had required initially. For example, after achieving the standard performance rating of 100, the operator no longer pushed the pin to its very end to check that the element "position" was completed, but decreased the required feedback to merely releasing the pin as a sign that the element "position" actually was completed.

b. Increase in the level of smoothness of performance, mainly between transport and stationary elements, resulting in rhythmical overall performance. Presumably this is because more than the first three quarters of the transport motion is performed in a smooth and automatic way, whereas less than the last quarter is completely controlled by the human. But this last quarter of the distance in transport elements (mainly "move") takes as long to perform as the first three quarters of the distance. Presumably this is because of the accuracy required at the end movement. The smoothness effect could occur because of partial takeover of the human manual operation by human automatic performance.

c. More effective utilization of fingers and thumb, generally resulting in fewer and simpler motion patterns. The operator grasps the object in the manner in which it will be moved and positioned, taking into consideration the least and most "efficient" muscular contraction of the

Exhibit 6.1.4 Changes in Histogram Performance Times of Elements and Cycle During the Performance of More Than 50,000 Cycles (855 Work Periods of 10 Min Each) on the One-Hole Test,[a] Showing Work Periods (*a*) No. 2, (*b*) No. 64, and (*c*) No. 8.

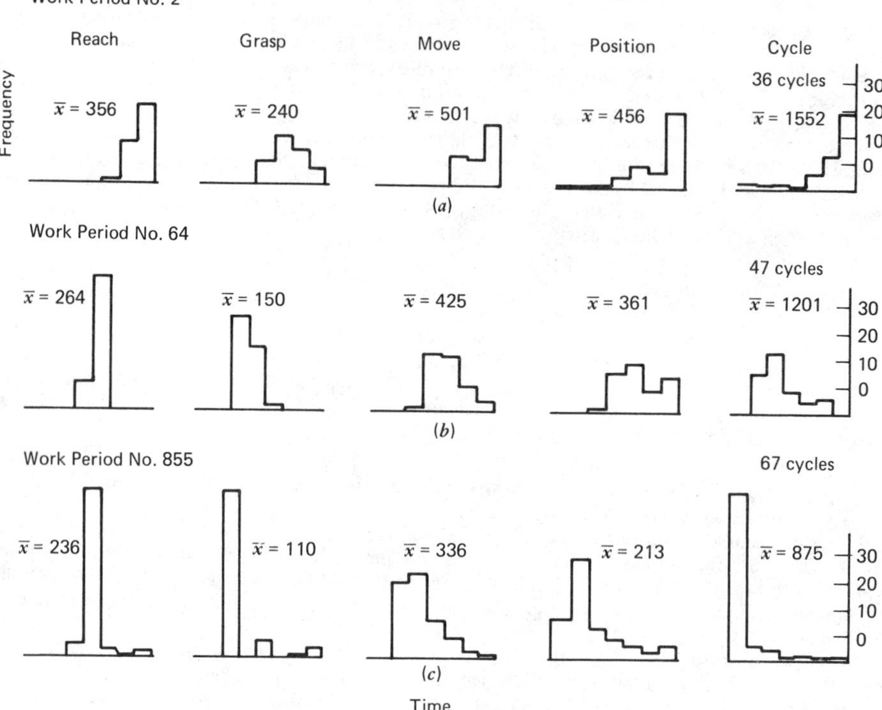

[a]Mean ($\bar{x}$) elemental and cycle times are presented in thousandths of seconds.

fingers and thumb. This is derived (as has been indicated by the operators) from unconscious changes in the work methods adopted by the operator.

d. Increase in the length of time between the termination of eye fixation on an operation and the termination of that operation. It is achieved presumably because of a kinesthetic sense that takes over from much of the vision.

e. Increased ability to cope with the "unexpected" because of previous trial-and-error learning. For example, at the 100 performance rating, fumbles lasted nearly twice as long as at the 133 performance rating. Furthermore, when a fumble occurred at the 100 performance rating, the operator's pattern of performance was markedly disturbed, but this effect had nearly disappeared at the 133 performance level.

f. Decrease in the frequency and length of mental blocking occurring in a certain element or elements of the work cycle. Initially, when mental blocking occurred, the element lasted 3.5 times as long as the average. This decreased to 2.5 times the average element duration at the 133 performance rating.

6.1.4 MENTAL AND INFORMATION PROCESSING ASPECTS OF PSYCHOMOTOR PERFORMANCE

Mental Activities

Even apparently simple movements carried out in psychomotor performance depend upon a usually complex sequence of mental activity. Crossman[11] has pointed out five predominantly mental activities or steps present in most psychomotor tasks: plan, initiate, control, end, and check.

In planning, the operator decides what is to be accomplished and selects an appropriate sequence of activity to achieve the goals. At the proper moment, neural signals to activate muscle groups are issued, and the activity sequence is initiated. Movements and other activity then must be controlled to make sure they are proceeding in agreement with plan, that they do not overshoot or undershoot the final objective, and that they are corrected when unexpected events occur. Finally, the operator must sense when movements should be ended. The overall effect of the activity sequence then is checked against the goals of the initial plan to determine if further action is required.

These five mental activities can be used to describe the operator's global task (e.g., meeting the daily production quota) as well as the elemental motions (e.g., reaching for a wire or other small part) that compose the overall task.

Information Processing Model of Psychomotor Performance

The operator's ability to perform these crucial mental activities, and therefore the ability to perform psychomotor tasks effectively, rests upon fundamental cognitive processes and functions. These basic mental functions and processes (or stages) appear in Exhibit 6.1.5, which represents an information processing model of the human operator.

In this model the operator is continuously presented with information about the task being performed. The operator uses (i.e., processes) this information to accomplish his or her work objectives. The operator is viewed as a channel through which information flows. In the model of Exhibit 6.1.5, three major information processing stages are shown: perception, decision making, and response control. Also shown are three memory systems (sensory, short-term, and long-term memory) for storing information needed for task performance. Overall psychomotor performance depends upon, and is limited by, the information processing capacities of these three major stages and by the storage characteristics of the three memory systems.

Information Overload and the Need for Selectivity

Limits of psychomotor performance arise from two characteristics of the major information processing stages: (1) they require a minimum time in which to perform their functions, and (2) they have limits as to the amount of information they can process per unit time. If information arrives too rapidly, a stage may become overloaded and unable to operate effectively. This limit to the rate at which a stage can handle (i.e., transmit) information is its channel capacity.

This limit can be reached in three ways. First, a task may be inherently difficult and present information to a particular stage at an excessive rate. Psychomotor performance can improve if the processing capacity of the affected stage(s) increases. Kalsbeck and Sykes[12] studied the task of handwriting and found evidence for increasing capacity limits of the response control stage.

Inexperienced operators are prone to a second source of stage overload. Typically, much of the information available to an operator is either irrelevant or redundant. A novice operator will fail

Exhibit 6.1.5 Information Processing Model of the Human Operator.

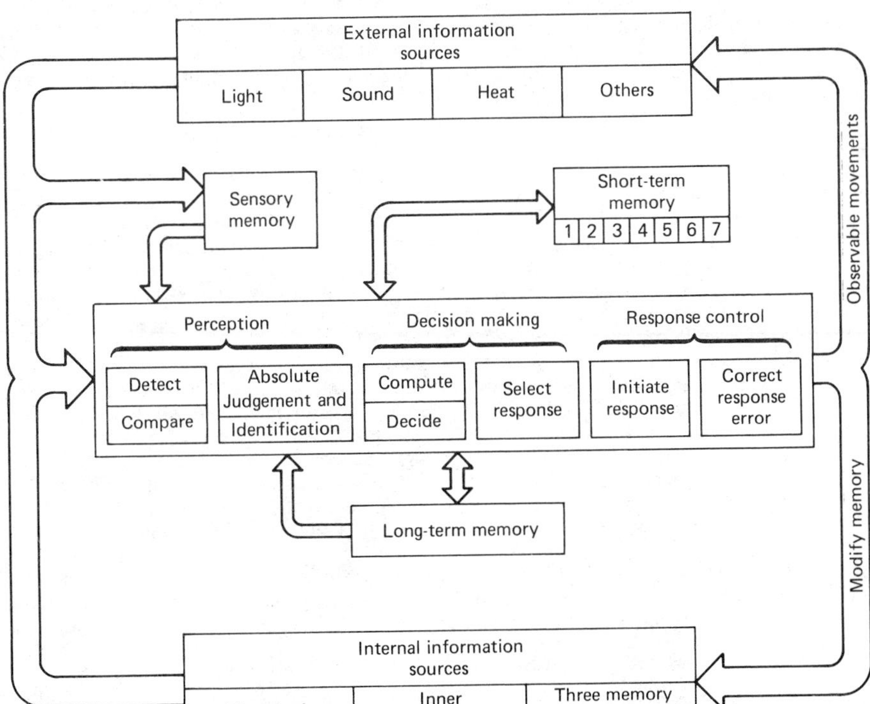

to recognize this and attempt to process more information than necessary. This results in overload and consequently low performance. For example, an operator may attend to (i.e., process) many small and irrelevant details in the appearance of a workpiece and thus fail to detect a critical flaw.

Similarly, an operator working with moving objects–for example, workpieces on a continuously moving conveyor–must keep track of constantly changing positions. This becomes more difficult if the moving objects follow a rapidly changing path. Even in this case, however, an experienced operator can take advantage of the redundancy in the object's motion; for some value, x, the position of the object at time t is perfectly predictable by its position at time $t - x$. The experienced operator need only observe the object's location every x time units. Actually, even experienced operators observe the location of moving objects about twice as often as necessary.[13] However, novice operators sample this information far more frequently, and this imposes an unnecessary load in various processing stages.

The process whereby an operator comes to attend to only essential information in a task is a critical mechanism underlying the development of psychomotor skill. To take advantage of redundancy, an internal model of the task being worked upon must be developed by the operator. This internal cognitive model uses available information to make predictions about future task requirements. Thus, in the preceding example, the operator, having acquired knowledge of path regularities in the movement of the workpieces, was able to predict future workpiece locations, thus avoiding processing of unnecessary information. The availability of an accurate internal model of the task is the most significant advantage enjoyed by a skilled operator over a novice counterpart.

The third way in which overload may occur is when two tasks compete for an operator's attention and simultaneously present information to the same limited-capacity stage. In this case the operator may choose to process the information from only one task, thus drastically degrading performance in the other task. Or, the operator may choose to process some information from each task, thus producing milder degradation in both cases.

The foregoing model of the industrial operator as a series of stages sensitive to information flow rate (i.e., transmission rate) emphasizes the need for information selection mechanisms to protect the operator from overload. An internal model enhances the ability to select properly only the essential information. The skilled operator is one who efficiently selects only needed information for processing during psychomotor activity.

Memory Systems

The information processing model contains three memory systems. These systems contribute several essential functions in psychomotor performance. They act as buffers to store temporarily (from 1 to 2 sec) rapidly arriving sensory information (sensory memory). They temporarily store up to seven "chunks" (words, names, digits, etc.) of information (short-term memory). Finally, they provide long-term storage that underlies learning and improvement in psychomotor performance (long-term memory).

6.1.5 TIME-SHARING

Industrial work often requires the operator to time-share, or simultaneously perform, several separate subtasks. This time-sharing demand occurs on three levels. First, in even simple tasks, the operator must receive information, make decisions, and control response movements. Efficient performance may require that these activities occur parallel. Second, more complex tasks often require the operator to make several separate responses simultaneously; for example, concurrent, but separate, hand motions. Third, the operator may be asked to perform two quite separate tasks at once. How efficiently can activities at each of these levels overlap?

At the first level, evidence suggests that information reception can efficiently overlap both decision making and response control. However, these latter two functions interfere with each other. More specifically, the initiation and correction of movements interferes with decision making. These response functions occur primarily in the second phase of movement control. Hence performance can be enhanced by eliminating or minimizing second-phase control. This can be done by terminating movements with mechanical stops rather than with closed-loop, operator guidance.

Time-sharing at the second level can be enhanced if the same mental function, information reception, decision making, or response control is not needed simultaneously by both activities. The refractory period* of the central decision making stage requires that successive inputs to this process be separated by at least 300 msec. For example, if the operator is required to identify and respond to two successive signals, those signals should not occur within 300 msec of each other.

The processes involved in closed-loop movement control, including monitoring, selecting an ap-

*The period during which the operator is unable to process any new information.

propriate corrective response, and initiating the correction, impose particularly high information processing demands. Hence, when these processes are required by two simultaneous subtasks (e.g., independently moving each hand), information overload and consequent interference between the subtasks can be expected. For example, elements "position" and "grasp" both impose high information processing loads because they require significant second-phase, closed-loop control. Therefore, they cannot be effectively time-shared. At the other extreme, elements "reach" and "move" (which do not generally involve precise, closed-loop movement control) generally can be carried out parallel with other elements.

Another critical factor in time-sharing efficiency is respone-response compatibility. Some combinations of responses can be performed more easily than others. In executing simultaneous movements, performance is best when the hands (or feet) move in the same direction (e.g., both forward). Next best is complementary movements (e.g., one forward, one backward). Performance is worst for perpendicular arrangements (one forward, one sideways). Similarly, responses that start at the same time are easier to time-share than those that do not. Selecting, initiating, or monitoring parallel (or successive) responses that have similar characteristics apparently requires less information processing than occurs in the case of unrelated movements. Symmetrical relationships between movements enhance this similarity effect even when the movements are made in opposite directions.

Time-sharing efficiency will be greatly enhanced when highly compatible stimulus-response (S-R) relationships are used. High S-R compatibility reduces the load on the decision making stage responsible for selecting responses. Responses may almost become "self-selecting" with the most compatible mappings. This most readily occurs with tactile signals. For example, a vibrating machine control provides a highly compatible signal for the response of grasping the control more firmly. The operator may do this almost immediately, with no disruption of other movement activities.

Finally, at the most complicated level, performing two separate tasks at once, performance depends on a wide variety of factors, including the priorities that the operator attaches to the competing tasks. Typically, when an easy task was combined with a more difficult one, a greater percentage decline in performance was found for the easier task.[14]

Time-sharing efficiency improves with task experience for a variety of reasons. First, there is evidence that time-sharing is a general ability that can be enhanced by training. Operators who effectively time-share one pair of tasks often are superior at time-sharing other task pairs. Second, as operators become well trained, tasks impose lower information processing loads and even appear to become "automatic." Several reasons for this have been considered, including the following:

An internal task model frees the operator from processing redundant information.

Kinesthetic information, which may be processed faster than visual information (Exhibit 6.1.6) and which often is highly S-R compatible, is gradually substituted for visual information.

Certain information processing steps (e.g., "check" operation) may be minimized or deleted entirely.

More efficient movement sequences involving less second-phase, closed-loop control are developed.

Because the "automatic" time-shared tasks each impose lower information processing demands upon the operator, there is less likelihood of overload, and efficient time-sharing is possible.

The preceding discussion has focused on time-sharing difficulties resulting from "central" (i.e., cognitive) interference between two activities. Obviously tasks also may interfere with each other because of "structural" interaction: If one task requires the operator to look to the right, while the other requires the operator to look left, the tasks will be mutually interfering. Such structural interference often is quite difficult to distinguish from central interference. This represents a primary difficulty in attempting to use "secondary-task" methods to assess mental work load.

Exhibit 6.1.6 Minimum Reaction Times (*Kp*) for Various Stimulation Modalities

Stimulation Modality	Reaction Time (msec)
Visual	150–225
Auditory	120–185
Tactual	115–190

6.1.6 CHARACTERISTICS OF INFORMATION PROCESSING CAPABILITIES

Perception

Detection of information is the first important cognitive function. Two systems most important to psychomotor performance are vision and kinesthesia. The latter system provides information on the position and motion of the operator's limbs.

Typically, perceptual information is conveyed by stimulus changes, for example, in luminous intensity. Sensory systems must be sensitive to such variation. Thus it is appropriate to consider the threshold of just noticeable difference (JND), the minimum detectable stimulus change. The visual system is typical of most other senses in that the JND increases in constant proportion to the background reference against which the change occurs. This proportionality yields the Weber fraction (see Exhibit 6.1.7).

The ability of the kinesthetic sense system to supply information appears comparable to that of other modalities. Marteniuk et al.[16] asked operators to reproduce level movements of 45, 90, and 125°. Their study yielded JND sensitivities of 1.95, 2.20, and 2.13°. While these values are impressively low, the important thing to note is that they yield Weber fractions of .043 (1.95/45), .024, and .017. In other words, the kinesthetic system seems to differ from most other sense modalities in that it exhibits constant absolute sensitivity rather than proportionally constant sensitivity.

This JND sensitivity involves the basic function of comparison. In such "side-by-side" comparisons, operators usually can distinguish hundreds of different objects. A much more difficult function is absolute judgment. This requires the operator to recognize or identify an object presented in isolation. Humans are surprisingly limited in this ability. There appears to be a limit of approximately 3 bits* (not bits/sec) on the amount of information that can be effectively transmitted in absolute judgment. This means that an operator can reliably recognize only about eight levels of a single sensory dimensions. For example, Pollack[17] trained operators to identify auditory stimuli differing only in loudness or only in pitch. His operators could distinguish only about 5 stimuli of different loudness and about 5.5 stimuli of different pitch.

This seems to contradict everyday experience. We can identify readily many different sounds. However, these usually differ simultaneously on many sensory dimensions. The human can utilize these simultaneous differences to improve his or her absolute judgment capabilities. However, as different dimensions are combined, their contributions to absolute judgment performances are not additive. For example, Pollack[17] found that his operators were able to identify only about 8 stimuli rather than 10.5 (5 based on loudness discrimination and 5.5 based on pitch discrimination) when stimuli differed in both pitch and loudness.

Absolute judgment ability in the kinesthetic sense modalities also is quite low. For example, Marteniuk[18] asked operators to identify movements of different length and found that only about six movements could be accurately identified.

The implication of these findings is that task designs requiring absolute judgment should be avoided whenever possible. Very skilled workers may be able to overcome this limitation somewhat by developing very good internal references or models that can be used as standards for comparison. An exerienced worker may have a very accurate idea of what an acceptable product should look like. Thus, his or her task, when presented with a test product for inspection, approximates

Exhibit 6.1.7 Weber Fractions for Various Sense Dimensions

Sense Dimension	Weber Fraction, W^a
Brightness	.016
Pitch	.003
Loudness	.088
Cutaneous Pressure	.136
Lifted Weights	.088

Source. Reference 15.

$^a W$ = JND/background reference level.

*A "bit" is a unit of information. One bit is enough information to allow an accurate decision between two equally likely alternatives.

a side-by-side comparison–a skill in which people excel. To the novice operator, however, the task requires absolute judgment and consequently is far more difficult.

Decision Making

Decision making refers to the processes whereby operators evaluate information made available by the initial perceptual processing. Decision making results in the selection of an intended course of action. Two decision making characteristics are especially important: how much time decision making requires and how accurate decisions are.

Decision delays stem from two sources, capacity limitations and refractory limitations. Capacity limitations arise because decision making stages can process information at only a limited rate. The amount of information transmission involved in a decision increases logarithmically with the number of possible stimuli that might be presented and the number of alternative responses from which the operator might select. In general, doubling the number of possible stimuli and responses increases the information transmitted in the decision by one bit. For example, if an operator must sort two different types of parts into two bins, each decision concerning a part involves one bit of information. However, when sorting among four parts, each decision involves two bits. More specifically,

$$Ht \text{ (information transmitted, bits)} = Hr + Hs - Hsr$$

where

Hr (response information, bits) = $-Pi \log Pi$, and Pi = probability of the ith response possibility

Hs (stimulus information, bits) = $-Pj \log Pj$, and Pj = probability of the ith possible stimulus

Hsr (joint S-R information, bits) = $-Pij \log Pij$, and Pij = probability of the joint occurrence of stimulus j and response i (This is a measure of decision consistency.)

For perfectly accurate decisions with N equally likely stimuli and responses, the formula for Ht simplifies to

$$Ht = \log N$$

Hick[18] determined that reaction time for simple decisions obeyed the function now called Hick's law:

$$RT = Kp + Cd \times Ht$$

where Kp = the sum of all delays not associated with decision making, and Cd = the time needed to process one bit of information. Therefore $1/Cd$ is a measure of the information handling capacity of the decision making stages. With visually presented information, Hick found values for Kp and Cd of 150 msec and 220 msec/bit, respectively.

Responses are fastest when no decision making is involved (i.e., when the operator knows in advance what and when information will be presented and what response should be made). Under these conditions, Exhibit 6.1.6 shows the minimum reaction times (values of Kp) that can be expected (the range of values shown accounts for effects of stimulus intensity) for information presented in various modalities.

The information transmitted in the decision making process depends on the average accuracy of the outcome. If average decision accuracy is reduced, each decision will involve less information processing and thus can occur faster. This is the speed-accuracy trade-off effect. An operator can speed the decision making if he or she is willing to tolerate more errors. Hick[19] showed that within wide margins increases in speed are compensated by losses in accuracy, so the *rate* of information flow per unit time remains constant at about 1 bit/220 msec. If the operator exceeds these margins by trying to go too fast, however, accuracy drops very rapidly, and the rate of information transmission will fall. This occurs when the operator tries to increase the speed more than about 20%.

The other source of decision making delay is a fixed delay of about 300 msec that must separate successive decisions. This is the so-called psychological refractory period. If information is presented to the decision making stage within 300 msec of a previous decision, decision making will be delayed until the psychological refractory period has elapsed. This refractory delay does not decline with practice.

Both decision delay and the slope of the Hick's law function sharply decrease as S-R compatibility becomes greater. S-R compatibility refers to the "naturalness" of the relationship between a stimulus and the response with which it is associated. Some examples of natural relationships

Exhibit 6.1.8 Some Human Biases That Affect Psychomotor Performance

Quantity Estimation	Bias
Horizontal distance	Underestimate
Height	Overestimate when looking down
	Underestimate when looking up
Speed	Overestimate if object accelerating
	Underestimate if object decelerating
Angle	Underestimate acute angles
	Overestimate obtuse angles
Temperature	Overestimate heat
	Underestimate cold
Weight	Overestimate if bulky
	Underestimate if compact
Numerosity	Consistently underestimate
Probability	Overestimate pleasant event likelihood
	Underestimate unpleasant event likelihood

reflect innate human characteristics. Thus illuminating a push button very effectively elicits the response of pushing it. Other relationships must be learned—for example, in the United States, light switches are lifted up to turn on room lights.

Decision accuracy depends not only on the operator's speed and accuracy strategy, but also on built-in biases. Some of these biases are listed in Exhibit 6.1.8 (see also Chapter 4.2).

Response Control

The precision of movements is limited largely by the response control stage. Fitts[20] defined an index of movement difficulty, ID, by analogy to information theory:

$$\text{ID (bits)} = \log\,(2A/W)$$

where A = the amplitude or distance an operator must move to complete a movement, and W = the width of the target area at which the operator is aiming.

Fitts found that the time needed to complete a movement once started could be predicted from the movement ID:

$$MT = Km + Cm \times \text{ID}$$

where

Km = a delay constant that depends on the body member being used for responding (i.e., the foot and hand yield different values for Km) (For hand movements, a typical value for Km is 0.177 sec.)

Cm (sec/bit) = a measure of the information handling ability of the response control stage ($1/Cm$ is its channel capacity. Typically, Cm is approximately 0.1 sec/bit or greater. For repetitive responses—such as moving back and forth between two target locations—Cm is slightly lower than for single, isolated movements.)

A more detailed study of movement reveals two phases: first, an initial, gross ballistic phase by which the operator moves to the general vicinity of the target (with an error of about 7% of the movement amplitude), followed by a second, closed-loop phase in which the operator makes a series of fine corrective control movements. Each movement correction requires about 300 msec and reduces error by 93%. Visual information is used to control second-phase movements. The second phase may be absent for movements made in less than 260 msec. At least visual guidance is not used for movements shorter than this duration. Movements involving rapid changes (i.e., corrections) in direction or speed impose high mental loads and therefore should be avoided. Smooth, continuous movements are more efficient.

Overall Performance Characteristics

Perceptual, decision making, and response control characteristics may be combined. For signals well above threshold, the minimum reaction times listed in Exhibit 6.1.7 may be used as an estimate of perceptual processing and other unavoidable delays. These factors are combined to yield Kp. Then, the time needed to receive information, select an appropriate response, and carry out the corresponding movement will be

$$\text{total time} = [\text{perceptual delays}] + [\text{decision making delay}] + [\text{movement time}]$$

$$\text{total time} = [Kp] + [Cd \times Ht] + [Km + Cm \times \log (2A/W)]$$

This formula predicts changes in the time needed for movements as a function of movement, decision, and perceptual variables.

Application Example

To illustrate application of the foregoing formula, consider a sorting task in which the operator reaches into a general parts bin containing color-coded parts, withdraws one, and then moves it to an appropriate specific bin and drops it in. Suppose there are N different types of parts (and therefore N specific parts bins). Also suppose the specific parts bins are a distance A from the general bin and that each specific bin is of width W. Assume that the time, Kw, needed to return from the specific bin and withdraw a part from the general bin is 1.0 sec.

Exhibit 6.1.9 shows, in idealized form, how cycle time would be expected to vary as a function of several important parameters of this task, N, W, and A. These calculations assume a perceptual delay of 150 msec for discriminating the part's color code.* The operator is assumed able to process 1 bit/220 msec when deciding into which bin the withdrawn part should be placed. Accuracy is stressed, so errors are assumed to be rare. The Km is taken as 0.177 sec, and the operator's movement control channel capacity is taken as 0.2 sec/bit. Notice from Exhibit 6.1.11 that changing the "target width" (i.e., the width, W, of the specific parts bins into which the operator drops each sorted part) can exactly compensate for increases in the extent of required transport distance, A.

Although the preceding discussion has focused on discrete movements, the concepts apply to

Exhibit 6.1.9 Idealized Cycle Time as a Function of Task Parameters N, A, and W

N^a	Decision Ht (bits)	A^b (cm)	W^c (cm) (Width of Bin)	Movement ID (bits)	Total Time (sec)
4	2	12	3	3	2.01
4	2	12	6	2	1.81
4	2	12	12	1	1.61
4	2	24	3	4	2.21
4	2	24	6	3	2.01
4	2	24	12	2	1.81
4	2	48	3	5	2.41
4	2	48	6	4	2.21
4	2	48	3	3	2.01
8	3	12	3	3	2.23
8	3	12	6	2	2.03
8	3	12	12	1	1.83
8	3	24	3	4	2.43
8	3	24	6	3	2.23
8	3	24	12	2	2.03
8	3	48	3	5	2.63
8	3	48	6	4	2.43
8	3	48	12	3	2.23

[a] N = number of different bins/parts.
[b] A = distance to bin.
[c] W = width of bin.

*In practical situations the operator may have to move his or her eyes and head, as well as refocus his or her vision, during the performance of a task. Although these times, which may total up to 0.75 sec, can vary with task parameters such as movement amplitude, they are not explicitly considered in this idealized example, which is restricted to delays associated with primarily mental operations.

continuous tasks (such as tracking) as well. Delays inherent in the human information processing system impose an upper limit on the frequency of continuous information to which the operator can respond. The operator can deal effectively only with continuous signal components below 1 Hz. In responding to continuous, time varying input signals, the operator can be modeled as an intermittent servomechanism making a series of discrete corrective movements about twice per second.

6.1.7. SUMMARY

In this chapter some of the basic capabilities and limitations of the industrial worker are discussed. In addition, an information processing model of the human has been presented. By utilizing the perspective provided by this model, the demands placed upon operators by their work tasks can be better understood. Consideration of these demands and the inherent psychomotor capabilities of the human to meet them can result in better prediction of job performance, more effective job design, and enhanced quality of operators' work lives.

REFERENCES

1. G. SALVENDY and W. D. SEYMOUR, *Prediction and Development of Industrial Work Performance*, Wiley, New York, 1973, pp. 105–125.
2. SALVENDY and SEYMOUR, *Prediction and Development of Industrial Work Performance*, pp. 195–197.
3. D. WECHSLER, *The Range of Human Capabilities*, 2nd ed., Williams and Wilkins, Baltimore, 1952.
4. S. A. MULAIK, *The Foundations of Factor Analysis*, McGraw-Hill, New York, 1972.
5. E. A. FLEISHMAN, "Toward a Taxonomy of Human Performance," *American Psychologist*, Vol. 30, 1975, pp. 1127–1149.
6. N. A. DUDLEY, *Work Measurement: Some Research Studies*, London, Macmillan, 1968.
7. E. A. FLEISHMAN and W. E. HEMPLE, JR., "Changes in Factor Structure of a Complex Psychomotor Test as a Function of Practice," *Psychometrika*, Vol. 19, 1954, pp. 239–252.
8. G. SALVENDY, "Selection of Industrial Operators: The One-Hole Test," *International Journal of Production Research*, Vol. 13, 1975, pp. 303–321.
9. H. SANFLEBER, "An Investigation Into Some Aspects of Predetermined Motion Time Systems," *International Journal of Production Research*, Vol. 6, 1967, pp. 25–45.
10. G. SALVENDY, "Learning Fundamental Skills–A Promise for the Future," *AIIE Transactions*, Vol. 1, No. 4 (1969), pp. 300–305.
11. SALVENDY and SEYMOUR, *Prediction and Development of Industrial Work Performance*, pp. 24–32.
12. J. W. H. KALSBECK and R. N. SYKES, "Objective Measurement of Mental Load," in A. F. Sanders, Ed., *Attention and Performance*, North-Holland, Amsterdam, 1970.
13. P. M. FITTS and M. I. POSNER, *Human Performance*, Brooks/Cole, Belmost, CA, 1967, p. 118.
14. B. H. KANTOWITZ and J. L. KNIGHT, "Testing Tapping Time-Sharing. II: Auditory Secondary Task," *Acta Psychologica*, Vol. 40, 1976, pp. 343–362.
15. J. W. KLING and L. A. RIGGS, Eds., *Woodworth/Schlosbergs Experimental Psychology*, 3rd ed., Holt, Rinehart & Winston, New York, 1971.
16. R. G. MARTENIUK, K. W. SHIELDS, and S. CAMPBELL, "Amplitude, Position, Timing, and Velocity as Cues in Reproduction of Movement," *Perceptual and Motor Skills*, Vol. 35, 1972, pp. 51–54.
17. I. POLLACK, "The Information in Elementary Auditory Displays. I," *Journal of the Acoustical Society of America*, Vol. 24, 1952, pp. 745–749.
18. R. G. MARTENIUK, *Information Processing in Motor Skills*, Holt, Rinehart & Winston, New York, 1976.
19. W. E. HICK, "On the Rate of Gain of Information," *Quarterly Journal of Experimental Psychology*, Vol. 4, 1952, pp. 11–26.
20. P. M. FITTS, "The Informational Capacity of the Human Motor System in Controlling the Amplitude of Movements," *Journal of Experimental Psychology*, Vol. 47, 1954, pp. 381–391.

BIBLIOGRAPHY

HOLDING, D. H., *Human Skills*, Wiley, New York, 1981.

KANTOWITZ, B. H., *Human Information Processing: Tutorials in Performance and Cognition*, Erlbaum, Hillsdale, NJ, 1974.

SALVENDY, G., and SEYMOUR, W. D., *Prediction and Development of Industrial Work Performance*, Wiley, New York, 1973.

WELFORD, A. T., *Fundamentals of Skill*, Methuen, London, 1968.

WELFORD, A. T., *Skilled Performance: Perceptual and Motor Skills*, Scott, Foresman, Glenview, IL, 1976.

CHAPTER 6.2

Reduction of Human Error

DAVID MEISTER

U.S. Navy Personnel Research and Development Center

6.2.1 OVERVIEW

Traditional approaches to reducing error in production rely heavily on personnel selection, placement, and training, supplemented by motivational campaigns to "eliminate production defects." These approaches must, however, be improved or modified if maximum production quality is to be achieved at acceptable cost.

The improvements referred to can be achieved if they are based on an understanding of the nature, frequency, and effects of production errors. Moreover, the attitude of workers will become more positive if the emphasis is changed from one of "motivating" them to try harder, for example, the "zero defects program," to one of securing their participation in identifying those job elements most responsible for errors. The assumption underlying this strategy,[1] termed the "work situation approach," is characteristic of the work situation that *predisposes* to error rather than from poor worker attitudes. Few workers consciously and deliberately attempt to make errors.

The work situation approach is aimed at reducing the probability of errors and resultant production defects by careful design of (1) the item to be produced, (2) the tooling, (3) drawings and instructions, (4) procedures, and (5) the work environment. A key aspect of this approach is the utilization of the worker as a member of the design team.

If the production job has already been designed, the work situation approach focuses on the identification of error predisposing conditions and their elimination or modification. In either case, the end result is a job that is more compatible with the capabilities and limitations of the worker, what Rook[2] has called "design for producibility."

The Nature of Human Error

Before one can talk about reducing human error, one must know what it is. This is because all errors are not alike: They vary in frequency and in terms of consequences, and most important, if they are to be eliminated, they vary in terms of their causation. For example, an error may or may not have an effect on the quality of the item being fabricated; the effect, if any, may be major or minor, and the error may occur at different phases of the fabrication process. The error may result from a variety of causes, for example, incorrect or unintelligible blueprints or instructions, inadequate tools, poor working environment, inappropriate human engineering design of production equipment, or unsatisfactory workplace layout.

Product defects, failures, and accidents are invariably the result of human error. If a material, component, or piece of equipment fails to meet its requirements, it is because it was erroneously designed, selected, applied, fabricated, accepted, installed, used, and/or maintained. If an error occurs because the worker is inadequate to the job, it is because the company's selection, job classification and training systems were improperly designed.

Since the worker is merely part of the production system, which has been consciously and deliberately designed, it stands to reason that *those who designed the system are responsible for any inadequacies occurring in it.* If error occurs because of inadequate system design, that error can be avoided or eliminated by better system design.

The author wishes to express his appreciation to Alan D. Swain, from whose book, *Design Techniques for Improving Human Performance in Production* (reference 1), much of the material in this chapter has been taken, and to Steven Konz of Kansas State University, for information on quality circles. Nonetheless, the opinions expressed in this chapter are those of the author alone.

Human Variability

To understand human error, it is first necessary to understand human variability, because error is a function of this factor. Nothing is more variable than a human being; no one does anything the same way twice. Hence each action is an opportunity for error.

Errors are inevitable unless there are no tolerance limits. Although errors can probably never be completely eliminated, those that are the result of inadequacies in the work situation can be reduced. We learn much about why errors occur and what can be done about them by knowing which of two types of variability occurs in any given work situation.

Consider a rifleman firing 10 shots at a target, and call any shot off the target an error. "Random variability," the first type of variability, is characterized by a dispersion pattern centered about a desired norm, the bull's-eye in this example (Exhibit 6.2.1). When the variability is large, some shots will miss the target. Notice that the pattern of shots is diffuse. These random errors are largely a result of the human's inherent variability and can be controlled by personnel selection, training, supervision, and quality control programs. Training, for example, will cause the rifleman to reduce his variability so that the dispersion pattern is tightened up.

"Systematic variability" is characterized by a different shot pattern in which there is, for example, a bias to the right-hand side of the target (Exhibit 6.2.2). Overall dispersion is small, but some factor, perhaps an incorrect sight adjustment, is causing the rifleman to pull his shots to the right. Once this factor is discovered, it can be corrected.

The point of the example is that certain errors are produced by features designed into the work situation, and it is this class of errors that the industrial engineer or human factors specialist seeks to uncover and eliminate. If the worker has been properly selected and trained, random errors should be reduced to a tolerable level. Under these circumstances an excessive production error rate is probably the consequence of some systematic factor in the situation which can be discovered and modified.

Where errors are random, efforts aimed at their reduction would be better focused on personnel selection and training. Where errors are systematic, they involve some aspect of the work situation that is causing a systematic bias, and efforts for remediation should be aimed at that aspect. In either case it is necessary to identify the type of error made if one expects to apply the proper remedy.

There are four general types of errors: (1) failure to perform a required action, (2) performance of an unnecessary action, (3) performance of a required action at an incorrect time, and (4) making a substandard response. (For a more detailed listing of error classes, see Altman.[3]) Nevertheless, a particular error is specific to the work situation in which it occurs. Unfortunately, error is seldom the result of any single cause, but rather of an often large number of contributing factors. Any factor that can influence work-related behavior can contribute to work-related errors.

6.2.2 THE FREQUENCY OF HUMAN ERROR

Rigby and Swain[4] point out that people are really quite reliable. In an industrial setting one can expect that, for discrete, work-related acts such as reading a five-digit number, moving a control, or putting a part into place, workers will average about one error per 1000 to 10,000 acts. With highly intensive inspection, about 80 to 98% of their errors will be detected and corrected in normal reviews and inspections; more frequently, inspectors catch only 70 to 80% of the defects. Fortunately, only 20 to 30% of the undetected errors will have a significant effect. Thus the prob-

Exhibit 6.2.1 Random Dispersion

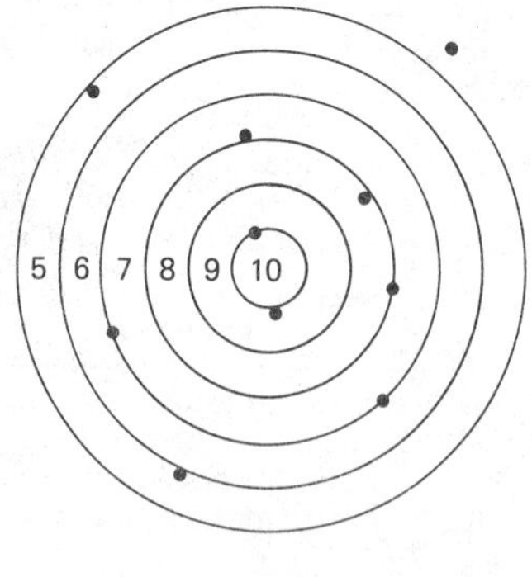

5 6 7 8 9 10

Exhibit 6.2.2 Systematic Dispersion

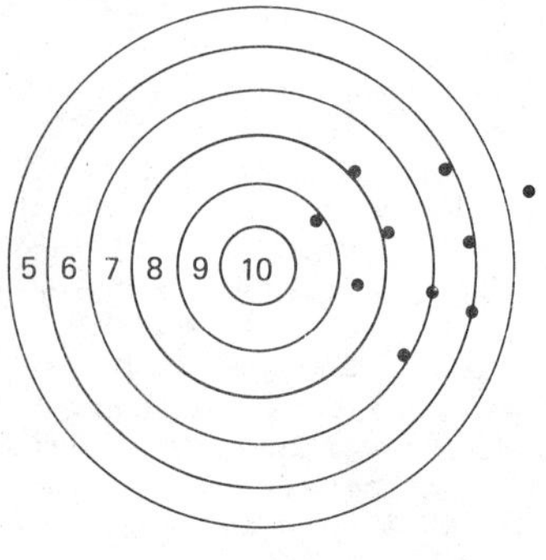

5 6 7 8 9 10

ability that an error will occur in any discrete act, will remain undetected, and will cause a significant effect, will often be only about .00006 to .0000004.

The reason that errors seem to be so frequent, and that they are actually a major problem, is that in almost any work situation a great many people are doing a great many things over a long period. The Ford Motor Company,[5] for example, estimates that it experiences about 3 billion opportunities for assembly error every day.

6.2.3 ERROR CHARACTERISTICS

An analysis performed by Meister[6] indicated that 40% of various kinds of equipment failures and accidents resulted from error of one sort or another. Presumably the remainder of these failures resulted from the normal wearing out of equipment, non-operator-related design deficiencies, or other causes that can be ascribed only remotely to personnel. The irreducible minimum of error to be expected even under optimal conditions is not known. However, error is not only frequent, but significant. Rook[7] reports that, of 23,000 production defects in nuclear weapons manufacturing, 82% were caused by human error.

The effect of an error is, however, also a variable. The effect of some errors is immediate, whereas that of others is delayed. Because of that immediacy, some errors are more visible than others. Many workmanship errors are difficult to discover even during inspection. Error visibility may also produce different error consequences. If the error is apparent to the worker who made it, the chances of its being rectified are increased. This suggests that some feedback to workers about the quality of their work is essential in an error reduction program.

Some errors are frequent, but minor, in their effect; others are infrequent, but critical. Making an error need not necessarily lead to disastrous consequences. Much equipment is so designed that, even if one makes an error, that error, once noted, can be rectified by performing the operation again. Unfortunately, this is not characteristic of production errors.

Manifestly, only those errors that have serious consequences will be remedied. Too many errors are made, and industry has too few industrial engineers and human factors specialists, to remedy all situations conducive to errors. Moreover, management must weigh error reduction against the cost of that error reduction. When the cost becomes excessive, it is unlikely that management will pursue a particular error situation.

6.2.4 WHY PEOPLE MAKE ERRORS

Idiosyncratic errors must be distinguished from the situation-caused errors discussed so far. Situation-caused errors are related to design of the work situation. Idiosyncratic errors are peculiar to the individual and his or her characteristics. Idiosyncratic factors include marital and other interpersonal relationships, emotional conflicts, and attitudes.

Situation-caused errors are the responsibility of management because management designs the work situation. Because management can control the work situation, but not the worker's home or personal problems, it should concentrate its efforts at reducing production error on situational rather than idiosyncratic factors. Motivational approaches to error reduction, for example, zero defects, are likely to be ineffective[2] because it is difficult to control private attitudes.

Swain[1] has listed a large number of what he calls "performance shaping factors" (PSFs). These predispose the worker to error. He divides them into three categories: (1) those external to the individual (situational and task and equipment characteristics), (2) those within the individual (idiosyncratic factors), and (3) physiological stresses that form a bridge between the other two (see Exhibit 6.2.3).

Those factors that are external can be readily modified. They include[6]:

Inadequate Work Space and Poor Work Layout. Highly precise motor manipulations require adequate work space and proper layout. Where containers for parts are not arranged in accordance with assembly procedures, for example, the probability of selecting an incorrect part increases.

Poor Environmental Conditions. Examples are inadequate lighting, high temperature, and high noise level. Inadequate lighting increases the difficulty of positioning and wiring small components properly; high temperature and noise level reduce work effort.

Inadequate Human Engineering Design. Includes inadequate design of machinery, handtools, and checkout equipment. This factor affects production equipment just as it does operational equipment; for example, in one piece of equipment used to check out autopilot amplifiers in the factory, investigators found test accessories that took up most of the working area, difficulty in hooking up the unit, and poorly laid out control panels.[8]

Inadequate Methods of Handling, Transportation, Storing, or Inspecting Equipment. One production department had an exceptional failure record for a highly expensive electronics compo-

Exhibit 6.2.3 Performance Shaping Factors

Situational Characteristics

Work environment (e.g., temperature, noise)
Cleanliness
Staffing
Work hours and breaks
Supplies
Actions by supervisors, peers, union representatives
Rewards, recognition, benefits
Organizational structure
Job instructions

Task and Equipment Characteristics

Job requirements
Task complexity
Task frequency and repetitiveness
Feedback (knowledge of results)
Task criticality and narrowness
Team structure
Man-machine interface factors (e.g., design of prime equipment, tools)

Psychological Stresses

Task speed and load
Fear of failure, job loss
Monotony
Sustained attention
Motivational conflicts
Physiological stresses
Fatigue, pain, discomfort
Hunger, thirst
Temperature extremes
Constricted movement
Lack of physical exercise

Idiosyncratic Factors

Previous training and/or experience
Present skill
Personality and intelligence
Motivation and attitudes
Physical condition
Social factors (family and friends)

---

nent until it was discovered that the components were being transported in carts that permitted them to slip off onto the floor. Redesigning the cart cut the failure rate to an acceptable level.

Inadequate Job Planning Information. Includes inadequate or unavailable operating instructions or blueprints. It is not unheard of to find components being fabricated according to out-of-date instructions because the information has been delayed in reaching the worker.

Poor Supervision. One production department had a very high defect rate until it was discovered that the supervisor refused to allow his people to sit at the benches at which they worked.

The effect of these factors is to create a work situation favorable to the commission of production errors; presumably the probability of error increases as the number of inadequate production characteristics increases.

6.2.5 METHODS OF REDUCING THE POTENTIAL FOR ERROR

Two Approaches

Two different, but related, approaches to improving the quality of industrial production exist. The traditional approach emphasizes changing the worker. The work situation approach emphasizes design of the production setting to support the worker. In the latter approach the emphasis is on human engineering the work situation so that the demands it creates are compatible with the capabilities, limitations, and needs of the worker.

The two approaches are not incompatible; the difference between them is a matter of emphasis. Both should be utilized as appropriate, but emphasis is given here to the work situation methodology because selection and training are usually one-shot affairs, and the options they present are either to fire the worker or to retrain him or her. The work situation methodology may be used in setting up the production job and repeatedly thereafter to dispose of error situations if they recur.

The most effective strategy is to prevent the occurrence of error by designing the work situation so that the potential for error is minimized. It is relatively inexpensive to make design changes when planning a system compared to making them in later stages. After the production system is operational, remedial changes are usually restricted to changes in work procedures and methods and to minor changes in fixtures and handling equipment.

It would take far more space than is available in this chapter to describe in detail the method required to anticipate and avoid error during the design of the production process. That method involves a man-machine system analysis in which jobs and tasks are analyzed to discover error-likely situations (ELS), and design changes are then recommended to avoid the errors. The reader is advised to consult Chapter 6.8 of this handbook, which describes man-machine system analysis in general.

Although avoidance of ELS by proper design of the production system is the most desirable strategy, realistically the industrial engineer must expect to be faced most often with the task of reducing excessive error in an already designed production area. The following material describes what Swain[1] terms an "error-cause removal" (ECR) program, which is somewhat idealistic because it depends on management-worker cooperation, which is not easily achieved.

The ECR Program

One of the main characteristics of the ECR program is its emphasis on preventive, rather than merely remedial, action. Therefore the identification and analysis of ELS is at least as important as the identification and analysis of errors that have occurred.

A requirement of an effective ECR program is the direct participation of production workers (inspectors, assemblers, suppliers, handlers, machinists, maintenance personnel, etc.) in the data collection, analysis, and design recommendation aspects of the program. The most effective program will be one that these production personnel see as their own.

The ECR program consists of teams of production workers with appropriate team coordinators, who might be supervisors or workers with special technical and group skills. The role of the coordinator is to keep the group's activities goal directed. The group should consist of no more than 8 or 12 people, and these whould be people who work together.

Reports of errors and ELS are made by individual workers at periodic meetings of an ECR team. These reports are discussed, and suggestions are made for preventive or remedial measures. Each team presents its own proposals, through the team coordinator, to management for appropriate evaluation and implementation. Human factors specialists and others are available to assist each team and also to help management evaluate proposed design solutions.

The ECR program should be restricted to identifying work situations that require redesign in order to reduce the potential for error. It should not be contaminated by tacking on other goals, such as increasing production quantity. An attempt to expand the program might well raise workers' suspicions that the program was merely a speedup ploy.

ECR Elements

The elements of the ECR program are:

1. Everyone involved is educated about the value of an ECR program.

2. Workers and team coordinators are trained in the data collection and analysis techniques to be employed.

3. Workers report errors and ELS; they analyze the reports to determine causes of errors and develop proposed design solutions to remove these causes.

4. Specialists and management evaluate the proposed design solutions in terms of worth and cost and implement the best of these or develop alternative solutions.
5. Management recognizes appropriately the efforts of production personnel in each ECR program.
6. Specialists evaluate the effects of redesign changes to the production process, aided by continuing inputs from the ECR program.

This last point is most important. The effects of instituting a remedial program for ELS must be evaluated, and the ECR program itself must be thought of as a continuing one.

The primary function of the ECR team coordinator is to stimulate participation. If the program is perceived by the members of any team as being threatening, coercive, or "another management gimmick," that team will not function well. The program is not another "cost savings" effort, and no monetary worth is assigned to any design solutions suggested and accepted. Management must make it clear that, even if a design solution cannot be accepted because of technical or cost considerations, the identification of a production problem that did or could lead to errors is a valuable input.

Workers should be allowed to set up their own teams. However, participation in the program must be mandatory because, although all workers will not be equally enthusiastic about it, those participating cannot be made to feel that they are performing an additional duty from which others are excused.

ECR Analysis

The data to be collected in the ECR program consist of errors, ELS, and accident-prone situations (APS). The latter are important because most accidents are the result of some work situation that encourages errors leading to accidents.

Most errors in an industrial setting will be related to defects, accidents, and near accidents. Identification of defects presents few problems in production operations where quality standards of the product are clearly defined. Disagreement among members of the ECR team as to whether or not a product is defective indicates a need for more careful definition of tolerance limits.

Errors are important even when no severe consequences have resulted. This is because the error may indicate a deficiency in the work process that could lead to a much more serious error later. This point should be emphasized to workers because their natural tendency will be to ignore what they consider nonconsequential errors. The same logic applies also to reports of accidents and near accidents.

The essence of an ELS and APS analysis is the recognition that certain characteristics of the situation are undesirable and make excessive demands on the worker. Questions to be asked include the following: Is the task within the worker's capability? Is there anything about the task that does or could cause the worker fatigue or discomfort? Is there sufficient information feedback? Does the task make demands that require too much precision or too many movements? Is the physical environment (temperature, noise, lighting) adequate? Questions should be asked about each of the situational PSFs listed in Exhibit 6.2.3.

Each ECR team should meet periodically (without management participation) and evaluate the reports turned in. The team would then comment on the suggestions made, and some consensus would be reached. This consensus would be added to the error, ELS, or APS report sent to management for evaluation. Every report must be passed upward; even if the suggestions for improvement are unworkable, at least a problem has been identified.

ECR Evaluation

Each suggestion made by the ECR team for a redesign of the work situation should be evaluated by a committee of specialists in terms of:

1. **Technical Worth.** Will the redesign really reduce errors and by what amount?
2. **Nontechnical Worth.** Will the redesign result in other improvements, such as greater job satisfaction? Some of these nontechnical factors are intangibles to which it is difficult to assign a monetary value. Nevertheless, they may reduce turnover, absenteeism, and grievances.
3. **Cost-Effectiveness.** Will the cost of the redesign be paid for by the presumed reduction in errors or by other considerations of worth?

For management to make an appropriate evaluation of errors, ELS, or APS, their importance must be ascertained. Importance is determined by three factors: consequences in terms of money and/or customer loss, likelihood of error occurrence, and cost of corrective action. Each factor can be rated on the following scales (from Swain,[1] as modified by the author):

1. **Potential ELS Consequences:**
   a. ⩾$100,000 scrap and 1000 customers lost.
   b. ⩾$10,000 scrap and 100 customers lost.
   c. ⩾$1000 scrap and 10 customers lost.
   d. ⩾$100 scrap and 1 customer lost.
   e. Negligible.
2. **Likelihood of Uncorrected Error:**
   a. Several times a month.
   b. Once a month.
   c. Several times a year.
   d. Once a year.
   e. Less than once a year.
3. **Cost of Corrective Action:**
   a. Less than $100.
   b. $100 to $1000.
   c. $1001 to $10,000.
   d. $10,001 to $100,000.
   e. More than $100,000.

Admittedly these scales are quite crude, and the resulting values cannot be combined; however, they present a means of semiquantifying the ELS that management must deal with. Obviously, an ELS or APS rated E on each scale could be dismissed, but one rated A on each scale should be seriously considered.

Quality Control Circles

A variation of the ECR approach is the concept of quality control (QC) circles, developed in Japan in 1963 and since then highly successful in that country in solving quality control problems. The essence of QC circles is participative problem solving. Groups of 8 to 10 people (e.g., supervisors, workers, production engineers) performing similar or interrelated work form a circle on a voluntary basis. They are given special training in statistical quality control techniques, training that, in the United States, is usually reserved for quality control engineers. The training includes use of Pareto diagrams to isolate the major problems producing a defect, cause-effect diagrams (something like a fault tree diagram), histograms, graphs, control charts, stratification, and binomial probability.

Meetings are held at least once a month, usually on a paid basis. With the help of a facilitator who may be a supervisor, the group selects a problem that is responsible for product defects and that appears amenable to solution. Using Pareto diagrams,[9,10] the circle analyzes the factors responsible for the greatest percentage of the defects found. The group then develops a cause-effect diagram to determine the specific cause of the problem. After establishing goals for reducing the defect rate, the circle recommends potential solutions, such as improved procedures or changes in design, and attempts to implement the solution. All of this is done with the active cooperation of management.

Certain elements of the ECR team and the QC circle are much the same, for example, the problem solving orientation, the concept of participative democracy, and the crossover among levels of management, engineering, and production. The QC circle differs from the ECR team in the emphasis on training in statistical quality control techniques, although the training is practical rather than academic; the formal way in which problems are investigated (i.e., through the use of Pareto and cause-effect diagrams); and, most particularly, the emphasis on teamwork, pride, and identification with the company, an emphasis that is viewed by American management as being peculiarly Japanese.

QC Circle Analytic Aids

The two major analytic aids used in training in the QC circle methodology are the Pareto and the cause-effect diagrams.

Pareto Analysis. The technique of arranging data according to priority or importance and tying them to a problem solving framework is called Pareto analysis. The following discussion is taken from the American Society for Quality Control's training manual for QC circles.[11]

The Pareto distribution is based on the order-of-measurement value for each element considered rather than on the quantitative value of the element. The steps involved are:

Exhibit 6.2.4

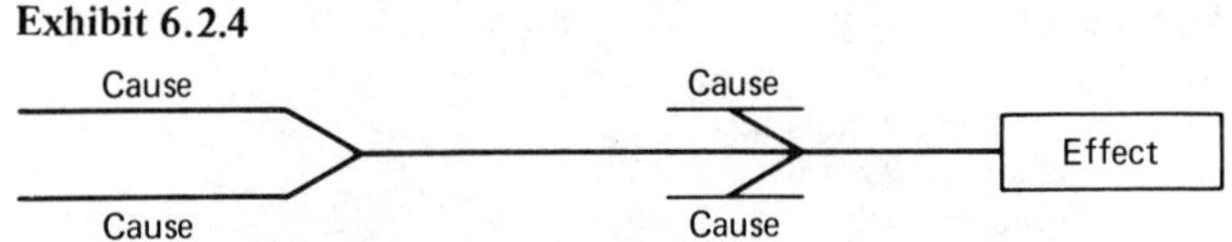

1. List all the elements of interest.
2. Measure the elements, using the same unit of measurement for each element.
3. Order the elements according to their measure, not their classification.
4. Create a cumulative distribution for the number of items and elements measured.

For example:

Type of Defect	Number of Elements	Percentage of Elements	Total Dollars	Percentage of Dollars
A	1	14.29	5865	79.58
B	2	28.57	6736	91.40
C	3	42.86	7005	95.05
D	4	57.14	7181	97.44
E	5	71.43	7275	98.71
F	6	85.71	7330	99.46
G	7	100.00	7370	100.00

The preceding distribution can be plotted as a curve, with one axis being percentage of value, and the other, percentage of items. This enables one to see that a few elements account for a disproportionate sum of all measurements and should therefore be the elements to attack. In the example given, it is obvious that defect type A is responsible for 80% of the problem and that major attention should therefore be given to that defect.

Cause-Effect Diagrams. The cause-effect diagram developed by Ishikawa in 1950 consists of defining an occurrence (effect) and then reducing it to its contributing factors (causes). The relationships among the contributing factors are illustrated in a "fishbone" arrangement, as shown in Exhibit 6.2.4.

In developing this diagram the principal factors or causes are first listed in terms of four categories (manpower, machine, methods, and material) and then iteratively reduced to their subcauses. The process is continued until all possible causes are listed. The factors are then critically analyzed in terms of their probable contribution to the effect or the problem.

The major objection voiced to the QC circle concept is that the procedures employed cannot be utilized in a non-Japanese society; in other words, it would be difficult to gain the necessary cooperation, teamwork, and motivation from westernized workers. The QC circle method has been applied primarily in Japan (where it has been enormously successful), and has been tried out in a number of American plants with positive results. Because of resistance on the part of American industrial management, however, the method has not been tried out on a large scale in this country.

REFERENCES

1. A. D. SWAIN, *Design Techniques for Improving Human Performance in Production*, author, 712 Sundown Place, S.E., Albuquerque, NM 87107, 1977.
2. L. W. ROOK, "ZD: Momentary or Momentous," *Quality Assurance*, Vol. 4, October 1965, pp. 24–28.
3. J. W. ALTMAN, "Classification of Human Error," in W. B. Askren, *Symposium on Reliability of Human Performance in Work*, Report AMRL-TR-67-88, Aerospace Medical Research Laboratories, Wright-Patterson Air Force Base, OH, May 1967.
4. L. V. RIGBY and A. D. SWAIN, "Effects of Assembly Error on Product Acceptability and Reliability," *Proceedings, 7th Annual Reliability and Maintainability Conference*, ASME, New York, July 1968.
5. I. METZ, "Building Better Cars," *The Wall Street Journal*, September 24, 1968.
6. D. MEISTER, *Human Factors: Theory and Practice*, Wiley, New York, 1971.
7. L. W. ROOK, "Reduction of Human Error in Industrial Production," Technical Memorandum SCTM 93-62(14), Sandia Corporation, Albuquerque, NM, June 1962.

8. R. E. URMSTON and C. M. CUTCHSHAW, *Human Engineering Principles Applied to the Design of Factory Test Equipment. I. TET-704*, Report AE60-0290, Convair/Astronautics, San Diego, April 11, 1960.

9. J. M. JURAN, "The QC Circle Phenomenon," *Industrial Quality Control*, January 1967, pp. 329–336.

10. S. KONZ, "Quality Circles: Japanese Success Story," *Industrial Engineering*, October 1979, pp. 24–27.

11. D. M. AMSDEN and R. T. AMSDEN, *QC Circles: Applications, Tools and Theory*, American Society for Quality Control, Milwaukee, WI, 1967.

CHAPTER 6.3

Engineering Anthropometry and Occupational Biomechanics

DON B. CHAFFIN

The University of Michigan

6.3.1 BACKGROUND

It is a tradition of industrial engineering to consider the worker an integral component in a complex production or service system. F. W. Taylor at the end of the nineteenth century and the Gilbreths in 1912 specifically made note of how a worker's physical capabilities had to be carefully considered when designing a job if total system performance were to be maximized. In fact, the Gilbreths made quantitative studies of reach capability that have formed the basis for many workplace layouts today.[1] Worker body dimensions were reported by LeGros and Weston[2] in 1926, and seat comfort was a topic discussed by Lay and Fisher[3] in 1940.

Today the study of human size, mobility, and shape for the purpose of designing products and our physical surroundings is referred to as "engineering anthropometry." The discipline for such studies arose from physical anthropology, wherein it was necessary to quantify such human physical attributes precisely for comparative purposes. The early industrial engineers realized the potential for the application of these data in the workplace, which motivated the engineering anthropometric field. A recent book by Roebuck et al.[4] reviews this development since the turn of the century.

Likewise, it became necessary to know how the use of force required in a job could affect a worker's capabilities and health. Anatomists in the late part of the nineteenth century began describing human motion and force loadings in kinematic terms. These early studies led to the multidisciplinary science referred to as "biomechanics," which is the study of mechanical reactions of the body to external or inertial loadings. Several distinct applications of biomechanics have emerged in the last two decades. "Impact biomechanics" addresses the problems resulting from sudden external forces acting on the body (e.g., a vehicle collision or fall from a height), which usually results in acute trauma (e.g., concussion, whiplash, laceration). "Occupational biomechanics" deals more with volitional acts (e.g., lifting of loads, pushing carts), wherein the person's musculoskeletal system may be loaded to the maximum. The problems of who can engage in such high exertion safely in industry are addressed in occupational biomechanical studies. It is this latter topic that is described further in the following subsections.

6.3.2 PROBLEMS OF WORKER-JOB MISMATCH

Acts that are nonrepetitive, but physical (e.g., an extreme reach to a seldom-used control or the occasional picking up of an unusually heavy object or tool), often can be overlooked in the traditional job description and evaluation process. Yet these same atypical acts can create substantial managerial problems and human suffering. What follows is a description of the types of problems managers must expect when they are unaware of the anthropometric and biomechanical basis for worker-job matching in this context. The effect of *repetitive* physical exertions is discussed in Chapter 6.4.

Worker Performance Problems

It should be obvious from careful inspection of a workplace layout that controls and objects that need to be moved about must be within easy reach of most workers. Yet the wide variation in worker reach capability that exists today, particularly with regard to the short woman, the older employee, and the physically limited worker, has greatly increased the complexity of the problem facing the job analyst or designer.

It was not long ago that a job designer would assume that, if an overhead reach to a control were required in a particular job, a man would be given that job. Today that is not the case, with more than 43% of the labor market's being composed of women. Where an easy overhead grasp type of reach of about 77 in. (195 cm) was given in design books with reference to men, now approximately 73 in. (185 cm) is recommended to order to accommodate 95% of women.[5] Such a change certainly makes it easy to reach the object of concern, but there are complications. For instance, a 73 in. high control, if not carefully placed in the work space, could hit the heads of more than 20% of fully clothed men walking under it, as discussed by Morgan et al.[6]

The anthropometric variations in the work force today dictate a great concern to ensure that the majority of people can physically reach *all* the objects necessary to perform a job. Overhead, side, and forward reach data must be scrutinized carefully and used to ensure that such is the case. Providing a step stool may be a solution worth considering in a particular overhead reach situation, but the falling and tripping hazards presented in a majority of situations make this solution questionable in industry where overhead reach or visibility problems exist.

Further, if an object to be moved requires *both* a forceful exertion and an extended arm posture overhead, to the side, or forward, then even greater care is necessary in designing the workplace layout. For instance, the arm force capability is greatly limited when the arm is extended and lifting an object up vertically in front of the body. This is illustrated in Exhibit 6.3.1 from Martin and Chaffin,[7] using average male strength data. As also depicted by Martin and Chaffin[7] and discussed by Laubach,[8] arm reach strength involving the shoulders is very limited for women, averaging between 40% and 50% of men's strength.

The general effect of producing any hand force perpendicular to the long axis of the forearm is similar to that depicted in Exhibit 6.3.1. That is, as the hands are located further from the torso, the perpendicular hand force requirement operating on the enlarged moment arm causes an increased torque, or turning moment, at various joints, particularly the shoulder. This moment effect is not well compensated for by the muscles. Thus the exerted hand force capability greatly decreases as the arm is extended. In this sense, such hand forces of a high magnitude should not be required when the arm is so extended. Some practical recommendations on avoiding the associated worker performance problems are presented in Section 6.3.4.

Worker Health and Safety

Anthropometric mismatching can also produce safety and health hazards. Various anthropometric survey reports, which have been available for more than 30 years, giving specific data regarding size, mobility, and shape of the population, can be used in assigning the cause of an accident.

Exhibit 6.3.1 Predicted Two-Handed Isometric Vertical Lifting Strength of a Fiftieth Percentile Male[a]

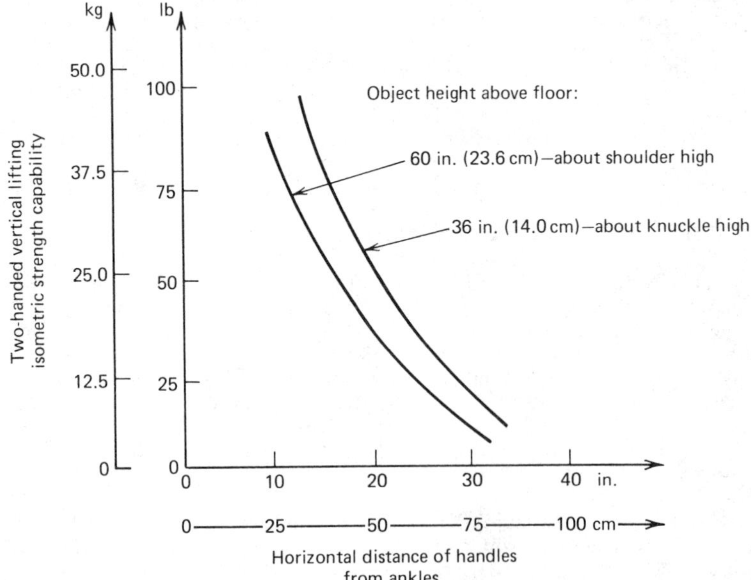

[a]Adapted from reference 7.

Exhibit 6.3.2 Incident Rate (*a*) and Severity Rate (*b*) of Low Back Pain for Workers on Jobs Requiring Various Proportions of Their Isometric Strengths[a,b]

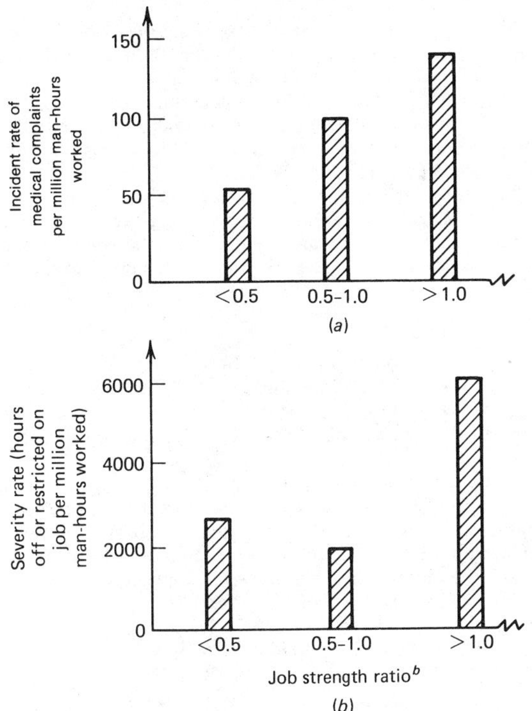

[a]Adapted from reference 9.
[b]Job strength ratio = maximum weight lifted on job/isometric strength in job simulation test.

The *Journal of Applied Ergonomics*,* which has been published for more than 10 years, has specific examples of anthropometric mismatches that caused or contributed to various accidents.

From a biomechanical standpoint, mismatching a worker's strength with the job requirement has been shown in several studies to increase greatly the probability of future musculoskeletal problems (i.e., strain and sprain injuries and episodes of low back pain). Exhibit 6.3.2 from Chaffin et al.[9] gives the effect of mismatching on the incidence of low back pain. Both the incidence rates (i.e., number of low back pain cases per million man-hours on the jobs) increased by almost three times when the person placed on the job did not have an isometric strength at least equivalent to that required for performing the most strenuous elements of the job. A similar result was reported by these authors for other musculoskeletal problems when the maximum physical exertions were repeated frequently during the day.

Since strength variability in the working population has increased over the last decade along with the concern for providing a safe and healthful workplace, it has become necessary to design workplaces that accommodate the low-strength individual. This is an explicit goal of various affirmative action programs for increasing the employment of women, older workers, and the physically handicapped. Thus it is wise today to coordinate the design of new or existing workplaces and/or machines with the plant personnel department in order to ensure that the policy regarding affirmative action for these groups of people is being considered.

In summary, from the standpoint of both worker performance and health and safety, anthropometric and biomechanical factors must be considered in job design. Simple assumptions about the reach and strength capability of the working population are not acceptable. Data are available on these human attributes, and they must be considered. The cost of mismatching a job and worker is growing. What follows are some suggested methods for controlling this situation.

**Journal of Applied Ergonomics*, IPC Science & Technology Press Limited, P.O. Box 63, Westbury House, Bury Street, Guilford, Surrey GU2 5BH England.

6.3.3 ANTHROPOMETRIC DESIGN CONSIDERATIONS

Need

As described in the preceding subsection, when designing for a population with widely varying anthropometric attributes, conflicts in design objectives must be recognized. Generally these conflicts will be based on the need to provide adequate clearance and comfort for the large person while providing an easy reach for the small person. Economics, safety, and worker well-being must be considered in meeting these conflicting objectives, and each case will require different approaches. It is always good, however, to have available the best depiction of the worker population attributes of concern when developing the design.

Engineering anthropometry provides the human size, shape, and mobility data necessary for many workplace designs. It provides a structure for acquiring the necessary data. Roebuck et al.[4] have reviewed the methods used to acquire anthropometric data for design purposes and should be read carefully before performing any type of worker anthropometric survey.

Data Sources—Static

Anthropometric data fall into two categories: static and dynamic. Static anthropometric data describe a person either at rest (sitting or standing) or while holding an extreme posture. Most of the data so reported depict a specific body segment (e.g., length, circumference, shape) or joint mobility. Thus the user must often combine the data to form a description of the population attribute of interest.

For instance, to estimate overhead reach while seated, the lengths of the upper extremity and torso can be combined to provide the necessary prediction for the population. However, the prediction may have considerable error, since the amount of upward rotation of the shoulder and torso erection (flattening) of the spine would not be depicted, and thus the estimate might be too short. Such an error may be quite acceptable, however, especially in the early stages of a design process when "ball park" estimates are needed.

Exhibit 6.3.3 Segment Lengths as Proportion of Stature[a]

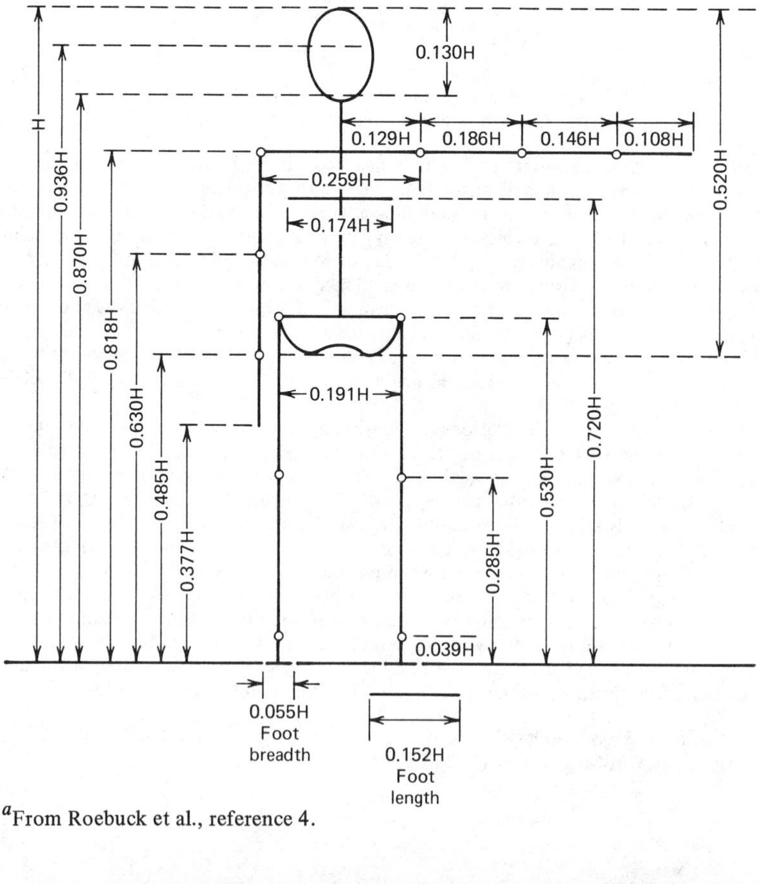

[a]From Roebuck et al., reference 4.

Exhibit 6.3.4 Stature of U.S. Adults From 1971 to 1974

	Mean		SD		Percentiles		
Sex and Age	In.	Cm	In.	Cm	5th	50th	95th
Men (18–74 years)	69.0	(175)	2.8	(7.11)	64.4 (164)	69.0 (175)	73.6 (187)
18–24 years	69.7	(177)	2.8	(7.11)	65.1 (165)	69.7 (177)	74.4 (189)
25–34 years	69.6	(177)	2.9	(7.34)	64.8 (165)	69.5 (177)	74.3 (189)
35–44 years	69.1	(176)	2.7	(6.86)	64.7 (164)	69.2 (176)	73.4 (186)
45–54 years	68.9	(175)	2.6	(6.60)	64.7 (164)	68.8 (175)	73.2 (186)
55–64 years	68.3	(173)	2.6	(6.60)	64.1 (163)	68.2 (173)	72.5 (184)
65–74 years	67.3	(171)	2.6	(6.60)	63.2 (161)	67.3 (171)	71.6 (182)
Women (18–74 years)	63.6	(162)	2.5	(6.35)	59.5 (151)	63.7 (162)	67.8 (172)
18–24 years	64.3	(163)	2.5	(6.35)	60.2 (153)	64.3 (163)	68.4 (174)
25–34 years	64.1	(163)	2.4	(6.10)	60.2 (153)	64.0 (163)	68.2 (173)
35–44 years	64.1	(163)	2.5	(6.35)	59.9 (152)	64.1 (163)	68.4 (174)
45–54 years	63.6	(162)	2.3	(5.84)	59.9 (152)	63.7 (162)	67.3 (171)
55–64 years	62.8	(160)	2.4	(6.10)	58.6 (149)	62.8 (160)	66.6 (169)
65–74 years	62.3	(158)	2.4	(6.10)	58.2 (148)	62.3 (158)	66.2 (168)

Source. Adapted from reference 11.

Various sources of static anthropometric data exist in tabular form. The most recent comprehensive tabulation is in three volumes. Edited by Webb Associates for NASA, it lists 59 anthropometric variables for 12 selected populations.[10]

The need to combine these data into a more useful form than a tabulation has resulted in the development of drawing board manikins. These are clear-plastic scale models of the human body, articulated at various major joints. Side, top, and front views are available. Kits can be acquired from Anthropometric Data Application Manikin, P.O. Box 2653, Santa Barbara, CA 93120.

The most recent development in design manikins allows the torso to flex and extend in a realistic fashion. Plans for this model can be acquired from 6570 Aerospace Medical Research Laboratory, Attention–Mr. Kenneth Kennedy, Wright-Patterson AFB, OH 45433. Layouts for more simplified fifth, fiftieth, and ninety-fifth percentile drawing board manikins also are presented in the *NASA Anthropometric Source Book.*[10]

Typical segment lengths in a kinematic depiction of the human body are given in Exhibit 6.3.3. When combined with the stature data presented in Exhibit 6.3.4, they can be used to easily describe some of the major static anthropometric data needed for design.

Data Sources—Dynamic

Using drawing board manikins or stature correlations in combining static data to predict a functional (dynamic) anthropometric approach does not reduce the error by much, but it certainly assists in easing the use of these data. Because the error may be unacceptable in many design situations, various functional anthropometric data have been developed for typical work situations (e.g., overhead reach, seated reach forward, reaches to each side). One method of depicting these types of functional data is to regress each variable onto the stature (standing height) of the person, such as was done in Exhibit 6.3.3 for static data. Though such coefficients of determination are relatively low (r^2 = .3 to .8), the results can be depicted usefully in a nomogram form, as done by Diffrient et al.[12] Thus, by choosing a stature percentile of interest, the nomogram provides a first-order prediction of many different functional dimensions of the population.

In utilizing any functional anthropometric data for design, it is evident that many different environmental and/or task factors can affect the data. Still, functional reach data do exist and provide design guidance. For instance, Exhibit 6.3.5 depicts the forward seated erect reach capability of the small (fifth percentile) woman, as obtained from an industrial population at Eastman Kodak by Faulkner and Day.[13] These data do not include shoulder or torso assistance, but, rather, represent a comfortable reach with a two-point thumb-to-finger grasp. They are presented as an example of the type of data available in various anthropometric source books.[4–6]

To develop more generality in the use of the data for widely varying design situations, computerized human form models have been developed. Though these are still research tools, some limited applications have been reported for advanced designs of aircraft cockpits and general work stations.[14–16] A typical human form model is depicted in Exhibit 6.3.6. As better techniques are developed to predict the postures that are preferred by a person when performing a reach task,

Exhibit 6.3.5 Maximum-Reach Curve Above Working Surface for Small (Fifth Percentile) Women Using Thumb-Finger Grasp[a]

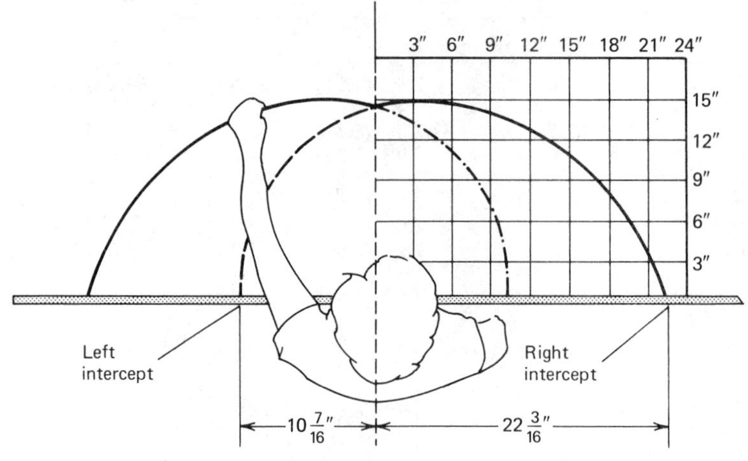

Reaches for right hand, fifth percentile (All dimensions are referenced from the front edge of the workplace.)

Distance Along Front Edge of Workplace	Height Above Working Surface (Elbow Height when Seated Erect)						
	1 in.	6 in.	11 in.	16 in.	21 in.	26 in.	31 in.
9 in. left	$7\frac{3}{4}$	$9\frac{13}{16}$	10	$9\frac{3}{8}$	$5\frac{3}{4}$	–	–
6 in. left	$11\frac{5}{8}$	$12\frac{13}{16}$	13	$12\frac{11}{16}$	$10\frac{1}{8}$	$6\frac{5}{8}$	–
3 in. left	$13\frac{5}{8}$	$14\frac{5}{8}$	$14\frac{7}{8}$	$14\frac{9}{16}$	$12\frac{3}{8}$	$9\frac{3}{8}$	3
0	$14\frac{1}{2}$	$15\frac{5}{8}$	$15\frac{7}{8}$	$13\frac{1}{2}$	$13\frac{1}{2}$	$10\frac{7}{8}$	$5\frac{7}{16}$
3 in. right	15	$15\frac{7}{8}$	$16\frac{1}{8}$	$15\frac{7}{8}$	$13\frac{15}{16}$	$11\frac{1}{8}$	$5\frac{7}{8}$
6 in. right	$14\frac{3}{4}$	$15\frac{3}{4}$	16	$15\frac{3}{4}$	14	$10\frac{15}{16}$	$5\frac{1}{4}$
9 in. right	14	$15\frac{1}{4}$	$15\frac{9}{16}$	$15\frac{5}{16}$	$13\frac{1}{2}$	$10\frac{1}{4}$	$3\frac{5}{8}$
12 in. right	$12\frac{7}{8}$	$14\frac{1}{4}$	$14\frac{1}{4}$	$14\frac{1}{4}$	$12\frac{1}{8}$	$8\frac{1}{2}$	$\frac{7}{8}$
15 in. right	$11\frac{3}{16}$	$12\frac{1}{2}$	$12\frac{5}{8}$	$12\frac{5}{8}$	$10\frac{1}{8}$	6	–
18 in. right	$8\frac{1}{2}$	$9\frac{7}{8}$	$9\frac{13}{16}$	$9\frac{13}{16}$	7	$\frac{3}{4}$	–
21 in. right	$3\frac{7}{8}$	$6\frac{1}{8}$	$5\frac{1}{2}$	$5\frac{1}{2}$	$1\frac{1}{2}$	–	–
Left intercept	$10\frac{7}{16}$	$11\frac{3}{8}$	$11\frac{1}{4}$	$11\frac{1}{8}$	$9\frac{3}{4}$	8	$4\frac{3}{16}$
Right intercept	$22\frac{3}{16}$	$23\frac{9}{16}$	$23\frac{7}{8}$	$23\frac{1}{4}$	$21\frac{9}{16}$	$18\frac{1}{4}$	$12\frac{5}{8}$

[a] Adapted from reference 13.

these models will provide the future means for matching either a person's or a population's anthropometry with a work space requirement.

6.3.4 BIOMECHANICAL DESIGN CONSIDERATIONS

Background

The human musculoskeletal system is a kinematic linkage. Thus, when a load is applied to the hands, the reactive forces are transmitted throughout the linkage. The ability to tolerate such forces varies greatly from individual to individual and from joint to joint.

The maximum occasional load (i.e., strength) that can be produced safely by a person performing a specific act has been assessed by several methods. Three common methods are:

1. In-plant epidemiology of overexertion injuries.
2. Biomechanical studies of load handling.
3. Isometric strength and psychophysical studies.

Exhibit 6.3.6 Typical Computerized Human Form Model[a]

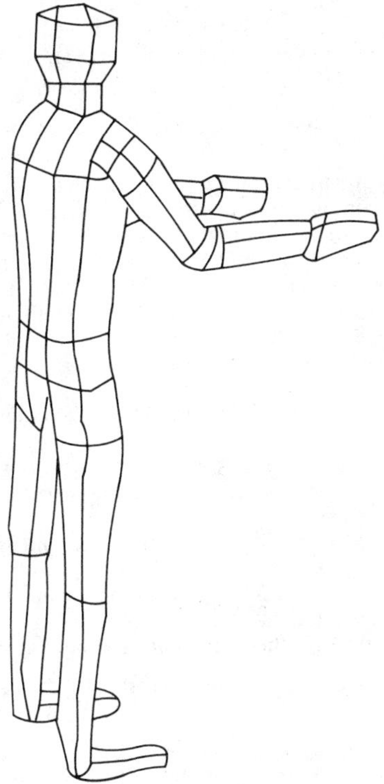

[a]From the Center for Ergonomics at the University of Michigan.

In-plant epidemiological studies of physical exertion have been used to determine the potential adverse effects on a worker's health when performing general types of physical acts. As an example, one study indicated that lifting compact (tote box) size loads of more than 120 lb in five geographically different plants resulted in about eight times more frequent complaints of low back pain per man-hour on the job than did lighter (less than 35 lb) lifts.[17] Unfortunately, such studies are difficult to control, and the results are often too general to provide design guidance for a specific task. Rather, these data indicate that there is some type of a problem related to the physical efforts required on a job and that more detailed evaluation by other means is needed.

Such detailed evaluations may rely on the second method, that of creating biomechanical models of the human musculoskeletal system. These produce more specific results, but the accuracy depends on the goodness of the model assumptions. Simple kinematic models have been shown to be effective in comparing various industrial manual materials handling tasks by E. R. Tichauer.[18] In this approach the body is considered a set of solid links articulated at the major body joints. Because epidemiology has indicated that low back pain develops when handling heavy loads, one major joint of concern in the kinematic linkage is the lumbosacral joint (or L_5/S_1 disc at the base of the lumbar spine). By comparing the turning moment about this joint created by lifting loads in various postures, Tichauer has demonstrated how severe the act of lifting a large bulky object can be in terms of low back stress. Both Tichauer and others have confirmed these model results using electromyography of the back extensor muscles, which provides an estimate of the muscle tension in a given act.[18]

Comparison of the forces acting to compress the lumbosacral joint during various physical acts with those maximum forces required to cause the disc of cadaver spines to fail has further confirmed the problems associated with the lifting of heavy loads. Chaffin[19] has shown that it is not unusual to expect 1400 lb (634 kg) of compressive force to develop at the lumbosacral disc, leading to chronic back problems. If the load were a bulky, heavy load, it could be expected that the compressive forces on the lumbosacral spine would exceed 2000 lb (906 kg).

The third method of predicting the safe load handling capability of the population relies on the cooperation of workers to demonstrate their capabilities to perform specific acts. If the demonstration is isometric (static), it is referred to as an "isometric strength test."[9] If the worker performs a dynamic task and adjusts the resistive load to the level that is tolerable for a given period (usually 8 hr), then it is referred to as a "psychophysical test." Snook of the Liberty Mutual Insurance Company has written extensively on the psychophysical basis for setting limits to manual materials handling tasks.[20]

Given these three methods, some general rules have emerged for specific physical acts that are nonrepetitive (i.e., for short exertions of a couple seconds performed infrequently during the 8 hr shift).

Lifting

As indicated by the preceding discussion on the biomechanics of lifting, the size (bulkiness) of an object to be lifted is a major consideration in establishing a "safe" limit. Also, a decision must be made regarding the strength of the population selected to perform such acts. The National Institute for Occupational Safety and Health (NIOSH), which is considering these matters, has suggested a preliminary guideline, as depicted in Exhibit 6.3.7. Though this is not now a formal proposal, it represents the most comprehensive attempt to combine various types of data into a uniform guideline for the design of a lifting task for both men and women. As such, it is believed to be the most reasonable design reference for occasional, two-handed symmetric lifts from near the floor to 30 in. (76 cm).

The following conditions for the NIOSH lifting guidelines should be noted, however:

1. If workers are selected for such work by passing specific strength and other related medical tests, then the upper values are recommended. If there is no selection or training program for such tasks, then the lower values apply.
2. If an object does not have to be lifted from the floor, but is at or above knee height, the limits would be higher by about 20%.
3. If an object is to be lifted above a 30 in. (76 cm) high table, the limits are reduced by about 1% for each inch above the table (0.4%/cm)
4. If the object is lifted frequently, then the limits are reduced in proportion to work rate (see Chapter 6.4).

Pushing and Pulling Carts

Two-handed pushing and pulling tasks (e.g., pushing a cart) have not been as extensively studied in industry as have lifting tasks. Clearly a major factor in such acts is the foot traction provided the worker. If it is assumed that very high traction exists (an assumption that must be carefully assessed), then the worker's strength becomes the dominant factor in short-duration pushes and

Exhibit 6.3.7 Recommended Weight and Size Limits for Occasional Lifts (Less Than Once Every 5 Min)

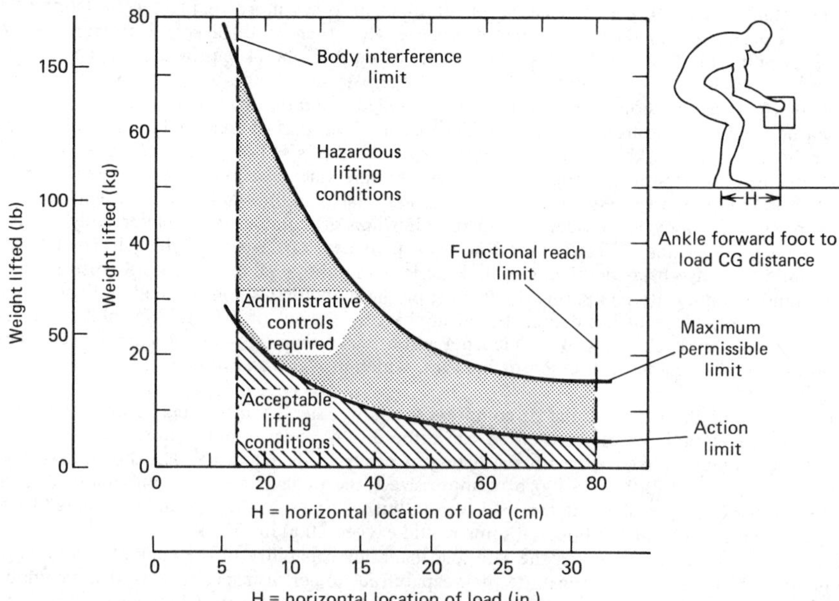

Exhibit 6.3.8 Pushing (*a*) and Pulling (*b*) Force Capability (in Pounds) for Male of Average Size and Strength, With Varied Hand Locations and Good Traction[a]

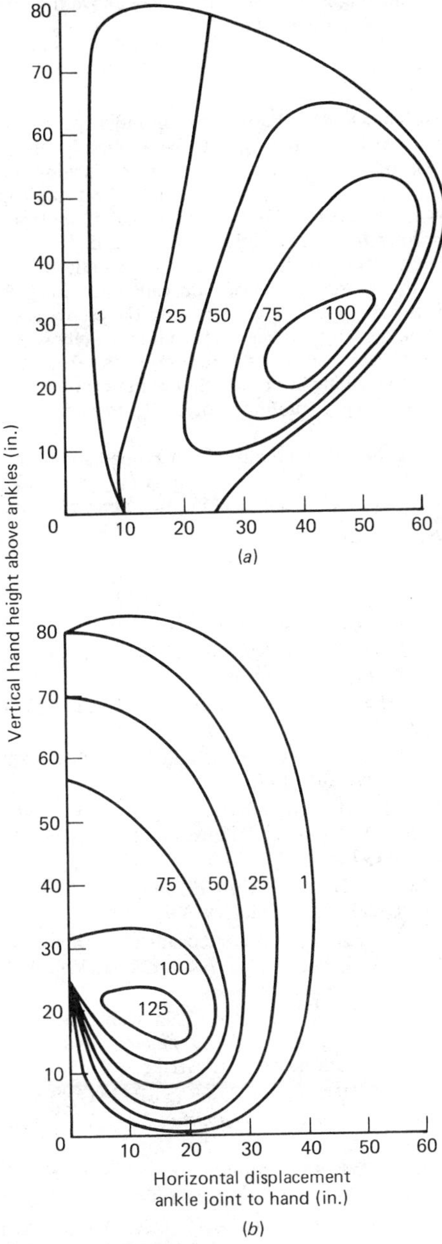

[a]Adapted from reference 7.

pulls. As was the case with lifting, strength depends upon postures that are allowed by the workplace and the object being pushed or pulled. Exhibit 6.3.8 depicts these trade-offs for the average-sized male. Once again, if the forces depicted must be sustained for more than a few seconds, the corrections discussed in Chapter 6.4 need to be applied to reduce the values accordingly.

The psychophysical pushing and pulling data of Snook[20] also suggest that the values in Exhibit 6.3.8 would be reduced to approximately 50% of the value shown if one were designing in order to accommodate 90% of the women. Also, it must be noted that often in industry the shoe-floor frictional coefficient can be expected to reach values of only .5, even on dry concrete. Thus, if a lightweight person (i.e., about 100 lb) is expected to perform a push or pull, hand forces of more than 50 lb (23 kg) could create a high slip potential, even in a good posture. In this regard good practice suggests that, when moving a cart on a ramp, stay above the cart. If a slip does occur, this will avoid the cart's coming down on the worker.

Control Movement

One-handed pushes and pulls may often be required in moving manual controls. As discussed at the beginning of this chapter, any time the force to be applied by the hand to an object acts perpendicular to the long axis of the forearm, it can create turning moments at the elbow and shoulder which overburden the muscle's capability. Thus the force capability decreases quickly with the amount of elbow extension and the distance the hands are from the shoulder, as depicted previously for lifting, in Exhibit 6.3.1. A simple rule for pushing and pulling forces is to keep the load forces on the hand from acting to extend or flex the shoulder when the elbow is already in the extended position. In other words, for maximum capability, keep the required load vector on the hand directed through the shoulder, especially when the elbow is extended (i.e., pull or push toward or away from the shoulder), by flexing or extending the elbow.

Also, if there is a choice between exerting a force across the front of the body (i.e., pushing to the left or right) as opposed to pushing in or out, choose in or out. Strength capability is far better (by about two times) when pushing or pulling in or out from the body than when pushing toward one side or the other.

Specific strength values are depicted graphically for one-handed activities in the *NASA Anthropometric Source Book*[10] and in many other human factors books. These values provide the basis for many of the specific workplace and machine guidelines presented later in this section (see Chapters 6.5 through 6.9).

REFERENCES

1. R. M. BARNES, *Motion and Time Study*, Wiley, New York, 1949.
2. L. A. LEGROS and H. C. WESTON, *On the Design of Machinery in Relation to the Operator*, Industrial Fatigue Research Board Report 36, London, 1926, p. 9.
3. W. E. LAY, and L. C. FISHER, "Riding Comfort and Cushions," *SAE Transactions*, Vol. 47, No. 5 (1940), pp. 482–96.
4. J. A. ROEBUCK, JR., K. H. E. KROEMER, and W. G. THOMPSON, *Engineering Anthropometry Methods*, Wiley, New York, 1975.
5. WEBB ASSOCIATES, Eds., *Anthropometric Source Book–Volume II: A Handbook of Anthropometric Data*, NASA Reference Publication 1024, Scientific and Technical Information Office, Clearlake, TX, 1978.
6. C. T. MORGAN, J. S. COOK, III, A. CHAPANIS, and M. W. LUND, Eds., *Human Engineering Guide to Equipment Design*, McGraw-Hill, New York, 1963.
7. J. B. MARTIN and D. B. CHAFFIN, "Biomechanical Computerized Simulation of Human Strength in Sagittal-Plane Activities," *AIIE Transactions*, Vol. 4, No. 1 (March 1972), pp. 19–28.
8. L. L. LAUBACH, "Human Muscular Strength," in Webb Associates, Eds., *Anthropometric Source Book–Volume I: Anthropometry for Designers*, NASA Reference Publication 1024, Scientific and Technical Information Office, Clearlake, TX, 1978.
9. D. B. CHAFFIN, G. D. HERRIN, and W. M. KEYSERLING, "Preemployment Strength Testing: An Updated Position," *Journal of Occupational Medicine*, Vol. 20, No. 6 (June 1978), pp. 403–408.
10. WEBB ASSOCIATES, Eds., *NASA Anthropometric Source Book*, 3 Vols., NASA Reference Publication 1024, Scientific and Technical Information Office, Clearlake, TX, 1978.
11. U.S. DEPARTMENT OF HEALTH, EDUCATION, AND WELFARE (DHEW), "Weight and Height of Adults 18–74 Years of Age–U.S., 1971–74," *Vital and Health Statistics* Series 11, No. 211, National Center for Health Statistics, Hyattsville, MD, 1979.
12. N. DIFFRIENT, A. R. TILLEY, and J. C. BARDAGJY, *Humanscale*, H. Dreyfuss Associates, New York, 1978.

13. T. W. FAULKNER and R. A. DAY, "The Maximum Functional Reach for the Female Operator," *AIIE Transactions*, Vol. 2, No. 2 (June 1970), pp. 126–131.

14. S. M. EVANS, *Updated Users Guide for the COMBIMAN Programs*, Technical Report AMRL-TR-78-31, University of Dayton Research Institute, OH, 1978.

15. M. C. BONNEY and K. CASE, "SAMMIE Computer Aided Work Place and Work Task Design System," *CAD/CAM*, February/March 1978, pp. 3, 4.

16. K. E. KILPATRICK, "A Biokinematic Model for Workplace Design," *Human Factors*, Vol. 14, No. 3 (1972), pp. 237–247.

17. D. B. CHAFFIN and K. S. PARK, "A Longitudinal Study of Low-Back Pain as Associated with Occupational Weight Lifting Factors," *American Industrial Hygiene Association Journal*, December 1973, pp. 513–525.

18. E. R. TICHAUER, *The Biomechanical Basis of Ergonomics: Anatomy Applied to the Design of Work Situations*, Wiley, New York, 1978.

19. D. B. CHAFFIN, "Low Back Stresses During Load Lifting," in D. Ghista, Ed., *Human Body Dynamics*, Oxford University Press, 1981.

20. S. H. SNOOK, "The Design of Manual Handling Tasks," *Ergonomics*, Vol. 21, No. 12 (1978), pp. 963–985.

CHAPTER 6.4

Physiological Basis for the Design of Work and Rest

ELIEZER E. KAMON

The Pennsylvania State University

6.4.1 INTRODUCTION

This chapter is concerned with physical work. Physical work is performed by the activation of the musculoskeletal system. The muscular contractions during work require the support of the respiratory and circulatory systems in order to transport oxygen to, and to carry metabolic by-products from, the muscles. Consequently, the responses of these supporting systems are closely correlated to the intensity of the work. In engineering terms the work can be considered the stress, and the physiological responses the ensuing strain. Therefore the physiological responses can be used to estimate performance and to design work. However, the nature of the physiological strain depends on the type of the muscular contraction involved. There are two types of muscular contractions:

1. Dynamic, involving rhythmical contractions of large muscle groups where the length of the muscles is changing (isotonic).
2. Static, involving prolonged contraction without a change in the length of the muscles (isometric).

The physiological responses to each type of muscular contraction are different, and therefore the criteria for the design of work and rest differ for each case.

6.4.2 DYNAMIC WORK

Dynamic muscular work is defined as rhythmical contractions of large muscle groups. When external work is performed, it is manifested by transformation of chemical energy in the muscles, which requires the oxidation of two primary food elements: carbohydrates and fats. The combustion of these elements yields about 5 kcal for each liter of O_2 consumed. On an hourly basis this is equivalent to 5.68 W.

The Supporting Systems

Since dynamic muscular work is sustained by oxidation, it depends on the capacity of the respiratory and circulatory systems to transfer air O_2 to the muscles. During work the activity of the respiratory and circulatory systems increases in proportion to the intensity of the muscular work. The relaxation periods between contractions allow adequate perfusion of the muscle with blood from which O_2 is extracted and into which CO_2 and other by-products are discarded. The gas exchange between muscles and blood is one of the major avenues of feedback to the central mechanism, or central nervous system (CNS), which coordinates the functions of the systems and keeps the process of muscle contractions operational.

The interaction among the contracting muscles, the supporting systems, and the feedback mechanisms involved is shown in Exhibit 6.4.1. It can be seen that, while CO_2 production is the basic feedback for the control of ventilation, the O_2 requirement is the major contributor to the cardiovascular responses. Whereas the first could be proven experimentally, the latter is more of an assumption, since the control mechanism of the cardiovascular system during muscular work is poorly understood. The neuronal feedback shown in the exhibit induces respiratory and circulatory responses by way of central sympathetic excitatory drive directly from the brain. This neuronal feedback is seen in the immediate transitory response to the onset of muscular action. The chemical feedback due to the muscle-blood exchanges is slower, but sustains the steady state response.

Transient Responses

During a change from rest to work, or from one level of work to another, the slow chemical feedback makes the full adjustments of the supporting system gradual (Exhibit 6.4.2). The O_2 uptake and the heart rate reach a steady state 2 to 3 min after the onset of the new work level. During the steady state the O_2 demand of the muscles is met. The transient period accrues O_2 deficit, and the muscles' energy is provided by anaerobic chemical processes that do not require immediate O_2 supply. The deficit, or O_2 debt, is repaid during the recovery rest period.

Steady State Responses

Oxygen Uptake

The steady state minute oxygen uptake ($\dot{V}_{O_2}$) is linearly related to the external work performed. This is shown in Exhibit 6.4.3 for activities involving external work of cycling against measurable resistance or climbing against gravity.

Efficiency of Dynamic Work

The efficiency of dynamic work is about 20% (Exhibit 6.4.1), unless excessive static work is involved. For example, shoveling is only 6% efficient, because of the static contraction needed for trunk stabilization. Efficiency of work (*Ef*) is expressed as the ratio

$$Ef = \frac{\text{external work}}{\text{net energy expenditure}} \tag{1}$$

The net energy can be derived in two ways: (1) as the difference ($\dot{V}_{O_2}w - \dot{V}_{O_2}r$), where $\dot{V}_{O_2}w$ and $\dot{V}_{O_2}r$ are the O_2 uptake in 1 min, respectively, for the work and the resting (sitting); and (2) as the difference ($\dot{V}_{O_2}w - \dot{V}_{O_2}z$), where $\dot{V}_{O_2}z$ represents the $\dot{V}_{O_2}$ for zero work (the *y* intercept of the regression of $\dot{V}_{O_2}$ on the external work in Exhibit 6.4.3). The $\dot{V}_{O_2}z$ is actually the value for the maintenance of the posture during the activity involved (such as standing still on a moving ladder or moving stairs for the climbing and pedaling against no resistance in cycling). The external work and the O_2 uptake ($\dot{V}_{O_2}$) are converted to energy units: 2.34 g cal/kpm (kilopondmeter) of external work and 5 g cal/ml O_2 consumed.

Exhibit 6.4.1 The Neural and Biochemical Feedbacks in the Interaction Between the Muscles and the Respirocirculatory Systems[a]

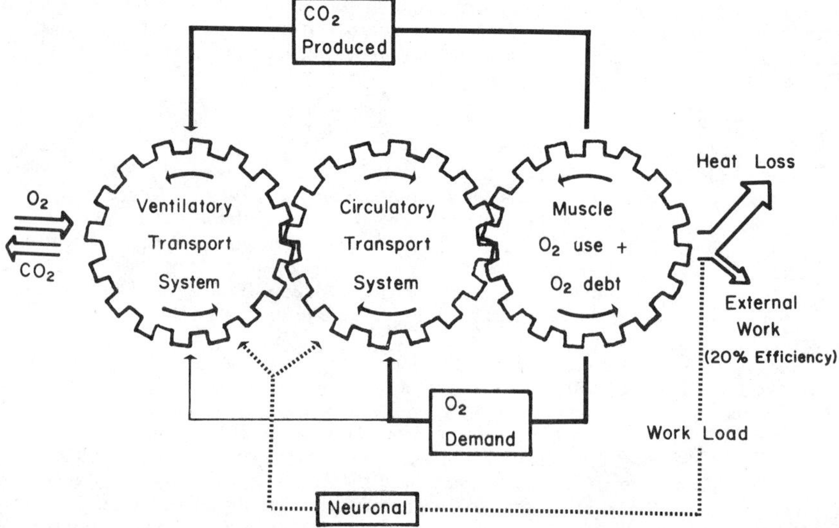

[a]Adapted from reference 1.

Exhibit 6.4.2 The Transient and Steady State Periods During Exercise and Recovery

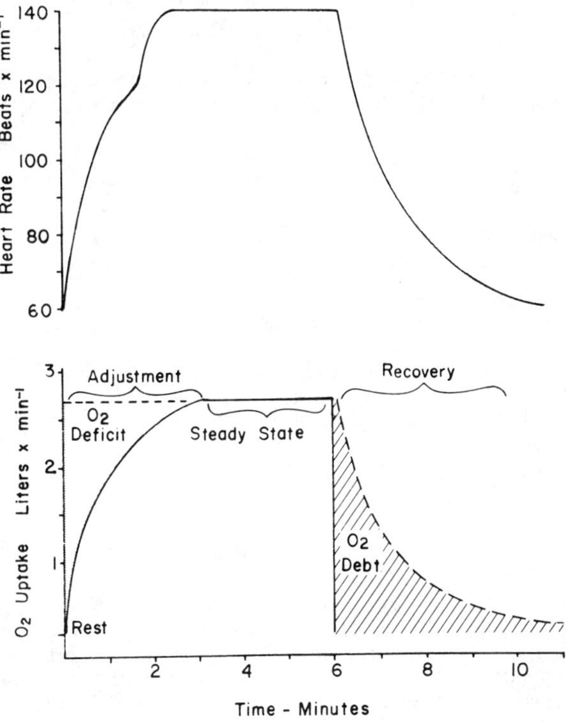

Exhibit 6.4.3 The Relationship Between Dynamic External Work and Oxygen Uptake

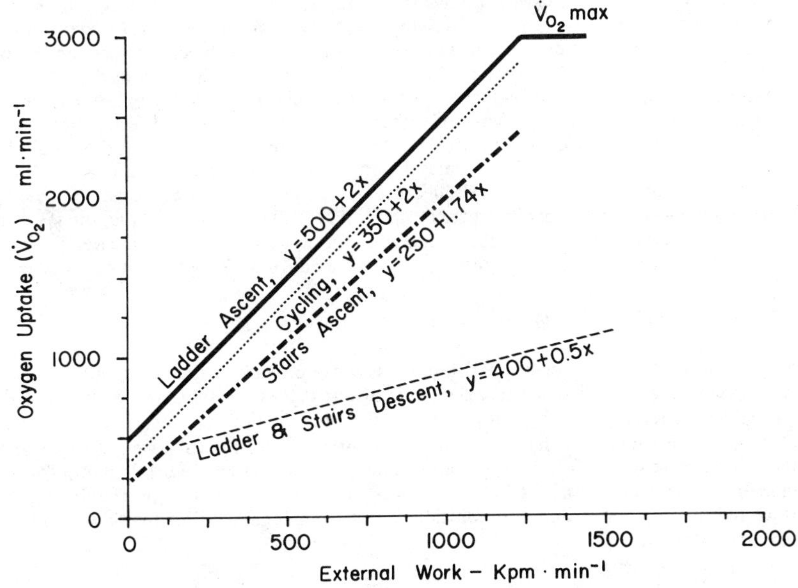

When $\dot{V}_{O_2}r$ is used, a mechanical efficiency (*Efm*) is derived as

$$Efm = \frac{\text{kp} \times \text{m/min} \times 2.34}{(\dot{V}_{O_2}w - \dot{V}_{O_2}r)\ \text{ml/min} \times 5} \qquad (2)$$

When $\dot{V}_{O_2}z$ is used, the denominator $(\dot{V}_{O_2}w - \dot{V}_{O_2}z)$ is the slope b of the $\dot{V}_{O_2}$ regression on the external work. Since $b = 1/\text{kpm}$, a physiological efficiency (*Efp*) is derived as

$$Efp = \frac{1 \times 2.34}{b \times 0.5} = \frac{0.47}{b} \qquad (3)$$

Since $\dot{V}_{O_2}z$ is usually larger than $\dot{V}_{O_2}r$, *Efp* is also larger than *Efm*, as shown in the following examples: Climbing at 12 m/min for a 70 kg person, the mechanical efficiency (using $\dot{V}_{O_2}r = 350$ ml/min) is

$$Efm = \frac{12 \times 70 \times 2.34}{(1680 - 350) \times 5} = 0.20$$

The physiological efficiency, using the slope of the regression (Exhibit 6.4.3), is

$$Efp = \frac{0.47}{2} = 0.24$$

Negative Work

The $\dot{V}_{O_2}$ regression for descent (ladder or stairs) is only 0.5 ml/kpm (Exhibit 6.4.3). Thus the physiological efficiency is $0.47/0.5 = 0.96$. Whereas descending involves resistance to gravity, ascending is work against gravity. While descending, the muscles are stretched during contraction (eccentric contraction); while ending, they shorten (concentric contraction). Dynamic work involving eccentric contraction sometimes is called "negative work." Negative work is about 100% efficient as compared to the average 20% efficiency of positive work (concentric contractions).

Efficiency of Different Tasks

Some tasks involve a mixture of concentric and eccentric muscle contractions: Their efficiency is higher than that for positive work. Walking and running involve mixed concentric and eccentric contraction for each step; the efficiency could be as high as 30%.

Efficiency can decrease because of excessive static muscle contractions and excessive trunk movements. Since static contraction requires O_2, yet no external work is involved, efficiency decreases. The additional external work due to excessive trunk motion can be taken into account by measuring the range of motion of the center of gravity of the trunk. With such considerations the efficiency will be close to the 20% expected for positive dynamic work. Examples are repeated shoveling and lifting. If only the external work of moving the load is considered, efficiency for both is about 6%. If the external work for moving the trunk in shoveling or moving the whole body in lifting is accounted for, the derived efficiency is more likely to be 15 to 20%. This is still less than the 20 to 25% efficiency for only leg work, because of static muscle contractions.

Walking on the level actually involves no external work. It is customary to predict the $\dot{V}_{O_2}$ from the speed of walking. However, uphill walking includes the external work for the vertical elevation of the body. The relationship between the $\dot{V}_{O_2}$ and the work of walking was found to be

$$\dot{V}_{O_2} = 7 + 0.001\,S^2 + 0.012\,S \times G \qquad (4)$$

where $\dot{V}_{O_2}$ is in ml/kg body weight per minute, S is speed of walking in m/min, and G is the uphill slope in percentage of the distance traveled. This means that for walking at a given speed the relationship between the $\dot{V}_{O_2}$ and the external work (elevating the body) is linear. Carrying weights of up to 30 kg on the trunk in such a way that the line of gravity of the total weight (body and extra weight) falls within the base of the feet requires no additional energy per kilogram. The $\dot{V}_{O_2}$ per kilogram total weight can be derived from equation 4. However, carrying on the arms in front of the trunk reduces efficiency and the additional cost per kilogram of load is 50% more than per kilogram body weight.[2]

Estimation of the Energy Cost of Dynamic Work

The metabolic cost of a task can be measured directly using the method of collecting expired air during work and deriving the $\dot{V}_{O_2}$ from the volume of the sample and the difference between the O_2 concentration in the sample and the 20.9% in the inspired ambient air. However, estimation, rather than direct measurement, is possible. The metabolic cost of a job can be derived either from equation 4 and Exhibit 6.4.3 or from the measured external work and the expected efficiency of the task. For convenience, some values of energy cost are given in Exhibit 6.4.4 for common industrial tasks.

Since energy expenditure depends on the capacity of the supporting system to meet the O_2 demand, measurements of pulmonary ventilation and the heart beats can serve as predictors for $\dot{V}_{O_2}$, but more important, they can serve as criteria for strain.

Ventilation

Reasonably good correlations were found between the volumes of expired air ($\dot{V}_E$) and of O_2 uptake ($\dot{V}_{O_2}$). An example is shown in Exhibit 6.4.5 for hundreds of measurements taken during walking, crawling, shoveling, and load carrying. Although the regression of $\dot{V}_{O_2}$ on $\dot{V}_E$ is curvilinear, linearity could be applied to the lower range of $\dot{V}_{O_2}$. At the higher range, as the maximal $\dot{V}_{O_2}$ is

Exhibit 6.4.4 Energy Expenditure (M)[a] for Some Industrial Tasks

Task	M (kcal/min)	(W)
Packaging		
Bag and pack paper rolls	2.5	(174)
Small item packing	3.5	(244)
Janitorial		
Using vacuum cleaner	3.3	(230)
Clean bathroom (walls, commode)	3.6	(251)
Dust mop and wet mop floors	4.9	(341)
Cafeteria: mop floors and tables	4.5	(313)
Window washing (inside and outside)	5.0	(348)
Wet mop lobby room and stairs (1 kg mop)	6.5	(452)
Steelmill		
Forging	6.5	(452)
Tending furnace	8.0	(557)
Hand rolling	9.0	(626)
Wire bundling	10.0	(696)
Slag removal	11.5	(800)
Aluminum Smelting		
Sweep bath	6.4	(445)
Sequence of anode change	6.5	(452)
Break crust and mock pot	7.0	(487)
Shovel	9.0	(626)
Skim bath off top of aluminum	10.5	(731)
Break bath off anode	15.0	(1044)
Coal Mining		
Brattice (plastic) hanging	7.0	(487)
Cut or set timber	7.4	(515)
Top off brattice	7.5	(522)
Shovel coal	8.0	(557)
Roofer: bolt trussing and tighten turnbolt	10.8	(752)
Hammer wedges	11.8	(821)

[a]Values were measured by the author except for some of the steelmill tasks which were given by J. V. G. A. DURNIN, and R. PASSMORE, *Energy, Work and Leisure*, Heinemann Books, London, 1967, p. 74. W = kcal/min/60/1.16.

Exhibit 6.4.5 The Relationship Between Oxygen Uptake ($\dot{V}_{O_2}$) and Pulmonary Ventilation (Volume of Expired Air $\dot{V}_E$) for Various Activities[a]

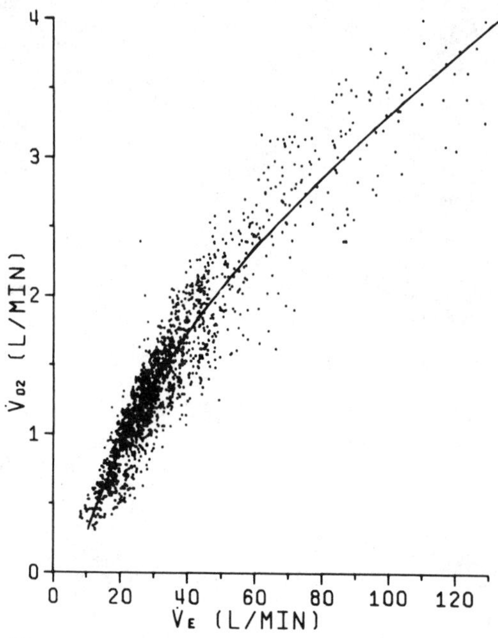

[a]From reference 3.

approached, $\dot{V}_E$ increases disproportionately because of excessive CO_2 production and the $\dot{V}_E$ dependency on CO_2 (Exhibit 6.4.1). Although ventilation is increasing rapidly with the increase in $\dot{V}_{O_2}$, the ability of the lungs to deliver the needed O_2 is not the limit to work.

Circulation

The circulatory system limits the O_2 delivery, probably because of the limited functional capacity of the heart. The factors that determine the pumping rate of the heart are the stroke volume (*SV*) at each beat and the rate of beating (*HR*). Thus the cardiac output (*CO*) is

$$CO = SV \times HR$$

where *CO* is in l/min, *SV* is in 1/min, and *HR* is in beats/min. The change in each factor depends on the work level. At work intensities of up to $\dot{V}_{O_2}$ of 1 l/min, the *CO* is first changed by increase in *SV*. Above $\dot{V}_{O_2}$ of about 1 l/min, *SV* reaches its maximal levels and the *CO* becomes *HR* dependent.[4] The relationship between $\dot{V}_{O_2}$ and *HR* is linear. However, there are substantial individual differences in the regression of *HR* on $\dot{V}_{O_2}$, mainly because of differences in physical fitness. Fit persons have lower *HR* at rest and a less steep slope of *HR* on $\dot{V}_{O_2}$ than nonfit persons.

Maximal Aerobic Capacity—Limits of Dynamic Work

The increase in $\dot{V}_{O_2}$ is limited to a maximal level beyond which the increase in work load does not yield any increase in $\dot{V}_{O_2}$. The expected maximal $\dot{V}_{O_2}$ level ($\dot{V}_{O_2,max}$) of an average male is shown in Exhibit 6.4.3 for ladder ascent.

Maximal aerobic capacity or maximal O_2 uptake ($\dot{V}_{O_2,max}$) is measured during the last minute of the highest work load that can be sustained for 2 to 3 min. Short periods of work loads above $\dot{V}_{O_2,max}$ can be sustained because the muscles resort to anaerobic processes of energy production.

Maximal Heart Rate

The increase in *HR* also is limited. The maximal *HR* (HR_{max}) is age dependent. Although there are individual differences in HR_{max}, an average value can be predicted from the formula HR = 220 –

age, ($s = 7$ beats/min) where age is in years. The HR_{max} usually accompanies the $\dot{V}_{O_2,max}$, which indicates that the limits of work represent the individual functional capacity of the heart.

Expected $\dot{V}_{O_2,max}$

Individual differences in HR_{max} on the one hand and the differences in the regression of $\dot{V}_{O_2}$ on HR on the other are reflected in individual differences in $\dot{V}_{O_2,max}$. These differences depend on age, sex, and fitness level. Age and sex play an important role in the design of work for a potential population of workers. Values of expected average $\dot{V}_{O_2,max}$ are given in Exhibit 6.4.6. However, fitness is classified according to $\dot{V}_{O_2,max}$. For example, compared to the mean value of 2.88 l/min given in the exhibit for men in the 30–39 age group, poor fitness will be $\dot{V}_{O_2,max}$ below 2.3 l/min, and high fitness $\dot{V}_{O_2,max}$ above 3.7 l/min. The values in Exhibit 6.4.6 are those expected for mixed arm and leg work typical to industrial tasks. Compared to the values in this exhibit, the $\dot{V}_{O_2,max}$ is 20% higher for leg work (walking and cycling), and it is 20% lower for arm work (cranking). In text $\dot{V}_{O_2,max}$ will be given for persons of average weight.

Relative $\dot{V}_{O_2}$ - $f\,\dot{V}_{O_2,max}$

The $\dot{V}_{O_2}$ - HR relationship can be normalized by using relative, rather than absolute, $\dot{V}_{O_2}$ values. This means using submaximal $\dot{V}_{O_2}$ as a fraction of $\dot{V}_{O_2,max}$ ($f\,\dot{V}_{O_2,max}$ or % $\dot{V}_{O_2,max}$). Take, for example, wire bundling (Exhibit 6.4.4), which requires $\dot{V}_{O_2}$ of 2 l/min (10 kcal/min). Compared to the $\dot{V}_{O_2,max}$ values in Exhibit 6.4.6 for the age group 20–29, $\dot{V}_{O_2}$ of 2 l/min is 0.63 $\dot{V}_{O_2,max}$ for men (2/3.16) and 0.93 $\dot{V}_{O_2,max}$ for women (2/2.14). Notice that the closer the demand of the task is to $\dot{V}_{O_2,max}$, the less is the reserve capacity of the heart (limited by HR_{max}). This consideration plays an important role when ambient stressing conditions, such as heat, are involved (see Section 6.4.4).

The expected linear regressions of $f\,\dot{V}_{O_2,max}$ on HR are shown in Exhibit 6.4.7. The age-divided lines start at 0.3 $\dot{V}_{O_2,max}$ and HR of 100 to 110 beats/min and end at HR_{max}, which drops with age. At 0.3 $\dot{V}_{O_2,max}$ the age reduction in HR is negligible. Notice the higher $f\,\dot{V}_{O_2,max}$ (% $\dot{V}_{O_2,max}$) for a given HR shown for older, as compared to younger, men and for women as compared to men. Actually, at 0.3 $\dot{V}_{O_2,max}$, women's HR is expected to be at about 110 beats/min, compared to men's 100 beats/min. Only the line for women of 20–29 years is shown in the exhibit. Comparing this line to that for the men, it seems that the slope of the regression for women is about the same as that for men who are one decade older.

Today there are devices for measuring HR without interfering with the worker's activities. The HR can be used to estimate the strain in terms of $f\,\dot{V}_{O_2,max}$ (Exhibit 6.4.7). Then the estimated $f\,\dot{V}_{O_2,max}$ can be used for the design of work and rest.

6.4.3 ENDURANCE AND RECOVERY

Endurance

Endurance is inversely related to work intensity. At maximal intensity an average fit person's endurance is limited to about 3 min. At submaximal intensities the endurance extends to 30 min at 0.8 $\dot{V}_{O_2,max}$, to 2 hr at 0.5 $\dot{V}_{O_2,max}$, and to 8 hr at 0.33 $\dot{V}_{O_2,max}$. For example, while skimming

Exhibit 6.4.6 Mean and SD of Maximal Aerobic Capacity ($\dot{V}_{O_2,max}$) During Mixed Arm and Leg Work, for Average Women and Men[a]

	$\dot{V}_{O_2,max}$ (l/min)	
Age (Years)	Women	Men
20–29	2.14 ± 0.25	3.16 ± 0.30
30–39	2.00 ± 0.23	2.88 ± 0.28
40–49	1.85 ± 0.25	2.60 ± 0.25
50–59	1.65 ± 0.15	2.32 ± 0.27

[a] Adapted from references 5 and 6. Average body weight is 70 kg for men and 58 kg for women.

Exhibit 6.4.7 The Regressions of Relative Oxygen Uptake (% $\dot{V}_{O_2,max}$) on Heart Rate for Men (——) and Women (- - -), (•) Maximal Values. The Slopes for Women Follow Those of One-Decade-Older men.

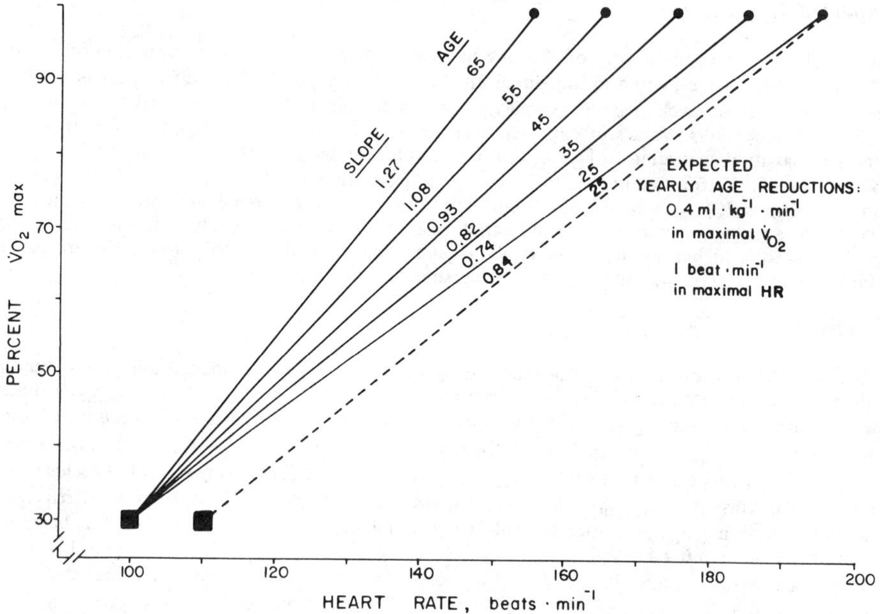

bath in aluminum smelting (10.5 kcal/min, Exhibit 6.4.4) demands from a woman effort close to or at her maximal capacity (Exhibit 6.4.6), it demands from an average man 0.63 $\dot{V}_{O_2,max}$ only. Whereas the woman will be limited to few minutes of work, the man can endure it for at least 30 min. Fitness improves endurance to the extent that highly fit persons can endure work at 0.8 $\dot{V}_{O_2,max}$ and 0.5 $\dot{V}_{O_2,max}$ for 2 and 8 hr, respectively.

Sound occupational practice calls for safe and acceptable working conditions. Working until fatigued provides neither. Therefore it is not desirable to assign working periods that require the worker to continue until exhausted. Few attempts have been made to find an optimal intermittent work schedule to prevent fatigue. In general it was found that, for high-intensity work loads, reducing the working period to less than half the time at which the worker became exhausted increased work time and reduced strain as indicated by the leveling off in the *HR* [7,8]. If a period of one third of the time to exhaustion is used for any given work intensity between 0.5 $\dot{V}_{O_2,max}$ to $\dot{V}_{O_2,max}$, a reasonable formula is

$$Tw = \frac{40}{f\,\dot{V}_{O_2,max}} - 39 \qquad (5)$$

where *Tw* is working time in minutes (Exhibit 6.4.8) and $f\,\dot{V}_{O_2,max}$ can be based on the $\dot{V}_{O_2,max}$ values in Exhibit 6.4.6, unless directly measured. For example, suppose that a 45-year-old man and a 35-year-old woman are candidates for a forging job demanding a $\dot{V}_{O_2}$ of 1.3 l/min. Their expected $\dot{V}_{O_2,max}$ are 2.6 l/min and 2 l/min, respectively (Exhibit 6.4.6). Therefore their respective relative work demand will be 0.50 $\dot{V}_{O_2,max}$ and 0.65 $\dot{V}_{O_2,max}$. Using equation 5, the acceptable safe working period will be 41 min and 22.5 min, respectively. The engineer responsible for the pace of the work should take this substantial difference into account.

Another example can be given for truss bolting in coal mines. Tightening the inserted rods to the roof requires 2.16 l/min (Exhibit 6.4.4). For a man and a woman in their mid-twenties, this task demands 0.68 $\dot{V}_{O_2,max}$ (2.16/3.16) and 1 $\dot{V}_{O_2,max}$ (2.16/2.14), respectively. The expected periods of acceptable and safe performance are, respectively, 19.8 min and 1 minute (equation 5). However, two points should be made: (1) The woman requires longer rest periods between each bolting, and thus the design of the pace of repetitive bolting becomes important for her; and (2) frequent repetition, to the extent that the average woman could find the task beyond her capacity, does not imply that all the potential women applicants could not perform the task. Suppose the $\dot{V}_{O_2,max}$ for the woman applicant is 2.41 l/min, which is 1 SD above the mean. The truss bolting task then will require an effort of 0.90 $\dot{V}_{O_2,max}$, which the woman can endure for

Exhibit 6.4.8 Working and Resting Time as a Function of Relative Dynamic Work ($s\ \dot{V}_{O_2}$, max).

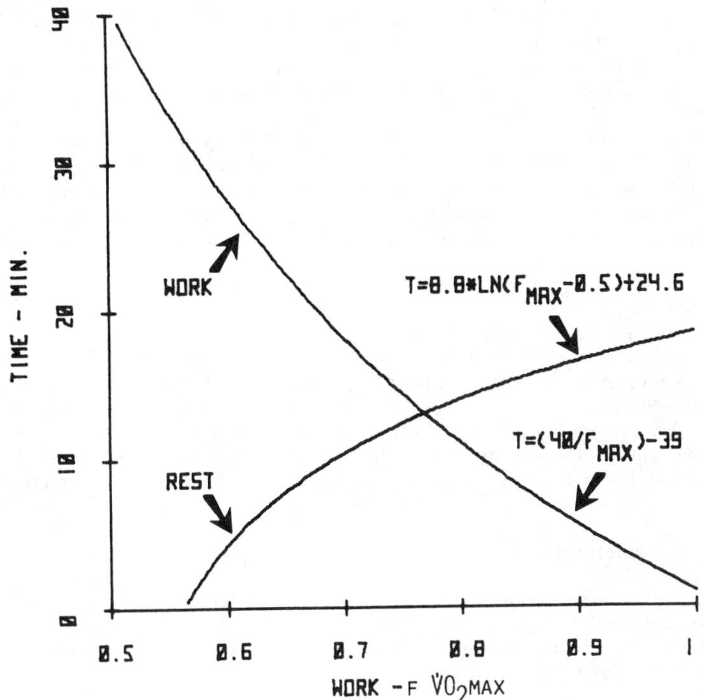

5.4 min, and she will require less recovery time. If the design of the job will permit a schedule of 5 min of bolting followed by adequate rest, then the 34% of the women (with $\dot{V}_{O_2,max}$ 1 SD or more above the mean) are expected to be able to perform the task. The next step is to define the required rest on the basis of the expected physiological recovery.

Recovery

Recovery is proportional to the intensity and the duration of the dynamic work. It is acceptable among exercise physiologists that an all-out effort at $\dot{V}_{O_2,max}$, which may last between 3 and 5 min, can be performed once or at the most twice a day. This means that recovery to the extent that the maximal effort can be repeated requires 4 or, for some, 8 hr of recovery. At submaximal work load of 0.40 $\dot{V}_{O_2,max}$, no special resting periods are required, provided that the conventional meal and two coffee breaks are available to the workers. Workers who are accustomed to their physical work most likely will be able to maintain a pace of 0.50 $\dot{V}_{O_2,max}$ during the shift. Observations of workers who chose their own pace of physical work seem to indicate that they indeed performed at 0.35 to 0.50 $\dot{V}_{O_2,max}$.

It should be mentioned that in the past the German school of occupational physiologists[9] suggested the following general formula for rest allowance:

$$RA = \left(\frac{M}{4.2} - 1\right) \times 100 \qquad (6)$$

where RA is rest allowance as percentage of the working time, M is the metabolic cost of the work in kcal/min, and the constant, 4.2, represents the basic cost of work that does not require rest allowance. The 4.2 kcal/min was derived from calculated energy requirements of workers whose nutritional needs were followed for a prolonged time during World Wars I and II. Since this value is close to the 0.30 $\dot{V}_{O_2,max}$ of an average male, the use of this formula for a rough estimate of rest allowance is acceptable for young men. However, our current knowledge of sex and age differences in working capacity provides for better design of resting periods. One possible adjustment of equation 6 is to substitute $f\ \dot{V}_{O_2,max}$ for M and 0.30 $\dot{V}_{O_2,max}$ for the 4.2. A more sophisticated approach is suggested next.

Work loads demanding more than the 0.50 $\dot{V}_{O_2,max}$ call for rest allowances. Physiologically, as the intensity increases above 0.5 $\dot{V}_{O_2,max}$, there is an increase in the anaerobic energy production, which results in an exponential increase in the lactic acid production. Acidity is considered an important factor in the impairment of tissue function, including that of the muscle. Rest allows elimination of lactic acid and restoration of normal muscle function. The use of lactic acid production as a criterion for rest can be based on the following factors: (1) the expected lactic acid production during the working period and (2) the rate of the lactic acid appearance in, and then disappearance from, the blood during the recovery period.

For example, the lactic acid produced during work at 0.70 $\dot{V}_{O_2,max}$ appears and then is eliminated from the blood during 10 min of recovery; that produced at 0.90 $\dot{V}_{O_2,max}$ appears and is eliminated in 15 min. The rate of lactic acid elimination is constant at about 3 mg %/min.[10] Based on the expected amount of lactic acid production in the muscles and on the rate of its elimination, the time required to reduce the blood lactic acid to about twice the level found during rest is

$$Tr = 8.8 \text{ Ln } (f\,\dot{V}_{O_2,max} - 0.5) + 24.6 \tag{7}$$

where Tr is resting period in minutes (Exhibit 6.4.8).

Light muscular activity at about or below 0.30 $\dot{V}_{O_2,max}$ was found to double the 3 mg %/min rate of elimination of lactic acid.[11] Although this means that, following a very demanding task, recovery with lighter work is more beneficial than full rest, it is not recommended. In addition to lactic acid, other factors (mostly neurological) play a role in fatigue. Therefore a well-designed rest schedule could be most effective with a combination of full rest and light work. For example, if a person works 1 min at his $\dot{V}_{O_2,max}$, the rest required is 18.5 min. However, after 8.5 min of rest, full recovery is expected by assigning light work (0.25 $\dot{V}_{O_2,max}$) for only 5 min.

Population Considerations

Since the energy demand of most tasks depends on the nature of the physical work, unless the pace can be individually designed, differences in performance must be expected. Therefore it is desirable to predict performance at least on the basis of the population's expected physical fitness. Such a prediction calls for the following steps:

1. Measure or estimate the $\dot{V}_{O_2}$ for the task or job.
2. Measure the work duration (Tw) for the task.
3. Derive the $f\,\dot{V}_{O_2,max}$ for Tw by rearranging equation 5; $f\,\dot{V}_{O_2,max} = 40/(Tw + 39)$.
4. Define the minimal $\dot{V}_{O_2,max}$ required to perform the task safely; $\dot{V}_{O_2,max}$ required = $\dot{V}_{O_2}$ measured/$f\,\dot{V}_{O_2,max}$.
5. Compare the required $\dot{V}_{O_2,max}$ to the appropriate mean and SD values in Exhibit 6.4.6 and derive the percentile of workers who could safely perform the task. The following example is given from observations made in a low seam coal mine.

Setting timber is a task of timbermen that is more demanding than usual timbering because of the stooping posture involved. The directly measured $\dot{V}_{O_2}$ was 1.48 l/min. The duration for one unit of continuous setting of 15 timbers was 6.08 min. Based on equation 5, this task can be performed safely if it demands 0.89 $\dot{V}_{O_2,max}$ or less, which also means that the performer has to have $\dot{V}_{O_2,max} > 1.66$ l/min. Since this is typical arm work, the $\dot{V}_{O_2,max}$ in Exhibit 6.4.6 should be reduced by a factor of 0.8. It is observed that, with this correction and the SD shown in Exhibit 6.4.6, all men in the age range of 20–40 could perform the task. The women's $\dot{V}_{O_2,max}$ is 1.71 ± 0.25 and 1.6 ± 0.23 for ages 20–29 and 30–39, respectively. Therefore the percentiles of women who could perform the task for the respective ages are 58 and 40. However, women in these percentiles could maintain their pace of the timbering only if there were adequate rest between the setting of each unit of timber.

A time study has shown that, on the average, eight units of timber are set in 2.1 hr, or one unit is set in 15.75 min. Work time of 6.08 min/unit allows 9.67 min of rest per unit. To perform at 1.48 l/min for 6.08 min, the job should demand 0.89 $\dot{V}_{O_2,max}$. However, at this level of strain the resting period should be 16.3 min, rather than the current 9.67 min. The options are either to reduce the work intensity to below 0.68 $\dot{V}_{O_2,max}$ or to decrease the work-rest ratio for the women. It is reasonable to reduce the setting to less than 15 timbers/unit and to prolong the time for the total timber setting, unless there is room for mechanization.

In summary, the options to the engineer are either to reduce the work load to match it to the potential employee or to select the appropriate percentile for the work load, but adjust the pace according to the required rest allowance.

6.4.4 HOT ENVIRONMENTS

Heat Stress

Heat stress results from excessive air temperature (*Ta*), humidity, and, in many industrial situations, radiant source temperatures (*Tr*). The humidity, or vapor pressure (*Pa*), plays a role in cooling by sweat evaporation. If the air and mean radiant temperatures are below the body's skin temperature, heat can be lost to the environment; otherwise, external heat is added to the body. These sources of heat stress and the heat balance under comfortable ambient conditions are discussed in Chapter 6.12.

Physiological Strain

The physiological strain due to the stress of hot ambient conditions is judged mainly by the apparent increments in *HR* and by rise in body temperature. The first responses to uncomfortable ambient conditions are cardiovascular, because of increased skin blood flow to facilitate heat dissipation. For work efficiency of 20% or less, the main heat load during muscular exercise is metabolic (*M*). Depending on the ambient temperature, radiative and convective $(R + C)$ heat either can be added or some of the metabolic heat can be lost through these avenues. The sum $M \pm (R + C)$ is the net heat load that is transported to the skin for dissipation and that is reflected by increase in *HR*. If $(R + C)$ is either inadequate or a source of heat load, cooling by sweat evaporation is required.

When the avenues of heat exchange and the cardiovascular responses are adequate, the core body temperature (*Tc*) can be maintained at the work-specific level. Core body temperature can be measured in the rectum, the ear canal, or orally. Normal oral temperature is 0.2 to 0.3°C lower than the other two. The *Tc* of about 37°C during rest will rise to a new equilibrium level in proportion to the intensity of the work. The expected rise is 2.5°C/1 $f\,\dot{V}_{O_2,max}$ for work in *Ta*, from 10°C to 50°C. As was shown previously, the *HR* also is expected to equilibrate at levels proportional to $f\,\dot{V}_{O_2,max}$, but for a narrower range of *Ta*. A summary of the expected *HR* and *Tc* for given $f\,\dot{V}_{O_2,max}$ is shown in Exhibit 6.4.9.

The Prescriptive Zone

The increase in *HR* and skin blood flow is expected when the *Ta* rises above 25°C. If the heat dissipation at the skin surface is adequate, *Tc* is maintained at the work-specific level. Well-trained and heat-acclimated persons do not reveal as large an *HR* increment as novices do because of better ability to redistribute blood from the internal organs (splanchnic area) to the skin. However, in evaluating heat strain, it is recommended that one assume *HR* increments in proportion to the rise in either air temperature or humidity, as is shown in Exhibit 6.4.10.

The exhibit represents expected increments in *HR* for work at 0.3 $\dot{V}_{O_2,max}$, within a zone of 25 to 50°C and vapor pressures (*Pa*) that become smaller as *Ta* increases. The increments in *HR* above the work-specific level are expected to maintain the work-specific *Tc* (Exhibit 6.4.9). Beyond this prescriptive zone, *Tc* either will equilibrate at above the work-specific level or will rise continuously. Notice that, in the zone of *Ta* between 38°C and 44°C, the *HR* increments are faster (2 beats/min × torr), but the work-specific *Tc* is maintained. In general, additional radiant heat will narrow this prescriptive zone.

Designing Work-Rest Periods in Heat

The expected *HR* increments due to heat stress result in a reduction in the cardiovascular reserve $(HR - HR_{max})$. Therefore the heat-induced increments in *HR* were equated with an increase in $f\,\dot{V}_{O_2,max}$ in the design of work-rest schedules.[12] In other words, the increments due to heat

Exhibit 6.4.9 Expected Core Body Temperatures (*Tc*) and Heart Rates (*HR*) at Steady State Work for Each Relative Work Load (% $\dot{V}_{O_2,max}$)

	Percentage $\dot{V}_{O_2,max}$				
	25	33	50	80	100
Tc (°C)	37.5	37.7	38.2	38.9	39.4
HR (beats/min)	90	110	125	160	180

Source. Reference 12.

Exhibit 6.4.10 Psychometric Chart Description for Expected Increase in Heart Rate in the Prescriptive Zone for Core Body Temperature[a]

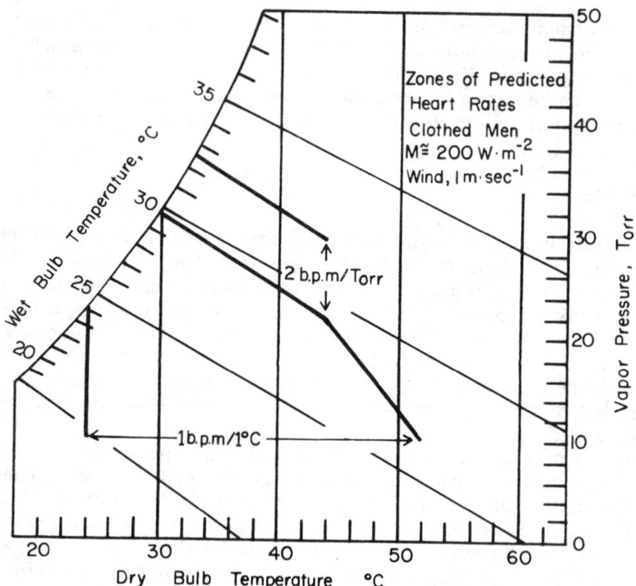

[a]Adapted from reference 13.

were added to the expected *HR* due to work; the total *HR* was used to derive a strain equivalent, $f\ \dot{V}_{O_2,max}$ (Exhibit 6.4.7). The strain equivalent $f\ \dot{V}_{O_2,max}$ served as the basis for the design of work and rest periods. Testing this procedure under laboratory experiments proved efficient in maintaining *Tc* at the expected levels. The following example demonstrates the use of this approach.

A forging job requires $\dot{V}_{O_2}$ of 1.3 l/min (6.5 kcal/min, Exhibit 6.4.4). The workers on the job are men in their twenties. From Exhibit 6.4.6, the job is expected to require 0.41 $\dot{V}_{O_2,max}$ (1.3/3.16). The ambient conditions in the shop are $Ta = 50°C$ and $Pa = 10$ torr. Thus the *HR* is expected to be 25 beats/min above the work-specific *HR* (Exhibit 6.4.10). The slope in Exhibit 6.4.7 is 0.74 $\dot{V}_{O_2,max}$ per 1 beat/min. Thus for 25 beats/min the equivalent increase in $f\ \dot{V}_{O_2,max}$ is 0.19 $\dot{V}_{O_2,max}$. The 0.41 $\dot{V}_{O_2,max}$ work load then can be considered as 0.60 $\dot{V}_{O_2,max}$ (0.41 + 0.19). Using equations 5 and 7, the safe design will be 27 min for the working period and 4 min for the rest period.

The work-rest for an average woman in her twenties should be considered as follows: $\dot{V}_{O_2}$ of 1.3 l/min is 0.61 $\dot{V}_{O_2,max}$ (Exhibit 6.4.6); the 25 beats/min heat-induced increments in *HR* are equivalent to 0.21 $\dot{V}_{O_2,max}$ (Exhibit 6.4.7); the strain can be considered 0.82 $\dot{V}_{O_2,max}$ (61 + 21), for which the respective work and rest are 10 and 35 min.

6.4.5 STATIC WORK

When muscles contract without observable change in their length, no external work is performed, yet energy is spent as long as the contraction is maintained. Such isometric muscular contraction can be found in two common work situations: (1) when forces are resisted either manually or by foot (pedal) or (2) when a posture is maintained with parts of the body (trunk, limbs) pulled by gravity.

A static work situation involves a balanced system of torques around a joint which serves as the fulcrum. Thus the muscular tension involved can be expressed in terms of either the resisted force or the torque; for example, 50 kg hanging on the wrist with the elbow flexed at 90°. The muscular tension for a person whose forearm length is 0.3 m can be defined as either 491 N or 147 Newton-meters (n × m). Expressing the isometric tension in torques is preferable because there is less variability resulting from differences in methodology and in anthropometric characteristics. Details on methods, standardization, and practical use of torques are given in Chapter 6.3.

Maximal Voluntary Contraction

The exertion of force by the muscles around a given joint is limited. The maximal voluntary contraction (MVC) is the highest tension the muscles can develop when rapidly contracting against resistance and maintaining the tension for at least 3 sec. The MVC values are not the same for the different joints and for each joint at different angles. The expected MVC values for angles at which exertion is most likely to be employed in industrial situations are summarized in Exhibit 6.4.11. The MVC drops with age at a rate of about 1%/year for men. No sufficient information is available for women.

The physiological responses to a given submaximal isometric contraction vary widely because of large individual differences in MVC. However, normalization of the responses is possible by the use of the fraction of MVC (f MVC) at which submaximal muscle tension is held.

Physiological Responses

Sustained isometric muscle contraction elicits dramatic increases in $\dot{V}_E$, blood pressure, and HR. Other apparent changes during isometric contraction are consistent electrical output and substantial lactic acid production. These responses are proportional to the tension developed (f MVC) and to some extent to the muscle mass involved.

Respiratory Responses

Isometric contraction causes a slight rise in $\dot{V}_{O_2}$ and a disproportionally marked increase in pulmonary ventilation ($\dot{V}_E$). At 0.40 MVC, $\dot{V}_E$ could increase up to 35 l/min, which is usually observed for dynamic work requiring 7.5 kcal/min (Exhibit 6.4.5).

Circulatory Responses

During sustained isometric muscle contraction there is an increase in the systolic and diastolic blood pressure (BP) and in HR. It has been shown[15] that sustained contraction of 0.4 MVC to the limit of endurance resulted in a continuous increase up to the following: systolic/diastolic of 180/120 mmHg and HR of 120 beats/min. The rise in BP has important clinical implications for workers with a diseased heart or with high resting BP. Such workers should avoid prolonged static work and intense isometric contractions.

Electrical Activity

Muscular contraction is accompanied by a change in the muscular membrane potential. This change spreads throughout the tissues and, if appropriately amplified, can be picked up at the skin surface.[16] There is an increase in frequency and amplitude of the recorded electromyograms (EMGs) which is proportional to the intensity and sustenance of the contractions. Moreover, the

Exhibit 6.4.11 Males (M) and Females (F) Expected Mean Torques (n × m)[a] for MVC of the Muscles Around Joints Flexed at Different Angles

	Angle							
	45°		90°		135°		180°	
Joint Action	M	F	M	F	M	F	M	F
Shoulder flexion	67	29	68	30	47	21		
Elbow flexion	52	24	85	43	60	23		
Back extension			240	130			250[b]	150[b]
Grip							470[c]	269[c]
Knee extension	135	93	196	130	174	136		
Foot plantar flexion	110	83	127	111	101	108		
Maximal lift strength[d]	166	55	293	155				
	553[c]	217[c]	687[c]	364[c]				

[a]Mean values for age below 40 years, the expected coefficient of variance 25%. Data are based on reports from the U.S.A., see reference 14.
[b]Back is actually hyperextended beyond 180°.
[c]Values are in newtons. Grip with wrist at 180°.
[d]Maximal lift strength (MLS) for angle at lower back, knees slightly bent.

nature of the EMG signals depends on the fatigue state of the muscle. Consequently, attempts have been made to quantify the EMG by integration and by power spectrum analysis of the signals and to use the values to determine strain and fatigue.

Lactic Acid

The contracting muscle tissue develops internal hydrostatic pressure which compresses the blood vessels. The throttle effect of the sustained contraction impairs blood flow, or at a given point of tension, blood supply is completely occluded. Such occlusion develops at different f MVC for different muscles. For example, 0.20 MVC, 0.50 MVC, and 0.70 MVC occludes the vessels, respectively, for the foot plantar flexors, elbow flexors, and forearm grip muscles. The restricted blood flow interferes with the gas and by-product exchanges between the muscle and the blood. This results in the reliance of the muscles on anaerobic energy resources, and consequently, the content of the muscle lactic acid is rising. Upon termination of the contraction, blood perfuses the muscles in a reactive increased flow, and the lactic acid is washed out. The impaired blood flow effects the functional capacity of the muscle tissue, and the acidity introduced by the lactic acid impairs normal biochemical processes. Therefore it is not surprising that static work is endured for very short periods and requires prolonged recovery periods.

Endurance

Similar to the approach taken for dynamic work, endurance during static work can be normalized in terms of the f MVC (Exhibit 6.4.13). Attempts to fit one equation to describe the endurance for data collected on different occasions were not too successful. One suggestion for T in minutes was[17]

$$T = \frac{0.19}{(f\ \text{MVC})^{2.42}} \tag{8}$$

The formula seemed to fit the data for sustained contraction at f MVC < 0.75, but not above 0.75 MVC. It seemed that between 0.75 MVC and MVC the data could better fit a linear regression of endurance over effort. Therefore predicting the expected endurance from the formulas shown in Exhibit 6.4.12 seems adequate.

Endurance represents limits of tolerance that are not common in industrial situations. Usually, isometric contractions at given f MVC last for a period t below T $[(t/T) < 1]$.

Recovery

Fatigue due to isometric contraction usually is evident in the inability to maintain the f MVC at the level and duration that existed before fatigue ensued. Two factors determine the recovery rate from static work: (1) the intensity of the contraction (f MVC) and (2) the duration t relative to the maximal tolerance time T (t/T). The closer each of them is to 1, the longer is the required recovery. In one extensive study[18] the rate of drop in the elevated HR due to the contraction was used as the criterion for recovery. The formula for rest allowance suggested was

$$RA = 18\ (t/T)^{1.4}\ (f\ \text{MVC} - 0.15)^{0.5} \tag{9}$$

where RA is in fraction of t, the contraction time, and T is the limit of endurance (equation 8 or Exhibit 6.4.12). At 0.15 MVC, fatigue occurs after prolonged contraction (Exhibit 6.4.12) with little increments in HR.

The following example demonstrates the use of the preceding information in designing pace of work involving isometric contraction. In a plant producing polyethylene fiber, the fibers were made into spools. Each spool was small enough to be grasped by one hand. The motion required placing the hand under the spool and, with a light grasp, transferring the spool from the spinning wheel into a box. The main effort was done by the elbow flexors at an elbow angle of 90°. The job was performed, and was expected to be performed, mainly by women. The following observations and derivations were made:

1. Each spool's weight was 10 kg. The time, t, for placing each spool into the packing box was 0.1 min.
2. Women's average lever arm (elbow to fist) is 0.3 m and their MVC is 43 n × m (Exhibit 6.4.11).
3. The elbow torque in transferring each spool is 29.4 n × m (10 kg × 9.81 × 0.3), which is 0.68 MVC (29.4/43).

Exhibit 6.4.12 Endurance as a Function of the Fraction of Maximal Isometric Voluntary Contraction (*f* MVC)

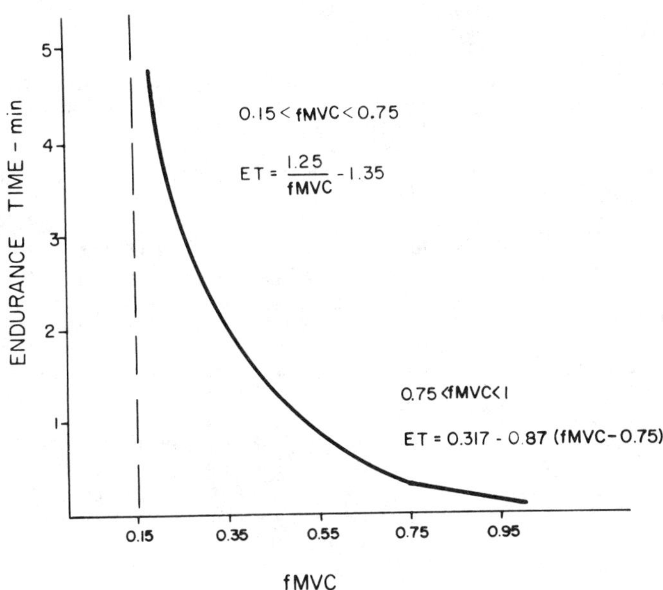

4. The T for 0.68 MVC is 0.48 min (Exhibit 6.4.12 or equation 8), and 6 sec of transfer is 0.21 of T ($t/T = 0.21$).

5. Using equation 9, the RA is 1.46 t, or 8.76 sec.

6. The total transfer time is 14.76 sec/spool (8.76 + 6), or about four spools/min.

6.4.6 COMBINED DYNAMIC AND STATIC WORK

Industrial tasks quite often involve a combination of dynamic and static muscular contractions as well as semidynamic contractions. Some jobs involve slow, rhythmic contractions of small muscle groups, such as in the forearm during packaging and hand precision operations or in the calf during fine pedal control. The rhythmic contractions do not increase the $\dot{V}_{O_2}$ substantially, but local fatigue develops rapidly (see Chapter 6.5). Such jobs also call for static contraction of the postural muscles. Usually, the physiological responses ($\dot{V}_E$, BP, HR) to the isometrically contracting muscles are superimposed on the responses to the dynamically contracting muscles, and fatigue develops in proportion to the intensity of the static components. However, when the dynamic and static components both involve large muscle groups, the extent to which the static components override the dynamic depends on the nature as well as the intensity of the contractions.

The $\dot{V}_{O_2}$ and HR were measured for tasks in which the dynamic component was walking at 2.5 km/hr and the static components were either holding (elbow at 90°), pulling, or pushing between 59 and 235 N (6 and 24 kg).[19] Extra $\dot{V}_{O_2}$ and HR costs were defined as the difference between the cost for the combined work and the sum of the cost when the dynamic and static components were performed separately. Although no extra cost was found for holding (carrying), extra cost of 50% to 100%, depending on the intensity of the static components, was found for all the pushing. Extra cost (50%) was found for pulling against the 235 N (24 kg) only.

Lifting

Lifting is the most common industrial task that includes combined static and dynamic muscular contractions. Lifting has received much attention because it is considered the major factor in musculoskeletal injuries, particularly of the lower back. The isometric component, mostly of the back extensors, is considered the crucial part which justifies a pseudostatic treatment for the posture assumed at the onset of lifting. Such a treatment is close to reality for a single heavy (close to or maximal) lift. Repetitive lifting of submaximal weights can be treated as dynamic work. Indeed, lifting light to heavy weights (6 to 36 kg) from floor to waist level at an increased rate of up to $\dot{V}_{O_2,max}$ showed that (1) the limit to frequent lightweight lifting was cardiovascular and that

(2) the limit to low-frequency lifting of heavy weights was the isometric fatigue of the forearm (grip) muscles.[20]

Floor Level Lifting

The static consideration of lifting from floor up to about 1 m is based on the isometric strength of the muscles and the torques, mostly around the lower back. The MVC of the back extensors or of the total MLS at 90° (Exhibit 6.4.11) can be, and indeed was, used[21] for such consideration (see also Chapter 6.3).

The dynamic consideration of lifting is based on the rate of lifting, the $\dot{V}_{O_2}$, and the specific $\dot{V}_{O_2,max}$ for lifting. The $\dot{V}_{O_2,max}$ for the weight and frequency of lifting expected in industry is about 0.85 of the $\dot{V}_{O_2,max}$ values shown in Exhibit 6.4.6 for combined arm-leg work.[20]

An example of the rate of lifting as a function of the weight is shown in Exhibit 6.4.13 for lifting rates that are expected to result in less than 0.50 $\dot{V}_{O_2,max}$ and that thus require no specific design for rest. It is not recommended that one lift weight of more than 0.70 MLS for repetitive lifting.[21] The MLS and the rate of lifting for women were found by European investigators to be 0.7 of those for men (Exhibit 6.4.13) and 0.4 of those for males in the United States (Exhibit 6.4.11). Psychophysical methods where the worker chooses the weight acceptable to him or her for prolonged work (see Chapter 5.1) seem to agree with the physiological results.[22] In laboratory experiments men preferred weights of 20 to 22 kg for lifting rates of 3 to 6/min and about 15 kg for lifting rates of 9 to 12/min. The $\dot{V}_{O_2}$ for lifting these subjectively selected weights seems to be at about 0.35 of the expected $\dot{V}_{O_2,max}$.[23] Moreover, judged by the $\dot{V}_{O_2}$ cost per total kilograms lifted, the economy of lifting improves sharply as the weight of each single lift is increasing and the lifting rate is decreasing, but only to a certain point. The economy of lifting at rates eliciting 0.35 to 0.50 $\dot{V}_{O_2,max}$ does not improve if the weight is more than 15 kg for men and most likely about 10 kg for women.

High Lifting

There are no sufficient data on the physiological responses to repetitive lifting from about knuckle height. Compared to lifting from floor level, the biomechanical stress and metabolic efficiency are much improved because of the reduction in the torques around the lower back and in the range of motion of the trunk. This does not mean that the MLS is larger; in fact, it is smaller (45° MLS in Exhibit 6.4.11) because the back and leg muscles are less involved. Biomechanical stresses do

Exhibit 6.4.13 Frequency of Lifting Versus Either Relative Lifting Strength (% MILS) or the Expected Weight During Work at Intenisty Below 0.5 of Aerobic Capacity[a]

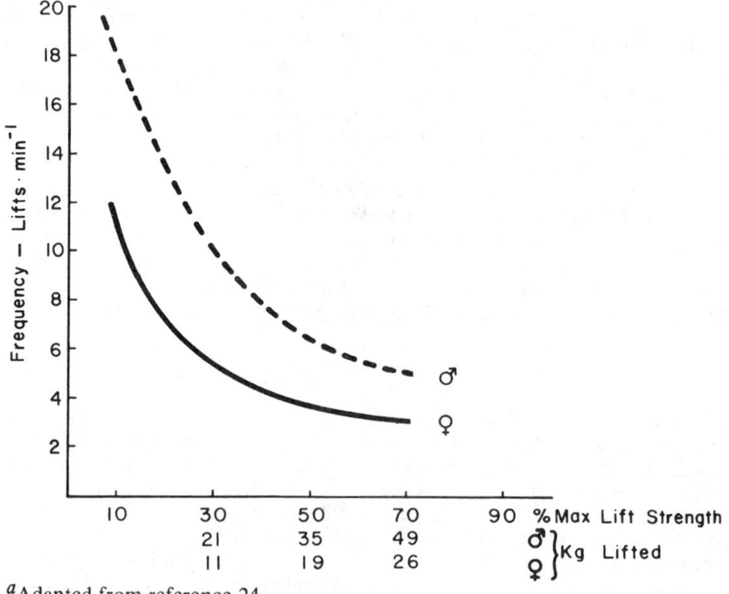

[a]Adapted from reference 24.

arise for lifting above shoulder height. There is a shift in the postural demand on the back, and the most active muscles are the extensors of the shoulders and the elbows. In general, the MLS from shoulder to arm reach is 0.6 to 0.8 of the strength of lifting from the floor.

REFERENCES

1. K. WASSERMAN, A. L. VAN-KESSEL, and G. G. BURTON, "Interaction of Physiological Mechanisms During Exercise," *Journal of Applied Physiology*, Vol. 22, 1967, pp. 71–85.
2. E. KAMON and H. S. BELDING, "The Physiological Cost of Carrying Loads in Temperate and Hot Environments," *Human Factors*, Vol. 13, 1971, pp. 153–161.
3. T. BERNARD, K. KAMON, and B. A. FRANKLIN, "Estimation of Oxygen Consumption From Pulmonary Ventilation During Exercise," *Human Factors*, Vol. 21, 1979, pp. 417–421.
4. P. O. ASTRAND and K. RODAHL, *Textbook of Work Physiology*, McGraw-Hill, New York, 1977, p. 398.
5. ASTRAND and RODAHL, *Work Physiology*, pp. 318–355.
6. F. J. NAGLE, "Physiological Assessment of Maximal Performance," in J. H. Wilmore, Ed., *Exercise and Sport Sciences Reviews*, Vol. 1, Academic Press, New York, 1973.
7. ASTRAND and RODAHL, *Work Physiology*, p. 398.
8. E. SIMONSON, *Physiology of Work Capacity and Fatigue*, Charles C Thomas, Springfield, IL, 1971, p. 451.
9. G. LEHMAN, *Praktische Arbeitsphysiologie*, Gerog Thieme Verlag, Stuttgart, 1962, p. 68.
10. L. HERMANSEN and I. STENSVOLD, "Production and Removal of Lactate During Exercise in Man," *Acta Physiologica Scandinavia*, Vol. 86, 1972, pp. 191–201.
11. A. WELTMAN, B. A. STAMFORD, and C. FULCO, "Recovery From Maximal Effort Exercise: Lactate Disappearance and Subsequent Performance," *Journal of Applied Physiology*, Vol. 47, 1979, pp. 677–682.
12. E. KAMON, "Scheduling Cycles of Work for Hot Ambient Conditions," *Ergonomics*, Vol. 22, 1979, pp. 427–439.
13. E. KAMON, B. AVELLINI, and J. KRAJEWSKI, "Physiological and Biophysical Limits to Work in the Heat for Clothed Men and Women," *Journal of Applied Physiology*, Vol. 44, 1978, pp. 918–925.
14. L. L. LAUBACH, "Comparative Muscular Strength of Men and Women: A Review of the Literature," *Aviation Space Environment Medicine*, Vol. 47, 1976, pp. 534–542.
15. J. S. PETROFSKY, R. L. BURSE, and A. R. LIND, "Comparison of Physiological Responses of Women and Men to Isometric Exercise," *Journal of Applied Physiology*, Vol. 38, 1975, pp. 863–868.
16. D. B. CHAFFIN, "Electromyography–A Method of Measuring Local Muscle Fatigue," *The Journal of Methods Time Motion*, Vol. 14, 1969, pp. 29–36.
17. H. MONOD and J. SHERRER, "Capacité de Travail Statique d'un Groupe Musculaire Synergque Chez l'Homme," *Comptes Rendues Société Biologie Paris*, Vol. 151, 1957, pp. 1358–1362.
18. W. ROHMERT, "Ermittlung von Erholungspausen fur Statische Arbeit des Menschen," *European Journal of Applied Physiology*, Vol. 18, 1960, pp. 123–164.
19. J. SANCHEZ, H. MONOD, and F. CHABAUD, "Effects of Dynamic, Static and Combined Work on Heart Rate and Oxygen Consumption," *Ergonomics*, Vol. 22, 1979, pp. 935–943.
20. J. S. PETROFSKY and A. R. LIND, "Metabolic, Cardiovascular and Respiratory Factors in the Development of Fatigue in Lifting Tasks," *Journal of Applied Physiology*, Vol. 45, 1978, pp. 64–68.
21. E. PAULSEN and K. JORGENSEN, "Back Muscle Strength, Lifting, and Stooped Working," *Applied Ergonomics*, Vol. 2, 1971, pp. 133–137.
22. S. H. SNOOK, "The Design of Manual Handling Tasks," *Ergonomics*, Vol. 21, 1978, pp. 963–965.
23. A. GARG and U. SAXENA, "Effects of Lifting Frequency and Technique on Physical Fatigue With Special Reference to Psychophysical Methodology and Metabolic Rate," *American Industrial Hygiene Association Journal*, Vol. 40, 1979, pp. 894–903.
24. K. JORGENSEN and E. PAULSEN, "Physiological Problems in Repetitive Lifting With Special Reference to Tolerance Limits to the Maximum Lifting Frequency," *Ergonomics*, Vol. 17, 1974, pp. 31–39.

CHAPTER 6.5

Bodily Discomfort

ILKKA KUORINKA

Institute of Occupational Health

6.5.1 INTRODUCTION

Bodily discomfort is a painful state which can follow unsuitable work stress, ill health, or psychological or sociological conflict. It is closely related to fatigue, both mental and physical.

Bodily discomfort is best defined with the help of its antithesis: It is the loss of bodily comfort. According to the World Health Organization's definition of health, bodily discomfort may be regarded as "a condition of ill health."[1]

Bodily discomfort is basically a warning signal. The organism is being warned of something that may be harmful or dangerous.

This chapter deals with the bodily discomfort caused by physical, physiological, and anatomic factors and with outcomes that are mainly physical in character. The important role of psychological factors is recognized here, but is dealt with in Chapters 5.1 and 6.6.

6.5.2 MANIFESTATION, LOCATION, AND CAUSES

A predominant sign of bodily discomfort is pain. Pain can manifest itself in many forms, from a vague feeling to acute pain preventing all activity. Sometimes discomfort is described as numbness, prickling, or some other form of abnormal sensation.

Two special characteristics of the manifestation of bodily discomfort deserve consideration. First, in manual tasks chronic, mild pain is common. Workers often regard it as "normal" and do not report it spontaneously. Even such mild pain may affect well-being and cause later pathological consequences.

Second, the manifestation of pain and discomfort in older workers differs from that in young ones. In social medicine investigations older workers have reported fewer subjective symptoms in comparison to the objective findings reported in a clinical examination. Young workers, on the other hand, report correspondingly more subjective symptoms.

The location of bodily discomfort is generally relative to the exposure in question; that is, it is specific to the task. However, when the exposure is more general in nature, such as in immobile sitting tasks or when one has poor posture in general, two body sites are predominantly affected: the back and the neck and shoulder area.

One investigation that illustrates the specificity of bodily discomfort involved 430 manual workers who were examined through palpation of the muscles.[2] The investigators found that hard and painful muscles were common and that, in addition, they were mainly those muscles used by workers the most in their tasks. Bodily discomfort thus seems to be job related, even though no compensable occupational disease has been recognized in connection with it.

Epidemiologic studies concerning the back indicate that 80% of the population will experience incapacitating back pain during its active life. Back pain and back disorders are one of the leading causes of work absenteeism in all industrialized countries. Many occupational factors have been proposed as causes of back pain, but none of them has yet been exactly proved. Reducing bodily discomfort can help prevent back pain.[3]

Manifestations of discomfort are also common in the neck and shoulder area. Three working populations (altogether 378 persons) were recently studied: packers on a production line, manual workers in a metal shop, and shop assistants.[4] One of the symptoms, the tension neck syndrome, appeared in 38, 58, and 54% of the workers in each group, respectively. It was concluded that this type of discomfort is common in many different occupations.

The primary causes of bodily discomfort are not always easily detected in everyday life. Multiple psychological and social variables may cause discomfort. It might be more appropriate to speak of factors related to discomfort than of its causal factors.

The following physical factors are known to be related to bodily discomfort:

Static muscle work (see Chapter 6.4).
Poor anthropometric design (see Chapter 6.3).
Work methods that immobilize the worker at the workplace.
Poor work postures.
Whole-body vibration.
Other environmental factors.

Static muscle work causes bodily discomfort—an experience familiar to everyone. But structures of the body other than muscles cannot tolerate static loading either; ligaments and joint surfaces, for example, need movement and variation in order to function well. The same applies to spinal discs, veins, and so on.

Poor anthropometric design, which does not allow for limb movement and comfort areas or does not correspond to a population's measures, causes poor postures or stressful movements that lead to discomfort.

Whole-body vibration causes definite discomfort that depends on the frequency range in a peculiar way. Different frequencies cause maximum discomfort in different areas of the body.

6.5.3 CONNECTION WITH DISORDERS AND INCAPACITY

Whether disease or some type of disorder follows bodily discomfort depends on many factors. Psychological aspects are very important in this respect. In addition, individual factors modify the effect of exposure to static work, bad posture, and so on.

In general, the longer the duration of, and the more intense the exposure to, discomfort causing factors, the greater the probability of the occurrence of a disease or disorder.

Neck and Shoulder Diseases

A neck and shoulder disease called "occupational cervicobrachial syndrome" is one of the discomfort-related diseases. In a Japanese investigation, 10,000 cases of this disease were found among 6 million workers in 1970–1971. Discomfort-related fatigue was also reported for 21% of production line workers, but for only 6% of sales personnel and 4% of management staff.[5] As has already been pointed out, the tension neck syndrome was found in about half of the investigated groups.[3]

Two features characterize neck and shoulder affections. First, discomfort and disease are common both among manual workers and in the general population. Second, neck and shoulder disorders are not easily definable or diagnosed. Thus health statistics do not give a true picture of the situation.

Back Pain

Although back pain is a common disorder, its exact cause is unknown. There is also the question of whether back pain or back discomfort predicts back disease. The answer is yes: Attacks of back pain predict at least certain types of back diseases.[6]

Knowledge about the relationship of occupational factors to back pain can be summarized as follows: If the relationship between the two is measured in terms of the time absent from work because of back trouble, certain features of a job will increase the absence. These features are hard physical work, static body postures, stooping postures, sudden back loads, and lifting. These characteristics also are definitely related to discomfort.

Psychosomatic Disorders

The importance of the psychological and psychosocial aspects of bodily discomfort has been stressed by Swedish psychosocial researchers. In their workplace research, Bolinder and Ohlström[7] demonstrated a clear relationship between psychosocial and physical factors and psychosomatic disorders, which often take the form of bodily discomfort. Other examples of related discomfort causing occupational situations and a certain disease might be sustained standing, yielding varicose veins, and long-term sitting, yielding back trouble.

6.5.4 INTERFERENCE WITH PRODUCTIVITY

Bodily discomfort also can be interpreted as a sign that an organism is being used for an unnecessary activity instead of for productive work or leisure-time activities. If, for example, a job is done in a poor posture, the circulatory and muscle capacity used to maintain the posture is not available for productive work. Furthermore, the fatigue caused by the poor posture may impede recovery and the enjoyment of leisure time.

Bodily discomfort also may have an additional effect: It prevents the mobilization of body energy to other activities. When an important part of the energy supply is used for maintaining alertness and arousal in an unfavorable situation, less capacity is left for creative activity on the job.

The interference of bodily discomfort with productivity has been studied in two contexts: effect of postural discomfort and static muscle work.

Postural Discomfort

Sämann[8] made a broad investigation of the effect of poor postures on total energy consumption and work performance. He divided total energy consumption into three parts: basic energy consumption (rest), energy used to maintain a poor posture, and energy available for sustained work.

Exhibit 6.5.1 compares the available energies in two work situations. The energy used for maintaining poor posture heavily taxes the capacity available for productive work.

In practice the different energy proportions are not easily distinguished, because work and posture are intimately connected. Often, total energy consumption simply is shown to be higher for a task done in a poor posture or worker fatigue is more apparent.

Pulse rate measurements, together with measurements of energy consumption, can be used to estimate the relative energy lost to the maintenance of poor posture. The following two criteria are relevant: (1) pulse rate rises in certain postures, and (2) the relationship Δ pulse rate$_{work}/\Delta$ energy consumption serves to estimate the task mode in a certain posture.

These theoretical considerations have been verified in practice. Energy consumption was measured in slag removing tasks.[9] Scholz compared the same heavy task done in two different postures; that is, the workpiece was at two different levels, one allowing free standing and the other demanding a stooping position. The work rate and other conditions were the same. He found that work done in an inconvenient posture was 25 to 34% more strenuous than that done in a convenient position.

Many such comparisons have been published. Their message could be summarized as follows: Even in innocent looking cases one fourth of the energy consumed may be used for maintaining a poor posture. The increase of activity metabolism may approach 100% when heavy work is done in a poor posture compared to the same job's being done in a convenient position. This conclusion is not surprising once one takes into consideration the physical energy needed to move or maintain the position of a torso weighing, say, 100 lb (45 kg).

Static Work

The nature and limits of static work, as well as recovery from it, have been presented in Chapter 6.4. Discomfort from static work strictly follows the limits of muscle performance. The rating of pain is linearly dependent on the percentage of the maximal endurance time of static work.[10] Pain from static work may be severe enough to make a person stop the work involved. If intense, it may lead to prolonged fatigue. Repeated or persisting static work may cause painful and hardened muscles.

As explained in Chapter 6.4, high-intensity, dynamic muscle work also contains a static (anaerobic) component. How much static muscle work can be permitted before the capacity used for productive work is taxed or persistent muscle pain appears? No clear-cut answer exists. The amount of static work tolerated depends on the work regime, number of pauses, and so on. Work

Exhibit 6.5.1 Relative Proportions of Physical Capacity Available for Productive Work in Connection With Good and Poor Posture

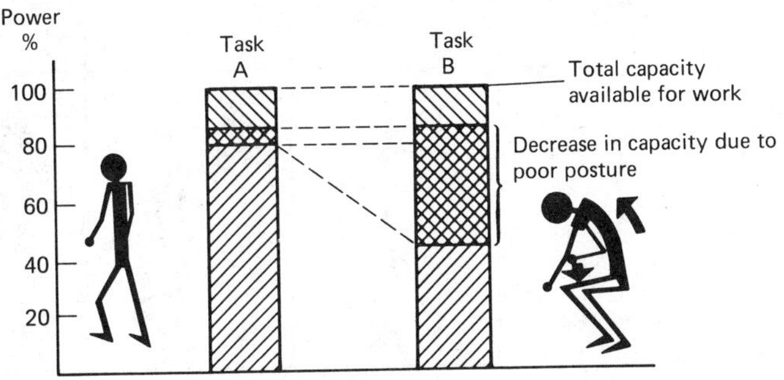

physiology offers one guideline: In continuous static work the force to be maintained should not exceed 10 to 15% of maximal voluntary force, maybe not even 5 to 10%. Some additional rules of thumb are: Design shorter work spells and more pauses, and vary the work intensity. A newly discovered old rule states that active pauses allow a more complete recovery than passive ones.[11]

6.5.5 ASSESSMENT OF BODILY DISCOMFORT

Because bodily discomfort is a strongly subjective phenomenon, it limits the methods available and favors the psychophysical methods' approach. In practice these methods can determine (1) whether occupational bodily discomfort exists, (2) how serious the discomfort is in comparison to some other situation, and (3) the location of the discomfort.

Another area of assessment is whether factors causing bodily discomfort exist in the workplace. This question can be answered through an ergonomic analysis of a workplace. This topic has also been outlined in other chapters of Section 6.

Methods of Assessment

None of the various methods available for assessing bodily discomfort cover all aspects of the problem. They always concentrate on specific points, and users must choose those they want to investigate.

Observation of Bodily Discomfort

Work study engineers and foremen always have observed work, sometimes for discomfort. For a competent observer, one glance at a person at work may be enough to provide useful information about the work, work load, and discomfort. When one sees a manual worker resting in a squatting position, he or she may conclude that the preceding work spell was strenuous. When a worker rises holding his back or a keypuncher stretches her neck and shoulders, it may be interpreted as an indication of discomfort. Unfortunately, this type of observation requires empathy and experience —both of which are not easily taught or acquired.

Systematic observation (e.g., work sampling), photographs, videotape, and so on, have been used to record behavior related to discomfort. For example, in passenger comfort studies the behavior reducing discomfort has been studied and registered.

A method that combines the work sampling techniques used in work study and a postural discomfort rating has been employed in a Finnish steel company.[12] This method is used by work study engineers in their daily routine and includes 72 postural combinations commonly found in the steel industry. The work study engineers are trained to observe, record, and analyze postures. Every posture is classified on a 4-point discomfort scale, on which the highest class implies an urgent need for corrective action. This classification is based on ratings made by experienced steelworkers, as weighted by existing medical, physiological, and ergonomic data. Exhibit 6.5.2 shows five postures needing urgent attention and five very easy ones.

Exhibit 6.5.2 Five Comfortable and Five Uncomfortable Postures, as Rated by a Panel of Steelworkers.[a]

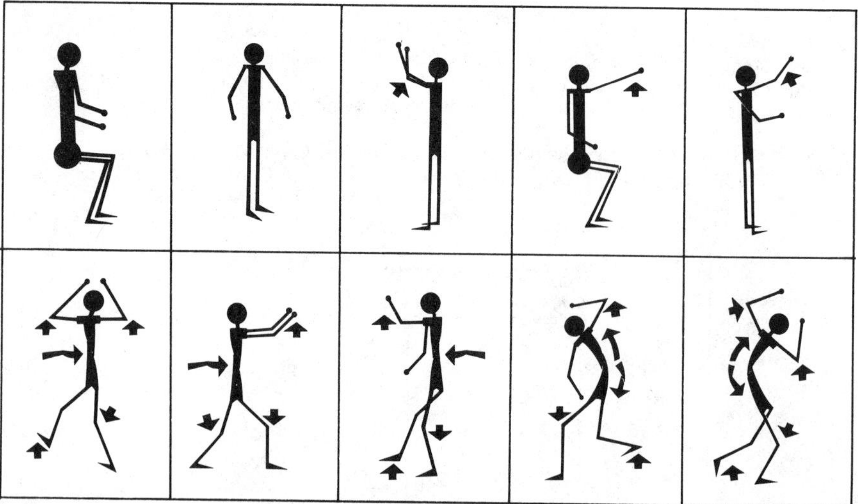

[a]From reference 11.

The advantage of this method is that it combines the routine of a work study engineer and the collection of discomfort data in a practical manner.

Rating and Locating Bodily Discomfort

A discomfort rating made by the subject is the most common method used to study discomfort. Several methods and procedures exist, from verbal and scalar types to paired comparisons and even more sophisticated procedures. Gathering reliable data requires basic knowledge of how scales and ratings work (see Chapter 6.4).

Two additional considerations are important with respect to discomfort ratings. First, the verbal expressions in rating scales limit the content and quality of the answers. Careful consideration of the items to be included and their unambiguous formulation are needed.

Second, semantic problems are greater than one tends to expect. The vocabulary and expressions of a worker and even his or her closest supervisor may differ considerably and lead to misunderstandings. Whenever possible, the rating questions should be validated for the target population.

Corlett and Bishop[13] have developed an illustrative and concrete method for rating discomfort. They applied it in the analysis of spot welding machines and operator discomfort and found it to work very satisfactorily. In their method, discomfort is rated on a rating scale and also on body diagrams. At given intervals the subjects are shown diagrams of the body and are asked to point out and mark the most painful areas. The next painful areas then are delineated, and the procedure is repeated until no new areas are indicated. The successive diagrams are scored to form a discomfort rating for body parts. A total discomfort index is obtained by asking the subjects to rate the discomfort on a 7-point scale.

The authors applied this method in analyzing the ergonomic qualities of spot welding machines and the discomfort of their users.[14] A number of different machines were analyzed, and the anthropometric dimensions of 60 operators were recorded. Clear conflict was found between many of the dimensions and qualities of the machines and the anthropometric measures of the users.

The dimensions and controls could be changed somewhat on three of the selected machines. This situation allowed the authors to study discomfort both before and after redesign. The results indicated that, in each of the three modified cases, better postures and easier operation improved the overall and body-part discomfort ratings. Back and neck regions showed the most marked improvements. The use of the machines increased accordingly, and idle time decreased.

The example of Corlett and Bishop's method shows how the assessment of discomfort can provide reliable and useful data for industrial design.

Other Procedures for Assessing Discomfort

Pictorial assessment of discomfort seems to be an easy and reliable way for subjects to express their discomfort. Whitham and Griffin[15] developed a pictorial method for rating vibration discomfort. After the subjects were exposed to vibration, they had to mark location of the worst discomfort on a human figure. As Exhibit 6.5.3 shows, the results clearly demonstrate the relationship of discomfort to vibration frequency.

Contribution of Occupational Health Personnel to Discomfort Assessment

Occupational health personnel are in a key position in the analysis and recording of painful states. Unfortunately, few working systems exist for providing systematic information to designers and method engineers from observations made in occupational health centers or other examinations.

The use of occupational health nurses or physiotherapists as links between designers and doctors could serve as a workable model. In the Institute of Occupational Health, Finland, a system for analyzing neck and upper limb disorders and discomfort has been developed. The procedure is based on a structured examination performed by a trained occupational physiotherapist.[4] In addition to epidemiologic data, the method yields a detailed analysis of the location and severity of symptoms and signs.

Broader Basis for Discomfort Assessment Through Position Analysis Methods

Discomfort assessment can be limited to the analysis of the manifestation of discomfort mostly in special cases, for example, by answering the question of whether certain design features cause discomfort.

If a more complete picture of the many factors linked to bodily discomfort is wanted, a position analysis for work profile must be made. A number of such methods have been proposed. Two of them may serve as examples, although a complete presentation is not possible.

Exhibit 6.5.3 Bodily Discomfort Reported in Vibration Exposure Experiments. Location of Maximal Discomfort Depends on the Frequency of the Vibration.

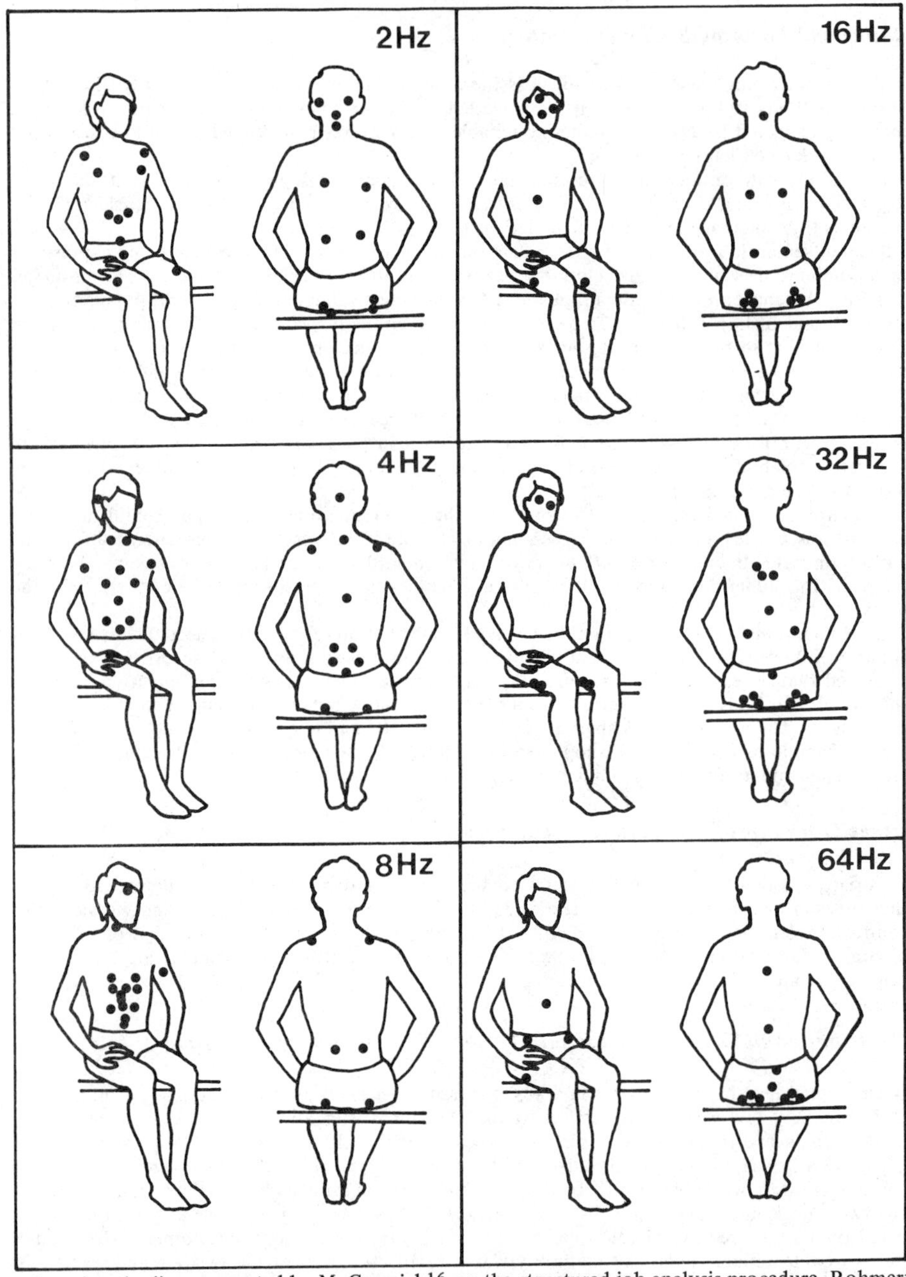

Pursuing the lines presented by McCormick[16] on the structured job analysis procedure, Rohmert and Landau developed a practical method with formulas, systems, and formalized education. Its name is AET analysis (Arbeitswissenschaftliche Erhebungsbogen für Tätigkeitsanalyse).[17]

In the AET analysis, the work system, task, and task demands are analyzed in a structured form. Observation and questioning are used for gathering answers to 216 structured items (work system–143 items; task–32 items; and task demands–41 items). The outcome is formulated as a work profile. A more sophisticated analysis produces information on work content, the main categories of which are muscle force production, sensorimotor coordination, information handling, and information production. Multiple items in the AET analysis provide information on factors producing bodily discomfort.

Lucas (from the Renault car company, France) has presented an analytical method for workplace analysis called "Les Profils de Postes" (profiles of workplaces).[18] The structure of the profiles of workplaces is even more practically oriented than the AET method. In the profiles 5 27 multiple items on the workplace are recorded according to a given procedure. The main groups of items in the analysis are workplace layout (4 items), work safety (1 item), ergonomic factors (14 items), and psychosocial factors (8 items). All the items are placed into (mainly) five categories. The profiles of workplaces also produces an analytical work profile as an end result.

Common features of both of these methods are their trend toward practical applicability and their multifaceted approach. The systems are offered complete with user education and, to some extent, analysis service.

6.5.6 REDUCTION AND PREVENTION OF BODILY DISCOMFORT

Two characteristics of bodily discomfort should be accounted for in the palnning of its reduction and prevention, that is, the multicausal nature of bodily discomfort and the lack of knowledge about many important details.

Not enough is known about the causal factors of bodily discomfort for prevention to be directed only toward known causal factors such as poor postures. Both of the characteristics important to prevention (multicausality and lack of details) imply the need for prevention on broad lines before prevention directed toward specific questions.

An analysis of and suggestions for prevention have been presented by Van Wely, who did a study on the design factors related to musculoskeletal diseases at work in the Philips Company.[19] Van Wely's starting point was the high absenteeism because of musculoskeletal disorders. A group of 50 patients was selected and carefully diagnosed. A list of the probable relationships between workplace design factors and musculoskeletal symptoms was made. The workplaces of the 50 patients were visited, and their ergonomic qualities were recorded. When diagnoses and workplace design factors were compared, the hypothesized relationship was confirmed in 39 cases and not confirmed in 8 cases; in the other 3 cases no ergonomic failures were found. Suggestions for prevention were:

Better equipment design.

Better instruction and training for newcomers.

Preemployment testing for determining low-tolerance people.

An intervention study on keypunch operators supports Van Wely's proposals for prevention.[20] Sixty keypunch operators participated in a study in which their neck and upper limb symptoms were analyzed before and after the intervention. The intervention consisted of adjusting the furniture, constructing supports for papers, providing advice on work pause activities, and so on. A substantial number of discomfort signs had disappeared by the time the second examination was made a half year later. A long-persisting effect points to biological effects in addition to the inevitable Hawthorne effect. The effects were gained through fairly minor modifications of the work and workplace.

Two more aspects of prevention can be added, based on experiences gathered during projects concerning musculoskeletal disorders and discomfort. Early diagnosis seems to be important (in the case of both individual clinical diagnosis and the diagnosis of workplace design). Good physical fitness seems to be related to individual insensitivity to work load. Immobilizing workplaces and methods promote poor physical fitness.

The following list summarizes the important aspects of the prevention and reduction of bodily discomfort.

Workplace layout and anthropometric design are of utmost importance in the prevention of bodily discomfort (see Chapter 6.9).

Hard physical work, poor postures, and immobilizing work methods cause discomfort and should be redesigned.

Environmental factors cause discomfort in addition to their specific detrimental effects (see Chapters 6.10 through 6.12).

6.5.7 CLOSING COMMENTS

Bodily discomfort is a common and complex phenomenon which is rooted in both social and individual factors inside and outside the workplace. Efforts have been made to prevent and reduce bodily discomfort, with very promising results. However, fast and dramatic changes cannot be expected. Small steps and continuous effort produce better results.

Not enough is known about the causes and mechanics of bodily discomfort. The singling out of one factor for the target of prevention repeatedly has been shown to be ineffective. Thus all factors potentially related to bodily discomfort must be controlled if possible. Bodily discomfort is a bird to be shot with a shotgun rather than a rifle.

REFERENCES

1. WORLD HEALTH ORGANIZATION, "Constitution," WHO, Geneva, Switzerland, April 7, 1948.
2. A. HILTUNEN, M. I. KARVONEN, J. KIHLBERG, and R. LAMMINPAA, "Muscle Spasm in Manual Laborers," *A.M.A. Archives of Industrial Hygiene and Occupational Medicine*, Vol. 9, June 1954, pp. 476–480.
3. F. DUKES-DUBOS, "What Is the Best Way to Lift and Carry?," *Occupational Health and Safety*, Vol. 46, No. 1 (1977), pp. 16–18.
4. I. KUORINKA and P. KOSKINEN, "Occupational Rheumatic Disease and Upper Limb Strain in Manual Jobs in Light Mechanical Industry," *Scandinavian Journal of Work Environment and Health*, Supplement No. 3, Vol. 5, (1979), pp. 39–47.
5. K. MAEDA, "Occupational Cervicobrachial Disorder and its Causative Factors," *Journals of Human Ergology*, Vol. 6, (1977), pp. 193–202.
6. A. NACHEMSON, "Critical Look at the Treatment for Low Back Pain," *Scandinavian Journal of Rehabilitation Medicine*, Vol. 11, No. 4 (1979), pp. 143–147.
7. E. BOLINDER and B. OHLSTRÖM, *Stress på Svenska Arbetsplatser*, Prisma, Lund, Sweden, 1971.
8. W. SÄMANN, "Charakteristische Merkmale und Auswirkungen ungünstiger Arbeitshaltungen," *Schriftenreihe "Arbeitswissenschaft und Praxis"*, Band 17, Beuth-Vertrieb GmbH, Berlin, Köln, Frankfurt/M, 1970.
9. H. SCHOLZ, *Die physische Arbeitsbelastung der Giessereiarbeiter.* Forsch. ber des Landes NRW, Nr. 1185, Verlag, Köln und Opladen, Westdt., 1963, p. 182.
10. N. S. KIRK and T. SADOYAMA, "A Relationship Between Endurance and Discomfort in Static Work," unpublished Master of Science report, Department of Human Sciences, University of Technology, Loughborough, England, 1973.
11. E. ASMUSSEN and B. MAZIN, "Recuperation After Muscular Fatigue by Diverting Activities," *European Journal of Applied Physiology and Occupational Physiology*, Vol. 38.1, 1978, pp. 9–15.
12. O. KARHU, P. KANSI, and I. KUORINKA, "Correcting Working Postures in Industry: A Practical Method for Analysis," *Applied Ergonomics*, Vol. 8, No. 4 (1977), pp. 199–201.
13. E. N. CORLETT and R. P. BISHOP, "A Technique for Assessing Postural Discomfort," *Ergonomics*, Vol. 19, No. 2 (1976), pp. 175–182.
14. E. N. CORLETT and R. P. BISHOP, "The Ergonomics of Spot Welders," *Applied Ergonomics*, Vol. 9, No. 1(1978), pp. 23–32.
15. E. M. WHITHAM and M. J. Griffin, "The Effects of Vibration Frequency and Direction on the Location of Areas of Discomfort Caused by Whole-Body Vibration," *Applied Ergonomics*, Vol. 9, No. 4 (1978), pp. 231–239.
16. E. J. MCCORMICK, P. R. JEANNERET, and R. C. MECHAM, "A Study of Job Characteristics and Job Dimensions as Based on the Position Analysis Questionnaire (PAQ)," *Journal of Applied Psychology Monograph*, Vol. 56, No. 4 (1969), pp. 347–368.
17. W. ROHMERT and K. LANDAU, *Das Arbeitswissenschaftliche Erhebungsverfahren zur Tätigkeitsanalyse (AET)*, Handbuch, Verlag Hans Huber, Bern Stuttgart Wien, 1979.
18. *Les Profils de Postes*, Méthode d'Analyse des Conditions de Travail, Collection Hommes et Savoirs, Sirtes et Masson, Paris, 1976.
19. P. VAN WELY, "Design and Disease," *Applied Ergonomics*, Vol. 1, No. 5 (1970), pp. 262–269.
20. T. LUOPAJÄRVI, R. KUKKONEN, and V. RIIHIMÄKI, *Prevention of Health Hazards of Key Punchers with Applications of Ergonomy and Occupational Physiotherapy*, Institute of Occupational Health, Helsinki, 1979 (in Finnish).

CHAPTER 6.6

Occupational Stress

GAVRIEL SALVENDY
JOSEPH SHARIT

Purdue University

The purpose of this chapter is to acquaint the practitioner with the field of occupational stress; a field which can have a profound effect on job design, productivity, and quality of working life. Specifically, this chapter discusses (1) characteristics of stress and factors related to it, (2) methodologies for measuring stress, (3) sources of occupational stress, and (4) strategies for stress management.

6.6.1 CHARACTERISTICS OF STRESS

Selye's[1,2] viewpoint on stress, although controversial, appears to be the one most frequently cited in the literature. According to Selye, the presence of stress can be inferred in an individual from a very generalized physiological response pattern (e.g., increases in adrenaline secretion, the dumping of sugar into the bloodstream, and other related physiological processes) whose elicitation can be provoked by a wide variety of agents and situations, such as drugs, fear, and job ambiguity.

This conceptualization of stress provides a basis for the development of various illnesses, most notably the psychosomatic, to which the debilitation of an ever-increasing sector of the work force can perhaps be attributed. Selye refers to these maladies as "diseases of adaption," since they are not a direct function of the agent or situation that elicited the response pattern, but a consequence of the body's faulty adaptive reaction. Such situations can presumably become facilitated through frequent repetition of the stress-response pattern. There is nothing wrong per se with escalations occurring in these response patterns; they are, in fact, essential to the individual's overall functional capabilities. Elicitation of these responses, however, often occurs where the stressors are such that an intensive mobilization of the body's defensive apparatus is inappropriate (as is often the case in white collar work). In such instances, these responses will most likely increase the *wear and tear* on the body, especially if this elicitation process becomes chronic.

Selye's conceptualization of stress is illustrated in Exhibit 6.6.1. Its most essential feature concerns the intensity of the demand for readjustment or adaptation brought about by the stressor, regardless of whether it is pleasant or unpleasant. This view deals only with the effects of the stressor on the physiological system, and the physiological response is assumed to be independent of the individual's *feeling* state. Also critical in this view is that some degree of stress is necessarily always present.

When shifting the emphasis from the general concept of stress to occupational stress in particular, demands on the worker for readjustment can be thought of as paralleling job and/or task demands. Carrying the analogy a step further, it would appear that discriminating these work-related demands as either very pleasant or unpleasant would make no difference in terms of the worker's physiological stress response. The minimal stress level would ideally correspond to the most "desirable" degree of demand imposed on the worker; that is, the level of demand that proves least harmful to the worker's state of health and that simultaneously results in both the most desirable work performance and worker satisfaction.

It is unlikely, however, to expect this level of demand to simultaneously result in the most desirable physiological responses, work performance, and worker satisfaction. Realistically, some type of trade-off needs to be arrived at, which would then correspond to some point along the curve in Exhibit 6.6.1. The particular level, however, would depend largely on the following individual factors:

The writing of this chapter was facilitated by financial supports from the National Institute of Occupational Safety and Health (contract numbers 210-800002 and 210-800034) and the National Science Foundation grant number APR7718695.

Exhibit 6.6.1 Theoretical Model Regarding the Relationship Between Physiological Stress, as Defined by Selye, and Pleasant, Indifferent, and Unpleasant Experiences of Various Environmental Stimuli, for Example, "Life Change." Note That the Physiological Stress Level is Lowest During Indifference, But Never Goes Down to Zero. Pleasant as Well as Unpleasant Emotional Arousal is Accompanied by an Increase in Physiological Stress (But Not Necessarily in Distress).

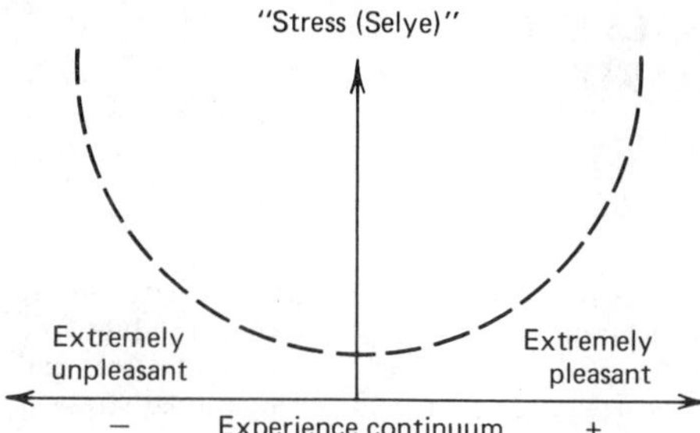

Source: Reference 2, p. 13, reproduced by permission.

1. Genetic predisposition, which would influence the individual's bodily response to stress and the potential ensuing development of diseases of adaption.
2. Early social experience, significantly affecting personality development and, in turn, affecting the perception of and response to stress.
3. A lifelong process of conditioning and cultural factors. These, in turn, would contribute to the development of the individual's coping behavior which, in effect, determines how the person will deal with stress.

Utilizing information regarding these individual factors would assist in matching the worker to the job and hence allow workers to operate in the region of the stress curve (see Exhibit 6.6.1) that would optimize the trade-off (undoubtedly a very subjective process) between physiological responses and worker performance and satisfaction.

6.6.2 FACTORS RELATED TO STRESS

Although there are a number of factors that relate to stress, the most closely associated ones include mental work load, fatigue, and the level of arousal present in an individual. This section attempts to distinguish these factors from each other and from the characteristics of stress while underscoring their relevance to the area of occupational stress.

Mental Work Load

Mental work load relates to that part of the task invoking predominantly decision making processes as distinct from those parts requiring primarily physical effort. Previously, interest focused on evaluating the stress of physical work loads through the use of physiological techniques both in order to improve work methods by determining how variations in the task could reduce the "cost" of these measures[3] and as a basis for determining rest allowances.[4] Interest is currently shifting to tasks consisting primarily of mental work load components. With the advent of computerized manufacturing systems, robotics, office automation, and of an overall reliance by many industries on computer technology, work processes have tended to become more a function of (1) activities in which the worker acts as a monitor, (2) the kinds of demands resulting from working with computerized visual display units, (3) decision making activity, and (4) responsibility factors. A side effect arising from this emphasis has been the problem for the worker in coping with increases in moment-to-moment uncertainty.

Evidence from a series of laboratory experiments has indicated that the variability (in time) between successive heart beats (sinus arrhythmia, (SA)) might be related to mental work load.[5] Specifically, it was demonstrated that as the mental work load was systematically increased, SA correspondingly decreased, despite no significant changes in the mean heart rate. In essence, these findings implied that the distribution of the heart beats and not the number (i.e., rate) were critical in evaluating mental work load. The importance of these findings is that they potentially augment the methods traditionally utilized by the ergonomist so that the assessment of mental as well as physical work is feasible. The usefulness of the SA measure for work design is discussed in the section on experiments on stressors in work situations.

When the work situation in the industrial as opposed to the laboratory environment is considered, factors such as the operator's skill and strategy and interactions between physical and mental work load may obscure, to some degree, attempts at evaluating the mental work load component. Still, the SA measure could prove useful[5]; however, it was more likely to do so by indicating to what extent the mental work load is acceptable. In addition, the SA measure could potentially detect moments of "peak load."[6] In this context the number of times situations of peak load arise within a given period could serve as a useful index of mental work load.

Another way by which the mental work load can be evaluated is through the method of distraction stress.[7] Basically, the method calls for manipulating the degree of difficulty in a worker's primary task while observing the deterioration of performance in some secondary task. Despite the popularity of using secondary tasks for evaluating mental work load, many interpretive problems emerge. A discussion of the methodological and theoretical issues involved in the use of secondary tasks in measuring mental work load can be found in Ogden et al.[8] Behavioral and physiological measures of mental work load are reviewed in Williges and Wierwille[9] and Wierwille,[10] respectively.

Fatigue

Factors influencing fatigue are presented in Exhibit 6.6.2.[11] The factor of fatigue has become associated with a variety of meanings. It is generally considered a "time-correlated" disorder of skill affecting skilled performance in the following ways: (1) responses are made too late or too early; (2) responses are made with an intensity greater than necessary; (3) occasional response omissions are typical; (4) there are tendencies toward increased variability in cycle time, with increased fatigue related to decreased performance[12]; and (5) there are decreases in mean performance times. Fatigue also has been related to strenuous and/or sustained physical work, as reflected in decreases in physiological efficiency (see Chapter 6.4). Subjective aspects of fatigue also have been investigated in terms of both the sensations experienced by the person and the ratings based on his or her appearance.

Industry traditionally has been interested in problems of *acute* fatigue. These situations are generally easier to identify, assess, and remedy. In some work situations the focus of interest may instead be on *chronic* fatigue. Research on civil aircrews[13,14] has interpreted this latter type of fatigue as a generalized response to stress over a period of time. In this context, cumulative effects of the work situation become important. This, in turn, would imply that the time required for recovery may be the appropriate criterion for assessing the severity of the fatigue. When fatigue is viewed in this way, complete recovery through the normal daily processes of rest, recreation, and

Exhibit 6.6.2 Cumulative Effect of Daily Causes of Fatigue. Fatigue Is Compared to the Level of a Liquid in a Container, and Recovery is Shown as the Outflow From the Container.

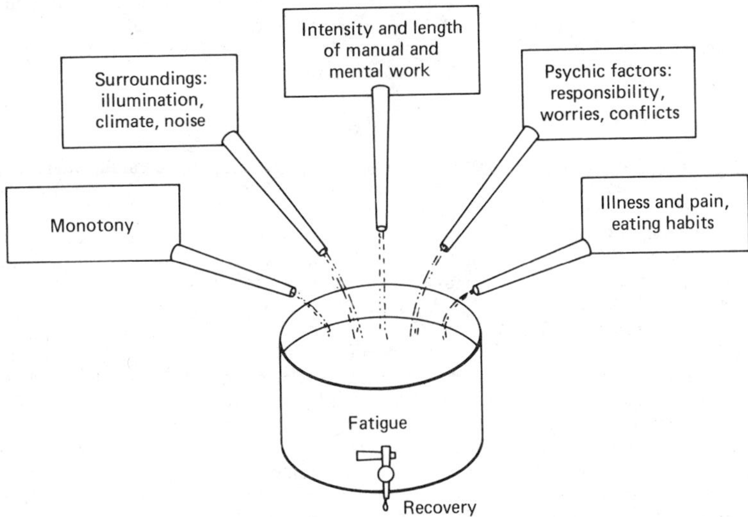

Source: Reference 11, p. xviii, reproduced by permission.

sleep between successive exposures to the work process is unlikely to occur. Physical fatigue (i.e., acute fatigue) in industry can be considered a diminishing problem, perhaps suggesting that industries need to revise their thinking of fatigue more along the lines on cumulative (e.g., over days, weeks, months, or even years) effects. The practical implications of the phenomenon of chronic fatigue for industry and some possible solutions for this problem are discussed in Cameron.[13]

Arousal

Arousal, or the general level of attentive behavior, has as its ultimate basis the activity in select nerve pathways in the brain. Its relationship to stress becomes more apparent if one visualizes a kind of set point, as in a thermostat, in operation in the brain, whose function is to maintain an optimum level of alertness in the individual. Tasks that can be considered to have either a lack (e.g., vigilance-type tasks with low signal rates) or an excess of stimulation would threaten this set point and hence provoke conditions of stress in the Selye sense (see Section 6.6.2 and Exhibit 6.6.1) of introducing a demand for readjustment. Often accompanying these demands are physiological and performance characteristics of the stress-response pattern. It appears then that both understimulation and overstimulation are stressful.

An elegant example that relates these concepts to stress in an actual work situation can be found in a study on sawmill workers in Sweden.[15] In this task, mechanical control, standardized motion patterns, and constant repetition of short-cycle operations were considered as examples of job underload (understimulation). Piece-rate rush and high demands on attention (an inspectional task component in this case) were, on the other hand, examples of overload (overstimulation). The stressful nature of this job can become clearer in the situations where skilled judgment, that is, the demands on attention, were required at short intervals (high rate of machine pace), creating a situation of time pressure and monotony. Since both of these states are producing strong demands for readjustment (although in opposite directions), the sum total will inevitably result in a high state of stress for the worker.

The effects of physiological response on performance have generally been characterized by an inverted-U function. Both low and high levels of physiological response are associated with submaximal performance, whereas optimal performance occurs at moderate levels. These latter levels would, in a sense, correspond to some optimal balance between understimulation and overstimulation and would likely be a function of both the individual and the situation. A more detailed treatment of this concept is given in Welford.[16]

6.6.3 MEASUREMENT OF STRESS

Because the causes and effects of stress are multidimensional, a multidimensional approach to the measurement of stress must be undertaken. Occupational stress can be assessed by utilizing some or all of the following four categories of measures: (1) physiological, (2) biochemical, (3) psychological, and (4) performance.

The task of selecting suitable measure(s) within each of these categories is often difficult. The choice typically is reflected in the specialized expertise of the investigator(s), the availability of equipment, the perceived degree of obtrusiveness of the equipment, and the unique information needed about the task under investigation. In general, it appears most useful to collect both subjective measures of stress, which serve to indicate the degree of stress perceived by the operator, and the more objective physiological and performance measures.

Physiological Measures

Physiological methods can be used for evaluating both physically and mentally demanding work. With respect to physically demanding work, these methods generally have been employed as a basis for determining rest allowances or for improving work methods.

Advances in physiological recording and processing technology have fueled a recent trend toward the return of on-the-job evaluations of the physiological cost associated with various tasks. These improvements include the use of magnetic cassette tape recorders, which are worn across the worker's belt[17] and allow for up to 24 hr of continuous recording; telemetry techniques (radio transmission of signals)[18]; and a respirometer for measuring oxygen consumption and ventilation volume relatively free from interference with normal activities.[19] As a result of these innovations, physiological measures can now be evaluated over rest and sleep in addition to the work cycles.[20]

Recording of physiological measures such as heart rate (HR), heart rate variability (HRV), blood pressure (BP), respiratory rate (RR), and electromyography (EMG) constitutes a continuous process over whatever period one collects the data. This aspect and the fact that these measures are associated with a physiological system that can be considered fast acting in response to internal (e.g., pathological worry) and external (e.g., threat from a supervisor, difficulty in solving a problem) stimulation contribute to the usefulness of these measures. In essence, they have the potential for conveying information that is more momentary in nature. As a result, these measures frequently enable one to isolate the effects of job components by evaluating minute-to-minute changes or changes at 10 min intervals. To realize this goal, one of the four input channels of the cassette tape recorders[17] can be used as an event marker so that physiological responses such as HR can be easily equated to work cycles. A study [21] of truck assembly line workers and medical nurses illustrates this strategy. The use of continuous HR recordings enabled a detailed breakdown of the work activities for these two groups. Results demonstrated that the HR responses for the nurses, in contrast to those for the assemblers, were sensitive to emotional stress. The HR responses for the assemblers were found to be more a function of their physical work activity. These findings emerged despite data on energy expenditure indicating that both occupations could be classified as light industrial work.

Another study further demonstrates how these measures can be applied in order to identify components of the task related to operator stress.[22-24] Utilizing modern computer and sensor technology, a methodology was developed for unobtrusively monitoring industrial assembly line workers for HR, RR, and BP. The methodology allows for on-line data collection simultaneously for six operators.

In one study,[25] HR data was transmitted from managers throughout the work day. This data was related to the activities experienced by the managers during the day, and illustrates the use of these measures for evaluating white collar work stress.

Biochemical Measures

Biochemical measures such as adrenaline and noradrenaline can be obtained from various fluids of the body, such as urine, blood, and saliva. Analysis of urine has been the most popular choice, largely because of its ready availability and the minimum of disturbance its collection causes in the work situation.

Justification for the use of these measures in the analysis of occupational stress stems from:

1. Clinical evidence that has implicated the chronic elevation of adrenaline and noradrenaline in functional disturbances in various organs and organ systems, which in turn may lead to psychosomatic and cardiovascular diseases.
2. The fact that the natural daily rhythms of these measures allow for evaluation of shift work in terms of the degree to which these measures change from their normal rhythmic pattern when changes in the work shift are initiated.
3. The availability of numerous automated processes for isolation (from urine) and subsequent measurement of the biochemical constituents of interest.
4. The ability to obtain the needed fluids easily during and after work periods.

Evidence linking chronic evaluations of adrenaline and noradrenaline with cardiac disease affords a meaningful criteria by which strategies in the job placement of workers with preclinical signs of cardiac disease can be evaluated. Other physiological measures could also conceivably serve this function. For example, it may be quite reasonable to speculate that repeated and sustained increases in BP, especially in occupations where the physical work loads are negligible, may ultimately result in chronic disturbances such as hypertension.

The biochemical measures, especially those that display daily rhythmic behavior (i.e., high and low levels of excretion throughout the day), are very appropriate for the examination of various shift work policies. In one study,[26] railroad workers followed 3 weeks of day work with 3 weeks on the night shift. On examining the pattern of adrenaline excretion in the workers, it was found that very little adjustment occurred after 3 days of night work. What this basically means is that adrenaline levels in the workers still remained high during the day, which is consistent with the normal pattern for adrenaline secretion. Unfortunately, the workers were now expected to sleep during this time, implying that the high adrenaline levels were one of the sources of sleep problems that are often reported by shift workers. In fact, the adrenaline excretion patterns for these workers still had not adjusted by the third week of night work.

The ability to obtain indexes of worker stress after work hours is a distinct advantage. At times there are reasons to believe that constraints in a particular work situation may impede the individual's response to stress. On the other hand, the worker may be more apt to exhibit these responses after work hours. The relative ease involved in urine collection could then enable one to validate whether work-related stress is indeed being manifested after work hours.

In contrast to measures such as HR and RR, the practical restriction on the collection procedures for biochemical measures necessarily dictates that they reflect responses to stress over fairly long periods (e.g., several hours or days). As a result, biochemical measures are relatively incapable of identifying which components of the task are most stressful. This factor alone is most critical in selecting between these two categories of measures for the purpose of evaluating occupational stress. Practical problems with biochemical measures usually involve the need to exercise rigid control in the kinds and quantity of food and liquid ingested by the worker. A more complete treatment of the considerations governing the use of biochemical measures is given in Levi.[27]

Psychological Measures

Many standardized measurement devices are available that can be used for evaluating occupational stress. Some of these are discussed below.

Drawing on Selye's notion that stress may be characterized by the intensity of the demand for readjustment or adaptation brought about by some agent or situation, two investigators[28] have developed a Social Readjustment Rating Scale. This scale attempts to quantify the stress potential of various life events that often follow a series of life changes. A linear relationship was found between the magnitude of the life crises and the risk of change in health. When applied to studies on occupational stress, the scale offers a method for recognizing and evaluating which types of responses (e.g., psychosomatic complaints) may have actually originated from factors extraneous to the workplace.

A popular method of validating physiological response criteria has been the use of mood checklists. These measurement devices serve to gauge the worker's feeling state at any given time and are relatively simple to administer. One such checklist[29] has been found to be capable of differentiating stress from arousal. The validity of this checklist has been supported by studies that have indicated a differential sensitivity of these two factors to a variety of environmental and task effects. As an example, after a prolonged and monotonous repetitive task, significant increases in self-reported stress were found together with significant decreases in self-reported arousal. Another mood checklist, the Multiple Affect Adjective Checklist,[30] has proved useful as a measure of immediate or daily anxiety levels and has been found to correlate with work performance over time.

Performance Measures

In utilizing performance measures in studies of occupational stress, the worker's particular task will necessarily govern which performance criteria will be selected. Quantity, quality, and variability of work performance are among the most frequent measures used.

Individual attributes often can significantly affect performance and the ensuing inferences regarding the severity of stress associated with the worker's task. Examples include: (1) the capacities of the worker and his or her strategies used to reconcile the job's demands with those capacities[31]; (2) the changes in strategy under stress; (3) the systematic changes that occur with age, for example, changes related to signal-to-noise ratio (i.e., decreased reliability in detecting a signal), memory, and speed[32]; and (4) the effects of the worker's personality.[33]

6.6.4 SOURCES OF OCCUPATIONAL STRESS

Merely listing sources of stress on the basis of intuition alone offers no sound framework for ultimately actuating job redesign policies aimed at improving the quality of working life. More important is assessing the quantitative and qualitative effects of these presumed sources, the kinds of people in which to expect these effects, the kinds of situations that can potentially modulate the effects, and the kinds of stress management procedures that would be most effective. Still, some kind of source classification scheme would seem to be a prerequisite.

Generally, stress in the work environment stems from the following three sources: (1) the individual (e.g., personality attributes), (2) the environment (social and physical environments at work and at leisure), and (3) the task (e.g., mental load, pacing). This breakdown is probably an oversimplification; a complex interaction between these sources is more realistic. However, for the sake of simplicity, each of these is discussed separately. Exhibit 6.6.3 illustrates the ways in which these sources and their components relate to mental health and cardiac risk factors.

Individual Sources

Three main sources of occupational stress that are related to the individual concern (1) health-related factors, (2) the degree of match between the job workers actually perform and their capabilities, likes, and dislikes as they relate to the work environment and (3) the personality makeup of the individual.

With respect to the worker-job mismatch, approaches used to locate and predict job stress consider the relationship between stress and the degree of fit between the person and the work environment. Within these so-called person-environment fit models,[35,36] the interaction between the characteristics of the person and the potential source of stress in the work environment determines the degree of either coping or maladaptive behavior and subsequent stress-related symptoms. The larger the mismatch, the greater the stress experienced by the worker.

Assembly line work has received much attention in terms of the delineation of the kinds of

Exhibit 6.6.3 A Model of Stress at Work

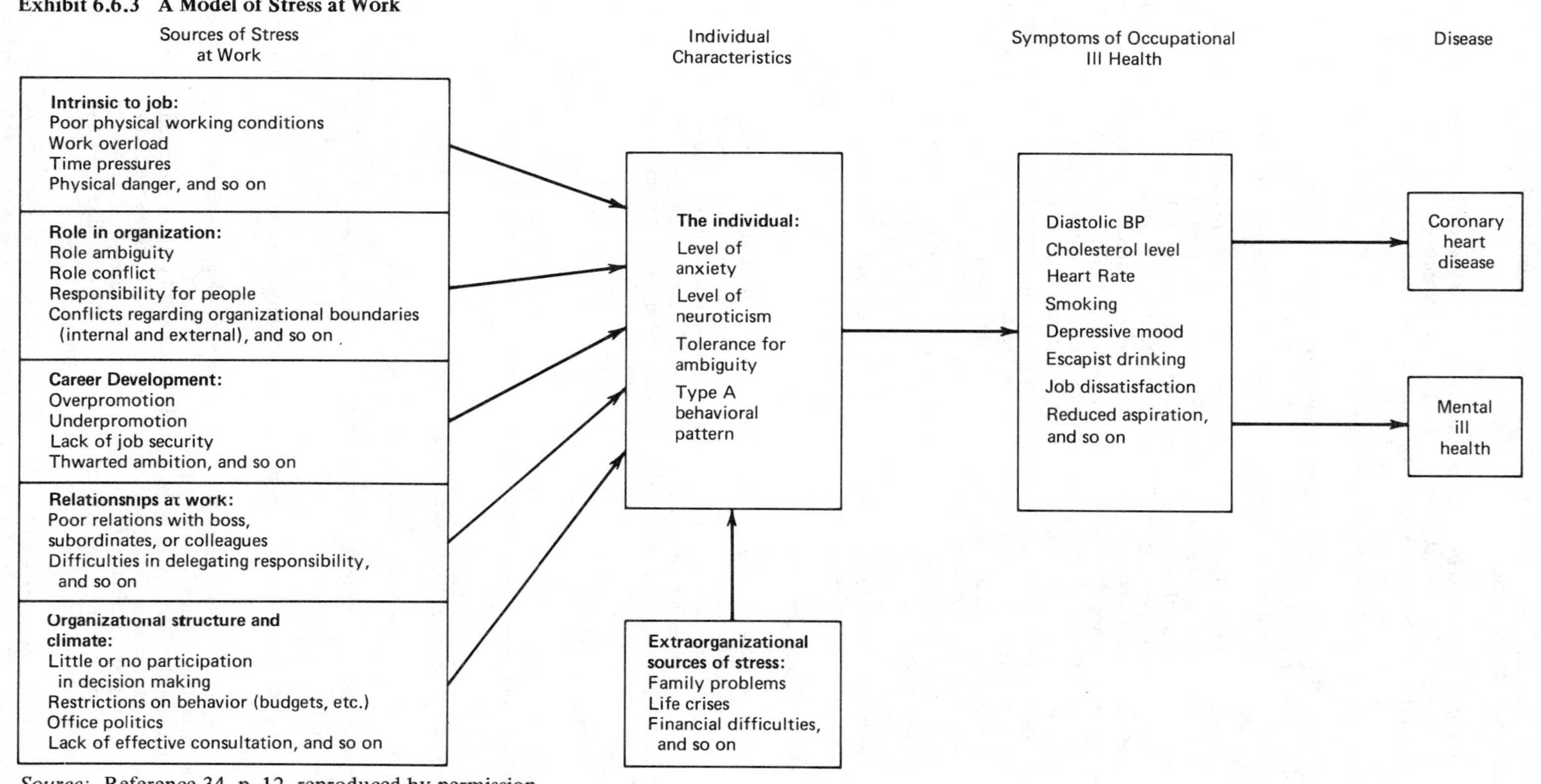

Source: Reference 34, p. 12, reproduced by permission.

personalities that may either conflict with or successfully adapt to this type of work. Existing evidence implies the following:

> *(1) anxious or autonomically labile individuals (i.e., individuals with very active physiological systems) will find assembly-line work especially frustrating; (2) persons with high ego strength will cope most successfully with paced operations; (3) persons high on sensation seeking will do poorly on the (assembly) line; and (4) persons high on authoritarianism (individuals resistant to changes in their duties) will adapt well to assembly-line jobs.*[37]

Other personality features that have been implicated as significant factors in modifying the individual's response to various work-related situations have been the degree of extroversion or neuroticism[38] (measured by the Eysenck Personality Inventory) and the Type A versus Type B personality.[39]

The Eysenck Personality Inventory[38] has been useful in predicting individual susceptibility to various jobs. For example, machine-paced work is often done under conditions preventing social interaction, a factor most likely to provoke discontent in extroverted, as opposed to introverted, workers. Scores on this personality inventory have also been associated with the degree of arousal manifest in the individual (see section on arousal). Since low arousal is frequently associated with fatigue, boredom, increased accidents, and decreased quality of performance, certain work situations may warrant the use of these personality measures in facilitating job placement policies.

The Type A and Type B personalities have been mostly associated with the development of coronary heart disease. Certain types of behavioral patterns, labeled Type A, have been characterized by "excessive drive, aggressiveness, ambition, involvement in competitive activities, frequent vocational deadlines . . . [and] an enhanced sense of time urgency. . . ."[40] The relative absence of these behavior patterns has been called Type B and is basically the converse of the Type A pattern. The Type A person, in contrast to the Type B, has been shown to be much more susceptible to coronary heart disease. The implications of these behavioral types would seem to be for worker selection, that is, in minimizing occupational stress and perhaps coronary heart disease by providing a better job-worker fit. However, for this to become practically feasible, reliable techniques for adequately discriminating the Type A person need to become available.

Environmental Sources

Social and physical factors that can potentially affect worker stress are fully discussed in other chapters of this handbook as follows: organizational factors (2.1); job structure (2.5); and physical work environment factors, such as noise (6.10), illumination (6.11), and the presence of toxic substances (6.13). Whether the source of the worker's stress on the job is related to problems arising from the home or personal life can potentially be established from the Social Readjustment Rating Scale (see section on psychological measures).

Task-Related Sources

Generally, two basic approaches have been utilized by investigators in studying task-related stresses, namely, surveys of stressors in work situations and statistically balanced experiments (see Chapter 13.4).

Surveys of Stressors in Work Situations

Within this category, establishing whether the stresses are person related, environment related, or task related is not feasible. However, this approach does provide an effective means of collecting data on large segments of the population engaged in various work-related activities.

An excellent example of these kinds of surveys is the NIOSH survey[35] in which 2010 men employed in 23 diversified jobs were given questionnaires and from which a subsample of 390 men representing eight job categories had detailed physiological measures taken. Some conclusions drawn were:

1. Personality variables had no direct effect on psychological and physiological strain (somewhat contradicting findings reported earlier).
2. Job dissatisfaction appears to be strongly influenced by underutilization of skills and abilities, simple and repetitive work, low participation in making decisions which affect one's work, job insecurity, and poor social support from one's immediate supervisor and from others at work.[41]
3. Assemblers and relief workers on machine-paced assembly lines have higher levels of stress and strain than any of the other workers studied in this survey.

Another survey examined a large cross section of Swedish and American workers.[42] Two sources of occupational stress were focused on: job demands (represented as work overload and time pres-

Exhibit 6.6.4 Effects of Job Demands and Job Decision Latitude on Stress. Note That Under Any Level of Job Demands, the Amount of Stress Will Be a Function of the Degree of Job Decision Latitude. This Effect is Most Pronounced Under High Job Demands, Where a Similar Increase in Decision Latitude Significantly Reduces the Probability of Developing Stress-Related Symptoms

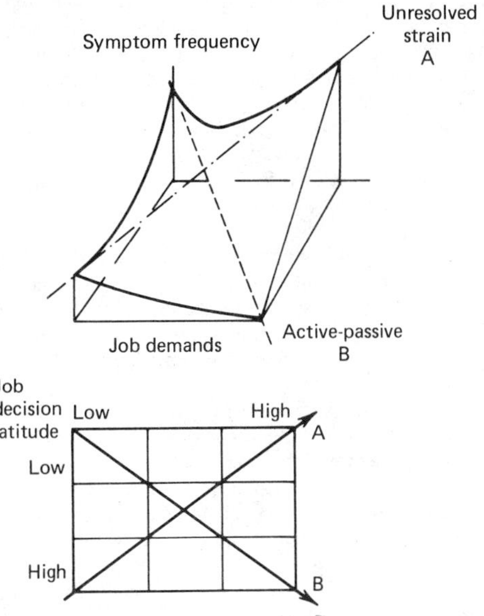

Source: Reference 42, p. 294, reproduced by permission.

sures in Exhibit 6.3.3) and job decision latitude (the decision discretion permitted the worker). Job demands were viewed as that which "places the individual in a motivated and energized state of stress," and job decision latitude as the "constraint which modulates the release as transformation of stress (potential energy) into the energy of action."[43] A model was developed which asserted that psychological strain results from the "joint effects of the demands of a work situation and the range of decision-making freedom (discretion) available to the worker facing those demands."[44]

Exhibit 6.6.4 illustrates the combinations of job demands and job decision latitude which are associated with probabilities (percentages) of experiencing relatively severe depression or exhaustion (the upper arrow being in the direction of unresolved strain). The author of this study warns that unclear relationships will emerge if results are examined for (linear) trends of either demands or decision latitude alone. The joint effects of job strain and job decision latitude led this investigator to the conclusion that "literature lamenting the stressful burden on executive decision making misses the mark . . . constraints on decision making, not decision making per se, are the major problem and this problem affects executives as well as workers in low status jobs with little freedom for decision."[45]

Experiments on Stressors in Work Situations

Much of the available evidence resulting from experiments relating the effects of task design configurations to stress have been conducted on comparative studies between machine-paced (M/P) and self-paced (S/P) work. The design of these studies was such that the total amount of work performed as well as the duration was identical under both pacing conditions. In addition, the same operators performed, in alternating order, in both conditions.

The tasks utilized in these studies were predominantly of the following types: light assembly operations, man-machine computer interactions, arm ergometer, and visual inspection. For heavy physical work, energy expenditure was utilized, whereas for tasks consisting of light physical work, SA was the criteria for evaluation. Some conclusions from these studies are summarized here.

The earlier studies comparing the effectiveness of M/P versus S/P work attended principally to the physiological consequences of these pacing systems. Evidence indicates[46] that for tasks requiring heavy physical work, the physiological efficiency of the younger operators (20 to 45 years of age) is highest during S/P work, whereas for the older operators, it is highest during M/P work. The higher the physiological efficiency of the operator, the longer he or she can be expected to sustain higher production and lower stress levels as compared to the operator with lower physiological efficiency.

Subsequent physiological studies have concentrated more on work environments requiring very light physical work but higher levels of decision making; a situation becoming increasingly more common in work situations. Utilizing SA measures of work stress, higher stress levels were found to be present during S/P, as compared to M/P, work situations.[47] The rationale given for this finding was that a timekeeping factor, referred to as an "internal pacing mechanism," is present during S/P work. In M/P work, on the other hand, this responsibility for timekeeping is removed from the worker and instead maintained externally by the machine. Assuming then that the SA measure is sensitive to changes in mental work load, these findings imply that S/P work imposes a higher mental work load on the worker. However it only *suggests* that such work is more stressful. To make a stronger argument for the worker's experiencing a significantly greater amount of stress in S/P work, additional data, as might be obtained from subjective measures, must necessarily be integrated.

In a later study,[48] these basic findings were supported for tasks requiring high decision making, but not for tasks where this requirement is low. Results also indicated that the error rates committed by the operator in the "low-stress" task were higher than those committed in the "high-stress" task—providing additional support for the notion of an optimum level of arousal in the individual (see section on arousal). When the effects associated with the availability of performance feedback are observed, it was shown[49] that as the accuracy of the performance feedback increases in S/P tasks, the stress level decreases. Specifically, for such tasks without performance feedback, the stress level is much higher than for M/P performance. This difference however, reverses itself as the knowledge of results systematically increases.

Finally, in an industrial study involving light electronic assembly operations,[50] it was concluded that M/P work had twice as many non-work-related movements as S/P work. Within each pacing condition, the number of non-work-related movements was greater for tasks requiring high decision making activity and lower for tasks requiring low decision making. Non-work-related movements are signs of potential stress and have an impact on the efficiency of human performance by directing the efforts of the worker to nonproductive, as opposed to productive, tasks. In the same study,[51] utilizing attitudinal, intelligence, and personality questionnaires, it was concluded that operators who were assertive, imaginative, shrewd, self-sufficient, and more intelligent preferred S/P rather than M/P work. In contrast, operators who were humble, practical, forthright, group dependent, and less intelligent preferred M/P to S/P work. This kind of information can potentially serve the function of guidance and placement of personnel.

Although the experiments on M/P and S/P work that were performed in statistically balanced experiments (see Chapter 13.4) had the advantage of fairly precise control, the artificially short durations of these experiments make their results, namely, that S/P work was more stressful, somewhat questionable for generalizing to actual industrial situations. Statistically unbalanced industrial studies (e.g., Frankenhaeuser and Gardell[15]) have, on the other hand, supported the notion that M/P work is more stressful. However, in these studies we often do not know if the observed differences were in fact due to the work pace, the people (the same workers were not studied under both work paces), or the job (the job content and job structure were not the same under the different work paces). In addition, different studies use different measurements of stress (see Section 6.6.3), and the possibility exists that these measures are not reflecting the same phenomena. However, a recent industrial study,[52] which utilized 33 experienced operators in a statistically balanced design (where the same operators performed the identical job under both M/P, and S/P conditions), did conclude that there were no differences between M/P and S/P work based on measures of systolic, diastolic, and mean BP, mean HR, SA, and RR. Hence it appears that no conclusive evidence currently exists as to which one of the two work paces is more stressful.

6.6.5 MANAGEMENT OF OCCUPATIONAL STRESS

Ideally, we should not have to consider strategies for the management of occupational stress. With a vast body of literature on industrial-organizational psychology, principles of worker selection and placement, consideration of the human factor in the design of the workplace, and task- and job-related stress responses of workers employed in numerous occupations, it would perhaps appear more realistic to set our goals on the *prevention* of occupational stress. This approach, in fact, encompasses a large aspect of the stress management program—namely, the preventive approach. Unfortunately, prediction, the necessary prerequisite, is often inaccurate. The long-term effects of

Exhibit 6.6.5 Coping Responses Most Frequently Used (in Descending Order)

1. Let people know exactly where they stand.
2. Consider a range of plans for handling the situation–set priorities.
3. Forget work when finished for the day.
4. Try to find out more about the situation–seek out additional information.
5. Try to reassure yourself that everything is going to work out all right.
6. Make sure people are aware you are doing your best.
7. Try to see the humor of the situation.
8. Follow proper channels of procedure to cover yourself.
9. Try to take some immediate action on the basis of your preset understanding of the situation.
10. Think objectively about the situation and keep your feelings under control.
11. Do not let the problem go until you have solved it or reconciled it satisfactorily.
12. Talk about the situation with someone else at work.

Source: Reference 53, reproduced by permission.

the design of workplaces and organizations are often not sufficiently understood, and thereby designers of workplaces are frequently incapable of accommodating the effects of random events into the interactive complexity of the particular work environment. The *curative* approach to stress management will therefore most likely need to be incorporated along with preventive strategies. This approach is frequently brought about through job redesign strategies that are based on some systematic evaluation of the worker in relation to the task and/or work environment.

The development and evaluation of individual coping strategies is rapidly becoming recognized as an effective means of reducing high levels of worker stress. Emphasis along these lines is justified on the basis that what is manifest and evaluated as stress by the worker is inextricably related to the short- and long-term consequences of the coping strategies the worker has available. Some effective coping strategies are listed in Exhibit 6.6.5.

Stress management strategies used predominantly for either blue or white collar workers are discussed in the next two subsections. It should be kept in mind, however, that these strategies are often applicable to both work sectors.

Stress Management in Blue Collar Work

Stress management in blue collar workers can be most effectively realized through job redesigning strategies. An illustration would be removing a worker from shift work or changing the duration of the day and/or night shift on the basis of biochemical and psychological (mood checklists) data. For example, a worker whose adrenaline rhythms exhibit relatively rapid changes that are in line with changes in the work shift, and whose social and domestic activities are not significantly upset by these changes, may be better suited for less prolonged periods on any one shift.

Another example might be in machine-paced operations. Fixed work paces do not account for changes in workers' abilities through the workday. By adjusting the work pace so that it is optimally matched to time varying operator capabilities as determined by physiological, performance, and psychometric measures, or alternatively, the worker's self-judgment on setting the pace, periods of potential underload and overload can be avoided. If in the evaluation of a work situation these measures are found to conflict, changes would need to be made that reflect the interests of both the worker and the company. That is, priorities or weights associated with each category of measures would need to be arrived at.

Use of reliable methods for identifying sources of worker stress is often critical when initiating the job design or redesign process. One such method is the ergonomic job analysis,[54] a systematic analysis of work situations whose underlying objective is to enhance the worker's well-being. This is generally accomplished by constructing a checklist to assist in the assessment of task components that constitute the overall work load. The ergonomic job analysis is especially useful in that it may be preventive or curative, depending on whether it is applied to new or long-established work methods.

A study examining work stress in telegraphists[55] concluded with a section on preventive and remedial measures, with extensive applications for the blue collar work sector. These measures include:

1. Use of previous work and sickness records and history of frank emotional disturbances to help in proper selection.
2. Instituting job enlargement; increased identification, involvement, and communication; and the training of management and supervisory staff in the social psychology of industry in order to increase their skill in personal relationships.

3. Decentralizing authority, increasing initiative and responsibility, and decreasing administrative rigidity.

4. The implementation of a mental health program as part of an occupational health service, assisting in matching personality characteristics with job demands and in predicting mental difficulty. Such a program would include education in personal and social psychology; marital adjustment; nutrition; the effects of smoking, drinking, and other drug taking; and the use of leisure time.

Other factors that often need to be attended to are the complexity of travel to work and back and its effect on absenteeism and turnover, the need for knowledge of results, and instilling in the worker a sense of perceived autonomy. This latter factor is especially important in assembly line work where operators prefer to exert control over both the pace and the methods. The study on sawmill workers discussed previously[15] illustrates this nicely. The investigators of that study concluded that the interest in and worthwhileness of making skilled and economically important decisions in an extremely short time is negated by the fact that the pace and methods are no longer controlled by the worker, but by the machine. The inevitable result is that the worker is not allowed the time necessary to do a good job relative to his or her skills.

Stress Management in White Collar Work

A comprehensive review of the literature on strategies for stress management applicable primarily to white collar workers can be found in Newman and Beehr.[56] The problem these investigators frequently encountered was the lack of evaluation research in this area, so that many of the strategies seemed suggestive at best. This situation can be attributed largely to the tendency for researchers in the area of occupational stress to assess the response of workers as compared to some control (neutral) condition. Little attention has been given to the logical next step—designing, implementing, and evaluating a stress management program.

The psychotherapeutic approach, as applied to individuals whose problems are exhibited in the work environment, is rapidly gaining acceptance. Reports have it[57] that large corporations are setting up in-house psychiatric facilities and that smaller companies are bringing in outside consultants to help employees with emotional problems stemming from the job. For example, the five following experiences have been found to be particularly stressful to managers: (1) the first job, (2) the first promotion, (3) relocation to a new area, (4) the first supervisory position, and (5) retirement. Companies can lessen the strain on managers by offering professional counsel to help them deal with these phases.

Many prescribed behavioral regimens aim primarily to reduce the presence of cardiac risk factors such as smoking, high BP, and obesity. A stress clinic in England[58] offers an array of sophisticated diagnostic, analytic, and predictive checks on a patient's predisposition to coronary heart disease, and a corrective and preventive treatment package which includes nutritional, gymnastic, and postural management. In terms of the ultimate savings in both human and economic cost, companies should perhaps consider incorporating these kinds of programs.

Finally, relaxation and biofeedback programs made available by companies may provide simple and relatively inexpensive ways in which conditions such as high BP could be controlled. A simple relaxation response based on a procedure similar to transcendental meditation has been shown to be very effective in decreasing BP.[59] This route may be especially desirable for individuals who want to avoid a continuous intake of medication in order to control BP. These authors have suggested that programs be established in companies where a "relaxation response break" (in which relaxation techniques specifically aimed at reducing muscular tension are employed) could be made available in place of a coffee break. Biofeedback approaches can allow an individual to learn to modify responses such as their BP by continuously providing information, in the form of a visual or auditory signal, about the momentary degree of presence of that physiological measure. More evaluative research with this technique, however, needs to be carried out within the context of the company.

6.6.6 CONCLUSION

Methods are available for measuring, identifying, and rectifying sources of occupational stress that could improve worker health, job satisfication, quality of work, and productivity. Hence, there is no need to maintain an environment in which workers are unduly stressed. Uncertainty is the single biggest factor influencing stress at the workplace. These factors include uncertainties associated with: inadequate performance feedback; variations in mental load associated with task performance; length and variability of cycle time in machine-paced work; nonspecified performance goals; job insecurity; inadequate training to cope with work situations; and anticipation in man-computer interactive work. Another factor that can reduce stress at the workplace includes work design that provides a balance between task complexity and the level of decision latitude that an employee can exercise. Worker-job mismatch increases occupational stress. This may be reduced or alleviated through effective personnel selection and placement strategies.

REFERENCES

1. H. SELYE, "The Evaluation of the Stress Concept," *American Scientist*, Vol. 61, 1973, pp. 692–699.
2. L. LEVI, "Introduction: Psychosocial Stimuli, Psychophysiological Reactions, and Disease," *Acta Medica Scandinavica*, Supplement 528, Vol. 191, 1972, pp. 11–27.
3. H. L. DAVIS, T. W. FAULKNER, and C. I. MILLER, "Work Physiology," *Human Factors*, Vol. 11, 1969, pp. 157–166.
4. S. KONZ, *Work Design*, Grid Publishing Company, Columbus, OH, 1979.
5. J. W. H. KALSBEEK, "Sinus Arrhythmia and the Dual Task Method in Measuring Mental Load," in W. T. SINGLETON, J. G. FOX, and D. WHITFIELD, Eds., *Measurement of Man at Work*, Taylor & Francis, London, 1971.
6. J. W. H. KALSBEEK, "Do You Believe in Sinus Arrhythmia?," *Ergonomics*, Vol. 16, 1973, pp. 99–104.
7. J. W. H. KALSBEEK, "Measurement of Mental Work and of Acceptable Work: Possible Applications in Industry," *International Journal of Production Research*, Vol. 7, 1968, pp. 33–45.
8. G. D. OGDEN, J. M. LEVINE, and E. J. EISNER, "Measurement of Workload by Secondary Tasks," *Human Factors*, Vol. 21, 1979, pp. 529–548.
9. R. C. WILLIGES and W. W. WIERWILLE, "Behavioral Measures of Aircrew Mental Workload," *Human Factors*, Vol. 21, 1979, pp. 549–574.
10. W. W. WIERWILLE, "Physiological Measures of Aircrew Mental Workload," *Human Factors*, Vol. 21, 1979, pp. 575–593.
11. E. GRANDJEAN and K. KOGI, "Introductory Remarks," in K. HASHIMOTO, K. KOGI, and E. GRANDJEAN, Eds., *Methodology in Human Fatigue Assessment*, Taylor & Francis, London 1971.
12. K. F. H. MURRELL and B. FORSAITH, "Laboratory Studies of Repetitive Work II: Progress Report on Results From Two Subjects," *International Journal of Production Research*, Vol. 2, 1963, pp. 247–263.
13. C. CAMERON, "Fatigue Problems in Modern Industry," *Ergonomics*, Vol. 14, 1971, pp. 713–720.
14. C. CAMERON, "A Theory of Fatigue," *Ergonomics*, Vol. 16, 1973, pp. 633–648.
15. M. FRANKENHAEUSER and B. GARDELL, "Underload and Overload in Working Life: Outline of a Multidisciplinary Approach," *Journal of Human Stress*, Vol. 2, 1976, pp. 35–46.
16. A. T. WELFORD, "Stress and Performance," *Ergonomics*, Vol. 16, 1973, pp. 567–580.
17. W. S. SMITH and C. O'BRIEN, "A System for Rapid Analysis of Long-Term Recordings of Heart Rate and Other Physiological Parameters," *Biomedical Engineering*, Vol. 11, 1976, pp. 128–131.
18. U. REISCHL, D. M. MARSCHALL, and P. REISCHL, "Radiotelemetry-Based Study of Occupational Heat Stress in a Steel Factory," *Biotelemetry*, Vol. 4, 1977, pp. 115–130.
19. C. ELEY, R. GOLDSMITH, D. LAMAN, and B. M. WRIGHT, "A Miniature Indicating and Sampling Respirometer (MISER)," *Journal of Physiology*, Vol. 256, 1976, pp. 59–60.
20. K. RODAHL, and Z. VOKAC, "Work Stress in Norwegian Trawler Fishermen," *Ergonomics*, Vol. 6, 1977, pp. 633–642.
21. C. O'BRIEN, W. S. SMITH, R. GOLDSMITH, M. FORDHAM, and G. L. E. TAN, "A Study of the Strains Associated with Medical Nursing and Vehicle Assembly," in C. MACKAY and T. COX, Eds., *Response to Stress: Occupational Aspects*, IPC Science and Technology Press, London, 1979.
22. J. L. KNIGHT, G. SALVENDY, L. A. GEDDES, J. JANS, and E. SMITH, "Monitoring Respiratory and Heart Rate of Assembly-Line Factory Workers," *Medical and Biological Engineering and Computing*, Vol. 18, 1980, pp. 797–798.
23. J. L. KNIGHT, G. SALVENDY, and L. A. GEDDES, "A Minicomputer System for Long-Term Automatic Blood Pressure Monitoring," *The Annals of Biomedical Engineering*, Vol. 7, 1979, pp. 369–374.
24. J. L. KNIGHT, L. A. GEDDES, and G. SALVENDY, "Continuous Unobstrusive Performance and Physiological Monitoring of Industrial Workers," *Ergonomics*, Vol. 23, 1980, pp. 501–506.

25. J. K. HENNIGAN and A. W. WORTHAM, "Analysis of Workday Stresses and Industrial Managers Using Heart Rate as a Criterion," *Ergonomics*, Vol. 18, 1975, pp. 675–681.

26. T. AKERSTEDT, "Inversion of the Sleep-Wakefulness Pattern: Effects on Circadian Variations in Psychophysiological Activation," *Ergonomics*, Vol. 20, 1977, pp. 459–474.

27. L. LEVI, "Methodological Considerations in Psychoendocrine Research," *Acta Medica Scandinavica*, Supplement 528, Vol. 191, 1972, pp. 28–54.

28. T. H. HOLMES and R. H. RAHE, "The Social Readjustment Rating Scale," *Journal of Psychosomatic Research*, Vol. 11, 1967, pp. 213–218.

29. C. MACKAY, T. COX, G. BURROWS, and T. LAZZERINI, "An Inventory for the Measurement of Self-Reported Stress and Arousal," *British Journal of Social and Clinical Psychology*, Vol. 17, 1978, pp. 283–284.

30. M. ZUCKERMAN, B. LUBIN, L. VOGEL, and E. VALERIUS, "Measurement of Experimentally Induced Affects," *Journal of Consulting Psychology*, Vol. 28, 1964, pp. 418–425.

31. C. D. WELFORD, "Mental Work-Load as a Function of Demand, Capacity, Strategy, and Skill," *Ergonomics*, Vol. 21, 1978, pp. 151–167.

32. A. T. WELFORD, "Thirty Years of Psychological Research on Age and Work," *Journal of Occupational Psychology*, Vol. 49, 1976, pp. 129–138.

33. R. STAGNER, "Boredom on the Assembly Line: Age and Personality Variables," *Industrial Gerontology*, Vol. 2, 1975, pp. 23–44.

34. C. L. COOPER and J. MARSHALL, "Occupational Sources of Stress: A Review of the Literature Relating to Coronary Heart Disease and Mental Ill Health," *Journal of Occupational Psychology*, Vol. 49, 1976, pp. 11–28.

35. R. D. CAPLAN, S. COBB, J. R. P. FRENCH, R. V. HARRISON, and S. R. PINNEAU, *Job Demands and Worker Health*, U.S. DHEW Publication No. (NIOSH) 75-160, Washington, DC, 1975.

36. D. COBURN, "Job-Worker Incongruence: Consequences for Health," *Journal of Health and Social Behavior*, Vol. 16, 1975, pp. 198–212.

37. R. STAGNER, "Boredom on the Assembly Line," p. 40.

38. H. J. EYSENCK, *Biological Basis of Personality*, Charles C Thomas: Springfield, IL, 1967.

39. C. D. JENKINS, R. H. ROSENMAN, and R. FRIEDMAN, "Development of an Objective Psychological Test for the Determination of the Coronary Prone Behavior Pattern," *Journal of Chronic Diseases*, Vol. 20, 1967, pp. 371–379.

40. C. D. JENKINS et al., "Development of an Objective Psychological Test for the Determination of the Coronary Prone Behavior Pattern," p. 371.

41. R. D. CAPLAN et al., *Job Demands and Worker Health*.

42. R. A. KARASEK, "Job Demands, Job Decision Latitude, and Mental Strain: Implications for Job Redesign," *Administrative Science Quarterly*, Vol. 24, 1979, pp. 285–308.

43. R. A. KARASEK, "Job Socialization and Job Strain: The Implications of Two Related Psychosocial Mechanisms for Job Design," unpublished report, Department of Industrial Engineering and Operations Research, Columbia University, New York, 1979.

44. R. A. KARASEK, "Job Socialization and Job Strain," p. 287.

45. R. A. KARASEK, "Job Socialization and Job Strain," p. 303.

46. G. SALVENDY and J. PILITSIS, "Psychophysiological Aspects of Paced and Unpaced Performance as Influenced by Age," *Ergonomics*, Vol. 14, 1971, pp. 703–711.

47. I. MANENICA, "Comparison of Some Physiological Indices During Paced and Unpaced Work," *International Journal of Production Research*, Vol. 15, 1977, pp. 261–275.

48. G. SALVENDY and A. P. HUMPHREYS, "Effects of Personality, Perceptual Difficulty and Pacing of a Task on Productivity, Job Satisfaction and Physiological Stress," *Perceptual and Motor Skills*, Vol 49, 1979, pp. 219–222.

49. J. KNIGHT, and G. SALVENDY, "Effects of Task Feedback and Stringency of External Pacing on Mental Load and Work Performance," *Ergonomics*, in press.

50. B. BASILA, S. SUOMINEN, G. SALVENDY, and G. P. MCCABE, "Non-Work Related Movements in Machine-Paced and in Self-Paced Work," *Proceedings of the Human Factors Society*, 23rd annual meeting, 1979, pp. 149–153.

51. S. SANDERS, G. SALVENDY, and J. L. KNIGHT, "Attitudinal, Personality, and Age Characteristics for Machine-Paced and Self-Paced Operations," *Proceedings of the Human Factors Society*, 23rd annual meeting, 1979, pp. 153–157.

52. G. SALVENDY, J. L. KNIGHT, and V. L. ANDERSON, "Physiological and Psychological Effects of Machine-Paced and Self-Paced Work: An Industrial Study," unpublished report, School of Industrial Engineering, Purdue University, West Lafayette, IN, 1981.

53. P. DEWE, D. GUEST, and R. WILLIAMS, "Methods of Coping With Work-Related Stress," in C. MACKAY and T. COX, Eds., *Response to Stress*; *Occupational Aspects*, IPC Science and Technology Press Limited, Surrey, England, 1979.

54. G. C. E. BURGER and J. R. DEJONG, "Evaluation of Work and Working Environment in Ergonomic Terms–Aspects of Ergonomic Job Analysis," *Ergonomics*, Vol. 5, 1962, pp. 185–193.

55. D. FERGUSON, "A Study of Occupational Stress and Health," *Ergonomics*, Vol. 15, 1973, pp. 649–663.

56. J. E. NEWMAN and T. A. BEEHR, "Personal and Organizational Strategies for Handling Job Stress: A Review of Research and Opinion," *Personnel Psychology*, Vol. 32, 1979, pp. 1–43.

57. D. ROBBINS, "Psychiatric Consultation in the World of Work," U.S. DHEW Publication No. (NIOSH) 78–140, Washington, DC, 1978.

58. "Putting a Stop to the Strain Drain," *Personnel Management*, August 1976, pp. 1273–1274.

59. R. K. PETERS and H. BENSON, "Time Out From Tension," *Harvard Business Review*, Vol. 56, 1978, pp. 120–124.

BIBLIOGRAPHY

COOPER, C. L., and R. PAYNE, Eds., *Stress at Work*, New York, 1979.

HASHIMOTO, K., K. KOGI, and E. GRANDJEAN, Eds., *Methodology in Human Fatigue Assessment*, Taylor & Francis, London, 1971.

LEVI, L., Ed., *Society Stress and Disease: Working Life*, Vol. 4, Oxford University Press, London, 1979.

MACKAY, C., and T. COX, Eds., *Response to Stress; Occupational Aspects*, IPC Science and Technology Press, Surrey, England, 1979.

MONAT, A., and R. S. LAZARUS, Eds., *Stress and Coping*, Columbia University Press, New York, 1977.

MORAY, N., Ed., *Mental Workload: Its Theory and Measurement*, Plenum, New York, 1979.

Reducing Occupational Stress, Report No. 78-140, U.S. DHEW (NIOSH), Washington, DC, 1978.

SALVENDY, G., and M. J. SMITH, Eds., *Machine-Pacing and Occupational Stress*, Taylor & Francis, London, 1981.

SELYE, H., *Stress in Health and Disease*, Butterworth, Boston, 1976.

SHARIT, J., and G. SALVENDY, "Occupational Stress: Review and Reappraisal," *Human Factors*, in press.

SHARIT, J., G. SALVENDY, and M. P. DEISENROTH, "External and Internal Attentional Environments: I. The Utilization of Cardiac Deceleratory and Acceleratory Response Data for Evaluating Differences in Mental Workload Between Machine-Paced and Self-Paced Work," *Ergonomics*, in press.

SINGLETON, W. T., J. G. FOX, and D. WHITFIELD, Eds., *Measurement of Man at Work*, Taylor & Francis, London, 1971.

CHAPTER 6.7

Work Performance and Handicapped Persons

THOMAS J. ARMSTRONG
DEV S. KOCHHAR
The University of Michigan

6.7.1 INTRODUCTION

Meaningful employment for the disabled can link them into the fabric of society and enhance their sense of worth. A physical or mental disability is recognized as a handicap to employment only when, in spite of reasonable accommodations, it results in an inability to perform a given job in a manner that is efficient and safe for the worker and those around him or her or when the job aggravates the worker's impairment.

Legislation designed to help integrate the disabled into society and the work force places the burden of proof of whether or not an impairment constitutes a handicap upon the employer rather than upon the employee or prospective employee.[1] Whether the impairment is congenital or the result of an injury or a disease, the emphasis is on accepting the worker's limitations and assessing the residual capabilities so as to utilize them optimally. Often, potential employers are unaware of the unique, but variable, residual characteristics of the physically or mentally impaired and of the techniques that would increase the disabled person's motivation, efficiency, and productivity.

This chapter provides information that will help engineers understand and deal with the placement and accommodation of persons with some of the most common handicapping conditions.

6.7.2 EMPLOYERS' CONCERNS

There are four major reasons why employers should be interested in job applicants with potentially handicapping conditions. First, employers have an ethical obligation to help qualified handicapped persons find employment. Work, employment, economic independence, self-fulfillment, and health are all inseparably related in our society.

Second, employers who do not consider applicants with potentially handicapping conditions are overlooking a tremendous resource. It has been estimated that as much as 30% of the U.S. population is afflicted with conditions that are potentially handicapping in one way or another.[2] Through minor accommodations, most of these persons become reliable and productive workers and an asset to their employers.

Third, under section 503 of the 1973 Vocational Rehabilitation Act, employers receiving federal contracts in excess of $2500 "shall take affirmative action to employ and advance in employment of qualified handicapped individuals." A "qualified handicapped individual" is defined broadly in the corresponding regulations[3] as an individual who is capable of performing a particular job with "reasonable accommodation," but who has an impairment that substantially limits one or more of his or her major life activities (including work). "Reasonable accommodation" is defined as:

> *(d) Accommodation to physical and mental limitations of employees. A contractor must make a reasonable accommodation to the physical and mental limitations of an employee or applicant unless the contractor can demonstrate that such an accommodation would impose an undue hardship on the conduct of the contractor's business. In determining the extent of a contractor's accommodation obligations, the following factors among others may be considered: (1) business necessity and (2) financial cost and expenses.*

Penalties for failure to comply with these regulations can include withholding of payments, termination of contracts, or debarment from future contracts. In addition to financial penalties for not hiring, financial incentives are provided for hiring qualified handicapped individuals. For current information, see your local State Bureau of Rehabilitation, Internal Revenue Service, or Department of Labor office.

The fourth reason deals with employers' responsibility to employees who are injured or who become ill because of their job. According to the National Safety Council,[4] there were an estimated 2.3 million disabling work injuries in 1977, which resulted in 80,000 permanent impairments. Based on the size of the 1975 labor force, this is about two injuries per 100 man-years of work. In spite of major efforts to improve worker safety and health, work-related injuries and illnesses probably will continue to be a problem. An employer's reaction to and management of disabled workers can have a major influence in the time and cost of recovery and return to work.

6.7.3 ANALYSIS OF WORK REQUIREMENTS

Since disabling conditions constitute handicaps to employment when they impair individuals' abilities to perform given work elements, handicaps must be discussed in reference to specified work requirements. Although job analysis is described in detail in Chapter 2.4, the importance of its applications in employment of the handicapped merits further discussion here.

The purpose of the job analysis is to identify all of the physical and mental work requirements that could limit successful and safe performance by a given applicant. This information then can be used in further evaluating or training applicants and in determining necessary accommodations.

Although job analysis procedures vary in thoroughness, the fidelity between the job description and the job is related to the analysis time and the skill of the analyst. The methods used for job appraisal should complement the methods used for appraising workers. The simplest and quickest analysis is to use a checklist of work requirements and workplace attributes such as the classic example used by Bridges in 1946.[5] Information for this and other methods of analysis is obtained most readily from five sources: (1) existing job descriptions, (2) interviews of supervisors, (3) interviews of incumbent workers, (4) observations of work activities, and (5) measurements of the workplace and workplace objects. The analyst simply checks each attribute on the checklist that corresponds to the job under study. Checklists, though requiring a minimum of analyst time and skill, are inflexible in their classification of work elements and do not provide information about frequencies of activities.

A job analysis can be made more thorough by dividing the job into functionally distinct tasks having both physical and mental components that cannot be differentiated. For example, the worker may perform the physical elements of grasp, lift, reach, transport, and so on, while simultaneously performing the mental elements of plan, estimate, compute, and so on. Thus it is possible to examine the physical and mental attributes associated with each task, although a correct sequence of activity sometimes is difficult to establish. As an example, an analysis of a sorter job is shown in Exhibit 6.7.1. The sorter job is divided into two tasks: get stock and sort washers. The second task is divided into a sequence of 10 elements. Because there are no hard-and-fast rules for defining elements, some trial and error will be necessary in order to find an acceptable level of detail.

The job analysis should include a sketch of the workplace layout; a description of possible chemical, physical, and biologic stresses in the environment; and a list of required tools and personal protective equipment.[5–7] This information will assist with the analysis of task elements and with the design of accommodations. A workplace sketch for the sorter job is shown in Exhibit 6.7.2.

Analyzing Physical and Mental Attributes

Physical and mental attributes are listed across the top of Exhibit 6.7.1. The exhibit thus can be used as a checklist to identify the attributes corresponding to each element and task. For some attributes, additional information can be obtained from measurements of the workplace and activities and can then be recorded under the corresponding attributes. The listed attributes include frequency and duration of task, gross body action, reach, gross strength, energy, hand action, hand strength, perception, sensation, and psychomotor, cognitive, and affective attributes.

Frequency and duration of task indicate how much of the time each element or task is performed. Measurement of frequency and times of work elements is a traditional industrial engineering function and is discussed in Chapters 3.2 and 3.3.

Gross body action describes how the body is used and its posture. Some examples include reach, lift, position, push, and carry. All observations of gross body action also should include additional notes whenever an operator leans on the edge of a bench, wall, or seat for support. Footnotes should be used for recording additional information. Information about gross body action for each element is readily obtained from interviews and observations.

Reach includes the horizontal, vertical, and lateral locations of the right and left hands with respect to some fixed frame of reference. For a mobile standing operator, the frame of reference might be the feet, as shown in Exhibit 6.7.3*a*, where the origin is the midpoint of a line between the heels of the feet. For immobile seated workers, the frame of reference might be the workbench, as shown in Exhibit 6.7.3*b*. Reach coordinates are measured easily with a ruler.

Gross strength is a measure of the load that is transferred through the body to the seat or feet of the worker. Force can be recorded as two numbers corresponding to the right and left hands or as one number corresponding to the resultant force. Force is easily measured with a hand scale.[8,9]

Energy is a measure of the metabolic cost of performing a given element or task. Energy is reported in kilocalories per minute, liters of oxygen per minute, watts, or mets (the ratio of working energy expenditure to resting energy), as well as in other units. In some cases energy costs are expressed per unit of worker body mass. Tables of energy expenditure, such as those in Chapter 6.4 or those published by the American College of Sports Medicine,[10] can be used to estimate the requirements for similar tasks under analysis. A set of empirical equations for estimating the energy cost of each element has been proposed by Garg.[11] Work energy requirements also can be assessed through measurements of oxygen consumption by incumbent workers.[12] Measurement of oxygen requires specialized equipment and skilled technicians.

Hand action describes how the hands are used[13] and requires two entries, one for each hand. Examples of hand action include power grip, hook grip, press, pulp pinch, and lateral pinch. Although hand postures can be determined by careful observations, films or videotapes that can be played in a slow motion are most helpful.

Hand strength is a measure of the resultant force on each hand and requires one entry for each hand. In most cases these forces are not directly measurable and must be estimated. Electromyography and films are sometimes useful for such measurements.[14]

The psychophysiological requirements of each job and each task element can be simplified to some extent to suit the capabilities of a worker. Every industrial or office task requires some aspects of the attributes of perception, sensation, psychomotor abilities, cognition, and the affective attributes. These attributes cannot be viewed as being independent of each other and are, in fact, quite interrelated, with overlapping perceptual, cognitive, and motor processes. However, each of these attributes has an underlying structure, described briefly here.

Perception implies the ability to note relevant detail, location, form, color, texture, and other features in objects and in written, graphic, or verbal material and to make visual comparisons and discriminations. A visual job analysis in which the visual requirements of each task element can be specified and related to the corresponding requirements in the worker is often a useful subset of the overall job analysis.[15]

Sensation is an attribute that refers to employing all sensory skills–seeing, tasting, hearing, touching, smelling, and the vestibular sense, or the sense of orientation and balance. It is difficult to distinguish between sensation and perception. Several sensory skills are required at the different steps in perceptual processing of information, such as planning (an action), initiating (movement of a limb), controlling, ending (an action), and checking (feedback) to determine if the end result is satisfactory. Visual acuity, including peripheral field of view, color perception, and depth perception, can be measured using commercially available vision testers. Hearing ability can be measured with an audiometer.[16] Touch and pressure sensitivity can be determined using methods described in Geldard.[17] Balance and orientation, indicating muscular tone and postural adjustment, or the "feel" of bodily position, are determined qualitatively by common tests such as walking a straight line or balancing when standing on one foot while, with eyes closed, extending the arms to the side or front or upward and then touching the tip of the nose with the forefinger of either hand. This ability to maintain balance and the sense of touch may often be lost in certain disabling conditions, as discussed in Section 6.7.4.

It is difficult to distinguish between cognition and perception because cognition is based on perception; cognitive abilities involve attaching meaning to and drawing inferences from sensory and perceptual phenomena. Cognition includes the higher mental activities involving specific areas of intelligence, thinking, remembering, and using symbols or language. Every task element requires cognition to some extent. For example, a task element such as "reach for washer in stock bin" (element 1 in Exhibit 6.7.1) warrants a different response than "position in jig" (element 4). In element 1, "reach" precedes an eventual pinch grip using the fingers. Element 4, "position," requires the use of sensory feedback for its successful completion. Both elements involve cognition, which implies intelligence, judgment, and an ability to define a procedural sequence.

Psychomotor attributes are related to the ability to coordinate exertion of limbs in response to visual, auditory, tactile, and other stimuli. This ability involves the kinesthetic sense and motor coordination of joint positions and movements. Psychomotor abilities most often are determined in terms of a score achieved on a psychomotor test, although no universal standard test exists.[18] Elements of control precision, multilimb coordination, response orientation, reaction time, speed of arm movement, rate control, manual dexterity, finger dexterity, arm-hand steadiness, wrist and finger speed, and aiming are tested. Other tests, such as the Wechsler test and the Bender visual motor test, also have been used.[19] A critical tracking task, as reported by Dott and McKelvey,[20] can provide an indication of psychomotor abilities.

Exhibit 6.7.1 Job Analysis of a Sorter Job[a]

	Frequency	Duration (min)	Gross Body Action	Reach Location (cm)		Gross Strength		Energy Cost (kcal/min)	Hand Action (kp)		Hand Force (kp)		Perception	Sensation				Psychomotor		Cognitive	Affective
														Vision	Hearing/Speech	Touch	Orientation/Balance	Motor Coordination	Feedback		
				R	L	R	L		R	L	R	L									
				Vert. Horiz. Lat.																	
Tasks																					
1. Get stock.	2/day																				
2. Sort washers.	1/day																				
Elements for Task #2																					
1. Reach for washer in stock bin.	400/day	–	Reach	83 25–30 ±13	–	–	–	↑	–	–	<1	–	×	•	•	×	×	×	×	×	×
2. Grasp washer in bin.	400/day	–	Grasp	83 25–50 ±13	–	–	–		Pulp pinch	–	<1	–	×	•	•	×	×	×	×	×	×

3. Transport to jig.	400/day	–	Move	81 1 0	–	–	–		Pulp pinch	–	<1	–	×	•	•	×	×	×	×	×	×
4. Position in jig.	400/day	–	Position	81 1 0	–	–	–		–	–	<1	–	×	•	•	×	×	×	×	×	×
5. Release washer.	400/day	–	Release	81 1 0	–	–	–	2.5	–	–	<1	–	×	•	•	×	×	×	×	×	×
6. Inspect. If washer falls through slot, go to 1.	–	–	–	–	–	–	–		–	–	<1	–	×	•	•	×	×	×	×	×	×
7. Grasp washer in jig.	400/day	–	Grasp	81 1 0	–	–	–		Pulp pinch	–	<1	–	×	•	•	×	×	×	×	×	×
8. Transport to reject bin.	400/day	–	Move	76 0 51	–	–	–		Pulp pinch	–	<1	–	×	•	•	×	×	×	×	×	×
9. Release washer in bin.	400/day	–	Release	76 0 51	–	–	–		Pulp pinch	–	<1	–	×	•	•	×	×	×	×	×	×
10. Go to 1.	400/day	–	–	–	–	–	–		–	–		–	×	•	•	×	×	×	×	×	×
Environment																					
Indoors Temperature: 60°F to 80°F WBGT Noise level: 70 to 80 dBA Emissions from forklift truck																					

[a] • = desirable attribute; × = essential attribute; – = attribute not required.

Exhibit 6.7.2 A Workplace Sketch for the Sorter Job

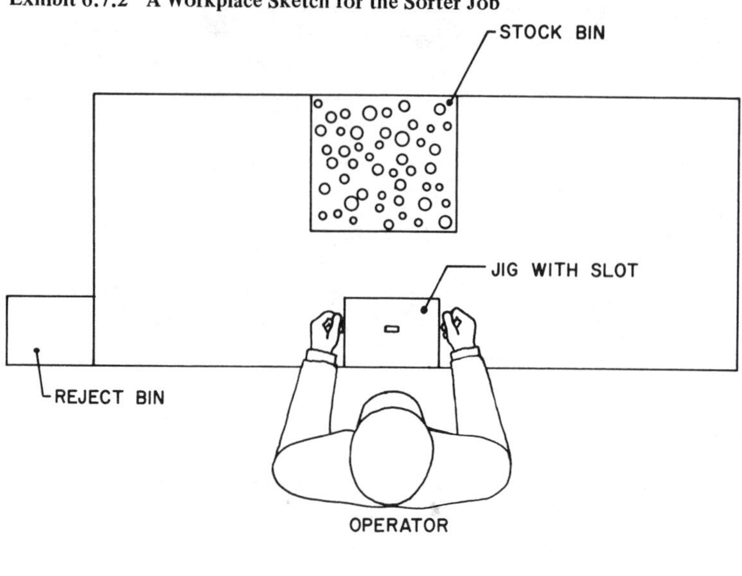

Affective attributes are factors such as motivation, emotional stability, self-confidence, personality, interests, persistence, impulsiveness, rigidity of opinions, adaptability, and reactions to peers or superiors. These have relevance to job design for the handicapped in several ways. For some disabling conditions, such as cerebrovascular or cerebellar dysfunction or trauma, brain tumor, or head trauma, and for some levels of mental retardation, a person may have lived with consistently low expectations and stimulation and thus may have little motivation or enthusiasm to understand

Exhibit 6.7.3 Measurement of Reach Coordinates With Respect To *(a)* a Midline Between the Heels of the Feet for a Standing Operator or *(b)* From the Edge of the Workbench and Floor for a Seated Operator

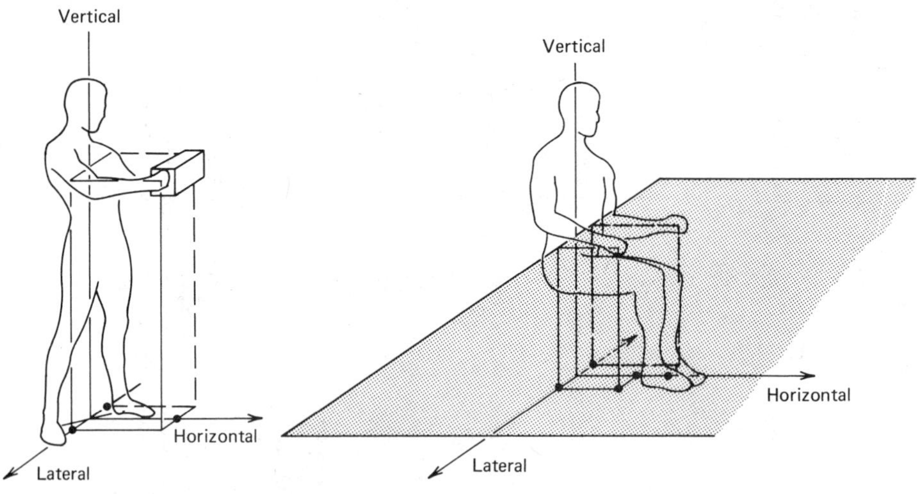

role performance in a social world. An understanding of a handicapped person's attitude toward and insight into his or her disabling condition is important to the job designer.

Several tests are available to evaluate qualitatively the affective attributes as they apply to certain disabled groups.[19] However, since the test results obtained can give rise to different interpretations, and since the tests do not consider all disabled groups, these tests should be used with qualifications.

6.7.4 DISABLING CONDITIONS

Data from the National Health Survey[21] show that in 1974 2.6% of the 122,546,000 U.S. men and women of working age (between 16 and 64 years) had chronic conditions making them unable to carry on a major activity such as working, housekeeping, or going to school; a total of 7.7% were limited in the amount or kind of major activity; and 3.9% were limited, but not in their major activity. A histogram of the most common causes of activity limitations (Exhibit 6.7.4) shows that about 40% involve the musculoskeletal system, 25% involve the heart, 13% involve the respiratory system, 6% involve special senses, 3% involve cancer, and 14% involve metabolic, gastrointestinal, and urinary disabilities.

For the conditions listed in Exhibit 6.7.4 to be considered handicapping, they must interfere with an individual's ability to perform a specified set of work requirements. The major performance aspects affected by each disorder that would be of concern when placing and accommodating an affected worker are listed in Exhibit 6.7.5. Before placing the potential worker, individual performance assessments must be performed by qualified medical or rehabilitation personnel, particularly when the individual's impairments could be life threatening. The employer's cooperation in assessing jobs, developing accommodations, and managing handicapped workers greatly facilitates return to work.

Perhaps one of the first considerations in placing a disabled person is to determine if the disabling condition is progressive or stable and if the job might aggravate the employee's condition. Progressive conditions generally are the result of disease processes such as arteriosclerosis, arthritis, tendonitis, and glaucoma; nonprogressive conditions are usually the result of an accident, a birth defect, or a past disease, such as amputations, phocomelia, and polio. The performance of persons with nonprogressive conditions tends to be stable or may even improve with therapy; the job designer should consider this when providing quantitative descriptions of work requirements to help the supervising medical personnel decide if there is risk of further injury.

Accommodations include either enhancement of worker performance by means of an adaptive device or modification of the work and workplace so as to reduce the performance requirements. A guiding principle is to remove completely or to reduce to a tolerable level the job demands that a disabled worker cannot meet. When such demands cannot be eliminated or reduced to a tolerable level, the work should be reorganized so that the more difficult activities are transferred to those workers who are capable of performing them.

Adaptive devices include wheelchairs, braces, artificial limbs, eyeglasses, and hearing aids; workplace modifications include special procedures, special jigs and tools, ramps for wheelchairs, and braille encoding of controls.

When physical disability is accompanied by perceptual, sensory, and mental disabilities, or when there is evidence of any of the latter alone, accommodation is more difficult and each case often is unique. For example, one of the most ubiquitous defects is in perception. There is a wide range of perceptual difficulties associated with brain tumors, cerebrovascular disease, and multiple sclerosis. Depending on the area of brain affected, these difficulties may or may not be apparent upon casual observation and may even be unknown to the individual. In general, mental disability by itself or in the presence of physical disability is evidenced by neurologic disorders resulting from traumatic or incipient diseases of the brain, spinal cord, and nervous system. Causes may vary, but they generally can be ascribed to intracranial, spinal cord, and peripheral nerve disorders and myopathies.

Physical Disabilities

The largest group of disorders involves the musculo-tendinous-osseous system and affects the ability of individuals to perform work elements that require movements and exertions of the body (see Exhibit 6.7.5). An inability to perform work elements involving the lower extremities sometimes can be accommodated by using specialized seating, wheelchairs, crutches, canes, walkers, and so on. Special considerations often are required to permit safe access to the workplace with a wheelchair or crutches. Some of the architectural guidelines for wheelchair accessibility recommended by ANSI[22] are shown in Exhibit 6.7.6.

It is estimated that 42% of the 680,000 wheelchair users in the United States include paraplegics, leg amputees, and persons with partial paralysis, advanced arthritis of the legs, and cardiac conditions.[23] These persons are capable of unimpaired seated reach (see Chapter 6.3). Reach limits

Exhibit 6.7.4 The Most Common Conditions Resulting in Activity Limitations in the U.S. Population Under 65 Years of Age[a]

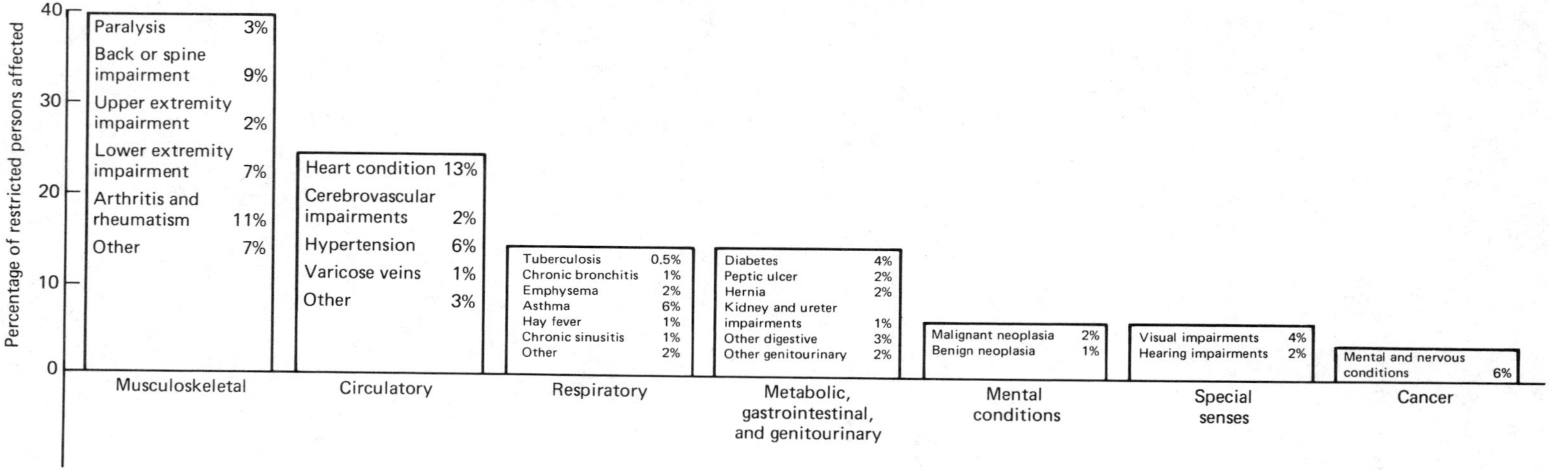

Percentage of restricted persons affected

40
30
20
10
0

Paralysis 3%
Back or spine impairment 9%
Upper extremity impairment 2%
Lower extremity impairment 7%
Arthritis and rheumatism 11%
Other 7%

Musculoskeletal

Heart condition 13%
Cerebrovascular impairments 2%
Hypertension 6%
Varicose veins 1%
Other 3%

Circulatory

Tuberculosis 0.5%
Chronic bronchitis 1%
Emphysema 2%
Asthma 6%
Hay fever 1%
Chronic sinusitis 1%
Other 2%

Respiratory

Diabetes 4%
Peptic ulcer 2%
Hernia 2%
Kidney and ureter impairments 1%
Other digestive 3%
Other genitourinary 2%

Metabolic, gastrointestinal, and genitourinary

Malignant neoplasia 2%
Benign neoplasia 1%

Mental conditions

Visual impairments 4%
Hearing impairments 2%

Special senses

Mental and nervous conditions 6%

Cancer

[a]Reference 23.

recommended by ANSI for wheelchair occupants are shown in Exhibit 6.7.6. Published reach and strength data for persons with specific disorders are rare.[24,25] Because the reach performance of persons with musculoskeletal disorders is extremely variable, strength and reach performance averages may greatly overpredict or underpredict the performance of a given individual. Performance of employees with potentially handicapping conditions should be evaluated on a case-by-case basis (see Section 6.7.3).

As discussed in Chapter 6.3, strength is related to the location of the load with respect to the body; as a general rule, strength decreases with an increase in distance of the load from the body. Strength tests should accurately reflect the posture and reach requirements of the job; this information can be obtained from analysis of overall physical work requirements as discussed in Section 6.7.3. Administration and interpretation of both strength tests and other tests of physical work capacity are regarded as medical procedures and should be performed under the supervision of qualified medical or rehabilitation professionals–particularly when conditions are potentially progressive.

Although isometric strength tests provide useful performance indicators for occasional lifting, allowances for fatigue must be applied when lifting is repetitive. Unfortunately, available fatigue allowance schedules (Chapter 4) do not take into consideration the effects of disabling conditions. Therefore interpretation of strength test data and application of allowances must involve qualified medical personnel.

Elements that limit reach and strength can be identified by comparing work requirements with the performance of potential employees. Some general accommodations include relocation of reach targets close to the body, relocation of the worker close to the reach target, reduction of force requirements, and use of power assistive devices. A "shadow" or "buddy" system is used in some work places. In this system workers who have no medical work restrictions are assigned to assist restricted workers for extreme exertions.

Cardiopulmonary Disabilities

The relationship between the cardiopulmonary system and physical work activities was discussed in Chapter 6.4. Suffice it to say here that an impairment at any level in the cardiopulmonary system not only can reduce the ability to perform physical work, but also can cause death. Although physical work requirements that are too great can result in fatigue or other adverse health effects, in some cases a certain amount of physical activity can improve the physical work capacity. Employment of persons with cardiopulmonary disorders requires careful management by *qualified medical personnel*; however, the employer can help by assisting with analyses of physical work requirements, development of accommodations, and management of work activities. Evaluation of such persons is described so that the employer can work more effectively with supervising medical personnel; it is not intended as a set of instructions for evaluating potential workers.

Stress tests typically consist of a task, such as stepping, walking, or bicycling, that is performed to steady state at successively greater intensities. The results are reported as the maximum oxygen consumption, energy expenditure, or mechanical work intensity that corresponds to the age-predicted maximum HR, excessive fatigue, EKG irregularities, chest pains, or shortness of breath. Oxygen consumption and energy expenditure rates are often reported per kilogram of body weight. Mets are used commonly in cardiac rehabilitation.[10]

Many parameters, including HR, BP, and oxygen consumption, are measured and recorded during tests of cardiopulmonary performance under stress. Unfortunately, BP and oxygen consumption are not measured easily under work conditions; therefore HR is used most often as an indicator of circulatory stress.[26]

The relationship between work intensity and HR for three levels of fitness is shown in Exhibit 6.7.7. It can be seen that a significantly greater burden is placed on the heart if the worker is in poor condition than if he or she is in good condition. Similar plots can be constructed to estimate the circulatory burden for tasks of given levels of energy requirements for given individuals and populations. Energy requirements for sample tasks shown in Exhibit 6.4.4, Chapter 6.4, can be used for rough estimates of energy requirements for similar tasks. Direct measurements of metabolic and circulatory responses of incumbent workers can be used to identify the most stressful elements of the job. Accommodations then can be designed to control these stresses.

Other important factors of cardiorespiratory performance include the type of work, the thermal environment, and chemical and physical agents. The type of work determines how the work load is distributed among the parts of the body and whether the muscles contract statically or dynamically. Laboratory studies have shown that higher HRs and BPs are attained for a given dynamic work intensity of the upper extremities than for a corresponding work intensity of the whole body.[28] It also has been shown that physical work capacity is less for arm work than it is for whole-body work.

Additional work has shown that significant circulatory burdens can be produced during static muscle contractions in seemingly sedentary persons at low levels of total energy expenditure.[29–32]

Exhibit 6.7.5 Major Performance Aspects Affected by Each Disorder

Disorder		Posture	Mobility	Reach	Strength	Dexterity	Work Capacity: Physical	Work Capacity: Mental	Endurance	Perception	Sensation: Vision	Sensation: Hearing/Speech	Sensation: Touch	Sensation: Orientation/Balance	Psychomotor: Motor Coordination	Psychomotor: Kinesthesia	Cognition	Affective	Progressive Disability
Musculoskeletal System																			
Amputations		•	•	•	•	•	•		•										
Arthritis	Osteoarthritis	•	•	•	•	•	•		•										•
	Rheumatoid arthritis	•	•	•	•	•	•		•										•
Burns			•	•	•	•							•						
Congenital abnormality		•	•	•	•	•	•								•				
Gout		•	•																
Muscular dystrophy		•	•		•	•	•		•					•					•
Myasthenia gravis		•			•	•	•		•	•	•								
Repetitive trauma disorders		•	•	•	•	•	•											•	
Spina bifida		•	•	•	•	•			•						•			•	
Central Nervous System/Spinal Cord																			
Brain tumor/head trauma		•	•	•	•	•	•	•		•	•	•	•	•	•	•	•	•	
Cerebral palsy		•	•	•	•	•	•	•	•	•				•	•	•		•	
Cerebrovascular disease/trauma		•	•	•	•	•	•	•	•	•	•	•	•	•	•	•	•	•	

Epilepsy[a]																		
Hemiplegia	•	•	•	•	•	•		•	•	•	•	•	•	•	•	•	•	
Multiple sclerosis	•	•	•	•	•	•	•	•	•	•	•	•	•	•	•	•	•	•
Paraplegia	•	•	•	•	•	•						•					•	
Parkinson's disease	•	•	•	•	•	•		•		•	•		•	•	•	•	•	•
Polio	•	•	•	•	•	•							•	•			•	
Quadraplegia	•	•	•	•	•	•	•	•	•	•	•	•	•	•	•	•	•	
Circulatory System																		
Arrhythmias						•		•										
Congenital defects				•		•		•									•	
Coronary artery disease				•		•		•										
Hypertension						•	•	•										
Valve defects				•		•		•										
Respiratory System																		
Asthma				•		•		•										
Chronic bronchitis				•		•		•										
Emphysema				•		•		•										
Pneumoconiosis				•		•		•										
Tuberculosis				•		•		•										
Other																		
Alcoholism				•	•	•	•	•	•			•	•	•	•	•	•	
Cancer		•	•	•	•	•		•									•	
Drug addiction				•	•	•	•	•	•			•	•	•	•	•	•	
Mental retardation	•				•	•	•	•	•	•	•		•	•	•	•	•	
Nonpsychotic disorders																•	•	
Psychotic disorders						•										•	•	

[a] All affected only during and/or after seizure.

Exhibit 6.7.6 Some ANSI Architectural Guidelines for Wheelchair Accessibility[a]

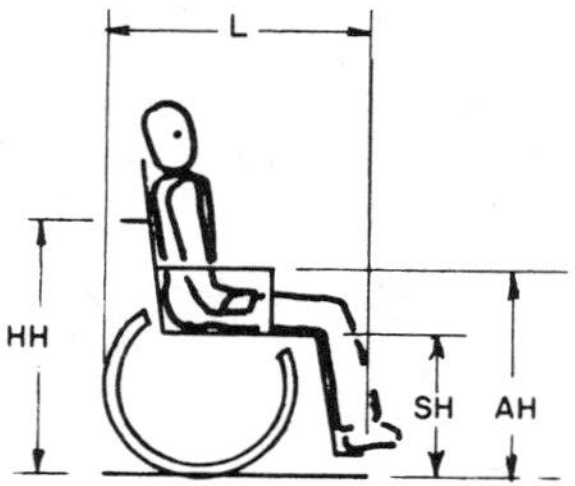

		AVERAGE (cm)	RANGE (cm)
AR	ARMREST HEIGHT	73.7	
HH	HANDLE HEIGHT	91.4	
L	LENGTH	106.7	
SH	SEAT HEIGHT	49.5	
W	WIDTH	63.5	
HR	HORIZONTAL REACH	81.8	68.6 - 90.2
TR	TABLE REACH	78.2	72.4 - 84.3
VR	VERTICAL REACH	152.4	137.2 - 198.1

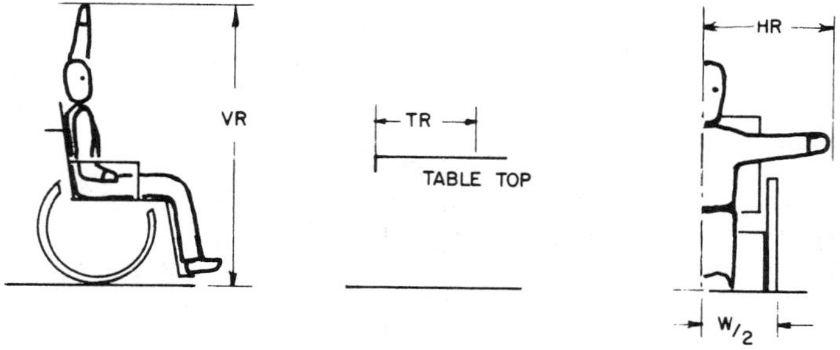

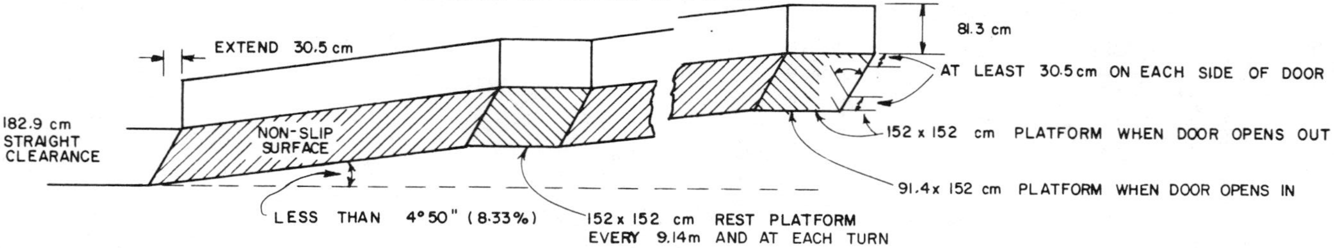

[a]Reference 24.

UM COLLEGE OF ENGINEERING

Exhibit 6.7.7 The Relationship Between Work Intensity (in kcal/min and mets) and HR for a 45-Year-Old 80 kg Male With Low, Average, and High Cardiac Fitness[a]

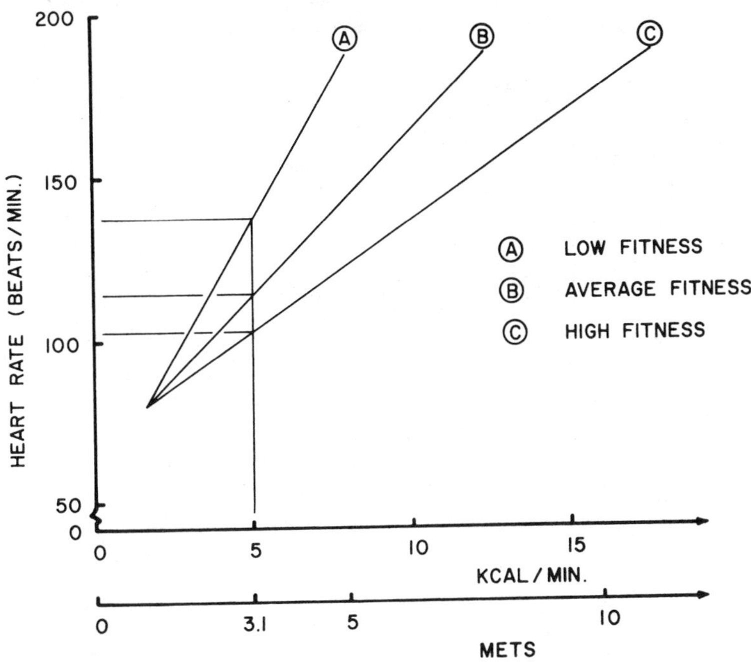

[a]Reference 29.

Jackson et al.[30] use the term "airport angina" to describe chest pains that persons with coronary artery disease experience during static work, such as standing while holding luggage. A plot of predicted circulatory responses to three intensities of a sustained static grip exertion is shown in Exhibit 6.7.8. Lind and McNicol[29] and Petrofsky and Lind[31] warn that persons predisposed to cardiovascular or cerebrovascular accidents should be aware of the risks of static work. Sustained static work elements, such as gripping, holding, and positioning, should be minimized for persons with cardiovascular impairments.

Additional circulatory burdens can result from work in hot or cold environments and from exposure to certain chemicals.[33] Work in hot environments can result in increased circulatory burden due to loss of fluid and electrolytes, increased peripheral blood flow, and energy cost of heavy protective equipment.[33,34] Important factors that affect burden and that should be assessed include dry air temperature, relative humidity, air velocity, radiant heat, and work intensity.[6]

Persons with cardiopulmonary disabilities may be especially sensitive to certain chemicals and to radio waves.[33] A few examples of such chemicals include ammonia, chlorine, carbon monoxide, nitroglycerin, carbon disulfide, azides, chlorinated solvents, and nitrobenzene.[6,33,35] Although it has been reported that excessive radio frequency energy may increase the incidence of heart disease, the greatest concern is for the effects of microwave radiation on heart pacemakers.[33,36] Measurement of chemicals and radio waves requires special equipment and trained personnel, but employers should be aware of chemical and radio wave presence.[7]

Finally, there should be plans regarding what to do if a disabled worker becomes incapacitated on the job. The plans should include control of equipment so that other workers are not injured, prevention of injury to the worker as a result of contact with the work surface or equipment, first aid, and transportation to an appropriate medical facility. A buddy system may be useful in which someone who is trained in emergency procedures supervises workers. The buddy also can help with stressful activities and provide important moral support to the worker.[33]

Perceptual, Sensory, and Other Mental Disabilities

Several tests exist that enable the determination of perceptual, sensory, and other mental attributes and the presence of any disabilities. Work evaluation is a process to determine what a person is capable of doing. It involves interviews with a rehabilitation counselor; the use of vocational tests,

Exhibit 6.7.8 Predicted HR (*a*) and Mean Arterial Responses (*b*) (in Percentage of Resting Values) to Sustain Static Grip Exertions at 25%, 50%, and 75% of Maximum Strength for a Young, Healthy Subject, Showing That Significant Circulatory Responses Can be Produced in Seemingly Sedentary Work[a]

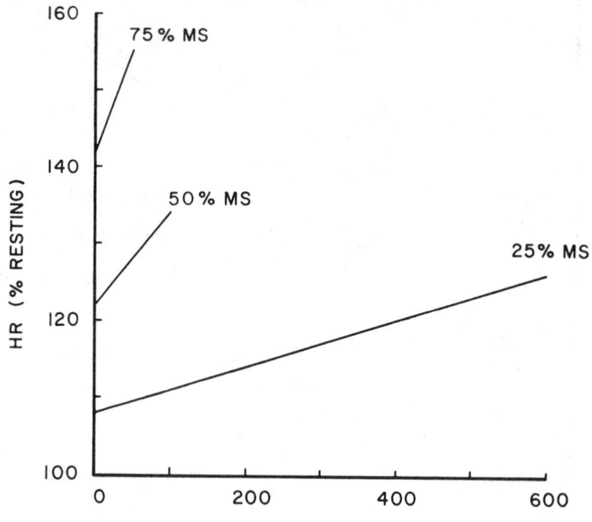

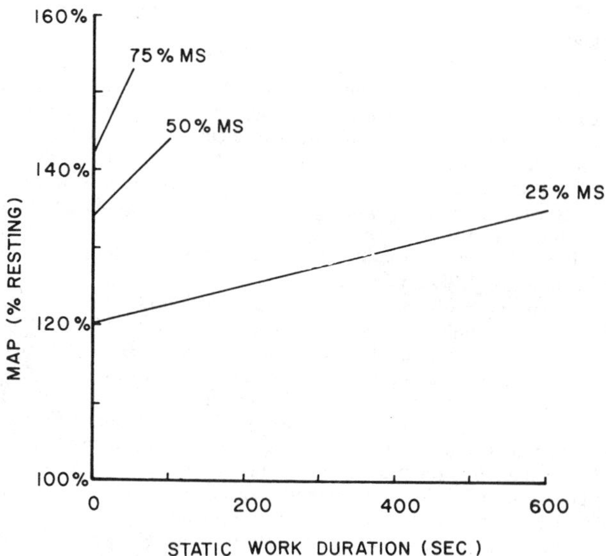

[a]Reference 34.

such as the Available Motions Inventory (AMI) or the General Aptitude Test Battery (GATB) of the USES; and assessment of the worker in an actual work situation and of his or her ability to perform a work or job sample. Evaluation should be a joint effort among the personnel department, the plant physician, and the industrial engineer. It helps to know the effects of the most commonly disabling conditions upon the sensory, perceptual, congnitive, and other attributes.

Decision making is a principal job activity. Indeed, several decisions are made in the conduct and execution of even the simplest task—the choice of which hand to use, which fingers, the form of grasp, and the procedural sequence. In most activities the worker derives direct nonsymbolic information from his or her task by way of the senses. This information is abstracted and translated by the worker into a form meaningful for the task. Unfortunately, perceptual and sensory disabilities are also the ones that occur most commonly. Limitation of sight or loss of visual acuity in one or both eyes, hemianopsia and impaired peripheral vision, defective color vision, and stereopsis can occur in endless combinations in disabling conditions such as cerebral palsy, glaucoma, trauma, or retinopathy. Specific visual defects, such as strabismus (squint), aphakia (absence of lens) in one or both eyes, ptosis (paralysis of accommodation), or immobility of the pupil or muscle paralysis leading to severe bilateral impairment, should be noted. Other perceptual problems, such as diplopia (double vision), perceptual distortion of geometric figures, agnosia (which is often a common accompaniment of cerebellar dysfunction, trauma, cerebral palsy, mental retardation, or other limitations), or sensation due to impaired nerve reception from spinal cord injury or multiple sclerosis, may be manifested. Sensory disabilities, such as difficulty in interpreting information, limitation of speech, susceptibility to fainting, dizziness, seizures, incoordination, or limitation of stamina, or perceptual deficits that can result in learning disabilities such as confusion of directionality, confused laterality, nonrecognition of symbols, aphasia where comprehension or expression of words is impaired, and apraxia, may not be readily apparent. Other cortical and sensory disorders, such as loss in learning, loss in interpretation of aural stimuli, and amnesia, also can be disabling.

Other sensory disturbances include anesthesias and a high stimulation threshold for pain, pressure, or thermal sensations. For purposes of job design, this translates into an extended delay in response, but subsequent performance may well be adequate.

In each case the job designer must assess the information processing capacity of the worker and match it with the information load of the task.[37] This can be done by reducing the information contained in each task element to a level where a disabled person can process it and respond to it successfully. A good principle is to provide assistance to bolster the impaired sense and/or to provide alternative means of getting the required information to the operator. The job designer should dispense with information that the handicapped individual is unable to receive normally. For example, considerations for work design require that those with poor vision should not engage in fine, close work. However, information to the worker need not always be visual or auditory; other sense modalities that are unaffected could be called into action. For example, even though true ability to judge distances is lost in a person with monocular vision, new methods are substituted for determining the relationships of objects in space.[38] However, job placement considerations require that these workers not be placed where accurate judgment of distance and speed is required in a short time. A job designer should remove those features of the environment with which the worker cannot cope. In all cases adequate individual protection also should be provided.

The mentally retarded are a special group of people with perceptual disabilities. Mental retardation is characterized by impairment in learning, maturation, and social functioning. There is evidence of a limited capacity for information processing, and hence placement considerations for manual industrial or office tasks are important. The social aspects of mental handicaps can have important considerations in work design, for there is often severe mental and social incapacity, evidenced by marked dependence upon others for personal needs.

The American Association on Mental Deficiency defines four levels of retardation: mild, moderate, severe, profound. Eighty-nine percent of those classified as retarded evidence mild retardation[39] showing minimal brain damage. Those moderately retarded have problems in the social awareness area, but because they have fair motor development, they have the potential to work in a semiindependent manner. The severely and profoundly retarded have very poor motor development. Placement proceeds from a consideration of a person's work personality and recognizing that, when properly placed, the majority of retarded persons perform tasks assigned to them as efficiently and rapidly as normal employees; in fact, they may perform routine repetitive tasks better and tire less quickly.[40] On routine jobs the mentally retarded workers may display a high degree of job satisfaction. When properly assisted by well-structured vocational training programs, the mentally retarded function adequately.[41]

Other Disabilities

When compared to perceptual learning disabilities, other disabilities such as arthritis, muscular dystrophy, myasthenia gravis, polio, hemiplegia, paraplegia, quadraplegia, Parkinson's disease, and brain tumor or head trauma can manifest themselves in motor learning difficulties, psychomotor impairment, and impaired kinesthesia. Difficulties in motor coordination for both gross and fine movement are related to perceptual difficulties. Other motor disorders are also associated with paralysis, athetosis, spasticity, intentional tremors, ataxia, and generalized motor retardation. It is important to appreciate, when designing a job, that such individuals are easily fatigued and distracted and may suffer from defects in concentration. Tests such as the Purdue Pegboard Test of Fine Manual Dexterity, the Bennet Hand Tool Dexterity Test, Crawford's Small Parts Dexterity Test, or the GATB can be used for evaluation.[19] There are certain basic skills that can be used as an indication of what a person could do: (1) recognition of coins and/or bills, and so on; (2) ability to read signs or forms containing simple instructions; (3) ability to write simple messages; (4) ability to measure (linear, weight measures); and (5) ability to arrange in gross order. Basic work-related skills include an ability to use the telephone, sorting by characteristics (such as color, size, shape), an ability to package and tie, and simple cleanliness and orderliness.

A myriad of deficiencies is associated with the affective attributes. Psychological assessment focuses on three specific areas of an individual's work personality, namely, intelligence, personality, and psychomotor tests of ability. Several standard tests for assessment of work motivation, mobility, maturity, organization, self-concept, productivity, learning ability, and so on, are described in Chapter 5.2. Indications should be obtained of ego strength to face obstacles, frustration, crises, and other occurrences.

Many cases of brain damage are accompanied by an inability to think or reason symbolically or in abstract terms, even though there may be no indication of a loss of intelligence. Symptoms of personality disturbances that are the result of brain damage can manifest themselves as disturbances of attention or interest and, in milder forms, as an inability to work in isolation or in large open workshops. Organically derived disorders and those associated with impaired cerebral tissue function include dementia, epilepsy, delirium and explosive outbursts, and temper tantrums. In dementia there is a slow disintegration of personality and intellect because of impaired insight and judgment. There is cognitive dysfunction and, in terms of job performance, the acquisition of new skills becomes difficult. Delirium is reversible cerebral insufficiency caused by organic factors and is manifested as a clouding of consciousness and impairment of recent memory. With regard to epileptic individuals, it is important for the job designer to appreciate the characteristics of the seizure and postseizure phases. For persons experiencing petit mal attacks lasting from 10 to 30 sec, there is a mild loss of consciousness after the seizure. There is sudden stoppage of activity in which the person is engaged, with subsequent resumption. Normally, no brain damage is indicated, and individuals are highly intelligent. For those having grand mal seizures that last longer, or status epilepticus, when seizures follow one another with no intervening periods of consciousness, activity should be restricted.

6.7.5 SUMMARY

Integration of handicapped persons in the workplace is a growing concern for ethical, production, legal, and rehabilitation concerns. Since physical and mental conditions are recognized as handicaps only when they impair individuals' abilities to perform given work elements, handicaps must be discussed in reference to specific work requirements.

Analysis of physical and mental work requirements is best performed by an industrial engineer familiar with the principles of work measurement and the performance limitations of common disabling conditions. Standard performance tests for physical and mental attributes should be administered by trained medical or rehabilitation professionals. Together, industrial engineers and the medical and rehabilitation professionals can identify limiting work elements that either could not be performed or that might be too stressful for given individuals. It is best to eliminate these limitations through workplace modifications, such as barrier-free construction, relocation of controls and stock, use of jigs and fixtures, or enhancement of controls and displays that benefit the productivity, safety, and health of all workers. Adaptive devices, braces, artificial limbs, specialized controls, and so on, should be used where workplace modifications are not feasible. The industrial engineer is a key person in the design and implementation of accommodations for handicapped persons.

REFERENCES

1. Vocational Rehabilitation Act of 1973, Public Law 93-112, 93rd Congress, H.R. 8070, September 26, 1976.
2. T. B. GRALL, "A Feasibility Study of Product Testing Reporting for Handicapped Consumers," *Human Factors Society Bulletin*, Vol. 23, No. 1 (1980), p. 7.
3. "Affirmative Action Obligations of Contractors and Subcontractors for Handicapped Workers," Part 60-741, *Federal Register*, Vol. 41, No. 75 (April 16, 1976), pp. 16147–16155.
4. *Accident Facts*, National Safety Council, Chicago, 1978.
5. C. D. BRIDGES, *Job Placement of the Physically Handicapped*, McGraw-Hill, New York, 1946.
6. J. OLISHIFSKI, Ed., *Fundamentals of Industrial Hygiene*, 2nd ed., National Safety Council, Chicago, 1979.
7. *Accident Prevention Manual for Industrial Operations*, 6th ed., National Safety Council, Chicago, 1969.

8. D. CHAFFIN, G. HERRIN, W. KEYSERLING, and A. GARG, "A Method for Evaluating the Biomechanical Stresses Resulting from Manual Material Handling Jobs," *American Industrial Hygiene Association Journal*, Vol. 38, No. 12 (December 1977), pp. 662–675.
9. S. SNOOK, "The Design of Manual Handling Tasks," *Ergonomics*, Vol. 21, No. 12 (December 1978), pp. 963–985.
10. AMERICAN COLLEGE OF SPORTS MEDICINE, *Guidelines for Graded Exercise Testing and Exercise Prescription*, Lea and Febiger, Philadelphia, 1976.
11. A. GARG, D. CHAFFIN, and G. HERRIN, "Prediction of Metabolic Rates for Manual Materials Handling Jobs," *American Industrial Hygiene Association Journal*, Vol. 39, No. 8 (1978), pp. 661–674.
12. P. ASTRAND and K. RODAHL, *Textbook of Work Physiology–Physiological Basis of Exercise*, McGraw-Hill, New York, 1977.
13. D. JACOBSON and L. SPERLING, "Classification of the Hand Grip, A Preliminary Study," *Journal of Occupational Medicine*, Vol. 18, No. 6, pp. 395–398.
14. T. ARMSTRONG, D. CHAFFIN, and J. FOULKE, "A Methodology for Documenting Hand Positions and Forces During Manual Work," *Journal of Biomechanics*, Vol. 12, 1978, pp. 131–133.
15. C. E. BURGER and J. R. DEJONG, "Aspects of Ergonomic Job Analysis," *Ergonomics*, Vol. 5, 1962, p. 185.
16. K. D. KRYTER, *The Effects of Noise on Man*, Academic Press, New York, 1970.
17. F. A. GELDARD, *The Human Senses*, Wiley, New York, 1972, pp. 290–300.
18. E. A. FLEISHMAN, "Towards a Taxonomy of Human Performance," *American Psychologist*, Vol. 30, No. 12 (1975), pp. 1127–1149.
19. C. H. PATTERSON, "Methods of Assessing the Vocational Adjustment Potential of the Mentally Handicapped," in L. K. Daniels, Ed., *Vocational Rehabilitation of the Mentally Retarded*, Charles C Thomas, Springfield, IL, 1974.
20. A. B. DOTT and R. K. MCKELVEY, "Influence of Ethyl Alcohol in Moderate Levels on Visual Stimulus Tracking," *Human Factors*, Vol. 19, No. 2 (1977), pp. 191–199.
21. *Limitation of Activity Due to Chronic Conditions, United States, 1974*, Data from the National Health Survey, Series 10, Number 111, DHEW Publication No. (HRA) 77-1537, U.S. DHEW, Public Health Service, Health Resources Administration, National Center for Health Statistics, Rockville, MD, June 1977.
22. *Specifications for Making Buildings and Facilities Accessible to and Usable by Physically Handicapped People*, ANSI A117.1-1961, ANSI, Inc., New York.
23. N. DIFFRIENT, A. TILLEY, and J. BARDAGJY, *Humanscale 1/2/3 Manual*, MIT Press, Cambridge, MA, 1974.
24. C. K. ROZIER, "Three-Dimensional Workspace of Amputee," *Human Factors*, Vol. 19, No. 6 (1977), pp. 525–533.
25. D. SMITH and L. GOEBEL, "Estimation of the Maximum Grasping Reach of Workers Possessing Functional Impairments of the Upper Extremities," in *Proceedings of the 1979 Spring AIIE Conference*, San Francisco, 1979.
26. G. C. E. BURGER, "Heart Rate and the Concept of Circulatory Load," *Ergonomics*, Vol. 12, 1969, p. 857.
27. *Exercise Testing and Training of Apparently Healthy Individuals: A Handbook for Physicians*, American Heart Association, New York, 1972.
28. T. REYBROUCK, G. HEIGENHAUSER, and J. FAULKNER, "Limitations to Maximum Oxygen Uptake in Arm, Leg, and Combined Arm-Leg Ergometry," *Journal of Applied Physiology*, Vol. 38, No. 5 (1975), pp. 774–779.
29. A. R. LIND and G. W. MCNICOL, "Circulatory Responses to Sustained Hand-Grip Contractions Performed During Other Exercise, Both Rhythmic and Static," *Journal of Physiology*, London, Vol. 192, 1967, p. 595.
30. D. H. JACKSON, T. J. REEVES, L. T. SHEFFIELD, and J. BURDESHAW, "Isometric Effects on Treadmill Exercise Response in Healthy Young Men," *The American Journal of Cardiology*, Vol. 31, 1973, p. 344.
31. J. S. PETROFSKY and A. R. LIND, "Aging, Isometric Strength and Endurance, and Cardiovascular Responses to Static Effort," *Journal of Applied Physiology*, Vol. 38, 1975, p. 91.
32. T. ARMSTRONG, D. CHAFFIN, J. FAULKNER, G. HERRIN, and R. SMITH, "Static Work Elements and Circulatory Diseases," *American Industrial Hygiene Association Journal*, Vol. 41, No. 4 (1979), pp. 254–260.

33. E. PLUNKETT, "Cardiac Work: Practical Aspects of Heart Disease on the Job," *The International Journal of Health and Safety*, Vol. 43, No. 5 (October 1974), pp. 20–22.

34. M. BATTIGELLI, "Determination of Fitness to Work," in C. Zenz, Ed., *Occupational Medicine–Principles and Practical Applications*, Year Book Publishers, Chicago, 1975.

35. F. W. MACKISON, R. S. STRICOFF, and L. PATRIDGE, Eds., *Pocket Guide to Chemical Hazards*, DHEW (NIOSH) Publication No. 78-210, Superintendent of Documents, Washington, DC, September 1978.

36. *Radio Frequency (RF) Sealers and Heaters. Potential Health Hazards and Control*, Current Intelligence Bulletin No. 33, NIOSH, Cincinnati, OH, December 4, 1979.

37. P. M. FITTS, "The Information Capacity of the Human Motor System in Controlling the Amplitude of Movement," *Journal of Experimental Psychology*, Vol. 47, 1954, pp. 381–391.

38. D. S. KOCHHAR and T. M. FRASER, "Monocular Peripheral Vision as a Factor in Flight Safety," *Aviation, Space and Environmental Medicine*, Vol. 49, No. 5, 1978, pp. 698–706.

39. D. E. BROLIN, *Vocational Preparation of Retarded Citizens*, Charles E. Merrill, Columbus, OH, 1976.

40. J. KELLEY and A. SIMON, "The Mentally Handicapped as Workers: A Survey of Company Experience," *Personnel*, Vol. 46, No. 5 (1969), pp. 58–66.

41. A. HALPERN, "General Unemployment and Vocational Opportunities for EMR Individuals," *American Journal for Mental Deficiency*, Vol. 78, 1973, pp. 123–127.

CHAPTER 6.8

Design of Man-Machine Systems

TARALD O. KVÅLSETH

University of Minnesota

6.8.1 INTRODUCTION

In a general sense, most human activities, whether they be associated with productive work or leisurely pursuits, involve interactions between men and machines. "Machine" is interpreted here in its broadest sense, that is, ranging from simple tools and equipment to complex physical processes. The design of such systems involving men and machines, known as man-machine systems, has traditionally been, and is still too often, focusing attention on the design of the hardware or machine component while failing to give proper consideration to the human component. The design relies instead on man's versatility and unique abilities to take care of any design problems and make the system work. The proper integration of man and machine, which benefits the human operator and enhances the overall system performance, is a primary aim of the human factors/ergonomics discipline. This chapter presents, within the limited space allotted, the human factors approach to man-machine systems design; it provides some specific design principles and guidelines.

As a point of departure, consider the schematic representation of a general man-machine system shown in Exhibit 6.8.1. Human operators receive information from various displays (dials, counters, lights, charts, buzzers, cathode-ray tubes, etc.). They respond by moving or exerting pressure on various controls (knobs, wheels, pedals, levers, keyboards, etc.) or by using their voice. The outputs of the control devices in turn provide the input to the controlled system (often referred to as the "machine" or the "plant"); it may be a relatively simple piece of equipment, a machine in the traditional sense (milling machine, lathe, etc.), a vehicle, a complex industrial process, and so on. This system interacts with its environment; it receives information about the goals or desired outputs of the system (generally referred to as the "reference input" or "forcing function") as well as possible noise (disturbances), which may affect the various system elements.

The main emphasis of this chapter is on the design of the man-machine interface, that is, the design of individual displays and controls as well as their spatial arrangement. The material presented has been compiled and summarized from various human factors textbooks[1–4] and other sources as indicated.

6.8.2 DISPLAYS

Types of Displays

The most common means of providing the human operator with information is through the use of visual displays. However, in some situations it may be desirable to use auditory displays (e.g., bells or buzzers for alarm signals). Although other sensory modalities such as kinesthesia (i.e., sensation of position, movement amplitude, velocity and acceleration/deceleration, and force generated by various body members), cutaneous senses (sensation of temperature, touch, and pain by nerve endings in the skin), and chemical senses (smell, taste) provide additional channels for information transmission, this section is concerned with visual displays, which are by far the most predominant source of information in man-machine systems.

The wide variety of visual displays used may be conveniently classified as follows:

1. *Quantitative displays* provide information about the numerical value of some variable, which may be either dynamic (i.e., changing with time, such as the temperature or pressure of a chemical process) or static. Such displays again may be divided into two classes: *analog displays*, for which the position of the pointer along the scale is analogous to the value it represents for the associated variable, and *digital displays*, which provide the same information in a direct numerical format.

Exhibit 6.8.1 The Basic Man-Machine System

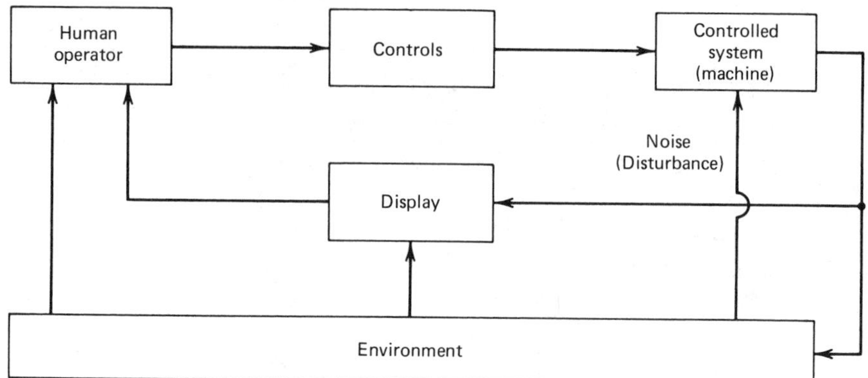

2. *Qualitative displays* provide information about a limited number of discrete states of some variable. Typical applications of such displays include check reading of whether or not a variable is within normal or acceptable operating range, on-off indicators (e.g., warning lights), and general indications of trends and rates of change of a variable.

3. *Representational displays* provide the user with a visual impression of the process or machine, its variables, and its environment. Included among the many varieties of such displays are those that present pictorial information or "mimic diagrams" (e.g., process flow diagrams, maps, wiring diagrams) and symbolic and graphic information (e.g., line and bar graphs, histograms).

4. *Alphanumeric displays* present information by means of alphanumeric characters.

This classification of visual displays may not be unique. In many cases individual displays provide the user (operator) with a mixture of the preceding types of information. However, the subsequent outline of design recommendations is based on the classification given here.

Principles for Display Design

Quantitative Displays

The three basic types of quantitative displays are given in Exhibit 6.8.2, and some of their relative advantages and disadvantages are outlined for various applications. This relative evaluation is based on the results of a number of experimental studies.[5] In general, a display with a moving pointer and a fixed scale is preferable to one with a fixed pointer and a moving scale. However, if the scale has to have a wide range of values and fine reading precision, a fixed scale may be inappropriate. Then a fixed-pointer, moving-scale design with a window (so that only part of the scale is visible at any given time) would be preferable. Thus, for example, the scale may be a long moving tape going in loops around several pulleys or drums (one of which is driving the tape). One of the primary advantages of a digital display over the two basic types of analog displays is that the digital display is more suitable for making quick and precise numerical readings; it does not require the user to make the types of interpolations needed for analog displays when determining the pointer position between adjacent scale markers. However, a digital display is unsuitable for situations in which the displayed values are changing rapidly, since each value may not be displayed long enough for the user to read it. Some specific design recommendations for digital displays are given in the section on alphanumeric displays.

Several variations of the preceding two basic types of analog displays are available: (1) fixed circular and semicircular (curved) scale with moving pointer, (2) fixed vertical and horizontal straight scale with moving pointer, (3) moving circular and semicircular scale with fixed pointer, and (4) moving vertical and horizontal straight scale with fixed pointer. See Van Cott and Kinkade[6] and McCormick[7] for further details about the relative suitability of these display alternatives. It generally may be inferred from available research studies that, irrespective of whether the scale is circular, semicircular, or straight, the moving-pointer, fixed-scale design tends to be preferable to the fixed-pointer, moving-scale design. An exception is when the scale needs to cover a wide range of values. Also, in most situations circular and semicircular scales are preferable to vertical and horizontal straight scales. Some experimental data, whose general validity perhaps may be questioned, have indicated that a straight horizontal scale is better than a vertical one because it minimizes the likelihood of reading errors.

A number of specific recommendations are summarized in Exhibit 6.8.3 for the design of scales and pointers. The various recommended dimensions are based on a normal viewing distance of

Exhibit 6.8.2 Ratings of Some Major Types of Displays

Characteristics	Type of Display: Moving Pointer, Fixed Scale	Moving Scale, Fixed Pointer	Digital (2954)
Reading speed and accuracy	Acceptable	Acceptable	Good
Check reading and trend and rate identification	Good	Poor	Poor
Continuous control (tracking)	Good	Acceptable	Poor
Control setting (when associated with a control device)	Good	Acceptable	Good
Economy of space use and illumination area	Acceptable	Good	Good

about 12 to 30 in. (30 to 75 cm). For a greater viewing distance, these dimensions should be enlarged proportionately. Scale numbers should increase in a clockwise, left-to-right, or upward direction for circular (and semicircular), horizontal, and vertical scales, respectively. The scale should be linear—that is, the separation between scale markers should be the same along the length of the scale; logarithmic or other nonlinear scales tend to increase the likelihood of reading errors. The scale markers, scale numbers, and pointer should contrast well in tone and color with the general display face—a design factor that should be combined with good display illumination and absence of glare or reflections.

Qualitative Displays

When the operator only needs to know which one of a few distinct states the process is in at any point, an excellent display arrangement may be to use a moving pointer and a fixed scale on which different colored lines or bands correspond to the different states of the process or variable. Thus, for example, whenever the pointer is within a green band, it signifies to the operator that the process is in a normal (satisfactory) operating condition. If the pointer is within a yellow or red band, it indicates that the process is in a caution or danger mode, respectively. Such type of color coding also may be superimposed on a quantitative analog display with numbered scale markers as long as the general principle that a display should present no more information than necessary is not violated. When color coding is not feasible (e.g., difficult illumination conditions or color-blind operators), alternative methods may be used for coding zones or operating regions on a display, for example, symbolic codes or different types of shading.

For providing the operator with warning signals, either visual or auditory indicators may be used. For very important signals, some redundancy, using both visual and auditory indicators, may be warranted. For very important visual warning displays, flashing lights are more effective than continuous ones. Although very limited and not entirely conclusive, experimental results[9] relevant to warning lights indicate that a flash rate between 1 and 10/sec seems to be appropriate. The recommended color for warning lights depends on the brightness of the signal and the background color and illumination. If the signal-to-background brightness contrast is low, a red signal light is recommended. For design guidelines concerning auditory displays, see Van Cott and Kinkade[10] or McCormick.[11]

Representational Displays

Although reference is made to McCormick[12] for specific design principles for representational displays, one general recommendation is offered relating to simplicity: The display should be as simple as possible and should omit any details that are not strictly relevant. The search for relevant

Exhibit 6.8.3 Specific Design Recommendations for Scales and Pointers on Analog Displays Used for Normal Viewing Distance[a]

Display Characteristics	Design Recommendations
Diameter of circular scale	
For general-purpose use	2.25 to 3 in. (55 to 75 mm)
For high-accuracy reading of small changes	4 to 6 in. (100 to 150 mm)
Length of linear scale	
For general-purpose use	2.5 to 3 in. (62 to 75 mm)
For high-accuracy reading of small changes	4 to 5 in. (100 to 125 mm)
Pointer characteristics	
Spacing between pointer tip and scale markings	0.02 to 0.06 in. (0.5 to 1.5 mm)
Width of pointer	0.03 to 0.09 in. (0.8 to 2.3 mm)
Approximate tip angle	20°
Number sequence of major scale markers	1 2 3 4 5 and so on 5 10 15 20 25 and so on 10 20 30 40 50 and so on
Major scale markers	
Height	0.22 in. (5.6 mm)
Width (high-low illumination)	0.013 in. (0.33 mm) to 0.035 in. (0.89 mm)
Intermediate scale markers	
Height	0.16 in. (4.06 mm)
Width (high-low illumination)	0.013 in. (0.33 mm) to 0.030 in (0.76 mm)
Minor scale markers	
Height (high-low illumination)	0.09 in. (2.29 mm) to 0.10 in. (2.54 mm)
Width (high-low illumination)	0.013 in. (0.33 mm) to 0.025 in. (0.64 mm)
Minimum separation between centers of minor scale markers (high-low illumination)	0.05 in. (1.27 mm) to 0.08 in. (2.03 mm)
Size of numbers/letters on the scale	
Height	0.19 in. (4.83 mm)
Stroke width	0.025 in. (0.64 mm)

[a]Adapted from references 5 and 8.

features of a display requires more time and is subject to higher error rates if the display is cluttered with irrelevant details.

A general trend in display technology is the increasing use of cathode-ray tube (CRT) displays, including television, for a variety of different applications, such as computer-generated representational displays. Further discussion of such displays is given in Section 12 of this handbook and in a number of other references relating to man-computer interactions.[13,14]

Alphanumeric Displays

Based on the experimental results summarized in various human factors textbooks,[1,4,15] the following design principles may be suggested for displays using alphanumeric characters:

1. The dimensions of both letters and numerals have to be related to the viewing distance between the eye and the display. Under favorable illumination conditions, the height H (in mm) of characters should be related to the viewing distance D (in mm) as follows:

$$H = 0.003D \text{ to } 0.005D$$

Additional recommended dimensions are as follows:

$$\text{stroke width} = \begin{cases} H/8 \text{ to } H/6 & \text{for black characters on white background} \\ H/10 \text{ to } H/8 & \text{for white characters on black background} \end{cases}$$

separation between letters = $H/5$ to $H/4$
width of letters = $2H/3$ to H
width of numerals = $2H/3$

2. When several numerals and/or letters are displayed simultaneously, both perception and short-term memory will be assisted by grouping the characters. Groups of from two to four characters appear to be the most advantageous. For example, the number 361524 ought to be displayed as 361 524.

3. Whether to use capital or lowercase letters or a mixture of both depends on a broad spectrum of independent variables. Very limited experimental evidence is available to aid the display designer. A couple of general principles perhaps may be offered. For labels such as those used for instrument identification, capital letters appear to be preferable, as also seems to be the case when visual search is required. For sentence reading, lowercase or a mixture of both seems to be preferable.

6.8.3 CONTROLS

Types and Choice of Controls

A wide variety of control devices is available for use in man-machine systems. The most commonly used controls are listed in Exhibit 6.8.4. The exhibit also provides a commonly accepted assessment of their suitability for different purposes or requirements. The majority of these controls are operated by means of arm, hand, and/or finger movements, although some are operated by foot movements requiring minimal force, and others by movements requiring force exertion.

The various controls given in Exhibit 6.8.4 are assessed with respect to four operational criteria: speed, accuracy, force, and range. Specifically, these refer to the speed with which an operator can make a control movement, the accuracy and range of control movements, and the amount of force that an operator can exert. Also shown are the control functions for which each type of control tends to be suitable. These alternative functions refer to activation (usually on–off), discrete settings (at distinct positions), and continuous control and settings; "continuous" refers to both continuous control movements and setting of a control at any position along a continuum (i.e., a quantitative setting).

In addition to the characteristics of speed, accuracy, force, and range, a number of other factors need to be taken into account when selecting controls. Some of the most important factors are outlined briefly in the subsequent sections, and some general recommendations are given.

Exhibit 6.8.4 Various Traditional Control Devices and Their Operational Characteristics and Control Functions

Type of Control	Operational Criteria and Control Ratings				Control Function
	Speed	Accuracy	Force	Range	
Cranks					
Small	Good	Poor	Unsuitable	Good	Continuous
Large	Poor	Unsuitable	Good	Good	Continuous
Handwheels	Poor	Good	Fair/poor	Fair	Continuous
Knobs	Unsuitable	Fair	Unsuitable	Fair	Discrete/ continuous
Levers					
Horizontal	Good	Poor	Poor	Poor	Continuous
Vertical (to–from body)	Good	Fair	Short–poor; Long–good	Poor	Continuous
Vertical (across body)	Fair	Fair	Fair	Unsuitable	Continuous
Joysticks	Good	Fair	Poor	Poor	Continuous
Pedals	Good	Poor	Good	Unsuitable	Continuous
Push buttons	Good	Unsuitable	Unsuitable	Unsuitable	Discrete
Rotary selector switch	Good	Good	Unsuitable	Unsuitable	Discrete
Joystick selector switch	Good	Good	Poor	Unsuitable	Discrete

Source. Reference 4.

Hand Versus Foot Controls

Traditionally there has been a clear preference for using hand-operated, as opposed to foot-operated, controls. However, there is little or no solid experimental evidence to justify such a general preference. In terms of speed of response, for example, the simple reaction time of the foot is only approximately 15% higher than that of the hand. Also, the apparently widely held belief that foot movements are much less accurate than hand movements appears to be neither supported nor discredited by published experimental results.[16] For design purposes, the following points may be considered as general guidelines:

1. Preferably, a system should not be so designed that it requires important control functions to be performed by the feet and hands simultaneously.
2. Foot controls may be used to provide some rest for the hands and some task variation, but should not be used to the extent that the feet do most of the control actions while the hands are idle. In general, the hands should be responsible for most of the control activity.
3. Foot-operated pressure controls should be used only when the operator is seated.
4. For foot controls requiring movements, that is, pedals, there appears to be no conclusive evidence indicating the optimal location of the fulcrum, although some data tend to favor a fulcrum location under the heel.
5. A pedal should return to the null position when pressure is removed; that is, elastic resistance (spring loading) should be used.
6. If a standing operator has to perform control functions with his or her leg, a knee-operated control device may be used, but not a foot control.

Control Resistance

Controls may be subject to four types of resistance: elastic (spring loading), friction (static and sliding), viscous damping, and inertia. These different types of resistance have various advantages and disadvantages, depending on the amount of resistance and on certain characteristics and requirements of the control task. Other sources[17,18] should be consulted for details, but a few general points are as follows:

1. All forms of control resistance reduce the possibility of accidental activation.
2. Some spring loading has the advantage of returning the control to the null position when the operator releases it and of providing the operator with a useful pressure cue, since he or she feels a pressure that is proportional to the distance of the control from its center (null position).
3. Frictional resistance has the advantage of holding a control setting in a fixed and selected position, but it may interfere with precise adjustments by causing the control to "jump."
4. Viscous damping, which causes a resistance that is directly proportional to the speed of the control movement, tends to smooth control action. It provides the operator with feedback information ("feel") about the speed of control movement; however, the operator is probably not able to interpret it as well as displacement feedback related to elastic resistance (spring loading).
5. Inertial resistance, which is directly proportional to the acceleration or deceleration of the control movement and independent of displacement and velocity, helps to smooth control movements, but increases the difficulty of making quick and precise control adjustments involving changing movement directions.

Control Sensitivity

The sensitivity (gain) of a control device (the inverse of the so-called control/display ratio), which relates any displacement of the device to the corresponding displacement of the moving element of the display (pointer, cursor, etc.), is a critical design factor affecting operator performance. However, since control sensitivity is affected by a number of factors (e.g., display size, tolerance requirements, system time lags), there are no specific guidelines or formulas for determining the optimal sensitivity for given circumstances. Instead, the optimal control sensitivity for any given system for which speed and accuracy are of critical importance should be established empirically.

Control Coding

Whenever an operator is responsible for a number of controls, it may be of utmost importance that he or she be able to identify each control device rapidly and correctly. Control coding is such an identification aid. The primary methods of coding include color, shape, texture, size, location, operational method, and labeling. Each of these methods has various desirable and undesirable features and principles of application.[19] Rather than using only one coding method, it may be advantageous to use combinations of codes. Such combinations, for example, identifying each control by a distinct color and a distinct shape, may be used to create some redundancy, which can be particularly useful whenever correct identification is especially critical.

Computer Input Devices

Many different types of control devices have been designed for providing inputs into computers. Such devices, frequently referred to as "data entry devices," include keyboards, joysticks (displacement or force operated), thumb wheel controls, rolling balls (trackballs), punches and readers for cards and tape, graphic tablets (digitizers), and light pens. Detailed descriptions of such devices and the factors influencing their selection are given elsewhere,[20] but a few points are made here regarding those devices that generate continuous data.

Of the continuous control devices, the joysticks and the trackballs have been used most extensively so far for such purposes as moving a cursor across a CRT for target acquisition and tracking tasks, text manipulation, and "zooming" onto particular sections of a representational display (see Section 6.8.2) in order to examine them in greater detail. The more recent graphic tablet (digitizer) and also the light pen have considerable potential as control devices. The graphic tablet requires the operator to move a stylus (pen) or cursor across a flat surface while the coordinates of the stylus are being measured at a preset sampling rate and recorded by the computer or displayed on the CRT as a point or in terms of numerical values. The light pen, which generally has a lower resolution level and is less accurate than the graphic tablet, is used for pointing directly on the face of the CRT itself while the computer detects its position coordinates.

6.8.4 LAYOUT OF DISPLAYS AND CONTROLS

General Guidelines

The location of individual controls and displays in relation to each other and to the operator is an important factor in man-machine system design, affecting system performance, safety, and the general job satisfaction of the operator. In theory, it may be possible to determine optimal locations for controls and displays in certain given situations. However, in practice, such optimal design of a man-machine interface is often difficult or impossible, and priorities have to be established and subjective judgment used. Nevertheless, general principles and specific guidelines are available to aid the designer. Following are four such general principles:

1. The *importance principle* is concerned with the operational importance of an instrument (display or control) in terms of the extent to which it may influence the total performance of the system. According to this principle, the most important instruments should be located in the optimal positions in terms of convenient access and good visibility.
2. The *frequency-of-use principle* stipulates that the most frequently used instruments should be placed in the optimal locations.
3. The *sequence-of-use principle* requires that, when instruments are used in a fixed sequence, they should be laid out in that order.
4. The *functional principle* recommends that instruments having related functions be grouped together.

The application of these principles generally will require the use of judgment and some compromise between the principles. In cases in which an existing system is to be modified, it may be possible to carry out activity analyses in order to obtain quantitative data relevant to the principles, especially to principles 2 and 3, as outlined in a succeeding section of this chapter. However, when a new system is to be designed, the designer has to rely on whatever relevant information on these principles can be pieced together.

Following are some additional general guidelines that should be taken into account when designing an instrument layout:

Displays that are most important and most frequently used ideally should be located so that they are within 30° below the standard horizontal line of sight (i.e., within ±15° of the normal sight line for a seated operator; for a standing operator, the normal sight line is 10° below the horizontal sight line) and horizontally within ±15° to each side of the standard line of sight. Certain recommendations regarding viewing distance are given in Section 6.8.2.

Controls that are (a) the most important and frequently used, (b) emergency controls, and (c) used for precise manipulations should be located within a vertical area extending from approximately 10 in. (25 cm) to 30 in. (76 cm) above the seat reference point (i.e., intersection between the middle of the seat and the back of chair) and 15 in. (38 cm) to its left and right. Other secondary controls should not be located more than about 40 in. (102 cm) above and 20 in. (51 cm) to each side of the seat reference point. For the placement of foot controls, see McCormick.[21]

Control-display compatibility dictates that corresponding controls and displays be arranged in corresponding patterns. If it is feasible to place a control close to its corresponding display, which is highly desirable, then the control should be placed either below or, if need be, to the right of the display. If some displays and controls are used in a certain sequence, then they should be arranged in that order, from left to right.

Control clearances between neighboring control devices need to be sufficiently large to prevent the possibility of accidental activation. Such minimum clearances, as well as preferred ones, which depend on the type of controls and the nature of their use, have been tabulated by McCormick[22] and Shackel.[2] The latter reference also provides data for control operations with gloved hands. For example, the recommended minimum and the preferred separation are 3 in. (7.6 cm) and 5 in. (12.7 cm), respectively, for two round control knobs that are operated with both hands simultaneously (without gloves) as compared to 1 in (2.5 cm) and 2 in. (5 cm) if only one hand is used at a time.

Angled side panels may be used to place displays and controls within a convenient area if there are too many instruments to be fitted onto a single panel.

Use of Anthropometric Data

As discussed in more detail in Chapter 6.3, anthropometric data relates to measures of various human traits, primarily physical dimensions of different parts of the human body, weight, range of movements, and muscular strength. A substantial amount of such data has been compiled for both sexes and for different percentiles of the population.[23,24] Such data include both structural (static) body dimensions, that is, measurements made when the body members are in fixed and standardized positions, and functional (dynamic) body dimensions, that is, measurements made while the body members are in motion.

Clearly such data are useful and necessary for the design of man-machine interfaces and work stations if these are to be compatible with the physical characteristics of the potential user population. For example, such data may be used to determine the area within which controls have to be placed so that they can be reached by sitting operators who are expected to fall within the range of the fifth to the ninety-fifth percentile for the arm reach of adult males. If the ninety-fifth percentile value of reach (and grasp) is 83.8 cm from the seat reference point, this means that 95% of the given population has a reach of 83.8 cm or less. Similarly, anthropometric data may be used to determine the best location of a pedal that requires a given static force for its activation, since data have been tabulated for the maximal forces that can be exerted by one foot of a seated operator for different foot positions.

Use of Activity Analyses Data

To obtain data with relevance to the preceding principles of frequency and sequence of use when modifying an existing system, activity analysis is a useful technique. Basically, this technique involves recording the control activities of an operator's hands and feet and his or her display monitoring behavior. Such control activity data will reveal the extent and frequency with which each control device is being used and the transition frequencies between the various controls. Similarly, display monitoring data indicate how often and for how long each display is being observed as well as the scanning patterns between the different displays. A number of different techniques may be used for data collection, such as direct observation and the use of motion picture and videotape equipment and eye-movement cameras.

The summary data from such an activity analysis may be presented in some matrix form (i.e., the so-called from-to chart) where the number in the (i, j)th cell denotes the transition frequency from the ith to the jth display and where the diagonal elements represent the dwell-time distribution between the displays, that is, the proportion of the total actual monitoring or observation time that the operator is effectively looking at a display. The same type of chart also may be used for recording control activities. However, such charts do not include any information about the actual relative locations of the different instruments. An alternative method of summarizing activity data is to superimpose the data for transition frequencies and dwell-time distribution onto an instrument layout diagram, which should be drawn to scale.

By analyzing such summary data, various improvements in the instrument layout may be immediately apparent. For example, if the transition frequencies between two remote instruments are relatively high, this would indicate that the two instruments ought to be located closer together. Also, if the dwell time for one display is relatively high, the display should be located in a central position on the display panel.

6.8.5 SUMMARY

This chapter has been concerned with the "nuts and bolts" of the human factors approach to man-machine systems design. The emphasis has been on the design of individual displays and control devices as well as their spatial layout. A number of specific design principles and guidelines have been presented as an aid to system designers. It has been claimed that a proper display-control design, that is, a man-machine interface design, is a prerequisite for reliable, safe, and efficient system operation and a strong determinant of an individual's feeling of job satisfaction and of his or her physical and mental well-being.

Although the nature of the interaction between man and machine is one of the most essential components of the man-machine system design process, it also must be recognized that such a system is not a closed one, but rather one that interacts with its environment. Other chapters in Section 6 of this handbook are devoted specifically to such environmental considerations.

REFERENCES

1. E. J. MCCORMICK, *Human Factors in Engineering and Design*, 4th ed., McGraw-Hill, New York, 1976.
2. B. SHACKEL, Ed., *Applied Ergonomics Handbook*, IPC Science and Technology Press, Guilford, United Kingdom, 1974.
3. H. P. VAN COTT and R. G. KINKADE, Eds., *Human Engineering Guide to Equipment Design*, rev. ed., U.S. Government Printing Office, Washington, DC, 1972.
4. K. F. H. MURRELL, *Ergonomics: Man in His Working Environment*, Chapman and Hall, London, 1965.
5. MCCORMICK, *Human Factors in Engineering and Design*, pp. 62–112.
6. VAN COTT and KINKADE, *Human Engineering Guide to Equipment Design*, pp. 81-84.
7. MCCORMICK, *Human Factors in Engineering and Design*, pp. 67–70.
8. SHACKEL, *Applied Ergonomics Handbook*, pp. 18-26.
9. MCCORMICK, *Human Factors in Engineering and Design*, pp. 78–81.
10. VAN COTT and KINKADE, *Human Engineering Guide to Equipment Design*, pp. 78–81.
11. MCCORMICK, *Human Factors in Engineering and Design*, pp. 113-141.
12. MCCORMICK, *Human Factors in Engineering and Design*, pp. 81–87.
13. E. EDWARDS and F. P. LEES, *Man and Computer in Process Control*, The Institution of Chemical Engineers, London, 1973.
14. T. B. SHERIDAN and G. JOHANNSEN, *Monitoring Behavior and Supervisory Control*, Plenum, New York, 1976.
15. E. GRANDJEAN, *Fitting the Task to the Man*, Taylor & Francis, London, 1980.
16. K. H. E. KROEMER, "Foot Operation of Controls," *Ergonomics*, Vol. 14, 1971, pp. 333-361.
17. VAN COTT and KINKADE, *Human Engineering Guide to Equipment Design*, pp. 350–352.
18. E. C. POULTON, *Tracking Skill and Manual Control*, Academic Press, New York, 1974, pp. 312–320.
19. MCCORMICK, *Human Factors in Engineering and Design*, pp. 240–244.
20. VAN COTT and KINKADE, *Human Engineering Guide to Equipment Design*, pp. 311-344.
21. MCCORMICK, *Human Factors in Engineering and Design*, pp. 303–304.
22. MCCORMICK, *Human Factors in Engineering and Design*, pp. 305–306.
23. VAN COTT and KINKADE, *Human Engineering Guide to Equipment Design*, pp. 467–584.
24. MCCORMICK, *Human Factors in Engineering and Design*, pp. 267–289.

CHAPTER 6.9

Design of Handtools, Machines, and Workplaces

E. NIGEL CORLETT

University of Nottingham, England

6.9.1 INTRODUCTION

This chapter is concerned with the design of the place in which work is done and with the machines installed therein. Handtools are hand-held machines, which may be used in a variety of workplaces and environments. Static machinery usually requires the worker's presence nearby during the work period if its productivity is to be maintained. Hence it introduces particular ergonomic problems related to the limited amount of movement that may be imposed on the operator. Workplaces, whether they are machine stations, assembly benches, or office desks, must be considered in conjunction with the machinery used in the workplace and the work the operator is required to carry out. Thus some consideration of seating must be incorporated in workplace design. Finally, the building, be it office or factory, will have an influence on the success of the worker at his or her job. Apart from those problems of environment that are discussed in the three following chapters, certain ergonomic factors must be presented here if the designer is to succeed in effecting an ergonomic match between the work and the worker.

6.9.2 HANDTOOLS

Although handtools are later divided into powered tools and those tools deriving power only from the operator, it is useful first to consider some aspects common to both types.

Weight

Except in particular instances, for example, large hand grinders used in some fettling jobs, the weight of a handtool contributes to limiting its use. The muscles of the shoulder and arm cannot support even 2 lb (1 kg) at three quarters of maximum reach for more than a few minutes; to repeat the gesture frequently during the day, perhaps while exerting a force, would be quite impossible.

Exhibit 6.9.1 gives relevant data for design guidance. A primary design consideration, therefore, should be lightness; where work conditions make it possible, heavier tools should be supported. A heavy handtool cannot be positioned precisely with any speed, and fatigue will make the performance worse.

Grip and Handles

Where a tool is to be gripped around its body and pushed, the diameter may be larger than where other actions are required. If the tool is to be wielded, such as a hammer, then a wraparound grip of the hand is needed. Since the user can be either male or female, a diameter of about $1\frac{1}{4}$ in. (30 mm) should preferably not be exceeded. This suits the lower end of the distribution of women's hand sizes. (See Chapter 6.3 for details.)

When holding a tool in the working position, the handles must be placed so that the hands fall on them, and grip them, without the axis of arm and hand deviating from a straight line. Any design that requires force or grip to be applied by means of an angled wrist is unacceptable, since it exposes the user to a high risk of tenosynovitis of the wrist.

Exhibit 6.9.1 The Holding Time to Unacceptable Levels of Fatigue. If an Object of the Given Weight Is To Be Held Repeatedly, It Should Be Held for 10% or Less of the "Time to Reach Significant Fatigue" on Each Occasion, With Intervals Between Each Event of at Least Six Times the Holding Time. Note That Graphs Show Values for Men. For Women, No More Than Two Thirds the Given Values Should Be Employed.[a]

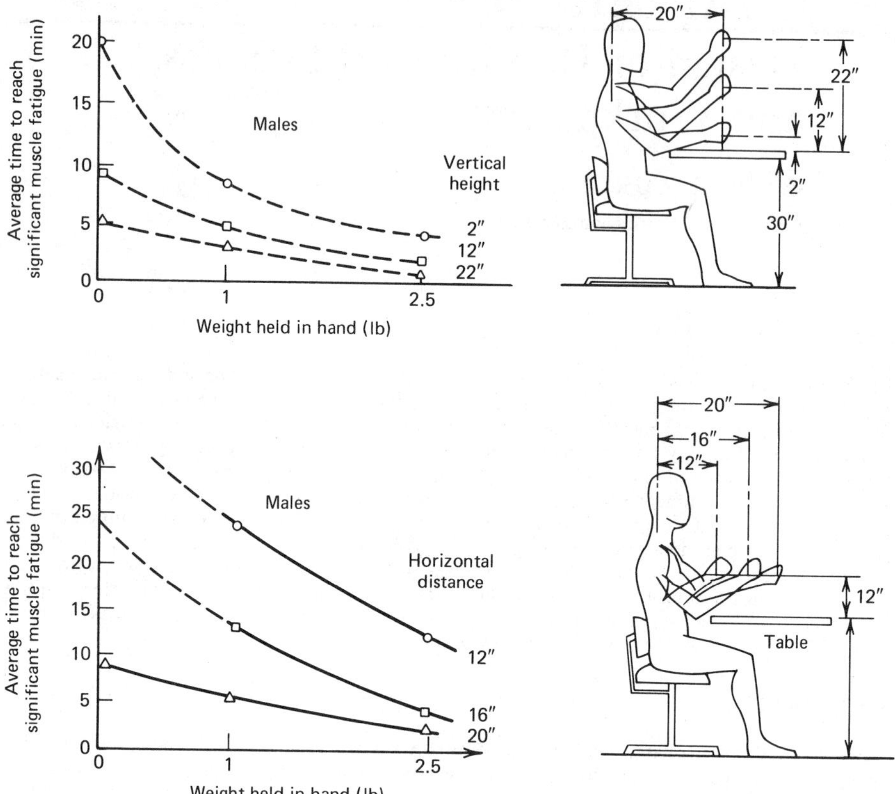

[a]With acknowledgments to L. GREENBERG and D. B. CHAFFIN, *Workers and Their Tools*, Pendell Publishing Company, Midland, MI, 1977.

Tremor

If the use of handtools is interspersed by periods when fine adjustments or skilled movements must be made, it is even more necessary to reduce the weight and the forces applied. On ceasing strong muscular activity, a worker will experience a tremor induced by way of the muscles which will prevent finely controlled movements from being successful. Similarly, the ability to recognize variations in smooth movements–for instance, by workers sanding furniture and testing the sliding fit of drawers–will be reduced in sensitivity after the performance of heavy work.

Operator-Powered Tools

As with any other machinery, the power supplied to handtools should be transferred to do its function as efficiently as possible. The use of power for doing anything other than the function should be reduced to a minimum. Thus hand punches, grips, or riveters should have springs or other aids to return them to the operating position; these should not have a stiffness that is a significant fraction of the operating load. Where this appears to be necessary, the tool or the technology is inadequate and needs redesigning. The forces must be applied by the operator at the most favorable biomechanic posture (see Chapter 6.3) and must be transmitted from the hands through broad pressure surfaces.

Exhibit 6.9.2 The Recommended Maximum Opening for a Handtool of the Form Shown Would be $3\frac{1}{2}$ in (90 mm) and the Maximum Force for Occasional Use Should Not Exceed About 15 lb (7 kg). For Frequent Use Where Force Is To Be Exerted, About 10% of This Loading Should Be Aimed for.[a]

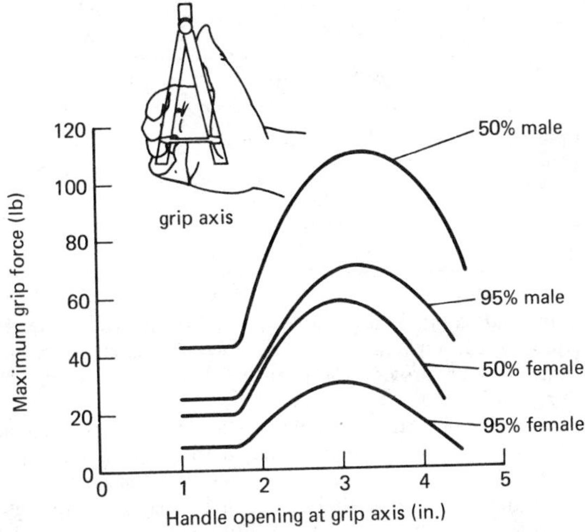

[a]With acknowledgments to L. GREENBERG and D. B. CHAFFIN, *Workers and Their Tools*, Pendell Publishing Company, Midland, MI 1977.

If force by means of the two hands is to be applied, forcing the hands together at chest level provides the strongest thrust. In any case force applied with the arms away from the body produces an unbalancing moment, which requires additional muscular effort to resist, as well as reducing the force that can be applied. Gripping with one hand, such as with a pair of pliers, has maximum values, as shown in Exhibit 6.9.2, but is not a suitable task for very frequent repetition. Powered tools should be introduced if production requires such an activity.

Powered Tools

Of great importance in powered tools is the minimizing of vibration transmitted to the hand. If this is not done, the fingers of the exposed hand begin to demonstrate a pallor and slow return of blood to the fingertips, which gets progressively worse. The disease is known as "vibration white finger," and after its early stages is irreversible.

It is also important not to use triggers or buttons that require continuous pressure while the tool is in use. Because the muscles operating the fingers are relatively small and easily fatigued, constant activation can lead to pains and cramp. In addition, the restriction on the blood supply to the finger is a health risk. A lever along the handle, which can be squeezed in the palm, is a better procedure, as is a twist grip or a two-position switch.

Powered tools are usually the heaviest of the handtools, and their support for long periods during the day will contribute to working fatigue. Wherever possible, they should be supported independently of the operator. Where this is impossible, some form of sling to transfer part of the load from the arms to the shoulders should be investigated. The closer a heavy tool is held to the body, the less the weight that has to be supported by the arms. Also, the less the elbow is bent, the longer it can be supported without high levels of discomfort. To allow the operator to maintain a straighter and more vertical arm position, the workplace for, say, fettling with a hand grinder will probably be lower than the bench height suitable for the use of a light sander.

6.9.3 MACHINES

The machines considered in this section are nonportable; hence the operator must approach them or station himself or herself in their vicinity in order to use them. This relative immobility of the

Exhibit 6.9.3 Minimum Requirements for Compensating for the Effects of Pacing, Using Either Tolerance Time, (i.e., Time for Which Work is Available) or Buffer Stocks. Values Apply to Short-Cycle Assembly or Machine Feeding Tasks.

Rate of Pacing	Tolerance Time (minimum)	Minimum Buffer Capacity: If Tolerance Time Equals Cycle Time	Minimum Buffer Capacity: If Tolerance Time Equals Zero
2.5% below to mean unpaced performance	Twice feed cycle time	1	2
Mean unpaced performance to 5% above mean	Three times feed cycle time	2	3
7½% above mean unpaced performance	Four times feed cycle time	3	4

operator should be *minimized* as much as possible. To be inactive for any length of time is physiologically bad. Opportunities for movement–for instance, by so designing the machine that its operator may sit or stand at will–provide the facility for recovery from a given posture without the necessity for stopping work.

Another important problem concerns the general relationship between machine and operator. If the machine *must* have attention at specific times before it can continue operating, then any inability in this respect on the part of its operator means that the machine is not productive. For machine feeding, assembly lines, and even the regular maintenance of equipment, it is desirable to decouple the operator from the cyclic needs of the equipment. It is more productive for an operator to load a magazine at his or her own speed, and for the machine to work from the magazine, than for the operator to feed the machine directly. This applies *even if the machine cycle when feeding from the magazine is no faster than the average feed cycle of the operator*. Thus the use of feeders, buffer stocks on assembly lines, and regular maintenance systems that can be operated when the equipment is still in use permit operators to work at their own speeds, while the machinery operates in its own cycle. What is more, the operator is in charge of the machine rather than the other way around. This is what is needed if good operators are to give good performance (see Chapter 2.5 and Exhibit 6.9.3).

The Machine Workplace

The general considerations for all workplaces are presented in the next section of this chapter. For the arrangement of a machine work station, these considerations must be incorporated into the design. Aspects particular to the operation of machinery are outlined in this section.

The most fundamental requirement for the machine user is that, in order to exercise control, he or she must see, reach, and exert whatever forces are necessary. Exerting these forces must not put the operator outside the reach or vision area for other aspects of the process that are concurrently important. Such simple rules frequently are not observed; it is common to see operators stretching, leaning, twisting, bending, standing on one foot, or adopting other unbalanced postures while working their machines. Indeed, this is so common that it usually goes unnoticed. We are not aware that all the effort and fatigue generated by operators in coping with the machinery are more of a hindrance than a help to them in making a product.

The second point of importance, already introduced, is the necessity to change posture at any time. If the work can be done only in certain positions, then of necessity there will be periods when it is not done. The opportunity to sit *or* stand while working will require the operator's shoulders to remain in the same relative position to the work in either posture. Hence the provision of footrests and, where necessary, the duplication of foot controls will ensure comfortable sitting.

The third important point in design is to equip the person-machine interface with controls and methods for showing the necessary information from the machine that are designed to suit human requirements. The sources listed in the bibliography provide detailed specifications for handle sizes, forces, visibility of numbers and letters at various distances, and so on, but the following general principles should be observed.

Principle 1. *Where possible, design so that, when the equipment gets old or dirty, boundary values will not be exceeded.*

With age the forces to move controls usually increase because of lack of maintenance, wear, or misuse. Dial markings or legend plates suffer reduced contrast between lettering and background because of abrasion or dirt, with a consequent increase in reading difficulty. Any force that must be frequently exerted should be as low as possible, preferably not more than 10% of the fifth percentile female maximum value for the particular direction under consideration. However, friction is not always a bad characteristic in controls, since a very freely moving control may make precise positioning difficult.

In view of the frequent use of handwheels and cranks on production machinery, the general principles defining the choice of dimensions for these controls are outlined in this section. The performance criteria relevant to the design should be clearly thought out. The maximum force that can be exerted, or the maximum turning speed that can be achieved, rarely is industrially relevant. Neither is a suitable criterion for selecting these types of control. The following two criteria are more useful:

Maximum Force To Be Overcome. Use a handwheel, setting it in the horizontal plane at chest height for the shortest operator (waist height for the tallest operator) and at the largest diameter feasible, but not greater than 20 in. (500 mm) in diameter.

Maximum Precision of Setting. This depends primarily on the dial or other display being set; the larger the display, the higher the precision. Provided that the pointer and dial marking are of equal thickness, mark width is unimportant. Handwheel or crank diameters have no recognizable effect on *precision* of setting (above 4 in. (100 mm) in diameter), but should be selected to overcome resisting torque or friction. Skilled workers will set with equal precision against a wide range of resisting torques if handwheel diameters are appropriately chosen. Exhibit 6.9.4 illustrates the relationship between dial diameter and setting precision. These values apply for resisting torques ranging between 3 and 60 in.-lb and for handwheel diameters between 4 and 12 in. (100 to 300 mm).

Principle 2. *Controls should be identifiable without error by their feel and position.*

Users will then not have to take their eyes off the process to be sure they have grasped the correct control (see Chapter 6.8). Controls should be arranged logically in relation to the processes being controlled, for example, in the same sequence as the process flow. Various functions should be identified by a unique and consistent control shape. For example, all amplitude controls might have mushroom-shaped heads, all frequency change switches be made as a knurled cylinder, and so on. Where various aspects of a process are controlled from one panel, each aspect should be clearly separated and marked off from the others. As an example, the group of controls dealing with feeding material into a hot forging machine should be distinct from those associated with the heating process, the ram movements, and so on.

Principle 3. *Controls, as well as displayed information, are best designed to behave in a manner similar to that of the equipment with which they are associated.*

Movement of a control to the right should not move its associated mechanism to the left. At the same time, common stereotypes should be adopted where possible. Red lights are for warning, and green ones indicate a safe situation; knobs turned clockwise are usually thought to increase the level or intensity of the items they control. Caution is sometimes necessary, however. In the United States a downward movement of an electric switch would be interpreted as turning the current off, whereas in Great Britain it would be interpreted as switching the equipment on. (Chapter 6.8 gives further data on controls.)

It is increasingly common to use symbols as an aid to identifying controls; many attractive and ingenious ones have been designed. Unfortunately, several investigations have shown that even widely used ones may not be understood. An interesting result from a study of symbols used in Dutch railway stations was the widespread ignorance of the meaning of the sign indicating the exit. Thus it is important to use international standards for symbols wherever possible rather than local standards.

The operator often is assumed to be the only person associated with machinery. We are inclined to forget that it must be assembled, transported, maintained, and often retooled. The requirements of the personnel engaged in each of these activities should be determined and provided for. Small punch presses sometimes have a bed height so low that a tool setter may have to crouch or even kneel to set tool clearances. Another ergonomic problem with presses arose in Great Britain, which requires a visual inspection of punch press clutch and brake units at statutory intervals. The introduction of welded steel press frames with a crankshaft fed into the frame through a large hole in one side, rather than lifted into half bearings with bearing caps, meant that a lengthy dismantling

Exhibit 6.9.4 The Range Within Which a Skilled Operator May Be Expected to Set a Dial of the Stated Diameter on 19 Occasions out of 20 Trials on the Average. This Value May Be Expected When Using Handwheels Ranging From 4 in. to 10 in. in Diameter Against Resisting Torques Between 3 in.-lb and 60 in.-lb.[a]

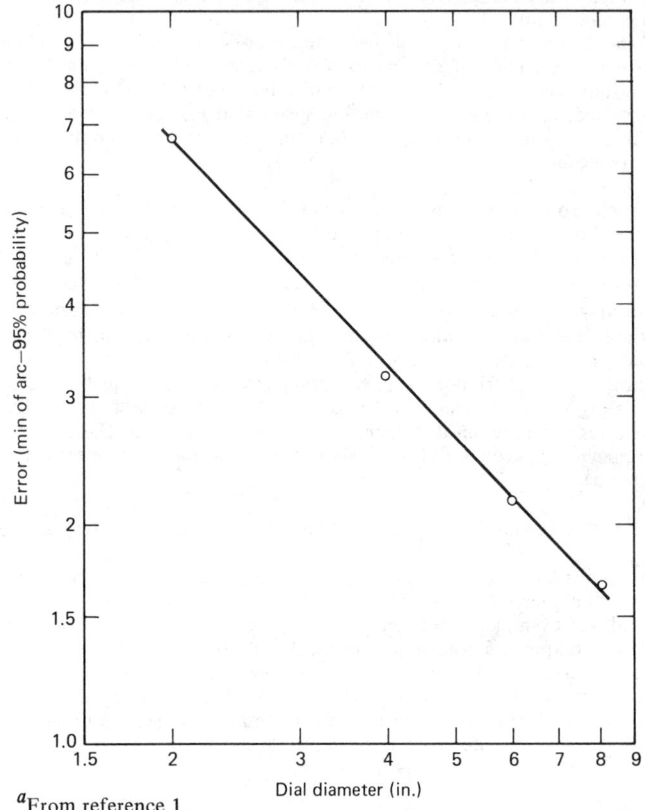

[a]From reference 1.

procedure was needed to inspect these units. Several cases of noninspection were revealed, the fitter having listened to and observed the unit in action and assessed it as satisfactory without visually inspecting the vital components.

To provide for the requirements of ancillary personnel, a task analysis (Chapter 2.4) should be carried out; all points where force has to be exerted, precise movements performed, or information obtained by means of the senses should be determined. The postures necessary to achieve the tasks then should be assessed, and the holding times and frequencies estimated. The worker's loading while performing his or her functions should be assessed, using the guidance provided in this chapter and the others in Section 6.

6.9.4 WORKPLACES

Of primary importance in practically every workplace is the necessity to see the task. Work should be arranged so that it can be seen without lowering the line of gaze more than 30° below the horizontal and without raising it more than 5° above the horizontal, with an optimum of 10° to 15° below horizontal. For fine detail, that is, anything equivalent to, or finer than, book or newspaper print, the work should be between 250 mm and 350 mm from the eyes. The illumination level and the direction from which the light comes in relation to the viewer affect contrast and glare and are crucial aspects of adequate seeing. Refer to Chapter 6.2 for more detailed information in this area.

Having decided on the positions in space appropriate for seeing the work, this visual area will be restricted further by the need to manipulate it. Chapters 3.2 and 6.3 give relevant data for specifying important aspects of manipulation, particularly the methods engineering contribution and the capacities and limitations of the human skeleton and musculature. Combining these aspects in the arrangement of a workplace will be facilitated by following the principles given here. These principles have been compiled with a clear recognition of the requirements of the human body for safe and healthy functioning when exposed to long-term work activities. The sequence of principles is deliberate—those higher up on the list should take precedence over those below them if a conflict arises during design.

The *principles for the arrangement of workplaces* are as follows:

1. The worker should be able to maintain an upright and forward-facing posture during work.
2. Where vision is a requirement of the task, the necessary work points must be adequately visible with the head and trunk upright or with just the head inclined slightly forward.
3. All work activities should permit the worker to adopt several different, but equally healthy and safe, postures without reducing the capability to do the work.
4. Work should be arranged so that it may be done, at the worker's choice, in either a seated or standing position. When seated, the worker should be able to use the backrest of the chair at will, without necessitating a change of movements.
5. The weight of the body, when standing, should be carried equally on both feet, and foot pedals should be designed accordingly.
6. Work should not be performed consistently at or above the level of the heart; even the occasional performance where force is exerted above heart level should be avoided. Where light hand work must be performed above heart level, rests for the upper arms are a requirement.
7. Rest pauses should allow for all loads experienced at work, including environmental and information loads and the time interval between successive rest periods.
8. Work activities should be performed with the joints at about the midpoint of their range of movement. This applies particularly to the head, trunk, and upper limbs.
9. Where muscular force has to be exerted, it should be by the largest appropriate muscle groups available and in a direction colinear with the limbs concerned.
10. Where a force has to be exerted repeatedly, it should be possible to exert it with either of the arms or either of the legs, without adjustment to the equipment.

How a workplace suitable for both sitting and standing work can be achieved has already been described. If the leg space permits a worker to cross his or her legs, this is an advantage; certainly the space should be wider than just the width across the worker's knees. The under-thigh pressure from long periods of sitting can be relieved if the sitter can move about on the seat; this necessitates considerable clearance on either side of the knees to allow for shifts in posture. Another factor in prolonged sitting is swelling of the lower leg. Leg movements, preferably a brief period of walking, will counteract such swelling. It can be reduced, but not prevented, if there is room to stretch the legs and activate the leg muscles during sitting.

Industrial seats are noted mostly for their being uncomfortable. It is simple common sense to provide comfortable seats if this reduces the discomfort or health risks associated with the job. If an operator cannot use a backrest when doing his or her job, and if the job is done for a major part of the working day, then the work situation should be examined in order to improve this aspect. However, there also must be opportunities for the worker not to have to use the backrest at times. The workspace design should not pin the operator into the chair.

Inevitably, industrial seats must be adjustable over a range of about 6 in. (150 mm). This adjustment must be simple, or it will not be used. It has been said that, if a seat is at an inappropriate height, one is alert enough to overcome the difficulty in the morning, but too tired to bother in the afternoon. Hence a simple ratchet or bayonet lock system is better than a screw thread or complex mechanism. Supporting the chair on five feet adds to stability, but for many factories puts a premium on flat floors. Even so, after providing for adjustment, stability is probably the next most important feature. The seat and back should have resilient coverings which do not get sticky with perspiration in hot weather and which can be readily wiped. Resilience, rather than softness, is the requirement; the seat should have a rounded front edge and a backrest that swivels to adjust to the curvature in the lumbar region of the back.

Within this general description, a wide range of opportunities in seat design exists. However, seats are not designed independently of the work or machinery. The seat is there not in its own right, but to help the person to do the work. Seat standardization is, of course, possible, but only after study and design of the machinery, workbenches, and so on, and what might become standard in one company would not necessarily be standard for others.

Fitting Trials

In designing a workplace, the use of material from Sections 3 and 6 of this handbook by itself is not sufficient to ensure a satisfactory situation. New designs should be tested in mock-up, even if this is a primitive arrangement. During testing the various movements and activities of workers should be simulated using tall, short, fat, and thin men and women from the fifth and ninety-fifth percentiles of the population (see Chapter 6.3).

If there is doubt as to whether it is necessary to build some adjustment into the equipment, the procedure, during each person's test, should be to adjust the dimension in question in increasing and then decreasing steps over the whole of its range, which is then reported as "acceptable" or "unacceptable" by the person. Different people will be found to have different ranges of acceptability. If all the ranges are drawn on the same graph, side by side, it will be clear immediately whether adjustment is necessary. If a line can be drawn through all the range graphs, then obviously no adjustment is necessary; if not, then the gap between the bottom of the highest range and the top of the lowest range illustrates the amount of adjustment that should be provided. Exhibits 6.9.5*a* and *b* illustrate how the results for fixed and adjustable requirements, respectively, will appear.

If several dimensions must be tested, then, after separate trials for each one on its own, all should be assembled on the mock-up at their chosen positions and the people brought back for a final test of effectiveness. This is desirable in the event that some unforeseen interaction between different dimensions of the workplace introduces difficulties.

Repetitive Work

A stressful, but common, work situation in the factory and increasingly in the office is work that is relatively short cycle and unvarying over long periods. Stress arises because there is a constantly

Exhibit 6.9.5 (*a*) Results From a Study to Determine the Working Height for the Bottom Electrode of a Spot Welder. The Bottom Electrode Tip Specified the Working Height. At a Height of 41½ in. (105 cm) the Bottom Electrode is Satisfactory for All Sizes of Operator. (*b*) The Range of Adjustment Needed for a Chair to Suit All Sizes of Operator When Working at the Electrode Height of 41½ in. (105 cm) Determined in the Previous Study[a]

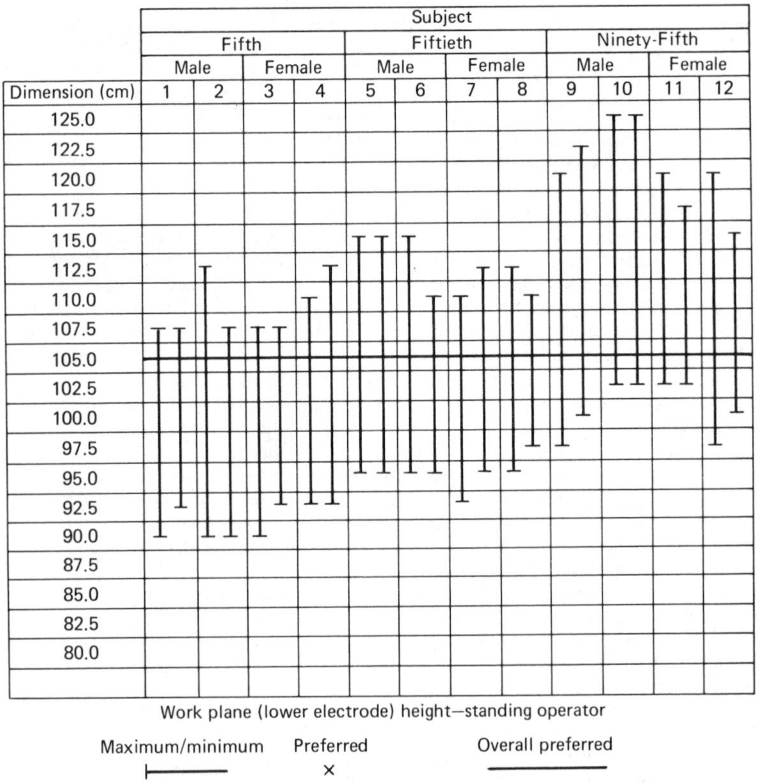

Subject

Fifth: Male (1, 2), Female (3, 4); Fiftieth: Male (5, 6), Female (7, 8); Ninety-Fifth: Male (9, 10), Female (11, 12)

Dimension (cm): 125.0, 122.5, 120.0, 117.5, 115.0, 112.5, 110.0, 107.5, 105.0, 102.5, 100.0, 97.5, 95.0, 92.5, 90.0, 87.5, 85.0, 82.5, 80.0

Work plane (lower electrode) height—standing operator

Maximum/minimum ⊢—— Preferred × Overall preferred ——

(*a*)

[a]From reference 2.

Exhibit 6.9.5 ***(Continued)***

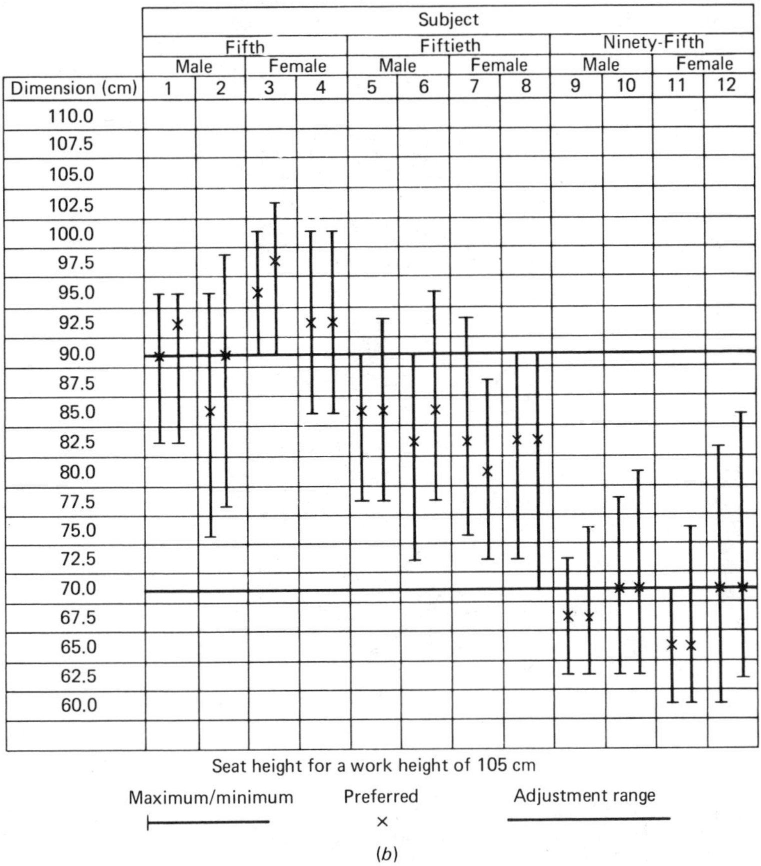

Subject
Fifth
Fiftieth
Ninety-Fifth
Male
Female
Male
Female
Male
Female
Dimension (cm)
1 2 3 4 5 6 7 8 9 10 11 12
110.0
107.5
105.0
102.5
100.0
97.5
95.0
92.5
90.0
87.5
85.0
82.5
80.0
77.5
75.0
72.5
70.0
67.5
65.0
62.5
60.0

Seat height for a work height of 105 cm

Maximum/minimum Preferred × Adjustment range

(*b*)

repeated load on the same muscle groups. Physiologically there may be little variation in energetic activity. Mentally there is little variety to provide interest and stimulus, so that the necessary attention to do the task must be maintained by willpower or stimuli gathered from outside the work situation (see chapter 2.5).

It is widely recognized today that pacing a worker's output by a machine cycle, computer response, or moving assembly line tends to be inefficient. The recognized variability in human performance times is such that, if pacing were set to cope with 95% of the unpaced operator's cycle times, the result would be grossly inefficient. It is preferable to decouple operators from the machine's working speed wherever possible; feeding material into buffer stores beside each worker is now more common. A buffer of four or five components for a short-cycle operation, and perhaps only one or two for long-cycle work, can allow variations in speed and attention to quality virtually uninfluenced by the rate of feed of the machine or line.

Exhibit 6.9.6 illustrates the effects of pacing on output for a single operator and clearly shows the trend of missed parts. This can be somewhat compensated for by increasing the time available for the part to be picked up before it is missed entirely. The tabulated data at the foot of the exhibit show the improvements obtained with small buffers. Where a buffer is shared by a small group of workers doing successive operations, suitable design can permit the group to work "up the line," emptying the upstream buffer and filling the downstream one, so that they may choose their times for rest and refreshment without in any way affecting the performance of preceding or succeeding operators.

6.9.5 OFFICES AND BUILDINGS

This part of the chapter deals only with particular features relevant to the match between people and their work. Environmental factors are discussed later—noise and vibration in Chapter 6.10, illumination in Chapter 6.11, and climate in Chapter 6.12.

Exhibit 6.9.6 The Losses Arising With Pacing and How Some Compensation May Be Obtained by Increasing the Tolerance Time or Buffer Stocks. Tolerance Time Is Specified in Seconds. Mean Unpaced Cycle Time Was 6 Sec for the Study.[a]

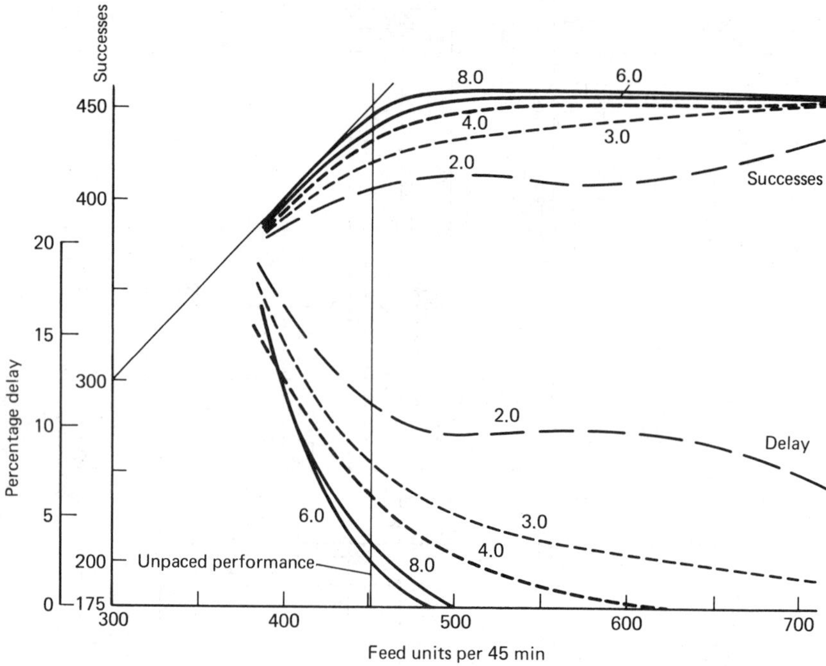

[a]From reference 3.

Apart from housing the hardware, the office or factory houses people. Broadly, people pursue two functions when within the building: the performance of their particular activities and communication with each other. The second activity is important in terms of the obvious need to give and receive orders, but also in terms of safety (seeing and being seen by others in hazardous work areas, for instance). Another important communication aspect is the social interaction that is necessary for human beings.

If it is necessary to shout in order to communicate, then people will do it as little as possible; hence the noise levels in a plant should not be set just against deafness criteria (see Chapter 6.10). Many machines give important information to their users from the noises associated with their operation; these should be audible some distance from the machine if confident control is to be assumed by the operator. If people in offices or plants are to be able to talk to each other from their workplaces, this presupposes certain noise levels related to the distances between people; Exhibit 6.9.7 gives specific values. Since people will always seek other human contacts, the workplace should be arranged so that this can be done at times without having to stop work. Human contact can also be maintained by seeing others; the isolated workplace is always to be avoided. Apart from it being dangerous for the isolated operator should he or she become ill or be injured, to most people loneliness quite quickly leads to unpleasant or disoriented sensations. Human presence and activity is stimulating and enlivening and can provide some evidence of normality when circumstances get difficult.

The point concerning normality also applies to being able to see outside the factory. It is generally preferable to have windows looking out on the world so that the stages of the day and the state of the weather can be seen. A very large factory building may require this less than a small plant or office, which would be claustrophobic without a visual link. Where the necessity is not so strong, it may still represent a valued amenity, however.

It frequently goes unrecognized that a factory or office indicates to its users much about how they are seen and valued by the company. There are not dramatic differences between the social lifestyles of senior managers and shop floor workers today in Western countries. A dirty building in which facilities are not cleaned or repaired and in which the housekeeping is neglected says to those employed that the management values them little as people and cares little for their comfort and sensibilities. Since good performance can arise only from the self-motivation of the workers

Exhibit 6.9.7 The Effects of Ambient Noise Levels on the Recognition of Speech, With the Distance Between Speakers for 90% Understanding of Conversation[a]

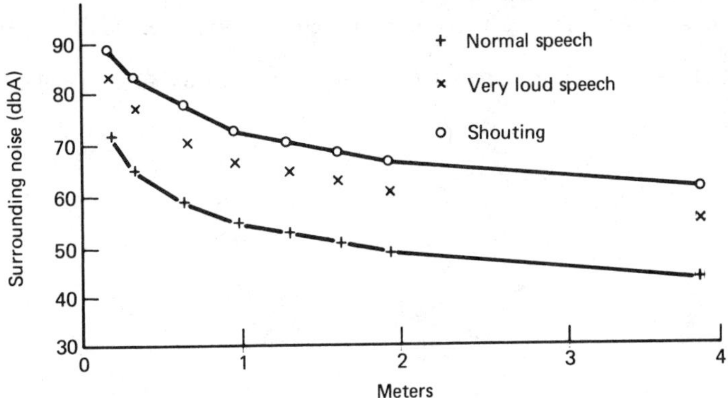

[a]Derived from data in reference 4.

concerned, it would seem little short of foolish to demonstrate that the company rated them as little more than a resource without higher human perceptions or feelings.

This perspective may, perhaps, be applied to the whole of this chapter and be a common one throughout many other chapters in this handbook. Industrial engineering is about people in industry, and it is belatedly recognized that industrial efficiency ultimately depends on those people. Technology is a resource that industry uses and it is not independent of the people themselves. To design the workplace and the equipment to suit people is therefore not just a kindness, but the obvious and only method to use in creating the essential conditions for efficient industry. Without this approach to work design, the industrial scene is littered with obstructions to efficient work, both physical and psychological, to which we give our time and effort to overcome–obstructions that should not have been there in the first place.

REFERENCES

1. E. N. CORLETT, "The Accuracy of Setting of Machine Tools by Means of Handwheels and Dials," *Ergonomics*, 1961, Vol. 4, pp. 53–62.
2. R. P. BISHOP, "The Ergonomics of Foot-Operated Spot Welders," Unpublished report, Department of Engineering Production, University of Birmingham, 1973.
3. R. J. SURY, "Allowing for the Human Factor in Designing Conveyor Systems," *Mechanical Handling*, January, 1967, pp. 25–29.
4. H. P. VAN COTT and R. G. KINKADE, *Human Engineering Guide to Equipment Design*, U.S. Government Printing Office, Washington, DC, 1972.

BIBLIOGRAPHY

General Texts on Industrial Ergonomics

MCCORMICK, E. J., *Human Factors in Engineering and Design*, McGraw-Hill, New York, 1970.

GRANDJEAN, E., *Fitting the Task to the Man*, Taylor & Francis, London, 1980.

Text Dealing Particularly With Handwork

GREENBERG, L., and D. B. CHAFFIN, *Workers and Their Tools*, Pendell Publishing Company, Midland, MI, 1977.

Sources of Ergonomic Data and Information

Ergonomics Abstracts and Data (four times per year), Taylor & Francis, 10-14 Macklin Street, London WC2B 5NF England.

VAN COTT, H. P., and R. G. KINKADE, *Human Engineering Guide to Equipment Design*, U.S. Government Printing Office, Washington, DC, 1972.

Journals With Relevant Articles

Applied Ergonomics (four times per year), IPC Science and Technology Press, IPC House, 32 High Street, Guildford, GU1 3EW England.

Human Factors (12 times per year), The Human Factors Society, Inc., Box 1369, Santa Monica, CA 90406.

Ergonomics (12 times per year), Taylor & Francis, 10-14 Macklin Street, London, WC2B 5NF, England.

CHAPTER 6.10

Noise and Vibration

MALCOLM J. CROCKER

Purdue University

6.10.1 INTRODUCTION

"Noise" is usually defined as unwanted sound. In the frequency range of about 15 to 16,000 Hz, sound is audible and is sensed mainly by the ear. At frequencies below about 15 Hz, sound is termed "infrasound" and, if sufficiently intense, can cause different body organs to vibrate with unpleasant sensations. Above about 16,000 Hz, sound is termed "ultrasound" and is no longer audible by people, although it can be detected by animals such as dogs and bats.

Vibration is undesirable for several reasons. After long periods, vibration can cause structural fatigue and eventual failure of mechanical systems. Vibration in machinery, vehicles, and airplanes can cause annoyance and disturbance. In addition, such structural vibration in machinery and vehicles can result in sound radiation and airborne noise.

This chapter is mainly concerned with control of noise and with vibration that results in noise. Noise has several undesirable effects. In industry the main effect is that intense noise experienced for long periods throughout a working life can result in permanent deafness. Other possible secondary effects, yet to be conclusively attributed to noise, include increased accidents and reduced efficiency and productivity. Industrial noise interferes with conversation, warning signals, telephone communication, and so on, at work. It can also be a source of community annoyance and complaints, even affecting sleep and other human activities in severe cases.

This chapter first reviews the way in which sound propagates and its main effects on people. The chapter concludes with a discussion of methods of measuring and controlling noise and some case histories.

6.10.2 NOISE PROPAGATION

Sound waves propagate rather like ripples on a pond: They travel out from a source at a constant speed. Water waves on a pond are circular and two-dimensional, whereas the sound waves in air are spherical and three-dimensional, although of course they cannot be seen.

Sound Pressure Level and Decibels

The ear responds to both very small sound pressures and very large sound pressures, up to a million times greater. Because of this large range, acousticians normally use a logarithmic sound pressure scale. The sound pressure level (SPL or L_p) of a sound is defined as[1]

$$\text{SPL} = L_p = 10 \log_{10} (p^2{}_{rms}/p^2{}_{ref}), \text{ dB} \quad (1)$$

where p_{rms} is the root mean square sound pressure and p_{ref} is 0.00002 Pa, a reference sound pressure* (approximately equal to the smallest sound an average young person can hear) The units of L_p in equation 1 are nondimensional and are referred to as "decibels" (dB). Exhibit 6.10.1 shows the SPL of some common sounds or noises.

Sound Intensity

The intensity I of a sound wave in air (in the region sufficiently far enough away from the source known as the "far field") is given by

$$I = p^2{}_{rms}/\rho_o c_o, \text{ W/m}^2 \quad (2)$$

*0.00002 pascal = 0.00002 N/m^2
≈ 0.0002 microbar

Exhibit 6.10.1 Typical Sound Pressure Levels

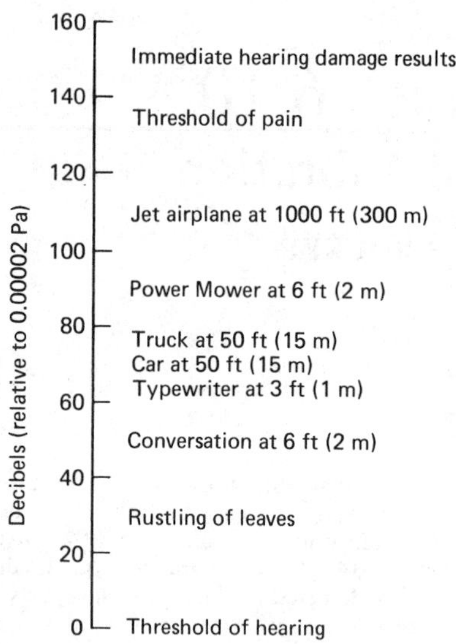

where

$\rho_o c_o$ = the characteristic impedance of air = 415 rayls (1 rayl (MKS) = 1 kg/m^{2}s)

ρ_o = the undisturbed air density (1.21 kg/m^3)

c_o = the speed of sound in air = 1125 ft/sec = 343 m/s

(All quantities given at a normal temperature of 20°C.)

The speed of sound in air depends only on the square root of the absolute temperature.

As sound waves propagate away from a sound source at the speed of sound, so does the sound energy. "Sound intensity" is defined as the sound energy that passes through unit area in unit time. Hence intensity is sound power per unit area. Sound intensity is seen to be rather like light intensity. Further analogous behavior between sound and light waves are discussed later.

An intensity level L_I may also be defined:

$$L_I = 10 \log (I/I_{ref}), \text{dB} \tag{3}$$

where I is the sound intensity (W/m^2) and I_{ref} is a reference intensity equal to 10^{-12} W/m^2. Note that, at 20°C, $\rho_o c_o$ = 415 rayls (MKS), and it is easily shown that $L_p \doteq L_I$. For other temperatures in air, $L_p \approx L_I$, and slightly greater differences can result.

Inverse Square Law

All the sound energy spreading out from a source has to pass through an imaginary sphere drawn around the source. Since the spherical area increases by four times each time the distance from the source is doubled, the power per unit area (i.e., the intensity) will reduce to one fourth of the original. Equations 1 and 2 show that the L_p correspondingly decreases by 10 $\log_{10}$ 4, or very nearly 6 dB. Hence, provided that the sound source is in free space (e.g., out-of-doors with no obstacles to cause reflections), L_p decreases by 6 dB for each doubling of distance. This is known as the "inverse square law."

Very close to the source of sound, the sound intensity is not simply related to the sound pressure. This region close to the source is called the "near field," and in this region there is a deviation from the inverse square law. Outside of this region lies the true sound field, the far field. In a building, as the distance increases from a machine, the reflections become greater in magnitude than the direct sound. The distance at which the direct and reflected sound fields are equal is called the "critical distance" (r_c) and defines the boundary between the free field and the reverberant field (see Exhibit 6.10.2).

Exhibit 6.10.2 Ideal and Practical Sound Sources Shown on Plot of SPL Against Distance r (on a Logarithmic Scale)

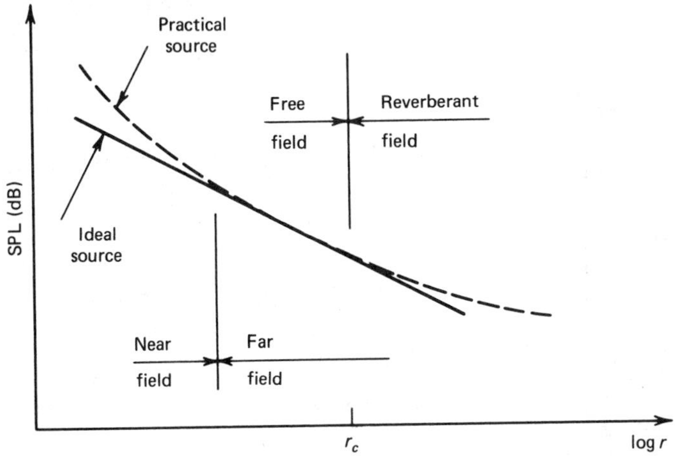

Reflections

The reverberant field is now examined more closely. If a sound wave is progressing in a medium, and the value of $\rho_o c_o$ (the characteristic impedance of the medium) changes, then there is a reflection. Solid obstacles have very different values of $\rho_o c_o$ from air and thus produce strong reflections. In such a case the inverse square law is violated. If strong pure tones, that is, discrete frequencies, are present in a noise spectrum, then the progressing wave and the reflected wave interfere to produce standing waves. There are regions in space of high sound pressure (known as "antinodes") and regions of low sound pressure (known as "nodes") for sound at these discrete frequencies. This fact is of considerable importance when the noise of a machine is measured in a room.

Plane Waves

If a source of low-frequency sound is situated in a hard-walled duct, then the sound field becomes one-dimensional, and there is no attenuation as the sound waves travel down the duct. Such waves are known as "plane waves."* If the wall of the duct is made absorbent, then the waves are attenuated as they travel down the duct. Absorption is discussed later in this chapter in Section 6.10.6.

Diffraction

When sound waves reach an obstacle, some energy is reflected and some diffracted (or bent) around the obstacle. The amount of sound energy that is diffracted depends upon the ratio of the size of the obstacle and the sound wavelength, λ. This phenomenon is also observed with light waves. In both sound and light, if the obstacle dimension is much greater than a wavelength, a strong shadow is cast. If the dimension is much less than a wavelength, a weak shadow is cast, and much energy is diffracted around the obstacle. The wavelength, λ, is given by

$$\lambda = c_o/f \tag{4}$$

Sound Power

If any imaginary closed surface is drawn around a noise source, then by summing the intensity I in a direction normal to the surface all over, the closed surface will yield the sound power of the source W, watts. For an omnidirectional source (one that radiates intensity equally in all direc-

*If the frequency of the sound is low enough so that $f < 0.59c_o/d$ where d is the duct diameter, then only plane waves can exist. At higher frequencies, cross modes can exist in addition to plane waves.

Exhibit 6.10.3 Typical Sound Power Levels

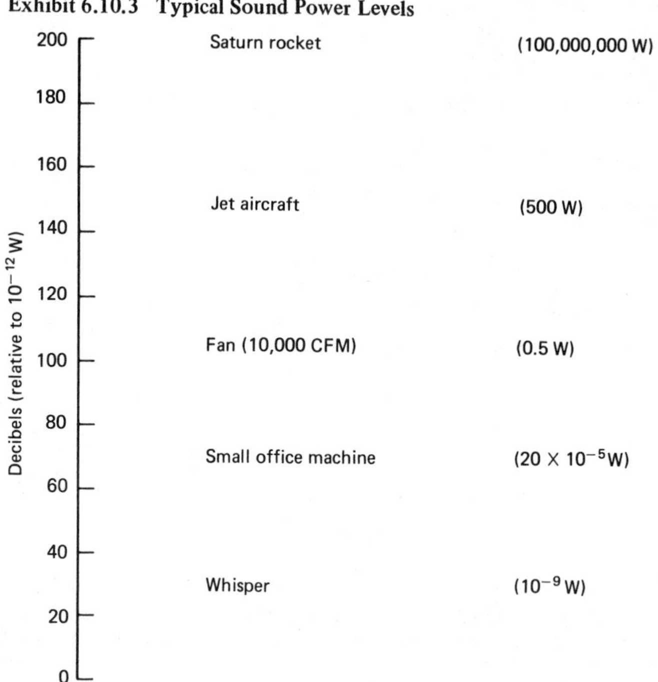

tions), $W = I4\pi r^2$, where I is the intensity, W/m^2, at a distance r meters. The sound power level (PWL) of a noise source is

$$\text{PWL} = L_W = 10 \log (W/W_{ref}), \text{ dB} \tag{5}$$

where W_{ref} is the reference sound power = 10^{-12} W. Exhibit 6.10.3 shows the sound power level of some well-known sources.

6.10.3 SUBJECTIVE RESPONSE TO NOISE

Construction of the Ear

Hearing is probably the most highly developed human sense. Exhibit 6.10.4*a* shows a cross section of the ear. The ear is normally divided into three regions: outer, middle, and inner. Sound waves are focused by the pinna into the auditory canal and travel along the canal to excite the eardrum into vibration. The eardrum (Exhibit 6.10.4*b*) is very thin–about the thickness of paper–and under tension. The eardrum vibration is transmitted by three small bones (auditory ossicles) to the inner ear (cochlea). The outer and middle ear are filled with air, whereas the inner ear is filled with liquid. Motion of the ossicles produces compressional waves in the fluid in the cochlea, which are sensed by thousands of microscopic hair cells. These hair cells transmit electrical signals through nerves to the brain, producing the sensation of hearing. Static pressure changes are equalized across the eardrum by swallowing, which opens the Eustachian tube to the back of the throat.

Loudness of Sounds

The ear is used mainly to listen to human speech, which is mostly in the range of about 250 to 4000 Hz, and is most sensitive in this frequency region. Exhibit 6.10.5 shows equal loudness contours for average young people. These contours were obtained by playing a pure tone (single frequency) sound to people at 1000 Hz and then having them adjust a tone at another frequency until it appeared equally as loud. For very intense sounds (about 100 dB), the contours are much flatter than for sounds at low levels. At low levels, the ear hears low-frequency sounds very poorly. (If it were more sensitive at low frequency, we would probably hear our own digestive processes and musculoskeletal movements!) The lowest contour shows the hearing threshold (the quietest sound that can be heard at any frequency). If the ear were only a little more sensitive (about 10 times) to quiet sounds, we would hear the random motion of air molecules (Brownian motion)!

Exhibit 6.10.4 (*a*) **Simplified Cross Section Through the Human Ear and** (*b*) **Eardrum and Three Auditory Ossicles**[a]

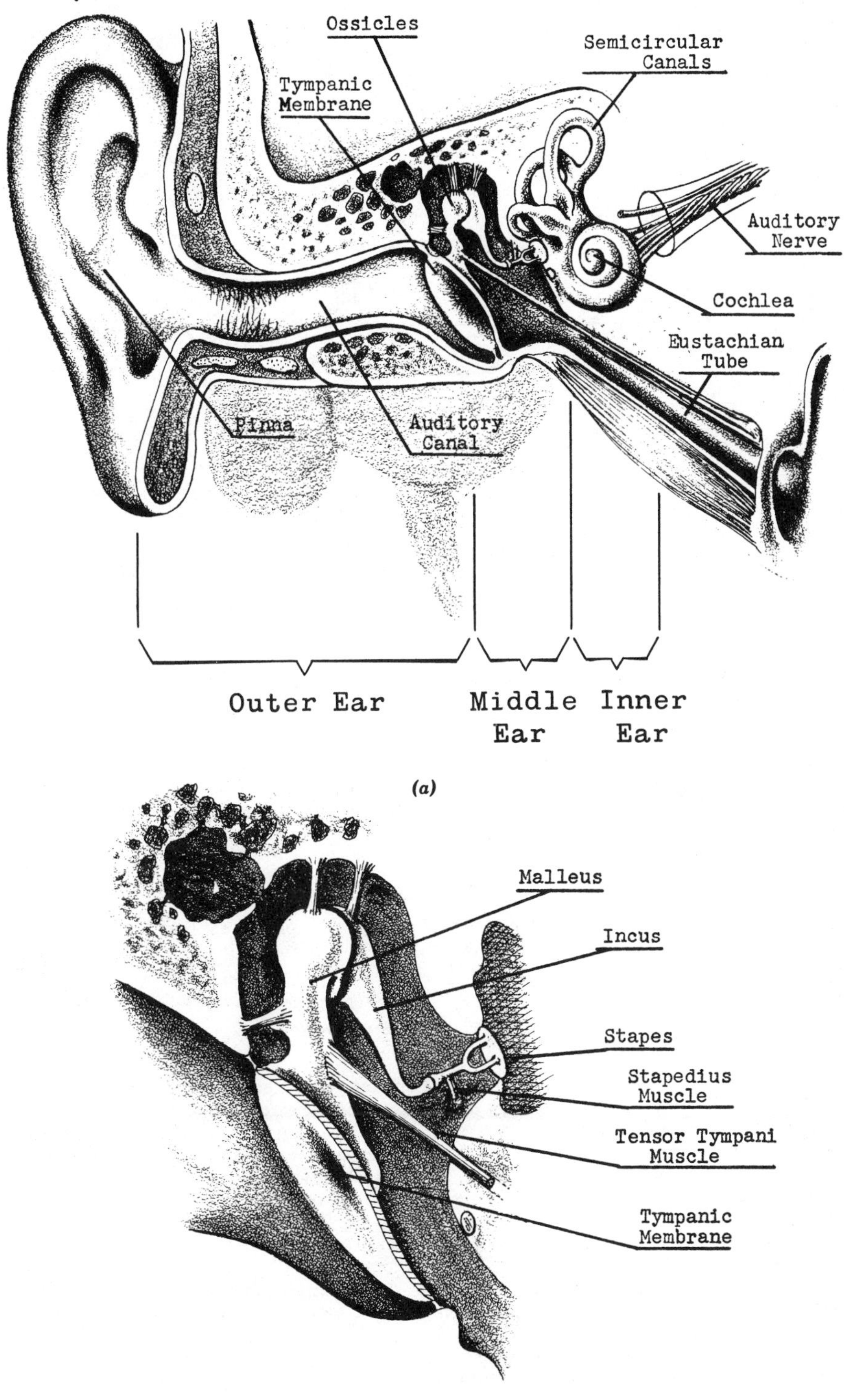

(*a*)

(*b*)

[a]From J. W. PALMER, *Anatomy for Speech and Hearing*, Harper & Row, New York, 1972. With permission.

Exhibit 6.10.5 Equal Loudness Contours for Pure Tones[a]

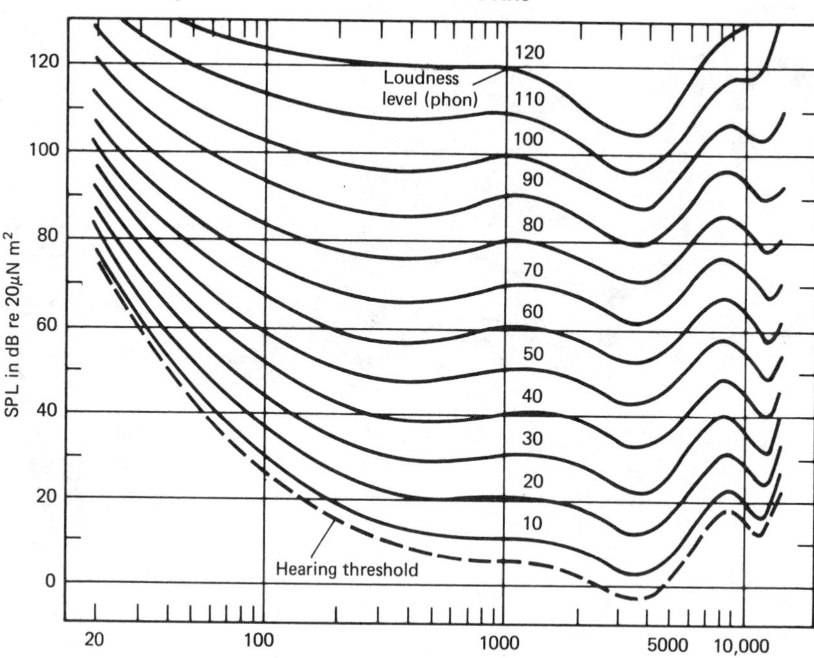

[a]From D. W. ROBINSON and R. S. DADSON, "A Re-determination of the Equal-Loudness Relations for Pure Tones," *British Journal of Applied Physiology*, Vol. 7, 1956, p. 166. With permission.

Each contour in Exhibit 6.10.5 is labeled with a number in phons. The loudness level of a sound in phons, P, is thus the SPL of a tone at 1000 Hz that seems equally as loud. Sometimes a linear loudness scale is more desirable, and the loudness S of a sound in sones is internationally agreed to be given by

$$S = 2^{(P-40)/10} \tag{6}$$

Although Exhibit 6.10.5 is for pure tones, experiments with people and bands of noise have resulted in a similar set of curves for equal loudness contours of bands of noise. With noise, equation 6 is again used internationally to obtain loudness. Thus a doubling of apparent loudness S occurs each time a noise increases by about 10 dB (when P is increased by 10 in equation 6).

Deafness and Hearing Damage

Extremely intense continuous or impulsive noise (such as gunfire or intense impacts) can cause immediate damage (see Exhibit 6.10.1). This usually occurs in the eardrum or ossicles (see Exhibits 6.10.4*a* and *b*). This damage may be permanent, although some partial healing can occur. Most people gradually become deaf by a so-called natural aging process, "presbycusis." Presbycusis affects all the population to varying degrees and men a little differently from women. Presbycusis is permanent. Usually high-frequency hearing is lost first, although the hearing loss progresses to lower frequencies later and generally becomes greater in magnitude with increasing age. Its causes are incompletely understood, but it is believed to be caused by deterioration of the inner ear, nerves to the brain, and possibly areas of the brain (cortex) also.

Industrial noise at levels of about 90 to 110 dB is insufficiently intense to cause immediate hearing loss. However, such noise, if experienced during work over a period of months or years, also causes permanent hearing loss. The damage occurs in the cochlea in the inner ear, where the microscopic hair cells are gradually destroyed. The hearing loss normally called "noise-induced hearing loss" usually first appears at about 4000 Hz, where the ear is most sensitive to sound. As the loss increases, it also progresses to higher and lower frequencies. Except for the characteristic loss at 4000 Hz, the hearing loss is similar to, and hard to distinguish from, presbycusis.

Other Effects of Noise

Noise is believed to act as a general stressor and possibly to have some effect on task concentration, efficiency, and productivity at work; absenteeism; and so on. Other adverse health effects sometimes attributable to noise are increased incidence of heart attacks, miscarriages, headaches, and so on. However, these additional effects are hard to isolate and attribute directly to noise. Indeed, some authorities insist that people adapt to noise and that, provided that noise is insufficiently intense to cause a noise-induced hearing loss, these other effects will not occur. This view is widely held in the United States, although more work should be done on such effects.

Noise does, of course, interfere with communication, sleep, and other human activities. Thus it is a problem both at work and in residential communities near to industry. Speech interference is caused by the masking effect of noise—a well-known phenomenon.

Noise Regulations and Legislation

To protect people working in noisy environments, federal noise regulations were first introduced in the United States in 1969. These are shown in Exhibit 6.10.6. In 1970 these were extended to cover almost all those working in noisy environments, totaling about 20 million workers, with the exception of farmers and construction workers. The exhibit shows that halving of exposure time allowed for 5 dB(A) increase in the noise level. In many other countries using energy arguments, noise regulations now exist which halve the exposure time allowed per day for each 3 dB(A) increase in noise level.

6.10.4 INSTRUMENTATION

Microphones

Just as the ear responds to sound pressure, the basic instrument available to measure sound, the microphone, produces a voltage signal proportional to the sound pressure. Microphones work on a variety of principles. The most common are the piezoelectric crystal microphone, the condenser microphone, and the recently introduced electret microphone. Piezoelectric microphones have several advantages—relative immunity to damage and humidity and high capacitance—but they have a rough and limited frequency response at high frequency (5000 to 15,000 Hz). Condenser microphones have a broader, smoother frequency response, but are more expensive and susceptible to damage and humidity. They also have a lower capacitance, which poses problems with the associated input to the sound level meter or preamplifier with which they are used. Electret microphones, which are quite new, are similar to condenser microphones, but less expensive; however, they are more rugged and immune to humidity, although they do suffer from a few disadvantages, such as unproven long-term stability.

Sound Level Meters

All microphones require some form of signal amplification. The sound level meter is a portable combination of microphone, amplifier, and meter. The meter is used to measure the signal, and normally the root mean square pressure, p_{rms}, is obtained. Most sound level meters have A, B, and

Exhibit 6.10.6 Permissible Noise Exposures for Occupational Noise Allowed by OSHA

Duration Per Day (Hr)	Sound Level in dB(A)
8	90
6	92
4	95
3	97
2	100
1.5	102
1	105
0.5	110
0.25 or less	115

Exhibit 6.10.7 Filter Response Curves for A, B, and C Weighting Filters

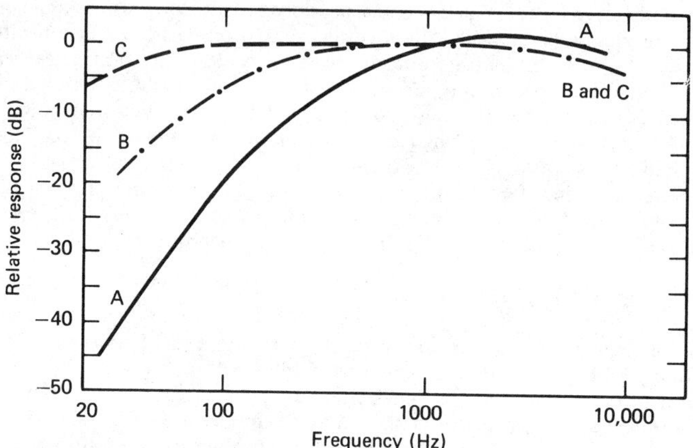

C weighting filters; some also have octave band filters. The A, B, and C weighting filters, shown in Exhibit 6.10.7, weight the sound frequency in much the same way as the human ear. For low-level sounds, the ear responds much as the A weighting filter. It is insensitive to low-frequency sound (below 1000 Hz), quite sensitive to middle- to high-frequency sound (between 1000 and 10,000 Hz), and then insensitive again above 10,000 Hz. The B weighting filter more nearly represents the ear response to middle-level sounds, and the C weighting filter to high-level sounds (where the ear frequency response is much more nearly flat). The sound level meter is sometimes used as a "loudness" meter. Most people give readings in dB(A) (decibels using the A weighting filter). Readings in dB(B) and dB(C) are not given so frequently.

Many people assume that microphones are nondirectional. This is true at low frequencies. However, at high frequencies (1000 Hz upwards), microphones are more sensitive to head-on sound than sound from the side. Most manufacturers alter the sound level meter design to modify this effect, and it is important to follow the manufacturer's instructions on whether the microphone should be pointed at the source or positioned sideways to the source.

The sound level meter response is affected considerably by reflections from the operator located only 2 or 3 ft away. Errors of up to 5 to 10 dB can be introduced. For more accurate measurements, the microphone can be located remotely from the sound level meter (see Exhibit 6.10.8).

Tape Recorders and Narrow-Band Analysis

If more narrow-band frequency information is needed, it is common practice to feed a signal from a sound level meter into a portable tape recorder and to analyze the signals later in the laboratory. Analyzers are available in several bandwidths (see Exhibit 6.10.9), including octave,* one half

Exhibit 6.10.8 Use of (*a*) Sound Level Meter and (*b*) Sound Level Meter With Remote Microphone

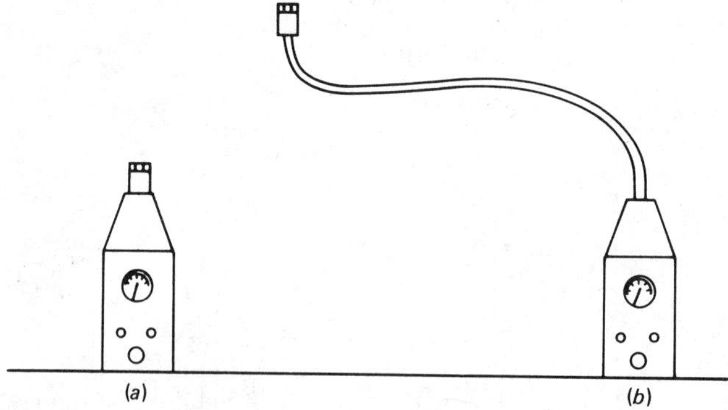

*An octave is a doubling of frequency.

Exhibit 6.10.9 Comparison Between Bandwidths of (*a*) Constant Percentage (One Third Octave) and (*b*) Constant Bandwidth Filters (20 Hz Bandwidth) at the Same Frequencies

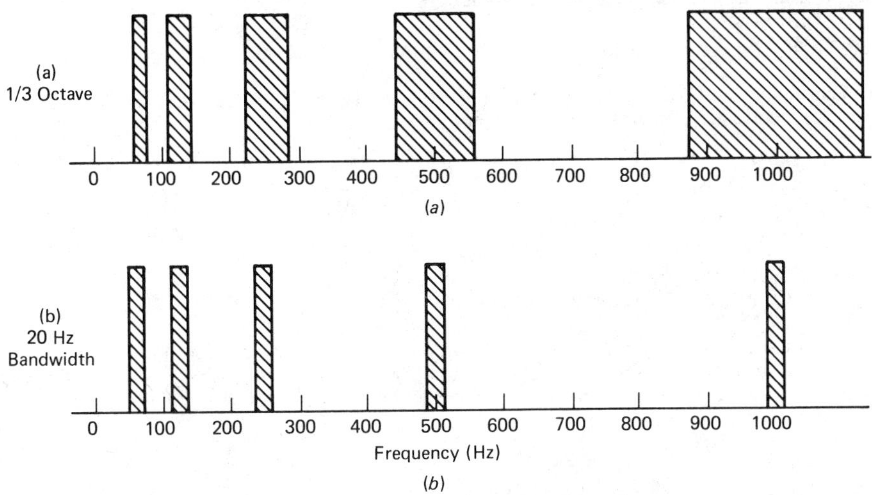

octave, one third octave, one tenth octave, and constant frequency bandwidth filters, for example, 5 Hz, 10 Hz, 20 Hz. Narrow-band equipment is more expensive, and narrow-band analog analysis is time consuming, but the results are useful in identifying the sources of noise. Real-time analyzers and fast Fourier transform (FFT) analyzers are expensive, but are capable of obtaining narrow-band frequency information very rapidly. The FFT analyzers have also recently been used with two microphones to measure the intensity I in a sound field. This new approach has become very useful in obtaining noise source information.

6.10.5 NOISE MEASUREMENT

We may want to measure the noise emitted by a machine for one or more of the following reasons:

1. To determine the noise at a certain distance (say, where an operator may be situated).
2. To verify that the noise produced by a machine lies within certain specified limits.
3. To make a comparison between the noise produced by different production models of the same design machine or between the noise produced by different machines.

Different methods will be used, depending upon the reason for the noise measurement. In case 1 it may be desired simply to measure the sound level (SL) in dB(A) at the operator's head position (when he or she is absent). In many cases the machine is fixed, or a laboratory is not available, and the noise must be measured with the machine in situ. In some cases the machine is sufficiently portable or small to take into a special laboratory where the noise can be measured more accurately. These two cases are now discussed.

Measurement of Noise from Machines *In Situ*

Standard methods are available for measuring machine noise.[2–5]

If one wants simply to measure SPL, this can be done as follows: The normal procedure is to mark a hemisphere around the machine, if possible, in the far field and in the free field. (These conditions may be checked by the 6 dB inverse square law already discussed.) The SL or the SPL is then measured at several positions on the hemisphere. For the SL, it is recommended that the A weighting setting be used. For the SPL, it is suggested that octave bands be used, although for some purposes one third octave bands may be more desirable.

It is desirable in some cases to determine the sound power (W) of the machine, since from the sound power it is possible to predict the SL or SPL of the machine in other surroundings. The sound power of a machine resting on a hard floor may be obtained by calculating the intensity W/m^2 at different points on the hemispherical surface and thus computing the total power. Use of the new two-microphone intensity technique allows the intensity to be measured directly at different points on the surface. This can then be summed over the hemisphere to obtain the total power W. However, if such sophisticated equipment is unavailable, then the classical approach must be used. In this approach the SPL is measured, and the intensity is assumed to be given by equation 2. Equation 2 assumes that measurements are in the far field. The procedure for determining W using the classical approach is given here.

Mean SL or SPL

The mean sound level (SL) or mean band sound pressure level (SPL) is calculated by averaging the results (after correcting the results for background noise):

$$\overline{L} = 10 \log_{10} \left\{ \frac{1}{n} \left[\text{antilog}_{10} \left(\frac{L_1}{10} \right) + \text{antilog}_{10} \left(\frac{L_2}{10} \right) + \cdots, \text{antilog}_{10} \left(\frac{L_n}{10} \right) \right] \right\} \tag{7}$$

where $\overline{L}$ = the mean SL or mean band SPL in dB
L_1 = the SL or the band SPL at first position and so on
n = the number of measuring positions

If the variation in the readings L_1, L_2, and so on, is less than 5 dB, an error of no more than 0.7 dB results if an arithmetic mean is taken of the dB readings instead of using equation 7. If the variation is less than 10 dB, the same procedure results in an error of less than 2.5 dB.

Sound Power Level

The sound power level (PWL) may be calculated as follows:

$$\text{PWL} = L_w = \overline{L} + 10 \log_{10} 2\pi r^2 \tag{8}$$

where PWL is the sound power level of the machine and $\overline{L}$ is the mean sound pressure level, SPL, determined above on a hemisphere of radius r, measured in meters.

It is necessary that r be great enough so that the hemisphere is in the far field. Normally, if r is greater than about twice the overall machine dimension, then this condition is satisfied (except for very low frequencies). The measurements should also be made in the free field (see Exhibit 6.10.2), and thus the effect of room reflections should be negligible. Any sound absorbing material close to the machine (inside the measuring hemisphere) will reduce the machine sound power and also should be removed. If the machine is mounted onto a floor or some reflecting plane, this might be regarded as an integral part of the machine noise source, and the machine should be tested in the same condition. If the machine is mounted in "free space," then a measuring sphere may be used instead of a hemisphere, and $2\pi r^2$ should be replaced by $4\pi r^2$ in equation 8.

Directivity Index

The directivity index, DI_θ, is given by

$$DI_\theta = L_\theta - \overline{L} + 3 \text{ dB} \tag{9}$$

where L_θ is the SPL measured at angle θ on the hemisphere at the radius r, and $\overline{L}$ is the mean of the L_θ measurements measured on a hemisphere of radius r determined from the preceding method.

If the machine is mounted in free space and the sound radiation is spherical, then the 3 dB in equation 9 is omitted.

Measurement of Noise in Anechoic and Reverberant Rooms

These rooms are available in laboratories for specialized measurements. Such facilities are expensive and their numbers somewhat limited.

Anechoic Rooms

An anechoic room is simply a means of providing a work space unaffected by atmospheric conditions; it is essentially a simulation of "free field" conditions. Anechoic rooms are of great use when the directionality of the noise source is an important feature to be determined.

The first problem is to ensure that there is low background noise in the room. This is normally achieved by building a room with fairly massive walls. If the room is built in an environment where there is a high noise level (e.g., in a noisy factory area or near to road traffic noise), it may be necessary to use a double wall construction.

It may also be necessary to isolate the room from vibration if there is a high level of vibration caused by road traffic or rail traffic or by machines, if machines such as presses are operating nearby. This is normally achieved by mounting the room on soft metal springs or rubber pads.

The second problem is to provide sufficient sound absorbing material on the walls of the room to keep reflections from the noise source under investigation to a minimum in the frequency range of interest. This is normally achieved by mounting absorbing wedges on the walls of the room. Typical wedges are made of fiberglass or foam rubber. Practical problems associated with the use of such rooms are discussed in Harris[5] and Crocker.[6]

Reverberant Rooms

A reverberant room is opposite to an anechoic room. Absorption is kept to a minimum, so that not far from the noise source the sound field is diffuse, and almost the same reading is obtained with a microphone anywhere in the room.

The room should be built with massive walls and isolated from the rest of the building (in a similar way to that described for an anechoic room) in order to keep the background noise level low. The walls of the room are made highly reflective by covering them with plaster and painting them with a gloss paint.

Reverberant rooms are ideal for measuring the sound power of a machine, provided that it does not produce much sound energy at discrete frequencies (usually called "tones"). Sound power can be much more rapidly determined in a reverberant room than in the hemispherical method described previously in this chapter.

6.10.6 NOISE CONTROL APPROACHES

Source-Path-Receiver

All noise control problems can be expressed as a *source*, *path*, and *receiver*, as shown in Exhibit 6.10.10.

This may appear to be an obvious statement, but it is nevertheless very useful to think of each noise control problem using this concept. In many noise control problems, there are several sources, several paths of energy flow from each source, and several receivers. In any noise control problem, it is the best approach to determine both the sources and the paths of energy flow for each noise source in order of importance. The best approach is to reduce the dominant noise source if this is possible. The secondary source will now probably become more important or annoying, and it may be necessary to reduce this, and so on.

Unfortunately, in many machine noise problems, it is not always practical to reduce the strength of the source, since it either may be very expensive or time consuming or may interfere with the operation of the machine. In such cases it may be more efficient to interfere with the paths of energy transmission. Examples of this are the use of enclosures, barriers, absorbing material, vibration isolators, and vibration damping material. Some of these techniques are discussed in succeeding sections; further discussion may be found in Crocker et al.[7]

In some cases even the use of enclosures, barriers, absorption, and so on, is not practical since the operator may need continuous access to a machine. In this case we must try to affect the receiver in Exhibit 6.10.10, the human ear. Using earplugs or earmuffs or making administrative changes to reduce exposure are examples of effecting changes at the receiver. Such approaches often are not satisfactory and should be considered as a last resort.

Enclosures

Enclosures are often the most effective means of reducing the acoustic energy path. Neglecting the stiffness and damping of a wall gives the transmission loss, TL, of an enclosure wall as

$$\mathrm{TL} = 20 \log_{10} (mf) - 34, \mathrm{dB} \tag{10}$$

where f is the frequency (Hz) and m is the mass per unit area (lb/ft^2).

The TL predicted by equation 10 agrees quite well with experiment, except at very low and high frequency. It is seen that enclosures are effective at high frequencies and also if they are made massive. Each time the frequency or the mass per unit area of the enclosure wall is doubled, the TL increases by 6 dB. Life is more complicated than this, however. At low frequencies we find

Exhibit 6.10.10 Source-Path-Receiver

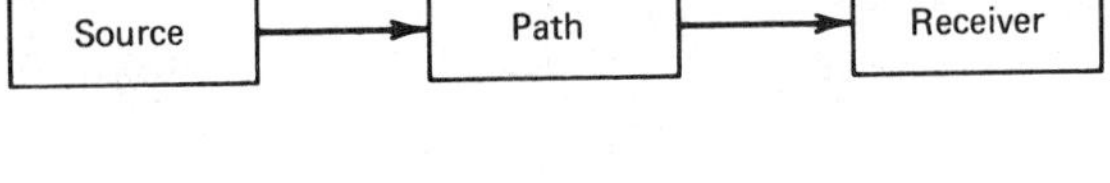

Source → Path → Receiver

Exhibit 6.10.11 Random Incidence Transmission Loss

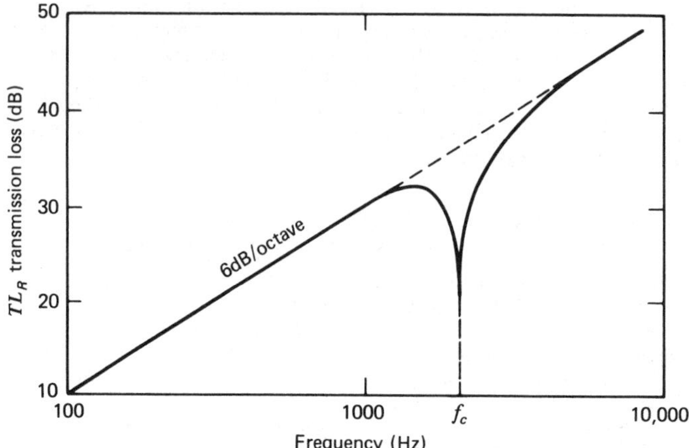

that the panel and air cavity stiffness no longer can be neglected, and the TL is decreased by resonances in the enclosure walls and the air cavity (particularly if the cavity is made small). There is also a reduction in TL at high frequencies because of the coincidence effect.[7] The critical coincidence frequency, f_c, is given by

$$f_c = 500/h, \text{ Hz} \tag{11}$$

where h is the panel thickness in inches for steel or aluminum panels. For other materials, the value 500 is replaced by another constant. The idealized TL of a typical wall is shown in Exhibit 6.10.11.

Vibration Isolation

Machines are often attached directly to the floor or to large metal surfaces. Such large surfaces are often efficient sound radiators at low frequencies. Careful vibration isolation can reduce this problem considerably. Vibration isolation can become particularly desirable when the air paths have been reduced by the use of enclosures. The theory of vibration isolation is well known and is discussed in most vibration textbooks. Briefly, the spring isolators must be chosen so that the resonance frequency is several times smaller than the frequency of the exciting force. For further discussion, see Crocker et al.[7,8]

Sound Absorbing Materials

The use of sound absorbing materials is very effective inside machine enclosures or in reverberant factory spaces in reducing the sound pressure caused by machines. The absorption coefficient α of a material is defined as the fraction of incident intensity I that is absorbed (see Exhibit 6.10.12). The absorption area A of a material in sabins is the product of its absorption coefficient α and its area S. Assuming that the sound field in the enclosure or space is diffuse (which is only approximated in the higher-frequency range), then the reduction in the sound pressure level, ΔSPL, caused by adding absorbing material with absorption area A_2 sabins is

$$\Delta\text{SPL} = L_1 - L_2 = 10 \log_{10} \left(\frac{A_1 + A_2}{A_1}\right), \text{ dB} \tag{12}$$

where L_1 = the original SPL
L_2 = the SPL after the absorption A_2 sabins are added
A_1 = the original absorption area (sabins)

Further discussion of absorption is found in Crocker and Price[9] or in other sources.[2–5]

Damping Materials

In some cases, vibration is controlled by resonances; in these cases the vibration amplitude is proportional to the structural damping present. Since the sound radiated is also proportional to the

Exhibit 6.10.12 Absorption Coefficient of a $\frac{3}{4}$ in. Thick Layer of Foam + $\frac{1}{2}$ lb/ft^2 Vinyl Coating. Noise Reduction Coefficient (NRC) is the Average of the Absorption Coefficients at 250, 500, 1000, and 2000 lb.

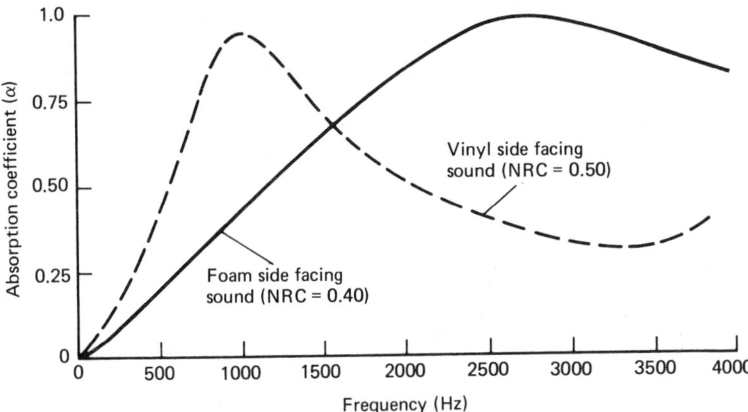

vibration amplitude, increasing the structural damping can reduce the sound radiated. Suitable damping materials made from viscoelastic substances are available. Preferably they should be applied so that their thicknesses are two or three times the metal thickness.

Leaks

Leaks in enclosures can provide paths through which sound energy can pass with no alteration. Leaks, of course, will become more important as the transmission loss of an enclosure is increased by making it more massive. In such cases even very small leaks are important and should be avoided.

Where air passages to enclosures are necessary for the machine operation (for instance, in cases where cooling is necessary), they should be lined with acoustic material, and the passages should be bent so that there is no direct "line of sight" along which the sound can travel.

Barriers

Diffraction around obstacles was discussed previously. Exhibit 6.10.13*a* shows the attenuation caused by a barrier in the case of a sound source placed on a hard floor. Exhibit 6.10.13*b* shows that barriers can be used with advantage to reduce sound when the sound wavelength is small compared with the obstacle, that is, for high frequencies. Such barriers are used along roads or railroads to shield houses or apartments. Smaller barriers can be used inside factory spaces. If the factory ceiling is low, absorbing material should be placed on it above the barrier to prevent reflections from the ceiling bypassing the barrier.

Comparison of Noise Reduction Methods

Exhibit 6.10.14 shows a comparison of noise reduction methods applied to a typical machine. Approximate A-weighted sound level reductions are given.

Personal Protective Equipment

In cases where noise reduction at the *source*(s) or along the *path*(s) is difficult or expensive, it may be necessary to consider the *receiver*(s) (usually the human ear). It is possible to enclose personnel completely with an acoustic booth. Such booths may need to be isolated from vibration from the floor and should have adequate transmission loss. Acoustic leaks should be minimized, and absorbing materials such as acoustic ceiling and wall tiles (and maybe carpet) should be used inside. Alternatively, earplugs, earmuffs, and even helmets may be used individually by personnel to reduce individual noise exposure. However, plugs, muffs, and helmets are uncomfortable to wear, particularly in a hot environment, and should be used as a last resort when engineering controls to reduce noise are too difficult or expensive.

Exhibit 6.10.13 (*a*) Noise Source and Effective Barrier Height Seen by Listener and (*b*) Noise Shielding Provided by a Barrier

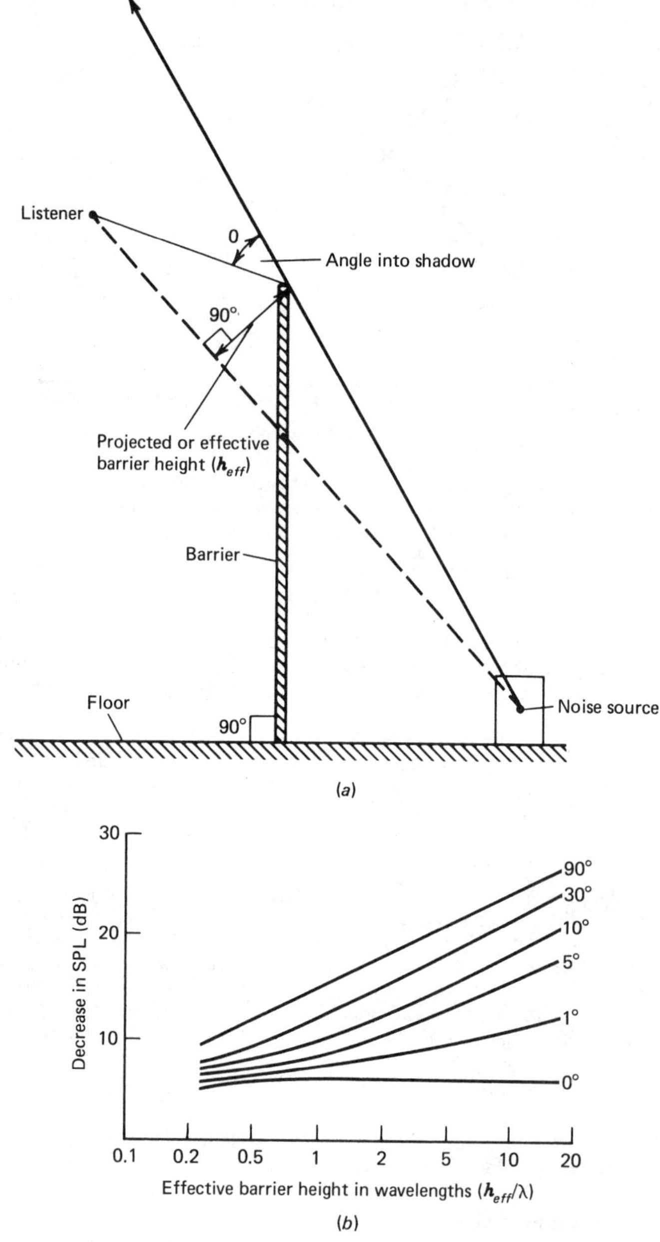

6.10.7 CASE HISTORIES OF NOISE REDUCTION

Folding Carton Manufacture

Exhibits 6.10.15*a* and *b* show top and side views, respectively, of the cutting press and automatic strippers for removal of waste material between cartons.[10]

The main noise problem in this type of machine does not usually come from the cutting press (if it is in good mechanical adjustment), but rather from the scrap disposal system. The noise is

Exhibit 6.10.14 Comparison of Noise Reduction Methods Applied to a Machine

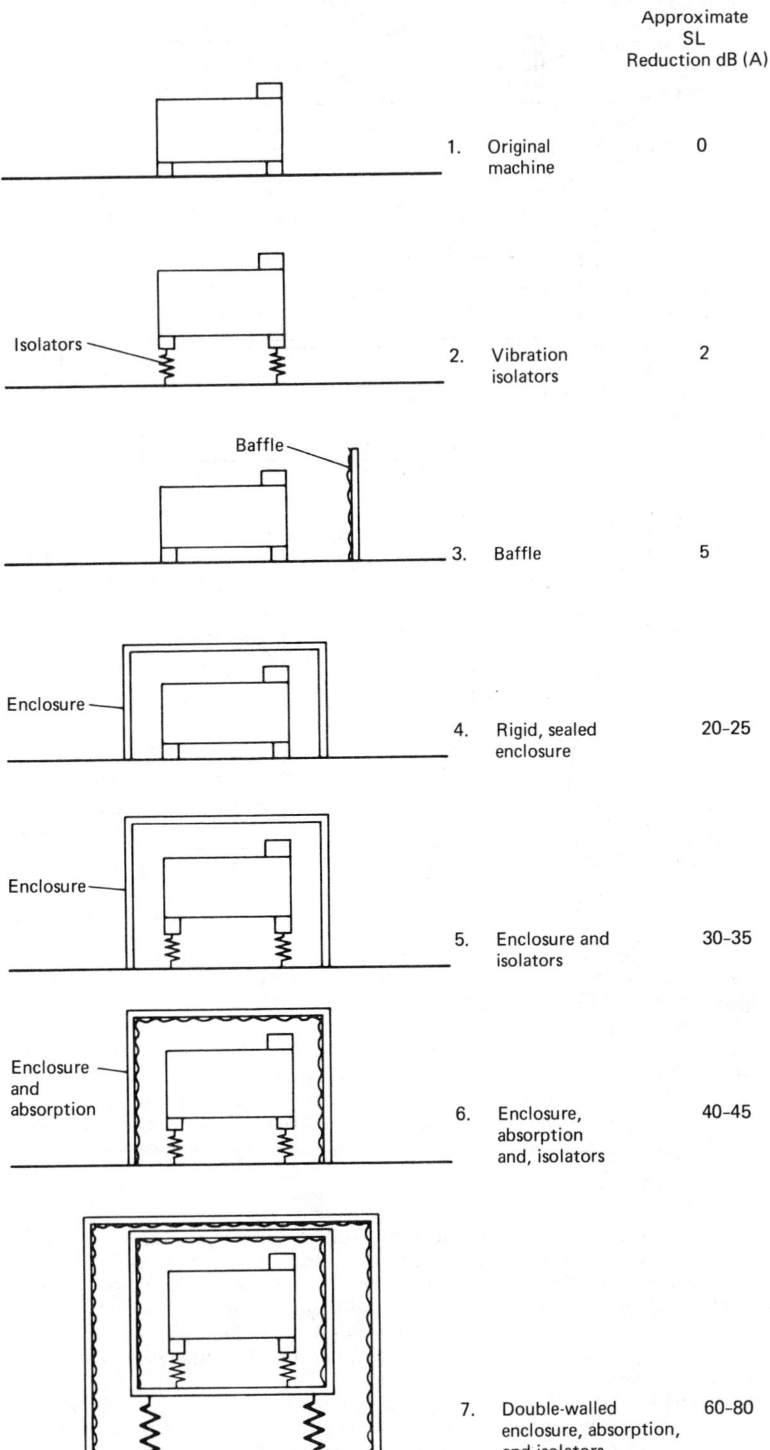

Exhibit 6.10.15 Scrap Handling System for Cutting Press, Showing (*a*) Top View and (*b*) Side View

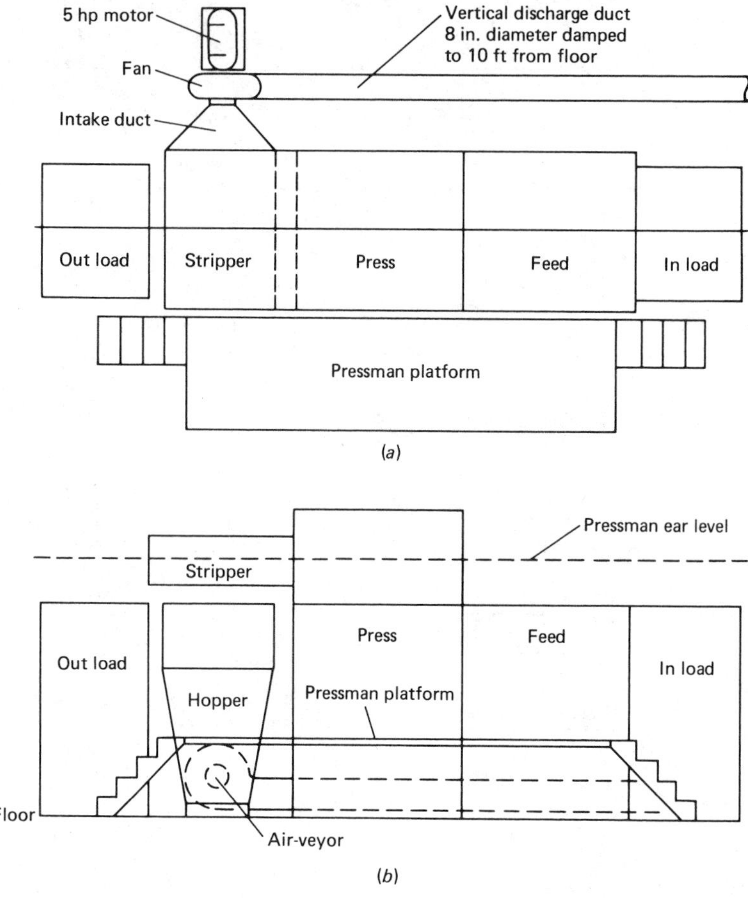

created when pieces of paper scrap strike the sides of the intake conveyor under the press stripper, the sides of the intake hood to the fan, and the fan and outlet ducts. The noise created by the scrap impacts at the pressman station reached 95 dB(A) with each stroke of the press, making the noise almost continuous.

The noise problem was reduced by gluing a layer of lead sheeting ($\frac{1}{32}$ in. thick and 2 lb/ft^2) to the outside of the surfaces mentioned previously. This increased the structural *damping* and also the *transmission loss.* This treatment reduced noise levels to about 88 to 90 dB(A) at the pressman station. Further noise reduction could be achieved by covering the duct with a lagging material.

Straight and Cut Machines

These machines straighten heavy gauge wire in an in-feed to cutoff unit which is set to cut repeated lengths. In this case the noise level was 92 dB(A) at the operator position.[10]

The noise control technique adopted was to install a *barrier* (see Exhibit 6.10.16). This was because the management had decided that minor redesign of the machine should not be attempted. The barrier materials used were $\frac{1}{4}$ in. plywood and $\frac{1}{8}$ to $\frac{1}{4}$ in. Plexiglas Ⓣ(perspex) were needed for viewing ports. The barrier was only about 6 to 8 in. from the cutter and extended about 26 in. past the ends of the region occupied by the cutter.

For an estimated $100, a 7 dB(A) noise reduction was achieved, with a final noise level of 85 dB(A) at the operator's ear position. It should be noted that such barriers can easily be removed by operators and other personnel, and thus their use should be supervised. Also bypassing can occur with low ceilings as already discussed in the section on barriers. This can be reduced by placing absorbing material on the ceiling above the barriers.

Exhibit 6.10.16 Barrier Wall for Straight and Cut Machine

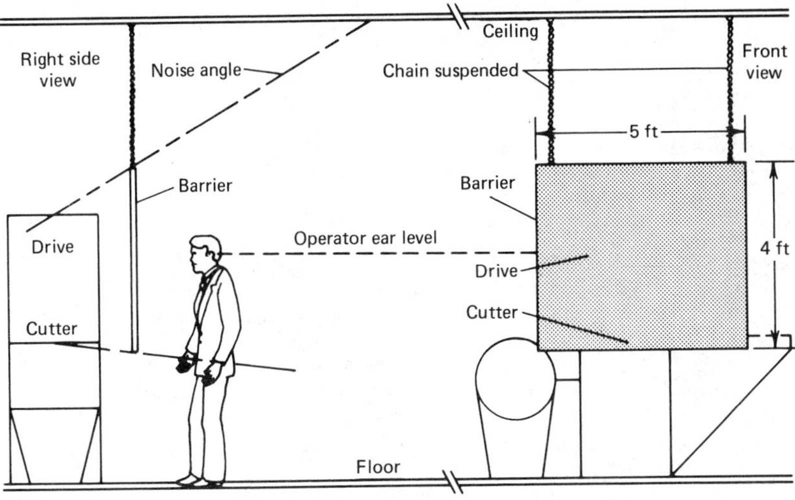

Parts Conveying Chute

Chutes are frequently used in industry to convey small parts. Much noise can be produced and radiated by the chutes when parts impact if the chutes are not properly damped.

Constrained-layer *damping* applied to the chute is normally very effective.[11] In this case history, 30 caliber cartridge cases were carried in the chute shown in Exhibit 6.10.17. The constrained layer can be placed either on the parts side or on the underside of the chute. If on the parts side, the metal layer must be wear resistant to the impacting parts. The application of the cardboard and the 20 gauge galvanized sheet to the chute shown in Exhibit 6.10.17 produced enough damping to reduce the noise from 88 to 78 dB(A) at 3 ft from the side of the chute (see Exhibit 6.10.18). Rubber deflector plates were also used to funnel parts to the center of the chute so that they did not strike the untreated sides.

Nail Making Machine

Ten nail making machines were mounted on a weak concrete floor.[8] The machines were operating at 300 strokes/min. The noise level at the operator position was 103.5 dB(A).

It was believed that the vibration caused by the impacts in the machine process was being transmitted to the concrete floor and that it was being radiated as noise. An octave band analysis of the machine noise before noise reduction was attempted is shown in Exhibit 6.10.19. It was decided to use *vibration isolators*. It was believed that the shock pulse duration was about 10 msec. The repetition period was 1/(300/60) sec or 200 msec.

Elastomeric isolators were selected to have a natural period of 100 msec corresponding to a machine natural frequency of 10 Hz and a static deflection of 0.1 in. under load. Since 10 msec < 100 msec < 200 msec, the design conditions were fulfilled. Exhibit 6.10.19 shows the octave band levels after the isolators were installed. The A-weighted sound level was reduced by 8.5 dB(A), from 103.5 dB(A) to 95 dB(A).

Since the noise level was still high, further noise reduction could have been achieved with the use of a barrier made from plywood and perspex. Absorbing materials should be used on low ceilings

Exhibit 6.10.17 Chute for Conveying Cartridge Cases

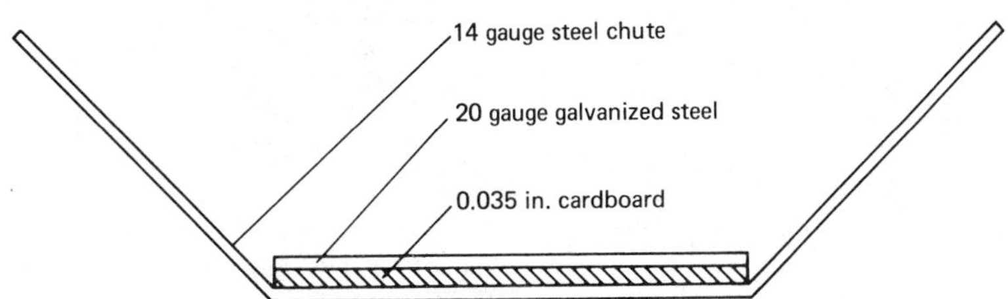

Exhibit 6.10.18 Noise Spectra Measured 3 Ft From Chute

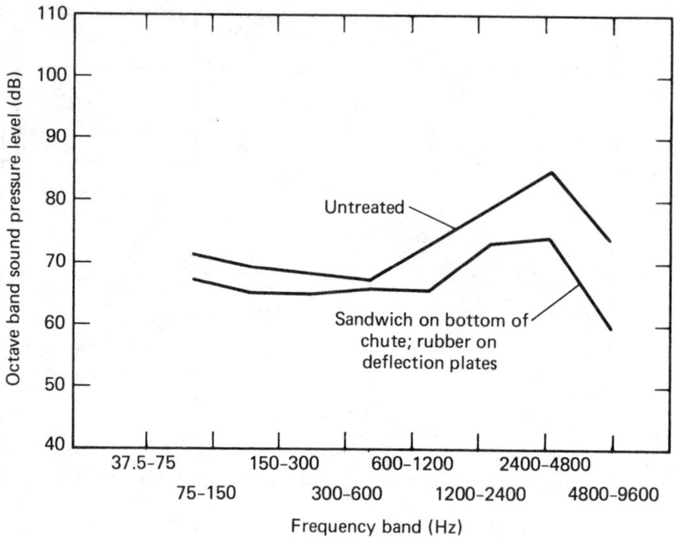

Octave band sound pressure level (dB)
110
100
90
80
70
60
50
40
Untreated
Sandwich on bottom of chute; rubber on deflection plates
37.5-75
75-150
150-300
300-600
600-1200
1200-2400
2400-4800
4800-9600
Frequency band (Hz)

above barriers to cut down reflections. Also, if several machines are used in one room, reverberation can be reduced by using sprayed-on absorbing materials or by attaching absorbing baffles to the walls and ceiling.

Metal Cutoff Saws

It is difficult to protect workers from machines that must be guided or manipulated directly. Saws are an example. Here the noise is radiated both by the saw blade and the workpiece. The noise of saws can be reduced by the use of vibration damping material on the blade or of close- or loose-fitting *enclosures*.

Handley[12] reports a loose-fitting enclosure that was built to cover the whole saw. Workpieces passed through slots in the enclosure. The workpiece is viewed through $\frac{1}{4}$ in. clear plastic. Exhibit 6.10.20 shows the noise reduction of 13 dB(A) that was achieved.

Exhibit 6.10.19 Operator Position Noise Levels for Nail Making Machine

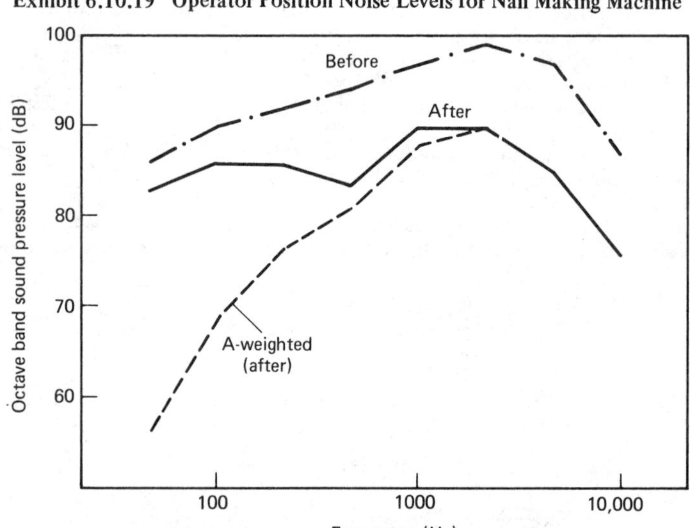

Exhibit 6.10.20 Operator's Exposure Before and After Enclosure of Metal Cutoff Saw

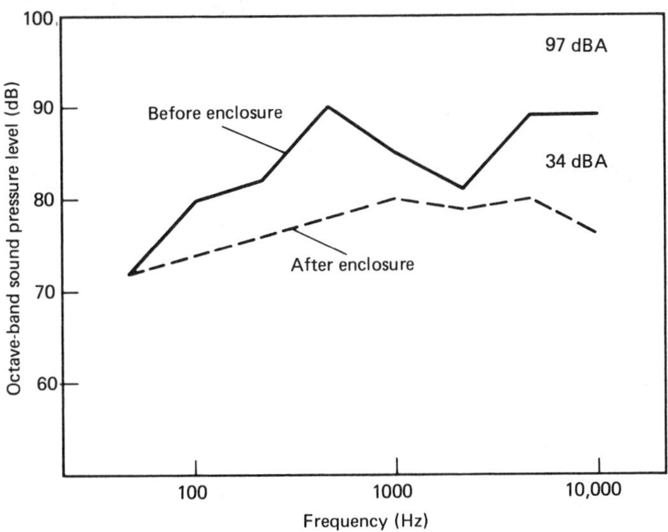

REFERENCES

1. M. J. CROCKER and A. J. PRICE, *Noise and Noise Control*, Vol. 1, CRC Press, Boca Raton, FL, 1975, p. 16.
2. L. L. FAULKNER, Ed., *Handbook of Industrial Noise Control*, Industrial Press, New York, 1976, pp. 67–112.
3. L. L. BERANEK, Ed., *Noise and Vibration Control*, McGraw-Hill, New York, 1971.
4. C. M. HARRIS, Ed., *Handbook of Noise Control*, McGraw-Hill, New York, 1958.
5. C. M. HARRIS, Ed., *Handbook of Noise Control*, 2nd ed., McGraw-Hill, New York, 1979.
6. M. J. CROCKER, "Use of Anechoic and Reverberant Rooms for Measurement of Noise from Machines," in M. J. CROCKER, Ed., *Tutorial Papers on Noise Control*, *Proceedings of the Inter-Noise Conference 72*, Noise Control Foundation, Poughkeepsie, NY, 1972, pp. 116–123.
7. M. J. CROCKER and F. M. KESSLER, *Noise and Noise Control*, Vol. 2, CRC Press, Boca Raton, FL, 1981.
8. M. J. CROCKER, J. F. HAMILTON, and A. J. PRICE, "Vibration Isolation for Machine Noise Reduction," *Sound and Vibration*, Vol. 5, No. 11 (1971), p. 30.
9. M. J. CROCKER and A. J. PRICE, *Noise and Noise Control*, Vol. 2, pp. 205–288.
10. V. SALMON, J. S. MILLS, and A. C. PETTERSON. *Industrial Noise Control Manual*, DHEW Publication No. (NIOSH) 75-183, U.S. Government Printing Office, Washington, DC, June 1975.
11. A. L. CUDWORTH, "Field and Laboratory Example of Noise Control," *Noise Control*, Vol. 5, No. 1 (1959), p. 39.
12. J. M. HANDLEY, *Noise–The Third Pollution*, Bulletin 6.0011.0, Industrial Acoustics Company, Bronx, NY, 1973.

CHAPTER 6.11

Lighting

CORWIN A. BENNETT

Kansas State University

6.11.1 LIGHTING DESIGN

Selection of lighting systems must take into account the physical characteristics of lighting, including surface materials; energy and dollar cost considerations; and human factors.

Relationships and Dimensions[1]

A light source produces a quantity of luminous flux (lumens, ϕ, lm–metric and English) which falls on a surface as illuminance (also illumination, E, footcandles–English; lux–metric; $fc = 10.761x$). The illuminance may be reflected from a surface (reflectance, ρ, dimensionless, 0 to 1) or transmitted through a surface (transmittance, τ, unitless, where zero is no transmittal and one is a perfect transmittal). The reflected or transmitted light is luminance (photometric brightness, L, footlamberts–English; candellas per square meter or nits, cd/m^2–metric; fL = 3.43 cd/m^2). Brightness or luminance is what people see, so that in terms of physical characteristics the effectiveness of the lighting system depends upon the flux of the source, the relationship of the surface to the source, and the character of the surface: $E = d\phi/dA$, where A is the area, ft^2. Also, $L = (E)(\rho)$ or $(E)(\tau)$.

Although luminance is what people see, quality electronic photometers, which measure luminance, may cost thousands of dollars. Although photometers based upon *human* brightness matching tend to be less expensive, they require some skill and give less precise and reliable measurements. Lightmeters, which measure illuminance, are 10 to 100 times lower in cost, and thus much light measurement and specification is in illuminance, with inferences made to luminance.

Energy and Cost Considerations

Although payback periods for capital investments in energy saving approaches must be considered, with high electricity costs money costs are closely associated with energy costs. Although there are many techniques for improving lighting efficiency, the principal ones include localized task lighting, use of high-efficacy sources, use of daylight, light switching and dimming, and maintenance.

By placing a source near a task, the dispersion of light can be restricted to the task area only rather than illuminating areas where light is not needed. Given sufficient general (room) lighting to provide safety of movement in nontask areas and to reduce adaptation problems in looking or moving from area to area, "task-ambient" lighting can be highly energy efficient. Furniture-mounted fixtures (task lights) can be used in both offices and factories. Further, resulting light variation may improve aesthetics compared to uniform room illumination. In manufacturing, task lighting may be an advantage in viewing three-dimensional objects where directional lighting can facilitate visibility by producing shades and shadows.

The amount of light output of a source per unit of electrical power is the source's efficacy, lm/W. For a particular source, say, metal halide, as the lamp wattage increases, the efficacy increases*: 175 W, 80 lm/W; 400 W, 85 lm/W; 1000 W, 110 lm/W. Greater differences may be found among source types. For high-wattage lamps: incandescent 22 lm/W, fluorescent 74 lm/W, high-pressure sodium 126 lm/W, low-pressure sodium 150 lm/W. Lamp life is correlated with efficacy also. Against high efficacy, color may be traded. Low-pressure sodium is monochromatic and unpleasant and permits no color discrimination (which might be unsafe for color-coded objects). All other types may be acceptable in the factory, except for color critical tasks where

*Does not include ballast losses.

the color acceptability of some mercury is negative and high-pressure sodium is questionable. In any case the most efficacious, acceptable source should be used.

Although intuitive reactions might lead to few, small, or no windows for energy saving, the lighting advantages of daylight plus seasonal heating and cooling may make windows advantageous. Since illuminance adds, light from windows on a task directly reduces artificial lighting requirements. With photosensor-controlled lighting systems, the energy saving can be readily effected. Control devices, such as blinds, must be provided to prevent glare. If windows are on room sides rather than in ceilings, then performance, comfort, and aesthetic advantages may result. Various calculational methods are available for daylighting.[1]

Light switching should be provided not only for turning room lights on and off, but for controlling lights in part of a large room. Dimmers, which save energy, are available not only for incandescent, but also fluorescent and high-intensity discharge (HID) lamps (mercury, metal halide, and high-pressure sodium).

Inadequate or infrequent maintenance may reduce light to less than 50% of its initial value. Most loss is due to aging lamps and dirt on lamps, luminaires, and room surfaces. With longer-life lamps, systematic ("group") relamping may be economical (see Chapter 10.5).

Criteria of Ergonomic Lighting

In order of importance, the ergonomic criteria applicable to lighting systems are provision for (1) health and safety, (2) performance, (3) (dis)comfort, and (4) aesthetics.[2]

Although workspace conditions such as viewing of the sun, lasers, or welding pose a threat to vision, interior lighting system intensity apparently is simply a matter of discomfort glare, although more research is needed.[3] Lighting should provide for emergency power outage.

Lighting for visual performance is as critical as the performance itself. Primarily this involves providing sufficient illuminance. However, in manufacturing, either the direction of the light or its color may be critical to portraying a particular task.

Discomfort glare may be a problem from windows, luminaires, or reflections off the task.

Lighting of offices and manufacturing facilities may involve aesthetic considerations.

6.11.2 HEALTH AND SAFETY

North American standards call for 1 ft-c (10 lux) on the floor of routes of escape and 5 ft-c (50 lux) on emergency signs for evacuation purposes in case of power failure. Power should be supplied by an independent source, such as local or central batteries or an emergency generator.[4]

For normal use, lighting for optimal task performance should prove to be suitable for safety.

6.11.3 PERFORMANCE

Primary factors in seeing are visual object size, luminance, contrast with object background, task duration, and seeing ability of viewer. The latter factor may be associated with individual differences regardless of age or may be due to universal visual deterioration associated with age. To some extent these factors are compensatory. For example, a person with good vision may see adequately despite somewhat poor task contrast. Generally, none of the factors except luminance is under control; therefore higher luminance is frequently used to compensate for other poor conditions. Lighting for performance should take all these factors into account at least implicitly.[5,6]

A further negative factor in seeing is disability glare. As a function of the illuminance and the closeness of the source to the line of sight, stray illuminance will mask the desired luminance. This commonly occurs as veiling reflectance where the light source is above and in front of the viewer (either from ceiling or task lights) and reflects off the task into the eyes, masking the task. Side lighting precludes this.

Determining Illuminance Requirements

In determining the illuminance required, one must take into account the visual task or type of space, the age of the viewers, the speed and accuracy required for the task, and the task background. The procedure for determining illuminance[7] is as follows:

1. Using the visual task (activity), refer to Exhibit 6.11.1 to determine the illuminance letter category.

2. Estimate the average age of people performing the activity. Refer to Exhibit 6.11.2 to determine the appropriate age weights.

Exhibit 6.11.1 Illuminance Letter Category[a]

Activity or Space Type	Letter Category
Exit stairways and landings, corridors, doorways	B
Administrative and lobby areas	D
Work table, coarse work	E
Work table, fine work	G
Recreation area	D
Dining area	D
Janitor's closet	C
Toilet and bathing facilities	C
Elevators	B
Detailed drafting and designing, cartography	G
Rough layout drafting	F
Reading poor reproductions, business machine operation, computer operation	F
Reading handwriting, reading fair reproductions, filing, mail sorting	E
Reading high-contrast or well-printed materials and conferring and interviewing	D
Conference rooms	E
Assembly or inspection	
Simple	D
Moderately difficult	E
Difficult	F
Very difficult	G
Exacting	H
Service garages: repairs	E
Active traffic	C
Write-up	D
Machining: rough bench or machine work	D
Medium bench or machine work, automatic machines, rough grinding, medium buffing and polishing	E–F
Fine bench or machine work, fine automatic machines, medium grinding, fine buffing and polishing	G
Extrafine bench or machine work, grinding, other fine work	H
Materials handling: wrapping, packing, labeling, stock picking, classifying	D
Loading inside truck bodies and freight cars	C
Paint shops: dipping, simple spraying, firing, rubbing, ordinary hand painting and finishing, stencil and special spraying	D
Fine hand painting and finishing	E
Exacting hand painting and finishing	G
Plating	D
Type foundries: matrix making, type dressing, casting	E
Font assembly sorting	D
Printing: color inspection and appraisal, proofreading	F
Machine composition, composing room, presses, electrotyping	E
Printing: blocking, trimming, electroplating, washing, and backing	D
Photoengraving: etching, staging, blocking	D
Routing, finishing, proofing, tint laying, masking	E
Inactive storage	B
Active storage: rough, bulky items	C
Small items	D
General testing	D
Exacting testing, extrafine instruments and scales, and so on	F
Welding: orientation	D
Precision manual arc welding	H

Exhibit 6.11.1 (*Continued*)

Activity or Space Type	Letter Category
Woodworking: rough sawing and bench work, sizing, planing, rough sanding, medium-quality machine and bench work, gluing, veneering, cooperage	D
Fine machine and bench work, fine sanding	E
Conferring and note taking during projection	D

[a]A more complete listing can be found in reference 1.

Exhibit 6.11.2 Weighting Factors

	Weight		
Task and Worker Characteristics	1	2	3
Average age of workers	Under 40 years	40 to 55 years	Over 55 years
Demand for speed and/or accuracy	Low (not important)	Average (important)	High (critical)
Reflectance of task background (average of walls and floor if there is no task)	Greater than 70%	30 to 70%	Less than 30%

3. Estimate the demand for speed and accuracy. Refer to Exhibit 6.11.2 to determine the appropriate demand weight.
4. Estimate the reflectance of the task background. Refer to Exhibit 6.11.2 to determine the average reflectance weight.
5. Add the weights from Exhibit 6.11.2, steps 2, 3, and 4.
6. Refer to Exhibit 6.11.3. Determine the illuminance level—"lowest," "middle," or "highest."
7. Using the letter category from step 1 and the illuminance level from step 6, refer to Exhibit 6.11.4. First find the range of the illuminances corresponding to the letter category. Then select the lowest, middle, or highest illuminance. This is the illuminance desired for the activity or space.

Procedures for Designing General Room Illuminance[1,8]

One critical characteristic of the lighting system is the directness of the luminaire. There are five standard categories of lighting: (1) direct lighting, where 90% or more of the flux is emitted below the horizontal down onto the task; (2) semidirect (90 to 60% downward); (3) general diffuse (60 to 40% downward); (4) semi-indirect (40 to 10% downward); and (5) indirect lighting (10% or less downward, that is, where 90% or more is emitted *above* the horizontal (onto the ceiling to be reflected back down into the room). Thus with direct lighting more of the light falls onto the task; however, glare problems may exist because of the contrast of the luminance of the luminaire or the task with its surroundings. With indirect lighting the entire ceiling would be more or less uniform, which would mean fewer shadows, and less glare.

A second major design feature is the room surface reflectance. To obtain efficient use of the light in the room, the various surfaces must be sufficiently reflective to produce satisfactory interreflectances, ultimately onto the task, rather than absorbing much of the light. Further, sufficient reflectances will help reduce contrasts with light sources and the task, thus reducing discomfort glare. Generally, maximum reflectances have been recommended for ceilings, modest reflectances

Exhibit 6.11.3 Illuminance Level Within a Range—Activities Case

Total Weight	Illuminance Level
3	Lowest
4	Lowest
5	Middle
6	Middle
7	Middle
8	Highest
9	Highest

Exhibit 6.11.4 Recommended Ranges of Illuminance

Category Letter	Range of Illuminances in Footcandles (lux)	Type of Activity or Area
A	2 to 3 to 5 (20 to 30 to 50)	Public areas with dark surroundings
B	5 to 7.5 to 10 (50 to 75 to 100)	Areas for brief visits
C	10 to 15 to 20 (100 to 150 to 200)	Working spaces where visual tasks are only occasionally performed
D	20 to 30 to 50 (200 to 300 to 500)	Performance of visual tasks of high contrast or large size–for example, reading printed material, typed originals, handwriting in ink, and good xerography; rough bench and machine work; ordinary inspection; rough assembly
E	50 to 75 to 100 (500 to 750 to 1000)	Performance of visual tasks of medium contrast or small size–for example, reading medium-pencil handwriting or poorly printed or reproduced material; medium bench and machine work; difficult inspection; medium assembly
F	100 to 150 to 200 (1000 to 1500 to 2000)	Performance of visual tasks of low contrast or very small size–for example, reading handwriting in hard pencil on poor-quality paper and very poorly reproduced material; highly difficult inspection
G	200 to 300 to 500 (2000 to 3000 to 5000)	Performance of visual tasks of low contrast and very small size over a prolonged period–for example, fine assembly; very difficult inspection; fine bench and machine work
H	500 to 750 to 1000 (5000 to 7500 to 10,000)	Performance of very prolonged and exacting visual tasks–for example, the most difficult inspection; extrafine bench and machine work; extrafine assembly
I	1000 to 1500 to 2000 (10,000 to 15,000 to 20,000)	Performance of very special visual tasks of extremely low contrast and small size–for example, surgical procedures

(20 to 40%) for floors, and intermediate reflectances for walls (40 to 60%) and furnishings (25 to 45%).

Whether general room lighting is to be the sole illumination source, is to be supplemented by task lighting, or is to supplement task lighting, it is desirable to determine how many lights are necessary to produce some level of illuminance in the space. In the first instance the goal is generally to provide uniform illuminance throughout the room, maintained over some period of time.

By definition, illuminance = luminous flux/area, such as E, ft-c = ϕ/A, lm/ft^2. *If there were no complications*, one could directly determine the number of lamps of a type needed to produce a desired illuminance in a room. If one wished 100 ft-c over a 100 ft × 100 ft room, then 1 million delivered lm would be required. Exhibit 6.11.5 gives lumens for sample lamps. Thus to light this

Exhibit 6.11.5 Lumens from Sample Lamps

Lamp Type	Number of Initial Lumens
60 W incandescent	860
100 W incandescent	1,740
1000 W incandescent	21,800
40 W fluorescent, cool white	3,150
215 W fluorescent, cool white	16,000
175 W metal halide	14,000
400 W metal halide	34,000
1000 W metal halide	110,000
250 W high pressure sodium	27,550
400 W high pressure sodium	50,000

room, for example, one would need 1200 60 W incandescents, 63 215 W fluorescents, 20 400 W high-pressure sodiums, or 10 1000 W metal halide lamps.

One complication is that not all the lumens reach the working plane. Thus the design of the luminaire, the less-than-perfect reflectance of room surfaces, and the geometry of the room must be taken into account. A coefficient of utilization (CU), which is the ratio of lumens reaching the work plane to lumens emitted from the lamps, may be determined from tables. In a method called the "zonal cavity" method, the space is divided into up to three vertical cavities–the ceiling cavity between the ceiling luminaires and the ceiling, the room cavity between the luminaires and the working plane, and the remaining floor cavity. These become entries in tables with other geometric characteristics and the reflectance and luminaire characteristics to determine the CU. The CU, which is a simple fraction, reduces the expected output luminous flux and is needed to calculate *initial illuminance*.*

Initial illuminance is multiplied by a "light loss factor" (LLF) to calculate *maintained* (over time) *illuminance.* The LLF is determined from four "unrecoverable" and four "recoverable" factors, although typically only two factors are used: (1) lamp lumen depreciation, which increases with time and is provided by lamp manufacturers, and (2) luminaire dirt depreciation, which depends on a maintenance category and space atmosphere dirtiness. All eight factors are coefficients between 0 and 1 and are multiplied together to produce LLF.

For rough estimation purposes, the formula defining illuminance can be modified by insertion of a "correction" factor of one half: Rearranging terms: area/fixture = (0.5) luminous flux/fixture/ desired illuminance. Total area can then be divided by area fixture to estimate the number of fixtures needed.

The layout procedure for desired uniform, general illumination[1] is as follows:

1. The calculated number of luminaires needed, say 45.9, can be rounded upward to a convenient value for layout, say 48 = 6 × 8.
2. The selected number of luminaires can then be layed out in some desired configuration, such as a checkerboard, with the spacing resulting from the room area and number of luminaires. If work stations are located at walls, then luminaires at the edges should be placed closer than one half of the interrow spacing to avoid inadequate illuminance near the walls. To improve uniformity further, central rows may be spaced farther than outer rows.
3. Special luminaires may be added for special requirements, such as for bulletin boards.
4. Where general illumination is the only or the primary lighting, procedures to provide uniform illumination create flexibility in furniture placement. Because of veiling reflectances, however, visibility may still vary considerably over the space.

Workplace Lighting

Task, supplementary, or workplace lighting may be appropriate in either the factory or the office. Supplementary luminaires are classified according to their distributional characteristics and vary from concentrating sources to uniform, diffuse types. For inspection of various materials as may occur in both assembly and inspection jobs, it will be desirable to position the light at a particular angle to enhance visibility of critical details through appropriate reflection off the task or through it. (A detailed set of recommendations is made in Kaufman.[8]) If task lighting is adjustable, instructions for use should be provided to workers.

It may be appropriate to calculate the illuminance that will be falling on a point, say a task, for either workplace or room lighting. Although calculation of illuminance at a point *p* usually would be on a horizontal surface, it might be more appropriate on a vertical or inclined surface. This calculation consists of two parts–a direct component and a reflected component–which may then be added.

The procedure for calculation[1] is as follows:

1. For a point source of light, a number of methods of calculating the direct component are available, including formulas, nomographs, and tables. One method directly uses the inverse square and cosine laws. In a convenient form, *Ep*, *fc* = (luminous flux, lm) × (cosine of angle of incidence of the light with the normal to the surface of interest)/area, ft^2.
2. For a long linear source, calculation can be made from *Ep*, *fc* = (*L*, *fL*) × (width of source, ft)/2 × (distance from source to point *p*, ft).

*Although bulky tables are needed for actual calculation, its general nature may be outlined here. A variety of complex and simple programs are available for lighting calculations on computers and hand calculators. These are generally considered proprietary and are available from commercial lighting laboratories and elsewhere. The Illumination Engineering Society (IES) (New York) has a Computer Subcommittee (of the Design Practice Committee). *Lighting Design and Application* publishes articles about computer programs.

3. For a very large area source, the illuminance, Ep, fc = source luminance, fL.

4. The problem of reflected components is similar to the calculation of general room illuminance where room geometry and surface reflectances must be taken into account. In addition, the location of p must be considered. Tables to determine the required coefficients show the advantage (greater illuminance) of central positions in the room.

6.11.4 COMFORT

Consideration of comfort in lighting has been concerned primarily with glare, although other facets, such as light color acceptability, may frequently be important.

Glare can produce disability effects or loss of visual performance, as in the case of veiling reflections. Glare may also cause discomfort, which may have the benefit of motivating a change of lighting, perhaps precluding performance or health threats.

Discomfort glare effects may result from fairly uniform but excessive, luminance or, more likely, especially indoors, from sources that have excessive contrast with their surroundings. These sources might be direct–windows or luminaires, or reflections in the field of view. The size of the glare source and its closeness to the line of sight increase the likelihood of discomfort. The most important determinant of discomfort is the sensitivity of the particular individual.[9]

Control of glare should include window shielding, luminaire enclosures, and lighting systems in which an appreciable component of the light is indirect. Major discomfort conditions may result from recessed ceiling fixtures where insufficient luminance is present on the surrounding ceiling and from any light sources in darkened (perhaps for aesthetics) spaces.

Lengthy calculations may be performed to determine the "visual comfort probability." This is the probability that observers will not exceed a criterion called "borderline between comfort and discomfort." This procedure is suitable for large offices lighted with certain types of fluorescent luminaires.[1]

Although lighting color acceptability has been of concern in certain applications for some time, low pressure sodium is unacceptable in any applications where others' complexions are observed. High-pressure sodium is used widely in manufacturing and less often in office spaces, with slightly less acceptability found in the latter applications.[10] Thus the type of space may be important. Energy conservation motivation may change acceptability standards over time.

6.11.5 AESTHETICS

In a narrow sense *pleasantness*, or the evaluative dimension of lighting, has been shown in research to be associated with having a variety of lighting sources such as ceiling and peripheral lighting.[11] It is not determined by the illumination level, although variation in lighting level–which is often associated with generally darkened spaces–is probably important.

More broadly, *spaciousness* is enhanced by peripheral rather than ceiling, lighting and also by amount of light.[11] *Visual clarity* is associated with the amount of light.[11] The question of a desirability of these qualities is one of appropriateness to the space. Other less well defined relationships would include lighting to provide orientation within a space (spatial definition), lighting to provide a sense of privacy (if people can see or hear others, they may feel that they can be seen or heard), or lighting to provide arousal (perhaps time changing lighting).

Common "white" lights have only minor or idiosyncratic effects on pleasantness. Apparent preferences for incandescent over fluorescent sources may be due to size, shape, and spacing rather than color.[12]

6.11.6 MOCK-UPS AND MODELS

Lighting effects are difficult to portray by graphic rendering and generally difficult to visualize. As design aids, full-scale lighting mock-ups and scale lighting models can be useful. Any economically important, unusual, or unusually lighted space should be mocked up or modeled.[13]

REFERENCES

1. J. E. KAUFMAN, Ed., *IES Lighting Handbook: 1981 Reference Volume,* Illuminating Engineering Society, New York, 1981.
2. C. A. BENNETT, *Spaces for People: Human Factors of Design*, Prentice-Hall, Englewood Cliffs, NJ, 1977.
3. D. H. SLINEY and B. C. FREASIER, "Evaluation of Optical Radiation Hazards," *Applied Optics*, Vol. 12, 1973, pp. 1–24.
4. J. A. SHARRY, Ed., *Life Safety Code Handbook*, National Fire Protection Association, Boston, 1978.

5. R. H. HOPKINSON and J. B. COLLINS, *The Ergonomics of Lighting*, MacDonald, London, 1970.
6. *Recommended Method for Evaluating Visual Performance Aspects of Lighting*, International Commission on Illumination, Report 19-2, 1979.
7. J. E. FLYNN, "The IES Approach to Recommendations Regarding Levels of Illumination," *Lighting Design and Application*, Vol. 9, No. 9 (1979), pp. 74–77.
8. J. E. KAUFMAN, Ed., *IES Lighting Handbook: 1981 Applications Volume*, Illuminating Engineering Society, New York, 1981.
9. C. A. BENNETT, "Discomfort Glare: Parametric Study of Angularly Small Sources," *Journal of the Illuminating Engineering Society*, Vol. 7, No. 1 (1977), pp. 2–15.
10. J. E. FLYNN and T. J. SPENCER, "The Effects of Light Source Color on User Impression and Satisfaction," *Journal of the Illuminating Engineering Society*, Vol. 6, No. 3 (1977), pp. 167–179.
11. J. E. FLYNN, C. HENDRICK, T. SPENCER, and O. MARTYNIUK, "A Guide to Methodology Procedures for Measuring Subjective Impressions in Lighting," *Journal of the Illuminating Engineering Society*, Vol. 8, No. 2 (1979), pp. 95–120.
12. C. A. BENNETT, P. A. ALI, A. PERECHERLA, and R. W. RUBISON, "Two Studies of Lighting Aesthetics," in *Proceedings of the Human Factors Society*, Detroit, October 1978.
13. T. M. LEMONS and R. B. MACLEOD, "Scale Models Used in Lighting System Design and Evaluation," *Lighting Design and Application*, Vol. 2, No. 2 (1972), pp. 30–38.

CHAPTER 6.12
Climate

STEPHAN KONZ
Kansas State University

6.12.1 AIR VOLUME AND QUALITY

Supply of oxygen and removal of carbon dioxide rarely are constraints for air volume; the most common constraints are odor removal or temperature control. The American Society of Heating, Refrigeration, and Air Conditioning Engineers' (ASHRAE) standards are given in air volume per person-min to allow for varying occupancies during a 24 hr period. Areas with smoking require more air changes.

It is expensive to bring supply air to a desired temperature, humidity, and quality; move it through a space; and then throw it away. Ventilation may use as much as 50% of the energy requirements of an office building. Reuse the air. The recycled air, processed to remove pollutants and odors, then is mixed with outside air (formerly called fresh air) and brought to the desired values of temperature and humidity. For information on air quality, see Chapter 6.13.

Filters and precipitators are used to remove contaminants and odors. Remove contaminants locally (such as through exhaust hoods) rather than letting them spread and then having to process many times the volume of air with general ventilation procedures. Be sure the worker's breathing zone is not positioned between the fume source and the hood. Exhaust air from "clean" areas (such as offices) can be used without processing as input for less critical areas (such as paint booths, storage). Warm exhaust air, when run through heat exchangers, can preheat input air and thus reduce heating loads. Heated air normally will be trapped (stratified) near the ceiling. In winter, use a heat inverter (fan at the top of a vertical duct) to bring this warm air down to the level of the people. In summer, let the heated air remain in a stagnant upper layer and keep the air conditioning down at the level of the people.

Avoid excess air movement in and out of a building. Doors should have "air locks" (two doors with an intervening space). Flexible plastic curtain strips can replace solid doors when traffic is frequent. Power shipping dock doors so they are kept closed as much as possible. Use dock seals to fill the gap between a truck and the wall.

Computerized building energy management saves in two ways: it reduces total load and it reduces peak load. To reduce total load, the computer may turn off the heat in a building after 5 p.m. and on weekends, turn ventilation fans off after 5 p.m., reduce hot water temperature when no one is working, and so on. To reduce peak loads, the computer may turn off ventilating fans 1, 3, and 4 from 3 to 3:05 p.m.; turn fans 1, 3, and 4 on and fans 2, 5, and 6 off from 3:05 to 3:10 p.m.; turn all fans off from 3:10 to 3:12 p.m., when hot water is brought up to temperature; permit all fans to be on at 3:12 p.m., when two big machine tools shut down for setup; and so on.

6.12.2 THERMAL COMFORT

Variables of Comfort

"Comfort" is defined by ASHRAE as "that state of mind which expresses satisfaction with the thermal environment."[1] Comfort is influenced by six major factors. The individual factors of metabolic rate and clothing usually are not controlled by the designer, who must work with the environmental factors of dry bulb air temperature (DBT), water vapor pressure (humidity), air velocity, and radiant temperature.

Exhibit 6.12.1, the psychometric chart, gives the relationship between DBT (horizontal axis) and humidity (vertical axis). The absolute humidity is given on the right-hand axis and the relative

---

The material in this chapter is a condensed version of Chapter 20, "Climate," from *Work Design*, by Stephan Konz, Copyright 1979, Grid Publishing, Inc., Columbus, OH, used with permission.

Exhibit 6.12.1 (See opposite page for legend.)

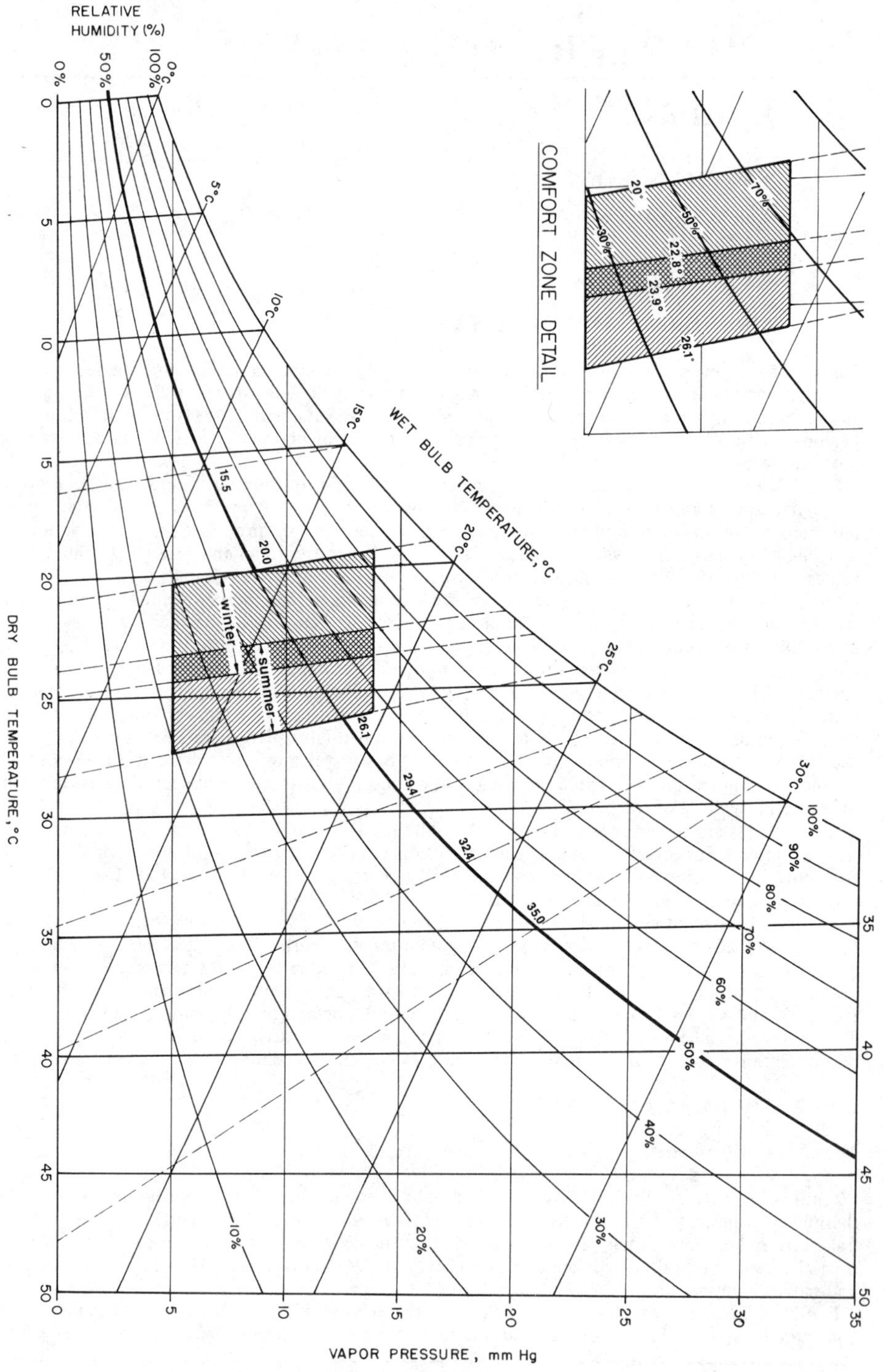

Exhibit 6.12.1 Psychometric Charts Primarily Show the Relationship Between Dry Bulb Temperature (Horizontal Axis) and Water Vapor Pressure (Vertical Axis). The Right Vertical Axis Is in Absolute Units (mm Hg, or torr), and the Left Vertical Axis Is in Relative Units (Relative Humidity). A Specific Point of 25 Dry Bulb Temperature and 15 mm Hg Has a Relative Humidity of 64%, Since the Maximum Vapor Pressure at 25 is 23.5 mm Hg. The Numbers Along the 100% Humidity Line Give Dew Point Temperature When Approached Horizontally. For Example, the Dew Point Temperature for 25 and 15 mm Hg Is 18. If the Numbers on the 100% Humidity Line Are Approached Following the Slanting Solid Lines, Then It Is Psychometric Wet Bulb Temperature; for 25 and 15 mm Hg, Psychometric Wet Bulb Is 20.2. The Numbers on the 100% Humidity Line Represent Effective Temperature (ET) When Approached Along the Dashed Lines; for 25 and 15 mm Hg, ET Is 24. These Dashed Lines Represent Conditions of Equal Skin Wettedness and Equivalent Comfort. In 1971 the Dashed Line Was Labeled With the Dry Bulb Temperature Where It Crossed the 50% Relative Humidity Line Instead of the 100% Line. The New Label Is Called New Effective Temperature (ET*). Thus the Dashed Line Formerly Labeled With 24 ET Now Is Labeled 26 ET*.

The ASHRAE Comfort Zone Is Shown for Winter and Summer Conditions. It Assumes That, Even Though the Interior Climate Is Relatively Constant, People Will Wear More Clothing in Winter and Less in Summer. It Is for People Sitting Quietly in Low Air Velocities.

---

humidity on the left-hand axis and the curves coming from it. The curved top of the exhibit (not the truncated horizontal portion), which indicates 100% humidity, has numbers that can be called three different things, depending upon how one got to the point. If reached by following the solid slanting lines, the number is called "psychrometric wet bulb temperature" or "wet bulb temperature." If determined by reading horizontally, it is "dew point temperature." If determined by the dashed line, it is "effective temperature."

Comfort Zone

The ASHRAE "comfort zone" is shown in Exhibit 6.12.1. It is the combination of DBT and humidity at which 80% of the people are comfortable. The zone is based on the responses of thousands of people who were exposed by Rohles and his colleagues to various combinations of temperature and humidity. It is for a metabolic rate of "sedentary sitting," "low" (0.15 m/s) air velocity, and 0.35 to 0.6 clo units of clothing in the summer and 0.8 to 1.2 in the winter.[2] Studies indicate that people dress according to the outdoor climate, even though the indoor conditions are constant.

Save energy by remembering that the comfort zone is a *zone*, not a point. Thus drift (maximum rate of 0.6 C/hr, maximum amount of ± 0.6 C beyond the zone) rather than slavishly fixating on a single point; this drifting zone is called the "zero energy band."

For an environment with sedentary sitting, low air velocity, and 0.4 to 0.6 clo, the mean thermal sensation (*TS*) and the percentage of people dissatisfied (*PPD*) can be predicted as follows:

$$\begin{array}{rl} ET^* < 20.7 & TS = -1.047 + 0.158\,ET^* \\ 20.7 < ET^* < 31.7 & TS = -4.444 + 0.326\,ET^* \\ ET^* > 31.7 & TS = 2.547 + 0.106\,ET^* \end{array} \quad (1)$$

where

TS = thermal sensation vote (1 = cold, 2 = cool, 3 = slightly cool, 4 = comfortable, 5 = slightly warm, 6 = warm, 7 = hot)

PPD = percentage of people dissatisfied (voting other than 3, 4, 5) corresponding to cumulative area from negative infinity for $CSIG$ or $HSIG$

$CSIG$ = number of SD from 50% for cold conditions (< 25.3 ET^*) $= 10.26 - 0.477\,(ET^*)$

$HSIG$ = number of SD from 50% for hot conditions (> 25.3 ET^*) $= -10.53 + 0.344\,(ET^*)$

ET^* = new effective temperature, C (2)

For example, for ET^* of 18 C, $CSIG = +1.67$; from a normal table, 90% are dissatisfed. For ET^* of 30 C, $HSIG = -0.21$, and 39% are dissatisfied. At 25.3 ET^*, the minimum (6%) are dissatisfied.

Although humidity has relatively little influence on comfort, low humidity in winter causes respiratory problems. Keep humidity above 12 mm Hg to prevent increases in colds and upper respiratory infections.

Adjustments for Nonstandard Conditions

The following are simple trade-offs against DBT to maintain comfort. For a detailed analysis, solve the Fanger comfort equations or use their graphic solutions.[3] For practical purposes, comfort is not influenced by the occupant's age or gender (men and women, if equally clothed and active, prefer the same temperature) or by the season of year or time of day, or even by geographic location (people in Detroit and San Antonio prefer the same temperature).

Air Velocity

For each 0.1 m/s increase in velocity above 0.15 m/s up to 0.6 m/s, increase DBT by 0.3 C; for each 0.1 m/s between 0.6 and 1.0 m/s, increase DBT by 0.15 C. (For a simple conversion from meters per second to feet per minute, multiply m/s × 200.) Stated another way, at low air velocities (< 0.6 m/s), you can increase DBT 1 C for every 0.33 m/s increase in air velocity with no loss in comfort. The "maximum" air velocity of 0.8 m/s is determined more by annoyance (blowing paperwork) than physiological effects, so that velocities over 0.8 m/s can be used for manual labor situations. Although clean air can be blown from any direction, in industrial practice blow away from the face (not into), setting the fans low and to the rear of the user.

Mean Radiant Temperature

When mean radiant temperature (MRT) differs from DBT, for each 1 C deviation of MRT from DBT, change DBT 1 C in the opposite direction. If MRT = 27 C, then to get comfort equivalent to DBT = 25 and MRT = 25, the DBT would have to be 23 C. Radiant heaters (up to 200 W) on the modesty panels of desks are quite effective in improving the comfort of seated workers; they should have a small light indicating when they are on. Radiant heaters also are used in exposed areas such as docks.

Clothing

Although the designer has little control over clothing worn, the worker can increase or decrease clothing over the "standard" clothing. The clothing insulation is expressed in clo units. For every increase in clo of 0.1, the comfortable DBT decreases 0.6 C if metabolic rate is less than 225 W and 1.2 C in the rate is over 225 W. For men, a woven shirt, sweater, and trousers has a clo of about 0.85. Females tend to wear clothing with less insulation value; a sweater and slacks has clo values from 0.6 to 0.9. There is about 0.33 clo/kg of clothing. For more on clothing, see ASHRAE[3] and Konz[4].

Metabolic Rate

For every 30 W increase in total metabolism above 115 W, decrease the comfort DBT by 1.7 C.

6.12.3 HEAT STRESS

Criterion and Limits

For more extreme environments, the criterion becomes not comfort, but the effect on performance (physical and mental) and on health.

Konz[5] presents curves from a number of studies, showing the physical and mental deterioration versus time in heat stress. Shoveling performance of acclimatized workers began to decline at about 32 ET*. The rate of decline depended upon air velocity; it declined 20% from the performance at 32 ET* at air velocity of 0.5 m/s and 40 ET*. Mental performance begins to decline at about 35 ET*. These studies show what highly motivated people can do in laboratory environments; performance probably will decline more rapidly for normal work situations.

For health, NIOSH limits the core temperature to a maximum of 38 C. "Maximum" means not exceeding 38 C for prolonged periods, such as 1 to 2 hr. "Core" temperature generally is measured by rectal temperature.

Since measuring rectal temperatures of workers on the job is not practical, researchers needed to be able to predict rectal temperature from specified environments. Lind developed the concept

Exhibit 6.12.2 Rectal Temperature Remains Constant (i.e., Stays Horizontal) With Increasing Environmental Temperature Until a Certain Critical Temperature Is Reached—the Prescriptive Zone. The Location of the Zone Depends Upon the Person's Metabolic Rate—Occurring at Lower Environmental Temperatures for Higher Metabolic Rates. In the Nonstress Zone, Rectal Temperature Can Be Estimated as 37.0 + 0.0038 (Metabolic Rate, W).

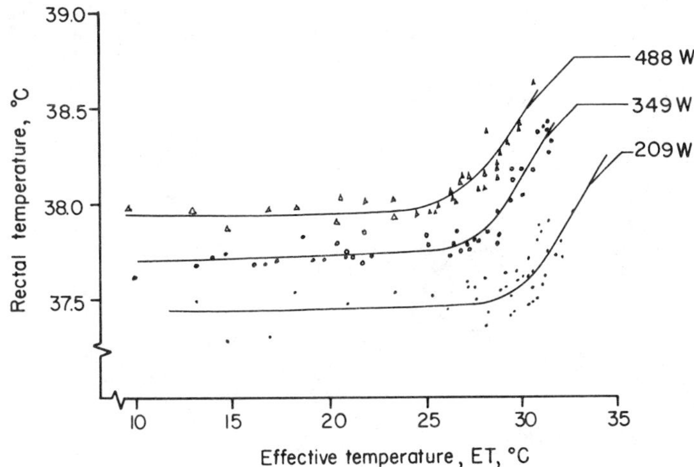

of the prescriptive zone; see Exhibit 6.12.2. At lower environmental temperatures, the body can maintain thermal equilibrium—rectal temperature stays constant for a metabolic rate. At some point—the start of the prescriptive zone—the body will not be able to maintain thermal equilibrium, and body temperature will begin to rise. The exact location of the zone depends upon the environment, the individual's acclimatization, and the metabolic rate.

Lind's concept and the 38 C core criterion was accepted by NIOSH, which developed the threshold values given in Exhibit 6.12.3. These values were designed to protect 95% of the population; it is assumed that the 5% of the population that is heat intolerant would not be working on hot jobs. These values are not maximums; they are the values at which precautions (provision of adequate drinking water, annual physical examinations, training in emergency first aid for heat stroke) should *begin* to be taken.

The recommendations in Exhibit 6.12.3 are given in an index called "wet bulb globe temperature" (WBGT). It attempts to combine into one number the effect on people of DBT, humidity, air velocity, and radiant temperature. The WBGT can be measured with a WBGT meter or calculated from its components. For an environment with radiant temperature close to air temperature,

$$WBGT = 0.7\, NWB + 0.3\, GT \qquad (3)$$

If radiant temperature is not close to air temperature,

$$WBGT = 0.7\, NWB + 0.2\, GT + 0.1\, DBT \qquad (4)$$

where $WBGT$ = wet bulb globe temperature
NWB = natural wet bulb temperature (The sensor with a wet wick is exposed to natural air currents. The NWB is not the same as psychometric wet bulb [WB], which has a

Exhibit 6.12.3 Wet Bulb Globe Temperatures (C) at Which Heat Stress Precautions Should Be Begun[a]

Metabolic Rate (Activity + Basal), W		Low Air Velocity (up to 1.5 m/s)	High Air Velocity (1.5 m/s or more)
Light	(up to 230 W)	30.0	32.0
Moderate	(up to 350 W)	27.8	30.5
Heavy	(over 350 W)	26.0	28.9

[a] A velocity of 1.5 m/s is a "noticeable breeze."

sensor with a wet wick exposed to air velocity of 3 m/s. For air velocity > 2.5 m/s, $NWB = WB$. For $0.15 < V < 2.5$, $NWB = 0.1\ DBT + 0.9\ WB$.)
GT = globe temperature
DBT = dry bulb temperature

Reduction of Heat Stress

To reduce heat stress, often it is useful to calculate what contributes to the heat load. The following equations treat the body as an open loop system in a constant environment and thus are not as accurate as computer models, which consider all the factors simultaneously in a dynamic closed loop system. However, they can give the engineer a good feel for the problem.

The key equation is the heat balance equation.

$$S = M - (\pm W) + (\pm R) + (\pm C) + (\pm E) + (\pm K) \qquad (5)$$

where S = heat storage rate, W
M = metabolic rate, W
W = mechanical work accomplished rate, W (walk up steps = +; down = –)
R = radiant heat rate, W (gain = +; loss = –)
C = convection rate, W (gain = +; loss = –)
E = evaporation rate, W (gain = condensation = +; loss = –)
K = conduction rate, W (gain = +; loss = –)

Storage

Storage can be calculated from equation 5 or from

$$S = 1.5\ m\ C_P(MBT_f - MBT_i)/t \qquad (6)$$

where S = storage gain (+) or loss (–), W
m = body weight, kg
C_P = specific heat of body = 0.83 kcal – kg/C
MBT_i = initial mean body temperature, C
MBT_f = final mean body temperature, C
t = time, hr

Mean body temperature is calculated by weighting skin temperature by one third and rectal temperature by two thirds.

Metabolism

Metabolic rates for various activities are given in Chapter 6.4. Reducing metabolic rate reduces the amount of heat storage. Working slower using the existing method or giving longer breaks reduces productivity; thus mechanization is the preferred way of reducing metabolic rate. To get metabolism in watts, multiply kcal/hr by 1.163; multiply mets by $58.2x\,(m^2)$; multiply btu/hr by 4.6.

Work

Mechanical work (W) must be subtracted from metabolic rate (M) to determine the net heat gain to the body; $W = 0$ for most activities.

Radiation

Radiant heat transfer is a function of many variables.

$$R = \sigma A\ f_{eff} f_{clr} F_{clr}\ e\ (T^4_{mrt} - T^4_{skin}) \qquad (7)$$

where R = radiant gain (+) or loss (–), W
σ = Stefan-Boltzmann constant = 5.67×10^{-8} W/($m^2 - K^4$)
A = skin surface area (usually figured from the DuBois formula), $m^2 = 0.007\ 184\ (HT)^{0.725}\ (WT)^{0.425}$
HT = height, cm
WT = weight, kg
f_{eff} = effective radiant area factor due to posture = 0.725 for standing, 0.696 for sitting
f_{clr} = $1 + 0.155\ I_{clo}$

I_{clo} = insulation value of clothing, clo
F_{clr} = multiplier to radiant heat transfer coefficient to adjust to clothing barrier
= $1/(1 + 0.55\ (5.2)\ I_{clo})$
e = emissitivity (bare skin = 0.99; clothing in nonvisible radiation = 0.7)
T_{mrt} = temperature of environment, K = C + 273
T_{skin} = temperature of the skin, K = C + 273

The really important factor in the equation is the mean radiant temperature raised to the fourth power since skin temperature does not change more than a few degrees. In general, skin temperature in the heat is about 35 C. Thus, if T_{mrt} is below 35 C, the body loses heat; above 35 C it gains–quickly. To reduce radiant temperature, the most important principle is to "work in the shade," since heat radiation behaves much as light radiation.

Clothing (a mobile heat shield) is the first line of defense. That is why experienced workers wear hats and long-sleeved shirts in the sun or near high-temperature radiant sources. Standing in the sun can add 170 W to a clothed person; it is considerably higher for people without a shirt and hat. The clothing color does not matter for nonvisible radiation; for work in the sun (and other visible radiation), use light colors.

A fixed shield between the person and the source is a second line of defense. Reflective heat shields are effective. If the operator must see the heat source, use a screen of chains or coated glass (which, however, tends to get broken or dirty). Cover conveyor entrances and exits of ovens with a screen of chains to reduce heat gain. Insulate the furnace walls to reduce heat gain (as well as to save energy).

Convection

Convective heat transfer is

$$C = h_c A\, f_{clc}\,(t_a - t_{sk}) \tag{8}$$

where C = convective gain (+) or loss (–), W
h_c = convective heat transfer coefficient, W/(m^2 – C)
= 4.5 for standing adults in "normal" air velocity
= $8.3\ V^{0.6}$ for seated adults
V = air velocity, m/s
A = skin surface area, m^2 (see equation 7)
f_{clc} = multiplier for clothing
= $1/(1 + 0.45\ I_{clo})$
I_{clo} = insulation value of clothing, clo
t_a = air temperature, C
t_{sk} = skin temperature, C

The key variable is t_a; as with radiant temperature, keep it below skin temperature of 35 C. The second variable is V–more air removes more heat. However the 0.6 exponent means that beyond 2 m/s on the *skin* there is little benefit. Air velocity drops rapidly with distance from a fan. Keep the fan close and pointed toward you. (The same applies for air from a duct.)

The primary effect of clothing is not obvious from the equation–it drops air velocity at the skin close to zero. Thus, to maximize convective loss, wear little clothing–remember, however, radiant heat, insects, and nudity customs. Do not restrict air circulation at the neck, arms, waist, or ankles.

Evaporation

Evaporative heat transfer is

$$E = h_e A\ W\ F_{pcl}\,(P_a - P_{sk}) \tag{9}$$

where E = evaporative gain (+) or loss (–), W
h_e = evaporative heat transfer coefficient = $2.2\ h_c$
A = skin surface area, m^2
W = fraction of skin that is wet (0.06 for no thermoregulatory sweating)
= $0.06 + 0.94\ W_{rsw}$
W_{rsw} = fraction of skin wet as a result of regulatory sweating (see Figure 14 in ASHRAE[3] for clothed, sedentary, low velocity situations)
F_{pcl} = decrease in evaporative efficiency due to permeability of clothing = $1/(1 + 0.41\ I_{clo})$
I_{clo} = insulation value of clothing, clo
P_a = vapor pressure of water in air, mm Hg
P_{sk} = vapor pressure of water on skin (45 mm Hg if t_{skin} = 35 C)

Evaporation can be limited by the environment or the body. From the environment, the maximum is

$$E_{max} = 12\ V^{0.6}\ (P_a - 45) \tag{10}$$

If evaporation is measured, 1 kg/h = 672 W.
The heat stress index (HSI) is:

$$HSI = E_{req}/E_{max} \tag{11}$$

where E_{req} = evaporation required for thermoneutrality (i.e., $S = 0$), W.

To improve evaporation, remove clothing and increase air velocity (up to 2 m/s on the skin). Clothing should be permeable to moisture (such as loosely woven cotton). Dehumidification of the air (i.e., reducing P_a), although expensive, gives good comfort since skin wetness is strongly related to comfort.

Evaporation can be limited by the body as well as the environment. An unacclimatized person can sweat 1.5 L/hr; within 10 days of acclimatization, this can reach 3 L/hr. Acclimatization does not come automatically by living in hot climates; it requires exercise in the heat. It can be lost in as little as 10 days, so workers returning from vacation must be careful.

The thirst drive is not sufficient to replace water loss from heavy sweating. Supervisors must insist that workers drink water; small amounts often are better than large amounts occasionally. Drink from a container, rather than a water fountain since more water is drunk. "Tank up" before heat exposure. Under normal circumstances the body has sufficient salt reserves for heavy sweating. Salt tablets should not be given; people take them excessively and get stomach problems and high blood pressure. If more salt is needed, salt the food or eat salty food such as potato chips or pickles.

Conduction

In most circumstances, conduction is small.

Heat Stress Allowances

Heat stress allowances are difficult to determine with scientific precision (also see Chapter 6.4). General guidelines are:

The amount of the allowance should be a function of the amount of heat storage, S. Greater S should give a greater allowance.

Because environmental conditions often change during the year, the allowance is quite likely to be much lower in the winter than in the summer, since the S value will change.

The amount of the allowance should be sufficient to keep average body temperature below the criterion level over the working shift. That is, short excursions to 38.5 C for a few minutes probably do no harm.

Body temperature due to heat stress should not exceed a ceiling level—probably about 39 C—for any period. This does not mean that a 39.2 C body temperature for a specific individual in an emergency such as a fire will cause harm, but the ceiling should not be exceeded in normal circumstances.

Since the goal of the allowance is to keep body temperature below a specified level, time for coffee breaks and meals should be included in the allowance time.

More accurate allowances probably will come from computer modeling of the heat balance equation, such as by Azer and Hsu,[6] than from general indexes such as WBGT.

6.12.4 COLD STRESS

Criterion and Limits

As we depart downward from the comfort zone shown in Exhibit 6.12.1, the first sensation is of discomfort; equation 1 indicates when people will sense the environment as slightly cool, cool, or cold. Next comes loss of manual dexterity when hand skin temperature reaches 15 to 20 C. (When the body core is warm, calm air temperature probably will be $< 18°C$.) When core temperature drops below 35.5 C (because heat storage, S, is negative too long), expect decreases in mental performance. At 35 C, dexterity is reduced so much you cannot light a match. Below 35 C, the mind becomes confused; at 32 C, there is loss of consciousness.

Exhibit 6.12.4 Windchill "Equivalent" Temperatures Predict the Effect of Air Velocity at Various Temperatures. The Number in the Table Gives the Temperature at 1.8 m/s, Which Has the Same Windchill as the DBT at an Air Velocity. The WCI Is Based on Cooling of a Water Container Rather Than a Person With Clothing. The Circles Show the –14 "Contour."

DBT \ Air Velocity (m/s): 0	2	4	6	8	10	12	14	16
+4	+3.3	+1.8	–5.0	–7.4	–9.1	–10.5	–11.5	–12.3
+2	+1.3	–4.2	–7.6	–10.2	–12.0	–13.5	–14.6	–15.4
0	+0.8	–6.5	–10.3	–12.9	–15.0	–16.5	–17.6	–18.5
–2	–2.8	–8.9	–12.9	–15.7	–17.9	–19.5	–20.7	–21.6
–4	–4.9	–11.3	–15.5	–18.5	–20.8	–22.5	–23.8	–24.8
–6	–6.9	–13.7	–18.1	–21.3	–23.7	–25.5	–26.9	–27.9
–8	–9.0	–16.1	–20.0	–24.1	–26.6	–28.5	–29.9	–31.0
–10	–11.0	–18.5	–23.4	–26.9	–29.5	–31.5	–33.0	–34.1
–12	–13.1	–20.9	–26.0	–29.7	–32.5	–34.5	–36.1	–37.2
–14	–15.1	–23.3	–28.6	–32.4	–35.3	–37.5	–39.1	–40.4
–16	–17.2	–25.7	–31.3	–35.2	–38.2	–40.5	–42.2	–43.5
–18	–19.2	–28.1	–33.9	–38.0	–41.1	–43.5	–45.3	–46.6
–20	–21.2	–30.5	–36.5	–40.8	–44.0	–46.5	–48.3	–49.7

What is the limiting cold environment? As with heat, people want one number that combines all factors. The index presently used is the windchill index, WCI, where WCI is in kcal/(m^2 – hr). When WCI = 400, the sensation is cool; when WCI = 600, the sensation is very cool; 800 = cold; 1000 = very cold; 1200 = bitterly cold; and 1400 = freezing of exposed flesh. Since the primary variables are air temperature and air velocity, Exhibit 6.12.4 gives the windchill equivalent temperatures. Be cautious about the exhibit since it was developed from cooling a bottle of water–not a clothed, warm human being.

Reduction of Cold Stress

Clothing is the primary defense. Air is the best insulator, so clothing that traps air is best–such as bulky sweaters and multiple layers of clothing. Windproof clothing helps.

With regard to parts of the body, especially protect the head, hands, and feet. The head skin does not vasoconstrict, so heat loss from an uncovered head may be 50% of resting metabolism. Wear a cap; stocking caps are excellent. In the wind use earmuffs and face masks. Hands should have gloves (if dexterity is needed) or mittens (for more protection). Feet need to be kept dry as well as warm–a multiple layer approach of wearing work shoes plus rubber galoshes works well.

A serious problem is exercise in cold, since exercise causes sweating; the sweat accumulates in the clothing and eventually freezes. Avoid sweat accumulation three ways: (1) remove excess garments (i.e., buy clothing that is easily removable); (2) wear clothing that has ventilation areas, gaps, and drawstrings so that trapped, moist air can escape; and (3) wear a fabric that breathes (wool is better than cotton). Avoid plastics, nylon, or tightly woven fabrics that do not breathe. A removable nylon outer layer (a Windbreaker Ⓣ) may be useful.

In the environment provide windbreaks for the workers. If moisture is a problem, be sure workers can change wet clothing. From a management viewpoint, someone should be in charge of "weatherization" to make sure that workers wear proper clothing, that equipment has proper lubrication oils, that first aid procedures are known, and so on.

REFERENCES

1. ASHRAE, "Physiological Principles, Comfort and Health," *Handbook of Fundamentals*, New York, 1977, p. 820.
2. L. BERGLUND, "New Horizons for 55–74: Implications for Energy Conservation and Comfort," *ASHRAE Transactions*, Part 1, 1980, pp. 507–515.
3. ASHRAE, "Physiological Principles, Comfort and Health."
4. S. KONZ, *Work Design*, Grid, Columbus, OH, 1979.
5. KONZ, *Work Design*, p. 436.
6. N. AZER and S. HSU, "OSHA Heat Stress Standards and the WBGT Index," *ASHRAE Transactions*, Vol. 83, No. 2 (1977), pp. 30–40.

CHAPTER 6.13
Toxicology

KARI LINDSTRÖM

Institute of Occupational Health, Helsinki, Finland

6.13.1 INTRODUCTION

Toxicology is a science that deals with toxic agents harmful to living organisms. Whether a chemical agent is toxic or not is relative, because every substance can produce adverse toxic effects when the exposure is significant enough. Industrial toxicology is one main branch of toxicology and includes the detection of toxic agents, analytics, affecting mechanisms, diagnostics, and therapy. The main purpose of industrial toxicology is the prevention of harmful toxic effects in the working environment.

The form of intoxication can be acute, delayed, or chronic. Acute intoxication follows short-term exposure when the chemical agent is absorbed rapidly. Delayed intoxication involves symptoms that appear days or weeks after the actual exposure. Chronic intoxication means the exposure has been long term and repeated. The clinical poisoning symptoms usually occur when the agent has accumulated in an organism or when the effects of repeated exposure are additive. The reaction of an organism depends on the physical and chemical characteristics of the agent and on the biological characteristics of the organism itself.[1]

Behavioral Toxicology

Behavioral toxicology involves assessing, by means of psychological test methods, impairments of the nervous system's functional capacity resulting from exposure to neurotoxic agents. In the context of occupational health, the role of behavioral toxicology is to provide the means for the detection of early and subclinical symptoms to complement the more severe clinical symptoms found with traditional methods. Behavioral toxicology applies test methods from clinical psychology and neuropsychology to measure cognitive functions, such as intellectual and memory performances, and perceptual and motor performances, such as reaction time and manual dexterity. Objective measurement of affective and personality reactions can also be included. The biggest group of neurotoxic agents with proven behavioral effects is the industrial solvents–aromatic, halogenated, and aliphatic hydrocarbons or their mixtures. Heavy metals, lead, mercury, and manganese also are known for their neurotoxic effects.

6.13.2 AFFECTING MECHANISMS OF NEUROTOXIC AGENTS

Chemical Agents

At work the absorption of chemical agents usually occurs through the lungs and skin, seldom orally. The most common means of industrial exposure is inhalation; such exposure is greatest when around aerosols, gases, fumes, and vapors. The concentration of the agent in air can be measured, but the amount of the agent absorbed in an organism seldom is known. Agents absorbed through the lungs can directly affect various organs, such as the brain and kidneys. Toxic agents have an affinity for specific tissues, which are not necessarily their primary site of influence. For example, some organophosphorus pesticides are strongly lipid soluble; those accumulate in fatty tissues and can remain there for years.

Organic substances change in organisms as a result of metabolic reactions. The metabolism of chemical agents also depends on the physiological factors characterizing the exposed workers, such as age, sex, nutritional state, pregnancy, and possible diseases. Also, environmental factors (such as the time of day), other possible stressors, and simultaneous exposure to more than one chemical agent can produce antagonistic or synergistic action in the metabolism and in the effects. The excretion of toxic agents occurs through urine, bile, exhaled air, sweat, stomach-intestine secretions, and so on.

Neurotoxic Agents

These agents can affect different parts of a living organism and produce a variety of effects, such as carcinogenic and mutagenic ones. Behavioral effects usually coincide with changes in the central and peripheral nervous systems, but are not necessarily identical with them. Mental functions, such as the perceptual and motor skills, and cognitive functions are based on the interaction of many brain areas and functions. No single corresponding area in the brain can be found for a certain behavioral system, although some correlations do exist. In particular, the effects of long-term exposure on the behavioral level include both the direct effects of an agent and the secondary effects modified by many intervening factors associated with the entire exposure situation.

6.13.3 MEASUREMENT OF CHEMICAL EXPOSURE

Threshold limit values (TLVs) refer to the amount of exposure that will not induce harmful health effects in an average individual. These hygienic limit values may vary among various countries for the same chemical agent because of the different theoretical framework and thinking behind the setting of the values. In the United States TLVs for different chemical agents are set by the American Conference of Governmental Industrial Hygienists (ACGIH) and by OSHA. These values refer to the concentration measured in the work atmosphere (i.e., environmental monitoring).

There are different kinds of TLVs (ACGIH). Usually mentioned is the "threshold limit value–Time Weighted Average" (TLV-TWA). This refers to the time-weighted average concentration for a normal 8 hr workday or 40 hr workweek. "Threshold limit value–short term exposure limit" (TLV-STEL) refers to the maximal concentration to which workers can be exposed for up to 15 min without suffering from, for example, irritation, chronic tissue change, or harmful effects on work safety or work efficiency. "Threshold limit value–ceiling" (TLV-C) is the concentration that should not be exceeded. When two or more substances are present, it is preferable to calculate their combined effect. These effects are usually seen as additive, but in some cases synergistic action also may occur.[2,3]

As a prelude to the TLVs and airborne concentrations for neurotoxic agents pointed out in Section 6.13.4, Exhibit 6.13.1 presents TLVs set by the ACGIH. Actual individual chemical exposure can also be determined by measuring the content of, for example, solvent in alveolar air samples. In *biological monitoring* the internal exposure is determined by measuring the metabolic end products of the agents in urine, blood, hair, and so on.

Exhibit 6.13.1 The TLV-TWAs for Some Neurotoxic or Potentially Neurotoxic Substances[a]

Substance	TLV-TWA[b]
Acetone	1000 ppm
Carbon monoxide	50 ppm
Carbon disulfide	20 ppm
Dichloromethane	200 ppm
Hexane (*n*-hexane)	100 ppm
Lead, inorganic	0.15 mg/m^3
Manganese and compounds	5 mg/m^3
Manganese, fumes	1 mg/m^3
Mercury (alkyl compounds)	0.01 mg/m^3
Mercury (except alkyl)	0.05 mg/m^3
Methylene chloride	200 ppm
Perchloroethylene	100 ppm
Stoddard solvent	100 ppm
Styrene	100 ppm
Toluene	100 ppm
Trichloroethylene	100 ppm
1,1,1-trichloroethane	350 ppm
Xylene	100 ppm

[a]Adapted from reference 2.

[b]These values may change. The reader is advised to consult the latest government publications and reference 3.

6.13.4 NEUROTOXIC AGENTS AND THEIR EFFECTS

The chemical agents included here are those most commonly used in work life and having behavioral effects. The main neurotoxic agents are mercury, lead, organic solvents, and pesticides. Some other single agents also are presented.

Mercury and Its Inorganic Compounds

Exposure to mercury can occur during metallurgic processes and in refineries. Mercury is used in chloralkali plants, in laboratories and laboratory equipment, in the electric industry, in producing amalgam, and so on. An organism usually picks up metallic mercury through inhalation of mercury vapors.

Acute intoxication begins quite slowly. The most common subjective symptoms found have been sleep disturbances, social shyness, nervousness, and loss of appetite, all of which have correlated with the time-weighted exposure. Tremor and psychomotor disturbances generally have been identified as the main effects on the nervous system.[4]

Psychomotor performances are slow, eye-hand coordination becomes poorer, and memory disturbances also have been found. Peak exposure concentrations have correlated highest with these findings. In behavioral studies the concentration of inorganic mercury in the work atmosphere or urinary mercury has been applied as the exposure index. The number of peak concentrations over 0.5 mg/l especially has proved to produce behavioral effects in chronic exposure. The amount of organic mercury in a worker's surroundings also can be determined from blood and hair. Exposure to inorganic mercury can be estimated from the mercury concentration in blood or urine. However, the excretion of alkylmercuric compounds in urine is very small, these compounds are therefore best analyzed in blood. Alkyl and inorganic mercury can even be analyzed independently.

Lead

The use of lead and its compounds in work life varies. The main sources of exposure are in the mining industry, metallurgy, battery production, and the paint industry. In addition, exposure outside of the workplace can be excessive. Such sources of exposure can be food, water, or traffic. An organism usually acquires inorganic lead through inhalation, particularly of lead vapors, fumes, and dust. The digestive system and skin can also be sites of entry.

Acute lead intoxication is seldom found in industry. Chronic intoxication can be divided into three phases: presaturnism, chronic intoxication, and sequel of chronic intoxication. The first phase refers only to biological change, not to a disease. The subjective symptoms characterizing mild, chronic intoxication due to low, long-term exposure are gastrointestinal symptoms, unusual fatigue, mood disturbances, and difficulties with the environment. These symptoms are neurotic manifestations. Neurological symptoms, such as pains in the arms and legs, are also typical. Cognitive functions, such as visuoconstructive intelligence and memory, and tasks demanding manual dexterity also were lowered because of low-level lead exposure (blood lead level never exceeded 70 mg/100 mg).[5] Such lead absorption indexes as the blood lead level and zinc protoporphyrin level have proved to be valid measures of exposure and useful in monitoring the behavioral effects of lead. Tetraethyllead is toxic to the central nervous system and leads to psychological disturbances. Exposure to alkyl lead can be estimated from the amount of lipid-soluble lead in the blood. Acute intoxication by alkyl lead is best monitored by monitoring the urinary excretion of lead.

Manganese

Manganese is highly toxic to the CNS. Serious emotional disturbances, fatigue, sleeping disturbances, and insecurity of manual dexterity are characteristic of chronic intoxication.

Carbon Monoxide

Common sources of carbon monoxide exposure are heating ovens, central heating units, motor exhaust, foundries, and smoking. Carbon monoxide reaches an organism through inhalation. Its behavioral effects result from the lack of oxygen (anoxia) in the CNS. Subjective symptoms such as unusual tiredness, lack of mental energy, irritability, and memory and concentration difficulties are common symptoms associated with carbon monoxide exposure. Under experimental conditions of carbon monoxide exposure, a level of carboxyhemoglobin as low as 5% adversely affected cognitive functions, such as arithmetic tasks. At a slightly higher level, manual dexterity and tasks demanding eye-hand coordination were impaired.[6] Actual occupations under investigation because of risks of carbon monoxide exposure have involved traffic or driving. Results have shown that, at a 5 to 10% carboxyhemoglobin level, reaction times, coordination, and occasionally the degree of vigilance have decreased. When dealing with chronic symptoms resulting from carbon monoxide exposure, it is more a question of the cumulation of effects rather than of the exposure.

Solvents

Solvents are the most common neurotoxic chemical agents in occupational use today. The number of single solvents and their mixtures is very high. Those focused upon here are the aromatic, halogenated, and aliphatic hydrocarbons; some others also are mentioned briefly. Solvents are absorbed mainly through inhalation and in part through the skin. They usually first induce acute narcotic effects or irritation.

Aromatic Hydrocarbons

Toluene, styrene, and xylene are the outstanding members of this group. Benzene, which also belongs to the aromatic hydrocarbons, has proved to be leukemogenic. However, because of its limited use, it has been excluded from this section.

Toluene. Toluene is used in paints, lacquers, glues, thinners, and cleaning liquids. Therefore painters and varnishers using solvent-based epoxy paints are exposed to toluene and its mixtures. In the photogravure printing industry toluene exposure is common. Toluene also is used as a psychotropic drug by the "glue sniffers" in the drug culture because it produces a euphoric state. During high experimental exposure it has produced acute effects such as drowsiness, unusual tiredness, headache, dizziness, and nausea. Under experimental conditions exposure to 300 ppm has lead to prolonged reaction times, and exposure to 700 ppm to impaired perceptual speed. Chronic exposure in photogravure printing work has shown to produce organic psychosyndrome, or wide-ranging mental decline, and, when the exposure is 60 to 200 ppm, lowered short-term memory.[7] Toluene exposure can be measured according to the airborne concentration, the concentration in blood or alveolar air, and the amount of hippuric acid or *o*-creosol in urine.[8]

Styrene. Styrene is used mostly in the production of reinforced polyester plastic products, such as boats and containers. Acute exposure symptoms are irritation and prenarcotic symptoms. The chronic symptoms include fatigue, difficulties in concentration, and irritation. Cognitive function decline has not been found among styrene-exposed workers, but visuomotor accuracy and reaction times have been found to be impaired after long-term exposure.[5] Reaction times have also been affected after short-term exposure to 350 ppm of styrene.[4] Both urinary mandelic and phenylclyoxylic acid concentrations have proved to be valid measures of the styrene exposure related to behavioral changes.[5] The behavioral changes listed here have occurred at lower levels than the current hygienic limit values for styrene in many countries.

Xylene. Xylene and its mixtures are used mostly as thinners in paints and varnishes and as solvents. They produce the same kind of acute symptoms as toluene and styrene. After short-term exposure to 90 to 200 ppm of xylene, lowered vigilance, prolonged reaction time, and changes in equilibrium have been found.[9] Also, under experimental conditions, the exposure of 300 ppm combined with physical strain caused the cognitive intellectual and memory functions to be disturbed. Long-term exposure to xylene has not been studied as such. However, the behavioral changes found among house and car painters exposed to solvent mixtures including xylene have indicated both cognitive disturbances and disturbances in manual dexterity.[5,10] Methylhippuric acid has been used as a biologic indicator of internal xylene exposure.

Aromatic Solvent Mixtures and Paint Solvents. Paints containing aromatic hydrocarbons were once quite common. Nowadays, however, water-based paints and paints containing mainly aliphatic hydrocarbons are more common. Nevertheless, older painters and varnishers who are still working have acquired remarkable past exposure to those aromatic solvent mixtures during their individual exposure histories. The spray painters who use epoxy paints have a history of heavy exposure, but so have repair painters, partly because the ventilation conditions in repair work are seldom arranged, and thus the exposure can be quite high.

The acute symptoms are usually the aforementioned prenarcotic symptoms, the chronic symptoms include tiredness, sleep disturbances, headache, concentration difficulties, irritability, and other types of mood changes. Cognitive functions, such as visuoconstructive abilities, have been found to be deteriorated, but the verbal abilities only occasionally. Short-term memory has proved to be quite sensitive to the adverse effects of paint solvent exposure. Impaired manual dexterity and lowered speed of perception have also been found among both poisoned and exposed workers.[5,10] The estimation of the actual and, especially, the total paint solvent exposure is usually difficult because of the variation in the type of paint used, in the manner of application, and, of course, in the working and ventilation conditions. It seems, however, that airborne and alveolar hygienic measures, combined with a detailed interview on past exposure conditions, can provide quite a valid measure of the total exposure relating to the behavioral changes.

Halogenated Hydrocarbons

Halogenated hydrocarbons are also neurotoxic and can affect the liver and kidney. Agents included in this group have mainly depressing effects on the central nervous system.

Dichloro-methane (Methylene Chloride). This agent is used as a solvent. One common use is as a paint remover. The carboxyhemoglobin increases as a result of dichloro-methane exposure and leads to the same kinds of effects that carbon monoxide exposure has. Attention lapses and decreased manual dexterity have been found already after short-term exposure to 500 ppm.[11] Practical problems at work are related to the drunken feelings and coordination difficulties.

Methyl Chloride. Methyl chloride is used as a refrigerant. It primarily affects the CNS and can produce ataxia, weakness, dizziness, difficulties in speech, visual disturbances, and so on. Cognitive tasks have been found to be clearly disturbed among workers exposed to methyl chloride.[12] These cognitive changes were related to the air concentrations.

1,1,1-Trichloroethane (Methylchloroform). This agent is usually used in the degreasing of metals. Under experimental exposure conditions, 350 ppm of methylchloroform has prolonged reaction times, lowered visuomotor speed, and also produced disturbances in manual dexterity.[13] Long-term exposure to this solvent has not yet been shown to produce any behavioral effects. Most methylchloroform taken in is excreted through exhaled air, and part is excreted in the form of trichlorocompounds in the urine.[8] Methylchloroform concentration in blood can be used as an index of exposure.

Trichloroethylene. In the degreasing of metals and in dry cleaning, the use of trichloroethylene has been common. Perceptual performances, manual dexterity, and cognitive memory tasks in workers have been found to be disturbed already at the exposure level of 110 ppm, but in some studies the changes have been first seen only after 300 ppm. In general, the acute effects on behavior are somewhat contradicting,[1] whereas chronic trichloroethylene exposure has produced quite clear behavioral effects. To avoid severe psycho-organic deterioration, it has been concluded that the long-term average exposure should be under 50 ppm. Subjective symptoms of chronic trichloroethylene exposure are unusual tiredness, headache, dizziness, nausea, sleep disturbances, and memory difficulties, addiction to trichloroethylene also is possible. Trichloroacetic acid and trichloroethanol in urine have been used as biological indicators of trichloroethylene exposure.[8]

Tetrachloroethylene (Perchloroethylene). This agent is also used in degreasing metals and in dry cleaning. Subjective acute symptoms are similar to those of trichloroethylene. No behavioral effects have been found after experimental exposure to 100 ppm, but a decline in coordination ability has appeared after short-term exposure to 100 to 150 ppm. Some case studies and clinical studies indicate that tetrachloroethylene can produce encephalopathy and pseudoneurotic disorders. Controlled field studies, however, leave these results unconfirmed.[1] Biologic monitoring of exposure to tetrachloroethylene is best achieved by measurement of the parent compounds in urine.

Aliphatic Hydrocarbons

Aliphatic hydrocarbons usually serve as the main components in various solvent mixtures, particularly in refined petroleum solvents such as petroleum ethers, rubber solvents, varnish makers' and painters' naphtha, mineral spirits, stoddard solvents, and kerosene. In these solvent mixtures the total aromatic hydrocarbon content is less than 20%.[14] Some amount of benzene is always present in refined petroleum solvents. These solvents are classified according to their boiling point and degree of evaporation. They are used for cleaning and degreasing purposes and especially in alkyd paints.

***n*-Hexane.** *n*-Hexane is an aliphatic hydrocarbon used, for example, in glues. Polyneuropathy has been found among women glue workers. High concentrations have produced nausea, headache, and eye and throat irritation.

Mineral Spirits. These are used as thinners in the paint and varnish industry and in dry cleaning. In experimental situations they have caused prolonged reaction times and disturbances in short-term memory at an exposure level of 700 ppm.[13]

Jet Fuel. This agent is an aviation turbine fuel. Acute symptoms found among exposed workers have been dizziness, headache, nausea, and respiratory tract symptoms. Chronic mental symptoms have been neurasthenic and neurotic symptoms. Psychological tests have indicated disturbances in attention and sensorimotor speed, whereas no signs of memory and manual dexterity impairments were found.

Other Solvents and Solvent Mixtures

Carbon Disulfide. This agent is used in the production of artificial fibers by the viscose method and in the cold cure of rubber. Carbon disulfide has anesthetic effects. In chronic intoxication the neuropsychiatric symptoms are predominantly sleeplessness, headache, dizziness, nausea, loss of memory, irritability, depressive moods, loss of initiative, and nightmares. In psychological tests deterioration has been found even before a real clinically diagnosable intoxication. In severe cases psycho-organic syndromes, indicating wide-range deterioration in cognitive, perceptual, and motor functions, have appeared, the most apparent disturbances being slowness in psychomotor performances and clumsy manual dexterity. Cognitive functions, especially visuoconstructive abilities, have also displayed deterioration. Findings on memory functions are scarce, but the depressive moods and lowered vigilance have negative effects on memory performances.[5]

Alcohols (Methanols, Ethanols, Butanols, etc.). These have narcotic and irritating effects in occupational use.

Ketones. These also have narcotic effects, but they are regarded as having low toxicity. However, disturbances in reaction time performances have been found after exposure to 200 ppm of acetone. Behavioral effects have also been found in animal experiments due to methyl *n*-butyl ketone.[1]

Anesthetic Gases. Anesthetic gases such as halothane, nitrous oxide, and alcohols have been studied among anesthetic nurses.[13] Behavioral impairments, however, were not found among them. Recovery after being anesthetized by one of these agents has shown lowered manual dexterity and perceptual speed among young volunteers.

Pesticides

The pesticides are divided into insecticides, rodenticides, herbicides, fungicides, and molluscicides. They are used mainly in farming and forestry, but also in health care systems.

Organochlorinated Insecticides

Compounds such as DDT stimulate the CNS and can lead to epileptic-type seizures. Such acute subjective symptoms as nausea, headache, imbalance, dizziness, and tremor are possible. Nowadays, however, organophosphorus compounds have replaced the DDT-type agents.

Organophosphorus Compounds

The adverse behavioral effects of these compounds are due to their inhibition of brain acetylcholinesterase, and the somatic symptoms largely originate peripherally in the autonomic nervous system. In acute intoxication the symptoms related to the CNS are anxiety, dizziness, headache, and tremor. In chronic intoxication depression is seen as one of the central symptoms, usually associated with sleep disturbances. In acute intoxication, vigilance and concentration disturbances are often found. Among exposed workers in farming and in industry memory disturbances also have been detected. Disturbances in manual dexterity and information processing have been present in persons with at least mild symptoms of clinical intoxication.[15]

6.13.5 PREVENTION OF ACUTE AND CHRONIC BEHAVIORAL EFFECTS

In preventing the adverse effects of industrial chemicals, best results are achieved when taking the risks into account in the planning phase of new industrial installations or work methods. Ventilation systems in particular are difficult to construct adequately in later phases. The use of personal protective equipment is not seen nowadays as a highly recommended alternative. In addition to being somewhat uncomfortable, such equipment also can disturb perceptual and motor performances and increase experienced strain. Adequate information at the initiation of work or introduction of a new chemical agent is important. Such safety and health education increases the workers' knowledge of the properties and dangers of the chemical agents and helps and motivates them to avoid the dangers.

The TLVs often are gauged so that they protect only average individuals against harmful effects. This is especially true regarding chronic behavioral effects, which seldom are included in the documentations of TLVs. Special groups, such as the young, the old, females, and people with limited working capacity, are not necessarily working safely under conditions fulfilling the hygienic standards. It also is known from clinical practice that people differ quite significantly in sensitivity and tolerance to harmful agents. The onset of symptoms typical of harmful chemical exposure can be very quick after beginning a new job.

Initial health examinations should take into account symptoms and disorders among applicants that are seen as contraindicators, for example, to solvent exposure work. Those symptoms that are related to chemical exposure are all quite common in the general population. If the symptom level indicating neurotic or neurasthenic disorders is high, it is probable that those symptoms tend to increase rapidly under exposure. Also, people having earlier neurological and psychiatric diseases are possibly more sensitive to harmful behavioral effects. Periodic health examinations of workers exposed to neurotoxic agents are motivated by the same reasons as for the initial health examinations. The behavioral changes can be seen as the early influence of harmful effects. Therefore both subjective psychological and neurological symptom surveys and behavioral test methods are tools in the prevention of more severe intoxication.

REFERENCES

1. R. R. LAUWERYS, *Précis de Toxicologie Industrielle et des Intoxications Professionelles*, Editions J. Duculot, Gembloux, 1972.
2. *TLVs®: Threshold Limit Values for Chemical Substances and Physical Agents in the Work Environment with Intended Changes for 1980*, American Conference on Governmental and Industrial Hygiene, Cincinnati, OH, 1980.
3. F. W. MACKINSON, R. S. STRICOFF, and L. J. PARTRIDGE, Jr., *NIOSH/OSHA Pocket Guide to Chemical Hazards*, DHEW (NIOSH) Pub. No. 78-210, U.S. Government Printing Office, Washington, DC, 1980.
4. G. D. LANGOLF, D. B. CHAFFIN, R. HENDERSON, and H. P. WHITTLE, "Evaluation of Workers Exposed to Elemental Mercury Using Quantitative Tests of Tremor and Neuromuscular Functions," *American Industrial Hygiene Association Journal*, Vol. 39, 1978, pp. 976–984.
5. H. HÄNNINEN, "Psychological Test Methods: Sensitivity to Long Term Chemical Exposure at Work," *Neurobehavioral Toxicology*, Vol. 1, Supplement 1, 1979, pp. 157–161.
6. R. R. BEARD and N. GRANDSTAFF, "Carbon Monoxide Exposure and Cerebral Functions," *Annals of the New York Academy of Sciences*, Vol. 174, 1970, pp. 385–395.
7. K. -H. COHR and J. STOCKHOLM, "Toluene. A Toxicological Review," *Scandinavian Journal of Work Environment and Health*, Vol. 5, 1979, pp. 71–90.
8. R. R. LAUWERYS, "Biological Criteria for Selected Industrial Toxic Chemicals: A Review," *Scandinavian Journal of Work Environment and Health*, Vol. 1, 1975, pp. 139–172.
9. K. SAVOLAINEN, V. RIIHIMÄKI, and M. LINNOILA, "Effects of Short-Term Xylene Exposure on Psychophysiological Changes in Man," *International Archives of Occupational Environmental Health*, Vol. 44, 1979, pp. 201–211.
10. M. HANE, O. AXELSON, J. BLUME, C. HOGSTEDT, I. SUNDELL, and B. YDREBORG, "Psychological Function Changes Among House Painters," *Scandinavian Journal of Work Environment and Health*, Vol. 3, 1977, pp. 91–99.
11. G. WINNEKE and G. G. FODOR, "Dichloromethane Produces Narcotic Effects," *International Journal of Occupational Health and Safety*, Vol. 45, 1976, pp. 34–35.
12. J. D. REPKO, P. D. JONES, L. S. GARCIA, E. J. SCHNEIDER, E. ROSEMAN and C. R. CORUM, *Behavioral and Neurological Effects of Methyl Chloride*, U.S. DHEW, Cincinnati, OH, 1976.
13. F. GAMBERALE, "Behavioral Effects of Exposure to Solvent Vapors," Arbete och Hälsa 14, Arbetarskyddsverket, Stockholm, 1975.
14. *Occupational Exposure to Refined Petroleum Solvents. Criteria for a Recommended Standard*, U.S. DHEW, NIOSH, 1977, Cincinnati, Ohio.
15. H. S. LEVIN and R. L. RODNITZKY, "Behavioral Effects of Organophosphate Pesticides in Man," *Clinical Toxicology*, Vol. 9, 1976, pp. 391–405.

CHAPTER 6.14

The Management of Occupational Safety

JAMES M. MILLER
The University of Michigan

6.14.1 THE SAFETY PROFESSIONALS

Assuring safe (and healthful) working conditions is becoming an increasingly important responsibility of an organization's industrial engineering personnel. The large organization may have a separate division within its industrial engineering department in which several people devote all of their time to some aspect of worker safety. It is becoming less common to find the management of the safety function assigned to personnel departments. This is because of the current importance placed on worker safety by mandated government requirements (OSHA) and the technical complexities involved in satisfying these requirements. Also, the sophisticated needs and demands of the modern worker have forced an organization's top management to rely on individuals who can apply both an engineering background and an understanding of human factors when managing safety efforts.

This individual is the safety engineer, or safety professional, who usually develops expertise in safety on top of an engineering discipline which includes fundamentals such as mathematics, physics, and chemistry. Engineering bachelor degree programs include these as well as electrical, mechanical, computer, and industrial engineering requirements. Since safety engineering concentrates on physical facilities within an organization, this type of course work is an essential prerequisite to becoming either a safety engineer or a safety manager.

The topics covered in this section on ergonomics/human factors may be introduced in a bachelor degree level program (particularly industrial engineering). They are becoming fundamental to safety professionals and, in particular, to persons with safety *management* responsibilities, who would be at a disadvantage if they did not have education in these areas. However, such education usually occurs at the graduate level or in continuing education courses. The importance of this advanced education is reinforced by the numerous graduate degree and extension-type programs in occupational safety and health engineering which have, since 1970, been supported by NIOSH.

With the increasing availability of formally trained safety professionals, organizations having several positions for occupational safety personnel are usually able to appoint as safety manager of the group a person who has some experience, a bachelor's degree, and post-bachelor's degree training in safety and health engineering. Working under the supervision of this manager could be a person with 2 year or 4 year "safety technology," "safety specialist," "safety education," or basic engineering training.

6.14.2 PHILOSOPHY OF SAFETY PROGRAMS

Top management often calls upon the safety manager to solve crises-level problems related to either a safety issue that is interfering with productivity or profitability or a requirement facing the organization because of government, unions, public opinion, or court mandate. Some examples of the types of problems that a safety manager's boss may present are:

"The community doesn't think this organization takes a humane view of the work force with respect to its safety and health. What should we do about it?"

"My costs for workers' compensation insurance and accident-related medical costs have doubled in the past 2 years. Why?"

"The workers and I are upset with the accident that happened to Joe Smith. He was a skilled worker and a friend who meant a lot to all of us. Can you find out what caused it and take corrective action?"

"I heard at a conference that OSHA is changing a significant number of its standards. How will this affect our operations?"

"I brought you into this organization because we need to start paying more attention to worker safety. Will you analyze our situation and report back to me with some recommendations regarding what we should be doing?"

The preceding questions, though politely presented, represent issues that may turn into organizational crises. Viewed as unfortunate to management, but beneficial to workers and the safety professionals, the Occupational Safety and Health Act of 1970 ranked high on the current list of leading causes of crises among American businesses. Required to give safety and health issues high priority, most organizations have been forced to increase the financial and manpower resources available to safety managers. It thus becomes imperative for the safety manager to be well organized for responses to safety and health crises, such as serious accidents or surprise OSHA inspections. A satisfactory handling of such crises may temporarily please top supervision. However, the long-range progress to reduce the losses generated by hazards can occur only with an effective day-to-day safety program. Herein lies the most challenging responsibility of the safety manager—that of planning, organizing, executing, and controlling a uniquely designed program that moves the organization toward safety and health objectives supported by both employer and employees.

The Safety Climate

It would be wishful thinking to believe that most safety programs have emerged after a well-funded and supported study by management or expert consultants. In spite of the popularly publicized slogan "safety pays," most top officials traditionally have had difficulty accepting this. Thus most programs have evolved slowly in response to the crises needs associated with governmental requirements. As a result, the "climate" for safety and health programs may not be ideal.

The relevance of this climate to designing and carrying out a "Safety by Objectives" program has been addressed by Petersen.[1] The organization whose safety program evolves in response to crises is labeled therein as "the negligent company." This is obviously not a healthy climate for safety (and probably not for productivity). The concept of organizational climate has played an important role in management theory and practice over the last 20 years. Authors such as MacGregor, Likert, and Argyris are behavioral scientists who have led this movement. More recently, Petersen has been a leader in applying their behavioral approaches to safety management. The concept of the "corporate climate" has been found by Petersen to be particularly relevant to understanding the "safety climate" in terms of determining the dimensions and potential success of a safety program.

What is "climate" in an organization? Not any single thing, but a number of factors that all come together to give the employees and first-level supervisors an intangible feeling or attitude toward their work environment. Climate factors include style of management leadership (from participative to authoritarian); management's demonstrated concern for the higher needs of employees (beyond job and financial security); and support or pressure by unions and/or employee peer groups to motivate workers toward a fair day's work using safe, healthful, and productive work methods. With a good climate, the people in the organization generally will have a feeling that top management, supervisors, and workers are part of a team. The team is striving to achieve a number of organizational and personal objectives, many of which are being attained at the same time team members are receiving positive reinforcement for their personal needs and objectives.

The kind of organizational climate must be recognized and taken into consideration as a safety program (or any organizational endeavor) is planned and executed. Otherwise, the program is not likely to succeed. Although climate is theoretically determined from the very top of an organization down, in practice there will probably be many types of climates within an organization, depending on individual managers and their independence to control their particular departments. Therefore, even though a department's primary goals may be determined by higher managers, individual managers may have the liberty to determine how these goals are to be achieved, what additional objectives are established, and what will be the climate in which employees work toward these objectives. It is anticipated that the safety manager will have this kind of liberty, at least with respect to the safety program.

Systematically Approaching Objectives

Professional management experts state that a safety program will most likely "succeed" if it is systematically approached using a method such as "Management by Objectives," or, as Petersen has called it, "Safety by Objectives."[1]

It is the fortunate organization that has a safety manager with experience in utilizing these systematic management approaches as well as expertise in safety and health engineering. Because each field is relatively new, the number of persons having capabilities in both is still small, and those who do probably practice mostly in a consulting capacity. Therefore many companies often decide to use the services of professional management consultants when initiating new programs. These individuals assist the organization's personnel in analyzing the needs and problems, designing objectives and goals to satisfy the needs or resolve the problems, organizing to meet the objectives, and executing a program. A considerable amount can be learned about these systematic management techniques through texts, seminars, and short courses. Eventually, an organization's own safety professionals will have the knowledge and experience to initiate and carry out such programs themselves.

Accident Prevention Philosophy

In the initial phases of a safety program, the safety manager must address some key questions. The answers will, of course, be organizationally specific and will provide both the philosophy and the direction of the program. The question about the organizational climate, already raised, concerns the prerequisite condition that determines the environment in which the safety program will operate. Next in importance are the most traditional questions:

1. What causes accidents—the unsafe acts of persons or the unsafe design of physical things (machines, processes, tools, equipment, workplaces, etc.)?
2. How can accident prevention be accomplished effectively?

Developing an understanding of accidents that would be applicable to designing accident prevention programs was first addressed by the father of modern industrial safety, H. W. Heinrich. As early as 1931, he formalized the philosophy and the approach in his text *Industrial Accident Prevention*, which is still available and an excellent reference in its newer editions.[2] Therein Heinrich advocated the philosophy that success in safety work depends upon a sound knowledge of *what* an accident is, *how* and *why* it occurs, the *reasons* and *incentives* for prevention, and the *opportunities* and *practical methods* of achievement. Several key questions are embodied in this philosophy, and in total they constitute nearly everything that not only Heinrich, but even contemporary safety researchers and practitioners have been addressing.

What Is an Accident?

The simplistic, most popular response to this question, particularly among government rule makers and many practitioners, is: An accident is an event that causes an *injury*. This limited definition has gained support because of the OSHA language defining "reportable accidents" as those that result in a fatality, hospitalization, lost workdays, medical treatment, job transfer or termination, or loss of consciousness. Minor incidents requiring only first aid treatment are not "reportable accidents."

It was found by Heinrich that about 90% of all accidents produce *no* injury. These mishaps might thus be labeled "near misses" rather than "accidents." What about mishaps that cause physical damage to property, equipment, or materials; are these accidents? It depends on one's relationship to the mishap. To the person or organization suffering the damages, the mishap certainly will be labeled an accident. Thus prevention of physical damage evolves as another reason for accident prevention by whomever suffers the damages.

Researchers and practitioners who are not satisfied with an accident definition that relates only to injury or physical damage are interested in searching for accident causes and predictors. They are not satisfied to learn only retrospectively from recorded accidents. They point out that the occurrence of injury or damage resulting from an accident is largely fortuitous. However, the occurrence of the accident that causes the injury or damage is largely preventable. Consequently, today's "near misses," or nonloss accidents, may be tomorrow's recordable losses if the causes are not determined and eliminated. Near misses can therefore be a useful source of information in reducing the potential for accidents.

In response, then, to "What is an accident?," we might say that it is an unplanned, uncontrolled, and unexpected event that resulted in, nearly resulted in, or had the potential to result in injury or other damages.

How Does an Accident Occur?

The most accepted general explanation of how an accident occurs comes again from Heinrich. It is called either the "sequential" or "domino" theory, which derives from Heinrich's original description of the occurrence of an accident as

> *. . . the natural culmination of a series of events or circumstances. . . . One is dependent on another and one follows another, thus constituting a sequence that may be compared with a row of dominoes placed on end and in such alignment in relation to one another that the fall of the first domino precipitates the fall of the entire row.*[3]

This theory also explains how and why accident prevention programs work, that is, that prevention stops the "accident sequence" by removing one of the factors that leads up to it (i.e., an unsafe act or unsafe condition). The factors in the Heinrich sequential depiction originally portrayed as dominoes were (1) ancestry and social environment, (2) the fault of the person, (3) an unsafe act and/or unsafe mechanical or physical environment, (4) the accident, and (5) the injury.

More recently, the accident analysis experts have drastically expanded the number and complexity of factors and steps in the sequence. The concept remains the same. In addition to this sequential model, numerous other conceptual models of the accident process have been proposed. Most of them provide additional insight into how and why accidents happen. Among these other approaches are the epidemiologic model, energy exchange model, behavioral models, systems model, Haddon model, multiple causation model, dynamic model, and Surry combined model. These are described in Surry's book, *Industrial Accident Research.*[4]

Why Does an Accident Occur?

The aforementioned models generally agree that the immediate causes of accidents relate to *unsafe acts* and/or *unsafe mechanical or physical* conditions. Many safety practitioners believe that 90% of accidents are related to unsafe acts. This leaves only about 10% that are due to unsafe mechanical or physical conditions. Accepting this belief often has led the safety practitioner to focus almost entirely on training people not to commit unsafe acts, accepting the reality that the unsafe conditions exist and will remain uncorrected.

The ergonomics philosophy emphasizes that *most unsafe acts are design induced and design preventable.* This suggests that machines, workplaces, tools, methods, tasks, and physical conditions must be engineered to be compatible with people's physical and psychological capabilities *and limitations.* (See other chapters in Section 6.) Among these limitations is the tendency for a person to behave in probabilistically predictable ways, given a certain background, mechanical and/or physical environment, and situation. Sometimes this normal behavior or reaction to a situation results in errors or unsafe acts–predictable ones. Education and training will help to condition behavior by reducing the probability of an error or unsafe act, but it never can eliminate entirely the possibility of unsafe acts and errors occurring. Thus attention to hazard elimination must be the first priority, with personal protective equipment and education and training methods used only where immediate hazard elimination is not feasible.

Hazard prevention by engineering design or redesign has been promoted in recent years by, among other things, product liability litigation and third-party liability actions in occupational accidents. Critical questions that must be asked in analyzing physical situations revolve around "reasonable foreseeability." For example, was it reasonably foreseeable that a person would find himself or herself in a given hazardous situation? Did the person respond in a reasonably foreseeable manner (even though it was a response that was unsafe or in error)? Did the equipment or other individuals also respond in reasonable ways? Could a different engineering design have prevented the accident and/or injury from occurring or have reduced the extent of damages?

In more and more accidents it is being recognized that the designs of the machines, equipment, processes, and workplaces have been more of a contributory cause than necessary based on the state of the art. The negligent party may be the equipment manufacturer, employer, job designer, or safety engineer. The price paid may be in the form of an OSHA penalty, medical costs, lost production, higher insurance premiums, or straight cash to an injured party. Blaming an accident on the "idiot" victim who committed the error or unsafe act is becoming an unpopular defense and a poor excuse. Nevertheless, management continues to fight the tendency, even among its own ranks, for "engineering" safety problems to be approached rationally, while "human error" related safety problems are accepted as unpreventable, since human behavior is often believed to be unchangeable.

A safety program structured to divide its emphasis between the direct causes–unsafe acts and unsafe conditions–only partially addresses what remedial actions might be taken. The indirect (or proximal) causes often play an equally important role in the sequence of events leading to the accident. These proximal causes include:

1. Error inducing or negligently designed or maintained workplace, tools, machinery, or other physical condition or aspect of the environment.
2. Incompatible match between worker and job (or task).
3. Failure of management to initiate, support, or provide a climate for a well-designed and well-executed safety program.

4. Noncompliance with recommended or required standards for safe work practices or safe working conditions.
5. Lack of (management or worker) knowledge or skill.
6. Improper management or worker attitude.

This list places a higher degree of responsibility for accident causes on management shortcomings than do traditional views. This, however, is justified because of (1) recent government mandates in the OSHA which make the employer responsible for providing safe and healthful working conditions, free from recognized hazards, and (2) the recognition that unsafe acts and unsafe conditions that cause accidents are symptoms of the management system's failure to manage with an understanding of employee needs and motivation.

The preceding discussion provides the philosophy and insight into what types of remedial activities will be necessary parts of an accident prevention program. Individual components of most safety programs are discussed in the rest of this section. They include government-mandated requirements (OSHA and others), industry and corporate standards, identifying past and future accident causes, physical hazard control, and unsafe behavior control.

6.14.3 GOVERNMENT-MANDATED REQUIREMENTS

Nothing in the history of industrial safety has affected the work load responsibilities of the safety department more than the OSHA of 1970. It required that "each employer shall furnish to each of his employees employment and a place of employment which are free from recognized hazards that are causing or are likely to cause death or serious physical harm to his employee."[5] This is known as the "General Duty Clause." The employer has this general duty mandatory responsibility not only to (1) protect his employees from "recognized hazards," but also to (2) "comply with occupational safety and health standards promulgated under the Act"[6] and to (3) comply with other requirements contained within the OSHA (i.e., record keeping, posting) or promulgated as regulations in accordance with the OSHA. These three categories, in addition to workers' compensation requirements, are the mandatory safety- and health-related employer responsibilities being enforced by federal and state governments.

Enforcement by Federal OSHA Versus State Plans

The OSHA gives umbrella authority for enforcement to the U.S. Department of Labor's Occupational Safety and Health Administration except for workers' compensation requirements, which have always been under direct state control. This administration has authority over all states, all federal agencies, and nearly all workplaces. However, a state wishing to assume responsibility for the requirements mandated under the OSHA of 1970 may do so by meeting certain requirements of Section 18 of the act. Namely, the state must demonstrate that it can administer an occupational safety and health plan "at least as effective" as standards and enforcement that would exist if complete control were provided through the administration. The primary issue is whether workplaces will be subject to enforcement through state or federal inspectors.

Safety practitioners will find it important to know whether the particular state in which they are practicing is under the primary authority of the federal or state government. The mandatory requirements for which an employer is responsible in state OSHA programs often will be identical with federal requirements since many states have adopted the federal OSHA standards. Thus employers in locations with state OSHA plans usually will have to comply with all the standards and requirements originating from the federal OSHA *plus* satisfy any additional OSHA standards and requirements originating at the state level. As of 1980 those states and territories having their own plans are Alaska, Arizona, California, Hawaii, Indiana, Iowa, Kentucky, Maryland, Michigan, Minnesota, Nevada, New Mexico, North Carolina, Oregon, Puerto Rico, South Carolina, Tennessee, Utah, Vermont, Virgin Islands, Virginia, Washington, and Wyoming.

None of the standards or requirements of any specific state plan is presented herein; only the federal act and the activities of the federal administration are discussed. The safety professional practicing in one of the locations listed must become familiar with any differences that exist between the federal and state standards and requirements.

Government Regulation Process

The OSHA of 1970 as passed by Congress said that there shall be safety and health standards. The act itself does not contain such standards, but provides authority for the federal administration to promulgate standards and regulations using procedures that give those affected by the standards a chance to participate, review, and critique the rule making. This rule making must be done in accordance with the Administrative Procedures Act,[7] which requires that proposed standards or rules be published in the *Federal Register* as public notice, that the public be given a chance to comment on the standards, and that public hearings be held if they are requested by any members of the public. The OSHA must take into account all the public evidence and not act "arbitrarily or capriciously" in arriving at a final standard that is mandatory for all. Additional "Rules of Procedures for Promulgating, Modifying, or Revoking Occupational Safety and Health Standards" have been promulgated and appear under Title 29, Part 1911 of the Code of Federal Regulations.

Often, employer or employee representatives believe that the administration has erred in its weighting of the facts and of public opinion and has decided upon an action or final standard in a manner that was arbitrary, capricious, in excess of statutory jurisdiction, unwarranted by facts, or contrary to constitutional rights, to name a few.[8] In such cases a suit usually is filed immediately against OSHA in a federal court. The federal, appeals, or Supreme Court judges must then decide the disputed issues.

The process of promulgating, modifying, and deleting procedures, regulations, and standards continues daily, not only within OSHA, but within nearly all government agencies. The Department of Labor alone administers more than 100 acts of Congress. Fortunately, there is a uniform system for an updated cataloging and publishing of all of these agencies' promulgated standards and regulations in one document, called the *Code of Federal Regulations.* This should not be confused with the daily publication, the *Federal Register.* The *Code* becomes the body of finalized government standards, regulations, procedures, and requirements which have the force of law. The *Code* is divided such that all regulations and standards issued and controlled by a given government department and agency usually appear under about the same major divisions, called "title" and "chapter." Title 29 is "Labor," and Chapter XVII is for the OSHA. The OSHA "General Industry Safety and Health Standards" has the official designation *Code of Federal Regulations*, Title 29, Chapter XVII, Part 1910, or for a short citation, "29 CFR 1910."

A complete and updated republishing of the *Code* occurs annually in a series having more than 50 volumes. Each volume can be ordered separately from the U.S. Government Printing Office. Most organizations have found that the best way to stay current with all aspects of what is happening within a given area of government control is to subscribe to a service that specializes in keeping its readers abreast. This is particularly important since regulations are changing continually. With regard to occupational safety and health, one such service is *The Occupational Safety and Health Reporter*, published by the Bureau of National Affairs (a private publishing company). A single subscription to this or a similar service is recommended as the most efficient way in which safety managers can have available nearly all related safety and health information current to within 2 weeks of significant happenings.

General Duty "Recognized Hazards"

For many of the recognized hazards within workplaces, standards were developed initially through consensual standards organizations such as ANSI and the National Fire Protection Association (NFPA). Many of these consensual standards have been adopted as mandatory government standards by OSHA, and these are now part of the *Code of Federal Regulations.* These standards have been criticized because, among other reasons, they probably address only about one fourth to one third of the hazardous situations in workplaces. Those accidents causing the most frequently occurring injuries are said not to be addressed by the standards (i.e., acts causing back injuries). Such criticism alerts the safety professional to the fact that most of the "recognized hazards" are not addressed by specific OSHA standards. Therefore complying with these specific standards will not entirely satisfy the administration. Such compliance also will not satisfy an organization's managers who wonder why accident and injury rates are not decreasing after all the money and manpower that have been devoted to complying with OSHA standards. The answer goes back to the "General Duty Clause." Both for the benefit of the employer and for compliance with the act, each organization must discover and take remedial action against its own "recognized hazards that are causing or are likely to cause death or serious physical harm to employees."

The general duty hazards that are dominant as accident causes will become recognizable through the collection and analysis of accident data, interviews with workers, accident investigation and analysis, exchange of information among organizations with similar manufacturing or service objectives, and information in research studies performed by government and private organizations. In the case of an OSHA inspection, these same sources will be what the government will use to determine those situations in a facility that are recognized hazards and thus possible violations of the act. This searching out of an accident's causes is the general duty of the employer. It is a much more difficult burden to the employer than complying with a specific set of standards, but will be more humanistically and financially rewarding because the employer will be addressing those hazards most responsible for the organization's accidents and costs.

Specific OSHA Requirements

Since the passage of the OSHA in 1970, part of the safety manager's responsibility has focused on those required activities that would bring an employer into compliance with the act and with the administration's regulations. In addition to the general duty activities already described, there are five other categories of required activities.

Before these other requirements are discussed individually, it should be noted that the safety manager must know that all these requirements are actively enforced by federal and state compliance occupational safety and health officers. These officers generally have the right to enter any workplace unannounced and to inspect that employer's records and workplace to determine violations, which, if found, may be followed by penalties and a time limit within which the employer must comply.

Posting and Employee Notification

Employees are to be kept informed about several aspects of safety and health in their workplace. This is to be done by posting notices at a prominent workplace location, that is, one to which employees are frequently exposed. The following are those items that must be posted:

1. "Safety and Health Protection on the Job" (OSHA Form 2003). This is a poster designed by and available from the government. It must be posted at a prominent location and describes for the employees the objective of the act, the responsibilities of the employer to provide a safe and healthful place of employment, and the rights of the employees to complain to the government without threat of recourse about conditions in their workplaces that they feel are hazardous and need corrective action by their employer.

2. Annual Summary of Accidents and/or Illnesses. The routine record keeping that is mandatory for the employer also must be summarized once a year. This annual summary must be posted in a prominent location so that employees can see for themselves their safety and health performance for the previous year. Using and posting OSHA Form 200 satisfies this requirement.

3. Citations, Penalties, Abatement Times, and Variances. If a federal or state OSHA compliance officer finds violations in a workplace and issues a citation to the employer, this citation is to be posted along with any notice of penalty and abatement time. If the employer contests the violation, penalty, or abatement time or seeks to obtain a "variance" from complying to a particular standard, information pertaining to any of these actions also must be posted.

4. Notification of Employee Toxic Exposure. Where employees have been or are being exposed to toxic materials or harmful physical agents in concentrations or at levels that exceed those prescribed by an OSHA standard, the individual employee will be so notified along with information regarding the corrective action being taken.[9]

Routine Record Keeping

Except for "minor injuries requiring only first aid treatment," the employer must take action in the event an injury or illness occurs in the workplace. This action may be as minimal as recording on a standard OSHA Form 200 that the incident occurred or may require notification of the local OSHA office within 48 hr of a serious incident.

However, thanks to the awareness generated by the OSHA, the Consumer Product Safety Act, the Motor Vehicle Safety Act, growing product liability litigation, and even professional liability claims, engineers are beginning to realize that people are not infinitely flexible and cannot accommodate to any engineering design. Nor are people perfectly reliable in repeatedly interfacing with the tools, equipment, and product in the workplace. Finally, people no longer can be considered an easily replaceable asset if an engineering design slowly or quickly destroys them physically or psychologically.

Specifics as to how to eliminate an identified hazard are probably not as taxing to the safety manager or safety engineer as identifying the hazards and error inducing designs. In both steps, however, the integration of traditional industrial engineering, ergonomics, and safety engineering methodologies usually proves to be most effective in achieving the goal of a safer and more productive workplace. One such methodology is called "job analysis."[15] It consists of steps similar to the following: (1) selecting from accident data a hazardous job to analyze, (2) breaking the job down into successive steps or elements, (3) identifying the hazardous elements and potential accidents, (4) determining ways to eliminate the hazards and prevent potential accidents, and (5) providing the coordination and leadership to implement the necessary changes.

The application of job analysis will be no better than the individuals performing it. Individuals with a broad engineering background are required if its full value is to be realized. Training in mechanical, electrical, industrial, human factors, management, and safety and health engineering would be useful.

In addition to the job analysis technique, one other methodology, "fault tree analysis," has proved to be particularly useful. This approach forces the analyst to review all possible failure or accident modes associated with a task, tool, or product. The entire analysis is depicted in a tree-like structure; hence its name. This technique forces the engineer analyst to disect a task, product use, or human behavior into small elements and then focus on all possible outcomes or aspects associated with these elements. The technique is useful for accident reconstruction, analyzing an existing task being performed, or predicting from conceptual drawings of a future system or product what the electrical, mechanical, and human behavior failure modes might be. Because of its broad capabilities, it has become extensively used by safety professionals in both the product and the occupational settings. Malasky[17] provides excellent guidance on using this technique.

Personal Protective Equipment

Certain types of hazards that have not yet been, or cannot feasibly be, eliminated may be guarded against by the use of personal protective equipment worn directly on the body. Helmets, safety glasses, gloves, aprons, ear coverage, and shoes are among the products available for this purpose. The general OSHA regulations call for personal protective equipment "where there is a reasonable probability of injury that can be prevented by such equipment."[18] Further requirements pertain to the type of construction of the personal protection equipment with respect to its ability to actually provide protection. The OSHA regulations generally require that protective equipment be in compliance with the appropriate ANSI standard dealing with the test that a particular piece of equipment must pass or how it is to be constructed.[19]

Several problems exist regarding personal protective equipment. First, when to use it is somewhat a discretionary decision, which becomes one person's opinion versus another's. Second, the ANSI standards for such equipment have been criticized for being neither strict enough nor broad enough. For example, early versions of the required ANSI tests for safety shoes dealt only with toe protection and not with shoe sole nonskid considerations. (The data have shown that slips and falls are a more frequent cause of accidents than crushed toes). Third, the purchaser of personal protective equipment has had to rely on the manufacturer's word that the equipment sold does comply with the appropriate ANSI standard. Recent government sample testing indicates that some of the equipment sold does not provide the required ANSI protection. Fourth, the use of such equipment is an "active" type of safety measure. It requires the worker to personally "act" by using the equipment provided. Often such equipment is not worn because it is forgotten, lost, dirty, inconvenient, uncomfortable, or work interferring. Whatever the reason, like seat belts in cars, if not used, there is no protection. It is this last issue that is most concerning, in that for most accidents that could have been prevented using personal protection, any protection would have helped even if it did not completely comply. This suggests that most accidents do not test equipment at its design and construction limit, but well below this. For example, in accidents causing eye injuries, standard glasses, though not recommended, may provide successful intervention against a projectile.

It is this frequent absence of any protection that makes the personal protection equipment preventive measure unpopular. Where no other immediate remedy exists, however, use of such equipment becomes the only answer. The tendencies of persons not to wear such equipment then must be overcome by such things as training, improved comfort and convenience, and, as a last resort, discipline. Where it is necessary, such discipline is legal and has the full support of both unions and government.

Administrative Controls of Exposure

Although any actions by management could be labeled "administrative," the types of controls usually included under this category include the limiting of an individual's exposure time to a given hazard (scheduling) and the matching of a worker to a job (personnel assignment). Some authors include programs to control unsafe acts under this category of administrative controls. Because of its importance, it will be addressed separately.

Scheduling. In the case of exposure to noise, radiation, and toxic substances in general, the detrimental effects to bodily health occur as a function of time of exposure and dosage while exposure occurs. Other types of hazards now being researched by NIOSH eventually may be found to be equally as time exposure related, such as vibration, repetitive lifting, and repetitive body motion trauma. In many cases it has not been feasible to eliminate the hazard or provide adequate personal protection equipment. Therefore control is exercised by assigning the worker to a different job for a part of the total workday. This alternative job(s) would not expose the worker to the same hazard and thus allow the individual TWA exposure to remain within some safe limits. Administratively, it is usually implemented by putting a group of workers on a rotation of several jobs, with the time of exposure recorded for each job.

Personnel Assignment. Certain tasks are more hazardous to some persons than others. Saying it another way, the propensity for a particular type of accident to occur in a given situation depends partly on the characteristics of the individual involved. At one time the concept of the "accident-prone" person was popular. Today the experts generally agree that it is certain vulnerable *characteristics* of persons subjected to particular hazards that ultimately determine the probability and severity of an accident. There is less agreement on what these characteristics are, but among the candidates are lifting and other strength capabilities, reaction time, visual acuity and perception, decision making, short- and long-term memory, age, size, mental stress, psychomotor skill, education, experience in given task, and training. (Note that sex has not been included.)

Because of the increasing costs of accidents through losses in productivity, equipment, workers' compensation payments, medical and rehabilitation costs, and training new employees, employers have begun researching their job selection-placement policies to determine how individuals can be better matched to jobs so that productivity will be higher and accidents lower. For example, pre-employment strength testing now is being done in several large companies to determine if there is a relationship between the strength of persons and their propensity to have lifting-related accidents in high strength requirement jobs. Such data will be used as justification for screening out those potential employees who would most likely experience what might be a permanent injury if they were allowed to perform those tasks.

Controlling Unsafe Behavior

A traditional opinion shared by many today is that stupid, careless, negligent people are the cause of 90% of all accidents and that the tools, machinery, or processes involved should not even be considered as being primary potential causes. These individuals believe that, regardless of how equipment is designed, the responsibility for operating it safely lies with the persons who interface with it.

At the other extreme are those who believe that the primary responsibility for accidents lies with the designer-manufacturer and that these persons are "strictly liable" for their product. This viewpoint supports products so designed that, even in the presence of untrained careless persons, the product will have human fail-safe features that will provide protection for persons from their own errors and negligence. This position is probably closer to the ideals of OSHA, labor representatives, plaintiff attorneys, and many engineers and ergonomists.

The middle-ground realities are, first, that tools, machinery, and systems are for the most part not designed to be human fail-safe; certain kinds of errors made by humans will lead to accidents and injury. The second reality is that unsafe acts resulting from human errors, carelessness, or negligence do occur frequently in spite of the fact that the average employee is an intelligent, careful, and conscientious person. These unsafe acts and errors occur because of such things as lack of knowledge, lack of skill, lack of recent experience, inattentiveness, fatigue, and mental-physical-environmental stressors.

Some program that strives to control unsafe employee behavior is essential in every operation. Activities to control unsafe behavior are limited in that they can only partially compensate for the hazard, which could have been *eliminated* from the job or the product had it been more safely designed. Such emphasis on eliminating hazards cannot cease. However, the presence of training and unsafe behavior control programs has been known to take energies away from eliminating the hazards at their origin! Unsafe behavior control also is not effective in reducing accident causing behavior in that portion of the population that does not respond to training, reminders, and motivation schemes. In these cases preventing accidents requires removing the hazard from the job or removing the job from the person.

In general, most work situations involve responsive, intelligent employees who want to and who will work productively and safely given an adequate program of training and motivation. It is the new employee in particular for whom training programs are most beneficial. Companies that emphasize new employee training have found it to be highly cost-effective. This is because a new employee does not make a production profit for the company until the knowledge, skills, and hazards of the job are learned. Without these, the new employee is also the most likely to be involved in a costly accident.

With respect to safety, such programs will consist of informing new employees of the known hazards in their work, providing them with information and demonstrations of safe operating procedures, and following up with a variety of regular on-the-job reminders and motivators for both new and experienced employees. Training, educational materials, safety contests, safe practice audits, feedback on accident records, and even discipline for unsafe practices are all signs to employees that management cares about their safety.

A reasonably new component in health and safety programs has been the *participation by employees* in training, controlling unsafe acts, and general safety program management. Many current collective bargaining contracts place requirements on labor and management to conduct such programs cooperatively. This approach has been quite successful in identifying new hazards, recommending remedial action for hazard abatement, developing safe work practice guides, and promoting the total work force cooperation that is necessary for any program to be successful. Where organizations are not unionized, participation through elected employee representatives can function in the same way.

The act defines reportable injuries and/or illnesses as those that involve "work related deaths, injuries, and illnesses (other than minor injuries requiring only first aid treatment), loss of consciousness, restriction of work or motion, or transfer to another job."[10] These reporting requirements have been formalized and expanded by the administration in its regulation 29 CFR 1904, "Recording and Reporting Occupational Injuries and Illnesses." These regulations pertain generally to every employer and, among other things, prescribe the format that must be used by the employer to satisfy the record keeping requirement (using OSHA Form 200). In addition to OSHA requirements, the employer will have to respond in accordance with state workers' compensation reporting requirements as noted later in this chapter.

Nondiscrimination Against Employees

Few employers realize that there are antidiscrimination provisions within the act. All employees under the act may complain to a local OSHA office if they believe the employer is violating an OSHA standard.[11] This complaint may bring an OSHA inspector to the workplace and result in a citation's being issued to the employer. Some employers have reacted negatively toward the employee(s) involved in such action only to find that they as employers may be even more severely penalized for having discriminated against employees, in further violation of the act–Section 11(c)–which prohibits reprisals in any form against employees who exercise rights under the act. The administration places great emphasis on the Section 11(c) employee rights. How this section is interpreted and enforced by OSHA is contained in 29 CFR 1977.

Immediate Reporting of Fatalities and Serious Accidents

Following the occurrence of an employment accident that is fatal to one or more employees or that results in the hospitalization of five or more employees, the employer shall, within 48 hr, report the accident orally or in writing to the nearest area director's office of OSHA (29 CFR 1904.8).

Compliance With Applicable Safety and Health Standards

Recall that the source of the previous OSHA requirements has been both the act itself and the regulations promulgated by the administration and cited as a *Code of Federal Regulations* number (i.e., the 29 CFR 1904.8 listed in the previous paragraph). Some of the regulations promulgated by the administration are known as "Safety and Health Standards." The most well known among these is that cited as 29 CFR 1910, "General Industry Safety and Health Standards." The safety manager must not get caught believing that there is only one set of OSHA standards; in fact, there are several. More than one set is likely to be applicable to a particular employer. These sets of safety and health standards are:

29 CFR 1910: Safety and Health Regulations for General Industry
29 CFR 1915: Safety and Health Regulations for Ship Repairing
29 CFR 1916: Safety and Health Regulations for Ship Building
29 CFR 1917: Safety and Health Regulations for Shipbreaking
29 CFR 1918: Safety and Health Regulations for Longshoring
29 CFR 1926: Safety and Health Regulations for Construction
29 CFR 1928: Safety and Health Regulations for Agriculture
29 CFR 1960: Safety and Health Provisions for Federal Employees

Each of these covers a particular scope of activity. The two that are encountered by most employers are Part 1910, General Industry, and Part 1926, Construction. Most general industry employers will have "construction work" activities occurring within their operations and thus must comply with Part 1926 for such activities. "Construction work" is defined as "work for construction, alteration, and/or repair, including painting and decorating."[12]

All of the preceding are, of course, *federal* OSHA standards. It has been indicated previously in this chapter that state OSHA standards may exist in lieu of the federal standards. The major responsibility of the safety manager will be to plan, organize, execute, and control those activities in an organization that will bring the operation into compliance with all the applicable federal and/or state OSHA safety and health standards.

Requirements of Workers' Compensation Legislation

The OSHA itself states that "nothing in this Act shall be construed to supersede or in any manner affect any workmen's compensation law."[13] Thus the safety manager must be responsive to the requirements of the workers' compensation legislation specific to each state within which an organization's operations are conducted. Such legislation specifies that insurance be provided for by an employer to the end that employees' injuries or illnesses "arising out of or in the course of employment," under a "no fault" policy, will be compensated for if the injury or illness resulted in more than a certain number of days lost (such as 5). Management activities related to workers' compensation most likely will include arranging for insurance, paying premiums, hosting visits by insurance company representatives, responding to insurance company suggestions for workplace hazard elimination, reporting and detailing the circumstances surrounding accidents, paying the medical costs of an accident or illness, paying for rehabilitation of workers, and making actual compensation payments to workers if an organization has in-house self-insurance. Aside from OSHA responsibilities, these workers' compensation activities amount to a considerable work load on a safety manger. Thus personnel departments usually share in these responsibilities.

6.14.4 INDUSTRY AND CORPORATE STANDARDS

The safety manager's responsibilities do not stop with efforts to comply with federal OSHA, state OSHA, and state workers' compensation requirements. Prior to all of these, and still important today, are the other "standards of practice" or "state of the art" norms recommended by non-government bodies. Professional, trade, and industry organizations have, for more than 50 years, been developing standards of practice relating to worker, product, equipment, process, and system safety. These are represented by such organizations as the ANSI, NFPA, Society of Automotive Engineers (SAE), ASME, Underwriters' Laboratories (UL), American Society of Testing Materials (ASTM), and ACGIH. Though not having the force of law, recommended standards developed by organizations such as these dictate the reasonable state of the art and standards of practice in applications that go beyond federal and state OSHA requirements. For example, product liability litigation and third-party liability, often encountered in work-related injuries, depend more on these organizations' standards than on OSHA standards. Therefore the safety manager must ensure that the employer complies with the consensual requirements in the non-government-mandated standards. The body of such standards applicable to an organization will usually exceed by at least 10-fold the volume of OSHA standards for which mandatory compliance is required.

Beyond mandatory government standards and recommended consensual organization standards are company-specific standards or policies, which add yet another dimension to the safety manager's responsibility. For example, the federal administration retracted the OSHA standard that would have made unlawful the placing of hands between dies for loading or unloading presses. Nevertheless, a number of companies have adopted "no hands in dies" as company policy. Many companies have "recognized hazards" within their operations for which no government or consensual standard exists. Because of the "General Duty Clause," it is necessary for a company to develop a program or standards that address these hazards.

6.14.5 ADMINISTRATIVE RESPONSE VERSUS ACCIDENT PREVENTION

The previous sections have emphasized the many mandatory and obligatory responsibilities of the safety manager that often are the source of crises situations to which the manager must respond. Managing rapid responses to "hot spots" that develop because of an accident, illness, or government inspection too often becomes the primary role of the safety manager. With no safety staff, the safety "manager" may in fact be a one-person department who spends too much time personally responding. The key question one must stop and ask is, "Do these administrative responses to satisfy government mandatory requirements reduce the accident losses to an organization?" The sad answer from most safety experts would be "Probably not!"

The safety manager's position thus becomes a frustrating one in that several years of accident and illness data for an organization may show little improvement. This occurs even though it appears to the safety manager and to other organization personnel that the safety management function is being well performed, since the organization is in compliance with all applicable federal and state OSHA standards. It is at this important discovery point that an organization is primed to move ahead to a safety program that focuses on the real causes of accidents and illnesses unique to that particular organization's people, processes, product, machines, methods, and materials. The real challenge for the safety professional involves identifying the past and future causes of accidents and making action plans to control the physical hazards and unsafe acts involved. These causes must be discovered and controlled because they are the real predators on profits and productivity.

It is the control of these that will allow the safety manager to show that indeed safety pays, and it is this objective to which the balance of this chapter is devoted.

6.14.6 IDENTIFYING PAST AND FUTURE ACCIDENT CAUSES

Frequency and Severity Rates

Since its founding in 1913, the National Safety Council (NSC) has played a major role in the collection of injury data, particularly for all types of accidents occurring within the United States that fall into the classes following: work related, motor vehicle, home, and public. Each year accident data for these classes are collected, summarized, and eventually printed in *Accident Facts*. This NSC publication, which is sold for only a few dollars, continues to be the most recognized profile of accidents occurring in the country.

With respect to the NSC data pertaining to work-related accidents, two types of "accident rates" have been adopted. "Frequency rate" is the number of lost-time accidents per million man-hours worked. "Severity rate" is: the number of lost workdays per million man-hours worked. (The approximate number of hours that 500 employees would work in 1 year is 1 million hr.) These are the rates that traditionally have been reported in *Accident Facts* for work-related accidents.

The frequency and severity rate concepts originated from ANSI Standard Z16.1, "Method of Recording and Measuring Work Injury Experience." The adoption of it by the NSC is largely responsible for its wide general use prior to the passage of OSHA.

OSHA Incidence Rate

A change occurred in the collecting and reporting of work-related accidents as a result of the OSHA. This became necessary because of an increased emphasis on occupational illnesses and because of statutory language in the act that defined when an accident was serious enough to be recorded and counted in OSHA's accident data base. The traditional NSC definition of lost-time accident frequency and severity rates did not coincide with this new, congressionally mandated definition. Therefore, to reduce the confusion that existed, OSHA originated a new injury and illness rate measure called the "incidence rate." This has been defined by the administration as the number of recordable injuries and illnesses per 200,000 man-hours worked. This rate differs in that (1) the "per million man-hours worked" used as the common denominator in the frequency and severity rate calculation has been changed to 200,000 man-hours worked and that (2) "lost workday" criteria have been changed to "recordable injuries and illnesses."[14]

Information is collected annually for OSHA from employers by the BLS. These incidence rates are calculated for the nation and for specific industries using the well-recognized Standard Industrial Classification (SIC) coding system. These data are published annually by the BLS in a number of different forms, allowing employers and OSHA to compare specific industries and companies to the national average. For 1978 this national average incidence rate was 9.2, that is, 9.2 reportable injuries and illnesses/200,000 man hr worked (see Exhibit 6.14.1). Since one person works about 2000 hr/year, this incidence rate can also be interpreted as 9 injuries and illnesses/100 full-time workers per year. The OSHA can estimate this incidence rate nationally and on a more localized geographical area. The administration can also obtain lists of companies in a particular geographical area within each SIC, which is useful to OSHA in setting priorities on how, in terms of enforcement, it can use its resources most effectively.

Workers' Compensation and BLS Supplemental Data

As previously noted, state workers' compensation requires employers to submit a "first report" of injury or illness whenever any employee has had an injury or illness "arising out of or in the course of employment" that exceeds a certain number of lost workdays. This information is collected statewide by the Workers' Compensation Division of state government and by insurance companies.

The BLS has recently initiated a program called the "Supplemental Data System." This program involves the national collection, summarization, and dissemination of workers' compensation data by state. Thirty states are participating in this program thus far. The advantage of accident and injury information being available from workers' compensation report sources is that considerably more detail is provided about the circumstances surrounding an accident or illness. Many of the states have tailored their accident reporting after part of the ANSI Z16.2 standard. If usefully summarized and interpreted, this can lead the safety manager closer to identifying the causes of accident or injury. Unfortunately, even these data usually do not lead to specific accident causes.

Exhibit 6.14.1[a] OSHA's Incidence Rate for Injuries and Illnesses for Different Segment Industries

Industry	Incidence Rate per 100 Full-Time Workers per Year	Total Employment (Thousands)
All private industry totals	9.2	71,533
All agriculture, forestry, and fishing	11.0	891
All mining	11.3	851
Anthracite mining	19.3	
All construction	15.8	4,271
All manufacturing	12.8	20,476
Durable goods	13.7	12,246
Lumber and wood products	22.3	
Furniture and fixtures	17.2	
Stone, clay, and glass products	16.4	
Primary metal products	16.5	
Fabricated metal products	18.8	
Nondurable goods	11.4	8,230
Food and kindred products	18.7	
Paper and allied products	13.3	
Rubber and plastic products	16.6	
Transportation and public utilities	9.9	4,927
Trucking and warehousing	16.1	
All wholesale and retail trade	7.8	19,499
Food stores	10.6	
All finance, insurance, and real estate	2.0	4,727
All services	5.3	15,891
Hotels and other lodging places	9.0	
Miscellaneous repair services	9.7	

[a]Abstracted by the author from BLS tables for 1978 occupational injury and illness incidence rates. The listing of major industry categories is complete. The subcategories listed were chosen to point out industries within a category that had high incidence rates.

Key Facts in Accidents (ANSI Z16.2)

The ANSI and the NSC have long had an interest in providing a system that would help safety managers everywhere identify past and future accident causes. This system, adopted as a consensual standard, is known as ANSI Z16.2, "Method of Recording Basic Facts Relating to Nature and Occurrence of Work Injuries." The purpose of the standard is to identify certain "key facts" about injuries and the accidents that produced them and to record these facts in a form that might show general patterns of injury and accident occurrence. The ANSI Z16.2 system requires that the following key facts be identified: (1) nature of injury, (2) part of body, (3) source of injury, (4) accident type, (5) hazardous condition, (6) agency of accident, (7) unsafe act, and (8) contributing factors. The ANSI Standard Z16.2 also suggests examples of specific sets of classifications for all of the preceding key facts. For example, for accident type the well-known classification includes struck against, struck by, caught in (on or between), fall on same level, fall on different level, overexertion, slip (not a fall), and so on. The use of the ANSI Z16.2 system by the NSC in their data collection and publication has made this system well known.

Any of these generalized approaches to collecting accident and illness data almost never will be sufficient to identify specific accident causes in an organization's operations. Many companies have thus developed their own system, tailored to the types of operations characteristic of their organization.

The ANSI Z16.2 approach certainly should be the starting point for the safety manager who wants to better identify past and future accident causes–but not the ending point. Substantial modifications to this approach will be necessary if it is to identify the accident causes unique to a specific organization.

Other Sources of Data and Assistance

As the safety manager concentrates on activities other than those that are government mandated, considerable assistance and insight can be gained from several organizations that also are particularly concerned with helping to prevent accidents in a particular organization. The assistance available may be in the form of detailed accident data, recommended standards of practice, direct consultation services, accident illness research capability, safety program audiovisual aids, or training programs. Among the sources providing such assistance are:

NSC.

Professional societies, such as American Society of Safety Engineers (ASSE), American Industrial Hygiene Association, Systems Safety Society, and ACGIH.

Trade associations, such as National Machine Tool Builders Association, Association of General Contractors, Industrial Safety Equipment Association, National Association of Manufacturers, Motor Vehicle Manufacturers Association.

Consensual organizations, such as ANSI, NFPA, UL.

Insurance associations, such as American Mutual Insurance Alliance, American Insurance Association.

Safety publications, such as *National Safety News*, *Professional Safety* (ASSE), *Journal of Safety Research*, *Journal of Systems Safety Society*, *Journal of Occupational Accidents*.

Government and public research organizations, such as NIOSH.

Universities with strong research and graduate programs in safety, such as Michigan, North Carolina, Ohio State, Texas Tech.

Federal and state consultation services—Congress has provided funds to pay for free safety and health consultation to companies. Contact local state department of labor or federal OSHA area office for information.

Private consultants and consulting firms—since 1970, a multitude of individuals and firms have evolved that specialize in health- and safety-related consultation services.

Addresses for most of the preceding sources are found in the NSC's *Accident Prevention Manual*,[15] and a few of the key addresses are given at the end of this chapter.

6.14.7 DIRECT ACCIDENT PREVENTION ACTIVITIES

Setting Priorities

Every safety program will require a substantial amount of activity to control physical hazards. Such control may be required in order to comply with OSHA standards, because of labor-management agreements, or in response to specific injury or illness causes identified through injury and illness data or research sources. Recognizing that all possible physical hazards identified cannot be controlled immediately, the safety manager must establish short-term and long-term priorities.

The most desirable priority system should be based on controlling those physical hazards for which the least amount of manpower and money is needed, the most number of injuries are prevented, and the greatest accident cost savings result. This is an oversimplified ideal, since the safety manager, even with complete information, is faced with multiobjective priority setting. For example, one might be forced to focus on complying with OSHA standards so that an OSHA compliance officer would be least likely to find those physical hazards for which there would be violations if an OSHA inspection occurred. This is not an unreasonable objective, since an organization's management would probably be quite pleased with a safety manager who could bring a facility into sufficient OSHA compliance to allow it to come out "clean" from an OSHA inspection. A strategy to achieve this might be to attack those physical hazards that compliance officers are most likely to give attention to while performing an inspection. What might these be? A source for such information is the set of statistics produced annually by OSHA, which ranks the total numbers of violations, by standard, found by all federal OSHA inspectors during the year. Exhibit 6.14.2 is provided on the theory that the violations found most frequently will point to the physical hazards OSHA inspectors most closely look for.

The threat of a federal or state OSHA inspection may be a consideration to some managers in allocating how much priority should go to safety. The probability that such an inspection will occur can be gleaned if one recognizes the priorities OSHA itself uses to determine which employers to visit with the manpower available. Those priorities for inspections are:

1. Response to an imminent danger.
2. Investigation of a fatality or catastrophic event (hospitalization of five or more employees).

Exhibit 6.14.2[a] Violations of Standards Found Most Frequently by OSHA Inspectors

Number of Violations Cited	Standard Violated (29 CFR)	Type of Hazard
16,348	1910.309A	1970–1971 National Electrical Code (all installations)
3,566	OSHA Sec. 5(a)(1)	General Duty Clause
3,347	1910.309(b)	1971 National Electrical Code (new installations)
3,062	1903 (a)(1)	OSHA poster not displayed
2,964	1910.212(a)(1)	Guarding for point of operation, nip points, rotating parts, flying chips and sparks inadequate
2,944	1910.219(d)(1)	Mechanical power transmission: pulley guarding
2,345	1910.252(a)(2)	Approval and marking of welding, cutting, and brazing cylinders and containers
2,343	1910.212(a)(3)	Point of operation machine guarding inadequate
2,209	1910.022(a)(1)	General housekeeping requirements
2,123	1910.219(e)(1)	Mechanical power transmission guarding vertical and inclined belts
1,945	1910.023(c)(1)	Guarding floor and wall openings
1,851	1910.132(a)	Personal protective equipment not used
1,771	1910.215(b)(9)	Abrasive wheels; work rests not provided
1,668	1904.2(a)	Log of occupational injuries and illnesses not maintained
1,596	1904.5(a)	Annual summary of injury and illness not compiled
1,438	1910.219(f)(3)	Mechanical power transmission: sprockets and chain guards
1,428	1910.106(e)(2)	Storage or use of flammable and combustible liquids
1,415	1910.212(a)(5)	Guarding fan blades within 7 ft of floor
1,316	1910.133(a)(1)	Eye and face protection not used

[a] Developed by the author from information provided by OSHA, Division of Compliance Programming, for Fiscal Year October 1, 1977–September 30, 1978.

3. Valid complaints received from employees.
4. Target industry program (industries identified as having particularly high accident rates: longshoring, lumber and wood products, roofing, and manufacturing of mobile homes and other transportation equipment).
5. General industry inspections (which are usually based on a "worst first" criterion, using the incidence rates obtained from the BLS Annual Survey Data (Exhibit 6.14.1).

Within the first four categories, more than 80% of the safety and health inspection resources are consumed. Therefore another strategy that a safety manager might use is to (1) identify and respond immediately to imminent dangers before OSHA does, (2) concentrate on those operations wherein the possibility of fatalities or multiple serious injuries is great, (3) provide a mechanism for employees to bring their safety and health complaints and grievances to management rather than to OSHA, and (4) recognize that the probability of a general OSHA inspection will be directly related to the incidence rate for the SIC within which a particular organization falls.

Another approach to identifying the physical hazards that should be given priority involves the utilization of accident and injury data such as those derived from workers' compensation reporting systems. The reader will recall that the Supplemental Data System of the BLS has been mentioned as such a compilation. Although the data are summarized across a broad base of organizations, they still provide good insight into what types of accident and injuries are the typical problem areas within given industries. Exhibits 6.14.3 and 6.14.4 are representative subsets of data from the BLS Supplemental Data System. These data happen to be for the year 1976 and represent the states Arkansas, Delaware, North Carolina, and Wisconsin.

It is obvious that the exhibits *do not* present a consistent pattern from which the safety manager can establish priorities. Namely, the most frequently cited standards by OSHA inspections (Exhibit 6.14.2) differ substantially from the types of accidents in which the largest number of injuries are occurring (Exhibit 6.14.3) and from the most frequent sources of injury and illness (Exhibit 6.14.4). This finding supports what was previously suggested concerning the dilemma faced by the safety manager as multiobjective program priority setting is considered (i.e., compliance with OSHA versus responding to accident causes).

Exhibit 6.14.3[a] Type of Accident As a Function of Percentage of Total Workers' Compensation Payments

Type of Accident	Percentage of Total Workers' Compensation Payments
Struck by	17.2
Overexertion	16.4
Caught in, under, between	13.6
Falls from elevation	11.6
Falls on same level	9.4
Motor vehicle	7.2
Body reaction	5.3
Struck against	5.3
Contact with temperature extremes	2.0
Contact with radiation, caustics, and so on	1.9
Others	10.1
	100.0

[a] Data were summarized by NIOSH, Division of Safety Research, Morgantown, WV, from 1976 BLS Supplemental Data System workers' compensation data for Arkansas, Delaware, North Carolina, and Wisconsin.

Exhibit 6.14.4[a] Sources of Injury or Illness As a Function of Percentage of Total Indemnity Compensation

Source of Injury or Illness	Percentage of Total Indemnity Compensation
Working surfaces	21.0
Vehicles	13.2
Machines	12.6
Metal items	9.0
Boxes, barrel, containers	6.8
Bodily motion	5.3
Wood items	3.5
Handtools, not powered	2.2
Electric apparatus	2.0
Hand-powered tools	1.5
Hoisting apparatus	1.4
Buildings and structures	1.4
Furniture, fixtures, and so on	1.4
Other person	1.3
Plants, trees, and vegetation	1.0
Conveyors	1.0
	84.6

[a] Data were summarized by NIOSH, Division of Safety Research, Morgantown, WV, from 1976 BLS Supplemental Data System workers' compensation data for Arkansas, Delaware, North Carolina, and Wisconsin.

Traditional Accident Control Measures

A hazard usually exists because (1) it was created by the engineering design of a process, tool, machine, or structure and because (2) the design required a worker to (a) use; (b) work in, on, between, or around; or (c) be exposed to the hazard through the environment or in transit. The obvious approach to accident prevention is not creating the hazards in the first place or eliminating them by reengineering if they already exist. For example, assume that the hazards of open manholes exist in walkways. The traditional remedial options include *warning* persons of the hazards with signs, *training* persons to be aware of open manholes and to walk around them, or *engineering* and installing protective covers for the manholes. The latter, of course, is preferred, because if there are no hazards, there will be no accidents; or "If there are no people (i.e., complete automation), there are neither accidents nor injuries." The common argument that 90% of all accidents are caused by an operator, a user, or the injured party suggests that *unsafe acts* committed in the presence of engineered hazards cause accidents. Does one therefore conclude that it is (1) the unsafe acts that must be prevented or (2) the unsafe condition (hazard) that must be eliminated by engineering? If "neither of the above," shall we protect the worker from himself or herself by allowing no or limited exposure to a hazard? Or shall we protect the worker from the hazard by using physical guarding between the worker and the hazard (i.e., machine guards, helmets, respirators, etc.)?

These questions have been answered within the NSC's *Accident Prevention Manual for Industrial Operations* as follows:

> *The basic measures for preventing accidental injury, in order of effectiveness and preference are:*
>
> *1.* Eliminate *the hazard from the machine, method, material, or plant structure.*
> *2.* Control *the hazard by enclosing or guarding it at its source.*
> *3.* Train *personnel to be aware of the hazard and to follow safe job procedures to avoid it.*
> *4. Prescribe* personal protective equipment *for personnel to shield them against the hazard.*[16]

The first two of these basic measures define what is meant by "engineering a safe workplace." Absent from this NSC list is a fifth measure, which should be added: Prevent or limit exposure by *administrative controls*. These approaches are discussed in the subsections to follow.

Physical Hazard Control

Engineering a Safe Workplace

The engineers who are responsible for tools, machines, and workplaces in general may have negligible experience and knowledge regarding human capabilities and limitations. They often do not realize that "death traps" are being built because their designs, while satisfying some mechanical, electrical, or process objectives, are inducing people to make predictable human errors, some of which may cause accidents and injuries. It is expected, then, that the safety manager will be in the middle of an unlimited number of hazards created by his or her peers. An unpleasant, but necessary, job of the safety manager is the influencing of these same peers to change the designs of which they are so proud.

REFERENCES

1. D. PETERSEN, *Safety by Objectives*, Aloray, River Vale, NJ, 1978.
2. H. W. HEINRICH, *Industrial Accident Prevention*, 4th ed., McGraw-Hill, New York, 1959. (5th ed. available in 1980.)
3. HEINRICH, *Industrial Accident Prevention*, p. 15.
4. J. SURRY, *Industrial Accident Research*, Labour Safety Council, Ontario Department of Labour, Toronto, 1971.
5. U.S., *Statutes at Large*, OSHA of 1970, Public Law 91-596, Section 5(a)(1).
6. U.S., *Statutes at Large*, OSHA, Section 5(a)(2).
7. *United States Code*, Administrative Procedures Act, Title 5, Public Law 89-554.
8. *United States Code*, Administrative Procedures Act, Chapter 7 (Judicial Review).
9. U.S., *Statutes at Large*, OSHA, Section 8(c)(3).
10. U.S., *Statutes at Large*, OSHA, Section 8(c)(2).
11. U.S., *Statutes at Large*, OSHA, Section 8(f)(1).
12. *Code of Federal Regulations*, Title 29, Part 1910.12(b).
13. U.S., *Statutes at Large*, OSHA, Section 4(b)(4).
14. *Code of Federal Regulations*, Title 29, Part 1904.
15. *Accident Prevention Manual for Industrial Operations*, 6th ed., National Safety Council, Chicago, 1969. (7th ed. now available.)
16. *Accident Prevention Manual.*
17. S. U. MALASKY, *System Safety*, Spartan Books, Rochelle Park, NJ, 1974.
18. *Code of Federal Regulations*, Title 29, Part 1910.133(a)(1).
19. *Code of Federal Regulations*, Title 29, Part 1910.133(b)(1).

BIBLIOGRAPHY

ANTON, T. J., *Occupational Safety and Health Management*, McGraw-Hill, New York, 1979.

FIRENZE, R. J., *Guide to Occupational Safety and Health Management*, Kendall-Hunt, Dubuque, IA, 1973.

HAMMER, W., *Occupational Safety Management and Engineering*, Prentice-Hall, Englewood Cliffs, NJ, 1976.

PETERSEN, D., *Safety Management*, Aloray, River Vale, NJ, 1975.

PETERSEN, D., *Techniques of Safety Management*, 2nd ed., McGraw-Hill, New York, 1978.

SIMONDS, R. H., and J. V. GRIMALDI, *Safety Management*, rev. ed., Irwin, Homewood, IL, 1963.

THYGERSON, A. L., *Safety–Principles, Instruction, and Readings*, Prentice-Hall, Englewood Cliffs, NJ, 1972.

KEY ADDRESSES

For ANSI Standards

American National Standards Institute
1430 Broadway
New York, NY 10018

For NFPA Standards

National Fire Protection Association
470 Atlantic Avenue
Boston, MA 02210

For OSHA Standards and OSHA Act

Directorate of Safety Standards
Occupational Safety and Health Administration
U.S. Department of Labor
200 Constitution Avenue
Washington, DC 20210 (or see local telephone directory under U.S. Government, Department of Labor)

For TLV (Other Than OSHA Standards)

American Conference of Governmental and Industrial Hygienists
P.O. Box 1937
Cincinnati, OH 45201

For NSC Publications and Services

National Safety Council
444 North Michigan Avenue
Chicago, IL 60611

For BNA Publications

Bureau of National Affairs, Inc.
1231 25th Street, NW
Washington, DC 20037

For NIOSH Safety Research Reports and Safety Technical Information Services

Division of Safety Research
National Institute for Occupational Safety and Health
Morgantown, WV 26505

SECTION 7

MANUFACTURING ENGINEERING

CHAPTER 7.1

Manufacturing Engineering

HAROLD N. BOGART

Ford Motor Company

7.1.1 MANUFACTURING ENGINEERING FUNCTION

Background

As a function, manufacturing engineering is still evolving toward total planning for manufacturing. The earliest engineering function in manufacturing that received appropriate attention was that which was known prior to World War II as tool engineering. The principal function of the tool engineer was to design jigs and fixtures for use with manually loaded and operated machines so that locating points, reference dimensions, and other critical manufacturing tolerances might be adhered to properly. At the same time the tool engineer was responsible for developing cutting tool technology, including specifying speeds and feeds used for cutting tools in machining operations. As concepts for mechanization took shape, the tool engineer reduced these concepts to practice. It is obvious that this use of tool engineering was not limited to the metal cutting industry, but extended into metal working, assembly, and other production areas.

In a slightly different context, within the foundry industry, the pattern making function was responsible for providing patterns, core boxes, fixtures, and related manufacturing aids. Other areas developed specialized expertise of this nature as their route to automated manufacturing.

At that time (before 1940), the function of planning for manufacturing was shared by the tool engineer, the plant master mechanic, and the production manager. The master mechanic, as the name suggests, had responsibility for all of the mechanical devices, from conception to retirement, including equipment selection, operation, maintenance, and new concept development. The production manager had an important voice in the planning for manufacturing, but this was limited since it was only part of the production manager's job, and he or she could not be expert in all of the fields for which he or she was responsible. Each of these individuals—the tool engineer, the master mechanic, and the production manager—depended on the timekeeper and the time standards analyst to provide him or her with information on the execution of the planned operation. As manufacturing operations and equipment have become more complicated and sophisticated, planning for manufacturing has been evolving toward a true engineering discipline.

Accompanying the evolution of a manufacturing engineering discipline were changes in procedures to support the new discipline. Routing sheets that conveyed to machine operators information on the movement of parts from raw material to finished assembly have been replaced by process sheets that serve the additional function of describing the details of each individual step in the process, including machine tool and automation requirements; thus the actual operating conditions for individual machine tools and process equipment are defined. Simple conveyable devices, such as bins, barrels, and other in-process storage containers, have successively been replaced and improved through conveyorization and, finally, in today's language, automation. The control of the individual machine tools has advanced from the manual switch, the manual gauging, the hand-operated advance and retract to sophisticated electronic controls and automatic gauges.

Definition

In short, manufacturing engineering has advanced from a relatively simple exercise of engineering logic to a highly sophisticated engineering field in which the whole spectrum of engineering skills is employed. In recognition of the complexities that have been introduced into the manufacturing engineering function, the American Society for Tool Engineers (ASTE) expanded its charter to become the Society of Manufacturing Engineers (SME). Formal recognition of the ever-enlarging role of the manufacturing engineer was made by SME in a brochure entitled *Manufacturing Engineering Defined.*[1] Part of the definition derived from an SME board of directors' action on May 8, 1978, wherein an official definition of manufacturing engineering was adopted.

Manufacturing Engineering is that specialty of professional engineering which requires such education and experience as is necessary to understand, apply and control engineering procedures in manufacturing processes and methods of production of industrial commodities and products; and requires the ability to plan the practices of manufacturing, to research and develop the tools, processes, machines and equipment, and to integrate the facilities and systems for producing quality products with optimal expenditure.

Functional Position in Manufacturing Hierarchy

As inferred in the preceding discussion of planning for manufacturing, and as specifically incorporated into the SME definition, a principal role of the manufacturing engineer is to provide, within a given manufacturing organization, the capability of producing capital goods of high quality at a low cost in a timely manner. This type of function places the manufacturing engineer in a pivotal position—his or her decisions weigh heavily not only on the company's ability to operate at a profit, but also on its ability to produce a quality product, thereby ultimately serving the consumer. Neither of these goals can be met unless the manufacturing engineer has an impact on the ultimate product design such that design and manufacturing optimization becomes part of new product development. To discharge this latter function effectively, the manufacturing engineer must occupy a place in the manufacturing hierarchy from which he or she can act on product information and intent from the product engineer and appropriately translate it to his or her colleagues in the manufacturing planning activities. Thus the manufacturing engineer will select and activate processes and equipment that will optimize the cost-benefit performance of the operating activity. This becomes the route of translating conceptual product designs into functional operating processes using new machines and equipment.

The pivotal position of the manufacturing engineer may be envisioned as in Exhibit 7.1.1, which describes the flow of information from the product engineer to and from the production operation. Upon receiving product information from product engineers, the manufacturing engineer interprets these data in terms of all the process requirements within the processing quadrant. This

Exhibit 7.1.1 Product Information Flow for Product Engineering

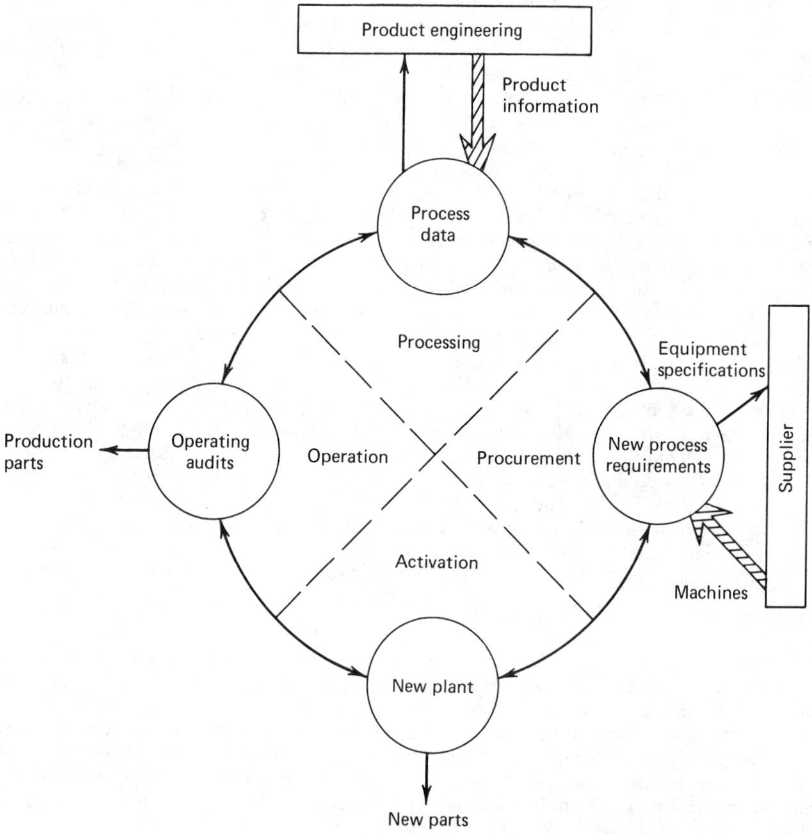

process information then flows to the manufacturing engineering function, where the machinery and equipment is identified and specified in terms communicable to the machinery supplier; this is the procurement quadrant.

The supplier's response (after appropriate interaction, as discussed later) is in terms of machines, tools, and equipment responsive to the original specification. This hardware, replete with instructions for its use, is installed and activated by the manufacturing function.

The communication flow is through the customer and a judgment of the adequacy of the planning is based on the performance of purchased equipment.

The communication loop continues in the operations quadrant, where production volumes are achieved and operating audits judge the adequacy of the entire planning and activation. The loop is closed as results of actual operation are available to the manufacturing process engineer and, through him or her, to the product engineer.

Accepted Practice

The overall function of the manufacturing engineer, then, is to translate product design into a highly replicated practice, the commonly known mass production. Henry Ford I was often quoted as saying, "If you can make one, you can make a million." This is a gross oversimplification of the type of travail to which the manufacturing engineer is regularly subjected. He or she must have a high confidence level, so that when the plant or an individual machine is activated, it will produce according to schedule, cost, and required quality on a consistent, day-by-day basis.

A high confidence level is achievable only if the sum total of proven experience in the field is available to the manufacturing engineer. Documentation of such experience is most useful. Even for small organizations, a full catalog of operating experience can help prevent "the reinventing of the wheel" or the compounding of mistakes (by replication). The use of manufacturing standards, wherein proven practices are identified, is highly recommended. A starting point for a manufacturing standards program is an accumulation of industry standards, such as joint industry standards, for major components and materials required for the machinery and equipment in common use by the particular operation. The second source of basic information is some assessment of performance for the various components and machines within existing manufacturing operations. Any major differences in performance among similar components or machines will become the basis for identifying preferred components or preferred features of components. These data, merged with the available industry standards, become the skeleton for building a manufacturing standards program. Historically, a mental catalog of this type was the justified bias of the previously referenced master mechanic. Many of the larger manufacturing companies have formal manufacturing and process standards, which are available, upon request, to smaller organizations.

Relationship to Other Manufacturing Functions

The definition of the manufacturing engineering function clearly shows its support nature. Within all of manufacturing, the production operation is the only operation that actually produces capital goods. Exhibit 7.1.2 shows this prime function for any manufacturing activity. The support functions other than manufacturing engineering include, among others, quality control and reliability, production control, finance, industrial relations, plant engineering and maintenance, material handling, toolroom, material supply, laboratory, and receiving and shipping. A typical communications network is shown in the exhibit to indicate the various inputs that the various manufacturing support functions provide to operations. It is obvious that two or more interrelating and interacting functions provide support for the same basic activity. The diversity of interest expressed by the different functions ensures a compromised, balanced position, minimizing the opportunity for unforeseen and possibly calamitous events. As is shown later, the manufacturing engineering function serves as a coordinator among technical functions and between technical and administrative functions.

Relationship to Industrial Engineering

The interaction between the manufacturing engineering function and the industrial engineering function is a good example of the intimate relationship between various segments of the manufacturing operations required for effective support. The manufacturing engineer has a particularly close relationship with the industrial engineer, since the performances of persons and machines in a production environment are so inextricably joined. Significant functions performed by the manufacturing engineer to support the industrial engineer are to communicate and coordinate new product information and new product programs and to translate these data into manufacturing-related information in light of existing manufacturing facilities and practices. This is of particular importance to the industrial engineer in that any early selection of processes by the manufacturing

Exhibit 7.1.2 Typical Plant Functions[a]

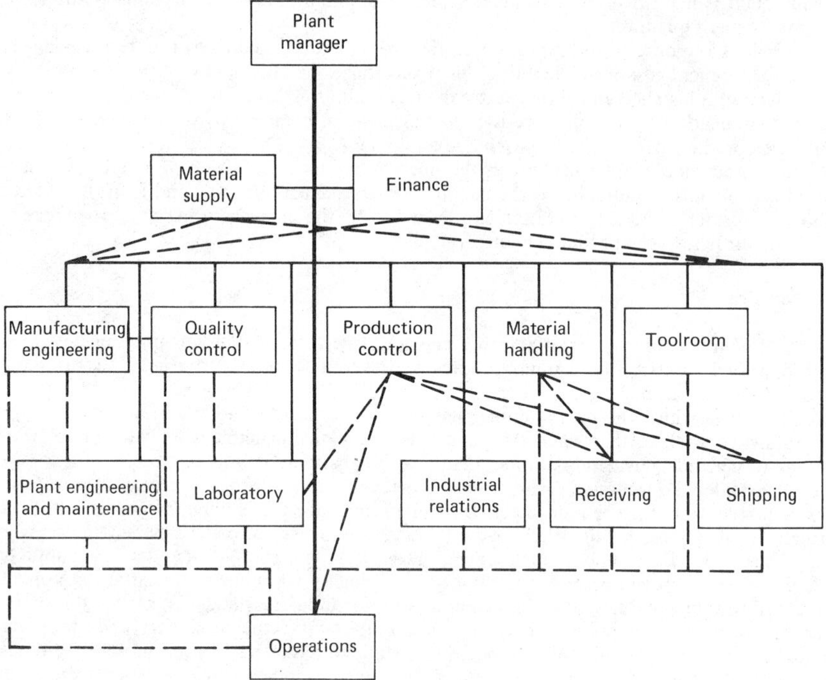

[a]Broken line indicates representative communication channels.

engineer should be reflected in cost estimates of the finished product. This information, generated by the industrial engineer, is forwarded through the manufacturing engineer to those responsible for developing corporate plans. In addition to the coordinating role that extends across the whole group engaged in planning for manufacturing, the manufacturing engineer provides active support to the industrial engineer in methods improvement programs and in the important line balancing function for assembly lines.

Line and Staff Variations

It is clear from the various manufacturing functions that are performed in the typical manufacturing plant that diverse hierarchical structures may be used for the manufacturing function. This may depend partly upon the personalities of the personnel in the various functions and certainly depends upon the size of the individual operation. There also are variations between line and staff in a given organization, where relationships at a staff level may be entirely different from those at the operational level. The basic principle prevails—communication between functions and between line and staffs must be enhanced in order to support top management decision making. Examples can be cited where functions, such as material handling and material handling engineering, occurring as separate functions within a plant or a division are combined at a corporate level.

7.1.2 MANAGEMENT REPORTING RELATIONSHIPS

Basic Principles

Although there is a wide variation in the management reporting relationships for the manufacturing engineer, there are basic principles that should be addressed in selecting the most appropriate reporting relationship for any given situation.

First, the *manufacturing engineering function must support the total manufacturing and technical activity* in the plant or company. It is possible that this function could actively manage and coordinate all technically based activities.

Second, the *manufacturing engineering function must be well represented in the top tier of plant supervision so that adequate recognition is given to planning for manufacturing.* Once the manufacturing process has been selected and activated, it is essential that the support services modify their activities based on the requirement of the manufacturing process rather than modifying the manufacturing processing solely to meet the real or imagined needs of a support activity.

Third, *the manufacturing engineer must recognize that he or she has a fundamental support role* and that variations in processing place first priority on actual manufacturing operations, with the convenience of the manufacturing engineer occupying a subservient role.

Fourth, *the span of control* within the manufacturing engineering operation and from the top management of the plant *must not be too broad.* The manufacturing engineering function should not be so compartmentalized as to have unnecessarily large numbers of people in the technical area reporting to any level of plant management.

Actual Practice

The most effective practice for a given operation will vary widely. This variation will be influenced by:

The characteristics of individuals within the total manufacturing system.

The nature of the processes themselves.

The underlying management and delegation philosophy.

Management skills.

Exhibit 7.1.3 illustrates some typical reporting relationships for the manufacturing engineering activities. Exhibit 7.1.3*a* and *b* represent potentially frustrating organizations for management and engineer alike. Basically the frustration can stem from too wide a span of control, not only reducing the effectiveness of communication, but also affecting a firm's ability to attract and compensate highly competent engineers. Whereas in a medium-size establishment the manufacturing engineer may report to the plant manager (Exhibit 7.1.3*c*), in larger establishments he or she would logically report to an operations manager or a plant staff manager (Exhibit 7.1.3*d*). In establishments with multiplant activities performing the same function, it is entirely possible that the principal manufacturing engineering planning will be the responsibility of a staff organization in the interest of uniformity from plant to plant (Exhibit 7.1.3*e*). In this case, of course, the manufacturing engineer in the plant is less responsible for the planning for manufacturing and more responsible for the day-to-day execution of the previously made manufacturing plants.

Recommended Reporting Relationship

All of the basic principles will be recognized if the manufacturing engineer reports as part of the top tier of the plant or company organization. This will ensure that the top management of the operation has cognizance of the manufacturing effort and recognizes the importance of giving corporate emphasis to machinery selection and maintenance. In a large plant it is recommended that a technical manager have the responsibility for plant manufacturing engineering, plant and maintenance engineering, quality control, and all of the important technical activities. This will reduce the span of control for the plant management and should provide a balanced voice during major decision making.

7.1.3 MANUFACTURING PLANNING AND UTILIZATION OF RESOURCES

The heart of the entire manufacturing engineering function is planning for manufacturing, which starts with the optimal utilization of all resources available to the manufacturing engineer and, through him or her, to the individual operation. This particular step in the manufacturing engineering function is the prelude to the actual process planning, that is, point-by-point detailed planning for the actual manufacturing operation. A major resource available to the manufacturing engineer is the operating experience with existing equipment that is performing operations equivalent to those required by the design under study. Three important steps must be followed in order to ensure the use of all resources available to the manufacturing engineer. These are:

Detailed engineering examination of the existing system.

Development of lists of engineering alternates.

Selection and promulgation of the best alternates.

Examining the Existing System

The analysis of the existing system should include a statement of generalized experience with known operating total systems and, where applicable, a mathematical model as an aid to the effective use of all resources. From the actual production floor will come detailed observations of

Exhibit 7.1.3 Typical Manufacturing Engineering Reporting Relationships[a]

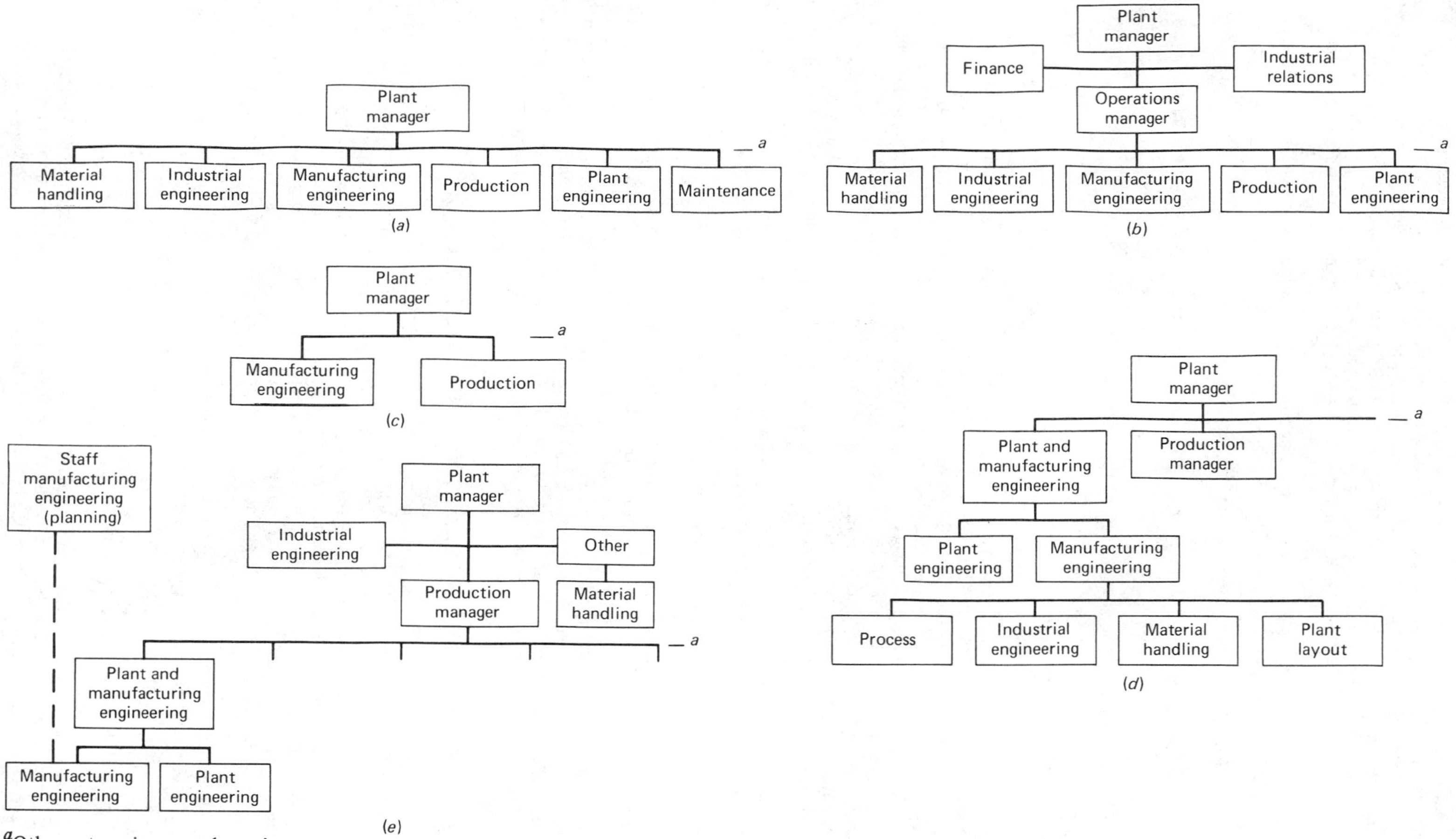

[a]Other categories may also exist.

the production machinery, the use of materials and processes, the efficiency of material handling, and the distribution and use of manpower, along with an analysis of the total logistics of the system.

Operating machinery will be examined for its entire function, that is, to produce parts to design intent at the planned production rate at the projected cost. This analysis will obviously result in identification of machine downtime, assignment of causal factors for the downtime, duration of the downtime, cost of repairs, and all factors that will assist in planning for new machines of similar type or function.

Direct materials used in producing the component parts and the efficacy of the related process will be examined in the same detail in which the machinery is examined. Such factors as scrap, rework, total cost, and cost variances will be detailed. Indirect materials, that is, those used in support of the production operation, will be subjected to an in-depth study of function and utility. Ascertaining energy effectiveness should be part of the resource analysis.

Another important resource is material handling devices. As is seen later, the ability to provide material at each step of a production operation and to remove the finished article from each operation is an important consideration. And in optimizing the use of materials, it is necessary to examine continually the material handling engineering and the exercise of this function.

The most important resource is the manpower involved in both direct and support operations. The entire system should be critiqued in terms of opportunities to use more effectively the skilled and engineering manpower.

Having examined all of the important details of the system, the manufacturing engineer must critique the total system and the interactions among the individual elements. Recently, new dimensions have been added to the required analyses, namely, the impact of regulation upon the total system planning. Thus beyond the normal cost-effectiveness judgment must be a critique of the ability of the system to respond to regulations for occupational safety and health, environmental protection, toxic substances and energy, among others.

Developing and Promulgating Engineering Alternates

The second step leading to the optimal utilization of resources is to develop a list of engineering alternates responsive to the opportunities identified as part of the analyses of the original system. These alternates must be explored in sufficient depth to allow the assessment of the cost and engineering improvements that can be realized from new or revised manufacturing operations. This includes construction of new facilities, modification of new facilities, or running change within existing machinery, materials, or processes.

The outgrowth of this engineering analysis, and the third step in the optimal utilization of resources, should be the promulgation of the selected alternates as recommended practices to those activities within a given organization that can either act on the listing of opportunities or provide guidance to those who can respond. Circulation of detailed and summary reports within the manufacturing planning activities by means of manufacturing standards and provisional improved processes is an important adjunct to the development of the engineering alternate. Equally important is providing summary reports to activities that carry out technological forecasting or product planning specifically aimed at future products and business opportunities. The manner in which these data are incorporated into a master corporate plan is discussed in Chapter 11.1.

7.1.4 PLANNING AREAS

The systematic approach to planning for manufacturing that has been outlined has found application in all branches of manufacturing. The exact steps may be modified to reflect operation size, unusually high or low production volumes, and specific processes. This treatment is recommended for use in discrete component manufacturing, in process industries, and in hybrid combinations such as heat treating operations in a machining and small-part assembly plant (e.g., an automotive transmission plant).

Exhibit 7.1.4 is a representative display of the entire spectrum of manufacturing processes with which the manufacturing engineer deals. Although the introductory paragraphs in this chapter focus on metal removal and working, enlarging the scope of manufacturing engineering to encompass all useful manufacturing processes will make the well-established engineering methodology available to any producing operation.

In logical sequence the exhibit illustrates (1) the flow of materials from the process industry through discrete part manufacturing to assembly and (2) the breadth of technical expertise demanded of today's manufacturing engineer.

Process industries are characterized as processors of bulk or nonspecific materials, often starting as ores, petroleum crudes, and similar basic materials. Even when incorporated into discrete part manufacturing, process equipment remains relatively nonflexible, with a limited identification with the final product.

Exhibit 7.1.4 Manufacturing Processes Spectrum

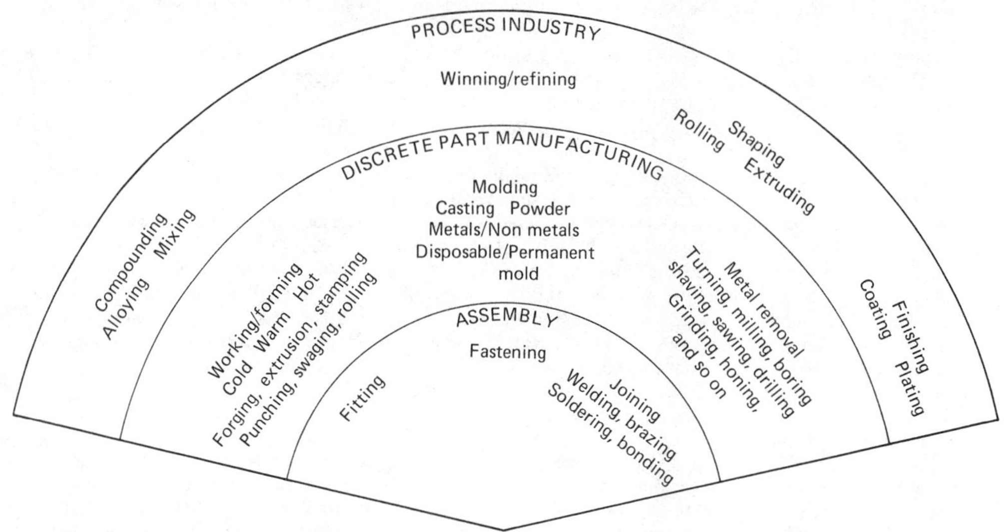

Discrete part manufacturing, on the other hand, produces functional components whose identity is retained through one or many operations from rough form to finished part or assembly.

In the production of most complex products, the processes used will include some combination of process industry and discrete part manufacturing. The diversity of skills required to coordinate, plan, activate, and control such a wide range of manufacturing methods emphasizes the need for a carefully organized manufacturing engineering organization.

7.1.5 MECHANICS AND TIMING OF THE MANUFACTURING PLANNING AND CONTROL FUNCTION

Manufacturing Feasibility

As has been shown in the preceding exhibits, the manufacturing engineer is the principal contact for the product engineer in ensuring the faithful translation of design intent from the blueprint to the production floor. Discussions between these two engineering groups broadly fall within the scope of what may be called manufacturing feasibility. Even though this particular subfunction is most important in large, highly structured operations, there is a need for a formal consideration of issues affecting manufacturing feasibility in operations of all sizes. "Manufacturing feasibility" is defined as an engineering step that examines each component and each assembly within a product design in light of the manufacturing capability to produce to design intent. It is obvious that manufacturing feasibility of the entire product or of individual components can be examined as a guide to the product engineer during the design phase. It should be emphasized that, in the case of an existing plant, manufacturing feasibility is affected by existing machinery, layout, and manpower.

Manufacturing feasibility can be established most effectively when the product designer and the manufacturing engineer work together closely to initiate and modify designs through the entire design sequence until final release to production operations. This same relationship should continue during the production life of the design and during all subsequent modifications to meet the demand of the marketplace or to improve the cost, quality, or reliability of the product. In some cases these modifications will be carried out during normal cost improvement efforts or in other more concentrated efforts, such as value engineering.

Formal response to issues of manufacturing feasibility remains the responsibility of the product engineering group within the individual operation. Considerations for manufacturability, producibility, and related terminologies are reflected in detail in Chapter 7.2. Certainly the starting point for effective manufacturing is effective design for manufacturing. Moreover, the dramatic growth in new materials and new processes places additional responsibility on the product engineer and the manufacturing engineer alike to ensure that adequate consideration is given to the new opportunities that arise from these new methods of doing business.

Another formalized, detailed, cooperative look for designing for manufacturing is carried out in a semidiscipline that has come to be known as value engineering. Discussions in Chapter 7.3 provide details of a formalized approach to design review of existing designs. Again, the need for a close working relationship between the product engineer and the manufacturing engineer is apparent. The success of all efforts to enhance the manufacturing feasibility of proposed or existing design will hinge upon the engineering skills of the participants and their flexibility in interpersonal relationships.

Process and Equipment Selection

The heart of the entire manufacturing engineering and planning for manufacturing function is the selection of process and equipment to be used to meet the design intent established by the product engineer. Even in the case of an existing product, the manufacturing engineer should examine options available to him or her to improve the quality of the product and/or to reduct its cost during the manufacturing operation. The formal steps in this process and equipment selection are:

1. Developing a general statement of the manufacturing operations to be performed.
2. Establishing a provisional process.
3. Developing a list of process alternates.
4. Selecting the production process.
5. Communicating the process selection to other affected activities.
6. Performing detailed processing.

Developing a General Statement

The general statement of a manufacturing procedure is system oriented in that the type of equipment required to provide the design feature is generically identified. Thus, if the preliminary engineering suggests that a transfer line would be required to perform a given manufacturing operation, a statement to this effect helps the overall manufacturing strategy to emerge.

Since in most cases there is an existing, operating model of the conceptualized facility, the first step in developing the general statement is to accumulate as much of the historical and experiential background data as is practical. Thus operating experience of an existing production facility is the logical starting point. The most useful assemblage of these data can be made when an identical product is referenced. This will allow a detailed examination of the production rate, both planned and actual; the machine performance in terms of quality and maintainability; the cost of maintenance; and the anticipated life of the machine. In cases where such facilities are not available in a particular operation, it may be necessary to estimate these same characteristics in order to provide a clear statement of the base from which the new process and equipment selection is made. Another source of sound engineering information is the analogical reasoning from sister industries performing similar types of work, but on different products, or, in many cases, information from industries that are quite dissimilar to the particular operations under study. At times it may be desirable to extract the general statement of the proposed facilities from a pilot operation that has been used to demonstrate the feasibility of a manufacturing process concept such that the confidence level of the manufacturing engineering and the manufacturing management is raised to the point where the new process can be recommended for production.

Establishing a Provisional Process

There may be cases where it is not necessary to develop a formal, general statement of the production system to be established, even though the discipline of such an exercise is an important adjunct to the total planning for manufacturing. In any case the second and mandatory step is to establish a provisional process to provide each individual feature identified by the product designer. Several additional inputs are necessary before beginning the selection of the provisional process. Specifically, one must do the following:

1. *Establish targets for facility and piece costs.* Accept or develop cost targets by which to judge the effectiveness of the manufacturing engineering.

2. *Specify the raw material.* The blueprint may or may not specify the exact form, size, and chemistry of the raw material. Thus it will be necessary for the manufacturing engineer to add to the simple material specification of hot rolled steel, casting, forging, or the like a more precise definition of the chemistry of the material; any chemical or thermal treatment to parts prior to delivery to the manufacturing operation; an exact identification of the form and size, in the case of standard shapes or preforms; or the amount of stock removal from castings, forgings, or other rough shapes.

3. *Determine hourly production volume preparatory to establishing machine capacity.* It will be necessary for the manufacturing engineer to determine the required net hourly production rates, based on anticipated usage and operational factors such as annual work hours. During the course

of this determination, the manufacturing engineer also must take into account established efficiency factors as they relate to both equipment and manpower.

4. *Establish timing.* Prior to selecting a provisional process, the manufacturing engineer must obtain an indication of the timing, that is, the required delivery date for the first production pieces and for full-scale production. Comparing timing requirements with timing commitments from machinery and equipment suppliers will allow an early assessment of potential trouble spots if the suppliers are unable to meet the timing required by the marketplace as interpreted by the manufacturers.

5. *Select the provisional process.* As part of the selection of the provisional process, the manufacturing engineer will estimate the number of steps and consequent stations necessary to provide all of the design features identified on the blueprint. This will require visualization of each individual sequence, an estimate of manpower required, and a rough approximation of the necessary layout provisions to accommodate each step of the process. It is at this point that reference to the previously developed experiential statements will be most helpful, as the exact steps required in the processing are identified. It is readily seen that actual process experience will assist in the judgmental selection of the provisional process. An integral part of selecting the provisional process is to identify the engineer's confidence level for each step of the process. Certainly, even for preliminary processing, no process should be selected with a confidence level below that acceptable to the particular management (0.92 is a suggested minimum).

6. *Develop judgmental costs.* Based on the provisional process, the manufacturing engineer will develop judgmental costs of facilities and materials and, with the assistance of the industrial engineering function, will develop the preliminary piece cost of the components for the final assemblies. These costs will be developed using the same broad methodology that is described under detailed processing.

7. *Compare with cost targets.* Prior to extensive planning and to commitment for facilities, it is essential that the manufacturing engineer compare cost projections of the provisional process analysis with the targeted facility cost and piece cost. A favorable comparison may support prompt decision making; at the very least, the comparison assists in subsequent analysis.

Developing a List of Process Alternates

Upon completing the provisional processing steps, the manufacturing engineer should develop a list of process alternatives, particularly for those areas where detailed analysis of the preliminary processing has shown high cost, questionable performance, or places where the confidence level of achieving the requirements of the individual operation is judged to be marginal. Of assistance in developing these opportunities are the historical data relative to similar operations. Thus the first source of process alternates for the manufacturing engineer is the identification of problem areas during the experiential and historical survey to be addressed in the course of the new process planning. Second, the new product and/or process should be examined in light of possible successful pilot operations that have been run on the same or similar applications. A third source for new approaches is manufacturing research and development directed at providing new processes and new materials for use in products similar to those under consideration. Finally, many innovations have been introduced into the various manufacturing operations through the use of analogical reasoning from similar industries and processes. This, of course, requires a clear identification of the similarities between observed and newly proposed operations. Often even less-than-complete knowledge of the analogical model can direct the engineer in a new, fruitful direction.

Selecting the Production Process

A careful step-by-step comparison between each phase of the provisional process with each phase of the alternative process will allow the manufacturing engineer to select the compromised position, which optimizes all the elements of cost, quality, flexibility, and inherent risk. Flexibility not only includes the ability to produce similar but also different parts, but also infers the possibility of increasing rates of production at a minimal cost. All engineering management and manufacturing considerations being equal, production processes will be chosen on the basis of the most favorable return on investment or other financial criteria.

Communicating the Process Selection

As noted previously, an ideal position for the manufacturing engineer is one that provides coordination and communication of engineering decisions to all appropriate affected areas. Thus, upon the completion of process selection, it is incumbent upon the manufacturing engineer to communicate the results of his or her analytic work to the product engineer as an additional insurance that the design intent is met. At the same time, the following areas, among others, should be notified of the process decision:

Industrial engineering
Plant and maintenance engineering
Industrial relations
Finance
Operations

Communication is particularly vital when new processes or machinery are involved. The interaction of all the elements previously cited is essential to the successful adaptation of new technology to existing plant and staff.

Performing Detailed Processing

When the process has been selected and communicated to all affected components, the final detailed processing upon which all actions depend is initiated. As the detailed processing proceeds, the manufacturing engineer will make extensive use of the body of engineering knowledge that resides with the machine and equipment suppliers. Chapters 7.4 through 7.11 illustrate the detailed engineering knowledge available within the vendor and user industries for the guidance of the manufacturing engineer. The logical flow from manually operated, transported, and controlled operations may be seen as the chapters' topics move from conventional machine tools through automation, numerical control machines, and group technology to computer-aided manufacturing, where machine operation and part movement are under computer control. The trend toward integration of all manufacturing functions (production, inspection, and part control) into computer-managed systems is apparent. Inspection of the subject matter does show a major emphasis on metal removal. Space does not allow a full treatment of other disciplines. Consequently, the individual engineer must recognize that a parallel body of knowledge exists for each major process technology, such as casting, metal working, joining, and assembly, as was shown in Exhibit 7.1.4.

As in the establishment of the provisional process, the starting point in establishing the detailed processing is to assemble the latest information relative to product design, production rate, facility cost, and part cost targets. The detailed processing follows the same format as did the provisional processing, except that each element of the process will be completely identified and documented through the use of process estimate sheets, such as the one shown in Exhibit 7.1.5. Information is included on the source of the material for each process from rough stock to finished piece. The subsequent operation should be identified on each process sheet so that there can be an orderly part flow through each of the manufacturing operations.

To examine the required amount of processing documentation, a particular example—a machining operation—will be used. The basic principle involved is that the instructions detailed in the manufacturing engineering document, that is, the process sheet, be sufficiently explicit that operational personnel can perform every function necessary to produce the finished component and that operations can establish staffing and piece cost from which to judge operation efficiency during and after physically launching the operation. The process sheet contains columns for recording the operation number, a description of the operation, numbers and types of machines required, effective operational rates, labor distribution, and minute costs; provision is also made for recording facility and tool costs. In large operations the responsibility for these individual entries will be divided up according to skill: The process engineer will fill out those portions of the process sheet that relate to establishing the process and selecting the machinery; the plant engineer will enter data relating to the installation cost of the machinery; and the industrial engineer will appropriately ascertain and register the direct labor minutes of each operation. In smaller operations the process and/or industrial engineer may fill out all of the portions of the process sheet. In any event, the finished document, or some document similar to the process sheet, should be used as a means of communicating with each part of the manufacturing planning and operating system.

The sample entry on the process sheet in Exhibit 7.1.6 shows the manner in which the manufacturing engineer describes each step or station in the manufacturing process. The engineer numbers and names the operation, namely, drill. He or she indicates the features, namely, the holes, and establishes their limits, that is, their size and depth. With this information, it is possible to identify the machine or machines required to carry out this specific step. In the example the machine is a single-spindle drilling machine; based on the features required in the part, the machine can be a single-spindle or multiple-spindle and can be part of additional automation, that is, a transfer line or a dial machine. All of this information is included on the process sheet. Numbers of machines required to meet the production volume and the space required complete the description of this individual process step. If fixtures or tools are required, they should be individually listed and costed as durable tools or special tools in the space provided. Each subsequent operation is identified in this exact detail until the part is finished. The initial data on the process sheet are extended to complete the analysis, providing a single document that contains all of the basic manufacturing engineering detail.

Exhibit 7.1.5 Typical Process Sheet

PLANT: ____________ **PROCESS ESTIMATE SHEET** DEPARTMENT: ____________

PROGRAM OR ECR NO.	PART NAME		ISSUE DATES	PART NO.	
FOR MODELS	MATERIAL	WT./LBS. RGH. FIN.		RELEASE	SHEET OF

OPER. NO.	OPERATION DESCRIPTION	TOOL – MACHINE – EQUIPMENT DESCRIPTION – TOOL OR B.T. NUMBER	MACH'S. REQ'D.	NET HOURLY CAPY.	EST. MINUTES	FACILITY AND DURABLE TOOL COST				SPECIAL TOOL COST				EXPENSE COST
						TOTAL	BASIC	FREIGHT	INSTAL-LATION	TOTAL	DESIGN	BUILD	INST. TRYOUT	
TOTALS														

REMARKS:

		PROCESS ENGR.	PLT. LAYOUT	AUTOMATION	DESIGN	MATL. MDLG. ENGR.	DAILY SERVICE	REQ'D. PER VEHICLE	NEXT ASSY:	OPER. NO.
Ford TCH 7745	MFG. ENG. JULY '62	INDUSTR. ENGR.	LAB.	QUAL. CONTR.	PLT. ENGR.	PRODN.	DAILY PLT. PLANNING VOLUME	REQMTS. PC/HR. HRS.	SUPERSEDES:	

With parts of complex configuration, or parts requiring a large number of operations, the processing specialist may find it advantageous to construct a tolerance chart to represent the interaction between operations graphically. By incorporating machine capability into this chart, tolerance stack-ups in excess of engineering requirements can be identified during the manufacturing engineering stage rather than after production start-up, when corrections and the accompanying time delay are so costly.

Information relative to the cost of the machinery is entered based on experience with previous purchases or on estimates available from the suppliers. Although estimates of machine and equipment delivery are not entered here, the manufacturing engineer is responsible for accumulating them so that the timing may be established and met. Analysis of these process sheets also allows the manufacturing engineer and the industrial engineer to develop a list of the manual tasks and to extend these data to provide minute cost and, ultimately, direct and indirect labor cost as a basis for developing piece costs.

The manufacturing engineer also documents on the process sheet the tools, dies, and fixtures required for each process step. These data are then assembled in such a way that listings of the tools, dies, fixtures, material handling, and support equipment can be generated as a basis for further engineering by specialists. Engineering specialists in large organizations will supervise the designing and building of special tools, dies, and fixtures, along with appropriate support and material handling equipment. Smaller organizations will depend on engineer generalists, suppliers, or contracted consultants to provide these special services.

Developing a Launch Plan

Upon complete extension of the process sheet to include all the elements of engineering that go into the planning for manufacturing, the responsible manufacturing planner (engineer) should develop an activation or launch plan. This plan is essential, irrespective of the size of the operation. Basically it identifies and provides the timing for each event that must take place from the inception of the planning until the plant is operating according to plan. The first elements of the launch plan begin to fit into place as the detailed process planning proceeds. The major elements, however, can be identified only as the process planning is well documented. The manufacturing engineer must coordinate the launch plan with each element of the manufacturing organization so that agreement is reached on all functions to be performed. This responsibility is not exclusively a manufacturing engineering function and can be performed by other segments of the organization as appropriate. The function, however, is essential for adequate manufacturing planning to meet target costs and timing.

Emerging directly from the activation planning is the overall program management, which monitors each element of the activation plan. The manufacturing engineer must continue to feed information to, and receive it from, each element of the planning system as the engineering becomes increasingly specific. Other functions, such as industrial engineering, plant engineering, material handling, quality control, and production control, continue to update their analyses and contributions to the process sheet as the planning advances. The assumption is made in this particular description of the manufacturing engineering function that there are at least some special machines or tools required to perform the function. Consequently, the engineering planning is done over a relatively prolonged period.

In many major industries this activity is carried on concomitantly with product development in such a way that the product engineering intent is constantly updated during the planning for manufacturing. Each segment of the manufacturing operation thus provides detailed planning for its engineering functions as the facility planning moves toward production. The plant engineering will provide complete planning for building design, construction, and delivery of the various utilities to the individual pieces of machinery and equipment in the plant. Detailed layouts must be developed within the plant engineering and the manufacturing engineering activities so that all preliminary work can be done in preparation for the actual plant construction.

The material handling function must provide for the transport of raw materials to the first operation and for the delivery of the finished product to the shipping location. In addition to identifying the flow of material, the machines and equipment required to handle the materials and parts effectively must be identified.

Quality control and reliability procedures must be established in sufficient detail so as to facilitate the procurement of inspection and analytic devices and to identify the cost impacts of the quality control procedures.

Production control must develop a plan for procuring and using direct materials (those materials that become the product) and indirect materials (those that support the machinery and equipment required to operate the facility). Indirect materials include lubricants, abrasives, cleaning materials, cutting tools, and maintenance parts.

Exhibit 7.1.6 Sample Process Sheet Entry

PROCESS ESTIMATE SHEET

PLANT: ____ DEPARTMENT: ____

PROGRAM OR ECR NO.	PART NAME		ISSUE DATES	PART NO.	
FOR MODELS	MATERIAL	WT./LBS. RGH. FIN.		RELEASE	SHEET OF

OPER. NO.	OPERATION DESCRIPTION	TOOL – MACHINE – EQUIPMENT DESCRIPTION – TOOL OR B.T. NUMBER	MACH'S. REQ'D.	NET HOURLY CAP	EST. MINUTES	FACILITY AND DURABLE TOOL COST				SPECIAL TOOL COST				EXPENSE COST
						TOTAL	BASIC	FREIGHT	INSTAL-LATION	TOTAL	DESIGN	BUILD	INST. TRYOUT	
10	DRILL (3) LOCATOR HOLES .500" D x 1.5" DEEP	SINGLE SPINDLE DRILL PRESS- LOCATING FIXTURE REQUIRED	1	16	3.75									
TOTALS														

REMARKS:

		PROCESS ENGR.	PLT. LAYOUT	AUTOMATION	DESIGN	MATL. MDLG. ENGR.	DAILY SERVICE	REQ'D. PER VEHICLE	NEXT ASSY:	OPER. NO.
		INDUS. ENGR.	LAB.	QUAL. CONTR.	PLT. ENGR.	PRODN.	DAILY PLT. PLANNING VOLUME	REQMTS. PC/HR. HRS.	SUPERSEDES:	

Ford TCH 7745 MFG. ENG. JULY '62

Initiating Procurement

Parallel to the activation and monitoring of the detailed planning, the responsible engineering activity should initiate procurement documents. This is a vital step in the planning for manufacturing and must be considered very early, even during the detailed processing, if items whose delivery times are excessively long are involved. For example, if large transfer lines are required to meet the system plan, time must be allocated for the design and building of these highly sophisticated special machine tools. The sequences leading to the procurement are vital to the successful launching of plants using other than general-purpose equipment. Although the manufacturing engineer must take the lead in initiating procurement of the production equipment, a parallel effort must be made by all organizational components requiring procurement action.

The procurement function, generally vested in a purchasing department, cannot be initiated until a clear definition of the particular equipment or machinery is developed. The manufacturing engineer and the related organizational components must develop specifications that clearly describe each individual item to be purchased. When a standard or catalog machine can be used without modification, specification is relatively simple. In a normal commercial situation, however, such directed procurement may not provide the advantages of competitive bidding by potential suppliers. Further, many manufacturing engineering organizations have found it desirable, and even necessary, to use industry standards, such as joint industry standards, to meet mandated machinery performance or to provide a more guaranteed performance.

Thus, in general, even for the procurement of standard or shelf items, many manufacturing engineers develop a sufficiently detailed specification that translates their own experience into the new machinery and equipment that is to be procured. In the case of the special machine tools, the specification must include:

Required delivery.

Production rate.

Machine configuration.

Preferred components, including such items as spindles, controls, and type.

Degree of automation to furnish parts to, and remove parts from, the special machine.

Standards of machine acceptance.

Mandated requirements, such as air and water pollution, noise.

In organizations where machinery and equipment procurement is done on a regular basis, a general specification may be available to be included as part of the specific requirements relative to the new equipment.

It is the responsibility of the manufacturing engineer as the coordinator for procurement of all production and support equipment to obtain appropriate authorization at all levels of management and to transmit the authorization and the specifications to the internal purchasing activity.

Upon the issuance of the appropriate purchasing document, the manufacturing engineer will continue to interpret the design intent and the process intent to the selected vendors. Without intimate day-by-day contact with the suppliers, the potential for misunderstanding is present. Failure to detect misunderstandings prior to the manufacture of the machine and equipment can result in extra cost and delay in delivery. Thus the supplier of machine and equipment becomes a part of the planning for manufacturing team. As a result of this communication, it may be necessary for the manufacturing engineer to perform a certain amount of reprocessing as problems are encountered in the final machine and equipment design.

Installation and Activation

The responsibility for installing and activating new equipment must be shared with the corporate function responsible for developing the building construction plan, including the delivery of utilities to each individual piece of equipment. This includes electricity, air, water, hydraulics, and other commonly shared utility-type materials. As the detailed machinery and equipment design becomes firm at the machine builders, these data are transmitted by the manufacturing engineer to the plant engineering organization. Equipment building proceeds under the continuous direction of the manufacturing engineer so that he or she may anticipate problems that may develop as the equipment is delivered and installed. As the equipment is completed at the vendor or supplier plant, it is inspected jointly by the vendor and the manufacturing engineer according to a previously agreed-upon testing procedure. Normally this requires a live cycling of the equipment to ensure that the pieces or components produced meet the design intent.

At this point, the approved or certified machinery and equipment is moved to the production floor and is installed according to the previously agreed-upon layout. Installation prints supplied by the vendor will spell out such items as foundations, pits, and location points for utility delivery. The vendor's responsibility, however, does not end until the machinery is installed and has produced, in the plant environment, individual parts to final design intent at the established rate.

The launching itself is a programmed, step-by-step procedure, moving from first sample to final production rate. Customarily this will occur over a period ranging from days to months, depending upon the complexities of the total system. The launching of the equipment is not independent of the need to develop special skills among the operators. The previously mentioned launch or activation plan must anticipate the availability or the lack of skilled manpower to carry out the planned operation; appropriate training programs may be required.

7.1.6 AUDIT

The responsibility of the manufacturing engineer does not end at the point of full-scale production. The function itself should encompass an adequate check, once production is established, on the adequacy of the planning and the execution of the plan. This will require detailed observations of the adherence to the original process sheets at each step of the manufacturing operation.

A quick assessment of the overall planning can be made by comparing the original manufacturing system targets with the gross statistics of production rate, component and assembly quality, projected unit costs, and actual costs. A more detailed audit in retrospect may be useful in identifying the degree of adherence to good engineering and cost-effective practice by the manufacturing planners. Of particular value will be the examination of the total manufacturing system in terms of new information gathered during the course of the total planning effort. This may result in an identification of opportunities for system optimization by modifications of processes, automation, or controls for use in existing facilities or in planning new facilities.

Audit of the plan's execution does require a more in-depth examination of individual machine or system performance and will probably be done on a random basis, with emphasis on the more costly and more troublesome operations. The standard of comparison remains the approved process sheets.

In any given system under audit, the use of all the resources of production should be examined. Thus the utilization of machines and equipment, basic processes, materials, and manpower will be brought under scrutiny.

7.1.7 FEEDBACK

Results from the planning and execution audits will allow an organized identification of opportunities for future planning. The feedback of this updated experiential data to the manufacturing planning activity becomes the starting point for new facility planning and closes the information flow loop shown in Exhibit 7.1.2.

Even more important, the listing of opportunities facilitates another vital responsibility of the manufacturing engineer, namely, to identify, support, sponsor, and/or execute advanced manufacturing engineering and manufacturing research and development (R & D) efforts in support of future manufacturing operations. This vital responsibility is the lifeline through which manufacturing will meet corporate and national goals of increased productivity. Research and development aimed at developing a clearer understanding of all the resources available to the manufacturing engineer is the starting point for effective R & D programs.

From this starting point, the manufacturing engineer should be encouraged to develop specific R & D programs in support of future manufacturing activities. The execution of such programs may rest with many other groups in industry or universities, but the manufacturing engineering input is vital. Since the basic premise of the R & D programs is that new technology, innovation, and sound engineering will reduce costs, improve quality, and assist in meeting delivery, the manufacturing engineer must also participate in the decision to bring the results of R & D effort to the production floor–using the same basic manufacturing engineering processing procedures described herein.

There is one basic caution–the timing from concept to production application must not be underestimated. Prudent planners estimate that completely new products and processes require 7 to 13 years to advance from concept to production.

REFERENCE

1. *Manufacturing Engineering Defined–Special Report to the Membership of SME, Society of Manufacturing Engineers*, Dearborn, MI, June 1978.

CHAPTER 7.2

Designing for Manufacturing

BENJAMIN W. NIEBEL

The Pennsylvania State University

7.2.1 INTRODUCTION

The objective of designing for manufacturing is to produce a design that will satisfy both functional and physical objectives at a cost that is compatible to the user. Thus the design itself must be producible. Designing for manufacturing as a discipline implies producibility, which involves adapting a design without altering its performance objectives so that it can be produced at the lowest cost with available materials in the shortest total time. Sound designing for manufacturing represents cost-effectiveness practiced by the design engineer by incorporating producibility during the conception, development, and production phases of the life cycle of the product. To produce designs of the highest quality at competitive costs, designing for manufacturing must take place in concert with functional design. By so doing, performance objectives will not be compromised or adversely affected by factors that could be introduced to maximize producibility. There must be an ongoing concern for producibility beginning at concept formulation and continuing throughout the entire design effort.

7.2.2 SPECIFIC OBJECTIVES

The incorporation of sound designing for manufacturing throughout the life cycle of a product is based on an effort to meet established delivery dates and to minimize unit costs, tooling costs, elaborate test systems, use of high-cost and/or critical processes, design changes in production, and use of limited availability items. Sound designing endeavors to maximize design simplicity, standardization of materials and components, product inspectability, product testing, safety in production, and competitive procurement. The larger the production quantity, the more important that creative design for manufacturing take place.

7.2.3 DRAWINGS

Drawings represent the heart of designing for manufacturing since they are the principal means of communication between the functional designer and the producer of the design. They alone control and completely delineate shape, form, fit, finish, function, and interchangeability requirements that lead to the most competitive procurement. An engineering drawing, when supplemented by reference specifications and standards, should permit a competent manufacturer to produce the part shown within the dimensional and surface tolerance specifications provided. It is the engineering drawing that should demonstrate the most creative design for manufacturing thinking.

Certain product specifications may not be included on the drawings in view of space constraints. Product specifications such as quality assurance checkpoints, inspection procedures, and general design criteria may be separately summarized, but should always be cross-referenced on the engineering drawing. At all times the design engineer must remember that the end product drawing is the communication medium between the design engineer and the producer. It is the basis for interchangeability for repair parts; it provides the form, fit, and function to the manufacturing function.

Too often the language of drawings is incomplete. For example, chamfers may be indicated, but are not dimensioned; worse yet, they may be desired, but are not even shown. Frequently the finish desired is omitted. Complex coring may be incorrectly shown. The principal errors common to many designs are as follows:

1. Design is not conducive to the application of economic processing.
2. Designer has not taken advantage of group technology and creates a new design for an already existing item.

3. Design exceeds the manufacturing state of the art.
4. Design and performance specifications are not compatible.
5. Critical location surfaces have not been established.
6. Design specifies the use of proprietary items.
7. Design specifications are not definitive.
8. Inadequate consideration has been given to measurement problems.
9. Tolerances are more restrictive than necessary.
10. Item has been overdesigned.

7.2.4 THE DESIGN PROCESS

The design process should be undertaken systematically. Although the procedure will vary somewhat from one concern to another, the sequence of design should include the following phases: initial conception and evaluation, analysis, general design, detailed design, hardware development, and production of the design. The design process is illustrated graphically in Exhibit 7.2.1.

Throughout the design process consideration should be given to producibility so that the end product will be designed for manufacture. Without a continued effort toward producibility, there will invariably be deficiencies in the design itself. Notable among these deficiencies is excessive complexity. For example, the design may be stronger than actually required or heavier than need be. It could call for complex locking or indexing mechanisms when much simpler ones would suffice. The more complex a design, the more the chance for design error. Simpler designs not

Exhibit 7.2.1 The Engineering Design Process

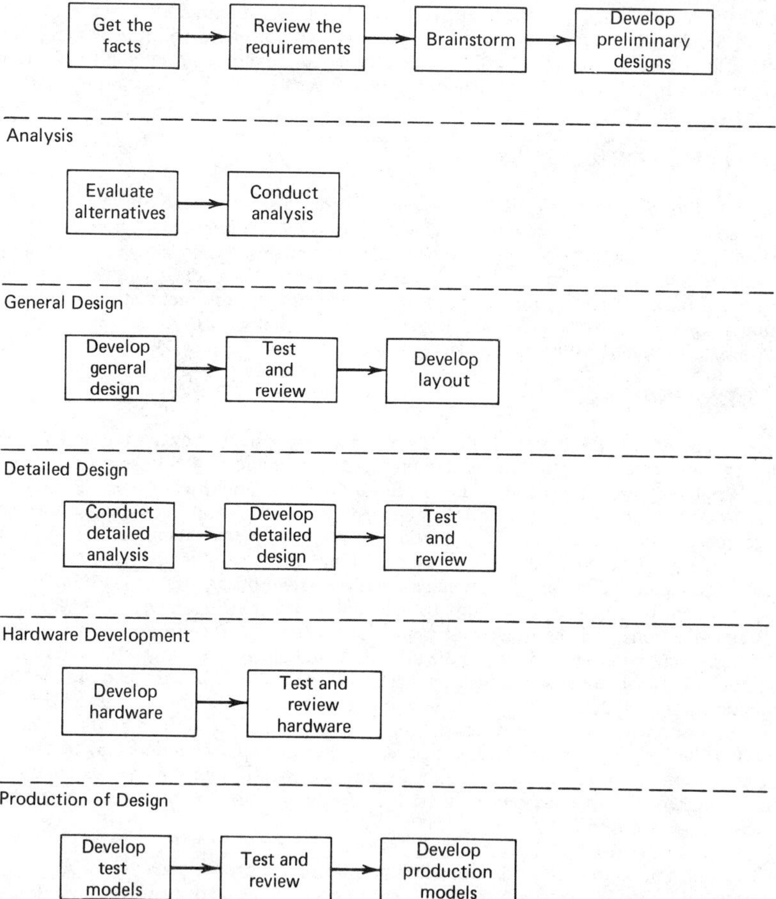

only reduce costs, but usually are more reliable and more easily maintained. The quality and performance of the simpler design will exceed the performance of the more complex design.

A second common deficiency in the design process that frequently occurs when producibility has not been exploited is production restrictiveness. The functional designer inadvertently dictates the method by which his or her design is to be produced. Because of inexperience or lack of knowledge, the designer may be unaware of alternative materials or processes, which may be detrimental to the design itself. The creative designer who works in concert with the manufacturing engineer interested in producibility will search for diverse ideas, explore new concepts, and seek design simplicity while avoiding limiting generalizations.

For example, the common errors of inconsistent double dimensioning, specifying a material that is not compatible with a desired process, specifying a threaded inside diameter to the bottom of a blind hole, and specifying a ground tolerance on an outside diameter to a shoulder represent the inattention to producibility.

7.2.5 PROCESSES AND MATERIALS FOR PRODUCING THE DESIGN

The selection of the ideal processes and materials with which to produce a given design cannot be an independent activity. It must be a continuing activity that takes place throughout the design life cycle–from initial conception to production. Material selection and process selection need to be considered together; they should not be considered independently.

In considering the selection of materials for an application, it usually is possible to rule out entire classes of materials because of cost or their obvious inability to satisfy specific operational requirements. But even so, with the acceleration of material development there are so many options for the functional design engineer that optimum selection is at best difficult. The suggested procedure for organizing data related to material selection is to divide it into three categories: properties, specifications, and data for ordering.

The property category usually will provide the information that suggests the most desirable material. A property profile is recommended, where all information, such as yield point, modulus of electricity, resistance to corrosion, and so on, is tabulated. Those materials that qualify because of their properties will stand out.

Each material will have its own specifications on the individual grades available and on their properties, applications, and comparative costs. The unique specifications of a material will distinguish it from all competing materials and will serve as the basis for quality control, planning, and inspection.

Finally, the data needed when physically placing an order needs to be maintained. This includes minimum order size, quantity breakpoints, sources of supply, and so on.

In the final selection of a material, cost of the processed material needs to be considered–hence the need for close association between material selection and process selection in connection with design. Design evaluation invariably is in terms of a processed material cost, which may be derived by analyzing the involved processing steps, including setup and lead time costs along with the preprocessed material cost.

7.2.6 DESIGNING FOR BASIC PROCESSES–METAL

Liquid State

Early in the planning of the functional design, one must decide whether to start with a basic process that uses material in the liquid state, such as a casting, or in the solid state, such as a forging. If the engineer decides a part should be cast, he or she will have to decide simultaneously which casting alloy and process can most nearly meet the required dimensional tolerance, mechanical properties, and production rate at the least cost.

Casting has several distinct assets: the ability to fill a complex shape; economy when a number of similar pieces are required; and a wide choice of alloys suitable for use in highly stressed parts, where light weight is important or where corrosion may be a problem. There are inherent problems, too, including internal porosity, dimensional variations caused by shrinkage, and solid or gaseous inclusions stemming from the molding operation. However, most of these problems can be minimized by sound design for manufacturing.

Casting processes are basically similar in that the metal being formed is in a liquid or highly viscous state and is poured or injected into a cavity of a desired shape.

The following design guidelines will prove helpful in reducing casting defects, improving their reliability, and assisting in their producibility:

1. When changes in sections are required, use smoothly tapered sections to reduce stress concentration. Where sections join, use generous fillets and blending radii.

2. Machining allowances should be detailed on the part drawing so as to ensure adequate stock and to avoid excessive differences in casting thickness.

3. Remember that, when designing castings to be produced in a metal mold or die, convex forms are easy to mill, but concave notches are both difficult and expensive.

4. Raised lettering is simple to cut into a metal mold or die; depressed lettering will cost considerably more.

5. Avoid the design of thin sections since they will be difficult to fill.

6. To facilitate the secondary operations of drilling and tapping, cored-through holes should have countersinking on both ends of the holes.

7. Avoid large, plain surfaces. Break up these areas with ribs or serrations to avoid warpage and distortion.

8. For maximum strength, keep material away from the neutral axis. Endeavor to keep plates in tension and ribs in compression.

Exhibit 7.2.2 identifies the important design parameters associated with the various casting processes and provides those limitations that should be incorporated by the functional designer to ensure producibility.

Solid State

A forging, as opposed to a casting, is usually used because of improved mechanical properties, which are a result of working metals into a desired configuration under impact or pressure loading. A refinement of the grain structure is another characteristic of the forging process. Hot forming breaks up the large dendritic grain structure characteristic of castings and gives the metal a refined structure, with all inclusions stretched out in the direction in which plastic flow occurs. A metal has greater load carrying ability in the direction of its flow lines than it does across the flow lines. Consequently, a hot-formed part should be designed so that the flow lines run in the direction of the greatest load during service.

An extension of conventional forging known as precision forging can be used to acquire geometric configurations very close to the final desired shape, thus minimizing secondary machining operations.

Guidelines that should be observed in the design of forging in order to simplify its manufacture and help ensure its reliability are as follows:

1. The maximum length of bar that can be upset in a single stroke is limited by possible buckling of the unsupported portion. The unsupported length should not be longer than three times the diameter of the bar or distance across the flats.

2. Recesses in depth up to their diameter can be easily incorporated in either or both sides of a section. Secondary piercing operations to remove the residual web should be utilized on through-hole designs.

3. Draft angle should be added to all surfaces perpendicular to the forging plane so as to permit easy removal of the forged part. Remember that outside draft angles can be smaller than inside angles since the outside surfaces will shrink away from the die walls, and the inside surfaces will shrink toward bosses in the die.

4. Deeper die cavities require more draft than shallow cavities. Draft angles for hard-to-forge materials, such as titanium and nickel base alloys, should be larger than when forging easy-to-forge materials.

5. Uniform draft results in lower-cost dies, so endeavor to specify one uniform draft on all outside surfaces and one larger draft on all inside surfaces.

6. Corner and fillet radii should be as large as possible to facilitate metal flow and minimize die wear. Usually 6 mm is the minimum radius for parts forged from high-temperature alloys, stainless steels, and titanium alloys.

7. Endeavor to keep the parting line in one plane since this will result in simpler and lower-cost dies.

8. Locate the parting lines along a central element of the part. This practice avoids deep impressions, reduces die wear, and helps ensure easy removal of the forged part from the dies.

Exhibit 7.2.3 provides important designing for manufacturing information for the major forging processes.

Other Basic Processes

In addition to casting and forging, there are several other processes that may be considered basic since they impart the approximate finished geometry to material that is in the powdered, sheet, or rod shape form. Notable among these processes are the following: powder metallurgy, cold head-

Exhibit 7.2.2 Important Design Parameters Associated With Various Casting Processes

	Casting Process							
	Sand Casting							
Design Parameter	Green	Dry/Cold Set	Shell	Plaster (Preheated Mold)	Investment (Preheated Mold)	Permanent Mold (Preheated Mold)	Die (Preheated Mold)	Centrifugal
Weight	100 g to 400 MT	100 g to 400 MT	100 g to 100 kg	100 g to 100 kg	Less than 1 g to 50 kg	100 g to 25 kg	Less than 1 g to 30 kg	Grams to 200 kg
Minimum section thickness	3 mm	3 mm	1.5 mm	1 mm	0.5 mm	3 mm	0.75 mm	6 mm
Allowance for machining	Ferrous–2.5 to 9.5 mm; non-ferrous–1.5 to 6.5 mm	Ferrous–2.5 to 9.5 mm; non-ferrous–1.5 to 6.5 mm	Often not required; when required, 2.5 to 6. 5 mm	0.75 mm	0.25 mm to 0.75 mm	0.80 mm to 3 mm	0.80 mm to 1.60 mm	2.50 mm to 6.5 mm
General tolerance	±0.4 ~ 6.4 mm	±0.4 ~ 6.4 mm	±0.08 ~ ±1.60 mm	±0.13 ~ ±0.26 mm	±0.05 mm to ±1.5 mm	±0.25 mm to ±1.5 mm	±0.025 mm to ±0.125 mm	±0.80 mm to ±3.5 mm
Surface finish (μrms)	6.0 to 24.0	6.0 to 24.0	1.25 to 6.35	0.8 to 1.3	0.5 to 2.2	2.5 to 6.35	0.8 to 2.25	2.5 to 13.0
Process reliability	90	90	90	90	90	90	95	90
Cored holes	Holes < 6 mm	Holes < 6 mm	Holes < 6 mm	Holes < 12 mm	Holes as small as 0.5 mm diameter	Holes as small as 5 mm diameter	Holes as small as 0.80 mm diameter	Holes as small as 25 mm diameter; no undercuts
Minimum lot size	1	1	100	1	20	1000	3000	100
Draft allowances	1 to 3°	1 to 3°	$\frac{1}{4}$ to 1°	$\frac{1}{2}$ to 2°	0 to $\frac{1}{2}$°	2 to 3°	2 to 5°	0 to 3°

Exhibit 7.2.3 Important Design Parameters Associated With Various Forging Processes

	Forging Process				
Design Parameter	Open Die	Conventional Utilizing Preblocked	Closed Die	Upset	Precision Die
Size or weight	500 g to 5000 kg	Grams to 20 kg	Grams to 20 kg	20 mm to 250 mm bar	Grams to 20 kg
Allowance for finish machining	2 to 10 mm	2 to 10 mm	1 to 5 mm	5 to 10 mm	0 to 3 mm
Thickness tolerance	+0.6 mm −0.2 mm to +3.00 mm. −1.00 mm	+0.4 mm −0.2 mm to +2.00 mm −0.75 mm	+0.3 mm −0.15 mm to +1.5 mm −0.5 mm	–	+0.2 mm −0.1 mm to + 1 mm 0.2 mm
Fillet and corners	5 to 7 mm	3 to 5 mm	2 to 4 mm	–	1 to 2 mm
Surface finish (μrms)	3.8 to 4.5	3.8 to 4.5	3.2 to 3.8	4.5 to 5.0	1.25 to 2.25
Process reliability	95	95	95	95	95
Minimum lot size	25	1000	1500	25	2000
Draft allowance	5 to 10°	3 to 5°	2 to 5°	–	0 to 3°
Die wear tolerance	±0.075 mm/kg weight of forging	±0.075 mm/kg weight of forging	±0.075 mm/kg weight of forging	–	±0.075 mm/kg weight of forging
Mismatching tolerance	±.25 mm ±0.01 mm/3 kg weight of forging	±0.25 mm ±0.01 mm/3 kg weight of forging	±0.25 mm ±0.01 mm/3 kg weight of forging	–	±0.25 mm ±0.01 mm/3 kg weight of forging
Shrinkage tolerance	±0.08 mm	±0.08 mm	±0.08 mm	–	±0.08 mm

ing, extrusion, roll forming, press forming, spinning, electroforming, and automatic screw machine work.

In powdered metallurgy powdered metal is placed in a die and compressed under high pressure. The resulting cold-formed part is then sintered in a furnace to a point below the melting point of its major constituent.

Cold heading involves striking a segment of cold material up to 25 mm in diameter in a die so that it is plastically deformed to the die configuration.

Extrusion is performed by forcing heated metal through a die having an aperture of the desired shape. The extruded lengths are then cut into the desired length. From the standpoint of producibility, the following design features should be observed:

1. Very thin sections with large circumscribing area should be avoided.
2. Any thick wedge section that tapers to a thin edge should be avoided.
3. Thin sections that have close space tolerance should be avoided.
4. Sharp corners should be avoided.
5. Semiclosed shapes that necessitate dies with long, thin projections should be avoided.
6. When a thin member is attached to a heavy section, the length of the thin member should not exceed 10 times its thickness.

In roll forming, strip metal is permanently deformed by stretching it beyond its yield point. The series of rolls progressively change the shape of the metal to the desired shape. In designing the extent of the bends in the rolls, allowance must be made for springback.

In press forming, as in roll forming, metal is stretched beyond its yield point. The original material remains about the same thickness or diameter, although it will be reduced slightly by drawing or ironing. Forming is based upon two principles:

1. Stretching and compressing material beyond the elastic limit on the outside and inside of a bend.

2. Stretching the material beyond the elastic limit without compression or compressing the material beyond the elastic limit without stretching.

Spinning is a metal forming process in which the work is formed over a pattern, usually made of hard wood or metal. As the mold and material are spun, a tool (resting on a steady rest) is forced against the material until the material contacts the mold. Only symmetrical shapes can be spun. The manufacturing engineer associated with this process is concerned primarily with blank development and proper feed pressure.

In electroforming a mandrel having the desired inside geometry of the part is placed in an electroplating bath. After the desired thickness of the part is achieved, the mandrel pattern is removed, thus leaving the formed piece.

Automatic screw machine forming involves the use of bar stock, which is fed and cut to the desired shape.

Exhibit 7.2.4 provides important design for manufacturing information for these basic processes.

7.2.7 DESIGNING FOR SECONDARY OPERATIONS

Just as there should be careful analysis in the selection of the ideal basic or primary process, so must there be sound planning in the specification of the secondary processes. The parameters associated with all process planning include the size of the part, the geometric configuration or shape required, the material, the tolerance and surface finish needed, the quantity to be produced, and of course the cost. Just as there are several alternatives in the selection of a basic process, so there are several alternatives in determining how a final configuration can be achieved.

With reference to secondary removal operations, several basic guidelines should be observed in connection with the design of the product in order to help ensure its producibility.

1. Provide flat surfaces for the entering of the drill on all holes that need to be drilled.
2. On long rods, design mating members so that male threads can be machined between centers, as opposed to female threads, where it would be difficult to support the work.
3. Always design so that gripping surfaces are provided for holding the work while machining is performed and ensure that the held piece is sufficiently rigid to withstand machining forces.
4. Avoid double fits in designing for mating parts. It is much easier to maintain close tolerance when a single fit is specified.
5. Avoid specifying contours that require special form tools.
6. In metal stamping avoid feather edges when shearing. Internal edges should be rounded, and corners along the edge of the strip stock should be sharp.
7. In metal stamping of parts that are to be subsequently press formed, straight edges should be specified, if possible, on the flat blanks.
8. In tapped blind holes the last thread should be at least 1.5 times the thread pitch from the bottom of the hole.
9. Blind drilled holes should end with a conical geometry to allow the use of standard drills.
10. Design the work so that diameters of external features increase from the exposed face and diameters of internal features decrease.
11. Internal corners should indicate a radius equal to the cutting tool radius.
12. Endeavor to simplify the design so that all secondary operations can be performed on one machine.
13. Design the work so that all secondary operations can be performed while holding the work in a single fixture or jig.

Exhibit 7.2.5 provides a comparison of the basic machining operations used in performing the majority of secondary operations.

7.2.8 DESIGNING FOR BASIC PROCESSES—PLASTICS

There are more than 30 distinct families of plastic from which evolve thousands of types and formulations that are available to the functional designer. However, in the fabrication of plastics, either thermoplastic or thermosetting, only a limited number of basic processes are available. These processes include compression molding, transfer molding, injection molding, extrusion, casting, cold molding, thermoforming, calendering, and blow molding. The functional designer usually gives little thought to how the part will be made. He or she is usually concerned primarily with the specific gravity; hardness; water absorption; outdoor weathering; coefficient of linear thermal expansion; elongation; flexural modulus; izod impact; deflect temperature under load; and flexural yield, tensile, shear, and compressive strengths.

Exhibit 7.2.4 **Important Design Parameters Associated With Manufacturing Information for Basic Processes**

Design Parameter	Powder Metallurgy	Cold Heading	Extrusion	Roll Forming	Press Forming	Spinning	Electroforming	Automatic Screw Machine
Size	Diameter–1.5 to 300 mm; length–3 to to 225 mm	Diameter–0.75 to 20 mm; length–1.50 to 250 mm	1.5 mm to 250 mm diameter	1 to 2000 mm	Up to 6 mm diameter	6 mm to 4000 mm	Limited to size of plating tanks	0.80 mm diameter by 1.50 mm; length to 200 mm diameter by 900 mm length
Minimum thickness	1 mm	–	1 mm	0.075 mm	0.075 mm	0.1 mm	0.0025 mm	–
Allowance for finish machining	To size	To size	To size	To size	To size	To size	To size	To size
Tolerance	Diameter– ±0.025 to 0.125 mm; length– ±0.25 to 0.50 mm	Diameter– ±0.05 to 0.125 mm; length– ±0.75 to 2.25 mm	Flatness– ±0.01 mm/in. of width; wall thickness– ±0.15 to ±0.25 mm; cross section– ±0.15 to ±0.20 mm	Cross section– ±0.050 to 0.35 mm; length– ±1.5 mm	±0.25 mm	Length– ±0.12 mm; thickness– ±0.05 mm	Wall thicknesses– ±0.025 mm; dimension– ±0.005 mm	Diameter– ±0.01 to 0.06 mm; length– ±0.04 to 0.10 mm; concentricity– ±0.06 mm
Surface finish (μrms)	0.125 to 1.25	2.2 to 2.6	2.5 to 3	2.2 to 2.6	2.2 to 4.0	0.4 to 2.2	0.125 to 0.250	0.30 to 2.5
Process reliability	95	99	99	99	99	90 to 95	99	98
Minimum lot size	1000	5000	500 ft	10,000 ft	1500	5	25	1000
Draft allowance	0	–	–	–	0 to $\frac{1}{4}°$	–	–	–
Bosses permitted	Yes	Yes	Yes	Yes	Yes	No	Yes	Yes
Undercuts permitted	No	Yes	Yes	Yes	Yes	Yes	Yes	Yes
Inserts permitted	Yes	No	No	No	No	No	No	No
Holes permitted	Yes	No	Yes	Yes	Yes	No	Yes	Yes

Exhibit 7.2.5 Machining Operations Used in Performing Secondary Operations

Process	Shape Produced	Machine	Cutting Tool	Tolerance	Surface Finish (μrms)	Relative Motion: Tool	Relative Motion: Work
Turning (external)	Surface of revolution (cylindrical)	Lathe, boring machine	Single point	±0.005 mm to ±0.025 mm	0.8 to 6.4		
Boring (internal)	Cylindrical (enlarges holes)	Boring machine	Single point	±0.005 mm to ±0.025 mm	0.4 to 5.0		
Shaping and planing	Flat surfaces or slots	Shaper, planer	Single point	±0.025 mm to ±0.050 mm	0.8 to 6.4		
Milling (end, form, slab)	Flat and contoured surfaces and slots	Milling machine—horizontal, vertical, bed-type	Multiple points	±0.025 mm to ±0.050 mm	0.8 to 6.4		
Drilling	Cylindrical (originating holes 0.1 to 100 mm diameter)	Drill press	Twin-edged drill	±0.050 mm to ±0.100 mm	2.5 to 6.4		Fixed
Grinding (cylindrical surface, plunge)	Cylindrical, flat, and formed	Grinding machine—cylindrical, surface, thread	Multiple points	±0.0025 mm to ±0.0075 mm	0.2 to 3.2		
Reaming	Cylindrical (enlarging and improving finish of holes)	Drill press, turret lathe	Multiple points	±0.0125 mm to ±0.0500 mm	0.8 to 2.5		Fixed
Broaching	Cylindrical, flat, slots	Broaching machine, press	Multiple points	±0.005 mm to ±0.0150 mm	0.8 to 2.5		Fixed

Exhibit 7.2.5 (*Continued*)

Process	Shape Produced	Machine	Cutting Tool	Tolerance	Surface Finish (μrms)	Relative Motion	
						Tool	Work
Electric discharge machining	Variety of shapes depending on shape of electrode	Electric discharge machine	Single-point electrode	±0.050 mm	0.8 to 5.0	↕	Fixed
Electro-chemical machining	Variety of shapes; usually odd-shaped cavities of hard material	Electrochemical machine	Dissolution process	±0.050 mm	0.3 to 1.5	Andoic dissolution; tool is cathode	Workpiece is anode
Chemical machining	Variety of shapes; usually blanking of intricate shapes, printed circuit etching, or shallow cavities	Chemical machining machine	Chemical attack of exposed surfaces	±0.050 mm	0.6 to 1.8	Chemical attack of exposed surfaces	Fixed
Laser machining	Cylindrical holes as small as 5 μ	Laser beam machine	Single wavelength beam of light	Holes are reproducible within ± 3%	0.6 to 2.5	Fixed	Fixed
Ultrasonic machining	Same shape as tool	Machine equipped with magnetostrictive transducer, generator power supply	Shaped tool and abrasive powder	±0.025 mm	0.3 to 0.9	↕	Fixed

Electron beam machining	Cylindrical slots	Electron beam machine equipped with vacuum of 10^{-4} mm of mercury	High-velocity electrons focus on workpiece	±0.025 mm	0.6 to 1.8	↕	Fixed
Gear generating	Eccentric cams, ratchets, gears	Gear shaper	Single-point reciprocating	±0.013 mm to ±0.025 mm	1.8 to 3.8	↕ ↻	↻
Hobbing	Any form that regularly repeats itself on periphery of circular part	Hobbing machine	Multiple points	±0.013 mm to ±0.025 mm	1.8 to 3.8	↻	↻
Trepanning	Large through holes, circular grooves	Lathe-like machine	One or more single-point cutters revolving around a center	±0.13 mm	2.5 to 6.4	↻	↻

Compression Molding

In compression molding an appropriate amount of plastic compound (usually in powder form) is introduced into a heated mold, which is subsequently closed under pressure. The molding material, either thermoplastic or thermosetting, is softened by the heat and formed into a continuous mass having the geometric configuration of the mold cavity. If the material is thermoplastic, hardening is accomplished by cooling the mold. If the material is thermosetting, further heating will result in the hardening of the material.

Compression molding offers the following desirable features:

1. Thin-walled parts (less than 1.5 mm) are readily molded with this process with little warpage or dimensional deviation.
2. There will be no gate markings, which is of particular importance on small parts.
3. Less and more uniform shrinkage is characteristic of this molding process.
4. It is especially economical for larger parts (those weighing more than 1 kg).
5. Initial costs are less, since it usually costs less to design and make a compression mold than a transfer or injection mold.
6. Reinforcing fibers are not broken up as they are in closed-mold methods such as transfer and injection. Therefore the fabricated parts under compression molding may be both stronger and tougher.

Transfer Molding

Under transfer molding the mold is first closed. The plastic material is then conveyed into the mold cavity under pressure from an auxiliary chamber. The molding compound is placed in the hot auxiliary chamber and subsequently is forced in a plastic state through an orifice into the mold cavities by pressure. The molded part and the residue (cull) are ejected upon opening the mold after the part has hardened. Under transfer molding there is no flash to trim; only the runner needs to be removed.

Injection Molding

In injection molding the raw material (pellets, grains, etc.) is placed into a hopper, called the "barrel," above a heated cylinder. The material is metered into the barrel every cycle so as to replenish the system for what has been forced into the mold. Pressure up to 1750 kg/cm^2 forces the plastic molding compound through the heating cylinder and into the mold cavities. Although this process is used primarily for the molding of thermoplastic materials, it can also be used for thermosetting polymers. When molding thermosets, such as phenolic resins, low barrel temperatures should be used (65 to 120°C). Thermoplastic barrel temperatures are much higher, usually in the range of 175 to 315°C.

Extrusion

Like the extrusion of metals, the extrusion of plastics involves the continuous forming of a shape by forcing softened plastic material through a die orifice that has approximately the geometric profile of the cross section of the work. The extruded form is subsequently hardened by cooling. With the continuous extrusion process, such products as rods, tubes, and shapes of uniform cross section can be economically produced. Extrusion to obtain a sleeve of the correct proportion almost always precedes the basic process of blow molding.

Casting

Much like the casting of metals, the casting of plastics involves introducing plastic materials in the liquid form into a mold that has been shaped to the contour of the piece to be formed. The material that is used for making the mold is often flexible, such as rubber latex. Molds may also be made of nonflexible materials such as plaster. Epoxies, phenolics, and polyesters are plastics that are frequently fabricated by the casting process.

Cold Molding

Cold molding takes place when thermosetting compounds are introduced into a room temperature steel mold that is closed under pressure. The mold is subsequently opened, and the formed article is transferred to a heating oven, where it is baked until it becomes hard.

Thermoforming

Thermoforming is restricted to thermoplastic materials. Here sheets of the plastic material are heated and drawn over a mold contour so that the work takes the shape of the mold. Thermoforming may also be done by passing the stock between a sequence of rolls that produce the desired contour. Most thermoplastic materials become soft enough for thermoforming between 135 and 220°C. The plastic sheet that was obtained by calendering or extrusion can be brought to the correct thermoforming temperature by infrared radiant heat, electrical resistance heating, or ovens using gas or fuel oil.

Calendering

Calendering is the continuous production of a thin sheet by passing thermoplastic compounds between a series of heated rolls. The thickness of the sheet is determined by adjusting the distance between the rolls. After passing between the final set of rolls, the thin plastic sheet is cooled before being wound into large rolls for storage.

Blow Molding

In blow molding a tube of molten plastic material, the "parison," is extruded over an apparatus called the "blow pipe" and is then encased in a split mold. Air is injected into this hot section of extruded stock through the blow pipe. The stock is then blown outward, where it follows the contour of the mold. The part is then cooled, the mold opened, and the molded part ejected. In very heavy sections, carbon dioxide or liquid nitrogen may be used to hasten the cooling. This process is widely used in molding high- and low-density polyethylene, nylon, polyvinyl chloride (PVC), polypropylene, polystyrene, and polycarbonates.

Parameters Affecting the Selection of the Optimum Basic Process

Selecting the optimum basic process in the production of a given plastic design will have a significant bearing on the success of that design. The principal parameters that should be considered in the selection decision include the plastic material to be used, the geometry or configuration of the part, the quantity to be produced, and the cost.

If the functional designer cannot identify the exact plastic material that is to be used, he or she should be able to indicate whether a thermoplastic or thermosetting resin is being considered. This information alone will be most helpful. Certainly both thermoforming and blow molding are largely restricted to thermoplastics. Then, too, on sizable orders compression molding is usually restricted to thermosetting resins, as is transfer molding. Injection molding is used primarily for producing large-volume thermoplastic moldings, and extrusion for large-volume thermoplastic continuous shapes.

Geometry or shape also has a major impact on process selection. Unless a part has a continuous cross section, it would not be extruded; unless it were relatively thin walled and bottle shaped, it would not be blow molded. Again, calendering is restricted to flat sheet or strip designs, and the use of inserts is restricted to the molding processes.

The quantity to be produced also has a major role in the selection decision. Most designs can be made by simple compression molding, yet this method would not be economical if the quantity were large and the design and material were suitable for injection molding.

The following designing for manufacturing points apply to the processing of plastics:

1. Holes less than 1.5 mm diameter should not be molded, but should be drilled after molding.
2. Depth of blind holes should be limited to twice their diameter.
3. Holes should be located perpendicular to the parting line to permit easy removal from the mold.
4. Undercuts should be avoided in molded parts since they require either a split mold or a removable core section.
5. The section thickness between any two holes should be greater than 3 mm.
6. Boss heights should not be more than twice their diameter.
7. Bosses should be designed with at least a 5° taper on each side for easy withdrawal from the mold.
8. Bosses should be designed with radii at both the top and the base.
9. Ribs should be designed with at least a 2 to 5° taper on each side.
10. Ribs should be designed with radii at both the top and the base.
11. Ribs should be designed at a height of $1\frac{1}{2}$ times the wall thickness. The rib width of the base should be half the wall thickness.

12. Outside edges at the parting line should be designed without a radius. Fillets should be specified at the base of ribs and bosses and on corners and should be not less than 0.8 mm.

13. Inserts should be at right angles to the parting line and of a design that allows both ends to be supported in the mold.

14. A draft or taper of 1 to 2° should be specified on the vertical surfaces or walls parallel with the direction of mold pressure.

15. Cavity numbers should be engraved in the mold. The letters should be 2.4 mm high and 0.18 mm deep.

16. Threading below 8 mm diameter should be cut after molding.

Exhibit 7.2.6 identifies the major parameters associated with the basic processes used to fabricate thermoplastic and thermosetting resins.

7.2.9 COMPUTER-AIDED MANUFACTURING PLANNING

If it were known at the functional design stage what basic and secondary processes were to be used to produce a given design, then it would be relatively simple to incorporate good producibility in the design. The planning of the manufacturing procedure to establish the best way to produce a part has always been an important step in the chronological sequence of events that take place between the functional design of a product and its distribution to the market. Good manufacturing planning will help ensure product quality and reliability, help achieve customer satisfaction, and help ensure a competitive product. Poor manufacturing planning may result in costs that are so prohibitive that the product will never be sold.

Since so many parameters need to be considered in deciding how best to produce a part, and since there is seldom enough time for the analyst to make a detailed study of all alternatives during the initial planning, invariably improvements can be introduced into almost every product on the market. Even if the present planned method is quite satisfactory today, it may not be in the immediate future. New materials and new processes are continually being developed, which may make the present product and its method of manufacture obsolete.

To do the best job of process planning, we must resort to the digital eomputer. The speed and memory bank of the modern computer make it economical to consider all practical alternative ways of producing a given design component. To develop a suitable equation for the computer to solve, it is first necessary to identify those parameters that are important in determining the best process to use for the design under study.

Process families that need to be identified may strategically be broken down into three main classes: basic processes, secondary processes, and finishing processes. Different parameters will enter into the selection process of each of these three families.

In connection with the assignment of the basic processes, the following parameters would need to be considered: size of part, microstructure resulting from the process, geometry or configuration, material being utilized, quantity to be produced, relative cost of secondary operations, and the cost of manufacture through the basic process.

Size and/or weight has a limiting effect on many of the basic processes. For example, die castings larger than 35 kg are seldom produced. Similarly, sand castings smaller than 50 g are usually not made. Extrusions with cross-sectional areas of more than 900 cm^2 are seldom, if ever, made.

Microstructure resulting from the process also has a restraining impact. Parts not having grain flow may not be suitable for end-use products that are impact loaded. If the grain flow cannot be controlled, the process may be unsuitable.

Geometry of the product is one of the most important parameters that acts as a limiting restraint. For example, very complex geometries cannot be forged. Nonsymmetrical bowl-shaped parts are not spun. Roll forming must include generous radii. Holes or recesses more than two thirds the diameter of the major axis are not forged when working with ferrous materials. Designs with undercuts or reentrant angles cannot be made of powdered metal except as a secondary operation.

The material, too, has a significant impact on the processes that should be considered. For example, it is generally conceded that only nonferrous metals can be economically die cast. Similarly, plaster mold casting is limited to nonferrous metals. Compression molding is the customary technique for producing thermoplastic plates less than 6 mm thick.

Of course the quantity to be produced has a significant bearing not only on the process to be employed, but also on the sophistication and extent of the tooling to be used in connection with the process. For example, 10 parts of a given design would never be produced on an automatic screw machine because of economics. Similarly, 100,000 bushings would never be made on an engine lathe. We would never make a special broach for broaching just one keyway–it would be milled. However, we seldom would mill 10,000 keyways–they would be broached.

Another important consideration in evaluating the various alternative ways of producing a part is the cost of subsequent operations. There are many instances in which one alternative is considerably more costly than another, but in view of the low cost of subsequent operations when com-

Exhibit 7.2.6 Basic Processes Used to Fabricate Plastics and Their Principal Parameters

	Parameter									
Process	Shape Produced	Machine	Mold or Tool	Material	Typical Tolerance	Minimum Wall Thickness	Ribs	Draft	Inserts	Minimum Quantity Requirements
Calendering	Continuous sheet or film	Multiple-roll calender	None	Thermoplastic	0.05 to 0.200 mm depending upon material		None	None	None	Low
Extrusion	Continuous form such as rods, tubes, filaments, and simple shapes	Extrusion press	Hardened steel die	Thermoplastic	0.10 to 0.30 mm depending upon material		None	None	Possible to extrude over or around wire insert	Low (tooling is inexpensive)
Compression molding	Simple outlines and plain cross sections	Compression press	Hardened tool steel mold	Thermoplastic or thermosetting	0.04 to 0.25 mm depending upon material	1.25 mm			Yes	Low
Transfer molding	Complex geometries possible	Transfer press	Hardened tool steel mold	Thermosetting	0.04 to 0.25 mm depending upon material	1.5 mm	3 to 5° taper; height < 3 times wall thickness	$\frac{1}{2}$ to 5°	Yes	High
Injection molding	Complex geometries possible	Injection press	Hardened tool steel mold	Thermoplastic or thermosetting	0.04 to 0.25 mm depending upon material	1.25 mm	2 to 5° taper; height = $1\frac{1}{2}$ times wall thickness; width = $\frac{1}{2}$ of wall thickness	$\frac{1}{4}$ to 4°	Yes	High
Casting	Simple outlines and plain cross sections	None	Metal mold or expoxy mold	Thermosetting	0.10 to 0.50 mm depending upon material	2.0 mm			Yes	Low to medium depending on mold
Cold molding	Simple outlines and plain cross sections	None	Mold of wood, plaster, or steel	Thermosetting	0.10 to 0.50 mm depending upon material	2.0 mm			Yes	Low
Thermoforming	Thin walled and cup shaped	Thermoforming machine	Suitable form	Thermoplastic					No	Low
Blow molding	Thin walled and bottle shaped	Pneumatic blow molding machine	Tool steel mold	Thermoplastic					No	High
Rotational molding	Full enclosures or semienclosures (hollow objects)	Rotomolding system	Cast aluminum or fabricated metal	Thermoplastic, limited thermosetting	0.30 to 0.60 mm	3.0 mm			No	Medium
Filament winding	Tubes, piping, tanks	Filament winding machine	Must have axis about which the filament can be wound	Single-end continuous strand glass fiber and thermoplastic	0.20 to 0.50 mm	3.0 mm				Medium

pared to the alternative processes' subsequent operation costs, the more expensive basic operational process may be the more economical. For example, a powdered metal ferrous gear might be much less costly than a similar gear produced as a machined casting.

Finally, cost must eventually be considered. Assuming equal quality and reliability, we will always want to choose the process that will give us the lowest unit cost if delivery can be made on schedule.

The preceding parameters can be incorporated in the memory bank of the high-speed digital computer. Thus the computer will be able to reject all processes that are unacceptable because of unique size, geometry, material, and so on. Those processes for producing a given design that are acceptable can be compared economically through the computer's solution of appropriately developed cost equations. In this manner the computer can materially assist the manufacturing engineer in favorable manufacturing planning.

BIBLIOGRAPHY

DOYLE, L. E., *Manufacturing Processes and Materials for Engineers*, Prentice-Hall, Englewood Cliffs, NJ, 1969.

Engineering Design Handbook: Design Guidance for Producibility, Headquarters, U.S. Army Material Command, Washington, DC, 1971.

GREENWOOD, D. C., *Product Engineering Design Manual*, McGraw-Hill, New York, 1959.

GREENWOOD, D. C., *Engineering Data for Product Design*, McGraw-Hill, New York, 1961.

GREENWOOD, D. C., *Mechanical Details for Product Design*, McGraw-Hill, New York, 1964.

LEGRAND, R., *Manufacturing Engineers' Manual*, McGraw-Hill, New York, 1971.

NIEBEL, B. W., and E. N. BALDWIN, *Designing for Production*, Irwin, Homewood, IL, 1963.

NIEBEL, B. W. and A. B. DRAPER, *Product Design and Process Engineering*, McGraw-Hill, New York, 1974.

TRUCKS, H. E., *Designing for Economical Production*, Society of Manufacturing Engineers, Dearborn, MI, 1974.

CHAPTER 7.3
Value Engineering

DAVID J. DEMARLE
M. LARRY SHILLITO

Eastman Kodak Company

7.3.1 BACKGROUND

In 1961 Lawrence D. Miles in his book *Techniques of Value Analysis and Engineering* defined "value analysis (VA)"* as "an organized creative approach which has for its purpose the efficient identification of unnecessary cost, i.e., cost which provides neither quality nor use nor life nor appearance nor customer features."[1] Numerous other definitions of value engineering (VE), including that of the Society of American Value Engineers (SAVE),[2] have been used. We perceive VE as the systematic application of recognized techniques to identify the functions of a product† or service and to provide those functions at the lowest total cost. The philosophy of VE is implemented through a systematic rational process consisting of a series of techniques, including (1) function analysis to define the reason for the existence of a product or its components, (2) creative and speculative techniques for generating new alternatives, and (3) measurement techniques for evaluating the value of present and future concepts. The techniques used in the VE process are not unique to VE. They are a collection of various techniques from many fields. As Ernst Bouey, president of SAVE, says, "VE . . . holds no respect for proprietary concepts that are merely ways of thinking. VE will adopt any technique or method (or piece thereof) for use in any of its procedural phases."[4]

Value engineering should not be confused with modern or traditional cost reduction analyses; it is more comprehensive. Based on function analysis, the process concentrates on a detailed examination of utility rather than on a simplistic examination of components and component cost. The improvement of value is attained without any sacrifice in quality, reliability, or maintainability. Collateral gains are often realized in performance, productivity, parts availability, lead time, and quality. Historically VE has returned between \$15 and \$30 for every \$1 expended in effort. It was originally developed and traditionally applied in the hardware area. However, recent years have seen a proliferation of its use in numerous new and nontraditional areas. It should be viewed as applicable to almost any subject. Its use is limited only by the user's imagination. The literature offers examples of applications in construction,[5] administration,[6,7] training,[8,9] nonhardware,[10] management,[11] systems and procedures,[12] venture analysis,[13] forecasting,[14] resource allocation,[15,16] and marketing.[17]

7.3.2 HISTORICAL PERSPECTIVE

Value engineering had its origin in studies of product changes resulting from material shortages during World War II. The substitution of materials in designs without the sacrifice in quality and performance caught the attention of Larry Miles and the purchasing people at the General Electric Company (GE). Mr. Miles, who is considered to be the inventor of VA and VE, organized a formal methodology in which teams of people examined the function of products (function analysis) manufactured by GE. Through team-oriented creative techniques, they made changes in products that lowered their cost without affecting their utility. This new methodology was VA. Although most of the original techniques that were used were not new, the philosophy of a functional approach was unique.

*The term "value analysis" is used interchangeably here with "value engineering" (VE). Traditionally, VE has been used to refer to the design stage or before the fact, whereas VA has been used to refer to an existing product or after the fact.

†The definition of "product" used in this chapter is the one developed by Arthur Mudge (reference 3): "Anything which is the result of someone's effort." Therefore "product" is not limited to hardware.

What emerged from the early experience of GE and other organizations was the discovery and development of some fundamental concepts that provided the foundation for the development of VA methodology. These basic concepts were (1) the use of interdisciplinary teams to effect change, (2) the development of change through the study of function, (3) a basic questioning logic, and (4) a job plan. Over the years the techniques of VA were expanded, as were the areas of its application. Today VA, or VE, is a widely recognized discipline for improving the value of products or services.

7.3.3 NATURE AND MEASUREMENT OF VALUE

It would be inappropriate to discuss VE and some of its techniques without a brief explanation of the nature and measurement of value. Value can be perceived as the ratio of the sum of the positive and negative aspects of an object. The following equation depicts this simplistic interpretation.:

$$\text{value} = \frac{\Sigma(+)}{\Sigma(-)} \tag{1}$$

In reality, this equation is more complex, since we are dealing with many variables of different magnitude. A more descriptive equation is

$$\text{value} = \frac{(mb_1 + mb_2 + \cdots + mb_n)}{(mc_1 + mc_2 + \cdots + mc_n)} \tag{2}$$

where m = the magnitude of a given factor or criterion
b = a specific benefit
c = a specific cost

Equations 1 and 2 are both general and can be used to measure the value of items and alternatives. The field of psychophysics has provided us the means with which to measure the parameters of equations 1 and 2. L. L. Thurstone,[18,19] S. S. Stevens,[20] and others[21] have discovered and proved the validity and utility of quantitative subjective estimation applied to both physical and non-physical stimuli. The principles of these psychophysical experiments have been used by ourselves and others[22–24] to develop several value measurement techniques. They can be used to quantify the importance of beneficial functions and characteristics present in objects and services. They can also be used to measure cost, difficulty, risk, and other negative factors present in new or old designs. They are very effective in evaluating alternatives. The quantification of these parameters allows value, as expressed in equations 1 and 2, to be measured. This psychophysical principle of subjectively quantifying parameters forms the basis for deriving value by means of the value measurement techniques discussed in a later subsection.

7.3.4 THE VE PROCESS

The VE process is a rational and structured process using an interdisciplinary team to (1) select the proper project or product for analysis in terms of time invested in the study; (2) display and measure the current (state 1) value of a product or its components in terms of functions that fulfill a user's needs, goals, or objectives; (3) develop and evaluate new alternatives to eliminate or improve component areas of low value; and (4) match the new alternatives with the best way of accomplishing them. Exhibit 7.3.1 displays the phases of this process. It also diagrams the basic questioning logic and lists the various functions, activities, milestones, and techniques for each phase.

The process begins with the *origination phase*, wherein a VE study team is formed and a project is selected and defined. The product and all its components are examined in detail in order to obtain a thorough understanding of their nature.

This familiarization leads to the *information phase*, where the function(s) of the product and/or its components are documented by function analysis techniques. Constraints that dictate an original design, materials, components, or procedures are challenged for validity. The importance and cost of functions are quantified by various value measurement techniques, which are discussed later. This analysis is referred to by the authors as a "state 1," or current state, analysis. A Value Index is established by computing the cost-benefit ratio for each function. The authors use value graphing as a very vivid method for displaying the relationship of importance to cost. The output of the information phase is an ordered list of functions or items, arrayed from highest to lowest relative value as they currently exist. The low-value items become candidates for value improvement.

These candidates are selected for the *innovation phase*, where various creative techniques are used to generate new alternatives for their replacement or improvement. This phase generally produces copius alternatives. The task then becomes one of reducing the large list of alternatives for development and recommendation. Obviously, the team cannot work on the hundred or more ideas that are usually produced.

The objective of the *evaluation phase* is to prescreen the large list using various information reduction techniques. The highest-scoring alternatives (usually 5 to 20 items) that emerge from the prescreening are subjected to further evaluation by more discriminating value measurement techniques. These are the same measurement techniques as those used in the state 1 analysis of the information phase. Value measurement in this phase is referred to as "state 2," or future state, measurement. The highest-valued alternatives that emerge, now numbering only two or three, are further examined for economic and technical feasibility, for their ability to perform the desired function satisfactorily, and for their ability to meet other standards, such as accuracy, quality, reliability, safety, ease of repair, and environmental impacts.

During the *implementation phase* a report is prepared to summarize the study, present conclusions, and specify proposals for the decision maker. Program and action plans are developed to produce and implement the alternatives that survive the analysis phase. Techniques used here fall in the realm of production and/or project management. The VE change proposals are monitored and followed up in order to provide assistance, clarify any misconceptions, and ensure that the recommended actions are achieved.

The following subsections detail the activities in each phase. A simple desk compass (Exhibit 7.3.2) is used to illustrate the use of the various methodologies.

Origination Phase

Organization

Experience has shown that management support is important to the initiation and success of a VE study.[27] The following questions should be addressed at the outset: What is the scope of the study? Define what the study is or is not. Who funds the study, and how much funding is required? Who approves the study? Who is the requestor? What are the study's start and completion dates? Who should be on the study team? Who are the intended users of the study results? What is the expected format? Oral? Written? What levels of the organization should be involved? What geographic areas? What are the expected time, manpower, and cost?

Once the study parameters have been reviewed, it is beneficial to have management endorse the VE study in a properly distributed letter. The letter should designate the team members, give them the time to participate in the study, pave the way for their access to needed information, and define the study's sponsor and scope.

Project Selection

Many VE studies arise out of necessity in a very specific, well-defined area. Consequently, a predefined project may obviate the need for formal project selection. However, the resources that can be allocated to a VE study are limited. Traditionally, the main criteria for a VE project have been high dollar volume and high total costs in relation to the function performed. There are other general criteria that should also be considered. The VE study should:

1. Solve a problem. The need should be real and should be supported by management.
2. Have a good probability of success and implementation.
3. Have objectives that are credible.
4. Be important to the people in the area being studied.
5. Have the commitment of the requestor and the team members.
6. Have receptivity. The sponsor or decision maker must be receptive to change.

A U. S. Department of Defense publication[22] lists additional project selection criteria.

The VE Team

A VE team normally comprises three to seven members. With more than seven members, interaction becomes complex, discussions become indistinct, and the group begins to fracture. An odd number is also helpful because it reduces chances for split decisions. The following characteristics are very important when assembling a team:

1. It should be interdisciplinary, incorporating a good balance of backgrounds, viewpoints, and disciplines, as well as good geographic representation.

2. Members should be from equivalent levels in the organization's hierarchy in order to minimize peer pressure and politics.[26]

Exhibit 7.3.1 The VE Process

	Process Phase				
	Origination			Information	Innovation
Functions	Organize Study Select Project Form Team	Define[a] Product	Define Function	Derive Value	Generate Alternatives
Basic questioning logic flow The VE process	What is our official charge? Team building How are we going to work? Project definition What do we work on? What is our mission (goal)?	Product[a] definition What is it?	Function analysis What does it do? Constraints analysis What causes it to be done that way?	Worth analysis Importance comparison What is it worth? What is its importance? Importance (I) Cost accounting Cost appraisal What does it cost? How much does it contribute to cost? Cost (C) I/C Value analysis 1 (Value Index) (Figure of merit) What is its value? $V_1 = I/C$ Value graphing I / C	High value Individual items or functions Low value Innovation/ speculation What else will do the job? Alternatives
Activity	Study definition Team building Project selection; project definition; mission (goal) definition	Systems analysis Data collection	Function analysis Constraints analysis Data collection	State 1 value analysis (current state) Current state value measurement Data collection	Speculation/ideation
Results, output, milestones	Project defined; team formed; study launched.	Product[a] and components documented.	Current/existing functions or functional families defined. Functional hierarchy of product or system diagrammed.	Worth/importance and cost quantified for each function or item. Ratio of I/C derives a Value Index for each current item or function. Functions/items now arrayed from highest to lowest value.	New alternatives for low-value functions/items generated.
Techniques used	Organization development; team building	Explosion diagrams	Two-word functional description; FAST diagrams; functional matrix	Worth analysis Importance comparison Ranking Paired comparison Direct magnitude estimation Cost accounting Cost appraisal	Brainstorm Nominal group technique Gordon technique Catalog technique Synectics Checklists Attribute listing Morphological synthesis

[a]Product = Anything that is the result of someone's effort (objectives, devices, services, procedures, items, components, subsystems, etc.).

3. It is sometimes helpful to include a decision maker on the team since acceptance of results may depend upon who is on the team. Caution should be taken if having a decision maker on the team may induce peer pressure and the "will of the boss."
4. It is necessary that one or more members be versed in the VE process. Alternatively, a third-party facilitator or outside consultant can supply the VE methodology.
5. One member should be an "expert" on the product or subject being studied.

The team members themselves should:

1. Be at least generally familiar with the product or area of study.
2. Know the sources of data for their area of expertise.
3. Have interest, motivation, and commitment to engage in the task.
4. Be able to get cooperation and assistance while representing their organizational area.
5. Have sufficient time to do the job and should be engaged long enough to provide continuity to the project.
6. "Be able to create, accept and be eager to exploit change."[27]
7. Have an open mind and able to work and communicate with others.

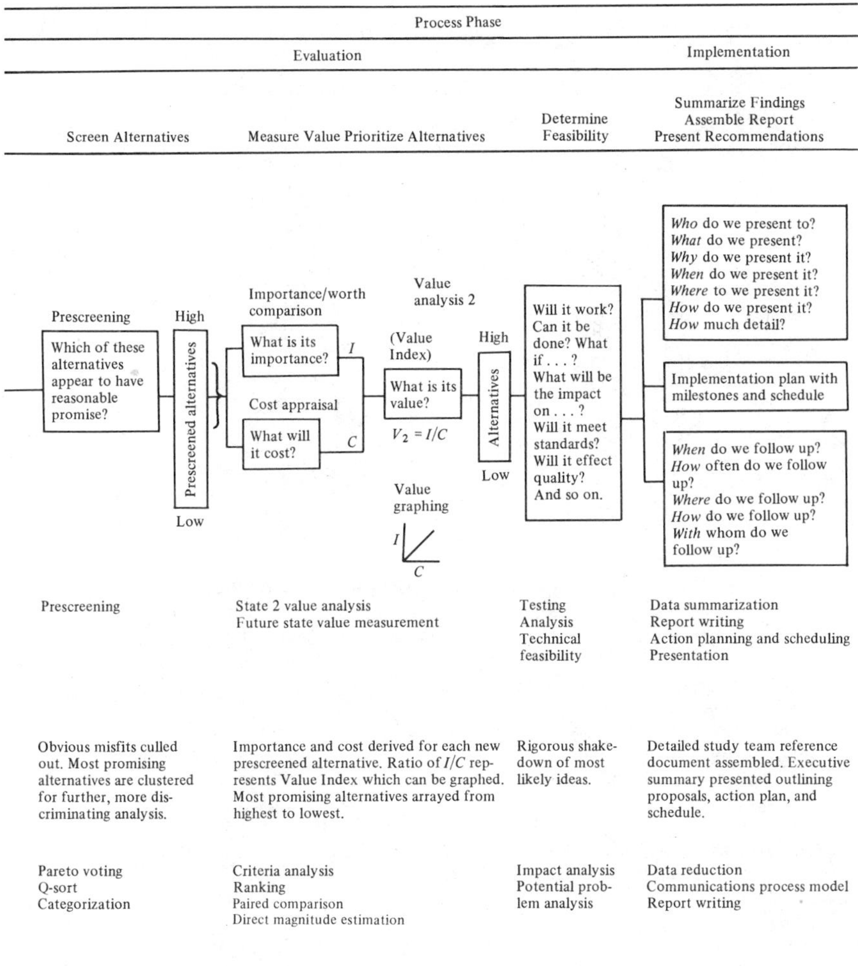

Project or Study Mission

Once the topic of the VE study has been selected and defined and the team formed, it is very helpful for the team to formulate a mission statement. This statement is a short, broad definition of what is to be accomplished by the project or study and why. It essentially creates a target or goal for the team. A sample mission statement is, "Manufacture component 'X' for $1.00 by the end of the year to meet requirements." Although the statement may at first appear to be unreasonable or impossible, it does help provide a challenge and engender a positive, creative attitude among the members. Similar approaches have been used by others, such as "design to cost"[28] and "life cycle costing."[29]

Product Definition and Documentation

When organized and ready, the VE team begins by collecting information on the subject being studied and defining the product and its components through a review of facts. This is equivalent to asking the VE question, "What is it?" All possible sources of information should be pursued. It is better to collect too much information than too little. The purpose of this product definition

Exhibit 7.3.2 Desk Compass: (*a*) Old Design and (*b*) New Design[a]

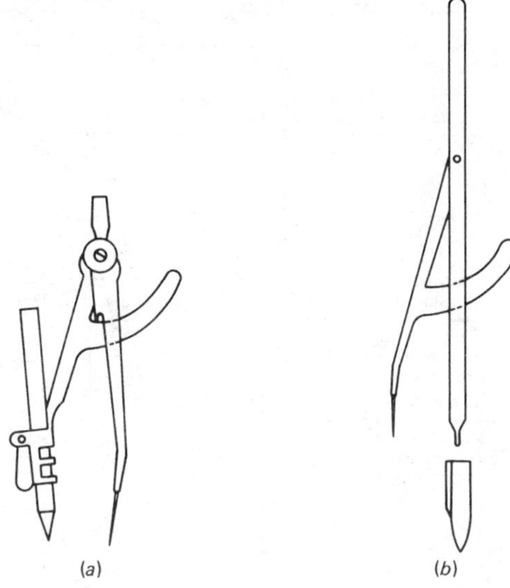

[a]Note that further functional improvement is still possible. For example, the pin arm may be molded into one piece containing an integral pin point, the radial arm can be improved, the ballpoint pen with cap can be simplified, and so on.

is to define the product as it is designed, manufactured, or used today. It is necessary to have sufficient factual data in order to minimize personal opinion and bias, which can adversely affect the direction of the study and the final results.

Explosion diagrams detailing the product and all its components are helpful at this stage. Hardware items can be brought to team meetings and disassembled into all of their parts. Exhibit 7.3.3 shows the component listing for the desk compass example.

Once the team is thoroughly familiar with the product and its components, a qualitative analysis of their value is performed.

Exhibit 7.3.3 Component Listing and Function Description for a Product (Compass) and a Function (Scribe Arc)

Component[a]	Function[b]	%I[c]	%C[d]	Value Index (%I/%C)[e]
A. Pencil leg	Contain marker	16	25	0.6
B. Pencil lock	Apply leverage	5	4	1.3
C. Lock rivet	Create fulcrum	1	1	1.0
D. Handle	Access assembly	0	9	0.0
E. Screw	Connect components (fasten legs, fasten handle)	12	7	1.7
F. Nut	Induce torque	7	3	2.3
G. Washer	Maintain friction	1	4	0.3
H. Pin leg	Hold pin	12	20	0.6
I. Pin	Anchor rotation axis of compass	21	4	5.3
J. Pencil[f]	Deposit graphite	25	23	1.1

[a]Components listed from product definition stage.
[b]Functions derived from function analysis stage.
[c]Derived by way of paired comparison (Exhibit 7.3.6) in value measurement stage.
[d]Derived in same manner as %*I*.
[e]Value Index for components derived by dividing %*I* by %*C*.
[f]Assumes pencil is part of compass assembly.

Information Phase

Qualitative Analysis of Value–Function Analysis

The product and all its components are studied to determine their functions (or purposes). Functions are noted for all of the components that were listed in the information phase. Function analysis consists of definitional and structural techniques that employ a semantic clarification of function. It provides the groundwork for deriving a quantitative analysis of value.

The method requires functions to be described with only two words, a verb and a noun. By so restricting functional specifications, clear descriptions of the functions are possible–descriptions that are not confounded by unnecessary modifying phrases, adjectives, and adverbs. Longer, more verbose function descriptions restrict creative approaches that may be required later in the process. They also can make it difficult to develop alternatives and increase the chances of directing the VE effort to the wrong subject or item. Concise function descriptions reduce the possibility of detailed semantic elaboration. They force a rational approach by eliminating superfluous frills and minimizing emotion. The rules of function description are:

1. Determine the user's needs for a product or service. What are the qualities, traits, or characteristics that define what the product *must* be able to do? Why is the product needed?

2. Use only one verb and one noun to describe a function. The verb should answer the question "What does it do?"; the noun should answer "What does it do it to or with?" Where possible, nouns should be measurable and verbs should be demonstrable or action oriented.

3. Avoid passive or indirect verbs, such as provides, supplies, gives, furnishes, is, prepares. Such verbs contain very little information.

4. Avoid goallike words or phrases, such as improve, maximize, minimize, optimize, prevent, least, most, 100%.

5. List a large number of two-word combinations and then select the best pair. Teams can be used to derive a group definition of function. Examples are:

a. Light bulb: "Emit light."

b. Coffee cup: "Hold liquid."

c. Screwdriver: "Transfer torque," or, if a painter uses it to open cans, it would be "transfer linear force." Function depends on the user's intended use.

Function descriptions should be derived for the product and all of its components. Occasionally items may legitimately have more than one function. When listing components and their functions, keep them at the same level of abstraction. For example, during the redesign of a wooden pencil, the components listed–wood, graphite, eraser, eraser holder, paint, glue, and so on–would be kept at one level of assembly. Components in the paint (pigment, solvent, carrier, etc.) are constituents of a main component. Such constituents are at a much lower level of abstraction and most likely would not be considered in designing a new pencil. Exhibit 7.3.3 illustrates a functional description for the compass and its components.

Functions are frequently classified as either basic or secondary. A basic function is the prime reason for the existence of the product. It describes the output of the product in terms of the user's need. A good question to determine the basic function is, "If you take this function from the product, will the purpose of the product still be fulfilled?" Secondary functions, on the other hand, support the basic function(s) and allow them to occur. They generally are present as a result of a specific chosen design. Secondary functions may improve dependability or convenience or may serve solely sensory or aesthetic functions. They are prime candidates for elimination, improvement, and innovation. Classifying functions as basic or secondary is very helpful for identifying redundant or unnecessary functions later on, as well as for identifying cost distribution within a product. For example, if more dollars are spent on unnecessary secondary functions than on basic or secondary functions needed to fulfill specifications and requirements, this should be a signal to a VE team for potential cost improvement.

Functional Analysis Systems Technique. A well-known and powerful structural technique of function analysis used by VE practitioners today is Functional Analysis Systems Technique (FAST) diagramming, developed by Charles Bytheway in 1965. The technique allows two-word function definitions to be ordered in a hierarchy based on cause and consequence. It expands upon the reason for the existence of the product, or the basic function. The FAST diagrams visually display the interrelationship of all functions that must be performed to accomplish a basic function. They should not be confused with time sequence flow charts or critical path charts. A FAST diagram is developed in the following manner:

1. All the functions performed by the product and its elements are defined by using the two-word function description. Each function is written on a separate small card to facilitate construction of the diagram. The cards are placed on a surface where they are visible and easily accessible and can be moved about. It is convenient to use a table covered with paper to which the cards can eventually be fastened with tape.

2. The card that best describes the basic function is selected from among the many cards. Included in the initial listing of functions will be cards that describe secondary functions. Although many secondary functions may be present, some may have questionable value and may add unnecessary cost. They are prime candidates for improvement and innovation.

3. A branching tree structure should be created from the basic function (Exhibit 7.3.4). This is best done by personal analogy by assuming that you are the item under analysis and asking the question "How do I (verb) (noun)?" A more depersonalized branching question would be, "How does (the product) do this?" In any case, answers to the question are placed to the immediate right of the basic function. The "how" question will result in branching and is repeated until branching has stopped and the function order is in a logical sequence. The individual cards are a convenient aid to arranging and laying out the logical order of the "how" questions.

4. The logic structure is verified in the reverse direction by asking the question "Why do I (verb) (noun)?" for each function in the logic sequence. The "how-why" questions are used to test the logic of the entire diagram. Answers to the "how" and "why" logic sequence questions must make sense in both directions; that is, answers to "how" questions must logically flow from left to right, and answers to "why" questions must read logically from right to left.

5. A "critical function path" may result from the logic sequence of the basic and secondary functions. It is composed only of those functions that must be performed to accomplish the basic function. Any function not on this path is a prime target for redesign, elimination, and cost reduction.

6. The FAST diagrams are usually bounded on both ends by "scope lines," which delineate the limits of responsibility of the study. For example, if one is value analyzing an overhead projector, the FAST diagram would be expanded up to the point where current is conducted to the device. "Generate electricity" is outside the scope of the study.

A FAST diagram by itself is not very useful and may appear confusing and formidable to those not involved in its construction. The main value of the diagram lies in the intensive questioning and penetrating analysis required for its development. In this respect the construction process serves as an excellent communications device. It is reemphasized that a FAST diagram is not a sequential flow diagram related to time. It is a tree diagram depicting the interrelationship of functions in a hierarchical order. The FAST diagram becomes even more valuable and useful in the value measurement phase if costs and importance are allocated and posted to functions on the diagram.

Exhibit 7.3.4 is an example of a completed FAST diagram for the compass example. For more detailed information on the construction and use of FAST diagrams, the reader is referred to other sources.[30–36]

Constraints Analysis. It is also helpful to challenge any constraints or reasons that dictate a particular component, material, design, or procedure that is currently being used. The authors define this procedure as "constraints analysis." Its purpose is to challenge whether or not any constraints are still valid today. Too often products become overdesigned because the original constraints that dictated the design are no longer valid or are misinterpreted. Questions to answer are, "Why do we use what we use?" and "Is this reason still valid?" Constraints analysis should not be underemphasized. In some cases such an analysis has even obviated the need for a VE study. Once identified, costs due to invalid constraints should be derived and combined with function cost.

Quantitative Analysis of Value–State 1 Value Measurement

Cost Derivation. After completing the function analysis of the product, one should (1) determine the cost of the functions, (2) determine their worth or importance, (3) derive a figure of merit (FOM), or Value Index,[37–39] for these functions, and (4) post these measures to the respective functions listed on the FAST diagram if one is used. Please note that, although a FAST diagram can be very useful, it may not always be used. Also, not using one does not preclude deriving a Value Index.

Function cost analysis answers the basic VE question "What does it cost?" Function costs are derived by determining the cost of the items, components, or labor necessary to provide respective functions. Costs may consist of actual, or "hard," costs, such as materials and labor dollars, or of subjective, or "soft," costs, such as difficulty, risk of failure, or even "guesstimated" material and labor dollars when real costs are not available. If hard costs are available, they should be used. They can be derived by customary cost accounting methods. However, for those occasions when they are not available, estimates can be derived by cost appraisal methods that produce subjective fig-

Exhibit 7.3.4 A FAST Diagram for a Desk Compass

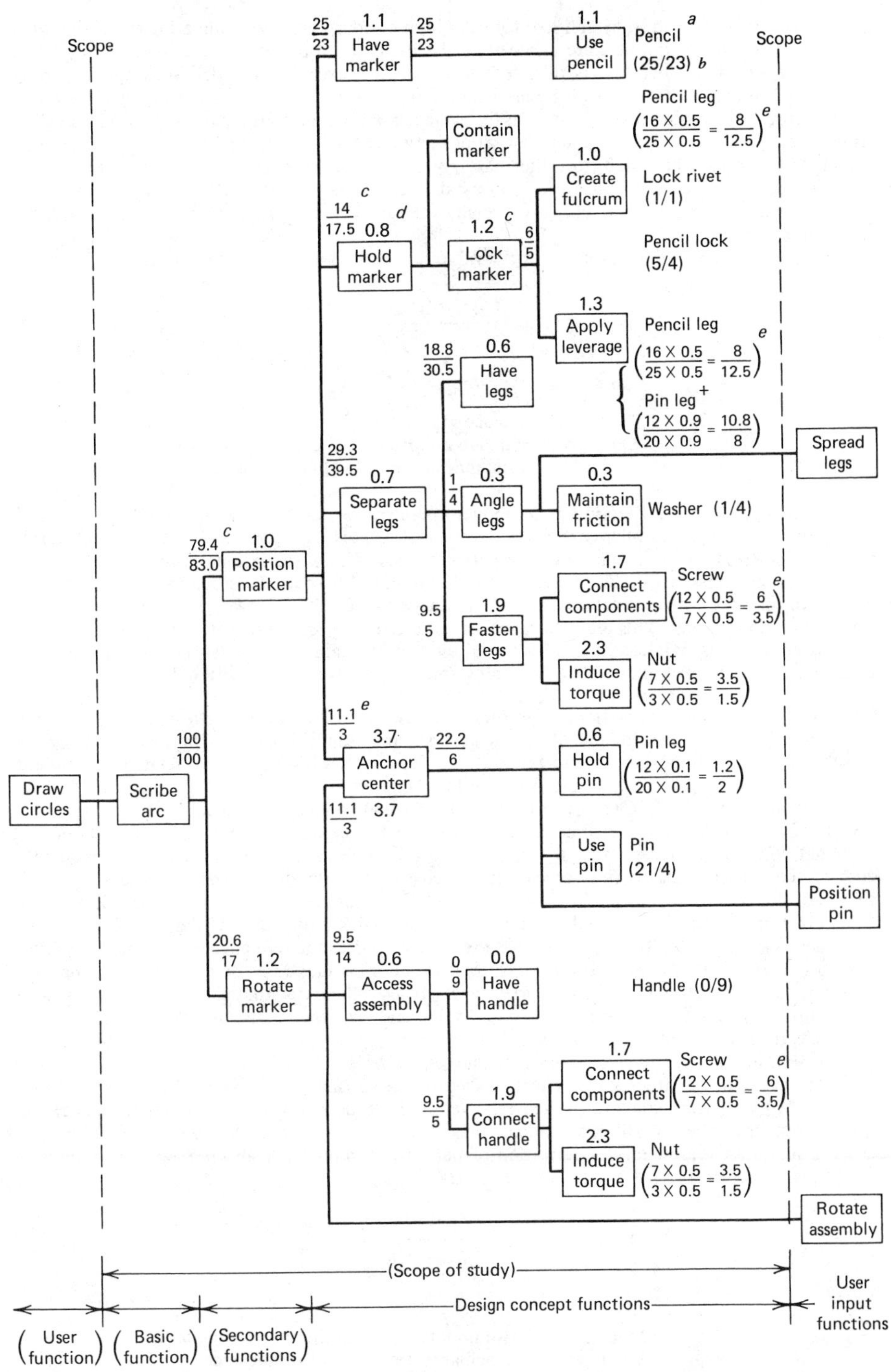

[a] Components posted to respective functions as listed on component listing of Exhibit 7.3.3.
[b] Numerator represents I, denominator C. Both I and C are derived by paired comparison (Exhibit 7.3.6) or direct magnitude estimation and posted from Exhibit 7.3.3.
[c] I and C equal the sum of all I and C to right in diagram (e.g., "hold marker," $I = 14 = 8 + 1 + 5$; "lock marker," $I = 6 = 1 + 5$; "position marker," $I = 79.4 = 25 + 14 + 29.3 + 11.1$).
[d] Numbers on top of functions are computed value of I/C (e.g., $14/17.5 = 0.8$).
[e] Components and functions are split. The I and C for multifunctional components factored according to perceived contribution to each function (e.g., pencil leg split 50/50, pin leg split 90/10).

ures. These soft costs can be established for each item or component by comparing and rating each relative to the others. Numbers are assigned to the items in proportion to their perceived relative costs. They are then converted (normalized) to individual percentages by dividing a specific rating by the total numerical rating of all items. When a group of people estimates costs, an average percentage cost can be obtained for an item by averaging across all of the participants' ratings. Percentages are used in order to standardize ratings across different raters if there is more than one participant. It should be noted, too, that hard costs may also be normalized to percentages. Normalized costs are often used since they show individual costs as a proportion of the total overall cost of the product or system. This is especially relevant when studying components of systems, procedures, or services.

Equation 3 provides a denominator for the value ratio expressed in equations 1 or 2.

$$\text{average } \%C = \frac{(\%C_1 + \%C_2 + \cdots + \%C_n)}{n} \tag{3}$$

where

$\%C$ = average relative cost of an item or component
$\%C_{(1,2,\ldots,n)}$ = individual participants' estimates of cost
n = total number of individuals rating cost

Occasionally, individual function or component percentages can be converted to dollars by multiplying them by the total dollar cost of the entire system. This is a very useful method for establishing target costs. It can also be used to derive target costs or goals for various components and tasks of projects and missions at the outset of a VE project. All costs, both real or relative, may be posted to the respective functions on FAST diagrams (Exhibit 7.3.4), explosion diagrams, or similar function listings. This allocation process translates the traditional item-cost viewpoint to a systems view of cost relationships, which is often quite revealing. The total function cost of basic or secondary functions can now be obtained by summing costs on the function diagram.

Worth or Importance Derivation. The functions are also studied for worth or importance. The term "worth" and the technique of worth analysis are usually used with hardware items. Conversely, the term "importance" and the methods for its derivation are often related to nonhardware items. The worth or importance of functions, like cost, is determined indirectly by deriving the worth and importance of the items or components that collectively provide those functions.

Worth is established by a technique of external cost comparison and is defined as the lowest cost that will reliably achieve the required function. It is determined by creatively comparing the cost of the function of the item or part to the cost of the function of analogous external items that can also reliably perform the same function. For example, the worth of a tie clasp whose function is to "anchor tie" could be derived by examining other objects that can also "anchor tie." The cost of the least costly reference item is selected as the reference cost. A paper clip that costs less than \$0.01 also provides the anchor tie function. It illustrates that the worth of the basic function could conceivably be in this price range. It also illustrates that the major worth of a tie clasp consists primarily of prestige or esteem value and that the total worth is a combination of the basic function (anchor tie) and aesthetic secondary functions. Worth analysis using creative comparisons of this type is covered in other sources.[22,40,41]

Importance, on the other hand, is established by an internal comparison of the items or components by comparing and rating each relative to the others. Evaluation of importance is performed by subjective quantification techniques in the same manner as was done to derive soft costs. Equation 4 represents the derivation of percentage importance and provides the numerator for the value ratio expressed in equations 1 and 2.

$$\text{average } \%I = \frac{(\%I_1 + \%I_2 + \cdots + \%I_n)}{n} \tag{4}$$

where

$\%I$ = average relative importance of an item or component
$\%I_{(1,2,\ldots,n)}$ = individual participants' estimates of importance
n = total number of individuals rating importance

As with soft cost derivation, the subjective numbers are normalized to a percentage. The relative importance of each item in a system is now available in terms of its proportion to the total overall importance attributed to the entire system. Both worth and importance may also be posted to their respective functions on the FAST diagram, as in Exhibit 7.3.4.

It is important to emphasize that function comparison, whether it be for worth, importance, or cost, be performed at the same hierarchical level. For example, all the functions listed in Exhibit 7.3.3 are at the same level and were used to derive importance and cost. An opposite example can be taken from the FAST diagram in Exhibit 7.3.4: That is, the two functions "induce torque" and "separate legs" cannot be directly compared for deriving importance or cost. "Separate legs" is at a higher level and comprises several lower-level functions, one of which is "induce torque." "Separate legs" can, however, be compared with "hold marker," "anchor center," and other functions at that same level.

The Value Index. Once cost and importance (or worth) are derived and posted for each function, they are used to compute the Value Index. There are several ways to compute the index. Either worth or importance may be used in the numerator; whichever is used is situational. Similarly, the denominator may consist of actual hard costs, relative (normalized) hard costs, or subjective estimated soft costs, although the former are preferred. For example,

$$\text{Value Index} = \frac{\text{absolute worth}}{\text{absolute cost}} \tag{5}$$

$$\text{Value Index} = \frac{\text{relative }\%I}{\text{relative }\%C} \tag{6}$$

It is important to note that, whichever parameters are used to derive the numerator or denominator of the Value Index, the numbers representing those parameters must be either all relative units or all absolute units. Consistency is necessary so that the units cancel out to give a dimensionless index.

The Value Index is a dimensionless number that allows one to array a system of functions (or items) in order of perceived value. Generally, a Value Index greater than 1.0 represents good value; an index less than 1.0 can indicate a function or component that needs attention and improvement. The index highlights areas of low value that may be improved in the speculation phase.

A vivid way to illustrate the index is to plot relative importance (or worth) versus relative cost for each function or item. The authors refer to the plotting procedure as "value graphing"[39,42] (see Exhibit 7.3.5). The 45° line on the graph has a Value Index of 1 and represents acceptable value. Areas above this line possess good value, whereas areas below exhibit poorer value. In the example, components A, H, and D appear to be prime candidates for improvement or elimination in the innovation phase. They also illustrate the point that value graphs more vividly depict relative value than does a tabular listing of numerical value indexes, as in Exhibit 7.3.3. That is, the graphs readily display the relative magnitude and relationship of the components that form the structure of the Value Index for each item rated.

Exhibit 7.3.5 Value Graph of Compass Components[a]

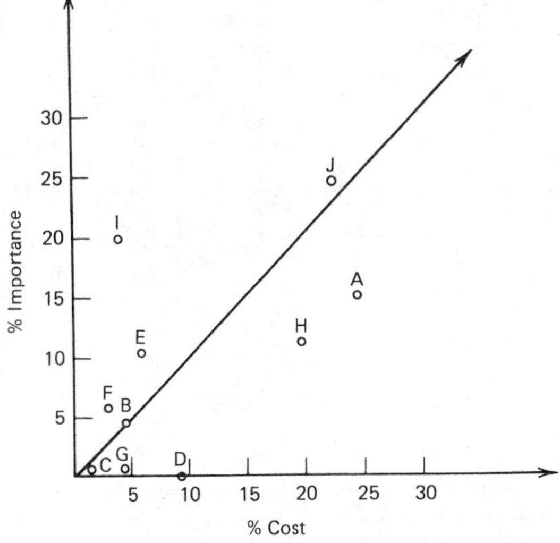

[a]Letters refer to components as listed in Exhibit 7.3.3.

Value Measurement Techniques. There are numerous subjective quantification techniques available to aid in measuring the parameters of the Value Index. The authors have found the value measurement techniques of paired comparison and direct magnitude estimation (DME) to be most useful and descriptive.

Paired comparison is used and described by Mudge.[3,43] The process employs a comparison of items wherein an individual chooses from a pair of functions or items the one that possesses a higher level of some specific characteristic (i.e., cost, importance, etc.). By iterating the procedure over all possible pairs and subsequently summing the numbers associated with each function or item, it is possible to quantify functional or item relationships. The matrix can be used to quantify positive and negative characteristics such as importance and cost. The summed figures for each rated item are normalized to a percentage of the total rating of all items.

Exhibit 7.3.6 is a paired comparison of the desk compass example. Although importance is illustrated in the exhibit, the method was also used to derive relative costs. Both parameters are shown in the component listing in Exhibit 7.3.3 and in the FAST diagram in Exhibit 7.3.4. Paired comparison may be used by a group of individuals by averaging the individual matrices across all participants. Paired comparison is highly discriminating and useful for prioritizing items that are difficult to separate and rank. The process can be cumbersome and impractical if there are a large number (more than 10) items to evaluate. In this case, DME is recommended.

Direct magnitude estimation[39,42] is a method in which people assign numbers to items in direct proportion to the magnitude of a characteristic that the items possess. It is very useful where it is necessary to discriminate among a large number of items. It is easy to use with groups that need to determine the importance of items. The instructions may be modeled on the following example:

> *You will be presented with a list of items in irregular order. Your task is to tell how important they seem by assigning numbers to them. Allot any number that seems to be appropriate to you to the first item. Then assign successive numbers to the subsequent items in such a way that they reflect your subjective impression of the importance of these items relative to the preceding items. For example, if the second statement seems 20 times as important, assign it a number 20 times greater, and so forth. Use positive fractions, whole numbers, or decimals, but make each assignment proportional to the importance as you perceive it.*

Magnitude estimates have been found to yield distributions that are log-normal.[20,44] Consequently, averaging is best done by taking the geometric mean of the estimates, averaging them

Exhibit 7.3.6 Paired Comparison Matrix for Desk Compass

	Items										
Items	B	C	D	E	F	G	H	I	J	Σ	%
A[a]	A-2[b]	A-3	A-3	A-1	A-2	A-3	0	I-1	J-3	14	16
B		B-1	B-2	E-1	0	B-1	H-2	I-3	J-3	4	5
C			C-1	E-3	F-2	0	H-2	I-3	J-3	1	1
D				E-3	F-3	G-1	H-2	I-3	J-3	0	0
E					E-1	E-2	0	I-1	J-2	10	12
F						F-1	H-1	I-2	J-2	6	7
G							H-3	I-3	J-3	1	1
H								I-2	J-1	10	12
I									J-1	18	21
J										21	25
Total										85	100%

[a] Letters refer to components as listed in Exhibit 7.3.2.
[b] Numerical rating scale: 0 = no difference in importance; 1 = minor difference in importance; 2 = medium difference in importance; 3 = major difference in importance.

across individuals, and finally normalizing them on a percentage basis as previously described. Computer programs that plot histograms of individual estimates and that calculate the geometric mean of the resulting distributions are useful. In addition to its simplicity, DME allows one to express the magnitude of the differences between items. Experience shows that it is most effectively used with less than 30 items.

Another version of assigning numbers to items in direct proportion to their perceived importance is category scaling. This procedure employs tailor-made descriptors assigned to ranges of numbers. Respondents first choose the category that best fits their overall perception and then assign a number within that category's numerical range that best matches their feeling of importance. A typical category scale is: 90–100, "most important"; 70–89, "very important"; 50–69, "moderately important"; 30–49, "slightly important"; and 10–29, "little or no importance." Category scaling, compared to DME, has the advantages of being less confusing, easier to understand, and more convenient to use. It promotes greater consistency among raters in terms of interpretation, perception, and quantification because everyone has the same reference set (scale). Also, since perceptions quantified by a category scale are not log-normal, and since all participants use the same scale, it is not necessary to use a geometric mean to calculate an average response. Scaling may, however, restrict a person's expression of perceived differences in importance between items because it may narrow the range of response.

The value measurement phase culminates in the quantification of functions and components according to value, importance (or worth), and cost. These parameters are used to select likely candidates for redesign, improvement, or elimination. Such candidates usually possess low value, high cost, low importance, or combinations thereof. They become the focus for improvement and speculation in the innovation phase. Referring to the compass example in Exhibit 7.3.4, soft cost, importance, and value of the components are posted to the respective functions on the FAST diagram. Components may serve several functions simultaneously, as in the case of the pencil leg, pin leg, screw, and nut. In this case importance and cost are allocated according to the perceived contribution of the components to the various functions. A similar allocation method is used by Ruggles.[45] Importance and cost for the secondary functions "position marker" and "rotate marker," as well as any design concept functions, were derived by summing the importance and cost of all the functions to the right that contribute to these functions. Value for all functions was individually computed from importance and cost summations at any point in the diagram.

Innovation Phase

Improvement of Value

The innovation phase is the creative part of the VE process and a vital step in the redesign process. Activities in this phase are geared toward creating alternative ways of accomplishing functions. Creative effort is aimed at eliminating or combining low-value secondary functions. The thrust of this phase is directed toward answering "What else will do the job?" and "How can we eliminate secondary functions and still perform the basic function(s)?"

There are many idea generating and collecting methods. Not all of the techniques may be appropriate to all situations. They can be classified along a continuum based on the quantity and uniqueness of ideas that their application produces. One must first decide what type of ideas he or she wants to accumulate. Do you want to collect currently existing ideas in order to search for a new solution? This is often sufficient. Do you want to take a quantum leap into the twilight zone above and beyond those ideas you already have? Experience shows that it is usually best to start with the simplest method and to progress to the more bizarre, mind-stretching methods. The simplest method is to collect ideas that already exist. This can be done by collecting from either documents or individuals by means of interviews, questionnaires, literature searches, and so on, or from groups by means of a meeting where people merely pool their ideas through the use of a facilitator. If no satisfactory ideas emerge from simple collection, then more unique ideation techniques may be tried. Many such techniques are available for use in developing new ideas. They are all designed to circumvent negative thinking and often uncover unexplored areas by stimulating creative thinking. They are designed to suppress the habitual perceptual, cultural, and emotional blocks that inhibit creativity.

Two basic precepts are common to all of the techniques. First, all criticism and evaluation are eliminated from the idea producing stage in which ideas are merely listed. This allows for a maximum production of ideas and prevents premature rejection of any potentially good ideas. Second, all ideas, even the apparently impractical ones, are considered at a later time. The reasoning here is to encourage everybody to explore even the seemingly impractical ideas and to feel free to express thoughts without fear of ridicule.

Exhibit 7.3.7 lists some creativity techniques in common use. References are also listed as a source of detailed information regarding their use and application.

Probably the most familiar technique is brainstorming. It has been widely used for a number of

Exhibit 7.3.7 List of Creativity Techniques and Their References

Technique	References
Brainstorming	22, 46–48
Gordon technique	22, 48, 49
Nominal group technique	70, 50–54
Checklists	22
Morphological synthesis	22, 48
Attribute listing	22
Catalog technique	70, 48
Synectics	46, 47, 49, 55, 56

years. The nominal group technique (NGT), developed by Delbecq and Van de Ven, is a relatively new technique which the authors have found to be more effective than traditional brainstorming. This effectiveness is supported by other research.[54,57] Brainstorming is performed by interacting groups as opposed to nominal groups. Interacting groups involve spontaneous group discussion. All communication takes place among members, with minimal controls and formal structuring. Nominal groups, on the other hand, involve individuals who work in the presence of others, but who do not interact verbally with one another except at specific designated and controlled periods. Various sources[46,50,51,53,54] discuss in detail the advantages and disadvantages of both techniques and give detailed instructions in the application of NGT. The authors highly recommend this technique.

Whatever techniques are used, the innovation phase often produces a multitude of ideas. The task now becomes one of selecting the most promising and practical ideas for further consideration and action. In the compass example, the handle, washer, pencil leg, and pin leg had low-value functions. They were the subjects of creativity sessions for improving, combining, or eliminating the functions they provided. Various combinations of function innovations produced several alternative designs. They were evaluated under the evaluation phase to produce the new design shown in Exhibit 7.3.2*b*.

Evaluation Phase

Prescreening–Qualitative Analysis of Value

The evaluation phase entails a selection process in which the ideas produced in the innovation phase are examined and a small number of ideas are selected. An extensive analysis of the value of all the ideas produced in the innovation phase is impractical since the time and effort involved would be excessive. Consequently, it is desirable to prescreen the large list in order to reduce it to manageable size.

Once screened, the higher-ranking items are subjected to a more discriminating quantitative evaluation. Two effective prescreening techniques are Pareto voting[58] and Q-sort.[59] Both methods are very fast and simple, yet provide a high degree of discrimination. The more ideas to be evaluated, the more useful these methods can be. They both furnish a starting point from which a more quantitative analysis can proceed. However, before these methods are used, the ideas should be examined for redundancy, which can be eliminated by combining highly similar ideas.

The principle behind Pareto voting is Pareto's law of maldistribution, which, when applied to VE, postulates that about 80% of the value of a list of items is invested in approximately 20% of the items. In application, participants vote (by secret ballot or overtly) on the value of the items on the list. Each voter is asked to select a certain number of items (about 20%) that he or she perceives as important. Voters are not asked to prioritize within the selected unit. Only one vote is allowed per item, and the full allotment of votes must be exercised. Votes by item are then postulated and tallied. The items receiving the most total votes are then discussed and considered for further investigation. Shillito[58] provides detailed instructions on the use of this technique.

In the Q-sort method each item is listed on an individual card. The rater is then required to sort the items physically into a hierarchy of value categories. Initially, they are sorted into two piles representing "high" value and "low" value (sort one). The second sort entails splitting the two piles into four, which represent "very high," "high," "low," and "very low" value. A third sort is then performed in order to obtain a middle category of "medium" value. It is obtained by sorting out from the two middle piles all the medium-valued items. This final sort produces five piles of cards representing items of "very high," "high," "medium," "low," and "very low" values. Candidates for further VE study can be selected from the highest-valued items.

Quantitative Analysis of Value–State 2 Value Measurement

The select group of items that emerges from the prescreening is subjected to a more rigorous and discriminating evaluation to arrive at two or three candidates for possible development. The same value measurement techniques discussed under the innovation phase–paired comparison and DME–can be used to evaluate the prescreened alternatives. However, a more effective and discriminating technique often used at this stage is criteria analysis.[38]

Criteria analysis is a matrix scoring technique especially designed for evaluating alternatives by judging their individual merit against a set of criteria important to their end use. A decision matrix is employed wherein alternatives to be judged are arrayed against these criteria. Weighting factors are assigned to the criteria in proportion to their perceived importance. The relative merit of each alternative is determined by scoring its performance against each criterion. Direct magnitude estimates or category scaling can be used for the scoring procedure. A weighted FOM for each alternative is derived by summing the products of the DME and the criteria weighting factors. The weighted FOM can then be normalized to a percentage.

Exhibit 7.3.8 depicts a completed criteria analysis matrix for the desk compass alternatives. The matrix illustrated was used to derive a weighted FOM and percentage importance for positive criteria only. A similar matrix was also used to derive a weighted FOM for negative criteria. A Value Index may be computed according to equation 6 by dividing the positive FOM, or percentage importance, for each alternative by the respective negative FOM (percentage cost). Value graphs as used in the state 1 value measurement are also useful at this stage for displaying alternatives according to importance and cost.

Criteria analysis has many advantages. The most important is that the process utilizes a common reference set of criteria. With paired comparison or DME, no common set of evaluation criteria is established. Each rater uses his or her own reference set and point of view. Criteria analysis also displays how each alternative satisfies each criterion. Decision rationale is documented and can be repeated in light of new information or challenge. Computer programs can be used to perform the computations involved in the process and to plot value graphs and histograms of individual ratings. A valuable side benefit of criteria analysis is the communication that ensues in the process of constructing the matrix. The technique is well suited to tasks in which one wants to select the best approaches from a small set of alternatives. Other similar matrix scoring techniques are also available, such as the weighted constraints method,[22,40] decision matrix,[25,26] Combinex,[23] Delphi,[16] and criteria function analysis.[24] Higher-order statistical techniques, such as correlation and cluster analysis,[60] can also be used for analysis of data. However, the reader is cautioned in using them with subjective data.

The most promising alternatives to evolve from the state 2 value measurement are now extensively reviewed for cost-effectiveness, technical feasibility, and possible implementation. Considerable spadework is done to answer questions such as the following: Can it be done? What if . . . ? Will it meet standards? What will it actually cost? Is it compatible with the system? What will be its impact on . . . ? The assistance of knowledgeable people external to the VE study team is often solicited in order to judge more fully the potential and feasibility of the most likely alternatives.

Exhibit 7.3.8 Criteria Analysis for Desk Compass Example

	Criteria						
Item (Item No.)	Ease of Use (15)[a]	Ease of Manufacturing (30)	Safety (20)	Quality (25)	Attractiveness (10)	Weighted FOM	% I
Regular compass (1)	100[b]	30	50	70	10	545	21.8
Drawing, Exhibit 7.3.2*b* (2)	80	100	90	50	30	845	33.9
Alternative 2[c] (3)	30	50	70	40	70	505	20.2
Alternative 3[c] (4)	50	60	80	50	60	600	24.1
Total						2495	100.0

[a] Weighted factor.
[b] Ratings derived by scaling. They may also be derived by DME, paired comparison, or similar techniques.
[c] Alternatives not illustrated in text.

New criteria may evolve that may warrant further analysis. Other techniques that are helpful in locating unforeseen problems are potential problem analysis[61] and impact analysis as described by Shillito.[62] Testing the alternatives may involve developing prototypes and so on as well as extensive economic analysis, including cash flow analysis. When final evaluation is complete, the remaining alternative is ready for implementation.

Implementation Phase

This phase concerns the preparation and presentation of the VE team's recommendations to management. A report is prepared that describes the proposal(s) and lists suggested action plans for implementation. To minimize rejection, careful consideration should be given to this report and its reception. Answering basic communications questions as listed in the implementation phase of Exhibit 7.3.1 can be helpful. If a written document is presented to a decision maker for implementation, it is often more effective to include a brief two or three page executive summary. A separate study team report containing detailed information can be prepared and used as a working document during implementation. An oral presentation is often an excellent supplement to a written document.

Actual implementation of the recommendations may be performed by the VE team, but is often carried out by others. The VE teams are often dissolved after their recommendations are acted upon, and the responsibility for implementation is given to line personnel. The study team provides the necessary input for the decision maker, and the team members, along with the detailed report, are available to assist the department or individuals responsible for implementation. Assistance is in the form of clarification or additional input. Monitoring should be performed to check on progress and to see that implementation takes place.

A major factor in obtaining acceptance and use of the recommendations involves making the decision maker feel comfortable with the recommended change(s). Having the decision maker as a team member is very helpful. Where this is not possible or practical, the decision maker should be periodically informed of the VE team's activities. It is helpful to solicit his or her input during the course of the team's work. Scheduling periodic reviews to solicit input is very helpful and alleviates anxiety over change or potentially threatening information. It allows the decision maker to contribute to the team and fosters participation in the direction and outcome of the study. Such participation increases the chances for acceptance and positive action. Implementation is more likely if the VE study team prepares a suggested plan of action and a timetable for accomplishment. Planning and scheduling methods such as PERT are very useful in actually implementing VE recommendations.[22,63]

7.3.5 USING THE PROCESS

Usually the VE process is followed in sequence, phase by phase. However, VE is not a rigid, one-way process in terms of sequence of activities or application of the specific techniques. The process is really cyclical and modular. In practice it is frequently necessary to return to a previously completed phase for more information or additional work prior to progressing to the next phase or making a decision. In this respect the process is iterative and looping.

It should be noted, too, that the techniques themselves are not restricted to a particular phase in the process flow. It is not necessary to use all the techniques on all the problems. Only those that yield useful information are used; those that are not applicable are eliminated. The techniques are also useful by themselves. They have wide application in many areas, some of which have little or nothing to do with VE. For example, technological forecasting, discussed in Chapter 11.1, employs many VE techniques, such as FAST diagramming, as well as many of the value measurement techniques. Because the techniques are self-containing, the VE process should also be perceived as modular. Depending on the nature of the request, it may not be necessary to use the entire VE process. Although it may sound like heresy to the long-standing proponents of the VE process, using the full process is not always mandatory or a good idea. Beware of overkill.

7.3.6 BEHAVIORAL AND ORGANIZATIONAL ASPECTS OF VE

Although the VE process is a rational, systematic, and structured process, its foundation is structured around the effective use of people in teams. Once we add people, teams, organizations, emotions, and power to the process, the process becomes more complex. Consider the following problems and roadblocks frequently encountered in the VE process.

1. Value engineering is usually performed with teams. They can waste time, be overly conservative, and avoid decisions.
2. Individuals involved in VE usually have other jobs and are already busy.
3. Strong parochial interests are common.

4. The output of a VE study may be threatening, especially to designers, planners, and current decision makers.
5. Emotional as well as rational conflict of interest is often generated.

The success of a VE study is enhanced if organizational and behavioral aspects of the study are considered early in the process. Roadblocks can be anticipated and planned for before they occur. Some factors to be considered are:

1. **Organizing for VE.** Organization of the VE team itself is just as important as selecting a project or using the process. There are several important questions to ask. How should a team be organized? How many people should be involved? Where should they come from? What level? Are the team members at comparable levels of the organization? Are the "decision makers" also the "doers"? Who should be the leader? What are the roles of the various members? Who does the team report to? How much time will they need? These questions and others need to be addressed at the inception of the VE process. Answers to these and similar questions are discussed in the section on the origination phase as well as in Lashutka and Lashutka.[64]
2. **Decision Making.** How are decisions made within the team? Who outside the team can influence, veto, or approve recommendations? What is the process for approval? How can the opinions of management be determined without risking a pocket veto? Can the decision makers on the team speak for or make decisions for their organizations? The key is to analyze early in the VE process how decisions will actually be made.
3. **Gaining Commitment.** A VE recommendation is only as good as the commitment to implement it. Who must budget to implement it? Who will be affected? Who gains and who loses? Who can delay or block the recommendation? How do we plan to gain commitment of key parties? Again, analyze the commitment issue early and have a plan for gaining the necessary commitment.
4. **Effective Meetings.** Teams can work effectively or can waste time and make poor decisions (or none at all). The structure and momentum of the team is maintained by the VE process, methodology, and techniques. They provide the vehicle for overcoming obstacles. They help encourage rational conflict and minimize emotional conflict. They help the team work at the right level of detail. However, questions that still must be considered are: How will you get a team started quickly? How will you check the health of the meetings? What will be done if members do not attend meetings or do not do their homework?

People and teams can complicate the VE process, but are a necessary part of it. Behavioral science aspects must be treated as carefully as any other part of the VE process. Look ahead, predict the problems, and carefully plan to deal with them. Team building workshops conducted prior to doing the actual VE project are very effective and well worth the time. The purposes of the workshop are (1) to help the members develop communications skills and learn to work together as a group and (2) to use the group skills effectively to analyze problems, make decisions, and work as a team. There are some excellent sources on team building and process consultation.[11,64–67] Reigle[27] also addresses the behavioral and organizational concerns quite well. For further information regarding behavioral science aspects of organizations, the reader is referred to Chapters 2.1 and 2.2.

7.3.7 RESULTS AND BENEFITS OF VE

Scharf[68] defines "effectiveness" as "doing the right thing" and "efficiency" as "doing things right." Value engineering provides assistance in both areas. Qualitative techniques such as function analysis guide VE into an effective domain of work on "the right thing," while the quantitative techniques provide a framework for being efficient within that domain. From a manager's perspective, VE provides a method to prioritize and put dollars on high-value items. There is a growing use of VE as a prioritization process to optimize resource allocation. Certainly there are many mathematical or operations research techniques for resource allocation (see Sections 9 and 14). They are efficiency techniques geared more to "after the fact," that is, after the resource recipients have been identified. In contrast, VE is an effectiveness technique that provides a means of identifying the proper recipients. From a generic point of view, VE

1. Enables people to pinpoint areas that need attention and improvement.
2. Provides a method of generating ideas and alternatives for possible solutions to a concern.
3. Provides a means for evaluating alternatives.
4. Allows one to evaluate and quantify intangibles and to compare apples with oranges.
5. Provides what Roper[69] describes as a "vehicle for dialogue." It does this by (a) allowing large amounts of data to be summarized in concise form, (b) allowing new and better questions to be asked, and (c) using numbers to communicate in an information searching mode. Numbers, in this context, replace semantics and give us a common language by making opinions on sometimes vague

concepts measurable. Through employment of the interdisciplinary team and measurement techniques, the process broadens the perspective, which, in turn, increases communication. The increased communication ultimately increases effectiveness.

6. Documents the rationale behind recommendations and decisions. The process can be repeated and explained as well as amended in light of new or different information or challenge.
7. Materially improves the value of goods and services.

It is important to reemphasize that VE is applicable to all areas: hardware, purchasing, products, services, systems, or procedures. The authors by no means imply that VE is a panacea for all problems or that its results are axiomatic inputs to decision making. We do believe that the process and its techniques can assist industrial engineers and managers in their common task of value improvement.

REFERENCES

1. L. D. MILES, *Techniques of Value Analysis and Engineering*, McGraw-Hill, New York, 1961, p. 1.
2. J. L. PIERCE, P. G. GOSCHER, E. D. SPARTZ, M. N ZABYCH, and E. D. JOHNSON, "Guidelines for Value Engineering (VE)," *Value World*, Vol. 2, No. 6 (March–April 1979), pp. 7–9.
3. A. E. MUDGE, *Value Engineering, A Systematic Approach*, McGraw-Hill, New York, 1971, p. 16.
4. E. BOUEY, "Value Engineering Is Unique," *Interactions*, Vol. 5, No. 8 (December 1979), p. 1.
5. A. J. DELL'ISOLA, *Guide for Application of Value Engineering to the Construction Industry*, 2nd ed., McKee-Berger-Mansueto, Inc., Washington, DC 1972.
6. F. FIFIELD, "Administrative Value Analysis," *Industrial Engineering*, November 1973, pp. 24–28.
7. T. C. FOWLER and B. HIGGINS, "Organization Value Analysis Made Easy," *Proceedings, SAVE Regional Conference*, Detroit 1974, pp. 3.1–3.7.
8. D. J. DEMARLE, "The Application of Subjective Value Analysis of Training," *Proceedings, SAVE Regional Conference*, Detroit 1971, pp. 5.1–5.6.
9. R. W. DOBLES, P. M. DROST, S. S. HAZEN, and R. C. HINKLEMAN, "If You're Not Doing It Already, Verify Your Training Objectives," *Training*, Vol. 16, No. 12 (December 1979), pp. 36–45.
10. P. E. ILLMAN, "Value Analysis in the Non-Hardware Field: A New Approach in Design and Training," *Proceedings, Society of American Value Engineers*, Vol. 6, May 1971, pp. 43–62.
11. W. J. RIDGE, *Value Analysis for Better Management*, American Management Association, New York, 1969.
12. R. F. VALENTINE, *Value Analysis for Better Systems and Procedures*, Prentice-Hall, New York, 1970.
13. C. RAND, "New Venture Value Search (Making Companies Well Through VE/VA)," *Value World*, Vol. 3, May–June 1979, pp. 17–23.
14. D. J. DEMARLE, "Use of Value Analysis in Forecasting," paper presented at James R. Bright's Technology Forecasting Workshop, Castine, ME, Industrial Management Center, Austin, TX, 1973.
15. G. D'ASCANIO, "Social Value Analysis," *Proceedings, Society of American Value Engineers*, Vol. 10, May 1975, pp. 174–178.
16. S. F. LOVE, "Resource Allocation by the Delphi Decision Process," *Optimum*, Vol. 6, No. 1 (1975), pp. 39–48.
17. L. GROENEVELD, "Value Analysis and the Marketing Concept," *Industrial Engineering*, February 1972, pp. 24–27.
18. L. L. THURSTONE, "A Law of Comparative Judgment," *Psychological Review*, Vol. 34, 1927, pp. 273–286.
19. L. L. THURSTONE, "The Measurement of Values," *Psychological Review*, Vol. 61, No. 1 (1954), pp. 47–58.
20. S. S. STEVENS, "A Metric for the Social Consensus," *Science*, Vol. 151, 1966, pp. 530–541.
21. E. GALANTER, "The Direct Measurement of Utility and Subjective Probability," *American Journal of Psychology*, Vol. 75, 1962, pp. 208–220.

22. U.S. DEPARTMENT OF DEFENSE, *"Principles and Applications of Value Engineering,"* Vol. 1, U.S. Government Printing Office, Washington, DC, 1968.

23. C. FALLON, *Value Analysis to Improve Productivity*, Wiley-Interscience, New York, 1961.

24. R. S. SCHERMERHORN and M. I. TAFT, "Measuring Design Intangibles," *Machine Design*, Vol. 40, December 1973, pp. 108–112.

25. D. BORCK, "Using Decision Theory in Value Analysis Studies," *Systems and Procedures Journal*, March–April 1968, pp. 28–31.

26. C. W. DILLARD, "Value Engineering Organization and Team Selection," *Proceedings, Society of American Value Engineers*, Vol. 10, May 1975, pp. 11–12.

27. J. REIGLE, "Value Engineering: A Management Overview," *Value World*, Vol. 3, No. 3 (October–December 1979), pp. 5–8.

28. W. G. BANCROFT, "Value Engineering Makes Design to Cost Easy," *Proceedings, Society of American Value Engineers*, Vol. 10, May 1975, pp. 124–127.

29. B. S. BLANCHARD, "Life Cycle Costing—A Review," *Terotechnica*, Vol. 1, 1979, pp. 9–15.

30. C. W. BYTHEWAY, "FAST Diagrams for Creative Function Analysis," *Journal of Value Engineering*, March 1971, pp. 6–10.

31. C. W. BYTHEWAY, "Innovation to FAST," *Proceedings, SAVE Regional Conference*, Detroit, 1972, pp. 6.1–6.7.

32. T. F. COOK, "Function Analysis Systems Technique Task Oriented FAST Diagram," *Value World*, Vol. 3, No. 2 (July–September 1979), pp. 24–28.

33. R. CREASEY, *FAST Manual*, Value Design Press, Fort Worth, TX, July 1973.

34. J. K. FOWLKES, W. F. RUGGLES, and J. D. GROOTHUIS, "Advanced FAST Diagramming," *Proceedings, Society of American Value Engineers*, Vol. 7, June 1972, pp. 45–52.

35. E. RONEN, "Functional Analysis of Procedures and Organizational Structures," *Performance*, Vol. 5, No. 6 (November-December 1975), pp. 12–17.

36. F. X. WOJCIECHOWSKI, "FAST Diagram—Its Many Uses," *Proceedings, SAVE Regional Conference*, Detroit, 1972, pp. 10.1–10.4.

37. D. J. DEMARLE, "A Metric for Value," *Proceedings, Society of American Value Engineers*, Vol. 5, April 1970, pp. 135–139.

38. D. J. DEMARLE, "Criteria Analysis of Consumer Products," *Proceedings, Society of American Value Engineers*, Vol. 6, May 1971, pp. 267–272.

39. D. J. DEMARLE, "The Nature and Measurement of Value," *Proceedings, 23rd Annual AIIE Conference*, May 1972, pp. 507–512.

40. R. F. BECKER, "A Study in Value Engineering Methodology," *Performance*, Vol. 4, No. 4 (July–August 1974), pp. 24–29.

41. R. L. CROUSE, "Function and Worth," *Proceedings, Society of American Value Engineers*, Vol. 10, May 1975, pp. 8–10.

42. D. M. MEYER, "Direct Magnitude Estimation: A Method of Quantifying the Value Index," *Proceedings, Society of American Value Engineers*, Vol. 6, May 1971, pp. 293–298.

43. A. E. MUDGE, "Numerical Evaluation of Functional Relationships," *Proceedings, Society of American Value Engineers*, Vol. 2, April 1967, pp. 111–123.

44. A. R. FUSFELD and R. N. FOSTER, "The Delphi Technique: Survey and Comment," *Business Horizons*, Vol. 14, June 1971, pp. 63–64.

45. W. F. RUGGLES, "Cost Function Relationships," *Proceedings, Society of American Value Engineers*, Vol. 10, May 1975, pp. 1–7.

46. T. RICKARDS, *Problem-Solving Through Creative Analysis*, Wiley, New York, 1974.

47. M. I. STEIN, *Stimulating Creativity, Part 2: Group Procedures*, Academic Press, New York, 1975.

48. C. WHITING, *Creative Thinking*, Reinhold, New York, 1958.

49. W. J. J. GORDON, *Synectics, The Development of Creative Capacity*, Harper & Row, New York, 1961.

50. A. L. DELBECQ and A. H. VAN DE VEN, "A Group Process Model for Problem Identification and Program Planning," *Journal of Applied Behavioral Science*, Vol. 7, No. 4 (1971), pp. 466–491.

51. A. L. DELBECQ, A. H. VAN DE VEN, and D. H. GUSTAFSON, *Group Techniques for Program Planning: A Guide to Nominal Group and Delphi Processes*, Scott, Foresman, Glenview, IL, 1975.

52. G. P. HUBER and A. DELBECQ, "Guidelines for Combining the Judgments of Individual

Members in Decision Conferences," *Academy of Management Journal*, Vol. 15, No. 2 (1972), pp. 161–174.

53. M. H. MELCHER, "Amplifying Group Decision-Making in the Project Management Environment," *Project Management Institute Proceedings*, Montreal, 1976, pp. 248–256.

54. A. H. VAN DE VEN and A. L. DELBECQ, "Nominal Versus Interacting Groups for Committee Decision-Making Effectiveness," *Journal of the Academy of Management*, Vol. 14, June 1971, pp. 201–212.

55. T. ALEXANDER, "Synectics: Inventing by the Madness Method," *Fortune*, August 1965, pp. 165–171.

56. E. RAUDSEPP, "Forcing Ideas With Synectics, A Creative Approach to Problem Solving," *Machine Design*, October 16, 1969, pp. 134–139.

57. T. RICKARDS, "Brainstorming: An Examination of Idea Production Rate and Level of Speculation in Real Managerial Situations," *R&D Management*, Vol. 6, No. 1 (1975), pp. 11–14.

58. M. L. SHILLITO, "Pareto Voting," *Proceedings, Society of American Value Engineers*, Vol. 8, May 1973, pp. 131–135.

59. A. F. HELIN and W. E. SOUDER, "Experimental Test of a Q-Sort Procedure for Prioritizing R & D Projects," *IEEE Transactions on Engineering Management*, Vol. EM-21, No. 4 (November 1974), pp. 159–164.

60. M. L. SHILLITO, "Cluster Analysis: Amplification of the Value Index?," *Proceedings, Society of American Value Engineers*, Vol. 9, April 1974, pp. 144–150.

61. C. H. KEPNER and B. B. TREGOE, *The Rational Manager*, McGraw-Hill, New York, 1965.

62. M. L. SHILLITO, "Impact Analysis," paper presented at James R. Bright's Technology Forecasting Workshop, Castine, ME, The Industrial Management Center, Austin, TX, June 1977.

63. J. W. GREVE and F. W. WILSON, Eds., *Value Engineering in Manufacturing*, Prentice-Hall, Englewood Cliffs, NJ, 1967.

64. S. LASHUTKA and S. C. LASHUTKA, "The Management of Team Development in Value Analysis/Engineering," *Value World*, Vol. 3, No. 4 (January–March 1980), pp. 5–10.

65. N. N. BARISH, "Evaluating Intangibles," in *Economic Analysis*, N. N. Barish (Ed.), McGraw-Hill, New York, 1962.

66. W. BOOTHE, *Developing Teamwork*, Golle and Holmes Corporation, Minneapolis, July 1974.

67. E. H. SCHEIN, *Process Consultation: Its Role in Organization Development*, Addison-Wesley, Reading, MA, 1969.

68. A. SCHARF, "Management by Emphasis," *Proceedings, 22nd Annual AIIE Conference*, May 1971, pp. 11–16.

69. A. T. ROPER, "Technology Assessment: A Vehicle for Dialogue," paper presented at Second International Congress On Technology Assessment, University of Michigan, Ann Arbor, October 1976.

70. A. M. BIONDI, Ed., *Have an Affair With Your Mind*, Creative Synergetic Associates, Ltd., Great Neck, NY, 1974.

CHAPTER 7.4

Conventional Machine Tools

KENNETH M. GETTELMAN

Modern Machine Shop Magazine

7.4.1 INTRODUCTION

The machine tool is the royalty of capital equipment. Only the machine tool can reproduce itself. Every manufactured item can be traced to capital equipment, and all capital equipment is the product of one or more machine tools. Thus the machine tool is the fountainhead of all industrial processes and all manufactured and processed items.

Although the potter's wheel, a basic turning tool, has been in existence for some 3000 or 4000 years, with historical records containing references to other basic machine tool concepts, it was John Wilkinson of Birmingham, England, working with James Watt, the inventor of the steam engine, who developed the first in the continuum of today's machine tools. The first machine tool, the Wilkinson boring mill, was instrumental in enabling Watt to develop the precision cylinder required for his steam engine. The year was 1776, and the event was historical–mechanical energy was harnessed and the machine tool developed, launching the modern industrial age. The first boring mill was water powered, but the steam engine it helped produce was utilized to power additional machine tools to produce more steam engines to launch the productive capital formation process.

7.4.2 DESCRIPTION OF MACHINE TOOLS

There are only two basic families of mechanical machine tools. The first is the group of chip removal machines that shape metal by removing individual chips from the workpiece. They include milling, drilling, boring, planing, shaping, broaching, turning, grinding, and sawing machines. The other family is the category of forming machine tools, including punch and forming presses, press brakes and shears, and casting and molding machines. Feeding and material handling equipment is not considered a machine tool, but functions as an accessory item.

The basic element of the machine tool is the frame, upon which are mounted the powered spindle and the work holding and toolholding components. Traditionally the frame has been a casting, but many machine tools today are constructed from heavy plate sections welded together. This is weldment-type construction.

The majority of machine tool frames are the basic "C" type, with the powered toolholding spindle mounted on the upper arm of the "C" and the work holding table mounted on the lower arm. Both the chip-type and the forming-type machine tools rely heavily upon the basic C-frame. The other alternative is a closed box, or "O," type of frame in which there are a bed that holds the workpiece and two uprights that support a cross member upon which is mounted the powered spindle containing the cutting tool. The single purpose of the frame is to provide a rigid support system for the cutting tool and workpiece so that they may effectively interact with each other to produce a machined workpiece to the specified shape and tolerance.

An essential element of all machine tools is the powered spindle, which holds either the cutting tool or, in the case of turning equipment, the workpiece chuck. Historically the drive spindle received its power from an overhead or jack shaft that ran through the plant. Each machine was powered by a belt deriving its power from the shaft. Today all machine tools have their individual motors (nearly all electric, but a few hydraulic) as the main driving force. From the main motor there may be shafts that provide motion for the driving of other machine elements such as the table, upon which is mounted the workpiece, or the feed screws, which control the rate of workpiece feed into the cutting tool.

One universal rule of machining says that the workpiece must be held rigidly on the machine tool. The spindle-type machine tools, such as milling, drilling, and boring machines, provide slotted tables upon which the workpiece may be clamped.

The same rule that applies to the workpiece also applies to the cutting tool–it must be rigidly held. For the spindle-type machines there are a number of standard spindle nose arrangements and toolholding systems. Some are held in tapered holders by friction. Others are held by powered draw bars, and still others are held in different types of chucks. This in itself is an extremely extensive subject. From an industrial engineering standpoint, many plants standardize, as much as possible, their toolholding systems to reduce both inventory and industrial processing requirements. Thus, with the workpiece and tool both rigidly held, it is the function of the machine tool to power the cutting tool, and usually the workpiece, in motions that will produce a machined part. With some very simple machine tools, it is the operator who powers the actual feeding of the workpiece or tool. An example is the very simple drill press. A motor-driven spindle powers the twist drill rotation, while the operator forces the advance of the drill into the workpiece.

7.4.3 CHIP-TYPE MACHINES

All chip-type machining can be broken down into the three basic elements of feed, speed, and depth of cut. These elements apply to the single cutting tool of the lathe, drill press, planer, or shaper or to the multiple-tooth cutting tools such as milling machines or saws.

Feed refers to the thickness of the bite of the cutting tool into the workpiece (Exhibit 7.4.1). For most chip-type machining work the feed usually ranges from 0.005 to 0.050 in. (0.127 to 1.27 mm) per tooth or cutting edge, depending upon the type of workpiece material, its toughness and hardness, the cutting tool material, and the inherent capabilities of the machine tool. Common feed rates are 0.005 to 0.015 ipt (0.187 to 0.371 mmpt).

Speed is a measure of how fast or at what rate the cutting tool traverses the workpiece. It is usually expressed in feet per minute or meters per minute. With today's machines and cutting tool materials, speed may range anywhere from 100 to 4000 fpm (30 to 1200 mpm), with 400 to 600 sfpm (120 to 150 mpm) being a common range.

Exhibit 7.4.1 All Metal Cutting, Regardless of the Tool, Involves Feed, Depth of Cut, and Speed. Feed Is Usually Expressed in Terms of the Distance That the Tool Advances Into the Work With Each Revolution, With Each Stroke, or Within a Given Time Period. The Depth of Cut Is How Far the Tool Penetrates Into the Workpiece, and the Speed Is How Fast the Cutting Tool Is Forced Through the Workpiece.[a]

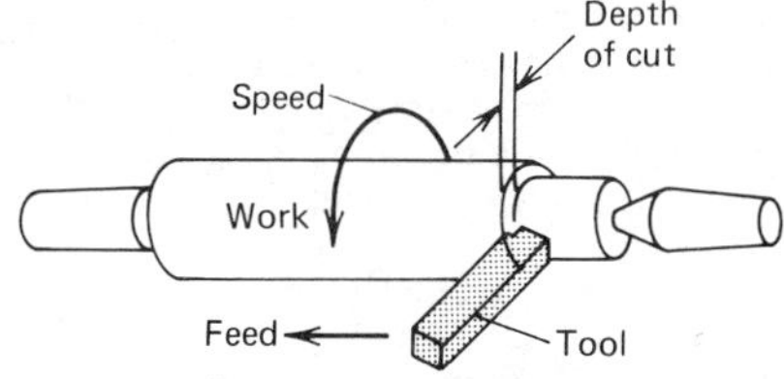

Turning–work rotates

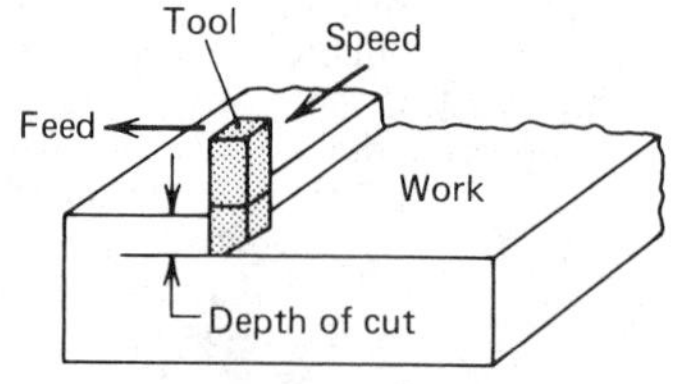

Single-point tool moves through work

[a]Courtesy of *Modern Machine Shop* magazine.

Depths of cut may range anywhere from 0.005 to 1 in. (0.127 to 25.4 mm), although most fall within a range of 0.125 to 0.375 in. (2.8 to 8.4 mm). The operation of any chip-generating-type machine tool involves the selection of a feed, speed, and depth of cut. The machining parameters may be selected by the machine operator or determined as a function of industrial engineering. In any machining operation there is an optimum combination of feed, speed, and depth of cut that yields the lowest-cost machining or the highest rate of production, which are usually not the same. For a comprehensive resource on recommended machining parameters for all kinds of materials, with the currently employed types of cutting tool materials and formulas for determining optimum rates, the reader is referred to *Machinability Data Handbook*, published by Metcut Research Associates, 3980 Rosslyn Drive, Cincinnati, OH 45209.

Milling Machines

Milling machines range from small 2 hp vertical spindle models used in the toolroom to massive 100 hp models used for aerospace or similar-type work. They share the common factor of employing a multiple-tooth milling cutter. It may be made from solid tool steel or may use inserted teeth. In most instances the milling machine table moves the workpiece past the rotating cutter. There are a few traveling-column-type milling machines in which the workpiece remains stationary and the cutter is passed through the workpiece.

The conventional milling machine is available in many different variations. The horizontal spindle models often are equipped with supported arbors upon which are mounted shell-type or circular cutters for milling top surfaces or milling slots in workpieces. The horizontals may also be tooled with face or end milling cutters for working the vertical surfaces of a workpiece.

Vertical spindle milling machines may also be tooled with a variety of different-type cutters, including face and end mills for working the top surface of a workpiece.

Except for the small, inexpensive models, the conventional milling machine will have a powered feed system running off the main spindle motor. There will be controls for selecting any one of a range of feeds and also clutches for engaging the power feed. Milling machines also have a range of available speeds.

Within the milling machine group are universal mills with tables that can be tilted or swiveled for special orientation of the workpieces. Milling machines may be equipped with special heads to do vertical-type work on a horizontal machine, and vice versa. It is also possible to mount milling heads on special frameworks for unique applications, such as construction or shipbuilding, where the machine must be taken to the work. In all applications of the multiple-tooth milling cutter, the proper feed, speed, and depth of cut are keys to successful milling operations.

Drilling and Boring Machines

The drilling and boring group of machine tools is very similar in concept to the milling family. The difference lies in the cutting tools. Drilling and boring are single-point-tool metal cutting operations.

For the most part, drilling machines have the basic C-frame configuration. They may range from small, fractional horsepower models used for fine handwork to massive 50 hp models used with full 3 in. diameter twist drills. The small hand or bench models are often called "sensitive drills" because the feed is determined by the operator as he or she forces the cutting tool into the workpiece. The larger drilling machines will have power feed. By its very nature the twist drill requires a heavy thrust to force its dead center into the workpiece. It is for this reason that drilling machines need a rigid framework to produce an accurate hole.

One important variation of the drilling machine is the radial drill. Instead of a fixed frame, the radial has a solid base and a round column. An arm is mounted on the column and may swing through as much as a 240° arc, while the spindle head may be shifted to any point along the arm. This allows the drilling spindle to reach all locations on large, flat workpieces.

Drilling machines are also available with multiple-spindle drilling heads. In some instances several hundred individual twist drills may be mounted in such heads. This type of equipment is usually employed in long production runs on workpieces with many holes. A variation is a gang drill arrangement, where individually powered drill spindles are clustered in a pattern for drilling multiple holes in a single workpiece.

Tapping is the process of machining threads in drilled holes. The key to successful tapping is the ability to advance the tapping tool into the hole at a rate matching the lead of the thread. The machine spindle must be reversed to extract the tap from the hole without damaging or stripping the threads.

Unlike the drilling machine, which produces the hole from the solid, the boring machine enlarges and finishes an already produced hole. The boring tool is also a single-point cutter. Its feed, speed, and depth of cut are often very light since the final objective is usually more of a fine machining to precision tolerance than massive metal removal.

The similarity of machine construction, the universality of operations, and the commonality of chip formation are such that some machines are built as drilling, boring, and milling machines–sometimes referred to as DBMs–to handle all three machining operations. In order to get the high precision often required of boring, it may mean the machine cannot be used for heavy milling, but the ability to accomplish three machining operations may make a combination machine worthwhile.

Planing and Shaping Machines

The planing and shaping machines are among the time-honored conventional machine tools that are found in many plants, but they are manufactured in very small numbers today because they use a single-point cutting tool. The multiple-tooth milling cutter has proved to be much more applicable and efficient in many machining situations formerly handled by planing or shaping.

In a shaping operation the workpiece remains rigid on the machine table while the single-point shaping tool is passed through the workpiece. Each tool pass at a determined feed, speed, and depth of cut removes a portion of the workpiece surface.

Planing is just the opposite. The cutting tool is rigidly held while the workpiece is carried past the cutting tool by the moving table. With each pass of the table, the feed of the cutting tool is advanced. Feeds, speeds, and depths of cut are very similar to those found in turning a round workpiece.

The typical toolroom shaper of past years had a stroke of 6 to 10 in. and was used to machine the surface of smaller workpieces. The planers, on the other hand, were often made with beds 50 ft or even longer. Often the cutting tool was supported by a bridge that straddled the table. For special oversize workpieces, some planers were made with an open side or C-frame concept.

Many of these older units have been reworked, and milling heads have been mounted on the cross arm. These units are known as planer mills.

Broaching Machines

For very fast rates of metal removal, the broaching machining method is unsurpassed. The broaching cutting tool is a multiple-tooth cutter which may be as long as 10 ft. Each outer tooth is located so that it advances into the workpiece the desired depth of cut from the preceding tooth. In the broaching process the broaching machine either pulls or pushes the cutter through or past the workpiece surface to be machined. The reverse may also be true. The workpiece may be forced through or past a stationary cutting tool.

A very common form of broaching is the generation of a spline on the inside of a ring. The workpiece is clamped to a table, and the broaching tool is pulled through the workpiece. In a single stroke the workpiece is machined.

Gears are often broached by pushing them through a "pot" broaching tool with its cutting teeth on the inside so that it machines the outside teeth of the gear blank.

Surfaces are broached by rigidly holding the workpiece and pulling a rigidly supported tool over the surface to be machined.

Although broaching is very fast in terms of its ability to remove metal from a workpiece, the method has several drawbacks. Broaching cutting tools are extremely expensive. The process requires solid fixturing to hold the workpiece, and the machines must be powerful and massive. Setup time is often lengthy. Thus broaching has found its greatest application in the automobile and mass production industries, where the machine, setup, and tooling costs can effectively be offset by the rapid machining rates offered by the method.

Turning Machines

The machine tool for machining the round workpiece is the lathe. The basic engine lathe has a C-frame construction, with the powered headstock at the left, as the operator faces the machine, and the tailstock at the right. The headstock is equipped with a chuck or other work holding and driving device, whereas the tailstock supports the outboard end of the workpiece. The single-point cutting tool is mounted on a toolholder, which is supported by a cross slide, so that the tool may be adjusted to the proper diameter of the workpiece.

One of the key elements of the lathe is the lead screw, which constantly feeds the cutting tool into the workpiece as it rotates. By setting the proper combination of workpiece rotation with screw advance, it is possible to thread a workpiece as well as machine its surface.

In the hands of a skilled operator, the basic engine lathe is a highly versatile machine tool. Engine lathes are available in sizes from tiny bench models used by jewelers to massive units with distances of more than 100 ft between centers used to turn large machine components.

A variation of the engine lathe is the turret lathe with a tool turret mounted on the cross slide.

The turret may hold four, five, six, or eight cutting tools. The center of the turret is referenced to a line through the centerline of the spindle. The cutting tools are usually employed on the face of a workpiece rather than on its periphery. Normally the workpiece is chucked without any outboard support; thus the turret lathe is ideal for workpieces with an emphasis on diameter rather than length.

The ram-type turret lathe has a special ram that advances the tool turret into the workpiece. The saddle type has the whole saddle or cross slide advance to the workpiece.

The automatic chucking machine is a logical outgrowth of the turret lathe and features an automatic sequence of operations. Instead of the operator's actuating tool selection and advancement into the workpiece, the automatic chucking machine sequences through a completely automatic cycle. One cutting tool is advanced and completes its operation. It then retracts, and the next cutting tool on the turret is rotated into place and advanced into the workpiece. This continues until the workpiece is finished.

Many of the automatic chucking machines have an inverted C-frame construction, with the mouth of the "C" at the bottom. This allows the chips generated by machining to fall into a pan or conveyor for easy removal from the machining area.

Another important member of the turning machine family is the automatic bar or screw machine. These units are fully automatic and are used to turn workpieces up to about 2 in. in diameter. They are similar to the automatic chucking machines in that the tool sequencing is automatic. They differ in that the workpiece is a piece of bar stock fed through the spindle of the machine rather than a blank mounted in the chuck. The bar is automatically advanced to a stop. The machining takes place, followed by an automatic cutoff. Once set up, these units can run for extended periods on their own. They may even be equipped with automatic bar feeders so that, when one bar is consumed, another is fed into place.

Another variation, used for long production runs, is the multiple-spindle automatic, which is available in a four-, six-, or eight-spindle variety. Each spindle is mounted with a section of bar stock, and the entire spindle carriage rotates from tooling station to tooling station. Thus, instead of the tools rotating on a turret, with each tool utilized only one sixth of the time on a six-tool turret, each tool on a multiple-spindle automatic is in constant use, and one workpiece is completed with each tool cycle of the machine. The disadvantages are that each spindle must rotate at the same speed so that the top speed is governed by the slowest operation. Likewise, the tool turret cycle is governed by the longest single machining operation. In addition, the multiple-spindle unit is a complex machine to set up and tool. It has proved itself over the years in volume production—especially in the automotive and appliance industries.

Grinding Machines

One important type of chip generating machine tool is the abrasive group that employs a grinding wheel or abrasive segment. To some it seems an anomaly, but the abrasive wheel or stone does remove metal by a chip process. Each abrasive grain, as it passes through the workpiece, removes a small particle of metal, which, when viewed under a magnifying glass, is seen to be a small chip.

As with most machine tools, most grinding machines have a basic C-frame construction. The common surface grinder employs a wheel that uses its periphery as the portion in contact with the workpiece. Generally the wheel is mounted on a horizontal spindle, and the workpiece is mounted on a reciprocating table that travels below the wheel. With each pass of the table, the wheel is fed progressively into the workpiece. Grinding feeds are light, with 0.002 in. (0.050 mm)/pass being a substantial feed rate. Surface grinders come in all sizes, from small toolroom models with a 6 by 12 in. (150 by 300 mm) table to a unit that may have a table 50 ft in length with 50 hp or more driving the wheel spindle.

Another important type of grinding machine is the cylindrical grinder, which mounts a workpiece between two centers, with the wheel head traversing on ways from one end of the workpiece to the other as the workpiece rotates.

A variation of the cylindrical grinder is the universal grinder, in which the wheel head, the worktable, or both may be set at different angles for grinding tapers, shoulders, and other variations of cylindrical paths.

The internal grinder employs a small wheel on the end of an arbor capable of reaching the internal surface of cylindrical shapes. The workpiece is mounted on a powered work holding spindle that rotates as the grinding wheel is both fed into and reciprocated the length of the surface to be ground.

The jig grinder is a specialized type of unit that employs a small abrasive wheel on the end of a vertical spindle noted for its precision. The machine is used primarily for very close tolerance toolroom work. The spindle is normally vertical, and surfaces such as bearing or other critical machine or tooling surfaces are normally ground with the jig grinder. The workpiece surface is passed by the grinding wheel.

One highly specialized abrasive-type machine that does not use wheels, but that employs abrasive stone segments, is the vertical spindle surface grinder with a large rotary table. The unit is employed to abrade the surface of large, flat plates and similar workpieces.

Another abrasive-type machine for large surfaces is the unit that employs abrasive belts. Some belt machines have as much as 150 hp drives and can remove metal at the rate of 15 in.^{3}/min or even faster.

With the exception of the abrasive-type machine tool, all machining is inherently degrading. Lathes or milling machines are incapable of machining workpieces to as high a degree of precision as found in the mechanical components of those machine tools. Abrasive machines are enhancing in that they can produce surfaces to a greater precision than is found in the machine itself. This is because of the thousands or even millions of individual abrasive grains in a wheel or belt that have the ability to prepare a surface to within one millionth of an inch, or one fortieth part of a micron deviation from the mean.

Sawing Machines

The final chip generating machine tools are the powered saws. There are two types: the band saw with its continuous blade and the hack saw which utilizes a reciprocating action. There are definite feed and speed rate recommendations for sawing as for other machining operations. These depend upon the material being cut, the number of teeth per inch of blade, and the material of the blade itself.

7.4.4 METAL FORMING MACHINES

Metal forming machines are the second family of metalworking equipment. The basic unit is the simple punch press. It is either constructed on the C-frame basis or has support on all four corners. Almost invariably the bed of the press supports the die section of the tooling, while the ram supports the punch half. When forming is the operation, the male half of the die is usually mounted on the ram, and the female on the press bed.

Exhibit 7.4.2 A Constantly Varying Tonnage Is Delivered as a Press Progresses Through Its Cycle. A Shaft Delivering 100 Ft-tons of Torque When the Crank Is Horizontal Delivers a Constantly Increasing Effective Tonnage as the Crank Approaches Bottom. The Lever-Ram Distance Constantly Becomes Smaller Until at Bottom It Is Zero. Remember That Zero Divided Into any Quantity Other Than Zero Is Infinity.[a]

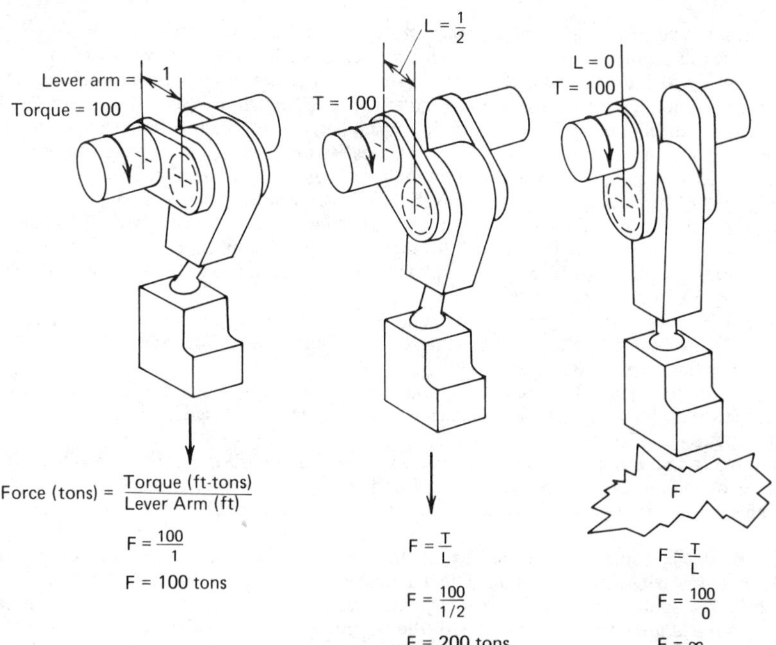

[a]Courtesy of *Modern Machine Shop* magazine.

If the press is hydraulically powered, it has a constant tonnage rating throughout its entire stroke. If it is a mechanical press, it is rated at a certain tonnage with the ram at a specified distance from the bottom of the stroke. The tonnage is a constantly changing factor throughout the entire press stroke because of the changing lever arm ratio. More problems occur with presses over this one factor than any other. The clutch has a constant torque at all times, but the tonnage the ram is able to deliver changes until it reaches infinity at the bottom of the stroke (Exhibit 7.4.2).

A most important factor of press operation is its tooling. Ideally it should include automatic feeding so that the operator does not need to move close to the danger area. If a press is hand-fed, safety considerations become a critical factor, and the operator must be prevented from allowing his or her hands to enter the pinch point.

Presses may also be used for hot or cold forging. Here the metal is simply squeezed between two die halves to produce a rough shape.

Another portion of the press family is the category of die casting and plastic forming presses which force molten metal or plastic into a cavity made of two die halves. When the material solidifies, the die halves are pulled apart, and the workpiece is ejected.

There are the multipurpose machines, such as index machines or transfer lines, where many different machining spindles are mounted in either a circular or a linear pattern, and the workpiece passes from one station to the next until all machining or working operations are completed. Both the index and the transfer types are used for very long production runs.

Another concept just now beginning to be employed is the variable or flexible manufacturing system. This concept is based on a cell or grouping of numerically controlled machining centers connected with powered conveyors under the control of a central computer. Pallets of different workpieces are placed within the system, and the computer controls their passage from one machine to the next and the sequence of operation at each machining center. The objective is a mass production efficiency for small-lot production work.

There are four "nontraditional" machining methods that have found general acceptance since World War II. They are thermal, chemical, electrical, and electrochemical. Thermal methods include the laser and electrical discharge machining (spark erosion), electron beam, and plasma. Chemical machining is the use of chemical agents to dissolve metal from exposed surfaces. Electrical machining is the direct application of current to remove metal by an ionization process and is the reverse of electroplating. Electrochemical machining combines the electric current with a chemical reaction. Any one of the four of these methods is used when there is either a material or a tooling condition that poses a problem for the mechanical machining processes.

Even the nontraditional machining processes are most often executed in a machine with the basic C-frame or four-post construction, for they involve the basic principle of a workpiece and tool's interacting with each other, and both must be supported and controlled throughout a series of machining moves.

BIBLIOGRAPHY

MOLTRECHT, K. H., *Machine Shop Practice*, Vols. I and II, Industrial Press, New York, 1981.

POLLACK, H. W., *Manufacturing and Machine Tool Operations*, 2nd Ed., Prentice-Hall, Englewood Cliffs, NJ, 1979.

PORTTER, H. W., O. D. LASCO, and C. A. NELSON, *Machine Shop Operations and Setups*, 3rd Ed. American Technical Society, Chicago, 1967.

CHAPTER 7.5

Automation

RALPH E. CROSS, SR.

Cross & Trecker Corporation

7.5.1 EARLY AUTOMATION

The automation of machine tools is simply the bringing together of three basic building blocks: machine tool, material handling, and controls.

Automation as it is known today became a major factor in mass production during World War II. But it had earlier beginnings. As population and markets expanded early in the twentieth century, the need became apparent for productivity improvement beyond that possible with conventional machine tools. "Special" machine tools were developed to answer that need.

The early special machines performed a single type of machining operation, such as turning, milling, or drilling. Each machine had a single spindle, performed one operation on one workpiece, and needed an operator. Performing all machining called for on a workpiece required an arrangement of many single-purpose machines in a line, usually served by a manually operated roller conveyor which transported work from one machine to the next.

Output from these early production lines was restricted by the time required to load, machine, and unload work and to convey it between the single-operation machines. This restriction was partially relieved by the development of the multiple-spindle machine. With this machine, a single motor driving several spindles through a gear train allowed multiple operations to be performed by one machine. Machining time cycles did not change, but more machining operations could be performed within each cycle. And several machining operations could be performed on one machine by a single operator.

A major advance in productivity was inaugurated by the multiple-unit, multiple-spindle, single-station machine. This concept allowed several sets of operations to be performed on different workpiece surfaces all on one machine by one operator. The number of machines required on a production line could be reduced by this concept, and work output per operator was increased. Total load-machine-unload time was not changed, but productivity was increased considerably because fewer machines and fewer operators were required.

The advent of workpiece indexing, or transfer, made it possible for one operator to control the work performed at several machining stations. In addition, the operator was able to load and unload at the load station while machining was taking place. This reduced total cycle time by combining the load-unload function with the machining function.

7.5.2 DIAL AND TRANSFER MACHINES

The breakthrough that led to automation as it is known today came with the development of automatic workpiece indexing and linear transfer devices. These resulted in the creation of the multiple-station rotary indexing, or "dial," machine and the multiple-station in-line indexing, or "transfer," machine, respectively. Both provided automatic movement of the workpieces from station to station and permitted more operations to be performed by a single machine operator.

The huge multiple-station in-line transfer machines that followed performed incredible machining feats. But they had one drawback: When any one of the many machining stations went down for repairs or tool change, the entire line had to be shut down.

7.5.3 SECTIONIZED AUTOMATION

Sectionized automation (Exhibit 7.5.1) was developed to prevent loss of production during such periods of downtime. Sectionized automation divides a transfer machine into sections, which permits shutting down some of the operations without interrupting the production of others. This is accomplished through a program in which transfer devices move workpieces through various

Exhibit 7.5.1 The Principle of Banking Used on Sectionized Machines Within a Linear Transfer Line. When Tools Need Changing, or an Unexpected Problem Results in the Shutdown of One of the Machine Sections, Parts Are Automatically Banked by the Transfer Device Preceding the Down Section, While Allowing the Other Machine Sections to Continue Operating at Capacity.

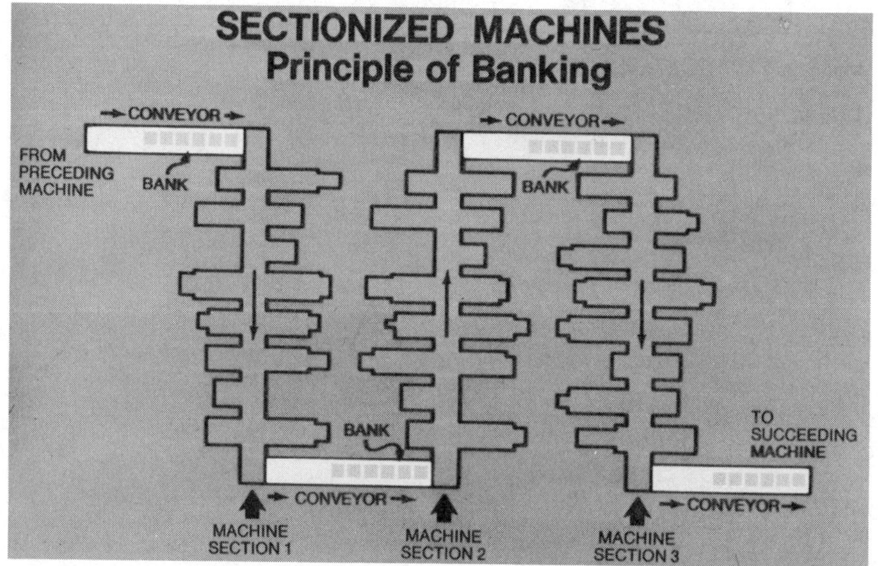

machining operations, while other transfer devices, working independently, move workpieces from machine to machine. Provision is made for banking parts between each of the sections.

When an automatic device signals the need for tool changing in one section, that section is shut down. But the other sections continue full automatic operation. The operator continues to feed parts into the first section, and work is processed by all but the down section, where work is banked up to wait for production to resume after the tool change. Sectionized automation overcame the obstacles that had previously handicapped automation.

To illustrate the quantum jump in productivity that was made possible by such automated machining, one need only consider the example of machining an automobile cylinder block, which once required on the order of 600 machining operations (Exhibit 7.5.2). Loading, unloading, and complete machining of the block on a line of single-spindle, single-station machines of the 1920s would have consumed 600 min of productive labor. Today's multiple-station in-line indexing transfer machine reduces that time to less than 1 min of productive labor, with far closer machining tolerances (Exhibit 7.5.3).

7.5.4 COLLATERAL DEVELOPMENTS

With the advancement of automated machining techniques, collateral developments have made significant contributions to productivity and have provided job enrichment and far lighter work for machine operations. Automatic loading and unloading devices have relieved workers of the necessity for heavy lifting. Work handling systems have been refined. The serious downtime factors caused by tool changing have been minimized, first with preset cutting tools, later by machine cycle counting units that signaled the need for tool changes, and finally by completely automatic tool wear inspection and compensation without interruption of the machining cycle.

The newest technological contribution to automation has been the increasing sophistication of control devices. Limit switches were used initially to control travel and machining cycles, and they are still in wide use to control many machine functions. The programmable controller came next, a solid-state switching device that can effect machine functional changes.

Present state of the art in controls is computer numerical control (CNC), which provides considerably more intelligence to the machine, allows greater flexibility so that families of parts can be run on the same machine, and supplies information regarding operating conditions. As a tool wears, for example, automatic tool gauging signals the machine that it must compensate for the tool wear.

Exhibit 7.5.2 Productivity Increases Possible Through Incorporation of the Various Stages of Automation Techniques. The Single-Spindle, Single-Purpose Machine Shown at the Left Can Drill and Ream the Holes in the Part Shown Above at the Rate of 5 Pieces/H. This Rate Includes Load and Unload Time and the Time Necessary to Change Tools From a First Pass to a Second Pass Drill and Then to a Reamer. The Multiple-Spindle Machine in the Center Performs Drilling and Reaming on the Part Simultaneously. By Providing a Multiple-Spindle Head, a Group of Four Fixtures, and a Rotary Indexing Table, Production Is Increased to 30 Pieces/H. This Machine Allows Concurrent First and Second Pass Drilling and Reaming of the Two Holes in Each Part and Loading at Each of the Four Fixture Positions, Respectively. When the Operations Are Complete, the Rotary Table Indexes, and Each Operation Is Repeated With a Finished Part Being Presented to the Load/Unload Station. The Machine at the Right Shows the Next Step in Production Increase. By Using a Larger Index Table, Doubling the Number of Drill and Ream Spindles and the Number of Fixtures, the Production Rate Is Doubled to 60 Pieces/H

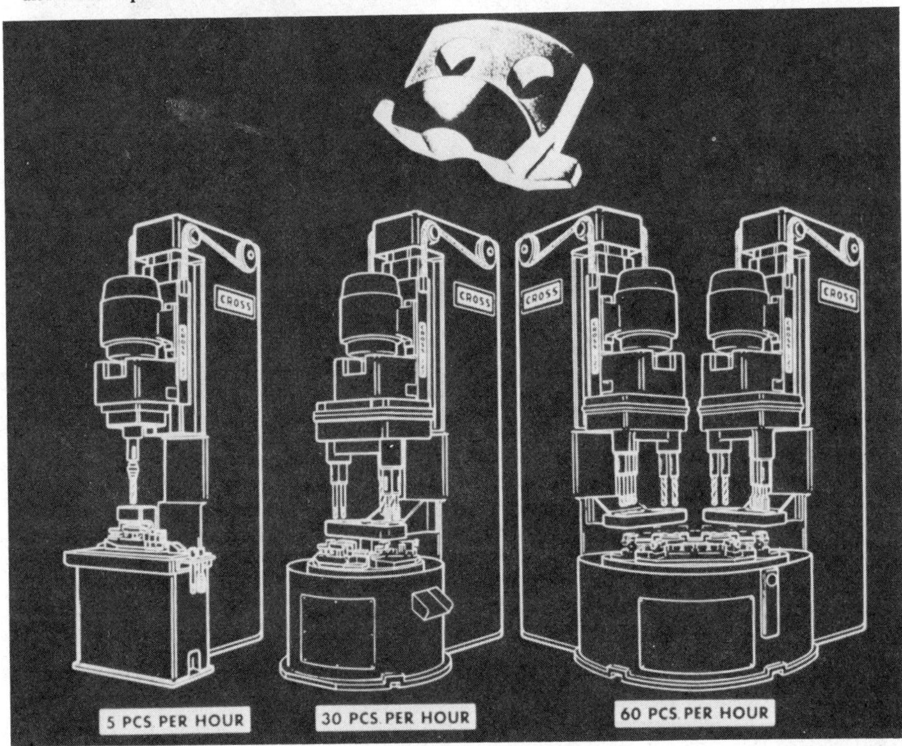

Leading builders of automated equipment now use a building block concept under which standard modules can be applied in varying arrangements to make up a variety of special station-type machines. Modules used include slide units, vertical column assemblies, single- and multiple-spindle heads, index tables, and so on. Use of such modular units simplifies machine design and construction, expedites maintenance, allows reuse of many units when the machine is converted to make a different part or group of parts, and reduces spare parts inventories. Further, the cost of each module is lower because of the economies inherent in building them in quantity.

7.5.5 INTEGRATED MANUFACTURING SYSTEM

Machines originally developed as special-purpose tools for production inspection and assembly operations have been integrated into automated machining systems, which has led to today's most advanced method of manufacturing—the totally integrated manufacturing system. This system uses groups of machines linked by data pathways to solid-state controls and vastly multiplies output of workers for mass production of similar parts or assemblies. One such system is composed of close to 100 metalworking, assembly, and test machines, along with a variety of automatic material handling and in-process storage equipment. The system produces completely machined and assembled truck wheel hub and drum assemblies from raw castings. It is totally automatic and requires no machine operators, only maintenance and service personnel.

Exhibit 7.5.3 Often One Section of Machines in a Line Cannot Perform Its Operations at a Rate Equal to That of the Preceding Machine Section. In This Case the Slower Section Is Divided Into Legs, With Banking Provisions Preceding the Section Legs. In This Way the Slower Cycle Time Is Compensated by the Addition of a Greater Number of Machines. Also, as in the Case of the Illustration, Either Leg of the Slower-Cycle Machine Section Can Be Shut Down for Tool Change Without Interrupting Production of the Other Leg.

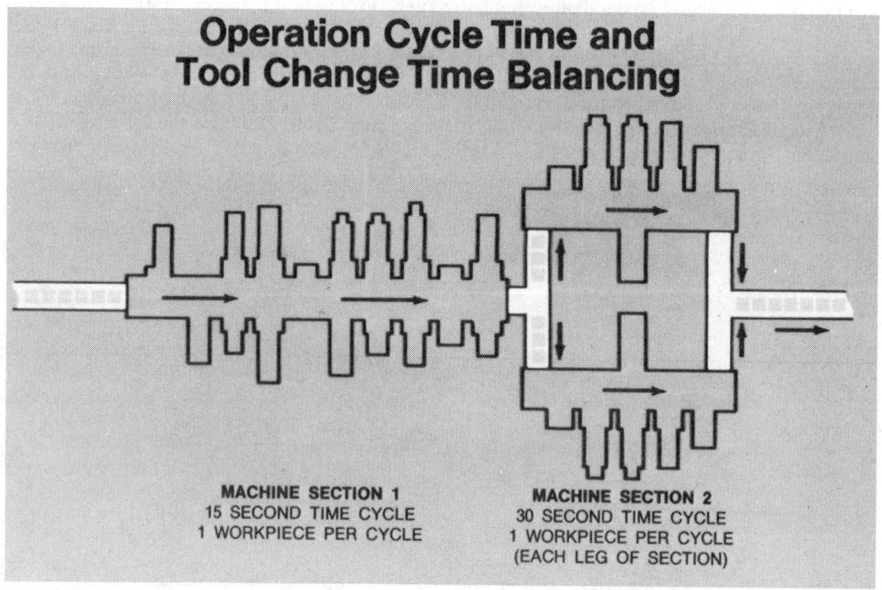

The integrated manufacturing system described also employs a variation of the transfer system. In certain parts of the system, where testing and assembling operations have variable cycle times, a nonsynchronous transfer system is employed to allow each station in the machine to cycle independently without affecting the other stations. In other cases synchronous transfer is employed where it is more advantageous to move work in this mode.

Early machine tools were able to perform tasks beyond the limits of human strength and manual skills. Today's automated machines have released mankind from onerous, repetitive burdens and have made possible time and energy for more creative endeavors. Automation tomorrow may help us to perceive a need in the marketplace, design a product to fill the need economically, functionally, and aesthetically—and automatically transform the design into a finished product without human participation beyond the function of creative thinking.

BIBLIOGRAPHY

"Automation, Numerical and Computer Control," in Dallas, D. B., Ed., *Tool and Manufacturing Engineers' Handbook*, 3rd ed., McGraw-Hill, New York, 1976.

LEONE, W. C., *Production, Automation, and Numerical Control*, Ronald, New York, 1967.

SYKORA, J., *An Automation Dictionary*, Adler, New York, 1976.

WEEKS, R. C., *Machines and the Man: A Sourcebook on Automation*, Irvington, New York, 1972.

CHAPTER 7.6

Industrial Robotics

RONALD L. TARVIN, MERTON D. CORWIN,
and **WAYNE E. MECHLIN**

Cincinnati Milacron, Inc.

7.6.1 WHAT IS AN INDUSTRIAL ROBOT?

An "industrial robot," as defined by the Robot Institute of America, is a programmable, multifunction manipulator designed to move material, parts, tools, or specialized devices through variable programmed motions for the performance of a variety of tasks. What separates an industrial robot from other types of automation is the fact that it can be reprogrammed for different applications; hence a robot falls under the heading of "flexible automation," as opposed to "hard" or dedicated, automation.

Industrial robots comprise three basic components:

1. The *manipulator* (or *arm*), which is a series of mechanical linkages and joints capable of movement in various directions to perform the work task.
2. The *controller*, which actually directs the movements and operations performed by the manipulator. The controller may be an integral part of the manipulator or be housed in a separate cabinet.
3. The *power source*, which provides energy to the actuators on the arm. The power source may be electrical, hydraulic, or pneumatic.

7.6.2 ROBOT ARM AND WRIST CONFIGURATIONS

Any discussion of robot arm configurations must entail a description of the type of "work envelope" generated by each configuration. The work envelope is a product of the robot arm's reach and for a stationary base device is defined in terms of three axes of arm movement: horizontal arm sweep, which is rotation about the center axis or linear travel along a horizontal axis; vertical motion; and arm extension. Arm configurations are usually defined in terms of the work envelope that results from a particular combination of axes.

Thus all industrial robot arms fall under one of the following configuration classifications: cylindrical, spherical, or jointed spherical.

Robot arms of the *cylindrical* type (sometimes called "post-type") are so termed because their work envelope is a portion of a cylinder. Such robots consist of a horizontal arm mounted on a vertical column, or post, which is in turn mounted on a base. The arm moves in and out and up and down on the column, and the column rotates on the base. Hence cylindrical robot arm configurations contain two linear and one rotary axes (see Exhibit 7.6.1).

A *spherical* (or "polar") configuration resembles that of a turret on a tank. The arm moves in and out, pivots on a horizontal axis, and rotates in a horizontal plane about the base. Thus such a configuration consists of one linear and two rotary axes, with the resulting work envelope being a portion of a sphere (see Exhibit 7.6.2).

The *jointed spherical* arm configuration, also called "anthropomorphic" or "articulated," operates in much the same manner as the human arm. In this configuration the arm extends from the base, or trunk, and is jointed at the "elbow" and at the "shoulder," where the arm and base meet. The base provides rotary motion. This configuration consists of three rotary axes and also yields a portion of a sphere as its work envelope (see Exhibit 7.6.3).

In cases in which the robot base itself travels on tracks, a rectangular dimension is added to the others, regardless of the type of arm configuration.

In each configuration the robot arm moves in three axes, or "degrees of freedom." These three

Exhibit 7.6.1 Cylindrical Arm Configuration

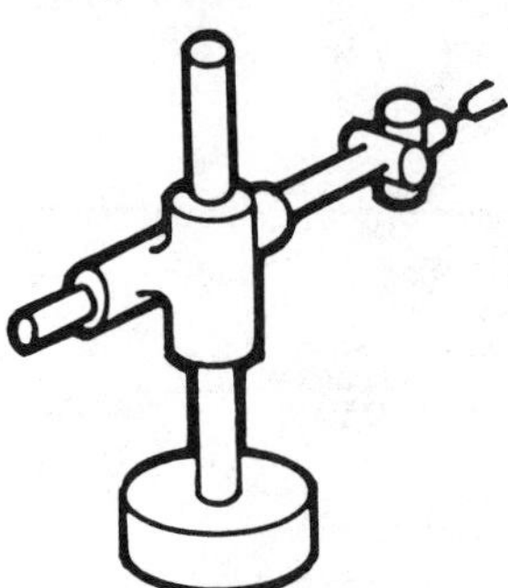

Exhibit 7.6.2 Spherical Arm Configuration

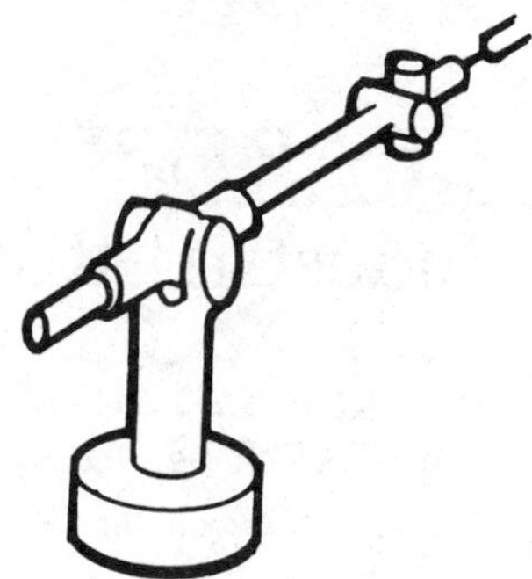

axes are sufficient for positioning the arm in X-Y-Z space. As many as three additional degrees of freedom are provided at the extremity of the robot arm in what is commonly called the "wrist."

Wrist axes include "roll" (rotation in a plane perpendicular to the end of the arm, "pitch" (rotation in a vertical plane through the arm), and "yaw" (rotation in a horizontal plane through the arm). These wrist axes make little contribution to the shape and size of the work envelope; their purpose is to provide orientation of the arm about an X-Y-Z *point* (see Exhibit 7.6.4).

Many wrist configurations are used in industrial robots, but their common function is to orientate the "end-effector," the tool or gripper with which the robot performs its task.

A particular wrist configuration is determined by the number of axes of movement (or, again, degrees of freedom) and also by whether an axis is of the roll or bend type and the combination and sequence of these rolls and bends.

7.6.3 END-EFFECTORS

A mounting surface is provided on the last axis of the wrist for installation of an end-effector, which, again, is the tool or gripper with which the robot performs its task. Though the end-effector is determined by the application (discussed later) and may in fact be unique to that application, several basic types of grippers and tools are commonly used.

Normally grippers are used for "pick-and-place" operations—that is, operations in which the robot arm must pick up an object and put it down in a specified location. Most grippers employ suction cups, magnets, or articulated handlike mechanisms (linkage designs) to hold the object. A variety of grippers has been developed, in most cases to meet the demands of particular applications. Sometimes the demands produce rather exotic solutions, such as a gripper consisting of tentacles filled with magnetic fluid that wrap around a workpiece, but most grippers are relatively simple and straightforward. A few examples of the more common types of grippers in use are shown in Exhibit 7.6.5.

Exhibit 7.6.3 Jointed Spherical Arm Configuration and Work Envelope

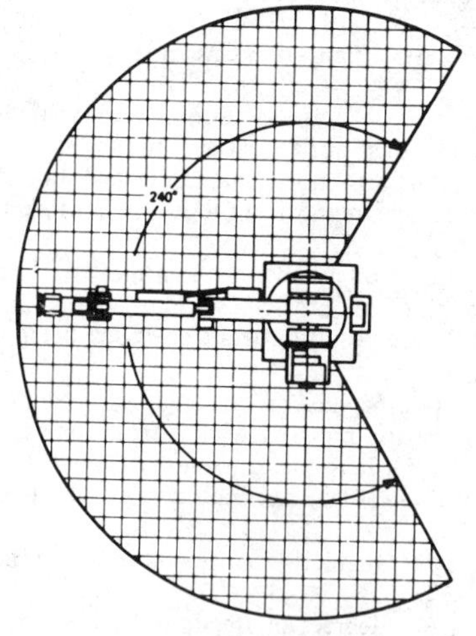

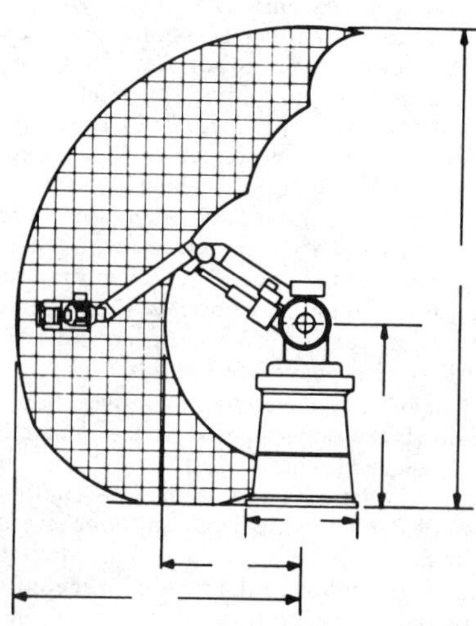

Exhibit 7.6.4 Axes of Movement in a Jointed-Arm Robot

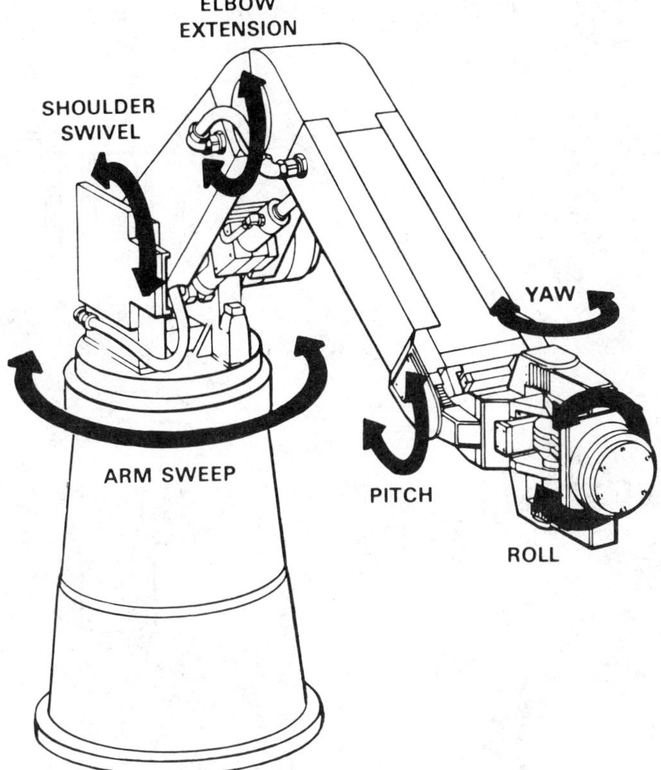

Tools currently in use in robot applications include welding guns, drills, boring tools, spray guns (primarily used in painting), and routers.

7.6.4 NON-SERVO-CONTROLLED VERSUS SERVO-CONTROLLED ROBOTS

From an operational viewpoint, robots may be classified as either non-servo-controlled or servo-controlled devices.

Non-Servo-Controlled Robots

Non-servo-controlled robots are characterized by high-speed motion and good repeatability. These robots are relatively low in cost and simple to operate, program, and maintain. However, each axis of a non-servo-controlled robot is limited to a small number of discrete programmable positions.

A non-servo-controlled robot's sequencer-controller initiates action by sending signals to control valves located on the axes to be moved. The valves open, admitting air or oil to the actuators, which drive the axes, and the axes then move until they are physically constrained by end stops. When the end stops are reached, limit switches signal the controller, which commands the control valves to close. The sequencer then indexes, and the controller sends out new command signals. These signals may go to the control valves or to some external device, such as a gripper. This process is repeated until the entire sequence of steps has been executed.

Programming of non-servo-controlled robots is accomplished by specifying the desired sequence of moves on the sequencer and by adjusting the end stops or switches of each axis. The sequencer thus provides the capability for several motions in a program, but only to the set points of each axis.

Servo-Controlled Robots

Servo-controlled robots, on the other hand, provide maximum positioning capability because of their ability to position each axis anywhere within its limits of travel. Such robots also permit

Exhibit 7.6.5 Common Types of Grippers: (*a*) Vacuum; (*b*) Four-Fingered External; (*c*) External, Small Diameters; (*d*) External, Large Diameters; (*e*) Three-Fingered Internal; and (*f*) Two-Handed External

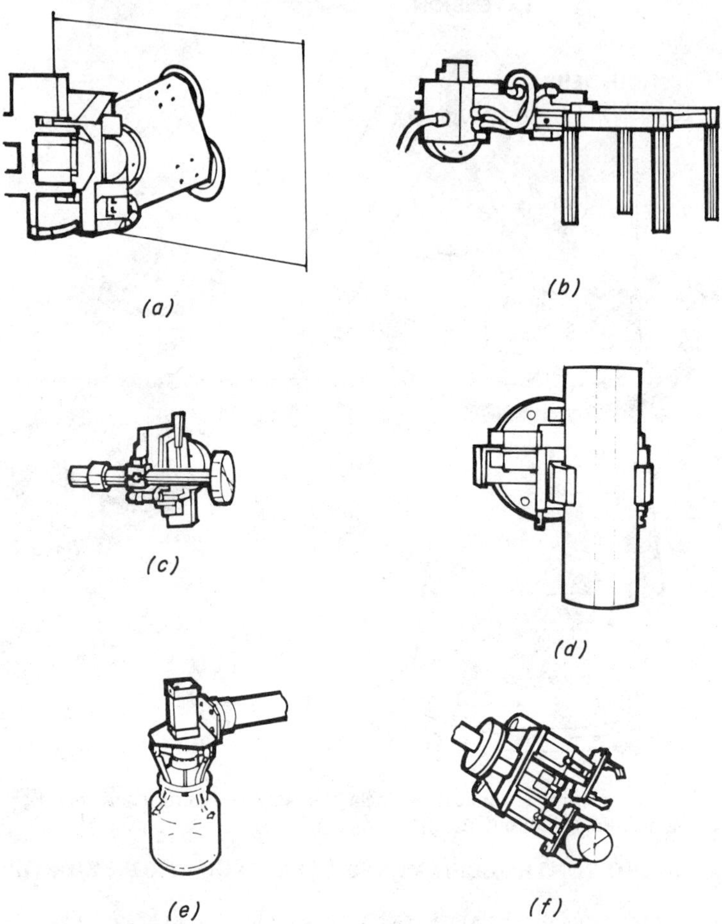

more control over the movement of heavy loads by letting the user specify the speed of movement and, in some cases, acceleration and deceleration rates. However, because of their complexity, servo-controlled robots are more expensive and sometimes less reliable.

Servo-controlled robots operate by recalling prerecorded positional data from memory, using these data to generate motion command signals for each arm axis. As the individual axes move, the feedback devices mounted on the axes are read continuously to determine the amount of error that exists between the actual and desired axis position. When the feedback indicates that the axes are approaching their destinations, the axes are brought to a controlled stop. Then any operations to be performed at the programmed location are carried out.

There are three basic modes of path control operation associated with servo-controlled robots: point-to-point, continuous path, and controlled path operation.

Point-to-Point

This type of path control is the simplest and most frequently used control method. Teaching is done by moving each axis of the robot individually until the combination of axis positions yields the desired position of the robot and end-effector. When this desired position or point is reached, it is programmed into memory, thereby storing the individual position of each robot axis. In replaying these stored points, each axis runs at its maximum or limited rate until it reaches its final position. Consequently, some axes reach their final value before others. And because there is no coordination of motion between axes, the path and velocity of the end-effector between points is not easily predictable. For this reason, point-to-point control is used for applications in which only the final position is of interest, and the path and velocity between points are not prime considerations.

Continuous Path

This type of control is used when the path of the end-effector is of primary importance to the application, such as is required for spray painting. The unit is generally not required to come to rest at unique positions and to perform functions as is common in applications employing a point-to-point control. Typically, robots using this type of control are taught by physically grasping the unit and leading it through the desired path in the exact manner and speed in which it is to repeat the motion.

Controlled Path

With this type of path control, the computational power of a minicomputer or microcomputer is used to provide coordinated control of the robot axes during both the teaching and the automatic modes of operation. Controlled path systems combine desired characteristics of both point-to-point and continuous path systems. During teaching, controlled path systems let the operator position and orient the robot to desired points without having to command individual axes. It also is not necessary to grasp and lead the unit physically.

For example, an operator can command the arm to move straight up or down, in or out, and left or right and can let the computer issue the necessary commands to the appropriate axes to accomplish this "coordinated axis" motion. Also, during teaching the operator is not required to generate the desired path, but only to identify unique end points. Then, during replay or automatic operation, the computer automatically generates a controlled path (usually a straight line) at a desired velocity between the designated end points. This type of path control can greatly reduce the number of data points that need to be programmed and can give the operator more control over the movement of his or her robot's arm.

7.6.5 HYDRAULIC, ELECTRIC, AND PNEUMATIC DRIVE

Robots may be hydraulically, electrically, or pneumatically driven. Briefly, the advantages of hydraulically driven robots include mechanical simplicity (few moving parts), higher load capacity, and high speed. However, hydraulically driven robots generally offer lower repeatability than their electrically driven counterparts.

In most cases electric robots are not as strong or as fast as hydraulic robots, but they generally show better accuracy because they tend to be stiffer. Electrically driven robots save floor space and decrease noise levels since no hydraulic power unit is necessary to their operation.

Some non-servocontrolled robots may be pneumatic, or airdriven. For applications involving low payloads and simple programs, such robots are often the best solution because of their simplicity of operation. However, the low-payload capabilities make pneumatically driven robots extremely limited in their range of application.

7.6.6 ADVANTAGES AND APPLICATIONS

Industrial robots are already in use in a wide variety of industries and applications, and the number of industries that are investigating robots and putting them to use is increasing every day. The reasons for this are as follows:

1. **Productivity.** With productivity an increasing concern, robots offer consistently excellent uptime and reliability. They can work 24 hr a day with no change in the speed or quality of their output, which means a manufacturing operation can be run on a continuous basis with little or no scrap generated. Also, in many cases the robot works faster and more consistently than a human, providing additional productivity increases.
2. **Adaptability.** Robots offer adaptability in manufacturing operations in two ways. First, they are adaptable to other production equipment; that is, they can be interfaced with other equipment to perform and coordinate a manufacturing operation. Second, they are adaptable to the task at hand and in most cases the environment as well. At present, industrial robots have generally been employed only in jobs that humans cannot or do not want to do: boring jobs, repetitive jobs, jobs involving the handling of heavy weights, jobs performed in a hazardous environment, and so on. Robots cannot feel bored or degraded. They have weight handling capacities ranging into the hundreds of pounds in some cases and are or can be made immune to the effects of fiberglass, asbestos, paint, and so on.
3. **Safety.** Existing and proposed worker safety rules require extensive modifications to, or replacement of, production line equipment. Restrictions on where a worker can place his or her

hands within a machine and on plant noise levels are just two examples. Where a large number of machines are involved, it may be less costly to convert to robot operation than to change the existing equipment.

4. Training. Since, by definition, industrial robots are reprogrammable devices, they can be programmed as many times as desired for use on different operations. Programming can be accomplished quickly and easily, and any program can be stored for later use.

5. Return on Investment. Although many advantages can be achieved through the use of industrial robots, economics is the ultimate criterion used to determine their justification and worth in industrial applications. It has generally been agreed that their use can be justified economically if they can be utilized to replace one person on a two-shift basis. The return on investment will usually be greater than 30% and will normally result in a payback period of between 1 and 2 years.

6. Reliability. An uptime of around 98% is normal for industrial robots. This figure is the result of the relatively simple mechanical design of the manipulator unit. When something does go wrong, trained in-house maintenance personnel generally can repair the robot quickly and get it back into production in minimal time. The use of many standard, readily available, and highly reliable commercial parts also helps to maximize the mean time between failures.

On the basis of considerations such as the preceding ones, robots are now in operation in welding and assembly, drilling and routing, inspection, material handling, machine loading, die casting, and a wide variety of other applications.

Certainly one of the most predominant operations to which the servo-controlled robot has been applied is spot welding, and the industry that has most firmly embraced the robot in this application as well as in others is the automotive industry. In some plants robots are spot welding car bodies as they pass by on moving conveyors. In such applications robots are often required to handle welding guns weighing 100 lb or more in order to place welds accurately without stopping the line. A servo-controlled robot with sophisticated computer control is required in this type of operation.

Another major application area is arc welding. This application involves sophisticated interfacing among various elements of the system, including the robot, the welding power supply, the wire reels and feeders, and the part positioning table. Though arc welding applications in the field are already numerous, widespread implementation has been limited because of the lack of a cost-effective sensor that would allow the welding gun to follow each seam accurately, regardless of variations in its width or location. Various types of sensors are undergoing research.

7.6.7 THE FUTURE

The industrial robot industry is currently experiencing phenomenal growth, and this growth is expected to continue for some time as robots are brought into new industries and adapted to new applications.

BIBLIOGRAPHY

CUNNINGHAM, C. S., "Robot Flexibility Through Software," paper presented at the Ninth International Symposium on Industrial Robots, sponsored by the Society of Manufacturing Engineers and the Robotic Institute of America, Washington, DC, 1979.

DAWSON, B., "Moving Line Applications With a Computer Controlled Robot," paper presented at the Robots II Conference, sponsored by the Society of Manufacturing Engineers and the Robotic Institute of America, Detroit, MI, 1977.

HEGINBOTHAM, W. B. et al., "Robot Application Simulation," *Industrial Robot*, Vol. 6, No. 2 (June 1979), pp. 76–80.

HOHN, R. E., "Application Flexibility of a Computer-Controlled Industrial Robot," SME Paper MR76-603, presented at the First Industrial Robot Conference, sponsored by the Society of Manufacturing Engineers and the Robotic Institute of America, Chicago, IL, 1976.

HOLMES, J. G., and B. J. RESNICK, "A Flexible Robot Arc Welding System," SME Paper MS79-790, presented at the Robots IV Conference, sponsored by the Society of Manufacturing Engineers and the Robotic Institute of America, Detroit, MI, 1979.

HOLT, H. R., "Robot Decision Making," SME Paper MS77-751, presented at the Robots II Conference, sponsored by the Society of Manufacturing Engineers and the Robotic Insitute of America, Detroit, MI, 1977.

KOROLIEV, V. A., S. M. SERGEEV, and S. P. AZAROV, "Point-to-Point Pneumatic Robots for Assembly," SME Paper MS79-255.

LOCKETT, J. H., "Small Batch Production of Aircraft Access Doors Using an Industrial Robot," SME Paper MS79-783, presented at the Robots II Conference, sponsored by the Society of Manufacturing Engineers and the Robotic Institute of America, Detroit, MI, 1977.

NEVINS, J. L., and D. E. WHITNEY, "Assembly Research," *Industrial Robot*, Vol. 7, No. 1 (March 1980), pp. 27-43.

NOF, S. Y., J. L. KNIGHT, and G. SALVENDY, "Effective Utilization of Industrial Robots: A Job and Skills Analysis Approach," *AIIE Transactions*, Vol. 12, No. 3 (September 1980), pp. 216-225.

PAUL, R. L., and S. Y. NOF, "Robot Work Measurement–A Comparison Between Robot and Human Task Performance," *International Journal of Production Research*, Vol. 17, No. 3 (May 1979), pp. 277-303.

ROTH, B., "Performance Evaluation of Manipulators from a Kinematic Viewpoint," in *Manipulators.* National Bureau of Standards Special Publication No. 459, 1975.

WARNECKE, H. J., and B. BRODBECK, "Analysis of Industrial Robots on a Test Stand," *Industrial Robot*, Vol. 4, No. 4 (December 1977), pp. 194-198.

CHAPTER 7.7

Numerical Control Machines

KENNETH M. GETTELMAN

Modern Machine Shop Magazine

7.7.1 INTRODUCTION

Numerical control (NC) has been a working machine tool controlling methodology since 1954. It had its genesis in the effort of John Parsons of Parsons Manufacturing Company, Traverse City, Michigan, to manufacture helicopter rotor blades to the complex airfoil shape called for in the mathematical design criteria. With the aid of digital computing equipment that was emerging in the 1950s, it was possible to define accurately the mesh of points that described the airfoil surface and to generate printouts of them. There was no manufacturing technology that would drive a cutting tool to machine closely the defined surface contour. All tracer and checking templates were made by hand to a few widely spaced points, with the intervening distance approximated by skilled toolmakers. The requirements of modern jet aircraft called for closer manufacturing control. Parsons realized that, if the emerging computer technology could quickly define the closely spaced points of a mathematical surface description, then it surely must be possible to utilize that same capability in a controlling mode to direct the machining of that described surface. Conceptually he was absolutely correct, but it took 6 years of effort to bring the concept to a working reality. The work was done at MIT's Servomechanisms Laboratory.

Thus it became possible to establish a numerical definition into a control mechanism and then to have the control mechanism execute a series of machine tool motions that were necessary to machine the workpiece. Hence the definition "numerical control." In the actual process of defining a workpiece, both numbers and symbols are actually used; a better definition, then, might have been "symbolic control," but NC it was and has continued to be by acceptance and use.[1]

7.7.2 DESCRIPTION

With NC the fundamental control of the machine tool has passed from the operator, who studied the workpiece drawing and then manually directed the machine, to a workpiece programmer, who studies the workpiece drawing and then lists the motions required of the machine tool to produce the part. The manuscript developed by the programmer is usually called a "source document." It must be converted to some medium that can be acted upon by the machine tool control unit that will direct the machine tool.

The traditional medium has been the punched tape with eight columns. The pattern of punches in each row across the tape corresponds to a letter or number code in the workpiece program. The punched tape became so prevalent as an input medium that NC often was referred to as "tape" control. It is still in common use and is bad practice since workpiece program data may be stored in other media, including magnetic tape, tabulating cards, some kind of computer memory, and diskette. The latter is commonly known as the "floppy disc" and is rapidly growing in popularity since a 7 in. diameter disc can store the same program information that would require as much as 3000 linear ft of punched tape. The information is usually entered into the program storage medium by some kind of keyboard device, although there are a few that will work with a limited number of spoken commands.

Once into the machine control unit by means of the tape, disc, and so on, the control unit then translates the data into controlling signals that direct the machine tool servomechanism drives, which are almost always hydraulic or electrical, with the latter growing in popularity.

7.7.3 COORDINATE MEASUREMENT SYSTEM

The foundation upon which the NC concept rests is the method of describing workpiece geometry and machine motion in terms of two and three axis Cartesian or rectangular coordinates with the occasional use of polar coordinates (see Exhibits 7.7.1 and 7.7.2).

The coordinate system handles any combination of positive and negative values. Polar descriptions are stated in terms of an angle starting in the positive X direction and sweeping toward the Y and a distance from the origin along the angle line. For example, 30° 5 would be an angle of 30° and five units along the angle line in the plus direction.

If the rectangular coordinate concept is transferred to the table of a machine tool, it can be seen that any location on the table can be described in terms of an X and Y coordinate location. If a flat workpiece is mounted upon the table, it can be given corresponding coordinate location values. Taking it one step further, the center of a series of holes to be drilled in the workpiece could be stated in terms of XY coordinate locations.

The third axis is Z, which refers to distance above or below the XY plane and perpendicular to it. The direction above the XY is positive, and Z values below are negative (Exhibit 7.7.2). Thus any point in space can be defined by the distance and direction along the three coordinate axes. Likewise, any point on a workpiece can be defined in the same three-axis manner, from an origin point with a fixed relationship to the workpiece.

The normal use of the coordinate measurement system is based on an absolute set of values from a fixed origin. In NC programming the incremental concept is often employed. An incremental coordinate location is not stated in terms of the fixed origin. Rather, it is expressed in terms of distance and direction from the preceding point. Thus an $X3$ $Y4$ location is not in the first quadrant 3 units along X and 4 along Y measured from the base reference or origin point. The incremental notation means it is 3 units along the X axis in a positive direction and 4 units along the Y, also positive, from the preceding program point wherever that point may be. The experienced

Exhibit 7.7.1 The Origin Point May Be the Corner of the Machine Table or May, in Machines With a Zero Shift, Be Relocated at a Point on the Workpiece Itself. If a Hole Is To Be Drilled at the $X3$ $Y4$ Location, the Center Point of the Drill Is Directed to the Programmed Point. If a Slot Represented by the Broken Line Is To Be Milled From the Origin Point to $X3$ $Y4$, the Center Point of the Cutting Tool Is Sent From the First to the Second Position. Slot Width Is Determined by the Cutting Tool Diameter. To Get This Straight-line Configuration, the Table Drive Screws or Spindlemoving Mechanisms Would Be Controlled by the Machine Control Unit so That Three Units of X Motion Occur Simultaneously With Four Units of Y Movement. Although There Is a Natural Tendency to Seek Setups That Will Allow Programming in the All-Plus Quadrant, Programmers Can Use Any Quadrant They Wish. The Machine Tool Spindle Could Not Care Less.[a]

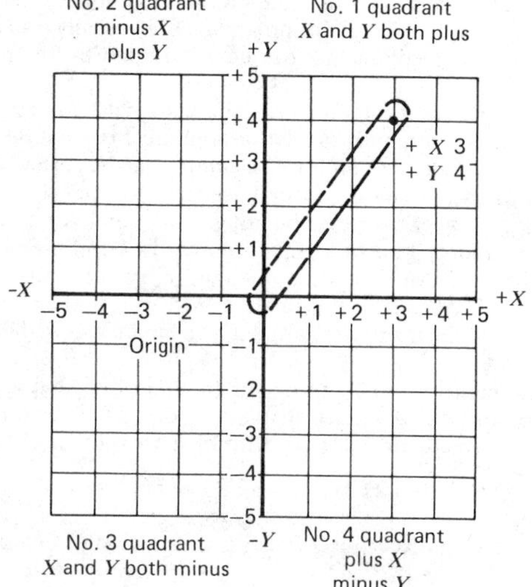

Exhibit 7.7.2 Three-Dimensional Coordinates Showing Depth as a Measurement Along the *Z* Axis. On Most Machines the *Z* Measurement Begins Where the Tool Spindle Is Fully Retracted Rather Than at the Origin Point. A Programmer Determines the Depth and Programs the Workpiece Accordingly, Using the Surface of the Workpiece as a Reference.[a]

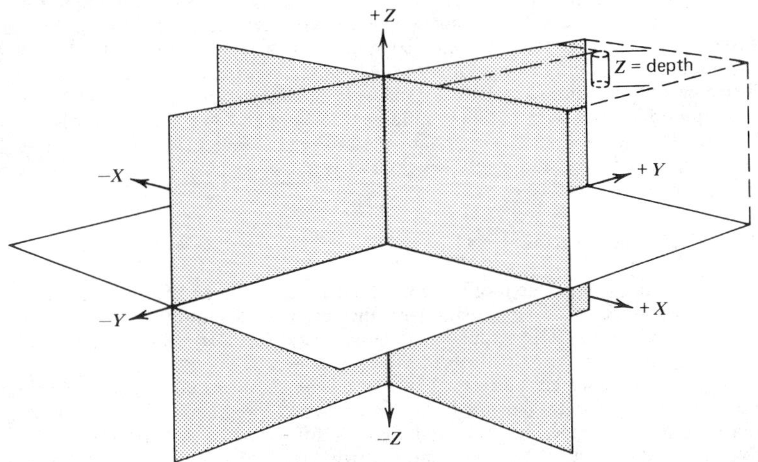

[a]Courtesy of *Modern Machine Shop 1980 NC/CAM Guidebook*, Cincinnati, OH.

NC programmer will feel free to use either absolute or incremental dimensioning as best suits the programming situation.

On the spindle-type numerical control machine tool, the *X* axis is normally the longest move of the machine table. The *Y* axis is normally shorter than the *X* move of the table, and the *Z* is the advance and retract of the spindle. There may be additional axes of motion, including a tilt or swivel of the spindle, table rotation, or secondary and tertiary moves of additional spindles or tables parallel to the principal axes of motion.

On turning equipment there are two primary axes of motion, the *X* and the *Z*. The *X* axis is the movement of the cutting tool parallel to the spindle line, with the positive direction away from the headstock. The *Z* axis is the tool motion toward or away from the workpiece centerline, with the positive direction away. Turning equipment may have a number of secondary motions, including rotation of a tool turret, moves of a second turret, or controlled rotation of the spindle itself (Exhibit 7.7.3).

By controlling the motions of the various machine tool axes, the workpiece programmer controls the relationship of workpiece to cutting tool, which in turn determines the nature of the machined workpiece. Mathematically, any geometric shape can be defined and programmed. The ultimate dimension and accuracy of the workpiece, however, depends upon the capabilities of the machine tool and its driving servomechanisms combined with workpiece holding and tooling. For example, the machine control unit might have a programming resolution of 0.0001 in. or 0.001 mm. Thus, although the programmer may program to the nearest 0.0001 in. or 0.001 mm, the action of the servomechanisms, the flex of the spindle and workpiece fixturing, the inherent accuracy limitations of the machine tool mechanical components, and such things as thermal expansion and so on may all contribute to a degradation of accuracy such that the machined workpiece may have only a 0.001 in. or 0.01 mm accuracy.

7.7.4 THE MACHINING CENTER

Machine tools, until after World War II, were designed for single-purpose use, such as drilling, milling, and boring. If a workpiece required several machining operations for its completion, it was taken from one machine to the next. This meant a separate setup, operator, and work handling for each machining function.

The original NC machine tools were also single-purpose units. The big breakthrough in their design came with the programmed automatic tool change, taking cutting tools from a magazine and inserting them into the spindle as needed. Thus the machining center came into being and rapidly grew in acceptance as it afforded the opportunity to accomplish many machining operations on a workpiece in one setup on one machine with one machine operator who may oversee several NC machine tools under the right circumstances.

The typical NC machining center may have a magazine of anywhere from 18 to 60 different cutting tools mounted in holders. As the machining operations change from milling to drilling, to boring, to tapping, and so on, the programmed arm removes the used tool from the spindle and exchanges it for the next one taken from the tool storage magazine.

The machining center concept is now growing to the point where pallets of different workpieces on conveyors may be transported into place for machining so that, as one workpiece is finished, the next one moves into place. Some machining centers have more than one tool magazine. As cutting tools become dull, a whole new magazine of sharpened tools is moved into the operating position. Sensors detect dull cutting tools through either in-process gauging of the workpiece or sensing increased torque required to drive a spindle with a dull cutting tool.

A similar situation exists with turning equipment. Although fewer cutting tools are normally required to machine a workpiece on a lathe, the automatic tool turret has become a standard feature of most NC turning equipment, whether it is a bar machine, a chucking unit, or a lathe that turns workpieces between centers.

7.7.5 MACHINE CONTROL UNIT

An integral part of an NC machine tool is the machine control unit. In 25 years it has gone through at least four generations of development, from the original vacuum-tube-type controls to those featuring transistors, on to those with solid-state circuitry, to those with integrated circuits, and finally to the computer numerical controls of the latest generation.

Three important functions are performed by the machine control unit. It must first accept the workpiece program input data generated by the workpiece programmer. At one time it was almost exclusively punched tape, but direct computer information or diskette-stored data are more frequently utilized. The control unit must store either the complete workpiece program or a large enough portion of it so that it can handle sufficient program data without hesitation. It must then process the workpiece program data to generate output signals that control the functioning of the servomechanism drives on the machine tool.

Virtually all machine control units produced today are electronic. There were a number of designs based on fluidics, or hydraulics, but with the lowering of cost and the reduction in sizing of electronic components along with the concurrent increasing of their capabilities, the other controlling methods have largely dropped from use.

Exhibit 7.7.3 The Ultimate of NC Lies in the Programmer's Ability to Control the Available Axes of Motion Simultaneously. The Diagrams Illustrate the Axes of Motion That Must Be Considered.[a]

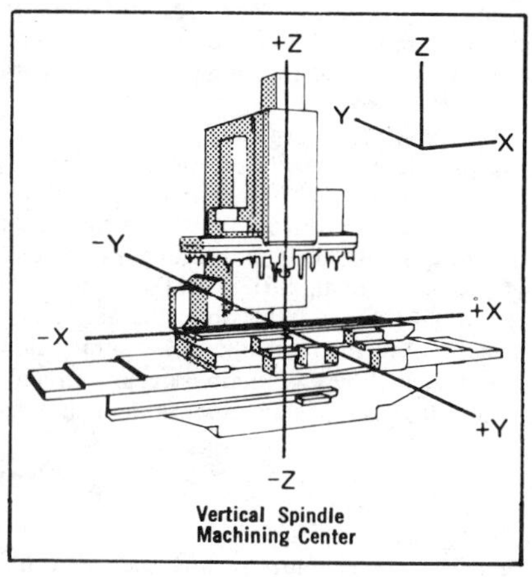

Vertical Spindle Machining Center

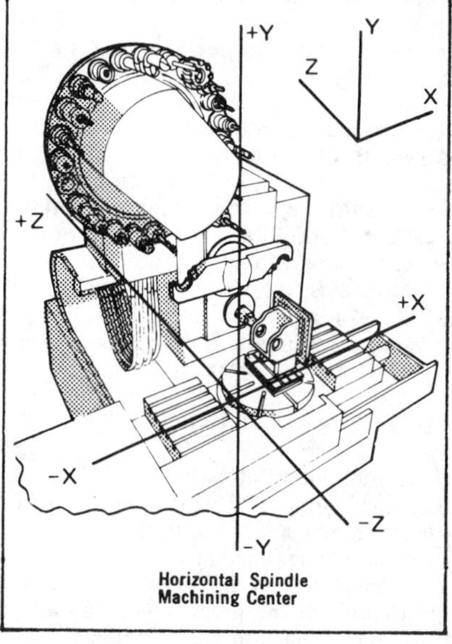

Horizontal Spindle Machining Center

[a]Courtesy of *Modern Machine Shop 1980 NC/CAM Guidebook*, Cincinnati, OH.

Exhibit 7.7.3 (*Continued*)

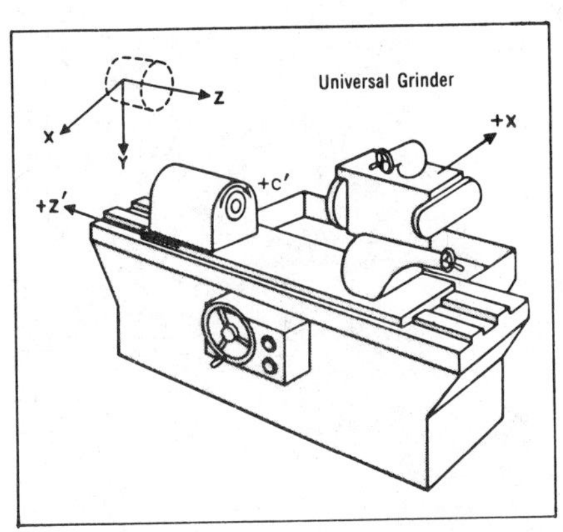

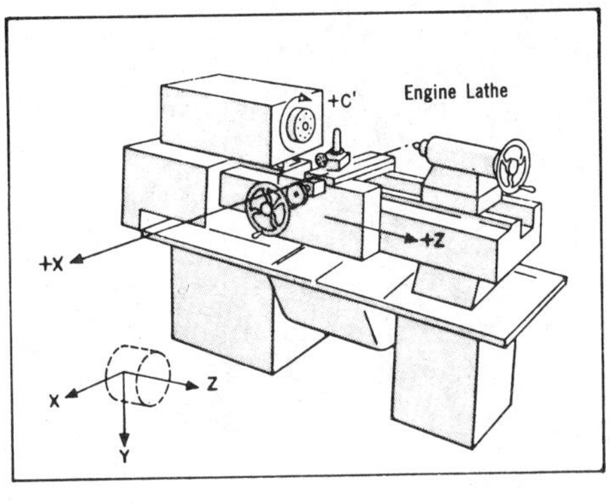

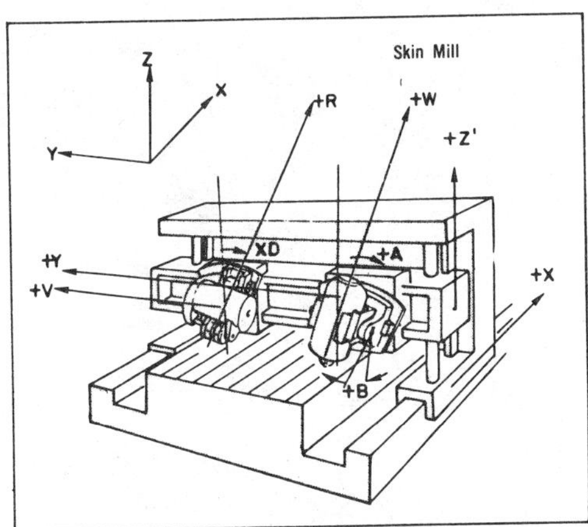

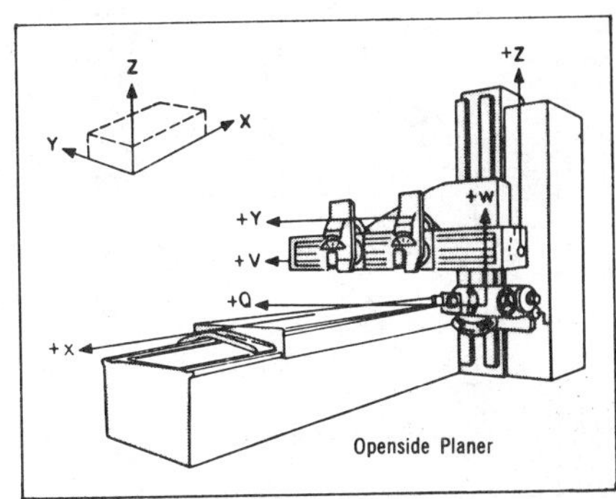

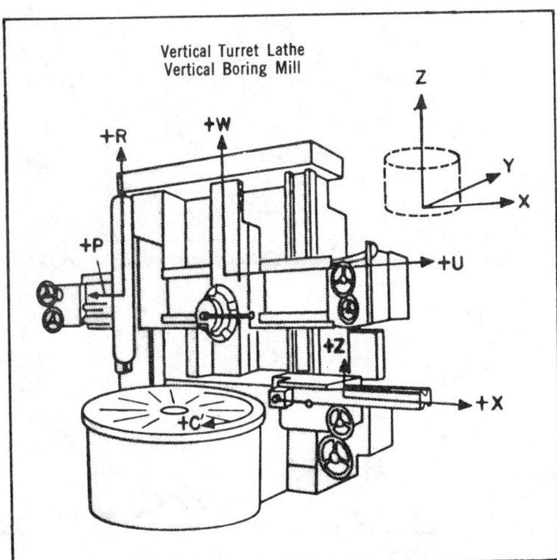

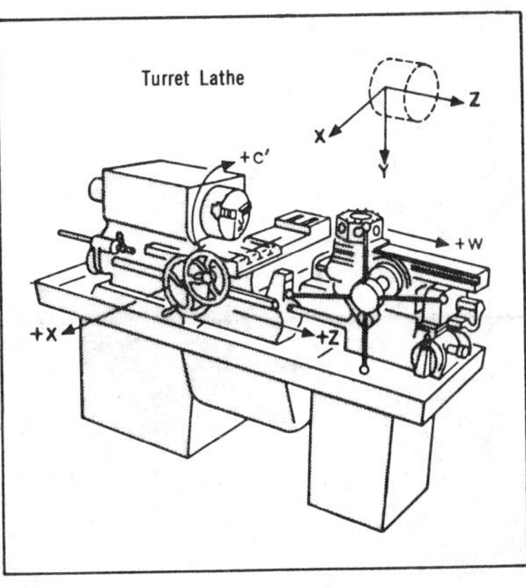

Exhibit 7.7.3 (*Continued*)

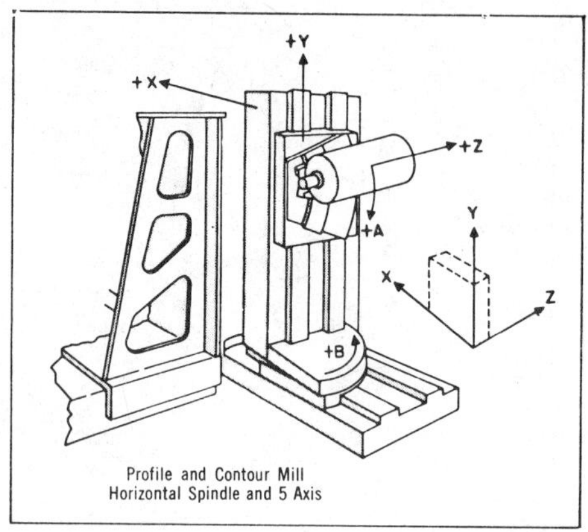

Profile and Contour Mill
Horizontal Spindle and 5 Axis

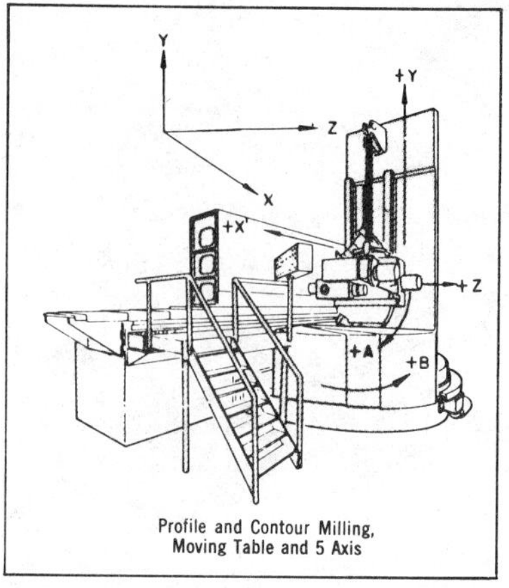

Profile and Contour Milling,
Moving Table and 5 Axis

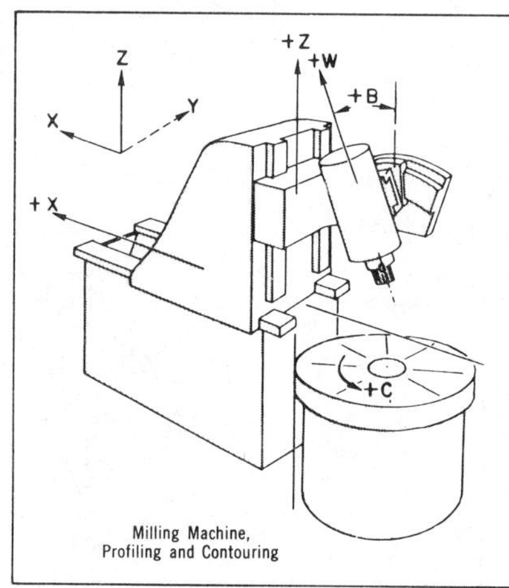

Milling Machine,
Profiling and Contouring

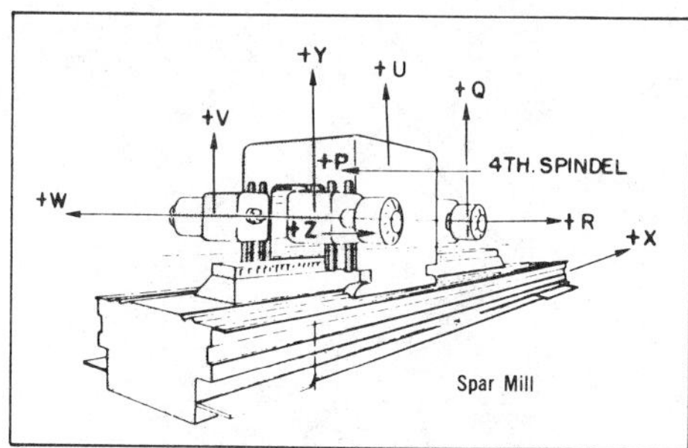

Spar Mill

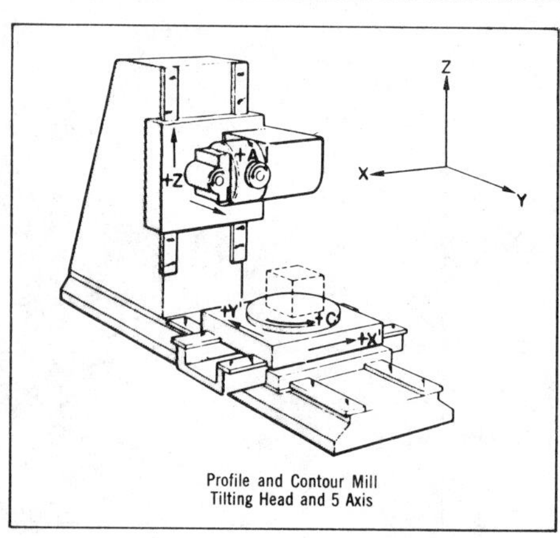

Profile and Contour Mill
Tilting Head and 5 Axis

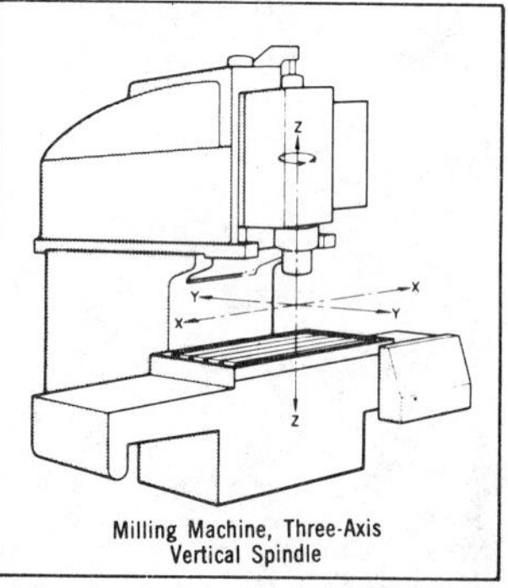
Milling Machine, Three-Axis
Vertical Spindle

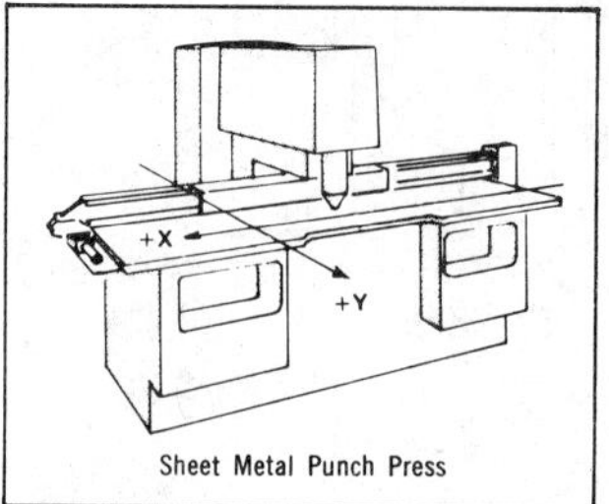

Sheet Metal Punch Press

Exhibit 7.7.3 (*Continued*)

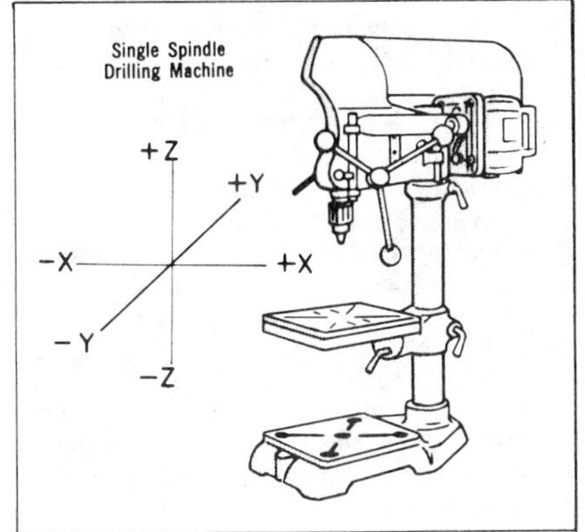
Single Spindle
Drilling Machine
+Z
+Y
−X
+X
−Y
−Z

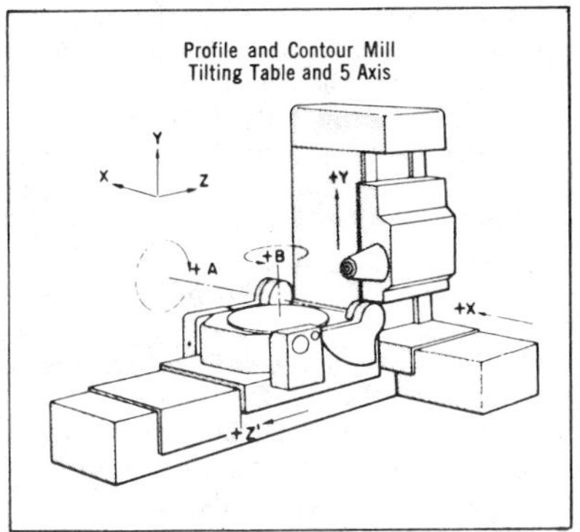
Profile and Contour Mill
Tilting Table and 5 Axis
Y
X
Z
+Y
+A
+B
+X
+Z'

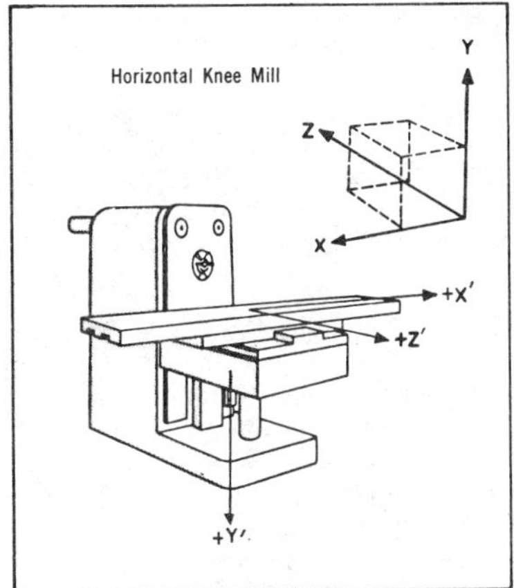
Horizontal Knee Mill
Y
Z
X
+X'
+Z'
+Y'

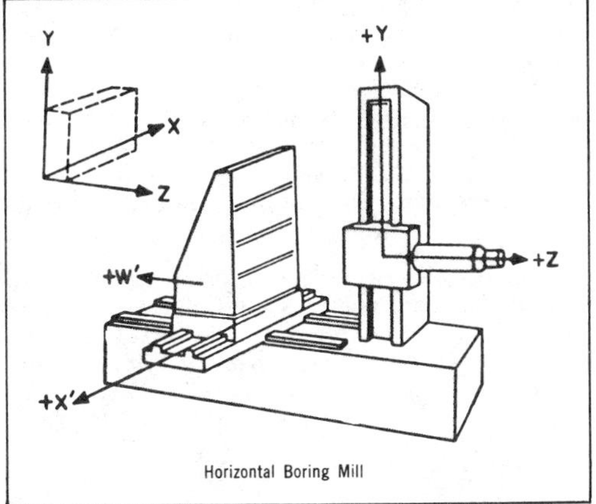
Y
+Y
X
Z
+Z
+W'
+X'
Horizontal Boring Mill

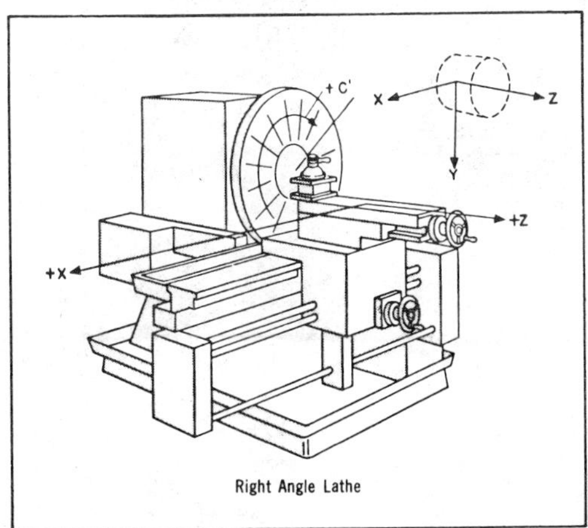
+C'
X
Z
Y
+Z
+X
Right Angle Lathe

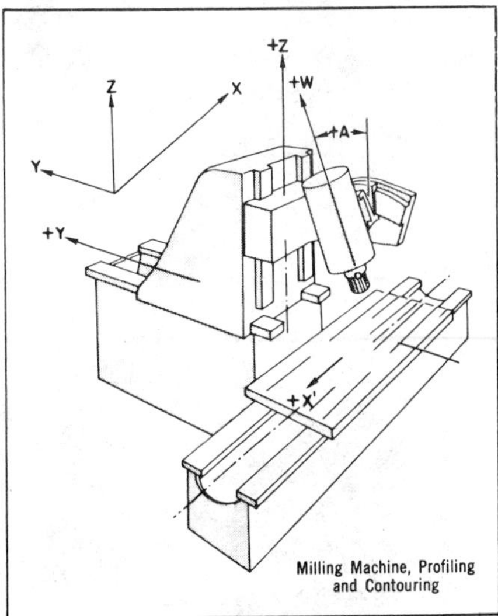
+Z
Z
X
+W
+A
Y
+Y
+X'
Milling Machine, Profiling
and Contouring

7.7.6 WORKPIECE PROGRAMMING

Workpiece programming is accomplished by one or two means: manually or with computer assistance. With manual programming the programmer must state every machine motion necessary to machine the workpiece. This is usually done on a manuscript form. After the form has been written and checked, it is then given to a clerk, who usually keyboards it to produce a tape, diskette, or some other medium in which the program is stored. The final proof comes when the first workpiece is machined and inspected. Manual programming may be feasible for very simple workpieces where nothing more than a few drilled holes or straight milling cuts are needed.

Experience has shown that, for some types of workpieces, especially those with a complex milled surface geometry, and those with many different machine locations with some commonality among them, computer-assisted programming is essential. With the assistance of a computer, it is possible to do in minutes what would otherwise require days or even weeks (see Exhibit 7.7.4).

The important part of computer-assisted workpiece programming is the particular processor language that will be used. Computers do not have any innate knowledge of either machine tools or machining processes. That knowledge rests in the processor language that has been developed to enable the computer to generate machining instructions when a workpiece description is entered. It still remains the programmer's responsibility to understand the workpiece, the logical sequence of matching operations, the basic machining processes, and how NC interfaces with the machining process. The computer can only speed the calculations that otherwise would have to be made by the workpiece programmer. The computer offers only speed and accuracy; it brings no creative thinking to the programming process.

An excellent example of how the computer speeds the programming process is the simple bolt hole circle. If done manually, the programmer would have to calculate the coordinate location of each hole and then specify the center drilling, chamfering, and tapping of each hole. It may take several hours to write and debug a program for 12 holes. If processed on the computer with an adequate NC processor language, it might take 5 min to state the essential facts and enter them

Exhibit 7.7.4 The Operating Efficiency of an NC Machine Tool Is Determined by the Part Programmer. His or Her Manuscript Contains the Basic Instructions That the Machine Tool Will Follow From One Part to the Next With No Deviation. With Manual Operation the Output Results From a Combination of the Operator's Efficiency and Capability.[a]

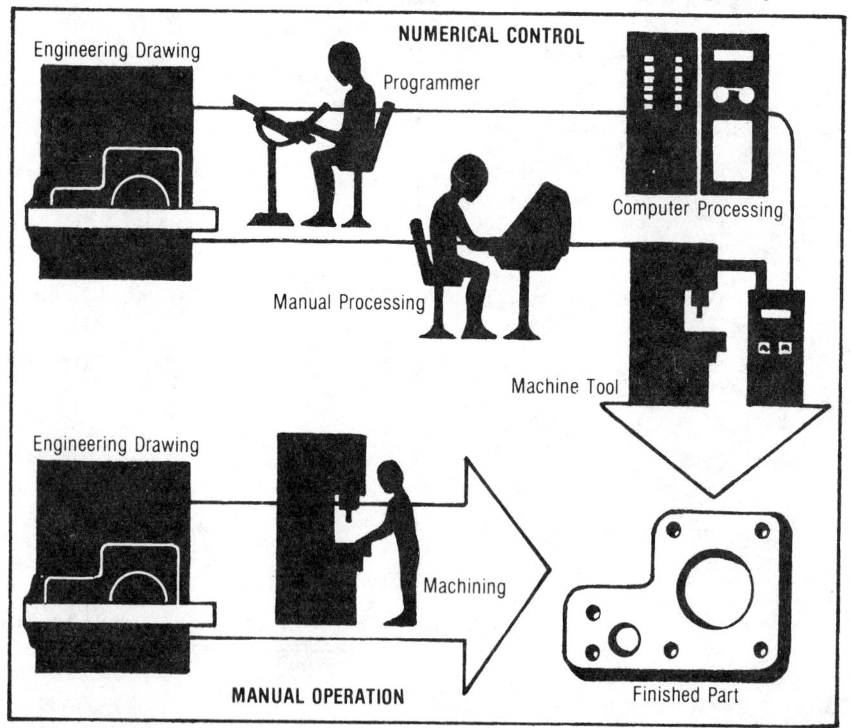

[a]Courtesy of *Modern Machine Shop 1980 NC/CAM Guidebook,* Cincinatti, OH.

into the computer, and it would then take less than a minute to process the data and another 2 or 3 min to generate a program tape or disc.

One of the oldest, most powerful, and most widely employed processor languages is Automatic Programming of Tools (APT). It was originally created by the Aerospace Industries Association to assist in developing workpiece programs for the first NC machine tools that went to the aerospace industry. The task was so massive and ongoing that the further development and enhancement of the processor was contracted out in 1961 to the Illinois Institute of Technology Research in Chicago. During the next 14 years several hundred man-years of additional development went into the processor. It was adapted to most mainframe computers, and its capabilities were further enhanced. Then APT was turned over to Computer Aided Manufacturing–International (CAM-I) of Arlington, Texas, where it is now maintained and documented. The language is in the public domain, and its complete documentation may be obtained from CAM-I, 611 Ryan Plaza Drive, Suite 1107, Arlington, TX 76012 for a modest fee. In addition to APT, there are more than 50 other NC processor languages commercially available either by purchasing or time-sharing. A complete listing is found in the *NC/CAM Guidebook.*[2]

7.7.7 APPLICATION OF NC

Justification of NC requires a good understanding of what it is and what it is not. First, it is a method of control; it is not a machining process. Numerical control will not magically endow a 5 hp spindle with 10 hp capabilities. Once the cutting tool is in the cut, the rate of metal removal is not a function of NC or manual machine control. The method merely offers the opportunity to control much more efficiently and accurately the operation of a machine tool. Thus it is particularly applicable to those machining areas where the traditional mass production approaches cannot be applied. Workpieces logical for the NC machine tool include:

Those requiring substantial tooling costs in relation to the total manufacturing costs by conventional means.

Those requiring lengthy setup times compared to the machine run time in conventional machining.

Those machined in small or variable lots.

A wide diversity of parts requiring frequent changes of the machine setup and a large tool inventory.

Those produced at intermittent times because of some cyclic demand.

Those with complex configurations requiring close tolerances and machined relationships, and those with complex, mathematically defined surfaces or contours.

Those which are very expensive and for which human error would be very costly as the part nears completion.

Machine operators are not removed from their traditional role because the actual machining sequences are under NC. Operators must still be present to load the program, mount the workpiece, and initiate the program sequence. They must observe and watch for any accidents, malfunctions, or untoward events that could occur. They are a valuable source of feeding back information to the programmer about the efficiency of the written workpiece program.

With the conventional machine tool, operators need perfected motor skills to coordinate the various machine tool motions. With NC, their role has changed to that of a supervisor and overseer of the whole operation. They must still know good machining practice, the sound of a cutting tool growing prematurely dull, and what to do if the unexpected does occur. Operators may even be allowed to make minor changes in feed or speed to optimize the functioning of a program.

Numerical control is significant for the optimization and efficiencies it can bring to the machining process. Equally as significant are the knowledge and experience gained in working with preplanned and documented data, often obtained with the assistance of a computer, which naturally leads to a better working understanding of other computer-aided manufacturing (CAM) applications, such as group technology, shop floor control, bill of materials organization, computer-assisted quality control, computer-assisted production planning, and finally the full integrated computer-assisted manufacturing facility.

REFERENCES

1. MODERN MACHINE SHOP, *Modern Machine Shop 1980 NC/CAM Guidebook*, Author, Cincinnati, OH, p. 269.
2. *1980 NC/CAM Guidebook*, p. 186.

CHAPTER 7.8
Group Technology

INYONG HAM

The Pennsylvania State University

7.8.1 INTRODUCTION

Basic Concept

More and more manufacturing industries involved with small lots and a variety of products are becoming interested in group technology, which is particularly applicable in the area of batch-type manufacturing. Group technology has also been recognized as an essential element of the foundation for the successful development and implementation of CAM through the application of the part family concept.

Group technology is generally considered to be a manufacturing philosophy or concept that identifies and exploits the sameness or similarity of parts and operation processes in design and manufacture. In batch-type manufacturing each part has traditionally been treated as being unique in design, process planning, production control, tooling, production, and so on. However, by grouping similar parts into part families, based on either their geometric shapes or operation processes, as shown in Exhibit 7.8.1, and also, if possible, forming machine groups or cells that process the designated part families, it is possible to reduce costs through more effective design rationalization and design data retrieval, fewer stocks and purchases, simplified and improved process planning and production control, reduction of tooling and setup times, semi-flow-line production by machine groups or cells, less in-process inventory, reduction of total throughput time, reduction of NC programming, more efficient utilization of expensive NC machines and machining centers, and so on.

Historical Background

The basic concept of group technology has been practiced for many years as part of "good engineering practice" or "scientific management." For example, a classification and coding system developed by F. W. Taylor[1] for formation of part families was used in manufacturing as early as the beginning of this century. Through the years many companies devised their own classification and coding systems and have been using them in various areas, such as design, materials, and tools. There are numerous examples of machine groups or cells, group tooling devices, part family groupings and programming, and so on, which have been used for many years in various sectors of industry. These practices and applicatons of group technology concepts were in many cases identified under different names and in various forms of engineering, manufacturing, and management functions.

Around the world, group technology has been practiced in various forms and degrees for many years. Many countries took an interest in it in the 1950s and 1960s. At that time various classification and coding systems were developed, machine cell concepts were practiced, and many excellent group tooling practices were reported. Until recently, group technology has not been formally recognized and rigorously practiced as a systematic scientific technology. In the recent years, advanced manufacturing industry appears to be undergoing a revolution in the area of improving its manufacturing productivity. This has led to an intensified effort in integrated CAM. These current trends have stimulated a strong renewed interest in group technology since it provides the essential means for higher manufacturing productivity and for CAM, for example, computer-aided process planning.

Major Areas of Application

One of the most important reasons for increasing manufacturing productivity is economic. Manufacturing contributes a major part of the gross national product (GNP) of modern industrialized

Exhibit 7.8.1 Examples of Part Families
(*a*) Similar in Shape and Geometry and
(*b*) Similar in Production Operation Processes

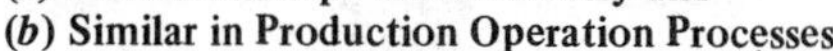

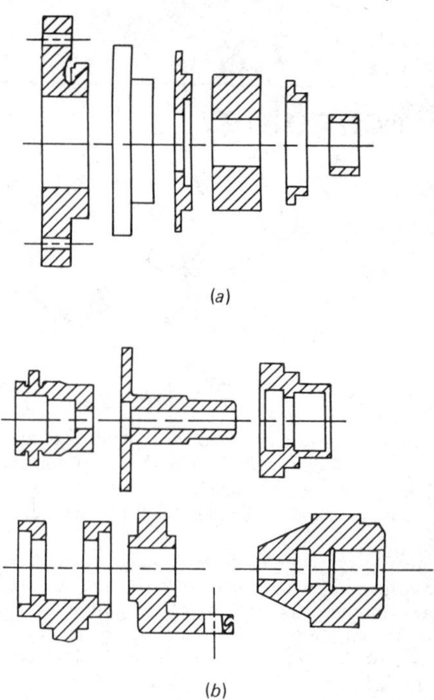

countries. Yet in spite of that, manufacturing, although normally thought of as a highly productive and efficient activity, generally can still be improved significantly. This is especially true in a batch-type manufacturing environment. Indeed, the potential for economic improvement of manufacturing by group technology not only is tremendous now, but will grow with time.

Rationalization of various engineering activities, such as design data retrieval, process selection, and process planning, can be readily achieved by effective implementation of the group technology concept. It has been recognized that in batch-type manufacturing major efforts should be made for continuous improvements in in-process inventory and efficient machine loading in order to achieve higher productivity. Again, group technology provides a key element in this effort.

Current Trend and Future Prospects

For various objectives in achieving higher productivity from design to manufacture, many manufacturing industries that are related primarily to batch-type manufacturing have become increasingly interested in group technology implementation to meet their needs. Many industrial companies have been applying group technology principles in their own way, although in some cases it was not identified as group technology, but simply as good engineering practice and effective scientific management. Group technology implementation in many instances has been limited primarily to cellular manufacturing. However, recently more companies are interested in applying group technology concepts as a part of the total system of the overall company operations through design to manufacture.

A forecast of the future of production technology advancement carried out by both the University of Michigan[2] and the International Institute for Production Engineering Research (CIRP)[3] predicted that approximately 50 to 75% of manufacturing industry will use group technology concepts in the period 1980–1990. This forecast also predicted that the computer-automated factory will be a full-blown reality in many industries well before the end of this century. It is evident that new technological innovations, such as direct numerical control (DNC), CNC, machining centers, industrial robots, and microprocessors, will lead the way toward more automated, computer-integrated manufacturing systems involving CAM and thus ensure more integrated applications of group technology for optimum manufacturing, resulting in higher productivity. The effort related to the integrated CAM is a positive approach to achieving those objects.[4]

A part classification system, which is an integral part of, and has been used as, an essential tool of group technology applications, can also be evolved as a means of describing parts in a form that can be readily integrated into a computer data base structure, which will link design and production. Furthermore, as evolution of integrated CAM leads to generative design and process planning, certain part classification and coding systems will become an integral part of the total generative system evolving with integrated CAM.

Group technology is a dynamic and evolutionary development which continues to expand its influence on manufacturing systems. It is evident that the role of group technology will certainly broaden with more innovative advancements in theory and application, not only for improving productivity in conventional batch-type manufacturing systems, but also for proper adaptation of CAM systems.

7.8.2 PART FAMILY FORMATION AND MACHINE GROUPING

Methods and Procedure

A "part family" may be defined as a group of related parts that have some specified sameness and similarities. They may have similar geometric shape or may share similar processing requirements, as shown in Exhibit 7.8.1. Parts may be dissimilar in shape, but could be grouped as a part family because of some common production operations, or vice versa. Parts are considered to be similar with respect to production operations when the same machines and processes are used and when the type, sequence, and tooling requirements are similar. In grouping part families the number of parts and their frequency of manufacture should be taken into consideration. The greater the similarity of processing requirements and lot frequency, the more effective it is to form the part family for practical applications of the group technology concept in forming machine groups or cells and in scheduling for optimum sequencing and machine loading.

The grouping of similar parts into part families is the key to group technology implementation. The problem that immediately arises is how the parts are to be efficiently grouped into these families. There are three basic methods used to form part families: (1) manual visual search, (2) production flow analysis, and (3) classification and coding systems.

The first method is obviously very simple, but is limited in its effectiveness when dealing with a large number of parts. In general, the other two methods are more commonly used in forming part families and machine groups or cells.

Production Flow Analysis

Production flow analysis[5] is a technique to analyze the operation sequence and the routing of the part through the machines and work stations in the plant. Parts with common operations and routes are grouped and identified as a part family. Similarly, the machines and work stations used to produce the part families can be grouped to form the machine group or cell.

An example of forming part families by this method is shown in Exhibit 7.8.2. To use this method successfully, one should ensure that a company has a reliable data source of routing or operation sheets. One of the advantages of this method is that part families can be formed with or without a classification and coding system, since the part families are formed using the data from operation or routing sheets. There are a number of disadvantages that stem from the method's reliance on existing production data and routing methods.

Classification and Coding Systems

A classification and coding system provides an effective means for sorting the coded parts in forming part families based on the specific parameters of the system, regardless of the origin or use of the parts. Especially for CAM applications, such a system becomes an essential requirement for effective implementation of group technology concepts.

"Classification" involves arranging items into groups according to some principle or system whereby like things are brought together by virtue of their similarities and are then separated according to a specific difference. A "code" can be a system of symbols used in information processing in which numbers or letters or a combination thereof are given a certain meaning.

Many varieties of classification and coding systems are being developed and used around the world. An example of a coded part using a publicly available system[6] is shown in Exhibit 7.8.3.

Classification and coding for group technology applications is a very complex problem, and although many systems have been developed, and countless efforts made to improve them, there is as yet no universally acclaimed system. Since each company has its own specific needs and conditions, it is necessary to search for a suitable system that can be adapted to the specific needs and requirements of the company. It is essential that an adapted system be usable by all concerned departments in the company, including design/engineering, planning/control, and manufacturing/tooling, as well as by management.

Exhibit 7.8.2 Part Family and Machine Grouping by Production Flow Analysis: (*a*) Before Grouping and (*b*) After Grouping

Machine \ Part No.	1	2	3	4	5	6	7	8	9	10	11	12	13	14	15	16	17	18	19	20
L[a]	✓	✓		✓	✓		✓	✓	✓		✓	✓		✓	✓		✓	✓	✓	✓
M_1[b]	✓	✓	✓		✓	✓	✓		✓		✓		✓	✓		✓				✓
M_2[c]			✓	✓				✓		✓		✓	✓		✓		✓	✓	✓	
D[d]	✓	✓	✓	✓		✓	✓	✓		✓	✓	✓	✓	✓		✓	✓	✓		✓
G[e]	✓	✓	✓	✓		✓			✓			✓	✓		✓			✓		✓

(*a*)

Machine \ Part No.	1	2	20	7	11	14	9	5	4	18	12	8	17	15	19	3	13	6	16	10
L	✓	✓	✓	✓	✓	✓	✓	✓												
M_1	✓	✓	✓	✓	✓	✓	✓	✓												
D	✓	✓	✓	✓	✓	✓														
G	✓	✓	✓				✓													
L									✓	✓	✓	✓	✓	✓	✓					
M_2									✓	✓	✓	✓	✓	✓	✓					
D									✓	✓	✓	✓	✓							
G									✓	✓	✓			✓						
M_1																✓	✓	✓	✓	
M_2																✓	✓			✓
D																✓	✓	✓	✓	✓
G																✓	✓	✓		

(*b*)

[a] L = Lathe.
[b] M_1 = Milling machine I.
[c] M_2 = Milling machine II.
[d] D = Drilling machine.
[e] G = Grinding machine.

For group technology applications, a well-designed classification and coding should be able to group part families as needed, based on specific parameters. An example of such part family grouping of the parts as was shown in Exhibit 7.8.1*a*, using a suitable coding system along with the related data need for the coding, is shown in Exhibit 7.8.4.

Cell Layout

In general, there are three basic types of plant layout: (1) mass production flow line layout, (2) functional layout, and (3) group layout. In the practice of group technology, a group of machines for producing one part family or more may be formed such that it can perform all the operations required by the family or families of parts. The machines themselves are arranged in a semi-flow-line to minimize transportation distances and waiting problems. The result is very similar to a modern NC machining center. If conditions warrant, a machining center may be used instead of a group of single-purpose machines. A conventional functional layout and a group layout of machine tools based on the group technology concept are shown in Exhibits 7.8.5*a* and 7.8.5*b*, respectively, Exhibit 7.8.5*b* illustrates the features of a group/cell layout.

Exhibit 7.8.3 Coded Example Using a Classification and Coding System (Refer to Exhibit 7.8.8)

Part name: Pin
Material: Mild steel (AISI-1020) forged round bar
Treatment: Surface hardening by carbonizing/fine finish
Operations: Turning of outside diameter and drilling hole

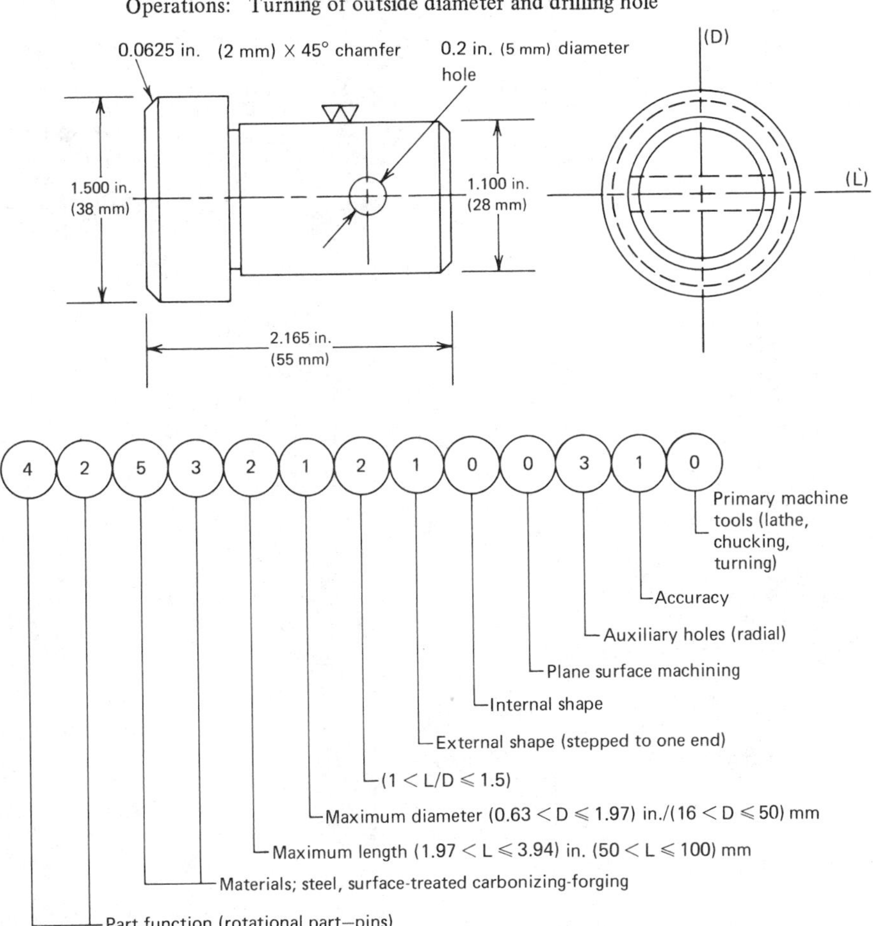

The formation of machine groups or cells to process part families is relatively easy if a well-designed classification and coding system has been introduced. It is also possible to form machine groups/cells or part families using the production flow analysis technique. Here the operation sequence and routing of the part through the machines in the plant is analyzed using the information obtained from the operation or routing sheets as illustrated in Exhibit 7.8.2. Each part family should have a certain group of operations associated with it. This group of operations indicates the types of machines and facilities needed to process each part within a family. The amount of time needed for each particular operation for each job in the family of parts can be determined if basic data such as lot sizes, setup times, and machining times are available. These times will be the basis of determining how much capacity is needed for each machine within the group or cell. The formation and grouping of part families permits the computation of the machine load in hours for each machine in the group cell.

In the practice of group technology, an effort is made to maximize the utilization of machines in the group by (1) extending basic part families by adding parts of a similar type or merging two or more subfamilies and by (2) machining two or more part families on the same machine group.

A simple example of machine loading computation for forming machine groups/cells for given part families is exhibited in Exhibits 7.8.6 and 7.8.7.

The machine loading analysis indicates that the part family #1 needs the total times of 2783 hr for turning, 1721 hr for milling, 1085 hr for drilling, and 3367 hr for grinding, thus requiring two lathes, one milling machine, one drill press, and two grinding machines, respectively.

Exhibit 7.8.4 Part Family Grouping Using a Classification and Coding System

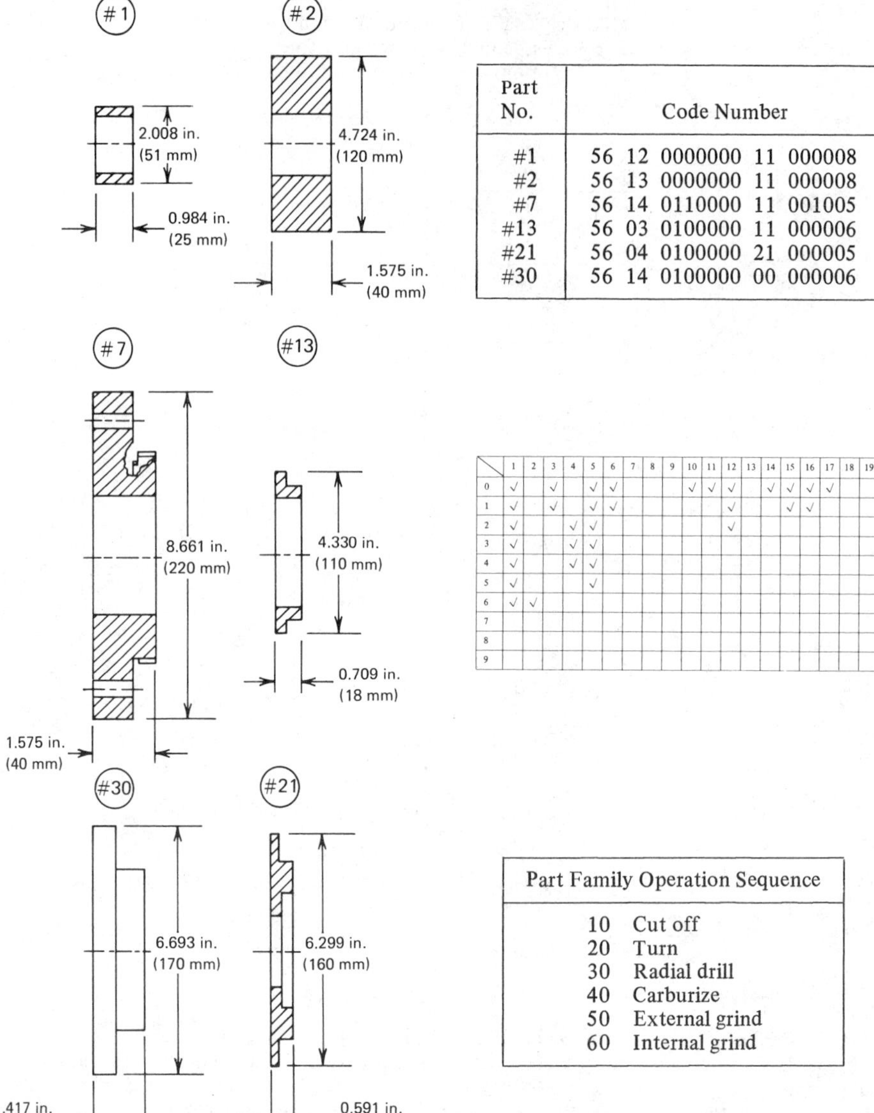

Part No.	Code Number
#1	56 12 0000000 11 000008
#2	56 13 0000000 11 000008
#7	56 14 0110000 11 001005
#13	56 03 0100000 11 000006
#21	56 04 0100000 21 000005
#30	56 14 0100000 00 000006

	1	2	3	4	5	6	7	8	9	10	11	12	13	14	15	16	17	18	19
0	√		√		√	√				√	√	√		√	√	√	√		
1	√		√		√	√						√			√	√			
2	√			√	√							√							
3	√			√	√														
4	√			√	√														
5	√				√														
6	√	√																	
7																			
8																			
9																			

Part Family Operation Sequence	
10	Cut off
20	Turn
30	Radial drill
40	Carburize
50	External grind
60	Internal grind

Although the group layout of machines for group technology applications has many advantages, it may also involve some problems. For example, it may be difficult to balance the labor and machine utilization. Also, there may be difficulty in finding suitable supervisory personnel.

7.8.3 CLASSIFICATION AND CODING SYSTEMS

Types and Features

There are many types of classification and coding systems for general purposes. Three basic forms for the current group technology applications are as follows:

1. Hierarchical structure (monocode).
2. Fixed-digit-type structure (polycode).
3. Combined structure (multicode).

Exhibit 7.8.5 Functional Layout (*a*) and Group/Cell Layout (*b*)

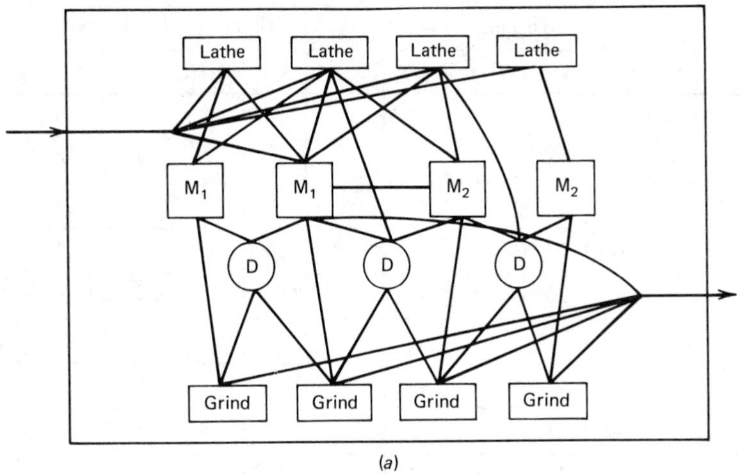

(*a*)

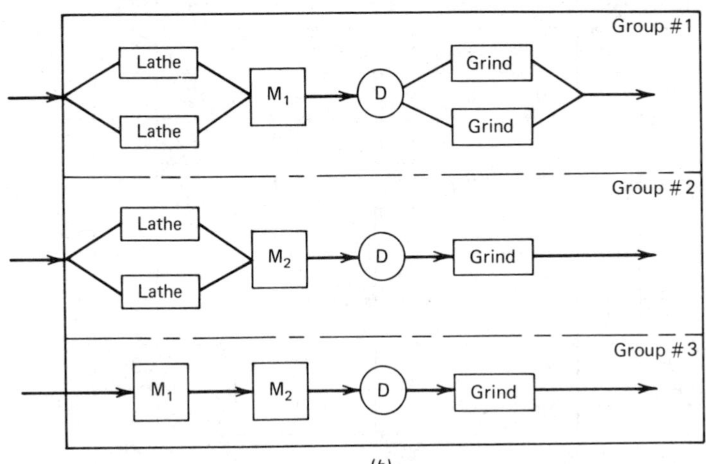

(*b*)

A well-designed classification and coding system for group technology implementation should meet several basic requirements. It can provide many benefits and facilitate group technology applications in many areas of company operations. The major benefits of a well-designed classification and coding system for group technology applications can be summarized as follows:

1. Formation of part families and machine groups (cells).
2. Effective retrieval of designs/drawings and process plans/routings.
3. Design rationalization and reduction of design costs.
4. Standardization of product design.
5. Securing of reliable workpiece statistics.
6. Accurate estimation of machine tool requirements, rationalized machine loading, and optimized capital expenditure.
7. Rationalization of tooling setup and reduction of setup time and overall production time.
8. Rationalization of tool design and reduction of time and cost for tool design and fabrication.
9. Standardization of process routings/tooling.
10. Rationalization of production planning and scheduling.
11. Accurate cost accounting and cost estimation.
12. Better utilization of machine tools, work holding devices, and manpower.
13. Improvement of NC programming and effective use of machine and machining centers.
14. Establishment of a master data base.

Exhibit 7.8.6 Basic Data for Machine Loading Analysis

Part No.	Code No.	Lot Size	Turning		Milling (1)		Milling (2)		Drilling		Grinding	
			S/T	M/T	S/T	M/T	S/T	M/T	S/T	M/T	S/T	M/T
1	010 120	115	20	9	13	5	–	–	15	3	6	20
2	010 124	30	15	12	15	7	–	–	23	6	7	15
3	002 123	55	–	–	13	5	16	8	23	12	6	10
4	111 121	105	33	14	–	–	15	6	25	8	8	9
5	010 120	25	23	10	10	8	–	–	–	–	–	–
6	002 123	10	–	–	15	8	–	–	18	10	8	23
7	011 123	5	15	9	11	9	–	–	15	15	–	–
8	110 124	12	15	15	–	–	10	9	24	11	–	–
9	011 124	18	23	12	10	8	–	–	–	–	5	23
10	002 120	35	–	–	–	–	13	6	30	11	–	–
11	011 123	10	23	14	12	6	–	–	24	7	–	–
12	112 122	10	30	8	–	–	12	10	18	10	7	15
13	001 123	21	–	–	11	7	10	9	25	11	6	15
14	011 123	61	15	8	10	5	–	–	24	4	–	–
15	111 121	4	15	17	–	–	12	9	–	–	5	18
16	002 120	5	–	–	10	6	–	–	20	13	–	–
17	110 120	24	23	10	–	–	10	8	20	11	–	–
18	111 121	46	30	11	–	–	11	17	30	7	5	10
19	111 120	61	20	9	–	–	13	5	–	–	–	–
20	012 124	10	15	10	11	9	–	–	20	5	5	18

Exhibit 7.8.7 Machine Loading Analysis for Turning Operations of Part Family #1 (Refer to Exhibits 7.8.5*b* and 7.8.6)

Operation: Lathe turning Part family #1

Part No.	Code No.	Lot size (N_l)	Setup Time (S_t)	Machining Time (M_t)	$(M_t) \times (N_l)$	$(M_t) \times (N_l) + (S_t)$
1	010 120	115	20	9	1035	1055
2	010 124	30	15	12	360	375
20	012 124	10	15	10	100	115
7	011 123	5	15	9	45	60
11	011 123	10	23	14	140	163
14	011 123	61	15	8	488	503
9	011 124	18	23	12	216	239
5	010 120	25	23	10	250	273
Total						2783 hr

Basic Requirements

For group technology applications, a classification and coding system should meet the following basic requirements:

1. Be all embracing.
2. Be mutually exclusive.
3. Be based on permanent characteristics.
4. Be specific to user needs.
5. Be adaptable to future changes.
6. Be adaptable to computer processing.
7. Offer companywide applications.

Some typical topics that should be addressed are:

1. Objective.
2. Scope of application.
3. Costs and time.
4. Adaptability to other systems.
5. Management problems.

7.8.4 DESIGN RATIONALIZATION

Design Data Retrieval

A classification and coding system facilitates a part reduction and standardization program, which can be valuable to both the company and its customers. When a well-designed classification and coding system is efficiently implemented in the design area, the system provides a simple, systematic, and efficient method for storing information in an organized manner. It will also use a computer data base if needed. The system provides for retrieving design data, for example, drawings, specifications, geometric data, and materials. A code makes it possible to recall all stored data relative to a specified part family grouped together based on their designated similarities. The design data retrieval system sorted by part family grouping provides the following important features, which assist significantly in design rationalization:

1. Part family grouping for design rationalization.
2. Retrieval of existing design information for new applications, modifications, and references.
3. Standardization of design features, specifications, and materials.
4. Improvement for better design.
5. Elimination of duplicate designs.
6. Simplified effective cost estimation.

The retrieval system may be processed manually or by computer. An example of a flow diagram[7] for a design data retrieval system using the group technology concept is shown in Exhibit 7.8.8.

A successful classification and coding system application in design leads to substantial economic gains. The most significant and immediate savings result from design rationalization through effective design data retrieval. An industrial survey[8] reports that an average design engineering cost per new part is approximately \$2000. Approximately a 15% reduction of new design activity will take place when a suitable effective retrieval method is utilized. Therefore, if a company releases approximately 2500 new parts annually, it can be estimated that the annual design cost of new parts released is approximately \$5 million. The savings from an effective retrieval system would be estimated at \$750,000 per annum.

In many cases more intangible savings are realized by various indirect benefits resulting from design rationalization, for example, design standardization, design improvement, and productivity improvement in design activity in general.

Standardization

When all active parts are classified and coded using a suitable system, it is possible to analyze the part population and the frequency of usage of specific parts in the part family. In practice, parts that belong to a specific part family can be sorted to identify standard designs that are most frequently used. It must be recognized that such standardization is possible only when a part family grouping is made to identify the standard features clearly. Usually, when parts are designed independently at different times without any means of grouping into part families, it is difficult to identify the nature and degree of unnecessary duplications and obvious similarities among the parts of the company.

Exhibit 7.8.8 Flow Diagram for Design Rationalization

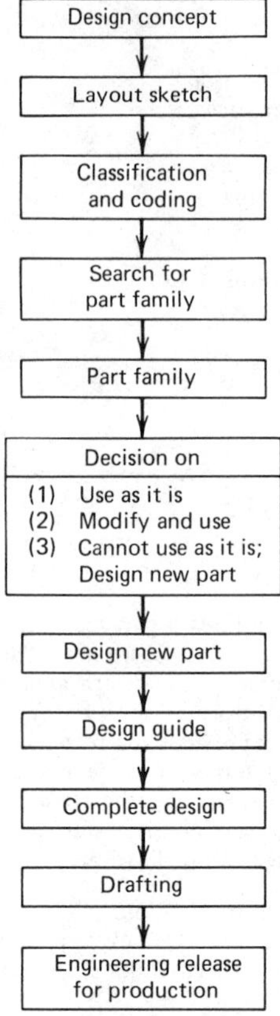

Exhibit 7.8.9 shows a part family of washers. One immediately recognizes the standard parts based on the usage frequency of this part family. In many cases many parts in the part family have a very minor variation in size, shape, and features. In such cases the part family grouping of a selected part assists not only in identifying such similarities, but also in preventing unnecessary variety through effective standardization. Although design personnel are knowledgeable as to the desirability of standardization and believe they have incorporated standard parts, they are not always aware that similar designs exist where standardization can be made. By part family groupings and design data retrieval, it is readily possible to incorporate a high level of standardization.

7.8.5 GROUP TOOLING

Composite Part

The composite part provides an aid for the application of group technology concepts in the standardization of parts, standardization of process planning, machine grouping, design of group jigs and fixtures, the planning of group tooling setups, NC part family programming, and so on. Exhibit 7.8.10 illustrates a group of parts represented by a composite part possessing all the shape characteristics and processing features of a part family that is illustrated in Exhibits 7.8.1*a* and 7.8.4. If process planning or tooling is developed for the composite part, then any part in the family can be processed with the same operations and tooling.

An example of such a group jig for drilling a part family is shown in Exhibit 7.8.11. To drill the holes of six different parts in this part family, it requires only one group jig (Exhibit 7.8.11*b*) and six different auxiliary adapters (Exhibit 7.8.11*a*) to accommodate some minor differences in sizes, numbers, and locations of the holes and in size and shape of parts. Therefore, instead of designing, fabricating, and using six individual drill jigs as is done in a conventional production method,

Exhibit 7.8.9 Standardization of Parts

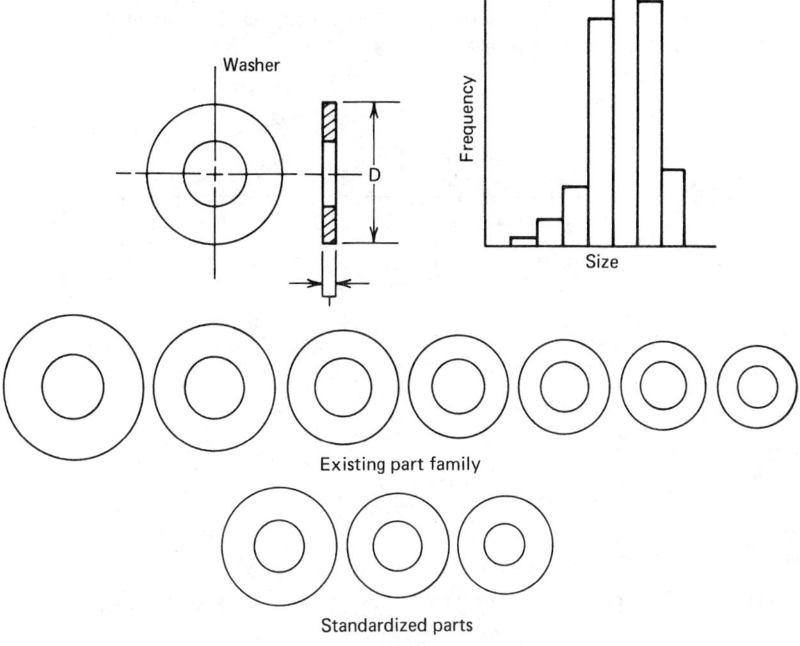

Exhibit 7.8.10 Composite Part

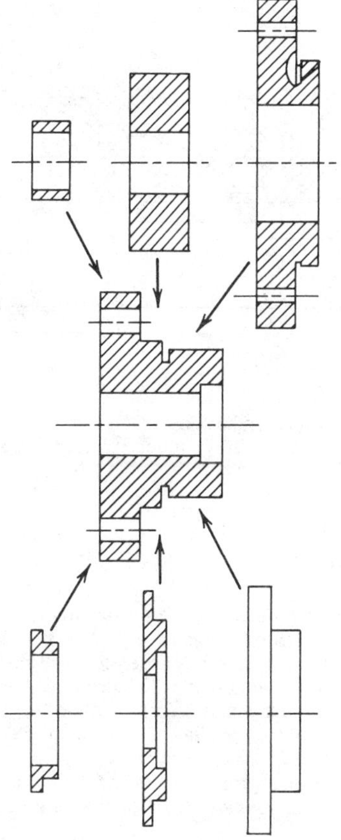

Exhibit 7.8.11 Group Tooling Design for Part Family: (*a*) Part Family of Round Plates and (*b*) Group Jig for Drilling

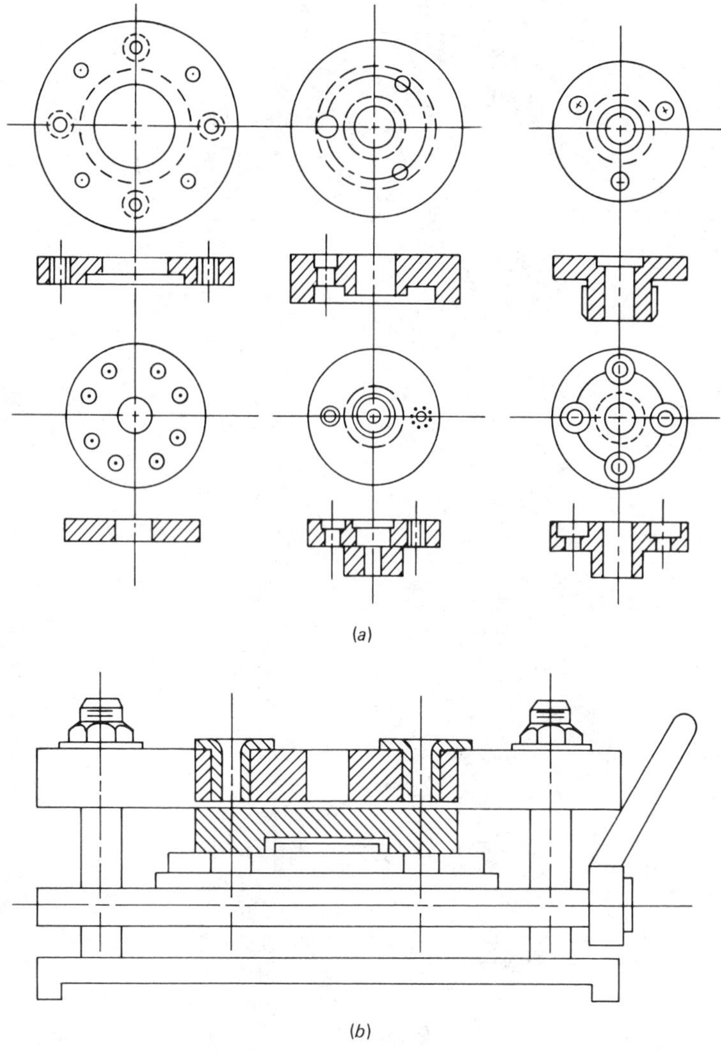

only one group jig and six adapters or bushing plates, which are essentially inexpensive, are required. Therefore it becomes evident how much tooling costs can be reduced using group technology.

7.8.6 GROUP SCHEDULING

Basic Concept

Production scheduling is greatly simplified by group technology. The scope of the problem is reduced from a large portion of the shop to a small group of machines. If the families of parts and groups of machines have been formed correctly, each job will indicate by its code number which group of machines will be used to process it. The group/cell layout and part family concepts ideally lend themselves to optimal sequencing efforts. Within the group of machines, the scheduling problem is again reduced to simply scheduling the given jobs through the machines in the cell. Even though a machine group/cell is not formed, the production scheduling could be very much simplified by use of part families in assigning jobs to the various machines in the shop. A computer program can be utilized to schedule jobs of a part family to a corresponding machine group/cell.

The jobs can be properly sequenced in the family and the families properly sequenced through the machine group/cell.

Proper scheduling is an integral part of group technology. Good scheduling, combined with reduced setup time and reduced transportation, will result in a significant cost reduction. The most obvious benefit is reduced total production time. With this reduction, production can more closely match demand so that inventories can be reduced and parts produced on schedule. Proper application of group technology concepts in production scheduling will:

1. Reduce setup times and costs.
2. Permit optimal sequences of group and job.
3. Permit flow line production.
4. Optimize group layout.
5. Provide overall economic advantages.

Group scheduling has some specific features that differ from conventional scheduling problems, as follows:

1. Optimization for group and job sequence and machine loading.
2. Possibility of flow shop pattern.
3. Reduction of setup times and cost.
4. Economic savings.

Algorithms for Group Scheduling

Optimal Sequencing Analysis

Group scheduling can be analyzed in a multistage manufacturing system. In the case of manufacturing multiple parts (jobs) grouped into several part families, both optimal group and optimal group sequences can be determined such that the total flow time (makespan) is minimized by means of various methods, for example, branch-and-bound method and a heuristic method.[9–11]

Machine Loading Analysis

Analysis for machine loading for group scheduling is a complex problem, and it is not simple to develop an adequate algorithm for practical applications. However, some mathematical models for the machine loading and product mix analysis problems for group technology applications are available.[12,13]

Integrated Applications With MRP

Group technology takes all of the component parts and attempts to classify them into a workable set of part families and/or assign them to machine groups/cells. Ideally, each part family has enough interpart similarity that the individual parts may be processed by means of a particular subset of the total production processes. Material requirements planning (MRP) is yet another tool for the alleviation of problems within the multistage, multiproduct plant. In its simplest form, MRP reduces each final product into its elementary parts and, using a forecast of the requirements for the final product, assigns the required quantities of each elementary part to a specific time period. An integrated use of MRP and group technology scheduling provides a viable system for effective production control.[14]

7.8.7 ECONOMICS OF GROUP TECHNOLOGY

Economic Benefit and Justification

Appropriate and successful implementation of group technology will lead to such improvements as more effective design, less stock and fewer purchases, simplified production planning and control, optimum sequencing and loading, reduced tooling and setup times, reduced in-process inventories, shorter throughput time, and more efficient utilization of expensive machines. Significant economic benefits will be achieved, as shown in Exhibit 7.8.12.

It is desirable to analyze these economic gains through cost analysis of specific applications by comparing the economic benefits of both the present conventional method and the proposed group technology method. Economic justification is a key to implementation of group technology.

Exhibit 7.8.12 Reduction of Manufacturing Costs Through Group Technology Application

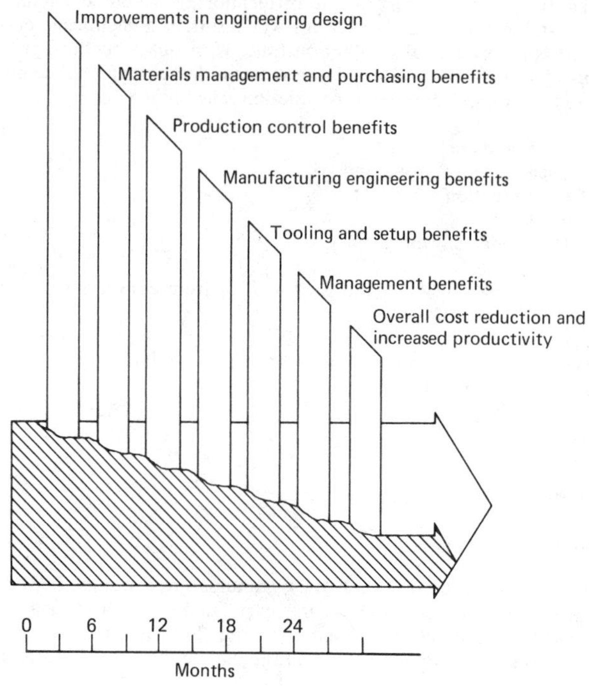

Various formulas and procedures have been developed for economic analysis, some examples of which are presented in the succeeding subsection.

The economic gains through successful group technology applications are great. However, it requires some time to gain new savings since a considerable cost is involved in maintaining the system, as indicated in the example shown in Exhibit 7.8.13.

Exhibit 7.8.13 Examples of Cost Savings, Expenditure, and Time Period for Group Technology Implementation

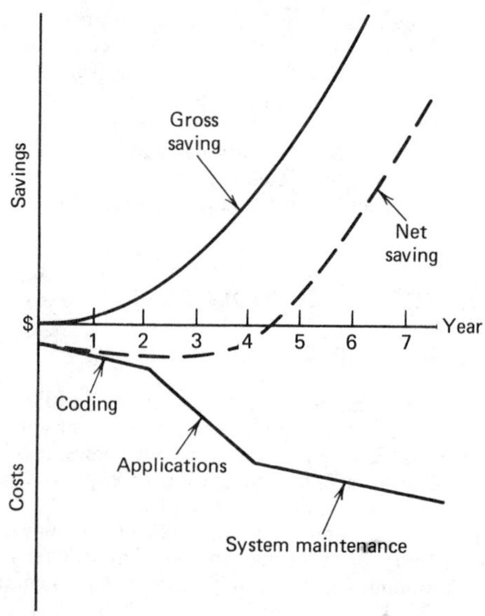

Comparative Cost Analysis

Group Tooling Costs

One of the advantages of group technology applications is the rationalization of tool designs and the reduction of tooling setups, which lead to reduction of tooling costs and production costs as a whole. The cost analysis of group tooling (group jigs and fixtures) in comparison with that of conventional tooling methods becomes essential for the justification of group technology applications in tooling.[15]

1. **Conventional Tooling Method**

$$C_{tw1} = \sum_{i=1}^{p} C_{w1}(i)$$

where C_{w1} = cost of a jig or fixture of conventional tooling method, \$
C_{tw1} = total tooling costs of conventional methods using p different jigs or fixtures, \$
p = number of different jigs or fixtures used (also, possibly, number of different parts to to be produced)

2. **Group Tooling Method**

$$C_{tw2} = \sum_{i=1}^{q} C_a(i) + C_{w2}$$

where C_{w2} = cost of a group jig or fixture, \$
C_{tw2} = total costs for the group tooling using a group jig or fixture with q different adapters, \$
C_a = cost of an adapter, \$
q = number of adapters used for the production of a family of parts

3. **Unit Tooling Cost**

a. Conventional Tooling Method

$$C_{u1} = \left[\frac{C_{tw1}}{N}\right] = \left[\frac{\sum_{i=1}^{p} C_{w1}(i)}{N}\right]$$

where C_{u1} = unit tooling cost for conventional tooling method, \$/piece
N = number of parts produced

b. Group Tooling Method

$$C_{u2} = \left[\frac{C_{tw2}}{N}\right] = \left[\frac{\sum_{i=1}^{q} C_{a(i)} + C_{w2}}{N}\right]$$

where C_{u2} = unit tooling cost for group tooling method, \$/piece

The data shown in Exhibit 7.8.14 are given for comparison of a conventional tooling method using conventional milling fixtures and a new group tooling method using a master group fixture

Exhibit 7.8.14 Cost Data for Comparative Analysis

Item	Conventional Tooling Method	Group Tooling Method
Cost of the drill jig	\$815	\$2208
Number of jigs required	6	1
Cost of an adapter	–	\$450
Number of adapters required	–	5
Number of pieces to be produced	240	240

Exhibit 7.8.15 Computed Example of Tooling Costs for Comparison

No. of parts in part family	Conventional Method		Group Tooling Method	
	C_{tw1}	C_{u1}	C_{tw2}	C_{u2}
1	$ 815	$3.40	$ 2,658	$11.08
2	1,630	3.40	3,108	6.48
3	2,445	3.40	3,558	4.94
4	3,260	3.40	4,008	4.18
5	4,075	3.40	4,458	3.72
6	4,890	3.40	4,908	3.41
7	5,705	3.40	5,358	3.19
8	6,520	3.40	5,808	3.03
9	7,335	3.40	6,258	2.90
10	8,150	3.40	6,708	2.80
11	8,965	3.40	7,158	2.71
12	9,780	3.40	7,608	2.64
13	10,595	3.40	8,058	2.58
14	11,410	3.40	8,508	2.53
15	12,225	3.40	8,958	2.49
20	16,300	3.40	11,208	2.34

and adapters. The total tooling costs (C_{tw}) and the unit tooling costs (C_u) of the conventional tooling method and the group tooling method in relation to the number of different parts in the part family or the group are computed and listed in Exhibit 7.8.15.

The total tooling costs (C_{tw}) and the unit tooling costs (C_u) as a function of the number of parts in the part family or group are plotted in Exhibits 7.8.16 and 7.8.17, respectively.

Exhibit 7.8.16 Total Tooling Costs of Conventional and Group Tooling Methods (Refer to Exhibits 7.8.14 and 7.8.15)

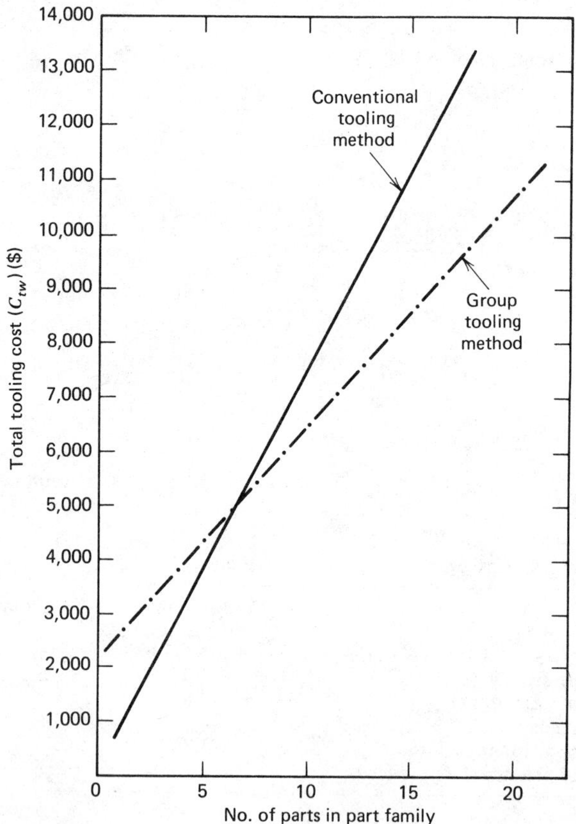

Exhibit 7.8.17 Unit Tooling Costs of Conventional and Group Tooling Methods (Refer to Exhibits 7.8.14 and 7.8.15)

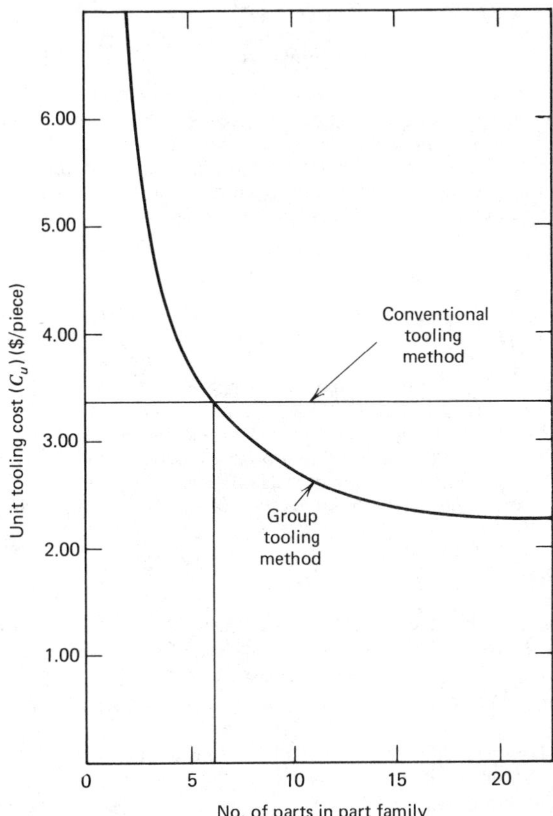

As shown in these exhibits, the rate of increase in the total tooling costs for the conventional tooling method is exceedingly larger than that for the group tooling method. From the standpoint of the unit tooling costs, as the number of parts in the part family increases, the unit tooling costs for the group tooling methods become far more economical compared to the conventional tooling method, which is not affected by the number of parts in the part family. However, the sharp decrease in the unit tooling costs levels off after a certain number of parts in the group. This indicates that there is a limit up to which reduction of the unit tooling cost is effective. Also, both graphs of the total tooling costs and the unit tooling costs indicate the break-even points at which the decision for selection of appropriate tooling method can be made.

Group Machining Costs

Group machining is one of the most important features of group technology applications. Although group machining is advantageous from various technical points of view, it is still desirable to confirm the advantages of the group machining method over the conventional machining method.

1. The total machining cost for a single lot of a part with a special individual tooling may be expressed as

$$C_{tm} = C_o (T_c N_\ell + T_s) + D_t$$

where C_{tm} = total machining cost, $
C_o = labor rate, $/min
T_c = unit machining time per piece, min/piece
N_ℓ = lot size, no. of pieces/lot
T_s = setup time per a lot, min/lot
D_t = depreciation of tooling per a lot, $/lot

2. The total machining costs for machining of n lots or n different parts in the part family for both conventional and group machining can be expressed as follows:

a. Conventional (Individual) Machining

$$C_{tm1} = C_o \left[\sum_{i=1}^{n} T_{c1}(i) N_{\ell 1}(i) + \sum_{i=1}^{n} T_{s1}(i) \right] + \sum_{i=1}^{n} D_{t1}(i)$$

where C_{tm1} = total machining cost for conventional machining, \$
n = number of lots or number of different parts to be produced
T_{c1} = unit machining time per piece by conventional machining, min/piece
T_{s1} = setup time per lot for conventional machining, min/lot or part
D_{t1} = average depreciation of tooling per lot for conventional machining, \$/lot or part

b. Group Machining

$$C_{tm2} = C_o \left[\sum_{i=1}^{n} T_{c2}(i) N_{\ell 2}(i) + T_{s2} + \sum_{i=1}^{n-1} T_{sa}(i) \right] + \left[D_{t2} + \sum_{i=1}^{n-1} D_{ta}(i) \right]$$

where C_{tm2} = total machining cost for group machining, \$
n = number of parts in the part family
T_{c2} = average unit machining time per piece by group machining, \$/pc
T_{s2} = setup time per lot (per a part family) for group machining, min/lot or part family
T_{sa} = setup time per adapter for group machining, min/adapter
D_{t2} = depreciation of tooling per lot or part family for group machining, \$/lot or part family

REFERENCES

1. H. D. HATHAWAY, "The Mnemonic Systems of Classification; As Used in the Taylor System of Management," *Industrial Management*, Vol. 60, No. 3 (September 1920), pp. 173–183.
2. L. EVANS, "Production Technology Advancements: A Forecast to 1988," Industrial Development Division, Institute of Science and Technology, University of Michigan, Ann Arbor, 1973.
3. M. E. MERCHANT, "Delphi-type Forecast of the Future of Production Engineering," *CIRP Annals*, Vol. 20, September 1971.
4. D. E. WISNOSKY, W. A. HARRIS, and O. L. SHUNK, "An Overview of the Air Force Program for Integrated Computer Aided Manufacturing (ICAM)," Technical Paper #MS77-254, SME, Dearborn, MI, 1977.
5. J. L. BURBIDGE, *Production Planning*, Heinemann, London, 1971.
6. JAPAN SOCIETY FOR PROMOTION OF MACHINE INDUSTRY, *Guide Book for Group Technology Implementation*, (Japanese), Tokyo, Japan, 1979.
7. A. R. THOMPSON, "Establishing a Classification and Coding System," Technical Paper No. MS76-276, SME, Dearborn, Michigan, 1976.
8. I. HAM and W. REED, "Preliminary Survey Results on Group Technology Applications in Metal-Working," *Machine Tool Blue Book*, May 1977, pp. 100–108. (Also Technical Paper No. MS77-328, SME, Detroit, MI, 1977.)
9. V. A. PETROV, *Flowline Group Production Planning*, Business Publications, Ltd., London, 1968.
10. I. HAM, R. J. DUTKOSKY, and K. HITOMI, "Production Scheduling in Group Technology Applications," Technical Paper No. MS76-275, SME, Dearborn, MI, 1976.
11. K. HITOMI and I. HAM, "Group Scheduling Techniques for Multi-production Multistage Manufacturing Systems," *ASME Transactions*, August 1977, pp. 419–422.
12. I. HAM and K. HITOMI, "Machine Loading for Group Technology Applications," *CIRP Annals*, Vol. 25, August 1977, pp. 279–281.
13. I. HAM and K. HITOMI, "Machine Loading and Product-mix Analysis for Group Technology," *ASME Transactions*, Vol. 100, August 1978, pp. 370-374.
14. I. HAM, J. IGNIZIO, and N. SATO, "Integrated Applications of Group Scheduling and Materials Requirement Planning (MRP)," *CIRP Annals*, Vol. 27, August 1978, pp. 471–473.
15. S. P. MITRAFANOV, *Scientific Principles of Group Technology* (English translation), J. Grayson, Ed., National Lending Library for Science and Technology, United Kingdom, 1966.

BIBLIOGRAPHY

BURBIDGE, J. L., *Proceedings of International Seminar on Group Technology*, International Centre for Advanced Technical and Vocational Training, Turin, Italy, 1969.

BURBIDGE, J. L., "A Study of the Effects of Group Production Methods on the Humanization of Work," final report International Labor Office, Geneva, Switzerland, June 1975.

BURBIDGE, J. L., *The Introduction of Group Technology*, Wiley, New York, 1975.

DEVRIES, M. F., S. M. HARVEY, and V. A. TIPNIS, *Group Technology: An Overview and Bibliography*, Machinability Data Center (MDC 76-601), Cincinnati, OH, 1976.

EDWARDS, G. A. B., *Readings in Group Technology*, Machinery Publishing Company, London, 1971.

GALLAGHER, C. C., and W. A. KNIGHT, *Group Technology*, Butterworths, London, 1973.

HAM, I., and D. T. ROSS, *Integrated Computer-Aided Manufacturing (ICAM) Task-II Final Report*, Vol. 1, Group Technology Classification and Coding, U.S. Air Force Technical Report AFML-TR-77-218, Wright Patterson Air Force Base, Dayton, OH, December 1977.

MITRAFANOV, S. P., *Scientific Principles of Group Technology* (Russian), Mashinostroyenie, Moscow, 1970.

MITRAFANOV, S. P., *Scientific Principles of Machine Building Production* (Russian), Mashinostroyenie, Moscow, 1976.

OPITZ, H., *A Classification System to Describe Workpieces*, Parts 1 and 2, Pergamon, London and New York, 1970.

CHAPTER 7.9

Computerized Manufacturing Systems for Discrete Products

MOSHE M. BARASH

Purdue University

7.9.1 PRODUCTION VOLUME AND MANUFACTURING MODE

The volume in which complex metal products (typically elements of machines and instruments) are being manufactured ranges from individual units to hundreds of thousands per order. The mode of production depends largely upon the volume, whether it be small, medium, or large batch manufacturing or "mass production." Typical equipment for each mode of production, order size, and relative cost per unit are shown in Exhibit 7.9.1. It is of economic significance that nonmass production contributes to more than 70% of income derived in the manufacture of complex metal products[1]; moreover, at least 50% of all such items are made in batches of less than 50 units.[2]

Numerically controlled machine tools (see Chapter 7.7) have significantly raised the productivity in small batch manufacturing compared to conventional machine tools, but do not provide automatic handling from machine to machine. An "ideal" mode of manufacturing would be that which combines the flexibility of general-purpose machine tools with the high productivity of an automatic transfer line (see Chapter 7.5). Ways for realizing such a concept were first outlined in the early 1960s; best known are System 24 by Molins,[4] the NC-Line by Sundstrand (today White-Sundstrand),[5] and the VARIABLE MISSION Manufacturing System®.*[6]

The Sundstrand line was the first integrated NC system to be commercially utilized. The parts manufactured on it are magnesium alloy housings for the speed control gear for aircraft electric generators. About 70 different types of housings (that fit into a cube of just over 1 ft, or 30 cm) are being made in job lot sizes of 25 to 300. The machining section of the system includes eight 5-axis NC machining centers and two automatic multispindle drills. These machine tools are arranged in two rows, and between these rows is a "conveyor loop," that is, two parallel powered roller conveyors with cross-connectors at the ends. Workpieces are manually attached to pallets, placed on the conveyor, and allowed to circulate in the system. Each pallet has a mechanical code strip that identifies the workpiece, and each machine has a code reader. When the machine reads a pallet code that "matches" it, the pallet is automatically stopped and transferred to the machine table for processing. The next step in the automatically controlled sequence is activation of the tape reader, with subsequent machining (or another appropriate processing) of the workpiece. After completion of the work programmed on the tape, the machine returns the pallet to the conveyor. The pallet then is stopped at the next process station as prescribed in the prepared sequence plan and continues so until all processes have been executed.

In addition to machining stations, the facility includes cleaning, fluorescent dye inspection, and anticorrosion treatment sections. An automatic storage facility located above the machining line holds more than 6000 parts which are delivered to pallet loading stations by an overhead monorail conveyor.

The NC-Line replaced a shop with approximately 100 conventional machine tools that required 125 persons to operate them. Only 25 persons are needed to perform all functions in the line, including tool care, tool preparation, inspection, maintenance, and supervision. The product quality is consistently better; in-process inventory was reduced by 40%. Other gains are shorter lead time, reduced floor space, fewer fixtures and tools, and a cleaner environment.

Although controlled by electronic relays and not a computer, this system has served as the prototype for many computerized systems that were built later.

*VARIABLE MISSION Manufacturing System is a registered trademark of Cincinnati Milacron.

Exhibit 7.9.1 Dependence of Relative Cost of Machining One Piece Upon Mode of Production and Equipment: A–Experimental and Prototype; B–Small and Medium Batch; C–Large Batch and High Volume; D–Mass Production; 1–Toolroom Machinery; 2–General-Purpose Machine Tools; 3–Special-Purpose Machines; 4–Automatic Transfer Lines; 5–NC and CNC Machine Tools; 6–Computerized Manufacturing Systems[a]

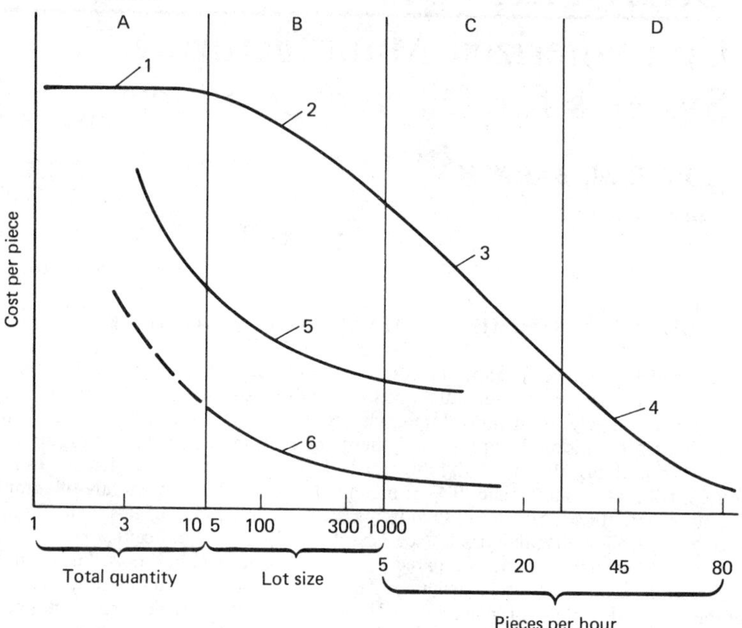

[a]Adapted from reference 3.

7.9.2 THE COMPUTERIZED MANUFACTURING SYSTEM

The time-sharing computer and the inexpensive minicomputer have made it possible to upgrade the concept of the NC-Line into computerized manufacturing systems (CMSs). A CMS (sometimes called a flexible manufacturing system, VARIABLE MISSION Manufacturing System, versatile manufacturing system, OMNILINE®,* and so on) is a production facility that consists of a group of process equipment units, such as machine tools or auxiliary equipment (inspection machines, washing stations, chip disposal devices, etc.), linked with an automatic materials handling system that reaches every process station, the entire facility being integrated under common computer control. Through the combination of the flexibility of NC machine tools with automatic materials handling and computer-controlled production management, the CMS achieves in batch manufacturing levels of efficiency approaching those of mass production.

Processes Included in CMS

Although most of the existing CMSs are intended for machining, a few have been built to perform forming or welding operations. An example is a computer-controlled flame cutting system[7] with four machines. On this system nine operators produce $2\frac{1}{2}$ times the work done earlier by 11 operators with eight machines. Improvements in predictability and control of these nonmachining processes will make more of them amenable to computerized automation.

Workpiece Classification

Workpieces processed on the machining systems may be conveniently classified as palletized and nonpalletized. "Palletized" workpieces include prismatic parts such as plates, brackets, gear cases, valve bodies, and most other nonrotational shapes. In most cases the workpiece is held in a fixture

*OMNILINE is a registered trade name of the White-Sundstrand Machine Tool Company, Division of White Consolidated Industries, Inc.

which is bolted to the pallet. Occasionally, rotational parts are being palletized, for example, gears with bores.

"Nonpalletized" parts are those that can be picked up by an industrial robot or a manipulator and properly located in a fixture or a chuck. Most bodies of rotation belong to this category. Some small nonrotational parts are being successfully handled by robots in CMSs.

Examples of CMS

Computerized manufacturing systems have been and are being built in several industrialized countries. The systems described here represent only a small fraction of the installed facilities. The machine tool builders who engaged in this field can and have put together CMSs of most diverse configurations and capabilities.

A CMS with six tool changing machining centers placed on both sides of a powered roller conveyor loop[2,8] is shown in Exhibit 7.9.2. Parts to be machined are placed on coded pallets in the four reload stations. From there the pallets are automatically transferred by the staging shuttle D to buffer conveyor K and to the loop. The machines pick up pallets after their code has been read. The system is used to machine a diversity of cast-iron parts of up to 450 lb (200 kg) mass and 3 ft (90 cm) cube size. Sixteen pallets can be in the system at the same time. In all, close to 180 part numbers are being produced in batch sizes from one up. Yearly production is from 12 to 20,000 parts, depending on part number. The system is controlled by an IBM 360/30 computer; three operators plus a foreman constitute the direct labor.

A very different method for transporting pallets with workpieces is employed in the type of system shown in Exhibit 7.9.3. The machine tools (machining centers and a digital inspection machine in this case) and the pallet loading stations are arranged in two rows; between them is laid a double-rail track on which two shuttle cars move under computer control to the appropriate stations, to pick up and deliver pallets with workpieces. Each car is driven by a pinion that engages with a rack attached to one of the rails. A feedback loop ensures high car positioning accuracy, even though it travels at a speed of 300 ft (91 m)/min. The pallet changer can swivel through 180°, to serve stations on either side of the track. The system can be easily extended through addition of machine tools (or other stations) and track modules that come in 20 ft (6 m) long sections.

This particular system machines 10 different parts, such as gear cases, the mainframe, and the clutch housing for two tractor models. The part mix varies considerably from one order to another. The average daily output suffices for 12 tractor assemblies. Systems of this type employ pallets of 32, 42, or 52 in (813, 1067, or 1321 mm) diameter, for a combined fixture and part mass of up to 20,000 lb (9000 kg). The method of control, which today is practically universal for CMS, is by a central computer in a direct numerical control (DNC) mode (see Chapter 7.7) interfaced with the CNC (see Chapter 7.7) controls of the individual machines. The CNC also provides backup if needed.

Geometric features of workpieces, such as the presence of groups of holes, make it sometimes economical to employ multiple-tool (spindle) heads. The feasibility of incorporating in a CMS a

Exhibit 7.9.2 Computerized Manufacturing System With Powered Roller Conveyor Loop: A–Four-Axis OD3 OMNIDRILL® Drilling Machine; B–Five-Axis OM3 OMNIMIL® Machining Center; C–Four-Axis OM3 OMNIMIL Machining Center; D–Staging Shuttle; E–Pallet Reader; F–Pallet Loader-Unloader; G–Pallet Indexer; K–Buffer Conveyor; L–Refixture Station; M–Compount Shuttle; 1, 2, 3, 4–Reload Stations[a]

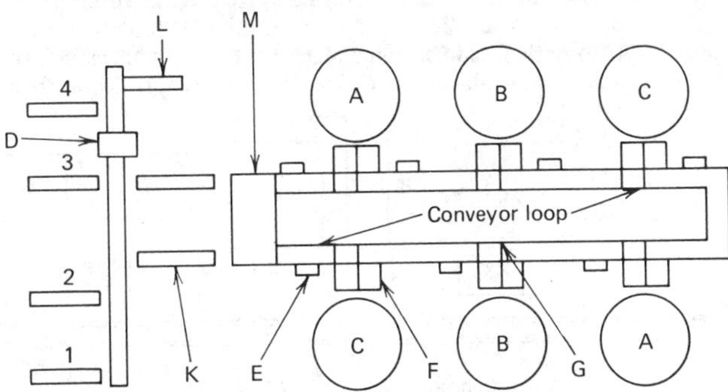

[a]Courtesy of White-Sundstrand Machine Tool Company, Division of White Consolidated Industries, Inc.

Exhibit 7.9.3 OMNILINE™ CMS with Series 80 OMNIMIL Machines: A–Fixed Head 40 hp OMNIMIL; B–Tilt Head 25 hp OMNIMIL; C–Inspection Machine; D–Wash Station; E–Shuttle Car; F–15 Load/Unload Stations[a]

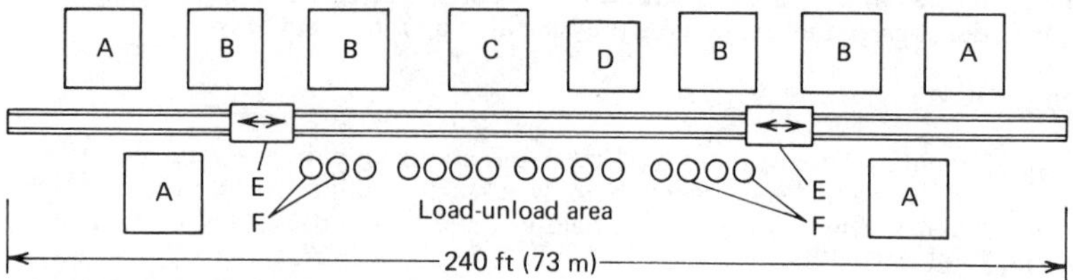

[a]Courtesy of White-Sundstrand Machine Tool Company, Division of White Consolidated Industries, Inc.

machining center that changes such heads was proved in the experimental VARIABLE MISSION Manufacturing System built by Cincinnati Milacron.[3] Today, multiple-tool heads of various sizes—up to 11,000 lb (5000 kg) mass—and head changers of different designs are being used, either as part of systems or in a stand-alone mode.

Machine tools and other process and auxiliary equipment of very diverse types can be integrated in a CMS. The system shown in Exhibit 7.9.4 includes tool changing vertical turning centers, a head changer, and single-tool changing machining centers. Its function is to machine a family of gear boxes and shaft housings. Cart-type transporters are used for moving pallets with workpieces.

Another method of transporting pallets that is used in CMSs, especially when the path is relatively complicated, is by towline carts. The towing chain is installed below floor level; a computer-activated mechanism located at each station disengages and engages the latch that connects the cart to the chain. A towline transport system can be extended if new machines or other stations are being added to the system.

The system shown in Exhibit 7.9.5 employs up to 23 towline carts. The seven part types machined in the system are various tractor gear box and transmission case castings. The part mix per order varies substantially, but the average daily output is 25 tractors. The parts have groups of holes which make the use of multispindle heads economical. Different heads are presented to the workpieces by numerically controlled indexing head carriers, the so-called head indexer modules. A module includes a single-axis NC slide. Modules are made with a single indexer (simplex module) or with two (duplex module). In the system shown, duplex modules are employed. Depending upon head size, up to 10 can be stored in each indexer. The heads are randomly accessible.

CMS With Nonrandom Travel

All the CMSs that have been described allow the workpiece to visit system stations in a sequence that has no resemblance to the "geographic" location of these stations in the system. Such "random" motion has advantages in terms of flexibility, but requires a more complex material handling facility and more involved control algorithms than does a "transfer line" type of system, in which the parts move through once. In some cases, especially when the parts are large and the number of part types is small, a system composed of NC machines, but operating in a transfer line mode, is the most economical solution. Such systems employ machining centers, head changers, head in-

Exhibit 7.9.4 Computerized Manufacturing System With Different Types of Machine Tools: A–36 in. Vertical Turning Center With Automatic Tool Changer; B–Automatic Head Changing Machine; C–Head Magazine; D–10HS Horizontal Spindle Machining Center With Automatic Tool Changer[a]

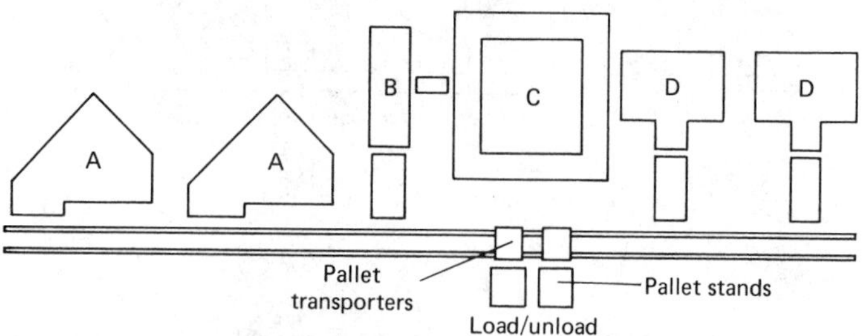

[a]Courtesy of Giddings & Lewis, Inc.

Exhibit 7.9.5 Flexible Manufacturing System With Towline Carts and Duplex Head Indexers[a]

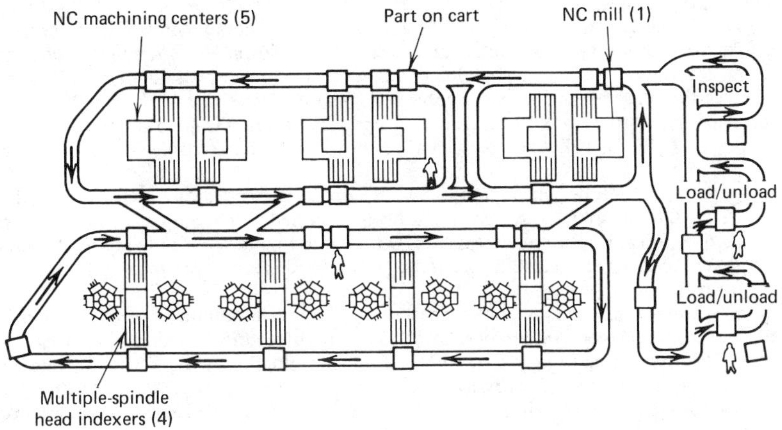

[a]Courtesy of Kearney & Trecker Corporation.

dexers, and special-purpose machines, some representing new concepts in machining (e.g., Kearney & Trecker Corporation[9]). The combination of multiple, selectable tooling and NC programmability makes these systems flexible and adaptable to a degree unattainable in a conventional transfer line.

As the cost of NC hardware comes down, not only of the electronic controls, but also of mechanical elements such as servomotors and drives, digitally controlled machines will continue to gradually displace the traditional, mechanically controlled units. There are already in operation CMSs that are dedicated to one product (e.g., Cincinnati Milacron[3]) and that in effect replace a special-purpose machine or a transfer line. They are economical for a small volume of production and are easily adapted to product modifications. An example of such a system, designed for a product of heroic proportions, is shown in Exhibit 7.9.6. The workpiece, a tank hull, is machined in succession on four stations. To handle the 40 ton (36,000 kg) mass of the hull and its pallet, water "bags" are used that are inflated with a fluid (actually the cutting coolant) and lift the pallet, allowing it to be pulled into position.

Status and Future of CMSs

The number of CMSs in operation is growing at an accelerating rate. The systems described here are a selection from the U.S. scene, but comparable developments are taking place abroad. Systems that have been well "tuned" show utilization levels of 75% and better, which is $1\frac{1}{2}$ to two times as high as for stand-alone NC machine tools that have to meet similar production demands. The numbers of direct labor personnel for a CMS are much lower than for stand-alone NC machines, amounting to approximately one person per two machine tools for systems manufacturing a mix of products; fewer personnel are needed for dedicated systems. Tool supervision and replacement are still a large part of the human activities in a CMS; automation of tool delivery and replacement and improvement in tool life (better tools, wider use of adaptive control, and more stringent quality control of work material) will further reduce the numbers of required operating personnel. (For adaptive control see Chapters 7.7 and 7.11). If one includes programming and other support functions, the number of total personnel is of course larger. Detailed information is difficult to obtain, but according to one foreign source,[10] the ratio of total personnel (operators, inspectors, super-

Exhibit 7.9.6 Dedicated System for Large Parts: A–Machining Center; B–Head Changing Machine; C–Special-Purpose Vertical Bridge-Type Turning Machine; D–Special-Purpose Vertical Gantry-Type Automatic Tool Changing Machine; E–Part on Pallet. The 40 Ton Part-Pallet Combination Is Supported by Fluid Pressure When Moved From Station to Station[a]

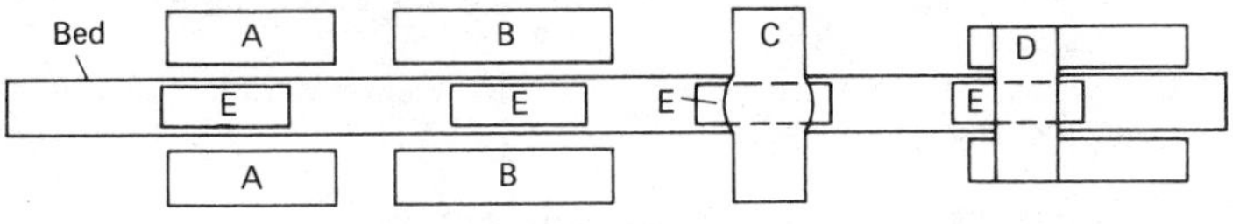

[a]Courtesy of Cincinnati Milacron.

visors, setup personnel, loading-unloading personnel, materials handling workers, programmers, NC technicians) for a CMS, stand-alone NC machines, and conventional machines, for the same production volume, is, respectively, 1:1.5:3.4.

Recently, industrial robots (see Chapter 7.6) have appeared as components of CMS, primarily for handling parts of rotation such as shafts, pinions, gears, and so on.[11,12] A single robot can serve two or three machines, the load and unload conveyors, and an inspection station. Systems with more machines that employ two or more robots are also in operation and their number is increasing. In terms of basic structure and fundamental operating rules, a CMS that employs robots does not differ from a CMS for palletized parts that employs computer-controlled shuttle cars (Exhibit 7.9.3).

The latest trends in CMS design point to a greater diversity of ideas in moving tools rather than workpieces (e.g., Jablonowski[11]). A system has been described[10] that employs a large, central tool magazine for 605 tools. Robots move cassettes with tools to and from individual machine magazines. The system has seven machining centers, with magazines for 60 tools, and an automatic warehouse for pallets with parts.

The overall prospect for CMSs is one of accelerating growth in numbers, sophistication, and diversity of processes included. Integration of CMSs into automatic factories may begin within a decade.[13,14] It is interesting to note that, although Japan is the first country to investigate seriously the concept of an "unmanned" factory for mechanical products, the first actual machining centers that can work for several shifts without attention were built in the United States and delivered to Sweden.[9,15]

7.9.3 PLANNING FOR A CMS

A CMS is a costly installation, and careful advance planning is required in order to avoid disappointment. The problems of such planning and of optimal operation of a CMS were investigated at Purdue University in a study supported by the National Science Foundation (NSF).[16] A number of design "tools" were developed, including simulation methods for CMSs[17] and a mathematical model[18] that requires only a limited amount of input information, but that identifies bottlenecks and underutilization in a proposed CMS and provides an overall productivity estimate. The time required to perform computer analysis of one proposed system configuration using the mathematical model is on the order of a few seconds. Various possible CMS configurations can be so analyzed, and the best few can then be simulated in great detail to identify possible productivity improvements, which are likely to constitute about 10%. In this way the final system configuration can be determined.

System scheduling and control rules were also analyzed, and it was found that no one rule is best for all system types. Such rules should be tested and selected through simulation. Probably the best solution is to have a simulation program included in the CMS software and delivered with it, making certain that adequate computer power is available—for operating the system and for performing simulation when required. Recent developments in microprocessor technology indicate that such computing ability can be achieved at relatively low cost.

REFERENCES

1. *Manufacturing Technology—A Changing Challenge to Improved Productivity*, Report to the Congress by the Comptroller General of the United States, U.S. General Accounting Office, Washington, DC, 1976.
2. N. H. COOK, "Computer-Managed Parts Manufacture," *Scientific American*, Vol. 232, No. 2 (1975), pp. 22–29.
3. CINCINNATI MILACRON, Product Literature (Manufacturing Systems), Publ. No. SP-145, Cincinnati, OH, June 1978.
4. D. T. N. WILLIAMSON, "Molins System 24—A New Concept of Manufacture," *Machinery and Production Engineering*, September 13, 1967, pp. 544–555; October 25, 1967, pp. 852–863.
5. B. C. BROSHEER, "The NC Plant Goes to Work," *American Machinist*, Vol. 112, October 23, 1967, pp. 41–47.
6. C. B. PERRY, "VARIABLE MISSION Manufacturing Systems," *Proceedings of International Conference on Product Development and Manufacturing Technology*, University of Strathclyde, Glasgow, Great Britain, September 5, 1969.
7. D. F. DOMINICK, "Manufacturing Productivity (A View From Caterpillar)," *Proceedings, Manufacturing Productivity Solutions Conference*, SME, Washington, DC, October 2 and 3, 1979.
8. WHITE-SUNDSTRAND MACHINE TOOL COMPANY, Division of White Consolidated Industries, Inc., Product Literature (OMNILINE), Belvidere, IL.

9. KEARNEY & TRECKER CORPORATION, "The Case for IMAGINEERING, Multi-Station Palletized N/C Manufacturing System," Supplement to Data Sheet 523–675, *Special Products News*, Vol. 2, No. 1 (1978), p. 6.

10. V. A. LESHCHENKO, Ed., (in Russian), *NC Machine Tools*. Mashinostroyeniye, Moscow, 1979, pp. 507–535.

11. J. JABLONOWSKI, "Aiming for Flexibility in Manufacturing Systems," *American Machinist*, Vol. 124, No. 3 (1980), Special Report 720, pp. 167–182.

12. CINCINNATI MILACRON, "Discover the Tomorrow's Tool Today (The T3 Robot)," Pat. No. K-259-2, Cincinnati Milacron, March 1979.

13. H. YOSHIKAWA, "The Japanese Project on the Automated Factory. PROLAMAT 76," *Third International Conference on Programming Languages for NC Machine Tools*, Vol. 1, International Federation for Information Processing and International Federation of Auto-Matic Control, Stirling, Scotland, 1976.

14. M. M. BARASH, "The Future of Numerical Controls," *Mechanical Engineering*, Vol. 101, No. 9 (1979), pp. 26–31.

15. R. SKOLE, "Unmanned Machining at Work," *American Machinist*, Vol. 123, No. 6, (1979), pp. 99–102.

16. NATIONAL SCIENCE FOUNDATION, Grant APR74-15256, "Optimal Planning of Computerized Manufacturing Systems, Summary (Final)," *Proceedings of the Eighth NSF Grantees' Conference*, Stanford University, Palo Alto, CA, January 1981. pp. J-1–J-10.

17. E. J. LENZ and J. J. TALAVAGE, *A Generalized Simulation Model for Computerized Manufacturing Systems*, Report No. 7, Optimal Planning of Computerized Manufacturing Systems, NSF Grant No. APR74-15256, Purdue University, School of Industrial Engineering, West Lafayette, IN, August 1976.

18. J. J. SOLBERG, "A Mathematical Model of Computerized Manufacturing Systems," *Production and Industrial Systems: Proceedings of the Fourth International Conference on Production Research*, Taylor & Francis, London, 1978.

BIBLIOGRAPHY

Understanding Manufacturing Systems, Vol. 1, Kearney & Trecker Corporation, Milwaukee, WI.

Automated Small-Batch Production, 3 vols., National Engineering Laboratory, East Kilbride, Glasgow, Great Britain, 1978.

HUTCHINSON, G. K., and B. E. WYNNE, "A Flexible Manufacturing System," *Industrial Engineering*, Vol. 5, No. 12 (1973), pp. 10–17.

CHAPTER 7.10

Computers in Continuous Process Control

EDWARD J. KOMPASS

Control Engineering

7.10.1 INTRODUCTION

Distributed control is the most important trend in control systems for continuous industrial processes. The digital computer, but particularly the microprocessor, is the technology at the heart of this trend. Better control of industrial processes will be the result.

7.10.2 PRIOR CONTROL PRACTICE

Prior to World War II most automatic control in the continuous process industries was truly distributed. The difference was that these automatic control systems were not integrated, but rather far-flung and disorganized sets of independent control loops. Thus a flow controller would be mounted on the process unit it controlled, perhaps on the very pipe in which it controlled the flow rate and very near the differential pressure taps sensing the flow rate and the valve being adjusted by the controller. Pressure gauges with scales to indicate flow rate or some other kind of flowmeter were generally mounted near the sensor and controller, along with a round-chart recorder. The recorder was important, because it provided an immediate means by which the operator could see that the control loop equipment was operating properly during periods when he or she was attending to some other part of the process. But it also provided, when analyzed with other recordings, a way to assess overall process performance and the controller adjustments or tuning that could account for the effect of one loop on another by way of the process itself.

Thus what these early automatic control loops needed most of all was a way in which the plant operator could see all of his or her real-time indicators at once. The process variable recorders would then become less important in operating the plant, though they would remain important as the best way to determine the cause of any upset in the steady-state behavior.

7.10.3 PNEUMATIC SIGNAL TRANSMISSION

These early control loops, remember, were generally mechanical, sensing the physical motion of a diaphragm, bourdon tube, or liquid column and amplifying this motion through pneumatic relays to move a diaphragm or piston-actuated valve stem. The development of pneumatic amplifiers of small mechanical motions that operated at low gauge pressures and the standardization of the input pressure range accepted by indicators, controllers, and recorders led to the widespread use of pneumatic transmission of industrial process control signals. This 3 to 15 psig pneumatic signal standard is still used and in fact was the basis of about half of all the control instrumentation sales worldwide in 1978.

Pneumatic transmission permitted the collection in one place, for the first time, of all the indicators required for safe and correct operation of an entire process unit. Because the indicators, controllers, and recorders all responded to the same 3 to 15 psig air pressure range, these mechanical instruments were often combined, or at least grouped. Because taking over manual control in the event of poor control due to mistuning or controller malfunction required manually supplying a simulated controller output, the manual controls also became part of the controller package. Controller output also was standardized at 3 to 15 psig, and pneumatic relaying to higher pressures for valve actuation was done at or near the valve.

Pneumatic control signal transmission, with its roots in the inherent safety of air and mechanical systems in flammable, potentially explosive, and otherwise hazardous environments, thus became the practical source of centralized industrial control systems. From an operational point of view, this centralization was essential to improve plant operation.

The unfortunate genius of the concept of combining indicator and controller (and sometimes recorder) all in one instrument also led to a 30 year period of stultified thinking about control system organization, in which all control equipment, except the sensor-transmitters and the actuators, seemed obviously to *belong* in the control room. The distributed control system as a concept was buried by the wisdom implicit in common practice.

7.10.4 ELECTRONICS AND COMPUTER CONTROL

The advent of electronics in the world of industrial control happened almost concurrently with the coming of the electronic digital computer to process control in the late 1950s.

The electronic controllers were simply analogs of the earlier pneumatic controllers, with the primary difference being the advantages of electrical signal transmission in place of the older pneumatic transmission. The obvious advantage was speed of transmission—the electronic speed of light versus the pneumatic speed of sound. Neither speed limit is quite achieved, of course, because of transmission line characteristics. But electrical speeds will not slow control responses, even over control loops stretched to many thousands of feet from the process unit under control. What you had to put up with instead was a new kind of sensitivity to extraneous electrical noise—a sensitivity that could be successfully coped with, at the expense of more complicated circuitry.

It is interesting in the light of present trends that there is hardly any record of suggestions to deal with the threat of electronics by speeding up the pneumatic control loop by separating the indicators from the controllers and putting the controllers back in the field. (Beckman did it in the 1950s to solve transmission lag problems on fast control loops.) Such is the power of the common wisdom.

The electronic controller, in the same rut, was destined to stay in the control room. But it offered easier interfacing to a variety of new indicators and recorders, and hence had some new room in which to grow. Pneumatics by this time was a highly evolved, mature technology, dependent for growth on new materials, for example. So 4 to 20 mA became the common electronic control signal standard, the analog of pneumatics' 3 to 15 psig. (Since voltages and currents are more easily scaled and converted than air pressures, some suppliers opted for 1 to 5 mA, 0 to 10 V, and other ranges, all easily converted one to another in any case.)

At first the indicators remained packaged together with the controller circuitry, as in the pneumatics contollers. This was done because of the force (or rut) of practice it seems, because they were much more easily separated in the electronic design. But the electronic controllers, built of early day transistor technology, were not rugged enough to move out of the control room into the field, in any case. The advantages in simpler, smaller panel board design made possible by the split architecture controller (so-called when the display and the controller circuitry are separated) were not obvious in the day of the miniature-case control system. (The increasing technical possibility of the shared display CRT terminal would put the emphasis on split architecture later.)

Meanwhile, the digital computer was developing apace, with continuing efforts at application by many control engineers. In the early days, the late 1950s and early 1960s, computers were both so expensive and so unreliable that they required complete time-sharing and complete backup. In other words, a single, central computer was used to control all loops, with the loops simulated one at a time in sequence and the calculated control output held until the next time the same loop was scanned for control. Each loop so controlled had a conventional electronic analog controller installed in parallel to control the process if the computer should fail.

Although control reliability had always been important, the all-your-eggs-in-one basket approach of direct digital control by computer made system reliability discussions the order of the day. The lessons learned are well applied today, as industrial control gets more and more complex, even though distributed system configurations provide more inherent reliability.

The main point to be noted is that the computer did not at first lead to any kind of distributed control system arrangement. In fact, the computer tended to solidify even further the dominance of centralized control designs. But the germ of distributed control was there, too.

7.10.5 DIGITAL CONTROL SIGNAL TRANSMISSION

It was evident from the earliest attempts to apply the computer in control systems that it would be of great benefit if the control signal transmissions were digital rather than analog. Digital signals were easier to protect against degrading by electrical noise sources, and digital signals could carry information or data encoded to any accuracy desired, whereas analog signals were limited by practical signal-to-noise ratios.

In the beginning of control computer applications, the analog-to-digital conversion of each and every control loop signal was a dream far beyond practical reach. In the last 20 years, however, this dream has become reality, along with other developments even more important. The effort to achieve universal digital transmission of control signals is partly responsible for the payoff of today's distributed control. Control engineers have long been ready for digital control signals.

The other developments involved in the sudden arrival of distributed control are large-scale integration (the most recent step in electronics integrated circuitry technology); its precocious child, the microprocessor; and time-shared CRT display terminals with graphics capability and the data highway concept, which allows extending the addressable computer bus outside of the computer. Thus computers could be in direct addressable contact with anything down to individual transducers and operator input devices, no matter how remote in the plant. So, by 1975, the stage was set for the beginnings of true distributed control systems.

7.10.6 DEFINITION OF DISTRIBUTED CONTROL

The most obvious form of distributed control is the system in which each loop controller is physically located on the process close to the control actuator and measurement sensor for that loop. Note that this is almost diametrically opposite to the traditional practice of the last 20 to 30 years in the flow processing industries, where the loop controllers have been brought together in central control rooms in order to centralize the loop display instruments integral to the controllers. This first and most obvious method of control distribution is generally what we mean by distributed control.

A less obvious method of "distributing" control follows from the more and more common practice of using multiloop digital controllers. When a single microprocessor-based controller is time-shared, that is, when a single piece of hardware is in fact controlling a number of different process variables simultaneously, it is considered prudent that these variables be chosen among a number of different process units. In this way it is possible that loss of that controller hardware will not cause dangerous operational conditions or plant shutdown. This kind of control distribution is known as "functional distribution." Since the trend in multiloop digital controllers is definitely to more and more loops rather than fewer, such controllers must automatically switch to a backup controller on failure, as in fact all of them now can. Functional distribution of control is thus an academic consideration.

Besides distribution of the controllers themselves, it is also possible to distribute only the inputs and outputs of the controllers (or of a data acquisition or data logging subsystem), in order to replace the individually hard-wired inputs/outputs with only a few wires by some form of signal multiplexing. Distributed input/output is an economic move in the direction of distributed control and was probably a necessary step in the development of distributed control. But it is not distributed control.

In distributed control the loop controllers must be optionally (at least) or inherently located remote from the central control room. The central control room houses the central operator's station, which is merely supervisory, but from which all loop variables can be viewed, and all loop controller tuning constants can be set or changed. A supervisory or hierarchical computer is entirely optional. *Distributed control does not mean computer control, although it can include computer control.*

In true distributed control each control loop is local to the process and operates independently of the supervisory control function of the central control room. Each loop thus can be optimally short, fast, and relatively safer from accidental damage and insensitive to noise. Yet all necessary information concerning process behavior and control instrument and controller integrity is immediately available to the operator at the central station. Control wiring is minimized, too.

7.10.7 THE CONFIGURATIONS OF PROCESS CONTROL

Integrated process control systems available today generally fit into three simplified categories based on system configuration,[1] that is, the philosophy according to which the system components are interconnected. These three categories are reduced to their barest essential descriptors in Exhibit 7.10.1. The order in which they are presented in the exhibit is that of their historical descendence and is not meant to suggest any sense of relative value of one system over another, even though such values will always in fact exist in their applications.

Before considering the characteristics of the three configuration types in detail, a word needs to be said about that word "integrated." An integrated process control system is any system that can be built from a set of system components, all available from the same manufacturer, that have been designed for easy assembly into a complete arrangement that tends to have a uniform, modular appearance from one final application embodiment to almost any other. Process control systems can thus be integrated or a pastiche of components from many manufacturers put together by an end user. An integrated system can be assembled and installed by the manufacturer, by the end user, or, as is common, by an engineering construction firm or engineer-architect.

Even integrated process control systems are rarely "packaged" in the sense that central control room panels or operator's consoles always look exactly alike. Yet they will always have that obvious family resemblance.

Exhibit 7.10.1 Early Field-Mounted, Automatic Feedback Control Loops Such as This Flow Controller (*a*) Required the Process Operator to Check Gauges and Recorder Charts Regularly to Monitor Loop Operation and Compare Charts of Other Loops to Estimate Process Performance. Centralized Control System Concept–Configuration Type 1 (*b*)–Brings All Loop Indicators Into the Control Room for Integrated Plant Operation. The Problem Is That Controllers Are in Central Locations With Their Indicators, So Loops Are Long, Vulnerable to Noise and Damage, and Total Wiring or Tubing Is Excessive and Very Costly. Optionally Distributed Control System Concept–Configuration Type 2 (*c*)–Permits Controllers To Be Physically Remote to Environmentally Controlled Local or Unit Operation Area. All Loop Information and Access to Controller Setpoints, Tuning Constants, and Manual Control of Each Loop Is Still Available in Central Control by Way of Data Highway. Control Loops and Plant Units Operate Independently If Highway Fails. Fully Distributed System–Configuration Type 3 (*d*)–Mounts Each Loop Controller in the Field and Reduces Plant Control Wiring to Absolute Minimum, While Still Maintaining Full Plant Information and Supervisory Control in the Central Control Room. Controllers Must Be Designed to Operate in Severe Environments and Be Serviceable in Field Location.

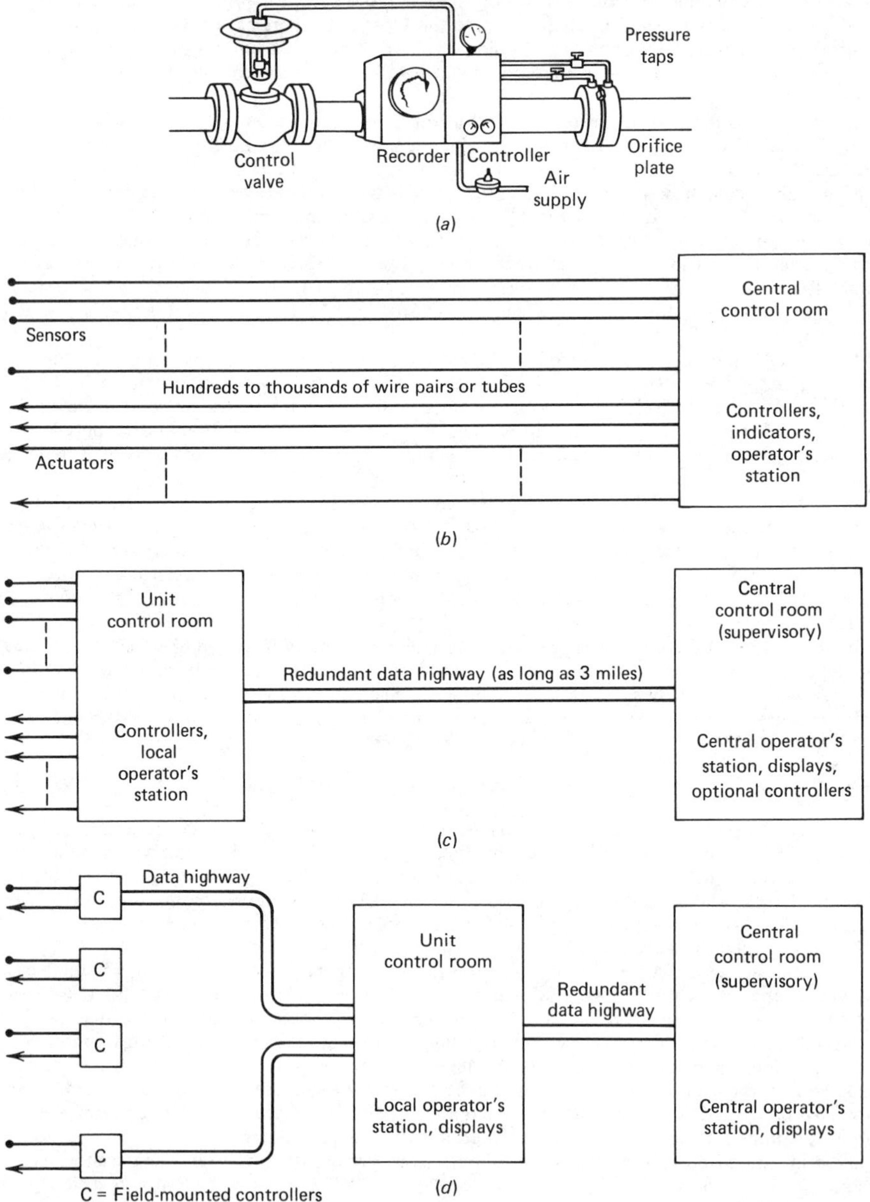

Configuration Type 1

Process control system type 1 is the centralized control system and can be considered the conventional configuration. It is certainly the most common configuration in use in all process industries today.

Basically, configuration is the simplest of "star-type" configurations, in which all of the decision functions of the control system are centralized, and individual signal paths are brought into the central control room from every sensor and sent out to every actuator.

Thus in type 1 systems all controllers for every control loop, all displays for operator decision, and all knobs, switches, keyboards, and so on, for operator action, are located in the central control room. The controllers themselves may be integrated with their displays of loop parameters or separated for easier maintenance in so-called split-architecture controller designs, but both the displays and the controllers will be in the control room area.

Controllers and other central control room equipment can be either analog or digital. Thus sophisticated displays such as digital panel meters and CRT terminals with microprocessor-driven graphics are commonly found in type 1 control systems.

Type 1 systems can also be pneumatic and in fact are likely to be so in about half of all process control systems. Type 2 and type 3 systems are all, so far, electronic.

Type 1 systems have evolved greatly since the early 1960s, and it is common to find analog loop controllers interfaced to digital CRT display terminals, and in some systems, to digital signal buses that permit admixtures of analog and digital controllers, terminals, printers, and supervisory computers in the system—all in the control room environment, of course. Such systems can and do include multiple-loop and single-loop microprocessor-based controllers and even multivariable advanced control capabilities without the hierarchical supervisory computer control that can be added.

In another variation that might be called configuration type 1A (not illustrated), a digital bus can be extended into the field to a remote multiplexer terminal, where a sizable number of field points, sensors, or actuators can be connected. The configuration is thus a star at the end of one or more of the lines radiating from the central control room. The objective is to reduce significantly the number of signal paths to save wire and installation costs.

Configuration Type 2

Process control system type 2 is a star or radial configuration that also provides the possibility of true distributed control; that is, the closed-loop controllers for each process loop can be removed from the control room location itself and placed closer to the field sensors and actuators. This is done by extending a digital bus or data highway connecting the controllers with the operator's console and displays. The control loops themselves can then become physically shorter (from possibly a mile, say, to a hundred feet or so) and therefore less vulnerable to noise or damage. Loss of the long, usually redundant, digital data bus or highway means loss of operator intelligence, but controllers continue to operate locally. Multiplexing is done at the controller location. Again, systems of this type could be implemented with either analog or digital controllers, or even with pneumatic controllers, but to date the multiple-loop digital controller prevails.

Configuration Type 3

The third type of process control system configuration is based on a digital data highway in a loop configuration with data transmission in both directions. If the loop is damaged at any point, full communication still can be maintained with all controllers, without a redundant highway.

7.10.8 HOW CONTROL IS DISTRIBUTED

The data highway, the data bus, the dataway, the control bus, call it what you will, is the key that opened the door to distributed control. The computer itself, in the form of a single-chip large-scale integration (LSI) processor, had to become cheap enough to fit into just about every addressable station for it all to come true. That is not exactly right, of course. Lots of data buses use just a chip called a UART (universal asynchronous receiver transmitter), which is a modulator-demodulator (MODEM) that serializes parallel data bits on one end and parallels the serial data on the other to look for a code match between a transmitted "word" and a device address register. But that UART is a small product of the same technology that gives us the microprocessor.

So picking the data bus that is used is a large part of designing a distributed control system and a large part of choosing between them.[2] The choices here are important. They include speeds, message protocols, highway traffic control methods, degrees of redundancy available, interfaces and retrofit capabilities, and relative amounts and kinds of data traffic (such as real-time control loop data or supervisory data only).

Even given the data bus, however, and the addressability of "stations" that might have all manner of inputs or outputs attached, why distributed control?

The beginnings of the trend came from the desire to reduce the number of wires connecting all the sensors and actuators in an industrial plant to that central control room (or controller room) that no one could conceive of giving up. The addressable station idea theoretically permitted all of these thousands or tens of thousands of long, long wires (perhaps thousands of miles of wire!) to be eliminated. Total wiring costs (which could run to hundreds of thousands of dollars in large, complex industrial plants) could be reduced 90% or more. A single loop of twisted-pair cable, or a couple of redundant straight runs using coaxial cable, would bring all the data into the control room and back to the actuators.

7.10.9 COMPUTERS ON THE PROCESS

The next step was the increasing economic capability to put a complete computer in the form of a microprocessor in each station if that were useful. This new wrinkle made the remote stations able to preprocess in any degree the data to be transmitted. In fact, *the continuous control information really didn't need to be sent at all.* The stations could be the controllers, and the data bus used only to update the operator on what was going on at the process unit, a plus permitting him or her to change set points or take control manually.

This arrangement, by the way, is true distributed control. The controller is no longer in the control room–it is back out near the process unit under its control, just like it was in the old mechanical days before central control. The big difference, of course, is that the plant operator is fully in touch with the controller and what is going on in his or her process.

One of the major benefits of putting the controller back out near the process unit is that the control loop itself is physically shortened. This reduces the possibility of physical damage due to unforeseen accident and thus increases system reliability to some extent. Since the control loop itself is most likely still analog, the shorter loop is also less susceptible to signal degradation by electrical noise. In fact, there is no real reason that the control loop cannot be pneumatic, either because an installed pneumatic system is retrofitted with a data bus or to make use of the special safety advantages pneumatics offer.

The image up to now still considers that the controllers, whether microprocessor-based digital controllers, electronic analog controllers, or pneumatic controllers, are still in some kind of control room–a small control room concerned only with a single process unit most likely, but still an environmentally controlled space. It is possible to go another step further. This step puts the controller right back out on the pipe itself–right back into the field environment. It takes a rugged controller, to be sure. The old pneumatic equipment has lived this way for years, and successfully. Electronic analog controllers have not been designed for it, in general, and can not be used this way. But by 1978 at least one microprocessor controller had been, and so had some data highway drop-station equipment.

The special microprocessor controller is found in a type 3 system[3] which mounts a single-loop controller directly on the valve to be controlled, sensing valve stem position directly. A color graphic CRT terminal communicates with up to 32 of these weather-proofed controllers, including adjustment of controller tuning constants.

Other distributed control systems on the market might still require that control room environment. But they also offer a flexibility about where that control room environment is located. The controllers can be remote from a process-unit local operator's station, but they can also just as well be handy to the central operator's station, if that is an overriding advantage in some plants. The distribution of the controllers to the field or not is optional in these systems. This option might make retrofitting a possibility.

7.10.10 SHARED DISPLAY GRAPHICS

The availability of shared display graphic CRT terminals is also an essential element of any distributed control system. The importance of centralized display of data from each control loop was the reason for bringing all the signals into the control room in the first place. It cannot be given up in a move to distributing control. In fact, the capability of making tuning adjustments to the controllers, or of taking over manual control at any time, is also necessary. It is unlikely that a user would be willing to send an operator to a field-mounted controller for adjustment of tuning constants any more, even though the plant control engineer can see his or her adjustments as they are made and communicate with him or her by means of a walkie-talkie. Controller tuning and remote controller configuration (selection of inputs, signal conditioning, outputs, etc.) should be available only to the engineer by way of a separate engineering terminal or a portion of the operator's terminal under lock and key. In any case, local unit operator's display terminals, with limited control and data capability, should be locatable anywhere on the data highway for maximum user flexibility.

7.10.11 COMPUTERS IN DISTRIBUTED CONTROL

It might be noticed that, once the capability of the computer was absorbed into the distributed control system for data communications and controller functions, nothing further was said about computer control as such. This is because a separate computer is not necessary to most of these distributed control systems.

Note the "most." For example, some systems require a computer to be on the data bus in order to accomplish controller configuration. Others require the computer for color graphics, whereas some provide even that without a separate computer. The control computer—generally a minicomputer of considerable capability—is used in all cases for advanced control algorithms such as optimizing control. The microprocessor controllers implemented in the distributed systems on the market implement anywhere from a dozen or so up to 180 control algorithms without the addition of the minicomputer. Microprocessor-based controllers, where used, usually are multiloop controllers, which should have redundant fallback capabilities for system reliability, but the trend is definitelv to single-loop microprocessor controllers.

The central minicomputer can, of course, be used to do great things with the CRT colorgraphic displays and all the data available from the distributed system by way of the data highway. Interactive displays are available in some systems, with light pens or oil-tight mylar switch panels used as entry means.

The old chart recorders are replaced in many distributed control systems by magnetic tape cassettes or disks, from which trend displays are generated on the color CRT terminals.

7.10.12 TRENDS IN CONTROL MEASUREMENT

In the two most important types of continuous process control measurement, temperature and pressure, practically no advancements at the primary sensor level are discernible over the past 25 years. Thermocouples and resistance bulbs, bourdon tubes, and diaphragms hold sway today as then. LSI technology is sneaking into the transducers in the form of integrated operational amplifiers, sputtered strain gauges, and infrared sensors. But the real trend here, again, is the use of microprocessors, which are beginning to move from the panel-mounted readout into the field-mounted transducer case. All this is part of the large trend to digital distributed control.

Flowmeters have shown more of a succession of new inventions than a recognizable trend. The orifice plate and differential pressure measurement are omnipresent. But magnetic flowmeters, turbine flowmeters, thermal mass flowmeters, swirl-type and bluff-body-type vortex flowmeters, and ultrasonic flowmeters have all appeared during the last 25 years. The message here, again, is the large trend to digital electronics.

Computer capability in the transducer is nowhere more evident than in analysis instruments, in which the composition percentage of a particular component in a complex stream might be calculated, for example, from the relationships of several types of measurements. This kind of smart sensor is a long-term trend, and some control engineers think that future advances in control will depend on computer-based sensors.

7.10.13 TRENDS IN CONTROL ACTUATION

Discussions surrounding process control valves today are much the same as 25 years ago. The subjects are valve position versus flow characteristics and their effects on system controllability and valve sizing. New designs, except for the digital valve of multiple binary-graded orifices operated by electronic selection of solenoids, have been mainly refinements of old principles. The activity has been largely in new actuators, for high stem forces, for electronic or computer compatibility, or for just modifying the position-flow curve. The Occupational Safety and Health Administration caused a considerable flurry of valve design activity around 1972 to reduce noise generated by valves that must drop high pressures.

But the process control valve may finally get a real jolt, as soaring energy costs make control engineers take a hard new look at speed control of pumps instead of throttling to control flow.

7.10.14 FUTURE POSSIBILITIES

Consideration might also be made during design or selection of a distributed control system for what might be coming up next. We have seen that most of these systems, in the unit operator's stations out on the data highway, still input analog transducer signals and output analog controller signals to the actuators. The day might well be soon that custom chips in transducer cases will permit all signals to be digital. Some actuators, such as stepping motors and certain digital valves, already accept serial digital signals. These developments might make it unnecessary to go all the way back to field mounting the controllers. Will it then be useful to have these digital end-elements directly match the data highway, for example?

REFERENCES

1. E. J. KOMPASS, "The Configurations of Process Control: 1979," *Control Engineering*, March 1979, pp. 43–57.
2. M. J. MCGOWAN, "From the S-100 to CAMAC: The Diversity of Digital Buses," *Control Engineering*, April 1979, pp. 31–34.
3. M. J. MCGOWAN, "Process Control System Is Distributed to the Valves," *Control Engineering*, December 1978, pp. 41, 42.

CHAPTER 7.11

Sensor Technology

RICHARD A. MATHIAS

Macotech Corporation

7.11.1 SENSOR CLASSIFICATION

The application of CAM to the instrumentation and control of material removal, material forming, and welding processes is paced by sensor technology. Basically there are two categories of sensors.

One category, common to all modern-day equipment, comprises sensors for basic machine position and velocity measurement as required for NC. These would include lead screw or slide position measurement; slide, tool, or workpiece proximity detection; lead screw revolutions per minute (RPM); slide velocity; and spindle RPM.

The other category of sensors comprises those uniquely suited to the processes to which they apply. Sensors currently used in material removal processes are classified with respect to their application and function in Exhibit 7.11.1. Those for material forming processes and arc welding are likewise classified in Exhibit 7.11.2.

7.11.2 BASIC DESCRIPTION OF CURRENTLY USED SENSORS

Lead Screw Angular Position

This is the most widely used method of slide position sensing. The two types of sensors are angular resolvers and photo-optic digital angular encoders. An angular resolver is a pair of transformers, one of which rotates relative to the other. The amplitude of the alternating current (AC) voltage produced across the stationary transformer is proportional to the cosine of its angle of orientation with the rotating transformer.[1]

A photo-optic digital angular encoder operates either by interruption or by reflection of light from a light emitting diode (LED). A disc with alternating opaque and transparent segments spins between an LED and a light pulse detector to produce a pulse train. There are two types of pulse train: incremental and absolute. The incremental type has a fixed spacing between pulses, with each pulse representing a fixed movement of the lead screw and machine slide. The absolute type has the opaque and transparent spacing arrayed to produce directly a digital coding of the shaft position.[2]

Slide Position

Direct slide position encoding is achieved by four different methods. An optical grating laid along a slideway bed is illuminated by an LED attached to the slide. Reflections are sensed by a photodetector to produce pulses whose pitch spacing represents a specific slide movement. A vernier effect, increasing sensor resolution, is achieved through a second grid mounted to the slide and having a pitch slightly different from that on the bed.

A linear coil mounted to the bed scanned by a moving coil on the slide is another method, called a "linear inductosyn." A third method, called the Sony Magne Scale®,* uses a fine magnetic grating along a nickel-steel rod that is scanned with a magnetic pickup head. Resolution is within 50 μ in.

The fourth method is the laser interferometer. Although used primarily as a master gauge to check machine accuracy, it has been implemented into the machine control loop in cases where extreme precision is required.[3] The interferometer method boosts resolution to within one half wavelength of the light source. This represents 12.5 μ infrared light. The principle of operation of the interferometer is the splitting of a coherent light beam (laser beam) into two paths, one of which changes with the slide movement to be measured. When the two beams are recombined, interfer-

*National Machine Systems, Tustin, CA, distributor for Sony Magne Scales.

Exhibit 7.11.1 Application and Function of Sensors Used in Material Removal Processes

	Sensor Application and Function		
Process Description	In-Process Monitor and Control	Postprocess Part Inspection	Postprocess Tool Inspection
Metal cutting with conventional machines; transfer lines; NC, CNC, DNC controlled machines; machining centers; flexible manufacturing systems; unmanned machining centers	Motor horsepower Spindle torque Spindle deflection Spindle force Spindle vibration Tool post force (for turning) Workpiece diameter (turning and grinding) Tool tip temperature	Workpiece diameter Bore diameter Hole depth Surface position Surface finish Surface integrity	Tool tip position Tool tip flank wear Tool diameter (drilling, milling, grinding) Broken tool Tool life
Electric discharge machining (EDM)	Current, voltage, electrode-workpiece shorting	Surface position Surface finish Surface integrity	Electrode wear
Electrochemical machining (ECM)	Tool-workpiece voltage drop	Surface position Surface finish Surface integrity	Tool wear

ence of the beams produces cyclic occurrence and absence of light at a photodetector for each one half wavelength of position change.

Proximity Detection

Detecting the proximity and small displacement of ferrous objects is simply achieved by use of inductive probes. There are three types.

One type is the eddy current probe, which detects changes in induced currents (eddy currents) in the surface of a ferrous object. The currents are induced by a pulsating magnetic field from a coil in the probe excited with an AC voltage. The eddy current strength changes with displacement of the object surface. The output signal can be filtered so as to produce a digital pulse upon proximity of the probe to the object.[4]

A second type measures the changing inductance between a probe, surrounded by an electrically

Exhibit 7.11.2 Application and Function of Sensors Used in Material Forming and Welding Processes

	Sensor Application and Function		
Process Description	In-Process Monitor and Control	Postprocess Part Inspection	Postprocess Tool Inspection
Die casting and injection molding	Die temperature Internal die pressure Cycle time Coolant flow rate Coolant temperature Plunger force Shot cylinder pressure Hydraulic fluid temperature	Successful part ejection Part weight	Die temperature
Bending of sheets, tubes, and beams	Part springback	Bend angle Bend radius	
Rolling	Sheet thickness Sheet edge location	Sheet thickness Surface finish	
Electric arc welding	Welding current Welding temperature	Weld hardness	

excited coil, and a ferrous object. For small displacements the inductance changes and resulting probe output signal changes are proportional.

A third type, which is becoming one of the lowest-cost means of proximity detection, is the Hall-effect sensor.[5] Analog types provide a direct current (DC) output proportional to magnetic field strength. Digital types provide digital voltages triggered by a specific magnetic field strength.

For detecting the proximity and small displacement of conductive and nonconductive surfaces, capacitance gauges and electropneumatic proximity probes are applied. Capacitance types produce a current proportional to the distance between two surfaces having a voltage drop across them.[6]

Electropneumatic types detect the change in orifice pressure as an orifice with constant air flow is brought into proximity with a surface.[7] For small displacements the effect is linear.

Lead Screw and Spindle RPM

There are three basic types of shaft-mounted tachometers. One type is the DC tachometer, which is actually a small DC generator whose output is proportional to RPM. A second type is a slotted disk that interrupts the path of a light source to a photodetector/pulse counter. This is an inherent feature of the incremental photo-optic angular position encoders discussed previously. The third and most widely used type is an inductive probe in proximity to the slotted periphery of a ring.[8] Output pulses may be shaped electrically or by design of the slots to suit digital data requirements or pulse counting logic.

Motor and Spindle Horsepower

The watts transducer is the most accurate and most widely used method. By monitoring the instantaneous motor voltage and current and multiplying these together, a watts transducer produces a signal proportional to motor horsepower. For AC motors, voltage is sensed by a voltage transformer connected across the motor inputs. Armature current is sensed by a current transformer which is primarily looped around one of the motor input power heads. For DC motors, armature voltage may be measured directly, and armature current is sensed as the voltage drop across a DC shunt placed in series with the motor power line. Multiplication of signals from voltage and current sensors is inexpensively achieved with integrated circuit multiplier chips.

A signal proportional to spindle horsepower is achieved by setting the watts transducer signal to zero when the spindle is at proper RPM and unloaded. The signal produced thereafter when the spindle is loaded is proportional to spindle horsepower.

Spindle Torque

There are three methods. The least accurate and least costly method is one of sensing motor current, as described previously, as a measure of spindle torque. Though motor current is roughly proportional to motor torque, motor power factor losses and drive train losses are included in the motor current sensor signal.

The second and most widely used method, particularly for adaptive control and torque-controlled drilling,[9] is one of extracting torque from the spindle horsepower sensor output discussed in the preceding section. Since power is the product of torque and spindle RPM, the watts transducer signal must be divided by a signal representing spindle speed to yield a signal proportional to spindle torque. Torque resolution achieved with this method is within 3% of the spindle torque available at the given RPM.

The third and most costly method, but one providing direct torque sensing nearest to the spindle load, is the spindle torque strain gauge dynamometer. Placing strain gauges on spindles and reliably bringing the signals out to external amplifiers are requirements for applying this method. Dynamometer components and engineering assistance in specific applications are available.[10]

Spindle Deflection and Spindle Force

The primary method used is that of sensing small displacements of the spindle relative to the spindle bearing housing using inductive noncontact probes mounted in or near the spindle dustcover.[11] These displacements are a measure of forces on the spindle. Force deflection calibration remains very consistent for properly designed and preloaded spindle bearings. Calibration is a function of the distance between the plane of loading and the spindle bearing. Resolution on 5 in. spindles varies between 20 and 200 lb, depending on spindle and bearing runout characteristics.

Spindle and Machine Vibration

This is commonly sensed using piezoelectric accelerometers. These use quartz crystals, which output small voltage signals when they are subjected to force, in this case an acceleration force.[12]

Tool Post Force (for Turning)

Tool post dynamometers are either the strain gauge type[13] or the piezoelectric crystal type.[14] The main drawback for production applications is that they must be custom designed for specific tooling requirements.

Workpiece and Bore Diameter

Traditionally these measurements have been achieved during constant diameter grinding operations using caliper-type gauges having pneumatic, electrical, or mechanical sensors on the fingers.[15] These sensing methods have short ranges of application and are suited for large-volume production of the same-size part. Where in-process measurement of several diameters or bores during turning or boring is required, optical methods have more range. Several laser-beam-type calipers have been developed.[16]

Hole Size, Depth Location, and Workpiece Surface Position

Spindle-mounted probes and associated programs for automatic interrogation of any surface's location are now in production use on machining centers.[17] Probe stylus deflections are sensed by either spring-loaded linear variable differential transducers (LVDTs) or by precision magnetic gratings* mounted in the probe head which allow ample deflection for overtravel and recovery upon probe contact with the workpiece. This type of probe has been designed for tool post and turrent mounting in turning and boring applications.†

Surface Finish

No convenient method for the in-process sensing of surface roughness exists, one of the main reasons being the lack of a uniform, reliable standard criterion.[18] A promising method is that of measuring the intensity of laser light reflected from the workpiece surface.[19] Postprocess surface finish inspection is performed with conventional surface finish equipment, which scans the surface at a fixed rate with a contact stylus containing a piezoelectric quartz crystal.

Surface Integrity

Surface integrity is the degree of improvement or damage to a surface resulting from a machining process. This would include minute cracks, burns, surface stresses, stress corrosion, and so on. Nondestructing testing methods include eddy current probes, ultrasonic testing, and magnetic particle and magnetic fluid penetration.[20]

Tool Tip Position, Tool Diameter, Broken Tool

Postprocess tool position and tool diameter sensing systems have been developed for design into machining centers[21] and NC turning centers[22] using electropneumatic- and electromechanical-type proximity probes, discussed previously. Broken tool detectors are more in demand with increased emphasis on unmanned machining centers. They are beginning to appear on new machine tools, where the cutter is brought in proximity with noncontact probes that signal the presence of a metallic surface.[23]

Tool Wear, Tool Tip Temperature, Tool Life

Extensive research has been conducted worldwide in pursuit of reliable tool wear[24] and tool tip temperature[25] sensing techniques. To date, few applications have been cost-effective, and these are restricted to special applications. On-line tool life measurement is the monitoring of the total time in cut between tool changes. These are desirable data for management information and feedback control purposes. The primary technique used for tool-in-cut detection has been electric contact or cutting force developed between tool and workpiece.

Electrode-Workpiece Shorting in EDM

Sudden current changes and the change in phase angle between current and applied voltage are good indicators of shorting.

*National Machine Systems, Tustin, CA, Distributor for Sony Magne Scales.
†Feed Back Gauge (FBG) Control Unit, Ikegai Tekko Company, Tokyo, 1979.

Electrode Wear by EDM

Graphite electrode wear can best be monitored postprocess with noncontact pneumatic probes or electromechanical contact probes.

Electrode Wear by ECM

Electrochemical machining electrodes are good conductors, but are generally nonferrous. Pneumatic capacitance or electromechanical probes are used for postprocess inspection.

Die Casting and Injection Molding Processes

A discourse has been written on the monitoring and instrumentation requirements for die casting,[26] and this is applicable to plastic injection molding also.

Die, coolant, and hydraulic temperature are sensed with thermocouples. Die pressure is most conveniently measured with "load washers."[27] Short cylinder pressure is monitored with conventional strain-gauge-type pressure transducers.[28] Plunger force is sensed with strain gauges mounted to the plunger. Coolant flow rate is monitored by a conventional flowmeter mounted at the coolant inlet or outlet. A cable attached to the plunger actuates a linear potentiometer and DC tachometer for respective monitoring of plunger position and velocity. Successful part ejection and part weight can be detected with force transducers or strain gauges mounted to the ejection tray.

Bend Angle, Bend Radius, and Springback

As NC tube bending, beam forming, and sheet roll forming become more computerized, sensors for bend angle, bend radius, and springback are required. Adaptive-controlled tube bending has been developed where tube springback is measured with an LVDT probe on an independent axis, and this information is used to set up the corrective bending cycle automatically.* Scanning of formed parts to measure bend radius and bend angle is best done with NC inspection machines using conventional electromechanical contact-type probes.

Sheet Thickness, Sheet Edge Location

In-process gauging is achieved with noncontact pneumatic, inductive, or capacitance-type calipers and probes, depending on the sheet material, tolerance, and surface condition. Postprocess thickness measurement of large metal sheets is traditionally obtained with the ultrasonic thickness tester.† This gauge measures thickness as the time duration for transmission of sound pulses through the sheet, reflection from the back side, and return to the front side.

Welding Temperature, Weld Defects, and Weld Hardness

Welding temperature is best measured with a noncontact radiation pyrometer whose principle of operation is to sense the power of infrared radiation emitted. Welds are usually scanned for defects with eddy-current-type proximity detectors, discussed previously. Nondestructive weld hardness testing can be time consuming using conventional Brinell or Rockwell equipment. A faster method and one having an electrical signal as a go-no go measure of hardness is the Grindo-Sonic principle.‡ This technique measures the hardness as the dissipation rate of sound pulses transmitted through the part.

7.11.3 FUTURE TRENDS IN SENSOR TECHNOLOGY

In 1978 the Machine Tool Task Force was formed from 115 international experts in all phases of manufacturing technology. Its purpose is to characterize the state of the art of machine tool technology and to identify promising future trends in this technology. A final report was released when the study ended in September 1980.[29] This report contains new advancements and requirements in sensor technology not included in this writing.

REFERENCES

1. J. R. MCDERRMOTT, "Simplifying Resolver Application," *Control Engineering*, January 1963, pp. 105–107.

*Microprocessor-Controlled NC Bending Machines, Teledyne Pines, Aurora, IL.
†Branson Vidigauge Ultrasonics Thickness Testers, Branson Instruments Inc., Stamford, CT.
‡Grindo-Sonic, J. W. Lemmens, Inc., Anaheim, CA.

2. L. TESCHLER, "Transducers for Digital Systems," *Machine Design*, July 1979, pp. 66–67.
3. R. W. SCHEDE, "Laser Interferometer Feedback for Numerical Control," Paper No. 69CP828IGA, *19th Annual Institute of Electrical and Electronics Engineers Machine Tool Conference*, Oak Ridge Y-12 Plant, Oak Ridge, TN, October 1969.
4. TESCHLER, "Transducers for Digital Systems," pp. 69–70.
5. TESCHLER, "Transducers for Digital Systems," p. 71.
6. L. MICHELSON, "Greater Precision for Noncontact Sensors," *Machine Design*, December 1979, pp. 117–121.
7. A. NOVAK, "Exploratory Study of the Non-Contact Pneumatic Gauge," *Adaptive Control as a Part of the Manufacturing System*, Stockholm Royal Institute of Technology, April 1976.
8. TESCHLER, "Transducers for Digital Systems," p. 68.
9. R. A. MATHIAS, "New Developments in Adaptive Control," Paper No. MS78-217, Computer Automated Association/SME Western Tool Exhibition and Technical Conference, Los Angeles, March 1978.
10. S. HIMMELSTEN AND COMPANY, *MCRT Torquemeters, Bulletins 760 and 761*, Elk Grove Village, IL, 1976.
11. R. A. MATHIAS, "Production Experience With Adaptively Controlled Milling," Paper No. MS74-724, SME Western Tool Exhibition and Technical Conference, Los Angeles, March 1974, p. 2.
12. KRISTAL INSTRUMENT CORPORATION, "Piezoelectric Measuring Systems," Catalog K2.002 8.77, Grand Island, NY, September 1978, p. 8.
13. S. HIMMELSTEIN AND COMPANY, *MCRT Torquemeters, Bulletin 510*, Elk Grove Village, IL, 1976.
14. KRISTAL INSTRUMENT CORPORATION, *Piezoelectric Measuring Systems, Bulletin 20.055e*, Grand Island, NY, September 1978.
15. NOVAK, *Adaptive Control*, pp. 64–65.
16. NOVAK, *Adaptive Control*, pp. 81–100.
17. KEARNEY AND TRECKER COMPANY, *KT Spindle Probe Unit, Publication 687*, Addendum 1, Milwaukee, WI, 1980.
18. M. O. NICOLLS, "Analyzing Surface Finish," *American Machinist*, May 13, 1974, pp. 67–71.
19. H. MURRAY, "Exploratory Investigation of Laser Methods For Grinding Research," *Annals of the CIRP*, Vol. 22, No. 1 (1973), p. 137.
20. METCUT RESEARCH ASSOCIATES, Section 5.3, *Machining Data Handbook*, Cincinnati, OH, 1972, p. 844.
21. K. ESSEL and W. HANSEL, "Development of Sensors For Process Control Systems in the Field of Production Engineering," Gesellschaft Fur KernForschung MbH, Karlsruhe, PDV-report KFK-PDV 41, RWTH Auchen, Germany, May 1975.
22. W. D. CAIN, *Automatic Tool Setters*, SME Paper No. MS72-629, Union Carbide Company, Detroit, MI, 1972.
23. KEARNEY AND TRECKER COMPANY, *Broken Tool Detector*, Addendum 2, Milwaukee, WI, 1980.
24. N. H. COOK, K. SUBZAMANIAN, and S. A. BASILA, "Survey of the State of the Art of Tool Wear Sensing Techniques," Materials Processing Laboratory, Department of Mechanical Engineering, MIT, September 1975.
25. NOVAK, *Adaptive Control*, pp. 150–155.
26. R. MOORE, *Monitoring the Die-Casting Process, Bulletin D2395*, Honeywell, Inc., Test Instruments Division, Denver, September 1969.
27. KRISTAL INSTRUMENT CORPORATION, *Data Bulletin K6.001* (5/72), Elk Grove Village, IL, 1972.
28. KRISTAL INSTRUMENT CORPORATION, *Data Bulletin 2.021e* (2.75), Elk Grove Village, IL, 1975.
29. S. K. BIRLA, "Sensors for Adaptive Control and Machine Diagnostic," in Machine Tool Task Force and Lawrence Livermore Laboratory, University of California, Livermore, California, Eds., *The Technology of Machine Tools–A Survey of the State of the Art*, Vol. 4, *Machine Tool Controls*, Chicago, October 1980, pp. 7.12-1–7.12-7.

SECTION 8
QUALITY ASSURANCE

CHAPTER 8.1

The Quality Assurance System

CHRISTIAN H. GUDNASON

Technical University of Denmark

8.1.1 CONCEPTS OF QUALITY AND QUALITY ASSURANCE

Concept of Quality

Although no concise definition of "quality" exists, it is generally agreed that it characterizes the degree to which products satisfy consumers' wishes and expectations. Typical of such definitions is that of the European Organization for Quality Control (EOQC): "the totality of features and characteristics of a product or service that bear on its ability to satisfy a given need."[1] Expressions such as "fitness for use" or "user's satisfaction"[2] often stand for product quality, but it is clearly the consumer's subjective evaluation that is central to the quality concept.

There are many problems attached to these definitions, particularly that of their relevance (in reality), since producers have only limited means of determining a new product's quality before it is marketed.

From the industrial organization's viewpoint, no general definition of product quality is operable; instead, other concepts are used, such as quality characteristics, quality parameters, and quality specifications. In setting such standards, the aim is to strike a balance between the costs of achieving a given quality level and the profits arising from increased sales, thus promoting the organization's own economic interests, although setting such standards may be claimed to be done for the consumer's benefit. The need to quantify product quality first arises when there is an information gap between producer and consumer, which it is in the producer's interests to bridge.

Quality–Consumers and Society

If product quality is taken to mean a concept adopted by industrial organizations to make their products more acceptable to consumers, such a concept would be desirable if industrial and societal goals were identical. However, recent experience has shown that *any definition of product quality devoted exclusively to organizational goals is not only unsatisfactory, but damaging to society.* Consequently, industry must bear in mind that both consumer and environment will be increasingly protected by legislation against the harmful effects of its products or production methods. A number of consumer protection organizations already exist.

Jurisdiction increasingly makes producers responsible for damage arising from their production. There are three concepts involved: negligence, warranty, and liability. They aim at ensuring that products are defect free, with producers increasingly being called upon to demonstrate that this is so, and that production and sales methods are sufficiently regulated.

Apart from governmental intervention, the most important of many other safeguarding mechanisms is *consumerism.* Here a special pressure group emerges, in the absence of other effective consumer protection, to influence both governmental and industrial policies. Those industrial organizations that introduced the quality concept for their own purposes only have had to come to terms with consumers' interests, which has increased the need for a dialogue. Herein the concept of quality becomes bound up with the producer's need to inform the consumer adequately so that the latter may choose objectively and be aware of the implications of his or her consumption.

This author believes that the concept of quality should be further broadened to cover far more than the gap between producer and consumer. There is another gap, which may prove to be far more serious: the gap between man and nature. It must be narrowed if terrible catastrophes are to be avoided. As the production of material goods increases, along with the consumption of raw materials, so, too, does pollution. This would not be dangerous were it not that the rates of increase of resource use and pollution production are greater than nature can withstand and that doubts exist whether human needs may be exclusively satisfied by material goods. Production involves the transformation of natural resources from one given state to another, not their destruction. With today's economic systems, all we can expect is transformation from a "natural" to a polluted state. Unfortunately, since national wealth is closely related to the production of material goods, and since this wealth is measured at a certain time, there is a tendency to postpone some costs to some future time. This is like asking about a person's assets without inquiring about his or her liabilities. If society expects to survive, it would do well to inquire into what is happening to nature.

Given our contemporary problems, it is urgent that we develop a comprehensive concept of quality embracing the needs of industry, consumers, and society in general. It should be a framework to facilitate communication among many parties, not merely a constrained optimization of the relationship between industry and consumers. The framework may be seen in a model (Exhibit 8.1.1) illustrating a dynamic mechanism in setting quality standards. This model has four elements: industry, consumers, society, and nature.

Industry must show an open-minded response, particularly to the environment. Feasible, not "optimal," solutions should be sought, since the latter are both short term and partial. This author's postulate is that industry's survival depends on seeking more understanding of society's needs and how to satisfy them. Only such a framework, embodying the four elements and taking them all seriously—rather than maximizing profit—can ensure the survival of the industrial organization as we know it.

Industrial Use of the Quality Concept

The many definitions of quality might be summarized as follows: Quality is the degree of consumer satisfaction. The needs and expectations of consumers are directed toward the properties of a product, and an application of the preceding—and other—definitions necessitates further elaboration of the concept of quality properties. Thus it seems appropriate to distinguish between quality parameters and quality characteristics.

"Quality parameters" describe, in general terms, the need for or expectation from a product imposed by consumers. It is a superior way of characterizing a product through appearance, suitability, reliability, and similar general characteristics. Properties constituting side effects of production and consumption, including pollution and resource consumption, should also be considered quality parameters.

"Quality characteristics" are product properties that determine to what extent a property mentioned under quality parameters is built into the product. It is a property that determines, for instance, the degree of suitability of a product. Quality characteristics are thus product properties that influence the degree of fulfillment of consumer expectations.

In dealing with the quality concept in relation to quality assurance, it is appropriate to divide the concept into three parts (Exhibit 8.1.2):

Quality of design.
Quality of manufacture.
Quality of marketing/service.

"Quality of design" is the planned quality. It is laid down primarily in the product specifications, but also in the fundamental decisions concerning market segments and specifications for use and service. Differences in quality of design refer to differences in specifications for the same functional use.

"Quality of manufacture" is the degree to which the manufactured product conforms to the product specifications. Differences in quality of manufacture refer to differences, originating at the manufacturing stage, in properties of products with the same specifications.

The first two aspects of quality are classical. More recent is the concept of "quality of marketing service," which may be defined as the degree to which a product in use conforms to the fundamental decisions on marketing/maintenance and product service. The argument for introducing this third category of quality is primarily that a product, by definition, is far more than the physical product. A number of elements associated with the product determine how a product is experienced (Exhibit 8.1.3).

The Concept of Quality Assurance

The term "quality assurance" has various meanings, and there seems to be a great deal of confusion about the concept. According to the EOQC,[3] quality assurance is:

> *A system of activities whose purpose is to provide assurance and show evidence that the over-all quality control job is in fact being effective.*

Exhibit 8.1.1 A Dynamic Mechanism in Setting Quality Standards

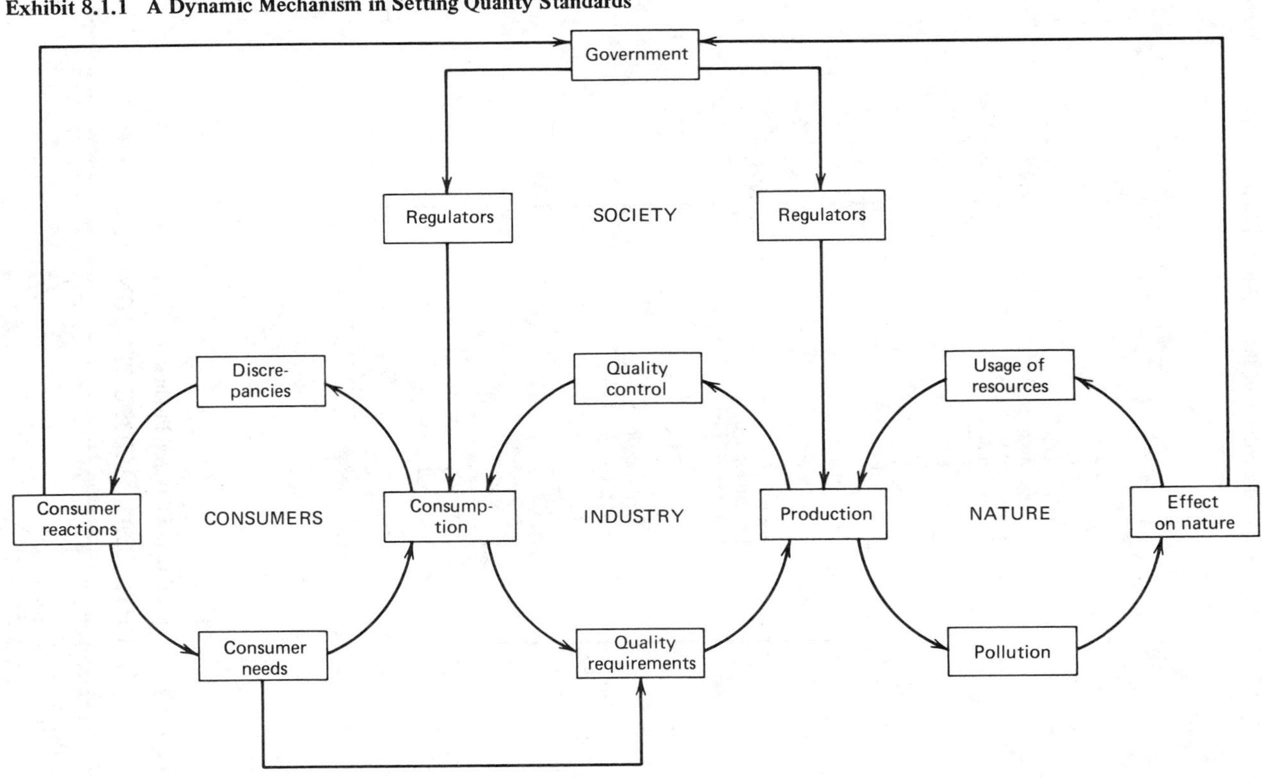

Exhibit 8.1.2 Breakdown of the Quality Concept

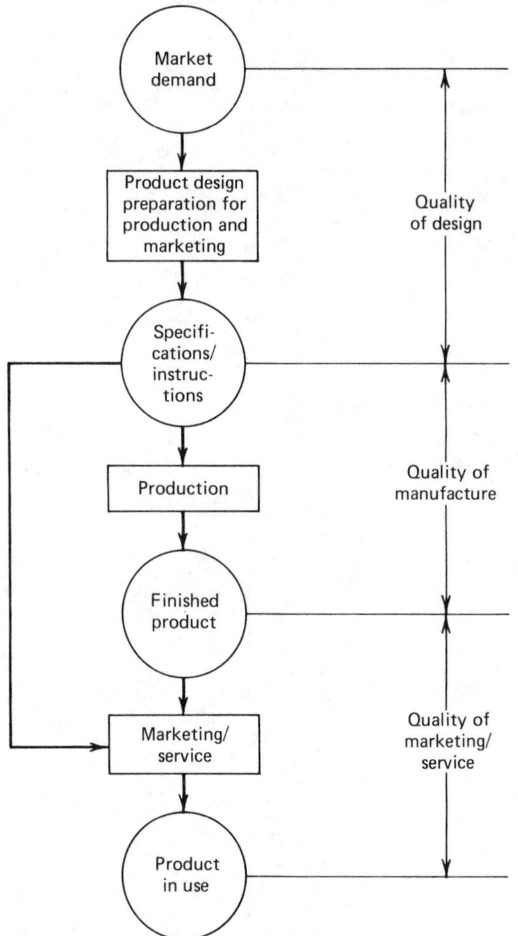

Exhibit 8.1.3 Definition of Physical Product

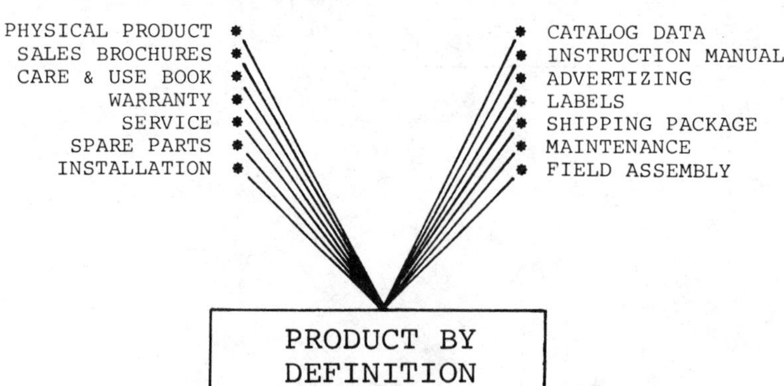

The system involves a continuing evaluation of the adequacy and effectiveness of the overall quality control program with a view to having corrective measures initiated where necessary. For a specific product or service, this involves verifications, audits and the evaluation of the quality factors that affect the specification, production, inspection and use of the product or service.

In this chapter "quality assurance" is used as a synonym for "quality control" or "total quality control," that is, all activities in the enterprise concerned with the attainment of quality. Consequently, the term "control" is used in a more narrow sense; that is, it is a part of the assurance function covering the function of comparing performance and results with plans and specifications. Evaluation of performance may result in a corrective action and/or in adjustment of plans and/or specifications (Exhibit 8.1.4).

8.1.2 MOTIVES AND MOTIVATION FOR QUALITY ASSURANCE

In viewing the motivational aspects of quality assurance, two questions arise. First, what are companies' motives in establishing a quality assurance program or system? Second, what motivates and stimulates people within the company to cooperate in the company's endeavors to improve quality? Some of the motivational aspects are mentioned in the following subsections.

Motives Related to Company Objectives

Market Share and Competition

With the rising costs in the industrialized world, quality becomes a sales parameter of ever greater importance. Experience as well as planned investigations have proved that there is a significant correlation between the quality level and the market shares of products. Correspondingly, a close connection has been shown between quality level and business economy.

The fixing of the quality level is contained in the activities of quality assurance. Whether a product has a high or low quality level is, among other things, a question of how it performs its function, how many functions (features) it has, its lifetime and reliability, its appearance, and so on. It is important that a company try to choose the right quality level. The quality level of a product may be set either too high or too low—both may be disastrous.

Exhibit 8.1.4 Quality Assurance as a Feedback Cycle

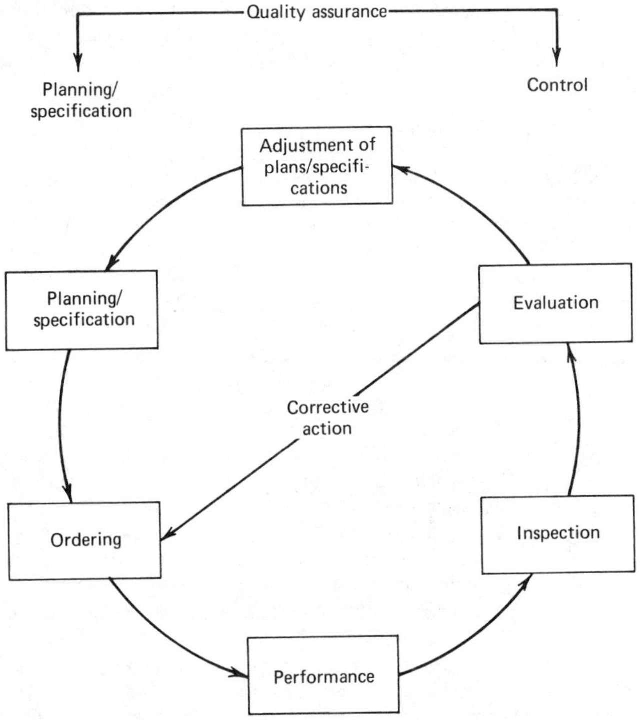

Exhibit 8.1.5 Quality Level and Profitability

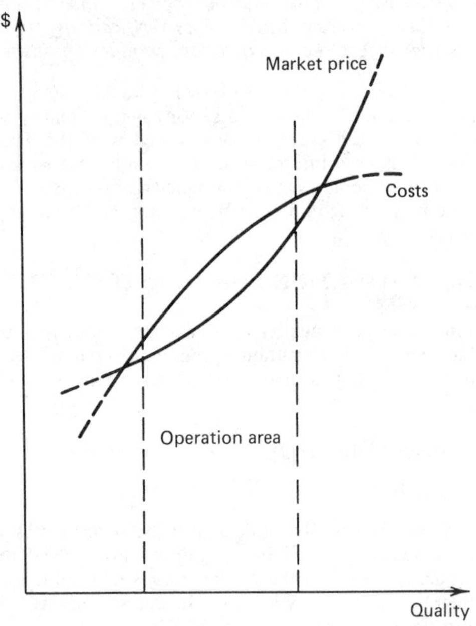

In many branches of industry, the scope for the selection of a quality level is rather narrow (see Exhibit 8.1.5). If the wrong quality level is chosen for a product, the contribution of the product to the total profitability will diminish. That the wrong quality level is assigned to a product through development and design (being either too high or too low) is in many cases due to there has not been, either before or at an early stage of the development phase, an explicit decision of the required quality level of new products. The lack of such a goal for the quality of new products, resulting from the lack of decisions concerning quality parameters, can destroy the entire course of development and in the long run ruin the company's image and goodwill.

Reduction of Quality Costs

The importance of the careful supervision of quality costs has to be estimated on the basis of the absolute order of these costs as well as in connection with the possibilities of obtaining savings through endeavors to reach an optimal distribution of the total costs of different categories of costs. The internationally accepted breakdown of total quality costs is shown in Exhibit 8.1.6.

It is extremely difficult to give precise indications as to which level of quality costs should be considered reasonable for given enterprises or branches of industry. This is partly because of the many different characteristics of enterprises that are of importance for the cost level and partly because it would demand an absolutely unambiguous definition of the different kinds of costs; this last point is especially difficult because of differences in the characteristics of enterprises.

Exhibit 8.1.6 Breakdown of Quality Costs

	Breakdown and Definition
Costs of quality assurance	Prevention costs: Costs for quality planning, design reviews, specification, acquisition and analyses of quality data, quality reporting, education and training, and so on.
	Appraisal costs: Costs for quality control, including incoming material, control, process control. Control of finished goods, tests and maintenance of test equipment, and so on.
Costs of failure	Internal failure costs: Scrap, sorting, reworking and retesting, idle facilities, troubleshooting, and so on.
	External failure costs: Costs related to complaints, warranty, returned products, allowances, and so on.

Exhibit 8.1.7 Quality Costs and Control Efforts

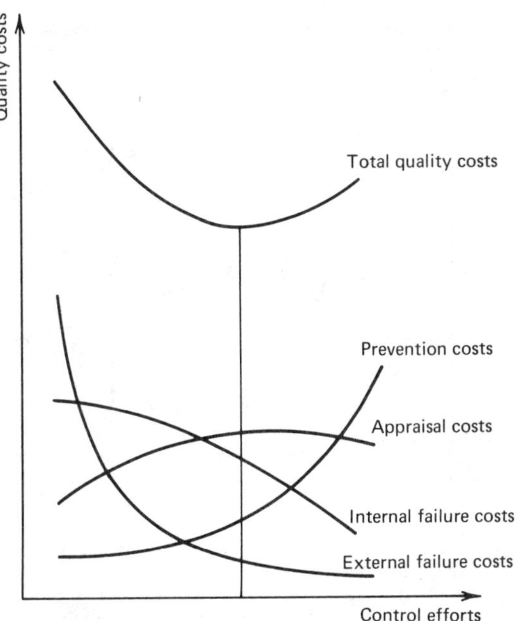

The statements of costs known from practice show that total quality costs often constitute 10 to 20% of the turnover of a company. From the literature one may get an impression of the order of magnitude of costs, which seems to depend very much on the particular branch of industry involved. The lowest quality costs are to be found in the heavy iron industry, whereas costs that are many times higher may be discovered in precision industry, that is, fine mechanics and electronics.[2]

One explanation of the big variations may be that the cost levels of enterprises are placed nearer to or further from an "optimal" value; in this connection it must be strongly stressed that most of the cost figures given in the literature are hardly optimal. That one often finds quality costs at about 10% of the turnover should not be perceived as a norm, that is, something one has to accept or be satisfied with. In most cases you ought to aim at a lower value. If the quality costs are depicted as a function of the degree of control (see Exhibit 8.1.7), one may conceive that, with such a given degree of control, that is, with a certain control effort, a minimal cost level will be found to which corresponds an optimal quality level with regard to costs.

It seems to be a common experience in enterprises that have been working systematically to improve their quality assurance that essential savings may be obtained by spending relatively more on planning and specification activities, that is, the preventive aspect of quality assurance. Usually companies are to the left of the minimum point for quality costs suggested in Exhibit 8.1.7.

An example of a cost development is shown in Exhibit 8.1.8. The figures are drawn from a company within the electromechanical industry, which has gone through a process of reorganization, including an improvement of the quality assurance system. In the period 1971–1979, the prevention costs have been increased from 5% to 15% of the total quality costs. During the same period, the total quality costs were reduced by about 2% of the turnover. It is normal that an effort to improve the quality assurance system results in a reduction of the total quality costs corresponding to 2 to 5% of the turnover. This will, in all essentials, be reached by an increase in the prevention costs.

Improving Communication

Good quality assurance practice has proved to be a very efficient means of improving communication both among sectors, departments, groups, and so on, within the company and between the company and its environment, that is, the market, the vendor, suppliers, authorities, and so on.

Examples of typical effects of poor communication within a company are changes in product design, production methods, and tools. Many companies suffer from such changes during the final phases of product design and during production preparation and start-up. The number of instructions, new directives, and so on, of alterations and changes made can never be zero, but in

Exhibit 8.1.8 Cost Development in an Electromechanical Plant

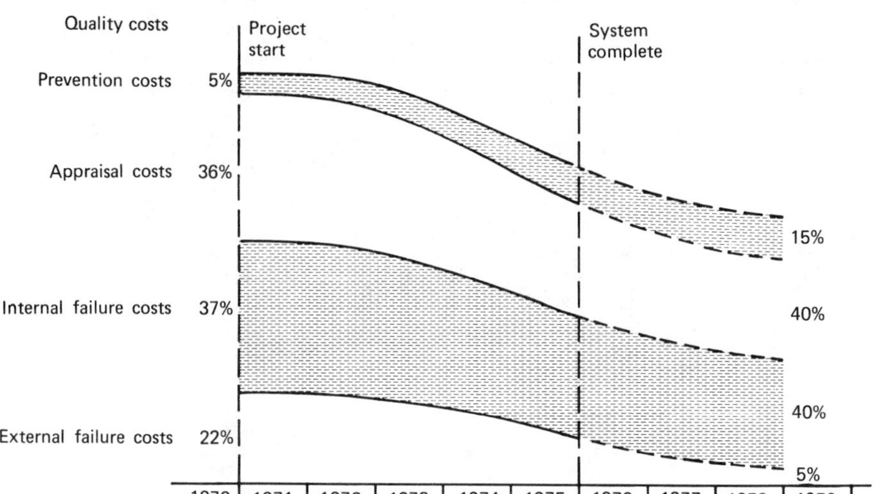

many companies it may be reduced to a more reasonable size by, for instance, systematic design reviews and other preventive quality measures.

It is a fundamental principle of quality assurance to try to identify and correct potential failures as early as possible. If a potential quality problem can be revealed at the outline stage, it is of course cheaper to correct the sketch on the drawing board than to have to correct design details and change tools and so on later.

A means of coordinating various engineering functions around the development of new products is formalized design reviews, where an interdepartmentally composed group, at different stages in the course of development from idea to preproduction, goes through and discusses the current results and problems of the development project. Design reviews are thus a means of communication whereby, for instance, experience and know-how from production, quality control, marketing and service, and, if necessary, purchasing, and so on, are channeled to the designer. In companies where there is a positive, constructively critical, and mutually open attitude to problems in the course of development, really good results may be obtained, one of which, among others, is a reduction in the number of changes in specifications.

Good and effective communication between the company and its environments has been increasingly important with the growth in international trade and with the tendency to increased industrial cooperation–nationally and internationally. The development of international standards has made communication between cooperating firms considerably better, but often a receiving company will require documented evidence, for instance, by means of a quality manual, that the supplier's quality assurance system is sufficiently well established. This applies particularly in connection with military supplies within the space industry, the nuclear power industry, the medical industry, and the like. In the case of military supplies, a strict observance of NATO specifications will often be required of a supplier's quality assurance system.[4] These NATO requirements have formed the basis for working out the general requirements for subcontractors' quality assurance systems in many private enterprises.

Reducing Risks of Product Liability

Because of the reaction to the harmful consequences of production and consumption, governments have started to play a more active part in regulating the relationship between industrial organizations and their environments. Of particular interest is the concept of "product liability." This designation covers the responsibility of suppliers for damage caused by their products.

In recent years it has been necessary for companies to increase their awareness of product liability. The increasing importance of this factor is attributable to, among other things, the internationalization that industrial companies have experienced, particularly with respect to marketing, and also, with the passing of time, to the geographic distribution of their products.

In the United States, companies have experienced a substantial increase in the number of cases where demands for compensation based on product liability have been put forward. Magazines and papers have stated that in 1975 more than 1 million claims were made, which reveals that product liability is not an unimportant area for some members of the American law profession.[5]

In Europe the suggestions of a European Economic Council (EEC) directive on product liability were passed in the Commission of the European Communities (CEC) in 1976; in 1977 a convention of the Council of Europe on product liability was passed by the Council.[6] A precise evaluation of the effects, however, demands a high degree of legal expertise.

The management of a company must, of course, consider what ways are available to meet the consequent risk of product liability. Four fundamentally different courses that the company's management may take are:

1. Regulation of risk through careful statement of conditions of sale and delivery.
2. Transfer of risk through the underwriting of product liability insurance.
3. Reduction of risk through the introduction of effective quality assurance.
4. Acceptance of risk through passivity.

Points 1 through 4 are not mutually independent. The reduction of the existing risk through introduction or improvement of quality assurance will give better conditions for regulation and transfer of risk.

Motivation for Quality

A primary condition for achieving intended benefits of quality assurance is the engagement and motivation of all individuals within the organization concerned with quality matters. Just as the organization should adjust its perception of quality and the content and spirit of the quality assurance function to the forces in the environment, it should be aware of the needs and expectations of the individual members of the organization itself. To achieve quality objectives, management must learn to understand the motivational forces behind the creation of quality work. Studying the work of motivation theorists may contribute to such understanding. (For references and more detailed studies, see Chapter 2.2 of this handbook.) Frederick Herzberg distinguishes between factors producing job satisfaction and those causing job dissatisfaction. Motivators are the primary cause of satisfaction, and hygiene factors the primary cause of unhappiness on the job. Among the motivator factors are achievement, recognition, responsibility, and the work itself. Among the hygiene factors is supervision.

Work satisfaction research brought Herzberg and others to the conclusion that the *content of jobs*, rather than the conditions surrounding them, had the significant influence on performance, satisfaction, and motivation. Job enlargement and job enrichment are essential concepts in the endeavor to develop job content. Experiments have proved that workers with enlarged jobs were concerned with the importance of their jobs and work methods, with responsibility for quality, and with performance of preparatory activities.

Other interesting developments are the concepts of sociotechnical systems and job design, according to which emphasis is placed more on *role content*. The results of organizational and job design research have lead to five categories of requirements: responsibility, autonomy, adaptability, variety, and participation. A long series of experiments in various forms of work organization has followed, with the purpose of combining various requirements, such as work satisfaction, flexibility, productivity, and quality. Decentralization of planning and control activities, together with an increased degree of responsibility, plays an important role in these endeavors.

An interesting development is the Japanese QC circles, which have come to play a major role in companywide quality control efforts. Quality control circles are small groups of workers and a leader who work voluntarily on the improvement of quality by diagnosing and remedying quality problems. Prior to and combined with the participation in QC circles, an intensive training and education is taking place. There is no doubt that QC circles are an essential reason for the success of Japanese industry in reversing its quality image totally in the last 20 years.

8.1.3 THE QUALITY ASSURANCE SYSTEM[7]

The Subsystems of Quality Assurance

The concept of quality was previously broken down into three parts: quality of design, quality of manufacture, and quality of marketing/service. Similarly, the concept of quality assurance may be broken down into three parts:

1. *Assurance of quality of design*, that is, the planning/specification and control of quality characteristics of products and production processes and the methods and basic decisions concerning means for securing intended use, maintenance, service, and so on.
2. *Assurance of quality of manufacture*, that is, planning/specifying and control of the degree of conformity of the final product to quality of design.
3. *Assurance of quality of marketing/service*, that is, planning/specifications and control of prin-

Exhibit 8.1.9 Subsystems of Quality Assurance

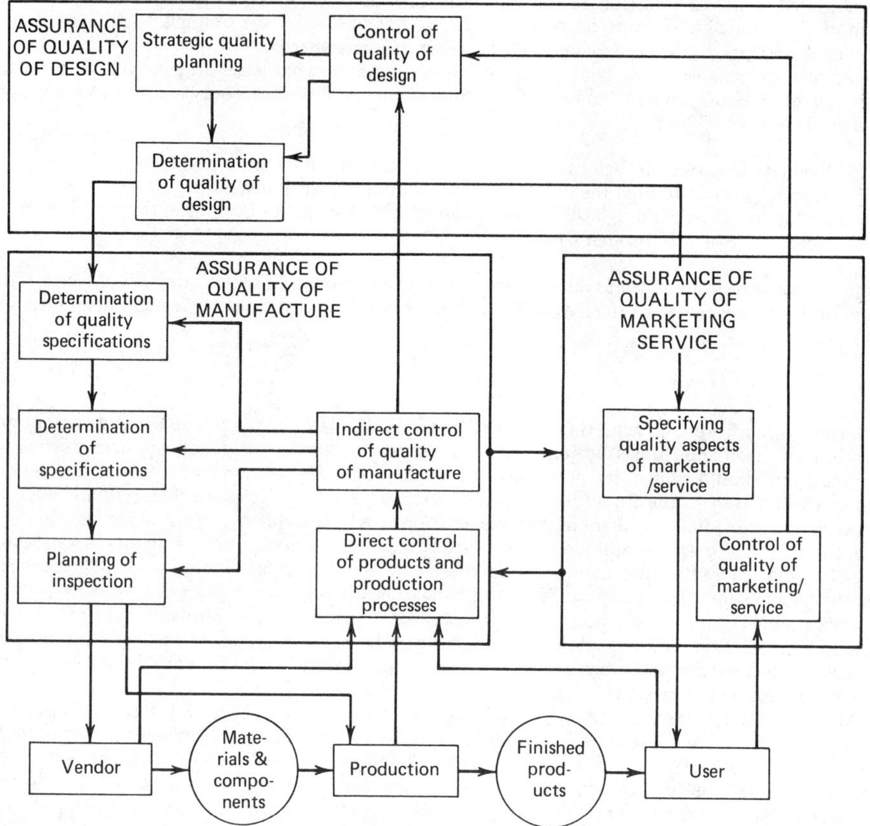

ciples, methods, and levels of marketing/maintenance, service, and similar aspects that concern consumers' impressions of and experience with products.

These three parts of quality assurance make up three main systems within the total system of quality assurance.

The three main systems can be further broken down into a number of subsystems, as illustrated in Exhibit 8.1.9. According to the concept of control systems as feedback mechanisms, the decision processes of the subsystems have the character either of planning and specification or of control or appraisal, that is, inspecting, analyzing, and comparing conformity and thus forming the basis of decisions for corrective action and/or adjustments to specifications.

Note that in this presentation a distinction is made between direct and indirect control. "Direct control" means the direct control of products and production processes in accordance with specifications. "Indirect control" means an appraisal of the appropriateness of prearranged and applied specifications.

The decisions made within the quality assurance systems at various levels in the organizational hierarchy differ in character as to time horizon, frequency, level of detail, and so on. At *the strategic level*, decisions are made on goals, objectives, and policies. Examples are decisions on quality objectives and policy, market needs to be statisfied, quality level of products, the weight of various quality parameters, resources allocated to quality assurance, and so on. At *the tactical level*, the main functions are to prepare procedures and rules by which goals and objectives can be fulfilled, subject to given restrictions. The decisions concern quality assurance activities in engineering (design and production), vendor appraisal and the like. Decisions at *the operative level* are of a more routine or programmed nature. They concern, for instance, inspection activities, maintenance of measuring equipment, and the like.

The Function of Quality Assurance

Quality assurance was previously described as a system broken down into a number of interacting subsystems. The quality assurance function may also be described by means of a number of general

Exhibit 8.1.10 The Functions of Quality Assurance

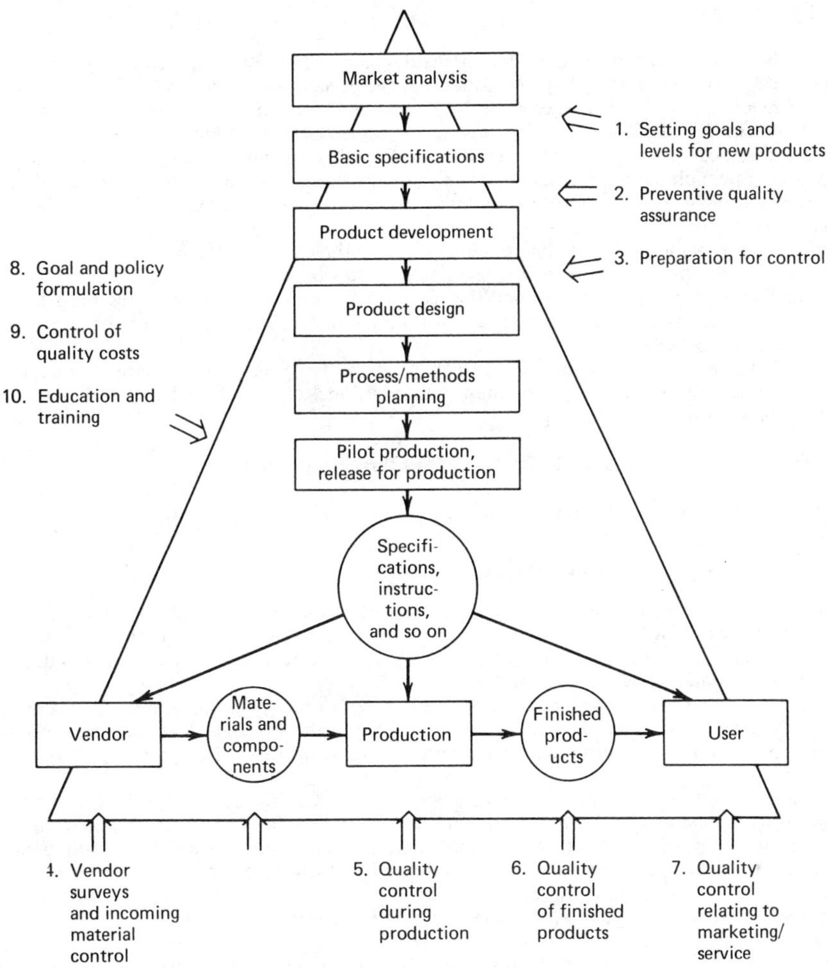

tasks or activities, each of which is essential to quality assurance. The importance of the various activities depends on the type and characteristics of the particular company.

The total number of activities may be grouped or classified according to various criteria. Exhibit 8.1.10 suggests a classification into the following three groups:

Activities relating to preparations for production and marketing.

Activities relating to the flow of materials and products.

General activities.

Activities Relating to Preparations for Production and Marketing

1. Setting Goals and Levels for New Products. This activity includes market analyses and identification of needs, which are the basis for determining quality of design. Included are decisions on quality levels for new products and the determination of quality parameters. It also includes decisions concerning quality aspects of principles for marketing and installation, use, maintenance, service, and appraisal of these principles.

2. Preventive Quality Assurance. This activity contains all the endeavors of building quality into the product through

Product development and design.

Production preparation, that is, process and methods planning.

Determination of principles and procedures for marketing and service.

Final product evaluation.

During the various stages of product development, analyses and tests must be carried out in order to concretize the various quality characteristics that the product must have to fulfill the customer's needs or requirements, such as design reviews, analyses of potential defects, and reliability of production. The function also covers endeavors to ensure the compatibility of the design characteristics and associated quality levels to actual manufacturing capability, including the location and elimination of possible sources of manufacturing troubles before the start of formal production.

The function includes the appraisal (i.e., indirect control) of production methods in relation to quality requirements.

Decisions on principles of marketing, including principles relating to the many elements causally associated with the product (advertising, maintenance, service, etc.), must be built up by working out detailed procedures, instructions, and the like, which are necessary to ensuring proper use or application. Appraisal of the appropriateness of the said guidelines is important in order to ensure continuous user satisfaction.

The final evaluation of the quality of the new product, based on data from previous design reviews and tests and possibly supplemented with additional tests, should be carried out before release for production and marketing.

3. **Preparation for Control.** This function covers three main activities:

Working out quality specifications.

Planning inspection.

Determining measurement techniques and equipment.

Determination of quality specifications is the basis of quality assurance of manufacture. A part of the activity is to classify and set quality levels for the various product properties. The classification includes an evaluation of the importance and risks of defects caused by materials at different stages from raw materials to finished products.

Specification and planning of inspection, together with quality specifications, form the total basis for inspection activities. The appraisal of quality specifications and the plans and specifications for inspection is a continuous evaluation regarding the need for alterations to existing specifications.

Good quality assurance depends, to a high degree, on quantifications of product and process characteristics. Preparation of control thus includes the determination of methods of testing and measurement, the purchase of measuring equipment, and the maintenance of such equipment. To maintain accuracy in measuring, a continuing system of calibration control is often required.

Activities Relating to the Flow of Materials and Products

These activities all belong to the control or appraisal part of the quality assurance function (see Exhibit 8.1.4).

4. **Vendor Quality Surveys and Quality Control by Incoming Inspection.** Vendor quality surveys cover the attempts to predict a vendor's ability to meet quality requirements. The evaluation concerns a vendor's quality assurance system, procedures, facilities, and equipment. The concept is increasingly applied by industry, defense, and public authorities. Vendor relations often extend to joint planning and cooperation between producer and vendor on quality assurance problems. The cooperation may include assistance to set up effective quality control programs.

Incoming material control is the current inspection of materials and components purchased. The purpose is to evaluate whether the goods received conform to specifications or whether they should be rejected. A suitable instrument might be the acceptance sampling plans, for instance, International Organization for Standardization 2859 (Military Standard 105D). In addition to this direct control of quality of incoming material, the activity may include an indirect quality control function through the registration and evaluation of the quality level of each supplier.

5. **Quality Control During Production.** This activity includes:

Approval of setups, which are the assembling of machines, tools, instruments, and material. The setup must be adjusted so that the product will conform to specifications. Approval may be based on first-piece inspection.

Process inspection, which is a current inspection of production processes. The aim of process inspection is to ensure that the process is "in control" in order to prevent production defects.

Statistical tools (control charts) are often used. Results from process inspection form an important basis for evaluating product specifications and production methods.

Inspection of lots of semiproduced items moving from one production area or department to another. Sampling plans may be used. Besides the current inspection activities, quality control during production covers various forms of analyses.

Troubleshooting, that is, analyses with the purpose of restoring the status quo to a process that has gone out of control.

Process capability analyses, which are important in selecting appropriate manufacturing equipment, renewal of parts in manufacturing equipment, in determining realistic quality specifications (tolerances), in using statistical control charts in process inspection, and so on.

6. Quality Control of Finished Products. The current inspection of finished products is the last chance to observe defects before marketing or delivery. A number of functional product properties may first be measured or observed after the final assembly, for instance, reliability and environmental testing. Acceptance sampling plans are used to a great extent. Control of certain product properties often requires 100% inspection, that is, inspection of each unit of production.

7. Quality Control Relating to Marketing/Service. The control function includes the surveillance of the direct occurrence of defects in the marketplace. It also includes a general observation of customer reactions, that is, whether the use of the product or service gives less user satisfaction than aimed at by the determination of quality level and quality parameters. Thus the surveillance concerns all three concepts of quality assurance–those of design, manufacture, and marketing/service. Sources of information on defect occurrence at the market level may be reports on users' complaints, service reports, and reports from dealers and market analyses.

The activity also includes the actual treatment of complaints and of cases concerning product liability, as well as the possible recall of products.

General Activities

To fulfill its function, the quality assurance system must include some additional activities that do not belong to the two groups of activities already mentioned. These activities are of a more general organizational character and include a substantial aspect of management.

8. Formulation of Quality Goals and Policies. Goals and policies form the basis of a company's quality assurance program. Quality goals and objectives include fundamental decisions on the quality levels of products, on fundamental combinations, and on the mutual weighing of quality parameters. Quality goals may also include a mutual weighing of the three concepts of quality: quality of design, quality of manufacture, and quality of marketing/service.

From quality goals may be derived a set of quality policies, which form guidelines according to which the main functions are to be carried out to attain quality goals. More specific policies may contain fundamental decisions concerning the quality assurance system. Among other things are decisions concerning the proportion of resources allocated to the preventive and planning parts and to the controlling part of quality assurance. Also, fundamental decisions on resources allocated to the organization and maintenance (auditing) of the quality assurance system may be included in the set of policies.

Goals and policies should be frequently reviewed in order to ensure that they are, at any given time, adapted to the expectations and requirements from markets and environments in general.

9. Control of Quality Costs/Quality Income. The order of magnitude of quality costs and the possibilities of reducing quality costs were discussed previously, along with the traditional classification of different cost categories. Activities concerning the proper control of a budget relating to quality assurance include various types of economic analyses, budgeting, and budget control.

Analyses cover, among other things, the evaluation of effects on market share/income from ineffective quality assurance, for example, lost orders and loss of income due to poor quality image or wrongly chosen quality levels for products. Analyses should also be carried out with the purpose of locating critical areas and causes of high costs and of giving background information for a selection of meaningful cost reduction projects.

As a budgeting tool, quality cost control permits coordinated cost optimization and quality assurance planning; most important, it contributes to ensuring performance in the light of prescribed goals and objectives. Thus quality costs should be carefully budgeted and periodically reported on and studied by management at various levels.

10. Education and Training Regarding Quality and Quality Control. Since quality and quality control affect most organizational units in a company, it is essential that education and training activities be carried out concurrently. Planning and implementation of these activities, which may differ in content and form, may often include all the organizational functions in a company and all levels in the organizational hierarchy. External as well as (company) internal educational activities may be included, ranging from seminars for top management to programmed education in specific techniques, for instance, inspection.

8.1.4 MANAGEMENT AND QUALITY ASSURANCE

Organization of Quality Assurance

Identification of Quality Assurance Activities

Organization of quality assurance includes, among other things, the identification of the quality assurance activities or tasks necessary to achieve given company goals and objectives.

The quality assurance system described in Section 8.1.3 aims at giving a general survey of activities that should be included in the total quality function. It ought to be considered as a *framework* and not as a standard system of quality assurance. A given company may not necessarily need all the activities listed in the system. Which activities are important depends on the type, size, and characteristics of the company. An R & D intensive company may allocate substantial resources to the activities of preventive quality assurance, whereas a company producing standard goods may need to place more emphasis on control activities, that is, activities relating to the flow of materials.

The framework should be considered as a tool to be used by the individual company for analyzing its own situation. For this purpose, the quality assurance functions are described partly as a number of subsystems and partly as a number of activities. The framework may be used as a checklist or as a template in order to locate weak points and to identify those activities that are most important to the company in relation to its objectives and strategy.

Responsibility for Performing the Activities

The framework describes the quality assurance function from a systems point of view. It does not give any directions as to which employees in the organization should be in charge of the various activities. Another important organizational task, therefore, is the assignment of responsibility for performing the quality assurance activities.

In view of the almost evolutionary development of the quality assurance function during recent decades, the organizational tasks have become correspondingly more difficult. Once primarily an inspection function, the quality assurance function is now a complex interaction between departments and persons within a company and with customers, suppliers, and other external groups or institutions. Consequently, it will normally not be possible to place all the activities within the total quality assurance system in one department of quality assurance. A number of activities may more appropriately be placed in other departments (product development, purchasing, sales, etc.), and certain activities and areas of responsibility are often best taken care of by interdepartmental groups or committees.

Organizational Structure

There is no such thing as a "best" organizational structure. Structures vary considerably with the size of the organization, the nature of its products, the category of customers, and so on; different organizational forms may, in principle, give the same results. Over the years the structure known as the "functional structure" has been expanded by a number of alternatives, such as divisional structure, multinational structure, matrix structure, and innovative structure.

As a result of these developments, the question of *centralization or decentralization* has attracted much attention. A centralized form of organization, where almost all activities are placed in one central department, should, in principle, secure a better coordination of many activities, but experience has shown that this organizational form does not promote cooperation between main departments. A decentralized form of organization seems to give better quality motivation within departments mainly because of the motivational aspects of responsibility. A disadvantage is that the responsibility for cooperation often becomes unclear. In most instances a mixed organizational structure will function best. New viewpoints on motivation, industrial democracy, joint responsibility, and so on, have strengthened the interest in and tendency toward decentralization. Gradually it has become more and more common to leave the control or inspection functions to operators. Another trend is toward changes in work organization, including the formation of autonomous production groups given responsibility for planning and controlling their own work.

At the same time, the increasing complexity of the total quality assurance function has necessitated the presence of a central organizational unit responsible for coordination and long-range

Exhibit 8.1.11 Quality Organization in a Divisional Concern

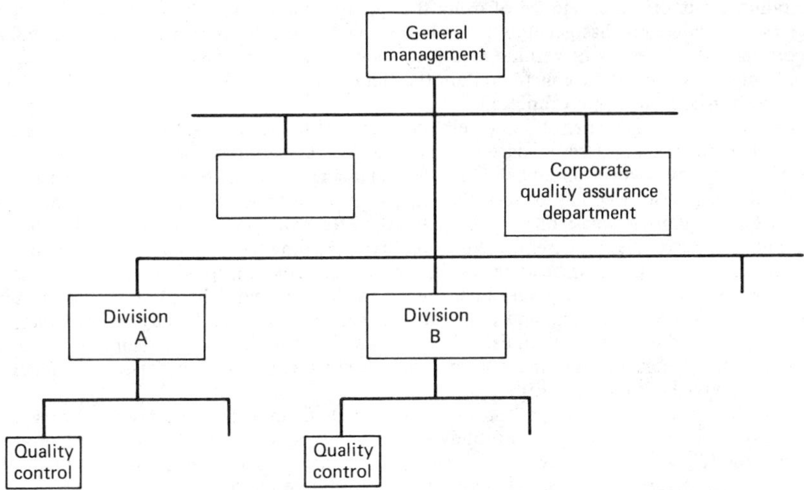

development. A central quality assurance department, assisted by interdepartmental groups, strongly supported by upper management, and with the authority and professional competence to coordinate a large number of decentralized quality functions, seems at present to be a basic element in an up-to-date organization form.

Exhibit 8.1.11 shows the basic structure of a quality organization in a divisional industrial concern.

The duties of the corporate quality managers are partly to assist general management and partly to assist divisional management. They may include the following:

Development of quality policies and procedures.
Auditing divisional quality performance.
Appraisal of divisional quality managers.
Quality systems auditing and development.
Solving interdivisional quality problems.
Training and education of quality personnel.

Quality and Upper Management

The engagement of upper management in the assurance of quality may be seen in the light of a general development of the roles and functions of management, which have become ever more comprehensive and complex. Until around 1950, managers were concerned primarily with *converting* potential profit into real profit. Management activities followed the functional structure of the productive process—for example, production management, sales management—and the task of general management was that of coordinating and integrating the functional activities. Since 1950 it has been increasingly important for management to explore and identify areas of new opportunities and to stimulate the development of new products and services to these areas. In other words, management is becoming more concerned with *creating* the profit potential for the company.

In the 1970s, societal forces and pressures necessitated the increased participation of management in the company's noncommercial environment, that is, with legislative, judicial, and regulatory authorities as well as with social groups, such as labor unions, consumers, and ecologists. As a result, a new classification of managerial activities has emerged which includes[8] (1) competitive management, (2) entrepreneurial management, and (3) societal or political management. A fourth type of management activity—administrative management—concerns itself with providing the skills and the systems required by each of the other three types.

The first three management activities place different demands on the skills of the company. In the course of development, priorities have shifted from competitive to entrepreneurial to societal management. This does not mean that competitive management is less important or less critical to the progress or survival of the company than it was years ago; rather, it means that the total management function has been expanded.

From a quality assurance point of view, it is interesting to note that the development of the management function can be seen as a parallel to the changing concept of quality as proposed earlier in this chapter and to a consequently expanded content of quality assurance. In that section it was pointed out that the practice of using the quality concept mainly to support the economic objectives of a company has generated trends that have gradually forced industrial organizations to incorporate the interests of various groups in society. A quality assurance system based on an enlarged concept of quality seems to contain the elements of a basic philosophy suitable as a guide to a balanced total management function.

Looking at the management roles and relating them to the activities of quality assurance, it becomes evident that competitive management should keep a careful watch on quality costs in order to benefit from the savings obtainable through balancing the costs of quality assurance and the costs of failure. It also becomes evident that entrepreneurial management should be aware of the increased quality consciousness of consumers, of the effect of good quality on market share, and so on, and should, to a larger extent, include quality as a prime factor in endeavors to explore new opportunities. Furthermore, it becomes evident that a quality concept that includes the interests of a number of social groups in society and quality assurance that is based on such a concept will be effective tools for the societal and political management function. Finally, it seems important that managers in the field of administrative management should involve themselves actively in systems auditing to ensure that the quality assurance systems, among others, are constantly adjusted to changing environments.

Thus quality assurance constitutes an essential ingredient in a firm's corporate strategy. It has been widely said that "quality is everybody's job"; conversely, it has been widely demonstrated that "everybody's job is nobody's job." Therefore, since a part of everybody's job influences product quality to varying degrees, someone must coordinate and pull it all together—the general manager.

REFERENCES

1. EUROPEAN ORGANIZATION FOR QUALITY CONTROL, *Glossary of Terms Used in Quality Control*, 4th ed., July 1976, p. 16.
2. J. M. JURAN, Ed., *Quality Control Handbook*, 3rd ed., McGraw-Hill, New York, 1974.
3. EOQC, *Glossary*, p. 26.
4. *NATO Quality Control Systems Requirements for Industry*, AQAP-1, The Military Agency for Standardization, December 1972.
5. R. M. JACOBS, "Liability Responsibility and a Method to Reduce Risk," *Proceedings of the 20th European Organization for Quality Control (EOQC) Conference on Quality Control*, Vol. A, p. 98.
6. COUNCIL OF EUROPE, *European Convention on Product Liability in Regard to Personal Injury and Death*, Pub. No. 91, Strassbourg, January 1977.
7. O. HARTZ, "Quality Control Systems: Elements of a Theory," in C. H. GUDNASON and E. N. CORLETT, Eds., *Development of Production Systems*, Taylor & Francis, London, 1974.
8. H. ANSOFF, "The State of Practice in Planning Systems," *Sloan Management Review*, Winter 1977, pp. 1–24.

BIBLIOGRAPHY

FEIGENBAUM, A. V., *Total Quality Control*, McGraw-Hill, New York, 1961. A pioneering book on principles, practices, and technologies included in the concept of total quality control.

GROOCOCH, J. M. *The Cost of Quality*, Pitman, London, 1971. The book describes how practical quality systems have developed and how quality costs are used to measure achievement.

Guide for Reducing Quality Costs, American Society for Quality Control (ASQC), Milwaukee, WI, 1977.

HARRIS, D. H., and F. B. CHANEY, *Human Factors in Quality Assurance*, Wiley, New York, 1969. The book provides the reader with an understanding of human behavior in industry—and with special reference to quality assurance objectives.

HAYES, G. E., *Quality Assurance, Management and Technology*, Charger Productions, Inc., Capistrano Beach, CA, 1975. The book is a textbook, describing the role of quality assurance in the manufacturing process.

JURAN, J. M. and F. M. GRYNA, *Quality Planning and Analysis*, McGraw-Hill, New York, 1970. The book—with the subtitle *From Product Development through Usage*—offers an excellent exposure to the entire spectrum of the quality function.

Job Reform in Sweden, Swedish Employers' Confederation, Stockholm, 1975.

KONDO, Y., "The Roles of Manager in QC Circle Movement," *Statistical Applied Research*, Japanese Union of Scientists and Engineers, Vol. 23, No. 2, June 1976.

NIXON, F., *Managing to Achieve Quality and Reliability*, McGraw-Hill, New York, 1971. A book for management, describing the interrelationship and managerial aspects of activities necessary for attaining quality and reliability.

Quality Costs–What and How, ASQC, Milwaukee, WI, 1971.

CHAPTER 8.2

Measurement Assurance

KARL F. SPEITEL

Eastman Kodak Company

8.2.1 INTRODUCTION

The Importance and Value of Useful Measurement

Measurement is recognized as one of the fundamental elements of any quality control system. The decisions made during product development, process development, and later in product and process control depend largely upon the quality of the data collected for these purposes. The ability to make sound decisions in each of these important areas is therefore directly related to (1) the availability of *adequate* measurement processes, (2) the proper selection of the *right measurement process* for each job, and (3) the *correct operation* of each measurement process under controlled conditions. Experience has proved that these essential requirements for good measurement are not achieved spontaneously. On the contrary, they are usually the result of carefully planned measurement assurance programs supported by a sympathetic, knowledgeable management. The importance and value of *useful* measurement cannot be overemphasized. Experienced quality engineers are quick to point out that a high percentage of industry's "quality problems" are solved by identifying and correcting the real problems–inaccurate data produced by inadequate measurement processes.

Walter A. Shewhart[1] referred to the significance of measurement to the control function by saying,

> *In any program of control we must start with observed data; yet data may be either good, bad, or indifferent. Of what value is the theory of control if the observed data going into that theory are bad? This is the question raised again and again by the practical man.*

These words leave no doubt concerning Dr. Shewhart's philosophy regarding the importance of good data produced by *adequate measurement processes.*

The main objectives of this chapter are to introduce many of the basic concepts of measurement and to explain how this knowledge is used to evaluate and improve the adequecy of one's measurements. It emphasizes the importance of useful measurements, presents methods to detect ineffective measurement processes, and provides proven techniques to determine the capability of measurement processes and the quality of the measurements they produce. Although no attempt is made to describe the equipment and procedures required to make specific types of measurements, several excellent references are presented to aid the reader seeking this type of information.

Indicators of Measurement Weakness

Experience has revealed a number of situations that frequently indicate measurement weakness. Examples of these indicators are as follows:

1. Frequent occurrences of "quality problems" that are attributed to unknown origins. This situation often indicates that measurement capability is not adequate.
2. Pronounced dissatisfaction by the supervisory or operating personnel with their lack of control of the production processes for which they are responsible. This state of affairs is also frequently caused by inadequate measurement capability.
3. Vague or negative responses are given to questions regarding measurement capability or the measurement specification relationship.

8.2.2 BASIC CONCEPTS AND TERMINOLOGY

As is true of any science, metrology (the science of measurement) has certain basic terms and concepts that must be defined and understood by those who wish to make use of it. These terms and concepts actually form a foundation upon which rests the succeeding discussion of measurement assurance.

Measurement as a Process

Measurement is essentially a *production process*, the product being numbers. A characteristic of a measurement process is variability, that is, that repeated measurements of the same item result in a series of nonidentical numbers. Although it is only natural to attribute a portion of this variability to the irregularity of the item being measured, this chapter is more concerned with the sources of variability that are inherent in the measurement process itself. These sources of variability become more apparent as one considers the principal elements of a measurement process, as shown in Exhibit 8.2.1.

In the exhibit the item to be measured enters the measurement process by way of an operator, who follows a specific procedure, using some sort of measurement equipment to compare the characteristic of the item being measured with a reference, to produce the measurement. This process is operating in some type of environment, which often can have a significant effect on the quality of the measurement produced. In fact, the quality of the measurement will be affected by each of the elements defining the measurement process: the knowledge and skill of the operator, the completeness and clarity of the operating procedures, and the care and maintenance of the reference and equipment. Each of these elements must be present in the proper amount in order to achieve satisfactory results. Treating measurement as a production process is valuable because it not only identifies several potential sources of variability, but also suggests two possible approaches to measurement assurance, that is, product control and process control. Before considering these concepts, however, let us define several other concepts and terms that are fundamental to all measurement discussions.

True Value

True value refers to the theoretically correct value of the characteristic being measured. Although this concept is expressed rather easily, it is not as easily applied in the real world because of its dependence upon the definition of the measured characteristic. Consider, for example, how differently the "true length" of a gauge block might be defined by a parts inspector using it to check the accuracy of his or her micrometer as compared with the definition of this same characteristic supplied by a laboratory technician using it as a master to calibrate other gauge blocks (see Eisenhart[2]).

Exhibit 8.2.1 Measurement as a Production Process Concept

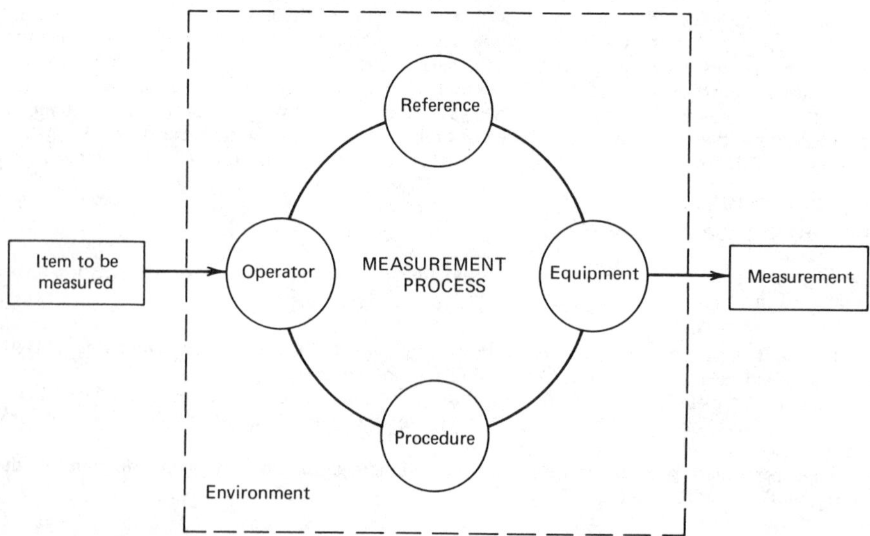

Precision

Precision of a measurement process is the degree of agreement among individual measurements of the same sample.

Relative to a test method, precision is the degree of mutual agreement among individual test results obtained under prescribed similar conditions when the number of individual observations in a single test result is specified in the method of test.[3]

The precision of a measurement process is ordinarily summarized by the SD of the process, which quantifies the characteristic disagreement of repeated measurements of a single quantity by the process concerned. It thus serves to indicate the extent to which a particular measurement is likely to differ from other values that might have been produced by the same measurement process even when the same sample is measured.

The term "repeatability" is commonly used to express precision estimates for a limited set of specifically defined conditions, such as those encountered within the same laboratory during a relatively short period. For example, one may refer to a one-day-operator-instrument repeatability. When a broader set of specifically defined conditions is involved, such as those involving several operators in different laboratories over a longer period, the term "reproducibility" is used to express the precision.

Bias

Bias, or systematic error, refers to any difference between the average of repeated measurements of an accepted reference and the true value of the reference being measured. The bias or systematic error associated with a measurement process may be composed of either constant or variable errors. Constant errors affect *all* measurements produced by the measurement process by the same amount and are generally associated with setup procedures, that is, inaccurate standards, faulty alignment, incorrect conversion factors, and so on. For example, a dial indicator that has been "zeroed out" on the minus side, that is, adjusted with its indicator pointer to the left of the zero line, will consistently produce negatively biased measurements until this setup error is corrected. Variable errors, on the other hand, are errors that display a change of magnitude within the usable range of the measurement process. These errors are frequently found to be related to the construction of the measurement equipment and may be reduced or eliminated completely through repair or reconditioning. An example of this type of error is a steel rule whose graduations were manufactured uniformly, but slightly closer together than specified. This defective rule would then produce measurements with variable biases, that is, the magnitude of error would be proportional to the length of the object measured.

Accuracy

Accuracy refers to the agreement of individual measurements with an accepted reference value.

Relative to a test method, accuracy is the degree of agreement of individual test results with an accepted reference value when the number of individual observations in a single test result is specified by the test method.[3]

It is especially important that we know and understand the distinction between the meanings of the terms "accuracy," "bias," and "precision" as applied to measurement processes. Stated in simplest form, accuracy deals with *closeness to the truth*, bias with the *predictable difference from the truth*, and precision with *closeness of replicate measurements.* The relationship among the three terms is illustrated in Exhibit 8.2.2. The precision of repeated measurements of the same item

Exhibit 8.2.2 The Accuracy Distribution, in Which Each Individual Observation Is Affected by the Precision and Bias of the Measurement Process

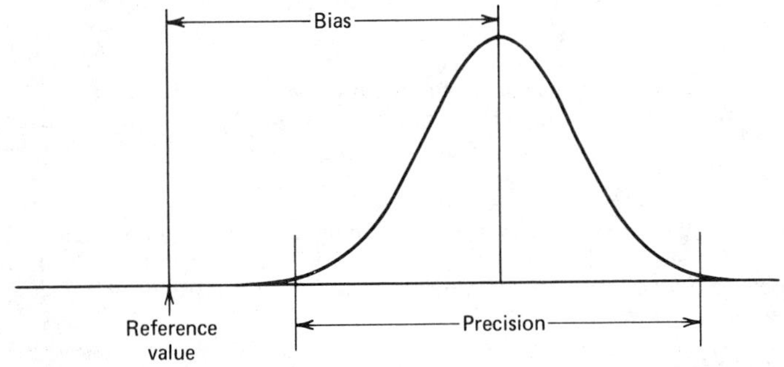

under similar conditions is depicted by the typical bell-shaped curve or distribution. The offset of the long-term average of this distribution from an accepted reference value indicates the amount of bias associated with the measurement process. The combined effect of both bias and precision represents the meaning of the term "accuracy." To characterize the accuracy of a measurement process properly, it is thus necessary to indicate (1) its precision, (2) any bias or systematic error that may be present, and (3) the form of the distribution of the individual measurements about the process average.

Traceability

Several definitions of this important concept currently are in use. The older, more traditional definition expresses traceability as "the ability to demonstrate conclusively that a particular instrument or artifact standard has either been calibrated by the National Bureau of Standards at accepted intervals, or has been calibrated against another standard in a chain or echelon of calibrations, ultimately leading to a calibration performed by NBS."

A newer definition, which stresses measurement traceability as opposed to standards traceability, states

> *Traceability to designated standards (national, international, or well-characterized reference standards based upon fundamental constants of nature) is an attribute of some measurements. Measurements have traceability to the designated standards if and only if scientifically rigorous evidence is produced on a continuing basis to show that the measurement process is producing measurement results (data) for which the total measurement uncertainty relative to national or other designated standards is quantified.*

These definitions are thoroughly discussed in Belanger.[4]

Calibration

Calibration is the comparison of a measurement system or device of unverified accuracy to a measurement system or device of known or greater accuracy in order to detect or correct any variation from required performance specifications of the unverified measurement system or device.

Discrimination

Discrimination refers to the fineness of the scale divisions of an instrument, that is, the smallest division of the scale that can be read reliably.

Sensitivity

Sensitivity is defined as the minimum input that will result in a discernible change in output.[5] This characteristic of measuring equipment is often referred to as an "instrument's threshold of response."

Resolution

Resolution is the ratio of the width of one scale division (one output unit) to the width of the hand (the readout element).[5] This is the characteristic of a dial indicator, for example, that allows one to clearly separate the divisions on its scale.

Stability

Stability is a measure of the dependability of a measurement process. It is best characterized by the absence of drift or other changes in the process that would be readily detected with control charts.

Maintainability

Maintainability is a characteristic of measurement equipment which indicates the probability that, under stated conditions of usage and maintenance, it will retain its capability to perform its required functions. Maintainability should be considered from the aspect of its effect on production as well as its impact on operating costs.

8.2.3 MEASUREMENT COSTS

Cost Considerations

There are few functions in the manufacturing environment that are more closely associated with costs and productivity than measurement. Measurements may be extremely expensive for several reasons. First, they may be inadequate because of "undermeasurement"; that is, wrong decisions are being made as a result of measurement error. Second, they may be superfluous because of "overmeasurement," that is, using measurement processes that are much more sophisticated than actually required.

A third reason for high measurement costs might stem from measuring the wrong characteristic. This problem is illustrated with an example involving a tapered valve for controlling fluid flow in a coating operation. To determine the functional quality of this valve, one might approach the problem indirectly, that is, by making many difficult measurements to ensure that the mating parts meet specifications. Alternatively, one could check the quality directly, that is, by measuring the flow rate at several settings *in situ*. This latter approach would undoubtedly be more efficient and effective, from the standpoint of both the cost and the reliability of the results.

The cost of producing any given measurement can usually be determined; however, the total cost of operating a measurement process is often obscured. This total cost of measurement has the following components:

Initial Costs

1. Capital cost of devices and equipment.
2. Capital cost of standards and reference materials.
3. Cost of setting up the process, including training the operators, developing and writing operating procedures, and debugging the entire process.

Operational Costs

1. Cost of producing the measurements, including recording and reporting the data.
2. Equipment maintenance costs.
3. Cost of verifying that the system is operating properly, including revalidating standards and references and retraining operators.

One must consider all aspects of measurement cost, both initial and operational, when making a choice between two or more *adequate* measurement processes. Experience has shown that the initial capital cost is often only a small fraction of the overall cost of producing a given type of measurement.

8.2.4 MEASUREMENT ASSURANCE

Measurement assurance is a program to establish, evaluate, and control the quality of measurement. Although special Measurement Assurance Programs (MAPs)* are being established by the National Bureau of Standards (NBS) for specific types of measurements, the term is used here in a much broader sense. That is, it refers to a general program designed to improve the quality of any measurement process to which it is applied. This program for measurement assurance is best described by discussing the following six subjects:

1. Identifying the measurement process.
2. Determining the precision of a measurement process.
3. The search for bias or systematic error.
4. Checking the stability of the measurement process.
5. The precision/tolerance (P/T) ratio.
6. The effect of the P/T ratio on decision making.

Identifying the Measurement Process

The need to control the manufacturing process generates the need to measure. However, this does not dictate the method of measurement that should be used. It therefore becomes the responsibility of the development engineers and quality control personnel to select or develop a satisfactory

*In 1980 the NBS offered nine MAP services in the following areas: mass, DC voltage, resistance, capacitance, DC voltage ratio, electric energy (watt-hour meters), laser power and energy, temperature, and gauge blocks.

method of measurement. This is an extremely important step in many measurement situations. Although it would be impossible to discuss the countless methods that have been developed to make both general and special measurements, the author would be remiss if the reader were not provided with some guidance on this important topic. Several excellent references[6–8] have been selected to supply this valuable information, and one reference[9] is a bibliography containing 3600 subject-classified methods pertaining to dimensional measurement.

Once the measurement method has been selected, one can then proceed to define the elements of the measurement process he or she wishes to evaluate. Questions such as the following must be considered:

1. What if different instruments of the same type and from the same manufacturer were used?
2. Does the type or brand of instrument affect the measurement process results?
3. Is the process sensitive to environmental conditions such as temperature, moisture, dust, or vibration?
4. What effect does operator experience have on the process?
5. How sensitive is the process to changes in the procedure?
6. How does sample preparation (cleaning, mixing, deburring, etc.) affect the measurement?

These are examples of the types of questions that should be answered when one wishes to ascertain the adequacy of the measurement process that has been selected.

Determining the Precision of a Measurement Process

One of the prime performance characteristics of any measurement process is its precision, that is, the uniformity of its *product*. Succeeding subsections explore the importance of the precision tolerance relationship and compare the impact of this relationship upon the decision making ability of measurement. In making these comparisons, one needs a quantitative expression rather than a qualitative, descriptive term for precision, such as "fairly good" or "very good." The best way to find out how precisely a given measurement process can perform is to allow it to repeat a given measurement a sufficient number of times and to examine the data to estimate its precision.

For example, in column 1 of Exhibit 8.2.3, 20 values are listed in the order in which they were

Exhibit 8.2.3 Measurement Data Used to Evaluate Precision

Value (Arbitrary Units)	Deviation (d) from Average Value[a]	d^2
58.34	0.02	0.0004
58.37	0.01	0.0001
58.36	0.00	0.0000
58.38	0.02	0.0004
58.34	0.02	0.0004
58.36	0.00	0.0000
58.39	0.03	0.0009
58.33	0.03	0.0009
58.37	0.01	0.0001
58.36	0.00	0.0000
58.32	0.04	0.0016
58.35	0.01	0.0001
58.37	0.01	0.0001
58.35	0.01	0.0001
58.34	0.02	0.0004
58.36	0.00	0.0000
58.37	0.01	0.0001
58.35	0.01	0.0001
58.36	0.00	0.0000
58.38	0.02	0.0004
Average value = 58.36	Mean deviation = 0.014	Sum = 0.0061

Highest value	58.39
Lowest value	58.32
Maximum variation (range)	0.07
Maximum deviation from average value	0.04
SD	0.018

[a] All deviations refer to the *absolute* values of the deviations from the average value.

obtained during the course of an investigation of a proposed measurement process. The easiest way to indicate the precision or "closeness together" of the measurements is to state the highest and lowest values, 58.39 and 58.32, respectively; the difference between them, 0.07; or perhaps the maximum deviation from the average value, 0.04.* None of these is a very good expression of precision, because the highest value may be unusually high, or the lowest abnormally low, for some extraordinary reason. This would give a false impression of the series as a whole.

There is always a temptation to discard a "wild" value from a series, even though there may be no apparent reason why it should have turned out to be so different from the others. The mean deviation from the mean provides a useful guide to determining whether or not rejection of a given value is justifiable. If the deviation of the doubtful value from the average computed without it is more than five times the mean deviation, the doubtful value should be rejected because there is little chance that such a value would occur under normal conditions; that is, some unnoticed abnormal condition probably caused the wide aberration. Applying this rule to the data in Exhibit 8.2.3, the maximum deviation, 0.04, is less than five times the mean deviation ($5 \times 0.014 = 0.070$); consequently, all of the values must be retained. Statisticians prefer other, more sophisticated criteria for the rejection or retention of values, but this simple rule serves this purpose reasonably well.

The most important measure of dispersion is the SD, which is calculated as follows:

$$\text{SD} = s = \sqrt{\frac{\Sigma d^2}{n - 1}}$$

In this formula, Σd^2 is the sum of the squares of the individual deviations from the mean, listed in column 3 of the exhibit, and n is the number of individual values in the series. Substituting the values of these quantities in the formula gives the following result:

$$s = \sqrt{\frac{0.0061}{19}} = \sqrt{0.000321} = 0.018$$

The SD always turns out to be a little greater than the mean deviation, partly because the divisor is $(n - 1)$ instead of n. This is a sort of built-in margin of safety to compensate for using a relatively small number of individual values.

There are two principal ways in which the SD can be useful: (1) as related to the precision of the measurement process and (2) to define the confidence one may place in a single measurement. It can be shown that, when the distribution of values is normal and the number of individual values is large, approximately two thirds of the values are expected to be within the range of ±1 SD from the average value; approximately 95% of the values will be within ±2 SD; and 99.7% will be within ±3 SD. These statements are based upon the areas of a normal curve, as shown in Exhibit 8.2.4.

If the measurement process used to obtain the 20 values listed in Exhibit 8.2.3 were applied to another unknown sample from the same process, and if one obtained 58.30 as a result of a single measurement, one could be reasonably certain (a 2 to 1 probability) that, if the measurement were repeated many times, the average value would lie between 58.28 and 58.32 (58.30 ±1 SD). Or one could state that the odds are 19 to 1 that the single result is not more than 0.036 units from the

Exhibit 8.2.4 Areas of the Normal Curve

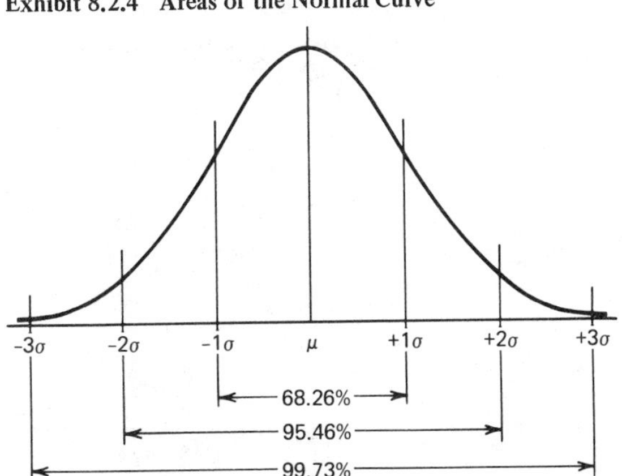

average of many measurements of the sample. In fact, one would be willing to bet 333 to 1 that the single measurement is within 0.054 units from the average. Statements such as these are based upon confidence intervals for the population mean (see Natrella[10]).

Furthermore, one's confidence increases with the square root of the number of times the same sample is measured; that is, one is twice as confident of the average of four measurements as one is of a single measurement. This statement is based upon the following equation:

$$\hat{\sigma}_{\overline{M}} = \frac{\hat{\sigma}_M}{\sqrt{n}}$$

where $\hat{\sigma}_{\overline{M}}$ = estimated SD of averages of n measurements
$\hat{\sigma}_M$ = estimated SD of individual measurements
n = number of measurements averaged

This relationship and its effect upon the variability of averages as compared to individual measurements is illustrated in Exhibit 8.2.5.

Thus, by using some of the simpler statistical concepts, one obtains both an estimate of the precision of the measurement process and a measure of the confidence one may place in the results obtained through that process.

The Search for Bias or Systematic Error

Bias, or systematic error, is defined as any difference between the average of repeated measurements of an accepted reference and the true value of the reference being measured. Although this definition clearly states the meaning of bias and even suggests how to evaluate it, this type of error is not easy to detect. As one might imagine, the source of the difficulty lies in locating reliable references whose true value is known. The task of establishing and calibrating references is normally handled by a standards laboratory. The importance of this work generally requires that it be accomplished by highly trained specialists using extremely fine equipment under carefully controlled environmental conditions. The standards laboratory unquestionably plays a major role in any measurement assurance program by establishing and maintaining reference standards as well as providing the vitally important link with national and international standards.

Once accurate references are available, the search for bias in a measurement process becomes a matter of determining how closely the average of repeated measurements of the reference agrees with its established true value. Here we solicit the aid of statistical tests of significance to control this decision making process. (See Mandel[11] and Natrella[10] for discussions of the t and F distribu-

Exhibit 8.2.5 Family of Normal Curves, Showing the Reduction in Variability of Averages of n Measurements as Compared to Single Measurements

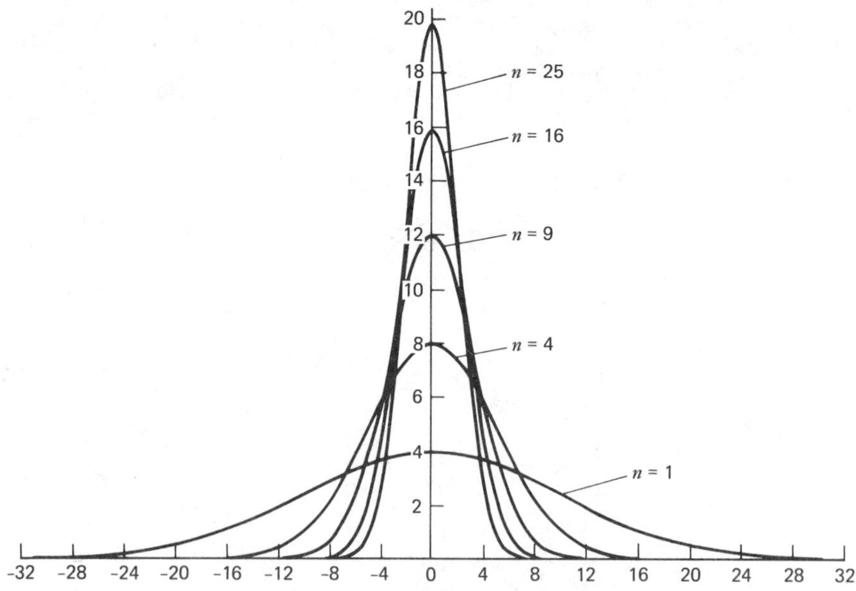

tions in metrology.) These tests must be used with caution, since they may indicate evidence of "statistical significance" that a real bias exists. However, this does not mean that this bias is of *practical* importance.

Finally, what does one do when no established reference is available? In this situation Mandel[12] states that "the evaluation of accuracy in the absence of an exactly known reference value is reduced to an educated guess." He then offers two methods for making such a guess. One method suggests estimating "upper bounds" for the various systematic errors and then combining these estimates into an upper bound for the total bias. The second method utilizes a comparison of results obtained by different and independent measurement processes.

Checking the Stability of the Measurement Process

The techniques for determining precision and bias presented in the preceding subsections are very useful for defining and debugging measurement processes; however, neither of these techniques addresses the question of the stability of a measurement process. This important characteristic can be determined only by studying its performance over time.

The standard statistical technique designed specifically for this purpose is the Shewhart control chart. The fundamental purpose of this technique is to ascertain whether there is a statistically stable process in operation. The technique performs this function by informing the user when some assignable cause has entered the process and shifted either the process level or the variability. In the absence of these assignable causes, the process producing the data is said to be in a "state of statistical control," which simply means that the process is operating in a predictable manner within set limits based on the inherent random variability. This inherent random variability, which is often referred to as "noise," is merely another name for the short-term measurement precision that was discussed previously. In fact, the control chart limits can be established by using either the previously determined variability estimate or an estimate based upon newly acquired data. The latter approach is used here, since it introduces the reader to a method of achieving an estimate of the precision as well as to a method of checking the stability of the measurement process.

The Shewhart control chart is based upon the premise that the data being generated are coming from a "constant cause system," which is characterized by a single distribution. If this premise is true, the variables, such as averages, ranges, or SDs of subgroups of samples selected from this distribution, will behave in a predictable manner. In fact, one can establish an estimate of the average and SD of each of these variables and thereby determine statistical limits to judge the continued existence of this state of affairs. By plotting these variables on a time scale, a graphic representation of the situation is created. The addition of the central line, representing the average, and the statistically determined limit lines transforms each graph into a control chart for each variable. These charts commonly use "three-sigma limits," which should include 99.7% of the plotted values when the process producing the data is normally distributed and "in control." Under these circumstances, a point falling outside these three-sigma limits would indicate that either a very unlikely event had occurred, that is, 1 chance in 333, or an assignable cause had altered the "constant chance cause system" and in fact the measurement process had shifted. Also note that the sensitivity of the control chart's alarm system is completely defined by the limits selected by the user.

The following subsection describes how to establish a control chart that will check the stability of a measurement process and simultaneously provide an estimate of the process variability.

Establishing a Control Chart

The data for this phase of measurement capability are collected and analyzed as follows:

1. Twenty or more samples are selected and sequentially numbered.
2. Each of these samples is measured three to five times in the designated order over an extended period. (It is extremely important that each of these measurements is performed independently; that is, the entire procedure must be completely repeated for each measurement, including any zeroing out operations, removal and repositioning of the object being measured, etc.)
3. The range of the values obtained for each sample, that is, the largest value minus the smallest value, is then determined and plotted in the order in which the items were measured.
4. The average range ($\overline{R}$) is then calculated using the following equation:

$$\overline{R} = \frac{\Sigma R}{n}$$

where ΣR = sum of R values
n = number of items measured

Exhibit 8.2.6 Factors for $\overline{X}$ and R Control Charts

No. in Sample	A_2	D_3	D_4	d_2
2	1.880	0	3.267	1.128
3	1.023	0	2.574	1.693
4	0.729	0	2.282	2.059
5	0.577	0	2.114	2.326

5. The upper (UCL) and the lower (LCL) control limits (three-sigma) for this range chart are determined using the following equations:

$$UCL = D_4\overline{R}$$

$$LCL = D_3\overline{R}$$

The values for the constants D_3 and D_4 are listed in Exhibit 8.2.6.

6. The control limits are added to the chart, and it is then checked for evidence of out-of-control conditions, that is, one or more points falling outside the control limits or, similarly unlikely, having seven or more consecutive points falling on one side of the average. The occurrence of either of these signs indicates a lack of control and the need to search for the assignable cause. A chart displaying none of these signs provides evidence of a statistically stable measurement process.

7. When the control chart for ranges is "in control," the average range can be used to estimate the variability of the measurement process as follows:

$$\hat{\sigma}_M = \frac{\overline{R}}{d_2}$$

The values for the constant d_2 are listed in Exhibit 8.2.6.

An example of this technique is illustrated in Exhibit 8.2.7. The values plotted on this chart were obtained by measuring the length of 20 film cartridges three times each. As is often the case with this type of measurement, the actual measurements were made with a special comparator gauge, which merely indicates the deviations of the part from the aim or nominal value of the specification. The ranges of the three repeated measurements of each of the 20 measured cartridges are listed in Exhibit 8.2.8.

The average range and three-sigma control limits calculated with these data are:

$$\begin{aligned}\overline{R} &= \Sigma R/n = 0.0105 \text{ in.}/20\\ &= 0.00052 \text{ in. (or } 5.2 \text{ in.} \times 10^{-4})\end{aligned}$$

$$\begin{aligned}LCL &= D_3\overline{R} = 0\ (5.2 \times 10^{-4})\\ &= 0\end{aligned}$$

$$\begin{aligned}UCL &= D_4\overline{R} = 2.574\ (5.2 \times 10^{-4})\\ &= 13.4 \text{ in.} \times 10^{-4}\end{aligned}$$

Upon reviewing the control chart for ranges in Exhibit 8.2.7, one finds no evidence to indicate an out-of-control condition. Therefore one can assume that the measurement process that produced these data is stable and can estimate its variability as follows:

$$\begin{aligned}\hat{\sigma}_M &= \overline{R}/d_2 = (5.2 \times 10^{-4})/1.693\\ &= 3.1 \text{ in.} \times 10^{-4}\end{aligned}$$

Once an "in control" range chart has been established for a base period, its limits can be used to monitor the variability of the measurement process until changes are made or noted in the process.

Finally, note that, if a reference or standard is substituted for the samples used to establish the stability of the measurement process, it is possible to create a second control chart for the averages ($\overline{X}$s)* of each subgroup of n measurements. This chart of the subgroup averages will provide a method of checking for a shift in the level of the measurement process. Therefore, by establishing both $\overline{X}$ and R control charts through the use of periodic repeat measurements of a reference, one

*The values for the constant A_2, which are used to calculate the limits for $\overline{X}$ control charts, are also listed in Exhibit 8.2.6.

Exhibit 8.2.7 A Control Chart for Ranges

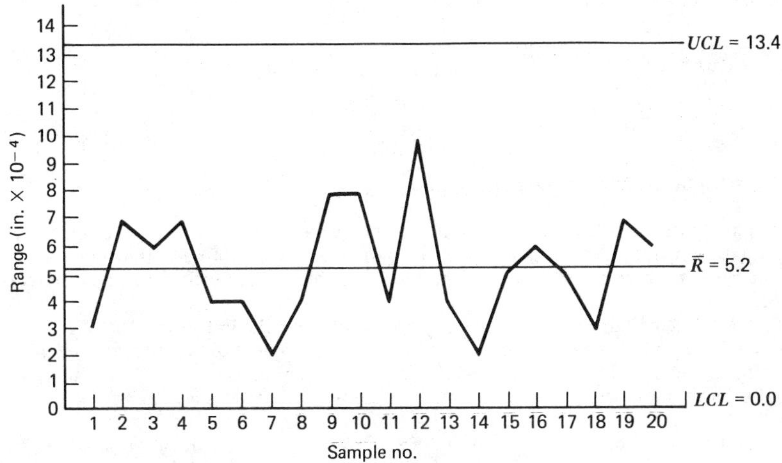

can monitor both the level and the variability of a measurement process. (See other sources[1,10,13,14] for additional information concerning the use of control charts.)

The Precision/Tolerance Ratio

A major concern in metrology involves the relationship between measurement precision (repeatability) and product tolerances. This precision/tolerance relationship must be considered for each new measurement application because it can adversely affect both the value and the cost of measurement. If ignored, one might, on the one hand, be unaware of the excessively high risks one is taking in some measurement applications or, on the other hand, waste valuable time, effort, and money to attain unnecessarily high levels of perfection. This general problem is the measurement requirement referred to earlier as "selecting the right measurement process for each job." To investigate this problem, one can study the ratio of the measurement precision estimate to the total

Exhibit 8.2.8 Ranges of Film Cartridge Length Data Used to Control the Measurement Process (Three Replicates per Sample)

Sample No.	R (in.)
1	0.0003
2	0.0007
3	0.0006
4	0.0007
5	0.0004
6	0.0004
7	0.0002
8	0.0004
9	0.0008
10	0.0008
11	0.0004
12	0.0010
13	0.0004
14	0.0002
15	0.0005
16	0.0006
17	0.0005
18	0.0003
19	0.0007
20	0.0006

tolerance (the difference between the minimum and maximum limits) of the characteristic being measured. This ratio may be expressed as follows:

$$\text{P/T ratio} = \frac{\pm 3\hat{\sigma}_M}{(\text{maximum} - \text{minimum})}$$

$$= \frac{6\hat{\sigma}_M}{\text{total tolerance}}$$

where $\hat{\sigma}_M$ = SD of the measurement distribution

Note: For the purpose of this discussion, it is assumed that the measurement errors are independent, normally distributed, and independent of part size. Furthermore, it is assumed that sources of bias or systematic error will be eliminated or that their effect will be removed by correction procedures.

The literature[6] generally suggests using the "tenth rule" for the P/T ratio, that is, $\pm 3\hat{\sigma}_M$ equal to one tenth of the total tolerance. However, you may find some authors who extend this number to one fifth or even more. The "tenth rule" has withstood the test of time, but this type of test is generally quite one-sided. Perhaps this general rule is too conservative! In any event, we should explore this area to better understand the consequences of these decisions.

To begin the discussion, consider Exhibit 8.2.9, which shows a pair of vertical lines to depict tolerance limits and a normal curve, which represents the effect of the measurement process. The arrow at the bottom indicates the actual "true size" of the part that is being measured. Note from the position of the curve with respect to the tolerance limits in this example that measurement capability will not be a factor in determining conformance to specifications.

Exhibit 8.2.10 displays two additional situations to indicate the effect of the measurement process when measuring two more parts, both smaller than the first. The sample identified as #2 is closer to the lower tolerance limit. In fact, there is now a clearly discernible probability that this sample might be rejected, that is, the shaded area under the curve beyond the specification limit. Sample #3 is so far outside specification that, once again, measurement capability is no longer a factor in determining conformance; that is, it has little chance of being accepted.

One can determine the exact nature of the measurement effect by constructing the operating characteristic (OC) curve for the measurement process. This curve for the process discussed here is illustrated in Exhibit 8.2.11.

To make this plot more generally useful, one can plot probability of acceptance versus percentage of total tolerance from aim. These changes, which are shown in Exhibit 8.2.12, convert our measurement curve for one set of conditions into a universal measurement curve for a particular P/T ratio. The ratio in Exhibit 8.2.12 is equal to 1:5.

With this type of plot one can now determine the probability of acceptance associated with any size part. One can then characterize the P/T ratio effect for this measurement process by stating

Exhibit 8.2.9 Measurement of a Single Part by a Measurement Process Without Bias

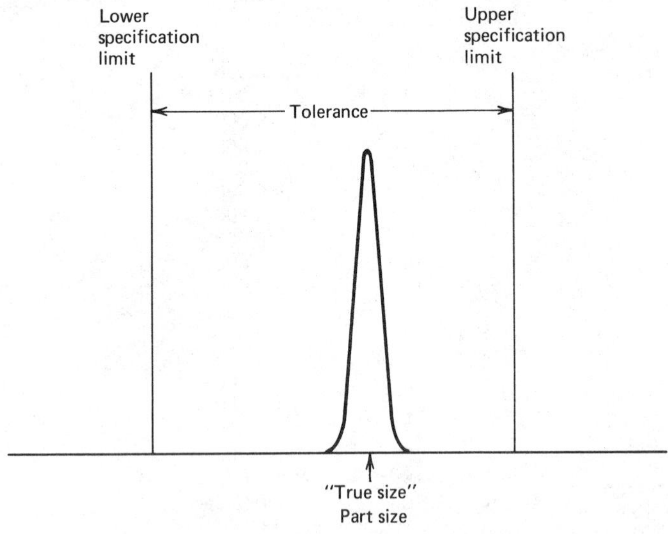

Exhibit 8.2.10 Measurement of Three Parts by the Same Measurement Process

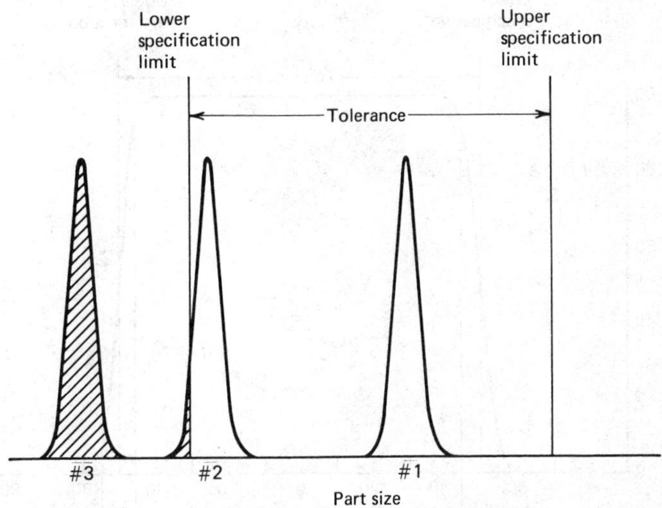

the size of its "critical region" (CR) associated with specified α and β risks. Each of these regions is defined when values are selected for α, the probability of rejecting a good part, and β, the probability of accepting a nonconforming part. These zones for $\alpha = \beta = .05$ are indicated and labeled "CR" in Exhibit 8.2.12.

If we wish to compare the P/T ratio effect of several measurement processes, we can calculate their operating characteristic curves and superimpose them on the same set of axes. An example of this procedure is seen in Exhibit 8.2.13, which compares the curves for processes with P/T ratios of 1:10, 3:10, and 1:2.

The Effect of the P/T Ratio on Decision Making

Although these operating characteristic curves are extremely useful in determining the relationship between measurement precision and specifications and in comparing different P/T ratios, the real question concerning the effect of measurement capability on the ability to make correct decisions still remains unanswered. This question can be answered only by determining how much product submitted to the measurement process will actually fall into the critical zones defined by a given P/T ratio and what percentage of the decisions will be correct.

Exhibit 8.2.11 An OC Curve for a Measurement Process Without Bias

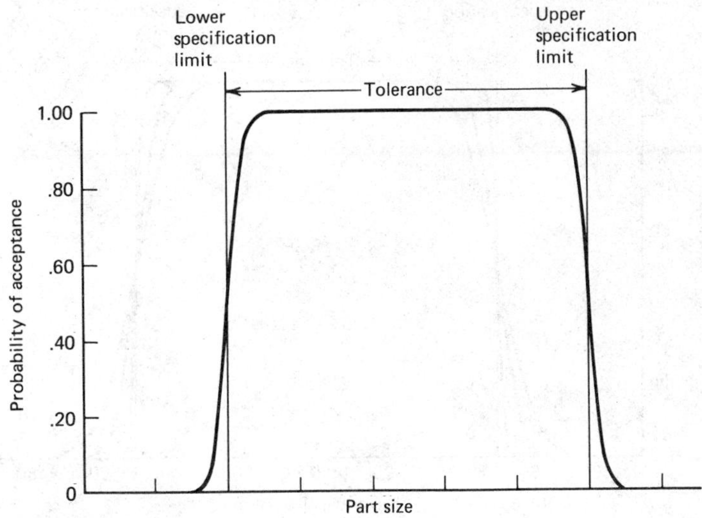

Exhibit 8.2.12 A Measurement OC Curve for a P/T Ratio of 1:5.

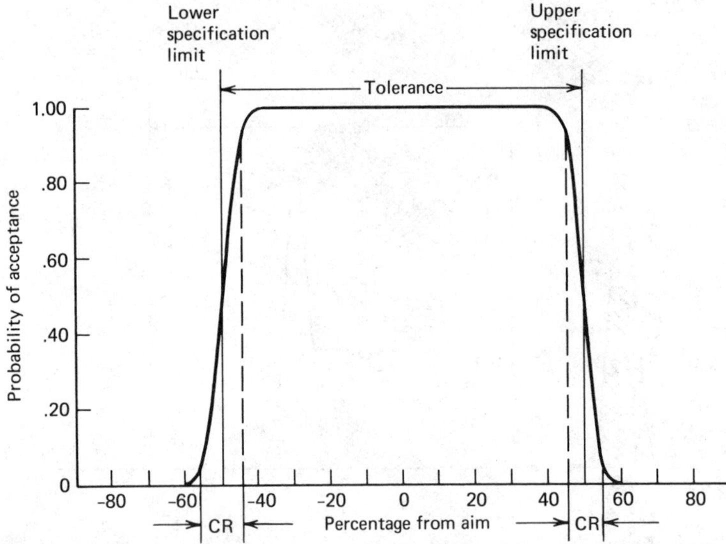

Exhibit 8.2.14 shows the joint effect of both the manufacturing and measurement process capabilities on the decision making process. The scale units of the horizontal axis represent equal percentage of tolerance from aim, with 0, +50, and -50 indicating the aim, the upper specification limit, and the lower specification limit, respectively. The values along the vertical axis indicate the percentage of correct decisions. The manufacturing process capability is represented by the bell-shaped curve, with variability ($\pm 3\sigma$) equal to the total tolerance. The three curves at the top of Exhibit 8.2.14 were produced by computer simulation and depict the percentage of correct decisions one can expect when the P/T ratio is 1:10, 3:10, or 1:2, respectively.

For example, when the manufacturing process is positioned on aim, as shown in Exhibit 8.2.14, only a small portion of the product will actually fall near either of the tolerance limits. The percentage of correct decisions is therefore very high, that is, greater than 99% for all three P/T ratios. This clearly indicates that there is little difference in the decision making capability of the three measurement processes when the parts distribution is positioned at the aim and has variability equal to the total tolerance. As the position of this parts distribution shifts away from the aim

Exhibit 8.2.13 Measurement OC Curves for P/T Ratios of 1:10, 3:10, and 1:2.

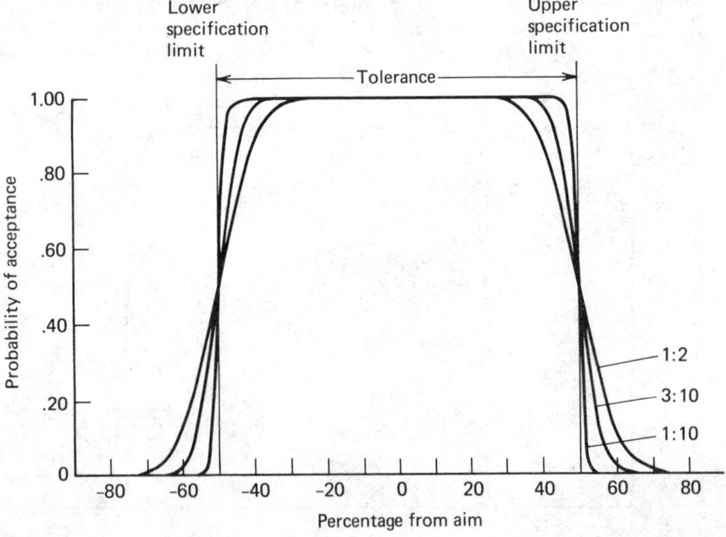

Exhibit 8.2.14 The Effect of the P/T Ratio on Decision Making

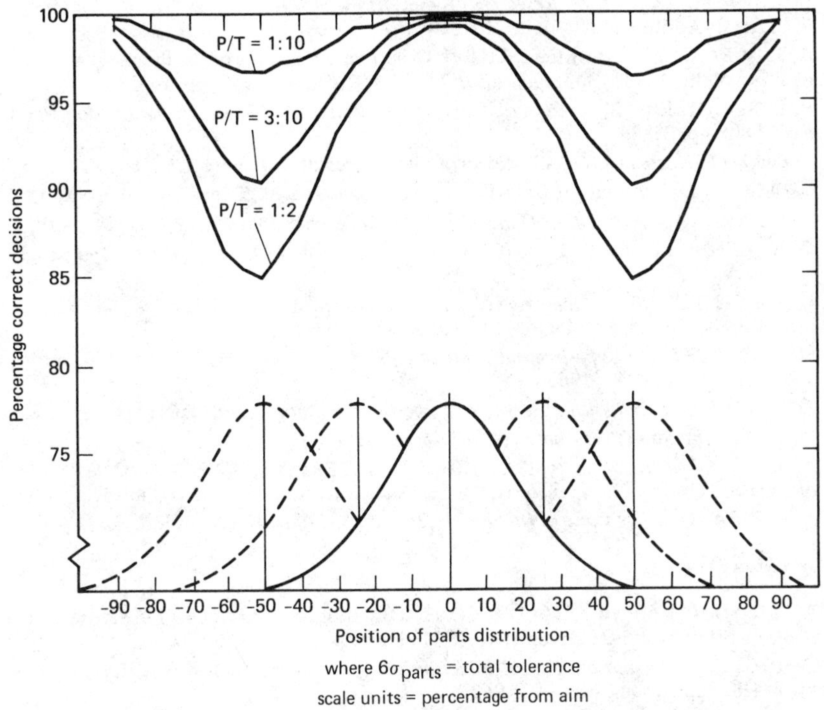

toward either of the specification limits, however, the differences among the selected P/T ratios become more pronounced. As one might expect, these differences reach a maximum when the parts distribution is positioned at either of the specification limits and gradually decreases again as the position of the parts distribution moves beyond these points.

By tying together all of the important factors that contribute to the effectiveness of the decision making function, this chart provides the following information:

1. The actual effect of measurement precision upon the decision making function is undefined until the capability of the manufacturing process that is producing the parts distribution is known.

2. Both the variability and the position of the parts distribution affect the decision making function, emphasizing the importance of setting up and controlling our manufacturing processes as close to aim as possible.

3. Under the conditions of this example, the percentage of correct decisions for all three selected P/T ratios is high, that is, greater than 99%, when the manufacturing process is positioned exactly on aim.

4. When the manufacturing process is positioned at either of the specification limits, however, the percentage of correct decisions for a P/T ratio of 1:2 drops to 85%, whereas the P/T ratio of 1:10 remains relatively high at 96%. This uniformly safe performance by the P/T ratio of 1:10 is undoubtedly one reason why the "tenth rule" has remained unchallenged for many years. By knowing the actual capability of both the measurement and the manufacturing processes, we realize that there are times and circumstances when this "tenth rule" can be ignored without sacrificing the effectiveness of one's decision making. This message is extremely important to achieving our goal of measurement assurance!

Finally, it should be noted that a change in the variability of the parts distribution will change the decision making curves for these P/T ratios. Experience has shown, however, that the conditions used for this example are fairly representative of those encountered in many industrial situations.

REFERENCES

1. W. A. SHEWHART, *Economic Control of Quality of Manufactured Product*, Van Nostrand, New York, 1931, p. 376.

2. C. EISENHART, "Realistic Evaluation of the Precision and Accuracy of Instrument Calibration Systems," *Journal of Research of the NBS-C. Engineering and Instrumentation*, Vol. 67C, No. 2 (April–June 1963), pp. 161–187.
3. A. J. DUNCAN, "Views of the E-11 Task Group on Statements of the Precision and Accuracy of a Test Method," *ASTM Standardization News*, December 1978, pp. 16–18.
4. B. C. BELANGER, "Traceability: An Evolving Concept," *ASTM Standardization News*, January 1980, pp. 22–28.
5. "Selecting Dimensional Gages," *Metalworking Magazine*, November 1960, p. 25.
6. T. BUSCH, *Fundamentals of Dimensional Metrology*, 2nd ed., Delmar, Albany, NY, 1965.
7. F. T. FARAGO, *Handbook of Dimensional Measurements*, Industrial Press, New York, 1968.
8. C. W. KENNEDY and D. E. ANDREWS, *Inspection and Gaging*, 5th ed., Industrial Press, New York, 1977.
9. I. H. FULMER, *Dimensional Metrology*, NBS Miscellaneous Publication 265, U.S. Government Printing Office, Washington, DC, 1966.
10. M. G. NATRELLA, *Experimental Statistics*, NBS Handbook 91, U.S. Government Printing Office, Washington, DC, 1963.
11. J. MANDEL, *The Statistical Analysis of Experimental Data*, Wiley, New York, 1964.
12. MANDEL, *The Statistical Analysis of Experimental Data*, p. 125.
13. AMERICAN SOCIETY FOR TESTING AND MATERIALS, *ASTM Manual on Presentation of Data and Control Chart Analysis*, Special Technical Publication 15-D, Philadelphia, 1976.
14. J. M. JURAN, *Quality Control Handbook*, 3rd ed., McGraw-Hill, New York, 1974.

BIBLIOGRAPHY

AMERICAN SOCIETY FOR QUALITY CONTROL, *Glossary and Tables for Statistical Quality Control*, Statistics Technical Committee, ASQC, Milwaukee, WI, 1973.

BAIRD, D. C., *Experimentation: An Introduction to Measurement Theory and Experiment Design*, Prentice-Hall, Englewood Cliffs, NJ, 1962.

CAMERON, J. M., "Measurement Assurance," NBS Internal Report 77-1240, National Bureau of Standards, Washington, DC, April 1977 (available from the Office of Measurement Services).

EAGLE, A. R., "A Method for Handling Errors in Testing and Measuring," *Industrial Quality Control*, March 1954, pp. 10–14.

EISENHART, C., "Expression of Uncertainties of Final Results," *Science*, Vol. 160, 1968, pp. 1201–1204.

GRUBBS, F. E., "Errors of Measurement, Precision, Accuracy and the Statistical Comparison of Measuring Instruments," *Technometrics*, Vol. 15, February 1973, pp. 53–66.

GRUBBS, F. E., "On Estimating Precision of Measuring Instruments and Product Variability," *Journal of the American Statistical Association*, Vol. 43, 1948, pp. 243–264.

KU, H. H., Ed., *Precision Measurement and Calibration*, NBS Special Publication 300, Vol. 1, U.S. Government Printing Office, Washington, DC, 1969.

MANDEL, J., "Repeatability and Reproducibility," *Journal of Quality Technology*, April 1972, pp. 74–85.

MURPHY, R. B., "On the Meaning of Precision and Accuracy," *Materials Research and Standards*, April 1961, pp. 264–267.

PROSCHAN, F., "Rejection of Outlying Observations," *American Journal of Physics*, Vol. 21, No. 7 (October 1953), pp. 520–525.

SOCIETY OF MANUFACTURING ENGINEERS, *Handbook of Industrial Metrology*, Prentice-Hall, Englewood Cliffs, NJ, 1967.

SPEITEL, K. F., "Making Your Measurement Effort Pay Off," *Transactions of the 36th Annual Quality Control Conference*, Rochester, NY, March 11, 1980, pp. 85–93.

TRAVER, R. W., "Measuring Equipment Repeatability–The Rubber Ruler," *1962 ASQC Convention Transactions*, Milwaukee, WI, pp. 25–32.

WERNIMONT, G., "Design and Interpretation of Interlaboratory Studies of Test Methods," *Analytical Chemistry*, Vol. 23, No. 11 (1951), pp. 1572–1576.

YOUDEN, W. J., "How to Evaluate Accuracy," *Materials Research and Standards*, April 1961, pp. 268–271.

CHAPTER 8.3
Quality Control

PHILIP S. BRUMBAUGH

Quality Assurance, Inc.

SYMBOLS

A_2	Factor for $\bar{x}$-chart trial control limits
α	Producer's risk
β	Consumer's risk
c	Number of nonconformities per inspection unit (control charts); acceptance number (acceptance sampling)
c'	Process average number of nonconformities per inspection unit
$\bar{c}$	Sample average number of nonconformities per inspection unit
d_2	Factor for estimating σ from $\bar{R}$
D_3	Factor for R-chart lower trial control limit
D_4	Factor for R-chart upper trial control limit
k	Number of subgroups (control charts); critical value of standardized $\bar{x}$ (acceptance sampling)
K	Statistical tolerance factor
L	Lower specification limit
μ	Process mean
n	Sample size
N	Lot size
p	Sample fraction nonconforming
p'	Process fraction nonconforming
p''	Specified value of process fraction nonconforming
$\bar{p}$	Mean of a number of values of p
p^*	Standardized p
P_a	Probability of acceptance
R	Sample range
$\bar{R}$	Mean of a number of values of R
s	Sample SD
s^2	Sample variance
s_c	Estimate of standard error of c
s_p	Estimate of standard error of p
$s_{\bar{x}}$	Estimate of standard error of $\bar{x}$
σ	Process SD
σ_c	Standard error of c
σ_p	Standard error of p
$\sigma_{\bar{x}}$	Standard error of $\bar{x}$
Σ	Summation
u	Average number of nonconformities per inspection unit
$\bar{u}$	Mean of a number of values of u
U	Upper specification limit
$\bar{x}$	Sample mean
$\bar{\bar{x}}$	Mean of a number of values of $\bar{x}$
z	Standardized normal variate

8.3.1 SPECIFICATIONS, TOLERANCES, AND ALLOWANCES

Specifications

Product specifications provide procedural instructions for the operator. These instructions should be clear and precise, but where they are not, or where there is conflict with, for example, past shop practice, resolution will have to be obtained at a higher level than that of the operator. Product specifications also provide standards for test and inspection. Again, clarity and precision are essential; where they are lacking, the inspection department will tend to "fill the void" and make decisions in this area, often with predictably unfortunate consequences for product reliability, safety, and fitness for use.

Tolerances

Tolerance is the maximum amount of variation that can be permitted from a nominal or stated specification. In general, close (or numerically small) tolerances must be maintained on parts that are to be assembled with other parts, and the closer the tolerances, the greater the ease of assembly. This is a particularly critical consideration in mass production situations, where interchangeability of parts is required. However, while close tolerances reduce assembly costs, they increase the cost of fabricating the individual parts. Hence a balance of the two types of costs must be achieved, with the objective of minimizing their sum. Expressed differently, it may at times be economically advantageous to incur the cost of some selective fitting of parts in order to realize the fabrication cost savings that result from a relaxation of tolerances.

Mandatory and Advisory Tolerances

Mandatory tolerances are tolerance requirements that must be met because they directly affect important quality characteristics of the end product. Advisory tolerances are provided for manufacturing convenience; they are designed to assist the operator in achieving the mandatory tolerance requirements. An example might be: "Add 10 oz ± 0.1 oz (283.49 g ± 283 g) of compound X to achieve a solution strength of 18% ± 0.5%." The weight tolerances are advisory; the strength tolerances are mandatory. In the absence of clear distinctions between the two types of tolerances, inspection departments will tend to interpret all tolerances as mandatory.

Unilateral and Bilateral Tolerances

Both unilateral and bilateral tolerances are encountered in practice. A unilateral tolerance permits variation in only one direction from a nominal value; a bilateral tolerance permits variation in both directions. For example, so-called 5% resistors carry ratings of X Ω ± 5%; this is a bilateral tolerance. Precision resistors, in contrast, carry ratings of X Ω $^{+1\%}_{-0\%}$; this is a unilateral tolerance. Unilateral tolerances have the advantage of making it relatively simple to maintain a given allowance for mating parts while changing the tolerances. (See the example in the next part of this chapter.)

Allowances

An allowance is defined as the minimum specified clearance between mating parts.

Example 8.3.1. If a piston whose diameter specification is 3.500 in. ± 0.002 in. (88.900 mm ± 0.051 mm) is to be assembled into a sleeve whose diameter specification is 3.505 in. ± 0.002 in. (89.027 mm ± 0.051 mm), then the allowance is the difference between the smallest permissible sleeve and the largest permissible piston:

allowance = 3.503 in. (88.976 mm) – 3.502 in.
(88.951 mm) = 0.001 in. (0.025 mm)

Suppose that one desires to relax both the piston and sleeve tolerances from ± 0.002 in. (± 0.051 mm) to ± 0.003 in. (± 0.076 mm) while maintaining the allowance at 0.001 in. (0.025 mm). Since the tolerances are bilateral, this can be achieved only by changing either or both of the nominal dimensions. Thus one solution would be to specify the piston diameter at 3.500 in. ± 0.003 in. (88.900 mm ± 0.076 mm) and the sleeve diameter at 3.507 in. ± 0.003 in. (89.078 mm ± 0.076 mm). We can then verify

allowance = 3.504 in. (89.002 mm) – 3.503 in.
(88.976 mm) = 0.001 in. (0.025 mm)

If the original specifications had incorporated unilateral tolerances, the piston diameter would have been specified as 3.502 in. $^{+0.000\text{ in.}}_{-0.004\text{ in.}}$ $\left(88.951\text{ mm}\ ^{+0.000\text{ mm}}_{-0.102\text{ mm}}\right)$, and the sleeve diameter as 3.503 in. $^{+0.004\text{ in.}}_{-0.000\text{ in.}}$ $\left(88.976\text{ mm}\ ^{+0.102\text{ mm}}_{-0.000\text{ mm}}\right)$. We can then change the tolerances to the same degree as before without altering either the allowance or the nominal dimensions. The new piston diameter specification will be 3.502 in. $^{+0.000\text{ in.}}_{-0.006\text{ in.}}$ $\left(88.951\text{ mm}\ ^{+0.000\text{ mm}}_{-0.152\text{ mm}}\right)$, and the new sleeve diameter specification will be 3.503 in. $^{+0.006\text{ in.}}_{-0.000\text{ in.}}$ $\left(88.976\text{ mm}\ ^{+0.152\text{ mm}}_{-0.000\text{ mm}}\right)$.

Design Review Teams

Historically, the product design department has had the dominant influence in setting product specifications. With the evolution of increasingly complex products, and the ever-larger capital commitments required for their manufacture, it has become necessary to broaden the responsibility for determining specifications. Establishment of a design review team is a prevalent response to this requirement. In addition to representation from product design, the design review team may comprise, but need not be limited to, the following:

Market research specialist—what specifications do the customers really require?

Manufacturing specialist—what are the production cost implications of specification decisions?

Reliability engineer—what are the product reliability and maintainability implications of specification decisions?

Product safety specialist/lawyer—what are the safety and product liability implications of specification decisions?

Quality control engineer—what are the inspection and quality cost implications of specification decisions?

8.3.2 NATURAL AND ENGINEERING TOLERANCE LIMITS

Natural Tolerance Limits

True natural tolerance limits for a normally distributed variable are located three process SDs from the process mean ($\mu \pm 3\sigma$). Numerical values for μ and σ, however, are rarely known with certainty. Even when calculations are based upon a large amount of data, the resulting values cannot be considered perfectly accurate because of the likely presence of sampling error.

Engineering Tolerance Limits

In practice, engineering tolerances can be established through the use of statistical tolerance limits. The term "statistical" indicates that an estimation procedure is involved, utilizing sample data. If it can be assumed that the manufacturing process yields dimensions that are normally distributed, then several alternative approaches are available.

A common procedure is to take a random sample of n items from the process, measure each item for the dimension in question, and then compute the sample mean $\bar{x}$ and the sample SD s. The statistical tolerance limits are set at $\bar{x} \pm Ks$, where K is a factor obtained from a standard table. Tables of K are available from most standard references on quality control, for example, Bowker and Lieberman[1] and Juran et al.[2] The numerical value of K is governed by the sample size employed, the confidence level chosen, and the proportion of the population to be included within the limits. For a discussion of the considerations involved in making these choices, see Chapters 13.4 and 13.6.

Example 8.3.2. Bottles of soda water filled by a certain machine are specified to contain 12.0 oz (0.355 liter) of product. One wants to obtain statistical tolerance limits that will include 99% of the bottles filled by this machine, with 95% confidence. A sample of size $n = 30$ filled bottles yields a mean quantity $\bar{x} = 12.08$ oz (0.357 liter) and an SD $s = 0.031$ oz (0.001 liter). For a sample size of $n = 30$, confidence level = 0.95, and proportion = 0.99, standard tables provide a value of $K = 3.350$. Thus the statistical tolerance limits are calculated as 12.08 oz (0.355 liter) ± 3.350 (0.031 oz [0.001 liter]), or 11.976 oz (0.354 liter) and 12.184 oz (0.360 liter).

Two alternative techniques for establishing statistical tolerance limits under the assumption of normality make use of the sample range R and the mean of several sample ranges $\bar{R}$; these approaches avoid the need to calculate s, the sample SD. When the assumption of normality is not justified, a distribution-free method of establishing tolerances is available. For a detailed discussion of all of these techniques, see Section 22 of Juran et al.[2]

8.3.3 MERGER OF TOLERANCES

Additive Parts

Merger of tolerances occurs when parts are assembled in an additive manner. The conventional method of calculating the tolerance on the assembled dimension is to add the individual part tolerances. Thus, if an assembly A is formed by additively combining parts X, Y, and Z, then the tolerances T are conventionally related by

$$TA = TX + TY + TZ$$

The statistical approach recognizes that simultaneous occurrences of extreme values are very unlikely. The merged tolerance is then computed as

$$TA = \sqrt{(TX)^2 + (TY)^2 + (TZ)^2}$$

This method is based upon the fact that, when random variables are added, their variances (which are squared quantities) are additive, but their SDs (which are first-power quantities) are not.

Example 8.3.3. A 200 Ω resistance is to be formed by placing in series three resistors of 40 Ω ± 2 Ω, 60 Ω ± 3 Ω, and 100 Ω ± 5 Ω. Conventionally, the tolerance on the assembled resistor is 2 Ω + 3 Ω + 5 Ω = 10 Ω. Statistically, the tolerance is $\sqrt{(2\ \Omega)^2 + (3\ \Omega)^2 + (5\ \Omega)^2} = 6.16\ \Omega$. The designer may, for product quality reasons, wish to stay with the indicated statistical tolerance. Alternatively, if a tolerance of 10 Ω is satisfactory, then a cost saving may possibly be achieved by relaxing any or all of the individual resistor tolerances, as long as the square root of the sum of the squared tolerances does not exceed 10 Ω.

Mating Parts

Application of the principle of merger of tolerances is not limited to situations involving additive parts. In cases where assembled parts are mating, the concept is equally valid.

Example 8.3.4. Refer to example 8.3.1 on allowances discussed earlier in this chapter. Suppose that, instead of the bilateral tolerances shown for the piston and the sleeve, an allowance had been specified as 0.0015 in. (0.000038 mm) ± 0.0004 in. (0.00001 mm). The conventional approach would be to divide the tolerance on the allowance into two equal portions—± 0.0002 in. (0.000005 mm)—and to assign one portion as tolerance on the piston and the other as tolerance on the sleeve. Maintaining equality of the tolerances on the two parts ($TP = TS$), the statistical approach would hold that the tolerance on the assembly, TA, is

$$TA = \sqrt{(TP)^2 + (TS)^2} = \sqrt{2T^2}$$

Numerically

$$0.0004\text{ in. } (0.00001\text{ mm}) = \sqrt{2T^2}$$

or

$$T = \pm\, 0.00028\text{ in. } (0.000007\text{ mm})$$

As a result, the piston and sleeve tolerances may each be relaxed from their conventional values.

8.3.4 DEGREES OF SERIOUSNESS

When a product or service is evaluated in terms of customer usage, safety and economic considerations are vital. Any nonconformity that occurs with a severity sufficient to cause a product or service not to satisfy intended normal or reasonably foreseeable usage requirements is termed a "defect."[3] It is frequently useful in evaluating the product or service to employ the following classification of defects by degree of seriousness:

Class 1 (Very Serious). Leads directly to severe injury or catastrophic economic loss.

Class 2 (Serious). Leads directly to significant injury or significant economic loss.

Class 3 (Major). Related to major problems with respect to intended normal or reasonably foreseeable use.

Class 4 (Minor). Related to minor problems with respect to intended normal or reasonably foreseeable use.

As part of such a rating system, it is sometimes advantageous to employ a modifier to describe the likelihood that the potential defect will be found in the operation of the product or service. The modifiers that are used form a likelihood continuum, with the terms "virtually certain" and "virtually no chance" at the extremes. The intermediate modifiers all relate to a 50% chance of occurrence and are termed "substantially above," "somewhat above," "around," "somewhat below," and "substantially below."

Other uses of classification systems are (1) to rank quality parameters according to their importance and (2) to determine an index of quality for purposes of process control. In conjunction with a rating or classification procedure, it is customary to use penalties or demerits, which are essentially arbitrary units that serve as weighting factors.

8.3.5 PROCESS CAPABILITY

Basic Concept

Process capability is the measured, inherent reproducibility of the product turned out by a process.[4] This statement refers to a specific manufacturing process and is concerned with the degree of product homogeneity resulting from a state of statistical control. A process is in a state of statistical control when all variation in the output variable(s) can be attributed to chance. Expressed differently, statistical control exists in the absence of assignable (specific and identifiable) causes of variation. The rationale for studying process capability is that it is essential to be able (1) to predict whether satisfactory products will be produced by a given process and (2) to diagnose the problems when a process fails to produce satisfactory products.

Process Capability Measurement

Process capability can be measured in several ways; the most common technique is through the use of control charts. In this approach at least 10 subgroups, each customarily containing five consecutively produced items, are selected. Appropriate measurements are made on each item, and for each subgroup the values of the mean $\overline{x}$ and the range R are calculated and are plotted on $\overline{x}$ and R charts (this is discussed later in this chapter). If statistical control appears to be present, then the process capability is computed as six SDs, which can be estimated from the average range $\overline{R}$ as $6\overline{R}/d_2$. The value of d_2 is a function of n, the number of items in each subgroup; see Exhibit 8.3.1. If statistical control does not appear to be present, then the assignable cause(s) of variation must be identified and eliminated and the process capability measured again. If the apparent lack of control is slight, it is often ignored.

Process Capability Analysis

Once process capability has been measured, it must be related to product tolerance; this procedure is known as "process capability analysis." One use of such analysis is predictive: Will process X be

Exhibit 8.3.1 Factors for Estimating σ' from $\overline{R}$ or $\overline{\sigma}$[a]

Number of Observations in Subgroup n	d_2 Factor for Estimating σ' from $\overline{R}$ ($\sigma' = \overline{R}/d_2$)	c_2 Factor for Estimating σ' from $\overline{\sigma}$ ($\sigma' = \overline{\sigma}/c_2$)	Number of Observations in Subgroup n	d_2 Factor for Estimating σ' from $\overline{R}$ ($\sigma' = \overline{R}/d_2$)	c_2 Factor for Estimating σ' from $\overline{\sigma}$ ($\sigma' = \overline{\sigma}/c_2$)
2	1.128	0.5642	21	3.778	0.9638
3	1.693	0.7236	22	3.819	0.9655
4	2.059	0.7979	23	3.858	0.9670
5	2.326	0.8407	24	3.895	0.9684
			25	3.931	0.9697
6	2.534	0.8686	30	4.086	0.9748
7	2.704	0.8882	35	4.213	0.9784
8	2.847	0.9027	40	4.322	0.9811
9	2.970	0.9139	45	4.415	0.9832
10	3.078	0.9227	50	4.498	0.9849
11	3.173	0.9300	55	4.572	0.9863
12	3.258	0.9359	60	4.639	0.9874
13	3.336	0.9410	65	4.699	0.9884
14	3.407	0.9453	70	4.755	0.9892
15	3.472	0.9490	75	4.806	0.9900
16	3.532	0.9523	80	4.854	0.9906
17	3.588	0.9551	85	4.898	0.9912
18	3.640	0.9577	90	4.939	0.9916
19	3.689	0.9599	95	4.978	0.9921
20	3.735	0.9619	100	5.015	0.9925

Reproduced by permission from *ASTM Manual on Presentation of Data*, American Society for Testing and Materials, Philadelphia, 1945.

[a]Estimate of $\sigma' = \overline{R}/d_2$ or $\overline{\sigma}/c_2$ where $\overline{R}$ is the average of sample ranges and $\overline{\sigma}$ is the average of sample SDs.

Exhibit 8.3.2 Illustration of (*a*) Adequate Process Capability and (*b*) Inadequate Capability

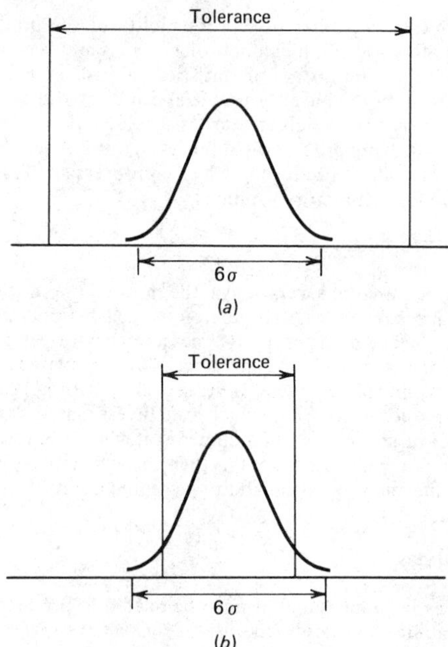

able to produce items that meet the required tolerance? The basic technique is to compare the computed six SD value to the tolerance. If this value is less than the tolerance, the process is capable of meeting the specifications; if not, the process is not capable (see Exhibit 8.3.2).

The other use of process capability analysis is diagnostic: Why is process X not able to produce items that meet tolerance requirements? There are many potential reasons, such as improper setup, faulty measurements, commingling of products, and the occurrence of sudden or assignable causes of variation. Detailed discussion of the most widely used techniques of analysis is available in Juran et al.[2]: graphic methods, Section 16; evolutionary operation, Section 27A; and response surface methodology, Section 28. In addition, experimental design is a very valuable technique (see Chapter 13.4). General statistical concepts, along with the considerations involved in the planning of tests, are discussed in Juran and Gryna.[5]

8.3.6 PROCESS CONTROL–FRACTION NONCONFORMING

The *p*-Charts

The statistic p is the fraction of nonconforming units contained in a sample; the p-chart "tracks" its value through time in order to control the fraction nonconforming p' in the process output. The sampling distribution of p has a standard error $\sigma_p = \sqrt{p'(1 - p')/n}$, where n is the size of the sample (or subgroup). In reality the value of p' is not likely to be known, and it is usually estimated by the statistic $\overline{p}$, which is the average value of p computed from a number of samples. The standard error σ_p is then estimated by $s_p = \sqrt{\overline{p}(1 - \overline{p})/n}$. In setting up a control chart, a number k of subgroups (usually $k = 20$), each containing n units, are selected in sequence and inspected for attributes. The values of $\overline{p}$ and s_p are calculated; the central line on the control chart is established at $\overline{p}$, and the trial control limits for p are set at $\overline{p} \pm 3s_p$. These "three-sigma" limits will not necessarily enclose 99.73% of the distribution of p, since p is not normally distributed (see Chapter 13.5). However, we may say that, if the process is in control with respect to fraction nonconforming, it is very unlikely that a computed value of p will fall outside of the limits. The principal reasons for a lower limit in controlling fraction nonconforming are to detect instances of careless inspection or the occurrence of a significant improvement in the manufacturing process.

The next step is to plot all k values of p in *proper sequence* on the chart. If any value of p falls outside a control limit, it is necessary to determine whether it can be explained by an assignable cause. If so, that value of p is discarded, $\overline{p}$ and s_p are recomputed on the basis of the remaining subgroups, and new trial control limits are established. If not, it is concluded that k is too small, and additional data are needed. If all plots of p are within the limits, we then check for randomness of the sequence. This can be done roughly by visual inspection of the pattern on the chart or statistically by means of a test for runs (see Chapter 13.5). If nonrandomness is present, we must identify and eliminate the causes and revise the control limits accordingly.

After the preceding steps have been accomplished, and assuming the average fraction $\overline{p}$ is satisfactory, the trial control limits can be tentatively accepted, and the chart can be used for controlling the process. Periodically, a subgroup of n items is inspected, and the value of p is computed and plotted on the chart. If a plot of p falls outside a control limit, or if there appears to be a drift toward a control limit, it is concluded that an assignable cause of variation has occurred; it is necessary to investigate and take appropriate corrective action. As long as the plots remain between the control limits and the sequence appears random, it is concluded that the process remains in a state of statistical control. As more data are accumulated, of course, the control limits can be further refined. The frequency of sampling is governed by the per unit cost of sampling and inspection, the cost of a "false alarm" (type 1 error), and the cost of operating out of control (type 2 error).[6]

Example 8.3.5. In a retail store, sales tickets may contain arithmetic errors; any such ticket is considered nonconforming. In a certain department, a subgroup (sample) containing $n = 200$ sales tickets was selected on each of 20 consecutive days. Each ticket was classified as conforming or nonconforming, with the following results:

Subgroup, i	Nonconforming, x_i	$p_i = x_i/200$
1	10	0.050
2	12	0.060
3	9	0.045
4	10	0.050
5	7	0.035
6	13	0.065
7	8	0.040
8	10	0.050
9	10	0.050
10	11	0.055
11	9	0.045
12	13	0.065
13	12	0.060
14	10	0.050
15	9	0.045
16	11	0.055
17	10	0.050
18	8	0.040
19	10	0.050
20	12	0.060

Values of $\overline{p}$ and s_p are calculated as follows:

$$\overline{p} = \sum_{i=1}^{20} P_i/20 = 0.0505$$

$$s_p = \sqrt{0.0505(1 - 0.0505)/200} = 0.0155$$

The central line (CL) and the upper and lower control limits (UCL and LCL) are computed as follows:

$$CL = \overline{p} = 0.0505$$

$$UCL = \overline{p} + 3s_p = 0.0505 + 3(0.0155) = 0.0970$$

$$LCL = \overline{p} - 3s_p = 0.0505 - 3(0.0155) = 0.0040$$

These results are plotted in Exhibit 8.3.3.

Since all of the plots of p are within the control limits and appear to be random, the limits may be tentatively accepted, provided that the mean fraction nonconforming (0.0505) is not considered excessive. If the sample size n varies from subgroup to subgroup, it is necessary to modify the preceding procedure because the control limits are a function of s_p, which in turn is a function of n.

Exhibit 8.3.3 A *p*-Chart

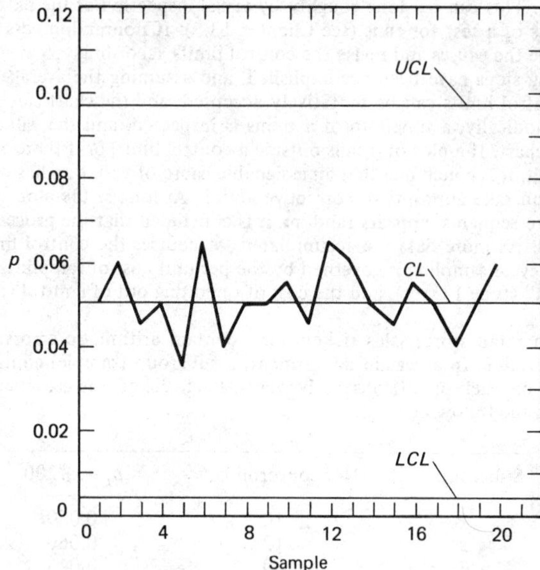

One approach is to compute variable control limits whose values change each time the sample size changes; this is a somewhat cumbersome procedure. Another approach is to achieve fixed control limits by employing the average sample size in computing s_p. However, care is necessary in interpreting plots that fall close to these limits, since they are approximate.[7] A third approach is the use of a stabilized p-chart, as discussed in the next subsection.

Stabilized *p*-Charts

When the sample or subgroup size varies, the control limits can be made constant by standardizing the sample fraction nonconforming p as follows:

$$p^* = (p - p'')/\sigma_p''$$

where p^* = standardized sample fraction nonconforming
p'' = specified or "goal" process fraction nonconforming
$\sigma_p'' = \sqrt{p''(1 - p'')/n}$

The net effect of this procedure is that the variable being plotted is expressed in SD units. A stabilized p-chart always has its central line set at zero, and the upper and lower control limits are at $+3$ and -3, respectively.

The *np*-Charts

In situations where the subgroup size is constant, the use of an np-chart may be appropriate. The statistic np, which is plotted on the chart, is simply the observed number of nonconforming units in the subgroup; hence the computational step in the p-chart procedure of dividing the number of nonconforming units by n is avoided. The CL and the UCL and LCL are computed as follows:

$$CL = n\bar{p}$$

$$UCL = n\bar{p} + 3s_{np}$$

$$LCL = n\bar{p} - 3s_{np}$$

where $s_{np} = \sqrt{n\bar{p}(1 - \bar{p})}$.

It is sometimes felt that the np-chart, in comparison to the p-chart, is more easily understood by shop personnel. However, this advantage may be more than offset by the added complexity of using both types of charts, since the np-chart has only limited applicability.

8.3.7 PROCESS CONTROL—NUMBER OF NONCONFORMITIES

The *c*-Charts

In some cases it may be difficult to identify discrete items for inspection. For example, electrical wire might be inspected for pinhole nonconformities in the insulation. It would be impossible to determine the number of locations on the insulation where a nonconformity might occur and to identify them. Therefore, instead of attempting to quantify the fraction of nonconforming locations, it is more logical to select some length of wire as an inspection unit and to count the number of nonconformities.

This variable, which is symbolized by c, tends to follow the form of the Poisson distribution.[8] Theoretically, the central line of a c-chart will be at c', the process average number of nonconformities per inspection unit, and the control limits will be at $c' \pm 3\sigma_c$, where $\sigma_c = \sqrt{c'}$. In reality, of course, the value of c' is usually not known; its value is estimated by $\bar{c}$, the sample average number of nonconformities per inspection unit (generally computed from 20 inspection units). The trial control limits are then set at $\bar{c} \pm 3s_c$, where $s_c = \sqrt{\bar{c}}$. The principal reason for a lower limit in controlling the number of nonconformities is to detect instances of careless inspection or the occurrence of a significant improvement in the manufacturing process.

The 20 values of c are plotted *in proper sequence* on the chart. If any value of c falls outside a control limit, it is necessary to determine whether there is an assignable cause. If not, it is an indication that further data are needed. If so, the value of c outside the limit can be discarded, and $\bar{c}$ and the control limits recomputed on the basis of the remaining data. When the control limits contain all the plots, it is necessary to determine whether the plots form a random sequence. This can be done roughly by visual examination of the pattern of the plots or statistically by a test for runs (see Chapter 13.4). After all of the preceding steps have been accomplished, the trial control limits may be tentatively accepted, and the chart used for controlling the process. As additional data are accumulated, of course, the control limits can be further refined.

Example 8.3.6. An inspection unit of 1000 ft (304,800 mm) is selected for the inspection of electrical wire for insulation nonconformities. For each inspection unit i, the number of nonconformities is recorded as follows:

i	c_i
1	27
2	23
3	29
4	37
5	30
6	26
7	29
8	32
9	29
10	40
11	27
12	28
13	33
14	35
15	14
16	29
17	27
18	46
19	37
20	31

$$\bar{c} = \sum_{i=1}^{20} c_i/20 = 30.45$$

$$s_c = \sqrt{\bar{c}} = 5.518$$

$$CL = \bar{c} = 30.45$$

$$UCL = \bar{c} + 3s_c = 30.45 + 3(5.518) = 47.0$$

$$LCL = \bar{c} - 3s_c = 30.45 - 3(5.518) = 13.9$$

These results are plotted in Exhibit 8.3.4.

Exhibit 8.3.4 A c-Chart

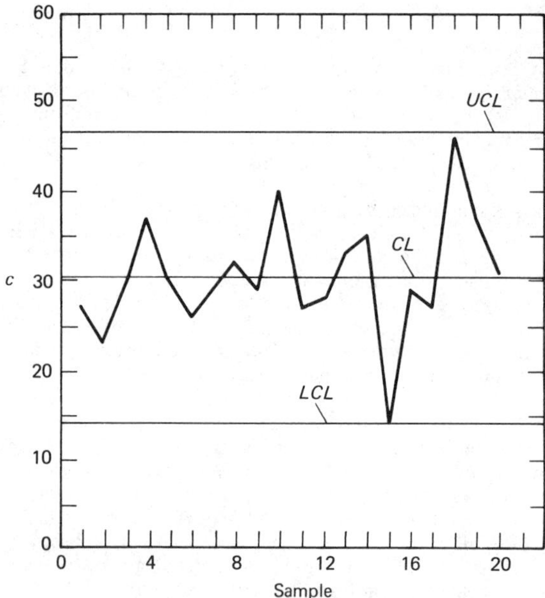

The u-Charts

At times it may not be possible to fix the quantity of product inspected at a constant amount. This would be true, for example, where 100% inspection of output is required, since the volume of production will undoubtedly vary from day to day. It is advantageous to treat the sample as being composed of k inspection units, each of fixed size and hence containing equal opportunity for nonconformities to occur. The number of nonconformities per sample c can then be converted to the average number of nonconformities per inspection unit $u = c/k$. This procedure permits the establishment of a chart with a CL of constant value, $\bar{u}$. The UCL and LCL will vary from sample to sample as the number of inspection units varies.

$$UCL = \bar{u} + 3\sqrt{\bar{u}/k}$$
$$LCL = \bar{u} - 3\sqrt{\bar{u}/k}$$

In setting up a u-chart, the same procedural considerations are involved as with a c-chart.

8.3.8 PROCESS CONTROL—VARIABLES

The $\bar{x}$-Charts

In many cases it is economically advantageous to control a process on the basis of variables rather than of fraction nonconforming. The reason is that, since a set of measurements contains more information than the same number of counted observations, the same degree of control over a process can be maintained with fewer measurements than counts. Partially offsetting this advantage, however, is that the process of obtaining measurements and performing calculations with them may be more time consuming (and more costly) than the simple classification of items as conforming or nonconforming.

When a process is to be controlled on the basis of variables, it is necessary to control both the mean and the variability; the $\bar{x}$-chart is the most widely used technique for controlling the mean. The design of an $\bar{x}$-chart is based upon the fact that the sample mean $\bar{x}$ is normally distributed and that it itself has a mean equal to the process mean μ and an SD (or standard error) $\sigma_{\bar{x}} = \sigma/\sqrt{n}$, where σ is the process standard deviation. When the process output is normally distributed, the normality of $\bar{x}$ is exact; when the process output is not normally distributed, the normality of $\bar{x}$ is approximate, with the degree of approximation improving as the sample size n increases. Ideally, if a process is in control with a specified mean μ'' and a specified SD σ'', a control chart utilizing the statistic $\bar{x}$ can be set up with CL at μ'' and three-sigma control limits at $\mu'' \pm 3\sigma_{\bar{x}}''$. This procedure will provide for a "false alarm" probability of 0.0027.

In reality, of course, when control over a process is yet to be established, the values of μ and σ are usually not known. It is necessary to take from the process a series of k samples or subgroups (usually $k = 20$), each containing n units. For each subgroup the mean $\bar{x}$ and the range R are calculated; the range is defined as the difference between the largest measurement and the smallest. The resulting k values of $\bar{x}$ are averaged to form $\bar{\bar{x}}$, and the k values of R are averaged to form $\bar{R}$. An $\bar{x}$-chart is then set up, with CL at $\bar{\bar{x}}$ and UCL and LCL as follows:

$$UCL = \bar{\bar{x}} + A_2\bar{R}$$

$$LCL = \bar{\bar{x}} - A_2\bar{R}$$

where A_2 has a value determined from Exhibit 8.2.5 according to the size n of the subgroups. The product $A_2\bar{R}$ is an estimate of $3\sigma_{\bar{x}}$; alternatively, a computed value of $s_{\bar{x}}$ could be used to estimate $\sigma_{\bar{x}}$, but this is a computationally more cumbersome procedure.

The limits just calculated are trial control limits; the next step is to plot the k values of $\bar{x}$ on the chart *in proper sequence.* At this point it is necessary to stress that control over the process variability is equally as important as control over the process mean. Therefore an R-chart should be de-

Exhibit 8.3.5 Factors for $\bar{x}$, R, σ, and x Control Charts–Trial Control Limits[a]

	Factors for $\bar{x}$-Chart[b]		Factors for R-Chart[c]		Factors for σ-Chart[d]		Factors for x-Chart[e]	
Number of Observations in Subgroup	From $\bar{R}$ A_2	From $\bar{\sigma}$ A_1	Lower D_3	Upper D_4	Lower B_3	Upper B_4	From $\bar{R}$ E_2	From $\bar{\sigma}$ E_1
2	1.880	3.759	0	3.268	0	3.267	2.660	5.318
3	1.023	2.394	0	2.574	0	2.568	1.772	4.146
4	0.729	1.880	0	2.282	0	2.266	1.457	3.760
5	0.577	1.596	0	2.114	0	2.089	1.290	3.568
6	0.483	1.410	0	2.004	0.030	1.970	1.184	3.454
7	0.419	1.277	0.076	1.924	0.118	1.882	1.109	3.378
8	0.373	1.175	0.136	1.864	0.185	1.815	1.054	3.323
9	0.337	1.094	0.184	1.816	0.239	1.761	0.010	3.283
10	0.308	1.028	0.223	1.777	0.284	1.716	0.975	3.251
11	0.285	0.973	0.256	1.744	0.321	1.679	0.946	3.226
12	0.266	0.925	0.284	1.717	0.354	1.646	0.921	3.205
13	0.249	0.884	0.308	1.692	0.382	1.618	0.899	3.188
14	0.235	0.848	0.329	1.671	0.406	1.594	0.881	3.174
15	0.223	0.817	0.348	1.652	0.428	1.572	0.864	3.161
16	0.212	0.788			0.448	1.552		
17	0.203	0.762			0.466	1.534		
18	0.194	0.738			0.482	1.518		
19	0.187	0.717			0.497	1.503		
20	0.180	0.698			0.510	1.490		
21	0.173	0.680			0.523	1.477		
22	0.167	0.662			0.534	1.466		
23	0.162	0.647			0.545	1.455		
24	0.157	0.632			0.555	1.445		
25	0.153	0.619			0.565	1.435		
Over 25	†	†			†	†		

[a] Factors reproduced from 1950 *ASTM Manual on Quality Control of Materials* by permission of the American Society for Testing and Materials, Philadelphia. All factors in exhibit are based on a normal distribution.

[b] $UCL_{\bar{x}} = \bar{\bar{x}} + A_2\bar{R}$ and $LCL_{\bar{x}} = \bar{\bar{x}} - A_2\bar{R}$ or $UCL_{\bar{x}} = \bar{\bar{x}} + A_1\bar{\sigma}$ and $LCL_{\bar{x}} = \bar{\bar{x}} - A_1\bar{\sigma}$.

[c] $UCL_R = D_4\bar{R}$ and $LCL_R = D_3\bar{R}$.

[d] $UCL_\sigma = B_4\bar{\sigma}$ and $LCL_\sigma = B_3\bar{\sigma}$.

[e] $\bar{x} + E_2\bar{R}$ and $\bar{x} - E_2\bar{R}$ or $\bar{x} + E_1\bar{\sigma}$ and $\bar{x} - E_1\bar{\sigma}$.

† Values of these constants may be determined for larger sample sizes from formulas given in J. M. Juran, F. M. Gryna, and R. S. Bingham, Jr., *Quality Control Handbook,* 3rd ed., McGraw-Hill, New York, Appendix I, pp. 1–7.

veloped concurrently with the $\bar{x}$-chart (see the next subsection). If any plotted value of $\bar{x}$ falls outside a control limit, it is necessary to determine whether there is an assignable cause. If not, it is an indication that further data are needed. If so, the value of $\bar{x}$ outside the limit can be discarded, and $\bar{\bar{x}}$, $\bar{R}$, and the control limits recomputed on the basis of the remaining data. When the control limits contain all the plots, it is then necessary to determine whether the plots form a random sequence. This can be done roughly by visual examination of the pattern of the plots or statistically by a test for runs (see Chapter 13.4). After all of the preceding steps have been accomplished, the trial control limits may be tentatively accepted, and the chart used for controlling the process mean. As additional data are accumulated, of course, the control limits can be further refined. For a discussion of the economic factors to be considered in designing an $\bar{x}$-chart, see Saniga.[9]

Example 8.3.7. Thermometers are tested in a solution known to be at 37.4°F (3.0°C). A series of 20 subgroups was tested, with means and ranges as follows (each subgroup i is a sample of size $n = 8$):

i	$\bar{x}_i$	R_i
1	3.16	1.31
2	3.36	1.26
3	3.10	2.02
4	3.36	1.07
5	3.06	1.13
6	2.88	2.58
7	3.44	1.64
8	2.70	2.39
9	3.38	2.55
10	3.32	0.70
11	3.24	1.39
12	3.06	0.74
13	2.62	1.98
14	3.26	0.33
15	2.90	1.27
16	3.32	1.72
17	2.88	0.56
18	3.10	1.51
19	2.94	2.25
20	3.22	0.30

$$\bar{\bar{x}} = \sum_{i=1}^{20} \bar{x}_i/20 = 3.115$$

$$\bar{R} = \sum_{i=1}^{20} R_i/20 = 1.44$$

$$CL = \bar{\bar{x}} = 3.115$$

$$UCL = \bar{\bar{x}} + A_2\bar{R} = 3.115 + (0.373)(1.44) = 3.652$$

$$LCL = \bar{\bar{x}} - A_2\bar{R} = 3.115 - (0.373)(1.44) = 2.578$$

These results are plotted in Exhibit 8.3.6.

These trial limits may be tentatively accepted, provided that the estimated process mean $\bar{\bar{x}}$ = 3.115°C is considered satisfactory. Note that the $\bar{x}$-chart of itself does not establish complete control over the process; it is necessary to set up an R-chart in conjunction with the $\bar{x}$-chart.

The *R*-Charts

The purpose of an R-chart is to maintain control over the variability of a process. A series of k subgroups (usually $k = 20$), each containing n units, is selected from the process. For each subgroup the range R is calculated; the range is defined as the difference between the largest measurement and the smallest. The resulting k values of R are averaged to form $\bar{R}$. An R-chart is then set up, with CL at $\bar{R}$, UCL at $D_4\bar{R}$, and LCL at $D_3\bar{R}$. The factors D_3 and D_4 have values determined from Exhibit 8.2.5 according to the size n of the subgroups. The product $D_4\bar{R}$ establishes a control limit that is approximately three standard errors of R above $\bar{R}$, and the product $D_3\bar{R}$ establishes a con-

Exhibit 8.3.6 An $\bar{x}$-Chart

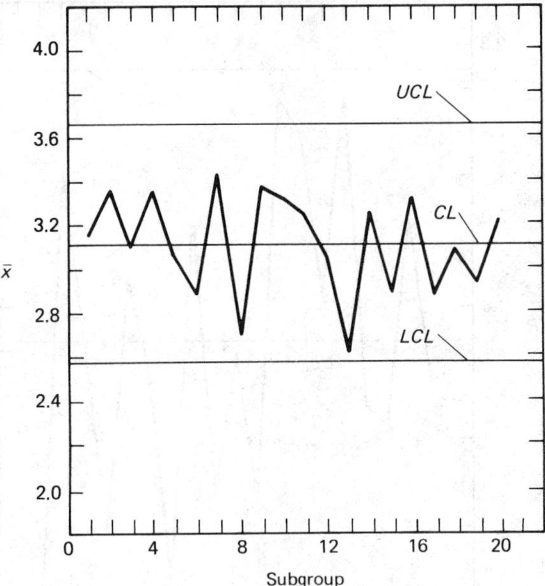

trol limit that is approximately three standard errors of R below $\bar{R}$. The principal reason for an LCL on the process variability is to detect instances of careless inspection or the occurrence of a significant improvement in the manufacturing process.

These limits are trial control limits; the next step is to plot the k values of R on the chart *in proper sequence.* At this point it is necessary to stress that control over the process mean is equally as important as control over the process variability. Therefore an $\bar{x}$-chart should be developed concurrently with the R-chart (see previous subsection). If any plotted value of R falls outside a control limit, it is necessary to determine whether there is an assignable cause. If not, it is an indication that further data are needed. If so, the value of R outside the limit can be discarded, and $\bar{R}$ and the control limits recomputed on the basis of the remaining data. When the control limits contain all the plots, it is necessary to determine whether the plots form a random sequence. This can be done roughly by visual examination of the pattern of the plots or statistically by a test for runs (see Chapter 13.4). After all of the preceding steps have been accomplished, the trial control limits may be tentatively accepted, and the chart used for controlling the process variability. As additional data are accumulated, of course, the control limits can be further refined. For a discussion of the economic factors to be considered in designing an R-chart, see Saniga.[9]

Example 8.3.8. Refer to the data in example 8.3.7. The 20 values of R are averaged to form

$$\bar{R} = \sum_{i=1}^{20} R_i/20 = 1.44$$

$$CL = \bar{R} = 1.44$$

$$UCL = D_4\bar{R} = (1.864)(1.44) = 2.684$$

$$LCL = D_3\bar{R} = (0.136)(1.44) = 0.196$$

These results are plotted in Exhibit 8.3.7.

These trial limits may be tentatively accepted, provided that the amount of process variability is satisfactory. Note that the R-chart of itself does not establish complete control over the process; it is necessary to set up an $\bar{x}$-chart in conjunction with the R-chart.

Other Variables Charts

Additional chart techniques have been developed for special situations. For example, when the subgroup size n exceeds 12, the sample range R tends to lose efficiency. In such cases the subgroup

Exhibit 8.3.7 An R-Chart

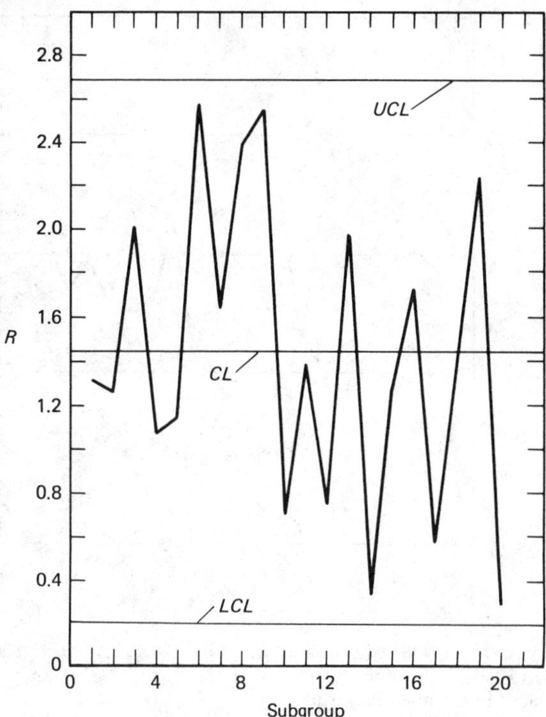

can be subdivided, and the mean range $\overline{R}$ for the subdivisions can be plotted on a chart. Further techniques for controlling process variability include the s-chart for sample SDs and the s^2-chart for sample variances. Obtaining values for s and s^2 is computationally tedious, however.

Alternatives to the $\bar{x}$-chart for controlling the process mean include charts for the median and the midrange. These statistics are easy to calculate, but are less efficient than the mean. (Larger sample sizes are needed to attain given risk levels.) For details concerning all of these chart techniques, see Duncan.[10]

8.3.9 PROCESS CONTROL–OTHER TECHNIQUES

A criticism of the standard (or Shewhart) charts just described is that they provide decision rules based only upon the most recent observations. The cumulative sum (cusum) chart permits decisions to be based upon more data. Cusum charts can be designed for attributes or for variables, for one-sided decision procedures or for two-sided procedures (see Duncan[11]). One study[12] has shown that, when both a cusum chart and a Shewhart chart are of minimum-cost design, there is little difference between them. However, when a Shewhart chart is nonoptimal, a comparable cusum chart will result in lower cost. A cusum chart, though, is more likely than a comparable Shewhart chart to react to minor process disturbances.

A technique that is useful when there is a natural "drift" in a process (e.g., due to tool wear), or when the inherent variability of a process is much narrower than the tolerance, is the acceptance control chart. Specific designs are treated in Duncan.[13] Adaptive control techniques are widely employed in the chemical industry, where the type and amount of corrective action must be predicted; see Juran et al.[14]

8.3.10 LOT ACCEPTANCE PLANS–ATTRIBUTES

Single-Sample Plans

In acceptance sampling a decision to accept or reject a collection of items–called a "lot"–is based upon the inspection of a sample of items taken from the lot. In many cases sampling is economically advantageous because inspection of the entire lot would be too costly. Sometimes, of course, sampling is necessary, as in those cases where testing is destructive.

A single-sample plan is defined by specifying values for n, the sample size, and c, the acceptance number. If the observed number of nonconforming units in the sample does not exceed c, the lot is deemed acceptable. If the observed number of nonconforming units does exceed c, then the lot is rejected. The properties of a given acceptance sampling plan are customarily shown by its operating characteristic (OC) curve. This is a graph in which the probability of accepting a lot is plotted against the lot or process fraction nonconforming. It is often useful to distinguish between a type A OC curve, in which we are concerned with an isolated lot, and a type B OC curve, in which we are concerned with a series of lots formed essentially at random from the output of some process.[15] No such distinction will be made in this chapter; all OC curves will be considered as Type B. An OC curve is shown in Exhibit 8.3.8.

In the exhibit AQL signifies the acceptable quality level; this is a product fraction nonconforming sufficiently small that a sampling plan should provide for acceptance of such a product a large percentage of the time. The percentage of the time (or probability) that such a product will be rejected by the sampling plan is called the producer's risk and is symbolized by α.

The symbol LQL signifies limiting quality level; this is a product fraction nonconforming sufficiently large that a sampling plan should provide for acceptance of such a product only a small percentage of the time. This percentage (or probability) is called the consumer's risk and is symbolized by β.

In designing a sampling plan, or selecting one from a set of plans, the standard procedure is to assign numerical values to α, β, AQL, and LQL on the basis of the cost of sampling and inspection, the cost of accepting a poor product, and the cost of rejecting a good product. Recent research has been directed toward optimization, that is, the design of sampling plans that will result in the best combinations of expected outgoing quality and expected cost of quality assurance.[16] The standard procedure is illustrated in example 9.

Example 8.3.9. In a certain application it is desired that, when product quality (i.e., fraction nonconforming) = 0.01 = AQL, the probability of acceptance should be 0.95 (α = 0.05). Also, when product quality = 0.06 = LQL, the probability of acceptance should be 0.10 (β = 0.10). Assume that a type B OC curve is applicable and thus that the sampling probabilities follow the binomial distribution. Since cumulative probabilities are of concern here (x *or fewer* nonconforming units in a sample), it is advantageous to use a cumulative Poisson table to approximate the exact binomial calculations, which are quite tedious.

First, form the ratio of LQL/AQL = 0.06/0.01 = 6.0. Next, from a cumulative Poisson table[17] find that, for α = 0.05 and β = 0.10, the LQL/AQL ratio is 6.50 for c = 2 and 4.89 for c = 3. Since the desired value of the ratio is 6.0, the value of the acceptance number can be set at either 2 or 3. Next, set c = 2 and find the value of n, with α held at 0.05, by dividing the 0.95 value of the Poisson parameter in the table by AQL; this yields a value of n = 0.818/0.01 = 81.8 ≐ 82. Thus one sampling plan, which we may call plan A, is defined as n = 82, c = 2. This means that, whenever a sample of 82 units taken from the incoming product contains 0, 1, or 2 defectives, accept the product; whenever the sample contains 3 or more defectives, reject the product.

Next, continue with c = 2 and find the value of n, with β held at 0.10, by dividing the 0.10 value of the Poisson parameter in the table by LQL; this yields a value of n = 5.32/0.06 = 88.7 ≐ 89. Thus plan B is defined as n = 89, c = 2. Then repeat the preceding two steps for c = 3. With α held at 0.05, plan C is defined as n = 136.6 ≐ 137, c = 3; with β held at 0.10, plan D is defined as n = 111.3 ≐ 112, c = 3.

Exhibit 8.3.8 An OC Curve

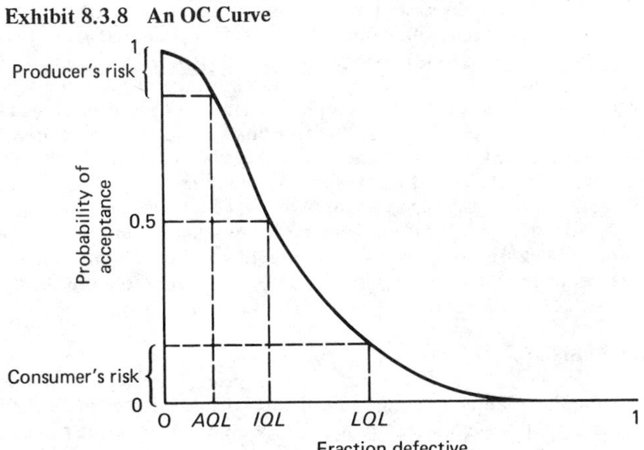

The four plans from which a choice is to be made can be summarized as follows:

Plan	α	β
A. $n = 82$, $c = 2$	0.05	0.14
B. $n = 89$, $c = 2$	0.06	0.10
C. $n = 137$, $c = 3$	0.05	0.04
D. $n = 112$, $c = 3$	0.03	0.10

Thus plans A and C have OC curves that pass through the desired point whose coordinates are AQL, $1 - \alpha$ (or 0.01, 0.95), but not through the desired point whose coordinates are LQL, β (or 0.06, 0.10). Plan C provides more protection to the consumer than required, whereas plan A provides less. Plans B and D have OC curves that pass through the desired point whose coordinates are 0.06, 0.10, but not through 0.01, 0.95. Plan D provides more protection to the producer than required, whereas plan B provides less. The reason that there is no plan whose OC curve passes through both of the desired points is that the values of n and c are required to be integers.

Military Standard 105D

Standard military procedures for acceptance sampling by attributes have been in existence for almost 40 years. The current version, designated MIL-STD-105D, was issued by the U.S. Department of Defense in 1963.[18] This standard is essentially a set of sampling plans indexed with respect to AQL. The selection of a specific plan from MIL-STD-105D depends upon the lot size, the AQL that has been chosen, and the inspection level that has been chosen.

There are three general levels and four special levels of inspection; the level to be used in a given application depends upon the degree of discrimination needed between good and bad lots, upon the cost of the item, and upon whether testing is destructive. There is a provision for a mandatory shift to tightened inspection when it appears that product quality has deteriorated significantly and a provision for an optional shift to reduced inspection when it appears that product quality is exceptionally good on a consistent basis. The standard recognizes the classification of defects into critical, major, and minor and also makes provision for double- and multiple-sample procedures if desired. Selected type B OC curves are provided, as are average sample size curves for the double and multiple plans without curtailed inspection. In its mode of operation, MIL-STD-105D pushes the manufacturer to supply a product at least as good as the specified AQL.

Double-Sample Plans

In the area of sampling and inspection costs, a double-sample plan can provide significant savings over a comparable single-sample plan. Such a plan is defined by specifying values for the five numbers n_1, n_2, c_1, c_2, and c_3, with $c_1 < c_2$ and $c_2 < c_3$. If in a sample of n_1 units taken from a lot the number of nonconforming units is less than or equal to c_1, the lot is accepted. If the number of nonconforming units exceeds c_2, the lot is rejected. If the number of nonconforming units exceeds c_1, but does not exceed c_2, a second sample of size n_2 is taken. If there are c_3 or fewer nonconforming units in the combined samples, the lot is accepted; if there are more than c_3 nonconforming units, the lot is rejected. In many cases the value of c_2 is set equal to that of c_3.

The cost savings with respect to a comparable single-sample plan arise from two sources. (Comparable plans are those that have essentially equal values for AQL, LQL, α, and β). First, n_1 for the double-sample plan is smaller than n for the single-sample plan, and therefore whenever the lot is accepted or rejected on the basis of the first sample, the amount of inspection will be less. Second, when inspecting the second sample, it is possible to utilize curtailed inspection when the lot is rejected. This means that inspection ceases as soon as c_3 has been exceeded. These relationships are shown graphically, without reference to any specific sampling plans, in the average sample size curves in Exhibit 8.3.9. The average sample size is the weighted average of n_1 and $n_1 + n_2$, where the weights are the probabilities of only one sample being required and of a second sample being required. These probabilities are functions of the product fraction nonconforming.

Tables of double-sampling plans are given in MIL-STD-105D, discussed previously. For a detailed discussion of the design of double-sample plans, see Duncan.[19]

Multiple-Sample Plans

A lot acceptance sampling plan containing more than two stages is known as a "multiple-sample plan" or, at times, as "group sequential sampling." There are typically seven or eight stages, at each

Exhibit 8.3.9 Average Sample Size Curves

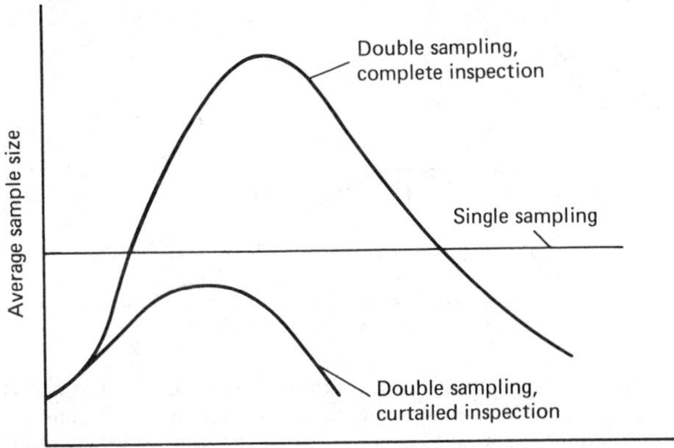

of which a sample of n units is taken for inspection, unless an accept or reject decision was reached at the preceding stage. Thus at each stage the cumulative sample size is n, $2n$, $3n$, and so on.

In addition, at each stage the plan provides an acceptance number and a rejection number whose values, along with the value of n, are determined by the choice of AQL, LQL, α, and β. The rejection number is always larger than the acceptance number. The decision rule at each stage is to accept the lot if the total number of nonconforming units in the combined samples through that stage is less than or equal to the acceptance number. A decision to reject the lot is made if the number of nonconforming units is equal to or greater than the rejection number. If the number of nonconforming units is greater than the acceptance number and less than the rejection number, proceed to the next stage. A decision is forced at the last stage because the rejection number is equal to the acceptance number plus 1.

The inspection cost associated with a multiple-sample plan can be significantly less than that with a comparable single-sample plan because the average sample size of the former tends to be less than the fixed sample size for the latter. (See the discussion of the average sample size curve in the previous subsection.) This holds true even without curtailed inspection in the second and subsequent stages. The inspection cost advantage will be offset to some degree by the greater administrative complexity. Tables of multiple-sample plans are provided in MIL-STD-105D. For details concerning the construction of multiple-sample plans see Duncan.[20]

Sequential Sampling Plans

When items are taken from a lot one at a time for inspection, the procedure is known as item-by-item sequential sampling. It was noted previously that, when the sample size n and the acceptance number c are fixed, it is not always possible to hold both α and β exactly at the desired levels because of the integer-value constraints on n and c. Sequential sampling permits α and β to be fixed very close to the desired values because the sample size n varies from one lot to another. At each sampling step there is an acceptance number and a rejection number whose values are determined by the values that have been assigned to AQL, LQL, α, and β. When the cumulative number of nonconforming units equals or falls below the acceptance number, the lot is accepted; when the cumulative number of nonconforming units equals or exceeds the rejection number, the lot is rejected. For values in between, the sampling is continued.

A major advantage of sequential sampling is that inspection economies can often be realized. It has been estimated, for example, that the average sample size for a sequential sampling plan may be as low as one-half of the fixed sample size required in a comparable single-sample plan.[21] For details concerning the construction of sequential sampling plans, see Duncan.[19]

8.3.11 LOT ACCEPTANCE PLANS—RECTIFICATION

Rectifying Inspection

In sampling inspection by attributes, a rejected lot may be subjected to screening or 100% inspection in order to remove all nonconforming units. This general procedure is called "rectifying inspection." The specific procedure employed in any given situation may follow any one of several variations on the general theme: If a lot passes sampling inspection, the nonconforming units in

Exhibit 8.3.10 An *AOQ* Curve

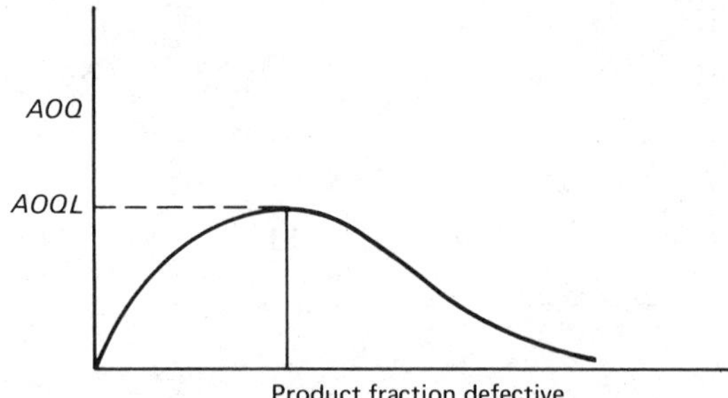

the sample may or may not be removed, and, if removed, may or may not be replaced by conforming units; when a rejected lot is screened, the removed nonconforming units may or may not be replaced by conforming units. The purpose of rectification is to establish quality assurance in the sense of setting an upper limit on the outgoing fraction nonconforming. This fraction is called the "average outgoing quality" (*AOQ*) and is a weighted average of the fraction nonconforming contained in the rectified and the unrectified lots. The fraction nonconforming in rectified lots is taken to be zero, and that in the unrectified lots is the average fraction nonconforming turned out by the process from which the lots are formed. The weights are the probabilities of acceptance as given by the OC curve for the plan that is being used; these probabilities, of course, are a function of the product fraction nonconforming.

The equation for computing *AOQ* depends upon which of the specific rectifying procedures described is being used. In all cases the *AOQ* function will have a maximum, called the "average outgoing quality limit" (*AOQL*), because when product quality is good, the unrectified lots will contain good quality, and when product quality is poor, most of the lots will be rectified. An *AOQ* curve, without reference to any specific sampling plan, is shown in Exhibit 8.3.10.

Inspection cost can be taken into account in selecting a plan through the calculation of the average total inspection (*ATI*).

$$ATI = n + (1 - P_a)(N - n)$$

where n = sample size
N = lot size
P_a = probability of acceptance for a given incoming quality level (from the OC curve)

In addition to the single-sample application, rectifying inspection can also be employed with double-sample and multiple-sample attributes plans and with variables plans (see Duncan[22]).

Dodge-Romig Plans

Tables have been developed that give sampling plans for rectifying inspection.[23] These are known as the Dodge-Romig tables and are indexed by *AOQL*. Both single-sample and double-sample plans are presented, and it is assumed that all rejected lots are screened. Unlike MIL-STD-105D, the Dodge-Romig tables do not provide for the classification of nonconformities and do not contain procedures for tightened and reduced inspection. The Dodge-Romig tables also provide sampling plans indexed by lot tolerance percent defective (*LTPD*) [limiting quality level (*LQL*)].

8.3.12 LOT ACCEPTANCE PLANS—VARIABLES

Single-Sample Plans

In situations where a quality determining characteristic of a product is continuous and has a known distribution, it is sometimes economically advantageous to employ an acceptance sampling plan that utilizes the actual measurements of the characteristic rather than attributes, that is, classification as conforming or nonconforming. The advantage is based upon the principle that a sample of n measurements contains more information than a sample of n counts, and hence a plan based on measurements (or variables) will require a smaller sample size than one based on attributes that has comparable properties (similar OC curve). Principal disadvantages are the cost of variables inspection compared to attributes inspection, the necessity of having a different plan for each characteristic, and the potential for serious inaccuracies when the distributions are not normal.

It will be assumed here that the measurements follow a normal distribution; this assumption will permit the use of normal variates in conjunction with AQL, LQL, α, and β. In a situation where the SD σ of the process from which lots are formed is assumed to be constant, then the fraction nonconforming contained in such lots will be determined by the mean of the process. If there is a single specfication limit, for example, a lower limit L, then the process mean μ_1, which yields a lot fraction nonconforming equal to AQL, can be transformed into a standard normal variate z_1 as follows:

$$\frac{(\mu_1 - L)}{\sigma} = z_1$$

Similarly, the process mean μ_2, which yields a fraction nonconforming equal to LQL, can be transformed as follows:

$$\frac{(\mu_2 - L)}{\sigma} = z_2$$

The statistic used for acceptance or rejection of a lot will be the sample mean $\overline{x}$, and if k represents the critical value of the standardized form of $\overline{x}$, then

$$(k - z_1)\sqrt{n} = -z_\alpha$$

and

$$(k - z_2)\sqrt{n} = z_\beta$$

The symbol $-z_\alpha$ represents a value of the standard normal variate z that establishes the producer's risk α under the sampling distribution of $\overline{x}$ when the process mean $\mu = \mu_1$. Similarly, the symbol z_β represents a value of z that establishes the consumer's risk β under the sampling distribution of $\overline{x}$ when the process mean $\mu = \mu_2$. The sampling plan can then be defined as

$$k = \frac{z_1 - z_\alpha}{\sqrt{n}}$$

or

$$k = \frac{z_2 + z_\beta}{\sqrt{n}}$$

and

$$n = \left[\frac{(z_\alpha + z_\beta)}{(z_1 - z_2)}\right]^2$$

The acceptance sampling procedure will be to take a random sample of n units from the lot, measure each for the appropriate characteristic, and compute the mean $\overline{x}$. If $(\overline{x} - L)/\sigma \geqslant k$, accept the lot; if $(\overline{x} - L)/\sigma < k$, reject the lot. This procedure is based upon the assumption that σ is constant and that its value is known. If σ is constant, but its value is not known, then the computations must be modified as follows:

$$k = \frac{(z_\alpha z_2 + z_\beta z_1)}{(z_\alpha + z_\beta)}$$

and

$$n = \left(\frac{1 + k^2}{2}\right)\left[\frac{(z_\alpha + z_\beta)}{(z_1 - z_2)}\right]^2$$

Then, letting the symbol s represent the sample SD, the decision rule is: If $(\bar{x} - L)/s \geqslant k$, accept the lot; if $(\bar{x} - L)/s < k$, reject the lot. Because of the symmetry of the normal distribution, it is a relatively simple matter to modify the preceding techniques to accommodate a situation requiring a single upper specification limit U. The OC curve will resemble that for an attributes plan (see Exhibit 8.3.8).

Example 8.3.10. A certain type of dry cell used in emergency lighting systems has a normally distributed operating life with a one-sided lower specification limit of 30 A-hr (108,000 A-sec). An acceptance sampling plan is desired that will accept with probability 0.95 ($\alpha = 0.05$) product whose fraction nonconforming is $AQL = 0.02$ and that will accept with probability 0.10 ($\beta = 0.10$) product whose fraction nonconforming is $LQL = 0.08$. The z-values associated with these requirements are $z_{0.02} = 2.0540$; $z_{0.08} = 1.4053$; $z_{0.05} = 1.6450$; and $z_{0.10} = 1.2817$.

a. Assume $\sigma = 1.5$ A-hr (5400 A-sec).

$$n = [(1.6450 + 1.2817)/(2.0540 - 1.4053)]^2 = 20.35 \doteq 21$$

$$k = 1.4053 + 1.2817/\sqrt{21} = 1.685$$

or

$$k = 2.0540 - 1.6450/\sqrt{21} = 1.695; \text{let } k = 1.690$$

Decision rule: If $(\bar{x} - 30[108{,}000])/1.5[5400] \geqslant 1.690$, accept the lot; otherwise, reject the lot.

b. Assume σ is unknown.

$$k = [(1.6450)\,(1.4053) + (1.2817)\,(2.0540)]/(1.6450 + 1.2817) = 1.689$$

$$n = [1 + (1.689)^2/2]\,[(1.6450 + 1.2817)/(2.0540 - 1.4053)]^2 = 49.38 \doteq 50$$

Decision rule: If $(\bar{x} - 30[108{,}000])/s \geqslant 1.689$, accept the lot; otherwise, reject the lot.

Variables acceptance sampling plans can be constructed for situations in which there are both upper and lower specification limits. The specific procedure to be used depends upon the relative magnitudes of the process SD and the distance between the specification limits. For a detailed discussion, see Duncan.[24]

Military Standard 414

Procedures for acceptance sampling by variables, known as MIL-STD-414, have been developed by the U.S. Department of Defense.[25] The plans contained in the standard are single-sample plans only and are indexed according to AQL. The specific plan to be employed depends upon the lot size, the AQL that has been selected, the level of inspection that has been selected, and whether single or double specification limits are to be utilized.

Plans are provided for situations in which the process SD is either known or unknown, and in the latter case a plan may be selected that is based upon either the sample SD or the sample range. Whatever specific plan is employed, the standard contains provisions for shifting to tightened or reduced inspection when appropriate. The rationale for the development of MIL-STD-414 was the possibility of a reduction in inspection costs through the use of smaller samples than would be required under MIL-STD-105D. However, the exercise of care in utilizing MIL-STD-414 is necessary for several reasons:

1. The risks associated with any given plan can be significantly altered if the assumption of normality is not valid.
2. The lot size classes are not the same as in MIL-STD-105D.
3. The rules for a shift to tightened inspection are not the same as in MIL-STD-105D.
4. The behavior of the producer's risk as a function of sample size is not the same as in MIL-STD-105D.

8.3.13 SPECIAL TECHNIQUES

Skip-Lot Sampling Plans

When an attributes-based lot acceptance sampling plan is being employed, it is possible under certain conditions to reduce the total amount of inspection through a skip-lot technique.[26] The procedure is to impose sampling inspection upon only a certain fraction of the lots, provided that a predetermined number of consecutive lots have been accepted. The skip-lot procedure remains in effect until one lot is rejected, at which time all subsequent lots are subjected to the acceptance sampling plan. Reinstatement of the skip-lot procedure is permissible when the consecutive-lot criterion has been satisfied again. Rectifying inspection (see earlier parts of this chapter) can be utilized in conjunction with a skip-lot technique.

Chain Sampling Plans

In situations where the per unit cost of inspection is high, as in the case of destructive testing, the sample size n is of necessity very small. The only meaningful acceptance number is $c = 0$. Every such plan has an OC curve whose second derivative is positive throughout; that is, there is no inflection point. In such cases the plan has very limited ability to distinguish between good and bad lots; in particular, the producer has little protection against chance rejection of lots formed from a satisfactory process.

The chain sampling plan was devised to correct this situation.[27] The procedure is to select an appropriate attributes-based single-sample plan with a small value of n and with $c = 0$. At the same time, select a value for the parameter i to be used in conjunction with the plan ($i = 1, 2, 3, \ldots$). When i consecutive lots have each yielded a sample of n containing no nonconforming units, set $c = 1$ and continue. As soon as a sample is taken that contains one or more nonconforming units, set $c = 0$ and repeat the procedure.

There are certain conditions necessary for proper use of chain sampling:

1. The application must be a continuing one, not an isolated one.
2. The process quality must be essentially stable.
3. The consumer should have no particular reason to suspect poor quality in any given lot.
4. There must be confidence that the supplier will not on occasion deliberately submit a poor lot in the hope that it will pass acceptance sampling when $c = 1$.

Discovery Sampling

On a lot-by-lot basis, not all quality levels are equally likely to occur. Discovery sampling[28] requires the estimation of the fraction of partially nonconforming lots in the total number of lots in order to adjust the sample size periodically. An *AOQL*-type protection (see earlier part of this chapter) can be achieved, often with smaller sample sizes than would be necessary with conventional lot acceptance sampling plans.

Continuous Sampling Plans

When units are produced in a continuous production system, the formation of inspection lots creates both conceptual and practical difficulties. For this reason, continuous sampling plans have been developed that differ from the lot-by-lot plans described in Sections 8.3.10 through 8.3.12. The earliest techniques date from World War II,[29] and the original plan is known as CSP-1.

Values must be selected for two parameters, i (a positive integer) and f ($0 < f < 1$). The operation of the plan begins with 100% inspection until i consecutive units are found to be free of nonconformities. At this point the inspection rate drops from 100% to f until a nonconforming unit is found. The 100% inspection rate is then reinstated, and the cycle begins again. All nonconforming units found are corrected or replaced by conforming units. For any CSP-1 plan, the OC curve, the *AOQL*, and the average fraction of units inspected (*AFI*) can be computed.

In addition, a system of classification of nonconformities can be employed (see Section 8.3.4). Since a process that is in control can produce a random nonconforming unit, modifications known as CSP-2 and CSP-3 require that two out of k nonconforming units be found before reinstating 100% inspection. Multilevel continuous plans also have been developed which provide for decreasing values of f as long as no nonconforming units have been found. Others have developed continuous sampling plans that differ from Dodge's in that (1) units are inspected in groups rather than individually, (2) inspection begins on a sampling basis rather than at the 100% rate, and (3) an *AOQL* is assured regardless of whether the process is in control. For details, see Girschick[30] and Wald and Wolfowitz.[31]

Compressed Limit Gauging

In using an attributes-based acceptance sampling plan, it is sometimes advantageous to employ tighter specification limits than those prescribed by engineering design. The purpose is to reduce the needed sample size and hence lower the total inspection cost. The usefulness of this technique is greatest when both *AQL* and *LQL* are small and when the cost of variables inspection is high. However, the risks associated with sampling can be precisely controlled only if there is no marked departure from normality and if the process standard deviation remains constant from lot to lot. The technique is sometimes known as narrow limit gauging or increased severity testing and is sometimes applicable to the construction of p-charts for process control. For further details, see Duncan.[32]

Bulk Sampling

The most common objective of bulk sampling is to determine the mean quality of the material where the material may be a gas, a liquid, or a solid. If the material is contained in packages, and if it can be assumed that it is uniform within each package, then techniques for discrete-product quality control can be used (see Sections 8.3.10 through 8.3.12). However, a different approach is required if the contents of the packages are commingled, if there is reason to suspect that the within-package contents are not uniform, or if the material is not packaged at all, that is, if it is contained in a pipeline or it is stored on the ground in a pile.

For each situation it is necessary to develop a model that will determine the sampling procedure and the acceptance criteria. The model, in turn, requires the establishment of assumptions concerning the statistical properties of the materials. In general, models fall into two groups: (1) models for distinctly segmented bulk material and (2) models for bulk material moving in the stream. See Juran et al.[33] for a comprehensive overview of these models, along with an extensive list of references.

REFERENCES

1. A. H. BOWKER and G. J. LIEBERMAN, *Engineering Statistics,* 2nd ed., Prentice-Hall, Englewood Cliffs, NJ, 1972.
2. J. M. JURAN, F. M. GRYNA, JR., and R. S. BINGHAM, JR., *Quality Control Handbook*, 3rd ed., McGraw-Hill, New York, 1974.
3. AMERICAN SOCIETY FOR QUALITY CONTROL, "ANSI/ASQC Standard A2," Milwaukee, WI, 1978.
4. JURAN et al., *Quality Control Handbook*, p. 16.
5. J. M. JURAN and F. M. GRYNA, JR., "Statistical Aids for Planning and Analysis Tests," *Quality Planning and Analysis*, McGraw-Hill, New York, 1970.
6. D. C. MONTGOMERY and R. G. HEIKES, "Process Failure Mechanism and Optimal Design of Fraction Defective Control Charts," *AIIE Transactions,* December 1976, pp. 467–472.
7. A. J. DUNCAN, *Quality Control and Industrial Statistics*, 4th ed., Irwin, Homewood, IL, 1974, p. 405.
8. DUNCAN, *Quality Control and Industrial Statistics,* p. 418.
9. E. M. SANIGA, "Joint Economically Optimal Design of $\bar{x}$ and R Control Charts," *Management Science*, Vol. 24, No. 4 (1977), pp. 420–431.
10. DUNCAN, *Quality Control and Industrial Statistics*, pp. 453–455.
11. DUNCAN, *Quality Control and Industrial Statistics*, pp. 464–484.
12. A. L. GOEL, "A Comparative and Economic Investigation of $\bar{x}$ and Cumulative Sum Control Charts," unpublished doctoral thesis, University of Wisconsin–Madison, 1968.
13. DUNCAN, *Quality Control and Industrial Statistics*, pp. 485–502.
14. JURAN et al., *Quality Control Handbook*, pp. 29-41–29-48.
15. DUNCAN, *Quality Control and Industrial Statistics*, pp. 157–161.
16. C. W. MORENO, "A Performance Approach to Attribute Sampling and Multiple Action Decisions," *AIIE Transactions*, September 1979, pp. 183–197.
17. DUNCAN, *Quality Control and Industrial Statistics,* p. 166.
18. U.S. DEPARTMENT OF DEFENSE, *Sampling Procedures and Tables for Inspection by Attributes*, MIL-STD-105D, Washington, DC, 1963.
19. DUNCAN, *Quality Control and Industrial Statistics*, pp. 179–198.
20. DUNCAN, *Quality Control and Industrial Statistics*, pp. 199–208.
21. A. WALD, *Sequential Analysis*, Wiley, New York, 1947, p. 57.
22. DUNCAN, *Quality Control and Industrial Statistics,* pp. 333–357.
23. H. F. DODGE and H. G. ROMIG, *Sampling Inspection Tables, Single and Double Sampling*, 2nd ed., Wiley, New York, 1959.
24. DUNCAN, *Quality Control and Industrial Statistics*, pp. 247–267; 268–282.
25. U.S. DEPARTMENT OF DEFENSE, *Sampling Procedures and Tables for Inspection by Variables for Percent Defective*, MIL-STD-414, Washington, DC, 1957.

26. H. F. DODGE, "Skip-Lot Sampling Plans," *Industrial Quality Control*, February 1955, pp. 3–5.
27. H. F. DODGE, "Chain Sampling Inspection Plan," *Industrial Quality Control*, January 1955, pp. 10–13.
28. E. F. TAYLOR, "Discovery Sampling," *Proceedings of the Ninth Annual ASQC Convention*, ASQC, Milwaukee, WI, 1955.
29. H. F. DODGE, "A Sampling Inspection Plan for Continuous Production," *Annals of Mathematical Statistics*, Vol. 14, 1943, pp. 264–279.
30. M. A. GIRSCHICK, *A Sequential Inspection Plan for Quality Control*, Technical Report No. 16, Applied Mathematics and Statistics Laboratory, Stanford University, Stanford, CA, July 1954.
31. A. WALD and J. WOLFOWITZ, "Sampling Inspection Plans for Continuous Production Which Insure a Prescribed Limit on the Outgoing Quality," *Annals of Mathematical Statistics*, Vol. 16, 1945, pp. 30–49.
32. DUNCAN, *Quality Control and Industrial Statistics*, pp. 264–265.
33. JURAN et al., *Quality Control Handbook*, pp. 25A-1–25A-14.

BIBLIOGRAPHY

FEIGENBAUM, A. V., *Total Quality Control–Engineering and Management*, McGraw-Hill, New York, 1961.

GRANT, E. L., and R. S. LEAVENWORTH, *Statistical Quality Control*, 4th ed., McGraw-Hill, New York, 1972.

CHAPTER 8.4

Improving Inspection Performance

COLIN G. DRURY

State University of New York at Buffalo

8.4.1 INTRODUCTION

Inspection is seen by the general public (and dictionary compilers) as a careful search for nonconformities or errors. As part of industrial quality control, it does much more, with three functions being of prime importance:

1. Preventing nonconforming goods or materials from proceeding further in the processing or from being sold to a customer. This is the customer protection function. Here the purpose is to remove nonconforming product.
2. Collecting data on specific characteristics of goods or materials for use in decisions regarding overall quality. This is the typical role of the inspector in a statistical quality control (SQC) scheme. Here the purpose is to identify imperfections whether or not they are severe enough to be considered nonconforming.
3. Collecting data on specific characteristics of goods or materials to give feedback to a manufacturing process. This is the process feedback function. As in function 2, the identification of imperfections can lead to trend analysis to prevent nonconfirmities.

The importance of all of these functions has increased rapidly over the past decade. First, the public is demanding greater accountability of industry for manufacturing and materials defects, leading to a wild increase in the costs of product liability litigation and in the corresponding insurance premiums. Second, it is now possible to design SQC schemes in which the inspection is assumed to have a known amount of error, rather than being done by a hypothetical perfect inspector. The increased savings from this mathematical treatment can be large,[1] but to obtain these savings a measure of inspection error is needed. Third, in the process feedback function, the enormous capital investment in complex production processes is putting an even greater premium on rapid, accurate inspection feedback to control the process near to its optimum working conditions.

The inspection function will continue to gain in importance, but its nature is constantly changing. The changes have always been to augment the human inspectors' senses so that he or she can be more discriminating between conforming and nonconforming products. These changes will continue, but they will not decrease the importance of the inspector. Ultimately it is the human inspector who makes the decision on an item of product, whether the inspector is performing unaided visual inspection, reading a gauge, or setting up and calibrating an automatic inspection device.

The three functions of inspection are often seen as different jobs within an organization and often have very different characteristics. The customer protection function is typically a 100% sorting of the product, often done as part of an assembly or packing process, and often without much augmentation of the human senses. An example is the examination of bakery products[2] or integrated circuit chips.[3] The SQC function usually involves the sampling of a product and measurement by more highly trained inspectors using gauges. For example, Evans[4] evaluated micrometer inspection of metal parts, and McKennal[5] examined the grading of wool quality on the basis of a small sample. Inspection process feedback is often closely associated with the process itself and indeed can be performed by the process operator rather than a separate inspector. The output of such a process inspection can be displayed to the process operator as a control chart. An extensive study in many industries by Stok[6] has shown that the use of control charts has a consistent, positive effect on quality.

In all of these applications, the inspection device, whether unaided human, human plus gauges, or automated, *will* make errors. It is the purpose of this chapter to show how to measure these errors, how to reduce them as much as possible, and finally how to build a reliable quality control scheme despite the errors.

Inspection error is a fact of life, and designing quality control schemes that ignore inspection error is attempting to force the inspection system to perform an impossible task and hence to fail in its designated task. It is instructive to note that inspection errors had been recognized as early as the 1920s and 1930s (e.g., by Juran in 1935[7]). Inspectors were found to be less than 100% accurate in 100% inspection, so alternatives to 100% inspection were sought. These were SQC schemes, however, that assumed perfect inspectors. The imperfect 100% inspector was supposed to become the perfect sample inspector! If an inspector examines only, say, 1% of a product instead of 100%, there is much that can be done to use the cost savings to improve the inspector's performance, for example, by using more expensive and accurate gauging or by allowing considerably more time per item of product inspected. However, under conditions where the inspection task remains the same, there is no evidence that the inspector behaves any differently in sampling inspection than in 100% inspection.[8]

8.4.2 DEFINITIONS OF INSPECTION ERROR

Statistical quality control distinguishes between two types of inspection, attributes and variables, and the same distinction must be made in reducing inspector error. For attributes inspection, the inspector must classify each item of product inspected into two or more categories. The most usual categories are just two: accept and reject (also referred to as "conforming" and "nonconforming" or by other names). Items can be truly conforming and nonconforming, and the inspector can classify them as accept or reject. We thus have the outcomes shown in Exhibit 8.4.1.

In this exhibit, A, B, C, and D refer to the number of items in each cell of the table. Obviously we should like to have all items classified correctly, with A and D as large as possible and B and C as small as possible. This goal will never be achieved perfectly because the two wrong decisions, B and C, are never zero. The two errors have special names, as follows:

B. Nonconforming item is accepted. Usually called a type 2 error or a missed nonconformity.

C. Conforming item is rejected. Usually called a type 1 error or a false alarm.

Corresponding to these two errors are the two measures of inspection performance.

A. Conforming item is accepted. The performance measure is probability of good item being accepted $A/(A + C) = p_1$.

D. Nonconforming item is rejected. The performance measure is probability of a nonconforming item being rejected $D/(B + D) = p_2$.

These two measures, p_1 and p_2, measure the effectiveness of an attributes inspector. The efficiency is measured by the time or cost used to classify each item. As will be shown later, p_1 and p_2 are closely related to each other and to the time and cost of inspection.

The effects of p_1 and p_2 on inspection performance can be seen by replacing the numbers in Exhibit 8.4.1 with probabilities. If the overall probability of a nonconforming product is p' in the usual SQC terminology, then we can rewrite the outcomes as shown in Exhibit 8.4.2.

Values of p_1 and p_2 have been measured in a variety of industries, from electronics, through glass and metal products, to textiles and even food. Sinclair[9] gives an extensive tabulation of values, which average around .90 to .99 for p_1 and .80 to .90 for p_2. Some values are much lower, with this author having found p_1 and p_2 as low as .5 in particularly poor industries.

From the measures p_1 and p_2, we can define an important system performance measure, the effective fraction nonconforming. This is the fraction of the total input that is rejected by the inspector and is what management thinks is the fraction nonconforming if it believes inspection is perfect. From Exhibit 8.4.2 this is

$$\text{effective fraction nonconforming} = p'_e = (1 - p_1) - p'(1 - p_1 - p_2)$$

Exhibit 8.4.1 Outcomes for Attributes Inspection

Decision of Inspector	True State of Item: Conforming	True State of Item: Nonconforming
Accept	A	B
Reject	C	D

Exhibit 8.4.2 Outcomes for Attributes Inspection Expressed as Probabilities

	True State of Item		
Decision of Inspector	Conforming	Nonconforming	Total[a]
Accept	$p_1(1-p')$·	$(1-p_2)p'$	$p_1+p'(1-p_1-p_2)$
Reject	$(1-p_1)(1-p')$	p_2p	$(1-p_1)-p'(1-p_1-p_2)$
Total	$1-p'$	p'	1

[a]Note that the totals have been rearranged for computational convenience.

Exhibit 8.4.3 shows that both p_1 and p_2 have large effects on p_e'. In particular, the farther away p_1 is from 1.0, the worse the quality looks to management. The usefulness of p_e' is that it can be used in place of the true fraction nonconforming p' in the design of SQC schemes to redesign sampling plans to behave correctly despite inspection error. Section 8.4.5 develops this further.

For variables inspection the inspector must take a measurement of an individual product characteristic, such as a dimension or a resistance. The value of the measurement will not be identical for all individual items of product, nor will an identical value be obtained if a specific item is remeasured. The distributions of both item-to-item and measurement-to-measurement variation tend to be reasonably unimodal and symmetrical and are often approximated by normal distributions. The *true* distribution of the measurement of the product has a mean value μ_p and a variance σ_p^2. This true distribution can be subjected to two types of error, called by Mei, et al.[10] "bias" and "imprecision." Their separate and combined effects are shown in Exhibit 8.4.4. Bias shifts the mean of the measurements, and imprecision increases the variability of measurements. The distribution of measured values, when affected by an inspector bias μ_c and an imprecision μ_c^2, can be seen to have a mean $\mu = \mu_p + \mu_c$ and a variance $\sigma^2 = \sigma_p^2 + \sigma_c^2$ if the inspection error is independent of the value of the product measurement, as is often the case.

Measurements of bias and imprecision are not often reported in the inspection literature since they tend to be specific to the product characteristic measured and to the instruments used for measurement. Imprecision is a real problem, however, with Lawshe and Tiffin[11] reporting the

Exhibit 8.4.3 Effect of Inspection Error on Effective Fraction Nonconforming

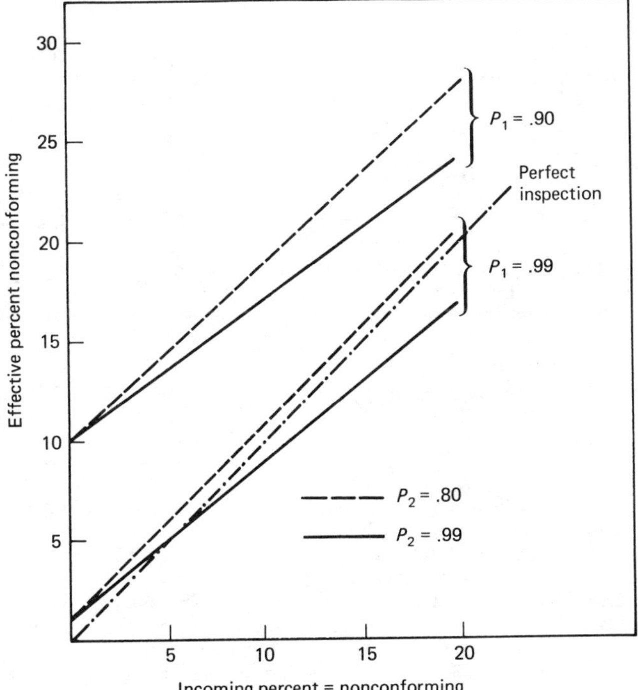

Exhibit 8.4.4 Effects of (*a*), Bias, (*b*) Imprecision, and (*c*) The Two Combined on Variables Inspection

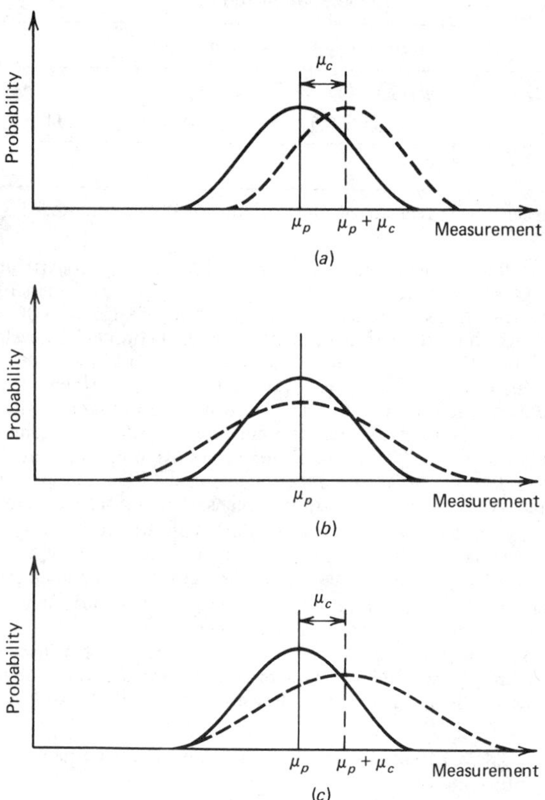

results of more than 100 inspectors using 11 different precision instruments, in which the imprecision was too high for 50% of the inspectors on most of the instruments. The trend in variables inspection is toward increasing both the precision and the reliability of the measurement tool. With digital readouts, go/no go lights, and other such improvements at the human/machine interface, the potential for error in variables inspection is greatly reduced.

8.4.3 ANATOMY OF AN INSPECTION JOB

To improve an inspection job, we must first be able to measure what needs improvement. Inspection is a complex job, despite the relatively simple measures of inspector performance (p_1, p_2, μ_c, σ_c^2) defined in Section 8.4.2. It is complex enough that many schemes have been proposed for suitable components of the job as an aid to analysis. For attributes inspection almost all inspection jobs involve the following four tasks[12]:

1. Present preselected items for inspection.
2. Search the item to locate possible nonconfirmity ("imperfections").
3. Decide whether each imperfection is sufficiently severe to be classified as a nonconformity.
4. Take the appropriate action of acceptance or rejection.

The first and last of these tasks are relatively reliable operations, whether done by human or machine. Human performance in these tasks involves mainly movements of the hands, so that the well-known predetermined motion time system models in Chapter 4.5 can be used to identify critical factors, suggest improvements, and estimate speed of performance.

It is the search and decision making tasks in inspection that take a large proportion of the time and contribute greatly to inspection error. Search and decision making are considered in turn to show how inspectors perform each of the two tasks and how to improve performance on each.

For any industrial job, only three changes are possible: change the person performing the job, change the job, or change the place of the job within the organization. For both search and decision components, the effective strategies for deciding which change to make are presented. In general, changing the person by selection and training has been found to be less effective in inspection jobs than changing the job itself or its organizational environment. In particular, selection for inspection was reviewed extensively by Wiener[13] who concluded that "Personnel Selection, long a favorite of industrial management, offers little promise at present."

8.4.4 IMPROVING INSPECTION PERFORMANCE

Search Tasks

The most usual search for nonconformity is visual search, for example, searching glass plates for inclusions[14] or trays of baked goods for malformed goods.[2] However, search can also be procedural, as in searching complex electronic chassis for electrical and visual nonconformity.[15] In either case, search proceeds as a series of functions linked by eye movements for visual search or hand or eye movements for procedural search. Procedural search can be dealt with using standard task analysis methods, and its reliability estimated from these task analyses.

In visual search almost all of the information taken in by the inspector is taken in during the fixations, which average about one third of a second in duration and account for more than 90% of the search time. Eye movements are extremely rapid, and the time they take is so short compared with the fixation time that they are usually ignored. This has led to models of visual search as a sequence of fixations. In each fixation an inspector can detect imperfections in an area of the item, called the "visual lobe," around the center of the fixation. This area varies with the conspicuity of the imperfection and is *not* simply the most sensitive part of the retina, called the "fovea centralis."

The area of this visual lobe is a function of the luminance of the object inspected, the contrast between the object and the nonconformity on the object, the size of the imperfection, and the distance of the imperfection from the inspector's eyes. It is possible to derive detailed mathematical expressions for how much a person can see in one fixation given exact values of all these factors (e.g., Greening[16]), but for inspection tasks it is sufficient to know that these factors affect the lobe size and to optimize them in designing inspection jobs.

Eye movement studies of inspectors show that they do not follow a simple pattern in searching an object.[17] Some tasks have very random appearing search patterns (e.g., circuit boards), whereas others show some systematic search components in addition to this random pattern (e.g., knitwear[17]). However, all who have studied eye movements agree that performance, measured by the probability of detecting an imperfection in a given time, is predictable assuming random search.[12] The equation relating probability (p_t) of detection of an imperfection in a time (t) to that time is

$$p_t = 1 - \exp\left(-\frac{t}{\bar{t}}\right)$$

where $\bar{t}$ is the mean search time. Further, it can be shown that this mean search time can be expressed as

$$\bar{t} = \frac{t_o A}{apn}$$

where t_o = average time for one fixation
A = area of object searched
a = area of the visual lobe
p = probability that an imperfection will be detected if it is fixated (This depends on how the lobe area (a) is defined. It is often defined such that $p = \frac{1}{2}$. That is an area with a 50% chance of detecting an imperfection.)
n = number of imperfections on the object

The relationship between p_t and t can be seen in Exhibit 8.4.5 found for inspectors searching for two types of imperfection in sheet glass. Obviously, as the inspector is given more time to search, he or she has a better chance of finding the imperfection. It is also evident from the exhibit that search times will be highly variable, ranging (in this example) from 0.7 to 2.5 sec for the easier imperfection and from 1.5 to 8.5 sec for the more difficult imperfection. Later we shall see that this extreme natural variability of search times has definite implications for design of inspection jobs.

The preceding search model means that to maximize search performance we must improve the *conspicuity* of an imperfection in relation to its background. This conspicuity is measured by the

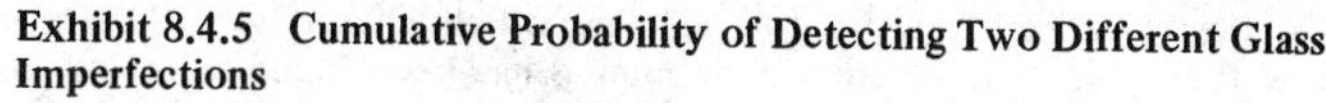

Exhibit 8.4.5 Cumulative Probability of Detecting Two Different Glass Imperfections

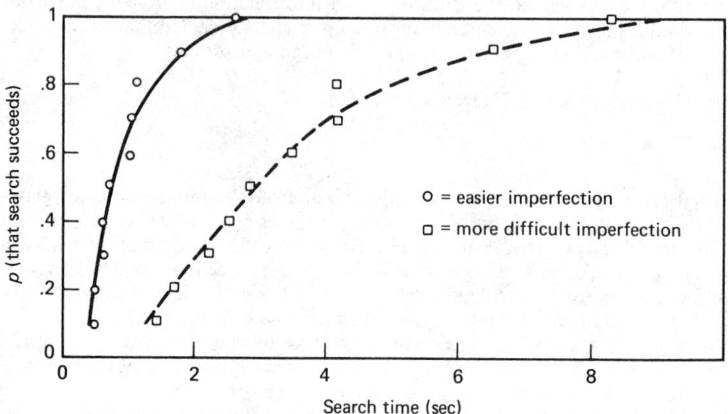

amount of product that can be searched per unit time, that is, by $(1/\bar{t})$. We can proceed through the expression for $\bar{t}$ and deduce the necessary conditions for rapid accurate search.

Improving conspicuity means:

Decreasing the area of object or objects to be searched in a given time. Examples of this are instructions and visual aids to allow the inspector to concentrate on those parts of the object that are most likely to contain nonconformities. The other obvious way to use this principle is to increase the time available for inspection, as in Exhibit 8.4.5. The optimum time per item for a search task is about 2 to 3 times the mean time needed to locate a single imperfection, if one is present. Morawski[18] shows that this intuitive time is reasonably constant over a number of different economic conditions of inspection.

Increasing the visual lobe size. This means increasing visual size of the nonconfirmity, for example, by magnification or shadowgraphing for transparent objects; increasing the contrast between the fault and its background by special lighting techniques[19]; decreasing the incidence of clutter in the background, which may be confused with an imperfection; or increasing the overall illumination, although this gives diminishing returns above about 1000 lux. Visual lobe size can also be increased by choosing inspectors with better visual acuity in the periphery (not the foveal acuity usually measured in eye tests). Ericson[44] found that peripheral visual acuity correlated with search performance. Special methods of improving search efficiency include overlays and blink inspection. Overlays are special patterns that are laid over (or projected onto) an object in such a way that any differences between the pattern and the object are enhanced. Teel et al.[20] used overlays to improve the detection of nonconformity in printed circuit board photomasks by 42%, at a savings of $10,000 per year in 1966 prices. Blink inspection carries this integration of a pattern and an object one stage further by showing the inspector a perfect object and the object to be inspected in rapid alternation on the same visual field. Any difference between object and perfect objects will appear to flash on and off with each alternation. A recent evaluation of this method[21] showed it to be more suited to some imperfections than others and to work best on complex products such as circuit boards.

Decreasing the duration of each fixation by choosing inspectors with faster fixations or training them to fixate more rapidly would seem like an obvious strategy, but it does not work. Schoonard et al.[3] found that the best inspectors were the ones who reached the target in fewest fixations, not the ones with more rapid fixations. What seems to happen is that better inspectors have a larger visual lobe. There is evidence that lobe size, and hence speed of inspection, can be trained by a properly designed training procedure.[22]

Automating search, as with flying-spot scanners for sheet products, which detect surface imperfections, can be used to mark any surface imperfection above a particular size, and the decision on each nonconformity or each object left to the human inspector or automated. As search is the main consumer of time, it is the subtask of inspection that is usually automated after the simpler materials handling subtasks. Exhibit 8.4.6 shows 10 examples of automating the search component of inspection. The advent of cheap, distributed computing power microprocessors has made it possible to utilize the large amounts of data generated by the detection system, so that 100% inspection again becomes a reality.

Exhibit 8.4.6 Automated Search in Inspection[a]

Object Searched	Imperfection Types	Method Used	Authors of Chapters Listed in Reference 23
Textile fabrics	Broken threads	Optical diffraction/ optical computing	Kasdan and Mead
Wood planks	Cracks, holes	Photodiode array/ microcomputer	Claridge and Purll
Metal, paper plastic, glass sheet, 3 ft wide	Surface defects, 0.04 in. diameter	Photodiode array/ microprocessor	Claridge and Purll
Turned metal products, for example, bolts, shafts	Damaged threads, missing features	Fiber optic or capacitance probe	Brooks and Bhattacharya
Plug-in digital modules	Electrical performance faults	Computer simulation	Kline
Machined metal parts	Surface faults, 10^{-4} in. diameter	Laser scanner/photodetector	Fadl and Parsons
Sheet products up to 13 ft wide	Surface faults, 10^{-3} in. diameter	Laser scanning/photomultiplier	Clarke and Bedford
Cylinder bores	Surface faults, 4×10^{-3} in. diameter	Laser, rotating probe, photomultiplier	West and West
Gun cartridges, glass reinforced plastic pipes	Thin spots; thickness change, 3×10^{-3} in.	X rays, scintillation counter	Basler
Shell projectiles, 15 in. long by 4 in. diameter	Cavities, cracks, 2×10^{-2} in.	Gamma ray scattering, photomultipliers	Karlin and Guttwein

[a]From reference 23.

Decision Making

At its simplest, the inspector's decision on an object is to decide whether it is to be accepted or rejected. If search is involved, he or she must first locate the imperfection; having located it, the decision about acceptability of the item may be more or less difficult. If the standards say that no imperfection, however small, is acceptable, then the inspection task involves search only: the decision making is trivial. Such a situation is characterized by an almost complete absence of type 1 errors, because there is no reason to reject a good object. At the opposite extreme are tasks in which search is trivial, as in grading of wool,[5] but decision making is complex and prone to error. In most tasks nontrivial search leads to an imperfection requiring a nontrivial decision, for example, the inspection of castings for chain saw housings for surface blemishes. Performance on such an inspection task was measured before and after lighting improvements, with the results shown in Exhibit 8.4.7. The results were typical of inspection tasks requiring both search and decision making in that more time per object gave an increase in imperfection detection performance, but was accompanied by an increase in false alarms (decrease in p_1). This is because, as more time is allowed, more imperfections are found in the search, but not all of them should cause rejection. Thus at long inspection times inspectors will have more opportunity to reject good items.

The pure decision component of inspection can be measured by concentrating on tasks that require no search. Many such tasks have been measured in the factory and in the laboratory, for example, the studies reviewed in Drury and Fox.[24] The general conclusion of these studies is that such tasks are those rare tasks where a human being behaves like a rational economic decision maker, balancing the costs and payoffs involved to arrive at an optimum performance. Unfortunately, what is optimum to the inspector may not be optimum to the company, so we must consider decision making in more detail.

Just as the four outcomes of a decision making inspection can have probabilities associated with them (Exhibit 8.4.2), they can have costs and rewards also: costs for errors and rewards for correct decisions. Exhibit 8.4.8 shows a general cost and reward structure, usually called a "payoff matrix," in which rewards are positive and costs negative. A rational economic maximizer would multiply the probabilities of Exhibit 8.4.2 by the corresponding payoffs in Exhibit 8.4.8 and sum them over the four outcomes to obtain the expected payoff. He or she would then adjust those factors under his or her control, p_1 and p_2, so as to maximize this expected payoff. Such an exercise can be

Exhibit 8.4.7 Effect of Time Allowed on p_1 and p_2 for Chain Saw Housing[a]

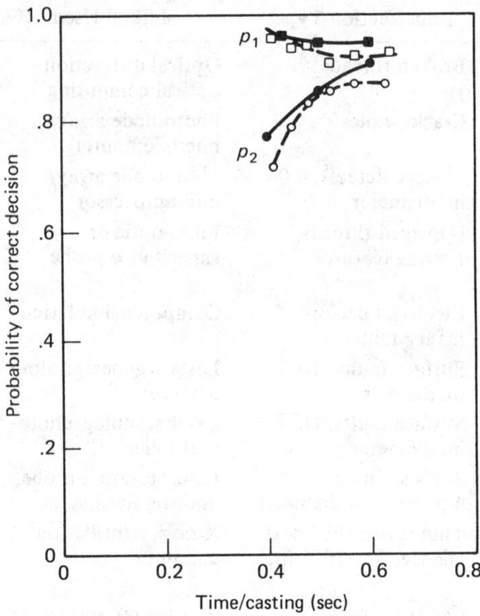

[a]Full lines represent performance after implementation of lighting improvements.

carried out,[12] given assumptions about how p_1 and p_2 can be changed by the inspector. The most often-tested set of assumptions comes from a body of knowledge known as the theory of signal detection, or TSD (e.g., McNichol[25]). This thoery has been used for numerous studies of inspection, for example, sheet glass,[26] electrical components,[27] and ceramic gas igniters,[28] and has been found to be a useful way of measuring and predicting performance.

Basically, TSD states that p_1 and p_2 vary in two ways. First, if the inspector and task are kept constant, then as p_1 increases, p_2 decreases, with the balance between p_1 and p_2 dependent upon the costs and payoffs (a, b, c, and d) and the nonconforming rate (p'). This balance point is known as the "criterion." Second, changes to the inspector and/or the task will increase p_1 and p_2 together by changing the discriminability for the inspector between acceptable and rejectable objects.

The objective in improving decision making is to reduce decision errors. These can arise directly from forgetting imperfections or standards in complex inspection tasks or indirectly from making an incorrect judgment about an imperfection's severity with respect to a standard. Ideally, the search process should be designed so as to improve the conspicuity of rejectable imperfections (nonconformities) only, but often the measures taken to improve conspicuity apply equally to nonrejectable imperfections. Reducing decision errors usually reduces to improving the discriminability between imperfection and a standard.

The first prerequisite is to have the standard available for the inspector, as close as possible in time and space to the imperfection severity. This can take the form of limit standards for each imperfection available at the workplace. For surface imperfections or misshapen parts, the limit standards can be photographs of imperfection severities or examples of nonconformities. If there are many imperfections, a random access slide projector can be used to enable the inspector to obtain a particular standard quickly at the inspection station. Limit standards change the decision making task from one of absolute judgment to the more accurate one of comparative judgment.

Exhibit 8.4.8 Payoff Matrix for Attributes Inspection

	True State of Item	
Decision of Inspector	Conforming	Nonconforming
Accept	a	−b
Reject	−c	d

Harris and Chaney[29] showed that limit standards for soldered joints gave a 100% improvement in inspector consistency for near-borderline cases. Kelley[30] showed that limit standards for TV screen inspection reduced undetected nonconformities from 2.3% to 1.4% and customer returns from 1.7% to 0.4%.

Collecting the limit samples and obtaining management and customer agreement on them is a necessary step in providing them for the inspector. It is also a step that is valuable in itself in translating inspection requirements directly into inspector actions. Often this step alone will reduce inconsistency between inspectors and, by forcing management and customers to define what they both mean by quality, will reduce management-customer conflicts.

Given the limit standards, then, any means of increasing the precision of comparative judgment will allow decision making to improve. Instruments, tools, and gauges all achieve this function, although as Evans[4] has shown, few skilled workers can use measuring tools consistently without regular retraining. The simplest tools to use are the go/no go gauges, which are usually employed to measure sizes of components, although they can be used to measure voltages, air pressures, and so on. The standard is in effect built into the measuring instrument, so that the range of the variable is dichotomized into "accept" and "reject" regions. However, even go/no go gauges are not guarantors of 100% inspection performance (see next part of this chapter).

The next step after installation of gauging or measuring instruments is to develop semiautomatic or fully automatic measuring equipment. Semiautomation shows a value of a variable (or set of variables) which is interpreted by an inspector, whereas fully automatic inspection does the interpretation and merely indicates the decision. The advent of cheap, distributed computing power is changing the field of automatic inspection rapidly: a sample of 10 recent applications is shown in Exhibit 8.4.9.

With any form of measuring instrument, whether manual or automatic, general principles of human factors in equipment design apply. Such methods have been used to reduce inspection costs for rubber seals by 25%[31] and to design less error-prone inspection drawings,[15] which improved inspection accuracy 42%.

There appear to be no selection tests available that will choose better decision makers, but training has been found to improve decision making performance. The methods available for training inspectors have been reviewed by Embrey[32] and include both on-line and off-line methods. As with most training schemes (e.g., Salvendy and Seymour[33]), off-line training is particularly cost-effective as both the cues given by the task and the feedback to the inspector can be controlled.

Exhibit 8.4.9 Automated Decision Making in Inspection[a]

Object Inspected	Parameter Measured	Transducer	Authors of Chapters Listed in Reference 23
Machine tool slideways	Straightness	Sensor coil to detect position	Crabtree, Goodhead, McGoldrick
Ground cylindrical parts	Surface finish	Tracer pin/magnet and coil	Dutschke and Eissler
Coal on conveyer	Volume of flow	Two-dimensional photodiode array	Claridge and Purll
Parts turned on engine lathe	Dimensions, measured during cutting	Laser/photodetector	Mergler and Sahajdak
Hypodermic needle point	Quality of point	Optical diffraction/ optical computing	Kasdan and Mead
Nuclear power fuel elements	Diameter of element	Capacitance transducer with digital readout	Boorman and Wilson
Machine tool slides	Flatness and straightness	Laser interferometry/ liquid surface	Burdekm, Cowley, and Fields
Cartridge cases	Four dimensions of case	Linear diode array	Coleman and Swinth
Various objects	Position of object	Limit switches: inductive, capacitative, magnetic, hall effect	Rasmussen
Automobile oil filter	Quality of seal	Air pressure/ diaphragm pressure transducer	Heginbotham et al.

[a]From reference 23.

For example, Parker and Perry[34] trained glass component inspectors by progressively introducing them to more and more difficult faults, with practice and feedback after each new level of difficulty. Errors were halved after substituting this progressive training program for a more normal "exposure on the job" approach. In another example on machined metal parts, a set of four 1 hr training sessions of lecture, demonstration, and discussion improved performance of highly experienced inspectors by 32%[15]. A recent study by Czaja and Drury[22] has shown that an active approach to training, which requires the trainee to make an overt decision at each step, leads to large increases in performance over a well-designed passive training scheme. Such an active method is particularly useful with older trainees.

Organizational Factors

The most general organizational factor is whether inspectors should be part of production or in a separate quality control department. It has been found[35] that inspectors see themselves as "leagued together against a hostile world" and rarely take the production side in production/inspection disputes. Jamieson[36] found that a separate quality control department made fewer errors than inspectors in production, but there could be an element of suboptimization. The objective should be to improve production and inspection together. Although *inspection* is improved if inspection and production are separate, to improve *production quality* the inspection should be linked to production as closely as possible in order to give the most rapid feedback of quality performance. Indeed, in an extensive study of the effect of visual quality feedback on production in more than 30 industries, Stok[6] found that visual feedback through control charts always improved product quality.

Given that inspectors tend to "stick together," it is natural that they develop group norms of accepted performance. McKenzie[37] reported that, in one plant where total production was doubled, rejects remained constant at the numbers the inspectors thought reasonable. Drury and Addison[26] also found that inspectors change their effective standards with incoming quality, in this case keeping output fraction nonconforming constant. Thomas and Seaborne[35] showed clearly the effects of group norms on a go/no go gauging task where experimental subjects all shifted their rejection rate to conform with the rejection rates of the other three members of the group.

The reward structure was shown earlier to be a determiner of the balance between type 1 and type 2 errors. It is only natural, therefore, to use this as a method of improving performance. Unfortunately, there is little evidence from industry that paying inspectors an incentive based on their type 1 and type 2 errors has any effect on the incidence of these errors. Of course their outputs of acceptable and rejectable parts must be sampled to derive p_1 and p_2, which means both reinspection costs and payments based upon highly variable small samples. (Incentive schemes based on throughput have been used to increase throughput, but since the objective of inspection is *not* throughput, they are irrelevant to improving inspection performance, i.e., p_1 and p_2). In laboratory tasks, rewards and costs have been demonstrated to change performance in the way predicted, but industrial inspectors are subject to far larger and more complex pressures and cannot be expected to change their performance in a simple manner. These pressures are from manufacturing (to pass items), from sales (to prevent nonconformities reaching the customer), and from individual production workers (to maintain consistent decisions on individual objects).

Incentive schemes not only provide rewards to inspectors, but also provide feedback on performance. It may well be that this feedback is a motivator, irrespective of the financial rewards. In their study of glass inspection, Drury and Addison[26] found that provision of rapid feedback from a small sample reinspection reduced type 1 and type 2 errors by half. Exhibit 8.4.10 shows how the discriminability changed after the provision of this feedback.

Feedforward of the types of imperfections likely to be encountered in a multiimperfection inspection task has also proved effective. Exhibit 8.4.11 shows how the provision of feedforward information to inspectors improved the TSD parameter of discriminability in a study of metal parts inspection.[38] The effect was to halve the errors and swamp a small age decrement in performance.

The problem of conveyor-paced inspection needs to be addressed. We have seen that, for tasks involving search, the time needed to inspect an object can vary over a wide range. Any conveyor or paced system must recognize this fact by allowing for extreme variability in inspection times. With imperfections' detection time ranging about a half to a third of the good object acceptance time, the mean pacing speed will have to vary with imperfection rate. But the choice of correct mean speed is not nearly so important as the allowance of a wide tolerance time about this mean speed. It is for this reason that most authors agree that conveyor-paced performance is worse than unpaced performance at the same speed. For example, Fox and Richardson[39] found only half as many imperfections detected in paced inspection as were detected in unpaced inspection. If conveyor-paced inspection is a practical necessity for production despite its problems, Buck[40] gives much good advice on how to organize the conveyor pacing.

Exhibit 8.4.10 Improvement in Discriminability of Nonconformity in Glass Sheets With a Change to Rapid Feedback

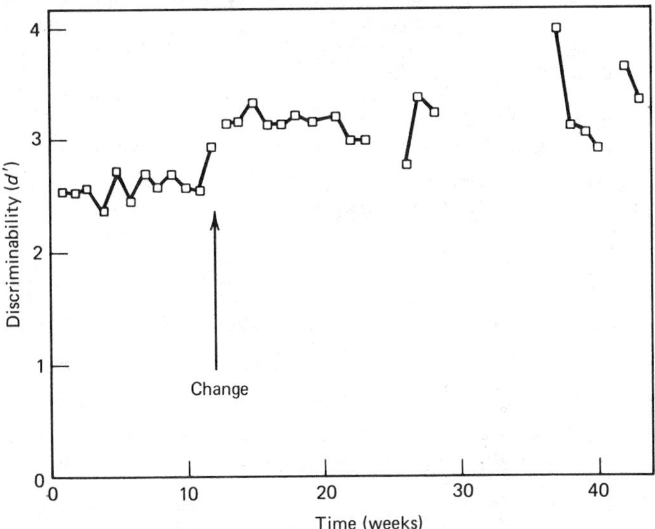

Finally, it should be remembered that all of this discussion of improvement is based on the inspection *task*—not on the whole inspection *job*. The *task* of inspection is what the inspector spends much of his or her time doing, but there is more to any *job* than the central *task*. A prime example is in inspection of the same object for prolonged periods. This is usually seen as a classic problem in trying to maintain a state of vigilance for long periods. In laboratory tests it is almost always found that performance starts to deteriorate after a few minutes on monotonous monitoring tasks, and that after 20 to 30 min performance is at quite a low level. The author has found only one documented "vigilance decrement" in an industrial inspection task, where Sinclair[2] found an initial warm-up in performance of chicken inspectors, followed by a gradual fall in performance over a 2 hr period. However, most textbooks continue to recommend that a single inspection task should be limited to 20 to 30 min in duration.

In many industrial jobs it has been found that both rotation and enrichment of jobs improve productivity and worker satisfaction. Rigby and Swain[45] recommend the formation of pairs of

Exhibit 8.4.11 Improvement in Discriminability of Nonconformity in Metal Hooks With Provision of Feedforward

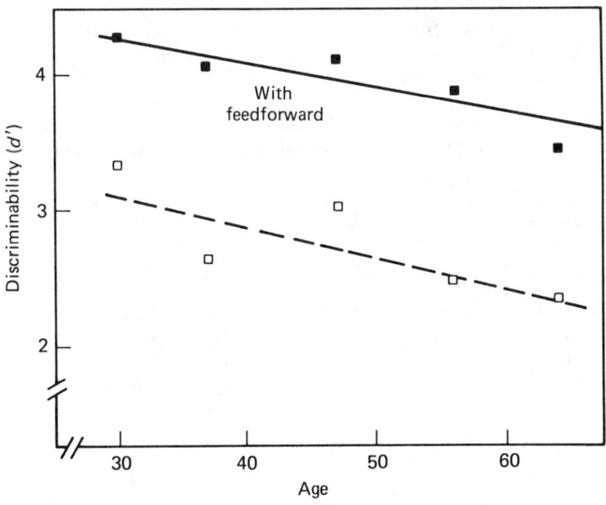

operators who alternate jobs in the assembly *and* inspection of pairs of components. In this do-check-do method, each operator inspects the previous section of assembly done by his or her partner and then assembles the next section. This cycle is repeated until the assembly is complete. In an example from a high technology industry, Maher et al.[41] enriched the jobs of 22 inspectors to that of quality analyst, with a much wider range of responsibilities and duties. Reject rate fell from 7% to less than 3%, time required for inspection fell by 50%, and there were measurable improvements in attitude. However, not all inspectors had their jobs enriched, and those who did received a promotion, so job enrichment may not be the whole story.

Hackman et al.[42] note that jobs should be enriched in terms of five components:

1. Skill variety–the worker needs the challenge of using several skills.
2. Task identity–the job should accomplish some meaningful whole.
3. Task significance–the worker should perceive that what he or she is doing benefits society at large.
4. Autonomy–the worker needs to have some freedom and independence rather than being preprogrammed.
5. Feedback–the worker needs timely and accurate knowledge of the effectiveness of his or her efforts.

Three major strategies to accomplish improvements in these components are:

1. *Establishing client relationships* by opening contacts between the worker and the ultimate user of the product. This is often done in a qulaity control context by having inspectors visit the plants of customers and/or suppliers.
2. *Vertical loading*, or allowing the worker to take responsibility for the planning, scheduling, and organizing of the job as well as just doing the job. In inspection jobs this may mean being decoupled from the production line and being allowed to help use the information the inspector is generating to improve production quality.
3. *Opening feedback channels*, which is notoriously difficult in quality control work, but, as Exhibit 8.4.10 shows, it can have a dramatic effect on inspector accuracy at relatively low cost.

As was stated previously, inspection is always of people: the people who produced the items the inspector inspects. It is thus a stressful job, and any steps that can be taken to make it more meaningful and less stressful will ultimately have benefits in employee health care costs as well as in direct performance improvement.

8.4.5 RELIABLE SYSTEMS WITH UNRELIABLE INSPECTORS

So far the emphasis has been on measuring and reducing inspection error. However, there will always remain a nonzero incidence of error in any human activity, and the design of the quality control system must take this into account explicitly. For 100% inspection, there is no alternative but reinspection, although this must also be less than perfect. For example, if 10% of the incoming objects are nonconforming and if an inspector catches 90% of the nonconformities ($p_2 = 0.9$) and makes 2% type 1 errors ($p_1 = 0.98$), then the objects after inspection will contain only 1.12% nonconformities. Bearing in mind the discussion of the effect of nonconformity rate in an earlier part of this chapter the detection of the remaining nonconformities will be much more difficult. Belbin[43] showed just this effect in the reinspection of ball bearings.

One problem is that, if there are many types of imperfection, both the original inspector and the reinspector will tend to miss the same type of nonconformity. A partial solution is to instruct the reinspector(s) to look for nonconformities different from those found by the original inspectors.

A partial solution to the problem in nonconformity rate in the reinspected batch may be to bias the original inspector toward rejection so heavily that the outgoing quality is almost perfect. The reinspection can then be done on the items *rejected* so as to find good items. For example, if our inspector ($p_1 = 0.98$, $p_2 = 0.90$) is encouraged to reject so that he or she has p_1 and p_2 reversed ($p_1 = 0.90$, $p_2 = 0.98$), then the outgoing percent nonconforming will only be 0.25%. The percentage of conforming items in the rejected items will have increased from 16.7% to 47.9%, making reinspection of the rejected items much easier.

For sampling inspection, both for attributes and variables, methods exist for designing sampling schemes to have known characteristics despite inspection errors. In variables inspection, Mei et al.[10] show that a bias, μ_c, can be added directly to a lower specification limit decision constant or subtracted from one for an upper specification limit, for imprecision. They define

$$h = \frac{\sigma^2}{\sigma_c^2}$$

and show that, for any variables plan, the sample size n should be multiplied by $(h + 1)/h$ to compensate for the inspector error.

In attributes inspection the compensation for human error is again quite simple. Wherever a fraction nonconforming, p', occurs in the equations defining the sampling plan, it is replaced by its effective value

$$p'_e = (1 - p_1) - p'(1 - p_1 - p_2)$$

Bennett[1] uses this method to compensate for various levels of inspection error in a single sampling plan, with and without rectification, and shows large savings compared with the usual practice of ignoring inspection error. He uses the same values of p_1 and p_2 at all values of p', which, as has been shown in an earlier part of this chapter may not be as realistic as allowing p_2 to increase and p_1 to decrease with increasing p'. Drury[12] shows how the search and decision making models of an earlier part of this chapter can be used to compensate a single-sampling plan for inspection error. He shows that, provided that enough time per object is allowed to complete searching, there need be no increase in sample size to compensate for inspectors decision errors.

Finally, it should be reemphasized that reliable inspection systems can be designed despite inspection error, but only if the error is faced and measured. Ignoring inspection error always leads to increased quality costs. And exhorting inspectors to be perfect, when both management and inspectors know that perfection is impossible, only increases the stress associated with the inherently stressful job of inspection.

REFERENCES

1. G. K. BENNETT, "Inspection Error: Its Influence on Quality Control Systems," in C. G. Drury and J. G. Fox, Eds., *Human Reliability in Quality Control*, Taylor & Francis, London, 1975.
2. D. E. CHAPMAN and N. A. SINCLAIR, "Ergonomics in Inspection Tasks in the Food Industry," in C. G. Drury and J. G. Fox, Eds., *Human Reliability in Quality Control*, Taylor & Francis, London, 1975.
3. J. W. SCHOONARD, J. D. GOULD, and L. A. MILLER, "Studies of Visual Inspection," *Ergonomics*, Vol. 16, 1973, pp. 365–380.
4. R. N. EVANS, "Training Improves Micrometer Accuracy," *Personnel Psychology*, Vol. 4, 1975, pp. 231–242.
5. A. C. MCKENNAL, "Wool Quality Assessment," *Occupational Psychology*, Vol. 32, No. 1 (1958), pp. 50–60; Vol. 32, No. 2 (1958), pp. 111–119.
6. T. L. STOK, *The Worker and Quality Control*, University of Michigan Press, Ann Arbor, 1905.
7. J. JURAN, "Inspector's Errors in Quality Control," *Mechanical Engineering*, Vol. 57, 1935, pp. 643–645.
8. Y-C TSAO, C. G. DRURY, and T. B. MORAWSKI, "Human Performance in Sampling Inspection," *Human Factors*, Vol. 21, No. 1 (1929), pp. 99–105.
9. N. A. SINCLAIR, "The Use of Performance Measures on Individual Examiners in Inspection Schemes," *Applied Ergonomics*, Vol. 10, No. 1 (1970), pp. 17–25.
10. W-H MEI, K. E. CASE and J. W. SCHMIDT, "Bias and Imprecision in Variables Acceptance Sampling: Effects and Compensation," *International Journal of Production Research*, Vol. 13, No. 4 (1975), pp. 327–340.
11. C. H. LAWSHE and J. TIFFIN, "The Accuracy of Precision Instrument Measurement in Industrial Inspection," *Journal of Applied Psychology*, Vol. 29, No. 6 (1945), pp. 413–419.
12. C. G. DRURY, "Integrating Human Factors Models Into Statistical Quality Control," *Human Factors*, Vol. 20, No. 5 (1978), pp. 561–572.
13. E. L. WIENER, "Individual and Group Differences in Inspection," in C. G. Drury and J. G. Fox, Eds., *Human Reliability in Quality Control*, Taylor & Francis, London, 1975, pp. 101–122.
14. C. G. DRURY, "Inspection of Sheet Materials: Model and Data," *Human Factors*, Vol. 17, 1975, pp. 257–265.
15. D. H. HARRIS and F. B. CHANEY, *Human Factors in Quality Assurance*, Wiley, New York, 1969, pp. 131–132, 138–140.
16. C. P. GREENING, "Mathematical Modelling of Air-to-Ground Target Aquisition," *Human Factors*, Vol. 18, 1975, pp. 111–148.
17. E. D. MEGAW and L. BELLAMY, "Variables That Affect Search Patterns," in J. N. Claire and N. A. Sinclair, Eds., *Search and the Human Observer*, Taylor & Francis, London, 1979, pp. 65–73.

18. T. B. MORAWSKI, "Economic Models of Industrial Inspection," unpublished master's thesis, State University of New York at Buffalo, 1979.
19. T. W. FAULKNER and T. J. MURPHY, "Lighting for Difficult Visual Tasks," in C. G. Drury and J. G. Fox, Eds., *Human Reliability in Quality Control*, Taylor & Francis, London, 1975.
20. K. S. TEEL, R. M. SPRINGER, and E. E. SADLER, "Assembly and Inspection in Microelectronic Systems," *Human Factors*, Vol. 10, 1968, pp. 217–224.
21. J. LUZZO and C. G. DRURY, "An Evaluation of Blink Inspection," *Human Factors*, Vol. 22, 1980, pp. 201–210.
22. S. J. CZAJA and C. G. DRURY, "Training Programs for Inspection," *Human Factors*, Vol. 23, in press.
23. *Proceedings of the Third International Conference in Automatic Inspection and Product Control*, International Fluidic Services Ltd., Kempston, England, 1979.
24. C. G. DRURY, and J. G. FOX, Eds., *Human Reliability in Quality Control*, Taylor & Francis, London, 1975.
25. D. MCNICHOL, *A Primer of Signal Detection Theory*, Allen and Unwin, London, 1972.
26. C. G. DRURY and J. L. ADDISON, "An Industrial Study of the Effects of Feedback and Fault Density on Inspection Performance," *Ergonomics*, Vol. 16, 1973, pp. 159–169.
27. P. M. WALLACK and S. K. ADAMS, "A Comparison of Inspector Performance Measures," *AIIE Transactions*, Vol. 2, 1970, pp. 97–105.
28. X. K. ZUNZANYIKA and C. G. DRURY, "Effects of Information on Industrial Inspection Performance," in C. G. Drury and J. G. Fox, Eds., *Human Reliability in Quality Control*, Taylor & Francis, London, 1975.
29. D. H. HARRIS and F. B. CHANEY, *Human Factors in Quality Assurance*, pp. 157–162.
30. M. L. KELLY, "A Study of Industrial Inspection by the Method of Paired Comparisons," *Psychological Monographs*, Vol. 69, No. 394 (1955), pp. 1–16.
31. R. W. ASTLEY and J. G. FOX, "The Analysis of an Inspection Task in the Rubber Industry," in C. G. Drury and J. G. Fox, Eds., *Human Reliability in Quality Control*, Taylor & Francis, London, 1975.
32. D. E. EMBREY, "Approaches to Training for Industrial Inspection," *Applied Ergonomics*, Vol. 10, No. 3 (1979), pp. 139–144.
33. G. SALVENDY and W. D. SEYMOUR, *Prediction and Development of Industrial Work Performance*, Wiley, New York, 1973.
34. G. C. PARKER and G. PERRY, *Lighting for Glassware Production*, Parts 1 and 2, British Glass Industry Research Association Technical Note 157, Sheffield, England, 1972.
35. L. F. THOMAS and A. E. M. SEABORNE, "The Sociotechnical Context of Industrial Inspection," *Occupational Psychology*, Vol. 35, 1961, pp. 36–43.
36. G. H. JAMIESON, "Inspection in the Telecommunications Industry: A Field Study of Age and Other Performance Variables," *Ergonomics*, Vol. 9, 1966, pp. 297–303.
37. R. M. MCKENZIE, "On the Accuracy of Inspectors," *Ergonomics*, Vol. 1, No. 3 (1958), pp. 258–272.
38. C. G. DRURY and J. J. SHEEHAN, "Ergonomic and Economic Factors in an Industrial Inspection Task," *International Journal of Production Research*, Vol. 7, 1969, pp. 333–341.
39. J. G. FOX and S. RICHARDSON, "The Complexity of the Signal in Visual Inspection Tasks," unpublished paper, University of Birmingham, Department of Engineering Production, unpublished report no. ERG/70/146, England, 1970.
40. J. R. BUCK, "Dynamic Visual Inspection: Task Factors, Theory and Economics," in C. G. Drury and J. G. Fox, Eds., *Human Reliability in Quality Control*, Taylor & Francis, London, 1975.
41. J. MAHER, W. OVERBAGH, G. PALMER, and D. PIERSOL, "Enriched Jobs Improve Inspection Performance," *Work Study and Management Services*, October 1970, pp. 821–824.
42. J. R. HACKMAN, G. OLDHAM, R. JANSON, and K. PURDY, "A New Strategy for Job Enrichment," *California Management Review*, Vol. 17, No. 4 (1975), pp. 57–71.
43. R. M. BELBIN, "New Fields for Quality Control," *British Management Review*, Vol. 15, 1957, pp. 79–89.
44. R. A. ERICKSON, "Relation Between Visual Search Time and Peripheral Acuity," *Human Factors*, Vol. 6, 1964, pp. 165–177.
45. L. V. RIGBY and A. D. SWAIN, "Some Human Factors Applications to Quality Control in a High Technology Industry," in C. G. Drury and J. G. Fox, Eds., *Human Reliability in Quality Control*, Taylor & Francis, London, 1975, pp. 201–216.

CHAPTER 8.5

Reliability and Maintainability

KAILASH C. KAPUR

Wayne State University

8.5.1 INTRODUCTION

The ultimate objective of any system is the performance of some intended function. This function is frequently called the "mission." The term often used to describe the overall capability of a system to accomplish its mission is "system effectiveness." System effectiveness is defined as the probability that the system can successfully meet an operational demand within a given time when operating under specified conditions. Effectiveness is influenced by the way the system is designed, manufactured, used, and maintained. Thus the effectiveness of a system is a function of several attributes, such as design adequacy, performance measures, safety, reliability, quality, producibility, and maintainability. The disciplines of assurance sciences help to increase the overall effectiveness of any system. The assurance sciences are engineering disciplines that have the common objective of attaining product integrity (product does what it says it is supposed to do).[1,2] The term "product assurance" is also used by some companies. This chapter is concerned with the reliability, maintainability, serviceability, and availability aspects of the product assurance system.

Reliability is one of the major attributes determining system effectiveness. It is generally defined as the probability that a given system will perform its intended function satisfactorily, for its intended life, under specified operating conditions. With this definition, the obvious problems are (1) the acceptance of the probabilistic notion of reliability; (2) the problems associated with defining adequate performance, particularly for system parameters that deteriorate slowly with time; and (3) the judgment required to determine the proper statement of operating conditions.

Reliability is an inherent attribute of a system resulting from design just as is the system's capacity, performance, or power rating. The reliability level is established at the design phase, and subsequent testing and production will not raise the reliability without a basic design change. Because reliability is an abstract concept that is difficult to grasp and to measure, many organizations find themselves unable to implement a comprehensive reliability program primarily because of the lack of understanding on the part of both management and technical system design personnel. This is not to say that the system designers or managers in the organization are not interested in a reliable product, but rather, the pressures on the design engineer, and very often on the organizational structure, impede the development of an effective reliability program.

With increasing system complexity, reliability becomes an elusive and difficult design parameter. It becomes more difficult not only to define and achieve as a design parameter, but also to control and demonstrate in production and thus to ensure as an operational characteristic under the projected environmental conditions of use. However, past history has demonstrated that, where reliability was recognized as a necessary program development component, with the practice of various reliability engineering methods throughout the evolutionary life cycle of the system, reliability can be quantified during the specification of design requirements, can be predicted by testing, can be controlled during production, and can be sustained in the field. The purpose of this chapter is to present some of the reliability and maintainability methodologies and philosophies that are applicable throughout the life cycle of a system.

Reliability and Maintainability Activities During the System Life Cycle

Reliability and maintainability activities should span the entire life cycle of the system. Exhibit 8.5.1 shows the major points of reliability practice in a typical system life cycle.[3] The activities given in the exhibit are briefly explained here. There is a continuous feedback, and designs go through several cycles of the reliability program activities.

Exhibit 8.5.1 Reliability and Maintainability Activities During System Life Cycle

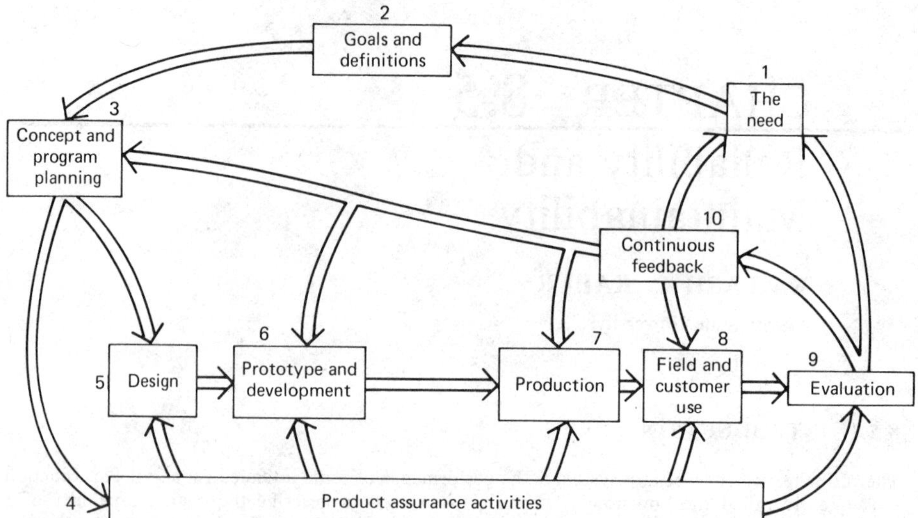

Step 1–The Need

The need for reliability and maintainability programs must be anticipated right from the beginning. The need for such programs cannot be overemphasized. These programs are justified based on specific system requirements in terms of life cycle cost and other operational requirements. As mentioned before, the effectiveness of a system is determined by its reliability and maintainability characteristics.

Step 2–Goals and Definitions

All the requirements must be specified in terms of well-defined and quantitative goals. The goals and requirements are defined by some of the following measures:

1. **Reliability Measures.** Mission reliability, a reliability function based on specified failure distribution, mean time to failure (MTTF) and failure rate.
2. **Maintainability Measures.** Maintainability function based on time to repair distribution, mean time to repair (MTTR), percentile of time to repair, and maintenance ratio.

In addition, there are other measures, and the relationship among them is given in Exhibit 8.5.2.[4]
The term "reliability" has already been defined; some of the other terms are defined as follows:

1. "Mission reliability" is the probability that the product and/or system will successfully complete a given mission with specified operating requirements and time duration.
2. "Maintainability" is defined as the probability that a failed system can be made operable in a specified interval of downtime. As shown in Exhibit 8.5.2, the downtime includes the failure detection time, the active repair time, the logistics time connected with the repairs of the product, and all the administrative time. The maintainability function describes probabilistically how long a system remains in the failed state.
3. "Repairability" deals only with the active repair time and can be defined by the time to actively repair random variable and the associated distribution. Repairability is defined as the probability that a failed system be restored to a satisfactory operating condition in a specified interval of active repair time. This measure is more valuable to the administration of the repair facility since it helps to quantify the workload for the facility and its workers.
4. "Serviceability" is defined as the "ease" with which a system can be repaired. Serviceability, just like reliability, is a characteristic of the system design and must be planned at the design stage. Serviceability is difficult to measure on a ratio scale; however, it can easily be measured on an ordinal scale by a specifically developed rating and/or ranking procedure, which requires that systems be compared and ranked according to the ease of serviceability.
5. "Operational readiness" is defined as the probability that a system either is operating or can operate satisfactorily when the system is used under stated conditions. Operational readiness deals with all the time elements, including storage time, free time, operating time, and downtime.

Exhibit 8.5.2 Relationship Among Various Product Assurance Measures

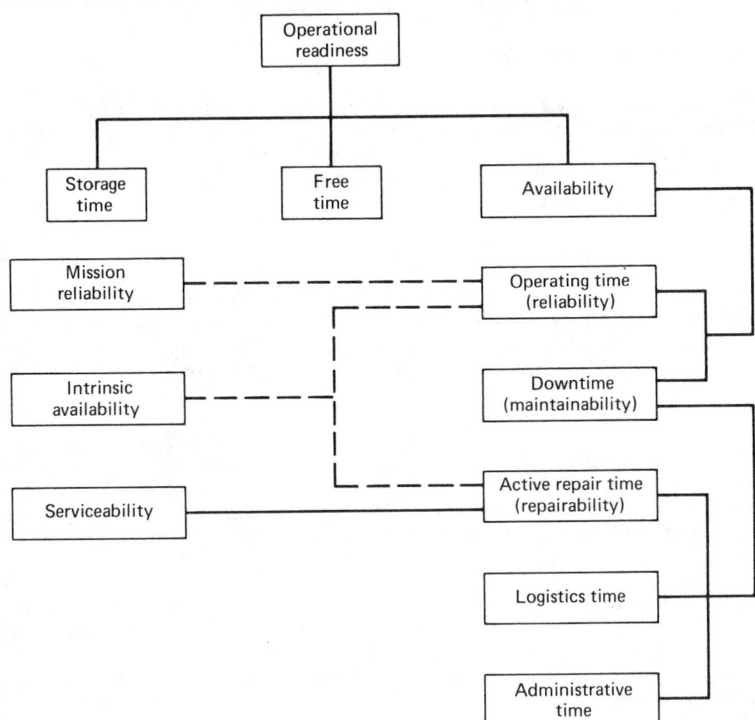

6. "Availability" is defined as the probability that a system is operating satisfactorily at any point in time. Availability considers only the operating time and downtime, thus excluding the idle time. Therefore it is a measure of the ratio of operating time of the system to the operating time plus the downtime. Availability is a function of both the reliability and maintainability of the system.

7. "Intrinsic availability" is more restrictive than availability because it is limited to operating and active repair time only.

These reliability, maintainability, and availability measures should be defined, and system requirements specified using these quantitative measures. The effectiveness of the total product assurance program depends on these definitions.

Step 3–Concept and Program Planning

Based on the reliability and other operational requirements, various concepts are developed that can potentially meet these requirements. Also, at this stage total product assurance program plans must be formulated, and responsibilities assigned to different groups. The conceptual stage is an important part of the system life cycle because it has a major impact on the future system. Studies done by the U.S. Department of Defense indicate that 70% of the system life cycle cost is determined by the decisions made at the concept stage. The detailed nature of the reliability programs will also determine the effectiveness of the total program.

Step 4–Product Assurance Activities

The plans developed in step 3 are implemented, and the total program is continuously monitored, as indicated in Exhibit 8.5.1. An organization for the implementation of these plans must exist, with well-defined responsibilities.

Step 5–Design

The conceptual system selected in step 3 is designed. Reliability and maintainability of this design are assessed. Various methodologies, such as design review, failure mode and effect analysis, fault tree analysis, and probabilistic design approach, can be applied at this step. Reliability is a design parameter and must be incorporated in the system at the design step.

Step 6–Prototype and Development

Prototypes are developed based on the design specifications. Reliability of the design is verified by testing. If the design has certain deficiencies, they are corrected by redesign. Reliability growth management plans must be developed for this step in order to monitor continuously the growth and progress of the program. After the system has achieved the required level of reliability, the design is released for production.

Step 7–Production

The system is manufactured based on design specifications. Quality control methodologies are essential during this step. All the parts, materials, and processes are controlled based on methodologies discussed in previous chapters in the area of quality assurance. One of the objectives of the quality control program is to make sure that the inherent reliability of the design is not degraded.

Step 8–Field and Customer Use

Before the system is actually used in the field by its customers, it is very important to develop all the service and maintenance instructions, which are well documented. Just like reliability, maintainability is considered throughout the life cycle, and its purpose is to sustain required levels of reliability and availability in the field. Maintainability program plans are developed at the planning step.

Step 9–System Evaluation

The system in the field is continuously evaluated to determine whether the original reliability and maintainability goals are met by the system. For this purpose a reliability monitoring program and field data collection program must be established.

Step 10–Continuous Feedback

There must be continuous feedback among all the steps in the system's life cycle. A proper communication system should be developed among all the groups responsible for the various steps. All the field deficiencies must be reported to the appropriate groups. This will help guide the system improvements.

The methodology related to some of the activities during the system life cycle is given in this chapter.

Reliability and Life Characteristic Curve

Reliability has sometimes been described as "quality in the time dimension."[5] The reliability characteristics of a product change with time. One of the characteristics is the concept of failure rate which is defined mathematically later in this chapter. The failure rate, or the hazard rate, changes with the age or life of a product and has three distinct periods, as shown in Exhibits 8.5.3*a* and 8.5.3*b*. These three periods are described here.

Infant Mortality Period

The total item population or a system generally exhibits a relatively high failure rate in the beginning, which decreases rapidly and stabilizes at some approximate time t_1. This initial period is generally called the "burn-in," "infant mortality," or "debugging" period. The item population has "weak" items, and these fail in the beginning. To understand the nature of these early failures, some of their causes are listed:

Substandard workmanship.

Poor quality control.

Substandard materials.

Insufficient debugging.

Poor manufacturing techniques.

Poor processes and handling techniques.

Problems due to assembly.

Contamination.

Exhibit 8.5.3 Failure Rate–Life Characteristic Curve (*a*) and Failure Rate Based on Components of Failure (*b*)

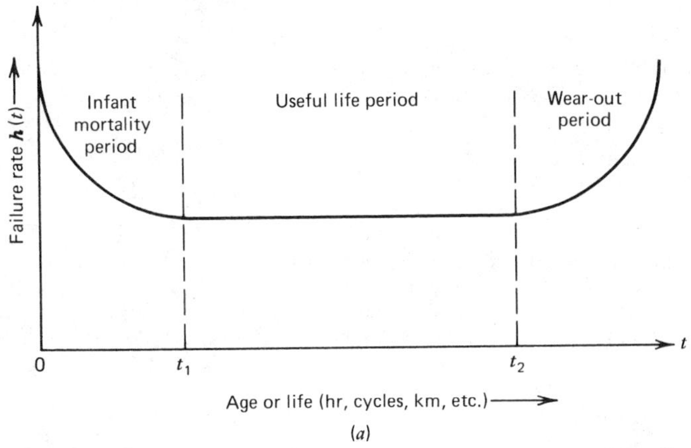

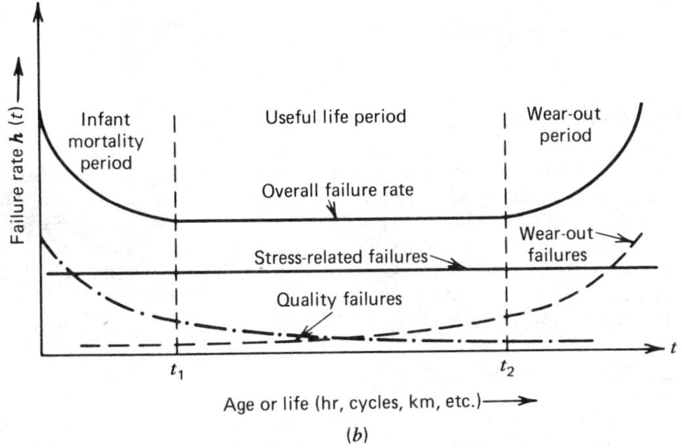

Improper installation.
Improper start-up.
Human error.
Parts failure in storage and transit.
Improper packaging and transportation practices.

Fundamentally, these failures reflect the "manufacturability" of the product, and many are due to quality control. Thus these early failures would show up during process audits, in-process or final tests, life tests, environmental tests, and so on. Most manufacturers provide a burn-in period for their products so that the early or infant mortality failures occur in the plant and are not experienced by the customer in the field, where it is much more costly to fix these failures. The duration of the burn-in period determines what portion of the early failures is eliminated.

Useful Life Period

After having "burned in," the item population reaches its lowest failure rate level, remaining relatively constant during this period. This failure rate is related to the inherent design reliability of the product and hence is given the most weight during the design reliability. This is also the most significant period for reliability prediction and assessment activities. Some of the causes of these failures are as follows:

Low safety factors.
Stress-related failures–higher-than-expected random loads.

Lower random strength than expected.
Defects that cannot be detected by the "best" available inspection techniques.
Abuse.
Human errors.
Failures that cannot be observed during debugging.
Failures that cannot be prevented by the "best" preventive maintenance practices.
Unexplainable causes.
"Act of God" failures.

Wear-Out Period

Most of the products are designed to last for a specified period of useful life. The time t_2 in Exhibits 8.5.3*a* and 8.5.3*b* indicates the end of useful life or the start of the wear-out period. After this point, the failure rate increases rapidly. The wear-out, or deterioration, results from a number of familiar chemical, physical, or other causes, some of which are as follows:

Corrosion or oxidation.
Frictional wear or fatigue.
Aging and degradation.
Creep.
Poor maintenance or service practices.
Improper overhaul practices.
Short designed-in life.
Shrinkage or cracking in plastics.

A reliability program must consider all three of these distinct periods. It must also be pointed out that not all products have these three periods. The importance of the periods depends on the magnitudes of time t_1 and t_2, where $0 \leqslant t_1 \leqslant t_2 < \infty$. Thus we can develop various types of life characteristic curves, depending upon the values of t_1 and t_2. Early failures can be eliminated by systematic procedures of controlled screening, quality control, and burn-in tests. Stress-related failures during the useful life can be minimized by providing adequate design or safety margins. Wear-out failures can be minimized by preventive maintenance and replacement policies.

8.5.2 RELIABILITY MEASURES

Reliability has been defined as the probability that a given system will perform satisfactorily its intended function for its intended life under specified operating conditions. Thus reliability is related to the probability of the successful performance of any system. It is clear that we must define what the successful performance of any system is or what we mean by the failure of the system; otherwise, it is not possible to predict when any system will fail to perform its intended function. The time to failure or "life" of a system cannot be deterministically defined, and hence it is a random variable. Thus we must quantify reliability by assigning a probability function to the time to failure random variable.

Mathematics of Reliability Measures

Let **t** denote the time to failure random variable. Then reliability at any time t, denoted by $R(t)$, is the probability that the system will not fail by time t, or mathematically

$$R(t) = P[\mathbf{t} > t] \qquad (1)$$

Let $f(t)$ be the probability density function for the failure random variable **t**. Then the cumulative distribution function $F(t)$ is given by

$$P[\mathbf{t} \leqq t] = F(t) = \int_0^t f(\tau)\, d\tau \qquad (2)$$

Hence from equations 1 and 2 we have some fundamental relationships between the reliability function, cumulative distribution, and probability density function

$$R(t) = 1 - P[\mathbf{t} \leqslant t] = 1 - F(t) = 1 - \int_0^t f(\tau)\, d\tau \tag{3}$$

Exhibits 8.5.4*a*, 8.5.4*b*, and 8.5.4*c* show, respectively, the failure probability density function, the cumulative distribution function, and the reliability function for the well-known case when the time to failure is exponentially distributed. Here we have

$$f(t) = \lambda e^{-\lambda t}, \quad t \geqslant 0, \quad \lambda > 0 \tag{4}$$

$$F(t) = \int_0^t \lambda e^{-\lambda t}\, d\tau = 1 - e^{-\lambda t}, \quad t \geqq 0 \tag{5}$$

and

$$R(t) = e^{-\lambda t}, \quad t \geqq 0 \tag{6}$$

These functions are all related, and selection of any one determines the shape of the others. This can be easily seen by studying equations 1, 2, and 3.

Obviously, the reliability function inherent in a system, by virtue of its design, dictates the probability of successful system operation during the system's life. A natural question is then, "How does one know the shape of a reliability function for a particular system?" There are basically three ways in which it can be determined:

1. Test many systems to failure using a mission profile that is identical to use conditions. This would allow one to develop empirically a curve such as that shown in Exhibit 8.5.4*c*.
2. Test many subsystems and components to failure where use conditions are recreated in the test environment. This allows one to develop empirically the component reliability functions and then to derive analytically the system reliability function.
3. Based on past experience with similar systems, the underlying failure distribution may be hypothesized. Then one can test fewer systems to determine the parameters needed to adapt the failure distribution to a particular situation. For example, the lifetime of many different kinds of electronic components follows the exponential distribution as previously given in equation 4. To apply this distribution one must know the value of the parameter λ for a particular situation. Elaborate studies have been done, so that for a given environment and mission the parameter λ can be determined for most electronic components.[6,7]
4. In some cases the failure physics involved in a particular situation may lead one to hypothesize a particular distribution. For example, fatigue of certain metals tends to follow either the lognormal or Weibull distributions.[8] Here again, once a distribution is selected, the parameters for a particular application must be ascertained.

Another measure that is frequently used as an indirect indicator of system reliability is the MTTF, which is the expected or mean value of the time to failure random variable. Thus, the MTTF is theoretically defined as

$$MTTF = E[\mathbf{t}] = \int_0^\infty tf(t)\, dt = \int_0^\infty R(t)\, dt \tag{7}$$

Sometimes the term "mean time between failures" (MTBF) is also used to denote $E[\mathbf{t}]$. The problem with using only the MTTF as an indicator of system reliability is that it uniquely determines reliability only if the underlying time to failure distribution is exponential. If the failure distribution is other than exponential, the MTTF can produce erroneous comparisons.

If we have a large population of the items whose reliability we are interested in studying, then for replacement and maintenance purposes we are interested in the rate at which the items in the population, which have survived at any specific time, will fail. This is the failure rate, or hazard rate, and is given by the following relationship:

$$h(t) = \frac{f(t)}{R(t)} \tag{8}$$

The failure rate for most components follows the curve shown in Exhibit 8.5.3*a*, which is also called the life characteristic curve.

Exhibit 8.5.4 Exponential Density Function (*a*), Exponential Distribution Function (*b*), and Reliability Function for Exponential Distribution (*c*)

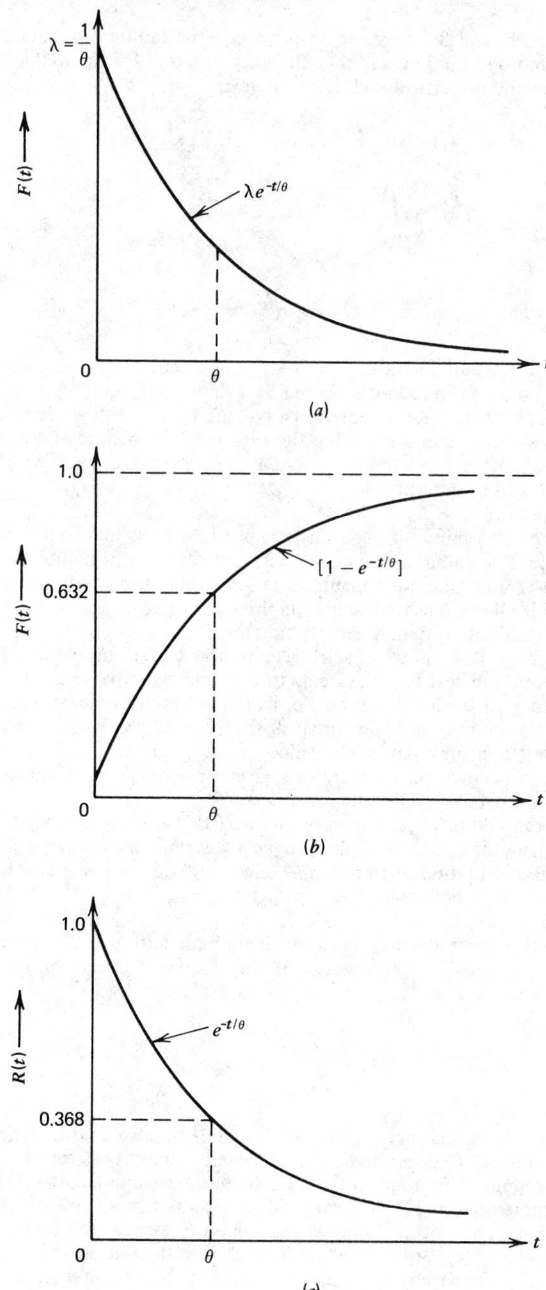

To understand the notion of failure rate or hazard rate, basic mathematical relations are given here.

The hazard rate is defined as the limit of the instantaneous failure rate given no failure up to time t and is given by

$$
\begin{aligned}
h(t) &= \lim_{\Delta t \to 0} \frac{P[t < \mathbf{t} \leqslant t + \Delta t \mid \mathbf{t} > t]}{\Delta t} \\
&= \lim_{\Delta t \to 0} \frac{R(t) - R(t + \Delta t)}{\Delta t \cdot R(t)} \\
&= \frac{1}{R(t)} \left[-\frac{d}{dt} R(t) \right] \\
&= \frac{f(t)}{R(t)} \qquad (9)
\end{aligned}
$$

Also

$$f(t) = h(t) \exp \left[-\int_0^t h(\tau)\, d\tau \right] \qquad (10)$$

and thus

$$R(t) = \exp \left[-\int_0^t h(\tau)\, d\tau \right] \qquad (11)$$

Various Life Distributions

The properties of some life distributions that are used in reliability and maintainability discipline are given in the next subsections.

Exponential Distribution

$$f(t) = \lambda e^{-\lambda t}, \qquad t \geqslant 0 \qquad (12)$$

$$F(t) = 1 - e^{-\lambda t}, \qquad t \geqq 0 \qquad (13)$$

$$R(t) = e^{-\lambda t}, \qquad t \geqq 0 \qquad (14)$$

$$h(t) = \lambda \qquad (15)$$

$$MTBF = \theta = \frac{1}{\lambda} \qquad (16)$$

Thus the failure rate for the exponential distribution is always constant.

Normal Distribution

$$f(t) = \frac{1}{\sigma \sqrt{2\pi}} \exp \left[-\frac{1}{2} \left(\frac{t - \mu}{\sigma} \right)^2 \right], \qquad -\infty < t < \infty \qquad (17)$$

$$F(t) = \Phi \left(\frac{t - \mu}{\sigma} \right) \qquad (18)$$

$$R(t) = 1 - \Phi \left(\frac{t - \mu}{\sigma} \right) \qquad (19)$$

$$h(t) = \frac{\phi[(t - \mu)/\sigma] / \sigma}{R(t)} \qquad (20)$$

$$MTBF = \mu \qquad (21)$$

Thus $\Phi(z)$ is the cumulative distribution function and $\phi(z)$ is the probability density function, respectively, for the standard normal variate z. The failure rate for the normal distribution is a monotonically increasing function. Normal distribution should be used as a life distribution when $\mu > 6\sigma$ because then the probability that t will be negative is exceedingly small. Otherwise, truncated normal distribution should be used.

Lognormal Distribution

$$f(t) = \frac{1}{\sigma t\sqrt{2\pi}} \exp\left[-\frac{1}{2}\left(\frac{\ln t - \mu}{\sigma}\right)^2\right], \qquad t \geqslant 0 \tag{22}$$

$$F(t) = \Phi\left(\frac{\ln t - \mu}{\sigma}\right) \tag{23}$$

$$R(t) = 1 - \Phi\left(\frac{\ln t - \mu}{\sigma}\right) \tag{24}$$

$$h(t) = \phi\left(\frac{\ln t - \mu}{\sigma}\right) \Big/ t\sigma R(t) \tag{25}$$

$$MTBF = \exp\left[\mu + \frac{\sigma^2}{2}\right] \tag{26}$$

The failure rate for the lognormal distribution is neither always increasing nor always decreasing. It takes different shapes, depending on the parameters μ and σ.

Weibull Distribution (Three Parameters $\theta > \delta$)

$$f(t) = \frac{\beta(t-\delta)^{\beta-1}}{(\theta-\delta)^{\beta}} \exp\left[-\left(\frac{t-\delta}{\theta-\delta}\right)^{\beta}\right], \qquad t \geqslant \delta \geqslant 0 \tag{27}$$

$$F(t) = 1 - \exp\left[-\left(\frac{t-\delta}{\theta-\delta}\right)^{\beta}\right] \tag{28}$$

$$R(t) = \exp\left[-\left(\frac{t-\delta}{\theta-\delta}\right)^{\beta}\right] \tag{29}$$

$$h(t) = \frac{\beta(t-\delta)^{\beta-1}}{(\theta-\delta)^{\beta}} \tag{30}$$

$$MTBF = \delta + (\theta - \delta)\,\Gamma\left(1 + \frac{1}{\beta}\right) \tag{31}$$

The failure rate for the Weibull distribution is decreasing when $\beta < 1$, is constant when $\beta = 1$ (same as the exponential distribution) and is increasing when $\beta > 1$.

Gamma Distribution

$$f(t) = \frac{\lambda^{\eta}}{\Gamma(\eta)} t^{\eta-1} e^{-\lambda t}, \qquad t \geqslant 0 \tag{32}$$

$$F(t) = \sum_{k=\eta}^{\infty} \frac{(\lambda t)^k \exp[-\lambda t]}{k!}, \qquad \text{when } \eta \text{ is integer} \tag{33}$$

$$R(t) = \sum_{k=0}^{\eta-1} \frac{(\lambda t)^k \exp[-\lambda t]}{k!}, \qquad \text{when } \eta \text{ is integer} \tag{34}$$

$$h(t) = \frac{f(t)}{R(t)} \text{ (using equations 32 and 34)} \quad (35)$$

$$MTBF = \frac{\eta}{\lambda} \quad (36)$$

The failure rate for the gamma distribution is decreasing when $\eta < 1$, is constant when $\eta = 1$, and is increasing when $\eta > 1$.

Summary

Properties of extreme value distributions are given by Gumbel.[9] Other mathematical properties of the preceding distributions and their use in the reliability field have been discussed extensively by several authors, and results are available in the literature.[10,11] To use these distributions in the product assurance field, one must understand their nature and properties and the conditions under which they are applicable to describe various physical phenomena.

8.5.3 SYSTEM RELIABILITY MODELS

To analyze and measure the reliability and maintainability characteristics of a system, there must be a mathematical model of the system that shows the functional relationships among all the components, the subsystems, and the overall system. The reliability of the system is a function of the reliabilities of its components. A system reliability model[12] consists of some combination of a reliability block diagram or cause-consequence chart, a definition of all equipment failure and repair distributions, and a statement of spare and repair strategies. All reliability analyses and optimizations are made on these conceptual mathematical models of the system.

Reliability Block Diagram

A reliability block diagram is obtained from a careful analysis of the manner in which the system operates. An analysis has to be done of the effects on overall system performance of failures of the various components; the support environment and constraints, including such factors as the number and assignment of spare parts and repairpersons; and the mission for the system.

Engineering analysis on the system has to be done in order to develop a reliability model. The engineering analysis consists of the following steps:

1. Develop a functional block diagram of the system based on physical principles governing the operations of the system.
2. Develop the logical and topological relationships between functional elements of the system.
3. Performance evaluation studies are used to determine the extent to which the system can operate in a degraded state.
4. Define the spares and repairs strategies (for maintained systems).

Based on the preceding analysis, a reliability block diagram is developed, which is used for calculating various measures of reliability and maintainability. The reliability block diagram is a pictorial way of showing all the success or failure combinations for the system. Some of the guidelines for drawing these diagrams are as follows:

1. A group of components that are essential for the performance of the system and/or its mission are drawn in series (Exhibit 8.5.5*a*).
2. Components that can substitute for other components are drawn in parallel (Exhibit 8.5.5*b*).
3. Each block in the diagram is like a switch: It is closed when the component it represents is working and is opened when the component has failed. Any closed path through the diagram is a success path.

The failure behavior of all the redundant components must be specified. Some of the common types of redundancies are:

1. *Active redundance* or *hot standby*, where the component has the same failure rate as if it were operating in the system.
2. *Passive redundancy*, spare or *cold standby*, where the standby component cannot fail. This is generally assumed of spare or shelf items.
3. *Warm standby*, where the standby component has a lower failure rate than the operating component. This is usually a realistic assumption.

Exhibit 8.5.5 Series Configuration (*a*), Parallel Configuration (*b*), and Bridge Structure (*c*)

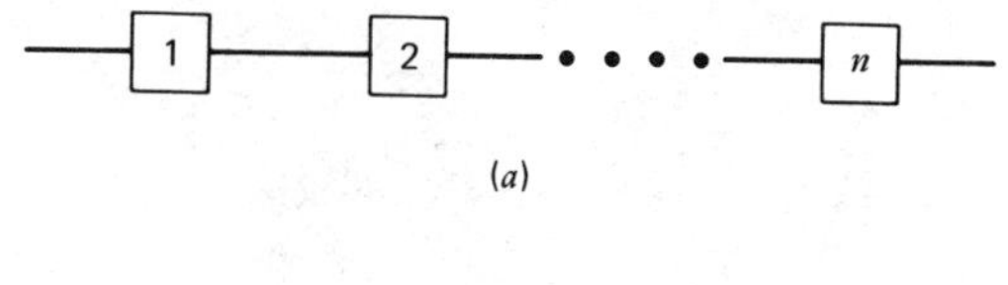

(*a*)

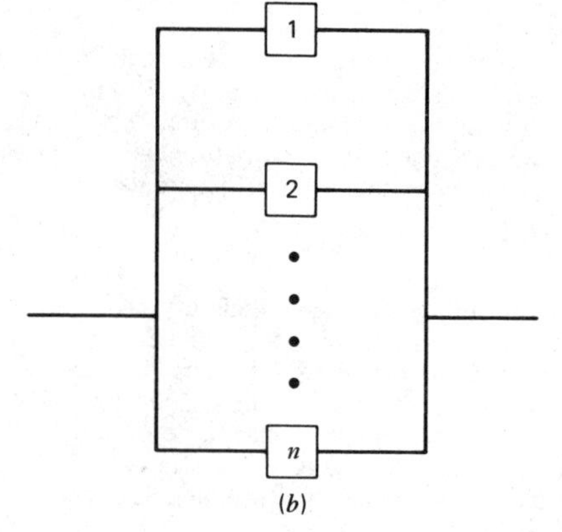

(*b*)

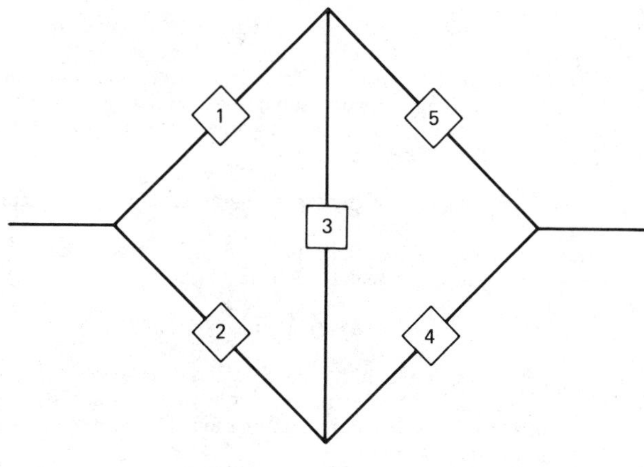

(*c*)

Some mathematical relationships between the system reliability and the reliabilities of its components are given in the next subsections. In the following, R_s denotes the reliability of the system, and R_i denotes the reliability of the ith component, where $i = 1, 2, \ldots, n$ and the system has n components. In addition, in the following relationships, it is also assumed that all the components work or fail independently of each other.

*Series Configuration**

For the static situation we have

$$R_s = \prod_{i=1}^{n} R_i \tag{37}$$

*See Exhibit 8.5.5*a*.

and for the dynamic situation

$$R_s(t) = \prod_{i=1}^{n} R_i(t) \tag{38}$$

The failure rate $h_s(t)$ for the system is given by

$$h_s(t) = \sum_{i=1}^{n} h_i(t) \tag{39}$$

where $h_i(t)$ is the failure rate of the ith component.

If all the components have an exponentially distributed time to failure, we have

$$h_s(t) = \sum_{i=1}^{n} \lambda_i \tag{40}$$

and *MTBF* for the system is given by

$$\frac{1}{\sum_{i=1}^{n} \lambda_i} \tag{41}$$

*Parallel Configuration**

For the static case, we have

$$R_s = 1 - \prod_{i=1}^{n} (1 - R_i) \tag{42}$$

and for the dynamic case

$$R_s(t) = 1 - \prod_{i=1}^{n} (1 - R_i(t)) \tag{43}$$

If the time to failures for all the components is exponentially distributed with *MTBF* θ, then the *MTBF* for the system is given by

$$\sum_{i=1}^{n} \frac{\theta}{i} \tag{44}$$

where θ = *MTBF* for every component.

The k-out-of-n Configuration

In this configuration the system works if and only if at least k components out of the n components work, $1 \leqslant k \leqslant n$. For this case, when $R_i = R(t)$ for all i, we have

$$R_s(t) = \sum_{i=k}^{n} \binom{n}{i} [R(t)]^i \, [1 - R(t)]^{n-i} \tag{45}$$

If $R(t) = e^{-t/\theta}$, for exponential case, *MTBF* for the system is given by

$$\sum_{i=k}^{n} \frac{\theta}{i} \tag{46}$$

*See Exhibit 8.5.5*b*.

Coherent Systems

The reliability block diagrams for many systems cannot be represented by the preceding three configurations. In general, the concept of coherent systems can be used to determine the reliability of any system. The performance of each of the n components in the system is represented by a binary indicator variable, x_i, which takes the value 1 if the ith component functions and 0 if the ith component fails. Similarly, the binary variable ϕ indicates the state of the system, and ϕ is a function of $x = (x_1, \ldots, x_n)$.

The function $\phi(x)$ is called the "structure function" of the system. The structure function is represented by using the concept of minimal path and minimal cut. A "minimal path" is a minimal set of components whose functioning ensures the functioning of the system. A "minimal cut" is a minimal set of components whose failures cause the system to fail. Let $\alpha_j(x)$ be the jth minimal path series structure for path A_j, $j = 1, \ldots, p$ and $\beta_k(x)$ be the kth minimal parallel cut structure for cut B_k, $k = 1, \ldots, s$. Then we have

$$\alpha_j(x) = \prod_{i \in A_j} x_i \tag{47}$$

$$\beta_k(x) = 1 - \prod_{i \in B_k} (1 - x_i) \tag{48}$$

and

$$\phi(x) = 1 - \prod_{j=1}^{p} [1 - \alpha_j(x)] \tag{49}$$

$$= \prod_{k=1}^{s} \beta_k(x) \tag{50}$$

For the bridge structure (Exhibit 8.5.5c) we have

$$\begin{aligned} \alpha_1 &= x_1 x_5 & \beta_1 &= 1 - (1 - x_1)(1 - x_2) \\ \alpha_2 &= x_2 x_4 & \beta_2 &= 1 - (1 - x_4)(1 - x_5) \\ \alpha_3 &= x_1 x_3 x_4 & \beta_3 &= 1 - (1 - x_1)(1 - x_3)(1 - x_4) \\ \alpha_4 &= x_2 x_3 x_5 & \beta_4 &= 1 - (1 - x_2)(1 - x_3)(1 - x_5) \end{aligned}$$

Then the reliability of the system is given by

$$R_S = P[\phi(x) = 1] = E[\phi(x)]$$

If R_i is the reliability of the ith component, we have for the bridge structure

$$\begin{aligned} R_S = {} & R_1R_5 + R_1R_3R_4 + R_2R_3R_5 + R_2R_4 \\ & - R_1R_3R_4R_5 - R_1R_2R_3R_5 - R_1R_2R_4R_5 \\ & - R_1R_2R_3R_4 - R_2R_3R_4R_5 + 2R_1R_2R_3R_4R_5 \end{aligned}$$

If all $R_i = R = 0.9$, we have

$$\begin{aligned} R_S &= 2R^2 + 2R^3 - 5R^4 + 2R^5 \\ &= 0.9785 \end{aligned}$$

The exact calculations for R_S are generally very tedious because the paths and the cuts are dependent, since they may contain a same component. Bounds on system reliability are given by

$$\prod_{k=1}^{s} P[\beta_k(x) = 1] \leqslant P[\phi(x) = 1] \leqslant 1 - \prod_{j=1}^{p} \{1 - P[\alpha_j(x) = 1]\}$$

Using these bounds for the bridge structure, we have, when $R_i = R = 0.9$,

$$\text{upper bound on } R_S = 1 - (1 - R^2)^2\,(1 - R^3)^2 = 0.9973$$

$$\text{lower bound on } R_S = (1 - (1 - R)^2)^2\,(1 - (1 - R)^3)^3 = 0.9781$$

The bounds on system reliability using the concepts of minimum paths and cuts can be improved.[10] Further details and derivations for coherent systems can also be found in Barlow and Proschan.[10]

Fault Tree Analysis

Fault tree analysis is one of the methods for system safety and reliability analysis.[13] The concept was originated by Bell Telephone Laboratories as a technique for safety evaluation of the Minuteman Launch Control System. Many reliability techniques are inductive and are concerned primarily with ensuring that hardware will accomplish its intended functions. Fault tree analysis is a detailed deductive analysis that usually requires considerable information about the system. It is concerned with ensuring that all critical aspects of a system are identified and controlled. It is a graphical representation of Boolean logic associated with the development of a particular system failure (consequence), called the "top event," to basic failures (causes), called "primary events." These top events can be broad, all-encompassing events, such as "release of radioactivity from a nuclear power plant" or "inadvertent launch of an ICBM missile," or they can be specific events, such as "failure to insert control rods" or "energizing power available to ordnance ignition line."

Fault tree analysis is of value in:

1. Providing options for qualitative and quantitative reliability analysis.
2. Helping the analyst to understand system failures deductively.
3. Pointing out the aspects of a system that are important with resepct to the failure of interest.
4. Providing the analyst an insight into system behavior.

A fault tree is a model that graphically and logically represents the various combinations of possible events, both fault and normal, occurring in a system that lead to the top event. A fault event is an abnormal system state. A normal event is an event that is expected to occur. The term "event" denotes a dynamic change of state that occurs in a system element. System elements include hardware, software, human, and environmental factors. Details about the construction of fault trees can be found in Barlow et al.[13]

8.5.4 ALLOCATION OF RELIABILITY REQUIREMENTS

Reliability and design engineers must translate overall system performance, including reliability, into component performance, including reliability. The process of assigning reliability requirements to individual components in order to attain the specified system reliability is called "reliability allocation." There are many different ways in which reliability can be allocated in order to achieve this end.

The allocation problem is complex for several reasons, among which are: (1) the role a component plays for the functioning of the system; (2) the methods available for accomplishing this function; (3) the complexity of the component; and (4) the reliability of the component, which may change with the type of function to be performed. The problem is further complicated by the lack of detailed information on many of these factors early in the system design phase. However, a tentative reliability allocation must be accomplished in order to guide the design engineer. The typical decision process from a reliability allocation standpoint is illustrated in Exhibit 8.5.6. A process such as this attempts to force all concerned to make decisions in an orderly and knowledgeable fashion rather than on an ad hoc basis.

Some of the advantages of the reliability allocation process are:

1. The process forces system design and development personnel to understand and develop the relationships among component, subsystem, and system reliabilities. This leads to an understanding of the basic reliability problems inherent in the design.
2. The design engineer is obliged to consider reliability equally with other system parameters, such as weight, cost, and performance characteristics.
3. Reliability allocation ensures adequate design manufacturing methods and testing procedures.

Exhibit 8.5.6 Reliability Allocation Process

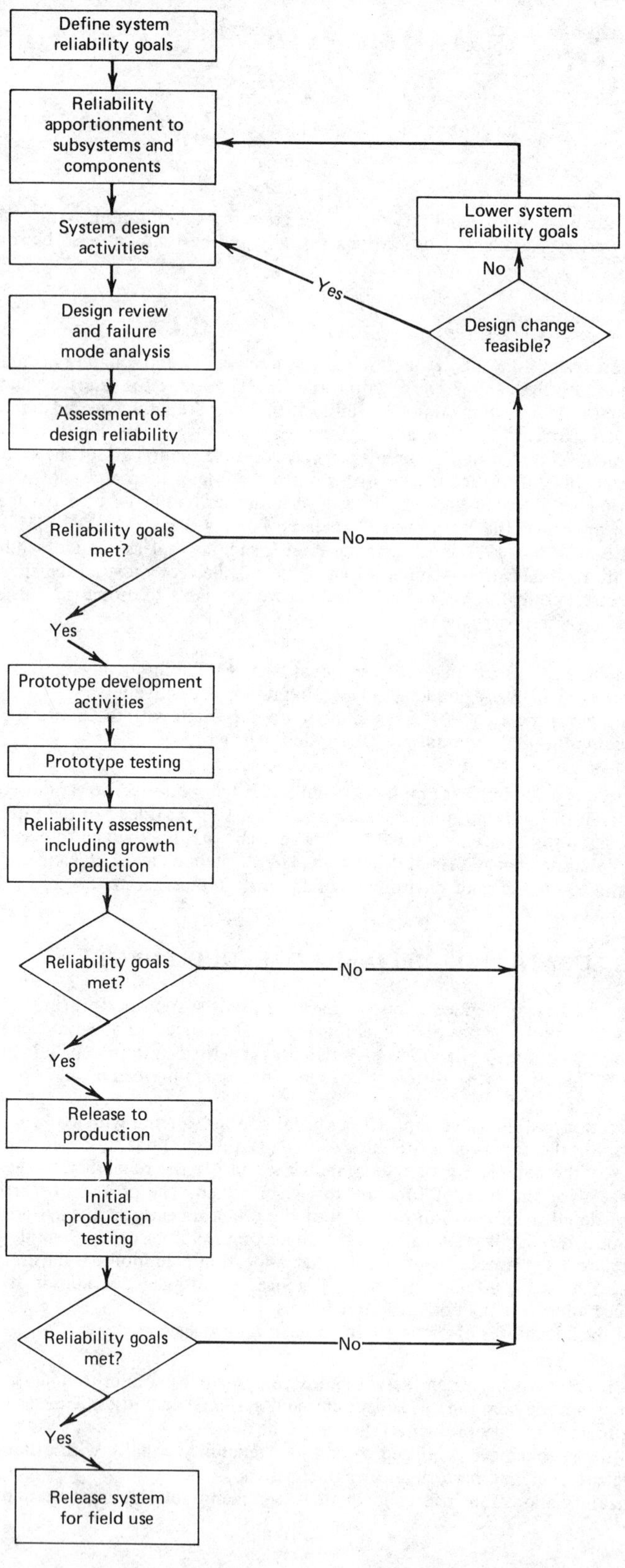

The allocation process is approximate, and the system effectiveness parameters, such as reliability and maintainability apportioned to the subsystems, are used as guidelines to determine design feasibility. If the allocated parameters for a system cannot be achieved using the current technology, then the system must be modified and the allocations reassigned. This procedure is repeated until an allocation is achieved that satisfies the system requirements.

Various allocation algorithms for reliability and availability requirements are available.[4,11,12]

8.5.5 DESIGN FOR RELIABILITY

Often the terms "reliability" and "quality control" are used interchangeably as if they both implied the same thing. This is an erroneous belief. Quality control ensures conformance to engineering design specifications during manufacturing of the system. However, after a design has been completed and released for manufacturing, the maximum reliability level of the system has been determined by virtue of its design. Essentially, the reliability effort is over once the design is released, and all that quality control can do is to ensure that this reliability level does not deteriorate during manufacturing. From a reliability standpoint, quality control is after the fact, and thus it is too late to consider reliability. So the reliability effort must be an integral part of system design and development, since this is where the reliability level is established. Reliability activities that can be performed during system design and development are described briefly in the remainder of this section.

To ensure most economically and effectively the production of a reliable product, the reliability activities must start early in the product development cycle. However, at this stage, the identification of reliability improvements depends heavily on the experience of the personnel studying the product from blueprints and preliminary system mock-ups, because no hard data are available for a quantitative assessment of reliability. To consider reliability early in the design cycle, one must rely on a formalized design review procedure, which is now briefly explained.

Design Review

The design review, a formal and documented review of a system design, is conducted by a committee of senior company personnel who are experienced in various pertinent aspects of product design, reliability, manufacturing, materials, stress analysis, human factors, safety, logistics, maintenance, and so on. The design review extends over all phases of product development from conception to production. In each phase previous work is updated, and the review is based on current information.

A mature design requires trade-offs between many conflicting factors, such as performance, manufacturability, reliability, and maintainability. These trade-offs depend heavily on experienced judgment and require continuous communication among experienced reviewers. The design review committee approach has been found to be extremely beneficial to this process. The committee adopts the system's point of view and considers all conceivable phases of design and system use, to ensure that the best trade-offs have been made for the particular situation.

A complete design review procedure must be multiphased in order to follow the design cycle until the system is released for production. A typical example of a review committee, including personnel and their responsibilities, is shown in Exhibit 8.5.7. Here the review process has been subdivided into three phases, and each phase is an updated or more detailed analysis based on the latest knowledge.

Ultimately the design engineer has the responsibility for investigating and incorporating the ideas and suggestions posed by the design review committee. The committee's chairperson is responsible for adequately reporting all suggestions by way of a formal and documented summary. The design engineer then can accept or reject various points in the summary; however, he or she must formally report back to the committee, stating reasons for his or her actions.

It should be recognized that considerably more thought and detail than the basic philosophy presented here must go into developing the management structure and procedures for conduct in order to have a successful review procedure. It should be noted that this review procedure considers not only reliability, but all important factors in order to ensure that a mature design will result from the design effort.

In the next subsection attention is focused on a technique that has been proved effective in identifying failure situations early in the design cycle and before product testing.

Failure Mode and Effects Analysis

Failure mode and effects analysis is a design evaluation procedure used to identify all conceivable and potential failure modes and to determine the effect of each failure mode on system performance. This procedure is accomplished by formal documentation, which serves (1) to standard-

Exhibit 8.5.7 Design Review Committee

Member	Review Phase 1	2	3	Responsibility
Chairperson	x	x	x	Ensure that review is conducted in an efficient fashion. Issue major reports and monitor follow-up.
Design engineer (of this product)	x			Prepare and present initial design with calculations and supporting data.
Design engineer (not of this product)	x	x	x	Review and verify adequacy of design.
Reliability engineer	x	x	x	Evaluate design for maximum reliability consistent with system goals.
Manufacturing engineer		x	x	Ensure manufacturability at reasonable cost. Check for tooling adequacy and assembly problems.
Materials engineer		x		Ensure optimum material usage considering application and environment.
Stress analyst		x		Review and verify stress calculations.
Quality control engineer		x	x	Review tolerancing problems, manufacturing capability, inspection, and testing problems.
Human factors engineer		x		Ensure adequate consideration to human operator, identification of potential human induced problems.
Safety engineer		x		Ensure safety to operating and auxiliary personnel.
Maintainability engineer		x	x	Analyze for ease of maintenance repair and field servicing problems.
Logistics engineer		x	x	Evaluate and specify logistic support. Identify logistics problems.

ize the procedure, (2) as a means of historical documentation, and (3) as a basis for future improvement.

The procedure consists of a sequence of logical steps, starting with the analysis of lower-level subsystems or components. The analysis assumes a failure point of view and identifies all potential modes of failure along with the causative agent, termed the "failure mechanism." The effect of each failure mode is then traced up to the systems level.[14]

A criticality rating is developed for each failure mode and resulting effect. The rating is based on the probability of occurrence, severity, and detectability. For failures scoring high on this rating, design changes to reduce criticality are recommended. This procedure is aimed at providing a better design from a reliability standpoint.[15]

Probabilistic Approach to Design

Reliability is basically a design parameter and has to be incorporated in the system at the design stage. One way to quantify reliability during design and to design for reliability is the probabilistic approach to design.[16,17] The design variables and parameters are random variables, and hence the design methodology must consider them as random variables. The reliability of any system is a function of the reliabilities of its components. To analyze the reliability of the system, we first have to understand how to compute the reliabilities of the components. The basic idea in reliability analysis from the probabilistic design methodology viewpoint is that a given component has certain strength, which, if exceeded, will result in the failure of the component. The factors that determine the strength of the component are random variables, as are the factors that determine the stresses or loading acting on the component. "Stress" is used to indicate any agency that tends to induce failure, whereas "strength" indicates any agency resisting failure. "Failure" is taken to mean failure to function as intended; it occurs when the actual stress exceeds the actual strength for the first time.

Let $f(x)$ and $g(y)$ be the probability density functions for the stress random variable x and the strength random variable y, respectively, for a certain mode of failure. Also, let $F(x)$ and $G(y)$ be the cumulative distribution functions for the random variables x and y, respectively. Then the reliability R of the component for the failure mode under consideration, with the assumption that the stress and the strength are independent random variables, is given by

$$R = P\{\mathbf{y} > \mathbf{x}\} \tag{51}$$

$$= \int_{-\infty}^{\infty} g(y) \left\{ \int_{-\infty}^{y} f(x)\, dx \right\} dy \tag{52}$$

$$= \int_{-\infty}^{\infty} g(y)\, F(y)\, dy \tag{53}$$

$$= \int_{-\infty}^{\infty} f(x) \left\{ \int_{x}^{\infty} g(y)\, dy \right\} dx \tag{54}$$

$$= \int_{-\infty}^{\infty} f(x)\, \{1 - G(x)\}\, dx \tag{55}$$

For example, suppose that the stress random variable **x** is normally distributed with mean value of μ_x and SD of σ_x and that the strength random variable **y** is also normally distributed with parameters μ_y and σ_y. The reliability R is then given by

$$R = \Phi \left[\frac{\mu_y - \mu_x}{\sqrt{\sigma_y^2 + \sigma_x^2}} \right] \tag{56}$$

where $\Phi(z)$ is the cumulative distribution function for the standard normal variate **z**.

The reliability computations for other distributions, such as exponential, lognormal, gamma, Weibull, and extreme value distributions, have also been developed.[11] In addition, the reliability analysis has been generalized when the stress and strength variables follow a known stochastic process. The references cited in this subsection also contain simple design examples illustrating the use of the probabilistic approach to design.

8.5.6 HUMAN FACTORS

All systems are of, by, and for humans. Human factors therefore become active participants in the system design process and consequently must be weighed against safety, reliability, maintainability, and other system parameters in order to obtain trade-offs to increase system effectiveness. Human interaction with the system of interest consists of:

1. Design and production of systems.
2. Operators and repairers of systems.
3. Operators and repairers as decision elements.

Man-machine interface consists of such aspects as allocation of functions (man versus machine), automation, accessibility, human tasks, stress characteristics, information presented to the human, and the reliability of inferences coupled with the decisions on the basis of such information. Both human and machine elements of a system can fail, and their failures have varying effects on the system's performance. Some human errors cause total system failure or increase the risk of such failure. Human factors exert a strong influence on the design and ultimate reliability of a system. A summary of results on human performance reliability predictive methods is given by Meister.[18]

Both reliability and human factors are concerned with predicting, measuring, and improving system effectiveness. When the man-machine interface is complex, the possibility of human error increases, which results in an increase in the probability of system failure. An interesting facet of the human factors-reliability-maintainability relationship is that the system's reliability-maintainability depends on the detection and correction of system malfunctions. This task is generally performed by humans. Thus the system performance can be enhanced or degraded depending upon the human response. The quantification of human reliability characteristics and the development of a methodology for quantifying human performance, error prediction, control, and measurement are discussed by Meister and O'Connell.[19,20]

Reliability of a system is affected by the allocation of system functions to man, machine, or both. Characteristics tending to favor humans are:

1. Ability to detect certain forms of energy.
2. Sensitivity to a wide variety of stimuli within a restricted range.

3. Ability to detect signals and patterns in high-noise environments.
4. Ability to store large amounts of information for long periods and to remember relevant facts.
5. Ability to profit from experience.
6. Ability to use judgment.
7. Ability to improvise and adopt flexible procedures.
8. Ability to arrive at new and completely different solutions to problems.
9. Ability to handle low-probability or unexpected events.
10. Ability to perform fine manipulations.
11. Ability to reason instinctively.

Characteristics tending to favor machines are:

1. Computing ability.
2. Performance of routine, repetitive, and precise tasks.
3. Quick response to control signals.
4. Ability to exert large amounts of force smoothly and precisely.
5. Ability to store and recall large amounts of data for short periods.
6. Ability to reason deductively.
7. Insensitivity to extraneous factors.
8. Ability to handle highly complex operations that involve doing several things at once.

A human performance reliability model developed at Sandia Laboratories, Technique for Human Error Rate Prediction (THERP), and is defined as follows:

> *THERP is a method to predict human error rates and to evaluate the degradation to a man-machine system likely to be caused by human errors in association with equipmenting, operational procedures and practices, and other system and human characteristics which influence system behavior.*[21]

8.5.7 RELIABILITY MEASUREMENT

Reliability measurement techniques provide a common discipline that can be used to measure, predict, and evaluate system reliability throughout the system life cycle.[22] The two major components of the reliability measurement system are the test program and the data system. Test programs have to be developed throughout the life cycle, and the test effort has to ensure that the reliability goals are met at different stages in the cycle. Procedures for gathering the data generated throughout all the phases must be documented in sufficient detail for complete identification and integration into the data processing system.

Test Programs

Exhibit 8.5.8 shows a sequence of different types of tests that may be used throughout the life cycle, consisting of design, development, production, and service/field use phases. Brief descriptions of these tests follow.

Design Support Tests

These tests are used to determine the need for parts, materials, and component evaluation or qualification to meet system performance and other reliability design criteria. Some of the objectives are:

Parts application data.
Parts evaluation.
Parts qualification.
Parts comparative evaluation.
Vendor control.

Design Verification Tests

These tests are used to verify the functional adequacy of design and to corroborate preliminary predictions and failure mode and effects analysis that disclose high-risk areas and reliability problems in the proposed design. Design verification tests fulfill the following essential design phase functions:

Exhibit 8.5.8 Integrated Test Flow Diagram

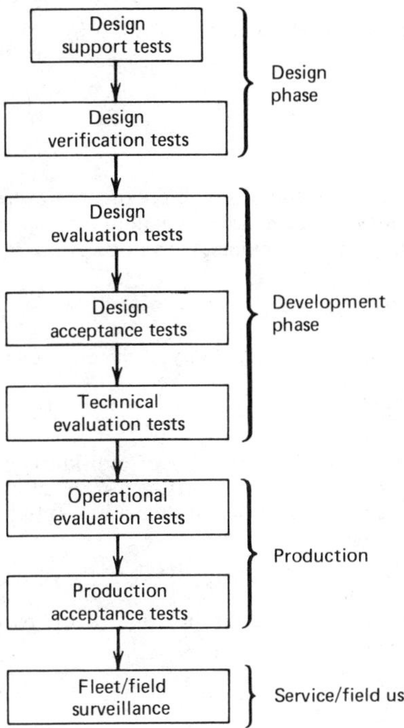

Analytical verification.
Functional evaluation.
Parts and materials definition.
Preliminary reliability verification.

Design Evaluation Tests

These tests are used to evaluate the design under environmental conditions, to verify the compatibility of subsystem interfaces, and to review the design from the maintainability point of view. Some of the tests under this category are:

Environmental (evaluation tests).
Longevity (failure rates) tests.
Operability tests.
Engineering change evaluation tests.

Design Acceptance Tests

These tests are used to demonstrate that design meets required levels of reliability. Thus a reliability demonstration test is considered mandatory for design acceptance. Definitions of test requirements are:

1. Define acceptable levels of reliability and decision risks (type 1 and type 2 errors, confidence limits).
2. Define test conditions.
3. Define the specific test plan.
   a. MTBF tests.
   b. Mission reliability test.
   c. Availability tests.

4. Define "failure" and scoring criteria.
   a. System failures.
   b. Mission failures.
   c. Maintenance actions (chargeable, nonchargeable).
   d. Criticality factors.
5. Determine sample size.

Technical Evaluation Tests

These tests are used to evaluate the technical suitability of a prototype or a preproduction model. It is sometimes practical to integrate technical evaluation with operational evaluation when the earlier design acceptance test demonstrates complete conformance.

Operational Evaluation Tests

These tests are used to evaluate operational suitability of the production model.

Production Acceptance Tests

These tests are used to determine the acceptability of individual production items in order to ensure production control of critical interfaces, parts, and material quality. Manufacturing operations that result in significant reliability degradation should be carefully studied.

Fleet/Field Surveillance Tests

These tests and evaluation programs during the field use of the product are for the continuing assessment of reliability and quality.

Test Procedures

Any test procedure must consider the following factors:

1. Purpose of test.
2. Test items–description and sample selection.
3. Test monitoring and review procedures.
4. Test equipment requirements.
5. Test equipment calibration procedures.
6. Test equipment proofing.
7. Environmental conditions to be applied.
8. Operating conditions.
9. Test-point identification.
10. Definition of failures and scoring criteria.
11. Procedure for conducting tests.
12. Test report procedures and documents.

Reliability Estimation

Reliability measurement tests are used to make estimates of the reliability of a system or a population of items. Parametric and nonparametric estimates are used. Parametric estimates are based on a known or assumed distribution of the system characteristic of interest. The parameters are the constants that describe the shape of the distribution. Nonparametric estimates are used without assuming the nature of the underlying probability distribution. Generally, nonparametric estimates are not as efficient as parametric estimates. Nonparametric reliability estimates apply only to a specific test interval and cannot be extrapolated. Parametric estimates are described in this section when the underlying distribution is exponential and Weibull. The three types of parametric estimates that are frequently used are:

1. Point estimate–a single-valued estimate of a reliability measure.
2. Interval estimate–an estimate of an interval that is believed to contain the true value of the parameter.
3. Distribution estimate–an estimate of the parameters of a reliability distribution.

Reliability Estimation: Exponential Distribution

The two types of test procedures considered here are:

1. Type 1 censored test—the items are tested for a specified time T, and then the testing is stopped.
2. Type 2 censored test—the test time is not specified, but the testing is stopped when a desired number of items fail.

Let us consider the situation where n items are placed and the test is stopped as soon as r failures are observed ($r \leqslant n$). This is type 2 censoring with nonreplacement of items. Let the observed failure times be, in order of magnitude,

$$0 = t_0 < t_1 < t_2 < \cdots < t_{r-1} < t_r \tag{57}$$

Then, making the transformation,

$$u_i = \begin{cases} nt_1, & i = 0 \\ (n-i)\,(t_{i+1} - t_i), & i = 1, 2, \ldots, r-1 \end{cases} \tag{58}$$

it is well known that the $\{u_i, i = 0, \ldots, r-1\}$ is independently and identically distributed with common density function $(1/\theta)\, e^{-u/\theta}$.[23]

It is clear that the total time on test is given by

$$\begin{aligned} V(t_r) &= \text{total time on test} \\ &= \sum_{i=0}^{r-1} u_i \\ &= \sum_{i=1}^{r} t_i + (n-r)\, t_r \end{aligned} \tag{59}$$

Then

$$\hat{\theta} = \frac{V(t_r)}{r} = \frac{1}{r}\left[\sum_{i=1}^{r} t_i + (n-r)\, t_t\right] \tag{60}$$

is the minimum variance unbiased estimator of θ. Since $V(t_r) = \sum_{i=0}^{r-1} u_i$ and the $\{u_i\}$ are independently distributed with a common exponential density function, then it follows that $V(t_r)$ has a gamma distribution with parameters (θ, r). Hence $2V(t_r)/\theta = 2\hat{\theta}r/\theta$ is distributed as χ^2_{2r}.

The 100 $(1 - \alpha)$% confidence limits on θ are given by

$$P\left[\chi^2_{1-(\alpha/2),\,2r} \leqslant \frac{2\hat{\theta}r}{\theta} < \chi^2_{\alpha/2,\,2r}\right] = 1 - \alpha$$

or

$$\frac{2\hat{\theta}r}{\chi^2_{\alpha/2,\,2r}} \leqslant \theta \leqslant \frac{2\theta r}{\chi^2_{1-(\alpha/2),\,2r}} \tag{61}$$

The life testing procedures are often used in a quality control context in which we wish to detect the deviations of θ below some desired level, say, θ_0. Then, for a significance level of α, the probability of accepting H_0 is

$$P_a = P\left[\frac{2r\hat{\theta}}{\theta_0} \leqslant \chi^2_{\alpha,\,2r} | \theta = \theta_0\right] = 1 - \alpha \tag{62}$$

The expected time to complete the test is given by

$$E[t_r] = \theta \sum_{i=1}^{r} \frac{1}{n-i+1} \tag{63}$$

Let

$$\begin{aligned}
\theta_0 &= \text{desired reliability goal for } MTBF \\
1-\alpha &= \text{probability of accepting items with true } MTBF \text{ of } \theta_0 \\
\theta_1 &= \text{alternative } MTBF\ (\theta_1 < \theta_0) \\
\beta &= \text{probability of accepting items with true } MTBF \text{ of } \theta_1
\end{aligned}$$

With this information, reliability testing consists of putting n items on test and stopping the test when the number of failures is given by the smallest integer satisfying

$$\frac{\theta_1}{\theta_0} \leqslant \frac{\chi^2_{\alpha,2r}}{\chi^2_{1-\beta,2r}} \tag{64}$$

Thus, when we know θ_0, θ_1, α, and β, we can compute the necessary value for r.

For the type 1 censored test, where r failures are observed on an interval of total test time T, the $100(1-\alpha)\%$ confidence limits on θ are given by (a modification of equation 61)

$$\frac{2T}{\chi^2_{\alpha/2,2(r+1)}} \leqslant \theta \leqslant \frac{2T}{\chi^2_{1-(\alpha/2),2r}} \tag{65}$$

Reliability Estimation: Weibull Distribution

Weibull distribution is probably one of the most widely used distributions in life testing applications. One of the reasons is the ease with which graphic procedures can be used to estimate the parameters of the Weibull distribution and thus the reliability of the product. Confidence limits can also be easily developed. In addition, various statistical estimation procedures have recently been developed, and these can also be easily used by reliability engineers.

The density function for the Weibull time to failure random variable is given by (see equation 27)

$$f(t) = \frac{\beta(t-\delta)^{\beta-1}}{(\theta-\delta)^{\beta}} \exp\left[-\left(\frac{t-\delta}{\theta-\delta}\right)^{\beta}\right], \qquad t \geqslant \delta > 0 \tag{66}$$

where

$$\begin{aligned}
\beta > 0 &= \text{shape parameter or the Weibull slope} \\
\theta &= \text{scale parameter or the characteristic life} \\
\delta &= \text{location parameter or the minimum life}
\end{aligned}$$

If the minimum life $\delta = 0$, the cumulative distribution is given by

$$F(t) = 1 - \exp\left[-\left(\frac{t}{\theta}\right)^{\beta}\right] \tag{67}$$

After rearranging and taking twice the natural logarithm we have

$$\ln\left[\ln \frac{1}{1-F(t)}\right] = \beta \ln t - \beta \ln \theta$$

or

$$\ln t = \frac{1}{\beta} \ln\left[\ln \frac{1}{1-F(t)}\right] + \ln\theta \tag{68}$$

Exhibit 8.5.9 Weibull Probability Paper

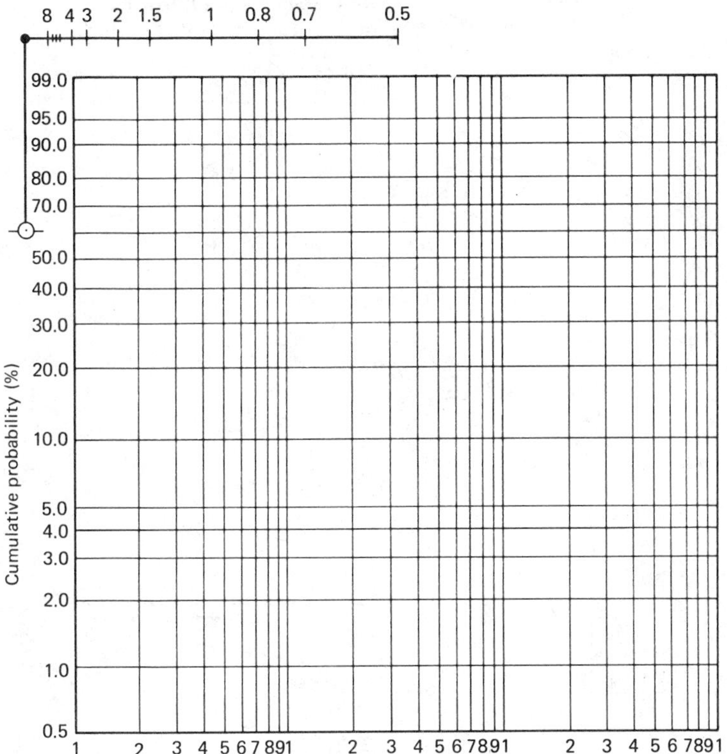

Weibull graph paper is constructed by plotting ln t as the horizontal axis versus ln [ln · 1/(1 − $F(t)$)] as the vertical axis, and then β is the slope of the straight-line plot. Exhibit 8.5.9 shows such Weibull paper. Various plotting procedures as well as statistical estimation methods with tables are available.[24] Mann et al.[24] offer point and interval estimation procedures for various other distributions and also tables that can be used for statistical estimation of the parameters of the Weibull distribution. This source is also a good reference for testing reliability hypothesis.

8.5.8 MAINTAINABILITY

Maintainability is one of the system design parameters that has a great impact on the effectiveness of the system. Failures will occur no matter how reliable a system is made to be. A system's ability to be maintained, that is, retained in or restored to effective usable condition, is often as important to system effectiveness as is its reliability. Maintainability is a characteristic of the system and its design just like reliability. It is concerned with such system attributes as accessibility to failed parts, diagnosis of failures, repairs, test points, test equipment and tools, maintenance manuals, displays, and safety. Maintainability may be defined as a characteristic of design and installation that imparts to a system a great inherent ability to be maintained, so as to lower the required maintenance man-hours, skill levels, tools, test equipment, facilities, and logistics costs and thus achieve greater availability.

Maintainability Measures

Maintainability is the probability that a system in need of maintenance will be retained in or restored to a specified operational condition within a given period. Thus the underlying random variable is the maintenance time. Let **t** be the repair time random variable. Then the maintainability function $M(t)$ is given by

$$M(t) = P[\mathbf{t} \leqslant t] \tag{69}$$

Exhibit 8.5.10 Lognormal Probability Density Functions (*a*) and Maintainability Functions *M*(*t*) Based on Lognormal Distribution with Median Equal to 15 min (*b*)

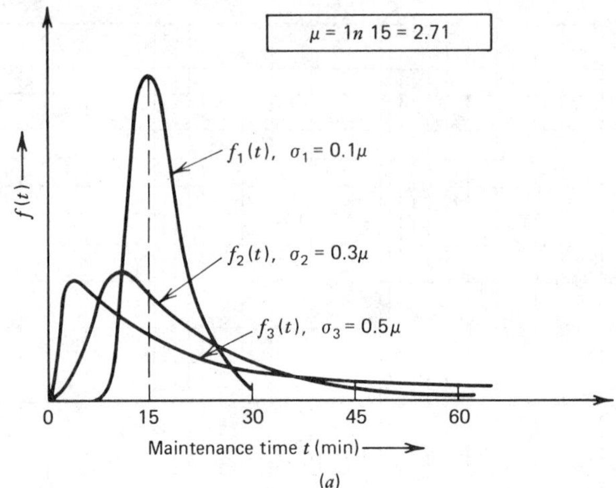

(*a*)

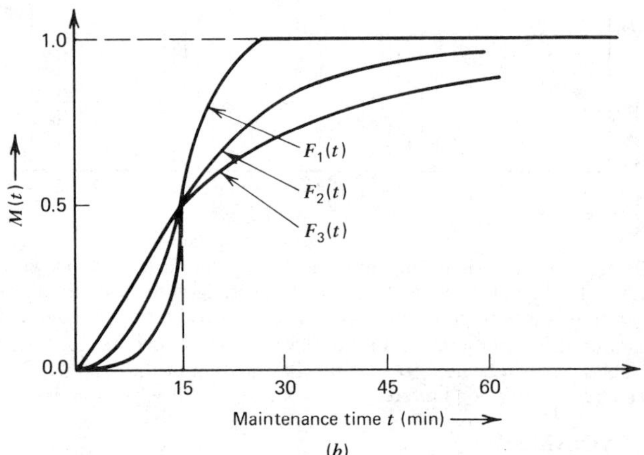

(*b*)

If the repair time **t** follows the exponential distribution with *MTTR* of $1/\mu$, where μ is the repair rate, then

$$M(t) = 1 - \exp\left(-\frac{t}{MTTR}\right) \tag{70}$$

Various other distributions, such as lognormal, Weibull, and normal, are used to model the repair time. In addition, other time-related indices, such as median (fiftieth percentile) and M_{max} (ninetieth or ninety-fifth percentile), are used as maintainability measures. The lognormal probability density functions with an *MTTR* of 15 min, but with different values for SD, are given in Exhibit 8.5.10*a*, whereas the associated maintainability functions are shown in Exhibit 8.5.10*b*. From the maintainability function plot, different percentiles, such as the ninetieth, can be easily read. In other instances the maintenance man-hours per system operating hours or maintenance ratio (*MR*) may be specified, and maintainability design goals then derived from such specifications.

The *MTTR*, which is the mean of the distribution of system repair time, may be evaluated by

$$MTTR = \frac{\sum_{i=1}^{n} \lambda_i t_i}{\sum_{i=1}^{n} \lambda_i} \tag{71}$$

where n = number of components in the system
λ_i = failure rate of the ith repairable component
t_i = time required to repair the system when the ith component fails

In addition, other measures, such as mean active corrective maintenance time and mean active preventive maintenance time, are used to measure maintainability. Some of the components of the corrective maintenance tasks are:

Localization. Determining the location of a failure to the extent possible, without using accessory equipment.

Isolation. Determining the location of a failure by the use of accessory test equipment.

Disassembly. Disassembling the equipment to gain access to the item being replaced.

Interchange. Removing the failed item and installing the replacement.

Alignment. Performing any alignment, testing, and adjustment made necessary by the repair action.

Checkout. Performing checks or tests to verify that the equipment has been restored to a satisfactory operating condition.

As an example of *MTTR* computation, assume that a communication system consists of five assemblies, with the data given in Exhibit 8.5.11. Column 2 gives the number of units n_i for assembly i. Column 3 indicates the failure rate per thousand hours for each unit. Thus column 4 gives us the total failure rate for an assembly i. Column 5 gives the average time to perform all the maintenance actions discussed previously. Then, *MTTR* is given by

$$MTTR = \frac{\Sigma n_i \lambda_i t_i}{\Sigma n_i \lambda_i} \tag{72}$$

$$= \frac{63.5}{161} = 0.394 \text{ hr}$$

8.5.9 AVAILABILITY

Availability is the vehicle that translates measures of reliability and maintainability into a combined index of effectiveness for a system. It is based on the question, "Is the equipment available in a working condition when it is needed?" Availability analysis can be used for trading between and establishing requirements for reliability and maintainability.

Availability Measures

By its very nature, availability measures are time related. The breakdown of total time upon which the availability analyses are based was briefly described earlier in this chapter. The time elements are:

1. Storage, free, and off time.
2. Operating time.
3. Standby time—availability for operations.
4. Downtime, which consists of corrective and preventive maintenance and also time due to administrative and logistics delays.

Exhibit 8.5.11 Worksheet for *MTTR* Prediction

Assemblies	n_i	λ_i ($\times 10^3$)	$n_i\lambda_i$ ($\times 10^3$)	t_i (hr)	Repair Time per 10^3 hr ($n_i\lambda_i t_i$)
1	4	10	40	0.10	4.0
2	6	5	30	0.20	6.0
3	2	8	16	1.00	16.0
4	1	15	15	0.50	7.5
5	5	12	60	0.50	30.0
			$\Sigma = 161$		$\Sigma = 63.5$

Measures for operational readiness are based on all the time elements. However, the availability measures do not consider the off time, including storage and free time. Achieved availability (A_a) is used for development and initial production testing where the system is not operating in its intended operational environment and is equal to operating test time divided by operating test time plus total preventive and corrective maintenance time (clock time). Excluded are operator before-and-after operating checks and supply, administration, and waiting times. Standby time is excluded both by definition and by environment. Thus

$$A_a = \frac{OT}{OT + TPM + TCM} \tag{73}$$

where OT = operating time
TPM = total preventive maintenance time
TCM = total corrective maintenance time

Operational availability (A_o) covers all segments of the time that the system should be operative. Thus we must consider standby time (ST) as well as administrative and logistics delay time ($ALDT$). Hence

$$A_o = \frac{OT + ST}{OT + ST + TPM + TCM + ALDT} \tag{74}$$

Sometimes there is a need to define the availability with respect to operating time and corrective maintenance when the system is operating in an ideal support environment. This form of availability, called "inherent availability" (A_I), is useful for determining certain figures of merit for the system per se, such as frequency and type of failure occurrence, repairability (active repair time), and analysis of maintenance actions. Thus A_I is given by

$$A_I = \frac{MTBF}{MTBF + MTTR} \tag{75}$$

Standby time, delay times associated with scheduled or preventive maintenance, and administrative and logistics downtime are excluded.

Reliability-Maintainability-Availability Trade-Off

The system available A_I is a function of variables of reliability ($MTBF$) and maintainability ($MTTR$) as given by equation 75. Since $MTBF = 1/\lambda$ where λ is the failure rate, and $MTTR = 1/\mu$ where μ is the repair rate (both valid when the underlying distribution is exponential), equation 75 may be rewritten as

$$A = \frac{\mu}{\mu + \lambda} \tag{76}$$

A generalized plot of equation 75 is given in Exhibit 8.5.12*a*. This shows that, to optimize availability, it is desirable to make the ratio of $MTBF$ to $MTTR$ as high as possible. Since increasing $MTBF$ and decreasing $MTTR$ is desirable, the equation for availability is plotted in terms of $MTBF$ and $\mu = 1/MTTR$, as shown in Exhibit 8.5.12*b*. Each of the curves representing the same availability is called an "isoavailability contour"; corresponding values of $MTBF$ and $MTTR$ give the same value of A, all other things being equal. Based on various physical, technological, and economic constraints, trade-off optimization models can be developed. There are practical limits as to how high a value for $MTBF$ can be achieved or how low $MTTR$ can be made. To increase $MTBF$ may require the redundancy level to be so high that the desired reliability could not be realistically achieved within the state of the art, or else the cost would be high. Low values for $MTTR$ would require extreme maintainability design features, such as complete built-in test features, automatic fault isolation, and automatic switchover from a failed to a standby item.

8.5.10 RELIABILITY GROWTH

As the product goes through the various steps in the life cycle, the reliability of the product should be estimated and predicted. These values, when plotted at selected points in the life cycle, result in a growth curve, as shown in Exhibit 8.5.13, which reflects the comparative levels of reliability.

Exhibit 8.5.12 Availability (*A*) as a Function of *MTBF* and *MTTR* (*a*) and Availability as a Function of *MTBF* and Repair Rate (*μ*) (*b*)

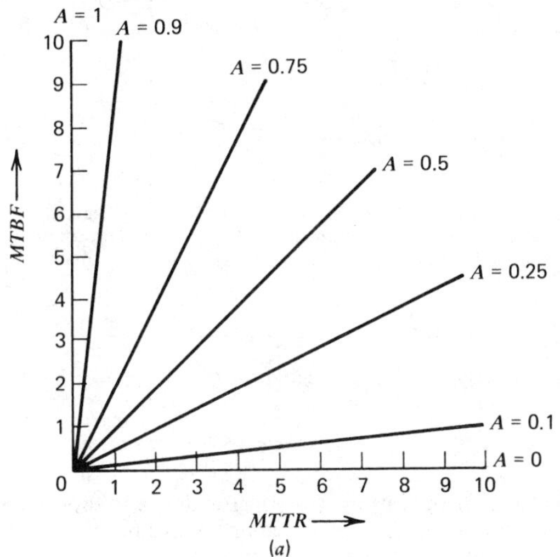

(*a*)

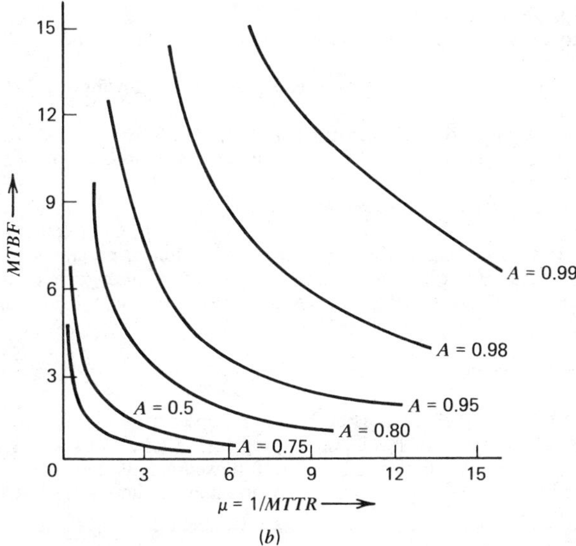

(*b*)

Reliability growth represents the effort spent to achieve the reliability potential either during design and development or during production or subsequently during field and operational use. During early development the achieved reliability of a prototype is much lower than its predicted reliability because of initial design and engineering deficiencies as well as manufacturing flaws. Also, the reliability of a fielded system is much lower than its inherent or potential reliability predicted during design and development because of the following reasons:

1. Reliability degradation due to manufacturing, assembly, and quality control errors as well as to ineffectiveness of some of the screening tests.

2. Reliability degradation due to interaction of man, machine, and environment. Degradation may be due to rough handling, extended duty cycles, or neglected maintenance.

3. There is degradation due to maintenance activities because excessive handling brought about by frequent preventive maintenance or poor maintenance practices reduces reliability.

Exhibit 8.5.13 Reliability Growth and Degradation During System Life

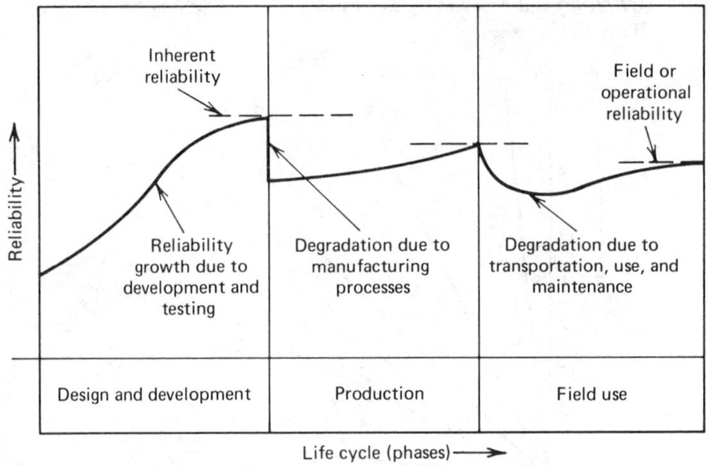

Inherent reliability

Field or operational reliability

Reliability

Reliability growth due to development and testing

Degradation due to manufacturing processes

Degradation due to transportation, use, and maintenance

Design and development | Production | Field use

Life cycle (phases)

However, during all phases there is also reliability growth due to the underlying learning process. There is reliability growth during design and development as a result of an iterative design process. The essential elements involved in achieving reliability growth are:

1. Detection of failure sources.
2. Feedback of problems identified.
3. Redesign effort based on problems identified that will potentially be corrected.

If the failure sources are detected by testing, then we have the following:

4. Manufacturing or fabrication of prototype hardware.
5. Redesign, detection, and analysis of failure sources serve as verification of redesign effort.

Some of the benefits of reliability growth methodology are:

1. It enables us to take advantage of experience gained in similar programs in the past.
2. It enables us to evaluate better the progress being made by an ongoing program.
3. It enables us to evaluate possible courses of corrective actions if an ongoing program experiences problems.

Reliability Growth Models

In 1964, Duane of GE published a report describing his observations on failure data for five different types of systems during their development programs at GE.[25] His analysis revealed that for these systems the observed cumulative failure rate followed a consistent pattern, approximately a straight line, when plotted on a log-log paper as a function of cumulative test hours. This can be expressed mathematically as

$$\Lambda(t) = kt^{-\beta} \tag{78}$$

where $\Lambda(t)$ = cumulative failure rate at time t
t = cumulative test time
k = a constant ($k > 0$)
β = growth factor ($\beta > 0$)

Let $N(t)$ be the total number of failures accumulated over the cumulative test time t. Then

$$\Lambda(t) = \frac{N(t)}{t} \tag{79}$$

Thus the cumulative failure rate $\Lambda(t)$ will decrease as reliability grows as a result of the development and corrective effort because fewer failures will be observed. If we take logarithms of both sides of equation 78, we have

$$\log \Lambda(t) = \log k - \beta \log t \tag{80}$$

The value of constant k depends on system complexity, design margins, and design objectives for reliability, whereas the value of growth factor β depends on the development effort.

Substituting equation 79 in equation 78, we have

$$N(t) = kt^{1-\beta} \tag{81}$$

$$= kt^{\beta'} \qquad (\beta' = 1 - \beta) \tag{82}$$

Taking the logarithm of both sides of equation 82, we have

$$\log N(t) = \log k + \beta' \log t \tag{83}$$

Thus, if we plot the cumulative number of failures with respect to cumulative test time t on a log-log paper, we get a straight line with slope β', which again represents the growth rate. It is clear from equation 78 that the higher the value of β, the more growth the system has.

The actual failure rate of the system if the design is released after test time t is given by

$$\lambda(t) = \frac{dN(t)}{dt}$$

$$= k(1-\beta)\, t^{-\beta} \tag{84}$$

Thus, if we are given some goal for failure rate $\lambda(t)$, we can compute the total test time required during the development effort using equation 84.

8.5.11 DESIGN AND MANAGEMENT OF RELIABILITY PROGRAMS

The establishment of an effective product assurance program throughout the system life cycle requires a management with exceptional perception of the assurance sciences. This is because of the many conflicting factors involved when making decisions on the delegation or redistribution of responsibility and authority and also because reliability is at best a poorly understood discipline without a universal approach that applies to every product and every organization. A reliability program extends far beyond the estimation of reliability numbers: It must create an attitude of anticipation of reliability problems and initiate the preplanning necessary to eliminate or reduce the effects of unreliability to an acceptable and planned-for level. Because of the broadness of the subject, a product assurance program must span the total system life cycle; therefore the program will infringe on several well-established technical management groups, such as design, testing, purchasing, quality control, manufacturing, sales, and service groups.

Unreliability is actually not the result of any one group, but is due to the complexity of today's systems and of the organization required to create these systems. During system design, development, and production, anything that can contribute to unreliability should be identified and reviewed. A reliability management structure must be created to accomplish this task. The question then is, "Who should do this?" In a "perfect" organization with complete flow of information, the unreliability problems might be taken care of by existing groups. However, this "perfect" organization rarely exists in practice; thus over the years many organizations have found it necessary and advantageous to spend the extra effort to create a reliability group and to integrate this group into the existing organization. The reliability group performs essentially an assurance function, overseeing the design and development process from a reliability point of view.

The establishment and maintenance of a viable reliability program must be done based on management foresight and intuition, since the payback is not readily measured in dollars. Unreliability problems uncovered by a reliability group tend to be taken care of outside of the formal organizational communication channels and in general will be attributed to design engineering. To force a reliability group to point up all unreliability problems will mean that management is forcing them into a position of accusation, and this will create an intractable climate that may well hinder the product development cycle and the reliability improvement effort. Thus it is very important to understand the role and the function of the reliability group in the total organization.

Elements of a Reliability Program

Management and control of system reliability must be based on a recognition of the system's life cycle, beginning at concept and extending through design, production, use, and discarding of the

Exhibit 8.5.14 Elements of a Reliability Program

	Life Cycle Phase[a]				
Elements	Concept of Planning	Design	Development	Production	Field
Requirements definition	V	V C	C L		
Reliability models	V	V	V L	L	L
Reliability allocation	V	V N	N		
Design review		V	N L		
Failure mode and effects analysis		V	N L		
Probabilistic design		D	L		
Parts selection	N	V N	N L		
Reliability measurement			CD	N	CV
Reliability growth			N D	L	L
Acceptance tests			V C	N L	L
Quality control			L V	C	D
Maintainability		V N	N L	L	V
Serviceability		V N	N L	L	C
Planning of a reliability program	V C	C L	L	L	
Failure analysis			V	V L	V
Field and Data Collection				L	V
Management of the program	V	V	V	V	V

[a]C = critical activity (errors usually disastrous); V = very important activity (errors often disastrous); N = necessary activity (integral part); D = desirable activity (for high success); and L = low key activity (update previous results).

system. One of the objectives is to achieve acceptable levels of operational reliability and maintainability. Achievement of this objective requires numerous tasks. Some of these are given in Exhibit 8.5.14, which also shows the importance of the tasks throughout the system life cycle. Thus the activities of a reliability program have applications throughout the system life cycle. Some of these applications are as follows:

Applications During Design

Develop safety margins from reliability viewpoint.

Predict component reliability from the data bank of the failure rates.

Compute system reliability from component reliability.

Determine amount of redundancy needed to achieve a reliability goal.

Provide input to human engineering.

Interact with value engineering.

Evaluate design changes.

Perform trade-off analysis.

Compare two or more designs.

Provide guidelines for design review.

Work with cost reduction programs.

Applications During Development of Testing

Establish reliability growth curves for the development and testing phase.

Develop guidelines for the amount of testing.

Develop bathtub curve based on the failure rate data and the test data.

Participate in the development of failure definition and scoring criteria document.

Participate in the scoring of the test failures.

Applications During Manufacturing

Provide guidelines for manufacturing processes.

Provide input to quality control.

Provide input to guidelines to evaluate the suppliers and vendors.

Develop product burn-in or debugging time.

Applications During Field Use

Establish warranty cost and help reduce it.
Optimize the length of warranty.
Reduce inventory costs.
Develop maintenance procedures, both corrective and preventive.
Provide input to the spare parts allocation models.
Participate in the collection and analysis of the field data.
Participate in the feedback process to report and correct the field failures.

The reliability group coordinates and directs the overall reliability effort to provide assurance that the optimum reliability has been achieved and that the consequences of unreliability have been considered in the overall plans. Obviously, the reliability group can be effective only if given proper authority, demanding formal sign-off during all critical stages of system development.

An acceptable reliability level for a system is really a many-faceted problem that requires many complex trade-offs. Considerations in design, cost, manufacturability, material availability, maintenance, and serviceability all enter into this problem. For example, a highly reliable design that cannot be manufactured effectively or that cannot be maintained may represent an unacceptable situation from the total system point of view. The important thing is to plan for adequate overall system effectiveness utilizing good knowledge on the actual reliability level of the system.

References

1. E. R. CARRUBBA, R. D. GORDON, and A. C. SPANN, *Assuring Product Integrity*, Lexington Books, Lexington, MA, 1975.
2. S. HALPERN, *The Assurance Sciences: An Introduction to Quality Control and Reliability,* Prentice-Hall, Englewood Cliffs, NJ, 1978.
3. *Quality Assurance–Reliability Handbook,* AMC Pamphlet No. 702-3, Headquarters, U.S. Army Materiel Command, Alexandria, VA, October 1968.
4. AERONAUTICAL RESEARCH INCORPORATED, Engineering and Statistical Staff, in W. H. Von Alven, Ed., *Reliability Engineering*, Prentice-Hall, Englewood Cliffs, NJ, 1964.
5. *Reliability Design Handbook*, Reliability Analysis Center, RDG-376, Griffiss Air Force Base, New York, March, 1964, p. 4.
6. MIL-HDBK-217A, "Reliability Stress and Failure Rate Data for Electronic Equipment," *Military Standardization Handbook*, U.S. Department of Defense, December 1, 1965.
7. MIL-HDBK-217B, "Reliability Prediction of Electronic Equipment," *Military Standardization Handbook*, U.S. Department of Defense, September 20, 1974.
8. W. WEIBULL, *Fatigue Testing and Analysis of Results*, Macmillan, New York, 1961.
9. E. J. GUMBEL, *Statistics of Extremes*, Columbia University Press, New York, 1958.
10. R. E. BARLOW and F. PROSCHAN, *Statistical Theory of Reliability and Life Testing*, Holt, Rinehart & Winston, New York, 1975.
11. K. C. KAPUR and L. R. LAMBERSON, *Reliability in Engineering Design*, Wiley, New York, 1977.
12. *Engineering Design Handbook*, Part 3, Reliability Prediction, AMC Pamphlet No. 706-197, U.S. Army Materiel Command, Alexandria, VA, January 1976.
13. R. E. BARLOW, J. B. FUSSEL and N. D. SINGURWALLA, Eds., *Reliability and Fault Tree Analysis*, Society of Industrial and Applied Mathematics, Pennsylvania, Philadelphia, 1975.
14. ARP-926, *Design Analysis Procedure for Failure Mode, Effects and Criticality Analysis (FMECA)*, Society of Automotive Engineers, New York, N.Y., September 1967.
15. MIL-STD-1629 (SHIPS), "Procedures for Performing a Failure Mode and Effects Analysis for Shipboard Equipment," U.S. Department of the Navy, Naval Ship Engineering Center, Hyattsville, MD, November 1, 1974.
16. E. B. HAUGEN, *Probabilistic Approach to Design,* Wiley, New York, 1968.
17. D. KECECIOGLU and D. CORMIER, "Designing a Specified Reliability Directly into a Component," *Proceedings Third Annual Aerospace Reliability and Maintainability Conference*, 1968.
18. D. MEISTER, "A Critical Review of Human Performance Reliability Predictive Methods," *IEEE Transactions on Reliability*, Vol. R-22, August 1973, pp. 116–123.
19. D. MEISTER, "Human Factors in Reliability," in W. G. Ireson, Ed., *Reliability Handbook*, McGraw-Hill, New York, 1966.

20 R. D. O'CONNELL, *Handbook of Human Performance Measures*, Space Biology Laboratory, University of California, Los Angeles, 1972.

21. A. D. SWAIN, "Short Cuts in Human Reliability Analysis," in E. J. Henly and L. W. Lynn, Eds., *Generic Techniques in System Reliability Assessment,* Noodhoff Publishing Company, Holland, 1974, p. 394.

22. *Engineering Design Handbook*, Part-4; AMC Pamphlet No. 706-198, Reliability Measurement, U.S. Army Materiel Command, Alexandria, VA, January 1976.

23. B. EPSTEIN, "Tests for the Validity of the Assumption That the Underlying Distribution of Life is Exponential," *Technometrics*, Vol. 2, No. 1 (February 1960), Part I, pp. 83–101 and Part II, pp. 163–184.

24. N. R. MANN, R. E. SHAFER, and N. D. SINGPURWALLA, *Methods for Statistical Analysis of Reliability and Life Data*, Wiley, New York, 1974.

25. J. T. DUANE, "Learning Curve Approach to Reliability Monitoring," *IEEE Transactions Aerospace*, Vol. 2, 1964, pp. 563–566.

SECTION 9
ENGINEERING ECONOMY

CHAPTER 9.1

Company and Cost Accounting

EVAN F. BORNHOLTZ

General Motors Institute

9.1.1 GENERAL ACCOUNTING

Purposes of the Accounting System

Accounting systems exist to meet the needs of their users. Users such as managers and owners need to know how profitable their firm has been and what assets and obligations it has today. Managers also must plan future operations and then control those operations to see that the firm runs efficiently. The necessary accounting activities and their sequence is shown in Exhibit 9.1.1.

For reporting purposes the life of the firm is divided into accounting periods. Items such as sales or profits are generated continuously and thus must be reported per period (annually, quarterly, or monthly). For other items, such as inventories carried or the amount of money borrowed, users need to know the current level or balance. These balances are determined and reported as of the end of the accounting period. Establishing consistent reporting periods allows users of financial data to compare the current period's results with those of past accounting periods. For example, sales for the year just completed may be compared to sales generated last year to see if they increased, decreased, or stayed the same. The inventory carried at the end of the current period may be compared to the inventory level at the end of the previous period. This comparison shows any change that occurred during the accounting period.

Basic Accounting Statements

General financial information is presented to managers and owners in the balance sheet and in the income statement. (Because many firms face shortages of cash, a third statement is becoming important. This statement details the flow of funds within the firm.)

Very briefly, the balance sheet is a *statement of position* showing the firm's investment in assets and its obligations to lenders and owners as of the end of the accounting period. The format of this statement is based on the accounting equation as shown in Exhibit 9.1.2. For examples of statements, refer to annual reports of your firm or any financial accounting textbook.

The income statement serves as a *score card* of revenues less expenses, showing how well the business did in generating profits in the accounting period. (See Exhibit 9.1.3 for the basic format.) To obtain an accurate profit figure, it is important that expenses be *matched* with the revenues they helped to generate. This matching is achieved by recognizing both revenues earned and expenses incurred (used) in that same accounting period. It should also be noted that whether or not a cash flow has occurred is ignored in the development of an income statement. For example, sales are recognized as revenues whether they are cash or credit sales. Inventories become expenses only when used and not when they are purchased. Finally, depreciation is an estimated cost and thus a noncash charge.

Exhibit 9.1.1 Sequence of Accounting Activities

Before Accounting Period	During Accounting Period	After Accounting Period
Plan firm's operations	Gather, digest, and report on 1. Profits earned during period. 2. Position at end of period.	Evaluate cost control and profit performance

Exhibit 9.1.2 The Balance Sheet and the Accounting Equation

Basic Format[a]

Things of value owned by business = borrowed funds + owner-supplied funds

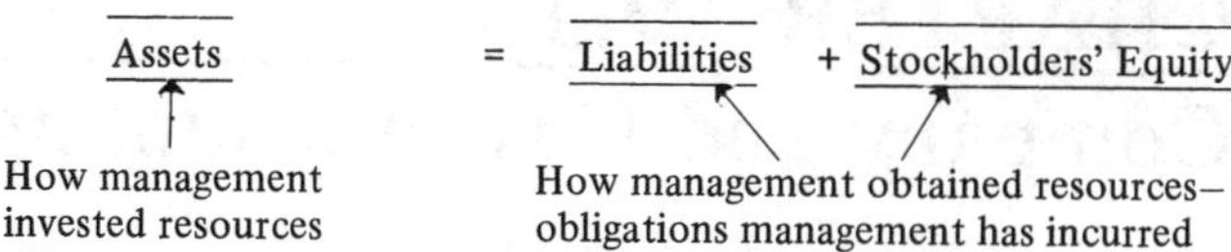

How management invested resources

How management obtained resources—obligations management has incurred

Subclassification of Assets, Liabilities, and Stockholders' Equity

Current assets: Things of value that will be used (turned into cash through sales) in the coming year. Listed as to how quickly they can or will be turned into cash, the items typically included are cash, short-term investments, receivables, inventories, and prepaid expenses.

Noncurrent assets: Things of value having useful lives longer than one year, such as buildings and equipment or investments in subsidiaries.

Current liabilities: Obligations that will come due and require payment of cash in the coming year. Items typically included are notes payable, accounts payable, and wages and taxes payable.

Noncurrent liabilities: Portions of long-term debt obligations such as mortgages or bond issues that will not require payments in the coming year.

Stockholders' equity: Funds from two sources:

1. Capital that was paid in by owners when the firm sold stock.
2. Retained earnings, which are profits that were not paid out to owners as dividends. Earnings reinvested to finance the growth of the firm.

[a]For a specific time—levels existing at end of period.

Exhibit 9.1.3 Income Statement

Basic Format[a]

Sales revenue	Sales revenue	
Less: Expenses ←	Less: Cost of goods sold	
Net profit	Gross margin ←	{Highlights "mark up" used in pricing
	Less: Operating expenses	
	Selling	
	Administration	
	Net profit	

[a]For a period of time—amounts accumulated during period.

Types of Firms and Their Costs

The types of costs incurred and their relative importance depend on the nature of the firm involved. For example, a service firm does not sell goods, so it has no cost of goods sold. Cost of goods sold for a merchandising firm is the cost to purchase, whereas for a manufacturing firm it is the cost to produce. All three types of firms have selling and promotional expenses, but they are typically very important to the retail merchandising firm. Overhead items such as administration, clerical, and depreciation costs are common to all types of businesses and are particularly important to service types of firms. Manufacturing firms have a special group of these costs, called "factory overhead," which is discussed in the following sections. Many of the comments and control techniques for factory overhead may also be applied to nonmanufacturing overhead. For example, the cost control techniques employed by manufacturing firms have been used by hospital management with excellent results.

9.1.2 DETERMINING AND REPORTING MANUFACTURING COSTS

Types of Manufacturing Costs

Manufacturing costs are divided into three basic classifications (Exhibit 9.1.4). The first two are direct costs in that they are directly involved in the production of the goods. Manufacturing a unit

Exhibit 9.1.4 Manufacturing Cost Classifications

Classification	Description
Direct material	Materials that become part of the product
Direct labor	Labor performed directly on product
Factory overhead	Manufacturing costs other than direct material and labor costs Indirect labor Operating supplies Expense tools Utilities Maintenance Employee fringe benefits Losses and errors Depreciation, property tax, insurance

requires that a certain amount of material and a certain amount of labor be used (constant amounts per unit). Manufacturing a thousand units requires a thousand times the material and a thousand times the labor. For example, assume that engineering design says that producing a product requires 100 lb (45 kg) of steel per unit and that purchasing says that this steel costs \$0.40/lb. The resulting direct material estimate is \$40/unit and \$40,000 for a thousand units.

Not all material and labor costs are direct. Manufacturing operations require a variety of supplies such as cutting oils and lubricants. Also required are indirect labor items such as supervision and clerks' wages. These indirect costs, which facilitate operations but do not become part of the product, are classified as overhead. Note that there is no precise definition of what is a direct cost and what is overhead. Although some costs clearly fall in one category, others such as glue could be treated either way. We could determine the amount of glue needed to manufacture an individual unit, or its cost could be determined by just checking the level of the "glue pot" at the end of the period. Determining the cost per unit and treating that cost as direct require more effort, but provide better control of that cost. Management must decide whether the added control is worth the added cost.

The factory overhead classification consists of a variety of types of costs (Exhibit 9.1.4). Some of these, such as supplies or losses and errors, are strongly related to production volume. Other overhead costs, such as supervision or depreciation, are independent of the level of operations. These latter costs relate to the production capacity the firm maintains rather than to the level at which that capacity is used.

Depreciation

One overhead cost requires special discussion because its nature is often misunderstood. This cost is depreciation, which is the estimated cost of using property, plant, and equipment. For example, assume a machine is purchased for \$100,000 and then sold 8 years later for \$20,000. Clearly, the cost of using this fixed asset is \$80,000 over its 8 year useful life. A problem, however, is that income statements are developed annually, so we need to know the cost of using the asset each year. One solution is to spread the \$80,000 cost evenly over the 8 year useful life, recognizing \$10,000 depreciation per year. Depreciation costs are also needed for income statements generated *during* the asset's life. Thus the calculation of annual depreciation must be based on estimates of both useful life and salvage value (often assumed to be zero). Exhibit 9.1.5 illustrates this calculation and graphs the value at which the fixed asset is carried on the accounting books (original cost less the sum of depreciation charges against the asset). This book value declines linearly over the asset's life, giving rise to the term "straight-line depreciation."

The purpose of depreciating a fixed asset is to have a fair cost of using the item to deduct from revenues on the income statement. This differs from the popular view of depreciation as the decline in market value during a period of time. (E.g., I drove the car off the lot and it depreciated \$1000.) This popular view leads to a misinterpretation of what asset values on the accounting books represent. Book values of assets are *not* current market values. In fact, market values are considered only when an asset is purchased and when it is sold. Interim market values are not considered because the fixed asset was purchased to be used and not to be resold. Likewise, book values do not indicate the asset's value to the firm. For example, an asset may be fully depreciated, showing zero value on the books, but still have years of serviceability left for the firm. Book values also do not consider replacement cost. A replacement machine may be twice as expensive to acquire. However, until the existing machine is actually replaced, depreciation expense and the book value are those of the old asset. (See Chapter 9.4 for an economic analysis of fixed asset replacement.)

Exhibit 9.1.5 Annual Depreciation–Straight-Line Method

Cost to acquire machine	$400,000
Estimated useful life	12 years
Estimated salvage value	$ 40,000

$$\text{Annual depreciation} = \frac{\text{cost to acquire} - \text{salvage value}}{\text{estimated useful life}}$$

$$= \frac{\$400{,}000 - \$40{,}000}{12 \text{ years}} = \underline{\underline{\$30{,}000/\text{year}}}$$

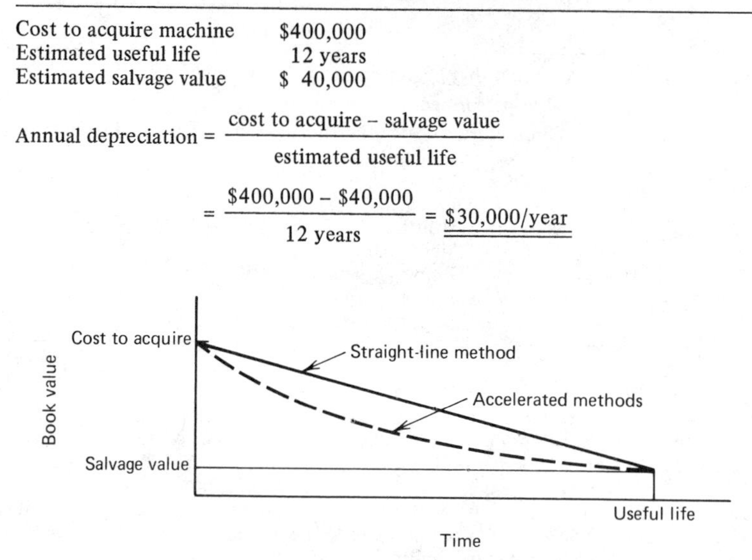

The straight-line method charges depreciation expense equally to each year of the asset's useful life. Other methods, termed "accelerated depreciation," provide a faster write-off early in the asset's life. This accelerated write-off causes the book value of the asset to decline more rapidly initially than the straight-line method. (See dashed line in Exhibit 9.1.5.) However, later in the asset's life the depreciation charge must be correspondingly reduced because the total write-off is limited to the cost to acquire less estimated salvage value. Accelerated depreciation methods simply redistribute depreciation charges forward in the asset's life. Exhibit 9.1.6 illustrates the calculations for two of these accelerated methods.

Many firms choose accelerated over straight-line depreciation for use on their accounting books or tax records. Accountants cite the fact that book values of plant and equipment are less with accelerated depreciation. If one of these assets has to be replaced before the end of its estimated life, an accounting loss must be recognized. With accelerated depreciation the loss is smaller, causing less distortion of the net income for the period. More compelling, the payment of taxes can be delayed if accelerated depreciation is used. The greater depreciation early in the asset's life reduces taxable income and thus taxes. Later in the asset's life the depreciation is less and the taxes more. Although the same total taxes are paid, the firm has the use of the funds during the time the tax payments are delayed. For further discussion of depreciation methods and their effect in aftertax discounted cash flow analysis, see Chapter 9.4.

Overhead Rates: Their Purpose, Calculation, and Application

Management must know the cost to manufacture a unit. This per unit information is needed as a basis for setting selling prices and in establishing standards for cost control purposes. Direct material and direct labor costs per unit can be developed by direct estimate. First, the quantity of material or labor required to manufacture a unit is estimated. Then the cost per quantity is developed. Finally, multiplying the quantity and the cost yields the estimated cost per unit. (See example given in Exhibit 9.1.7.)

Factory overhead, in contrast, is composed of a variety of items, including some costs not directly related to the manufacture of a unit. For example, how much supervision or depreciation cost is required to produce a single unit? Such a question cannot be answered directly, and thus some indirect means must be employed to determine overhead cost on a per unit basis. The indirect means employed is called an "overhead rate."

Overhead rates are developed by estimating total factory overhead for the period and comparing it to the total expected for some overhead base, such as direct labor cost. (See Exhibit 9.1.7 for formula and example of calculation.) The resulting overall rate for overhead can then be multiplied times the per unit amount of the base to estimate the overhead cost of manufacturing an individual unit.

The function of the overhead base is to act as a link or bridge between total overhead and the units produced. Establishing a good link with units is easily achieved by using items that can be

Exhibit 9.1.6 Accelerated Depreciation Methods

Example: Cost to acquire = \$100,000
Estimated useful life = 5 years
Estimated salvage value = \$10,000
Straight-line annual depreciation = \$18,000/year

Sum-of-the-Years-Digits Method[a]

$$\text{Sum of the years digits} = 1 + 2 + 3 + 4 + 5 = \underline{\underline{15}} = \frac{(n+1)n}{2} = \frac{6(5)}{2}$$

Year	Calculation	Annual Depreciation	Book Value
1	(5/15) × \$100,000 – \$10,000)	\$30,000	\$70,000
2	(4/15) × (\$90,000)	24,000	46,000
3	(3/15) × (\$90,000)	18,000	28,000
4	(2/15) × (\$90,000)	12,000	16,000
5	(1/15) × (\$90,000)	6,000	10,000

Double Declining Balance Method[b]

Annual depreciation = 2 × (straight-line depreciation rate assuming zero salvage value) (book value)

$$\text{Depreciation rate} = \frac{1}{\text{useful life}} = \frac{1}{5 \text{ years}} = 20\%/\text{year}$$

Year	Calculation	Annual Depreciation	Book Value
1	(2 × 20%) × (\$100,000)	\$40,000	\$60,000
2	(40%) × (\$60,000)	24,000	36,000
3	(40%) × (\$36,000)	14,400	21,600
4	(40%) × (\$21,600)	8,640	12,960
5	(40%) × (\$12,960)	2,960[c]	10,000

[a] Reversing fractions method.
[b] Built-in salvage value.
[c] Book value cannot decrease below estimated salvage value.

Exhibit 9.1.7 Estimating the Cost to Manufacture a Unit

Item	Estimate	Source
Direct material quantity	100 lb/unit	Product design
Material cost	\$0.40/lb	Purchasing
Direct labor hours	$\frac{1}{2}$ hr/unit	Methods
Labor rate	\$10.00/hr	Payroll
Total overhead for period	\$25,000	Budgeting group
Production volume	1000 units	Budgeting group

Estimated Cost to Manufacture a Unit

Direct material	100 lb at \$0.40/lb =	\$40.00
Direct labor	$\frac{1}{2}$ hr at \$10.00/hr =	5.00
Overhead	500%[a] × \$5.00 =	25.00
Total cost per unit		\$70.00

[a] Overhead rate (using direct labor cost as the base) = estimated total overhead/(overhead base) (expected volume) = \$25,000/(\$5.00/unit) (1000 units) = \$25,000/\$5,000 = 5; 500% of direct labor cost.

Exhibit 9.1.8 Applying Overhead

Overhead rate = 500% of direct labor cost

Costs Attached to Goods as Manufactured During Period

Actual direct material = 95,000 lb at \$0.42/lb = \$39,000
Actual direct labor = 510 hr at \$10.00/hr = 5100
Applied overhead = 500% × 5100 = 25,500

End-of-Period Comparison of Applied to Actual Overhead

Actual overhead	\$26,000	
Applied overhead	25,500	Added to cost of goods sold to correct to actual cost
Underabsorption	\$ 500	

readily stated on a per unit basis, such as direct material cost, direct labor hours or cost, or machine hours. The adequacy of the link to total overhead depends on what types of overhead costs are dominant in a particular operation. For example, labor-intensive operations should use direct labor hours or cost. A highly automated operation should use machine hours as its overhead base. Also, if a firm manufactures more than one product, the number of units produced will not make a good overhead base. Using units as the base will assign each product the same overhead per unit, regardless of how difficult it is to build.

The indirect nature of overhead also causes problems in assigning this cost to goods as they are manufactured. Actual direct material costs can be determined from material requisitions, and actual direct labor costs from time tickets. Total actual overhead, however, will not be known until after the period is over. Thus an estimate must be employed if all costs are to be attached to the goods as they are completed. This estimate is called "applied overhead" and is calculated as the overhead rate times the actual amount of the overhead base. (See Exhibit 9.1.8 for an example of the calculation.)

Since applied overhead is an estimate, it will differ from the actual cost incurred, making an end-of-period correction necessary. The adjustment required is called an "overabsorption" if too much has been applied, and an "underabsorption" if the amount applied is less than actual overhead. Generally the correction is made to the cost of goods sold for the period.

The Manufacturing Cost Cycle

The record keeping for manufacturing costs follows the physical flow of the goods being produced from the purchase of raw materials through the sale of finished inventories to customers. Four distinct types of inventories are used in recording these costs. Raw material records are kept either as supplies or direct material inventories, depending on the types of items involved. Costs of work in process and finished goods are recorded in inventory accounts of the same name. The flow of these manufacturing costs through the inventory accounts is detailed in Exhibit 9.1.9. Exhibit 9.1.10 is a numerical example that also illustrates the recording of costs through the manufacturing cost cycle.

Manufacturing costs of the period are reported internally in a statement of goods manufactured and sold (Exhibit 9.1.11). This statement is also designed to follow the physical flow of goods through the inventory accounts. The basic calculation employed in this statement is worthy of note in itself. It is often called an "input-output analysis" and considers four things about an inventory: beginning levels, ending levels, flows in, and flows out. If three of these factors are known, the fourth can be determined. Note that this basic calculation is used three times in the statement, with each output leading into the next inventory. It is first used for direct material inventory. The cost of direct material used in the period is calculated, which is one of the additions to work in process. The second calculation covers work-in-process inventory, yielding the cost of the goods completed during the period. This cost is, in turn, the input to finished goods inventory. Finally, the output of the finished goods calculation is the cost of the goods sold for the period.

Variable and Fixed Overhead Rates

Direct material, direct labor, and a portion of overhead costs (such as supplies, small tools, losses, and errors) relate strongly to the manufacture of a unit. These costs are described as "variable" in that they tend to be a certain amount for each unit, and their totals increase linearly with the number of units produced (see Exhibit 9.1.12). The remainder of overhead costs (such as supervision, depreciation, and property taxes) relate more to plant capacity than to production volume. These costs are described as "fixed" in that they tend to be independent of volume on a total basis. On a per unit basis, however, these costs vary inversely with volume. (Note that the preceding

Exhibit 9.1.9 Manufacturing Cost Cycle

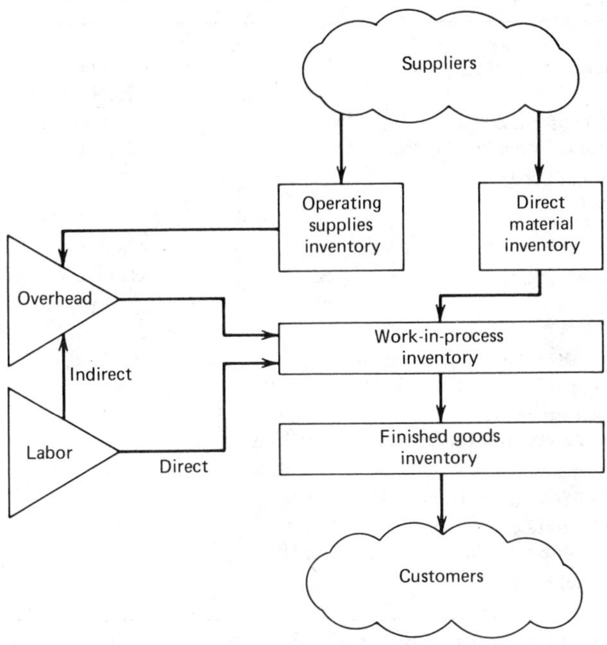

Exhibit 9.1.10 Illustration of the Recording of Manufacturing Costs

Events of the Period

1. $55,000 in direct material and $8000 in supplies are purchased.
2. $50,000 in direct material and $6000 in supplies are requisitioned to production.
3. $40,000 of direct labor cost and $10,000 of supervision and clerical costs are incurred.
4. Overhead is applied at a rate of 200% of direct labor cost.
5. Other actual overhead costs totaling $60,000 are incurred.
6. All work in process is completed and transferred to finished goods.
7. Finished goods that cost $160,000 to manufacture are sold.
8. Costs of goods sold is adjusted for the $4000 overabsorption of overhead.

Recording of the Events in the Accounting Records

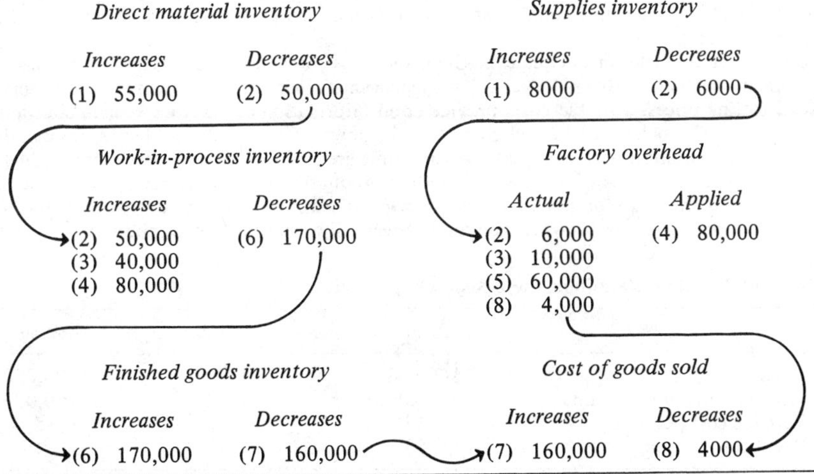

Exhibit 9.1.11 Statement of Goods Manufactured and Sold for the Year Ended December 31, 19__[a]

Direct material		
Inventory, January 1, 19__	$15,000	
Purchases	55,000	
Total available for use	$70,000	
Less: Inventory, December 31, 19__	20,000	
Direct Material Used		$ 50,000
Direct labor cost		40,000
Actual overhead		
Supplies	$ 6,000	
Supervision, clerical	10,000	
Other	60,000	
Total overhead		76,000
Manufacturing costs originating during the year		$166,000
Work-in-process inventory, January 1, 19__		30,000
Total manufacturing costs		$196,000
Less: Work-in-process inventory, December 31, 19__		30,000
Cost of goods manufactured		$166,000
Finished goods inventory, January 1, 19__		32,000
Total available for sale		$198,000
Less: Finished goods inventory, December 31, 19__		42,000
Cost of goods sold		$156,000

[a]Basic form: Additions in period (purchases, added costs, completions) + beginning inventory = total available in period (production costs, completed inventory) − ending inventory = cost of inventory manufactured or sold in period.

Exhibit 9.1.12 Cost Benavior With Changes in Production Volume

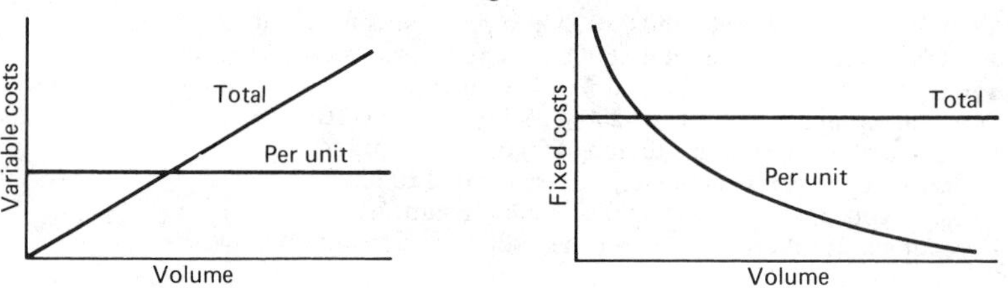

statements are true only for a limited range of production volumes. If volume were to differ significantly, variable costs per unit could change. Also, management might change capacity, resulting in a new level of fixed costs.)

Prices are set to yield an adequate mark-up over cost, and once established, it is difficult to change them greatly from year to year. Thus management needs stable per unit costs to serve as a basis for setting prices. Variable costs provide good information in that they remain constant on a per unit basis. The behavior of fixed cost per unit, however, creates a problem. Consider the example shown in Exhibit 9.1.13. Fixed costs per unit are low in year 1 when volume is high, but they rise drastically in year 2 because of the low production volume. If management establishes prices based on the costs of the first year, then profits will be less than adequate in the second year. Raising prices in year 2 when volume is already low may not be a good idea. A better ap-

Exhibit 9.1.13 Determining the Fixed Cost per Unit

	Year 1	Year 2	Average
Total fixed cost	$600,000	$600,000	$600,000
Production volume in units	200,000	100,000	150,000
Variable cost per unit	$5	$5	$5
Fixed cost per unit	$3	$6	$4

Exhibit 9.1.14 Fixed and Variable Overhead Rates

$$\text{Fixed overhead rate} = \frac{\text{estimated fixed overhead}}{\text{(overhead base per unit) (normal volume)}}$$

where normal volume is the average volume expected in the next few years.

$$\text{Variable overhead rate} = \frac{\text{estimated variable overhead}}{\text{(overhead base per unit) (expected volume)}}$$

where expected volume is the volume forecast for the coming year.

proach would be to build in the highs and the lows by using average volume in developing fixed cost data for pricing purposes. This average volume is termed "normal volume."

Cyclical firms such as those selling consumer durables experience significant swings in production volume. These swings cause a stability problem in their fixed cost per unit data. To overcome this difficulty, these firms must first estimate (budget) their overhead for the coming period in two parts. The fixed portion of estimated overhead is then combined with normal volume, yielding a fixed overhead rate which is independent of the level of production. (See Exhibits 9.1.14 and 9.1.15 for equation and example of calculation of the fixed overhead rate, respectively.) Since there is no such stability problem with the variable portion of overhead, the variable rate is calculated using the volume expected in the coming year.

Exhibit 9.1.15 Calculation of Fixed and Variable Overhead Rates

	Projected Volumes					
	Year 1	Year 2	Year 3	Year 4	Year 5	Average
Product A	6,000	6,700	7,200	7,500	7,600	7,000
Product B	10,000	12,000	13,000	15,000	15,000	13,000

Other projections for year 1:
Variable overhead = $ 6,840
Fixed overhead = $14,100

Estimated Cost to Manufacture

	Product A		Product B	
Direct material	40 ft at $0.02 =	$0.80	30 ft at $0.02 =	$0.60
Direct labor	0.3 hr at $10.00 =	3.00	0.2 hr at $10.00 =	2.00
Variable overhead	$1.80 × 0.3 hr =	0.54	$1.80 × 0.2 hr =	0.36
Fixed overhead	$3.00 × 0.3 hr =	0.90	$3.00 × 0.2 hr =	0.60
Total cost per unit		$5.24		$3.56

Calculation of Overhead Rates

For normal volume use the average of the projected volumes. Use direct labor hours as the overhead base.

$$\text{Variable overhead rate} = \frac{\$6840}{(0.3 \text{ hr})(6000 \text{ units}) + (0.2 \text{ hr})(10{,}000 \text{ units})}$$

$$= \frac{\$6840}{3800 \text{ hr}} = \$1.80/\text{hr}$$

$$\text{Fixed overhead rate} = \frac{\$14{,}100}{(0.3 \text{ hr})(7000 \text{ units}) + (0.2 \text{ hr})(13{,}000 \text{ units})}$$

$$= \frac{\$14{,}100}{4700 \text{ hr}} = \$3.00/\text{hr}$$

Absorption Versus Variable Costing Statements

Two different types of income statements are used internally by manufacturing firms. The major difference between them is how fixed production costs are treated. The absorption costing statement, which is the traditional approach and must be used for external reporting, attaches all costs, including fixed overhead, to the goods as they are produced. Thus this type of statement is also known as a "full costing statement." The variable or direct costing statement differs in that it attaches only the variable costs to the goods as they are manufactured and placed in finished goods inventory. Fixed overhead is treated as a period cost and is charged off as an expense each accounting period.

One advantage of variable costing is that fixed overhead does not have to be applied to work-in-process inventory. Thus no overabsorption or underabsorption of overhead need be recognized. Another advantage is that the *contribution margin* is highlighted on a variable costing statement. Contribution margin is the difference between selling price and variable cost and thus is the amount available from each unit sold to cover fixed costs and then yield profits. For further discussion, see section in Chapter 9.4 on break-even assessment.

Costs Centers and Departmental Overhead Rates

Most manufacturing operations can be divided into departments such as foundry, machinery, and painting. If these departments are viewed as separate *cost* or *overhead centers*, having their own costs and their own overhead rates, several advantages result. First, for the purpose of better cost control, having a defined area for cost supervision means one person can be made responsible for all costs incurred by that area. In addition, separate cost centers provide a more detailed breakdown of manufacturing costs. For example, direct labor cost would be known for each department rather than just for the whole operation. Such a breakdown makes for easier location of trouble spots.

Second, using separate overhead rates for each cost center improves the allocation of overhead costs among products. The result is better information for setting prices. The example in Exhibit 9.1.16 serves to illustrate this point. Note that products A and B both require the same total

Exhibit 9.1.16 Departmental Overhead Rates

	Foundry	Machining	Painting	Total Plant
Overhead	$1,500,000	$1,000,000	$500,000	$3,000,000
Direct labor cost (DLC)	500,000	200,000	50,000	750,000
Overhead rate	300% of DLC	500% of DLC	1000% of DLC	400% of DLC

Product A

Product B

Product C

		Product A		Product B		Product C	
Department	Rate	DLC	Overhead	DLC	Overhead	DLC	Overhead
		Application of Departmental Overhead Rates					
Foundry	300%	$ 80/unit	$240	$ 70/unit	$210	–	–
Machining	500%	$ 30/unit	$150	$ 30/unit	$150	–	–
Painting	1000%	$ 5/unit	$ 50	$ 15/unit	$150	$10/unit	$100
		$115/unit	$440	$115/unit	$510	$10/unit	$100
		Application of Overall Overhead Rate					
Total plant	400%	$115/unit	$460	$115/unit	$460	$10/unit	$ 40

direct labor cost per unit. Using one overhead rate for the whole operation results in the same amount of overhead being assigned to each product. Actually, however, a unit of product A requires less overhead cost to manufacture. This is true because it spends more time in the foundry, which is a relatively low overhead cost area. Using the overall rate, which is an average rate, will cause the firm to overprice product A and unnecessarily reduce the number of units sold. Now note that product C passes only through the painting cost center. Therefore it should be assigned overhead at the higher rate of that department. Using the overall rate understates product C's overhead cost, and the firm will probably price the product too low to make an adequate profit. The point of this illustration is that using departmental overhead rates better reflects the actual cost of manufacturing the product.

Deciding on the number of cost centers to be established requires a compromise between data precision and the cost of generating the data. Having a cost center for each operation would yield the most precise cost data, but would be very expensive. The person making this judgment should keep in mind that problems arise when a cost center includes operations that would have significantly different overhead rates if viewed alone. In this case the departmental overhead rate is an average, assigning an average amount of overhead to each product the firm manufactures.

Allocation of General Plant Overhead and Service Department Costs

Manufacturing operations consist of more than just the producing departments discussed in the previous material. The manufacture of goods also requires a variety of service departments, from maintenance to plant protection, and general plant overhead costs such as administration and building depreciation and insurance.

If all manufacturing costs are to be attached to the goods as they are produced, then these service department and general plant overhead costs must be transferred to the producing departments. Once these costs are in the producing departments, they can be built into the departmental overhead rates and applied to the goods as they are manufactured. This transfer requires an allocation of cost among the various departments. For example, general plant overhead might be assigned to departments based on the proportion of plant square footage they use (as in Exhibit 9.1.17).

The allocation of general plant overhead and service department cost can be viewed as a three-step process:

Step 1. *Assign all costs to a producing or a service department.*

Assign costs directly, whenever possible, to minimize the need to allocate costs.

General plant overhead is allocated in this step.

Step 2. *Allocate all service department costs to the producing departments.*

Choose some logical basis for type of service involved, such as machine hours for maintenance department costs.

All costs will now be in the producing departments.

Step 3. *Determine the overhead rates for the producing departments* (if allocating budgeted data).

Job Versus Process Costing

Although the recording of manufacturing costs incurred always involves the same three basic cost categories, the *focus* for accumulation of these costs differs. In job costing these costs are accumulated by the individual job or item being produced. That is, each job will have its own work-in-process inventory where the material, labor, and factory overhead costs incurred in its manufacture are recorded. Job costing is used by firms that manufacture products having unique specifications, such as machine tool manufacturers. In contrast, process costing is used by manufacturers who produce large quantities of similar or standardized items. These items are produced by passing them through a series of manufacturing processes. Costs are accumulated by department, with each department being a separate cost center having its own work-in-process inventory.

Equivalent Units

Management needs to know what it costs to manufacture a unit. This information is accumulated separately for each item produced under job costing. However, in process costing, where cost accumulation is by department, manufacturing costs must be *spread* over the units produced. Because production is a continuous process, some units are in a partial state of completion at the end of the period. A simple division of costs incurred by units completed cannot be made because these costs are for both the completed and uncompleted units. The difficulty is handled by using *equivalent units* rather than completed units. For example, if at the end of the period there are 20,000

Exhibit 9.1.17 Allocation of General Plant and Service Department Cost

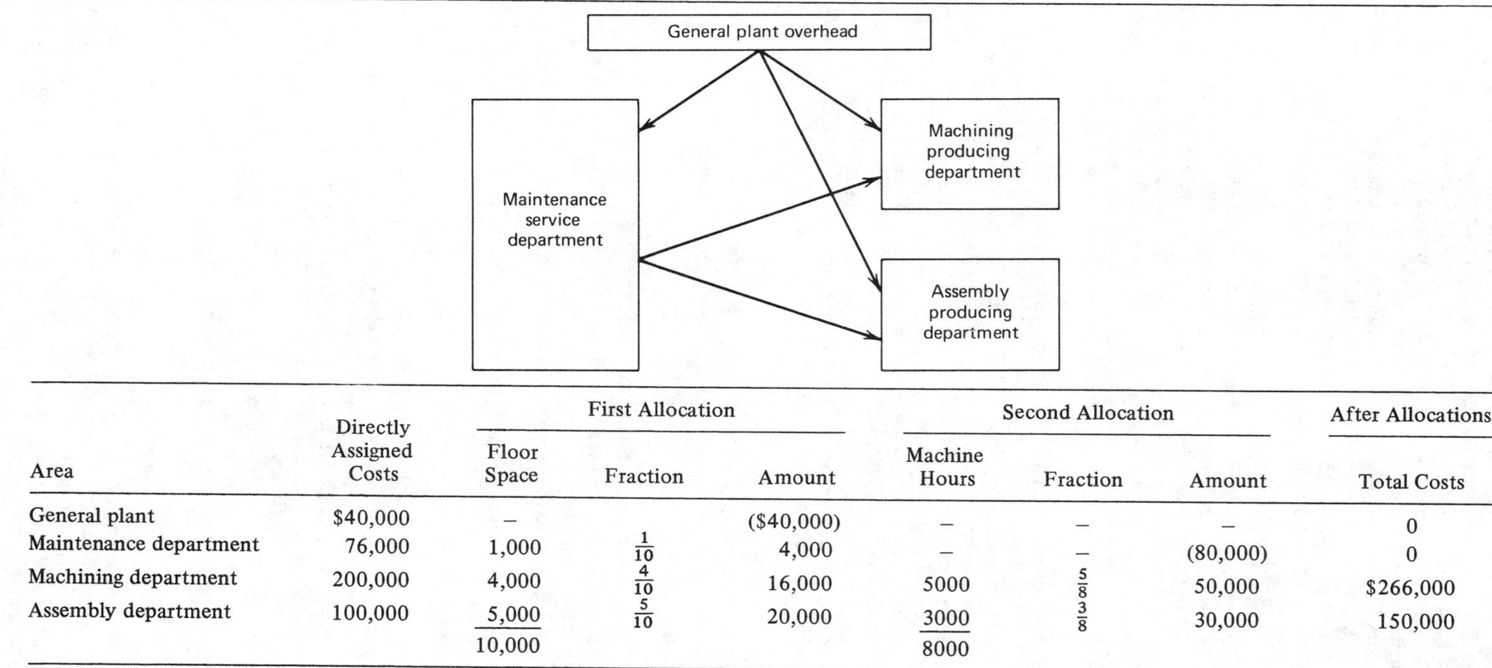

Area	Directly Assigned Costs	First Allocation: Floor Space	First Allocation: Fraction	First Allocation: Amount	Second Allocation: Machine Hours	Second Allocation: Fraction	Second Allocation: Amount	After Allocations: Total Costs
General plant	$40,000	–		($40,000)	–	–	–	0
Maintenance department	76,000	1,000	$\frac{1}{10}$	4,000	–	–	(80,000)	0
Machining department	200,000	4,000	$\frac{4}{10}$	16,000	5000	$\frac{5}{8}$	50,000	$266,000
Assembly department	100,000	5,000	$\frac{5}{10}$	20,000	3000	$\frac{3}{8}$	30,000	150,000
		10,000			8000			

Exhibit 9.1.18 Equivalent Units

Manufacturing costs incurred in period = \$54,000
Units completed in period = 100,000
Ending work-in-process inventory = 20,000 units (40% completed)
Beginning work-in-process inventory = 0

$$\text{Equivalent units} = \text{units completed} + \left(\begin{array}{c}\text{units in ending}\\ \text{work-in-process inventory}\end{array}\right)\left(\begin{array}{c}\text{estimated \%}\\ \text{completion}\end{array}\right)$$

$$= 100 + 20{,}000\ (40\%) = \underline{\underline{108{,}000 \text{ units}}}$$

$$\text{Cost per equivalent unit} = \frac{\$54{,}000}{108{,}000 \text{ units}} = \underline{\underline{\$0.50/\text{unit}}}$$

$$\text{Cost of completed units} = (\$0.50)\ (100{,}000 \text{ units}) = \underline{\underline{\$50{,}000}}$$

$$\text{Cost of ending work-in-process inventory} = (\$0.50)\ (8000 \text{ units}) = \underline{\underline{\$4{,}000}}$$

units estimated to be 40% complete, then this is considered equal to having completed 8000 units. (See Exhibit 9.1.18 for illustration of this per unit calculation.) Note that many firms simply assume beginning and ending work-in-process inventories to be equal and thus avoid the complexity of using equivalent units.

9.1.3 ANALYSIS OF COST CONTROL AND PROFIT PERFORMANCE

Standard Costs

The control of costs is very important to a firm and its management because costs have a direct impact on profits (a penny saved is a penny earned, and vice versa). To achieve this control, individuals are made responsible for the costs incurred in the areas they manage. Thus, to do their jobs, managers need information on how well they have been controlling their costs. The method used for evaluating the control of costs involves establishing standard costs to serve as *yardsticks*. At the end of the period, the actual costs incurred are measured against these established standards. If actual cost exceeds what the standard allows (negative variance), then this is considered an indication of poor cost control. If actual cost is less than standard cost (positive variance), good cost control is assumed.

The key factor in making a fair and accurate cost control analysis is the standard used. Standard costs are not actual costs, but rather, they are *model* or representative costs that are developed prior to the period. If these standards are set very loose or very tight, any motivational effects will be lost. Thus the standards established should represent a very good cost control performance, but still be attainable under normal operating conditions. There will be some differences in actual costs from standard as a result of small changes in prices or quantities used. What management is concerned with are the large variations, because concentrating efforts there should yield the maximum improvements in cost control.

Variance Analysis of Direct Costs

Exhibit 9.1.19 presents an example of cost control analysis complete with the necessary equations. The same three equations are used to evaluate the control of both direct material and direct labor because of the similar nature of these costs. First, the total variance establishes a standard that is calculated as the standard cost per unit times *actual volume*. In the example the direct labor standard allows \$5.00/unit for each of the 40,000 units actually produced. Actual costs are then subtracted from this allowable amount, so that a negative result represents an unfavorable variance or poor cost control. In the example the total labor variance is unfavorable because the actual cost exceeded the allowable by \$9000.

Next, the total variance is split into a *price* and a *quantity* variance. Refer to the standard costs per unit developed prior to the year and note that they consist of a standard quantity per unit multiplied by a standard price. The price variance calculation determines any variation caused by the actual price's differing from the standard price. The quantity variance calculation determines any variation caused by the quantity used per unit's differing from what the standard allows. Calculating these two variances provides additional information about what caused any differences from standard. In the example the unfavorable total labor variance is due completely to a change in the wage rate. Assuming that production supervisors have no control over the rate paid labor, then they are not responsible for the unfavorable price variance. Supervisors are, however, responsible for seeing that labor performs efficiently. The favorable labor quantity variance of $10,000 shows that labor was effectively used, and supervisors should receive proper credit for this performance.

Variance Analysis of Overhead Costs

The variance analysis for overhead uses a *flexible* standard. This is necessary because overhead costs are part variable and part fixed. Note that in the variance calculation of Exhibit 9.1.19 the allowable variable overhead is determined as the standard of $12.00/unit times the actual volume of 40,000 units. As with direct material and direct labor, a standard variable amount is allowed for each unit actually produced. Fixed overhead, being independent of volume, requires different treatment. The standard established for fixed overhead is simply the budgeted fixed overhead with no adjustment for volume. The sum of the adjusted variable and the budgeted fixed overhead becomes the standard (here called "total allowable overhead"). Now to judge the control of overhead costs, actual overhead is subtracted from allowable. The difference is called the "spending variance."

Exhibit 9.1.19 Variance Analysis of Cost Control

Prior to the Year

Standard cost per unit

Direct material	22 lb at $0.40 =	$ 8.80
Direct labor	0.5 hr at $10.00 =	5.00
Variable overhead	240% × $5.00 =	12.00
Fixed overhead	160% × $5.00 =	8.00
		$33.80

Expected volume = 38,000 units
Normal volume = 50,000 units

Budgeted variance overhead = $456,000
Budgeted fixed overhead = 400,000

Actual Results for the Year

Direct material = 900,000 lb at $0.43 = $387,000; units produced = 40,000 units
Direct labor = 19,000 hr at $11.00 = 209,000; overhead = $860,000

Variance Analysis of the Direct Costs

Total variance = (standard cost) (actual volume) – actual cost

Total material variance = ($8.80) (40,000) – $387,000
= $352,000 – $387,000
= ($35,000) (unfavorable)

Total labor variance = ($5.00) (40,000) – $209,000
= $200,000 – $209,000
= ($9000) (unfavorable)

Price variance = (actual quantity) (standard price – actual price)

Material price variance = 900,000 lb ($0.40 – $0.43)
= 900,000 (–$0.03)
= ($27,000) (unfavorable)

Labor price variance = 19,000 hr ($10.00 – $11.00)
= 19,000 (–$1.00)
= ($19,000) (unfavorable)

Quantity variance = Standard price [(standard quantity) (actual volume) – actual quantity]

Material quantity variance = $0.40 [(22 lb) (40,000) – 900,000 lb]
= $0.40 (880,000 – 900,000)
= $0.40 (–20,000 lb)
= ($8000) (unfavorable)

Exhibit 9.1.19 (*Continued*)

Actual Results for the Year (Continued)

Variance Analysis of the Direct Costs (Continued)

Labor quantity variance = \$10.00 [(0.5 hr) (40,000) – 19,000 hr]
= \$10.00 (20,000 – 19,000)
= \$10.00 (+1000 hr)
= \$10,000 (favorable)

Flexible Budget Standard for Overhead Variance Analysis

Flexible budget standard		
Variable overhead adjusted to actual volume		
= (240%) (\$5.00) (40,000) = \$12.00 (40,000) =	\$480,000	
Budgeted fixed overhead =	400,000	
Total allowable overhead	\$880,000	
Actual overhead	860,000	
Overhead spending variance	\$ 20,000	(favorable)

Three-Variance Method–Evaluating Overhead Absorption

Applied overhead = (composite overhead rate) (actual direct labor)
= (240% + 160%) (\$209,000)
= \$836,000

Underabsorbed overhead = actual overhead – applied overhead
= \$860,000 – \$836,000
= \$24,000

Spending variance (favorable) = \$20,000 (overabsorption)

Capacity variance = (budgeted fixed overhead) – (fixed overhead rate) (standard direct labor at actual volume)
= \$400,000 – (160%) (\$200,000)
= \$400,000 – \$320,000
= \$80,000 (underabsorption)

Efficiency variance = (Composite overhead rate) (actual direct labor – standard direct labor at actual volume)
= (240% + 160%) (\$209,000 – 200,000)
= 400% (\$9,000)
= \$36,000 (overabsorption)

Underabsorbed overhead = – \$20,000 + \$80,000 – \$36,000
= \$24,000 (check)

The Three Variance Method for Overhead

In addition to the spending variance, two other overhead variances can be calculated, yielding either a two- or three-variance approach. (See Exhibit 9.1.19 for an example of the calculation.) These other two variances are not directed toward the evaluation of cost control; rather, they serve to explain why an overabsorption or underabsorption of overhead resulted.

Overhead is applied to the goods as they are manufactured by multiplying the sum of the variable and fixed overhead rates by the actual amount of the overhead base. Since the base is variable depending on production volume, the applied overhead will also be variable. This is appropriate for the variable portion of overhead. However, the correct amount of fixed overhead will not be applied unless the firm just happens to operate at normal volume where the fixed rate was determined. The misapplication of fixed overhead due to volume is measured by the capacity variance. Overhead costs can also be overabsorbed or underabsorbed as a result of the actual amount of the overhead base. If the base is 4.5% higher because of poor cost control, then the applied overhead will be increased by 4.5%. This misapplication of overhead is measured by the efficiency variance.

Return on Investment Ratios

The concept of return on investment (ROI) is often used to measure *how well* a business has been managed. Among management's duties are deciding how best to invest the firm's resources in assets, setting prices that will result in an acceptable number of units sold, and controlling costs so that adequate profits are earned from sales. The management of a small business has invested in a small amount of assets and generates a small amount of sales and profits. A large business has a large amount of assets, and its management is expected to generate a correspondingly large amount of sales and profits. Both large and small businesses are trying to earn an adequate ROI.

In addition, the ROI ratio is often factored into profit margin and asset turnover. These two supporting ratios provide a partial answer to the question of *why* a certain ROI was earned. Exhibit 9.1.20 diagrams this analysis and outlines what these two ratios measure. Exhibit 9.1.21 illustrates the calculation of these ratios.

Performance Analysis

Management's performance in operating the business is often evaluated using ROI ratios. Actual ROI, profit margin, and asset turnover ratios are calculated using the period's sales, profits, and level of assets. These ratios of actual performance are then compared to ratios calculated from budgeted data. Before this comparison is made, however, the budgeted data must be adjusted to the actual level of operations. Using the flexible budget concept, variable items, such as direct materials or current assets, are adjusted to the actual level of sales. Fixed items are included at the budgeted amount.

Note that in the example in Exhibit 9.1.22 all three ratios show improvement when compared to the original budget. However, since actual volume is higher than expected, there should be improvement as a result of better utilization of fixed costs and fixed assets. Comparing the actual to the adjusted budget reveals that management's performance is really poorer than what should be expected. Although asset turnover is slightly better than the adjusted budget, the profit margin is significantly off, causing the lower ROI.

Exhibit 9.1.20 Performance Ratios

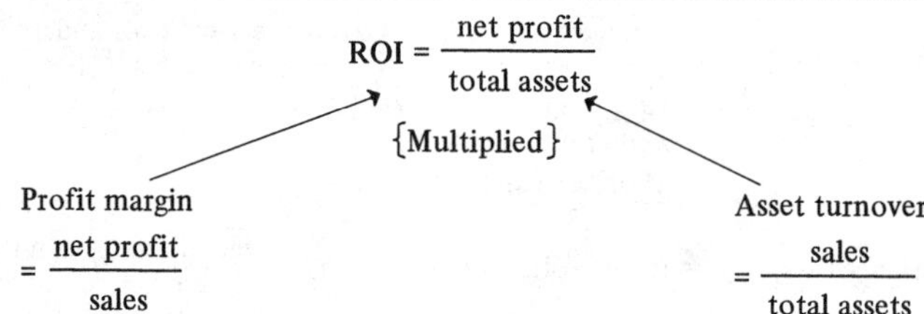

$\text{ROI} = \dfrac{\text{net profit}}{\text{total assets}}$
{Multiplied}
Profit margin
$= \dfrac{\text{net profit}}{\text{sales}}$
Asset turnover
$= \dfrac{\text{sales}}{\text{total assets}}$

Profit margin measures the (1) efficiency with which costs were controlled and the (2) mark-up used in pricing policy.

Asset turnover measures the utilization of assets in generating sales.

Operating cycle

Profit margin represents the pennies of profit left from a typical sales dollar that flows through the operating cycle.

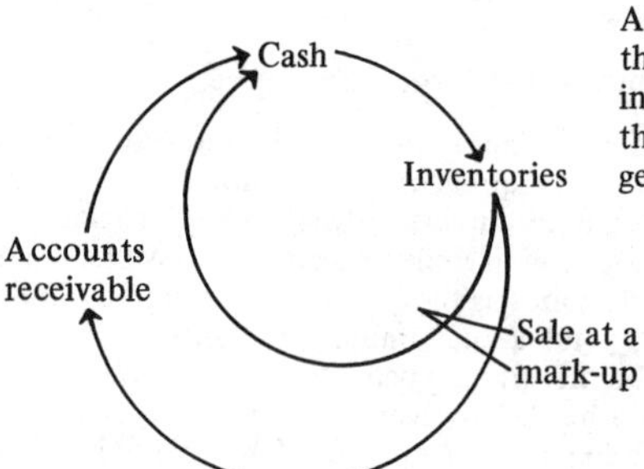

Asset turnover concerns the volume of sales dollars in the operating cycle and the assets employed in generating those sales.

Exhibit 9.1.21 Illustration of ROI Calculations

Income Statement	Auto Manufacturer	Auto Dealer
Sales revenue	$28,240	$4800
Less: Cost and expenses	26,320	4709
Net income	$ 1,920	$ 91
Total assets	$16,230	$ 800
Profit margin	$\frac{\$1,920}{\$28,240} = 6.8\%$	$\frac{\$91}{\$4800} = 1.9\%$
Asset turnover	$\frac{\$28,240}{\$16,230} = 1.74$	$\frac{\$4800}{\$800} = 6.00$
ROI	$\frac{\$1,920}{\$16,230} = 11.8\%$	$\frac{\$91}{\$800} = 11.4\%$

Comments:

Manufacturer is forced into a low turnover because of all the assets required to produce autos.

Dealer uses price competition to attract customers, relying on volume to generate an adequate ROI.

Exhibit 9.1.22 Performance Analysis Report

	Budget (400)		Adjusted Budget (460)	Actual (460)	Variance
	Per Unit	Total			
Income Statement					
Volume in knits		400	460	460	
Sales	$30	$12,000	$13,800	$13,800	
Less: Direct material	$ 3	$ 1,200	$ 1,380	1,360	20
Direct labor	4	1,600	1,840	1,870	(30)
Variable overhead	8	3,200	3,680	3,930	(250)
Fixed overhead		3,600	3,600	3,600	0
Factory profit		$ 2,400	$ 3,300	$ 3,040	
Investment Data					
Current assets (25% of sales)		$ 3,000	$ 3,450	$ 3,300	
Plant and equipment		3,000	3,000	3,000	
Total assets		$ 6,000	$ 6,450	$ 6,300	
Profit margin		20.0%	23.9%	22.0%	
Asset turnover		2.00	2.14	2.19	
ROI		40.0%	51.2%	48.3%	

9.1.4 THE OPERATING BUDGET

Purposes

The operating budget or profit plan is a detailed plan of how the firm is to be operated in the near future. The basic purpose of developing such a plan is to ensure that the business is run efficiently and thus achieves its profit goals. If these goals are to be reached, managers must make good decisions. These decisions, of course, are made for their impact on future performance. (The past is past and only serves as a base for making projections.) Thus another purpose of budgeting is to force managers to be forward-looking in their decisions. Since the operation of most firms requires more than one manager, the decisions of the various managers must be coordinated. Managers are justifiably most concerned with their own areas or departments. However, the decisions they make must serve to improve the performance of the total firm. Combining individual manager's projections into an overall budget reveals inconsistencies in their plans. Finally, the budget is a statement of how the firm should be operated. After the period, actual costs and profits can be compared to the budget in order to measure management's performance in operating the business.

Time Span and the Type of Budget

Many firms develop their operating budgets annually for the coming year. These annual plans of operation are further broken down into budgets for each month or quarter. Other firms choose to prepare a new budget each quarter, covering the next four quarters. This *rollover* process, though more costly, causes budgets to be reviewed automatically. Periodic review and revision are very important if the budget is to contain the best possible projections.

In contrast, strategic budgets are for much longer time spans and do not have to concern the total operation. Instead they often focus on specific management questions such as "What will the impact of mass transit be in the next 20 years?" Thus strategic budgets serve to assist top management in making major policy decisions. After these decisions are made, the operating budget serves to ensure that the policies are carried out in the firm's operations.

Developing the Operating Budget

The budgeting process typically involves all parts of the firm. First, a budget committee composed of top management is formed. This group then appoints a budget director to oversee the development of the budget. Next, the committee issues guidelines to the firm's various departments. Following these guidelines, the departments develop their individual budgets and then submit them to the budget director. The director combines the individual plans into a budget for the total firm. Where departmental budgets are in conflict or otherwise unsatisfactory, the director asks for revisions. If the departments feel that these revisions are not appropriate, they must justify their positions. These questions are thus settled through a negotiation process between the departments and the budget director.

After a coordinated total budget has been developed, it is reviewed by the budget committee. If this top management group does not feel that the budget is satisfactory, it will be sent down for further revisions. When approval is finally granted, the budget is returned to the individual departments as the operating plan they will follow in the coming period.

Budgeting for a Manufacturing Firm

The operating plan for a manufacturing firm is divided into four subbudgets. These individual budgets follow the flow of goods from sales backward through the inventories carried to purchases from suppliers. Exhibit 9.1.23 provides a description of and sample calculation for each budget. After establishing how many units will be produced and sold, costs can be determined and assembled into a projected statement of goods manufactured and sold and a projected income statement. With the addition of these pro forma statements, a complete profit plan is developed.

Exhibit 9.1.23 Example of a Manufacturing Budget[a]

	Standard Costs	
	Pro	Sport
Direct material		
A	20 ft at $0.40 = $8.00	15 ft at $0.40 = $6.00
B	10 lb at $0.50 = 5.00	–
C	–	8 lb at $0.30 = 2.40

	Projected Inventories	
	Beginning	Ending
Direct material		
A	32,000 ft	19,000 ft
B	3,800 lb	2,400 lb
C	11,000 lb	8,000 lb

Exhibit 9.1.23 (*Continued*)

	Projected Inventories	
	Beginning	Ending
Finished goods		
Pro	200 units	400 units
Sport	800 units	1,600 units

Sales Budget[b]

Model	Units	Selling Price	Sales Revenue
Pro	1000	$42	$ 42,000
Sport	4000	29	116,000

Production Budget[c]

	Pro	Sport
Sales in units	1000	4000
Ending finished goods inventory	400	1600
Total required	1400	5600
Less: Beginning finished goods inventory	200	800
Required production	1200	4800

Material Requirements Budget[d]

		Material A	Material B	Material C
Per Unit				
Pro		20 ft	10 lb	–
Sport		15 ft	–	8 lb
Required Production				
Pro	1200 units	24,000 ft	12,000 lb	
Sport	4800 units	72,000		38,400 lb
Total material required		96,000 ft	12,000 lb	38,400 lb

Purchases Budget[e]

	Material A	Material B	Material C
Material required	96,000 ft	12,000 lb	38,400 lb
Ending direct material inventory	19,000	2,400	8,000
Total required	115,000	14,400	46,400
Less: Beginning direct material inventory	32,000	3,800	11,000
Required purchases	83,000 ft	10,600 lb	35,400 lb
Purchase Price	$ 0.40	$ 0.50	$ 0.30
Cost of purchases	$33,200	$5,300	$10,620

[a] This firm manufactures two products, in one department, from three materials.

[b] Specifies the units to be sold from finished goods inventory.

[c] Input-output analysis of finished goods inventory that determines how many units must be produced.

[d] Determines the direct material needed to produce the required units.

[e] Input-output analysis of direct material inventory to determine the amount and cost of material that must be purchased.

BIBLIOGRAPHIC COMMENT

College textbooks provide additional illustrations and discussions of the material presented in this chapter. General accounting is discussed in detail in financial accounting texts, while managerial accounting and cost accounting texts cover the remainder of the material. General financial information for large firms is provided by their annual reports or by data services such as Moody's or Standard and Poor's, which are found in most public libraries.

CHAPTER 9.2
Cost Estimating

PHILLIP F. OSTWALD

University of Colorado, Boulder

9.2.1 ESTIMATING—AN EVERYDAY, EVERYBODY PROBLEM

Cost estimating is a popular activity within industrial engineering. Even though the professional person may be titled a cost estimator, cost engineer, cost analyst, labor estimator, material planner, and so on, the emphasis remains the same. He or she is required to answer a familiar question: "How much will it cost?" Although the purposes that underlie this question are varied, we find that businesses, government, and not-for-profit organizations desire timely and reliable measures of economic want. For it is an estimator who does the appraisal, analysis, forecasting, and compiling of a pro forma document that extends from the basic cost ingredients to the bottom line of an estimate. Using this evaluation other management people may determine a price, make versus buy decision, ROI, or public fiscal-year budget. Thus the estimator finds a future value that responds to a specified need.

This historical trail of development of cost estimating is intimately tied to industrial engineering. The original concept of a labor "standard" was seminal in the development of "standard" cost plans found widely throughout business. The genesis of estimating predates the 1900s since it was connected closely with manufacturing and construction. Cost estimating is a long-established job and an everyday occurrence for many industrial engineers.

9.2.2 DESIGN—THE STIMULUS

Cost estimating is an activity done for engineering and management. Initiation of cost estimating work usually arises from some sort of a design. Loosely, and for the purposes of cost estimating, the following definition is used: Every design is a new combination of preexisting knowledge that satisfies an economic want. This definition includes bridges, cars, chemical plants, highways, machine tools, radios, piece parts, service work, and systems of machines and people as well as plans, technical reports, and models that are a part of the engineering job. With design as the focal point, *cost estimating is the body of theory and practice that provides a measure of the economic want of the design*. Notice that the cost estimating activity is not involved predominantly with the "satisfaction" part of the design. Satisfaction is determined externally by the marketplace; internally the firm's management may stop an undesirable design or continue it. Politics may also enrich or abrogate the satisfaction requirement of designs that deal with society.

Design does not mean board work per se, but is the work of the many creative engineering talents in widely dispersed activities. Design precedes the estimating and spending of money, considering the costs of design to be excluded. Design could also imply methods design (see Chapter 3.1), line balancing (Chapter 6.8), plant layout (Chapter 10.2) and many other industrial engineering designs as described within this handbook. In view of the importance of design, the estimator has to be able to understand and interpret many kinds of designs.

For purposes of the cost estimating specialization, four kinds of design are defined: operation, product, project, and system. Exhibit 9.2.1 shows a classification of design for cost estimating purposes. Concurrently, there are four kinds of estimates associated with these designs.

9.2.3 FOUR KINDS OF ESTIMATES

Operation Estimate

The process of producing a change in value or a way of working establishes the content of an operation estimate. Frequently called "direct cost," an operation estimate is a forecast of direct labor and direct material required for a design. If the design is a toy, radio, or building, there is work, a

Exhibit 9.2.1 A Classification of Design for Purposes of Cost Estimating

Design	Fundamental Characteristics	Symbolic Measure of Economic Want	Examples
Operation	Man and tool	Cost	Assembler and handtools, secretary in office, crew workers, driver and transportation vehicle
Product	Replication	Price	Toys, radios, houses, typewriters, bridges, NC machine tools, transportation vehicle
Project	Uniqueness	Return	Bridge, plant addition, refinery, factory prototype, capital tooling for product, right-of-way structure for transportation vehicles
System	Configuration	Public effectiveness	Weapon system, hospital, rapid transit system

worker, and a tool, however simple or complicated that may be. This notion includes one worker with one tool, one worker with multiple machines, or crew work. It can be applied to the factory, construction site, office, service work, maintenance work, hospital, or government. The worker can be skilled, unskilled, craft, apprentice, journeyman, or professional. The work is classified as direct if it can be clearly traced to the function of the design. The work that engineers do for a weapon system, for example, can be classified as direct or overhead; the classification is optional. A clear-cut example is a turret lathe operator making parts that are identified clearly in a product or an electrician wiring a commercial building under construction. Techniques used for operation estimating are different from those required for other types of estimates. Furthermore, operation estimates are used as input information for other kinds of estimates.

Product Estimate

In a product estimate an entire product, rather than a subassembly or a component part, is estimated. Product estimating includes the essential characteristic of replication. Replication here means a likeness or reproduction, not repetition. For estimating, "replication" is defined as a deliberate change in the design. Model 2 differs somewhat in design from model 1, for instance. For many reasons these intentional design differences cause changes in the estimate. But the estimator does not do an estimate of model 2 from scratch. Instead, by several ways the estimator adds and substracts for the replicated distinctions in design.

Note that a product estimate does not depend solely upon production quantity. A product quantity may be anywhere from a few units to many. Traditionally, there has been a mistaken belief that it is necessary to have large quantities of consumer goods for a product estimate and that the production time should be reasonably brief in relationship to the quantity produced. However, bridges, turbines, and airplanes, as well as consumer products, have the quality of being replicated designs. Homes and apartments being constructed in a development have different facades, interiors, and so on, but there is similarity. In dealing with replicated product designs, the estimator uses many similar techniques whether the product estimate is for $2 toys or $200 million turbines. Techniques used for product estimating are usually different from those required for other types of estimates, and they depend upon operation estimates.

Project Estimate

A project estimate, whether it is a plan, a plant, equipment, capital tooling for a product, or a prototype is one of a thing. In a project design the emphasis is on oneness. The design is custom, perhaps, and there will be only one manufactured or constructed. Usually the dollar amount is considered capital rather than expense. Examples include a refinery, plant, turbine, bridge, prototype, and airplane. Although some of these examples illustrated the product estimating category, project estimating is done for single, rather than multiple, goods. The techniques used in project estimating are essentially different from those used for other types of estimates; however, they require operation and product estimates as information.

Another distinction between product and project estimating occurs in the buyer-seller relationship. If a factory producing NC machine tools were manufacturing several or more units, the factory cost estimator would connect the problem with product design and use techniques of product estimating. The engineer representing his or her company as an estimator in evaluating the NC machine for a purchase would approach the problem as a project design and apply methods of project estimating.

System Estimate

A system design involves designs of operations, products, and projects in any arrangement. The descriptor term of a system is "configuration." Thus, "system" is defined differently for the needs of cost estimating. A system design is complex, and the elements of the system estimate include operation, product, and project estimates. In this lexicon the system design deals in the public, government, or not-for-profit domain of enterprise. This arena involves political, altruistic, and defensive factors and providing for the general needs and goals of society. One example of a system is a rapid transit system. This would involve, for example, initial nonrecurring costs, such as securing right of way and constructing the road structure for the vehicles, which would involve project estimates. Transit vehicles, not all similar, would be necessary, and product estimates would thus be required. A driver and the vehicle, or the man and tool, is the grist for an operation estimate and is necessary as information. Thus a system estimate uses other estimates in arriving at a value for the measure of the economic want.

9.2.4 MEASURES OF ECONOMIC WANT

The task facing the estimator is to provide a fact or number that represents the economic want of the design. A "want" is a value exchanged between competing and selfish interests. The price a consumer is willing to pay for an item stocked on the grocery shelf, a contractor-owner agreement on the bid value of a building project, and the fiscal-year budget value for a weapon system that the U.S Department of Defense proposes and Congress accepts are typical examples of wants exchanged. Exhibit 9.2.1 provides symbolic measures of wants. Although dollar cost is the usual measure for operation designs, man-hours or man-days are also used. In projects the rate of return is commensurate with present value, future sum, and annual cost; methods provided in Chapter 9.3 and 9.4 are useful. In system designs public effectiveness may imply a measure of benefit-cost or a budget fiscal-period total that sums cost streams throughout the system life cycle. However desirable other dimensions for the evaluation of designs may be, the cost estimator deals principally in one term—dollars.

9.2.5 PROCEDURES

Estimate requests are initiated outside the cost estimating function or department, for cost estimators seldom start the cycle. A request for quotation (RFQ) or request for proposal (RFP) is received by marketing and sales or engineering design from potential buyers. A customer generally communicates with sales rather than cost estimating; a request for estimate (RFE) begins the estimating work and stems from an RFQ, an RFP, or a production inquiry.

The RFE has many forms, one of which is shown in Exhibit 9.2.2. Although the information varies with the nature of the design, the estimator needs information about the status of the design, quality specifications, quantity and rate, time of effort, location, delivery, legal requirements, and many other details. It is not the estimator's responsibility to provide this information, and if overlooked, the quality of the estimate suffers. Information specific to the nature of the design and required before estimating can begin is shown in Exhibit 9.2.3.

Sources of estimating information are internal and external to the organization. Product estimating data, since they pertain to industrial origin, are usually internal. Project data, since they deal with capital types of designs, are primarily external to the organization. Union halls, consultants, published and private indexes, and commercial data are gathered and used for project estimates.

Although much of the preceding information is historical, remember that it is future costs that are desired. Information is raw and needs to be recast into appropriate values to match the estimating requirements.

Before estimating is started, it is fundamental that analysis of elements of cost be undertaken or known. Analysis is directed toward labor, material, and overhead costs of various kinds. If anything is clear about successful estimating, it is that wise analysis precedes it. This analysis needs to be timely, economical in itself, and accurate.

9.2.6 LABOR ANALYSIS

Labor constitutes one of the most important items of operation designs. Labor has received intensive study, and many recording, measuring, and controlling schemes have been developed in an effort to manage it. Labor can be classified in a number of ways, including direct-indirect, recur-

Exhibit 9.2.2 Request for Estimate

TO: Estimating Department

REQUEST NO.: 1048

FROM: [X] O. E. M. Sales Dept.
[ ] Export Sales Dept.
[ ] A. M. Sales Dept.

DATE: June 9

DUE DATE: June 30

CUSTOMER: Eastwood Company

ADDRESS: 6682 Whaley Drive, Central City, Michigan

[X] YES, [ ] NO Customer blueprints (and, or) specifications sent to Engineering.

ITEM NO.	PART NO.	CUSTOMER'S PART NO.	DESCRIPTION	TOTAL QUANTITY	SCHEDULED QUANTITY	F.O.B.
1	X 60371	A-4189	Sheet metal chassis	15,000	1,200/mon.	Home

PACK REQUESTED: 15 unit pack

SPECIAL INSTRUCTIONS: If we win the bid, production will begin December 1

R. F. E. ISSUED BY: Bill Smith

REFERENCE: Requested by their Mr. Harrison per letter June 3

ENGINEERING DEPT.: [ ] Production items, no additional information to be issued.
[X] Cost Estimate No. CE 4262 issued to cover the above.

BY: J. Wilkon DATE: 6-12

ring-nonrecurring, designated-nondesignated, exempt-nonexempt, wage-salary, blue collar–management, and union-nonunion. Other ways in which to classify labor are according to social, political, and educational divisions and type of work. Payment for wages may be based upon attendance or performance. For cost estimating operation designs, the direct-indirect classification is the most appropriate.

For operation designs there is an unquestioned dependence upon the simple qualitative formula

$$\text{labor cost} = (\text{time}) \times (\text{wage}) \tag{1}$$

The selection of "time" matches the requirements of the operation design. Time is expressed relative to a unit of measure, which is denoted by terms such as piece, bag, bundle, container, 100 units, or 1000 board ft. The usual ways to measure labor are by (1) time study (see Chapter 4.4), (2) predetermined motion time systems (see Chapter 4.5), (3) work sampling (see Chapter 4.6), and (4) man-hour reports. Job tickets, especially for smaller organizations, are analyzed and allocated to units of work. For instance, a job ticket may state "136 units turned of part number 8671" and list "6 man hr." Simple analysis would show 0.044 hr/unit. The estimator would use 0.044 hr the next time this part were run. Although hardly accurate because of the nature of historical work reports, man-hour reports are used because of their simplicity. Man-hour estimating data are especially popular in construction work.

Exhibit 9.2.3 Sources of Information for Cost Estimating

Design	Information Required for Estimating	Sources
Operation	Labor time, wages, fringes Material shape and policy cost Material losses Variable overhead rates	Industrial engineering, methods and standards, personnel Engineering, industrial engineering Estimating Accounting
Product	Bill of materials, drawings Special tests, packaging, shipping Quantity and rate of production Operation estimates	Engineering Engineering Marketing, scheduling Cost estimating
Project	Fixed costs of equipment Life periods Direct operating costs, cash flows Indirect costs, working capital Operation and product estimates Project design specifications and drawings	Industrial engineering Industrial engineering Accounting, estimating Accounting Estimating Engineering
System	Operation, product, and project estimates Budgets, wages Fiscal money available Reimbursable and nonreimbursable costs Taxation, sources, amounts Eminent domain, legal requirements	Cost estimating Public authority Public authority Public authority Public authority Legal officer

Direct observation and measurement of labor are of little use to the estimator except for reestimates of similar work or reruns of the same work. Although the cost estimator may not be directly involved with the measurement of labor, he or she does depend on work measurement. The estimator is satisfied if such labor measurements are objective, as far as that is possible, and is willing to use the information, provided that engineering techniques were used in the determination of time. Although the time measurements are of value, it is immensely more important that work measurement data be transformed into information that can be applied prior to the time of the operation design. The time measurements are more valuable when expressed as standard time data (see Chapter 4.8) and presented in a table or computer format (see Chapter 4.7). The estimating data may be described in terms of elements, which are the subwork descriptors of operations, or be expressed as time estimating relationships (TERs) for operations. Standard data expressed at the predetermined motion time level are too detailed for much cost estimating work. A typical TER for a drill press operation of sheet metal parts is for setup 0.2 + 0.05/tool hr and for run time 0.015 + 0.003/tool + 0.001/hole hr per unit. Thus, if a sheet metal part requires two different countersinks for 22 holes, setup would be 0.3 hr, and run time 0.043 hr/unit. The TERs as preliminary estimating information are perhaps less accurate than detailed estimating information, but are faster. Speed of estimating versus accuracy of the estimate is always a necessary trade-off.

In some situations the estimate of time may be done from a "guesstimate" and be unrelated to measured, referenced, and analyzed data. A guesstimate is based upon the estimator's observational experience. There are circumstances where these judgmental numbers are unavoidable.

The second part of equation 1, "wage," is defined in the context of the operation design that is being estimating. The operation design may be for one worker and one machine, for a crew with one machine, or for a crew with several machines or processes. In the simplest case, one on one, the job description (see Chapter 2.4) and job design (see Chapter 2.5) are specifications available to the estimator. The number used for the wage corresponds to the time period of work and is money out of pocket. Regression methods (see Chapter 13.6), labor contract (see Chapter 5.6), or personnel planning (see Chapter 5.3) are sources for wage trend information.

The units for the wage are dimensionally compatible to the time estimate. If the time as estimated is in units of hours, then the wage is expressed in dollars per hour. The wage may be the amount that the worker sees or may include all or part of the fringe costs that can be associated with the wage. The practice of what is included in the wage amount is coordinated with the finding of the overhead (see Chapter 9.1). Fringe additions could include effects of paid holidays and vacations, health insurance and retirement benefits, Federal Insurance Contributions Act (FICA) benefits, workers compensation, bonuses, gifts, uniforms, special benefits, profit sharing costs, education, and so on.

Forecasting the estimated time accurately, but multiplying by an inaccurate average wage rate, creates inaccurate labor estimates. Some effects to watch for are amount of overtime hours, merit increases, promotions, labor contracts, and unskilled to skilled labor mix. These effects are significant for operation designs involving crew work. For crew work the forecasted, weighed average is the central focus of the schedule for the operation design. Overescalation is as erroneous as underescalation.

9.2.7 MATERIAL ANALYSIS

The term "direct materials" includes raw materials, purchased parts, standard commercial items, interdivisional transfers, and subcontracted items required for the design. Direct material cost is the cost of material used in the design. The cost should be significant enough to warrant the cost of estimating it as a direct cost (see Chapter 9.1). Some material, by virtue of the difficulty of computation and estimating, may be classified as either indirect or direct costs. The latter estimates are preferred. Paint material of irregularly shaped objects is an example of material that can be classified either way.

The estimator begins by calculating the final exact quantity or shape required for a design. To this quantity, losses for scrap, waste, and shrinkage are added. The general model for cost of direct material is

$$C_{dm} = S(1 + L_1 + L_2 + L_3)P_s - R \tag{2}$$

where S = shape expressed in various engineering units, pound, volume, length, area, and so on, in compatible dimensions to price P_s and as required for design
L_1 = losses due to scrap, decimal
L_2 = losses due to waste, decimal
L_3 = losses due to shirnkage, decimal
P_s = price per pound of material, price per linear foot, price per volume, and so on
R = unit price of anticipated material salvage, dollars per unit

Scrap is material that is lost because of human mistakes, whereas waste is necessary because of the design. Shrinkage losses are due to theft, or physical law deterioration. In estimating of foodstuffs, if direct material is not processed at the appropriate time, or if it is mishandled, shrinkage of the quantity will result. It is required that these three losses be estimated and that their percentages be added to the theoretical finished requirement.

An example of material estimating is given by the 12 fluid oz beverage can, which is composed of the body, top, and pull ring. The container body is blanked from 3004 H19 aluminum coils, with the layout given in the Exhibit 9.2.4. An intermediate cup is formed, without any significant change in thickness. The cup is drawn in a horizontal drawing machine, and metal is squeezed to sidewall thickness of 0.0055 in. (0.140 mm), while bottom thickness remains unchanged. The can is trimmed to final height to give an even edge for later rolling to the lid. Calculations are given in Exhibit 9.2.4.

9.2.8 NEED FOR ACCOUNTING DATA

Cost accounting has always been important to the performance of diverse estimating functions. As colleagues in the gathering, analysis, and reporting of business data, accountants provide overhead rates, standard costs, and budgeting data. The estimator reciprocates with manpower and material estimates for the several designs. In many situations the estimate can serve as a mini profit and loss statement for special products. Thus there is interdependence between these two professions. The construction of the cost accounting information is not demonstrated here (see Chapter 9.1); the estimator is less interested in balance sheets, profit and loss statements, and the intimate details of the structure of accounts. Overhead rates are vital for the estimating functions, however, since the estimator may apply these rates in the estimate.

By definition, overhead is that portion of the cost that cannot be clearly associated with particular operations, products, projects, or systems designs and that must be distributed among the cost units in some arbitrary way. The overhead rate is simply

$$\text{overhead rate} = \frac{\text{predicted nondirect costs associated with a design}}{\text{predicted direct costs associated with a design}}$$

A classification of overhead methods would include:

1. Whether the rate includes fixed costs, as in absorption costing, or not, as in direct costing.
2. The base used to distribute overhead, such as direct labor dollars, direct labor hours, or machine hours.

Exhibit 9.2.4 Calculation of Direct Material Cost

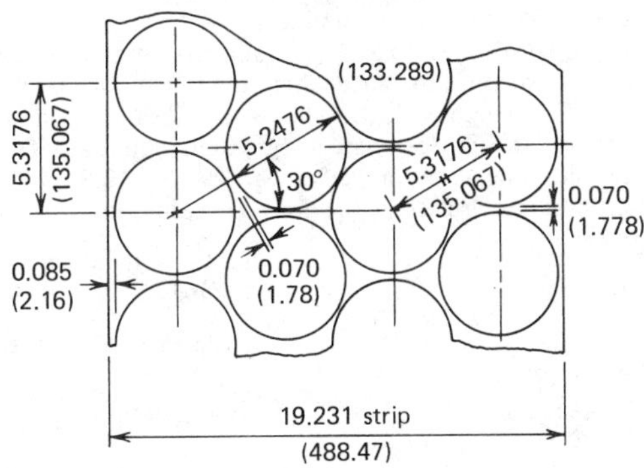

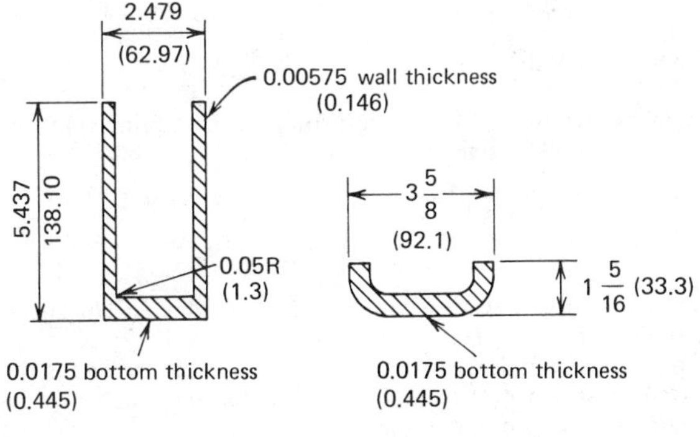

Final metal volume in 12 oz can body = 0.3266 in.^{3} (5355 mm^3)
Strip volume per can body = 5.3176 × 19.231 × 0.0175/4 = 0.4474 in.^{3} (7328 mm^3)
Metal efficiency = 0.3266/0.4474 = 73.0%
Metal volume in 5.2746 (133.975) blank = 0.3785 in.^{3}
Can body to blank efficiency = 0.3266/0.3785 = 86.3%
Cost of metal = $1.0017/lb ($2.226/kg)
Cost of can body = 0.4474 × 0.0982 lb/in.^{3} × 1.0017 = $0.0441
Waste recovery value = $0.4830/lb ($1.073/kg)
Recovery of waster per can body = (19.231) (5.3176) (0.0175) (1 – 0.73) (00982) (0.4830) = $0.0057
Net cost per can body in strip = 0.0441 – 0.0057 = $0.0384
Metal cost in can body = (0.3266) (0.0982) (1.0017) = $0.0321

3. The scope of the application of the rate—whether it is for the plant, cost center, machine, or design.

4. Whether the rate applies to all designs (such as product lines) or to one line of the design.

9.2.9 FORECASTING

Many forecasting techniques have been developed to handle a variety of problems (see Chapter 13.9). Each has its special advantage, and care is necessary in choosing techniques for cost estimating. Selection of a method depends on the context of the forecast, availability of historical data, accuracy desired, time period to be forecast, and value to the company. The estimator should adopt a technique that makes the best use of the data. He or she should initially use the simplest technique and not gold plate the effort with a more advanced technique than is justified. For estimating requirements, we are concerned with data about labor, material overhead, and their quantities and cost. The forecasts should reflect those values under the proposed actions of the company and environment. It is necessary to recall that forecasting is a future prediction about line elements of the estimate. Forecasts should not deal in overall or grand average cost, time, and quantities, but should be matched to line items required by the pro forma estimate. "Forecasting" is not "estimating," as the term is used here, since forecasting takes data and frames it in a new picture, and judgment is suppressed as much as possible.

Methods in forecasting begin with the simplest: a graphic plot of an independent variable against cost, time, or an economic want measure. As a first step, the linear approximation of $y = ax + b$ on Cartesian grids is satisfactory. The graphic plotting helps to give a vicarious sense of the situation. Once plotting has been attempted, single-variable regression equations (see Chapter 13.6) are applied, or moving average and time series models (see Chapter 13.9) or indexing are adopted. Some of the more important nonlinear relationships for cost estimating are

$$y = ae^{bx} \qquad \text{semilog fit} \qquad (4)$$

$$y = ax^{b} \qquad \text{log-log fit} \qquad (5)$$

$$y = a + \frac{b}{x} \qquad \text{reciprocal } x \text{ fit} \qquad (6)$$

$$y = \frac{1}{a + bx} \qquad \text{reciprocal } y \text{ fit} \qquad (7)$$

$$y = \frac{x}{(a + bx)} \qquad \text{hyperbolic fit} \qquad (8)$$

$$y = a + b_1x + b_2x^2 + \ldots \qquad \text{polynominal fit} \qquad (9)$$

9.2.10 INDEXES

Cost estimating indexes are useful for a variety of purposes. Principally they are multipliers to update an old cost to a new cost, using the formula

$$C = C_r \frac{I}{I_r} \qquad (10)$$

where C_r is the reference cost associated with a reference index I_r. The cost C to be determined is linked in terms of time to the index I.

Indexes are prepared and published by the government, private industry, banks, consultants, associations, and trade magazines. It is important to determine one's own indexes, especially for materials or labor not charted by other groups.

A cost index is a dimensionless number representing the change in cost of material or labor or both over a period. Prices, which are the input of an index, must relate to specific material or labor. An index for lumber is based upon the price changes of a specific quantity and type of lumber, such as a board foot of 2 by 8 in. (50 by 200 millimeters) utility-grade pine. Quantity and quality must remain constant over the period so that price movements represent a true price change rather than a change in quality or quantity. This is difficult for indexes that are charted over many periods.

A cost index is meaningful only in that it expresses a change in price level between two specific times. A cost index for lumber in 1985 alone is meaningless. An index for material A has no relationship to the index for material B. Similarly, the cost indexes for material A in two geographical areas may not be directly comparable.

To compute a price index for a single material, a series of prices must be gathered covering a period for a specific quantity and quality of the material. Index numbers are usually computed on a periodic basis. The federal government gathers data and calculates and divulges index numbers for periods as short as a month. The prices gathered for the material may be average for the period (month, quarter, half year, or year) or may be a single observed value as found on invoice records for one purchase.

Assume that the following prices have been collected for a standardized unit of silicon laser glass material:

	Period					
	0	1	2	3[a]	4	5
Price	$43.75	$44.25	$45.00	$46.10	$47.15	$49.25
Index	94.9	96.0	97.6	100.0	102.3	106.8

[a]Benchmark period.

Index numbers are computed by relating each period price to one of the prices that has been selected as the base. If period 3 is the benchmark period, period 2 price divided by period 3 price = \$45.00/\$46.10 = \$0.976. When period 3 price is expressed as 100.0, period 2 price can be expressed as 97.6.

Movements of indexes from one period to another are expressed as percent changes rather than as changes in index points.

Current index	106.8
Less previous index	102.3
Index point change	+4.5
Divided by previous index	102.3
Equals	+0.044 = +4.4%

Periods 6, 7, and so on can be projected, and if a reference price is known, a future price can be calculated. For instance, if C_2 = \$3700, and if we want to know C_7, then projecting C_7 = \$3700 × (110.0/97.6) = \$4170. The use of indexes for updating material is straightforward. Distortion effects are easily identified and corrected for single-commodity indexes.

A composite index is often required, say, for learning curve projections, purchased materials, or adjustment of purchase price for "quote-or-price-in-effect" types of purchase contracts. Equally important is the updating of estimates of complicated assemblies, buildings, and plants. Although updating past estimates might be handled by the compound growth factor of the Consumer Price Index (CPI) for inflation, labor, and material elements of an assembly, do not follow the CPI where technology progress is active.

Assume that a product called "10 cm disk aperture laser amplifier" is selected for a composite index. Although the 10 cm disk amplifier was produced only during period 0, tracking of selected cost items has continued. To worry about all amplifier components is too involved, so major items were picked for individual tracking, and spot prices have been gathered for 4 years. The quantity of each of the five materials is in proportion to the initial one-time cost of the material to the total cost. Some materials have declined in price, whereas others have increased. Prices for each material have been gathered (or imputed for periods where no information was available) and are shown in Exhibit 9.2.5. The prices conform to a quantity and quality specification. With the index at 100 for benchmark period 0, the following indexes are calculated as 94.1, 89.6, 87.6, and 93.3. If the unit cost is \$43,650 during period 0, the estimated cost is equal to \$37,953 at period 5.

One may argue that cost facts, materials, quantities, and qualities are not consistent as given in Exhibit 9.2.5. Indeed, if technology is active, a decline in the cost and index is possible. But indexes should reflect basic price movements alone. Index creep results from changes in quality, quantity, and the mix of materials or labor. Exhibit 9.2.5 is an example of a product index. The components in this case are selected on the basis of their contribution to the product value. Selection of components could be 100%, random, or stratified in accordance with the needs of cost estimating. Quantity is determined proportional to the design requirements. Specifications provided by engineering are used to fix quality characteristics. But product indexes can be maintained by noting the changes when they occur, inputting all previous data, and recalculating the previous year's indexes. Every so often it may be necessary to reset the benchmark year whenever delicate effects are influencing the index and are not being removed.

Exhibit 9.2.5 Calculation of Index for Material, Quantity, and Quality Specifications

			Period				
Material	Quantity	Quality	0[a]	1	2	3	4
Laser glass	3 to 10 cm disk	Silicate	\$26,117	\$24,027	\$22,345	\$21,228	\$22,713
Stainless steel turnings	18 kg	AISI 304	1,913	2,008	2,129	2,278	2,460
Aluminum extrusion	4 kg	3004	418	426	439	456	479
Fittings	3 kg	MIL-STD-713	637	643	656	657	689
Harness cable	4 braid 4 m	MIL-STD-503	2,103	2,124	2,134	2,305	2,466
Annular glass tube	12 m	Tempered $\frac{3}{16}$ in. wall PPG-27	4,317	4,187	4,103	4,185	4,311
Total index			\$35,505	\$33,415	\$31,806	\$31,109	\$33,118
percentage			100.0	94.1	89.6	87.6	93.3

[a]Benchmark period = 100.

The several kinds of indexes are:

1. Material.
2. Labor.
3. Material and labor.
4. Regional effects.
5. Design type.

9.2.11 PRELIMINARY AND DETAILED METHODS OF ESTIMATING

Although applications and designs may differ, the methods used in estimating are remarkably similar. These methods are intended to develop either a preliminary or a detailed estimate. A preliminary method is used in the formative stages of design. Other terms used in practice may be "battery limit," "order of magnitude," "quickie," and so on. Preliminary methods are fast and not as accurate as those used to prepare detailed estimates. At the other extreme we use detailed methods to set a price, make a bid, or take economic action. Detailed methods are more quantitative, and arbitrary and excessive judgmental factors are suppressed, although they are never eliminated. The techniques presented here progress from preliminary to detailed methods and are as follows:

1. Judgment and conference.
2. Comparison.
3. Unit method.
4. Cost and time estimating relationships.
5. Probability approaches.
6. Standard time data.
7. Factor method.

If broadly defined, these methods can be used for the four types of designs.

Judgment and Conference Method

The use of personal judgment is an integral part of the estimating process. It is easy to be critical of guesstimates, but in the absence of data and with the shortage of time, there may be no other way but to use judgment in the evaluation of designs. The estimator is selected for the job because of his or her observational experience, common sense, and knowledge about the designs. In judging the economic want of a design, the mettle of the estimator is tested. Time, cost, or quantities with regard to minor or major line elements are picked using this inner experience; the estimator must be objective in attempting to measure all future factors that affect the out-of-pocket cost. Judgment estimating is alse done collectively.

The conference method is a nonquantitative consensus method of estimation. It provides a single value or estimate made on the basis of experience. In addition to cost or price, other information such as savings potential, marginal revenue, and the like can be estimated. The conference method relies on the collective judgment of contrasting differences between previously determined estimates and their associated designs and an unknown, but to-be-determined, estimate with its design. The method follows many rituals, but usually involves representatives from various departments conferring with the estimating department in a round table fashion and jointly estimating cost or price of line items or a lump sum. Sometimes labor and material are isolated and estimated, with overhead, distribution, selling expenses, and profit added later through various formula methods by the estimator. The conference method may be used within the cost estimating department. Estimators having specialized knowledge confer on a design and determine a cost figure without counsel from other departments.

The way in which the method is managed depends on the available information. Various techniques are used to sharpen judgment. Major drawbacks of the conference method are the lack of analysis and a trail of verifiable facts leading from the estimate to the governing situation. Although little faith in the accuracy is given to a conference estimate, the lack of analysis and procedural rigor seldom deters the use of this method.

Comparison Method

The comparison method is similar to the previous method, except that it attaches a formal logic. If we are confronted with an unsolvable or excessively difficult design and estimating problem, we designate it problem A and construct a simpler design problem for which an estimate can be found. The simpler problem is called problem B. This simpler problem might arise from a clever manipulation of the original design or a relaxation of the technical constraints on the original problem. Thus we gain information by branching to B, since various facts may already exist about B. Estimate B may be in final form, or portions may exist and there need only be a minor restructuring of data to allow comparison. The alternative design problem B must be selected to bound the original problem A in the following way:

$$C_A(D_A) \leqslant C_B(D_B) \tag{11}$$

where $C_{A,B}$ is the cost value of the estimate for designs A or B and $D_{A,B}$ is design A or design B. Also, D_B must approach D_A as nearly as possible. We adopt the cost value C_A of our estimate as something under C_B. The sense of the inequality in equation 11 is for a conservative posture. It may be management policy to estimate cost slightly higher at first, and then once the detailed estimate is completed with D_A thoroughly explored, we comfortably find that $C_A(D_A)$ is less than the original comparison estimate.

An additional lower bound is possible. Assume a similar circumstance for a known or nearly known design *C*, and a logic can be expanded to have

$$C_C(D_C) \leqslant C_A(D_A) \leqslant C_B(D_B) \tag{12}$$

We assume that designs *B* and *C* satisfy the technical requirements, but not the economic estimate, as nearly as possible.

Although the foregoing is a formalized approach, in practice a great deal of estimating uses the comparison logic. Standard cost plans (see Chapter 9.1) provide "similar to" approaches, and analogy plans and computer retrieval schemes use this essential logic.

Unit Method

The unit method is the most popular of the preliminary estimating methods. Many other terms are used to describe this method—"order of magnitude," "lump sum," "module estimating," "flat rates"—and involve various refinements. Extensions of this method lead to the factor estimating method discussed later. Examples of unit estimates such as the following are found in all activities:

Cost of house construction per square foot of livable space.
Cost of fabricated components per pound of casting.
Cost of electrical central power station per kilowatt generated.
National norm cost of university education per student year.
Chemical plant cost per barrel of oil capacity.
Factory cost per machine shop man-hour.

Although these estimates are typically vague in these contexts, the strongest assumption necessary for their application is that the design to be estimated be similar to the composition of the parameter used to determine the estimate. Notice that the estimate is "per" something. Data for these estimates are collected from technical literature, government, banks, or the files of cost engineering or accounting. If accounting data are used, they must be recast to be useful for estimating costs.

Unit estimates are easily figured. Consider the manufacturing operation of turning, for example. Using job tickets, the total time for several jobs and many parts for a lathe can be ascertained. Divide this quantity by the number of inches (centimeters) turned. The result is a unit estimate, or hours per inches (centimeters) of length. Even though this estimate is simplistic expressed in this context, unit estimating methods are widely used. We see that the unit estimate implies an average value based upon observed time study or man-hour data.

Cost and Time Estimating Relationships

Cost estimating relationships (CERs) and TERs are mathematical or graphic models that estimate cost or time. Although statistical regression was previously discussed as forecasting, its contribution to estimating also develops CERs and TERs. Forecasting tends to deal with time serial (see Chapter 13.9) and minor analytical problems. However, CERs and TERs are formulated to give estimates in either final or line item form of a pro forma cost estimate. Rules of thumb such as unit estimates are not classified as CERs or TERs.

The learning curve is a TER or CER and is also discussed in Chapter 4.3. A typical approach is given by

$$T_U \text{ or } T_{AC} = KN^S \tag{13}$$

where T_U = time, cost, or value per unit of production, such as man-hours or dollars, required to produce Nth unit
or T_{AC} = average cumulative time, cost, or value for N units
and N = unit number, 1, 2, 3, . . .
K = constant or estimate for $N = 1$, expressed in dimensions compatible with T
S = slope parameter of the improvement rate, equal to log ϕ/log 2 where ϕ = learning (S is usually negative.)

If learning of 93% is proposed, $S = \log 0.93/\log 2 = -0.1047$.

The learning curve can have either the unit or the cumulative average line as the straight one adopted. One theory suggests that, on log-log graph paper, the cumulative average line is straight and the unit line curves under from unit 1 until 10 or 20 units from which point it parallels the cumulative line. The other theory holds that the individual unit line is straight when plotted on log-log graph paper and that the cumulative average line, though starting together with the unit line at unit 1, will curve above the unit line, and that by 10 to 20 units the two curves will run parallel. Either practice is acceptable, but it is important that the estimator identify the practice to avoid misunderstanding. The two approaches are shown in Exhibits 9.2.6 and 9.2.7.

Historical records are examined, and either the unit or the average cumulative values are regressed against unit number N. With knowledge of these slopes or other learning experiences, the estimator determines the appropriate factor for the job to be estimated.

The learning curve is discussed extensively in Ostwald,[1,2] and tables have been codified and are supplied in Gallagher.[3] Follow-on estimation is described in Exhibit 9.2.8. The assumption of this application is that the cumulative average is a straight line.

The power law and sizing model is frequently used for estimating equipment or parts as a lump sum. This model is concerned with designs varying in size, but similar in type. The unknown costs of a 200 gal (0.75 m^3) kettle can be estimated from data for a 100 gal (0.40 m^3) kettle provided that both are similar in design. No one would expect that the 200 gal kettle would be twice as costly as the smaller one. The law of economy of scale ensures that. The power law and sizing model is

$$C = C_r\left(\frac{Q_c}{Q_r}\right)^m \tag{14}$$

Exhibit 9.2.6 Calculations for a Learning of 93% Using Unit Factor

N	Unit Factor	Total Cumulative Unit Factor	Average Cumulative Unit Factor
1	K = 1.0000	1.0000	1.0000
2	0.9300	1.9300	0.9650
3	0.8913	2.8213	0.9404
4	0.8649	3.6862	0.9216
5	0.8449	4.5312	0.9062
10	0.7858	8.5604	0.8560
15	0.7531	12.3856	0.8257
	$T_U = KN^{-S}$	$\sum_1^N T_U$	$\sum_1^N T_U/N$

Exhibit 9.2.7 Calculations for a Learning of 93% Using Average Cumulative Factor

N	Average Cumulative Factor	Total Average Cumulative Factor	Unit Average Cumulative Factor
1	K = 1.0000	1.0000	1.0000
2	0.9300	1.8600	0.8600
3	0.8913	2.6739	0.8140
4	0.8649	3.4596	0.7856
5	0.8449	4.2246	0.7650
10	0.7858	7.8578	0.7073
15	0.7531	11.2969	0.6767
	$T_{AC} = KN^{-S}$	NT_{AC}	$KN^{1+S} - K(N-1)^{1+S}$

Exhibit 9.2.8 Follow-On Estimating Using the Learning Curve (Log-Log Graph)

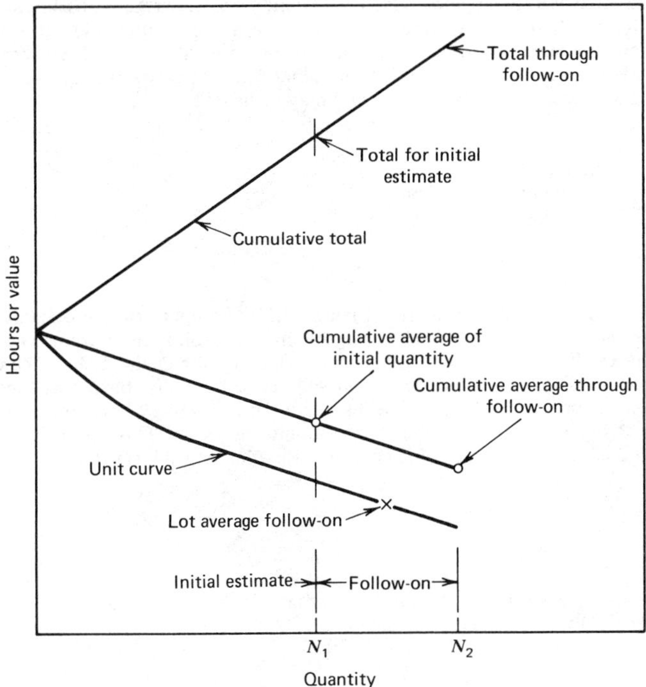

where C = total cost sought for design size Q_c
C_r = known cost for a reference size Q_r
Q_c = design size expressed in engineering units
Q_r = reference design size expressed in engineering units
m = correlating exponent, $0 < m \leqslant 1$

An equation expressing unit cost C/Q_c can be used as

$$\frac{C}{Q_c} = \left(\frac{C_r}{Q_r}\right)\left(\frac{Q_c}{Q_r}\right)^{m-1} \tag{15}$$

As total cost varies as the mth power of capacity, C/Q_c will vary as the $(m - 1)$ power of the capacity ratio. If we let $m = 1$, we have a strictly linear relationship and deny the law of economy of scale. For chemical processing equipment, m is frequently near 0.6, and for this reason the model is sometimes called the "sixth-tenth model." The units on Q are required to be consistent since it enters only as a ratio. The model can be altered to consider changes in price due to inflation or deflation and effects C_I independent of size, or

$$C = C_r\left(\frac{Q_c}{Q_r}\right)^m\left(\frac{I_c}{I_r}\right) + C_I \tag{16}$$

Another CER is $C = KQ^m$ where K is a constant for a project and may be a processing plant or equipment. The concept of economy of scale is derived from this CER where capital cost per unit produced reduces as the plant size increases. The scale factor m is not constant for all project designs. General scale-up or scale-down by more than a factor of 10 should be avoided.

For instance, where symbols have been previously supplied, a multivariable CER is also possible.

$$C = KQ^m \quad N^{S_1}$$

Probability Approaches

Cost is usually treated as a single point value under conditions of certainty. Estimators knowing the weaknesses of information and techniques recognize that there are probable errors. The indication that cost is a random variable opens up the topic of probability estimating. A "random variable" in statistical parlance is a numerically valued function of the outcomes of a sample of data. Various approaches deal with probabilistic outcomes and may be listed as:

1. Expected value.
2. Computer simulation.
3. The 0.1, 0.5, 0.9 percentile method.
4. The PERT-based beta distribution.

Expected Value

For the most part estimators have preferred to deal with the simplest case of certainty. Despite this widespread practice, it seldom exists. The category involving risk is appropriate whenever it is possible to estimate the likelihood of occurrence for each condition of the design. These probabilities describe the likelihood that the predicted event will occur. Formally, the method incorporates the effect of risk on potential outcomes by means of a weighted average. Each outcome of an alternative is multiplied by the probability that the outcome will occur. This sum of products for each alternative is entered in an expected value column, or mathematically for the discrete case

$$C(i) = \sum_{j}^{n} p_j x_{ij} \tag{17}$$

where $C(i)$ = expected cost of the estimate for alternative i
p_j = probability that x takes on value x_j
x_{ij} = design event

The p_j's represent the independent probabilities that their associative x_{ij}'s will occur with $\Sigma p_j = 1$.

Computer Simulation.

Widely employed in engineering, science, and business, simulation is used to estimate the cost of systems. "Simulation" is defined as the manipulation and observation of a synthetic model representative of a real design, which, for technical or economic reasons, is not susceptible to direct experimentation. This synthetic model ideally represents the essential characteristics of the real system with the frills excluded. The computer is mandatory in this type of analysis. Simulation is discussed in Chapter 13.11, and its use in economic risk analysis is described in Chapter 9.5.

Percentile Method

Estimates reflecting uncertainty may be specified by three values representing the tenth, fiftieth, and ninetieth percentiles of an unstated probability distribution. The estimator's best value is the fiftieth percentile. The tenth percentile cost is one for which there is a one tenth chance that the actual value will be lower, and for the ninetieth percentile cost, there is a one tenth chance that the actual value will be greater. The tenth and ninetieth percentile costs are roughly equivalent to best- and worst-case estimates. The method is illustrated as follows:

	Percentile			Difference	
Item	10th	50th	90th	(50 – 10)	(90 – 50)
1	\$36	\$42	\$53	\$6	\$11
2	7	12	15	5	3
3	9	10	14	1	4

These costs can be assumed to combine independently, that is, a low cost with a mid-cost with another low cost, and the number of combinations for any product is vast. After estimating, we express the tenth and ninetieth percentile as differences from the mid-value. Next we square the differences and sum.

	$(50-10)^2$	Mid-value	$(90-50)^2$
	\$36.00	\$42.00	\$121.00
	15.00	12.00	9.00
	1.00	10.00	16.00
Total	62.00	64.00	146.00
Square root	7.87		12.08

Total estimate at tenth percentile = \$64.00 − 7.87 = \$54.13
Total estimate at fiftieth percentile = \$64.00
Total estimate at ninetieth percentile = \$64.00 + 12.08 = \$76.08

Sensitivity can be found using this approach or

Item	Contribution to Low Uncertainty	Contribution to Total Cost	Contribution to High Uncertainty
1	58% (36/62 × 100)	66% (42/64 × 100)	83% (121/146 × 100)
2	40%	19%	6%
3	2%	16%	11%

This rough sensitivity analysis will identify candidates for cost reduction and watching.

The PERT-Based Beta Distribution

The following procedure is based on a method developed for PERT. It involves making a most likely cost estimate, an optimistic estimate (lowest cost), and a pessimistic estimate (highest cost). These estimates are assumed to correspond to the beta distribution, which can be symmetrical or skewed left or right. With the three estimates made, a mean and a variance for the cost element can be calculated as

$$E(C_i) = \frac{L + 4M + H}{6} \tag{18}$$

$$\text{var}(C_i) = \left(\frac{H - L}{6}\right)^2 \tag{19}$$

where $E(C_i)$ = expected cost for element i
L = lowest cost, dollars
M = modal value of cost distribution, dollars
H = highest cost, dollars
$\text{var}(C_i)$ = variance of cost for element i

If several elements are estimated this way, and if their costs are assumed to be independent of each other and are added together, the distribution of the total cost is approximately normal. This follows from the central limit theorem. Exhibit 9.2.9 illustrates how this method is used to find contingency effects for a project design. There must be several or more elements to satisfy the conditions of the central limit theorem.

$$E(C_T) = E(C_1) + E(C_2) + \cdots + E(C_n) \tag{20}$$

and

$$\text{var}(C_T) = \text{var}(C_1) + \text{var}(C_2) + \cdots + \text{var}(C_n) \tag{21}$$

where $E(C_T)$ is the expected total cost in dollars and $\text{var}(C_T)$ is the variance of total cost in dollars.

Standard Time Data

The most precise way to estimate direct labor is with standard time data. (For more discussion on this topic, see Chapter 4.8.) Time data as raw observations are not readily usable by the estimator. Frequently these data include bad methods, unwarranted conditions, or nonaverage operators.

Exhibit 9.2.9 Estimating for Risk Contingency

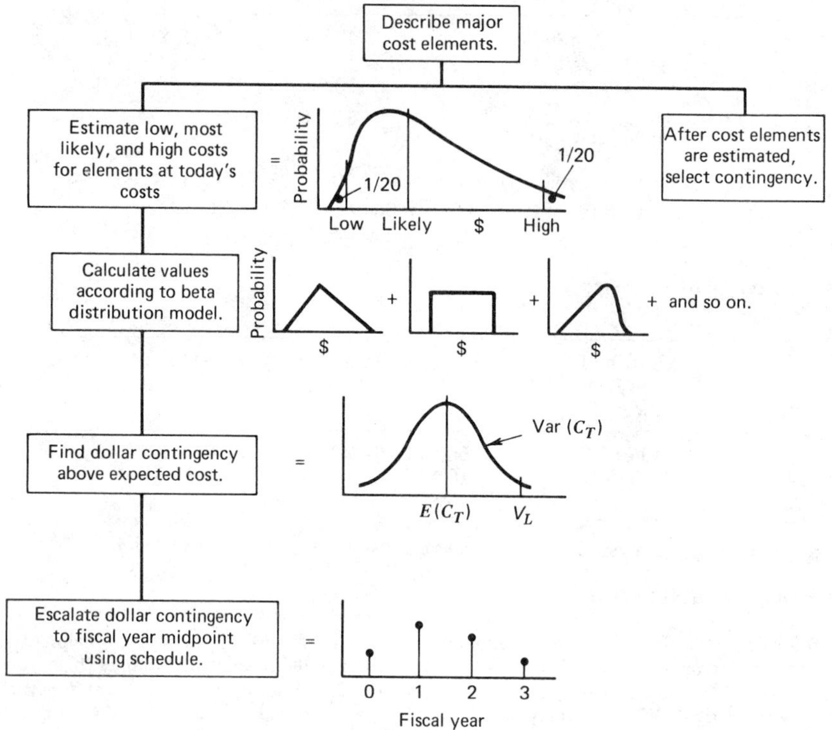

Sometimes they are lacking because of the limitations of a narrow range of observed work. The industrial engineer may use regression analysis or other techniques to extend these raw observed data into a more digestible form. It is not the original time measurements that the estimator ultimately desires; rather, it is a set of engineering performance data, or standard time data, or, more briefly, standard data, that he or she uses for estimating.

Standard time data may be defined as a catalog of standard tasks that are used for performing a given class of work. Parenthetically, it should be indicated that such data are arranged in a systematic order and are used over and over again. The advantages over direct observation methods, such as time study, predetermined motion time data, work sampling, and man-hour reports, are lower cost and greater consistency. Description is provided in advance of the need for the data, and more estimators can use standard time data.

Standard time data may be divided into preliminary or detailed data. The estimator is more likely to be concerned with preliminary standard data early in estimating; later, detailed data become more important. Standard time data are ordinarily determined from any of the various methods of observing work. In manufacturing, time study and predetermined motion time data are the major sources. In construction and hospitals, work sampling and man-hour reports are the principal means of information. In certain government work, such as that of the post office or armed forces, work sampling and man-hour methods are used. Whatever the source of work data, the development of standard data is similar.

Curves or formulas are not preferred as the final expression for standard time data since there is a tendency toward incorrect interpolation, and since they are time consuming to read and subject to faulty extension. Consequently, charts replace curves as the preferred final expression of standard time data because they are faster and consistently applied when compared to curves. Charts can be prepared for machines, processes, and bench work. Machine time tables have been codified,[4] and production information can be found in Nordhoff.[5] Construction work standard time data are found in various references.

The standard time data are used for a designated work design for direct labor. Estimating when using standard time data is considered accurate and relatively inexpensive.

Factor Method

The factor method is a basic and important method for project estimates. Other methods, such as ratio, parameter, and percentage methods, are about the same thing. The unit method, such as hour per inch of turned diameter, is one factor used to find composite cost. The unit cost estimating method was limited to a single factor for calculating overall costs. A natural extension of the unit method achieves improved accuracy by adopting separate factors for different cost items. For example, the approximate cost of a plant can be estimated by multiplying the area by an appropriate unit estimate such as the dollars per square foot (dollars per square meter) factor. As an improvement, individual cost per unit area factors can be used for heating, lighting, electrical, and the like, and their value summed for the separate design requirements.

$$C = \left[C_e + \sum_i f_i \, C_e \right] \left(1 + f_I \right) \tag{22}$$

where C = cost of design being evaluated
C_e = cost driver or subdesign used as base
f_i = factor for estimating instrumentations, structures, site clearing, and so on
f_I = factor for estimating indirect expense such as engineering, contractor's profit, and contingency
$i = 1, \ldots, n$ factor index

There are several variants to this problem, but the basic idea is that a cost driver, such as a significant functional unit in a chemical plant, is chosen as C_e. If C_e is a high-pressure–high-temperature reactor handling caustic materials, for example, the piping, structure, instrumentation, and site cost would be more expensive than if the reactor were of low pressure and temperature. These factors are correlated, and historical data, design, indexes, and business data are necessary to conducting a

Exhibit 9.2.10 Estimate for Manufacturing a Part

Part number 672	Part name Sheet metal chassis	Material 5052-H34
Quantity 250	General notes Becomes a	0.040 in. (1.02 mm)
Estimator PFO	part of a two-piece	Sheet size, 48 x 96 in.
Date 3/9	subassembly	(12 x 24 m)
Estimate expires on 6/9		

Work Center	Op. No.	Description of Operation (List tools and gauges)	Setup Hours	H/100 Units	Unit Estimate	Labor Rate	Labor and Overhead Rate	Material	Labor and Overhead
Stockroom	-	Raw material	-	-			-	0.45	-
7100	10	Shear	0.1	0.100	0.0014	17.25	1.9		0.05
81300	20	Tape, piece, and notch NC tape	0.5	3.835	0.040	18.50	3.15		2.35
47103	30	Countersink 18-0.12000 holes 4-0.12500 holes	0.3	4.327	0.043	18.41	2.5		2.00
853	40	Degrease	0.1	0.118	0.0016	16.45	1.9		0.05
71	50	Brake 2 lips	0.3	0.528	0.007	18.20	2.6		0.31
71	60	Brake 2 lips	0.3	0.528	0.007	18.20	2.6		0.31
47103	70	Ream 24-0.125 in, 00 holes & 4-0.18900 holes	0.3	2.774	0.028	18.41	2.5		1.28
853	80	Degrease	0.1	0.653	0.007	16.45	1.9		0.22
Bench	90	Deburr	0.05	0.094	0.001	15.37	1.3		0.02
950	100	Heliarc 4 corners	0.2	3.509	0.036	19.18	1.7		1.17
2342	110	Grind welds	0.1	2.050	0.021	17.68	1.4		.52
							Component total	$0.45	$8.28

factor estimate. Indeed, it is also possible to estimate direct labor costs on the basis of a factor relating to capital assets.

9.2.12 OPERATION ESTIMATING

The heart of operation estimating is, in a substantive way, a talent for breaking down a task into essential elements. The subdivision of the design task into large portions of direct labor and material comes first, but this is followed by progressive finer detail until a description of labor and materials is very broad. At this point, dollar extensions of labor and material are made to reflect the cost of the design.

Exhibit 9.2.10 is an example of a manufacturing operations sheet. The estimate is for sheet metal work. Exhibit 9.2.11 is an example of a construction estimate.

9.2.13 PRODUCT ESTIMATING

The product estimate determines the full cost of the product; operation estimates, various overhead rates, and other companion estimates are at hand for the detailed estimate. A preliminary estimate may have preceded the product estimate, but that effort served the purposes of screening or feasibility. The pro forma (a Latin word meaning "for sake of form") estimate is the company's recognized and official cost estimating form, which summarizes the total direct labor, material, and

Exhibit 9.2.11 Construction Cost Estimate

Estimate Summary

Customer *A-B-C Oil Company* Prop. No. *93500*
Location *Central USA* Job No. *–*
Project *Typical Refinery Unit* Date *Nov. 3*
By *WRW*
Rev. No. Rev. Date By

Act	Description	Craft Hours	Labor	Material	Other	Total	Total Field Labor Cost	Direct Field Labor Cost
A	Earthwork *1300 MH*	6 540	28 400	13 700	31 500	73 600	1 5	
B	Concrete *1100 MH, 2250 CY*	22 000	96 800	72 900	18 700	188 400	3 8	
C	Buildings and structures *1400 MH*	9 010	54 200	124 300	36 700	215 200	4 3	
D	Process equipment *17,500 MH*	12 400	75 300	994 900	862 900	1 933 100	38 8	
E	Piping	54 480	351 200	530 900	122 300	1 004 400	20 2	
F	Electrical *9700 MH*	100	600	64 400	93 900	158 900	3 2	
G	Painting *4300 MH*	–	–	–	48 500	48 500	1 0	
L	Plant items	100	600	1 000	–	1 600	–	
N	Instruments and controls	5 970	38 500	240 600	–	279 100	5 6	
P	Insulation *6500 MH*	–	–		134 600	134 600	2 7	
	Avg. Rate = $5.84/MH							
	Direct field cost	110 600	645 600	2 042 700	1 349 100	4 037 400	81 1	(2.26)
H	Field expense	5 550	139 900	3 200	24 300	167 400		25 9
H	All risk, pr tax, bond	–	–	–	103 500	103 500		16 1
K	Construction supplies	–	–	21 300	–	21 300		3 3
M	Start-up	1 100	6 600	–	–	6 600		1 0
S	Temporary facilities	7 600	28 600	6 400	17 200	52 200		8 1
V	Craft benefits	–	–		51 900	51 900		8 0
V	Construction camp.	–	–	–	–	–		–
W	Construction equip.	1 250	7 600	16 100	127 200	150 900		23 4
	Indirect field cost	15 500	182 700	47 000	324 100	553 800	11 1	85 8
	Total field cost	126 100	828 300	2 089 700	1 673 200	4 591 200	92 2	
J	Engineering and base salaries *× 1.08 × 1.45 (Base O.H.)*					387 300	7 8	60 0
	Total field and eng. cost					4 978 500	100 0	
Q	Sales tax					66 500	1 3	
R	Premium pay					20 000	4	3 1
	Escalation - *Included Above*					–		
	Contingency					237 000	4 8	
	Subtotal					5 302 000		
Y	Fee					190 000	3 9	
	Total					5 492 000		
	Process Equip. Ratio = 5492 M / [illegible] M = 3.10							

Exhibit 9.2.12 Operation Method of Product Estimating

Part number 492 7151

Quantity 250

Estimator PFO

Date 3/12

Estimate expires on 6/9

Part name Final sheet metal radio chassis assembly

General notes Has four subassemblies

Subassembly Part Number	Material Cost	Labor and Overhead Cost
7150	$18.36	$109.38
7149	27.46	42.15
7148	102.18	87.33
7149	7.47	17.26

Work Center	Op. No.	Description of Operation (List tools and gauges)	Setup Hours	H/100 Units	Unit Estimate	Labor Rate	Labor and Overhead Rate	Cost of Material	Cost of Labor and Overhead
Stockroom	–	Supply subassemblies							
181-3	10	Finish box-pack	0.2	1.955	0.020	16.65	1.45	0.87	0.48
2700	20	Pack shipping documents	0.1	1.056	0.016	16.45	1.40	0.26	0.39

1. Cost of goods manufactured: total	$156.60	$256.99
2. General and administrative costs at 90% of 1		372.23
3. Engineering cost prorated		33.83
4. Contingencies		–
5. Selling costs at 40% of 1		165.44
Total unit cost of manufacturing, development, and sales		$985.09

Exhibit 9.2.13 Department Method of Product Estimating

Manufacturing estimate for *laser* products

Description *Speed decreaser - L H*

Customer *Endicott* Quote no. *7/8* Date *3/24*

Base on quantity of *480 units* During period of *6/1 – 10/1*

Dept.	Description	Hr/Lot	Rate	Labor
103	Finishing	33.60	$15.60	$ 524.16
105	Machine shop	143.20	$18.00	$2577.60
106	Miscellaneous machines	118.20	$14.10	$1666.62
108	Precision assembly	28.80	$17.00	$ 481.10
131	Electronic assembly	68.05	$15.65	$1064.98
125	Cable assembly	—	—	—
233	Inspection	28.80	$17.20	$ 495.36
24	Stockroom	48.00	$14.20	$ 681.60
	Subtotal	468.65		$7491.42
	Learning curve factor - 8%			$ 599.31
	Subtotal - manufacturing			$6892.11
	Fringe labor cost +28%			$1929.79
	Total manufacturing labor			$8821.90

Item	Description	Unit	Lot
1	Manufacturing labor (per above)	$18.379	
2	Manufacturing overhead @ 75% of item 1	$13.784	
3	Total	$32.163	
4	General & Administrative @ 20% of item 3	$24.122	
5	Parts & materials per estimate	$23.210	
6	Overhead on parts & materials @ 15%	$ 3.482	
7	Custom engineering labor per estimate	$18.330	
8	Custom engineering overhead @ 40%	$ 7.332	
9	Research & development per estimate	—	—
10	Contingencies per estimate	—	—
11		$108.639	$52146.57
12	Selling at 25% of item 11	$ 27.160	$13036.64
13	Total cost	$135.299	$65183.21
14	Profit @ 13%	$ 17.654	$8473.82
15	Selling price	$153.453	$73657.03

Cost approvals		Price approvals	
Estimated by	WEJ	Accounting	XLM
Approved by	JXL	Product manager	PEO
Division head	MSP	Director of service engineering	F. Haller
		General manager	

overheads of various kinds. It is the single most important document of the product design estimating efforts, since it is often approved by the owner, president, or senior executive officer. There are varieties of pro forma estimates, and three are shown in Exhibits 9.2.12, 9.2.13, and 9.2.14.

Exhibit 9.2.12 is an extension of Exhibit 9.2.10, which estimated part costs. Following the indentured bill of material part numbering scheme, the parts, subassemblies, and major assemblies are finally rolled up into a final assembly operation process sheet. Because of the dependence upon the operation process sheet, the method is called the "operation method."

The department method is simpler and less costly to prepare, although it is not as accurate. The initial information is a direct labor estimate for lot, given a specific department. Operation process sheets could be canvassed for this information, or other means could be used. Exhibit 9.2.13 shows how the departmental method is developed.

The variable cost method is appropriate for high-volume consumer goods. In the consumer goods

Exhibit 9.2.14 Variable Cost Method of Product Estimating

PRODUCT COST ESTIMATE SUMMARY

Preliminary or (detail) ________

Product line ________ Retail price $0.50 Retail mark-up 44.5%

ITEM	AMOUNT	PER CENT
LIST PRICE	$0.2775	100.00
VARIABLE MANUFACTURING COSTS:		
Direct material	0.0741	26.70
Direct labor	0.0243	8.76
Variable manufacturing overhead	0.0550	19.82
Freight to warehouses	0.0042	1.51
Irregular merchandise allowance	0.0022	0.80
Freight to customer	0.0021	0.77
TOTAL VARIABLE MANUFACTURING COSTS	0.1619	58.36
VARIABLE MARKETING COSTS:		
Trade discount allowed	0.0028	1.01
Field allowances	0.0001	0.04
Promoting allowances	0.0035	1.26
Royalties	—	—
Variable selling expenses	0.0031	1.11
TOTAL VARIABLE MARKETING COSTS	0.0095	3.42
VARIABLE ADMINISTRATIVE COSTS:		
Cash discounts allowed	0.0050	1.80
Variable administrative expenses	0.0040	1.44
TOTAL VARIABLE ADMINISTRATIVE COSTS	0.0090	3.24
TOTAL VARIABLE COSTS	0.1804	65.00
STANDARD PROFIT CONTRIBUTION	0.0971	35.00
STANDBY FIXED COSTS:		
Manufacturing	0.0208	7.50
Selling and marketing	0.0077	2.77
Advertising	0.0002	0.07
Administrative	0.0069	2.48
TOTAL STANDBY	0.0356	12.82
PRODUCT FIXED COSTS:		
Manufacturing	0.0066	2.38
Selling & marketing	0.0039	1.40
Advertising	0.0066	2.38
Administrative	0.0200	7.24
TOTAL PRODUCT	0.0371	13.37
STANDARD EARNINGS	0.0244	8.79

Exhibit 9.2.15 Work Breakdown Structure to Fifth Level

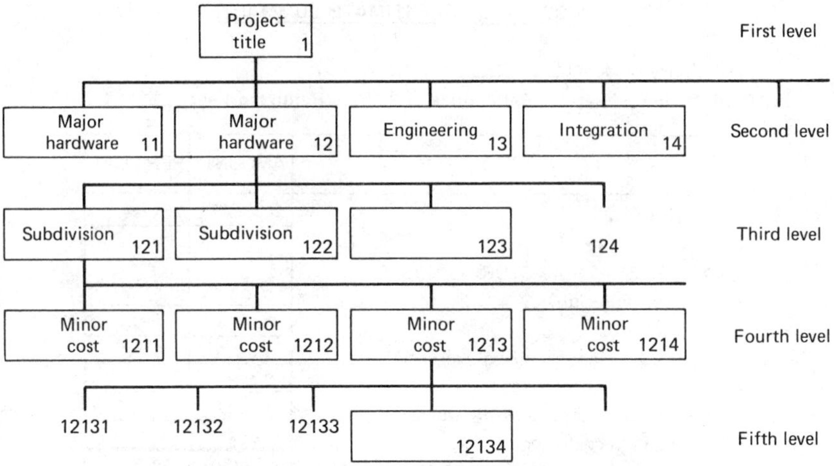

Project title 1 — First level

Major hardware 11 | Major hardware 12 | Engineering 13 | Integration 14 — Second level

Subdivision 121 | Subdivision 122 | 123 | 124 — Third level

Minor cost 1211 | Minor cost 1212 | Minor cost 1213 | Minor cost 1214 — Fourth level

12131 | 12132 | 12133 | 12134 — Fifth level

category, product estimates are developed to greater detail and accuracy. Exhibit 9.2.14 is an example.

9.2.14 PROJECT AND SYSTEM ESTIMATING

Project estimates provide the costs where basic economic considerations are the design, capital, expense, income, and time. System estimates, since they deal with governmental requirements, use the same pro forma estimate as found in operation, product, and project estimates.

The required information is found in a work package, which includes the work package definition, procurement estimate, labor estimate, supplies and expenses estimate, and finally the contract pricing proposal. The bibliography provides additional readings for this information. The work package definition identifies what is to be done and when it is to be done. It relates to the work breakdown structure, an example of which is shown in Exhibit 9.2.15. The work breakdown structure includes more tasks or equipment than a product, for instance. The bill of material, which is critical to product estimates, is only one part of the structure. There are additional major elements of cost that are not hardware, and the work breakdown structure provides for their inclusion. The structure can be organized, scheduled, and described on the work package definition, or as shown in Exhibit 9.2.16.

Appendix Exhibit 9.2.A1 is an example of a project pro forma estimate which summarizes other estimates. The exact nature and kinds of costs vary somewhat from form to form, and instructions from offerer to offerer. The offerer must explain how the package is put together and how the estimate uses factual data and also must discuss the protection against future uncertainties.

9.2.15 ESTIMATE ASSURANCE

As estimates are made prior to the spending of money, there is the natural inclination to evaluate the success of estimating efforts. Too often the post-estimate analysis is limited to cost control measurements or variances or crude comparisons of actual to estimates using the formula

$$\text{error percentage} = \frac{C_e}{C_a} - 1 \times 100 \tag{23}$$

where C_e is the estimate and C_a is actual value. Often an analysis such as that in equation 23 is unobtainable because a long period may have elapsed between the estimate and the realization or because the data were incompatible. The notion that cost control satisfies the need for estimate assurances misses the point. Cost estimators estimate, and they have little influence on cost control activities. Usually, operation estimates can be audited in a simple fashion; however, for product estimates there is little in the way of accurate data useful for comparisons. Where successful estimating departments are operating, it is not surprising to find the estimated cost adopted as the true cost and the efforts in estimating assurance maturing into analysis for data base improvements.

Exhibit 9.2.16 Work Package Definition

See reverse side for procedures and instructions

① Proposal/project	Work package ②	
Nova	N12	

③ Page 1 of 57

④ Title Laser-Nova System

⑤ W.A. No. ____ W.O. No. ____

⑥ No.	⑦ Description
01	Concept definition
02	Conceptual design
03	Detailed design
04	Parts procurement
05	Fabrication
06	Test
07	Integration

⑩ Proposal/project ____ Work package ____ Page ____ of ____

Milestone schedule ⑧ Start date:

No.	Title/forecast period	5	6	7	8	9	10	11	12	13	1	2	3	4	5	6	7	8	9	10
01	Conceptual definition	Δ	—	—	Δ															
02	Conceptual design				Δ	—	—	Δ												
03	Detailed design							Δ	—	—	Δ									
04	Parts procurement								Δ	—	—	Δ								
05	Fabrication										Δ	—	—	Δ						
06	Test													Δ	Δ					
07	Integration														Δ	—	—	Δ		

(The "Year" heading spans the forecast-period columns.)

⑨ Prepared by PFO 3/8 Date Reviewed by ____ Date Approved by ____ Date

Exhibit 9.2.A1 Example of Federal Pricing Form

DEPARTMENT OF DEFENSE **CONTRACT PRICING PROPOSAL** *(RESEARCH AND DEVELOPMENT)*		*Form Approved* *Budget Bureau No. 22-R100*	
This form is for use when *(i)* submission of cost or pricing data *(see ASPR 3-807.3)* is required and *(ii)* substitution for the DD Form 633 is authorized by the contracting officer.		PAGE NO.	NO. OF PAGES
NAME OF OFFEROR	SUPPLIES AND/OR SERVICES TO BE FURNISHED		
HOME OFFICE ADDRESS *(Include ZIP Code)*			
DIVISION(S) AND LOCATION(S) WHERE WORK IS TO BE PERFORMED	TOTAL AMOUNT OF PROPOSAL $	GOVT SOLICITATION NO.	

DETAIL DESCRIPTION OF COST ELEMENTS					
1. DIRECT MATERIAL *(Itemize on Exhibit A)*			EST COST ($)	TOTAL EST COST[1]	REFERENCE[2]
a. PURCHASED PARTS					
b. SUBCONTRACTED ITEMS					
c. OTHER - (1) RAW MATERIAL					
(2) YOUR STANDARD COMMERCIAL ITEMS					
(3) INTERDIVISIONAL TRANSFERS *(At other than cost)*					
TOTAL DIRECT MATERIAL					
2. MATERIAL OVERHEAD[3] *(Rate %X $ base =)*					
3. DIRECT LABOR *(Specify)*	ESTIMATED HOURS	RATE/ HOUR	EST COST ($)		
TOTAL DIRECT LABOR					
4. LABOR OVERHEAD *(Specify department or cost center)*[3]	O.H. RATE	X BASE =	EST COST ($)		
TOTAL LABOR OVERHEAD					

Exhibit 9.2.A1 *(Continued)*

5. SPECIAL TESTING *(Including field work at Government Installations)*	EST COST ($)		
TOTAL SPECIAL TESTING			
6. SPECIAL EQUIPMENT *(If direct charge) (Itemise on Exhibit A)*			
7. TRAVEL *(If direct charge) (Give details on attached Schedule)*	EST COST ($)		
a. TRANSPORTATION			
b. PER DIEM OR SUBSISTENCE			
TOTAL TRAVEL			
8. CONSULTANTS *(Identify - purpose - rate)*	EST COST ($)		
TOTAL CONSULTANTS			
9. OTHER DIRECT COSTS *(Itemise on Exhibit A)*			
10. TOTAL DIRECT COST AND OVERHEAD			
11. GENERAL AND ADMINISTRATIVE EXPENSE *(Rate % of cost element Nos.*			
12. ROYALTIES 4			
13. TOTAL ESTIMATED COST			
14. FEE OR PROFIT			
15. TOTAL ESTIMATED COST AND FEE OR PROFIT			

This proposal is submitted for use in connection with and in response to *(Describe RFP, etc.)*

and reflects our best estimates as of this date, in accordance with the instructions to offerors and the footnotes which follow.

TYPED NAME AND TITLE	SIGNATURE
NAME OF FIRM	DATE OF SUBMISSION

DD FORM 1 APR 68 633-4

Exhibit 9.2.A1 (*Continued*)

EXHIBIT A - SUPPORTING SCHEDULE *(Specify. If more space is needed, use blank sheets)*

COST EL NO.	ITEM DESCRIPTION *(See footnote 5)*	EST COST ($)

I. HAVE THE DEPARTMENT OF DEFENSE, NATIONAL AERONAUTICS AND SPACE ADMINISTRATION, OR THE ATOMIC ENERGY COMMISSION PERFORMED ANY REVIEW OF YOUR ACCOUNTS OR RECORDS IN CONNECTION WITH ANY OTHER GOVERNMENT PRIME CONTRACT OR SUBCONTRACT WITHIN THE PAST TWELVE MONTHS?

☐ YES ☐ NO *If yes, identify below.*

NAME AND ADDRESS OF REVIEWING OFFICE *(Include ZIP Code)*	TELEPHONE NUMBER/EXTENSION

II. WILL YOU REQUIRE THE USE OF ANY GOVERNMENT PROPERTY IN THE PERFORMANCE OF THIS PROPOSED CONTRACT?

☐ YES ☐ NO *If yes, identify on a separate page.*

III. DO YOU REQUIRE GOVERNMENT CONTRACT FINANCING TO PERFORM THIS PROPOSED CONTRACT?

☐ YES ☐ NO *If yes, identify:* ☐ ADVANCE PAYMENTS ☐ PROGRESS PAYMENTS *OR* ☐ GUARANTEED LOANS

IV. DO YOU NOW HOLD ANY CONTRACT *(or, do you have any independently financed (IR & D) projects)* FOR THE SAME OR SIMILAR WORK CALLED FOR BY THIS PROPOSED CONTRACT?

☐ YES ☐ NO *If yes, identify*

V. DOES THIS COST SUMMARY CONFORM WITH THE COST PRINCIPLES SET FORTH IN ASPR, SECTION XV *(See 3-807.2 (c) (2))*?

☐ YES ☐ NO *If no, explain on a separate page.*

Exhibit 9.2.A1 (*Continued*)

INSTRUCTIONS TO OFFERORS

1. The purpose of this form is to provide a standard format by which the offeror submits to the Government a summary of incurred and estimated cost (*and attached supporting information*) suitable for detailed review and analysis. Prior to the award of a contract resulting from this proposal the offeror shall, under the conditions stated in ASPR 3-807.3, be required to submit a Certificate of Current Cost or Pricing Data (see ASPR 3-807.3(e) and 3-807.4).

2. As part of the specific information required by this form, the offeror must submit with this form, and clearly identify as such, cost or pricing data (that is, data which is verifiable and factual and otherwise as defined in ASPR 3-807.3(e)). In addition, he must submit with this form any information reasonably required to explain the offeror's estimating process, including:

a. the judgmental factors applied and the mathematical or other methods used in the estimate including those used in projecting from known data, and

b. the contingencies used by offeror in his proposed price.

3. When attachment of supporting cost or pricing data to this form is impracticable, the data will be specifically identified and described (*with schedules as appropriate*), and made available to the contracting officer or his representative upon request.

4. The format for the "Cost Elements" is not intended as rigid requirements. These may be presented in different format with the prior approval of the contracting officer if required for more effective and efficient presentation. In all other respects this form will be completed and submitted without change.

5. By submission of this proposal, offeror, if selected for negotiation, grants to the contracting officer, or his authorized representative, the right to examine, for the purpose of verifying the cost or pricing data submitted, those books, records, documents and other supporting data which will permit adequate evaluation of such cost or pricing data, along with the computations and projections used therein. This right may be exercised in connection with any negotiations prior to contract award.

FOOTNOTES

1 Enter in this column those necessary and reasonable costs which in the judgment of the offeror will properly be incurred in the efficient performance of the contract. When any of the costs in this column have already been incurred (e.g., on a letter contract or change order), describe them on an attached supporting schedule. Identify all sales and transfers between your plants, divisions, or organizations under a common control, which are included at other than the lower of cost to the original transferror or current market price.

2 When space in addition to that available in Exhibit A is required, attach separate pages as necessary and identify in this "Reference" column the attachment in which information supporting the specific cost element may be found. No standard format is prescribed; however, the cost or pricing data must be accurate, complete and current, and the judgment factors used in projecting from the data to the estimates must be stated in sufficient detail to enable the contracting officer to evaluate the proposal. For example, provide the basis used for pricing materials such as by vendor quotations, shop estimates, or invoice prices; the reason for use of overhead rates which depart significantly from experienced rates (reduced volume, a planned major rearrangement, etc.): or justification for an increase in labor rates (anticipated wage and salary increases, etc.). Identify and explain any contingencies which are included in the proposed price, such as anticipated costs of rejects and defective work, or anticipated technical difficulties.

3 Indicate the rates used and provide an appropriate explanation. Where agreement has been reached with Government representatives on the use of forward pricing rates, describe the nature of the agreement. Provide the method of computation and application of your overhead expense, including cost breakdown and showing trends and budgetary data as necessary to provide a basis for evaluation of the reasonableness of proposed rates

4 If the total royalty cost entered here is in excess of $250 provide on a separate page (or on DD Form 783, Royalty Report) the following information on each separate item of royalty or license fee: name and address of licensor; date of license agreement; patent numbers, patent application serial numbers, or other basis on which the royalty is payable; brief description, including any part or model numbers of each contract item or component on which the royalty is payable; percentage or dollar rate of royalty per unit; unit price of contract item; number of units; and total dollar amount of royalties. In addition, if specifically requested by the contracting officer, a copy of the current license agreement and identification of applicable claims of specific patents shall be provided.

5 Provide a list of principal items within each category indicating known or anticipated source, quantity, unit price, competition obtained, and basis of establishing source and reasonableness of cost.

GPO 864-232

REFERENCES

1. P. F. OSTWALD, *Cost Estimating for Engineering and Management*, Prentice-Hall, Englewood Cliffs, NJ, 1974.
2. P. F. OSTWALD, Ed., *Manufacturing Cost Estimating*, SME, Dearborn, MI, 1980.
3. P. F. GALLAGHER, *Project Estimating by Engineering Methods*, Hayden, New York, 1965.
4. *Machining Data Handbook*, 2nd ed., Metcut Research Associates Inc., Cincinnati, OH, 1972.
5. W. A. NORDHOFF, *Machine Shop Estimating*, McGraw-Hill, New York, 1960.

BIBLIOGRAPHY

AMSTEAD, B. H., P. F. OSTWALD, and M. L. BEGEMAN, *Manufacturing Processes*, SI version, 7th ed., Wiley, New York, 1979.

BLS Handbook of Methods, U.S. Department of Labor, Bureau of Labor Statistics, Bulletin 1920, 1976.

CALDER, G., *The Principles and Techniques of Engineering Estimating*, Pergamon Press, Oxford, England, 1976.

CLARK, F. D., and A. B. LORENZONIA, *Applied Cost Engineering*, 2nd ed., Dekker, New York, 1978.

JELEN, F. C., *Cost and Optimization Engineering*, McGraw-Hill, New York, 1970.

KHARBANDA, O. P., *Process Plant and Equipment Cost Estimating*, Craftsman, Solana Beach, CA, 1979.

PATRASEU, A., *Construction Cost Engineering*, Craftsman, Solana Beach, CA, 1978.

PETERS, M. S., and K. TIMMERHAUS, *Plant Design and Economics for Chemical Engineers*, 3rd ed., McGraw-Hill, New York, 1980.

Periodical literature is also available as follows:

Building Cost File, Construction Publishing Company, Inc., New York (annual volume).

BUREAU OF LABOR STATISTICS, *Monthly Labor Review*.

BUREAU OF LABOR STATISTICS, *Current Wage Developments* (monthly).

BUREAU OF LABOR STATISTICS, *Producer Prices and Price Indexes* (monthly).

BUREAU OF LABOR STATISTICS, *Chart Book on Prices, Wages, and Productivity* (monthly).

GODFREY, R. S., Ed., *Building Construction Cost Data*, Robert Snow Means Company, Duxbury, MA (annual).

CHAPTER 9.3

Discounted Cash Flow Techniques

RAYMOND P. LUTZ

The University of Texas at Dallas

9.3.1 DEFINING ALTERNATIVE SOLUTIONS

A capital expenditure opportunity entails a cash outlay with the expectation of future benefits over 2 or more years. These expenditures include those for machinery, equipment, and buildings, as well as less tangible ones for research and advertising. The analysis of the feasibility of these expenditures requires the engineer to estimate the future costs and benefits through forecasts or predictions. The uncertainties of the future dictate that the engineer consider a set of alternative capital expenditures so that the best solution to the problem can be discovered and implemented.

To determine the worth of any alternative capital expenditure, four variables must be defined: (1) all relevant costs and the time when these expenditures will occur, (2) all benefits or revenues attributable to the expenditures and the time when they will occur, (3) the economic life of the alternative, and (4) the prevailing interest rate at the time at which costs and revenues accrue and which accommodates the risk of the uncertain future.

Determining the Economic Life

The economic life of a capital expenditure is the number of years that this item will make a positive economic contribution to the firm. The equipment used for a project will be retired when management observes (1) unsatisfactory functional characteristics, such as wear or deterioration; (2) unsatisfactory economic characteristics; (3) termination of need; and (4) obsolescence due to changes in policy regulations of technology. The economic life is estimated by considering when the preceding conditions will occur, consulting historical service records of equivalent assets when these records are available. The economic service life is used for tax depreciation. The average service life of an asset for taxes is often a function of legislative economic policy.

Developing Cash Flow Profiles

The cash flow profile should include all cash items flowing into or out of a project. Each cash flow item must be identified specifically according to the time at which this flow occurs. The major items included in an engineering economy study are:

1. First cost, P, is the sum of the costs of engineering, construction, purchasing, installation, and so on, to bring the asset into service. This cost is considered to occur at the initiation of the project, $t = 0$. However, if the project requires a number of years to construct, then the major expenditures to be made each year during construction should be identified at the time the cash flow occurs.

2. Salvage, V, is the net sum realized from the disposal of the project or asset at the termination of its useful economic life.

3. Income (revenues), other benefits, and expenses are identified according to their type of flow over time.

a. Periodic cash flow items occur at specific times, such as an overhaul of an engine every 3 years.

b. Uniform series revenues or expenses are equal periodic amounts, such as property taxes, leasing costs, sales, interest on debt, and so on.

c. Continuous flows of revenues or expenses occur continuously and uniformly over the life of the project, such as the savings realized from a new assembly technique.

To ensure that all positive and negative cash flow items are included in the analysis, and to better visualize these cash flows, it is useful to construct a *cash flow diagram* or *table.* A generalized illustration of a cash flow diagram for n periods of time is shown here. In the diagram P is the present value of the first cost; A is the value of an annuity, a uniform series of equal cash savings occurring at the end of the period; and V is the salvage value received for the asset upon termination of the project.

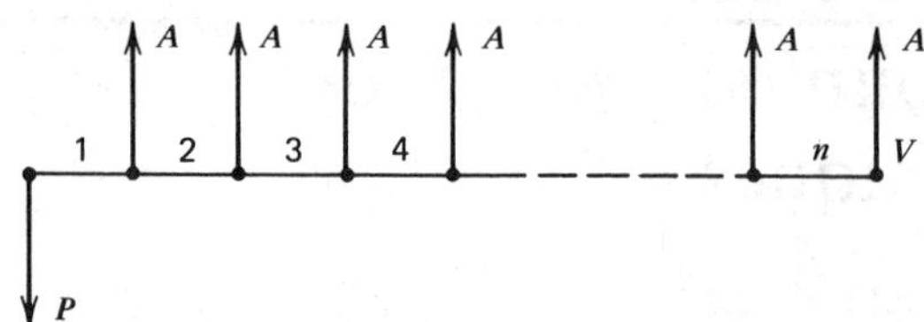

Convention in engineering economy studies dictates that all discrete cash flows are considered to occur at the end of the period, which is normally at the end of the year. However, the period may be quarterly, monthly, or even daily.

To describe a specific cash flow, consider the following discrete items, with the cash flow developed using a tabular format.

End of Year	Receipts	Disbursements	F_{jt}
0	$0	-$6000	-$6000
1	2000	- 500	1500
2	3500	- 1000	2500
3	5000	- 2000	3000
4	5000	- 1000	4000

In this table F_{jt} = net cash flow for investment j at time t. If $F_{jt} < 0$, then F_{jt} represents a net cash disbursement or expense. If $f_{jt} > 0$, then F_{jt} represents a net gain or revenue.

Selecting the Interest Rate

When analyzing the feasibility of any investment, two principles must be observed. First, the value of a sum of money is a function of the time span between the base point of reference and the date an expenditure or the receipt of revenue will occur. For example, the value of $100 one year from today is not the same as $100 today. Consequently, money has a *time value*, which is measured by the interest rate. Second, because of this time value of money, all comparisons of receipts and disbursements must be made at a common specific time. Compound interest factors (Section 9.3.2) provide the means of shifting cash flows to this common time.

The minimum interest rate a company should earn on its invested capital is determined by the capital structure of the firm. The firm must earn an adequate return both to support the long-term debt and to compensate the stockholders adequately for their equity investment. This minimum interest rate, C_c, is calculated using the capital asset pricing model and is often referred to as the weighted cost of capital.

$$C_c = C_d + C_e$$

where C_c = weighted cost of capital, %
C_d = weighted cost of long-term debt, %
C_e = weighted cost of equity, %

The weighted cost of long-term debt, C_d, can be determined by

$$C_d = k_i \times W_{ltd}$$

where k_i is the return necessary to support debt, stated as a percentage, and W_{ltd} is the percentage of the firm's capital structure represented by long-term debt. The values of k_i and W_{ltd} are determined by examining the financial statements of the firm. The value of k_i would represent the average interest rate charged on the firm's long-term obligations.

$$W_{ltd} = \frac{\text{long-term debt}}{\text{long-term debt + equity}} = \frac{LTD}{LTD + E}$$

The weighted cost of equity, C_e, would be determined by

$$C_e = k_e \times W_e$$

where k_e is the return necessary to support the firm's equity, stated as a percentage, and W_e is the percentage of the firm's capital structure represented by equity. As would be expected, W_e is the complement of W_{ltd} and would be calculated by

$$W_e = \frac{\text{equity}}{\text{long-term debt} + \text{equity}} = \frac{E}{LTD + E}$$

The before-tax rate of return required to support the firm's equity structure would reflect both the risk attributable to equity instruments and the risk associated with the firm's particular business. The rate of return required to support equity, k_e, is

$$k_e = \frac{(\text{risk-free rate}) + (\text{risk premium})}{1 - \text{tax rate}}$$

$$k_e = \frac{r^* + \beta(R_m - r^*)}{1 - t}$$

where r^* = risk-free rate of return
R_m = risk attributable to the general equity market
t = tax rate
β = systematic risk of a stock due to underlying movements in security prices

The value of the risk-free rate of return, r^*, can be estimated as being equal to the interest rate paid on U.S. Treasury Bills or guaranteed savings accounts. The interest rate equivalent to the general equity market, R_m, was found to be approximately 9% by Fisher and Lorie in the *Journal of Business*, July 1968. The tax rate, t, can be calculated from data obtained from the firm's annual financial reports.

$$t = \frac{\text{taxes paid}}{\text{profit before taxes}}$$

The final term to determine, β, reflects systematic risk of a stock due to underlying movements in security prices. Values of β can be obtained for the majority of firms traded on major stock exchanges from references such as *Value Line Investment Survey.*

As an example of determining the minimum rate of return that must be earned to maintain the integrity of the capital structure of the firm, consider Texas Instruments for 1977.

R_m = 9% from Fisher and Lorie, *Journal of Business*, July 1968
r^* = 7% assumed risk-free interest rate
t = 44.7% from firm's financial reports
β = 1.2 from *Value Line Investment Survey*
E = \$714,900,000
LTD = \$29,700,000

$$k_e = \frac{r^* + \beta(R_m - r^*)}{1 - t}$$

$$= \frac{7 + 1.2(9 - 7)}{1 - 0.447}$$

$$k_e = 17\%$$

$$W_e = \frac{E}{LTD + E}$$

$$= \frac{\$714.90}{\$29.70 + \$714.90}$$

$$W_e = 96\%$$

$$C_e = k_e \times W_e$$

$$= (0.17) \times (0.96)$$

$$C_e = 16\%$$

$$k_i = \frac{\text{interest paid on long-term debt}}{\text{long-term debt}}$$

$$= \frac{\$2{,}700{,}000}{\$29{,}700{,}000}$$

$$k_i = 9\%$$

$$W_{ltd} = \frac{LTD}{LTD + E}$$

$$= \frac{\$29.70}{\$29.70 + \$714.90}$$

$$W_{ltd} = 4\%$$

$$C_d = k_i \times W_{ltd}$$

$$= (0.09) \times (0.04)$$

$$C_d = 3.6\%$$

$$C_c = C_e + C_d$$

$$= 16.0 + 3.6$$

$$C_c = 19.6\%$$

The preceding calculations will determine the *minimum* interest rate a company should earn on its invested capital. However, a firm may adopt differing interest rates in order to evaluate the feasibility of specific capital expenditures. These interest rates may vary from year to year according to economic conditions, demand for funds, or management policy related to the risk of the endeavor. The following set of before-tax minimum project interest rates established by one firm is a typical example.

1. Safety, quality, or legal requirement: Return on funds projection is not required. The *immediate* need for the project and the lack or inadequacy of alternative solutions must be clearly demonstrated.

2. Increased profit

a. Cost reduction: Projects with a rate of return of at least 20% over a 10 year period are worthy of consideration.

b. Increase production capacity for an existing product: Projects for a proven product where the risk of production termination for that product is small, the rate of return should be at least 20% over a 10 year project life.

c. Provide facilities to manufacture and distribute a new product or product line: Because of greater risk, the rate of return should be at least 40% over a maximum 10 year life.

Note that these rates of return would apply only to expenditures within the United States and Canada. Profit increasing capital projects in other countries in categories 2a and 2b should offer a rate of return of at least 25% for the maximum 10 year period. Projects associated with new products outside the United States and Canada must offer at least a 50% rate of return over the maximum 10-year period.

The interest rate used to evaluate capital investment alternatives should be greater than the minimum rate of return. However, the specific rate to be used in capital budgeting and project feasibility analysis must be a management decision, depending on the type of activity, the risk, and, the opportunity cost.

9.3.2 USING INTEREST FACTORS

Interest formulas provide the mechanism for converting a cash flow at one specific time into an equivalent cash flow at another time. The tables on pages 9.3.18 to 9.3.25 provide this information for discrete compound interest factors; continuous compounding interest factors; geometric series factors; and continuous compounding, continuous flow factors.

Simple Interest

The interest payment each year is found by multiplying the interest rate times the principle, $I = Pi$. After any n time periods, the accumulated value of money owed under simple interest, F_n, would be

$$F_n = P(1 + in)$$

For example, \$100 invested now at 9% simple interest for 8 years would yield

$$F_8 = \$100[1 + 0.09(8)] = \$172$$

Simple interest is rarely used in engineering economy analyses.

Compound Interest

The interest payment each year, or each period, is found by multiplying the interest rate by the accumulated value of money, both principle and interest.

End of Period (EOP)	Accumulated EOP Value or Amount Owed (1)	Interest for Period (2)	Amount Owed or Value Accumulated Next Period (3) = (1) + (2)
0	P	Pi	$P + Pi = P(1 + i)$
1	$P(1 + i)^1$	$[P(1 + i)^1]i$	$P(1 + i) + P(1 + i)\,i = P(1 + i)^2$
2	$P(1 + i)^2$	$[P(1 + i)^2]i$	$P(1 + i)^2 + P(1 + i)^2\,i = P(1 + i)^3$
3	$P(1 + i)^3$	$[P(1 + i)^3]i$	$P(1 + i)^3 + P(1 + i)^3\,i = P(1 + i)^4$
...	...	...	...

Consequently, the value for an amount P invested for n periods at i rate of interest using compound interest calculations would be

$$F_n = P(1 + i)^n$$

For example, \$100 invested now at 9% compound interest for 8 years would yield F_8 = \$100 $(1 + 0.09)^8$ = 100(1.9926) = \$199. Compound interest is the basis for practically all monetary transactions.

Nominal and Effective Interest Rates

For many engineering economy studies it is appropriate to consider interest periods of one year. However, financial agreements can call for interest to be compounded or paid more frequently, say quarterly, monthly, or even daily. Interest rates associated with compounding more frequent than annually are usually stated as "8% compounded quarterly," for example, in the case of quarterly interest periods. The *nominal interest rate* is expressed as an annual rate, without considering the effect of any compounding. It is obtained by multiplying the effective rate per interest period times the number of interest periods per year. The *effective annual interest rate* is then the true or actual annual interest rate, taking into account the effect of compounding during the year.

The nominal interest rate per year, r, is

$$r = im$$

where i is the effective interest rate per compounding period and m is the number of compounding periods per year.

The effective interest rate per year, i_a, is

$$i_a = (1 + i)^m - 1 \text{ when } m < \infty$$

For example, when the effective interest rate is 2%/month, the nominal interest rate per year is

$$r = (0.02)\,(12) = 24\%$$

and the effective interest rate per year is

$$i_a = (1 + 0.02)^{12} - 1 = 26.8\%$$

Compound Interest Factors: Discrete Cash Flow, Discrete Compounding

The notation used in this chapter is

i = effective interest rate per interest period
n = number of compounding periods
A = end-of-period cash flows (or equivalent end-of-period values) in a uniform series continuing for a specified number of periods (The letter A implies annual or annuity.)
F = future sum of money (The letter F implies future or equivalent future value.)
P = present sum of money (The letter P implies present or equivalent present value.)

The compound interest factors described in this section are used for discrete cash flows compounded discretely at the end of each interest period. All of these factors can be found in Exhibit 9.3.1, including their algebraic and functional formats. Numerical values for each factor for a number of interest rates can be found in tables in the sources listed in the bibliography at the end of the chapter. As mentioned previously, compound calculations form the basis for analyzing capital expenditure proposals.

Compound Amount Factor (Single Payment)

This factor finds the equivalent future worth, F, of a present investment, P, held for n periods at i rate of interest. For example, what is the value in 9 years of \$1200 invested now at 10% interest?

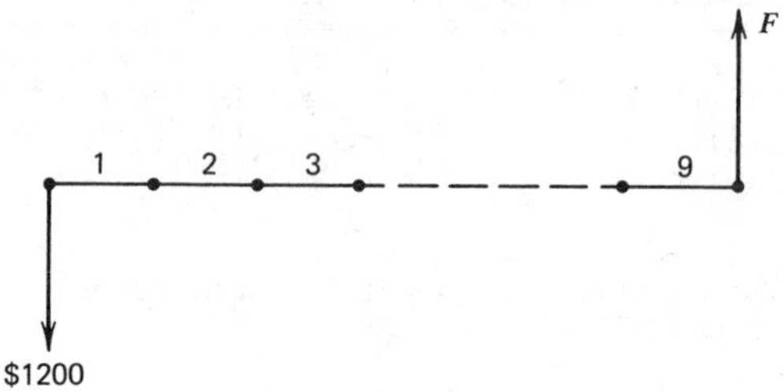

Exhibit 9.3.1 Compound Interest Factors: Discrete Cash Flow, Discrete Compounding

To Find	Given	Name of Factor	Format: Algebraic	Format: Functional
F	P	Compound amount factor (single payment)	$(1+i)^n$	$(F/P, i\%, N)$
P	F	Present worth factor (single payment)	$(1+i)^{-n}$	$(P/F, i\%, N)$
F	A	Compound amount factor (uniform series)	$\frac{(1+i)^n - 1}{i}$	$(F/A, i\%, N)$
A	F	Sinking fund factor	$\frac{i}{(1+i)^n - 1}$	$(A/F, i\%, N)$
A	P	Capital recovery factor	$\frac{i(1+i)^n}{(1+i)^n - 1}$	$(A/P, i\%, N)$
P	A	Present worth factor (uniform series)	$\frac{(1+i)^n - 1}{i(1+i)^n}$	$(P/A, i\%, N)$
A	G	Arithmetic gradient conversion factor (to uniform series)	$\frac{(1+i)^n - (1+ni)}{i[(1+i)^n - 1]}$	$(A/G, i\%, N)$
P	G	Arithmetic gradient conversion factor (to present value)	$\frac{1 - (1+ni)(1+i)^{-n}}{i^2}$	$(P/G, i\%, N)$

Using the algebraic format,

$$\begin{aligned} F &= P(1+i)^n \\ &= \$1200(1+0.10)^9 \\ &= \$1200(2.3579) \\ &= \$2829 \end{aligned}$$

Using the functional format,

$$\begin{aligned} F &= P(F/P, i\%, N) \\ &= \$1200(F/P, 10\%, 9) \\ &= \$1200(2.3579) \\ &= \$2829 \end{aligned}$$

Throughout the remainder of this chapter, only the functional format will be used. The reader can quickly obtain the algebraic format, if needed, from Exhibit 9.3.1.

Present Worth Factor (Single Payment)

This factor finds the equivalent present value, P, of a single future cash flow, F, occurring at n periods in the future when the interest rate is i percent per period. (Note that this factor is the reciprocal of the compound amount factor–single payment.) For example, what amount would you have to invest now to yield \$2829 in 9 years if the interest rate per year were 10%?

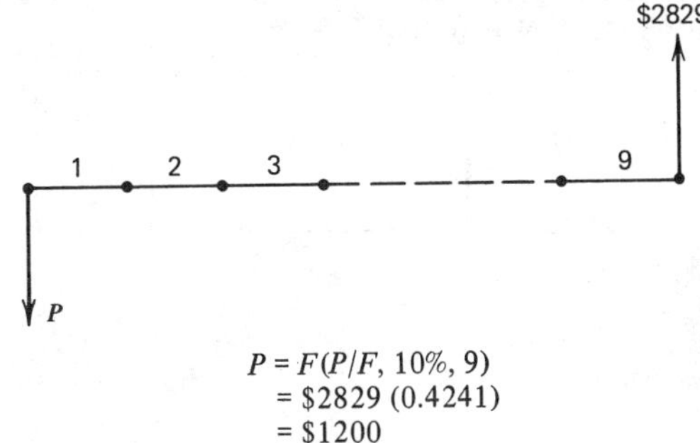

$$\begin{aligned} P &= F(P/F, 10\%, 9) \\ &= \$2829\ (0.4241) \\ &= \$1200 \end{aligned}$$

Compound Amount Factor (Uniform Series)

This factor finds the equivalent future value, F, of the accumulation of a uniform series of equal annual payments, A, occurring over n periods at i rate of interest per period. For example, what would be the future worth of an annual year-end cash flow of \$800 for 6 years at 12% interest per year?

F
1 2 3 4 5 6
\$800 \$800 \$800 \$800 \$800 \$800

$$\begin{aligned} F &= A(F/A, 12\%, 6) \\ &= \$800\ (12.2997) \\ &= \$9840 \end{aligned}$$

Sinking Fund Factor

This factor determines how much must be deposited each period in a uniform series, A, for n periods at i percent interest per period to yield a *specified* future sum, F. For example, if a \$1.2

million bond issue is to be retired at the end of 20 years, how much must be deposited annually into a sinking fund at 7% interest per year?

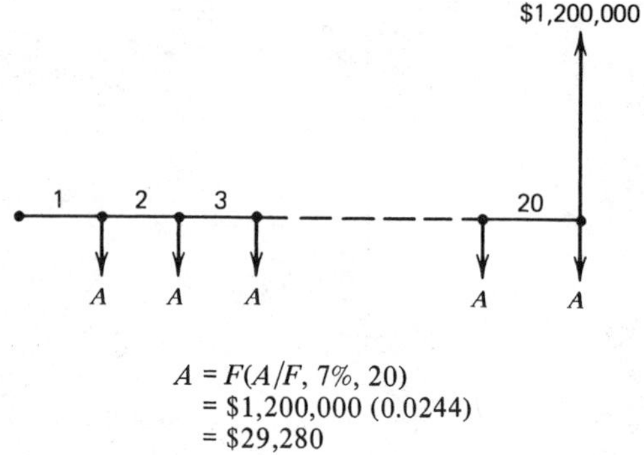

$$\begin{aligned} A &= F(A/F, 7\%, 20) \\ &= \$1{,}200{,}000\ (0.0244) \\ &= \$29{,}280 \end{aligned}$$

This factor was historically used to find the required annual payments that must be made into a "sinking fund" to retire a bond issue by a particular date.

Capital Recovery Factor

This factor finds an annuity, or uniform series of payments, over n periods at i percent interest per period that is equivalent to a present value, P. For example, what savings in annual manufacturing costs over an 8 year period would justify the purchase of a $120,000 machine if the firm's minimum attractive rate of return (MARR) were 25%?

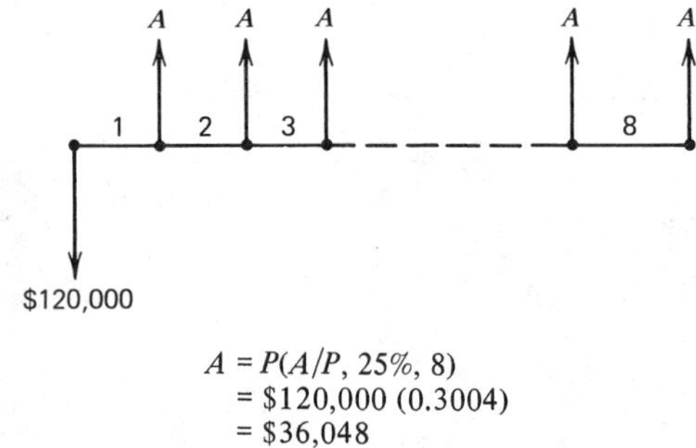

$$\begin{aligned} A &= P(A/P, 25\%, 8) \\ &= \$120{,}000\ (0.3004) \\ &= \$36{,}048 \end{aligned}$$

This problem of finding the revenue that must be generated each period to justify a capital expenditure is one of the most common facing the engineer.

Note that the uniform series of cash flows of $36,048 in this example would accomplish two purposes. First, it would return the initial investment of $120,000 to the firm. Second, it would earn a rate of return of 25%/year on the unrecovered invested capital each year, accounting for the remaining $168,384 in total earnings over the 8 years.

$$\begin{aligned} 8 \times \$36{,}048 = {} & \$288{,}384 \text{ total cash flow} \\ & \underline{-\$120{,}000} \text{ return of capital} \\ & \$168{,}384 \text{ interest earned on unrecovered invested capital} \end{aligned}$$

Present Worth Factor (Uniform Series)

This factor finds the equivalent present value, P, of a series of end-of-period payments, A, for n periods at i percent interest per period. For example, a donor has offered to give the Hospital Authority a new wing for the treatment of allergy cases. However, since the operation and maintenance of the existing facility requires all of the funds available under the Authority's taxing limits, how much additional endowment would be required to provide $50,000/year over the 30 year life of the structure if the endowment will be invested at 9% interest?

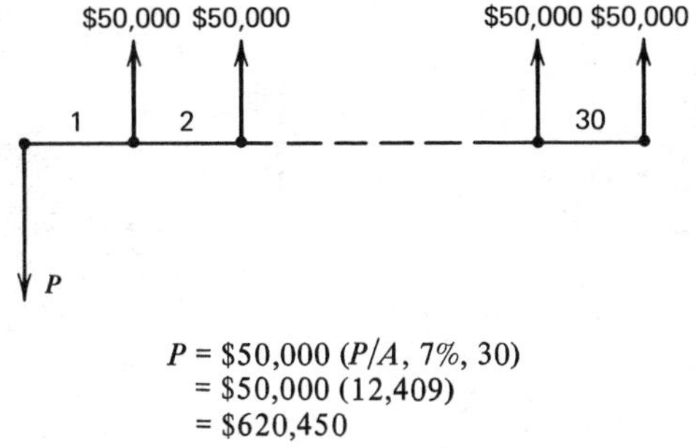

$$\begin{aligned} P &= \$50{,}000\ (P/A,\ 7\%,\ 30) \\ &= \$50{,}000\ (12{,}409) \\ &= \$620{,}450 \end{aligned}$$

Arithmetic Gradient Conversion Factor (to Uniform Series)

Many times annual payments do not occur in equal amounts. Inflation causes annual increases in operating costs, and maintenance costs often increase with the age of the equipment. If a series of payments increases by an equal amount or gradient, G, each year, then a special compound interest factor can be used to reduce the gradient series to an equivalent equal-payment series. The following illustration shows a four-period gradient series that increases by G each period.

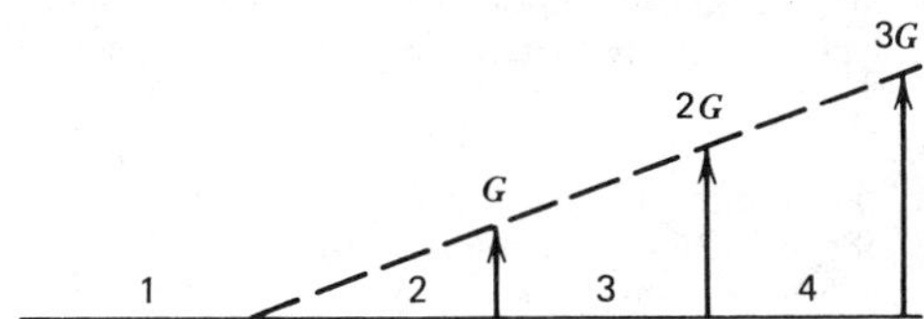

The arithmetic gradient conversion factor (to uniform series) is used when it is necessary to convert a gradient series into a uniform series of equal payments. For example, what would be the equal annual series, A, that would have the same net present value (i.e., be equivalent) at 20% interest per year to a 5 year gradient series that started at $1000 the first year and increased $150 every year thereafter?

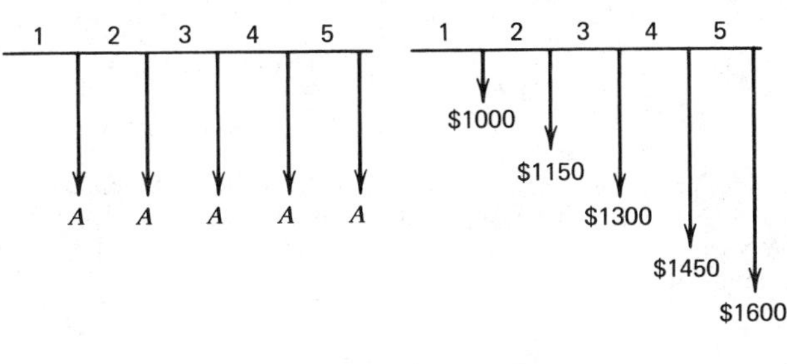

$$\begin{aligned} A &= A_g + G(A/G,\ 20\%,\ 5) \\ &= \$1000 + \$150\ (1.6405) \\ &= \$1246 \end{aligned}$$

where A_g is the uniform base value of the gradient series. In the previous four-period gradient series illustration, $A_g = 0$.

Arithmetic Gradient Conversion Factor (to Present Value)

This factor converts a series of cash amounts increasing by a gradient value, G, each period to an equivalent present value at i percent interest per period. For example, a machine will require $1000 in maintenance the first year of its 5 year operating life. Further, the cost of maintenance will increase by $150 each year. What is the present worth of this series of maintenance costs if the firm's MARR is 20%?

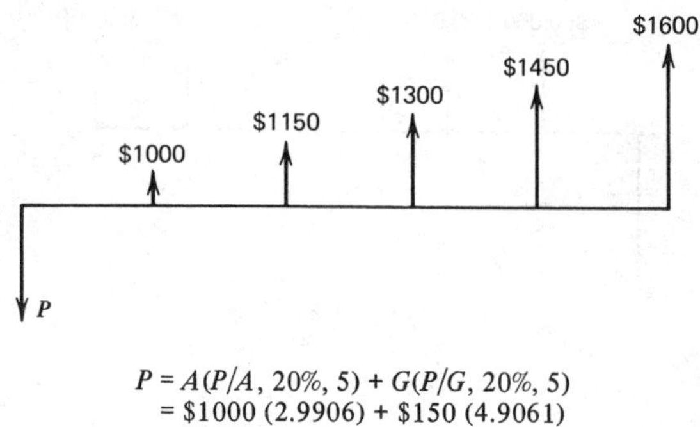

$$\begin{aligned} P &= A(P/A, 20\%, 5) + G(P/G, 20\%, 5) \\ &= \$1000\,(2.9906) + \$150\,(4.9061) \\ &= \$3727 \end{aligned}$$

Compound Interest Factors: Discrete Cash Flow, Continuous Compounding

Monetary institutions and firms alike strive to keep their funds working at all times. Techniques of cash management, such as electronic transfer of funds and the immediate profitable use of cash when available, as well as increasing competitive pressures, have resulted in the shortening of compounding periods until a sizable number of institutions are in effect using continuous compounding. In the concept of discrete cash flows with continuous compounding, it is assumed that the cash flows occur once per year, but that compounding is continuous throughout the year. Thus, if

$$\begin{aligned} r &= \text{nominal interest rate per year} \\ M &= \text{number of compounding periods per year} \\ n &= \text{number of years,} \end{aligned}$$

then at the end of 1 year one unit of principal will equal

$$\left[1 + \left(\frac{r}{m}\right)\right]^M \tag{1}$$

Let $k = M/r$, equation 1 then becomes

$$\left[1 + \frac{1}{k}\right]^{rk} = \left[\left(1 + \frac{1}{k}\right)^k\right]^r \tag{2}$$

The limit of $(1 + 1/k)^k$ as k approaches infinity is e. Thus the equation 2 can be written as e^r, and the single payment continuous compounding amount factor at r percent nominal annual interest rate for N years is e^{rN}. Also, since e^{rN} (for continuous compounding) corresponds to $(1 + i)^N$ for discrete compounding,

$$e^r = 1 + i$$

or

$$i = e^r - 1$$

By use of this relationship, the compound interest factors for discrete cash flows compounded continuously shown in Exhibit 9.3.2 can be derived from the discrete compounding factors in Exhibit 9.3.1.

Continuous Compounding Compound Amount Factor (Single Payment)

This factor is used to find the equivalent future worth, F, of a present value, P, when interest is continuously compounded at the nominal annual rate of $r\%$. For example, consider the problem of finding the future worth in 6 years of \$5000 invested now at 9% nominal interest rate compounded continuously.

$$F = Pe^{rn}$$

Exhibit 9.3.2 Compound Interest Factors: Discrete Cash Flow, Continuous Compounding

			Format	
To Find	Given	Name of Factor	Algebraic	Functional
F	P	Continuous compounding compound amount factor (single payment)	e^{rn}	$(F/P, r\%, N)$
P	F	Continuous compounding present worth factor (single payment)	e^{-rn}	$(P/F, r\%, N)$
A	F	Continuous compounding sinking fund factor	$\frac{e^r - 1}{e^{rn} - 1}$	$(A/F, r\%, N)$
A	P	Continuous compounding capital recovery factor	$\frac{e^{rn}(e^r - 1)}{e^{rn} - 1}$	$(A/P, r\%, N)$
F	A	Continuous compounding compound amount factor (uniform series)	$\frac{e^{rn} - 1}{e^r - 1}$	$(F/A, r\%, N)$
P	A	Continuous compounding present worth factor (uniform series)	$\frac{e^{rn} - 1}{e^{rn}(e^r - 1)}$	$(P/A, r\%, N)$
P	G	Continuous compounding arithmetic gradient conversion factor (to present value)	$\frac{e^{rn} - 1 - n(e^r - 1)}{e^{rn}(e^r - 1)^2}$	$(P/A, r\%, N)$
A	G	Continuous compounding arithmetic gradient conversion factor (to uniform series)	$\frac{1}{e^r - 1} - \frac{n}{e^{rn} - 1}$	$(A/G, r\%, N)$
P	A_1, c	Continuous compounding geometric gradient conversion factor (to present value)	$\frac{1 - e^{(c-r)n}}{e^r - e^c}$	$(P/A, r\%, c\%, N)$ $r \neq c$
F	A_1, c	Continuous compounding geometric gradient conversion factor (to uniform series)	$\frac{e^{rn} - e^{cn}}{e^r - e^c}$	$(F/A, r\%, c\%, N)$ $r \neq c$

or equivalently

$$\begin{aligned} F &= P(F/P, r\%, N) \\ &= P(F/P, 9\%, 6) \\ &= \$5000\ (1.7160) \\ &= \$8580 \end{aligned}$$

Note that the only difference with continuous compounding and discrete amounts in finding equivalent values of F, P, A, and G is the interest factor used (r, the nominal annual interest rate). Consequently, to solve discrete cash flow continuous compounding problems, use the same procedures illustrated for discrete compounding.

Continuous Compounding Geometric Gradient Conversion Factor (to Present Value)

When conditions such as inflation cause a gradient that increases at a fixed percent per period, c, then geometric gradient conversion factors should be used. For example, find the present worth of a series of costs that increase 10%/year from the initial first-year cost of \$1000 for 5 years when the firm's MARR is 15%.

$$\begin{aligned} P &= A(P/A, r\%, c\%, N) \\ &= \$1000\ (P/A, 15\%, 10\%, 5) \\ &= \$1000\ (3.9037) \\ &= \$3904 \end{aligned}$$

Compound Interest Factors: Continuous Uniform Cash Flows, Continuous Compounding

Many cash flows that engineers must consider can be assumed to occur continuously, such as accounts receivable, the cost savings resulting from productivity improvements, or the costs of carrying inventories. The following cash flow diagrams serve to illustrate the differences between discrete and continuous cash flows.

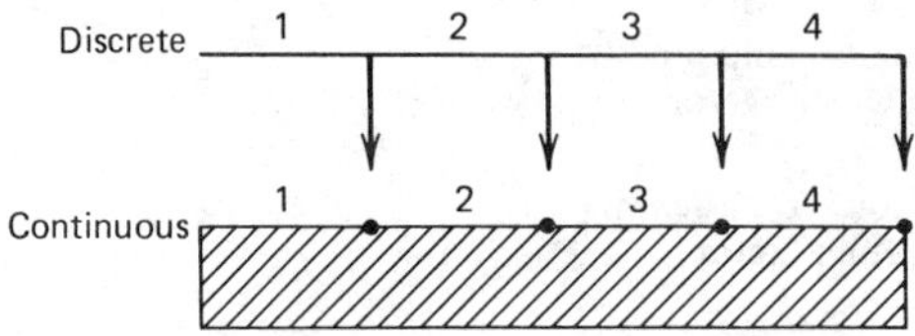

Again, as with discrete cash flows, to solve for one variable given another, it is only necessary to select the proper interest factor for continuous cash flow, continuous compounding. A listing of these factors can be found in Exhibit 9.3.3.

As an example, consider the equivalent future worth of a uniform series of continuous cash flows totaling \$2000/year for 10 years compounded continuously at 15% nominal annual rate of interest.

$$\begin{aligned} F &= \bar{A}(F/\bar{A}, r\%, N) \\ &= \$2000\,(F/\bar{A}, 15\%, 10) \\ &= \$2000\,(23.2113) \\ &= \$46{,}423 \end{aligned}$$

9.3.3 COMPARING ALTERNATIVES

In comparing alternatives to meet a need or an objective, plans (1) should provide the same quality and quantity (or level) of service and (2) should provide that service over the same period of time. Competing plans should be alternative ways to accomplish the same end. Any differences in expected revenue or other benefits must be credited to the plan providing the additional services. The analysis is concerned only with the *differences* in the cash flows between the alternatives.

Present Worth

The present worth method compares all of a project's estimated expenditures to all of its estimated revenues and other benefits at a reference time called the "present" ($t = 0$). For a particular inter-

Exhibit 9.3.3 Compound Interest Factors: Continuous, Uniform Cash Flows, Continuous Compounding

To Find	Given	Name of Factor	Format	
			Algebraic	Functional
$\bar{A}$	F	Continuous compounding sinking fund factor (continuous, uniform payments)	$\dfrac{r}{e^{rn}-1}$	$(\bar{A}/F, r\%, N)$
$\bar{A}$	P	Continuous compounding capital recovery factor (continuous, uniform payments)	$\dfrac{re^{rn}}{e^{rn}-1}$	$(\bar{A}/P, r\%, N)$
F	$\bar{A}$	Continuous compounding compound amount factor (continuous, uniform payments)	$\dfrac{e^{rn}-1}{r}$	$(F/\bar{A}, r\%, N)$
P	$\bar{A}$	Continuous compounding present worth factor (continuous, uniform payments)	$\dfrac{e^{rn}-1}{re^{rn}}$	$(P/\bar{A}, r\%, N)$

est rate, if the present value of the revenues and other benefits exceeds the present value of the expenses, the project is considered acceptable. The present worth of alternative j with cash flows that last for n periods of time at i percent interest per period is

$$PW(i)_j = \sum_{t=0}^{N} F_{jt}(P/F, i\%, t)$$

If two or more alternatives are being compared, the alternative with the greatest present worth or net present value is recommended. To compare alternatives fairly using the present worth method, it is necessary that all alternatives have a common retirement date.

For example, two pieces of equipment are being considered by a hospital to perform a particular service. Brand A will cost $30,000 and will have an annual operating and maintenance cost of $5000 over its 8 year economic life, with a salvage value of $3000. Brand B will cost $15,000, will have an annual operating and maintenance cost of $8000 over the first 3 years and $10,000 over the last 3 years of its economic life, and will have a negligible salvage value. Which brand of equipment would you recommend, using the present worth comparison with an interest rate of 10%/year?

Cash Flow Item	Brand A	Brand B
First cost	–$30,000	–$15,000
Operating and maintenance		
A: –$5000 (P/A, 10%, 6)		
= –$5000 (4.3553)	–21,776	
B: (1 – 3) – $8000 (P/A, 10%, 3)		
= –$8000 (2.4869)		–19,895
(4 – 6) – $10,000 [(P/A, 10%, 3)		
(P/F, 10%, 3)]		
= –$10,000 [(2.4869) (0.7513)]		–18,684
Salvage value		
A: +$3000 (P/F, 10%, 6)		
= +$3000 (0.5645)	+1,694	
B: 0		0
Present worth	–$50,082	–$53,579

Brand A would be the recommended alternative since the present worth of its total cost is smaller, or its present worth is greater.

Annual Worth

The annual worth method converts all cash flows to an equivalent uniform series of equal annual payments. As in the present worth method, if the annual worth of the revenues is greater than the annual worth of the costs for the specified interest rate, then the project is acceptable. The annual worth of alternative j at i percent rate of interest per period, which lasts for n periods, is

$$AW(i)_j = PW(i)_j(A/P, i\%, N)$$

It is usually necessary to calculate the present worth of all cash flows, $PW(i)$, first, since these cash flows are rarely a uniform series that can be summed directly to find $AW(i)$.

If two or more alternatives are being compared, the alternative with the greatest annual worth (cash receipts are positive and disbursements are negative) is the recommended alternative.

If you must compare alternatives with differing economic lives, the annual worth method is preferred when the "repeatability assumption" is valid for the analysis. (See Chapter 9.4 for a detailed discussion of the comparison of alternatives with unequal service lives.) If this assumption is valid, then the annual worth at the time of renewal of the asset is exactly the same as before. Therefore you are actually comparing the annual worth of two infinite series.

Using the example from the preceding subchapter, the following calculations are made.

Cash Flow Item	Brand A	Brand B
First cost		
A: –$30,000 (A/P, 10%, 6)		
= –$30,000 (0.2296)	–$6,888	

Cash Flow Item	Brand A	Brand B
B: $-\$15,000$ $(A/P, 10\%, 6)$		
$= -15,000$ (0.2296)		$-\$3,444$
Operating and maintenance		
A	$-5,000$	
B: (1 – 3) – \$8000 $(P/A, 10\%, 3)$		
$(A/P, 10\%, 6)$		
$= -8000$ (2.4869) (0.2296)		$-4,568$
(4 – 6) – \$10,000 $(P/A, 10\%, 3)$		
$(P/F, 10\%, 3)$ $(A/P, 10\%, 6)$		
$= -\$10,000$ (2.4869) (0.7513)		
(0.2296)		$-4,290$
Salvage value		
A: +\$3000 $(A/F, 10\%, 6)$		
= +\$3000 (0.1296)	389	
B		0
Annual worth	$-\$11,499$	$-\$12,302$

Brand A is the recommended alternative since it has the greatest annual worth.

Note that $AW(i)_j(P/A, i\%, N) = PW(i)_j$. Also note that either of these methods of comparison recommends the same alternative.

A common method of finding the annual worth of an alternative is

$$AW(i)_j = R_j - FC(A/P, i\%, N) - (O \;\&\; M) + V(A/F, i\%, N)$$

where R = revenues per year (uniform series)
FC = first cost
V = salvage value
$O \;\&\; M$ = operating and maintenance cost (uniform series)

This is exactly what was done in the preceding tabular format.

Future Worth

The future worth method is comparable to the present worth method, except that the comparison between the project's estimated expenditures and benefits occurs at a reference time called the "future" ($t = F$). As was the case in present worth analysis, in future worth analysis a project is acceptable at a particular interest rate if the future value of the revenues and other benefits exceeds the future value of the expenses. Likewise, the preferred alternative, given equal future benefits, would be the alternative with the lowest future costs.

For example, if the estimated future worth of a stream of revenues and other benefits from proposed materials handling equipment at the end of 10 years is \$1,200,000, should the new equipment be purchased? The firm's MARR before taxes is 25%. The initial cost would be \$125,000, and the annual maintenance cost would be \$750/year for the 10 year life.

$$\begin{aligned} FW(25\%) &= \$125,000\ (F/P, 25\%, 10) + \$750\ (F/A, 25\%, 10) \\ &= \$125,000\ (9.313) + \$750\ (33.253) \\ &= \$1,164,125 + \$25,940 \\ &= \$1,189,065 \end{aligned}$$

Since the future worth of the benefits exceeds the future worth of the costs, the purchase of the equipment would be justified.

Rate of Return

The rate of return method finds the interest rate that equates the cash flows of receipts and disbursements. That is, an alternative's rate of return is the interest rate at which the present worth of the cash flows is equal to 0.

$$0 = \sum_{t=0}^{N} F_{jt}(1 + i)^{-k}$$

Thus, for alternative j the rate of return is the "break-even" interest rate between incomes and expenses.

For example, what rate of return would be earned from a \$64,000 investment in a testing device if the savings were to be \$16,000/year for 8 years?

$$-\$64{,}000 = \$16{,}000\,(P/A, i\%, 8)$$
$$i = 18.62\%$$

When more than one uniform cash flow is involved, an iterative method is necessary to solve for the rate of return. See Newnan[1] for a discussion of these calculations and the potential problem of multiple rates of return.

If it is necessary to compare mutually exclusive alternatives using the rate of return method, Chapter 9.4 should be consulted regarding *incremental analysis* for mutually exclusive alternatives.

A more complex example would consider the investment yielding the 4 year stream of cash flows illustrated in the subchapter on cash flow profiles. What rate of return would equate the –\$6000 disbursement at $t = 0$ with the positive cash flows of \$1500, \$2500, \$3000, and \$4000 at the end of years 1, 2, 3, and 4, respectively?

$$0 = -5000 + \$1500\,(P/F, i\%, 1) + \$2500\,(P/F, i\%, 2) + \$3000\,(P/F, i\%, 3) + \$4000\,(P/F, i\%, 4)$$

The rate of return for this set of cash flows is approximately 25%/year and would be determined using an iterative trial-and-error solution method.

Payback Period

The payback period method determines the length of time required to recover the initial investment, or first cost, at a zero rate of interest. This is the most common general definition of payback period. Using this definition, the payback period for alternative j is

$$PP_j = \frac{\text{first cost of project}}{\text{uniform } net \text{ benefits per period}}$$

$$PP_j = \frac{FC_j}{R - D}$$

where R equals the equivalent uniform benefits per period and D equals the equivalent uniform costs per period. For example, find the payback period for a \$10,000 investment that will return net uniform benefits of \$1250/year.

$$PP = \frac{\$10{,}000}{\$1250} = 8 \text{ years}$$

The alternative with the shortest payback period would be the preferred alternative.

The payback period method has two serious shortcomings. First, it does not consider the timing of cash flows. It weighs cash flows 10 years from now the same as cash flows occurring today. Second, it ignores the duration of the cash flows. These weaknesses in the payback period method render it less desirable than the other measures of merit presented in this section.

Payback period is still widely used for comparing alternatives. Its use precludes the necessity of specifying an interest rate or performing interest rate calculations. However, its widest use is as a surrogate measure of risk.

Benefit-Cost Analysis

The benefit-cost method is often utilized to determine the feasibility of public sector expenditures. The benefit-cost criterion for the jth alternative, B/C_j, can be expressed as

$$B/C_j = \frac{\Sigma_{t=1}^{N} B_{jt}(1+i)^{-t}}{\Sigma_{t=1}^{N} C_{jt}(1+i)^{-t}}$$

where B_{jt} equals the public benefits accruing to alternative project j during year t and C_{jt} equals the governmental costs associated with alternative project j during year t. A project is deemed to be acceptable if $B/C_j \geqslant 1.0$, that is, if the project's benefits equal or exceed its costs.

The calculation of the benefit-cost ratio is not different in principle from the other methods of comparing alternatives. However, determining the monetary value of the costs and benefits associated with projects in the public sector is difficult. This difficulty arises because of the often subjective nature of the costs and benefits and because they often occur far into the future. A detailed discussion of the problems of defining public costs and benefits and of their use in calculating the benefit-cost ratio can be found in other sources, such as Smith[2] and White et al.[3]

An example of the use of the benefit-cost method is as follows: A government bridge project requires an initial investment of \$10 million and operation and maintenance costs of \$250,000/year for the 20 year life of the project. The annual user benefits of \$1,950,000/year are estimated to arise from savings in travel distance and time. At an annual interest rate of 7%, the benefit-cost ratio is

$$B/C = \frac{\$1{,}950{,}000\ (P/A, 7\%, 20)}{\$10{,}000{,}000 + \$250{,}000\ (P/A, 7\%, 20)}$$

$$= \frac{\$1{,}950{,}000\ (10.5940)}{\$10{,}000{,}000 + \$250{,}000\ (10.5940)}$$

$$= \frac{\$20{,}658{,}300}{\$10{,}000{,}000 + \$2{,}648{,}500}$$

$$= 1.63$$

Thus the project benefits would exceed its costs.

9.3.4 NONUNIFORM SERVICE LIVES

Engineers are often required to purchase or install a number of items at one time in order to provide a service or manufacture a product. The group of such items is called a "vintage group." For example, you may install 2000 telephone poles or 50 automobiles. Assuming that all these items must be in working order in order to provide the required level of service, whenever an item fails, it must be replaced. Thus, if 10 poles are destroyed or fail during the first year of service, they must be replaced in order to maintain the desired level of service. The result of having to replace items to maintain the desired level of service means that the true cost will always be higher than that estimated by conventional interest factors. Therefore the error is always one of *underestimation.* The following figure illustrates the effect of assuming the rectangular distribution, as opposed to a survival distribution, representing the true failure rate of the items in question. The shaded portion of the distribution indicates the items that must be replaced to maintain a given level of service.

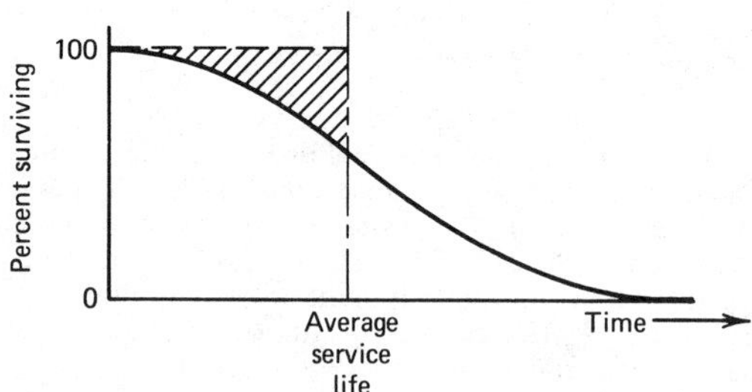

It is the cost of these replacement items that provides the error of underestimation. See Smith[2] for details on how to correct for this error in group properties that is due to nonuniform service lives.

9.3.5 PERPETUITIES AND CAPITALIZED COSTS

In some major public works projects, such as dams, locks, or bridges, the life of the investment is considered to be infinite. In the case of an infinite life asset, the amount needed to construct or acquire that asset initially plus the amount to provide for the perpetual maintenance and replacement of that asset is referred to as the "capitalized cost." A "perpetuity" is a uniform series of payments that continues indefinitely, such as one would find from the conversion of a capitalized cost to an annuity.

DISCRETE COMPOUND INTEREST FACTORS
INTEREST RATE 12.0 PER CENT

N	SINGLE PAYMENT		UNIFORM SERIES				GRADIENT SERIES		N
	COMPOUND AMOUNT FACTOR	PRESENT WORTH FACTOR	CAPITAL RECOVERY FACTOR	PRESENT WORTH FACTOR	COMPOUND AMOUNT FACTOR	SINKING FUND FACTOR	UNIFORM SERIES FACTOR	PRESENT WORTH FACTOR	
	FIND F GIVEN P F|P,I,N	FIND P GIVEN F P|F,I,N	FIND A GIVEN P A|P,I,N	FIND P GIVEN A P|A,I,N	FIND F GIVEN A F|A,I,N	FIND A GIVEN F A|F,I,N	FIND A GIVEN G A|G,I,N	FIND P GIVEN G P|G,I,N	
1	1.1200	0.8929	1.1200	0.8929	1.0000	1.0000	7.4405	0.0	1
2	1.2544	0.7972	0.5917	1.6900	2.1200	0.4717	6.7389	0.7972	2
3	1.4049	0.7118	0.4164	2.4018	3.3744	0.2964	6.1980	2.2207	3
4	1.5735	0.6355	0.3292	3.0373	4.7793	0.2092	5.7913	4.1272	4
5	1.7623	0.5674	0.2774	3.6048	6.3528	0.1574	5.4962	6.3969	5
6	1.9738	0.5066	0.2432	4.1114	8.1151	0.1232	5.2935	8.9300	6
7	2.2107	0.4524	0.2191	4.5637	10.0890	0.0991	5.1669	11.6441	7
8	2.4760	0.4039	0.2013	4.9676	12.2996	0.0813	5.1023	14.4713	8
9	2.7731	0.3606	0.1877	5.3282	14.7756	0.0677	5.0878	17.3561	9
10	3.1058	0.3220	0.1770	5.6502	17.5486	0.0570	5.1136	20.2539	10
11	3.4785	0.2875	0.1684	5.9377	20.6544	0.0484	5.1711	23.1286	11
12	3.8960	0.2567	0.1614	6.1944	24.1330	0.0414	5.2532	25.9521	12
13	4.3635	0.2292	0.1557	6.4235	28.0289	0.0357	5.3541	28.7021	13
14	4.8871	0.2046	0.1509	6.6282	32.3924	0.0309	5.4686	31.3622	14
15	5.4735	0.1827	0.1468	6.8109	37.2794	0.0268	5.5929	33.9199	15
16	6.1304	0.1631	0.1434	6.9740	42.7529	0.0234	5.7234	36.3668	16
17	6.8660	0.1456	0.1405	7.1196	48.8833	0.0205	5.8574	38.6971	17
18	7.6899	0.1300	0.1379	7.2497	55.7492	0.0179	5.9926	40.9078	18
19	8.6127	0.1161	0.1358	7.3658	63.4391	0.0158	6.1273	42.9977	19
20	9.6462	0.1037	0.1339	7.4694	72.0518	0.0139	6.2600	44.9674	20
21	10.8038	0.0926	0.1322	7.5620	81.6980	0.0122	6.3896	46.8186	21
22	12.1002	0.0826	0.1308	7.6446	92.5016	0.0108	6.5152	48.5541	22
23	13.5522	0.0738	0.1296	7.7184	104.6018	0.0096	6.6362	50.1774	23
24	15.1785	0.0659	0.1285	7.7843	118.1540	0.0085	6.7521	51.6927	24
25	16.9999	0.0588	0.1275	7.8431	133.3323	0.0075	6.8627	53.1045	25
26	19.0399	0.0525	0.1267	7.8957	150.3321	0.0067	6.9678	54.4175	26
27	21.3246	0.0469	0.1259	7.9425	169.3719	0.0059	7.0672	55.6367	27
28	23.8836	0.0419	0.1252	7.9844	190.6964	0.0052	7.1610	56.7672	28
29	26.7496	0.0374	0.1247	8.0218	214.5800	0.0047	7.2492	57.8139	29
30	29.9595	0.0334	0.1241	8.0552	241.3296	0.0041	7.3320	58.7819	30
31	33.5547	0.0298	0.1237	8.0850	271.2886	0.0037	7.4095	59.6759	31
32	37.5812	0.0266	0.1233	8.1116	304.8435	0.0033	7.4818	60.5008	32
33	42.0910	0.0238	0.1229	8.1353	342.4246	0.0029	7.5493	61.2611	33
34	47.1418	0.0212	0.1226	8.1566	384.5151	0.0026	7.6121	61.9611	34
35	52.7988	0.0189	0.1223	8.1755	431.6573	0.0023	7.6704	62.6050	35
36	59.1347	0.0169	0.1221	8.1924	484.4556	0.0021	7.7246	63.1969	36
37	66.2308	0.0151	0.1218	8.2075	543.5901	0.0018	7.7747	63.7405	37
38	74.1785	0.0135	0.1216	8.2210	609.8206	0.0016	7.8211	64.2393	38
39	83.0798	0.0120	0.1215	8.2330	683.9985	0.0015	7.8639	64.6967	39
40	93.0494	0.0107	0.1213	8.2438	767.0784	0.0013	7.9035	65.1158	40
41	104.2153	0.0096	0.1212	8.2534	860.1274	0.0012	7.9399	65.4996	41
42	116.7211	0.0086	0.1210	8.2619	964.3423	0.0010	7.9735	65.8509	42
43	130.7275	0.0076	0.1209	8.2696	1081.0627	0.0009	8.0044	66.1721	43
44	146.4148	0.0068	0.1208	8.2764	1211.7896	0.0008	8.0328	66.4659	44
45	163.9846	0.0061	0.1207	8.2825	1358.2045	0.0007	8.0589	66.7341	45
46	183.6626	0.0054	0.1207	8.2880	1522.1880	0.0007	8.0829	66.9792	46
47	205.7019	0.0049	0.1206	8.2928	1705.8494	0.0006	8.1048	67.2028	47
48	230.3862	0.0043	0.1205	8.2972	1911.5510	0.0005	8.1250	67.4068	48
49	258.0322	0.0039	0.1205	8.3010	2141.9351	0.0005	8.1434	67.5928	49
50	288.9961	0.0035	0.1204	8.3045	2399.9673	0.0004	8.1603	67.7624	50

DISCRETE COMPOUND INTEREST FACTORS
INTEREST RATE 15.0 PER CENT

N	SINGLE PAYMENT: COMPOUND AMOUNT FACTOR, FIND F GIVEN P, F|P,I,N	SINGLE PAYMENT: PRESENT WORTH FACTOR, FIND P GIVEN F, P|F,I,N	UNIFORM SERIES: CAPITAL RECOVERY FACTOR, FIND A GIVEN P, A|P,I,N	UNIFORM SERIES: PRESENT WORTH FACTOR, FIND P GIVEN A, P|A,I,N	UNIFORM SERIES: COMPOUND AMOUNT FACTOR, FIND F GIVEN A, F|A,I,N	UNIFORM SERIES: SINKING FUND FACTOR, FIND A GIVEN F, A|F,I,N	GRADIENT SERIES: UNIFORM SERIES FACTOR, FIND A GIVEN G, A|G,I,N	GRADIENT SERIES: PRESENT WORTH FACTOR, FIND P GIVEN G, P|G,I,N	N
1	1.1500	0.8696	1.1500	0.8696	1.0000	1.0000	5.7971	0.0	1
2	1.3225	0.7561	0.6151	1.6257	2.1500	0.4651	5.1544	0.7561	2
3	1.5209	0.6575	0.4380	2.2832	3.4725	0.2880	4.6941	2.0711	3
4	1.7490	0.5718	0.3503	2.8550	4.9933	0.2003	4.3796	3.7863	4
5	2.0114	0.4972	0.2983	3.3521	6.7423	0.1483	4.1808	5.7750	5
6	2.3131	0.4323	0.2642	3.7845	8.7537	0.1142	4.0727	7.9366	6
7	2.6600	0.3759	0.2404	4.1604	11.0667	0.0904	4.0351	10.1923	7
8	3.0590	0.3269	0.2229	4.4873	13.7267	0.0729	4.0514	12.4805	8
9	3.5179	0.2843	0.2096	4.7716	16.7857	0.0596	4.1083	14.7546	9
10	4.0455	0.2472	0.1993	5.0188	20.3035	0.0493	4.1948	16.9793	10
11	4.6524	0.2149	0.1911	5.2337	24.3490	0.0411	4.3023	19.1287	11
12	5.3502	0.1869	0.1845	5.4206	29.0014	0.0345	4.4238	21.1847	12
13	6.1527	0.1625	0.1791	5.5831	34.3515	0.0291	4.5538	23.1350	13
14	7.0756	0.1413	0.1747	5.7245	40.5042	0.0247	4.6880	24.9723	14
15	8.1370	0.1229	0.1710	5.8474	47.5798	0.0210	4.8232	26.6929	15
16	9.3575	0.1069	0.1679	5.9542	55.7168	0.0179	4.9568	28.2958	16
17	10.7611	0.0929	0.1654	6.0472	65.0743	0.0154	5.0869	29.7827	17
18	12.3753	0.0808	0.1632	6.1280	75.8353	0.0132	5.2122	31.1563	18
19	14.2316	0.0703	0.1613	6.1982	88.2106	0.0113	5.3316	32.4211	19
20	16.3663	0.0611	0.1598	6.2593	102.4420	0.0098	5.4446	33.5820	20
21	18.8212	0.0531	0.1584	6.3125	118.8083	0.0084	5.5509	34.6447	21
22	21.6444	0.0462	0.1573	6.3587	137.6294	0.0073	5.6502	35.6149	22
23	24.8911	0.0402	0.1563	6.3988	159.2737	0.0063	5.7426	36.4987	23
24	28.6247	0.0349	0.1554	6.4338	184.1646	0.0054	5.8282	37.3022	24
25	32.9184	0.0304	0.1547	6.4641	212.7891	0.0047	5.9072	38.0313	25
26	37.8561	0.0264	0.1541	6.4906	245.7073	0.0041	5.9799	38.6917	26
27	43.5345	0.0230	0.1535	6.5135	283.5632	0.0035	6.0465	39.2889	27
28	50.0646	0.0200	0.1531	6.5335	327.0974	0.0031	6.1074	39.8282	28
29	57.5743	0.0174	0.1527	6.5509	377.1619	0.0027	6.1630	40.3145	29
30	66.2104	0.0151	0.1523	6.5660	434.7358	0.0023	6.2136	40.7525	30
31	76.1419	0.0131	0.1520	6.5791	500.9458	0.0020	6.2595	41.1465	31
32	87.5631	0.0114	0.1517	6.5905	577.0872	0.0017	6.3012	41.5005	32
33	100.6975	0.0099	0.1515	6.6005	664.6499	0.0015	6.3390	41.8163	33
34	115.8020	0.0086	0.1513	6.6091	765.3467	0.0013	6.3731	42.1033	34
35	133.1722	0.0075	0.1511	6.6166	881.1482	0.0011	6.4038	42.3586	35
36	153.1479	0.0065	0.1510	6.6231	1014.3196	0.0010	6.4316	42.5871	36
37	176.1200	0.0057	0.1509	6.6288	1167.4670	0.0009	6.4566	42.7915	37
38	202.5379	0.0049	0.1507	6.6338	1343.5859	0.0007	6.4790	42.9742	38
39	232.9184	0.0043	0.1506	6.6380	1546.1230	0.0006	6.4992	43.1374	39
40	267.8560	0.0037	0.1506	6.6418	1779.0398	0.0006	6.5173	43.2830	40
41	308.0342	0.0032	0.1505	6.6450	2046.8948	0.0005	6.5336	43.4128	41
42	354.2388	0.0028	0.1504	6.6478	2354.9253	0.0004	6.5481	43.5285	42
43	407.3745	0.0025	0.1504	6.6503	2709.1638	0.0004	6.5611	43.6317	43
44	468.4802	0.0021	0.1503	6.6524	3116.5352	0.0003	6.5727	43.7234	44
45	538.7522	0.0019	0.1503	6.6543	3585.0151	0.0003	6.5831	43.8051	45
46	619.5645	0.0016	0.1502	6.6559	4123.7617	0.0002	6.5924	43.8777	46
47	712.4988	0.0014	0.1502	6.6573	4743.3242	0.0002	6.6007	43.9423	47
48	819.3728	0.0012	0.1502	6.6585	5455.8164	0.0002	6.6081	43.9997	48
49	942.2783	0.0011	0.1502	6.6596	6275.1875	0.0002	6.6147	44.0506	49
50	1083.6191	0.0009	0.1501	6.6605	7217.4609	0.0001	6.6205	44.0958	50

DISCRETE COMPOUND INTEREST FACTORS
INTEREST RATE 20.0 PER CENT

	SINGLE PAYMENT		UNIFORM SERIES				GRADIENT SERIES		
	COMPOUND AMOUNT FACTOR	PRESENT WORTH FACTOR	CAPITAL RECOVERY FACTOR	PRESENT WORTH FACTOR	COMPOUND AMOUNT FACTOR	SINKING FUND FACTOR	UNIFORM SERIES FACTOR	PRESENT WORTH FACTOR	
N	FIND F GIVEN P F|P,I,N	FIND P GIVEN F P|F,I,N	FIND A GIVEN P A|P,I,N	FIND P GIVEN A P|A,I,N	FIND F GIVEN A F|A,I,N	FIND A GIVEN F A|F,I,N	FIND A GIVEN G A|G,I,N	FIND P GIVEN G P|G,I,N	N
1	1.2000	0.8333	1.2000	0.8333	1.0000	1.0000	4.1667	0.0	1
2	1.4400	0.6944	0.6545	1.5278	2.2000	0.4545	3.6111	0.6944	2
3	1.7280	0.5787	0.4747	2.1065	3.6400	0.2747	3.2639	1.8518	3
4	2.0736	0.4823	0.3863	2.5887	5.3680	0.1863	3.0710	3.2986	4
5	2.4883	0.4019	0.3344	2.9906	7.4416	0.1344	2.9906	4.9061	5
6	2.9860	0.3349	0.3007	3.3255	9.9299	0.1007	2.9906	6.5806	6
7	3.5832	0.2791	0.2774	3.6046	12.9158	0.0774	3.0464	8.2550	7
8	4.2998	0.2326	0.2606	3.8372	16.4990	0.0606	3.1394	9.8830	8
9	5.1598	0.1938	0.2481	4.0310	20.7988	0.0481	3.2557	11.4335	9
10	6.1917	0.1615	0.2385	4.1925	25.9585	0.0385	3.3849	12.8870	10
11	7.4300	0.1346	0.2311	4.3271	32.1502	0.0311	3.5195	14.2329	11
12	8.9160	0.1122	0.2253	4.4392	39.5802	0.0253	3.6541	15.4666	12
13	10.6993	0.0935	0.2206	4.5327	48.4963	0.0206	3.7850	16.5682	13
14	12.8391	0.0779	0.2169	4.6106	59.1955	0.0169	3.9096	17.6007	14
15	15.4069	0.0649	0.2139	4.6755	72.0345	0.0139	4.0264	18.5094	15
16	18.4883	0.0541	0.2114	4.7296	87.4413	0.0114	4.1346	19.3207	16
17	22.1859	0.0451	0.2094	4.7746	105.9296	0.0094	4.2337	20.0419	17
18	26.6231	0.0376	0.2078	4.8122	128.1154	0.0078	4.3239	20.6804	18
19	31.9477	0.0313	0.2065	4.8435	154.7384	0.0065	4.4053	21.2439	19
20	38.3372	0.0261	0.2054	4.8696	186.6859	0.0054	4.4783	21.7394	20
21	46.0046	0.0217	0.2044	4.8913	225.0230	0.0044	4.5435	22.1742	21
22	55.2055	0.0181	0.2037	4.9094	271.0273	0.0037	4.6015	22.5546	22
23	66.2466	0.0151	0.2031	4.9245	326.2327	0.0031	4.6528	22.8867	23
24	79.4958	0.0126	0.2025	4.9371	392.4790	0.0025	4.6981	23.1760	24
25	95.3950	0.0105	0.2021	4.9476	471.9749	0.0021	4.7379	23.4276	25
26	114.4739	0.0087	0.2018	4.9563	567.3694	0.0018	4.7729	23.6460	26
27	137.3686	0.0073	0.2015	4.9636	681.8430	0.0015	4.8034	23.8353	27
28	164.8422	0.0061	0.2012	4.9697	819.2112	0.0012	4.8301	23.9995	28
29	197.8107	0.0051	0.2010	4.9747	984.0532	0.0010	4.8534	24.1406	29
30	237.3726	0.0042	0.2008	4.9789	1181.8630	0.0008	4.8736	24.2628	30
31	284.8469	0.0035	0.2007	4.9824	1419.2346	0.0007	4.8912	24.3681	31
32	341.8159	0.0029	0.2006	4.9854	1704.0796	0.0006	4.9064	24.4588	32
33	410.1790	0.0024	0.2005	4.9878	2045.8948	0.0005	4.9195	24.5368	33
34	492.2144	0.0020	0.2004	4.9898	2456.0716	0.0004	4.9309	24.6038	34
35	590.6570	0.0017	0.2003	4.9915	2948.2849	0.0003	4.9407	24.6614	35
36	708.7881	0.0014	0.2003	4.9929	3538.9404	0.0003	4.9492	24.7108	36
37	850.5454	0.0012	0.2002	4.9941	4247.7266	0.0002	4.9565	24.7531	37
38	1020.6538	0.0010	0.2002	4.9951	5098.2656	0.0002	4.9628	24.7893	38
39	1224.7842	0.0008	0.2002	4.9959	6118.9180	0.0002	4.9682	24.8204	39
40	1469.7400	0.0007	0.2001	4.9966	7343.6992	0.0001	4.9728	24.8469	40
41	1763.6875	0.0006	0.2001	4.9972	8813.4375	0.0001	4.9768	24.8696	41
42	2116.4233	0.0005	0.2001	4.9976	10577.1172	0.0001	4.9802	24.8890	42
43	2539.7073	0.0004	0.2001	4.9980	12693.5352	0.0001	4.9831	24.9055	43
44	3047.6467	0.0003	0.2001	4.9984	15233.2344	0.0001	4.9856	24.9196	44
45	3657.1755	0.0003	0.2001	4.9986	18280.8750	0.0001	4.9877	24.9316	45
46	4388.6055	0.0002	0.2000	4.9989	21938.0273	0.0000	4.9895	24.9419	46
47	5266.3242	0.0002	0.2000	4.9991	26326.6211	0.0000	4.9911	24.9506	47
48	6319.5820	0.0002	0.2000	4.9992	31592.9102	0.0000	4.9924	24.9581	48
49	7583.4961	0.0001	0.2000	4.9993	37912.4805	0.0000	4.9935	24.9644	49
50	9100.1914	0.0001	0.2000	4.9995	45495.9570	0.0000	4.9945	24.9698	50

CONTINUOUS COMPOUNDING INTEREST FACTORS
INTEREST RATE 9.0 PER CENT

	SINGLE PAYMENT		UNIFORM SERIES				GRADIENT SERIES		
	COMPOUND AMOUNT FACTOR	PRESENT WORTH FACTOR	CAPITAL RECOVERY FACTOR	PRESENT WORTH FACTOR	COMPOUND AMOUNT FACTOR	SINKING FUND FACTOR	UNIFORM SERIES FACTOR	PRESENT WORTH FACTOR	
N	FIND F GIVEN P F|P,I,N	FIND P GIVEN F P|F,I,N	FIND A GIVEN P A|P,I,N	FIND P GIVEN A P|A,I,N	FIND F GIVEN A F|A,I,N	FIND A GIVEN F A|F,I,N	FIND A GIVEN G A|G,I,N	FIND P GIVEN G P|G,I,N	N
1	1.0942	0.9139	1.0942	0.9139	1.0000	1.0000	0.0	0.0	1
2	1.1972	0.8353	0.5717	1.7492	2.0942	0.4775	0.4775	0.8352	2
3	1.3100	0.7634	0.3980	2.5126	3.2914	0.3038	0.9401	2.3620	3
4	1.4333	0.6977	0.3115	3.2102	4.6013	0.2173	1.3877	4.4550	4
5	1.5683	0.6376	0.2599	3.8479	6.0347	0.1657	1.8206	7.0055	5
6	1.7160	0.5827	0.2257	4.4306	7.6030	0.1315	2.2388	9.9192	6
7	1.8776	0.5326	0.2015	4.9632	9.3190	0.1073	2.6424	13.1147	7
8	2.0544	0.4868	0.1835	5.4500	11.1966	0.0893	3.0316	16.5220	8
9	2.2479	0.4449	0.1696	5.8948	13.2510	0.0755	3.4065	20.0808	9
10	2.4596	0.4066	0.1587	6.3014	15.4989	0.0645	3.7674	23.7400	10
11	2.6912	0.3716	0.1499	6.6730	17.9585	0.0557	4.1145	27.4557	11
12	2.9447	0.3396	0.1426	7.0126	20.6497	0.0484	4.4479	31.1912	12
13	3.2220	0.3104	0.1366	7.3229	23.5944	0.0424	4.7680	34.9156	13
14	3.5254	0.2837	0.1315	7.6066	26.8164	0.0373	5.0750	38.6031	14
15	3.8574	0.2592	0.1271	7.8658	30.3418	0.0330	5.3691	42.2325	15
16	4.2207	0.2369	0.1234	8.1028	34.1992	0.0292	5.6507	45.7864	16
17	4.6182	0.2165	0.1202	8.3193	38.4198	0.0260	5.9201	49.2508	17
18	5.0531	0.1979	0.1174	8.5172	43.0380	0.0232	6.1775	52.6151	18
19	5.5289	0.1809	0.1150	8.6981	48.0910	0.0208	6.4234	55.8707	19
20	6.0496	0.1653	0.1128	8.8634	53.6200	0.0186	6.6579	59.0114	20
21	6.6194	0.1511	0.1109	9.0144	59.6696	0.0168	6.8815	62.0329	21
22	7.2427	0.1381	0.1093	9.1525	66.2889	0.0151	7.0945	64.9323	22
23	7.9248	0.1262	0.1078	9.2787	73.5316	0.0136	7.2972	67.7084	23
24	8.6711	0.1153	0.1065	9.3940	81.4564	0.0123	7.4900	70.3609	24
25	9.4877	0.1054	0.1053	9.4994	90.1276	0.0111	7.6732	72.8905	25
26	10.3812	0.0963	0.1042	9.5957	99.6153	0.0100	7.8471	75.2987	26
27	11.3588	0.0880	0.1033	9.6838	109.9964	0.0091	8.0121	77.5876	27
28	12.4286	0.0805	0.1024	9.7642	121.3552	0.0082	8.1686	79.7600	28
29	13.5990	0.0735	0.1016	9.8378	133.7838	0.0075	8.3168	81.8190	29
30	14.8797	0.0672	0.1010	9.9050	147.3828	0.0068	8.4572	83.7680	30
31	16.2810	0.0614	0.1003	9.9664	162.2625	0.0062	8.5899	85.6106	31
32	17.8142	0.0561	0.0998	10.0225	178.5435	0.0056	8.7154	87.3508	32
33	19.4918	0.0513	0.0993	10.0738	196.3574	0.0051	8.8340	88.9925	33
34	21.3275	0.0469	0.0988	10.1207	215.8492	0.0046	8.9460	90.5397	34
35	23.3360	0.0429	0.0984	10.1636	237.1767	0.0042	9.0516	91.9967	35
36	25.5336	0.0392	0.0980	10.2027	260.5125	0.0038	9.1512	93.3675	36
37	27.9382	0.0358	0.0977	10.2385	286.0461	0.0035	9.2451	94.6560	37
38	30.5693	0.0327	0.0974	10.2712	313.9841	0.0032	9.3335	95.8664	38
39	33.4481	0.0299	0.0971	10.3011	344.5535	0.0029	9.4167	97.0025	39
40	36.5981	0.0273	0.0968	10.3285	378.0015	0.0026	9.4949	98.0681	40
41	40.0447	0.0250	0.0966	10.3534	414.5996	0.0024	9.5685	99.0670	41
42	43.8159	0.0228	0.0964	10.3763	454.6443	0.0022	9.6377	100.0027	42
43	47.9422	0.0209	0.0962	10.3971	498.4597	0.0020	9.7026	100.8788	43
44	52.4571	0.0191	0.0960	10.4162	546.4019	0.0018	9.7635	101.6985	44
45	57.3972	0.0174	0.0958	10.4336	598.8591	0.0017	9.8207	102.4651	45
46	62.8026	0.0159	0.0957	10.4495	656.2563	0.0015	9.8743	103.1816	46
47	68.7169	0.0146	0.0956	10.4641	719.0588	0.0014	9.9245	103.8510	47
48	75.1883	0.0133	0.0954	10.4774	787.7759	0.0013	9.9716	104.4761	48
49	82.2690	0.0122	0.0953	10.4895	862.9626	0.0012	10.0157	105.0596	49
50	90.0166	0.0111	0.0952	10.5006	945.2317	0.0011	10.0569	105.6039	50

GEOMETRIC SERIES FACTORS: DISCRETE COMPOUNDING
FUTURE WORTH FACTOR F|A
INTEREST RATE 15.0 PER CENT

N	C=4 0/0	C=5 0/0	C=6 0/0	C=8 0/0	C=10 0/0	C=12 0/0	C=15 0/0	C=20 0/0	N
1	1.0000	1.0000	1.0000	1.0000	1.0000	1.0000	1.0000	1.0000	1
2	2.1900	2.2000	2.2100	2.2300	2.2500	2.2700	2.3000	2.3500	2
3	3.6001	3.6325	3.6651	3.7309	3.7974	3.8649	3.9675	4.1425	3
4	5.2650	5.3350	5.4058	5.5502	5.6980	5.8495	6.0835	6.4919	4
5	7.2246	7.3507	7.4792	7.7432	8.0168	8.3004	8.7450	9.5393	5
6	9.5249	9.7296	9.9392	10.3740	10.8298	11.3078	12.0681	13.4585	6
7	12.2189	12.5291	12.8486	13.5170	14.2258	14.9778	16.1913	18.4632	7
8	15.3677	15.8155	16.2795	17.2583	18.3084	19.4350	21.2801	24.8159	8
9	19.0414	19.6653	20.3153	21.6980	23.1982	24.8262	27.5311	32.8381	9
10	23.3208	24.1664	25.0520	26.9517	29.0358	31.3231	35.1785	42.9235	10
11	28.2992	29.4202	30.6007	33.1533	35.9848	39.1274	44.5008	55.5537	11
12	34.0835	35.5435	37.0890	40.4579	44.2356	48.4750	55.8283	71.3168	12
13	40.7970	42.6709	44.6645	49.0447	54.0093	59.6421	69.5527	90.9304	13
14	48.5816	50.9571	53.4971	59.1210	65.5629	72.9518	86.1383	115.2691	14
15	57.6005	60.5806	63.7825	70.9262	79.1947	88.7817	106.1346	145.3986	15
16	68.0414	71.7465	75.7463	84.7372	95.2510	107.5722	130.1916	182.6150	16
17	80.1206	84.6913	89.6485	100.8737	114.1335	129.8383	159.0778	228.4954	17
18	94.0864	99.6869	105.7885	119.7046	136.3078	156.1797	193.7005	284.9553	18
19	110.2251	117.0465	124.5110	141.6561	162.3137	187.2965	235.1306	354.3213	19
20	128.8656	137.1302	146.2130	167.2200	192.7763	224.0032	284.6316	439.4175	20
21	150.3864	160.3528	171.3521	196.9639	228.4202	267.2498	343.6921	543.6670	21
22	175.2230	187.1915	200.4541	231.5419	270.0830	318.1404	414.0671	671.2214	22
23	203.8761	218.1954	234.1257	271.7092	318.7354	377.9614	497.8213	827.1099	23
24	236.9219	253.9959	273.0637	318.3369	375.4993	448.2070	597.3853	1017.4229	24
25	275.0232	295.3201	318.0720	372.4282	441.6736	530.6167	715.6169	1249.5317	25
26	318.9424	343.0044	370.0747	435.1406	518.7590	627.2085	855.8774	1532.3555	26
27	369.5559	398.0105	430.1350	507.8076	608.4905	740.3291	1022.1145	1876.6821	27
28	427.8726	461.4451	499.4771	591.9663	712.8733	872.7021	1218.9656	2295.5518	28
29	495.0518	534.5818	579.5100	689.3879	834.2246	1027.4900	1451.8738	2804.7271	29
30	572.4277	618.8845	671.8545	802.1125	975.2209	1208.3618	1727.2283	3423.2439	30
31	661.5347	716.0388	778.3755	932.4915	1138.9524	1419.5752	2052.5215	4174.0977	31
32	764.1372	827.9817	901.2190	1083.2317	1328.9880	1666.0635	2436.5400	5085.0547	32
33	882.2656	956.9436	1042.8547	1257.4526	1549.4490	1953.5530	2889.5815	6189.6250	33
34	1018.2527	1105.4871	1206.1223	1458.7451	1805.0896	2288.6746	3423.7146	7528.2383	34
35	1174.7842	1276.5627	1394.2910	1691.2456	2101.3989	2679.1157	4053.0706	9149.6875	35
36	1354.9468	1473.5620	1611.1191	1959.7161	2444.7090	3133.7788	4794.1992	11112.7969	36
37	1562.2917	1700.3872	1860.9333	2269.6401	2842.3259	3662.9768	5666.4727	13488.5000	37
38	1800.9021	1961.5251	2148.7078	2627.3293	3302.6755	4278.6484	6692.5586	16362.3086	38
39	2075.4751	2262.1379	2480.1667	3040.0520	3835.4783	4994.6250	7898.9766	19837.3086	39
40	2391.4094	2608.1599	2861.8904	3516.1711	4451.9375	5826.8867	9316.7344	24037.6719	40
41	2754.9194	3006.4219	3301.4602	4065.3184	5164.9844	6793.9687	10982.0937	29113.0586	41
42	3173.1458	3464.7717	3807.5747	4698.5703	5989.5039	7917.2617	12937.4336	35243.6797	42
43	3654.3115	3992.2498	4390.2656	5428.6914	6942.6953	9221.5742	15232.2656	42646.6445	43
44	4207.8516	4599.2305	5061.0508	6270.3516	8044.3320	10735.5234	17924.4766	51583.3164	44
45	4844.6484	5297.6719	5833.1953	7240.4648	9317.2422	12492.2617	21081.6094	62368.4492	45
46	5577.1797	6101.3008	6721.9336	8358.4453	10787.7070	14530.0742	24782.5977	75380.7500	46
47	6419.8281	7025.9297	7744.8086	9646.6797	12486.0352	16893.2344	29119.5273	91076.4375	47
48	7389.1133	8089.7148	8921.9883	11130.9023	14447.1250	19632.9023	34199.9414	110004.062	48
49	8504.0469	9313.5703	10276.6758	12840.7422	16711.2070	22808.2227	40149.2656	132824.250	49
50	9786.4805	10721.5195	11835.5469	14810.2656	19324.5859	26487.4570	47113.9141	160331.312	50

GEOMETRIC SERIES FACTORS: DISCRETE COMPOUNDING
PRESENT WORTH FACTOR P|A
INTEREST RATE 15.0 PER CENT

N	C=4 0/0	C=5 0/0	C=6 0/0	C=8 0/0	C=10 0/0	C=12 0/0	C=15 0/0	C=20 0/0	N
1	0.8696	0.8696	0.8696	0.8696	0.8696	0.8696	0.8696	0.8696	1
2	1.6560	1.6635	1.6711	1.6862	1.7013	1.7164	1.7391	1.7769	2
3	2.3671	2.3884	2.4098	2.4531	2.4969	2.5412	2.6087	2.7238	3
4	3.0103	3.0503	3.0908	3.1734	3.2579	3.3445	3.4783	3.7118	4
5	3.5919	3.6546	3.7185	3.8498	3.9858	4.1268	4.3478	4.7427	5
6	4.1179	4.2064	4.2970	4.4850	4.6821	4.8887	5.2174	5.8185	6
7	4.5936	4.7102	4.8303	5.0816	5.3480	5.6307	6.0870	6.9410	7
8	5.0237	5.1702	5.3218	5.6418	5.9851	6.3534	6.9565	8.1124	8
9	5.4128	5.5901	5.7749	6.1680	6.5944	7.0572	7.8261	9.3347	9
10	5.7646	5.9736	6.1925	6.6621	7.1773	7.7427	8.6957	10.6101	10
11	6.0828	6.3237	6.5775	7.1261	7.7348	8.4102	9.5652	11.9410	11
12	6.3705	6.6434	6.9323	7.5619	8.2680	9.0604	10.4348	13.3297	12
13	6.6307	6.9353	7.2593	7.9712	8.7781	9.6936	11.3044	14.7788	13
14	6.8660	7.2018	7.5607	8.3556	9.2660	10.3103	12.1739	16.2910	14
15	7.0789	7.4451	7.8386	8.7165	9.7327	10.9109	13.0435	17.8689	15
16	7.2713	7.6673	8.0947	9.0555	10.1791	11.4958	13.9130	19.5153	16
17	7.4454	7.8701	8.3308	9.3739	10.6061	12.0655	14.7826	21.2334	17
18	7.6028	8.0553	8.5484	9.6729	11.0145	12.6203	15.6522	23.0261	18
19	7.7451	8.2244	8.7489	9.9536	11.4052	13.1606	16.5217	24.8968	19
20	7.8738	8.3788	8.9338	10.2173	11.7789	13.6869	17.3913	26.8489	20
21	7.9903	8.5198	9.1042	10.4650	12.1363	14.1994	18.2609	28.8858	21
22	8.0955	8.6485	9.2612	10.6975	12.4782	14.6985	19.1304	31.0113	22
23	8.1907	8.7660	9.4060	10.9160	12.8052	15.1846	20.0000	33.2292	23
24	8.2768	8.8733	9.5395	11.1211	13.1180	15.6581	20.8696	35.5435	24
25	8.3547	8.9713	9.6625	11.3137	13.4172	16.1192	21.7391	37.9585	25
26	8.4251	9.0607	9.7758	11.4946	13.7035	16.5682	22.6087	40.4784	26
27	8.4888	9.1424	9.8803	11.6645	13.9772	17.0056	23.4783	43.1079	27
28	8.5464	9.2170	9.9767	11.8241	14.2391	17.4315	24.3478	45.8517	28
29	8.5985	9.2851	10.0654	11.9739	14.4895	17.8463	25.2174	48.7149	29
30	8.6456	9.3472	10.1473	12.1146	14.7291	18.2503	26.0870	51.7025	30
31	8.6882	9.4040	10.2227	12.2468	14.9583	18.6438	26.9565	54.8200	31
32	8.7267	9.4558	10.2922	12.3709	15.1775	19.0270	27.8261	58.0730	32
33	8.7615	9.5032	10.3563	12.4874	15.3872	19.4002	28.6956	61.4675	33
34	8.7930	9.5464	10.4154	12.5969	15.5877	19.7637	29.5652	65.0096	34
35	8.8215	9.5858	10.4698	12.6997	15.7796	20.1177	30.4348	68.7057	35
36	8.8473	9.6218	10.5200	12.7962	15.9631	20.4624	31.3044	72.5625	36
37	8.8706	9.6547	10.5663	12.8869	16.1386	20.7982	32.1739	76.5869	37
38	8.8917	9.6847	10.6089	12.9720	16.3065	21.1252	33.0435	80.7864	38
39	8.9107	9.7121	10.6482	13.0520	16.4670	21.4437	33.9130	85.1684	39
40	8.9280	9.7372	10.6844	13.1271	16.6207	21.7538	34.7826	89.7410	40
41	8.9436	9.7600	10.7178	13.1976	16.7676	22.0559	35.6522	94.5124	41
42	8.9577	9.7809	10.7486	13.2639	16.9081	22.3501	36.5217	99.4913	42
43	8.9704	9.8000	10.7770	13.3261	17.0425	22.6366	37.3913	104.6866	43
44	8.9819	9.8173	10.8031	13.3845	17.1711	22.9156	38.2609	110.1078	44
45	8.9924	9.8332	10.8272	13.4393	17.2941	23.1874	39.1304	115.7646	45
46	9.0018	9.8477	10.8495	13.4908	17.4118	23.4521	40.0000	121.6674	46
47	9.0103	9.8610	10.8699	13.5392	17.5243	23.7099	40.8696	127.8268	47
48	9.0180	9.8731	10.8888	13.5847	17.6319	23.9609	41.7391	134.2541	48
49	9.0250	9.8841	10.9062	13.6273	17.7349	24.2054	42.6087	140.9608	49
50	9.0313	9.8942	10.9222	13.6674	17.8334	24.4435	43.4783	147.9592	50

CONTINUOUS COMPOUNDING, CONTINUOUS FLOW FACTORS
INTEREST RATE 15.0 PER CENT

N	CAPITAL RECOVERY FACTOR FIND $\bar{A}$ GIVEN P $\bar{A}$|P,R,N	PRESENT WORTH FACTOR FIND P GIVEN $\bar{A}$ P|$\bar{A}$,R,N	SINKING FUND FACTOR FIND $\bar{A}$ GIVEN F $\bar{A}$|F,R,N	COMPOUND AMOUNT FACTOR FIND F GIVEN $\bar{A}$ F|$\bar{A}$,R,N	N
1	1.0769	0.9286	0.9269	1.0789	1
2	0.5787	1.7279	0.4287	2.3324	2
3	0.4139	2.4158	0.2639	3.7887	3
4	0.3325	3.0079	0.1825	5.4808	4
5	0.2843	3.5176	0.1343	7.4467	5
6	0.2528	3.9562	0.1028	9.7307	6
7	0.2307	4.3337	0.0807	12.3843	7
8	0.2147	4.6587	0.0647	15.4674	8
9	0.2025	4.9384	0.0525	19.0494	9
10	0.1931	5.1791	0.0431	23.2112	10
11	0.1857	5.3863	0.0357	28.0464	11
12	0.1797	5.5647	0.0297	33.6642	12
13	0.1749	5.7182	0.0249	40.1911	13
14	0.1709	5.8503	0.0209	47.7743	14
15	0.1677	5.9640	0.0177	56.5847	15
16	0.1650	6.0619	0.0150	66.8210	16
17	0.1627	6.1461	0.0127	78.7137	17
18	0.1608	6.2186	0.0108	92.5312	18
19	0.1592	6.2810	0.0092	108.5847	19
20	0.1579	6.3348	0.0079	127.2363	20
21	0.1567	6.3810	0.0067	148.9063	21
22	0.1557	6.4208	0.0057	174.0833	22
23	0.1549	6.4550	0.0049	203.3350	23
24	0.1542	6.4845	0.0042	237.3204	24
25	0.1536	6.5099	0.0036	276.8057	25
26	0.1531	6.5317	0.0031	322.6814	26
27	0.1527	6.5505	0.0027	375.9810	27
28	0.1523	6.5667	0.0023	437.9065	28
29	0.1520	6.5806	0.0020	509.8538	29
30	0.1517	6.5926	0.0017	593.4436	30
31	0.1514	6.6029	0.0014	690.5618	31
32	0.1512	6.6118	0.0012	803.3979	32
33	0.1511	6.6194	0.0011	934.4937	33
34	0.1509	6.6260	0.0009	1086.8052	34
35	0.1508	6.6317	0.0008	1263.7661	35
36	0.1507	6.6366	0.0007	1469.3665	36
37	0.1506	6.6408	0.0006	1708.2380	37
38	0.1505	6.6444	0.0005	1985.7676	38
39	0.1504	6.6475	0.0004	2308.2107	39
40	0.1504	6.6501	0.0004	2682.8367	40
41	0.1503	6.6524	0.0003	3118.0891	41
42	0.1503	6.6544	0.0003	3623.7844	42
43	0.1502	6.6561	0.0002	4211.3125	43
44	0.1502	6.6576	0.0002	4893.9219	44
45	0.1502	6.6589	0.0002	5687.0117	45
46	0.1502	6.6600	0.0002	6608.4414	46
47	0.1501	6.6609	0.0001	7678.9883	47
48	0.1501	6.6617	0.0001	8922.7969	48
49	0.1501	6.6624	0.0001	10367.8750	49
50	0.1501	6.6630	0.0001	12046.8281	50

CHAPTER 9.4

Project Selection and Analysis

GERALD J. THUESEN

Georgia Institute of Technology

9.4.1 DECISION RULES FOR PROJECT SELECTION

Incremental Analysis for Mutually Exclusive Alternatives

When comparing mutually exclusive alternatives, it is the future *difference* between the alternatives that is relevant for determining the economic desirability of one compared to the other. It is this fundamental concept that is the basis for the discussion concerning decision criteria in this chapter.

The reason why the difference between alternatives is so fundamental in the comparison of alternatives is demonstrated by the comparison of alternatives $A1$ and $A2$ shown in Exhibit 9.4.1. To compare the two alternatives described in the exhibit, it is sufficient to examine the cash flow that represents the difference between $A1$ and $A2$ because the advantage or disadvantage of alternative $A2$ over alternative $A1$ is completely described by the cash flow representing $A1$ subtracted from $A2$. The cash flow representing alternative $A2$ can be viewed as being the sum of two separate and distinct cash flows. One of these cash flows is identical to the cash flow of alternative $A1$. The other is the cash flow representing the difference between alternative $A1$ and alternative $A2$. To decide which of the two alternatives is economically superior, it is sufficient to utilize the following simple decision rule: If cash flow $(A2 - A1)$ is economically *desirable*, alternative $A2$ is preferred to alternative $A1$. If cash flow $(A2 - A1)$ is economically *undesirable*, alternative $A1$ is preferred to alternative $A2$.

For the example cited in Exhibit 9.4.1, the decision to undertake alternative $A2$ rather than alternative $A1$ requires an additional or incremental investment of \$500 now and \$100 one year hence. The extra receipts expected from the extra investment are \$500 at the end of years 2 and 3. Do the extra receipts justify the extra investment? This is the question that must be answered to determine which of the two alternatives is economically more desirable.

Present Worth on Incremental Investment

When making decisions between mutually exclusive alternatives, it is the differences between alternatives that are relevant for decision making purposes. The present worth on incremental investment criterion provides an example of this rule since it requires that the incremental differences between alternative cash flows actually be calculated.

Exhibit 9.4.1 Differences Between Mutually Exclusive Alternatives

End of Year	Alternative $A1$	Alternative $A2$	Cash Flow Difference $(A2 - A1)$
0	-\$1000	-\$1500	-\$500
1	800	700	- 100
2	800	1300	500
3	800	1300	500

Material in this chapter is based on H. G. THUESEN, W. J. FABRYCKY, and G. J. THUESEN, *Engineering Economy*, 5th ed., Prentice-Hall, Englewood Cliffs, NJ, 1977, and W. J. FABRYCKY and G. J. THUESEN, *Economic Decision Analysis*, 2nd ed., Prentice-Hall, Englewood Cliffs, NJ, 1980.

When comparing one alternative to another, the first task is to determine the cash flow representing the difference between the two cash flows. Then the decision whether to select a particular alternative rests on the determination of the economic desirability of the additional increment of investment required by one alternative over the other. The incremental investment is considered to be desirable if it yields a return that exceeds the MARR. In other words, if the present worth amount for the incremental investment is greater than zero, the increment is considered desirable, and the alternative requiring this additional investment is deemed best.

To apply this decision criterion to a set of mutually exclusive alternatives such as shown in the following, certain steps must be utilized:

1. List the alternatives in ascending order of their equivalent first cost or initial disbursements.

	Alternatives			
End of Year	Do Nothing	$A1$	$A2$	$A3$
0	$0	-$5000	-$8000	-$10,000
1–10	0	1400	1900	2,500

2. Select as the initial "current best" alternative the one that requires the smallest first cost. In most cases the initial current best alternative will be the alternative "do nothing," as it is in this example. All too frequently investment alternatives are compared without including the possibility of not undertaking the project at all. The exclusion of the do nothing alternative can lead to the investment of a scarce resource, money, in unproductive activities, that is, activities that yield a return that is less than the MARR.

3. Compare the initial current best alternative and the first "challenging" alternative. The challenger is always the next highest alternative in order of first cost that has not been previously involved in a comparison. The comparison is accomplished by examining the differences between the two cash flows. If the present worth of the incremental cash flow evaluated at the MARR is greater than zero, the challenger becomes the new current best alternative. If the present worth is less than or equal to zero, the current best alternative remains unchanged, and the challenger in the comparison is eliminated from consideration. The new challenger is the next alternative in order of first cost that has not been a challenger previously. Then the next comparison is made between the alternative that is the current best and the alternative that is currently the challenger.

4. Repeat the comparisons of the challengers to the current best alternative as described in step 3. These comparisons are continued until every alternative other than the initial current best alternative has been a challenger. The alternative that maximizes present worth and provides a rate of return that exceeds the MARR is the last current best alternative.

Step 3 and 4 lead to the following calculations for the alternatives being considered. Assume that the MARR is equal to 15%.

The first comparison to be made in this example is between alternative $A1$ (the first challenger) and the do nothing alternative (the initial current best alternative). The subscript notation in $PW(15)_{A1-0}$ indicates that the present worth amount is for the cash flow representing the difference between alternative $A1$ and do nothing. The do nothing alternative is signified by zero.

$$PW(15)_{A1-0} = -\$5000 + \$1400(\overset{P/A,15,10}{5.0188}) = \$2026.32$$

Note that, when comparing an alternative to the do nothing alternative, the cash flow representing the incremental investment is the same as the cash flow on the total investment.

Because the present worth amount of the differences between the cash flows is greater than zero ($2026.32), alternative $A1$ becomes the new current best alternative as dictated by step 3. The second challenger becomes alternative $A2$. Alternative $A2$ is then compared to alternative $A1$ on an incremental basis as follows:

$$PW(15)_{A2-A1} = -\$3000 + \$500(\overset{P/A,15,10}{5.0188}) = -\$490.60$$

Since this value is negative, alternative $A2$ is dropped from further consideration, and alternative $A1$ remains the current best alternative. The third challenger is alternative $A3$. Comparing the current best with the next challenger yields

$$PW(15)_{A3-A1} = -\$5000 + \$1100(\overset{P/A,15,10}{5.0188}) = \$520.68$$

The present worth on the additional investment required by alternative $A3$ over alternative $A1$ is positive, and therefore that increment is economically desirable. Thus alternative $A3$ becomes the current best alternative, and the list of alternatives has been exhausted so that there is no new challenger possible. According to step 4, when all challengers have been considered, the current best alternative is the alternative that maximizes present worth and provides a return greater than the MARR. Therefore alternative $A3$ is the optimum selection from the set of alternatives shown.

The substitution of the annual equivalent (annual worth) amount or the future worth amount for the present worth amount as the basis for comparison for incremental decision making will lead to consistent solutions. The following relationships confirm this fact and can be proved in a manner similar to that used for the present worth amount.

$$AE(i)_B - AE(i)_A = AE(i)_{B-A}$$

and

$$FW(i)_B - FW(i)_A = FW(i)_{B-A}$$

Using the incremental approach and the annual equivalent amount requires the following calculations for the set of alternatives being considered.

$$AE(15)_{A1-0} = -\$5000(\overset{A/P,15,10}{0.1993}) + \$1400 = \$403.75$$

$$AE(15)_{A2-A1} = -\$3000(\overset{A/P,15,10}{0.1993}) + \$500 = -\$97.75$$

$$AE(15)_{A3-A1} = -\$5000(\overset{A/P,15,10}{0.1993}) + \$1100 = \$103.75$$

By following the decision rules for an incremental analysis, the optimum solution is to select alternative $A3$. This is the same decision given by the incremental analysis using present worth.

Rate of Return on Incremental Investment

This particular decision criterion is based on the same type of incremental analysis applied to the previously discussed criterion, present worth on incremental investment. The only difference in the decision rules between these two criteria is the decision rule in step 3 that determines whether an increment of investment is economically desirable. For the rate of return on incremental investment criterion, the increment of investment is considered desirable if the rate of return resulting from the increment is greater than the MARR ($i^*_{B-A} > MARR$).

To apply rate of return on an incremental basis, it is first necessary to rank the alternatives in order by increasing equivalent first cost and then to select the initial current best alternative. Using the previous example, steps 3 and 4 of the incremental analysis procedure require the following calculations. Find the value i^* of i so that the equation representing the present worth of the incremental cash flow is set equal to zero. That is, find the rate of return (i^*) on the increment between the current best alternative and the challenger. Again, the MARR is assumed to be 15%. For increment $A1 - 0$

$$0 = -\$5000 + \$1400(\overset{P/A,i,10}{\quad\quad})$$

$$i^*_{A1-0} = 25.0\%$$

Because the rate of return on the increment is greater than the MARR, alternative $A1$ becomes the initial current best alternative, and the do nothing alternative is dropped from further consideration. Next, compare alternative $A2$ to alternative $A1$. For the increment ($A2 - A1$)

$$0 = -\$3000 + \$500(\overset{P/A,i,10}{\quad\quad})$$

$$i^*_{A2-A1} = 10.5\%$$

Because the rate of return of this increment is less than the MARR, alternative $A1$ remains the current best and alternative $A2$ is rejected. Then compare alternative $A3$ to alternative $A1$, the current best alternative. For increment $A3 - A1$

$$0 = -\$5000 + \$1100(\overset{P/A,i,10}{\quad\quad})$$

$$i^*_{A3-A1} = 17.6\%$$

Alternative $A3$ becomes the current best alternative, and alternative $A1$ is removed from consideration. Since all the alternatives have been compared, alternative $A3$, the last current best alternative, is the optimum solution. This is the same solution given by the present worth on incremental investment criterion.

One very important point to recognize from this example is that selecting the alternative with the highest rate of return on its total cash flow may *not* lead to the alternative that will maximize the total present worth at the MARR. The rates of return for the *total* cash flows of the alternatives in Exhibit 9.4.2 are

$$i_0^* = 15\%, \quad i_{A1}^* = 25\%, \quad i_{A2}^* = 19.9\%, \quad i_{A3}^* = 21.9\%$$

If alternative $A1$ is selected because it has the maximum rate of return, the total present worth will *not* be maximized for a MARR = 15%.

Benefit-Cost and Incremental Investment

Suppose that four mutually exclusive alternatives have been identified for providing recreational facilities in a certain urban area. The equivalent annual benefits, equivalent annual costs, and benefit-cost ratios are given in Exhibit 9.4.2.

Inspection of the BC ratios might lead one to select alternative B because the BC ratio is a maximum. Actually, this choice is *not* correct. The correct alternative can be selected by applying the principle of incremental analysis as previously described. In this instance the additional increment of outlay is economically desirable if the incremental benefit realized exceeds the incremental outlay. Thus, when comparing mutually exclusive alternatives $A1$ and $A2$, the decision rule is as follows:

If $BC(i)_{A2-A1} > 1$, accept alternative $A2$.

If $BC(i)_{A2-A1} \leq 1$, reject alternative $A2$ and retain alternative $A1$.

Just as described in the subchapter on present worth on incremental investment, the alternatives should be arranged in order of increasing value of the denominator. Thus the alternative with the lowest denominator should be first, the alternative with the next lowest denominator second, and so on.

If the do nothing alternative is to be considered, assume that the cash flow associated with that alternative is zero. When comparing an alternative to the do nothing option, the incremental benefit-cost ratio is computed using this assumption, and the decision rules just described are applied.

The sequence of calculations required to produce the results presented in Exhibit 9.4.2 follows. For example, the do nothing alternative is considered to be a feasible alternative. To compare the alternative with the smallest initial investment to do nothing, compute the benefit-cost ratio using the total benefits and total costs for alternative D as follows:

$$BC(i)_{D-0} = \frac{\$95{,}000}{\$50{,}000} = 1.90$$

This procedure is identical to an incremental comparison where the cash flows for the do nothing alternative are considered to be zero. Since this benefit-cost ratio is greater than 1, alternative D is seen to be preferred to the do nothing alternative. Therefore the do nothing alternative (the initial current best alternative) is rejected, and alternative D becomes the new current best alternative. It is the alternative with the lowest investment of the four alternatives being considered (exclusive of the do nothing alternative) that is acceptable.

Next, it is necessary to determine whether the incremental benefits that would be realized if alternative C were undertaken would justify the additional expenditure. Therefore compare

Exhibit 9.4.2 Benefit-Cost Ratios for Four Alternatives

Alternative	Equivalent Annual Benefits	Equivalent Annual Costs	*BC* Ratio
A	$182,000	$91,500	1.99
B	167,000	79,500	2.10
C	115,000	78,500	1.46
D	95,000	50,000	1.90

alternative C to alternative D as follows:

$$BC(i)_{C-D} = \frac{\$115{,}000 - \$95{,}000}{\$78{,}500 - \$50{,}000} = \frac{\$20{,}000}{\$28{,}500} = 0.70$$

The incremental benefit-cost ratio is less than 1, and therefore alternative C is rejected, and alternative D remains as the current best alternative.

Next, compare alternative B to alternative D as follows:

$$BC(i)_{B-D} = \frac{\$167{,}000 - \$95{,}000}{\$79{,}500 - \$50{,}000} = \frac{\$72{,}000}{\$29{,}500} = 2.44$$

The incremental benefit-cost ratio in this instance exceeds 1, and therefore alternative B is preferred to alternative D. Alternative B becomes the new current best alternative.

Alternative A is now compared to alternative B as follows:

$$BC(i)_{A-B} = \frac{\$182{,}000 - \$167{,}000}{\$91{,}500 - \$79{,}500} = \frac{\$15{,}000}{\$12{,}000} = 1.25$$

Since the incremental benefit-cost ratio for this comparison is greater than 1, alternative A is preferred to alternative B. Alternative A becomes the current best alternative, and there are no more comparisons to be made. The alternative that should be selected is the current best alternative that remains after the final comparison. Therefore alternative A is the preferred of the four alternatives. Selection of this alternative will ensure that the equivalent annual benefits less the equivalent annual costs are maximized and that its BC ratio is greater than 1.

Total Investment Analysis for Mutually Exclusive Alternatives

Present Worth on Total Investment

This criterion is one of the most frequently used criteria for selecting an investment alternative from a set of mutually exclusive alternatives. Since the stated objective of the selection of alternatives is to choose the alternative with the maximum present worth amount, the rules for this criterion are rather simple. All that is required is to calculate the present worth amount for the cash flow representing each alternative. Then select the alternative that has the maximum present worth amount, provided that this amount is positive. The present worth amount must be positive to ensure that the alternative yields a return that is greater than the MARR.

To see the computational simplicity of this criterion, it is applied to the mutually exclusive alternatives described in the subchapter on present worth on incremental investment. Using a MARR = 15%, the calculations of the present worth amounts give

$$PW(15)_0 = \$\ 0.00$$

$$PW(15)_{A1} = -\$\ 5000 + \$1400(\overset{P/A,15,10}{5.0188}) = \$2026.32$$

$$PW(15)_{A2} = -\$\ 8000 + \$1900(\overset{P/A,15,10}{5.0188}) = \$1535.72$$

$$PW(15)_{A3} = -\$10{,}000 + \$2500(\overset{P/A,15,10}{5.0188}) = \$2547.00$$

It is seen that the maximum value of the present worth amounts for these four alternatives is \$2547.00, the present worth of alternative $A3$. Although for this example the alternative selected happened to have the largest first cost, it is certainly possible for alternatives with the smaller first costs to have present worths greater than those alternatives with the larger first costs. For example, if alternative $A3$ is excluded from consideration, it is seen that alternative $A1$ has a larger present worth than alternative $A2$, even though it requires less initial outlay.

When the receipts from a number of alternatives are assumed to be equal, it is common to describe the cash flows of the alternatives by showing only their costs. If the costs are shown as positive numbers, then the decision rule for this criterion is to select the alternative that minimizes the present worth amount of the costs.

Annual Equivalent on Total Investment

If either the annual equivalent amount or the future worth amount is substituted for the present worth amount as the basis for comparison in this criterion, the same conclusion will result. By applying the annual equivalent on total investment criterion or the future worth on total investment criterion to the alternatives being considered, alternative $A3$ is again selected as expected.

$$AE(15)_0 = \$\ 0.00$$

$$AE(15)_{A1} = -\$\ 5000\left(\overset{A/P,15,10}{0.1993}\right) + \$1400 = \$403.50$$

$$AE(15)_{A2} = -\$\ 8000\left(\overset{A/P,15,10}{0.1993}\right) + \$1900 = \$305.60$$

$$AE(15)_{A3} = -\$10{,}000\left(\overset{A/P,15,10}{0.1993}\right) + \$2500 = \$507.00$$

or

$$FW(15)_0 = \$\ 0.00$$

$$FW(15)_{A1} = -\$\ 5000\left(\overset{F/P,15,10}{4.046}\right) + \$1400\left(\overset{F/A,15,10}{20.304}\right) = \$\ 8195.60$$

$$FW(15)_{A2} = -\$\ 8000\left(\overset{F/P,15,10}{4.046}\right) + \$1900\left(\overset{F/A,15,10}{20.304}\right) = \$\ 6209.60$$

$$FW(15)_{A3} = -\$10{,}000\left(\overset{F/P,15,10}{4.046}\right) + \$2500\left(\overset{F/A,15,10}{20.304}\right) = \$10{,}300.00$$

Comparison of Alternatives With Unequal Service Lives

Often it is necessary to compare alternatives for which the time span of service will not be equal. In such situations it is necessary to make certain assumptions about the service interval so that the techniques of decision making just discussed are applicable.

When comparing alternatives with unequal lives, the principle that all alternatives under consideration must be compared over the same time span is basic to sound decision making. The time span over which alternatives are considered must be equal so that the effect of undertaking one alternative can be considered identical to the effect of undertaking any of the other alternatives. Clearly, the direct comparison of alternative A with a 5 year life and alternative B with an 11 year life fails to consider the possible investments that could be undertaken during the 6 years following alternative A's termination. There are two basic approaches that can be used so that alternatives with different lives can be compared over an equal time span.

Study Period Approach

This approach confines the consideration of the effects of the alternatives being evaluated to some study period that is usually the life of the shortest-lived alternative. To illustrate this approach, suppose a decision must be made as to which alternative should be selected from the two alternatives described in Exhibit 9.4.3. It is assumed that these two alternatives provide the same service for each year that they are in existence.

Exhibit 9.4.3 Two Alternatives With Unequal Lives

End of Year	Alternative $A1$	Alternative $A2$
0	−$15,000	−$20,000
1	−7,000	−2,000
2	−7,000	−2,000
3	−7,000	−2,000
4	−7,000	—
5	−7,000	—

The study period chosen for this example is 3 years, the life of alternative $A2$. Using the annual equivalent on total investment for an interest rate of 7% yields

$$AE(7)_{A1} = -\$15{,}000(\overset{A/P,7,5}{0.2439}) - \$7000 = -\$10{,}659/\text{year}$$

The \$15,000 first cost of alternative $A1$ is distributed over its entire life to find its equivalent cost per year.

For alternative $A2$

$$AE(7)_{A2} = -\$20{,}000(\overset{A/P,7,3}{0.3811}) - \$2000 = -\$9622/\text{year}$$

The cost advantage of alternative $A2$ over alternative $A1$ is \$1037/year for the first 3 years. For years 4 and 5, alternative $A1$ costs \$10,659 more than alternative $A2$, which provides no service for those last 2 years. Since the study period is 3 years, the cost advantage of alternative $A2$ over alternative $A1$ is stated as \$1037/year for 3 years. The costs occurring after the study period are disregarded since the equivalent costs are being compared only for the period indicated.

The costs occurring after the study period would be considered when alternative $A2$'s successor is to be compared to continuing with alternative $A1$. The decision about $A2$'s successor is assumed to be separable from the original decision when the study period approach is used. An implication of this approach is that, for any alternatives with a life longer than the study period, the unrecovered balance of their first cost at the end of the study period is the assumed salvage value for these alternatives. For the example just discussed, the assumed salvage for alternative $A1$ after 3 years would have to be

$$\$15{,}000(\overset{A/P,7,5}{0.2439})(\overset{P/A,7,2}{1.8080}) = \$6615$$

Estimating Future Alternatives

The second approach to the problem of unequal lives is to estimate the future sequence of events that are anticipated for each alternative being considered so that the time span is the same for each alternative. Two methods that are frequently used to accomplish this end are:

1. The explicit consideration of future alternatives over the same time span.
2. The assumption that an investment opportunity will be replaced by an identical alternative until a common multiple of lives is reached.

To illustrate the first method, suppose it is anticipated that, after alternative $A2$ in Exhibit 9.4.3 is terminated, the service it was providing is continued by incurring costs of \$15,000 at the end of years 4 and 5. Now the service is provided over equal time spans of 5 years, and the annual equivalent costs for alternative $A2$ and the additional expenditures required in years 4 and 5 are

$$AE(7) = \left[-\$20{,}000 - \$2000(\overset{P/A,7,3}{2.6243})\right](\overset{A/P,7,5}{0.2439}) - \$15{,}000(\overset{F/A,7,2}{2.070})(\overset{A/F,7,5}{0.1739}) = -\$11{,}558$$

The annual equivalent cost for alternative $A1$ has been computed for a life of 5 years to be \$10,659/year. Now alternative $A1$ has an annual cost advantage of \$11,558 less \$10,659 over alternative $A2$ and its replacement. This advantage is stated as \$899/year for 5 years.

The second method that can be used to equate alternatives with unequal lives is to assume that each opportunity will be replaced by itself until a common multiple of lives is reached. For the alternatives described in Exhibit 9.4.3, this assumption produces the cash flows presented in Exhibit 9.4.4.

The annual equivalent comparison should be applied when such an assumption is made since it is computationally the most efficient approach. Because the cash flows for each alternative consist of identical repeated cash flows, it is only necessary to calculate the annual equivalent for the original alternative. That is, the 5 year equivalent annual cost for alternative $A1$ described in Exhibit 9.4.3 equals the 15 year equivalent annual cost for alternative $A1$ presented in Exhibit 9.4.4. Thus, under the assumption of repeated replacements, the annual equivalents for the two alternatives in Exhibit 9.4.4 are

$$AE(7)_{A1} = -\$15{,}000(\overset{A/P,7,5}{0.2439}) - \$7000 = -\$10{,}659/\text{year}$$

Exhibit 9.4.4 Two Alternatives With Identical Replacements for a Common Multiple of Lives

End of Year	Alternative $A1$		Alternative $A2$	
0	-$15,000		-$20,000	
1	-7,000		-2,000	
2	-7,000		-2,000	
3	-7,000		-2,000	-20,000
4	-7,000		-2,000	
5	-7,000	-15,000	-2,000	
6	-7,000		-2,000	-20,000
7	-7,000		-2,000	
8	-7,000		-2,000	
9	-7,000		-2,000	-20,000
10	-7,000	-15,000	-2,000	
11	-7,000		-2,000	
12	-7,000		-2,000	-20,000
13	-7,000		-2,000	
14	-7,000		-2,000	
15	-7,000		-2,000	

and

$$AE(7)_{A2} = -\$20{,}000(\overset{A/P,7,3}{0.3811}) - \$2000 = -\$9622/\text{year}$$

The lowest common multiple of years for these two alternatives is 15 years. Therefore, when using this method of examining alternatives over equal time spans, the cost advantage of alternative $A2$ over alternative $A1$ is stated as $1037/year for 15 years. If, in fact, the alternatives are replaced with similar alternatives as assumed, this approach is sound. However, it is infrequent that a sequence of alternatives will repeat itself since technological progress can lead to improved alternatives in the future. This method of comparing alternatives tends to overstate the differences between the alternatives when it assumes that the differences will occur over a time span that exceeds the service lives of the current alternatives.

To use present worth calculations for the method just discussed requires additional computation. The annual equivalent can be calculated for the life of each alternative and is then converted to a present worth amount over the same period.

$$PW(7)_{A1} = -\$10{,}659(\overset{P/A,7,15}{9.1079}) = -\$97{,}081$$

$$PW(7)_{A2} = -\$\ 9622(\overset{P/A,7,15}{9.1079}) = -\$87{,}636$$

It should be clear that *to calculate the present worth for cash flows of unequal duration is incorrect*. That is, for the example just presented, the following calculations are incorrect for comparing alternatives $A1$ and $A2$ of Exhibit 9.4.3.

$$PW(7)_{A1} = -\$15{,}000 - \$7000(\overset{P/A,7,5}{4.1002}) = -\$43{,}701$$

$$PW(7)_{A2} = -\$20{,}000 - \$2000(\overset{P/A,7,3}{2.6243}) = -\$25{,}249$$

Such a calculation and comparison imply that, for years 4 and 5, alternative $A2$ will provide at no cost a service or income equal to that of alternative $A1$. Thus, when present worth comparisons are made, it is essential that the alternatives be compared over the same time span. This same principle holds when making rate of return comparisons on an incremental basis.

Methods for Considering Capital Constraints

To illustrate the technique required to incorporate a budget constraint in the decision process, the present worth on total investment criterion is applied to the proposed investments described in Exhibit 9.4.5. Each set of proposals with the same letter designation ($A1, A2$) is considered to be

Exhibit 9.4.5 Cash Flows for Five Investment Proposals

Proposal	First Cost	Net Income (Years 1–10)	Salvage Value (Year 10)	$PW(8)_j$
$A1$	–\$10,000	\$2000	\$1,000	\$3,883
$A2$	–12,000	2100	2,000	3,018
$B1$	–20,000	3100	5,000	3,117
$B2$	–30,000	5000	8,000	7,255
$C1$	–35,000	4500	10,000	–173

independent of the other sets having different letter designations ($B1$, $B2$). Proposals identified by the same letter are assumed to be mutually exclusive. Therefore no more than one proposal with the same letter can be accepted, but proposals having different letters can be accepted together. The problem is to select the proposal or proposals that maximize total present worth if the amount of money available for investment is \$35,000 and the MARR equals 8%.

Method of Enumeration

No changes at all are required in order to include the budget constraint in the decision process. All that is required is that the proposals be rearranged into mutually exclusive alternatives. To reduce the number of mutually exclusive combinations that must be considered, it is usually worthwhile to first calculate the present worth for the cash flows of each proposal. If any proposal has a present worth that is not positive, it can be eliminated from consideration immediately.

The present worth amounts at 8% for each proposal are shown in the far right-hand column of Exhibit 9.4.5. All the present worths are positive, except for proposal $C1$, which can be immediately dismissed from further consideration. The remaining proposals are arranged into mutually exclusive alternatives, and these alternatives, along with their cash flows and present worth amounts, are shown in Exhibit 9.4.6.

Next, those alternatives that have a first cost that exceeds the budget amount of \$35,000 must be dropped from consideration. Thus the alternative to accept $A1$ and $B2$ and the alternative to accept $A2$ and $B2$ are eliminated since they require more funds than are available.

The last step is to select the remaining alternative that maximizes the present worth. It is seen that alternative 5, representing the acceptance of proposal $B2$, meets this objective with a present worth equal to \$7255. Thus the optimum solution to this problem is to select proposal $B2$ and to reject all the other proposals. It is assumed that the \$5000 that is still available for investment after alternative $B2$ is undertaken will be invested at the MARR.

Usually, the present worth on total investment criterion is the most efficient criterion to use when solving this type of problem by enumerating all the mutually exclusive alternatives. It is important to realize that it would be just as easy and just as correct to use annual equivalent or future worth on the basis of total investment.

Linear Integer Programming Formulation

It is obvious that, when there are large numbers of proposals under consideration, the number of mutually exclusive alternatives that can be formed is quite large. For instance, the number of mutually exclusive alternatives that can be formed from 100 independent proposals is approximately 1.268×10^{30}. Therefore, to solve many problems of practical significance, it is necessary

Exhibit 9.4.6 Cash Flows for the Mutually Exclusive Alternatives From Proposal in Exhibit 9.4.5

Alternatives	Proposals X_{A1}	X_{A2}	X_{B1}	X_{B2}	Proposals Accepted	First Cost	Net Income Years 1–10	Salvage Value Year 10	$PW(8)_j$
1	0	0	0	0	None	\$ 0	\$ 0	\$ 0	\$ 0
2	1	0	0	0	$A1$	–10,000	2000	1,000	3,883
3	0	1	0	0	$A2$	–12,000	2100	2,000	3,018
4	0	0	1	0	$B1$	–20,000	3100	5,000	3,117
5	0	0	0	1	$B2$	–30,000	5000	8,000	7,255[a]
6	1	0	1	0	$A1, B1$	–30,000	5100	6,000	7,000
7	1	0	0	1	$A1, B2$	–40,000	7000	9,000	11,138[b]
8	0	1	1	0	$A2, B1$	–32,000	5200	7,000	6,135
9	0	1	0	1	$A2, B2$	–42,000	7100	10,000	10,273[b]

[a]Alternative that maximizes present worth for a limited budget.
[b]Infeasible because the alternative requires more money than is available for investment (\$35,000).

to be able to utilize mathematical techniques that can consider all possible alternatives without having to make calculations for each alternative.

One such technique is linear integer programming. This technique requires that the problem under consideration be formulated according to a particular format. Once the problem has been properly formulated, there are a number of solution procedures or algorithms that are available to solve problems of such a structure. These algorithms usually converge to the optimum solution in a highly efficient manner. Many of these algorithms are available as computer programs, and the problem solutions generally require only a few minutes of computer time.

The general format of the linear integer programming problem is as follows:

$$\text{Maximize } Z = c_1x_1 + c_2x_2 + \cdots + c_nx_n$$

subject to

$$a_{11}x_1 + a_{12}x_2 + \cdots + a_{1n}x_n \leqslant b_1$$
$$a_{21}x_1 + a_{22}x_2 + \cdots + a_{2n}x_n \leqslant b_2$$
$$\vdots \qquad \vdots \qquad \vdots$$
$$a_{m1}x_1 + a_{m2}x_2 + \cdots + a_{mn}x_n \leqslant b_m$$
$$x_i \text{ for } i = 1, 2, \ldots, n \text{ must be an integer} \geqslant 0$$

The c's, a's, and b's are constants, and the x's are the decision variables representing the values to be determined. The solution to this type of problem is given by the values of the x's such that Z will be maximized and all the constraints (the equations under "subject to") are satisfied.

The problem of making decisions between alternatives is easily converted to the linear integer programming format. The decision variables X_j, rather than being any integer greater than or equal to zero, are confined to the integers 0 (reject proposal j) or 1 (accept proposal j). The value of c_j is the present worth of proposal j at the MARR, $PW(\text{MARR})_j$. The constraints for this type of problem are the result of two different types of relationships. One type of constraint reflects the limitation of the amount of money available for investment. That is, the total first cost of the proposals undertaken must not exceed the amount of the budget. The second type of constraint reflects the relationships between proposals, such as whether the proposals are mutually exclusive, independent, or contingent. Thus, for the decision problem related to the proposals in Exhibit 9.4.5, the integer programming formulation is as follows:

$$\text{Maximize } Z = \$3883X_{A1} + \$3018X_{A2} + \$3117X_{B1} + \$7255X_{B2} - \$173X_{C1}$$

subject to

(budget constraint)

$$\$10{,}000X_{A1} + \$12{,}000X_{A2} + \$20{,}000X_{B1} + \$30{,}000X_{B2} + \$35{,}000X_{C1} \leqslant \$35{,}000$$

(mutually exclusive) constraint

$$X_{A1} + X_{A2} \leqslant 1$$

(mutually exclusive) constraint

$$X_{B1} + X_{B2} \leqslant 1$$

(zero–one constraint) $X_j = 0$ or 1 for all proposals

Solving this problem with existing techniques yields the solution*

$$X_{A1} = 0, X_{A2} = 0, X_{B1} = 0, X_{B2} = 1, X_{C1} = 0$$

Thus proposal $B2$ is accepted, and all the other proposals are rejected as previously shown in Exhibit 9.4.6. The value of Z for this solution is \$7255.

*Techniques for solving integer linear programming problems are presented in Chapter 14.2.

Depreciation and Tax Considerations

In economic analysis the primary importance of depreciation is its effect on estimated cash flows resulting from the payment of income taxes. Depreciation as an amortized cost influences profits as shown on a company's profit and loss statement since depreciation appears as an expense to be deducted from gross income. Income taxes are paid on the resulting net income figure, and these taxes do represent actual cash flows, although the depreciation charges are bookkeeping entries.

In general, an asset must be used for the purpose of producing an income, whether or not an income actually results from its use, in order that a deduction may be made for its depreciation. In cases where an asset such as an automobile is used both as a means for earning income that is taxable and for personal use, a proportional deduction is allowable for depreciation. Intangible property, such as patents, designs, drawings, models, copyrights, licenses, and franchises, may be depreciated.

Straight-Line Method of Depreciation

The straight-line depreciation method assumes that the value of an asset decreases by a uniform amount each year. Thus, if an asset has a first cost of $5000 and an estimated salvage value of $1000, the total depreciation over its life will be $4000. If the estimated life is 5 years, the depreciation per year will be $4000/5 = $800. This is equivalent to a depreciation rate of 1/5, or 20%/year. For this example, the annual depreciation and the book value for each year are given in Exhibit 9.4.7.

Under the straight-line method, the cost of an asset less its estimated salvage value is deducted as an expense in equal annual installments over the asset's useful life. Although the use of straight-line depreciation is widespread, the liberalization of depreciation practices by the Internal Revenue Code of 1954 has favored the adoption of other methods for computing depreciation. Many assets experience more rapid decrease in value during the early portion of their lives, and straight-line calculations do not reflect this condition. However, there are numerous firms, especially public utilities, that continue to account for capital consumption using straight-line depreciation methods.

Declining Balance Method of Depreciation

The declining balance method of depreciation assumes that an asset decreases in value by a greater amount in the early portion of its service life than in the latter portion of its life. For this method a fixed percentage is multiplied by the book value of the asset at the beginning of each year to determine the depreciation charge for that year. Thus, as the book value of the asset decreases through time, so does the size of the depreciation charge. For an asset with a $5000 first cost, a $1000 estimated salvage value, an estimated life of 5 years, and a depreciation rate of 30%/year, the depreciation charge per year is shown in Exhibit 9.4.8.

Exhibit 9.4.7 Straight-Line Method

End of Year t	Depreciation Charge During Year t	Book Value at End of Year t
0	–	$5000
1	$800	4200
2	800	3400
3	800	2600
4	800	1800
5	800	1000

Exhibit 9.4.8 Declining Balance Method

End of Year t	Depreciation Charge During Year t	Book Value at End of Year t
0	–	$5000
1	(0.30)($5000) = $1500	3500
2	(0.30)(3500) = 1050	2450
3	(0.30)(2450) = 735	1715
4	(0.30)(1715) = 515	1200
5	(0.30)(1200) = 360	840

If the declining balance method of depreciation is utilized for income tax purposes, the maximum rate that may be used is double the straight-line rate that would be allowed for a particular asset or group of assets being depreciated. Thus, for an asset with an estimated life of n years, the maximum rate that may be used with this method is $2(1/n)$. Many firms and individuals choose to depreciate their assets using declining balance depreciation with the maximum allowable rate. Such a method of depreciation is commonly referred to as the "double declining balance" method of depreciation.

In Exhibit 9.4.8 the book value at the end of year 5 is \$840, which is less than the estimated salvage of \$1000. If the salvage value for this example had been zero, it is observed that for this method of depreciation the book value would never reach zero, regardless of the length of the time span over which the asset is depreciated. Thus adjustments are necessary in order to reconcile the differences between the estimated and calculated book value of the asset. In most situations these adjustments are made at the time of disposal of the asset, when accounting entries are made to account for the difference between the asset's actual value and its calculated book value.

Sum-of-the-Years-Digits Method of Depreciation

The sum-of-the-years-digits depreciation method assumes that the value of an asset decreases at a decreasing rate. If an asset has an estimated life of 5 years, the sum of the years will be $1 + 2 + 3 + 4 + 5 = 15$. Thus, if the first cost of the asset is \$5000 and the estimated salvage value is \$1000, the depreciation during the first year will be $(\$5000 - \$1000)\frac{5}{15} = \$1333.33$. During the second year, the depreciation will be $(\$5000 - \$1000)\frac{4}{15} = \$1066.67$, and so on. These values are given in Exhibit 9.4.9.

The sum of the years digits for any number of years n can be computed from the expression

$$\sum_{j=1}^{n} j = 1 + 2 + 3 + \ldots + n - 1 + n = \frac{n(n+1)}{2}$$

Thus, for an asset with a 5 year life, the sum of the years digits is $(5)(5 + 1)/2 = 15$, as seen earlier.

The sum-of-the-years-digits method of depreciation produces larger depreciation charges in the asset's early life, with the smaller charges occurring later in the asset's life. In this method the depreciation rate decreases through time, and this decreasing rate is multiplied times a fixed amount, the first cost minus the salvage value. This calculation is in contrast to the declining balance method, which multiplies a fixed rate times a decreasing book value. Both methods are similar in that the depreciation charges in the early portion of an asset's life are greater than during the later part.

Aftertax Economic Analyses

The introduction of income taxes into economic analysis requires special consideration because there are a number of factors that affect the amount of taxes that must be paid for a particular venture. These factors include the net income generated by the project, any interest charges on borrowings associated with the project, the method of depreciation to be applied, and whether any investment credit is appropriate.

The important fact to recognize is that taxes are an actual cash flow, and they must be reflected in any analysis that is intended to represent the true economic impact of an investment. To ignore the effect of taxes on an investment is to disregard some of the most significant costs incurred by those doing business in the private sector of our economy.

Exhibit 9.4.9 Example of the Sum-of-the-Years-Digits Method of Depreciation

End of Year t	Depreciation Charge During Year t	Book Value at the End of Year t
0	–	\$5000
1	$\frac{5}{15}$ (\$4000) = \$1333	3667
2	$\frac{4}{15}$ (4000) = 1067	2600
3	$\frac{3}{15}$ (4000) = 800	1800
4	$\frac{2}{15}$ (4000) = 533	1267
5	$\frac{1}{15}$ (4000) = 267	1000

Exhibit 9.4.10 Tabular Method for Alternative *A*1

End of Year A	Before-Tax Cash Flow B	Depreciation Charges C	Taxable Income[a] (B + C) D	Taxes (−0.4 × D) E	Aftertax Cash Flow (B + E) F
0	−$30,000				−$30,000
1	10,000	−$10,000	$ 0	$ 0	10,000
2	10,000	−8,000	2000	−800	9,200
3	10,000	−6,000	4000	−1600	8,400
4	10,000	−4,000	6000	−2400	7,600
5	10,000	−2,000	8000	−3200	6,800

[a]Consider only those revenues and costs that affect taxable income. (The initial cost of investment has no effect on taxable income. Salvage value affects taxable income only if the actual salvage differs from the estimated salvage.)

The cash flows that represent the actual receipts and disbursements associated with an investment alternative can be either before-tax or aftertax cash flows. Since taxes constitute a substantial portion of the disbursements that are related to an alternative, it is sound decision making to compare the aftertax cash flows of investment alternatives.

To determine the aftertax cash flows of an investment alternative, a tabular method may be used. Tabular methods have the advantage that they can be made to reflect complex situations with simple mathematics. They are also easy for the layperson to understand.

Suppose that a firm has $30,000 available for investment. Three possible alternatives have been suggested, and they are considered to be mutually exclusive. Alternative *A*1 and alternative *A*2 differ only in the method of depreciation used in the calculation of their aftertax cash flows. Alternative *A*1 requires sum-of-the-years-digits depreciation, whereas alternative *A*2 uses the straight-line method. The net before-tax cash flows for alternatives *A*1 and *A*2 are presented in column B of Exhibits 9.4.10 and 9.4.11, respectively. The lives of these alternatives are 5 years, and the estimated salvage value is zero. The effective tax rate is 40%, and the MARR *after taxes* is considered to be 10%.

Alternative *A*3 is somewhat different from *A*1 and *A*2 in that it requires an initial investment of $50,000. Since the firm has only $30,000 available, it must borrow the additional $20,000 required at an interest rate of 8%/year in order to undertake alternative *A*3. This alternative has a life of 5 years, similar to *A*1 and *A*2, but the estimated salvage at the end of its life is expected to be $25,000. Column B of Exhibit 9.4.12 presents the before-tax cash flow of the alternative exclusive of the loan and its associated interest of payments. The cash flow related to the loan is shown in column C of the exhibit. Now that the aftertax cash flows are available for each of the three alternatives, the computation of the present worth amounts for these alternatives is straightforward. For an aftertax MARR of 10%, the present worth on total investment for each alternative is computed.

$$PW(10)_{A1} = -\$30{,}000 + [\$10{,}000 - \$800(\overset{A/G,10,5}{1.8101})](\overset{P/A,10,5}{3.7908}) = \$2419$$

$$PW(10)_{A2} = -\$30{,}000 + \$8400(\overset{P/A,10,5}{3.7908}) = \$1843$$

$$PW(10)_{A3} = -\$30{,}000 + \$8240(\overset{P/A,10,5}{3.7908}) + \$5000(\overset{P/F,10,5}{0.6209}) = \$4341$$

Exhibit 9.4.11 Tabular Method for Alternative *A*2

End of Year A	Before-Tax Cash Flow B	Depreciation Charges C	Taxable Income (B + C) D	Taxes (−0.4 × D) E	Aftertax Cash Flow (B + E) F
0	−$30,000				−$30,000
1	10,000	−$6000	$4000	−$1600	8,400
2	10,000	−6000	4000	−1600	8,400
3	10,000	−6000	4000	−1600	8,400
4	10,000	−6000	4000	−1600	8,400
5	10,000	−6000	4000	−1600	8,400

Exhibit 9.4.12 Tabular Method for Alternative *A*3

End of Year A	Before-Tax Cash Flow B	Cash Flow for Loan C	Cash Flow After Loan Payment D	Depreciation Charges E	Taxable Income (D + E) F	Taxes (−0.4 × F) G	Aftertax Cash Flow (D + G) H
0	−$50,000	$20,000	−$30,000				−$30,000
1	12,000	−1,600	10,400	−$5,000	$5,400	−$2,160	8,240
2	12,000	−1,600	10,400	−5,000	5,400	−2,160	8,240
3	12,000	−1,600	10,400	−5,000	5,400	−2,160	8,240
4	12,000	−1,600	10,400	−5,000	5,400	−2,160	8,240
5	12,000	−1,600	10,400	−5,000	5,400	−2,160	8,240
5	25,000[a]	−20,000	5,000				5,000

[a]Salvage value.

Economically, the most desirable alternative is *A*3. Although alternatives *A*1 and *A*2 are economically the same on a before-tax basis, alternative *A*1 is favored over alternative *A*2 on an aftertax basis. This difference between these two alternatives arises because of the effect the different methods of depreciation have on income taxes.

There are many instances in the evaluation of economic alternatives where only the costs of the alternatives are considered. Most often this occurs where alternatives that are intended to provide the same service are to be compared. Since the benefits to be derived are equal for each alternative, it is common practice to eliminate the estimation of associated revenues to reduce the evaluation effort.

It might appear that ignoring revenues for alternatives would prevent the comparison of these alternatives on an aftertax basis. Fortunately, this is not the case. It can be seen that the direct application of the procedures presented in Exhibits 9.4.10, 9.4.11, and 9.4.12 will permit the incorporation of tax effects in the comparison cash flows having equal revenue streams. By following the sign convention just used, (+) revenues, (−) costs, the tabular method will produce tax-adjusted cash flows that can be directly compared as long as (1) their revenues are assumed to be equal and (2) the firm on the whole is realizing a profit. To see how the procedure operates, examine the cost-only cash flow in Exhibit 9.4.13.

For this example, straight-line depreciation is used. The estimated salvage for the investment is $5000, and the effective tax rate is 42%. Since there are no annual operating revenues being considered, the taxable income appears as $25,000 in costs each year. That is, the effect of this project is to reduce the firm's profits by $25,000/year (if revenues are not considered). Assuming that the firm has sufficient earnings from other activities to offset these costs, there will be a "savings" in taxes of $10,500. By following the sign convention utilized in the previous exhibits, the taxes in column E are positive, indicating that $10,500 in taxes would be avoided yearly if the costs in columns B and C are incurred.

The aftertax cash flow is found by adding these tax savings to the before-tax cash flow shown in column B. By comparing this aftertax cash flow with a similarly calculated aftertax cash flow for a competing alternative, the alternative with the minimum equivalent aftertax cost can be identified. Thus, without any change in procedure, an aftertax comparison can be made for projects having the same revenue stream. Since this comparison does not require consideration of the possible revenue effects of the alternatives, much time and effort may be saved in performing the analysis.

The tabular method for developing aftertax cash flows allows the analyst the flexibility to include many other tax considerations not discussed here. For example, tax considerations such as capital gains and losses and the investment credit are easily accounted for when using this approach.

Exhibit 9.4.13 Tax Effects for a Cost Cash Flow

End of Year A	Before-Tax Cash Flow B	Depreciation Charges C	Taxable Income (B + C) D	Taxes (Savings) (−0.42 × D) E	Aftertax Cash Flow (B + E) F
0	−$45,000				−$40,000
1	−15,000	−$10,000	−$25,000	$10,500	−4,500
2	−15,000	−10,000	−25,000	10,500	−4,500
3	−15,000	−10,000	−25,000	10,500	−4,500
4	−15,000	−10,000	−25,000	10,500	−4,500
4	5,000[a]				5,000

[a]Salvage value.

If the tax considerations are complex, it is suggested that the analyst rely on tax experts to determine the actual taxes that will be incurred if the project is undertaken. These tax cash flows are then inserted in the column D, and the analyst proceeds as previously described.

9.4.2 METHODS OF ANALYSIS

Replacement Analysis

To facilitate the discussion of the principles involved in replacement analysis, it is necessary to introduce some important terms commonly used in replacement studies. Two terms representing interpretations that are widely accepted by practitioners involved in replacement analyses are "defender," or the existing old asset being considered as the asset to be replaced, and "challenger," or the asset proposed to be the replacement.

Because the economic characteristics of the defender and the challenger are usually so dissimilar, special attention is required when these two options are compared. One obvious feature of replacement alternatives is that the duration and the magnitude of cash flows for old existing assets and new assets are quite different. New assets characteristically have high capital costs and low operating costs. The reverse is usually true for assets that are being considered for retirement. Thus capital costs for an asset to be replaced may be expected to be low and decreasing, whereas operating costs are usually high and increasing.

In addition, the remaining life of an asset being considered for replacement is usually short, and the future of the asset can be estimated with relative certainty. There is also the advantage that a decision not to replace it now may be reversed at any time in the future. Thus a decision may be made on the basis of next year's cost of the old asset, and if it is not replaced, a new decision can be made on the basis of next year's cost a year later, and so on.

Evaluation of Replacements Involving Sunk Costs

The method of treating data relative to an existing asset should be the same as that used in treating data relative to a possible replacement. In both cases, only the future of the assets should be considered, and *sunk costs should be disregarded*. Thus the value of the defender that should be used in a study of replacement is not what it cost when originally purchased, but what it is worth at the present time.

The following example will be used to illustrate correct and incorrect methods of evaluating replacements where sunk costs are involved. Suppose that machine A was purchased 4 years ago for $2200. It was estimated to have a life of 10 years and a salvage value of $200 at the end of its life. Its operating expense had been found to be $700/year, and it appeared that the machine would serve satisfactorily for the balance of its estimated life. Currently, a salesperson is offering machine B for $2400. Its life is estimated at 10 years, and its salvage value at the end of its life is estimated to be $300. Operating costs are estimated at $400/year.

The operation for which these machines are used will be carried on for many years in the future. Equipment investments are expected to justify a 15% MARR, in accordance with the policy of the company concerned. The salesperson offers to take the old machine in on trade for $600. This appears low to the company, but the best offer received elsewhere is $450. All estimates relative to both machines have been carefully reviewed and are considered sound.

To make a proper comparison of alternatives, the analysis may be undertaken from the standpoint of a person who has a need for the service that machine A or machine B will provide, but who owns neither. In attempts to purchase a machine, the person finds that he or she can purchase machine A for $600 and machine B for $2400. This analysis of which to buy will not be biased by the past since he or she was not part of the original transaction for machine A and therefore will not be forced to admit a sunk cost. With this *outsider viewpoint*, the appropriate cash flows are presented in Exhibit 9.4.14. The important effect of using the outsider viewpoint is that the old machine's present market value is identified as the investment required to continue its use. This method of analysis is correct, even though the retention of the old machine requires no actual disbursement at the present.

If machine A is retired, its original investment of $2000 should be ignored. If an outsider were to consider this problem, he or she would have to anticipate paying $600 for machine A because this figure represents its worth at the present. That is, the $600 amount that would be received if machine B were purchased and machine A were "sold" represents the present best estimate of its worth. If the outsider were to purchase machine B, he or she would pay its present market price of $2400, not having an asset to trade in. Thus the logical alternatives are (1) to consider machine A to have a value of $600 and to continue with it for 6 years and (2) to purchase machine B for $2400 and use it for 10 years. Because these alternatives have different service lives, the study period approach discussed previously is applicable. A study period of 6 years is assumed.

Exhibit 9.4.14 Outsider Viewpoint for a Replacement Problem

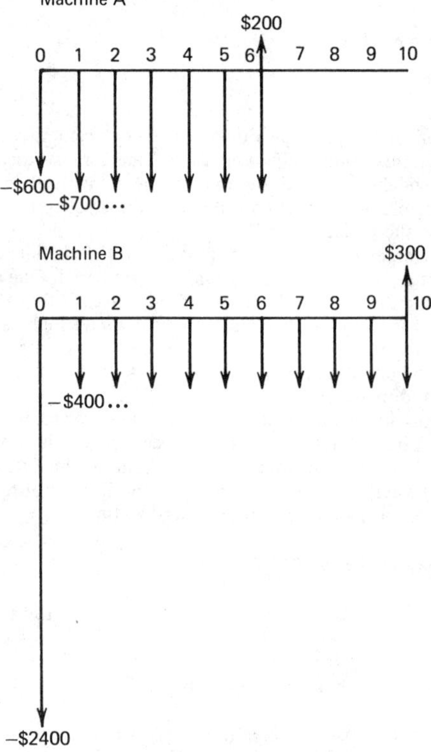

The equivalent annual cost to continue with machine A for 6 years is calculated as follows:

$$\text{Annual capital recovery with return, } (\$600 - \$200)\overset{A/P,15,6}{(0.2642)} + \$200(0.15) = \$136$$
$$\text{Annual operating cost} = \underline{700}$$
$$\$836$$

The equivalent annual cost to dispose of machine A, purchase machine B, and use B for 10 years is calculated as follows:

$$\text{Annual capital recovery with return, } (\$2400 - \$300)\overset{A/P,15,10}{(0.1993)} + \$300(0.15) = \$464$$
$$\text{Annual operating cost} = \underline{400}$$
$$\$864$$

If the alternative to continue with machine A is adopted, the annual saving prospect for the next 6 years is $864 – $836 = $28. For the next 4 years after that time, the amount of savings will depend upon the characteristics of the machine that might have been purchased 6 years from the present to replace machine A. If it is assumed that machine A will be replaced after 6 years by a machine identical to machine B, the equivalent annual costs of the two alternatives will be the same after the first 6 years.

The Economic Life of an Asset

The preceding section has discussed the types of analyses that may be applied when the service life is known. However, there are many instances when the length of time a particular asset will be retained is only conjecture. Since replacement analyses are usually sensitive to the lives assumed, it is prudent to consider each alternative in its most favorable circumstances. Thus, *when comparing an existing asset and its possible replacement, the lives that should be assumed are the lives that*

Exhibit 9.4.15 Tabular Calculation of Economic Life

End of Year n	Salvage Value When Asset Retired at Year n	Operating Costs During Year n	Annual Equivalent Costs of Capital Asset When Retired at Year n	Annual Equivalent Cost of Operating for n Years	Total Annual Equivalent Cost When Asset Retired at Year n
1	\$1500	\$1000	\$1860	\$1000	\$2860
2	1000	1700	1303	1330	2633[a]
3	500	2400	1101	1647	2748
4	0	3100	987	1951	2938

[a]Economic life.

are most favorable to each asset. In other words, the comparison should be made on the basis of each alternative's *economic life*.

The "economic life" of an asset is the time interval that minimizes the asset's total equivalent annual costs or maximizes its equivalent annual net income. The economic life is also referred to as the "minimum cost life" or the "optimum replacement interval."

Because in replacement analysis the defender and challengers should be compared on the basis of the lives most favorable to each, some time needs to be devoted to describing how to calculate the economic life of an asset. If the future could be predicted with certainty, it would be possible to predict accurately the economic life for an asset at the time of its purchase. The analysis would simply involve the calculation of the total equivalent annual cost at the end of each year in the life of the asset. Selection of the total equivalent annual cost that is a minimum would specify a minimum cost life for the asset. The application of this approach is demonstrated by the following example.

The economic future of an asset whose first cost is \$3000, with decreasing salvage values, and operating costs beginning at \$1000 and increasing by \$700 each year for an interest rate of 12% is shown in Exhibit 9.4.15. To find this asset's economic life, it is necessary to identify the relevant cash flows associated with retaining the asset 1, 2, 3, or 4 years. These cash flows are depicted in Exhibit 9.4.16, and they are the basis for the annual equivalent calculations shown in Exhibit 9.4.15. The economic life for this asset is 2 years. If the asset were sold after 2 years, it would have a minimum annual equivalent cost of \$2633/year, and that is the life that is most favorable for comparison purposes.

Replacement Analysis Based on Economic Life

Two assets must be evaluated at the time replacement is being considered: the defender and its challenger. However, because of the cost patterns normally associated with these assets, there may

Exhibit 9.4.16 Cash Flows of Asset Retained 1, 2, 3, or 4 Years

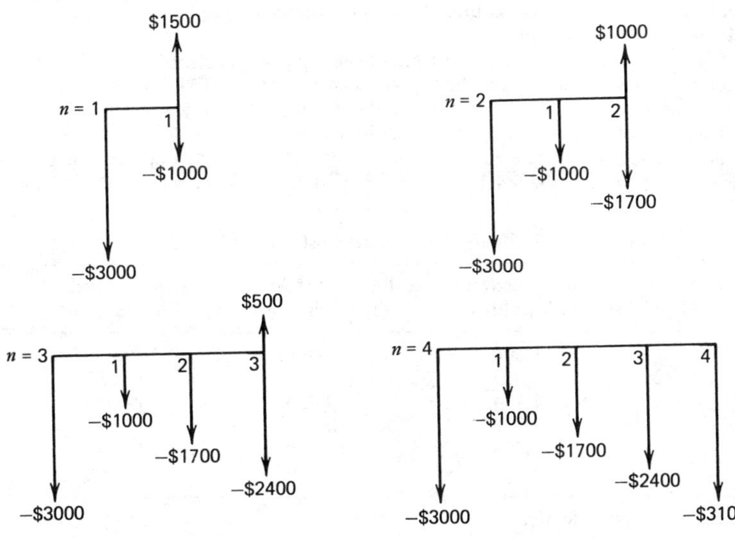

be a number of alternatives that must be considered. For each asset, it may be necessary to evaluate the equivalent annual cost of keeping each asset for 1, 2, 3, . . . years. Having computed the equivalent costs associated with each of these alternatives, the economic lives for the defender and the challenger are easily identified. Then the comparison is reduced to just two alternatives, the defender and the challenger retained for their respective economic lives.

The basic elements that should be considered when undertaking a replacement study can be summarized as follows:

1. Sunk costs should be ignored. (Use the outsider viewpoint.)
2. Find the economic lives of the assets under consideration. (Use the lives most favorable to each asset.)
3. Compare the replacement alternatives. (Use the principles of comparison for assets with unequal lives presented in Section 9.4.1.)

To illustrate the comparison of alternatives on the basis of their economic life, consider the following example. Three years ago a chemical processing plant installed a system at a cost of \$20,000 to remove pollutants from the wastewater that is discharged into a nearby river. The present system has no present salvage value and will cost \$14,500 to operate next year, with operating costs expected to increase at the rate of \$500/year thereafter. A new system has been designed to replace the existing system, and its installed cost is expected to be \$10,000. The new system is expected to have first-year operating costs of \$9000, with these costs increasing at a rate of \$1000/year. The new system is estimated to have a useful life of 12 years. Because the original system and the new system are specially designed for this particular chemical process, their salvage values at any future time are expected to be equal to zero. Should the company replace the existing pollution control system if their MARR is 12%?

First, the \$20,000 initial cost of the existing plant is ignored as a sunk cost. Second, the economic life of the old system must be found. However, since the present and future salvage value for the old system are zero, and the operating expenses are increasing, the economic life is 1 year.

The total equivalent annual costs that will be incurred if the old system is retained for n more years are as follows:

$$\text{Total equivalent annual costs} = \text{capital recovery with return} - \text{equivalent annual operating costs}$$

$$\text{Total equivalent annual costs for } n \text{ years} = [(P - F)(\overset{A/P,12,n}{\quad}) + F(0.12)] + [\$14{,}500 + \$500(\overset{A/G,12,n}{\quad})]$$

Since $P = F = \$0$, the capital recovery with return for any n is equal to zero. Thus, for the pollution control system currently in service, the annual operating costs are the total costs to be incurred if the system is retained. Since the annual operating costs are increasing each year, the equivalent annual operating costs are also increasing for each additional year the present system is retained. With this pattern of increasing costs, the life for which the total equivalent annual costs will be minimized is the shortest possible life, 1 year. The total equivalent annual costs for the old system retained 1 more year is \$14,500.

Next, it is necessary to find the economic life for the new system. In this case the total equivalent annual costs are calculated for the first 6 years, as shown in Exhibit 9.4.17. These costs would continue to increase if the new system is operated for more than 6 years. The economic life for the new system is 5 years, and the total annual equivalent cost is \$13,544.

These alternatives can now be compared on a basis that is most favorable to each alternative. The methods presented in Section 9.4.1 for comparing assets with unequal lives should now be applied.

Exhibit 9.4.17 Equivalent Annual Cost for the New System

n	Capital Recovery With Return	Equivalent Annual Operating Costs	Total Equivalent Annual Cost
1	\$11,200	\$ 9,000	\$20,200
2	5,917	9,470	15,387
3	4,164	9,920	14,084
4	3,292	10,360	13,652
5	2,774	10,770	13,544[a]
6	2,432	11,170	13,602

[a]Economic life.

The conclusion is that the new system, if kept for its economic life, is preferred to operating the old system 1 more year. If the new system is adopted, its cost pattern should be reviewed with regard to any future systems that might be an improvement over the existing system. This review procedure should occur periodically.

Project Evaluation Considering Inflation

To incorporate the effects of inflation in economic studies, it is necessary to use the interest factors so that the inflationary effects on dollars occurring at different times can be recognized. The usual procedure for dealing with the loss in buying power that accompanies inflation is to follow these steps:

1. Estimate all the costs associated with a project in terms of today's dollars. (Today's dollars, or constant dollars, are the dollars in terms of purchasing power relative to some fixed time, usually the beginning of the project, that are required to purchase future goods or services associated with that project.)
2. Modify the costs estimated in step 1 so that at each future date they represent the cost at that date in terms of the future dollars that must be expended at that time. (Future dollars, or actual dollars, are the actual out-of-pocket dollars that would be expended to acquire the goods or services required by the project at future times.)
3. Calculate the present worth of the cash flow based on the future dollars resulting from step 2 by using the MARR. (The MARR is considered here to reflect the market rate of interest, i, which includes inflation effects.)

Suppose the cash flows for the four alternatives presented in Exhibit 9.4.18 have been estimated in terms of today's dollars. If the inflation rate is 9%, the cash flows shown in Exhibit 9.4.19 represent the investments' cash flows after they have been transformed to future dollars at the 9% rate. For example, the actual cost at the end of year 2 for alternative $B1$ will be

$$-\$2500(\overset{F/P,9,2}{1.188}) = -\$2970$$

To compare these four alternatives, compute the present worth of each alternative using the MARR, which for this problem is 10%.

$$PW(10)_{B1} = -\$10{,}000 - \$2725(\overset{P/F,10,1}{0.9091}) - \$2970(\overset{P/F,10,2}{0.8265}) + \$1295(\overset{P/F,10,3}{0.7513}) = -\$13{,}959$$

$$PW(10)_{B2} = -\$12{,}000 - \$1635(\overset{P/F,10,1}{0.9091}) - \$1782(\overset{P/F,10,2}{0.8265}) + \$1943(\overset{P/F,10,3}{0.7513}) = -\$13{,}499$$

$$PW(10)_{B3} = -\$12{,}000 - \$1308(\overset{P/F,10,1}{0.9091}) - \$1426(\overset{P/F,10,2}{0.8265}) + \$1943(\overset{P/F,10,3}{0.7513}) = -\$12{,}908$$

$$PW(10)_{B4} = -\$15{,}000 - \$436(\overset{P/F,10,1}{0.9091}) - \$475(\overset{P/F,10,2}{0.8265}) + \$3885(\overset{P/F,10,3}{0.7513}) = -\$12{,}870$$

For an inflation rate of 9%, it is observed that alternative $B4$ is the least-cost alternative. In contrast, for no inflation the least-cost alternative was $B3$.

The procedure just described requires considerable computational effort, but it produces inter-

Exhibit 9.4.18 Net Cash Flows for Four Alternatives Providing the Same Service (Today's Dollars)

End of Year	Alternatives			
	*B*1	*B*2	*B*3	*B*4
0	-\$10,000	-\$12,000	-\$12,000	-\$15,000
1	-2,500	-1,500	-1,200	-400
2	-2,500	-1,500	-1,200	-400
3	1,000[a]	1,500	1,500	3,000

[a] Positive values can arise as the result of the salvage value received when the asset is sold at the end of its life.

Exhibit 9.4.19 Inflated Cash Flows for Alternatives in Exhibit 9.4.18 (Future Dollars)

End of Year	Alternatives			
	*B*1	*B*2	*B*3	*B*4
0	–$10,000	–$12,000	–$12,000	–$15,000
1	–2,725	–1,635	–1,308	–436
2	–2,970	–1,782	–1,426	–475
3	1,295	1,943	1,943	3,885

mediate values that can be easily related to the investor's actual experience. When computer calculation is possible, this method is the one most analysts adopt.

Another method that provides the same results with less calculation is to find i', the interest rate without inflation. Then the estimates in today's dollars are converted directly to the present worth amount. An example of this calculation is as follows:

$$(1 + i') = \frac{(1 + i)}{(1 + \lambda)} = \frac{(1.10)}{(1.09)} = (1.009174)$$

$$i' = 0.9174\%$$

For alternative $B1$, the calculation of present worth for $i' = 0.9174\%$ gives

$$PW_{B1} = -\$10{,}000 - \$2500(\overset{P/A,0.9174,2}{1.97281}) + \$1000(\overset{P/F,0.9174,3}{0.97297}) = -13.959$$

This amount is identical to the present worth of alternative $B1$ calculated at 10% when the cash flows were inflated at 9%.

Applying the approach just discussed to the alternatives in the subchapter on present worth on incremental investment if the inflation rate is 10% and the MARR is 15% yields the following results. (It is assumed that the cash flows are stated in terms of today's dollars.)

$$i' = \frac{(1 + i)}{(1 + \lambda)} - 1 = \frac{(1.15)}{(1.10)} - 1 = 4.55\%$$

$$PW(4.55)_{A1} = -\$\ 5000 + \$1400(\overset{P/A,4.55,10}{7.8933}) = \$6051$$

$$PW(4.55)_{A2} = -\$\ 8000 + \$1900(\overset{P/A,4.55,10}{7.8933}) = \$6997$$

$$PW(4.55)_{A3} = -\$10{,}000 + \$2500(\overset{P/A,4.55,10}{7.8933}) = 9733$$

With inflation considered, alternative $A3$ is preferred over the other two alternatives. (This same conclusion was reached when no inflation was assumed.) However, it is observed that alternative $A2$ is preferred to alternative $A1$ for a 9% rate of inflation, while, as seen earlier, assuming no inflation caused the preference to be reversed. Therefore, when comparing economic alternatives, it is essential that the effects of inflation be included in the analysis.

Lease or Purchase

Most physical assets are either owned or leased by the firm that utilizes them. The economic advantage of owning or leasing can be determined by evaluating the aftertax cash flow that is associated with each of these options.

When the assets are to be purchased outright, there are three primary sources of capital. The purchaser can borrow the funds, use retained earnings, sell stock to raise the capital, or use some combination of these approaches. Because the effect on income taxes of utilizing borrowed funds is different from investing equity funds, an aftertax analysis is performed so that these different effects are accurately reflected. The following example will demonstrate, for a single asset, the economic effects of obtaining that asset with equity funds, with borrowed funds, or by leasing.

The asset to be considered has a first cost of $10,000, and it is estimated that its annual operating expenses for the next 4 years will be $3000/year. The asset's salvage value at the end of 4 years is estimated to be \$2000, and straight-line depreciation will be applied. The effective tax rate for this firm is 45%, and the MARR after taxes is 15%.

Equity Funds

The effect of financing investments from retained earnings is essentially the same as utilizing equity funds. Therefore the calculation of the aftertax cash flow shown in Exhibit 9.4.20 would be the same whether stock were sold or funds accumulated from previous investments were utilized to finance the purchase.

Since no funds are borrowed, there is no loan cash flow in this example. The column "cash flow after loan" is the before-tax cash flow reflecting the out-of-pocket costs experienced by the firm. Once the taxes are calculated, they (in this case "tax savings") are combined with the "cash flow after loan" column to obtain the "aftertax cash flow" column. The present worth of the aftertax costs is

$$PW(15) = -\$10{,}000 - 750(\overset{P/A,15\%,4}{2.8550}) + \$2000(\overset{P/F,15\%,4}{0.5718}) = -\$10{,}997$$

Borrowed Funds

When the funds are borrowed, the interest to be repaid by the borrower is deducted as an expense before taxes are calculated. In the example shown in Exhibit 9.4.21, the loan is for \$10,000, with an 18% annual rate of interest. For the first 3 years of the loan, interest *only* will be repaid with the principal, and the last year's interest will be paid at the end of the fourth year.

The present worth of the aftertax cash flow for the borrowing option yields

$$PW(15) = -\$1740(\overset{P/A,15,4}{2.855}) - \$8000(\overset{P/F,15,4}{0.5418}) = -\$9542$$

Because the interest payments are tax deductible, it is less costly to finance this asset by borrowing. Even though the before-tax cost of borrowing is 18%, which is greater than the aftertax MARR of 15%, it is advantageous to borrow. This advantage will continue to exist as long as the aftertax cost of borrowing, (1 – 0.45)(18%) = 9.9%, remains less than the aftertax MARR.

Leasing

When a lease arrangement is utilized to obtain an asset, it must be recognized that the lessee cannot depreciate the asset since it is not owned by him or her. Suppose that, for the example being considered, the lease requires yearly payments of \$2900, with the lessee continuing to pay the \$3000 annual operating cost. The aftertax cash flow is computed as shown in Exhibit 9.4.22. Based on

Exhibit 9.4.20 Calculation of Aftertax Cash Flow for Equity Funds

End of Year	Before-Tax Cash Flow	Loan Cash Flow	Cash Flow After Loan	Depreciation	Taxable Income	Taxes	Aftertax Cash Flow
0	–\$10,000	\$0	–\$10,000	–	–	–	–\$10,000
1–4	–3,000	0	–3,000	–2000	–5000	+2250	–750
4	2,000	0	2,000	–	–	–	+2,000

Exhibit 9.4.21 Calculation of Aftertax Cash Flow for Borrowed Funds

End of Year	Before-Tax Cash Flow	Loan Cash Flow	Cash Flow After Loan	Depreciation	Taxable Income	Taxes	Aftertax Cash Flow
0	–\$10,000	\$10,000	\$ 0	–	–	–	\$ 0
1–4	–3,000	–1,800	–4800	–2000	–6800	+3060	–1740
4	+2,000	–10,000	–8000	–	–	–	–8000

Exhibit 9.4.22 Calculation of the Aftertax Cash Flow for Leasing

End of Year	Before-Tax Cash Flow	Depreciation	Taxable Income	Taxes	Aftertax Cash Flow
0	\$ 0	–	–	–	\$ 0
1–4	–5900	\$0	–\$5900	+\$2655	–3245
4	–	–	–	–	–

this particular lease agreement, the present worth on an aftertax basis gives

$$PW(15) = -\$3245\overset{P/A,15,4}{(2.8550)} = -\$9264$$

This amount indicates that, of the three options of financing available in the example, the least costly approach is to lease, followed by borrowing, with the use of equity funds being least desirable. It should be recognized that each method of financing must be evaluated on its own merits and that the conclusion reached for the example presented should not be generalized.

Break-Even Analysis of Alternatives

When the cost of two alternatives is a function of the same variable, it is usually useful to find the value of the variable for which the alternatives incur equal cost. The first step is to express the cost of each alternative as a function of the common decision variable as follows:

$$TC_1 = f_1(x) \quad \text{and} \quad TC_2 = f_2(x)$$

where TC_1 = fixed plus variable cost for alternative 1
TC_2 = fixed plus variable cost for alternative 2
x = common decision variable affecting alternatives 1 and 2

Next, the value of x resulting in equal cost for alternatives 1 and 2 is sought. This is accomplished by setting $TC_1 = TC_2$, giving $f_1(x) = f_2(x)$. Solution for x will give equal cost for the alternatives. It is called the "break-even point."

Equipment Selection

Suppose that a fully automatic attachment for a machine tool can be fabricated for \$1400 and that it will have an estimated salvage value of \$200 at the end of 4 years. Maintenance cost will be \$120/year, and the cost of operation will be \$0.85/hr.

As an alternative, a semiautomatic attachment can be fabricated for \$550. This device will have no salvage value at the end of a 4 year service life. The cost of operation and maintenance is estimated to be \$1.40/hr.

With an interest rate of 10%, the annual equivalent total cost for the automatic attachment as a function of the number of hours of use per year is

$$\begin{aligned} TC_A &= (\$1400 - \$200)\overset{A/P,10,4}{(0.3155)} + \$200(0.10) + \$120 + \$0.85N \\ &= \$378 + \$20 + \$120 + \$0.85N \\ &= \$518 + \$0.85N \end{aligned}$$

and the annual equivalent total cost for the semiautomatic attachment as a function of the number of hours of use per year is

$$\begin{aligned} TC_S &= (\$550)\overset{A/P,10,4}{(0.3155)} + \$1.40N \\ &= \$174 + \$1.40N \end{aligned}$$

Break-even occurs when $TC_A = TC_S$, or

$$\begin{aligned} \$518 + \$0.85N &= \$174 + \$1.40N \\ \$0.55N &= \$344 \\ N &= 625 \text{ hr} \end{aligned}$$

Exhibit 9.4.23 An Equipment Selection Decision

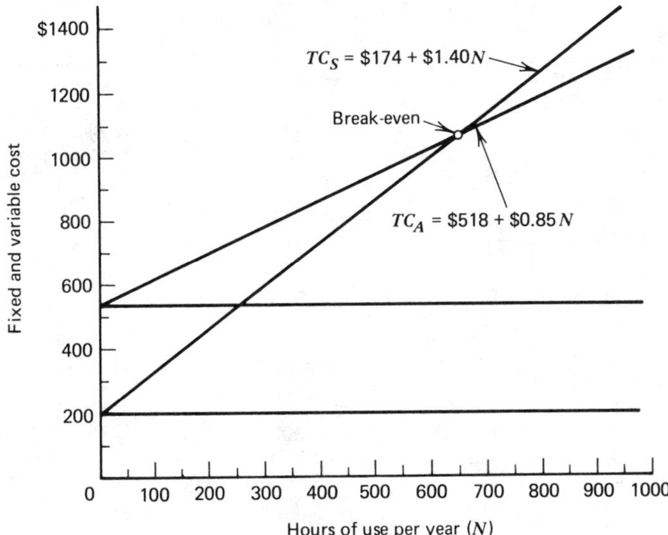

Exhibit 9.4.23 shows the two cost functions and the break-even point. For rates of use exceeding 625 hr/year, the automatic attachment would be more economical. However, if it is anticipated that the rate of use will be less than 625 hr/year, the semiautomatic device should be used.

Equipment Selection for Expanding Operations

In the early stages of an enterprise, when production volume is low, it will usually prove best to purchase equipment whose fixed costs are low. In the latter stages, when sales are approaching the ultimate level, high fixed-cost equipment permitting low variable production costs may be most economical. Consider the following example.

Suppose it is estimated that annual sales of a new product will begin at 1000 units the first year and increase by increments of 1000 units/year until 4000 units are sold during the fourth and subsequent years. Two proposals for equipment to manufacture the product are under consideration.

Proposal A involves equipment requiring an investment of approximately $10,000. Annual fixed cost with this equipment is calculated to be $2000, and the variable cost per unit of product will be $0.90. The life of the equipment is estimated at 4 years.

Proposal B involves equipment requiring an investment of approximately $20,000. Fixed cost of this equipment is estimated at $3800/year, and variable cost per unit of product will be $0.30. The life of this equipment is also estimated at 4 years.

On the basis of the ultimate annual production of 4000 units, the cost per unit for proposal A will be

$$\frac{\$2000 + (4000 \times \$0.90)}{4000} = \$1.40$$

and the cost per unit for proposal B will be

$$\frac{\$3800 + (4000 \times \$0.30)}{4000} = \$1.25$$

On the basis of the ultimate rate of production, proposal B is superior to proposal A. On the basis of the total production during the life of the equipment, the analysis in Exhibits 9.4.24 and 9.4.25 applies.

The calculated advantage of proposal A over proposal B would have been increased by considering the time value of money. But perhaps even more important than the difference in cost per unit is the lesser investment required by proposal A. This is particularly important for a new enterprise that must conserve its funds or where there is considerable uncertainty regarding the sales volume.

Exhibit 9.4.24 Expanding Operations (Proposal A)

Year of Life	No. of Units Made	Fixed Cost	Variable Cost
1	1,000	$2000	1000 × $0.90 = $ 900
2	2,000	2000	2000 × $0.90 = 1800
3	3,000	2000	3000 × $0.90 = 2700
4	4,000	2000	4000 × $0.90 = 3600
	10,000	$8000	$9000

Cost per unit = ($8000 + $9000)/10,000 = $1.70

Exhibit 9.4.25 Expanding Operations (Proposal B)

Year of Life	No. of Units Made	Fixed Cost	Variable Cost
1	1,000	$ 3,800	1000 × $0.30 = $ 300
2	2,000	3,800	2000 × $0.30 = 600
3	3,000	3,800	3000 × $0.30 = 900
4	4,000	3,800	4000 × $0.30 = 1200
	10,000	$15,200	$3000

Cost per unit = ($15,200 + $3000)/10,000 = $1.82

Profit Analysis

There are two aspects of production operations. One consists of assembling production facilities, material, and labor for the production of goods or services. The other consists of the sale of the goods or services. The economic success of an enterprise depends upon its ability to carry on these activities to the end that there may be a net difference between receipts and the cost of production.

If receipts and costs are assumed to be linear functions of the quantity of product to be made and sold, analysis of their relationship to profit is greatly simplified. In this section both mathematical and graphic break-even models are presented for the analysis of profit.

Formulation of the Linear Break-Even Model. Under the assumption of linearity, the patterns of income and costs will appear as in Exhibit 9.4.26. Fixed production costs are represented by the line HL. The sum of variable production cost and fixed production cost is represented by the line HK. Income from sales is represented by the line OJ.

Analysis of existing or proposed production operations can be made mathematically if the linear condition exists. Let

N = number of units of product made and sold per year
R = amount received per unit of product in dollars (R is equal to the slope of OJ)

Exhibit 9.4.26 General Graphic Representation for Income, Cost, Units of Output per Year

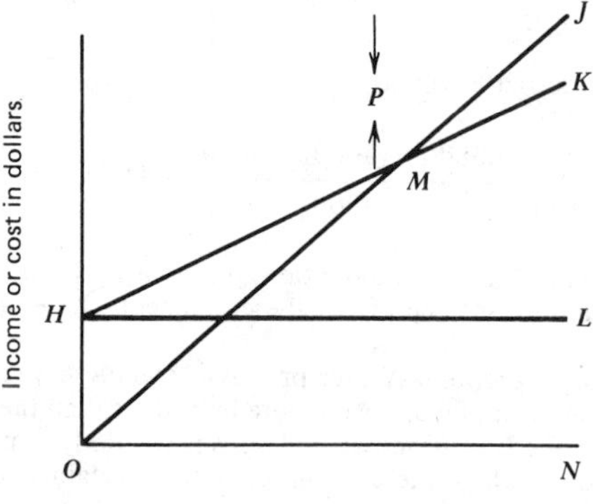

$I = RN$, the annual income from sales in dollars ($I = RN$ is the equation of line OJ)
F = fixed cost in dollars per year, represented by OH and HL
V = variable cost per unit of product (V is equal to the slope of HK)
TC = sum of fixed and variable cost of N units of product, $F + VN$ ($TC = F + VN$ is the equation of line HK)
P = annual profit in dollars per year ($P = I - TC$; negative values of P represent loss)
M = break-even point (At this point $P = O$)

The break-even point occurs when income is equal to cost. In Exhibit 9.4.26 this occurs where lines OJ and HK intersect. At this point $I = C$ and $RN = F + VN$. Solving for N,

$$N = \frac{F}{R - V}$$

If $F/(R - V)$ is substituted for N in $I = RN$, the income at the break-even point may be found. It is

$$I = R\left(\frac{F}{R - V}\right)$$

Likewise, if $F/(R - V)$ is substituted for N in $TC = F + VN$, the cost at the break-even point may be found. It is

$$TC = F + \frac{VF}{R - V}$$

Since P is the annual profit, it is often desirable to have a relationship that expresses P as a function of the number of units made and sold, N. This relationship may be derived as follows:

$$\begin{aligned} P &= I - TC \\ &= RN - (F + VN) \\ &= (R - V)N - F \end{aligned}$$

Profit Improvement With Break-Even Analysis. Any change in the selling price or the production cost will affect the break-even point. Changes that will lower the break-even point are usually desirable in that the firm will be able to meet costs at a lower level of output.

Suppose that a firm is currently operating with an annual fixed cost of \$400,000, a revenue of \$11/unit, and a variable cost of \$6/unit. The break-even point under these conditions is

$$N = \frac{F}{R - V} = \frac{\$400{,}000}{\$11 - \$6} = 80{,}000 \text{ units/year}$$

If 100,000 units are being produced per year, the annual profit is

$$\begin{aligned} P &= (R - V)N - F \\ &= (\$11 - \$6)\ 100{,}000 - \$400{,}000 \\ &= \$500{,}000 - \$400{,}000 \\ &= \$100{,}000 \end{aligned}$$

If a worker training program is implemented that will reduce the variable cost to \$5.75/unit because of a reduction in the cost of direct labor, the break-even point will be

$$N = \frac{\$400{,}000}{\$11 - \$5.75} = 76{,}200 \text{ units/year}$$

At 100,000 units of production per year, the annual profit will now be

$$\begin{aligned} P &= (\$11 - \$5.75)\ 100{,}000 - \$400{,}000 \\ &= \$525{,}000 - \$400{,}000 \\ &= \$125{,}000 \end{aligned}$$

Thus up to \$125,000 – \$100,000 = \$25,000 could be spent on the training program to recover its cost in 1 year.

Often it is possible to reduce the break-even point and increase profit by reducing fixed cost. For example, suppose that the firm is considering a consolidation of certain production equipment that will save \$30,000/year in supervision and related fixed costs. With a revenue of \$11/unit and a variable cost of \$6/unit, the break-even point will be

$$N = \frac{\$370{,}000}{\$11 - \$6} = 74{,}000 \text{ units/year}$$

At 100,000 units of production per year, the annual profit will now be

$$\begin{aligned} P &= (\$11 - \$6)\ 100{,}000 - \$370{,}000 \\ &= \$500{,}000 - \$370{,}000 \\ &= \$130{,}000 \end{aligned}$$

As with the training program example, two benefits will be experienced. First, the profit is improved at the current level of production. In addition, the level of production may fall below the original break-even value before the firm begins to experience a loss.

As a third method for reducing the break-even point, consider an advertising campaign that will make it possible to sell units for \$11.25. The break-even point would be

$$N = \frac{\$400{,}000}{\$11.25 - \$6} = 76{,}190 \text{ units/year}$$

At a sales volume of 100,000 units/year, the annual profit will now be

$$\begin{aligned} P &= (\$11.25 - \$6)\ 100{,}000 - \$400{,}000 \\ &= \$525{,}000 - 400{,}000 \\ &= \$125{,}000 \end{aligned}$$

Thus up to \$125,000 – \$100,000 = \$25,000 could be spent on the advertising campaign per year.

BIBLIOGRAPHY

AMERICAN TELEPHONE & TELEGRAPH COMPANY, *Engineering Economy*, 3rd ed., McGraw-Hill, New York, 1977.

BARISH, N. N., and S. KAPLAN, *Economic Analysis for Engineering and Managerial Decision Making*, 2nd ed., McGraw-Hill, New York, 1978.

BIERMAN, H., and S. SMIDT, *The Capital Budgeting Decision*, 3rd ed., Macmillan, New York, 1971.

BLANCHARD, B. S., *Life Cycle Cost*, M/A Press, Portland, OR, 1978.

BUSSEY, L. E., *The Economic Analysis of Industrial Projects*, Prentice-Hall, Englewood Cliffs, NJ, 1978.

CANADA, J. R., and J. A. WHITE, Jr., *Capital Investment Decision Analysis for Management and Engineering*, Prentice-Hall, Englewood Cliffs, NJ, 1980.

CHERNOFF, H., and L. E. MOSES, *Elementary Decision Theory*, Wiley, New York, 1959.

CLIFTON, D. S., and D. E FYFFE, *Project Feasibility Analysis*, Wiley, New York, 1977.

COHN, E., *Public Expenditure Analysis*, Heath, Lexington, MA, 1972.

COUGHLAN, J. D., and W. K. STRAND, *Depreciation*, Ronald, New York, 1969.

DEAN, J., *Managerial Economics*, Prentice-Hall, Englewood Cliffs, NJ, 1961.

DEGARMO, E. P., J. R. CANADA, and W. G. SULLIVAN, *Engineering Economy*, 6th ed., Macmillan, New York, 1979.

ENGLISH, J. M., *Cost Effectiveness*, Wiley, New York, 1968.

FABRYCKY, W. J., and G. J. THUESEN, *Economic Decision Analysis*, 2nd ed., Prentice-Hall, Englewood Cliffs, NJ, 1980.

FLEISCHER, G. A., *Capital Allocation Theory*, Appleton-Century-Crofts, New York, 1969.

GILLIS, F. E., *Managerial Economics: Decision Making Under Certainty for Business and Engineering*, Addison-Wesley, Reading, MA, 1969.

GRANT, E. L., W. G. IRESON, and R. S. LEAVENWORTH, *Principles of Engineering Economy*, 7th ed., Ronald, New York, 1982.

HIRSLEIFER, J., *Investment, Interest, and Capital*, Prentice-Hall, Englewood Cliffs, NJ, 1970.

HOLLOWAY, C. A., *Decision Making Under Uncertainty Models and Choices*, Prentice-Hall, Englewood Cliffs, NJ, 1979.

JEYNES, P. H., *Profitability and Economic Choice*, The Iowa State University Press, Ames, 1968.

MAO, J. C. T., *Quantitative Analysis of Financial Decisions*, Macmillan, New York, 1969.

MORRIS, W. T., *Engineering Economic Analysis*, Reston Publishing, Reston, VA, 1976.

NEWNAN, D. G., *Engineering Economic Analysis*, Engineering Press, San Jose, CA, 1976.

OAKFORD, R. V., *Capital Budgeting*, Ronald, New York, 1970.

OSTWALD, P. F., *Cost Estimating for Engineering and Management*, Prentice-Hall, Englewood Cliffs, NJ, 1974.

RADFORD, K. J., *Managerial Decision Making*, Reston Publishing, Reston, VA, 1975.

RAIFFA, H., *Decision Analysis: Introductory Lectures on Choice Under Uncertainty*, Addison-Wesley, Reading, MA, 1968.

REISMAN, A., *Managerial and Engineering Economics*, Allyn and Bacon, Boston, 1971.

RIGGS, J. L., *Economic Decision Models*, McGraw-Hill, New York, 1968.

SCHELLENBERGER, R. E., *Managerial Analysis*, Irwin, Homewood, IL, 1969.

SMITH, G. W., *Engineering Economy*, 2nd ed., Iowa State University Press, Ames, 1973.

SPECTHRIE, S. W., *Industrial Accounting*, 2nd ed., Prentice-Hall, Englewood Cliffs, NJ, 1959.

SPENCER, M. H., *Managerial Economics*, 3rd ed., Irwin, Homewood, IL, 1968.

STEINER, H. M., *Public and Private Investments: Socioeconomic Analysis*, Wiley, New York, 1980.

STERMOLE, F. J., *Economic Evaluation and Investment Decision Methods*, Investment Evaluations Corporation, Golden, CO, 1974.

STOKES, C. J., *Managerial Economics*, Random House, New York, 1969.

TARQUIN, A. J., and L. T. BLANK, *Engineering Economy*, McGraw-Hill, New York, 1976.

TAYLOR, G. A., *Managerial and Engineering Economy*, 2nd ed., Van Nostrand Reinhold, New York, 1975.

THUESEN, H. G., W. J. FABRYCKY, and G. J. THUESEN, *Engineering Economy*, 5th ed., Prentice-Hall, Englewood Cliffs, NJ, 1977.

VAN HORNE, J. C., *Financial Management and Policy*, 3rd ed., Prentice-Hall, Englewood Cliffs, NJ, 1974.

WESTON, J. F., and E. F. BRIGHAM, *Managerial Finance*, Holt, Rinehart & Winston, New York, 1969.

WHITE, J. A., M. H. AGEE, and K. E. CASE, *Principles of Engineering Economic Analysis*, Wiley, New York, 1977.

WINFREY, R., *Economic Analysis for Highways*, International Textbook Company, Scranton, PA, 1969.

Selected Periodicals

Accounting Review
American Economic Review
AIIE Transactions
American Journal of Agricultural Economics
Appraisal Journal
Bell Journal of Economics
California Management Review
Decision Sciences
Economic Journal
Engineering Economist
Financial Analyst's Journal
Financial Management
Harvard Business Review
Industrial Engineering
Journal of Accountancy
Journal of Business
Journal of Finance
Journal of Financial and Quantitative Analysis
Journal of Taxation
Management Science
Public Utilities Fortnightly
Quarterly Review of Economics and Business

CHAPTER 9.5
Economic Risk Analysis

JAMES R. BUCK
The University of Iowa

JOSE M. A. TANCHOCO
Virginia Polytechnic Institute and State University

9.5.1 INTRODUCTION

Decisions on prospective projects require economic justification, as the other chapters in this section of the handbook have stated. Various methods of analysis for economic justification are shown in these chapters based on the assumption that all of the component cash flows for the project (or alternative) are known and certain. However, in most cases the amount and timing of these cash flows are estimated, and uncertainties exist in the estimation process. Furthermore, there is usually more uncertainty with some component cash flows than others, and some of these component flows affect the economic criteria more than others. Thus additional methodologies and concepts are needed for economic analysis when explicit information on the effects of uncertainties in the timing and amounts of the cash flows is important. These methodologies and concepts are the focus of this chapter.

Numerous factors contribute to the uncertainties in the estimates of the amount and timing of component cash flows. Delivery or construction delays, unexpected bottlenecks in new projects, inflationary or recessionary pressures, labor negotiations, and problems in R&D are but a few examples of changes that can and do occur to alter the amounts and timing of disbursements and receipts of monies. Although these possibilities are usually recognized during the early planning phases of a project, the actual cash flows are uncertain, and there is a risk associated with the resulting project's present worth, benefit-cost ratio, or other measure of economic merit being used. Since this economic risk is as important to the decision maker as the other aspects of economic analysis, explicit information regarding the risk should be developed as part of the analysis. Approaches to this form of analysis and some of the relevant techniques are described in this chapter.

9.5.2 PROCESSES IN ENGINEERING ECONOMIC ANALYSIS

There are three different processes to address in project planning and economic risk analysis: (1) the cash flow estimation process, where the various component disbursements and receipts of money are developed; (2) the discounting process, where the future cash flow estimates are discounted to account for the time value of money; and (3) the "risk-uncertainty" description process, where the uncertainties in the amount of timing of the component cash flows are made explicit. These processes are different and should be separated in the analysis thinking in order to maintain clarity for subsequent decision making.

In the cash flow estimation process, the elements of the project where component cash flows will occur are identified, and the expected amount and timing of each component cash flow is estimated. Discussions of many features of this cash flow estimating process are given in Chapter 9.2. These discussions primarily address single future cash flows or a uniform series of flows over time. A much wider variety of cash flow time patterns exists, which is considerably useful to practitioners because it provides more faithful descriptions of economic situations and can help them compute the discounted cash flows of projects. Some cash flows that occur very frequently can be better described as a continuous flow over time than as a single end-of-the-year flow. In other cases the cash flow is best described as a discrete series over time. With either the continuous or series flows, the amount of flow may increase with time (e.g., maintenance) or decrease over

Exhibit 9.5.1 Some Functional Cash Flow Series and Continuous Functions With the Total (Undiscounted) Cash Flow

Series or Functions of Cash Flow (and Parameters)	Description	Total Cash Flow (Undiscounted)	Discounted Cash Flow–Present Worth[a]
Uniform series $(\overline{A}, k)$	\$, A, k, Time	$\sum_{n=1}^{k} A = kA$	$\frac{A}{i}[1-(1+i)^{-k}]$
Uniform continuous $(\overline{A}, k)$	\$, A, k, Time	$\int_{O}^{k} \overline{A}\, dt = \overline{A}t$	$\frac{\overline{A}}{j}(1-e^{-jk})$
Gradient series (G, k)	\$, G, 1, k, Time	$\sum^{k} Gn = G(k+1)\frac{k}{2}$	$\frac{G}{i}[1-(1+i)^{-k}]$ $+\frac{G}{i}[1-(k+1)(1+i)^{-k}]$

Ramp continuous $(\overline{G}, k)$	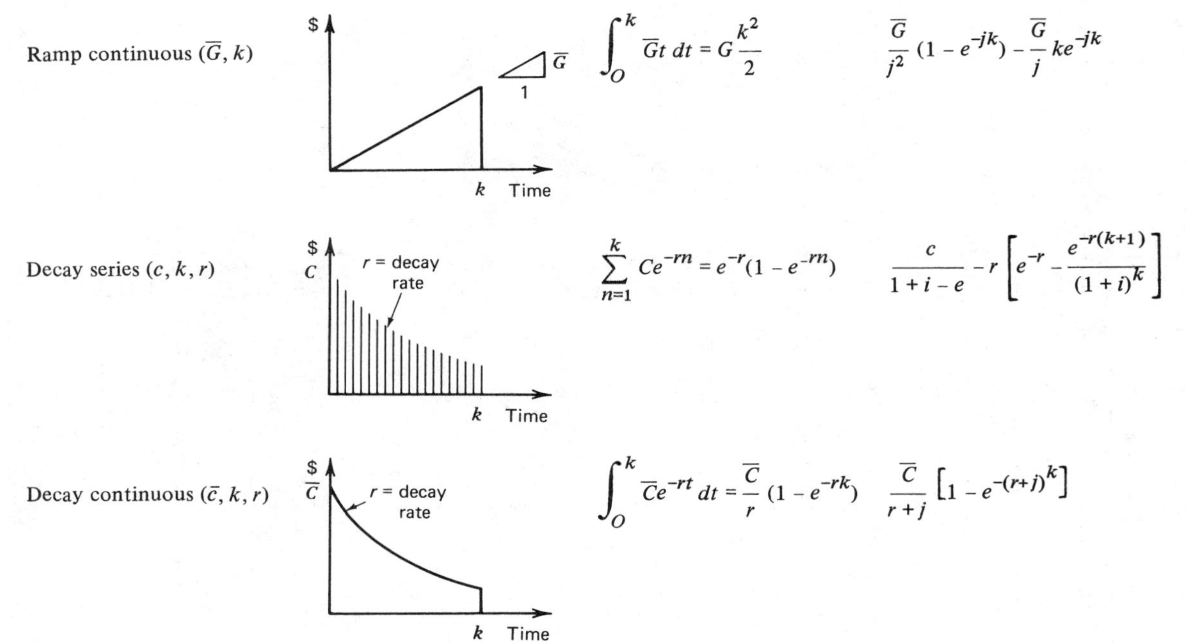	$\int_O^k \overline{G}t\,dt = G\frac{k^2}{2}$	$\frac{\overline{G}}{j^2}(1 - e^{-jk}) - \frac{\overline{G}}{j} ke^{-jk}$
Decay series (c, k, r)		$\sum_{n=1}^{k} Ce^{-rn} = e^{-r}(1 - e^{-rn})$	$\frac{c}{1 + i - e} - r\left[e^{-r} - \frac{e^{-r(k+1)}}{(1+i)^k}\right]$
Decay continuous $(\bar{c}, k, r)$		$\int_O^k \overline{C}e^{-rt}\,dt = \frac{\overline{C}}{r}(1 - e^{-rk})$	$\frac{\overline{C}}{r+j}\left[1 - e^{-(r+j)^k}\right]$

[a] i = discrete discount rate and j = continuous discount rate.

time (e.g., cost savings), and the form of these changing amounts with time needs to be described. Parameters of the model need to be estimated as part of the cash flow estimation process.[1]

Exhibit 9.5.1 illustrates a few of the basic forms of continuous and serial cash flow forms that may be used. Formulas for computing the sum of the undiscounted cash flow quantities over time are shown as an aid to the estimation process. These forms of cash flow serve to describe components flows of a project. Whatever the actual estimates of the component cash flows, these forms of serial and continuous cash flows, or combinations thereof, can be used to describe the expected receipts and disbursements associated with a proposed project.

The second process in engineering economic analysis is the discounting of the cash flows with respect to their future timing and amount in order to give an equivalent present worth or other criterion value. Chapter 9.3 describes the fundamentals of the discounted cash flow analysis and other economic criteria. In addition, Chapter 9.4 describes the application of these techniques in the choice between alternative projects.

The identification and description of *uncertainty* constitutes the third process in engineering economic analysis. Estimates are needed about the probable variation in component cash flow amounts and the timing of these flows. An explicit structuring of the uncertainty information may be critical to the making of a sound decision. Thus the correct handling of this area to meet the needs of a specific analysis and decision maker is essential.

The term "uncertainty" is used in this chapter to mean a *lack* of precise knowledge regarding the amount of cash flow, its timing, or the resulting values of economic criteria. Other definitions of the term appear in the decision making or information theory literature, and these alternative definitions should not be confused with the meaning intended here. Although uncertainty decisions are sometimes defined where probability estimates are not available, it is assumed here that the lack of knowledge of the amount of cash flow can be described by a probability (density) function with estimated parameter values.

Chapter 9.3 addresses the final decision making in the general sense, where some of the underpinnings are described and illustrated. The focus here is on concepts and techniques for handling engineering economic risk analysis. These concepts and techniques are outlined from the simpler, more traditional approaches to the advanced techniques. It is not possible in a single chapter to cover the details of this diverse and expanding area. Consequently, emphasis is placed on the basic methodology, which is illustrated through several examples. For more detailed pursuits of specific topics, references guide the reader to specific topics in the literature.

9.5.3 CONCEPTS OF ECONOMIC RISK ANALYSIS

Risk is the failure to meet the forecasted return on the investments. These failures may be due to inadequacies in the cash flow estimation process of engineering economic analysis. Component cash flows may be overlooked. Magnitudes of cost or return or their timing may be inaccurately estimated. These and other causes of inadequacies occur because forecasting of the future is an imperfect art. The result may be an undesirable underreturn or a desirable overreturn on investments as the risk. Experienced decision makers recognize these imperfections and attempt to adjust their analysis and thinking accordingly. In some cases the risks are perceived as insignificant, and so the risk is ignored. Cash flow estimates are then treated as if they were known with complete certainty. For other cases the "assumed certainty" approach is recognized to be inappropriate. The concepts and techniques presented here address this latter case.

Over the years numerous methods have been used for evaluating the risk associated with one or more prospective projects. Many of these methods were devised on "rules of thumb" bases and for computational simplicity. One basis that has been employed is that more future cash flows are usually more difficult to estimate and therefore are more prone to risk. This rule of thumb led to the use of the related finite time horizon and payback period methods of analysis as described here. Two other methods, based on this rule of thumb and on the concept that estimate adjustments are needed, include the discounting for risk method and the conservative adjustment method, also described here. These approaches to economic risk analysis and the concept that the ROI is more affected by some estimates than others led to the notion of sensitivity analysis. These approaches to risk analysis generally predate the probabilistic methods that are advocated in the contemporary literature and described later in this chapter. Although the nonprobabilistic techniques have a number of drawbacks, which are noted, they are often computationally simpler and serve well in some circumstances as a primary or secondary criterion.

Finite-Time Horizon and Payback Methods

In the finite time horizon method of risk analysis, a predetermined period of time from the start of a project is established for purposes of analysis. All cash flows that occur after this time horizon are then ignored on the basis of unreliable estimates. It remains, then, to compare the cumulative net cash flows of a project within the time horizon; the project with the greatest cumulative net cash flow (i.e., subtracting cash outflows from inflows) would be perceived as the project with least risk. Although the more typical practice with this method is also to ignore the time value of money, the future cash flows within the time horizon may be discounted for a more accurate analysis. In either the nondiscounted or the discounted form, the finite time horizon method of risk analysis has been criticized because the planning horizon is arbitrary, and unreliabilities in estimates of component cash flows are not necessarily related to the more future time periods. Uncertainties in the *timing* of some component cash flows may make some project appear almost riskless because a large cash outflow may be estimated to occur just beyond the time horizon.

The payback period method, as described in Chapter 9.4, avoids the arbitrariness of the predetermined time period because the criterion is the time period in which the early investment is recovered by the accumulated net cash flow. Since shorter payback periods are generally preferred, this method provided an intuitively acceptable criterion. However, the payback period method ignores cash flows beyond the payback period. This oversight can lead to the selection of safe, but economically poor, projects. Also, the payback period method does not account for the time value of money or the economic life of the physical assets. Perhaps these deficiencies have been the cause of a much lower use of the payback period than the discounting techniques as a primary criterion. The payback period method is often used as a secondary criterion—that is, as a measure of how quickly initial capital investments are projected to be recovered and thus as one indicator of project risk.[1–3]

Discounting for Risk and Conservative Adjustment Methods

In the discounting for risk method of risk analysis, it is assumed that there is a progressive riskiness because of the more and more future component cash flows.[1,3] So instead of selecting a single point of future time as a cutoff point, as in the finite time horizon and payback period methods, it is argued that the progressive effect of unreliability in estimation can be made by *inflating* the discount rate. Hence the riskless discount rate may be increased by a factor that describes the analyst's perception of risk associated with the estimations of a particular project. There is, of course, the obvious difficulty of determining a single factor that captures the overall risk of a project and the subjectivity in making this estimate. Some firms have attempted to reduce, but not eliminate, these difficulties by creating classes of project riskiness by fiat and then having the analyst choose a class where the inflating discount rate factor is prescribed. Although this practice reduces the variability of analysts, it in no way overcomes the criticisms leveled. Moreover, this method confounds the discounting and uncertainty description processes of economic risk analysis so that the time value of money and the uncertainties in component cash flow amounts and timing are unclear.

The method of conservative adjustment for risk analysis avoids some of the difficulties of the discounting for risk method. In the conservative adjustment method, magnitudes or timings of one or more component cash flows are scaled to reflect a lower cash inflow, higher cash outflows, or shorter project lives.[3] That is, component cash flows or other parameters of the projects that would normally produce a higher return on investment are scaled to reduce the return in proportion to the analyst's perception of the associated uncertainty. Discount rates are kept at their riskless level. The resulting discounted cash flow of a project is therefore a conservative estimate, which may be used to compare with those of other projects for decision making. Sometimes two or three levels of conservative adjustments are performed, where mildly, moderately, and strongly conservative adjustments may be made and alternative projects may be compared at each level.

Although this method of risk analysis does keep the discounting process separate from the description of uncertainty process, there is little guidance offered on the amount of adjustment to be made or on the means of maintaining consistency in this adjustment over alternative projects. Nevertheless, this adjustment method does provide an indication of the effects of some of the component cash flow estimates and a *means* for investigating the sensitivity of the economic attractiveness to particular estimates. This latter indication is best described under sensitivity analysis, the next topic.

Sensitivity Analysis

There are two principal approaches to *economic* risk situations: (1) reduce the estimation uncertainty to a minimum and (2) quantify the amount of risk. Sensitivity analysis serves to help in reducing the uncertainty by identifying the relative importance of the component cash flow variables to the principal economic criterion. The implication is straightforward in that those variables that exert greater influence on the criterion require greater estimation resources. It follows, but dimly, that greater resources in estimation process will reduce the uncertainty of the cash flow variables. Although this last point may easily be contested, most people would agree that identifying the key variables and reducing the amount of uncertainty associated with them to the minimum is an important heuristic approach to decision making.

The procedure of sensitivity analysis is simplistic. First, identify a reference situation for a project; typically the most likely case is selected for this reference. Second, make perturbations on each of the cash flow model variables (one at a time) and determine the change in the outcome criterion associated with each variable.[1] These perturbations should consist of the expected variation of each variable. Resulting changes in the outcome criterion indicate the relative effects of the variables, assuming all else remains constant. For a more complete sensitivity analysis, which gets around this tenuous assumption that only one variable can change at a time, one should theoretically investigate all combinations of change; however, this procedure is not feasible if there are many factors to consider. Also, perturbations should be made at several steps around the reference situation over the range of uncertain values of the variables when nonlinearities may result.

Example 9.5.1 illustrates a sensitivity analysis for a simple project where three variables of the cash flow model are uncertain. The results of the analysis from this example are shown in Exhibit 9.5.2 and 9.5.3, which demonstrates that the annual worth of the project is most sensitive to the annual revenue and less sensitive, respectively, to the annual operating and maintenance costs and the MARR.

Example 9.5.1. Consider a project characterized by the following cash flow items:

Item	Most Likely Estimate
Initial investment	$10,000
Project life	10 years
Annual revenue	$3000
Annual operating and maintenance cost	$1000
Minimum attractive rate of return, i	10%

The annual worth (AW) method is used as the basis for establishing the profitability of the project. It is assumed here that the initial investment and the project life are known with certainty. However, the annual revenue, the annual operating and maintenance cost, and the MARR are not

Exhibit 9.5.2 Expected Minimal Worth of a Project (with Baseline Values of i = 10%) for the Problem Illustrated in Example 9.5.1

Percent Change in Revenue	Percent Change in Cost	Percent Change in Minimum Attractive Rate of Return				
		-25	-10	0	+10	+25
	-25	43	-58	-127	-198	-306
	-10	-107	-208	-277	-348	-456
-25	0	-207	-308	-377	-448	-556
	+10	-307	-408	-477	-548	-656
	+25	-457	-558	-627	-698	-806
	-25	493	392	323	252	144
	-10	343	242	173	102	-6
-10	0	243	142	73	2	-106
	+10	143	42	-27	-98	-206
	+25	-7	-108	-177	-248	-356
	-25	793	692	623	552	444
	-10	643	542	473	402	294
0	0	543	442	373 (expected AW)	302	194
	+10	443	342	273	202	94
	+25	293	192	123	52	-56
	-25	1093	992	923	852	744
	-10	943	842	773	702	594
+10	0	843	742	673	602	494
	+10	743	642	573	502	394
	+25	593	492	423	352	244
	-25	1543	1442	1373	1302	1194
	-10	1393	1292	1223	1152	1044
+25	0	1293	1192	1123	1052	944
	+10	1193	1092	1023	952	844
	+25	1043	942	873	802	694

Exhibit 9.5.3 Results of Sensitivity Analysis Example

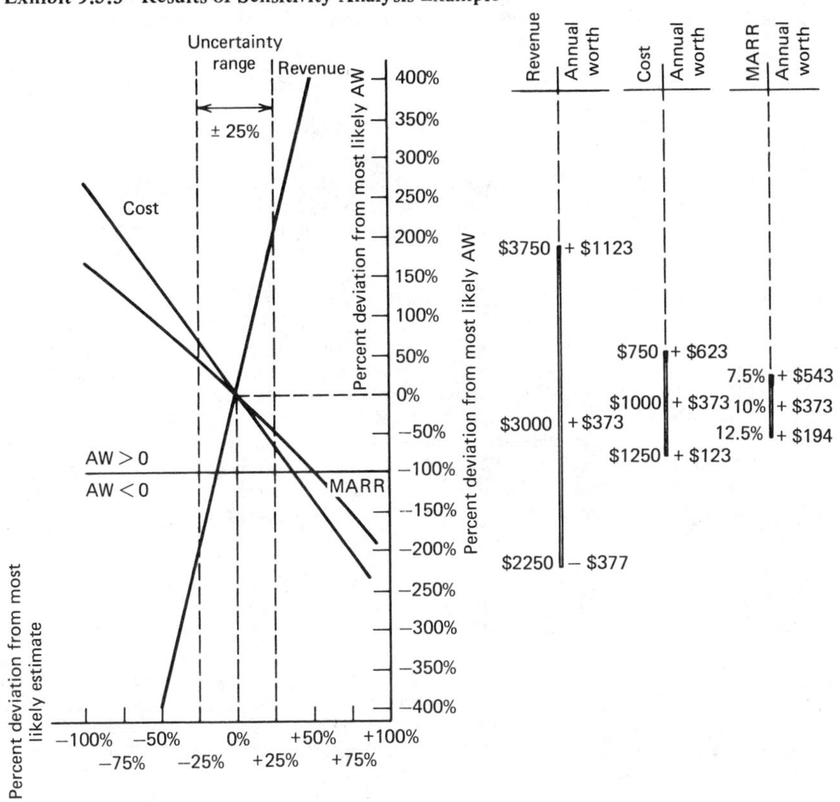

known with certainty. For a decision under assumed certainty, the most likely estimates are used as though these values were deterministic. Thus the annual worth of the project is calculated as

$$AW(i = 10\%) = 3000 - 1000 - 10{,}000\ (A/P, 10\%, 10) = \$373$$

where the symbol $(A/P, 10\%, 10)$ represents the capital recovery interest factor. Therefore the project is acceptable under assumed certainty since it would most likely generate returns at a rate greater than the MARR. However, since the revenues, costs, and MARR are subject to uncertainty, the economic performance of the project would be greatly affected by the actual values that these three variables assume. A sensitivity analysis is then in order. A −25%, −10%, +10%, and +25% variation from the best estimates are made for the three variables in question. The annual worths are then calculated for each triple variation (revenue, cost, and MARR) and are tabulated (Exhibit 9.5.2). The interpretation of the results shown in this exhibit is not an easy task for a full factorial analysis. If it could be reasonably assumed that the variable deviations occur one at a time, then a chart such that the one shown in Exhibit 9.5.3 can be utilized. Since revenue is observed as the most sensitive variable, it would be worthwhile to obtain more accurate estimates of the annual revenue. It should be noted that all the factors of uncertainty were viewed here by the same percentage variations. When greater uncertainties are associated with some factors over others, then greater variations should be made for the more uncertain factors.

9.5.4 PROBABILISTIC METHODS OF ECONOMIC RISK ANALYSIS

Chapter 13.3 is about decision making principles and analysis. That chapter describes many of the basic concepts and techniques of the probabilistic methods of general risk analysis. In essence, these methods of risk analysis view the uncertain and uncontrolled variables as a set of "events." The outcomes of not selecting the project are evaluated as no change in the economic situation or as a zero change in present worth or whatever criterion is used. If the project is selected, however, then the economic criterion changes with the events. When the events are exclusive (i.e., only one can surely happen), and probability measures can be assigned to each event, then the average economical criterion value can be determined. For m exclusive events that are indexed by $j = 1, 2,$

$\ldots, m$, with p_j as the probability assigned to each event and u_j as the economic criterion for the project given the event, the expected (over the events) criteria value of the project is

$$E(U) = \sum_{j=1}^{m} p_j u_j \tag{1}$$

Thus a project with an expected present worth that is positive would be a good risk in the sense that the average economic worth considering all events would yield a return on that investment. Example 9.5.2 illustrates a project decision involving uncertainties of the amount and duration of cost savings.

Example 9.5.2. A proposed project requires a \$200 investment with returns as cost savings over the future. It is uncertain whether these cost savings will be \$100, \$200, or \$300/year and whether these levels of savings will be realized over 1, 2, or 3 years, but the savings are expected to be uniform regardless of the duration. Accordingly, nine exclusive events can be identified as an exhaustive set (i.e., all possible events are accounted for) with the combinations of savings amounts and durations. If a 20% MARR is required, then the net present worth can be found using the analytical method shown in previous chapters in this section. Since the three amounts of savings may be represented by the variable x and the durations by the variable t, then the probabilities may be assigned subjectively or by other means as $p(x \,\&\, t)$. Arbitrary values of these joint probabilities are shown here. Assume that the following probabilities are estimated for the various amounts x and durations t of these cost saving cash flows:

$$p(x = 100) = 0.20 \qquad p(t = 1) = 0.30$$
$$p(x = 200) = 0.40 \qquad p(t = 2) = 0.50$$
$$p(x = 300) = 0.40 \qquad p(t = 3) = 0.20$$

The joint events consist of all combinations of x and t values, yielding nine joint events. If the x values are *not* statistically independent of the t values, then the joint probabilities are *not* products of those just shown, and the case shown below reflects the dependence. Net present worth (NPW) values (i.e., the future returns discounted less the \$200 investment) are also shown to the nearest dollar value in the following:

	Events								
	$x = 100$			$x = 200$			$x = 300$		
	$t = 1$	$t = 2$	$t = 3$	$t = 1$	$t = 2$	$t = 3$	$t = 1$	$t = 2$	$t = 3$
$p(x \,\&\, t) =$	0.20	0	0	0.07	0.25	0.08	0.07	0.25	0.08
NPW; $u_j =$	−117	−47	11	−38	106	221	50	258	432

The expected NPW of this project is

$$E(U) = -117(0.2) - 47(0) + 11(0) + \cdots + 50(0.05) + 258(0.25) + 432(0.08) = \$121$$

where the positive expected result indicates that one would expect to have a return surplus of \$121 beyond a 20% return on the investment of \$200.

9.5.5 PRESENT WORTH EXPECTATIONS OF CONTINUOUS EVENTS

Perhaps the most used and accepted method of economic risk analysis is the expected discounted cash flow technique.[1,3,4] Example 9.5.2 illustrated this method with discrete events and present worth as the discounted cash flow criterion. In that example one of the uncertain variables was the occurrence time or duration of cash flow, a time variable. An alternative and more efficient approach to viewing *discrete* events on timing is to let time be a continuous random variable t and to view discounting as a continuous operation. The continuous equivalent to a year of discounting is

$$e^j = 1 + i \tag{2}$$

where i is the annual riskless MARR and j is the equivalent continuous discount rate. If the left- and right-hand sides of equation 2 are raised to the power t for time in years, then equivalence still holds.

With a single future cash flow of F dollars, the expected present worth is

$$E(PW) = \int_0^{\infty} F\, p(t)\, e^{-jt}\, dt = F \int_0^{\infty} p(t)\, e^{-jt}\, dt \qquad (3)$$

where $p(t)$ is the probability (density) for the occurrence of this single cash flow. In this case $p(t)$ is a continuous probability function of the single cash flow timing, and the integral on the far right-hand side of equation 3 depends upon the probability density function and its parameter values. In fact, this integral can be directly determined for specific functions $p(t)$ as

$$E(pw) = FE(e^{-jt}) \qquad (4)$$

where the final factor on the right-hand side is shown in column 2 of Exhibit 9.5.4 for various $p(t)$ functions of timing.[5] Example 9.5.3 illustrates the simplicity of this calculation of present worth expectation when a single future cash flow has uncertain timing.

Example 9.5.3. A single cash inflow of $10,000 was known to occur, but the timing of it was uncertain. It was expected with equal likelihood anytime between 2 and 4 years. The MARR in this case is set at 22%, so that the continuous discount rate is

$$e^j = 1 + 0.22$$

By taking the natural logarithm of both sides, it is seen that j is approximately 20%. Since the timing is a continuous random variable with a uniform probability density between $a = 2$ years and $b = 4$ years, the value of $E(e^{-jt})$ can be found from Exhibit 9.5.4 to be

$$E(e^{-jt}) = [e^{-(0.2)2} - e^{-(0.2)4}]/0.2(4 - 2) = 0.552478$$

The expected present worth, from equation 4 is

$$E(PW) = \$10{,}000\ (0.552478) = \$5524.78$$

When the amount of the single future cash flow F is also a random variable that is *statistically independent* of the timing random variable, then $E(F)$ can replace F in equation 4. Hence an uncertain single future amount can be handled with these formulas, so long as the mean value of F can be determined and independence can be assured, as Example 9.5.4 illustrates.

Example 9.5.4. Suppose that a single future cash flow has an uncertain *amount*, which is described as a uniform random variable of at least $8000 and at most $12,000, and that the *timing* of this flow is distributed as a negative exponential with an expected time of 3 years. The expected F is (8000 + 12,000)/2, or $10,000, and the discounted timing expectation is found from the Exhibit 9.5.4 formula and j = 20% to be

$$E(e^{-jt}) = \frac{1}{1 + 0.2(3)} = 0.625$$

Accordingly, the expected present worth is

$$E(PW) = 10{,}000\ (0.625) = \$6250$$

Although the expected F and the expected t of this example are the same as in the previous example, the present worth is much greater because the functional form of the negative exponential distribution has a greater probability of earlier occurrence times than the uniform timing distribution.

Uniform continuous cash flows of $\overline{A}$ dollars per year have a present worth that is determined in part by the duration t of the cash flow (see Exhibit 9.5.1). If the duration is uncertain, but this random variable is described with a probability density function with specified parameters, then the expected present worth is

$$E(PW) = \overline{A}\ \{[1 - E(e^{-jt})]/j\} \qquad (5)$$

Formulas for computing $E(e^{-jt})$, as given in column 2 of Exhibit 9.5.4, could be substituted directly into equation 5 for computation of present worth. However, the formulas for the entire factor

Exhibit 9.5.4. Present Worth Expectation Due to Uncertain Timing

$p(t)$ Probability Density Function	$E(e^{-jt})$ Single Cash Flow Occurrence Time	$[1 - E(e^{-jt})]/j$ Continuous Step Duration	$E(te^{-jt})$
Uniform probability $V(t) = (b-a)^2/12$ a = earliest time b = latest time $E(t) = (a+b)/2$ $V(t) = (b-a)^2/12$	$\dfrac{e^{-ja} - e^{-jb}}{j(b-a)}$	$\dfrac{1}{j} - \dfrac{e^{-ja} - e^{-jb}}{j^2\,(b-a)}$	$\dfrac{(ja+1)\,e^{-ja} - (jb+1)\,e^{-jb}}{j^2\,(b-a)}$
Gaussian (normal) $p(t) = \dfrac{1}{\sigma\sqrt{2\pi}}\,e^{-(t-\mu)^2/2\sigma^2}$ μ_2 = mean time σ = time variance $E(t) = \mu_2$ $V(t) = \sigma^2$	$e^{-j\mu + (j^2\sigma^2/2)}$	$(1 - e^{-j\mu + j^2\sigma^2/2})/j$	$(\mu - j\sigma^2)\,e^{-j\mu + \sigma^2 j^2/2}$
Gamma $p(t) = \dfrac{t^{a-1}e^{-t/b}}{\Gamma(a)\,b^a}$ $E(t) = ab_2$ $V(t) = ab$	$(1+jb)^{-a}$	$\dfrac{1}{j}[1 - (1+jb)^{-a}]$	$ab(1+jb)^{-(a+1)}$
Negative exponential gamma with $a = 1$, $\Gamma(a) = 1$	$(1+jb)^{-1}$	$b/(1+jb)$	$b/(1+jb)^2$

shown in braces on the right-hand side of equation 5 are given in column 3 of this exhibit for reader convenience. These uniform cash flow present worth factors are used in the same manner as were the formulas for $E(e^{-jt})$ with single cash flows. It also follows that $\overline{A}$ can be treated as a random variable where $E(\overline{A})$ replaces $\overline{A}$ in equation 5 when $\overline{A}$ is statistically independent of the duration of the uniform cash flow.[6]

Example 9.5.5. A uniform cash flow is expected to amount to \$5000 each year, but this amount is uncertain. The Gaussian (normal) distribution provides a reasonable description of this uncertainty when the SD is set at σ = \$1000. Also, the duration of this uniform continuous cash flow is expected to be 5 years, but variance of a uniformly distributed uncertainty is estimated as 1 year. Since the duration uncertainty is uniform with an expectation and variance of 5 and 1, respectively, then from column 1 of Exhibit 9.5.4 it is seen that

$$\frac{a+b}{2} = 5 \qquad \frac{(b-a)^2}{12} = 1$$

where the parameter values are solved as $a = 3.27$ and $b = 6.73$. The expected present worth with $j = 20\%$ is computed using equation 5 and the third column of Exhibit 9.5.4 to be

$$E(PW) = \$5000 \left(\frac{1}{0.2} - \frac{e^{-0.2(3.27)} - e^{-0.2(6.73)}}{0.04\,(6.73 - 3.27)} \right) = \$15{,}617.97$$

The uniform continuous cash flow shown in equation 5 is assumed to start immediately and then go on for an uncertain duration. If the duration of this flow *after the flow start* were as probabilistically described, but the *start* were delayed exactly s time periods (years), then the expected present worth would be precisely e^{-js} times that value computed in equation 5. When the start of the uniform cash flow s is statistically independent of both the duration t and the amount per time period $\overline{A}$, then the right-hand side of equation 5 can be multiplied by $E(e^{-js})$, which is computed using the formulas of $E(e^{-jt})$ for the uncertainty form of the delayed starts (see the formulas of Exhibit 9.5.4). Example 9.5.6 illustrates the uniform continuous cash flows with a delayed start.

Example 9.5.6. A uniform continuous cash flow is expected, with an average of \$5000/year, but the actual annual amount is felt to vary according to the normal distribution with a variance of \$1000. Once this cash flow component starts, it is expected to continue for 5 years, but the duration of flow is viewed as gamma distributed with a variance of 1 year. Delays in the start of this cash flow are uniformly distributed from now to 3 years hence. By temporarily ignoring cash flow delay, the expected present worth at $j = 20\%$ may be computed using equation 5 where the factor in braces is computed from Exhibit 9.5.4 using the gamma function and column 3. The expected duration and duration variance of this cash flow are

$$ab = 5$$

and

$$ab^2 = 1$$

so that $a = 25$ and $b = \frac{1}{5}$. Substituting these parameters into the formula for $[1 - E(e^{-jt})]/j$ yields

$$\{1 - [1 + (0.2)(\tfrac{1}{5})]^{-25}\}/0.20 = 3.124416$$

The value 3.124416 times the expected \$5000/year yields the expected present worth if starting is not delayed, or \$15,622.08. A uniformly distributed delay from $a = 0$ to $b = 3$ years has a present worth factor of

$$E(e^{-jt}) = [e^{-0.2(0)} - e^{-0.2(3)}]/[0.2(3-0)] = 0.7520$$

so that the expected present worth is

$$E(PW) = 0.7520\,(15{,}622.08) = \$11{,}747.50$$

for the delayed starting case.

It should be stated that annual or future worth could be used as a criterion rather than present worth. However, the annual and future worth must be set to some exact and prespecified time duration or future time k. Since these future durations or points in future time are *random variables* in economic risk analysis rather than the known present time, present worth analysis is easier to use and is therefore recommended.

Present Worth Variances

If no uncertainties exist in a component cash flow, then deterministic equations could be written as shown in Exhibit 9.5.1. In these equations the occurrence times of future cash flows and the amount of the cash flow or the time durations and amounts of annual flow become products in the equations. These times and amounts are the random variables in the uncertain cases. When these random variables are statistically independent, then the expectation of the product of random variables is the product of the individual expectations. However, that is not true for the variance of products. Rather, the variance of the products of random variables F and e^{-jt}, for the future single cash flow case,[7] is

$$V(PW) = V(F)\,V(e^{-jt}) + V(F)\,E(e^{-jt})^2 + V(e^{-jt})\,E(F)^2 \tag{6}$$

Therefore all that is needed to compute these present worth variances of single future cash flow components with uncertain timing is $V(e^{-jt})$ for various probability density functions of t. In the case of uniform continuous cash flows of uncertain duration t and uncertain amounts per year $\overline{A}$, the present worth variance is

$$V(PW) = V(\overline{A})\,[1 - V(e^{-jt})]/j + V(\overline{A})\,\{[1 - E(e^{-jt})^2]/j\} + \{[1 - V(e^{-jt})]/j\}\,E(\overline{A})^2$$

Here again, $V(e^{-jt})$ is needed for computation of equation 6. Formulas for computing $V(e^{-jt})$ are given in Exhibit 9.5.5 for the uniform, Gaussian (normal), and gamma distributions of the random variable t.[8] Example 9.5.7 illustrates the use of these formulas in conjunction with equation 6.

Example 9.5.7. Consider the cash flow component that was described in Example 9.5.4. The amount of the single future cash flow F is uniformly distributed from a = \$8000 to b = \$12,000, so that

$$V(F) = \frac{(b-a)^2}{12} = \frac{(12{,}000 - 8000)^2}{12} = 1.333 \times 10^6$$

from column one of Exhibit 9.5.4. Since the timing of F is described as negative exponentially distributed with a mean of 3 years, then the parameter b in this timing distribution is 3. It then follows from the Exhibit 9.5.5 formula for the gamma with $a = 1$ that

$$V(e^{-jt}) = [1 + 2(0.2)3] - [1 + 0.2(3)]^{-2} = 0.06392$$

It was also shown in Example 9.5.4 that $E(F)$ = \$10,000 and $E(e^{-jt}) = 0.625$. The present worth variance at the discount rate of 20% follows from equation 6 as

$$V(PW) = (1.333\ldots \times 10^6)(0.06392) + (1.333\ldots \times 10^6)(0.625)^2 + (0.06392)(10{,}000)^2 = \$6{,}998{,}060$$

and the resulting SD of present worth is \$2645.39.

The preceding situations represent single component cash flows rather than total project cash flow models, and the type of component flows are restricted to either single future cash amounts or to uniform continuous flows over time. However, the procedure for expansion to the project case is straightforward so long as the individual components of cash flow are statistically independent of each other because the total expected present worth is simply the algebraic sum of the component expectations where inflows are signed oppositely to outflows. Total *variances* (not SDs) of present worth are simply the sum of the component present worth variances. Therefore the development of project present worth expectations and variances poses no theoretical difficulties so long as independence is assured.

Other Functional Cash Flow Models

The preceding development allows for the combining of expectations and variances of single future cash flows or continuous uniform cash flows. However, cash flows that follow other functional forms such as ramp or decay forms cannot be so treated. Exhibit 9.5.6 provides formulas

Exhibit 9.5.5 Some $V(e^{-jt})$ and $V(te^{-jt})$ Formulas

Probability Density Function	$V(e^{-jt})$	$V(te^{-jt})$
Uniform continuous (parameters a, b)	$\frac{e^{-2ja} - e^{-2jb}}{2j(b-a)} - \frac{e^{-ja} - e^{-jb^2}}{j(b-a)}$	$\frac{e^{-2ja}(2ja+1) - e^{-2jb}(2jb+1)}{4j^2(b-a)} - \frac{e^{-ja}(ja+1) - e^{-jb}(jb+1)}{j^2(b-a)}$
Gamma (negative exponential $a = 1$)	$(1+2jb)^{-a} - (1+jb)^{-2a}$	$\frac{ab}{(1+2jb)^{a+1}} - \left(\frac{ab}{(1+jb)^{a+1}}\right)^2$
Gaussian (normal)	$e^{-2j\mu+2\sigma^2 j^2} - e^{-2j\mu+\sigma^2 j^2}$	$(\mu - 2j\sigma^2)\, e^{-2j\mu+2j^2\sigma^2} - [(\mu - j\sigma^2)\, e^{-j\mu+j^2\sigma^2/2}]^2$

Exhibit 9.5.6 Formulas for Computing Present Worth Expectations for Three Functions of Continuous Cash Flow

Continuous Cash Flow Function Form	Present Worth Expectations
Uniform continuous $E(\overline{A})$ dollars/year	$E(\overline{A})\,[1 - E(e^{-jt})]/j$
Ramp continuous	$E(\overline{G})\left[\frac{1}{j^2} - \frac{E(e^{-jt})}{j^2} - \frac{E(te^{-jt})}{j}\right]$
Decay continuous $E(\overline{C})$ dollars/year at start-decay rate r	$E(\overline{C})\left[\frac{1}{j+r}\right]\left\{1 - E[e^{-(j+r)t}]\right\}$

for present worth expectations for the continuous uniform, ramp, and decay cash flow functions. Expected present worth can be found for combinations of these formulas, as illustrated in Example 9.5.8.

Example 9.5.8. A *growth form* of cash flow function may be described as continuously increasing cash amounts with a diminishing rate of increase. This cash flow function may be formed by combining the uniform continuous and the decay continuous forms with $E(\overline{A}) = E(\overline{C})$. If the latter form is subtracted from the former, then the cash flow function rises at a diminishing rate toward the eventual value of $E(\overline{C})$. When the value of the steady state cash flow is Gaussian with $\mu = \$1000$ and $\sigma^2 = 400$, the duration after the start is gamma with $a = 2$ and $b = 3$ (giving an expected duration of 6 years), and the growth and discount rates are, respectively, 0.3 and 0.2, the present worth formula (Exhibit 9.5.6) is

$$E(\overline{A})\ \{[1 - E(e^{-jt})]/j - \left[\frac{1}{j+r}\right]\ [1 - E(e^{-(j+r)t})]\}$$

The values of the expectations for the $E(e^{-jt})$ and $E(e^{-(j+r)t})$ are found in the second column of Exhibit 9.5.4 to be

$$E(e^{-jt}) = [1 + 0.2\ (3)]^{-2} = 0.390625$$

$$E(e^{-(j+r)t}) = [1 + (0.2 + 0.3)\ 3]^{-2} = 0.25$$

If the cash flow is initiated immediately, then the expected present worth is found from the preceding formula to be

$$1000\ \{[1 - 0.390625]/0.2 - [1/0.5]\ [1 - 0.25]\} = \$1546.88$$

It may be observed that the formulas in Exhibit 9.5.6 are the same as the present worth formulas of the deterministic case shown in Exhibit 9.5.1, except for $E(e^{-jt})$ replacing e^{-jk}. Values of $E(e^{-jt})$ may be computed from formulas given in Exhibit 9.5.4. However, one of the components of the continuous ramp present worth expectations shown in Exhibit 9.5.6 is $E(te^{-jt})$ rather than $E(e^{-jt})$. Depending upon the probability density function of the variable t, the value of $E(te^{-jt})$ can be found through the formulas given in the last column of Exhibit 9.5.4, as developed in Zinn et al.[9] Formulas for $V(te^{-jt})$ are shown at the bottom of Exhibit 9.5.5, as developed in Hillier,[10] Rosenthal,[8] and Wagle.[11] These added formulas extend the methods of computing present worth expectations and variances to a wider variety of continuous cash flow functions over time. Example 9.5.9 demonstrates how these formulas can be used to compute the expectations and variances of present worth.

Present Worth Expectation-Variance Criterion

Few people argue against the present worth expectation as a necessary criterion, but many feel that this criterion alone is not sufficient. Ignoring all the noneconomic aspects of a decision or the pattern of cash inflows, projects with high present worth values can also possess high variability, thereby exposing the firm to potential losses. This observation led to the inclusion of the variance or SD of present worth along with the expectation. There are numerous proponents of this expectation-variance (E-V) criterion.[6,10,12–15] Chapter 13.3 shows some variations of this criterion and some of the background reasoning.

Once the present worth expectations and variances of the *component* cash flows are computed as previously described, these expectations and variances may be combined separately as previously discussed. If the resulting *project* present worth can be considered normally distributed (under the central limit theorem[10]), then a confidence interval about the expected present worth can be estimated by multiplying the SD (i.e., square root of the variance) by a standardized normal random deviate. Normal random deviates for 90%, 95%, and 99% confidence limits are, respectively, $Z = 1.645$, 1.960, and 2.575. This normal confidence bound approximation is also illustrated in Example 9.5.9. Use of both expectation and variance criteria, or, equivalently, the confidence bounds, is simply variations of the E-V criterion.

Example 9.5.9. A continuous stream of cost savings is described as exponentially decreasing (decay) with $r = 0.4$ from an expected initial cost of \$100 that is Gaussian (normally) distributed with $\sigma = \$5$. The duration of this stream is expected to be 5 years, but the uncertainty of the duration is also Gaussian distributed with $\sigma = 1$ year. However, the start of this cost stream is uniformly uncertain with $a = 0$ and $b = 2$ years. Decay cash flow functions have the equation of present worth for an immediate start as (from Exhibit 9.5.6)

$$E(PW) = E(\bar{C})\left[\frac{1}{j+r}\right]\ [1 - E(e^{-(j+r)t})]$$

where j is given as 10%. Expectation and variance of the uncertain duration effect are (see Exhibits 9.5.4 and 9.5.5), respectively,

$$1 - E[e^{-(j+r)t}] = [1 - e^{-0.5(5)+0.25(1)/2}]/0.5 = 1.813971$$

and

$$V[e^{-(j+r)t}] = e^{-2(0.5)5+2(1)0.25} - e^{-2(0.5)5+1(0.25)} = 0.002457$$

The expected present worth and the variance for an immediately starting cash flow are then, respectively,

$$E(PW) = 100[1/(0.1+4)]\ [1.813971] = \$362.79$$

$$V(PW) = 25(0.002457) + 25(1.813971)^2 + (0.002457)200^2 = 106.90$$

Since the cash flow start is uncertain, which is distributed, then

$$E(e^{-jt}) = \frac{[e^{-0.1(0)} - e^{-0.1(2)}]}{0.1\,(2-0)} = 0.906346$$

$$V(e^{-jt}) = \frac{e^{-2(0.1)0} - e^{-2(0.1)2}}{2(0.1)\,(2-0)} = \left[\frac{e^{-0.1(0)} - e^{-0.1(2)}}{0.1\,(2-0)}\right] = 0.002736$$

The resulting expected present worth for the cash flow with the uncertain start is

$$E(PW) = 0.906346(362.79) = \$328.81$$

and the variance of present worth is

$$V(PW) = (0.002736)\,(106.90) + (0.002736)\,(362.79)^2 + (106.90)\,(0.906346) = \$448.26$$

If the distribution of present worth were approximated by a Gaussian (normal) distribution with the expectation and variance statistics just shown, then a 95% confidence interval would be

$$328.81 \pm 1.96\sqrt{448.26} = \$287.31 \text{ to } \$370.31$$

If the required investment for this project were equal to the lower confidence limit of \$287.31, then there is a risk of 0.025 that the prospective project would not yield the 10% MARR.

As an alternative to the E-V criterion, it has been proposed[16] that the substitution of semivariance for the variance provides a better measure of risk. Semivariances are measurements of dispersion from a central statistic separately either upward or downward. Discounted costs would be measured upward and returns downward. The idea behind the use of semivariances is that the asymmetry of variation (or skewness) is accounted for by the directional semivariance measurement. Still others[17–19] have recommended the use of still higher moments of present worth to account for the dispersion of economic values associated with projects or groups of potential projects. Ideally, one should examine the entire distribution of present worths associated with one or more projects. The methods shown here are restricted to computing only the first two central statistics.

Correlation Effects on Present Worth Statistics

The preceding techniques are based on independent random variables. Although true statistical independence is seldom encountered in practice, moderate departures from independence exhibit only minor effects on the statistics. However, if the dependency factor is considered important, then the variance of sums or products of nonindependent random variables x and y is found as

$$\begin{aligned} \sigma^2_{x+y} &= \sigma^2_x + \sigma^2_y + 2\gamma_{x:y}\sigma_x\sigma_y \\ \sigma^2_{x\cdot y} &= E(x)^2\sigma^2_y + E(y)^2\sigma^2_x + 2\gamma_{x:y}E(x)\,E(y) \end{aligned} \tag{7}$$

where $\gamma_{x:y}$ is the covariance of these random variables. (See Tanchoco and Buck[7] for the use of these formulas.) Since $\gamma_{x:y}$ divided by the product of $\sigma_x\sigma_y$ is the well-known correlation coefficient, the correlations and covariances give similar meanings. In particular, positively correlated random variables increase the variance of a sum or product, whereas negatively correlated random variables decrease the variance. These correlation and discounting effects on present worth are illustrated in Example 9.5.10. (See Canada[3] and Canada and Wadsworth[13] for further details.)

Example 9.5.10. Consider the nine events shown in Example 9.5.2, where the probabilities of \$100, \$200, and \$300 cash flow savings are, respectively, .2, .4, and .4, and the durations of these cost savings for 1, 2, and 3 years are, respectively, 0.3, 0.5, and 0.2. If the amounts and durations of cost savings were independent random variables, then the joint probabilities would be the product of the marginal probabilities. Joint probabilities in Example 9.5.2 clearly are not independent, and the expected present worth was \$121. A computation of correlation in that example yields a correlation coefficient of $\rho = .52$. If these random variables had been independent, then the expected present worth would have been \$116.00, a reduction of about \$5. The present worth SD in the correlated case is 161.4, whereas independence would reduce this SD to 150.00.

Mutually Exclusive Versus Independent Projects

Mutually exclusive projects imply, by definition, a decision in which at most one project will be selected. Typically this situation occurs when several different ways of accomplishing the same objective are being considered. Two questions are usually posed in this class: (1) Which prospective project in this collection is economically best? and (2) Is the best project economically sound? In answering the first question under the maximum expected present worth principle, one would only need to compute the expected present worths of each prospective project and then select that project with the greatest expectation. This approach clearly ignores the variance of present worth and the risk that this variance implies. If the E-V criterion were used, then the project with the greatest *lower* confidence bound may be selected. In each case the second question can be answered for either criterion by seeing whether the criterial value (i.e., the expected present worth or the lower confidence bound) is greater than zero. As an alternative measure of risk with the E-V criterion, one could find the confidence level α where the lower confidence bound is zero and show the value of $\alpha/2$ as the probability that the project will fail to generate the MARR specified, as shown in Example 9.5.11.

Example 9.5.11. Consider the situation shown in Example 9.5.9, where $E(PW)$ = \$328.81 and $V(PW)$ = \$448.26 for the cost savings stream. If the investment required to achieve these cost savings is \$300, then there is some normal random deviate Z that will make the lower confidence bound equal to zero as

$$328.81 - 300.00 - Z\sqrt{448.21} = 0$$

where $Z = 1.36$. The probability of a Z value being 1.36 or more is .9131. Therefore there is about a 9% chance of failing to reach the 10% MARR value if this project is selected.

In the case of an independent set of prospective projects, then, two or more projects may be selected, depending upon the available investment *funds*. When two or more projects are selected, the statistics of present worth resulting from the combination of selected projects depend upon how the two or more projects in the combination are related statistically. If the projects are statistically independent, then the expectations and variances of NPW for the combination simply add, and the lower confidence bound changes accordingly. Combinations of projects with high ratios of expected present worth to variances will result in lower-risk decisions. The converse also results. It also follows that combinations of independent projects can be selected in order to balance a high-risk project with those of low risk to achieve an appropriate mixture of risk. Such a mixture depends upon the trade-off one is willing to accept between expectations of present worth and acceptable risks.

If the independent projects are not statistically independent, then the risk depends upon the nature of the statistical dependence between the projects. A procedure for this situation is indicated by Buck,[20] where variance adjustments can be made if the correlation coefficient can be estimated. A technique is shown in Canada[3] and Canada and Wadsworth.[13] Here again, mixtures of projects can be obtained for desired levels of risks and expected present worths. This type of problem, which occurs in capital budgeting and rationing, is known as the "portfolio problem."[2,15]

Utility Models

Utility theory is briefly described in Chapter 13.3, along with some difficulties with the theory. In essence, utility is a single metric on the unit interval denoting the degree of desirability of an item or a quantity of items with respect to a completely defined collection of such items. Thus an item or group of items with the greatest desirability would have a utility of 1, and a least desirable item, a zero utility. All items and groups within the collection range between these extremes in an ordered fashion. Amounts of monetary receipts and disbursements would provide a utility function from zero to unity. A monetary gamble would be reviewed in this theory as a linear combination of the amount won and lost in the gamble, with the expected utility as probabilistically proportional between the utility of a loss along the straight line to the utility associated with winning. Once a person's utility function is derived, the theory of utility denotes how one should act in order to remain consistent with his or her denoted goals. Accordingly, utility theory is a description of normative economic behavior based on several stated axioms.

A number of advocates of this theory have therefore recommended that utility functions be established and economic risk analysis conducted with respect to this theory. That is, projects with the greatest expected utility should be selected by rational economic decision makers. There are many compelling features to this approach. However, it also involves the required development of the utility function, which is not a simple task; the question of whose utility function should represent the firm; and other perplexing problems.[21] Also, it has been shown[22] that current methods of risk cash flow analysis do represent a reasonable and rational approximation of the utility theory approach. There are also challenges to the axioms of existing theories of utility.[23] Because of these and other detractions, the utility theory approach has not enjoyed popularity among many practitioners.

Decision Trees

Chapter 13.3 also describes the concept of the decision tree as an explicit description of a decision in network form. Since most decisions constitute a choice of an action by the decision maker from among alternative actions, (a choice node) followed by the possibility of one or more uncontrolled events (a chance node), the decision network usually expands from node to node in a tree form of graph; hence the name. When a sequence of two or more decisions is associated with a project, then the decreasing tree provides a way of structuring the decision.

Exhibit 9.5.7 illustrates a single two-stage research and development decision tree. It is clear from this exhibit that decision trees provide an intuitive medium for describing the analytical structure to higher management, and this feature adds to the appeal of this methodology.

To find the statistics of monetary returns associated with the decision to start R & D in Exhibit 9.5.7, one would have to estimate the probabilities along the branches from the circular choice nodes, the various profits and losses at the termini of the end branches, and the timing for the costs and returns. Each distinct action in the decision tree is the specific combination of all choices within the tree as a strategy (see Chapter 13.3). Distinctive actions could each be evaluated for a decision. Also, the different branches from chance nodes in this decision tree can be viewed as random variables with present worth evaluations in the manner shown previously.[1,3,24,25]

A principal difficulty with multiple-stage decision trees is in projecting the downstream decisions, such as the second decision point in Exhibit 9.5.7. The typical procedure for handling this type of difficulty is to assume a decision rule based on the type of success and failure evaluation in the earlier decision stages (e.g., in the R & D testing) and then to show the evaluation effects over variations in the success and failure testing. This approach is a form of sensitivity testing as described earlier. Frequently this sensitivity testing provides a means for determining a sequential testing cutoff point in the early testing process of an R & D decision.

Exhibit 9.5.7 A Decision Tree

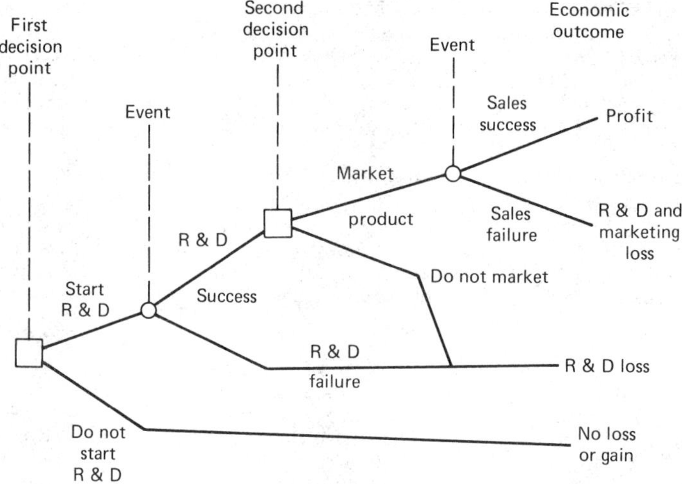

Expected Loss Integrals

When there is a single random variable that affects the profits or costs of a project, then there is some value of this variable that is optimum. Other values of this random variable result in higher costs or lower profits, and such increases in cost or decreases in profit create a loss in prospective opportunity. If the random variable x represents a source of uncertainty where the opportunity loss is a function of x, $f(x)$, which is optimum at x', and where $p(x)$ is the probability density function, then the upper expected loss integral is

$$ELI_u = \int_{x'}^{\infty} f(x)\, p(x)\, dx \tag{8}$$

A similar lower expected loss integral exists with integral limits from ∞ to x'. Since these two directional expected loss integrals represent opportunities that are foregone if $x \neq x'$ and if the value of x is uncertain and uncontrolled, the sum of these expected loss integrals represents the expected cost of uncertainty and/or lack of control. Accordingly, information that changes $p(x)$ has an expected value equal to the *change* in the sum of the expected loss integrals (see Chapter 13.3 for details).

In the situation where a number of prospective projects are under consideration and are all affected economically by the random variable x, there is an optimum economic value for each value of x. For a given value of x, a project that has a lower profit or a higher cost than that of the optimum project yield results in an opportunity loss associated with that project. This definition of loss is identical to the concept of regret as described in Chapter 13.3. When the loss function $f(x)$ is described in this manner, and x' is the best possible value of the random variable, the project with the least expected loss integral (or least expected regret) will also have the greatest expected worth or the least expected cost. This viewpoint of expected loss integrals merely provides a different manner of representing the economic values, but it will yield the same decisions.

A more common use of expected loss integrals is to set x' in equation 8 at a central statistic, or at a break-even point of x, for direct project comparisons. With two projects and x' as a break-even point, the difference between the costs or profits with $x \geqslant x'$ represents the loss function $f(x)$ associated with the more costly or lower-profit project. A similar loss function for $x \leqslant x'$ exists for the other project. It then follows that one may compare the upper and lower expected loss integrals to identify the better project. When three or more projects are being compared, there are usually multiple break-even points, and so simultaneous comparisons cannot be made in this manner. In the case where x' is set at the statistic $E(x)$, the upper and lower expected loss integrals are directional measures of asymmetric variation in a similar, but not identical, manner to the concept of semivariance, described previously. Accordingly, the directional expected loss integrals may be used in lieu of the semivariance as a further variation of the E-V criterion.

Computation of the expected loss integral has been shown for a linear $f(x)$ when $p(x)$ is Gaussian.[4,25] An augmentation of this technique has been given[20] when $f(x)$ is approximated by piecewise linear segments. It also follows that numerical integration techniques[26] can be used to approximate any nonlinear form of $f(x)$ for virtually any probability density function.

9.5.6 SIMULATION TECHNIQUES

Closed-form techniques for computing the expectations and variances of present worth are useful analytical procedures when the cash flow parameters are statistically independent and the first two statistical moments are sufficient. However, in more complex situations, such as those involving correlated cash flow parameters or changes in the discount rate, difficult computational problems ensue, and computer simulation techniques are usually recommended. Simulation techniques provide a whole distribution of results, rather than just the statistical moments, allowing for embedding of internal decision rules into the logic of the program. These added capabilities of digital computer simulation models make this procedure a most viable analytical approach.

However, the simulation technique has some detractions. Typically, computer simulation requires more analytical effort and cost than the closed-form techniques, and it requires the availability of an adequate digital computer with appropriate software support. However, newer computer languages and lower computer costs are making these detractions less and less important. Another detraction of the simulation technique is that the assumptions that must be made in the simulation model are not easily identified. These assumptions are often very difficult to detect, and misleading results may be accepted without adequate review. Therefore the documented program logic should be carefully reviewed before application if this detraction is to be minimized.

Modeling of a simulation program often starts with a decision tree, as previously described, or some description of a process. Various modeling and analysis concepts and techniques are available.[3,24,27] The simulation approach to economic risk analysis is rapidly gaining in popularity, particularly with intermediate and larger decisions.

Transform and Other Techniques

Some time ago present worth computations were observed to be direct applications of transform theory. It has been shown[28] that a continuous cash flow function of time $f(t)$ had a present worth under the discount rate j as

$$PW = \int_0^\infty f(t)\, e^{-jt}\, dt = P(_j) \tag{9}$$

where the present worth is observed as the Laplace transform of the time function $f(t)$ into the state function $P(_j)$. Although present worth calculations in this form and in the probabilistic cases occurred earlier,[13,29] the Laplace transform of probabilistic cash flow timing of single future cash flows and uniform continuous cash flows with uncertain durations was extended to expected present worths[6] as

$$E(PW) = \int_0^\infty p(t)\, Fe^{-jt}\, dt = P(_j) \tag{10}$$

where $p(t)$ is the probability density function of the occurrence time of the future cash flow F. In essence, equation (10) is the Laplace transform of the probability density function $p(t)$ to the state function $P(_j)$. It was later shown[9] that the cash flow that followed various functional forms over time and had probabilistic and statistically independent starting and stopping times could be shown as Laplace transforms of both functions (i.e., the changing amount of flow over time and the probability of a timely start and stop). The present worth variance calculations were extended[8] by taking advantage of statistical properties and properties of the Laplace transform. These theoretical developments on the continuous case provided the basis of the closed-form economic risk analysis previously shown.

Developments in a discrete cash flow analysis proceeded later, but in a similar sequence. First came the discovery that the present worths of discrete cash flow time series were equivalent to the zeta transform of the functional form of the series.[30] Then it was discovered that these discrete present worth functions could be partitioned into terms of a single timing variable so that statistics of present worth could be computed in closed form for statistically independent timing variables.[7] Consequently, discrete time series can be treated in a similar fashion to the continuous cash flows as previously described.

A development associated with these discrete series present worth algorithms is that of the alpha/beta series. This particular series is the solution of a first-order constant-coefficient difference equation where the alpha and beta values are the coefficients.[31] It has been shown that this series can be used to describe learning phenomena,[32] all current depreciation methods,[33] and other economic situations and that zeta transforms exist for this series in order to compute present worth values directly.[34] As a result, there are wide varieties of engineering economic analyses served efficiently through the transform methods. Also, the use of discrete series analysis allows one to study parametrically the series coefficients in sensitivity analysis.

These developments in applying transform theory to engineering economy problems in general and to economic risk analysis in particular have made it possible to compute statistical moments of present worth through the use of available formulas. With the use of inexpensive electronic calculators and these algorithms, economic risk analysis can now be done quickly and easily.

9.5.7 SUMMARY

Project investment decisions require analyses for economic justification. The focus of this chapter is the analytical techniques and interpretations of criteria used to provide economic risk justification. Since there are three distinct processes in the analysis of risky cash flows (i.e., cash flow estimation, discounting, and describing uncertainties), the recommended forms of analysis provide the means of separating these processes for clear thinking during the analysis and evaluation of risk. Sensitivity analysis is shown as a means of identifying the economic importance of component cash flows. Break-even analysis is a special case of sensitivity analysis and is usefully connected to loss integral analysis. The principal emphasis of this chapter, however, is on the methods for doing economic risk analysis through present worth expectations and variances. Closed-form equations for computing these discounted cash flow statistics are discussed. While only continuous cash flow models and techniques are discussed in this chapter, the theory and techniques used for the analysis of discrete cash flow series follow similar concepts. Both forms of analysis provide statistics of present worth.

Expectations and variances of present worth are criticized as being necessary, but not sufficient, measures of economic risk. Alternative criteria, such as semivariances and expected loss integrals, are shown as ways to overcome some of the insufficiency. Some of the alternative methods of economic risk are also shown and criticized. Also, computer simulation is suggested for use in analyzing the more important economic decisions, especially when the cash flows are complex. Finally, the notion and usefulness of transform techniques in economic risk analysis are discussed.

REFERENCES

1. H. G. THUESEN, W. D. FABRYCKY, and G. J. THUESEN, *Engineering Economy*, 4th ed., Prentice-Hall, Englewood Cliffs, NJ, 1971.
2. L. E. BUSSEY, *The Economic Analysis of Industrial Projects*, Prentice-Hall, Englewood Cliffs, NJ, 1978.
3. J. R. CANADA, *Intermediate Economic Analysis for Management and Engineering*, Prentice-Hall, Englewood Cliffs, NJ, 1971.
4. W. T. MORRIS, *Management Science: A Bayesian Approach*, Prentice-Hall, Englewood Cliffs, NJ, 1968.
5. H. L. BEENHAKKER, "Sensitivity Analysis of the Present Value of a Project," *The Engineering Economist*, Vol. 20, No. 2 (1975), pp. 123–150.
6. D. YOUNG and L. CONTRERAS, "Expected Present Worths of Cash Flows Under Uncertain Timing," *The Engineering Economist*, Vol. 20, No. 4 (1975), pp. 257–268.
7. J. M. A. TANCHOCO and J. R. BUCK, "A Closed-Form Methodology for Computing Present Worths of Risky Discrete Cash Flows," *AIIE Transactions*, Vol. 9, No. 3 (1977), pp. 278–287.
8. R. E. ROSENTHAL, "The Variance of Present Worth of Cash Flows Under Uncertain Timing," *The Engineering Economist*, Vol. 23, No. 3 (1978), pp. 163–170.
9. C. D. ZINN, W. G. LESSO, and B. MOTAZED, "A Probabilistic Approach to Risk Analysis in Capital Investment Projects," *The Engineering Economist*, Vol. 22, No. 4 (1977), pp. 239–260.
10. F. S. HILLIER, "The Deviation of Probabilistic Information for the Evaluation of Risky Investments," *Management Science*, Vol. 9, No. 3 (1963), pp. 443–457.
11. B. WAGLE, "A Statistical Analysis of Risk in Capital Investment Projects," *Operational Research Quarterly*, Vol. 18, No. 1 (1967), pp. 13–33.
12. W. J. BAUMOL, "An Expected Gain-Confidence Limit Criterion for Portfolio Selection," *Management Science*, Vol. 10, October 1963, pp. 174–182.
13. J. R. CANADA and H. M. WADSWORTH, "Methods of Quantifying Risk in Economic Analysis of Capital Projects," *The Journal of Industrial Engineering*, Vol. 19, No. 1 (1968), pp. 32–37.

14. D. E. FARRAR, *The Investment Decision Under Uncertainty*, Prentice-Hall, Englewood Cliffs, NJ, 1962.
15. H. M. MARKOWITZ, *Portfolio Selection: Efficient Diversification of Investments*, Cowles Commission Monograph 16, Wiley, New York, 1953.
16. J. C. T. MAO, "Models of Capital Budgeting, E-V vs. E-S," *Journal of Financial and Quantitative Analysis*, January 1970, pp. 657–675.
17. F. D. ARDITTI, "Risk and the Required Return on Equity," *The Journal of Finance*, Vol. 22, No. 1 (1967), pp. 19–36.
18. W. H. JEAN, "The Extension of Portfolio Analysis to Three or More Parameters," *Journal of Financial and Quantitative Analysis*, Vol. 6, No. 1 (1971), pp. 505–515.
19. H. LEVY, "A Utility Function Depending on the First Three Moments," *The Journal of Finance*, Vol. 24, No. 4 (1969), pp. 715–720.
20. J. R. BUCK, "Truncated Linear Gaussian Loss Integrals," *American AIIE Transactions*, Vol. 9, No. 2 (1977), pp. 170–175.
21. B. GOETZ, "Perplexing Problems in Decision Theory," *The Engineering Economist*, Vol. 14, No. 3 (1969), pp. 129–140.
22. J. HIRSCHLEIFER, "Risk, the Discount Rate and Investment Decisions," *American Economic Review*, Vol. 51, No. 2 (1961), pp. 112–130.
23. H. RAIFFA, *Decision Analysis, Introductory Lectures on Choices and Uncertainty*, Addison-Wesley, Reading, MA, 1968.
24. D. HERTZ, "Risk Analysis in Capital Investment," *Harvard Business Review*, January–February 1964, pp. 95–106.
25. R. SCHLAIFER, *Probability and Statistics for Business Decisions*, McGraw-Hill, New York, 1967.
26. F. SHIED, *Theory and Problems of Numerical Analysis*, Schaums Outline Series, McGraw-Hill, New York, 1968.
27. J. R. BUCK, G. E. PECK, R. M. FRANZ, and G. S. BANKERS, "Tabletting Production Times, Cost, and Risks Through Simulation," *Drug Development and Industrial Pharmacy*, Vol. 6, No. 3 (1980), pp. 237–253.
28. J. R. BUCK and T. W. HILL, JR., "Laplace Transforms for the Economic Analysis of Deterministic Problems in Engineering," *The Engineering Economist*, Vol. 16, No. 4 (1971), pp. 247–263.
29. G. A. FLEISCHER, Ed., *Risk and Uncertainty: Non-Deterministic Decision Making in Engineering Economy*, Publication No. 2, Monograph Series (Engineering Economy Division), AIIE, Norcross, GA, 1975.
30. T. W. HILL, JR. and J. R. BUCK, "Zeta Transforms, Present Value, and Economic Analysis," *AIIE Transactions*, Vol. 6, No. 2 (1974), pp. 120–125.
31. S. GOLDBERG, *Introduction to Difference Equations*, Wiley (Science Editions, Inc.), New York, 1961.
32. J. R. BUCK, J. M. A. TANCHOCO, and A. L. SWEET, "Parameter Estimation for Discrete Exponential Learning Curves," *AIIE Transactions*, Vol. 8, No. 2 (1976), pp. 184–194.
33. J. R. BUCK and T. W. HILL, JR., "Generalized Depreciation Methods and Their Economic Evaluation," *The Engineering Economist*, Vol. 22, No. 2 (1977), pp. 79–96.
34. J. R. BUCK and T. W. HILL, JR., "Alpha/Beta Difference Equations and Their Zeta Transforms in Economic Analysis," *AIIE Transactions*, Vol. 7, No. 3 (1975), pp. 330–358.

CHAPTER 9.6

Production Economics

FERDINAND F. LEIMKUHLER

Purdue University

9.6.1 INTRODUCTION

Production economics is the study of how producers combine costly inputs to make profitable outputs. For analytic purposes, production planning can be divided into three stages of decision making as follows:

1. *Input decisions*, or how many of the variable factors of production to use so as to produce a specified output at minimum cost.
2. *Output decisions*, or how much output to make (and input to use) so as to maximize net profit or to achieve some other production objective.
3. *Capacity decisions*, or how to schedule long-term investments in production facilities so as to optimize the return on investment.

These decisions must take into account both the physical capabilities of the production processes used and the market conditions associated with the purchase of inputs, the sale of outputs, and the acquisition of capital funds.

The analysis begins with a determination of the minimum-cost way of producing an output in the short run by considering the productivity characteristics of the process and the cost of the inputs. This provides a basis for determining how the producer responds to market demand for the output in order to achieve sales objectives and also how the producer responds to changes in the cost of inputs. Finally, by taking a long-run view of these input-output relationships, a framework is developed for making optimal investment decisions in production facilities. In the first stage of the analysis, the element of time is treated as an implicit dimension by assuming that input and output activities occur at steady rates and can be adjusted instantaneously. However, when making capacity decisions, the element of timing becomes an important explicit aspect of the problem.

The viewpoint taken in this analysis is that of an optimizer, that is, a decision maker who acts so as to achieve a stated objective to the greatest degree possible. Although this goal may not be realized in practice for various reasons, it provides a consistent basis for making comparisons and developing a comprehensive theory of production. This survey of production theory is limited to the more general concepts and models that have been developed. The references listed at the end of the chapter give a thorough treatment of these ideas and include important extensions and applications of the models. The decision rules that emerge from the theory identify the important elements of production decisions, the kinds of data that are needed for decisions, and available methods for processing production information to guide managerial decisions. Simplified numerical examples are used throughout this chapter to demonstrate how analytic methods might be developed and applied in practical situations.

9.6.2 PRODUCTIVITY OF INPUTS

The output rate of a production process depends on the input rates of the factors used. The relative efficiency with which a process uses an input is measured by the input's "average productivity," that is, output rate divided by input rate, and by its "marginal productivity," that is, the rate of change in output rate with a change in input rate. The ratio of marginal and average productivity is the "output elasticity" of an input, which measures the percentage change in output to a percentage change in input.[1,2]

In the analysis of production systems, the marginal productivity of an input is equal to the partial derivative of a "production function," which defines the output rate of a process as a function of the input rates. For example, a production function often used in economic analysis is the Cobb-Douglas function,[3] which has the form

$$q = a(x_1^{b_1})(x_2^{b_2}) \ldots (x_m^{b_m})$$

$$\log q = \log a + b_1 \log x_1 + \cdots + b_m \log x_m$$

$$q'_i = \frac{\partial q}{\partial x_i} = \frac{b_i q}{x_i} = b_i \bar{q}_i \qquad (1)$$

where q = output rate of a process with input rates x_i for each of the m types of input
q'_i = marginal productivity of input i
$\bar{q}_i$ = average productivity of input i

Note that the logarithmic form of this production function is useful for statistical estimation by means of linear regression methods (see Chapter 13.6). The regression coefficient b_i is an estimate of the output elasticity of input i.

A production process is "scaled up" when all of the inputs are increased in a fixed proportion to one another. The percentage change in output with a percentage change in all inputs is called the "scale elasticity" of the process, and it is equal to the sum of the individual output elasticities of the inputs. A process has decreasing, constant, or increasing "returns to scale," depending on whether the scale elasticity is less than, equal to, or greater than unity. A class of production functions that exhibit constant returns to scale are "linear homogeneous" functions, which can be written in the following form:

$$q = q'_1 x_1 + q'_2 x_2 + \cdots + q'_m x_m \qquad (2)$$

An example of such a process is the following Cobb-Douglas production function:

$$q = 2x_1^{1/2} x_2^{1/2} = q'_1 x_1 + q'_2 x_2 = \left(\frac{\frac{1}{2}q}{x_1}\right) x_1 + \left(\frac{\frac{1}{2}q}{x_2}\right) x_2$$

$$x_2 = \frac{q^2}{4x_1}; \qquad \frac{dx_2}{dx_1} = \frac{-q^2}{4x_1^2} = \frac{-x_2}{x_1} \qquad (3)$$

The productivity characteristics of this process are plotted in Exhibits 9.6.1 through 9.6.4. Note that the marginal productivity falls as the input rate is raised. This is an example of the economic "law of diminishing returns." In linear homogeneous processes, inputs are used efficiently when the output elasticity of each input is between 0 and 1. This is called the "normal production region."

Exhibit 9.6.1 Productivity Curve A-B When Input x_1 Changes and Input x_2 Is Constant Along Expansion Path A-C

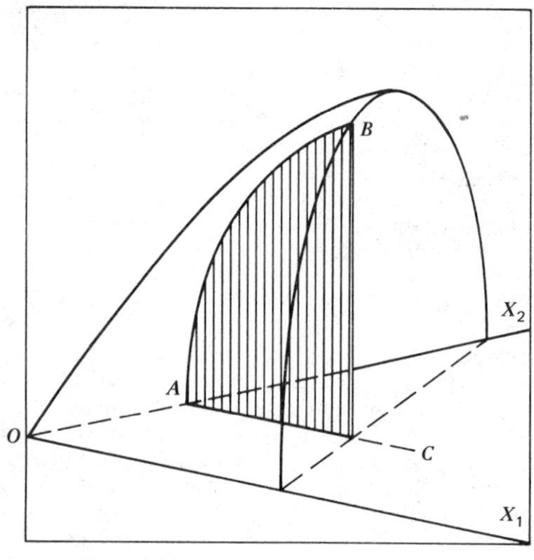

Exhibit 9.6.2 Returns to Scale Along Output Curve *0-A* When Inputs x_1 and x_2 Change Proportionally Along Expansion Path *0-B*

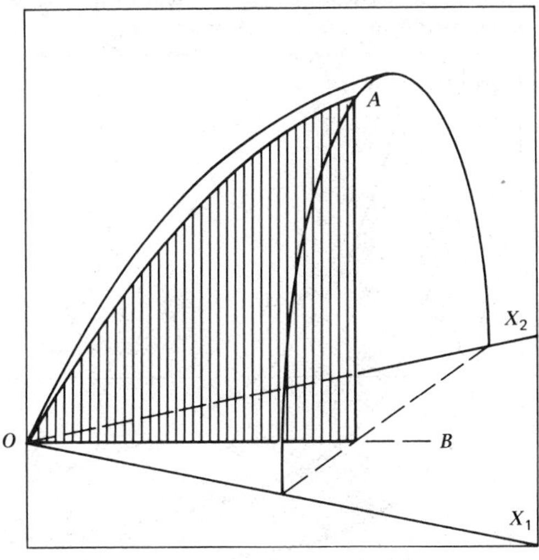

9.6.3 SUBSTITUTION OF INPUTS

Those combinations of input rates that yield the same output rate define a constant output curve, or "isoquant," of a process, as shown in Exhibit 9.6.3. The ratio of the marginal productivities of two inputs measures their rate of technical substitution and is equal to the slope of the isoquant times –1. For the previous example, in equation 3, the marginal productivity ratio q'_1/q'_2 equals x_2/x_1, which is the rate of technical substitution of input 2 for input 1. For example, when the process is using 100 units of x_1 and 400 units of x_2, the output level of 400 units of q can be maintained by substituting 4 units of x_2 for 1 unit of x_1. Note that the rate of technical substitution changes as the input ratio changes. When the rate of technical substitution is a positive constant, the inputs are said to have "perfect substitution," and when it equals zero, there is "no substitution."

An important example of no substitution is in processes where inputs are used in fixed proportions. Such a process is called a "linear process with fixed technical coefficients,"[1,4] where a_i

Exhibit 9.6.3 Isoquant Curve *A-B* When Inputs x_1 and x_2 Change So As To Keep the Output Constant

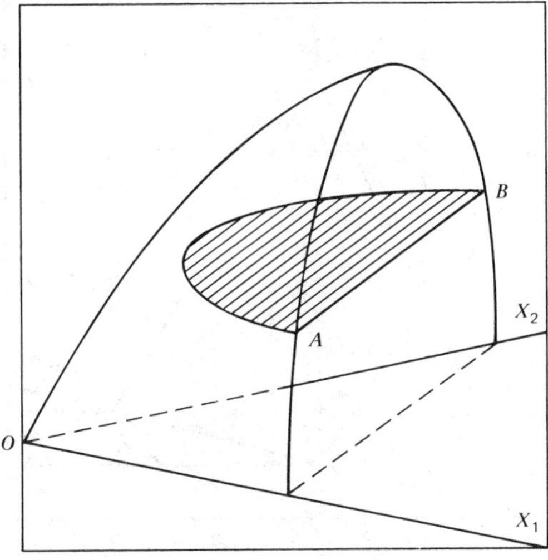

Exhibit 9.6.4 Isoquant Curves *Q-Q* for a Nonlinear Production System (Equations 22–24) With a Typical Budget Line *B-B*, Expansion Path *0-A* When Inputs Are Unlimited, and Expansion Path *0-C* When Input x_2 Is Limited to 16 Units

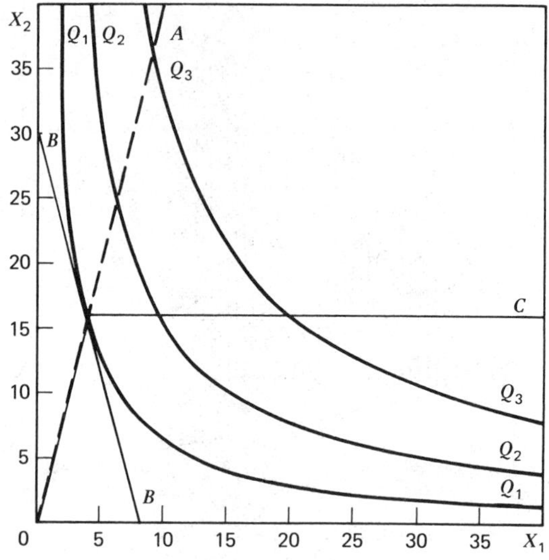

denotes the fixed amount of input i required per unit of output q. The production function of a linear process can be written as follows:

$$q = \text{minimum}\left(\frac{x_1}{a_1}, \frac{x_2}{a_2}, \ldots, \frac{x_m}{a_m}\right) \tag{4}$$

An example of a linear process with only two inputs is shown in Exhibit 9.6.5, where the iso-

Exhibit 9.6.5 Isoquant Lines *Q-Q* for a Linear Production System (Equation 7) With Two Processes Having Expansion Paths *0-P_1* and *0-P_2* and Used Either Simultaneously (Solid Line) or One at a Time (Dashed Line)

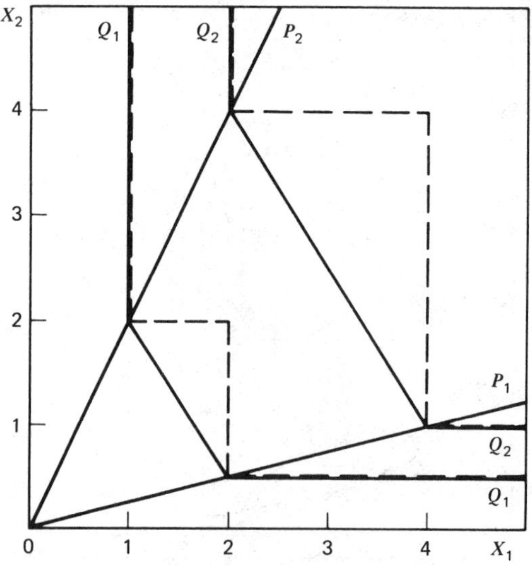

quants are L-shaped with corner points on a process vector, that is, the diagonal line from the origin with slope a_2/a_1. Along the process vector the input rates x_1 equal a_iq, which is the minimum amount of x_i needed to make q, and the marginal products are zero. If there is an excess amount of one input, its marginal product is zero, and if there is a shortage of one input, its marginal product is equal to $1/a_i$. In the latter case the deficient input has an output elasticity of one, but otherwise it is zero.

When two or more linear processes are used to make the same output, they form a linear production system with coefficients a_{ij} denoting the minimum amount of input i required to make one unit of q by process j. Substitution of processes may or may not be possible. When there is no substitution and processes are used one at a time, the production function has the form

$$q = \text{maximum}_j \ \text{minimum}_i \left(\frac{x_i}{a_{ij}} \right) \tag{5}$$

That is, the best output obtainable from each process is determined first and then the best of these is used. The isoquant for such a system has a step pattern consisting of segments of the L-shaped isoquants of the processes as shown in Exhibit 9.6.5. The case where process substitution can occur is discussed in the next section.

9.6.4 LINEAR PRODUCTION SYSTEMS

Where the same output is produced using two or more distinct linear processes, input substitution occurs when different combinations of the processes are used. If q_j denotes the output rate of process j, the sum of the q_j's is the total output (q) of the system, and a_{ij} is the amount of input i per unit of output using process j, then the production function for a linear system is defined by the following linear programming formulation:

$$q = \text{maximum} \ (q_i + q_2 + \cdots + q_n)$$

where

$$q_j \geqslant 0 \qquad \text{for} \ \ j = 1, 2, \ldots, n$$

$$x_1 \geqslant a_{11}q_1 + a_{12}q_2 + \cdots + a_{1n}q_n$$
$$\ldots \quad \ldots \quad \ldots \quad \ldots \quad \ldots$$
$$x_m \geqslant a_{m1}q_1 + a_{m2}q_2 + \cdots + a_{mn}q_n \tag{6}$$

No more than m of the n processes are used at the same time in the linear programming solution (see Chapter 14.2 and other sources[1,4,5]).

An example of a linear production system with one product, two resource inputs, and two linear processes, each of which produce the output, is one with the following production function:

$$q = \text{maximum} \ (q_1 + q_2); q_1 \geqslant 0; q_2 \geqslant 0$$
$$x_1 \geqslant 1.00q_1 + 0.52q_2$$
$$x_2 \geqslant 0.25q_1 + 1.00q_2 \tag{7}$$

For this system, an isoquant with constant output level q consists of three line segments as shown in Exhibit 9.6.5 and as defined by the following equations:

$$x_1 = a_{12}q = 0.5q, \quad \text{when} \quad x_2 \geqslant a_{22}q = q \tag{8}$$

$$x_2 = a_{21}q = 0.25q, \ \text{when} \quad x_1 \geqslant a_{11}q = q \tag{9}$$

$$x_2 = \frac{(a_{11}a_{22}q - a_{12}a_{21}q - a_{22}x_1 + a_{21}x_1)}{(a_{11} - a_{12})} \tag{10}$$

$$x_2 = 1.75q - 1.5x_1, \ \text{when} \quad x_1 \ \text{and} \ x_2 \ \text{are otherwise} \tag{11}$$

The latter equation defines the isoquants in the triangular region between the two process vectors P_1 and P_2 in Exhibit 9.6.5, that is, where the input ratio x_2/x_1 is between $a_{21}/a_{11} = 0.25$ and $a_{22}/a_{12} = 2$ for this example. This is the normal production region of a linear system, since otherwise there are excess amounts of input.

In the normal production region, the marginal products and the technical rate of substitution are as follows:

$$q_1' = \frac{a_{22} - a_{21}}{(a_{11}a_{22} - a_{12}a_{21})} = \frac{6}{7} \tag{12}$$

$$q_2' = \frac{a_{11} - a_{12}}{(a_{11}a_{22} - a_{12}a_{21})} = \frac{4}{7} \tag{13}$$

$$\frac{q_1'}{q_2'} = \frac{a_{22} - a_{21}}{a_{11} - a_{12}} = 1.5 \tag{14}$$

The constant rate of technical substitution indicates perfect substitution along a linear isoquant. Also, in the normal production region, the output rates are given by the following equations:

$$q = \frac{(a_{22}x_1 - a_{21}x_1 + a_{11}x_2 - a_{12}x_2)}{(a_{11}a_{22} - a_{12}a_{21})} \tag{15}$$

$$= a_1'x_1 + q_2'x_2 = \frac{6x_1 + 4x_2}{7}$$

$$q_1 = \frac{(a_{22}x_1 - a_{12}x_2)}{(a_{11}a_{22} - a_{12}a_{21})} = \frac{8x_1 - 4x_2}{7} \tag{16}$$

$$q_2 = \frac{(a_{11}x_2 - a_{21}x_1)}{(a_{11}a_{22} - a_{12}a_{21})} = \frac{8x_2 - 2x_1}{7} \tag{17}$$

These equations are the general solutions to the linear programming problem of equation 7 in the normal region. Equations 12 and 13 for the marginal products are the optimal values of the corresponding dual variables of the linear programming problem (see Chapter 14.2 and other sources[1,4,5]). This analysis can be extended to linear systems with more than two inputs and more than two linear processes.

9.6.5 PRODUCTION COST FOR LINEAR SYSTEMS

The cost of production depends on the technology used and the cost of the inputs. In linear systems, if the cost per unit of each input is fixed, then the cost per unit of output of each process is also fixed under normal conditions. If there is no limitation on the use of inputs and processes, the cheapest process is used, and production cost is equal to the minimum process cost. However, when there are limitations on the amount of input or on the use of a process, increased output requires the use of more expensive processes. The marginal and the average costs of production go up. Under these conditions, a general production cost model is the following linear program, where k_i denotes the cost per unit on input i, c_j denotes the cost per unit of output q_j from process j, and x_i are the available amounts of input:

$$\begin{aligned}
c_q &= \text{minimum } (c_1q_1 + c_2q_2 + \cdots + c_nq_n) \\
c_j &= k_1a_{1j} + k_2a_{2j} + \cdots + k_ma_{mj} \\
q &\leqslant q_1 + q_2 + \cdots + q_n; q_i \geqslant 0 \\
x_1 &\geqslant a_{11}q_1 + a_{12}q_2 + \cdots + a_{1n}q_n \\
x_2 &\geqslant a_{21}q_1 + a_{22}q_2 + \cdots + a_{2n}q_n \\
\cdots & \quad \cdots \quad \cdots \quad \cdots \quad \cdots \\
x_m &\geqslant a_{m1}q_1 + a_{m2}q_2 + \cdots + a_{mn}q_n
\end{aligned} \tag{18}$$

The cost of producing output q can be found by solving the linear program for all values of q from 0 to the maximum output defined by equation 6. This cost equation will consist of a series of line segments as shown in Exhibit 9.6.6.

In the previous example of a linear system with the production function of equation 7, if inputs x_1 and x_2 both cost \$4/unit, then the output from process 1 costs \$5/unit, and the output from process 2, \$6/unit. Furthermore, if there are 40 units of x_1 and 45 units of x_2 available, then the

Exhibit 9.6.6 Total Production Cost Curve *0-A*, Marginal Cost *B-C*, and Average Cost *B-D* for a Linear Production System (Equation 19) With Input Limitations

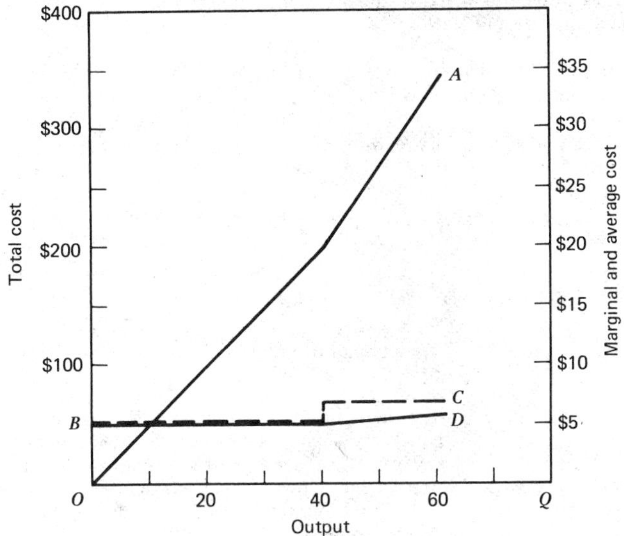

production cost can be found by solving the linear program

$$
\begin{aligned}
c_q &= \text{minimum } (5q_1 + 6q_2) \\
q &\leqslant q_1 + q_2; q_1 \geqslant 0; q_2 \geqslant 0 \\
40 &\geqslant 1.00q_1 + 0.50q_2 \\
45 &\geqslant 25q_1 + 1.00q_2
\end{aligned} \tag{19}
$$

for output rates from 0 to 60. The resulting cost curves are plotted in Exhibit 9.6.6 and defined by the following equations:

When $0 \leqslant q \leqslant 40$	When $40 \leqslant q \leqslant 60$
$c_q = 5q$	$c_q = 7q - 80$
$\bar{c}_q = 5$	$\bar{c}_q = 7 - 80/q$
$c'_q = 5$	$c'_q = 7$
$e_c = 1$	$e_c = 1/(1 - 11.43/q)$

where c_q = cost of producing q units of output
$\bar{c}_q$ = average cost of production
c'_q = marginal cost
e_c = elasticity of production cost (i.e., the percentage change in cost for a percentage change in output)

For output rates up to 40 units, the percentage change in cost is the same as the percentage change in output, but when output is between 40 and 60 units, the percentage change in cost is greater than the percentage change in output.

9.6.6 PRODUCTION COST FOR NONLINEAR SYSTEMS

A general method of finding a minimum production cost is the Lagrangian method used in constrained optimization problems (see Chapter 14.2 and other sources[1,2]). The Lagrangian has the form

$$L(x, v) = k(x) + v[q - Q(x)] \tag{20}$$

where x = set of variable input rates, $(x_1, \ldots, x_m)$
v = Lagrangian variable
$k(x)$ = cost as a function of x
$Q(x)$ = output as a function of x (i.e., the production function)
q = some required output level

Those values of the variables that minimize cost usually are found by setting the partial derivatives of the Lagrangian function with respect to the x_i's and v equal to zero and then solving these equations simultaneously.

When the cost function is linear, the *solution* of the Lagrangian function reduces to the following optimality conditions:

$$L(x, v) = k_1x_1 + k_2x_2 + \cdots + k_mx_m + v[q - Q(x)]$$

$$v = \frac{k_1}{q'_1} = \frac{k_2}{q'_2} = \cdots = \frac{k_m}{q'_m}, \text{ when } \frac{\partial L}{\partial x_i} = 0$$

$$\frac{k_h}{k_i} = \frac{q'_h}{q'_i} \text{ for any pair of inputs} \tag{21}$$

At optimum the marginal input cost is the same for all inputs and is equal to v, which is the marginal cost of production; that is, v equals c'_q when q is being produced at minimum cost. The rate of technical substitution q'_h/q'_i for any pair of inputs along the q isoquant is equal to the ratio of their units' costs k_h/k_i; that is, the isocost line is just tangent to the isoquant, as shown in Exhibit 9.6.4. The locus of the tangency points for all output levels defines the "expansion path" of the system, chosing the optimal mix of inputs to use.

Additional requirements or limitations on the system can be included in the analysis by adding terms to the Lagrangian function. For example, let a production system with a Cobb-Douglas-type production function as defined by equation 3 have an upper limit on the input rate x_2 of 16 units, and let there be a fixed cost rate of \$48 and a cost of \$4/unit of x_1 and \$1/unit of x_2. Then, the Lagrangian function and its solution are as follows:

$$L(x, v) = 4x_1 + x_2 + v_1(q - 2x_1^{1/2}x_2^{1/2}) + v_2(16 - x_2)$$

$$q \leqslant 16: x_2 = 4x_1 = q; x_1 = \frac{q}{4}; v_1 = 2; v_2 = 0 \tag{22}$$

$$c_q = 48 + 2q; c'_q = 2; \bar{c}_q = 2 + \frac{48}{q}$$

$$q \geqslant 16; x_2 = 16; x_1 = \frac{q^2}{64}; v_1 = \frac{q}{8}; v_2 = 1 - \frac{q^2}{64} \tag{23}$$

$$c_q = 64 + \frac{q^2}{16}; c'_q = \frac{q}{8}; \bar{c}_q = \frac{64}{q} + \frac{q}{16} \tag{24}$$

The cost functions are plotted in Exhibits 9.6.7 and 9.6.8. Note that at optimum, v_1 is equal to the marginal cost of production, and v_2 is equal to the marginal cost of the constraint on x_2; that is, by increasing the x_2 limit, the cost of production would be decreased marginally by an amount equal to v_2. The Lagrangian variables are equivalent to the "shadow prices" of the constraints. In Exhibit 9.6.4 the expansion path of the system follows the boundary of the x_2 constraint for outputs of 16 units or more.

9.6.7 PRODUCTION COST AND OUTPUT DECISIONS

Production output decisions depend on the output cost function, the sales revenue function, and the objectives of the managers of the production system. Five output levels of special interest to decision makers are:

1. Minimum break-even output level.
2. Minimum average cost output level.
3. Maximum profit output level.
4. Maximum break-even output level.
5. Maximum sales revenue level.

Exhibit 9.6.7 Production Cost Curve *K-K* for a Nonlinear Production System (Equations 22-24), Total Revenue Curve *R-R* at $5/unit, and Four Typical Output Levels (Equations 25-29)

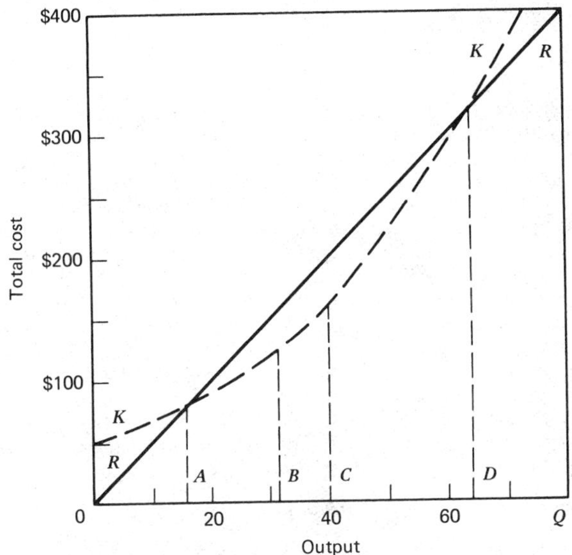

Level 1 is important for planning purposes, level 2 is a usual production objective at the plant level, and level 3 normally is assumed to be the objective at corporate level. Levels 4 and 5 are alternate and appropriate objectives for some production systems. These output decisions are shown graphically in Exhibits 9.6.7 and 9.6.8 for the production system in the previous example, with cost curves defined in equation 22 and with a fixed selling price of $5/unit of output q.

In this example, levels 4 and 5 are the same. The conditions for the five output levels can be

Exhibit 9.6.8 Average Cost Curve *K-K* and Marginal Cost Curve *M-M* for a Nonlinear Production System (Equations 22-24), With Marginal and Average Revenue Curve *R-R* and Four Typical Output Levels (Equations 25-29)

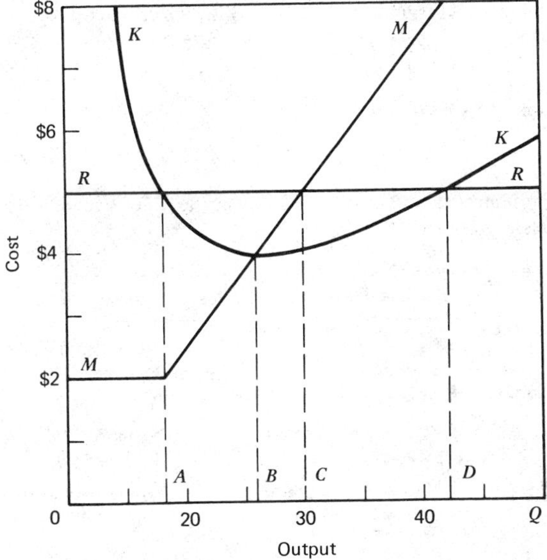

summarized as follows:

1. $\bar{c}_q = \bar{r}_q; \bar{c}'_q \leq \bar{r}'_q$ (25)
2. $\bar{c}_q = c'_q; c''_q \geq 0$ (26)
3. $\bar{c}_q \leq \bar{r}_q; c'_q = r'_q; c''_q \geq r''_q$ (27)
4. $\bar{c}_q = \bar{r}_q; \bar{c}_q \geq \bar{r}_q$ (28)
5. $\bar{c}_q \leq \bar{r}_q; \bar{r}_q = 0; r''_q < 0$ (29)

At break-even level 1, average cost $\bar{c}_q$ equals average revenue $\bar{r}_q$, and average cost is changing at a smaller rate $\bar{c}'_q$ than that of the average revenue $\bar{r}'_q$. Similar conditions hold at the maximum break-even level 4, except the latter condition is reversed. At the minimum average cost level 2, average cost equals marginal cost, and the marginal cost is increasing. At the maximum profit level 3, the marginal cost c'_q and marginal revenue r'_q are equal, and marginal cost is changing at a higher rate c''_q than that of marginal revenue r''_q. Also, to exclude a loss situation, the average cost is not greater than average revenue. In level 5 marginal revenue equals zero and is decreasing.

With a fixed selling price, as in the preceding example, the marginal revenue is equal to the selling price, and a profit maximizing producer supplies that output where selling price equals the marginal cost. For this reason, the marginal cost curve also defines the producer's "short-run supply curve," that is, the amount supplied for a given fixed price. When the selling price depends on the amount sold, the demand curve, that is, average revenue as a function of output or $\bar{r}_q$, is not a horizontal line. In this case the profit maximizing producer still equates marginal cost with marginal revenue, but the supply curve as such is not readily defined. In highly competitive markets, there is a long-run trend for the surviving producers to be producing at the minimum average cost in the industry, with no excess revenue over full cost.

9.6.8 DEMAND AND EXPENDITURE FOR INPUTS

A producer's demand and expenditure for inputs depends on the input prices, the output cost and revenue functions, and the producer's objective, which can be one of the following:

1. Minimize the cost of a given output level.
2. Maximize output with a given cost budget.
3. Maximize profit.
4. Maximize the break-even output rate.
5. Maximize sales revenue and avoid a loss.

Objectives 3, 4, and 5 correspond to output levels 3, 4, and 5 in the previous section, but not 1 and 2. Input demand curves showing the amount of input needed as a function of its price per unit and the producer's objectives are shown in Exhibit 9.6.9. Expenditure curves are shown in Exhibit 9.6.10.

The demand curves in Exhibit 9.6.9 are derived for a process with the production function of equation 3, a cost of \$1/unit for input x_2, and the following cost and revenue functions:

$$q = 2x_1^{1/2}x_2^{1/2}; c_x = k_1x_1 + x_2 \tag{30}$$

$$c_q = k_1^{1/2}q; c'_q = \bar{c}_q = k_1^{1/2} \tag{31}$$

$$r_q = 4q - 0.02q^2; r'_q = 4 - 0.04q; \bar{r}'_q = 4 - 0.02_q \tag{32}$$

For this system, the input demand curves $x_1(k_1)$ and the input expenditure curves k_1x_1 for each of the preceding objectives are:

Objective	Demand Curve $x_1(k_1)$	Expenditure k_1x_1	
1	$x_1 = q/2k_1^{1/2}$	$k_1x_1 = qk_1^{1/2}/2$	(33)
2	$x_1 = c/2k_1$	$k_1x_1 = c/2$	(34)
3	$x_1 = 50k_1^{-1/2} - 12.5$	$k_1x_1 = 50k_1^{1/2} - 12.5k_1$	(35)
4	$x_1 = 100k_1^{-1/2} - 25$	$k_1x_1 = 100k_1^{1/2} - 25k_1$	(36)
5	If $k_1 \leq 4$, $x_1 = 50\,k_1^{-1/2}$	$k_1x_1 = 50\,k_1^{1/2}$	(37)
	If $k_1 \geq 4$, same as for objective 4	k_1x_1 same as for objective 4	(38)

Exhibit 9.6.9 Input Price–Demand Curves for a Nonlinear Production System Under Five Different Production Objectives (Equations 33–38)

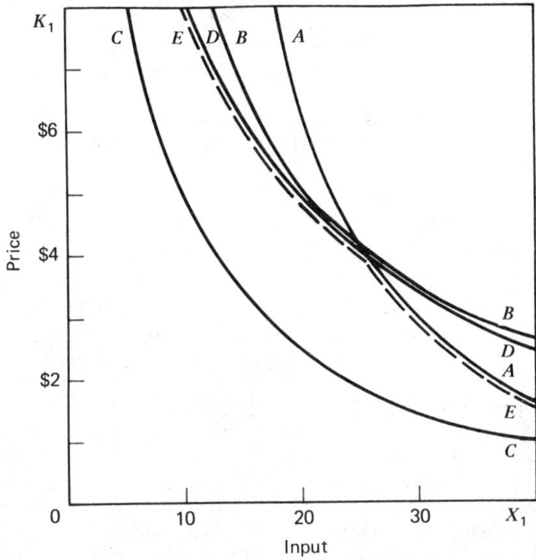

The expenditure curves are plotted in Exhibit 9.6.10. In Exhibit 9.6.9, demand curve 1 assumes q is 100, and demand curve 2 assumes c is $200. Demand curve 1 shows only a substitution effect of a price change since the output is constant, but the other curves show both a substitution and an output effect since the change in price also causes a change in output.

When a production process has fixed technical coefficients, that is, when inputs are a fixed proportion of the output, there is no substitution, and price changes can only have an output effect on the input demand. Input demand curves for such a linear process are shown in Exhibit 9.6.11 for each of the preceding objectives. In this example the revenue function and cost of $1/unit of x_2 are the same as the last example, but the production function and output cost function are as

Exhibit 9.6.10 Input Expenditure Curves for Input x_1 in a Nonlinear Production System as a Function of the Input Price k_1 and the Production Objective (Equations 33–38)

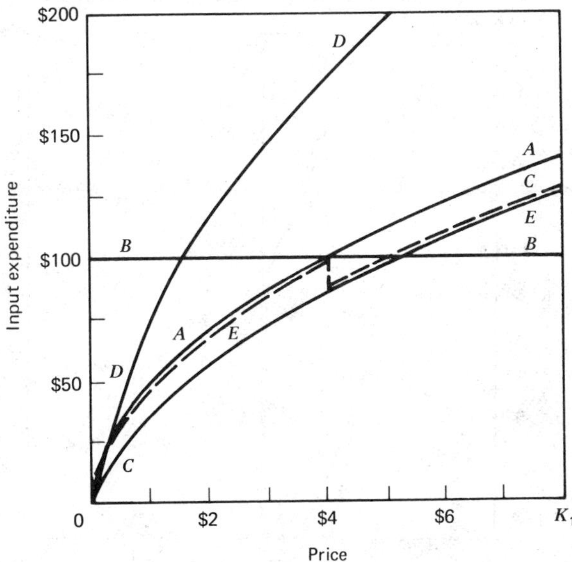

follows:

$$q = \text{minimum}\ (x_1/0.2, x_2/1.2); c_x = k_1x_1 + x_2 \tag{39}$$

$$c_q = (0.2k_1 + 1.2)\ q; c_q' = \bar{c}_q = 0.2k_1 + 1.2 \tag{40}$$

$$r_q = 4q - 0.02q^2; r_q' = 4 - 0.04q; \bar{r}_q = 4 - 0.02q \tag{41}$$

For this system the input demand curves $x_1(k_1)$ and the input expenditure curves k_1x_1 are as follows:

Objective	Demand Curve $x_1(k_1)$	Expenditure k_1x_1	
1	$x_1 = 0.2q$	$k_1x_1 = 0.2qk_1$	(42)
2	$x_1 = c/k_1 + 6$	$k_1x_1 = c/1 + 6/k_1$	(43)
3	$x_1 = 14 - k_1$	$k_1x_1 = 14k_1 - k_1^2$	(44)
4	$x_1 = 28 - 2k_1$	$k_1x_1 = 28k_1 - 2k_1^2$	(45)
5	If $k_1 \leqslant 4, x_1 = 20$	$k_1x_1 = 20k_1$	(46)
	If $k_1 \geqslant 4$, same as for objective 4	k_1x_1 same as for objective 4	(47)

In Exhibit 9.6.11 demand curve 1 assumes q is 100, and curve 2 assumes c is \$100. The expenditure curves for the input x_1 as a function of price and the different objectives are shown in Exhibit 9.6.12.

The "price elasticity of demand" is the ratio of the percentage decrease in demand to a percentage increase in price; the "price elasticity of expenditure" is the ratio of the percentage change in expenditure to a percentage change in price. As with other elasticity measures, they are both equal to the ratio of the marginal and average values of the functions. The elasticities for the previous example processes are as follows:

Process and Elasticity	Objective 1	Objective 2	Objectives 3 and 4	
Nonlinear				
Demand	0.5	1.0	$2/(4 - k_1^{1/2})$	(48)
Expenditure	0.5	0.0	$2/(4 - k_1^{1/2})$	(49)
Linear				
Demand	0.0	$k_1/6 + k_1$	$k_1/14 - k_1$	(50)
Expenditure	1.0	$6/6 + k_1$	$7 - k_1/7 - \frac{1}{2}k_1$	(51)

Exhibit 9.6.11 Input Price–Demand Curves for a Linear Production Process for Different Production Objectives (Equations 42–47)

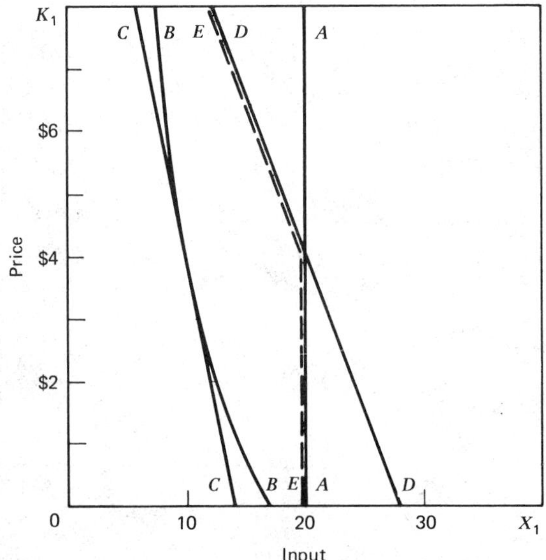

Exhibit 9.6.12 Input Expenditure Curves for Input x_1 in a Linear Production Process as a Function of the Input Price k_1 and the Production Objective (Equations 42–47)

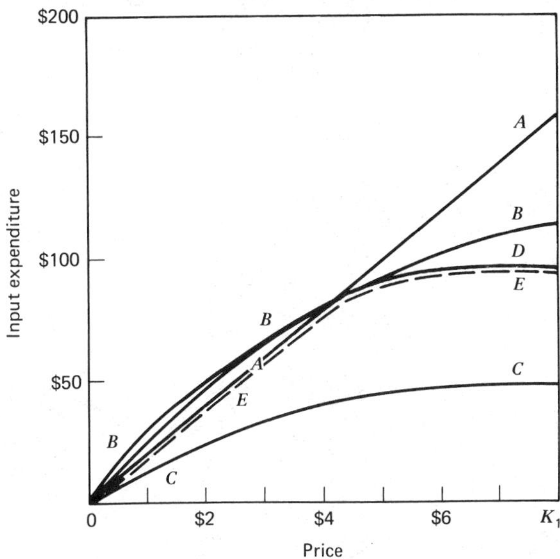

For example, under objective 1, minimum cost with fixed output, a 20% rise in the price of x_1 causes a 10% drop in the use of x_1 and a 10% increase in the expenditure on x_1 for the nonlinear system, but in the linear process the use of x_1 is not affected by the price increase, and expenditure on x_1 rises 20%.

Of more general interest is whether the demand is "elastic" or "inelastic," that is, whether the percentage decrease in use is greater or less than the percentage increase in price. Under objective 1, demand is inelastic for the nonlinear case and perfectly inelastic for the linear process. Under objective 2, maximum output with fixed budget, demand has unit elasticity for the nonlinear system and is inelastic for the linear process. Under objective 3, maximum profit, and objective 4, maximum break-even output, demand is inelastic for low prices of x_1 and elastic for high prices of x_1.

The expenditure on x_1 is inelastic in all cases, except for the nonlinear system under objectives 3 and 4, where expenditure is elastic for high prices of x_1 in both the linear and nonlinear cases. Inelasticity means that the percentage increase in expenditure on x_1 is somewhat less than the percentage increase in the price of x_1, because of adjustments made in other inputs and output. Demand and expenditure patterns for linear systems with more than one linear process approach those of the nonlinear system.

9.6.9 SHORT-RUN AND LONG-RUN PRODUCTION COSTS

In the short run, a producer seeks to optimize an existing system that has fixed input limitations and cost characteristics. But in planning for the long run, the producer can specify the system's capacity by considering the "long-run production cost curve," which defines the minimum cost of output under optimal capacity conditions. The long-run cost curve is the lower boundary, or envelope, of the various possible short-run cost curves.

For example, in Section 9.6.6 and equations 22, 23, and 24, a short-run cost function is derived for a nonlinear production system with a limit of 16 units of input x_2. Equations 23 and 24 have the general form

$$q \leqslant s,\ c(q, s) = k_s s + 2q;\ c_q' = 2;\ \bar{c}_q = 2 + \frac{k_s s}{q} \tag{52}$$

$$q \geqslant s,\ c(q, s) = (k_s + 1)\, s + \frac{q^2}{s};\ c_q' = \frac{2q}{s}; \tag{53}$$

$$\bar{c}_q = \frac{(k_s + 1)\, s}{q + q/s} \tag{54}$$

Exhibit 9.6.13 Short-Run Total Cost Curves for a Nonlinear Production System With Three Capacity Levels at (*A*) 16 Units, (*B*) 25 Units, and (*C*) 40 Units of Input x_2 (Equations 23), and the Long-Run Total Cost Curve (Dotted Line) for the Same System (Equations 55–57)

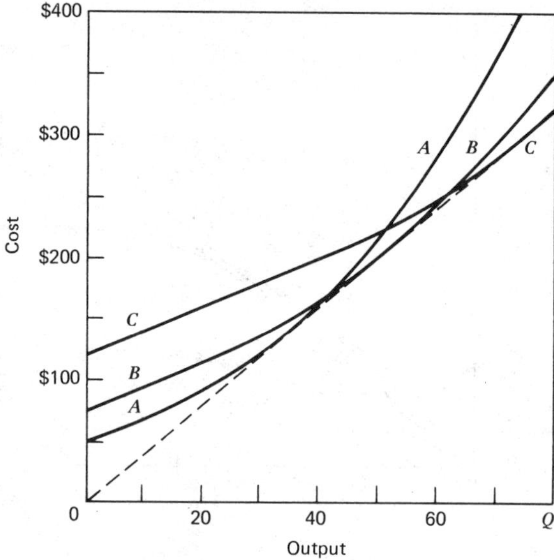

where s denotes the upper limit for input x_2 and k_s is the cost per unit time for one unit of "capacity" level s.* For this example, k_s is equal to $3, as it is in equations 23 and 24. All other parameters in equations 52, 53, and 54 are the same as in equations 23 and 24. The short-run total, average, and marginal costs defined by equations 52, 53, and 54 are plotted in Exhibits 9.6.13 and 9.6.14 for three different values of s: 16, 25, and 40 units of x_2.

The long-run cost curve can be derived by taking the derivative of the short-run cost function with respect to the capacity variable, setting this equal to zero, and solving for the capacity value that minimizes the short-run cost function. Substituting this optimal value of capacity into the short-run cost function will yield the long-run cost function. When this procedure is followed with equations 52, 53 and 54, the optimal capacity level is found to be equal to half of the desired output level. At this value of s, the long-run total, average, and marginal cost functions are as follows:

$$c_q = 4q \tag{55}$$

$$c'_q = \bar{c}_q = 4 \tag{56}$$

$$s = \frac{q}{2} \tag{57}$$

These long-run cost functions are plotted in Exhibits 9.6.13 and 9.6.14 along with the short-run curves.

The specification of an optimal capacity level† depends on the long-run cost function, the revenue function, and the producer's objective. For the preceding example, the optimal output levels and the corresponding optimal capacity levels are derived in order to (1) maximize profit or (2) maximize the break-even output level, when the revenue function is defined as follows:

$$r_q = 6q - 0.02q^2; r'_q = 6 - 0.04q \tag{58}$$

$$r'_q = c'_q; q = 50; \quad s_o = 25 \text{ at maximum profit} \tag{59}$$

$$r_q = c_q; q = 100; s_o = 50 \text{ at break-even} \tag{60}$$

*The capacity cost rate is equal to the capital investment multiplied by the continuous capital recovery factor $(\overline{A}/P, i, n) = i/(1 - e^{-in})$, where i is the nominal interest rate and n is the life of the investment (see Chapter 9.3).

†See Chapter 11.5 for a discussion of capacity measurement and management methods.

Exhibit 9.6.14 Average Cost Curves (Solid Lines) and Marginal Cost Curves (Dashed Lines) for a Nonlinear Production System With Three Capacity Levels at (*A*) 16 Units, (*B*) 25 Units, and (*C*) 40 Units of Input x_2 (Equations 23), and the Long-Run Marginal and Average Cost Curve (Dotted Line) for the Same System (Equations 55–57

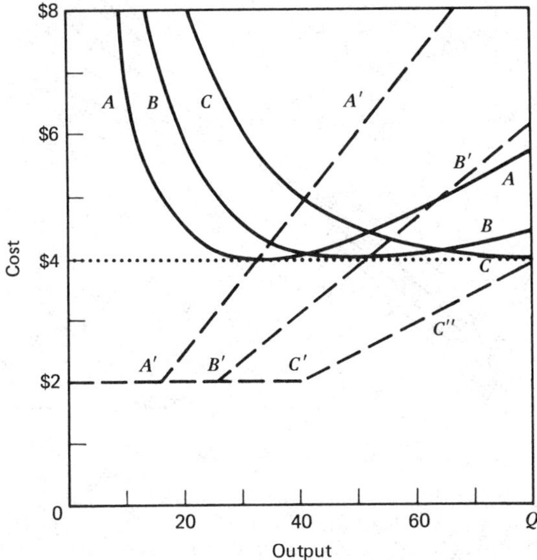

Thus, in this example, the profit maximizing producer would prefer a system capacity of 25 units of x_2 and an output rate of 50 units of q to obtain a net profit rate of \$50. If the producer has to work with a system having a capacity of only 16 units of x_2, the optimal output rate would be about 36 units of q, and the maximum obtainable profit rate would be \$45, or 10% less than optimal. If the capacity were already fixed at 40 units of x_2, the best output rate would be about 67 units of q, with a net profit rate of only \$40, which is 20% less than is obtainable with an optimal capacity level of 25 units of x_2.

9.6.10 EXPANSION OF PRODUCTION CAPACITY

Planning for the expansion of production capacity involves a trade-off between the temporary loss of return on excess production capacity and the economies of scale in building larger production facilities. The initial investment needed for a given level of production capacity frequently can be approximated by a power function[6,7] of the form $k_s q_s^b$, as shown in Exhibit 9.6.15. The power b is called the "economy of scale factor" and is equal to the elasticity of the investment cost function, that is, the ratio of the percentage change in investment cost to a percentage change in capacity q_s. The scale factor 0.6 is frequently used in the chemical industry and is called the "six tenths rule." A likely physical basis for the 0.6 rule is the relation between the area (cost) and the volume (capacity) of cylinders (equipment).

When planning for the expansion of production capacity in order to satisfy demand, which is expected to grow at a steady annual rate over a fairly long planning horizon, an optimal plan consists of a uniform series of investments at regularly spaced intervals,[6,7] as shown in Exhibit 9.6.16. The amount of capacity added at each interval is equal to the total growth of demand in the interval. When demand is growing at a rate of g units per year, then for each interval of length t years, capacity is increased by qt units at a cost of $k_s(gt)^b$. The equivalent uniform annual cost during the interval t equals $k_s(gt)^b$ multiplied by the capital recovery factor $(\overline{A}/P, i, t) = i/(1 - e^{it})$, and the equivalent present value of capacity cost over an indefinitely long planning horizon is found by dividing the equivalent annual cost by the nominal interest rate i (see Chapter 9.3). Finally, the optimal expansion interval t_o is found by minimizing the present value of the expansion costs, that is,

$$p(t) = \frac{k_s(gt)^b}{1 - e^{-it}}$$

$$p(t) = 0;\ it_o = b(e^{it_o} - 1) = 0.95 \text{ if } b = 0.6 \qquad (61)$$

Exhibit 9.6.15 Relative Cost of Expanding Production Capacity as a Function of Four Different Scale Factors

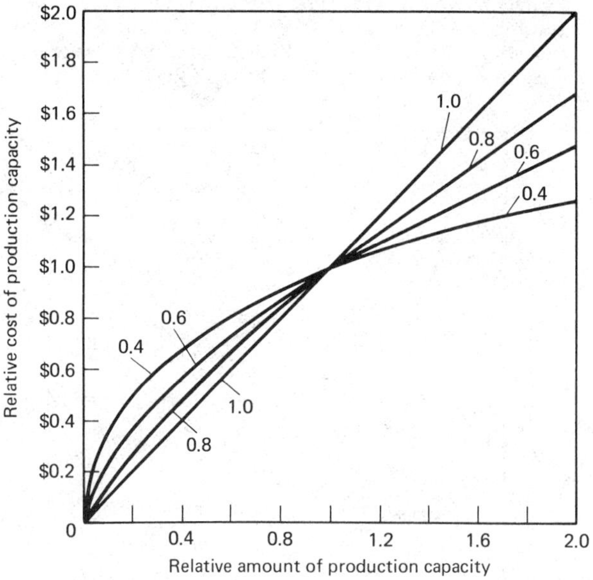

When the scale factor b equals 0.5, 0.7, 0.8, and 0.9, the optimal value of it_o decreases with the following approximate respective values: 1.230, 0.676, 0.431, and 0.207. If b equals one, $t_o = 0$, indicating that capacity should be increased *continuously* as needed.

Note that the optimal expansion interval does not depend on the cost factor k_s nor on the growth rate g. However, it is inversely proportional to the interest rate; that is, a doubling of the interest rate will cut the optimal interval in half, which implies about 33% reduction in the amount of capital invested in each of the successive capacity expansion projects. It has been shown[6] that the present value of cost is not very sensitive to small changes from the optimal values of the expansion interval. Various applications and extensions of this analysis have been discussed.[6,7]

Exhibit 9.6.16 Optimal Step Patterns for Expanding Production Capacity When the Scale Factor Is 0.6, the Interest Rate Is 25%, and the Demand for Production Increases at (*A*) 4 Units/Year or (*B*) 2 Units/Year

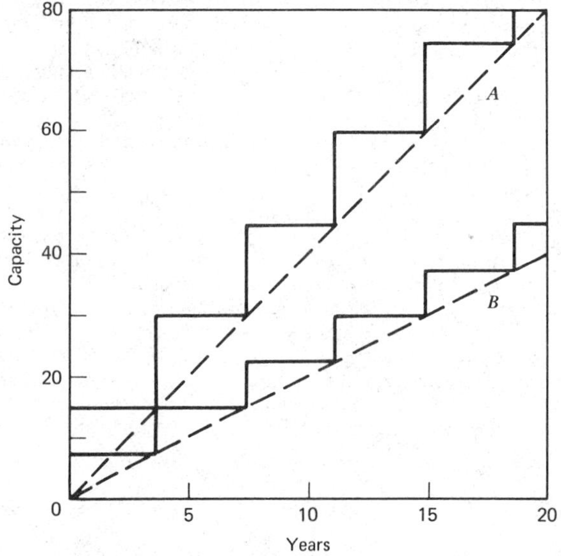

REFERENCES

1. W. J. BAUMOL, *Economic Theory and Operations Analysis*, 4th ed., Prentice-Hall, Englewood Cliffs, NJ, 1977.
2. J. M. HENDERSON and R. E. QUANDT, *Microeconomic Theory*, 3rd ed., McGraw-Hill, New York, 1980.
3. M. NERLOVE, *Estimation and Identification of Cobb-Douglas Production Functions*, Rand McNally, Chicago, 1965.
4. R. DORFMAN, P. A. SAMUELSON, and R. SOLOW, *Linear Programming and Economic Analysis*, McGraw-Hill, New York, 1958.
5. K. E. BOULDING and A. W. SPIVEY, Eds., *Linear Programming and the Theory of the Firm*, Macmillan, New York, 1960.
6. A. S. MANNE, Ed., *Investments for Capacity Expansion: Size, Location, and Time-Phasing*, MIT Press, Cambridge, MA, 1967.
7. L. M. ROSE, *Engineering Investment Decisions: Planning Under Uncertainty*, Elsevier Scientific Publishing, Amsterdam, 1976.

SECTION 10

FACILITIES DESIGN

CHAPTER 10.1

Location: Single and Multiple Facilities

GEORGE M. PARKS

Emory University

10.1.1 IMPORTANCE, DEFINITIONS, AND SCOPE

Central to the profitability of any organization is the location of its manufacturing plants, warehouses, retail establishments, service centers, and other units of economic activity. Decisions regarding location rest squarely in the middle of a complex planning process, where they are affected by several other corporate planning activities and in turn dictate policy for still others. They are typically long-term decisions, affecting corporate profitability for 25 years or more.

Required as input and central to the entire corporate planning activity is a system of short- and long-term demand forecasts, which will drive the entire decision process. Following this, there must be a strategic plan, which will, in turn, dictate a capacity plan, which itself must be based on economies of scale and other scale of operations considerations. The facility location plan results directly from these issues and in turn will determine optimum transportation and distribution logistics activities, will dictate inventory requirements and policies, will control customer service, will have important implications for productivity of the enterprise, and, in sum, will have a major impact on the "bottom line."

This chapter describes qualitative and quantitative methods and techniques for the location of economic activities. The decision regarding a single or multiple facility is analyzed. The focus is on optimizing the number, size, and location of such facilities, ordinarily (but not exclusively) from a cost-profitability viewpoint. Three levels of the selection process are examined. First, the region or regions must be selected. Assuming the organization seeks nationwide or even worldwide distribution of its products or services, a region is considered here to be a fairly large geographic area, such as the West Coast, the Southeast, or New England. With the region determined, areas or locations must be selected. In this chapter a location means a geographic area, such as southern California, greater Chicago, northern Indiana, or Atlanta. When the location has been determined, specific sites, meaning a particular piece of ground, must be chosen within the area. Numerous specific actual case studies are described throughout the chapter.

Greatly increased energy costs, high interest rates, and severe competition for limited investment capital require that progressive organizations take a fresh look at their plant-warehouse network to see if they have the proper number of each of the correct size in the right geographic locations. The objective is to minimize total costs, including inbound and outbound freight, labor, energy, local taxes, and any other location-dependent expenses. Although it may once have been possible to build a company, an industry, or even an entire economy on the concepts of cheap labor, cheap energy, and cheap money, those days are clearly gone forever. With transportation bills rising rapidly, forward-looking companies are more motivated than ever to assess the impact of a great many alternative configurations on the bottom line, not only for present product lines and demand patterns, but also for contemplated changes.

Emphasis here is placed on the location of manufacturing and warehousing facilities, but many of the same issues are equally relevant to retail establishments and a great variety of service centers.

10.1.2 FORECASTING

The issue is not whether to forecast, but how to forecast. Every business decision concerning location or capacity involves a demand forecast because such decisions are concerned with the future. They must be based on a set of assumptions about both short- and long-term demand patterns and market shares. It is vital to make these forecasts as explicit as possible, since the long-term effects of the decisions are crucial for the enterprise.

Forecasting techniques are not covered here, since they are considered elsewhere in this volume. They are highlighted here to indicate their importance and to caution against proceeding on the basis of implicit assumptions that cannot be justified. There are three principal classes of forecasting techniques, all of which apply to location decisions. These are: subjective methods, of which Delphi techniques are the best known; time series analysis; and econometric methods.

A new company, or a company producing a new product, may have to rely on subjective techniques. This, however, does not mean pure guesswork. Delphi techniques, which are an organized method of obtaining forecasts from a consensus of experts, have been successfully applied by many organizations. It may be possible to draw careful analogies with other new products or services introduced by the organization or its competition.

When past sales and demand data are available, more rigorous forecasting techniques should be employed. Time series forecasting (see Chapter 13.9), which depends solely on historical demand patterns, including trends, seasonal factors, and business cycle influences, is most commonly employed, with perhaps the best known of these analyses being some form of exponential smoothing. Econometric methods, which depend not only on past demand patterns, but also on relevant environmental data, are particularly appropriate for longer-term aggregate forecasts of the type required for location decisions. Perhaps the best known of these is regression and correlation analysis (see Chapter 13.6), which is employed later in this chapter for specific site selection of retail establishments.

It is equally important to assess the variability in the forecast and to consider explicitly the associated risks. Decision makers should not accept any single value forecast as the unalterable truth. Nor should they rely exclusively on that expectation. They must be prepared to deal with the "What would happen if . . . ?" questions that might arise because of uncertainties in the forecasts and other effects of the competitive environment. Any location decision contains such forecasts, either implicitly or explicitly, and decision makers should carefully analyze this area as the first step in location decisions.

10.1.3 STRATEGIC PLANNING

If an organization plans to conduct all its activities in a single location, then it may proceed with the regional, area, and site decisions discussed in this chapter. When, however, expansion is considered, certain strategic decisions must first be made. One possibility, of course, is merely to expand in the existing location. But it may be preferable to move to a new location, where economies of scale, production efficiencies, better transportation arrangements, higher productivity, lower taxes, or a host of other benefits may significantly increase profitability. The considerations dictating such choice are discussed in more detail later. But as before, only a single facility is being considered. More challenging and potentially more rewarding are multifacility strategies. Significant cost and managerial control considerations will have an impact on this choice.

Schmenner,[1] writing in the *Harvard Business Review*, identifies five general types of multiplant manufacturing strategies. The following is adapted from his discussion. Though it is concerned with manufacturing facilities, the same issues apply to warehouses, service centers, or retail establishments.

1. **Product Plant Strategy.** Under this plan, individual plants manufacture only a subset of the total company products or product lines and supply the company's entire market needs for these products. This allows each facility to concentrate on only a limited set of activities, and certain efficiencies regarding technology, equipment, and organization may result. Certain complexities may be avoided, and the benefits of economies of scale may be maximized. Duplication of expensive specialized processing and test equipment can be minimized. There may also be transportation advantages where raw materials are found only in specific areas or where the company wishes to operate under a highly decentralized or venture management concept.
2. **Market Area Plant Strategy.** Under this plan, a particular plant will manufacture the entire product line of a company, but will serve only a specifically defined market area. This is the standard notion of a branch plant and would be particularly appropriate when transportation costs are large relative to other variable facility-dependent expenses. It also gives management the flexibility to minimize the effects of work stoppages, supply interruptions, or natural disasters such as fire, flood, or tornado. This strategy also places manufacturing facilities close to specific customers, which may have important revenue producing and servicing effects. Economies of scale considerations are particularly important here, since a company must decide when it is better to expand a given network of plants or to add additional ones with the attendant increase in overhead and coordination costs.
3. **Product-Market Plant Strategy.** If a company is large enough, or if its product line is limited enough, it may be best to combine the two previous strategies. This might also be appropriate if scale of operations considerations allowed relatively small manufacturing units. Under this strategy, a particular plant would produce only a subset of the products or product lines and would supply only a specific market area. Although this strategy may allow a company to benefit from the

advantages of both previous strategies, coordination and management problems might be greater. However, the opportunity for a particular plant to develop highly specialized knowledge of particular products and/or particular market areas may provide a flexibility and a benefit that will outweigh these considerations.

4. **Process Plant Strategy.** Vertically integrated companies or those with several stages in their manufacturing process may elect the strategy of having specific plants conduct separate stages in the manufacturing process, perhaps feeding their output to one or more final assembly plants. As with the product plants, each stage may involve different technology, different material supply sources, different managerial control systems, and the like, and separating them can lead to significant benefits from specialization and decoupling. Economies of scale may play a major role here also, in particular, as opposed to the market area plant strategy. Production planning and inventory control problems may be more complex, and coordination may be more difficult. This will require a more highly skilled and technically trained management control team to ensure that the successive stages do not become seriously out of balance.

5. **General-Purpose Plant Strategy.** A fifth approach, which is in part a combination of the previous four, is to operate plants with flexible enough planning and control systems that they may continuously adapt to changing market needs. This obviously requires considerably higher levels of flexibility on the part of both labor and management. Control systems must be such that they can be adapted to cover all of the previous situations. If the company's products are not particularly complex, and if scale economies allow relatively small units, such plants may be extremely efficient and competitive in reacting quickly to changing customer demands, product designs, and market opportunities. The development and introduction of new products may be facilitated, and more of a spirit of entrepreneurship can be engendered. But coordination of the various units and avoidance of unnecessary duplication may be serious problems.

These are all important issues to consider carefully before undertaking economic analysis concerned with locating a particular activity in a particular place. Issues of organization design, technological development, market strategy, and management control may override strictly economic considerations. Of course all of these considerations have an impact on the bottom line as well, but it is usually extremely difficult, if not impossible, to cost them out directly. Strategic planning relating to the issues discussed here may prove to be far more beneficial to the long-term competitive position of the organization than the most carefully conducted location analysis of the type described later in the chapter.

Location analysis will be especially important if the company has grown by a process of merger and acquisition. It would be most fortuitous and extremely unlikely if the separate planning activities of the previously independent companies would result in anything like an optimum strategic plan for the combined organization. This plan would result in a series of facilities, perhaps well located for their original missions, but not well placed when these missions change as a result of the strategic issues just described. Facility proliferation may well be the result of such mergers and acquisitions, with the attendant excess overhead and control costs. Integration and consolidation issues, though having an obvious economic dimension, must precede and be combined with conventional location analysis.

10.1.4 SCALE OF OPERATIONS

One of the most persistent problems affecting productivity is low capacity utilization. Production economics were thought to dictate high capacity facilities, which then could not be fully utilized because of high raw material and finished good transport costs or sluggish market demands. Capacity and strategic planning were stressed previously, and this area clearly requires careful analysis in any location study.

Some form of break-even analysis (discussed fully in Chapters 9.6 and 11.6) is the principal analytical tool. The methodology is not repeated here, but is highlighted because of its importance in any location analysis. All fixed and variable costs must be carefully considered with regard to technological alternatives associated with any production process, with special reference to the familiar problem of the relative intensity of labor and capital in the production process. This is not purely an issue of costs, but also involves the availability of the various inputs and the desired flexibility to adjust to future uncertainty in both supply and demand markets.

Economies of scale will have a major impact not only on the profitability of a single facility, but on decisions concerning the economic desirability of the various multiplant strategies. When these considerations have been clearly understood and carefully analyzed, the analyst is in a position to proceed with specific location decisions.

10.1.5 REGIONAL ANALYSIS

Many of the issues affecting facility locations discussed in this section do not apply to the regional firm that has made a conscious decision to restrict its activities to a narrowly defined specific region. Such companies might skip this part of the decision process and go directly to the location (general area) analysis described in the next section. The distinction between a region and a general area, as the terms are used in this chapter, is not always clear, and some companies might choose to consider some of the factors presented here in an area analysis. Discussed here are what are often called the "nonquantitative" factors in a location decision; specific cost factors are considered in the next section of the chapter. Methods for handling these nonquantitative factors, the trade-offs between them, and the development of summary measures also are discussed in the next section. Presented here are the factors and issues, even though they cannot be reduced to specific dollar terms, as can inbound and outbound transportation costs, direct and indirect labor, utilities, taxes, insurance, land, construction, and overheads.

Demographics

There has been and continues to be a significant shift of both population and industry from the Snow Belt to the Sun Belt. This has caused a redefinition of areas of job concentrations and markets, and it appears that these trends will continue. Infrastructure development in the various regions affected has not always kept pace with the influx of industry and people. The distribution of income levels, buying patterns, and industry concentrations is continually changing, and the location analyst must take into account not only the present situation, but projections on future trends. Much of this analysis relates to demand forecasting, but there are other important specific questions to consider. For example, is the labor supply sufficient in both quantity and quality, and what is the competition for it? What is the availability of water, fuel, power, and other utilities? Can the region attract sufficient labor and management resources, either from other locations within the company or from outside? What is the ethnic and racial mix? Might there be language problems?

Incentives

The federal government, certain regions, and some states provide special incentives for an industry locating in or relocating to particular areas. These include the availability of attractive terms on industrial development financing, subsidized training of the local work force for the specialized tasks of the company, favorable tax and zoning treatment, the provision of working capital at favorable terms, assistance in marketing and contract procurement, and a myriad of other inducements that are limited only by the imagination (and laws) of the region or state in question. These are particularly prevalent in surplus labor areas or areas designated for urban redevelopment. Although these special incentives may be extremely important to a particular firm in the establishment of its facility and in quickly obtaining a relative stability, future growth and profitability are likely to depend more on the hard cost items discussed later. More than one company has been attracted to a particular location by these special inducements only to find that its long-term viability and profitability are seriously jeopardized by the location thus selected.

Constraints

Some industries find their regional choice severely constrained by market concentration, resource availability, and/or special transportation and processing requirements. For example, a supplier of coated coiled steel for the automobile industry would look toward Detroit. A manufacturer of pulp for the paper industry would seek regions where the appropriate wood resources are available, and a manufacturer or user of semiconductor technology might seek a location in or near Silicon Valley. A company providing services to the motion picture industry could not ignore southern California, and a firm servicing stock markets and commodity exchanges would have to consider New York, Chicago, and San Francisco. Other regional constraints might include the need for (relatively) low-cost electrical power, for processed water of a certain purity, for barge (waterway) transportation, or for proximity to a well-equipped seaport in order to meet export requirements.

International Considerations

Many organizations have obtained major benefits from locating facilities outside the United States, especially when international markets are involved, and these possibilities merit serious consideration by the location analyst. However, though these locations may be intuitively appealing, the pitfalls are many. Enticingly lower labor costs may be more than offset by concomitant lower productivity—not because the workers are not as skilled or industrious, but because of the lower levels of capital equipment and technology utilized. The extra costs of transportation, at least to supply domestic markets, may also offset any potential labor savings. In addition, wage rates in many of the more industrialized parts of the world are now equal to or even greater than those in the United States because of the higher social costs incurred in many of these countries. Language and cultural differences can present almost insurmountable problems in motivation, management, and control of the work force, the supplier network, and other foreign constituencies with which the company must deal.

A number of countries have local sourcing requirements that restrict freedom in the selection of raw materials, parts, and components, and this in turn may lead to cost inefficiencies and difficult quality control problems. Special financial requirements, difficulties with the repatriation of funds, taxation, the availability of appropriate management talent and labor skills, and the level of technological development may cause unfamiliar and costly problems to the company locating outside the United States. Political stability and the dangers of expropriation must also be assessed.

None of the preceding pitfalls is meant to deny the excellent opportunities available to the company willing to make careful and systematic analysis of both the risks and the rewards associated with offshore development. Caution is expressed, however, against a "me too" policy or responding too quickly to what may appear at first impression to be an overriding consideration. Communication, coordination, and control problems are magnified, but for the right company in the right situation at the right time, exceptional opportunities do still exist. The ancient homily "if you can't do it in your own backyard, you probably can't do it anywhere else" merits attention.

Business Climate

Although "climate" means many different things to many different people, it has come to be used as a measure of attractiveness for business operations in a given state. A 1979 report, prepared by Alexander Grant and Company and commissioned by the Conference of State Manufacturers Association, ranked the 48 contiguous states of America by a summary measure of those aspects of the business environment that are directly controllable by the actions of the state governments. Eighteen categories were analyzed, including labor union membership as a percentage of the total nonagricultural work force, the energy cost per million btu's, the average weekly manufacturer's wage, the man-hours lost as a result of work stoppage per worker, the state and local taxes per capita, the private pollution abatement expenses compared to value of industrial shipments, the unemployment compensation benefits paid per covered worker, worker's compensation per hundred dollars of manufacturing salary, state debt and spending per capita, and a number of other issues relating to doing business in the state. The eight states ranking best in this study were, in order, Mississippi, North Carolina, South Carolina, Utah, New Mexico, Arkansas, Georgia, and Oklahoma. All of these states had net gains in manufacturing jobs for the 1968–1978 period. The bottom eight states in the ranking, listed from best to worst, were Illinois, Delaware, Washington, Pennsylvania, Ohio, New York, New Jersey, and Michigan. All of these states, with the exception of Washington, suffered a net loss in manufacturing jobs for the same period.

Business climate also refers to the productivity of the local or regional work force, the willingness of local communities to accept new industries of various types, the availability of capital resources from the private sector, and the general attitude of the local population and regulatory agencies toward the corporate community. One oft-quoted and important item is the existence of right-to-work legislation. For example, from 1968 to 1978, the 20 right-to-work states showed a net gain of 1,073,000 manufacturing jobs, whereas those states that allow compulsory unionism reflected a loss of 481,600 jobs.

Personal Preferences

Many executives and company owners express strong personal preferences in making location decisions, either for production facilities or for corporate offices. These can include climate considerations; proximity to favorite pastimes; the desire for a large metropolitan location; specific antagonisms toward certain states or regions; the location of competitor's facilities; the ease of travel by management, sales, and vendor personnel; and a whole host of other personal issues. Although there is no doubt that these issues can and do have a major impact on location decisions, it may be argued that an economic analysis of the type described later in this chapter should be performed in any case to determine their economic impact when stated in terms of dollars and cents. It may be that the decision maker does not hold these preferences as dearly as he or she originally indicated.

Government Regulations

There is a myriad of federal, regional, state, and local laws and codes affecting facility location decisions. Probably the best known of these regulations are those of the Environmental Protection Agency (EPA) and OSHA. These obviously need to be carefully considered by the location analyst, but there are many others at all levels of government that can seriously affect the profitability of a facility. State, county, or city building codes; "freeport" legislation; rulings of the Interstate Commerce Commission; regulations covering waste disposal and smoke abatement; laws limiting the scope and hours of work permitted; costs for unemployment insurance and workers' compensation;

and taxation legislation are but a few of the many issues that must be understood, analyzed, faced, and dealt with. In fact, three bills were introduced into the 96th Congress (1979), which, while they have not yet been passed, may represent a future trend and merit the attention of all concerned with location decisions. They are referred to as "plant relocation" or "hostage" bills and would place burdens on business and industry to justify any plant relocation activity. Requirements to be placed on a business considering relocation would include prior notification to the community, employees, and union; the preparation of an economic impact statement; an investigation by the Secretary of Labor; the payment of former employees after a plant relocates; and reimbursement to local governments for taxes lost by relocation. Under these proposals, the federal government would also offer financial assistance to any organization that wanted to buy the business or obtain controlling interest. Although some companies have done some of these things voluntarily, they are now matters that have attracted the attention of the U.S. Congress.

Attractions

The availability of appropriate cultural, educational, and recreational activities may be an important determinant of a company's ability to attract a work force of sufficient quality and quantity. This includes management personnel as well as labor, especially with the increased emphasis on quality of working life considerations.

Climate

Although climate is certainly a matter of personal preference, there are many direct cost consequences as well. Construction costs will be affected, since the facility must be built to withstand the forces of nature, including extreme temperature variations, snow, earthquakes, winds, tornados, and hail.

The probabilities of damage and production interruptions as well as employee absenteeism caused by any of these forces of nature must also be assessed. Direct costs associated with heating or air conditioning for both personnel and production processing must be included. There are also weather-caused costs of maintenance and costs for the protection of products and raw materials to avoid deterioration in their value.

Summary

Many of the considerations raised in this section apply also to a location (area) analysis, and there are other not directly quantifiable issues that apply specifically to more narrowly defined geographic areas.

10.1.6 LOCATION (AREA) ANALYSIS

A particular region may rank relatively poorly in regard to one or more of the nonquantitative issues raised in the previous section, but may well contain locations within the region that would rate relatively high. Therefore a location analysis should include at least a reassessment of the issues raised in the preceding section so that the rankings reflect a true assessment of the considered locations. Other factors more specific to a given location are listed next.

Nonquantifiable Factors

In addition to all the factors listed in the preceding section under regional analysis, there are others relating to employees, suppliers, professional services, and customers that should be considered. The availability of educational opportunities from primary school through graduate work, appropriate cultural and recreational activities, access to professional sports teams, the adequacy of available libraries, and other quality of life issues bear serious attention. Not only must they be available within a particular location, but they should also be accessible. The facility itself should be accessible to employees and other persons having reason to visit it, such as suppliers, bankers, accountants, attorneys, and visitors from other corporate facilities. Involved here are questions of airport availability and service, the availability of public transportation, the necessity for and provision of employee and visitor parking spaces, and a consideration of commuting distances and times for the various kinds of personnel required by the facility. Careful study must be made of the number and types of suppliers of goods and services readily available in the location under consideration. Their quality and reliability must be assessed. Closeness to banking and other professional relationships may prove particularly important, since the availability of these resources in the immediate area will enable personal relationships to develop between those communities and company personnel that are important to the smooth operation of any business.

There are also income producing factors relating to the proximity of competitors and customers that should be assessed. Customers may require the close personal contact, the ease of access to the producing facility, and the quick service that can be developed only when producer and consumer are located in the same area. Locating near competitor facilities may enhance a company's abilities in head-to-head competition, but it may be preferable to locate away from competitors, thereby giving up a smaller market share of a larger total market, but improving the chances of obtaining a larger market share of a smaller total market area.

The relevant issues from this discussion and from the previous discussion of regional issues would then be selected and become candidates for inclusion in one or more of the composite ranking procedures discussed later.

Measurable Costs

Inbound Transportation

All raw materials, supplies, purchased parts, components, and other goods necessary in the production process bear freight and delivery costs that are clearly variable, depending on location. This is true even if the amounts are not billed separately and regardless of whether the supply is from an outside source or another facility of the same company. Even where such costs are included in the supplier's price, they should be estimated, since it may be possible to negotiate discount arrangements when locations are close.

Labor

Direct and indirect labor costs for the contemplated production volume must be assessed. Unless national labor contracts are involved, these costs will also vary from location to location because of unionization, localized prevailing rates, and the impact of competition. Not only must labor rates be ascertained, but an analysis should be made of expected productivity, so that realistic estimates of expected per unit costs can be developed.

Outbound Transportation

Costs associated with the delivery of the finished product to the market often constitute the most significant location-dependent cost. Consideration must be given to distances, commodity classifications, rates, and alternative transportation modes. Customer pickup and back-haul possibilities should be explored.

Utilities

Costs of electric power, fuel, water, and waste disposal may vary significantly according to location and should be estimated in total and on a per production unit basis.

Raw Materials

The cost of raw materials may vary from location to location for reasons other than inbound transportation. Such reasons might include zone pricing, quantity discounts, opportunities for alternative modes of transportation, and the availability of special materials handling equipment.

Inventory

Different locations, because of different distances from customers and/or suppliers, will require different levels of inventory to meet customer service specifications and to fill pipeline needs. Determining proper levels of inventory for multifacility situations is especially complex.

Cost of Space

This includes the cost of purchasing land and building, land and construction costs, or facility rental. While land costs may be viewed as an investment, purchase or construction costs for buildings should be converted to an equivalent annual cost, ordinarily through the capital recovery factor.

Taxes and Insurance

Real estate, inventory, payroll, and other taxes must be estimated, and costs must be included for workers' compensation, liability, fire, comprehensive, business interruption, and any other necessary insurance. Of increasing importance are costs associated with security.

Overhead

Costs of supervision, banking, travel, sales, administration, and other services are location-dependent and must be estimated as a function of different potential output levels.

Methods and Techniques

Ranking Procedures

Although ranking, or rating, procedures are never totally satisfactory and in fact may even be misleading by giving a false sense of rigor and objectivity to the analysis, they do have an important advantage over pure cost analysis in that the nonquantitative or intangible factors can be taken into account. These methods may be used to compare regions, locations, or individual sites against each other.

The first step is to develop a list of the relevant factors, to each of which a scale is then assigned. The factors are weighted relative to each other, and each potential location is scored according to the scale defined. Scores and weights are then multiplied together and totaled for each potential location. If quantitative factors are included in the list, it may then be possible merely to chose the location with the highest total score. On the other hand, it would ordinarily be preferable to perform an economic analysis (see the next subsection) for those factors that are quantifiable and to use ranking methods for those that are not. The two analyses would then be used together to determine the best location, which of course might not be the one with the lowest costs for the quantified factors, nor would it necessarily be the one with the highest total score for the nonquantified factors. Explicit trade-offs between the two are not possible, but at least the nonquantifiable factors can be "costed out" by comparing the differences in the ranked totals with the differences in the computed costs for the various alternatives.

A factor list can be developed from consideration of the various factors listed earlier in this chapter or reference can be made to available checklists (e.g., see Table 2-1 in Monks[2] or Table 6.1 in Ireson and Grant[3]). Reed[4] defines four general ranking methods found in the literature and gives examples of each. They are:

1. Assign equal weights to all factors and evaluate each location along the factor's scale. For instance, assume that 30 factors were selected by which to rate each of 10 sites. A scale of 0–10 might be defined for each factor, and then each site would be assigned a value on this scale for each factor. An overall site rating would be the sum of the assigned points on the scale for each factor, and this would be compared with the sums for each of the other nine sites.
2. Assign variable weights to each of the factors and evaluate each location along the factor scale. Then proceed as in method 1.
3. Assign variable weights to each factor. Rate each location by a common scale for each factor. Multiply the location rating for each factor by the factor weight to obtain the point assignment for that factor for each location. The overall measure is then the sum of these point assignments for all factors.
4. Establish a subjective scale common to all factors. For example, a 6-point scale ranging from very poor through poor, satisfactory, good, very good, and excellent might be established. Assign points against this scale for each factor, and assign the factor points. Again, the total for each location over all factors will provide a relative ranking for each alternative location.

Quantitative Methods

A great many quantitative methods, varying in degree of complexity and computer processing requirements, have been developed and applied to location problems. Two commonly applied ones are described here, with brief descriptions of actual applications. Others may be found in the references at the end of this chapter.

"Center-of-Gravity" Approach. The first of these models is based on a "center-of-gravity" theory, which says that the optimum location for a producing facility or facilities is that point or points where the total sum of the market demand in each sales area times the Euclidean distance from the facility to the market area plus a factor multiplied by the raw materials required times the distances from their sources is minimized. This approach does not explicitly consider dollar costs of production and distribution in the solution, but can be extremely useful in situations where there is complete freedom in facility location and where the available cost data are fragmentary or based on unreliable estimates. This might be the case particularly where new products are concerned. The general theory of this approach and the mathematical development for the basic model are presented in Francis and White[5] and are not repeated here.

The term "center of gravity" is used in quotation marks because while it has intuitive appeal and is easily explainable, the model actually employed is not precisely the physical definition of the center of gravity. Rather, the Weber model is used, which requires one to find x and y such that

$$\sum_{i=1}^{m} w_i \times [(x - a_i)^2 + (y - b_i)^2]^{1/2}$$

is minimized, where m = number of market areas to be served by a new plant
(x,y) = coordinate location of new plant
(a_i, b_i) = coordinate location of market area i
w_i = market demand for area i

The preceding equation considers only outbound freight; in many applications, however, inbound transportation is a significant factor. Thus raw material sources should also be included in the summation to be minimized, and the corresponding w_i's adjusted to reflect scrap rates, differing freight rates due to bulk commodity shipping factors, and conversion to common shipment units (e.g., equivalent cases).

As indicated, the basic goal is to locate the center (or centers) of gravity, where the pull is generated by market demands and raw material requirements. Required inputs are:

1. Forecasts of product demand by sales region, state, city, Standard Metropolitan Statistical Area (SMSA), warehouse, specific customer, or any other convenient breakdown.
2. Latitude and longitude coordinates for each demand point—taken at a specific point or at the principal city in a region.
3. Ratio of each raw material used to final product.
4. Location of suppliers or raw materials, with coordinates.
5. Amount of each raw material supplied by each supplier as a percentage of the total required.

The following parameters are ordinarily varied to perform sensitivity analyses:

1. Demand, to reflect uncertainties.
2. Relative contribution made by each raw material supply point.
3. Ratio of raw material to final product.
4. Number of raw materials.
5. Number of center-of-gravity areas (proposed plants).
6. Organization of regions about centers of gravity (proposed plants) to allow varying plant capacities.

If the inputs are completely specified, and if there is only one facility to be located, this model provides answers to the classic "What would happen if . . . ?" question, where varying assumptions are made to reflect either uncertainty or unresolved issues. In such cases the output will be in the form of an optimum region, rather than a single point, within which probability measures can be assigned or secondary criteria can be brought to bear. In multifacility situations, regional boundaries can be changed to reflect variations in demand patterns and/or plant capacities.

For single-facility problems, costs of production, utilities, taxes, insurance, overheads, and so on are then added to obtain a total cost for this computed location. This may be compared with other alternatives by forcing the (x,y) coordinates to any proposed location and using the model to compute a comparable transportation sum, to which these other costs for the proposed location would be added for comparison purposes.

Linear Programming Approach. For multifacility problems, it is usually desirable to consider an explicit minimization of the sum of the various quantifiable costs described earlier. This can be accomplished by utilizing an approach based on the well-known transportation model of linear programming, wherein the total costs of all facilities can be minimized, the optimum number of facilities determined, and capacities computed for each. Optimum distribution boundaries for each facility are also determined as output of this analysis.

Again, the basic mathematical theory and the model are described here, since they are well covered in many other sources (e.g., see Dantzig,[6] Hillier and Lieberman,[7] and Chapter 14.2 of this volume). The basic objective is to compute the economic consequences of a great variety of proposed facility locations, plant capacities, and distribution requirements and to explore the sensitivity of the solutions so generated to varying assumptions concerning the costs of production and distribution.

The mathematical procedure is to minimize

$$\sum_{i=1}^{m} \sum_{j=1}^{n} C_{ij} X_{ij}$$

subject to

$$\sum_{i=1}^{m} X_{ij} = b_j$$

$$\sum_{j=1}^{n} X_{ij} = a_i$$

and all $X_{ij} \geqslant 0$

where m = number of factories (including a "dummy" to account for excess–i.e., unsatisfied–total demand, if appropriate)
n = number of sales regions (including a "dummy" to account for excess total production capacity, if appropriate)
a_i = capacity of factory i
b_j = demand in sales region j
C_{ij} = total cost of producing a unit in factory i and shipping it to sales region j
X_{ij}, to be determined = amounts of each sales region's demand to be supplied from each factory

The solution procedure must utilize a proposed facility configuration as input data and will then generate a solution that will allocate market demands to production facilities in such a way that total costs are minimized. This initial proposed configuration may be generated by the center-of-gravity approach or may be specified by the analyst, based in part on the present facilities and on his or her "feel" for the problem at hand. The user may thus evaluate as many proposed configurations as desired, and this methodology will find the best of these proposals and will indicate the sensitivity of the configurations to changes–either real or simulated to reflect uncertainty–in the input data assumptions.

The input data required are the demand data mentioned previously and the following:

1. Production costs at each present and proposed manufacturing location.
   a. Variable unit costs of production, including such items as labor, direct materials, power, other utilities, and local taxes and insurance.
   b. Plant overhead costs as a function of plant capacity.
   c. Freight rates for each inbound raw material from proposed sources of supply.
2. Shipping and distribution costs per unit from each manufacturing facility (present and proposed) to each warehouse or sales region. These costs may be obtained directly from current actual freight charges from present facilities, from quotations from shippers for proposed facilities, or from published rate books.
3. Construction (or purchase) cost estimates for each proposed new facility and applicable charges for capital (capital recovery factors).

The model will then generate the minimum cost solution for each proposed configuration by allocating production requirements to producing facilities and by determining optimum shipping and distribution patterns. The analyst will usually want to vary much of the input data in order to explore sensitivities, perhaps the major ones being the number, size, and location of the production facilities and the patterns of future demand.

Overhead and construction (or purchase) costs appropriate for each proposed configuration are then added to the total obtained from the model to enable a total cost comparison for each alternative.

Applications of Methods. A major manufacturer of pulp and paper products was faced with the problem of determining the optimum number, location, and capacity of manufacturing facilities for a new consumer product being introduced for nationwide distribution. Pulp, the principal raw material, was to be supplied from existing company mills. Scale of operations considerations dictated that the number of facilities would be between two and five. For each of these possibilities, the center-of-gravity model was used to find preliminary locations for further, more detailed analysis using linear programming. Various alternative configurations for each of the feasible number of plants were explored, and the optimal locations for each feasible number were determined. Overhead and construction costs were then added to the computed costs resulting from the best locations for each number of facilities, thereby enabling the determination of an overall optimum configuration. This overall optimization resulted in a conclusion that 40% of the total nationwide demand should be produced in a facility near Wilmington, Delaware; another 40% near St. Louis, Missouri; and the remaining 20% near Portland, Oregon. Optimum distribution patterns (geographic boundaries) were computed as well. This case study is described in considerably greater detail in Parks.[8]

A food manufacturer, with manufacturing facilities currently in New Jersey, Illinois, and Arizona, wished to evaluate the possibility of either consolidating all U.S. manufacturing under one roof at a site to be determined or combining the New Jersey and Illinois facilities in a new plant and operating a two-plant system. The major differences between this application and the one just described were that five different raw materials, available from a variety of sources, rather than one, were to be considered and that three plants were already in existence and operating. There were also certain differences in distribution policies to be considered. Preliminary analysis utilizing the preceding logic eliminated the two-plant alternative as economically infeasible, so the center-of-gravity model was utilized independently to analyze the location for the possible consolidation of all three plants. The described methodology pinpointed a location in central Pennsylvania.

A building products manufacturer with existing plants throughout the country was seeking a location for a new manufacturing facility in the Southeast to relieve pressure on existing plants and to serve the regional market better. Because this product is highly freight-intensive–both inbound and outbound–and since competitors' plants were already located at various points throughout the region, a special analysis was developed to project market share potential by specific marketing area within the region as a function of distance from all plants within the region and their installed capacities. The center-of-gravity model was utilized, and a specific site was selected in southern Virginia. Linear programming could be used to determine appropriate capacity to install, since this is a multiplant environment, and the company's present plants now serving the Southeast are operating at or near capacity.

Both models were utilized by a consumer appliance manufacturer with nationwide distribution whose market demand was exceeding the capacity of its single manufacturing facility located in New Jersey. Rather than expand capacity in New Jersey on land the company has already purchased for this purpose, a three-plant configuration was shown to be best. Considerable cost reduction was possible by locating new facilities in the general areas of western Kentucky and southern California and allocating to these new locations not only the necessary capacity expansion, but also some of the existing production.

A multiplant manufacturer, also of consumer electrical appliances, was contemplating decentralizing its warehousing system by setting up a number of new regional warehouses. The number and locations were to be determined. Primarily because of the high capital and inventory carrying costs, this methodology clearly indicated that the best solution was to remain centralized and concentrate on improved internal control systems, rather than attempt to solve the problems by warehouse proliferation.

Other Methods. There are a number of more advanced analytical techniques of both theoretical and practical interest. Procedures involving mixed-integer programming or Monte Carlo methods have been successfully applied to multifacility problems.[9–12] Heuristic programming methods have also been employed (e.g., see Kuehn and Hamburger[13]). As pointed out in Johnson et al.,[14] heuristic programming methods are typically employed when absolute optimum solutions could theoretically be obtained by complete enumeration of all alternatives, but the combinatorial nature of such problems renders this approach prohibitively expensive and time consuming. Thus an organized, commonsense search procedure is utilized. Typically, such procedures contain provisions to:

1. Eliminate initially all alternatives that by some simple logic are extremely unlikely to be optimal.
2. Eliminate those costs not likely to be important during preliminary analysis stages and consider only the major ones.
3. "Divide and conquer" the complex problem by breaking it into simpler component parts and devise procedures to bring them back together. Johnson et al.[14] discuss such procedures in detail.

The Computer's Role

The complexity of the calculations involved in these analyses clearly requires the use of a computer. Computer organization and processing are covered in Chapters 12.1 and 13.11. Suffice it to note here that the simulation capability thus provided to deal with the "What would happen if . . . ?" question is at least as important as the capability of determining optimal solutions. Such sensitivity analyses enable the analyst to predict in advance the effects of either planned or contemplated changes in any of the cost and environmental variables or in changes that will be forced upon the organization by competitive actions, government regulations, or other environmental influence. Sensitivity of solutions to particular data points can be assessed in order to determine how much effort should be expended to obtain accurate data and when rough approximations will be satisfactory. Another important use of such sensitivity analyses is to assess the impact of uncertainty in the underlying data. Specific points of leverage and critical factors of operating success can be identified and then subjected to more intense analysis, while the less relevant ones can be considered with approximations, if not ignored completely. A wide variety of computer software to deal with facility location problems is readily available from computer manufacturers, software houses, and consulting organizations, and the interested analyst would be well advised to consult these sources before embarking on a project to "reinvent the wheel."

Also, many city, state, and regional governments, as well as private real estate and consulting organizations, maintain computerized data bases of available sites. The analyst may specify certain characteristics required or desired for his facility, and conduct a rapid and efficient search for feasible locations.

Combining the Qualitative and Quantitative

The importance of location decisions to both short- and long-term profitability of an enterprise requires the location analyst to consider carefully all relevant factors. As previously noted, explicit dollar trade-offs between the qualitative and quantitative are not possible, but the rigor and discipline imposed on the decision process by the methodologies described should lead to improved location decisions. An imaginative blending of the measurable and unmeasurable will enable the organization to compete effectively and to meet successfully competitive challenges of the future.

10.1.7 SITE SELECTION

After a general location has been found, a specific site must be selected. This will involve a rigorous examination of local real estate and business tax laws; the availability of plant services such as sewage disposal, fire protection, and public transportation; the availability of local development funds; the rates of county, local, or municipality income taxation; county and city building codes; and precise land and building requirements. These and other factors will dictate whether the preferred site should be urban, suburban, rural, and/or located in an industrial park. Plant design and layout (see Chapter 10.2) will affect this choice, as will the need for parking areas and for space for future expansion possibilities. Many local communities offer attractive inducements to obtain new industry and offices, and these should be checked carefully in all the local political subdivisions within the area selected.

The buy, build, or lease decision must be made, ordinarily by means of an engineering economy study (see Section 9). The decision will depend upon the availability of suitable existing facilities, local costs of construction, and the availability of capital in relation to other company investment alternatives.

Alternative sites may be compared by the ranking methods described in the preceding section, but ordinarily more precise estimates of the costs that will differ from site to site can be readily obtained. This enables one to conduct reasonably straightforward cost comparisons among the various alternative sites.

Depending on the particular situation, other specific considerations might include the availability of railroad sidings, interstate highway access, airport availability, ease of travel for suppliers and other visitors, and specific access and egress traffic patterns. The availability of suitable housing within reasonable commuting distances and/or a suitable resident work force with relatively easy access to the site may be important. The answers to such seemingly inconsequential questions as "What do employees do on their lunch break?" may have significant effects on morale and productivity. Plant security and the safety of employees traveling to and from the site are becoming increasingly important. Attractive surroundings, whether urban or rural, can have a significant impact on how employees feel about coming to work.

Although manufacturing plants, warehouses, and retail establishments should consider all these factors, the latter face a more difficult additional issue. For retail businesses the specific site will generally have a major effect on revenues, a situation not usually faced by plants and warehouses. Thus the retail site selection process requires some methodology to forecast sales as a function of the various environmental variables surrounding a particular proposed site. A method that has been successfully employed by supermarkets, banks, fast-food franchises, convenience stores, chain motels, and the like to accomplish this is the statistical forecasting technique of regression analysis. Specifics of the technique are not described here (see Chapter 13.6), but the following example will illustrate its use.

A large supermarket chain with some 800 existing sites sought a better methodology for evaluating both new site proposals and the performance of existing units. The linear regression model was employed, wherein

$$y = a_1x_1 + a_2x_2 + \cdots + a_nx_n + b$$

It was decided that the dependent variable y would be sales. (Actually, several different analyses were conducted to estimate separately sales for the different product categories, e.g., groceries, dairy, frozen foods, nonfood items, produce, and meat.) Some 80 different independent variables ($x_1, x_2, \ldots, x_{80}$) were originally identified by company management as being important determinants of sales at a particular site. These ranged from various demographic measures, the quantity and quality of competition in the trading area, and relative pricing practices to the ease of left-hand turns into the parking lot from neighboring highways. Further management discussion reduced the number of independent variables to 20 after it was observed that many of them would be closely correlated with each other and that others would be extremely unlikely to have a significant impact on sales. Subsequent statistical analysis reduced this number still further, to 11. These included population density, income levels, traffic density, the square footage of competitive selling space within a 5 mile radius, the number of parking spaces, and others.

Data were then collected from a representative sample of existing stores, which included the actual sales (by category) and the actual values for the 11 environmental variables (x_1 through x_{11}) that had been chosen for inclusion in the study. The regression analysis was then conducted to determine statistically the values for the weighting coefficients (a_1 through a_{11}) and the constant term b.

Sales could then be forecast for a proposed site by ascertaining the specific values for the 11 independent variables particular to that site, multiplying each by its appropriate weighting value as determined from the regression analysis, computing the sum of these products, and adding the constant term. Confidence intervals were also computed to obtain a measure of risk. Relevant costs specific to the site were then determined, so that profitability could be forecast and desirability of the site determined.

10.1.8 CONCLUSION

Increasingly complex business environments and the rapid change in cost structures present a major challenge to organizations of all sizes and all types in the determination of facility locations. Rigorous analysis and a careful blending of the economic and noneconomic factors are necessary to avoid costly mistakes and to ensure the long-term profitability of the enterprise. Increasingly sophisticated statistical and mathematical techniques can be a major source of assistance in the decision making process and should be employed wherever feasible and practical. Although these can be effective extensions of management's capabilities, the highest levels of creativity and imagination are required to deal effectively with the many noneconomic and intangible factors. Manufacturing, service, government, nonprofit, and financial organizations can all benefit from increased examination and reexamination of these important decisions. The future will surely bring more complex problems and better methodology to deal with them, and the competitive organization will be continuously seeking improved profit opportunities through better location decisions.

REFERENCES

1. R. W. SCHMENNER, "Look Beyond the Obvious in Plant Location," *Harvard Business Review*, January–February, 1979, pp. 126–132.
2. J. G. MONKS, *Operations Management: Theory and Problems*, McGraw-Hill, New York, 1977, p. 54.
3. W. G. IRESON and E. L. GRANT, *Handbook of Industrial Engineering and Management*, 2nd ed., Prentice-Hall, Englewood Cliffs, NJ, 1971.
4. R. REED, JR., *Plant Location, Layout, and Maintenance*, Irwin, Homewood, IL, 1967.
5. R. L. FRANCIS and J. A. WHITE, *Facility Layout and Location–An Analytical Approach*, Prentice-Hall, Englewood Cliffs, NJ, 1974.
6. G. B. DANTZIG, *Linear Programming and Extensions*, Princeton University Press, Princeton, NJ, 1963.
7. F. S. HILLIER and G. J. LIEBERMAN, *Introduction to Operations Research*, Holden-Day, San Francisco, 1967.
8. G. M. PARKS, "Determining the Number, Location and Capacity of Manufacturing Facilities–One Detailed Case Study and Three Quickies," *Proceedings*, Spring Annual Conference, AIIE, Dallas, 1977.
9. J. A. WHITE, "A Quadratic Facility Location Problem," *AIIE Transactions*, June 1971, pp. 156–157.
10. A. A. ALY and D. W. LITWHILER, JR., "Police Briefing Stations: A Location Problem," *AIIE Transactions*, March 1979, pp. 12–22.
11. L. F. MCGINNIS, "A Survey of Recent Results for a Class of Facilities Location Problems," *AIIE Transactions*, March 1977, pp. 11–18.

12. A. GEOFFRION and R. MCBRIDE, "Langrangean Relaxation Applied to Capacitated Facility Location Problems," *AIIE Transactions*, March 1978, pp. 40–47.

13. A. A. KUEHN and M. J. HAMBURGER, "A Heuristic Program for Locating Warehouses," *Management Science*, July 1963, pp. 643–666.

14. R. A. JOHNSON, W. T. NEWELL, and R. C. VERGIN, *Operations Management–A Systems Concept*, Houghton Mifflin, 1972, pp. 149–155.

CHAPTER 10.2

Plant Layout

JAMES A. TOMPKINS

Tompkins Associates, Inc.

10.2.1 INTRODUCTION

Plant layout is the field of selecting the most effective arrangement of physical facilities to allow the greatest efficiency in the combination of resources to produce a product or service. Plant layout applies to the selection of the arrangement of physical facilities not only for manufacturing plants, but also for offices, hospitals, airports, shopping centers, and all other types of facilities. A more accurate title for the field of plant layout would be "facilities layout." Nevertheless, the field has traditionally been referred to as plant layout, and for the purposes of this chapter will continue to be.

Significance of Plant Layout

Since 1955 approximately 8% of GNP has been spent annually on new facilities in the United States.[1] Exhibit 10.2.1 indicates the typical expenditures, in percentage of GNP, for the major industry groupings. The task of establishing plant layouts for this volume of new facilities in itself makes the field of plant layout of the utmost importance. In addition to these new facilities, a significant percentage of previously purchased facilities are altered each year and require alterations in the layout.

Although the scope of plant layout is indicated by the dollar volume of the facilities laid out each year, the impact of effective plant layout does not necessarily follow. An inkling of the importance of effective plant layout may be gained by considering the following questions:

1. What impact does plant layout have on handling and maintenance costs?
2. What impact does plant layout have on employee morale, and how does employee morale affect operating costs?
3. In what do organizations invest the majority of their capital, and how convertible is their capital once invested?
4. What impact does plant layout have on the management of a facility?
5. What impact does plant layout have on a facility's capability to adapt to change and satisfy future requirements?

Although these questions may not be unilaterally answerable, they clearly tend to highlight the importance of effective plant layout. As an example, consider the first question. It has been estimated that between 20 and 50% of the total operating expenses within manufacturing are attributed to material handling. It is generally agreed that effective plant layout can reduce these costs by at least 10 to 30%. If effective plant layout were thus applied, the annual manufacturing productivity in the United States would increase approximately three times more than it has in any year in the last decade. Although it is difficult to relate these projections to other sectors of the economy, a concluding statement that plant layout is one of the most significant fields of the future, and in fact one of the most crucial to increasing the rate of productivity improvement, certainly would appear to be appropriate and accurate.

Objectives of Plant Layout

Plant layout may be likened to painting a picture or playing a musical instrument. There are general guidelines, principles, and techniques which, if followed, may lead to an effective plant

---

Portions of this chapter are adapted from J. A. Tompkins and J. A. White, *Facilities Planning*, Wiley, New York, 1982.

Exhibit 10.2.1 Percentage of GNP Typically Expended on New Facilities Between 1955 and the Present, by Industry Grouping

Industry	Percentage GNP
Manufacturing	3.2
Mining	0.2
Railroad	0.2
Air and other transportation	0.3
Public utilities	1.6
Communication	1.0
Commercial and other	1.5
All industry	8.0

Source: Reference 1.

layout, a beautiful piece of art, or a splendid performance. However, one must go beyond an intellectual understanding of these guidelines, principles, and techniques and develop a feel for the interrelated, often conflicting, objectives that affect the overall results. Plant layout, then, although becoming more scientific, remains today as an art.

An acceptable overall objective for painting a picture or playing a musical instrument is that the results are attractive and pleasant. In a similar manner, an acceptable overall objective for plant layout is that the result allows an enterprise to maximize profits for the service provided. The overall objectives of attractiveness, pleasantness, and maximum profits or service are certainly admirable, but give little practical guidance in the painting of a picture, the playing of an instrument, or the laying out of a facility. If clear, precise, exacting, measurable, and harmonious objectives could be set forth for these activities, they would be referred to as sciences and not as arts.

Meaningful objectives for plant layout are difficult to set forth. Such objectives may include the following:

1. To minimize backtracking, delays, and handling.
2. To maintain flexibility.
3. To utilize manpower and space effectively.
4. To promote high employee morale.
5. To provide for good housekeeping and ease of maintenance.

At times these objectives conflict with one another. Therefore they must be applied cautiously. For example, the objective of minimizing backtracking, if not realistically applied, may lead to a design contrary to the flexibility objective. Additionally, if not applied discreetly, the objective of minimizing equipment delays may lead to a design that is inconsistent with the objective of effectively utilizing manpower. None of these objectives should be relied upon without considering the effect on the other objectives.

Steps in Establishing a Plant Layout

The remainder of this chapter deals with the steps required to establish a plant layout. An overview of this procedure is as follows:

1. Define the objective of the facility to be laid out.
2. Specify the primary activities that must be performed to accomplish the objective.
3. Specify the related activities required to support the primary activities.
4. Determine the space requirements for all activities.
5. Determine the interrelationships among all activities.
6. Generate alternative layouts.
7. Evaluate alternative layouts.
8. Finalize and implement the layout.

10.2.2 DEFINING THE OBJECTIVE OF THE FACILITY

The first step in laying out a facility is to determine the objective of the facility. To specify this objective, one must document what is to take place in the facility once it is placed in operation. The defining of the objective of different types of facilities is illustrated in Exhibit 10.2.2. A review of the exhibit indicates that the questions that must be answered to define the objectives of a facility are (1) What is to be produced? and (2) How much is to be produced? The next two portions of this section describe how to answer these two questions.

Exhibit 10.2.2 Information Required to Define the Objective of Different Types of Facilities

Type of Facility	Information Required
Manufacturing	The products to be produced The market forecast
Hospital	The services to be offered The level of various types of patient activity
Airport	The types of aircraft to be serviced The volume of travelers to be handled
Warehouse	The types of products to be stored The volume of each type of product to be handled and stored

Product Design

The description of what products are to be produced is referred to as product design. Product design requires answers to the questions (1) What products are to be produced? and (2) What are the details of the products to be produced?

The first question should be answered by management and should not be altered throughout the layout planning process. For example, the management of an appliance manufacturing facility should try to determine the specific appliances that are to be produced within the facility being laid out. Similarly, the management of a hospital should attempt to determine the specific services (the product of a hospital) the hospital is to offer.

Specification of the details of the products to be produced by a facility may be viewed from operational and economic viewpoints. For facilities that provide various services, such as airports, libraries, hospitals, and banks, the operational viewpoint is concerned with the need for providing various services; the economic viewpoint is concerned with the costs of providing the services. For example, management may have decided that a hospital should offer x-ray services. The specification of the services requires a description of the types of x rays the hospital should have the capability of making. The decision should be made by trading off the needs for various types of x-ray services with the costs of providing the services.

For facilities that manufacture a product, the operational and economic viewpoints are described in Exhibit 10.2.3. The aesthetic considerations of product design are concerned with consumer acceptance. The objective is to have a product that is unique, eye-catching, and attractive. The functional considerations of product design are concerned with product liability and reliability. The objective is to design products that will function, even if not properly used, in a safe manner over their useful life. The output from the aesthetic and functional product design considerations are detailed operational specifications and a pictorial representation of the product.

An often-used pictorial representation for plant layout purposes is an exploded assembly drawing (Exhibit 10.2.4). The encircled numbers on such drawings refer to part numbers. Although the drawings are made to scale, specifications and dimensions are typically omitted.

The economic viewpoint of product design for manufactured components, given in Exhibit 10.2.3, includes material and manufacturing considerations. These considerations have as an objective the accomplishment of the operational requirements of the product at the least cost. Often, the techniques of value analysis or value engineering (Chapter 7.3) and group technology (Chapter 7.8) are used to achieve these objectives.

Value analysis or value engineering is a method of improving product value by improving the relationship between the function of each component or product and its cost. Group technology is a method of grouping various parts to be produced within a facility and utilizing similar approaches to produce these parts to obtain the least total materials and manufacturing costs for a family of parts.

Exhibit 10.2.3 Product Design Viewpoints

Viewpoint	Considerations	Responsibility
Operational	Aesthetics Function	Marketing and product engineering Human factors and product engineering
Economic	Materials Manufacturing	Purchasing and product engineering Manufacturing and product engineering

Exhibit 10.2.4 Exploded Assembly Drawing of a Tap Wrench

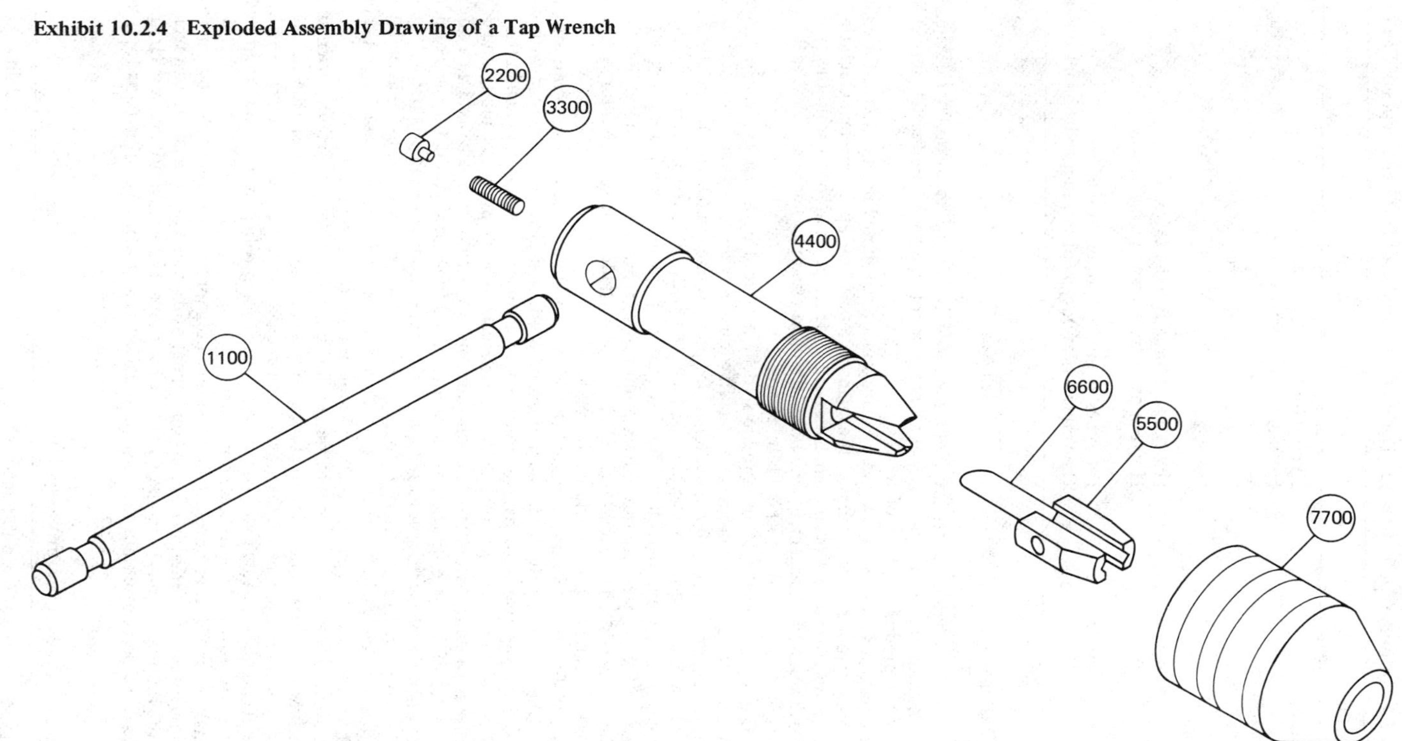

Exhibit 10.2.5 Component Part Drawing

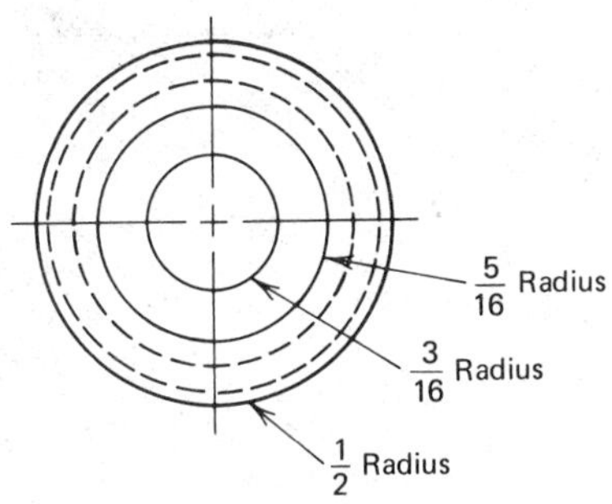

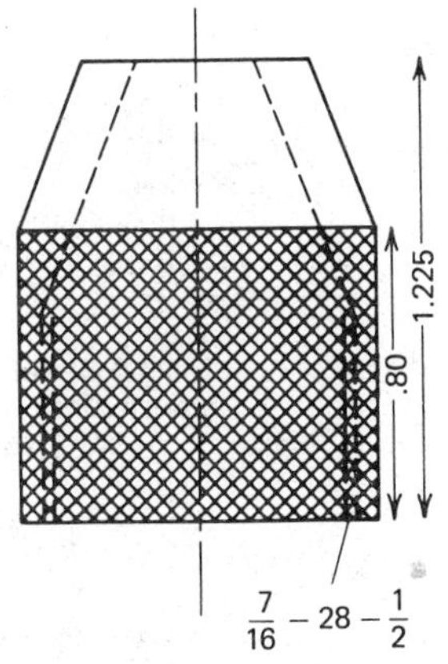

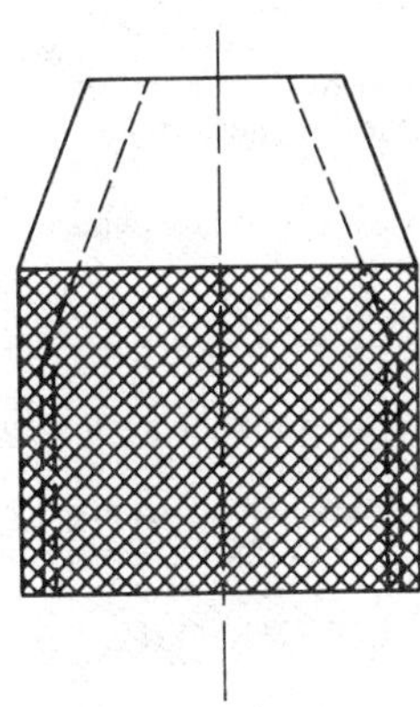

The output from the material and manufacturing product design considerations is a set of detailed component part drawings for every component of every part (Exhibit 10.2.5). These drawings should contain part specifications and dimensions in sufficient detail to allow fabrication of the part. The composite of the exploded assembly drawings and component part drawings fully documents the design of the products to result from a manufacturing facility.

Market Analysis

An effective plant layout process requires information describing the volumes, product mix, trends, and predictability of future demands for the various products or services produced by the facility. Such information should be obtained by means of a market analysis conducted by the marketing department and approved by management.

As a minimum, a market analysis should consist of the information given in Exhibit 10.2.6. The difficulty with market analysis information such as that given in the exhibit is that it assumes that all information is deterministic and known with certainty. A more desirable market analysis would consist of probabilistic estimates of the likelihood of a range of production volumes. Unfortunately, such stochastic information is often unavailable; therefore the plant layout must be based

Exhibit 10.2.6 Market Analysis Indicating Minimal Information Requirements for Plant Layout

Product or Service	First-Year Volume	Second-Year Volume	Fifth-Year Volume	Tenth-Year Volume
A	5000	5000	8000	10,000
B	8000	7500	3000	0
C	3500	3500	3500	4,000
D	0	2000	3000	8,000

Exhibit 10.2.7 Valuable Information To Be Obtained From a Market Analysis and How a Layout Planner May Use This Information

Information to be Obtained From a Market Analysis	Plant Layout Issues Affected by This Information
Who are the consumers of the product?	Packaging Susceptibility to product changes Susceptibility to changes in marketing strategies
Where are the consumers located?	Facilities location Method of shipping Warehousing systems design
Why will the consumer purchase the product?	Seasonality Variability in sales Packaging
Where will the consumer purchase the product?	Unit load sizes Order processing Packaging
What percentage of the market does the product attract and who is competition?	Future trends Growth potential Need for flexibility
What is the trend in product changes?	Space allocations Material handling methods Need for flexibility

on the type of information given in Exhibit 10.2.6. Constant care must be taken not to get overly confident or precise with the deterministic and assumed-to-be-certain information. In addition to the volume, product mix, trend, and predictability of future demands, the information given in Exhibit 10.2.7 should be obtained from the market analysis.

10.2.3 SPECIFYING THE FACILITY'S PRIMARY ACTIVITIES

The conversion of a facility's objectives into the primary functions required to accomplish the objectives is called "process specification." Process specification occurs in three steps:

1. **Process Design.** Determine the types of equipment required to accomplish the objectives of a facility.
2. **Process Requirements.** Determine the number of each type of equipment required to accomplish the objectives of a facility.
3. **Departmentalization.** Combine into departments the required number of each type of equipment and specify the departmental area and service requirements.

Process Design

Process design takes place in two phases. The first phase involves the identification of the types of processes to be performed, and the second involves the determination of the specific processes to be performed.

The scope of a facility will dictate to a large extent the types of processes that must be performed. The extremes for a manufacturing facility may range from a firm that purchases raw materials and proceeds through a multitude of refining, processing, and assembly steps to obtain a finished product to a firm that simply purchases components and assembles finished products. Decisions must be made concerning the condition of the materials to be purchased. These decisions are referred to as "make or buy" decisions, and the input to the plant layout procedure from this decision process is a parts list indicating the items to be made and those to be purchased.

The selection of the specific processes required is based upon past experiences, related requirements, available equipment, production rates, and future expectations. Consequently, it is not uncommon for different processes to be selected in different facilities to perform identical operations. However, the procedure used to arrive at these different conclusions should be the same. The process selection procedure consists of the following steps:

1. Define the elemental operations.
2. Identify alternative processes for each operation.

Exhibit 10.2.8 Route Sheet Data Requirements

Data	Production Example	Service Example
Component name and number	Plunger housing–3254	Wash dishes–WD1
Operation description and number	Shape, drill, and cut off–0104	Prepare water–0104
Equipment requirements	Automatic screw machine and appropriate tooling	Wash sink, rise sink and drainage rack
Location of where operation is to take place	Turning department	Kitchen
Unit times	Setup time–5 hr; operation time–0.0057 hr/component	Setup time–0.10 hr; operation time–0.0055 hr/dish
Raw material requirements	1 in. diameter by 12 ft aluminum bar/80 components	15 gal water/setup; 15 gal soap/setup

3. Analyze the alternative processes.
4. Standardize the processes.
5. Evaluate the alternative processes.
6. Select the best processes.

Inputs into the process selection procedure for a manufactured product would include a parts list, component part drawings, and the quantities of parts to be produced. The outputs from the procedure are the processes, equipment, and raw materials required to produce the components of the products specified in the objectives of the facility. This output may best be recorded on a route sheet, which should contain at least the data given in Exhibit 10.2.8. The method of combining the components into products to accomplish the objectives of the facility is documented in an assembly chart (Exhibit 10.2.9). By superimposing the route sheets, which provide information on the production methods, onto the assembly chart, which indicates how components are combined, an overview of the flow within the facility can be obtained in the form of an operation process chart (Exhibit 10.2.10).

Process Requirements

Once the specific process and equipment types required have been defined, the quantity of each process or equipment type must be determined. The first step in specifying the process requirements involves the determination of the quantities of components that must be produced. The estimated annual volume requirements obtained from the market analysis must be adjusted by appropriate scrap production factors for each operation in order to obtain the total annual volume requirements. The volume of components to start into an operation is defined as

$$Tn = \frac{Pn}{1 - Xn}$$

Where Tn = number of components flowing into operation n
Pn = desired production out of operation n
Xn = percent scrap produced on operation n

The next step in specifying the process requirements involves the calculation of the quantity of equipment required for each operation. This quantity, called the "equipment fraction," is determined by dividing the total time required to perform an operation by the time available to complete the operation. Mathematically, the equipment fraction of an operation is defined as

$$Fn = \frac{(Sn)\,(Tn)}{(En)\,(Hn)\,(Rn)}$$

where Fn = equipment fraction for operation n
Sn = standard time for performing operation n
Tn = number of times of performing operation n
En = historical efficiency of performing operation n
Hn = time available to perform operation n
Rn = historical reliability factor of the equipment

Exhibit 10.2.9 Assembly Chart for an Air Flow Regulator

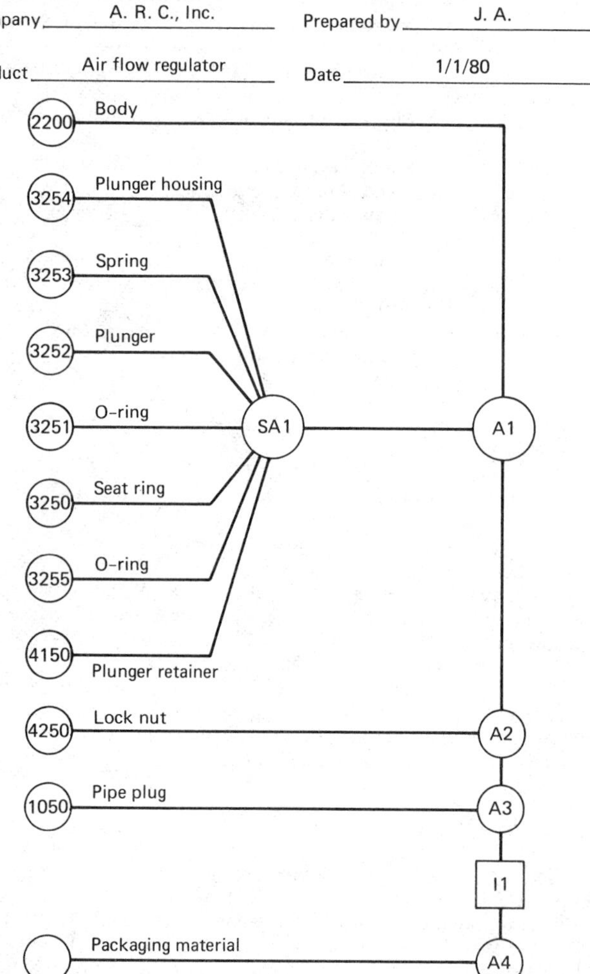

The final step in specifying the process requirements involves combining the equipment fractions for identical equipment types in order to determine the total equipment requirements. This procedure, however, is not necessarily straightforward. Strictly interpreting the sum of the equipment fractions for an equipment type as the total requirements for that equipment type will often result in production shortages or overtime due to scheduling or setup problems. Information on the cost of the equipment, the length of machine setups, the cost of in-process inventories, the cost and feasibility of setting up on overtime, the expected future growth, and several other qualitative factors must be analyzed in determining the quantity of equipment required.

Departmentalization

Departmentalization involves the conversion of process requirements into departmental specifications. The first phase of departmentalization determines the area requirements for each equipment type by considering all the interactions among the equipment, materials, and personnel and developing a detailed description of all work stations.

A work station must include areas for machinery, materials, and personnel. The machinery areas for a work station consist of space for (1) the machine, (2) machine travel, (3) machine maintenance, and (4) plant services. This information should be readily available from machinery data sheets. If machinery data sheets are not available, a physical inventory should be performed to determine at least the following:

1. Machine manufacturer and type.
2. Machine model and serial number.

Exhibit 10.2.10 Operation Process Chart for an Air Flow Regulator

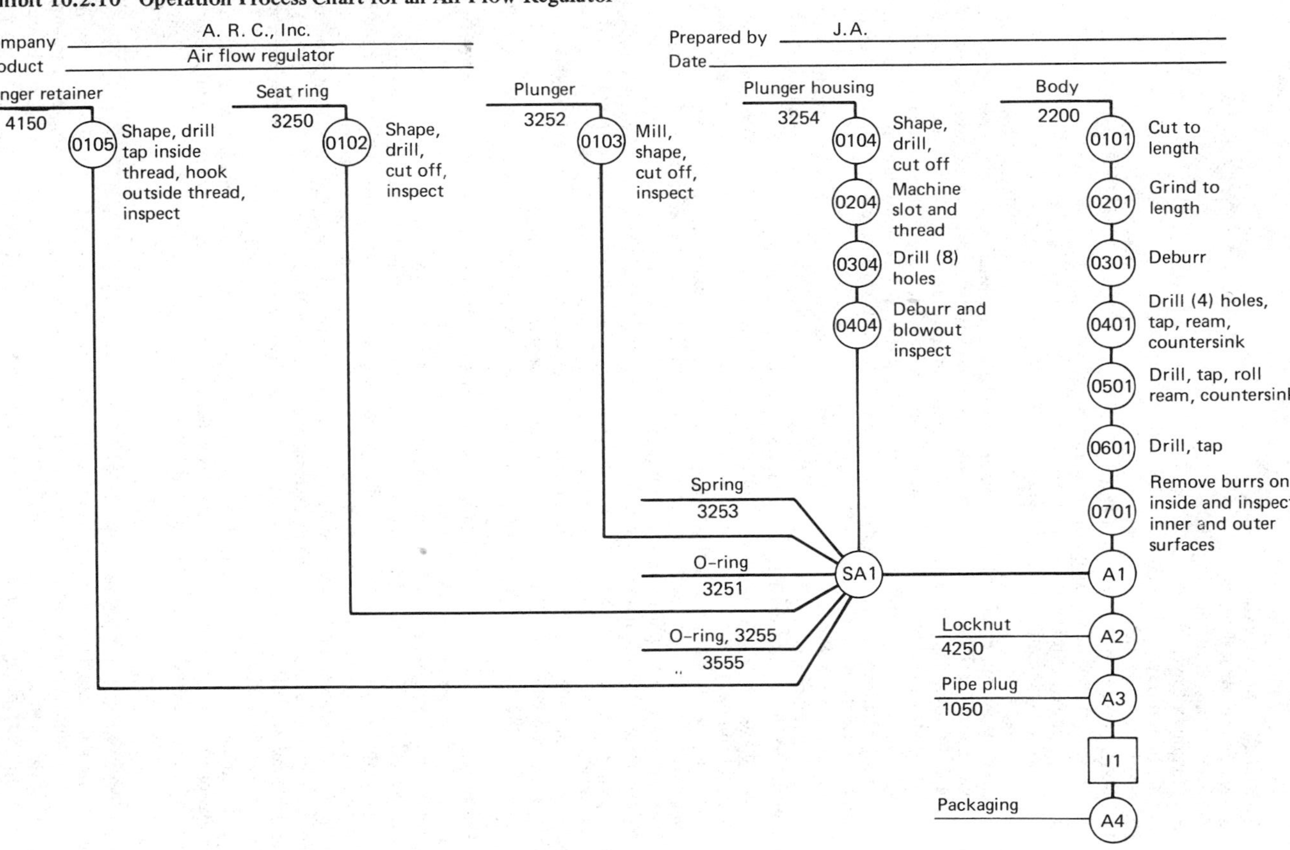

3. Location of machine safety stops.
4. Floor loading requirement.
5. Static height at maximum point.
6. Maximum vertical travel.
7. Static width at maximum point.
8. Maximum travel to the left.
9. Maximum travel to the right.
10. Static depth at maximum point.
11. Maximum travel toward the operator.
12. Maximum travel away from the operator.
13. Maintenance requirements and areas.
14. Plant service requirements and areas.

The floor area requirement of the machine, to include machine travel, may then be determined by calculating the product of the total width (static width plus maximum travel to the left and right) and the total depth (static depth plus maximum travel toward and away from the operator). To the floor area requirement of the machine should be added the maintenance and plant service area requirements. The resulting sum will represent the total machinery area for a machine. The sum of the machinery areas for all machines within a work station results in the machinery area requirement for the work station.

The materials areas for a work station consist of space for (1) receiving and storing materials; (2) in-process materials; (3) storing and shipping materials; (4) storing and shipping waste and scrap; and (5) tools, fixtures, jigs, dies, and maintenance materials.

To determine the area requirements for receiving and storing materials, in-process materials, and storing and shipping materials, the unit loads to be handled and the flow of the material through the machine must be known. Sufficient space should be allowed for the number of inbound and outbound unit loads typically stored at the machine. Additional space may be needed to allow for in-process materials to be placed into the machine, for the material to project beyond the machine, and for the removal of material from the machine. Space for the removal of waste (chips, trimmings, etc.) and scrap (defective parts) from the machine and for storage prior to removal from the work station must be provided. Space is also required for tools, fixtures, jigs, dies, and maintenance materials. A decision with respect to the storage of these items at the work station or in a central storage location will have a direct bearing on the area requirement. At the very least, space must be provided for the accumulation of tools, fixtures, jigs, dies and maintenance materials required while altering the machine setup.

The personnel area for a work station consists of space for (1) the operator, (2) material handling, and (3) operator ingress and egress. The specification of the space requirements for the operator and for material handling can be obtained directly from the method of performing the operation. The method should be determined using a motion study of the task (Chapter 3.2) and an ergonomics study (Chapter 6.9) of the operator. In addition to the space required for the operator and for material handling, space must be allowed for operator ingress and egress. A minimum 30 in. (76 c) aisle should be allowed for operator passage past stationary objects. If the operator is passing between a stationary object and an operating machine, a minimum 36 in. (92 c) aisle should be allowed. If the operator is passing between two operating machines, a minimum 42 in. (107 c) aisle should be allowed.

A sketch should be drawn for each work station in order to visualize the operators' activities. A simulation of the operator's daily activities will ensure the adequacy of the space allocation and may aid in significantly improving the efficiency, effectiveness, productivity, safety, and satisfaction of the operator.

Once all work stations are specified, the layout planner must determine which work stations should be combined to form departments. The departments to be formed at this point in the layout planning process are not necessarily the ultimate managerial departments that will exist in the facility. The departments to be formed are *planning departments*. These planning departments are groups of work stations that will be laid out during the plant layout process. After the layout has been completed, managerial departmentalization should be considered.

The combination of work stations into planning departments follows two basic philosophies:

1. **Product Planning Department.** Combine work stations performing operations on similar products or components.
2. **Process Planning Department.** Combine work stations performing similar processes.

Product planning departments can be further subdivided into the following:

1. **Production Line Departments.** All work stations required to produce a standardized product with large, stable demand are grouped together.

2. **Fixed Material Location Equipment.** All work stations required to produce a physically large and awkward-to-move product with low, sporadic demand are combined with a product staging area.
3. **Product Family Department.** The work stations required to produce a group of similar products with low individual demand are combined.

Most facilities will consist of a mixture of product and process planning departments. For example, in a facility consisting of mainly process planning departments producing a large variety of rather unrelated products, it would not be surprising to find paint, plate, heat treat, or assembly departments consisting of work stations based on a product planning department philosophy. Conversely, in a facility consisting mainly of product planning departments producing a few high-volume, standard products, it would not be surprising to find several "specialized" components produced in process planning departments. The approach to combining work stations into planning departments should follow a procedure of evaluating each product and component and determining the best approach.

The only remaining step in fully documenting the required processes for a facility is to summarize the departmental area requirements and to set forth the departmental service requirements. The summation of departmental areas is not simply the sum of the areas of the individual work stations to be included within the department. It is quite possible that machine maintenance, plant services, incoming and outgoing materials, and operator ingress and egress areas for various work stations may be combined to save space. Care must be taken that operational interference is not created by attempting to combine areas that are needed by individual work stations. Additional space is required within each department for material handling within the department. At this point in the layout planning process, it is impossible to determine aisle space requirements accurately since the department configurations, the work station alignments, and the material handling system have not been defined. The factors known at this point may be useful in approximating the percentage of the department needed for aisles are the relative sizes of the loads to be handled. Exhibit 10.2.11 sets forth a guide that may be used to obtain estimates of the aisle space requirements.

The departmental service requirements are given by the sum of the previously determined service requirements for the individual work stations that are to be included within a department. These requirements, as well as the departmental area requirements, should be recorded on a departmental service and area requirements sheet such as the one shown in Exhibit 10.2.12.

10.2.4 SPECIFYING THE RELATED ACTIVITIES

Several secondary activities are normally required to support the primary activities involved in producing goods or services. The most universal related activities include receiving, shipping, storage, and maintenance functions. The layout of storage functions is discussed in Chapter 10.4.

Receiving and Shipping Departments

The receiving department should be viewed as the gateway to a facility through which all materials, supplies, and equipment must pass. The shipping department should be viewed as a funnel through which the finished and sold product must pass. Deficiencies in the planning and layout of either of these activities can result in productivity losses and poor customer service.

The first step in the layout planning process for a receiving or shipping department is to determine what is to be received or shipped. A chart that is useful for defining what is to be received and shipped is a receiving and shipping analysis chart, as shown in Exhibit 10.2.13. The first seven columns of the chart include information on the what, how much, and when of items received or shipped. For an existing facility, or for one that is to have similar objectives to an existing facility, this information may be obtained from past receiving reports or shipping releases. For a new facility, the parts lists and the market analysis information for all products must be analyzed to deter-

Exhibit 10.2.11 Aisle Allowance Estimates

Size of Largest Load	Aisle Allowance Percentage
Less than 6 ft^2 (0.56 m^2)	5–10
Between 6 and 12 ft^2 (0.56 and 1.11 m^2)	10–20
Between 12 and 18 ft^2 (1.11 and 1.67 m^2)	20–30
Greater than 18 ft^2 (1.67 m^2)	30–40

Exhibit 10.2.12 Departmental Service and Area Requirement Sheet

Company A. B. C. Prepared by J. A.

Department Turning Date ______ Sheet 1 of 1

Work Station	Quantity	Service Requirements			Floor Loading	Ceiling Height	Area in Square Feet			
		Power	Compressed Air	Other			Equipment	Material	Personnel	Total
Turret lathe	5	440 V AC	10 CFM @ 100 psi		150 psf	4 ft	240	100	100	440
Screw machine	6	440 V AC	10 CFM @ 100 psi		190 psf	4 ft	280	240	100	620
Chucker	2	440 V AC	10 CFM @ 100 psi		150 psf	5 ft	60	100	40	200

Net area required 1260

10% aisle allowance 130

Total area required 1390

mine reasonable unit loads and order quantities. Once this information is obtained, the first seven columns of the receiving and shipping analysis chart may be completed.

The eighth and ninth columns of the chart require the identification of the types of carriers to be utilized for receiving and shipping materials. At a minimum, one should specify the type of carrier; its overall length, width, and height; and the height to the carrier's deck. This information may be obtained by investigating the types of carriers utilized by the transportation companies to carry the types and quantities of material to be received or shipped.

The last two columns of the chart concern the handling of materials off of or onto a carrier. If the chart is being completed for an existing receiving or shipping area, the current methods of unloading or loading materials should be evaluated and, if acceptable, recorded on the chart. If the chart is being completed for a new operation, alternative methods of handling material should be identified and evaluated to determine the best method. The time required to unload or load a carrier may be determined for an existing operation by means of historical data, work sampling, or time study (Chapters 4.4 and 4.6). The accuracy and time required to determine time standards using these approaches, however, typically results in the facility planner's using predetermined time elements. These elements are also used for new receiving or shipping operations (Chapter 4.5).

Once all the information required on the receiving and shipping analysis chart is completed for all materials to be received or shipped, the receiving or shipping function is fully defined. The next step of the layout planning process involves the determination of the number and type of docks required in the receiving or shipping department.

The determination of the number of docks required will depend upon a number of variables, including the average time interval between truck arrivals and the average time required to service a truck. Two analytical tools that may prove useful in determining the number of docks required are queuing analysis (Chapter 13.7) and simulation (Chapter 13.11).

Once the number of docks is determined, the proper dock configuration must be determined. The first consideration is the flow of the carriers about the facility. For rail docks, the location and configuration of the railroad spur will typically dictate the flow of railroad cars and the configuration of the rail dock. For truck docks, the truck traffic patterns must be analyzed. Truck access to the property should be planned such that trucks need not back into the dock area. Truck guidelines that should be taken into consideration are:

1. Two-directional service roads should be 24 ft (7.32 m) wide.
2. One-way service roads should be 12 ft (3.66 m) wide.
3. If pedestrian travel is to be along service roads, a 4 ft (1.22 m) wide walk physically separated from the service road should be included.
4. Gate openings for two-directional travel should be 28 ft (8.53 m) wide.
5. Gate openings for one-way travel should be 16 ft (4.88 m) wide.
6. Gate openings should be 6 ft (1.83 m) wider if pedestrians will also use the gate.
7. All right-angle intersections must have a minimum 50 ft (15.24 m) radius.
8. If possible, all traffic should circulate counterclockwise since left turns are easier and safer to make than right turns.
9. Truck waiting areas should be allocated adjacent to the dock apron and should be of sufficient magnitude to hold the maximum expected number of trucks waiting at any given time.

Once these guidelines have been taken into consideration, the overall flow of trucks about a facility may be determined and a decision made between 90° docks or finger docks. As can be seen in Exhibit 10.2.14, 90° docks require greater apron depth, but less bay width; consequently, 90° docks require greater outside turning area, and finger docks require greater inside maneuvering area. Since outside space costs considerably less to construct and maintain than inside space, 90° docks should be used whenever space exists. The space requirements for 90° docks are given in Exhibit 10.2.15. If adequate apron depth does not exist for a 90° dock, a finger dock must be utilized. When finger docks are utilized, the largest angle finger dock should be employed. The space requirements for finger docks are given in Exhibit 10.2.16.

Dock widths of 10 ft (3.05 m) are adequate for spotting trucks. The potential for accidents and increase in maneuvering time, however, have made 10 ft (3.05 m) widths virtually obsolete. The accepted width is now 12 ft (3.66 m), and for some docks where traffic is to be very high, 14 ft (4.27 m) dock widths are often recommended.

The last step in determining the receiving or shipping department space requirements deals with the space requirements within the facility. The two most important internal space requirements are allocations for buffer and staging areas and for material handling equipment maneuvering.

Buffer areas are areas within receiving where material removed from carriers may be placed until it is dispatched. If the operating procedure is to deliver all merchandise to stores, inspection, or the department requesting the merchandise immediately upon receipt, a buffer area is not required. However, if the procedure is to remove material from the carrier and place it in a holding area prior

Exhibit 10.2.13 Receiving and Shipping Analysis Chart

Company ______________ Date ______________ Raw materials ______ Finished goods

Prepared by ______________ Sheet ____ of ______________ Plant supplies

Description	Unit Loads				Size of Shipment (Unit Loads)	Frequency of Shipment	Transportation		Materials Handling	
	Type	Capacity	Size	Weight			Mode	Specifications	Method	Time

Exhibit 10.2.14 Trade-Offs Between Bay Width and Apron of (*a*) 90° Dock and (*b*) 45° Finger Dock

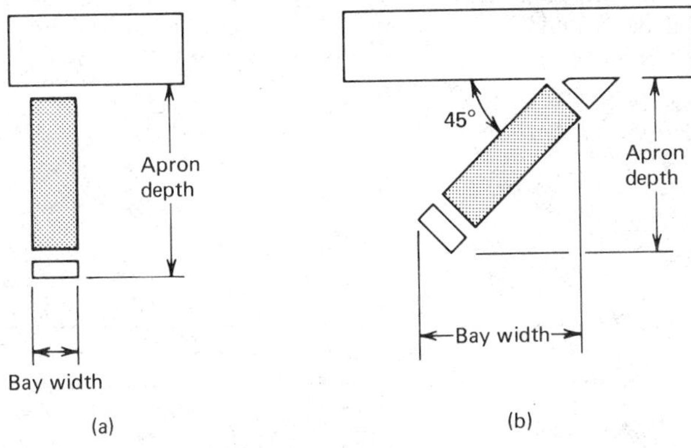

Exhibit 10.2.15 Space Requirements for 90° Docks

Truck Length in Feet (Meters)	Dock Width in Feet (Meters)	Apron Depth in Feet (Meters)
40 (12.19)	10 (3.05)	46 (14.02)
	12 (3.66)	43 (13.11)
	14 (4.27)	39 (11.89)
45 (13.72)	10 (3.05)	52 (15.85)
	12 (3.66)	49 (14.94)
	14 (4.27)	46 (14.02)
50 (15.24)	10 (3.05)	60 (18.29)
	12 (3.66)	57 (17.37)
	14 (4.27)	54 (16.46)
55 (16.76)	10 (3.05)	65 (19.81)
	12 (3.66)	63 (19.20)
	14 (4.27)	58 (17.68)
60 (18.29)	10 (3.05)	72 (21.95)
	12 (3.66)	63 (19.20)
	14 (4.27)	60 (18.29)

to dispatching, space must be allocated in which to store this merchandise. In a similar manner, staging areas are areas within shipping where merchandise is placed and checked prior to being loaded into a carrier. If merchandise is loaded into carriers directly from being withdrawn from the warehouse, no staging area is required.

The space required for buffer or staging areas may be determined by considering the number of carriers of merchandise to be stored in these areas and the area required to store a carrier of merchandise. Typically, when buffer and staging areas are utilized, sufficient space should be allocated

Exhibit 10.2.16 Finger Dock Space Requirements for a 65 ft (19.81 m) Trailer

Finger Angle	Dock Width in Feet (Meters)	Apron Depth in Feet (Meters)	Bay Width in Feet (Meters)
10°	10 (3.05)	50 (15.24)	65 (19.81)
	12 (3.66)	49 (14.94)	66 (20.12)
	14 (4.27)	47 (14.33)	67 (20.42)
30°	10 (3.05)	76 (23.16)	61 (18.59)
	12 (3.66)	74 (22.56)	62 (18.90)
	14 (4.27)	70 (21.34)	64 (19.51)
45°	10 (3.05)	95 (28.96)	53 (16.15)
	12 (3.66)	92 (28.04)	54 (16.46)
	14 (4.27)	87 (26.52)	56 (17.07)

Exhibit 10.2.17 Minimum Maneuvering Allowances for Receiving and Shipping Areas

Material Handling Equipment Utilized	Minimum Maneuvering Allowance in Feet (Meters)
Tractor	14 (4.27)
Platform truck	12 (3.66)
Fork lift	12 (3.66)
Narrow aisle truck	10 (3.05)
Hand lift (jack)	8 (2.44)
Four-wheel hand truck	8 (2.44)
Two-wheel hand truck	6 (1.83)
Manual	5 (1.52)

for one full carrier for each dock. When fluctuations in hourly unloading or loading rates become pronounced, however, space for storing two or more carriers of merchandise in buffer or staging areas should be considered. The space required to store a carrier of merchandise depends upon the size of the carrier, the cube utilization in the carrier, and the cube utilization in the buffer or staging area. Merchandise can typically be stored at least as high and often much higher in a buffer area than in the carrier. The inverse is true for staging areas where order checking must take place.

The maneuvering space required for material handling equipment is the area between the backside of the dock leveler and the beginning of the buffer or staging areas. The amount of maneuvering space required depends upon the type of material handling equipment utilized (Exhibit 10.2.17).

The final result of the receiving and shipping planning analysis will be a set of departmental service and area requirement sheets (Exhibit 10.2.12) for the receiving and shipping departments. For additional information on the receiving and shipping function, consult Chapter 10.4.

Maintenance Department

The first step in the layout planning process for a maintenance department is to define the objectives and scope of the department just as the objectives and scope of the overall facility were defined. Such a definition will include a description of the types and quantities of maintenance services to be provided.

Once the types and quantities of services are defined, the activities and space required to provide these services can be determined in the same manner used to specify the primary activities of the overall facility. The end result of the maintenance planning function will be a set of departmental service and area requirement sheets (Exhibit 10.2.12) for the maintenance department.

10.2.5 DETERMINING SPACE REQUIREMENTS FOR ALL ACTIVITIES

In addition to the space requirements for the primary activities and the related support activities of a facility, space must be provided for a number of other necessary activities. Foremost among these are offices, food services, medical services, lavatories, and locker rooms. Office layout is described in detail in Chapter 10.6 of this handbook. The space requirements of the other activities listed are discussed here.

Food Services

The OSHA of 1970 requires that, in all places of employment where employees are permitted to lunch on the premises, an adequate space suitable for that purpose be provided for the maximum number of employees who may use such space at one time. Such space shall be physically separate from any location where there is exposure to toxic materials.

The most common food services areas and their space requirements per person using the area at one time are shown in Exhibit 10.2.18. A departmental service and area requirement sheet (Exhibit 10.2.12) should be completed for the food services area.

Medical Services

Depending upon the occupational hazards within a facility and the number of employees, the medical services required may vary from a first-aid room to a plant hospital. A first-aid room should consist of at least a first-aid cabinet, bed, and two chairs and should not be smaller than 100

Exhibit 10.2.18 Space Requirements for Food Services Areas for Each Anticipated Employee at Peak Usage

Type of Facility	Square Feet (Square Meters) Required per Person
Full kitchen and cafeteria	20 (1.86)
Serving line and cafeteria	17 (1.58)
Vending machines and cafeteria	13 (1.21) for 25 or fewer persons
	12 (1.11) for 26 to 74 persons
	11 (1.02) for 75 to 149 persons
	10 (0.93) for 150 or more persons

ft^2 (9.29 m^2). An estimate of the space required for a first-aid room may be obtained by approximating the size of the medical staff assigned to the medical service. For each 500 employees, an industrial nurse should be hired. One physician should be employed if between 800 and 2000 employees work in the firm. For each additional 1000 employees, an additional physician should be employed. To estimate the space requirements, approximately 250 ft^2 (23.23 m^2) should be added to the minimum of 100 ft^2 (9.29 m^2) for each nurse, and 400 ft^2 (37.16 m^2) should be added for each physician. A departmental service and area requirement sheet (Exhibit 10.2.12) should be completed for the medical services area.

Lavatories

The OSHA defines very clearly what is required as far as toilet facilities and sinks (subpart J, paragraph 1910.141, c–f). From a plant layout point of view, the important initial considerations are:

1. Separate facilities shall be provided for each sex not to be further than 200 ft (60.96 m) from the location where workers are regularly employed.
2. The number of water closets for each sex shall be determined by Exhibit 10.2.19.
3. Where the number of men employed is greater than 10, one water closet less than given in Exhibit 10.2.19 may be provided for each urinal. The number of water closets must remain at least at two thirds of the number specified in Exhibit 10.2.19.
4. At least one sink with adequate hot and cold water shall be provided for every 10 employees, or portion thereof, up to 100 persons; for more than 100 persons, one sink shall be provided for each additional 15 persons. A sink shall be at least 24 in. (60.96 cm) wide and shall have an individual faucet.
5. Where 10 or more women are employed, at least one resting area is required. Two beds are required if between 100 and 250 women are employed, and one additional bed is required for each additional 250 women employees.

For space planning purposes, 12 ft^2 (1.11 m^2) should be allowed for each water closet, 5 ft^2 (0.46 m^2) for each urinal, and 6 ft^2 (0.56 m^2) for each sink. Entrance doorways should be designed such that the interior of the lavatory is not visible from outside of the room when the door is open. An allowance of 14 ft^2 (1.3 m^2) should be made for this entrance. For each bed required in the women's lavatory, a 60 ft^2 (5.56 m^2) allowance should be made.

Locker Rooms

The OSHA requires that separate locker rooms be provided for each sex whenever it is the practice to change from street clothes to working clothes. If locker rooms are not provided, OSHA requires facilities to be provided for hanging outer garments. For planning purposes, 6 ft^2 (0.56 m^2) should be allocated for each person using the locker room. Often it is convenient to combine a lavatory

Exhibit 10.2.19 The OSHA Standard for Number of Toilet Facilities Required

Number of Persons	Minimum No. of Water Closets
1 to 9	1
10 to 24	2
25 to 49	3
50 to 74	4
75 to 100	5
1 additional water closet for each additional 30 persons	

and the locker room. Although it is desirable to locate this facility in a convenient place for the employees, if possible it should be located away from the primary flow of materials and in a location that provides good ventilation. Mezzanines or locations along an outside wall are therefore often effectively utilized for lavatories and locker rooms. A departmental service and area requirement sheet (Exhibit 10.2.12) should be completed for the locker room.

Summarizing Space Requirements

Departmental service and area requirement sheets (Exhibit 10.2.12) should be prepared for any other activities requiring space within the facility. The space requirements for the individual activities should then be summarized in tabular form. This information will be a valuable input into the generation of alternative plant layouts.

10.2.6 DETERMINING THE INTERRELATIONSHIPS AMONG ALL ACTIVITIES

Arranging departments within a facility is based upon the flow of materials, individuals, and information among departments. To evaluate alternative arrangements, a measure of flow must be established. Flows may be specified in a quantitative or qualitative manner. Qualitative measures may include the absolute necessity that two departments be close to each other, the importance of two departments being close to each other, or the preference that two departments not be close to each other. In facilities having large volumes of material, information, and people moving between departments, a quantitative measure of flow will typically be the basis for the arrangement of departments. In contrast, in facilities having very little actual movement of material, information, and people between departments, but having significant communication and organizational interrelationships, a qualitative measure of flow will typically be the basis for the arrangement of departments. Most often a facility will have a need for both quantitative and qualitative measure of flow, and both measures should be used.

Quantitative Interrelationships

Departmental interrelationships may be quantitatively measured in terms of the amount of material, individuals, or information moved between departments. A common method used to record these quantitative interrelationships is a from-to chart (Exhibit 10.2.20). A from-to chart is a square matrix having the departments listed down the rows and across the columns in the same order. The chart is not symmetrical about the diagonal; that is, the flow volumes from department A to department B are not normally the same as the flow volumes from department B to department A.

A from-to chart may be constructed by the following steps:

1. Include all departments for which departmental service and area requirement sheets have been prepared.
2. List the departments down the row and across the column in such a way that the overall flow pattern is followed.

Exhibit 10.2.20 From-To Chart

From \ To	Stores	Milling	Turning	Press	Plate	Assembly	Warehouse
Stores		12[a]	6	9	1	4	
Milling					7	2	
Turning		3			4		
Press					3	1	
Plate		3	1			3	
Assembly	1						7
Warehouse							

[a]Units = trips per day.

3. Establish a measure of flow for the facility that will accurately indicate equivalency of flow volumes. If the items moved are equivalent with respect to ease of movement, the number of trips may be recorded on the from-to chart. If the items moved vary in size, weight, value, risk of damage, shape, or any other characteristic, however, an equivalency among items must be established so that the quantities recorded on the chart represent the proper relationships among the volumes of movement.

4. Based on the flow paths for the items to be moved and the established measure of flow, record the flow volumes on the from-to chart.

Qualitative Interrelationships

Qualitative departmental interrelationships may be recorded by using closeness relationship values in a relationship chart (Exhibit 10.2.21). A relationship chart may be constructed according to the following steps:

1. List all departments for which departmental service and area requirement sheets have been prepared on the relationship chart.

Exhibit 10.2.21 Relationship Chart[a]

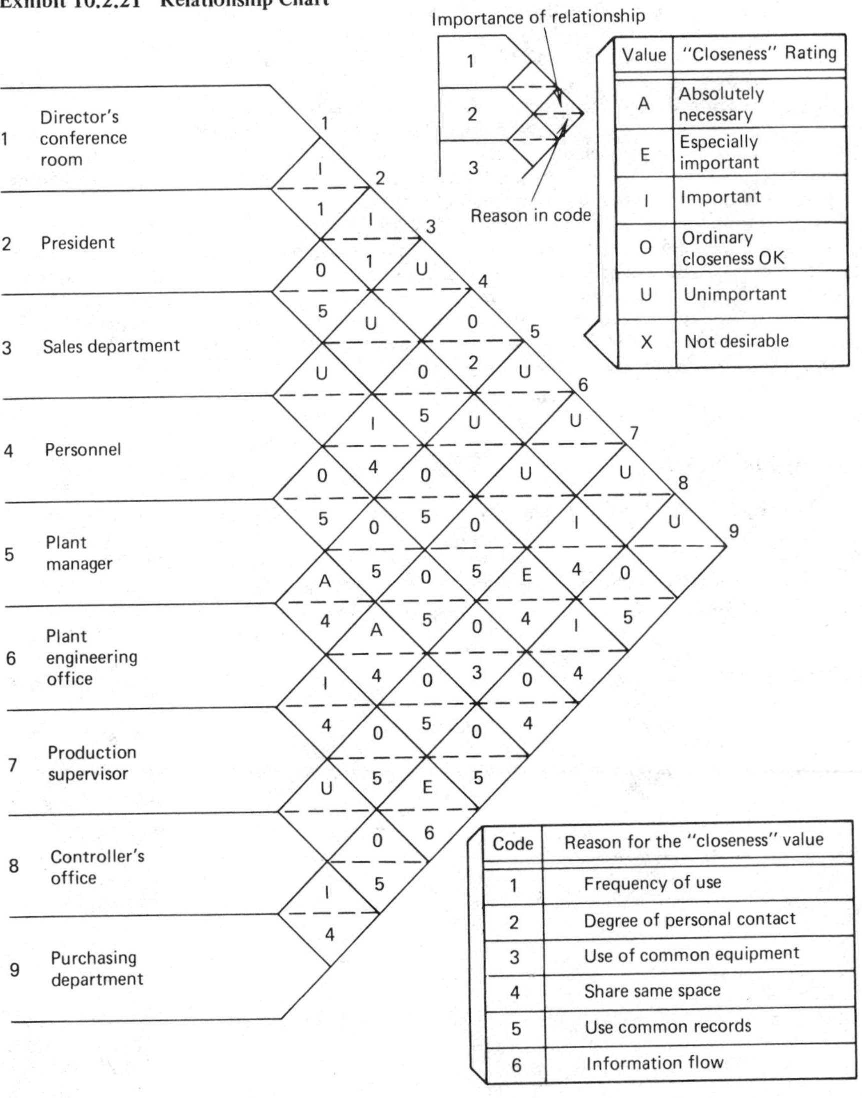

[a]Adapted from reference 2.

2. Conduct interviews or surveys with the department heads for the departments listed on the relationship chart and with the management responsible for all departments.
3. Define the criteria for assigning closeness relationships and itemize and record the criteria as the reasons for relationship values on the relationship chart.
4. Establish the relationship value and the reason for the value for all pairs of departments.
5. Allow everyone having input to the development of the relationship chart to have an opportunity to evaluate and discuss changes in the chart.

It is of the utmost importance that these steps be followed in developing a relationship chart. If, instead of the layout planner's synthesizing the relationship among departments as described previously, the department heads are allowed to assign the closeness relationships with other departments, inconsistencies may develop. The inconsistencies follow from the form of the chart. The relationship chart, by definition, requires that the relationship value between departments A and B be the same as the relationship value between departments B and A. If individual relationship values were assigned by department heads, and if the head of department A said the relationship with B was unimportant (U), whereas the head of department B said the relationship with A was of ordinary importance (O), an inconsistency exists. It is best to avoid these inconsistencies by having the layout planner assign relationship values based on the inputs from the important parties and then have the same parties evaluate the final result.

10.2.7 GENERATING ALTERNATIVE LAYOUTS

During the specification of the primary, related support, and other activities of a facility, the space requirements for each activity are determined and preliminary departmental layouts developed. The next step in the layout planning process is to mold the various departmental layouts into an overall facility layout. The individual departmental layouts should then be reviewed and altered as required to obtain a fully integrated facility layout.

Layout planning prior to 1950 consisted of facilities planners' drawing schematics of department locations and mentally evaluating the advantages and disadvantages of each drawing. This approach worked well, and still works well, for very small facilities having only a few interactions. However, if one is considering the layout of a more complex facility where several interactions take place, it is quite possible that a more rigorous approach to layout planning will be required. Several graphic and computer-aided layout techniques have been developed since 1950 to provide more methodical approaches to layout planning.

Graphic Layout Techniques

Data Requirements

The data requirements for the graphic layout techniques presented here consist of the department areas and the departmental relationships. Information on the department areas should be available from the department service and area requirement sheets (Exhibit 10.2.12). The relationships among departments may be recorded either quantitatively on a from-to chart or qualitatively on a relationship chart. Each of the graphic layout techniques is demonstrated here by being applied to the design of a facility having the requirements given in Exhibits 10.2.22, 10.2.23, and 10.2.24.

Spiral Technique

The objective of the spiral technique is to arrange departments so that the volume of flow between adjacent departments is maximized. The first step of the spiral technique involves the ranking of the flow volumes in descending order. For the example problem given in Exhibit 10.2.23, the rank-

Exhibit 10.2.22 Department Areas for the Example Problem

Code	Function	Area in Square Feet (Square Meters)
A	Receiving	12,000 (1114.84)
B	Milling	8,000 (743.22)
C	Press	6,000 (557.42)
D	Screw machine	12,000 (1114.84)
E	Assembly	8,000 (743.22)
F	Plating	12,000 (1114.84)
G	Shipping	12,000 (1114.84)

Exhibit 10.2.23 From-To Chart Indicating the Number of Trips per Week Made by an Electric Platform Truck

	A	B	C	D	E	F	G
A		45	15	25	10	5	
B				30	25	15	
C					5	10	
D		20			35		
E						65	35
F		5			25		65
G							

ing is F–G, E–F, A–B, E–G, D–E, B–D, A–D, B–E, F–E, D–B, A–C, B–F, A–E, C–F, A–F, F–B, and C–E.

The next step of the spiral technique involves the actual arrangement of departments. The arrangement is performed in two phases. The first phase is performed without regard to area and results in only a schematic diagram of the facility; it is conducted by bringing the departments into the layout in the order indicated by the flow rankings. The departments are positioned in a schematic diagram as indicated by the graphic representation. For the example problem, the first two departments to enter the layout are departments F and G; they are placed adjacent to each other as shown in Exhibit 10.2.25*a*. The second highest flow ranking is between departments E and F; therefore the next department to enter the layout is department E. A review of the material flow

Exhibit 10.2.24 Relationship Chart for the Example Problem

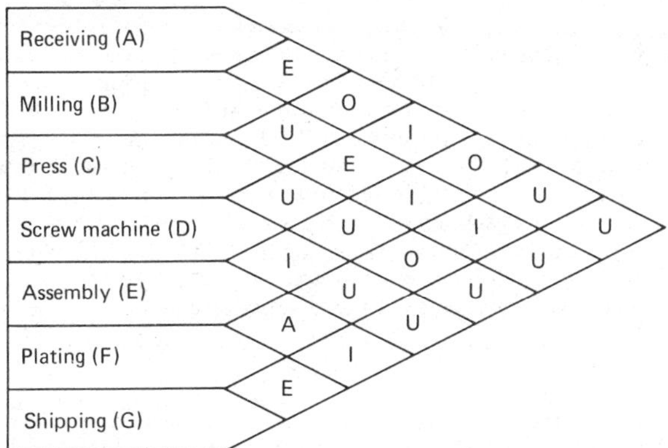

Exhibit 10.2.25 Schematic Development of Layout for the Example Problem Using the Spiral Technique

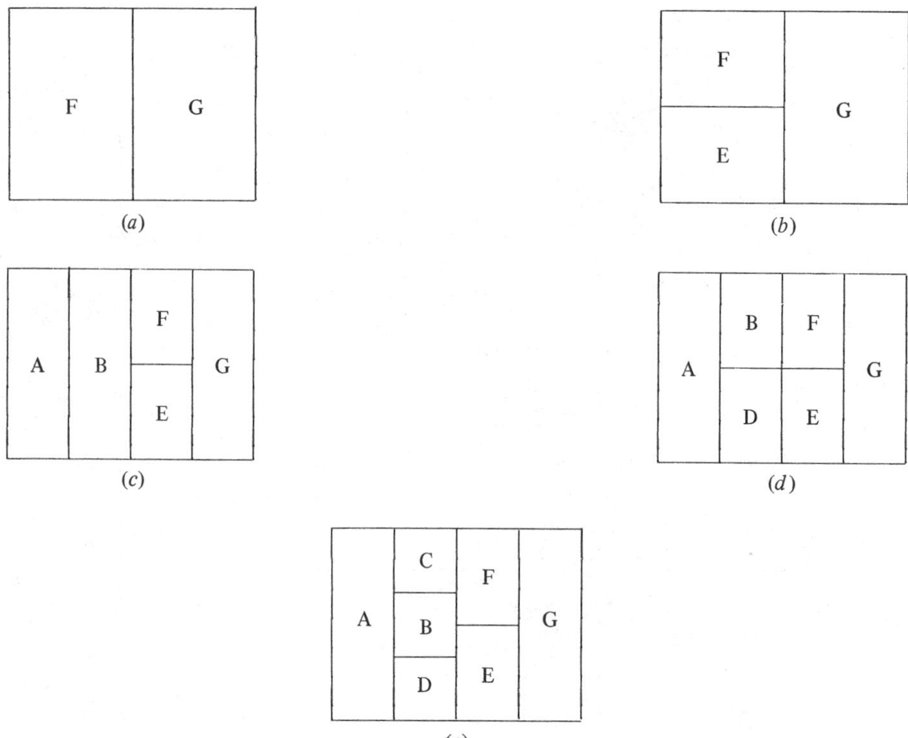

rankings indicates E should be associated with both F and G; therefore the resulting schematic diagram is as shown in Exhibit 10.2.25*b*.

The next most important flow is between A and B; therefore these departments will be next to enter the layout. A review of the from-to chart indicates that department A has only a slight interaction with departments E and F–10 and 5 trips/week, respectively–as compared to department B; therefore the schematic diagram shown in Exhibit 10.2.25*c* is chosen. Proceeding down the rankings, department D is next to enter the layout, leaving department C to enter last. Exhibit 10.2.25*d* indicates the placement of department D so that it is adjacent to A, B, and E as indicated in the from-to chart. The placement of department C in the final schematic diagram is as shown in Exhibit 10.2.25*e*.

The second phase in arranging the departments may now be completed by considering the actual department area requirements and including the areas in the schematic diagram. The layout for the example problem may be obtained by adding the department area requirements given in Exhibit 10.2.22 to the final schematic diagram given in Exhibit 10.2.25*e*. The layout given in Exhibit 10.2.26 results.

Several alternative layouts may be developed using the spiral technique while still adhering to the material flow rankings. The alternative layouts may be rated by dividing the flows for which departments are not adjacent by the total flow within the facility. The flows that are not between adjacent departments for the solution given in Exhibit 10.2.26 are A–E, A–F, and C–E, for a total of 20 trips/week. The total of all flows is determined by adding the values listed in the from-to chart; for the example problem, the total is 425 trips/week. Therefore the inefficiency rating for the solution given in Exhibit 10.2.26 is 5% (20 trips per week/435 trips per week).

The spiral technique is a good method of visualizing the flow within a facility. However, there is no systematic procedure leading to improved solutions; therefore the quality of the final solution depends upon the ingenuity and persistence of the facilities planner.

Travel Charting

The objective of travel charting is to minimize the product of the quantities moved and the length of the moves. This product is referred to as the "volume-distance product" since it is the product

Exhibit 10.2.26 Resulting Layout for the Example Problem Using the Spiral Technique

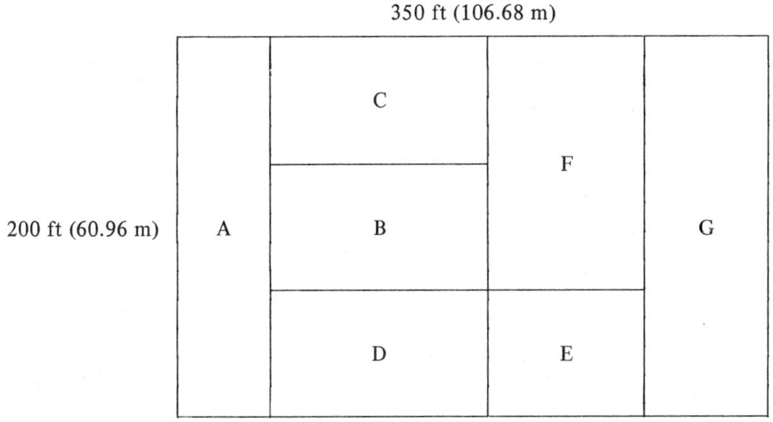

350 ft (106.68 m)

200 ft (60.96 m)

A C B D F E G

of the flow volumes given in a from-to chart and the distance between the origination and destination of the move given in a distance matrix.

The first step in travel charting is to establish an initial layout. This layout may be based upon the spiral technique or upon intuition. The first layout for the example problem is given in Exhibit 10.2.27. Once the initial layout is established, the lengths of all flow paths must be determined. These lengths may be difficult to determine since the actual origin and destination of the moves may not be known. The assumption typically employed to overcome this difficulty is that all flows originate and terminate at the centers, or centroids, of the departments. It is further typically assumed that the flow path between these origins and destinations will proceed along perpendicular aisles. Therefore the lengths of the flow paths are assumed to be the rectilinear flow path between the centroids of the departments. Hence the next travel charting step is the determination of the centroids of all departments and the calculation of the rectilinear distances between the centroids for all moves. The resulting distances are recorded in a distance matrix. For the example problem, the distance matrix for the initial layout is given in Exhibit 10.2.28.

The next travel charting step involves multiplying the flow volumes (from the from-to chart) by the flow paths (from the distance matrix) to obtain the travel chart. The sum of the elements of the travel chart is the volume-distance product of the initial layout. The product of the from-to chart and the distance matrix for the example problem results in the travel chart given in Exhibit 10.2.29.

The final travel charting step involves the altering of the layout so as to reduce the volume-distance product. One procedure that attempts to reduce the volume-distance product is to reduce the rectilinear distance between the centroids of the departments having the largest elements in the travel chart. This step can be repeated until no alteration in the layout can be found that reduces

Exhibit 10.2.27 Initial Layout for Travel Charting for the Example Problem

350 ft (106.68 m)

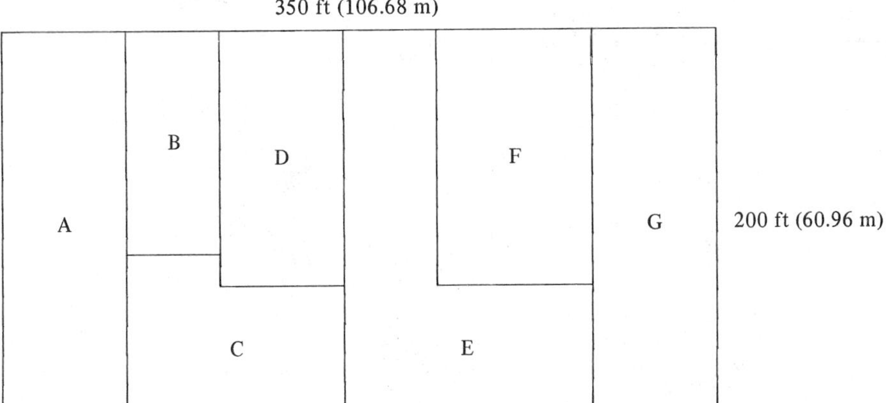

200 ft (60.96 m)

Exhibit 10.2.28 Distance Matrix for the Initial Layout of the Example Problem, in Feet (Meters)

	A	B	C	D	E	F	G
A		105 (32.00)	130 (39.62)	140 (42.67)	260 (79.25)	240 (73.15)	
B				85 (25.91)	235 (71.63)	185 (56.39)	
C					130 (39.62)	250 (76.20)	
D		85 (25.91)			150 (45.72)		
E						150 (45.72)	100 (30.48)
F		185 (56.39)			150 (45.72)		80 (24.38)
G							

Exhibit 10.2.29 Travel Chart for the Initial Layout of the Example Problem

	A	B	C	D	E	F	G	Row Totals
A		4725	1950	3500	2600	1200		13,975
B				2550	5865	2775		11,200
C					650	2500		3,150
D		1700			5250			6,950
E						9750	7000	16,750
F		925			3750		5200	9,875
G								0

Volume-distance product = 61,900

the volume-distance product, at which point the layout having the lowest volume-distance product would be selected. An evaluation of the travel chart for the example problem (Exhibit 10.2.29) results in the decision to alter the layout given in Exhibit 10.2.27 by moving the centroid of department E closer to the centroids of departments F and G. It is anticipated that the altered layout will reduce the largest travel chart elements E to F and F to G. This iteration and others proceed in a manner similar to the initial iteration and therefore are not described here.

Travel charting is a good technique for evaluating alternative layouts and determining methods of improving a layout. The assumption that all flows originate and terminate at department centroids is limiting and must be continually considered in order to ensure the development of realistic layouts. Travel charting depends upon the designer's ingenuity and can be extremely tedious for a realistically sized facility.

Relationship Diagramming

Relationship diagramming is no more than an organized approach to working manually with various layouts in an attempt to maximize the closeness relationship requirements as specified on a relationship chart. Relationship diagramming consists of two phases. The first phase determines the relative location of departments, and the second phase develops the actual layout.

Phase One. Determining the relative location of departments is performed without regard to the departmental areas. All departments are represented by block templates having the same shape and size. On each of these templates are recorded the department name, the department code, and the relationships with all other departments. Based on the relationship chart in Exhibit 10.2.24, the block templates for the example problem may be developed as shown in Exhibit 10.2.30.

The process begins with the selection of the template having the greatest number of "A" relationships. If two or more templates have the greatest number of "A" relationships, they should be subjected to the following tie breaking hierarchy: the greatest number of "E" relationships; the

Exhibit 10.2.30 Block Templates for the Example Problem

A– X–
A
Receiving
E–B I–D
O–C, E U–F, G

A– X–
B
Milling
E–A, D I–E, F
O– U–C, G

A– X–
C
Press
I–B, D,
E– U–
O–A, F E, G

A– X–
D
Screw machine
E–B I–A, E
O– U–C, F, G

A–F X–
E
Assembly
E– I–B, D, G
O–A U–C

A–E X–
F
Plating
E–G O–B
O–C U–A, D

A– X–
G
Shipping
I–E
E–F U–A, B
O– C, D

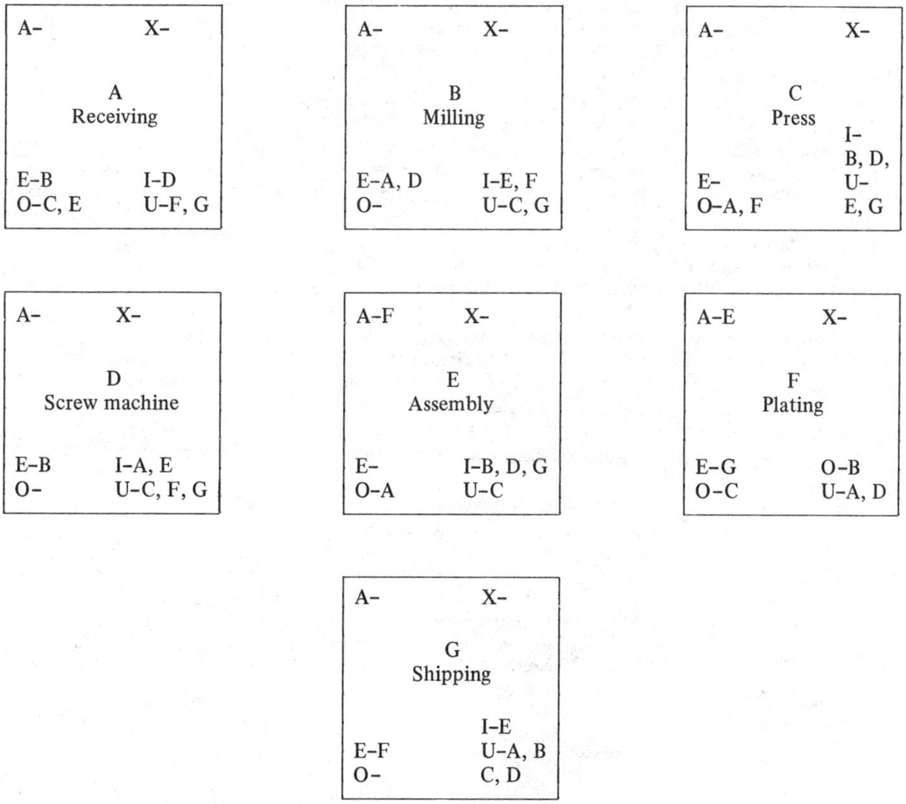

Exhibit 10.2.31 Relative Location of Block Templates for the Example Problem

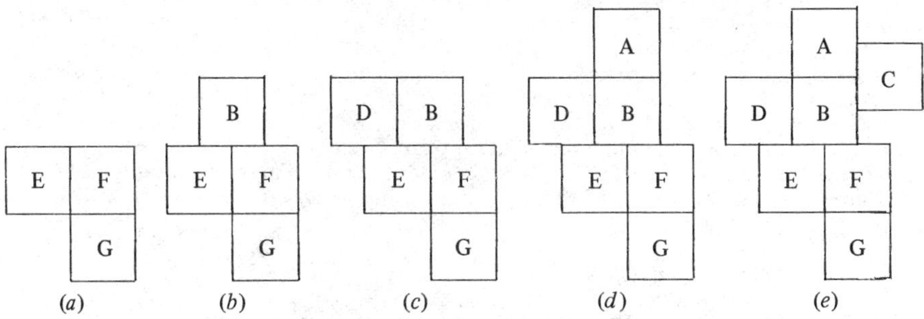

greatest number of "I" relationships; the fewest number of "X" relationships; and, finally, the random selection of one of the remaining templates. The selected template should be placed in the center of the layout. For the example problem, the templates with the greatest number of "A" relationships are templates E and F. Since template F has more "E" relationships, it is selected and placed in the center of the layout.

The next template to enter the layout should have an "A" relationship with the template already selected and the greatest number of other "A" relationships. If a tie exists, the tie breaking hierarchy should be utilized. The template selected second should be placed next to the first template. For the example problem, only template E has an "A" relationship with the selected template F; therefore it is selected and placed next to template F.

The next template selected should have the highest combined relationship with the templates already selected, where the highest possible combined relationship would be an "A" relationship with each template already selected. Next highest would be all "A's," except for one "E" relationship. The combined relationship hierarchy continues in this manner until either a template is selected or a tie exists, at which point the tie breaking hierarchy should be applied. The selected template should be placed as close as possible to the templates with which it has the closest relationship. For the example problem, the template having the highest relationship with templates E and F is template G, which has an "I" and "E" relationship, respectively. Since template G and F have an "E" relationship, and G and E have only an "I" relationship, when G is brought into the layout, it is positioned as shown in Exhibit 10.2.31*a*.

The next template selected would be the one whose combined relationship was the highest with those already selected. The procedure continues in this manner until all templates are included in the layout. Template B in the example problem has relationships "I," "I," and "U" with selected templates E, F, and G, respectively; therefore it is next to enter the layout. Exhibit 10.2.31*b* shows the positioning of template B as a result of the "I" relationships with E and F. Template B is followed by template D and positioned as shown in Exhibit 10.2.31*c*. Template D is followed by template A, leaving template C to enter last. Exhibits 10.2.31*d* and 10.2.31*e* show the positioning of templates A and C, respectively.

Phase Two. The second phase of relationship diagramming takes into consideration departmental areas and uses the unit area templates. The first step is to select a unit area that can be approximately divided into the department areas an integer number of times. The division results in the number of unit area templates required for each department. The templates should be labeled with the proper department codes and cut out. Utilizing the templates, the final relative location of block templates, and some common sense, a final layout may be developed. A quick look at

Exhibit 10.2.32 Conversion From Department Areas to Number of Unit Area Templates for the Example Problem

Code	Department	Number of Unit Area Templates
A	Receiving	6
B	Milling	4
C	Press	3
D	Screw machine	6
E	Assembly	4
F	Plating	6
G	Shipping	6

Exhibit 10.2.33 Final Layout for the Example Problem Using the Relationship Diagramming Technique

	D	A	A
D	D	A	A
D	D	A	A
D	C	C	C
F	F	B	B
F	F	B	B
F	F	G	G
E	E	G	G
E	E	G	G

Exhibit 10.2.22 indicates that the appropriate unit area for the example problem is 2000 ft^2 (185.81 m^2). The number of unit area templates resulting from 2000 ft^2 (185.81 m^2) unit squares is as shown in Exhibit 10.2.32.

The final layout may then be determined by attempting to mold the unit templates into the general configuration established with the block templates while still considering the practical constraints on the facility. One attempt at a final layout is as shown in Exhibit 10.2.33. Because of the subjective nature in which block and unit templates are located, several block template relative location layouts and final layouts should be developed. A rating of the layouts is possible by assigning a point value for "A," "E," "I," "O," "U," and "X" relationships and giving credit for each of these relationships when departments are adjacent. For example, using the point values given in Exhibit 10.2.34, the rating of the layout shown in Exhibit 10.2.33 may be calculated as shown in Exhibit 10.2.35.

Summary. Relationship diagramming is an organized approach to converting a relationship chart and departmental area requirements into a layout. However, it does not provide significant aid to the facilities planner while arranging the departments; therefore the quality of the solutions depends upon the designer's ingenuity and persistence. For large problems, the effort required to generate each alternative detracts from the usefulness of the technique.

Computer-Aided Layout Techniques

Computer-aided layout techniques may be classified by the method of recording flows among departments and by the method of generating layouts. Flows among departments may be recorded

Exhibit 10.2.34 Point Values for the Relationships Given in a Relationship Chart

Relationship	Point Value
A	8
E	4
I	2
O	1
U	0
X	–8

Exhibit 10.2.35 Rating of the Final Layout for the Example Problem Given in Exhibit 10.2.33

Adjacent Departments	Relationship	Point Value
D–A	I	2
D–C	U	0
D–F	U	0
F–C	O	1
F–B	I	2
F–G	E	4
F–E	A	8
A–C	O	1
C–B	U	0
B–G	U	0
		18

either quantitatively on a from-to chart or qualitatively on a relationship chart. Of the two routines described here, one–Computerized Relative Allocation of Facilities Technique (CRAFT)–requires quantitative flow inputs, and the other–Computerized Relationship Layout Planning (CORELAP) –requires qualitative flow inputs.

Layouts may be generated by improving upon an existing layout (an improvement routine) or by constructing the layout "from scratch" (a construction routine). The CRAFT routine is an improvement routine, whereas CORELAP is a construction routine.

The discussion of these computer-aided layout techniques will be brief. The user's manual for each of these routines should be consulted, or, if unavailable, Tompkins and Moore[3] should be consulted for details and instructions on how to obtain the routines. In Addition, Tompkins and Moore should be consulted for three other routines: Computerized Facilities Design (COFAD),[4,5] Computerized Plant Layout Analysis and Evaluation Technique (PLANET),[6] and Automated Layout Design Program (ALDEP).[7]

CRAFT

The CRAFT[8,9] is a heuristic routine whose objective is to develop a layout that will approach the minimal transportation cost, where "transportation cost" is defined as the product of the volume-distance product and the initial input move costs. This routine requires move costs to be input in the form of a cost per unit distance. To input move costs in this manner requires the acceptance of the following assumptions:

1. Move costs are independent of the utilization of the equipment.
2. Move costs are linearly related to the length of the move.

In many situations these assumptions cannot be justified, and therefore move costs are assigned a value of unity. By negating move costs in this manner, the objective of CRAFT reverts to the objective of minimizing the volume-distance product. For this reason, CRAFT may be referred to as "computerized travel charting."[10]

The CRAFT procedure begins by determining the centroids of the departments in the initial layout. It then calculates the rectilinear distance between department centroids and stores these distances in a distance matrix. The transportation cost (volume-distance product) for the initial layout is determined by calculating the product of the from-to chart, move cost matrix, and distance matrix.

The procedure next considers departmental interchanges for departments that are of equal area or that have a common border in an effort to reduce the transportation cost. The following types of interchanges can be considered:

1. Pairwise interchanges.
2. Three-way interchanges.
3. Pairwise followed by three-way interchanges.
4. Three-way followed by pairwise interchanges.
5. The best of pairwise or three-way interchanges.

The transportation cost is approximated for each departmental interchange. The departmental interchange offering the greatest reduction in the transportation cost is added to the layout. The

CRAFT continues by considering departmental interchanges to the new layout, approximating the transportation costs for these interchanges, and once again making the interchange that offers the greatest estimated reduction in the transportation cost. The procedure continues until no interchanges in the layout can be found that reduce the transportation cost; the model is then terminated.

A user of CRAFT should be aware of the path orientation of the routine as well as the use of "dummy" departments. Since CRAFT develops a final layout by following a path of improvement from the initial layout through several iterations to the final layout, it follows that the final layout depends upon the initial layout. Several different initial layouts should be utilized to overcome any path orientation problems.

"Dummy" departments are departments having no flows with other departments, but encompassing a specific area. Dummy departments may be used to serve one of the following purposes:

1. To fill building irregularities.
2. To represent fixed areas in the facility where departments may not be located, that is, stairways, elevators, plant services, rest rooms, and so on.
3. To aid in evaluating aisle locations in the final layout.

The first two uses of dummy departments follow from the CRAFT requirement that all facilities be rectangular or square and not have interior void spaces. The third use of dummy departments allows the facilities planner to *use* CRAFT to evaluate practically usable layouts. By including in the initial layout several dummy departments of various areas that are fixed to an exterior wall, these dummies may be used in later runs to represent and therefore determine the impact of various aisle locations. Rarely will the CRAFT-generated department configuration allow for the inclusion of straight, noninterrupted aisles as is desired in the final layout. By altering the department configurations, fixing departments to specific areas, and placing the dummies in the layout to represent the aisles, CRAFT may be used to calculate the effects of molding the computer-generated layout into a practical layout.

CORELAP

Relationship diagramming, described previously, is a procedure that converts relationship chart information into a layout. For problems involving 10 departments, only 45 pairs of relationships exist, and thus relationship diagramming may be effectively utilized. For problems involving 45 departments, however, there are approximately 1000 pairs of relationships, and relationship diagramming becomes extremely difficult to use. The CORELAP procedure may be used to address the same types of problems as relationship diagramming and easily handles the larger problems.[11]

The basic inputs required by CORELAP are the same as those required by relationship diagramming. The CORELAP procedure begins the process of constructing a layout for a facility by calculating the total closeness rating (TCR) for each department, where the TCR is the sum of the numerical values assigned to the closeness relationships ("A" = 6, "E" = 5, "I" = 4, "O" = 3, "U" = 2, "X" = 1) between a department and all other departments. The department having the highest TCR is then placed in the center of the layout. If there is a tie for the highest TCR, the following tie breaking rule is applied: the department having the largest area and then the department having the lowest department number.

Next, the relationship chart is scanned, and if a department is found that has an "A" relationship with the selected department, it is brought into the layout. If none exists, the relationship chart is scanned for an "E" relationship, then an "I", and so on. If two or more departments are found to have the same relationship with the selected department, the department having the highest TCR is selected; if a tie still exists, the tie breaking rule is utilized.

The third department to enter the layout is determined by scanning the relationship chart to see if an unassigned department exists that has an "A" relationship with the first department selected. If so, this department is brought into the layout. If a tie exists, the TCR and then the tie breaking rule are utilized. If no unassigned department exists that has an "A" relationship with the second department, this procedure is repeated considering "E" relationships, then "I" relationships, and so on. If a tie occurs, the TCR and then the tie breaking hierarchy are utilized. The same procedure is repeated for the fourth department to enter the layout, except that the three departments previously selected are included in the search. The procedure continues until all departments have been selected to enter the layout.

Once a department is selected to enter the layout, a decision must be made as to where to place the department. This decision is made by calculating the placing rating for the available locations of the department, where the placing rating is the sum of the weighted closeness ratings between the department to enter the layout and its neighbors. For example, consider the layout consisting of departments 1 and 7 in Exhibit 10.2.36*a*. If department 2 is next to enter the layout, if it has an

Exhibit 10.2.36 Illustration of CORELAP Placing Rating

1	1	1				
1	1	1	7	7	7	
		(*a*)				
	2	2				
1	1	1				
1	1	1	7	7	7	Placing rating = 64
		(*b*)				
1	1	1		2	2	
1	1	1	7	7	7	Placing rating = 16
		(*c*)				
1	1	1	2	2		
1	1	1	7	7	7	Placing rating = 80
		(*d*)				

"A" relationship with department 1 and an "E" relationship with department 7, and if an "A" relationship is weighted 64 and an "E" 16, the placing ratings are as shown in Exhibit 10.2.36*b* through *d*.

If a tie occurs for the placing rating, the boundary lengths of the tied locations are compared. The boundary length is the number of unit square sides that the department to enter the layout has in common with its neighbors. In Exhibit 10.2.36*b* and *c* the boundary lengths are 2, and in Exhibit 10.2.36*d* the boundary length is 3. By placing the departments in the layout, CORELAP develops layouts by growing like a crystal out from the center.

After the final CORELAP layout has been prepared, CORELAP evaluates it by calculating the layout score, which is defined as

$$\text{layout score} = \sum_{\text{all departments}} \text{numerical closeness rating} \times \text{length of shortest path}$$

It should be noted that CORELAP utilizes the shortest rectilinear path between departments as opposed to the distance between department centroids as in CRAFT. The shortest rectilinear path is used because it is assumed that each department will have a dispatch area and a receiving area on that side of its layout nearest its neighbor.

The problems associated with attempting to use CORELAP-generated layouts are similar to those associated with several construction-type computer-aided layout techniques. The CORELAP procedure does not restrict the final layout to a uniform building shape, nor does it have the capability of fixing departments to certain locations. The CORELAP-generated layouts therefore often take on unrealistic shapes, and little may be done to use the routine to evaluate the effects of manually adjusting the layout. Therefore CORELAP must be utilized primarily as an initial layout generator and not as a final solution. In addition, case must also be taken in interpreting the layout scores, since the shortest rectilinear path between departments may not always be a realistic measure.

10.2.8 EVALUATING ALTERNATIVE LAYOUTS

As stated in section 10.2.1, plant layout is the field of selecting the most effective arrangement of physical facilities to allow the greatest efficiency in the combination of resources to produce a product or service. The criterion utilized to evaluate alternative layouts is the flow of materials, individuals, or information. To be used as a criterion for evaluation, flow must be measurable; indeed, Section 10.2.6 presented methods of measuring departmental flows both quantitatively and qualitatively. The graphic and computer-aided layout techniques discussed in Section 10.2.7 use these quantitative and qualitative measures of flow to generate alternative overall facility layouts, and each routine has an associated layout rating technique that can be used to compare the alternative layouts.

In addition to considering the layout ratings generated by the graphic and computer-aided layout techniques, the alternative facility layouts should be evaluated by how well the flow within each layout would conform to several basic flow principles. Flow principles are rules that, when properly applied, will typically result in effective flow. The following are flow principles:

1. Maximize the use of directed flow paths.
2. Minimize flow.
3. Minimize the costs of flow.

A directed flow path is an uninterrupted flow path progressing directly from origin to destination. An uninterrupted flow path is one that does not intersect with other paths. A flow path progressing directly from origin to destination is a flow path with no backtracking. Backtracking significantly increases the length of a flow path.

The second flow principle of minimizing flow represents the work simplification approach to material flow:

1. **Eliminate Flow.** Plan for the delivery of materials, information, or people directly to the point of ultimate use and eliminate intermediate steps.
2. **Minimize Multiple Handlings.** Plan for the flow between two consecutive points of use to take place in as few movements as possible, preferably one.
3. **Combine Operations Whenever Possible.** Plan for the movement of materials, information, or people to be combined with a processing step.

The third flow principle of minimizing the costs of flow may be viewed from either of the following two perspectives:

1. **Minimize Manual Handling.** Minimize walking, manual travel distances, and motions.
2. **Eliminate Manual Handling.** Mechanize or automate flow to allow workers to spend full time on their assigned tasks.

For each situation both perspectives should be applied, and the one resulting in the least total cost should be selected.

Alternative facility layouts that do not allow, or that in fact encourage, the application of these basic flow principles should be revised or eliminated from further consideration. The layout that scores well on the measurable quantitative and qualitative criteria and that allows effective application of the basic flow principles should be selected as the best facility layout.

10.2.9 FINALIZING AND IMPLEMENTING THE LAYOUT

Layout Plans

Once the overall facility layout has been selected, the details of the layout must be recorded in layout plans. Layout plans may be two- or three-dimensional representations.

Two-dimensional representations include handmade drawings, template and tape layouts, and computer graphics layouts. Handmade drawings may be the best approach for layout plans of small areas; however, they are much too expensive to produce and alter to be used for final layout plans for large areas. The most common method of creating layout plans for large facilities is to use templates and tapes. Templates may be homemade or purchased and can be either block or contour templates. A block template is simply a labeled rectangle representing the maximum length and width of equipment. A contour template illustrates the contour and the clearances for the movable portions of the machine. A computer graphics layout is generated using an input terminal, a central processor, and output peripherals. The input terminal typically will consist of a keyboard, a light pen, and a CRT. The keyboard and light pen are used by the layout planner to provide instructions to the computer graphics system. The CRT is used to display graphically the layout being developed. A layout is developed by calling for specific templates and symbols from the central processor and then, with the use of the light pen, positioning the templates and symbols to obtain a layout. Once the layout is complete, it may be reproduced by any one of many output peripherals.

Three-dimensional models are the clearest, most easily understood method of developing layout plans. However, because of the difficulty of duplicating a three-dimensional model, a two-dimensional model is still required. Therefore the costs of three-dimensional models are often prohibitive. The two types of three-dimensional models are:

1. **Modular Block Models.** These models consist of modular building blocks used to represent equipment. The cost of modular block models is reduced since special models of each machine are not required.
2. **Scale Models.** Scale models consist of special models representing each piece of equipment and, like two-dimensional templates, may be either block or contour models.

No matter what type of representation is utilized, the same procedure should be followed to create the layout plan. The systematic procedure for developing a manufacturing facility layout plan is as follows:

1. Select the scale. If possible, the same scale should be used as is being used by the architect, construction engineer, or other professionals working on the facility plan.

2. Decide on the method of representation. In general, the selection of the method of representation should be based upon a combination of clarity and economics.
3. Obtain layout plan supplies and/or equipment.
4. For an existing facility, locate all permanent facilities on the layout plan. All columns, windows, doors, walls, ramps, stairs, elevators, sewers, cranes, and other permanent facilities should be initially located on the layout plan.
5. Locate the exterior wall that includes the receiving function.
6. For nonexisting facilities, locate all columns. The size, span, and location of columns must be among the initial decisions for a new facility layout.
7. Locate all manufacturing departments and equipment. Beginning with the receiving department, each department should be tentatively located in the layout plan in accordance with the department layout.
8. Locate all personnel and plant services. Alterations should be made to the manufacturing layout to include all personnel and plant services.
9. Audit the layout plan. The layout plan should be reviewed from both a material flow and a personnel perspective. The material flow audit should involve tracing all materials through the facility. The personnel audit should involve the mental tracing of the tasks to be performed by every person to be employed within the facility.
10. Finalize the layout plan. Permanently locate everything on the layout plan and record any appropriate headings, clarifying notations, and a detailed legend on the plan.

Selling the Layout

If the finalized layout is not properly sold, the probability of acceptance is very low, no matter how high the quality of the layout. The three components required to sell a successful layout are (1) A high-quality layout plan, (2) a written report, and (3) an oral report. The previous sections of this chapter have been concerned with developing a high-quality layout plan. The other two components are discussed now.

The objective of the written report is to describe the benefits of the layout and document why the particular layout selected is the best layout, not to describe the procedures required to establish the layout. The process that should be followed to produce a quality layout planning written report includes the following steps:

1. Define the objective of the report. The objective should be written and should state what conclusions should be reached by anyone reading the report. These objectives should be frequently referred to while writing the report.
2. Establish a table of contents. The table of contents serves as an outline of a written report. It forces one to organize the material to be placed within the report and to verify that a logical train of thought evolves from the introduction to the conclusions.
3. Determine who is going to read the report. The style, level of detail, and length of a report will be dictated by the person(s) for whom the report is written. Clearly, a report written for the chief facilities planner would differ from that written for a chief executive officer.
4. Write the report. The report should be brief, but should accurately communicate the reasons for implementing the facilities plan. It should be assumed that the reader will believe the results presented and that therefore he or she should not be forced to read through detailed justifications or explanations. These details should be placed in tables or figures, and full documentation of these tables and figures should be placed in appendixes.
5. Document the report. Prepare appendixes that define the sources of data and the assumptions included in the report. Include within the appendixes data and procedural details describing exactly how the conclusions reached in the report were determined. Provide sufficient information so that the reader can recreate the conclusions reached in the body of the report.
6. Edit the report. Prior to finalizing, allow a few days to pass and then critically review the report. Focus your attention on the objective of the report and on brevity and clarity.

The written report should be given to all persons who will be attending the oral presentation at least 2 days prior to the presentation. After distributing the report, it may be wise to check with a few key persons who will attend the presentation and to learn their reactions. These reactions may be very helpful in preparing the oral presentation.

The first step in preparing an oral presentation is to determine to whom the presentation will be made. The perspective of each person who will hear the presentation should be analyzed, and it should be determined what each person expects to hear. Fully understanding the members of the audience and viewing the presentation from their perspective will allow the presenter to predict what questions will be asked and what benefits to emphasize.

Once such an understanding is obtained, a decision needs to be made concerning what should be presented. The two critical factors that should be kept in mind while deciding what to present are

(1) brevity and (2) progressing from what to why to how. The oral presentation should be brief. For most projects a 1 hr planned presentation is all that is necessary. If it may be done effectively in 15 min, then plan for only a 15 min presentation.

Oral presentations should begin by stating *what* should be done, then describing *why* these things should be done, and then describing *how* to do the things recommended. One should not attempt in the oral presentation to train the attendee in how to establish a layout plan. To the contrary, an oral presentation should dwell on what plans should be implemented, why these plans should be implemented, and how these plans should be implementea. Helpful guidelines to preparing and presenting an oral presentation include the following:

1. Be certain not to oversell. Be realistic and make certain that all statements may be proved.
2. Be certain to refer to other systems similar to the one being recommended. Be certain the audience knows that what is being recommended has been successful elsewhere.
3. Just prior to concluding the oral presentation, describe the implementation plan. Explain how to accomplish what is being recommended.
4. Conclude the presentation with a tabulation of the results. Show cost savings, improvements in space utilization, and so on.
5. Practice the oral presentation. This breeds self-confidence and makes the presentation easier to accept.
6. Prepare quality visuals. Do not try to place too much information on any one visual. Do not use reproductions from the report.
7. Be appropriately dressed for the presentation. Look successful and your presentation will be much more believable.
8. Begin the presentation on time. Have visual aid equipment ready and focused.
9. Address all objections and questions. Be aware and communicate to the audience that objections and questions are good to get out in the open. Listen carefully and allow all objections and questions to be fully voiced. Be open-minded.
10. If you do not know the answer to a question, respond, "I do not know, but I will find out."

Implementing the Layout

The facility layout, once approved, will be given to a plant engineering group or a contractor to implement. The layout planner should work with whomever is responsible for installing the layout so that if and when alterations are required, they may be made while considering the overall effects of such a change. Often what may appear to be a slight change from an installation point of view may significantly affect the operation of the facility. Once the layout is implemented, the layout planner should follow up with the people responsible for the operation of the facility on procedures, methods, and utilization of the layout as designed.

10.2.10 CONCLUSIONS

The layout planning process is never completed; it is a continuous process. The information developed to construct a layout and the operation of a facility should be continually reviewed and updated as conditions change within the facility. Plant layout is an art as much as it is a science. The ingenuity, persistence, and experience of the layout planner will have a tremendous bearing upon the quality of the resulting layout. This chapter has described some of the available techniques and systematic procedures that can guide the layout planner to an effective plant layout. Sources that provide a more in-depth treatment of the material presented in this chapter are given in the reference list.[2,12–16]

REFERENCES

1. U.S. BUREAU OF THE CENSUS, *Statistical Abstract of the United States*, 97th ed., Washington, DC, 1976.
2. R. MUTHER, *Systematic Layout Planning,* Cahners, Boston, MA, 1976.
3. J. A. TOMPKINS and J. M. MOORE, *Computer Aided Layout: A User's Guide,* AIIE, Norcross, GA, 1977.
4. J. A. TOMPKINS and R. REED, JR., "An Applied Model for the Facilities Design Problem," *International Journal of Production Research,* September 1976, pp. 583–595.
5. J. A. TOMPKINS and R. REED, "Computerized Facilities Design," *Technical Papers 1973,* AIIE, Norcross, GA, 1973.
6. M. P. DEISENROTH and J. M. APPLE, "A Computerized Plant Layout Analysis and Evaluation Technique (PLANET)," *Technical Papers 1962*, AIIE, Norcross, GA, 1972.

7. J. M. SEEHOF and W. O. EVANS, "Automated Layout Design Program," *Journal of Industrial Engineering,* May–June 1958, pp. 13–15.

8. G. C. ARMOUR, E. S. BUFFA, and T. E. VOLLMAN, "Allocating Facilities with CRAFT," *Harvard Business Review*, March–April 1964, pp. 136–158.

9. G. C. ARMOUR and E. S. BUFFA, "A Heuristic Algorithm and Simulation Approach to Relative Location of Facilities," *Management Science*, January 1963, pp. 294–309.

10. J. A. TOMPKINS, *Facilities Design*, North Carolina State University, Raleigh, 1975.

11. R. C. LEE and J. M. MOORE, "CORELAP–Computerized Relationship Layout Planning," *Journal of Industrial Engineering*, March 1967, pp. 195–200.

12. J. A. TOMPKINS and J. A. WHITE, *Facilities Planning*, Wiley, New York, 1982.

13. R. L. FRANCIS and J. A. WHITE, *Facility Layout and Location: An Analytical Approach*, Prentice-Hall, Englewood Cliffs, NJ, 1974.

14. J. M. MOORE, *Plant Layout and Design*, Macmillan, New York, 1962.

15. R. REED, JR., *Plant Layout: Factors, Principles and Techniques*, Irwin, Homewood, IL, 1961.

16. J. M. APPLE, *Plant Layout and Material Handling,* Ronald, New York, 1977.

CHAPTER 10.3

Materials Handling Systems

E. RALPH SIMS, JR.

E. Ralph Sims, Jr. & Associates, Incorporated

10.3.1 DEFINITIONS IN MATERIALS HANDLING

"Materials handling" is that portion of the business and economic system which affects the physical relationship of materials, products, and packaging to the product, process, facility, geography, or customer without adding usable worth or changing the nature of the products. From an engineering point of view, materials handling is defined as the art and science involved in the moving, packaging, and storing of substances in any form. In its broadest definition, materials handling includes the movement of liquids, bulk solids, pieces, packages, unit loads, bulk containers, vehicles, and vessels. Because of this wide range of coverage, most authorities limit the definition of materials handling to bulk solids, slurries, pieces, packages, unit loads, and bulk containers.

A "materials handling system" is defined as a series of related equipment elements or devices designed to work in concert or sequence in the movement, storage, and control of materials in a process or logistics activity. Each system must be custom designed to serve in a specific operating environment and with designated materials. The nature of the materials handling system and equipment is determined by the characteristics of the product and the type of movement. The system concept can be applied to the design of the workplace, the manufacturing or processing operation, a department or entire plant, or the logistics functions of a whole industry. The basic principles apply at all levels; however, the optimum economy of a materials handling system is based on the concept that *the best handling is no handling at all.*

10.3.2 PRINCIPLES IN SYSTEM DEVELOPMENT

To optimize the materials handling function, one must first define the objective of the movements and apply the fundamental principles to the simplification and elimination of moves. A primary principle in this approach states that one must move the largest possible volume, quantity, or unit of material to the next point of use before releasing it or breaking it down into smaller units.

Some of the commonsense rules of the game can be stated as follows:

Handling costs money and adds no product value.

The best handling is none at all.

Materials in transit should move as close as possible to the next point of use before being halted.

Straighten and shorten moves whenever possible.

Pre-position for the next operation whenever possible before depositing the materials being handled.

Combine or eliminate moves and handling operations whenever possible.

Consider moving workers rather than materials.

Mechanical handling should replace manual labor where practicable.

Use "air rights," or overhead space, whenever possible. Integrate handling and material control systems, but do not let administrative or control procedures dominate material flow patterns.

Use document systems to preclude moves, preassemble orders, and sequence operations.

Remember that material flow is cash flow, that stopped or stored materials are inventory, and that it costs money to keep inventory stopped.

10.3.3 EQUIPMENT CLASSIFICATION

In general, materials handling equipment is divided into (1) package or unit handling and (2) bulk handling. There is a great deal of overlap between these two general classifications. Bulk material is often handled in bags, sacks, barrels, or other containers, and in some cases, package or piece-type items (loose materials, castings, machine parts, etc.) are handled over belt conveyors or in tubs or hoppers in the same manner as bulk materials.

Materials handling equipment can also be classified on the basis of the nature of the move to be made. The movement classifications may be:

Fixed path versus flexible route
Intermittent versus continuous
Long distance versus short haul
Indoors versus outdoors
Vertical versus horizontal

Various authorities have classified materials handling equipment in many different ways. For the purpose of analyzing production materials handling equipment, which is in itself a major segment of the materials handling field, the following general breakdown is often useful:

Fixed Path Handling Equipment. This group includes conveyor equipment of all types, monorail and railroad systems, elevators, skip hoists, piping, duct systems, and other permanently installed materials handling devices.

Limited Area Handling Equipment. Included in this group are bridge and jib cranes, cable and boom systems, gantry cranes, and various materials handling devices that are flexible within a permanently restricted operating area.

Mobile Materials Handling Equipment. This group includes forklift trucks, skid trucks, tractors, and trailers, pedestrian power trucks, and other industrial vehicles designed for indoor use. Yard vehicles, including cranes, straddle carriers, side loaders, power shovels, front-end loaders, bulldozers, dump trucks, highway trucks, and other outdoor vehicles are also part of this group.

Materials Handling Tools and Storage Equipment. This group includes hand trucks, hand jacks, casters, dollies, rollers, chain hoists, power pullers, dock plates, pallets, skids, scales, racks, shelves, bins, and so on.

Classification methods are of value to the industrial engineer during the selection of handling equipment because they simplify the operation of reviewing the market and determining the requirements of the project. After defining the basic problem, the engineer must establish a general range of equipment that can aid him or her in developing a solution. The use of classification systems can aid in the statement of requirements to vendors and purchasing agents.

10.3.4 FACTORS INVOLVED IN EQUIPMENT SELECTION

Many factors are involved in the selection of materials handling equipment. They include the proper application or classification of equipment, the reliability of the product selected, economics, management, financing, labor relations, safety, plant characteristics and environment, and many others.

It has been found that the cost of production usually varies in direct proportion to the indirect cost of manufacturing. A large portion of these indirect costs are included in the cost of materials handling. Because of this, selection of the correct materials handling system must be based upon an economic analysis comparing one type of system to another. The selection of mechanical handling equipment for industrial or distribution facilities is usually based upon the ability of the equipment to:

Reduce handling costs.
Shorten work cycles.
Reduce inventory requirements.
Expedite shipments or deliveries.
Improve space utilization.
Simplify flow and increase operating efficiency.
Reduce damage and waste.
Improve safety.

Ultimately, the economics of the situation controls the choice of equipment. Handling equipment applications and costs can be compared in terms of man-hours required to move materials, return on capital investment, direct operating expenses, and/or the indirect system effects of the proposed method. Techniques for calculating ROI and project economics are discussed in Section 9 on engineering economy.

The handling operations may be repeated frequently and with many variations. Handling may be integrated in the fabrication, assembly, picking, packing, or storage operations. In every instance, although the magnitude of the move or the volume of material may vary, materials handling consists of only three basic operations: picking up, transporting, and depositing.

It is inconceivable that a materials handling engineer would select an expensive type of machinery that would not appreciably reduce the cost of operation or fit within the environment of the facility. He or she would check floor loadings, door dimensions, clear ceiling heights, structural strengths, OSHA and fire regulations, environmental conditions, fume problems, traffic safety, power supply, and all of the many engineering characteristics that are factors in the selection of the particular equipment item. Thus, when selecting equipment for existing facilities, many installation characteristics must be considered before a particular piece of equipment is chosen for the job. Quite frequently, several methods or types of machinery can perform the required work, and the facility consideration is the final determining factor in choosing between the various methods.

If a list of selection factors were made, the following might be included:

Applicability of the equipment to the solution of the problem.
Reliability of the equipment.
Adaptability of the equipment to the operating environment.
Capital investment.
Cost recovery based upon savings.
Safety hazards.
Flexibility of equipment if the operation is changed.
Worker acceptance.
Supervisor acceptance.
Complexity of training.
Maintenance requirements.
Fuel or power supply.
Availability of spare parts and service.
Application materials requirements.

Many other factors might also be applied in special cases. These could include atmospheric conditions, explosion risk, and so on.

10.3.5 RELIABILITY OF EQUIPMENT

Materials handling equipment, like any other manufactured product, exhibits varying degrees of ruggedness, quality, and design characteristics. Each manufacturer has a different set of criteria upon which to base the design and manufacturing standards. Some of the criteria that are applicable to reliability analysis might be as follows:

The use of reputable standard components (bearings, rollers, belting, structural members, etc.).
Sound connections, whether welded, bolted, riveted, or using special fasteners.
Adequately rated power equipment.
Adequately designed transmission equipment.
Suitable lubrication systems or sealed-in lubricants.
The backing of a good service organization.
The reputation of the vendor.

The principles of reliability engineering and related decision factors that are discussed in Chapter 8.5 under quality assurance also apply to materials handling equipment.

10.3.6 COMPARISON OF COMPETITIVE EQUIPMENT

Optimum cost does not necessarily mean least cost. If two items of materials handling equipment are comparable in design, are equal in their applicability to the problem, and have adequate service support from competent vendor service organizations, the selection is usually based upon price. In this case the less expensive item is usually purchased. If any of these factors is lacking, price should be secondary to the best interest of the using company.

Many engineers include all factors, such as maintenance costs, spare parts, anticipated length of service, depreciation, labor savings, and power, in their analysis. The more common approach is to take straight-line depreciation on the cost of equipment and its installation and to compare it to the direct labor savings resulting from the installation. In many cases an even simpler approach is used, wherein the capital investment for the equipment and installation is divided by the dollars saved each year through reduction of direct labor to develop the number of years to recover the investment. In this case the criterion of 3 years or less for capital recovery is often used.

When complex systems are being considered, it is sometimes difficult to define the net performance impact of the proposed concept. In such cases simulation techniques are often employed to project the behavior and compare the capabilities of the alternative proposals. Such mathematical tools as queuing, simulation, and linear programming are often applied in these studies. These techniques are discussed in Section 13 of this handbook.

A general statement of the relationship among design, application, service, and price might be restated as follows: When competitive equipment items develop similar economics from the point of view of operating savings, and their comparable design features are both suitable to the job application and supported by competent service organizations, the selection should be made on the basis of price. If any of these factors is lacking in one or the other of the products, price should become a secondary factor, and operating characteristics, serviceability, and life expectancy of the product should be the primary basis for selection.

Using The Transaction/Inventory Ratio Method of Equipment Comparison[1]

Many methods have been devised to assemble the four primary system cost elements of building, storage aids, labor, and equipment into comparison tables to permit the evaluation of alternative system configurations. The transaction/inventory (T/I) ratio method is a means of categorizing inventory and handling-related costs and comparing systems on the basis of the ratio of transactions to inventory. To perform this comparison, transactions and inventory must be expressed in the same units, such as pallets, boxes, or bins.

After all other factors have been accounted for in the comparison of equipment, the final decision must be based on the evaluation of the relative merits of performance and cost. The true cost of a materials handling system is composed of both inventory and handling-related costs. Inventory costs can be separated from handling-related (transaction) costs and are influenced by such factors as storage height, aisle size, required clearances, and the type of storage hardware (pallet racks, shelving, etc.). Environmental costs (security, refrigeration, air conditioning, etc.) are additional cost factors that are not directly system dependent, but that will be affected by the particular system selected and must be included in any evaluation.

Handling-related or transaction costs are influenced to some extent by inventory and system characteristics such as travel distance and the time required to access the highest level of storage, both of which are related to the equipment. It can be said that, in general, handling costs are composed primarily of operator labor and equipment amortization and operating expenses.

By using conventional cost analyses to define the differing costs for the transaction and inventory elements, it is possible to devise a cost equation that will express the annual cost associated with a given inventory level and transaction rate. By comparing the equations for two systems, it is possible in most cases to determine a level of transactions and inventory at which the annual cost for the two systems will be equal.

In making the comparison, a set of two equations with two unknowns is developed. If a value for either inventory or transaction is inserted, the required value for the variable can be determined. Without inserting specific values, an infinite number of transaction-inventory combinations, producing equality of the two annual costs, is possible.

By solving the equations for a ratio of the variables, a fixed ratio of transactions to inventory (T/I) can be found that will maintain annual cost equality. A ratio of transactions to inventory is used to produce a direct relationship between the factors since an increase or decrease in the transaction rate will produce a corresponding change in the T/I ratio. This approach is the preferred method since the inventory is generally the fixed quantity, and system performance can be evaluated under conditions of changing throughput. The absolute value of a T/I ratio is determined entirely by the choice of measure for transactions. Transactions may be measured on a daily, weekly, monthly, or yearly basis. The more important factors in using the T/I method can be summarized as follows:

Transactions and inventory must be expressed in the same units, such as pallets, cubic feet, or pounds. The preferred measure should be volume related (cubic feet) since the materials handling operation is concerned with load movement, not product value or density.

The transaction period may be expressed in a measure convenient to the user. This might be, for example, pallets per day or pallets per month. The inventory must be expressed in the same units, in this case, pallets.

The T/I ratio *must not* be confused with inventory turnover. Only under particular conditions will the T/I ratio actually equal the inventory turns. The T/I ratio is a measure of material movement relative to the size of the facility.

All T/I ratios used in a comparison must be computed on the same base. A T/I comparison cannot be made using different periods for the transactions, that is, daily versus weekly.

A sample calculation of T/I ratios is shown in Exhibit 10.3.1. In computing these ratios, three possibilities are evident:

1. The transaction or inventory costs for the two systems are equal, and system selection is determined by the remaining variable. No T/I ratio is possible.

2. Both the transaction and inventroy costs in one equation are less than those in the other equation. One system is always less costly than the other, and there is no possibility for equality and therefore no T/I ratio.

3. The transaction and inventory costs are such that equality is possible and a T/I ratio can be determined.

A graphic concept of the T/I ratio is shown in Exhibit 10.3.2. The system cost lines T_1I_1 and T_2I_2 represent a total system cost composed of inventory and transaction-related components for systems A and B, respectively. The lines represent the various combinations of transactions and inventory required to produce a range of costs. The range of transactions and inventory is selected to the values needed by both systems to produce equality. At only one combination of transactions and inventory (T_0 and I_0) will the lines cross (point X). This represents the point of equality at which the annual cost of system A equals the annual cost of system B. The slope of the line from the origin (point 0) through the intersection of the cost lines (point X) represents the T/I ratio or the relationship of T_0 to I_0. If the slope of this line is steeper (closer to the transaction axis), the ratio is increasing and is dominated more by transactions costs. If the slope of the line is less steep (closer to the inventory axis), the ratio is decreasing and is dominated more by inventory costs.

By applying this analysis to several systems with various heights, a ranking chart can be developed to show the most desirable system for a given inventory and transaction. By using this method, the less desirable systems can be eliminated early in the analysis, and effort can be placed on evaluating the remaining systems. The various T/I ratios provide checkpoints that indicate the relationship between transactions and inventory at which annual costs are equal and what effect changes in inventory or activity will have on the desirability of given systems.

Exhibit 10.3.1 Sample Calculation of T/I Ratios

Example: Assume the following cost equations for various systems operating at a given height:

system A $\$10.04I + \$63.54T = \text{annual cost } (A_1)$
system B $\$10.61I + \$65.17T = A_2$
system C $\$\ 9.58I + \$52.48T = A_3$
system D $\$\ 9.35I + \$64.83T = A_4$

There are six comparisons that can be made among these four systems:

A/B, A/C, A/D, B/C, B/D, C/D

By examination, three of these comparisons (A/B, A/C, B/C) reveal that there is no equality point since one system is always less costly than the one it is being compared to. Therefore there is no T/I ratio for these three situations and for all T/I ratios the following relationships apply:

system A less than system B
system C less than system A
system C less than system B

Comparison of systems A and D reveals

$$\$10.04I + \$63.54T = \$9.35I + \$65.83T$$
$$0.69I = 2.29T$$
$$0.30l = \text{T/I}$$

At a T/I ratio of 0.30*l*, the annual costs are equal.
For T/I greater than 0.30*l*, the sytem with the lower transaction cost is less expensive.
For T/I less than 0.30*l*, the system with the lower inventory cost is less expensive.

Similarly, for systems B and D, T/I = 1.91 and for systems C and D, T/I = 0.017.

Exhibit 10.3.2 Transaction/Inventory Ratio Comparison Concept

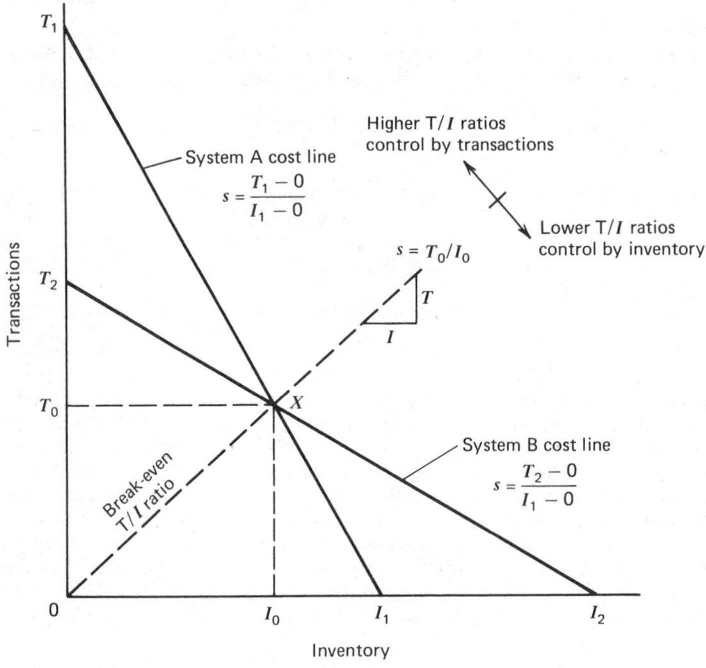

10.3.7 BULK MATERIALS HANDLING EQUIPMENT

By definition, bulk materials mean any loose, powdery, granular, or lumpy substance, such as wheat, sand, or coal. These materials may be transported, elevated, and stored by a variety of apparatus known as bulk materials handling equipment. Some typical classes of bulk materials handling equipment (Exhibit 10.3.3) are chutes, conveyors (belt, flight, apron, screw, etc.), bucket elevators, pneumatic pressure and vacuum systems, vibrating conveyors, hydraulic systems, hoists, and vehicular handling equipment.

Gravity chutes, screw conveyors, and belt conveyors are the most common bulk handling devices. The other types of bulk handling equipment are usually found in engineered systems, whereas the chutes, screws, and belts are also available as unit machines and devices in portable or "package" design form.

One of the critical factors in the application of bulk handling equipment is the flowability of the material being handled. Such factors as moisture content, fragility, angle of repose, stickiness, abrasiveness, and density are major elements in the design and selection of bulk handling equipment. It is essential to study the product characteristics carefully before attempting to specify the system components.

Another issue involves precision of control. In general, vibratory and screw conveyors provide the most precise metering capabilities and are normally found in such applications as weighing, mixing, and process feeding systems.

Belt conveyors are most often selected for long-run operations such as mining and power plant coal systems. Flight conveyor is favored for hot applications in foundries and in the handling of heavy, lumpy, and/or rough materials.

There are many types of bucket elevators, and each application must be designed for the specific situation. Different pickup, discharge, and bucket configurations are needed for each material and process combination.

Pneumatic systems are normally used with light and flowable materials such as grain, sawdust, and granular plastics. Abrasion and lumping are the critical product characteristics in the application of such equipment.

Air fluidization of dry materials to provide for free flow is another common means of moving light granular products. In some cases the same objective is achieved by making a water or oil slurry and using pumps and piping instead of conveyor.

Exhibit 10.3.3 Typical Bulk Conveyor Equipment

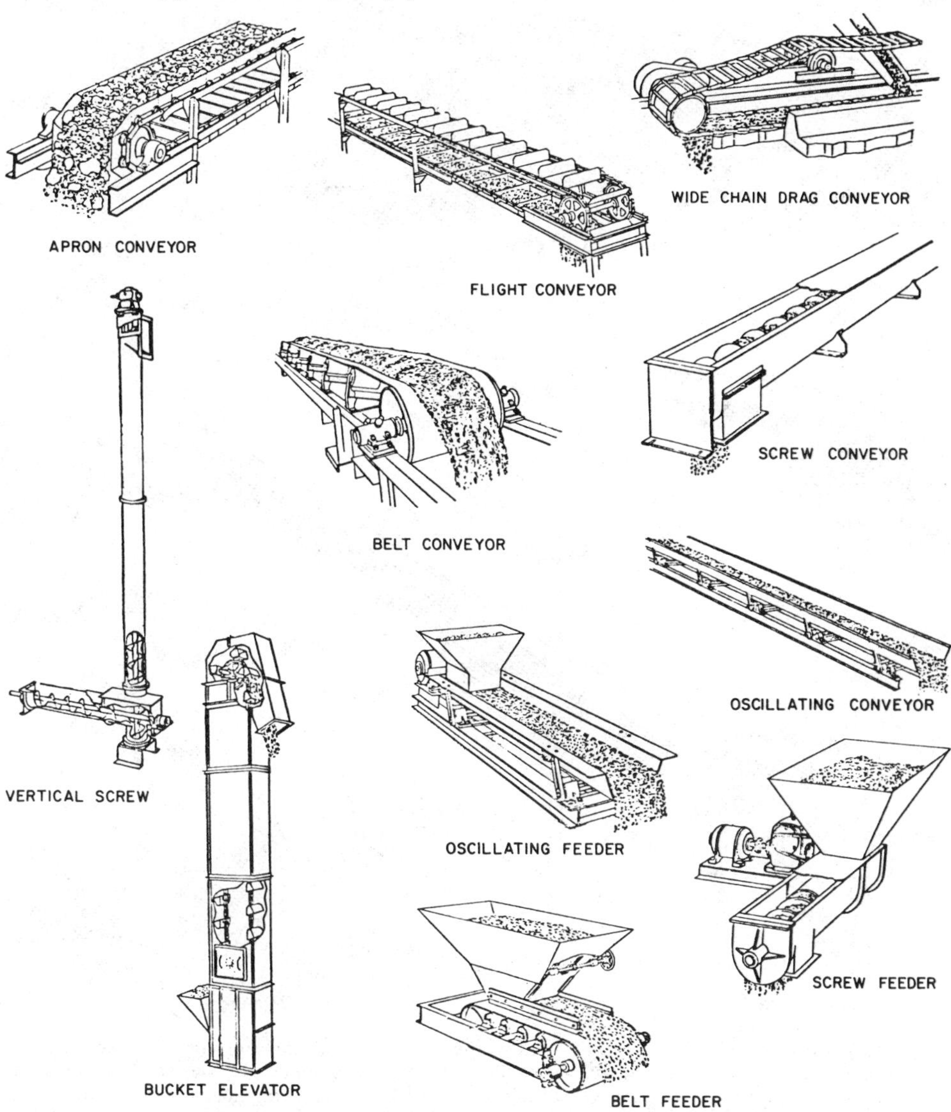

These are the more common types of bulk handling equipment. In view of the infinite variations of materials, applications, and equipment designs, it is essential to engineer in detail each application of bulk materials handling equipment.

10.3.8 FIXED PATH, PACKAGE, PARTS, OR UNIT HANDLING

The more typical applications of fixed materials handling equipment are found where transportation requirements exist for high-volume movement of packages and unit loads between two or more fixed points on a repetitive cycle basis. Accordingly, the equipment considered for this purpose includes conveyors, elevators, cranes, cableways, and pallet loaders. All of these are suited to materials handling situations involving high volumes, unit loads, packaged products, fixed points, and repetitive cycles. The basic types of package conveyors are gravity, belt, and live roller (Exhibit 10.3.4).

A large portion of the conveyors used in industry are the gravity type. There are many types of

Exhibit 10.3.4 Typical Package Conveyor Equipment

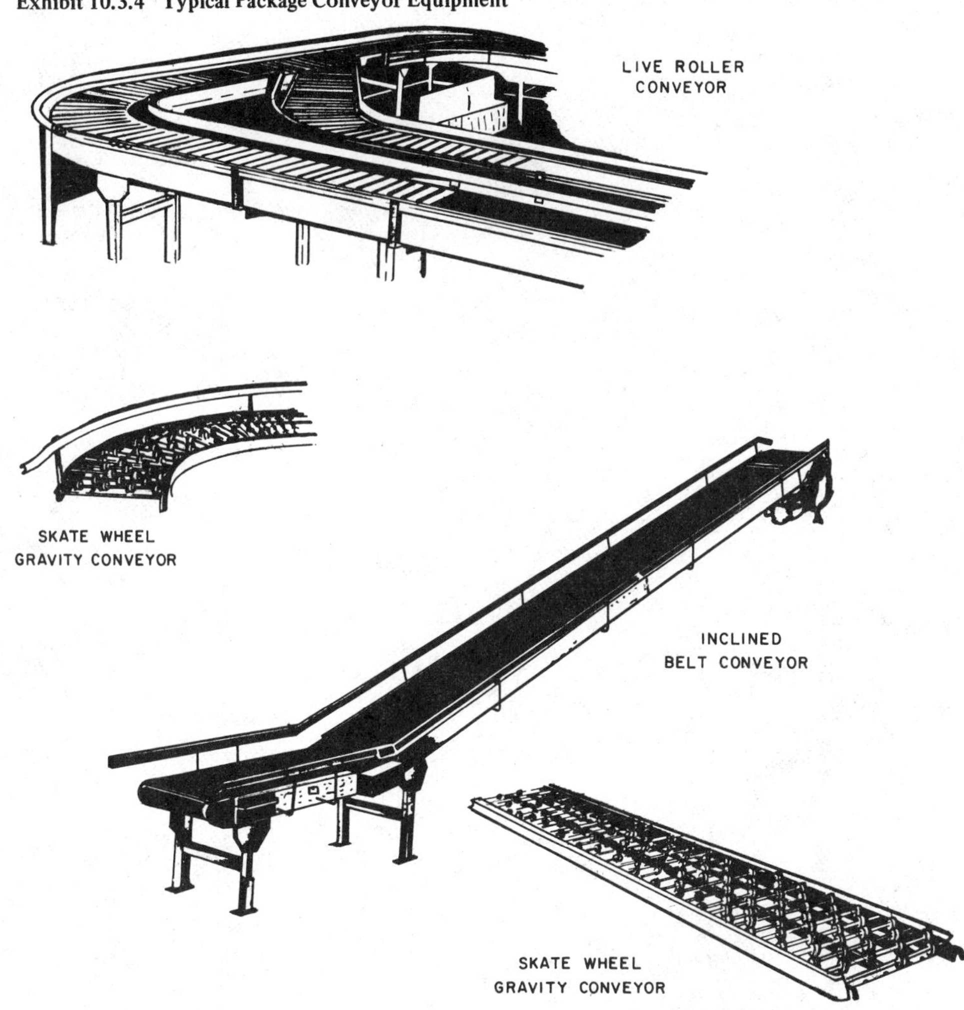

gravity conveyors and chutes. Gravity conveyors provide material transportation without energy cost and offer a simple and economical handling means when suitable. The investment in gravity conveyors is relatively low compared with other materials handling equipment, and maintenance expense is negligible. Gravity conveyors can be portable or permanently installed.

The major classes of gravity conveyors are skate wheel, roller, adjustable gravity, transfer tables, slides, and chutes.

Skate wheel conveyor is usually best applied to smooth, hard bottom items. Roller conveyor can handle similar products in heavier units and with rougher or more irregular bottoms. The rollers can be also formed with V- or U-shaped troughs to handle cylindrical items. Roller conveyor can better absorb impact and cross-slide loading.

Transfer tables are usually constructed of a series of steel or plastic balls or swivel caster wheels to permit the multidirectional movement and positioning of products. They can be used to position heavy workpieces or unit loads or can be treated as junctions in conveyor systems. In some cases transfer balls or casters are mounted on widely spaced posts to permit workers to move heavy plate or workpieces in and out of machines while walking between the posts.

Gravity conveyor can be equipped with switches and curves to route packages. Belt boosters are often inserted in gravity systems to recover height, and power switches are used to alter travel routes.

The more common types of power conveyor are belt, live roller, slat, apron, chain, and flight conveyors. Slat, apron, chain, and flight types are normally used in heavy handling applications such as foundries and machine shops. Belt conveyor is found in a wide range of sizes, speeds, and applications throughout industry.

Belt conveyor can operate on inclines of 30° or more, depending on the package size and shape and the type of belt surface. Belt curves and spirals are also available. Belts can be made of rubber, plastic, metal mesh, and other materials, depending on the application. Magnetizable materials can move vertically on belts with magnetic backup plates, and metal mesh belts can operate in ovens, furnaces, or chemical baths. In some cases the return strand of the belt is also used to move cargo in the opposite direction. One major disadvantage of belt conveyor is its inability, in most cases, to accumulate, because the product remains in a stable relationship with the belt surface. This is desirable in constant timing applications, but reduces the effectiveness of the belt conveyor as a storage cushion between operations. An advantage of belt conveyor is its ability to handle almost any shape that will not overhang or roll off en route. Portable belt conveyors are often used for loading and unloading vehicles. Some of these are designed to telescope into the vehicle.

Life roller conveyor is often driven by a conveyor belt pressed against the underside of the rollers with snubbing rolls. Later versions use V-belts, round belts, or chain drive systems. In any case the live roller has the advantage of permitting accumulation of packages and the varying of package spacing and speed along the route. Live roller conveyor has a much more limited incline-decline capability than belt conveyor and requires a relatively flat bottom on the package to be conveyed.

Exhibit 10.3.5 Mass Movement Unit Handling Systems

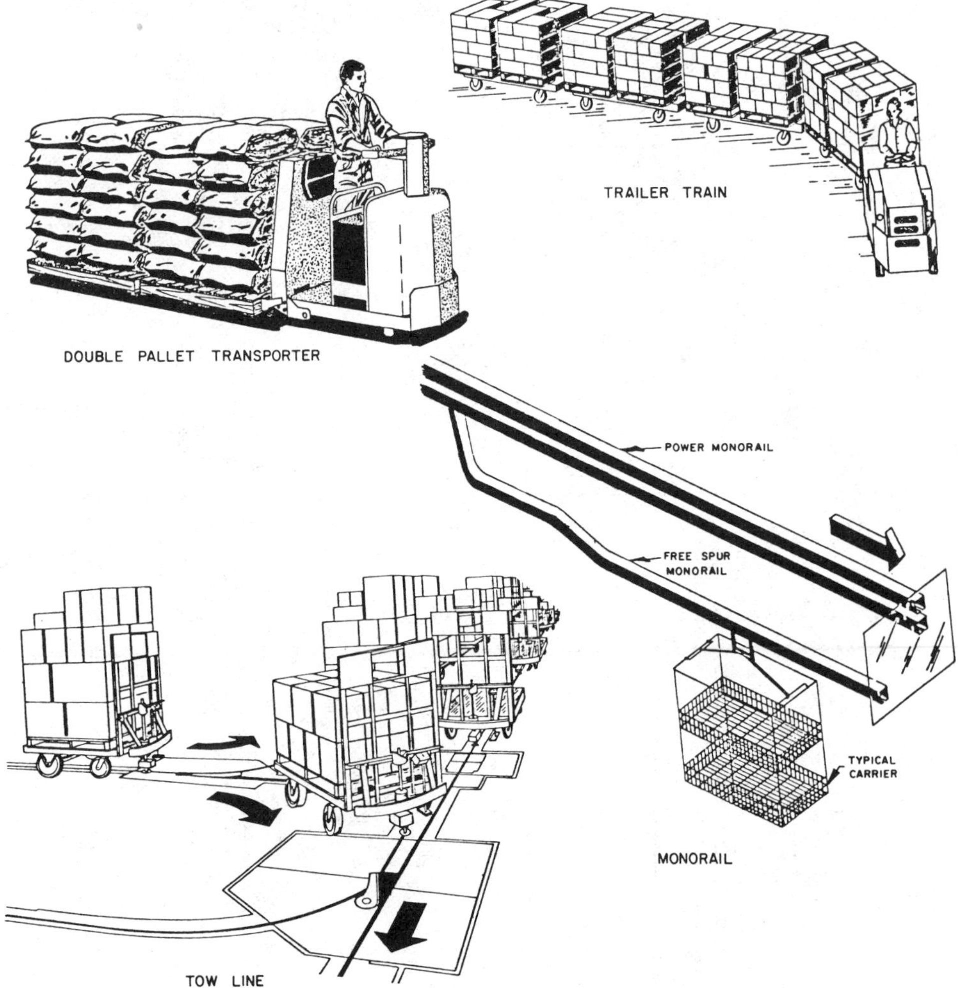

Live roller curves, switches, and brakes are normal parts of most systems. Telescoping live roller units are also available for vehicle loading operations.

Steep inclines are sometimes dealt with by using push bar conveyors. In these devices a slide plate on a roller conveyor bed is straddled by two chains connected together with a series of space bars or rods. Packages are pushed up the incline by the bars. This equipment can also be used for controlled descents.

Reciprocating incline elevators, scissor lifts, and elevators can also be used as portions of conveyor systems to move loads vertically between floors and work levels.

Towlines and monorail conveyors (Exhibit 10.3.5) are another general class of fixed path equipment. In each case the movement is in discrete handling units over a fixed and continuously powered path. The monorail can be of the power and free type and can move units at random by connecting to and releasing from the prime movement chain or cable by automatic preset command. The towline cart functions in a similar manner through a pin connection to an underfloor chain drive. Both of these systems offer inexpensive long-haul movement, random pickup and discharge, automatic unmanned route control, en route storage, recycling capability, and the elimination of cross-travel blockages. Monorail also offers vertical route flexibility, and towlines

Exhibit 10.3.6 Typical Counterbalanced Forklift Trucks

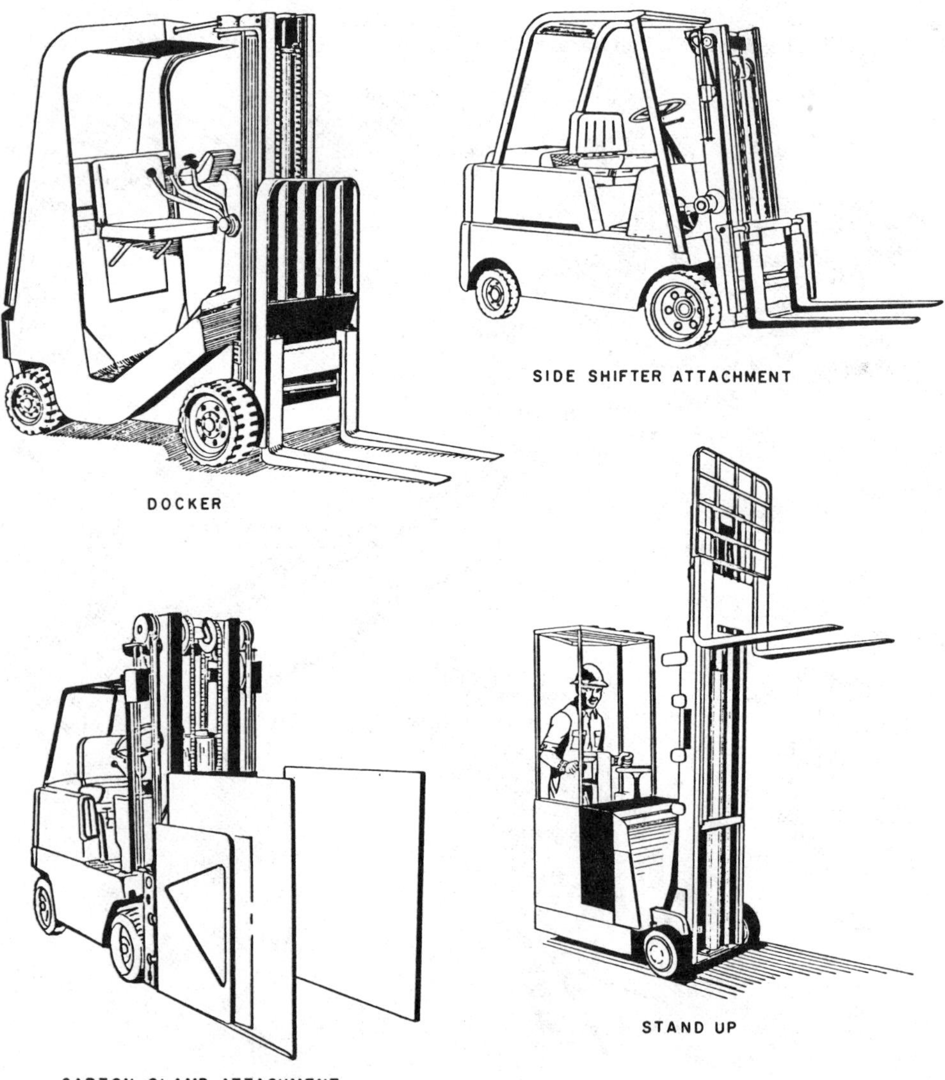

can negotiate ramps with a slope of up to 13°. Towline carts can also be equipped for manual handling and/or alternative trailer train operations.

10.3.9 POWER VEHICLE SYSTEMS

There is an almost infinite variety of powered vehicular devices used in materials handling operations. The well-known power shovels, bulldozers, front-end loaders, and so on, of the construction industry are often used in bulk materials handling operations in manufacturing. More conventionally, industrial vehicles are designed around the movement of pallets or large units. The warehouse tractor, pallet transporter (Exhibit 10.3.5) and forklift truck (Exhibit 10.3.6) have developed rapidly in recent years to a point of near optimum design versatility. At the same time, the forklift truck has reached a capability plateau.

Exhibit 10.3.7 Front Loading, Narrow Aisle Forklift Trucks

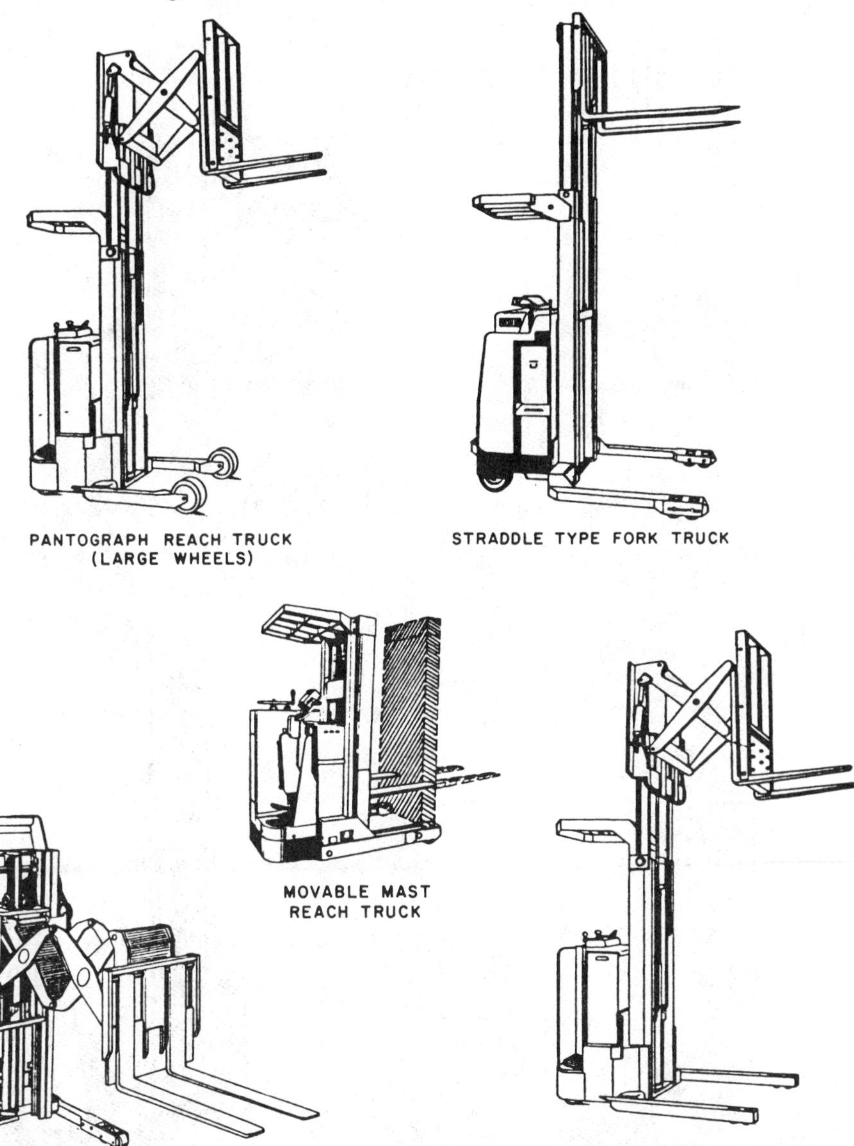

There are essentially three basic classes of forklift trucks available to the user. Each has an effective set of operating parameters (a performance envelope) that overlaps, but does not compete, with the others. In each case, aisle width, storage height, and equipment cost factors provide distinct application breakpoints.

Counterbalanced forklift trucks represent the original design concept (Exhibit 10.3.6). These machines require wide aisles–10 to 16 ft (3 to 5 m) with 4 ft (1.2 m) load lengths–and normally operate to a maximum storage height of about 24 ft (7.3 m) with 2000 to 6000 lb (900 to 2700 kg) loads. Typical building ceilings for these machines are in the 26 to 28 ft (8 to 8.5 m) clear height range. These machines are available with battery, gasoline, propane, or diesel power plants.

The modern conventional forklift truck design (Exhibit 10.3.7) for warehousing and manufacturing operations is a reach truck unit. These machines require smaller aisles–7 ft 6 in. to 9 ft 6 in. (2.3 to 3.0 m) for a 4 ft (1.2 m) load length–and can handle 2000 to 4000 lb (900 to 1800 kg)

Exhibit 10.3.8 Typical Narrow Aisle Forklift Trucks

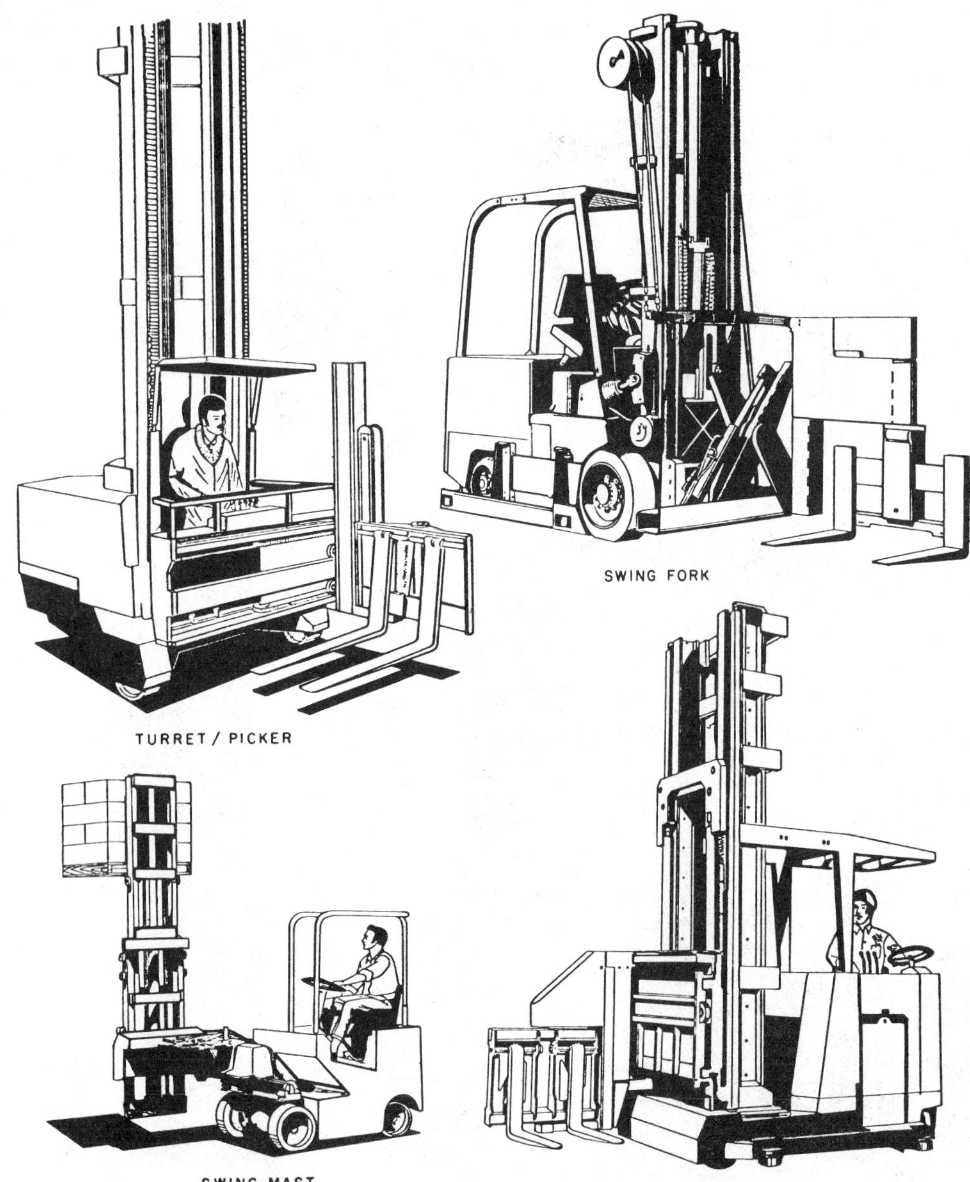

loads to about 24 ft (7.3 m) with a similar 26 to 28 ft (8 to 8.5 m) clear height ceiling. From a performance point of view, the reach truck has the same operating characteristics as the counterbalanced machine, with the benefits of better aisle economy, lower "footprint" pressures on the floor, and an all-electric power system.

The more recent developments in forklift trucks (Exhibit 10.3.8) provide models that are capable of storing to 40 ft (12 m) heights in aisles less than 6 ft (1.8 m) wide. These machines require electronic wire or mechanical rail guidance and are electric powered. They can be computer directed, and in the larger units are very costly. The payoff for their use is found in high-ceiling buildings–44 ft (13.5 m) clear–which require less land cover and provide a lower total facility cost. When using these machines, it is usually necessary to design the floors to very close tolerances–$\frac{1}{8}$ in. (3 mm) in 20 ft (6 m)–and column spacings must be matched to the vehicle and rack configuration.

In general, forklift trucks should not be used for long-haul transportation. They should be limited to repetitive cycle operations of 150 to 200 ft (46 to 60 m) one way loaded and should be inter-

Exhibit 10.3.9 High Cube Mechanized Storage Equipment

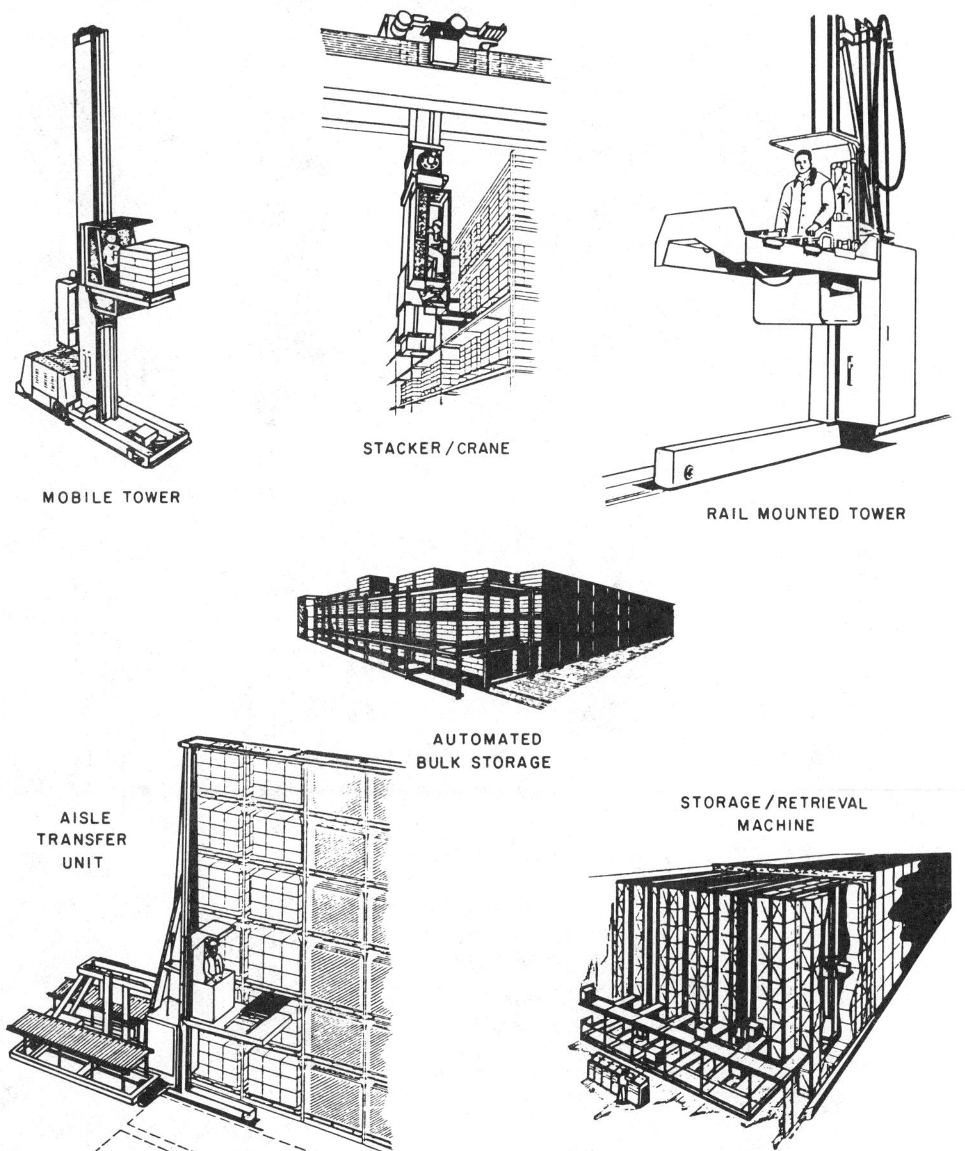

faced with long-haul transport equipment such as towlines, tractor trailer trains, and pallet transporters (Exhibit 10.3.5) for longer hauls.

10.3.10 FIXED PATH RAIL-MOUNTED SYSTEMS

These systems use a blend of forklift and automated storage and retrieval technology (Exhibit 10.3.9). These systems are specifically engineered to each application. They are discussed in Chapter 10.4.

10.3.11 ORDER PICKING

Order picking operations are usually people oriented and vehicle based (Exhibit 10.3.10). Recent developments in automated package picking and computer-directed vehicles have been applied in high transaction rate operations. In general, most order filling activities depend on the management of people through control documents and the mechanization of movement through the use of vehicles and conveyors.

Exhibit 10.3.10 Order Picking Equipment

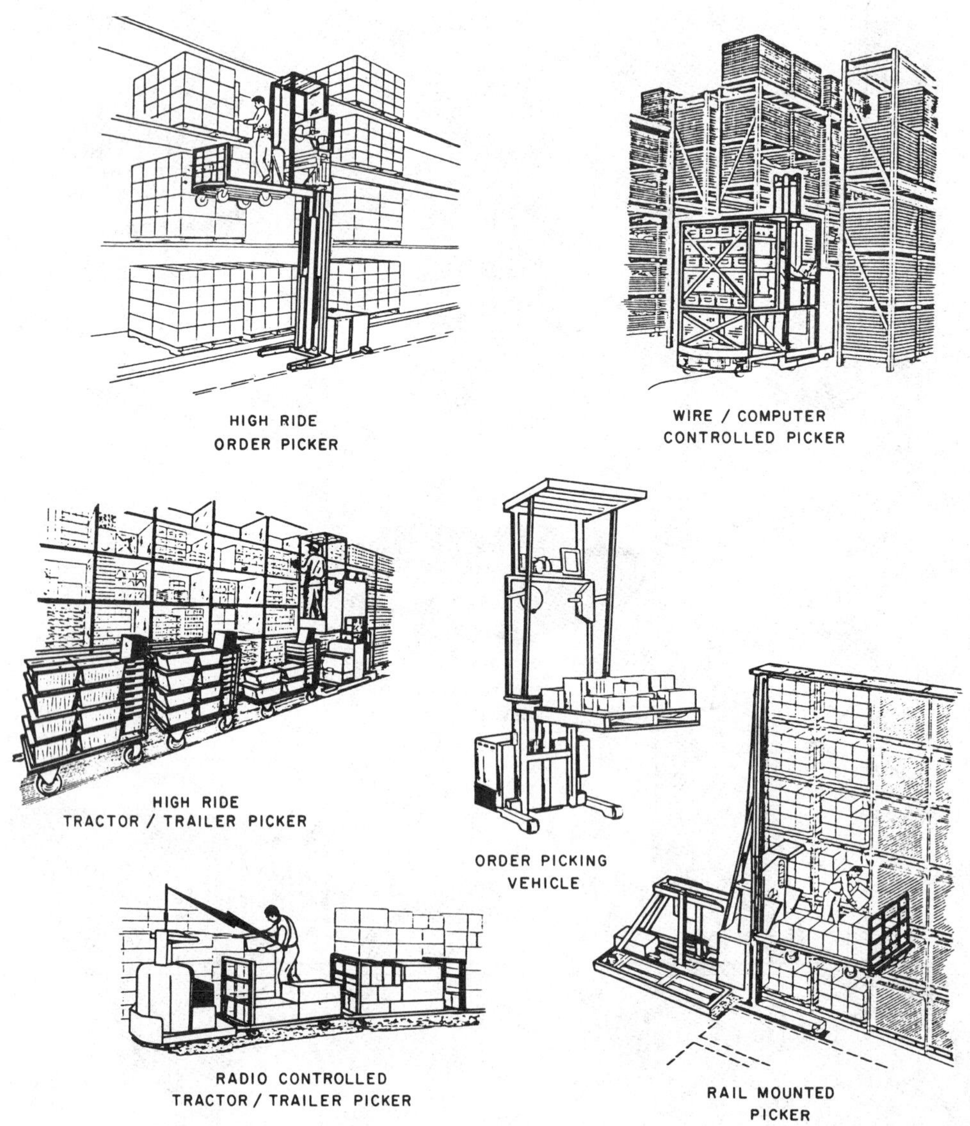

Exhibit 10.3.11 Materials Handling Tools and Storage Furniture

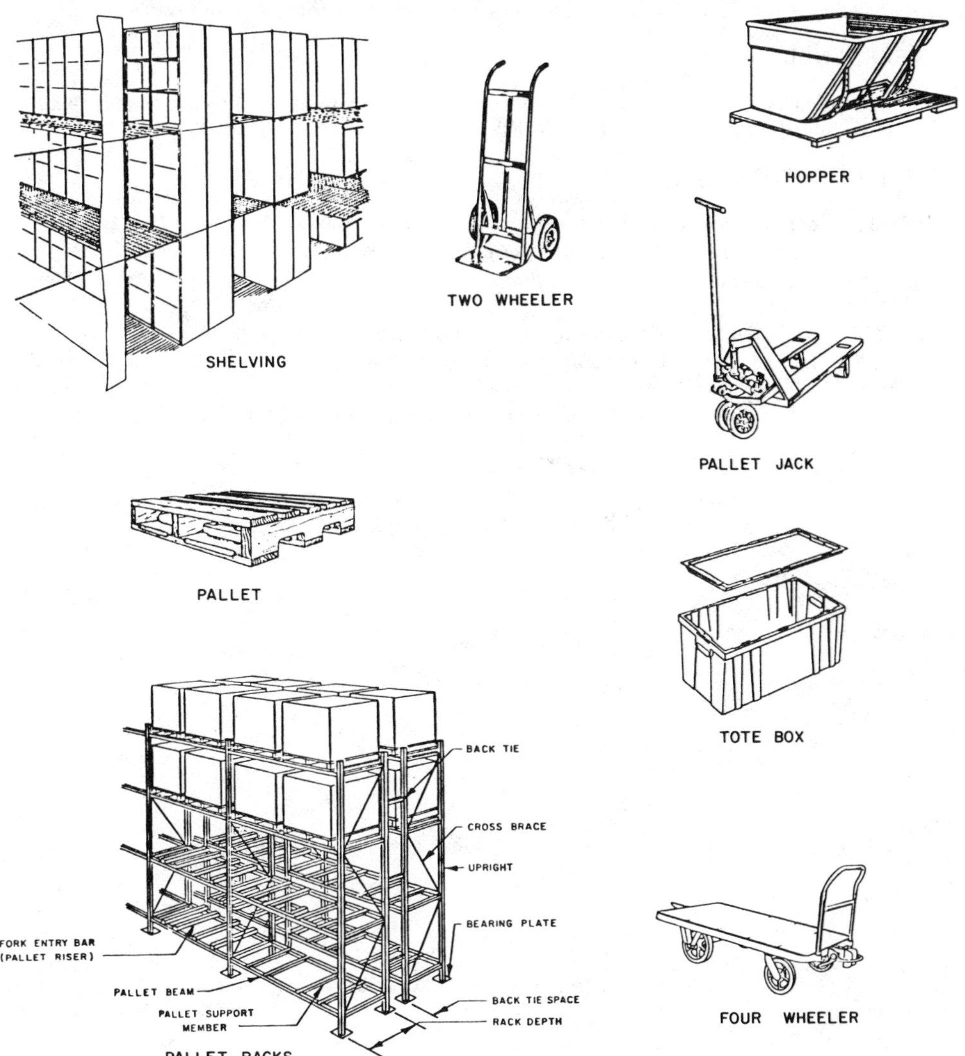

10.3.12 MATERIALS HANDLING TOOLS

Materials handling tools and storage furniture are required to bridge the gap between manual and mechanized operations and to provide the common denominators of the system, as shown in Exhibit 10.3.11. These tools include pallets, tote boxes, hand pallet jacks, two-wheel carts, four-wheel carts, pallet racks, shelving, and hoppers. Each of these items must be properly designed for the specific application to ensure capacity and dimensional compatibility.

10.3.12 SUMMARY

In the final analysis, the design of a materials handling system is the interpretation of a business function into a configuration of carefully engineered equipment applications. The controlling criteria are the size, shape, weight, and physical characteristics of the products to be moved and/or stored; the volume of movement in terms of total volume and volume per time period; and the permanence or expected service life of the system. Fully automated systems (which are discussed in Chapters 7.5 and 10.4) are applicable when volume and transaction rates are large and operating stability is predictable.

Each application must be custom designed; "cookbook" solutions are seldom applicable. Materials handling is an "all cost" function, and every penny saved is a penny of profit earned.

REFERENCE

1. U.S. DEPARTMENT OF THE NAVY, *Warehouse Modernization and Layout Planning Guide*, Naval Supply Systems Command, NAVSUP PUB 529, Washington, DC, 1978.

BIBLIOGRAPHY

APPLE, J. M., *Plant Layout and Materials Handling*, 3rd ed., Ronald, New York, 1977.

APPLE, J. M., *Material Handling Systems Design*, Wiley-Interscience, New York, 1972.

BLANDING, W., *Blanding's Practical Physical Distribution*, Traffic Services Corporation, Washington, DC, 1978.

BOLTZ, H. A., *Materials Handling Handbook*, Wiley-Interscience, New York, 1958.

MUTHER, R. and K. HAGANAS, *Systematic Hcndling Analysis*, Management and Industrial Research Publications, Kansas City, MO, 1969.

SIMS, E. R., JR., *Planning and Managing Materials Flow*, Industrial Education Institute, Boston, 1968.

CHAPTER 10.4

Storage and Warehousing

JOHN A. WHITE

Georgia Institute of Technology

HUGH D. KINNEY

SysteCon, Inc.

10.4.1 INTRODUCTION

In designing and/or improving manufacturing, distribution, and service-related systems, the approach used in storing materials and supplies can have a significant impact on the productivity of the total system. Requirements for the storage of materials exist in practically any organization. For example, hospitals, banks, manufacturing plants, insurance firms, distribution centers, assembly plants, maintenance centers, and ships all have materials, supplies, records, furniture, equipment, and/or tools that must be stored.

The terms "storage" and "warehousing" are often used to mean different things. For example, some prefer to distinguish between the two by referring to the storage of finished goods as warehousing and letting storage refer to raw materials, supplies, and work in process. For the purposes of this discussion, the terms will be used somewhat interchangeably.

In considering storage and warehousing from the point of view of the industrial engineer, it is important to focus on "the design, improvement, and installation" of storage and warehousing systems. Since storage and warehousing systems are examples of "integrated systems of people, material, equipment, and energy," industrial engineers can draw "upon specialized knowledge and skills in the mathematical, physical, and social sciences together with the principles and methods of engineering analysis and design to specify, predict, and evaluate the results to be obtained from such systems."*

Even though storage and warehousing systems are ideally suited for attention from the industrial engineer, the evidence to date indicates that little attention has been directed toward the area. One study indicated that less than 20% of the firms surveyed had reached a meaningful level of ongoing measurement to support productivity improvement in warehousing and distribution. Furthermore, 85% of those responding to the survey indicated that they were not using standards for warehousing activities.[1]

10.4.2 FUNCTIONS

The basic warehousing functions traditionally have been considered to be as follows:

1. Receiving.
2. Identification and sorting.
3. Dispatching to storage.
4. Placing in storage.
5. Storage.
6. Removing from storage.
7. Order accumulation.
8. Packing.

*The quotations were taken from the official definition of industrial engineering adopted by the AIIE, 1978.

9. Shipping.
10. Record keeping.

However, there is a difference in *warehousing functions* and *functions performed in the warehouse*. Depending upon the particular needs of an organization, the warehousing functions listed might be decentralized among several buildings or building modules. Additionally, other "nonwarehousing functions" might be assigned to the warehouse facility. Examples of functions that might be located in the warehouse are inbound inspection, parts preparation, kitting, and packaging.

Inbound inspection is generally a quality control function, but because of its relationship to receiving is often located in the warehouse. Depending upon the particular situation, special storage areas might need to be provided for inbound materials until released for issue by quality control. The inspection requirements can have a significant impact on the material handling, storage, and control systems in the warehouse.

Parts preparation consists of activities performed on materials prior to manufacturing to facilitate the manufacturing operations. In the electronics industry, wire leads for components might be bent and cut to length for subsequent insertion in circuit boards. In the aerospace industry, first-cut operations are performed in the warehouse to reduce the handling requirement to and from manufacturing; by cutting blanks from the sheet metal in the warehouse, smaller parts are delivered to manufacturing, and less scrap must be removed from manufacturing.

Another example is detrashing inbound materials at receiving. By removing vendor pack materials not required for parts protection, the corrugated trash is not taken into manufacturing to be removed subsequently; furthermore, potential quality problems due to contamination from the corrugated trash are reduced by detrashing.

Kitting occurs when predetermined quantities of parts for a manufacturing or assembly lot are removed from storage and placed in one or more containers for subsequent release to manufacturing or assembly. The kits are stored until needed and then dispatched to the appropriate work station(s).

Packaging of individual items can occur on-line in assembly or manufacturing or can be performed at some other location. If packaging is not performed on-line, it is frequently performed in the warehouse. Parts are either conveyed individually or delivered in bulk to packaging stations located in the warehouse; depending upon the quantities involved and packaging requirements, automatic packaging and palletizing equipment might be justified.

10.4.3 OBJECTIVES

The basic resources of *people*, *equipment*, and *space* must be integrated to obtain effective and economical methods and systems for *handling*, *storing*, and *controlling* materials and supplies. In designing, improving, and installing storage and warehousing systems, a number of objectives exist. Typically, the following criteria are considered:

1. Maximize personnel utilization.
2. Maximize equipment utilization.
3. Maximize space utilization.
4. Maximize energy utilization.
5. Maximize throughput.
6. Maximize loss control.
7. Maximize customer service.
8. Maximize productivity.
9. Minimize costs.

No design can possibly satisfy all of these objectives. For example, decisions to enhance, say, space utilization might have an adverse effect on throughput, and vice versa. What is needed is a multicriteria optimization model that explicitly incorporates all of the criteria. Unfortunately, the state of the art in multicriteria optimization has not advanced to such a point that its successful implementation in the context of warehousing is likely in the near term.

The approach generally used in evaluating alternative designs is first to define minimum acceptable levels of performance for each criterion and eliminate from consideration those that are unsatisfactory. Next, the remaining alternatives are evaluated on the basis of cost performance. Subjectivity is still involved to some degree in the final selection when issues such as flexibility are considered. A weighted factor comparison method can be used to combine quantifiable and nonquantifiable criteria.

10.4.4 PRINCIPLES

Principles or rules of thumb are commonly used by engineers in designing systems, and warehousing systems are no exception. One popular rule of thumb used in designing storage systems is the "85% rule of thumb." It takes various forms, including the following:

1. Plan on a maximum of 85% of slots being occupied in pallet storage racks.
2. Plan on a maximum of 85% of the storage cube in the slot being utilized.
3. Plan on a maximum equipment utilization of 85% for input/output conveyor systems for automated storage and retrieval systems.
4. Plan on 85% of the receipts being generated by 15% of the suppliers.
5. Plan on space for randomized storage equaling two times 85%, or 170%, of the sum of the space requirements for the average inventory levels.[2]

The danger in using principles or rules of thumb developed by others on the basis of their experience is that you do not know how they were derived, how good they are, and whether or not they apply to your situation. For this reason, another rule of thumb is that there are no accurate rules of thumb for warehousing.

As noted by Morris,[3] "a principle is simply a loose statement of something which has been noticed to be sometimes, but not always true." He goes on to state that "principles, then, must be applied carefully, in the light of what we like to call common sense, for they represent tremendous condensations of extensive and complicated experience."

Having cautioned against blindly applying warehousing principles and rules of thumb, the following principles are provided as general guidelines, as indicators of areas for consideration and concentration. A number of the principles are adopted from the "Twenty Principles of Material Handling" developed by the College-Industry Committee on Material Handling Education.

1. **Planning.** Develop a strategic plan for handling, storing, and controlling materials that is supportive of strategic plans for manufacturing, marketing, and distribution.
2. **Hybrid System.** Plan a system that integrates handling, storage, and control of materials; handle different things differently, store different things differently, and control different things differently.
3. **Material Flow.** Develop the layout for the warehouse on the basis of inbound, outbound, and internal flow of material.
4. **Control.** Plan a system that provides real-time physical, fiscal, inventory, and management control of material.
5. **Simplification.** Simplify the handling, storage, and control of material.
6. **Throughput Capacity.** Plan a system that maximizes the throughput for the warehouse.
7. **Space Capacity.** Plan a system that maximizes the utilization of the storage capacity of the warehouse, based on cubic space.
8. **Unit Size.** Increase the quantity, size, and weight of the load handled and stored.
9. **Automation/Mechanization.** Automate handling, storage, and control functions.
10. **Equipment Selection.** Select equipment on the basis of material characteristics and flow requirements, including handling, storage, and control requirements.
11. **Standardization.** Standardize handling, storage, and control methods and types and sizes of equipment.
12. **Adaptability/Flexibility.** Plan a building and a system for handling, storing, and controlling material that can respond to changing requirements.
13. **Layout/Aisle.** Develop the layout and determine lengths, widths, heights, and placement of aisles on the basis of the handling, storage, and control requirements.
14. **Utilization.** Maximize the utilization of personnel and equipment.
15. **Maintenance.** Plan for preventive maintenance and scheduled repairs for all handling and storage equipment.
16. **Obsolescence.** Periodically review the handling, storage, and control system and make needed replacements to increase productivity and/or reduce costs.
17. **Performance.** Develop performance measures to be used periodically to evaluate and provide feedback on warehouse productivity.
18. **Audit.** Design the audit program for the handling, storage, and control system when the system is designed, and perform the systems audit periodically.
19. **Facility.** Design the facility to house the handling, storage, and control system; base ceiling heights and column spacing on the requirements of the warehouse system.
20. **Safety.** Plan for safe handling, storage, and control of material.

10.4.5 ALTERNATIVE STORAGE METHODS

In designing the material storage system for a warehouse, numerous alternatives are available for consideration. Storage methods can be categorized in a number of ways. One approach is to categorize them on the basis of the type of product stored and retrieved, for example, item, cases, and pallets. A second approach is based on the location of items, for example, randomized versus dedicated storage methods. Another basis for categorizing alternative storage methods is the equipment used, for example, bins, totes, miniload, carousel, shelving, block stacking, case flow rack, selective pallet rack, drive-in rack, automated storage and retrieval systems, deep lane storage systems, and pallet flow rack. A fourth category is based on the storage and retrieval method, for example, in-aisle versus end-of-aisle order picking methods.

Hybrid Systems

Interestingly, many firms recognize the need for using different storage methods because of significantly different *material* characteristics. However, not nearly as many recognize the need for using different storage methods because of differences in the *move* or flow characteristics of the material being stored. As an example, depending upon the relative flow rates for products, fast movers might be picked from carousels or bins in a dedicated picking area, whereas, slow movers might be picked from pallet rack in the reserve storage area.

Hybrid storage systems, as well as hybrid storage and retrieval systems, can provide the correct solution to storage problems involving significantly different material and/or move characteristics. Such systems also can provide an added degree of flexibility in accommodating changes in the mix of either material or move characteristics.

Type of Product

The storage method to be used is a function not only of the type of product being stored, but also of the storage and retrieval pattern used. In particular, the following storage and retrieval combinations typically occur: pallet-pallet, pallet-case, case-case, case-item, and item-item. Occasionally pallet-put-away and item-pick occurs, especially with large welded-wire containers of unpackaged items being stored. However, case–put-away and pallet-pick seldom occurs.

Pallet storage can involve block stacking or the use of conveyors or racking. Cases or totes can be stored on shelves, conveyors, or storage racks. Items can be stored in bins or totes on shelves, conveyors, or storage racks.

Location Method

The two storage location methods that in some sense represent extreme points of view are randomized storage and dedicated storage. Randomized storage is used when an individual stockkeeping unit (SKU) can be stored in any available storage location. The term "floating slot" is used to describe randomized storage. Two variations of randomized storage exist. One method of randomized pallet storage is to store an inbound pallet load in an empty slot, with the slot selected randomly from among all empty slots. Similarly, pallet retrieval would occur by randomly selecting the pallet from among all slots occupied by the desired item. Although such a storage and retrieval policy is truly *random* storage, it is not the policy generally referred to in practice as *randomized* storage.

The most common operational definition of randomized storage appears to be the following: When an inbound load arrives for storage, the closest available slot is designated as the storage location; retrievals are performed on a first-in, first-out (FIFO) basis. The FIFO retrieval policy provides for a uniform rotation of stocks and, in the long run, is equivalent to "pure" randomized retrieval. The closest-available-slot storage policy can yield results that are similar to the "pure" randomized storage policy *if the storage level remains fairly constant and at a high level of utilization*; otherwise, there will be differences in the throughput rates obtained.

Dedicated storage is used when an SKU is assigned to a specific storage location or set of locations. The term "fixed slot" is used to describe dedicated storage. Two variations of dedicated storage are commonly used. Storage of items in part number sequence is one form of dedicated storage; another approach is to determine the storage location for an SKU based on its activity and inventory levels. The latter method is preferred when there are significant differences in either the activity level or the inventory level for SKUs.

The number of openings assigned to an SKU must accommodate its maximum inventory level. Hence the total number of openings required for dedicated storage is equal to the sum of the openings required for each SKU. With randomized storage, however, the total number of openings required in the system is the number of openings required to store *all* SKUs. Since, typically, all SKUs will not be at maximum inventory levels at the same time, randomized storage usually requires fewer openings than dedicated storage.

Dedicated storage, if based on activity, will maximize throughput, but will also maximize storage space requirements. On the other hand, randomized storage will minimize storage space, but will reduce throughput. The selection of the appropriate storage method depends upon the weight given to storage space versus that given to throughput.

There are two reasons for randomized storage resulting in less storage space than required for dedicated storage. First, if an "out-of-stock" condition exists for a given SKU, the empty slot continues to remain "active" with dedicated storage; whereas it would not with randomized storage. Second, if there are multiple slots for a given SKU, then as the inventory level decreases, empty slots will develop.

When inventory shortages seldom occur, and single slots are assigned to SKUs, then there are no differences in the storage space requirements for randomized and dedicated storage. Many carousel and miniload systems meet these conditions.

To maximize throughput when using dedicated storage, assign SKUs to storage locations based on the ratio of their activity to the number of openings or slots assigned to the SKU. The SKU having the highest ranking is assigned to the preferred openings, and so on, with the lowest-ranking SKU assigned to the least-preferred openings. Since "fast movers" are up front and "slow movers" are in back, throughput is maximized.

It is important in ranking SKUs to define "activity" as the number of storages and/or retrievals per unit time, not the quantity of material moved. It also is important to think of "part families" as well. "Items that are ordered together should be stored together" is a maxim of activity-based storage.

If storage locations are determined on the basis of activity and inventory levels, studies have shown that dedicated storage can yield savings in travel time or increases in throughput of from 15% to 50% when compared with "pure" randomized storage. Simulation studies have shown that dedicated storage can require from 20% to 60% more slots than required for randomized storage.

Despite the greater throughput of dedicated storage, it is not used as often as it should be. One reason is that it requires more *information* to plan the system for maximum efficiency. Very careful estimates of activity levels and space requirements must be made. It also requires more *management* to continue to realize the benefits of dedicated storage after the system is installed. When conditions change significantly, items must be relocated. Hence randomized storage is often used under highly seasonal and dynamic conditions.

When you have many SKUs, dedicated storage based on each SKU may not be practical. Instead, SKUs can be assigned to classes based on their activity-to-space ratios. Class-based dedicated storage, with randomized storage within the class, can yield the throughput benefits of dedicated SKU storage and the space benefits of randomized SKU storage. Depending upon the activity-to-space ratios, three to five classes might be defined.

The effect on space and throughput of the storage method used can be demonstrated with an example. Suppose the storage area for the warehouse is designed as shown in Exhibit 10.4.1. A single input/output (I/O) point serves the storage area. All movement is in full pallet load quantities. The storage area is subdivided into 10 ft by 10 ft storage bays. Three classes of products (A, B, and C) are to be stored. Class A items represent 80% of the I/O activity and have a dedicated storage requirement of 40 storage bays, or 20% of the total storage; class B items generate 15% of the I/O activity and have a dedicated storage requirement of 60 storage bays, or 30% of the total; and class C items account for only 5% of the throughput for the system, but represent 50% of the storage requirement.

Assuming that lift truck travel between the I/O point and individual storage bays can be approximated by rectilinear distances between the dock and the centroid of each bay, the distances are as shown in Exhibit 10.4.2. Based on the ratio of I/O activity to dedicated storage requirement, the product classes will be placed in the layout in the rank order A, B, and C to obtain the product layout given in Exhibit 10.4.3.

The expected distance traveled for the dedicated storage layout shown in Exhibit 10.4.3 is

Exhibit 10.4.1 Warehouse Layout Example

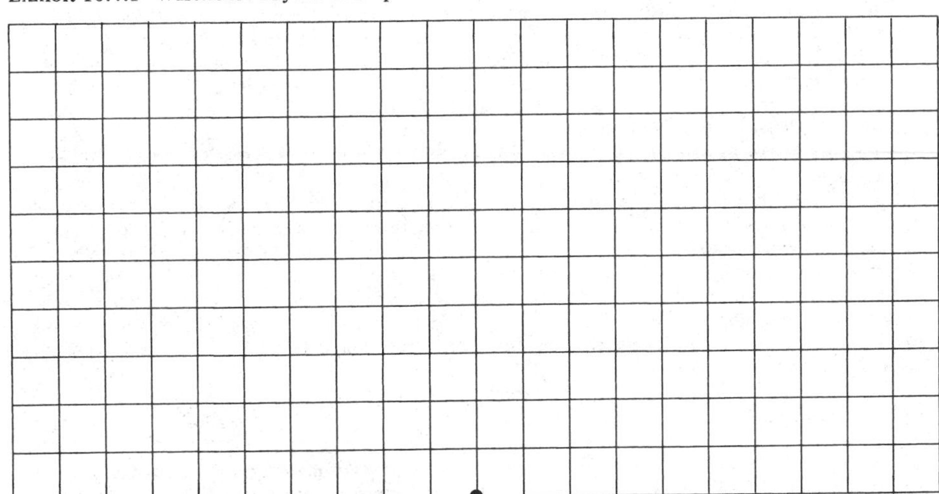

Exhibit 10.4.2 Average Distances for Warehouse Layout Example

190	180	170	160	150	140	130	120	110	100	100	110	120	130	140	150	160	170	180	190
180	170	160	150	140	130	120	110	100	90	90	100	110	120	130	140	150	160	170	180
170	160	150	140	130	120	110	100	90	80	80	90	100	110	120	130	140	150	160	170
160	150	140	130	120	110	100	90	80	70	70	80	90	100	110	120	130	140	150	160
150	140	130	120	110	100	90	80	70	60	60	70	80	90	100	110	120	130	140	150
140	130	120	110	100	90	80	70	60	50	50	60	70	80	90	100	110	120	130	140
130	120	110	100	90	80	70	60	50	40	40	50	60	70	80	90	100	110	120	130
120	110	100	90	80	70	60	50	40	30	30	40	50	60	70	80	90	100	110	120
110	100	90	80	70	60	50	40	30	20	20	30	40	50	60	70	80	90	100	110
100	90	80	70	60	50	40	30	20	10	10	20	30	40	50	60	70	80	90	100

Dock

Exhibit 10.4.3 "Optimum" Dedicated Storage Layout

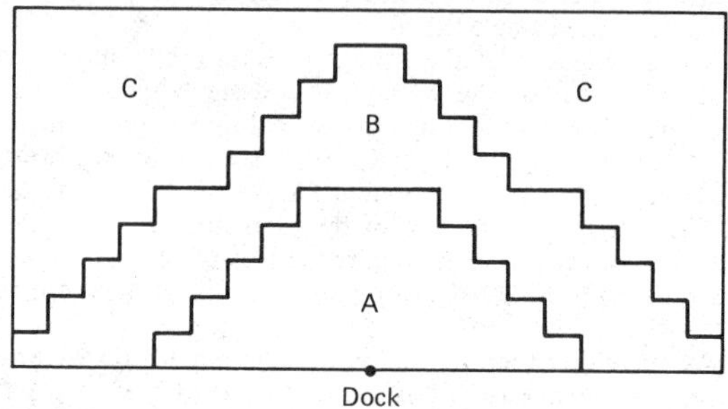

53.15 ft. If randomized storage is used such that each bay is equally likely to be used for storage, then the expected distance traveled will be 100 ft. However, with randomized storage a total storage requirement of less than 200 bays is anticipated. The exact storage requirement will depend upon the demand and replenishment patterns for the three product classes.

Even though the storage requirement for randomized storage is not known, it is possible to compute a storage bound that will yield an expected distance traveled equal to or less than that for dedicated storage. Storage bays can be eliminated from consideration in reverse order of their distances from the I/O point, and expected distance values can be computed until a sufficient number have been eliminated. For the example, 138 storage bays, or 69%, must be eliminated for randomized storage to yield an expected distance traveled equal to 52.90 ft. The resulting randomized storage layout is shown in Exhibit 10.4.4.

If storage space is to be rectangularly shaped, and if 10 ft by 10 ft storage bays are to be used, then the randomized storage layout having minimum expected distance will be as shown in Exhibit 10.4.5. The resulting expected distance traveled for this exhibit is 50 ft. However, only 50 storage bays can be required for storage, rather than the 62 required in Exhibit 10.4.4. (The comparison also serves to demonstrate the effect that building design can have on expected distance traveled.)

From the bound obtained for randomized storage from the example, it is seen that it is not likely that randomized storage will yield a reduction in space sufficient to obtain throughput values comparable to those obtained from dedicated storage. However, if space costs are significantly greater than handling or throughput costs, then randomized storage might be preferred, regardless of the impact on throughput.

Equipment Used

In considering the effect of the equipment used on the storage of material, only palletized storage will be considered. However, similar analyses can be performed for alternative case, tote, and item storage systems.

Exhibit 10.4.4 Randomized Storage Layout

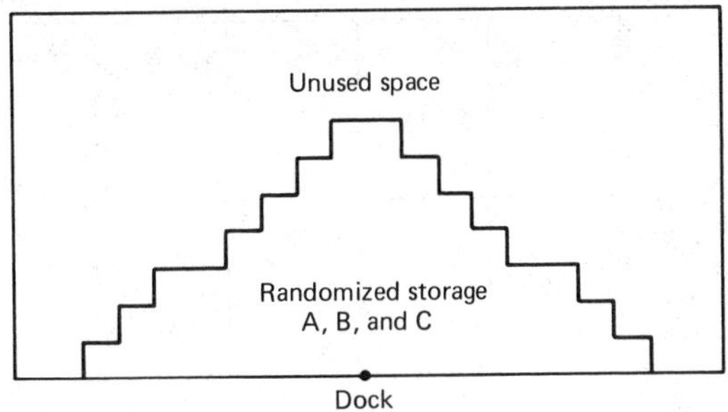

Consider three methods of storing pallet loads: block stacking, drive-in rack, and selective pallet rack. Basically, block stacking and drive-in rack are quite similar in terms of the impact on space and throughput.

Block stacking provides an apparent high utilization of floor space at a low cost; selective pallet rack provides maximum selectivity, but with reduced utilization of space and at increased cost. Drive-in rack would be used to achieve the benefits of block stacking when the product either is not stackable or is limited in the stacking height.

Both block stacking and drive-in rack suffer from "honeycombing." The empty space resulting from pallet retrieval is referred to as honeycomb loss since it is not usable for the storage of additional pallets if stock rotation is to be achieved. With block stacking, both horizontal and vertical honeycombing occur.

To demonstrate the effect of honeycombing, consider the situation in which 15 pallet loads are to be block stacked. Three levels of storage are allowed. As shown in Exhibit 10.4.6, different bay or lane depths can be used. Horizontal honeycomb loss occurs with four-deep, three-deep, and two-deep storage because of partially filled bays. For the two-deep situation, Exhibit 10.4.7 illustrates the vertical honeycomb loss due to partially filled columns.

Given a lot size of Q pallets to be stored and a uniform depletion of inventory, DeMars, Matson, and White[5] give the average floor space requirement for various block stacking designs as

$$S = y\frac{(W + c)}{288Q}(xL + 0.5A)\,(2Q - xyT + xT)$$

where the following notation, illustrated in Exhibit 10.4.8, is used:

S = average floor space requirement (ft^2)
x = depth of lane or bay, measured in columns
y = number of lanes or bays
L = pallet length or depth (inches)
W = pallet width (inches)
A = aisle width (inches)
c = clearance between lanes or bays (inches)
T = number of tiers or levels in a column
Q = lot size (pallets)
$Q \leqslant xyT$

Exhibt 10.4.5 Rectangularly Shaped Randomized Storage Layout

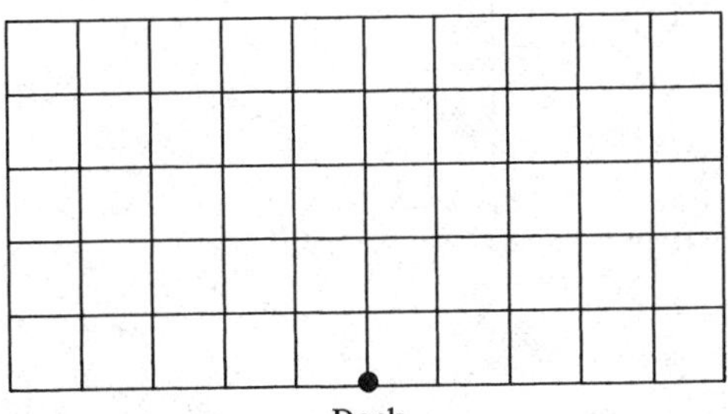

Exhibit 10.4.6 Block Stacking Storage Bay Options[a]

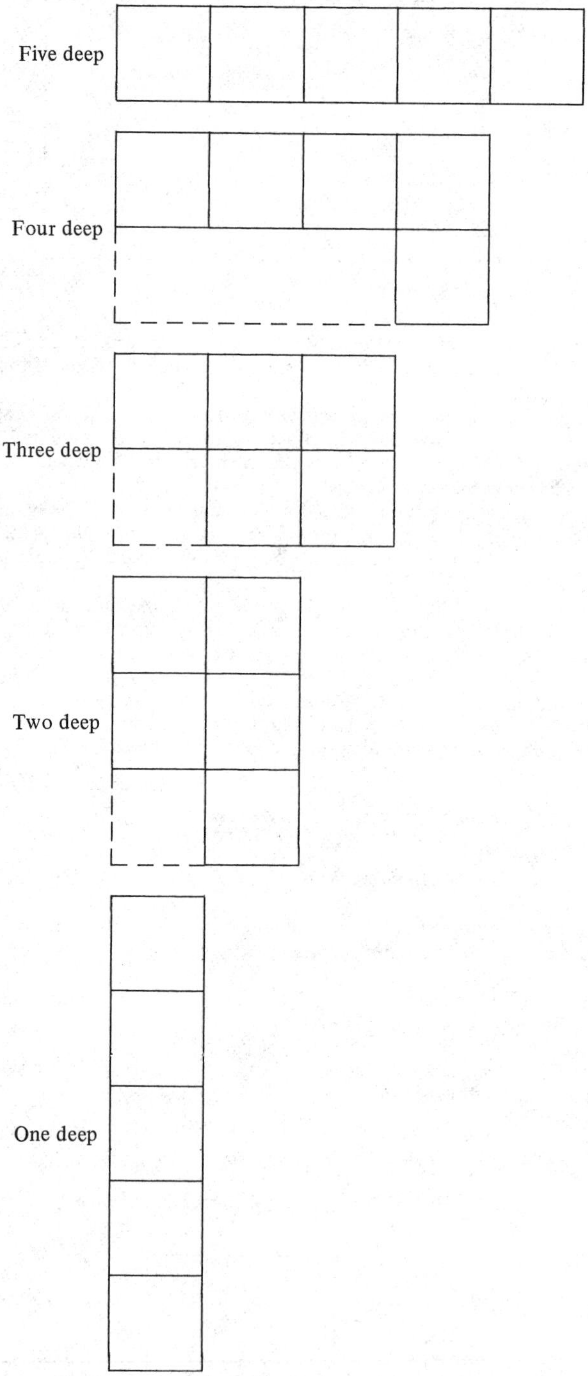

Five deep

Four deep

Three deep

Two deep

One deep

[a]Taken from reference 4 with permission.

Exhibit 10.4.7 Block Stacking Storage Profile[a]

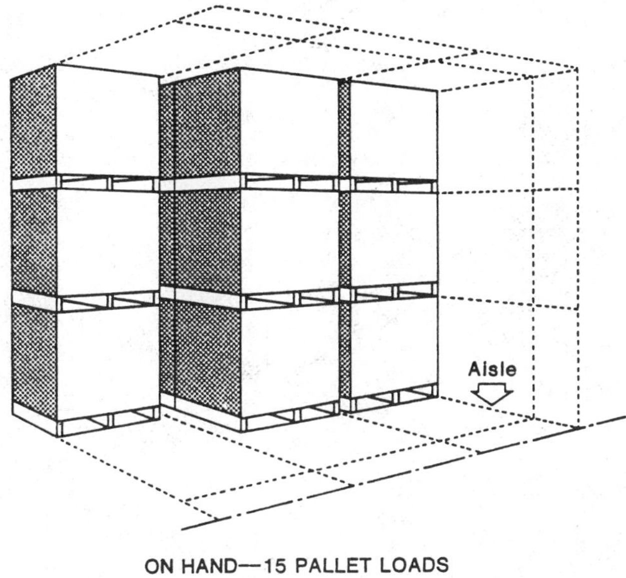

ON HAND—15 PALLET LOADS

(a)

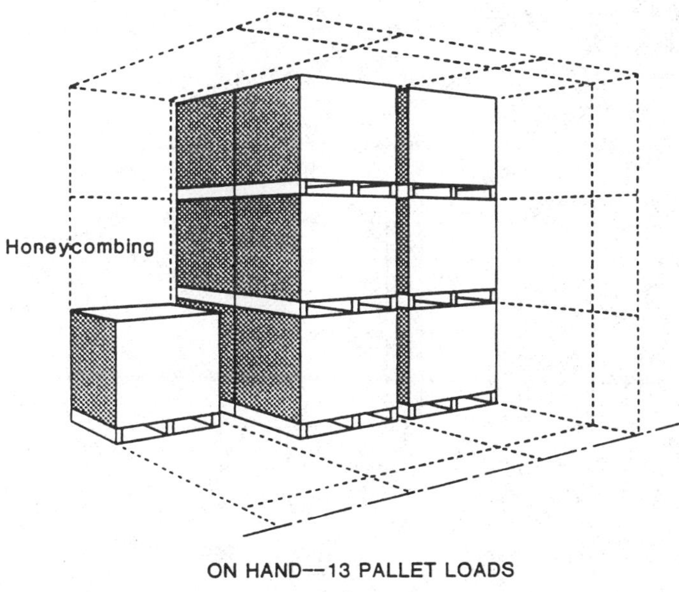

ON HAND—13 PALLET LOADS

(b)

[a]Taken from reference 4 with permission.

Exhibit 10.4.7 (*Continued*)

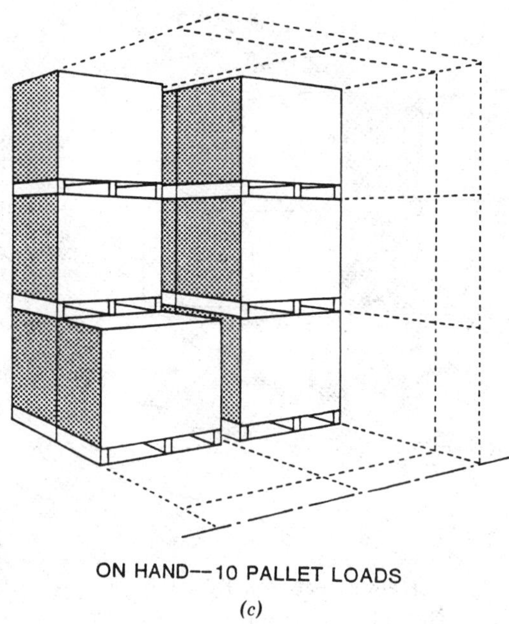

ON HAND—10 PALLET LOADS

(*c*)

If space is to be minimized, then values of x, y, and T would be selected such that S is minimized subject to $xyT \geqslant Q$. As an example, suppose a lot of 15 pallets is to be block stacked three high. Pallet dimensions are 50 in. by 42 in.; a 12 ft aisle and a 10 in. clearance are to be used. Hence L = 50 in., W = 42 in., A = 144 in., c = 10 in., T = 3, and Q = 15. Thus it is desired to minimize

$$S = 0.012(50xy + 72y)\,(30 - 3xy + 3x)$$

Exhibit 10.4.8 Plan View of Block Stacking (Floor Stacks)[a]

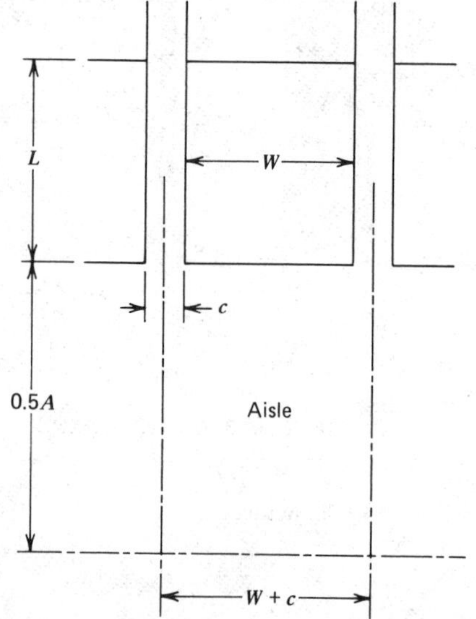

[a]Taken from reference 4 with permission.

Exhibit 10.4.9 Block Stacking Space Requirements for the Example

Lane Depth (x)	Number of Lanes (y)	Floor Space (S)
5	1	116.28
4	2	117.87
3	2	112.33
2	3	111.80
1	5	132.17

subject to $xy \geqslant 5$. Values of S are given in Exhibit 10.4.9 for various values of x and y. As can be seen, a two-deep configuration minimizes S.

To compare space requirements for block stacking with single-deep pallet storage rack, consider Exhibit 10.4.10. No honeycomb loss occurs with single-deep pallet storage rack. However, the following space losses occur for rack storage:

1. Vertical clearance between the top of a load and the load beam above it, plus the load beam dimension.
2. Horizontal clearance between a load and the upright truss, plus one half the dimension of the upright truss.
3. Horizontal clearance between loads on the same load beam.
4. Flue spacing between loads (back to back).

The following new notation (illustrated in Exhibit 10.4.11) is added to that established for block stacking:

r = width of upright truss rack member (inches)
f = flue spacing, load to load (inches)
c_1 = rack-to-load horizontal clearance (inches)
c_2 = load-to-load horizontal clearance (inches)

Typically, $c_1 = c_2$, and for convenience c will denote the value for c_1 and c_2.

The resulting expression for S is given in DeMars, Matson, and White[5] to be

$$S = \frac{[W + 0.5\,(r + 3c)\,]\,[L + 0.5\,(A + f)\,]\,(Q + 1)}{288T}$$

Exhibit 10.4.10 Rack Space Allowances[a]

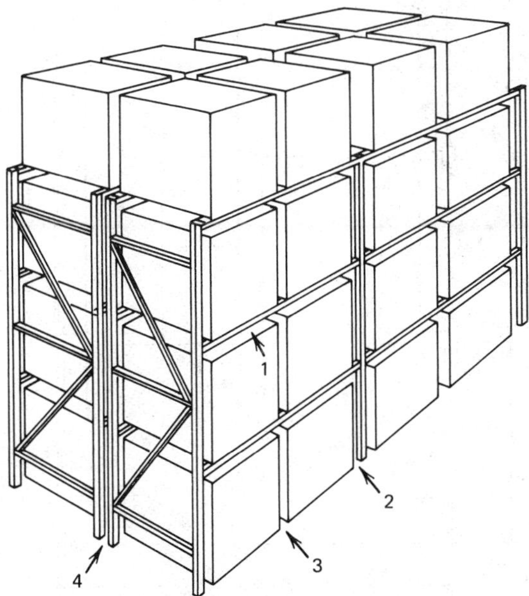

[a]Taken from reference 4 with permission.

Exhibit 10.4.11 Plan View of Single-Deep Rack[a]

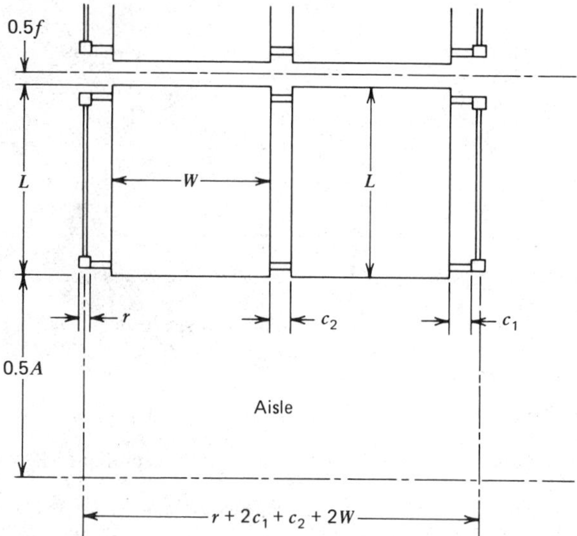

[a]Taken from reference 4 with permission.

For the case of L = 50 in., W = 42 in., A = 144 in., T = 4 in., Q = 15, r = 3 in., f = 6 in., and c = 4 in., a calculation establishes that S equals 85.94 ft^2.

To compare double-deep pallet storage rack with single-deep pallet storage rack and block stacking see Exhibit 10.4.12 and note that the only honeycomb loss is due to the two-deep configuration; no vertical honeycomb loss occurs because of the rack members. The assumption is made that both pallets in a two-deep "lane" are for the same product.

Exhibit 10.4.12 Plan View of Double-Deep Rack[a]

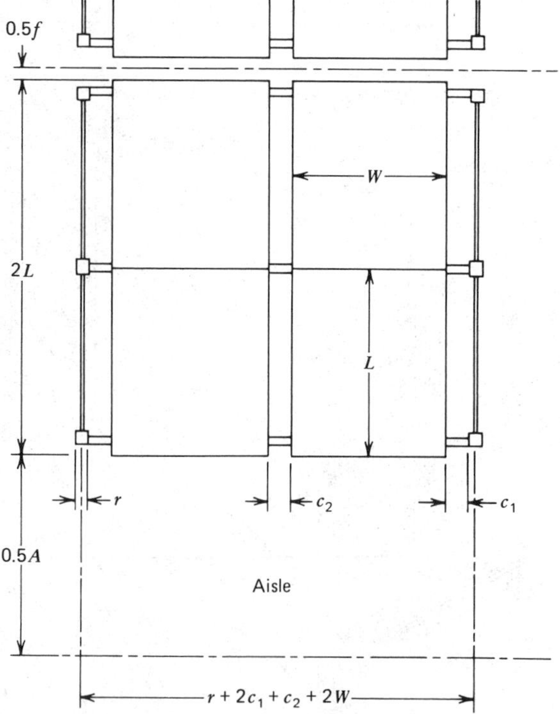

[a]Taken from reference 4 with permission.

Exhibit 10.4.13 Deep-Lane Storage

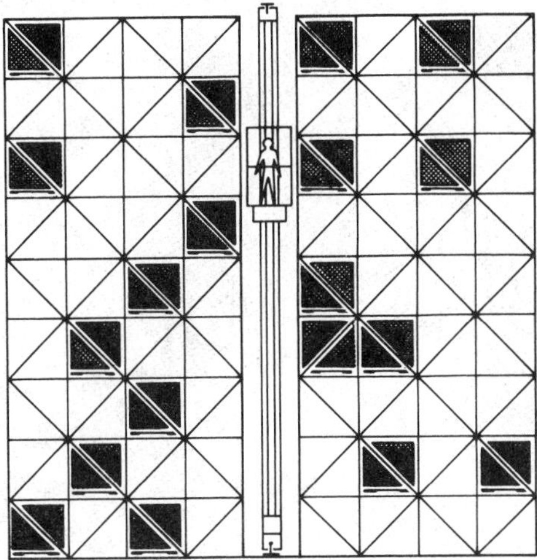

DeMars, Matson, and White[5] show that the space requirement for a double-deep configuration can be given as

$$S = \frac{[W + 0.5\,(r + 3c)]\,[2L + 0.5(A + f)\,]\,[2Q^2 + 4Q + [1 - (-1)^Q]]}{1152QT}$$

For this case, S equals 64.17 ft^2.

The analysis for drive-in rack is not as direct as for the storage methods already considered. With drive-in rack, vertical honeycombing can occur; however, the same product need not be stored at each tier in a bay. The results for block stacking closely approximate drive-in rack applications, with appropriate adjustments for space losses due to vertical rack members.

Deep-lane storage systems, unlike drive-in rack, do not have vertical honeycomb losses. A deep-lane storage system is depicted in Exhibit 10.4.13. With a last-in, first-out (LIFO) discipline for a lane, the resulting expression for S is given for deep-lane storage to be[5]

$$S = \frac{z\,(W + r + 2c)[0.5(A + f) + xL](2Q - xz + x)}{288QT}$$

where $xz \geqslant Q$ and z *is defined to be the smallest integer greater than or equal to* Q/x; z represents the number of lanes x deep required for storage of Q units. For the case of L = 50 in., W = 42 in., A = 144 in., T = 4, Q = 15, f = 6 in., c = 3 in., r = 3 in., and x = 2 (such that z = 3), a calculation establishes that S equals 57.55 ft^2.

In Aisle Versus End-of-Aisle Order Picking*

Among the various storage methods available, one of the most interesting concerns order picking. The choice is between moving the person to the material storage location and performing in-aisle order picking (Exhibit 10.4.14*a*) and moving the material to the person's work station and performing end-of-aisle order picking (Exhibit 10.4.14*b*). The question is, "Picker to the picking face, or picking face to the picker?"

The choice of moving the person or the material is not a new one. Years ago the questioning attitude of work simplification focused attention on the issue of direct labor personnel traveling to get their own material versus having *something* or *someone* (indirect labor) bring the material to the worker.

The material-to-person versus person-to-material issue was also suggested by Sir Francis Bacon in his *Essays of Boldness* when he wrote, "If the hill will not come to Mahomet, Mahomet will go to the hill." The message is clear—alternatives exist in bringing together people and material!

Included in the list of possible ways of moving the order picker to the material are counterbalanced trucks, reach trucks, order picker trucks, person-aboard automatic storage/retrieval machines, and walking, among others. Alternative methods of moving the material to the order

*This section is based on reference 6.

Exhibit 10.4.14 Order Picking Alternatives of (*a*) In Aisle and (*b*) End of Aisle[a]

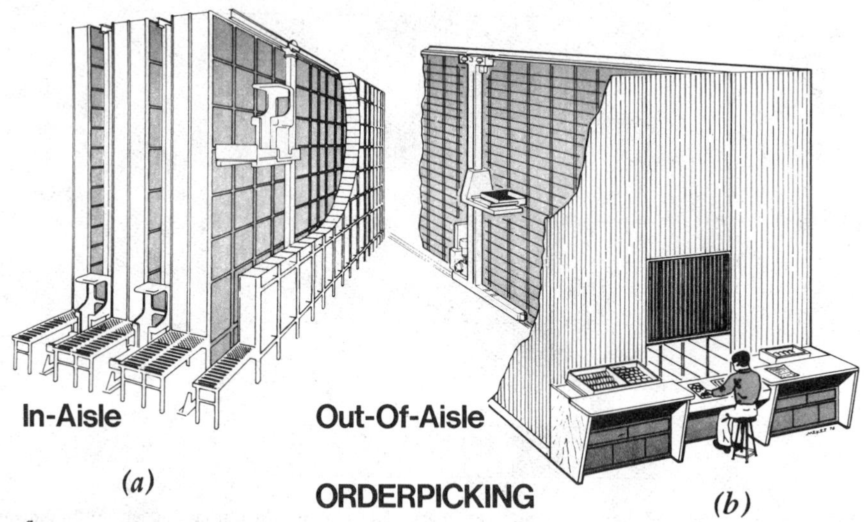

[a]Courtesy of the Munck System, Inc.

picker's work station include storage/retrieval carousels and miniload and unit load automatic storage/retrieval machines, as well as conveyor-based systems for automatic case and item retrieval.

Despite the alternatives of moving the material to the person and moving the person to the material, major handling systems continue to be designed without serious consideration being given to *both* alternatives.

The best alternative(s) for a specific application will depend on the material characteristics, flow requirements, economics, and organizational environment. Experience indicates that each alternative method will be preferred to the others under a particular set of circumstances. Among the factors that will influence the selection are size and weight of the material, storage height, hazardous and fragile material conditions, pilferable material, energy requirements, throughput or activity levels, facility limitations, and the desired level of computer control. Additionally, order characteristics, such as the number of line items per order and the amount ordered per line item, must be considered.

In making the selection, the *total system* must be considered. System productivity, rather than component productivity, and the system cost per unit picked should provide the basis for the selection.

The material-to-person versus person-to-material issue also exists in manufacturing areas. Material-to-person examples include pneumatic conveyors delivering tools and supplies to work stations from a tool crib, transporter conveyors feeding material from flow racks or carousel storage conveyors to individual electronic assembly stations, and driverless carts supplying material to machining centers. In each case the material could have been obtained by having the person travel to the material storage point. As work-in-process storage and control systems develop, increased consideration must be given to material-to-person alternatives.

10.4.6 AUTOMATED STORAGE AND RETRIEVAL SYSTEMS

The automated storage and retrieval system (AS/RS) has had a dramatic impact on storage and warehousing. Through the use of computer control, handling and storage systems have been integrated into manufacturing and distribution processes. Although the primary emphasis had been on storing and retrieving finished goods, more recently the focus has shifted to work in process, raw materials, and supplies.

The AS/RS has also affected the design of the warehouse. Rack-supported buildings, for example, are frequently constructed to house the AS/RS. A typical rack-supported building is shown in Exhibit 10.4.15. Such structures are usually more economical to construct than conventional buildings; additionally, they frequently receive favorable income tax treatment because of their special-purpose construction.

The AS/RS consists of storage racks, storage/retrieval (S/R) machines, and I/O or pick-up/deposit (P/D) stations. A typical AS/RS layout is given in Exhibit 10.4.16. The basic function of the S/R machine is to store and retrieve loads. Typically, each S/R machine operates in a single aisle with storage racks on either side. The typical S/R machine consists of a single or double mast frame, a carriage, a shuttle, and a drive and guidance mechanism(s). The frame is designed to add stability

Exhibit 10.4.15 Rack-Supported Building[a]

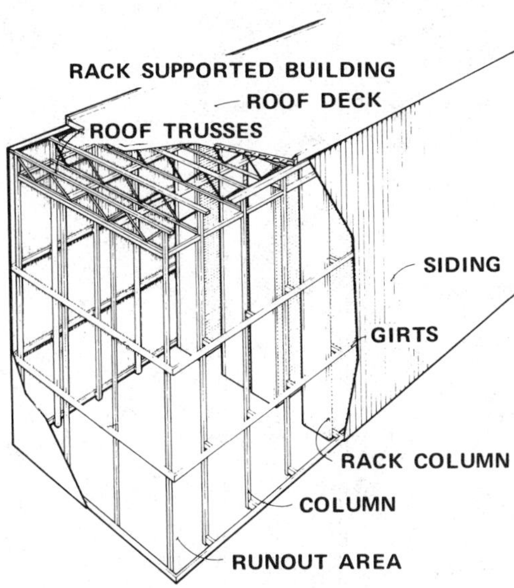

to the machine while moving at reasonably high speeds. The lower portion of the frame has wheels, which normally run on a single floor rail or two rails. It is guided from above by a support rail that runs along the rack structure. The frame is also used to guide the carriage. Input to and output from the AS/RS can be by conveyor, industrial truck, automatic guided vehicles, and so on.

The carriage carries the shuttle, which is the mechanism that transfers the load from the S/R machine to the rack opening (or to the P/D station), and vice versa. Typically, an S/R machine has

Exhibit 10.4.16 Typical AS/RS Layout[a]

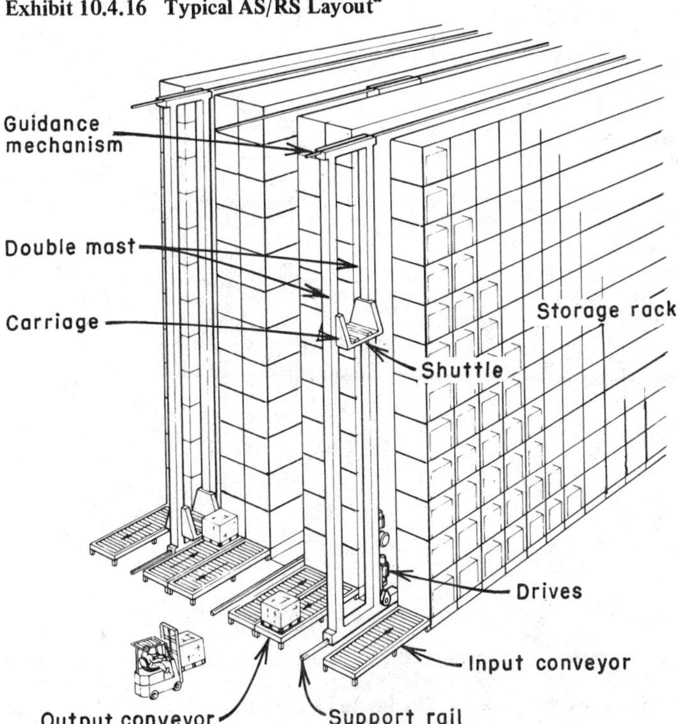

Exhibit 10.4.17 Unit Load S/R Machine[a]

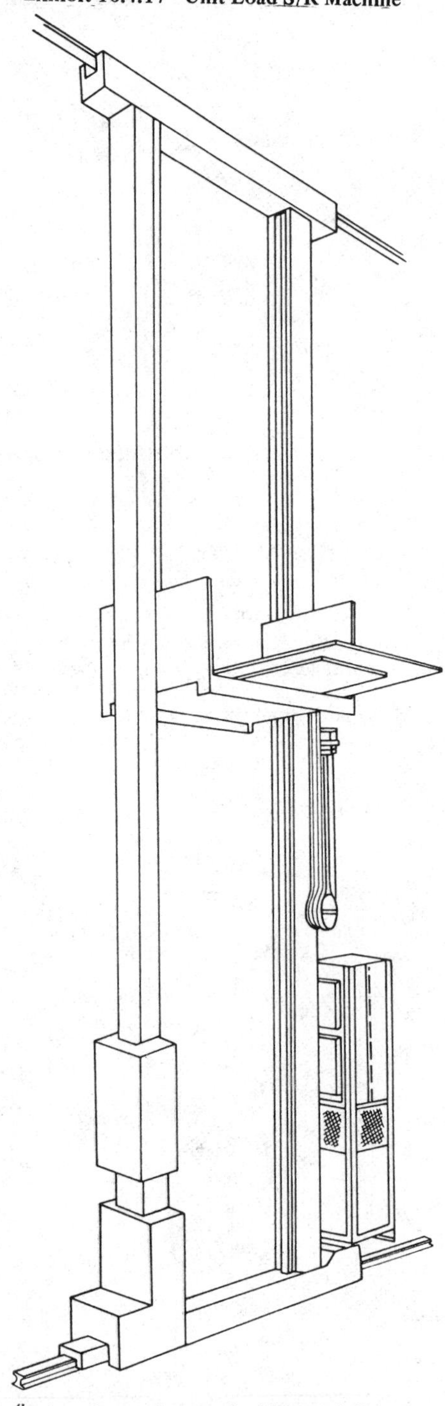

[a]Taken from reference 7 with permission.

three mechanical drives: the horizontal drive, which moves the frame back and forth along the aisle; the vertical drive (hoist drive), which raises and lowers the carriage; and the shuttle drive, which transfers the load between the S/R machine and either side of the aisle. The horizontal and vertical drives can both operate at the same time, so that the machine moves diagonally in order to reduce travel time.

The S/R machines have numerous technical specifications, depending upon the particular system under consideration. In general, the system can be considered to be one of the following: unit load AS/RS, person-on-board AS/RS, or miniload AS/RS.

Unit Load AS/RS

Unit load S/R systems are most commonly used for warehousing finished goods or raw materials in situations where a high volume of material needs to be stored and retrieved. Typically, the loads are palletized or placed in a tote or a container. For light loads the S/R machine capacity generally varies between 300 and 700 lb, with a lifting height of up to 40 ft. The handling mechanism usually consists of a shuttle table or a mechanical clamp. However, it can also be a vacuum or magnet-based mechanism for handling sheet metal and coils. A typical unit load S/R machine is shown in Exhibit 10.4.17.

For heavier loads the S/R machine capacity generally varies between 700 and 8800 lb, with a lifting height of 100 ft and beyond. Typically, the horizontal travel speed is between 250 and 500 fpm, whereas the vertical travel speed is between 60 and 100 fpm. Shuttle travel is generally around 88 fpm.

When throughput requirements are low, it might not be justified to dedicate an S/R machine to each aisle. In such cases transfer cars are used to transfer the S/R machine between aisles. The transfer car is located at the end of the aisle; it typically has a travel speed from 100 to 160 fpm.

Unit load systems can store loads in single-deep or double-deep storage racks. Occasionally, with small loads, three-deep storage might be performed. Multishuttle S/R machines have been used to facilitate double-deep operations; additionally, double-shuttle S/R machines have been used to accommodate peak periods of storage or retrieval.

Deep-Lane AS/RS

A variation of the unit load AS/RS is the deep-lane AS/RS. As shown in Exhibit 10.4.18, a deep-lane storage system is similar to a unit load AS/RS, with the exception that a single S/R machine handles multiple pallet depths on both sides of the aisle. The primary benefit of the deep-lane AS/RS is the savings in aisle space. Person-on-board systems also enable order picking to be combined with storage in order to increase efficiency. However, they require a slightly different rack structure and a special-purpose vehicle to interface the rack opening with the S/R machine. Such a vehicle is typically a moving platform that carries the load into the lane, senses proper load location, deposits load, and returns to the cube face for the next load. On FIFO operations, two vehicles are required, one at each end of the lane. The vehicle at one end of the lane performs storages, and the vehicle at the other end of the lane performs retrievals. On LIFO operations, a single vehicle stores and retrieves from the same end of the lane.

The S/R machine is typically capable of handling loads up to 4000 lb. The horizontal and vertical speeds of the machine can generally go up to 400 fpm and 90 fpm, respectively. The vehicle that interfaces the rack opening with the S/R machine generally weighs around 850 lb. It has a

Exhibit 10.4.18 Deep-Lane Storage[a]

[a]Taken from reference 7 with permission.

load capacity of up to 4000 lb and a horizontal speed of up to 120 fpm. It typically measures between 34 in. and 43 in. in width and 44 in. and 52 in. in length and is 7 in. in height.

Person-on-Board AS/RS

There are several ways of storing and retrieving materials in less than unit load quantities. As noted in the preceding section of this chapter, one approach is to store/retrieve at the storage location or in-aisle. For in-aisle picking, an aisle-captive, person-on-board S/R machine such as that shown in Exhibit 10.4.19 is sometimes used. The operator picks from shelves, bins, or drawers within the

Exhibit 10.4.19 Person-on-Board S/R Machine[a]

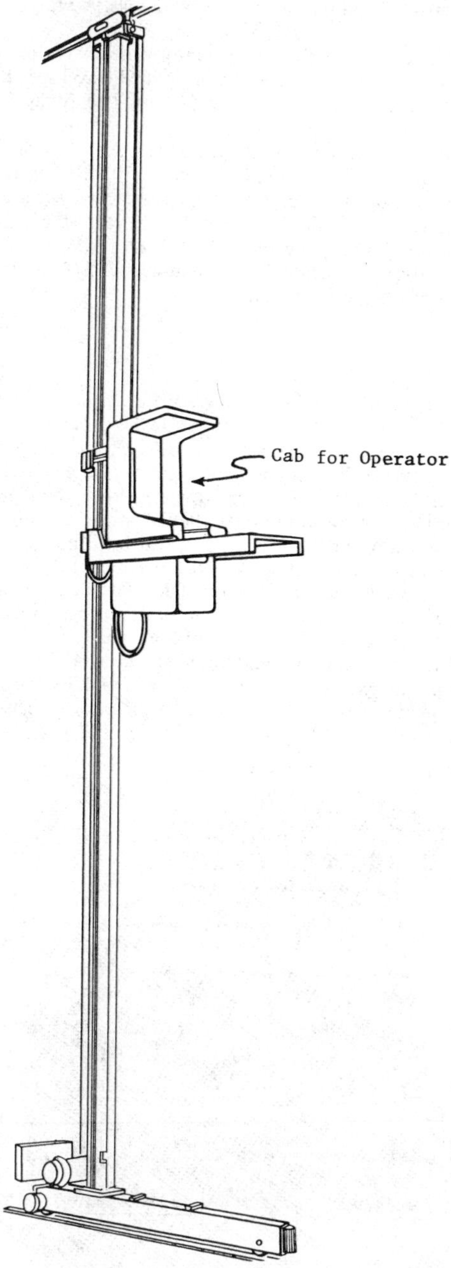

Cab for Operator

[a]Taken from reference 7 with permission.

storage structure and places the picked items into totes or modules, which are then carried by the S/R machine either to the end of the aisle or to an intermediate conveyor in the aisle for dispatch. The operator's platform may contain auxiliary devices to facilitate retrieving heavy items. The machine can be operated manually. However, efficiency can be increased by utilizing automatic controls, including an on-board computer terminal to provide the operator with such information as the location of the item and the amount to be picked. The S/R machine capacity generally ranges up to 1500 lb, with a lifting height between 40 ft and 80 ft. Vertical and horizontal travel speeds are similar to those of the unit load AS/RS. Some S/R machines will handle up to 4000 lb, including the operator.

In-aisle picking also can be performed using more conventional methods. Walking to bin shelving and riding from aisle to aisle on order picker trucks, storage/retrieval trucks, and so on, are alternatives to the aisle-captive, person-on-board S/R machine.

Miniload AS/RS

As was also noted in the preceding section, an alternative to in-aisle storage/retrieval of less than unit load quantities is end-of-aisle order picking and replenishment. One method of performing end-of-aisle order picking and replenishment is the miniload AS/RS. The miniload AS/RS is appropriate for handling small parts that are stored in containers, tote boxes, and so on. Such systems are generally space efficient and provide real-time inventory control. A typical miniload AS/RS installation is depicted in Exhibit 10.4.20.

The S/R machine for the miniload system is similar in structure to that of the unit load system. However, it is generally equipped with additional or more sophisticated extractors and guidance mechanisms in order to meet the requirements imposed by tighter clearances. (Handling smaller units requires increased accuracy.) Load weights generally vary between 300 and 700 lb. The lifting height is generally between 10 ft and 40 ft. Horizontal travel speed is generally 350 fpm, whereas vertical travel speed is generally between 60 fpm and 80 fpm. As with the unit load version, the miniload S/R machine can be fully automated.

Carousels

As noted previously, another method of performing end-of-aisle order picking is with carousel conveyors. As depicted in Exhibit 10.4.21, the carousel concept consists of a horizontal revolving bin that brings the bin contents to the pick station. It consists of a set of carriers, and each a carrier consists of vertical rows of multishelf bins. The drive mechanism rotates the carriers in order to bring the appropriate bin to the picker. The drive mechanism typically is a conveyor with overhead or floor-mounted (for heavier loads) drive.

Carousels are used for order picking, progressive assembly, production kitting, staging, sorting, test accumulation, and maintenance stores. The number of carriers generally varies between 20 and

Exhibit 10.4.20 Miniload S/R Machine[a]

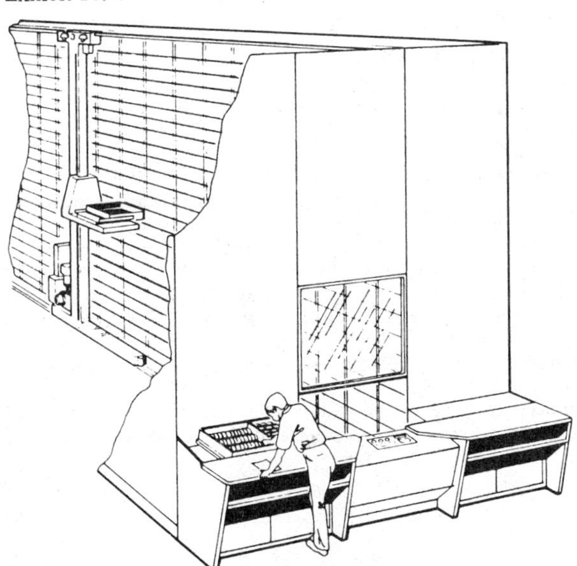

[a]Taken from reference 7 with permission.

Exhibit 10.4.21 Carousel S/R Machines[a]

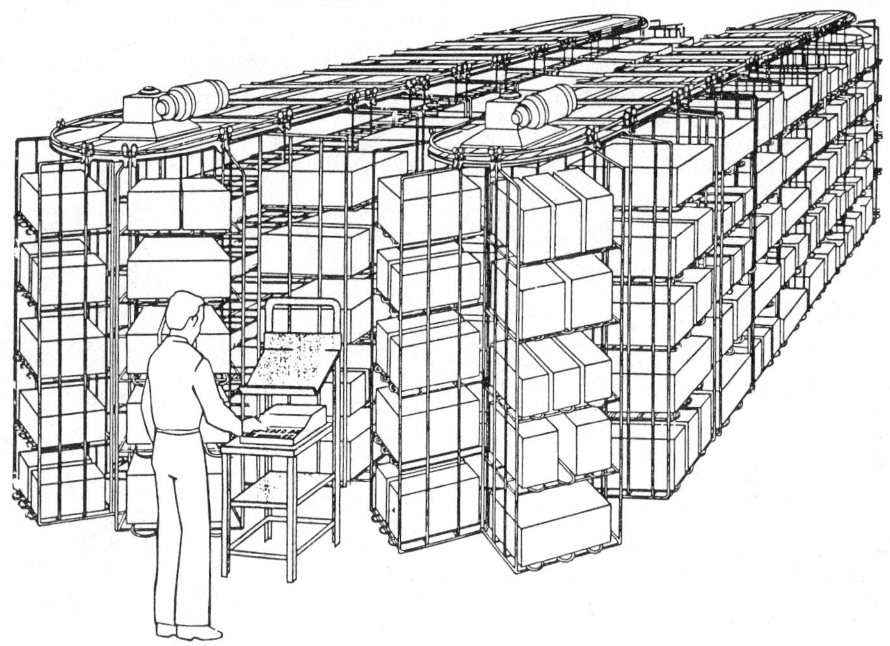

[a]Courtesy of Saratoga Conveyor Corporation.

70 carriers. Carriers can handle up to 5000 lb and have a depth of 12 in. to 22 in., a width of 3 ft 8 in. to 5 ft 4 in., and a height of 6 ft to 10 ft. The carousel speed is usually 60 to 80 fpm. The rated capacity of a carousel depends upon the length, bin size, load distribution, usage factor, and the desired speed. They can handle units up to 125 ft long. Furthermore, special bin and shelf designs are also made available.

Automated storage/retrieval carousels have increased the performance characteristics of the carousel system. One such system consists of multiple levels of carousels in which parts are stored in tote boxes or storage bins. The bins are automatically retrieved by a robot device; the bin is dispatched by means of a conveyor to a remote station for order picking; the bin is returned to the carousel system by conveyor; and the robot device automatically stores the bin on the carousel.

Cycle Time Calculations

The unit load S/R machine is capable of traveling in the aisle both vertically and horizontally simultaneously. Hence the time required to travel from the P/D station to a storage or retrieval location is the maximum of the horizontal and vertical travel times.

Because of the importance of the cycle time in developing the throughput for the system, considerable emphasis is placed on the time for the S/R machine to perform both *single-command* and *dual-command* cycles. A single-command cycle consists of either a storage or a retrieval, but not both, whereas a dual-command cycle involves both a storage and a retrieval.

A single-command storage cycle generally begins with the S/R at the P/D station; it picks up a load, travels to the storage location, deposits the load, and returns empty to the P/D station. A single-command retrieval cycle also generally begins with the S/R at the P/D station; it travels empty to the retrieval location, picks up the load, travels to the P/D station, and deposits the load.

A dual-command cycle generally begins with the S/R at the P/D station; it picks up a load, travels to the storage location, deposits the load, travels empty to the retrieval location, retrieves the load, travels to the P/D station, and deposits the load. A total of two pickups and two deposits is performed with a dual-command cycle.

Assuming randomized storage, an end-of-aisle P/D station located at the base of the rack, constant horizontal and vertical velocities, and single-sized rack openings, cycle times for the S/R are given by Bozer and White[8] as

$$T_{SC} = \alpha \left(1 + \frac{\beta^2}{3}\right) + 2T_{P/D}$$

$$T_{DC} = \frac{\alpha}{30} (40 + 15\beta^2 - \beta^3) + 4T_{P/D}$$

where

$\alpha = \max(t_h, t_v)$

$\beta = \min(t_h/t_v, t_v/t_h)$

with

T_{SC} = single-command cycle time

T_{DC} = dual-command cycle time

$T_{P/D}$ = time to perform either a pickup or a deposit

t_h = time required to travel horizontally from the P/D station to the furthest location in the aisle

t_v = time required to travel vertically from the P/D station to the furthest location in the aisle

If aisle-captive equipment is used that performs horizontal and vertical travel sequentially rather than simultaneously, then Bozer and White[8] provide the following expressions for cycle times:

$$T_{SC} = t_h + t_v + 2T_{P/D}$$

$$T_{DC} = \frac{4}{3}(t_h + t_v) + 4T_{P/D}$$

Design of AS/RS

In designing an AS/RS a number of decisions must be made. Depending upon the particular situation, some of the decisions will be made by the system supplier, and others will be made by the user of the system. Information must be collected on a number of aspects, and values for many design parameters must be established. Among some of the most important considerations are the following[8]:

1. Load size(s) and opening size(s).
2. Number and location of P/D stations.
3. Building construction—rack supported or conventional.
4. Land availability, condition, cost, and zoning restrictions.
5. Number, height, and length of storage aisles.
6. Percentage of the operations (storages and retrievals) to be performed on a dual-command basis.
7. Applicability of transfer cars.
8. Randomized or dedicated storage or some combination.
9. Dwell point for the S/R machine when it is idle; e.g., in-aisle or at P/D station.
10. Level of automation.
11. Level of computer control.
12. Requirement for physical inventory.
13. Replenishment requirement.
14. Requirement for maintenance.
15. Use of slave pallets versus vendor pallets.
16. Mode of I/O.
17. Plan for evolution and change.
18. Throughput requirement—peak and average.
19. Level of specifications to be developed for hardware and software.
20. Provision for interrupt and priority storage/retrieval.
21. Storage depth—single, double, or deep lane.
22. Provision for mixed loads on a pallet.
23. Use of automatic identification systems.
24. Use of simulations to support design decisions.
25. Amount of queue space required.
26. Impact of randomness versus scheduled operation on the requirements for the system.
27. Impact of energy and utilities.
28. Impact of inflation and income taxes.
29. Sprinkler requirements.
30. Plan for start-up, debugging, and postauditing.

10.4.7 DESIGNING WAREHOUSING SYSTEMS

The previous discussion of AS/RS design did not adequately cover the broader subject of designing warehousing systems, since the vast majority of warehousing requirements do not justify AS/RS designs. Second, the emphasis in the preceding section was on storage and retrieval; such functions as receiving, dispatching to and from storage, order picking and accumulation, and shipping were not treated.

In analyzing warehousing requirements and in designing new or improved systems for handling, storing, and controlling materials and supplies, a questioning attitude must prevail. The basic questions of *why*, *what*, *where*, *when*, *how*, *who*, and *which* should be asked constantly throughout the planning process. In particular, the following questions should be addressed, as a minimum:

1. Why?

a. Why are handling, storage, and control required?
b. Why are the operations performed as they are?
c. Why are the operations performed in the current sequence?
d. Why is material received as it is currently?
e. Why is material shipped as it is currently?
f. Why is material packaged as it is currently?
g. Why is material stored as it is currently?
h. Why is material controlled as it is currently?

2. What?

a. What is to be moved, stored, and controlled?
b. What data are available and required?
c. What alternatives are available?
d. What are the benefits and costs for each alternative?
e. What is the planning horizon for the system?
f. What should be mechanized and/or automated?
g. What should be done manually?
h. What should not be done at all?
i. What other firms have related problems?
j. What criteria will be used to evaluate alternative designs?

3. Where?

a. Where are material handling, storage, and control required?
b. Where do material handling (storage or control) problems exist?
c. Where should material handling, storage, and control equipment be used?
d. Where should material handling, storage, and control responsibility exist in the organization?
e. Where will future changes occur?
f. Where can operations be eliminated, combined, simplified?
g. Where can assistance be obtained?

4. When?

a. When should material be moved and stored?
b. When should I automate?
c. When should I consolidate?
d. When should I eliminate?
e. When should I expand (contract)?
f. When should I contact consultants?
g. When should I consult vendors?

5. How?

a. How should material be moved, stored, and controlled?
b. How do I analyze the material handling, storage, and control problem?
c. How do I sell everyone involved?
d. How do I learn more about material handling, storage, and control?
e. How do I choose from among the alternatives available?

6. Who?

a. Who should be handling, storing, and controlling material?
b. Who should be involved in designing the system?
c. Who should be involved in evaluating the system?
d. Who should be involved in installing the system?
e. Who should be invited to submit equipment quotes?
f. Who has faced a similar problem in the past?

7. Which?

a. Which operations are necessary?
b. Which problems should be studied first?
c. Which type of equipment (if any) should be considered?
d. Which alternative is preferred?

Exhibit 10.4.22 Material Handling System Equation

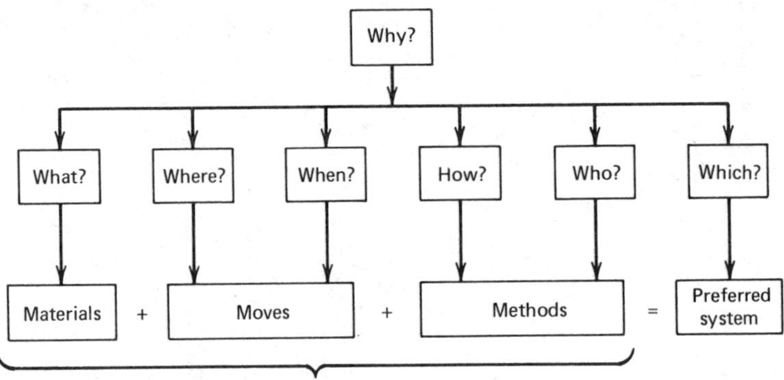

As shown in Exhibit 10.4.22, liberal use of the question "Why?" is essential to separating what must be from what has been; asking what and why defines the correct materials to be handled, stored, and controlled; asking where, when, and why identifies the necessary moves to be performed; asking how, who, and why establishes the correct methods to be used; and asking which and why yields the preferred design. Factors to be considered in designing material handling, storage, and control systems include the types of materials, their physical characteristics, the quantities to be moved, the sources and destinations for each move, the frequencies or rates at which moves must be made, the equipment alternatives, and the units to be handled, among others.

Some forms, given by Tompkins,[9] that can be used to facilitate the warehouse planning process are provided in Exhibit 10.4.23 and 10.4.24. As noted, information concerning the material, move, and method is required.

Receiving and Shipping

In designing receiving and shipping operations, it is necessary to consider the *handling*, *storage*, and *control* requirements of the activity and to provide the proper combination of *space*, *equipment*, and *people*. The number, location, and design of docks for both inbound and outbound shipments is an important consideration.

Docks are among the first requirements at a site and are vital for smoothly functioning operations. Additionally, plans for expansion must incorporate receiving and shipping. A rule of thumb concerning expansion is to expand the warehouse without disrupting dock operations. As shown in Exhibit 10.4.25, numerous expansion alternatives exist for a distribution center. The small arrows in the exhibit represent material flows, and the large, open arrows represent feasible directions for expansion.

Requirements for people, equipment, and space in receiving and shipping depend on the effectiveness of programs to incorporate prereceiving and postshipping considerations. For example, by working with vendors and suppliers, peak loads at receiving can be reduced. Scheduling inbound shipments is one method of reducing the impact of randomness on the receiving work load.

Another reason for being concerned with prereceiving activities is the opportunity to influence the unit load configuration of inbound material. If, for example, cases are hand stacked in the carrier by the vendor, then they will probably have to be unloaded by hand. Likewise, if slip sheets or clamp trucks were used by the vendor to load the carrier, and if receiving does not have similar material handling equipment, the shipment probably will have to be unloaded by hand. Finally, if the material is received in unit loads not compatible with your material handling system, then additional loading, unloading, or both might be necessary.

A third reason for trying to influence prereceiving activities is to provide a smooth interface between the vendor's and receiver's information system. Where automatic identification systems are used in receiving, some firms supply their vendors with the appropriate labels to be placed on inbound material to facilitate the receiving activity. The faster and more accurately receiving is performed, the more both the vendor and the buyer benefit: Information is provided to accounting and payments are made sooner, and material is available for use sooner, rather than sitting idle.

Just as the receiver wishes to influence the vendor, the customer wishes to influence the shipper. Hence postshipping activities must be considered. In addition to the reasons cited for considering prereceiving activities, the following are of concern in postshipping activities: returnable containers, returned goods, returning carriers, and shipping schedules.

Exhibit 10.4.23 Shipping and Receiving Analysis Chart

Company________ Date________ ________Raw materials ________Finished goods

Prepared by________ Sheet________of________ ________Plant supplies

Description	Unit Loads				Size of Shipment (Unit Loads)	Frequency of Shipment	Transportation		Materials Handling	
	Type	Capacity	Size	Weight			Mode	Specifications	Method	Time

Exhibit 10.4.24 Shipping Analysis Chart

Company________ Date________ ________Raw materials ________In-process goods

Prepared by________ Sheet______of______ ________Plant supplies ________Finished goods

Description	Unit Loads				Quantity of Unit Loads Stored			Storage Space			
	Type	Capacity	Size	Weight	Maximum	Average	Planned	Method	Specifi-cations	Area (ft^2)	Ceiling Height Required

Exhibit 10.4.25 Distribution Center Flow and Expansion Concepts

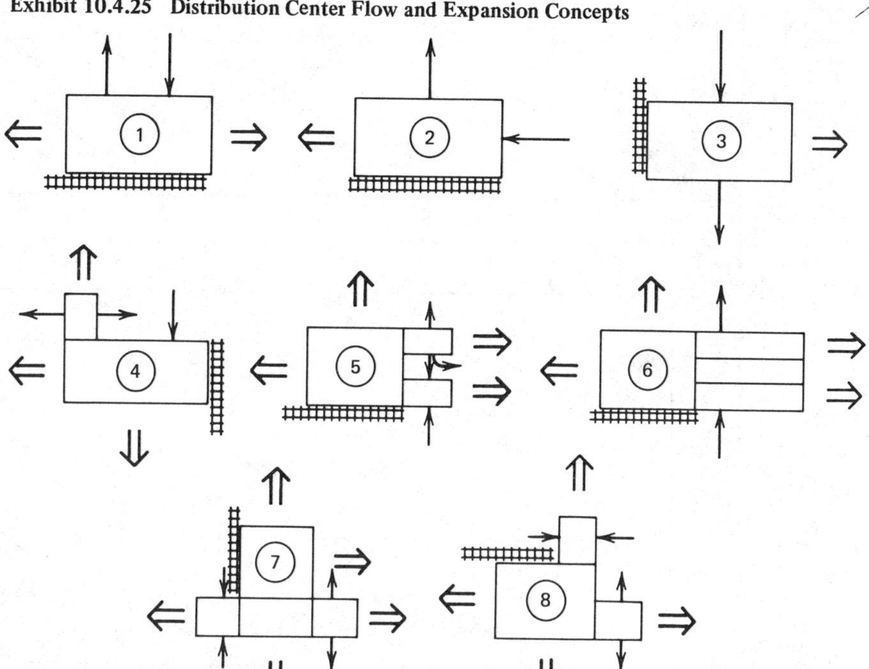

If goods are shipped to customers in returnable containers, then a system must be developed to keep track of the containers and to ensure that they are returned. Additionally, whether or not the shipping container or support is returnable, a natural attrition will occur, and replacements must be planned for.

Goods are returned because they fail to meet the customer's quality specifications, because mistakes are made in the type and amount of material shipped, or because the customer just simply decides not to accept the material. Regardless of the reason, returned goods must be *handled*, and an appropriate system for handling the material must be designed.

If supplier-owned equipment is used to deliver material to customers, then consideration should be given to the utilization of the capacity of the carrier on the return trip. The "backhaul" of the carrier could be used for returning the returnable containers; additionally, it could be used for other transportation purposes.

Shipping schedules can have a significant impact on the resource requirements for shipping. Hence close coordination is required between the shipper and the shipping department. The shipper might be the customer's carrier, its own carrier, a contract hauler, or a commercial carrier. If shipping activities are to be planned, then shipping schedules must be accurate and reliable.

In addition to the need for closely coordinating vendor and receiving activities and shipping and customer activities, it is equally important to coordinate the activities of receiving and production, production and shipping, and receiving and shipping.

The natural sequence for the flow of material is vendor, receiving, storage, production, warehousing, shipping, customer. However, in some cases material might go directly from receiving to production and from production to shipping. Hence such possibilities should be included in the system design.

Why should receiving and shipping be coordinated? Common space, equipment, and/or personnel might be used to perform receiving and shipping. Additionally, when, say, slave pallets are used in a manufacturing or warehousing activity, empty pallets will accumulate at shipping and must be returned to the loading point at either receiving or production.

To improve productivity in receiving and shipping, the following areas and topics are suggested for possible application[10]:

1. Use automatic loading and unloading trailers.
2. Use automatic identification systems, label printers, and scanner/readers to facilitate data entry in both receiving and shipping.
3. Use automatic palletizers and depalletizers, shrink-wrap/stretch wrap/strapping for unitizing/stabilizing loads.

4. Use special attachments on industrial trucks, for example, slip sheet, clamp, barrel, and appliance attachments.
5. Use extendable conveyors that go into trailers to facilitate loading/unloading trailers.
6. Use dock levelers or sturdy dock boards.
7. Schedule arrival of both outbound and inbound carriers and preassign spots for loading/unloading.
8. Centralize receiving and shipping; receive and ship at different times of the day or on different days of the week.
9. Require inbound shipments to be unitized and, possibly, palletized.
10. Provide an ample number of dock doors.
11. Eliminate floor stacking materials and provide racks for staging materials.
12. For inbound shipments requiring inspection, pull samples for quality control, dispatch remainder to storage, and "lock" material in storage until released by quality control.
13. Preassign receiving numbers and provide computer terminals at the dock area for expediting receiving.
14. Evaluate side-loading and top-loading trailers to facilitate automated loading.
15. Consider unpacking small parts at receiving and placing in standard containers for movement throughout the system; compact and bale waste packing materials.

Order Picking

When an order is received to retrieve or pick materials, the quantity to be picked might be one or more full pallet loads, one or more full cases, or one or more individual items. Alternative storage methods for facilitating the order picking process were described previously.

A number of different approaches are used in picking the items on customers' orders. For example, some warehouses assign an order picker to an individual order; some batch several orders and assign them to an order picker; and some assign order pickers to zones within the warehouse. When "zone picking" is used, picking lists are developed for each zone by batching orders and separating the items by zone.

Picking single orders at a time maintains the integrity of the order, allows packing as picking occurs, and often results in travel time being a large percentage of the total order picking cycle time. With batch or zone picking, the orders must be accumulated following the picking operation; however, reductions generally occur in the percentage of the picking cycle devoted to travel time.

To determine the preferred method of order picking (single, batch, or zone), an analysis should be performed, based on the orders received and on the layout of the warehouse. Rankin[11] presented the results of an analysis of the order picking activities of a replacement parts warehouse for Deere and Company. The alternatives considered were single order picking and zone order picking. Using computer simulations of the order picking activities, the optimum number of orders to be batched for zone picking was determined.

Order picking times are affected not only by the storage and handling equipment and the order picking method, but also by the layout of the order picking area. For example, suppose one area in the warehouse is devoted to order picking, and another area is used for reserve storage. The amount of inventory in the picking area affects the size of the picking face (space devoted to an individual item along a picking aisle) and the frequency of replenishments. Hence, as the quantity stored at the pick location increases, (1) the size of the container for picking stock increases, (2) the travel time to perform a picking cycle increases, and (3) the number of required replenishments decreases. As the quantity stored at the pick location decreases, (1) the container size decreases, (2) the travel time to perform orderpicking decreases, and (3) the number of required replenishments increases.

Trade-offs exist between order picking costs and replenishment costs as the size of the picking face changes. Thus for any given situation there exists some optimum size for the picking face. Unfortunately, few picking areas are designed on the basis of an analysis of the pick-replenishment trade-offs. Rather, rules of thumb such as "store a week's supply" and "store as much as will fit in our standard bin" are used.

A second layout consideration that affects order picking productivity is the length of storage racks in a rack storage area. Suppose a warehouse has the layout shown in Exhibit 10.4.26*a*. To pick an order consisting of items located in slots A, B, C, and D requires more travel distance than is required for the layout shown in Exhibit 10.4.26*b*. Thus trade-offs between space and picking time occur as changes are made in the number of slots per storage rack module.

An analysis of the effect of rack length on storage space and order picking productivity was performed at Deere and Company.[11] The results of the analysis are provided in Exhibit 10.4.27. As can be seen, order picking productivity was maximized with a storage rack module 62.5 ft long; the associated number of line items picked per hour was 56, with a storage area of 169,000 ft^2. It should be noted that order picking productivity decreases with decreasing rack lengths below 62.5 ft. The analysis considered throughput only; reducing rack lengths increases the number of cross-

Exhibit 10.4.26 Rack Length Effect on Space and Order Picking

aisles and thus increases space. To determine an optimum rack length requires that both space and throughput be considered. Depending upon the cost of storage space versus the cost of order picking, an optimum value can be obtained for the length of a storage rack module. The lesson to be learned from the analysis is that decisions concerning the length of the storage rack module should not be made casually.

Exhibit 10.4.27 Effect of Storage Rack Length on Order Picking Productivity and Floor Space

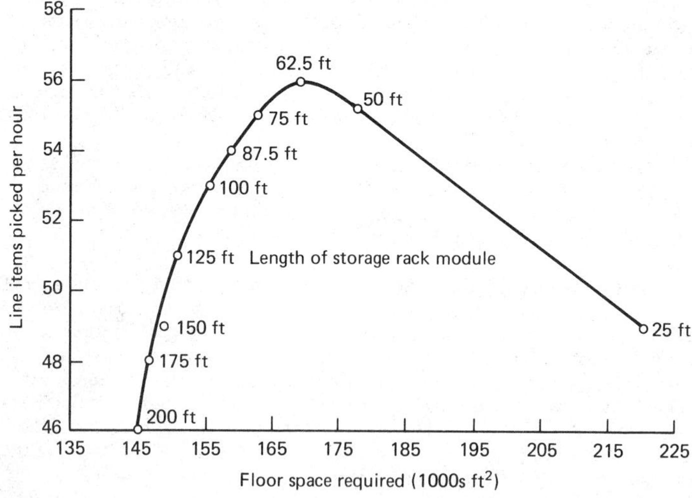

10.4.8 AUDITING THE WAREHOUSING SYSTEM AFTER INSTALLATION*

Once the automated material handling, storage, and control system has been installed, what next? Since conditions change and estimates are not exact, the system must be monitored or audited periodically.

This section addresses the need to plan carefully for a postaudit program. Accountability on the part of the user and supplier is addressed, and the responsibilities of each are identified. User accountability is explored for management, engineering, and operating personnel.

For the purposes of this discussion, an "audit" may be defined as a comprehensive, systematic examination of a material handling system: its use of space, equipment, personnel, energy, and financial resources; its handling, storage, and control components; its performance under current conditions; and its capacity to meet future requirements.

The audit is more than a financial audit, it is more than an operations audit, and it is more than an audit of the economic justifications performed. The audit being recommended is truly a *systems audit,* including audits of the planning, designing, building, installing, commissioning, operating, and maintaining phases for the automated material handling and storage system.

In performing the systems audit, the following areas should be examined:

1. Installation, to verify that the hardware and software systems installed were in fact what were designed and/or purchased.
2. System performance, to verify that satisfactory performance is being provided, focusing on throughput, personnel required, software, uptime, return on investment, costs, and productivity.
3. Requirements, to verify that sufficient capacity exists to meet requirements for space, equipment, and personnel.
4. Management of the system, to verify that the policies, procedures, controls, and reporting channels support effective management of the system.

Some reasons for performing the systems audit are to learn from the past, to assess the need for change, to develop a data base, and to establish or improve credibility within the organization. Despite the fact that many benefits can be obtained from formal system audits, very few firms appear to perform such audits.

Some of the more common reasons for firms' failing to conduct audits are:

1. **Fear.** Being afraid that promises were not kept, benefits were not realized, and resources were wasted.
2. **Mobility.** The lead times for planning, designing, and installing the system are such that the persons responsible have been promoted on transferred or have left the company.
3. **Lack of Accountability.** Typically, the persons responsible for planning, designing, and selling the system are not responsible for installing, operating, and maintaining the system.
4. **Time Pressures.** Other needs exist, along with the attitude "there's never time to do it right."
5. **Changing Requirements.** Because of internal and external factors, by the time the system is installed, the need has changed.
6. **Lack of Baseline.** The current status of the system cannot be compared with anything; neither relative nor absolute comparisons can be made.
7. **Hindsight Attitude.** Believing that "anyone can look back, but the real pros in the business are those who only look ahead" discourages audits.

As with many situations, success has many fathers, but failure is an orphan. Unfortunately, audits are associated with the binary judgments of failure and success. Yet, the terms "success" and "failure" are generally related to *outcomes*, not *decisions*. It is important to recognize the differences between decisions and outcomes. Good decisions can have good or bad outcomes; likewise, bad decisions can have good or bad outcomes. The merit of the decision is often related to the extent to which it was objective, analytical, comprehensive, repeatable, and explainable.

It is essential that the audit be performed on site. Too often quick judgments are made concerning systems; BOLD (based on little data) conclusions are drawn from comments made by persons lacking objectivity. Disgruntled suppliers, jealous users, and politically motivated employees can, through innuendo and rumor, cast doubts on a system that are difficult to overcome.

The audit must be performed at a level in the organization sufficient to allow it to be performed objectively. The report should be submitted at least at the funding approval level. Whether it

*This section is taken from reference 12, with permission of the Materials Handling Institute, Inc.

should be performed by someone at the site level, division level, or corporate level depends on the magnitude of the activity and the likely impact on the organization.

The audit should be designed when the system is designed. Too often audits are designed *after* the system is installed. Such an approach is similar to playing a game and *then* deciding how the scoring will be performed, or submitting bid packages, receiving vendor responses, and *then* deciding what factors will be used to select the supplier. One should know what things are important from an audit point of view when the system is being designed.

The initial audit is best performed approximately 6 months after it has been accepted by the user. To do it sooner is to invite the natural bias of "buyer's remorse." Immediately following start-up, and during debugging, the user typically undergoes a period of discouragement that can bias attempts to assess the system.

The audit not only can be performed too soon, but also can be performed too late. If the initial audit is delayed, then bad habits will have developed, firm judgments will have been made, and opportunities for improvements will have been foregone.

After the initial audit, annual audits are appropriate for some installations. For others, periodic, but indefinite, timing is preferred.

It is recommended that the audit be performed by someone who is objective; knowledgeable of automated material handling, storage, and control systems; tactful; and perceptive. The individual might be from plant, division, or corporate staff groups; someone from another plant or division might be used; or an outside consultant might be employed.

To be effective, the audit must be comprehensive. Hence managers, operators, maintenance engineers, designers, system suppliers, and functions being served by the system should be included in the audit. System suppliers can frequently provide recommendations for change or modification.

It is not necessary to perform formal audits of all material handling, storage, and control installations. However, it is advisable to audit those installations involving new technology (new for your firm, as well as new for the industry), radical changes from present methods, and complex (and expensive) systems. Dollar limits could be used as guides. However, frequency of occurrence in the company is another major consideration.

It is important that the audit be performed formally and that comparisons be made of the following:

1. Promises versus deliveries.
2. Expectations versus realizations.
3. Forecasts versus actual occurrences.
4. Assumptions versus outcomes.
5. Concepts versus installations.

Where differences exist, they must be explained.

The results of the audit should be reported formally; they should be reviewed with those affected before being submitted. It is important that feedback occur to ensure that opinions or perceptions are minimized and that facts are maximized. Actions for change or continuation must be recommended, and the impact of the recommendations must be assessed.

Accountability: Users' Responsibilities

Too often the user succumbs to the pressures of time and does not spend enough time planning and designing the automated material handling and storage system. The attitude seems to be "Let's just get something installed; we'll make the necessary modifications later!" Unfortunately, such an approach can be extremely costly.

As the life of an installation is viewed, it undergoes a cycle that includes planning, designing, building, installing, commissioning, maintaining, and modifying. The opportunities for cost savings decrease exponentially, and the cost of making changes in the system increase exponentially over the cycle. Hence it is important that more emphasis be placed on planning and designing.

As planning and designing are improved, the need for frequent postaudits will decline. In the meantime, how can users do a better job of fulfilling their responsibilities and become more accountable? Consider first the role of management.

Management must provide the right climate for audits to be productive. An adversary relationship must not exist; otherwise the audit process will be doomed to failure. Differences in decisions and outcomes must be recognized.

Management must be patient. Complex systems must be debugged. The period of time following start-up will tend to be a frustrating period; management support during this period is crucial.

Engineers need to be more accountable for their designs; they must avoid "pride of authorship" and recognize that changes in the design might be needed. Assumptions, objectives, alternatives considered, and rationale used must be documented throughout the design process for the audit process to be effective. Operating personnel must be involved in the design process. To the extent it is possible to do so, the design must be based on "facts" rather than guesses; it must be rooted in analysis, not hypotheses.

Operating personnel also need to be more accountable. It is important that they follow through on commitments. If reductions in personnel, equipment, and/or space were promised, then they should be made. Operating personnel need to be a part of the solution, not a part of the problem. Resistance to change, lack of confidence in the new system, and a lack of patience with the system do not create the proper climate for an effective audit.

Accountability: Suppliers' Responsibilities

The degree of responsibility shared by the supplier depends upon the role the supplier played in the design of the system. If the supplier designed the system, then the same responsibilities apply as were listed for the engineer.

The supplier will need to participate in the audit. Although there will be little, if any, legal responsibility during the audit, there will be a professional responsibility to ensure that the system meets the needs of the user.

Few, if any, suppliers can afford not to participate in the audit. The supplier probably has more to gain, as well as more to lose, from the audit than any individual within the user organization. Because of the mobility of the user, at the time the audit is performed the only person available who "was around at the beginning" may well be the system supplier. Such a situation lends further support for having a formalized audit program.

REFERENCES

1. *Measuring Productivity in Physical Distribution*, National Council of Physical Distribution Management, Chicago, 1978.
2. J. A. WHITE, "The 85% Rule of Thumb," *Modern Materials Handling*, July 1980.
3. W. T. MORRIS, *Analysis for Materials Handling Management*, Irwin, Homewood, IL, 1962, p. 16.
4. N. A. DEMARS, "Computer-Aided Warehouse Layout Design," 30th Annual Material Handling Short Course, Georgia Institute of Technology, Atlanta, February 1980.
5. N. A. DEMARS, J. O. MATSON, and J. A. WHITE, "Optimizing Storage System Selection," Proceedings of the 4th International Conference on Automation in Warehousing, Tokyo, Japan, 1981.
6. J. A. WHITE, "Picker to the Picking Face or Picking Face to the Picker?," *Modern Material Handling*, September 1979, p. 19.
7. *Consideration for Planning and Implementing an Automated Storage/Retrieval System*, Material Handling Institute, Pittsburgh, PA, 1977.
8. Y. A. BOZER and J. A. WHITE, "Optimum Designs of Automated Storage/Retrieval Systems," The Information Management Society/Operation Research Society of America Joint National Meeting, Washington, DC, May 1980.
9. J. A. TOMPKINS, *Facilities Design*, North Carolina State University, Raleigh, 1975.
10. J. A. WHITE, *Yale Management Guide to Productivity*, Eaton Corporation, Yale Industrial Truck Division, Cleveland, OH, 1979.
11. W. C. RANKIN, "Applications of Simulation in Distribution," *Proceedings*, 28th Annual Material Handling Short Course, Georgia Institute of Technology, Atlanta, 1978.
12. J. A. WHITE, "Auditing the System after Installation," *Proceedings 1980 Automated Material Handling and Storage Systems Conference*, Material Handling Institute, Pittsburgh, PA, April 1980.

BIBLIOGRAPHY

APPLE, J. M., *Material Handling Systems Design*, Wiley, New York, 1972.

BERNARD, C., II, "In Retrieval–Move the Man or Move the Load," *Production Engineering*, Vol. 26, No. 3 (March 1979), pp. 51–53.

FRANCIS, R. L., and J. A. WHITE, *Facility Layout and Location: An Analytical Approach*, Prentice-Hall, Englewood Cliffs, NJ, 1974.

KINNEY, H. D., "Planning for Storage and Warehousing," Material Handling Seminar, SME, Atlanta, January 1978.

KINNEY, H. D., "How to Size the Warehouse," Material Handling Management Course, AIIE, Norcross, GA, June 1979.

KINNEY, H. D., "A Total Integrated Material Management System," 1980 Annual Conference of the International Material Management Society, Cincinnati, OH, June 1980.

RYGH, O. G., "Orderpicking Alternatives," 1979 AIIE/MHI Seminars, Long Beach, CA, March 27, 1979.

TOMPKINS, J. A., "Problem Solving Techniques in Material Handling," *Proceedings 1976 Material Handling and the Industrial Engineer Seminar*, AIIE, Norcross, GA, 1976.

TOMPKINS, J. A., and J. A. WHITE, *Facilities Planning*, Wiley, New York, forthcoming.

WHITE, J. A., "Randomized Storage or Dedicated Storage?," *Modern Materials Handling*, January 1980, p. 19.

WHITE, J. A., "Aisle Lore–and More!," *Modern Materials Handling*, March 1980, p. 35.

ZOLLINGER, H. A., "Do It Yourself Guide to Costing Stacker Systems," *Automation*, September 1974.

CHAPTER 10.5

Energy Management

WAYNE C. TURNER
Oklahoma State University

WILLIAM J. KENNEDY, JR.
University of Utah

10.5.1 INTRODUCTION

Definition

Energy management is simply the judicious use of energy to maximize profits (minimize costs) and enhance competitive positions. Therefore any technique that encourages or leads to the prudent use of scarce and expensive energy for profit maximization is part of energy management. Energy management is energy conservation, but it is also much more.

For example, the use of dual fire capability with standby fuel utilization is certainly profit improvement, since equipment, process lines, and perhaps even entire plants can be run during periods of curtailments or brownouts. Since the secondary fuel almost always burns less efficiently than the primary fuel, more energy is required when it is utilized; so dual firing is not energy conservation. As another example, demand leveling and power factor improvement save few calories (btu) at the plant, but can save large amounts of money. This is energy management.

Objectives

Objectives vary somewhat among organizations, but some typical areas for an energy management program are:

1. Profit improvement.
2. Enhancement of competitive positions.
3. Energy conservation.
4. Reduction of curtailment or brownout impacts.
5. Good plantwide communications on energy matters at all levels.
6. Good energy reporting and monitoring system.
7. Minimization of employee discomforts due to energy programs.
8. Continued steady supply of goods and services during curtailments.
9. Incorporation of new products or services in the company's offerings to aid others in the energy crisis.

Results

It is difficult to summarize results, but as early as 1977, the aluminum industry had achieved a 5% reduction in energy consumption across all companies; the American paper industry, 4.1%; meat packing, 7%; and glass manufacturing, 12 to 15%–all in a 2 year period. Individual company savings increased a great deal, with some having substantial returns. A pattern seems to be emerging, as shown in Exhibit 10.5.1.

The literature abounds with reported savings of 40 to 60%, and in one recent case an office building reduced its consumption by 76%. The General Services Administration (GSA) operates an office building utilizing only 20% of the normal usage.

With savings like this, it is no wonder energy management is so popular. There are other reasons for involvement, however.

Exhibit 10.5.1 Typical Results of Energy Management Programs

Cumulative Savings	Action
5 to 10%	Low-cost–no-cost changes
25 to 30%	Capital-intensive engineering design (goal for most)
40 to 50%	Dedicated programs

Other Reasons

The United States is faced with a dramatic problem of rising energy costs and dwindling energy supplies. Consider the following:

1. Natural gas prices rose from $0.50/million btu in 1971 to $1.60 in 1976 to about $2.55 in 1979. For the same period, fuel oil rose from $0.50 to $2.00 to $4.10. (All figures are approximate national averages.)
2. The United States faces a severe balance-of-payments problem, yet the total cost of petroleum imports is more than the trade deficit. In other words, without petroleum imports, there would be no trade deficit.
3. With only one twentieth of the world's population, the United States consumes about one third of its energy. It also produces one third of its GNP, but countries such as Sweden, West Germany, and Japan require substantially less energy per unit of GNP than the United States. They have similar standards of living.

Relationship to Industrial Engineering

Energy management is good industrial engineering. Management is no longer concerned only with materials, labor, and equipment; energy must now be added to the list. Energy costs are now somewhere around 10% of total operating costs, and this figure is rising. Some predict that energy costs will escalate 3 to 10% faster than the underlying rate of inflation for the general economy.

The prudent industrial engineer will study energy management. The problem and subsequent potential returns are simply too large to ignore.

This chapter is presented in the section on facilities design because many of the activities of energy management are facility oriented: layout, caulking, insulating, modifying, and so on. It could also be in any of several other sections, however.

10.5.2 PROGRAM START-UP

Program Design

The actual program design must be done at the top management level, but that does not mean the concept must start there. Good industrial engineers at any level can plot energy costs and show forecasted costs and supply problems. A simple "what if" plot showing savings if energy costs were reduced 10, 15, 20%, and so on, is very impressive and will convince most progressive management of the real problem.

Organizationally, there should be one person who has complete responsibility for the energy management program. This person should be talented, have a technical base, be a good manager, and be a self-starter. To avoid problems with conflicting objectives, this person should report well up into the organizational structure of the company. This person is often called the "energy management coordinator."

Because of their systems orientation, industrial engineers are good candidates for this position, but so are plant engineers, talented maintenance personnel, good line managers, and other highly motivated persons. The key is to select a person who believes in the program and who will work for its success. (Perhaps the person who initiated the discussion is a logical choice.)

Unfortunately, the technical talent required for success does not normally lie in one person or discipline. Backup talent is needed and is often provided by some type of task force. This task force, often called the "energy management committee," should consist of the energy management coordinator as chairperson, plant engineers, mechanical engineers, chemical engineers, financial experts, and other talent deemed necessary.

In this same committee, or perhaps in a parallel effort, there needs to be substantial line involvement. The program needs this involvement in order to (1) generate communication, resulting in an upward flow of ideas, and (2) encourage participation, resulting in more cooperation.

Candidates for this committee include first-line supervisors, hourly employees (probably on a rotating basis), personnel department managers, and union officials. This broad spectrum of interest should help encourage unification or alignment of goals.

Program Initiation

Undoubtedly, one of the most important elements for a successful program is top management commitment to it. Without such commitment, the program will likely fail, but with it, the program has a fighting chance. This commitment must be strong and evident to all involved. Some ways in which managers can demonstrate their commitment include letters to employees, plant or department meetings at which they are present, newsletters, posters, billboards, and news releases—all of which are designed to attract and hold employees' attention. There are two other ways that are vitally important:

1. **Leadership by Doing.** Top management must be lead by showing the way. Thermostat settings must be consistent throughout the plant and everyone must learn to switch off lights and equipment.
2. **Resource Allocation.** Management must provide adequate resources. This includes, but is not limited to, funding cost-effective proposals.

Objectives must be stated early in the program and must be communicated to all involved. As soon as possible, specific objectives should be stated for all levels of the hierarchy. These objectives should be, for obvious reasons, tough, specific, and measurable.

The selection of early projects is critical and should be done with long-range goals in mind. For example, it is often wise to select the early projects carefully, based on guaranteed success with high returns. These golden opportunities exist for all organizations, and careful auditing and thought can lead to their selection.

For example, one company replaced mercury vapor lights in one of two refrigerated warehouses with high-pressure sodium. The system was carefully designed to improve visibility. Savings were large in that high-pressure sodium requires less energy to operate and also produces less waste heat that must be offset by the refrigeration. The project was a success, to the point that the employees in the other refrigerated warehouse asked for high-pressure sodium lights in their area. Projects such as this one are in your plant if you will seek them.

Program Management

A high level of creativity is required for a successful program; thus management should do whatever is necessary to encourage this development. Management by objectives (MBO) is one way that has worked well for other companies. Here, the energy management coordinator would meet with his or her boss and subordinates to define goals. Progress toward these goals would be monitored, and control exercised only as needed.

Management should recognize that the motivation for involvement in energy management takes many forms, but must be cultivated periodically. For example, a simple pat on the back, letter in an employee's file, or other form of recognition is often quite important to the type of person who is active in energy management.

Energy accounting through good monitoring and reporting is vitally important. Management should design such systems to evolve toward cost center metering and reporting. Someday energy will be part of the budgetary process instead of being overhead as it is today. This means that department heads will be allocated an energy budget, and their departments' usage metered. Excessive off-standard usages will have to be explained. Prudent management will anticipate this and be working toward it. The Energy Utilization Index (EUI) developed in the next section on energy audits is one good monitoring measure.

General Motors, 3M Company, and Carborundum all have designed good monitoring systems. Seidman and Seidman, a consulting firm, is also working on such systems.

10.5.3 ENERGY AUDITING

Purpose

One of the first steps in most energy management programs is the conducting of one or more energy audits. There seems to be no agreement in the literature as to what an energy audit is, but it is possible to list the purposes. From this list, the company can choose what it needs to accomplish and subsequently what type of audit it needs. The list is:

1. How much energy is being consumed and of what type? For these purposes, conduct a *gross audit*.

2. What processes are consuming the energy? For this purpose, conduct a *detailed audit* and a corrected load study.

3. What changes can be made to improve operations? For this purpose, conduct a *walk-through audit*.

a. Operating and maintenance (O & M) or low to no cost.
b. Energy management opportunities (EMOs).

Most plants find it necessary to conduct the first type (gross audit), whereas many do very little toward the second (detailed audit). All organizations with energy management programs do the third type (walk-through audit) in one form or another.

Gross Audit

The gross audit is a relatively simple audit designed to allow a firm to determine its total energy consumption and bill for each fuel type utilized. While doing this, it is possible to obtain some information on how well the energy is being used (e.g., power factor and demand billing for electrical utilization).

Forms have been designed to facilitate the conducting of the audit. A sample, rather comprehensive one is given in Exhibit 10.5.2, but a word of warning is in order. This form contains a great deal of information–perhaps more or less than you really need. The form should be designed for the specific organization being audited. It may contain more or less information than the one shown. *Instructions For Energy Auditors*[1] breaks this one form into about 10 separate forms. The auditor then simply chooses the forms needed.

Before going through the form, it is necessary to discuss quickly typical billing scheduling. The discussion here applies to electrical billing schedules. Schedules for most other forms (gas, gasoline, fuel oil, propane, and coal) are simpler and usually involve consumption billing only (tons of coal, cubic feet of gas, gallons of gasoline, etc.). Purchased steam sometimes includes a demand component similar to electricity.

The first level of electrical billing schedules is for small commercial and residential units. It involves a charge per kilowatt hour (measure of electrical usage equal to 8.6×10^5 cal or 3412 btu for 1 hr) usually on a declining block basis; that is, the cost per kilowatt hour decreases as the consumption goes up. The next schedule, for larger commercial concerns and small industry, involves the same type of consumption charge plus a demand charge. To understand this, consider Exhibit 10.5.3.

In Exhibit 10.5.3, the demand on the power company in kilowatts is plotted against time over 1 week. In customer A's case the demand fluctuates dramatically both daily and between days. Note that the utility has to provide the peak demand capability shown even though all of it is used only once that month. Customer B consumes the same number of kilowatt-hours, but has the same demand all the time. Naturally, the utility would rather have customer B than customer A, and so the rate schedule includes a demand charge component based on the peak demand. This means customer B can use the same kilowatt-hours, but pay less for them. Customer A thus has a motivation to "flatten" his or her curve to match customer B, saving money. Peak demand is usually based on the "average" demand over a set period (5, 15, or 30 min), so instantaneous surging of motor start-up does not drive the peak up much as is commonly thought.

The next level of billing maturity adds another component–power factor. Inductive loads, such as inductive motors, furnaces, and fluorescent lights, develop a reactive current that is out of phase with the working current. This reactive current does no real work, yet the distribution system must carry it, and the utility must be sized for it. This is illustrated in Exhibit 10.5.4.

The company can supply its own reactive power through capacitors (which are simple devices capable of storing energy), synchronous motors, or synchronous condensers. Consequently, large industries are usually on a rate schedule penalizing them (in dollar cost) for low power factors (often 75% or lower) and sometimes rewarding them for high power factors (often 85% or higher). The actual method of billing varies widely, but the net result is an encouragement to raise the power factor.

A summary of required calculations is as follows:

$$\text{kW} = (\text{hp})\left(0.746\ \frac{\text{kW}}{\text{hp}}\right)$$

$$\text{kW} = (\text{volts})(\text{amps})(\sqrt{3})^{*}\ (\text{power factor})^{\dagger}$$

(See Exhibit 10.5.4. Volts and amps are measured quantities.)

$$\text{kVA} = \text{kilovolt-amperes} = (\text{volts})(\text{amps})(\sqrt{3})^{*}$$

*For a three-phase motor. For a single phase, simply replace $\sqrt{3}$ with 1.
†Power factor is defined in Exhibit 10.5.4.

Exhibit 10.5.2 Gross Energy Audit Form

FORM A

GROSS AUDIT OF ALL ENERGY RESOURCES

DATE OF AUDIT ____________________

DATE OF LAST AUDIT ____________________

YEAR / MONTH	UNITS OF PRODUCTION (1)	ELECTRICITY: KWH (2)	ELECTRICITY: KW DEMAND ACTUAL (3)	ELECTRICITY: KW DEMAND BILLED (4)	ELECTRICITY: POWER FACTOR (5)	ELECTRICITY: TOTAL COST (6)	ELECTRICITY: (2) X A* BTU/MO. (7)	NATURAL GAS: KCF (8)	NATURAL GAS: TOTAL COST (9)	NATURAL GAS: (8) X B BTU/MO. (10)	GASOLINE: GALLON (11)	GASOLINE: COST (12)	GASOLINE: (11) X C BTU/MO. (13)	OTHER FUELS: # UNITS (GAL., FT^3) TON, ETC.) SPECIFY (14)	OTHER FUELS: COST (15)	OTHER FUELS: BTU # UNITS X BTU/UNIT (16)	MONTHLY TOTALS: (6)+(9)+(12)+(15) COST (17)	MONTHLY TOTALS: (7)+(10)+(13)+(16) BTU (18)	MONTHLY TOTALS: (18) ÷ (1) BTU/UNIT (19)	COMMENTS
JAN	200,000 ft^2	122800	407	407	.80	3716.16	12.28×10^8	10813	19520.79	1.0813×10^{10}	0	0	0	0	0	0	23,236.95	1.2041×10^{10}	6.02×10^4	No standard Unit
FEB																				of production,
MAR																				therefore, use
APR																				square footage of
MAY																				business.
JUN																				
JUL																				
AUG																				
SEP																				
OCT																				
NOV																				
DEC																				
TOTAL																				
CONVERSION FACTORS		A* = 10,000 BTU/KWH OR A* = 3412 BTU/KWH						B = 1,000,000 BTU/KCF			C = 130,000 BTU/GAL.			D = ______ SPECIFY BTU/UNIT						

*10,000 BTU/KWH IS CONVERSION TO ELECTRICITY AT SOURCE, 3412 BTU/KWH IS THE BTU EQUIVALENT OF 1 KWH. WHICH EVER ONE YOU DECIDE TO USE BE SURE AND BE CONSISTENT WHEN CALCULATING CONSUMPTION AND SAVINGS.

OG&E
ELECTRIC SERVICE BILL

KEEP THIS PART OF BILL

6516 0360 2
ACCOUNT NUMBER

RATE	SERVICE TO	CONSTANT	DEMAND	READING	KWH	AMOUNT
PL	02 20 TO 03 21	400	407	822	122800	

Power Factor* .90 — CURRENT BILL — 3716.16

*(This may or may not appear on your billing depending upon billing schedule and utility company.)

AMOUNT INCLUDES	
SALES TAX	14293
FRAN TAX **	
FUEL ADJ. AMT.	51195
FUEL ADJ. FACT.	.004169

A B C CORP
1000 N. Main
Stillwater, OK
74074

AMOUNT NOW DUE: 3716.16

DUE DATE	AFTER DUE DATE PAY THIS AMOUNT
04 15 79	3771.90

ADDRESS WHERE SERVICE WAS USED
I-40 & HWY 16

**AMOUNT IN EXCESS OF %

ONG
OKLAHOMA NATURAL GAS COMPANY
SALES TAX INCLUDED IF APPLICABLE

ACCOUNT NUMBER	SERVICE TO	RATE	RATES AVAILABLE UPON REQUEST
00 000 00 0000 0	3-26	2	

PRESENT READING	PREVIOUS READING	MCF	AMOUNT
		10813 0	19520.79

CODE: C CREDIT; ESTIMATED USE; ESTIMATED READING; FINAL BILL; MISC. CHARGES; PREVIOUS BALANCE; TRANSFER BALANCE; METER EXCHANGE

Rate Adjustments: PGA Purchased Gas Adjustment; EGSC Emergency Gas Service Charge

PGA per Mcf $.37963 — EGSC per Mcf $.03173

TOTAL — PAST DUE AFTER 4-19-79 — 19520.79

KEEP THIS STUB FOR YOUR RECORD

PLEASE RETURN THIS STUB WITH PAYMENT TO
OKLAHOMA NATURAL GAS CO.
P.O. BOX 1234
TULSA, OK 74186

ABC Corporation
1000 North Main
Stillwater, OK 74074

Copy

00 00000 0000 0 19520.79

Exhibit 10.5.3 Demand Billing

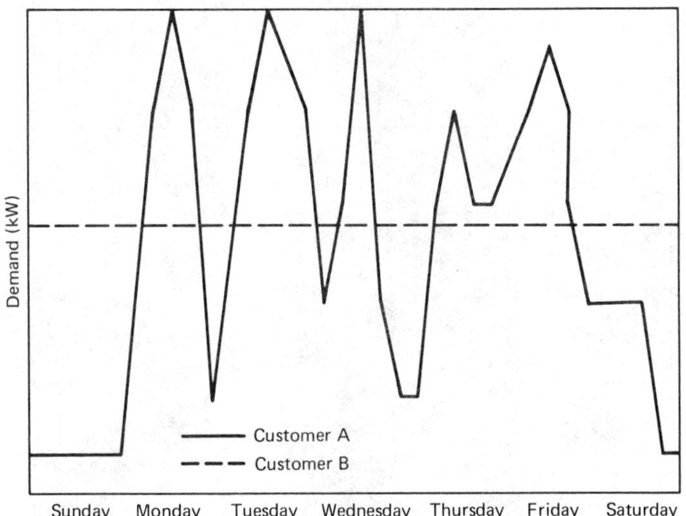

Typically, an electric utility will have three or more rate schedules progressing as just discussed. The choice of rate schedule determines the gross audit form section for electricity.

Often a company can save money by going to a higher or lower rate schedule. Also, utilities can make mistakes (they do not very often), so the energy management coordinator or someone should check each bill.

To fill in the gross audit form or forms, utility records for all energy supplies should be made available for 1 or 2 years past. Two is best, but one is acceptable. Then the data can be entered. The form is self-explanatory, except for two additional comments.

1. Billed demand may differ from actual demand if the company is on a ratchet clause where previous peaks can override actual peaks or when power factor billing involves a modification of demand.

2. The kilowatt-hour is 8.6×10^5 cal or 3412 btu, but the form lists 10,000 as the conversion to btu. This simply takes into account generating efficiency.

Detailed Audit

Detailed audits are necessary to determine what processes or equipment are using the energy and how much. Forms are given for electricity, natural gas, and gasoline in Exhibits 10.5.5 and 10.5.6.

Comprehensive detailed audits are not done very often for two main reasons:

1. They are expensive and time consuming, equipment for measuring is expensive, and recording the data is time consuming.

2. Energy management programs can be conducted without them. (Most managers have an idea of where the energy is going without conducting these audits.)

Exhibit 10.5.4 Power Factor

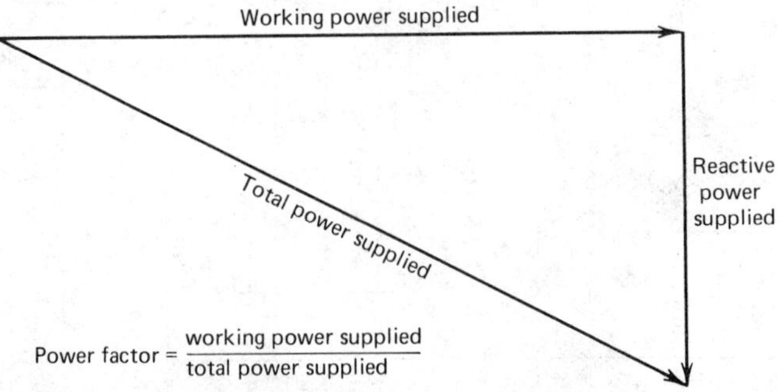

Exhibit 10.5.5 Detailed Audit for Electrical Units

Date of audit ______________ ___ page of ___ pages

Date of last audit ______________ Form B

Description of Unit	Rated Horsepower[a] (1)	Wattage[b] (2)	Operating Hours (3)	Average[c] Load (4)	(2) × (3) × (4) W-hr[d] (5)	(5) + 1000 Kw-hr (6)	(6) × 10,000 btu (7)	Comments
Total								

[a]One horsepower equals 746 W.
[b]This can be taken off of the nameplate of equipment.
[c]For motors: Average load is usually 70%; therefore insert 0.70. If wattage has been measured, use 1.0. For fluorescent tubes: Add 20% for ballast; that is, insert 1.20.
[d]If three-phase unit, multiply by 1.732; otherwise use 1 and proceed.

Exhibit 10.5.6 Detailed Audit for Natural Gas and/or Gasoline

Description of Unit	Natural Gas				Gasoline			Comments
	1000 cubic feet/hr (1)	(1) $\times 10^6$ btu/hr (2)	Operating Time (3)	(2) $\times$ (3) btu (4)	Miles or Hours (5)	Gallons (6)	(6) $\times 0.3 \times 10^5$) btu (7)	
Total								

Certainly, one should not expect a detailed audit to account for 100% of the energy shown in the gross audit. A good study will yield about 80% of the total quite quickly. The Pareto or ABC principle applies in that a small amount of equipment accounts for most of the energy. Small motors or gas burners can be ignored (there are many of them), but larger motors and large furnaces should be included.

Whether or not to do a detailed audit is a decision that must be made in specific cases. The program should probably at least begin to implement actual changes so results can be shown before spending much time on the detailed audit.

Walk-Through Audit

The walk-through audit is the first real step toward saving energy dollars. The audit involves a tour(s) of the facility with a checklist and the delineation of a number of energy management ideas.

There are basically two types of energy management ideas: O & M charges (often called "low-cost–no-cost ideas"), which involve little or no investment or engineering design, and EMOs, which require investment and/or engineering design. Naturally, the system should be fine tuned by implementing all applicable O & M charges before instituting many EMOs. The O & M charges, however, are long-lived devices and will always be around; thus their implementation never stops.

Some believe the checklist for the walk-through should include only O & M ideas, whereas others believe it should contain EMO ideas also. The authors believe some combination is desirable, since ideas do not separate according to this classification, and a walk-through will yield numerous ideas in both categories.

A walk-through audit is not a one-time occurrence. It should be repeated periodically and at different times. Likely times are as follows:

1. During operating hours alone.
2. During operating hours with others (management, employees, plant engineer, etc.).
3. During nonoperating hours (night–2 or 3 a.m.) to look for air handlers or other equipment left running, light left on, air or steam leaks (they can be heard), and so on.
4. During nonoperating hours (weekend) for essentially the same reasons as in item 3.

There is no shortage of forms or checklists. Several are referenced, and a sample is provided in Section 10.5.5.

10.5.4 OPERATING AND MAINTENANCE CHANGES

Why O & M Problems Exist

Two situations cause O & M problems that can cost in high energy consumption. First, most systems were designed and operating procedures developed when energy was inexpensive. Consequently, procedures were often developed to save personnel time, to create more comfortable environmental conditions than necessary, and for other purposes that today are not considered cost-effective.

Second, as time progresses, operating procedures evolve—usually toward the more chaotic. Most engineers have had the experience of designing good engineering systems only to come back later and find that the procedures and control thereof have deteriorated dramatically. Periodically, then, O & M checks need to be made. This dictates that *O & M checks are not a one-time occurrence*. They need to be repeated periodically.

How and When

The establishment of an O & M inspection procedure is relatively simple since many checklists have been developed, but maintaining such a system is difficult and requires discipline. The checklist should be utilized extensively at the beginning of the program and revised periodically to bring systems back into line.

As mentioned earlier, the timing of the O & M inspection should vary in order to be sure all phases are covered. (See Section 10.5.3 for a discussion of timing). The night and weekend inspections are particularly enlightening since they probably have not been done before, and interruptions caused by external noise and inquisitive people are minimized.

Checklist

Several checklists exist, and most are quite good. Actually, a checklist may not be necessary if the auditor(s) has a very inquisitive, open mind while making the audit. The checklist in the *Energy*

Conservation Program Guide for Industry and Commerce (EPIC) series[2] is quite good, although not too comprehensive. The checklist in *Total Energy Management*[3] is very good and is broken down by region.

Given in Section 10.5.5 is a checklist taken from *Instructions for Energy Auditors*,[1] published by the U.S. Department of Energy. The complete checklist is given in the reference and is much larger. This checklist presents O & M changes followed by EMOs.

10.5.5 ENERGY MANAGEMENT OPPORTUNITIES

The second basic type of energy opportunity is that which requires some capital investment and usually good engineering design efforts.

It is most important in EMOs to be sure that the engineering design is done well and that good economic analyses are done, since substantial amounts of money are often involved. Using the principles of engineering economy and life cycle costing as presented in Chapter 9.3, good decisions can be made.

It is very difficult to provide a checklist of EMOs, but the total energy management checklist mentioned earlier contains both O & M changes and EMOs. *Instructions for Energy Auditors*[1] also contains a long separate list of EMOs. The following checklist shows, first, O & M changes and then EMOs.

Checklist: Maintenance and Operational Changes

Space Heating

1. Lower the thermostats during the heating season, and raise them during the cooling season. If your building is like most, you can save about 8% of your heating fuel bill by lowering the thermostat(s) a mere 5°.

2. Use night setback on heating systems with zone controls. Maintaining 55° or 60°F at night will reduce energy consumption by 5 to 6% during the night hours.

3. Check burner firing period. If it is improper, it could be a sign of faulty controls.

4. Check automatic temperature control system and related control valves and accessory equipment to ensure that they are regulating the system properly.

5. Check flue gas analysis on a periodic basis: The efficient combustion of fuel in a boiler requires burner adjustment to achieve proper stack temperature of no more than 150° above steam or water temperature. There should be no carbon monoxide. For a gas fire unit, carbon dioxide should be present at 9 or 10%; for #2 oil, 11.5 to 12.8%; for #6 oil, 13 to 13.8%.

6. Check boiler stack temperature. If it is too high (more than 150°F above steam or water temperature), clean tubes and adjust fuel burner.

7. Adjust air/fuel ratios of firing equipment; the air-to-fuel ratio must be maintained properly. If there is insufficient air, the fire will smoke, causing tubes to become covered with soot and carbon, and thus lose heat, wasting energy. Most fuel service companies will test your units for a token fee and provide you with specific recommendations.

8. Examine operating procedures when more than one boiler is involved. It is far better to operate one boiler at 90% capacity than two at 45% capacity each. The more boilers used, the greater the heat loss.

9. Lower steam pressure to the minimum pressure that will just satisfy needs.

10. Turn off the boiler natural gas standing pilot during the summer months when the boiler is off.

11. Institute an operating procedure for multiple-boiler plants to ensure maximum loading of one boiler before a second boiler is put into service.

12. Do not heat parking garages, docks, and platform areas.

13. Shut off boilers in the spring and fall when the air conditioning machine is on and temperature control is not needed.

14. Feel the pipe on the downstream side of steam traps. If it is excessively hot, the trap is probably passing steam. This can be caused by dirt in the trap, excessive steam pressure, or worn trap parts (especially valve and seats). If it is moderately hot—as hot as a hot water pipe, for example—it probably is passing condensate, which it should do. If it is cold, the trap is not working at all. Replace or repair.

15. Keep a daily log of pressure, temperature, and other data obtained from instrumentation. This is the best method available for determining the need for tube and nozzle cleaning, pressure or linkage adjustments, and related measures. Variations from normal can be spotted quickly, enabling immediate action to avoid serious trouble. On an oil-fired unit, indications of problems include an oil pressure drop, which may indicate the following: a plugged strainer, a faulty regulator valve, or an air leak in the suction line. An oil temperature drop can indicate temperature control malfunction or a fouled heating element. On a gas-fires unit, a drop in gas supply pressure can indicate a malfunctioning regulator.

16. Reduce blowdown losses.

Ventilation

1. Reduce fresh air to legal limits. Ventilation systems often draw excess fresh air. This requires warming air to inside temperature at the cost of energy.
2. Reduce cubic feet per minute per occupant outdoor air requirements to the minimum, considering the tasks they are performing, room volume, and periods of occupancy.
3. Reduce exhaust air quantities to practical limits.
4. Establish a ventilation operation schedule so that exhaust system operates only when it is needed.
5. Reduce outdoor air to the minimum required to balance the exhaust requirements and maintain a slight positive pressure to retard infiltration.
6. If possible, use permanently sealed windows to reduce infiltration in climatic zones where this is a large energy user.
7. Operate the ventilation system only when the building is occupied. Also consider shutting off the air handling units on normal heating days before building is emptied. If the radiators are located properly, they should be able to maintain space temperature above freezing.
8. Increase the mixed air temperature setting on units to 65°F. If this is not practical, consider adjusting the fresh air linkage so that the mixed air temperature cannot go below 55°F.
9. In summer, when outdoor air temperature at night is lower than indoor temperature, use full outdoor air ventilation to remove excess heat and precool the structure to reduce air conditioning load.
10. Adjust the time clock day-night settings to operate ventilation units fewer hours during the day cycle.
11. Adjust all variable air volume (VAV) boxes so they operate precisely. This will prevent overheating or overcooling, both of which waste energy.
12. In noncritical areas reduce the fresh air drawn into the ventilation system to 25%. If a 50% situation is reduced to 25%, there will be a 50% saving in heat required to raise the incoming air to indoor conditions.

Air Conditioning

1. Set room temperature at 78°F in summer.
2. Use water-cooled refrigeration units rather than air-cooled ones since the former are up to 20% more efficient.
3. Turn off the cooling system during the night. Use ventilation air to cool the building at night.

Operating Procedures

1. Whenever possible, use outdoor air for cooling rather than using mechanical refrigeration. Use the economizer cycle where installed.

Hot Water

1. Reduce generating and storage temperature levels to the minimum required for washing hands, usually about 110°F. Boost hot water temperature locally for kitchens and other areas where it is needed, rather than provide higher than necessary temperatures for the entire building.
2. De-energize booster heaters in kitchens at night.
3. Consider replacing existing hot water faucets with spray-type faucets with flow restrictors where practical.
4. If water pressure exceeds 40 to 50 lb, install a pressure reducing valve on the main service to restrict the amount of hot water that flows from the tap.
5. Operate only one of the domestic hot water heaters. If one unit carries the load, leave the other off for standby.
6. Inspect insulation on storage tanks and piping. Repair or replace as needed.
7. Reduce laundry room hot water supplies to 160°F. This temperature will achieve the removal of soils, blood, and so on, which is the need of a hospital laundry. Sterilization should be done on a separate basis. Hot water reduction from 180° to 160°F will save 160 btu/gallon of water heated or 2000 btu/day per patient.
8. Use improved cake pan coatings to minimize washing water. If the washing cycle can be reduced by 25%, then the energy and dollars for washing go down accordingly.

Domestic Water

1. If water pressure is in excess of 40 psi, use a pressure reducing valve to lower the pressure.
2. Test hot water heater controls.

Lighting

1. Remove unnecessary lamps and/or replace present lamps with higher-efficiency units.
2. Use energy conserving fluorescent lamps. These lamps save about 15% of the energy for the same illumination.
3. Remove lamps or fixtures. If only lamps are removed, disconnect ballasts since a ballast accounts for 10 to 30% of the lamp's power drain.
4. Control exterior lighting. If a photocell is used to turn on the lamps and a timer to turn them off, there may be savings of as much as a third of present consumption.
5. Shut off lights in unoccupied rooms.
6. Reduce area lighting in all intense lighting areas and substitute task lighting.
7. Use color-coded light switches to avoid nonessential lights being put on during nonoperating hours; designate those needed for cleaning, register reading, and security.
8. Set a limit of 10 min after closing for the bulk of floor lights to be off; full night schedule to be in effect within 30 min.

General Buildings

1. Prepare an energy profile of building in as much detail as possible.
2. Conduct a survey of the total building on a space-by-space basis to determine the actual user needs.
3. A low power factor on an electrical system within a building will increase the losses in the electric utility system and reduce the system's capacity. Many electric utility companies have a penalty charge for low power factor. Correcting power factor can provide for more efficient use of energy as well as a reduction in the cost of electricity. Electrical devices known as capacitors can be installed to correct low power factor.
4. Put each apartment on separate electric meters. Lease without electric utility or pay rewards for usage below experiential standards.
5. Make the monthly energy consumption and cost data available to the manager and chief operating engineer so that they can evaluate and compare against previous months and normal budget.
6. Involve total building staff with energy conservation measures so that each individual has responsibility.
7. Provide a temperature control training program for the operating engineers that will give them a thorough understanding of how the heating and ventilating system was designed to operate. Include optimization of energy by means of temperature control.
8. Install storm windows or double glaze windows throughout. A single 36 ft^2 window will save about 3.5 million btu/year with storm windows added.
9. Examine the entire building for air leaks around windows, doors, and any place else that leaks might occur. Seal up the leaks. Open windows and outside doors during the heating or cooling seasons are criminal if the heaters or air conditioners are running.
10. Provide proper insulation for all equipment that is heated or refrigerated. This includes tanks, ovens, dryers, washers, steam lines, boilers, refrigerators, and so on.
11. Add insulation to the roof whenever the roof is going to be resurfaced or repaired. If insulation is not already in place, do not wait—put insulation on immediately.
12. Close off unused areas and rooms. Where possible, be certain that blinds or other shading devices are drawn, registers closed, and so on.
13. Utilize solid-state motor drives instead of motor generator sets for elevators.

Laundry Department

1. Use cold water detergents where permitted and satisfactory for your purposes.
2. Repair all steam leaks promptly.
3. Do not take make-up air from air conditioned spaces. Laundry make-up air should be taken from the outdoors. Stealing air from the conditioned spaces results in infiltration into them of outside air. Such air has to be cooled and dehumidified, a needless waste of energy.
4. Set schedules to reduce peak use and demand charges. Whenever possible, make arrangements to do laundry work during periods when the least amount of energy is being used at the property (nonpeak demand hours). If practical, try to limit the amount of equipment being used at the same time.

Food Preparation and Storage Equipment

1. Cook in largest volumes possible.
2. Cook meat slowly at low temperatures. Cooking one roast for 5 hr at 250°F could save 25 to 50% of the energy that would be used in cooking for 3 hr at 350°F.
3. Use the minimum amount of compressed air at the minimum required pressure. Have an outside air intake for the compressor. Compressed air is very expensive and should be avoided if at all possible.
4. Provide ovens, fryers, and washers with loads all of the time they are heated and on. An oven not baking 1 hr out of 7 is an oven wasting 14% of its energy.

Checklist: Energy Management Opportunities

Space Heating

1. Preheat oil to increase efficiency.
2. Replace existing boilers with modular boilers.
3. Use waste heat from pan driers to preheat make-up water. Each gallon of water each degree Fahrenheit raised saves 8 btu.
4. Preheat combustion air to increase boiler efficiency.
5. Reduce blowdown losses.
6. Replace old, inefficient burners with new, efficient ones. Changing a 60% unit for a 75% one with a 25,000 gal/year experience will save 5000 gal or more than $4000/year. Payback, about 1 year.
7. Install loading dock door seals. Although it may not be possible to completely seal off such an opening, the effort to do so is rewarding in energy savings.
8. Insulate hot, bare heating pipes. A 160°F bare hot water pipe, $1\frac{1}{2}$ in. in diameter, not insulated, will lose 13 million btu/year for each 10 ft length.
9. Insulate hot, bare heating pipes. Economic thicknesses can be supplied by contractors.
10. Replace all damaged insulation on heating pipes, including those in steam, condensate, hot water supply, and hot water return systems.

Ventilation

1. Consider installing economizer/enthalpy controls on air handling units in noncritical areas to minimize cooling energy required by using proper amounts of outdoor and return air to permit "free cooling" by outside air when possible.
2. When more than 10,000 CFM are involved, and when building configuration permits, consider installation of heat recovery devices such as a rotary heat exchanger.
3. Consider utilizing revolving doors for main access in addition to swinging doors needed by those in wheelchairs or on crutches.
4. In locations where strong winds occur for long duration, consider installing wind screens to protect external doors from direct blast of prevailing winds.
5. Install an automation system to operate the ventilation units so that supply air temperature and return air/fresh air dampers can be adjusted to maintain the desired spaces temperature in the room.
6. Consider installation of an air curtain, expecially in delivery areas.

Air Conditioning

1. Replace inefficient air conditioners. Newer units may save as much as 25% or more on the energy consumed for the same cooling.

Domestic Hot Water

1. Install shower head restrictors. This may save up to 50% of the hot water consumed.
2. Change the shower heads from the existing GPM rating to 2 GPM.
3. Install a small domestic hot water heater to maintain the desired temperature in the water storage tank to eliminate the need for running one of the large space heater boilers at a very low efficiency during the summer months.

Lighting

1. Remove unnecessary lamps if removal will still provide illumination levels required. When lamps are removed from a fluorescent luminaire, all lamps controlled by a given ballast should be

removed to prevent ballast failure or reduced lamp life. Also consider disconnecting ballasts that otherwise would continue to consume energy.

2. Consider replacing present lamps with those of lower wattage that provide the same amount of illumination or (if acceptable in light of tasks involved) a lower level of illumination.

3. Replace all incandescent parking lighting with mercury vapor high-intensity discharge or sodium lamps.

4. When existing circuitry makes it impossible to utilize less than 25% of the light in a given large space whenever light is needed, and when persons work during normally unoccupied periods, consider development of a desk lamp issuance program that enables persons working during unoccupied periods to use a simple desk lamp or two instead of a large bank of luminaires.

5. When natural light is available in a building, consider the use of photocell switching to turn off banks of lighting in areas where the natural light is sufficient for the task.

6. Move desks and other work surfaces to a position and orientation that will use luminaires to their greatest advantage (instead of moving luminaires).

General Building

1. Add storm windows. Consider adding to every other window of the building. Storms on 50% of the windows would still leave enough windows for ventilation during spring and fall schedules.

2. Insulate all roofs, walls, and floors having exterior exposures.

3. Use heat pumps in place of electrical resistance heating and take advantage of the favorable coefficient of performance.

10.5.6 INDUSTRIAL ENGINEERING, ENERGY MANAGEMENT, AND MAINTENANCE

A good maintenance program aimed at systems that consume energy can save a company large amounts of money in electricity, gas, or purchased steam costs and can enhance production capability by decreasing lost production caused by preventable breakdowns of major energy consuming systems. Other benefits of a good program include general cleanliness, improved employee morale, and increased safety.

An efficient and complete system for the maintenance of energy consuming systems can be developed in a three-step procedure. Step one is the examination of the present facility to determine immediate corrective maintenance needs. The outputs from this step include (1) a list of major equipment units, with location, operating hours, and condition of each; (2) a list of the locations of all relevant operating manuals; and (3) a priority listing of work to be performed to bring the most critical items to operating condition. The next section gives a guide to each major system and indicates possible trouble symptons. Step two is the development of a list showing what items must be maintained on a routine basis, how often each should be maintained, and approximately how long each routine maintenance task will take. In step three the data from step two are used to prepare a schedule for the routine program—checking to see that the scheduled work has been performed, revising the maintenance intervals or task time estimates, and adding or deleting tasks from the list.

Systems To Be Maintained

A major problem is determining what to maintain. As a guide, the following sections describe each of the major energy consuming systems, with lists of the most prevalent problems in each.

Building Envelope

The building envelope consists of the roof, exterior walls and doors, and any other part of the building in direct contact with the outside. The inspection of the building envelope should include at least these areas:

1. **Windows.** Look for missing windows and feel for drafts around window frames.

2. **Doors.** These should be reasonably tight and should not be forced inward or outward by an imbalance in air pressure between the inside and the outside.

3. **Walls.** Any cracks should be caulked.

4. **Vents.** Their dampers should be connected to the appropriate control mechanisms and not blocked open, nor should the linkages be disconnected.

5. **Roofs.** Roofs should be checked for holes and leaks around such openings as chimneys.

Boiler and Steam Distribution System

This system consists of the boiler, the steam distribution system, and the condensate return system. Improvements to this system require technical expertise, but frequently have immediate payback. A boiler system that has not been adjusted or maintained within the last 2 years generally represents an opportunity for a company to decrease its fuel consumption by that boiler by 30% or more. The boiler itself consists of many components, and these should be checked to see if they are functioning as intended: safety interlocks and boiler trip circuits, burners, coal boiler components (grates, stokers, air dampers, cinder treatment system, etc.), combustion controls, the furnace, the water treatment system, and all controls. Any professional inspection of a boiler should include a statement about each one of these components regarding both its present condition and the cost of restoring it to a workable state.

An inspection of the steam distribution system should pay special attention to steam traps—devices whose purpose is to keep live steam contained and to enable condensate and dissolved gases to escape from the steam distribution lines. Leaking steam traps can cost a great deal of money. To learn how to recognize leaking steam traps, consult a local distributor or see the *Energy Management Handbook.*[4] Also, vendors are often willing to test steam traps free of charge.

The condensate return system should be free from leaks, and the water treatment system should be working properly if a buildup of scale is to be prevented. In examining both the steam distribution system and the condensate return system, look for steam leaks and for leaks of hot water. Both are expensive.

Heating, Ventilating, and Air Conditioning System

This system is intended to provide occupants of an area with clean air of a reasonable temperature. Although there are many different system configurations, most systems have three main components: (1) plenums, dampers, and ductwork for air intake and mixing; (2) fans, chillers, and heating units to heat, cool, and distribute air; and (3) the control system. Check the plenums and ductwork for loose insulation, air leakage, and damage caused by the insulation of other systems; check the dampers to see that they close properly and that the control linkages are properly connected. Grillwork should also be examined to see that it is not plugged with lint or obstructed by furniture or other material. In checking fans be sure that fan motors are connected to the fans and that the belts are aligned properly. Also check any motors for excess noise that may indicate potential bearing failure, and feel pipes and pumps to see, first, that hot water is being pumped in hot water pipes and, second, whether it may be profitable to install insulation. (It may well be cost-effective to insulate a pipe that is too hot to touch or that is very cold.)

The detailed inspection of cooling towers and chillers is best left to someone who has done it before, but manuals are available from vendors that outline inspection procedures. A general rule is that the cooling in the summer should not be so vigorous that the inspector is cold in shirt sleeves, nor should the heating be so warm in the winter that the inspector is uncomfortable in a jacket. Checking the control system is also a job for an expert, but some preliminary checks can be performed—if the person doing the auditing knows what gauge readings should be. If he or she does not know, it is still worthwhile to look for gauges that are not working, for valves whose moving parts have been cannibalized, and for thermostats whose temperatures do not agree with those obtained by the thermometer carried by the energy auditor. A check should also be made to see that air is not necessarily preheated during the summer and that cooling and heating temperatures are kept to reasonable maxima and minima, respectively, to avoid unnecessary cooling and heating.

Electrical System

The electrical system can cause a waste of energy dollars if the power factor is too low, if voltage unbalances are present, or if some of the components are faulty. The power factor is the ratio of the resistive component of electrical power to the total delivered power. Most electrical meters read only the resistive load, so a low power factor indicates that the electrical utility is delivering more electricity than it is being paid for. The difference is made up in a charge for a low power factor, and the size of the charge is a good indication of the amount of emphasis that power factor increase should get. Typically, a lower power factor is caused by inductive loads, such as motors, transformers, and welders; it can be corrected by the installation of capacitors or by having self-contained power supplies for individual units.

Voltage imbalances can be detected by a detailed load analysis of the facility. Such imbalances can cause two problems—wires can become overloaded by equipment installed beyond that intended by the original designers of the electrical system, or the voltages at different sides of a three-phase motor can be different. In the first case too much overload can create a fire hazard; in the second case a difference of more than 1% can cause serious operating inefficiencies and heat buildup in the motor.

Faulty electrical components that should be checked include connections, which should be tight; switches, which should be free from arcing; and transformers, which should be properly sized, with unobstructed heat transfer surfaces and neither leaking nor in the region of water leaks.

Lighting

The amount of light used in an office is the most conspicuous indicator of the attention being given to energy management, and reducing that lighting often leads to employees' reducing energy consumption in many other ways. In addition, reducing the amount of energy used in lighting reduces the amount of heat that must be removed from an area.

In looking at the lighting system, two possibilities should be considered carefully. First is the possibility of removing existing lights or replacing those currently in use with lower-wattage lights. Either can be done subtly, by not replacing lights as they burn out, or directly, with fanfare. If fluorescent lights are being modified, the ballasts should also be removed if lights are being removed, because ballasts use energy and contribute to a poor power factor even without the lights; furthermore, spare ballasts are a good source for replacements.

The second major area for lighting improvement is in the examination of ways to make the existing system more effective. Luminaires can be cleaned, light giving windows can be washed, floors can be maintained, and walls can be painted light colors. All of these improve the light within a building and make it a more pleasant place to work.

Hot Water Distribution System

A typical industrial facility uses hot water for washing, for industrial cleaning, and as a source of heat. In each case two questions need to be answered: (1) Is there waste in the system? and (2) Is the water temperature too hot for the actual need?

Waste can occur in the form of leaks (very expensive, but easy to fix) or in practices that use large quantities of heated water unnecessarily. It is also possible that the water temperature is too hot for actual needs (105°F for washing hands) and that the extra energy put into heating is simply wasted. Two other measures can often help reduce this part of the energy bill: (1) putting a timer on the hot water heater and operating it only when it is needed and (2) checking hot water pipes to eliminate scale buildup. Find out when the hot water is actually needed and how long it takes the heater to bring water to this temperature, and then install a timer with the appropriate settings to eliminate hot water heating at night, on weekends, and otherwise when not needed. But know the piping system—frozen pipes are remembered. If scale buildup is a problem, consider the installation of water treatment equipment.

Air Compressors and the Air Distribution System

Compressed air leaks and poor compressor performance costs money. If the air is being used to power equipment, more power must be used if air leaks are present; if it is being used as a control medium for an air conditioning system, low air pressure will generally cause the vent system to shunt only warm air into rooms, with consequent discomfort; if the air is being used to blow away particules or for other cleaning, leaks will cause the compressor to run unnecessarily. Air lines should be checked for air leaks and for oil or water in the lines. If leaks are found, they should be repaired; if oil or water is found in any air lines, the compressor should be checked for leaks immediately before the liquid in the lines wrecks the controls (if the air is used to power pneumatic controls).

Manufacturing Equipment System

Most manufacturing systems have components that have not been described here. A general approach to such systems is to ask these questions:

1. Does the system need all the extra heating (or refrigeration) that it is currently using? For example, many present computer systems do not need the elaborate temperature protection that systems required 15 years ago.
2. Is the system (or its components) operating when it does not need to? It is possible that equipment can be turned off if it is not in productive use, with no loss to anyone.
3. Is it possible to reschedule the use of some equipment so as to reduce peak electrical demand? Electricity used during the peak can cost 5 to 10 times as much as that used in off-peak periods, or, put another way, reducing the peak by 1 kW can save $6.00 or more per month. The basic rule is: *If you don't need it on, turn it off—but don't freeze the pipes.*

10.5.7 FACILITIES AND ENERGY MANAGEMENT

Many of the changes (both O & M and EMOs) are actually facilities oriented through building changes such as caulking, weatherizing windows and doors, and insulation. Also, the layout of the facility can dramatically affect energy consumption. Therefore this chapter covers changes that aid in energy management by affecting the facility in a matter controlled by industrial engineers.

Energy Management Outside the Facility

The following ideas involve steps to take outside the facility to aid in managing energy:

1. **Locate Energy-Efficient Property.** Avoid energy-intensive spots such as wind areas or sunny areas with no trees. (If alternative energy sources are being considered, these spots may be more attractive.) The facility should be near energy supplies and good transportation services.
2. **Place Facility Carefully on Property.** Again, avoid windy or sunny areas and bottoms. Hillsides facing south are usually best. Utilize hills as berms and take advantage of natural shading.
3. **Landscape Well.** Avoid concrete or asphalt areas close to the building since grasses and shrubs are much cooler. Deciduous trees offer shading in the summer, yet allow winter sun through for warming. Shrubs, hedges, and vines can be used for additional shading. Evergreen windbreaks in the northwest corner divert winter winds while channeling summer breezes for cooling (see Exhibit 10.5.7). The heating load for an unprotected building can be four times that for a protected one.
4. **Orient the Facility Properly.** To minimize heat gain in the summer while maximizing solar input in the winter, locate the facility so the longer dimension runs east to west (see Exhibit 10.5.7). Place glass areas on southern exposures.

Exhibit 10.5.7 Energy-Efficient Landscaping

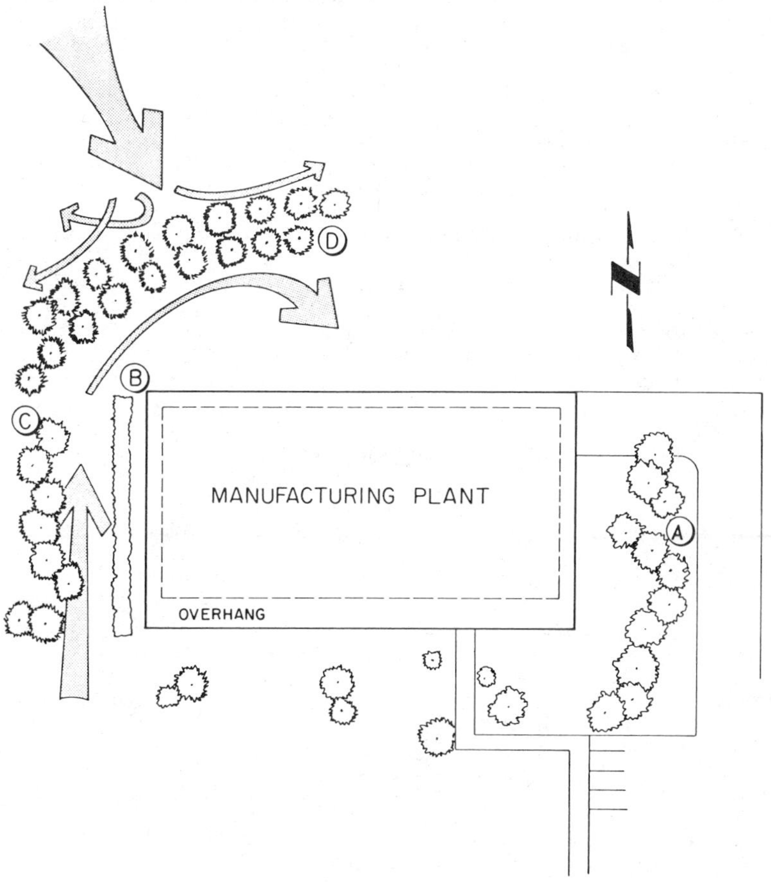

Exhibit 10.5.8 Temperature Distribution for Underground Construction[a]

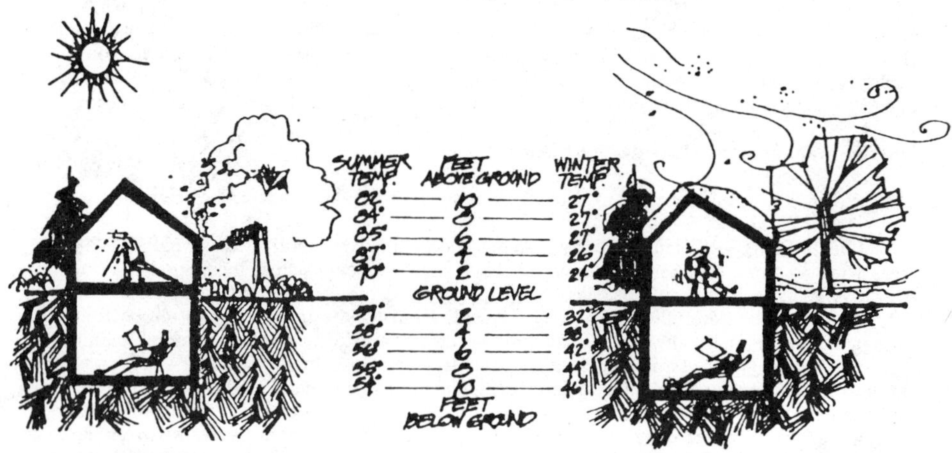

[a]*Source.* Reference 5.

5. Consider the Use of Underground Structures. Underground structures are becoming popular as warehouses or even manufacturing facilities. Berms can be used to obtain partial benefits of underground structures (see Exhibits 10.5.8 and 10.5.9).

6. Engineer Outside Lighting. Provide $\frac{1}{2}$ to 2 ft-c illumination for roadways and about 3 ft-c for parking areas. Install timers or photosensors. Avoid commercial lighting. Consider high-intensity discharge lights, such as high- or low-pressure sodium, since they are very energy efficient.

Energy Management in the Facility Envelope

In general, design to a good energy building code, such as ASHRAE 90-75 or one of the new codes.

1. Insulate well according to local needs. It is cheaper to insulate now than later.

2. Minimize wall perimeter area. Larger wall areas mean larger heat losses. Minimizing the perimeter through square or other regularly shaped buildings minimizes the wall area and subsequent heating and cooling bills.

3. Consider storm windows and doors and/or thermopane construction. Good engineering economic analysis is necessary to determine feasibility.

4. Use a minimum amount of glass (10% of wall area or so). Southern walls with proper overhangs or awnings are best (see Exhibit 10.5.10).

5. Use dark roofing material in cold climates and light material in warm ones. Consider solar-reflective coatings where air conditioning loads are large.

6. Engineer wall openings.

a. Caulk and weatherstrip.

Exhibit 10.5.9 Underground Manufacturing

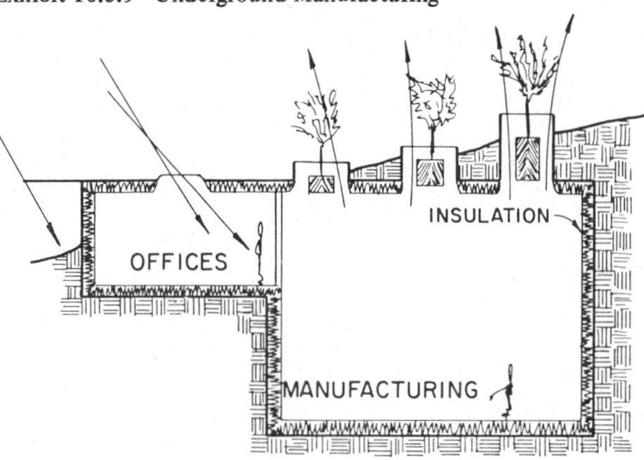

Exhibit 10.5.10 Use of Overhangs and Awnings

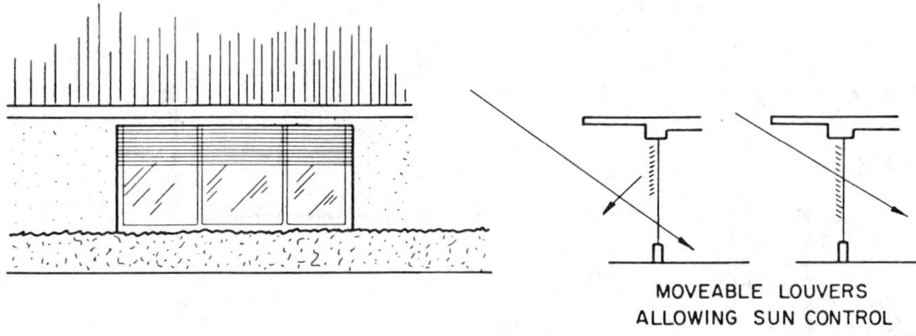

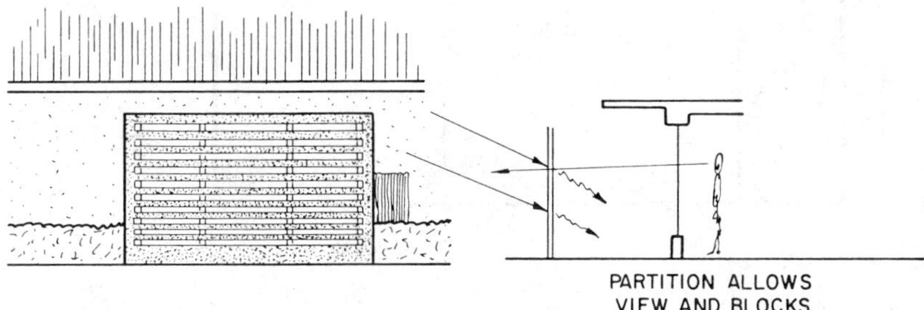

b. Insulate dock doors.
c. Use adjustable dock cushions.
d. Use good pedestrian doors for personnel travel.
e. Utilize vestibules or revolving doors to cut air infiltration in half or more over normal pedestrian doors.

7. Consider the use of windows that can be opened and/or tinted glass.
8. Consider economical cycles that utilize unconditioned outside air for cooling when temperature and humidity permit.

Energy Management Inside the Building

The lists of ideas in the preceding subsections are mostly ideas for use inside the facility and adequately cover this, except for the following ideas:

1. Design for cost center metering. If the distribution systems are designed properly for metering, energy can be measured and charged directly to departments rather than as overhead.
2. Locate boilers, air compressors, and so on, to minimize travel distances and thus transmission losses. Design steam systems to facilitate condensate return and/or reflashing.
3. Correctly size all equipment. Overdesign usually leads to poor energy efficiency and, in the case of electric motors, power factor problems.
4. Use light-colored furniture and walls to minimize lighting requirements. Avoid glare problems, however.
5. Lay out to minimize energy requirements and facilitate energy management (see Exhibit 10.5.11).

a. Place high personnel density areas in southern areas of the plant.
b. Utilize buffer zones of large machines or warehousing areas along outside walls.
c. Place exhaust and make-up air units close together to facilitate waste heat recovery.
d. *Avoid* northern and western walls for dock or other doors and railroad entry.
e. Design energy monitoring and management systems into the layout (computerized energy management systems).

6. Wire exhaust fans, lights, equipment, and so on, together to facilitate turnoff.
7. Wire lighting in banks so that small areas can be turned on or off instead of large spans.
8. Consider use of wood, waste oil, or other by-products as potential fuel. Plan storage areas accordingly.
9. Plan for standby or alternative fuel utilization with necessary storage and security.

Exhibit 10.5.11 Layout for Energy Management

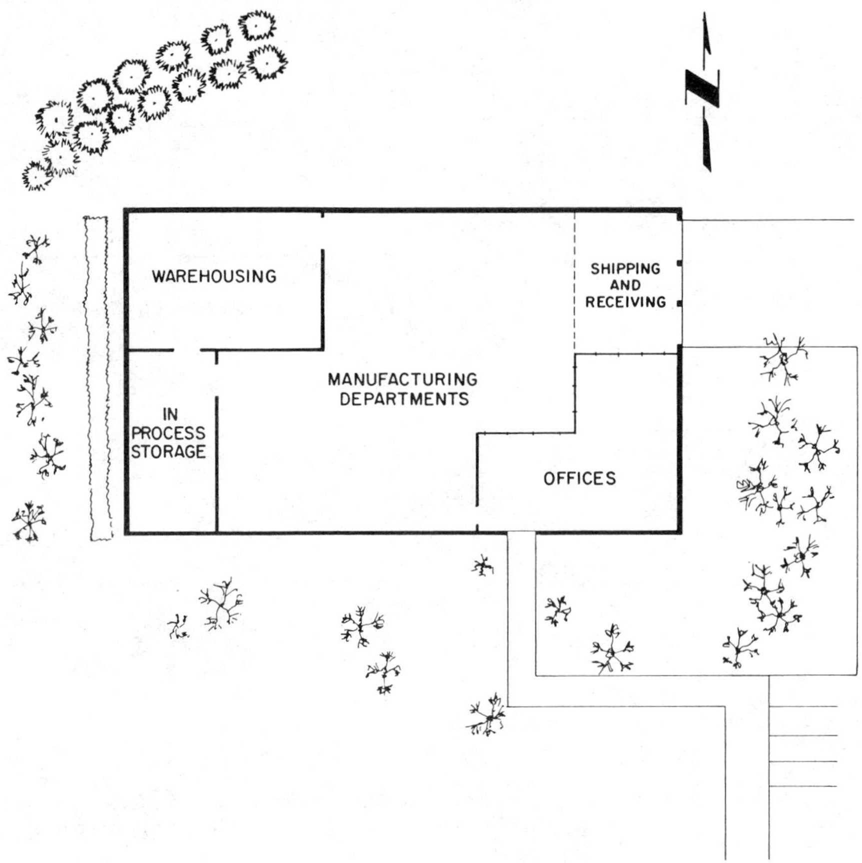

Energy Management and Materials Handling

Materials handling represents 10 to 90% of manufacturing cost, and industrial engineers are very active in materials handling. So any section on energy management for industrial engineering would not be complete without a section on energy management and materials handling. In general, the real savings for energy management lies not in the materials handling equipment itself, but in the interface between that equipment and other energy consumption systems. For example, the ventilation and subsequent reheating or recooling of make-up air as a result of the presence of fork trucks often exceeds the energy cost to run the truck itself. Even disregarding the cost to heat or cool air, the minimum exhaust requirements recommended by the American Conference of Governmental Hygienists are 5000 CFM/propane-fueled truck and 8000 CFM/gasoline-powered truck.[6] Replacing propane or gasoline-powered lift trucks with battery-powered trucks does away with this requirement, although it will still be necessary to ventilate the room where batteries are charged. Carrying the example further, it may be possible to do away with the heating and cooling as well as the ventilation if the storage system is automated.

In general, energy is used for three purposes in materials handling and storage: (1) to move material; (2) to condition spaces for material; and (3) to heat, cool, or ventilate spaces for people who are moving, storing, or removing the material. (These uses may interact—when, for example, waste heat from a refrigeration process is used to keep an office space warm.) To analyze a materials handling area for possible cost saving changes in energy management, consider each of the three purposes in the light of these questions: (1) How much time does relevant equipment *have* to be running? (2) Is the temperature being maintained either too hot or too cold for actual needs? and (3) Is all the lighting necessary? Consider these for each of the three purposes listed.

Equipment that is idling is using energy without doing useful work, and one useful question is whether all the idling is necessary. An unpublished study on underground mines, for example, indicated that conveyor belt systems generally run continuously, even when they are not carrying material. This same study reported that diesel trucks were idling 60% of the time they were on in a given shift. The manufacturers of diesel engines universally condemn idling for more than 5 to 10 min as injurious to the motor (not to mention the unnecessary pollution created). In other situations it has proven possible to have equipment on only when material is being moved.

Another possibility for saving energy dollars in the movement of material comes from examining the temperatures where the movement–loading, unloading, or retrieval–is taking place. If the material does not have any critical temperature requirements, then the reason for air conditioning is to keep people comfortable, and the amount of air conditioning can be reduced if the people are somehow protected from the natural environment of the movement area. (One way to achieve this is to have a small, comfortable enclosed area within the warehouse and to supply coats for use in warehouse areas.) Another possibility is to regulate the temperature of the warehouse within broad limits–say 55 to 85°F.

A third major area is that of lighting. Much warehouse lighting is in the wrong place and too high to be of much help. Lights that are over high storage racks should be eliminated. Where possible, fixed lights should be replaced by lights mounted on equipment. A lighting system should be set up so that the lights are off when no one is in a given area. Where lights must be on continuously and must be mounted on a high ceiling, sodium or mercury vapor lights should be considered.

If energy is used primarily to condition spaces for material, the approach used previously can help define problems for further investigation. If air is being cooled all day and heated all night to condition an area, why not turn on the intake fans very early in the morning and cool down the building with free outside air, with possibly the same idea operating to keep the building from cooling down in the evening? Companies that have tried this have achieved as much as 70% energy savings in air conditioning bills. Here again, if material is not sensitive to the amount of heat or cold in its environment, why cool it or heat it? And keep lighting to a minimum. The Illuminating Engineering Society recommends that 5 to 50 ft-c be used in a warehouse, depending upon the visual detail of the tasks that need to be performed there.[7]

The third area of savings concerns the people who are involved in the materials handling function. They can be provided with a comfortable area separate from the materials handling space, or they can be issued coats. (This is usually done in meat lockers or other refrigerated areas–why not make it routine in other places?) Or, schedules can be changed so that either (1) people come to work when the natural environment is comfortable–in the evening in summer and during the day in winter–or (2) schedules are changed so that the air conditioning load is uniform and avoids peak electrical demands. The use of conditioned air should also be studied carefully to see that it is not lost unnecessarily. For example, if the only convenient way to a concessionaire's truck during a coffee break is through a 12 ft by 20 ft door, that door will be open during the entire time of the coffee break. In this situation self-closing personnel doors pay for themselves rapidly.

Some other specific measures that have been found useful are given in Exhibit 10.5.12.

10.5.8 ALTERNATIVE ENERGY SOURCES

Much is being said today about future energy sources. This subject is a source of controversy among the experts, so no wonder it is confusing to the average person.

In general, it seems safe to say that for the professional lifetime of most practicing industrial engineers, conventional energy sources will still be prevalent, in fact, most likely dominant, for 20 years or so. Then, as conventional sources become more expensive and supplies less certain or more difficult to obtain, nonconventional sources will likely take over. (Here the word "nonconventional" implies any source not playing a *dominant* role today.)

Exhibit 10.5.12 Possibilities for Energy Saving in Materials Building

Problem	Solution
No barrier between conditioned and unconditioned areas	Try strip curtain doors unless there is a significant pressure difference between the areas.
Truck bays leak cold or hot air	Install loading cushions.
Excessive industrial truck fuel consumption	Reduce idling time or use smaller trucks. Check daily and weekly maintenance procedures. Lock up fuel supplies.
Hoists and cranes idling too much	Use special switches to turn off when not in use. Turn off during nonworking hours as part of regular procedures.
Conveyors and associated equipment run when not in use	Wire conveyors to light switch; install load-operated switch on conveyor.

10.5.9 SUMMARY

The energy future is laden with opportunity. First, the United States has the opportunity to pursue a course of energy decisions that can propel the country ahead in international competition and yet save our environment and economy. Second, and more germane to this discussion, the industrial engineer has the opportunity to lead the way in industrial energy management. Profit improvement options abound, and professional enhancement or visibility is available. The enterprising industrial engineer will now, and in the future, be practicing energy management.

The scenario to be followed in establishing good energy management programs must be unique to the company, but certain ingredients seem to be necessary. They are:

1. Appointment of energy management coordinator.
2. Appointment of energy management committee.
3. Securing and displaying of top management commitment.
4. Successful and visible projects first.
5. Recognition of service.
6. Establishment of a reporting and monitoring system.
7. Management for creativity.
8. Good energy audits.

a. Gross audit—historical consumption and cost survey.
b. Detailed audit—connected load survey.
c. Walk-through audit, including O & M changes and EMOs.

Industrial engineers bring many talents to energy management, but the following seem to offer the most to the success of an energy management program:

1. Engineering economy.
2. Maintenance program experience.
3. Facility location.
4. Facility layout.
5. Facility design.
6. Materials handling.

REFERENCES

1. U.S. DEPARTMENT OF ENERGY, *Instructions for Energy Auditors*, U.S. Government Printing Office, Washington, DC, 1979.
2. U.S. DEPARTMENT OF COMMERCE, *Energy Conservation Program Guide for Industry and Commerce*, NBS Handbook 115, U.S. Government Printing Office, Washington, DC, 1974.
3. NATIONAL ELECTRICAL CONTRACTORS ASSOCIATION and NATIONAL ELECTRICAL MANUFACTURERS ASSOCIATION, *Total Energy Management—In Existing Buildings*, NECA and NEMA, Washington, DC, 1977.
4. TURNER W. C., Ed., *Energy Management Handbook*, Wiley-Interscience, New York, 1982.
5. U.S. DEPARTMENT OF COMMERCE, *Options for Passive Energy Conservation in Site Design*, HCP/M5037-01, National Technical Information Service, Washington, DC, 1979.
6. L. G. HAUSER, *Industrial Ventilation*, American Conference of Governmental Industrial Hygienists, Lansing, MI, 1976, pp. 5–105.
7. R. D. HARDY, *IES Lighting Handbook*, Illuminating Engineering Society, New York, 1980.

BIBLIOGRAPHY

AMERICAN SOCIETY OF HEATING, REFRIGERATING, AND AIR CONDITIONING ENGINEERS, *ASHRAE Standard 90-75*, Ashrae, New York, 1975.

BYRER, T. G., *Energy Conservation in the Metalworking Industry*, SME Publication #MM75-129, Dearborn, MI, 1975.

CASSEL, R. T., *Energy Conservation at the Rocketdyne Division of Rockwell International*, SME Publication #E76-109, Dearborn, MI, 1976.

CLARK, G. W., *Good Lighting With Energy Conservation*, SME Publication #M76-105, Dearborn, MI, 1976.

COGGINS, J. L., *Techniques of Energy Management*, SME Publication #EM76-114, Dearborn, MI, 1976.

CONNELLY, R. R. *Conserve Energy by Controlling the Demand for Electric Power*, SME Publications #EM76-101, Dearborn, MI, 1976.

DALE, J. C., *The Energy Audit–First Step in an Energy Conservation Audit Program*, SME Publication #EM76-104, Dearborn, MI, 1976.

DEKOKER, N., *GM Energy Management–Organization and Results*, SME Publication #MM75-132, Dearborn, MI, 1975.

DOOLITTLE, J. S., *Energy*, Matrix, Champaign, IL, 1977.

DUBIN, F. S. H. L. MINDELL, and S. BLOOME, *How to Save Energy and Cut Costs in Existing Industrial and Commercial Buildings*, Noyes Data Corporation, Park Ridge, NJ, 1976.

ECKLER, N. H., *Energy Conservation in a Research and Development Facility*, SME Publication #MM75-916, Dearborn, MI, 1975.

GENERAL MOTORS CORPORATION, Energy Management Section, *Industrial Energy Conservation–101 Ideas at Work*, Detroit, MI, January, 1977.

ENERGY TASK FORCE OF THE AMERICAN COUNCIL ON EDUCATION, *Notes on Energy Management*, National Association of College and University Business Offices, Washington, DC, 1976.

HARDY, R. D. *Energy Savings With Powder Coatings*, SME Publication #FC74-189, Dearborn, MI, 1974.

HAUSER, L. G., *New Energy Usage Patterns in Manufacturing*, SME Publication #76-100, Dearborn, MI, 1976.

HAUSER, L. G., *Increasing Energy Efficiency–A Program of Industrial Workshops*, participant's workbook, Federal Energy Administration, Dallas, 1976.

INSTITUTE OF REAL ESTATE MANAGEMENT OF THE NATIONAL ASSOCIATION OF REALTORS, *Energy Cost Reduction for Apartment Owners and Managers*, Department of Energy, Washington, DC, 1977.

JONES, T., *Conserving Energy in Pretreatment Processes*, SME Publication #FC74-190, Dearborn, MI, 1974.

KLISIEWICZ, P. E., *Electrical Supply Capacities and Limitations*, SME Publication #EM76-112, Dearborn, MI, 1976.

LINDER, E. E., "The Consultant's Role in Energy Management for Production," SME Publication #EM76-113, Dearborn, MI, 1976.

Managing the Energy Dilemma–Technical Reference Manual, participant's workbook, Federal Energy Administration, Dallas, 1976.

NEAL, G. W., *Evaluating On-Site Power Generation for Industrial Plants*, SME Publication #EM76-110, Dearborn, MI, 1976.

NATIONAL ELECTRICAL CONTRACTORS ASSOCIATION AND NATIONAL ELECTRICAL MANUFACTURERS ASSOCIATION, *Total Energy Management–A Practical Handbook on Energy Conservation and Management*, NECA and NEMA, Washington, DC, 1976.

NATIONAL INSULATION CONTRACTORS ASSOCIATION, *Principles of Heat Transfer and Introduction to ETI*, NICA, June 1976.

OSTRANDER, B. W., *Fuel Savings Electronic Air Cleaners Applied to Welding Smoke*, SME Publication #76-102, Dearborn, MI, 1976.

OSTRANDER, B. W., *The Ozarks Regional Commission–Regional Energy Alternative Study*, Mathtech, Princeton, NJ, 1977.

OSTRANDER, B. W., *The Potential for Energy Conservation in Nine Selected Industries*, Vol. 2., Federal Energy Administration, Washington DC, 1975.

POWERS, H. R., *Economic and Energy Savings Waterborne Coatings*, SME Publication #FC75-561, Dearborn, MI, 1975.

PRESLEY M. and W. TURNER, *Save $$$–Conserve Energy*, School of Industrial Engineering and Management, Oklahoma State University, Stillwater, 1977.

RELICK, W. J. *Survey Your Plant's Electrical Power Needs and Save*, SME Publication #MM75-130, Dearborn, MI, 1975.

ROBNETT, J. D., *Energy Conservation at DuPont*, SME Publication #EM76-106, Dearborn, MI, 1976.

SHAFFER, E. W., *Computerized Approach to Energy Conservation*, SME Publication #EM76-107, Dearborn, MI, 1976.

SPENCER, R. S. and G. L. DECKER, *Energy and Capital Conservation Through Exploitation of the Industrial Steam Base*, SME Publication #EM76-103, Dearborn, MI, 1976.

STAFFORD, J., *Energy Conservation in Appliance Manufacturing*, SME Publication #EM76-108, Dearborn, MI, 1976.

THEKDI, A. C., *Airless Paint Drying for Energy Savings in Finishing Industry*, SME Publication #FC76-221, Dearborn, MI, 1976.

U.S. DEPT. OF COMMERCE, *Building Energy Handbook*, Vol. 1–Methodology for Energy Survey and Appraisal, U.S. Government Printing Office, ERDA-76/163/1, Washington, DC, 1976.

U.S. DEPT. OF COMMERCE, *Building Energy Handbook*, Vol. 2, ERDA-76/163/2, U.S. Government Printing Office, Washington, DC, 1976.

CHAPTER 10.6

Office Layout

JOHN J. MARIOTTI

General Motors Institute

10.6.1 INTRODUCTION

Most office interiors have been designed using a combination of private offices, partitioned areas for small groups, and large bull pen areas. The office systems are people-intensive, and typical interiors include desks, filing cabinets, and typewriters. People are grouped according to the organizational chart, and the position of an employee is reflected not only in the location and size of the office, but also in the perceived luxury of the furnishings.

In the 1930s the large open office was a rarity in most countries. However, in the United States companies were rapidly increasing in size, and with the concentration of plant, divisional, and corporate staffs in centralized office buildings, the open office was the custom.

A rapid increase in the construction of new office buildings occurred in the United States after World War II. The custom was to continue with the open office for the lower levels of workers and supervisors and to concentrate upper-level supervisors and managers along the perimeter window walls in private offices. As with most polar solutions, a continuum existed, and the age, size, and shape of the office building modified the open plan designs to include in one office not only groups of 100 or more, but also smaller groups of 10 to 30 people. The striking characteristic of these offices was the lack of visual and acoustic privacy. They were referred to as "bull pens," an appropriate but denigrating term.

The literature contained useful checklists that recommended rules for designing offices. A representative checklist that is still very useful is shown in Exhibit 10.6.1.

In the decade of the 1960s, modular office furniture was developed, and this mitigated the bull pen effect in large offices by reducing the visual signals that impinged upon workers. In Europe the Quickborner team (originators of the office landscaping concept) were experimenting with variations on modular furniture systems. Concurrently, in the 1960s and early 1970s, the published work of scientists stressing concepts of crowding, territoriality, invasion, and aggression began to influence the design of offices.[2] The results of some of this published work, which is pertinent to the design of office layouts, are included in this chapter.

10.6.2 POWER, COMMUNICATION, AND POSITION*

One's spatial relationship to other people can maximize or minimize one's communications with those people and influence the power one has over them. There is evidence that the shape and arrangement of furniture and the physical position of an individual in relation to others influences the perception held of one by others. For example, the end position on a rectangular table is invested with status, and a practical result is that persons selecting the end position in a jury room have twice as many chances of being elected foreman as persons selecting seats along the sides of the table.[3]

10.6.3 TERRITORIALITY, INVASION, AND OVERCROWDING

The territorial needs of humans are said[4] to be an inherited characteristic from our animal ancestry. Part of one's idea of onself and one's individuality is based upon the territory one considers one's own. The cultural aspect of territoriality is reflected in folk sayings such as "No stove is

*Portions of the sections on power, territoriality, noise, interviewing, and offices were abridged from J. J. MARIOTTI, "Proxemics the Spatial Interaction of Man With Man," *Proceedings*, 1978 Spring Conference, AIIE, Toronto, Ontario, Canada.

Exhibit 10.6.1 Representative Checklist for Traditional Design of Large Offices[a]

Desks should face the same general direction.

In open areas desks should be placed in rows of two.

For desks in rows of one, there should be 6 ft (1.83 m) from the front of a desk to the front of the desk behind it.

For desks in rows of two or more, and where ingress and egress is confined to one side, 7 ft (2.13 m) should be allowed from the front of a desk to the front of the desk behind it.

If employees are back to back, allow a minimum of 4 ft (1.22 m) between chairs.

Inside aisles within desk areas should be 3 to 5 ft (0.91 to 1.52 m) wide.

Intermediate aisles should be 4 ft (1.22 m) wide.

Main aisles should be at least 5 ft (1.52 m) wide.

Natural lighting should come over the left shoulder or the back of the employee.

From 50 to 75 ft^2 (4.65 to 6.97 m^2) are required for a work space consisting of a desk, shelf space, a chair, and a 2 ft (0.61 m) space allowance on a length and a width.

Desks should not face high-activity aisles and areas.

Desks of employees doing confidential work should not be near entrances.

Desks of employees having much visitor contact should be near entrances, and extra space should be provided.

Desk of the receptionist should be near the visitors' entrance.

Supervisors should be positioned adjacent to the secretaries.

Access to the supervisor's work station should not be through the work area.

Supervisors in open areas should be separated from their group by 3.3 ft (1 m).

The flow of work should take the shortest distance (see Exhibit 10.6.2).

People who have frequent face-to-face conferences should be located near each other.

Employees should be adjacent to those files and references that they use frequently.

Employees should be placed near their supervisors.

Five-drawer file cabinets should be considered in lieu of four-drawer cabinets.

Open shelf filing or lateral file cabinets should be considered in lieu of standard file cabinets (see Exhibit 10.6.3*a* and 10.6.3*b*).

Four- or five-drawer file cabinets should be considered as a substitute for 2 two-drawer cabinets.

The reception area should create a good impression on visitors, and an allowance of 10 ft^2 (0.93 m^2) should be used per visitor if more than one arrives at a given time.

The layout should have a minimum of offsets and angles.

Large open areas should be used instead of several small areas.

Open areas for more than 50 persons should be subdivided by use of file cabinets, shelving, railings, or low "bank-type" partitions.

Office space should not be used for bulk storage or for storage of inactive files.

Conference space should be provided in rooms rather than in private offices.

Conference and training rooms should be pooled.

The size of a private office will often be determined by existing partitions.

Private offices should have a minimum of 100 ft^2 (9.3 m^2) to a maximum of 300 ft^2 (27.9 m^2).

A 300 ft^2 (27.9 m^2) private office should be used only if the occupant will confer with groups of eight or more people at least once per day.

Related groups and departments should be placed near each other.

Minor activities should be grouped around major ones.

Work should come to the employees.

Water fountains should be in plain view.

Layouts should be arranged to control traffic flow.

Heavy equipment generally should be placed against walls or columns.

Noise producing work stations should be grouped together.

Access to exits, corridors, stairways, and fire extinguishers should not be obstructed.

All governmental safety codes should be followed.

In planning the office, consider the floor load, columns, window spacing, heating, air conditioning and ventilation ducts, electrical outlets, and lighting and sound.

The scale of the layout should be either $\frac{1}{4}$ in. = 1 ft (1 cm = 50 cm) or $\frac{1}{8}$ in. = 1 ft (1 cm = 100 cm).

Plastic reproducible grid sheets and plastic self-adhesive templates should be considered.

---

[a] Abridged from reference 1.

Exhibit 10.6.2 Planning Work Flow[a]

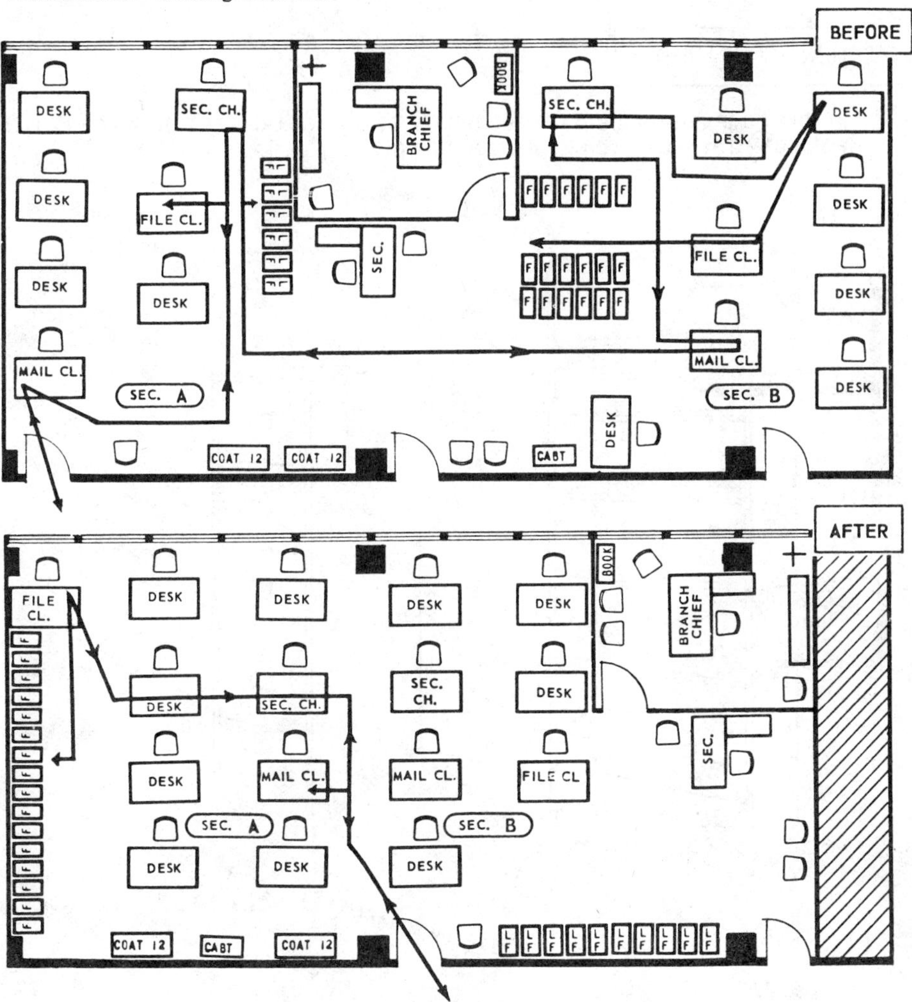

[a]Reference 1.

big enough for two women to cook on." What we have learned about crowding should include how one's expectations concerning densities affect one's perceptions about being crowded. For example, at a party we expect to be crowded and enjoy being crowded. At work we prefer more space, and at a religious retreat large amounts of space are needed.

It appears obvious that we sit closer to cooperative cohorts, and this is confirmed by studies showing that individuals who possess similar personality variables also share similar personal distance characteristics. For example, people who feel most comfortable sitting 4 ft away from another person will have characteristics in common with other people who feel comfortable at a distance of 4 ft.

10.6.4 NOISE

Noise is now perceived as an invasion of an individual's personal space. Noise increases one's sense of feeling crowded. For example, one study showed that (1) increasing environmental noise increased anxiety and that (2) with the subject under the stress of noise, other people were perceived as more disagreeable, more disorganized, and more threatening.

The reading achievement of children increased as their living quarters got farther away from street noise, a New York City study indicated. Noise is more liable to cause disturbance while

Exhibit 10.6.3 File Versus Shelves (*a*) and Space Comparison (*b*)[a]

EQUIPMENT COMPARISON

IN THE COMPARISION ILLUSTRATED BELOW, SEVEN SQUARE FEET OF SPACE IS USED FOR BOTH THE FILE CABINET AND SHELF UNIT.

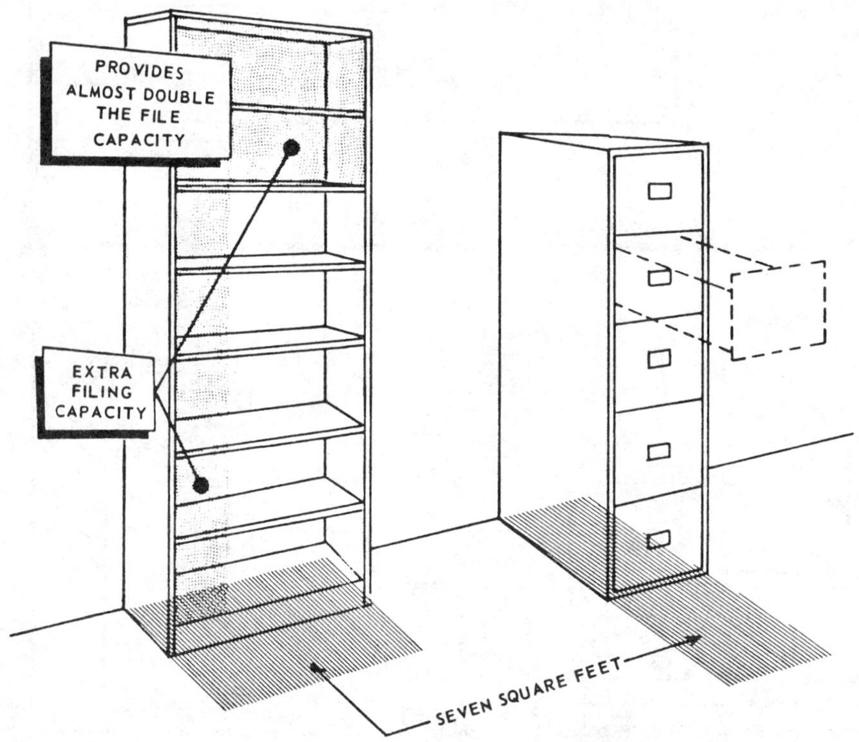

SAVINGS:

THE USE OF A SINGLE SHELF FILING UNIT (Seven Shelves High) PROVIDES ALMOST DOUBLE THE FILE CAPACITY OF THE CONVENTIONAL FIVE DRAWER FILE CABINET.

THE USE OF OPEN SHELF EQUIPMENT WOULD NET SAVINGS TO BOTH GENERAL SERVICES ADMINISTRATION AND THE AGENCIES.

(*a*)

[a]Reference 1.

people are at home than when they are outdoors or working. One study found that 56% of those who responded are distrubed by noise at home, 27% by noise outdoors, and only 20% by noise at work. On serial-spaced repetitive work that required complete attention, noise adversely affected the performance–the faster the pace, the poorer the performance. A Japanese study indicates that we complain more when we are the passive recipients of noise from other work than when we participate in the outbreak of noise.

Exhibit 10.6.3 (*Continued*)

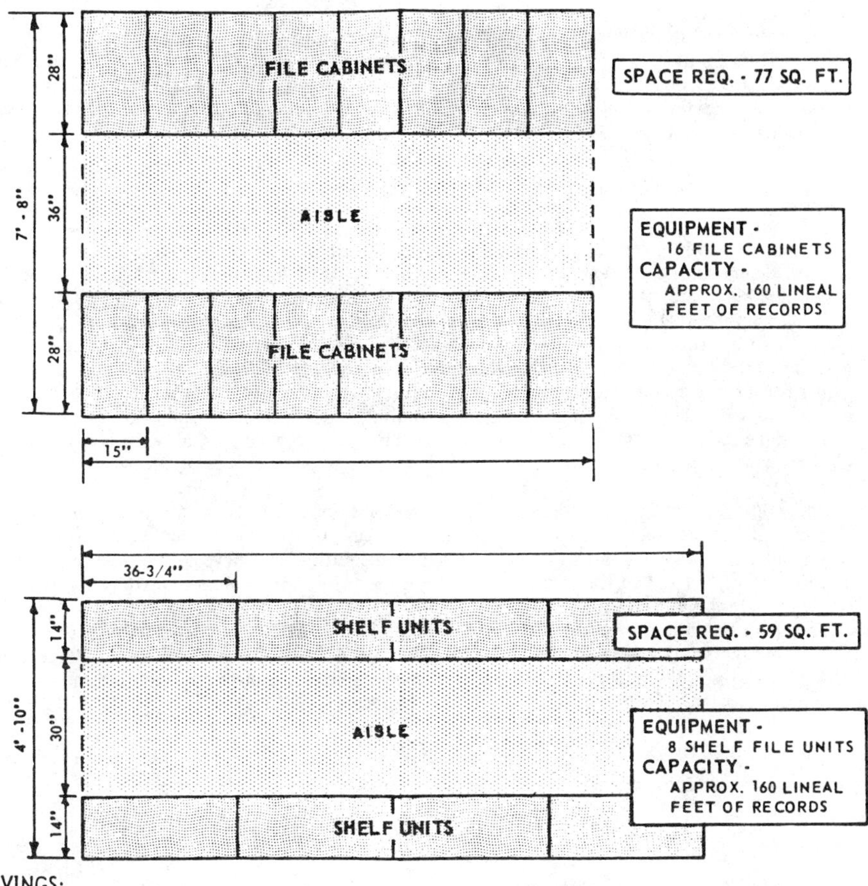

SAVINGS:

SHELF FILING RESULTS IN SPACE SAVINGS OF APPROX. 23% OVER FILE CABINETS (Letter or Legal Size).

[a]Reference 1.

(b)

10.6.5 EFFECT OF SPATIAL RELATIONSHIPS ON INTERVIEWING

Increasing the amount of eye contact, the feeling of comfort, and the amount of verbal response by the interviewee have been goals of investigators in designing interview situations. There is less eye contact at closer distances and when the inverview is conducted in a large room. In one study, when the interview was performed at distances between the interviewer and interviewee of 2 ft, $4\frac{1}{2}$ ft, and 9 ft, the greatest amount of talking by the interviewee occurred at 9 ft. In a psychiatric interview, of three distances (3 ft, 6 ft, and 9 ft), the most talking by the patient and the most eye contact occurred at 6 ft. In another doctor-patient study, more patients felt at ease when a desk was placed between the doctor and patient.

Factors in the interview design can influence how the interviewer is viewed by the interviewee. For example, an introvert interviewee with a frequently gazing interviewer will give the interviewer a more negative rating. If a desk is positioned between a highly anxious interviewee and the interviewer, the interviewee will rank the interviewer very high on credibility. Performing the interview in a pleasantly furnished room ensures that the interviewee will rank the interviewer very high on maturity.

Offices

It is clear that work stations are poorly designed for most secretaries and stenographers. Items such as files, pencil sharpeners, and paper punches, which are used not only by the secretary but also by other personnel, are frequently placed in what Hall refers to as the "secretary's personal space."[5] Better to duplicate facilities than to subject the secretary to the repeated invasions of the secretary's territory, most of which are accompanied by what has been positively described as "verbal strokings" and negatively described as the "idiot response."

In spite of the recent attention to open planning, modular equipment, and office landscaping, most of us still inhabit what many describe as bull pens (see Exhibit 10.6.4). Minimal space allocation should include sufficient distance so that the person can comfortably push his or her chair back from the desk without touching either the materials or the desk of the work station behind one.

Telephone systems may greatly affect how one feels about an office. A separate phone even for low users might be an economical decision. The incessant ringing of telephones, however, is the major noise producer for most bull pens. A standard solution is to have all the calls routed through a secretary in the front of the room. This brings on a curious phenomenon in which many an office worker will look up each time the secretary's phone rings, to see if the call is for him or her.

Time may be wasted in other ways, which are perhaps inevitable and irreducible. For example, a work sampling study by Serge Birn Associates indicated that the average male in a large bull pen office looked up 60 times per shift because of the presence of miniskirted women, and each of the distractions lost approximately 1 min.

Exhibit 10.6.4 Example of a Small Bull Pen[a]

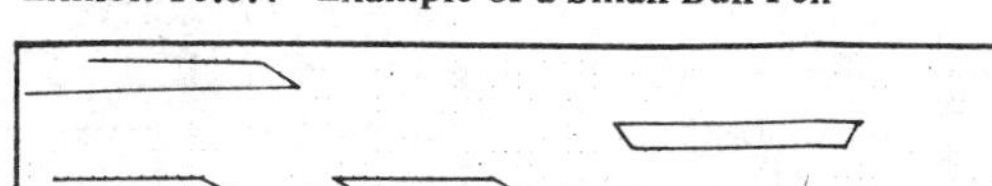

[a]Reference 1.

A long-standing question has been whether or not windowless buildings are more efficient than buildings with windows. One study found that the error rate of computer programmers increased when they were placed in a windowless room.

An experiment in which the subjects using office calculators were positioned 4 ft, 8 ft, and 12 ft apart showed the most work performed and the least errors made at the 8 ft distance.

Proxemics in the Design of Work Areas* †

Consider the design importance that can arise from the understanding that a taboo exists against people standing within each other's close personal zone. This taboo exists everywhere, unless people are related or riding in public conveyances such as subways, buses, or elevators. In assigning space and equipment in bull pen areas, shouldn't the territorial rights of individuals be considered and buffer zones be established? If some people have great negative feelings about being touched by others, should they have more space? In a bull pen, who owns the rights to the aisle? Should aisles be wide enough so that workers who border the aisle can assume ownership rights to the near part of the aisle?

Most of us would intuitively agree with the statement that, as stress increases, sensitivity to overcrowding increases and more space is needed. Perhaps our choice of conference rooms for meetings and the amount of space allocated to workers in bull pen offices should be related to the amount of stress expected to be reached. Many readers would agree that aggression can be controlled by establishing hierarchy and by proper spacing, but can that statement be extrapolated to the idea that workers with greater buffer zones between them have less need for close supervision? The design of work areas should reflect our subjective feelings about being cramped.

Since a hot room increases one's sense of feeling crowded, temperature control is a necessity. A more dramatic and much less practical solution is to vary the density of people according to changes in temperature. Noise control is also necessary, since noisy rooms increase the feeling of being crowded and possibly increase one's feeling of warmth. Libraries, by using carrels to reduce the amount of movement observed through peripheral vision, are able to handle a high density of people. Libraries are using table and desk arrangements to reduce the amount of eye contact and conversation. In reception rooms, if visitors are seated more than 12 ft apart, there is little or no conversation between the visitor and the receptionist.

In the remainder of the chapter, a planning process is applied to the development of an office layout. This is followed by a discussion of the cost-effectiveness of open plan offices and of demountable and portable offices. The chapter concludes with a list of references and bibliographic items.

10.6.6 PLANNING THE STUDY

In the early 1980s, planning for office layout is done in an environment in which there is an increasing awareness not only of the impact of territoriality, invasion, and population density, but also of the quality of work life of employees. In a survey sponsored by Steelcase, Inc., questionnaire results reflected a feeling among office workers that environmental conditions do affect their output.[7] Offices designed in the traditional rectangular grid were said to make for the sensory overload that distracts employees.

Employees expressed a desire for a heating and air conditioning system that avoided extremes in temperature and air movement. The greatest challenge to office design at this time is the challenge of individualizing ventilation and temperatures. This problem may be more difficult to solve for the open office than it is for the closed private office, but this is a task for the creative designer.

The employees responding to the questionnaire also wanted each work station and office system to provide access to required tools, equipment, and materials. In the Steelcase survey 74% said that an improved working environment would increase their productivity and their sense of worth. Of the workers surveyed, 54% use or operate data processing, word processing, telecommunications, or other electronic equipment.

Forces that create the climate for making changes in the office environment are:

1. Expansion of the work force.
2. Changes in personnel assignments.
3. Desire to increase work output.

*"Proxemics" is the term coined by Hall (reference 6) for the interrelated observations and theories of humankind's use of space as a specialized elaboration of culture.

†Portions of this section were abridged from J. J. MARIOTTI, "How Close Is Your Neighbor," *Industrial Engineering*, October 1969, pp. 14–18.

4. Rising construction costs.
5. Desire to improve the quality of work life.
6. Need to conserve energy.
7. Introduction of word processing and electronic information handling systems.

Many of these affect the planning objectives on specific projects.

Objectives

In developing specific objectives, the history of the organization, future expectations, and long-range plans must be considered. If the future can be extrapolated from the past, the change rate (i.e., the number of times the office environment was changed, for example, in the past 5 years) is a useful measure. If it is considered a high change (churn) rate, then the focus of attention will be on systems that allow the changes to be made quickly and inexpensively. Specific questions that must be answered are:

1. What is the duration of the planning period?
2. How many people will the system have to accommodate throughout the period?
3. What communication system will be used?
4. Is the organization's structure expected to change over the period in a manner that might cause fundamental changes in the layout?
5. Which subgroups are expected to expand or contract?
6. Will the management style change such that the layout might be significantly changed?[8]

The list can be endless, and the planning team has to probe those areas that might signal the need for a major change in the systems objectives. The resulting overall objectives, when completed and approved by managers and directors, form a structure within which detailed objectives can be developed for each department.

A Departmental Study

Since most office layouts are re-layouts of an existing office space, the focus is usually on overcoming existing problems, with the assumption that the desirable parts of the existing layout will remain. These assumptions and guides form constraints on the design of the layout and must be clarified. In one recent project an important and traditional constraint was to ensure that each of four people should be in the line of sight of the manager. Additional constraints—that interior walls and the overall supervisor's private office could not be moved—were lifted after several layouts proved inadequate. As more information was gathered about the overall organization and detailed work methods, further constraints developed.

Most managers accept the need to do a full-scale study of office organizations before arriving at an actual layout, but rarely is this done. Many layout engineers rely heavily on the wants of the customer department head, even to the extent of asking the department personnel to develop an initial layout. The problems involved in office layout are dramatized by a recent case: The headquarters staff of a division recently moved into a new contemporary structure. In this division it is customary to get signatures showing approval of the layout from each department head. After more than 3 years of work the move was finally made, but none of the department heads were willing to sign the layout.

If a full study is to be done with the expectation of cost reduction or cost avoidance, then the work simplification approach is useful. Work simplification in office systems has as its nucleus a quantitative and qualitative analysis of the work content. Interviews supplemented by techniques such as work log, delay studies, questionnaires, input-output analysis, flow charting, standard data, time study, work sampling, multiple regression, and short-interval scheduling are used to develop appropriate job assignments.

The major functions that are performed in an office area are subjected to a critical review to see if they are fulfilling office objectives. In a recent study of a mass mailing operation, the objective of which was to get more qualified applicants, there were indications that some applicants were receiving more than one mailing. Attempts to assess the extent of the duplications led to three possible actions. The first was to search for records of those applicants who communicated that they had received duplicates. None of these records had been maintained. The second alternative was to conduct a telephone and mail survey of applicants, but this was discarded because of the high cost. The third alternative was to compare lists. When this seemingly endless and tedious task was started, however, the source of the duplications was quickly found. Further analysis of the lists cut the mailing list by 65%.

The study dramatized the value of saying "Don't do things right, but do the right things."

For detailed information on performing a systems study, see Chapters 3.1, 3.2, and 3.3. Section 2, on organization and job design, is also pertinent.

Individual Workplace

Once the work simplification phase is complete, and it is clear the procedures will assist in fulfilling the objectives, then the space and equipment requirements of each workplace and each function are gathered. This is most efficiently done by using a questionnaire on which each person specifies the amount of space required for his or her storage and work surfaces. In addition, the type of conference facilities required and the frequency and duration of use are recorded. The perceived need for acoustic and standing and seated visual privacy is also recorded, along with the frequency of person-to-person interaction. The need to be adjacent to, or far away from, other individuals is one more required piece of information.

Storage and Surface Area

To supply adequate storage for a work station, the number of linear feet of shelving or drawers required for letter-, legal-, and computer-sized input and output cards and for paper, drawings, books, magazines, personal items, and miscellaneous items is recorded. Total surface space required considers the use of telephones, calculators, typewriters, computer input and output devices, and other information processing equipment.

Since the analyst is searching for the maximum required surface area, it is necessary to record the frequency with which large items, such as printouts, will be simultaneously in use. An individual may satisfy an occasional need for a large surface area by moving to a conference table or drawing board.

Privacy

The perceived need for individual privacy may, on further investigation, be less important than the questionnaire returns may suggest. It is important to isolate an individual's infrequent need for telephone privacy from another's need to have privacy for 20% of the calls. The need for private conferences among two or three people as well as the need for privacy for an occasional telephone call can be met by setting aside a small conference area and a separate telephone area, each designed for acoustical privacy.

Visual privacy needs vary, not only among individuals, but also among management levels. For many, only seated privacy is needed, and this gives layout designers flexibility (Exhibit 10.6.5). For middle and upper management, the perceived need is for complete privacy.[9] In the new forms of organization that center on maximizing employee involvement in decision making, the perceived need for complete privacy is negotiable. This is so since in these organizations there is an avoidance of status symbols, including private offices, throughout many layers of management.

Group Needs

Getting data for a group may be done concurrently with the data gathering for individual work stations. The data for storage, equipment, work surface area, territoriality, conference facilities, reception areas, lounge areas, acoustic and visual privacy, and relative closeness needs are gathered.

If an open office is the solution, then the heights of panels and screens become an important consideration in establishing the area for the group. An important factor that influences selection of panel height is whether or not each individual is to have seated visual privacy. A better solution is to isolate subgroups. In this case individuals sharing the open space would have little visual and acoustic privacy.

Standardization for Flexibility

With the available building block furniture systems, it is possible to have a wide variety of individualized work stations. In many of the systems, floor-supported furniture can be mixed with a basic panel-supported furniture system. The merits of the open area systems include the ability to rearrange work stations. A wide variety of individualized stations would inhibit this flexibility. For example, if each subgroup had an individualized color for work stations, an additional work station, transferred from another subgroup, would have another color.

Standardization in heights of panels is accomplished by using panels of approximately 4 ft (1.22 m) to 5 ft (1.52 m) height to provide seated visual privacy; heights of approximately 80 in. (2.03 m) provide standing visual privacy. The higher panels can be used to dramatize territorial boundaries, shield confidential areas, and add to visual variety. Individualization of every work station is impractical; therefore those workers doing the same work will have the same type of work station.

Exhibit 10.6.5 Example of Open Planning With Single and Two-Person Work Stations[a]

[a]Courtesy of Steelcase, Inc.

Power and Communications

In most existing office buildings, the power and communication network is distributed through wall conduits or through a floor distribution system. Recent years have brought very flexible grid systems installed in the ceiling plenum. Power poles deliver the wires to clusters of panel-connected work stations. The panels can be factory wired, and standard connectors are used to connect panels together. The power poles may be hard wired in the ceiling plenum or may be designed for a quick release.

The panels have raceways in the baseboards to service not only the baseboard outlets, but also the outlets at the top of the panel connector poles, which are designed specifically for ambient light fixtures.

Lighting

The conventional office relies on overhead direct lighting to illuminate all areas uniformly. The new office furniture systems allow a mixture of direct task lighting, where the fixture is supported on the panel above the desk, and an ambient source, which is also an integral part of the furniture system. There is a significant reduction in energy use for lighting because less power is used. Since light sources create heat, reducing light results in lower air conditioning requirements and further reduction in energy use. By replacing ceiling lighting panels, the sound absorbency area of the ceiling is increased 10% to 15%. For detailed coverage of lighting, refer to Chapter 6.11.

Acoustics[10]

The sound in typical offices is produced by typewriters, telephones, air conditioning and heating systems, duplicators, and serious and casual conversations. Both too much and too little sound may create distractions, which in turn may result in less productivity. A sound-absorbent ceiling prevents sound waves from ricocheting in all directions. Carpeting will reduce noise, especially foot noise, since carpets absorb about 10 times more sound than other floor surfaces. Screening panels obstruct sound and prevent it from traveling to areas where it is not wanted. Panels can be used to form enclosures that trap noise and keep it contained.

These measures may make the office so quiet that every sound is magnified and becomes a distraction, which may interfere with privacy of conversation. To prevent that from happening, masking sound is introduced. Masking sound is a nonirritating electronic sound projected from speakers in the ceiling or from other areas in the environment. It raises the background noise level to mask or cover up other sounds without offending the eardrums or hearing. For a detailed coverage of acoustics, refer to Chapter 6.10.

Color[11]

Color is a tool used to relieve monotony and create variety. The psychological effects of color have been studied, and several general comments can be made. Red and yellow bring things nearer and increase the feeling of warmth. Blue, gray, and violet, make a room larger and suggest coolness. Red, yellow, and orange are exciting colors and should be considered when the work itself is monotonous. Blue, green, and purple are comfortable and soothing and should be considered if the work itself is interesting. The practical considerations of maintenance and whether or not there is little or much dirt affect the choice of colors. For example, darker colors, which hide dirt, should be considered for floors.

Colors can be used to reduce the sense of disorientation. In open plan offices in which panels are used to reduce visual stimuli, color can be used to provide orientation and territorial delineation and to highlight the location of aisles, entrances, and exits.

Whether or not color affects productivity in offices has not been determined, since so many variables influence productivity.

10.6.7 RELATIVE CLOSENESS NEEDS*

During the data gathering process, each person is asked to record his or her relative closeness need to:

1. Others in the work group.
2. Specific functional areas, such as a file room, a copy machine, or an executive office area.
3. Fixed building features, such as a front lobby and warehouse entrance.

These needs are expressed with relative closeness values such as:

1. Must be adjacent.
2. Important–within easy walking distance.
3. Ordinary–within a reasonable distance.
4. Unimportant.

*This section on relative closeness needs and space estimating is abridged from *The Procedures, Office Survey and Space Planning Program*, Steelcase, Inc., Grand Rapids, MI, sections F, G, and H, n.d.

This is also done for groups. The example will consider solely the relative closeness needs among groups. The values will be given by the managers or supervisors of the departments or groups involved. If two department heads give different values for the closeness needs of their departments with each other, the difference must be resolved. The values are entered on the relative closeness matrix, and a reason is recorded for each (see Exhibit 10.6.6).

Using the completed matrix, a closeness chart is prepared. Starting with group 1, all "A" relationships (3 lines) are entered, and each group entered is represented by a circle (Exhibit 10.6.7*a*). The circles are all of equal size. When all the "A" relationships in the matrix have been entered, the process is repeated for the "B" relationships (2 lines, Exhibit 10.6.7*b*). The "B" lines are drawn 25% longer than the "A" lines. The process is again repeated for the "C" lines, which are drawn 25% longer than the "B" lines (Exhibit 10.6.7*c*).

It may be advantageous to redraw the chart to reduce the crossing of lines after the "B," and

Exhibit 10.6.6 Relative Closeness Matrix[a]

	Group or unit	1	2	3	4	5	6	7	8	9	10	11
1	Credit and loan manager	1										
2	Analytical	B/2	2									
3	Investigation	C/4	C/4	3								
4	Loan control	C/2	C/3	A/1	4							
5	Credit policy	B/2	C/2	A/	A/2	5						
6	Consumer credit	A/2	C/2	C/3	B/2	C/2	6					
7	Loan review	B/2	C/2	C/3	C/5	C/4	A/2	7				
8	Repossession	C/2	C/5	C/5	C/5	C/5	A/13	A/11	8			
9	Loan adjustment	C/5	A/1	C/2	C/5	C/2	D/6	D/	D/	9		
10	Copy machine	D/	C/3	C/2	C/5	C/3	C/2	C/5	C/3	D/	10	
11	File room	D/	D/	A/3	A/3	D/3	D/	D/	D/	D/	D/	11
	Group or Unit	Credit and loan manager	Analytical	Investigation	Loan control	Credit policy	Consumer credit	Loan review	Repossession	Loan adjustment	Copy machine	File room

Code	Reasons
1	Extensive face-to-face contact
2	Work on joint tasks/projects
3	Share files
4	Share equipment
5	Paper flow
6	Noise level

Value	Relative Closeness
A	Must be adjacent
B	Important
C	Ordinary
D	Unimportant

[a]Courtesy of Steelcase, Inc.

Exhibit 10.6.7 Closeness Chart Showing (*a*) A Relationships, (*b*) A and B Relationships,[a] and (*c*) A, B, and C Relationships

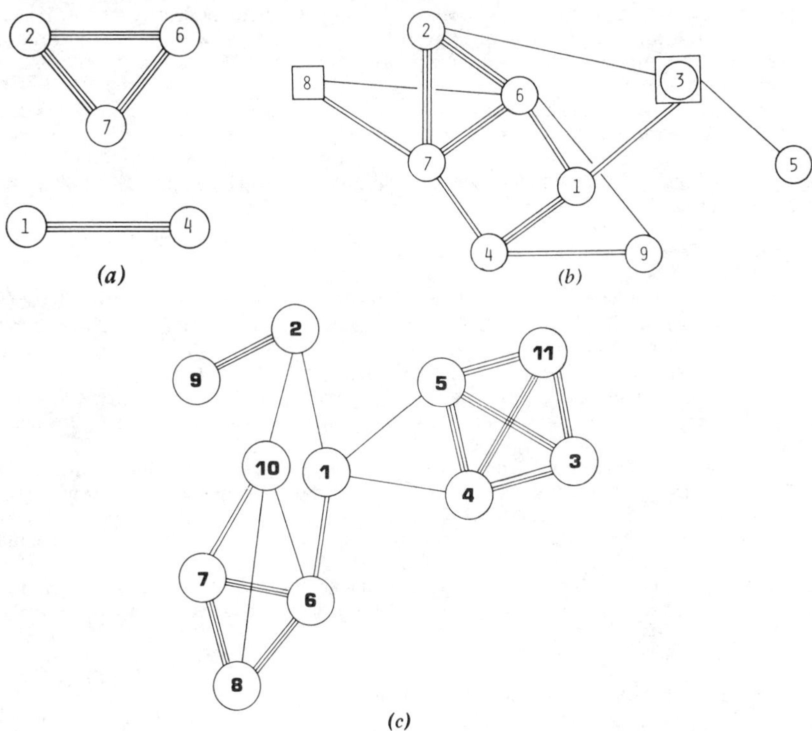

[a]Courtesy of Steelcase, Inc.

again after the "C," symbols are added. The objective is to keep the "A" relationships close together, and, if possible, the "B" relationships are also kept close together. It is not as important to keep the "C" relationships close together. The "D" relationships are ignored.

Figuring Total Floor Space Required

The next phase is to estimate the space required for each work station, group, and department until an estimate has been made of the total floor space required. To the normal dimensions of individual work stations must be added an allowance for aisles and access. A rule of thumb is to add 2 ft (0.61 m) to both the length and width dimensions of the work station. The floor space for a workstation requiring 60 ft^2 (5.6 m^2) would be increased to 95 ft^2 (8.83 m^2) by adding the aisles and access allowance in Exhibit 10.6.8.

Exhibit 10.6.8 Space for a Single Work Station

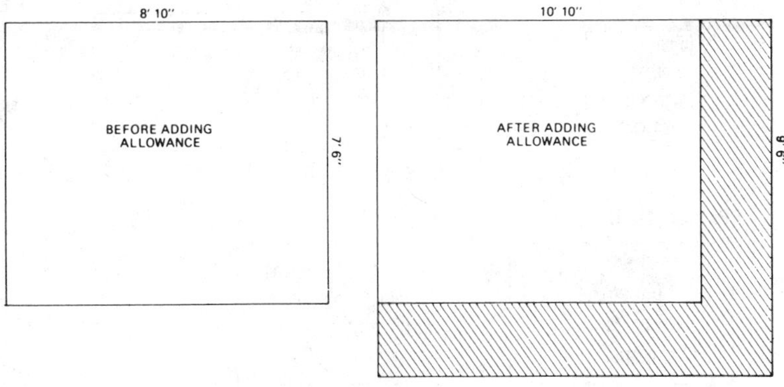

Density Considerations

The 2 ft added to each dimension in Exhibit 10.6.8 provides a medium-density layout. Decreasing the 2 ft (0.46 m) down to a minimum of $1\frac{1}{2}$ ft (0.46 m) increases the density. Enlarging the allowance over 2 ft (0.61 m)–generally required for adequate separation of work stations in a "pure landscape" layout–decreases the density.

The square footage needed for a single work station is multiplied by the number of work stations to arrive at total square footage required for that type of standard work station. This procedure is repeated for each standard work station in that particular group or unit. Adding up the total square footage for each type of work station will give the total floor space required for the individual work stations in that group or unit.

Group and Conference Space Requirements

The same procedure is used for group equipment. However, if group equipment requirements call for files or counters, the aisle and access allowance figure is increased by an additional 2 ft (0.61 m) to a total of 4 ft (1.22 m). This additional allowance provides space for drawers to open or for employees to stand at a counter or work surface, as illustrated in Exhibit 10.6.9.

Conference areas and similar spaces that are self-contained within a fixed perimeter, such as lounge or reception areas, labs, and workrooms, are computed in the same way as individual work stations.

Adding up the space requirements for group equipment, conference, and similar areas and combining them with the space required for individual work stations provides the total space required for the overall office layout (Exhibit 10.6.10).

Total floor space required for all groups must be equal to or less than the net available building floor space, which is calculated next (Exhibit 10.6.11).

Space occupied by columns is considered part of available floor space and need not be subtracted.

For an example on how to start with gross total floor space and arrive at the space available per work station, refer to Exhibit 10.6.12.

Exhibit 10.6.9 Space for Files and Counters

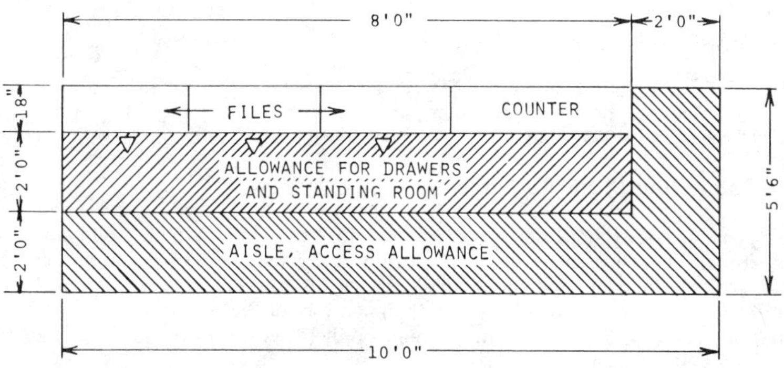

Exhibit 10.6.10 Space Required for Group/Unit

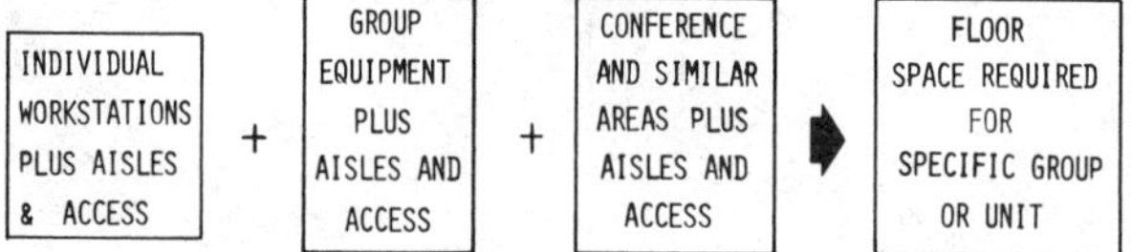

Exhibit 10.6.11 Net Space Available

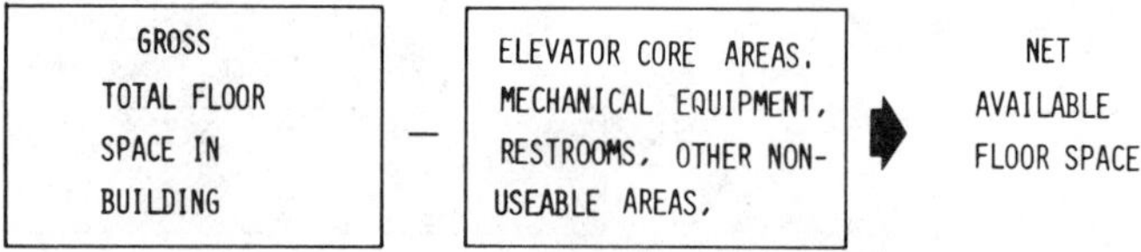

Exhibit 10.6.12 Estimating Square Feet per Work Station for an Existing Area: An Example Problem

Assume that 4000 ft^2 (371.6 m^2) of space is available.
Net = gross area minus corridors, elevators, and rest rooms.
Net = 4000 (371.6 m^2) – 800 ft^2 (74.3 m^2) (corridor)
– 100 ft^2 (9.3 m^2) (elevator)
– 200 ft^2 (18.6 m^2) (rest room)
= 2900 ft^2 (269.4 m^2)
Area for work stations = 2900 ft^2 (269.4 m^2) – 30% for interior aisles
= 2030 ft^2 (188.6 m^2)
Area for engineers and secretaries
= 2030 ft^2 (188.6 m^2) – 140 ft^2 (13 m^2) (general supervisor)
– 240 ft^2 (22.3 m^2) (2 supervisors)
– 150 ft^2 (13.9 m^2) (conference area for 8 people)
– 59 ft^2 (5.5 m^2) for 160 linear ft (48.8 m) of records in the shelf units
– 48 ft^2 (4.5 m^2) reproduction center
– 120 ft^2 (11.1 m^2) word processor station
= 1273 ft^2 (118.3 m^2)

$$\text{Net area per work station} = \frac{1273\ \text{ft}^2\ (118.3\ \text{m}^2)}{19\ \text{work stations}}$$

= 67 ft^2 (5.2 m^2)

The 30% for interior aisles may give a congested area. If more aisle space is required, then the options listed previously in this chapter under "Reducing Space Required" should be considered.

Reducing Space Required

If the total floor space required exceeds the net available building space, then the total floor space must be reduced. In reducing space required, the following options are available:

1. Decrease the individual work station.
2. Decrease group equipment aisle.
3. Decrease access allowance spaces.
4. Eliminate provisions for growth.
5. Eliminate space for future employees.
6. Reduce work station size.
7. Reduce group equipment by transferring or combining storage equipment.

All such modifications have to be taken with consideration of the impact on worker performance. After the space has been estimated for each group, a scaled block chart is developed.

The Scaled Block Chart

The scaled block chart is developed by adding to the closeness chart the space requirements of each group in square feet. Using a consistent scale, square blocks are created, which represent the square footage requirements of each group, and are substituted for the circles of the original closeness chart. The dimensions of each square block are scaled as the square root of the square footage requirement of that group. For example, Exhibit 10.6.13 would replace the circle for group 7, resulting in the block chart in Exhibit 10.6.14.

The next phase is to create the block plan by placing the scaled space requirements of each group in the layout area while retaining the closeness relationships represented by the block chart (Exhibit 10.6.15).

Evaluating the Block Plans

Several block plans are prepared and evaluated against a set of criteria. The criteria are developed from the planning objectives, and weights are assigned to each of the criteria. Then each of the plans is given a rating for each of the criteria.

The individual ratings are then multiplied by the assigned weight for the respective criteria. The results are summed for each plan, and the plan with the highest total is presented to the committee. Occasionally the committee may request that additional plans be developed and evaluated.

Exhibit 10.6.13 The Scaled Block

7
THIS, ON THE CLOSENESS CHART,
BECOMES THIS ON THE
SCALED BLOCK CHART
7
C. S. CLERICAL
625 SQ. FT.
25′ x 25′

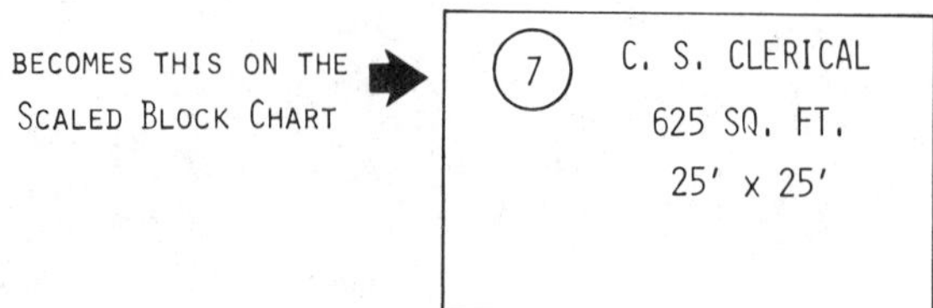

Exhibit 10.6.14 Block Chart[a]

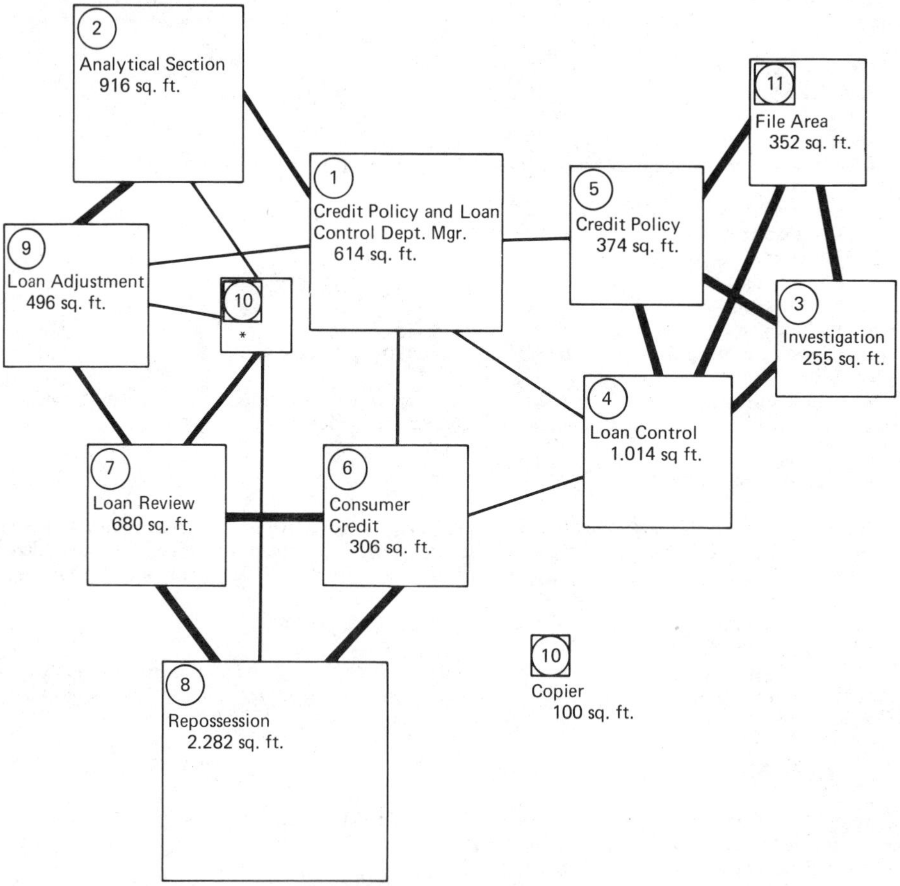

2
Analytical Section
916 sq. ft.
11
File Area
352 sq. ft.
1
Credit Policy and Loan
Control Dept. Mgr.
614 sq. ft.
5
Credit Policy
374 sq. ft.
9
Loan Adjustment
496 sq. ft.
10
*
3
Investigation
255 sq. ft.
4
Loan Control
1.014 sq ft.
7
Loan Review
680 sq. ft.
6
Consumer
Credit
306 sq. ft.
10
Copier
100 sq. ft.
8
Repossession
2.282 sq. ft.

[a]Narrow width line = 1 line; medium width = 2 lines; wide width = 3 lines.

Exhibit 10.6.15 Block Plan[a]

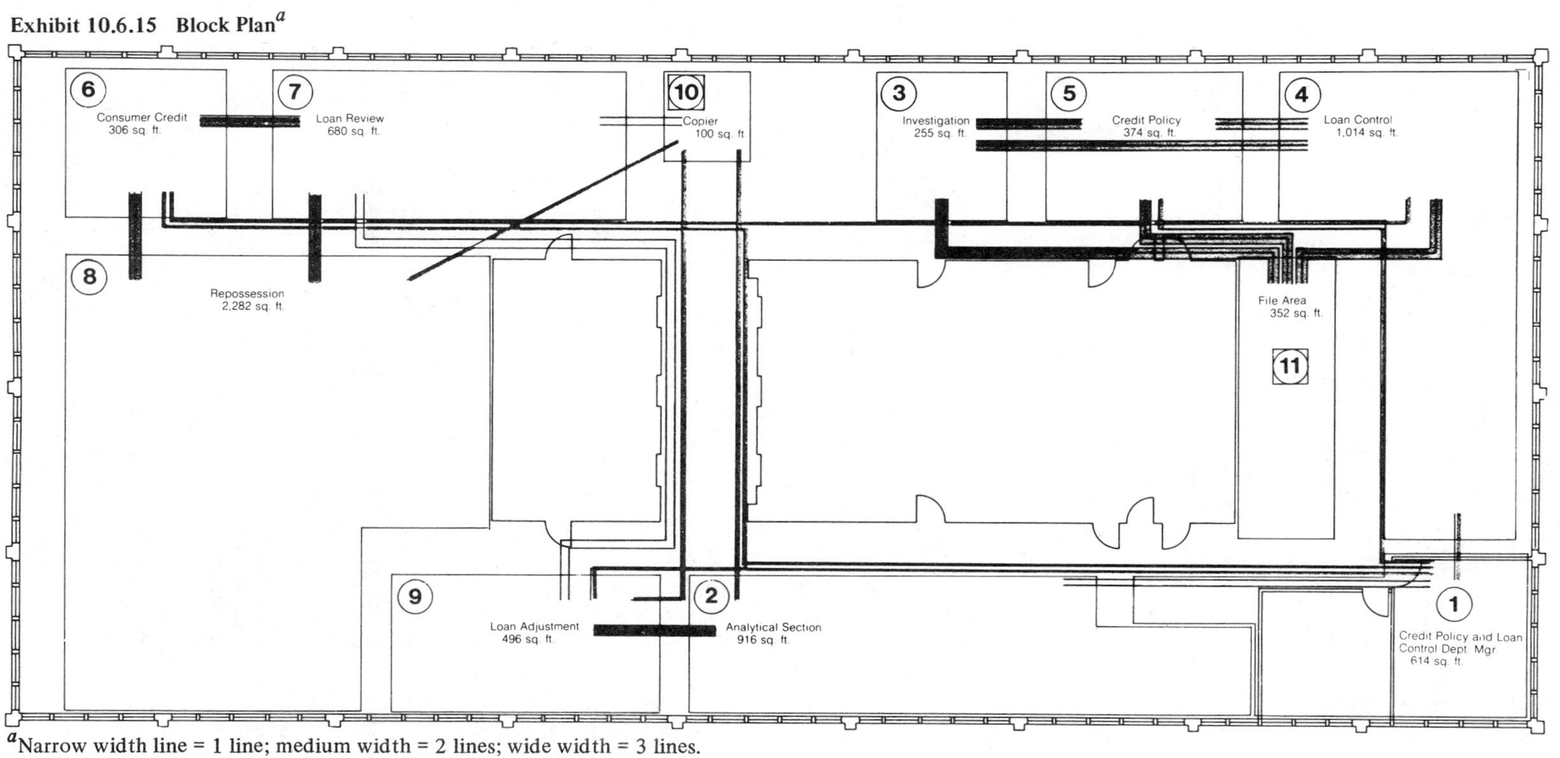

[a]Narrow width line = 1 line; medium width = 2 lines; wide width = 3 lines.

Exhibit 10.6.16 Evaluating the Block Plans

Office Layout–Evaluation

Office: Household goods loan Date: 8-8-80

Project: Q W L Analyst: JJM

Plan: A. Top floor B. Ground floor

Criteria	Weights	Plan A Rating	Plan A Rating X Weight	Plan B Rating	Plan B Rating X Weight
1. Noise distraction	8	9	72	7	56
2. Visual distraction	6	8	48	7	42
3. Reception	5	8	40	10	50
4. Flow; people, supplies	7	7	49	9	63
5. Convenience of personnel	8	9	72	8	64
6. Flexibility	10	10	100	6	60
7. Aesthetics	7	10	70	6	42
8. Area per employee	3	8	24	7	21
9. Energy costs	4	10	40	5	20
10. Cleaning costs	9	8	72	8	72
11. Maintenance costs	9	8	72	7	70
12. Project cost	10	9	90	6	60
13. Expected productivity	10	10	100	7	70
Total			749		690

Exhibit 10.6.16 shows two plans compared against a list of 12 criteria. The ratings and the weights each were selected from a scale of 1 to 10, with 1 as a low score and 10 as a high score. The weights and the scales are usually selected by the advisory committee.

Frequently, criteria such as safety, the cost of energy, cleaning, maintenance, and the initial cost of the project are classified as tangible criteria. They may be evaluated separately from the remaining criteria, such as criteria 1 through 8 from Exhibit 10.6.16. Criteria 1 through 8 would then be classified as intangible criteria and evaluated as in Exhibit 10.6.16.

The next step is to make a detailed layout (Exhibit 10.6.17) of the plan that best satisfies the objectives of the project.

Cost-Effectiveness of Open Plan Offices

The integrated furniture panel systems, when installed in existing structures, represent a significant investment per employee. Top management, when presented with the cost of implementation, may balk at approving the project and suggest instead a traditional layout consisting of many private offices and small open areas. Frequently voiced arguments for the integrated unified-panel system are presented in the following paragraphs.

In an extensive study by Axel Boje[12] comparing a large open office to a series of four-person offices, the estimated net gain for each employee in the open office was 2000 min/month. Over half the gain came from fewer distractions because of noise and from a rise in output because of optimum control of temperature and moisture content. Approximately 25% of the gain was because of a reduction in use of time for employee's private interests and an increase in sharing of imbalances in work load. The remaining savings resulted from elimination of approximately 32 door openings/employee per day, a reduction in length of trips to visit other employees, a reduction in wrong calls, fewer primary distractions from telephone calls and visitors, and a reduction in absenteeism.

Secondary distractions, traffic, and loud telephoning on the periphery of the group decrease the effectiveness of workers in open offices. In spite of the disadvantages of open offices, there is an estimated net gain in time per employee of 18.7%, according to Boje.

Exhibit 10.6.17 Final Layout[a]

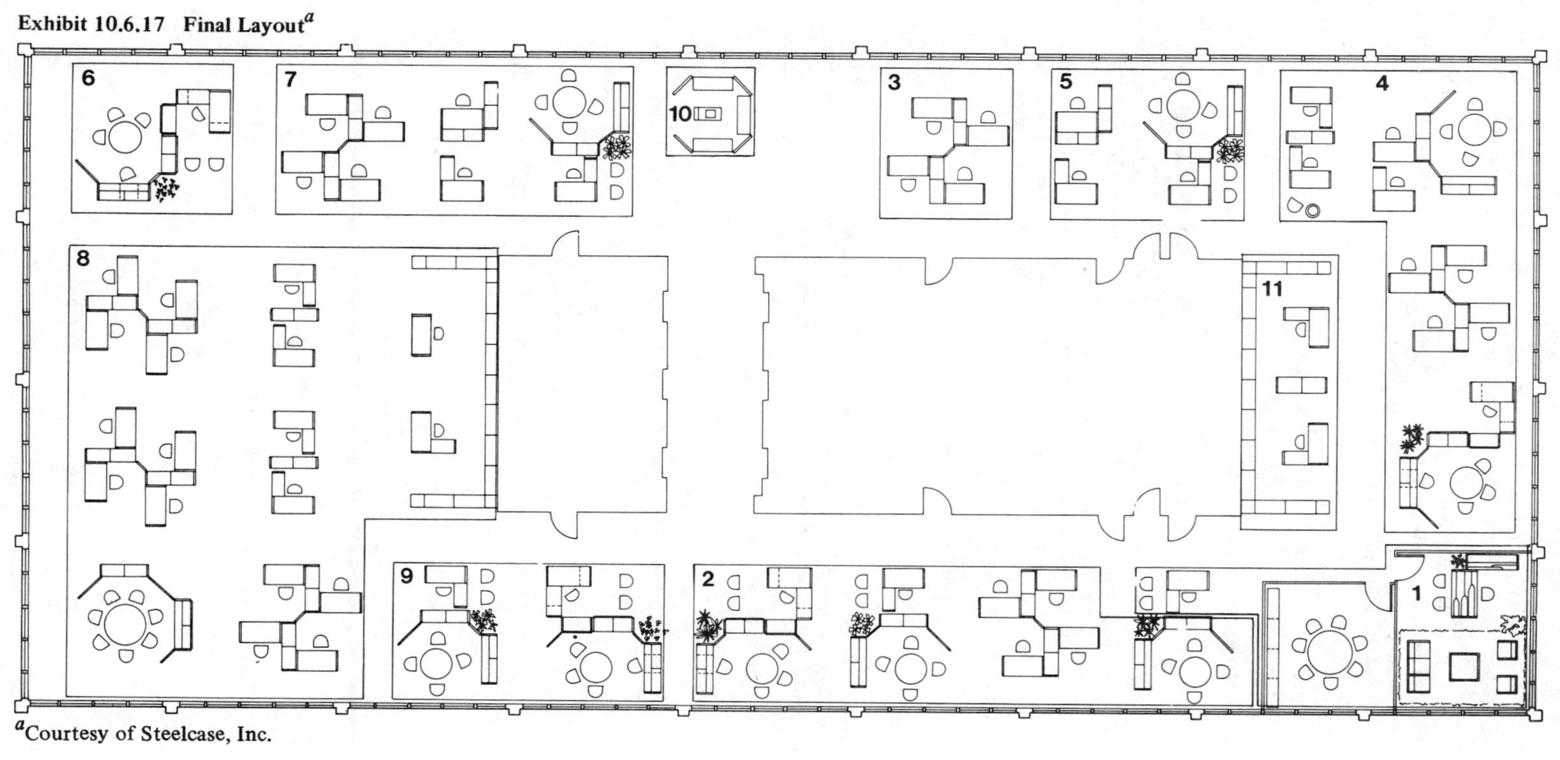

[a]Courtesy of Steelcase, Inc.

In addition, there are many testimonials from organizations, some of which are documented by confidential studies, which report an increase in output per employee from 9% to 20% arising from open offices.

Open offices reduce cleaning and maintenance costs and provide an estimated 80% reduction in the cost of rearrangement. Companies have experienced a 20% reduction in construction costs for new buildings with open offices.

Open office lower building costs come from:

1. A reduction in gross floor space.
2. A reduction in the number of interiors, walls, and doors.
3. A reduction in the complexity of the heating and air conditioning system.

Demountable Office Enclosures

In recent years there has been a move toward providing prefabricated office modules for use inside or outside factory and warehouse buildings. The modular design permits relocation and expansion to facilitate changing conditions. Standard modules may include offices fully wired with air conditioning and sound deadening, all of which are preassembled in the vendor's factory. The preassembled units are designed to be carried by a fork truck.

Some of the uses of the offices are for supervisors, utility rooms, lunch rooms, security control, waiting rooms, first aid stations, meeting rooms, and rooms for union representatives. Since both the preassembled and the prefabricated offices are modularly designed, they may be grouped together (with interior walls deleted) to form large areas.

The use of portable offices has been particularly successful in noisy factory environments. They provide a private, quiet place for production supervisors to confer with employees and management.

CONCLUSION

The study of office systems, like the study of manufacturing systems, is colored by the technical environment. The introduction of word processing and other computerized information systems into a typewriter environment frequently results in a reallocation of functions.[13] The 1980s should produce the paperless office. The office systems developed in the 1970s will provide the flexibility suitable to the electronic paperless office.

REFERENCES

1. *Guide for Space Planning and Layout*, Federal Stock No. 7610-145-0168, Superintendent of Documents, Washington, DC, n.d.
2. L. SHOSHKES, *Space Planning, Designing the Office Environment*, Architectural Record Books, New York, 1976, p. 2.
3. R. SOMMER, *Personal Space*, Prentice-Hall, Englewood Cliffs, NJ, 1969, p. 20.
4. K. LORENZ, *On Aggression*, Harcourt Brace Jovanovich, New York, 1966.
5. E. T. HALL, *The Hidden Dimension*, Doubleday, Garden City, NY, 1966.
6. HALL, *The Hidden Dimension*, p. 114.
7. LOUIS HARRIS AND ASSOCIATES, *The Steelcase National Survey of Office Environments*, Steelcase, Inc., Grand Rapids, MI, 1978.
8. M. SAPHIER, *Office Planning and Design*, McGraw-Hill, New York, 1968.
9. A. BAUM and Y. EPSTEIN, *Human Response to Crowding*, Erlbaum, Hillsdale, NJ, 1978, p. 174.
10. *Planning Guide*, Architectural Systems Division, Westinghouse Corporation, Grand Rapids, MI, 1975.
11. C. BENNETT, *Human Factors in Design*, Prentice-Hall, Englewood Cliffs, NJ, 1977.
12. A. BOJE, *Open Plan Offices*, Business Books Limited, London, 1971, p. 57.
13. H. MCCABE and E. POPHAM, *Word Processing*, Harcourt Brace Jovanovich, New York, 1977, p. 151.

BIBLIOGRAPHY

MONTAGUE, A. M. F., *The Nature of Human Aggression*, New York, Oxford University Press, 1976.

OIS Layout Planning Guide, E. F. Hauserman, Cleveland, OH, 1973.

SECTION 11

PLANNING AND CONTROL

CHAPTER 11.1

Technological Forecasting

DAVID J. DEMARLE
M. LARRY SHILLITO
Eastman Kodak Company

11.1.1 INTRODUCTION

Planning for the future is an everyday practice. In a world in which change is occurring exponentially, it is essential that industrial engineers avail themselves of modern tools to anticipate change. Frequently, industrial engineers and managers face a common problem—how to participate constructively in the change process. In this chapter we examine some modern forecasting techniques that are basic to subsequent planning and control processes.

11.1.2 FORECASTING MODELS

When people forecast the future, they do so by formulating speculative constructs of it. The word "construct" is used here to describe a mental model of the future that can be communicated to others. Mental constructs of the future allow us to sense the future and to anticipate and plan for it. Constructs of this type are fundamental to our adaptive ability and are extensions of our creative imagination. Such models are often intuitive, based on subconcious data. Experienced managers often perform like skilled chess players and act instinctively to optimize future success. Constructs of this type are a basis for progress and are involved in a significant amount of management practice. However, to err is human, and cases can be sited where faulty models of the future led managers and organizations to undesirable ends. Hence the constant search for improved forecasting models.

11.1.3 RATIONAL VERSUS INTUITIVE FORECASTS

Several factors at work in any forecasting process are shown in Exhibit 11.1.1. Three vectors are shown: vector A depicts the subject to be forecast; B depicts the information used in forecasting; and C depicts various methods used in forecasting. All vectors originate from a central point and move outward to levels of increasing complexity. The origin represents a position from which future states of increasing complexity originate.

These vectors are interrelated. As the subject to be forecast becomes more generic, more information is needed. To accommodate this, trend constructs replace event constructs, and a degradation occurs in the specificity of the data used in the forecast. Near the origin, factual data and knowledge provide a rich source from which reasonably reliable constructs can be made. Knowledge of cause-effect relationships allow factual information to be interrelated and structured to predict the near future with a large degree of accuracy. As the subject to be forecast is enlarged, the data required grow exponentially and make construction of detailed models impractical, if not impossible.

To accommodate these demands, models based on trends and vectors and their resulting interactions come into play. They are based on information that is frequently more opinionated and less factual. Here rational mathematic models begin to fail, and models based on intuition and expressed in language begin to prevail. As the scope of the subject enlarges, the complexity of the subject becomes so great that it defies rational comprehension, and managers use highly speculative constructs based on their past experience for forecasting and planning. Constructs of this type are largely intuitive and are the venue of planners and seers alike. At the end of this chapter, we have included an example of a technological forecast of future water supplies in a major water district. In identifying the scope of this forecast, we offer further elaboration of Exhibit 11.1.1.

From this brief discussion, we conclude that the scope of a forecasting effort controls the informational and modeling aspects of the study. Even the most elegant and sophisticated forecasting

Exhibit 11.1.1 Forecasting Parameters

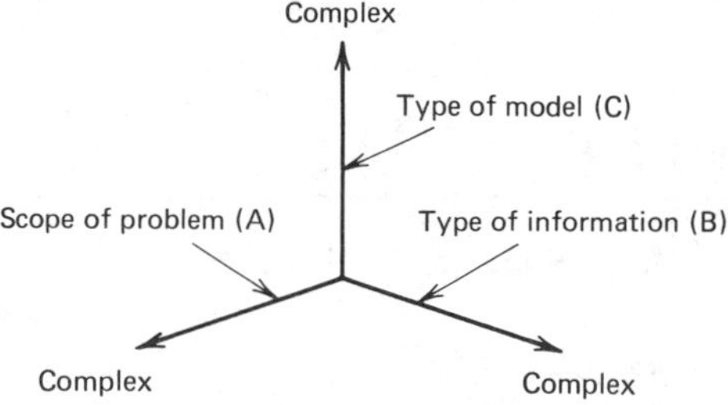

techniques cannot cope with the scope of some forecasting demands. Yet technological forecasts can aid planners in many areas, and we now examine this forecasting process.

11.1.4 DEFINITION AND BACKGROUND

"Technology forecasting" has been defined as a process in which data are gathered and analyzed in order to predict the future characteristics of useful machines, procedures, or techniques.[1] It grew out of the need of the military to predict the characteristics of future weapons systems. Technology forecasting has had widespread use in defense planning, and this use led to the spread of technology forecasting in industry.

In 1961 Professor James Bright developed a course on technology forecasting and planning at Harvard which was offered to R & D managers. In Europe Dr. Erich Jantsch published the results of a study of some 100 technological forecasting and planning activities conducted in 13 different European nations.[2] These events served to stimulate interest in this field and to attract new practitioners. Since the late 1960s, the field has seen the formation of consulting firms specializing in this type of forecasting; the publication of numerous articles, books, and journals on the subject; and the development of university courses devoted to the subject.

As the technology forecasting discipline grew, a large array of modern forecasting methods was developed and came into play, which allowed the forecaster to broaden the scope of the forecasts. Many of these used the calculating and word processing capabilities of the computer. These new tools have been used in market research and analysis, as well as in social, political, and environmental forecasting. Today, future studies using the technology forecasting techniques are used to forecast highly subjective areas. This broad type of forecasting, which uses the technical tools developed to predict technology, has been called technological forecasting. Thus technological forecasting, using constructs developed from mathematics, language, and art, is a modern methodology for forecasting the future.

11.1.5 RATIONALE FOR FORECASTING

Many forecasts never materialize. What, then, is the rationale for technological forecasting? What phenomenon leads us to believe that we can predict the future?

Inspection of the past reveals that goods and services come into being as replacements for prior goods and services. As in nature, there is a process of survival, where functionally superior items replace their predecessors. Thus incandescent lights replaced gas lanterns, and corporations replaced feudal estates. Time after time, value-based innovations produced better ways to accomplish fundamental needs. As time progressed, the "fittest" evolved as replacements for predecessors, which then died out.

As in natural evolution, this process of technological and social evolution is heavily influenced by the environment. Frequently, substitutions are brought about by shortages in key raw materials. This need-driven, valued-based substitution process lies at the heart of technological forecasting. Since substitutions take time to unfold, the forecaster can use substitution phenomena as one basis for logical predictions of the future. By examining historical information, forecasters derive parameters and relationships that portend the future. Models derived from the data allow the forecaster to develop a set of futures that are "likelier than not" to occur. These forecasts form a basis for planning and decision making. Many of these forecasts have an innate tendency toward self-fulfillment and are a basic ingredient in management planning. Technological forecasting, based as it is on a study of value-based substitution phenomena, can be a cornerstone of this planning effort.

11.1.6 THE TECHNOLOGICAL FORECASTING PROCESS

Technological forecasting involves the application of various forecasting techniques to a subject in an orderly and structured process. The process itself involves six phases, outlined as follows:

1. **Origination**
   a. Select subject for study.
   b. Bound the study.
      (1) Determine time frame of forecast.
      (2) Define scope of study.
      (3) Define forecasting mission.
   c. Form a forecasting team.

2. **Data Gathering**
   a. Identify data needs (subjects).
   b. Identify data sources.
   c. Plan data collection.
   d. Collect data.
   e. Establish data file.

3. **Data Organization and Analysis**
   a. Categorize data.
   b. Establish data chronology.
   c. Determine trends.
   d. Determine interrelationships.
   e. Interpret data for relevance.

4. **Projection**
   a. Establish optimistic, pessimistic, and realistic reference frames.
   b. Project likely futures.

5. **Situation Analysis**
   a. Assess impact of projections on current situation.
      (1) External impacts.
      (2) Internal impacts.
   b. Project these impacts into the future.
   c. Recommend modifications of current situation to improve future success.

6. **Monitoring**
   a. Determine what to monitor.
   b. Establish monitoring process.
   c. Upgrade data file.

Origination

The origination phase is of special importance in technological forecasting because it establishes the overall nature and depth of the effort. The scope of the forecast (vector A in Exhibit 11.1.1) controls the information gathering process (vector B) as well as the construct of the forecast itself (vector C). It is important that sufficient attention be given to this phase at the inception of a forecast. Often this is not done, and as a result, forecasts are often changed as second thoughts are given to the relevance and scope of the study.

Many of the questions posed in Chapter 7.3 of this handbook regarding the origination of a VE study are also relevant to a technological forecasting study. At this phase of a forecasting effort, it is essential to define in detail the scope of the forecast. By asking questions about the intended study, a clearer understanding of its purpose and mission will emerge. A short, well-written mission statement will allow the forecaster to communicate to others what the study's goals and objectives are. Why is the study being done? What will the forecast be used for? Why? What subjects will the study include and exclude? Who is the client for the study? Who is sponsoring the forecast? Why? When should the study be completed? Will a team be used to forecast? Who should be on this team? Answers to these questions serve to define the scope of the study and clarify the mission of the forecast effort.

Before forecasting, considerable attention should be devoted to the time frame of the forecast. Is a 1, 5, or 15 year projection desired? Why? Answers to these questions are very important because they dictate the approach to the study and the type of data needed. Often organizations set 5 year goals and objectives and develop plans to accomplish these. In many technological areas this time frame may be too short to accommodate innovative approaches. Longer time frames allow an organization to plan for major innovations, yet at the same time introduce an element of risk into planning that stems from the speculative nature of this longer time horizon. It is possible to create credible forecasts over a spectrum of distant horizons, but these forecasts take added time and effort, which need to be balanced against the value of the modified forecast.

Once sufficient attention has been given to the time frame and scope of the forecast, people should be assigned to do the forecast. Forecasting teams can be formed to study areas of interest to an organization. Chapter 7.3 contains useful information on VA teams that can also be applied to the formation of forecasting teams. A team should be made up of people from various backgrounds who have access to the specialized information that the study will require. Forecasters are not born, they are made, and a forecasting team should be led by a person skilled in forecasting and with a basic understanding of forecasting techniques. Since these techniques often involve some form of mathematical modeling, forecasters should have knowledge of this field, although they may rely on others for mathematical model building. They will also need to work with people whose input will seldom be couched in mathematical terms, people who convey their constructs of the future in words and drawings. Forecasters will need to interpret these data and work with them in mathematical and nonmathematical forms. Forecasters need to be objective and capable of gathering and sorting data in an unbiased manner. They should be imaginative and capable of drawing upon both fact and fiction to produce new constructs of the future. They should have had experience in working with groups of people and should be able to focus a group effort on the task to be accomplished. Forecasters should have a heuristic approach to data that allows them to identify data needs and sources. A gestalt view of the subject will aid forecasters in recognizing patterns and trends in the data.

At the inception of a study, it is helpful to educate the team in the technological forecasting process. Here the forecaster may assume the temporary role of teacher. Knowledge of the forecasting process will improve the efficiency of the team and is quite cost-effective. Training should focus on the subject to be forecast and may include elements of team building and goal setting. If proper attention is paid to these topics, the subsequent forecasting effort will be more productive.

Data Gathering

Identification of the forecasting mission should lead directly to an identification of information that will be needed in the study. This information may come from many different sources and often may consist of several different types of data.

Hart described three different types of information gathered in a forecasting study.[3] He included recorded information contained in published records or files, observational data such as data determined in laboratory or operational studies, and subjective data obtained by interviewing or questioning people. Some of the commercial computer data banks are excellent sources of recorded information. Searching these files for information is an easy way to start the data gathering process.[4] Often this recorded information will need to be augmented with interview and observational data as the study progresses. Both the data gathered and the data gathering process will influence the remainder of the study. Properly conducted interviews will enlist the aid of knowledgeable participants and users in the study. People being interviewed often will point to other information sources that should be consulted or will suggest observations that should be made. A study of recorded information may lead the forecaster to other data. The abstracts recorded in electronic data files can provide excellent data of an unbiased nature (different people recorded the abstracts at different times) and often can lead to articles that offer rich detail on elements of the problem at hand.

As data are gathered, they can be assembled in a central forecasting file for subsequent manipulation and study. It is important at this stage of a forecast to search far and wide for information and not to bias the data gathering process by excluding data as irrelevant because they do not fit preconceived concepts of what the future will be.

Forecasting Techniques of Value in Data Gathering

Several techniques used in technological forecasting are very useful in the data gathering phase of a forecast. By interviewing, constructing Delphi questionnaires, and searching through electronic data bases, the forecaster can rapidly collect a large amount of relevant data.

Interviewing. Interviewing is used as a data gathering mechanism. It generally involves qualitative future-oriented data, which are subjective and "soft" in nature. Proper planning for and conducting of interviews are essential. Questions to address in preinterview planning are: Who should be the interviewer and interviewees? Are all organizations/areas represented? Are the right disciplines represented? How many interviews? Usually interviews are interdisciplinary, representing many areas and backgrounds. The questions are generally open-ended and unstructured to produce stream-of-consciousness-type answers. Generic examples are: What do you see lying ahead in the next X years that could affect the future of . . . ? or What do you see as significant opportunities (or threats, etc.) for . . . ? Time horizons can be specific (eg., next 10 years) or completely open, to encourage creativity and imagination. Individual interviews work well. Group interviews should be used with caution so as to minimize peer pressure and personality conflicts. One interviewer has generally been used. However, experience indicates that two interviewers are more effective.

Questionnaires have not proven to be effective in gathering imaginative, qualitative-type data. See other sources[5–8] for further information.

Delphi. Delphi is an opinion taking, consensus forecasting technique employing experts or knowledgeable people to locate, analyze, evaluate, and forecast parameters of future events. It was originally designed to minimize the personality and psychological shortcomings of committees. The process is performed by conducting several carefully designed individual interrogations. Participants are questioned privately, without face-to-face confrontation with other participants. Each questioning round is followed by an anonymous information and opinion feedback from the previous round in order to strive for group consensus. According to Coates,[9] the process is "... applicable to any situation to which quantitative values may be assigned, whether these are dates, weightings or scalings, etc." The entire process is usually performed through a director or mediator.

Electronic Interactive Data Retrieval Services. Several commercial interactive data retrieval services are available that are excellent sources of information. Such electronic data banks should not be the sole source of data, but should be considered as supplementing the other data gathering methods. They are generally used to search for certain pieces of information in order to answer specific questions. They may also be used to look at aggregate information and rates of change of information availability.[4,10] This, in turn, provides an indicator of trends and amount of activity for a given subject. Searching through abstracts may also reveal new ideas and subjects to be searched and refined. "The procedure is often iterative and may lead into many areas not dreamed of when the search began."[4] Among the commercially available information services are:

The Information Bank, One World Trade Center, New York, NY 10048 (a subsidiary of the New York Times Company).

Lockheed Information Retrieval Service, "Dialog-Chronology," Lockheed Information Systems (D52-08, B/201), Lockheed Palo Alto Research Laboratory, 3251 Hanover Street, Palo Alto, CA.

SDC Search Service, Systems Development Corporation, 2500 Colorado Avenue, Santa Monica, CA 90406.

Data Organization and Analysis

As data are gathered in a forecasting study, they need to be organized and analyzed. Data that have been organized and analyzed take on a new character—the character of information. Hence, in this phase of a forecasting study, data are categorized and a data chronology is established. The data are examined for interrelationships as part of the categorization process, and trends are established as a result of establishing a data chronology. In this phase of the study, interpretations of the relevancy of data are made, which in turn influence the weighting given to data sets within the data on hand.

Categorizing Information

The subject under study often dictates the categorization of data that should take place. Numerous data organizing methods are available, and several are listed here.

FAST Diagrams. Chapter 7.3 discussed at some length a functional categorization technique called FAST diagramming. The authors find this process very useful in categorizing forecast information. Diagrams of this type can be constructed to depict both supply and demand functions at work. The example forecast at the end of this chapter illustrates the use of FAST diagrams as data organizing devices.

Relevance Trees. Lanford[11] describes a relevance tree as a "way of setting forth interrelating variables in a coherent graphic form." Relevance trees may be qualitative, also referred to by Swager[12] as "perspective" or "objective trees," or quantitative, employing the use of relevance numbers described by Bright[13] and others.[1,14] They may be visualized as a vertical FAST diagram, based on how-why logic, branching out from a single unique function, objective, or parameter to form a hierarchical structure. Each level of the tree represents a certain level of complexity that increases at successively lower levels. It is a useful device for displaying large amounts of information in concise structured form, searching for unforeseen relationships, and discovering new ways to accomplish objectives and fulfill needs. The process of constructing the tree also serves as an excellent communications device. The tree itself, because it contains a wealth of information, may be considered what Lanford[11] terms "miniinformation centers." Whatever scheme is used to categorize data, it should accommodate data on market "pull" and technology "push." It should also integrate the data in a logical way and identify by its subject headings the main parameters at work in the subject area.

Other Techniques. Many other categorization techniques can be used in this phase of the forecasting effort. Cause-effect diagrams,[15] morphological analysis,[16] numerical taxonomy,[17] conceptual maps,[18] and mission flow analysis[19] are all useful techniques that can be helpful at this point in a forecasting study.

Establishing a Data Chronology

Data take on new meaning when they are arrayed chronologically. When written descriptions of events occurring in a subject area are arranged chronologically, they frequently reveal patterns that tell the forecaster much about the level of development of a technology. Product life cycles are uncovered that reveal the level of development of an idea. One idea may be in a speculative phase, whereas another may have passed through much of the research and development process and may soon be introduced as a new product.

An important step in establishing data chronology involves segregating event data from trend data. News items are events in the development of a technology, whereas numerical data arranged chronologically often depict trends. Mathematically, future events are usually treated in terms of probability theory, whereas trends are often treated as material flows in a control system or as vectors that interact with one another to produce resultant forces.

Establishing Interrelationships

When the data collected in the data gathering step have been aggregated into distinct subject categories, it is necessary to determine interactions among the categories. To determine these interactions, matrices are often constructed in which subject parameters are arrayed against themselves (cross-impact analysis) or against external factors (impact analysis). Both of these procedures are described and illustrated in the technology forecast at the end of this chapter.

Cross-Impact Analysis.

> *Cross impact analysis as developed by T. J. Gordon and Olaf Helmer, is a computer simulation technique that uses a cross-impact matrix for input. Raters assign to each cell of the matrix, (1) the mode (direction of impact, positive, negative, or no effect), (2) the strength of impact between pairs of items, and (3) the time lag, how long it takes until the impact is realized. A quadratic equation relating the above three variables is programmed into a computer simulation program that computes the resulting probabilities of each pair-wise interaction of items, based on the matrix input. The output is a listing of the items, the initial probabilities, the probability shifts which occurred when the items were correlated, and the final probabilities. Items can then be ranked (by computer) by initial or final probability or by probability shift each of which produces a different scenario. The process permits an elementary analysis of the potential interactions between items.*[22]
>
> See also references 20 and 21.

Impact Analysis.

> *Impact analysis is a subjective quantification technique for evaluating the impact of external factors (influences that impinge on a set of items). It is a matrix scoring method and is similar to a force-field type of analysis wherein raters use (+)'s and (−)'s and zeros to quantify their perceptions of interactions of items. It is an effective technique for screening influences, identifying attributes and locating sensitive variables in a system. It can be used in any situation where it is desirable to measure both the positive and negative forces that impinge on a system or set of alternatives.*[23]

Other Techniques. The scenario technique is also useful in ascertaining the interrelationships and impacts of a forecast. This method, which is described under the projection phase of technological forecasting, is an excellent method for showing cause-effect relationships. Similarly, PERT diagrams, so often used in planning and project management, are also useful tools for ferreting out detailed interactions.

Projection

At this point, all of the necessary steps have been taken to allow the forecaster to project historical data into the future. If a mathematical model has been constructed, it is now run to simulate what may happen in the future. Before discussing mathematical models, however, we again point out that many intuitive methods offer constructs of the future where the base of the construct is never shown. Thus an artist's model of a future desalination plant, for example, is an intuitive construct in which the artist depicts what a future plant may look like. A newspaper report describing this same plant is similar in construct. Investigation of either of these projections invariably reveals that artist and writer alike do not project from a vacuum, but have in mind prior constructs that they have modified to meet interrelationships that they feel will affect the future. Managers often act in a similar instinctive manner. Such constructs can be improved by attending to the forecasting process described here.

Described here are several methods used by forecasters to project the future. In using any of these methods, it is desirable to consider three different types of projections: optimistic, pessimistic, and realistic. These projections should be made so as to provide a range of estimates of the future. In trend extrapolation this might represent plus or minus deviations from the normal projection; in scenario writing separate scenarios are written; and in cross-impact or impact analysis the model can be changed to simulate these different futures. The realistic future is, by definition, the most probable, yet all three projections should be made. These different futures will have impacts on current activities, which need to be assessed in the situation appraisal phase of technological forecasting.

Trend Analysis

Trend analysis consists of several similar statistical techniques to evaluate historical data in terms of time relationship and correlation between various parameters and events. Trend extrapolation entails plotting historical data and statistically extrapolating them into the future. Trend correlation consists of statistically quantifying the relationship between the trends.

In cases involving highly correlated trends, one variable or technology may be a precursor to another. In these cases the precursor can be monitored as a leading indicator of the follower. Martino[1] elaborates in depth on trend analysis methods. Trend analysis also involves the time series analysis and curve fitting methods as well as the traditional econometric forecasting methods. More elaborate trend extrapolation may involve envelope curve fitting and extrapolation, which involves analysis and extrapolation of a succession of several different technologies over time. The combined "envelope" curve serves as an indicator of capability improvement and technological progress. Many models can be used to fit historical data and provide valuable projections of the future. One point should be made, however. In using this technique it is important to consider any limits that may cause the trend to deviate from its historical path. Many times these limits can be determined and used to modify the trend line.

Substitution Analysis

The time-related techniques of trend analyses mainly involve quantifying the progress of past events. Substitution analysis,[24] on the other hand, entails quantifying the rate of which a new technology will expand into the economy and/or take over and replace an existing one. The technique is used mainly to predict the rate and the amount of substitution of one technology for another in order to estimate the market growth of new products as well as the rate of phaseout of existing ones. This method is finding increasing use in technology forecasting and often produces useful predictions where a new technology is replacing an older one. Such replacements can be categorized using the FAST diagramming technique described earlier, and substitution analysis provides a way of projecting these functionally based replacements.

Scenarios

Another method for projecting the future involves scenario writing. Scenario writing is a qualitative forecasting technique that is used to project various futures from present conditions based on stated assumptions. It is often used in conjunction with the more quantitative trend analysis models. Often these latter models cannot be constructed because numerical data simply do not exist. In this case scenarios are very useful. "It is a major technique for exploring future implications."[25]

Scenarios may be written in many different forms, lengths, formats, and combinations.[26] Generally, a set of several alternative scenarios is constructed as opposed to a single most-likely scenario. These scenarios must be believable and relevant and are usually based on data collected during the technological forecasting study. According to Hart,[3] scenarios can be a powerful means of forcing people to look at alternative futures because scenarios presume that events, especially unpleasant ones, have occurred. This, in turn, helps reduce bias. Vanston et al.[27] and MacNulty[28] relate the various steps in constructing alternative scenarios.

Situation Analysis

Once a series of projections has been made, the task of appraising the impact of these futures on current activities and future plans can begin. Often this calls for a recycling of the forecasting effort, which is now directed toward assessing the impact of alternative futures on current practice. In performing this analysis, it is advisable to list a large number of different aspects of the present situation and sort through these to select several major areas that should be studied further. Brainstorming is an excellent technique for listing these areas, and the value measurement techniques described in Chapter 7.3 can be used to select appropriate areas from a long list of subjects. Functional diagrams of a business can be used to select areas on which the future projections will have an impact. It is often useful to consider the impact of the forecast on such areas as research and development, capital spending, manufacturing, sales and distribution, and future manpower needs. This type of analysis should bring to light the consequences of the forecast to the internal composition and health of an organization.

The appraisal should also assess the impact of the forecast on agencies outside of the parent organization. What influences will the technological forecast have on other manufacturers, on municipalities, and on government services? Will these set up second-order effects on the organization itself? The extent to which this type of an analysis is done is highly situational. In practice, most assessments are restricted to first-order effects on the business itself, with further attention given only to major secondary effects. Because of the complexity of such an analysis, many times it is carried on as intuitive speculation on the part of the decision maker.

As the potential impacts of the technological forecast are enumerated, they should be itemized and aggregated. Kennell and Linneman use a factorial design to assess the impact of major trends on current practice.[29] In their "shirt-sleeve" approach to planning, they test the robustness of current plans to varying futures and set up a strategy for monitoring the future to allow an organization to better deal with its changing environment. Out of this type of situation analysis they select "best chance" opportunities, while at the same time they devise plans to alleviate the impact of negative forces on the organization. Recommendations to modify plans and activities are the major output of the technological forecasting process.

Monitoring

The last phase of technological forecasting involves the establishment of a process for monitoring developments in the field. Since forecasts are often unreliable, it is important to monitor developments continuously as they occur. The forecast itself should provide knowledge of sensitive areas to monitor. Often a critical path can be described that a technology will follow. Life cycle analysis provides insights into the stage of development of many emerging technologies and is a useful tool to use when following emerging technologies.

A monitoring system should be established to follow the future as it unfolds. The basic need here is to establish a file, or an information storage and retrieval system, and to assign responsibility for maintenance of this file to a clearly identified person. Using monitoring, the technological forecast can be followed for deviations from its predicted path. Is demand holding constant? Are breakthroughs occurring in any of the new technologies? Monitoring entails "keeping current tabs" on significant signals in order to assess change. In addition to employing literature searches and environment scanning, monitoring also includes choosing the right parameters for observations, evaluating new information for significance, and presenting the information in an effective and timely manner. Data are collected from general literature, but also from annual reports and patents and by personal observations or contacts with people in key strategic positions. Some thought should be given to data storage and retrieval because data once collected should be stored for subsequent rapid access. Simple notebooks may quickly become impractical. Monitoring is important because it carries on and updates important information uncovered by other parts of the technological forecasting process. It involves what Bright calls "assessing events in-being."[13]

From time to time, it is wise to recycle the entire technological forecasting effort. Depending on the complexity of the area, it may be necessary to study some subjects in more detail, especially if they appear to be in a rapid state of flux. The results of such studies should then be incorporated into any updated technical forecasts.

11.1.7 CONCLUSION AND SUMMARY

The technological forecasting process just described has proved of value to a large number of organizations. The forecasting techniques enable organizations to break old patterns and to create new ones by keeping an open mind. The techniques used in constructing models, whatever their form, are effective ways to do this. Yet they cannot do the job alone. Communication throughout the organization is essential in order to reduce bias. Communication is enhanced and can form the basis for "cooperative models," models built by the cooperation of many people in the organization. As Licklider and Taylor state, "society rightly distrusts modeling done by a single mind."[30]

In order for the technological forecasting process to effect "cooperative modeling" it is important that it do the following:

1. **Be interdisciplinary and participative.** Forecasting encompasses information from many areas and many viewpoints. The term "cooperative modeling" itself implies interaction. Without interdisciplinary participation, one can develop erroneous time horizons and narrow perceptions.
2. **Be systematic and structured.** People can too easily become blinded with favorite ideas, which can cause them to overlook important issues involved in complex, ill-defined problems. A systematic and structured approach can reduce this possibility and deter people from jumping to solutions.

3. **Be iterative.** It is important that the process be iterative. Parts will have to be revised and updated. Gaps in information will appear that will necessitate going back to earlier portions of the process.

4. **Be dynamic.** Forecasting is a dynamic process requiring review and revision whenever change occurs. All aspects of forecasting should be under almost constant review in order to act in anticipation of change, rather than to be reactive once change has occurred.

5. **Utilize both "hard" and "soft" data.** It is important to use judgmental and subjective data along with objective and hard data because, according to Enzer,[31] much of the insight necessary for anticipating change "cannot be obtained by any other means."

6. **Document assumptions.** It is important that all assumptions be documented. This also helps document the rationale behind decisions and focuses everyone's attention to a common inference point, which in turn enhances communication.

If these steps are taken, technological forecasting can be a process that allows an organization to anticipate and plan for change. Used properly it can help organizations maintain an adaptive and flexible posture that will allow them to grow and prosper in our ever-changing world.

11.1.8 EXAMPLE: TECHNOLOGICAL FORECAST OF FUTURE WATER SUPPLIES

The literature of technological forecasting is noticeably void of detailed examples of forecasts that illustrate the forecasting process. This is perhaps due to two factors: the relative newness of this process and the proprietary nature of many forecasts. In teaching technological forecasting to graduate students at the Rochester Institute of Technology, the authors became aware of the need for examples of technological forecasts. Accordingly, we illustrate the technological forecasting process here with an example of a forecast of the future water supply of a large water district. Because everyone has some familiarity with this subject, we hope it will be a useful and easily understood example of the forecasting process.

Origination

Exhibit 11.1.1 illustrated three basic parameters at work in any forecasting effort. Exhibit 11.1.2 is an elaboration of this figure, which illustrates two forecasts that could be made for a water district. In this drawing these two forecasts are illustrated by the two spheres. Sphere 1 represents a potential forecast of the completion date for a municipal reservoir. In this forecast factual data could be gathered and used to construct a critical path model for completion of the project. The technology is well known and highly predictable, and a close interplay of manpower allocation and technology makes the date of completion of this project highly predictable.

Exhibit 11.1.2 Illustration of Two Forecasts That Could Be Made for a Water District

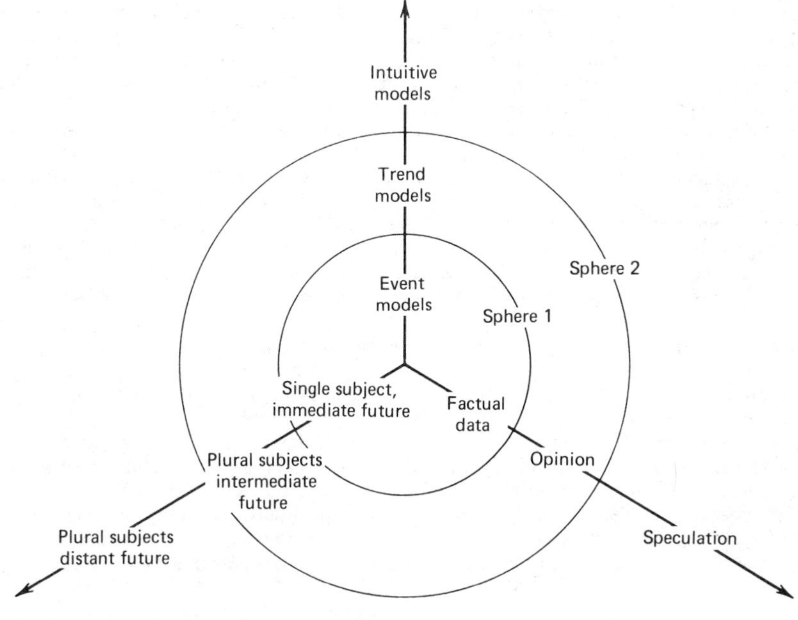

Sphere 2 contains a potential forecast of future municipal water systems in a large water district. Here a study of demand factors, such as population, industrial expansion, and per capita water use, might form the basis for a systems model that would interrelate these demand trends with supply factors such as reservoir construction, rainfall, hydrology, and future water recycling capabilities. Adding complexity to this forecast would be studies of the influence of new technologies such as desalination, cloud seeding, and others that may dramatically alter future water availability. Technical, economic, political, and social factors make this forecast much more complex than the first.

These are but two examples of a large number of forecasts that might be performed in a water district. In the origination phase of technological forecasting, it is important to develop a clear definition for a forecasting study.

In ascertaining the scope of this forecast, assume the study sponsor was asked the following questions:

Q. "Why are you interested in forecasting the future?"
A. "I'd like to gain more feeling for what we have to do to ensure an adequate water supply."
Q. "Who is we?"
A. "This agency–the Department of Water Affairs."
Q. "What water supplies are you concerned with?"
A. "All of them. Well, no, that is not really true. We're only interested in the municipal water supplies of our major cities. Agricultural water supplies can be excluded from this study."

This sort of questioning is essential and should lead to a written mission statement that the client and the forecaster agree upon. Let us assume that the following mission statement emerged.

Forecasting Mission Statement

Forecast the demand for municipal water in the major metropolitan areas of water district #1 over the next 20 years. Assess the role of new water delivery technologies in meeting this demand and examine the type of manpower the Water Department will need to meet its future water needs.

Such a statement is essential to a forecasting effort. It describes the scope of the forecast and lists major accomplishments that must be completed. The origination phase of technological forecasting requires a clear understanding of what subject is to be forecast and of the forecast time frame. The forecasting mission should be agreed upon by the forecast sponsor and the forecaster. In most cases the process will uncover data resources to be scrutinized and people who should be involved in the forecasting process.

Data Gathering

In conducting the district water forecast, records were searched for data on water supply and demand in the region. A large number of references were found that provided data on the historical consumption of water in the area. These searches provided information on population growth, industrial growth, and municipal water consumption in the district. The searches also yielded information about the water recycling and desalination, which are potential sources of new water for the district. These data, coupled with observational data on rainfall and river flows, provided a sound data base for the technological forecast.

These data allowed trends to be constructed that related the district's water demand to the district's future water supply. In most cases literature searches of this type would be complemented with interviews of key officials and planners in the area. Questionnaires and Delphi forecasts would be fashioned to elaborate on key issues uncovered in the data gathering process. This was not done in this particular study because of the remote location of the water district and the high quality of information uncovered in the literature searches.

Data Organization and Analysis

In organizing and analyzing data, extensive use was made of FAST diagrams. Exhibit 11.1.3 is a FAST diagram that was used to categorize information.

Using the breakdown shown in the diagram, historical data on reservoir construction and well drilling in the area were collected. Data on work in the region on water recycling and desalination were also collected, as were data on the extensive importation of water to the cities of the region through interbasin water transfer. In this latter region major tunnels and aqueducts had been built to divert water from adjacent water sheds to the region. Data on residential, industrial, and agricultural water demands were also collected and categorized as shown in Exhibit 11.1.3.

Exhibit 11.1.3 A FAST Diagram Utilized to Categorize Information

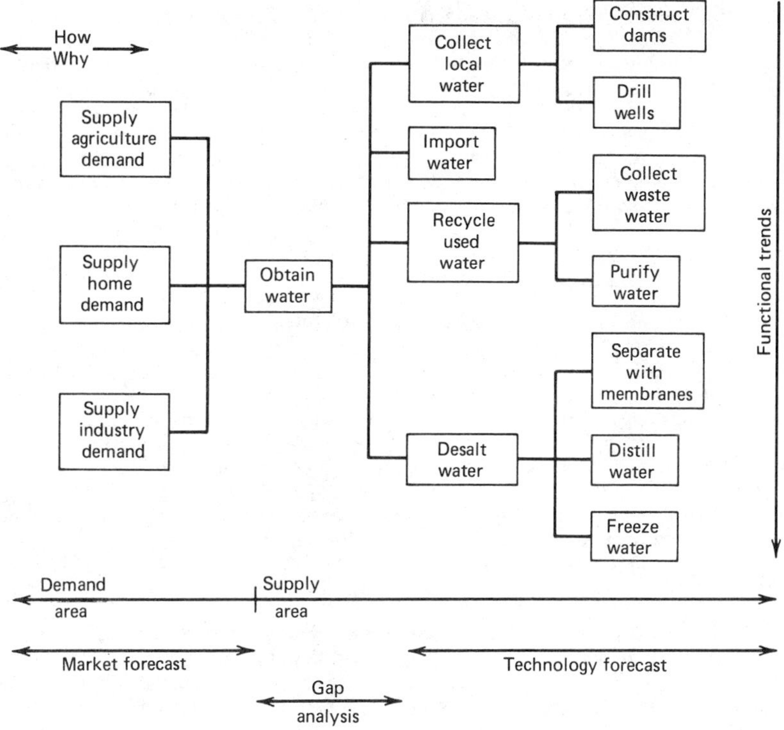

As the data were gathered, the functional trends shown on the diagram became apparent. It was then possible to relate mathematically the historic supply and demand nature of the area's water system and to project these factors into the future. This was done after establishing a data chronology in each of the areas of the FAST diagram.

As data were gathered and compiled, an extensive file emerged. Data from this file were then interrelated using an impact matrix, which evaluated the impact of several new technologies on the future water needs of metropolitan areas in the region. In this example the water shortage that five metropolitan areas may experience is listed across the top of Exhibit 11.1.4. The figures represent

Exhibit 11.1.4 Impact Matrix of New Water Technologies on Future Water Needs

		Cities				
		A	B	C	D	E
		Future Water Needs				
Water Technologies	Σ	287	274	253	112	92
Interbasin transfer	154	60	0	56	21	17
Cloud seeding	129	37	16	48	13	15
Water reuse	305	103	36	91	50	25
Desalination by distillation	64	23	25	8	8	0
Desalination by membranes	139	43	47	20	19	10
Desalination by freezing	94	20	63	3	0	8
Iceberg utilization	133	0	88	28	0	17

a 15 year projection of future water shortages. Various water supply technologies are listed down the left-hand side of the matrix. They will influence some cities more than others. Cities A and D are inland cities where ocean desalination is not feasible, whereas cities B, C, and E are coastal cities where this technology may be useful. In all cases the cities have grown beyond the capacity of nearby water resources and currently rely on interbasin water transport from distant drainage areas. The potential of each supply methodology to alleviate a city's water shortages was derived by the pair comparison method (see Chapter 7.3). The numbers entered on the matrix are estimates of the relative ability of the technologies to meet a city's future water needs. When these estimates are summed, they indicate the overall ability of the technologies to meet the district's water needs. The number for water reuse (305) indicates that this technology can be expected to play a major role in the future of the water district, whereas desalination by distillation will have much less significance.

Projection

Completion of the impact matrix provides a projection of the ability of new technologies to meet future water supply needs of the district. The projection of such futures from models constructed in the data organization and analysis phase of technological forecasting is a natural sequence of that phase. In forecasting future water consumption in this area, an upper limit of regionally available water was apparent from a study of historical rainfall patterns in the area. Several pearl curves[1] were derived to predict the amount of water that might be available from surface water, underground water, and water recycling in the area.[32] The total of all water from these sources is compared to a historical 7% annual growth in water demand for the area in Exhibit 11.1.5. Water shortages that each of the five metropolitan areas could experience were also ascertained and used to construct and run the impact matrix shown in Exhibit 11.1.6.

Situation Analysis

These projections clearly established that the water district will face major water shortfalls unless new technologies are found that can furnish large quantities of fresh water to the district. Information relating future water supplies to future demand often necessitates further forecasting studies that extend the forecast into areas requiring additional data gathering and analysis. In this example

Exhibit 11.1.5 Total of All Water Sources Compared to Historical 7% Annual Growth in Water Demand

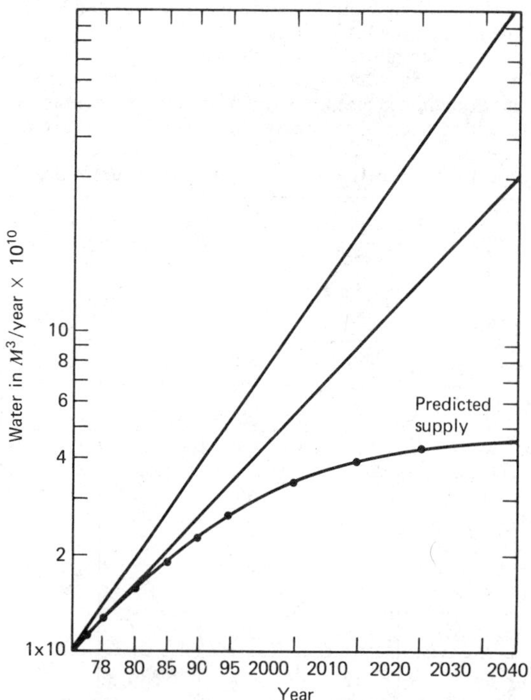

Exhibit 11.1.6 Impact Matrix of Professions and Water Technologies

		Water Technologies						
		Water Reuse	Interbasin Transfer	Membrane Desalination	Iceberg Utilization	Cloud Seeding	Freeze Desalination	Flash Desalination
		Impact Weightings for Water District						
Σ	Professions	305	154	139	133	129	94	64
74	Civil engineers	12	39	–	15	8	–	–
128	Chemical engineers	37	8	25	5	–	29	24
79	Electrical engineers	–	15	17	15	–	8	24
61	Refrigeration engineers	–	–	–	25	–	36	–
112	Chemists	46	–	40	5	8	8	5
101	Bacteriologists	58	8	25	5	–	–	5
97	Ecologists	58	8	8	15	8	–	–
131	Hydrologists	37	31	–	5	25	8	–
53	Meteorologists	–	8	–	5	40	–	–
165	Lawyers	58	39	–	15	40	8	5
25	Naval personnel	–	–	–	25	–	–	–

it was necessary to study the effect of water shortages on the growth of new technologies and to relate this growth to the future manpower requirements of the district.

As a consequence, the initial forecasting effort produced several new studies of the potential for several of the technologies to meet water demands in the district. New models were constructed to meet these needs. A trend cross-impact model was constructed that simulated the growth of several of the new technologies in one of the major coastal cities in the district. One of the technologies was the subject of additional studies, which led to a project proposal including detailed project parameters and cost estimates.[33]

Secondary forecasts of future manpower demands were performed using further impact models. Here a second impact matrix was used in which the growth of technologies examined in the original forecast was arrayed with several professional disciplines that could furnish the expertise needed to implement the new technologies. Exhibit 11.1.6 is a replica of this new impact matrix.

The numbers on the matrix are estimates of the ability of each profession to assist in the growth of each new technology. By summing across each row of the matrix, overall figures were obtained that depict the "need" for each professional category. The higher the number, the greater the need for each profession. Thus lawyers, hydrologists, chemical engineers, and chemists will be needed to provide expertise in future water supply projects. These needs represent a deviation from current practice. Continuation of the current staffing practice would add more civil engineers to the staff and would not prepare the district to adapt to new technological demands.

This example of the impact of the future on manpower needs represents one impact on the district. Many other areas will be affected by the technological forecast, including capital funding and R & D. It is important for the district to identify the impact of the future on these and other activities. In a comprehensive situation appraisal, it is important to look at these additional impacts. The manpower needs should also be quantified by estimating future manpower needs as a function of increasing water needs.

Monitoring

Several factors will need to be closely monitored by the water district. Overall water demand will need to be followed closely, as will demographic factors such as population growth, per capita water consumption, and industrial expansion. In addition, various new desalination technologies, such as reverse osmosis, electrophoresis, freezing, and distillation processes, will need to be closely monitored. Even exotic proposals, such as iceberg transport and processing, should be followed since any of them may play an important role in the future.

To monitor these factors, the district should establish an information data base that can be easily updated. Modern electronic data bases that provide for continuous monitoring of new abstracts make the task of following developments in this field much easier. Coupled with a management awareness program, they can alert the district to important events in time for it to adapt to them.

REFERENCES

1. J. R. MARTINO, *Technological Forecasting for Decisionmaking,* Elsevier North-Holland, New York, 1972.
2. E. JANISCH, *Technological Forecasting in Perspective,* Organization for Economic Cooperation and Developments, Paris, 1966.
3. J. L. HART, "Technological Forecasting, From Board Room to Drawing Board," *Machine Design,* Vol. 48, No. 3 (February 12, 1976), pp. 90–93.
4. D. J. DEMARLE and J. L. HART, "Searching for Signals Through Interactive Data Banks," paper presented at James R. Bright's Technology Forecasting Workshop, Castine, ME, The Industrial Management Center, Austin, TX, June 1977, p. 5.
5. J. K. FORDYCE and R. WEIL, *Managing With People,* Addison-Wesley, Reading, MA, 1971, pp. 137–156.
6. J. L. HART, "The Futures Study Applied to Manufacturing," paper presented at James R. Bright's Technology Forecasting Workshop, Castine, ME, The Industrial Management Center, Austin, TX, June 1977.
7. R. JOHANSON and J. A. FERGUSON, Mapping Views of the Future in a Small Group," *Futures,* Vol. 8, No. 2 (April 1976), pp. 163–169.
8. J. E. JONES, "The Sensing Interview," *The 1973 Handbook for Group Facilitators,* University Associates, San Diego, 1973, pp. 213–214.
9. J. F. COATES, "Some Methods and Techniques for Comprehensive Impact Assessment," *Technological Forecasting and Social Change,* Vol. 6, 1974, p. 349.
10. J. BUTLER, D. F. BALL, and A. W. PEARSON, "The Analysis of Technological Activity Using Abstracting Services," *R and D Management,* Vol. 7, No. 1 (1976), pp. 33–40.

11. H. W. LANFORD, *Technological Forecasting Methodologies*, American Management Association, New York, 1969, p. 120.
12. W. SWAGER, "Strategic Planning III: Objectives and Program Options," *Technological Forecasting and Social Change*, Vol. 4, 1973, pp. 283–300.
13. J. R. BRIGHT, *Practical Technological Forecasting, Concepts and Excercises,* The Industrial Management Center, Austin, TX, 1978.
14. J. H. BICK, *Planning and Forecasting Using a Combined Relevance Analysis and Cross-Impact Matrix Method,* Report RM-1757, General Research Corporation, McLean, VA, June 1974.
15. M. S. INOUE and J. L. RIGGS, "Describe Your System With Cause and Effect Diagrams," *Industrial Engineering,* April 1971, pp. 26–31.
16. A. V. BRIDGEWATER, "Morphological Methods–Principles and Practice," in R. V. Arnfield, Ed., *Technological Forecasting*, Edinburgh University Press, Edinburgh, 1969.
17. R. R. SOKOL and J. SNEATH, *Numerical Taxonomy,* Freeman, San Francisco, 1963.
18. H. JONES and B. C. TWISS, *Forecasting Technology for Planning Decisions,* PBI–Petrocilli Books, New York, 1978, pp. 103–111.
19. H. A. LINSTONE, "Mission Taxonomy," in J. R. Bright, Ed., *Technological Forecasting for Industry and Goverment,* Prentice-Hall, Englewood Cliffs NJ, 1968.
20. J. KANE, "A Primer for a New Cross-Impact Language–KSIM," *Technological Forecasting and Social Change*, Vol. 4, 1972, pp. 129–142.
21. M. TUROFF, "An Alternative Approach to Cross Impact Analysis," *Technology Forecasting and Social Change*, Vol. 3, 1972, pp. 303–339.
22. M. L. SHILLITO, "Impact Analysis," paper presented at James R. Bright's Technology Forecasting Workshop, Castine, ME, The Industrial Management Center, Austin, TX, June 1977, p. 30.
23. M. L. SHILLITO, "Impact Analysis," p. 1.
24. J. C. FISCHER and R. H. PRY, "A Simple Substitution Model of Technological Change," *Technological Forecasting and Social Change*, Vol. 3, 1971, pp. 75–88.
25. COATES, "Comprehensive Imapct Assessment," p. 353.
26. R. D. ZENTNER, "Scenarios in Forecasting," *Chemical and Engineering News,* October 6, 1975, pp. 22–34.
27. J. H. VANSTON, W. P. FRISBEE, S. C. LOPREATO, and D. L. PASTON, "Alternate to Scenario Planning," *Technological Forecasting and Social Change,* Vol. 10, No. 2 (1977), pp. 159–180.
28. C. A. R. MACNULTY, "Scenario Development for Corporate Planning," *Futures*, Vol. 9, No. 2 (April 1977), pp. 138–147.
29. J. D. KENNELL and R. E. LINNEMAN, "Shirt-Sleeve Approach to Long Range Plans," *Harvard Business Review,* March-April 1977, pp. 141–150.
30. J. C. R. LICKLIDER and R. W. TAYLOR, "The Computer as a Communications Device," *Science and Technology,* April 1968, pp. 21–31.
31. S. ENZER, "Cross-Impact Techniques in Technology Assessment," *Futures,* March 1972, pp. 30–51.
32. D. J. DEMARLE, "South African Iceberg Utilization: A Proposal for Fresh Water Production at Saldanha Bay," *Proceedings of the Value in Energy and Conservation 1979 International Conference,* Value Engineering and Management Society of South Africa, Republic of South Africa, October 29–31, 1979.
33. D. J. DEMARLE, "Design Parameters for a South African Iceberg Power and Water Project," *Proceedings of the Second Conference on the Use of Icebergs,* International Glaciological Society, Cambridge, England, 1980.

CHAPTER 11.2

Planning and Control for Manufacturing Systems and Projects

RANDALL P. SADOWSKI
Purdue University

DEBORAH J. MEDEIROS
The Pennsylvania State University

11.2.1 INTRODUCTION

Both planning and control have a critical impact on the successful operation of manufacturing systems. Shop loading and project planning are preproduction activities that define what is to be produced at an aggregate level. The quality of these planning activities significantly affects the ability to control actual production at the shop floor through the use of scheduling techniques.

Available scheduling techniques vary greatly in their ability to model existing systems accurately. Most of these procedures have been developed for a small class of production systems under a restrictive set of assumptions. Often, those procedures that are capable of producing acceptable schedules are very cumbersome to implement and prohibitively expensive.

To date, no generalized scheduling methodology has been found. The primary reason for this lack of a unified theory of scheduling is the great diversity of application areas. Manufacturing systems may be divided into three basic categories: continuous processes, batch production, and unique projects.

Continuous processes are commonly found in the chemical industry. Oil refining is one example. Batch production involves intermittent manufacture of a variety of related products in a wide range of quantities; this category includes the majority of manufacturing systems. Unique projects are characterized by their large size and extended duration. The construction of a ship is a typical example.

This chapter addresses the basic scheduling and control function. Since most production facilities are very complex, it is often necessary to implement numerous other planning and control tools prior to considering the use of effective scheduling techniques. Thus the first section provides a brief overview of manufacturing systems. This is followed by a discussion of sequencing and scheduling for batch production systems and a treatment of project planning and control. Continuous production systems are covered in Chapters 7.10 and 13.10.

11.2.2 MANUFACTURING SYSTEMS OVERVIEW

Manufacturing planning and control have traditionally involved five basic tasks: forecasting, routing, scheduling, dispatching, and expediting. Forecasting and routing are planning tasks, which determine what is to be produced and where it is to be produced. Scheduling and dispatching are control tasks, which determine when a product is to be produced and the action of releasing the order to the shop floor. Expediting is normally considered to be a compliance task, which provides a follow-up to determine if the performance of the system is adequate. These five basic tasks provide the overall framework for any manufacturing planning and control system.

These fundamental definitions provided the basic concepts for the planning and controlling of early industrial systems. In recent years manufacturing systems have become increasingly complex. These systems now contain large numbers of complex machine tools and produce a much wider variety of products than before. In addition, the need to control more closely the cost of capital expended for production machinery and in-process inventory has placed a much greater importance on the performance of these systems. The introduction of the computer into the industrial environment has greatly increased the ability to control such manufacturing systems. However, it has also required a much more precise definition of the control tasks as well as of their critical interactions. The initial development in this area was concentrated in management information systems (Chapter 12.3). These systems primarily served the function of automating existing tasks, mainly in the accounting area, to provide middle- and upper-level management with current data regarding the performance of their system. More recently, attempts have been made to integrate control and planning strategies into these management information systems.

Planning and Control Levels

The development and implementation of a total planning and control system has become a very complex task. Consideration must be given to the level of planning or control as well as to the relative time frame for each. It is often useful to think in terms of three general types of planning and control activities.

Strategic. Primarily a long-term planning activity at the end-product level, covering the time frame from 1 to 5 years.

Tactical. Both a short-term planning and a long-term control activity that might consider both end products and individual parts, covering the time frame from 0 to 2 years.

Operational. Primarily a control activity at the individual part level, covering the time frame from 0 to 2 months.

The design of any system should provide the means to perform all of these activities and to collect the necessary data. There are a large number of elements in such a system; numerous attempts have been made to integrate these elements into actual operating systems. The basic elements and their general linkages are shown in Exhibit 11.2.1.

Exhibit 11.2.1 Manufacturing System Elements

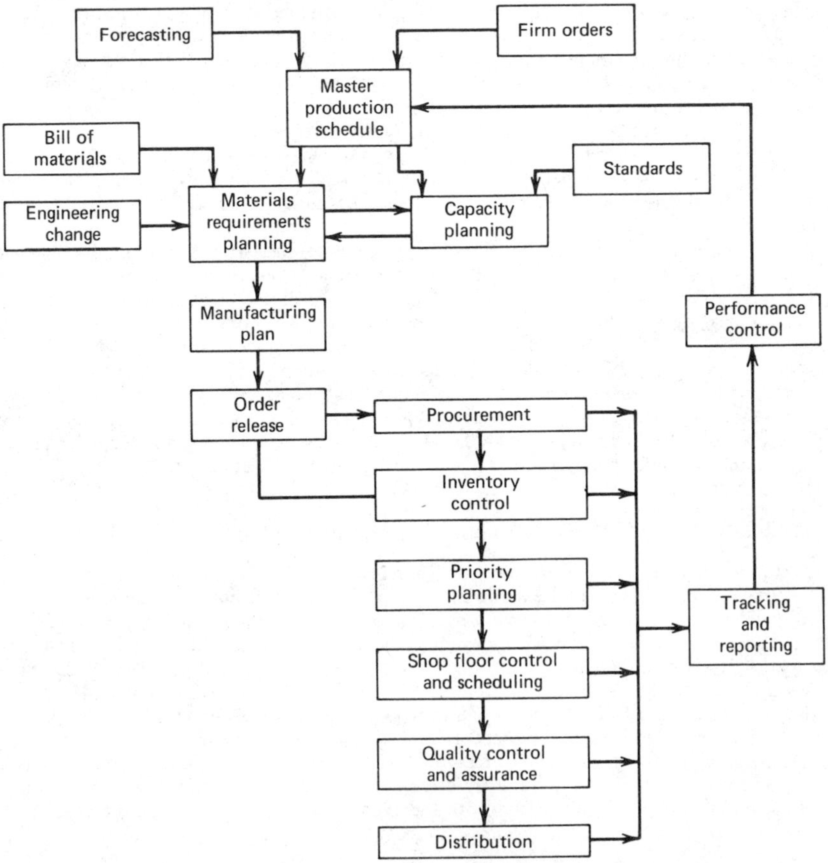

Manufacturing System Elements

The operational system begins with the forecasting of future demands at the end-product level (Chapter 11.1). These forecasts are combined with the firm or known demand to create the master production schedule (Chapter 11.4). This aggregate schedule is accessed by the material requirements planning element (Chapter 11.6) and the capacity planning element (Chapter 11.5), which utilize information from the bill of materials, engineering change, and standards files to create the manufacturing plan to be utilized by the system. The manufacturing plan provides the order release to the procurement (Chapter 11.10) and inventory control (Chapter 11.4) elements. When the material becomes available, the priority planning element assigns relative priority and releases the order to the manufacturing floor where the shop floor control and scheduling element guides it through completion. During its production, various quality control and assurance procedures (Section 8) may be employed to ensure that the resulting product is acceptable. The finished product is then forwarded to the distribution phase (Chapter 11.11) for storage or shipment to the customer. The manufacturing process is normally monitored by a tracking and reporting element, which records the progress of orders and provides current information and summary reports to the system users. This information is utilized by the performance control element to determine the overall system status and provide a feedback mechanism to the master production schedule for possible action.

The system described in Exhibit 11.2.1 illustrates the basic elements of a manufacturing planning and control system and the general flow of action and information throughout the system. Depending on the requirements of the system user, there are numerous additional linkages and interactions that may be added. The system described provides the basic elements and the primary interactions and flow of information required for the fundamental concept.

Implementation

The advent of computers has made the development and implementation of such a total system feasible. To define and create a complete system successfully, each of the basic elements would have to be further detailed, along with the necessary interactions with all other elements that depend on the specific system to be designed. As previously indicated, many of these elements are discussed in greater detail in other sections.

The original work on the design of such systems was performed by IBM in the late 1960s and published in an eight-volume set in 1972. Entitled "The COPICS (Communications Oriented Production Information and Control System) Series," it presented the basic concepts for the development of a complete computer-based system. The concepts expounded in this series are still valid today.

Many attempts have been made to develop such a total system for general use with very few true success stories. The tremendous complexity of the required system, combined with the prevalent inaccuracies in the available data, has resulted in many system failures. However, the knowledge gained from these attempts has provided some valuable insight into the development of such systems.

It has become clear that, prior to the implementation of any system, it is absolutely necessary that current data be accurate. This requires some type of accountability to ensure accuracy. If collected data are never audited, they will eventually decrease in quality. If the quality is not important, then the data are probably not important and should not be collected. In many instances the development of new systems has been based fundamentally on expansions of existing systems and their associated data bases. If the existing data base contains inaccurate data, the system will never function properly.

The overall objective of a manufacturing system is profitability. This objective must be translated into a specific set of performance measures that are used to evaluate the system functions. Many of these performance measures deal with the status of jobs on the shop floor. An example is percentage of delivery dates met. Thus it is necessary to control the processing of jobs through the shop, which necessitates the use of sequencing and scheduling procedures at the shop floor level.

11.2.3 SEQUENCING AND SCHEDULING

The priority planning and shop floor control and scheduling elements ultimately determine the performance of the production system. If capacity is available, if orders are released at the proper time, and if material is available, it might seem that scheduling is a simple task. However, this is not the case. The complexity of scheduling problems can be illustrated by considering the simple one-machine case. If N jobs are currently available to be processed on a single machine, there are $N!$ ways in which they can be sequenced. For 5 jobs, this results in 120 sequences; for 10 jobs, 3,628,800 sequences. As the number of jobs and the number of machines are increased, the number of combinations becomes extremely large, and only a fraction of these can be generated or evaluated with existing computers. If one also considers the time dynamics and the changing priorities of existing systems, the determination of the best schedule is often an impossible task.

The Single-Machine Problem

The single-machine deterministic problem has had an extensive analytical treatment. This work has provided numerous sequencing rules, which have been tested for more complicated situations, and has been the basis for many existing scheduling systems. The single-machine results for three of the most-cited rules follow.

Shortest Processing Time Rule

Sequence the jobs according to the shortest processing time (SPT), the job with the smallest time being first. This sequence will:

1. Minimize the average job flow time (where flow time is the difference between job completion time and arrival time).
2. Minimize the average job waiting time.
3. Minimize the average job lateness (where lateness is the difference between the completion time and due date).

The SPT is one of the most often used sequencing rules because of these properties. If each job is assigned a weight or priority, with larger weights indicating more important jobs, a generalization of SPT, called weighted SPT (WSPT), may be used to minimize the weighted average flow time. Jobs are sequenced in nondecreasing order of the ratio of processing time to weight or priority.

The SPT sequencing has a disadvantage in a very heavily loaded shop. If new jobs are arriving frequently, those jobs with very long processing times will be pushed back to the end of the queue and may remain in the system for an unacceptably long period. This property has led to an interest in rules that consider job due dates, such as the two that follow.

Due Date Rule

Sequence the jobs according to the earliest due date. This sequence will minimize the *maximum* job lateness. It will not minimize the *average* lateness. Interestingly, SPT, which utilizes no due date information, has this property. One might suspect that better results could be achieved by combining due date and processing time into a single rule. This line of reasoning led to the development of the slack rule.

Slack Rule

Sequence the jobs according to the smallest slack time, where slack time is the amount of time remaining before a job must be started if it is to be completed by its due date. The sequence will maximize the minimum job lateness.

Although intuitively appealing, this rule exhibits no useful properties for the single-machine problem. However, it is useful in problems involving multiple machines where due dates are a primary measure of performance.

An example of the application of these rules is given in Exhibit 11.2.2. These rules may have other beneficial properties, but the lists include only those that can be mathematically proved.

Variations on the Single-Machine Problem

Other objectives are often of interest in the single-machine case. One of these is minimization of the number of tardy jobs, that is, those jobs whose completion times exceed their due dates. No simple sequencing rule is known for this problem. However, a procedure known as Hodgson's algorithm will provide the minimum number of tardy jobs. The steps in the procedure are as follows:

1. Sequence the jobs using the due date rule.
2. If no jobs in the sequence are late, the optimal solution has been found; go to step 4. Otherwise, find the first late job in the sequence. Suppose this is the ith job.
3. Select the longest of the first i jobs; remove it from the sequence. Update the job completion times and then go to step 2.
4. Add the unsequenced jobs, if any, to the end of the existing sequence in any order.

Exhibit 11.2.2 Single-Machine Rules

Job	Processing Time	Due Date	Slack
A	3	16	13
B	4	11	7
C	5	17	12
D	6	12	6

SPT Schedule

A	B	C	D

0 3 7 12 18

Job	Wait Time	Flow Time	Lateness
A	0	3	−13
B	3	7	− 4
C	7	12	− 5
D	12	18	6
Mean	5.5[a]	10[a]	− 4[a]

Due Date Schedule

B	D	A	C

0 4 10 13 18

Job	Wait Time	Flow Time	Lateness
A	10	13	−3
B	0	4	−7
C	13	18	1[b]
D	3	10	−2
Mean	7.25	11.25	−2.75

Slack Schedule

D	B	C	A

0 6 10 15 18

Job	Wait Time	Flow Time	Lateness
A	15	18	2
B	6	10	−1
C	10	15	2
D	0	6	−6[a]
Mean	7.75	12.25	−0.75

[a]Minimum.
[b]Maximum.

An example of the procedure follows.

Job	Processing Time	Due Date
1	5	8
2	2	9
3	3	18
4	4	10
5	6	11

STEP 1. Sequence 1 2 4 5 3
Completion time 5 7 11 17 20

STEP 2. Job 4 is the first late job; $i = 3$.

STEP 3. Remove job 1.
Sequence 2 4 5 3
Completion time 2 6 12 15

STEP 2. Job 5 is the first late job; $i = 3$.

STEP 3. Remove job 5.
Sequence 2 4 3
Completion time 2 6 9

STEP 2. There are no late jobs.

STEP 4. Sequence 2 4 3 1 5
Completion time 2 6 9 14 20

Two jobs are tardy.

Such sequencing rules as SPT and due date provide optimal results for the single-machine problem with little computational effort. For other measures of performance or with extensions to the basic model, no such simple procedures have been found. An example is minimization of mean tardiness. Procedures such as dynamic programming and branch and bound can be used to meet this objective, as long as the problem is reasonably small.

An interesting extension to the basic problem is the inclusion of setup times that depend upon the order in which the jobs are processed. This problem can occur in such industries as paint manufacturing, where a changeover from a darker color to a lighter one takes more time than the opposite changeover because of the cleaning required. This introduces a new potential objective function: minimization of "makespan," defined as the difference between the completion time and start time of the entire sequence of jobs. Without setup times, makespan is constant for the single-machine problem and equals the sum of the processing times.

For the one-machine problem with setup, minimization of makespan is equivalent to minimization of setup time. This is because makespan is the sum of the setup times and the processing times, and the sum of processing times is a constant. This is the classic traveling salesperson problem (Chapter 14.2), where the machine corresponds to the salesperson, the jobs to the cities, and the setup times to the distances or costs of travel between cities. Little's algorithm may be used to determine the optimal solution.

Another extension of the basic problem is the dynamic shop, in which jobs arrive at random times, with the assumption of the preempt-resume discipline. This assumption states that a job, once it has begun processing, may be interrupted and then continued at a later time without any increase in total processing time. Because of this assumption, the rules for the static case may be applied with little change. An example is the shortest remaining processing time (SRPT) rule, which minimizes average flow time. This rule states that the machine should always be processing the job with the shortest remaining processing time. As each job enters the system, its processing time is compared with the remaining processing time of the job on the machine. If the new job has a smaller processing time, it preempts the current job. The current job is placed in the queue, and its processing time is decreased by the amount of time it was on the machine. No general results are known for the dynamic shop when the preempt-resume assumption does not hold.

The Parallel Machine Problem

A more complex scheduling problem occurs when each job has one operation, as before, but several machines are available to process the jobs. This problem involves both allocation of each job to a machine and sequencing the jobs on each machine. Because there are multiple machines, the makespan is again of interest. The longest possible makespan would result if all the jobs were processed on a single machine, whereas distributing the jobs among machines decreases the makespan. An integer programming formulation of the problem follows.

Let

x_{ij} = 1, if job i is processed on machine j; 0, otherwise
t_i = processing time of job i
n = number of jobs
m = number of machines

Minimize y subject to

$$\sum_{j=1}^{m} x_{ij} = 1 \qquad i = 1, \ldots, n$$

$$y - \sum_{i=1}^{n} t_i x_{ij} \geqslant 0 \qquad j = 1, \ldots, m$$

$$x_{ij} \geqslant 0 \text{ and integer}$$

The first constraint requires that each job be processed once and only once. The variable y represents the makespan. The second constraint requires y to at least equal the amount of processing on the busiest machine; because it is minimized, y will equal the sum of processing times on this machine, which is the makespan.

For small problems, it is possible to obtain a solution using integer programming. However, computation time increases exponentially with problem size, so this approach is not very useful for realistic problems. Instead, procedures have been developed that do not guarantee an optimal solution, but (we hope) produce good results. One such procedure that attempts to find a minimum makespan uses the longest processing time (LPT) rule. This rule is the opposite of SPT; jobs are sequenced by longest processing time. The procedure is as follows:

1. Sequence the jobs using the LPT rule.
2. Assign the jobs, in order, to the machine with the smallest amount of processing time currently allocated.

An example problem is solved in Exhibit 11.2.3. Because the LPT rule is the opposite of the SPT rule, it will tend to maximize average flow time for this problem. A variation of LPT, called reverse LPT (RLPT), adds a third step to the LPT procedure:

3. Reverse the order of jobs on each machine.

This does not change the makespan, but improves the average flow time, as shown in Exhibit 11.2.3.

If the objective is to minimize average flow time rather than makespan, a very simple optimal procedure, based on the SPT rule, may be used.

1. Sequence the jobs using the SPT rule.
2. Assign the jobs, in order, to the machine with the smallest amount of processing time currently allocated.

The schedule produced by this procedure is contained in Exhibit 11.2.3.

For more complex parallel-machine objectives, such as minimizing tardiness, weighted average flow time, or setup, no simple optimal procedures are known. Dynamic and integer programming have been used for the tardiness and weighted average flow time problems. The setup problem can be formulated as a multisalesperson "traveling salesman" problem or, if maximum work load constraints are imposed on the processors, as a vehicle delivery problem, where machines correspond to vehicles, maximum work load to capacity, and setup times to distances. All these approaches

Exhibit 11.2.3 Parallel-Machine Rules

Job	Processing Time
A	3
B	2
C	10
D	6

LPT Schedule

Machine 1: C (0–10)

Machine 2: D (0–6), A (6–9), B (9–11)

Makespan = 11

Average flow time = 9

RLPT Schedule

Machine 1: C (0–10)

Machine 2: B (0–2), A (2–5), D (5–11)

Makespan = 11

Average flow time = 7

SPT Schedule

Machine 1: B (0–2), D (2–8)

Machine 2: A (0–3), C (3–13)

Makespan = 13

Average flow time = 6.5

require a great deal of computation time for realistic problems. The suboptimal approach of allocating jobs to machines to balance work load, or due dates, and then sequencing the jobs on the machines will provide a quick and easily computed solution.

The Flow Shop and Job Shop Problems

Both the flow shop and the job shop contain multiple machines and process jobs consisting of a series of operations that must be performed in a specific order. In the job shop, each job may require a different sequence of machines. The flow shop has unidirectional flow, although each job does not necessarily require processing on each machine.

Exhibit 11.2.4 Johnson's Two-Machine Flow Shop Rule

Problem

Job	Processing Times	
	Machine 1	Machine 2
A	3	4
B	5	4
C	7	2
D	5	6

Procedure

Step	Unscheduled Jobs	Minimum Time/Job	Partial Schedule
1	A, B, C, D	2/C	XXXC
2	A, B, D	3/A	AXXC
3	B, D	4/B	AXBC
4	D	5/D	ADBC

Solution

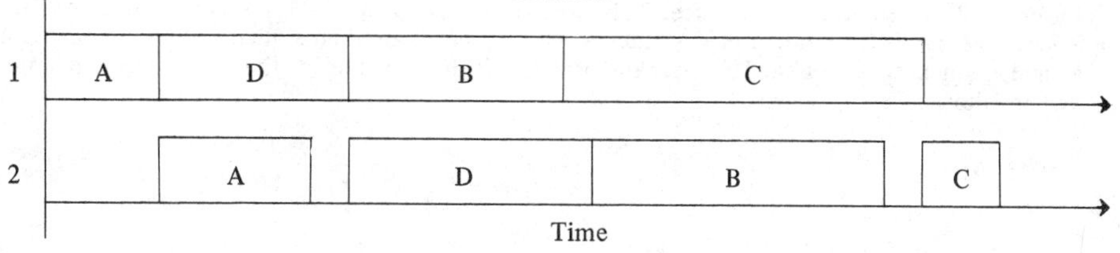

An optimal procedure for minimizing the makespan for the two-machine flow shop was developed by S. M. Johnson. The procedure is summarized as follows: Select the smallest available processing time from the unscheduled jobs. Schedule that job in the first available position if the processing time is for the first machine; otherwise, schedule that job in the last available position. Repeat until all jobs are scheduled. An example of this procedure is given in Exhibit 11.2.4.

Johnson's procedure was applied to the two-machine job shop by Jackson. Jobs are divided into four sets as follows:

{A}	Contains those jobs that require machine 1 only.
{B}	Contains those jobs that require machine 2 only.
{AB}	Contains those jobs that are first processed on machine 1 and then on machine 2.
{BA}	Contains those jobs that are processed on machine 2 and then on machine 1.

The steps of the procedure are:

1. Sequence the jobs in {AB} using Johnson's rule.
2. Sequence the jobs in {BA} using Johnson's rule.
3. Sequence the jobs in {A} and {B} in any order.
4. Construct the schedule as follows:

Machine 1	{AB}	{A}	{BA}
Machine 2	{BA}	{B}	{AB}

This procedure minimizes the makespan by minimizing the amount of time each of the machines is idle while waiting for work from the other.

Johnson's rule can also be used for flow shops with more than two machines. For the three-machine flow shop, two pseudomachines are created. The processing times for the pseudomachines

($\hat{t}_{ij}$) are calculated as follows:

$$\hat{t}_{i1} = t_{i1} + t_{i2} \qquad i = 1, \ldots, n$$

$$\hat{t}_{i2} = t_{i2} + t_{i3} \qquad i = 1, \ldots, n$$

where t_{ij} = processing time for job i on machine j
n = number of jobs
$\hat{t}_{ij}$ = processing time for job i on pseudomachine j

Johnson's rule is then applied to the two pseudomachines to create the job sequence. This procedure will produce the optimal makespan if either of the following conditions are met:

1. $\min_i \{t_{i1}\} \geqslant \max_i \{t_{i2}\}$
2. $\min_i \{t_{i3}\} \geqslant \max_i \{t_{i2}\}$

This implies that machine 2 is dominated by either or both machine 1 and 3 and thus is not a bottleneck. Also, if the optimal sequence, found by Johnson's rule, for machines 1 and 2 only is identical to the sequence for machines 2 and 3 only, then this sequence is also optimal for the three-machine problem.

For flow shops with more than three machines, a procedure proposed by Campbell, Dudek, and Smith may be used to find a good (though not necessarily optimal) solution to the makespan problem. It is a multiple-stage process, which creates two pseudomachines at each stage and uses Johnson's rule to produce a sequence. This procedure results in $m - 1$ potentially different sequences, where m is the number of machines. These sequences are then evaluated, and the one with the smallest makespan is used. Times for the pseudomachines for each of the $m - 1$ stages are computed as follows:

Stage 1

$$\hat{t}_{i1} = t_{i1}$$
$$\hat{t}_{i2} = t_{im}$$

Stage 2

$$\hat{t}_{i1} = t_{i1} + t_{i2}$$
$$\hat{t}_{i2} = t_{i,m-1} + t_{im}$$

Stage k, $3 \leqslant k \leqslant m - 1$

$$\hat{t}_{i1} = \sum_{j=1}^{k} t_{ij}$$

$$\hat{t}_{i2} = \sum_{j=m-k+1}^{m} t_{ij}$$

This procedure reduces the m machine flow shop to a series of two-machine flow shops and then utilizes the best sequence on all of the machines. For the 2 or 3 machine flowshop, the schedule which yields the minimum makespan has the same sequence of jobs on all machines. However, with more than three machines, the optimal schedule may require that the sequence of jobs be different for different machines. The Campbell, Dudek, and Smith procedure will not find schedules of this type; a methodology such as integer programming must be used to find such an optimal schedule.

Integer programming formulations have been developed for the job shop and flow shop problems; many of these have been published. Branch and bound procedures have also been applied to the job shop problem. Neither of these approaches is useful for realistic problems because they require excessive computer time and/or storage space.

One method of controlling a large shop is the use of dispatching procedures. Whenever a machine becomes idle, the queue of jobs awaiting that machine is examined, and a priority rule is used to select the next job to be processed. Some often-used priority rules are to select the job with:

Shortest operation time.
Longest waiting time in that queue.

Largest number of following operations.

Least work remaining.

Most work remaining.

Least slack.

Closest due date.

Smallest value of slack divided by the number of remaining operations.

It appears that the SPT rule is among the best of the priority rules. However, this depends upon both the performance measure used and the nature of the shop. The performance of different priority rules for a shop may be evaluated by computer simulation. Multiple runs of the simulation for each rule may be performed, and then the best rule with respect to mean and variance of the performance measure may be selected for implementation.

In addition to sequencing, several alternate approaches to this class of problems have been pursued. Much attention has been given to the determination of optimal buffers between machines for the flow shop. However, these procedures are principally applicable to those situations where intermediate storage is design limited, such as conveyor systems. For these situations, some rather dramatic effects have been obtained, but the procedures are generally very difficult to apply, and the results good only as long as the product mix does not change.

An alternative procedure is to control the flow by the selective assignment of manpower (Chapter 11.9). This type of procedure can produce good results, but a general method that can be easily used is still not available. Many procedures are available that provide good or optimal solutions in limited situations. The setup problem, the process control problem (Chapter 13.10), the transportation problem (Section 14), and the assembly line balancing problem (Chapter 3.4) are examples.

Performance Measures

The performance control element, which measures the quality of the production system, can also have a profound effect on the control function. The principal performance measures can be divided into three general objectives:

1. **Maximize Customer Satisfaction.** This implies that all orders should be completed by their due date. Generally a 90 to 95% rate is strived for, with a much lower rate achieved.
2. **Minimize Work in Process.** This implies that the system be controlled such that the minimum amount of inventory is in process, thus minimizing the inventory investment cost.
3. **Maximize Resource Utilization.** This implies that the resources (manpower, machines, etc.) be utilized to the maximum extent possible.

The ideal system would achieve all of these objectives, whereas in reality the achievement of any one tends to have a negative effect on the other two. Thus each user must rank order these objectives and attempt to find a weighted objective function that is acceptable.

In applying performance measures, one should ensure that the system objectives match the shop floor objectives. For example, the performance of the shop floor foreman is frequently measured by the output per month in pounds or dollars. This results in the foreman's trying vainly to produce the critical or expedited products in the first 2 or 3 weeks and then producing only those orders with a large number of pounds, or high dollar value, in the last week or two in order to meet his or her quota. The resulting system performance provides poor customer satisfaction and frequently results in poor machine utilization. It may reduce the department's work in process, but the high pound or dollar parts may have to wait for a future delivery date or for assembly with overdue low pound or dollar parts from other departments. The true work in process for the entire system may in fact increase under these circumstances.

The First Steps

Fundamentally, the inability to solve the general scheduling problem for real-world situations is due to the extraordinarily large number of combinations that one has to consider. Until adequate solution techniques are developed or computer capabilities allow direct computation of the solution, the practioner will have to rely on simple tools that already exist. The performance of most systems can be increased by concentrating efforts in those areas that will have the greatest impact.

Capacity planning (Chapter 11.5) is one of the most important and yet most neglected areas. A production system that is overloaded can never meet the imposed due dates and will continue to fall further and further behind regardless of any expediting effort. Conversely, an unintentionally underloaded system will yield a poor utilization of resources.

Basic department-level loading concepts should be applied to smooth the flow of parts through the system by establishing realistic order release dates. Orders that require large blocks of processing time and have a large number of operations obviously take longer to work their way through the system than orders with minimal operations and time requirements. The consideration of waiting time, material handling time, and the number of operations as well as the processing time can greatly reduce the congestion on the shop floor.

If capacity is available and the orders are released on time, dispatching based on priority assignment can provide a basic scheduling tool. The foreman can judge the relative importance of each order at each phase of its manufacture and take into consideration the status of the shop floor in determining the final schedule of jobs.

Machine scheduling problems occur when a group of usually unreleated jobs must be processed through a system. Occasionally, the manufacturing process consists of one very large job with many interrelationships among the operations of that job. An example is the manufacture of airplanes. Many thousands of operations and several months are required to complete a single plane. Further, the relationships among the operations are complex. Some groups of operations can proceed simultaneously, whereas others must wait for prior operations. The completion of a single operation may allow the start of many or no following operations, depending upon the status of the system. A small delay in one operation may have no effect on the final schedule or may require complete rescheduling of succeeding operations. In such circumstances, different scheduling techniques are required; these fall under the heading of project planning and control.

11.2.4 PROJECT PLANNING AND CONTROL

The initial concepts of project planning and control emerged during the 1940's in response to the need to better manage and control the complex projects confronted during and after World War II. These concepts were initially applied to defense systems. Since that time, they have been applied to numerous nondefense systems in the construction, aerospace, and shipbuilding industries, as well as to other R & D projects.

The Gantt chart was probably the first technique to be applied to project planning and control. This chart is a simple way to show graphically both the anticipated and completed portions of a project. The horizontal axis of the chart represents time. Activities are scheduled by plotting them as bars on the chart, observing all precedence relationships. The percentage of each activity that is complete is indicated either by shading the appropriate portion of the bar or by placing a caret on the bar. By drawing a vertical line through the current date, it can be easily determined if any activities are ahead of or behind schedule. The Gantt chart can be machine or project based. An example of each type is shown in Exhibit 11.2.5. The machine-based chart (Exhibit 11.2.5*a*) also includes a repair or maintenance activity, which is indicated by crossing out the time period in which the planned downtime will occur.

There are several advantages to using such a technique: It forces a plan to be made; work planned and work accomplished are easily compared; and the chart is easy to produce and is dynamic in nature. This type of chart is frequently used by federal agencies in the form of a milestone chart. Although it provides an excellent graphic display showing project progress, it does not always fully describe the dependency or interaction among various project tasks. Furthermore, though it can be an effective tool in project planning and control, it is not readily adaptable as an analytical tool.

Critical Path Method and PERT

The formalized concepts of project planning and control emerged in the late 1950's with the appearance of the Critical Path Method (CPM) and PERT. Although the techniques were developed independently, they were both based on the project network concept.

In CPM the activities or operations necessary to complete the project are shown in a graph called a "network," which also shows the order in which activities must be completed. Included with each activity is a single time estimate of how long it will take to complete the activity. Using rather simple computations, one can determine the longest route through the network—the critical path. The technique also provides early and late start times for all activities. Knowing the critical path, and hence the critical activities, can be of great value in the planning and controlling of a project. If an activity on the critical path is delayed, the entire project will be delayed. A delay in noncritical activities will not necessarily cause the project to be late, unless the delay exceeds the available slack. The CPM method also allows for the consideration of time-cost trade-offs.

Program Evaluation and Review Technique is very similar to CPM, except that it allows for uncertainty in the time estimates of activities. Each activity is given three time estimates: optimistic, most likely, and pessimistic. These three estimates are used to compute an expected time, which in turn is used to compute the critical path. The initial estimates are used to calculate deviations, which can then be used in the computation of statistical estimates of completion times.

Sucessful completion of any project requires careful integration of three key interdependent

Exhibit 11.2.5 Example (*a*) Machine-Based Gantt Chart and (*b*) Project-Based Gantt Chart

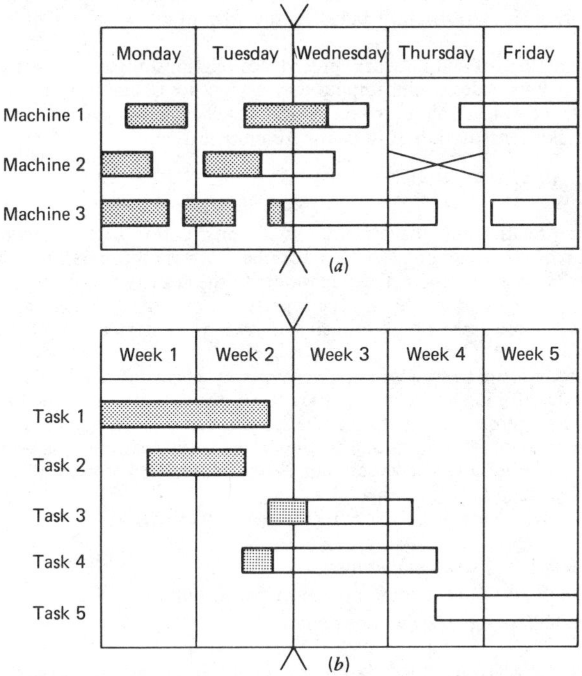

elements: objectives, resources, and timing. If these key elements are independently predetermined, project failure is likely. Therefore the first step in project planning is a clear statement of what objective is to be achieved and the allowable time horizon. Project success is facilitated by having a single main objective. Resources for a project include personnel, funding, data, technology, equipment, and products, some of which may not be readily available or fully developed at the needed time; therefore project planning is at best an interactive process, which starts with a rough plan and proceeds to a final detailed plan. Good managers estimate the effects of delays or constraints and modify resourses and/or time accordingly.

Developing the Network

There are several basic steps that must be performed before PERT or CPM can be applied. The first is to define the project objectives and to determine the personnel who will be responsible for accomplishing these objectives. The next step is to define the tasks or activities that must be completed in order to achieve the objectives. In this step the planner must establish the time increment for the tasks or activities, that is, months, days. If the increment is too large, the project may be difficult to control and monitor. If it is too small, accurate time estimates may be difficult to obtain. As the planning proceeds, it may become necessary to consolidate several activities into one or, conversely, to split one activity into several.

The third step is to develop time estimates for each activity. These may consist of deterministic estimates, as in CPM, or probabilistic estimates, as in PERT. Finally, the planner must develop the project network by defining all of the interactions or precedences among activities.

Defining Precedence Relationships

The precedence relationships must be specified carefully to provide the maximum flexibility in scheduling. It is best to include only technological restrictions, at least for the first network. For example, the foundations of a house must be excavated before the pouring of cement can occur. Thus a precedence relationship exists between these activities. However, activities such as wiring and plumbing can be done at the same time, so no precedence relationship should be included, even if the plumbing is usually done before the wiring.

A precedence relationship means that an activity may not begin until all its predecessors are complete. For some activities, this relationship is overly restrictive; a part of the former activity may need to be completed, rather than the entire activity, before the successor may begin. For example, only part of the excavation may need to be done before pouring of the cement can begin. This relationship may be expressed by breaking the former activity into two or more subactivities and including a precedence restriction between the first of these subactivities and the original successor activity.

Because of the structure of the network, precedence relationships are transitive. Thus, if activity A precedes B, and if B precedes C, this implies that A precedes C, and it is not necessary to include an explicit precedence restriction between A and C. The inclusion of unnecessary precedence relationships will make construction of the network more difficult.

Constructing the Network

Once all precedence relationships have been stated, construction of the network may begin. A network is made up of arrows and nodes. Two different network representations for projects are in common use: activity on arrow and activity on node. In the activity-on-arrow format, the activities are represented by arrows, and the nodes represent "events," defined as the initiation or completion of one or more activities. Before an activity may be started, all activities that terminate at its start node must be completed.

The activity-on-node format represents the activities by means of nodes. The arrows are used to show precedence relationships between activities. As in the activity-on-arrow format, all predecessors must be complete before an activity may be started. The two different formats are equivalent, and it is not difficult to convert from one to the other. Using the activity-on-node representation makes construction of the network easier, but many people find it more difficult to visualize a project when it is described in this manner.

For an activity-on-arrow network, three rules are generally observed:

No two events should have the same number.

No two events should be connected by more than one activity.

The network should have one start and one end node.

The first and third rules are included to prevent confusion in analysis. The first rule does not restrict the structure of the network. The second and third rules appear to limit the types of projects that can be represented. The limitation is removed by the use of dummy activities. Dummy activities have no time delay associated with them; they are used to express precedence relationships only. If the project requires that more than one activity exist between the same two nodes, an additional node, coupled with a dummy activity, can be used to avoid violating the second rule. To conform to the third rule, dummy activities can be used to connect all activities with no successors to a single end node and all activities with no predecessors to a start node. Dummy activities can also be used to describe other precedence constraints.

The activity-on-node representation is more compact than the activity-on-arrow format; dummy activities are not required. In essence, all the arrows are dummies because their only function is to express precedence.

Dummy nodes may be required in this representation if it is desired to have one start and one end node for the project. Examples of the use of dummies for both representations are shown in Exhibit 11.2.6. Exhibit 11.2.6*a* shows the use of a dummy activity to prevent activities B and C from having the same start and end node. Exhibit 11.2.6*b* uses a dummy activity to show a precedence relationship between activities C and E. The activity-on-node format has a dummy node so that the project will have a single end node.

The general concepts and procedure can best be illustrated by considering a simple example of a contractor who is going to develop a wooded area into a small park. The objective is to complete the park in 22 working days with the contractor's existing equipment and work force. The activities to accomplish this project are defined as follows:

A–Clear. Knock down unwanted trees, remove stumps, and make initial road cuts.

B–Cut trees. Cut up the usable firewood from the clearing activity and stack for removal.

C–Wood pickup. Pick up and remove usable firewood and dispose of branches and so on.

D–Roads. Make final road cuts and put down gravel.

E–Parking. Grade parking areas and put down gravel.

F–Utilities. Install water and electric service.

G–Shelter. Construct basic shelters and buildings.

H–Landscape. Landscape required areas with plantings and grass.

I–Finish. Connect utilities and perform general cleanup.

Exhibit 11.2.6 Examples of Dummy Activities

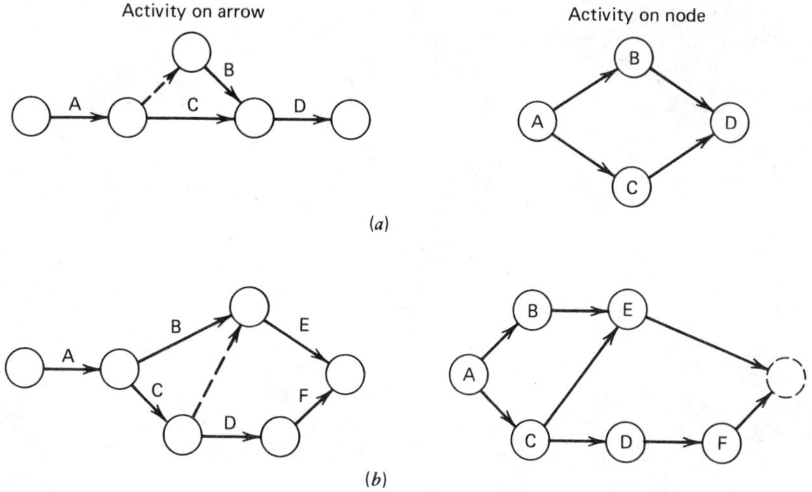

The time estimates and resulting project network are shown in Exhibit 11.2.7. Times for both a CPM and PERT analysis have been included. The project network is displayed as an activity-on-arrow network. Exhibit 11.2.8 contains an activity-on-node representation of the example problem.

The PERT times are given as three estimates:

t_o = optimistic time (The most optimistic time in which the activity can be completed.)

Exhibit 11.2.7 Example Project Network

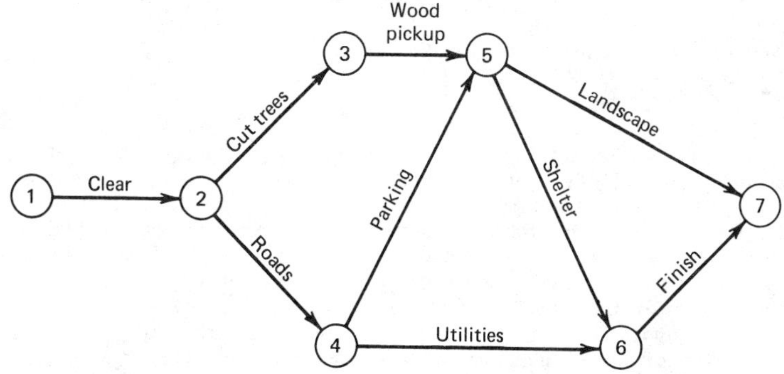

Activity	Arrow	CPM Time	PERT Times t_o	t_m	t_p	t_e	σ
A–Clear	1-2	4	2	4	6	4	0.66
B–Cut trees	2-3	4	3	4	5	4	0.33
C–Wood pickup	3-5	5	3	5	7	5	0.66
D–Roads	2-4	3	2	3	4	3	0.33
E–Parking	4-5	3	1	2	9	3	1.33
F–Utilities	4-6	4	3	4	5	4	0.33
G–Shelter	5-6	5	3	5	7	5	0.66
H–Landscape	5-7	8	7	8	9	8	0.33
I–Finish	6-7	2	1	2	3	2	0.33

Exhibit 11.2.8 Activity-on-Node Representation of Example Project Network

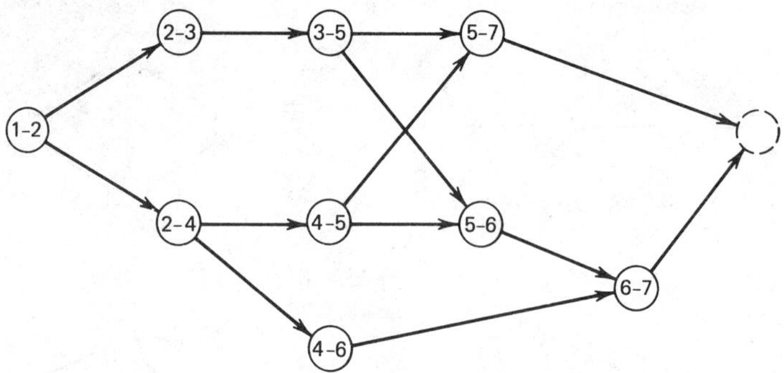

t_m = most likely time (The most likely or best guess of the time in which the activity will be completed.)

t_p = pessimistic time (The most pessimistic time that the activity will take.)

From these estimates, the expected time, t_e, can be computed as follows:

$$t_e = \frac{t_o + 4t_m + t_p}{6}$$

This provides an estimate of the mean completion time of each activity. For convenience, these times have been made equal to the deterministic times for the CPM analysis. The standard deviation can also be computed for each activity as follows:

$$\sigma = \frac{t_p - t_o}{6}$$

These values have also been included in Exhibit 11.2.7.

Critical Path Calculations

The initial analysis calls for the determination of the critical path and is essentially the same for both techniques. This analysis requires a forward pass through the network to establish the earliest possible start and finish times for each activity and a backward pass to establish the latest possible start and finish times. The computations for this analysis are shown in Exhibit 11.2.9. This exhibit gives the results for both the graphic method, which is suitable for small networks, and the tabular method, which is normally performed on a computer for large networks.

The computations for the forward pass of the graphic method are given by the circled numbers at each event or node. These circled numbers represent the early finish time for the previous activity; the largest of these values at any node represents the early start time for activities starting at that node. The backward pass values are found in the squares. These values represent the latest start time for activities following this event; the smallest of these values at any node represents the latest ending time for all activities that end at that node. Those values contained in both a circle and a square indicate the critical path times.

The equivalent values can be found by the tabular method. The table is constructed by entering in the first two columns the activities and their times. The order of the activities is such that all predecessors of an activity are listed before the actual activity is entered. If the network has been constructed such that the start node of each activity has a lower number than its end node, as in the example, then the activities can be entered into the table in increasing order based on the start nodes. Once these values have been entered, all activities that have no predecessors are assigned the same early start time. This value is normally zero, but could be a time representing the actual start of the project. The difference between the start and finish times, for both the early and late times, is the activity duration. In the example there is a single starting activity, 1–2, which was assigned a start time of zero and a finish time of 4, the activity duration. The early start time of the remaining activities depends on the early finish times of all immediate predecessors, that is, those activities that end at the start node of the activity being examined.

Exhibit 11.2.9 Forward and Backward Pass Analysis

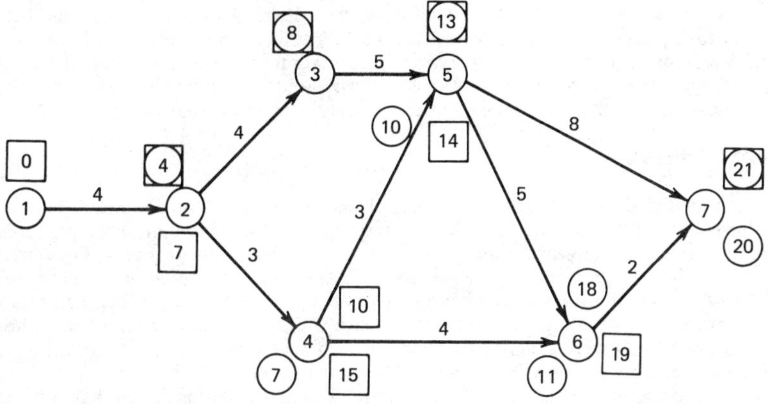

Activity	Time	Early Times Start	Early Times Finish	Late Times Start	Late Times Finish	Slack
1-2	4	0	4	0	4	0
2-3	4	4	8	4	8	0
2-4	3	4	7	7	10	3
3-5	5	8	13	8	13	0
4-5	3	7	10	10	13	3
4-6	4	7	11	15	19	8
5-6	5	13	18	14	19	1
5-7	8	13	21	13	21	0
6-7	2	18	20	19	21	1

The early start time for an activity is equal to the maximum early finish time of its immediate predecessors. If the activity has a single predecessor, as do activities 2-3 and 2-4, its early start time is equal to the early finish time of the predecessor, in this case activity 1-2, with an early finish of 4. Activities 5-6 and 5-7 each have 2 predecessors (3-5 and 4-5), so their early start times are equal to the maximum of the early finish times, 13 and 10. Thus the early start time for activities 5-6 and 5-7 is 13.

This procedure is continued until all the early times have been determined. The maximum early finish time determines the minimum project duration or the critical path time. Thus, for the example project, a minimum of 21 days is required.

The late times are found by a backward pass starting at the bottom of the table, or at the terminating node of the project, node 7. This backward pass is the reverse of the forward pass; the activities are examined in the reverse order, and the late finish time is determined before the late start time.

The first entries are the finish times of all activities that end at the terminating node (5-7 and 6-7). Normally these times are the same as the critical path time, 21, but a larger value could be used to represent the required completion date of the project. The late start times are then determined by subtracting the respective activity durations from the late finish times. The order in which the remaining late times are calculated depends upon the ending nodes of the activities. If the nodes are numbered as previously described, activities are examined in decreasing order of their end nodes. The late finish time for an activity is equal to the minimum late start time of its successors. Activities 5-6 and 4-6 have a single successor (activity 6-7); their late finish times are equal to 19, the late start time of activity 6-7. Activities 4-5 and 3-5 have two successors: activities 5-6 and 5-7, with late start times of 14 and 13, respectively. Thus the late finish time for activities 4-5 and 3-5 is 13. This procedure is continued until all late finish and start times are found.

The slack or float values are then found by computing the difference between the late and early start times (or the late and early finish times). These values represent the total amount of slack for the entire path on which the activity lies, not the slack for the individual activities. The critical path is a connected set of activities from the beginning to the end of the network, each of which has the minimum slack value. More than one critical path may exist. If the early finish time of the project is equal to the late finish time, as in the example, the slack for the critical path (or paths) will be equal to zero.

These slack values can be utilized to determine the criticality of paths other than the critical path. The critical path for the example was 1-2-3-5-7. The next most critical path was 5-6-7, with a slack of 1, compared to path 5-7. This type of information can be useful in tracking the progress of the project. As the project progresses, the analysis should be updated because the amount of slack may change drastically if certain activities are delayed or completed ahead of time.

Statistical Estimates

The times computed in the forward and backward analysis are based on deterministic or expected times, which will seldom occur. A PERT analysis allows the computation of probabilities of completion on schedule for selected paths. The initial objective of the project was to complete the park in 22 working days, whereas the critical path time was computed as 21 days, or a 50% probability of completing the critical path in 21 days. By utilizing the SDs calculated in Exhibit 11.2.7, the probability of completing the critical path in 22 days can be computed. Although the individual activity times were assumed to be distributed according to a beta distribution, the path time can be assumed to be normally distributed becasue of the central limit theorem. The variance of any path can be computed as the sum of the activity variances that make up that path. Thus the SD of the critical path can be computed as follows:

$$\sigma_{cp} = \sqrt{\sigma_{1-2}^2 + \sigma_{2-3}^2 + \sigma_{3-5}^2 + \sigma_{5-7}^2}$$

$$= \sqrt{0.66^2 + 0.33^2 + 0.66^2 + 0.33^2}$$

$$= \sqrt{1.108}$$

$$= 1.05$$

Utilizing standard statistical concepts (Chapter 13.2), the probabilities of completing the critical path in 22 days can be computed as follows:

$$Z = \frac{T_c - T_e}{\sigma}$$

where T_c = desired completion time (22)
T_e = expected path completion time (21)
σ = SD of the path (1.05)
Z = number of standard normal deviations that T_c is from T_e

$$Z = \frac{22 - 21}{1.05}$$

$$= 0.952$$

Using the standard normal tables, the probability of completing the critical path in 22 days or less is determined to be approximately .83. These results are illustrated in Exhibit 11.2.10. The probability of completing any other path or partial path in a given time can also be computed using the previous procedure. In addition, it is possible to determine the path duration that would be required for a given probability. For example, if the contractor wanted to be 95% sure of

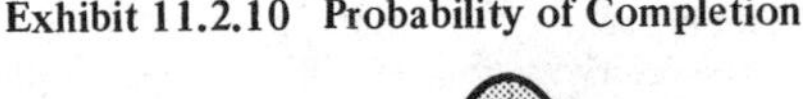

Exhibit 11.2.10 Probability of Completion

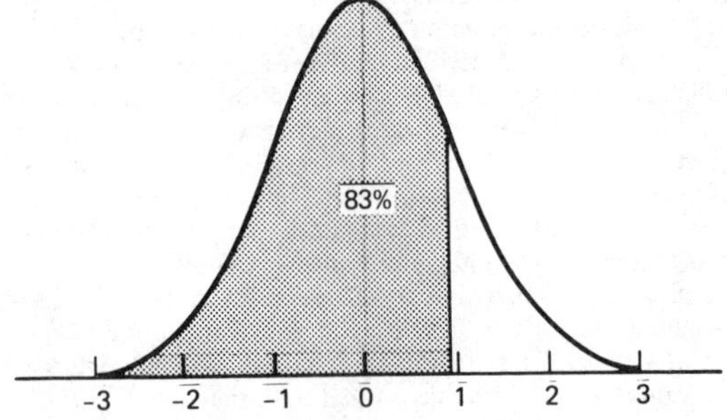

completing the critical path, he or she would have to allow 22.73 days ($Z = 1.65$ from the standard normal tables).

One criticism of the PERT statistical analysis is that probability values are calculated for completion of the critical path rather than the project. Unless the critical path is significantly longer than the other paths in the network, it is a poor assumption that the probabilities of completing the critical path and the project are equal. For example, the large variance associated with activity 4-5 could result in a lower probability of completion in 22 days for path 1-2-4-5-7 than the critical path if there were less slack available. Furthermore, the PERT assumptions result in overly optimistic estimates of the probability of completion of the critical path by a given time. Consequently, these values should be used with caution. One alternative is to employ simulation techniques (Chapter 13.11). With simulation, the user is not restricted to the modified beta distribution for modeling activity durations; any desired distribution type can be employed without increasing the difficulty of analysis. Using the time estimates, the project could be simulated several times. The resulting estimates of duration mean and variance will be much more accurate, and the distribution of the mean can be easily obtained. In addition, results apply to the project duration rather than the critical path duration. A criticality index for each activity may also be calculated; this is the probability that the activity is on the critical path.

Time-Cost Trade-Offs

The CPM approach, with deterministic activity times, allows the user to easily consider time-cost trade-offs. The objective of this procedure is to reduce project duration by compressing selected activities such that a minimum increase in cost occurs. This procedure assumes that activity times are based on a normal cost. The user may be able to "crash," or shorten the duration of, selected activities at an increase in the cost. To apply this procedure, the user must first identify those activities that can be crashed and the minimum time, or crash time, in which the activities can be completed.

At this point the cost associated with these reductions in time must be estimated. Two methods are available. The actual cost for each increment of reduction may be estimated, or the normal and crash time costs can be estimated, with all intermediate values interpolated by a linear approximation. The two methods are shown in Exhibit 11.2.11 for activity 5-7. Once all the cost estimates for all the activities have been made, the actual project compression can proceed.

If nonlinear cost curves are used, the problem becomes more complex. One method of alleviating the difficulty is to segment each activity into two or more pseudoactivities, each of which has a linear cost curve. The procedure for time-cost trade-off that follows is not optimal. If an optimal solution is desired, a network flow algorithm or linear programming may be used.

The basic procedure when the linear approximation is assumed is to identify that activity on the

Exhibit 11.2.11 Time-Cost Tradeoff Curve

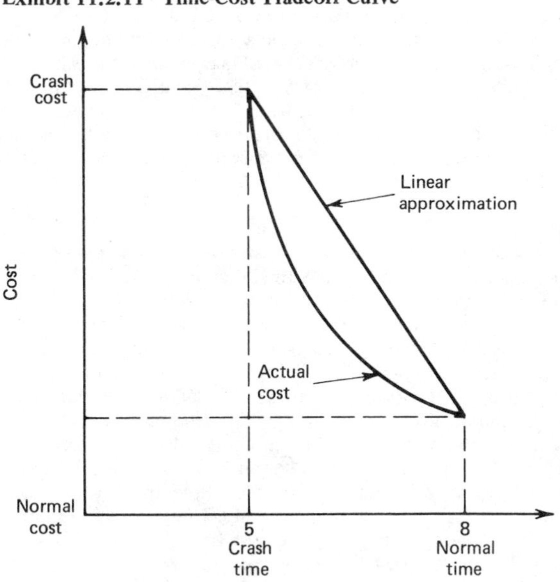

critical path which can be compressed or crashed at the minimum cost per time unit. This activity is then compressed until the slack on an alternative path becomes zero. In the example, activity 5-7 can be compressed by only one time unit, from 8 to 7, because at that point the time duration of path 5-6-7 has a zero slack, and further reduction of activity 5-7 would not reduce the total project duration. This procedure is repeated until the desired project duration is achieved. At some point in the procedure, it may become necessary to consider the reduction of several activity times at once in order to achieve a reduction in the project duration at the minimum cost.

This occurs when the original project has multiple critical paths or when previous crashing creates additional critical paths. Under these circumstances, the minimum cost reduction may involve simultaneous crashing of two or more activities. At least one activity from each critical path must be crashed for project duration to decrease; some candidate activities may be common to several critical paths. The cost of crashing is the sum of the costs for each activity. For simple problems with few critical paths, the minimum cost activity or set of activities may be found by inspection.

For complex problems more formalized procedures are required. One such approach is the use of a linear programming formulation as follows:

$$\min \sum_{(i,j)} \sum (a_{ij} - b_{ij} t_{ij})$$

subject to

$$e_i + t_{ij} - e_j \leqslant 0 \qquad \text{for } (i, j) \text{ in network}$$

$$t_{ij} \leqslant u_{ij} \qquad \text{for } (i, j) \text{ in network}$$

$$t_{ij} \geqslant l_{ij} \qquad \text{for } (i, j) \text{ in network}$$

$$e_n \leqslant T$$

where t_{ij} = duration of the activity between nodes i and j
e_i = occurrence time of event i
$-b_{ij}$ = slope of cost line for activity ij
a_{ij} = intercept of cost line for activity ij
u_{ij} = longest duration for activity ij
l_{ij} = shortest (crash) duration for activity ij
T = duration of project
n = index for last node in network

The objective function is minimization of total cost. The first constraint prohibits the occurrence of an event until all activities directly preceding that event are complete. The second and third constraints restrict activity duration between a lower bound (the crash time) and an upper bound. The upper bound may be the normal time for the activity or may be longer than the normal time, allowing the activity duration to be extended, at a savings in cost, if it is not critical.

If desired, a project cost curve may be produced that graphically illustrates the increased cost of successive reductions. If a project is crashed using the basic procedure described earlier, the project cost curve can be easily created by plotting the increased cost versus time saved after each reduction. In practice, it is more efficient to use a network flow algorithm to solve the problem and develop the project cost curve.

Resource Allocation

The CPM and PERT concepts, including the time-cost trade-offs and probability estimates, are relatively straightforward and easy to apply. Even large projects can be handled with the aid of a computer. Unfortunately, these techniques do not consider the availability of resources such as manpower and equipment. With the recent advances in technology and subsequent specialization of skills and equipment, the allocation of resources has become a rather complex problem. It is not uncommon for a large project to require as many as 40 to 50 different resources.

There are two basic approaches to the resource allocation problem. The first approach, resource leveling, assumes a project completion date with unlimited available resources and attempts to level the required resources. The second approach, resource constrained, assumes a fixed amount of available resources and attempts to minimize the project completion date subject to the resource constraints. The real-world problem requires a combination of both approaches because there is normally a fixed amount of resources available, but it is often possible to access additional key resources for short durations by renting equipment, temporary help, subcontracting, and so on. The total problem is often further complicated by the fact that several projects may be ongoing at the same time.

Exhibit 11.2.12 Estimated Activity Resource Requirements

Activity	Duration	Resource Requirements		
		Truck	Excavator	Crew
1-2 (clear)	4	1	1	2
2-3 (cut trees)	4	–	–	1
2-4 (roads)	3	1	1	2
3-5 (wood pickup)	5	1	–	2
4-5 (parking)	3	1	1	3
4-6 (utilities)	4	–	1	2
5-6 (shelter)	5	–	–	2
5-7 (landscape)	8	1	–	2
6-7 (finish)	2	–	–	1

The initial step of resource allocation is the determination of the types and number of resources required by each activity. These resource requirements depend on the perceived duration of the activity since the requirements are assumed to be constant for the duration. As with the time-cost trade-off, it is often possible to shorten an activity duration by increasing the resource requirements. Most resource allocation procedures assume the resource requirements and activity durations are fixed.

To illustrate the basic concepts, the previous example of the construction of a small park will be utilized. Three resource types are required: trucks, excavators, and work crews. The estimated resource requirements for each activity are given in Exhibit 11.2.12. The resource units are given in the number of units per time period, or day. Thus the first activity (clear) requires one truck, one excavator, and two crews for 4 days.

Prior to proceeding with the development of an actual schedule, it is often useful to determine the relative resource requirements of the project. Utilizing the data from Exhibit 11.2.12, one can calculate the total project requirements as 73 crew days, 14 excavation days, and 23 truck days. If the project is to be completed in 22 days, this requires an average of 3.32 crews, 0.63 excavators, and 1.05 trucks/day. These values can be very misleading because they do not take into account the actual timing of the resource requirements; they do, however, provide a general measure of the project needs.

If the project is small, a Gantt chart of the resource requirements can be constructed to gauge the initial effects of the timing. Utilizing the results from the previous forward and backward analysis and scheduling all activities at their earliest start times yields the resource requirements shown in Exhibit 11.2.13. The numbers contained within the resource blocks in the exhibit reference the activity that requires the resource. This schedule requires a maximum of seven crews, two excavators, and two trucks. The greatest deviation occurs in the crew schedule, with a range in requirements from two to seven crews. Further examination reveals that the simultaneous scheduling of activities 4-5 and 4-6 results in the maximum requirements for both the crew and the excavator resources. If there are no maximum upper limits on the availability of resources, or if the project duration is fixed, resource leveling concepts may be applied in an attempt to distribute the resource requirements more evenly over time.

Resource Leveling

If the project duration is equal to the critical path time, the early and late times determine the general bounds on where each activity can be scheduled. The slack time provides an indication of the maximum amount of movement for each activity. It should be recalled that the slack value is for a path in the network and not for each activity. For example, if the start of activity 2-4 is delayed until time 6, it will not be completed until time 9, and two units of slack would have been consumed. This would reduce the available slack for activity 4-5 from three units to one unit. If the project is to be scheduled for a duration that is longer than the critical path time, the late times and slack must be adjusted. This adjustment is simply a constant, equal to the difference between critical path and the desired duration, added to all the start and finish late times and the slack values.

In applying the concepts of resource leveling to projects with multiple resources, it is not uncommon to find that the benefit from the leveling of one resource is negated by the effect of the rescheduling of another resource. If this occurs, it may be necessary to establish priorities for the resources or create a weighted performance measure to allow consistent trade-offs. Obviously, as the number of resources and activities increases, the complexity of the problem grows at a much faster rate. The application of resource leveling to the example problem for a project duration of 21 days yields the results shown in Exhibit 11.2.14. This rescheduling of activities results in a

Exhibit 11.2.13 Resource Requirements for an Early Start Schedule

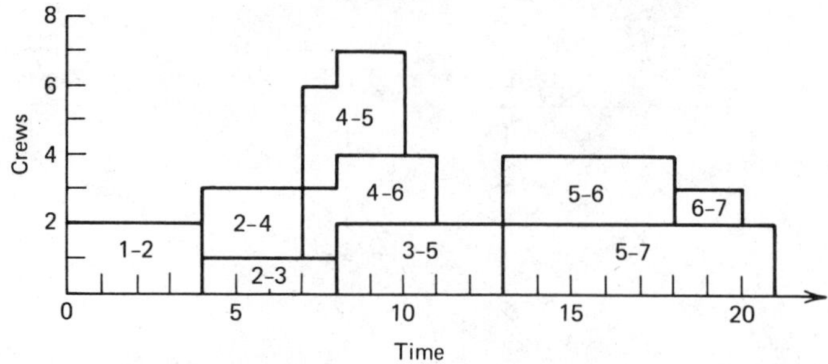

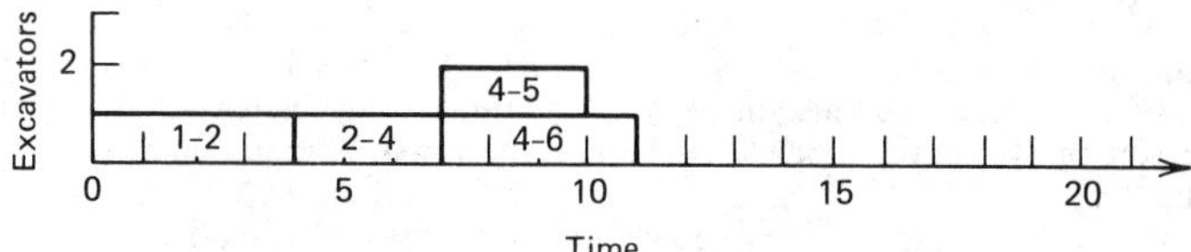

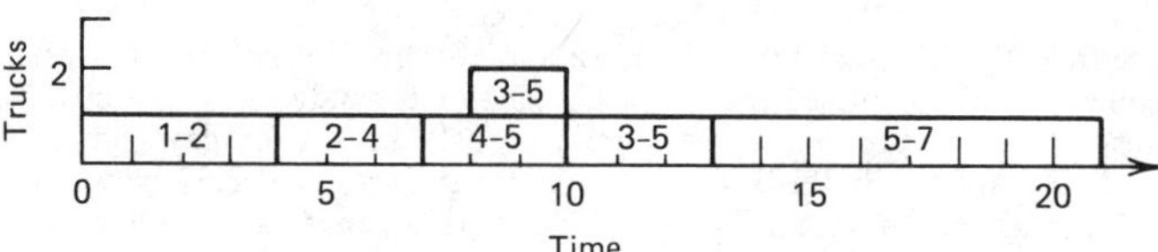

Exhibit 11.2.14 Resource Leveling Schedule

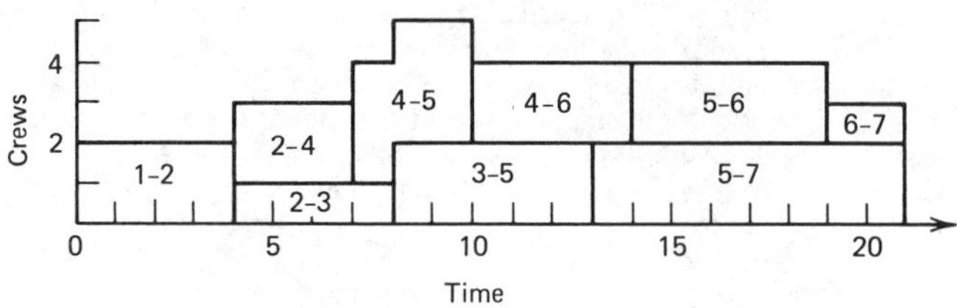

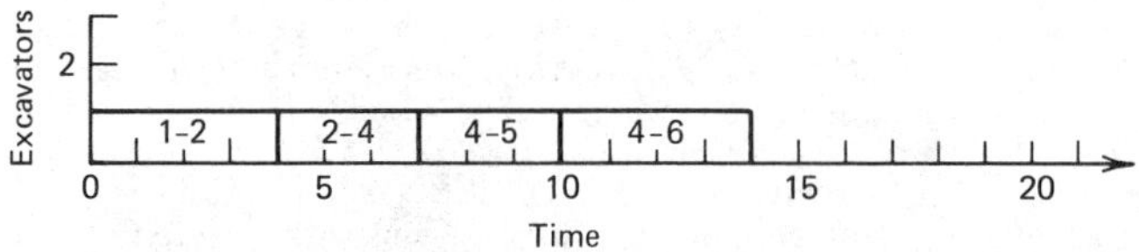

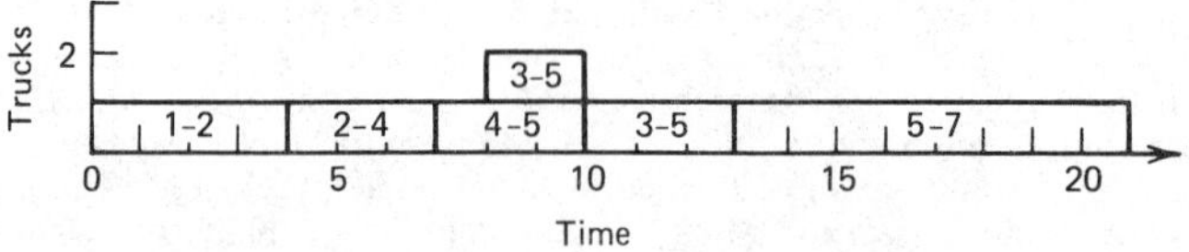

reduction of the maximum crew requirements from seven to five and the number of excavators from two to one. The resulting schedule still requires two trucks for days 9 and 10.

Constrained Resources

If the contractor is limited to only one truck for the project, then leveling the resources will not resolve the problem, and the resource constrained approach must be taken. Since 23 truck days are required, the minimum feasible schedule would be for 23 days. Applying this constraint to the problem and leveling the crew requirements as much as possible yields the schedule shown in Exhibit 11.2.15. This schedule allows the project to be completed with one truck and one excavator and reduces the maximum crew requirements from five to four. If only three crews were available, the project duration would have to be further extended from 23 to 33 days.

The basic concepts for the resource leveling and the resource constrained problems are relatively simple and easy to apply to small projects. Even if an acceptable schedule is developed, it must be continually updated if activities are completed ahead of time or if delays occur. An early activity completion may free critical resources and greatly reduce the expected project completion time; a delay in an activity completion time may well have the reverse effect. In either case this type of analysis does provide the project manager with information to make appropriate trade-offs and intelligent decisions.

Many mathematical programming formulations of the resource constrained and resource leveling problems have been created. They are not useful for most realistic projects because of the combinatorial nature of the problem. A great many heuristic, that is, nonoptimizing, procedures have been proposed for the generation of good project schedules.

One class of heuristics ranks the activities and attempts to schedule that activity with the highest priority first. If precedence or resource constraints are violated, the ranked list of activities is examined until an activity can be scheduled, or the list is exhausted. The procedure then steps to the first scheduled activity termination and repeats itself until all activities are scheduled. Following are some of the many priority rules that have been suggested.

Least slack.
Greatest total resource demand.
Shortest activity duration.
Greatest demand of a "key" resource.
Minimum late finish time.
Longest chain of following activities.

Exhibit 11.2.15 Resource Constrained Schedule

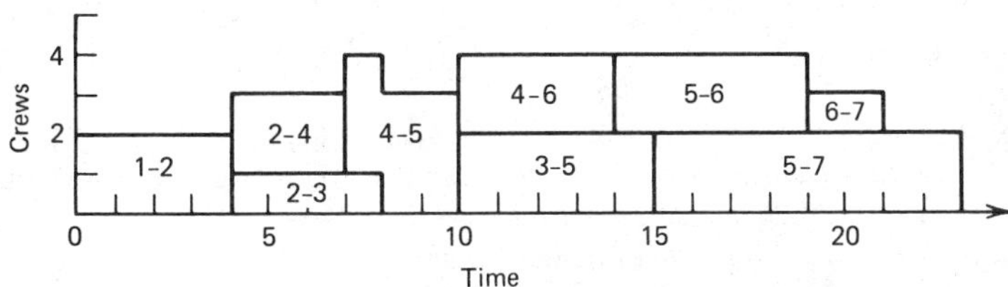

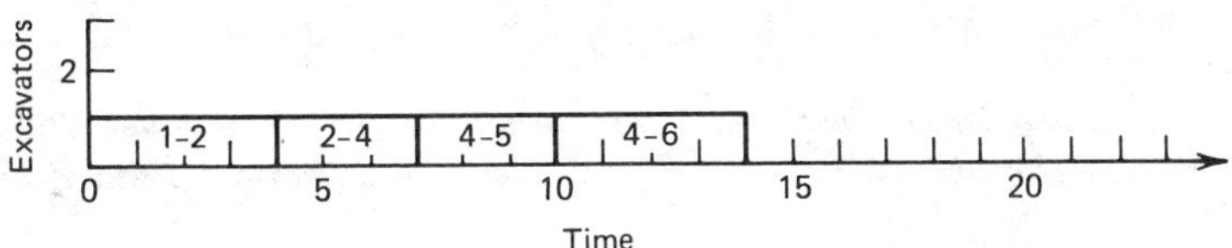

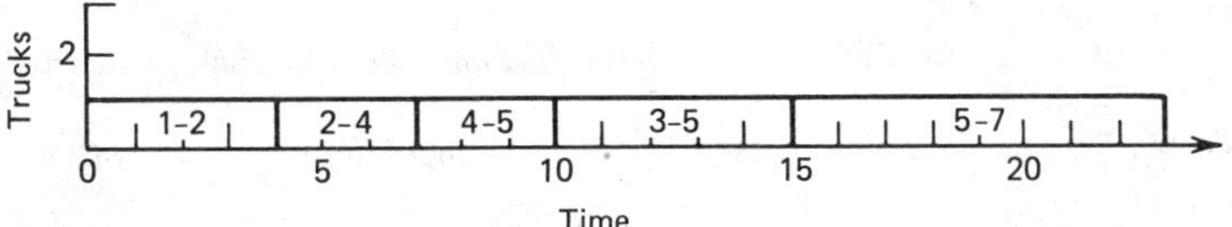

Several of these rules may be implemented in two ways. The priority could be calculated at the beginning of the procedure and never changed or could be recalculated after each time an activity is scheduled, thereby changing the ordering of the list. Also, these simple heuristics can be combined to produce more complex decision rules.

Numerous computer packages incorporating these or other heuristics are currently being marketed, although little is known about the procedures that they utilize for activity scheduling. These packages are generally quite expensive and require large computers. In many instances this expense is fully justified and provides the only reasonable way to manage and control large projects. An alternative is to use several heuristics to generate schedules and choose the best among those schedules. Simulation (Chapter 13.11) can be used to evaluate the performance of different heuristics for a particular project when activity times (and possibly some resource requirements and availabilities) are random variables.

11.2.5 SUMMARY

Scheduling problems are combinatorial in nature. With the exception of a few cases, problem size grows rapidly with only a small increase in the number of input variables (jobs, machines, activities, resources). Thus optimization involving partial or complete enumeration of the possible sequences or schedules is not feasible. Heuristic procedures, which seek good rather than optimal solutions, appear to be the only viable alternative unless optimization procedures are improved dramatically.

A great deal of research effort has been devoted to scheduling problems. These efforts fall into two classes: optimal algorithms and heuristic procedures. In most cases the problems that can be solved optimally are unrealistic, either in size or assumptions, or are of limited applicability. For most heuristic procedures there is no means of determining the goodness of the solution except by comparison with the solutions produced by other heuristics.

Conway, Maxwell, and Miller, in discussing the general job shop, summarized the difficulties inherent in most scheduling problems.

> *Although it is easy to state, and to visualize what is required, it is extremely difficult to make any progress whatever toward a solution. Many proficient people have considered the problem, and all have come away essentially empty handed. Since this frustration is not reported in the literature, the problem continues to attract investigators, who just cannot believe that a problem so simply structured can be so difficult, until they have tried it.*[1]

REFERENCE

1. R. W. CONWAY, W. L. MAXWELL, and L. W. MILLER, *Theory of Scheduling*, Addison-Wesley, Reading, MA, 1967, p. 103.

BIBLIOGRAPHY

BAKER, K. R., *Introduction to Sequencing and Scheduling*, Wiley, New York, 1974.

BEDWORTH, D. D., *Industrial Systems: Planning, Analysis, Control*, Ronald, New York, 1973.

BUFFA, E. S., *Modern Production Management*, Wiley, New York, 1973.

Communications Oriented Production Information and Control Systems, IBM, White Plains, NY, 1973.

DAVIS, E. W., "Project Scheduling Under Resource Constraints–Historical Review and Categorization of Procedures," *AIIE Transactions*, Vol. 5, No. 4, pp. 297–313.

DAVIS, E. W., *Project Management: Techniques, Applications and Managerial Issues*, AIIE, Norcross, GA, 1976.

GREENE, J. H., *Production and Inventory Control Handbook*, McGraw-Hill, New York, 1970.

GREENE, J. H., *Production and Inventory Control: Systems and Decisions*, Irwin, Homewood, IL, 1974.

JOHNSON, L. A., and D. C. MONTGOMERY, *Operations Research in Production Planning, Scheduling, and Inventory Control*, Wiley, New York, 1974.

MODER, J. J., and C. R. PHILLIPS, *Project Management with CPM and PERT*, Van Nostrand Reinhold, New York, 1970.

MUTH, J. F., and G. L. THOMPSON, *Industrial Scheduling*, Prentice-Hall, Englewood Cliffs, NJ, 1963.

WIGHT, O. W., *Production and Inventory Management in the Computer Age*, Cohners, Boston, 1974.

CHAPTER 11.3

Aggregate Production Scheduling

SAMUEL EILON

Imperial College of Science and Technology, London

11.3.1 INTRODUCTION

Aggregate production scheduling is involved when planning is undertaken at a broad level without consideration of individual products or activities. The purpose of the aggregate schedule is to map out a program of output to meet future demand under given constraints, thus utilizing the available resources so as to attain high levels of performance criteria.

Several issues need to be examined:

1. The objective or objectives by which the schedule is to be evaluated.
2. The constraints on the system and on the scheduler.
3. The type of production involved.
4. The control or decision variables.
5. The aggregation and disaggregation process.

Let us analyze these issues with respect to aggregate production scheduling.

Objectives

An examination of the literature and various case studies suggests that numerous objectives are often considered by managers and analysts, some of which are as follows:

Minimizing cost.
Maximizing profit.
Maximizing return or rate of return on investment (ROI).
Maximizing net present value (NPV) or internal rate of return (IRR).
Minimizing time span.
Minimizing throughput time or its variance.
Minimizing work in process.
Minimizing machine idle time.
Feasibility.
Maximizing regularity.

This is obviously not an exhaustive list, but it serves to illustrate the very wide divergence that is to be found in published models and in practice. Although some of the objectives mentioned are compatible with each other, many are not. For example, cost minimization and profit maximization—perhaps the two most prominent objectives—often lead to different solutions (a result known from inventory problems[1]), as do the objectives of maximum ROI (measured as the ratio of profit to cost) compared with maximum rate of return (the ratio of return per unit time). Similarly, the objective of maximizing NPV does not coincide with that of maximizing IRR.

Of these various objectives, cost minimization is probably the most prevalent, first because it embraces narrower criteria, such as minimization of work in process and maximization of production capacity utilization, and, second, because it pertains directly to decisions in the production environment (whereas profit, rate of return, NPV, and IRR are greatly affected by pricing and marketing policies and/or by cash flow considerations).

But the last two criteria in the preceding list of objectives are also important in practice. The one regarding feasibility (unlike the others listed) does not involve optimization; it is a satisficing objective, which merely requires that an acceptable solution be found. When the constraints imposed on the system are so severe that a feasible solution does not exist or is difficult to find, then the objective may be to violate these constraints as little as possible. The method that springs to mind is that of goal programming, which has been advocated as a tool for financial and corporate planning, but so far does not seem to have made a significant impact on the production area. The methodology of goal programming calls for the setting up of an objective function, which is the weighted sum of all the deviations from given constraints; the problem may then center on how trade-offs between different types of constraints can and should be determined.[2]

The last objective in the list reflects the desire of production managers to minimize disturbances to the smooth running of facilities, and this is greatly enhanced by determining a production cycle with recurring regularity. In the case of continuous production, regularity is achieved by adopting a constant production rate, but this must be tempered by the need to reduce fluctuations in inventory, so that the objective function may take the form $a\sigma_I + b\sigma_Q$, where σ_I and σ_Q are the SDs of the inventory and the production level, respectively, and a and b are weighting constants.[3] In one sense this is a cost minimization approach, where a and b may be interpreted as cost parameters, except that it obviates the need to determine a multitude of cost data, which would normally be required for a conventional cost model. An example for the use of this criterion for aggregate production scheduling is discussed later.

Constraints

Any system in which resources have to be allocated to perform an array of tasks is subject to constraints of various kinds.

Product specifications for physical, chemical, and quality characteristics.

Technical constraints on the capabilities of machines and people to perform certain tasks.

Market requirements regarding due dates and the need to avoid running out of stock.

Operational considerations concerning the capacities and availability of facilities, problems of flow and storage dictated by layout and the physical environment, limitations on overtime and/or subcontracting, the speed with which the system can adjust to changes in production rate and product mix, flexibility in the use of manpower.

There may, of course, be many others, but the list serves to bring out two important points. First, a distinction can be made between external constraints, imposed on the system from the outside (such as legal requirements, the needs of the customer, and restrictions dictated by trade unions), and internal constraints, imposed by a higher authority on lower echelons of the hierarchy (e.g., on the mode of operations and conditions in the plant). This distinction is of some consequence, since external constraints are more difficult—and sometimes impossible—to violate, whereas internal constraints may be negotiable. Admittedly, a rigid line of demarcation between the two categories is not always easy to draw: For example, delivery due dates are often considered in production models as inviolable, whereas in reality there is some room for negotiation between customer and supplier. Similarly, in cases of stock runouts there is a measure of product substitution that needs to be recognized, particularly when it is encouraged by price discounts. Thus, even in considering external constraints, it may be wise to bear in mind that some flexibility in definitions and determination of bounds is often possible.

The second point[2] is that there is no fundamental difference between objectives and internal constraints, particularly when the former are of the satisficing kind. This realization can greatly affect not only the choice of model and the details of its structure, but also the solutions that consequently emerge. The adage that it is not what is in a model that is important, but what is behind it, is certainly pertinent in relation to goals and constraints.

Types of Production

To understand the type of models that aggregate production scheduling can draw upon, we need to identify the various types of production activities that may be encountered. It has been found convenient to identify five main types for the purpose[4]:

1. *Continuous production*, where the demand for a product requires production on a continuous basis, but because of fluctuating demand from one period to another over a given horizon, the level of production needs to be adjusted from time to time.
2. *Batch production*, where the rate of demand for a product is well below the rate at which it can be produced, so that production is carried out intermittently to avoid excessive stockpiling.
3. *Job production*, where various jobs, each with its own array of processing requirements, need to be loaded in some sequence on a given set of production facilities.

4. *Project construction*, where complex one-off jobs can be controlled with the use of network analysis to assist in the timing of activities and the use of resources.

5. *Timetabling*, where a set of activities needs to be performed through the use of given resources and a timetable is required to assign the activities to time slots and marry resources for the purpose.

Perhaps a more important distinction is between problems concerned primarily with sequencing, those concerned with assignment, and those involving production levels, as schematically illustrated in Exhibit 11.3.1. When jobs need to be processed on several machines in some given sequence, as in Exhibit 11.3.1*a*, the production system consists of several queues in tandem, where a job proceeds from one queue to another until processing is complete and the job leaves the system. Here the question of production levels does not arise, since the processing requirements for each job are specified in advance. The problem is to steer the job through the system where alternative routes exist and to determine the queueing discipline at each queue, namely, to order the jobs in the sequence in which they are to be processed by each machine.

In Exhibit 11.3.1*b* each product proceeds through a single processing facility and then leaves the production system. The problem of sequencing encountered in Exhibit 11.3.1*a* does not arise or is considered irrelevant. If a product can be processed on alternative facilities, as in the example shown in Exhibit 11.3.1*b*, the question of allocating facilities to products needs to be resolved, as does the question of the production level for each product.

Exhibit 11.3.1 Decision Variables in Scheduling, Including (*a*) Sequential Processing–the Decision on the Sequence at Each Machine; (*b*) Alternative Processing–the Decision on Assignment of Products to Machines; and (*c*) Aggregate Production Scheduling–the Decision on the Level of Production

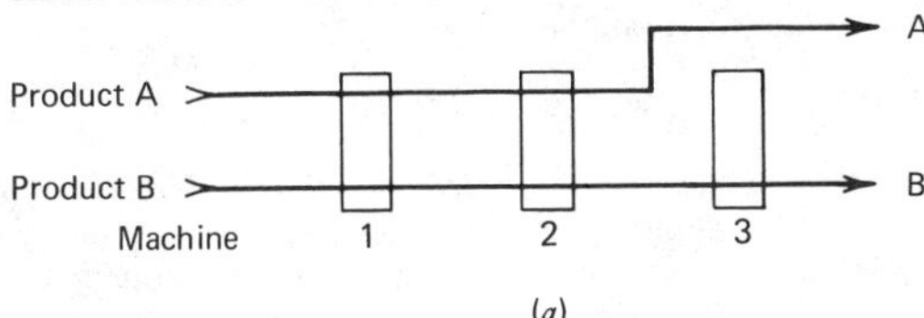

(*a*)

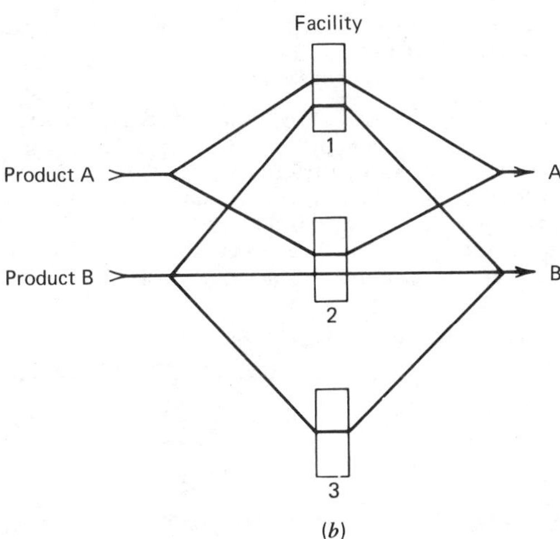

(*b*)

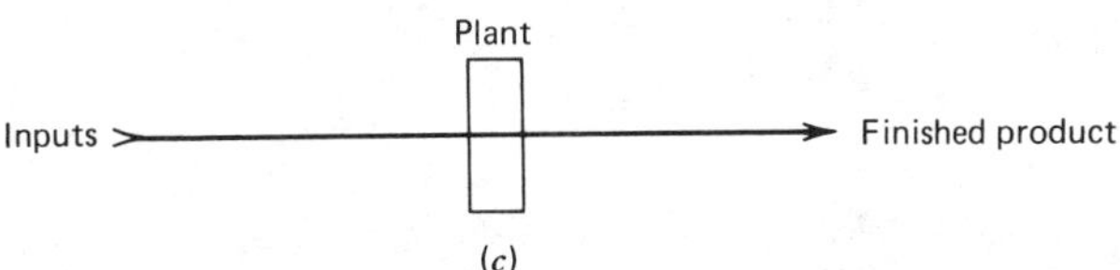

(*c*)

In Exhibit 11.3.1*c* it is assumed that all the products and activities can be represented by a single product and the production facilities by a single plant. Here the main problem for the scheduler is to determine the production level, and this is essentially *the aggregate production scheduling problem*.

Decision Variables

In the study of production scheduling problems, the decision variables boil down to determining how much to produce, in what order to produce (which in turn determines the timing of production), and which facilities should be involved. These three variables–*production level, sequencing*, and *assignment*–indicate a wide array of models for the analysis of the 10 scheduling problems summarized in Exhibit 11.3.2.[4]

The first three problems are concerned with continuous production for one or several products and for one or several machines. Problem 1 is the well-known production smoothing problem (except that it usually involves not only decisions for the level of production, but also for the level of the work force). Problem 2 envisages alternative production lines with different cost characteristics for production and for changing its level, so that, in addition to finding the total production rate, it is necessary to allocate it to the machines (involving a facility-mix problem, in interesting contrast to the more commonly found product-mix problem). Problem 3 widens the scope to the multiproduct case. Problems 4 and 5 relate to batch production, whereas problems 6 and 7 are concerned with job production, and problem 10 with timetables.

Exhibit 11.3.2 attempts to highlight the decision variables that predominate. Whereas in problems 6 through 10 the tasks to be performed are prescribed, and the production level therefore ceases to be a decision variable, it is a central issue in problems 1 through 5. Of these, the first has become perhaps the most prominent, both for its relative simplicity (it involves a single product processed in a single plant and requires essentially one decision variable) and for constituting the basic ingredients relevant to aggregate production planning.

Aggregation and Disaggregation

Thus *aggregate production scheduling is primarily concerned with determining the production level for each period over a given horizon* (e.g., monthly production volumes for 12 months ahead), and the results of this planning exercise may be crucial in any production environment described under problems 1 through 5 in Exhibit 11.3.2. The question for the scheduler is what modeling approach to adopt. In problem 1 the scheduler needs to consider only one decision variable, that of the global production level, but in the other problems he or she has to determine the production level for each product (except for problem 2, which, like problem 1, is a single-product case), and sequencing and assignment decisions also have to be made.

Basically, there are three approaches that may be considered:

1. Complete Disaggregation. A comprehensive array of products and tasks is constructed, together with a list of all the available facilities, the resource requirements for each task, the technological sequence in which tasks have to be performed, and so on. In the case of multiple products–for example, in engineering plants where products consist of many components, each involving a sequence of operations on several machines–the result is a massive data matrix, which can then be fed into a linear programming model to determine the production volumes for each product for each time period (say, a week or a month) and the activities to be performed on each of the available facilities.

Exhibit 11.3.2 Types of Scheduling Problems

		No. of Products		No. of Machines		Decision Variables		
Type of Production		One	Several	One	Several	Production Level	Sequencing	Assignment
Continuous	1	x		x		x		
	2	x			x	x		x
	3		x		x	x		x
Batch	4		x	x		x		
	5		x		x	x	x	x
Job	6		x	x			x	
	7		x		x		x	
Project	8	x			x		x	
	9		x		x		x	x
Timetables	10		x		x		x	x

2. Assignment First. Here the scheduler assigns the products or the activities to the machines, thereby converting the multimachine problem to several single-machine problems, before turning his or her attention to solving each problem. This process of decomposition reduces the size and complexity of the original model posed in approach 1, so that the resultant problems are much easier to handle, and each can be singled out for a more detailed study without directly affecting the rest of the schedule. The scheduler may then examine the results of the analysis and consider whether to amend the original assignment decision and proceed with an alternative decomposition.

3. Aggregation First. The scheduler visualizes the plant as if it were a single-product, single-machine case, namely, problem 1 in Exhibit 11.3.2, or Exhibit 11.3.1*c*. The first task, therefore, is to determine the global production volume for each period of a given horizon, or at least for the next period in the horizon. The scheduler then needs to disaggregate this global figure for each period by making assignment decisions (and sequencing decisions, where appropriate) to determine the work load on each of the facilities and the ensuing volumes for each product.

The advantage of complete disaggregation in approach 1 is that none of the decision variables is subjugated by the modeling approach, whereas the decomposition in approach 2 and the aggregation in approach 3 involve modeling in tandem, seeking a solution to one variable first before turning to the other(s). In this sense approach 1 may be regarded as the purest of the three, and in cases where the number of products and machines is relatively small, it would be the analyst's natural choice. However, when hundreds of products are involved, with thousands of components processed on numerous facilities in a volatile production environment, the demands on accurate information and the inflexibility that is often embedded in the programming method in approach 1 make implementation difficult. In comparison with approach 1, both 2 and 3 may be regarded as suboptimal philosophies, but they offer the scheduler the practical advantage of handling smaller and more manageable problems with greater flexibility of adjusting the production programs in the light of changing circumstances. There are cases where machines are so clearly suitable for processing certain products that assignment first, and hence approach 2, is called for; the treatment of each of the subproblems as problem 1 in Exhibit 11.3.2 can then follow. In other cases approach 3 is appropriate, where the total production is handled as for problem 1, with a suitable output measure defined as a surrogate to represent the production volumes of the individual products.

What follows is a more detailed description of the aggregate production scheduling problem (sometimes known as the "production smoothing problem") and the various methods that may be employed to tackle it.

11.3.2 THE PROBLEM

The production system is considered as a single facility that handles a single product. Time is divided into discrete periods, and demand forecasts for future periods are given. Demand is stochastic, so that discrepancies between forecasts and actual demand values may occur.

Exhibit 11.3.3 shows cumulative and production demand curves. The vertical difference between them denotes inventory (which turns into a "runout" or shortage when cumulative demand exceeds cumulative production), and the horizontal difference denotes storage time (assuming depletion on an FIFO basis).

Exhibit 11.3.3 Production and Demand

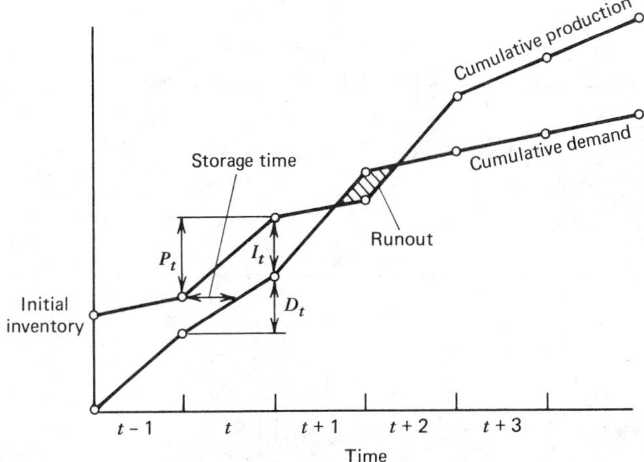

Three possible strategies may be considered for production planning to handle this case[1]:

1. *Have a static production program*, coupled with an inventory large enough to satisfy the fluctuating demand. The inventory level would fluctuate according to the demand pattern, with replenishment being provided by a constant flow from the plant. This method is greatly favored by the production department, since it simplifies planning, ensures higher machine utilization, allows better supervision and control, and promotes a sense of security among the workers. Average stock level is high, however, thus tying up capital and involving high carrying costs.
2. *Have a fluctuating production program*, to cater to the changing demand, and keep a constant inventory level. The purpose of the inventory in this case is to provide a safety cushion between production and marketing. Any change in the demand pattern requires a certain time lag before production can follow suit, and the safety stock enables management to satisfy demand in the interim period. The stock level does not, strictly speaking, remain constant, but the fluctuations and the average stock level are fairly low, compared with the previous method.
3. *Have a combination of the two systems*, so as to bring the total costs to a minimum. The problem is, therefore, to achieve a proper balance between the amount of fluctuations in the production program and those in the stock level.

This problem of production for fluctuating demand may be illustrated graphically as a network flow.[5] Sales forecasting and production planning are broken down into periods. The network consists of a chain connecting two lines, one representing production and the other representing inventory. The flow in the network is as indicated by the arrows in Exhibit 11.3.4 (in which the ith time period is shown), and the flow must always be positive. A certain amount of production capacity is carried over through the production flow line from the preceding period. At the beginning of the period, one can plan to increase the capacity (by adding labor, machines, or overtime or by subcontracting) or to reduce it (by lowering the labor force, etc.). The output flows into the inventory line, from which a certain quantity is tapped for sale. For the sake of simplicity it has been assumed in Exhibit 11.3.4 that the output of the ith period reaches the inventory flow line only at the end of the period and that the sales volume of the period also leaves at the end of the period, but modifications to this graphic model can easily be incorporated whenever necessary.

At any point of intersection in the network, we must have equilibrium of flow; namely, the flow into the intersection must be equal to the flow coming out of it. On the production flow line, previous capacity + any increase − any decrease = output = capacity carried over to the next period. On the inventory line, previous inventory + output = sales + inventory carried over the next cycle. Bearing in mind these conventions, we can omit the arrows in the network (leaving the quantities in each link), and represent the three strategies, listed earlier in graphic form, as shown in Exhibit 11.3.5.

The problem before us can be formulated as follows: Given the demand forecast F_t for any period t in the planning horizon, which extends over T periods, determine the production level P_t and the work force W_t for time t (where $t = 1, \ldots, T$).

Exhibit 11.3.4 A Network Presentation of Production and Inventory[a]

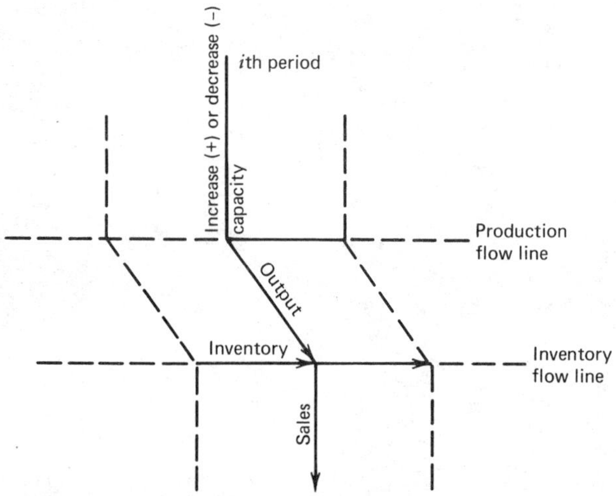

[a]Reference 5.

Exhibit 11.3.5 Policies in Analyzing Production Smoothing Problems: (*a*) Static Production Program, (*b*) Static Inventory Program, and (*c*) Combination of (*a*) and (*b*)

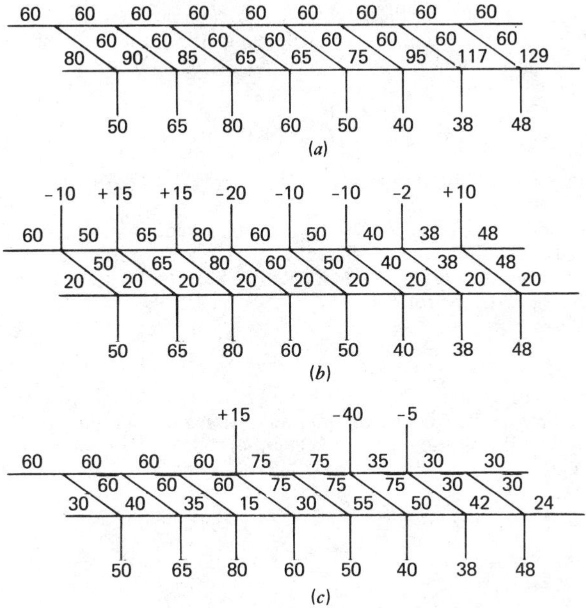

The following methods of solution are reviewed here:

1. The Holt, Modigliani, Muth, and Simon (HMMS) decision rule.
2. The Deziel and Eilon (DE) decision rule.
3. The management coefficients method.
4. Mathematical programming.
5. Production switching.

There are two reasons for selecting these methods for consideration. The first is to illustrate the differences in approach and in the assumptions employed, and hence to indicate their consequences in practical situations. The second is that all these methods emerged from real case studies: Method 1 was devised as a result of a production study of a paint factory in the United States; method 2 arose from a study of production of light bulbs in Britain and in Holland; method 3 was developed by its originator as a result of several studies in the United States; method 4 has been widely employed in a variety of circumstances; and method 5 was developed in conjunction with production planning in a food canning plant in Britain.

11.3.3 THE HMMS DECISION RULE

The Model

The smoothing problem was studied extensively by Holt, Modigliani, Muth, and Simon, who suggested that four cost factors should be accounted for: regular production costs, costs of hiring and firing, costs of overtime working, and inventory costs. They proposed the following cost structure[6,7]:

1. **Regular Production Costs**

$$c_t(1) = c_1 W_t \tag{1}$$

where c_1 is a constant. The assumption here is that costs are linearly related to the size of the work force, W_t. An additional fixed-cost term can be added to equation 1, but since it would add a constant amount to the total cost, it would not affect the solution.

2. **Hiring and Firing Costs.** The cost of increasing or decreasing the work force is assumed to

take the form of a quadratic function, so that the cost in period t is

$$c_t(2) = c_2(W_t - W_{t-1})^2 \tag{2}$$

where $W_t - W_{t-1}$ is the change in the level of the work force from interval $t - 1$ to t and c_2 is a constant. In equation 2 the cost is assumed to be symmetrical; that is, an increase or a decrease in the work force by a given amount incurs the same cost. Asymmetry in the cost function can be introduced, for example, by

$$c_t(2) = c_2(W_t - W_{t-1} - c_{11})^2 \tag{3}$$

but the authors state[8] that this additional constant "proves to be irrelevant in obtaining optimal decisions."

3. Cost of Overtime. It is assumed that, for a given production level, there is a corresponding desirable level of labor requirements. If there is a surplus of labor, then extra costs are incurred in idle time; if there is a shortage of labor, then overtime needs to be used. The cost of overtime and undertime is assumed to take the form

$$c_t(3) = c_3(P_t - c_4W_t)^2 + c_5P_t - c_6W_t \tag{4}$$

where $c_3 \ldots c_6$ are constants. This function has a point of minimum where the production level matches the work force, and also at that point $c_t(3) = 0$. A further term, $c_{12}P_tW_t$, can be added to equation 3 to improve the approximation. The authors add that c_5 turns out to be irrelevant in making the scheduling decisions because in the long term the cumulative production is unaffected by the decision rule.[8]

4. Cost of Inventory. If the inventory level at the end of period t is I_t, then

$$I_t = I_{t-1} + P_t - S_t \tag{5}$$

where I_{t-1} = inventory level at the end of period $t - 1$
P_t = production in period t
S_t = sales or shipment from the manufacturing plant in period t

The minimum-cost inventory level is assumed to be linearly related to the demand, namely, to take the form

$$c_8 + c_9F_t$$

where F_t is the forecast demand for period t, and c_8 and c_9 are constants. In fact, it is known from inventory theory that the optimal inventory level is proportional not to demand, but to the square root of demand. In the HMMS model, however, it is assumed that the linear relationship is an adequate approximation.

The total cost of inventory, including holding costs and runout costs, are then assumed to take the quadratic form

$$c_t(4) = c_7[I_t - (c_8 + c_9F_t)]^2 \tag{6}$$

where c_7, c_8, and c_9 are constants. The expression in brackets is the amount by which inventory deviates from the optimal level, and the penalty for such a deviation is assumed to be the same for equal positive and negative deviations.

The Solution

The total cost for period t is then

$$c_t = c_t(1) + c_t(2) + c_t(3) + c_t(4) \tag{7}$$

and over T periods

$$C_T = \sum c_t \tag{8}$$

where henceforth Σ stands for $\Sigma_{t=1}^{T}$ unless stated otherwise. The problem posed by the HMMS model is as follows: Find the values of P_t and W_t that will minimize the total cost function in equation 8. When derivatives of the quadratic cost function are taken, linear expressions are obtained, and the solution can finally be reduced to the following form:

$$P_t = a_0F_t + a_1F_{t+1} + a_2F_{t+2} + \cdots + g_1W_{t-1} - h_1I_{t-1} + e_1 \tag{9}$$

$$W_t = b_0F_t + b_1F_{t+1} + b_2F_{t+2} + \cdots + g_2W_{t-1} - h_2I_{t-1} + e_2 \tag{10}$$

These are the production and employment *linear decision rules*, where all the lower-case coefficients are constants. Each expression consists of a series of terms that include the forecasts for a given number of future periods, and each rule takes account of the present (i.e., at the end of period $t - 1$) levels of employment and inventory.

In the example of a paint factory, which the authors quote in connection with their study, production planning was carried out on a monthly basis, and forecasts were available for 12 months. The results were as follows:*

$$P_t = \left\{ \begin{array}{l} 0.463\ F_t \\ +0.234\ F_{t+1} \\ +0.111\ F_{t+2} \\ +0.046\ F_{t+3} \\ +0.013\ F_{t+4} \\ -0.002\ F_{t+5} \\ -0.008\ F_{t+6} \\ -0.010\ F_{t+7} \\ -0.009\ F_{t+8} \\ -0.008\ F_{t+9} \\ -0.007\ F_{t+10} \\ -0.005\ F_{t+11} \end{array} \right\} + 0.993\ W_{t-1} - 0.464\ I_{t-1} + 153 \tag{11}$$

$$W_t = \left\{ \begin{array}{l} 0.0101\ F_t \\ +0.0088\ F_{t+1} \\ +0.0071\ F_{t+2} \\ +0.0054\ F_{t+3} \\ +0.0042\ F_{t+4} \\ +0.0031\ F_{t+5} \\ +0.0023\ F_{t+6} \\ +0.0016\ F_{t+7} \\ +0.0012\ F_{t+8} \\ +0.0009\ F_{t+9} \\ +0.0006\ F_{t+10} \\ +0.0005\ F_{t+11} \end{array} \right\} + 0.743\ W_{t-1} - 0.010\ I_{t-1} + 2.09 \tag{12}$$

It should be noted that:

All the a's and all the b's are smaller than 1.

The a's and the b's decline rapidly, so that forecasts for the immediate future are weighted more heavily than those for the more distant future; in the example quoted in equations 11 and 12, half the forecasts (for the latter periods) could be ignored without any significant effect on the results.

$g_1, g_2 > 0$, that is, the higher the present level of the work force, the higher the future level of activity.

$h_1, h_2 > 0$, that is, a high level of inventory has the effect of reducing the values of P_t and W_t, but too low a level of inventory has the opposite effect because of the positive value of the last term in these decision rules.

The authors studied the effect of errors in estimating the cost parameters and say: "We conclude that an estimating accuracy of, say, ±50% is probably adequate for practical purposes. This accuracy will yield decision rules whose cost performance is tolerably close to the minimum possible."[9] This conclusion is supported by other writers.[10]

*These results are cited in reference 6 and are only marginally different from those given in reference 7.

The HMMS decision rules are based on demand forecasts, unbiased forecasts being treated exactly as if they were perfect forecasts[7]; the variance of the demand distribution has no effect on the decisions. Actual orders (as opposed to forecasts) are accounted for only indirectly by the effect they have on the inventory level I_{t-1}, which is included in the decision rules.

A comparison of the HMMS rules for alternative forecasting procedures with the hypothetical case of perfect forecasts provides an indication of the desirability to improve forecasting in any given situation.

Comments on the HMMS Rule

Criticisms of the HMMS model fall into several categories. The first relates to the assumptions of the cost functions. The quadratic form of these functions is clearly a matter of mathematical convenience (in that the derivatives of these functions lead to the linear decision rules in equations 9 and 10, and this may be justifiable over a narrow range; nevertheless, it is arbitrary and may be a source of serious errors.

What is the justification, for example, for assuming that the cost of hiring and firing is a quadratic function? Holt et al. suggest[8] that the plausibility of this assumption rests on the argument that "reorganization costs are more than proportionately larger for large layoffs than for small layoffs; and similarly the efficiency of hiring, measured in terms of the quality of the employees hired, may fall when a large number of people are hired at one time," but why should this increase costs? Such an argument defeats the notion of economies of scale, which is intuitively far more plausible.

The quadratic form of overtime costs in equation 4 is, in some ways, even more baffling. In practice, overtime rates are paid in direct proportion to the amount of overtime; sometimes a further higher rate applies above a certain level of overtime, for example, in the case of third shifts or weekend working. A piecewise linear cost function is perhaps a better description of such a situation, so that a quadratic form can be regarded only as an approximation. Another objection is that equation 4 also expresses the cost of undertime, namely, the cost of the work force being idle when the scheduled production rate is too low. It may be argued that the cost of the labor force, whether it is gainfully employed or not, is already included in the regular payroll in equation 1 and that to impose a further penalty for undertime is not appropriate.

Similarly, the justification for a quadratic function for inventory costs is dubious. Since inventory is a major factor in designing an aggregate production schedule, alternative strategies over a wide range need to be considered, and this implies that the use of a quadratic function, which may provide a reasonable fit over a short range, may generate wide margins of error.

It is interesting to quote in this context some results (see Exhibit 11.3.6) from the case study of the paint factory reported by Holt et al.[7] The company's performance for 1949–1953 was compared with the linear decision rule, first based on a moving average forecast and second on the assumption of having perfect forecasts.

Several interesting features of these results should be noted:

1. The regular payroll is not greatly affected by the introduction of the HMMS rule.
2. The cost of hiring and firing is so small (well under 1% of the total costs) that the advantage of including this cost term in the model is rather questionable.
3. The reduction in costs envisaged by the HMMS rule is *almost entirely due to reduction in*

Exhibit 11.3.6 Cost Comparisons (× 10^3 dollars)

	Company Performance		HMMS Moving Average Forecast		HMMS Perfect Forecast	
Regular payroll	1940		1834		1888	
Overtime	196		296		167	
Inventory holding	361	1927	451	1067	454	854
Back orders	1566		616		400	
Hiring and layoffs	22		25		20	
Total	4085		3222		2929	
Total excluding back orders	2519		2606		2529	
Hiring and layoffs as % of total	0.5		0.8		0.7	

total inventory costs, and this is achieved in the main by drastically reducing the cost of back orders through an increase in the average inventory level. In fact, if the cost of back orders were ignored altogether, the results for the HMMS rule would prove to be no better than the company performance (Exhibit 11.3.6). Bearing in mind the somewhat arbitrary manner in which runout costs are sometimes evaluated in practice, this result is certainly one that management is likely to view with some reservation.

The HMMS rule triggered off several investigations into the production smoothing problem, and many of these studies are reviewed by Silver[11] and Buffa.[12] Some writers merely concentrate on discussing the merits and the practical aspects of the HMMS rule, whereas others suggest some modifications and attend to the problem of disaggregation of the production plan in the case of a multiproduct system.

An interesting point is made by Peterson[13] regarding the basic assumption in the HMMS model that the amount, S_t, shipped by the manufacturer to the wholesaler in period t is the same as the amount ordered, D_t (i.e., $S_t \equiv D_t$). He suggests that if the manufacturer were to disregard this identity, he would have three decision variables at his disposal (P_t, W_t, and S_t) instead of the two considered by the HMMS rule (P_t and W_t). In this way the manufacturer introduces an additional degree of freedom into the decision matrix, and it may sometimes be possible to determine cheaper production plans than those obtained by the HMMS rule, even though the manufacturer must be prepared to pay some penalty (in the form of compensation to the wholesaler) for not meeting the precise amount specified in the order. Peterson suggests that a term be added to the cost function in equation 8, so that the total costs become

$$C_{\text{TOT}} = \sum (c_t + k Z_t^2) \tag{13}$$

where

$$Z_t = \sum_{i=1}^{t} (D_t - S_t) \tag{14}$$

is the cumulative imbalance between demand and shipment during the first t periods and k is a cost parameter. He goes on to show that a linear decision rule for S_t, similar in structure to the rules for P_t and W_t, can be derived.

Peterson puts forward this idea as an alternative to the one indicated by Abramovitz,[14] who suggests a tax on inventories in order to encourage the smoothing of orders. Another possibility is for the manufacturer to give the wholesaler a positive incentive to keep orders constant,[15] but there are some doubts as to whether the idea of an incentive or Peterson's idea of a deliberate imbalance, Z_t, between orders and shipments can be implemented in practice without changing fundamentally the structure of the production smoothing problem.[16]

Rather than introduce a new decision variable, Z_t, it may be pertinent to consider a model where W_t is not independent of P_t. In such a model a penalty would be imposed on increasing or decreasing the rate of production from P_{t-1} to P_t, and this penalty would absorb the adjustments that need to be made in this level of the work force. (This is discussed later in Section 11.3.7.)

11.3.4 THE DE DECISION RULE

The Model

A different approach to the production smoothing problem suggests that it is not the minimization of an overall cost function that should be the planner's objective, but the minimization of the fluctuations in production and inventory levels. This approach is exemplified by the method investigated by Deziel and Eilon[17], the DE rule. Their model is based on the following assumptions:

1. The decision variable is the production quantity Q_t, and this decision is made at the beginning of period t. There is no separate decision for the work force level.

2. There is a lead time of L periods for implementing the production decision, so that the production level P_t in period t is

$$P_t = Q_{t-L} \tag{15}$$

3. Production orders already in the pipeline cannot be altered.

4. Orders that cannot be filled at the end of a period are backlogged.

The decision rule takes the form

$$Q_t = k\left[R - I_{t-1} - \sum_{i=t-L}^{t-1} (Q_i - \bar{D})\right] + \bar{D} \tag{16}$$

where Q_t = decision made in period t for reorder quantity
R = safety stock
I_{t-1} = stock level at end of period $t - 1$
$\bar{D}$ = expected demand level
k = a smoothing constant ($0 \leqslant k \leqslant 1$)

$R - I_{t-1}$ describes the amount by which the stock level falls below the safety stock requirement, and the third term in the brackets represents the cumulative excess of production over demand during the lead time. The smoothing factor k gives a weight to the total inventory balance in the brackets. If $k = 0$ is taken, then the decision rule is reduced to ordering an amount equivalent to the expected demand.

If in place of the expected demand $\bar{D}$ the forecast F_t is substituted into equation 16, the decision rule becomes

$$Q_t = k\left[R - I_{t-1} - \sum_{i=t-L}^{t-1} Q_i\right] + (1 + kL)F_t \tag{17}$$

Here F_t represents the forecast for demand per period, and it may be derived from some forecasting procedure; for example, if simple exponential smoothing is used, then

$$F_t = \alpha D_{t-1} + (1 - \alpha) F_{t-1} \tag{18}$$

where F_t = updated demand forecast
F_{t-1} = previous demand forecast
D_{t-1} = actual demand in period $t - 1$
α = the smoothing constant in forecasting ($0 \leqslant \alpha \leqslant 1$)

Thus the decision involves two smoothing parameters, k and α. The performance of the system can be described by several measures, such as the following:

1. Fluctuations in the inventory level, measured by the SD σ_I.
2. Fluctuations in the reorder level, measured by the SD σ_Q.
3. If a sudden increase in demand occurs, the level of stock runout increases. The *additional* amount of stock that cannot be supplied as a result of this sudden impulse in the demand is defined as

$$\gamma = q' - q$$

where q is the expected level of future runouts when demand is stationary and q' is the level of future runouts when the mean demand is subject to a sudden increase. (Both q and q' are measured over a given horizon.) Thus γ expresses (in terms of runouts) the consequence of a "disturbance" in the demand pattern.

4. If a disturbance occurs (e.g., a discrete increase in the demand level), the system reacts by supplying the demand from stock and by issuing orders to increase the production level (see Exhibit 11.3.7, derived by simulating the system on an analog computer). After a while (called the time rise T_r) the stock is replenished sufficiently for the production level to settle down to the expected demand level.

Thus the purpose of the proposed control procedure is to provide a mechanism that will respond quickly enough to abrupt changes in the mean demand level and yet protect the production rate from being affected by spurious demand fluctuations. This is the essence of the smoothing problem.

Three alternative objectives are considered in the DE model:

1. Minimize

$$C = a\sigma_I + b\sigma_Q \tag{19}$$

where a and b are constants.

Exhibit 11.3.7 Linear System Response to a Demand Step (for $\alpha = 0.5$, $k = 1.0$, $L = 3$)[a]

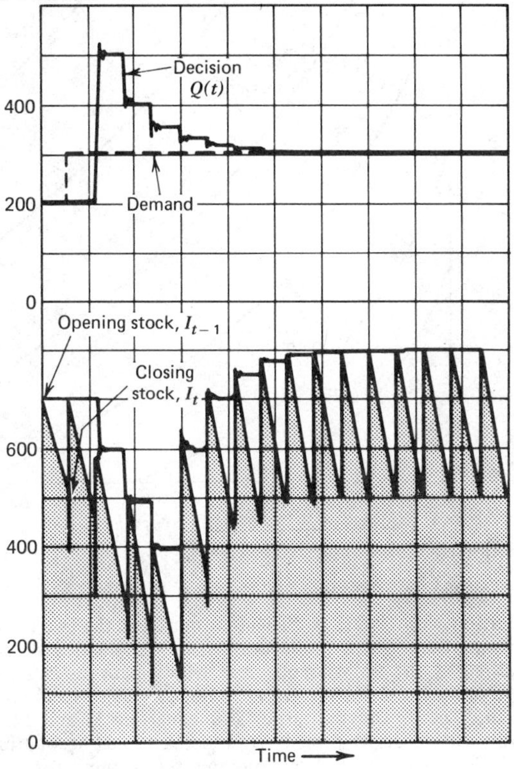

[a]Reference 17.

2. Minimize

$$C = a\sigma_I + b\sigma_Q + c_\gamma \tag{20}$$

where a, b, and c are constants.

3. Same as objective 1 and subject to a given maximum value of T_r.

Objective 1 is concerned mainly with the case where the demand distribution is essentially stationary and is therefore a special case of 2 or 3, which represent two alternative approaches. Objective 2 looks at the incidence of runouts and requires a penalty cost for allowing runouts to exceed an acceptable level, whereas objective 3 attempts to ensure that the system recovers from an abrupt disturbance within a reasonable period.

Comments on the DE Model

One interesting result from this study[17] is the symmetry of results with respect to a and k. Exhibits 11.3.8 and 11.3.9 show examples of isomers for σ_Q and σ_I, respectively, the latter being a saddle-shaped space with the saddle point at $a = k = 0.4$. An example of a cost function that follows objective 3 is shown in Exhibit 11.3.10 (where a sudden increase, ΔD, in demand is considered, the step being equal to the SD of demand σ_D), and in general one finds that the optimal solution for any of the three alternative objectives listed lies either at $a = k$ or sometimes at $k = 0$ (or alternative $a = 0$). This result substantially reduces the search (which often follows a simulation approach) required to establish the optimal value of the decision parameters a and k, since only a range of values for one parameter needs to be scanned.

The fact that the DE model is essentially reduced to determining a single decision variable is an obvious advantage for the sake of simplicity; also, the model takes account of the lead time L (and furthermore can be used to assess the effect of changing its value), a feature that is missing from the HMMS model.

Exhibit 11.3.8 Isomers of σ_Q (where L = 1)[a]

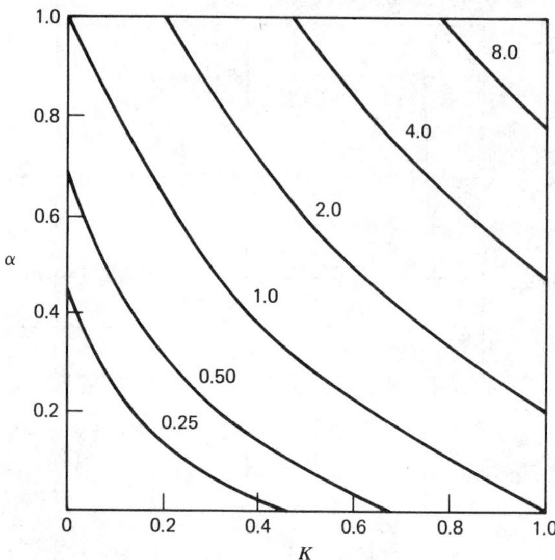

[a]Reference 17.

But a direct comparison with HMMS is difficult to make, as the two models are based on different structures and assumptions. The DE model dispenses with the need to determine the numerous cost parameters required by the HMMS model and avoids many of its problems, which were discussed earlier. On the other hand, for the DE model one needs to specify the two or three weighting parameters used in objective functions, such as in equations 19 and 20; these parameters imply costs, and yet are difficult to derive directly from conventional accounting procedures. Another shortcoming of the DE model is that it incorporates the demand forecast for the forthcoming period, but not for subsequent periods.

Exhibit 11.3.9 Isomers of σ_I (where L = 1)[a]

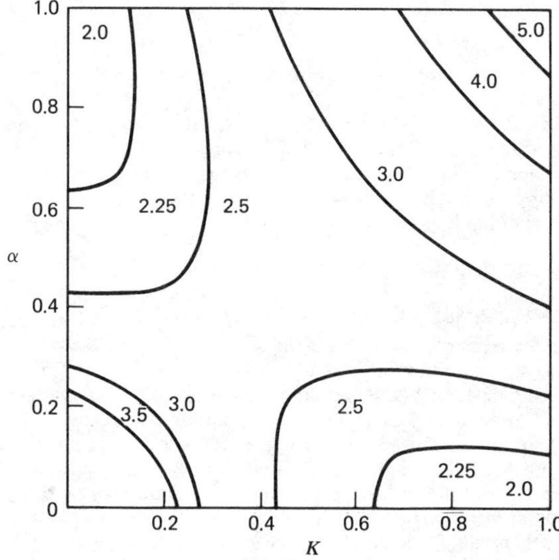

[a]Reference 17.

Exhibit 11.3.10 The Cost Function $C = a\sigma_I + b\sigma_Q + c_\gamma$ (for $a = 2, b = 1, c = 0.5, L = 3, \Delta D = \sigma_D$)[a]

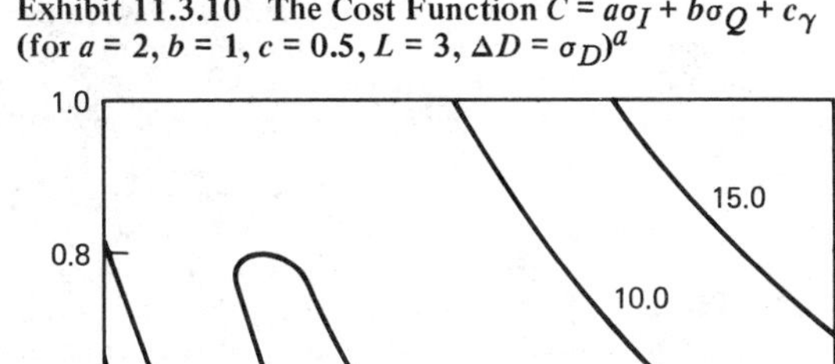

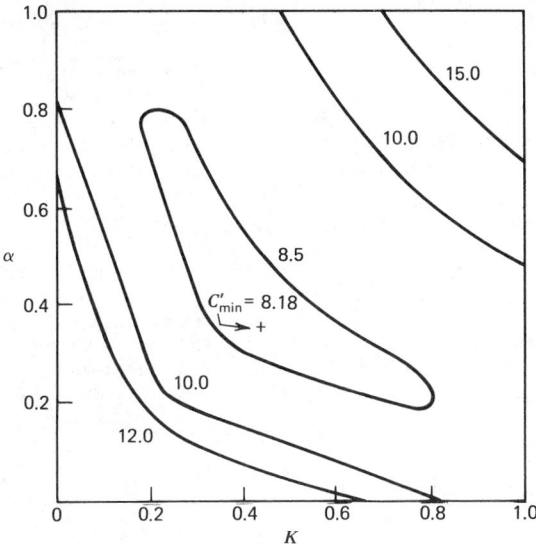

[a]Reference 17.

11.3.5 MANAGEMENT COEFFICIENTS APPROACH

The Model

This method, which is attributed to Bowman,[18] seeks to establish coefficients that describe management's decision making behavior in a given environment. Using statistical regression analysis, the scheduling rules are fitted to simple expressions, such as

$$P_t = a_1 F_t + a_2 W_{t-1} - a_3 I_{t-1} + a_4 \qquad (21)$$

$$W_t = b_1 F_t + b_2 W_{t-1} - b_3 I_{t-1} + b_4 \qquad (22)$$

where the a's and the b's are derived from the regressions. The assumption here is that management's decisions are in the main governed by the current work force, by the forecast for demand in period t, and by the inventory level.

There are, of course, many alternative multiple regression models that may be examined. For example, an attempt to account for the forecast in period $t + 1$ as well may take the form

$$P_t = a_1 F_t + a_2 F_{t+1} + a_3 W_{t-1} - a_4 I_{t-1} + a_5 \qquad (23)$$

Indeed, extended expressions such as equations 9 and 10 may be tried, although Bowman remarks that, in the cases he studied, such regressions gave poor results because of the high correlation between the forecast estimates.

The axiomatic notion of this approach is that, as Bowman puts it,[18] "experienced managers are quite aware of and sensitive to the criteria of a system" and that managerial decisions are "more erratic than biased." He proceeds to argue that managerial decisions are basically sound and that what is needed is to eliminate the "erratic" element by making them more consistent.

Comments

Criticisms of this approach are manifold:

1. The form of the multiple regression function is arbitrary, and regression of past decisions over a narrow range may lead to erroneous conclusions.

2. The test of goodness of fit is self-defeating. If the fit is poor, the regression model is rejected, because it does not sufficiently describe the behavior pattern of the manager; if the fit is good, the model is accepted. But the better the fit and the more confidence we have in the model, the smaller is the erratic element in managerial decisions and hence the smaller the potential benefit from removing inconsistencies through the use of this approach.

3. The regression model relies on decisions made by a manager or by a group of managers. Changes in personnel may render the model invalid.

4. Last, but not least: The fundamental assumption that managers are good decision makers and that what is needed is just to eliminate inconsistencies in their behavior is rather questionable. Such a philosophy is tantamount to suggesting that analysts should endeavor to make contributions only of a second-order magnitude, whereas many believe that management scientists should pose more penetrating questions about the performance of an industrial enterprise.

11.3.6 MATHEMATICAL PROGRAMMING

In the mathematical programming approach, the demand forecasts are assumed to be accurate, so that a production plan for the complete horizon of T periods may be determined at the outset. The plan may then be updated period by period as further information becomes available on the effect of actual demand on the inventory level and as forecasts for future demand are updated. Several alternative mathematical programming models can be constructed, depending on the complexity of the assumptions that are made, and a few examples are discussed here.[3]

Model 1: Production and Inventory Costs

In addition to notation introduced later, let

P_t = production rate in period t (where $t = 1, \ldots, T$) $= x_t + y_t + z_t$ (24)
x_t = production rate in regular time in period t
y_t = production rate in overtime in period t
z_t = production rate from subcontracting (or from third shift working) in period t
c_1 = cost per unit in regular time
c_2 = cost per unit produced in overtime
c_3 = cost per unit subcontracted
I_t = inventory level at end of period t
c_4 = cost per unit of closing inventory level
F_t = forecast demand for period t
S_t = shipment or sales in period t

Thus the costs of production are piecewise linear, and if

$$c_3 > c_2 > c_1$$

the solution will ensure that regular time is fully used before overtime is employed and, similarly, that no subcontracting takes place before all available overtime is used. The cost parameters c_1, c_2, c_3 are assumed here to be time independent, but the model can equally handle the case where each of these parameters has different values in different periods.

The inventory level at the end of period t is computed by equation 5, and an initial inventory I_0 at the beginning of the planning horizon can be accounted for. Thus

$$I_t = I_0 + \sum_{i=1}^{t} (P_i - S_i) \tag{25}$$

in the case of backlogged demand, and the inventory holding costs are $c_4 I_t$ for period t.

The total cost function then becomes

$$C = \sum (c_1 x_t + c_t y_2 + c_3 z_t + c_4 I_t) \tag{26}$$

Several constraints may have to be observed, such as

$$A_t^x \leqslant x_t \leqslant A_t \tag{27}$$

$$B_t^x \leqslant y_t \leqslant B_t \tag{28}$$

$$C_t^x \leqslant z_t \leqslant C_t \tag{29}$$

where A_t^x and A_t are the lower and upper bounds, respectively, for the production rate in regular time in period t and the other limits, B_t^x, B_t, and so on, are similarly defined. The lower bounds may often be zero, but sometimes certain prior commitments in terms of a minimum level of employment or a minimum undertaking with contractors may result in lower bounds having given positive values. The upper bounds simply reflect the production capacities of the available facilities, and account can be taken of the changing level of the work force by making the upper bounds period dependent.

Stock Runouts

So far in this linear programming formulation no reference has been made to stock runouts, and it is not difficult to see that the minimum cost solutions for equation 26 are achieved by producing quantities equal to the lower bounds. There are two ways to guard against excessive runouts:

1. **Model 1.1.** A constraint is imposed on the inventory level, for example, in the form

$$I_t \geqslant I^x \tag{30}$$

where I^x is the absolute lowest level that must be maintained. (In this context I^x may even be negative; also, if it is desirable to allow I^x to vary from period to period, then I_t^x will substitute I^x in equation 30.) Alternatively

$$I_t \geqslant k\,F_{t+1} \tag{31}$$

where k is a constant. This expression suggests that the closing stock of period t must be at least a given proportion of the forecast demand in the next period. The linear programming model may therefore be defined as follows: Find the values of x_t, y_t, and z_t that will minimize the objective function in equation 26, subject to the constraints and requirements expressed in 25, 27 through 30, or 31.

2. **Model 1.2.** A penalty is imposed on runouts, say c_5 per unit. The inventory level I_t is replaced by

$$I_t = u_t - v_t \tag{32}$$

where u_t is the closing physical stock at the end of period t and v_t is the amount of runout; both u_t and v_t are nonnegative. Thus, if $u_t > 0$, then $v_t = 0$, and if $v_t > 0$, then $u_t = 0$. The inventory cost term in the cost equation 26 is now replaced by

$$\text{inventory costs} = c_4 u_t + c_5 v_t \tag{33}$$

and we now regard u_t and v_t as decision variables. Since $c_4 \neq c_5$, and since there are no upper bounds to u_t and v_t, the solution to the linear program will include either u_t or v_t for each time period.

Model 2: The Work Force

In model 1 the work force does not appear as a decision variable. If an assumption is made that a certain relationship must be maintained between the available work force and the possible rate of production in regular time, then the model needs to be modified accordingly. Let W_t be the work force in period t and let

$$x_t \leqslant \alpha_1 W_t - \alpha_0 \tag{34}$$

where α_0 and α_1 are nonnegative parameters; α_1 is the output per worker, and α_0 is proportional to some minimum level of staffing that may be required before production can commence. The right-hand side of expression 34 describes the maximum possible rate of production with the available work force W_t.

New let

c_1 = cost per worker per period for regular time
c_6 = cost for hiring a worker
c_7 = cost for layoff per worker when the work force is reduced

Similar to what was done in equation 32, we introduce

$$W_t - W_{t-1} = U_t - V_t \tag{35}$$

where U_t is the number of people hired to increase the work force to W_t and V_t is the number of people fired to reduce the work force to W_t. Here, too, U_t and V_t are nonnegative, and if $U_t > 0$, then $V_t = 0$, whereas if $V_t > 0$, then $U_t = 0$. The total cost function now becomes

$$C = \sum (c_1 W_t + c_2 y_t + c_3 z_t + c_4 u_t + c_5 v_t + c_6 U_t + c_7 V_t) \tag{36}$$

where the penalty for runouts is assumed as in model 1.2, and the constraints are given in 27 through 29, 32, 34, and 35. A variation of model 2 may be constructed to account for costs incurred by increasing or decreasing *production* (rather than the work force), where an expression similar to equation 35 is introduced, and the total cost function is then similar to equation 36.

Comments on Linear Programming Models for Production Smoothing

The fundamental assumption in these models that cost functions are linear may be challenged, and where such an assumption cannot be sustained, there is a need to construct nonlinear programming models, which are more intricate and take much longer to solve than linear programming problems.

A more serious issue is the tacit assumption that we are dealing with a deterministic problem: All demand forecasts are treated as if they were accurate, and they all carry equal weight. Admittedly, a procedure for updating the solution to the linear programming model, say every period, helps to take account of new information, but the fact remains that the production level for the forthcoming period can be significantly affected by demand forecasts for the more distant future, even though such forecasts are far less reliable than the forecast for the immediate future. This difficulty can be alleviated to a certain extent by reducing the planning horizon, but the fundmental criticism of this point remains valid.

The planning horizon also raises the problem of the end conditions. If none are stated, the model will produce a solution where the final inventory will be zero at the end of the planning period; it may also drastically reduce resources (such as manpower) toward the end of the horizon in an attempt to minimize overall costs. This "end effect" can be eliminated either by specifying minimum end conditions for inventory and production that must be met or by planning for a longer horizon than that considered for implementation, so that the end effect is thereby diffused.

It is, of course, possible to subject expected costs for future periods to a discounting procedure, so that both the problems of equal weighting for forecasts and the end effect become less significant, but in that case the final results may well depend on the value of the discounting factor.

The advantage of linear programming for solving the production smoothing problem lies in the simplicity of the model and in the comparative ease with which it can handle constraints on availability of resources. Also, it provides some useful information about shadow prices to indicate profitable ways in which limited resources may be expanded.

It is precisely because of these advantages that, in cases where cost parameters are not linear, attempts are often made at linearization. There are many forms that such attempts can take (e.g., see Hanssmann and Hess[19]); one interesting approach is to aim at minimizing the value of nonlinear cost parameters by a goal programming formulation.[20] It is impossible, however, to make general statements about the effectiveness of such methods, whose performance appears to depend on the degree of nonlinearity involved, so that the justification for adopting any particular approach must rest with a comparative examination in any given case.

11.3.7 PRODUCTION SWITCHING

The Switching Model

In many production systems, adjustment of the production level incurs two types of costs: a discrete fixed cost associated with the fact that a change in level is to take place and a variable cost proportional to the amount by which the level is adjusted. If any production level may be considered (with upper constraints reflecting plant capacity), a mathematical model can be formulated with zero-one variables, where in each period such a variable takes the value of 0 when no change in production is introduced and the value 1 when a change is made.

Another approach to solving the problem for the stationary demand case is based on dynamic programming,[21,22] resulting in a solution illustrated in Exhibit 11.3.11, where P and I are the production level and inventory, respectively. The shaded area shows the domain where no action is called for, but when a point representing the current production and inventory lies outside the shaded area, the production is adjusted by the amount needed to bring the point to the envelope of this area. The computations involved in this method are usually time consuming, and it is not amenable to handling nonstationary demand distributions.

If production is confined to a relatively small number of prescribed levels (so that adjustment in production is achieved by given discrete steps), experience of performance and scheduled activities at each level provide good opportunities for controlling costs and minimizing the effects of change.

Exhibit 11.3.11 Dynamic Programming Solution to the Production Smoothing Problem (Beckman's Method)[a]

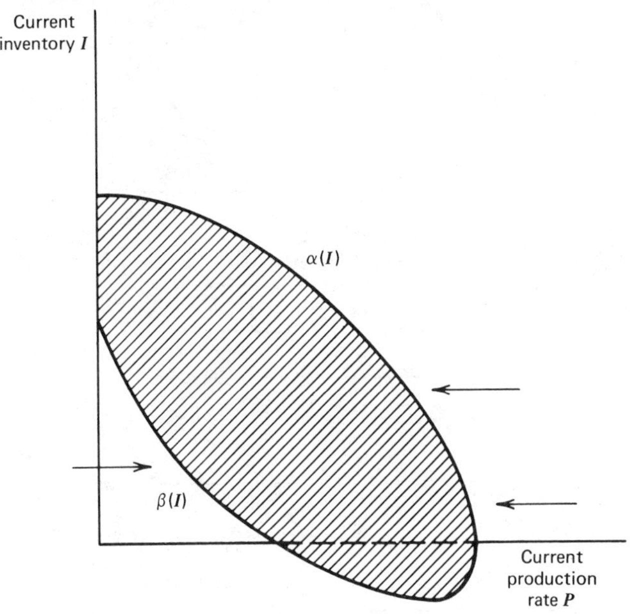

[a]Reference 21.

The total cost model is then expressed as[23]

$$C = \sum [c_1 P_t + c_2 z_t + c_3 I_t \, (I_t > 0) + c_4 I_t \, (I_t < 0)] \tag{37}$$

where P_t and I_t are the production rate and inventory level in period t and

c_1 = unit cost of production
c_2 = production changeover cost
c_3 = unit carrying costs for leftover inventory (when $I_t > 0$)
c_4 = unit runout cost (when $I_t < 0$)
z_t = number of step changes made in period t

The control mechanism is illustrated by the simple two-production-level case, described in Exhibit 11.3.12. The levels H and L represent the high and low production rates at which the system can operate. The inventory level is monitored, and when it crosses a control level a, this determines when production is to switch from H to L, and vice versa. A more elaborate control mechanism would involve two control levels, a and b (when $a > b$), and switching from L to H will take place when the inventory level crosses the control limit a from below. The rationale for such switching policies is similar to the two-bin or (s, S) inventory control system. In the three-production-level case, where three production rates, H, N, and L (with $H > N > L$), are possible, three control limits, a, b, and c, may be defined (where $a > b > c$), and the production rule will then follow the operating instructions

$$P_t = \begin{cases} H & \text{if } I_t \text{ passes } c \\ N & \text{if } I_t \text{ passes } b \\ L & \text{if } I_t \text{ passes } a \end{cases}$$

Clearly, the introduction of multiple production/control levels provides more degrees of freedom and added protection to the system when demand is very volatile or unpredictable.

An effective way of analyzing such a model for any given demand pattern is by simulation. An example for results obtained in one case with two production levels in a food canning plant[24] is given in Exhibit 11.3.13. Each of the points shown is based on 1000 simulation runs. In Exhibit 11.3.13a the lower production level L = 0, and the three curves correspond to the upper production level H = 300, 400, and 500, respectively. The control level a is shown on the abscissa scan-

Exhibit 11.3.12 The Two-Production-Level Case: (*a*) Single Control Level and (*b*) Two Control Levels

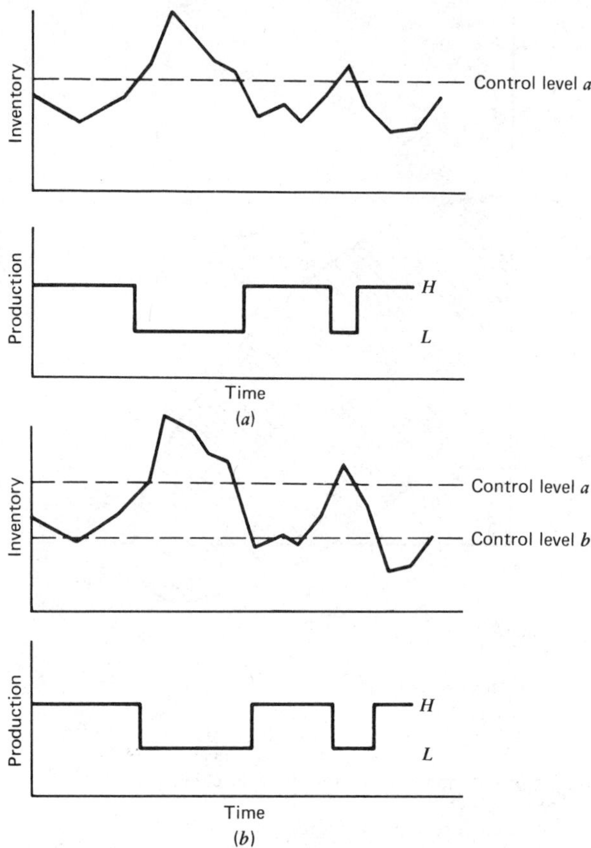

ning the range from L to H. The costs for changing the production level consist of a fixed cost per change plus a cost proportional to the magnitude of change, but the costs of inventory holding are ignored here. The demand is sampled from a normal distribution (truncated at zero), with lead times assumed for processing as well as for the supply of raw materials. A horizontal dotted line in the exhibit shows the costs incurred when no production capacity restrictions are imposed and when the quantity scheduled for production can be adjusted to attain an optimal solution for the inventory system on its own (designed to meet demand with minimum inventory, provided that no runouts are allowed). Such a policy would incur frequent adjustments in production levels (and thereby high changeover costs), and clearly these costs can be reduced by a switching method. The results in this case also suggest that costs decline when the control level a approaches either L or H. Exhibits 11.3.13*b* through 11.3.13*d* show similar results for $L = 50$, 100, and 150.

Comments on the Switching Model

When production operations can preferably be carried out at certain perdetermined levels (such as the opening or shutting of a production line), it is not appropriate to treat the production level as a continuous variable, and the model described here is a convenient way for analyzing alternative values of control parameters for switching purposes. In addition, the model can be used to determine desirable fixed production levels (if a decision to that effect is open to the management) and to carry out sensitivity analyses.

Since this is a simulation model, there are no constraints on the type of demand pattern that can be handled (it may involve a series of deterministic demand forecasts or sampling from a stationary or nonstationary distribution), and the objective cost function given in equation 37 may be easily extended as appropriate, for example, to include inventory holding costs of raw materials. Alternative cost functions may also be considered. For example, a function similar to equation 19 may be constructed to incorporate the SDs of inventory and work in process, respectively; simulation

Exhibit 11.3.13 Production Changeover Costs Against Control Levels, Where (*a*) $L = 0$, (*b*) $L = 50$, (*c*) $L = 100$, and (*d*) $L = 150$[a]

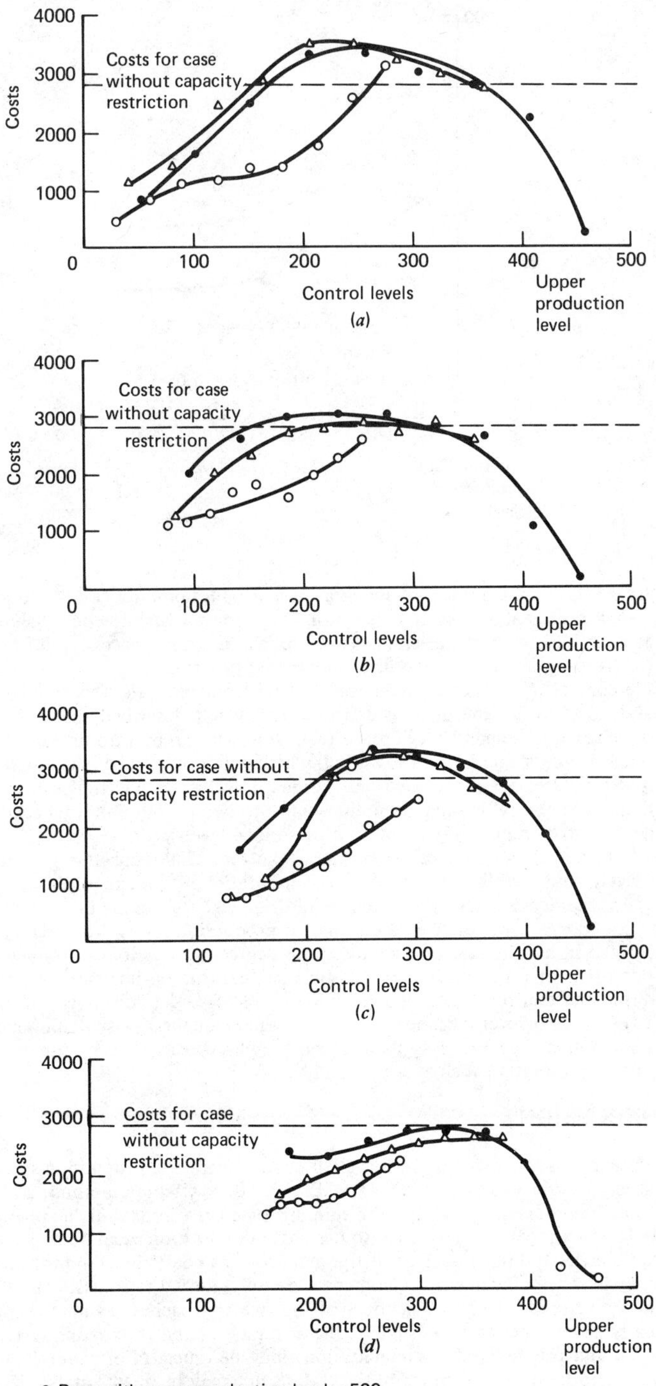

[a]Reference 24.

Exhibit 11.3.14 Standard Deviation of Inventory and Work in Progress, Where $L = 100, N = 300$, and $H = 500$[a]

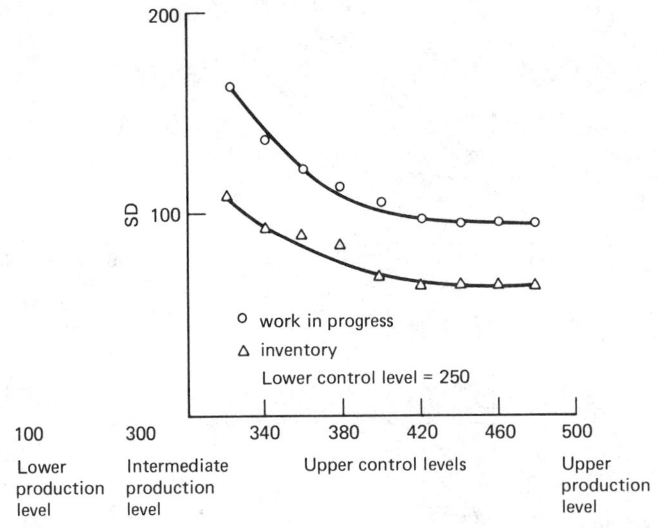

[a]Reference 24.

results for these SDs in an example involving a three-production-level case[24] are shown in Exhibit 11.3.14, and such results allow the cost function to be computed for various values of the control parameter. Mean values for inventories of raw materials, work in process, and finished goods may also be derived from simulation and fed into similar cost functions.

For a simple control procedure involving one or two control levels, this simulation model is an effective analytical tool, but the obvious danger in the system described in Exhibit 11.3.12 is that, when a drastic change in demand takes place (e.g., following an upward or downward trend), production will lock into a single high or low level, which may prove to be inadequate in dealing with a persistently depleted or soaring stock level. To guard against such extreme situations, additional monitoring is needed for telling management when the two-production-level system ceases to be effective and needs to be replaced (e.g., by new prescribed levels).

Such an elaboration of the control system may take the form of specifying multiple production and control levels, and clearly the amount of computational work required for such simulation models rises very rapidly, particularly when, in addition to determining the optimal control levels, investigations are carried out for several alternative production levels. The greater the number of levels incorporated in a switching system, the more degrees of freedom introduced and hence the greater the potential for finding a lower-cost solution than the optimal solution for the two-level case. However, as the control system increases in complexity, the advantages of simplicity imbedded in the two- or three-level methods are eroded, the frequency of switching is expected to increase, and management becomes less able to gear the plant to an effective use of resources during a relatively shorter production spell at any given level.

11.3.8 CONCLUSION

Each of the five approaches reviewed in this chapter for solving the production smoothing problem has certain strengths and weaknesses. The HMMS rule, the switching method, and the linear programming model are concerned mainly with minimizing costs (or maximizing profit in some linear programming formulations); the DE rule also has cost minimization as an objective (but of a different structure than the HMMS model); and the management coefficients approach strives to maintain consistency in scheduling decisions, and costs are only quoted as a justification. Because of the differences in the assumptions and objectives of the five approaches, a direct comparison is difficult, particularly since other factors–such as the accuracy of the forecasting system employed by the organization and problems of implementation–may be crucial. For example, if cost functions are indeed found to be quadratic in form, the HMMS rule may be most suitable; if the objective of the production scheduler is to strike a balance between fluctuations in production and fluctuations in inventory, while avoiding a direct evaluation of many of the cost parameters considered in the HMMS model, then the DE rule or the switching method may be effective and simple to use; the management coefficients method appears to be the least disruptive to implement and also perhaps the least rewarding; and linear programming models are powerful, for the most part quite satisfactory in spite of the linear cost assumptions, but vulnerable when demand forecasts are relatively inaccurate. The switching method is flexible in the type of assumptions and cost functions that can be entertained and is based on a simple control mechanism whose performance can be repeatedly and closely monitored by simulation. But this method is effective only when the number of production and control levels is relatively small; otherwise the amount of computation involved rises very rapidly.

The five approaches described here have been developed in response to real problems, and they have all been successful in their own way in terms of the particular production environment in which each was considered. They demonstrate that aggregate production scheduling can be approached from different angles and that some approaches involve reliance on elaborate data banks, whereas others attempt to reduce the scheduling problem to a small number of significant variables in the knowledge that detailed scheduling takes place at the subsequent disaggregation exercise. The first task of the scheduler is to make a judgment as to which approach is most appropriate to adopt for his or her own system.

REFERENCES

1. S. EILON, *Elements of Production Planning and Control*, Macmillan, New York, 1962.
2. S. EILON, *Aspects of Management*, Pergamon, Oxford, England, 1977.
3. S. EILON, "Five Approaches to Aggregate Production Planning," *AIIE Transactions*, Vol. 7, 1975, pp. 118–131.
4. S. EILON, "Production Scheduling," in K. B. Haley, Ed., *OR '78*, North-Holland, 1979, pp. 237–266.
5. T. C. HU and W. PRAGER, "Network Analysis of Production Smoothing," *Naval Research Logistics Quarterly*, Vol. 6, 1959, pp. 17–23.
6. M. ANSHEN, C. C. HOLT, F. MODIGLIANI, J. F. MUTH, and H. A. SIMON, "Mathematics for Production Scheduling," *Harvard Business Review*, Vol. 36, 1958, pp. 51–58.
7. C. C. HOLT, F. MODIGLIANI, J. F. MUTH, and H. A. SIMON, *Planning Production, Inventories and Work Force*, Prentice-Hall, Englewood Cliffs, NJ, 1960.
8. HOLT et al., *Planning Production, Inventories and Work Force*, p. 53.
9. HOLT et al., *Planning Production, Inventories and Work Force*, p. 165.
10. C. VAN DE PANNE and P. BOSE, "Sensitivity Analysis of Cost Coefficient Estimates: The Case of Linear Decisions Rules for Employment and Production," *Management Science*, Vol. 9, 1962, pp. 82–107.
11. E. A. SILVER, "A Tutorial on Production Smoothing and Work Force Balancing," *Operations Research*, Vol. 15, 1967, pp. 985–1010.
12. E. S. BUFFA, *Operations Management: The Management of Production Systems*, Wiley, New York, 1976.
13. R. PETERSON, "Optimal Smoothing of Shipments in Response to Orders," *Management Science*, Vol. 17, 1971, pp. 597–607.
14. M. ABRAMOVITZ, *Inventories and Business Cycles With Special Reference to Manufacturer's Inventories*, National Bureau of Economic Research, New York, 1950.
15. C. C. HOLT, "Dynamic Pricing and Economic Instability," *Management Science*, Vol. 17, 1971, pp. 609–611.
16. S. EILON, "On Smoothing Shipments—A Comment," *Management Science*, Vol. 17, 1971, pp. 608–609.
17. D. P. DEZIEL and S. EILON, "A Linear Production-Inventory Control Rule," *The Production Engineer*, Vol. 46, 1967, pp. 93–104.
18. E. H. BOWMAN, "Consistency and Optimality in Managerial Decision Making," *Management Science*, Vol. 9, 1963, pp. 310–321.
19. F. HANSSMANN and W. HESS, "A Linear Programming Approach to Production and Employment Scheduling," *Management Technology*, Vol. 1, 1960, pp. 46–51.
20. D. A. GOODMAN, "A New Approach to Scheduling Aggregate Production and Work Force," *AIIE Transactions*, Vol. 5, 1973, pp. 135–141.
21. M. J. BECKMAN, "Production Smoothing and Inventory Control," *Operations Research*, Vol. 9, 1961, pp. 456–467.
22. E. S. MILLS, "The Theory of Inventory Decisions," *Econometrica*, Vol. 23, 1955, pp. 46–66.
23. D. ORR, "A Random Walk Production-Inventory Policy: Rationale and Implementation," *Management Science*, Vol. 9, 1962, pp. 108–122.
24. J. ELMALEH and S. EILON, "A New Approach to Production Smoothing," *International Journal of Production Research*, Vol. 12, 1974, pp. 673–681.

CHAPTER 11.4

Inventory Management and Control

URBAN WEMMERLÖV

University of Wisconsin

11.4.1 INTRODUCTION

This chapter covers inventory management and control from two different perspectives. One is general and deals with the purpose, function, and consequences related to keeping inventories. In this context inventories in both producing and nonproducing companies are considered from a dual organizational and financial viewpoint. The other perspective is more narrow and concentrates on systems and procedures suitable for use in controlling inventory items subject primarily to "independent demand." For the purpose of inventory control, an independent demand item is considered to be unrelated to any other item in the system. This situation occurs basically only for organizations where no manufacturing takes place, such as for retailers and distributors. (To emphasize this, traditional inventory management is often called "distribution" inventory management.) For control of material subject to "dependent demand," for example, a component that derives its demand from an assembled product of which it is a part, the reader is referred to Chapter 11.6. As is seen below, however, the dependent/independent demand principle* for selecting inventory control systems is not always followed in practice.

11.4.2 IMPACT ON COMPANY OPERATIONS

The Financial Significance of Inventories

Inventories exist in all kinds of companies, with the exception of extreme service producing organizations where no product is a part of the service package. (One example would be a barber who offered nothing but haircuts.) They play a major role in determining both a company's internal effectiveness, that is, how efficiently the manufacturing operations can be carried out and production goals met,† and its external effectiveness, that is, how well the company serves the marketplace. This contribution, however, does not come without sacrifices (i.e., costs). The operational impact of inventories is discussed in the next section. The financial aspects are looked at here.

The annual income statement and the balance sheet are two prime sources of information for judging the economic performance of a company (see Exhibits 11.4.1 and 11.4.2). The income statement displays the company profit for the last year. Inventories do not appear explicitly on this report, but various costs related to keeping inventory are included under "cost of goods sold." (More details on these costs are discussed later in this chapter under "The Management of Inventories.") Profit, in isolation, is not a completely satisfying performance measure. It is also of interest to know the amount of capital the company needed, on the average, to generate the reported profit. This information can be found in the balance sheet. That a company generated $200,000 in profits in a particular year might seem impressive by itself, but the knowledge that this was the return on an average investment of $40 million will change the judgment (ROI = 0.5%). It is obvious that a company that generates the same amount of profit using only $5 million worth of assets is far better off (ROI = 4%).

Contrary to what is the case for the income statement, inventories (i.e., the estimated value of the total ending inventory for the year) appear on the balance sheet. They usually make up a substantial portion of a company's total assets, ranging from 10 up to 45%, depending on what type of industry the company represents. Thus, the effect of reducing inventories results in decreases in the cost of goods sold and in lower total capital investment. If an inventory reduction can be

*Originally formulated by J. A. Orlicky (see reference 1).

†When illustrating general concepts related to inventory management in this chapter, the most complex situation (i.e., inventories in a manufacturing environment) will often be chosen.

Exhibit 11.4.1 Simplified Income Statement (in $1000s)

Net sales	$5000	
Cost of goods sold	4437	
Gross profit		$563
Operating expenses	403	
Depreciation	24	
Interest	14	
		$441
Operating income		$122
Federal taxes		61
Net income after taxes		$61

achieved *without* sacrificing other performance measures such as customer service and production efficiency, then a company can foresee both increased profits and a higher ROI. A comprehensive picture of the relationship among inventory, profit, and ROI can be found in the Du Pont chart (see Exhibit 11.4.3), a well-known tool for financial analysis.[2]

Because of increased competition, shorter product lifecycles, increased product variety, a higher cost of capital, and, not the least, the emergence of high-speed computers, more and more firms in the last 10 years have come to realize that inventory management represents an avenue with substantial potential for increasing profits, lowering costs, and releasing scarce capital. The change has been especially dramatic for manufacturing companies with the emergence of a powerful technique, called material requirements planning, used to schedule production and control inventories (see Chapter 11.6). It should be pointed out that inventory management traditionally has not been viewed as being very important by top management, and crucial decisions have often been delegated to lower levels in the organization. Top management can and should,

Exhibit 11.4.2 Simplified Balance Sheet (in $1000s)

Assets	
Cash	$62
Accounts receivable	314
Inventory	341
Total current assets	$717
Land and buildings	288
Machinery	342
Other fixed assets	107
	$737
Less accumulated depreciation	126
Net fixed assets	$611
Total assets	$1328
Liabilities and Equity	
Accounts payable	$284
Other current liabilities	26
Total current liabilities	$310
Mortgage	184
Long-term debt	$184
Common stock	676
Retained earnings	158
Total stockholders' equity	$834
Total liabilities and equity	$1328

Exhibit 11.4.3 Du Pont Chart[a]

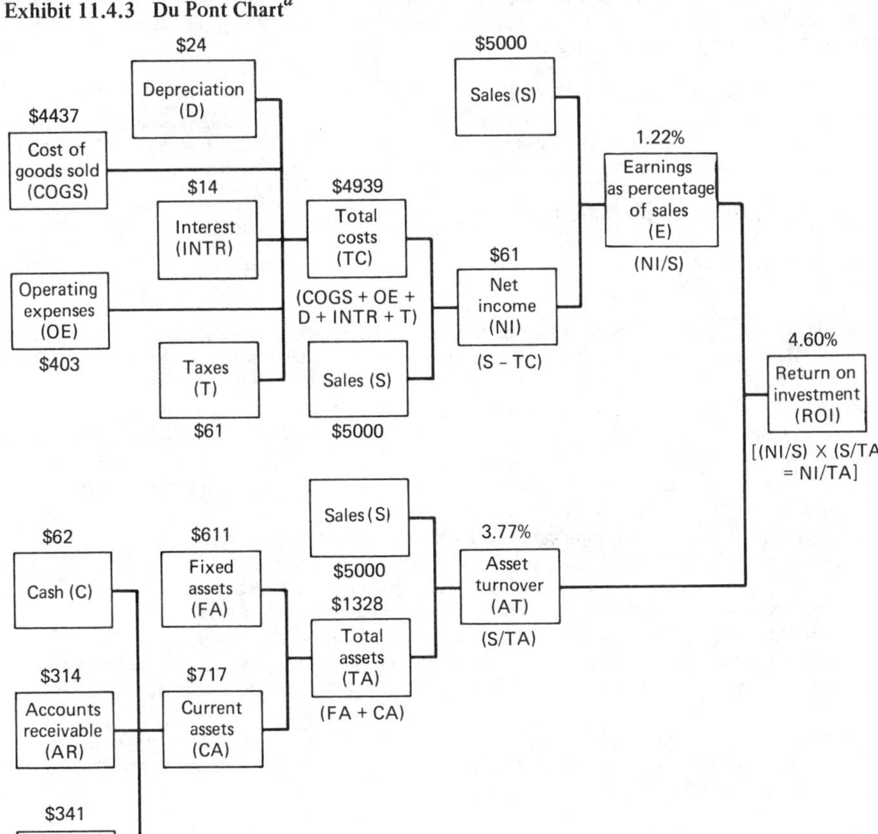

[a]Amounts in $1000s.

however, take an active interest in this area. The result will show up on the income statement and the balance sheet.

The Operational Significance of Inventories

Irrespective of whether a certain inventory is planned or not, it always serves this basic function: it *decouples* supply from demand. This general function is always performed by inventory, even if the purpose of its creation differed from this. Since reasons for keeping inventory are treated in more detail in the next section, one example will suffice here.

It is common for a company to buy large amounts of material at a time, if, in this way, it can lower the costs through price discounts, less expensive transportation, and so on. With the material on hand at the company, the buyer and the seller are decoupled from each other. That is, since the inventory is larger than what the company planned to use for the immediate future, the use of the material can now deviate from what was planned, without making the company *directly* dependent on its supplier. The inventory that was created out of economical reason will act as a buffer and give the company time to place a new order with its vendor.

It should be stressed that inventories in manufacturing environments, both planned and unplanned, are primarily a function of how the production is planned and carried out. This has led one of the experts in the field of MRP to state the following[3]:

> *I have banished the term "inventory management" . . . altogether, because I now reject the very notion that manufacturing inventories . . .* are *managed,* should *be managed,* or can *be managed in their own right.*

What the author points out is that inventory management might be a function in itself in a non-producing organization, whereas in a manufacturing company it is very much subjugated to the planning and scheduling of the production.

11.4.3 TYPES OF INVENTORIES

Inventories can be classified in various ways, depending on their specific purpose or their location in the production-distribution process. Some of the more useful concepts are introduced here.

A Classification Based on Materials Flow

The first classification is based on the state that the inventoried material is in. Bicycle manufacturing is used as an example.

Raw Materials. Refers to purchased material that needs further fabrication at the company before it can be used in assembly or sold to customers. Includes semifinished purchased components. Example: iron and aluminum bars to be used for bicycle frames.

Finished Parts. Encompasses raw material that has undergone some kind of manufacturing process at the company and is ready for use in assembly operations. Example: cut bars ready to be put together to make the bicycle frames.

Component Parts. This class covers purchased material that needs no further processing before it is used in assembly. Examples: saddles, tires, light fixtures.

Subassemblies. Refers to assembled material that represents a fairly large or important part of the complete product. Many companies make subassemblies out of parts and components and store them before final assembly takes place. This gives them a larger flexibility in responding to the market in a short time, since often many different products can be made from a limited number of subassemblies. Example: complete, unpainted bicycle frames.

Work in Process. Any material that is being worked in the factory, waiting to be worked on, waiting for transportation after being processed, or being transported inside the plant. Example: a bicycle being assembled at a work station.

Finished Goods. Includes the completed products that are ready for delivery to customers. Note that spare parts or other special incomplete items come from the finished parts, component parts, or subassemblies inventory.

These categories are, of course, relevant only for manufacturing companies and are not applicable to retailers or other types of businesses where no conversion of material takes place. Further, the categories deal only with material that, in one form or another, is directly included in the final product. Material needed to support the manufacturing process, such as lubricants and cutting tools, is usually included in a general category called supplies. Material used for packaging and distributing the final products to the customers can also represent a separate category.

Many times companies keep track of only three categories: raw materials, work in process, and finished goods. Included in the first class, then, is all purchased material, whereas work in process represents everything else that is not finished goods. The classification scheme is often used for accounting purposes as a guide to tracking down the total inventory a company is holding. The distribution of the inventory investment over these three classes depends on the manufacturing process and the strategic decisions made by the company. Companies that are very much directed toward serving the market place, usually consumer goods industries, might have a large inventory of finished goods ready to be shipped. Companies that make complex products to order, with long manufacturing times, will have their inventory investment dominated by work in process, and so on.

Distribution Versus Manufacturing Inventory

Another distinction can be made between manufacturing and distribution inventories. A distribution inventory consists of products that are sold to customers. A finished goods inventory at a company thus represents a distribution inventory. If the company has an extensive distribution system, covering a large geographical area, goods can also be stored in distribution centers and field warehouses. It should be noted that a distribution inventory can contain material that for the manufacturer represents finished goods, but for the buyer simply represents input material for the manufacturing process (i.e., raw material or components). All inventory kept by a nonmanufacturer, such as a supermarket or a bookstore, can in the same way be called distribution inventory.

Manufacturing inventory, on the other hand, is the collective name for all other types of inventoried goods in the production-distribution chain. It therefore includes everything from raw materials to subassemblies. The reason for separating distribution and manufacturing inventories is that different techniques should be applied when managing them. As has been pointed out before, this chapter concentrates on techniques most suitable for distribution inventories where the demand for one item is independent of the demand for the next. Chapter 11.6 deals explicitly with planning techniques for manufacturing inventories.

Transit and Organizational Inventory

Inventory can further be classified as "transit" inventory (also called "movement" or "pipeline" inventory) and "organizational" inventory.* The first class simply contains all material being transported either inside or outside the company. The significance of transit inventories should be apparent. Consider a company that sells its products on the European market. If it chooses to transport the goods by ship, the total customer delivery time might be 5 weeks. During this time the goods under shipment (in the pipeline) represent an inventory investment that cannot be utilized and that costs the company money. If, on the other hand, the option of air freight is available, the delivery time can be reduced to, say, 3 days. The cost of transportation will, of course, increase this way, and management has to decide whether this increase is outweighed by the speedier delivery and the reduced cost related to the now much smaller inventory investment. Transit inventories also exist in manufacturing, for example, when material is being moved from one work station to the next. Efficient materials handling, such as using conveyor lines instead of forklifts, can reduce these kinds of transit inventories.

Organizational inventories are created and further classified out from the specific purpose they serve. As pointed out earlier, inventories always decouple supply and demand. A finished goods inventory will, for example, make production less dependent on variations in customer demand. One can therefore say that inventories reduce the organizational effort needed to balance supply and demand exactly. Or, put another way, a reduction in inventory often requires more efficient planning and control to make the operation function as smoothly as it did before.

Organizational inventories can be subdivided into three categories: cycle stock, safety stock, and anticipation stock.

Cycle Stock

This type of inventory, also called "lot size inventory," is created whenever material is purchased or produced in a quantity larger than what is needed for immediate consumption. The name "cycle" implies that the inventories are replenished at regular intervals. The average size of the inventory over time will be determined by the demand and by how much is ordered each time.

Safety Stock

Sometimes labeled "buffer" or "fluctuation" stock, this type of inventory is created to act as a reserve stock in situations of unanticipated events. Fluctuations can occur for both expected supply and expected demand at any point in the production-distribution chain. Safety stocks can, for example, be built up to handle deviations from expected sales of a product, to act as a buffer between two work stations in cases of machine breakdown, to smooth the effect of operator variability along an assembly line, or to reduce the possibility of production stoppage due to excessive scrap in supplier deliveries.

Anticipation Stock

This type refers to inventory that is built up ahead of the time of demand because of limited or reduced supply. For example, if a company is selling a seasonal product, such as Christmas cards or ski equipment, the demand for production capacity can be smoothed if the product is not produced over a short time span immediately preceding the season, but over a longer period. (Note that inventory often is called "stored capacity.") A retailer can, in the same way, accumulate inventory ahead of a predicted sales increase. A company can further create anticipation stock in advance of internal or external labor disputes that could reduce production or stop shipments. Another reason for having anticipation stock is simply the restricted supply of raw materials during a short time period. Harvesting of ripe green peas, for example, must take place during a period of a few weeks, whereas the demand for the product might be evenly spread over the year.

11.4.4 THE MANAGEMENT OF INVENTORIES

The Nature of the Inventory Planning Problem

The size and complexity of the task of managing inventories will vary from company to company. Supermarkets and department stores keep a multitude of items in inventory. Keeping track of what

*Originally suggested in reference 4.

is coming in and what is going out is obviously more complicated for these types of operations than for a local hamburger stand. Manufacturing companies can range from small, one-product companies to large manufacturers of complex, high-technology products. The number of different products produced and their complexity in terms of number of components will determine the total number of items in the company's inventory system. For an average manufacturing company, this number will range between 10,000 and 50,000.[5] Items will vary further in terms of size, weight, and perishability, factors that will influence materials handling, ways of storing the goods, and the time it is desirable to stock them.

The managing of inventories is not simply a question of keeping track of the material, however. This function is only a necessary, but not sufficient, part of the inventory management problem. There must also be goals for desired inventory performance, so that actions can be taken to achieve these goals. It is the responsibility of top management to formulate company policies for inventory management and control. These policies would be statements of target inventory investments for a certain planning horizon, say 3 months, in each of several inventory categories. The categories can be either raw materials, work in process, and finished goods, or cycle stock, safety stock, and so on.

Managers often ask the question "How much inventory is enough?" This issue cannot be addressed in isolation, but must be solved together with other important policy variables. There are clear relationships among inventory levels, customer service, and, for a manufacturing company, production efficiency. Managers must therefore make some trade-offs and consider how much they are willing to invest in inventory of various categories in order to achieve satisfying results for the other performance measures mentioned.

The problem of deciding on investment levels for various categories of inventory is obviously not easy. First, top management cannot consider individual items, which would take too much time and effort, but must instead concentrate on aggregate inventories. This means that broad material groupings, such as similar products in a product line, are looked at, and in many cases the unit of measure is in dollars instead of in units. The decisions on inventory allocation are usually based on experience, recent inventory levels, and a desire to change these levels in a way that better satisfies the company's goals. For a manufacturing company, the decision to change the level of finished goods, for example, by building safety or anticipation stock, will have far-reaching effects on the inventory system as a whole. The levels of raw material and work in process that are needed to support the changes in finished goods can be estimated from historical data and managerial experience and judgment. With advanced MRP systems, simulations of future production and demand levels can be done to estimate the resulting inventory levels over time. Simulation is, of course, a useful technique for retailers as well. The cost and time involved, however, prohibit most companies from employing this tool on a regular basis.

Planning aggregate inventory levels will help the company to predict profits and ROI levels and to set the framework for future activities in production and distribution. Top management decisions on aggregate inventories will, for a manufacturing company, affect the making of the master schedule (see Chapter 11.3) and also detailed inventory control. It can be argued that it is at the single-item level that the real task of inventory management and control begins. (It is also toward this level that the emphasis of the remaining part of the chapter is directed.) As mentioned earlier, one reason for this is the sheer magnitude of the control problem. Another reason involves accuracy.

Before introducing techniques for inventory control, objectives and performance measures of inventory management and control are further discussed. Without setting objectives and having ways of measuring the performance, no control can be executed.

At the aggregate level, top management decides on inventory levels after a joint consideration of predicted demand, customer service levels, financial restrictions, capacity, and material availability. Decisions are often made without using formal decision models, but the underlying aim is to enhance company profits and the ROI. At the single-item level, where decision making is more "automated" and is handled routinely by inventory control models, the most common criterion is cost. A simplified objective then, is to minimize the cost related to inventory after certain restrictions have been fulfilled. Inventory-related costs actually serve three functions. They represent a way of expressing a goal, they represent one way of measuring performance, and they can be incorporated into inventory models whose output, if acted upon, ultimately will decide the inventory performance from a cost standpoint.

Inventory-Related Costs

Three types of costs are usually associated with inventory: the cost of ordering material, the cost of carrying the material in inventory, and the cost related to being unable to satisfy demand directly from inventory. In the following descriptions of each type, costs pertinent to both manufacturing and distribution inventories are dealt with. For more detailed information on the determination of inventory costs, the reader is referred to other sources.[4,6–8]

The Ordering Cost

The cost of ordering is related to the activities performed in connection with issuing an order in manufacturing or sending an order to an outside supplier. Since the activities involved are very different, separate ordering costs should be established for purchased and manufactured parts. However, for practical reasons it is often not possible to determine individual ordering costs for each single item in the inventory system. In these cases the costs actually chosen should at least be representative of a number of item families that, from an order preparation standpoint, can be considered similar.

For *purchased* items, the following activities might be performed in connection with material ordering:

Search for a supplier.

Selection of a supplier.

Negotiations with a supplier.

Issuing of orders.

Receiving of goods.

Inspection of goods.

Returning of defective items.

Transferring of goods to stockroom.

Handling of information system transactions related to the activities above.

The activities involved will be somewhat different for a distributor and a manufacturer. They will also differ substantially depending on whether it is a repeated buy or a first-time purchase of a new good. The cost of ordering stems from three different sources: personnel, equipment, and material. The last two categories can include such items as stationery, stamps, telephone calls, duplicating equipment, typewriters, word processors, and computers.

The exact cost of preparing a single purchase order can be difficult to assess because both fixed and variable costs are involved. A common approach in industry is to estimate the total annual cost for the purchasing department, including manpower, material, and equipment, and to divide this cost by the total number of orders processed in a year. (A preferable and more exact approach is to distribute the costs over approximate classes of items, so that each class will receive a unique ordering cost.) The ordering cost estimated this way will be an average cost and not a marginal cost. That is, because of the consideration of both fixed and variable costs, the ordering cost will not reflect the cost incurred if one extra order were placed or the cost saved if one less order were issued.

It is, however, the marginal cost that is of interest when deciding how many orders should be handled over time. For certain situations, for example, mailing a computer-printed material requisition back to the supplier, this cost is very low and almost negligible. From an economic standpoint this means that it is desirable to issue more orders, since the cost of each is so low. However, an increased number of orders would necessitate more resources in the form of personnel and equipment, which would raise the average cost per order. The marginal cost per order after hiring more people or acquiring more equipment can still be the same. This phenomenon occurs because the cost of processing orders basically follows a function that increases in steps when the processing capability of the system has reached a limit and more capacity must be added (see Exhibit 11.4.4). A company can therefore be advised to use the average cost approach, but it should also be aware that in most cases this approach will overstate the ordering cost. What is of interest to the company is obviously the aggregate impact of ordering and not a change that occurs for an individual item. Particular care must therefore be taken when changes are made to the ordering costs for all items in the inventory system.[9]

For *manufactured* material, the following cost components together will make up the ordering cost (which in this context often is called "setup cost"):

Cost of setup–adjustment of machine tools and equipment in preparation for next operation.

Cost of supporting material, for example, fixtures, tools.

Cost of materials handling–locating and handling material inside stockroom and transportation to place of operation.

Cost of productivity–this is a penalty cost in order to achieve longer series, which can yield decreased unit costs as a result of learning effects.

Cost of inspection–to be included only if the first and the last pieces produced in a batch are inspected.

Exhibit 11.4.4 The Ordering Cost as a Function of Number of Orders

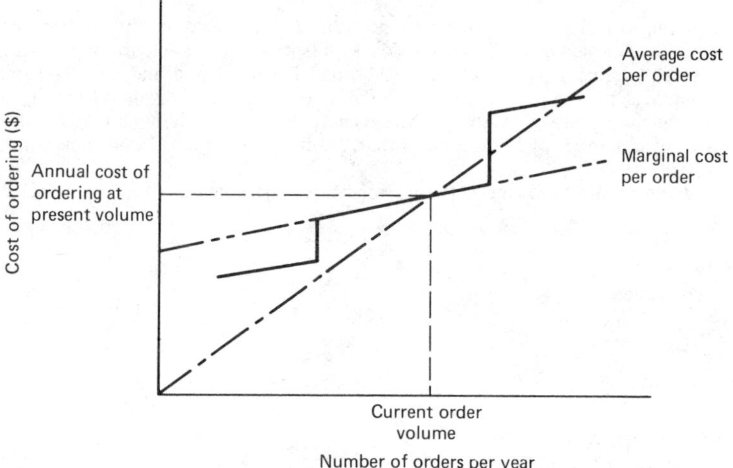

Cost of scrap–to be included if the number or cost of scrapped units is fairly constant and not related to the order size.

Cost of idle time–this is the cost of lost production for the time during which the manufacturing equipment is being prepared for the next task.

Cost of processing and printing information, for example, the making of a manufacturing order package.

It must be stressed that each cost element listed is relevant only if the cost is constant per order and is not related to other factors, such as the order size. The discussion concerning marginal and average costs is also pertinent to manufactured material. It can, for example, be argued that there is no cost involved in preparing one more machine if personnel hired to do this job are already on the payroll. Most companies, however, do include in the ordering cost the standard time for a setup multiplied by the standard labor hourly cost. Often this is also the only cost they consider, with the result that their ordering costs are underestimated. In practice, however, many of the costs mentioned here, such as the costs related to productivity and idle time, are difficult to assess. By including only labor cost a company might capture the majority of the actual costs involved. This must be investigated from case to case, however.

The Inventory Carrying Cost

That there is a cost associated with holding inventory can be attributed to several different factors. The following cost components will together, whenever relevant and related to the level of inventory, make up the inventory carrying cost:

Cost of capital invested in inventory.

Cost of stockrooms and warehouses and their operation and maintenance.

Cost of materials handling, including equipment.

Cost of insurance for goods and warehouses.

Cost of taxes on inventory.

Cost of spoilage, damage, and pilferage.

Cost of obsolescence.

Cost of counting inventory.

Cost of processing and printing information related to holding inventory.

The carrying cost per time unit is often separated into two multiplicative factors, I and C, where C represents the cost (value) of the stocked item in dollars, and I is the carrying charge factor expressed in percent per unit of time. The reasons for doing this are as follows:

1. The cost of carrying inventory per unit of time is assumed to be in direct proportion to the unit cost of the item. (Note that an item should be valued at either its purchasing cost or its variable manufacturing cost.)

2. If I is constant, then $I \times C$ implies that the carrying cost becomes adjusted to each individual item.

3. One can assume that I is constant for all items since the cost of capital is constant and this component dominates the other costs listed here.

4. A change in the carrying cost can easily be implemented for all items by changing the carrying charge I.

Not all the cost elements mentioned previously might be applicable to all materials. They may also vary substantially from situation to situation. The cost of pilferage, for example, is much higher in a drugstore than in a metal sheet producing plant. The rate of obsolescence is high in the fashion industry and in industries with a high rate of technological change, such as the electronics industry, but might be very low in other types of businesses. Many of the elements in the carrying cost are not directly proportional to the unit cost, nor are they always constant over time. As a matter of fact, the only factor that is likely to have this characteristic is the cost of capital. A recent study reported by Lambert and La Londe[10] showed that the cost of capital accounted for around 84% of the carrying charge. Even if a very small sample of companies were studied, it would confirm the earlier, widely held contention that the cost of capital dominates the holding cost.

Just as was the case for the ordering cost, the holding cost might vary between items or groups of items. (That is, the holding cost might not be proportional to the unit cost for all items.) Thus each company must consider its own situation and decide whether a uniform carrying charge should be applied or whether it should vary. A simplified way of finding the carrying charge is to sum up the total estimated cost per year for materials handling, stockrooms, insurance, taxes, and so on, and to divide this sum by the average inventory investment over the year. The cost of capital, expressed in percent per year, is then added to this figure to make up the total holding cost. A numerical example of this procedure is given in Exhibit 11.4.5.

The cost of capital is the most subjective factor in the holding cost. It is a policy variable that should be set by management to express sacrifices made when investing capital. It can therefore be looked upon as an opportunity cost, reflecting the return foregone by investing in a certain project. Since the return from the best alternative investment will keep changing all the time, the cost of capital should theoretically keep changing as well. A more pragmatic use of the cost of capital, then, is as a capital rationing instrument. For example, if management decides that the cost of capital is 25%/year, then no investment proposals will be accepted that yield a return less than that. The inclusion of this factor in the inventory carrying charge will lead to the most cost-efficient level of inventory investment.

A large study conducted by the American Production and Inventory Control Society (APICS) in 1973 indicated that three out of four companies used inventory carrying costs.[11] It further showed that 50% of the companies used a carrying charge of more than 21%, while the median value was 20%. A survey of the literature between 1951 and 1974, reported in Lambert,[7] showed that the carrying charge estimate ranged between 12 and 35%, with a mean around 25%. It is interesting to note that this estimate has remained fairly stable over the years. With the recent sharp increases in interest rates, however, the average inventory carrying cost is probably higher now than ever before.

The Shortage Cost

If a company is approached by a potential customer who asks for delivery of a certain product and the company cannot deliver, two possibilities exist. The first is that the customer accepts

Exhibit 11.4.5 Example of Inventory Carrying Charge Calculation

Costs Related to Inventory	Expenditure (Dollars per Year)
Rented warehouse and materials handling	$12,000
Insurance	5,400
Taxes	10,600
Spoilage, damage, and pilferage	2,320
Obsolescence	5,200
Administrative costs (variable)	5,350
	$40,870
Average annual inventory	$500,000
Cost of capital	24%/year
Annual inventory carrying charge	$24 + \frac{40{,}870}{500{,}000} \times 100 = 32\%$

what he or she can get, if anything, and agrees to have the remaining quantity delivered at a later date. The second possibility is that the customer accepts a partial delivery, but turns to a competitor for the rest, or simply cancels the order and has it filled by another company. In either of the cases described, shortages occur. The first situation, where the customer stays with the company, can be called a "backorder situation." The second situation implies a shortage that leads to lost sales.

The cost of a lost sale cannot simply be equated with the contribution the company could have made through the sale. As a result of the company's not being able to deliver, its reputation might suffer, and a loss of goodwill might occur. If the customer does not return for more purchases in the future, then all potential future sales should theoretically be included in the estimated shortage cost. If the customer accepts a backorder, that is, agrees to wait for the goods, the cost for the supplier might be much lower. There can still be a loss of goodwill, however, and the cost of the backorder can be quite steep if special arrangements, such as changing a production schedule, are made to accommodate the customer.

As is the case with ordering and holding costs, shortage costs can be difficult to estimate. It should also be noted that standard accounting data are of little help in these assessments.[12] In practice, many managers avoid the issue completely or turn to surrogate measures such as stock run-out probabilities or service levels. Because of the relative complexity of the inventory models that incorporate shortage costs in their structures, they are not further discussed in this chapter. Interested readers are referred to, for example, Hadley and Whitin[13] or Peterson and Silver.[14]

The Inventory Turnover Rate

At the single-item level, the inventory management objective can be formulated as minimizing the total inventory-related cost per unit of time with respect to such restrictions as demand and desired service level. Models that will help to achieve this objective are treated in the next section. To measure the performance of an inventory management and control system, attention is usually shifted from individual items to aggregates. Furthermore, costs, especially ordering and shortage costs, are no longer as widely used to measure performance as is the value of the inventory. The most common inventory performance measure at the aggregate level, the inventory turnover (IT) rate, also reflects this variable. The rate can be defined as follows:

$$\text{IT} = \frac{\text{annual usage (\$)}}{\text{average inventory investment (\$)}} = \frac{\text{annual usage (units)}}{\text{average inventory (units)}}$$

It must be emphasized that, if inventory is measured at cost, the annual usage, or sales, must be measured on the same basis. For example, if net sales for one product were \$6 million one year, measured at cost of goods sold, and the average value of the finished goods inventory were \$1.5 million, then the turnover rate would have been 6/1.5, or 4 turns/year. The turnover rate tells how many times, on the average, the inventory is filled up and emptied during a year in order to satisfy the demand during the same period. If a company can reduce its inventory and still maintain the same sales level, then the turnover rate will increase. A reduction in the inventory investment just mentioned, from \$1.5 to \$0.5 million, will, for example, increase the inventory turnover to 12. Turnover rates can be calculated for single items, product families, and different classes of material, such as raw materials and finished goods. An overall turnover rate can also be found for relating annual sales to the total inventory value of the company.

Turnover rates can be used as measures of relative performance if calculated and compared from period to period. An increased number of turns is thereby a good trend, whereas a decreasing rate can cause alarm and trigger extra managerial attention. Companies also compare their turnover rates to those of their competitors or of the industry as a whole, if these statistics are known. When doing this, the turnover measure often stops being simply an "after-the-fact" measure and becomes a goal variable. Management might then stipulate that, say, six turns per year should be achieved for all materials because that is what the competitors manage to do. This policy on an aggregate level will then affect the inventory decisions on an individual basis. It has been found that successful companies can demonstrate higher inventory turnover rates than not so successful competitors.[14] It does, therefore, seem logical that a company should use the best value achieved by its competitors as an *objective* for inventory management.

There is, however, an inherent danger in relying too much on the turnover ratio as a goal variable. There are several reasons for this (see also Bonsack[15] and Krupp[16]). One is that single items, or groups of items, might have actual turnover rates that differ, and should differ, substantially from the set goal. It can thus be bad policy to stipulate the same rate for all materials. Another reason is that the inventory turnover rate is only a surrogate measure, indirectly related to the holding cost through the inventory investment in the denominator. The measure completely disregards other inventory-related costs, such as the ordering cost and the cost of being out of stock. A strict enforcement of a high turnover rate, seemingly desirable in itself, might therefore lead to an increased cost of ordering (because of more turns) and a higher rate of stock run-outs (due to lower inventory), which together with the inventory holding cost might yield a higher overall cost than was the case before. Finally, one of the single most important elements in successful inventory management is top management support and involvement in aggregate inventory planning.[14] Sophisticated systems or ambitious turnover goals will not help a company achieve what the best of its competition has achieved unless this vital ingredient is present.

11.4.5 TECHNIQUES IN INVENTORY CONTROL

Dependent Versus Independent Demand

The nature of the demand an item is facing is an important factor to take into account when deciding upon an inventory control technique. Consider the manufacturing of bicycles. The number of bicycles produced will have to be based on a forecast of future sales. The number of saddles the manufacturer must purchase, or the number of wheels that must be made in order to assemble complete bicycles, need not, however, be forecast. The reason is that, once the number of bicycles scheduled to be produced in the future is determined, the amount of each component making up the bicycles will be known for the product structure. The dependent/independent demand principle[17] simply says, then, that end items that can be considered independent of other items in the inventory system should be controlled by statistical order point techniques (as outlined later in this section), whereas all parts that make up the end items (actually all manufacturing inventories) and are thus subject to dependent demand should be controlled by MRP systems.

There are, of course, always exceptions to the rule. A spare part, for example, can have both dependent and independent demand if the part is a component of an end item currently in production. Some of its demand must therefore be forecast, while some can be directly derived from the production schedule. For these types of items, MRP systems will be suitable. Spare parts can also be manufactured for products no longer sold by the company. In this case only an independent demand must be forecast; the techniques presented in this chapter are applicable here. Another case, quite the opposite, deals with items that clearly are subject to dependent demand, but whose value is so low and whose use is so regular that a less sophisticated system can be used to control the inventory. Finally, the statistical order point procedures discussed in this section can be, have been, and still are applied to all or a majority of a manufacturing company's inventoried parts, even if the techniques are best suited for distribution inventories. The MRP system is, after all, a fairly new planning and control device for production.

Forecasting and Ordering Systems

The three basic questions regarding item inventory control are as follows:

1. How much is needed over time?
2. How much should be ordered each time?
3. When should it be ordered?

The first question relates to demand estimation for an item. This is handled by a forecasting system. The system itself can range from an unstructured, informal, manual procedure to sophisticated computer-based methods. The following section outlines some simple, yet effective and widely used, short-term forecasting techniques. The second and third questions are answered by the ordering system, which decides on lot sizes and when to launch the orders.* These decisions are addressed following the discussion of forecasting.

Basic Forecasting Techniques

The Moving Average Technique

Most forecasting techniques are based on the often-correct assumption that the future may not be exactly like the past, but will be close to it. Thus an intuitive approach to estimating what will happen in the future is to find the average of past events and use that as a forecast. The simplest forecast in this sense is to state that what happened last period will also happen in the future. For example, if sales last quarter were 12,000 units, then sales next quarter will be the same. A way to stabilize the forecasts is to increase the number of data in the average, since this factor will affect the responsiveness of the forecast.

*Only ordering systems based on the fixed order quantity/reorder point principle are discussed here. For other ordering systems, see, for example, reference 18.

Exhibit 11.4.6 Sales of Product XYZ During the Last 6 Months

Month	Sales (Units)
January	41
February	60
March	57
April	52
May	49
June	63

The principle behind the moving average forecast technique can be illustrated as follows. If actual sales for a product over the last 6 months were as shown in Exhibit 11.4.6, then a three-period moving average, calculated at the end of June, is $(52 + 49 + 63)/3 = 54.7$. This number, rounded off, can then be used to forecast sales for July. When actual sales for July are recorded, let us say—58 units, a new three-period average can be found—$(49 + 63 + 58)/3 = 56.7$—and thereby a forecast for August, and so on. As a matter of fact, the last moving average calculated will be the forecast for any period in the future until new sales data have been recorded and the average updated.

The disadvantages commonly associated with moving averages are (1) that as many old data as there are periods in the average must be stored at any time and (2) that the average that is calculated and used as a forecast gives equal weight to old and new data. (The weight is $1/n$, if an n-period moving average is used.) It seems reasonable to put greater emphasis on more recent data and less emphasis on data the older they get. The exponential smoothing technique does exactly this.

The Exponential Smoothing Technique

Assume that a forecast has been made for a future period and that later, at the end of this period, the actual demand has been recorded. The difference between the actual demand in period $t(AD_t)$ and the forecast for the same period (FC_t) is called the "forecast error" (FE_t), that is, $FE_t = AD_t - FC_t$. If the forecast was incorrect, that is, there was a positive or negative forecast error, something should be learned from it. One way to do that is to take a portion of the error and add that to the old forecast in order to get a new forecast for next period.

The exponential smoothing technique, just as the moving average procedure, is based on the principle of averaging, or smoothing out, old data and using them as a forecast. If the fraction of the forecast error that is added to the old forecast (the old average) is α, where α is a number larger than zero and less than one, the technique can be formulated as follows (also see Exhibit 11.4.7):

1. $EA_t = \alpha \times (AD_t - FC_t) + EA_{t-1}$ where EA_t is the exponential average calculated at the end of period t.

Exhibit 11.4.7 A Graphic Illustration of the Exponential Smoothing Technique

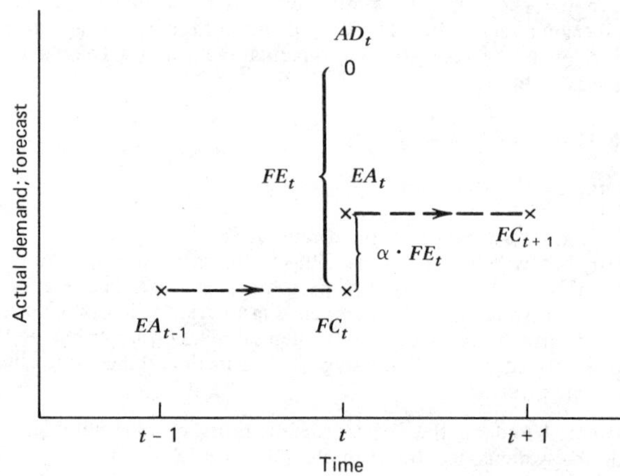

2. $FC_{t+n} = EA_t \quad n = 1, 2, \ldots$ that is, the forecast for period $t + 1$, $t + 2$, and so on, is equal to the most recent average.

3. Since $FC_t = EA_{t-1}$, the first equation is usually written as

$$EA_t = \alpha \times AD_t + (1 - \alpha) \times EA_{t-1}$$

Tracing this equation backward in time, it can be seen that the weight applied to AD_t is α, and the weight associated with AD_{t-1} is $\alpha \times (1 - \alpha)$, whereas the weight for AD_{t-2} is $\alpha(1 - \alpha)^2$, and so on. Since α is less than one, the weights get successively smaller.

If exponential smoothing is used to forecast the sales for the item referred to in Exhibit 11.4.6, then a starting average and a smoothing constant, α, are needed. The most commonly used α value in industry is 0.1. The starting average can, for example, be estimated as the average of the first 5 months, or 51.8. The exponential average at the end of June can then be found as

$$EA_6 = 0.1 \times 63 + 0.9 \times 51.8 = 52.9$$

The forecast for July is therefore 53 units.

Exponential smoothing is a simple and efficient technique (note that only the last average needs to be stored from period to period) for short-term forecasting. In the form described here, it is best suited for data that fluctuate around a constant mean. The method can, however, be easily expanded to consider situations involving a definite growth pattern or seasonal variations. (See, for example, Makridakis and Wheelwright[19].)

Forecasting future demand is always very difficult, cannot be exact (the customers do not read your forecast), and is therefore a constant source of frustration to managers.* It is, however, a necessary and fundamental activity for each business organization, and the manager should be aware of the strengths and weaknesses associated with the wide range of available techniques. (For an excellent overview, see Chambers et al.[21]; also see Chapter 13.9.)

Determination of Order Quantities

Assume now that demand over a certain horizon has been estimated, either by using exponential smoothing or simply by looking at past inventory records and assuming the future planning horizon will meet the same demand. The next problem is to determine how much should be ordered each time an order is placed over this horizon. To do that, the inventory-related costs discussed previously can be utilized. The common models discussed here consider only the holding and the ordering cost. For more complex models, which also include shortage costs, the reader is referred to Hadley and Whitin.[13]

Based on constant demand over time, instantaneous replenishments of inventory, and no stock-cuts, the inventory levels over time associated with two different order quantities are illustrated in Exhibit 11.4.8. It is clear from the exhibit that the larger the order quantity (Q), the fewer the number of orders placed per unit of time. Further, the larger the order quantity, the larger the average inventory over time. Let

D = average demand per unit of time
S = cost for ordering
C = unit cost per item
I = inventory carrying charge measured in percent per year

The following relationships can then be set up (the time unit is assumed to be 1 year):

$$\frac{D}{Q} = \text{total number of orders per year}$$

$$\frac{D}{Q} S = \text{annual ordering cost}$$

*In a recent study by Fogarty (reference 20) on the use of computers in industry, he found that managers were most dissatisfied with forecasting software. He concluded: "This is consistent with the inherent difficulties of forecasting and perhaps indicates that further work is required in improving the application of forecasting techniques and software packages. It also may indicate that expectations of forecasting results are not realistic."

Exhibit 11.4.8 Inventory Levels Over Time for Two Different Lot Sizes

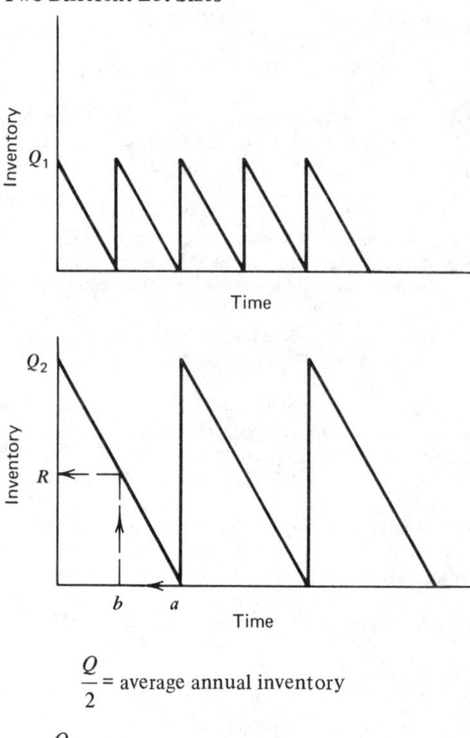

$$\frac{Q}{2} = \text{average annual inventory}$$

$$\frac{Q}{2} IC = \text{annual inventory holding cost}$$

The annual total cost, *TC*, for ordering and holding inventory is thus

$$TC = \frac{D}{Q} S + \frac{Q}{2} IC \tag{1}$$

The *TC* is a function of *Q*, and the best order quantity—called the economic order quantity (*EOQ*)—is the one that yields the lowest cost.

The solution can be found graphically, as in Exhibit 11.4.9. Analytically, the solution is found by taking the derivatives of *TC* with respect to *Q*.

Exhibit 11.4.9 A Graphic Solution to the Problem of Finding the Best Order Quantity

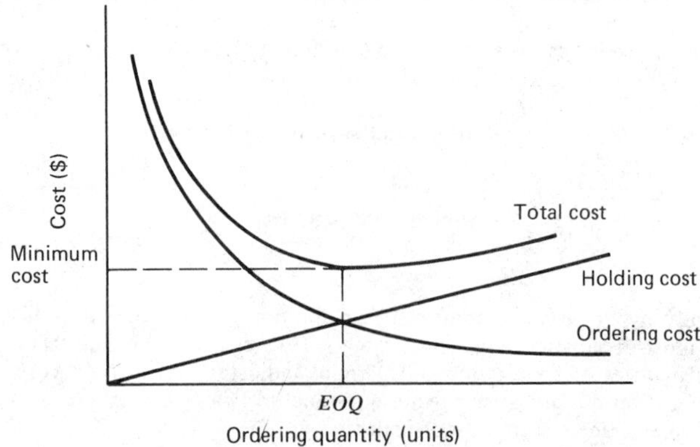

$$\frac{dTC}{dQ} = -\frac{DS}{Q^2} + \frac{IC}{2} \tag{2}$$

$$\frac{dTC}{dQ} = 0$$

yields

$$Q^* = EOQ = \sqrt{\frac{2DS}{IC}}$$

This result can be used to find the total cost when the order quantity is *EOQ*. The cost can always be found by inserting *EOQ* in equation 1, but is easier found as follows:

$$TC^* = \frac{DS}{EOQ} + \frac{EOQ}{2} \times IC = \text{since the two terms are equal} = 2 \times \frac{EOQ}{2} \times IC$$

$$= EOQ \times IC \tag{3}$$

The *TC* curve is relatively flat around its minimum, indicating that the penalty for deviating from *EOQ* is small. The following formulas can be used to investigate this[22]:

$$p = 0.5\left(\frac{1}{q} + q\right) \tag{4}$$

or

$$q = p \pm \sqrt{p^2 - 1} \tag{5}$$

where p is the relationship between the total cost for using an order quantity of Q units and the optimal total cost achieved when using *EOQ*, that is, $p = TC/TC^*$, and q represents the ratio between the order quantity Q and the optimal order quantity *EOQ*, that is, $q = Q/EOQ$.

Consider an item with the following data: The annual demand is 1200 units, and the ordering cost is $14. The unit cost is $1.25, and the carrying charge is 25%/year. The economic order quantity and the total cost for this item are then

$$EOQ = \sqrt{\frac{2 \times 1200 \times 14}{0.25 \times 1.25}} = 327.9 \approx 328 \text{ units}$$

$$TC = EOQ \times IC = 327.9 \times 0.25 \times 1.25 = \$102.47/\text{year}$$

Using equation 4, questions such as the following can be answered: "What are the consequences of increasing the ordering quantity by 50%?" Since in that case $q = 1.5$, we get $p = 1.0833$. Therefore the total annual cost related to an order quantity of 492 (327.9 × 1.5) is $111.01 (102.45 × 1.0833). With the help of equation 5, we can ask such questions as, "If a 1% increase above the optimal cost is tolerated, by how much can the order quantity vary?" Now $p = 1.01$, which yields $q_1 = 0.868$ and $q_2 = 1.152$. Thus the answer is that Q can vary between 285 and 378 (327.9 times 0.868 and 1.152, respectively) and still not exceed the total cost by more than $1.02.

It is obvious that the total cost is very insensitive to variations in the order quantity. This is, of course, a desirable quality, since it gives leeway in the actual use of order quantities. Sensitivity analyses such as that used here can also be carried out for the cost parameters S and IC, with similar results. It must be pointed out, however, that the statements about *EOQ* are, strictly speaking, valid only if the assumptions behind the model are fulfilled. The more the actual demand per period deviates from the average demand, the less accurate are the estimates.

The Reorder Point

The last question raised previously with regard to inventory control relates to the time when the lot size should be ordered. Consider again Exhibit 11.4.8. At point a the inventory is empty and must be replenished since stock-outs are not allowed. If the time elapsing between the placement of an order and its arrival in inventory (the lead time) is L time units, then ordering must, at the latest, take place at time b ($a - b = L$). This corresponds to an inventory level of R units. This particular inventory level is called the "reorder point" and is, in situations with no uncertainties, equal to the demand during lead time.*

For example, let $D = 1200$ units/year and $L = 4$ weeks. Using 50 weeks in a year, the demand per week is 24 units, and the reorder point is thus 4×24, or 96 units. If EOQ is 328 units, the decision rule will be to order 328 units whenever inventory falls below 96 units. This will happen about 3.7 (1200/328) times/year.

The Service Level Concept

A situation where demand is constant over time is rarely, if ever, found in real life. Most likely, the demand varies from period to period. Nor is the lead time constant, but can vary from one order to the next. There are further uncertainties as to whether all the ordered units will arrive, whether they are in good condition, or whether the recorded inventory level actually corresponds with the physical inventory. Ways to handle *demand uncertainty* will be treated in this and the following section. It will then be assumed that lead time does not vary. (For models considering lead time variation, see Hadley and Whitin.[13]) The last two sources of uncertainty have to be attached in other ways specific to the situation, for example, by increasing the order quantity or by establishing safety stock.

If actual demand varies over time and the reorder point is determined from the average demand, it is clear that stock-outs will occur fairly often. Since stock-outs can happen only when demand during lead time is larger than the reorder point (see Exhibit 11.4.8), they can be reduced by increasing the reorder point R. Stock-outs can further occur only when the reorder point has been passed, that is, during the time between the placement and the arrival of an order. An increased order quantity will lead to fewer orders being placed and thereby to a lower number of shortages.

The determination of order quantities and reorder points can be based on trade-offs between the ordering cost, the holding cost, and the shortage cost. As mentioned previously, however, the shortage cost is not very often used in practice, probably because of reluctance among managers to estimate this cost. Another reason can be that the inventory models become more complicated. A way to get around this problem is to use the concept of "service level," a measure most managers feel comfortable with. The level of service can be defined in several ways. One of the most common definitions is the following: The service level is the percentage of demand over time that is satisfied directly from inventory. That is, if 1200 units are demanded in a year and 1152 of them can be delivered immediately, the service level is 1152/1200, or 96%. The service level, β, can be mathematically formulated as follows:

$$\beta = 100 \left(1 - \frac{AUS(R)}{Q}\right) \tag{6}$$

where $AUS(R)$ is the average number of units short per order cycle. Since the order quantity, Q, is the average demand per order cycle, expression (6) follows directly from the previous definition of the service level.

Exhibit 11.4.10 shows the discrete distribution of demand during lead time for a stocked item. The average demand during lead time, D_L, can be found as

$$D_L = \sum_{D=0}^{D\max} D \times P(D) \tag{7}$$

Therefore

$$\begin{aligned} D_L &= 0 \times .1 + 1 \times .15 + 2 \times .15 + 3 \times .25 + 4 \times .15 + 5 \times .1 + 6 \times .05 + 7 \times .05 \\ &= 2.95 \text{ units.} \end{aligned}$$

From the same distribution, the average number of units short per order cycle can be determined. Shortages occur when demand during lead time exceeds the reorder point. Therefore

$$AUS(R) = \sum_{D=R+1}^{D\max} (D - R) \times P(D) \tag{8}$$

*In most systems a new order is triggered when the inventory position (i.e., on-hand inventory plus any outstanding orders) falls below the reorder point. This will take care of situations where lead time is longer than the order cycle.

Exhibit 11.4.10 Distribution of Demand During Lead Time

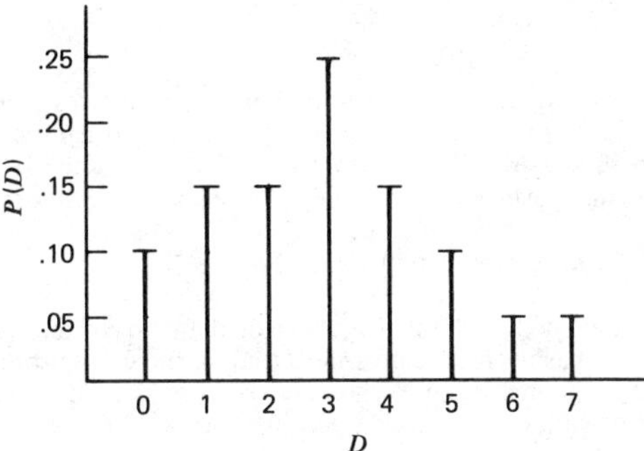

where $(D - R)$ represents the size of the shortage and $P(D)$ is the probability that this stock-out will occur. Various types of problems can now be addressed:

1. What reorder point is necessary to achieve a service level of at least 97%? Present order quantity is 20 units.
Solution:

$$97 = 100\left(1 - \frac{AUS(R)}{20}\right) \Rightarrow AUS(R) = 0.60 \text{ units/order cycle}$$

The reorder point must be found through trial and error. Try $R = 4$.

$$AUS(4) = \sum_{D=5}^{7} (D-4) \times P(D) = P(5) + 2 \times P(6) + 3 \times P(7)$$

$$= 0.1 + 2 \times .05 + 3 \times .05 = .35 < .60$$

This reorder point gives too much protection. Try $R = 3$.

$$AUS(3) = \sum_{D=4}^{7} (D-3) \times P(D) = P(4) + 2 \times P(5) + 3 \times P(6) + 4 \times P(7)$$

$$= .70 > .60$$

Since the objective is to achieve at least 97% service level, the reorder point must be 4 units. The concomitant service level will be

$$\beta_{R=4,Q=20} = 100\left(1 - \frac{.35}{20}\right) = 98.25\%$$

2. If the present order quantity is 20 units and the reorder point is 3 units, how is the service level affected by a doubling of the order quantity?
Solution:

a. $$\beta_{R=3,Q=20} = 100\left(1 - \frac{.70}{20}\right) = 96.50\%$$

b. $$\beta_{R=3,Q=40} = 100\left(1 - \frac{.70}{40}\right) = 98.25\%$$

The number of units short per order cycle has not changed, but the number of order cycles has. If annual demand is 150 units, the average number of units short *per year*, can be computed as:

a. $(150/20) \times .70 = 5.25$ units/year. Check with the service level definition. $\beta = 100(150 - 5.25)/150 = 96.50\%$.

b. (150/40) × .70 = 2.625 units/year. β = 100(150 – 2.625)/150 = 98.25%.

3. What is the average inventory level over time if a lot size of 20 units and a reorder point of 4 units are used?

Solution:

The expected inventory level at the time a new replenishment arrives is usually called the safety stock, *SS*. The safety stock is a function of the reorder point and can be written as $SS = R - D_L$ (see Exhibit 11.4.11). In this case *SS* is equal to 4 – 2.95, or 1.05 units. The average cycle stock is $Q/2$, or 10 units. Total average inventory over time is therefore 11.05 units.

The Use of Forecast Errors in Inventory Control

Up to now it has been assumed that the distribution of demand during lead time, such as the one depicted in Exhibit 11.4.10, has been time invariant; that is, the distribution of demand is constant over time. If a forecasting system is used to estimate the same demand, it is clear that this variable now will take on different values every time a new forecast is made. It is not the distribution of demand during lead time that is of interest, however, but *the distribution of the forecast errors.* The reason is that inventory related decisions are now based on the forecast and the distribution of the errors, given an indication of how far off the real outcome might be. Imagine, for example, a procedure that could forecast the future perfectly. Even if the actual demand showed large variations over time, the forecasts would always be accurate, with forecast errors equal to zero. Protection in the form of safety stock would therefore not be needed.

A good forecasting system automatically and continually estimates the standard deviation (SD) of the forecast errors, many times in the form of the mean absolute deviation (MAD). It is often assumed that the average forecast error is zero and that the distribution of errors around this mean follows the normal distribution.[23]* For this case there is a direct relationship between MAD, the average absolute error, and the SD of the forecast errors, σ_e, in that $\sigma_e = 1.25 \times MAD$.

Assume now that there is a distribution of forecast errors, $f(u)$, around a mean, $\bar{x}$, and that the SD is σ_e. The mean, $\bar{x}$, is equal to the forecast of the demand during lead time. The average number of units short during an order cycle for a continuous distribution can be expressed as

$$AUS(R) = \int_R^{\infty} (x - R) f(u)\, du \tag{9}$$

The ratio $AUS(R)/\sigma_e$, also called the "service function," can be tabulated for each value of the safety factor Z.

$$Z = \frac{R - \bar{x}}{\sigma_e} \tag{10}$$

Exhibit 11.4.11 Cycle Stock and Safety Stock

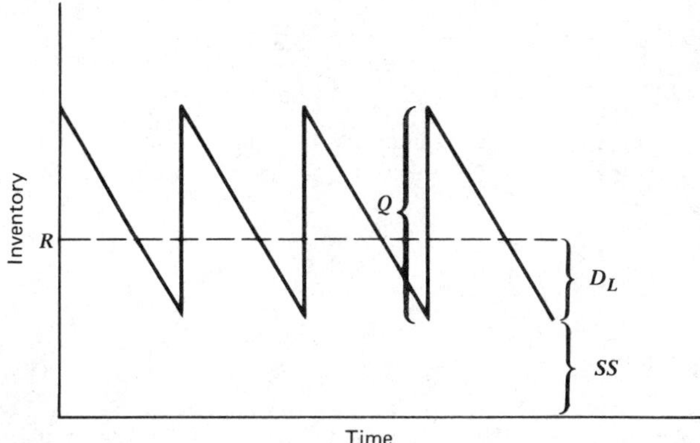

*If the average demand during lead time is less than 10 units, it has been suggested that Laplace-distributed forecast errors be assumed. For these cases, see Peterson and Silver, reference 14.

Exhibit 11.4.12 The Service Function as a Function of the Safety Factor

Z	$AUS(R)/\sigma_e$
0.00	0.3989
0.10	0.3509
0.20	0.3069
0.30	0.2668
0.40	0.2304
0.50	0.1978
0.60	0.1687
0.70	0.1429
0.80	0.1202
0.90	0.1004
1.00	0.0833
1.10	0.0686
1.20	0.0561
1.30	0.0455
1.40	0.0367
1.50	0.0293
1.60	0.0232
1.70	0.0183
1.80	0.0143
1.90	0.0111
2.00	0.0085
2.10	0.0065
2.20	0.0049
2.30	0.0037
2.40	0.0027
2.50	0.0020
2.60	0.0015
2.70	0.0011
2.80	0.0008
2.90	0.0005
3.00	0.0004

A limited table is given in Exhibit 11.4.12. (For complete tables, see, for example, Johnson and Montgomery.[24]) Recall that the safety stock was previously defined as $SS = R - D_L$. Since $\bar{x}$ is the estimate of D_L, the safety factor Z expresses how many SDs of the forecast error should make up the safety stock. It should be stressed that σ_e is based on the forecast errors during lead time. If the forecast interval, F, is different from the lead time, L, the SD should be multiplied by the factor $0.659 + 0.341 \times (L/F)$.[23]

Example 11.4.1 Assume that for an item the demand during a 2-week period is 700 units, the SD of the forecast errors during the same period is 42.1 units, the order quantity is 2500 units, and that lead time is 4 weeks.

1. What safety stock is needed to achieve a service level of 90%?
Solution:

$$90 = 100\left(1 - \frac{AUS(R)}{2500}\right) \Rightarrow AUS(R) = 250 \text{ units/order cycle}$$

$$\sigma_e = (0.659 + 0.341 \times 4/2) \times 42.1 = 56.5 \text{ units}$$

Therefore $AUS(R)/\sigma_e = 4.42$. This value cannot be found in Exhibit 11.4.12. As a matter of fact, Z must be negative to achieve a value of that size. This indicates that the protection, even without safety stock, is fairly high and that the reorder point must be set lower than the expected demand during lead time to meet a service level of 90%. For $Z = 0$, that is, $R = \bar{x} = 1400$ units, $AUS(R)/\sigma_e = 0.399$, and $AUS(R)$ is therefore 22.5 units. The resulting service level, using equation 6, is thus 99.1%.

In general, the service level without using safety stock ($Z = 0$) is

$$\beta = 1 - 0.4 \times \frac{\sigma_e}{Q} \tag{11}$$

2. What happens to the service level if the SD of forecast errors triples, the reorder point is set to 1500 units, and the order quantity is halved?
Solution:

$$Z = \frac{R - \bar{x}}{\sigma_e} = \frac{1500 - 1400}{3 \times 56.5} = 0.59$$

From Exhibit 11.4.12 we get $AUS(R)/\sigma_e = 0.1716$ use interpretation. This gives $AUS(R) = 169.5 \times 0.1716 = 29.09$ units/order cycle. The service level is therefore

$$\beta = 100 \left(1 - \frac{29.09}{1250}\right) = 97.67$$

By changing various variables in equations 6, 9, and 10 and observing the results, Exhibit 11.4.13 can be derived. This matrix can be interpreted as follows: An increase in the SD, coupled with an unchanged order quantity, necessitates an increased safety stock (i.e., an increased reorder point) if the service level is to remain constant. Likewise, an increased service level can be achieved from an increased order quantity, assuming that the variability in the forecast errors does not change.

The typical relationship between the service level and the level of safety stock is displayed in Exhibit 11.4.14. It is clear from the exhibit that more and more units in the safety stock are required to achieve the last percentage of service level. For example, if σ_e/Q is 0.25, an increase in β from 97 to 98% means that SS must increase by approximately 25%. If β is to be increased further, from 98 to 99%, the necessary increase in SS is now around 35%.

The service level is a somewhat vague measure in the sense that the consequences associated with various levels of service are not readily apparent. Based on the definition of the service level, $D(1 - \beta)$ represents the total number of units short in a year, if D is the annual demand. The ratio between the number of units short when the service level is changed from β_1 to β_2 is therefore $(1 - \beta_2)/(1 - \beta_1)$. If, for example, β is increased from 90 to 95%, the number of units short is cut in half. On the other hand, if β is lowered from 99 to 98%, that is, just by 1%, the average number

Exhibit 11.4.13 Relationships Among Service Level, Safety Stock, Order Quantity, and SD of the Forecast Errors

Service Level, β	σ_e/Q	Safety Stock, SS
Constant	+	+
+	Constant	+
+	–	Constant

Exhibit 11.4.14 Relationship Between Service Level and Safety Stock

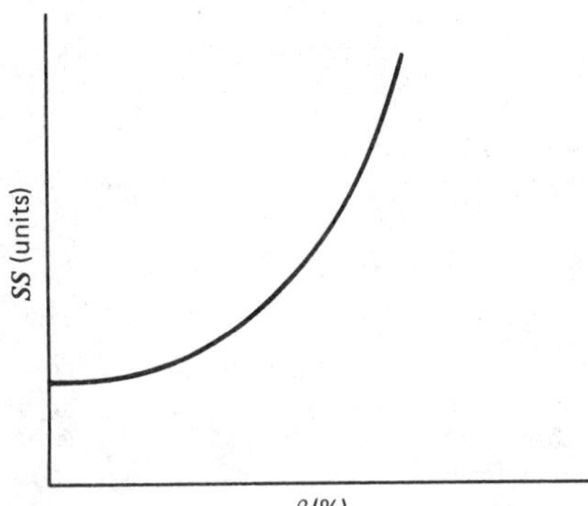

of units short will double. Another aspect is that it cannot be inferred from the measure itself how much it will cost to achieve a specific service level. As was pointed out in connection with Exhibit 11.4.14, the marginal cost of a percent of service is steadily increasing, and management must not decide on a service level in isolation, without also considering the consequences in the form of inventory investment and carrying cost.

Determination of Order Quantities When Quantity Discount Is Offered

Suppliers will commonly offer reduced prices to their customers if they purchase more than a certain amount each time. Of interest to the buyer in a price discount situation is not only the cost associated with ordering and holding inventory, but also the cost of purchasing the material. A formula for the total cost of ordering, holding inventory, and purchasing the goods can be set up, using the symbols as before.

$$TC = \frac{D}{Q}S + \frac{Q}{2}IC + DC \tag{12}$$

The unit cost, C, is now a variable that depends on Q, that is, $C = f(Q)$. To find the lowest cost, the following algorithm can be used:

1. Find EOQ for the lowest price offered. If it is valid, stop. Otherwise continue the search for a valid EOQ, successively increasing the price. A valid EOQ falls within the quantity range for which the particular price is offered.
2. Calculate the total cost for the valid EOQ.

$$TC_{EOQ} = EOQ \times IC + DC$$

3. Find the total cost for each price break associated with a quantity that is *larger* than the valid EOQ, if any. The unit costs associated with these price breaks will be *lower* than the one used to find the valid EOQ.

$$TC_Q = \frac{D}{Q} S + \frac{Q}{2} IC_Q + DC_Q$$

where C_Q is the price associated with each price break.
4. Find the lowest cost from steps 2 and 3.

Example 11.4.2 A supplier offers the following price discount schedule:

Order Quantity	Price per Unit
$0 \leqslant Q < 600$	$2.20
$600 \leqslant Q$	$2.10

The annual demand is 5000 units, the ordering cost is $20, and the carrying charge is 20%/year. There is one price break at 600 units. Which is the best ordering policy?
Solution:

1. Find the valid EOQ. Start with the lowest price.

$$EOQ\ (2.10) = \sqrt{\frac{2 \times 5000 \times 20}{0.2 \times 2.10}} = 690 \text{ units}$$

Since the EOQ is larger than 600 units, the solution is valid.
2. $TC_{EOQ} = 690 \times 0.2 \times 2.10 + 5000 \times 2.10$
$= \$10{,}789.80/\text{year}$
3. Since the valid EOQ was found for the lowest price, there is no price break related to a quantity larger than EOQ.
4. The best ordering policy is therefore to order 690 units each time, at an annual cost of $10,789.80/year.

Assume now that the price break is at 800 units instead of 600. The solution in this case would look as follows:

1. $EOQ\,(2.10) = 690 < 800$. Not valid.

$$EOQ\,(2.20) = \sqrt{\frac{2 \times 5000 \times 20}{0.2 \times 2.20}} = 674$$

This EOQ is valid, since if 674 units are purchased each time, the unit price of \$2.20 is applicable.

2. $TC_{EOQ} = 674 \times 0.2 \times 2.20 + 5000 \times 2.20$
$= \$11{,}296.56\text{/year}$

3. There is only one price break, and it is for a quantity larger than the valid EOQ. Therefore

$$TC_{800} = \frac{5000}{800} \times 20 + \frac{800}{2} \times 0.2 \times 2.10 + 5000 \times 2.10$$
$$= \$10{,}793.00\text{/year}$$

4. The lowest cost is \$10,793.00/year. Thus the best ordering policy is to order 800 units each time.

Time-Phased Demand Data

The parameters in the inventory control system that have been discussed so far, that is, the order quantity and the reorder point, should be updated and adjusted regularly or whenever changes in the demand or the lead time take place. In situations where a more detailed knowledge of the future demand exists, this information should be considered in an ordering system. One way to do this is to use what is called "time-phased records." An example of such a record is given in Exhibit 11.4.15.

The second row of the exhibit specifies the expected demand per period over a certain horizon. The third row indicates the current customer orders that have arrived. The demand row gives the demand that is used in the calculations. In this case the maximum of the forecast and the orders for each period have been chosen. This approach will lead to an overprotection should the total forecast be correct over the planning horizon. The fifth row indicates if there are any outstanding orders right now, and when they are expected to arrive. In this case a batch of 40 units will arrive in the beginning of period 4. The "on-hand" row shows the projected available inventory balance over time. Since the opening inventory for period 7 is 3 units, and the expected demand is 20 units, 17 extra units are needed to cover the demand. In period 8 the cumulative number of units needed is 37. Since the lead time is 2 periods, an order must be placed in period 5 to cover the shortage. With a fixed order quantity of 40 units, the "planned order" row will show 40 in period 5. If these 40 units arrive on time in period 7, the planned inventory balance will change from −17 to 23 units. The ending inventory for period 8, then, is 3 units.

The record in Exhibit 11.4.15 utilizes what is called a "time-phased order point."[25] Similar records are used in MRP systems for controlling manufacturing inventories (Chapter 11.6). While

Exhibit 11.4.15 Time-Phased Inventory Record

Period		3	4	5	6	7	8
Forecast		10	10	15	15	20	20
Orders		3	7	18	4	0	0
Demand		10	10	18	15	20	20
Open orders			40				
On hand	16	6	36	18	3	−17 23	−37 3
Net requirements						17	20
Planned order releases				40			

Lead time = 2

Q = 40 units

Exhibit 11.4.16 Replanning Because of a New Order

Period		3	4	5	6	7	8
Forecast		10	10	15	15	20	20
Orders		3	7	18	104	0	0
Demand		10	10	18	104	20	20
Open orders			40				
On hand	16	6	36	18	−86 0	−100 0	−126 0
Net requirements					86	20	20
Planned order releases			86	40			

Lead time = 2

Q = 40 units

the statistical order point procedures, as previously discussed, rely on actual inventory level, the time-phased order point relies on planned data. It is therefore possible to plan for the future and thus get a more precise idea of what is going to happen. It is possible, for example, to predict when orders will be placed simply by looking at the "planned order" row. If there are any delays in the released orders, these can be updated in the record and thereby, if necessary, change the planned orders. By foreseeing and acting upon changes, better protection can be achieved. Statistical reorder point systems are especially deficient when large deviations in demand occur. With a time-phased order point, a large order can be added to the "order" rows, and actions can be taken before a shortage occurs. Assume, for example, that a customer wants a delivery of 100 units in period 6. This will lead to a net requirement in the same period of 86 units. A normal lot size will not cover this situation. The solution taken in Exhibit 11.4.16 is to plan one order for 86 units to be released in period 4 and a normal lot size to be released in the period after.

Lot Sizing in Connection With Time-Phased Order Point

The net requirements row in the time-phased order point record indicates the amount of material that is needed to cover the expected demand. The order quantities that are issued should cover these net requirements. Because of the way the demand is specified in the records, several options are open.

One is to use the EOQ formula. To give the best fit, the average demand in the formula should be based on the average demand calculated from the data in the record. The simplest lot sizing technique is, no doubt, just to order what is needed in each period. In Exhibit 11.4.15, for example, the planned orders using this "lot-for-lot" technique would be 17 and 20 units, respectively. The rules that perform the best in a time-phased environment are economic lot sizing techniques that also are discrete. The latter means that each batch that is formed is made up from an integral number of consecutive net requirements. The expected ending inventory when such a batch is consumed is therefore zero. Several discrete rules are suggested in the literature.[1] One good and simple rule is presented here.

This marginal cost rule was recently proposed by Groff.[26] Letting D_t indicate the net requirement in period t and T be a pointer, the algorithm can be formulated as follows: (This algorithm is simplified, and will produce only the first lot size. It can easily be extended for repeated use.)

1. Let $T = 1$.
2. Compute $T^* = T(T + 1)$.
3. If $T^* \times D_{T+1} < 2S/IC$, let $T = T + 1$ and go to step 2. Otherwise, go to step 4.
4. Let the lot size be $\Sigma_{t=1}^{T} D_t$

Example 11.4.3 Assume that the ordering cost is \$20, that the holding cost is \$1/item and period, and that the following net requirements have been determined:

Period	1	2	3	4	5	6
Net requirements	10	5	15	0	25	5

The first lot size is then found as follows:

1. $T = 1$.
2. $T^* = 1(1 + 1) = 2$.
3. $2S/IC = 2.20/1 = 40$. $T^* \times D_2 = 2 \times 5 = 10 < 40$. Raise T to 2.
4. $T^* = 2(2 + 1) = 6$.
5. $T^* \times D_3 = 6 \times 15 = 90 > 40$.
6. The first lot size is therefore made up from D_1 and D_2 and will be equal to 15 units.

A repeated use of the algorithm will yield the following solution: (The last batch cannot be found by the algorithm.)

Period	1	2	3	4	5	6
Net requirements	10	5	15	0	25	20
Lot size	15		15		25	20

11.4.6 INVENTORY SYSTEMS

An inventory control system can be viewed as a set of procedures that executes the principles behind inventory control. Inventory systems vary as to what degree of information processing is involved, whether and how uncertainty is accounted for, how often system parameters are updated, and so on. The simplest system is the "two-bin," or "bin reserve system." Other types of systems based on reorder points are collectively called "continuous review systems." Although this chapter has presented only the order point/order quantity system, there are several other principles (e.g., see Peterson and Silver[27]).

The Two-Bin System

This system is especially suited for items with a high and steady demand, short lead time, and low item value, such as nuts, bolts, and washers. As the name implies, there are two bins of material, one large and one small. When the large bin is empty, the small bin is opened, and at the same time an order is placed to replenish the large bin. Since items controlled in this way are subject to routine buys, premade purchasing order documents can be placed on the top of the small bin, ready to be mailed to the supplier. The opening of the small bin represents the passing of a reorder point, and the content of the bin should be large enough to cover the demand until next replenishment arrives. Items controlled by two-bin systems are usually of the kind such that stock-outs cannot be tolerated. Therefore, overstocking of these items is common. This will have only a modest financial impact, since the inventory investment these items represent is relatively low.

Two-bin systems are easy to administer. The disadvantages are that, since no records are kept, the on-hand balance at each particular time is unknown. Further, because of the same reason, it is not possible to get a specific knowledge of demand over time. Recording the time that elapses between the opening and the emptying of a bin can, however, give one estimate of the demand.

The two-bin principle can be executed without using two separate bins. A colored line inside a large bin can just as well serve as a reorder point. Similarly, lines can be painted on iron bars and metal sheets to represent reorder points. Finally, there are several products that we use that have reorder points built in by the supplier. In a desk calendar, for example, there can be a sheet inserted at the beginning of October, indicating that it is time to order a new one. In containers for spices and holders for tape, there are usually small holes so that the user can see when the product is running low and it is time to replenish.

The Continuous Review System

A continuous review system, also called a "perpetual inventory system," differs from a two-bin system primarily when it comes to information processing. The name implies a constant watch over inventory levels. This is, of course, not the case. It is sufficient to check the inventory status each time a disbursement is made. If the reorder point has been passed, a new order must be issued. In a continuous review system, however, all inventory transactions are recorded, that is, not only disbursements, but also replenishments and returns of material. With a proper record keeping system, a lot of valuable information can be extracted for use in inventory control. The inventory transactions can be handled within a manual or a computerized system. The use of index cards and punched cards for recording transactions was common in industry not long ago. Index card files still represent excellent and easy-to-access data bases unless the number of items is too high.

The emergence in recent years of high-speed computers that can process lots of information at a fairly low cost has made it possible even for small businesses to use computers for inventory control purposes. Along with the hardware revolution there has been a significant increase in the availability of software for production planning and inventory control. This means that it is no longer necessary to "reinvent the wheel"; that is, a company can purchase computer packages instead of developing them in-house. Even if there always will be problems in adapting commercial software to a company's specific operations, the time and money saved might be well worth it.

ABC Analysis

The nineteenth-century Italian economist and sociologist Vilfredo Pareto was probably the first to state what has later come to be considered almost a universal law, namely, that in any group only a few of its members will be of real significance. For example, a small fraction of all items sold at a supermarket might generate the majority of the sales. Likewise, in any inventory a fraction, let us say 20%, of the stocked items might be significant in that they account for up to 80% of the total inventory investment. The figures just mentioned give this principle a special name, the 80/20 rule.

The contribution of the rule is that it helps management to concentrate on what is important. That is, it is better to spend some effort on the items that account for 80% of the cost ("the vital few") instead of wasting time analyzing the 80% of the items that cause only 20% of the cost ("the trivial many"). This would mean both a larger effort and a smaller reward. When investigating inventories, it is not the unit cost of each stocked item that is of interest, nor the volume of goods that passes through the inventory. Instead it is the combination of both. Specifically, an item's unit cost multiplied by its annual volume represents a figure that is related to the potential inventory investment.* Using the "annual cost volume" as a basis, items in an inventory system can be ranked and classified as belonging to class A, B, or C. Class A, which is normally composed of 10 to 25% of all items, usually accounts for 65 to 90% of the total cost volume. Class C items, making up around 45 to 55% of the items, might account for 5 to 10% of the cost volume. The remaining items belong to class B.

An example of an ABC analysis is shown in Exhibit 11.4.17 and in graphic form (called a Lorentz diagram) in Exhibit 11.4.18.[28] The division of items into separate classes is obviously subjective and is more difficult the flatter the Lorentz curve is. This analysis can have various uses. It can help decide on the frequency with which items should be cycle counted (see "Inventory Records Accuracy" below). It can also be used in connection with lot sizing in time-phased records. A common technique here is the fixed period requirement procedure.[29] The procedure adds together net requirements over a certain number of weeks. Together with ABC analysis, the procedure can, for example, look as follows: A items are ordered every 3 weeks, B items every 6 weeks, and C items every 12 weeks. The primary use of ABC analysis, however, is to help point out items that can be controlled in less rigorous ways.[30]

All modern inventory management software packages will perform ABC analyses. It can be argued, however, that with the advent of inexpensive data processing, the usefulness of this technique has diminished. The reason would be that all items can easily be controlled in the same way using the computer. Extensive record keeping and parameter calculation can therefore be done for all items, regardless of annual cost volume. There is some merit to that argument. However, if a company controls a large amount of items, ABC analysis is still a valid way to cut down the number of items that need close managerial attention. Furthermore, parameter estimation, such as the determination of statistical parameters used in selling reorder points, requires a lot of effort that might not be worthwhile for class C items. Instead, more approximate methods can be used.[31] Finally, there are still companies where the degree of computerization is relatively low and where these arguments are especially relevant. It should, by the way, be pointed out in this context that a strict use of ABC analysis in such a way that, for example, all C items should be controlled by two-bin systems is dangerous. There is an obvious difference between two items with the same annual cost volume if one has a high unit cost, but low volume, and the other item represents the opposite situation. Most likely, they should be treated differently. Furthermore, a C item in a manufacturing situation might be of equal importance to an A item. (Production cannot stop because of a missing bolt.) The ABC analysis should therefore be restricted to independent-demand items. For further aspects of ABC analysis, see Zimmerman.[32]

Inventory Records Accuracy

It is of utmost importance that the amount stated in an inventory record actually correspond with the physical inventory level in the stockroom. The information read from the file forms the basis

*Just as is the case with the inventory turnover concept, ABC analysis is based on the idea that inventory investment, or the holding cost, is the most critical variable in inventory management. Also see section on the inventory turnover rate.

Exhibit 11.4.17 Example of an ABC Analysis

Product	Cumulative Percentage of All Products	Unit Cost ($)	Annual Usage (Units)	Annual Cost Volume ($)	Cumulative Annual Usage ($)	Cumulative Percentage of Total Annual Usage	Value Class
B321	10.0	4.24	29,387	124,600	124,600	31.3	A
X474	20.0	1.50	62,333	93,500	218,100	54.7	A
X462	30.0	6.38	11,945	76,210	294,310	73.9	A
A124	40.0	15.98	2,550	40,750	335,060	84.1	B
C676	50.0	0.88	32,045	28,200	363,260	91.1	B
K131	60.0	1.08	15,370	16,600	379,860	95.3	C
L143	70.0	2.32	3,621	8,400	388,260	97.5	C
B068	80.0	8.10	753	6,100	394,360	99.0	C
C111	90.0	0.26	11,730	3,050	397,410	99.7	C
K832	100.0	3.27	312	1,020	398,430	100.0	C

Exhibit 11.4.18 Lorentz Diagram

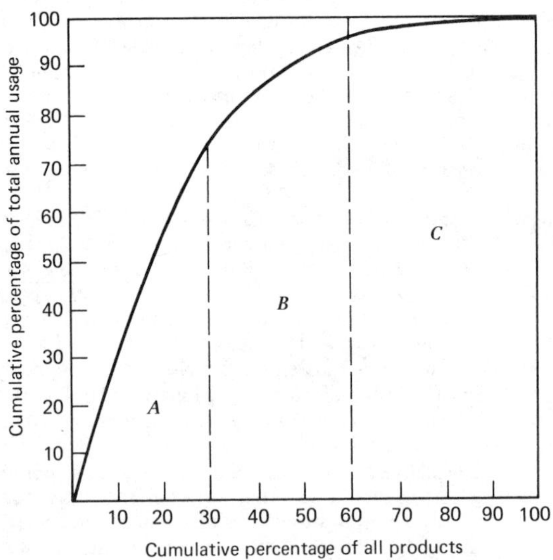

for customer order promising, production scheduling, and issuing of vendor orders. If the information is incorrect, it can lead to shortages and production inefficiencies or excessive holding of inventory.

There are many reasons why records and actual inventory levels are not always in accordance. In most cases it can simply be reduced to the fact that a physical handling of the goods was accompanied by an incorrect inventory transaction, or by no transaction at all. For example, goods can be miscounted in the receiving area; material can be moved to an incorrect storage area; personnel might read incorrectly from, or write incorrectly on, a materials requisition; personnel can remove items from the stockroom without notifying whoever is responsible; and so on.* Discrepan-

*Note the following sign that appeared on a stockroom wall: "If you must steal, leave a transaction card."

cies are especially common, and serious, in situations where companies keep separate inventory files without having a centralized data base.

Thus inventories must be checked now and then to see whether records and physical inventory are in agreement. This is necessary not only from an operational standpoint, but also because of auditing reasons. The inventory balance must be correct at least once a year, when the annual balance sheet is prepared. An annual inventory count forces a manufacturing company to practically shut down for a couple of days. Many people will be involved in the counting, several of whom have no experience or interest in the task. (For useful advice in connection with taking inventory, see Janson.[33]) The cycle counting technique,[34] on the other hand, is based on the concept of counting all items in the inventory system throughout the year, with an item frequency based on ABC analysis. There are several advantages to using this approach, which explain its increased use in recent years. Corrections to the inventory records will have less of an impact since not all errors are corrected at the same time. Special personnel can handle the counting all year around. Items can be counted when it is most beneficial to count them, that is, when the inventory record indicates a low or zero inventory balance. A 100% accuracy should, of course, be strived for, but can hardly be set up as a realistic goal. Instead, a 95 to 98% level of accuracy should be targeted.

Slow-Moving Items and Obsolete Inventory

One of the advantages with a good record keeping system is that it not only can tell current on-hand balances, but also can supply information on when the material in question was last issued. A modern computer-based system can, on request, produce a listing of all items that have not been disbursed from inventory for, let us say, 12 months. Items with such a low turnaround time should be looked at more closely. The reason might be that the products are not selling anymore; that the material, because of engineering changes, is no longer used in production; and so on. Whatever the reason, it must be decided whether the items should remain in the system or whether they should be considered obsolete and be written off. There is no use in occupying space in the storerooms and in the information system for items no longer in circulation.

REFERENCES

1. J. A. ORLICKY, *Material Requirements Planning*, McGraw-Hill, New York, 1975.
2. J. F. WESTON and E. F. BRIGHAM, *Essentials of Managerial Finance*, 3rd ed., Dryden Press, Hinsdale, IL, 1974, p. 57.
3. J. A. ORLICKY, *Net Change Material Requirements Planning*, G320-8170-0, IBM, 1975, (reprinted from IBM Systems Journal, 1973).
4. J. F. MAGEE, "Guide to Inventory Policy–I. Functions and Lot Size," *Harvard Business Review*, January–February 1956, pp. 49–60.
5. W. W. CHAMBERLAIN, "Is There an EOQ for All Seasons or Can We Make Our Current System More Responsive?," *Production and Inventory Management*, Vol. 18, No. 1 (1977), pp. 25–34.
6. L. K. ANDERSON, "The Framework of an Information Monitoring System for an Economic Order Quantity (EOQ) Inventory Model," unpublished doctoral dissertation, University of Wisconsin, Madison, 1970.
7. D. M. LAMBERT, *The Development of an Inventory Costing Methodology: A Study of the Costs Associated With Holding Inventory*, National Council Of Physical Distribution Management, Chicago, 1975.
8. R. E. ZIEGLER, *Criteria for Measurement of the Cost Parameters of an Economic Order Quantity Inventory Model*, unpublished doctoral dissertation, University of North Carolina, Chapel Hill, 1973.
9. T. E. VOLLMANN, *Operations Management*, Addison-Wesley, Reading, MA, 1973, p. 534f.
10. D. M. LAMBERT and B. J. LA LONDE, "Inventory Carrying Costs," *Management Accounting*, August 1976, pp. 31–35.
11. "State-of-the-Art Survey," *Production and Inventory Management*, Vol. 15, 1974, p. 8.
12. G. W. PLOSSL and O. W. WIGHT, *Production and Inventory Control*, Prentice-Hall, Englewood Cliffs, NJ, 1967, p. 90.
13. G. HADLEY and T. M. WHITIN, *Analysis of Inventory Systems*, Prentice-Hall, Englewood Cliffs, NJ, 1963.
14. R. PETERSON and E. A. SILVER, *Decision Systems for Inventory Management and Production Planning*, Wiley, New York, 1979.
15. R. A. BONSACK, "Inventory Ratios–Reader Beware," *APICS Conference Proceedings*, American Production and Inventory Control Society, Washington, DC, 1975.

16. I. A. G. KRUPP, "Inventory Turn Ratio–Still A Valid Concept," *Production and Inventory Management*, Vol. 18, No. 1 (1977), pp. 10–24.
17. ORLICKY, *Material Requirements Planning*, p. 22.
18. PETERSON and SILVER, *Decision Systems*, p. 214.
19. S. MAKRIDAKIS and S. C. WHEELWRIGHT, *Forecasting–Methods and Applications*, Wiley, New York, 1978, p. 72.
20. D. W. FOGARTY, "Utilization and Effectiveness of EDP in PIC: An Industrial Survey," *Production and Inventory Management*, 2nd quarter, 1977, p. 19.
21. J. S. CHAMBERS, S. K. MULLICK, and A. A. SMITH, "How to Choose the Right Forecasting Technique," *Harvard Business Review*, July–August 1971, pp. 55–64.
22. S. EILON, *Elements of Production Planning and Control*, Macmillan, New York, 1962, p. 244.
23. R. G. BROWN, *Smoothing, Forecasting and Prediction*, Prentice-Hall, Englewood Cliffs, NJ, 1963.
24. L. A. JOHNSON and D. C. MONTGOMERY, *Operations Research in Production Planning, Scheduling, and Inventory Control*, Wiley, New York, 1974.
25. ORLICKY, *Material Requirements Planning*, p. 90.
26. G. K. GROFF, "A Lot Sizing Rule for Time-Phased Component Demand," *Production and Inventory Management*, Vol. 20, No. 1 (1979), pp. 47–53.
27. PETERSON and SILVER, *Decision Systems*, p. 214.
28. K. G. LOCKYER, *Factory Management*, 2nd ed., Pitman, London, 1969, p. 322.
29. ORLICKY, *Materials Requirements Planning*, p. 124.
30. PETERSON and SILVER, *Decision Systems*, p. 78.
31. PETERSON and SILVER, *Decision Systems*, pp. 424–445.
32. G. ZIMMERMAN, "The ABC's of Vilfredo Pareto," *Production and Inventory Management*, Vol. 16, No. 3 (1975), pp. 1–9.
33. R. L. JANSON, *Production Control Desk Book*, Prentice-Hall, Englewood Cliffs, NJ, 1975, p. 66.
34. M. J. HABLEWITZ, "Cycle Counting," *APICS Conference Proceedings*, American Production and Inventory Control Society, Washington, DC, 1977.

BIBLIOGRAPHY

BUCHAN, J., and E. KOENIGSBERG, *Scientific Inventory Management*, Prentice-Hall, Englewood Cliffs, NJ, 1963.

LEWIS, C. D., *Demand Analysis and Inventory Control*, Saxon House, Westmead, Farnborough, England, 1975.

TERSINE; R. J., *Production/Operations Management: Concepts, Structure, and Analysis*, North Holland, New York, 1980.

The reader is also referred to the following journals covering the field of inventory management and production planning:

Harvard Business Review
Interfaces
Internnational Journal of Operations and Production Management
International Journal of Production Research
Journal of Purchasing and Materials Management
Management Science
Production and Inventory Management

CHAPTER 11.5

Capacity: Its Measurement and Management

ROBERT A. BOEHMER

General Motors Corporation

11.5.1 INTRODUCTION

Historical Concepts of Capacity

American industry and governmental agencies have struggled with the concept definition and measurement of capacity for many decades. Yet an honest appraisal of the results of the effort brings one, at best, to a rather bewildered state of confusion. It appears there are as many definitions of capacity as there are means of measuring it.

The reasons for this condition are apparent. Personnel engaged in the capacity affecting processes are interested in arriving at definitions and measurements that suit their particular purpose within their specific sphere of influence. Thus definitions used by individual firms are found with qualifiers such as "normal," "effective," "rated," "boom," "recession," "planned," and "actual." These definitions are also aggregated for divisions within various corporate enterprises so as to arrive at total company capacity. Additional aggregations are made to include companies within an entire industry and for larger divisions of the total economy.

Objections to specific definitions are commonplace because the definitions lack universal applicability. For example, capacity may be defined under conditions of a two-shift operation with equipment downtime of 20% for a given single-product firm. Another firm producing a variety of products may define its capacity under conditions of a three-shift operation with 35% equipment downtime. It is argued that the operational policies and procedures of these firms, as well as their respective product lines and market conditions, are justification for the differing bases of defining capacity.

The term "capacity" itself has caused much confusion at times. It is often used synonymously with "productivity," another term that needs consistent definition. But these terms are not the same. Productivity relates to the *actual* rate of output, sometimes using time as a base, sometimes implying cost as a base. Capacity, on the other hand, is "the plan." It is output *potential*, under given resource conditions. It is the *utilization* of capacity that results in productivity.

There is general agreement that the concept of capacity involves the relationship of output, time, and cost under given investment conditions. Broadly speaking, therefore, capacity is concerned with limitations placed on output over some period of time. These limitations may be physical, economic, or a combination of both. It is common practice to identify capacity as output that is limited by the slowest, or bottleneck, operation over some defined time span. Thus the focus of capacity studies normally centers on the relationship of the limiting piece(s) of equipment and the people operating it.

There is no question of the vital importance of establishing accurate levels of capacity at the level of the firm, the industry, sectors of the economy, and the economy as a whole. Yet the pursuit of a unique, consistent, comparable, and accurate definition and measure of capacity continues to be a very elusive quarry at all levels.

The author wishes to express his thanks to Irvin Otis, PE, AIIE fellow, Manager, Industrial Engineering, Corporate Manufacturing Staff, American Motors Corporation, for his reading of an earlier draft and for offering constructive editorial criticism. Also, the efforts of Eleanor Varga in typing and reading the manuscript are gratefully acknowledged.

The Need to Improve Capacity Measures and Management

Perhaps at no time in the history of the American economy have the measurement and the resultant management of capacity been more crucially important. This stems primarily from the present and projected future scarcity of capital in the U.S. economy. The rising cost of investment dollars for domestic use, combined with increasingly effective competition from foreign firms, makes improved utilization of existing plant and equipment a short- to medium-term imperative.

It has been a combination of underutilization of capacity and reduced rates of productivity that has fostered the inflationary spirals that further jeopardize our economic well-being. This is not to downplay the inflationary role of basic energy costs during the middle to late 1970s. It is a recognition of the impact of idle capital equipment on the standard of living. As a consequence, new machinery and equipment costs, along with labor rates, continue to escalate. Unskilled and semiskilled labor, replaced by more capital-intensive processes, are in turn replaced by more highly paid technicians and maintenance personnel to support the new equipment. Direct and indirect material costs continue to rise, thus increasing the cost impact of scrap generated by automated processes. And finally, in increasing numbers, labor is winning high levels of economic security. Direct labor costs are becoming more and more fixed, as workers are paid for being sent home for "company reasons," for example, machine breakdowns and stock shortages. Funds are created for supplemental unemployment benefits for longer-term layoffs, and when these private sector resources are depleted, the public sector takes over with various unemployment and welfare programs. All of these are social costs, whether funded by the firm or the government. And as the fixed cost of labor becomes more apparent, the alternate costs of the use of labor versus equipment will likely take on a different look.

Capacity—"the plan"—is a statement of the expected results of combinations of people, plant, and equipment, measured in terms of output, time, and cost. Improved concepts and measures of capacity are needed to improve its management and results in terms of returns to all people involved.

Method of Approach

The concepts of capacity measurement and management in this chapter are approached under four major topics:

1. A study of the factors involved in capacity management.
2. A universal definition of the concept of capacity.
3. A suggested technique for use in defining the capacity plan.
4. Extended use of the technique in supporting capacity management.

11.5.2 THE FACTORS INVOLVED IN CAPACITY MANAGEMENT

The principal factors involved in capacity management are the same as those concerning management of any enterprise or business. They are as follows:

1. **Providing the Answer to the Question "What?"** This vital first question is really an offshoot of the core mission, or "purpose for getting out of bed and going to work in the morning," of the firm. What products, parts, or services is the firm going to provide, and in what quantity, in what market? Based on answers to these questions, the cost and price structure can also be addressed.

2. **Providing the Answer to the Question "How?"** This second question, in combination with direction from the first, forms the short-, medium-, and long-term goals and objectives of the firm. How is the product, whether durable or service oriented, going to be provided? How much of it will be purchased complete or provided by some outside resource under lease or contract? How much will be made or provided in-house? And by what methods, processes, and support systems will the product be produced or made available for sale?

3. **Providing Resources.** Given the limits established by answers to the first two questions, operational resources can then be provided. This may entail acquisition of capital resources, depending upon the position of the firm at the time of capacity decision making. Other traditional resources include building, equipment, process, and support materials; people; and the systems to integrate all of these.

4. **Maintaining Resources.** This stage of capacity management assumes movement into some position of maturity of the business activity. All of the resources, capital, and people included must be maintained at levels commensurate with the mission, goals, and objectives of the firm. Of the remaining resources, the easiest to maintain are building, equipment, and materials. This is because they are tangible—measurable in both quantity and quality. The qualitative aspects of people and systems are not so easily measured and thus suffer the most from lack of maintenance.

5. **Measuring Results.** This is simply a matter of comparing the results of sales, output, and cost to the goals and objectives over determined periods of time.

6. **Improvement.** This last factor includes such concepts as "invent and innovate." It also includes "borrow and copy." History abounds with businesses that succeeded without "inventing here."

The interrelationship of these six factors forms the capacity management cycle for the firm.

11.5.3 THE CONCEPT OF CAPACITY

Definition

Capacity is a statement of the expected results of combining people and facilities. This predictive plan is normally stated in terms of rate, for example, output per unit of time. This is the case because capacity, by its nature, is the end result of resource expenditure put into action. It is believed that the concept of capacity has suffered because of this almost parenthetical reference to the base cost factor.

The concept of capacity is thus defined here as *maximum sustainable output* (of goods or services) *over a stated period of time, at a cost*. Three items require further explanation:

1. **Maximum Sustainable Output.** The key word is "sustainable." Capacity involves taking a look at all items affecting output, from the absolute maximum output rate, for example, gross pieces per hour, to all of the interferences that result in losses to that potential maximum. Thus the term "sustainable" recognizes losses that the firm comes to accept within its given resource plan.

2. **Period of Time.** Capacity is normally stated in annual terms. This has come about primarily because of the role of capacity in establishing annual financial budgets. There is no reason why the period cannot be stated to be shorter or longer than a year. What is vital is that the period *is* stated.

3. **Cost.** Cost refers to and includes *all* resources judged necessary to operating and supporting the operation in order to achieve maximum sustainable output at a cost acceptable to the profit goals of the firm (see Exhibit 11.5.1).

Major Conceptual Problems

In virtually every capacity study, there are two definitional hurdles to overcome. They stem from establishing bases of measurement for (1) the type of output rate and (2) the time frame to be used. It is reasonable to conclude that the major reason for this problem revolves around the establishment of a capacity base for the firm, which, when measured for utilization, offers some reasonable opportunity for "success." It is, after all, common to accept that continuous-process industries operate on three shifts per day, 7 days/week; that assembly lines operate two shifts per day, 5 days/week (with provisions for overtime); that major automated machining processes operate at 60 to 65% of gross output rates, while so-called conventional machines operate at 75 to 80% of their gross potential; and so on. These assumptions often result in output rate and time frame bases that are less than that which can actually be achieved. Utilization performance is thus given reasonable assurance of being graded satisfactory and, given a strong demand, of being judged outstand-

Exhibit 11.5.1 Examples of Cost Factors

Operational Costs

Gross machine pieces per hour (machine cycle, no interferences)

Standard pieces per hour × labor cost per hour (man assignment method plus labor allowances)

Worker allocation—one worker, one machine; one worker, two machines; and so on

Budgeted labor efficiency (assumed losses over and above those included in standards)

Support Costs

Normal quality and supply of materials

Product mix } Affect changeovers
Inventory management }

Tool quality

Maintenance procedures

"Presenteeism" of the labor force

The accuracy of sales forecasts versus actual schedules

Other

ing. This means capacity utilization percentages ranging from 80 to 100 and above. These are the standards that have come to be known as "doing well." (Historically, A's have not been given for grades of 40 to 60%).

Unfortunately, these same assumptions of rate and time virtually ensure a discontinuity in the capacity utilization series for the firm over its life span. Divisions within the same company that produce different products are not capacity comparable. And quite obviously, there is no chance to aggregate capacity accurately on the same base for an industry, a sector of the economy, or the economy as a whole.

Another serious problem affecting the establishment of capacity levels is the absence of, or extreme difficulty in obtaining, data concerning losses to optimum (gross) output at the bottleneck. The idea of equating capacity to maximum *sustainable* output is generally accepted. The problem in many firms is in documenting the type and extent of all bottleneck output losses. (Refer to Exhibit 11.5.1 for a typical listing of such losses.) Scrap losses have traditionally been measured for reasons of inventory control. Other loss areas, such as maintenance, changeover, and product mix, are becoming increasingly measureable with the aid of the computer.

Apart from getting data that affect the establishment of capacity targets, these measurements form a solid basis for determining the effectiveness of support staffs to production. This accountability factor is essential in order to enlist the wholehearted support of production in the establishment and operation of the capacity plan.

The "Ultimate" Problem

A review of the preceding material in this section reveals the following highlights:

1. Determining capacity involves the identification of the output limiting operations(s) of the firm.
2. Determining capacity involves the quantification of the level of maximum sustainable output over an extended period, typically a year.
3. Determining maximum sustainable output involves identification and quantification of all losses to gross (maximum) bottleneck output.

These concepts do not appear to cause undue concern in practical circles at the level of the firm, nor in circles that aggregate capacity utilization on more macro levels, such as the Commerce Department, McGraw-Hill, the National Industrial Conference Board, the Wharton School, or the Federal Reserve Board. In virtually all cases, it is recognized that maximum sustainable output will be less than gross (maximum, no loss) output. It will also be *less than standard* output, or that which is established from work standards as a base for labor productivity. This latter concept is easily accepted, since standard allowances normally do not recognize such nonrouted losses as machine breakdowns, scrap, lack of stock, deviated stock, defective tooling, changeovers, lunch breaks, absenteeism, deviation from the established manual method, power failures, and other "acts of God" and the like. Yet all of these interferences (losses) to maximum potential (gross) output do occur regularly. It is clear, then, that the capacity target will be some "net of standard," or "net of net," of gross.

Exhibit 11.5.2 describes the "ultimate" problem. The basic question is, how *much* less than standard output is capacity output? What are the average (normal) losses incurred (tolerated) by the process design and level of support (budgeted or actual) as evidenced by the dollars spent for those services?

Baselines for Establishing and Managing Capacity

There are three essential baselines for establishing and managing capacity upon which the remainder of the chapter focuses. They are as follows:

1. Output rate and time frame base data.

a. Output rate *base data* will be taken as maximum possible (gross output, no losses).

b. The time frame base will be 24 hr/day. The days per week will be consistent with the norms of the specific industry. For example, durable goods industry, 5 days/week with provisions for overtime; continuous-process industries, 7 days/week; and so on. *Note*: The percentage of capacity output to gross potential will be (much) lower than traditionally seen as a result of using maximum output rate and time frame as bases. However, the consistency, measurability, and comparability of data over time will more than compensate for the effort in changing the interpretation of relative values.

2. Improved systems must be developed to measure and report actual performance of all pertinent operational *and* operations support factors. Computer-assisted programs must be developed

Exhibit 11.5.2 Capacity Leveling–The Ultimate Problem

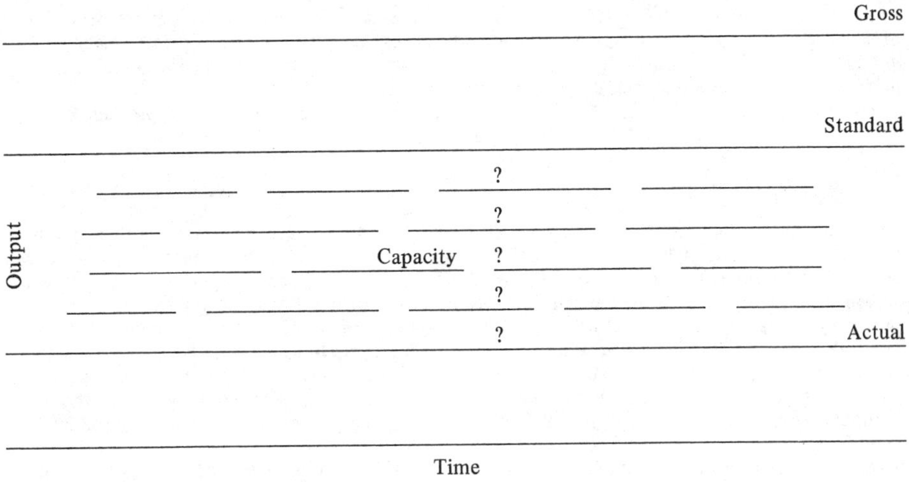

Gross

Standard

?

?

Output

Capacity ?

?

? Actual

Time

(a) to perform mundane weighed average calculations of long lists of values and (b) to aid in loss-to-gross output calculations for such things as maintenance, scrap, and changeover. Many larger industries already have these tools. With the rapid reduction in computer hardware costs, it is hoped that shared-time computer resource help will soon be economically available to all units of industry. Without these aids, *maintenance* of an effective capacity management system is very difficult.

3. A technique called "performance analysis" will be used to develop the capacity plan and manage its results. This empirical approach assumes that a firm already exists and that it has performance to *measure*. It should not, however, be difficult to make effective use of the methodology even if the firm is just in the original planning stage. Additional help in quantifying capacity from a theoretical standpoint can be found in Chapters 1.6, 9.6, 13.9, and 13.11.

11.5.4 ESTABLISHING THE CAPACITY TARGET

Three major approaches are normally used in varying combination to set the capacity target for a firm.

1. The engineering approach, based on design limitations at the bottleneck operation(s).
2. The economic approach, based on financial limits, marginal analyses, and optimization models.
3. The empirical approach, based on measuring what is actually taking place.

The empirical approach is the primary one used in this chapter. The reason for highlighting a study of "what is actually taking place" is based on a number of points. First, the engineering and economic approaches can define only what is expected to occur. The sustained long-term accuracy of these predictions is only as good as the data that are being fed back in some *consistent* and *uniform* manner. (The emphasis here is to underscore again the reason for using gross (no loss) output over a maximum time frame as a base for comparison.) Second, short-term business success (e.g., for annual fiscal periods) is likewise predicted on accurate performance feedback. Since capacity involves not just output over time, but also the effectiveness of all *support* factors on output, the need for a measure that accounts for the performance of all factors over the total time frame is essential. Third, it is known that plans are effective only to the extent that the personnel who carry them out understand and agree with them.

This approach to measuring what is and to establishing capacity targets involves the line organization as well as staff planners. The process involves five steps:

1. Defining the bottleneck operation.
2. Gathering data.
3. Establishing the performance analysis.
4. Establishing the capacity target.
5. Developing a program to continue utilization measurement.

Defining the Bottleneck Operation

Extreme care must be taken in this step. The temptation is to assume that the engineering estimate of slowest process or slowest operation is the bottleneck. In actuality, the "slowest" machine may *not* have the lowest effective output rate. The relationship of output rate per hour and the actual hours the job is able to run each day will determine the true bottleneck.

It is advised that bottlenecks be identified for every component of the total product and that each component be ranked relative to its output limitation on the total product.

Gathering Data

Once the true bottlenecks have been identified and ranked, the gross potential (no loss, no interference) output capabilities are determined, usually being expressed in terms of units (pieces or assemblies), volumes (cubic measures), or weights (pounds or kilograms) per hour. It is generally agreed that gross potential output will include the effect of labor only, to the extent that no output would happen without it. For example, in a manually operated machining operation, the load and unload of the part would be added to the machine cycle in order to arrive at the gross operational time. In an automated machine process, only the part transfer would be added to the machine cycle. In a manual assembly operation, only the work elements contributing to the direct assembly would be included. Other elements such as gauging, operator relief, or personal time would *not* be included.

Interferences to gross output are the next to be identified. There are two basic categories: routed (planned and regular) and nonrouted (not planned or regular). Some examples of interferences are noncyclical allowances, such as tool change, stock handling, and operator personal time, which are added to the repetitive manual elements to form a work standard; repetitive manual work elements that interfere with equipment output, such as gauging outside the machine cycle and multiple machine-man work assignments that result in machine wait time; lunch time; scrap; setup or changeover; equipment maintenance, whether planned or unplanned; and so on. The losses should be classified as routed or nonrouted. The particular choice of classification is left to the discretion of the firm. It is suggested, however, that the so-called standard attainable output, or output defined by the labor work standard, be identified. A comparison of standard attainable with gross attainable output defines the extent of loss incurred through the process *plan*.

Actual output of good quality product is the last major piece of data to be established.

Establishing the Performance Analysis

The term "performance analysis" indicates the process of identifying all of the pertinent data described in the preceding step and relating their significance to actual and potential output over some period of time. Exhibit 11.5.3 describes data for department ABC, a multiproduct (E and E′) department, over a 6 month time frame.

This data table is typical of that which might be expected from a department of a manufacturing firm producing multiple products. It defines base data concerning the bottleneck operations for the products made. This series of data, extending over a 6 month period, would likely represent true operating conditions.

Columns 11 and 12 are of interest. Note that the actual utilization of the bottleneck machines is only 43.4% of their gross attainable. Although this is only an example of a data analysis, durable goods manufacturers are coming to realize that highly automated transfer equipment really averages closer to 50% utilization than previous expectations of 60 to 65% actual utilization, for example, performing their designed, functional output process.

Column 12 indicates that the so-called standard cost utilization is 72.5% over the 6 months. This is interpreted to mean that the *best* utilization of the process would result in output slightly less than three quarters of the time. This level of output over a sustained period is virtually impossible, however, since it precludes losses not included in typical operational work standards, such as maintenance, scrap, changeover, and lunch.

Exhibit 11.5.4 plots the monthly U_A and U_{SC} data from Exhibit 11.5.3 and allows for a graphic demonstration of deviation from gross attainable. Gross, since it is the base, will always be 100% and thus will always be a horizontal line. The U_{SC} moves up and/or down on a monthly basis because the respective bottleneck standards on the two products differ. Thus the mix of the two products affects the weighted average standard attainable output. The movement of U_{SC} directly affects U_A to some degree, as it must. But the largest contributors to the ultimate position of U_A are the so-called nonstandard, or nonrouted, losses. These nonrouted losses are all taken as percentages of gross attainable. Exhibit 11.5.4 may thus be interpreted as follows: The U_A, which averages 43.4% for the 6 month period, would have been 46.4% had it not been for the 3% loss for scrap; an additional 9% gain in gross attainable would have been possible if changeover had not interrupted output. The U_A would now have been 55.4%; losses for maintenance (9.9%) and lunch

Exhibit 11.5.3 Performance Analysis

Department ABC	Product E	Product E′
Gross output:	20.0 pieces/hr	30.3 pieces/hr
Standard output:	15.4 pieces/hr	20.7 pieces/hr
Bottleneck:	Operation 10 (mill)	Operations 10, 12, 14 (mill and drill)

1	2	3	4	5	6	7	8	9	10	11	12	13
					Gross Attainable		Standard Attainable					
Month	Bottleneck days[a]	Total Average Shifts[b]	Changeovers[c]	Schedule[d]	Pieces per Hour[e]	Pieces per Month[f]	Pieces per Hour[g]	Pieces per Month[h]	Actual Output	U_A (%)[i]	U_{SC}[j] (%)	Standard Allowance Loss (%)[k]
Sept.	30	2.75	5	19,330	27.1	19,512	19.2	13,824	6,316	32.4	70.8	13.3
Oct.	31	2.50	3	16,855	23.7	17,632	17.4	12,946	7,479	42.4	73.4	15.4
Nov.	29	2.50	3	15,310	24.0	16,704	17.5	12,180	8,208	49.1	72.9	18.2
Dec.	23	2.75	2	11,209	28.8	15,898	20.0	11,040	6,118	38.5	69.4	17.0
Jan.	26	2.50	2	10,299	21.0	13,104	16.0	9,984	6,915	52.8	76.2	16.5
Feb.	23	2.63	1	7,025	24.1	13,303	17.7	9,770	6,670	50.1	73.4	18.1
Year to date	162	2.61	16	80,028	24.7	96,153	17.9	69,744	41,706	43.4	72.5	16.1

[a]Number of days per month worked at the bottleneck operation.
[b]Average staffing of the total department over three shifts.
[c]The number of line changeovers per month.
[d]The monthly schedule in total pieces (product E plus product E′).
[e]Weighted gross output pieces per hour average of actual output of products E and E′ per month.
[f]Column 2 × (colume 6 × 24).
[g]Weighted standard outpieces per hour average of actual output of products E and E′ per month.
[h]Column 2 × (column 8 × 24).
[i](Column 10 ÷ column 7) 100 = actual utilization (U_A).
[j](Column 9 ÷ column 7) 100 = standard cost utilization (U_{SC}).
[k][(1 ÷ column 8) − (1 ÷ column 6)] [column 10) ÷ (column 2 × 24)] 100.

Exhibit 11.5.4 Performance Analysis of Department ABC

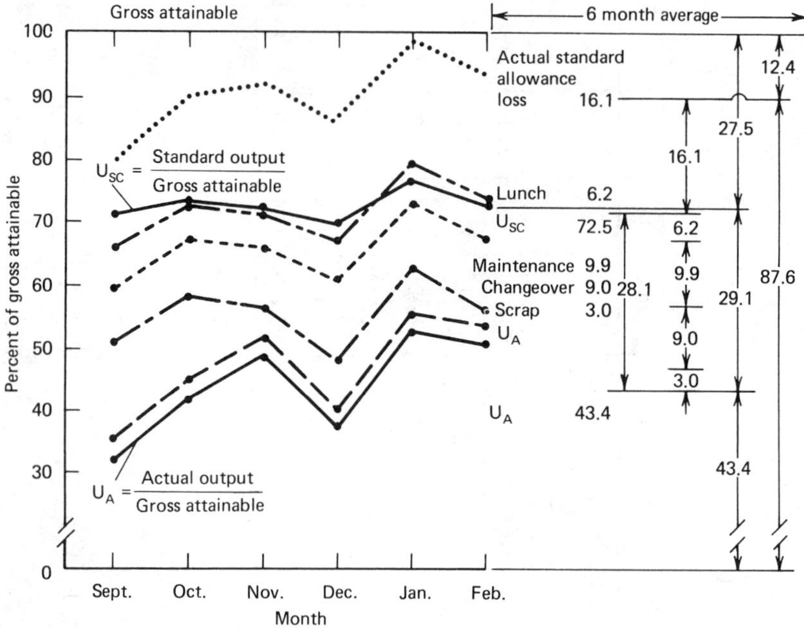

(6.2% at 3 half hour periods per day) account for another potential gain to gross attainable of 17.1%. Thus 28.1% of the 29.1% spread between U_{SC} and U_A has been accounted for. These performance measurements were made possible because the firm in the example had computer-assisted scrap control programs and maintenance dispatch and feedback systems. Changeover and lunch losses were maintained by a clerical person in the production office.

The final data line to be explained is the actual standard allowance loss. This is an expression of the *actual* losses to gross caused by allowances in the labor work standard extended by actual pieces *produced*. These data are shown added to the position of lunch loss.

Two important points are highlighted by this information. First, the losses implied by the difference in U_{SC} and gross (27.5%) will occur only if output is actually equal to standard (rarely, if ever). Second, only 87.6% of the activities composing the total gross have been identified. From an economic standpoint, this level of identification is probably sufficient. The unidentified losses (12.4%) comprise such things as inefficient methods deviations by the operator, use of materials that deviate from acceptable specifications, defective cutting tools, absenteeism, and new operator training and break-in periods—items that are difficult, if not impossible, to measure accurately on a sustained basis. In addition, it is the primary job of the foreman and his or her staff support to minimize these losses on an ongoing basis. It will be normal to expect a 10 to 15% void in the total explanation of 100% gross attainable bottleneck output.

There is one important loss that has not yet been explained. That loss is shifts not staffed. The bottleneck machinery in the example was staffed three shifts per day, so no loss for staffing was incurred. The procedure to follow on a two-shift operation is to identify a 33% loss to gross for the shift not staffed. Relate all other losses, initially, as percentages to a *16 hr* base gross attainable output. Then assume all losses (except lunch) would be the same percentage if the operation were to be run three shifts and plot on the 24 hr base graph. (Lunch loss on a two-shift basis typically would be included in the 33% shift-not-staffed loss.)

Establishing the Capacity Target

The output and loss data presented in the performance analysis in Exhibits 11.5.3 and 11.5.4 identify areas for potential improvement. Referring to a point established previously in this section, a team of people representing both staff and line organization units should now be called in by the data gatherers (typically, the industrial engineering department) to obtain agreement on potentials for output improvement and to assign responsibility for performance change. For example, examination of changeover and lunch loss data suggests improvement possibilities. Reduction in change-

Exhibit 11.5.5 Manufacturing Capacity

Department ABC

1. Part name Products E and E′

2. Bottleneck facility description Operation 10 (mill) for product E; Operations 10, 12, 14 (mill and drill) for product E′

	Present	24 Hr Present	24 Hr Proposed
3. Gross used time per day			
Less delays:			
Unplanned equipment downtime[a]		2.4	2.4
Planned equipment downtime[a]		0	0
Setup (not accounted for in work standard allowance)[b]		2.2	1.4
Material shortages[a]		0	0
Tool or fixture breakdown (unplanned)[a]		0	0
Manpower not available[a]		0	0
Lunch periods[c]		1.5	1.0
Miscellaneous (describe) ________		2.8	2.8
Nonidentified losses			
Scrap (not accounted for in work standard allowance)		0.7	0.7
4. Net productive time available per day		14.4	15.7
5. Gross production rate (pieces per hour)		24.7	24.7
Less:			
Operational work standard allowances		6.8	6.8
Group work standard allowances			
6. Net production rate (pieces per hour)		17.9	17.9
7. Daily capacity (item 4 multiplied by item 6)		257	282
8. Annual 3-8-5 capacity (item 7 multiplied by 240 days/year)		61,680	67,680
9. Effective date			

Prepared by: ____________________
Industrial Engineering

Approved by: ____________________
General Superintendent Production

____
Manager Production and Plant Engineering

____
Plant Manager

____
Manager Production Planning

[a] Daily average.
[b] Setup loss depends on reducing changeovers to an average of two per month.
[c] Lunch loss will be reduced by changing to three 20 min paid lunch periods.

Exhibit 11.5.6 Department ABC Capacity

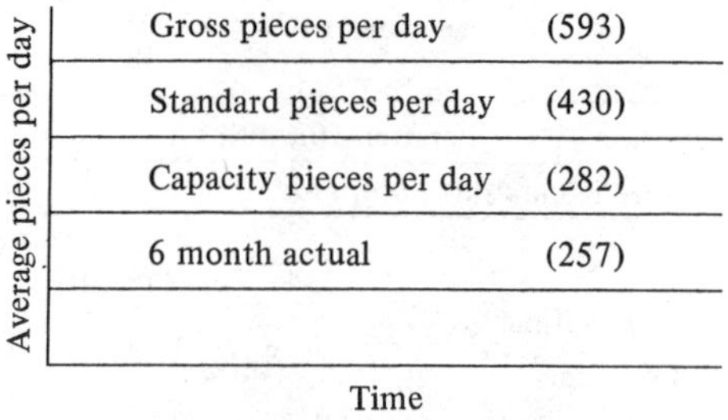

Average pieces per day	
Gross pieces per day	(593)
Standard pieces per day	(430)
Capacity pieces per day	(282)
6 month actual	(257)

Time

over from the experienced 2.67/month (16 ÷ 6) to 2.0/month will reduce loss to gross attainable by 2.1%.

$$162 \text{ days} \times 24 \text{ hr/day} \div 6 \text{ months} = 648 \text{ hr/month}$$

$$648 \text{ hr/month} \times 9\% \text{ changeover} = 58.3 \text{ changeover hr/month}$$

$$\frac{58.3}{2.67} = \frac{X}{2.0}$$

$$X = 43.6 \text{ changeover hr/month}$$

$$[(58.3 - 43.6) \div 648 \text{ hr/month}]100 = 2.1\%$$

Reduction in lunch loss from $1\frac{1}{2}$ hr/day to 1 hr/day will result in 2% improvement in gross attainable. (Any and all of the other losses can be examined in similar ways.)

Once agreement is reached by the responsible parties (production control and manufacturing in this case), a sign-off sheet similar to that shown in Exhibit 11.5.5 is attested to and the plan goes into effect. The agreement of all parties concerned on a new capacity level of an average of 282 pieces/day was based on high-credibility data combined with effective analysis. Thus the solution to the ultimate problem, the level of maximum sustainable output (capacity) over time at a cost, is arrived at.

If this technique had been in use during a previous capacity analysis, it would now be possible to compare results. Exhibit 11.5.7 presents a summary of (1) the results of the actual performance, (2) the plan to reduce the loss to gross attainable, (3) the new resultant capacity, and (4) the former capacity target.

One very important conclusion can be reached at this stage. It has been demonstrated that capacity can be increased with zero investment. In the example the firm paid an operating cost penalty for the 20 min paid lunch provision. Assuming a \$9.00/hr wage rate, \$9.00/hr × $\frac{1}{3}$ hr/day × 3 workers/day × 162 days ÷ 6 months = \$243/month or \$2916/year. This cost increase was incurred to achieve a 2% increase in capacity, or approximately 10 pieces/day or 2400/year. It was judged that the marginal profit increase from the sale of the additional 2400 pieces more than compensated for the \$2916 annual cost increase. A similar analysis could be made of the effect of reduced changeover (increased average lot sizes) on possible inventory costs. But that is not the purpose of this chapter.

What this example has shown is that it is possible to take a measured, rational look at the concept of capacity, to take it out of the realm of "black magic," and to make some sound business judgments based on the facts and on a group consensus that the decisions will work.

Developing a Program to Continue Utilization Measurement

Having gone through all the trouble of gathering the data for the performance analysis, it would be advisable to design them in a form that a computer could use to print out monthly capacity utilization and loss analysis reports. The report should contain the information shown in Exhibit 11.5.8.

Section A and section C data generally have high potential for direct computer-administered input from data-based files. Section B data depend largely on manual accumulation and input. A wide range of losses is listed (and perhaps an even wider range is *not* listed). They are intended as thought starters. Only those that are significant to the condition of the particular firm would be used.

Use of the data is the key, of course. The follow-up portion of almost any plan is perhaps the least-attended, least-used part. ("If it's going good, who needs it? If it's going bad, who wants it?")

Exhibit 11.5.7 Performance Analysis: Establishment of New Capacity, Department ABC

6 Month actual							New capacity target					
U_{SC} (%)	U_{SC} Loss to Gross	U_A (%)	U_A Loss to Gross	Average Gross Pieces per Hour	Average Shifts per Day	Actual Average Pieces per Day	Target Loss Reduction	Target % U_A	Reduced Loss Hours per Day	Increased Pieces per Day	Capacity Pieces per Day	Previous Capacity Target
72.5	27.5	43.4	29.1	25	2.6	257	4.1	47.5	1.0	25	282	260

Exhibit 11.5.8 Capacity Utilization and Loss Analysis Report

A. Part or Product Level Data

Identification numbers
Department producing part
Bottleneck operation
Current month schedule
Current month actual
Year-to-date schedule
Year-to-date actual output
Gross pieces per hour
Gross pieces per month
Standard pieces per month
Target (daily) capacity

B. Routed and Nonrouted Losses[a]

Scrap
Reworks
Changeovers
Downtime–breakdown
Downtime–preventive maintenance
Shifts not staffed
Shifts staffed
Days worked
Lunch
Routed allowances
Direct material not available
Indirect material not available
Manpower not available
Direct material quality
Feeds and speeds
Other

C. Capacity Utilization Performance Analysis[b]

Standard cost utilization
Actual utilization
Capacity utilization (actual output ÷ capacity output)
Percent schedule attained
Labor efficiency

[a]List all in terms of quantity and percentage of gross, monthly, and year-to-date.
[b]All data in percentages, monthly, and year-to-date.

Yet it is also understood by every competent manager that performance measurement and tracking is an absolute necessity for success. It is, perhaps, the single most important part.

11.5.5 USE OF THE TECHNIQUE TO SUPPORT CAPACITY MANAGEMENT

In addition to providing a potential for arriving at the most accurate and effective current capacity plan, this methodology also provides the experience base from which even more accurate estimates can be made for future plans. The time-to-date data accumulated in the follow-up performance analysis reports provides an easy-to-interpret and accurate statement of the condition of capacity utilization and the reasons for it. All of the output data and all of the factor losses are identified. The availability of a broad spectrum of product-related component capacities and capacity utilizations will permit the establishment not only of more effective annual budgets, but also of much more effective capital budgets.

11.5.6 SUMMARY AND CONCLUSION

In summary, the major topics covered were:

1. The factors involved in capacity management: What? How? Provide resources, maintain resources, measure results, and improve. The essential nature of total organizational involvement was stressed.

2. A presentation of the concept definition of capacity; the plan for sustained output level over a stated period of time at a cost.

3. A step-by-step demonstration of a performance analysis technique for measuring and identifying capacity utilization, using gross output capability and 24 hr/day as baselines.

4. An explanation of how and by whom the performance analysis data are used to establish annual capacity targets and longer-range capital expenditure plans.

To repeat a previous statement, the method of approach used in this work was slanted toward the "factory" practitioner. Its efforts were highly pragmatic and its data source highly empirical. The reader should not be led to conclude, however, that there is any feeling that the engineering or economic approaches are less worthy or have less potential for benefit to the firm. At some time, plans for investment-related capacity increases will need to be formulated. Engineering and economic tools will be called upon to form the vital plans and lead to the crucial decisions. It is firmly believed, however, that source data for engineering and economic capacity studies that result from programs and organizational methods recommended here will be superior to any experienced by traditional rules of thumb employed in the past.

With capital rapidly becoming one of this country's scarcest resources, businesses can ill afford to reach investment decisions with anything less than the optimum plan, from the most reliable data, generated and agreed upon by those who are responsible for carrying it out. This is the purpose for the measurement and management of capacity.

CHAPTER 11.6

Material Requirements Planning

GENE J. D'OVIDIO
RICHARD L. BEHLING

Booz, Allen & Hamilton, Inc.

11.6.1 OVERVIEW

Materials planning is a production and inventory control function found in virtually all manufacturing companies. Its objective is to manage and control inventories of the components and purchased parts or raw materials used in the manufacture or assembly of finished products made for stock or made to order. Control of these inventories is typically the job of material planners, who place component manufacturing or purchase orders to bring required components or raw materials into stock to meet assembly dates.

Techniques used to control these inventories must recognize two key characteristics of the demand for such components. First, since the products that require these components are usually made in lot quantities, demand for their components will occur in lumps at intervals. Second, the required quantities of these components can be calculated based on a forecast or prediction of the number of finished stock items required and when they must be assembled or manufactured. Material requirements planning is an inventory control technique specifically designed for use under these conditions of demand. Since the use of MRP typically requires the processing of significant amounts of data, the application of computers to manufacturing beginning in the late 1950s has greatly facilitated the application of this approach to inventory control. Today MRP enjoys wide use in a variety of manufacturing environments—particularly job shops or plants that employ discrete manufacturing processes.

In contrast to conventional statistical order point techniques used to control inventory (see Chapter 11.5), MRP is product rather than part oriented and is based on projected demand for the end item rather than on the historical behavior of its components. Beginning with a forecast of demand for an end item, the MRP approach requires the generation of time-phased material requirements based on the relationship between the end item and its components as expressed in a bill of materials. The major advantages of MRP over statistical order point techniques are its ability to handle intermittent, or "lumpy," demand effectively and its ability to project accurately anticipated changes in component requirements on a time-phased basis.

Selection of an Appropriate Approach

The nature of demand for an item is a determining factor in the selection of the appropriate inventory management approach. When the demand for an item is unrelated to the demand for other items, it is considered independent and is typically forecast. For example, finished products or spare parts sold unassembled are generally considered independent demand items. On the other hand, when the demand for an item or subassembly is directly related to or derived from the demand for another item or subassembly, it is considered dependent and can be calculated based on the demand for those items of which it is a component. The MRP approach to inventory control is particularly applicable to items with dependent demand. (It should be noted that some items may be either dependent or independent, depending on the way they are used. For example, in the case of service parts that are sold to customers and also used in production, service demand is independent, while production demand is dependent.)

The pattern of demand is another important factor in the choice of an inventory control technique. Statistical order point techniques assume relatively uniform, or continuous, demand, which is often the case for a finished product. Components used in the manufacture of end items, on the other hand, tend to be needed intermittently—that is, in discrete quantities—to satisfy lot sizing requirements at the end-item level. By recognizing the timing implications of intermittent demand,

MRP has the capability to delay replenishment until material is actually required. Under an order point system, in contrast, replenishment is automatically called for as soon as an order point is reached. In addition, MRP enables material planners to handle the lumpy demand that arises when several end items concurrently require the same component by aggregating the component requirements of all finished products, taking into account the points in the manufacturing process at which end items require their various components.

The Bill of Materials as a Key Data Source

To provide material planners with the information they need to effectively manage inventories of components with dependent demand, MRP requires basic data on how each end item provides data on the parts required, the quantity needed, the sequence in which the parts are required, and the necessary timing. In essence, the MRP approach calls for the calculation of future demand for all components on the basis of forecast demand for end items and the relationships between finished products and their components described in bills of material.

The Role of MRP In Production and Inventory Control

In the larger context of production and inventory control systems, MRP computer application systems play a central role in the overall flow of materials, as illustrated in Exhibit 11.6.1. The primary purpose of an MRP system is to support material planners in the timely and efficient execution of the master schedule by processing information and generating reports, which enable planners to ensure that both raw materials and manufactured components are available when needed and in the right quantities to support intermediate operations and final assembly. The key inputs that enable the system to carry out this task include the master schedule, or overall plan for production (Chapter 11.4); inventory records, which indicate balances on hand or on order plus the required lead times for purchasing raw materials or manufacturing components (Chapter 11.5); and bills of material for each product to be manufactured (Chapter 7.2). The output of the MRP system is a report containing a set of directives that enable material planners to execute the overall production plan by providing the purchasing department with schedules and instructions for obtaining raw materials and purchased parts and by directing the shop floor in the fabrication, manufacture, or assembly of work-in-process inventories.

11.6.2 BASIC MRP LOGIC

In essence, the MRP technique simulates, typically through the use of the computer, a plant's normal manufacturing processes. By simulating the flow of materials through the manufacturing process, the MRP system is able to determine:

What materials are required.
How much of each material is required.
When each material is required.

Beginning with the forecast (or supplied) demand for each independent demand item, the MRP technique derives the demand for the item's component parts using the item's bill of materials.

Preparation of the Master Schedule

The master schedule, which is a time-phased schedule of the production required to maintain desired levels of finished stock (Chapter 11.4), is the most important input to MRP. The master schedule is usually prepared on the basis of the following inputs:

Forecasts of demand.
Customer orders.
Finished goods stock requirements.
Service parts requirements.
Safety stock requirements.
Stock orders for production leveling or stabilization.

In a make-to-order plant, the master schedule may be derived solely from customer orders. In a make-to-stock plant, it may be based entirely on forecasts. In most plants, however, all of the inputs listed here are considered in the development of the master schedule.

The master schedule shows the quantity of required production in each period over a series of

Exhibit 11.6.1 The Role of MRP in Production and Inventory Control

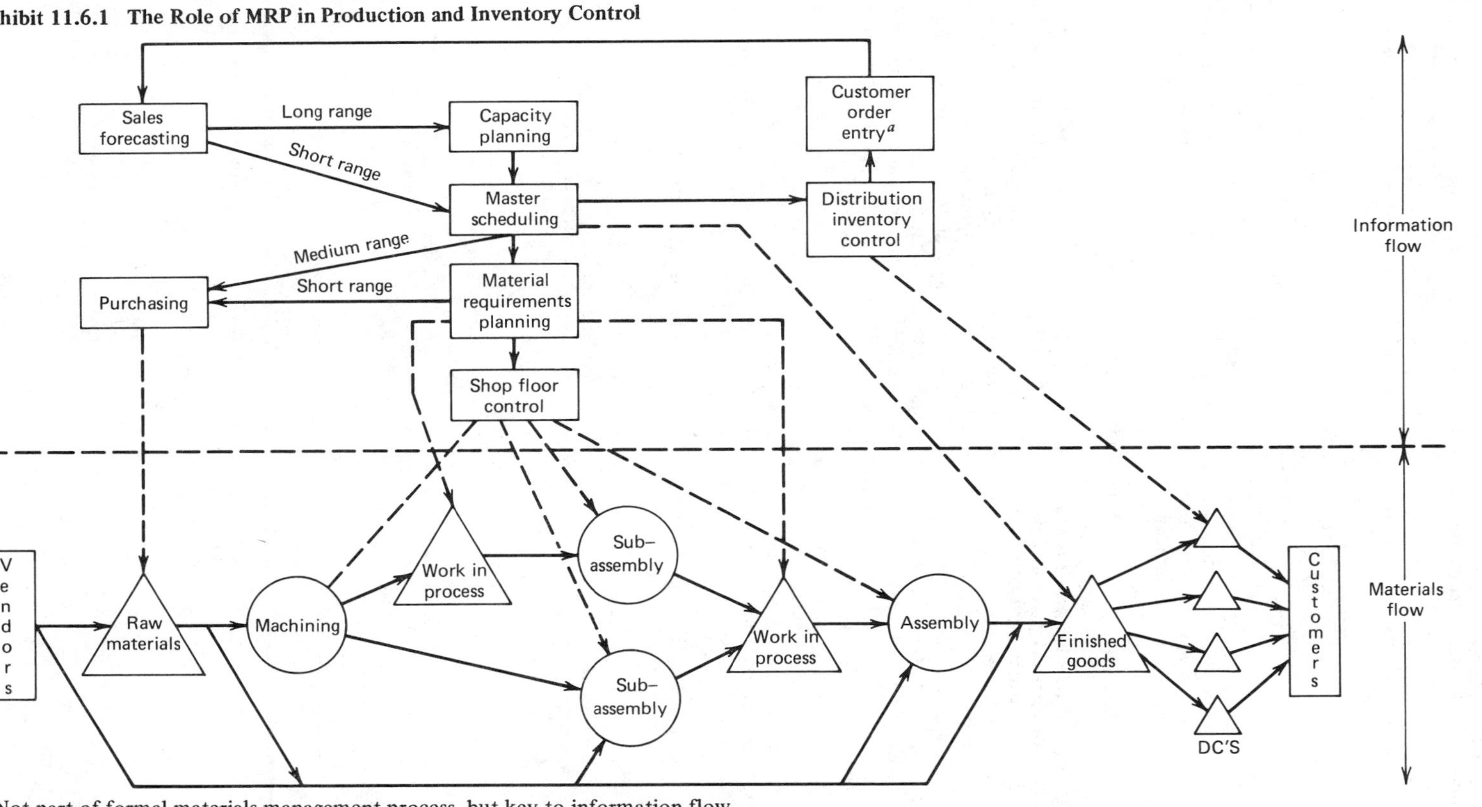

[a]Not part of formal materials management process, but key to information flow.

Exhibit 11.6.2 Master Schedule for Ajax Office Supply Company

	Period									
	1	2	3	4	5	6	7	8	9	10
Product A					80					100
Product X				60				40		
Product Z		100				100				100

continuous periods of fixed length, such as days, weeks, or months. The span of time covered by the entire schedule is called the "planning horizon." Exhibit 11.6.2 is a sample master schedule for the Ajax Office Supply Company.

The Bill of Materials: The Key Document Input to MRP

The bill of materials is the key document input to the MRP system because the collection of these bills–one for each end item and one for each end item's components that are themselves assemblies or manufactured parts–describes the flow of materials through the manufacturing process. It should be noted that purchased, as opposed to manufactured, parts do not require bills of material since, in the logic of a plant's MRP system, they do not have components. As a group, these bills show all of the components (both purchased and manufactured), at every level of the manufacturing process, that are needed to make a product and the required quantity of each.

The single-level bill of materials for a manufactured item lists the part number and quantity of all of the components that directly make up the product (referred to as "first-level components"). A single-level bill of materials, such as the parts list for product A shown in Exhibit 11.6.3, in essence describes one material transforming stage of the manufacturing process and one material flow stage of the manufacturing plant. Exhibit 11.6.4 shows the collection of bills of material for end item A.

When each level of an end item's multilevel bill is diagrammed and all the diagrams are joined together, as in Exhibit 11.6.5*b*, the resulting illustration, called a "product structure tree," clearly shows how material flows from purchased parts to the end item A. In the example the purchased

Exhibit 11.6.3 Single-Level Bill of Material for Product A–Desk Set

Parts List			
Item Number	Quantity	Part Number	Part Name
1	1	B	Holder assembly
2	1	G	Pen
3	1	H	Pencil

Exhibit 11.6.4 Collection of Bills for Product A–Desk Set

Parts List			
Item Number	Quantity	Part Number	Part Name
		Product A–Desk Set	
1	1	B	Holder assembly
2	1	G	Pen
3	1	H	Pencil
		Part B–Holder Assembly	
1	1	C	Base
2	2	E	Holder
3	2	F	Holder bolt
		Part C–Base	
1	1	D	Base blank

Exhibit 11.6.5 Product A–Desk Set (*a*) and Product Structure Tree (*b*)

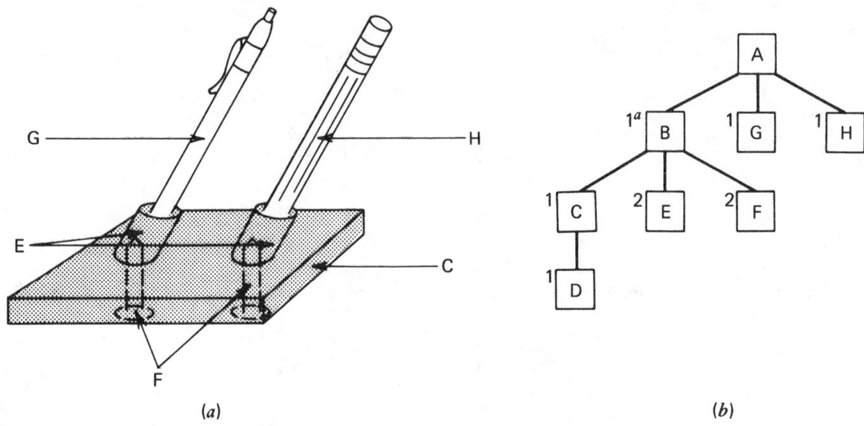

(*a*) (*b*)

[a]Quantity required per unit of A.

parts are D, E, F, G, and H. During the manufacturing process, D is transformed into manufactured part C, which in turn is combined with purchased parts E and F to produce assembly B. Then B, G, and H are transformed into the end item A. It is easy to tell by looking at the product structure tree that A has the direct component B, which in turn has the direct components C, E, and F. Components E and F are purchased rather than manufactured parts, but C has the component D. End item A's direct components, B, G, and H, are considered higher-level components of A than its second-level components because they are higher in the product structure tree.

Another way of expressing the composition of product A is through a multilevel bill of materials. One such multilevel bill, called the "indented bill of materials," highlights the product structure complexity of an assembly by showing concisely how many levels of components it has and the number of different components on each level. The level number for each component is indented for emphasis. Exhibit 11.6.6 is an indented bill of materials for product A.

If a plant produced only end item A, Exhibit 11.6.5 would portray the plant's entire material flow and completely specify its material requirements–what parts are needed, how much of each, and in what order. However, since most plants produce many end items, a collection of illustrations such as the one in Exhibit 11.6.5–one for each end item–served as a set of blueprints for the plant's overall material flow.

Determination of Time-Phased Gross Requirements

The MRP system determines gross component requirements for each end item by exploding the master schedule using the item's bill of materials. Exhibit 11.6.7 shows the product structure tree for product A, along with the lead times to procure and/or manufacture each component. Exhibit 11.6.8 shows the master schedule for product A exploded to first-level components B, G, and H. Since the lead time for each product being exploded determines the time periods in which its components are needed, components B, G, and H must be available one period before product A is required. Introducing the dimension of time into the master schedule during the process of explosion–that is, segmenting material requirements data by time period–is called "time phasing."

The planning horizon for the master schedule should be long enough to cover the longest cumulative lead time in the overall manufacturing process for master-scheduled items so that all materials

Exhibit 11.6.6 Indented Bill of Material for Product A–Desk Set

Level of Part Number	Quantity	Description of Part Number	Description
.1	1	B	Holder assembly
. .2	1	C	Base
. . .3	1	D	Base blank
. .2	2	E	Holder
. .2	2	F	Holder bolt
.1	1	G	Pen
.1	1	H	Pencil

Exhibit 11.6.7 Product Structure Tree for Product A, With Lead Times

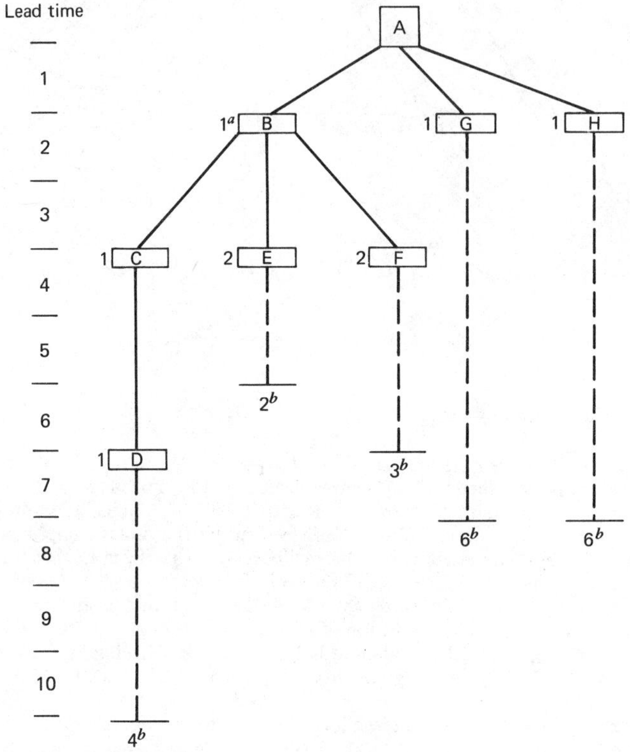

[a]Quantity required per unit of A.
[b]Purchased lead time.

Exhibit 11.6.8 Master Schedule for Product A Exploded to First-Level Components

				Period									
Part	Lead Time	End Item	Past Due	1	2	3	4	5	6	7	8	9	10
Product A	1	Master schedule						80					100
Components													
B	2	Required					80					100	
G	6	Required					80					100	
H	6	Required					80					100	

required can be planned for and made available within their lead times. In Exhibit 11.6.7, for example, the longest cumulative lead time path indicated by the multilevel bill of materials for product A is 10 periods, since the lead times for A, B, C, and D are 1, 2, 3, and 4 periods, respectively. The cumulative lead time determines the earliest point by which a given end item can be manufactured or, in the event an end-item schedule date has already been set, the latest date the lowest-level component order can be started. The bill of material explosion and time phasing logic of MRP are similar to the generalized theory of activity network planning discussed in Chapter 11.2. Exhibit 11.6.9 illustrates the cumulative lead time concept using the component requirements for desk set A.

Determination of Time-Phased Net Requirements

Gross component requirements derived from an end item's master schedule and exploded bill of materials may exceed actural requirements since, in practice, some needed components are already available—either on hand or on order. Accordingly, net requirements are determined by subtracting these available quantities from gross requirements.

Exhibit 11.6.9 Illustration of Cumulative Lead Time

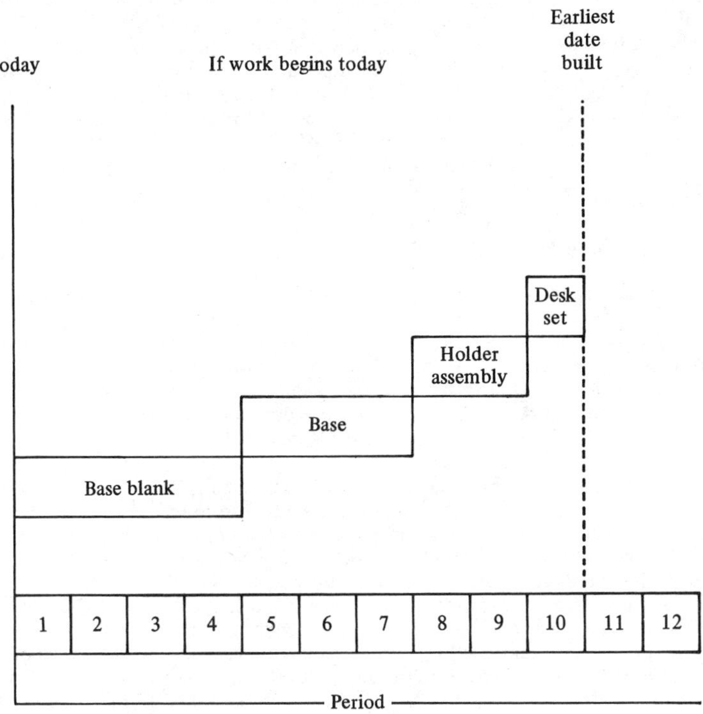

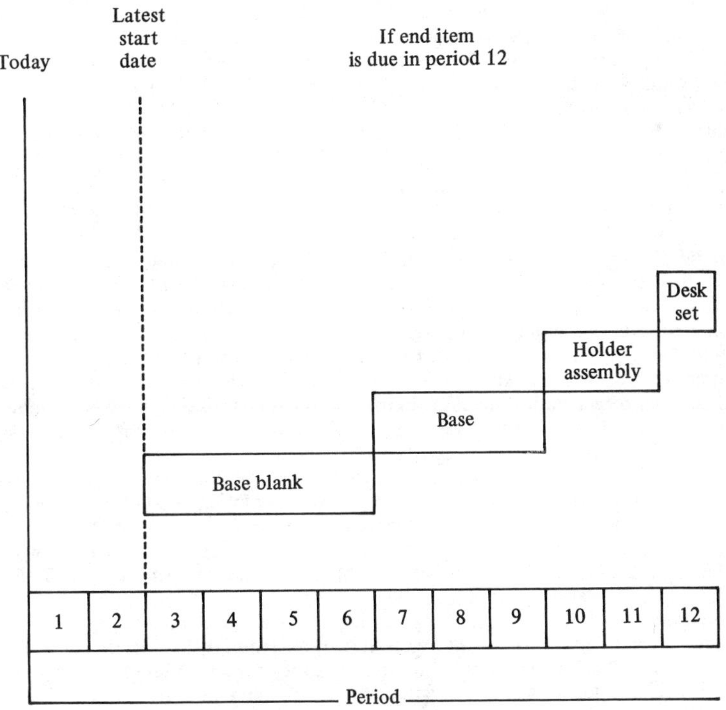

Exhibit 11.6.10 shows the net requirements for components B, G, and H derived from the master schedule and exploded bill of materials for end item A and adjusted for the on-hand and on-order quantities of B, G, and H. If a quantity on order is not acceptably scheduled, the MRP system will suggest rescheduling or expediting. For example, the 95 B components on order in period 7 are not required until period 9 and thus could be rescheduled to potentially obtain a two-period reduction in the cost of carrying this inventory. Similarly, the 110 G components on order in period 5 must be "expedited" to period 4 to satisfy the net requirement for 45 of these components in period 4. This expediting and rescheduling indicated by the MRP system enables material planners to keep actual "need dates" current and thereby maintain proper priorities for meeting component requirements.

Order Generation and Release Instructions

The MRP system also generates planned orders, which material planners use to ensure that material is available to satisfy net component requirements. In Exhibit 11.6.10 the system generates a planned order for part G in period 9 since otherwise there would be an insufficient quantity of part G to satisfy requirements. In this case the planned order is for 110 of part G because 110 is the lot size of this component. (Lot sizing using MRP is explained later in this chapter.) Planned orders are then offset by required lead time to determine when orders must be released–that is, changed from "planned" to "firm." Thus the planned order for 110 of part G is offset six periods to period 3 for release.

The release instruction shown on the net requirements plan constitutes one of the key outputs of the MRP system. Each release instruction represents a directive to issue orders for the manufacture or purchase of a specified quantity of a specified component or raw material in the period shown. The set of these directives for action tells the user what, how much, and when material is required to ultimately satisfy production needs specified in the master schedule.

The determination of net requirements continues through each level of the bill of materials. Planned orders are exploded to generate gross component requirements, which in turn are adjusted for on-hand and on-order quantities to yield net requirements. Additional planned orders are then generated as required. Under special circumstances, firm orders may also be exploded through the bill of materials. This occurs when firm orders have not been pulled from stock, or "kitted." Although these unkitted orders–often called "firm planned orders"–are not generated during the MRP process, they must be exploded through the bill of materials to generate component requirements in the same manner as planned orders. Firm planned orders are typically used in conjunction with MRP to override routine conditions, such as normal lot sizes or lead times. For example, in Exhibit 11.6.10, the 95 on-order B components in period 7 represent a firm planned order to override the normal B lot size of 100, since 5 B parts are still available after the period 4 requirement for 80 parts has been satisfied.

Exhibit 11.6.11 illustrates the complete net requirements explosion of the master schedule for part A.

Aggregating Gross Requirements

The sample net requirements explosion in Exhibit 11.6.11 indicated the quantities of the components required to produce only one independent demand item. The MRP technique aggregates the time-phased requirements of all independent demand items to calculate total gross requirements for each first-level component. These requirements are adjusted for on-hand and on-order quantities, and the appropriate planned orders are generated to satisfy time-phased net requirements. The computer system then explodes the net requirements for first-level components through the bills of material to generate time-phased gross requirements for second-level components and then again nets requirements and generates planned orders. This process is repeated, level by level, through all bills of material down to purchased components.

For components that occur in multiple levels of bills of material, the gross-to-net calculation process must be performed at the lowest level at which the component appears to ensure that all gross requirements are identified and aggregated prior to explosion. A low-level code assigned to each item indicates the lowest level at which it occurs in any bill of materials. The low-level code indicates that aggregation and explosion of components should be postponed until all gross requirements from higher levels have been identified.

The failure to postpone explosion of gross requirements results in the generation of understated gross requirements and, subsequently, understated net requirements at each level of the exploded bill of materials. This, in turn, results in erroneous lot sizes and inappropriate safety stock quantities. This cycle of calculation, adjustment, and recalculation continues through all the levels of bills of material. Each time the multiple-use item is encountered in a bill of materials, all previously completed gross-to-net calculations must be repeated, all lot sizes adjusted, and all safety stocks re-

Exhibit 11.6.10 Net Requirements for First-Level Components of Product A

Part	Release of Orders	Instructions for Release of Orders	Period										
			Past Due	1	2	3	4	5	6	7	8	9	10
A	Lead time = 1	Master schedule						80					100
B	On hand = 85	Required					80					100	
	Lead time = 2	On order								95 (future plan)			
	Lot size = 100	Net	85	85	85	85	5	5	5	100	100	0	0
		Planned											
		Available	85	85	85	85	5	5	5	100	100	0	0
		Release											
G	On hand = 35	Required					80					100	
	Lead time = 6	On order						110					
	Lot size = 110	Net	35	35	35	35	⟨45⟩	65	65	65	65	⟨35⟩	⟨35⟩
		Planned										110	
		Available	35	35	35	35	⟨45⟩	65	65	65	65	75	75
		Release				110							
H	On hand = 200	Required					80					100	
	Lead time = 6	On order											
	Lot size = 200	Net	200	200	200	200	120	120	120	120	120	20	20
		Planned											
		Available	200	200	200	200	120	120	120	120	120	20	20
		Release											

Exhibit 11.6.11 Master Schedule for Product A With Complete Net Requirements Explosions

Part			Period										
			Past Due	1	2	3	4	5	6	7	8	9	10
A	Lead time = 1	Master schedule						80					100
B	On hand = 85	Required					80					100	
	Lead time = 2	On order								95 (future plan)			
	Lot size = 100	Net	85	85	85	85	5	5	5	100	100	0	0
		Planned											
		Available	85	85	85	85	5	5	5	100	100	0	0
		Release											
C	On hand = 0	Required						95					
	Lead time = 3	On order											
	Lot size = 100	Net	0	0	0	0	0	95	95	95	95	95	95
		Planned						100					
		Available	0	0	0	0	0	5	5	5	5	5	5
		Release			100								
D	On hand = 60	Required			100								
	Lead time = 4	On order											
	Lot size = 200	Net	60	60	40	40	40	40	40	40	40	40	40
		Planned			200								
		Available	60	60	160	160	160	160	160	160	160	160	160
		Release	200										

E	On hand = 400	Required						190					
	Lead time = 2	On order											
	Lot size = 500	Net	400	400	400	400	400	210	210	210	210	210	210
		Planned											
		Available	400	400	400	400	400	210	210	210	210	210	210
		Release											
F	On hand = 50	Required						190					
	Lead time = 3	On order						500					
	Lot size = 500	Net	50	50	50	50	50	360	360	360	360	360	360
		Planned											
		Available	50	50	50	50	50	360	360	360	360	360	360
		Release											
G	On hand = 35	Required					80					100	
	Lead time = 6	On order						110					
	Lot size = 110	Net	35	35	35	35	45	65	65	65	65	35	35
		Planned										110	
		Available	35	35	35	35	45	65	65	65	65	75	75
		Release				110							
H	On hand = 200	Required					80					100	
	Lead time = 6	On order											
	Lot size = 220	Net	200	200	200	200	120	120	120	120	120	20	20
		Planned											
		Available	200	200	200	200	120	120	120	120	120	20	20
		Release											

calculated. In addition, since the net requirement for the item has changed, the gross demand for all lower-level components dependent on that item's net requirements must be recalculated.

11.6.3 APPLICATION OF MRP

The basic output of MRP is a time-phased schedule of lot-sized orders designed to economically satisfy net requirements for components and raw materials over a fixed planning horizon. On an ongoing basis, the output of MRP is a set of directives for action, which could include any of the following:

Release planned orders for production/purchase.

Expedite existing orders.

Reschedule existing orders.

Change quantity on existing orders.

Cancel existing orders.

These recommendations essentially reflect sets of factors: actions necessary to satisfy material requirements (what, how much, and when), adjusted for system/management policies, which serve to tailor recommendations issued by the system to the needs of management. This latter set of inputs to the MRP process may include the following:

The number of periods in advance of the release period, in which management desires the system to show directives for action.

The number of periods considered by management to represent a significant need to expedite or reschedule.

Parameters for an "acceptable" order quantity value when the order quantity is not equal to the lot size.

Guidelines for determining when "cancel," rather than "reschedule," is the appropriate action.

Policies such as these describe the operating environment of the MRP system. Excessively tight policy rules tend to make the MRP system "nervous" and overreactive. Rules that lack specificity, however, often result in an unresponsive system. Realistic, yet effective, policies are one of the keys to successful application of MRP systems.

Regeneration Versus Net Change MRP Systems

The MRP systems fall into two general categories on the basis of the way they update net requirements plans. A schedule regeneration MRP system periodically reexplodes the entire master schedule using the multilevel bills of material for all end items to regenerate new planned orders and directives for action. Under a net change approach, only changes to the master schedule or to the status of specific components are exploded through the bills of material.

Exhibit 11.6.12 compares the characteristics of schedule regeneration and net change system operation. The principal outputs of the two approaches are the same. The important differences are the frequency of replanning (high for net change versus limited for regeneration) and the planning trigger (a change in status for net change versus input of the entire master schedule for regeneration); the other differences between the two approaches are a function of these two key features. The net change approach is generally favored by users of MRP because it provides a quick response to change. However, in practice, net change systems usually require backup procedures to periodically purge the system of bad data through regeneration. In addition, limitations in a company's data processing capacity or capabilities may prevent it from adopting a net change MRP system.

Lot Sizing in an MRP Environment*

The quantities indicated on planned orders generated by the MRP system are not solely a function of net material requirements, but must also take into account the specified lot sizes of each item or the lot size rules established to help material planners calculate appropriate order quantities. The establishment of a standard lot size for each item may be influenced by such factors as the availability of resources, storage space constraints, or packaging requirements. However, the most important concern typically is to minimize the costs associated with ordering and/or setup and carrying inventory.

*See Chapter 11.5.

Exhibit 11.6.12 Schedule Regeneration Versus Net Change MRP Systems

Key System Characteristics	Schedule Regeneration	Net Change
Frequency of replanning	Limited—weekly or less	High—daily or continuous
Planning trigger	The entire master schedule on a regular basis	Changes in the status of the master schedule or specific parts
Extent of explosion	Every item in the master schedule	Only items with status changes
Processing mode	Batch	On-line or batch
Validity of requirements data over time	Deteriorates between batch processes	No deterioration because of continuous file updating
Data processing efficiency	Highly efficient	Relatively inefficient
Response time to change	Limited by infrequency of replanning	Quick because of frequent replanning or on-line updating
Ability to purge inaccurate requirements planning	Yes	No
Files that can be updated	Inventory data only	Inventory data and requirements data
Number of operating phases	Two—periodic requirements planning and intraperiod file updating	One—combined updating and requirements planning

Exhibit 11.6.13 provides brief descriptions of the most commonly used approaches to lot sizing, along with an indication of their applicability in an MRP environment. The first two techniques—EOQ and fixed order quantity—are based on the assumption of continuous demand, whereas the remaining approaches are specifically designed to handle the intermittent, lumpy demand usually associated with the use of MRP systems.

A number of additional factors must be considered in selecting an appropriate lot sizing technique, of which the following are the most important:

The variability of demand.

The length of the planning horizon.

The size of the planning period.

The ratio of setup to unit manufacturing costs.

For most components or groups of parts, there is not one lot sizing technique clearly superior to the rest. However, in general, lot sizing approaches designed to handle discrete, lumpy demand are more effective in an MRP environment. By specifying order quantities that are integer multiples of the net requirements of consecutive planning periods, these techniques minimize the generation of "odd" quantities, which only partially satisfy a specific period's requirements.

The Use of Safety Stocks Under an MRP Approach

Safety stocks of each component can be introduced into the MRP system either by subtracting the safety stock quantity from the stock on hand or by adding the safety stock quantity to the gross requirements. The two approaches produce the same effect—an increase in net requirements. However, caution should be exercised in the use of safety stocks in an MRP environment. Safety stocks cause the MRP system to generate overstated requirements, which in turn can invalidate order priorities and timing.

The primary purpose of safety stock is to compensate for unanticipated fluctuations in demand. However, because component demand is calculated, rather than forecast, under the MRP approach, safety stocks for components are not usually required. Rather, they should be one of the inputs considered in the establishment of the master schedule for independent demand items.

The only components with dependent demand that may require safety stocks are those whose supply is highly uncertain—usually restricted to a small percentage of a plant's purchased parts. In these cases safety stock can be introduced to the MRP system as an early delivery offset, which is the number of periods in advance of the due period in which the system should generate directives for the release of orders. Under this approach, the normally calculated order quantity is ordered, but the order may be received before it is actually needed. When this occurs, the balance on hand will include a safety stock equal to the number of periods prior to the period in which it is required multiplied by the average use per period.

In practice, however, many MRP users employ safety stocks to protect inventories from stock runouts due to delays in production and delivery or insufficient replenishment caused by spoiled lots or "holds" placed on production by quality control.

11.6.4 IMPLEMENTATION OF MRP

Implementation of an MRP system is far more than a simple modification made by a programmer to a company's existing computer systems—it represents a comprehensive change in the firm's approach to manufacturing. Successful implementation requires a thorough understanding on the part of users and management of both MRP and the manufacturing operation to which it is applied. The critical tasks involved in implementing an MRP system fall into three major categories: preplanning, planning, and installation.

Preplanning

Preplanning activities include tasks designed to identify areas of weakness within the current manufacturing operation and to establish the need for MRP. These preplanning tasks usually necessitate an operations diagnostic, which measures the performance of existing operations and identifies the causes of malfunction. The major areas to be examined in the operations diagnostic are as follows:

Operating policy and company background—for example, organization and staffing, sales and profit profile, customer service policies and goals.

Materials management systems—capacity planning, master scheduling, requirements planning, shop floor control.

Inventory—number of stocking points, investment targets, record accuracy.

Manufacturing operations—bills of material, process flow, process layout, labor efficiency, performance to schedule.

Customer demand—variability, concentration, forecast accuracy, lead time for change.

Customer service—service goals, order frequency, service performance.

Purchasing—volumes, lead times, performance to schedule.

The results of the diagnostic should be interpreted by persons familiar with both materials management systems in general and the existing manufacturing system in order to identify opportunities to apply the MRP technique. Because the skills required to diagnose and prescribe corrective action are not maintained by most companies, and because it is generally difficult for a company's manufacturing systems analysts to remain totally objective when measuring the performance of their own system, many firms find it valuable to assign this step to a high-level task force or to a manufacturing systems expert drawn from outside the firm.

The results of the diagnostic enable the expert in manufacturing systems to pinpoint the areas of manufacturing operation that must be fixed before planning for an MRP system can begin. Typically, deficiencies are discovered in four aeas:

Inventory record accuracy.

Bill-of-material structure.

Bill-of-material accuracy.

Master schedules.

Accurate inventory records are a fundamental requirement for successful application of MRP. The MRP system will not correct inventory inaccuracies—in fact, it will amplify the negative effect of erroneous records. Inaccurate inventory records will result in the failure to order required material and, conversely, orders for unneeded materials. Incorrect inventory records will result in inaccurate gross-to-net calculations and subsequent system malfunction. Furthermore, if records are not accurate, informal subsystems will almost invariably be set up by expediters and inventory controllers to handle parts shortages, and the integrity of the MRP system will be lost. A properly performed diagnostic not only should measure record accuracy for a broad sample of parts and materials, but also should identify the causes of inaccuracies and indicate each cause's contribution to the total problem. For items counted by weighing scales, errors of less than 2% are generally considered to be insignificant; low-value commercial hardware items are considered under control when actual quantities are within 5% of record figures. Inventory records of "A" items are significantly in error when records differ from actual counts by more than a few pieces. If existing inventory record keeping systems cannot meet these minimum requirements, steps must be taken to bring performance up to these levels. The following shortcomings are often found to be the causes of record inaccuracy:

Unlocked storerooms.

Lack of formal receiving procedures.

Poorly designed receiving and storage areas.

Inappropriate forms for the receipt and release of material.

Lack of formal procedures to check records—for example, check digits or cycle counting.

The bill-of-material structure is a second common problem area often requiring correction prior to the installation of an MRP system. Effective performance of MRP requires accurate descriptions of all products through their bills of material or combinations of modular bills (see Chapter 7.2). In manufacturing environments in which a very large number of end items are produced, methods for restructuring the bill of materials should be considered. For example, it would be impractical for a hydraulic cylinder manufacturer to maintain a bill of materials for every cylinder a customer could possibly specify. Requirements for most major components can be easily specified, but the cylinder body and piston rod can be produced in an infinite variety of lengths. In this case bills of material can be assigned to the more manageable number of modules that are first-level components of every final product—such as the head, cap, and piston. The final manufacture and assembly to specified length can be scheduled after the receipt of the customer's order. The use of modular bills of material also simplifies handling of engineering changes (i.e., additions, deletions, and other changes to the structure; see Chapter 7.1) and thereby reduces the chances for inaccuracies.

The bill-of-material structure should also provide the ability to identify components used in more than one level of the production process. A low-level code for each item can be used to indicate to

Exhibit 11.6.13 Lot Sizing in an MRP Environment

Technique	Description	Applicability in an MRP Environment[a]	Comment
Economic order quantity	The lot size that minimizes the total cost of ordering and/or setup and carrying inventory. The EOQ is based on the assumption of continuous, steady-rate demand.	○	Generally ineffective for an MRP system since the demand associated with the use of MRP systems typically involves discrete lumps of material required intermittently
Fixed order quantity	A lot size prespecified to the system. The quantity may be determined arbitrarily or based on intuition or empirical factors.	◒	Usually applied only to those parts whose characteristics (such as tool life, storage constraints, or pending product modification) are not adequately taken into account by other lot sizing techniques
Lot for lot	Components are ordered period by period in the exact quantities specified by the net requirements. This technique minimizes inventory carrying costs, but does not consider expenses associated with ordering and/or setup.	●	Can be highly effective when ordering and/or setup costs are low
Fixed-period requirements	The user specifies in advance how many periods of coverage each planned order should provide.	◒	Useful when requirements beyond a specified period are uncertain
Period order quantity	As in the fixed-period requirements approach, the user orders a specified number of periods of coverage. However, the number of periods is calculated by first determining the EOQ and then the number of orders per year based on a forecast of annual demand. The number of planning periods in one year is then divided by the number of orders per year to determine the ordering interval.	●	Employs classic EOQ logic modified to handle discrete demand

Least unit cost	The lot size that achieves the lowest setup plus inventory carrying cost per unit. First, the total cost (setup plus inventory carrying cost) is calculated for a quantity covering one period's requirements. Then the next period's requirements are added, and the total cost is recalculated. This process is repeated to include period $X + 2$, $X + 3$, and so on. Each total cost is divided by the number of units to determine the quantity with the lowest cost per unit.	○	Does not consider the entire planning horizon
Least total cost	The lot size for which the setup cost is equal or nearly equal to the inventory carrying cost. This approach is based on the assumption that the sum of all setup and inventory carrying costs will be minimized if these costs are equal.	◒	Incorporates analysis of all lot sizes required in the planning horizon
Part-period balancing	Employs the same logic as least total cost, but attempts to reduce the total cost further by considering the effects of wide variations in demand in the periods immediately before or after the period in which a tentative lot size is planned.	◒	Most effective for cyclical demand patterns
Wagner-Whitin algorithm	Uses a dynamic programming model to determine the optimal ordering strategy for the entire net requirements schedule.	○	Somewhat complex for most MRP applications; requires voluminous calculations

[a]This assessment applies to commonly found manufacturing environments. In practice, each of these techniques—alone or in combination—has useful applications. Key: ●—high degree of applicability; ◒—medium degree of applicability; ○—low degree of applicability.

the MRP system the lowest level at which the item is found in any bill of materials in a plant's files. Use of a low-level code greatly reduces processing time and minimizes the chances for error during the explosion for regeneration or the net change procedure by eliminating the repeated explosion and order planning of a multiple-use component or subassembly.

If bills of material are found to be insufficiently modularized or improperly coded, they must be restructured during planning to better support the use of an MRP system.

Accurate bills of material are essential to effective MRP system operation. Errors will cause the system to indicate a need for parts that are not needed or fail to plan orders for required components. Bill-of-material inaccuracies can often be traced to two causes:

Incomplete bill-of-material files–files missing formal bills of material for some products.

Insufficient control over engineering change–often due to a lack of formal cutoff procedures or poor communication systems.

A realistic master schedule is a critical input to an MRP system since it essentially drives the system. The master schedule must be accurate and complete and should fit the manufacturing environment. The following problems are typically identified during the operations diagnostic:

"Off"-schedule items–end items being produced, but not included on the master schedule.

Unrealistic master schedules–schedules that call for the production of unneeded goods or for finished goods in quantities that exceed the actual capacity of the plant.

Improper planning horizons–planning horizons too short to allow proper planning for the purchase or manufacture of the materials or parts with the longest cumulative lead time.

Inadequate control over changes to the master schedule–particularly troublesome when changes cannot easily be handled within required lead times.

Because any of these deficiencies in master scheduling would lead to problems in an MRP environment, corrective steps must be taken prior to MRP system installation.

In addition to identifying weaknesses in the existing system, the preplanning operations diagnostic should help to establish the need for an MRP system. The system cannot cure all manufacturing ills. In essence, an MRP system offers potential benefits if performance of the existing manufacturing operation is unacceptable in functions that can be improved through the use of such a system. To clearly establish and quantify the need for an MRP system, the following performance indicators should be examined during the course of the preplanning diagnostic:

Component inventory investment characteristics and trends.

Customer delivery performance.

Stock runout frequency.

Direct labor costs.

Indirect labor costs.

On the basis of a comparison of actual performance levels against levels desired by management and a comprehensive examination of manufacturing operations, a manufacturing systems analyst familiar with MRP can estimate the magnitude of the benefit that could be derived through the installation of an MRP system.

By the end of the preplanning stage, the manufacturing systems analyst should be able to specify the following:

Current levels of manufacturing performance and the magnitude of the opportunity to improve performance through the application of an MRP system.

The value of the benefits anticipated from installation of an MRP system, such as inventory reductions, improvements in customer service, reductions in manufacturing costs, and reductions in indirect labor costs.

The areas of operation that require correction prior to implementation of an MRP system.

Planning

Once the corrective actions specified during the preplanning are under way, planning for MRP system installation can begin. This stage typically involves three steps:

1. Development and execution of a comprehensive companywide training program.
2. A software package make-or-buy analysis–and selection of a package if the buy option is chosen.
3. Development of an action plan for installation.

Development and Execution of Training Program

The first and most critical step in planning for an MRP system installation is to prepare employees for the new ways they will be required to perform their jobs. This will typically require comprehensive training programs designed to educate all company personnel, from the president down to stockroom clerks, in the fundamentals of production and inventory control and in the use of MRP systems. These programs should provide all potential users of the system with a thorough knowledge of the concepts and techniques on which the system is based and should instill a sense of responsibility for the system's success. Many MRP experts cite training and education as representing more than half the effort required for successful installation.

Make-or-Buy Analysis

The second step of planning for an MRP installation involves a make-or-buy analysis to determine whether a system should be purchased or developed. The process of installing an MRP system on a computer can be speeded up significantly through the use of software programs available from most computer manufacturers and many application software vendors. In addition, these software packages provide a variety of analysis routines that enable the user to simulate and study the effects of changing the system of ordering. The alternative to purchasing an outside software package is to develop a system using in-house personnel. The decision to make or buy a package for MRP installation involves five basic steps, the first four of which provide the basis for comparing the best-suited commercially available packages to an internally generated system.

1. *Study the underlying issues* that must be addressed when considering the introduction of an MRP system. For example:

a. How are existing information systems linked to the company's manufacturing strategy?

b. Are current systems meeting the company's needs? Are they being used as originally intended?

c. What are the potential change triggers in the company's business, and how might the business need to evolve in the years ahead?

d. What is the company's historical record for managing change? (This assessment should consider the abilities of both internal and outside resources to cope with change.)

e. What issues are involved in interfacing manufacturing systems with other functional areas of the business?

2. *Specify application requirements*, including:

a. Functional features needed to capitalize on improvement opportunities.

b. Basic hardware and application and control software requirements.

c. Compatibility with the company's existing operational practices.

d. Modifications required to adapt an outside package to fit unusual in-house mainframe of terminal configurations or unique sets of coding conventions, product codes, or supplier codes.

3. *Evaluate source and buy options*. The following considerations are particularly critical:

a. The vendor's true motivation–is he or she trying to sell a piece of hardware? Software? Bodies? A total solution?

b. The depth of the vendor's manufacturing knowledge.

c. The vendor's sensitivity to the company's needs–is he or she pushing a largely canned approach?

d. The depth of the vendor's resources–are individuals stretched so thin that they are unable to support a major implementation?

e. The vendor's likely ability to modify and update the package to keep pace with the latest thinking.

f. The quality of the documentation.

g. The track records of vendors that pass a preliminary screening.

h. The experiences of manufacturers who installed similar packages.

4. *Assess the resource implications of buying a package.*

a. Human resources.

b. Technical resources.

c. Information resources.

d. Financial resources over the life cycle of a system.

5. *Perform a make-or-buy analysis.*

a. Summarize the total costs, advantages, and disadvantages associated with both make and buy options.

b. Using discounted cash flows (see Chapter 9.3), develop equivalent financial expressions to compare the long-term impact of faster installation of a commercial package against slower installation of a possibly better-designed internally generated system.

If it is decided that the buy option makes more economic sense, the next step is to evaluate the software package finalists and select one. A number of consulting firms have developed services to help companies evaluate software packages for MRP systems. One such firm, Manufacturing Software Systems, Inc., recommends that selection of the appropriate package be based on the following key criteria[1].

1. Functions included in the software, such as:

a. The MRP logic—for example, netting and exception, order planning and explosion, pegging, reporting, master production scheduling, entry into the processing sequence, firm planned orders.
b. Bill-of-material subsystem.
c. Inventory transaction subsystem.
d. Scheduled receipt subsystem.
e. Shop floor control.
f. Capacity requirements planning.
g. Input and output control.
h. Purchasing.
i. Distribution requirements planning.

2. Technical considerations:

a. The implications of each package's fit with existing hardware and software.
b. Implications of the programming language—for example, one-time or ongoing modifications, flexibility, documentation.
c. Compatibility and capabilities of data base management systems.
d. The advantages, disadvantages, and costs associated with on-line versus batch operating modes.

3. User considerations:

a. The history of the package—for example, number of successful installations, the package's stage of development.
b. Actual performance of the package in the field.
c. The extent of package "debugging."
d. The demonstrated degree of vendor support.

Development of Plan for Installation

Once the software package has been selected, an action plan for installing the MRP system should be developed. Such a plan should include the following critical components:

1. A campaign to gain the formal commitment and active support of top management. Such a program should include the following features:

a. A realistic cost-benefit analysis.
b. A clear definition of the system's objectives and an indication of expected results.
c. A schedule of periodic management reviews.

2. Establishment of an installation team and appointment of a full-time project manager.
3. A set of programs designed to:

a. Achieve at least 95% inventory record accuracy.
b. Eliminate bill-of-material inaccuracies through verification of all bills and restructuring where appropriate.
c. Ensure that inventory records contain the correct lead times, ordering quantities, and safety stocks for all components used in the manufacturing process.
d. Ensure that the master schedule is complete, accurate, and realistic.

4. An evaluation of the appropriateness of the production and inventory control department's organization in an MRP environment, and modifications if needed.
5. A formal installation schedule (PERT or Gantt chart), showing the timing of critical tasks and the project team personnel responsible for each task.

Installation

When all planning tasks have been completed, installation can begin. A pilot approach can facilitate a smooth transition to full MRP system operation. Under this approach, a single product line with as few components as possible in common with other products should be successfully converted to the MRP system before other product lines are installed on the system. The following key activities should be completed for the pilot product line prior to system startup:

A clear set of work center (Chapter 10.2) identifiers should be assigned, and capacities by work center established.

Programs should be implemented to ensure that the sequence of operations involved in the manufacture of each product is reflected in the most current routing (see Chapter 7.2).

Formal capacity planning and input and output control (Chapter 11.5) for the pilot product line should be installed if they are not already in place.

Formal shop floor controls for the pilot line should be established if they do not already exist.

Once the pilot product line is up on the system and is running, related groups of products should be cut over according to the same procedure.

Installation of the MRP system is more likely to be successful if engineering changes can be held to a minimum during startup. In addition, even if directives generated by the system call for a drastic reduction in orders to be sent to vendors, material planning should only gradually reduce the number of orders in order to minimize disruption of critical lines of supply. Finally, throughout the entire startup phase, bills of material should be checked and rechecked for accuracy.

Ongoing System Support

Even after the MRP system is fully operational, an ongoing effort is required at all levels of the manufacturing organization to ensure that the system performs to its full potential.

Effective Customizing Techniques

All MRP systems employ essentially the same basic techniques: The initial specification of design parameters rarely results in an unusable or unworkable MRP system. However, once the system is in operation, techniques often must be customized to adapt the system to its unique operating environment or to correct errors made during its installation and to ensure that it generates output that satisfies the needs of users. Lot sizing rules or exception reporting parameters in many cases must be modified to reflect changing conditions—such as supply shortages or rising inventory carrying costs—or to capitalize on opportunities to improve system operation. Although it is generally easy to change techniques, modifications should be undertaken only after they are thoroughly understood.

Ongoing User Training

Regular refresher courses should be provided to keep system users up to date on modifications made to the installed system. In addition, new company personnel who use the system or who are involved in its operation should be thoroughly versed in the design of the system and the objectives established by management.

Continued Top Management Support

Without the ongoing active support of top management, even the most knowledgeable system users equipped with the most accurate data records and the best techniques are unlikely to realize the potential benefits of an MRP system. Strong backing by management not only ensures that sufficient resources continue to be made available, but, even more important, reinforces the continuing commitment of system users.

REFERENCE

1. *MRP Software Evaluations*, Manufacturing Software Systems, Williston, VT, 1978.

BIBLIOGRAPHY

An Introduction To MRP, Mitrol, Lexington, MA, 1978.

Communications Oriented Production Information System (COPICS), IBM Publications, New York, 1972.

GREENE, J. H., *Production and Inventory Control Handbook*, McGraw-Hill, New York, 1970.

MRP Implementation Plan, Manufacturing Software Systems, Williston, VT, 1978.

ORLICKY, J., *Material Requirements Planning*, McGraw-Hill, New York, 1975.

ORLICKY, J. A., G. W. PLOSSL, and O. W. WIGHT, *Material Requirements Planning Systems*, IBM Publications, White Plains, NY, 1971.

PLOSSL, G. W., and O. W. WIGHT, *Production and Inventory Control*, Prentice-Hall, Englewood Cliffs, NJ, 1967.

TERSINE, R. J., *Materials Management and Inventory Systems*, Elsevier North-Holland, New York, 1976.

CHAPTER 11.7

Maintenance Management and Control

RINTARO MURAMATSU
YOSHIHIKO TANAKA
Wasada University, Japan

SEIICHI NAKAJIMA
Japan Institute of Plant Engineers

11.7.1 THE ROLE OF MAINTENANCE MANAGEMENT

Not only manufacturing, but all kinds of industries, are making efforts to increase labor productivity, to elevate the quality of products and services, and to improve the working environment by means of mechanization, automation, and the speeding up of the operation of production and service. Mechanization not only has increased the growth and profits of firms, but has stimulated the development of national economies and a rise in living standards.

However, when a breakdown occurs in equipment and facilities, the more mechanization progresses, the greater are the resultant loss and damage. For example, in the event of air or water pollution caused by an explosion in an atomic reactor or an accident in a chemical plant, or in the event of a breakdown of communication or service equipment and facilities, there will be severe physical, economic, and spiritual effects on human society. Plants and service firms are also affected in the following ways by a breakdown of equipment and facilities:

1. An outbreak of defective and degraded products.
2. A waste of production resources such as material, energy, and labor.
3. Increased useless repair costs.
4. Confusion in production and business planning and a delay in delivery dates.
5. A deterioration in the motivation of employees.
6. The occurrence of a public nuisance.

At a time when equipment and facilities used in plants and service firms are becoming extremely large, extremely small, or super high speed and moreover have to be operated under severe temperature and pressure conditions, it has become increasingly impossible for human senses and physical ability to cope with the discovery of an unusual condition of equipment and facilities while they are in operation, with the diagnosis of a failure, or with maintenance and repair operations.

Today, two of the efforts being made in maintenance management are to reduce to a minimum the effect of trouble on such equipment and facilities and to feed information regarding experience and knowledge obtained from maintenance work back to equipment and facilities design and manufacturing departments in order to reduce maintenance work. The necessary items for maintenance management in any industry, irrespective of the scale of business and the goods produced, are as follows:

1. Economic and availability analysis, and evaluation of maintenance management.
2. Function, organization, system, and procedure for maintenance management.
3. Maintenance diagnostic system.
4. Improvement of workers' skill and motivation in production and maintenance activities.

Exhibit 11.7.1 Factors of Downtime in Total Time

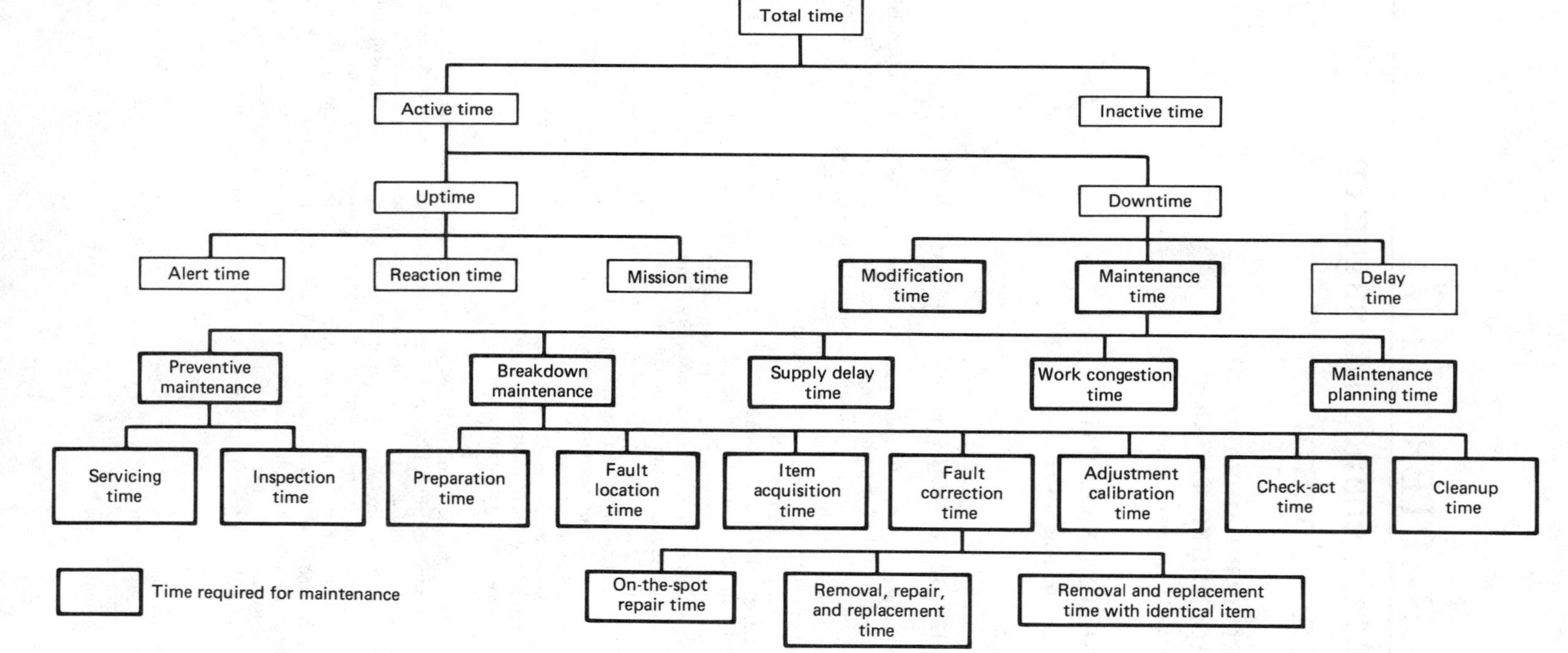

11.7.2 ANALYSIS OF AVAILABILITY, ECONOMY, FAILURE, AND DATA OF MAINTENANCE MANAGEMENT

Availability and Economic Analysis

Operation without any downtime during their total life cycles would be desirable for purchased equipment and facilities. However, in actuality, the downtime caused by failure or deterioration is liable to occur. Exhibit 11.7.1 shows the factors of downtime during the total time.

If there are no failures and deterioration, maintenance activity is unnecessary, and as a result there are neither loss nor maintenance costs caused by failure. On the other hand, the purchasing cost of equipment and facilities is extremely expensive. Therefore the economy of maintenance management must be evaluated as a total of the maintenance cost, opportunity costs, purchasing cost, and installation cost.

Availability is defined as follows:

$$\text{availability} = \frac{\text{uptime}}{\text{uptime} + \text{maintenance time}}$$

Availability improves when uptime increases by developing the reliability of equipment and facilities and also when the maintenance time is shortened by progressive maintenance techniques. Exhibit 11.7.2 shows the relationship between availability and costs.

The improvement of maintenance management in the case of equipment purchased on the basis of an evaluation of its economy represents an attempt to shorten the maintenance time, improve availability, and reduce the opportunity cost and maintenance cost.

Failure Characteristics

Exhibit 11.7.3 shows the change of failure rate of equipment according to usage time. Period 1 is called the "early failure period." In this period failure is caused mainly by misdesign, processing error, or rough handling. In period 2, the "chance failure period," failure is caused mainly by erroneous operation, with routine maintenance and corrective operation being required in such cases. In period 3, the "wear-out failure period," failure is caused mainly by wear on the machine, corrosion, or physical changes. This time preventive maintenance or corrective maintenance is effective. If preventive maintenance cost or corrective maintenance cost is extremely expensive, replacement is required.

Failure analysis means the systematic research and study for clarifying the effects of actual or potential failure causes, failure mechanisms, and occurrence probability on the job.

Methods shown in Exhibit 11.7.4 are used for reliability analysis during the design stage. Failure mode and effects analysis and fault tree analysis are the most popular methods. The practical procedures of these methods are described briefly.

Exhibit 11.7.2 Relationship Between Availability and Cost

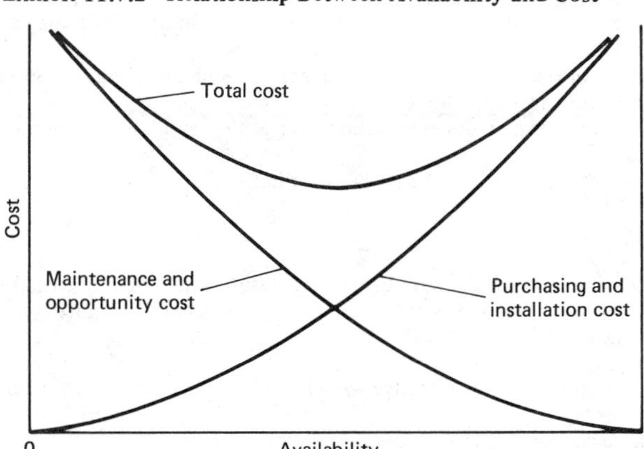

Exhibit 11.7.3 Relationship Between Usage Time and Failure Rate

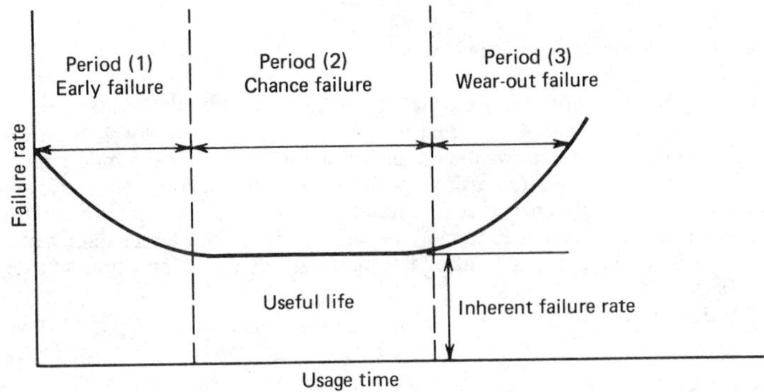

Exhibit 11.7.4 The Objectives and Methods of Deterioration and Failure Analysis

Objective	Method
Forecasting for variation of characteristics and deterioration	Worst-case analysis methods Statistical analysis methods
Decomposition of failure causes	Fault tree analysis
Sequence analysis of failure occurrences	Event tree analysis
Failure effects	Failure mode and effects analysis

Failure Mode and Effects Analysis

The purpose of this method is to estimate the potential failure modes and major failure causes and to evaluate the effects of the objective system. The practical procedures of this method are outlined as follows:

1. Analyze the structure of the objective system and decompose the system into subsystem, components, and parts.
2. Define the mission of subsystems, components, and parts on the basis of the system specifications.
3. Determine the decomposition level that is analyzed in step 1.
4. Classify subsystems, components, and parts by the functions of the system.
5. Establish the final mission of the objective system and compose the functional reliability block diagrams that express the functional relationships among the system, subsystems, components, and parts according to the classification of step 4.
6. Enumerate potential failure modes by functional reliability block diagrams through brainstorming method.
7. Select, under coordination with design, reliability, and test engineer, some effective failure modes among those enumerated in step 6.
8. Refer to the test or error documentation concerning similar systems in order to estimate major failure causes.
9. Record the preceding results in the analysis format.

Fault Tree Analysis

To eliminate "undesirable events" that disturb the safety and reliability of the system, the real causes of the events are to be found by using event symbols and logic gate symbols. The practical procedures of this method are as follows:

1. Define the structure of the objective system, missions, functions, reliability requirements, and safety requirements.
2. Select the top event among many "undesirable events."
3. Enumerate cause events for the top event.
4. Describe the relationship between top event and its cause events with logic gate symbols.

Exhibit 11.7.5 The Stages of Data Analysis for Reliability and Maintainability

Stages	Variable Studied	Contents
Collecting of failure data	Sampling interval Sampling size	(1) Sampling theory; (2) time to failure or time between failure; (3) time to repair
Data analysis	Precision Man-hour	(1) Classification (by importance or by mode; (2) statistical analysis (Pareto's diagram, histogram, hypothesis test and estimate, MTBF, MTTF, mean time to repair–MTTR); (3) failure analysis and failure mode and effects analysis (fault tree analysis, simulation)
Evaluation and counter measures	Cost to performance	(1) Evaluation of importance (MTBF, MTTR, cost effectiveness, ranking of failure mode and effects analysis); (2) determination of countermeasures (design improvement, process improvement, environment improvement, maintenance management improvement, test method improvement, education or training); (3) confirmation of effectiveness

5. Explore until the base events cause the events listed in step 3.
6. Summarize the results of steps 3, 4, and 5 in a fault tree diagram.
7. Obtain the occurrence probability of each event and calculate the top event's occurrence probability.
8. Using the results of step 7, improve the cause events that have an important effect upon the top event's occurrence probability.
9. If the occurrence probability of each event is not obtainable, select and improve some cause events that are considered to have an important effect.

The causes of failure are classified into the following factors: functional factors, environmental factors, and time factors.

Data Analysis Stages for Reliability and Maintainability

Exhibit 11.7.5 shows the stages of data analysis for reliability and maintainability

11.7.3 FUNCTION, ORGANIZATION, AND SYSTEMS OF MAINTENANCE MANAGEMENT

Maintenance Activities and Their Development

At the first stage, maintenance activities were to replace or repair equipment or tools that had deteriorated or broken down. At the second stage, when skill in repair work was required in proportion to the increase in mechanical complexity, workers with special repair skills were assigned or groups of such workers were formed to reduce other workers' direct downtime. At the third stage, in order to further reduce downtime in breakdown repair cases, substitute tools and machines and control methods that were to improve work methods and control inventory of spare parts were developed. The functions of periodic inspection and repair also were developed in order to reduce repair cost. At the fourth stage, as mechanization and automation have progressed, maintenance and opportunity costs due to maintenance deficiency have increased, and as a result, reliability, availability, maintainability, and economic analyses in maintenance have been devised.

Recently, in order to make maintenance management more efficient, activities of maintenance management have expanded. Also appearing is the concept that maintenance functions should not belong merely to the maintenance management department, but that portions of the functions should be allotted to the departments of production, R & D, design, engineering, and purchasing and finance, and to vendors, top management, and operators. Under such a concept, the British Department of Trade and Industry has emphasized the importance of "terotechnology," which is defined as a combination of management, financial, engineering, and other practices applied to physical assets in pursuit of economic life cycle cost. The practice of terotechnology is concerned with the specification for design of reliable and sound structures, including their installation, commissioning, maintenance, modification, and replacement, and with feedback of information on design, performance, and costs.

Functions

Maintenance functions can be categorized into management functions, technical functions, and operational functions as follows:

1. **Management functions:** maintenance policy determination; establishment of an organization and system for maintenance; planning, scheduling, and control of maintenance activities; economic analysis and evaluation; labor skills improvement and motivation elevation; subcontractor management; budgetary control; keeping of maintenance records and reports; maintenance effectiveness measurement; control of spare parts and substitute tools and equipment.
2. **Technical functions:** equipment performance analysis, failure causes analysis, preparation of standards and instructions for inspection, sweeping and repair, replacement analysis.
3. **Operational functions:** inspection (routine, periodic, and acceptance inspection), preparation operation (lubrication, adjustment, repair), engineering work (machining, can manufacturing, welding, finishing, etc.).

Organization

There are two types of organizations that perform maintenance functions. One is a centralized maintenance shop and the other a decentralized maintenance shop. The advantages of the centralized shop are:

1. Easier and earlier dispatch from each craft group to higher level within the maintenance system.
2. Justification of adequate maintenance budgets and manpower, suitable priority and maintenance technology, and higher-quality and more available equipment from the point of view of the company as a whole.
3. Saving and effective use of craft groups and maintenance staff.

Advantages of the decentralized shop are:

1. Reduction in travel time to and from job.
2. Improvement in knowledge of operative equipment condition.
3. Improvement in production-oriented relationships between production and maintenance.

Types of Maintenance

The types of maintenance are (1) breakdown maintenance, (2) routine maintenance, (3) corrective maintenance, (4) preventive maintenance, and (5) maintenance prevention. Exhibit 11.7.6 shows the types of maintenance and their features.

Exhibit 11.7.6 Types of Maintenance and Their Features

Type of Maintenance	Purpose	Features
Breakdown maintenance	Repair of equipment after failure	No preventive maintenance cost needed; suitable when there is spare equipment
Routine maintenance	Daily or weekly inspections, sweeping, adjustment, oiling, replacement	Aiming at protection of deterioration and reduction of repair cost
Corrective maintenance	Improvement of equipment, in addition to repair	Elevating of equipment productivity
Preventive maintenance	Periodic inspection and upkeep against early failure	Effective when inspection cycle is adequate
Maintenance prevention	Attempt at a maintenance-free system	No maintenance cost when condition is ideal; however, equipment cost is very expensive

Systems for Maintenance Management

A system for maintenance management implies a system that is connected organically with the previously cited activities so as to realize the maintenance management policy and target. This system consists of several subsystems, such as the maintenance control system, maintenance operation system, maintenance information system, computer-aided maintenance management system, scheduling and control system for maintenance, maintenance diagnostic system, and so on. The general model used to design various subsystems of maintenance management embodying maintenance policy is shown in Exhibit 11.7.7.

The maintenance control system includes the maintenance work control system, maintenance condition control system, and so on. The maintenance work control system includes setting standard times, scheduling and control using network methods and job shop scheduling methods regarding scheduled preventive maintenance, and breakdown maintenance and repairing operations. The maintenance cost and resources control system plans and controls the resources for maintenance activities. The maintenance condition control system consists of various failure informations.

The maintenance operation procedure is shown in Exhibit 11.7.8.

Parts of the subsystem of maintenance management, such as the diagnostic and information systems, which are shown in Exhibits 11.7.9 and 11.7.10, are aided by computer. However, in designing a computerized maintenance system, the following considerations are required:

1. Cooperation between a person familiar with maintenance operations and a person with knowledge of computer features and constraints.

2. Setting up of a person-computer interface.

11.7.4 MAINTENANCE DIAGNOSTIC SYSTEM

Even if scheduled maintenance and repair intervals are established, the optimum interval will fluctuate because of operating conditions and other reasons, resulting in overmaintenance or causing catastrophic failure. A maintenance diagnostic system is a system for diagnosing the condition of equipment while it is operating, with a view to improving operating efficiency and cutting maintenance cost by means of an accurate assessment of maintenance timing, carried out by utilizing sensing and measuring methods and signal processing techniques. This is also called a "running inspection system." The concept shown in Exhibit 11.7.11 is the same as that of the diagnosis of a human condition.[1]

Following is an example of the composition of a maintenance diagnostic system that has been applied to complicated precision equipment and facilities used in process industries such as iron and steel plants and chemical plants.

1. The technique for diagnosing the condition of equipment and facilities is as follows:

a. Divide the equipment and facilities into rational clusters according to the diagnostic characteristics.

b. Analyze the failure and characteristics of each cluster according to the second stage–data analysis–in Exhibit 11.7.5.

c. Develop and determine sensing, measuring, and discrimination methods and signal processing techniques, taking into account the object of diagnosis, the characteristics of equipment and facilities, and the observability, reliability, and economy of diagnosis.

2. Equipment and facilities testers are made according to the diagnostic technique developed for each cluster; these should preferably be portable.

3. A computer-aided maintenance diagnostic system is made in order to detect unusual equipment and facility conditions in the early stage and to indicate the location and the cause of failure together with the remedy, automatically.

11.7.5 WORKERS' SKILL AND MOTIVATION IN MAINTENANCE MANAGEMENT

Following the progress of mechanization and automation in production and service industries, workers' operations become simpler, and the number of jobs in which workers simply have to watch instruments increases. As a consequence, workers begin to lose interest in their jobs and in self-improvement, and their attention is distracted. Under such circumstances, many troubles in equipment and facilities arise from workers' carelessness in carrying out operations, and workers are affected by accidents at the same time at which productivity decreases.

As equipment and facilities have become more precise and complex, a higher level of maintenance skill has come to be required. In the past there was a tendency for management to separate not only repair jobs, but also jobs involving the checking of equipment and facilities, from the

Exhibit 11.7.7 General Model for Maintenance Management System

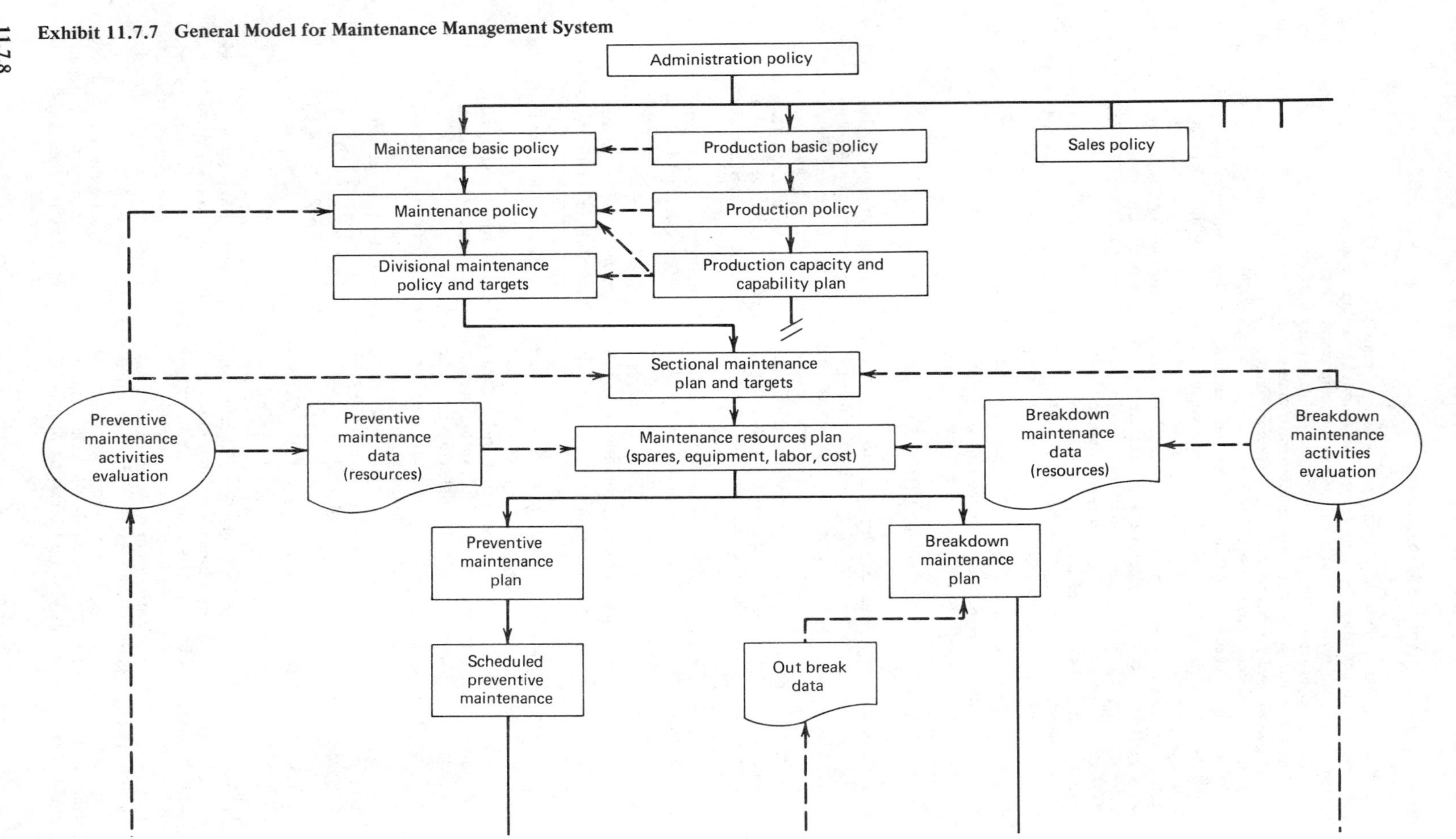

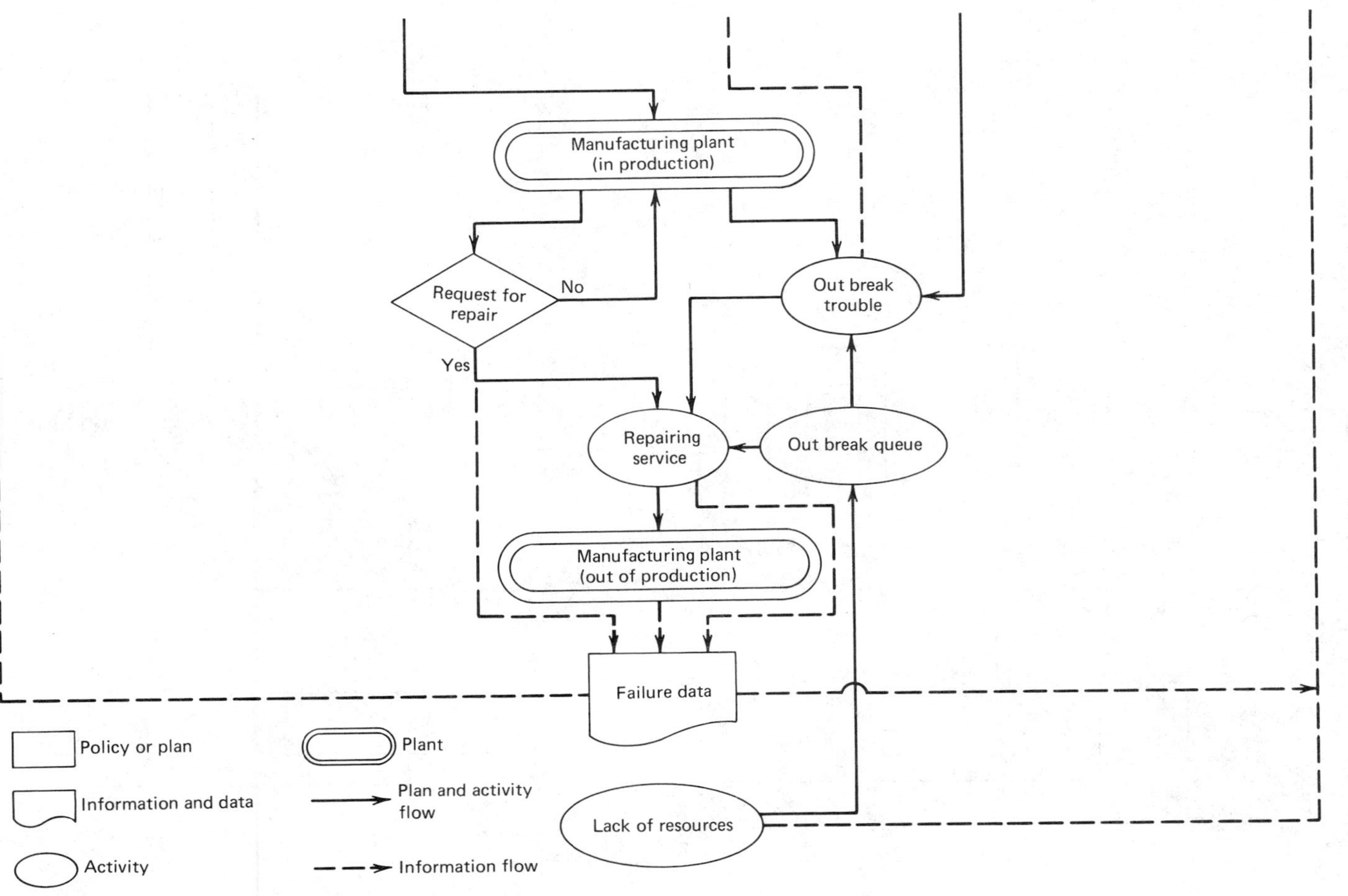
Manufacturing plant (in production)
Request for repair
No
Yes
Out break trouble
Repairing service
Out break queue
Manufacturing plant (out of production)
Failure data
Lack of resources
Policy or plan
Information and data
Activity
Plant
Plan and activity flow
Information flow

Exhibit 11.7.8 Maintenance Operation Procedure

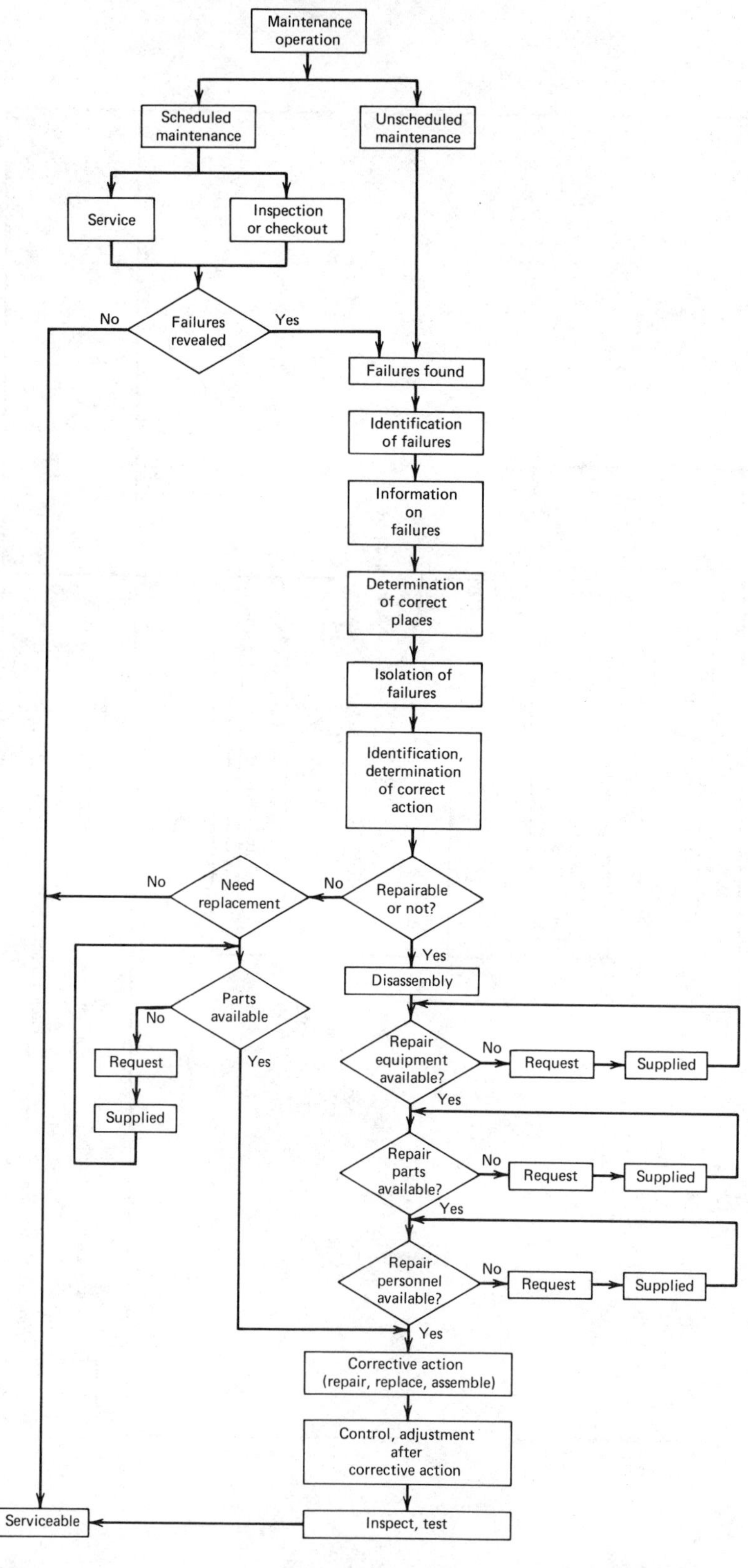

Exhibit 11.7.9 Maintenance Information System

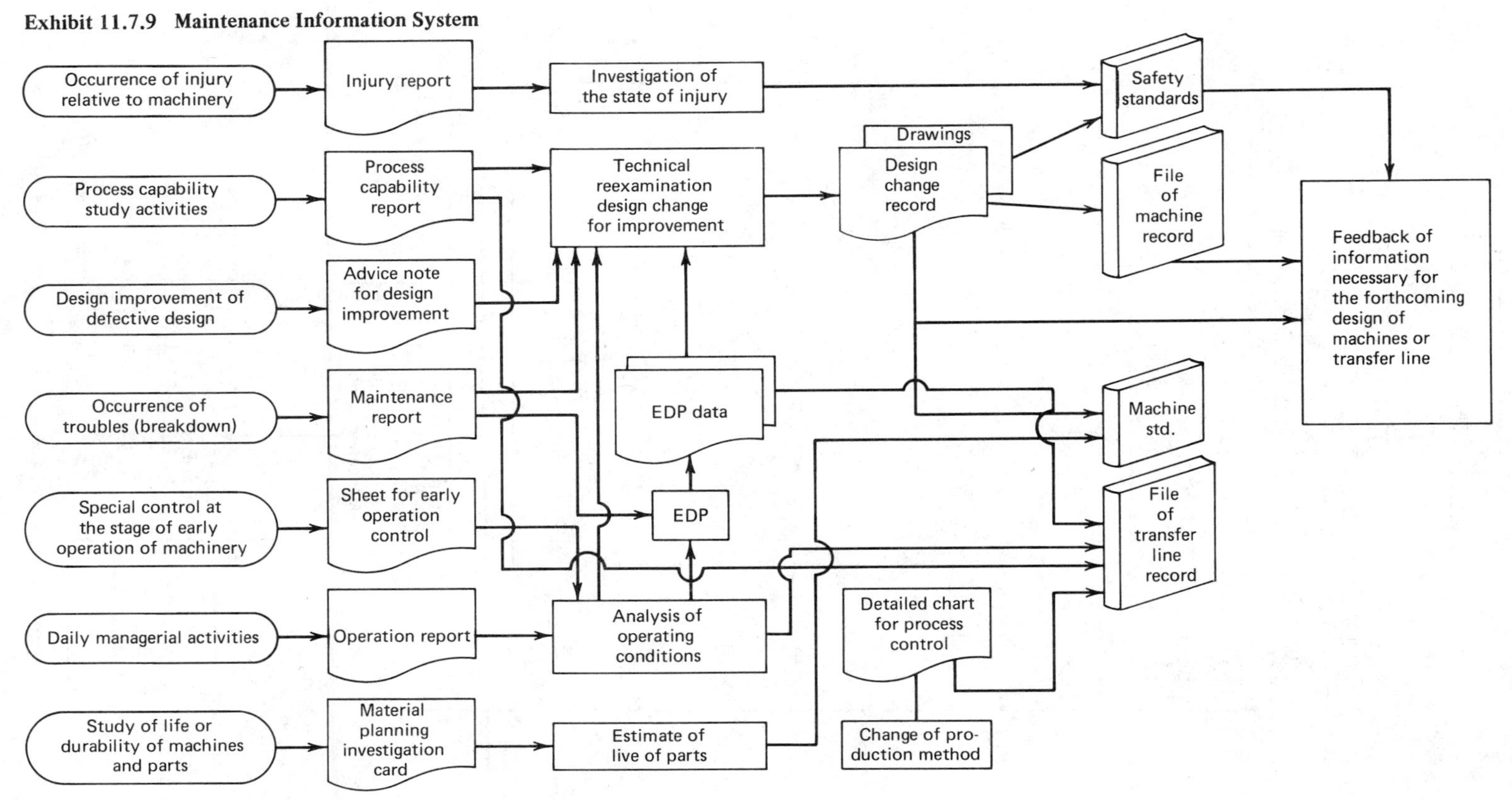

Exhibit 11.7.10 Information Flow for Maintenance Operation

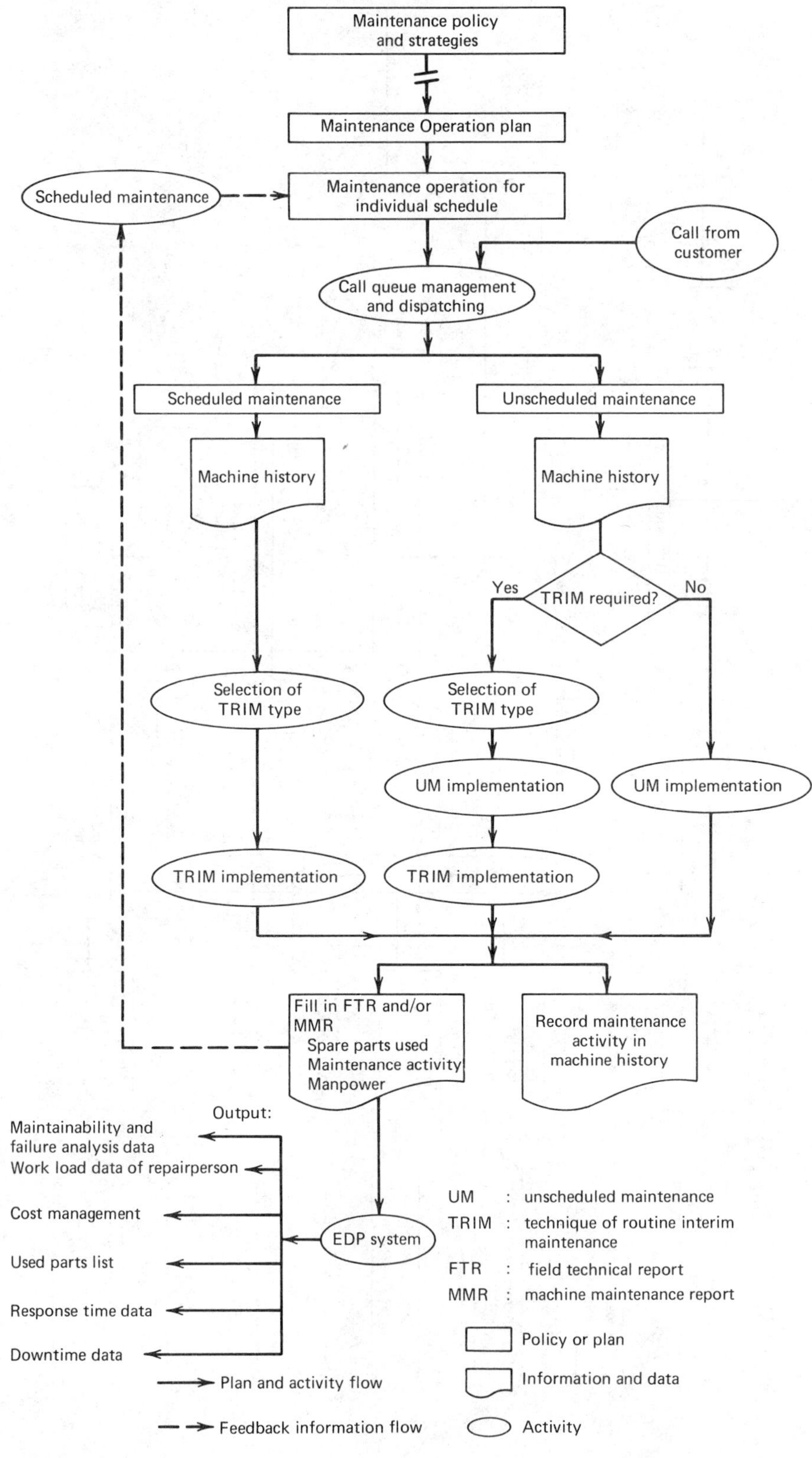

Exhibit 11.7.11 Comparison of the Concept of a Maintenance Diagnostic System With a System for Diagnosing People

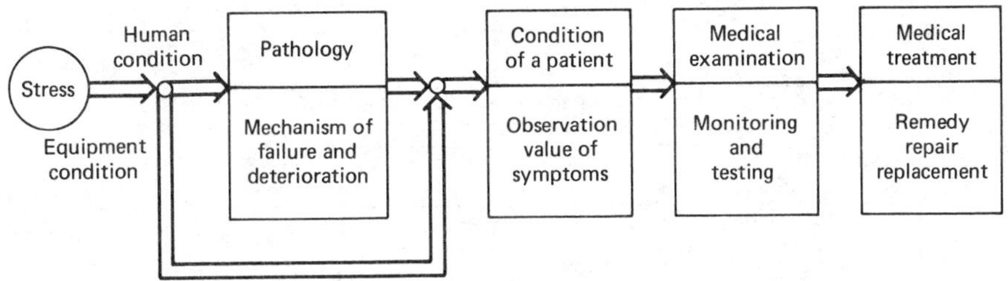

jobs of line operators and to refer these to maintenance specialists. As a result, workers have become subject to increasing stress arising from monotony and dullness.

When trouble in equipment and facilities is located at an early stage, maintenance work becomes easy, downtime is decreased, and maintenance and opportunity costs also are reduced. Moreover, troubles and failures arise at random, and it is a fact that the operators who are directly in charge of operating and watching the equipment can locate such troubles at an early stage.

On the basis of these facts, various companies have recently developed and introduced ways of harmonizing workers' needs with organizational needs. For example, a new job is designed in which the operator has the responsibility for finding an unusual equipment condition at an early stage and for checking the trouble and failure of equipment; additional training and education are given so as to improve the operator's skill and ability. In this way the company expects to satisfy the workers' needs and to relieve their monotony. Following from such an approach, many cases and methods of reducing monotony and increasing the motivation of workers and maintenance efficiency can be seen in various companies.

11.7.6 COMPANYWIDE MAINTENANCE ACTIVITY IN AN ELECTRICAL EQUIPMENT INDUSTRY

In the Nippon Denso Company, a Japanese company that makes electrical equipment, heaters, coolers, and many other products for automotive vehicles and which has 24,000 employees, a system of "total productive maintenance" (TPM) activities involving all the employees of the company has been in practice since 1969. The aims of the activities are as follows:

To keep all production equipment and facilities at peak operating efficiency and to guarantee the quality of products.

To establish a total system of productive maintenance for the whole life of all equipment and facilities.

To involve all personnel from top management down to those employees working on the shop floor.

To mobilize the personnel of those departments engaged in the work of planning, operating, and maintaining equipment and facilities.

To motivate and control all the personnel in small groups through the principle of autonomy and self-motivation.

By evaluating and implementing TPM activities based on its own evaluation methods, the company has achieved working efficiency in terms of equipment and facilities, has improved the quality of its products, has reduced production costs, and has made numerous other improvements.

Determination of TPM Policy and Its Embodiment

The basic policy of TPM, indicated in Exhibit 11.7.12, has been worked out by the managing director in charge of maintenance management for all equipment and facilities with a view to "firmly establishing the TPM system through the participation of all employees." On the basis of this policy, the company establishes annually various administrative measures and target values, determined by each of the functions according to the degree of control and administrative level.

In 1971 TPM QC safety circle activities aimed at improving self-motivation activity by small groups and organized by the employees themselves were started with the aim of positively handling quality, productive maintenance (PM), and safety. As a consequence, these circle activities have been of great value and significance for all personnel of the company in increasing PM knowledge

Exhibit 11.7.12 Embodiment of TPM Policy

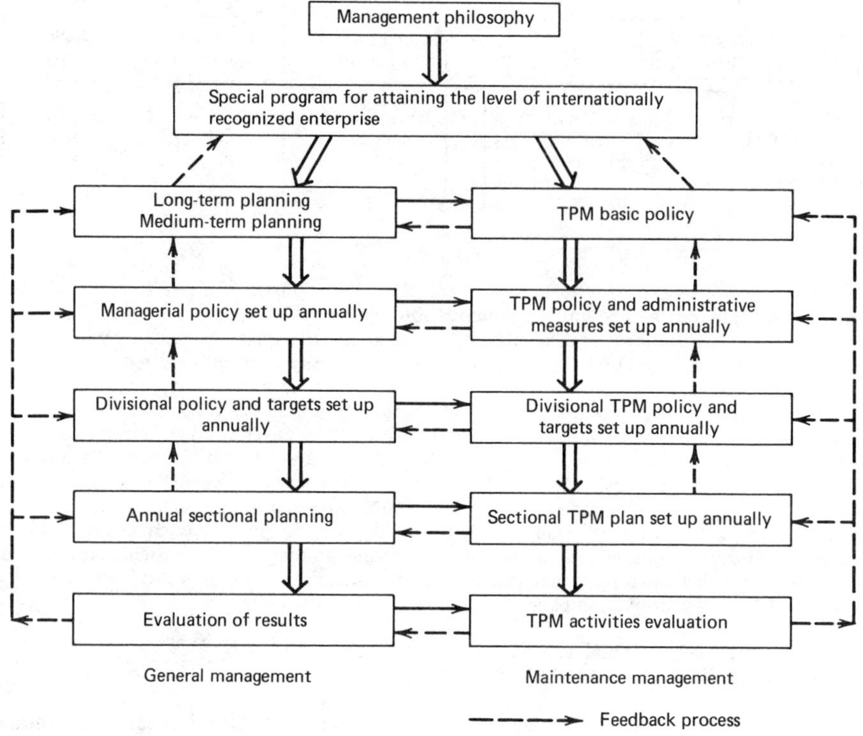

and morale, ensuring steadiness of daily maintenance practice, and further enhancing PM improvements.

PM Diagnostic System

In 1969 the company started a "PM diagnostic system" whereby it is possible to check and confirm the embodiment of PM policy and also to prompt feedback of the results of its implementation.

Under this diagnostic system, there are two categories of diagnosis: One is "self-diagnosis," carried on inside the division itself, and the other is "mutual diagnosis," conducted between one division and another. This mutual diagnosis constituted "total diagnosis" when examining all aspects of PM activities and "project diagnosis" when examining a specific PM function or a specific project currently taken up. These different types of diagnosis are conducted in an integrally linked manner.

Training and Education of PM Personnel

Production personnel are given such educational materials as a PM glossary and explanation, a lubrication control manual and PM slides, and so on, in systematically organized education schedules; further emphasis is also widely placed on on-the-job training for workers.

Maintenance personnel are given special education and training in connection with hydraulics, electricity, legal matters of safety, and so on, thus enabling each individual to become a specially skilled PM person.

In addition, the company practices the "Einsteller system" and the "rotation system," both of which are rather unique in personnel management concerning PM. The Einsteller system aims at educating and training specially designated workers in the workshops. Einsteller workers are in charge not only of production operations and daily maintenance work, but also of scheduling and making preparations for operations, of simple adjustments, of routine repairs, and of the training of unskilled workers. The merits of the Einsteller system are becoming increasingly clear as more high-precision production machines are introduced and as the number of unskilled workers increases. For maintenance or Einsteller operations the rotation system aims at educating machine

Exhibit 11.7.13 General Aspects of PM Evaluation System

	Result evaluation (mainly quantitative)		Activities evaluation (mainly qualitative)
Overall evaluation	Net operating rate Overall maintenance cost rate Facilities productivity Overall maintenance evaluation point		PM diagnosis (total diagnosis) Evaluation of the level of PM activities
Individual evaluation	Maintenance Scheduled maintenance rate Achievement rate Net maintenance rate Cost Maintenance cost rate Spare parts inventory Quality Process capability	Breakdown Frequency rate Severity rate Others Cost Maintenance loss and so on Quality Percentage defective Safety mInjury frequency rate relative to machinery	PM mutual diagnosis (project diagnosis) PM function diagnosis Maintenance function diagnosis Facilities function diagnosis Specific diagnosis PM self-diagnosis Evaluation of the level of MP activities
PM background	Education	Standardization and so on Morale (PM circle, suggestion system)	

manufacturing technicians from the machine manufacturing department, which is responsible for producing internal-use machines and is capable of installing, adjusting, and overhauling machines and turning them over to divisions according to a predetermined staffing plan.

Evaluation of PM and Its Effects

In evaluating the effects of PM activities, the company evaluates not only the results of maintenance itself, but also whether the PM policy set up by top management is observed among PM personnel in the workshop, whether PM morale is further deepened through all the personnel, whether prevention of early machine troubles and standardization are well advanced, and so on. The general aspects of the PM evaluation system are shown in Exhibit 11.7.13.

To carry out overall evaluation, PM managerial items are set up according to the functions and the graduation of managerial posts. Moreover, to clarify the relationships among important managerial items, a specially designed table is worked out whereby PM activities are graphically controlled or evaluated by the comparison of target with activity results.

11.7.7 A MAINTENANCE AND PRODUCTION SYSTEM FOR IMPROVING THE MOTIVATION OF WORKERS AND AVAILABILITY OF EQUIPMENT IN AN IRON AND STEEL PLANT

As a result of mechanization, automation, and rationalization, the number of workers required in Japan's Nippon Steel Corporation has been reduced by about 12,000 during the 6 years since 1970, and workers' jobs that used to demand skill and experience have been transferred to machines and controllers, so that more workers simply have the job of watching machines. Most of the workers in the plant have graduated from high school and have a fairly high intelligence level.

Considering the importance of terotechnology, the stabilization of employment, and the need to improve workers' motivation and skill, the plant has planned the development of a new maintenance management system. Maintenance diagnostic systems have been developed and installed by the maintenance department. The maintenance and production departments have jointly investigated the actual state of the jobs of production workers. As a result of the investigation, the following points were clarified:

1. There were few skilled jobs that needed more than 5 years experience, and there was only a slight correlation between the length of time it took to become skilled in a job and the length of job experience.

Exhibit 11.7.14 Comparison of a New Production and Maintenance System With an Old One

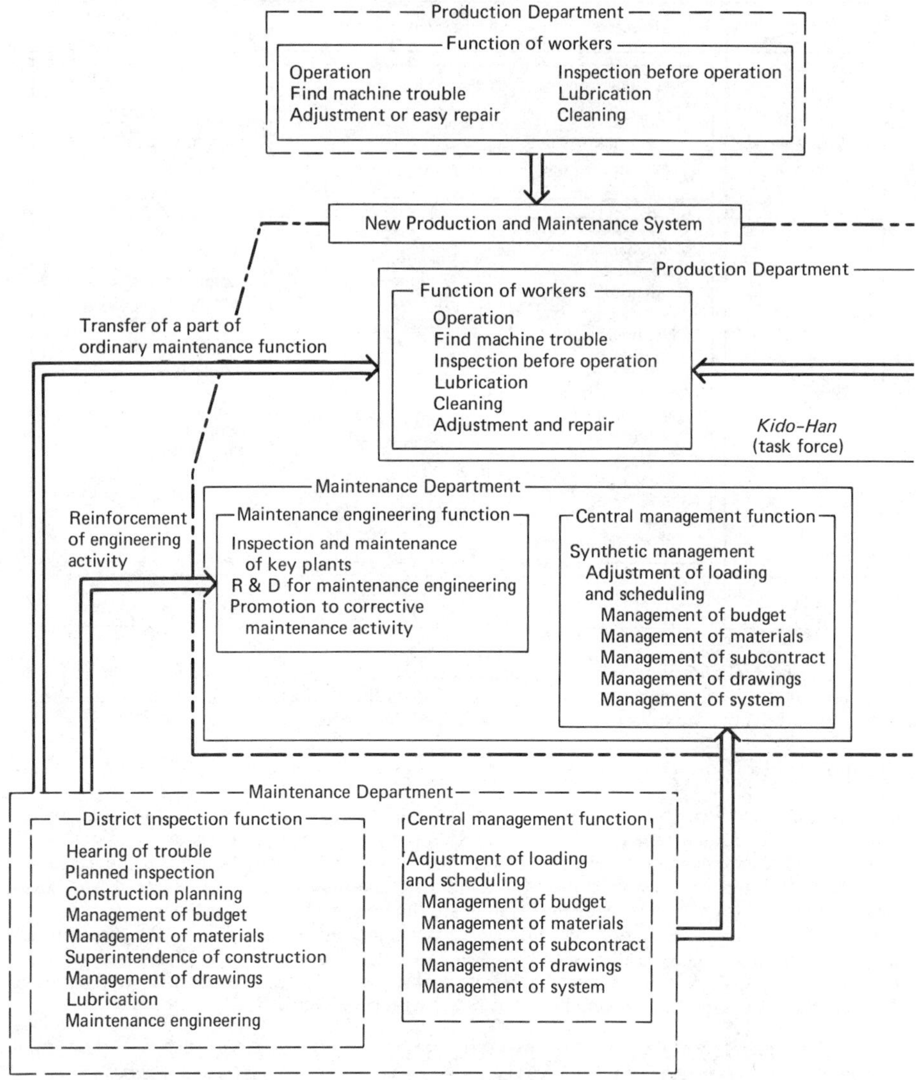

2. Job rotation has encouraged the ability to generate ideas in the case of a large number of workers.

3. Regarding items of maintenance trouble, 7% were caused by operating mistakes, 75% were troubles preventable by operators at an early stage, and 18% were independent of operators.

4. It was estimated that 70% of the ability to repair and discover faults at an early stage could be transferred to production workers after 3 or 4 years' training and that 30% of the ability to repair would remain as the specialized job of maintenance worker.

5. Since the fluctuation of the work load of repair jobs caused by the outbreak of troubles was large, many relief workers were required in the maintenance department.

6. A large number of workers feel various kinds of dissatisfaction with their present jobs.

On the basis of these facts, a new production and maintenance system has been developed and installed. The new system compared with the old one is shown in Exhibit 11.7.14.

The major characteristics of this system with regard to improvement in maintenance management are as follows:

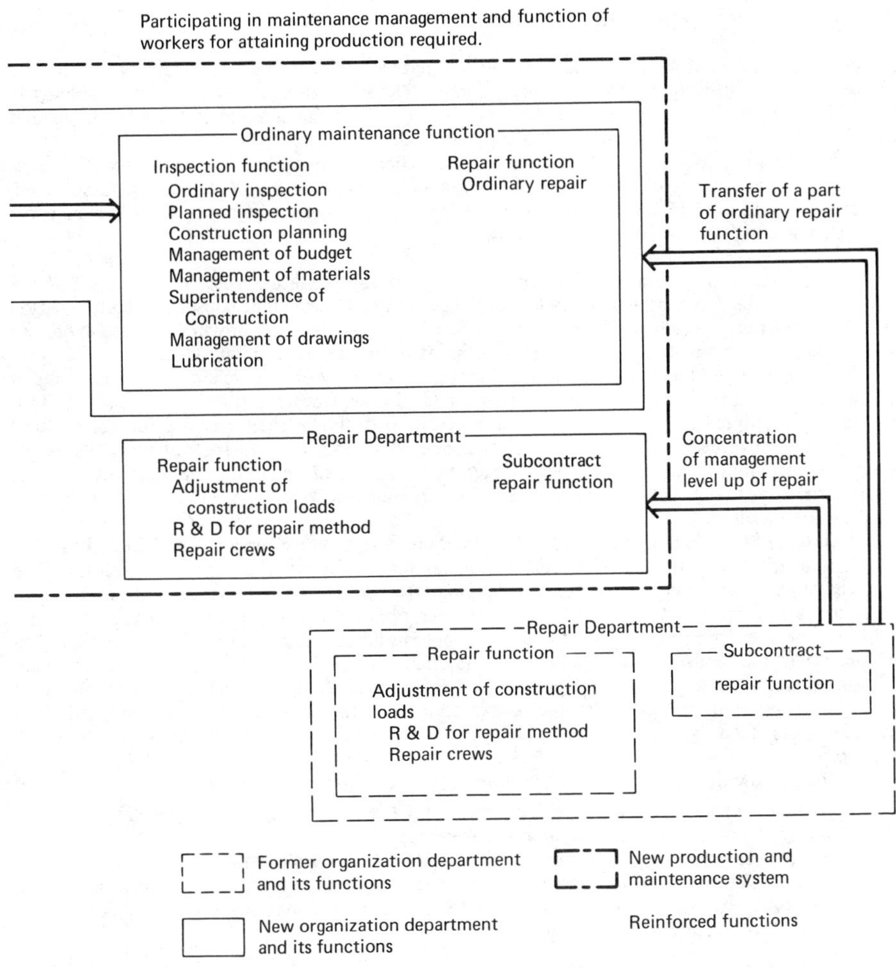

1. A broader application of the maintenance diagnostic system to both production and maintenance departments.
2. Enrichment and enlargement of the jobs to improve the motivation and ability of workers, and introduction of a training program on maintenance management.
3. Organization of a task force (*kido-han*) made up of both production department and maintenance department workers, so as to improve their maintenance skill by on-the-job training and to balance the fluctuation in the work load.

11.7.8 MAINTENANCE MANAGEMENT IN OFFICE WORK

Japan's Xerox-Fuji is a manufacturing and selling company of reprographic products. The selling of products is largely based on a rental system. The company's policy is to offer highly reliable products and high availability through maintenance service considering total life cost.

Features of Office Machines and Maintenance

Mechanization of office work has progressed rapidly, made possible by the use of technologies such as electronics. This brought about today's adaptability toward user needs to process a gigantic amount of information quickly. With the introduction of mechanization, the poor condition of office machines resulting from improper maintenance service not only lowers productivity, but also increases opportunity cost because of delayed decision making through lack of information.

Types of Maintenance for Office Machines

Maintenance activities for office machines such as copiers, duplicators, and electronic computers are conducted exclusively by manufacturers today. Therefore users expect to gain efficiency in using office machines, leaving maintenance responsibility with the maker, because of complex and precise mechanisms that require higher maintenance skills.

There are several ways to provide maintenance to office machines, based on their characteristics, their complexity, their frequency of use, the number of machines installed, the installation environment, the skill level of the operator or repair person, and the maintenance cost. Typical types of maintenance and their features are as follows:

1. **Maintenance by Operator.** The operator repairs the machine when it fails to function properly. This type of maintenance can be conducted only when the mechanism and fault isolation are simple and the parts replacement is easy. The advantages are short downtime and inexpensive repair cost, but operator training and identification of maintenance are required.
2. **Maintenance by Repair Persons Employed by the User.** A repair person or a group of repair persons provides failure diagnosis, does repairs, and also replaces parts when machine failure is reported by the operator. If such persons cannot solve a problem, they may ask for maintenance service from the outside. This type of maintenance is suitable to reducing downtime when the number of machines installed is relatively large, when the machines are located at remote areas or installed on board ships, or when the makers' maintenance service is not available. Inventory of space parts is required.
3. **Maintenance by Maker.** There are two types of maintenance service: unscheduled maintenance and scheduled maintenance. Unscheduled maintenance starts when the user requests repair from the maker's maintenance service department. Repair persons of the maker visit the user site to provide diagnosis and fault isolation, and then either make repairs or take office machines back to the maker for repair. Preventive maintenance is suitable when office machines are complex and precise in mechanism. The advantages are (a) that there is less possibility of incorrect repair or maintenance procedures by the operator and (b) that identical replacement is possible when maintenance or repair time is estimated to exceed a limit, thus reducing the user's downtime. A disadvantage is the longer downtime resulting from the repair person's travel time to the user's location.

To minimize downtime, the maker has to emphasize the following activities:

a. Determining properly the number, location, and manpower allocation of service stations.
b. Planning and controlling smooth and balanced work load.
c. Controlling spare parts by echelon level and better distribution system.
d. Feeding back information of maintenance activity analysis to design department in order to improve reliability and maintainability for future products and to maintanance service department in order to improve maintenance manuals and methods.

The user can select one of these three types of maintenance or use a combination of them based on the advantages of the respective maintenance types.

ACKNOWLEDGMENT

Three cases were presented to the authors from three companies. A case of companywide maintenance management in the electrical equipment industry (Section 11.7.6) was presented by Nippon Denso Company. A case of a new maintenance and production system for improving both workers' motivation and availability of equipment in an iron and steel plant (Section 11.7.7) was presented by Nippon Steel Company, Nagoya plant. A case of maintenance management in office work (Section 11.7.8) was presented by Zerox-Fuji Company. The authors rearranged and extracted from the original papers presented by each company. The authors appreciate the courtesy of each company.

REFERENCE

1. TAKADA, T., "Engineering System for PM," *Journal of Plant Engineering*, JIPE, December 1975.

BIBLIOGRAPHY

BAKER, J. T., "Automated Preventive Maintenance Program for Service Industries and Public Institutions," *Industrial Engineering*, Vol. 12, No. 2 (1980), pp. 18–32.

KELLY, A., and M. J. HARRIS, *Management of Industrial Maintenance*, Newnes-Butter Worth & Company Ltd., London, 1978.

MORROW, L. C., *Maintenance Engineering Handbook*, 3rd ed., McGraw-Hill, New York, 1973.

MURAMATSU, R., Ed., *Industrial Engineering and Management Handbook*, Maruzen, Tokyo, 1975.

CHAPTER 11.8

Work Schedules

STANLEY D. NOLLEN

Georgetown University

11.8.1 INTRODUCTION

Work schedules are the time patterns that employees follow on the job. Work schedules are one aspect of the deployment of labor in the production process. They are a way to match the production needs of the enterprise with the human needs and availabilities of workers. Work schedules are at the nexus of the man-machine interface. They are part of the sociotechnical system of the enterprise.

There are three dimensions along which work schedules vary: (1) length of work time (e.g., 40 hr/week), (2) allocation of work time (e.g., 8 or 10 hr/day), and (3) control over work time (e.g., by managers or employees).

Traditionally, a fixed work schedule is set by management for most employees, consisting of five 8 hr days/week starting at 8 a.m. and ending at 5 p.m. Factories in some industries regularly use two or three fixed 8 hr shifts of workers. For some workers, overtime hours are often available or ordered by management. (See Chapter 11.9 for a discussion of shift work.)

Alternatives to traditional work schedules are available. The length of time worked can be varied by using part-time employment as well as by changing the standard full-time work schedule. Part-time employment options include temporary employment, permanent part-time employment, job sharing, work sharing, phased retirement, sabbaticals, and work year contracts.

The allocation of a predetermined length of work time can be changed by using staggered work hours, flexible work schedules, or compressed workweeks. Among these, flexible work schedules transfer some control over work schedules from managers to employees.

The objectives of this chapter are to (1) describe the usage patterns of traditional and alternative work schedules, (2) summarize the economic and social effects that alternative work schedules have on employers and workers, and (3) sketch the key management issues, problems, and opportunities that these work schedules pose.

11.8.2 TRADITIONAL AND ALTERNATIVE WORK SCHEDULES: WHAT THEY ARE AND WHO USES THEM

This section describes the usage patterns and trends of traditional work schedules and alternatives to them. These work schedules include the standard workweek, shift work, overtime, part-time employment (including temporary and permanent part-time employment, job sharing, and work sharing), staggered work hours, flexible work schedules, and compressed workweeks.

The Standard 5 Day, 40 Hr Workweek

By far the most common work schedule is the standard 5 day, 40 hr workweek. Among all nonfarm workers, more than 80% work 5 days a week. By far the largest single percentage of workers—41%—work exactly 40 hr (see Exhibit 11.8.1). Nearly half work 35 to 40 hr. Morning arrival at work is predominantly at 8 a.m., and evening departure is usually at 5 p.m. (see Exhibit 11.8.2).* Although the multitude of work schedules prevents any one of them from achieving a majority use, the standard 5 day, 40 hr workweek outdistances all the rest by a wide margin.

The average weekly hours of work for all nonfarm workers was 38.6 in 1979; this number falls below 40 because of part-time employment. However, there is considerable variation in working

*The days per week estimate is based on unpublished government statistics for full-time wage and salary workers and on the author's estimate for part-time employees. Other estimates are from the U.S. Bureau of Labor Statistics (BLS) (see Exhibits 11.8.1 and 11.8.2).

Exhibit 11.8.1 Distribution of Weekly Work Hours, May 1978[a]

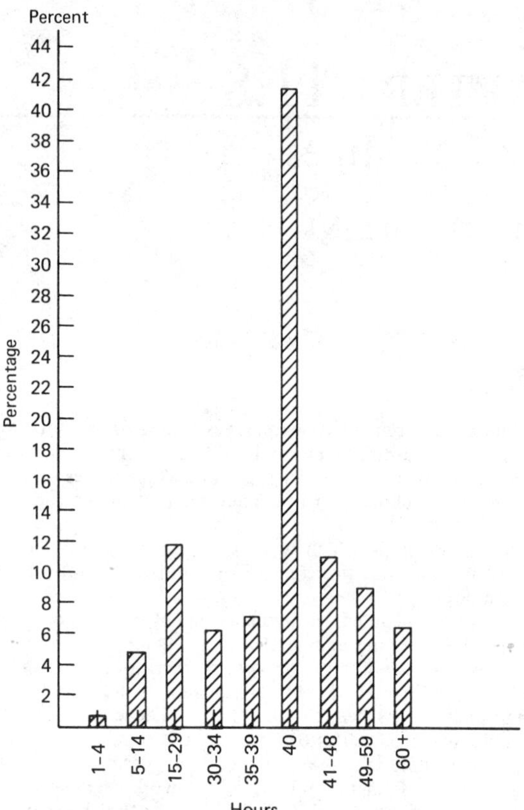

[a]Reference 1.

hours by age and sex of employees and by industry. For example, men aged 25 to 44 years worked 44 hr/week on the average in May 1978, whereas women of the same age worked 36 hr. Wage and salary workers in manufacturing worked 40.9 hr in 1979, whereas service industry workers put in 36 hours on the average.*

The standard 40 hr workweek has remained virtually unchanged in the post-World War II period. Although the 1978 figure of 38.5 hr is slightly below the 30-years-ago figure of 40.9, this is due mainly to changes in the composition of the labor force—more women and students—and to more part-time employment. Looking at men who were not students (in order to get a more homogeneous group for overtime comparisons), hours of work were unchanged from 1948 to 1978 at roughly 42.5 hr/week. There were, however, increases in vacation and holiday time over this period (see Exhibit 11.8.3).

Overtime

Overtime hours are hours worked beyond 40 in a week for which premium pay is required under the Fair Labor Standards Act. For employees working directly on federal government contracts, and for federal government employees themselves, hours worked in excess of 8/day are also overtime hours requiring premium pay. (The legislation that requires this is the Walsh-Healy Public Contracts Act and the Contract Work Hours and Safety Standards Act.) However, many employees are exempt from these requirements—managers and administrators who have supervisory responsibilities, professionals, and outside salespersons.

In May 1977 about 11% of all full-time employees received premium pay. This proportion was higher for manufacturing (13%) and lower for service industries (5%). Among blue collar workers, 20% of all full-time employees received premium pay, but only 6% of white collar workers did.

*The sources of Exhibits 11.8.1 through 11.8.3, as well as further details about them, are Chapter 3 of reference 1 and reference 2.

Exhibit 11.8.2 Distribution of Daily Work Beginning and Ending Times, May 1978[a]

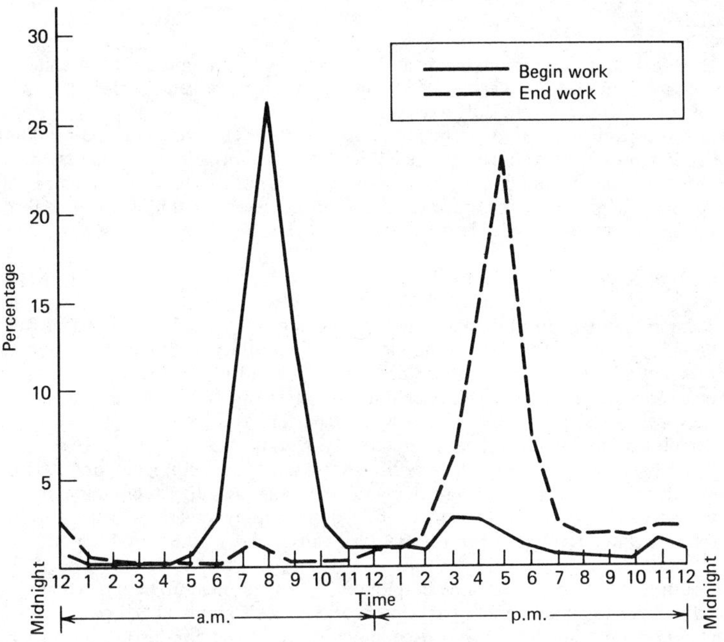

[a]Reference 1.

Exhibit 11.8.3 Average Weekly Hours of Work Over Time for Selected Industries[a]

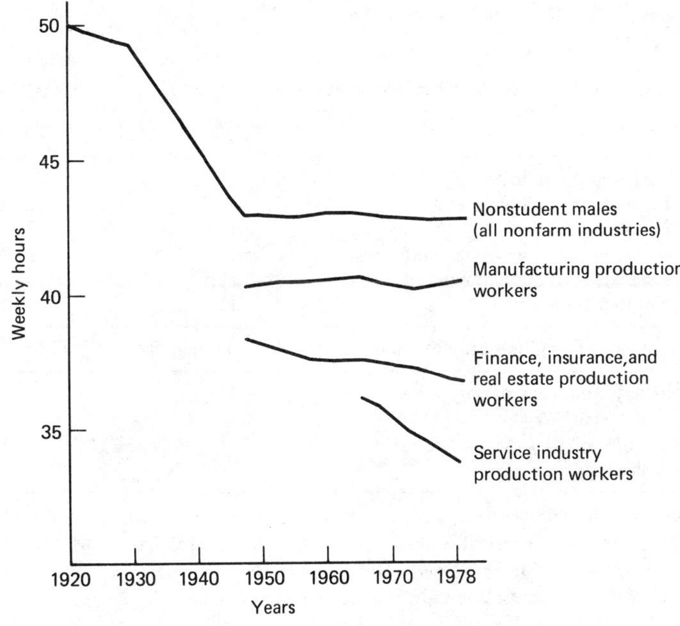

[a]Reference 1.

The number of overtime hours worked fluctuates with the business cycle. On the average, manufacturing industries use 3 to $3\frac{1}{2}$ weekly overtime hours per production employee.[1]

Shift Work

Shift work refers to work periods that fall largely after 4 p.m. and before 8 a.m. Evening shifts accounted for 4.9 million workers in 1978, and night shifts tallied another 2.1 million workers. Overall, 16% of all workers were shift workers.

Shift work is especially common in manufacturing industries, where nearly 3 production workers in 10 are shift workers, as well as in service industries. Use of shift work increased steadily in the 1960s as a proportion of total employment. However, that proportion has remained stable through the 1970s. (See Chapter 11.9 for more complete information on shift work; see also Hedges and Sekscenski[3] and *Employment and Training Report of the President.*[1])

Part-Time Employment

Part-time employment is an umbrella term that includes all work less than usual full-time hours. Part-time employment may be long term, stable, and voluntary, or it may be temporary, either by choice or as a work sharing response to economic downturn in lieu of layoffs. Some part-time employees are job sharers: two half-time people who share one full-time job.

Part-timers average about 20 hr/week. The time pattern is most often part of a day, but many part-time employees also work full days for part of a week.

In 1978, 22.1% of all wage and salary workers were part-time employees, or 17.6 million people. This covers all those who worked less than 35 hr/week, including temporary, intermittent, and involuntary part-time workers. Permanent part-time employees (those who usually and voluntarily work part-time) accounted for 13.9% of all those who worked, or 11.1 million people (see Exhibit 11.8.4).

Women outnumber men as part-time employees by a large margin; people who are young or old are especially likely to be permanent part-time workers (see Exhibit 11.8.5).

Permanent part-time employment is used much more in some industries and occupations than others. Both wholesale and retail trade as well as service industries use large numbers of permanent part-time workers both absolutely and relative to their total employment. Mining and manufacturing are especially low use industries (see Exhibit 11.8.6).

A large number of firms (somewhere between 55 and 75%) have permanent part-time employees, especially in office and clerical jobs, but they have only a few such employees (between 2 and 7% in most cases). Part-time minishifts and job sharing are used by a fifth or more of these firms.

Permanent part-time employment increased during the 1950s and 1960s, expanding roughly twice as fast as the overall labor force. But there has been no relative growth since then (see Exhibit 11.8.7). This pattern holds for both women and men.

Exhibit 11.8.4 Part-Time Employment Models Used, 1977

Type	Total[a]	Men[a]	Women[a]
All part-time employees			
Percentage of all wage and salary workers	22.1	14.5	32.5
Number (millions)	17.6	6.7	10.9
Usual voluntary part-time employees			
Percentage of all who worked	13.9	7.6	22.7
Number (millions)	11.1	3.5	7.6

Permanent Part-Time Employment Models	Percentage of User Firms[b]
Part day, full week or part week	75
Full day, part week or part month	49
Minishift	23
Job sharing	22

Sources. References 4–6.

[a]Data refer to nonagricultural industries. Part-time means 1 to 34 hr/week.

[b]The total percentage exceeds 100 because the categories are not mutually exclusive and because some firms use more than one model.

Exhibit 11.8.5 Characteristics of Voluntary Part-Time and Full-Time Workers[a]

Characteristic	Voluntary Part-Time (percent)	Full-Time (percent)
Marital status and sex—total	100	100
Married men	26	55
Single men	20	11
Women with children ≤15 years old	27	12
Other women	27	23
Age[b] and sex—total	100	100
Young men	21	10
Prime-age men	17	39
Older men	8	17
Young women	19	7
Prime-age women	23	19
Older women	13	9

Source. Reference 7.

[a]Data refer to nonagricultural wage and salary workers in 1973.

[b]Young = age < 25; prime-age = 25–49; older = age ≥ 50.

Exhibit 11.8.6 Part-Time Employment Usage by Industry and Occupation, 1977[a]

	Usual Voluntary Part-Time	
	Number (1000s)	Percent
Industry		
All nonagricultural industries	10,433	13.3
Mining	17	2.2
Construction	225	5.0
Manufacturing	687	3.4
Transportation, public utilities	334	6.2
Wholesale, retail trade	4,168	25.4
Finance, insurance, real estate	410	9.1
Service industries	4,294	19.9
Public administration	257	5.2
Self-employed	972	16.4
Occupation		
All nonagricultural workers	10,665	13.5
White collar	5,427	13.5
Professional, technical	1,341	10.9
Managerial	328	3.3
Sales	1,166	23.8
Clerical	2,645	17.2
Blue collar	1,981	7.0
Skilled	397	2.9
Operatives, except transport	556	5.5
Transport operatives	300	9.1
Laborers	813	18.9
Service workers, except private household	3,255	31.0

Sources. References 4 and 8.

[a]Data refers to wage and salary workers (except self-employed line).

Exhibit 11.8.7 Growth of Part-Time Employment[a, b]

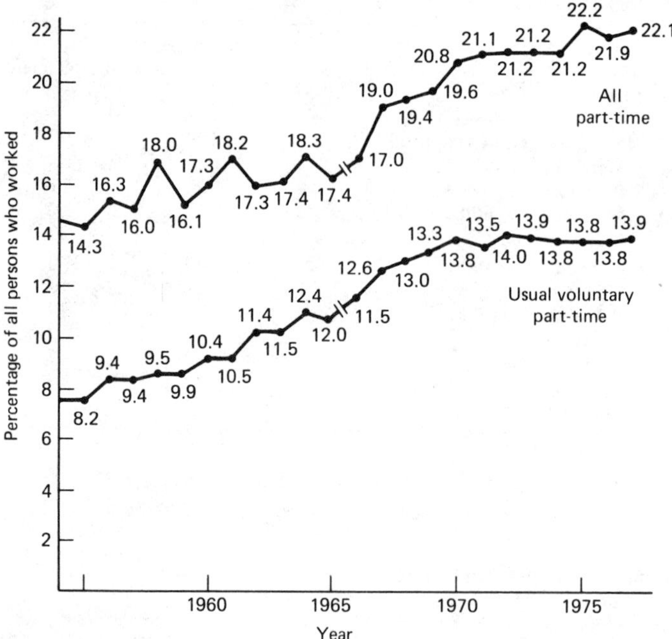

[a]Reference 4.

[b]Beginning in 1966, persons aged 14–15 were excluded, thus making data noncomparable between the 1954–1965 and 1966–1977 periods; the trend lines have been roughly adjusted for comparability, although the numbers have not.

Flexible and Staggered Work Hours

A schedule of flexible work hours, or "flexitime" for short,* is one under which employees choose their starting and quitting times within limits set by management. Flexitime schedules differ along three dimensions: (1) daily versus periodic (e.g., weekly or monthly) variation in starting and quitting times, (2) variable versus constant length of working day (whether credit and debit hours are allowed), and (3) core time—the hours of the day when all employees are required to be present.

Flexitime that requires employees to choose a starting and quitting time, to stick with that schedule for a period, and to work 8 hr every day is called "flexitour." If daily variation in starting and quitting times is possible, the schedule is called "gliding time." If credit and debit hours are allowed (e.g., working 10 hr one day and 6 hr another day), the term "variable day" is used.

Core time can be designed in several ways. Most user companies (88% in one survey) have some core hours requirement. Among these companies, more than half use a single core period (e.g., 10 a.m. to 3 p.m. or 9 a.m. to 4 p.m.). Others use two core periods, leaving a flexible lunch period (e.g., 9:30 to 11:30 a.m. and 1:30 to 3:30 p.m.).

In 1980 it was estimated that about 11% of all organizations and 9% of all workers, or 7 to 9 million people, were using flexitime. If professionals, managers, salespeople, and self-employed workers (who have long set their own hours without calling it flexitime) are excluded, the usage rate is 81%. It was used in all major industry groups, but with a somewhat heavier concentration in finance and insurance than in manufacturing. Among the three major flexitime models, there was a roughly equal distribution of use, but with perhaps a slightly more common use of gliding time (see Exhibit 11.8.8). Flexitime is more common in Europe than in the United States, especially in Germany and Switzerland where one third or more of the work force has flexible hours.

A schedule of staggered hours is one in which groups of employees regularly arrive at and leave from work at different times established by management. For example, one department's em-

*Many other short names aside from flexitime are in use, such as flex-hours. However, the word "Flextime" is a registered trademark applied to a time accumulating machine and should not be used to describe a work schedule.

Exhibit 11.8.8 Use of Flexitime in the United States by Occupation and Industry, 1980[a]

Occupation and Industry	Number (1000s)	Percent
All occupations	7638	11.9
Professional and technical workers	1914	15.8
Managers and administrators	1622	20.2
Sales workers	378	26.5
Clerical workers	1296	9.8
Craft workers	753	7.4
Operatives, except transport equipment	387	4.4
Transport equipment operatives	388	14.3
Laborers	214	7.3
Service workers	569	8.7
Occupations excluding professional and technical workers, managers and administrators, and sales workers	3608	8.1
All industries	7922	11.9
Mining	83	10.6
Construction	439	10.1
Manufacturing	1516	7.9
Transportation, public utilities	620	11.7
Wholesale, retail trade	1633	14.7
Finance, insurance, real estate	725	17.1
Professional services	1555	11.4
Other services	696	16.9
Federal public administration, except postal service	404	24.9
Postal service	47	7.6
State public administration	125	14.4
Local public administration	148	8.9

Source. Reference 9.

[a]All figures refer to nonfarm wage and salary workers who work full time.

ployees may begin work at 8:00 a.m. and leave 8 hr later, whereas another department's employees arrive at 8:15 and leave 8 hr later. Staggered hours differ from the most conservative flexitime model (flexitour) in that management sets the schedule rather than employees and entire groups of workers follow the same schedule.

Although there are no data on the overall incidence of the use of staggered hours, several large cities such as New York have encouraged staggered hours programs as a means of relieving rush-hour traffic congestion.

Compressed Workweeks

Compressed workweeks refer to full-time employment accomplished in less than 5 days/week. The schedules that are used include (1) 4 day workweeks with 10 hr days, (2) 3 day workweeks with 12 hr days, (3) $4\frac{1}{2}$ day workweeks with four 9 hr days and one 4 hr day (usually Friday), and (4) the 5/4-9 plan of alternating 5 day and 4 day workweeks with 9 hr days.

In 1980, 2.7% of all full-time nonfarm wage and salary workers, or 1.7 million people, were on compressed workweeks. Of this number, two thirds were working 4 day weeks. Compressed workweeks were used more in some industries than in others. Their heaviest use was in local public administration (especially police departments) and in small manufacturing firms (see Exhibit 11.8.9).

Compressed workweeks began in earnest in the United States in the early 1970s (there has never been noticeable use in Europe) and increased to 2.2% of the full-time work force by 1975. There was little change in use until 1980. However, the 4 day workweek has increased its share of all compressed workweeks at the expense of the $4\frac{1}{2}$ day week (see Exhibit 11.8.10).

11.8.3 EFFECTS OF ALTERNATIVE WORK SCHEDULES ON EMPLOYERS AND WORKERS

Work schedules result in a variety of economic benefits and costs for both employers and workers. In addition, work schedules have several social impacts on workers and may cause organizational

Exhibit 11.8.9 Compressed Workweek Use in the United States by Industry and Occupation in 1980[a]

	Employees on Workweeks Less Than Five Days	
	Number (1000s)	Percent
Industry		
Mining	17	2.0
Construction	148	3.4
Manufacturing	422	2.2
Transportation, public utilities	143	2.7
Wholesale, retail trade	267	2.4
Finance, insurance, real estate	76	1.8
Professional services	382	2.8
Other services	148	3.6
Federal public administration, except postal service	34	2.1
State public administration	31	3.6
Local public administration	180	10.8
Occupation		
Professional and technical personnel	303	2.5
Managers and administrators	83	1.1
Sales workers	63	1.9
Clerical workers	225	1.7
Craft workers	244	2.4
Operatives	290	3.3
Laborers	73	2.5
Service workers	439	6.7

Source. Reference 9.

[a]All figures refer to nonfarm wage and salary workers who usually work full time.

changes in the companies that use them. There are also several macro effects, both economic and social, that affect the society at large. In this section the ways in which the main alternative work schedules—flexitime, permanent part-time employment, and compressed workweeks—potentially affect employers, workers, and society are sketched. Experiences reported to date are then summarized.

Economic Effects on Employers and Workers

In general, alternative work schedules can affect the economic outcomes experienced by the user firm either by increasing the output of the firm (without increasing inputs) or by reducing the costs

Exhibit 11.8.10 Growth of Compressed Workweeks[a]

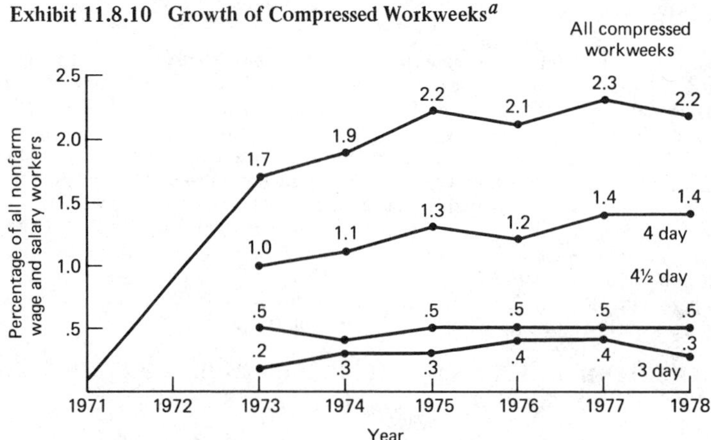

[a]References 11 and 12.

of production (without reducing output). One framework for analyzing the output increasing or cost reducing potential of new work patterns is to examine their effects on three main factors of production: labor, capital and production operations, and management and administration.

Flexitime

Individual labor productivity may be improved under flexitime in three ways. First, flexitime can increase *effective* labor input. More hours of actual work may be accomplished from the same number of workers and clock hours because of a reduction in paid absence (personal business, sick leave) and idle time while employees are at work. (They chat less after arriving and before departing because other employees are already working or have gone home.)

Second, flexitime can improve the organization of work. Work tasks may be more efficiently allocated over the day because meetings, telephone calls, and visits are concentrated in core hours, leaving work requiring thinking and concentration to be done at the beginning and end of the working day when it is more quiet and there are fewer distractions. Workers can finish jobs once they have started and thus avoid inefficient start-ups and shutdowns.

Third, flexitime aids workers by accommodating their different biological clocks. Morning people can come to work early in the day, whereas night people can come in later; each can be more efficient. In addition, if employee morale and job satisfaction increase with flexitime, that may bring about increases in productivity. This is the "happy worker is a productive worker" theory. If there are productivity gains from any of these sources, then there may also be a reduction in overtime payments.

Capital and production operations may be affected by flexitime in three different ways. First, flexitime may cause utilities and other overhead costs to go up. If the flexibility in arrival and departure time of employees requires buildings to be open more hours than before, then heating, cooling, and lighting costs will surely increase. If services such as the mail room, switchboard, or security need to be staffed for longer hours, then overhead costs also will go up.

Second, flexitime may improve the utilization of equipment and facilities. If the same number of employees are spread out over more hours of the day, then the demand for bottlenecks such as long distance telephone lines and photocopiers may be spread out. This means that capital costs can be reduced.

Third, scheduling, coverage, and communication can be either improved or worsened. When not all workers are required to be present at all times during the working day, there can be obvious problems. Disruptions in production operations can cause output losses. On the other hand, the availability of some employees over more hours of the day gives more scheduling flexibility, more communication opportunities (especially across time zones), and permits hours of customer service to be expanded.

Management and administration costs and effort will likely be affected by flexitime. First, supervision costs may rise because of flexitime. When workers' schedules are flexible, supervisors may have to spend more time ensuring that work stations are staffed. Perhaps supervision will need to be provided for a longer period to cover early arrivers and late leavers. Both of these result in higher supervisory cost. If new timekeeping equipment is installed (e.g., machinery that accumulates elapsed time put in by workers), another cost is added.

Second, personnel administration costs may go either up or down because of flexitime. Because employees desire flexible schedules, recruiting new employees and retaining old ones might be easier and cheaper. But if cross-training is required to teach one employee how to do another's job in order to solve coverage problems, or if new supervisory training is undertaken, then costs go up.

Empirical evidence on the economic effects of flexitime comes from several surveys and a large number of case studies covering several hundred companies and several thousand workers. Several of the studies have some more-or-less hard data, although a larger number report subjective opinion data. The general conclusions that are drawn from these data must remain tentative, both because of the questionable quality of the data and because several contingencies apply. Not all firms will have the same experiences suggested by the weight of this evidence.

In brief, the findings for flexitime are that the net economic effect of flexitime on the firms that use it is clearly favorable. Half or more of all users report outcomes that translate into dollars and cents gains; few encounter net losses. The size of these net gains when they occur is significant. It appears that output is increased or costs are reduced on the average by 5 to 20% overall.

The two chief sources of economic gains are increases in individual labor productivity and improvement in management practices coupled with elimination of excess supervision. Two additional sources of economic gains are easier recruiting and increased flexibility in scheduling production operations, including opportunities to extend customer service.

The chief sources of economic losses from flexitime, which are partial offsets to the gains, are increased supervisory effort, greater difficulty in managing production operations (scheduling, coverage, and communications), and increased utilities and overhead costs. Additional timekeeping and training costs are minor. Details of these econimc effects are presented in Exhibit 11.8.11.

Exhibit 11.8.11 Economic Effects of Flexitime on User Firms[a]

Effect	Direction of Effect	Frequency and Size of Effect
	Labor Performance and Costs	
Productivity	Positive	One third to one half of all users; 5 to 15% gain in output per worker.[a]
Absence and lateness	Positive	One half to three fourths of all users; 7 to 50% less absence; lateness virtually eliminated.[b]
Turnover	Positive	One half of all users; degree of reduction unknown, but likely small.[c]
Overtime pay	Positive	One third to three fifths of all users; degree of saving unknown, but can be large.[d]
	Capital and Production Operations	
Utilities and overhead	Negative	From 20 to 25% of all users; size of cost increase estimated to be only a few percent.[b]
Equipment and facilities utilization	Positive	No systematic data; case studies suggest gains are frequent but small.
Scheduling, coverage, communication	Often negative, sometimes positive	From 35 to 40% of all users report these outcomes to be worse under flexitime, but about 25% report them to be better; dollar impact indirect and unknown.[d]
	Management and Personnel Administration	
Supervision	Negative	About 20% of all supervisors feel adversely affected in terms of control and scheduling, but translation into dollar costs unknown.[e]
Timekeeping	Negative	Of all firms, 13% use flexitime accumulators, costing minimum $50/employee; for others, no change in timekeeping methods; cheating is infrequent (a problem for 1 in 10 users).[c]
Recruiting	Positive	Easier for 65% of all users; dollar impact unknown, but probably small.[d]
Training	Negative	Cross-training occasionally done, but with little or no out-of-pocket costs.

Source. Reference 10.

[a]The balance of the users, with few exceptions, reported "no change;" this result also holds for the other results reported here. Sources are 13 case studies with 21 "hard" measurements of the size of the productivity increase measured as output per worker before and after flexitime, and 8 surveys of user firms who subjectively reported the direction of productivity changes.

[b]Sources include some "hard" data and several subjective measures from case studies and surveys.

[c]Source is one survey.

[d]Source is a few surveys.

[e]Source is a few case studies.

Permanent Part-Time Employment

Increases in output are likely with part-time employment because of increased individual labor productivity that arises from three sources. First, as with flexitime, effective labor input can increase because absences can decrease. Part-time workers can more easily take care of personal business on their own time. In addition, many part-time workers are mature, stable women with good absenteeism records. (However, the outside interests and youthfulness of other part-timers may be an offset.) Part-day workers may have a feeling of greater urgency and spend less time taking breaks.

Second, for some jobs part-time scheduling is more rational than full-time scheduling. Both tedious work and mentally taxing work can be done more efficiently in time blocks of less than a day in length.

Third, if morale and job satisfaction are higher for part-time workers (because these jobs are highly sought after and meet their job holders' needs), then productivity may also be higher. For this and the other reasons, overtime costs can be reduced.

Labor costs can be higher for part-time employers, however, and perhaps can offset productivity gains if employees are paid the full range of fringe benefits. Although prorating benefits to time worked and offering cafeteria-style benefit packages minimize these added costs, administrative costs will still be higher.

Capital and production operations can be more efficient under part-time scheduling because the labor input can be better matched to the size of the work load. When work loads fluctuate predictably (e.g., heavy midday customer traffic in banks), part-time employment enables the firm to meet business needs without excess labor at off-peak hours. In some cases existing office or plant facilities can be utilized for more hours by adding a part-time minishift. The result is more output with a less-than-proportionate cost increase.

On the other side of the coin, some users of part-time employment may have coverage and communication problems—part-timers are not present all the time unless they are job sharers. Disruptions in production and output losses would then result.

Both supervision and personnel administration may be extra costs of part-time employment. Supervisors will need to solve problems of coverage and communication that arise because part-time workers are not always present. Personnel administration costs can go up if record keeping is complicated (more employees or different records for part-timers), if training effort increases (more trainees who have a longer payback period), or if recruiting is harder (nontraditional channels may have to be used). However, if there is an excess labor supply of part-time workers, then recruiting may be easier.

The empirical evidence on the potential economic impact of permanent part-time employment on the firms that use it is that the net effect is clearly positive. From two thirds to three fourths of all users experience gains in the form of cost savings or output increases whose magnitude may be in the range of 10 to 25% overall. However, these gains occur in only a small part of the firm's total operations (where part-timers are used), and so the economic impact firmwide will be much less.

The chief source of economic gain is the reduced labor cost from a better match between the size of the labor input and the size of the work load made possible by part-time employment. Higher individual productivity of part-time workers is a second major source of gain.

The largest economic cost of part-time employment, which is a partial offset to the gains, is higher fringe benefit costs, if they are offered. Additional supervisory effort and recordkeeping costs are smaller (see Exhibit 11.8.12).

Compressed Workweek

Individual labor productivity can increase because of compressed workweeks for the same reason that applies to other new work patterns: more effective labor input because of less paid absence. Compressed workweek employees have an extra weekday to use to handle personal business. However, the longer workday of compressed workweeks may not be a rational schedule for some jobs where fatigue is a factor. Productivity outcomes are thus in doubt.

Employee morale, if it increases, may also provide productivity dividends. Here the question is whether workers like compressed schedules better than standard schedules—seeing them as shorter weeks versus longer days.

From capital and production operations, there are three sources of economic gains from compressed workweeks. First, utilities and overhead costs can be reduced if the days of operation are reduced (e.g., from 5 to 4). Of course not all plants and offices will be able to do that. Second, work scheduling can become more efficient for some firms. Fewer start-ups and shutdowns associated with longer working days can increase output. Third, utilization of facilities can be increased in some cases, for example, two 12 hr shifts each working 3 days to provide around-the-clock operations 6 days a week.

Economic losses are also possible. Interfaces among work units will be hampered if not all work units in the firm have the same compressed schedule. Customer service may also be damaged if full staffing is not available during all business hours.

Neither management nor personnel administration should be much affected by compressed workweeks. In principle, supervision could be adversely affected by the need to extend supervision over more hours of the day and to coordinate work units on different schedules. Recruiting could be easier and cheaper if the availability of compressed workweeks increases the number of job applicants to the firm.

The empirical evidence is that the economic effects of compressed workweeks on user firms are mixed and variable, depending on the work technology of the firm and on the personal characteristics of workers. No general statement about net economic impact can be made. At best, about half of all successful continuing users experience significant economic gains; at worst, losses compel

Exhibit 11.8.12 Economic Effects of Part-Time Employment on User Firms[a]

Effect	Direction of Effect	Frequency and Size of Effect
	Labor Performance and Costs	
Productivity	Positive	One fourth to one half of all users; size of effect not documented.
Absence and lateness	Positive	From 40 to 50% of all users; size of effect unknown.
Turnover	Mixed	From 20 to 40% of all users have reduced turnover, but one third may have worse turnover; economic impact small.
Wages	Positive	Wage rates paid to part-timers are lower (versus full-timers) in 15 to 30% of user firms; gap is 8 to 30%, because of lower job levels.
Overtime pay	Positive	Of all users, 70% report less overtime; size of effect unknown.
Fringe benefits	Mixed	Of all users, 20% have proportionately higher costs; 60% have lower costs because not all benefits are offered; size of effect is hypothetically \$150 to \$1500/employee/year.
	Capital and Production Operations	
Scheduling, coverage, communication	Mixed	One half or more of all users solve scheduling problems with part-timers with large, but undocumented, cost savings; one third of all users have more difficult production operations.
	Management and Personnel Administration	
Supervision	Negative	From 35 to 50% of all supervisors have more difficult jobs because of part-time employment; cost impact unknown.
Record keeping	Negative	One third of all users; size of cost increase not documented, but likely small.
Recruiting	Mixed	One third of all users report each outcome; size of effect unknown.
Training	Negative	From 20 to 30% of all users have marginally higher training costs.

[a]Sources of quantitative data for most effects are one or two surveys supplemented by qualitative impressions from a small number of case studies. See footnotes to Exhibit 11.8.11.

discontinuance. This has occurred among 28% of all one-time users, according to the most recent survey.[6]

The chief source of economic gains is the opportunity compressed workweeks give for some firms to rationalize their production operations. Secondarily, the preference of many workers for longer blocks of leisure time results in increased morale and in reduced absence, which translate into higher labor productivity.

The main economic costs of compressed workweeks are (1) increased overtime pay required by labor law or collective agreements (or the accommodations firms make to avoid these costs, such as reducing the length of the workweek); (2) loss of productivity in some cases because of fatigue; (3) additional supervision, which is sometimes required because of the longer day or because of coordination problems; and (4) the frequent adverse effect on family life and household operations, which levies an unknown, but surely sizable, indirect economic cost on the firm (see Exhibit 11.8.13).

Social and Organizational Effects

Alternative work schedules can have several noneconomic effects on user companies, workers, and society at large. These effects potentially include (1) employee morale, job satisfaction, and the

Exhibit 11.8.13 Economic Effects of Compressed Workweeks on User Firms

Effect	Direction of Effect	Frequency and Size of Effect
	Labor Performance and Costs	
Productivity	Mixed	Gains for one third to one half of all continuing users, but losses for one fourth of all one-time users; size of effects unknown.[a]
Absence and lateness	Positive	Absence reduced by a modest amount (perhaps 10%) for one half or more of all users; lateness reduced less frequently.[a]
Turnover	Positive	One third to one half of all users; size of effect unknown.[b]
Overtime pay	Mixed	One third of all users pay less, but one fifth pay more; size of effects unknown, but likely sizable.[b]
	Capital and Production Operations	
Utilities and overhead	Positive	One fourth to one third of all users operate fewer days per week; size of effect not documented; may be partly offset by operating more hours per day.[b]
Equipment and facilities utilization	Positive	Gains for 10 to 20% of all users; size of effect unknown.
Scheduling, coverage, communication	Mixed	One fourth of all continuing users report gains, but one third report losses; dollar impact unknown, but likely substantial.[b]
	Management and Personnel Administration	
Supervision	Negative	One third of all users report management job is more difficult; size of effect unknown.
Recruiting	Positive	Two thirds of all users report better results; dollar impact unknown.
Planning, coordination, control	Negative	One fifth of all users have coordination problems; effect on dollar costs unknown.

Note: See footnotes to Exhibit 11.8.11
[a]Sources are a few surveys and case studies.
[b]Source is basically two surveys.

quality of work life; (2) family life; (3) labor-management relations; (4) the practice of management; (5) labor market conditions; and (6) transportation and energy.

Flexitime

Because flexitime is basically a proworker innovation that shifts some control and responsibility from supervisors to workers, it should result in an increase in employee morale and job satisfaction. When discontent with jobs and workplaces is rising,[11] a work schedule that implicitly places trust in workers and expands their range of choice should improve the perceived quality of work life.

Family life can be aided by flexitime if it permits work schedules and home schedules to be meshed more easily. Conflicts of this nature, focusing especially on child care, are quite common as women's labor force participation and economic status change.[11] However, the potential for flexitime may be limited unless both parents in two-parent families have quite liberal flexitime schedules.

Labor-management relations could be affected by flexitime if labor unions have a strong position for or against it. Management practices generally would be altered by flexitime because it changes the relationship between supervisors and workers.

Labor market conditions may be changed by flexitime on both demand and supply sides of the market. Labor supply may be increased if flexitime enables people to work who otherwise (with fixed schedules) could not. Labor demand may be increased if labor productivity rises and labor is substituted for capital; demand may fall if output cannot be expanded, substitution does not

Exhibit 11.8.14 Social and Organizational Effects of Flexitime

Effect	Nature and Direction of Effect
Employee morale, job satisfaction, and quality of work life	Improvements in morale reported by 85% or more of all workers stemming from easier commuting (reduced time and stress reported by three quarters of all workers) and more choice about work.
Family life	Better family life reported by 20 to 60% of all workers; stress and work-home conflicts reduced; results on role equality and time spent with spouse and children are mixed.
Labor-management relations	Labor unions usually support flexitour and gliding time, but object to variable days; less than 10% of flexitime workers are union members; it is seldom a bargaining issue.
Management practices	Changes sometimes reported toward less supervision, more self-management, and more short-term planning; companies with one alternative work schedule are likely to have other innovative human resource programs.
Labor market conditions	No direct evidence on effects of flexitime on employment or unemployment; labor supply likely to increase (since recruiting easier); no evidence of increases in moonlighting.
Transportation and energy	Rush-hour peaks are spread out, with savings in commuting time and cost for many; building energy costs for utilities went up for one in four flexitime users.

occur, and fewer of the more productive workers are needed. The critical outcomes to examine are the effects on unemployment and the composition of the labor force.

Transportation is likely to be improved by flexitime since workers are able to choose commuting times to avoid rush-hour peaks and perhaps use public transportation better, thus saving energy. But these energy savings may be offset if buildings are open longer to accommodate early arrivals and late leavers.

The evidence on these potential effects, summarized in Exhibit 11.8.14, is that employee morale is usually better because of flexitime and constitutes one of its chief advantages. Family life effects are uncertain; there is too little evidence. Labor unions are reluctant about those flexitime programs that permit variable day lengths because they erode the concept of the 8 hr day and may lead to abuses by management. Management practices change insofar as supervision is altered. There is little evidence on how flexitime affects labor market conditions. Transportation is improved, but energy effects are not documented.

Permanent Part-Time Employment

Part-time employment may aid family life. This work schedule permits people to achieve a better balance between labor market work and another major life interest such as child care. It also permits phased retirement and alternation between work and education.

Labor-management relations may be damaged by part-time employment if management adopts it over union objections. These objections may arise from concern that part-timers may take jobs away from workers who need full-time earnings. Management practices need not undergo changes when part-time employment is used, except for minor additional record keeping and scheduling tasks for supervisors.

Labor market conditions are likely to be affected by part-time employment. Labor supply is likely to be increased because this schedule permits some people to work who otherwise could not (e.g., parents with young children at home). However, others may decrease their work hours if good part-time opportunities are available (e.g., near-retirees). The net supply effect, if it is large, could affect unemployment rates and equity outcomes.

Transportation and energy effects of part-time employment would be negative if part-time employment means more workers are making commuting trips. This would be the case if full-time jobs are converted to part-time jobs.

Empirical evidence on these potential social and organizational effects of part-time employment is summarized in Exhibit 11.8.15.

Compressed Workweeks

Employee morale may be increased by compressed workweeks because blocks of leisure time can be met. (Every weekend is a 3 day weekend.) On the other hand, this schedule also means longer

Exhibit 11.8.15 Social and Organizational Effects of Permanent Part-Time Employment

Effect	Nature and Direction of Effect
Family life	Improvements in balancing work life with family life and other interests are indirectly suggested by evidence that most part-timers are also caring for a household or pursuing education.
Labor-management relations	Most labor unions are opposed to part-time employment except where necessary because of the nature of the work; it is seldom a bargaining issue; less than one in six part-timers is a union member.
Labor market conditions	Moonlighting is higher for part-timers, with a rate of 11 to 23% versus 5% for full-timers; sizable numbers of currently full-time workers (perhaps 8 to 20%) say they would prefer to work less; labor force entrance due to part-time employment unknown; net labor supply effect unknown.
Transportation and energy	Extra commuting trips are few since few part-time jobs result from conversion of full-time jobs, except for job sharing.

workdays and perhaps more fatigue. There is no other improvement in the relationship of the worker to the workplace.

Family life may be affected in two ways: Long work days may upset family schedules (especially child care and evening activities), but more time can be spent at home on the fifth weekday.

Labor-management relations are likely to be impaired by compressed workweeks. Unions usually oppose this schedule because it erodes the concept of the 8 hr day and raises concerns about health and safety. Management practices would not be expected to undergo any fundamental change because of compressed workweeks.

Labor market conditions could be affected by compressed workweeks if moonlighting increases, but no other labor force entry or exit would be expected.

Transportation and energy savings are potentially major advantages of compressed workweeks. Both commuting travel and building operating costs could be reduced if no work is done 1 day/week. However, personal travel could increase as a partial offset, and not all workplaces will be able to shut down 1 day/week.

The empirical evidence on these effects of compressed workweeks is summarized in Exhibit 11.8.16.

Exhibit 11.8.16 Social and Organizational Effects of Compressed Workweeks

Effect	Nature and Direction of Effect
Employee morale, job satisfaction, and quality of work life	Higher morale experienced by more than half of all compressed workweek workers on the average, but employee acceptance is highly variable; young workers and men favor compressed workweeks the most, whereas women with children and old workers often disapprove; greater fatigue reported by one third to one half of all compressed workweek employees.
Family life	One quarter of all workers report compressed workweeks upset family life, but some male workers spend more time in child care and household work.
Labor-management relations	Labor unions usually oppose compressed workweeks; collective agreements often require overtime pay beyond 8 hr in a day; compressed workweek workers are union represented in only 16% of all user firms.
Labor market conditions	Moonlighting increases in some cases (e.g., from 5% to 17% moonlighters), thus aggravating unemployment; no other labor supply or demand effects.
Transportation and energy	Reductions in building utilities costs reported by one third of all user firms (other users do not reduce their days of operation); commuting trips reduced, but personal travel increased on the day off for 70% of workers in one study.

11.8.4 MANAGEMENT ISSUES AND PROBLEMS

The objectives of this section are to sketch the issues and problems that management is likely to face when making the decision to adopt an alternative work schedule and to suggest several implementation steps to take. The topics to be discussed are (1) legal limitations to the use of alternative work schedules, (2) the role of labor unions, (3) matching the work schedule to the work technology (who can use alternative work schedules), (4) supervision and control, and (5) fringe benefits for part-time employees.

Legal Limitations

United States statutes in force in 1982 affect alternative work schedules by governing overtime pay and fringe benefits. Federal legislation requires overtime pay for all covered workers for hours worked beyond 40/week and for government employees and private sector employees working on government contracts for hours worked beyond 8/day. These provisions discourage the use of flexitime and compressed workweeks in some cases by increasing wage costs.

Legislation affects fringe benefits by requiring certain benefits—Social Security, pensions, and unemployment compensation—to be paid. These regulations can increase the cost of part-time employment, as is discussed later in this section.

The key labor laws and their effects on alternative work schedules are summarized as follows (see Nollen and Martin[6] for more detail):

1. The Fair Labor Standards Act covers all employees in interstate commerce and public administration. It requires payment of overtime at time-and-a-half rates after 40 hr/week for nonexempt employees. Nonexempt employees are blue collar and lower-level white collar workers, whereas exempt employees are professionals, managers, and administrators with supervisory responsibilities, and outside salespersons. (Some workers, such as farm workers, railroad and airline workers, and interstate truck drivers, are not covered.) A "workweek" is defined to be a consecutive 7 day period.

   This law increases the wage cost of variable day flexitime that permits credit and debit hours to be carried over across weeks, unless the standard workweek is 35 hr and no more than 5 hr are carried over. It also increases the wage cost of the 5/4–9 version of compressed workweeks.

2. The Walsh-Healy Public Contracts Act sets basic labor standards for nonexempt employees working directly on U.S. government contracts to manufacture or furnish more than $10,000 worth of goods. Its overtime provisions require payment at time-and-a-half rates for hours worked in excess of 40/week or 8/day. The 8 hr provision drastically increases the wage cost of the variable day model of flexitime and all compressed workweeks and effectively bars their use.

3. The Contract Work Hours and Safety Standards Act applies to U.S. government construction contracts exceeding $2000, service contracts exceeding $2500, and supply contracts exceeding $2500 but less than $10,000. It, too, specifies payment at time-and-a-half rates after 8 hr/day for nonexempt workers.

4. The Federal Pay Act establishes a basic 40 hr workweek for full-time U.S. government employees and provides payment at time-and-a-half rates for hours worked in excess of 8/day. Unless an agency would be seriously handicapped in carrying out its function or unless costs would be substantially increased, tours of duty must be scheduled not less than 1 week in advance—and, whenever possible, on 5 consecutive days of equal length Monday through Friday. Breaks of more than 1 hr may not be scheduled on a basic workday. These provisions limit the use of flexitime and preclude the use of compressed workweeks.

Two federal laws passed in 1978 are designed to encourage the use of alternative work schedules among federal government employees. The Flexible and Compressed Work Schedules Act of 1978 suspends the overtime pay provisions of both the Fair Labor Standards Act and the Federal Pay Act for a 3 year period (March 31, 1979, to March 31, 1982). This temporary change permits all forms of flexitime and compressed workweeks to be adopted by federal employees without requiring overtime pay if workers themselves choose to work more than 8 hr some days or 40 hr some weeks. However, overtime work ordered in advance by management still must be compensated at premium rates, and terms and conditions must be negotiated by labor and management where workers are organized. The new act also mandates experiments with flexible and compressed schedules in federal agencies and requires an evaluation of them.

The Federal Employees Part-Time Career Employment Act of 1978 encourages part-time employment in the federal government by requiring federal agencies to establish part-time employment programs with goals for increasing such employment. The Office of Personnel Management is charged with conducting demonstration programs, including job sharing. A new regulation in this act is the counting of part-time employees against total personnel ceilings according to the fraction of full time that they work. (Two half-time employees count as one, rather than two, personnel slots.)

Labor Unions

Many labor union leaders are opposed to part-time employment and compressed workweeks. There is also considerable union opposition to variable day flexitime, although flexitour and gliding time versions of flexitime often have at least tacit union approval. Local union people are sometimes more favorable to alternative work schedules than national officials, and there are several examples of locals having initiated and bargained for flexible and part-time schedules.

The key statistics on union representation among workers on alternative schedules are these:[6]

Only 7% of all flexitime firms had half or more of their flexitime employees in a labor union in 1977.

About 6% of all permanent part-time employees were union members in 1977.

Of compressed workweek firms, 16% had labor union representation among their compressed schedule workers in 1977.

These figures contrast with the 25% union representation among all workers nationwide in the late 1970s. Of course the smaller union representation for alternative work schedules is traceable in part to their relatively greater use in industries and occupations where there is less union activity even among full-time standard work schedule employees.

The reason why labor unions are often opposed to alternative work schedules is that they raise some legitimate threats to labor unions and collective bargaining. The main union objections are these:

1. Some flexitime and all compressed schedules erode the hard-fought concept of premium pay for days longer than 8 hr.
2. Exploitation of workers can occur if management denies fringe benefits and career status to part-time workers or subtly pressures flexitime workers to choose hours that meet the company's needs rather than their own preferences.
3. The long days of compressed workweeks may endanger workers' health and safety.
4. If full-time jobs are converted to part-time jobs, then workers who need full-time earnings will be hurt, and the drive toward shorter workweeks for all workers may be blunted.
5. Part-time workers may be hard to organize because they have other main interests.
6. All of these new work patterns could have adverse equity effects if they are not available to all workers. In addition, part-time employment in particular is usually low level with low pay; these jobs are usually held by women and marginal workers where inferior status would be further institutionalized.
7. New patterns of work could increase competition for jobs if they stimulate new labor force entry.

In contrast to these problems, alternative work schedules offer several advantages to labor unions, as follows:

1. They often increase workers' morale and job satisfaction and thus are a way for unions to improve the quality of work life.
2. If gains in labor productivity are traceable to new work patterns, then unions have a new tool to use to seek wage increases.
3. Voluntary cutbacks in working time by some full-time workers to permanent part-time hours may spread the work available to more people.
4. Union initiatives in new work patterns may be a source of new union membership.

Companies that consider the use of alternative work schedules will need to include labor unions in their planning so that industrial relations can be enhanced instead of damaged. A constructive and supportive union role will need to be fashioned.

Matching the Work Schedule to the Work Setting

Who can successfully use alternative work schedules? Which companies and which workers will find them beneficial? The answers to these questions depend on finding a match between the work schedule and the work technology of the company. Successful use also requires that the social system inside the company be compatible with the work schedule and that the external environment be receptive. For flexitime these matches are frequent and easy; for part-time employment the matches are somewhat more limited; for compressed workweeks, they can be hard to come by.

Flexible, part-time, and compressed schedules affect both the production operations of companies and the personal lives of employees. Sociotechnical systems theory provides a framework for incorporating these diverse effects into a decision whether or not to use an alternative work

Exhibit 11.8.17 Sociotechnical Systems Variables for Selection of Alternative Work Schedules

Technical System

Work Process

Job independence–extent to which job tasks can be done independently of other job tasks.

Work unit independence–extent to which work units depend on inputs from other work units.

External interface–how work units relate to customers, suppliers.

Continuity–importance of continuous coverage of work station or job tasks and of knowledge of preceding events.

Shift work–presence or absence and shift interface requirements.

Capital intensity and utilization–cost of start-up and shutdown and the degree of slack in utilization.

Repetitiveness, stress–degree to which work is either tedious or mentally taxing.

Labor Characteristics

Interchangeability–degree to which workers can substitute for each other.

Skill level–amount of specialized skill and supervision required.

Training–amount of on-the-job training required.

Output Demand Characteristics

Predictable cycles–degree to which work load varies regularly.

Criticality–degree to which output must be provided on demand rather than managed by inventories and advance scheduling.

Nonstandard size–output required more or less than 8 hr/day or 40 hr/week.

Social System

Employee Characteristics

Age–young or old versus middle-aged.

Sex.

Marital and family status.

Holds second job or not.

Organizational Climate

Management style and attitudes–degree to which management is participative, receptive to change, and oriented toward employees.

Supervisors' attitudes–level of trust in employees and willingness to change methods of supervision.

Employees' preferences and needs.

Labor-management relations–presence of a labor union, union leaders' attitudes, and health of industrial relations.

Organizational structure–mechanistic and bureaucratic versus flexible and decentralized.

Environment

Government Policy

Labor law–degree to which overtime pay requirements impinge on the work schedule.

Fringe benefits–degree to which statutory benefits affect workers on alternative work schedules.

Labor Markets

Tightness–overall labor supply-demand balance and availability of particular skills.

Functioning–quality and quantity of information about jobs and job seekers; recruiting aids.

Transportation and Energy

Public transportation–availability and quality.

Roads and parking–adequacy, accessibility, and cost.

Car pools, van pools, ride sharing–availability and flexibility.

Building operation policy–opening and closing hours and decentralization of heating, cooling, and lighting controls.

Source. Reference 6.

schedule. The technical system is the work technology: work processes, labor characteristics, and output demand characteristics. The social system includes demographic characteristics of employees, their attitudes, and the organizational climate. Environmental features to consider are government policies, labor market conditions, and other relevant institutional features.

For alternative work schedules, the technical, social, and environmental variables that need to be assessed are summarized in Exhibit 11.8.17.

Good Uses for Flexitime

Several favorable work settings for the use of flexible work schedules are suggested here, using a sociotechnical systems framework:

Independent tasks—one worker does not depend temporally on others (e.g., project work, batch process work, and such jobs as claims examiner, laboratory technician, repairperson, market analyst).

Interchangeable workers—one worker can do another worker's job with mimimum cross-training (where either jobs are quite simple or workers have multiple skills).

Output timeliness over the work day is not critical—jobs can be done from an "in-basket" or a stock of parts, and output can be produced for an "out-basket" or a buffer inventory (e.g., jobs such as copy editor, subassembler).

Irregular size of work load, which requires more hours to be done some days than other days (e.g., legislative assistant, printer, paramedic).

Heavy customer traffic with peak periods for which better service could be provided using longer hours of opening and variable staffing levels (e.g., doctor's office staff, local government offices).

Routine intracompany telephone communications across time zones (e.g., field sales office calls to headquarters credit department).

Mentally demanding jobs requiring concentration, creativity, quiet, and no interruptions (e.g., advertising copywriter, manuscript typist).

Medium-size or larger work units where there is more than one person available for coverage.

Workers are skilled and require little supervision.

Employee morale and job performance are poor.

Many parents with young children are employed.

Management style is participative and change oriented.

Supervisors trust employees and can change their supervision methods.

Commuting is difficult because of traffic congestion.

Good Uses for Part-Time Employment

A sample of conditions that favor the use of part-time employment includes:

Jobs that have discrete tasks with clear beginnings and endings (as opposed to continuous flow) and that can be done quite independently of close supervision (e.g., internal auditor, credit analyst, newsletter editor, safety inspector).

Repetitive work that can be done with more care and enthusiasm for short periods (e.g., copy typist, records clerk, keypunch operator, assembler).

Mentally stressful, tense, and energy draining work that can be done with more intensity for shorter periods (e.g., social caseworker, customer service worker, interviewer).

Cyclical demand for the work unit's output whose peaks can be predicted (e.g., bank teller, seasonal products or services, accounts payable clerk).

Extended hours of operation—evening or weekend hours, as in department stores and other retail outlets.

Nonstandard size of job—where the work to be done takes either more or less than 8 hr a day to accomplish and where overtime is common (e.g., legal assistant, laboratory technician, ticket agent, shipping department supervisor).

Work units where capacity constraints are binding and where increasing equipment, capital stock, plant space, or office facilities is not wise (e.g., insurance offices with limited desk space and computing capacity or a subassembly shop that is a bottleneck); part-time minishifts are suggested here.

Jobs that benefit especially from diverse skills, creativity, and cross-fertilization of ideas—they can be staffed with two job sharers (e.g., playground director, librarian).

Workers are already trained or jobs require little training.

Workers are students, old persons, women, or women with children.

Cafeteria fringe benefits plan is in effect.

Labor market is tight—skills are unavailable or workers are of low quality.

Good Uses for Compressed Workweeks

Several sociotechnical features of the work setting that are good for compressed workweek use are:

Jobs that require long start-up and shutdown times or travel time to a work site and that therefore are more efficiently scheduled for a longer work day or shift (e.g., road repair crews, telephone lineperson, chemical plant operative).

Capital-intensive production operations in which continuous utilization is needed and worker scheduling must meet equipment needs.

Work units requiring coverage longer than 8 hr a day (e.g., shipping and receiving departments, government offices dealing with the public).

Functions that benefit from heavier staffing at some times of the day or week (e.g., police departments, work units with heavy midweek demands).

Employees are young and without families.

No government contract work is done, and no collective bargaining agreement requires overtime pay after 8 hr/day.

Facilities can be shut down on the fifth day of the week.

Ill-Suited Work Settings

There are several features of work settings that can make the use of alternative work schedules difficult. For flexitime these unfavorable features are (1) continuous-process assembly lines requiring all work stations to be staffed simultaneously; (2) shift work in which corresponding workers on each shift mesh their departure and arrival times; (3) continuous coverage service or support jobs such as receptionist, switchboard operator, retail sales clerk, broadcaster, bus driver, receiving room clerk, nurses in nursing unit, and security guard; (4) jobs requiring close supervision or teamwork so that workers must have identical schedules, such as executive secretaries and computer operators; and (5) centralized, hierarchical, mechanistic organizational structures with traditional supervisors.

For part-time employment, jobs and work settings that are likely to be problems include (1) jobs with continuity requirements, such as management and supervisory jobs in which knowledge of preceding events and continuous availability is important; (2) jobs with high company-provided training requirements, such as management jobs, and where experience is very important to job performance; (3) labor markets and employees where moonlighting is common; (4) settings where management holds stereotypes of part-time employment as only temporary and low-level employment; (5) settings where fringe benefits are generous and offered in total to part-timers; and (6) setting where a labor union is present and is opposed to part-time employment.

For compressed workweeks the chief features of work settings that discourage use are as follows: (1) interfacing of work units both with other work units in the firm and with customers or suppliers is important; (2) partial staffing at some times or closing down at other times is not feasible; (3) many workers are women with household responsibilities or older people; and (4) government contract work is done or labor union agreements call for overtime pay after 8 hr/day.

Supervision and Control

Supervision is a critical factor in the management of flexible and part-time work schedules. In the case of flexitime, the supervisor's functions change, and new skills are required. Timekeeping and control methods need to be established. Some retraining and job redesign will be necessary.

The Supervisory Function Under Flexitime

When a company adopts flexitime, the basic role of the supervisor as the link or intermediary between the firm and the worker becomes more difficult. Flexitime changes the relationship between the worker and the workplace. The twin added supervisory responsibilities under flexitime are (1) to deliver good business results despite the potential for disruption and chaos following the removal of fixed worker's schedules and (2) to ensure that workers have genuine flexibility and free choice and are not subtly constrained to conform to management's wishes.

In particular, there are five changes in the supervisors' job.[12] First, the nature of control changes.

There is a shift in control away from supervisors and toward employees. The control practiced by supervisors is less negative (monitoring, checking up) and more positive (evaluation against goals and more facilitation). The feeling of "losing control" in the sense of losing power or the inability to "keep track of things" need not happen, despite fears.

Second, the supervisor's planning effort increases, especially short-term planning, caused by the increased need to coordinate work schedules with workers' schedules.

Third, the organizing function changes. There are fewer decisions to make because control is shifted to employees. But there is more responsibility because the job level is upgraded by virtue of the change in the nature of control and the increase in planning and because coordination of work activities to ensure good business results becomes more critical.

Fourth, many staffing activities can become easier, but some require more attention. If turnover and absenteeism are reduced and job satisfaction is increased, recruiting and maintaining a quality work force should be easier. Performance appraisals become more important and are utilized more often (because rewards must, under flexitime, be based on the job that is done rather than on the time worked). Cross-training may be required for some workers, and more supervisory coverage may be needed if working hours are extended.

Fifth, directing and leading become more important because formal rules as guides for behavior are relaxed. There is a shift toward participative management, emphasizing leadership skills for supervisors. More communication is required among workers and between supervisor and workers because of the lack of fixed schedules. Less supervisory effort is needed to motivate employees since increased morale is a usual consequence of flexitime.

Timekeeping and Control for Flexitime

Several methods are available for keeping track of employees' working time under flexitime:

The honor system—rely on employees' honesty to work the required number of hours.

Manually kept time sheets—sign-in-sign-out system.

Traditional time clock punch-in-punch-out system.

Flexitime accumulator—an electronic machine that records accumulated working time when activated by a key card, but that does not display arrival or departure times.

Computer-based time and payroll systems—remote terminals connected to a central computer that performs time accumulating and other payroll functions.

In practice, the timekeeping method used before flextime was adopted coninues to be used after implementation in most firms. The honor system is used in half of the user firms for exempt employees and in 3 out of 10 firms for nonexempt employees. Flexitime accumulation or computer-based systems are used by 13% of all flexitime firms.[6]

When establishing control methods, the guiding principle is not to impose new or additional controls in a misguided attempt to counter the loss of fixed workers' schedules. That is contrary to the spirit of flexitime, which is one of trust and free choice. To get results, establish minimum performance and output standards, charge employees with meeting those standards as a condition of flexitime, and evaluate results against those standards. In timekeeping be wary of moving from a de facto honor system to a formal time recording system—regard the latter as a convenience to all, not as a control against cheating.

Training and Job Redesign for Flexitime Supervisors

Because the supervisor's functions change under flexitime, training is necessary to make sure the needed skills change. The supervisor's job change is one of implicit upgrading and movement toward higher-level management, consisting of more planning, organizing responsibility, evaluation, leadership, and facilitation and less checking up and personnel administration.

Supervisor's skills to be strengthened via training include higher-level conceptual skills, ability to cope with uncertainty, understanding of power and control, interpersonal relations and communications, leadership, planning, evaluation methods, and ability to train workers.

The importance of including supervisors in the process of implementing flexitime is demonstrated by the fact that the critical outcomes of coverage, scheduling, and communication were significantly improved when meetings were held with supervisors.

Supervision for Part-Time Employment

Supervision is made more difficult by part-time employment in some cases, but it is not a major problem. There are three potential trouble spots. First, according to a third of the user firms, the scheduling of work and employees and internal communications are harder because part-time employees are not always present. Second, record keeping is a bigger job in some cases because part-time employment means more records (because of more employees or manual record keeping, such as for absenteeism) or special records (e.g., fringe benefits, pensions). Third, training tasks are potentially larger and less rewarding because part-time employees work less. In fact, training is usually not increased because part-time employees are fully trained already or the job requires little or no training.

In most cases (three quarters or more), part-time employment helps supervisors do their job; it is a staffing option they usually have the decision making power to use. Part-time employment often solves operating (usually scheduling) problems. It eases staffing when higher-quality workers are available part time in excess supply.

Job sharing eliminates some supervision problems of part-time employment, such as scheduling and communication, because these tasks are accomplished by the job sharers themselves.

Training for supervisors of part-time employees is principally a matter of (1) devising methods for meshing work schedules with workers' schedules and ensuring adequate communications, (2) education about fringe benefit administration, and (3) education to counter stereotypes of part-time employees as temporary and uninterested in career advancement.

Fringe Benefits for Part-Time Employees

Permanent part-time employees are often not paid the full range of fringe benefits that are offered to their full-time counterparts. In the majority of cases, this result is due to the proportionately higher cost of some fringe benefits for part-time employees. In this section the actual fringe benefit practices of employers are reported, and policies that are equitable to employees and of reasonable cost to employers are suggested.

Fringe Benefit Practices

Vacation leave is by far the fringe benefit most frequently offered to permanent part-time employees–80% of all organizations made it available to their part-time employees in 1977, almost always on a prorated basis. (Half-time employees, for example, got half the vacation days per year that full-time employees received.) Sick leave for part-time employees is also usually prorated to time worked, but in total only 55% of all organizations made it available to their part-time workers. (see Exhibit 11.8.18). In some instances benefits may be limited to employees who work at least half time.

Life and health insurance plans were offered to permanent part-time workers by just over half the organizations employing them. (Almost all full-time employees received these benefits.) These benefits were prorated in some fashion about a third to a half of the time; in the remainder of cases, part-timers received the same benefits as full-timers. Pension benefits were offered to part-time employees by 59% of their employers (prorated in more than two thirds of the cases), and some form of profit sharing was offered by 28% of the user organizations. (A total of 61% offered this benefit to full-time workers.)

Social Security

In the past, Social Security payments were potentially, if not actually, more expensive to employers for part-time compared to full-time workers. Because of the ceiling on the employer's contribution per worker, two half-time workers could have resulted in higher total contributions from the em-

Exhibit 11.8.18 Fringe Benefits Paid to Permanent Part-Time Employees

Fringe Benefit	Offered to Full-Timers[a]	Offered to Part-Timers[a]: Total	Prorated	Same as Full-Time
	(percent of all users of part-time employment)			
Vacation	99	80	75	5
Sick leave	95	55	49	6
Life insurance	96	51	27	24
Health insurance	97	52	19	33
Pension	93	59	42	17
Profit sharing	61	28	20	8

Source. Reference 6.

[a]Percentage of all users of part-time employment.

ployer, but only if they were paid salaries considerably above average. For example, given the 1981 wage ceiling for Social Security taxes, two half-time workers would have to be paid total salaries exceeding $29,700 (or $14,850 each) before this fringe benefit became more costly for part-time than full-time employment. These salaries for part-timers would be unusual. As the wage ceiling rises to $40,200 by 1986, the chances for higher Social Security costs for part-timers diminishes further.

Unemployment Compensation

Both federal and state unemployment tax is proportionally more expensive for part-time than for full-time employees, but the difference is small in magnitude since both the tax rates and the ceilings against which they are levied are low. For example, using current federal and state rates and ceilings, two half-time workers would cost an employer $134 more per year than one full-time worker if both half-time workers earned more than $4200.

Disability Insurance and Workers' Compensation

Disability insurance is regulated by states. Both rates and extent of coverage vary widely, but differences for part-time as opposed to full-time workers are small.

Holidays, Vacations, and Sick Leave

These benefits are easy to provide in a prorated fashion. For example, if full-time employees who work 40 hr a week get 10 days of vacation and 6 holidays, then part-time employees who work 20 hr a week can be offered 5 days of vacation and 3 holidays. Sick leave can be treated in a similar way.

Health Insurance

Health insurance is an expensive fringe benefit to the employer, and it is sometimes difficult to prorate. If part-time employees are offered the same health insurance benefits as full-time employees, that is, if the employer pays the same premium or percentage of premium for full- and part-time employees, the cost of providing this fixed-cost benefit is proportionally higher for part-time workers. For example, hiring two part-timers in lieu of one full-timer might increase the company's health insurance cost by $500 to $1000/year per part-time employee. When a firm employs only a few part-time workers, this proportionately greater cost may be ignored, but if large numbers of part-timers are employed, the dollar value of the difference becomes substantial.

In practice, the employer's total health insurance cost may not increase when coverage is provided for part-time employees. This is because many part-time workers who are already covered by another family member's plan will not choose to participate in their own employer's health insurance plan if it is one to which employees must contribute. Savings from nonparticipation for these part-time workers would offset the extra cost of contributions for other part-time employees.

In the future, if career part-time employment increases, more part-timers could be expected to want health coverage. One way to provide it without increasing the employer's cost would be to require part-time workers to pay a larger share of their premiums. Alternatives would be to provide some components of coverage but not others, limit coverage to the individuals and not extend it to the family, or reduce other benefits by the additional cost of health insurance.

*Retirement**

Since the amount an employer contributes to the pension plan for an employee is based on that person's earnings, it is proportionally the same for comparably paid full- and part-time employees. However, the Employee Retirement Income Security Act (ERISA) of 1974 has concerned some employers who previously excluded part-time workers. The act stipulates that employees who work 1000 or more hr/year (about half time or more) must all be treated alike—that is, included in an organization's pension plan (if it has one), made eligible to receive employer contributions based on annual earnings, and eventually made eligible for full vesting (100% ownership of employer contributions as well as any employee contributions).

Some employers exclude part-time workers simply by limiting the number of hours these employees may work to fewer than 1000/year. However, the law itself recognizes the legitimacy of concentrating retirement benefits on long-term employees and allows employers considerable latitude in setting participation standards, contribution formulas, and vesting schedules that exclude short-term (full-time and part-time) employees. For example, participation may be limited

*Based on reference 8.

to those who meet a 1 year minimum service requirement, may be postponed until an employee is 25 years old, or may be limited to those exceeding a minimum compensation requirement.

Cafeteria Benefits

The employer who wants to provide fringe benefits to part-time employees enjoys a wide range of options for doing so through manipulation of compensatory and supplementary benefits. Decisions about which benefits or what proportion of benefits to offer, and under what conditions they are to be offered, are within the purview of the employer.

The "cafeteria" or "market basket" system takes into consideration both the employer's need to control costs and the benefit preferences of individuals. It attaches a monetary value to each benefit provided by the employer. An employee is entitled to spend a certain percentage of salary on benefits and may choose the combination best suited to his or her own needs. Older workers, for example, might forgo maternity coverage in favor of retirement benefits, whereas married people can avoid duplication of benefits already received through their spouses' employers.

REFERENCES

1. *Employment and Training Report of the President*, U.S. Government Printing Office, Washington, DC, 1979.
2. U.S. BUREAU OF LABOR STATISTICS, *Employment and Unemployment During 1979*, Special Labor Force Report No. 234, U.S. Government Printing Office, Washington, DC, 1980.
3. J. HEDGES and E. SEKSCENSKI, "Workers on Late Shifts in a Changing Economy," *Monthly Labor Review*, Vol. 102, No. 9 (September 1979), pp. 14–22.
4. W. V. DUETERMANN and S. C. BROWN, "Voluntary Part-Time Workers: A Growing Part of the Labor Force," *Monthly Labor Review*, Vol. 101, June 1978, pp. 3–10.
5. U.S. BUREAU OF LABOR STATISTICS, *Work Experience of the Population, 1978*, Special Labor Force Report No. 236, U.S. Government Printing Office, Washington, DC, 1980.
6. S. D. NOLLEN and V. H. MARTIN, *Alternative Work Schedules*, AMACOM, New York, 1978.
7. J. OWEN, "Why Part-Time Workers Tend to Be in Low-Wage Jobs," *Monthly Labor Review*, Vol. 101, June 1978, pp. 11–14.
8. S. D. NOLLEN, B. B. EDDY, and V. H. MARTIN, *Permanent Part-Time Employment: The Manager's Perspective*, Praeger, New York, 1978.
9. U.S. BUREAU OF LABOR STATISTICS, *News Release*, February 24, 1981.
10. S. D. NOLLEN, *New Patterns of Work*, Work in America Institute, Scarsdale, NY, 1979.
11. R. P. QUINN and G. C. STAINES, *The 1977 Quality of Employment Survey*, University of Michigan Institute for Social Research, Ann Arbor, 1978.
12. L. A. GRAF, "An Analysis of the Effect of Flexible Working Hours on the Management Functions of the First-Line Supervisor," D. B. A. dissertation, Mississippi State University, Mississippi State, 1976.

BIBLIOGRAPHY

Flexitime

BERNARD, K. E., "Flexitime's Potential for Management," *Personnel Administrator*, October 1979, pp. 51–56.

BISHOP, A. L., "Flexitime in Manufacturing," *American Machinist*, April 1979.

EVANS, A. A., "Alternative Work Schedules: A European Overview," *Journal of the College and University Personnel Association*, Vol. 28, Summer 1977, pp. 30–42.

GOLEMBIEWSKI, R. T., and C. W. PROEHL, JR., "A Survey of the Empirical Literature in Flexible Workhours: Character and Consequences of a Major Innovation," *Academy of Management Review*, Vol. 3, October 1978, pp. 837–853.

GOLEMBIEWSKI, R. T., R. HILLES, and M. S. KAGNO, "A Longitudinal Study of Flexi-Time Effects: Some Consequences of an OD Structural Intervention," *Journal of Applied Behavioral Science*, Vol. 10, 1974, pp. 503–532.

GOMEZ-MEJIA, L. R., M. A. HOPP, and C. R. SOMMERSTAD, "Implementation and Evaluation of Flexible Work Hours: A Case Study," *Personnel Administrator*, Vol. 23, January 1978, pp. 39–41.

HARVEY, B. H., and F. LUTHANS, "Flexitime: Its Real Meaning and Impact," *MSU Business Topics*, Summer 1979, pp. 31–36.

HEDGES, J. N., "Flexible Schedules: Problems and Issues," *Monthly Labor Review*, Vol. 100, February 1977, pp. 62–65.

KRAMER, O. P., "Flexible Working Hours," *Journal of Systems Management*, December 1978, pp. 17–21.

LEE, R. A., and W. M. YOUNG, "A Contingency Approach to Work Week Restructuring," *Personnel Review*, Vol. 6, Spring 1977, pp. 45–55.

LEE, R. A., and W. M. YOUNG, "The Factor Method of Calculating Discertion in a Flexible Work Hour Schedule," *Journal of Management Studies*, Vol. 15, October 1978, pp. 265–284.

MARTIN, V. H., *Hours of Work When Workers Can Choose*, Business and Professional Women's Foundation, Washington, DC, 1975.

MORGAN, F. T., "Your (Flexi) Time May Come," *Personnel Journal*, Vol. 56, February 1977, pp. 82–85, 96.

NOLLEN, S. D., "Does Flexitime Increase Productivity?," *Harvard Business Review*, Vol. 57, No. 5 (September–October 1979), pp. 12, 16–18.

OWEN, J. D., "Flexitime: Some Problems and Solutions," *Industrial and Labor Relations Review*, Vol. 30, January 1977, pp. 152–160.

ROBERTS, K. H., and C. L. HULIN, *The Influence of Variable Work Schedules on Worker Responses to Their Jobs*, Employment and Training Administration, U.S. Department of Labor, Washington, DC, 1980.

SCHEIN, V. E., E. H. MAURER, and J. F. NOVAK, "Impact of Flexible Working Hours on Productivity," *Journal of Applied Psychology*, Vol. 62, August 1977, pp. 463–465.

SCHEIN, V. E., E. H. MAURER, and J. F. NOVAK, "Supervisors' Reactions to Flexible Working Hours," *Journal of Occupational Psychology*, Vol. 51, No. 4, pp. 333–337.

SWART, J. C., *A Flexible Approach to Working Hours*, AMACOM, New York, 1978.

SWART, J. C., "Flexitime's Debit and Credit Option," *Harvard Business Review*, Vol. 57, No. 1 (January-February 1979).

WINETT, R. A., and M. S. NEALE, "Modifying Settings as a Strategy for Permanent, Preventive Behavior Change: Flexible Work Schedules and Family Life as a Case in Point," in P. Kardy and J. Steffen, Eds., *Toward a Psychology of Therapeutic Maintenance*, Gardner Press, New York, 1979.

YOUNG, W. M., "Applying Flexible Working Hours in Production Areas," *Production Engineer*, April 1976, pp. 187–190.

YOUNG, W. M., "Application of Flexible Working Hours to Continuous Shift Production," *Personnel Review*, Vol. 7, Summer 1978, pp. 12–19.

ZAWACKI, R. A., and J. S. JOHNSON, "Alternative Workweek Schedules: One Company's Experience With Flexitime," *Supervisory Management*, Vol. 21, June 1976, pp. 15–19.

Part-Time Employment

BEDNARZIK, R. W., *A Micro Model of Labor Supply for Part-Time Workers Using Matched CPS Data*, BLS Staff Paper No. 10, U.S. Bureau of Labor Statistics, Washington, DC, 1979.

BUREAU OF NATIONAL AFFAIRS, "ASPA-BNA Survey No. 25–Part-time and Temporary Employees," Bulletin to Management No. 1295, Washington, DC, 1974.

CLARK, R. L., Ed., *Work Time and Employment*, Special Report No. 28, National Commission for Manpower Policy, Washington, DC, October 1978.

FREASE, M., and R. A. ZAWACKI, "Job Sharing: An Answer to Productivity Problems," *Personnel Administrator*, October 1979, pp. 35–38.

GREENWALD, C. S., and J. LISS, "Part-Time Workers Can Bring Higher Productivity," *Harvard Business Review*, September–October 1973, pp. 20, 22, 166.

MEIER, G. S., *Job Sharing: A New Pattern for Quality of Work and Life*, W. E. Upjohn Institute for Employment Research, Kalamazoo, MI, 1978.

MILLER, H. E., and J. R. TERBORG, "Job Attitudes of Part-Time and Full-Time Employees," *Journal of Applied Psychology*, Vol. 64, No. 4 (1979), pp. 380–386.

OLMSTEAD, B., "Job Sharing: An Emerging Work Style," *International Labor Review*, Vol. 118, No. 3 (May-June 1979).

SINGER, J. W., "Sharing Layoffs and Jobless Benefits–A New Approach is Attracting Interest," *National Journal*, February 9, 1980.

WERTHER, W. B., "Mini-Shifts: An Alternative to Overtime," *Personnel Journal*, Vol. 55, March 1976, pp. 130–133.

Compressed Workweek

ALLEN, R. E., and D. K. HAWES, "Attitudes Toward Work, Leisure, and the Four-Day Workweek," *Human Resource Mangement*, Vol. 18, Spring 1979, pp. 5–10.

ATWOOD, C. S., "A Work Schedule to Increase Productivity," *Personnel Administrator*, October 1979, pp. 29–33.

CALVASINA, E. J., and W. R. BOXX, "Efficiency of Workers on the Four-Day Workweek," *Academy of Management Journal*, Vol. 18, 1975, pp. 604–610.

DAVIS, H. J., R. O. BLALOCK, and K. M. WEAVER, "A Quantitative Decision-Making Guide to Four-Day Workweek Conversion," *Personnel Administrator*, Vol. 21, February 1976, pp. 45–50.

DICKINSON, T. L., and J. P. WIJTING, "An Analysis of Workers' Attitudes Toward the 4-Day, 40-Hour Workweek," *Psychological Reports*, Vol. 37, October 1970, pp. 383–390.

DOBELIS, M. C., "The Three-Day-Week–Offshoot of an EDP Operation," *Personnel*, Vol. 49, January–February 1972, pp. 24–33.

DUNHAM, R. B., and D. L. HAWK, "The Four-Day Week: Who Wants It?," *Academy of Management Journal*, Vol. 20, 1977, pp. 644–655.

FOTTLER, M. D., "Employee Acceptance of a Four Day Workweek," *Academy of Management Journal*, Vol. 20, December 1977, pp. 656–668.

HARTMAN, R. I., and L. M. WEAVER, "Four Factors Influencing Conversion to a Four-Day Work Week," *Human Resource Management*, Vol. 16, Spring 1977, pp. 24–27.

HEDGES, J. N., "How Many Days Make a Workweek?," *Monthly Labor Review*, Vol. 98, April 1975, pp. 29–36.

HODGE, B. J., and R. D. TELLIER, "Employee Reactions to the Four-Day Week," *California Management Review*, Vol. 18, Fall 1975, pp. 25–30.

IVANCEVICH, J. M., "Effects of the Shorter Workweek on Selected Satisfaction and Performance Measures," *Journal of Applied Psychology*, Vol. 59, December 1974, pp. 717–721.

IVANCEVICH, J. M., and H. L. LYON, "The Shortened Workweek: A Field Experiment," *Journal of Applied Psychology*, Vol. 62, February 1977, pp. 34–37.

LEWIS, R., "Engineers and the Four-Day Workweek," *Chemical Engineering*, Vol. 80, February 1973, p. 136.

MAKLAN, D. M., *The Four-Day Workweek: Blue-Collar Adjustment to a Nonconventional Arrangement of Work and Leisure Time*, Praeger, New York, 1977.

ROBISON, D., "Dupont's 12-Hour Shift Improves QWL and Employee Self-Esteem at Six Continuous-Process Plants," *World of Work Report*, Vol. 3, February 1978, pp. 13–19.

STEWARD, A. V., and J. M. LARSEN, "Four-Day/Three-Day Per Week Application to a Continuous Production Operation," *Management of Personnel Quarterly*, Vol. 10, Winter 1971, pp. 13–20.

STEWART, G. M., and A. GUTHRIE, "Alternative Workweek Schedules: Which One Fits Your Operation?," *Supervisory Management*, Vol. 21, June 1976, pp. 2–14.

SWERDLOFF, S., "The Revised Workweek: Results of a Pilot Study of 16 Firms," U.S. Bureau of Labor Statistics Bulletin 1846, U.S. Government Printing Office, Washington, DC, 1975.

WERTHER, W. B., and J. W. NEWSTROM, "Administrative Implications of the Four-Day Week: It Could Mean More Work for Managers," *Adminstrative Management*, Vol. 33, December 1972, pp. 18–19.

General

BEST, F., "Preferences on Worklife Scheduling and Work-Leisure Tradeoffs," *Monthly Labor Review*, Vol. 101, June 1978, pp. 31–37.

BEST, F., "The Future of Retirement and Lifetime Distribution of Work," *Aging and Work*, Summer 1979.

BEST, F., *Exchanging Earnings for Leisure*, U.S. Government Printing Office, Washington, DC, 1980.

BEST, F., *Flexible Life Scheduling*, Praeger, New York, 1980.

COHEN, A. R. and H. GADON, *Alternative Work Schedules: Integrating Individual and Organizational Needs*, Addison-Wesley, Reading, MA, 1978.

GLUECK, W. F., "Changing Hours and Work: A Review and Analysis of the Research," *Personnel Administrator*, Vol. 24, March 1979, pp. 44–67.

MAHONEY, T. A., "The Rearranged Workweek: Evaluations of Different Work Schedules," *California Management Review*, Vol. 20, Summer 1978, pp. 31–39.

MARIC, D., *Adapting Work Hours to Modern Needs*, International Labor Office, Geneva, Switzerland, 1977.

MCCARTHY, M., "Trends in the Development of Alternative Work Patterns," *Personnel Administrator*, October 1979, pp. 25–27, 33.

NATIONAL COUNCIL FOR ALTERNATIVE WORK PATTERNS. *Alternative Work Schedule Directory*: *First Edition*, NCAWP, Washington, DC, 1978.

NEWSTROM, J. W. and J. L. PIERCE, "Alternative Work Schedules: The State of the Art," *Personnel Administrator*, October 1979, pp. 19–23.

OWEN, J. D., *Working Hours*, Lexington Books, Lexington, MA, 1979.

U.S. COMPTROLLER GENERAL, "Contractors' Use of Altered Work Schedules for Their Employees—How It Is Working," U.S. General Accounting Office, Washington, DC, 1976.

WEINSTEIN, H. G., "A Comparison of Three Alternative Work Schedules: Flexible Work Hours, Compact Work Week, and Staggered Work Hours," Industrial Research Unit, The Wharton School, University of Pennsylvania, Philadelphia, 1975.

ZALUSKY, J. L., "Alternative Work Schedules: A Labor Perspective," *Journal of the College and University Personnel Association*, Vol. 28, Summer 1977, pp. 53–56.

CHAPTER 11.9

Personnel Scheduling

WILLIAM J. BURGESS
ROBERT E. BUSBY
Tennessee Eastman Company

11.9.1 INTRODUCTION

The general problem of determining how best to match available personnel to personnel requirements of an organization is one to which managers and industrial engineers devote much time. The problem is usually separated into smaller segments. Parts of the problem are treated under such headings as (1) personnel planning (manpower planning), (2) work schedules, (3) personnel scheduling, and (4) personnel work assignments (man-machine assignments). Personnel planning deals with determining long-term labor needs (numbers and skills) to meet the organization's goals and the planning necessary to ensure availability (see Chapters 5.2, 5.5, and 9.1). Work schedules concern work patterns as a function of work demands, employee acceptance, legal limitations, economics, and so on (see Chapter 9.1). Personnel scheduling deals with the assignment of available personnel to handle job demands that vary over time. Personnel work assignments involve the assignment of personnel to tasks and work stations (see Chapters 3.4 and 3.5).

The objective of this chapter is to provide the reader with an overview of personnel scheduling and the associated concepts and methodologies for use when scheduling personnel in various types of organizations and work situations. In many cases it is impossible to determine the best personnel schedule without considering alternate work schedules. Therefore the determination of work schedules is discussed in relation to optimizing personnel schedules.

11.9.2 THE PERSONNEL SCHEDULING PROBLEM—GENERAL APPROACH

The personnel scheduling problem is one of optimally matching available labor resources to labor needs of an organization considering all applicable constraints. The solution requires that staffing requirements, availability of labor, and the associated determinants (i.e., factors that determine labor requirements) be known. Exhibit 11.9.1 illustrates the general steps involved in analyzing and solving a personnel scheduling problem.

Determining Quantity of Work to be Done

The first step is to define the quantity of work to be done and when it needs to be done (i.e., time pattern of work demand by hour, day, month, etc.). A graph or chart of production- and/or service-level data should be developed. Time increments for data collection should be chosen as a function of the scheduling problem. If production and/or service demands are fairly level and constant within any workweek, scheduling and work demands should be tabulated by shifts. If work load and/or service rates vary within a shift, such as for toll gate operators, service line attendants, and telephone operators, then the problem is one of work shift scheduling, in which case work demand data should be tabulated by hour or some other subportion of a work shift, for example, 15 min intervals.

Work demand variability as related to any time unit can be divided into two components—a "within" time unit component and a "between" time unit component (see Exhibit 11.9.2). The analyst must decide which demand values to use or service level to provide for the time intervals being studied in order to minimize adverse cost effects. The expected values or expected values plus or minus some selected SD might be chosen based on economics. The work demand data and the operations from which they are collected should be analyzed to identify variability and its sources (determinants). Examples of determinants include production level, service level, waste

Exhibit 11.9.1 Steps in Analyzing and Solving a Personnel Scheduling Problem

STEP 1. *Determine quantity of work to be done.*
STEP 2. *Determine staffing required to do the work.*
STEP 3. *Determine personnel availability.*
STEP 4. *Match personnel to staffing requirements.*

a. Determine if labor needs and availability mismatches are significant.
b. Investigate changing the work demand pattern.
c. Investigate altering the time availability of personnel.
d. Develop a work schedule.
e. Develop a personnel scheduling management system.

level, service demand, and environmental conditions. This information will be useful later in solving the personnel scheduling problem.

Determining Staffing Required to Do the Work

The labor needed to satisfy the work demands during each time unit must be determined. These labor requirements should be based on the best work methods and procedures and should consider the following factors: work methods, auxiliary work, personal allowances, sickness, vacations, deferrability of tasks, and so on. Up-to-date work standards are an ideal basis for determining labor requirements (see Chapter 4.1).

The total staffing requirement for an operation is a function of the personnel availability. Staffing levels and personnel schedules will be affected by expected patterns of lost work time due to vacations, sicknesses, and so on. This is discussed in the following section.

Determining Personnel Availability

The next step is to determine numbers and time availability of personnel to perform the required work (work shifts, days per week of work, work hours per day, days off, etc.). Determine sources of personnel, such as part-time employees, labor pools, and temporary summer employees, that can be utilized to increase labor availability and flexibility. Personnel availability is also a function of time off because of vacations, sicknesses, and so on. The expected pattern of personnel time off because of various causes should be determined. In the case of vacations, management may either completely schedule employees' vacation times or allow varying degrees of freedom of employee choice. In any case, a pattern of time off will probably exist and should be determined. Lost time due to sickness is not scheduled, but frequently time patterns do exist.

Next, determine the availability of labor through a personnel relief system. This labor will be used to cover shortages of regular personnel due to such things as breaks, sickness, and vacation. Personnel relief can be provided in many ways–extra relief personnel assigned equally to each work group, relief labor pools, cross-trained personnel, overtime, and so on. Ingenuity used in developing a personnel relief system will make the personnel scheduling problem much easier because it relaxes constraints on scheduling.

Exhibit 11.9.2 Work Demand Pattern for Maintenance Service

Time (Hours)	Average Arrival Rate of Service Demands[a]	Range of Within-Hour Demand
8–9	20	15–25
9–10	15	11–18
10–11	10	7–12
11–12	12	9–15
12–1	5	2–8
1–2	15	11–18
2–3	12	8–15
3–4	10	8–13
4–5	6	2–9

[a]Range of between-hour demand: 5–20.

Matching Personnel to Staffing Requirements

Given known staffing requirements and personnel availability, the degree of mismatch and related "costs" should be determined. Consider all cost and intangible factors such as overtime, idle labor, training, equipment downtime, lost sales, poor service, employee morale, turnover, waste, off-quality product, productivity, and safety. If the cost of the mismatch is not excessive, there is little problem. Also, mismatches between labor availability and needs are a problem only when one or both are inflexible to change without incurring significant costs. Therefore, as a general approach, first try to remove restrictions. Try to eliminate the problem by changing the demand pattern for staffing. Depending on the nature of the business, this can be done in several ways. The basis for doing this should come from analysis of work variability and its determinants in step 1 (determining quantity of work to be done). Some examples are:

1. Level work demands by means of production planning and scheduling.
2. Reduce equipment failures and breakdowns by replacement or maintenance programs.
3. Offer inducements to customers to smooth service peaks and valleys; for example, reduce rates or prices during low-demand periods such as is done by utility companies (off-hour reduction in rates) and grocery stores (midweek sales).
4. Control scheduling of vacations to avoid fluctuations.
5. Redesign jobs to reduce absenteeism due to undesirable jobs.
6. Reassign deferrable work.

After consideration of changing the demand pattern, the next question concerns trying to alter the time availability of personnel. If time availability is flexible, then a dynamic personnel scheduling system should be developed. Ways of achieving flexibility include labor pools, part-time employees, and cross-training. Changes in work schedules may also need to be developed to facilitate personnel scheduling.

A dynamic personnel scheduling system is a management system where staffing needs are routinely determined and personnel are scheduled from a flexible labor force, such as a labor pool, to meet the needs. Such a system can be as simple or elaborate as needed and as cost justified. The development of dynamic personnel scheduling systems are covered later in the chapter.

If the time availability of personnel is basically inflexible, the problem is to develop a fixed work schedule that best matches personnel to staffing requirements during each time interval. A good example of this type of situation is the scheduling of toll gate operators. Some, but not much, alteration can be made to the work demands that affect staffing requirements by letting queues build up. The time availability of personnel is basically inflexible. There is one job to be done; hence labor pools or cross-training on multiple jobs are not practical. The number of work hours per shift and week must be within minimum and maximum limits to satisfy employee expectations. Mathematical and empirical approaches to developing work schedules when time availability is inflexible are discussed later in the chapter.

Once work schedules are developed, a supporting management system for scheduling personnel within the bounds of the schedule should be developed. Even where a work schedule is the primary solution to the personnel scheduling problem, there usually will be a need for some dynamic personnel scheduling, for example, to handle lost time due to vacations or sickness.

In actual practice, whether to concentrate on developing a work schedule or a personnel scheduling system is not always obvious. Both avenues should be considered and an overall solution developed that incorporates the appropriate parts of both. After both have been considered, reevaluation should be performed to determine whether changes in work demand patterns can be cost justified. For example, if work schedules and/or personnel scheduling management systems can be developed that handle a wide variance in staffing requirements, the company may not want to offer financial inducements to customers to alter the timing of their demand.

The remainder of this chapter is devoted to a more in-depth discussion of work schedules (as they relate to personnel scheduling) and the development of dynamic personnel scheduling management systems.

11.9.3 WORK SCHEDULES

In practice, a great portion of workweek schedules for production, sales, and service operations can be developed using straightforward empirical approaches. Other operations will require a mathematical approach for solving. The empirical-type problems are exemplified by the following characteristics: (1) an integer number of personnel are scheduled (usually a fixed number), (2) meal and break scheduling are ignored when developing the work schedule, (3) the shifts are nonoverlapping, and (4) the staffing requirements are relatively constant each shift and day, or can be made so by utilizing deferrable activities. The objective in these cases is not to match the personnel needs and availability optimally, but to develop a personnel workweek schedule that would satisfy the nonfinancial parameters (number of work stations available, days of work, etc.) of the work situation. The empirical approach is based on the assumption that the optimum or predetermined number of employees has already been defined.

The mathematical approach includes determining the optimum number of personnel and the optimum work schedule to match the personnel needs and availability. Allocation approaches to optimum personnel scheduling using linear programming are discussed in the section on mathematical approaches.

Empirical Approaches to Personnel Workweek Schedules

A workweek schedule should define weekdays on which work will be performed, shifts on which work will be performed, beginning and ending hours of the work shift, scheduled rest days, sequence of shift rotation (if required), and the repetitiveness of the schedule. For most operations there will be many alternative workweek schedules from which to choose that will fit the operation.

These alternatives include different work shift durations, one crew or multicrew schedules, schedules with and without weekend work, different shift rotational sequences, and many variations of duration of scheduled days of rest. Therefore it is important that the basic information and assumptions for a workweek schedule and the criteria for evaluating the different alternatives be clearly defined at the outset of the study.

Information and Evaluation Criteria for Empirical Approaches

Information necessary for developing personnel workweek schedules includes:

1. What work is to be performed (amount and variability).
2. When the work can be performed (e.g., Is work restricted to the day, afternoon, or night shifts? Can work be performed Monday through Sunday or should weekends be scheduled off?).
3. Labor staffing requirements.
4. Equipment or workplace limitations (e.g., size, number of workplaces or units).
5. Alternative shift configurations (e.g., hours per shift available, number of shifts per week, number of shifts per day).
6. Scheduling constraints (e.g., Should rest days be consecutive? Should the rest days rotate among personnel? Are the shifts to be fixed?).

Criteria that should be considered for evaluating the alternative workweek schedules include:

1. Ease of the employees' learning and retaining the workweek sequence.
2. Consecutive workdays scheduled.
3. Weekend time off.
4. Cost considerations (e.g., optimum use of personnel, optimal use of equipment).
5. Limitations (e.g., equipment, workplace, physiological and psychological nature of the work).
6. Wage and hour laws.
7. Compatibility with other company and community workweek schedules.

Using these criteria, the alternative schedules can be evaluated and an appropriate schedule chosen. Chapter 11.8, on work schedules, also discusses many of the factors that a manager should consider in choosing a workweek schedule.

If the operation for which alternative schedules are being developed is not a continuous (7 day, 24 hr) operation, workweek schedules are relatively easy to develop by trial and error. If the operation is continuous, the development of workweek schedules is more complex. For that reason the following section describes a general approach to developing 8 and 12 hr shift schedules for continuous operations. Examples are presented that should be beneficial as guides to developing other schedules for specific operations. All examples assume fixed shifts, equal rotation of rest days, and consecutive rest days.

Schedules for Continuous Operations

Most schedules for continuous operations are set up so that crews rotate through the different work shifts; therefore in the following discussion workweek schedules will be referred to as "rotating shift schedules."

Rotating shift schedules can be developed to rotate forward sequentially—for example, from the 7 a.m. to 3 p.m. (day) shift, to the 3 p.m. to 11 p.m. (afternoon) shift, to the 11 p.m. to 7 a.m.

(night) shift (A to B to C)–or rotate backward from the 11 p.m. to 7 a.m. shift, to the 3 p.m. to 11 p.m. shift, to the 7 a.m. to 3 p.m. shift (C to B to A). The preferred rotation is the backward rotation, based on psychological studies of sleep habits of shift workers which indicated that workers will lose less sleep and rotate easier on the C-to-B-to-A shift rotation.

Scheduling rest days for rotating shift schedules can be performed in a variety of ways. These include sequential rest days (e.g., Tuesday and Wednesday off one week, Thursday and Friday off the next week, etc.), repeating rest days (Saturday and Sunday off every week), and nonsequential rest days (Sunday and Monday off this week, and Monday and Tuesday off the next week, etc.).

Eight Hour Shift Schedules With Sequential Rest Days. The sequential rest day schedules have an interesting repeating pattern which makes them easier to schedule. The work schedule will repeat itself in a minimum of n number of weeks, where n is the number of crews scheduled to work. This is illustrated as follows:

	Crew 1	Crew 2	Crew 3	. . .	Crew n
Week 1	A Sch.[a]	B Sch.	C Sch.	. . .	n Sch.
Week 2	B Sch.	C Sch.	D Sch.	. . .	A Sch.
Week 3	C Sch.	D Sch.	E Sch.	. . .	B Sch.
.	.	.	.	. . .	.
.	.	.	.	. . .	.
.	.	.	n Sch.	. . .	.
.	n - 1 Sch.	n Sch.	A Sch.	. . .	.
Week n	n Sch.	A Sch.	B Sch.	. . .	n - 1 Sch.

[a]Sch. = schedule.

This is true for 8 and 12 hr schedules as long as the rest days are sequential.

With this information, a specific crew size, for example, four crews, can be taken, and one work week can be scheduled based on the work required. Then the repeating pattern can be applied to fill in the remaining weeks, for example, three. For example:

Week No.	Day	Crew 1 Sch. A	Crew 2 Sch. B	Crew 3 Sch. C	Crew 4 Sch. D
	Monday	11–7	3–11	7–3	R
	Tuesday	R	3–11	7–3	11–7
	Wednesday	R	3–11	7–3	11--7
1	Thursday	3–11	R	7–3	11–7
	Friday	3–11	R	7–3	11–7
	Saturday	3–11	7–3	R	11–7
	Sunday	3–11	7–3	R	11–7
2		Sch. B	Sch. C	Sch. D	Sch. A
3		Sch. C	Sch. D	Sch. A	Sch. B
4		Sch. D	Sch. A	Sch. B	Sch. C

All schedules presented will omit the a.m. and p.m. designation, and R's will indicate days of rest. The 11–7, 3–11, and 7–3 notations will represent 11 p.m. to 7 a.m., 3 p.m. to 11 p.m., and 7 a.m. to 3 p.m., respectively. Any 11–7 shift indicated will start at 11 p.m. of the day indicated. (For example, crew 1 would begin work on Monday night at 11 p.m.) This complete schedule is presented in Exhibit 11.9.3.

Another potential 8 hr sequential rest day work schedule is a widely accepted 24 hr shift schedule (see Exhibit 11.9.4), which is being used mainly in Europe, but is gaining acceptance in the United States. It is a "rapid rotation" shift schedule. This schedule allows the employees to rotate through all of the three shifts (e.g., 11–7, 3–11, and 7–3) within a 7 day work period. Experiments with this schedule in the United States have indicated that, when compared directly to a conventional shift schedule (such as in Exhibit 11.9.3), 70 to 80% of the employees prefer the rapid rotation system over the conventional schedule. The advantages cited for these schedules are that they accommodate social activities better, are less fatiguing, and provide more time with the family.

Although there are many variations of the rapid rotation shift schedule, Exhibit 11.9.4 depicts a schedule that has been used successfully.

Exhibit 11.9.3 Four Crews, 8 hr/Shift, Backward Rotation; Work 7–Rest 2, Work 7–Rest 3, Work 7–Rest 2 Pattern

Week No.	Day	Crew 1	Crew 2	Crew 3	Crew 4
	Monday	11–7	3–11	7–3	R
	Tuesday	R	3–11	7–3	11–7
	Wednesday	R	3–11	7–3	11–7
1	Thursday	3–11	R	7–3	11–7
	Friday	3–11	R	7–3	11–7
	Saturday	3–11	7–3	R	11–7
	Sunday	3–11	7–3	R	11–7
	Monday	3–11	7–3	R	11–7
	Tuesday	3–11	7–3	11–7	R
	Wednesday	3–11	7–3	11–7	R
2	Thursday	R	7–3	11–7	3–11
	Friday	R	7–3	11–7	3–11
	Saturday	7–3	R	11–7	3–11
	Sunday	7–3	R	11–7	3–11
	Monday	7–3	R	11–7	3–11
	Tuesday	7–3	11–7	R	3–11
	Wednesday	7–3	11–7	R	3–11
3	Thursday	7–3	11–7	3–11	R
	Friday	7–3	11–7	3–11	R
	Saturday	R	11–7	3–11	7–3
	Sunday	R	11–7	3–11	7–3
	Monday	R	11–7	3–11	7–3
	Tuesday	11–7	R	3–11	7–3
	Wednesday	11–7	R	3–11	7–3
4	Thursday	11–7	3–11	R	7–3
	Friday	11–7	3–11	R	7–3
	Saturday	11–7	3–11	7–3	R
	Sunday	11–7	3–11	7–3	R

Exhibit 11.9.4 Rapid Rotation Shift Schedule, Four Crews, 8 hr/Shift, Backward Rotation; Work 7–Rest 2, Work 7–Rest 2, Work 7–Rest 3 Pattern

Week No.	Day	Crew 1 Sch. A	Crew 2 Sch. B	Crew 3 Sch. C	Crew 4 Sch. D
	Monday	7–3	R	11–7	3–11
	Tuesday	3–11	7–3	R	11–7
	Wednesday	3–11	7–3	R	11–7
1	Thursday	11–7	3–11	7–3	R
	Friday	11–7	3–11	7–3	R
	Saturday	R	11–7	3–11	7–3
	Sunday	R	11–7	3–11	7–3
2		Sch. B	Sch. C	Sch. D	Sch. A
3		Sch. C	Sch. D	Sch. A	Sch. B
4		Sch. D	Sch. A	Sch. B	Sch. C

Eight Hour Shift Schedule With Nonsequential Rest Days. For nonsequential rest days, Exhibit 11.9.5 is an example of an 8 hr shift schedule that is in common use in industry. Under this schedule, employees would work 6 consecutive days and then rest for 2 days on a continuing basis. This schedule would require that the rest days be scheduled such that the last rest day in any given week would be the first rest day in the next week for the employee to work 6 consecutive days. As can be seen in this example, the rest days' sequence repeats in 8 weeks, whereas the entire shift and rest day sequence would eventually repeat in 24 weeks.

Exhibit 11.9.5 Four Crews, 8 hr/Shift, Backward Rotation, Work 6–Rest 2 Pattern

Week	Day	Crew 1	Crew 2	Crew 3	Crew 4
	Monday	7–3	3–11	11–7	R
	Tuesday	7–3	3–11	11–7	R
	Wednesday	7–3	3–11	R	11–7
1	Thursday	7–3	3–11	R	11–7
	Friday	7–3	R	3–11	11–7
	Saturday	7–3	R	3–11	11–7
	Sunday	R	7–3	3–11	11–7
	Monday	R	7–3	3–11	11–7
	Tuesday	11–7	7–3	3–11	R
	Wednesday	11–7	7–3	3–11	R
2	Thursday	11–7	7–3	R	3–11
	Friday	11–7	7–3	R	3–11
	Saturday	11–7	R	7–3	3–11
	Sunday	11–7	R	7–3	3–11
	Monday	R	11–7	7–3	3–11
	Tuesday	R	11–7	7–3	3–11
	Wednesday	3–11	11–7	7–3	R
3	Thursday	3–11	11–7	7–3	R
	Friday	3–11	11–7	R	7–3
	Saturday	3–11	11–7	R	7–3
	Sunday	3–11	R	11–7	7–3
	Monday	3–11	R	11–7	7–3
	Tuesday	R	3–11	11–7	7–3
	Wednesday	R	3–11	11–7	7–3
4	Thursday	7–3	3–11	11–7	R
	Friday	7–3	3–11	11–7	R
	Saturday	7–3	3–11	R	11–7
	Sunday	7–3	3–11	R	11–7
	Monday	7–3	R	3–11	11–7
	Tuesday	7–3	R	3–11	11–7
	Wednesday	R	7–3	3–11	11–7
5	Thursday	R	7–3	3–11	11–7
	Friday	11–7	7–3	3–11	R
	Saturday	11–7	7–3	3–11	R
	Sunday	11–7	7–3	R	3–11
	Monday	11–7	7–3	R	3–11
	Tuesday	11–7	R	7–3	3–11
	Wednesday	11–7	R	7–3	3–11
6	Thursday	R	11–7	7–3	3–11
	Friday	R	11–7	7–3	3–11
	Saturday	3–11	11–7	7–3	R
	Sunday	3–11	11–7	7–3	R
	Monday	3–11	11–7	R	7–3
	Tuesday	3–11	11–7	R	7–3
	Wednesday	3–11	R	11–7	7–3
7	Thursday	3–11	R	11–7	7–3
	Friday	R	3–11	11–7	7–3
	Saturday	R	3–11	11–7	7–3
	Sunday	7–3	3–11	11–7	R
	Monday	7–3	3–11	11–7	R
	Tuesday	7–3	3–11	R	11–7
	Wednesday	7–3	3–11	R	11–7
8	Thursday	7–3	R	3–11	11–7
	Friday	7–3	R	3–11	11–7
	Saturday	R	7–3	3–11	11–7
	Sunday	R	7–3	3–11	11–7

Days off repeat, but shift sequence will change:

Crew 1	Crew 2	Crew 3	Crew 4
to	to	to	to
11–7	7–3	3–11	R

Exhibit 11.9.6 Modified EOWEO, Four Crews, 12 hr/Shift; Work 2–Off 2, Work 3–Off 2, Work 2–Off 3 Pattern[a]

Week No.	Day	Crew 1 Sch. A	Crew 2 Sch. B	Crew 3 Sch. C	Crew 4 Sch. D
	Monday	R	N	R	D
	Tuesday	D	R	N	R
	Wednesday	D	R	N	R
1	Thursday	R	D	R	N
	Friday	R	D	R	N
	Saturday	N	R	D	R
	Sunday	N	R	D	R
2		Sch. B	Sch. C	Sch. D	Sch. A
3		Sch. C	Sch. D	Sch. A	Sch. B
4		Sch. D	Sch. A	Sch. B	Sch. C

[a]The D or N notation indicates whether the work shift is a day (D) shift (e.g., 6 a.m. to 6 p.m.) or a night (N) shift (e.g., 6 p.m. to 6 a.m.).

The advantages of this type of schedule are that it is easy for the employees to remember, relief employees are easy to schedule for use in the system, and there are always 2 days of rest after working 6 consecutive work shifts.

Twelve Hour Schedule With Sequential Rest Days. Twelve hour shift schedules with sequential rest days also follow the repeating pattern presented earlier. Twelve hour shift schedules are best utilized where the physical or mental demands of the shift work are not great. Studies of 12 hr shift work have indicated that fatigue begins to overtake the workers on physically or mentally demanding jobs after 8 hr.

There are many 12 hr shift schedules available for use in industry. Some schedule a maximum of 3 straight working days (modified EOWEO–every-other-weekend-off–schedule) with a work 2–rest 2, work 3–rest 2, work 2–rest 3 pattern; others schedule more straight working days (e.g., a work 4–rest 4 pattern schedule and a work 4–rest 8, work 4–rest 3, work 3–rest 1, work 3–rest 3 pattern repeating schedule). An example of the modified EOWEO schedule is presented in Exhibit 11.9.6.

This schedule alternatively schedules the crews for 48 and 36 hr of work per week for a total of 168 hr of work per 4 week period, or an average of 42 hr of work per week. Its major advantages are that it schedules every other weekend off for the employees, limits consecutive days of work to a maximum of 3, and provides more days off for the employees.

Mathematical Approaches to Personnel Work Schedules

The development of personnel work schedules involves commitment of the labor resource to satisfy staffing requirements. Consequently, problems of this nature are suitable for solution using an allocation linear programming model. This is the mathematical approach most frequently used.

Mathematical modeling should be considered if the following conditions exist:

1. Work time increments (work shifts or workweeks) being scheduled are not mutually exclusive, that is, overlapping is permitted. Nonoverlapping configurations make the allocation problem simple and relatively easy to solve. (See previous secton on empirical approaches.)
2. Changes in labor requirements occur within a time interval smaller (e.g., each hour) than the work increment being scheduled, for example, 8 hr shifts.

However, when these conditions do not exist, other pragmatic approaches, such as a dynamic management personnel scheduling system, might be more applicable. This approach is discussed later in the chapter.

For purposes of discussion, work schedules can be classed and treated according to:

1. Nature of demand, that is, cyclic versus noncyclic.
2. Work time increment scheduled, that is, work shift versus workweek.

Work Shift Scheduling–Cyclic Demand

The basic formulation of a work shift scheduling problem involves meeting personnel requirements R during time interval T using alternative work shift schedules. Personnel requirements are allowed to vary between intervals, but are assumed constant within intervals. The problem should be formulated such that the interval of time chosen is sufficiently small that the assumption of constant labor requirements is reasonable. In general, the problem can be mathematically formulated as follows:

$$\text{minimize} \quad \sum_{j=1}^{S} c_j X_j$$

$$\text{such that} \quad \sum_{j=1}^{S} a_{ij} X_j \geqslant R_i \qquad i = 1, 2, 3, \ldots, n$$

$$X_j \geqslant 0$$

where R_i = personnel requirements for time period $i, 1 \leqslant i \leqslant n$
n = hours in work cycle/T (T is time interval expressed in hours.)
X_j = personnel allocated to work shift pattern j, $1 \leqslant j \leqslant S$
S = total number of alternative work shift patterns
c_j = cost or undesirability of assigning one person to the shift pattern j
a_{ij} = 1 if shift pattern j assigns work to a person during period i; 0 otherwise

This problem is formulated so that each j vector of this a_{ij} matrix defines an alternative shift pattern. If the model is not too large, this problem can be solved using integer linear programming to determine the optimum schedule from the alternative shift patterns defined in a_{ij}. If the model is too large for integer linear programs because of the number of alternative shift patterns and/or the number of time intervals, linear programming techniques coupled with a heuristic rounding procedure can be used to obtain integer results. If the shift patterns consist of consecutive time periods, and if the R_i values are integers, then linear programming solutions will always be integers. Also, there are special cases of this problem that can be solved efficiently by other means.[1,2]

For example, consider the scheduling of personnel to staff a retail store open for business from 10 a.m. to 9 p.m. Based on available labor, management defines the "full-time" shift alternatives (shift patterns 1, 2, 3) and two part-time shift alternatives (shift patterns 4, 5) to be considered in determining an optimum daily schedule.

Shift Pattern	Hours of Work	Hours Worked	Employee Classification	Cost per Employee per Shift
1	10 a.m. to 6 p.m.	8	Full time	$25.0
2	1 p.m. to 9 p.m.	8	Full time	30.0
3	12 p.m. to 6 p.m.	6	Full time	15.0
4	10 a.m. to 1 p.m.	3	Part time	7.5
5	6 p.m. to 9 p.m.	3	Part time	8.0

Personnel requirements vary from hour to hour; thus an hour is chosen as the appropriate time interval T. The number of personnel needed during each time interval is as follows:

Hour	Personnel Requirement
10 a.m. to 11 a.m.	3
11 a.m. to 12 a.m.	4
12 a.m. to 1 p.m.	6
1 p.m. to 2 p.m.	4
2 p.m. to 3 p.m.	7
3 p.m. to 4 p.m.	8
4 p.m. to 5 p.m.	7
5 p.m. to 6 p.m.	6
6 p.m. to 7 p.m.	4
7 p.m. to 8 p.m.	7
8 p.m. to 9 p.m.	8

Determine the number of employees to work each shift pattern such that minimum staff requirements are met while total labor costs are minimized.

The problem can be formulated as a linear program as follows:

$$\text{minimize} \quad 25X_1 + 30X_2 + 15X_3 + 7.5X_4 + 8X_5$$

where X_i is the number assigned to the shift pattern.

Such that

$$\begin{bmatrix}1\\1\\1\\1\\1\\1\\1\\1\\0\\0\\0\end{bmatrix} X_1 + \begin{bmatrix}0\\0\\0\\1\\1\\1\\1\\1\\1\\1\\1\end{bmatrix} X_2 + \begin{bmatrix}0\\0\\1\\1\\1\\1\\1\\1\\0\\0\\0\end{bmatrix} X_3 + \begin{bmatrix}1\\1\\1\\0\\0\\0\\0\\0\\0\\0\\0\end{bmatrix} X_4 + \begin{bmatrix}0\\0\\0\\0\\0\\0\\0\\0\\1\\1\\1\end{bmatrix} X_5 \geqslant \begin{bmatrix}3\\4\\6\\4\\7\\8\\7\\6\\4\\7\\8\end{bmatrix}$$

$$X_j \geqslant 0 \qquad j = 1, 5$$

Solution of this linear program yields the solution $(X_1, X_2, X_3, X_4, X_5) = (0, 0, 8, 4, 8)$.

Sensitivity analysis can be used to analyze the costs and/or benefits associated with an increase or decrease in staffing during an interval of time or on any given shift pattern. (See Chapter 14.2 for discussion of sensitivity analysis and linear programming.) For example, the shadow price associated with the sixth constraint (time interval 6, or 3 p.m. to 4 p.m.) is 15. Working with integer amounts, a \$15 savings can be realized for each unit of labor reduced on shift pattern 3 at time interval 6, as long as the schedules in the solution stay optimum. The reduced cost associated with shift pattern 1 (X_1) is \$2.50/day. The current optimal staffing for shift pattern 1 is zero ($X_1 = 0$). If it is desired to staff at least one employee on shift pattern 1, costs must be decreased by \$2.50/day for shift pattern 1 to enter the solution as an equal cost alternative.

The preceding model and example do not make any specific allowances for the scheduling of meals or breaks. This consideration can be handled in two ways. First, such scheduling can be handled explicitly by defining these periods in the shift pattern. For example, if a 30 min meal period is desired in shift pattern 1 of the example, the shift pattern would have to be defined in 30 min, rather than 60 min, intervals. The a_{i1} vector associated with shift pattern 1 and a meal period from 12 p.m. to 12:30 p.m. would be

$$a_{i1} = \underbrace{1\ \ 1}_{10\ \ 11} \quad \underbrace{1\ \ 1}_{11\ \ 12} \quad \underbrace{0\ \ 1}_{12\ \ 1} \quad \underbrace{1\ \ 1}_{1\ \ 2} \quad \underbrace{1\ \ 1}_{2\ \ 3} \quad \underbrace{1\ \ 1}_{3\ \ 4}$$

$$= \underbrace{1\ \ 1}_{4\ \ 5} \quad \underbrace{1\ \ 1}_{5\ \ 6} \quad \underbrace{0\ \ 0}_{6\ \ 7} \quad \underbrace{0\ \ 0}_{7\ \ 8} \quad \underbrace{0\ \ 0}_{8\ \ 9}$$

Any number of alternate meal period schedules can be considered by definition of corresponding a_{ij} column vectors. If n different meal period schedules for shift pattern 1 are desired, n, rather than one, a_{ij} column vectors and shift variables (X_j's) would have to be placed in the linear program.

Alternatively, the slack (excess staffing) associated with each constraint (time interval) can be examined to determine when to schedule employees for meals or breaks. For example, in the sample problem the slack associated with the fifth constraint (2 p.m. to 3 p.m.) is one employee. Therefore, during this interval of time, employees can be scheduled for breaks or meal periods such that no more than one employee is off the job at any given time. Analysis of slack times can also be done to determine time available for deferrable duties as well as for breaks or meals.

Thus far, discussion has centered around the deterministic problem. Personnel requirements during any given interval of time are more likely to vary according to some probability distribution around a mean requirement. Some insight into the sensitivity of the solution to variability about the mean requirement can be obtained by examination of the slack associated with each constraint. Also, the staffing needed to meet a given service level or demand for personnel a given percent of the time can be calculated outside the linear program. The linear program can be reformulated by changing the requirements (R_i's) and can be solved, and a cost curve for various service levels can be determined.

A further refinement of the preceding model can be used to gain insight into trade-off shortage costs (cost associated with not meeting demand) and surplus costs (cost associated with overstaffing). In practice, unit shortage costs and surplus costs are difficult to identify. For example, shortage costs might include cost of customer ill will. As seen in the sensitivity analysis here, the marginal cost of raising one of the requirements is an output of the linear program (shadow price on that constraint in the optimal solution).

Even though the costs associated with staff shortage or staff surplus (relative to requirements) cannot be easily determined, considerable insight into the scheduling problem can be gained by estimating these costs and including them in the objective function of a linear program as follows:

$$\text{minimize} \quad \sum_{j=1}^{S} c_j X_j + \sum_{i=1}^{N} b_i Y_i + \sum_{i=1}^{N} d_i Z_i$$

$$\text{such that} \quad \sum_{j=1}^{S} a_{ij} X_j + Y_i - Z_i = R_i$$

$$Y_i \leqslant S_i$$

$$Z_i \leqslant T_i$$

$$X_j, Y_i, Z_i \geqslant 0$$

where Y_i = staff shortage in interval i
Z_i = staff surplus in interval i
b_i = cost of being one unit short in staffing interval i
d_i = cost of having one unit surplus staffing in interval i
S_i = maximum staff shortage permitted in interval i
T_i = maximum staff surplus permitted in interval i

The solution to this model may have to be rounded. Through careful selection of S_i and T_i, valuable insight into the impact of estimated shortage and surplus costs by means of postoptimality analysis of objective function coefficients can be gained.

Variability about a mean staff requirement has been discussed. What if the mean requirement varies from day to day? Either of two approaches can be used. First, a linear program can be formulated and solved for each workday. Once staff requirements have been determined for each workday, this information can be used to develop weekly work schedules by using another linear program. The second approach, and possibly the best approach, is to formulate the linear program to encompass the entire work cycle. In many problems the work cycle will be a week or less. Of course, formulation in this manner yields potentially very large linear programs (large a_{ij} matrix).

From the preceding discussion, we can see that the time value in formulating the personnel scheduling problem as a linear program is much more than just setting an optimum solution. Significant insights can be gained about the costs of various alternatives and options available to the manager from sensitivity analyses.

Workweek Scheduling–Cyclic Demand

The workweek scheduling problem occurs when the length of an employee's workweek is different from the length of the business week, for example, 5 day workweeks and 6 day or 7 day business operation, such as with nurses, toll collectors, or retail salespersons. The question many times is to determine how to schedule days off for employees. This problem can be formulated and handled using the same model as in work shift scheduling; the difference is one of scaling the time intervals. The j vectors of the a_{ij} matrix are defined to represent workweek patterns–c_j is the associated cost of work pattern j, and X_j is the staff allocated to work pattern j.

If both types of scheduling problems (work shift and workweek) exist, they can be handled sequentially by solving the work shift problem first. The two problems can be solved simultaneously by using very large shift pattern vectors that describe optimum terms of shift schedules and workweek schedules (accounting for days off). However, the mathematical formulation becomes very complex and large.

Work Shift, Workweek Scheduling–Noncyclic Demand

An example of a scheduling problem where the demand does not cycle (repeat) is a construction or a maintenance overhaul project. Scheduling of personnel for one-time projects can be done using the basic model discussed previously for cyclic demand. The following example is given to illustrate the formulation of the problem.

Example 11.9.1 The Libra Company is a contract maintenance and construction company specializing in overhauling and building coal-fired power plants. A typical job is located in a town distant to the company's home base and lasts 2 to 4 months. The number of crews needed fluctuates each week during the project duration and can be estimated accurately. The company uses semiskilled workers who are hired for each specific project. There are one-time costs associated with hiring, transporting, and training crews, as well as the normal time-related wage and benefits costs. Therefore it is sometimes economical to keep crews in excess of a given week's labor requirements. To develop a minimum-cost labor schedule for a specific project, the following formulation can be used:

$$\text{minimize} \quad \sum_{j=1}^{S} c_j X_j$$

$$\text{such that} \quad \sum_{j=1}^{S} a_{ij} X_j \geqslant R_i \qquad i = 1, 2, 3, \ldots, n$$

$$X_j \geqslant 0$$

where R_i = number of crews required for week i, $1 \leqslant 0 \leqslant n$
n = total number of weeks in project
X_j = number of crews allocated to a project work pattern j, $1 \leqslant j \leqslant S$
S = total number of alternative project work patterns
c_j = total costs of assigning one crew to a project work pattern (includes one-time start-up costs plus weekly costs)
a_{ij} = 1 if project work pattern assigns work to a crew during week i; 0 otherwise

If there are limitations on the minimum number of weeks a crew must work once it is employed, this restriction can be handled by appropriate formulation of the a_{ij} vectors.

11.9.4 DYNAMIC PERSONNEL SCHEDULING MANAGEMENT SYSTEM

In work situations where considerable variability in personnel requirements occurs and where flexibility in obtaining personnel can be established, many times the best solution to the personnel scehduling problem is the development of a dynamic personnel scheduling management system. After work schedules are established, inherent variability in staffing requirements and personnel availability can result from sickness, emergency vacations, shutdowns, unexpected production fluctuations, environmental changes, or other unanticipated circumstances.

The following discussion describes an approach to the development of a personnel scheduling management system. An ongoing personnel scheduling system is desirable for routinely matching available personnel to personnel requirements since both vary over time.

There are four basic steps in the development of a dynamic personnel scheduling management system: (1) defining "labor determinants," (2) defining the functional relationship of determinants and labor needs, (3) determining alternative means of matching personnel to labor needs, and (4) developing a management information system to balance labor needs and availability.

Defining Labor Determinants

Labor needs are a function of production levels, product mixes, service levels, labor standards, sickness and lost-time patterns, vacation patterns, personnel skill proficiencies, waste and machine interruption levels, and so on. A basic understanding of the labor determinants for an operation is a must in designing a dynamic personnel scheduling system. Analysis of available production records and the operation should result in identification of the labor determinants for any operation.

Defining the Functional Relationship of Determinants and Labor Needs

Once the determinants of labor needs are known, the associated labor needs for each possible determinant state must be defined. For example, define the labor needs associated with each possible product mix, service level, personnel skill proficiency, and so on. The formal definition of many of these relationships occurs in labor standards.

Optimal staffing is based on operating at the lowest total cost, including the costs for the labor itself, the equipment and/or service downtime costs, waste-produced costs, and other costs. The cost of downtime and waste and possibly some other factors depends on whether the operation is at full capacity or not. This should be considered in the analysis.

Determining Alternative Means of Matching Personnel to Labor Needs

Several means of matching personnel to labor needs are available that enhance flexibility in personnel availability. These include relief labor pools, cross-trained employees, temporary employees, and part-time employees. All of these should be considered for use in the personnel scheduling system to provide dynamic management of the system.

Relief Labor Pools

Relief pools are groups of personnel established to provide labor flexibility between and within job classification and/or organization units. To be able to use the concept, work must exist that can be deferred. The personnel normally doing the deferrable work can then be reassigned as needed to fill needs in other parts of the organization. Several factors affect the extent to which pools can be used: (1) the amount of deferrable work, (2) the complexity of the work, and (3) the extent to which the personnel needs and labor pool personnel work schedules coincide.

The amount of deferrable work determines the number of people that can be maintained to meet fluctuating needs. To determine the number of people to assign to the pool, total relief requirements must also be determined.

Labor pools work best where job complexity is not too great, such as the less complex entrance jobs in an organization.

In some situations, special shift schedules will need to be developed for people working in the labor pools on deferrable work so that they will be available for reassignment at the required times. For example, a company with an around-the-clock operation may find that there are more unscheduled absences on the afternoon shift (3 p.m. to 11 p.m.) and night shift (11 p.m. to 7 a.m.) than on the day shift (7 a.m. to 3 p.m.). Therefore an oscillating shift pool on the afternoon and night shifts might be established to have relief personnel available.

Cross-Trained Employees

Cross-training of employees to relieve others on comparable level jobs within a work unit is another excellent method of providing labor flexibility. In work units where jobs are sequenced by job grade, another use of cross-training is to train employees at one job grade to be able to relieve on the next highest grade in the promotion sequences. This cross-training of operators to relieve upward in their promotional sequences provides an economical method of providing employee relief at all levels. Since the employees at any level should eventually progress and be trained on the next highest-level jobs, the cross-training will not be completely lost and can be utilized immediately. It also results in benefits for areas of fluctuating production and/or service demands by enabling movement both up and down in staffing levels to meet demands without interrupting production or service.

Cross-trained employees can also be used to gain labor flexibility in two other ways. If jobs consisting primarily of deferrable work exist, then the employees can be cross-trained to fill personnel needs on jobs of nondeferrable work. Also, employees assigned to relief on nondeferrable work in one operation can be cross-trained to fill in on jobs in other operations when not relieving on their primary job.

Expected relief needs for each job or group of jobs can be determined, and the number of employees needed for relief can be assigned full time to the job or group of jobs. Since the actual shift-to-shift needs will vary considerably, the relief personnel should be cross-trained on several jobs for increased scheduling flexibility.

Temporary Employees

In many job environments, students or teachers are available during the summer months and can be used for vacation relief. The training and use of these employees is another excellent method for increasing labor flexibility. This also provides for vacations for personnel in the prime vacation months. Analysis of the vacation records should indicate if more vacations are preferred and taken in the summer months. If they are, then analysis of the job structure should be performed to determine if there are any available nonconfidential and appropriate jobs that these employees could perform. These employees can also be used to reduce overtime in the summer months if business demand is seasonal.

An extension of such a use of temporary employees is to use them anytime a peak temporary demand occurs that coincides with readily available short-term labor, for example, students helping retail businesses at Christmas or resort employees splitting time between summer resorts and winter resorts.

Part-Time Employees

Part-time employees may frequently provide a means of meeting irregular demands during work shifts and workweeks. For example, high school students can be used 4 to 5 hr on Friday and Saturday evenings to handle peak demands at a popular restaurant or grocery store. The main criteria for determining the potential for use of part-time employees is job complexity and availability of labor. Sources of part-time help include students, housewives, and retired former employees. This concept can be used in combination with the labor pool concept, for example, use of part-time employees in a secretarial pool. Refer to Chapter 9.8 for additional discussion on the use of part-time employees.

Developing a Management Information System to Balance Labor Needs and Availability

After defining the determinants of labor needs, how labor needs vary in relation to the determinants, and how personnel can be adjusted to meet varying needs, the next step is to develop a management system to ensure optimum movement of personnel to match labor needs. The system should be designed to be responsive to meeting monthly, weekly, daily, and shift-to-shift labor fluctuations according to their cost significance.

Someone will need to be designated to administer the system. The amount of time required to do so will depend on the size of the labor force, the complexity of the labor scheduling problem, and the organization units involved (coordination required). In complex operations with rotating shift work, someone may need to be designated on each shift to be responsible for coordinating personnel scheduling.

Once responsibility is assigned, an information system must be developed and implemented for routinely determining changes in personnel needs and assigning trained personnel to minimize labor excesses and shortages. Also, contingency plans for handling labor shortages and excesses should be developed.

Each information system must be individually tailored to the specific organization using it. The following example of a basic system is presented to acquaint the reader with what to consider in developing an effective information system. The operation for which it was developed is a continuous 24 hr, 7 day operation with rotating shift crews. The components of this system include an overall labor coordinator, shift labor coordinators, labor pools, cross-trained employees, and temporary summer employees.

Example 11.9.2 Basically, the system involves the first-line supervisors' filling out a form (no. 1) on Wednesday or Thursday of the present week, indicating the number of employees *available* to work on their shifts (based on accounting for known vacations, sicknessses, and other lost time) for the upcoming week. This form is then sent to their supervisors for summarizing available employees on all shifts. The supervisors calculate the employee *needs* on each shift based on the labor determinants and standards and forecast production levels for the upcoming week and summarize the employee needs and availability for each shift and crew on a second form (no. 2). A place is provided on this form for the supervisor to indicate employee shortages or excesses by subtracting personnel needs from available personnel for the upcoming week.

This form (no. 2) is then sent to the overall labor coordinator, who summarizes all information received from all work units on a master working sheet (form 3) and schedules the relief people (from labor pools, temporary summer employees, or other) to meet the identified needs. If an excess or a shortage of people exists after making all available labor assignments, the labor coordinator should take the following action: Excess labor is scheduled for deferrable work, training, or voluntary days off (if an option). On shifts where not enough labor is available for the needs, several alternatives should be considered, including shutting down less critical equipment or services to provide extra labor, scheduling overtime, working short, and balancing labor between shifts.

To assist in the decision making process, a priority listing (form 4) of possible actions to take when there is an excess or shortage of available labor should be generated and updated regularly. This listing should indicate any equipment or services that might be shut down or discontinued each day or week, the number of employees made available, and the maximum number of shifts or days the equipment or service can be shut down or discontinued without affecting production and/or service. Also, a priority listing should be developed identifying any deferrable work or training that should be performed when excess labor is available. Another option is to allow employees to take voluntary time off if no deferrable work is available.

A copy of this listing is also given to each shift labor coordinator to use to make adjustments in labor as required on each shift during the next week.

The overall labor coordinator will then distribute the master labor schedule to all areas on Friday morning, indicating where each pool employee is to work or train for the upcoming week. The shift coordinator also gets a copy.

With these copies, each area supervisor should know the labor situation for the entire unit. If changes needed to be made during the scheduled week, the area supervisors should contact the designated shift labor coordinator as soon as possible to ask for additional labor or to inform the coordinator of any excess labor. This daily information sharing provides the dynamic, ongoing movement of labor to optimize the use of the available labor.

Long-range scheduling for labor is accomplished by monthly and quarterly reviews with production planning, forecasting, and marketing people to determine future labor needs and availability. Information is then passed on to the personnel department concerning the need for hiring more personnel in the future or the need for placing excess personnel.

This information provides for a dynamic personnel scheduling system which should be able to react to all varying operating conditions to optimize the use of the labor for the unit.

Additional work that must be done to complete the development of the dynamic personnel scheduling management system includes (1) designing a system for maintaining trained and cross-trained personnel, (2) developing a feedback and monitoring system to determine the effectiveness of the system, and (3) implementing and evaluating the system.

11.9.5 CONCLUSION

This chapter has presented an overview of the basics of personnel scheduling. There is great variety in the nature of personnel scheduling problems. Considerable opportunity exists for ingenuity and creativity in the development of personnel schedules and dynamic personnel scheduling management systems. The best solution techniques for a specific problem depend on the characteristics of the work demands and the available personnel sources. The bibliography lists additional information sources that should be helpful in solving a variety of problems.

REFERENCES

1. J. J. BARTHOLDI, III and H. D. RATLIFF, "Unnetworks, With Application to Idle Time Scheduling," *Management Science*, Vol. 24, No. 8 (April 1978), pp. 850–858.
2. J. J. BARTHOLDI, III and H. D. RATLIFF, "Cyclic Scheduling Via Integer Programs With Circular Ones," *Operations Research*, Vol. 28, No. 5 (September–October 1980), pp. 1074–1085.

BIBLIOGRAPHY

BAKER, K. R., "Scheduling a Full-Time Workforce to Meet Cyclic Staffing Requirements," *Management Science*, Vol. 20, No. 12 (August 1974), pp. 1561–1568.

BAKER, K. R., "Workforce Allocations in Cyclical Scheduling Problems: A Survey," *Operational Research Quarterly*, Vol. 27, No. 1 (1976), pp. 155–167.

CHADWICK-JONES, J., "Shift Working: Physiological Effects and Social Behavior," *British Journal of Industrial Relations*, Vol. 5, 1965, pp. 237–243.

EUSTACE, V. F., "Shift Work," *Industrial Medicine and Surgery*, (now *International Journal of Occupational Health and Safety*), Vol. 34, No. 11 (November 1965), pp. 857–859.

HEALY, W. C., "Shift Scheduling Made Easy," *Factory*, October 1959, pp. 87–91.

KEITH, E. G., "Operator Scheduling," *AIIE Transactions*, Vol. 14, No. 1 (March 1979), pp. 37–41.

MONRON, G., "Scheduling Manpower for Service Operations," *Industrial Engineering*, August 1970, pp. 10–17.

MOTT, P. E., et al., *Shift Work–The Social, Psychological and Physical Consequences*, University of Michigan Press, Ann Arbor, 1965.

MURRELL, K. F. H., "Shift Work," in *Human Performance in Industry*, Van Nostrand Reinhold, New York, 1965.

ROTHSTEIN, M., "Scheduling Manpower With Mathematical Programming," *Industrial Engineering*, Vol. 4, No. 4 (1972), pp. 29–33.

SEGAL, M., "The Operator–Scheduling Problem: A Network Flow Approach," *Operations Research*, Vol. 22, No. 4 (1974), pp. 808–823.

SWENSON, A., Ed., "On Night and Shift Work," *Proceedings of an International Symposium*, National Institute of Occupational Health, Stockholm, Sweden, 1969.

TRIBREWALA, R., D. PHILLIPPE, and J. BROWNE, "Optimal Scheduling of Two Idle Periods," *Management Science*, Vol. 19, No. 1 (1972), pp. 71–75.

WAGNER, H. M., "Principles of Operations Research," 2nd ed., Prentice-Hall, Englewood Cliffs, NJ, 1975.

CHAPTER 11.10

Purchasing: A Strategic Force

LEROY H. WULFMEIER
MICHAEL R. HOTTINGER

General Motors Corporation

11.10.1 PURCHASING'S ROLE IN CORPORATE STRATEGY AND PERFORMANCE

The purchasing function's purpose is to buy all materials, goods, and services required by the business enterprise in a manner that maximizes profitability. This drive for profit must be conducted in a fair and ethical manner in order to establish and preserve valuable supplier relationships founded upon competitive pricing procedures, reliable quality standards, and a free exchange of developing technology.

The growing influence of the purchasing function on overall corporate strategy and performance stems from the increasingly critical nature of business variables controlled through procurement: material availability, realizable quality levels, lead time for new product development, and the cost of materials and services. In any economy characterized by supply restrictions or constrained production capacity, effective purchasing is the competitive edge vital to the achievement of overall business objectives. Since material costs account for an already substantial and growing share of each sales dollar, as noted in Exhibit 11,10.1, the probability of achieving cost savings through redoubled purchasing acumen far exceeds that of any other functional area. This "profit leverage" inherent in the purchasing function is further emphasized by the fact that American manufacturers reinvest nearly five times more capital in inventory accumulation than in plant and equipment assets combined.

Recognition of purchasing as a key profession, tantamount to marketing, finance, engineering, and personnel, is growing rapidly. Establishment of top corporate posts specializing in the procurement area are becoming the rule rather than the exception. This new awareness of purchasing and its pervasive role in corporate strategy and performance demands refined skills and advanced management techniques from those aspiring to the field.

Purchasing must necessarily be limited to the purchasing department. If responsibility for purchase commitments is diffused casually throughout various user departments, serious management problems will result. Suppliers become confused not knowing who in fact has the authority to buy. Material and service costs will rapidly inflate because no single control point exists. Investment and administrative costs soar in the absence of cohesive purchasing control. Legal entanglements

Exhibit 11.10.1 Material Costs as a Share of Total Sales in (*a*) 1968 Sales Dollars and (*b*) 1978 Sales Dollars Among Typical Manufacturing Firms

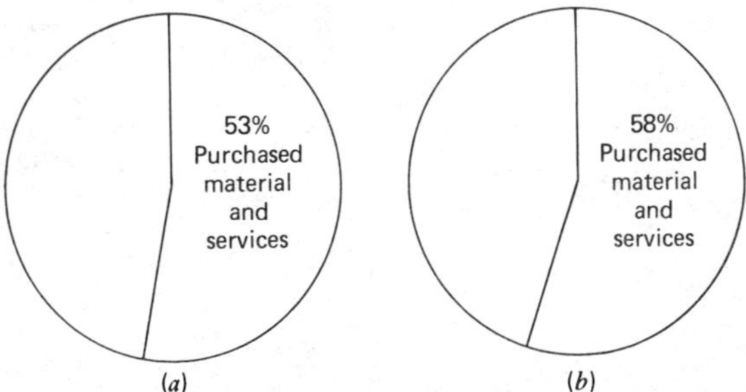

may appear as suppliers react to informal, implied commitments instead of comprehensive formal purchase agreements. The public image of the buying firm may be permanently damaged by conflicts of interest and unethical business relationships that inherently exist without a strong purchasing organization.

11.10.2 THE PROCUREMENT PROCESS

Classify Purchases

A taxonomy of purchased commodities should parallel the system of accounting for material, goods, and services. Generally, "direct materials" include steel, raw materials, and components destined to become part of the final product. From a financial perspective, direct materials are accounted for as a part of the cost of goods sold. "Indirect materials and services" include all items required for the business that do not become part of the finished product. Maintenance, repair, and operating supplies (MRO items) represent typical indirect material purchases. Financially, indirect items are recorded in burden or expense accounts. "Capital investment purchases" embrace machinery, equipment, construction, and other fixed assets. Accordingly, these purchases are accounted for as depreciable assets and are usually funded through a capital allocation system (appropriations).

Exhibit 11.10.2 provides a detailed list of examples for each category of purchased items.

Exhibit 11.10.2 Taxonomy of Purchases

Category	Description	Examples
Direct materials	Become part of the finished product	Raw materials Steel Aluminum Polyvinyl chloride Component parts Fasteners Stampings Injection moldings
Indirect items	Required for operations, but not part of the finished product	MRO Supplies Stationery Lubricants Machine repair Light bulbs Gloves Utilities Perishable tools Drills Reamers Taps Fixtures Services Advertising Cleaning
Capital goods	Long-term assets, investments	Machinery and Equipment Stamping presses Cranes Transfer machines Conveyor systems Computers Construction Plant additions Buildings Storage facilities Other fixed assets Cooling towers Air conditioner systems Plant security systems

Regardless of the category of goods purchased, certain fundamentals of the purchasing process provide the key to effective procurement.

Defining Authority to Purchase

Purchasing activity is initiated by a formal request (requisition) to seek out and establish sourcing for requirements. Routinely, definition of requirements and engineering specifications must be finalized *prior* to commitment of funds for purchase. Only in unusual circumstances involving highly specialized commodities (advanced computer systems, etc.) should the authority to purchase be granted prior to the availability of finalized specifications. Even when confronted with a "black box" requirement, purchasing must insist that basic performance criteria be established in the absence of detailed specifications. Requests for purchase that stipulate brand names or a specific single supplier require increased control by purchasing. In each case the reasons for a non-competitive requisition must be explained to purchasing's satisfaction and be thoroughly investigated.

Requests for purchase must be documented in writing. This document (see Exhibit 11.10.3) should require signed approval from authorized corporate officials and should contain a concise statement of requirements and specifications. Specifications that limit the number of potential suppliers (brand names, etc.) must be avoided in order to maintain a competitive purchasing climate, obtain maximum cost advantage, and preclude the specter of unethical conduct.

Identifying and Evaluating Potential Suppliers

Effective purchasing organizations are in constant contact with the supplier community. This contact must involve an aggressive search for new, untried potential suppliers. An attitude of open-mindedness and creativity will promote the development of new suppliers to foster competition and provide an influx of new technology and cost-effectiveness. Potential suppliers should be evaluated carefully and objectively along the following parameters: financial condition, management profile, technical capabilities, quality control systems, market position and external factors, and historical performance.

Financial Condition

Identify the form of ownership, extent of capitalization, degree of indebtedness, and general health. Prospective suppliers should be requested to submit audited financial statements for the buyer's review. Financial information may also be available through an independent credit analysis firm, such as Dun & Bradstreet or TRW. A supplier judged to be in poor financial condition cannot be depended upon for consistent, long-term supply of requirements.

Exhibit 11.10.3 Authority-to-Purchase Document

CP - 17B
Rev. 1-67

VENDOR:

Purchase Requisition

No. 161671

1

Date 12/18/79 | Date Wanted 1/2/80
Ship to Plant No. | Prev. Source Ric
Acct. or Appr. No. 700135531 H | Prev. P.O. No.

Inquiry No.	Date Required	Tax Status	Terms	P.O. No.
F.O.B. Buyers Plt, Unless Otherwise Indicated		Via		Buyer No.

Quantity	Code and Description	Price	Per
12	WD-4517 Carbide Buttons - Det. #8 in accordance with Blueprint dated August 10, 1975, revision dated January 5, 1980.		
	Ref: S/R-75337		

Purpose
Pt. #14011 Upper Control

Issuing Dept. M.M. Department	Charge to Or Required by Dept. M.M. Department	Deliver To Dept. or Room Bldg. #7
Signed John S. Doe	Approved	Signed
Dept. Delegate	Managers Delegate	Buyer / Purchasing Agent

Sources To Be Inquired

Management Profile

Purchasing should become throughly familiar with the management and organization of potential suppliers. This familiarity must encompass the operations management as well as normal sales and marketing contacts. Buyers should analyze and understand the corporate strategy and organizational structure of prospective suppliers, seeking an acute awareness of their marketing techniques, decision making procedures, and effective communication channels. This knowledge will facilitate problem solving efforts and aid the buyer in predicting supplier performance and behavior when a contractual relationship is established. A complete awareness of this dimension places the buyer at a distinct negotiating advantage.

Technical Capabilities

Determine the types and capacities of various production processes that are utilized by the prospective supplier. Evaluate these processes relative to the state of the art of the industry. Identify labor- versus capital-intensive processes. If a specific process is labor intensive, are the supplier's employees unionized? What is the state of the supplier's union-management relationship? Are work stoppages likely to occur that might curtail availability of the buyer's requirements? For capital-intensive processes, how will the supplier cycle production among various customers? Will the supplier conduct overtime operations to protect customer requirements? Buyers must also evaluate the degree of vertical and horizontal integration of each supplier. Subcontracting of critical processes by the primary supplier may be highly undesirable under certain procurement circumstances.

Quality Control Systems

Carefully evaluate each prospective supplier's commitment to quality. Documentation of receiving, in-process, and final inspection procedures should be surveyed during buyer visits to the supplier. If source quality systems are not acceptable to the buyer, it may be beneficial to provide assistance directly from the buyer's own quality control organization. The supplier's policy on product liability, acceptance of rejected material, and cost responsibility should also be assessed before a contractual relationship is established.

Market Position and External Factors

Analyze the supplier's market and industry. Determine his or her market share and competitive position. Review the supplier's historical growth pattern relative to the industry. A potential supplier who is losing market share in a declining industry may be headed for financial instability. Determine the supplier's role as a price leader in the industry. Identify external pressures on the supplier, such as government regulations, foreign competition, or labor supply, which may have an adverse impact on performance. Knowledge of the external issues affecting the supplier will allow the buyer to understand source behavior and accurately predict long-term performances.

Historical Performance

Independent evaluation of the various parameters already discussed may still fall short of revealing important attributes of potential suppliers. A review of the source's historical performance with other customers may prove to be the most critical element of the buyer's evaluation. This review should include information collected informally through the buyer's own professional contacts as well as the supplier's marketing and sales personnel.

Analysis of Suppliers

This list of parameters for potential supplier evaluation is not exhaustive. However, the message should be clear; it is purchasing's responsibility to analyze comprehensively all *supply risks* associated with a particular supplier or purchase requirement. Evaluation of suppliers should not be limited to potential suppliers. Established sources should be periodically reviewed along these same parameters to ensure the buyer's continuing awareness of changes, potential problems, and industry trends. Dedication to effective supplier evaluation through meaningful sales interviews, source visits, and professional collaboration will ensure a mobilized reserve of qualified suppliers for future requirements.

Soliciting Bids

Competitive selection of suppliers is essential if the optimum cost benefit is to be gained and if the buyer's image of fairness and honesty is to be preserved. Certain aspects of the bid solicitation process (inquiry) are critical, and these are discussed here.

Consistent Information

The description of requirements and all information provided to each bidder must be the same. No supplier should be afforded a competitive advantage by means of access to information restricted from his or her cobidders.

Confidentiality

Bidders should not be announced to the collective body of competitors. Bid responses from potential suppliers must be accepted under strict confidentiality. Moreover, open, simultaneous review of bid specifications with multiple competitors should be strictly avoided. Such joint supplier reviews discourage innovative competitors from submitting cost reduction ideas. Undesirable (and illegal) collaboration among bidders may also result.

Complete Specifications

The request for quotation should minimize the uncertainty of requirements. Specifications, terms, and conditions should be concisely stated (in writing) so that each bidder may have a thorough understanding of the buyer's requirements. It is a key purchasing responsibility to ensure that user organizations provide accurate specifications for the inquiry process. Vague specifications in the RFQ will ultimately increase the administrative cost of the procurement process.

Initiation of the RFQ

Requests for quotation must be issued by the purchasing department only. Failure to adhere to this rule may result in implied commitments tendered by nonpurchasing personnel who are *not* legal agents for the customer firm. Moreover, RFQs originating outside the purchasing department will fail to include appropriate business terms and conditions that are essential to the subsequent purchase agreement.

Bid Due Dates

A uniform bid due date must be established for all competitors involved in an inquiry. The deadline should allow enough time for thorough evaluation of the buyer's requirements and for preparation of a bid, yet should be consistent with the overall timing necessitated by the customer's needs. Bids received after the due date should normally be returned to the bidder unopened. *Late bids should not be included in the evaluation process in order to preserve the buyer's integrity and ethical reputation.* Buyers may agree to extend the bid due date at the legitimate request of bidders. However, the extended bid due date should be communicated to all bidders *in writing*. Bids may be accepted prior to the due date within the guidelines for confidentiality previously discussed.

Sealed Bid Procedure

Buyers may elect to impose a sealed bid procedure. This special procedure may be required for purchases of construction, machinery and equipment, or other major items involving sensitive procurement issues. Using a sealed bid procedure, the buyer provides a special self-addressed return envelope for each bidder. Sealed bids are returned to a *third party* within the customer organization, but outside the purchasing department. At a preestablished and announced date, place, and time, the sealed bids are opened and reviewed by the third party with the buyer. Bids received after the bid closing due date and time should be rejected and returned to the delinquent bidder unopened.

Analyzing Quotations

Selection of the supplier(s) to be awarded the buyer's requirements must be based on a comprehensive analysis of all quotations. This analysis must penetrate beyond a simple comparison of quoted selling prices. The buyer must seek to maximize *purchase value* by balancing the trade-offs between direct purchase cost and these criteria:

Expected quality.
Expected service.
Expected delivery performance.
Business terms and conditions required.
Hidden cost factors.

The effective buyer will determine the cost-benefit posture of each bidder, considering both economic and noneconomic variables. For example, the low bidder on an inquiry (low quoted selling price) may quote a price escalation contingency, may require advance or progress payments, or may refuse to accept important terms and conditions required by the buyer. The buyer must determine if the economic differential justifies acceptance of such contingencies and/or sacrifice of specified terms and conditions.

Similarly, the economic analysis of quotations must involve the total product cost. In addition to the quoted selling price, the cost of delivery, special inventory carrying costs (if required), and the use cost must be included.

The buyer's analysis of these trade-offs, along with the final selection of supplier, should be documented on a bid summary form. This summary should be maintained for reference and management audit.

Purchase Commitment

Once the final supplier has been selected, it is purchasing's sole responsibility to consummate the agreement between buyer and seller with a legally binding contract. A legal contract must consist of the buyer's *offer* to purchase and the seller's *acceptance* of the offer. Purchase orders or contracts represent only the buyer's offer until they are formally accepted by the seller. The seller's acceptance must usually be confirmed by written acknowledgment on the buyer's document within a specified period. Failure to secure acknowledgment may prevent the buyer from pursuing legal action against nonperforming suppliers. In some industries, trade practice allows delivery or supplier performance to constitute the seller's acceptance of the offer. Matters of this nature should be reviewed with the buyer's legal counsel.

The buyer's purchase document (offer) must carefully establish the commitments between buyer and seller, including:

Price.

Quantity.

Delivery schedule.

Method of delivery.

Responsibility for delivery costs.

Performance requirements of the seller's product.

Consequences of performance failure.

Method of payment.

Payment schedule.

Packaging requirements.

Limits of seller's product liability.

Compensation for breach of contract by buyer and/or seller.

Terms in the event of cancellation.

Exhibit 11.10.4 presents detailed examples of usual delivery and payment terms.

Most buying organizations have standardized purchase order and contract documents that outline general terms and conditions. Definition and control of the individual terms and conditions of specific purchases are key responsibilities of the procurement department.

Performance Follow-up

Since the buyer establishes the terms of payment and controls the opportunities for future business, purchasing is in the strongest position to influence supplier performance. For this reason purchasing should aggressively monitor supplier performance to ensure that the commitments of each purchase agreement are satisfied. Attention to performance follow-up will enhance purchasing's credibility within the customer organization and will gain the respect of the supplier community. Ultimately, experience with supplier performance will result in increasingly intelligent source selection decisions.

11.10.3 SOURCING STRATEGY AND SUPPLY RISK ANALYSIS

A sourcing strategy is a set of basic purchasing choices.

Should the requirement be purchased from outside suppliers or be made by the firm (integration)? (*make versus buy*)

Exhibit 11.10.4 Payment and Delivery Terms

Term	Definition
	Payment Terminology[a]
Net 25th instant	Payment will be made within 25 calendar days after receipt of an approved invoice. Invoices are "approved" only after satisfaction of the buyer's requirements.
Net 10 and 25th instant	Approved invoices received during the first 15 calendar days of the month will be paid on the 25th of that month. Approved invoices received during the last 15 days of the month will be paid on the 10th of the following month.
Net 25th proximo	Approved invoices will be paid on the 25th of the month following their receipt.
	Freight Terminology
f.o.b.: shipping point (seller's plant)	Buyer pays the transportation cost from the point of origin. Buyer designates freight mode and routing. Buyer assumes all rights and liabilities connected with possession during transit.
f.o.b.: buyer's plant (shipping destination)	Seller pays the transportation cost from the point of origin. Seller designates freight mode and routing. Seller assumes all rights and liabilities connected with possession during transit.
f.o.b.: shipping point equalized with destination	Buyer assumes all rights and liabilities of possession during transit. Buyer designates freight mode and routing. However, seller pays the cost of transportation.

[a]Buyers should carefully consider the cash flow impact of payment terms upon both the customer and the seller.

If purchased, *how many suppliers* are needed to satisfy the buyer's requirement?

What *form of purchase agreement* will best suit the procurement situation?

What *duration of commitment* is desirable?

Optimum sourcing strategies will (1) maximize the buyer's economic position, considering total costs, and (2) minimize the *supply risk* of the purchase. An effective sourcing strategy is the product of thorough financial analysis and systematic assessment of supply risk. Exhibit 11.10.5 graphically illustrates the relationship of financial and supply risk analysis to the sourcing strategy.

Substantial academic work has been devoted to financial analysis of procurement decisions. Unfortunately, little theory has been developed to address adequately the important subject of supply risk analysis. To this point, supply risk analysis has been treated as a qualitative process

Exhibit 11.10.5 Sourcing Strategy Development Process

Financial analysis
Material price
Investment
Total cost review

Supply risk analysis
Probability of source failure to perform

Sourcing strategy
Make versus buy
Number of suppliers
Mode of purchase
Duration of commitment

totally dependent upon the experienced judgment of seasoned purchasing professionals. Although experience is useful in supply risk analysis, it is the position of this chapter that the process can be structured for improved effectiveness. Future academic effort should concentrate on refinement of the supply risk process. This chapter proposes a model for supply risk analysis.

Multidimensional Model for Supply Risk Analysis

A model for supply risk analysis should identify the interaction of *critical dimensions* related to the specific purchase situation. Common dimensions that influence general procurement situations include:

- Volume (quantity) to be purchased.
- Expected forecast error for volume.
- Financial value of the purchase.
- Capital intensity of the manufacturing process involved.
- Number of operations required in the manufacturing process.
- Lead time from order placement to initial production.
- Level of technology involved.
- Structure of the supplier industry (many competitors, oligopoly, or monopoly).
- Structure of the buyer's industry (many buyers, oligopsony, or monopsony).

These nine dimensions of risk are not inclusive, but represent an effective starting point for supply risk analysis. The buyer must review each unique procurement situation for peculiar elements of risk. *Moreover, no single dimension alone is necessarily "high risk." It is the intersection of several dimensions that may portend a high-risk situation.*

Consider a buyer contemplating the purchase of an engine oil filter. Suppose the predicted volume (demand) for the filter is 10,000 units/day, which is double the rate required by any other type of filter currently supplied by the industry. Further, suppose the quoted price of the filter is $25/unit. No conclusion about supply risk can be established at this point.

Now consider the alternative scenarios defined in Exhibit 11.10.6. In each case (A, B, and C) a different description of the general risk dimensions results in a different supply risk outcome.

Case A: A Low-Risk Scenario

In case A (Exhibit 11.10.6) a single, powerful buyer (monopsony) confronts a supplier market consisting of many competitors. The technology required to manufacture the oil filter is lower than the current state of the art. Moreover, the lead time necessary to bring a new supplier into production is a nominal 4 weeks. These conditions, coupled with low expected forecast error, result in the conclusion that supply risk in case A is low.

Case B: A High-Risk Scenario

Case B (exhibit 11.10.6), on the other hand, requires one of many buyers to compete for the capacity of a limited supplier industry composed of only a few powerful suppliers (oligopoly).

Exhibit 11.10.6 Oil Filter Supply Risk Analysis Scenarios

General Risk Dimensions	Scenario A	Scenario B	Scenario C
Volume (estimated)	10,000/day	10,000/day	10,000/day
Financial value	$25/unit	$25/unit	$25/unit
Capital intensity	Low	High	High
Expected forecast error	Low	High	High
Processing operations	6	10	31
Lead time for new suppliers	4 weeks	40 weeks	64 weeks
Level of technology	Proven—below state of the art	State of the art	Innovative, totally new technology
Supplier structure	Many competitors	Oligopoly	Monopoly
Buyer structure	Monopsony	Many customers	Oligopsony
Risk outcome	Low risk	High risk	Moderate risk

In addition, the technology required to produce the part is at the current state of the art, and the expected error of the buyer's volume forecast is high. This demand uncertainty, in concert with the extended lead time of 40 weeks, increases the supply risk inherent in scenario B.

Case C: A Complex Scenario

Case C (Exhibit 11.10.6) involves several elements that might imply excessive supply risk. The level of technology is innovative, that is, beyond the state of the art. The lead time is great, 64 weeks. The processing required is complex (64 operations), and the buyer's expected demand forecast error is high. However, this scenario involves an influential buyer from an oligopsony (an industry with *few* buyers) and a single powerful supplier (monopoly). In this circumstance the supplier may face the threat of vertical integration by the buyer if the former's performance fails to satisfy requirements. Furthermore, the buyer industry, consisting of few firms, will focus special attention on a capital-intensive, technological innovation of major proportions. It should be noted that the risk dimension profile outlined in scenario C would result in extremely high risk with another buyer and supplier structure.

Summary

This multidimensional model of supply risk analysis is intended to promote a basic understanding. Obviously, supply risk analysis is a qualitative process based upon the professional judgment of the buyer. It is precisely this element of judgment that makes procurement an extremely creative and challenging field.

Make Versus Buy Analysis

Make or buy analysis considers whether the purchase requirements should be manufactured in-house or purchased from outside suppliers. This decision involves other key functional areas: finance, engineering, manufacturing management, business planning, and general management. Representatives from each area should meet regularly to review future requirements with purchasing management and to evaluate the advantages and disadvantages of making or buying. Sound make or buy decisions consider economic and business variables.

Economic Variables

Outside purchase prices should be compared with internal costs. It is essential that total cost factors be reviewed. For example, a make or buy analysis of steel stamping operations should consider the loss of scrap sale revenue, idled facilities, or reduction of work force if the buy alternative is selected. When substantial investment is involved, the NPV (net present value)–discounted cash flow analysis–of each make and buy alternative must be evaluated. If the loss of in-house business to an outside supplier would seriously burden internal operations, outside prices may be compared to internal variable costs, excluding fixed burden rates.

Business Variables

Compatibility of make alternatives with the firm's vertical integration strategy is vital. A poor fit with existing management capabilities should overrule apparently favorable economics. The long- and short-range impact upon the supplier community must be assessed. Cannibalizing business previously supplied outside may have serious ethical consequences. Conversely, purchasing may aggressively recommend the make option if excessive supply risk is inherent to the buy alternatives.

The make versus buy activity should include frequent review of existing operations (both purchased and made). Effective make or buy analysis will contribute to material cost control and to the solution of long-range supply problems.

Number of Suppliers Required

After the decision to purchase has been made, the buyer must determine how many suppliers are required to satisfy a specific purchase. In high supply risk situations, a multiple-source strategy will diversify and reduce risk. Intuitively, the more suppliers on contract for a requirement, the lower the supply risk. However, efforts to minimize supply risk must be balanced against the economic aspects of the sourcing strategy. For example, multiple sourcing may require substantial capital investment to be paid by the buyer. In addition, the buyer's administrative costs increase proportionately to the number of suppliers contracted. A multisource strategy also often requires the buyer to pay a unit cost penalty above the bid of the lowest bidder. The number of suppliers must be decided within the context of a trade-off between purchase economics and supply risk.

Form of Purchase Agreement

The purchasing department must also decide the form of purchase agreement to be used for each buying situation. This section briefly discusses the common forms and their impact on purchase economics and supply risk.

Requirements Contract

Requirements contracts are most often used for purchase of production components and direct materials. The supplier is given a commitment for a specified period (see next section on duration of commitment), with a firm price and designated delivery terms for a percentage share of the buyer's requirement. *Requirements contracts do not specify a fixed quantity to be purchased.* The buyer issues periodic delivery schedules for specific quantities during the term of the contract. Using this form, the buyer may transfer the cost of inventory investment to the seller. Therefore the requirements contract is suited to procurement situations where the buyer's volume demand is uncertain. Requirements contracts complement production control systems based on MRP (material requirement planning) principles.

Blanket Order

Blanket orders are analogous to requirements contracts, but are used for the purchase of repeatedly consumed nonproduct (indirect material) goods and services.

Spot Purchase Orders

Spot purchase orders provide the supplier with a commitment, specifying the quantity, delivery date, and technical specifications for an item or group of items. Spot orders are used routinely for capital equipment and nonrepetitive indirect goods and services.

Letters of Intent

Letters of intent are a promise to eventually issue a contract or purchase order if specific conditions are met by the supplier. Letters of intent may be required to reserve supplier capacity for long lead time requirements, pending completion of final specifications and delivery requirements. Letters of intent should be the exception, not the rule, in normal purchasing practice. Letters of intent should be carefully controlled by top purchasing mangement.

Forward Commitments

Forward commitments differ from letters of intent because the buyer's quantity, delivery, and specification requirements are completely defined. Forward commitments may be required in cases where the seller needs lead time beyond the buyer's normal commitment lead time for raw material, production processing, and the like. For example, a buyer may issue a requirements contract for a 1 year period and issue delivery schedules 12 weeks in advance of the delivery deadline, with specified quantities. The seller may actually need 24 week delivery commitments to allow necessary procurement of special raw materials. Forward commitments should also be carefully controlled by top purchasing management.

Duration of Commitment

Purchase agreements for repetitively consumed items (requirements contracts or blanket orders) should not exceed 1 year. An annual review of sourcing strategy will allow the buyer to review the competitive pricing and take advantage of favorable changes in the marketplace. Unique aspects of the buyer's or supplier's industry may influence the duration of commitment. For example, contracts for original equipment manufacturers in the automobile industry usually coincide with the annual model year, traditional in that industry.

Long-term contracts should be considered only when supply risk dictates or when extremely favorable economics are involved. In rare circumstances it may be desirable to establish a purchase commitment for more than 1 year. Caution must be exercised to avoid inflexible terms and conditions that may "lock" the buyer into an unfavorable position. Conversely, long-term commitments may provide an opportunity to reduce purchase prices in conjunction with a learning curve function. Also, multiyear agreements may be required for purchases involving extraordinary investment by the seller. Multiyear agreements should be reviewed carefully by top purchasing management.

11.10.4 PURCHASE COST CONTROL

This section discusses the key policies that affect purchasing's ability to control costs. No other aspect of the purchasing function influences the profit and loss statement of the firm more than cost control policies.

Payment Policies

Purchasing must strive to establish favorable payment policies for all commitments. The elements of a sound payment policy include (1) fair consideration of buyer's and seller's cash flow requirements and (2) incentive for the supplier to satisfy the full requirements of the purchase agreement. In sum, payment should be made *only* after the seller has satisfied the purchase agreement. Some policies that violate these principles are progress payments, advance payments, and time and material contracts.

Progress Payments

Suppliers may request a progression of payments that does not coincide with delivery. Such requests are sometimes connected with orders requiring extraordinary investment by the supplier, such as in capital equipment. In this situation the supplier may indeed feel a strain on cash flow. However, the buyer must evaluate his or her own cash flow circumstance and negotiate payment consistent with supplier delivery. Progress payments should always be avoided.

Advance Payments

Requests for payment in advance of delivery must be strictly rejected. Payment without delivery may result in forfeited payment if the goods are never received or in a lack of satisfactory performance if they are received.

Time and Material Contracts

In time and material contracts, the seller requests payment as he or she invests in labor and material to satisfy the buyer's requirement. In a construction contract, for example, the seller may be required to finance foundation and structural work many months before the overall contract is completed. However, full payment should be rendered only when the job is complete and when all of the buyer's requirements included in the purchase agreement have been satisfied. It is customary for the buyer to retain a share of each invoice amount submitted by the contractor until the purchase agreement has been satisfied.

Pricing Policies

The most favorable pricing policy possible under the procurement circumstances should be established. Typically prices should be established through competitive quotations and should be firm for the duration of the commitment. Exceptions to these guidelines should be managed within the context of fair and equitable dealings of the highest integrity. The most common forms of price policy deviations are cost-plus pricing, cost escalation agreements, and target pricing.

Cost-Plus Pricing

Suppliers may attempt to establish final prices on the basis of a cost-plus formula. This approach has several major disadvantages that must be avoided. First, the buyer will not know the final price prior to completion of the purchase agreement. Moreover, the supplier will have no real incentive to control costs since he or she is guaranteed a profit margin in the cost-plus formula. For these reasons, cost-plus pricing is rarely desirable for normal purchase agreements. Cost-plus pricing provides the seller with a blank check.

Cost Escalation Agreements

In the face of raw material cost increases that are expected to occur during the term of commitment, the supplier may propose an escalation formula. For example, the supplier may insist that the price be adjusted to coincide with price changes in raw material, labor, energy, and so on. Again, this type of policy reduces the supplier's incentive to increase productivity and control costs. Unfortunately, during periods of rampant inflation, it may be both fair and ethical to grant such agreements. Under these circumstances, interim price adjustments should be carefully *nego-*

tiated by the buyer to ensure that the amount of price increase accurately reflects the supplier's actual cost increase *only*. There should be no adjustment for overhead or profit.

Target Pricing

Buyers may attempt to evaluate the processing costs of a particular requirement and to establish a target price without supplier input. Although some suppliers may actually agree to provide requirements at the target price, this method may lead to serious procurement problems. First, the target price may not reflect competitive pressures in the supplier's industry, and the buyer will overprice the requirement. Conversely, the target price may be unrealistically low and actually force the supplier into financial difficulty.

Value Engineering

Purchase costs should be constantly reviewed to identify ways to reduce supplier costs and prices. Value engineering is a system of analyzing the functional characteristics of purchase requirements in an effort to determine alternate designs and materials for potential cost reduction.

Suppliers should be strongly encouraged to conduct their own independent reviews of the buyer's requirements. The buyer must be prepared to recognize supplier contributions to cost reduction through an increased share of the business or more frequent opportunity to bid competitively for the buyer's future requirements. The buyer and supplier are partners in the VE process.

Measuring Purchasing Performance

An optimum system of performance measurement for purchasing would comprehensively account for the qualitative and quantitative criteria discussed in this chapter. Obviously, such a comprehensive system could not omit cost control. Purchasing should be the focal point of responsibility for the direct and indirect material budgets of the firm.

Budget formulation should be based upon purchasing input, and comparison of actual cost versus budget should be reported to purchasing frequently.

Responsibility for cost control performance must be delegated to the *buying level* of the organization. Individual buying units in the purchasing organization should be held accountable for their specific procurement budget. Feedback concerning actual cost performance versus budgeted levels should be communicated not only to purchasing management, but to the individual buying units as well.

11.10.5 CURRENT TRENDS IN PROCUREMENT

Several recent trends in the purchasing field deserve brief review.

Materials Management

Many firms are reorganizing their procurement and logistics functions to integrate purchasing with traffic, production control, material scheduling, and inventory management. This concept is based upon the premise that increased interaction and cooperation among these supply organizations will improve performance. Increased career mobility and functional cross-training for personnel are also key advantages of the concept.

Centralized and Coordinated Purchases

Firms with multiple facilities in various geographic locations are also developing centralized or coordinated purchasing programs. For example, a company with five plant locations may have five individual buyers for the same basic indirect material commodity. The potential for common ordering at the combined volume level for price reduction is obvious. Moreover, personnel requirements and administrative costs may also be reduced if procurement responsibility can be combined at a single location.

The decision to centralize should be predicated on the opportunity to purchase *commonly used articles* in *greater volume* or to *standardize* similar articles for common use at all locations at reduced cost, with fewer people.

International Purchasing

Transnational firms may face an interesting set of constraints that dictate international purchasing transactions. Many countries require counterpurchasing agreements, which stipulate the transnational's imports into the foreign country as a function of its exports (purchases) from the local economy.

As the range of transportation systems increases, opportunities for worldwide sourcing at favorable costs will increase dramatically.

11.10.6 SUMMARY

This chapter has presented a comprehensive view of a rapidly growing and increasingly important business profession–purchasing. Control of purchase commitments within a single organizational entity promotes effective cost control and maintenance of ethical supplier relationships.

The purchasing process demands professional buyers who are skilled business specialists capable of identifying and analyzing potential suppliers in economic and qualitative terms. Technical knowledge of purchasing fundamentals is essential to effective procurement systems.

Supply risk analysis, along with financial evaluation, is the key to development of optimal sourcing strategies. Supply risk is a function of several dimensions of the procurement situation. These risk dimensions can be systematically modeled to facilitate analysis.

The focus of purchasing is cost control. It is purchasing's job to establish and manage the firm's direct and indirect material budgets. Unfavorable payment and pricing policies should be resisted by the buyer. Value engineering should be an ongoing search for cost reduction, with active supplier participation.

Traditional purchasing concepts of procedure and organization are changing to meet the challenge of a dynamic business environment. Effective procurement is a strategic force in any business enterprise.

BIBLIOGRAPHY

BRANSON, W. H., *Macroeconomic Theory and Policy*, Harper & Row, New York, 1972.

COREY , E. R., "Should Companies Centralize Procurement?," *Harvard Business Review*, November–December 1978, pp. 102–110.

D'ARCY, A. J., "Planning for Buying," *Journal of Purchasing*, Vol. 7, No. 3 (August 1971), pp. 24–32.

ELLIOTT, B., "At GM: A New Mark of Excellence," *Purchasing World*, August 1976.

ELLIOTT, B., "Detroit–A Purchasing Composite," *Purchasing World*, December 1976, p. 32.

HUMPHREYS, J., "New Concepts in Purchasing," *Michigan Purchasing Management*, May 1976, pp. 5, 18.

KARASS, C. L., "Effective Negotiating," *Purchasing World*, April 1979, p. 22.

KING A. S., "A Systematic Approach for the Analysis and Evaluation of Purchasing Policy," *Journal of Purchasing*, Vol. 9, No. 3 (August 1973), pp. 73–82.

LUSK, H. F., C. M. HEWITT, J. D. DONNELL, and A. J. BARVES, *Business Law: Principles and Cases*, Irwin, Homewood, IL, 1970.

MEITZ, A. A., and B. B. CASTLEMAN, "How to Cope With Supply Shortages," *Harvard Business Review*, January–February 1975, pp. 91–96.

MILLER, J. G., and P. GILMOUR, "Materials Managers: "Who needs Them?," *Harvard Business Review*, July–August 1979, pp. 143–153.

SHETH, J., "A Model of Industrial Buyer Behavior," *Journal of Marketing*, Vol. 37, No. 4 (October 1973), pp. 50–56.

WESTING, J. H., I. V. FINE, G. J. ZENZ, and MEMBERS OF THE MILWAUKEE ASSOCIATION OF PURCHASING MANAGERS, *Purchasing Management*, 3rd. ed., Wiley, New York, 1969.

CHAPTER 11.11

Distribution and Logistics

JOHN J. JARVIS
H. DONALD RATLIFF

Georgia Institute of Technology

11.11.1 INTRODUCTION

Distribution and logistics functions are concerned with the storage and transportation of people, goods, and services. Although the term "logistics" has often been given a military connotation (i.e., the transportation, quartering, and supply of troops), we use distribution and logistics interchangeably.

To simplify discussion, the quantities being moved or stored are referred to as "items." "Supply point" denotes any point from which items are shipped, and "demand point" denotes any point to which items are shipped. A facility such as a warehouse may be a demand point with respect to a manufacturing plant and a supply point with respect to the customers it supports.

11.11.2 SYSTEM FUNCTIONS AND PARAMETERS

Analysis of distribution systems can be divided into three major components: (1) facility location and design, (2) allocation of supply and demand points to facilities, and (3) transportation between supply and demand points. Clearly these components are not independent of one another. For example, a major consideration in locating and designing a facility is the cost of transportation to and from the facility. This, in turn, depends on which supply points are to serve the facility and which demand points are to be served by the facility. The allocation of supply and demand points is influenced by the location and design of the facility.

Within these components is a variety of planning and control functions. Since most of these functions are covered in other chapters of Section 11, and facility design is covered in Section 10, this chapter concentrates on analysis of the allocation and transportation components. However, the reader should also become familiar with Sections 10 and 11 before designing or analyzing distribution systems.

In allocating demand points to supply points and configuring the transportation system, the objective is to provide an acceptable level of service at minimum cost. Assuming that the facilities are already located and sized, the relevant costs include the purchase or manufacturing costs, the storage costs, the delay costs associated with getting the item from the supply point to the demand point, and the transportation costs. Since cost estimation is covered in Chapter 9.2, this chapter considers only those elements of these costs that are particularly important to decisions regarding the allocation and transportation subsystems.

The specification of an acceptable level of service implies the imposition of certain constraints on the system. These constraints may be the result of outside influences, such as government regulations, or of physical characteristics of the system, such as building size. However, they are usually imposed as an alternative to estimating hard-to-quantify costs, such as the cost of "lost opportunity." Hence it will be convenient for us to discuss costs and constraints simultaneously in considering the various system parameters.

Demand Variations

The changes in demand over time may be seasonal, in which case the demand is reasonably stable during each "season," or they may vary from day to day or week to week.

Seasonal changes may require some changes in scheduling and routing from one season to the next, but their major impact is on the capacity planning for the distribution system. If the capacity (e.g., warehouse size, delivery fleet, labor force) is to be the same for all seasons, then a trade-off must be made between the cost due to low utilization during off-seasons and the cost of not being able to satisfy demand during the peak season. In general, the space component of capacity is more difficult to vary than either the fleet or labor components. Hence permanent space is more likely to be based on peak demand, whereas leased vehicles and overtime may be used to augment the permanent fleet and labor force. If the capacity can be varied from season to season, then the fundamental problem is to determine how much permanent capacity to maintain. Since we would expect the additional capacity to be more expensive per unit than the permanent capacity, it may be cost-effective to have the permanent capacity exceed the off-season demand.

The day-to-day or short-term variations in demand have a major impact on routing considerations for situations where vehicles make more than one stop with a load. If demand is stable, then it is often possible to use "fixed" delivery routes where a vehicle makes the same stops along the route. If, on the other hand, the amount of vehicle capacity required to service a delivery point varies from one delivery to the next, either the routes must be altered on a day-to-day basis or we must accept a lower utilization of vehicle capacity.

These day-to-day demand variations also may have a significant impact on the required buffer inventory at each supply point. Buffer inventory levels are frequently based on mean demand plus some multiple of the SD of demand. As a result, when customer demands are reasonably independent, the required buffer inventory generally increases as the inventory is dispersed among more supply points. As an illustration, suppose we have two demand points, each with a mean demand of 20 and an SD of 5. Suppose also that we require a buffer stock equal to the mean demand plus 3 SDs. If we supply the demand points from two different supply points, the buffer inventory required is $20 + 3(5) = 35$ at each supply point. However, if we supply both from the same point, the combined demand distribution has an SD equal to $(5^2 + 5^2)^{1/2} = 7$. Hence the total buffer stock is $40 + 3(7) = 61$ units if both points are served from the same supply point versus 70 units if they are served from separate points.

This phenomenon is a result of the fact that the demands are independent. If demands are independent, we would expect that on a given day we would have some customers with a higher-than-average demand and others with a lower-than-average demand. Hence we would not generally have to stock enough buffer inventory to satisfy the maximum demand of each customer if the customers are supplied from the same inventory location.

On the other hand, if demands are very dependent (i.e., when one customer's demand increases, all other customer demands increase) as is often the case with seasonal demand variations, there is little advantage in terms of buffer inventory associated with having fewer inventory locations. Hence there are some inventory level advantages associated with having fewer locations from which customers are supplied, but only if customer demands are reasonably independent.

Source Constraints

It is common practice to put some restrictions on the quantities that are to be provided from different supply points. These restrictions are typically stated in terms of the fraction of demand to be supplied by a given supply point. The intent of most such restrictions is to limit the number of suppliers. The extreme case is what is called "single sourcing," where each demand point must be served by only one supply point. Although there are costs associated with dealing with each supplier (e.g., paperwork, time to develop good working relationships, specification of points of responsibility), these costs are very difficult to quantify. What we would like to do is balance these costs against the increases that will occur in purchase and transportation costs as a result of using fewer sources. However, as a result of the difficulties in assessing all of these costs, decisions determining the sourcing constraints are generally based on "management judgment."

In a few situations, particularly in purchasing for military use, there is the desire to use more, rather than fewer, suppliers. This is motivated by the need to maintain adequate lines of supply in case of an emergency need. Here the constraint is often posed in terms of the maximum fraction of demand that can come from a given supplier.

Delay Cost

The cost associated with delay in moving the item from the supply point to the demand point depends very much on the nature of the item. For items such as military ammunition, computer spare parts, or fresh blood, the cost can be very high. For other, less critical items, the cost of delay is insignificant, provided that the delay is within reasonable limits. Some of these costs, such as the cost of lost production or the cost to rent the item until it can be supplied, can be estimated in a straightforward fashion. Others, such as the cost of lost goodwill or the cost of death or injury, are almost impossible to assess reasonably. For this latter case, we often make a decision as to the maximum acceptable delay and then constrain the delay to be within that limit.

Route Considerations

The transportation cost varies appreciably, depending on whether shipment is in large lots, such as truck or boxcar loads, or small lots, in which case the vehicle must make many stops to disperse a load. The former most often occurs at the "trunking" level, where the transportation is from a manufacturing facility to a warehouse or distribution center. The latter occurs more frequently at the "local delivery" level, where delivery is made from the warehouse or distribution center to individual customers.

A restriction that delivery must be made during some specified time frame is common for many local delivery problems. This is a particularly important consideration if the time frames are small and the vehicle is making multiple stops. If a vehicle fails to get to a delivery point during its time frame, then the vehicle typically either has to wait at the location until the next time frame or has to make a second stop at the location. In either case the delivery cost can be increased substantially.

Usually we would like the elapsed time for a route or route segment to fall within a specified range (e.g., not more than 8 hr). In some cases failure to stay within the range results in increased cost, such as overtime and lodging expenses. For other cases where the range is set by union contract or government regulation, the consequences resulting from exceeding the range are much more serious. In such cases we would generally treat the elapsed time range as a constraint on the system rather than try to assess the cost.

There are at least two measures of vehicle capacity: volume and weight. In some situations only one measure will be critical, whereas in other instances we have to be concerned about both. A supplier of automobile parts may have weight as the critical factor in the winter when supplying large quantities of batteries, and volume as the critical factor in the summer when supplying large quantities of tires.

Capacity is more difficult to deal with when items and/or deliveries cannot be divided into small quantities. For example, suppose that we have a number of deliveries that cannot be split, and each requires approximately one half of the vehicle capacity. Then we must take care in matching the deliveries for each route so that the capacities are not exceeded and the trip cost is not excessive. Capacity is also more difficult to handle when it is not the same for all of the vehicles being used. In this case the routes must be vehicle specific in order to ensure that the capacities are not exceeded. If the vehicles all have the same capacity, then the routes can all be generated first and then assigned to the vehicles.

Loading and unloading restrictions depend on the type of equipment being used. The most common restriction is that the last item loaded must be the first item unloaded. If the items are all identical, then this restriction causes no problem. However, if the items are such that they are grouped on the vehicle by delivery point, this can be a major cause of concern. In this case, once the vehicle is loaded, it is frequently committed to a particular sequence of stops. If for any reason there is a deviation from this sequence (e.g., if a delivery point cannot take delivery), then there may be a major disruption in the delivery schedule for the vehicle.

For situations where both pickups and deliveries occur on the same route, care must be taken to ensure that the vehicle capacity is not exceeded and that picking up items at one point does not prevent the delivery of items at another. This problem is greatly simplified if we require the vehicle to make all deliveries before beginning pickups. This may result in substantial increases in cost.

The supply profile has a major impact on routes when delivery is made directly from production rather than from buffer inventory. In delivering newspapers, the system frequently is designed so that papers are loaded directly from the presses into the delivery trucks. This requires a truck at the loading dock while the presses are running. The truck not only must wait for production to occur before beginning a route, but also must return in time to receive additional production.

There is a variety of situations where we desire some measure of the work content of the vehicle drivers to be reasonably close for each route. If the driver must normally unload or unpack the items in the delivery, we may desire the unpacking time to be the same for all routes. If the driver writes orders and works on a commission, we may want the anticipated commission from each route to be the same. We may also want the expected duration of each route to be the same. All of the conditions are the result of a need for "fairness" in the distribution system design.

11.11.3 SYSTEMS MODELING CONCEPTS

To obtain maximum utilization of distribution system models, it is imperative that one have a clear understanding at the outset of the uses for the models, the usefulness of the models (themselves), the tractability of the models (i.e., their ability to yield solutions), and the capability to generate alternatives. This section briefly describes several important concepts that should be considered in selecting appropriate systems concepts.

Planning Versus Operations

In developing distribution systems models, one of the immediate questions that must be answered is, "What role will these models play?" Two important uses for systems models are planning and operation.

"Planning" connotes all of the analysis and design studies undertaken prior to system implementation. In distribution planning this would include (1) location and sizing of facilities, (2) fleet selection and sizing, (3) analysis of production and/or inventory capabilities and needs, and (4) financing requirements. Planning also includes an understanding of how the system is expected to operate after system implementation and/or modification, although not in as much detail as required during actual operation. Planning usually occurs over a longer time span, often months or years.

Quantities such as demand, costs, supply, and time are not as well known in the planning phase. However, this is not meant to infer that planning models are less complex than operating models. Because planning models attempt to measure total system effects and interactions, they tend to be more complex and larger in scope. Also, since more time is available for planning, models that take more time to analyze—both computer and analyst time—can and usually are employed. Depending on distribution system size, it is not unusual to find planning models that take tens or even hundreds of hours of computer time to run. At $1000 an hour, we can afford to spend a lot of computer hours to obtain even a 1% improvement in a $100 million distribution system.

Operation models for distribution systems are usually quite different. Operation often requires immediate decisions. Little time is available for generating and testing alternatives. As a result, operating models tend to be limited in scope, to be less sophisticated in methodology employed, and to take less time to analyze. They often involve policies and rules of thumb developed through experience or planning. Operating models invariably require greater detail of the specific system function involved, for example, knowledge of individual streets or hours of operation of individual customers.

If more care would be taken during the planning phase, we might anticipate, analyze, and suggest solutions for many of the operating problems that arise. This could save firms thousands or even millions of dollars in poor decisions.

Planning does not end when operation begins. Planning should continue throughout the life of any distribution system. Demand patterns change, vehicle capabilities change, and so on. New opportunities and alternatives become available, and system modifications should be made to take advantage of these opportunities.

Two other terms often used in system evaluation are "strategic" decisions versus "tactical" decisions. Strategic decisions are akin to planning; they involve that system implementation and/or modification that occurs over a longer time horizon and that may require facilities or vehicles. Tactical decisions involve immediate decisions (e.g., vehicle scheduling, routing, and loading or the determination of production overtime requirements).

Prior to any distribution system analysis and/or design effort, one should always ask the fundamental question, "How much time and money do we (practically) have to make the decision(s)?"

The Roles of Simulation and Optimization

Optimization models are effective in implicitly generating and evaluating large numbers of alternative system configurations. Such models are useful in developing candidate system configurations for further analysis and testing. However, optimization models tend to be deterministic in structure and thus cannot capture many of the system effects that are due to variability in operation.

Simulation models are excellent for analyzing the interaction effects due to system variability. Such models can identify queuing effects, bottleneck conditions, and delays. All of these phenomena result in higher system operating cost and usually necessitate the design of greater system capability (and higher capital cost). Unfortunately, simulation models cannot generate their own alternatives for evaluation. They must be provided with alternatives by the analyst.

As a result of the capabilities of each kind of model, both optimization and simulation models are useful in the design and analysis of distribution systems. Optimization models are best utilized as "first cut" models, to generate several promising candidate system configurations for further evaluation. These candidates can then be subjected to simulation analyses to test their capabilities under variability. It may be possible to iterate between the optimization and simulation models several times in order to "tune" the final system configuration.

Model Aggregation and Decomposition

In systems modeling and analysis, we must continually trade off *model detail* with *analysis capability*. The greater the detail of the model, the more accurate the model representation of the distribution system. Unfortunately, with greater model detail also comes greater model complexity and increased analysis effort (often to the point of intractability). We must be careful to select the appropriate level of detail for the distribution system models.

There are two important methods of achieving reasonable detail without overburdening the analysis function. One method is through model aggregation, whereas the other encompasses model decomposition. Each method is useful in its own way.

Aggregation can be employed to reduce model detail. Here we attempt to combine quantities with similar characteristics into a single quantity. For example, we are usually forced, because of sheer numbers, to combine individual customer locations into areas or zones of demand, with total customer demand for an area (zone) assigned to its centroid. Other examples of aggregation include treating time as if it were a single period, combining similar vehicles into a single vehicle type, and aggregating inventoried items into a few major categories.*

Another approach to achieving reasonable model analysis effort is to employ system decomposition. In this circumstance we attempt to separate (decouple) the overall system model into a series of subsystem models. Distribution systems models are commonly decomposed into location models, allocation models, and vehicle routing and scheduling models. Each of these models is analyzed independently, or else the analyst iterates among them to achieve a reasonable system configuration. The next section discusses these and other subsystem models in greater detail.

Alternatives Generation and Evaluation

Systems modeling, analysis, and synthesis does not usually account for the total evaluation and final selection of the desired distribution system configuration. In addition to the constraints and considerations represented by the model, there are often nonquantifiable constraints and socioeconomic and political considerations that are not incorporated into the original model, yet are extremely important to the final selection. Therefore it is important for the analysis-synthesis methodology to facilitate the generation of desirable alternatives to the "best" indicated system configuration resulting from the system model. This group of generated alternatives can be subjected to further analysis based on these other considerations, resulting in a final solution.

One commonly used technique for generating reasonable alternatives or evaluating the alternatives under study is sensitivity analysis. In a sensitivity analysis we would vary the "important" parameters of the system over some reasonable range and examine the effects on the various output measures (total cost, manpower, etc.).

11.11.4 CUSTOMER ASSIGNMENT MODELS

This and the next section describe some of the methodology used in distribution analysis. Because of similarity of structure, ease of understanding, and convenience of exposition, the methodology has been divided into two major areas of consideration: customer assignment models and delivery models.

The function of the customer assignment models is to organize the customer locations into logical subgroups that will be served by a single facility or vehicle. By relaxing or imposing various assumptions, we are able to generate a variety of models that incorporate a wide range of problem applications.

Constraints and Objective

The most basic question underlying customer assignment models is, "Which facilities (vehicles) will serve which customers?" To begin answering this question, we must clearly understand what constraints exist on the assignment results and what objective is sought.

There are a host of constraints that may exist with regard to the assignment of customers. A fundamental constraint is limited facility (vehicle) capacity. Others include time limit considerations, commodity mixing considerations, customer preference considerations, and various routing feasibilities.

Although the overall objective in distribution modeling is usually the minimization of total cost, most distribution studies concentrate on the two major components–facility cost and delivery cost. Each of these costs may include fixed and variable components, depending on the particular application.

Customer Allocation Models Involving Fixed Supply Points

The most straightforward customer assignment model is one involving allocation to fixed supply points. In this model the locations of the supply points are given, and one must decide which demand points should be allocated to which supply points. Exhibit 11.11.1 illustrates this allocation process for 4 supply points (enclosed in boxes) and 10 demand points enclosed in circles. Several important assumptions accompany the model depicted in the exhibit. First, we have assumed that multiple sourcing is allowed. This occurs when a particular demand point's requests are met by more than one supply point. Second, without explicitly developing the vehicle routes, we must

*For one study, the U.S. Air Force aggregated approximately 1 million different parts into about 500 categories.

Exhibit 11.11.1 An Example of a Customer Allocation Problem

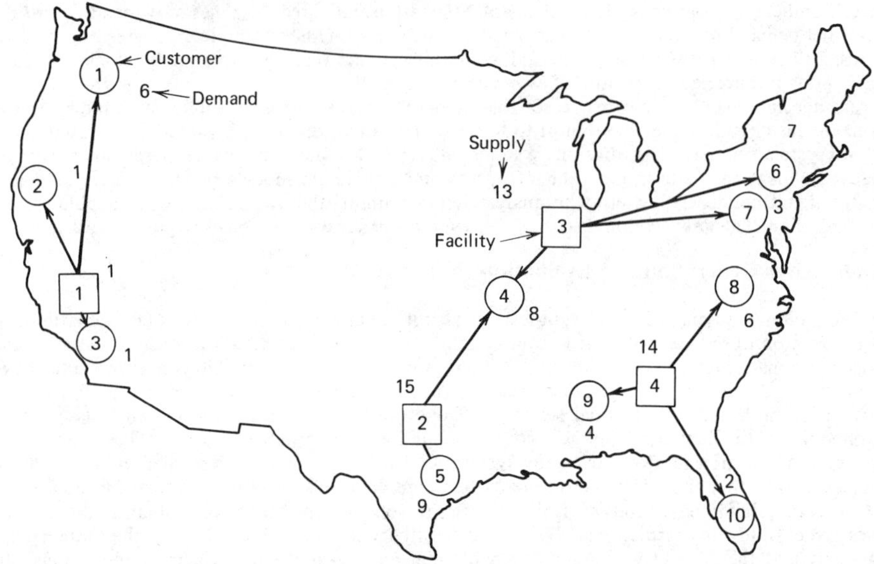

assume that some matrix of assignment costs is provided or that such costs are proportional to direct distance between supply point and demand point.

With the previous assumptions, we can develop a mathematical formulation for the fixed supply point assignment model as follows: Let c_{ij} be the cost of allocating one unit of demand point j's requirements to supply point i. Let x_{ij} be the amount of demand point j's requirements allocated to supply point i. Finally, let a_i be the supply point availability and b_j the demand point requirement. Then the model becomes

$$\text{minimize} \sum_i \sum_j c_{ij}x_{ij} \quad \text{(minimize allocation cost)} \tag{1}$$

$$\text{such that} \quad \sum_j x_{ij} \leqslant a_i \quad \text{(supply point availability constraints)} \tag{2}$$

$$\sum_i x_{ij} = b_j \quad \text{(demand point requirements)} \tag{3}$$

$$x_{ij} \geqslant 0 \tag{4}$$

This is an example of the classical transportation model. The solution of this model will most likely result in multiple sourcing. We can utilize the efficient transportation procedure to determine the best allocation. We may also include constraints on lower amounts of supply used at any supply point by employing network flow methods.

Next, suppose single sourcing is required. Then to the previous model we add the additional constraints

$$x_{ij} = 0 \text{ or } b_j \tag{5}$$

which transform the linear programming model into an integer programming model. This is a variant of the "generalized assignment problem" (see Ross and Soland[1]). In this case we most probably would combine a branch-and-bound technique with the classical transportation algorithm to achieve the desired allocation. Exhibit 11.11.2 illustrates a single-sourcing solution. As depicted in the exhibit, significant increases in distribution costs may result from additional constraint. Unfortunately, in many circumstances single sourcing is in fact a real requirement, usually as a result of customer considerations.

There will frequently be short-supply or no-stock situations, which make the single-sourcing assumption invalid. To anticipate and measure the effects of such situations, we could employ the transportation model as a "first cut" optimization model, and then a simulation model could be brought to bear to determine the effects of stock runouts, and so on.

Exhibit 11.11.2 A Solution Without Load Splitting

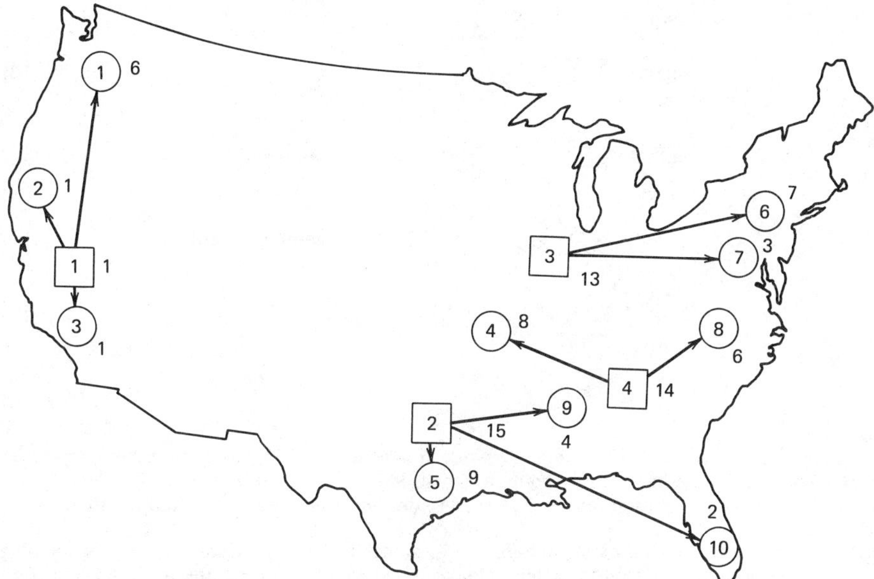

Multiple-Level Distribution Models Involving Fixed Facilities

Distribution systems most often involve multiple levels of goods transfer. Commodities are shipped from supply points to transfer or distribution points, from which they are dispersed to individual demand point locations. We can account for the additional level in the modeling process in one of several ways, depending on the specific assumptions involved.

If we assume that there are one set of costs associated with the transportation from supply points to distribution points and a separate set of costs from distribution points to demand points, then we can develop the following model:

$$\text{minimize} \quad \sum_i \sum_k s_{ik} x_{ik} + \sum_k \sum_j d_{kj} y_{kj} \tag{6}$$

$$\text{such that} \quad \sum_k x_{ik} \leqslant a_i \qquad \text{(supply point availability)} \tag{7}$$

$$\sum_i x_{ik} - \sum_j y_{kj} = 0 \quad \text{(flow balance through distribution points)} \tag{8}$$

$$\sum_i x_{ik} \leqslant f_k \qquad \text{(distribution point capacity)} \tag{9}$$

$$\sum_k y_{kj} = b_j \qquad \text{(demand point requirements)} \tag{10}$$

$$x_{ik} \geqslant 0, \qquad y_{kj} \geqslant 0 \tag{11}$$

where

s_{ik} = supply point–distribution point cost (and x_{ik} its associated flow)
d_{kj} = distribution point–demand point cost (and y_{kj} its associated flow)
f_k = distribution point capacity

The preceding model is directly solvable by simple and efficient network flow methods. By incorporating other considerations, we can generate more complex models. For example, suppose that we assume that transportation cost is a function of the entire route by which the item will be delivered. Also, suppose that multiple sourcing is permitted. In this case we can develop a straightforward extension to the model defined by equations 1 through 4 by adding an extra subscript to the allocation variable.

Specifically, let x_{ijk} be the amount of demand point j's requirement originating at supply point i and passing through distribution point k. Let c_{ijk} be the associated distribution cost and the transfer point capacity. The model then becomes

$$\text{minimize} \quad \sum_i \sum_j \sum_k c_{ijk} x_{ijk} \tag{12}$$

$$\text{such that} \quad \sum_j \sum_k x_{ijk} \leqslant a_i \quad \text{(supply point availability)} \tag{13}$$

$$\sum_i \sum_j x_{ijk} \leqslant f_k \quad \text{(distribution point capacity)} \tag{14}$$

$$\sum_i \sum_k x_{ijk} = b_j \quad \text{(demand point requirement)} \tag{15}$$

$$x_{ijk} \geqslant 0 \tag{16}$$

This model cannot be handled by simple network flow techniques. However, we can employ linear program methods, most likely using column generation to keep down its size. Exhibit 11.11.3 illustrates one possible solution for a multilevel distribution system with five intermediate distribution points.

When single-source constraints are required, the model becomes a mixed integer programming model. In such a case we would probably employ an enumeration procedure such as in Geoffrion and Graves[2] or Mairs et al.[3] Customers are first assigned to distribution points. Then a simple transportation problem is solved for the supply points and distribution points, with the demand point requirements allocated to the appropriate distribution points.

Thus far all of the models have been discussed in the context of the distribution of a single commodity. When multiple commodities are involved, the models become increasingly complex. In most cases the models can be updated by addition of another index (for commodity type) in the allocation variable (flow) and its associated cost. However, this usually means that the efficient solution techniques are no longer applicable, and one usually resorts to linear programming with column generation. Geoffrion and Graves[2] report very good success with a Bender's decomposition approach to the multicommodity problem, posed as a mixed integer program.

Exhibit 11.11.3 A Multilevel Distribution System

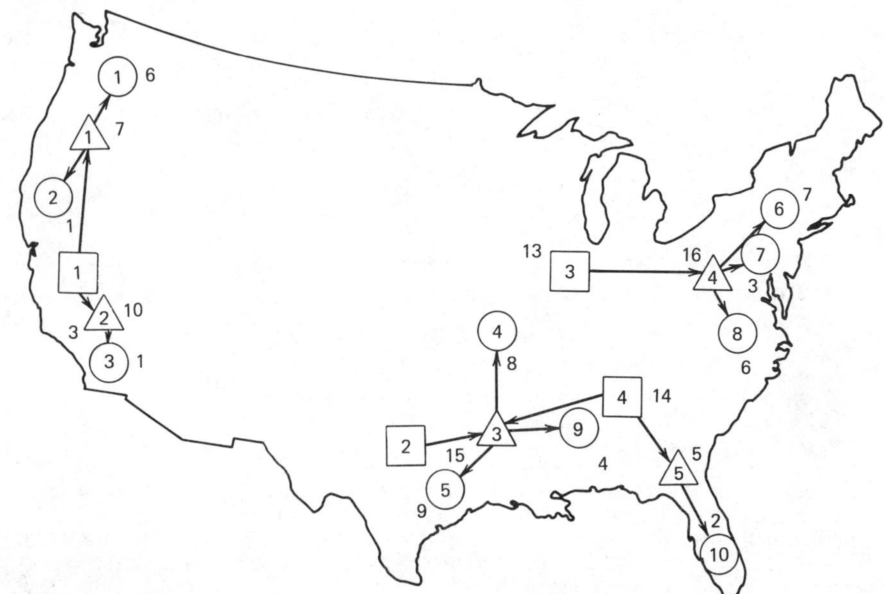

11.11.5 DELIVERY MODELS

By delivery models we mean those operational models that are concerned with the allocation of deliveries to vehicles (i.e., clustering) and the routing of the vehicles. There are three fundamental approaches to this problem: cluster first and then route, route first and then cluster, and cluster and route simultaneously. These approaches are all concerned with generating sets of feasible routes. The problem of determining an optimum set of feasible routes is a very difficult one. The class of covering and partitioning models seems the most attractive mathematical programming methodology to use in approaching this problem. This approach was first proposed for routing problems by Charnes and Miller[4] and for delivery problems by Balinski and Quandt.[5]

The essence of this approach is to represent each feasible route as a column in a matrix. The optimum set of columns that corresponds to the available vehicles is then selected. Obviously, for many delivery problems the number of feasible routes is much too large to actually enumerate and evaluate. However, the approach can still be effectively used by generating only a part of the possible routes and then choosing the best subset of those generated. If the routes are cleverly generated,[6] this approach tends to yield good, if not optimum, sets.

First the covering and partitioning models are discussed in more detail, followed by some of the approaches for generating vehicle routes. The latter can be used either in conjunction with the covering and partitioning models or as stand-alone heuristics.

Covering and Partitioning

The covering problem is an integer program defined by

$$\text{minimize} \sum_j c_j x_j \tag{17}$$

$$\text{such that} \sum_j a_{ij} x_j \geq 1 \qquad i = 1, \ldots, m \tag{18}$$

$$x_j = 0 \text{ or } 1 \tag{19}$$

where all of the a_{ij} are constants with the value of either 0 or 1. Since the $a_{ij} = 0$ or 1, and the x_j must be 0 or 1, then the constraints (equation 18) effectively specify that at least one of the variables x_j contained in constraint i must be 1. Partitioning models are similar to covering models, except that the constraints (equation 18) are required to be equalities.

Exact solutions to covering and partitioning problems require considerable computational effort.[2,7,8] However, because of their special structure, a number of heuristics are available that produce good solutions (e.g., Cullen et al.[6]). Covering and partitioning models permit us to determine the "best" simultaneous assignment of customers to facilities and to routes. This is because assignment routes implicitly fix assignment to facility.

To understand how the modeling process operates, consider the 4-facility, 10-customer problem given in Exhibit 11.11.4. The underlying travel network is indicated by the links. We shall, for convenience, assume that all of the customers require unit loads. If they do not, then we can split any such customer into several customers without loss in modeling power.

Associate with each customer a row of the covering problem; let each column represent a feasible route. The covering problem for Exhibit 11.11.4 has 10 rows (one for each customer) and a large number of columns (one for each feasible route). An example of the covering problem, with part of the routes enumerated, is given in Exhibit 11.11.5. Column x_j has a 1 in row i if route j will deliver to customer i. The cost of column j is the associated route cost.

To understand the covering model in Exhibit 11.11.5, consider column x_1. This column is scheduled to deliver only to customer 1, with a total route cost of 24 (the out-and-back distance from facility 1). Note that even though route x_1 went through the node associated with customer 2 (see Exhibit 11.11.4), no delivery was scheduled for that customer on this route. This could be a result of several considerations, one of which might be vehicle load limitations. Note that we also have defined a route, x_{11}, which delivers to customers 1 and 2 at the same cost as route x_1. Obviously, route x_{11} is preferable, and column x_{11} "dominates" column x_1. Techniques exist that identify undesirable columns as well as redundant rows in covering models. And, of course, any optimization scheme would never select column x_1 over x_{11}.

Variable $x_j = 1$ indicates that route x_j was selected in the final distribution solution. Variable $x_j = 0$ indicates that route x_j was not selected. A "cover" is the set of selected columns in the solution. An example of a cover (solution) for the problem in Exhibit 11.11.5 is $x_{20} = 1, x_{13} = 1, x_{15} = 1, x_9 = 1, x_{10} = 1$, with cost 111. This solution is also called a "partition," since each constraint of the covering model is satisfied as an equality.

Exhibit 11.11.4 An Example Problem for the Covering and Partitioning Formulation

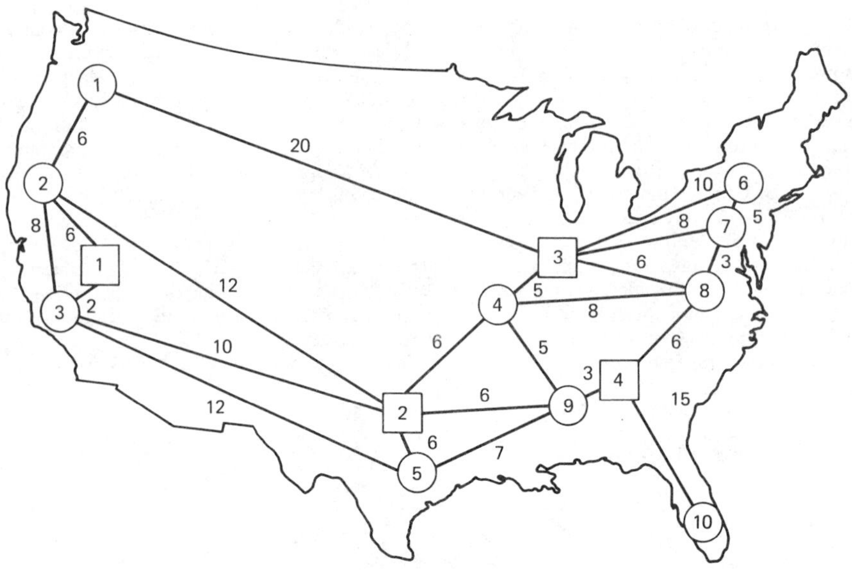

Cullen et al.[6] describe a method whereby prices, similar to dual variables in linear programming, can be defined for the rows of the covering model when a feasible solution (cover) is specified. Like dual variables, the covering prices satisfy three properties: (1) they are nonnegative; (2) they must be zero if the constraint is not exactly satisfied; and (3) when multiplied by a column of the cover, they must add up to the cost of the column. These prices can be applied to the columns of the covering model to determine whether good candidates exist to improve the covering solution. The procedure is similar to computing the shadow prices in linear programming. (See Chapter 14.2 for a discussion of shadow prices.) These prices can also be applied to the network in Exhibit 11.11.4 to identify good candidate columns to add to the covering model.

A variety of routing constraints can be handled by simply generating only those routes that satisfy the constraints. Also, the only restriction on the type of cost that can be handled is that one must be able to determine the cost once the route is generated. By adding additional constraints to the covering and partitioning model, we can include other considerations such as multiple vehicle types and facility capacities.

Cluster-First-and-Then-Route Approach

This approach is most appropriate when the vehicle capacity constraints are the major consideration. The idea is to first group or cluster the delivery points that are to be assigned to each vehicle and then to determine a route through the cluster. This approach is not generally applicable when there are binding route constraints such as time frame requirements since the routes are not generated until after the delivery points on the route are determined.

The basic idea in this approach is to first obtain a fairly gross estimate of the routing cost. This "pseudocost" is then used as a basis for clustering the points to satisfy vehicle capacity constraints. Two variations of this approach are presented here.

The first approach is a location-allocation approach.[9] We assume that travel cost is proportional to the Euclidean distance traveled.* Since we do not know the actual routing distance, we approximate it as follows: Assume that the vehicle will first travel from the warehouse to some cluster point and then to the delivery points, returning to the cluster point after each delivery. Finally, the vehicle returns from the cluster point to the warehouse. This idea is illustrated in Exhibit 11.11.6. We count twice the distance of each assigned customer to its cluster point plus the distance from the cluster point to the facility. This is an attempt to approximate the distance a vehicle might travel in servicing the customer group (cluster). Sometimes a multiplier, $0 < \lambda \leq 2$, is applied to each customer-to-cluster-point distance instead of twice the distance to reflect the fact that there will be some more efficient routing scheme that "out and back" within the cluster.

*The Euclidean (straight-line) distance between two points (x_1, y_1) and (x_2, y_2) is $d = [(x_1 - x_2)^2 + (y_1 - y_2)^2]^{1/2}$.

Exhibit 11.11.5 The Partial Covering Matrix for Exhibit 11.11.4

	x_1	x_2	x_3	x_4	x_5	x_6	x_7	x_8	x_9	x_{10}	x_{11}	x_{12}	x_{13}	x_{14}	x_{15}	x_{16}	x_{17}	x_{18}	x_{19}	x_{20}
1	1										1									1
2		1									1	1								1
3			1									1								1
4				1									1	1		1	1			
5					1								1						1	
6						1									1					
7							1								1			1		
8								1						1	1	1		1		
9									1				1	1			1		1	
10										1										
c_j	24	12	2	10	12	20	16	12	6	30	24	15	24	22	24	19	16	18	20	27

Exhibit 11.11.6 The Clustering Concept

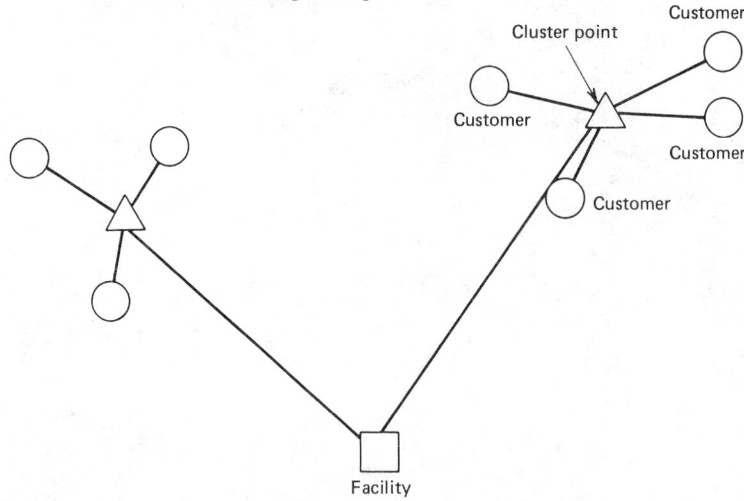

We can develop a mathematical model for this type of clustering. Let $(\bar{x}_0, \bar{y}_0)$ be the coordinates of the fixed facility location, $(\hat{x}_j, \hat{y}_j)$ the coordinates of the fixed demand points, and $(\bar{x}_i, \bar{y}_i)$ the coordinates of the unknown cluster points. Letting a_i be the capacity of vehicle i and b_j the requirement of customer j, the following model locates the cluster points and assigns delivery points to cluster points:

$$\text{minimize} \sum_i \sum_j \lambda [(\hat{x}_j - \bar{x}_i)^2 + (\hat{y}_j - \bar{y}_i)^2]^{1/2} x_{ij}$$

$$+ 2 \sum_i \sum_j [(\bar{x}_i - \bar{x}_0)^2 + (\bar{y}_i - \bar{y}_0)^2]^{1/2} x_{ij} \quad (20)$$

$$\text{such that} \sum_j x_{ij} \leqslant a_i \quad (21)$$

$$\sum_i x_{ij} = b_j \quad (22)$$

$$x_{ij} \geqslant 0 \quad (23)$$

where x_{ij} is the amount of demand point j's requirements supplied by supply point i.

It is not practical to develop optimal solutions to the preceding model. However, Cooper[9] has presented an iterative method for locating locally optimal solutions to the model. His procedure makes note of the fact that (1) if a set of m cluster points $(\bar{x}_i, \bar{y}_i)$, $i = 1, \ldots, m$, is fixed at some specified coordinates, then the preceding model is a simple transportation model, and (2) if x_{ij} is specified, then the preceding model is an unconstrained facility location model. The procedure is as follows:

STEP 0. Begin with any set of fixed cluster points $(\bar{x}_i, \bar{y}_i)$, $i = 1, \ldots, m$.

STEP 1. With the $(\bar{x}_i, \bar{y}_i)$ given, solve the resulting transportation problem for the values of x_{ij}.

STEP 2. With the specified values of x_{ij}, solve the resulting unconstrained facility location problem for a new set of cluster points $(\bar{x}_i, \bar{y}_i)$, $i = 1, \ldots, m$.*

STEP 3. Repeat steps 1 and 2 until some convergence criterion is satisfied (e.g., the same cluster points repeat).

By initiating the preceding model with different starting cluster point sets $(\bar{x}_i, \bar{y}_i)$, $i = 1, \ldots, m$, we can generate different locally optimal points on termination. It is also possible to initiate the

*The iterative procedure for solving this problem is given in Chapter 10.1.

procedure with specified values of x_{ij} and to proceed directly to step 2. The model is usually referred to as the location-allocation model because of the two problems encountered in its solution.

Once the model has been solved, we are not really interested in the cluster points; they were only a mechanism for determining the cluster. The information that we need is the delivery points assigned to each cluster. These points are then to be carried on (assigned to) the same vehicle.

By starting with different initial cluster points, different sets of clusters can be obtained. Hence this approach can be readily adopted for use with the covering and partitioning model.

A second method of clustering utilizes a tree growing approach. A tree is a connected network with no cycles. The network depicted in Exhibit 11.11.6 is a special case of a tree. In the tree growing approach, again an initial set of cluster points is selected. Then trees are constructed, or "grown," by adding links that do not create cycles or construct trees with more delivery points than can be placed on a single vehicle. Here the implicit assumption is that the cost is proportional to the length of the tree.

There are a variety of methods for selecting the links to be added to the trees. Perhaps the most common is to pick the cheapest link adjacent to the current partial tree. If only one tree were grown, this would yield the minimum-length spanning tree (i.e., the minimum-length tree containing all of the delivery points).

Traveling Salesman Problem

Once the points to be assigned to a vehicle have been determined by a clustering procedure, one must still determine a sequence in which the points are to be visited. If the only constraint on the sequence is that every point be included, then the problem of finding the minimum-length route is the famous "traveling salesman problem." Optimization procedures are available for solving such problems.[10] However, the computation time can be excessive, even for moderate-size problems (e.g., 20 to 50 delivery points). As a result, some sophisticated, but efficient, heuristics have been developed for solving traveling salesmen problems. One of the most widely used is called the "r-optimal method" and is due to Lin and Kernighan.[11]

A route is determined to be r-optimal if no better route can be obtained by replacing any r of its links by any other set of r links. When r is small, it is a simple matter to find an r-optimal route. As an illustration, consider the example in Exhibit 11.11.7*a*. Suppose we begin with the route in Exhibit 11.11.7*b* and consider any pair of links, say (B, C) and (F, D). We check to see if this pair of links can be replaced by another pair, in this case (B, F) and (C, D), to yield a better route. In this case the answer is yes, and the new route is given in Exhibit 11.11.7*c*. We continue considering pairs of links until no further improvement can be formed. The resulting route is 2-optimal. We must take care to ensure that after the replacement we still have a route. For example, replacing (B, C) and (F, D) by (B, F) and (E, F) would reduce the length and would not be acceptable since the result is not a route.

Since the time required to check for improvement increases rapidly with r, we do not normally look beyond 3-optimal routes. For many problems such routes have been found to be very good.

The idea of r-optimal routes can also be applied when there are constraints on the routes. The only modification required is that only interchanges of r links that yield feasible routes are allowed.

It should be noted that the r-optimal procedures always maintain a route with the same number of links as the route used to initiate the procedure. Hence, if no arc is used more than once in the initial tour, then no arc is used more than once in the final tour, even though it may be advantageous to do so. A recent survey of heuristics for the traveling salesman problem is given in Golden et al.[12]

Route-First-and-Then-Cluster Approach

Another approach to the vehicle routing problem is to first generate a single route (i.e., a traveling salesman tour) that visits all of the points. Then the route is divided into segments, each corresponding to a single vehicle route. There has been some success with this approach, particularly for problems such as garbage collection, where each link has many points to be visited.

This approach, like the cluster-first approach, is very limited in the kinds of route constraints that can be handled. It can handle capacity constraints and route-length constraints by simply making the segments sufficiently small so that these constraints are satisfied. However, unless individual deliveries are small compared to the vehicle capacity, we would not expect this approach to work as well as the cluster-first approach when vehicle capacity is a major consideration.

Another major drawback to the approach is that it is not possible to generate optimum traveling salesman tours if the number of delivery points is large. Therefore we must resort to applying a heuristic to generate the tour through all of the points and then apply another heuristic to separate the tour into segments. For the special case where each link must be visited (e.g., the garbage collection problem), there are efficient methods[13] for generating an optimum route through all of the points.

Exhibit 11.11.7 Finding an *r*-Optimal Route: (*a*) An Example, (*b*) Beginning Routing, and (*c*) Improved Routing

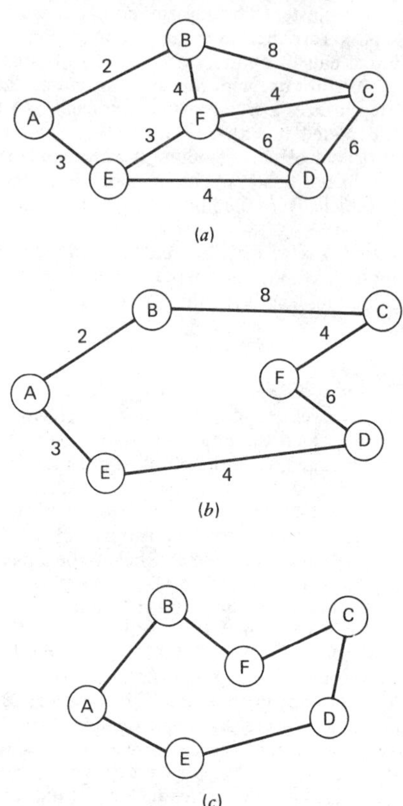

Simultaneous Cluster-and-Route Approach

To handle complex route constraints adequately, we must generally resort to heuristics that cluster and route simultaneously. This generally involves building routes by separately adding delivery points while maintaining feasibility at each step. We proceed by constructing routes one at a time or by constructing several in parallel. Constructing several routes in parallel often results in better routes, but requires more computational effort. Most of the heuristics proposed for building routes can be thought of as a "pricing" approach, a "geometric configuration" approach, or some combination of the two. These concepts are discussed next.

Pricing

The pricing idea can be used either to construct an initial set of routes or to try to improve on an existing set. Suppose that we have an existing set of feasible routes. For each delivery point i, let p_i be an estimate of the price or cost of serving the point using the existing set of routes. Now suppose that we start constructing a new set of routes. Let C_i be an estimate of the incremental cost of putting point i into a new route r. The quantity $p_i - C_i$ is then an estimate of the cost change from the current set of routes if delivery point i is put into r. With this estimate we can construct new routes by including points that indicate a cost improvement.

The best-known variation of this approach is the "savings" method of Clarke and Wright.[14] They start with an initial set of routes, each containing a single delivery point. The price for each delivery point is then simply the cost of a round trip from the warehouse to the delivery point (i.e., $p_i = 2q_{io}$ where q_{io} is the cost of a trip between the warehouse and the delivery point). Now suppose that point j is the only point in route r. The incremental cost of adding point i to route r is the new cost of route r minus the old cost of route r, or $C_{ir} = q_{oj} + q_{ij} + q_{oi} - 2q_{oj}$ and $p_i = 2q_{oi}$ and hence $p_i - C_{ir} = q_{oi} + q_{oj} - q_{ij} = s_{ij}$. The quantity s_{ij} is called the "savings" from combining points i and j in the same route. These savings values are put in a list ordered from largest to smallest. One way in which new routes can be created is by beginning at the top of the list and linking the two points i and j corresponding to the first s_{ij} if the resulting route satisfies all of the routing constraints. Then the next pair in the list is considered, and so on until all elements in the list have been considered.

Another pricing variant is to actually calculate the incremental cost of inserting a point between each pair of existing points in route r. The best location at which to insert the point is determined, and the corresponding cost is taken as the estimate of C_{ir}. The routes can then be constructed by adding points that result in a feasible route and that have the most attractive $p_i - C_{ir}$. This procedure is used iteratively by Cullen et al.[6] utilizing prices generated from the previous solution. If, rather than starting with a particular solution, we simply assign each delivery point a constant price (i.e., $p_i = m$), then this procedure amounts to adding to the route the point that yields the smallest incremental increase in the route cost. These procedures can be repeated using the new set of routes as the initial set; however, estimating the prices is more complex when the routes contain more than one delivery point. A good rule of thumb is to have the prices reflect, as nearly as possible, the actual cost to serve the point in the current route.

All of the routes generated can be incorporated into a set partitioning problem, which then selects from these the best set of routes. This can yield an appreciably better selection of routes. It may also be desirable to apply the r-optimal procedures discussed in the previous section to further improve the final route selection. For a discussion of other pricing strategies, see Cullen et al.[6]

Geometric Configuration

The basic approach is to divide the region into nonoverlapping geometric areas and then build a route through those points in the same area. The best-known of these methods is the sweep method of Gillet[15] and Gillet and Miller.[16] This approach utilizes the fact that the routes begin and end at the facility and that we would often like them to form a petal arrangement (see Exhibit 11.11.8). Essentially, this amounts to considering the region as if it were a clock centered at the warehouse. The hand of the clock is given an initial location, and then as it "sweeps" around, the points it encounters are put into the same route if the resulting route is feasible. When a certain amount of vehicle capacity has been used up, a new route is initiated. The geometric areas are then shaped like slices of a pie. Exhibit 11.11.9*a* shows four wedges forming four customer groups; Exhibit 11.11.9*b* indicates the routing of the four vehicles in these wedges of customers. By altering the edges of the slices in a systematic fashion, the routes may be improved.

Interactive Color Graphics

The most promising new techniques for solving routing problems are those involving color graphics. The capability of color graphics allows the use of all of the mathematical methods available for

Exhibit 11.11.8 Typical Routes Out of a Distribution Facility

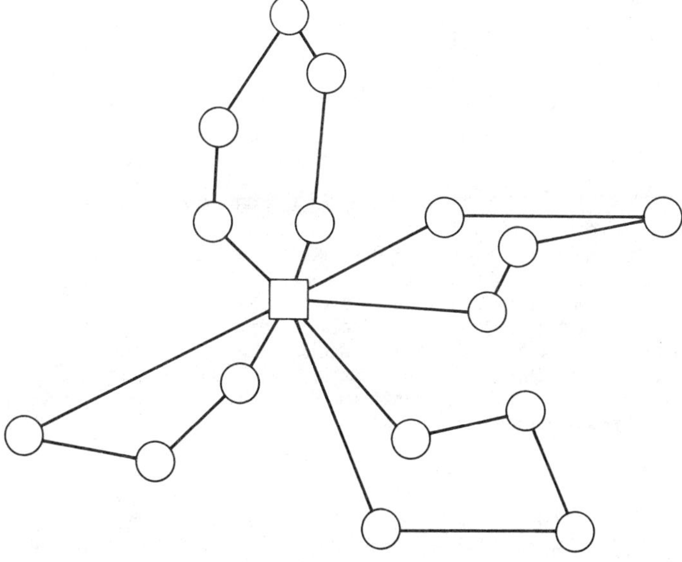

Exhibit 11.11.9 Four Customer Groups (*a*) and Four Associated Routes (*b*)

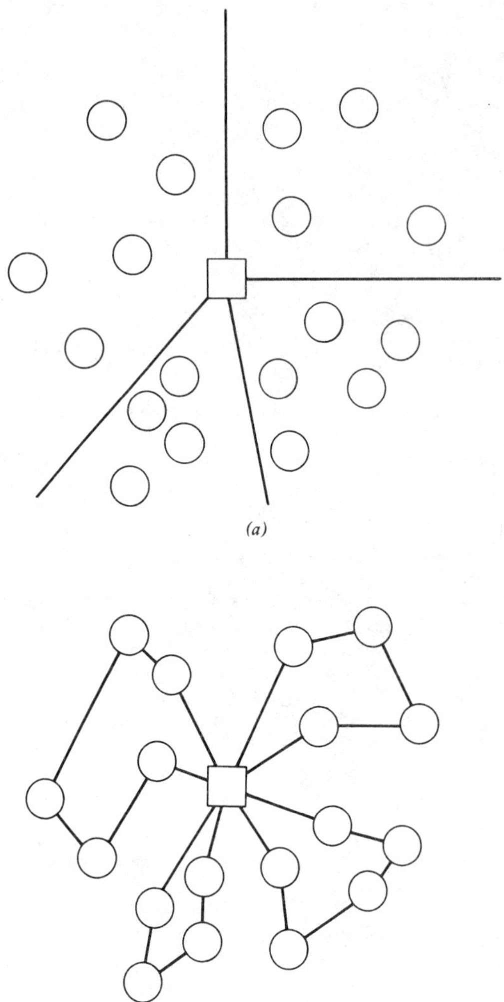

handling routing problems. It allows a human to guide the process and to take advantage of spatial information that is difficult to handle in a completely automatic procedure. Instead of trying to have the computer duplicate what a human would do, it helps the human in the process. The computer's role is to solve models, which provides the human with information to help build routes and to display the routes.

The system of Cullen et al.[6] discussed earlier lends itself to this computer-assisted approach. The computer develops the prices and potential savings associated with inserting delivery points into various routes and displays this information on a color graphics display. The human makes the decisions as to where the points are to be inserted in the routes. This procedure continues until a reasonable number of routes have been generated; the computer then solves the set partitioning problem to pick the best selection of routes from those generated. This process can be repeated until there is no improvement in the routes. The actual display utilizes seven colors, with all human input done by the use of a light pen.

11.11.6 METHODS

One of the often-overlooked steps in the design of a distribution system is its implementation. Implementation of many distribution system designs is difficult, if not impossible, because the analyst failed to answer one of the following basic questions: First, have we adequately considered the data requirements, including quality and quantity, for implementation of the distribution system models? Second, are the computer system and the associated models sufficiently responsive? (That is, are they able to give reasonable answers in a reasonable amount of time?) This second question may not be serious during the planning phase; however, it is critical during operation. Third, are the results understandable and believable to the user? Does he or she have enough information, but not too much? Finally, has the design been checked for reasonableness, and is it well documented for implementation? We briefly touch on each of these issues.

Data Requirements and Sources

If we could give one piece of advice on distribution system design, it would be this: *Don't forget the data requirements*! Too many otherwise beautiful models and designs have been discarded because inadequate attention was paid to the nature, quantity, and quality of the data required.

Ask yourself, "Are the data available, and, if so, where are they located?" Some distribution system models require an army of cost accountants just to supply the cost data. Before you get too far into model design, check out the data sources. If the data are not now available, how much effort (time and money) will it take to get them? Unless you are really fortunate, you will spend a lot more money on data collection than you will on model design. Also, do not forget that you still have to get the data into the model. If the model is a computer model, then the data must be keypunched and verified; if they already exist on a computer, they usually must be processed into a usable form for the current model.

We are not suggesting that the designer ignore significant model considerations simply because the data are not available in the form desired or because the quantity of data is excessive. It may simply be the case that some alternative data source might yield an adequate surrogate. For example, take the case of a distribution system being designed to serve 1000 customers. The travel cost matrix among these customers will have $(1000)^2 = 1{,}000{,}000$ entries. Where are all of these data going to come from, and who is going to keypunch them into the computer? Freight rate tables might be available, and you might even be lucky enough to find them on computer tape. But this might represent only part of the cost.

Another approach to this same cost matrix data problem might simply assume that cost is "related" to Euclidean (straight-line) distance. If this is the case, then we would need to develop and enter only the X-Y coordinate for the 1000 locations, (i.e., 2000 data entries). We could develop the "relationship" by sampling some points and applying regression analysis. Whenever possible, try to reduce the quantity of data required without significantly affecting model results.

Finally, careful attention must be paid to data quality needs. How accurate must the data be? At present it is hard to answer this question; however, some determination of the data quality required must be made before model building gets too far along.

Some answers about data quality can be obtained by approaching distribution system design as a multistage process. During the first stage the designer and/or analyst might choose to develop macro-level models of the distribution system. These models could then be subjected to robustness tests and sensitivity analyses to determine where the next level of effort must be concentrated.

During a spare parts distribution study for a large computer firm, a question was raised over the "actual" mileage cost for company vehicles. This debate might have thwarted the entire project had it not been for the fact that a sensitivity analysis indicated that the final system configuration was insensitive to changes in mileage rates over the range of debate.

Computability and Computer Resources

With regard to model execution, the designer and/or analyst must be concerned with the availability of computers or other computational capability. Careful attention must be given to computational effort, regardless of the medium used.

Computability refers to the effort to execute the distribution model. If little effort is required, then manual computation may suffice; otherwise some kind of computer resources will be required. If computer resources are not available, then the distribution models must be designed with manual computation in mind. Because distribution often involves points (customer locations) on a map, and because humans are good at spatial processing, a certain amount of manual computation (especially in model preprocessing) is possible. Except for boundary effects, humans are generally good at service area allocation, such as allocating customers to service districts on a map. Once this allocation is accomplished, a smaller distribution problem can be solved on each separate service area.

Even when computer resources are available, we must still be cognizant of the time required to execute the model on the particular computer. A computer model for a dial-a-ride transportation system was discarded by a northeastern city because it took 8 hr to execute. It was not practical for daily planning, much less for immediate dispatching needs during operation.

Problem size affects the solution method. What performs well in one range size may not do well in another. An interactive dispatching system developed for a large railroad employed an optimization subsystem for switching and delay. One approach tested utilized a linear program with a shortest-path calculation for column generation. On the minicomputer available (in the field, throughout the railroad system) the model took too long to execute. This linear programming approach was finally discarded in favor of an enumeration model, which was far superior in the particular-sized dispatching problem anticipated. Had the problem been twice as large, the first model would have been clearly superior. This example points out that we must always be aware of the problem dimensions. These dimensions very often dictate the solution approach, even when computers are involved.

Display and Presentation of Model Output

All of the modeling and analysis effort is wasted if the results are not understandable and useful to the user. All too often, executives are swamped by computer output. In one particular case, an executive received a computer output (approximately 1 in. thick) monthly on the operation of his system. He felt that the output contained too much information and that he could not find the information he needed for decision making purposes. The result was that the output ended up in the trash can.

The display and presentation of the distribution model results must be keyed directly to the user. Recommended choices should be accompanied by the analysis of reasonable alternatives so that the decision maker can interject nonquantifiable constraints and objectives into the final determination.

Presentation of the results of some sort of sensitivity analysis is probably one of the most useful pieces of information any decision maker can have. Many of the values of the model parameters will be open to question. Thus the decision maker will be more comfortable with the final outcome if he or she perceives a degree of robustness of the recommended system over some reasonable range of parameter values. Such a sensitivity analysis also helps to isolate the "critical" system parameters.

Model Validation and Documentation of Results

Few designers and/or analysts bother to validate their models adequately. Even fewer bother to document their models and results adequately.

An engineering firm was designing a rapid transit system for a large city. A complex computer model was being run to develop an overall system design. One day (after hundreds of hours of computer time using the final numbers) someone decided to plot patronage ingress and egress by station. An unusually low number for one particular station led to the discovery that parking costs at the station had been mispunched into the computer. Since several reports had already gone out, the firm had a very ticklish task of explaining why the results suddenly changed.

It behooves the designer to take more time in validating the model. This can be done in several ways. First, the model should be checked against historical information. Any discrepancies should be fully explained before further testing or implementation takes place. Second, various common-sense checks should be applied. For example, is the average distribution cost per item reasonable? Third, are any of the design results verified by independent checks? This can often be accomplished by applying very gross models to obtain "ball park" estimates for certain quantities that the distribution model will also yield.

For most firms, documentation of the model and results occurs at the end of the distribution system design. After people have been working on the design for months or years, they are tired. They do not want to document anything beyond the recommended system. However, let this recommendation sit on someone's desk for another year and then have the decision maker raise an important question or else have a condition change. If the analyst has left the firm, there is probably no hope of recovering the information. Even if the analyst is still around, he or she has probably forgotten most of what the model does and how it does it.

The only way we have found to force a reasonable job of model documentation is to require milestone reports, have presentations on these reports, and edit them carefully. Label them "drafts" so that they can easily be changed when someone remembers something else. Good documentation will be well worth the effort later on.

REFERENCES

1. G. T. ROSS, and R. M. SOLAND, "Branch and Bound Algorithm for the Generalized Assignment Problem," *Mathematical Programming*, Vol. 8, No. 1 (1975), pp. 91–103.
2. A. M. GEOFFRION, and G. W. GRAVES, "Multicommodity Distribution System Design by Bender's Decomposition," *Management Science*, Vol. 20, No. 5 (1974), pp. 822–844.

3. T. G. MAIRS, G. W. WAKEFIELD, E. L. JOHNSON, and K. SPIELBERG, "On A Production Allocation and Distribution Problem," *Management Science*, Vol. 24, 1978, pp. 1622–1630.

4. A. CHARNES, and M. H. MILLER, "A Model for the Optimal Programming of Railway Freight Train Movements," *Management Science*, Vol. 3, No. 1 (1956), pp. 74–92.

5. M. L. BALINSKI, and R. E. QUANDT, "On An Integer Program for a Delivery Problem," *Operations Research*, Vol. 12, 1964, pp. 300-304.

6. F. H. CULLEN, J. J. JARVIS, and H. D. RATLIFF, "Set Partitioning Based Heuristics for Interactive Routing," Report #J-80-1, ISyE, Georgia Institute of Technology, Atlanta, 1980.

7. E. BALAS, and M. W. PADBERG, "Set Partitioning: A Survey," *SIAM Review*, Vol. 18, 1976, pp. 710–760.

8. R. S. GARFINKEL, and G. L. NEMHAUSER, *Integer Programming*, Wiley, New York, 1972.

9. L. COOPER, "N-Dimensional Location/Allocation Used for Cluster Analysis," Report #C00-1493-23, Washington University, St. Louis, MO, 1969.

10. M. BELLMORE, and G. L. NEMHAUSER, "The Traveling Salesman Problem: A Survey," *Operations Research*, Vol. 16, 1968, pp. 538–558.

11. S. LIN, and B. W. KERNIGHAN, "An Effective Heuristic Algorithm for the Traveling Salesman Problem," *Operations Research*, Vol. 21, 1973, pp. 498–516.

12. B. GOLDEN, L. BODIN, T. DOYLE, and W. STEWART, "Approximate Traveling Salesman Algorithms," *Operations Research*, Vol. 28, No. 3 (1980), pp. 694–711.

13. N. CHRISTOFIDES, *Graph Theory*, Academic Press, New York, 1975.

14. G. CLARKE, and J. W. WRIGHT, "Scheduling of Vehicles From a Central Depot to a Number of Delivery Points," *Operations Research*, Vol. 12, 1964, pp. 569–581.

15. B. E. GILLET, "Vehicle Dispatching: Sweep Algorithm and Extensions," ORSA/TIMS National Meeting, Miami, 1976.

16. B. E. GILLET, and L. R. MILLER, "A Heuristic Algorithm for the Vehicle Dispatch Problem," *Operations Research*, Vol. 22, 1974. p. 340.

BIBLIOGRAPHY

CHRISTOFIDES, N., A. M. MINGOZZI, and P. TOTH, "The Vehicle Routing Problem," Imperial College Report, Louder, England, 1979.

EILON, S., C. D. T. WATSON-GANDY, and N. CHRISTOFIDES, *Distribution Management*, Hatner, New York, 1971.

KROLAK, P., W. FELTS, and J. NELSON, "A Man-Machine Approach Toward Solving the Generalized Truck-Dispatching Problem," *Transportation Science*, Vol. 6, 1972, pp. 149–170.

KROLAK, P., and J. H. NELSON, "A Family of Truck Clustering Heuristics for Solving Vehicle Scheduling Problems," Report 78-2, Computer Science, Vanderbilt University, Nashville, TN, 1978.

SALKIN, H. M., *Integer Programming*, Addison-Wesley, Reading, MA, 1974.

SECTION 12

COMPUTERS AND INFORMATION PROCESSING SYSTEMS

CHAPTER 12.1

Computer Fundamentals

HERBERT D. SCHWETMAN

Purdue University

12.1.1 INTRODUCTION

In the late 1940s, computers could be described as automatic calculators. Today that analogy can serve as an aid to understanding electronic computers, but it is an enormous oversimplification, ignoring the wide range of applications of modern computers. Because alphabetic information can be encoded as numbers, and because almost any sequence of computations can be performed accurately and at very high speed, computers have become essential tools in every segment of modern industry, government, and education.

This chapter provides concepts and background that are required for the following chapters on the uses of computers in industrial engineering. This chapter introduces both the electronic components (the hardware) of computer systems and the programs or sequences of instructions (the software) that are required to make use of these components. The total assembly of components—hardware and software—defines a system and determines how we, the human customers (users), make use of the system. In modern systems it is often true that the software component represents a larger investment than the hardware, in some cases by orders of magnitude. Furthermore, most errors, limitations, and inflexibilities are the result of shortcomings in the initial design of the software and not shortcomings of the hardware. Today, placing the blame for failure on the computer usually means blaming the software or the operator.

12.1.2 COMPUTER SYSTEM HARDWARE

Basic Concepts

This section presents some preliminary concepts, using the analogy of an automatic calculating machine. It can be skipped by readers who have at least a basic understanding of computers.

Consider a computational task in the days before computers were available. A scientist or engineer might formulate a function, which must be numerically evaluated for a variety of values. One approach would be to write down a sequence of steps, which are really keystrokes on a calculator, and give it, along with the set of data values, to a skilled operator. The operator, in turn, would repeat the sequence of steps for each data value and would write down a table of function values for each data value. This procedure, though slow and often inaccurate, would give the scientist the needed information. Many people can remember when such procedures were the only way of accomplishing this type of calculation. In fact, it was because of the limitations of this manual approach to preparing ballistics tables for artillery pieces in World War II that electronic computers were first developed on a wartime priority.

If we wanted to build a machine to perform this computational task, we would need the following:

A way of specifying a formula to be evaluated.

A way of specifying a list of data values.

A device for performing sequences of mathematical operations, possibly with sequences of intermediate results.

A device for presenting the results to human observers.

A computer system includes components that perform each of these functions. The basic unit that performs these functions is called a "central processing unit" (CPU). In a CPU there is an arithmetic unit, which can perform arithmetic operations on operands to produce a result; a memory, which holds both sequences of instructions and data values; an input/output (I/O) unit, which can receive data from input devices and send data to output devices; and a sequencer or

Exhibit 12.1.1 Schematic of CPU

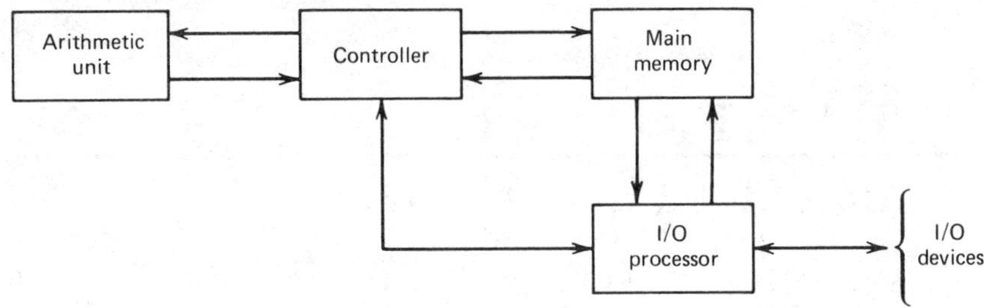

controller, which coordinates all operations of the CPU. A typical CPU is shown in schematic form in Exhibit 12.1.1.

The operation of the CPU can be described in terms of the basic fetch and execute cycle. Although this cycle is explained in greater detail in a subsequent section, it can be outlined as follows:

FETCH	Fetch next instruction (operation) from memory.
EXECUTE	Execute current instruction.

The execute phase usually involves locating two operands, sending these operands to the arithmetic unit, and then storing the result. We can compare these operations to the human operating the calculator as follows:

FETCH	Perceive next instruction in the procedure.
EXECUTE	Locate the operands, key them in, key in operation, and write down result.

As an example, suppose we are to evaluate the functions

$$f(x) = x + 9 \quad \text{and} \quad g(x) = \frac{f(x)}{2}$$

for $x = 1$, 3, and 5, using a calculator. Depending on the type of calculator, several sequences are possible. One possible sequence of operations would be:

1. Fetch $x + 9$ as next operation.
2. Execute $x + 9$ (note that x is 1), producing 10.
3. Fetch $f(x)/2$ as next operation.
4. Execute 10/2, producing 5 as first result.
5. Fetch $x + 9$ as next operation.
6. Execute 3 + 9, producing 12.
7. Fetch $f(x)/2$, execute 12/2, producing 6 as next result, and so on.

If the human operator had very little experience, the statement "Evaluate $f(x) = x + 9$ and $g(x) = f(x)/2$ for $x = 1$, 3, and 5" might not be explicit enough. In this case we might be forced to write something like

1. Key in 1 + 9 = and write down result on scratch paper.
2. Key in result /2 = and write down result in table.
3. Key in 3 + 9 = and write down result on scratch paper.
4. Key in result /2 = and write down result in table, and so on.

The point of this is that in a CPU the controller serves the function of the human operator; the calculator, the function of the arithmetic unit; the list of directions, the function of the computer instructions; and the paper as the memory device, where we can both save results and retrieve values. The list of the instructions is similar to a computer program. If we assume that we are dealing with an inexperienced operator, we see that we have to specify every operation in great detail. However, if the operator is experienced, then we can use more concise forms for expressing the steps to be followed. As we will see, we also have different levels for expressing computational procedures, depending on the computer system that is being used. Because humans have eyes, ears, and hands as well as very inventive and adaptive brains, dealing with them can be quite effortless. Computers, on the other hand, have none of these and thus require equipment and software to fulfill the functions needed. Everything must be correctly and completely specified; this usually requires great effort.

If we can visualize the CPU as a fully automatic equivalent of the human operator, with calculator and pencil and paper, and can envision all of the steps involved, we are beginning to understand computers. Of course the analogy breaks down in many ways. For one thing, computers are able to perform arithmetic operations at the rate of a million or more per second, whereas the human, with calculator, requires several seconds per operation. Also, computers, given correct sequences of instructions and values, can be expected to operate for days without making errors, whereas the human error rate is probably much higher. On the other hand, the human operator can perform a quality control function on both the sequence of instruction and values and can often either correct or at least reject erroneous sequences. Computers are quite content to perform irrational operations on nonsensical sequences of values, producing completely erroneous results. The expression "garbage in–garbage out" was probably invented to describe human use of computers.

Representing Data

An additional basic concept is the representation of data–numbers and characters–in a computer. All data in a computer are built up from one fundamental unit of information, called a "bit" (binary digit). Data in memory are stored as a sequence of bits, and all operations are implemented as transformations on sequences of bits. A bit can assume only one of two possible values; we normally write these two values as "0" and "1." A sequence of four bits could be written as "0101." If we are using four bits to represent some numbers, then a standard technique is to have the sequence of four bits "0000" represent the number 0 (zero), "0001" to represent 1, "0010" to represent 2, "0011" to represent 3, and so on. If we recall that in normal (decimal or base 10) notation the number 123 stands for the quantity $(1 \times 10^2) + (2 \times 10^1) + (3 \times 10^0)$, then we can see that the encoding technique for representing numbers using sequences of bits corresponds to representing numbers in binary (or base 2) notation; that is, "0011" corresponds to $(0 \times 2^3) + (0 \times 2^2) + (1 \times 2^1) + (1 \times 2^0)$.

Computer memories are usually implemented as a collection of fixed-length sequences of bits; each group of bits is called a "word." Many smaller computers have either 8 or 16 bits/word, whereas current larger systems have either 32, 36, 48, or 60 bits/word. A system with 16 bits/word could represent numbers in the range 0 to $2^{16} - 1 = 65535$. Actually, because it is necessary to deal with both positive and negative numbers, most computers appropriate the left-most bit in each word to designate the sign of the number, thus shifting the range to something like +−32767. If a larger range is needed, then either a system with more bits per word or some scheme for using multiple words to represent single numbers is required. This finite representation of numbers introduces problems while performing arithmetic operations on computers and can be a limitation to the types of computations that can be carried out.

In actual systems, signed integers are represented in one of three ways: signed magnitude, two's complement, or one's complement notation. Also, many systems have an alternative scheme for approximating decimal fractions, based on representing a number as a pair, consisting of a coefficient of some base number (usually 2, 8, or 16) and an exponent of that base. This representation is called "floating point" notation and is different for each type of computer. With special circuitry to operate on floating point numbers, fractions are automatically aligned and manipulated, providing a dramatic decrease in the programming effort required to implement many types of calculations. It should be pointed out that this type of arithmetic can lead to situations in which the numerical accuracy of some sequences of calculations is lost. The subject called "numerical analysis" is the study of such errors and the development of methods for reducing or eliminating them.

The introduction stated that the ability to handle alphabetic information within a computer was very important. Since all data are represented as sequences of bits, some technique for encoding characters (letters, numerals, punctuation marks, etc.) is required. Today most systems have adopted the convention that a character will be represented by a group of 8 bits, called a "byte." Although many encodings are possible, two specific ones are widely used; these are named American Standard Code for Information Interchange (ASCII) and Extended Binary Coded Decimal Interchange Code (EBCDIC). The second, EBCDIC, was developed by IBM, whereas ASCII is an international standard. It is unfortunate that there is more than one standard technique, because this complicates the interchange of data between systems that use different techniques.

In an ASCII-based system, the character "A" is represented by the sequence of bits "01000001," "B" by "01000010," and so on. Although 8 bits allow 256 different characters, ASCII has only 128 characters defined. It should be noted that for many applications the users of the system are not affected by the encoding technique used.

Mainframes

The main component of a computer is the CPU. However, as systems have become more complex, to the point of encompassing multiple CPUs, a broader term has come into use. "Mainframe" is used to include one or more CPUs, together with a block of main memory and an I/O processor. Each of these components is discussed in more detail.

The function of a processing unit is to successively fetch and then execute instructions. These instructions are commands that govern the operation of the processor. For example, one instruction can cause two numbers to be added together. Sequences of these instructions are called "programs." The operands of instructions are called "data." Both instructions and data must be in main memory if they are to be used by a processor. The I/O processor can cause data and instructions to be brought into main memory from an external or peripheral device; equivalently, data and instructions can be sent from main memory to an external device, all under the control of a program.

As stated earlier, memory consists of a collection of words. Each word has a unique name, called its "address" or "location." Typically, the first word in memory has the address 0, the second word 1, and so on. When a processor needs the number stored in a particular word (the contents of a word), it issues a read operation to the memory unit, specifying the address of the word. The memory unit locates the addressed word, fetches the contents of the word, and presents the fetched contents back to the processor. The contents of the word in memory are not altered by a read operation. Conversely, when a processor needs to store a number in memory, it presents both the address of the destination word and the contents (number) to be stored as the operands of a write operation to the memory unit. The memory unit then writes the contents into the specified location in the memory. Obviously, writing a word in memory destroys the previous contents at the destination word.

Most processing units have a set of special memory words, called "registers," internal to the processor. Although these registers operate like a small block of memory, they have much faster access rates and can be used to hold operands of instructions temporarily. A typical sequence of instructions consists of fetching a word from memory into a register, performing a series of transformations on the word, and then storing the result back in memory. Not all instructions require that their operands be stored in registers; some are able to manipulate words or sequences of words that are in memory, placing the results in memory, as well.

Often it is convenient to specify that the actual address (called the "effective address") of an operand of an instruction is formed by combining the contents of one or more registers with the address found in the instruction. This dynamic address calculation allows us to use one group of instructions to transform, for example, each word in a list of words. If we can specify the address of the first word in a linear list of words plus the contents of a register as the actual address of an operand, we can then repeatedly execute a group of instructions that manipulate each word in the list and then increment the register (called an "index register") so that each word in the list is transformed in the desired way.

Another important feature of a processing unit is the order in which instructions are executed. Instructions are stored in successive bytes or words in memory. The processor fetches and then executes each successive instruction until an instruction that alters this sequence is executed. Instructions that alter the order of execution are called "branches" or "jumps." This instruction sequencing (or flow of control) is implemented by having a special register, called the "program counter," which always contains the address of the next instruction to be fetched. Normally this register is incremented as each successive instruction is fetched from memory. Branch or jump instructions, as they execute, place a new address into the program counter, causing instruction fetching to begin at a different address in memory. We can summarize the operation of a processing unit by this basic fetch and execute cycle as follows:

1. Fetch instruction whose address is in program counter from memory.
2. Increment program counter by length of fetched instruction.
3. Calculate effective address(es) of operand(s), if required.
4. Execute instruction, using calculated address(es) of operands.
5. Repeat steps 1 through 4 until . . .

The I/O processor operates under the direction of a CPU to send or retrieve data from external or peripheral devices. A CPU can execute some special I/O instructions that cause the I/O processor to perform these functions. When doing these, the I/O processor accesses main memory either to store data that have been read from an I/O device or to fetch data that are to be written to a device. At the end of each operation, the I/O processor must signal the CPU that the operation has completed.

In modern systems the CPU can execute instructions at the rate of between 1 and 10 million instructions/sec (i.e., between 1×10^{-6} and 0.1×10^{-6} sec/instruction). However, the I/O processor can require from 20×10^{-3} to 100×10^{-3} sec or longer to read or write data at an external device. Because of this speed mismatch, most systems have provisions for allowing I/O operations to proceed while the CPU continues to execute instructions, thus achieving concurrent operation of both the I/O processor and the CPU. In such a system the I/O processor signals the CPU when an operation is complete by "interrupting" the CPU. Such an interrupt causes the CPU to stop processing instructions at one place and to begin processing them at another, the "new" instructions being those needed when an I/O operation is complete. When such an interrupt occurs, the program counter of the interrupted program is saved, so that subsequent execution can be continued at the point of interruption.

The function of the I/O processor is implemented in many different ways on different systems. In some systems there are several "data channels," where each channel can be processing one (or more) I/O operations. In other systems a collection of peripheral processors is responsible for managing I/O operations. In all cases the function is the same, and normally the I/O processor can operate concurrently with, but under the control of, a CPU.

Writing programs to perform I/O in this concurrent (asynchronous) mode is very complicated; thus systems with these facilities come equipped with programs and special hardware features that facilitate use of the I/O equipment. These I/O programs, plus many others, are all grouped together under the term "operating system," which is discussed in a later section. The special hardware features allow several programs to be making simultaneous use of the system's facilities (multiprogramming) and to ensure that these programs cannot interfere with each other. The result is that people writing normal programs can make efficient use of facilities of the system with little extra programming effort.

Main Memory

In the section on representing data, the organization of main memory was discussed with respect to bits, bytes, and words. It was pointed out that all information is represented in a computer as sequences of bits. In computers, bits are grouped into fixed-length words capable of storing a number or a collection of characters. A byte is a group of 8 bits.

Today the size of main memory is usually stated in terms of bytes or kilobytes (1 kilobyte is 1024 bytes). Small systems may come equipped with 16, 32, or 64 kilobytes of memory. Large systems may have up to 16 megabytes (1600 kilobytes). Most memories today are made up of integrated circuits, invariably called "chips," with from 4000 to 16,000 bits/chip. Densities of 64,000 bits/chip are starting to appear in commercial products. The speeds of these memories are on the order of 100 to 500 nsec/access (1nsec is 1×10^{-9} sec). The memory of the CRAY-1 supercomputer operates at the speed of 50 nsec/access. Memories fabricated from integrated circuits are said to be volatile because the contents of memory are lost if the power fails. If this is an important consideration, then either a battery backup power source is provided or another memory technology is used. The cost of main memory can range upward from about $4,000/megabyte. Prices of memory have been declining rapidly, as fabrication techniques have become more efficient.

Even at the speeds just given, memories are still slower than processors. To try to balance these component speeds, some systems have a high-speed memory buffer, called a "cache," which tries to ensure that data and instructions that are in current use can be accessed in times that are faster than the speed of main memory. A cache may be able to hold perhaps 8000 bytes and operate at a speed that is 10 times the speed of main memory (e.g., 50 nsec/access). When a reference is made to main memory, the addressed word and a block of words surrounding that word are brought into the cache. Then, if any of these words are referenced again, they can be fetched from the cache rather than from main memory. Of course the words in the cache will eventually be overwritten by more recently fetched words. Most manufacturers claim that their caches typically achieve a 95% hit rate; that is, 95% of references to main memory can be satisifed from the cache. Of course writes to memory cause a problem, since the data must eventually be stored in memory; also I/O transfers cause problems. Needless to say, the CRAY-1 does not have a cache, because all of its memory operates at speeds equal to or exceeding those of most caches.

In a multiprogrammed system, that is, a system that can have more than one program in main memory at the same time, it is necessary to provide features that guarantee that one program cannot interfere with another program. One such feature is called "memory protect," which makes certain that a program accesses only those portions of main memory that have been assigned to the program. Although there are a variety of ways of doing this, the most common way uses special registers to control all references to memory made by a program. This ensures that no program can generate addresses of memory words that are outside of the program's assigned area.

In a multiprogrammed system the management of memory that is being dynamically assigned and then reclaimed from programs becomes a problem. Also, many people want to write programs that need more memory than is available on a system. In response to these and other problems, many systems now have a memory system called a "virtual memory system" (VMS). The basic idea is that the main memory is divided into equal- and fixed-size blocks, called "page frames," typically 512 to 4096 bytes in size. A program, both instructions and data, is divided into pages, which are exactly the same size as a page frame. When a program is brought into main memory, the pages can be loaded into any available page frames. As this is done, an addressing problem is introduced, because the pages are probably not loaded into contiguous page frames. To solve this problem, a mapping table for each program is constructed by the operating system; this table is used to convert "virtual addresses" generated by the program into the actual (physical) addresses in main memory. When a program tries to access a word in memory, it uses the address of where the data should be (would be if the program were loaded into contiguous locations starting at word "0"). The hardware then looks this virtual address up in the mapping table for the program and uses the physical address that corresponds to the virtual address. It is possible that the page that is referenced is not currently in memory; if this happens, an interrupt (called a "page fault") is generated to the operating system. The operating system then locates the missing page, loads it into an available page frame, updates the mapping table, and then restarts the program, which will find the necessary word of memory. Such an addressing technique means that a program can be written as if there were more memory available than is actually present on the system; the operating system then juggles pages and page frames so that the program can run even though not all of the required memory is simultaneously present.

Needless to say, an operating system that can manage this type of memory addressing scheme in an efficient manner can become very complicated. There is a lot more to the subject of VMSs than has been presented in this very brief discussion. For example, if a page frame is needed and there are no available (empty) page frames, then the system must take some frames away from some active program. Bad or unfortunate decisions can lead to severely degraded system performance.

Rotating Mass Storage

Most computer systems must have access to a large amount of on-line storage in order to hold programs and data required by both the system and the users of the system. The most commonly used on-line storage devices are disk drives. The amount of storage attached to a system can range from perhaps 1 million to 2 billion bytes or more. Only very small systems, performing a limited set of tasks, can exist without some on-line storage.

A disk drive is a device that holds a disk pack and that can transfer data between the pack and the I/O controller or vice versa. A disk pack consists of a stack of one or more platters, which are metal disks coated with a magnetic recording material (iron oxide). Most disk drives have a single stack of movable read/write heads, one for each side of each platter. Information in the form of sequences of bits can be written (recorded) by the head on the recording material. Later the head can read the previously recorded information back. Writing new information destroys previously written information. The stack of platters revolves as a single unit, typically at speeds of up to 3600 RPM. As a platter revolves, the surface under a single read/write head is a track of bits or information. All of the tracks under a single position of the stack of heads is called a "cylinder." In most disk drives, a track is divided into fixed-size sectors. The sectors are where information is stored.

One currently available disk drive is the RM-03 drive from the Digital Equipment Corporation. This drive has five recording surfaces (and one servo surface, used to generate timing information). There are 823 concentric tracks per surface on an RM-03 disk pack; equivalently, there are 823 concentric cylinders per pack. Each track is divided into 32 sectors, and each sector contains 512 bytes. Thus the total capacity of a pack is $512 \times 32 \times 5 \times 823 = 67{,}420{,}160$ bytes. A more modern version of this drive, using a larger pack, has about 300,000,000 bytes/pack. The RM-03 can move the stack of five read/write heads (a seek operation) in an average of 0.030 sec/move (minimum of 0.006 sec, maximum of 0.055 sec); it can transfer nearly 1 million bytes of information per second.

On most disk drives, including the RM-03, a pack can be easily removed from the drive and another pack mounted in its place. Also, a pack can be used on other drives, providing a convenient way for preserving information and for carrying information between systems. An RM-03 drive costs about $20,000, and the packs cost about $500 each. The cost of the main memory ranges up from about 1¢/byte compared to the cost of an RM-03 drive, which is about 0.03 ¢/byte.

The least expensive disk unit is a floppy disk drive. It uses a diskette, which is a flexible, plastic recording surface with a capacity of about 500,000 bytes. Floppy disk drives operate at low speeds, but considering their low cost, are useful in some applications. Disk drives with capacities ranging up to 600 megabytes/pack and transfer speeds of up to 34 megabytes/sec are also available.

Because the stack of heads in a disk drive can be moved directly to any cylinder, any sector on the pack can be accessed very rapidly (less than 0.1 sec on the RM-03). Input/output devices with this type of accessing mechanism are called "direct access storage devices" (DASD); data stored on such a device are sometimes said to be "randomly accessible." This can be compared to a reel of magnetic tape (see next section) in which the data are stored sequentially. To access data at the end of the tape, all of the data preceding the needed data must be passed over, an operation that can require between 2 and 5 min, or longer.

Standard I/O Devices

In addition to disk drives, there are several other types of I/O devices available for use with computer systems. One of the most common devices (in terms of numbers) is a remote terminal used by people communicating with the computer. A terminal has a keyboard (similar to a typewriter keyboard) and a display device, either a printer (on paper) or a CRT, which resembles a TV set. A terminal is attached to a system by means of a communications line and a line controller (at the system). Users can send data by typing on the keyboard and can receive data on the display device. Printers can normally print at up to 30 characters/sec, whereas CRT units can receive data at up to 960 characters/sec. Normally the speed is limited by the type of communications line being used. It is possible to connect a terminal to a system by means of a normal telephone line; this arrangement can greatly enhance the availability of systems to people over wide geographic areas. Prices of CRT-based terminals begin at about $900 per terminal; printer-based terminals can be purchased for about $1200 per terminal on up.

Another common way for users to send data to a system is by using the familiar punched cards. A punched-card reader can read data from these cards and send it to the system. Similarly, data can be sent from a system to a printer and/or to a card punching device. This once was the most common way of using a computer: submitting decks of punched cards through a card reader and receiving printed output from a printer. Readers can read cards at rates of up to 1000 cards/min, and printers can print lines at rates ranging from 300 to 2000 lines/min.

Other commonly used devices for presenting output to people include digital plotters, electrostatic printer/plotters, and microfilm devices. Plotting devices are very useful for presenting data in graphic form. A computer output microfilm (COM) device is used when large volumes of printed output are needed in a compact form.

Reels of $\frac{1}{2}$ in. wide magnetic tape are one of the most commonly used media for storing large quantities of information in machine-readable form. The standard reel of tape is 2400 ft long and can record 800, 1600, or 6250 characters/linear inch. However, because there must be an inter-record gap between groups of information, the entire length of tape cannot be used. A reel of tape costs about $20. Tape drives, the devices that read and write on magnetic tape, can move the tape at speeds of up to 200 in./sec, with fairly inexpensive drives operating at speeds of around 45 in./sec. Tapes are often used as a medium for transferring large quantities of information between systems and for archival storage of information when the capacity of the disk-based storage units is exceeded.

An inexpensive variation of magnetic tape involves use of tape cassettes (normally used for audio recording) and tape cartridges. Drives for cassettes or cartridges can be easily and inexpensively attached to small systems.

The oldest technique for sending data to a system uses punched paper tape. This is also an inexpensive medium for recording data in machine-readable form. However, it is a fairly bulky and slow medium and is not in common use today.

Other I/O Devices

Another important class of I/O devices allows computers to monitor and control industrial and laboratory equipment. In general, these devices have a capability either of sampling and then digitizing a voltage or of sensing the state of a switch (input). For output, such devices can set a switch or generate voltage levels. The devices that monitor varying voltages are called "analog-to-digital" (A/D) converters; the devices that generate voltage levels are "digital-to-analog" (D/A) converters. These devices operate under the control of a program and can greatly extend the range of applications of computers. For example, if a thermocouple generates a voltage that is proportional to the temperature of a furnace, then an A/D converter can allow a computer to monitor this temperature. If the computer can also generate outputs that can set or reset some control devices, then the computer can be used to control automatically a process in the furnace.

There are other devices that can scan pictures, producing streams of numbers corresponding to the gray scale values in the picture, allowing computers to be used in analyzing maps and photographs. Other devices allow human operators to digitize coordinates from diagrams and maps, again allowing computers to be used to process graphical information. When coupled with digital plotters or plotting CRTs (e.g., to make motion pictures), computers have become powerful aids in the design process.

Computers have been interfaced with almost every kind of equipment in use. At airports, surveillance radar data are fed directly into computers to assist air traffic controllers. Computers have been used to control both cameras and equipment to create new kinds of special effects in motion pictures. In hospitals, computers have been attached to patient monitoring equipment in order to monitor the status of critically ill patients and to sound an alarm in case of trouble. In many industrial plants, badge readers connected to computers control entry into restricted areas. In stores, scanners read product identification codes in order to tabulate purchases. In oil refineries, computers monitor and control catalytic cracking units, adapting the process to produce a near-optimal mix of output fuels. All of these applications and thousands more depend on special I/O equipment, usually of the form just described.

Technology

The materials and techniques used to construct computer hardware have changed rapidly over the entire history of computers. The most radical changes have been the result of applications of new techniques for fabricating integrated circuits. Today CPUs are available as a single silicon chip, which is about $\frac{1}{4}$ in. on each side. This single chip replaces components that required several racks of equipment in earlier systems. The result is the microprocessor, which can be found in pocket calculators and thousands of industrial and consumer products. These microprocessors are revolutionizing the way electronic equipment is being designed and fabricated. Intelligence in the form of a computer can be integrated into almost any piece of equipment, from automobile carburetors to children's games to complicated testing equipment. The trend is toward greater computational power in each processor. Main memories have also benefited from application of new technologies. As was mentioned earlier, currently available memories are much faster and much cheaper than their predecessors.

Input/output devices have also changed. The bit densities on magnetic recording surfaces, particularly disks, have increased greatly. The capacities of currently available packs are an order of magnitude greater than those of devices available only 10 years ago. Also, these devices have become much more reliable and much less expensive. New devices based on high-density recording on plastic sheets or wide tapes are appearing. These mass storage devices have very great capacities and are designed to replace entire libraries of reels of magnetic tapes.

The logical organization of computers is, in some sense, undergoing many changes. The instruction sets of CPUs are undergoing almost no changes. The reason for this is that established manufacturers are very reluctant to introduce any systems that cannot be used by existing customers. The costs of producing and/or modifying programs have become so great that few customers are willing to buy new systems that will not execute existing programs. In fact, today new companies are building systems that are compatible with existing systems. The most notable example of this is the Amdahl computer system, which can execute programs written for the IBM S/370 systems with no changes required.

Exceptions to this lack of changes are a few new systems designed for special applications. For example, the CRAY-1 system is specially designed to perform arithmetic computations on vectors at very rapid rates, approaching 80 million operations/sec in some cases. Because there were no existing CRAY customers, this company was able to design a new system that was not compatible with any existing system.

Architecture: How Systems Differ

Although all digital computers operate in the same general way (a stored program, the fetch and execute cycle, I/O operations, etc.), there are differences between different models of systems and between systems from different manufacturers. These differences were alluded to in the presentation of ranges of word sizes, capacities, and processing rates. The choice of organization, the functional units, and the instructions are a few of the factors that make up the architecture of a computer system; an abbreviated list of architectural features found in modern systems could include:

Number of bits per word.
Number of registers per CPU.
Number of CPUs per mainframe.
Choice of instructions implemented, and how they operate.
Instruction formats.
Instruction execution rate.
Interrupt mechanism.
Addressing scheme.
Organization, size, and speed of main memory.
Organization of I/O processor.
Speed and bandwidth of I/O processor.
Method of connecting I/O equipment to the I/O processor.

Computer systems range in size from the very small to the very large. For purposes of classification, the following ranges of size are in common use (the prices listed are only rough estimates):

Microprocessor-based systems—a system using a "computer on a chip"; usually occupies one or two printed circuit boards (price—$100 to $1000, excluding I/O devices).

Minicomputers—a small system; usually occupies a single cabinet (price—$2500 to $100,000 or more).

Midicomputer—a large minicomputer (faster, more memory); usually occupies a few cabinets (price—$75,000 to $400,000).

Computer system (*small, medium, and large*)—computer system in traditional sense; occupies several cabinets (price—$300,000 to $10,000,000).

In general, the larger, more expensive systems have more main memory and faster instruction processing rates. Also, larger systems have several I/O processors, with greater potential for concurrent operations, and many I/O devices.

A modern architectural trend is to have multiple systems connected together, forming a network of computer systems. Although there are many ways of doing this, two basic techniques are appearing: (1) all systems are physically close together and connected by means of high-speed, high-bandwidth channels, and (2) some systems are at a distance from each other and are connected by means of a communications channel (lower speeds, lower bandwidth). The purpose of a network is increased reliability (availability) and increased performance when compared to a single-mainframe system. Many issues central to the concept of networks, such as communications channels and protocols and the type of system control (master-slave or decentralized), are not currently resolved. These are topics of current research and are also being tried out in some pioneering implementations.

In many cases the anticipated uses of a system may dictate some of the choices made when designing a new system. For example, the size of the address portion of instructions will determine the number of locations in main memory that can be directly addressed. If the address portion is small, then either only small amounts of memory can be attached to the system or some auxiliary registers must be used in generating memory addresses. The choice of word size has a direct bearing on the types of numerical calculations that can be efficiently carried out. If words are too small, then it becomes difficult to process numbers of large magnitude. Also, the presence of floating point instructions may be required if the system is to perform certain types of computations efficiently.

Systems that process characters frequently, as opposed to numbers, can benefit from the addition of special instructions for manipulating bytes within words. If a great deal of I/O activity is anticipated, then the capability for having several I/O operations simultaneously in progress may be crucial to satisfactory performance of the system. All of these considerations and hundreds more go into the design of computer systems. As purchasers of these systems, we must be aware of both the anticipated use of a system and the architectural features of prospective systems.

12.1.3 COMPUTER SYSTEM SOFTWARE

Computer Programs and Programming Languages

So far, we have said that programs were sequences of computer instructions and have implied that I/O operations are under program control. We have stated that the group of programs that actually control the I/O processor is the operating system. Most important, we have suggested that the costs of software are very great and are often the dominant factors in both the cost and the usefulness of the entire system. The next sections focus on computer software, both user provided and system (manufacturer) provided.

A computer program is really a detailed description of a computational task. If the program is written as actual machine instructions (or in a form close to machine instructions), then it is said to be written at a low level. Normally there is a system-provided program, called an "assembler," which allows users to write machine instructions in a convenient form. This is discussed in a subsequent section. However, because writing programs at this level is very tedious and error prone, other methods of writing programs have been developed. These methods make use of high-level programming languages such as FORTRAN, COBOL, PL/1, BASIC, and Pascal.

To provide an example of a program, consider the task of reading a list of numbers and calculating the mean and variance of the elements of the list. If we refer to a book on statistics, we can discover that, for a set of values $\{x_i \mid 1 \leq i \leq n\}$, the mean is given by

$$\overline{x} = \frac{1}{n} \sum_{i=1}^{n} x_i$$

and the variance by

$$\text{var}(x) = \frac{1}{n-1} \sum_{i=1}^{n} (\overline{x} - x_i)^2$$

Almost anyone with experience in calculating such statistics knows that the formula given for the variance is not very practical because it requires that the entire list of numbers be processed twice: once to calculate the mean and again to calculate the variance. Some simple algebraic manipulations will lead to an equivalent formula for the variance which requires the list to be processed only one time.

$$\operatorname{var}(x) = \frac{1}{n-1}\left[\sum_{i=1}^{n} x_i^2 - n\overline{x}^2\right]$$

If we are trying to write a program to accomplish this task, we might begin by giving a simple description of the steps required:

1. Form the sum and the sum of the squares of each element in the list; also, record n, the number of elements processed.
2. Calculate $\overline{x}$ as the sum divided by n.
3. Calculate var (x) as given by the formula.

If we want to express this procedure in terms that are like those used to write computer programs, we might have something like the following:

1. Let SUM be zero, let SUMSQ be zero, and let N be zero.
2. Read the next value; if end of list, proceed to statement 5.
3. Let N be N + 1 (increment count of number of values), let SUM be SUM + value-read (accumulate sum of values), let SUMSQ be SUMSQ + square of value-read (accumulate sum of squares).
4. Go back to statement 2.
5. Calculate mean and variance as follows:
   a. Let MEAN be SUM divided by N (calculate mean).
   b. Let VARIANCE be SUMSQ – (N times MEAN) all divided by (N – 1) (calculate variance).
6. Print "for," print N, " values, the mean is ", print MEAN, print ", and the variance is ", print VARIANCE.
7. End of program.

A programming language is both a language and a language processing program (called a "compiler") used in writing computer programs. Such languages are designed to allow people to express procedures in terms that are fairly easy to learn and understand. The compiler, a system-provided program, then translates expressions in this language into computer instructions. This approach of having a human-oriented language for writing programs and having the computer automatically convert programs in the language into sequences of computer instructions has proven to be the best way of preparing programs. If people have to express programs directly in computer-oriented terms, then preparing programs becomes an expensive chore.

One popular programming language is FORTRAN (Formula Translator). In Exhibit 12.1.2 a FORTRAN program that implements the task just described is given.

The program in Exhibit 12.1.2 is a real FORTRAN program and, when compiled and then executed, will give the desired results. For example, if the list of values were 1, 2, 3, –1, then the program would print:

for 3 values, the mean is 2.000, and the variance is 1.000

Exhibit 12.1.3 presents a program written in the Pascal programming language, which accomplishes the same computational task, namely, computing the mean and variance from a list of numbers. The purpose of presenting these two examples is not to teach either FORTRAN or Pascal, but rather just to give the "flavor" of these programming languages. Interested readers are invited to look into either of these languages and to acquire a knowledge of one or both of them.

Such a program as the one presented here is really of little value for calculating these statistics for three numbers; however, for 3000 numbers, it may be of great value. It should also be pointed out that this is a very simple program. Programs that are in production use are often very complex and very long, perhaps on the order of from 1000 to 10,000 statements or more. In fact, this is one problem that is encountered when trying to teach programming to students: Programming exercises are generally short and simple, whereas in industrial and research situations, programs are of much greater size and complexity. One does not write computer programs to solve simple problems; it is too expensive.

There are many programming languages in use today; Exhibit 12.1.4 lists some of them. Not all of the languages in the exhibit are available on all systems, but some are available on many differ-

Exhibit 12.1.2 FORTRAN Program to Compute Mean and Variance

```
              integer n
              real sum, sumsq, mean, var, value

              n = 0
              sum = 0
              sumsq = 0

1             read 1000, value
1000          format(f10.0)
              if(value .1t. 0) goto 10
              n = n + 1
              sum = sum + value
              sumsq = sumsq + value**2
              goto 1

10            mean = sum/n
              var = (sumsq - n*mean**2)/(n - 1)
              print 1001,n,mean,var
1001          format(" for ",i3," values, the mean is ",f6.3,
        1     ", and the variance is ",f6.3)
              end
```

ent systems, with FORTRAN probably the most prevalent. The most widely used language is COBOL. Aside from easing the task of implementing programs, use of one of these languages usually means that programs written for one system can be transported to another kind of system with only a modest additional effort. However, if a program makes use of some specific features of a particular system, for example, the number of bits per word, then transporting a program to another system may require extensive changes or may even be impossible.

Assembly Language

Assembly language denotes a class of language in which each statement is converted directly into a single computer instruction; the program that does this is an assembler. Each language in this class of languages is specific to a particular type of computer; that is, there is a different assembly language for each family of computers. This means that programs written in assembly language for one kind of computer cannot be transported to other kinds, unless that other kind is compatible at the instruction level with the original system.

Exhibit 12.1.3 Pascal Program to Compute Mean and Variance

```
program stats(output,input);
        var
                n : integer;
                sum,sumsq,mean,variance,value : real;

        begin
                n := 0;
                sum := 0;
                sumsq := 0;
                while not eof(input) do begin
                        readln(value);
                        n := n + 1;
                        sum := sum + value;
                        sumsq := sumsq + sqr(value)
                end;

                mean := sum/n;
                variance := (sumsq - n*sqr(mean))/(n - 1);
                write ('for ',n:3,'values, ');
                write('the mean is ',mean:8:3);
                writeln(' and the variance is ',variance:8:3);
        end.
```

Exhibit 12.1.4 Some Common Programming Languages

Name	Area of Application
COBOL	Business
FORTRAN	Scientific, general purpose
Pascal	General purpose, research
PL/1	General purpose, business
BASIC	Instruction, time-sharing
APL	Time-sharing, scientific
ADA	Real-time, military (new)
GPSS	Simulation

The major use of assembly language is to write programs that have either critical time and/or memory space requirements. Occasionally there are systems for which no high-level language is available, but this occurs less and less frequently. Because the programmer makes direct use of the system by writing machine instructions, the results are often programs that are smaller and faster running than those generated by a compiler. Also, all of the features of the system can be used by the programmer, features such as special instructions. Some of these factors may lead to the implementation of certain highly used programs (such as components of the operating system, compilers, etc.) in assembly language.

The major problems of using assembly language are the concomitant high costs and high programming error rates. Programs written in assembly language tend to have many errors because the programmer has to write many statements, even in simple programs; because little error checking can be performed by the assembler; and because the statements in the language are machine oriented rather than human oriented. Equivalent programs written in a high-level language have fewer statements (but they may be translated into more machine instructions), extensive error checking can be performed by the compiler, and the statements tend to be easy to understand. As hardware has become faster and less expensive, and as programming costs have risen, the justification for programs written in assembly language has diminished. Today even many operating systems and compilers are written in high-level languages.

Subroutines, Libraries, and Loaders

Often when writing a program, the writer may have within the program the same repeated group of statements. It would save time for the programmer if such a group of statements could be written just one time and then be "called" whenever it is needed. All programming languages have mechanisms for defining groups of such statements and for then passing control to them as needed. Such groups are called "subroutines" and/or "functions" in some languages and "procedures" in other languages. Exhibit 12.1.5 gives a trivial example of a FORTRAN program that calls a subroutine, in this case a subroutine that prints a line of text every time it is called.

Although the example in Exhibit 12.1.5 is trivial, notice that the subroutine had an input vari-

Exhibit 12.1.5 Example of FORTRAN Subroutine

```
      .
      .
      x = 15.0
      call pline (x)
      x = 20.0
      call pline (x)
      .
      .
      subroutine pline (a)
      real a

      print 1000,a
1000  format ("the answer is ", f10.3)
      return
      end
```

able, a, and that is was called with different values of this argument (e.g., 15.0 and 20.0). Subroutines can be written with several input arguments and can also return calculated values. Thus subroutines provide a convenient way of structuring programs into independent sections, each of which can be called as needed. Such an approach is a great aid to writing programs that are more nearly correct and that are easy to understand.

Most systems have a system-provided program, called a "linking loader" (or a "linkage editor"), which can combine a program and several subroutines together to obtain a single load module or program in executable form. In such a system the compilers for all available languages and the assembler all produce "object code" or object modules of machine instructions in a form that is required by the loader. In particular, a program can call subroutines that are not supplied by the programmer, but rather are available in a library of subroutines on the system. The loader is able to detect these calls and to load object code for these called subroutines from a library along with the object code produced by the compiler. A uniform technique for passing arguments to called subroutines and for passing values back to the calling program is essential to this process.

This ability to have libraries of precompiled subroutines is of enormous benefit to people writing programs. Most systems have several libraries of such subroutines, including subroutines for performing common mathematical functions such as solving systems of simultaneous linear equations, integrating a function over an interval, or solving a linear programming problem. Other examples of these include subroutines for applying statistical tests to a set of data and subroutines for plotting data on a digital plotter. Also, many system-provided functions, such as printing formatted lines and reading input lines, are implemented as subroutines in libraries.

The result is a saving in terms of time and reduced programming errors. There are even companies that sell libraries of subroutines containing efficient procedures for performing certain functions. This approach, of using previously written subroutines to perform well-defined computing tasks, is very important, for it allows programmers to benefit from other people's results rather than "reinventing the wheel" every time a wheel is needed.

Files

A file is a named sequence of data. The attributes of a file include its name, its location, and its structure. On some systems, a file is called a "data set." Most large-scale systems are said to be file-based or file-oriented systems. In such systems, programs and data are stored as files, usually on rotating mass storage devices. To keep track of the large number of files in a system, the system has directories of files, where a directory is a list of files and, often, other directories (subdirectories). A directory is used when a file is accessed; therefore entries in a directory include the attributes of a file required to locate that file. The collection of directories and files is called a "file system."

On most systems, some files are said to be permanent files, in the sense that they persist in time. Other files are temporary; they disappear at the termination of the job that created them. A system that is used by people at remote terminals must have some permanent file storage capability. On this type of system, when a user gains access to the system, he or she also gains access to the collection of files. Reliable storage of these files is essential to productive use of the system. Most systems have facilities for making copies of permanent files in the unlikely event of accidental loss of the files. Such losses could severely affect the users of the system.

On a file-based system, some files hold system-provided programs, such as compilers, loaders, and so on. Some files hold jobs submitted by users that are waiting to be executed; other files hold output generated by programs that is waiting to be printed on a printer; and yet other files hold intermediate output from one program in a job, to be used by a subsequent program in that job. All of these uses are in addition to the files of programs, data, and text created and saved by users of the system. The point is that the file system is vital to the operation of the system.

Utility Programs and Editors

Many commonly needed functions are fulfilled by a collection of programs that are part of the system library of programs. Compilers and loaders, which are examples of such programs, have already been mentioned. There is a variety of other programs, each of which usually performs a single function and is classified as a utility program. Examples of utility programs include a program to copy the contents of one file to another, a program to format text for printing, a program to save files on magnetic tape, and a program to maintain a library of subroutines, to name just a few. A complete set of utility programs can make a system easy to use.

One program that is essential to users at remote terminals is an interactive editor. Such an editor accepts commands and text from a terminal and creates and modifies files in response. For terminal users, this editor is one of the primary means of communicating with the system and of using the system. These users can spend a great deal of time creating and modifying programs, data files, and text files, so, to them, the editor is an important part of the system software.

Operating Systems

A system can be viewed as a collection of hardware resources. The software (programs) required to use these resources and to allocate them to the users of the system is the operating system. For example, the actual instructions that transfer data between main memory and a peripheral device (called a "device driver") would be part of the operating system. Because almost every system provides some sort of device drivers, almost every system has at least a rudimentary operating system. As systems become larger and more complex, the operating system grows as well.

Most larger systems provide a multiprogrammed environment; that is, several user programs can reside simultaneously in main memory, each competing with the others for use of the resources of the system (the CPUs, I/O processors, etc.). In this type of system, the allocation functions of the operating system become very important, to ensure not only correct use, but also fair use, of the resources. One user program should not be able to monopolize use of the resources to the extreme detriment of the other programs. Also, since files for several users may be stored on the disk drives, the operating system may be required to prohibit illegal access by other users to files belonging to one user. We can see that the operating system has to fulfill many different functions.

When computers were first designed, the I/O devices operated in a synchronized fashion and were quite simple to use; each user program executed its own I/O instructions. In an attempt to decrease the running times of programs, separate I/O processors were introduced; these permitted asynchronous (overlapped) I/O operations. Although these did reduce program run times, they also added to the complexity of writing I/O programs. Operating systems were introduced to ease this burden. Multiprogramming was also added to further increase utilization of system resources, the idea being that if one program could not use the CPU because it was waiting for an I/O operation to complete, then perhaps another program could use the CPU.

Asynchronous operation and multiprogramming had one major consequence: a source of indeterminancy was introduced. If programs are not carefully constructed, it is possible for one program to affect another inadvertently. If two programs share access to the same resource, for example, a word in memory or a file, then it is possible for the results of each program to be in error only because of the sequence in which operations in each program occur, introducing an undesirable indeterminancy into the execution of these programs. To remedy this, access to sharable resources was brought under the control of the operating system; the mechanisms for doing this are called "interlocks." Interlocks guarantee that only one program at a time has access to a critical system resource. Interlocks added another problem, namely, deadlocks. Suppose that program A owns resource X (i.e., has interlocked resource X so that no other program can access it) and needs resource Y. If, at the same time, program B owns resource Y and needs X, then deadlock occurs, because neither program can proceed. Deadlock problems are solvable, but the solution adds still more to the complexity of the system.

Users communicate with the operating system in two ways: by system calls within their programs (e.g., a request by a program to read data from a disk resident file) and by a system command language (either "control cards" or commands from a remote terminal). Statements in the command language may request the execution of a program or access to a file, to name two of many possibilities. The type of user interface provided by the operating system can have an impact on the productivity of people who are using the system. If the commands are inconsistent and confusing, then the users have an additional source of potential errors and must search for procedures that will solve their computational tasks. On the other hand, if the command language is consistent and easy to use, then the user's contact with the system is productive and even pleasant. Today the quality of the user's interface (the editor, the file system, and the command language) is often a greater factor in determining the usefulness of a system than the speed of the hardware resources.

In summary, the operating system is a collection of programs that manage the resources of the system and that provide an environment in which user-defined application programs operate. The operating system fulfills many functions, including:

1. Transferring data between programs and files.
2. Initiating program execution (user or system provided).
3. Prohibiting illegal access to stored information.
4. Ensuring correct and noninterfacing execution of programs.
5. Granting access to resources in a fair and efficient manner.

Because the operating system is in total control of the resources of the sytem, the performance of the entire system usually depends on the performance of the operating system. All systems now have special hardware features to assist the operating systems. These include memory protection features, supervisor/user CPU modes (e.g., all I/O instructions may be executable only in supervisor mode), special context switching instructions (to allow the operating system to switch rapidly the CPU between programs), and timers and timer interrupts (to guarantee a maximum interval that a program has had possession of the CPU). Today when we use a system, we are really using the system defined by the operating system.

12.1.4 COMPUTER SYSTEMS

Function and Operation

Computer systems are normally used in one of two distinct modes: dedicated to a single task (or group of related tasks) or available for general-purpose use by members of a community of users. Dedicated systems are used in many situations, some of which include:

Laboratories, where minicomputers monitor and control experimental chemical processes.

Manufacturing companies, where computers control machine tools and automatic assembly and test equipment.

Airlines, where large systems handle millions of seat reservations.

Armed services, where computers simulate combat environments.

Government, where computers monitor air and water quality.

Education, where computer-assisted instruction drills students in basic skills.

Business, where desktop systems aid in planning and forecasting.

These are in addition to the applications mentioned in prior sections.

General-purpose systems are found in computing centers and are provided as a service to groups of users. These centers are usually located in large organizations, such as corporations, research laboratories, government agencies, and universities. These systems provide a wide range of services to their users, including several programming languages and extensive libraries of subroutines. A full complement of I/O devices is usually available.

Most of these systems were once accessed by means of card readers and line printers. The system was said to be operating in a "batch processing" mode—user programs were handled in batches. Today most systems can be accessed in several ways: In addition to directly connected (local) card readers and printers, there are remotely located card readers and printers, remote terminals and other computers, all connected by means of communication lines.

For users of remote terminals, there are different modes of using the system. In the most limited mode, terminal users can create and modify program and data files, can submit these to the batch system for processing, and can retrieve the output generated by these programs. In an expanded mode, terminal users can actually interact with executing programs, including programs they have written themselves. Originally the term "time-sharing" denoted this type of user access; today "interactive access" is the more commonly used term.

In many organizations, people who were formerly users of a centralized, general-purpose system are starting to acquire their own, smaller systems and to use them in a dedicated fashion, to process their own programs. However, many of these people still need to share information with other users and to have access to the facilities of a central system. A developing trend is the use of computer networking technology to connect these small satellite systems to one or more central sites.

In computing centers there are often three groups of people involved with the delivery of service: the operations group, the systems and systems programming group, and the user services group. In addition, there are people who are responsible for interfacing with the system vendor, with communications suppliers, and so on. There is a planning function, which is trying to predict future needs. Managing the computing function in a large organization has become a challenging job.

System Performance

All systems must eventually produce results, either in the form of a response to a request for service made by a human or in the form of a control signal for some attached equipment. In almost every case, there is some requirement or expectation for the time that elapses between the time of the request and the time of the response. The interval between these two events is called "response time." The term "real time" has been applied to systems with critical response time requirements.

The life of a system usually begins with two phases: In the first phase, the main concern is with producing correct responses; in the second phase, the main concern becomes meeting the requirements or expectations about response times. Put in other terms, these are the system correctness and system performance phases. It is performance criteria that dictate the sizes and speeds of systems.

The topics of performance evaluation and system capacity are now of great interest to people in charge of systems. Many systems have provisions for gathering data on the demands for service by users and the response of the system to these demands. Such data can be of great assistance in trying to determine causes of inadequate performance. Another development has been the introduction of systems models to aid in the prediction of system performance when the operating environment or the system configuration is changed. One type of model is based on discrete event simulation, the other on queuing theory. The goal of these models is to assist management in making decisions about the need to procure new equipment in response to changing demands for service.

12.1.5 SUMMARY

Today computers are tools of industry, government, and education. They are responsible for many changes in the way things are done in each of these areas. Computers are now even coming into the home. Most disciplines within engineering make use of computers in many ways. Needless to say, many quantitative techniques of industrial engineering depend on the use of computers for their usefulness.

This chapter has introduced the digital computer. The emphasis has been on the hardware and software components of computer systems and on the relationship between these two. It is impossible to conceive of a computer without considering both of these components. The software now is often the more expensive of these two components, and certainly the quality of the software has a major impact on the quality of the entire system.

This introduction has been brief. Many important concepts have only been mentioned. Fortunately, the remaining chapters cover most of these in more detail. The purpose here has been to give the reader a glimpse into the area of computers and to provide a vocabulary. There are many sources of information for those who wish to pursue this area. A few books and articles on computer systems are listed in the bibliography. Almost all colleges and universities have a department of computer science or a computer-oriented group within the mathematics department, the school of business, or the school of engineering. Within these, there are courses at all levels of interest. Many magazines routinely feature articles on computers and their applications in many areas. There are two professional organizations, the Association for Computing Machinery (ACM) and the Institute for Electronic and Electrical Engineers Computer Society (IEEE/CS), that have several publications and that sponsor meetings and seminars. Also, many groups offer professional development seminars. Given the importance of computers in all technically oriented areas, there is a great deal of motivation to learn about them.

BIBLIOGRAPHY

BELL, C., J. MUDGE, and J. MCNAMARA, *Computer Engineering–A DEC View of Hardware Systems*, Digital Press, Bedford, MA, 1978.

Communications of the ACM, Vol. 21, No. 1 (January 1978), special issue on computer architecture.

HABERMANN, A., *Introduction to Operating Systems Design*, SRA, Chicago, 1976.

LEVY, H., and R. ECKHOUSE, JR., *Computer Programming and Architecture–The VAX 11*, Digital Press, Bedford, MA, 1980.

STONE, H., Ed., *Introduction to Computer Architecture*, SRA, Chicago, 1980.

CHAPTER 12.2

Fundamentals of Information Processing Systems: Classification and Characteristics

DANIEL TEICHROEW

The University of Michigan

12.2.1 TERMINOLOGY

Purpose of the Chapter

The industrial engineer frequently encounters terms such as "data," "data processing," "data processing system," "computer system," and "information processing systems." Although definitions appear in most of the standard encyclopedias and dictionaries, for example, Ralston and Meek[1] and Belzer et al.,[2] they frequently do not agree. The purpose of this chapter is to summarize the most commonly accepted terminology and to describe characteristics that are most relevant in industrial engineering.

Systems

A "system" is defined by Webster as "a set or arrangement of things so related or connected as to form a united or organic whole; as, a solar system, irrigation system, supply system." The type of system of concern here is one in which the unifying concept is "data." Chapter 12.1 describes information processing systems (IPSs) that include computer systems as subsystems.

Individual IPSs are usually identified by a name, for example, the production control system, or more commonly by an acronym, for example, PCS. In addition to the computing system, any such system usually includes the following:

Functions or operations to be performed, which are incorporated into a set of computer programs, usually called "application software," which perform the function.

Outputs (documents, reports, messages, etc.), which are produced by the system.

Inputs, which are accepted and processed by the system.

Data, usually called a "data base," which is stored by the system.

Noncomputerized procedures that must be followed.

Data Versus Information

The terms "data" and "information" are sometimes used interchangeably, although some consider them to be different concepts. In general, data are language, mathematical, or symbolic representations of people, places, things, and events, whereas information results from the filtering, processing, and formatting of data in a way that increases the level of pertinent knowledge for a recipient. Data are objective, whereas information is subjective and exists only when it is relevant to a recipient.

Data Operations

Data processing, or information processing, systems are systems that manipulate data to produce other data or information. In such systems different types of operations may be performed, such as those in the following list, which is adapted from Voich et al.[3]:

1. **Recording.** Identifying and transcribing data about events or transactions, for example, sale of a product, purchase of a material, payment of a bill, and production of goods.
2. **Classifying.** Encoding data on events or transactions by major types, for example, sales by salesperson and material purchases by type of material.
3. **Sorting.** Screening, segmenting, and compiling encoded data on events and transactions into categories, for example, identifying and compiling sales made by each salesperson and purchases by types of material.
4. **Calculating.** Adding, subtracting, multiplying, dividing, comparing, arranging, and so on, transaction data, for example, computing total sales volume for each salesperson during the month or total purchases or types of materials each month.
5. **Summarizing.** Arranging transaction data and related calculations into report format, for example, developing a sales performance report for each salesperson or a material status report for each type of material.
6. **Storing.** Holding transaction data or summarized data temporarily or permanently, as needed, for example, as a matter of official record or for use in later reporting periods.
7. **Retrieving.** File searching or scanning of stored data to perform additional calculations or to update data on file, for example, identification of all sales transactions recorded during a period.
8. **Reproducing.** Printing or developing reports, for example, sales performance reports or materials status reports.
9. **Distributing.** Moving reports to users of information, for example, to sales manager or procurement manager.

Levels of Mechanization

The mechanisms used to perform the data processing operations may be classified in three general categories:

1. Manual data processing operations are performed by a human being mentally or using paper and pencil.
2. Electromechanical data processing operations are performed with the aid of electrical and mechanical equipment, such as typewriters, cash registers, duplicating machines, and telephones.
3. Electronic data processing operations are performed by a computer system. The system consists of input devices, a CPU, and output devices. Most data processing or information processing systems in fact include all three types of mechanisms.

System Characteristics

Information processing systems have the following characteristics:

1. The systems are manmade; that is, they have to be designed, constructed, operated, and maintained. This is a nontrivial task and has led to the need for methods of system development, operation, and maintenance. An introduction to the topic is given in Chapter 12.4.
2. In the development and operation of information systems, both the programs and the data base are important.
3. Because of the large cost involved in developing information systems, there is an economic need for systems to share hardware, software, and data bases.
4. The systems tend to be large and costly to develop, operate, and maintain. This arises because of the economies of scale involved in larger hardware and in operation and maintenance of systems.
5. The systems involve man-machine communication at various levels, and problems of design and operation include problems of communication among individuals, of communication with the machine, and of communication among the various units of the machine. Therefore, documentation is an important aspect.
6. The uses of the systems and the technology on which the systems are developed are continuously changing, as are the organizations using them; consequently, the systems themselves are seldom, if ever, static.

Information systems are expensive to develop and to operate; consequently, analyses to determine whether they are serving the desired needs of users and the measurement of their performance are receiving considerable attention. Performance evaluation must be considered at a number of levels. At the top level, the value of the output of the system to the orgainzation that supports it must be determined. Once these specific outputs have been justified, the performance of the physical system in achieving these outputs must be measured. This performance is a combination of the performance of programs, software, and the hardware equipment itself. The process of designing and implementing systems is discussed in Chapter 12.4.

12.2.2 CLASSIFICATION OF INFORMATION PROCESSING SYSTEMS

Information processing systems of the type just defined may be classified in various ways. Some classification methods concerned with the functions performed by these systems in organizations are described in the next section. In this section some of the classification methods based on the physical organization and structure of the system are discussed. Systems are frequently categorized by the type of service performed and by response determined by the processing organization. In none of these categories is there any standard terminology, and the following is based on the most commonly used terms.

Type of Service

Systems may be classified by the type of service they are designed to provide, as follows:

1. **Computing Service.** These systems accept demands for computing in the form of computer programs and carry out their instructions. Normally these systems are intended for engineering and scientific users, but they are also used for one-of-a-kind administrative tasks.
2. **Information Storage and Retrieval.** These systems routinely maintain a file of documents and data and respond to inquiries. For example, a system might maintain data about current research projects.
3. **Command and Control.** These systems are extensions of the previous type in that they are programmed to issue notices when certain events occur. For example, the system mentioned in classification might be instructed to produce a notice to personnel who may be interested in a new report.
4. **Transaction Processing.** These systems accept (predefined) inputs and produce (predefined) outputs according to a given schedule. For example, a system may accept notices of deposits and withdrawals and produce a monthly account statement or a notice when the account reaches a negative balance.
5. **Message Switching.** These systems accept an input at a given geographical point and produce an output at some other specified point.
6. **Process Control.** These systems accept inputs from machines and produce signals that can alter the operation of the machine.

Each of the types of systems listed has a certain capability. In practice, systems being built today have a combination of these abilities and, in some cases, all the capabilities needed.

Response

One of the characteristics of interest, particularly to users, is the response of the system to a demand for service or an event. There are basically four different ways in which resources can be allocated to affect the response:

1. **Batch Processing.** The system resources are used to operate certain processes at certain times. The user gets a response when the last process needed is completed.
2. **Store and Forward.** The system resources are used as in a job shop. Each resource has a queue of items needing attention, and these are processed according to some priority rule. The user again receives a response when the last process needed has been completed. However, the user cannot be certain that it will be accomplished by a certain time, as in batch processing.
3. **Parallel operation.** If the system has more than one resource of a particular kind, provision can be made for parallel processing. In general, for a given total system cost, the response time will be decreased.
4. **Time-sharing.** In such a system each task requiring a resource receives a certain amount of time rather than waiting until all previous tasks have been completed, as in "store and forward." A user requiring only a little of each of a number of resources will receive a faster response, whereas the user requiring a large amount of one resource will receive a slower response.

12.2.3 ORGANIZATIONAL IPS

As has been indicated, IPSs may be designed and built to serve different purposes. This section outlines some of the methods used to classify organizational IPSs from the point of view of the using organization.

Application Area

"Application" refers to the major function performed by the system, for example, airline reservations system, inventory control system. Frequently these functions correspond to the functional areas of the organization; in a manufacturing company they might be marketing, production, finance, accounting, and planning.

Level of System*

There is no one theory of information systems; however, the user or designer of a system needs some conceptual model of systems. Because of the ill-structured nature of the systems field and its interdisciplinary origins, the best one can expect is to develop an intuitive model of information systems.

Anthony[4] has suggested several different types of decisions that are made in organizations: (1) strategic planning, which is the process of deciding on organizational objectives and means for achieving them; (2) managerial control decisions, which involve the manager's ensuring that resources are used efficiently and effectively to achieve objectives stated during strategic planning; and (3) operational control decisions, which involve specific tasks that must be completed efficiently and effectively on a day-to-day basis.

The Anthony framework stresses that different types of decisions require different types of information. For example, strategic planning decisions require infrequently updated, predictive, and aggregated data often from external sources. The major contribution of this framework is its distinction among types of decisions and their information requirements.

Systems are sometimes categorized in accordance with the Anthony framework by the level of the organization for which they are primarily intended.

Operational or "Routine" Systems

These systems carry out routine data processing that is closely related to the operational activities in the organization. For example, in a manufacturing operation the routine systems would process time cards and produce paychecks, process customer orders and produce invoices, and so on. Such operations are usually characterized as high volume and routine.

Middle Management Systems

These systems are intended to serve middle management or, in some cases, to replace some of middle management decision making by automated or semiautomated methods. For example, a system may automatically determine the quantity of goods to be purchased or select the appropriate vendor. These systems are also used for control, for example, checking amounts spent against the budget and determining variances.

Top Management or Strategic Planning Systems

These systems are designed to aid management. They generally include not only summary data from lower-level systems, but also data from outside the organization. Such systems are sometimes referred to as "management information systems" (MISs). This category is characterized by the generation of strategic decision aids that supply information to users, thereby facilitating fundamental organizational planning functions. Management information systems range widely in complexity and size, depending on requirements set forth by organization and/or functional users. The systems provide for the generation of information essential to financial, technological, marketing, and human resources analysis and planning. For example, a typical MIS concerned with human resources analysis and planning might be called the human resources information system. This system might generate decision aids in conjunction with:

Selection—recruiting and placing persons in appropriate jobs.

Utilization—distributing personnel according to current and future needs.

Maintenance—provision of services in terms of employee benefits, career pathing, salary administration, absence profiles, and so on.

Accounting—maintaining an ongoing evaluation of the usefulness of the human resources to the organization.

Planning—providing for future personnel needs according to existing and future skill and ability requirements.

In terms of scope an MIS may cover a wide area of data. New data sources may be incorporated as they become available. The users of MISs are customarily looking toward the future and contemplating some fundamental change. Often they are involved in the "development of innovation."

*The three paragraphs that follow are reprinted by permission of the author from a draft copy of Chapter 12.6 of this *Handbook*, "Systems Design to Encourage Use," by H. C. Lucas.

Type of Decision Supported*

Simon[5] has suggested two types of decisions: programmed and nonprogrammed. Programmed decisions are routine and repetitive–some specified procedure may be applied to reach a decision each time the situation arises. On the other hand, decisions that are nonprogrammed are novel and unstructured. There is no one solution to these nonprogrammed decisions since the problem has probably not appeared before.

Different types of decision making technology are suitable for attacking each type of decision. Programmed decisions have been made traditionally through habit, by clerical procedures or with other tools. More modern techniques for solving programmed decisions involve operations research, mathematical analysis, modeling, and simulation. Nonprogrammed decisions tend to be solved through judgment, intuition, and rules of thumb. Modern approaches to nonprogrammed decisions include special data analysis programs on computers, training for decision makers in heuristic techniques, and heuristic computer programs.

The operational support systems (OSSs) defined here may be divided, following the preceding dicotemization, into those that generate tactical decision aids (tactical OSSs) and those that generate programmed decision aids (programmed OSSs).

Tactical OSSs are concerned with the implementation and control of ongoing processes. They tend to operate in the short term and in "real" time. To a considerable extent they are an integral part of the system for which they act as the information processing subsystem (e.g., like the auto pilot subsystem of an aircraft). The operation of tactical OSSs is therefore characteristically paced by the system operation. The criteria for defining their effectiveness are defined in terms of the overall system performance within which they operate. Typical examples of systems generating tactical decision aids include control operations in petrochemical plants, air traffic control stations, and nuclear power control rooms. Tactical decision aids are closely bounded and have well-defined limits. Information acquisition, throughput, and output are specified and routinized. Repetition and prescribed performance are characteristic of tactical decision aids. Some of the typical operations in tactical OSSs are the description of current events; the monitoring of current performance; short-term prediction; and the analysis and simulation of new circumstances, contingencies, and so on.

Programmed OSSs employ thoroughly routinized and prescribed information flows. Automatic and electronic data processing systems are examples of programmed OSSs. Often such systems are concerned primarily with the transmission of data, with some minimal amount of data processing going on. For example, banking money transfer systems are set up to transfer accounts (dollar quantities) from one site to another, with the data processing activity being confined to sorting the items in terms of origin and destination. The overriding requirement of such systems is accuracy. Hence error detection routines (such as parity checks–two or more processes done in parallel to see if they yield similar results) are usually built into programmed OSSs. Programmed OSSs are commonly used for transferring data, consolidating data at a central point, updating and justifying records, and accessing data banks.

Combination of Level and Type of Decision*

Gorry and Scott-Morton[6] synthesized the framework of Anthony and Simon to develop a model of information systems. They cast the Anthony decision types as the columns of a matrix and Simon's programmed and nonprogrammed decisions as rows. Most existing information systems have attacked problems in the structured operational control cell of this matrix. These problems are similar in many organizations and are among the most easily understood. It is easier to mechanize these decisions and to predict and achieve cost savings than it is for less structured decisions or for strategic planning decisions. Since these operational systems play an important role in the daily functioning of the firm, they are high-priority applications.

Many experts in the systems field believe that decisions with the greatest payoff for the organizations are unstructured in nature. The development of systems for unstructured problems is a major challenge and undoubtedly more risky than the development of systems for structured problems. In the structured case the goal of an information system is usually to improve the processing of information. In an unstructured application the goal of the system is more likely to be improving the organization and presenting information input for the decision maker.

Degree of Integration

Systems are sometimes classified by the extent to which they cover the whole organization. Single-function systems perform one function. Any communication with other systems usually requires

*The two paragraphs that follow are reprinted by permission of the author from a draft copy of Chapter 12.6 of this *Handbook*, "Systems Design to Encourage Use," by H. C. Lucas.

that outputs of one system be transformed into inputs that are in a form acceptable to the next system.

Multifunctional systems are ones that accomplish two or more functions in the organization. Frequently such systems are built by combining the single-function systems by developing common interfaces.

Integrated systems are ones that serve all the needs of the organization. Implied in this definition is the concept of a single entry of information: A piece of data enters the system only once and is then used wherever needed. Similarly, a piece of data is stored in only one place.

The degree of integration is important because of the trade-offs involved. Single-funtion systems can be optimized for the individuals in that function. However, the same data may be entered or stored in several single-function systems. Similarly, it is much easier to obtain management information from an integrated system than from a number of single-function systems.

REFERENCES

1. A. RALSTON and C. L. MEEK, Eds., *Encyclopedia of Computer Science*, Petrocelli/Charter, New York, 1976.
2. J. BELZER, A. G. HOLZMAN, and A. KENT, *Encyclopedia of Computer Science and Technology,* Dekker, New York, 1975.
3. D. VOICH, H. MOTTICE, and W. SHROEDE, *Information System for Operations and Management,* Southwestern Company, Pelham Manor, NY, 1975.
4. R. ANTHONY, *Planning and Control Systems: A Framework for Analysis,* Division of Research Graduate School of Business Administration, Harvard University, Cambridge, MA, 1965.
5. H. SIMON, *The Shape of Automation for Men and Management,* Harper & Row, New York, 1965.
6. G. A. GORRY and M. S. SCOTT-MORTON, "A Framework for Management Information Systems," *Sloan Management Review,* Vol. 13, No. 1 (1971), pp. 55–70.

CHAPTER 12.3

Data Base Management

CLYDE HOLSAPPLE
University of Illinois

SHELDON SHEN
Ohio State University

ANDREW B. WHINSTON
Purdue University

12.3.1 INTRODUCTION

Spanning the series of revolutionary changes in computer technology and hardware has been an evolution of software and methodology. Of interest here are information handling methods and associated software. The evolution of information handling has progressed from primitive file management to sophisticated data base management. Although file management is still predominant, the use of data base management is spreading because of its inherent advantages over file management. Not the least of these advantages is the increased productivity afforded to those who develop, maintain, and use information systems, particularly where the application being modeled by the information system is of a complex nature.

A principal impediment to the more widespread use of data base management, particularly on minicomputers and microcomputers, has been the absence of a widespread understanding of just what a data base system is in the technical sense. Coupled with this is the shortage of persons trained in the design, development, and administration of information systems that utilize data base management software. In its loose, popular sense, "data base" simply refers to a collection of data. The way in which this collection of data is stored and processed is not implied in this loose usage of the term. It may involve the noncomputerized storage and manual processing of the data collection, the use of file management software, or the use of data base management software.

In this chapter "data base" is used in the strict, technical (and more informative) sense of the term. Whereas a file consists of many records of some given record type, a data base consists of many records of many different record types integrated into a single, organized structure in which all important record interrelationships are represented and maintained. The facilities available for defining a file (be it sequential, direct, indexed-sequential, or inverted in organization) differ substantially from the facilities available for defining a data base. An information system using file management software will employ many separate files. File processing centers on the activity of merging separate files based upon common (i.e., redundant) data values that exist in the separate files. In data base processing there is no merging of separate files; rather, the central activity is one of finding desired records in the data base on the basis of their interconnections with other records. This chapter offers a comparative study of the major features of three different kinds of data base systems.

A data base system is a software package. By itself, this software is not an information system, but it can be used to build an information system for some application (e.g., order processing, inventory management, payroll). A data base system is general in the sense that it is not oriented

Preparation of this chapter was supported in part by Army Research Office Grant Number DA 79C0154. The views and conclusions contained herein are those of the authors and should not be interpreted as necessarily representing the official policies, either expressed or implied, of Purdue University or any agency of the U.S. government. The authors are indebted to Jim Gray of Tandem Corporation for his helpful comments on an earlier draft.

toward any specific application. It provides facilities for defining the logical structure according to which records will be organized in a data base. It also provides facilities for manipulating data in the data base; "manipulation" includes the activities of creating, deleting, modifying, and finding records and extracting data.

Data base systems can be classified on the basis of the logical structuring facilities that they provide for building information systems. Some support full *network* logical structures, some support only *hierarchical* structures, and some allow only disjoint tables (i.e., *relations*) for modeling an information system's data. Each of the three data models is discussed in the next section. Following this discussion, a representative data base system is examined for each of the three data models: IBM's Information Management System (IMS) for the hierarchical model, System R for the relational model, and the Micro Data Base System (MDBS) for the network model. Each system is examined from the standpoints of its data base definition and data manipulation facilities. A simple supplier-parts application is used to illustrate the features of each system.

12.3.2 THE THREE DATA MODELS

The hierarchical, relational, and network data models employ different terminologies for defining the logical structure of a data base. Moreover, two data base systems subscribing to the same data model (e.g., two hierarchical systems) may utilize different terminologies. This presentation is based on the following definitions contained in the landmark CODASYL *Data Base Task Group Report* of 1971:[1]

1. **Data Item Type.** A data item is the smallest unit of named data. For example, supplier name (SNAME) and supplier address (ADDRESS) are data item types.
2. **Item Occurrence.** An actual occurrence or instance of a data item is an item occurrence. An example of an item occurrence for SNAME would be Smith.
3. **Record Type.** A record type is an aggregation of data item types. For example, a record type called SUPPLIER might be defined as consisting of the data item types SNAME and ADDRESS.
4. **Record Occurrence (or Simply: Record).** A record occurrence is an instance or occurrence of a specified record type. An example occurrence of the record type SUPPLIER might consist of the item occurrences Smith, 740 Main Street.
5. **Set Type (or Simply: Set).** A set is a named relationship between record types. One record type is called the set's owner, and another record type is called the set's member. A set represents a one-to-many (1 to N, where $N \geq 0$) relationship in which any *one* occurrence of the owner record type may be associated with *many* (N) occurrences of the member record type, but no member occurrence can be associated with more than one occurrence of the owner record type. For example, a customer can place many orders, but no order is placed by more than one customer. Thus we can define a set (call it PLACES) having CUSTOMER record type as its owner and ORDER record type as its member.
6. **Set Occurrence.** A set occurrence consists of an occurrence of the owner record type plus all associated member record occurrences (e.g., all of the order records associated with a particular customer record occurrence).

The foregoing definitions serve as a basis for understanding the terminologies of the hierarchical, relational, and network systems. The correspondences among the three terminologies are illustrated in Exhibit 12.3.1*a*, clearly indicating certain commonalities among the data models. The distinctions among the three kinds of logical structures are evident from examining the example logical structures of Exhibits 12.3.1*b* through 12.3.1*d*. A rectangle represents a record type (segment type, relation schema). A directed arc represents a set (parent-child relationship) pointing from the owner record type (parent segment type) to the member record type (child segment type).

In the network model, any record type can be the member as well as the owner of many sets. In the hierarchical model, a segment type can be the child in at most one parent-child relationship. With the relational model, the designer of a logical structure for some application is not permitted to define explicitly relationships between two record types. However, a relationship can be implicitly defined by repeating the same attribute in two different record types (i.e., two different relation schemata). This is illustrated later in the discussion of System R.

Each of these three approaches furnishes a different way of organizing one's thoughts about the nature of the application being modeled within an information system. They provide different ways for viewing or modeling the real world: as a network, as a tree(s), or as a collection of tables. These fundamental differences in specifying a data base's logical structure result in differing data manipulation facilities. Since the hierarchical and network models explicitly represent relationships among record types, data manipulation is accomplished in terms of those relationships (i.e., sets). With the relational model, there are only tables. Data manipulation is therefore accomplished in terms of combining tables on the basis of common attributes and extracting data from tables with operators such as SELECT, PROJECT, and JOIN.

Exhibit 12.3.1 Three Data Models: (*a*) Terminological Correspondences, (*b*) Example of Network Model, (*c*) Example of Hierarchy Model, and (*d*) Example of Relational Model

Network (MDBS)	Hierarchical (IMS)	Tabular/Relational (System R)
Data item type	Field	Attribute
Item occurrence	Field value	Attribute value
Record type	Segment type	Relation schema
Record occurrence (record)	Segment	Tuple
Set type (set)	Parent-child relationship	–
Set occurrence	–	–

(*a*)

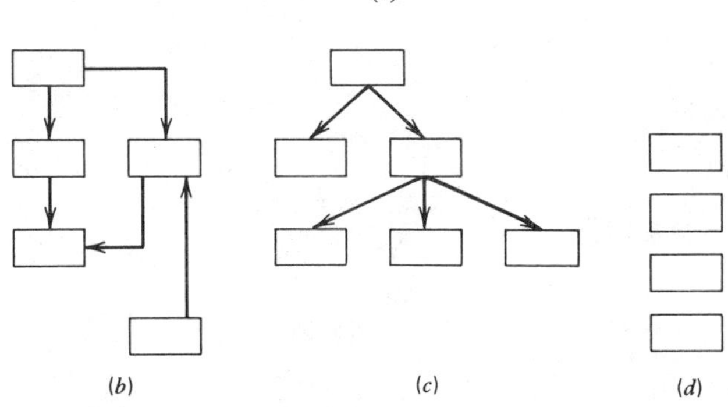
(*b*) (*c*) (*d*)

The sections that follow present the data description and manipulation facilities of IMS, System R, and MDBS. Certain distinctive features are emphasized. For example, IMS requires the specification of a physical storage organization in addition to a logical structure. The other two systems are less concerned with the physical placement of data base records, offering limited control over the clustering of records. The physical storage organizations available through IMS are discussed; clustering facilities in other systems are not addressed. With respect to System R, we stress SEQUEL, its data manipulation language. In the case of MDBS, its innovative logical structuring facilities (which go beyond the facilities of other network systems) are emphasized.

12.3.3 INFORMATION MANAGEMENT SYSTEM

Overview

International Business Machine's IMS is based on the hierarchical data model.[2] In IMS, one must differentiate between two kinds of data bases: a physical data base and a logical data base. The logical structure of a physical data base is a hierarchy of segment (i.e., record) types. The logical structure of a logical data base is also a hierarchy of segment types, each of which has already been defined to exist in the logical structure of a physical data base. Thus in IMS we can have physical data bases without logical data bases, but not vice versa.

An IMS data base, be it physical or logical, consists of segment (i.e., record) occurrences of the segment types specified in a hierarchy. The major difference between a physical data base and a logical data base is that, for the former, one of several physical storage techniques must be selected. The technique chosen determines how occurrences of segment types in the hierarchy will be physically stored relative to each other. Since a logical data base's hierarchy is defined in terms of segment types from one or more physical data base hierarchies, there is no choice of relative physical placements of segment occurrences in a logical data base.

The parent-child relationship between two segment types in a physical data base hierarchy is said to be a physical parent-child relationship, having a physical parent and a physical child. A logical data base hierarchy can contain physical parent-child relationships, but is also defined in terms of logical parent-child relationships. Unlike a physical parent and its physical child, pointers are used to keep track of the segment occurrence that is the logical parent of segment occurrences of the logical child segment type. A segment type can have at most one logical parent and one physical parent.

The foregoing notions are illustrated in the following example: Exhibits 12.3.2*a* and 12.3.2*b*

Exhibit 12.3.2 Physical Data Bases: (*a*) the SUPPDB Data Base, (*b*) the PARTDSDB Data Base, (*c*) the Logical Relationship Between Two Segment Types, and (*d*) User View Based on Logical Relationship

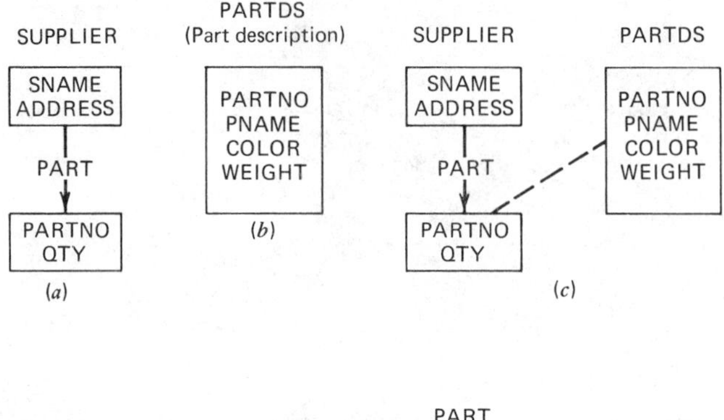

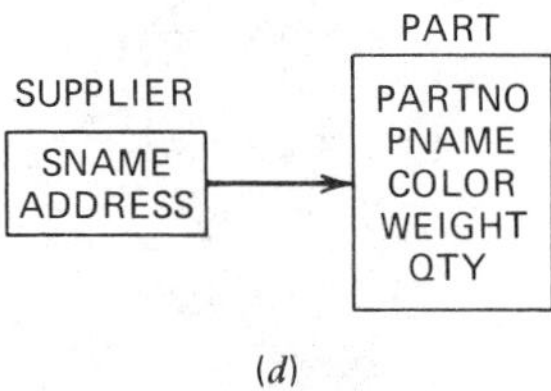

depict the logical structures (i.e., "hierarchy charts") of two physical data bases. The broken line in Exhibit 12.3.2*c* denotes a logical relationship between PART and PARTDS in which PARTDS is the logical parent of PART; SUPPLIER is the physical parent of PART. The logical relationship can be unidirectional (each occurrence of PART having a pointer to the related occurrence of PARTDS) or bidirectional (also having pointers from each occurrence of PARTDS to its related PART occurrences). This logical relationship can be used to declare the logical data base whose hierarchy chart is shown in Exhibit 12.3.2*d*, the logical relationship serving as a basis for the concatenation of the PART and PARTDS segment types. Data manipulation can proceed on the basis of the hierarchy chart in Exhibit 12.3.2*d*.

The rationale for permitting logical parent-child relationships in addition to physical parent-child relationships is to allow relationships to be established between physical data bases. This allows IMS to support limited forms of networking. It also leads to a reduction in redundancy. If the hierarchy chart of Exhibit 12.3.2*d* was for a physical, rather than a logical, data base, there would be greater redundancy. Since a particular part can be supplied by many suppliers, there would be many occurrences of the PART segment type for that part. Each of these occurrences would have the same value for their PNAME field. The same is true for their PARTNO, COLOR, and WEIGHT fields. The logical parent-child relationship shown in Exhibit 12.3.2*c* mitigates these redundancies by utilizing a pointer to point to the PARTDS occurrence associated with a PART occurrence.

Physical Storage Organizations

In IMS a user is required to specify a physical storage organization for each physical data base. Four main types of storage organizations are allowed: hierarchical sequential access method

Exhibit 12.3.3 Hierarchy Chart (*a*) and Two Data Base Records (*b*)

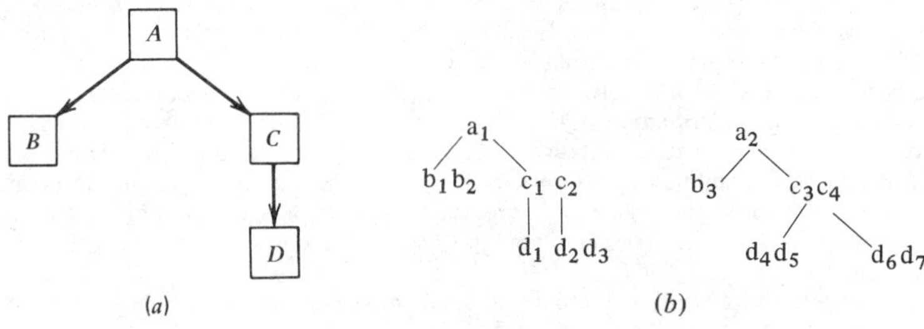

Exhibit 12.3.4 Structure of HSAM File

$a_1b_1b_2c_1d_1c_2d_2d_3a_2b_3c_3d_4d_5c_4d_6d_7\cdots$

(HSAM), hierarchical indexed sequential access method (HISAM), hierarchical direct access method (HDAM), and hierarchical indexed direct access method (HIDAM). Each denotes a different method for mapping segments to storage and a different method for accessing a "data base record." An IMS data base record must not be confused with the CODASYL notion of a record described previously. A CODASYL record is the same as an IMS segment. An IMS data base record consists of a root segment (an occurrence of the highest segment type in a hierarchy) plus all of its descendant segments. For example, a data base record for the hierarchy of Exhibit 12.3.2*a* consists of a SUPPLIER occurrence and all of its associated PART occurrences. Exhibit 12.3.3*b* shows two data base records for the hierarchy chart of Exhibit 12.3.3*a*. (Each lower-case letter denotes a segment.)

With HSAM, the segments of a data base record are related by physical adjacency, and the data base records are sequentially organized. Exhibit 12.3.4 shows the physical positioning of segments for the two data base records of Exhibit 12.3.3*b*. With HSAM, data base records are sequentially accessed.

The HISAM utilizes two storage areas: the primary and overflow areas. An attempt is made to relate the segments of a data base record by physical adjacency within the primary area. However, the space allocated for a data base record in the primary area is constrained. If all of its segments do not fit into the allocated primary space, the remaining segments are stored in the overflow area. A pointer is used to link these overflow segments with the appropriate data base record in the primary area. Within the primary area, data base records are stored sequentially (on the basis of a key field in the root segment type). A data base record is accessed through an index of the key values. Exhibit 12.3.5 shows the HISAM organization for the two data base records of Exhibit 12.3.3*b*.

In the physical organization used by HDAM, segments within a data base record are related by pointers. Two major options (with several variants) are available: hierarchical chaining and physical child/physical twin chaining. Hierarchical chaining provides a depth-first linkage of segments in a data base record, as shown in Exhibit 12.3.6*a*. Physical child/physical twin chaining relates all the segment occurrences under the same physical parent occurrence together by pointers. In addition, the left-most physical child occurrence is also pointed to by its physical parent occurrence, as shown in Exhibit 12.3.6*b*. Some segments of large data base records may be physically located in an overflow area. The HDAM allows a data base record to be directly accessed (by way of a hashing function) on the basis of a key field in its root segment.

Exhibit 12.3.5 The HISAM Organization for the Two Data Base Records of Exhibit 12.3.3b

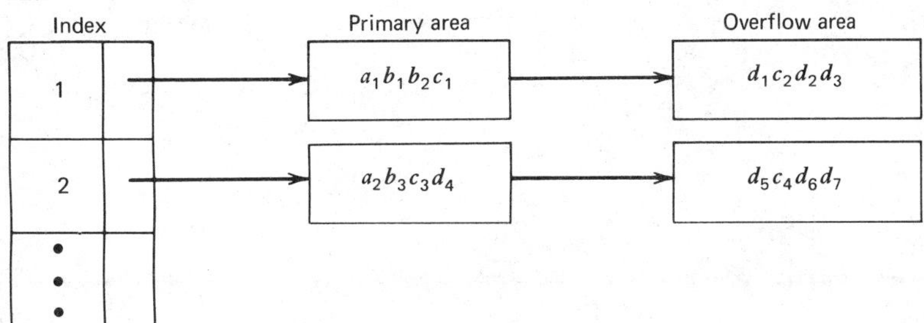

Exhibit 12.3.6 Hierarchical Chaining (*a*) and Physical Child/Physical Twin Chaining (*b*)

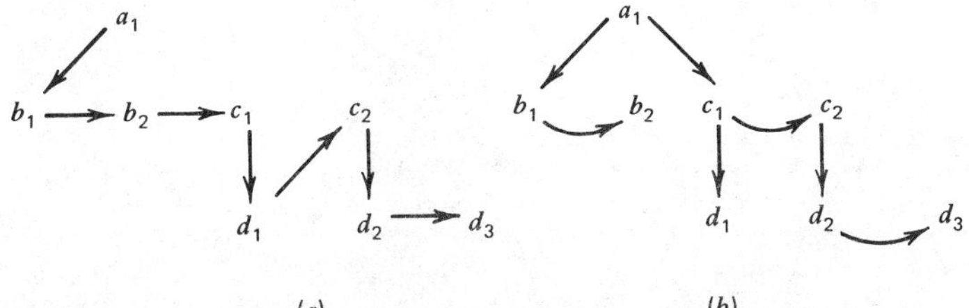

The HIDAM utilizes nearly the same approach to physical storage organization as HDAM. However, HIDAM uses an index to accomplish direct accessing of data base records. The index is based upon the key field of the hierarchy's root segment type.

Formal Specification of a Data Base

The critical data base design issues for IMS users involve decisions regarding the constitution of the hierarchy charts for physical data bases, where to use logical parent-child relationships, the hierarchy charts for logical data bases, which physical storage organizations to adopt for a physical data base, and the fields to be used as sequencing keys for each segment type. After these issues have been resolved, logical and physical data bases are formally specified using the IMS Data Base Description Language. The major kinds of statements in this language are described and illustrated as follows:

1. **DBD.** This statement names the data base and specifies the access method. The access method could be HSAM, HISAM, HDAM, or HIDAM, indicating one of the four storage organizations.
2. **DATASET.** This statement names the physical file that will contain the data base and specifies the physical storage device on which that file resides.
3. **SEGM.** This statement defines a segment type by giving its name and the name of its parent segment type(s). The parent clause has the following form: PARENT = (⟨physical parent⟩), (⟨logical parent⟩, ⟨data base⟩), indicating the physical and logical parents of the segment type. ⟨Data base⟩ is the physical data base to which the logical parent belongs.
4. **FIELD.** This statement defines a data item type in a segment type. The NAME parameter gives the name of the field. Other parameters specify physical length in bytes, starting position, and data type of the field. One field in a segment type can be declared to be a *sequence* (i.e., key) *field* for use in ordering segment occurrences. A sequence field can be unique or nonunique. A unique key field in a child segment type means that no parent segment occurrence can have two children segments with the same value for their sequence field.
5. **LCHILD.** This statement is used to define logical relationships, indicating a logical child.
6. **DBDGEN, FINISH, END.** These statements close the data base definition.

The following are the data base descriptions for the physical data bases shown in Exhibits 12.3.2*a* and 12.3.2*b*. Notice that the logical parent-child relationship, where PARTDS is the logical parent and PART is the logical child, is defined within the physical data base specifications.

```
DBD       NAME = SUPPDB, ACCESS = HDAM.
DATASET   NAME = SUPPHDAM,
SEGM      NAME = SUPPLIER, BYTES = 30
FIELD     NAME = (SNAME, SEQ, U), BYTES = 10, START = 1
FIELD     NAME = ADDRESS, BYTES = 20, START = 11
SEGM      NAME = PART, PARENT = ((SUPPLIER), (PARTDS, PARTDSDB)), BYTES = 5,
          POINTER = LPARENT
FIELD     NAME = (PARTNO, SEQ, U), BYTES = 5, START = 1
FIELD     NAME = QTY, BYTES = 10, START = 11
DBDGEN
FINISH
END
DBD       NAME = PARTDSDB, ACCESS = HDAM
DATASET   NAME = PDSDBSAM,
SEGM      NAME = PARTDS, BYTES = 50,
LCHILD    NAME = (PART, SUPPHDAM)
FIELD     NAME = (PARTNO, SEQ, U), BYTES = 5, START = 1
FIELD     NAME = PNAME, BYTES = 15, START = 6
FIELD     NAME = COLOR, BYTES = 10, START = 21
FIELD     NAME = WEIGHT, BYTES = 10, START = 31
DBDGEN
FINISH
END
```

A unique sequence field is denoted by (NAME, SEQ, U). An M in place of U indicates a nonunique sequence field. The second argument of the PARENT clause for the PART segment type indicates that PARTDS (in the PARTDSDB data base hierarchy) is the logical parent of PART. Logical data base hierarchies are defined in a fashion similar to that of physical data base hierarchies.

Once data base hierarchies have been defined, subhierarchies must be formally defined. Each subhierarchy is called a "program communication block" (PCB). It must contain the root segment type of a physical or logical data base. It may consist of an entire physical or logical data base. An application program can interface with a data base only in terms of a "program specification block" (PSB), which is a named collection of PCBs. A program's PSB effectively defines that program's view of the data bases. Information Management System uses a PSB to allocate and manage a buffer area between the data bases and an executing application program.

A program specification block is defined in terms of the following kinds of statements:

1. **PCB.** This statement defines a program communication block. DBDNAME specifies the name of the data base concerned. This statement is followed by a description of some subhierarchy within this data base.
2. **SENSEG.** This statement specifies the segment types to be included in the PCB view. NAME gives the name of the segment type. PARENT indicates the parent of this segment type. The parameter PROCOPT gives the processing options available for occurrences dealing with this segment type. These include G (get), I (insert), D (delete), R (replace), and K (key sensitive). A key sensitive option means that the segment type is not part of the PCB view and is mentioned only because some of its children are in the view.
3. **PSBGEN.** This statement specifies the host language (i.e., the language from which the data bases will be accessed) and names the PSB block. There may be several PCB statements per PSBGEN statement.
4. **END.** This statement terminates the PSB definition.

The following is an example of a simple PSB (named REPORT) consisting of one PCB. This PCB includes the entire SUPPDB data base hierarchy. The host language is PL/1.

```
PCB     DBDNAME = SUPPDB
SENSEG  NAME = SUPPLIER, PROCOPT = GI
SENSEG  NAME = PART, PARENT = SUPPLIER, PROCOPT = G
PSBGEN  LANG = PL/1, PSBNAME = REPORT
END
```

Data Manipulation Language

The data manipulation language (DML) of IMS is called Data Language/1 (DL/1). The DL/1 commands are made in the guise of procedure calls from application programs written in PL/1, COBOL, or assembler host languages. A DL/1 command is executed with respect to a particular PCB hierarchy existing in the program's PSB. Arguments for a DL/1 call can include an argument count, the name of a DL/1 command, the name of the PCB it is using, the name of an input/output (I/O) area, and segment search arguments (SSAs). An I/O area is simply a contiguous group of host language variables that serves as a "loading/unloading zone." If a DL/1 command entails the transfer of data from the data base to the program, those data are deposited in the variables of an indicated I/O area. Conversely, data to be inserted from the program to the data base must reside in an indicated I/O area prior to invoking a DL/1 command that performs the insertion.

One or more SSAs can be used to qualify (i.e., condition) a DL/1 call. An SSA may consist merely of a segment type name. Such an SSA is satisfied as soon as a segment of that type is found. A more complex SSA has not only a segment type name, but also the name of a field within that segment type, a relational operator (e.g., =, ≠, <), and a reference quantity. This SSA is satisfied when a segment of the indicated type is found whose value for the indicated field stands in the specified relation to the indicated reference quantity. Other, more complex SSAs can be constructed by an application programmer. These would involve Boolean connectors.

Following are descriptions of some of the basic commands provided by DL/1. The detailed format of procedure calls is not described in this brief examination of IMS.

1. **GET UNIQUE (GU).** This command retrieves the first-segment satisfying conditions specified by the SSAs. If there is more than one segment that satisfies the SSA conditions, the one that is "first" depends upon its sort key values and the data base's physical storage organization.
2. **GET NEXT (GN).** Starting from the most recently retrieved segment, GN finds the next segment satisfying the SSA conditions.
3. **GET NEXT WITHIN PARENT (GNP).** The segment retrieved by way of GU or GN becomes the "current parent" for a GNP call that follows. The GNP command visits descendant segments of the current parent segment until a segment is found that satisfies the specified SSAs.
4. **GET HOLD (GHU, GHN, GHNP).** The GET HOLD commands are like the three preceding commands, except that they do not allow the retrieved segment to be modified by other DL/1 commands in the application program.
5. **INSERT (ISRT).** The ISRT command is a semantically complex command that is used to add a new segment to a data base.

6. DELETE (DLET) and REPLACE (REPL). The DLET command deletes a segment that has been found and held and all of its physical descendant segments from a data base. The REPL command is used to modify a segment that has been found and held. The value of that segment's sequence field cannot be altered.

12.3.4 SYSTEM R

The Relational Data Model

The relational data model was proposed by Codd in 1970.[3] Before examining System R, we present the rudiments of the relational data model. A relational data base consists of two-dimensional tables called "relations." A relation has n columns, one for each of the n attributes (i.e., fields) that constitute that relation's schema. A row in the table is called a "tuple" and consists of one data value for each attribute. Thus a tuple is identical to the notion of a record occurrence, whereas a relation's group of attributes corresponds to a record type. Thus the logical structure of a relational data base is conveniently expressed as a collection of disjoint record types.

A logical relationship between two record types is represented by repeating an attribute(s) in both record types. This is illustrated later in the System R example (Exhibit 12.3.8). One or more attributes of each record type must be declared to be a key. The key must be selected such that each tuple in the relation has a key value that uniquely identifies that tuple within the relation.

Codd proposed two languages for manipulating data that have been organized into tables: relational algebra and relational calculus. In general, relational algebra commands take one or more tables as operands and produce a derivative table (i.e., a file). Two fundamental commands in relational algebra (among others) are projection and join. "Projection" strips specified columns out of a table to give a smaller table. (All duplicate tuples are eliminated from this new table.) "Join" can be performed on two tables having a common domain (i.e., a *repetitious* attribute). This is a matching operation that results in a new relation whose tuples are concatenations of tuples in the two original tables, based upon matches in specified attribute values.

The relational calculus is a mathematically oriented notation used to define a relation that is to be derived from existing relations. Upon receiving a relational calculus expression, a processor (i.e., a software component of the relational data base system) is supposed to determine automatically the sequence of relational algebra commands needed to derive the desired table. The processor must have the capacity to generate the algebraic steps corresponding to every calculus command. The language SEQUEL, discussed in the next subsection, is the System R version of a relational calculus from which the mathematical notation has been largely eliminated.

Overview of System R

System R[4,5] is a data base management system that supports a relational data base view and that gives its users a calculus-oriented language (SEQUEL) for data manipulation. Since System R has not yet been commercially released, this description is based on various prerelease specifications. The system includes two basic components: the relational data system (RDS) and the relational storage system (RSS). The RDS compiles SEQUEL statements and determines how the answer to a given request will be generated in terms of calls from a machine-language access module. It also maintains a catalog of external names (synonyms for attributes) and provides security. The RSS consists of the software to perform simple, record-at-a-time operations. It also provides data recovery, transaction management, and data definition facilities. It is called at run time by the access module generated by RDS. To facilitate retrieval, RSS utilizes access paths called "indexes and links" as well as sequential search. This allows a tuple in one relation that is associated with (i.e., has a matching attribute value with) a tuple in another relation to be accessed quickly, without a potentially exhaustive search of multiple files. The system structure is shown in Exhibit 12.3.7.

Since the relational model is based on a table view of the world, allowing no interrecord linkages to be explicitly specified and used by a user, data retrieval operations could be very expensive. System R allows a user to specify linkages through the CONNECT command in the RDS. The CONNECT command is similar to the notion of a set type in the data base task group (DBTG) specifications.[4] However, users still cannot explicitly use these linkages in their application pro-

Exhibit 12.3.7 System Structure of RSS and RDS

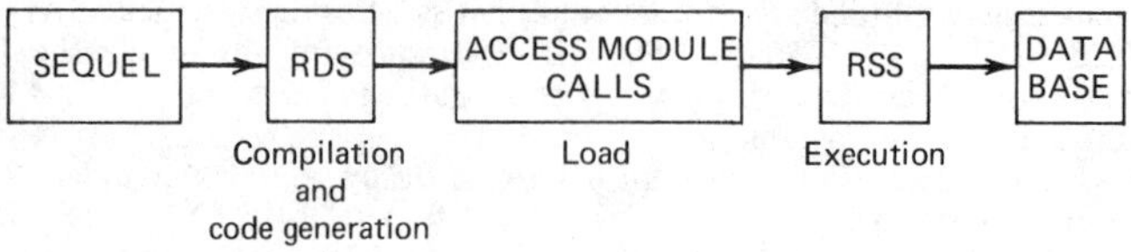

SEQUEL → RDS → ACCESS MODULE CALLS → RSS → DATA BASE

RDS: Compilation and code generation; ACCESS MODULE CALLS: Load; RSS: Execution

Exhibit 12.3.8 Schemata of Tables for the Supplier-Part Example of System R

SUPPLIER	PART	PARTDS
SNAME ADDRESS	SNAME PARTNO QTY	PARTNO PNAME COLOR WEIGHT

grams. Users of SEQUEL specify interrelation relationships in terms of the repeated attributes. During the compilation, RDS can use interrelation linkages for efficiency.

To illustrate System R, we use the three tables for the supplier-part example. The schemas of these tables are shown in Exhibit 12.3.8. Note that attributes are purposely repeated to indicate relationships among tables. In addition to repetition, there may also be linkages defined among tables.

Table Definition

A relation is defined or dropped from a relational schema through the SEQUEL commands CREATE TABLE or DROP TABLE, respectively. Two kinds of accessing methods can be specified with CREATE or DROP commands: index and link.

Indexing allows a tuple to be either directly accessed (by way of an index) or sequentially accessed on the basis of one or many attributes of the relation. The link access method provides access path links (i.e., pointers) from tuples of one relation to related tuples of another relation. For example, we can instruct System R to CONNECT (link) SUPPLIER and PART by matching SNAME. This linkage is not explicitly used in making retrieval statements. A user who wants to retrieve the parts for each supplier makes the request by stating that the SNAME field of PART should be matched with the SNAME field of SUPPLIER.

Data Manipulation

Data manipulation for a System R data base is accomplished with SEQUEL, which can be used as a stand-alone language or embedded in PL/1 or COBOL programs. The basic structure of a SEQUEL fetch statement involves SELECT, FROM, and WHERE clauses. A WHERE clause uses attributes within the context of arithmetic comparisons and operations, Boolean operators (AND, OR, NOT), set operations (UNION, INTERSECT, MINUS), set membership (X IN S, S CONTAINS X, X NOT IN S, and S DOES NOT CONTAIN X, where S is a table column and X is either a table column or an attribute value), or another SELECT-FROM-WHERE clause (representing a nested query). The main features of SEQUEL are demonstrated here by several examples, using the two tables in Exhibit 12.3.9 and 12.3.10.

The examples are as follows:

1. To print the suppliers that supply part 101, use the following:

```
SELECT    NAME
FROM      PART
WHERE     PARTNO = '101'
[ORDER BY NAME]
```

"ORDER BY NAME" is optional and is used when the output supplier names need to be sorted.

Exhibit 12.3.9 Part

SNAME	PARTNO	QTY
SA	# 101	200
SB	# 102	300
SC	# 103	500

Exhibit 12.3.10 PARTDS (Part Description)

PARTNO	PNAME	WEIGHT	COLOR
# 101	NUT	10	WHITE
# 102	BOLT	20	RED
# 103	WASHER	15	BLUE

2. Print the suppliers that supply at least one white part.

```
SELECT [UNIQUE] SNAME
FROM                PART
WHERE  PARTNO IN
    SELECT      PARTNO
    FROM        PARTDS
    WHERE       COLOR = 'WHITE'
                 or
SELECT  UNIQUE  SNAME
FROM    PART,    PARTDS
WHERE   PART.PARTNO = PARTDS.PARTNO
   AND COLOR = 'WHITE'
```

"UNIQUE" is used to eliminate duplicate supplier names.

3. Print the suppliers that supply all the white parts.

```
SELECT      SNAME
FROM        PART T
WHERE       (SELECT PARTNO
            FROM    PART
            WHERE   SNAME = T.SNAME)
CONTAINS
(SELECT     PARTNO
FROM        PARTDS
WHERE       COLOR = 'WHITE')
```

The tuple selected from PART for an SNAME is given a temporary name, "T." This temporary name is then used to refer to this tuple inside the WHERE clause.

4. Print the parts grouped by COLOR.

```
SELECT  PNAME,  COLOR
FROM    PARTDS
GROUP BY COLOR.
```

5. Insert a new supplier in the PART relation, and then delete all tuples in PART for the "SA" supplier.

```
INSERT INTO PART (SNAME, PARTNO)
<SD, #104>;
DELETE  PART
WHERE   SNAME = 'SA';
```

6. Update the weight of part 101 in PARTDS.

```
UPDATE  PARTDS
SET     WEIGHT = 50
WHERE   PART = '101'.
```

7. In a PL/1 or COBOL program, a SEQUEL statement must be prefixed with a dollar sign. Temporary tables, called "active sets," are defined with the LET command. Tuples within a temporary table are accessed (sequentially) one at a time on the basis of a "cursor." A cursor is used to maintain a position on one tuple of a temporary table; it keeps track of which tuple is currently being used by the program. (Note that a currency indicator that corresponds to the concept of a cursor is maintained on a data base rather than on a user-defined temporary table; see MDBS in the next section.)

The generation of tuples for a temporary table is accomplished by an OPEN statement. The resultant table conforms to the specifications of the LET command. A FETCH command transfers attribute values of the current tuple (i.e., as indicated by the cursor) into program variables as specified in the LET command. The cursor then "moves" to the next tuple of the temporary table. This tuple can then be fetched. The following example illustrates how SEQUEL is used within a host language:

```
$ LET     S  BE
    SELECT  SNAME
    INTO    $X
    FROM    SUPPLIER
    WHERE   PARTNO = $Y
$ OPEN           S;
$ FETCH          S;
$ CLOSE          S;
```

The letters X and Y are variables defined in the PL/1 or COBOL host language program. They are prefixed by a dollar sign to indicate that they are host language variables and to make writing a preprocessor easy. This code generates a table of supplier names for those suppliers that supply a particular part number (as indicated by the value of Y). The first supplier name is then fetched into X. The LET command is a declaration that indicates how the temporary table S is to be constructed (i.e., on the basis of a given PARTNO). Prior to the execution of the LET command, the program must assign some part number to the variable Y. The LET command also binds the value of SNAME (for the current tuple) to the variable X.

12.3.5 MICRO DATA BASE SYSTEM

Overview

The MDBS[6] is the preeminent data base system available on microcomputers. Although it is based on the network model, MDBS supports additional logical structuring features that are unavailable in any other data base system. Like other network systems, MDBS allows data bases to be defined in terms of data item types, record types, and set types (recall Exhibits 12.3.1*a* and 12.3.1*b*). A supplier-part network schema is shown in Exhibit 12.3.11*a*. The ORDERED set indicates that many quantities of a particular part may be on order. The SUPPLY set indicates that each of these quantities is associated with a particular supplier and that many quantities may be associated with a given supplier.

The Micro Data Base System also supports schemas that contain $N{:}M$ set types, recursive usage of set types, and multiple owner/member set types. Each of these additional logical structuring facilities is examined briefly.

When examining an application area with the intent of designing a data base structure for it, many-to-many relationships are frequently encountered. For instance, conceptually there is a many-to-many (N-to-M) relationship between suppliers and parts. That is, a part may be on order from *many* ($N \geqslant 0$) suppliers, and *many* ($M \geqslant 0$) parts may be on order from a supplier. In the schema of Exhibit 12.3.11*a*, this many-to-many relationship between SUPPLIER and PARTDS is captured by the SUPPLY and ORDERED sets, which have QUANTITY as their common member record type.

Since each supplier can be associated with many quantities (by way of SUPPLY), and since each quantity is associated with at most one part (by way of ORDERED), it follows that this logical structure represents the fact that many parts may be on order from any supplier. Conversely, since each part can be associated with many quantities (by way of ORDERED), and since each quantity is associated with at most one supplier (by way of SUPPLY), the schema also represents the fact that any part may be on order from many suppliers. This network approach to representing a many-to-many relationship incurs no redundancy of part or supplier data.

There is another many-to-many relationship that exists between suppliers and parts aside from which parts *are on order* from which suppliers: namely, which parts *can be ordered* from which suppliers. No mediating quantities are involved in this direct many-to-many relationship. The standard network approach for representing a direct many-to-many relationship is to use an artificial record type (a record type having no data item types), as shown in Exhibit 12.3.11*b*. Since each supplier can be associated with many occurrences of X (by way of the S1 set), and since each X occurrence is associated with at most one part (by way of the S2 set), it follows that this schema represents the fact that a supplier can supply many parts. Conversely, since each part can be associated with many X occurrences (by way of S2), and since each occurrence of X is associated with at most one supplier (by way of S1), the schema also represents the fact that a part can be supplied by many suppliers. There is still no redundancy in part or supplier data.

The $N{:}M$ set is an MDBS innovation that allows a many-to-many relationship to be represented without the use of an artificial record type and without incurring any redundancy. An $N{:}M$ set is like a traditional set, except that the restriction of at most one owner record occurrence per member record occurrence is eliminated. Thus the direct many-to-many relationship between suppliers and parts can be handled by an $N{:}M$ set (CAN-SUPPLY), as shown in Exhibit 12.3.11*c*. The decision as to which record type is the owner of an $N{:}M$ set is arbitrary. All owner record occurrences associated with a given member record occurrence can be accessed easily, just as all member occurrences associated with a given owner occurrence can be easily accessed. The advantages of an $N{:}M$ set over the traditional artificial record type approach are (1) greater structural simplicity and clarity, (2) greater storage and processing efficiency, and (3) less DML programming effort.

A multiple owner/member set type is a set type having more than one owner record type or more than one member record type. For example, salaried employees and hourly employees in the same department require different data item types. In Exhibit 12.3.11*d* the EMPLOY set has two member record types. Most network systems permit multiple member record types, but (aside from MDBS) they do not allow multiple owner record types.

The Micro Data Base System permits a set to be used recursively. This means that the set has the same record type as both owner and member. In the example of Exhibit 12.3.11*e*, the MANAGE set has the same record type (EMPLOYEE) as owner and member. As the set indicates, an em-

Exhibit 12.3.11 Supplier-Part Network Schema (*a*), Representation of a Many-to-Many Relationship With an Artificial Record Type (*b*), Representation of a Many-to-Many Relationship With an *N*:*M* Set (*c*), Set With Two Member Record Types (*d*), Recursive Use of a Set (*e*), and Schema With SYSTEM (*f*)

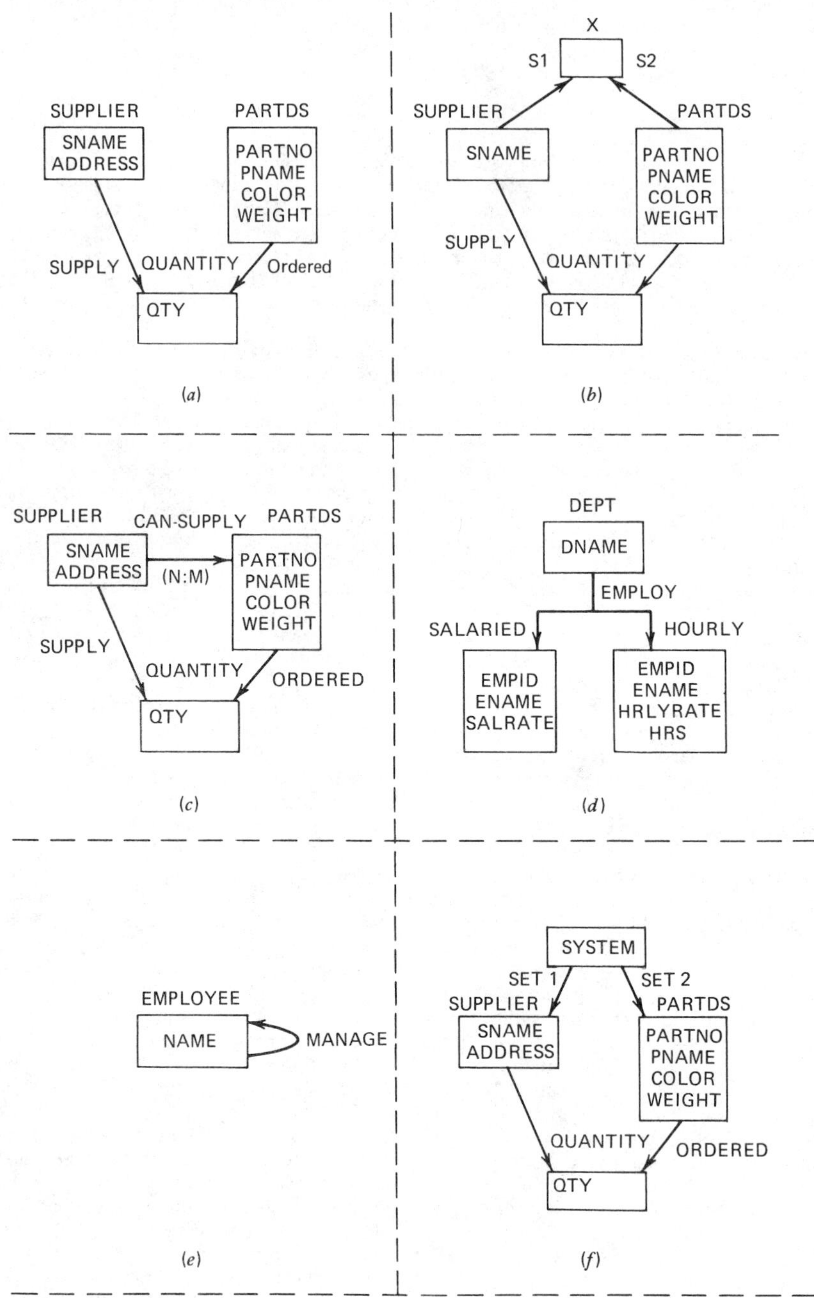

ployee can manage many other employees, but an employee is managed by at most one other employee. The $N:M$ sets can also be used recursively.

Data Description Language

The MDBS data description language (DDL) is used to specify formally the data item types, record types, and set types in an MDBS logical structure. Its statements are as follows:

1. **RECORD.** A RECORD statement gives a name of a record type.
2. **ITEM.** The ITEM statements following a RECORD statement specify names, types, and sizes of data items in the record type. Valid types are integer, real, binary, logical, and character. The size refers to the number of bytes allocated in a record occurrence for holding a value of this data item.
3. **SET.** A SET statement specifies a set type name, a set, and an indication of whether this is a traditional ($1:N$) set or an $N:M$ set. There are two storage classes: AUTO and MAN. The AUTO storage class indicates that, whenever an occurrence of the member record type for this set is created, that record occurrence will be automatically added to a set occurrence. If MAN (manual) is specified, then a user must explicitly add the new member record to the appropriate set occurrence. Traditional network systems permit a set order for member records only. In MDBS set order is used to establish the order of member records associated with an owner record and also (in the case of an $N:M$ set) the order of owner records associated with a member record.

Some set orders are explained here. Others, not described here, are also supported. As an example, CAN-SUPPLY in Exhibit 12.3.11*c* could have its members sorted on PARTNO and its owners sorted on SNAME. This means that the parts associated with any supplier can be accessed in order by part number. Also, the suppliers associated with any part can be accessed in order by supplier name.

a. FIFO. When a record occurrence is added to the set, it is logically placed as the last record occurrence in that set.

b. LIFO. When a record occurrence is added to the set, it is logically placed as the first record occurrence in that set.

c. SORTED. When a record occurrence is added to the set, it is placed in a sorted order in the set. The record occurrence with the smallest sort key value is logically first in the set. When a set order is SORTED, a data item name may be specified as the sort key for the set; otherwise the full record is used as a sort key.

d. IMMAT. The user is not concerned with the order of the record occurrences in the set. Use of the IMMATerial set order signals that the MDBS may insert records into the set to maximize access efficiency.

4. **OWNER, MEMBER.** These two statements are used to define the owner(s) and member(s) of a set type.

Every MDBS schema must contain a special record type called SYSTEM. The SYSTEM must be the owner of at least one set. The Micro Data Base System automatically creates one occurrence of SYSTEM, which serves as an entry point into a data base. All data manipulation uses the SYSTEM occurrence as a starting point. In Exhibit 12.3.11*f*, SYSTEM is the owner of two sets. The MDBS DDL specification for the supplier-part schema of Exhibit 12.3.11*f* is shown in Exhibit 12.3.12.

Data Manipulation

The MDBS DML provides the interface that allows a programmer to utilize a data base. The programmer can write application programs containing DML commands to manipulate data of any data base whose schema has been formally defined with the DDL. A DML command is performed by a subroutine call. The MDBS DML routines can be called from BASIC, FORTRAN, COBOL, PL/1, Pascal, and machine language host programs. The generic form of a DML call is as follows (the exact form varies across host languages):

EO = CALL (A, "routine name, arguments," host language variables)

where A = a DMS entry point address
routine name = name of a DML routine (command)
arguments = the list of arguments, separated by commas, gives data item, record type, set, or data block names as required by the DML routines
Host language variables = Host language variables, separated by commas, are used only in the DEFINE and EXTEND DML commands (The DEFINE command is used to give a name to a list

Exhibit 12.3.12 Data Definition Language in MDBS

```
RECORD   SUPPLIER
ITEM     SNAME      CHAR 20
ITEM     ADDRESS    CHAR 30
RECORDS  PARTDS
ITEM     PARTNO     INT
ITEM     PNAME      CHAR 20
ITEM     COLOR      CHAR 10
ITEM     WEIGHT     REAL
RECORD   QUANTITY
ITEM     QUANTITY   REAL
SET      SET1       AUTO     1:N
                                   SORTED SNAME
OWNER    SYSTEM
MEMBER   SUPPLIER
SET      SET2       AUTO     1:N
                                   SORTED PARTNO
OWNER    SYSTEM
MEMBER   PARTDS
SET      SUPPLY     MAN      1:N
                                   FIFO
OWNER    SUPPLIER
MEMBER   QUANTITY
SET      ORDERED    MAN      1:N
                                   IMMAT
OWNER    PARTDS
MEMBER   QUANTITY
END
```

of one or more program variables to form a data block. A data block is used in the transfer of data from a data base to program variables, and vice versa. The EXTEND command is used to extend the size of an existing data block so that it includes additional host language variables.)

EO = a variable in the application program (The DML routine sets the value of EO to indicate what has happened during the attempted execution of the DML command. A value of zero is returned for a successful execution. Other values indicate either an error in the programmer's usage of the DML or an attempt to find a record occurrence that does not exist in the data base.

In network systems, data manipulation is based on *currency indicators*. The MDBS uses four kinds of currency indicators:

1. The current record occurrence (CRO) of a record type.
2. The current owner (CO) of a set or N:M set.
3. The current member (CM) of a set or N:M set.
4. The current of the run unit (CRU).

At any instant during the execution of an application program, one of the occurrences of a record type is called its CRO. The programmer can use the DML commands to control which of a record type's occurrences is current at any given instant. There is one CRO indicator for each record type. Thus at any instant MDBS "knows" the location of one occurrence of each record type. Since the location is currently known, any of these occurrences is subject to immediate processing (e.g., transferral of data from the occurrence to program variables). The DML commands are used to make a different occurrence current (i.e., to make its location "known" to MDBS).

At any instant during the execution of an application program, one occurrence of the owner record type of a set is called the set's CO. The DML commands are used to control which owner occurrence is current (i.e., which is currently "known" or "seen" by MDBS). There is one CO indicator for every set and N:M set. The current owner of any set or N:M set is subject to immediate processing. Every set and N:M set also has a current member at any instant. Which record occurrence is the CM of a given set is controlled by the programmer through DML commands. Finally, the CRU is simply the record that was most recently accessed by the application program.

Several of the most commonly used MDBS DML commands are described here. Other commands are detailed in the MDBS *User's Manual.*[6]

1. DEFINE. EO = CALL (A1, "DEFINE, data block name," variables). A data block with the user-specified name is defined to consist of the indicated host language variables.

2. CRS (Create Record and Store data). EO = CALL (A0, "CRS, record type, data block"). A new occurrence of the specified record type is created and initialized with the contents of the variables that constitute the specified data block. If the record type is a member of an AUTO set, then the new record automatically becomes a member of (i.e., becomes linked to) the record occurrence that is the current owner of the set.

3. AMS (Add Member to Set). EO = CALL (A0, "AMS, record type, set type"). The CRO of the specified record type is added to the set occurrence identified by the CO of the specified set. That is, it becomes a member of the record occurrence that is the CO of that set.)

4. FMSK (Find Member based on Sort Key). EO = CALL (A0, "FMSK, set type, data block"). The members of the CO of the specified set are searched for the logically first member with a sort key value equal to the value of the data block's variable. The record that is found becomes the CM of the set.

5. FFM (Find First Member). EO = CALL (A0, "FFM, set type"). The first member of the specified set's current owner is made the CM of that set.

6. FNM (Find Next Member). EO = CALL (A0, "FNM, set type"). The logically next member following the specified set's CM becomes the new CM of that set. If a next member was found, EO is returned with a value of 0.

7. GETM (GET data from current Member). EO = CALL (A0, "GETM, set type, data block"). All data item values in the CM of the indicated set are returned in the variables of the specified data block.

8. GFO (Get Field from current Owner). EO = CALL (A0, "GFO, data item type, set type, data block"). The value of the specified data item type is extracted from the CO of the specified set. This value is returned in the data block's variable.

9. SOM (Set current Owner based on current Member). EO = CALL (A0, "SOM, set-1, set-2"). The record that is the CM of set-2 is made the CO of set-1. The first member of set-1 that is associated with this new CO becomes the CM of set-1.

10. SMM (Set current Member based on current Member). EO = CALL (A0, "SMM, set-1, set-2"). The record that is the CM of set-2 becomes the CM of set-1 as well. The logically first owner of set-1 that is associated with this new CM becomes the new CO of set-1.

Exhibit 12.3.13 shows a sample application program demonstrating how the DML commands can be used in conjunction with each other. This program generates a list of all the parts and quantities supplied by the supplier LEHR.

Query Interface

The user of a DML must be a programmer. In addition to a DML, MDBS supports a nonprocedural, English-like query language that allows nonprogrammers to interrogate any MDBS data base on an ad hoc basis. An MDBS query has the following generic format:

<COMMAND> <FIND CLAUSE> <CONDITIONAL CLAUSE> <PATH CLAUSE>

One command is LIST, which simply results in the automatic generation of a table (i.e., a file, a relation, a list). The content of a table is determined by the remaining clauses. The find clause con-

Exhibit 12.3.13 Data Manipulation Language in MDBS[a]

```
     EO = CALL (A1, "DEFINE, SUP", A2, A3)
     EO = CALL (A1, "DEFINE, QTY", N)
     EO = CALL (A1, "DEFINE, NAME", A4)
     NAME = 'LEHR'
     EO = CALL (A0, "FMSK, SET1, NAME")
     EO = CALL (A0, "GETM, SET1, SUP")
     PRINT A2, A3
     EO = CALL (A0, "SOM, SUPPLY, SET1")
10   EO = CALL (A0, "GETM, SUPPLY, QTY")
     EO = CALL (A0, "SMM, ORDERED, SUPPLY")
     EO = CALL (A0, "GFO, PNAME, ORDERED, NAME")
     PRINT A4, N
     EO = CALL (A0, "FNM, SUPPLY")
     IF EO.EQ.0 GO TO 10
```

[a] A0, A1, A2, A3, A4, N, and EO are program variables.

sists of a list of data item types whose values are to be retrieved subject to conditions indicated in the conditional clause. The path clause indicates which sets are to be used in answering the query.

To obtain a list of the suppliers from whom part number 101 has been ordered, the following query is used:

LIST SNAME FOR PARTNO = '101'
THRU SET1, SUPPLY, > ORDERED

In this case the find clause is SNAME, the conditional clause (which always begins with FOR) is FOR PARTNO = '101', and the path clause (which always begins with THRU) is THRU SET1, SUPPLY, > ORDERED. A path always begins with a SYSTEM-owned set. The specified path must lead to or through all record types containing data item types cited in the find and conditional clauses. In this case, the path must lead from SYSTEM through record types containing SNAME and PARTNO. Referring to Exhibit 12.3.11*f*, it is clear that SET1, SUPPLY, > ORDERED is just such a path. The symbol > preceding a set name indicates that the path uses the set by going from its member record type to its owner record type (e.g., from QUANTITY to PARTDS goes "upstream" through the ORDERED set).

The path clause is vital because there may be alternative paths that visit the query's data item types. Suppose that the schema of Exhibit 12.3.11*f* is augmented to include the $N{:}M$ set (CAN-SUPPLY) shown in Exhibit 12.3.11*c*. Without a path clause, LIST SNAME FOR PARTNO = '101' is ambiguous. It is not clear whether the user desires a list of suppliers from whom part 101 has been ordered or from whom part 101 can be ordered. The latter is handled by

LIST SNAME FOR PARTNO = '101' THRU SET1, CAN-SUPPLY

Notice that the SEQUEL analog to a path clause is the matching of fields from different relations in the SEQUEL WHERE clause.

Complex find and conditional clauses are supported, which involves arithmetic expressions and Boolean expressions. The following query results in a table showing supplier name and one tenth of the quantity ordered for any part that is white or whose weight is less than 39/2:

LIST SNAME, QTY/10 FOR COLOR = 'WHITE' OR (WEIGHT *2)
≤ 39 THRU SET1, SUPPLY, > ORDERED

To obtain a file of the part number, color, weight, and quantity of all parts ordered from LEHR in a quantity of more than 43, the query is

LIST PARTNO, COLOR, WEIGHT, QTY FOR SUPPLIER = 'LEHR'
AND QTY > 43 THRU SET1, SUPPLY, > ORDERED

Clearly, use of the MDBS query system relieves one from the programming effort such as that involved in Exhibit 12.3.13. In effect, the query system automatically constructs and executes the DML program needed to generate a desired table. Moreover, the MDBS query system supports other commands, such as STAT and CHANGE. Rather than listing data values, STAT outputs statistics on those values. The CHANGE command is an interactive command that allows selected data values to be modified. The query system also supports a number of report writer features, including control breaks, title definitions, and table lookups. The table generated by a query can be routed to a console, printer, and/or disc file for subsequent use as input to canned software packages. These and other features of the MDBS query system are detailed in the MDBS literature.[7]

Dynamic Restructuring of a Schema

The dynamic restructuring system (DRS)[8] allows any MDBS schema to be altered after data have already been loaded into the data base. In the absence of DRS, all data in the data base would have to be dumped onto files, the schema respecified with the DDL, and the data loaded into the new data base. This cumbersome process is avoided with DRS, which automatically reorganizes the data base records to be consistent with a revised schema. This is accomplished without any dumping or reloading of data.

For instance, to add a new data item type to an existing record type, one simply gives the interactive command ADI (Add Data Item). The DRS prompts for the name of this new data item type, its size, the record type to which it belongs, and so on. It alters the data base's schema and allocates space for this data item type in every existing occurrence of its record type. There are many kinds of schema alterations that can be performed with DRS. Aside from ADI, the most commonly used DRS commands include ART (Add Record Type), AST (Add Set Type), DDI (Delete Data Item), DRT (Delete Record Type), and DST (Delete Set Type).

12.3.6 DATA BASE SCHEMA DESIGN

Overview

We have seen that a data base system can be classified on the basis of the variety of logical structuring that it furnishes. Systems that exemplify the relational, hierarchical, and network varieties of logical structuring have been described. The nature of a system's data manipulation facilities is heavily influenced by the variety of logical structuring supported. The manipulation commands are stated in terms of the data structuring constructs (e.g., record type, set) that a system allows. Regardless of whether a relational, hierarchical, or network system is being used, the schema that is designed for a given application area establishes (and constrains) a user's view of the data base and serves as the basis for data manipulation. Thus the design of data base schematas is a very important activity. It is also a nontrivial activity.

A data base schema for an application area must be designed to model or simulate the structure of information pertinent to that application. It is a formal description of the real world. The different data models offer different descriptive tools. The choice of a data model to use in simulating the real world hinges largely on how one conceives of the real world. For one who considers the world to be a network, the network data model provides the most powerful simulation tool. With network systems it is easy to explicitly describe and give the semantics of (by way of a set or $N{:}M$ set name) relationships that exist in the real world.

On the other hand, if one views the world as being a collection of trees, the hierarchical model would be the most convenient simulation tool. However, unlike the network model, the hierarchical model generally forces a designer to incorporate redundancy into the design as a basis for establishing interconnections among trees. This problem is somewhat relieved, although not eliminated, by the use of IMS's logical relationships (yielding what amounts to limited networking capabilities). Finally, if one views the world in terms of a group of disjoint tables, then the relational model would provide a convenient tool for simulating that world. Relationships between the disjoint tables (i.e., files) are captured by redundancy. Semantics of these relationships are not explicitly represented; this can be unwieldy in schematas that have dozens of relationships (all implicitly represented).

Regardless of which view is adopted, there is the design task of transforming knowledge about the application world into a data base schema. Prior to presenting design approaches for the relational and network views, we examine a couple of issues of which a schema designer should be aware. These are data relationship preservation and structural anomalies.

The designer of a schema must preserve real-world data relationships through whatever mechanisms are available in the data model being used. Two of the four kinds of relationships have already been described: the many-to-one and the many-to-many. A one-to-many relationship is the complement of the many-to-one. A one-to-one relationship between two concepts exists when each instance of one concept uniquely identifies an instance of the other concept, and vice versa. Although these relationships are represented differently in the three data models, they all must be preserved in a schema. If relationships are not explicitly represented, it is advisable to document the nature of implicit relationships to allow the user of a data base to understand the semantics of its schema.

Whereas relationship preservation is concerned with a schema's representativeness, structural anomalies are concerned with consistency and redundancy problems that can arise from certain schema designs. Suppose that we have a record type with four data item types: SNAME, ADDRESS, PARTNO, and QTY. This design has the following four anomaly problems:

1. **Redundancy Anomalies.** The name and the address of a supplier are repeated for each part supplied.
2. **Update Anomalies.** As a result of redundancy, updates to a supplier name or address must be repeated in many places. Failure to update all repeated values of a data item results in inconsistency in the guise of update anomalies.
3. **Insertion Anomalies.** If a data base system prohibits null values in a record occurrence, then we cannot store a supplier's name and address if it does not currently supply at least one item.
4. **Deletion Anomalies.** If we delete all the items supplied by one supplier, we necessarily lose the name and address of this supplier. This is the converse of the insertion anomaly.

The foregoing awkward situations apparently result from grouping non-one-to-one related data items into one record type. Anomalies can be completely eliminated in the network model since data item linkages are through set types and no data items have to be repeated. In the relational model, data items are repeated for linkage purposes. Therefore anomalies are present across record types, although they can be removed from within any individual record type. With a data base whose schema is shown in Exhibit 12.3.8, changing a supplier's name involves changing a value of

Exhibit 12.3.14 The Seven-Step Procedure for Network Schema Design

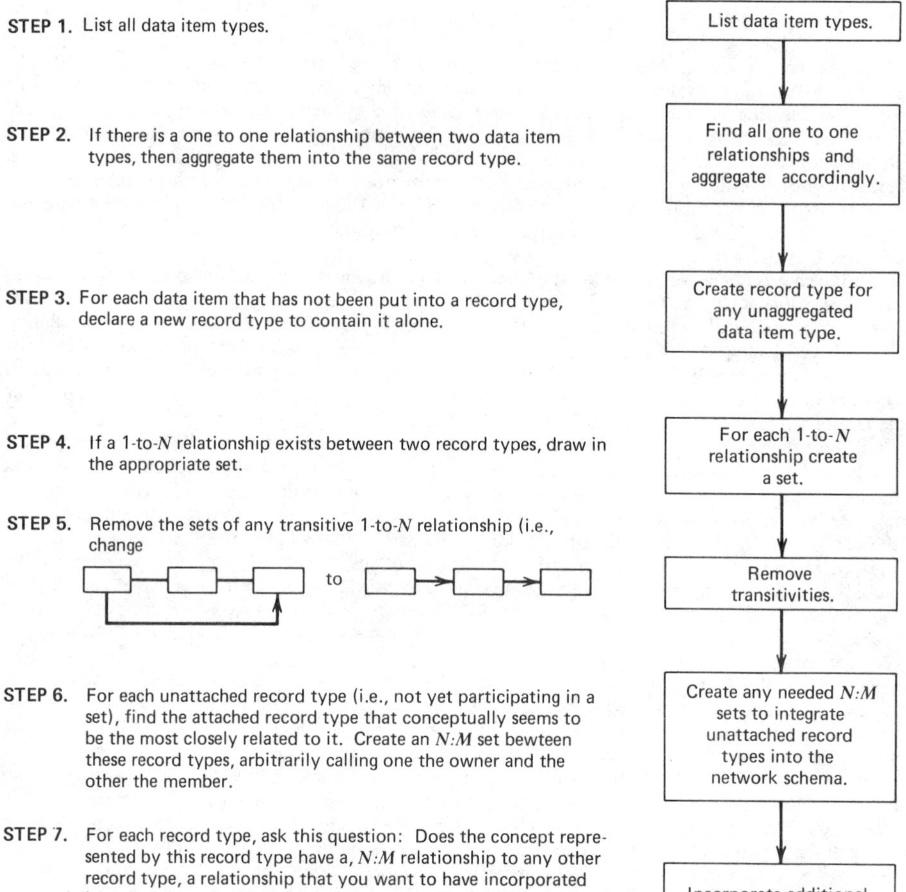

STEP 1. List all data item types.
STEP 2. If there is a one to one relationship between two data item types, then aggregate them into the same record type.
STEP 3. For each data item that has not been put into a record type, declare a new record type to contain it alone.
STEP 4. If a 1-to-N relationship exists between two record types, draw in the appropriate set.
STEP 5. Remove the sets of any transitive 1-to-N relationship (i.e., change
to
STEP 6. For each unattached record type (i.e., not yet participating in a set), find the attached record type that conceptually seems to be the most closely related to it. Create an N:M set bewteen these record types, arbitrarily calling one the owner and the other the member.
STEP 7. For each record type, ask this question: Does the concept represented by this record type have a, N:M relationship to any other record type, a relationship that you want to have incorporated into the schema? If it does, and if that relationship is not already represented in the schema, then add an N:M set to represent this relationship. Begin by considering those record types that are not owners of sets; then recursively consider those record types that own already-considered record types.
List data item types.
Find all one to one relationships and aggregate accordingly.
Create record type for any unaggregated data item type.
For each 1-to-N relationship create a set.
Remove transitivities.
Create any needed N:M sets to integrate unattached record types into the network schema.
Incorporate additional N:M sets into the schema, if needed.

SNAME not only in the SUPPLIER relation, but also everywhere that it is repeated in the PART relation.

An Approach to Relational Schema Design

In the relational model, data base design is based on the notion of "normal forms." The procedure used to derive normal forms is called "normalization." There are many kinds of normal forms. Following some preliminary definitions, three types of normal forms are discussed.

A data item (or data items) B is said to be functionally dependent upon a data item (or data items) A (denoted by $A \to B$) if each occurrence of A is associated with one and only one occurrence of B, although several occurrences of A may have the same occurrence in B. For example, if $A = \{\text{SNAME,PARTNO}\}$ and $B = \{\text{QTY}\}$, then $A \to B$ means that, given SNAME and PARTNO, QTY is uniquely determined. If $A \to B$ and $B \to C$, then $A \to C$ is true, and C is transitively dependent upon A. Functional dependency denotes a many-to-one relationship.

A key for a record type R is a data item or collection of data items K in R such that K is a unique identifier for R. There could be several keys for a given record type. We will call a data item A a "prime data item" if A is a member of any key for R. If A is not a member of any key, then A is nonprime. A data item (or data items) A in R is partially dependent upon a key K for R if, besides $K \to A$, A is also functionally dependent upon a proper subset of K. Data item A is fully dependent on K if A is not partially dependent on K and $K \to A$.

Three normal forms are as follows:

1. **First Normal Form (1NF).** A record type is in first normal form if none of its data items are themselves record types.
2. **Second Normal Form (2NF).** A record type R is in 2NF if R is in 1NF and if each nonprime data item in R is fully dependent upon every key. If R is not in 2NF, then for a nonprime data item X in R, $K \rightarrow X$ and $K_1 \rightarrow X$ are true where K is a key for R and $K = K_1 \cup K_2, K_2 \neq \emptyset$. Note that data items K_1 and X must be repeated for each different occurrence of K_2, yielding redundancies. For example, let record type R consist of four data items, K_1, K_2, X, and Y. If data items K_1, K_2 form a key and $K_1 \rightarrow X$ is true, then $k_1k_2xy_2$ and $k_1k_2'xy_2$ are two possible record occurrences in R. Data values k_1x are repeated for k_2 and k_2'. If R is changed into two record types, $R_1 = K_1K_2Y$ and $R_2 = K_1X$, then R_1 and R_2 are in 2NF, and redundancies of the preceding variety are eliminated.
3. **Third Normal Form (3NF).** A record type R is in 3NF if R is in 2NF and if no nonprime data item of R is transitively dependent on any key of R. If K is a key, if A is a nonprime data item, and if $K \rightarrow X \rightarrow A$ is true, then we cannot store an X occurrence with a K occurrence unless there is an A occurrence associated with that X occurrence. If we delete A, then the $K \rightarrow X$ association is also lost. Therefore a record type not in 3NF would result in insertion and deletion anomalies. So 3NF eliminates intrarecord anomalies. Additionally, a data base whose relations are in 3NF preserves all many-to-one relationships.

An Approach to Network Schema Design

There is very little published work on how to go about designing a network schema for a given application area. However, one useful procedure is summarized in Exhibit 12.3.14. This design strategy has been adapted from Holsapple.[9] To use this procedure, one must have a knowledge of the application area. One-to-one, one-to-many, and many-to-many relationships are automatically preserved by adhering to the procedure's seven steps. Anomalies do not arise on either an intra-record-type or interrecord-type basis. This is because redundancy is not used in relationship representation. It should be noted that some redundancy may be desirable for reasons of storage and/or processing efficiency. A network schema resulting from the design strategy could be modified to permit varying degrees of redundancy (e.g., by collapsing two record types into one). Users of the schema should be made aware of possible anomalies that result from such modifications.

12.3.7 CONCLUSION

Commencing with a description of the notion of data base management, we examined three data models. Each was illustrated by means of a specific data base management system. Commonalities and differences in data structuring and data manipulation were noted. Data base design issues were then considered. We conclude with several observations about selection of a data base management system and the future of data base management.

It is important to realize that there can be substantial differences among the data base management systems that stem from a given data model. For instance, there are numerous network data base systems. These differ in terms of the schema design features supported (e.g., MDBS has the N:M set construct, whereas others do not), the DDL syntax, the DML syntax, the host language(s) supported, the query interface, the method of implementing logical constructs, and the host hardware environment.

Differences among hierarchical systems tend to be somewhat more pronounced than differences among network systems, because the latter have a common CODASYL heritage. Hierarchical systems do not stem from a set of proposals comparable to the 1971 CODASYL *Data Base Task Group Report.* Relational systems, when they eventually appear, will doubtless have differences such as those just noted. However, these systems, like the network systems, have a common heritage in the work of Codd.[3]

As indicated previously, the choice of a particular data base system is influenced by the modeler's view of the world: the file or table view versus the hierarchical view versus the network view. Also important is a system's performance. There are few generalizations that can be made about the performance of systems adhering to one view versus the performance of systems adhering to another view. One generalization is Ullman's contention that relational systems tend to be less efficient than network systems.[10]

The data base management field has yet to reach the limits of its development. Adoption of a single data base standard is probably premature. One might reasonably expect to witness a merging of the various data models. This hierarchical model is conceptually a special case of the network model and the proposed manipulation language for System R is quite similar to the MDBS query language. The latter may be regarded as generating tables, or as a means for producing virtual relations, from a network data base. Conversely, tables (i.e., files) are used to load network data bases.

Another area of development is the automation of schema design. For instance, the desired output reports for a given application area might be used to drive a schema design process. This would necessitate a formal language for specifying the nature of output reports and a mechanism for converting a set of report specifications and their implied data relationships into a workable (or even optimal) schema specification.

Finally, it must be pointed out that file management is still widely used for constructing MISs. File management is not inappropriate for applications where there are few types of information to be managed and where interrelationships among the types of information are few and simple. The trend toward using data base management, rather than the older file management techniques, for MIS development is growing. Obstacles to this growth have been ignorance of what data base management is and the inaccessibility of data base systems.

The second obstacle is being rapidly overcome. With the advent of MDBS, data base management has become affordable for the masses of small applications (e.g., small businesses). It is priced at a fraction (one tenth to one hundredth) of the cost of data base systems for minicomputers and mainframe computers. Moreover, data base systems are becoming available on an increasing number of machines at the minicomputer, mainframe, and microcomputer levels.

The increased accessibility of data base systems will increasingly diminish the first obstacle cited. As data base systems become more widely used as the basis for MIS development, one would expect that they will also become a cornerstone in the construction of decision support systems.[11] Decision support systems will be built that interface a data base system with various procedural models (e.g., simulation models) in order to respond to nonprogramming users' requests, stated in an English-like, nonprocedural language.

REFERENCES

1. CODASYL Systems Committee, *Data Base Task Group Report*, ACM, New York, April 1971.
2. IBM, IMS/VS publications, *General Information* (GH20-1260), *System/Application Design Guide* (SH20-9025), *Application Programming Reference Manual* (SH20-9026), and *Systems Programming Reference Manual* (SH20-9027), IBM, White Plains, NY, 1978.
3. E. F. CODD, "A Relational Model for Large Shared Data Banks," *Communications of ACM*, Vol. 13, No. 6 (June 1970), pp. 377–387.
4. M. M. ASTRAHAN, M. W. BLASGEN, D. D. CHAMBERLIN, K. P. ESWARAN, J. N. GRAY, P. P. GRIFFITHS, W. F. KING, R. A. LORIE, P. R. MCJONES, J. W. MEHL, G. R. PUTZOLU, I. L. TRAIGER, B. W. WADE, and V. WATSON, "System R: A Relational Approach to Data Management," *ACM Transactions on Database Systems*, Vol. 1, No. 2 (June 1976), pp. 97–137.
5. M. W. BLASGEN, M. M. ASTRAHAN, D. D. CHAMBERLIN, J. N. GRAY, W. F. KING, B. G. LINDSAY, R. A. LORIE, J. W. MEHL, T. G. PRICE, G. R. PUTZOLU, M. SCHKOLNICK, P. G. SELINGER, D. R. SLUTZ, H. R. STRONG, I. L. TRAIGER, R. W. WADE, and R. A. YOST, *System R: An Architectural Update*, IBM Research Report RJ2581 (No. 33481), IBM, White Plains, NY, July 17, 1978.
6. MDBS, *User's Manual MDBS.DMS, MDBS.DDL*, Micro Data Base System, Inc., Lafayette, IN, 1979.
7. MDBS, *MDBS.QRS Query System/Report Writer*, Micro Data Base System, Inc., Lafayette, IN, 1980.
8. MDBS, *MDBS.DRS User's Guide*, Micro Data Base System, Inc., Lafayette, IN, 1980.
9. C. W. HOLSAPPLE, "Data Description and Manipulation Languages for Microcomputer Data Bases," *Computer Age*, December 1980.
10. J. D. ULLMAN, *Principles of Database Systems*, Computer Science Press, Potomac, MD, 1980.
11. R. H. BONCZEK, C. W. HOLSAPPLE, and A. B. WHINSTON, *Foundations of Decision Support Systems*, Academic Press, New York, 1981.

BIBLIOGRAPHY

CODASYL Systems Committee, *A Survey of Generalized Data Base Management Systems*, ACM, May 1967.

CODASYL Systems Committee, *Feature Analysis of Generalized Data Base Management System*, ACM, May 1971.

CODASYL Systems Committee, *COBOL Data Base Facility Proposal, Data Manipulation Language Proposal* DB/TG-73001.00. Technical Services Branch, Department of Supply and Service, Ottawa, Canada, January 1973.

CODASYL Systems Committee, *COBOL Journal of Development*, Canadian Government Specification Board, Ottawa, Canada, 1973.

CODD, E. F., "Further Normalization of the Data Base Relational Model," in R. Rustin, Ed., *Data Base Systems*, Prentice-Hall, Englewood Cliffs, NJ, 1972.

DATE, C. J., and E. F. CODD, *The Relational and Network Approaches: Comparison of the Application Program Interfaces*, IBM Research Report RJ 1401 (No. 21706), IBM, White Plains, NY, June 6, 1974.

HASEMAN, W. D., and A. B. WHINSTON, *Introduction to Data Management*, Irwin, Homewood, IL, 1977.

HOLSAPPLE, C. W., and A. B. WHINSTON, "The Significance of Data Base Management for Micro Computers," *Datamation*, April 1981, pp. 165–167.

CHAPTER 12.4

Analysis and Design Techniques for Information Processing Systems

JOHN V. PILITSIS

American Telephone and Telegraph Company

12.4.1 OVERVIEW

Outputs of basic work functions, whether physical or mental in nature, are related to decision making strategies undertaken by the person performing them. Burch et al.[1] and others emphasize that decisions are directly related to the change in the state of "knowledge" by the decision maker. In the work situation, this change is facilitated by designing basic work and organizational structures to provide the "decision maker" (this term is synonymous with employee, job incumbent, and consumer) with information essential to performing a given work function. If the clusters of work functions assigned to the decision maker are systematically designed into jobs and organizations through industrial engineering and behavioral science concepts (Chapter 2.1), the decision maker will most likely receive more *information* and little data, whereas if the job and organizational structure are poorly designed, the decision maker receives more *data* and little information. Conceptually, if organizational structures are defined on the premise of sound mission statements, strategic and tactical plans, and optimum work flow, they facilitate horizontal and vertical information transmission, resulting in high-quality decision making and corresponding overall performance.

Since IPSs do not in themselves perform work functions, make a decision, or act, for an IPS to be effective it must provide for decision aids that facilitate quality performance by the decision maker within a given organizational structure. As such, in computer-based IPSs the designer must focus on the development of three fundamental components: human performance (user needs), hardware subsystems, and software subsystems. Since the development of the earliest computers, the design process was dominated by systematic hardware development, whereas the other two components were developed on an ad hoc basis, resulting in a significant percentage of IPS failures during the 1960s and early 1970s.[2] Proliferation of these failures made it apparent that a primary contributor was the lack of a software development methodology. Attempts have been made, with varying amounts of success, to apply to the problems of IPS software development the engineering principles that have been reasonably successful in other disciplines,[2,3] namely, planning, life cycle management, and structured design philosophy.

The purpose of this chapter is to provide the reader with an overview of the methodologies essential to analysis and design of IPS software. (The human performance design issues are addressed, in general, in Chapter 6.8 and are implicit in the analysis and design methods discussed in this section.) The specific objectives of this chapter are:

To review the role of organizational planning in IPS development.

To provide descriptive and functional definition of system life cycle management concepts.

To identify and discuss major analysis and design techniques for IPSs.

12.4.2 THE ROLE OF ORGANIZATIONAL PLANNING IN IPS DEVELOPMENT

In recent years numerous organizations have instituted various types of organizational planning processes to cope with sociopolitical and technological change.[4] In these organizations the need for IPS development has become a natural outgrowth of such processes. Cleland and King,[5] Grindlay,[6] and others have shown that, when IPSs are conceived as part of organizational planning, the success rate is significantly greater than that of other similar development efforts. Concep-

Exhibit 12.4.1 Organizational Planning and IPS Development

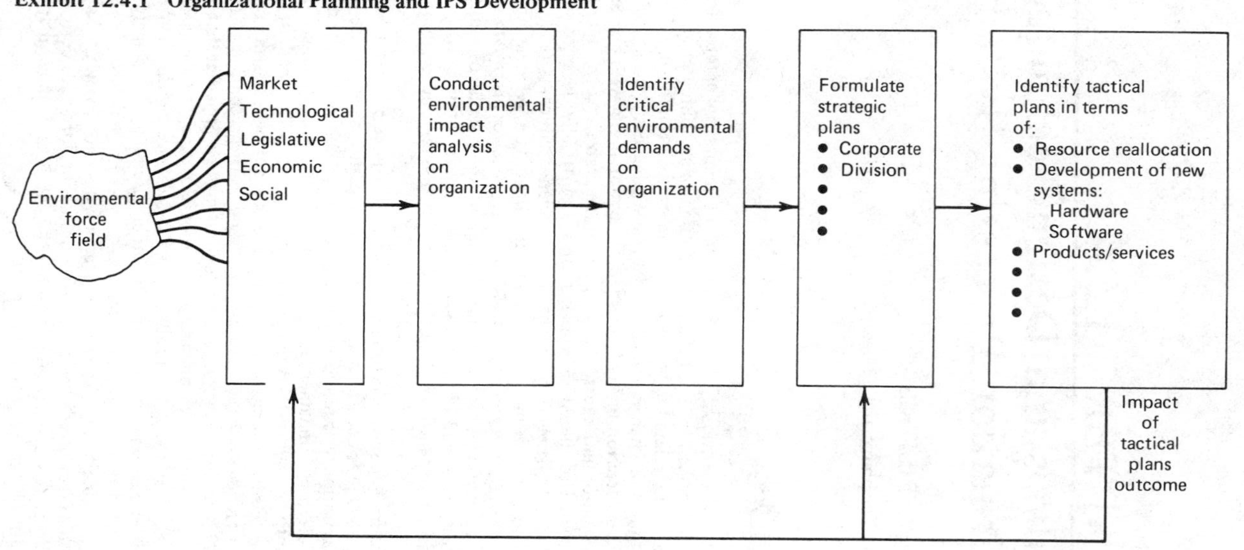

tually, organizational planning processes and their relationship to IPS development are reflected in Exhibit 12.4.1.

As illustrated in this exhibit, the overall planning process should be largely responsive to environmental demands.[7] These demands relate to current and anticipated market, technological, legislative, economic, and social issues, among others. Typically an environmental impact analysis is conducted, which leads to the identification of critical issues to be addressed by the organization and to the formulation of strategic plans for the various organizational areas. Corresponding tactical plans are developed to ensure execution of strategies. It is within the tactical plans that the need for each IPS development is conceptualized and is assigned to specific manager(s) to ensure that it comes to fruition. In organizations with clearly defined IPS development methodologies, each IPS is developed through life cycle management concepts.[5]

12.4.3 SYSTEM LIFE CYCLE MANAGEMENT CONCEPTS

Many of the IPS failures during the 1960s have been linked to top management's abdication of management and control of IPS development to system specialists.[8] Managers, often thinking that the introduction of sophisticated computers into the IPS necessarily warranted this abdication, apparently overlooked one important issue: System specialists are not as familiar as managers with the myriad complexities of decision making and organizational demands. In the recognition that neither the manager nor the specialists alone could develop and implement effective IPS, system life cycle management (Exhibit 12.4.2) was introduced as an effective resolution to this dilemma. (Although one system life cycle method is presented here, the reader may be interested in reviewing other sources[9-11]) Developed in the 1970s, system life cycle management is aimed at alleviating problems relating to:

Managerial lack of control and direction.[12]
Poorly designed IPS.
Costly and ineffective IPS.
Unstructured system design processes.

The focus of this section is on project organizing (appointing a project manager and formulating a project team), system development, and system maintenance processes. Discussions on project planning and control can be found elsewhere (Chapter 11.2).

Project Organization

Project organization consists of two key activities: appointing a project manager and formulating a project team.

Exhibit 12.4.2 System Life Cycle Management Process

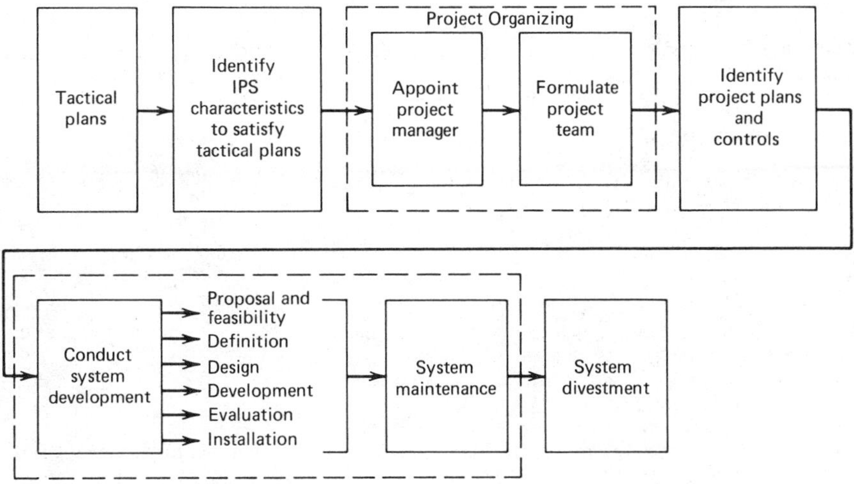

Project Manager

A project manager, typically a member of the user organization (for consumer-oriented IPS, this individual should be the functional product manager), is assigned after a proposed system has been identified by the tactical planners of a given organization. The project manager has direct day-to-day responsibility for overall project planning in terms of preestablished work plans, activities, costs, budgets, schedules, and quality control. The project manager is also held accountable for the continual tracking of the cost-benefit relationship of the IPS.

Project Team

If an IPS is to meet essential criteria, the application of a variety of technical skills needs to be integrated. To operationalize this integration, a project management team is formed on a matrix organization basis. The team may consist of the project manager, systems analysts, organizational design specialists, industrial engineers, and/or human factors specialists.

Within an IPS project team, a set of interfaces between the project manager and the project team members must be clearly defined. The team members are responsible for leading the various development efforts surrounding a specific IPS. The leaders of these efforts are referred to as "functional managers." (For example, the project team member who is responsible for the software architecture of a computer-based IPS is its functional manager.) This interface has been illustrated by Cleland and King[5] (see Exhibit 12.4.3).

System Development Stages

Since the early 1970s, effective and efficient IPSs have been developed by project teams that have followed structured system development processes. These processes are characterized by a series of work stages. Although the number of stages and the terminology used in the system development literature often vary, the basic work activities and related sequences remain constant. The following stages are used in this chapter's discussion of system development activities:

1. Proposal and feasibility.
2. Definition.
3. Design.
4. Development.
5. Evaluation.
6. Installation.

Each system development stage has been formatted* into the following parts:

Objective. What is to be accomplished in the stage.

Management Effort. Decisions that must be made by the project team are identified, and examples are provided.

Major Points to Consider. Issues are identified on which the project manager should focus attention.

Proposal and Feasibility

Objective. To evaluate whether a system should be built to do something faster, easier, better, cheaper, more safely, or with fewer people.

Management Effort. Assign personnel to collect more information about the proposed system from:

The potential users, that is, anyone involved in providing inputs, operating, maintaining, or using the system and/or its outputs.

Experts in areas relating to the system or its technology.

*The author wishes to acknowledge L. T. Martin, AT & T, for assistance in formatting the system development stages.

Exhibit 12.4.3 The Project-Functional Interface

Project Manager	Functional Manager
What is to be done?	How will the task be done?
When will the task be done?	Where will the task be done?
Why will the task be done?	Who will do the task?
How much money is available to do the task?	How well has the functional input been integrated into the project?
How well has the total project been done?	

Major Points to Consider. The following areas must be analyzed and are normally documented in a feasibility report:

What the system must do (system objectives).

What the system cannot do (system constraints).

Possible design approaches to meeting the system objectives (e.g., modularity—see details in the section entitled "Analysis and Design Methods for IPSs").

Estimates of the number and kind of people needed for system development.

Estimates of cost benefits (i.e., the anticipated savings or earnings versus the time and money required for system development).

Definition

Objective. To collect more detailed information from potential users and experts in relevant areas in order to further define the system.

Management Effort. Divide the system into subsystems, if it is a large and/or complex system.

Major Points to Consider. The following areas must be analyzed (for the system and subsystems) and are normally documented in a definition report:

Refined system objectives.

Inputs and outputs required to meet the objectives (as a result of I/O analysis—see details in the section entitled "Analysis and Design Methods for IPSs").

Refined constraints and performance requirements, for example, the outputs must be available within 2 hr of input.

The functions required to change inputs to outputs.

The relationships between functions and subsystems.

One or more design approaches that will satisfy the objectives (Wasserman[2]).

Refined estimates of the number of people needed for system development.

Refined cost-benefit estimates.

Design

Objective. To make trade-off decisions to identify the design that will meet the system objectives and provide maximum cost benefits, within the design constraints.

Management Effort. Analyze various design approaches using mathematical models, computer simulation, mock-ups, group discussions, and so on.

Obtain additional consultation from experts on potential hardware and software problems (see software design tools in the section entitled "Analysis and Design Methods for IPSs").

Identify design features that may degrade employee performance, consulting with human performance specialists.

Analyze system functions to determine which should be assigned to the users and which to the software. Again, trade-off decisions are made based on human and machine capabilities, time and cost factors, potential error rates, and so on.

Major Points to Consider. Detailed design activities for the user and the software begin along serial parallel paths. (Usually, discrete design efforts are established, composed of human performance specialists and software specialists.)

User	Software
All user functions are analyzed and documented to identify what tasks the system personnel must perform (providing inputs, operating, maintaining, or using the system and/or its outputs).	All software functions are analyzed and documented to identify what performance or production is required.
Initial performance specifications are developed, for example, quality and quantity standards.	Initial performance specifications are developed, for example, capacity or processing rates.

Interface design requirements are developed by analyzing the interaction of the user and software functions. For example, certain system outputs may have to be converted into notation understood by employees in order to avoid unnecessary, costly training and potential errors.

User	Software
Tasks that users must perform are grouped into units of work using criteria such as common inputs, work location, and skill level.	Software functions may be divided into smaller, more manageable units for further design, for example, software subsystems (referred to as "software modularization").
Each unit of work is analyzed to determine how required performance can best be achieved, for example, through personnel selection, training, or job aids.	Each component or subsystem is analyzed to determine how the required performance can be achieved.
Design specifications are developed, for example, work designs, personnel skill and knowledge requirements, staffing estimates, procedures, guidelines, displays, forms, training requirements, job aids (manuals, checklists, etc.), and work space layout.	Design specifications are developed, for example, materials for hardware manufacture or software language.

Both user and software specifications are refined and modified as the design reaches a greater level of detail. A change in either set of specifications normally affects the other; therefore constant communication between design efforts is coordinated and ensured by the project manager.

Development

Objective. To develop prototypes (first versions) for user hardware and software components (software prototypes are referred to as "software blueprints").

Management Effort. Prioritize subsystems, user work design, hardware, and software components based on criticality, logical relationships, resources, and so on.

Develop work activity schedules based on estimates of development time and priorities.

Refine software and user specifications (e.g., work designs, personnel skill and knowledge requirements, and staffing estimates) based on prototype development.

Major Points to Consider. As during the design stage, constant communication between development groups is required. Changes are identified during prototype development that affect both the user and the software prototypes.

Installation guidelines are developed to facilitate the integration of the system in the organization's environment. These guidelines include overviews of the system and subsystems and their functions; human-machine interfaces; work flow; personnel selection criteria; staffing requirements; performance data; and any information obtained during testing that may affect decisions in the preceding areas.

Evaluation

Objective. To conduct tests on site, using participants who meet the personnel skill and knowledge requirements.

Management Effort. Formulate a testing schedule based on the levels of testing to be completed and the development schedule. Decide on levels of testing to be completed based on some of the following criteria:

1. The more accurately testing simulates the ultimate operating environment, the greater the likelihood that:
   a. Problems will be identified and corrected prior to installation.
   b. Performance and staffing data will provide accurate estimates for installation planning.
2. Various levels of testing may be used to build up to a system test that integrates and tests all system components (software test and evaluation as discussed by Wasserman[2]).

Develop test plans based on the objectives of the subsystem or system being tested. These plans may be complex, requiring the development of simulated inputs.

Analyze test results and participant feedback.

Make trade-off decisions to prioritize system modifications.

Design, develop, and retest high-priority system modifications. Repeat this cycle as necessary.

Major Points to Consider. Did the subsystem or system meet its objectives? If not, what software and personnel subsystem designs must be changed?

How well were the objectives met? Do performance data indicate the need to modify performance specifications, personnel skill and knowledge requirements, staffing estimates, work flows, and so on?

What design inconsistencies, errors, or omissions were identified? What modifications can be made to correct the problems?

What interface problems were identified? What changes should be made?

Installation

Objective. To cut over to the new system at the user organization.

Management Effort. The cutover team, composed of the system designers and user personnel, uses the installation guidelines to plan and implement the following:

Modifications to jobs and methods (both supervisory and nonsupervisory) to meet new work requirements.

Changes in organizational structure resulting from job modifications.

Adjustments to work force.

Development of supervisor/employee ratios.

Development of performance measures.

Selection of clerical, craft, and supervisory personnel.

Obtaining and design of work space and layouts.

Modification to procedures, guidelines, displays, forms, training, performance aids, and so on.

Training of personnel.

Cutover to system.

Evaluation of cutover.

Major Points to Consider. Based on initial cutover experience, user personnel provide feedback to system design staff. This information is used to identify (1) needed system modifications and (2) areas in the system design process that should be changed or improved. Modifications that are to be delayed are documented, and potential system users are informed of these problem areas. Interim solutions for installation are jointly developed by the project team.

System Maintenance

System maintenance consists of three principal types of maintenance activities:

1. **Corrective.** Modifications that must be made to correct a system failure or malfunction.
2. **Preventive.** Routine activities performed on a preestablished and/or scheduled basis to prevent system performance degradation.
3. **Rearrangements.** Modifications necessary to accommodate revised user and/or environmental requirements.

System maintenance includes the annual review and evaluation of system effectiveness. It also facilitates the making of decisions pertaining to system divestment, replacement, and/or redevelopment. This responsibility is typically assigned to members of the information systems design staff.

12.4.4 ANALYSIS AND DESIGN METHODS FOR IPSs

The terms "analysis" and "design," as used here, refer to all the activities associated with the first three system life cycle stages (described in Section 12.4.3) or everything that goes on before coding of IPS software is initiated. The analysis and design methods presented in this section do not represent an exhaustive list of what is currently available. (For a comprehensive review of analysis and design methods, the reader is referred to Freeman and Wasserman,[13] Bergland and Gordon,[14] and Cougar.[15]) It does, however, identify major methods for IPS analysis and design used frequently by IPS developers. First, the IPS analysis methods are reviewed, followed by the IPS design methods. Exhibit 12.4.4 summarizes the methods and the corresponding applicability of each during the system life cycle stages.

Analysis Methods

Each of the seven methods shown in Exhibit 12.4.4 is discussed as follows:

Definition. A brief summary statement of the method.

Applicability. Discussion of when to employ each method, the utility of its output, and its limitations.

Approach. Task list of actual procedure.

Methods Products. Specific documents reflecting the results obtained after using the method.

Input/Output Analysis

Definition. Input/output analysis is a method of task analysis. It is the process by which inputs to and outputs from an activity are identified and compared. Lower-level activities that are needed to convert the inputs to the outputs are identified. Inputs and outputs at the lower level are also identified. This process is continued until a level of activity is reached where the user behaviors can be characterized in terms of performance requirements (i.e., knowledge, skills, experience).

Applicability. Input/output analysis is used primarily in the analysis of human functions. It is particularly useful when there are no existing similar systems from which to obtain data. Input/output analysis may be used to estimate skill and ability requirements for performing an activity, appropriate experience levels for a given activity, and the necessity for training or special job aids. It is usually used during the system definition stage.

Approach. The general procedure for I/O analysis may be summarized as follows:

1. Identify the required outputs of the human function and describe their characteristics. (Outputs are the required end products of an activity.)
2. Identify the available and required inputs to the function and describe their characteristics. (Inputs are information items available and required before the activity occurs.)
3. Identify the differences between inputs and outputs. This helps derive the activities necessary to change inputs to outputs.
4. Logically derive the lower-level activities necessary to convert the available inputs into required outputs.
5. Determine the optimum or required sequence of the lower-level activities. (Work flowcharts may be useful here.)
6. Check to see that all activities composing a higher-level activity are mutually exclusive and exhaustive.

Exhibit 12.4.4 Summary of Relationships Between IPS Methods and System Life Cycle Stages

	System Life Cycle Stages					
Methods	Proposal and Feasibility	Definition	Design	Development	Evaluation	Installation
Analysis						
Input/output analysis		X				
Decision analysis	X	X				
Contingency analysis			X			
Critical incident technique	X	X			X	
Field survey techniques	X	X			X	
Function flow logic diagrams	X	X				
Link analysis		X				
Design						
Structured design			X			
Structured systems analysis	X	X				
Hierarchy input-process output			X			
Program design languages			X			
Jackson design methodology			X			
Documentation standards			X	X	X	

7. Identify, if possible, the level of each activity in the activity hierarchy.

8. Identify, if possible, the human performance characteristics and requirements of the activities pertinent to each task. If the behavioral characteristics and requirements cannot be identified, go to the next step; otherwise the analysis is complete.

9. Identify the inputs and outputs of each activity that require further analysis and repeat from step 1.

Methods Products. Input/output analysis is carried out using a series of documentation forms, the most common of which is flow charting. Two forms of flow charting are particularly useful. Vertical working flowcharts emphasize the breakdown of human activities into lower-level activities. Horizontal working flowcharts are useful in identifying sequences between activities at or near the same level. Examples of both types of flowcharts are shown in Exhibit 12.4.5.

A second common type of documentation for I/O analysis is reflected in Exhibit 12.4.6. A function/task analysis record is a product that is used whenever the level of analysis is sufficient to characterize an activity in terms of its human behavior requirements. Prerequisite knowledge, skills, and abilities for the activity are then defined.

For further details on I/O analysis, see Amerman.[16]

Decision Analysis

Definition. Decision analysis is analysis of decisions made on the basis of information transmitted and the subsequent action that is taken. It is also the process by which alternative actions and conditions are identified.

Applicability. Decision analysis may be used in both proposal and feasibility and definition stages. It is most commonly used in IPSs because it specifies the information inputs required at any given point, indicates the sequence of decision making activities, and shows relationships between elements of a system. The need for decision analysis is generally found during the course of I/O

Exhibit 12.4.5 Input/Output Analysis Flowcharts: (*a*) Vertical and (*b*) Horizontal[a]

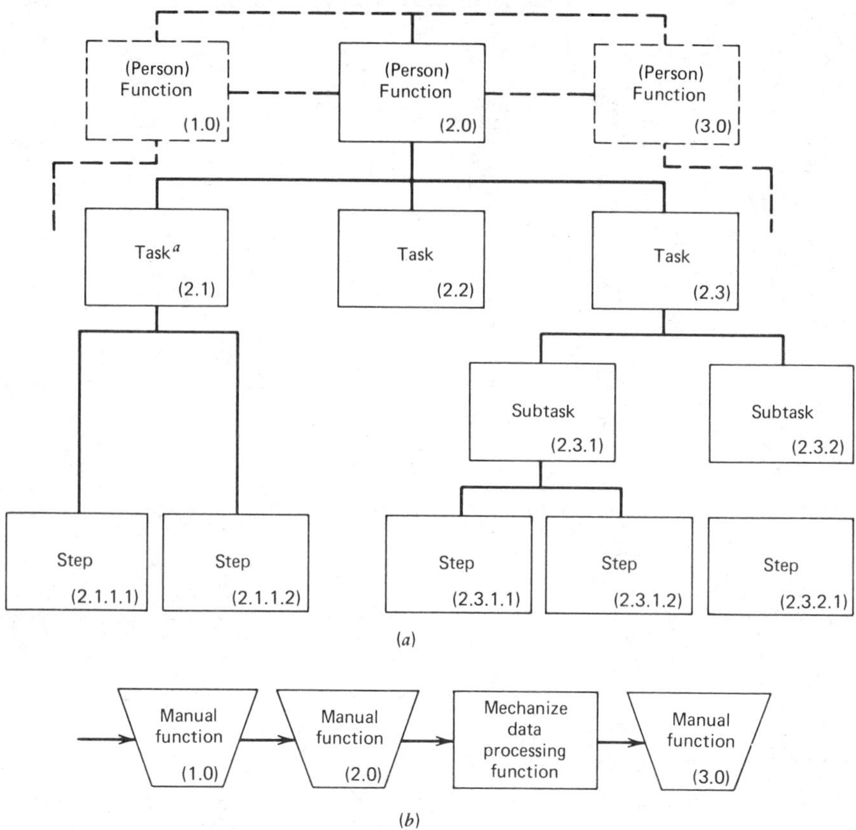

[a]Activity descriptions are entered inside boxes and symbols rather than "task" and so on. These descriptions begin with action verbs, for example, "prepare completion notices," "request credit references."

analysis. The results of decision analysis are usually documented for use as performance aids when systems become operational.

Approach. Decision analysis is conducted as follows:

1. From I/O analysis, identify activities that constitute sets of alternatives.
2. Identify the conditions that would cause the user to decide on each alternative within the set. (It may help to list the conditions in one column and the activities associated with each condition in an adjacent column.)
3. Identify the specific information that would reveal the existence of each condition. (Use flowcharts and decision tables, as the ones shown in Exhibit 12.4.7, to document information.)
4. Design and document the relationships between the conditions and the activities. Results of the decision analysis must be recorded in conjunction with the given task. A flowchart and a description of the decisions and the requirements for making them are generally used.
5. Develop the supporting performance aids to help users make correct decisions and take appropriate actions. (These aids should contain the same information depicted in the flowchart and/or decision tables.)

Methods Products. Decision analysis documentation includes flowcharts and decision tables. For further details on decision analysis, see Raiffa[17] and Morgan et al.[18]

Contingency Analysis

Definition. Contingency analysis is a method of identifying, analyzing, and developing procedures for handling potential errors, malfunctions, and other situations that decrease system effec-

Exhibit 12.4.6 Function/Task Analysis Record

Function name ____________________

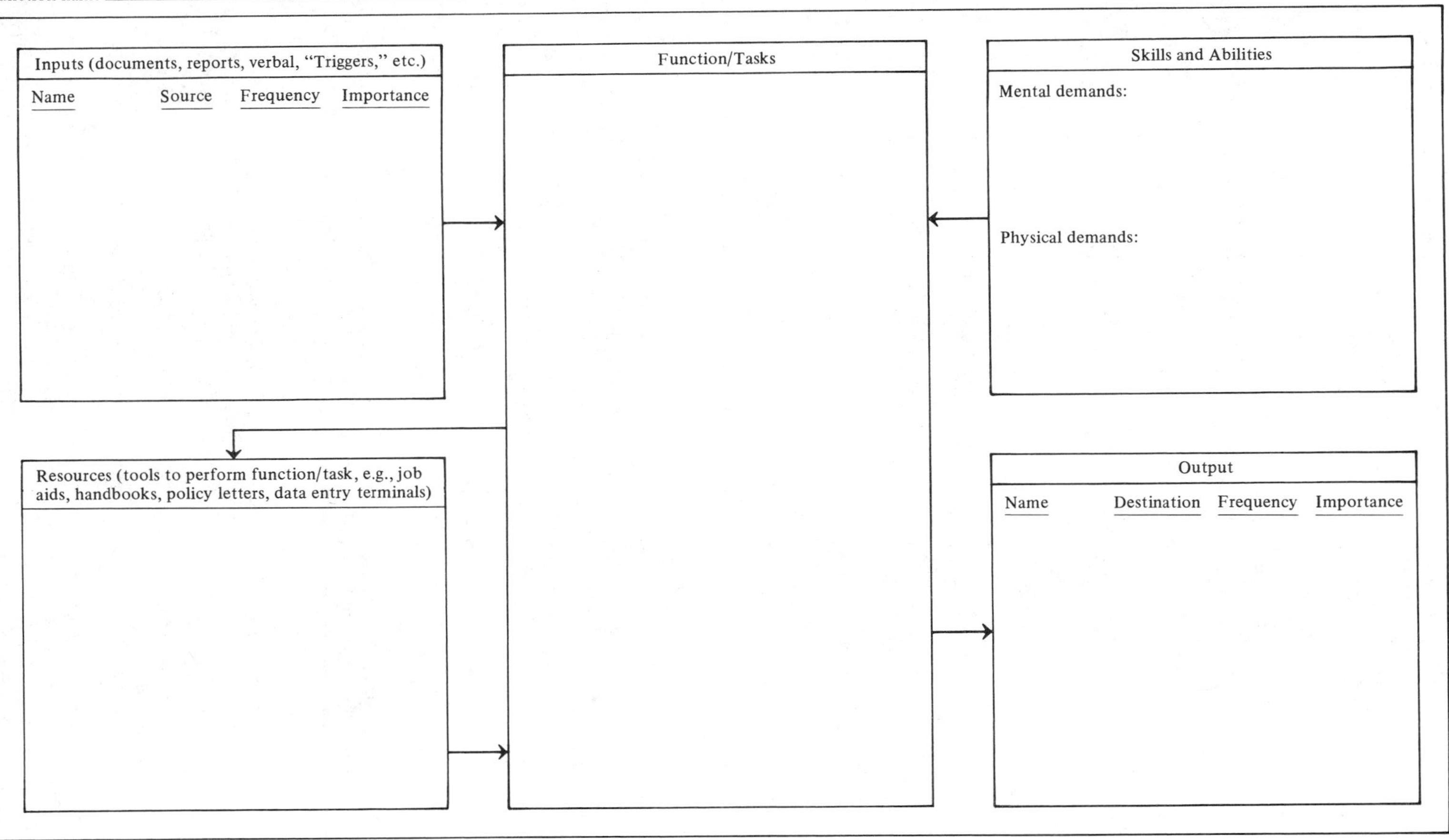

Exhibit 12.4.7 Examples of (*a*) Information Flowchart and (*b*) Decision Table

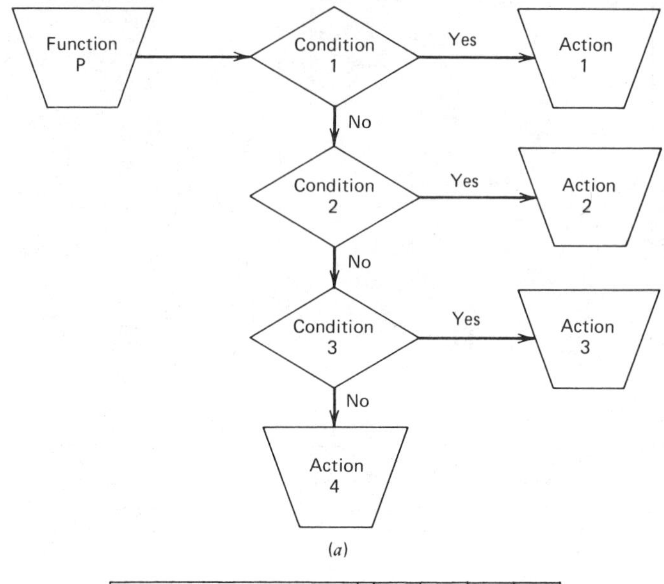

(*a*)

Condition 1	Y	N	N	N
Condition 2	N	Y	N	N
Condition 3	N	N	Y	N
Action 1	X			
Action 2		X		
Action 3			X	
Action 4				X

(*b*)

tiveness. The objective of contingency analysis is to design for minimal acceptable system errors, malfunctions, or problems. Fail-safe systems are ones in which all contingency is prohibited from developing into a situation threatening the integrity of the systems or having hazardous consequences.

Applicability. Contingency analysis is generally applied when system functions have been broken down into the step level by other forms of analysis (i.e., I/O analysis, decision analysis). Contingency analysis is usually done during the system design stage.

Approach. Contingency analysis is carried out as follows:

1. Determine for each task element what could possibly go wrong.
2. For each possible contingency (error or malfunction), determine the frequency of occurrence.
3. Determine the criticality of each contingency (if it occurred) in terms of system degradation. Consider the system and its objectives.
4. Identify possible design considerations that would minimize the frequency of the contingency's occurring.
5. Identify potential corrective procedures that would accommodate the contingency if it were not prevented.
6. Determine the impact on operations if preventive or corrective procedures were implemented (impact in terms of increased time, cost, training, etc.).
7. Based on the frequency and criticality of the contingency, determine and recommend an appropriate action.

Methods Products. The results of contingency analysis are documented (1) as part of user instructions and (2) as a corrective procedure if the corrective activities are lengthy and/or relatively infrequently required.

For further details on contingency analysis, see Pilitsis[19] and Nadler et al.[20]

Critical Incident Technique

Definition. The critical incident technique involves asking each of a large number of system users to describe extremely successful or extremely unsuccessful incidents that were experienced while using a particular system. These incidents are studied to identify highly significant features or elements. Checklists of task behaviors may be developed that represent the critical characteristics for effective or ineffective performance.

Applicability. Used during the proposal and feasibility and/or definition stages, the critical incident technique must be performed on an existing system since it uses occurrences of observed behaviors. The behavioral characteristics isolated by the critical incident technique may be used to:

1. Identify required behaviors for the successful or unsuccessful performance of a job.
2. Identify jobs that require similar behavior characteristics for their successful or unsuccessful performance (i.e., job families).
3. Identify both the static and dynamic characteristics of jobs.
4. Assist in the development of personnel selection tools.
5. Evaluate personnel selection procedures (i.e., the fewer negative critical incidents reported, the more successful the personnel selection tool).

Approach. In the critical incident technique, users are asked to recall examples of extremely poor task-related behaviors that they have observed in themselves or other users. Critical incidents may be collected in verbal or written form and may be administered to individuals or groups. Information on each incident should include what led up to the incident and the general circumstances under which it occurred, exactly what was so effective or ineffective, and why it seemed so helpful or detrimental.

The following steps are observed in carrying out the technique:

1. Determine the target population to be used in the collection of critical incidents.
2. Decide whether to administer questions in written or verbal form, in groups or individually.
3. Develop the format for data collection.
4. Isolate the critical behaviors from the critical incidents collected and organize them into categories.
5. Use the result as a comprehensive profile of critical behaviors pertinent to the job.

Methods Products. Data are collected through interviews and/or questionnaires. Data are summarized through a variety of methods.

For further details on the technique, see DeGreen,[21] Dunnette,[22] and Flanagan.[23]

Field Survey Techniques

Definition. Interviews and questionnaires are techniques for obtaining information using a question-and-answer approach. Interviews involve person-to-person interaction, either in a physical setting or over the telephone. They can be highly structured or loosely organized. Questionnaires are largely self-administered. They also can be open ended or closed ended. Some methods covered earlier (critical incident technique) represent special applications of the interview/questionnaire method.

Applicability. Used during the proposal and feasibility, definition, and evaluation stages, interviews and questionnaires facilitate collection of quantitative information or information on user needs, attitudes, values, and decision making models. In-person interviews, telephone interviews, and questionnaires are appropriate under different circumstances. Some of the advantages and disadvantages of each technique are summarized in Exhibit 12.4.8.

Approach. Gathering useful information by either the interview or questionnaire technique involves the following steps:

1. Define the purpose of the study.
2. Determine what information will be needed to meet this purpose; state this need clearly in terms of objectives.

Exhibit 12.4.8 Advantages and Disadvantages of Survey Techniques

Type of Survey	Advantages	Disadvantages
In-person interviews	Highest response rates More information More accurate information	More costly (time, dollars) Interviewers need training
Telephone interviews	Inexpensive Conducted more rapidly Supervised easily	Less sensitive Must be kept short Less information can be obtained
Questionnaires	Can generate large amount of information Trained interviewers are not required Minimizes "halo" effects	Instructions must be very explicit Returns tend to be low Require extensive coordination and preparation

3. Decide which type of interview or questionnaire technique is most appropriate in light of the objectives of the study and the existing constraints.

4. Develop questions that will facilitate the fulfillment of objectives defined in step 1.

5. Consider the motivation of the respondents as a critical variable; the idea is to get the person to respond fully and conscientiously to the questions. (That is, the purpose of the technique should be apparent, the instructions should be clear, and responding should require a minimum of time.)

6. Pretest the interview or questionnaire on a small group first. (The pretest group should be similar in as many ways as possible to the target group.)

Methods Products. Documentation of interview and questionnaire techniques may take many forms, depending on the kind of study undertaken, its objectives, and the use to which the results will be put. Documentation may be in the form of a narrative, tables, charts, and so on. There is no standard format for documentation of the results of an interview or questionnaire survey.

For further details on these techniques, see DeGreene,[21] Fetridge and Minor,[24] and Salvendy and Seymour.[25]

Functional Flow Logic Diagrams

Definition. The functional flow logic diagram is a technique used to repeatedly break down and analyze a system in terms of the functions or activities necessary to convert inputs to outputs. It is a method of identifying what is needed to achieve system objectives. The diagrams utilize boxes to represent functions or activities, arrows to represent inputs to and outputs from the boxes, and other special symbols (Exhibit 12.4.9).

Applicability. Functional flow logic diagrams are used to analyze systems critically in terms of their component functions or activities. They may be used to create a system model or to simulate a system or subsystem; they are general-purpose analytic tools, typically utilized during the proposal and feasibility and/or definition stages.

Approach. Functional flow logic diagrams are generated in the following manner:

1. Begin with a simple representation of the system or subsystem. Identify initial inputs and outputs. Use a box to represent the activity or function that transforms the inputs into outputs. Use arrows to show directionality. Use nouns rather than verbs to describe the inputs and outputs. Functions or activities should begin with a verb.

2. Consider the next lower level of detail. Identify the intermediate products or results that occur during transformation of inputs into outputs. Number the resulting activities in the lower right-hand corner of the boxes. Additional symbols are needed to indicate multiple inputs and outputs. Activity boxes are never intersected by arrows from the top or bottom.

3. Repeat step 2 until the desired level of detail in the analytic process is reached. A finer level of detail results with each repetition.

Methods Products. The diagram itself suffices as a documentation tool for communicating results of the analysis. Narrative materials may also accompany a diagram to further characterize activities or to present conclusions or implications that result from the study of a functional flow logic diagram.

For further details on these diagrams, see Burch et al.,[1] Behan,[26] and Passen.[27]

Exhibit 12.4.9 Symbols and Examples Used in Function Flow Logic Diagrams

Symbol	Meaning	Diagram	Explanation
⓪	"And"	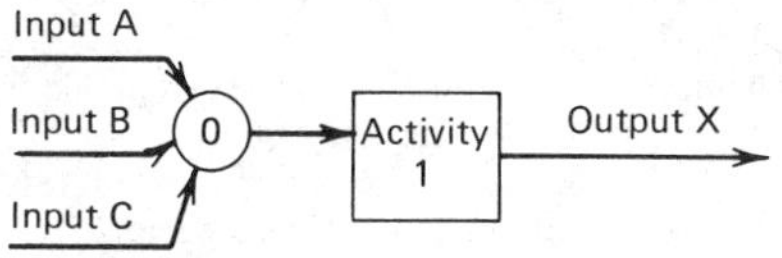 Input A, Input B, Input C → (0) → Activity 1 → Output X	A, B, and C are all required before activity 1 can result in X.
		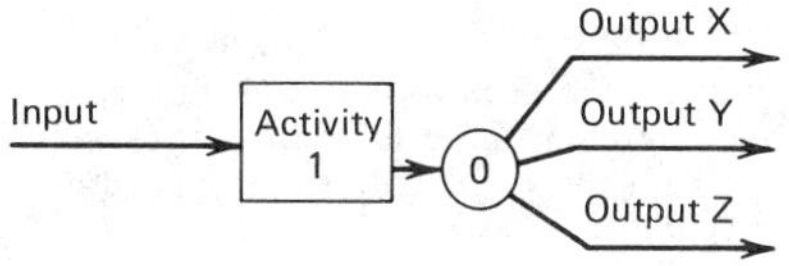 Input → Activity 1 → (0) → Output X, Output Y, Output Z	Given input A, activity 1 will result in X, Y, and Z.
①	"Either . . . or"	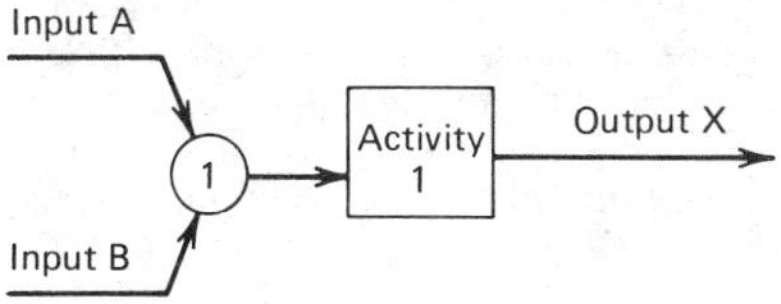 Input A, Input B → (1) → Activity 1 → Output X	Either A or B (but not both) is necessary before activity 1 results in X.
		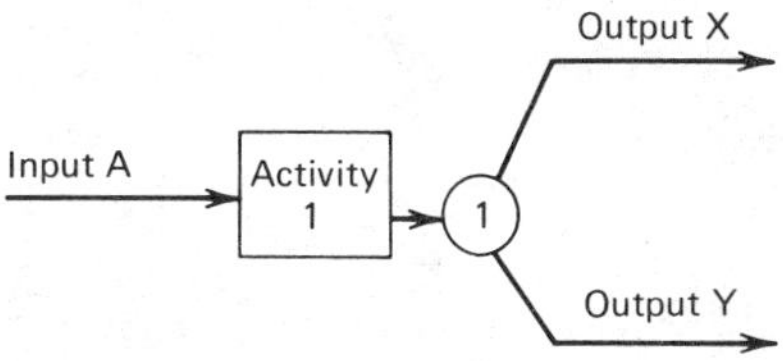 Input A → Activity 1 → (1) → Output X, Output Y	Given input A, activity 1 will result in X or Y (but not both).
⊕	"Any or all"	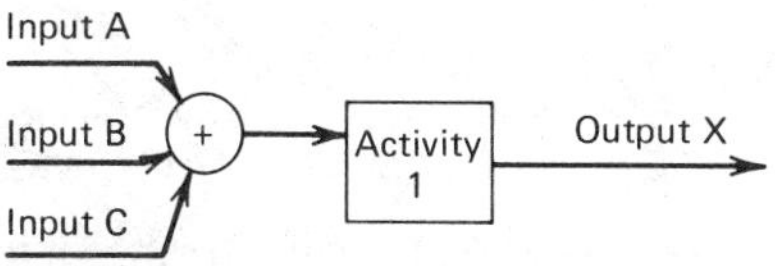 Input A, Input B, Input C → (+) → Activity 1 → Output X	If A, B, C, AB, BC, AC, or ABC occurs, activity 1 will result in X.
		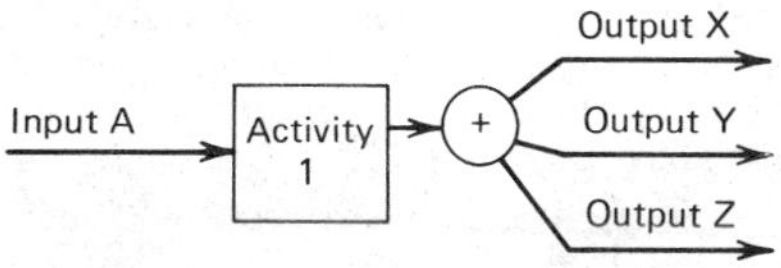 Input A → Activity 1 → (+) → Output X, Output Y, Output Z	Given input A, activity 1 will result in X, Y, Z, XY, YZ, XZ, or XYZ.

Link Analysis

Definition. Link or interface analysis is the process by which interactions between system components are identified and the components arranged so that these interactions will proceed effectively and efficiently. The system components may be humans, machines, or items of information. Links may be visual or auditory communication or physical movement.

Applicability. Link analysis is usually not appropriate until the detail substage of the design stage of system development. Before link analysis can be implemented, allocation of functions among humans and machines must be complete, data on human-machine configurations must be available, and results of a task analysis must be available. Link analysis may also be used to redesign existing systems to improve the working efficiency of system components and the overall system.

Approach. The procedure for conducting a link analysis of a human-machine system is as follows:

1. Draw a circle for each person in the system. Code each circle with a number to identify the person's function (i.e., 1 = clerk, 2 = supervisor).
2. Draw a square for each piece of equipment in the system. Code each square with a letter for further identification (i.e., A = computer, B = drawing board).
3. Draw connecting lines (links) between the humans who interact directly in the system.
4. Draw links between each human and machine that interact.
5. Redraw the resulting diagram, limiting the number of crossing links to obtain the simplest configuration.
6. Evaluate each link by one or more of the following criteria:

a. *Importance*—Have an experienced person rank each link according to its relative importance to the function objectives. (Low numbers indicate unimportant links.)

b. *Frequency*—Obtain data from the simulated or operational use of the system. Rank each link according to the frequency with which it is used.

c. *Frequency and Importance*—Sometimes both frequency of use and importance must be considered. Experienced employees can decide on relative weights to be given so as to assign a single combined value to each link.

7. Redraw the diagram so that the links with higher values are shorter than those with lower values. Reduce the number of crossing links. This is the optimal link diagram.
8. Redraw the diagram to fit into the available space, or even better, redesign the space to fit the shape of the link diagram.
9. As a final step, make a scale drawing of the actual positions of the humans, machines, and spaces in the system. This will serve as an aid to the design engineer.

Methods Products. A scale diagram with accompanying narrative description is generally used to summarize results.

For further details on link analysis, see Pilitsis,[28] Van Colt and Kinkade,[29] and McCormick.[30]

Design Methods

Each of the six methods shown in Exhibit 12.4.4 is presented here in terms of an overview, or summary statement of the utility and limitations of the method. Other sources are cited that provide further details on method and application.

Structured Design

Structured design aims at developing systems with properly "balanced" structures by facilitating comparison of design alternatives, determining relative quality, and transforming data flow diagrams into program structures.

Structured design is driven by the concept of "modularity." In this case modularity is measured by "cohesion" and "coupling." Cohesion is the extent to which module components are internally cohesive; coupling refers to the way in which modules are related and the ways in which data are communicated among modules. The overall mission of structured design is, therefore, to create system structures in which modules have high cohesion and low coupling. Under this condition, modules are relatively autonomous, exhibiting high I/O visibility and minimum data interdependency. By measuring cohesion and coupling, the designer is able to select system structures objectively. This method is used during the design stage. To be effective, it must be used in conjunction with structured system analysis outputs.

For further details on link analysis, see Pilitsis,[28] Van Colt and Kinkade,[29] and McCormick.[30]

Structured Systems Analysis

Structured systems analysis is a method for use with structured design. Structured systems analysis encompasses the following four concepts in creating IPS software specifications during the proposal and feasibility and/or definition stages:

Data Flow Diagrams. Symbolic representation of data flow among processing units, related processes, and places where data are stored.

Decision Tables or "Structured English" (Program Design Language). Utilized to depict the logic for each process box.

Data Dictionary. A data store that describes the nature of each piece of data used in the system, including process descriptions, glossary entries, and other items.

Immediate Access Analysis. Related data base and the types of operations performed.

For further details on this method, see Gane and Sarson,[33] DeMarco,[34] and Teichroew and Hershey.[35]

Hierarchy plus Input-Process Output

Hierarchy plus Input-Process Output (HIPO) has been used for software specification and design. It is a graphic technique that facilitates hierarchical organization of diagrams in which each module has specific input and output items. A HIPO process box (module) contains a nonprocedural description of the steps to be performed within the process or a more specific sequence of steps showing the actual process logic. The technique can be used during detail design.

For further details on HIPO, see Stay[36] and HIPO–A Design Aid and Documentation Technique.[37]

Program Design Languages

Program design languages is a method for specifying program logic in English-like readable form without conforming to the syntactical rules of any particular programming language. Typically used during detailed design, the method facilitates the illustration of a specific algorithm being used within a module.

Program design languages have been seen to be extremely effective replacements to traditional flowcharts. They facilitate definition of interfaces between modules encountered during program execution, of functions performed by a module, and of the logic used in realizing the function.

For further details on the method, see Gane and Sarson[33] and Caine and Gordon.[38]

Jackson Design Methodology

Jackson design methodology is a constructive method of program design. The initial step in applying it is to describe precisely the structure of the input and output data. A static model of the data files is constructed through graphic notation. The data structure diagrams are combined to form the program structure. The various executable operations that must be performed are listed and allocated within the resulting structure. This composition of I/O structure and executable operations lists facilitates translation into conventional programming languages for implementation.

The aim of Jackson design methodology is to facilitate creation of workable designs (not "best" designs) by "average" designers. It is used strictly during the design stage and requires as its input clearly defined I/O structures. Accordingly, resulting program structures are extremely sensitive to major changes in the I/O structures.

For further details on this methodology, see Jackson.[39,40]

Documentation Standards

Although not a specific design technique, the concept of identifying documentation standards for each of the three IPS subsystems (human performance, software, and hardware) is critical to effective design, redesign, operation, and maintenance of an IPS. In general, there are five types of documentation categories:

Analytical Documentation. All records and reports produced during the various IPS development stages (e.g., proposal and feasibility, definition).

Systems Documentation. All information needed to define proposed computer-based IPS to a level at which it can be programmed, tested, and implemented (e.g., system definition report).

Program Documentation. Records of detailed logic and coding of the constituent programs of a system. These records, prepared by the programmer, aid in development, acceptance, and maintenance and provide for overall continuity in case of programming staff changes.

Operations Documentation. Reflects procedures required for running the system by operations personnel. It presents the general sequence of events for performing the job and identifies precise procedures for data preparation, control and security, output dispersal, and so on.

User Aids. Reflect all descriptive and instructive material necessary for the user to facilitate running of the operational IPS, including details on the interpretation of the output information.

Every IPS development organization should establish documentation standards and ensure that these documents are a natural outgrowth of the design and development process rather than a discrete parallel activity to be addressed at any point in the IPS life cycle.

For further details on documentation standards, see Tausworthe[41] and London.[42]

REFERENCES

1. J. G. BURCH, F. R. STRATER, and G. GRUDNITSKI, *Information Systems: Theory and Practice*, Wiley, New York, 1979, pp. 4–5.
2. A. I. WASSERMAN, "Information System Design Methodology," *Journal of American Society of Information Science,* Vol. 31, No. 1 (1980).
3. J. M. BUXTON, P. NAYR, and B. RANDELL, Eds., *Software Engineering Concepts and Techniques*, Petrocelli/Charter, New York, 1976.
4. R. L. ACKOFF, *Redesigning the Future: A Systems Approach to Societal Problems*, Wiley, New York, 1974.
5. D. I. CLELAND and W. R. KING, *Systems Analysis and Project Management*, McGraw-Hill, New York, 1975.
6. A. GRINDLAY, "Managing the Information Systems Function: Ten Useful Guidelines," *The Business Quarterly*, February 1980, pp. 90–94.
7. C. W. HOFER and D. E. SCHENDEL, *Strategy Formulation: Analytical Concepts*, West Publishing, St. Paul, MN, 1978.
8. J. DIEBOLD, "Bad Decisions on Computer Use," *Harvard Business Review*, January–February 1969, pp. 64–68.
9. C. L. BIGGS, E. G. BIRKS, and W. A. ATKINS, *Managing the Systems Development Process*, Prentice-Hall, Englewood Cliffs, NJ, 1980.
10. D. J. REIFER, Ed., *Tutorial: Software Management*, IEEE/CS, Long Beach, CA, 1979.
11. *Second Software Life Cycle Management Workshop*, Publication 78C1390-4C, IEEE/CS, Long Beach, CA, 1978.
12. C. F. GIBSON and R. L. NOLAN, "Managing the Four Stages of EDP Growth," *Harvard Business Review*, Vol. 52, No. 1 (January–February 1974), pp. 76–88.
13. P. FREEMAN and A. I. WASSERMAN, *Tutorial on Software Design Techniques*, EHO 161-0, IEEE/CS, Long Beach, CA, 1980.
14. G. D. BERGLAND and R. G. GORDON, *Tutorial: Software Design Strategies*, EHO 149-5, IEEE/CS, Long Beach, CA, 1979.
15. J. D. COUGAR, "Evaluation of Business Systems Analysis Techniques," *Computing Surveys*, Vol. 5, No. 3 (September 1973), pp. 167–198.
16. H. L. AMERMAN, *Performance Content for Job Training*, Vol. 2, The Center for Vocational Education, Ohio State University, Columbus, 1977.
17. H. RAIFFA, *Decision Analysis*, Addison-Wesley, Reading, MA, 1968.
18. C. T. MORGAN, J. S. COOK, A. CHAPANIS, and M. W. LUNDT, Eds., *Human Engineering Guide to Equipment Design*, McGraw-Hill, New York, 1963.
19. J. V. PILITSIS, *Formation of Center Objectives: A Methodology*, AT & T New Technologies and Standards Document, AT & T, Basking Ridge, NJ, 1980.
20. D. A. NADLER, J. R. HACKMAN, and E. E. LAWLER, *Organizational Behavior*, Little, Brown, Boston, 1979.
21. K. DEGREEN, Ed., *Systems Psychology*, McGraw-Hill, New York, 1970.
22. M. D. DUNNETTE, *Personnel Selection and Placement*, Wadsworth, Belmont, CA, 1966.
23. J. C. FLANAGAN, "The Critical Incident Technique," *Psychological Bulletin*, Vol. 51, 1954, pp. 327–358.
24. C. FETRIDGE and R. S. MINOR, Eds., *Office Administration Handbook*, The Dartnell Corporation, Chicago, 1977.
25. G. SALVENDY and W. D. SEYMOUR, *Prediction and Development of Industrial Work Performance*, Wiley, New York, 1973.
26. R. A. BEHAN, *The State Change Approach to Systems Analysis*, Serendipity Associates, Boston, 1969.
27. B. J. PASSEN, *Program Flowcharting for Business Data Processing*, Wiley, New York, 1978.
28. J. V. PILITSIS, *System for Physical Arrangement of Centers (SPACE)*, New Technology and Standards Document, AT & T, Basking Ridge, NJ, 1980.

29. H. P. VAN COLT and R. G. KINKADE, Eds., *Human Engineering Guide to Equipment Design*, American Institute of Research, Washington, DC, 1972.

30. E. MCCORMICK, *Human Factors Engineering*, McGraw-Hill, New York, 1976.

31. E. YOURDEN and L. L. CONSTANTINE, *Structured Design*, Prentice-Hall, Englewood Cliffs, NJ, 1979.

32. C. J. MYERS, *Composite/Structural Design*, Van Nostrand Reinhold, New York, 1978.

33. C. GANE and T. SARSON, *Structured Systems Analysis: Tools and Techniques*, Prentice-Hall, Englewood Cliffs, NJ, 1979.

34. T. DEMARCO, *Structured Analysis and Systems Specification*, Yourdon, New York, 1978.

35. D. TEICHROEW and E. HERSHEY, "PSL/PSA: A Computer-Aided Technique for Structured Documentation and Analysis of Information Processing Systems," *IEEE Transactions on Software Engineering*, Vol. SE-3, January 1977, pp. 41–48.

36. J. F. STAY, "HIPO and Integrated Program Design," *IBM Systems Journal*, Vol. 15, No. 2 1976), pp. 143–154.

37. *HIPO–A Design Aid and Documentation Technique*, Document GC 20-1851, Data Processing Division, IBM, White Plains, NY, 1976.

38. S. H. CAINE and E. K. GORDON, "PDL a Tool for Software Design," *Proceedings of the American Federation of Information Processing Societies*, NCC 44, 1975.

39. M. A. JACKSON, "Constructive Methods of Program Design," in K. Samelson, Ed., *Lecture Notes in Computer Science*, Springer-Verlag, New York, 1976.

40. M. A. JACKSON, *Principles of Program Design*, Academic Press, London, 1975.

41. R. C. TAUSWORTHE, *Standardized Development of Computer Software*, Part II Standards, Prentice-Hall, Englewood Cliffs, NJ, 1979.

42. K. LONDON, *Documentation Standards*, Auerbach, Pennsauken, NJ, 1973.

CHAPTER 12.5

Information Processing Systems and Computers in Industrial Engineering

HAMED K. ELDIN

Oklahoma State University

12.5.1 INTRODUCTION

Increasingly, industrial engineers use the computer in performing many of their activities. Computer programs have been developed in almost every area of the industrial engineering profession, from job standards to mathematical optimizations. The objective of this chapter is to review the utilization of computers in the practice of industrial engineering over the years and to discuss future trends of computer applications.

The chapter starts with a brief introduction to the evolution of computer technology and the utilization of computers by the industrial engineering profession. This is followed by a description of some of the present computer applications in industrial engineering, with emphasis on the computing tools and techniques appropriate for the different areas of the discipline.

The last part of the chapter is intended to help the reader obtain a better understanding of the future trends in computer technology and their effect on the effective utilization of computers by industrial engineers in the future.

12.5.2 COMPUTER UTILIZATION

Industrial Engineering Evolution

Over the years both management and the industrial engineer have been hard pressed to define the exact boundaries of the industrial engineering discipline. One could be safe in saying that few will challenge the role of the industrial engineer in the traditional areas of quality control, plant layout, job evaluation, work simplification and measurement, and time standards. More recently, the systems analysis and operations research fields are credited with reviving the scientific worth of this discipline. The adoption of these two fields by industrial engineers should not deemphasize the venerable foundation of the profession; the industrial engineer can still make significant contributions to the basic techniques with the aid of computer technology.

With the complexity of industrial organizations today, an industrial engineer cannot hope to apply intuition and experience to the voluminous number of variables without the aid of a computer. His or her dependence on timely and accurate information on a multitude of organizational functions promises to make the industrial engineer one of the most frequent users of a computer system.

It is felt that the new industrial engineer should more appropriately fill the role of a member of the management team specifically charged with the unique functions of integrating the human factors, the physical contributions, and the new management information function essential for more effective utilization of the organizational resources.

The effective utilization of computer applications is one of the greatest challenges to the industrial engineer. To utilize the computer effectively, he or she should have adequate knowledge of information systems concepts (Chapter 12.2), solution techniques, and computing tools (Chapters 12.1, 12.2, and 13.11) and should be able to apply them to the different industrial engineering areas.

Methodology and Solution Techniques

Methodology

The "methodology" of industrial engineering is simply the systematic integration of activities toward the accomplishment of an objective. It is the process of establishing objectives, allocating resources, and organizing activity so that all aspects of a problem may be known as exactly as possible at any given time, and the process of providing the coordination necessary to achieve goals according to a predetermined plan and schedule. This process provides a bridge between what is needed and what is technically feasible and practical.

Methodology is distinguished from the tools and techniques that are used to derive a solution. (This difference is mainly one of philosophy; the two cannot be divorced in actual application.) The tools and techniques include mathematics, probability and statistics, operations research, modeling, simulation, decision theory, and related areas. One supporting tool worthy of individual recognition is the electronic computer. Its capabilities and the more advanced analysis techniques it makes possible can be used to manage the resources committed to the systems development and to improve implementation. Centralized control and rapid and accurate data management are possible when the computer is incorporated into the systems methodology. By use of the computer, vast amounts of data can be processed rapidly, and simulation models can be built to test new ideas. The really significant functions of systems methodology are the selection of the variables and relationships involved, the structuring of the problem, the establishment of the criteria, and the handling of the risks and uncertainties involved. This approach to decision making would be extremely difficult, if not impossible, without the availability of computers and mathematical techniques.

In addition to the use of systematic approach, methodology should follow a scientific procedure. The steps of scientific procedure can be summarized as follows:

1. Formulation of the problem.
2. Construction of a model to represent the system under study.
3. Derivation of a solution from the model.
4. Testing of the model and the derived solution.
5. Establishment of controls over the solution.
6. Implementation of the solution.

This methodology is also multidisciplinary, encompassing all science-based bodies of knowledge required in the solution of any given problem.

The aim is to determine, from all the available alternatives, the most favorable course of action for any given set of alternatives. The basic method is to study a problem by considering the total system (all significant factors, both human and material) and to analyze the resulting model (usually mathematical) by means of operations research techniques. Thus the basic tools and techniques are systems analysis and operations research. And since lengthy and complicated calculations are often necessary, the computer is usually employed.

Solution Techniques

The problems that face management arise from the different sectors of the organizational structure and are problems in content. They are synonymous with the functions of the organization and are classified as problems related to major systems, such as financial, personnel, and logistics, and to minor systems, such as R & D and marketing. It is obvious that not all of management's difficulties will fall neatly into the outlined categories, many will tend to overlap among the categories. Some may even have an implication for the whole structure.

The industrial engineer whose task is to apply the scientific knowledge to the solution of these management content problems has found from experience that the majority of these problems are composed of one or more basic structures, or forms. These forms into which we can classify most management problems are inventory, allocation, queuing, sequencing, routing, replacement, competition, and search. Again, a problem of one form, or structure, may also be formulated into a problem of another form. For example, some inventory problems can be reformulated as either allocation or queuing problems.

After the problem has been classified as a particular form, a model of the specific situation can be developed. The symbolic or mathematical model is the type of model that is most used and preferred. Aside from presenting complex situations in the easiest and most economical manner, the mathematical model provides the quantification of the problem and allows for clearer understanding and interpretation of the solution. To derive a solution from these developed models, certain tools and techniques must be applied.

In general, "techniques" involve the application of certain fundamental operations to solve a management problem. These basic operations to be employed in the solution are known as "tools," and some are found in all the various techniques.

The techniques utilized in solving the problems by form may be broken down into deterministic and stochastic techniques. The difference between these two is that deterministic techniques use data that are assumed to be correct as they stand, whereas stochastic techniques apply statistics and probability to give relative answers. In all these techniques, a computer is necessary if an in-depth study is desired.

The techniques and tools of particular value to the industrial engineer are summarized in the remainder of this section.

Mathematical, Probability, and Statistical Tools. Generally, the deterministic techniques use vast groups of mathematical tools. For example, linear programming has the matrix algebra as the main tool in its use. By a matrix representation of the data and the use of the identity matrix, the simplex method can be used to solve linear programming problems. In graphical linear programming solutions, the use of algebra in solving simultaneous equations is also valuable.

The stochastic techniques naturally rely more on statistics, probability, and distributions than do the deterministic techniques. Sensitivity analysis uses mostly the tools of defferential calculus to find the minimum point on a cost curve, probability to determine variations from this point, and expected value to evaluate any deviation from the optimum point.

Modeling, Simulation, and Optimization Techniques. First, there are deterministic techniques, which include linear programming, EOQ, economic level quantity, break-even analysis, improvement curve, critical path method, time series analysis, dynamic programming, dimensional analysis, and symbolic logic. The stochastic techniques include heuristic modeling, sensitivity analysis, decision theory, competitive modeling, queuing theory, statistical quality control, PERT, Monte Carlo theory, behavioral modeling, Markov process, and simulation.

Of course there are other modeling simulation and optimization techniques that lend themselves to the solution of managers' problems.

A working knowledge of these tools and techniques is essential for the industrial engineer. In most situations the use of computers will greatly aid in applying the various tools and techniques mentioned.

12.5.3 INDUSTRIAL ENGINEERING COMPUTER APPLICATIONS

Computer programs have been developed in almost every area of the industrial engineering profession, from time standards to mathematical optimizations.

This section provides industrial engineers with a broad view of computer utilization in the different areas of their activities. Within each area of activity, the following concepts are briefly discussed:

1. The type of information system and the quality and quantity of data involved.
2. The purpose of using the computer.
3. Software development (This will cover the computer language to be used, as well as the computing system configuration.).

In the early 1970s a survey was conducted in an effort to provide industrial engineers with information on sources from which to obtain available commercial applications.[1] This survey was recently updated, and the results are included in this section.

In 1974 another survey was conducted to quantify the need for industrial engineering computer software. For this purpose, a questionnaire was distributed to a list of 1500 industrial engineers who had been chosen in a geographically oriented random sample. The results of this survey were published in the AIIE magazine.[2] Of special significance were the respondents' answers to the following two questions:

Question and Answers	Percent
What type of software programs would be of value to you?	
Financial	7
Logistics	85
Personnel	8
Specifically mentioned:	
Applied mathematics	27
Production planning and control	26
Work measurement	10
Facility planning	9
Costs and cost control	7
Engineering economics	5
Organization planning	5
Applied psychology	4
Data processing systems design	4
Industrial engineering education, materials processing, and methods	3

Question and Answers	Percent
Which one of the following do your comments represent?	
Private industry	77
Government	9
University	6
Professional society	1
Others	7

This section also discusses the conventional computer applications in several industrial engineering areas. Each area provides an appreciation of the present and future prospects. References are given to sources of more detailed information.

Work Measurement and Time Standards

The innate value of computer applications in the work measurement and time standards area is the speed by which data can be retrieved and manipulated and the ability to program inductive sequences. All this is designed to relieve the industrial engineer of the drudgery of repetitive tasks. It also improves accuracy by eliminating errors in numerous calculations. The industrial engineer will then be free to make a more comprehensive evaluation of the work situation. The success of computerized work methods analysis in the direct labor areas permits the industrial engineer to expand this scientific analysis to that of the indirect labor and other minor tasks of a temporary nature, which otherwise would have been time consuming and costly.

The perspective that results from integrating such an application into the organization is better estimates and management information data. When standards are set prior to the work's beginning rather than after, the involved training period, production lines, labor relations, plant layout, and cost estimates can reflect a more realistic alternative on which to base a management decision. With labor costs becoming greater than those of plant equipment, time standards are one of the most critical elements in maintaining the competitive position.

Computer Software for Academic Instruction

Most universities providing instruction in the work simplification and time standard areas usually have computer subroutines for statistical analysis. Some universities may have a work sampling (ratio delay) simulator which could be used from an on-line terminal, depending upon the scope and size of the designed program.

In the human factors area, there are some specialized programs. The Bio-Mechanical Model® is one example of such a program. It is a large FORTRAN program that is used to perform statistical analysis of stresses on the human body. Input data consist of limb positions during stress applications and other parameters to describe the overall relationships. This is a major batch program. It can and should be compiled and stored in a software library during periods of frequent use.

Students are encouraged to write their own programs. A typical simple computer program for setting a time standard can be briefly described as follows:

1. **Input.** Each element is assigned an identifying activity code number, a pace rating, and an allowance for personal fatigue and delay (PFD). Some digits are used to indicate interruptions occurring during these elements, and the time values for the like digits must be combined in order to calculate the actual time for the element.

2. **Processing.** Computer applications for most of the work measurement and time standards area involve a large number of data items and relatively simple calculations. The main purpose of using the computer is storage and retrieval of information. Either batch or on-line processing is used, depending on the size of the job studied.

3. **Output.** The printout shows the summary of the time study and includes the following calculations:

a. The actual time for each activity and the total time.

b. The rating and PFD columns are totaled and divided by the number of occurrences of each to arrive at the average rating and PFD allowance.

c. The units studied are printed, and the time study standard is given.

Significant Contributions

Some of the significant contributions in the area of work simplification and time standards have been developed in the last decade; however, they are still used by industrial engineers. One is briefly described here to show its contribution to computerization in a particular area.

Work Activity Sampling Program (WASP) is a technique for continuous work sampling based on randomized, intermittent, and short observation periods. When integrated with other production and labor reporting systems, this method provides a low-cost, engineered statistical time standard. The unique feature of this method is a computer randomization of the time observations, the areas to be observed, the cost center observed, and the individual employee to be observed. The elimination of any employee anticipation lends more credence to the accuracy of the data.[3]

For detailed discussion related to computerized work measurement, see Chapter 4.7.

Process Control

The process industries, such as steel, power, and chemical, have long enjoyed computer process control. The combination of quick response to sensing devices and direct communication with the process under dangerous environmental conditions has proved highly successful. The basic functions of process control, then, are the collection of input sensing values, the monitoring of the control program, and, when necessary, the feeding back of control commands. Industrial engineers have a greater interest in, and better qualifications for application of process control computers to discrete item manufacturing. Process computer control is also utilized in the warehouse and in material handling and quality control activities.

Pike outlined the general nature of software systems for process control.[4] His thesis is that prepackaged software provides the most economical attack for supervisory process control requiring a minimum of special programming service. It is in this supervisory application that the industrial engineer can make the greatest contribution.

The process control computer affords a more consistent level of operation with tighter limits. It can respond to normal conditions as well as emergency conditions more quickly than the human operator. Some of the hidden benefits include more complete and accurate operating data, safer and smoother operation, better control through automatic equipment regulation, and more engineering data for analysis.

The accelerated growth of process control applications provides management with numerous benefits. Among these are improved quality, reduced labor costs, greater safety of equipment and for personnel, and improved throughput. One of the major drawbacks, however, has been the high cost of the software programs. At present, process control is mostly custom tailored to the given process. For this reason, a survey of specific computer applications by industry would not contribute to this general picture of the industry.

Production Planning and Control

The essence of production control is the efficient coordination of manufacturing activities by routing, scheduling, dispatching, and expediting individual tasks within the overall system. Undoubtedly, there are almost as many different applications as there are manufacturing firms. For this reason, there is no utility in expounding upon specific algorithms. Rather than emphasizing named programs, this section will deal with the general system approaches to the crucial activity of scheduling.

The computer is the ideal tool in determining the most economical way of scheduling production jobs. It is able to apply heuristic logic to determine the best of a multitude of alternatives in a short period of time.

Significant Contributions

Some of the computer programs developed in the last decade contributed to computerization in this area. Following are brief descriptions of such programs.

Lee Grossman Associates developed a general program that furnishes a daily plant schedule based on actual plant performance. It estimates the current and future manpower levels, establishes a priority to include rush orders, and provides accurate status reports to management. The basic concept is the pooling or grouping of similar activities on a job priority basis. This concept takes advantage of the cost savings inherent in reducing equipment and employee setup costs.[5]

In the area of maintenance, a well-programmed system should have a direct bearing on the overall efficiency of the production control system. The routine administrative burden of recording modifications, repairs, and spare parts involves a large labor expense and creates vast files of information which are far too expensive to evaluate for increased efficiency. Few organizations can justify economically a computerized maintenance effort as a major function. However, it can be developed within the system, as an integral part of a subsystem combining production control, sales, and inventory. The major functions of the maintenance system would be to schedule the present maintenance work, to schedule the preventive maintenance program, to optimize the department resources, and to provide management with special reports.

H. B. Maynard and Company developed a simple computer program to carry out the scheduling, routing, and reporting of maintenance profiles. It performs the three essential major functions mentioned with a minimum of input. Expanding this one application into a control system that integrates the manufacturing operation into a systems concept, H. B. Maynard and Company developed a customizing approach called Operations Control Systems. This system is tailored to present needs as well as future expansion. It alerts management to a more effective method of integrating manpower, machines, and raw materials to meet the consumer demand.

Future Trends

As indicated thus far, the more significant contributions of these computer applications are achieved with an integrated approach that quickly provides all levels with accurate data. The objective, then, would be the balancing of all organizational activities that affect the production.

In the future we will see increasing emphasis on the production applications. The direction will be to include quality control, purchasing, material handling, and accounting applications in the management information system.[6]

Inventory Control

For a large number of companies, the capital investment in inventories represents a substantial percentage of the organization's working capital. The efficiency or return on investment in inventories is generally determined by turnover rates. By their very nature, inventories influence storage space, handling costs, accounting procedures, production schedules, and so on. Essentially, the dynamic nature demands a simulation or forecasting technique that can provide accurate demand data that will minimize the quantity on hand after considering shipping costs, stock runout costs, and handling costs. Excess inventory is a result of poor forecasting and work load scheduling. The supervisor equates efficient utilization of his or her resources with having work lined up at all machines to eliminate idle machines or operators. The carrying costs of the inventory, attributed in part by early release of orders, are not traded off with scheduling and dispatching products. The true success of inventory control rests in shorter flow times.

Significant Contributions

Following are some of the features in inventory computer programs that were developed in the last decade. They are outlined here because of their contribution to computerization in this area.

Chrysler developed the Dynamic Inventory Analysis System (DIAS) as an answer to a dynamic and flexible material management computer application. Eliminating the horrendous amount of paperwork associated with the annual handling of more than a quarter of a million parts, the computer continuously scans and updates records while detecting impending problems with low levels and long delivery times. When a nonstandard condition arises, the computer generates exception reports for management. This on-line system forecasts and reports shortages and overages; it has the flexibility to plan for accepting new parts while phasing out the old, and it provides reports for management by exception.

Carrying the warehouse concept a step further to the automated shortage and parts delivery stage, Westinghouse has developed a computerized storage retrieval delivery system. This system supplies assembly lines according to the production schedule, or upon verified demand, in a matter of seconds. More important than the 15% cut in the inventory were the elimination of production delays, the 35% cut in storage costs, and the 40% reduction in damage.[7]

Control of Engineering Material Acquisition Storage and Transport (CEMAST) is a comprehensive total system used by BBC for inventory control and forecasting. The nucleus of the system is a software package known as Stock Control and Analysis (SCAN). This adaptive inventory control system forecasts and monitors, acquires and processes, records and orders, and accounts for the monetary value on the inventory. The SCAN package is used in the overall system to assess the total demand of a widely dispersed inventory and to determine what should be maintained in the central depot.

Available Utility Programs

Today there are a few utility inventory control programs that integrate several functional organizational areas. These programs calculate the EOQ, minimum reorder point, purchasing lead time, and group purchases and prepare the orders. There are also computer applications to develop and control warehousing. The basic concept is the consideration of the sales forecasts by segregating bulk and then ordering by descending weight classifications. Production plans are determined by the raw material inventory, which is similarly classified. The computer calculates the floor pallet spaces required and allocates building, zone, and slot spaces.

Some companies integrate their inventory control into their management information system.[6]

Quality Control

Data analysis is an extremely important aspect of a quality engineer's job and is frequently a difficult, complex, and time-consuming task.

Computers are invaluable to the industrial engineer, not only to save time, but also to reduce computational errors. The most basic computer program in data analysis is one that summarizes the data by computing the basic statistics of the sample, such as the mean and variance, and that plots a histogram of the data.

In the quality control field, the best and most natural way of summarization of any time dependencies in the data is by means of a control chart, which is essentially an analytical picture of the distribution of the mean and of the variation associated with a process over a period of time. The manipulation of data to determine if the system is within the inherent variability of the specifications is the quality control engineer's most important function. Computer applications that manipulate the data many times faster than the manual plotting of control charts are a most significant contribution to the cost reduction program. These programs greatly reduce the tedium of repetitive numerical shuffling on a calculator and significantly reduce the computational errors. With the use of automated machinery, human error is reduced, and specifications can be verified throughout the process. This process reduces the quality control function from the routine acceptance sampling to data recording and system improvement. The use of computer applications as an adaptive system quality control activity has been highly successful in reducing costs as well as in boosting customer confidence.

Recent Developments

In the last few years a series of computerized quality control programs has been developed by Merle D. Schmid of the University of Dayton. This series represents a timely landmark contribution to modern quality practices. Industry and government practitioners in quality assurance and professors teaching the subject should take advantage of the hundreds of hours of effort by Dr. Schmid that culminated in a series of programs for attributes and variables acceptance sampling.

The beauty of this series lies in the fact that existing technical aspects of acceptance sampling, often requiring a strong statistical background, have literally been placed at the fingertips of an inexperienced user. The basic concepts of attributes and variables acceptance sampling are well integrated into interactive, matching independent computer programs. The user-terminal interface is quite simple, with appropriate prompting of the user through brief, easily understood comments or questions.

In this series, 16 computer programs, each written in FORTRAN IV with excellent documentation, are presented. For attributes acceptance sampling, the programs permit designing single sampling plans to meet stated producer and consumer risks for either large or small lots. Existing sampling plans may be evaluated in terms of their operating characteristic curves using the Poisson, binomial, or hypergeometric distributions. For variables acceptance sampling, the programs permit designing sampling plans for single or double specification limits to meet stated producer and consumer risks. As in MIL-STD 414, the SD may be either known or unknown. Existing sampling plans may be evaluated in terms of their operating characteristic curves, showing probability of acceptance versus either fraction-defective or mean dimension.

In short, Dr. Schmid has taken a large step toward modernizing the use of existing quality control techniques. Fortunately, his contributions embrace those tools that are in frequent use in both industry and education. This series of programs is published in the journal *Computers and Industrial Engineering*, and the following are brief abstracts of these articles:

"Computer Design and Evaluation of Single-Sampling Plans."[8] This article describes two programs useful in designing and evaluating attribute single-sampling plans. The first program can be used to design single-sampling plans, given two points. The second program can be used to evaluate simultaneously up to five single-sampling plans. Both programs are written in FORTRAN IV, are interactive, are as machine independent as possible, and require a minimum hardware configuration of IBM 360/40, OS(64K). The two programs are used to illustrate design and evaluation criteria important to the successful operation of schemes that depend upon attribute single-sampling plans for lot acceptance or rejection.

"Design of Sampling Plans for Small Isolated Lots."[9] This article describes a series of interactive computer programs that solve most of the traditional and many special quality control problems. It shows practical applications for six programs. Since each program is written in FORTRAN IV and is as computer independent as possible, each is easily compilable on most companies' computers. Further, since all programs are interactive, they may be used by any quality control person who has access to a teletype terminal. The article also discusses how to design and evaluate sampling plans for special conditions not covered in standard sampling schemes. Actual examples are given, showing step-by-step procedures for designing and/or evaluating sampling plans that meet defined producer and consumer risks for isolated small lots, as well as the condition when lots are received in a steady stream from a given supplier.

"Computer Design of Optimum Variable Sampling Plans."[10] This paper shows practical applications for eight programs, all associated with variable sampling plans. Since each program is written in FORTRAN IV and is as computer independent as possible, each is easily compilable on most companies' computers. Further, since all programs are interactive, they may be used by any quality control person who has access to a teletype terminal. The article also discusses how to design and evaluate variable sampling plans for special conditions not covered in standard sampling schemes. Actual examples are given, showing step-by-step procedures for designing and/or evaluating sampling plans that meet defined producer and consumer risks for both fraction-defective and $\overline{X}$ plans for either single or double specification limits and with or without known sigmas.

Plant Layout

One of the uncontested roles of the industrial engineer is to optimize the plant layout for the most efficient utilization of the organizational resources. A poorly arranged flow of materials can result in congestion and high material handling costs, which will price the product right out of competition. A complex production operation can involve many departments and hundreds of interrelationships. As the number of departments increases and as the flow of materials branch, the time required to perform the analysis manually increases manyfold. A simple combinatorial problem would attest to the difficulty encountered. Several computer applications are available that attempt to reduce this complex interrelationship to a mathematical solution. These utility programs significantly reduce the designing function to a fast analysis of numerous alternatives.

Conventional Computer Programs

Computerized Relationship Layout Planning (CORELAP) is a path-oriented systematic analysis that applies heuristic logic to build methodically one department after another until the layout is completed. Although this program is not really optimizing in a mathematical sense, it seems to have overcome the limitations of the other programs that require the building configuration.

Computerized Relative Allocation of Facilities Technique (CRAFT) is unique in that it uses material flow data as the sole basis for the development of the closeness relationships. It is path oriented and specifically tied to production without regard to service facilities. Successive interchanges compare initial costs and savings to develop a suboptimum layout.[11]

Automated Layout Design Program (ALDEP) scores layouts on the basis of an interorganizational preference table considering as many as three levels. Alternate layouts are generated and scored according to the number desired and the minimum score requested.[11]

Richard Muther Associates Computer 1 (RMA COMP 1) considers the total closeness rating of those activities not yet formatted as well as those activities already placed on the layout. In effect, it is the computer version of the systematic layout planning. It places the largest total closeness rating in the center of the layout and reserves space in which related activities may be placed later.[11]

Computerized Drafting and Updating of Plant Layout (CDUPL) concentrates on the updating of the layout plans by incorporating the modifications desired and redrafting the layout. It can be utilized to simulate different plant alternatives. Its economic value seems to apply to the continuous updating of layouts with drafting accuracy and without additional cost.[12]

Office Layout

With the administrative overhead organic to most large organizations, a computerized model for the office layout is desirable. A heuristic adaptation emphasizing communications costs and minimizing personnel travel generates cost reduction data by exchanging personnel and equipment among the numerous possible locations.

What contribution can the industrial engineer make to the facilities modernization management planning program within the organization? H. B. Maynard and Company developed a computer program that combines product mix, work load capabilities, and elemental times to evaluate the future manpower, space, and equipment alternatives. Having obtained management's decision as to what equipment will be procured, it would be logical to integrate this information into a plant layout diagram for the future when the equipment would be delivered.

Future Trends

Although the preceding programs satisfy the needs of most industrial engineers, numerous efforts are being made by commercial software companies in the area of plant layout computer applications. The recent software programs incorporate the salient features of the preceding programs into a proven algorithm with a comparative economic advantage.

Engineering Economy

The fundamental purpose of engineering economic analysis is to determine the relative economic worths of engineering proposals. Economic feasibility is an essential prerequisite of successful engineering application.

Computer programs are utilized in a number of problem areas. On-line processing is utilized for small-to medium-size problems, either by direct programming or by use of stored preprogrammed routines.

The following are some conventional engineering economy problems that lend themselves to computerization:

Computing solutions for six different types of present and future value problems: present value of a single compounded amount, future value of a single compounded amount, present value of a uniform annual payment, future value of a uniform annual payment, required uniform annual payment into a sinking fund, required annual payment needed to pay off a loan in N years at an interest rate of R.

Comparing straight-line, sum-of-the-years-digits, and double declining balance depreciation methods calculates a depreciation schedule based on the service hours expected from an asset.

Computing a depreciation schedule for a given sum of money, fixed monthly payments, and simple annual interest rate.

Computing the optimal replacement policy for a purchased machine whose annual operating expenses increase and salvage value decreases.

Simulation models are also utilized in the study of cash flow situations. The size of the problems involved will determine the type of computer to be utilized. Large problems should be done in batch processing.

12.5.4 FUTURE TRENDS

Industrial engineers must keep abreast of the latest computational techniques. Herein lies a major problem. As the quantity of industrial engineering software increases, it becomes increasingly difficult to know what is available for specific applications. This leads to duplication in research and computer programming with attendant loss of time and increased costs. In industry a common resource of industrial engineering software is seldom available outside the individual corporation. Even worse, in many large corporations with multiple facilities, and in most governmental agencies, there is little transfer of software capabilities within the organization.[13]

The situation is potentially more serious in academic organizations because of increased constraints on manpower and funding. These constraints seriously restrict the search for existing software and limit the development of new software. As a result, the academic researcher in industrial engineering is quite often limited to the basic software available in his or her university computer center.

A comprehensive industrial engineering software library[14] with software descriptions is needed to solve these problems. The appropriate talent to establish and maintain such a library should be provided. It is also necessary to study detailed specifications for program submittal to the library and techniques for screening and verification must be developed. In addition, detailed library user procedures and a catalog of library programs should be published periodically.

Impact on Decision Making

The computer is currently an important aid to the decision maker, and it would appear that its importance will continue to increase in the future. We will continue to see added flexibility in software systems, computers conversing with computers, and quantification of what we once believed to be intangible. We will see expansion of supervisory applications to optimize total systems. The reduction in paperwork and human intervention will centralize decision making.

We will also see more acceptance of the minicomputer as an ideal approach for smaller company industrial applications whose size precludes larger operations, but favors the building block concept of smaller and more economical hardware. This concept, when added to the innovations of microprogramming and teleprocessing, allows low-cost information systems and interfaces with larger computers. Technology will make these computers mandatory for competitive businesses.

12.5.5 CONCLUSIONS AND RECOMMENDATIONS

Industrial engineering has clearly demonstrated its contribution to organizational effectiveness by applying the scientific analytical approach in integrating the physical, human, and informational activities. This new expansion of the boundaries of the field requires the broad perspective of management-oriented industrial engineers rather than highly specialized staff members. Accepting this larger, more encompassing responsibility, industrial engineers must utilize all the tools at their command in order to provide managers with timely, accurate, and meaningful data on which to base their decisions. The most obvious tool is the computer, with its capability of manipulating vast quantities of data and retrieving information, which used to take many hours of laborious effort. The effective use of this tool is one of the greatest challenges to the industrial engineer.

The effective use of the computer and its unlimited number of time saving, cost reducing applications will determine the future of the industrial engineering profession. The potential of new computer applications is limited only by the aggressiveness and initiative of the professional industrial engineer.

The benefits to be attained from computer applications do not rest in the simple conversion of a manual system to an automated system; conventional industrial engineering applications must be redesigned to take maximum advantage of the computer's unique capability. If industrial engineers can maintain their balance between overselling the computer and underutilizing the computer's capability, they will revolutionize management information systems and their profession.

Industrial engineers with the initiative and desire to utilize this new tool are frustrated only by the expense for a yet-to-be-proven application in their field. Where can these aggressive individuals turn for information with which to justify the use of this tool? It is recommended that industrial engineers initiate a program that would wholeheartedly support a single organization charged with the responsibility of developing and promoting the use of computer applications in industrial engineering tasks.[15] With contributions from educational agencies and the users, this organization would soon perform the most valuable function of providing all industrial engineers with a focal point from which to obtain actual computer applications or information on where computer software might be obtained.

REFERENCES

1. H. K. ELDIN and J. W. WOOD, "A Survey of Basic Industrial Engineering Computer Applications," *Industrial Engineering*, May 1971, pp.
2. H. K. ELDIN, "Software Library Survey," *Industrial Engineering*, May 1974, pp. 40–41.
3. D. GIBSON, "Work Sampling Monitors Job-Shop Productivity," *Industrial Engineering*, June 1970, pp. 12–19.
4. H. PIKE, "Process Control Software," *Proceedings of the IEEE*, Vol. 58, No. 1 (January 1970).
5. L. GROSSMAN, "Using a Computer to Control Production," *Automation*, October 1968, pp. 60–65.
6. E. S. BUFFA and W. H. TAUBERT, *Production Inventory System: Planning and Control*, rev. ed., Homewood, IL, R. D. Irwin, 1972.
7. "Take Charge Computer Dispatches Parts to Lines," *Modern Material Handling*, February 1968, pp. 56–61.
8. M. D. SCHMID, "Computer Design and Evaluation of Single-Sampling Plans," *Computers and Industrial Engineering*, Vol. 1, No. 4 (1977), pp. 217–233.
9. M. D. SCHMID, "Design of Sampling Plans for Small Isolated Lots," *Computers and Industrial Engineering*, Vol. 3, No. 1 (1979), pp. 1–39.
10. M. D. SCHMID, "Computer Design of Optimum Variable Sampling Plans," *Computers and Industrial Engineering*, Vol. 3, No. 2 (1979), pp. 101–165.
11. R. MUTHER and K. MCPHERSON, "Four Approaches to Computerized Layout Planning," *Industrial Engineering*, February 1970, pp. 39–42.
12. J. JANSARI and I. GUPTA, "A Program for Plotting Plant Layout," *Industrial Engineering*, March 1969, pp. 35–37.
13. H. K. ELDIN, "Planning an Industrial Engineering Software Library," *AIIE Proceedings*, 1973, pp. 3–12.
14. *Industrial Engineering Software Library*, report on the feasibility of establishing the library, NSF Grant No. GK-37354, National Science Foundation, Washington, DC, 1974.
15. H. K. ELDIN, "The Need for an Industrial Engineering Software Library," *Computers and Industrial Engineering*, Vol. 1, No. 4 (November 1977), pp. 199–206.

CHAPTER 12.6

Systems Design to Encourage Use

HENRY C. LUCAS, JR.

New York University

12.6.1 INTRODUCTION

Why should one be concerned with the effective use of information processing systems (IPSs)? Most organizations with computer-based processing have a large investment in equipment, software, and individuals to staff the information services function. Information systems are generally developed because of the anticipated benefits of applying a computer to solve an information processing problem in the organization. If information systems are not used effectively, the investment in these systems will not yield a return, and the full benefits of systems will not be realized.

Unfortunately, the issue is often not one of effective use, but of use at all. Examples from numerous organizations suggest that there have been failures to use systems. In other instances systems with a high potential for improving the effectiveness of the user and the organization are used only to the degree required, for example, to process simple transactions. The purpose of this chapter is to discuss what is known about the use of information systems and to offer guidelines on how to develop systems that will be used.

12.6.2 TYPES OF USE

The effective use of an IPS is essentially a behavioral phenomenon. For individuals to adopt and integrate an information system into decision making or control in the organization requires a change in behavior. Thus the study of the effective use of IPSs falls into the realm of the social sciences rather than engineering. The organizational and behavioral variables involved in system use tend to be less precise and more difficult to define than the variables involved in the construction of a model contained in an information system.

There are many different types of uses of an information system. One helpful distinction is between a voluntary system and one whose use is involuntary. With a voluntary system the user has discretion in employing it. Most decision-oriented systems are voluntary; the user can decide whether or not to work with them.

For transaction processing and operational control systems, those responsible for input usually have no option; they must provide input for the system as part of their jobs. However, a manager may choose to ignore or to use a report generated by one of these systems; thus there may be both voluntary and required use of the same system.

Systems designers need to consider how to encourage voluntary use. There is no guarantee that because designers or some managers rate a system highly that all potential users will take advantage of it. Even when the use of a system can be required as part of someone's job, one must still be careful since forced use of a poorly designed system can lead to alienation and even sabotage of the system.

12.6.3 IMPACT ON THE ORGANIZATION

An information system can have an impact on the structure of an organization, on work groups, on individuals themselves, and on the distribution of power. An organization's structure may be altered because of a new system or because of changes made possible by a new system. One firm with several different departments concerned with customer service is developing a new computer-based system that will make information available in one place. The firm is planning to create a new department responsible for all customer service, which represents a major change in the organizational structure of the firm. In this firm, work group composition will also be altered; different groups will be moved to new departments, and it is likely that some individuals will become members of new work groups. Restructuring of group membership occurs frequently when systems are developed.

For almost every information system, users must learn something new. They have to contend with new input in terms of content, format, and/or medium of entry; output is also frequently different after the installation of a new system.

These impacts on the organization have been known to affect the distribution of power in the firm. Power, or the ability to influence the behavior of others, may be altered by new information since information reduces uncertainty. The reduction of uncertainty for others is a sign of high power. Systems can also create new levels of interdependence in an organization, often between user and computer departments. High levels of dependence are associated with having low levels of power.

A common theme of these impacts is change. New information systems are developed because the organization wishes to change existing information processing procedures.

12.6.4 TYPES OF IMPLEMENTATION RESEARCH

There are three major types of research on information systems: theories, factor research, and process studies (see Exhibit 12.6.1). Most of the theories and proposals about implementation are not really scientific. It is difficult to develop testable hypotheses from the theories. The theorists provide insights and ideas to be used in implementing computer-based systems, but there is no guarantee that the suggestions will work.

Factor research seeks to identify variables and their relationship to implementation. A large number of factors have been identified through research on implementation, for example, attitudes toward a system and management support of it. Factors may be constructed from a single variable, such as age, or from a series of variables, such as a group of questionnaire items that assess user attitudes toward a system. Factors tend to be static, and this type of research usually suggests that factor A is associated with factor B—for example, management support may be related to the users' intentions to work with a new system. Factor studies are analytical and often quite sterile.

There is also research describing the use of information systems in the form of case studies that examine the process of design. "Process" refers to how a system is developed, for example, a description of the relationship among individuals working on a system. This research, unlike factor studies, attempts to probe the way in which a system has been developed and to examine various human relationships. These studies may appear unfocused at times and less rigorous than factor research; however, process research contributes a great deal to one's understanding of how information systems are used.

This chapter discusses factor research and process studies and makes suggestions for developing systems that will demonstrate high levels of use. Because of the rather limited scope of theories of systems use, they are not discussed in detail here. For an in-depth presentation of theories, factor research, and process studies, see Lucas.[1]

12.6.5 FACTORS ASSOCIATED WITH SYSTEMS USE

The studies reported in this section are concerned primarily with factors related to implementation success. The studies have been divided into four groups: the Northwestern University studies, the Lucas studies, research on decision style, and other studies. These groups are summarized in Exhibit 12.6.2.

The Northwestern University Studies

Researchers at Northwestern University have conducted a number of studies that include practical experience and implementation research. For the most part, this research concentrates on the use

Exhibit 12.6.1 Implementation Research

Type	Characteristics
Theories	Lack of scientific rigor
Factor studies	Identification of a large number of variables Cross-sectional research Problems of causal inference
Process studies	Emphasis on relationships Case studies predominate Problems of generalization

Exhibit 12.6.2 Summary of Factor Studies

Studies	Focus and Major Findings
Northwestern University	Focus on use of operations research models. Factors associated with use: management support, client receptivity, personal and situational variables, client-researcher relations, model quality.
Lucas	Focus on use of information systems. Factors associated with use: attitudes, system quality ratings, decision style, personal and situational variables, management support.
Decision style	Focus on ways of problem solving. Findings: decision style influences acceptance and use of information systems; difficult to predict relationship between decision style and use.
Others	Findings: involvement and use associated with each other; information systems may be perceived as highly rational.

of operations research or management science models; however, there is a great deal of similarity between model building and information systems. Information systems tend to be less complex mathematically and more complex technologically than operations research or management science models. However, since there is generally a paucity of research on systems use, it is helpful to include research on the use of models and to learn whatever possible from this work.

Through interviews, Rubenstein et al.[2] conducted a study of the effectiveness of operations research groups in 66 companies. The factors that appeared important in rating the effectiveness of the groups were management support and client receptivity, and the effectiveness of the groups seemed to differ in importance by stage of group development, for example, prebirth versus mature operations research groups.

Radnor et al.[3] examined the integration of management science groups. The variables highlighted in this study were management support and personal and situational factors. The authors found that management support was important in developing operations research and management science groups. In another project Radnor et al.[4] studied implementation success through an examination of client relations and management support. Their conclusion was that a good client-researcher relationship and management support are important to the successful implementation of operations research models. Neal and Radnor[5] looked at the successful use of operations research models. They examined the presence of formal procedures and a charter for the operations research group. The researchers found that the presence of formal procedures was associated with ratings of successful implementation for operations research models.

Maher and Rubenstein[6] examined the adoption of a particular model for project selection. The output quality of this model was associated with stated willingness to adopt; also, the researchers found that changes occurred in communications channel characteristics and that these were associated with willingness to adopt. Rather than focusing on general models of several companies, this particular study looked in depth at the adoption of a single model.

Bean et al.[7] also studied implementation success. Their research focused on structural and behavioral variables that were found to be associated with success. The authors pointed out, however, that few of these variables could be controlled by the operations research management science group.

For a good summary of the results of the Northwestern University studies, see Radnor.[8]

The Lucas Studies

Most of the studies in the series undertaken by Lucas involve only one system. The studies all deal with computer-based systems, and only one included an operations research model. Generally, level of use was the major dependent variable where use was voluntary. In cases where use was mandatory, satisfaction was employed as an indicator of systems success. Although most of the Lucas studies were set within single firms, they generally included a large number of respondents, making it possible to examine many different variables.

A study in a major university examined attitudes toward administrative systems, ratings of system quality, and levels of system use.[9] Use was positively related to ratings of systems quality and attitudes toward systems.

A study of a major sales information system examined attitudes toward the system, system quality, the decision style of users, personal and situational variables, and several different types of uses of information systems.[9] All of the independent variables were associated with different kinds of uses. Favorable attitudes were good predictors of use, but results were hard to forecast for decision style and personal and situational factors.

Another study was conducted in a branch bank.[9] This study was concerned with the decision style of the user, personal and situational variables, and the likelihood of taking action based on computer-generated reports. The results of the study indicated that the decision style and personal and situational variables were related to the use of reports and the likelihood of taking action.

A 1976 study was conducted of planning models developed by companies using a proprietary language on a time-sharing system.[10] All of the firms used the language and its modeling features to develop models of their organizations. The study found that user attitudes, management support, personal and situational factors, and decision style were all associated with high levels of use of the information systems. This study also found a high degree of correlation among different measures of system use, including number of hours of connect time on the time-sharing system and self-report measures of use.

A brokerage firm developed a large operations research model that made information available to account executives in the field.[11] This model could be used voluntarily by the sales force. A study in 1979 examined attitudes toward the model, personal and situational factors, management support, and decision style. The results indicated that personal and situational factors and decision style were associated with attitudes and use. Management support and decision style were also related to taking action from the model. The research demonstrated significant regional differences in the use of the model.

In a controlled field experiment, Lucas[12] examined the introduction of a new order entry form. A control group was employed to see that changes observed in the experimental group were due to the introduction of the new form. Satisfaction measures showed significant reductions for the group using the new optical character input system. There was evidence that benefits were not realized from the system and that the personal costs of using it were too high for many users.

The final study in this group focused on the level of use of a medical information storage and retrieval system. Use was measured through a monitor built into the retrieval system, and attitudes and other variables were assessed using a questionnaire. The results of the study offered strong evidence of the importance of management support and leadership in encouraging the use of an information system.

Studies of Decision Style

"Decision style" refers to the characteristic way individuals have of approaching a problem. Some individuals are analytic; that is, they tend to reason from models or analogies of the problem to be solved. Other individuals tend to be more intuitive or heuristic; they approach a problem without a set format for solution. It is thought that these different kinds of individuals require different types of information in order to make decisions.

The first study that pointed out the likelihood that decision style could be a constraint in the implementation of operations research models was conducted by Huysmans in 1970.[13] This rather complex laboratory experiment showed that cognitive style affected the choice and acceptance of operations research recommendations.

Doktor and Hamilton[14] conducted an experiment on cognitive style and found that subject population differences also affected the acceptance of management science recommendations. Extreme differences appeared between a group of students and a group of managers, each participating in different parts of the experiment.

In a 1979 study Larreche[15] examined a slightly different aspect of decision style, a theory known as "integrative complexity." This construct refers to the ability of an individual to perceive and interpret data and to integrate it with existing knowledge. He found that integrative complexity was associated with the extent and efficiency of information search behavior in an experiment employing the use of marketing models available on a time-sharing computer system.

In a 1976 study Kilmann and Mitroff[16] found that individuals with different cognitive styles had different ideal organizations that corresponded closely with their cognitive styles. In 1974 Mitroff et al.[17] were able to develop an information system in the laboratory that actually affected an individual's cognitive style, especially the preference for learning about both sides of a problem.

The studies of decision style have identified several psychological variables that appear to affect the use of information systems. However, sometimes the findings are counterintuitive, and it remains difficult to make predictions based upon decision style.

Other Factor Studies

Swanson[18] conducted a study of a computer-based retrieval system used for engineering project management. His study focused on user attitudes, involvement in the design of the system, and levels of system use. The results showed that high levels of involvement and favorable attitudes were associated with information systems use. The author suggests that involvement in the development of the system influences favorable attitudes and that favorable attitudes lead to use. Successful experience in using the system then improves attitudes.

Argyris[19] examined the functioning of a management science operations research group in a large organization. His study included the level of rationality and emotionality in the system and interpersonal relations. Argyris concluded that the introduction of information systems and management science requires a knowledge of the emotional aspects of the organization.

Conclusions

The factor studies discussed in this section have identified a large number of variables associated with the use of information systems. These factors can be grouped into various generic classes of variables to develop a model of the implementation process (see Exhibit 12.6.3).

The classes of variables extracted from the studies include the technical characteristics of the system, such as the accuracy of input and output, the reliability of the system, and completion of processing on schedule, and the interface between the system and the user. A class of variables called "client actions" includes the level of management support and the type of user involvement and influence in the design of systems.

Attitudes toward systems reflect the disposition of the individual toward the system. An individual with a certain attitude is likely to behave in a particular way. In the context of information systems, knowing a user's attitude toward a specific system should make it possible to predict the likelihood of his or her using the system.

Decision style is the characteristic way one has of approaching a problem. As mentioned previously, research has shown that this class of variables is important, but making specific predictions has proven difficult.

The final class of variables identified in factor research includes personal and situational variables, such as age, education, and length of time with the company. Unfortunately, these variables, too, have proven difficult to use as a basis for prediction.

Exhibit 12.6.3 is a model of the possible relationships among these various factors. Technical characteristics are expected to influence successful implementation as defined by measures of systems use and user satisfaction with the system. A system must be of sufficient technical quality so that it can be used. Technical characteristics also are expected to influence user attitudes since users are directly affected when working with the system. Client actions should influence the use of a system directly and through their impact on user attitudes; for example, management support encourages users to work with a system. Decision style and personal and situational variables also influence the use of a system.

The designer of a system has relatively little influence over most of the factors displayed in Exhibit 12.6.3. The designer is largely able to control the technical characteristics of a system because

Exhibit 12.6.3 Hypothesized Relationships Among Implementation Factors

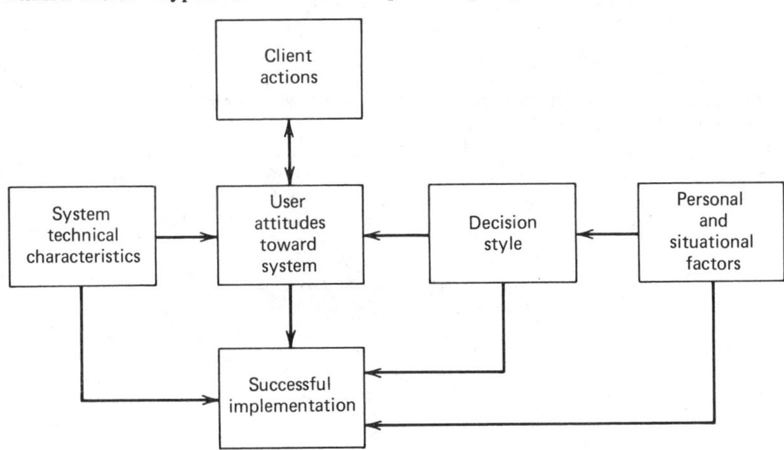

there is a wide range of options available and the designer is considered to be the technical expert. The designer can also often influence client actions to solicit more management support and to try to arrange more user input into the design process. However, it is extremely difficult to influence attitudes toward a system, at least in the short run. Decision style and personal and situational factors are also difficult for the designer to influence.

There is one major factor, however, that has been omitted from the factor studies: the process of design.

12.6.6 THE PROCESS OF DESIGN

The research discussed in this section generally consists of case studies. Instead of examining factors alone, this research tries to explain the relationship among variables and generally focuses on the relationship between the designer and user as a system is developed.

Among the important issues in the process of design are the following: What is the impact of a system on an organizational structure, on work groups, and on individuals? How is the design process to be controlled? How are users motivated to accept a new system?

One of the earliest studies of the development of a computer-based information system was conducted in an electric utility by Mann and Williams.[20] Their study examined the entire process of installation and the significant organizational restructuring that occurred. Their conclusions offered many insights into the impact of the first large batch processing system adopted by the utility.

Mumford and Banks[21] examined the implementation of the first computer systems at several English banks. They studied the impact on each individual. Mumford and Henshall[22] worked as consultants in the development of a new clerical information system at Rolls Royce in England. They applied the sociotechnical approach to participative design in their efforts. This approach emphasizes that, even though a new technology is being employed, there is a social system that forms the context for the new system.

Lucas and Plimpton[23] used the stages in one process model for systems development, the Kolb-Frohman model,[24] to analyze a system designed with the United Farm Workers' organizing committee. They found that careful attention to the impact of the system on users and concern over their ability to terminate the relationship with the union were extremely important. In particular, in order to have the system used, it was necessary for the designers to see that the members of the union became the psychological owners of the system. Ginzberg[25] also adopted the Kolb-Frohman model for a study of 29 projects. His results indicated that the termination stage in that model was extremely important in successful projects.

There have been a number of other process studies, each of which has reported the relationship between the user organization and the designer of the system. However, the work reviewed here is a good sample of the literature and offers many insights into the design process.

A particularly useful model mentioned in the studies here is the Kolb-Frohman approach to organizational development. This model was not designed specifically for the development of information systems, yet it can be applied to their design (see Exhibit 12.6.4). The model stresses the relationship between a change agent and the organization. Because of the changes created by information systems, designers do function as change agents whether it is realized or not.

The model consists of a series of stages. In the first stage, scouting, the activities of the designer and the user are characterized by a lack of commitment; each party tries to assess the motives of the other. During entry, the initial relationship between the implementer and the user is extended to develop a primary contact, or user, of the system. The implementer begins to develop a contract, a psychological, rather than a legal, document; agreement is needed on the goals of the project, the resources available, the method of approaching design, the benefits from the final product, and the nature of the relationship between the designer and the user.

During diagnosis, the objective is to define the information system in more detail. A feasibility study would be a typical output of the diagnosis stage of the model. The planning stage includes development of both the technical details of the system and the relationship between the user and the implementer.

The action stage encompasses the largest part of the project. During this stage the designer must

Exhibit 12.6.4 The Kolb-Frohman Model

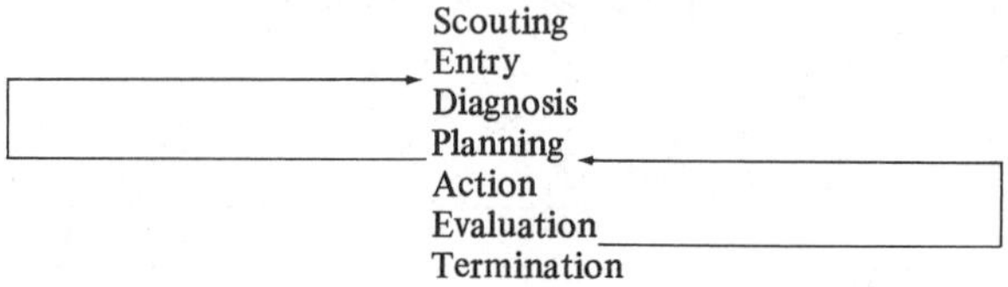

Scouting
Entry
Diagnosis
Planning
Action
Evaluation
Termination

be particularly concerned with the ultimate acceptance of the system after termination of the relationship with the user. During the action stage the actual change process is begun for users of the system. The systems design team works in conjunction with users to develop the specifications of the system.

Most researchers in the field argue for greater user influence and involvement in the design process, even to the extent of user control over design. The reason for greater user input and involvement is that users become better trained and more knowledgeable in how to use the system. Users surrender less power to the information services department because of their influence on the design, and the system itself will be better since users will be the ones who are ultimately responsible for it. Participation is also ego enhancing; it is thought to create greater commitment to a system.

During the evaluation stage, the results of the systems development process should be examined honestly. Evaluation also contributes to the planning of future systems.

The last stage in implementation, and possibly the most important, is termination. The objective here is to end successfully what has been a temporary relationship between the designer and the user of the system. The most successful termination of an information systems development project occurs when the user has developed a psychological ownership of the system. Termination, rather than being considered only as the final stage of design, must be continuously considered from the initiation and scouting phases through the very end of the development process.

The stages in the Kolb-Frohman model are really not distinct; rather, as the feedback loops imply, the relationship will move from one stage to another and back again during the development process. The main advantage of the model is that it focuses attention on the process of design and on the importance of considering the relationship between the designer and the user as well as the task of designing the system itself.

12.6.7 DESIGNING FOR USE

As the preceding review of the research indicates, the key task facing the designer is to blend concern with the task of developing a system with the development and maintenance of a relationship with the user. Consider the model from factor research that has been used to identify key variables such as the technical quality of systems, client actions, attitudes, decision style, and personal and situational variables. Then adopt a process model, something like the Kolb-Frohman model, and look for leverage points in the organization.

The following are some recommendations for improving the chances for success in systems analysis and design:

1. Be sure to design a technically adequate system. A technically adequate system is one that works and that is reliable; it has an easy-to-use interface with the user and includes the functions the user deems necessary to solve the problem at hand.
2. View implementation as a planned change activity in the organization. An information system is designed to change existing information processing procedures; plan for the changes and recognize that they have an impact on the organization, on work groups, and on individuals. Concern about the impact of a system on users and the organization should begin when the system is first suggested.
3. Seek management support, guidance, and resources in the development process. Management plays a key role in the design process described previously under client actions. Managers should indicate their support for the project orally, by attending design review meetings, and by providing adequate resources for development.
4. Stress meaningful user involvement and influence in design. One approach to user influence is user control or user codesign. The idea is that the user is given an opportunity to make suggestions and that those suggestions influence the design. User task groups, design review meetings, prototypes, and similar devices can be employed to help users contribute to a system.
5. Identify and deal with the key variables in the factor model. Consider the technical quality of the system, client actions, user attitudes, decision style, and personal and situational variables at each stage in a process model such as the one suggested by Kolb and Frohman. Seek appropriate client actions at each stage and develop a design that is contingent on decision style and personal and situational factors.
6. Concentrate on the task and also on the process of systems analysis and design. Too often the role of the analyst is defined only with respect to the technical characteristics of a system. The author hopes that the research review has demonstrated that the role is a much broader one; the designer must consider the relationship among the client, the system, and the organization along with the technical aspects of the design.

Although there is no guarantee for success, if these recommendations are followed, they should help encourage the effective use of computer-based information systems.

REFERENCES

1. H. C. LUCAS, JR., *Implementation: The Key to Sucessful Information Systems,* Columbia University Press, New York, 1981.
2. A. H. RUBENSTEIN, M. A. RADNOR, N. BAKER, D. HEIMAN, and J. MCCOLLEY, "Some Organizational Factors Related to the Effectiveness of Management Science Groups in Industry," *Management Science,* Vol. 13, No. 8 (1967), pp. B508–B518.
3. M. A. RADNOR, A. H. RUBENSTEIN, and A. S. BEAN, "Integration and Utilization of Management Science Activities in Organizations," *Operations Research Quarterly*, Vol. 19, No. 2 (1968), pp. 117-141.
4. M. A. RADNOR, A. H. RUBENSTEIN, and D. A. TANSIK, "Implementation of Operations Research and R and D in Government and Business Organization," *Operations Research,* Vol. 18, No. 6 (1970), pp. 967–991.
5. R. D. NEAL and M. A. RADNOR, "The Relation Between Formal Procedures for Pursuing OR/MS Activities and OR/MS Group Success," *Operations Research*, Vol. 21, No. 2 (1973), pp. 451–474.
6. P. M. MAHER and A. H. RUBENSTEIN, "Factors Affecting Adoption of a Quantitative Method for R and D Project Selection," *Management Science*, Vol. 21, No. 2 (1974), pp. 119–129.
7. A. S. BEAN, R. D. NEAL, M. A. RADNOR, and D. A. TANSIK, "Structural and Behavioral Correlates of Implementation in the U.S. Business Organizations," in R. L. Schultz and D. P. Slevin, Eds., *Implementing Operations Research/Management Science,* American Elsevier, New York, 1975.
8. M. A. RADNOR, "The Context of OR/MS Implementation," *TIMS Studies in the Management Sciences*, Vol. 13, 1979, pp. 17–34.
9. H. C. LUCAS, JR., *Why Information Systems Fail*, Columbia University Press, New York, 1975.
10. H. C. LUCAS, JR., "Methodologies for Research on the Implementation of Computer-Based Decision Aids," in P. Keen, Ed., *The Implementation of Computer-Based Decision Aids*, MIT Press, Cambridge, MA, 1976.
11. H. C. LUCAS, JR., "The Implementation of an Operations Research Model in the Brokerage Industry," *TIMS Studies in the Management Sciences*, Vol. 13, 1979, pp. 139–154.
12. H. C. LUCAS, JR., "Unsuccessful Implementation: The Case of a Computer-Based Order-Entry System," *Decision Sciences*, Vol. 9, No. 2 (1978), pp. 68–79.
13. J. H. HUYSMANS, "The Effectiveness of the Cognitive Style Constraint in Implementing Operations Research Proposals," *Management Science*, Vol. 17, No. 1 (1970), pp. 92–104.
14. R. DOKTOR and W. F. HAMILTON, "Cognitive Style and the Acceptance of Management Science Recommendations," *Management Science*, Vol. 19, No. 8 (1973), pp. 884–894.
15. J. C. LARRECHE, "Integrative Complexity and the Use of Marketing Models," *TIMS Studies in the Management Sciences*, Vol. 13, 1979, pp. 171–188.
16. R. H. KILMANN and I. MITROFF, "Qualitative Versus Quantitative Analysis for Management Science: Different Forms for Different Psychological Types," *Interfaces*, February 1976, pp. 17–25.
17. I. MITROFF, J. NELSON, and R. O. MASON, "On Management Myth-Information Systems," *Management Science*, Vol. 21, No. 4 (1974), pp. 371–382.
18. E. B. SWANSON, "Management Information Systems: Appreciation and Involvement," *Management Science,* Vol. 21, No. 2 (1974), pp. 178–188.
19. C. ARGYRIS, "Management Information Systems: The Challenge to Rationality and Emotionality," *Management Science*, Vol. 17, No. 6 (1971), pp. B275–B292.
20. F. MANN and L. WILLIAMS, "Observations on the Dynamics of the Change to Electronic Data Processing Equipment," *Administrative Science Quarterly*, Vol. 5, No. 2 (1960), pp. 217–256.
21. E. MUMFORD and O. BANKS, *The Computer and the Clerk*, Routledge and Kegan Paul, London, 1967.
22. E. MUMFORD and D. HENSHALL, *A Participative Approach to Computer Systems Design,* Associated Business Press, London, 1979.
23. H. C. LUCAS, JR., and R. B. PLIMPTON, "Technological Consulting in a Grassroots, Action-Oriented Organization," *Sloan Management Review,* Vol. 14, No. 1 (1972), pp. 17–36.

24. D. A. KOLB and A. L. FROHMAN, "An Organizational Development Approach to Consulting," *Sloan Management Review,* Vol. 12, No. 1 (1970), pp. 51–65.

25. M. J. GINZBERG, "A Study of the Implementation Process," *TIMS Studies in the Management Sciences,* Vol. 13, 1979, pp. 85–102.

BIBLIOGRAPHY

HAMMOND, J. S., "The Roles of the Manager and Management Scientist in Successful Implementation," *Sloan Management Review,* Vol. 15, No. 2 (1974), pp. 1–24.

LUCAS, H. C., JR., *The Implementation of Computer-Based Models,* National Association of Accountants, New York, 1976.

CHAPTER 12.7
Office Automation

ANDREW D. BAILEY, JR.

University of Minnesota

JAMES GERLACH, R. PRESTON McAFEE, ANDREW B. WHINSTON

Purdue University

12.7.1 INTRODUCTION

Office automation is an evolving trend centered around the automation of office procedures and document manipulation by means of computer technology. The purpose of Section 12.7.2 is to acquaint the reader with the generally accepted concepts and capabilities of office information systems (OISs), the issues surrounding their development, a real-life prototype (Citibank's Management Work Stations), and some of the implications that office automation will have for business and society.

Section 12.7.3 presents two research OIS prototypes: Officetalk-Zero® and System for Computerization of Office Processes (SCOOP). Officetalk-Zero is an electronic aid to performing routine office tasks, similar in manner to the traditional way of performing the equivalent clerical tasks; SCOOP stresses the automation not only of office devices, but also of office procedures. The need for nontechnical office personnel to implement automated office procedures leads to a discussion of the office worker–machine interface in Section 12.7.4. Two potential languages for describing automated office procedures are presented: the Office Procedures by Example language and the Business Definition Language. Each language is examined according to design philosophy, objectives, prospective uses, and limitations.

Automated office procedures and devices are effective means by which the office worker can manipulate, store, and send electronic documents. Such capabilities should be adequate for supporting formal communications. However, informal communications, serving important organizational needs, may not be adequately accounted for by the document-oriented OIS. Thus Section 12.7.4 concludes with a discussion of formal and informal communications in an automated office setting.

Section 12.7.5 deals with the modeling of OISs in terms of formal office activities and formal lines of communication. A model for describing and analyzing information flows within offices is presented. The model lends itself to automatic analysis and transformations intended to automate, streamline, and reorganize the OIS model. The outlook for OIS simulation as an analytic tool and a personnel training tool follows. In addition to office efficiency and productivity, the office designer and manager must incorporate internal control mechanisms in the automated office. A computer-assisted internal control verification system, TICOM II, is introduced as an effective means of analyzing OIS models for violations of sound internal control. Finally, Section 12.7.6 treats the problem of preparing for the implementation of OISs. The chapter ends with some thoughts on the implications of OISs for future office activities.

This project was funded in part by a grant from the Peat, Marwick, Mitchell Foundation through its research opportunities in auditing program. The views expressed herein are those of the authors and do not necessarily reflect the views of the Peat, Marwick, Mitchell Foundation. The research was supported in part by Army Contract No. DA79C0154. We gratefully acknowledge the fellowship support made possible by a grant from IBM.

12.7.2 OFFICE INFORMATION SYSTEMS IN PERSPECTIVE

With the introduction of the computer in the business environment, business functions such as inventory control, accounts receivable, accounts payable, and payroll accounting have been successfully automated. The benefits have included faster and more accurate reporting, more effective storage and management of large amounts of data, and greater reliability and consistency of system performance. However, despite the merits of automation, traditional office work has remained relatively unchanged. The goal of the automated OIS is to facilitate office workers' performance of their routine duties by providing them a totally office-oriented system that exploits the processing speed, precision, and generally infallible memory of computing machines.

The OIS is differentiated from an information processing system (IPS), a word processor (WP), or a decision support system (DSS) by its emphasis on the integration of all of the components necessary to support office personnel. The IPS, WP, and DSS are either not oriented toward office functions or address only a limited set of office activities. The basic functions of office work include text editing, document preparation and filing, copying, simple numerical processing, communication, and information analysis and verification. With respect to office procedures, a WP and overlapping areas of an IPS and DSS are only isolated elements or subcomponents of an OIS. The primary advantage of an OIS over a system in which each of the basic office functions is automated independently is the reduction in the complexity of the interface between the office personnel and the machine. Unless this interface is kept simple and is easy to learn and use, an OIS may be of little value except to the office specialist.

A second benefit obtained from an OIS as compared to a WP, IPS, or DSS is the result of the office structure the OIS is expected to automate. An office organization is designed to accommodate office workers performing highly interactive, autonomous, and concurrent business tasks. Fixed programs are inflexible, and simplistic interactive systems are deficient in meeting the needs of the office worker as required by the dynamic office environment.

General OIS Capabilities

As mentioned previously, an OIS is an integrated computer system comprising facilities for performing office work. In particular, office automation aids office personnel in the preparation of documents, information management, and decision making.[1] The types of documents to be prepared, processed, and managed by an OIS are memos, letters, reports, and typical business forms such as requisition and purchase order forms. The documents, still visually depicted in their regular format, are stored electronically in an OIS. The methods of capturing the information provided to the office workers produce documents with the standard headings and lettering appropriate to each particular form.

The office worker interacts with the OIS through work stations. A work station is a programmable microcomputer equipped with office-oriented devices and software. The work station will enable the office worker to see and manipulate the electronic documents, communicate electronically with other work stations, access data base information, and stepwise direct the work station to perform designated tasks on a one-time basis or automatically. Such a one-time task might be preparing a memo incorporating only sections of a previous memo and a previously prepared summary report.

An example of a task performed automatically by the work station might be searching an inventory data base to determine whether items listed on any purchase requisition are in stock or on order or whether they need to be special ordered. The work station operator could then fill out a blank purchase order, using the information found on the requisition form and in the inventory data base. Additional accesses to other system information could be made as the situation warranted. The completed purchase order form could then be transferred to another work station determined by the nature of further processing, such as managerial approval of the purchase.

In addition to electronic documents and the traditional data types, OISs will support video data such as pictures and graphs. Graphic packages are currently available and can be integrated into the OIS. Speech recognition would be a very desirable feature of an OIS, since some managers are opposed to keyboard interaction with the work station and prefer dictation and recordings in communicating with office personnel. Currently, speech recognition systems are in a very primitive stage of development. An experimental speech understanding system, Hearsay-II, recognizes utterances in a 1000-word vocabulary, with a correct interpretation rate of approximately 90%.[2] Other alternatives to the keyboard are touch screens and light pens.

Communications between work stations can be instantaneously carried out by means of electronic mail or other electronic-type messages sent over telephone lines or by way of a satellite communication network. The electronic mail system will contain known addresses and other pertinent information such as telephone numbers. The sending and receiving of mail is handled automatically with little interaction by the work station operator. A "mailbox" is maintained for each work station, the contents of which may be surveyed upon request. Immediate notification of newly arrived mail is possible if desired, and an indicator light that shines when the mailbox is not empty is feasible.

The physical configuration of an OIS could be as simple as a single processing unit supporting several work stations or as complicated as a distributed network of large, interconnected computers, each of which supports a cluster of work stations. The degree to which each work station can operate independently of other work stations and network processing devices determines the degree to which the OIS is immune to hardware malfunctions and downtime. Each work station will have local data bases and data objects and indirect access to information local to another work station. The design of a particular OIS will determine which data bases are available to each work station at any given moment and how that information may be accessed and used.

Multiple tasks may be performed in parallel at individual work stations. This will occur most often when the office worker describes entire tasks or parts of tasks so that the procedures may be handled automatically by the work station. When the work station encounters an unanticipated discrepancy or situation, it will notify the office worker, describe the problem, and await further instructions.

In addition to the concurrent processing at individual work stations, parallel processing may occur when multiple workstations perform operations on a single transaction. For example, consider the purchase order that requires the independent approval of two managers. Once the purchase order is ready for the approvals, it would arrive simultaneously at the two managerial work stations. Each manager could then approve the purchase, put it into a "wait state" for future action, or block the purchase by denying approval. The main point is that the two approval processes do not necessarily have to be staged serially; instead they can be performed concurrently. If one manager should disapprove the purchase, the other could be notified, and the purchase order handled accordingly.

Summary

An OIS can be viewed as a distributed network of programmable (intelligent), office-oriented work stations capable of communication with other work stations. The bulk of the work supported by an OIS focuses upon preparing and managing electronic documents. These documents are sent to the work stations where work is performed. The route of each document is denoted by the chain of actions required to process the document. The tasks associated with document processing may be autonomous and interactively performed. In addition, other facilities may be integrated into an OIS to automate office procedures. The OIS supports the concurrent processing of tasks assigned to an individual work station and of independent tasks associated with the processing of documents for a particular transaction. The Citibank system discussed next is an early attempt at creating an automated OIS.

Citibank's Management Work Stations

In 1976, the 12 management work stations (MWSs) installed at Citibank were intended to alleviate office workers' burdensome problem of managing and processing the paperwork needed to handle customer transactions.[3] The MWS system serves as an interesting introduction to OIS prototypes for a number of reasons: first, it is in a real-life business environment; second, it was designed to solve a business problem and not simply treat its symptoms; and third, it integrated several business functions into a single system with the end user, that is, the office worker, in mind.

The MWS system eliminated many of the inefficiencies of the traditional paperwork system by providing (1) automated internal correspondence, (2) an electronic mail system, (3) memo processing, and (4) financial and other report processing. The MWS system has built-in capabilities for word processing, document creation, editing and filing, business planning, distributed calendar maintenance, forms development, cost-benefit analysis, financial reporting, hard copy production, and the transmission of documents and messages from one work station to another. Each management work station consists of a PDP-8A processor, two keyboard-display units, an impact printer, and floppy disk drives. The MWSs are connected by telephone lines and shared by a manager and a secretary.

The MWS functions were designed to be consistent with the manual paper-based counterparts, so as to minimize the adverse effects of the transition and thus increase the acceptability of the system. Citibank observed secretaries and managers trained to use the new MWS system in order to gauge the effect of its introduction. Citibank concluded that the secretaries have showed "more flexibility and adventurousness in using the system than managers have."[4]

Citibank officials consider the primary advantages of the automated office over traditional office methods to be its impact on control, security, expense, and productivity issues. Electronic files were found to be relatively secure, and electronic forms were shareable and safe from accidental loss. Speedy filing and retrieval of office information and the other capabilities of the MWS system improved worker productivity and customer service. Together these advantages seem to justify the costs.

However, the limitations of the MWS system are many. Reliance on basic security and control mechanisms may be overly simplistic. The electronic mail system, to be more effective, needs to be able to handle external documents and correspondence as well as those generated internally. The MWS capabilities need to be extended to encompass the full spectrum of business operations and to permit the user to "program" the MWS to perform some tasks automatically. Clearly shown, however, is that an automated office, even this rather basic model, is practical and beneficial and will help set the trend for the OIS of tomorrow.

OIS: An Evolving Trend

As demonstrated by the Citibank example, a progressive and experimental project for its time, OISs are still in the developmental stage. Much of the technology required to make an OIS a reality is available. The degree to which the unsolved technological and psychological issues are resolved will greatly determine the usefulness and acceptability of OISs.

The technological issues remaining to be answered span several disciplines and are interrelated. The computer scientists' developments in natural office languages, communications, and hardware suitable for the office will strongly influence the roles and responsibilities of office workers and the physical organization of the office. Accountants and auditors, concerned with the possibilities of electronic fraud, need to ensure that internal controls are implemented. This will require adequate hardware and software support as well as managerial control of sensitive information.

Research into the psychological impact of the OIS on office workers is required so that the office workers' capabilities and views support its successful implementation. Sociologists may be able to provide recommendations based on their research into man-machine relationships and their effects on the social structure. The prospect of large, interconnected OISs will undoubtedly send reverberations through Washington, initiating, for example, the further propagation of privacy legislation regulating their use. Managing and integrating these essential advancements, findings, and technologies into a total OIS are themselves complex problems without immediate solutions.

The current limitations of OIS implementations are vast, but certainly not insurmountable. As work continues, answers to current questions will be found, and OISs will evolve, incorporating each new finding and experience. Since most of the formative work is done in laboratories, examining state of the art prototypes is appropriate.

12.7.3 OIS PROTOTYPES

Two OIS prototypes are discussed in this section. One, Officetalk-Zero, comes from an industrial setting, whereas the other, SCOOP, was developed in an academic setting.

Officetalk-Zero

Officetalk-Zero is a prototype OIS that was designed and implemented by William Newman, Tim Mott, and employees of the Office Research Group at Xerox Palo Alto Research Center (PARC).[5] The Officetalk-Zero project began in 1976, and the prototype was in operation by June 1977. Its goals are consistent with those designed into Citibank's MWS system. Officetalk-Zero is an electronic aid to performing routine office tasks, similar in manner to the traditional way of performing the equivalent clerical tasks. The fundamental object in Officetalk-Zero is the electronic document. The Officetalk-Zero commands are for the management and preparation of documents and for their transfer between work stations.

A major portion of the Officetalk-Zero development effort was invested in constructing a man-machine interface that would enable the office worker to use all of the system's facilities simply and uniformly. These goals were achieved by developing a work station supporting desktop activities and through access to an integrated system providing the user with a large set of office functions.

Configuration of Officetalk-Zero

Officetalk-Zero is configured as a network of interconnected minicomputers, each supporting a single work station. The minicomputer employed is the Xerox Alto with 128K 16-bit words of main memory and a 2.5 megabyte disk for external storage. The system supports both personal and system data bases for the storage of electronic forms, electronic mail, and information concerning each authorized user of Officetalk-Zero. Communications between work stations are conducted by means of the electronic mail system and the transfer of electronic documents. The work station is an advanced CRT device that electronically simulates the office worker's desktop. The user manipulates forms on the CRT device by employing a pointing device and a keyboard.

Work Station Description of Officetalk-Zero

The contents of the electronic desktop are represented by the display of forms on which work is currently being performed and four file indexes of forms available at the work station, for example, incoming mail, outgoing mail, retained forms, and blank forms, as shown in Exhibit 12.7.1. The system automatically updates indexes whenever the work station operator issues a command causing a form to be moved from one file to another. Each entry in the index consists of an action field specifying which Officetalk-Zero command may be used to process the form and pertinent descriptive information about the form.

Documents and the repository lists are displayed in rectangular windows on the CRT device. To start work on a form, the user indicates the form and the action to be performed by pointing the cursor to the action field of the appropriate entry of any index. If the specified action is to make a form available for processing, then the system will display the form in a new window. Each window includes a "menu" of Officetalk-Zero commands applicable to the form. Pointing the cursor to the desired function displayed on the menu causes the work station to enter the appropriate mode for interactive processing or prompts the work station to execute the selected function automatically. The user can enter data on a form by first pointing the cursor to the field that is to contain the data and then typing in the data on the keyboard. Data type checking is automatically performed by the system so that the entered information is in accordance with the forms specification; for example, a quantity field may contain only numerical data. A provided forms editor describes the standard letterings and format of the form and the style of each field on the form.

Opened windows on the CRT displaying forms or indexes may be adjusted by the office worker

Exhibit 12.7.1 Officetalk-Zero's Work Station Display

File Index F1

Personal Letter		Tom Smith
Memo		Richard Doe
Requisition		Pete Peels

Out-basket

	Recipient	Subject	Sent
Trace	Dick Tims	Purchase Request	7/15/80
Trace	Sally Hawkins	Transfer Request	7/16/80
			7/16/80

In-basket

	Sender	Subject	Received
Pull	Dick Tims	Purchase Order	7/16/80
Trace	Tom Smith	Employee Referral	7/16/80
			Date
			Date
			Date

Blank Stock Index

Purchase Request

Receive Print Copy File Send

American Acme Company
1616 Hudson Avenue
Camelot, IN 47903

July 16, 1980

International Shipping Company
1303 42nd Street
New York, NY 10079

Dear Sirs:

I have reviewed your latest contract proposal and found it to be so that the form can be seen in its entirety or in part. A form may be moved around under the window for viewing various parts, or the window may be relocated on the screen. Overlapping windows will produce the same effect as overlapping pages on a desktop; the last window moved is wholly visible and hides those portions of the windows beneath it. The work station user controls the movement of a window and the form within a window by pointing a cursor to the appropriate area(s) of the screen. When finished with a document, the user may file or mail it, freeing the window and causing the document to disappear from the screen.

The work station possesses another interesting capability: It allows the user to afix signatures or freehand illustrations to a form. The freehand markings can later be removed without altering the form's original condition.

Capabilities and Limitations of Officetalk-Zero

Officetalk-Zero provides the office worker with the basic functions required to process documents effectively. These capabilities include a text editor, electronic mail with a trace function that enables the sender to monitor the location of a mailed document, data entry capabilities, and functions for copying and filing documents. The limitations of Officetalk-Zero appear not to be the result of shortcomings in the implementation of its existing capabilities, but rather the omission of other desirable capabilities. The nonprogramming environment of Officetalk-Zero does not allow the user the flexibility of defining personalized office procedures and functions to enhance the effectiveness of the work station.

System for Computerization of Office Processes

Zisman developed SCOOP at the University of Pennsylvania as part of his doctoral research. The system stresses the automation of office procedures as well as of office devices.[5,6] The heart of SCOOP is a model-driven monitor that traces the progress of office procedures and automatically executes portions of them at the appropriate times. By so doing, the computer and the work station operator interact in a joint undertaking. The system description for driving SCOOP's monitor comprises document definitions and office activities expressed in a nonprocedural language. The underlying formalism of the system description is the Petri net augmented with simple and compound predicates whose truth initiates the execution of corresponding actions. The predicates can be used to detect certain events such as the existence of a record, a value set by a transaction, or the passing of a set amount of time. Like Officetalk-Zero, SCOOP has a single, uniform interface for integrating the special-purpose systems into a unified system. The overall system supports electronic message passing, document manipulation, file services, and other functions. The creative idea of automatically monitoring office procedures to assist the office worker is extendable to monitoring office work for compliance with security and internal control specifications. This issue is explored in Section 12.7.5.

12.7.4 THE OIS USER INTERFACES

Earlier this chapter stated the essential properties of an OIS: It must be flexible and powerful enough to meet the user's needs, and yet simple enough that the office worker, a nonprogrammer, can use it effectively. Historically, interactive application systems were developed under a systems analyst's supervision and management. The systems analyst's task was to survey the user's applications to establish the necessary system capabilities and to design a system meeting those needs. The system design was a blueprint from which the computer technicians programmed the system. Thus the systems analyst served as an intermediary, bridging the communication gap between the user and the programmer. If failures in communication were kept to a "reasonable" level, the system design would indeed meet most, if not all, of the user's needs and would be implemented as specified.

However, often the net result of these weeks or months of implementation effort would be an inflexible system of parameterized programs that were invoked through a kind of menu selection that only temporarily satisfied the user's requirements. Changes in such systems were often difficult and costly. Yet changes in the way business is conducted are inevitable and will require such alterations in the existing systems.

A high-level office language that office workers can easily learn and use would substantially improve the potential for efficient and effective OIS development and maintenance. However, because of the complexities inherent in office work itself and in natural language development, the attainment of such a language will not be easy. Nevertheless, the current state of the art makes it reasonable to expect the development of high-level office languages enabling the user to set up and maintain a wide variety of office application systems, and yet requiring only minimal training.

The languages will combine WP and data base management facilities, enabling the user to define, maintain, and query data bases and to incorporate the selected and processed data base information into reports, documents, memos, and so on. Micro Data Base Systems is a good example of a sophisticated data base management system for the control of complex data structures. It is commercially available and operational on microcomputers.[7] Word processor and data base management capabilities will make it possible for office workers to develop software for common office applications such as those in a doctor's office.

Two proposed high-level languages for use in OISs are presented next, followed by a brief discussion of the controversial issues surrounding the problem of the user interface with automated OIS.

The Office Procedures by Example Language

The Office Procedures by Example (OBE) language is a nonprocedural language intended for use in automating office and business applications.[8] The result of a research project of IBM, OBE is an extended version of the IBM-marketed Query-by-Example data base management language. The OBE language makes it possible for office personnel to maintain data bases and program work stations to create tables, forms, and other documents with information already available in a data base. Various mathematical calculations are also possible. Office Procedures by Example also maintains facilities to save programs for reuse or automatic invocation whenever certain critical events occur within the operation. Communication in OBE is by means of an electronic mailing system.

The fundamental object in OBE is the two-dimensional form. The user specifies how the data base information is to be mapped onto the form by filling in the form with sample data. The sample form specifications constitute an automated procedure, which, when executed, produces a document. This document can be edited further or can be sent to another work station in the network.

Exhibit 12.7.2 exemplifies an automated procedure written in the OBE style. The purpose of the procedure is to prepare a list of new employees arranged by manager. The procedure is set up to run automatically once a month without user interaction. The reports are sent automatically to the corresponding managers through the electronic mailing system. The procedure specification consists of (1) references to two data bases, (2) query criteria, (3) a template for creating the desired form, (4) a trigger command, and (5) a send command.

Exhibit 12.7.2 OBE Procedure for Personnel Reporting to Management

APPLICANT	NAME	H/R/W	POSITION	DATE	DEPT
TR1 (MONTHLY)	N	H	P	DT	D

DEPARTMENT	DEPT	MANAGER
	D	M

CONDITIONS
1/01/80 ≤ DT ≤ 12/31/80

A

Dear M

This is the current listing of new employees who are scheduled to work for you this year, 1980.

Juney Bird

NAME	POSITION	DATE
N	P	DT

SEND (TR1) A TO M

The first data base of the procedure description, APPLICANT, contains the name, status (H = hired, R = rejected, W = hired and working), job position, hiring date (implies status is an H or W), and hiring department for each applicant referral. The DEPARTMENT data base contains the title and manager's name for each department. The underlined character strings, N, P, DT, D, and M, are example elements. The presence of each example element in the data base description and in the body of the sample form indicates a mapping of the data base information onto the form. The APPLICANT and DEPARTMENT data bases are linked in this procedure by the department title field, DEPT. The department title field is common to both data bases. The result of this cross-referencing is that M is a place holder for the manager's name for each department. The character H in the status field of the APPLICANT data base is a constant and states that only applicants with a status of H are to be accessed by the procedure. The condition box places a further constraint as to which applicants are to be reported. Specifically, those applicants who have a hiring date in 1980 will be a part of the report.

The automatic scheduling facilities of OBE maintain a list of procedures that are to be run routinely. It automatically schedules them for execution at the appropriate times. The trigger command, TR1 (MONTHLY), and the SEND (TR1) A to M command specify that each monthly report is to be sent automatically to the manager that the letter A addresses. Whenever this procedure is executed, reports similar to the ones shown in Exhibit 12.7.3 are created. The exhibit also presents the contents of the data bases used to create the forms.

The OBE language provides means for defining, updating, querying, and delegating user access rights to data bases. Data base integrity can be preserved through the use of integrity constraints. Potential integrity constraints include the editing of the update data for reasonableness of validity

Exhibit 12.7.3 Sample Input and Output for OBE Procedure for Personnel Reporting

APPLICANT	NAME	H/R/W	POSITION	DATE	DEPT
	Sandy Wilson	H	Secretary	03/20/80	ACCT
	Burt Williams	R	Accountant		
	Sue Fields	H	Accountant	01/01/81	ACCT
	Tom Smith	W	Mgr-Asst	01/15/80	ACCT
	Joe Peels	H	Mgr-Asst	09/01/80	PAYR
	Gerald Farrel	H	Accountant	07/15/80	ACCT

DEPARTMENT	DEPT	MANAGER
	ACCT	Kathleen Roberts
	ENGR	Joe Doe
	PAYR	Peter Davis

Dear Kathleen Roberts

This is the current listing of new employees who are scheduled to work for you this year, 1980.

Juney Bird

Name	Position	Date
Sandy Wilson	Secretary	03/20/80
Gerald Farrel	Accountant	07/15/80

Dear Peter Davis

This is the current listing of new employees who are scheduled to work for you this year, 1980.

Juney Bird

Name	Position	Date
Joe Peels	Mgr-Asst	09/01/80

of values and for interdata and data base consistency. The integrity constraints may be established such that they are unconditionally enforced or enforced for only particular data base operations. As noted in the preceding discussion, in addition to executing procedures automatically based on specified timing criteria, procedures and other OBE actions may also be executed automatically whenever specified events occur, for example, an overdrawn account. The conditions controlling the automatic execution of OBE actions and commands may be simple or compound.

The Business Definition Language

The Business Definition Language (BDL) developed by IBM at its Thomas J. Watson Research Center is a high-level language designed specifically for business data processing.[9] The view of business data processing manifested in BDL centers on the fundamental observation that document processing is the primary operation in business data processing. Documents are prepared and transmitted singly and in groups between departments and divisions as a means of communication. These same documents are later used to support recording and cross-referencing transactions. The numerical operations involved in the preparation of these documents are usually very low level.

An objective in designing BDL was to create a language office workers could use to write programs that would be structured, modularized, self-documenting, and easily maintainable. In pursuing these goals, one means employed by the designers was providing the BDL programmer with a language in which whole algorithms are written in meaningful problem solving terms. Even though BDL was designed for general business data processing, the design philosophy employed in its development is a blueprint suitable for the design of future OIS languages.

The basic objects in the BDL are documents, steps, paths, and files. Documents serve as input and output mediums for steps and programs. A step is a sequence of actions for the manipulation of documents, and a program is a sequence of steps. Each step action is either irreducible, that is, a primitive step, or a composite of more primitive steps. An irreducible step represents a standard, predetermined document transformation or a user-written routine describing the input and output behavior of the step. Composite steps are reducible to primitive steps. Each step corresponds to an organizational unit such as a department or work station. The nesting of the composite steps is intended to reflect the hierarchical structure of the organization. A path is the representation of the flow of documents between steps. Files are used for long-term storage of documents.

Major Components

The three major components of a BDL system description are the form definition component (FDC), the document flow component (DFC), and the document transformation component (DTC). The FDC enables the user to define templates for creation of blank documents. Each document's design is developed by drawing it on a graphic device. Fields within a document are described by the type of data the field is to contain and by its format. Groups of fields can be defined as single entities, and a sort facility can be used to order the entities. Provisions also exist for describing floating fields and for handling form overflow (the extraneous information spills onto a user-designed overflow form).

The DFC application system is graphically defined in terms of the organizational units involved in the processing and flow of documents. The basic diagram constructs are rectangles, circles, and arrows, used to represent steps, files, and paths, respectively. Forked data flows indicate that a copy of the document is to be sent along each path identified. Exhibit 12.7.4 presents a DFC representation of a sales processing system.

The sales processing system description has three major composite steps: customer service, shipping, and accounts receivable. The internal structure of the major composite steps is further defined in terms of more elementary steps and data flows. These constituent steps may themselves be composite steps which could be further refined in other DFC diagrams. Previously unspecified file accesses may be introduced in the more refined DFC diagrams. The system is fully described by the DFC when each stop is irreducible, or a primitive step.

The sales processing example is especially interesting in that it combines aspects of both OIS and data processing applications. The customer service step prepares the invoice and approves credit terms. The invoice preparation step is performed by a secretary at the work station. The secretary obtains a blank invoice and fills in the customer's name, address, product numbers, and quantities. A procedure written by the secretary then accesses the price file and enters the product unit price, the total price by product number, and the sales invoice total price. Finally, the secretary reviews the invoice, makes any necessary changes, and sends it to the appropriate work station for credit approval action.

The credit officer, having specified general approval policies in granting credit terms on routine orders, only affixes a specific approval to requests outside the general approval policy bounds. For these special cases, the work station is programmed to display the sales invoice in question along with the customer's credit history. In either case, the approved invoices are forwarded to shipping.

Exhibit 12.7.4 Sales Processing System

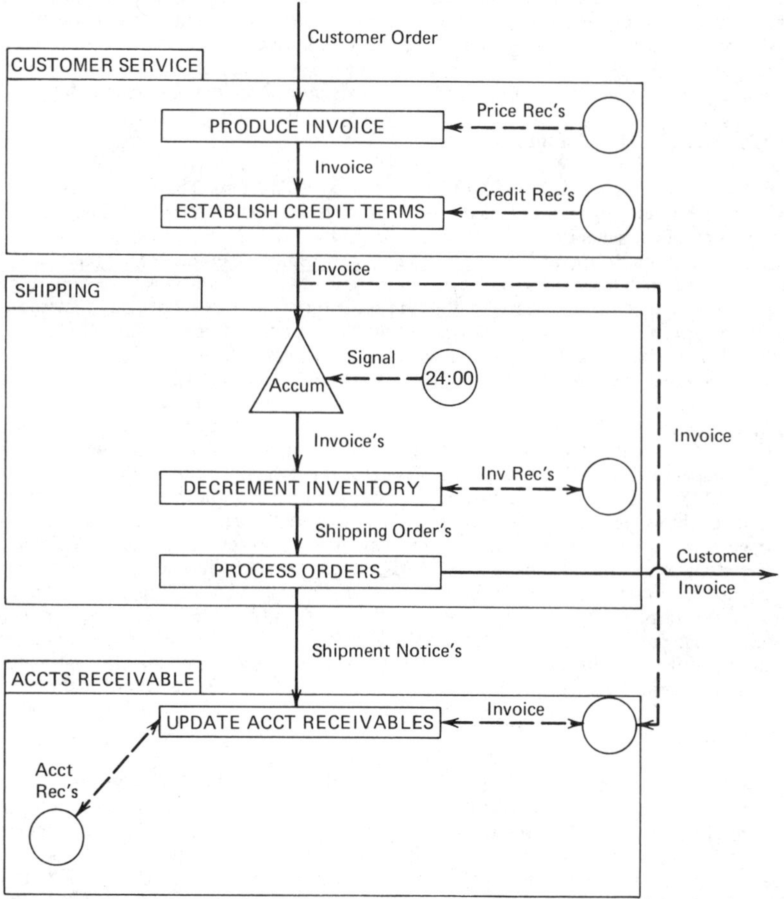

Shipping and accounts receivable are data processing systems with routines differing from the OIS procedures in that they need no autonomous procedures, but can be fixed algorithms requiring very little human interaction. The Accum function queues the invoices until midnight, at which time the function releases the day's entire group of invoices to the shipping data processing system. The shipping data processing system fills the orders according to a fixed priority scheme, updates inventory, and schedules shipping. In an automated warehouse, workers then supervise the packing of the orders and notify the system as each order is filled. The system then generates a shipment notice document destined for accounts receivable. Similarly, the accounts receivable data processing system accepts the shipment notices as input and updates the accounts receivable and invoice files.

The DTC of the system description is used to define the actual computation of each irreducible step in the DFC. Theoretically, any programming language or marriage of programming languages could be used in place of the DTC language. Thus, as in the previous example, an OIS language and a business data processing language could both be used to implement mixed systems, taking advantage of the office worker's programming skills and those of the application specialists. The very high level DTC language developed assumes a general structure for all steps and thus gives the programmer the advantage of a built-in control structure oriented toward business data processing.

In exchange for the resulting programming simplicity, we accept a loss in language flexibility. The programmer utilizes the resulting control structure in defining transformations of the input documents to create output documents. The computations appropriate to each field included on the output document are described using normal arithmetic operators, aggregate arithmetic operators, and logical and relational operators applied to available fields on the input and output documents. The DFC's run-time mechanism automatically readies the step for execution whenever documents are available on all its input paths, when prior system activity is complete to that step.

Office Workers and Machine Interface Issues

In any organization, including the business office, formal and informal communication lines exist between individuals and organizational units. Formal communications are designed to meet certain anticipated, critical information needs, with written documents usually serving as the medium. Preparation of the necessary forms is generally detailed as a part of the system description.

This type of communication closely corresponds to the document-oriented view of an OIS. The well-defined procedures lend themselves to being algorithmically specified. The documents can be structured and formatted and electronically or mechanically manipulated. Moreover, a mailing system is a natural and efficient means of transferring documents between the organizational units. Office information systems, properly complete with user-matched capabilities, are capable of efficiently supporting formal office communications.

The critical issue concerning formal communication in an OIS environment is the office worker-machine interface. As formal communications are machine representable, the question becomes whether clerical users can get the necessary support to perform office tasks with standard office capabilities such as WP and user-written routines.

Informal communications—conversations, gestures, jokes, and unwritten guidelines—serve an important organizational need often not adequately accounted for by the system. Dialogue is used by senior employees to provide on-the-job training for junior employees. Stimulating conversations about business problems, sharing ideas and knowledge, can give problem solvers new insights. Conversation can encourage personal involvement, concern, and a feeling of group membership. These forms of social contact are essential to maintaining an effective office organization and exemplify information communications that cannot be adequately supported by a purely document-oriented OIS. The integration of information communications into an OIS is a major concern in interfacing people and machines.

Office Worker–Machine Interface

The implications of a particular design philosophy and the nature of the resulting OIS interface are important to office personnel staffing and training. Clerical workers and managers cannot be expected to possess expert programming skills, and yet they need to control their own office procedures. An OIS will generally incorporate only specific problem-oriented functions and capabilities commonly needed in an office setting. For example, Odyssey is a knowledge-based system for aiding an individual user with business trip preparations.[10] Knowledge of trip preparation is built into Odyssey so that it can *intelligently assist* the user.

Specialized office assistance packages such as WP, DDS, and Odyssey can provide interactive procedures to aid the office worker in performing office tasks. These procedures enable the worker to perform effectively the particular task for which the system was designed. Users will require little training since most of the systems will have simple interfaces using basic problem-oriented commands or prompting techniques. However, they cannot cover the wide range of autonomous office applications currently required by business. Depending upon the nature of the task, office workers will be expected to specify and correct programs and to maintain their own automated office procedures.

This section addresses a few of the many issues concerning the design and implementation of OIS languages that will ease the office workers' burden in this respect. A number of these issues have been addressed by the computing community and correspond to questions regarding conventional programming languages: structured programming, concurrent languages, program and system documentation, data base languages, and the elimination of technicalities such as memory dumps from the programming environment.

Programming languages are defined by their syntactic and semantic rules. For a programming language to be properly interpreted by the computer, it must adhere to a set of rules that can be encoded as an algorithm. This need to structure and interpret the languages rigorously puts even limited natural language processing outside current programming technology. Clearly, the styles of conventional programming languages such as COBOL, FORTRAN, or Pascal are inappropriate for unsophisticated OIS users. For this reason, languages more appropriate for OIS have been described and are reviewed here.

An example of a special-purpose OIS language is OBE. Its simple, mimicking style makes OBE very easy for even unsophisticated clerical personnel to master. It is well suited to programming work stations called upon to prepare simple forms formatted according to the OBE standards. The automatic task initiation feature is also easy to use and very practical. The problem with special-purpose languages like OBE is that their application span is very narrow. Tasks coinciding with the abstract model on which the language was designed can be easily encoded; however, any deviation from the language's original problem formulation assumption will greatly increase the programming problem, if not make it impossible to express in the language.

At the other end of the language continuum are general-purpose languages. For a language to be general-purpose, it must consist of fundamental operators, data types, control structures, and data structures with which problem-oriented routines can be constructed. Translating a conceptualized algorithm into such a language is difficult; however, the more general-purpose the language, the wider the range of problems it can be used to solve. This benefit is obtained at the cost of increasing the conceptual difficulty in the programming task. Matching the right OIS language to the clerical user's or application specialist's needs and abilities is essential to effectively increasing productivity. A mismatch will reduce an OIS's effective use and could cause the system to fail.

Problem-oriented BDL is about midway between a general-purpose and a special-purpose language. Variations on BDL may provide an almost general system framework permitting the automatic interface of two or more languages. As the sales processing example showed, the DFC served as the system framework in depicting the irreducible steps and their relationships. Office-oriented steps could be programmed in the OIS language, and data processing steps could be programmed in a more appropriate language. If each step complied with the same document input and output concept, then the interface between the languages could be the document. On the surface this style of integrating languages is appealing because it permits the selection of the proper language for the task and the automatic interfacing of OIS with data processing systems.

The introduction of the *intelligent form* could help to lessen individual tasks by embedding controls within the form. The intelligent form would automatically perform complex edit and data consistency checks on itself and restrict the individual user's rights to view, update, or add information, depending upon the user and/or the form content. The intelligent form would schedule itself for work station processing according to specified priorities and current work station loads and could become persistent if neglected too long. Once the form is so described, office personnel would need be concerned only with primary document transformations. This would guarantee a measure of consistency in processing the document and would improve data security.

Another noteworthy issue includes the possibility of standardizing OIS languages to eliminate manufacturer dependency and problems involving constant, necessary training and retraining of personnel and to increase the portability of office procedures as well as data and procedure security.

Informal Communications

As previously mentioned, informal communications are necessary to performing group tasks in the business office. In a pure document-oriented environment, casual memos delivered by means of an electronic mail system cannot replace social contact and information exchange. An OIS that could support video and audio transmissions would heighten social interaction. Whether this would be adequate is an unanswered question. Office organization designs aimed at increasing social contact are a potential solution to the problem, but perhaps at the cost of some lost machine-based effectiveness. As hardware technology advances, work station portability will increase the prospect of the office worker's home becoming the office—a situation with far-reaching social and business implications.

As first-generation OISs are introduced into the business environment, there is likely to be little loss to the informal communication system. Implementation will probably not be complete, and initially clerical personnel will rely on each other while they learn how to use the system effectively. However, as OIS users become more proficient and OISs more powerful and complete, the task requirements for social contact will decrease.

12.7.5 OFFICE INFORMATION SYSTEM MODELING

Office information systems developed around the formal lines of communication and formal work station activities are subject to precise modeling. The advantages in modeling OISs are several. First, an OIS described in a formal manner can be checked for descriptive consistency and completeness. Second, a validated OIS model constitutes a standard and uniform documentation of the system. Third, the model can be used to further various types of analysis. For example, office efficiency specialists might use the OIS model to predict system performance, or auditors concerned with internal control issues would be able to analyze the control and processing structure. Fourth, managers, once convinced that the OIS model is secure, adequately controlled, and efficient as modeled, will require assurances that the implemented OIS complies with the model. Given the complexity of a full-scale office, it is infeasible to perform such analyses manually. What is needed is an OIS modeling process capturing these advantages as well as providing managers with the capability to restructure the organizational units and tasks under their authority as circumstances dictate.

Information Control Nets

Information Control Nets (ICN) is a model developed at Xerox PARC to describe and analyze office information flows.[11] The ICN model defines an office as a set of uninterpreted *activities* that access and update information stored in *repositories*. Possible execution sequences of activities are defined by "precedence constraints." A precedence constraint is a simple rule stipulating an immediate successor activity for the parent activity. Upon completion of an activity, one of its immediate successor activities can proceed. As with most office data flow models, the execution of a subsequent activity need not immediately follow the termination of its predecessor. If the work station operator is busy or away from the work station, then the activity is temporarily suspended.

The office models presented here share a fundamental perspective on the processing environment: that offices are transaction oriented and that, even though many transactions are performed concurrently, the activities required to process each type of transaction can be adequately described without explicitly considering the concurrent processing of like transactions. The models are designed to depict all preconceived processing for a particular type of transaction. Specific knowledge of how each type of transaction is individually processed can be used to infer knowledge of the highly parallel office environment.

An ICN can be represented graphically or as mappings defined over a set of activities and a set of repositories. In ICN diagrams, rectangles represent repositories; triangles represent temporary repositories; labeled circles denote activities; small, hollow circles denote conditional branches (choice nodes); solid lines connecting activities denote precedence constraints; broken lines denote repository access and updates; and a small, solid circle denotes the start (end) of parallel activity sequences.

Exhibit 12.7.5 is an example ICN graph. In the exhibit, activity a_1 begins upon the arrival of a departmental request. Activity a_1 simply logs the request in a logbook. Upon completion of a_1, a_2 is initiated, resulting in a typed requisition, which is then stored in the requisition form repository. Activity a_3 prepares a purchase order form. Choice node a_4 follows a_3 and establishes whether activity a_6 or activity a_5 is to follow. The choice depends upon whether the requested items are in stock. The branch associated with the in-stock choice is assigned a probability of occurrence of .75. If the out-of-stock branch is taken, activities a_6, a_7, and a_8 designate that the purchase order is to be sent to another worker for verification and approval; otherwise, activity a_5 completes a work order form for inventory. Note that the activities are only vaguely defined. The designers

Exhibit 12.7.5 Purchasing System

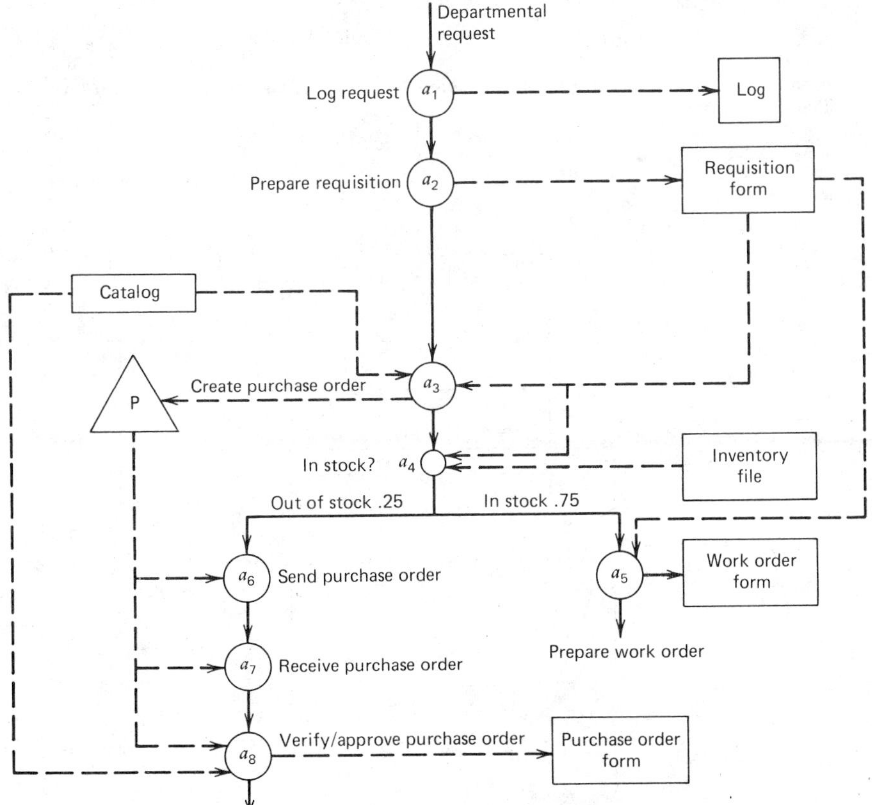

of the ICN suggest that a vagueness is advantageous since the model can illustrate information flows without obscuring the basic flows with overcomplicated operational details.

Ellis and Morris have developed techniques to automate the analysis and optimization of ICN models.[12] The ICN models are intended to be optimized according to three kinds of office transformations: *automation*, *reorganization*, and *streamlining*. Automation transformations replace manual activities with corresponding automated activities. Reorganization transformations shift activities to enhance some particular property of the model, such as concurrent processing. Streamlining transformations are used to simplify the ICN by eliminating useless activities and reducing communications overhead.

Exhibit 12.7.6 is a result that could be obtained by automated optimization transformations to the purchasing system of Exhibit 12.7.5. Activities a_6 and a_7 on Exhibit 12.7.5 are automated on Exhibit 12.7.6 within activities a_3 and a_8. The OIS communications system is utilized to accomplish this end. Shifting the prepared requisition activity (a_2) with respect to the log request activity (a_1) permits parallel processing of activities a_4 and a_1, the in-stock check and log request, respectively. This latter transformation belongs to the reorganization category. The primary advantage of parallel processing is to reduce the elapsed time requirement to complete a transaction. Activity a_3 has been rolled forward through decision node a_4 since the catalog needs to be accessed only if the requested items are out of stock. Since activities a_3 and a_8 access the catalog, and activity a_8 is an immediate successor to activity a_3, the catalog information obtained by activity a_3 can be stored with the purchase order form in the temporary repository p for use at a_8. This transformation will reduce competition for access to the catalog and will make the catalog information readily available to activity a_8. These preceding transformations were designed to reduce file access overhead and are examples of streamlining transformations.

By augmenting the ICN diagram with branching probabilities and average service times, it is possible to calculate the expected service time to process a single transaction. Consider Exhibit 12.7.7, which is an example of such an augmented ICN diagram. (Files are omitted for simplicity.) The precedence arrows are labeled with the branching probabilities and the expected service times, s, of each specified activity. For example, the service time for activity a_3 is 1.5 min, and the probabilities of activity a_3 initiating activity a_4 or a_5 are .3 and .7, respectively. Thus the expected service time required to complete activities a_3 and a_4 or activities a_3 and a_5 is $1.5 + .3(5) + .7(10) = 10$ min. Similarly, we could calculate the expected service time of the entire graph.[11] Using matrix notation, the computation is performed as follows:

$$\text{expected service time of the graph} = W[\Sigma(V\pi^i)]$$

Exhibit 12.7.6 Purchasing System of Exhibit 12.7.5, With Automated Optimization Transformations

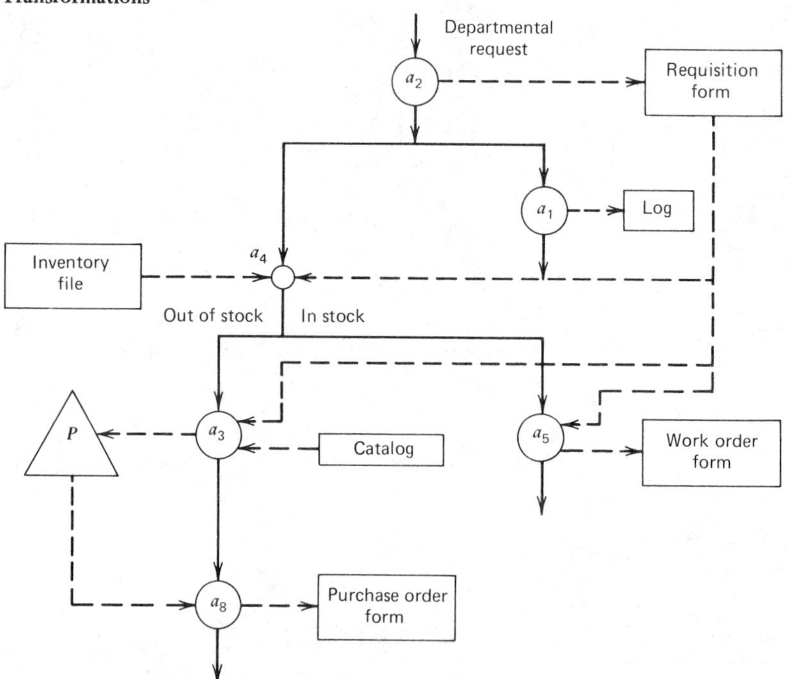

Exhibit 12.7.7. Augmented ICN Diagram and Expected Service Time Calculations

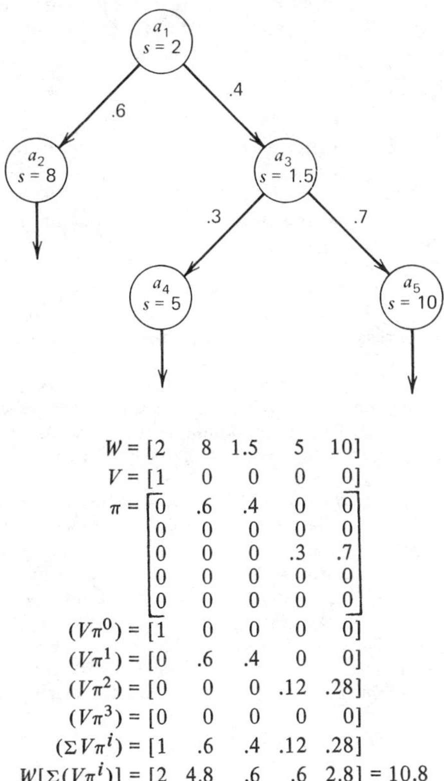

$$W = [2 \quad 8 \quad 1.5 \quad 5 \quad 10]$$

$$V = [1 \quad 0 \quad 0 \quad 0 \quad 0]$$

$$\pi = \begin{bmatrix} 0 & .6 & .4 & 0 & 0 \\ 0 & 0 & 0 & 0 & 0 \\ 0 & 0 & 0 & .3 & .7 \\ 0 & 0 & 0 & 0 & 0 \\ 0 & 0 & 0 & 0 & 0 \end{bmatrix}$$

$$(V\pi^0) = [1 \quad 0 \quad 0 \quad 0 \quad 0]$$

$$(V\pi^1) = [0 \quad .6 \quad .4 \quad 0 \quad 0]$$

$$(V\pi^2) = [0 \quad 0 \quad 0 \quad .12 \quad .28]$$

$$(V\pi^3) = [0 \quad 0 \quad 0 \quad 0 \quad 0]$$

$$(\Sigma V\pi^i) = [1 \quad .6 \quad .4 \quad .12 \quad .28]$$

$$W[\Sigma(V\pi^i)] = [2 \quad 4.8 \quad .6 \quad .6 \quad 2.8] = 10.8$$

where $W_i = s_i$

V_i = probability of entering the graph at activity i

π_{ij} = probability of a_j directly following a_i

This type of analysis may be useful in analyzing office models' characteristics. It is possible to ascertain the effects of introducing parallelism into a model, to answer questions concerning individual and group work loads and processing times, and to study system throughput characteristics. Such elementary analysis is useful, but very limited. For example, consider an office where order transactions are processed on a priority basis. The transaction's priority may be based upon simple criteria, such as a customer's credit rating, or on more complex, multiple criteria, such as the purchase amount, the customer's purchasing history, and/or current inventory levels. Now, complicate the problem by adding arrival rates for seasonal or sporadic transactions. Calculating the expected elapsed time to process a particular order transaction becomes extremely difficult and its meaning questionable. Currently, simulation techniques are employed to analyze such models.

OIS Simulation

System simulation is an interdisciplinary technique widely used to assist in measuring system performance. The ability to simulate presumes the ability to model the system. Industrial engineers and operations researchers use simulation as a tool for analyzing such systems as machine networks. Accounting researchers use simulation techniques to study transaction processing systems and to measure the effectiveness of the error detection and correction facilities incorporated in the system design. System designers also rely heavily upon simulation in fine-tuning system configurations. Common to most simulation applications of this type is the need to measure a system's performance and its components using measurements such as utilization, work load, throughput, and response time.

Simulation models employ descriptions of system procedures, queuing strategies, and probability distributions for such attributes as arrival rates, service times, system resource needs, and routing probabilities. Since the automated office is a system similar to those currently being simulated, office designers can also employ simulation techniques as a design tool. Detection of bottlenecks and underutilized work station capabilities can give the office designer crucial insights into reorganizing office and work station tasks. The statistical results of two simulation models operating under the same work load provide comparative measurements for evaluating one against the other.

Work Station Simulators

Nutt and Ellis propose the use of an office environment simulator not only for observing the real-time performance of an OIS, but also as a personnel training tool.[13] The personnel training tool is a work station simulator that can give the office trainee hands-on, real-time experience. Flight and driving simulators have already proved the effectiveness of simulators for learning purposes. The advantages of training on a work station simulator are as follows:

1. *Events and situations can be manufactured to specifications.* Using a real OIS, the trainee would repeatedly perform the same ordinary tasks, spending very little time handling important exceptional situations. At a work station simulator, all preconceived events and situations would occur according to specified frequency rates.
2. *Measurement of office worker performance.* Measurement of office worker performance is an extension of advantage 1. By presenting office workers with identical work loads and collecting statistical information as the office worker manipulates the work station, measurements are obtained that are comparative between office workers, between individual sessions, and against prespecified standards.
3. *A practicing environment for developing automated office procedures.* The work station simulator would permit the definition and execution of office procedures by the operator of the work station. This would enable the trainee to gain valuable programming and debugging experience without incurring the costs associated with mismanaged informations.
4. *A controlled environment (laboratory) for studying human performance in relation to changes in the work station design or capabilities.*
5. *A systems development tool.* As work station operators develop and refine automated procedures, these procedures can be used to simulate unmanned work stations. In this way the procedures' correctness and completeness can be tested in a real OIS environment.

A facility for real-time interactive simulation of a distributed office system has been developed at Xerox PARC. The facility, Backtalk, comprises a network of nodes, each of which is a work station or a dedicated device such as a printer or a data base server. One or more of the work stations are operated by automatic procedures that model the would-be operator in a realistic fashion. The remaining unautomated work stations serve as simulators that interact with the entire system.

The realism of the simulation depends upon the automated work stations' ability to reply to the simulator work stations' specific demands. Since the construction of the reply is contingent upon the nature of the demand, it is impossible for automated procedures to reply realistically to information communication events such as memos. However, a certain degree of realism can be achieved by identifying some basic characteristics of the demand and answering with prestored replies or replies taken from a data base of historical interactions between real work station operators.

The Backtalk facility possesses many of the abilities and properties presented in the list of advantages above. In a single work station simulator mode, it enables the work station operator to program procedures; access data bases; manipulate forms; transfer them to other automated work stations, which in turn can generate new forms; transfer forms to other work stations, including the work station simulator; and destroy forms.

Theoretic Internal Control Model—TICOM II

An important aspect of any OIS is the auditability, accountability, and internal control protections of the system. The Foreign Corrupt Practices Act of 1977[14] requires firms' procedures to satisfy certain internal control criteria that lessen the possibility of misallocation of corporate assets. As a result, every OIS operation is subject to external legal criteria. Consequently, significant design criteria are imposed on any OIS in addition to the usual engineering and convenience constraints.

The legal restrictions on office procedures, whether automated or not, are not the only criteria an OIS must satisfy. Firm managers also will wish to verify that only authorized individuals access company assets for valid purposes. When document processing is executed by machine rather than manually, the difficulty in verifying internal control procedures becomes more complex, as is corroborated by increasing "computer theft." As corporations grow more complex, successfully auditing firms in the traditional fashion—with involved, laborious flowcharting and narrative descriptions of document and asset flows—becomes more difficult.

Therefore satisfying legal constraints and management desires will soon require more extensive computer-assisted auditing. There are two important advantages to automated auditing. First, the speed, precision, and cost of computer processing compare favorably with their human counterparts. Second, interfacing an automated auditing and internal control system with an OIS will result in a nearly continuous audit, considered to be ideal in auditing literature.[15] In this way the OIS may be modified to satisfy legal constraints and permit economical internal control verification.

TICOM II[16,17] is an example of a computer-assisted internal control verification system. In a sense it proceeds in much the same way a human auditor does in establishing that certain internal control conditions are satisfied. However, the analytic capabilities are automated and therefore can be interfaced effectively with an OIS. The result of this conjunction is continual proof that the OIS does not violate sound management principles of internal control.

Intuitively, when firm employees use the OIS to process documents, the result of this processing will be examined by TICOM II. The latter will then consider whether the specified processing implies a violation of sound internal control. For example, it is considered good practice for the purchase order sent to the receiving department to contain only quantity data. The receiving department must enter the actual quantity received from the vendor and not merely copy from the quantity listed on the purchase order when completing a receiving report. Thus, if the receiving department attempts to access quantity data, TICOM II will immediately flag and prohibit the access as a violation of internal control. This prohibition exemplifies the interaction between TICOM II and an OIS in thwarting internal control violations.

Mechanics of TICOM II Modeling

The TICOM II modeling procedure is a generalization and semiautomation of the flowchart and narrative procedures auditors use to describe and evaluate internal control systems. The modeling input is a description of the duties and tasks of individuals within the firm. Instead of the flowchart, however, this modeling is expressed in the TICOM II modeling language.[16] The procedure is similar to flowcharting these duties just as writing a flowchart for an algorithm is similar to writing a computer program. An example of the TICOM II modeling is given in Exhibit 12.7.8. It shows the relationship of the flowcharted boxes representing document processing and the commands of the TICOM II modeling language. A complete description of the language and an extended example can be found in Bailey et al.[16]

A unique aspect of the language is embedded in the WAIT expression illustrated in Exhibit 12.7.8. The modeling procedure currently used for internal control descriptions involves the simultaneous construction of a flowchart for the entire firm. In the TICOM II modeling procedure, individual tasks are described separately, and these are integrated into a complete model of the firm by an automated algorithm. The links between individual tasks are determined by the WAIT expressions. In Exhibit 12.7.8 we see clerk 1 waiting for requisition approval. This informs the computer that the TRANSFER of approval fits into clerk 1's duties. Thus, by matching TRANSFER to WAIT expressions, a model of the firm's internal control system can be recovered from the individual descriptions.

Once the firm model is pieced together, TICOM II then analyzes it for consistency and completeness. This processing phase involves checking various items. First, it establishes whether each command performs a valid task, whether the document processing involves possible actions. For example, if an item on a purchase order, such as "vendor," is altered, it is established that this item really exists on a purchase order. Second, it is determined that every WAIT expression has a corresponding TRANSFER that satisfies it. Finally, each document in the system is examined to see that it is well defined. Consistency and completeness checking is a large part of the auditor's task, and thus the machine algorithm easily assists the auditor with automatic model construction and the initial testing.

The major value of the TICOM II system, however, is its ability to solve automatically queries about the internal control model. This is executed by means of a query language. The query language is expressed in a predicate calculus format, with all predicates (e.g., purchasing (p) means p is in the purchasing department) and terms defined in the model context. The types of queries the system can answer include the following: (1) Can an individual or department access a set of documents and assets? (2) Can a member of a given set of commands follow (precede) another set of commands? and (3) Can a set of conditions characterized by the predicate calculus query language ever be true in the firm model? These queries respond to most internal control issues currently examined in the literature. The mechanics of query answering and the computational complexity are examined in Bailey et al.[18]

Although the modeling procedure, consistency and completeness checking, and query processing have been extensively examined, the interface to an OIS is still under investigation. Intuitively, the procedure is to preprogram the TICOM II system to answer internal control queries that concern auditors. Then, every time the OIS is faced with document processing, it will be established that

Exhibit 12.7.8 Example of TICOM II Modeling

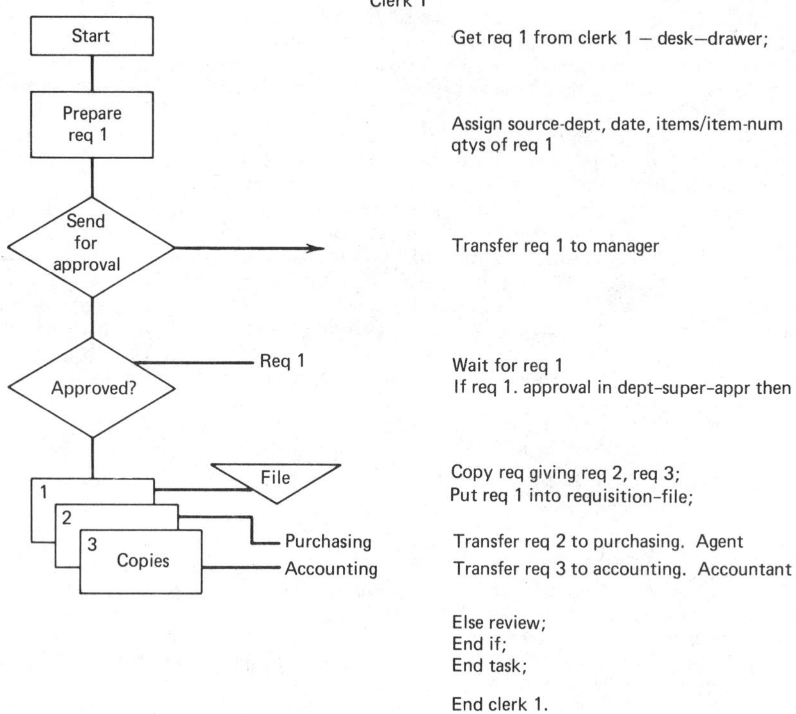
Clerk 1
Start
Get req 1 from clerk 1 – desk–drawer;
Prepare req 1
Assign source-dept, date, items/item-num qtys of req 1
Send for approval
Transfer req 1 to manager
Req 1
Approved?
Wait for req 1
If req 1. approval in dept-super-appr then
File
1
2
3
Copies
Copy req giving req 2, req 3;
Put req 1 into requisition-file;
Purchasing
Accounting
Transfer req 2 to purchasing. Agent
Transfer req 3 to accounting. Accountant
Else review;
End if;
End task;
End clerk 1.

this processing is consistent with the preprogrammed security issue. Thus every document processing request may be actively shown to satisfy accepted accounting practices.

Another important use of TICOM II involves the design of the OIS. Once an OIS is proposed, it may be modeled in the TICOM II modeling language. Relevant internal control issues may then be evaluated for the proposed OIS. If violations of accepted accounting principles are discovered, additional security precautions can be implemented until the OIS is deemed acceptable. In this way TICOM II functions not only as a security enforcer for an operational system, but also as a design tool for a proposed OIS.

Accounting principles form an important part of any operating management system. In an OIS, the links between electronic and manual document processing compound the difficulty of establishing satisfactory internal controls. Verifying that such controls exist is the purpose of TICOM II. In addition, TICOM II may be used as a design tool, by detecting failures in the internal control system. Finally, it may be interfaced with an operating OIS to provide continuous updating of internal control criteria satisfaction and a desirable, continuous audit.

12.7.6 INTRODUCING AN OIS INTO THE OFFICE SETTING

Implementation of an OIS is a much-discussed subject among business and office managers. Its particularly interesting problems are (1) the magnitude of the automation task; (2) the shortness of the implementation lead time; (3) the magnitude of the clerical training; (4) the acquisition of unavailable technicians for technical support; (5) the need to interface OIS applications with EDP applications; (6) the need to develop auditability criteria and techniques; and (7) the limited managerial, personnel, and technical experience in these areas. It would be folly for us to suggest step-by-step procedures for successful OIS implementation; these will be derived only through actual experience. However, careful planning should eliminate many possible pitfalls and lessen the consequences of inevitable, unforeseen obstacles.

One of the most positive steps an office manager could take would be to formalize job description specifications. A fundamental objective of an OIS is to permit the office worker to conduct business as before, but with the aid of electronic processing capabilities. Automating office work is not primarily intended to alter individual task assignments or responsibilities significantly, although this may result. It may be true that increased productivity brought on by automation will free office workers to perform more important tasks, but this will only add to their current roles.

Therefore accurate, precise, and complete job descriptions are an essential input to the automation cycle. It is strongly suggested that the job descriptions be represented in a uniform and formal manner so that the office worker and application specialist can automate directly from the procedural specifications.

A second desirable property of the job description specification would be its machine acceptability, using a language similar in style to a computer programming language. The TICOM II language may be particularly useful in this respect. Such a representation would provide not only an operational statement of the procedures, but also a systemwide control of data flow relationships among the tasks. Compilerlike programs could check the descriptions for inconsistencies, and models of the office could be mechanically constructed from the procedural descriptions. Using the computer, office analysts could then apply similar techniques to rigorously evaluate and measure the system's performance on efficiency and internal control scales. The validated office model and the system and job descriptions are the blueprints from which an OIS can be implemented.

The cost of such a project will be expensive, but cost beneficial, over the implementation cycle. It would seriously address and partially solve some of the aforementioned problems of OIS implementation, particularly, problems 1, 2, 3, 5, and 6. In addition, rigorous analysis of current office procedures could reap immediate gains from the discovery of inefficient and inadequate internal controls.

A managerial task force of representatives from all relevant disciplines should be formed to prepare specific long-term plans for OIS implementation. Suggested task force members include computer and EDP technicians, personnel and behavioral experts, accountants and auditors, departmental representatives, and office designers. Open communications between task forces to share ideas and identify problems and solutions would be beneficial. Throughout the planning and development stages, it must be remembered that an OIS is not in itself a labor saving device. It is by no means guaranteed to solve today's office problems automatically. The philosophy and design of the OIS must take into account the needs, abilities, and limitations of the organization and its individual members. However, if the organization is ready, the automated office will be implemented successfully and beneficially.

12.7.7 IMPLICATIONS OF OIS

As the world becomes increasingly interdependent, reliance on office-type communications, information processing, and knowledge storage and retrieval will grow. The advent of conglomerate firms has required business offices to handle information supplies and requests from all markets and all parts of the globe. This has strained traditionally labor-intensive office processing to the breaking point in some industries, for example, the stock exchanges. In banking and the stock market, electronic processing and office automation have proved necessary, simply to deal with the growing volume of transactions. Other industries have not yet been quite so strained.

The effects of office automation will be startling and pervasive. As computers direct more office procedures, office workers will increasingly find their job focusing on machine interactions. As for effects on the office job itself, office workers will be offered more direction from machines providing documents to be processed and comments on the processing. Consequently, the office worker will face a corresponding decrease in the need for self-reliance in job performance. The machine also will actually perform many of the simpler tasks now done by the office worker. However, as OIS concepts permit job requirements to be designed to fit the worker, job enrichment is a potential variable available in developing systems for maximum production.

Throughout this work, we have described and discussed OIS issues we feel will greatly affect the lives of office employees and the manner in which office business will be conducted. Conceptual knowledge of these OIS design issues should give the reader, the office manager, and the office researcher insights into the problems as well as the benefits of office automation. Although we could not report on all the aspects of OISs in this short space, we have surveyed the breadth of a new and exciting field.

REFERENCES

1. M. D. ZISMAN, "Office Automation: Revolution or Evolution," *Sloan Management Review*, Vol. 19, No. 3 (Spring 1978), pp. 1–16.
2. L. D. ERMAN, H. FREDERICK, V. R. LESSER, and D. R. REDDY, "The Hearsay-II Speech-Understanding System: Integrating Knowledge to Resolve Uncertainty," *ACM Computing Surveys*, Vol. 12, No. 2 (1980), pp. 213–253.
3. R. B. WHITE, "A Prototype for the Automated Office," *Datamation*, April 1977, pp. 83–90.
4. WHITE, "A Prototype for the Automated Office," p. 85.
5. C. A. ELLIS and G. J. NUTT, "Office Information Systems and Computer Science," *ACM Computing Surveys*, Vol. 12, No. 1 (March 1980), pp. 27–60.

6. M. D. ZISMAN, "Representation, Specification and Automation of Office Procedures," unpublished doctoral dissertation, Wharton School, University of Pennsylvania, Philadelphia, 1977.
7. MICRO DATA BASE SYSTEMS, INC., *User's Manual* (MDBS.DMS and MDBS.DDL), Author, Lafayette, IN, 1979.
8. M. M. ZLOOF, *A Language for Office and Business Automation*, IBM Research Report #35086, IBM, Cambridge, MA, January 1980.
9. M. HAMMER, W. G. HOWE, V. J. KRUSKAL, and I. WLADAWSKI, "A Very High Level Programming Language for Data Processing Applications," *Communications of ACM*, Vol. 20, No. 11 (November 1977), pp. 832–840.
10. R. E. FIKES, "Odyssey: A Knowledge-Based Personal Assistant," *Artificial Intelligence*, in press.
11. C. A. ELLIS, "Information Control Nets: A Mathematical Model of Office Information Flow," *ACM Proceedings of the Conference on Simulation, Modeling and Measurement of Computer Systems*, Boulder, CO, August 1979.
12. C. A. ELLIS and P. MORRIS, "Office Streamlining," IRIA, Rocquencourt, France, November 6–9, 1979.
13. G. J. NUTT and C. A. ELLIS, "Backtalk: An Office Environment Simulator," *Proceedings of the 1979 International Conference Communications*, Vol. 2, June 1979, pp. 22.3.1–22.3.5.
14. DELOITTE, HASKINS, and SELLS COMPANY, "Internal Accounting Control: An Overview of the DH&S Study and Evaluation Techniques," *DH&S*, 1979.
15. J. I. CASH, A. D. BAILEY, and A. B. WHINSTON, "A Survey of Techniques for Auditing EDP-Based Accounting Information Systems," *The Accounting Review*, October 1977, pp. 813–832.
16. A. D. BAILEY, J. H. GERLACH, R. P. MCAFEE, and A. B. WHINSTON, "TICOM II–The Internal Control Language," in A. R. Abdel-khalik, Ed., *Proceedings of the Florida Symposium on Internal Controls*, March 1981, in press.
17. A. D. BAILEY, J. H. GERLACH, R. P. MCAFEE, and A. B. WHINSTON, "Internal Controls in the Office of the Future," *IEEE Computer*, May 1981, pp. 59–70.
18. A. D. BAILEY, J. H. GERLACH, R. P. MCAFEE, and A. B. WHINSTON, "TICOM II: Theory," working paper, Krannert Graduate School of Management, Purdue Univeristy, West Lafayette, IN, 1980.

BIBLIOGRAPHY

KLING, R. W., "Social Analyses of Computing: Theoretical Perspective in Recent Empirical Research," *ACM Computing Surveys*, Vol. 12, No. 1 (March 1980), pp. 61–110.

WOHL, A. D., "A Review of Office Automation," *Datamation*, February 1980, pp. 117–119.

WOHL, A. D., "Replacing the Pad and the Pencil," *Datamation*, June 1980, pp. 169–176.

SECTION 13

QUANTITATIVE METHODOLOGIES FOR INDUSTRIAL ENGINEERS

CHAPTER 13.1

Mathematics for Industrial Engineers

WILLIAM E. BILES

Louisiana State University

13.1.1 INTRODUCTION

This chapter surveys the techniques of mathematical analysis that are useful to the industrial engineer in solving the kinds of problems confronted in designing, analyzing, and controlling industrial systems and processes. It covers those aspects of mathematics that are needed to understand the methodologies presented in other chapters of this handbook, especially those in Sections 13 and 14. Topics include linear algebra and matrices, calculus and differential equations (from the standpoint of continuous mathematics), difference equations and series analysis (from the standpoint of discrete mathematics), and complex variables and transform methods. More basic topics, such as algebra, geometry, and trigonometry, are omitted on the assumption that they are readily available to the industrial engineer in various mathematical handbooks.

13.1.2 LINEAR ALGEBRA AND MATRICES

Matrices

A "matrix" is defined as a rectangular array of numbers. It is denoted by a capital letter, and the elements of the matrix are denoted by lowercase letters. Let A be a matrix and a_{ij} the element in row i and column j. Therefore

$$A = \|a_{ij}\| = \begin{bmatrix} a_{11} & a_{12} & \dots a_{1j} & \dots a_{1n} \\ a_{21} & a_{22} & \dots a_{2j} & \dots a_{2n} \\ \vdots & \vdots & \vdots & \vdots \\ a_{i1} & a_{i2} & \dots a_{ij} & \dots a_{in} \\ \vdots & \vdots & \vdots & \vdots \\ a_{m1} & a_{m2} & \dots a_{mj} & \dots a_{mn} \end{bmatrix}$$

is an m by n matrix.

Example 13.1.1 The following matrix

$$A = \begin{pmatrix} 2 & 0.5 & 7 \\ 4 & -3 & 1 \end{pmatrix}$$

is of order (2, 3).

Operations and Properties

1. An mxn matrix is said to be of order (m, n).
2. Two matrices, A and B, are equal if and only if both are of the same order and $a_{ij} = b_{ij}$ for all i and j.

3. Matrices can be added or subtracted, provided that they are of same order. Then they are said to be conformable for addition (subtraction). Given matrices A and B,

$$C = A \pm B$$

defines a matrix C with elements

$$c_{ij} = a_{ij} \pm b_{ij}$$

Example 13.1.2

$$A = \begin{pmatrix} 1 & 3 \\ 5 & 1 \\ 8 & 2 \end{pmatrix}, \quad B = \begin{pmatrix} 7 & 4 \\ -1 & 3 \\ 5 & 0 \end{pmatrix}$$

$$C = A \pm B = \begin{pmatrix} 1+7 & 3+4 \\ 5-1 & 1+3 \\ 8+5 & 2+0 \end{pmatrix} = \begin{pmatrix} 8 & 7 \\ 4 & 4 \\ 13 & 2 \end{pmatrix}$$

4. Matrix addition is commutative.

$$A + B = B + A$$

5. Matrix addition is associative.

$$A + (B + C) = (A + B) + C = A + B + C$$

6. Given a matrix A and a scalar k, then scalar multiplication is defined as

$$kA = \|ka_{ij}\|$$

That is, the ijth element in matrix kA is the product of the scalar k and the element a_{ij} in matrix A.

Example 13.1.3

$$k = 1.2, \quad A = \begin{pmatrix} 2 & 0.5 & 7 \\ 4 & -3 & 1 \end{pmatrix}$$

$$kA = \begin{pmatrix} 2.4 & 0.6 & 8.4 \\ 4.8 & -3.6 & 1.2 \end{pmatrix}$$

7. Scalar multiplication is commutative.

$$kA = Ak$$

8. The product C of two matrices A and B is defined if and only if the number of columns in A is equal to the number of rows in B. Then A and B are said to be conformable for multiplication. The product

$$C = AB$$

has as its elements c_{ij} such that

$$c_{ij} = \sum_{k=1}^{n} a_{ik} b_{kj}, \qquad i = 1, \ldots, m \text{ and } j = 1, \ldots, r$$

Matrix A is of order (m, n), and B is of order (n, r). Then matrix C is of order (m, r).

Example 13.1.4

$$A = \begin{pmatrix} 2 & 0.5 & 7 \\ 4 & -3 & 1 \end{pmatrix}, \quad B = \begin{pmatrix} -2 & 5 \\ 0.5 & -3 \\ 4 & 0 \end{pmatrix}$$

$$C = AB = \begin{pmatrix} 24.25 & 8.5 \\ -5.5 & 29.0 \end{pmatrix}$$

9. Matrix multiplication is not necessarily commutative. That is, in general $AB \neq BA$.
10. Matrix multiplication is associative.

$$A(BC) = (AB)\,C = ABC$$

11. Matrix multiplication is distributive.

$$A(B + C) = AB + AC$$

12. The transpose of a matrix A is A^T, where A^T is obtained by interchanging the rows and columns of A. If A is of order (m, n), then A^T is of order (n, m). That is, if

$$A = \begin{bmatrix} a_{11} & a_{12} & \cdots a_{1n} \\ a_{21} & a_{22} & \cdots a_{2n} \\ \vdots & \vdots & \vdots \\ a_{m1} & a_{m2} & \cdots a_{mn} \end{bmatrix}$$

then

$$A^T = \begin{bmatrix} a_{11} & a_{21} & \cdots a_{m1} \\ a_{12} & a_{22} & \cdots a_{m2} \\ \vdots & \vdots & \vdots \\ a_{1n} & a_{2n} & \cdots a_{mn} \end{bmatrix}$$

Example 13.1.5

$$A = \begin{pmatrix} 2 & 0.5 & 7 \\ 4 & -3 & 1 \end{pmatrix}, \quad A^T = \begin{pmatrix} 2 & 4 \\ 0.5 & -3 \\ 7 & 1 \end{pmatrix}$$

13. If matrices A and B are conformable for addition,

$$(A + B)^T = A^T + B^T$$

14. If matrices A and B are conformable for multiplication, then

$$(AB)^T = B^T A^T$$

15. A matrix A is said to be symmetric if $A = A^T$.
16. A matrix A is said to be skew symmetric if $A = -A^T$.
17. A matrix of order (n, n) is said to be a square matrix of order n.
18. A scalar matrix is a square matrix having all elements along the main diagonal from upper left to lower right equal to the scalar k. That is,

$$A = \begin{bmatrix} k & 0 & 0 \ldots 0 \\ 0 & k & 0 \ldots 0 \\ 0 & 0 & k \ldots 0 \\ \vdots & \vdots & \vdots \quad \vdots \\ 0 & 0 & 0 \ldots k \end{bmatrix}$$

If A is a scalar matrix with scalar k, then

$$AB = kB$$

19. The identity matrix, denoted by I, is a scalar matrix having $k = 1$. Properties associated with the identity matrix include the following:

$$I_m A = A I_n = A$$
$$I^T = I$$

where I_m and I_n are of order m and n, respectively.

20. The elements in the identity matrix

$$I = \|\delta_{ij}\|$$

are called the "Kroneker delta." That is,

$$\delta_{ij} = \begin{cases} 1 & i = j \\ 0 & i \neq j \end{cases}$$

21. The null matrix, denoted by 0, has all zeros as its elements. Properties associated with the null matrix include the following:

$$A + 0 = 0 + A = A$$
$$A - A = 0$$
$$A0 = 0A = 0$$

Determinant of a Matrix

Given any square matrix A, there is a number called the "determinant," which is denoted by $|A|$. If A is of order 2, so that

$$A = \begin{bmatrix} a_{11} & a_{12} \\ a_{21} & a_{22} \end{bmatrix}$$

then the determinant is given by

$$|A| = a_{11}a_{22} - a_{21}a_{12}$$

Define the cofactor A_{ij} of the element a_{ij} of matrix A as the determinant of the matrix that is formed by deleting row i and column j of A, which is called the "minor," multiplied by the quantity $(-1)^{i+j}$. Then, for a square matrix of order 3, the determinant can be computed as follows:

$$A = \begin{bmatrix} a_{11} & a_{12} & a_{13} \\ a_{21} & a_{22} & a_{23} \\ a_{31} & a_{32} & a_{33} \end{bmatrix}$$

We compute the cofactors A_{ij} of the elements a_{ij} in any row (say, row i), multiply them by the corresponding a_{ij}, and add. For example, using row 1,

$$\begin{aligned} |A| &= a_{11}A_{11} + a_{12}A_{12} + a_{13}A_{13} \\ &= a_{11}(-1)^{1+1} \begin{bmatrix} a_{22} & a_{23} \\ a_{32} & a_{33} \end{bmatrix} + a_{12}(-1)^{1+2} \begin{bmatrix} a_{21} & a_{23} \\ a_{31} & a_{33} \end{bmatrix} \\ &\quad + a_{13}(-1)^{1+3} \begin{bmatrix} a_{21} & a_{22} \\ a_{31} & a_{32} \end{bmatrix} \end{aligned}$$

This technique is called "cofactor expansion." The determinant $|A|$ can be computed by the cofactor expansion of any row or column of the matrix A. For any square matrix of order n, the determinant $|A|$ can be computed by the cofactor expansion of any row or column in A. That is, for any row i

$$|A| = \sum_{j=1}^{n} a_{ij}A_{ij}$$

where the A_{ij} are determinants of order $n - 1$, which are computed in the same manner. That is, $n - 1$ successive reductions are needed until there are only matrices of order 2 from which the simple determinant $(a_{11}a_{22} - a_{21}a_{12})$ is computed.

Example 13.1.6 Find the determinant of the matrix A by cofactor expansion.

$$A = \begin{pmatrix} 1 & 3 & 6 \\ 5 & 1 & -2 \\ 2 & 6 & 8 \end{pmatrix}$$

$$\begin{aligned}
|A| &= (1)\,A_{11} + (3)\,A_{12} + (6)\,A_{13} \\
&= (1)(-1)^{1+1}\begin{vmatrix} 1 & -2 \\ 6 & 8 \end{vmatrix} + (3)(-1)^{1+2}\begin{vmatrix} 5 & -2 \\ 2 & 8 \end{vmatrix} + (6)(-1)^{1+3}\begin{vmatrix} 5 & 1 \\ 2 & 6 \end{vmatrix} \\
&= (1)(1)[(1)(8) - (-2)(6)] + (3)(-1)[(5)(8) - (-2)(2)] + (6)(1)[(5)(6) - (1)(2)] \\
&= (1)(1)(20) + (3)(-1)(44) + (6)(1)(28) \\
&= 20 - 132 + 168 \\
&= 56
\end{aligned}$$

Nonsingular Matrix

A matrix A is said to be singular if its determinant is zero and nonsingular if $|A| \neq 0$.

Rank of a Matrix

Let A be a matrix of order (m, n), where $m < n$. Then A is said to be of rank k if there exists at least one nonsingular square submatrix of A of order k and if all square submatrices of order greater than k are singular. Therefore the rank of A is found by evaluating the determinants of all square submatrices within A. If A is of order (m, n), then $k \leqslant m$ and $k \leqslant n$.

Inverse of a Square Matrix

Let A be a square matrix of order m. The inverse matrix of A, denoted by A^{-1}, is a square matrix of order m such that

$$A^{-1}A = AA^{-1} = I$$

where I is the identity matrix of order m. The A^{-1} exists only if A is nonsingular. The inverse of A is given by

$$A^{-1} = \frac{\text{adj}\,(A)}{|A|}$$

where adj (A) is called the "adjoint" of the matrix A. Clearly A^{-1} is not defined if $|A| = 0$. The following properties are associated with the inverse of a matrix:

1. If A is a nonsingular square matrix of order m with off-diagonal elements equal to zero, then

$$A^{-1} = \begin{bmatrix} 1/a_{11} & 0 & \dots & 0 & \dots & 0 \\ 0 & 1/a_{22} & \dots & 0 & \dots & 0 \\ \vdots & \vdots & & \vdots & & \vdots \\ 0 & 0 & \dots & 1/a_{ii} & \dots & 0 \\ \vdots & \vdots & & \vdots & & \vdots \\ 0 & 0 & & 0 & \dots & 1/a_{mm} \end{bmatrix}$$

2. If a nonsingular matrix A is symmetric, then A^{-1} is symmetric.
3. If A is a nonsingular square matrix, then

$$(A^{-1})^{-1} = A$$

4. If A is a nonsingular square matrix, then

$$(A^T)^{-1} = (A^{-1})^T$$

5. If A and B are nonsingular square matrices of the same order, then

$$(AB)^{-1} = B^{-1}A^{-1}$$

Adjoint Matrix

The adjoint of a square matrix A, denoted adj (A), is defined as the transpose of the matrix of co-factors. Therefore

$$\text{adj}(A) = \begin{bmatrix} A_{11} & A_{21} \ldots A_{m1} \\ A_{12} & A_{22} \ldots A_{m2} \\ \vdots & \vdots \quad\quad \vdots \\ A_{1n} & A_{2n} \ldots A_{mn} \end{bmatrix}$$

The following properties are associated with the adjoint of a matrix:

1. The adjoint of a scalar matrix is a scalar matrix.
2. If A is a symmetric matrix, adj (A) is also a symmetric matrix.
3. If A is a square matrix of order m, then

$$A[\text{adj}(A)] = |A|\,I$$

$$A[\text{adj}(A)] = [\text{adj}(A)]\,A$$

4. If A and B are nonsingular matrices, then

$$\text{adj}(AB) = \text{adj}(B) \times \text{adj}(A)$$

5. If A is a square matrix of order m, then

$$|\text{adj}(A)| = |A|^{m-1}$$

Example 13.1.7 Find the adjoint of the matrix

$$A = \begin{pmatrix} 6 & -1 & 2 \\ 4 & 5 & -3 \\ 2 & 0.5 & 1 \end{pmatrix}$$

$$\text{adj}(A) = \begin{pmatrix} A_{11} & A_{12} & A_{13} \\ A_{21} & A_{22} & A_{23} \\ A_{31} & A_{32} & A_{33} \end{pmatrix}^T = \begin{pmatrix} A_{11} & A_{21} & A_{31} \\ A_{12} & A_{22} & A_{32} \\ A_{13} & A_{23} & A_{33} \end{pmatrix}$$

$$A_{11} = (-1)^{1+1} \begin{vmatrix} 5 & -3 \\ 0.5 & 1 \end{vmatrix} = (1)[(5)(1) - (-3)(0.5)] = 6.5$$

$$A_{12} = (-1)^{1+2} \begin{vmatrix} 4 & -3 \\ 2 & 1 \end{vmatrix} = (-1)[(4)(1) - (-3)(2)] = -10$$

$$A_{13} = 8, \quad A_{21} = 2, \quad A_{22} = 2, \quad A_{23} = -5, \quad A_{31} = -7,$$

$$A_{32} = -2, \quad A_{33} = 34$$

Therefore

$$\text{adj}\,(A) = \begin{pmatrix} 6.5 & 2 & -7 \\ -10 & 2 & -2 \\ 8 & -5 & 34 \end{pmatrix}$$

Example 13.1.8 Find the inverse of the matrix A in Example 13.1.7.

$$A^{-1} = \frac{\text{adj}\,(A)}{|A|}$$

$$|A| = (6)\,A_{11} + (-1)\,A_{12} + (2)\,A_{13} = 65$$

$$A^{-1} = \frac{1}{65}\begin{pmatrix} 6.5 & 2 & -7 \\ -10 & 2 & -2 \\ 8 & -5 & 34 \end{pmatrix} = \begin{pmatrix} 0.1000 & 0.0308 & -0.1080 \\ -0.1539 & 0.0308 & -0.0308 \\ 0.1231 & -0.0769 & 0.5230 \end{pmatrix}$$

Vectors

A matrix with one row or one column is called a "vector." Let $\mathbf{x}$ denote an n-dimensional column vector. Therefore

$$\mathbf{x} = \begin{bmatrix} x_1 \\ x_2 \\ \vdots \\ x_n \end{bmatrix}$$

where the quantities x_i are real numbers. Vectors are matrices of order $(n, 1)$, so that all the properties previously stated for matrices relative to addition (subtraction) and multiplication apply.

Linear Combination of Vectors

If $\mathbf{x}_1, \mathbf{x}_2, \ldots, \mathbf{x}_m$ are n-dimensional vectors, and if $\alpha_1, \alpha_2, \ldots, \alpha_m$ are scalars, then the linear combination given by

$$\mathbf{x}_{m+1} = \alpha_1 \mathbf{x}_1 + \alpha_2 \mathbf{x}_2 + \cdots + \alpha_m \mathbf{x}_m$$

is also an n-dimensional vector.

Length of a Vector

If the origin is defined by 0, where

$$0 = \begin{bmatrix} 0 \\ 0 \\ \vdots \\ 0 \end{bmatrix}$$

then the length of a vector $\mathbf{x}$, denoted $\|\mathbf{x}\|$, is given by

$$\|\mathbf{x}\| = \sqrt{x_1^2 + x_2^2 + \cdots + x_n^2} = \sqrt{\mathbf{x}^T\mathbf{x}}$$

where the positive value of the square root is taken.

Example 13.1.9

$$\mathbf{x} = \begin{pmatrix} 2 \\ 3 \\ -1 \end{pmatrix}$$

$$\|\mathbf{x}\| = \sqrt{(2)^2 + (3)^2 + (-1)^2} = \sqrt{14} = 3.7417$$

Angle Between Vectors

Let $\mathbf{x}_1$ and $\mathbf{x}_2$ be two n-dimensional vectors. The angle θ between $\mathbf{x}_1$ and $\mathbf{x}_2$ is given by

$$\cos(\theta) = \frac{\mathbf{x}_1^T \mathbf{x}_2}{\|\mathbf{x}_1\| \, \|\mathbf{x}_2\|}$$

If $\cos(\theta) = 0$, then $\theta = 90°$ and $\mathbf{x}_1$ and $\mathbf{x}_2$ are said to be orthogonal.

Example 13.1.10

$$\mathbf{x}_1 = \begin{pmatrix} 2 \\ 3 \\ -1 \end{pmatrix} \quad \mathbf{x}_2 = \begin{pmatrix} -0.5 \\ 2 \\ 1 \end{pmatrix}$$

$$\cos\theta = \frac{(2, 3, -1)\begin{pmatrix} -0.5 \\ 2 \\ 1 \end{pmatrix}}{\sqrt{(2)^2 + (3)^2 + (-1)^2} \times \sqrt{(-0.5)^2 + (2)^2 + (1)^2}} = \frac{4}{(3.7417)(2.2910)} = 0.467$$

$$\theta = 1.085 \text{ radians} = 62.15°$$

Distance Between Vectors

Let $\mathbf{x}_1$ and $\mathbf{x}_2$ be two n-dimensional vectors. The distance between $\mathbf{x}_1$ and $\mathbf{x}_2$ is defined to be the length of the vector $\mathbf{x}_1 - \mathbf{x}_2$, or

$$\|\mathbf{x}_1 - \mathbf{x}_2\| = \sqrt{(\mathbf{x}_1 - \mathbf{x}_2)^T(\mathbf{x}_1 - \mathbf{x}_2)}$$

Example 13.1.11

$$\mathbf{x}_1 = \begin{pmatrix} 2 \\ 3 \\ -1 \end{pmatrix} \quad \mathbf{x}_2 = \begin{pmatrix} -0.5 \\ 2 \\ 1 \end{pmatrix}.$$

$$\|\mathbf{x}_1 - \mathbf{x}_2\| = \sqrt{(2 + 0.5)^2 + (3 - 2)^2 + (-1 - 1)^2} = \sqrt{11.25} = 3.354$$

Vector Spaces

Let V_n be a set of n-dimensional vectors $\mathbf{x}_i$ such that the following operations are defined: (1) scalar multiplication $k\mathbf{x}_i$, where k is a scalar, and (2) vector addition $\mathbf{x}_i + \mathbf{x}_j$. Then V_n is called a "vector space" over the field of real numbers. Important properties of vector spaces are as follows:

1. A subspace S_n of V_n is defined as a subset of V_n that is itself a vector space.
2. Every vector space V_n contains the zero vector.
3. Let V_n be a vector space containing the vectors $\mathbf{x}_1, \mathbf{x}_2, \ldots, \mathbf{x}_m$. If any vector $\mathbf{x}_i$ belonging to V_n can be expressed as a linear combination of $\mathbf{x}_1, \mathbf{x}_2, \ldots, \mathbf{x}_m$

$$\mathbf{x}_i = \sum_{j=1}^{m} k_j \mathbf{x}_j$$

where the k_j's are scalars, then the vectors $\mathbf{x}_1, \mathbf{x}_2, \ldots, \mathbf{x}_m$ are said to span V_n, or that V_n is spanned by the vectors $\mathbf{x}_1, \mathbf{x}_2, \ldots, \mathbf{x}_m$.

4. The space spanned by any set of n-dimensional vectors is a vector space.

5. If $\mathbf{x}_1, \mathbf{x}_2, \ldots, \mathbf{x}_m$ is a set of distinct vectors belonging to V_n such that

$$\sum_{j=1}^{m} k_j \mathbf{x}_j = 0$$

if and only if $k_j = 0$ for all j, then the vectors $\mathbf{x}_1, \mathbf{x}_2, \ldots, \mathbf{x}_m$ are said to be linearly independent.

6. Although linearly independent vectors are not necessarily orthogonal, orthogonal vectors are necessarily linearly independent.

7. If $\mathbf{x}_1, \mathbf{x}_2, \ldots, \mathbf{x}_m$ are a set of linearly independent vectors, then every subset of these vectors is also linearly independent.

8. If $\mathbf{x}_1, \mathbf{x}_2, \ldots, \mathbf{x}_n$ are linearly independent and span V_n, then $\mathbf{x}_1, \mathbf{x}_2, \ldots, \mathbf{x}_n$ are said to form a basis for V_n.

Convex Sets

Let $\mathbf{x}_1, \mathbf{x}_2, \ldots, \mathbf{x}_m$ be a set of vectors belonging to the vector space V_n. If

$$\mathbf{y}_1 = \sum_{j=1}^{m} \lambda_j \mathbf{x}_j$$

where λ_j is a nonnegative scalar for all j and

$$\sum_{j=1}^{m} \lambda_j = 1$$

then $\mathbf{y}_1$ is said to be formed by a convex combination of $\mathbf{x}_1, \mathbf{x}_2, \ldots, \mathbf{x}_m$. The set of vectors comprising a space $\mathbf{X}_n$ (not necessarily a vector space) is a convex set if every convex combination of vectors in $\mathbf{X}_n$ yields another vector in $\mathbf{X}_n$. A vector space V_n is necessarily convex, but a convex set of vectors is not necessarily a vector space.

Hyperplane

Let $\mathbf{x}_j$ be an n-dimensional column vector and $\mathbf{b}_j$ an n-dimensional column vector. Then

$$\mathbf{b}_j^T \mathbf{x}_j = k$$

defines a hyperplane in n-dimensional space, where k is a scalar. A hyperplane in two-dimensional space is a straight line, whereas that in three-dimensional space is a plane. A hyperplane divides a space into two half spaces,

$$\mathbf{b}_j \mathbf{x}_j \leqslant k$$

and

$$\mathbf{b}_j \mathbf{x}_j > k$$

A half space is convex.

Let B be a matrix of the m row vectors $\mathbf{b}_j$ and $\mathbf{k}$ a column vector of the m scalars k_j. Then

$$B\mathbf{x}_i \leqslant \mathbf{k}$$

defines the intersection of m half spaces. If this intersection is nonempty, then it is a convex set.

Eigenvectors and Eigenvalues

Let A be a symmetric matrix of order n, $\mathbf{x}$ be an n-dimensional column vector, and λ be a scalar such that

$$A\mathbf{x} = \lambda \mathbf{x}$$

for some $\mathbf{x} \neq 0$. The nonzero vectors $\mathbf{x}_i$ satisfying the stated condition are called "eigenvectors," "characteristic vectors," "latent vectors," or "invariant vectors." The scalars λ_i satisfying the condition are called "eigenvalues," "characteristic roots," or "latent roots."

Expressing the preceding equation in the form

$$(\lambda I - A)\,\mathbf{x} = 0$$

the solution vector (eigenvector) is nonzero if and only if

$$|\lambda I - A| = 0$$

The evaluation of the determinant $|\lambda I - A|$ results in a polynomial in λ of degree n and is known as the "characteristic equation" of the matrix A.

Example 13.1.12 Determine the eigenvalues of the following matrix:

$$A = \begin{pmatrix} -2 & 2 & -3 \\ 2 & 1 & -6 \\ -1 & -2 & 0 \end{pmatrix}$$

$$|\lambda I - A| = \begin{vmatrix} \lambda + 2 & -2 & 3 \\ -2 & \lambda - 1 & 6 \\ 1 & 2 & \lambda \end{vmatrix}$$

$$= (\lambda + 2)(-1)^{1+1} \begin{vmatrix} \lambda - 1 & 6 \\ 2 & \lambda \end{vmatrix} + (-2)(-1)^{1+2} \begin{vmatrix} -2 & 6 \\ 1 & \lambda \end{vmatrix} + (3)(-1)^{1+3} \begin{vmatrix} -2 & \lambda - 1 \\ 1 & 2 \end{vmatrix}.$$

$$= (\lambda + 2)(1)[(\lambda - 1)(\lambda) - (6)(2)] + (-2)(-1)[(-2)(\lambda) - (6)(1)]$$
$$+ (3)(1)[(-2)(2) - (\lambda - 1)(1)]$$
$$= (\lambda + 2)(1)[\lambda^2 - \lambda - 12] + (2)[-2\lambda - 6] + (3)[-4 - \lambda + 1]$$
$$= \lambda^3 + \lambda^2 - 14\lambda - 24 - 4\lambda - 12 - 9 - 3\lambda$$
$$= \lambda^3 + \lambda^2 - 21\lambda - 45$$

Setting $|\lambda I - A| = 0$, we get

$$\lambda^3 + \lambda^2 - 21\lambda - 45 = 0$$
$$(\lambda - 5)(\lambda + 3)^2 = 0$$

Thus the eigenvalues of A are

$$\lambda_1 = 5$$
$$\lambda_2 = \lambda_3 = -3$$

Example 13.1.13 In Example 13.1.12 find the eigenvector associated with the eigenvalue $\lambda_1 = 5$.

$$A\mathbf{x} = \lambda\mathbf{x}$$

For $\lambda = 5$

$$-2x_1 + 2x_2 - 3x_3 = 5x_1$$
$$2x_1 + x_2 - 6x_3 = 5x_2$$
$$-x_1 - 2x_2 = 5x_3$$

Rearranging, we get

$$-7x_1 + 2x_2 - 3x_3 = 0$$

$$2x_1 - 4x_2 - 6x_3 = 0$$

$$-x_1 - 2x_2 - 5x_3 = 0$$

The solution of these equations yields the eigenvector

$$\mathbf{x}_1 = \begin{pmatrix} 1 \\ 2 \\ -1 \end{pmatrix}$$

The following properties are associated with eigenvectors and eigenvalues:

1. If $\mathbf{x}_i$ and $\mathbf{x}_j$ are two eigenvectors corresponding to the distinct eigenvalues λ_i and λ_j for a symmetric matrix A, then $\mathbf{x}_i$ and $\mathbf{x}_j$ are mutually orthogonal.

2. If A is a square matrix such that

$$A^T = A^{-1}$$

then A is said to be an orthogonal matrix.

3. If A is a symmetric matrix of order n with $k \leqslant n$ identical eigenvalues each equal to the eigenvalue λ_j, then λ_j is said to have multiplicity k.

4. If A is a symmetric matrix of order n with the eigenvalues $\lambda_1, \lambda_2, \ldots, \lambda_n$, which need not be distinct, then there exists a set of n mutually orthogonal eigenvectors associated with $\lambda_1, \lambda_2, \ldots, \lambda_n$.

5. Two square matrices, A and R, are said to be congruent if there exists a nonsingular matrix B such that

$$R = B^T AB$$

6. Two matrices, A and R, are said to be similar if there exists a nonsingular matrix B such that

$$R = B^{-1}AB$$

7. If A is a symmetric matrix and B an orthogonal matrix such that

$$R = B^T AB$$

then R and A have the same eigenvalues.

8. Let A be a symmetric matrix of order n with the eigenvalues $\lambda_1, \lambda_2, \ldots, \lambda_n$, and let B be a matrix, the columns of which form a set of mutually orthogonal eigenvectors $\mathbf{x}_i$ associated with the eigenvalues λ_i. Then the matrix R is a diagonal matrix where

$$R = B^T AB$$

Quadratic Forms

Let $\mathbf{x}$ be an n-dimensional column vector and A a square matrix of order n. The polynomial equation given by

$$q = \mathbf{x}^T A\mathbf{x}$$

is called a "quadratic form." In terms of the elements x_i in $\mathbf{x}$ and a_{ij} in A, we have

$$q = \sum_{i=1}^{n} \sum_{j=1}^{n} a_{ij}x_i x_j$$

The term "quadratic" refers to the fact that q as defined in the preceding equations is a homogeneous quadratic equation in the variables $x_1, x_2, \ldots, x_n$. The matrix A is called the "matrix of the quadratic form." If $|A|$ is zero, the quadratic form is singular; it is nonsingular if $|A| \neq 0$. If A is a symmetric matrix, then q becomes

$$q = \sum_{i=1}^{n} a_{ii}x_i^2 + 2\sum_{i=1}^{n-1}\sum_{j=i+1}^{n} a_{ij}x_ix_j$$

The following properties are associated with quadratic forms:

1. Let A be a symmetric matrix of order n and $\mathbf{x}$ an n-dimensional column vector. Then the quadratic form

$$q = \mathbf{x}^T A\mathbf{x}$$

can be reduced a sum of squares by the transformation

$$\mathbf{x} = B\mathbf{y}$$

where B is a matrix of order n, the columns of which form a set of mutually orthogonal eigenvectors in A, and $\mathbf{y}$ is an n-dimensional vector.

2. If A is a square matrix of order n, then $|A_i|$ is called the "leading principal minor" of order $i \leqslant n$, where

$$A_i = \begin{bmatrix} a_{11} & a_{12} & a_{13} \cdots a_{1i} \\ a_{21} & a_{22} & a_{23} \cdots a_{2i} \\ a_{31} & a_{32} & a_{33} \cdots a_{3i} \\ \vdots & \vdots & \vdots \quad \vdots \\ a_{i1} & a_{i2} & a_{i3} \cdots a_{ii} \end{bmatrix}$$

The leading principal minor of order i of a square matrix of order n is obtained by deleting the last $(n - i)$ rows and columns of A and taking the determinant of the resulting matrix. The leading principal minor of order zero is defined to be unity.

$$|A_0| = 1$$

3. If A is a nonsingular symmetric matrix of the quadratic form q, with nonsingular leading principal minors,

$$q = \mathbf{x}^T A\mathbf{x}$$

then p_{nii} in the sum-of-squares expression

$$q = \sum_{i=1}^{n} p_{nii}z_{ni}^2$$

can be expressed as

$$p_{nii} = \frac{|A_i|}{|A_{i-1}|}$$

where

$$A_i = \begin{bmatrix} a_{11} & a_{12} \cdots a_{1i} \\ a_{12} & a_{22} \cdots a_{2i} \\ \vdots & \vdots \quad \vdots \\ a_{1i} & a_{2i} \cdots a_{ii} \end{bmatrix}$$

and

$$|A_0| = 1$$

4. A quadratic form $\mathbf{x}^T A\mathbf{x}$ is said to be

a. Positive definite if and only if $\mathbf{x}^T A\mathbf{x} > \mathbf{0}$ for all vectors $\mathbf{x} \neq \mathbf{0}$.

b. Positive semidefinite if and only if $\mathbf{x}^T A\mathbf{x} \geqslant \mathbf{0}$ for all vectors $\mathbf{x} \neq \mathbf{0}$ and $\mathbf{x}^T A\mathbf{x} = \mathbf{0}$ for at least one $\mathbf{x} \neq \mathbf{0}$.

c. Negative definite if and only if $-\mathbf{x}^T A\mathbf{x}$ is positive definite.

d. Negative semidefinite if and only if $-\mathbf{x}^T A\mathbf{x}$ is positive semidefinite.

5. If A is the matrix of a quadratic form, then A is positive definite, positive semidefinite, negative definite, or negative semidefinite if the quadratic form is positive definite, positive semidefinite, negative definite, or negative semidefinite.

6. The quadratic form that is not positive definite, negative definite, positive semidefinite, or negative semidefinite is said to be indefinite.

7. The quadratic form $\mathbf{x}^T A\mathbf{x}$ is positive definite if and only if

$$|A_i| > 0, \qquad i = 1, 2, \ldots, n$$

where A is a symmetric matrix of order n.

8. The quadratic form $\mathbf{x}^T A\mathbf{x}$ is negative definite if and only if

$$(-1)^i|A_i| > 0, \qquad i = 1, 2, \ldots, n$$

where A is a symmetric matrix of order n.

9. The quadratic form $\mathbf{x}^T A\mathbf{x}$, where A is a symmetric matrix, is

a. Positive definite if and only if every eigenvalue for A is positive.

b. Negative definite if and only if every eigenvalue for A is negative.

c. Positive semidefinite if and only if every eigenvalue for A is nonnegative and at least one eigenvalue is zero.

d. Negative semidefinite if and only if every eigenvalue for A is nonpositive and at least one eigenvalue is zero.

Simultaneous Linear Equations

Let A be a matrix of order (m, n) consisting of the known coefficients a_{ij}, $\mathbf{x}$ an n-dimensional vector of unknowns, and $\mathbf{b}$ an n-dimensional vector of known constants. Then

$$A\mathbf{x} = \mathbf{b}$$

represents a set of n simultaneous linear equations in n unknowns. In the special case where $m = n$, the solution is simply

$$\mathbf{x} = A^{-1}\mathbf{b}$$

if $|A| \neq 0$. If A is square, but $|A| \neq 0$, then Cramer's rule can be employed to find $\mathbf{x}$, where

$$x_i = \frac{|A_i|}{|A|}$$

where A_i is the matrix formed by replacing the ith column in A by the vector $\mathbf{b}$.

Another method for solving a set of n simultaneous equations in n unknowns is called "Gaussian reduction." Given the system of equations

$$\begin{array}{ccc} a_{11}x_1 + \cdots + a_{1n}x_n = b_1 \\ \vdots \quad \vdots \quad \vdots \\ a_{n1}x_1 + \cdots + a_{nn}x_n = b_n \end{array}$$

where $a_{11} \neq 0$ (the columns in A can always be rearranged so that $a_{11} \neq 0$), divide the first equation by a_{11} and use it to eliminate x_1 from the remaining equations $2, \ldots, n$. This produces the set of equations

$$\begin{array}{r} x_1 + a'_{12}x_2 + \cdots + a'_{1n}x_n = b'_1 \\ a'_{22}x_2 + \cdots + a'_{2n}x_n = b'_2 \\ \vdots \quad \vdots \quad \vdots \\ a'_{n2}x_2 + \cdots + a'_{nn}x_n = b'_n \end{array}$$

where

$$a'_{1j} = \frac{a_{1j}}{a_{11}}, \qquad j = 2, \ldots, n.$$

$$a'_{ij} = a_{ij} - \frac{a_{1j}}{a_{11}} a_{i1}, \qquad i = 2, \ldots, n \text{ and } j = 2, \ldots, n$$

$$b'_1 = \frac{b_1}{a_{11}} \text{ and } b'_i = b_i - \frac{a_{i1}}{a_{11}} b_1, \qquad i = 2, \ldots, n$$

In this new system of equations, divide equation 2 by a'_{22} and eliminate x_2 in equations $3, \ldots, n$. Repeat this process for each of the n equations until finally

$$\begin{aligned} x_1 + g_{12}x_2 + \cdots + g_{1n}x_n &= k_1 \\ x_2 + \cdots + g_{2n}x_n &= k_2 \\ \cdots\cdots & \\ x_n &= k_n \end{aligned}$$

Then $x_n = k_n$ is a solution for x_n. Substitute this result into the $(n - 1)$th equation and solve for x_{n-1}. Repeat this process until a solution for x_1 is found.

If we have a set of m equations in n unknowns with $m < n$, then there are $n - m$ arbitrary values of x_i. It is necessary to set $n - m$ of the x_i equal to constant values and solve for the remaining m values of x_i. If the constant values are zero, the solutions are called "basic solutions."

Example 13.1.14 Solve the following system of equations by Gaussian reduction:

$$\begin{aligned} 2x_1 + 6x_2 - x_3 &= 4 \\ 3x_1 - 2x_2 - x_3 &= 1 \\ 5x_1 + 9x_2 - 2x_3 &= 12 \end{aligned}$$

Multiply the first equation by $\frac{1}{2}$; then multiply the new first equation by -3 and add it to the second equation. This yields the reduced equations

$$\begin{aligned} x_1 + 3x_2 - \tfrac{1}{2} x_3 &= 2 \\ -11x_2 + \tfrac{1}{2} x_3 &= -5 \\ 5x_1 + 9x_2 - 2x_3 &= 12 \end{aligned}$$

Multiply the new first equation by -5 and add it to the third equation, yielding

$$\begin{aligned} x_1 + 3x_2 - \tfrac{1}{2} x_3 &= 2 \\ -11x_2 + \tfrac{1}{2} x_3 &= -5 \\ -6x_2 + \tfrac{1}{2} x_3 &= 2 \end{aligned}$$

Now multiply the new second equation by $-\frac{1}{11}$; then multiply that result by 6 and add it to the reduced third equation. This operation yields the further reduced equations

$$\begin{aligned} x_1 + 3x_2 - \tfrac{1}{2} x_3 &= 2 \\ x_2 - \tfrac{1}{22}x_3 &= \tfrac{5}{11} \\ 5x_3 &= 104 \end{aligned}$$

Now multiply this reduced third equation by $\frac{1}{5}$. This yields the solution $x_3 = \frac{104}{5}$. Substituting this result into the reduced second equation,

$$x_2 - \frac{1}{22}\left(\frac{104}{5}\right) = \frac{5}{11}$$

$$x_2 = \frac{5}{11} + \frac{104}{110} = \frac{50}{110} + \frac{104}{110} = \frac{154}{110}$$

Substituting these solutions for x_2 and x_3 into the reduced first equation,

$$x_1 + 3\left(\frac{154}{110}\right) - \frac{1}{2}\left(\frac{104}{5}\right) = 2$$

$$x_1 = 2 - \frac{462}{110} + \frac{104}{10} = \frac{220}{110} - \frac{462}{110} + \frac{1144}{110} = \frac{902}{110}$$

One can easily verify that the system of equations is satisfied by this solution.

13.1.3 CALCULUS AND DIFFERENTIAL EQUATIONS

Basic Definitions

Variables

A "variable" is a symbol, such as x, that may take on any value in some specified set of numbers. The set of values over which x may vary is called the "domain" of x. In most engineering applications, we refer to variables in some interval of numbers, such as the following:

1. Open interval:

$$(a, b) = \{x : a < x < b\}$$

2. Half-open interval:

$$[a, b) = \{x : a \leqslant x < b\}$$

or

$$(a, b] = \{x : a < x \leqslant b\}$$

3. Closed interval:

$$[a, b] = \{x : a \leqslant x \leqslant b\}$$

Function

Let X and Y be nonempty sets of variables. Let f be a collection of ordered pairs (x, y) with $x \in X$ and $y \in Y$ such that for every $x \in X$ there is a unique $y \in Y$. Then f is said to be a function. The notation

$$y = f(x)$$

is read "y is a function f of x."

Example 13.1.15 Let X be the set of real numbers. Then

$$y = \tfrac{1}{3}x - 3$$

is a function.

Independent and Dependent Variables

The first variable, x, in the ordered (x, y) is usually called the "independent variable," or "argument," of the function f. The second variable, y, is called the "dependent variable."

Functions of Several Independent Variables

Often we refer to a function that depends on several independent variables. In general, such a function is written as

$$y = f(x_1, x_2, \ldots, x_n)$$

Example 13.1.16 The volume v of a right circular cylinder is

$$v = \pi r^2 h$$

where r is the radius of the cross-sectional area and h is the length of the cylinder.

Limit

Suppose a function f is defined for values of x near a. We say that L is the limit of $f(x)$ as x approaches a, and we write this statement as

$$L = \lim_{x \to a} f(x)$$

The following properties are associated with limits (where all limits are taken as $x \to a$):

1. $\lim (mx + b) = ma + b$
2. $\lim [f(x) + g(x)] = [\lim f(x)] + [\lim g(x)]$
3. $\lim [f(x)\, g(x)] = [\lim f(x)]\, [\lim g(x)]$
4. $\lim [kf(x)] = k\, [\lim f(x)]$
5. $\lim \left[\dfrac{f(x)}{g(x)}\right] = \dfrac{[\lim f(x)]}{[\lim g(x)]}$
6. $\lim x^n = a^n$

There are several special results when a is equal to infinity, written as $x \to \infty$ In particular, if $\lim_{x \to \infty} f(x) = \infty$, then $\lim_{x \to \infty} \dfrac{1}{f(x)} = 0$.

Differential Calculus

Derivative of a Function

Let $y = f(x)$ define a function f. If the limit

$$\frac{dy}{dx} = \lim_{\Delta x \to 0} \frac{\Delta y}{\Delta x}$$

meaning

$$\frac{dy}{dx} = f'(x) = \lim_{\Delta x \to 0} \frac{f(x + \Delta x) - f(x)}{\Delta x}$$

exists and is finite, we call this limit the "derivative" of y with respect to x and say that y is "differentiable" at x. The symbol dy/dx is called the "derivative with respect to x," and Δx and Δy define changes in x and y, respectively.

The following operations are associated with derivatives:

1. The derivative of a constant is zero; that is, if $y = c$, $dy/dx = 0$.
2. If n is a positive integer, the derivative of x^n is nx^{n-1}; that is, if $y = x^n$

$$\frac{dy}{dx} = nx^{n-1}$$

3. $\dfrac{d(u \pm v)}{dx} = \dfrac{du}{dx} \pm \dfrac{dv}{dx}$
4. $\dfrac{d(uv)}{dx} = u\dfrac{dv}{dx} + v\dfrac{du}{dx}$
5. $\dfrac{d(u^n)}{dx} = nu^{n-1}\dfrac{du}{dx}$

6. $\dfrac{d(u/v)}{dx} = \dfrac{v(du/dx) - u(dv/dx)}{v^2}$

7. $\dfrac{d(\sin u)}{dx} = \cos u \dfrac{du}{dx}$

8. $\dfrac{d(\cos u)}{dx} = -\sin u \dfrac{du}{dx}$

9. $\dfrac{d(\tan u)}{dx} = \sec^2 u \dfrac{du}{dx}$

10. $\dfrac{d(\cot u)}{dx} = -\csc^2 u \dfrac{du}{dx}$

11. $\dfrac{d(\sec u)}{dx} = \sec u \tan u \dfrac{du}{dx}$

12. $\dfrac{d(\csc u)}{dx} = -\csc u \cot u \dfrac{du}{dx}$

13. $\dfrac{d(e^u)}{dx} = e^u \dfrac{du}{dx}$

14. $\dfrac{d(\ell n\, u)}{dx} = \dfrac{1}{u} \dfrac{du}{dx}$

15. $\dfrac{d(a^u)}{dx} = a^u \ell n\, u \dfrac{du}{dx}$

The second derivative of a function f is given by

$$\frac{d^2f(x)}{dx^2} = f''(x) = \frac{d}{dx} f'(x)$$

Differentials

The differential of x, written dx, is an independent variable such that

$$-\infty < dx < +\infty$$

Similarly, the differential of y is dy, and it represents the function

$$dy = f'(x)\, dx$$

where $f'(x)$ is the derivative at x of the function

$$y = f(x)$$

Example 13.1.17 If $y = x^3$,

$$f'(x) = 3x^2$$
$$dy = f'(x)\, dx = 3x^2\, dx$$

The formulas for differentials may be obtained by multiplying the derivative by dx. For instance,

1. $dc = 0$
2. $d(cu) = c\, du$
3. $d(u \pm v) = du \pm dv$
4. $d(uv) = u\, dv + v\, du$

5. $d\left(\frac{u}{v}\right) = \frac{v\,du - u\,dv}{v^2}$

6. $d(u^n) = nu^{n-1}\,du$

Continuity

If the function f has a finite derivative at $x = c$, then f is continuous at $x = c$. Suppose f and g are two functions that are continuous at $x = c$. Then the following functions are also continuous at $x = c$.

$$F_1 = f(x) + g(x)$$

$$F_2 = f(x) \cdot g(x)$$

$$F_3 = k\,f(x), \qquad \text{where } k \text{ is any number}$$

If $g(c)$ is not zero, then

$$F_4 = \frac{f(x)}{g(x)}$$

is also continuous at x.

Maxima and Minima

A function f is said to be convex in the interval $a \leqslant x \leqslant b$ if for $0 < \lambda < 1$

$$f(a) \geqslant f(a + h)$$

for all positive and negative values of h near zero. For a local minimum at $x = b$,

$$f(b) \leqslant f(b + h)$$

for values of h close to zero. If $f(a) \geqslant f(x)$ for all values of x, a is an absolute maximum. If $f(b) \leqslant f(x)$ for all values of x, b is an absolute minimum.

Let the function f be defined for $a \leqslant x \leqslant b$, and have a relative maximum or minimum at $x = c$, where $a < c < b$. If the derivative $f'(x)$ exists at c, then

$$f'(c) = 0$$

If f has a local maximum at c, then

$$f''(c) < 0$$

If f has a local minimum at c, then

$$f''(c) > 0$$

Example 13.1.18 Find the maximum and minimum of the function

$$f(x) = 2x^3 - 3x^2 + x - 5$$

Then

$$f'(x) = 6x^2 - 6x + 1 = 0$$

$$x = \frac{-(-6) \pm \sqrt{(-6)^2 - (4)(6)(1)}}{(2)(6)}$$

$$x = (0.2113, 0.7887)$$

Now $f''(x) = 12x - 6$. At $x = 0.2113$, $f''(0.2113) = -3.464$. Thus $x = 0.2113$ is a relative maximum. At $x = 0.7887$, $f''(0.7887) = 3.464$. Thus $x = 0.7887$ is a relative minimum.

Exhibit 13.1.1 Convex (*a*) and Concave (*b*) Functions

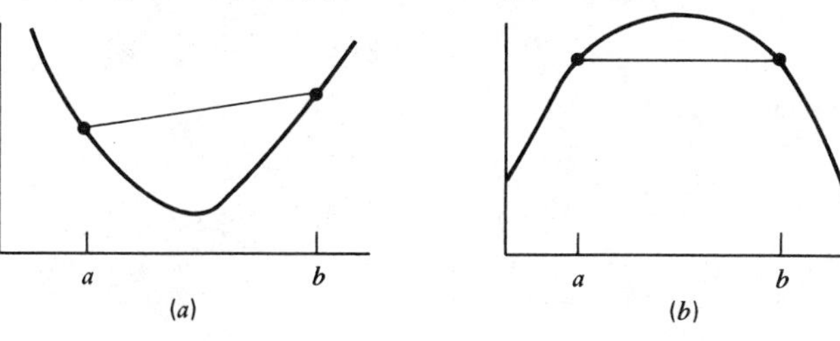

Convexity and Concavity

A function f is said to be convex in the interval $a \leqslant x \leqslant b$ if for $0 < \lambda < 1$

$$f[\lambda a + (1 - \lambda) b] \leqslant \lambda f(a) + (1 - \lambda) f(b)$$

and concave in the interval $a \leqslant x \leqslant b$ if

$$f[\lambda a + (1 - \lambda) b] \geqslant \lambda f(a) + (1 - \lambda) f(b)$$

The figures in Exhibit 13.1.1*a* and 13.1.1*b* illustrate convex and concave functions, respectively.

Newton's Method

Also called the "Newton-Raphson method," this technique is an iterative method for solving equations $f(x) = 0$ where f is differentiable. The idea is to approximate the graph of f by suitable tangents. Using a value x_0 obtained from the graph of f, we let x_1 be the point of intersection of the x axis and the tangent to the curve of f at x_0, as shown in Exhibit 13.1.2. Then

$$\tan \beta = f'(x_0) = \frac{f(x_0)}{x_0 - x_1}$$

Hence

$$x_1 = x_0 - \frac{f(x_0)}{f'(x_0)}$$

The general formula is

$$x_{n+1} = x_n - \frac{f(x_n)}{f'(x_n)}$$

Exhibit 13.1.2 Newton-Raphson Interactive Method for Solving Equations

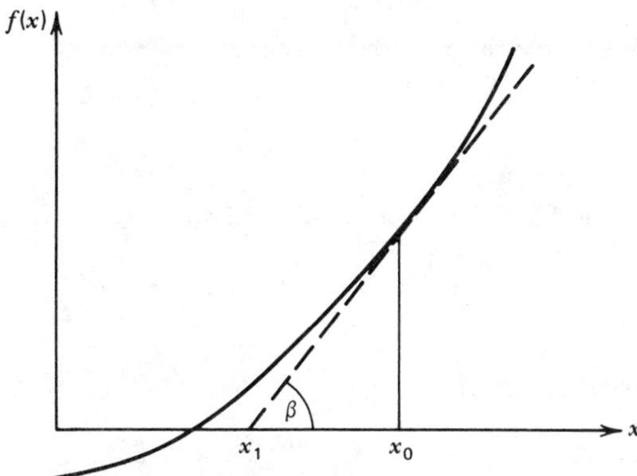

Example 13.1.19 Find a root of the following function:

$$f(x) = 6x^2 - 6x + 1$$

from the starting value $x_0 = 1$. Now, $f'(x) = 12x - 6$

$$x_1 = x_0 - \frac{f(x_0)}{f'(x_0)}$$

$$= 1 - \frac{[6(1)^2 - 6(1) + 1]}{[12(1) - 6]}$$

$$= 1 - \frac{1}{6}$$

$$= 0.8333$$

$$x_2 = x_1 - \frac{f(x_1)}{f'(x_1)}$$

$$= 0.8333 - \frac{[6(0.8333)^2 - 6(0.8333) + 1]}{[12(0.8333) - 6]}$$

$$= 0.8333 - \frac{(0.1653)}{4}$$

$$= 0.7917$$

$$x_3 = 0.7917 - \frac{[6(0.7917)^2 - 6(0.7917) + 1]}{[12(0.7917) - 6]}$$

$$= 0.7917 - 0.0030$$

$$= 0.7887$$

$$x_4 = 0.7887 - \frac{[6(0.7887)^2 - 6(0.7887) + 1]}{[12(0.7887) - 6]}$$

$$= 0.7887 - 0$$

$$= 0.7887$$

Thus we have converged to a root at $x = 0.7887$.

Integral Calculus

Integration

An equation $f(x)$ that specifies the derivative as a function of x (or as a function of x and y) is called a "differential equation." For example,

$$\frac{dy}{dx} = 2xy^2$$

A function $y = F(x)$ is called a "solution of a differential equation" if, over the interval $a < x < b$, $F(x)$ is differentiable and

$$\frac{d\,F(x)}{dx} = f(x)$$

$F(x)$ is called the "integral of $f(x)$ with respect to x." The symbol

$$\int \ldots dx$$

means "integral with respect to x." Thus integration is the inverse of differentiation.

The indefinite integral of $f(x)$ is written as

$$\int f(x)\,dx = F(x) + C$$

where C is an arbitrary constant. For example,

$$\int 3x^2\,dx = x^3 + C$$

Clearly then

$$\int f(x)\,dx = \int dF(x)$$

The following formulas are associated with integration:

1. $\int du = u + C$
2. $\int a\,du = a \int du$
3. $\int (du + dv) = \int du + \int dv$
4. $\int u^n\,du = \dfrac{u^{n+1}}{n+1} + C$
5. $\int u\,dv = uv - \int v\,du$
6. $\int a^u\,du = \dfrac{a^u}{\ell n\, a} + C, \quad a \neq 1$ and $a > 0$
7. $\int \cos u\,du = \sin u + C$
8. $\int \sin u\,du = -\cos u + C$

The definite integral is found by integrating a function $f(x)$ over some specified interval $a \leqslant x \leqslant b$ and is given by

$$\int_a^b f(x)\,dx = F(b) - F(a)$$

For example,

$$\int_1^2 3x^2\,dx = x^3 \Big|_1^2 = (2)^3 - (1)^3 = 8 - 1 = 7$$

Numerical Integration

The definite integral can be thought of as an area. When the indefinite integral form is known or can be found in a table of integrals, the definite integral can be evaluated. But often the indefinite integral is not known. In such cases, if the function to be integrated is continuous, the definite integral can be evaluated numerically. One of the simplest of these numerical methods is called the "trapezoidal rule," and, as seen in Exhibit 13.1.3, it consists of summing the areas of a series of trapezoids.

Exhibit 13.1.3 The Trapezoidal Rule

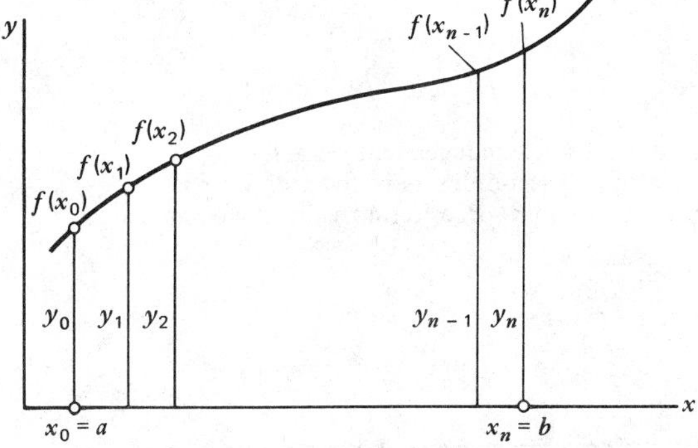

The formula for the trapezoidal approximation is

$$T = (\tfrac{1}{2}\, y_0 + y_1 + y_2 + \cdots + y_{n-1} + \tfrac{1}{2}\, y_n)\, \Delta x$$

where

$$y_0 = f(x)_0, \quad y_1 = f(x_1), \ldots, y_n = f(x_n).$$

$$\Delta x = x_i - x_{i-1}$$

Whereas the trapezoidal rule approximates the function $f(x)$ by a series of straight lines connecting the values $f(x_i)$ and $f(x_{i+1})$ for all i, Simpson's rule achieves greater accuracy by employing a series of parabolic arcs instead of straight lines. The formula for Simpson's rule is

$$T = \frac{h}{3}(y_0 + 4y_1 + 2y_2 + 4y_3 + 2y_4 + \cdots + 2y_{n-2} + 4y_{n-1} + y_n)$$

where n is the even number of subintervals of width $h = (b - a)/n$. The quantities a and b are values of x coinciding with y_0 and y_n, respectively.

Example 13.1.20 Use Simpson's rule with $n = 4$ to approximate $\ln 2$ from

$$\ln 2 = \int_1^2 \left(\frac{1}{x}\right) dx$$

For $n = 4$, $h = 0.25$. For $y = 1/x$

$$\begin{aligned} T &= \frac{h}{3}\,[y_0 + 4y_1 + 2y_2 + 4y_3 + y_4] \\ &= \frac{0.25}{3}\left[\left(\frac{1}{1}\right) + 4\left(\frac{1}{1.25}\right) + 2\left(\frac{1}{1.50}\right) + 4\left(\frac{1}{1.75}\right) + \left(\frac{1}{2}\right)\right] \\ &= \frac{0.25}{3}\,[1.00 + 3.20 + 1.33 + 2.29 + 0.50] \\ &= 0.693 \end{aligned}$$

Differential Equations

A differential equation is an equation that involves one or more derivatives, or differentials. Differential equations are classified by:

1. Type: ordinary or partial.
2. Order: the order of the highest-order derivative that occurs in the equation.
3. Degree: the exponent of the highest power of the highest-order derivative, after the equation has been cleared of fractions and radicals in the dependent variable and its derivatives.

Example 13.1.21

$$\left(\frac{d^3 y^2}{dx^3}\right)^2 + \left(\frac{d^2 y}{dx^2}\right)^5 + \frac{y}{x^2 + 1} = e^x$$

1. Ordinary, because it has a single independent variable x.
2. Third order, because the highest-order derivative is d^3y/dx^3.
3. Second degree, because the third-order term has the exponent 2.

The differential equation

$$\frac{\partial^2 y}{\partial x_1^2} = a^2 \frac{\partial^2 y}{\partial x_2^2}$$

is a partial differential equation of order two and degree one.

A function $y = f(x)$ is said to be a solution of a differential equation if the latter is identically satisfied when y and its derivatives are replaced by $f(x)$ and its corresponding derivatives. For example, if C_1 and C_2 are arbitrary constants, then

$$y = C_1 \cos x + C_2 \sin x$$

is a solution of the differential equation

$$\frac{d^2y}{dx^2} + y = 0$$

A differential equation of order n will, in general, have a solution involving n arbitrary constants. Ordinary differential equations can be solved in the following ways:

1. First-order equations with separable variables.

$$f(y)\,dy + g(x)\,dx = 0$$

The general solution is

$$\int f(y)\,dy + \int g(x)\,dx = C$$

where C is an arbitrary constant.

Example 13.1.22

$$(x + 1)\frac{dy}{dx} = x(y^2 + 1)$$

Rearranging,

$$\frac{dy}{(y^2 + 1)} = \frac{x\,dx}{x + 1}$$

and the solution is

$$\tan^{-1} y = x - \ln|x + 1| + C$$

2. First-order homogeneous equations.

$$\frac{dy}{dx} = F\left(\frac{y}{x}\right) \qquad \text{(homogeneous)}$$

Let $v = y/x$; then $y = vx$ and

$$\frac{dy}{dx} = v + x\frac{dv}{dx}$$

Therefore

$$v + x\frac{dv}{dx} = F(v)$$

so that by separation of variables the solution is

$$\frac{dx}{x} + \frac{dv}{v - F(v)} = 0$$

3. First-order linear equations.

$$\frac{dy}{dx} + Py = Q$$

where P and Q are functions of x.

Let $\rho = e^{\int P\,dx}$ be an integrating factor for the preceding equation, so that by multiplying the equation by ρ yields

$$\rho \frac{dy}{dx} + \rho P y = \rho Q$$

which becomes

$$\frac{d}{dx}(\rho y) = \rho Q$$

the solution of which is

$$\rho y = \int \rho Q\,dx + c$$

Example 13.1.23

$$\frac{dy}{dx} + y = e^x$$

Now

$$\rho = e^{\int P\,dx} = e^{\int dx} = e^x$$

Therefore we have

$$\frac{d}{dx}(e^x y) = e^{2x}$$

the solution of which is

$$e^x y = \tfrac{1}{2} e^{2x} + c$$

or

$$y = \tfrac{1}{2} e^x + ce^{-x}$$

4. First-order equations with exact differentials. An equation that can be written in the form

$$M(x, y)\,dx + N(x, y)\,dy = 0$$

and having the property such that

$$\frac{\partial M}{\partial y} = \frac{\partial N}{\partial x}$$

is said to be exact, and is solved by finding a function $f(x, y)$ such that

$$df = M\,dx + N\,dy$$

5. Special types of second-order equations, such as

$$F\left(x, y, \frac{dy}{dx}, \frac{d^2y}{dx^2}\right) = 0$$

can be reduced to first-order equations by a suitable change of variables.

6. Homogeneous linear second-order differential equations with constant coefficients, such as

$$\frac{d^2y}{dx^2} + 2a\frac{dy}{dx} + by = 0$$

where a and b are constants. In operator notation with $D = dy/dx$, this equation becomes

$$(D^2 + 2aD + b)\,y = 0$$

Calling the expression $(D^2 + 2aD + b)$ the characteristic equation, we see that this equation has two roots, r_1 and r_2, so that we can write the differential equation as

$$(D - r_1)(D - r_2)\,y = 0$$

Now let

$$(D - r_2)\,y = u$$

$$(D - r_1)\,u = 0$$

The second of these two equations is separable, so that

$$u = C_1 e^{r_1 x}$$

Substituting this into the first equation

$$(D - r_2)\,y = C_1 e^{r_1 x}$$

or

$$\frac{dy}{dx} - r_2\,y = C_1 e^{r_1 x}$$

which is linear and has as its integrating factor

$$\rho = e^{-r_2 x}$$

so that the solution is

$$e^{-r_2 x} y = C_1 \int e^{(r_1 - r_2)x}\,dx + C_2$$

7. Nonhomogeneous linear second-order differential equations with constant coefficients have the form

$$\frac{d^2y}{dx^2} + 2a\,\frac{dy}{dx} + by = F(x)$$

A homogeneous equation is obtained by letting $F(x) = 0$ and solving as in the preceding paragraph. The solution to this homogeneous equation is $y_h(x)$,

$$y_h = c_1 u_1(x) + c_2 u_2(x)$$

The solution to the nonhomogeneous equation has the form

$$y = y_h(x) + y_p(x)$$

where $y_p(x)$ is the particular function that satisfies the nonhomogeneous equation here. The method of solving for the nonhomogeneous equation once $y_h(x)$ is known is called the "method of variation of parameters." It consists of replacing the constants C_1 and C_2 in the expression for y_h with $v_1 = v_1(x)$ and $v_2 = v_2(x)$ such that the following two conditions are satisfied:

$$v_1' u_1 + v_2' u_2 = 0$$

$$v_1' u_1' + v_2' u_2' = F(x)$$

Solving these two equations for v_1' and v_2' and integrating these two functions to obtain v_1 and v_2 yields the solution to the nonhomogeneous equation.

Example 13.1.24

$$\frac{d^2y}{dx^2} + 2\,\frac{dy}{dx} - 3y = 6$$

Setting the left side to zero and solving yields

$$y_h = C_1 e^{-3x} + C_2 e^{x}$$

which gives

$$v_1' e^{-3x} + v_2' e^{x} = 0$$
$$v_1'(-3e^{-3x}) + v_2' e^{x} = 6$$

By Cramer's rule

$$v_1' = -\tfrac{3}{2} e^{3x}$$
$$v_2' = \tfrac{3}{2} e^{-x}$$

Hence

$$v_1 = \int -\tfrac{3}{2} e^{3x}\, dx = -\tfrac{1}{2} e^{3x} + C_1$$
$$v_2 = \int \tfrac{3}{2} e^{-x}\, dx = -\tfrac{3}{2} e^{-x} + C_2$$

Therefore

$$\begin{aligned} y &= v_1 u_1 + v_2 u_2 \\ &= (-\tfrac{1}{2} e^{3x} + C_2)\, e^{-3x} + (-\tfrac{3}{2} e^{-x} + C_2)\, e^{x} \\ &= -2 + C_1 e^{-3x} + C_2 e^{x} \end{aligned}$$

13.1.4 Calculus of Finite Differences

The previous section on calculus and differential equations assumed all functions and variables to be continuous. However, many problems in industrial engineering involve variables that are inherently discrete. Moreover, the solution of continuous problems using a digital computer often involves approximation methods that replace continuous variables by their discrete equivalents; that is, they progress in finite increments. The calculus of finite differences plays an important role in these operations.

Discrete Variable

If variable x is continuous on some open interval (a, b), then on any finite subinterval x may assume an infinite number of values. For example, the furnace temperature for heat treating metal castings may vary over a range (t_1, t_2). If x is discrete, it may assume only a finite number of values over a given interval. For example, the number of castings in the furnace may vary over the range $(0, N)$ where N is the capacity of the furnace.

Function of a Discrete Variable

Let $f(x)$ be a function of a discrete variable x, and assume that $f(x)$ is defined for $x = x_0, x_1, \ldots, x_n$. Therefore $f(x)$ is defined as

$$f(x) = \begin{cases} f(x), & x = x_0, x_1, \ldots, x_n \\ 0, & \text{otherwise} \end{cases}$$

Divided Difference

The discrete analog of the derivative is the divided difference, defined by

$$\Delta f(x_i) = \frac{f(x_{i+1}) - f(x_i)}{x_{i+1} - x_i}$$

If $\Delta x = x_{i+1} - x_i$ for all i, this expression becomes

$$\Delta f(x_i) = \frac{f(x_i + \Delta x) - f(x_i)}{\Delta x}$$

If $\Delta x = 1$, we have simply the difference

$$\Delta f(x) = f(x+1) - f(x)$$

Example 13.1.25 Let $f(x)$ be defined by

$$f(x) = \frac{x}{x+1}, \qquad x = 1, \frac{3}{2}, 2, \frac{5}{2}, \ldots$$

Evaluate $\Delta f(x)$ at $x = 1, 2, 3$. Since $\Delta x = 0.5$,

$$\Delta f(x) = f\,\frac{(x+0.5) - f(x)}{0.5} = 2\left[\frac{x+0.5}{x+1.5} - \frac{x}{x+1}\right]$$

Then

$$\Delta f(1) = 2\left[\frac{1.5}{2.5} - \frac{1}{2}\right] = 2\left[\frac{3-2.5}{21}\right] = \frac{1}{5}$$

$$\Delta f(2) = 2\left[\frac{2.5}{3.5} - \frac{2}{3}\right] = 2\left[\frac{15-14}{21}\right] = \frac{2}{21}$$

$$\Delta f(3) = 2\left[\frac{3.5}{4.5} - \frac{3}{4}\right] = 2\left[\frac{28-27}{36}\right] = \frac{1}{18}$$

Properties of the Difference Operator

Let $h(x)$ and $g(x)$ be functions of the discrete variable $x = x_1, x_2, \ldots$. Then the following properties apply to the difference operator Δ:

1. $\Delta[h(x_i) \pm g(x_i)] = \Delta h(x_i) \pm \Delta g(x_i)$
2. $\Delta[h(x_i)\,g(x_i)] = h(x_{i+1})\,\Delta g(x_i) + g(x_i)\,\Delta h(x_i)$
3. $\Delta\left[\dfrac{h(x_i)}{g(x_i)}\right] = \dfrac{1}{g(x_i)\,g(x_{i+1})}\,[g(x_i)\,\Delta h(x_i) - h(x_i)\,\Delta g(x_i)]$
4. Let $C(x) = \sum_{y=a(x)}^{b(x)} f(x, y)$

$$= f[x, a(x)] + f[x, a(x) + \Delta y] + f[x, a(x) + 2\Delta y] + \cdots + f[x, b(x) - \Delta y] + f[x, b(x)]$$

where x and y are discrete variables with constant increments Δx and Δy. Then

$$\Delta C(x) = \sum_{y=a(x)}^{b(x)} \Delta_x f(x, y) + \sum_{y=b(x)+\Delta y}^{b(x+\Delta x)} \frac{f(x+\Delta x, y)}{\Delta x} - \sum_{y=a(x)}^{a(x+\Delta x)-\Delta y} \frac{f(x+\Delta x, y)}{\Delta x}$$

where $\Delta_x f(x, y)$ is the divided difference of $f(x, y)$ with respect to x.

5. Let $C(x)$ be defined by the double summation

$$C(x) = \sum_{y=a(x)}^{b(x)} \sum_{z=g(x,y)}^{h(x,y)} f(x, y, z)$$

where x, y, and z have constant increments Δx, Δy, and Δz, respectively, such that $\Delta x = \Delta y = \Delta z$, where $a(x)$ and $b(x)$ are permissible values of y for all x, and where $g(x, y)$ and $h(x, y)$ are permissible values of z for all x and y. Then

$$\Delta C(x) = \sum_{y=a(x)}^{b(x)} \sum_{z=g(x,y)}^{h(x,y)} \Delta x\, f(x, y, z)$$

$$+ \sum_{y=a(x)}^{b(x)} \left[\sum_{z=h(x,y)+\Delta z}^{h(x+\Delta x, y)} \frac{f(x+\Delta x, y, z)}{\Delta x} - \sum_{z=g(x,y)}^{g(x+\Delta x, y)-\Delta z} \frac{f(x+\Delta x, y, z)}{\Delta x} \right]$$

$$+ \sum_{y=b(x)+\Delta y}^{b(x+\Delta x)} \sum_{z=g(x+\Delta x, y)}^{h(x+\Delta x, y)} \frac{f(x+\Delta x, y, z)}{\Delta x} - \sum_{y=a(x)}^{a(x+\Delta x)-\Delta y} \sum_{z=g(x+\Delta x, y)}^{h(x+\Delta x, y)} \frac{f(x+\Delta x, y, z)}{\Delta x}$$

Optimization of Discrete Functions

In the case of continuous functions, the derivative $f'(x)$ was used to find a local optimum, and the second derivative $f''(x)$ was used to determine whether it was a maximum or a minimum. The discrete function has its counterpart operation. It is first necessary to define a convex function of a discrete variable x.

Let the open interval (a, b) contain the permissible values of the discrete variable $x, x_0, x_1, \ldots, x_n$, and let λ be a scalar on the interval $[0, 1]$ such that $\lambda x_i + (1 - \lambda) x_j$ is a permissible value of x for $i, j = 0, 1, \ldots, n$. Then $f(x)$ is said to be convex from above on (a, b) if for $0 < \lambda < 1$

$$f[\lambda x_i + (1 - \lambda) x_j] \leqslant \lambda f(x_i) + (1 - \lambda) f(x_j)$$

and convex from below on (a, b) if

$$f[\lambda x_i + (1 - \lambda) f(x_j)] \geqslant \lambda f(x_i) + (1 - \lambda) f(x_j)$$

If $f(x)$ is convex on the open interval (a, b) containing the permissible values of $x, x_0, x_1, \ldots, x_n$, and x_i is an interior extreme point for $f(x)$ on (a, b), then either

$$f(x_{i-1}) \geqslant f(x_i) \leqslant f(x_{i+1})$$

or

$$f(x_{i-1}) \leqslant f(x_i) \geqslant f(x_{i+1})$$

where $i = 1, 2, \ldots, n - 1$. This definition leads to the necessary conditions for an interior extreme point for a function of a discrete variable, as follows: Let $f(x)$ be a function of the discrete variable x on the open interval (a, b). If x_i is an interior extreme point for $f(x)$, then either

$$\Delta f(x_{i-1}) \leqslant 0 \leqslant \Delta f(x_i)$$

or

$$\Delta f(x_{i-1}) \geqslant 0 \geqslant f(x_i)$$

These are necessary conditions for an extreme point at x_i, but a point satisfying these conditions is not necessarily an extreme point—it may be a point of inflection. To ascertain the nature of an interior extreme point x_i, it is necessary to compare $f(x_i)$ to $f(x_{i-1})$ and $f(x_{i+1})$. If

$$f(x_{i-1}) < f(x_i) > f(x_{i+1})$$

then x_i is a local maximum for $f(x)$, and

$$\Delta f(x_{i-1}) > 0 > \Delta f(x_i)$$

is a sufficient condition for a local maximum at x_i. Similarly, if

$$f(x_{i-1}) > f(x_i) < f(x_{i+1})$$

the x_i is a local minimum for $f(x)$, and

$$\Delta f(x_{i-1}) < 0 < \Delta f(x_i)$$

is a sufficient condition for a local minimum at x_i.

Antidifference

If $S[f(x)]$ is the antidifference of the function $f(x)$, then

$$\Delta S[f(x)] = f(x)$$

where S is the antidifference operator.

Example 13.1.26 Find the antidifference of $f(x)$ where

$$f(x) = x, \qquad x = 0, 1, 2, \ldots$$

$\Delta x = 1$. To determine $S[f(x)]$, it is necessary to find a function $F(x)$ such that

$$\Delta F(x) = f(x)$$

From calculus,

$$\int x\, dx = \frac{x^2}{2} + C_1$$

Therefore try

$$F(x) = \frac{x^2}{2} + C_1$$

But

$$\Delta F(x) = \frac{(x+1)^2 - x^2}{2} = x + \frac{1}{2}$$

which is not $f(x)$. Let $G(x)$ be defined such that

$$\Delta G(x) = \tfrac{1}{2}$$

so that

$$\Delta[F(x) - G(x)] = f(x) = x$$

Thus

$$G(x) = \frac{x}{2} + C_2 \quad \text{then} \quad \Delta G(x) = \frac{(x+1) - x}{2} = \frac{1}{2}$$

Therefore let us redefine $F(x)$ as

$$F(x) = \frac{x^2 - x}{2} + C_3$$

so that

$$F(x) = \frac{(x+1)^2 - (x+1) - (x^2 - x)}{2} = x$$

The antidifference is therefore

$$S[f(x)] = \frac{x^2 - x}{2} + C_3$$

Summation of Series

If $f(x)$ is a function of the integer-valued variable x, then

$$\sum_{x=a}^{b} f(x) = S[f(b+1)] - S[f(a)]$$

where $a \leqslant b$.

Summation by Parts

Let $f(x)$, $h(x)$, and $g(x)$ be functions of the integer-valued variable x such that

$$f(x) = h(x)\,\Delta g(x)$$

Then through summation by parts,

$$\sum_{x=a}^{b} f(x) = h(b+1)\,g(b+1) - h(a)\,g(a) - \sum_{x=a}^{b} g(x+1)\,\Delta h(x)$$

where $a \leqslant b$.

Higher-Order Differences

In calculus, the second derivative can be expressed as the first derivative of the first derivative. That is,

$$\frac{d^2}{dx^2} f(x) = \frac{d}{dx}\left[\frac{d}{dx} f(x)\right]$$

For functions of discrete variables, the second divided difference at x_i is defined as

$$\Delta^2 f(x_i) = \frac{1}{x_{i+2} - x_i}\,[\Delta f(x_{i+1}) - \Delta f(x_i)]$$

In general, the kth divided difference at x_i is given by

$$\Delta^n f(x_i) = \frac{1}{x_{i+n} - x_i}\,[\Delta^{n-1} f(x_{i+1}) - \Delta f(x_i)]$$

Example 13.1.27 Evaluate the first three differences of $f(x)$ at $x = 1, 2, 3, 4$ where

$$f(x) = (x+1)^2, \qquad x = 0, 1, 2, \ldots$$

The first three differences are as follows:

$$\Delta f(x) = \frac{f(x+1) - f(x)}{\Delta x} = (x+2)^2 - (x+1)^2 = 2x + 3$$

$$\Delta^2 f(x) = \frac{1}{2}\,[2(x+1) + 3 - (2x+3)] = 1$$

$$\Delta^3 f(x) = \frac{1}{3}\,[1 - 1] = 0$$

Evaluating these differences at $x = 1, 2, 3, 4$ yields the following results:

x	$f(x)$	$\Delta f(x)$	$\Delta^2 f(x)$	$\Delta^3 f(x)$
1	4	5	1	0
2	9	7	1	0
3	16	9	1	0
4	25	11	1	0

Difference Equations

In calculus we defined the differential equation as an equation involving the derivatives or the differentials of functions of continuous variables x, y. The difference equation is the discrete variable counterpart to the differential equation. For example, the nonhomogeneous linear second-order differential equation with constant coefficients was earlier seen to have the form

$$\frac{d^2y}{dx^2} + 2a\frac{dy}{dx} + by = F(x)$$

Its discrete analog, the nonhomogeneous linear second-order difference equation with constant coefficients has the form

$$\Delta^2 f(x) + 2a\,\Delta f(x) + b\,f(x) = F(x)$$

Thus we have the following definition of a difference equation: Let $f(x)$ be a function of the integer-valued variable x. A difference equation is an equation relating $f(x)$ and any of its differences $\Delta^i f(x)$, over a given range of values of x.

Example 13.1.28 Solve the difference equation

$$2f(x+2) - f(x) = 2, \qquad x = 0, 1, 2, \ldots$$

Rearranging, we get $f(x + 2)$ in terms of $f(x)$, or

$$f(x+2) = \frac{f(x)}{2} + 1$$

For even values of x, $x = 2, 4, 6, 8, \ldots$

$$f(2) = \frac{f(0)}{2} + 1$$

$$f(4) = \frac{f(2)}{2} + 1 = \frac{f(0)}{4} + \frac{3}{2}$$

$$f(6) = \frac{f(4)}{2} + 1 = \frac{f(0)}{8} + \frac{7}{4}$$

$$f(8) = \frac{f(6)}{2} + 1 = \frac{f(0)}{16} + \frac{15}{8}$$

By induction, for even x

$$f(x) = \frac{f(0)}{2^{x/2}} + \sum_{x=0}^{(x/2)-1} \frac{1}{2^i}$$

Then

$$f(x+2) = \frac{f(0)}{2^{(x/2)+1}} + \sum_{x=0}^{x/2} \frac{1}{2^i}$$

For odd x

$$f(3) = \frac{f(1)}{2} + 1$$

$$f(5) = \frac{f(3)}{2} + 1 = \frac{f(1)}{4} + \frac{3}{2}$$

$$f(7) = \frac{f(5)}{2} + 1 = \frac{f(1)}{8} + \frac{7}{4}$$

$$f(9) = \frac{f(7)}{2} + 1 = \frac{f(1)}{16} + \frac{15}{8}$$

which by induction leads to

$$f(x) = \frac{f(1)}{2^{(x-1)/2}} + \sum_{i=0}^{(x-3)/2} \frac{1}{2^i}$$

for odd x.

A particular solution for a difference equation is any function $f(x) \neq 0$ that reduces the difference equation to an identity. One such particular solution is that satisfying the initial conditions for the problem, although any function that satisfies the difference equation is a particular solution.

The linearly independent functions $f_i(x)$, $i = 1, 2, \ldots, n$ among the set of particular solutions to a difference equation form a fundamental set of solutions for the difference equation. This fundamental set of solutions has the property that any particular solution in the set cannot be represented as a linear combination of the remaining particular solutions in the set. However, any particular solution not belonging to the fundamental set can be represented as a linear combination of particular solutions in the set.

A general solution to a difference equation is one that is a linear combination of particular solutions in the fundamental set.

Linear Difference Equations

A difference equation defined over some set of values x is said to be linear if it can be expressed in the form

$$a_n(x)\, f(x+n) + a_{n-1}(x)\, f(x+n-1) + \cdots + a_0(x) f(x) = F(x)$$

where $a_i(x)$, $i = 0, 1, \ldots, n$ are functions of x, but not of $f(x)$. If $a_n(x) \neq 0$ and $a_0(x) \neq 0$ for all values of x, then the difference equation is an nth order linear difference equation. If $a_i(x)$ is constant for all i, then the difference equation is a linear difference equation with constant coefficients. If $F(x) = 0$, the difference equation is a homogeneous linear difference equation. If $F(x) \neq 0$, then it is a nonhomogeneous linear difference equation.

Example 13.1.29 The difference equation

$$xf(x+1) - f(x) = 0$$

is a homogeneous linear difference equation of order one, but does not have constant coefficients.

13.1.5 COMPLEX VARIABLES AND TRANSFORM METHODS

Complex Variables

The quantity $i = \sqrt{-1}$ is called an "imaginary number." Therefore the quantity s,

$$s = x + iy$$

which is composed of a real part, x, and an imaginary part, iy, is called a "complex number." Such numbers can be regarded as being in the complex plane, as shown in Exhibit 13.1.4.

Exhibit 13.1.4 Numbers in the Complex Plane

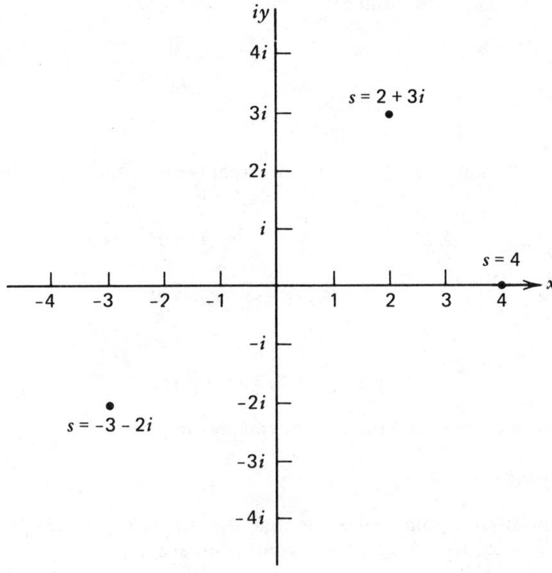

Operations with Complex Numbers

The following operations with complex numbers are carried out using the conventional rules of algebra, where $i^2 = -1$:

1. Addition (subtraction):

$$(x + iy) \pm (v + iw) = (x \pm v) + i(y \pm w)$$

2. Multiplication:

$$(x + iy)\,(v + iw) = (xv - yw) + i(xw + yv)$$

3. Division:

$$\frac{(x + iy)}{(v + iw)} = \frac{(xv + yw) + i(yv - xw)}{(v^2 + w^2)}$$

Let the point $s = x + iy$ be represented as shown in Exhibit 13.1.5, where r is the length of the line from the origin to s and θ is the angle formed between this line and the real axis. The squared distance from the origin to s is given by

$$r^2 = (x + iy)\,(x - iy)$$

Exhibit 13.1.5 Solving Complex Numbers with Conventional Rules of Algebra

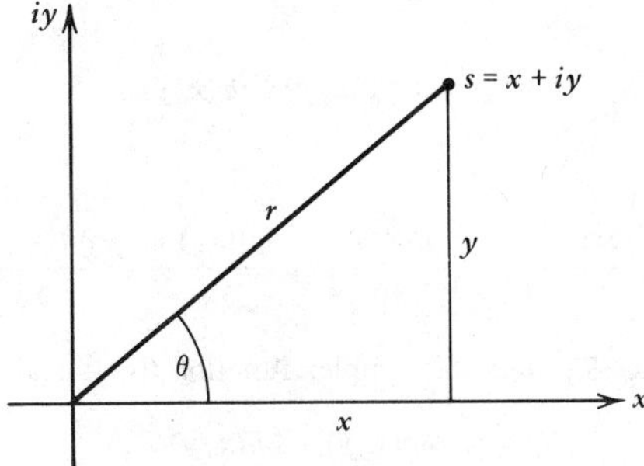

where $(x - iy)$ is referred to as the "complement conjugate" of s and is denoted by s^*. Therefore the distance from the origin to the point s is

$$r = \sqrt{ss^*}$$

or

$$r = \sqrt{x^2 + y^2}$$

Since the coordinates of s are located x and y units from the origin, respectively, then

$$\sin(\theta) = \frac{y}{r} \qquad \cos(\theta) = \frac{x}{r}$$

$$y = r \sin(\theta) \qquad x = r \cos(\theta)$$

Therefore

$$s = r\,[\cos(\theta) + i \sin(\theta)]$$

which is a polar coordinate representation of the complex number s.

Functions of Complex Variables

Let $f(s)$ denote a single-valued function of the complex variable s. The limit of $f(s)$ as s approaches s_o is $\lim_{s \to s_o} f(s)$, which is equal to $f(s_o)$ if $f(s)$ is continuous at s_o.

Derivative of a Complex Function

The derivative of $f(s)$ at s_o is given by

$$f'(s_o) = \lim_{\Delta s \to o} \frac{f(s_o + \Delta s) - f(s_o)}{\Delta s}$$

if the limit exists. Noting that $f(s)$ can be expressed as a function of x and iy, let

$$s_o = x_o + i\, y_o$$

and

$$\Delta s = \Delta x + i\, \Delta y$$

All the rules of differentiation in calculus hold for complex functions.

Analytic Complex Function

A complex function $f(s)$ is said to be analytic on the space S_o if and only if $f'(s)$ exists at all points in S_o.

Cauchy-Riemann Equations

Let

$$f(s) = g(x, y) + ih(x, y).$$

If the partial derivatives

$$\frac{\partial g(x, y)}{\partial x}, \quad \frac{\partial g(x, y)}{\partial y}, \quad \frac{\partial h(x, y)}{\partial x}, \quad \frac{\partial h(x, y)}{\partial y}$$

are continuous on the space S_o, then the complex function $f(s)$ is analytic on S_o if and only if

$$\frac{\partial g(x, y)}{\partial x} = \frac{\partial h(x, y)}{\partial y}$$

and

$$\frac{\partial g(x,y)}{\partial y} = \frac{-\partial h(x,y)}{\partial x}$$

for all s in S_o.

Singular Point

A point s_o at which $f(s)$ fails to be analytic is called a "singular point" or "singularity." If s_o is a singular point for $f(s)$ such that there exists a neighborhood of s_o in which the complex function $f(s)$ is analytic, then s_o is called an "isolated singularity."

Complex Integration

The definite integral, or, as it is called, the "complex line integral," of a complex function $f(s)$,

$$s = x + iy$$

is illustrated in Exhibit 13.1.6.

Let C be a smooth curve in the complex plane. The line interval of $f(s)$ along the curve C is denoted by

$$\int_C f(s)\, ds$$

Curve C is called the "path of integration." If C is divided into a series of arcs where the length of the ith arc is Δs_i and

$$\Delta s_i = |s_i - s_{i-1}|$$

let

$$M_i = \max_{s_{i-1} \leqslant s \leqslant s_i} f(s)$$

$$m_i = \min_{s_{i-1} \leqslant s \leqslant s_i} f(s)$$

then the upper and lower Riemann sums are

$$\overline{A}_{a,b,n}[f(s)] = \sum_{i=1}^{n} M_i \,\Delta s_i$$

$$\underline{A}_{a,b,n}[f(s)] = \sum_{i=1}^{n} m_i \,\Delta s_i$$

Exhibit 13.1.6 Complex Line Integration

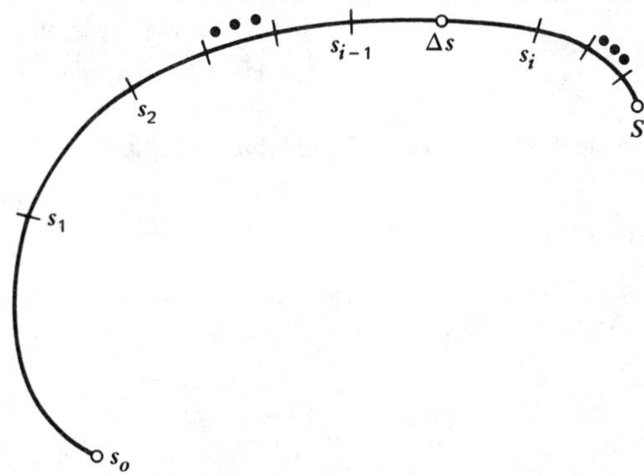

and then the line integral of $f(s)$ on C is

$$\int_C f(s)\,ds = \lim_{n\to\infty} \overline{A}_{a,b,n}[f(s)] = \lim_{n\to\infty} \underline{A}_{a,b,n}[f(s)]$$

if the limit exists. The projections of $\int_C f(s)\,ds$ on the x and y axes can be defined by defining the curve C by

$$y = g(x)$$

$$x = h(y)$$

The projection on the x axis is

$$\int_{a_x}^{b_x} f[x, g(x)]\,dx = \lim_{n\to\infty} \sum_{i=1}^{n} M_i\,\Delta x_i$$

and the projection on the y axis is

$$\int_{a_y}^{b_y} f[h(y), y]\,dy = \lim_{n\to\infty} \sum_{i=1}^{n} M_i\,\Delta y_i$$

To simplify the line integral, let

$$f(s) = f(x, y)$$

and if $y = g(x)$

$$ds = \sqrt{1 + [g'(x)]^2}\,dx$$

Then

$$\int_C f(s)\,ds = \int_{a_x}^{b_x} f[x, g(x)]\,\sqrt{1 + [g'(x)]^2}\,dx$$

Similarly, if $x = h(y)$

$$\int_C f(s)\,ds = \int_{a_y}^{b_y} f[h(y), y]\,\sqrt{1 + [h'(y)]^2}\,dy$$

Green's Theorem

Let R be a closed region bounded by the simple closed complex curve C. If $g(x, y)$, $h(x, y)$, $\partial g(x, y)/\partial y$, and $\partial h(x, y)/\partial x$ are single valued and continuous on R, then the integral over R is given by

$$\oint [g(x, y)\,dx + h(x, y)\,dy] = \iint_R \left[\frac{\partial h(x, y)}{\partial x} - \frac{\partial g(x, y)}{\partial y}\right] dx\,dy$$

where $\iint_R$ denotes the integral over the region R. Exhibit 13.1.7 illustrates this concept.

Cauchy's Integral Theorem

Let S_o be a closed region bounded by the simple closed complex curve C, and let $f(s)$ be analytic on S_o. If s_o is any interior point for S_o, then

$$f(s_o) = \frac{1}{2\pi i} \oint \frac{f(s)}{s - s_o}\,ds$$

where integration on C is in the positive direction with respect to S_o.

Exhibit 13.1.7 Green's Theorem

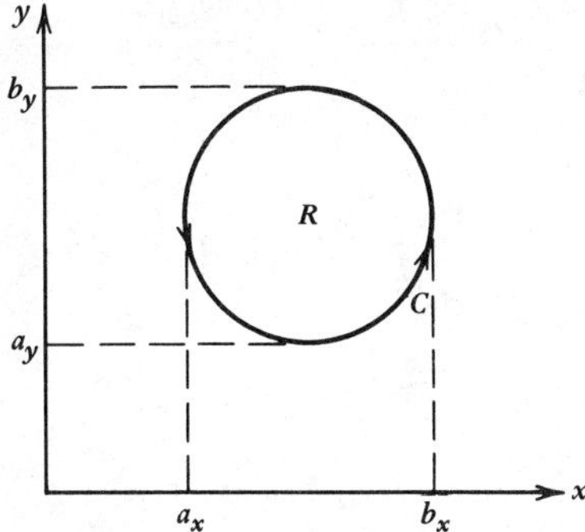

Transforms

Fourier Transforms

Let $f(t)$ be a function of the real variable t. Then

$$F_t(u) = \int_{-\infty}^{\infty} f(t) \exp[-iut]\, dt$$

is the Fourier transform or the Fourier integral of $f(t)$. The functions $F_t(u)$ and $f(t)$ form a transform pair. The Fourier transform is especially useful in probability theory in obtaining the moments of a function $f(t)$. Let the nth moment of $f(t)$ about zero be denoted by m_n. Then

$$m_n = \int_{-\infty}^{\infty} t^n f(t)\, dt$$

If the Fourier transform of $f(t)$ exists, then the nth moment of $f(t)$ about zero is given by

$$m_n = \frac{1}{(-i)^n} \frac{d^n}{du^n} F_t(u) \Bigg|_{u=0}$$

Example 13.1.30 Let

$$f(t) = \begin{cases} \exp[-at] & t \geqslant 0 \\ 0 & t \leqslant 0 \end{cases}$$

where $a > 0$. The Fourier transform of $f(t)$ is

$$F_t(u) = \frac{1}{a + iu}$$

The first and second moments about zero are given by

$$m_1 = \frac{1}{-i} \frac{d}{du} [(a + iu)^{-1}] \Bigg|_{u=0} = \frac{1}{a^2}$$

$$m_2 = \frac{1}{(-i)^2} \frac{d^2}{du^2} [(a + iu)^{-1}] \Bigg|_{u=0} = \frac{2}{a^3}$$

Convolution Theorem

Let $f(t_1)$ and $h(t_2)$ be bounded and continuous at all but a finite number of points on every closed interval $[a, b]$. Assume that

$$\int_{-\infty}^{\infty} |f(t_1)|\, dt_1$$

and

$$\int_{-\infty}^{\infty} |h(t_2)|\, dt_2$$

exist. If

$$g(t) = \int_{-\infty}^{\infty} f(t_1)\, h(t - t_1)\, dt_1$$

and if $F_{t_1}(u)$ and $F_{t_2}(u)$ are the Fourier transforms of $f(t_1)$ and $h(t_2)$, then the Fourier transform of $g(t)$ is given by

$$F_t(u) = F_{t_1}(u)\, F_{t_2}(u)$$

Extending this result to an arbitrary number of functions, let $f_1(t_1), f_2(t_2), \ldots, f_n(t_n)$ have the Fourier transforms $F_1(u), F_2(u), \ldots, F_n(u)$ and let

$$g(t) = \int_{-\infty}^{\infty} \int_{-\infty}^{\infty} \cdots \int_{-\infty}^{\infty} g(t - u_{n-1})\, g(u_{n-1} - u_{n-2}) \ldots g(u_2 - u_1)\, du_1 \ldots du_n$$

where

$$u_k = \sum_{j=1}^{k} t_j, \quad k < n$$

$$t = \sum_{j=1}^{n} t_j$$

Then the Fourier transform of $g(t)$ is given by

$$F_t(u) = \prod_{j=1}^{n} F_{t_j}(u)$$

Inverting the Fourier Transform

Let $F_t(u)$ be the Fourier transform of $f(t)$. If

$$\int_{-\infty}^{\infty} |F_t(u)|\, du < \infty$$

then

$$f(t) = \frac{1}{2\pi} \int_{-\infty}^{\infty} F_t(u) \exp\,[iut]\, du$$

for all values of t at which $f(t)$ is continuous.

Selby[1] gives the more useful Fourier transforms.

Laplace Transforms

The Laplace transform is given by

$$\mathcal{L}[f(t)] = \int_0^\infty f(t) \exp[-st]\, dt$$

Example 13.1.31 The Laplace transform of $t^n \exp[ct]$ is

$$\mathcal{L}[t^n \exp[ct]] = \int_0^\infty t^n \exp[ct] \exp[-st]\, dt$$

$$= \int_0^\infty t^n \exp[-(s-c)\,t]\, dt$$

$$= \frac{\Gamma(n+1)}{(s-c)^{n+1}}$$

where $\Gamma(n + 1)$ is the gamma function evaluated at $n + 1$, which is $n!$

Exhibit 13.1.8 gives the more useful Laplace tranforms.

Characteristic Function

Let x be a continuous random variable with probability density function $f(x)$, where

$$\int_{-\infty}^{\infty} f(x)\, dx = 1$$

The characteristic function of X is given by $\phi_X(u)$ where

$$\phi_X(u) = \int_{-\infty}^{\infty} \exp[iux]\, f(x)\, dx$$

If x is a discrete, integer-valued random variable with probability mass function $p(x)$, where

$$\sum_{x=-\infty}^{\infty} p(x) = 1$$

then the characteristic function of x is given by

$$\phi_X(u) = \sum_{x=-\infty}^{\infty} \exp[iux]\, p(x)$$

The characteristic function plays an important role in probability theory, in generating the moments of a random variable.

The Z Transform

When dealing with discrete variable x, the z transform is very useful. Let $f(x)$ be a function of a discrete variable x. The z transform of $f(x)$, denoted $\Psi_x(Z)$, is defined by

$$\Psi_x(Z) = \sum_{x=-\infty}^{\infty} Z^x f(x)$$

Exhibit 13.1.8 Table of Selected Laplace Transforms

$F(s) = \mathcal{L}\{f(t)\}$	$f(t)$
$1/s$	1
$1/s^2$	t
$1/s^n, \quad (n = 1, 2, \ldots)$	$t^{n-1}/(n-1)!$
$1/\sqrt{s}$	$1/\sqrt{\pi t}$
$1/s^{3/2}$	$2\sqrt{t/\pi}$
$1/s^a \quad (a > 0)$	$t^{a-1}/\Gamma(a)$
$\dfrac{1}{s-a}$	e^{at}
$\dfrac{1}{(s-a)^2}$	te^{at}
$\dfrac{1}{(s-a)^n} \quad (n = 1, 2, \ldots)$	$\dfrac{1}{(n-1)!} t^{n-1} e^{at}$
$\dfrac{1}{(s-a)^k} \quad (k > 0)$	$\dfrac{1}{\Gamma(k)} t^{k-1} e^{at}$
$\dfrac{1}{(s-a)(s-b)} \quad (a \neq b)$	$\dfrac{1}{(a-b)}(e^{at} - e^{bt})$
$\dfrac{s}{(s-a)(s-b)} \quad (a \neq b)$	$\dfrac{1}{(a-b)}(ae^{at} - be^{bt})$
$\dfrac{1}{s^2 + \omega^2}$	$\dfrac{1}{\omega} \sin \omega t$
$\dfrac{s}{s^2 + \omega^2}$	$\cos \omega t$
$\dfrac{1}{s^2 - a^2}$	$\dfrac{1}{a} \sinh at$
$\dfrac{s}{s^2 - a^2}$	$\cosh at$
$\dfrac{1}{(s-a)^2 + \omega^2}$	$\dfrac{1}{\omega} e^{at} \sin \omega t$
$\dfrac{s-a}{(s-a)^2 + \omega^2}$	$e^{at} \cos \omega t$
$\dfrac{1}{s(s^2 + \omega^2)}$	$\dfrac{1}{\omega^2}(1 - \cos \omega t)$
$\dfrac{1}{s^2(s^2 + \omega^2)}$	$\dfrac{1}{\omega^3}(\omega t - \sin \omega t)$
$\dfrac{1}{(s^2 + \omega^2)^2}$	$\dfrac{1}{2\omega^3}(\sin \omega t - \omega t \cos \omega t)$
$\dfrac{s}{(s^2 + \omega^2)^2}$	$\dfrac{t}{2\omega} \sin \omega t$

Exhibit 13.1.8 ***(Continued)***

$F(s) = \mathcal{L}\{f(t)\}$		$f(t)$
$\dfrac{s^2}{(s^2+\omega^2)^2}$		$\dfrac{1}{2\omega}(\sin \omega t + \omega t \cos \omega t)$
$\dfrac{s}{(s^2+a^2)(s^2+b^2)}$	$(a^2 \neq b^2)$	$\dfrac{1}{b^2-a^2}(\cos at - \cos bt)$
$\dfrac{1}{s^4+4a^4}$		$\dfrac{1}{4a^3}(\sin at \cosh at - \cos at \sinh at)$
$\dfrac{s}{s^4+4a^4}$		$\dfrac{1}{2a^2}\sin at \sinh at$
$\dfrac{1}{s^4-a^4}$		$\dfrac{1}{2a^3}(\sinh at - \sin at)$
$\dfrac{s}{s^4-a^4}$		$\dfrac{1}{2a^2}(\cosh at - \cos at)$
$\sqrt{s-a} - \sqrt{s-b}$		$\dfrac{1}{2\sqrt{\pi t^3}}(e^{bt} - e^{at})$

if the sum converges. Like the characteristic function, the z transform finds its primary use in developing the moments of probability distributions. Its applicability is limited to the discrete random variable, however.

REFERENCE

1. S. M. SELBY, *Standard Mathematical Tables*, The Chemical Rubber Company, Cleveland, 1972.

BIBLIOGRAPHY

KREYSZIG, E., *Advanced Engineering Mathematics*, Wiley, New York, 1972.

SCHMIDT, J. W., *Mathematical Foundations for Management Science and Systems Analysis*, Academic Press, New York, 1974.

THOMAS, G. B., *Calculus and Analytic Geometry*, Addison-Wesley, Reading, MA, 1972.

CHAPTER 13.2

Concepts of Probability

ROBERT G. MORRIS

General Motors Institute

13.2.1 INTRODUCTION

The objective of this chapter is to provide a basic review of the concepts and principles of probability and statistics to assist the industrial engineer in skillfully applying statistical methods to experimental and operational data.

Section 13.2.2 begins with definitions of terms associated with probability and some fundamental concepts of set notation used in studying probability. Section 13.2.3 is concerned directly with the concepts and definitions of probability together with the rules for computing probabilities. It also deals with the principles of counting as applied to computing probabilities. Section 13.2.4 follows the probability principles in introducing the concept of a random variable. Sections 13.2.5 and 13.2.6 develop the association of random variables and their probability distributions. Probability distributions provide the basis for a mathematically derived measure of uncertainty in the application of statistical methods.

13.2.2 DEFINITION AND NOTATION

In the ordinary sense, the word "probability" implies the relative certainty or chance with which we expect the occurrence of an event, as when the weather forecaster says, "The probability of rain tomorrow is 20%." This means that weather conditions like those currently prevailing produce rain on the average of once out of every five times they occur, or 20% of the time. Other examples are the chance of a manufacturing process producing a defective part and the probability that a construction job will be finished on time.

In these and similar cases, we are interested in the probability, or likelihood, that an event will occur. To be meaningful, the predictions must be made prior to the event. In many industrial applications, it is not known which outcome will occur, only that one of several possible outcomes will occur. Probability is the means by which the certainty or uncertainty of these outcomes can be quantified and interpreted. To the engineer, the methods of probability analysis offer an opportunity to make decisions with a limited amount of information.

Probability is a number, between 0 and 1, associated with the specified event. A probability of 1 indicates an absolute certainty, and a probability of 0 an impossibility. Generally, improbable events are assigned a 0 probability. Most events have a probability that lies between these two extremes. The probability of an event is a measure of how likely (probability near 1) or unlikely (probability near 0) the event is to occur. If the probability of a manufacturing process producing a defective part is one in a hundred, (.01), the probability of the next part produced being defective is considered to be .01. Conversely, the probability of a good part is .99. The value of an event's probability is determined by using the theory of probability, which is described next.

13.2.3 PROBABILITY THEORY

Probability theory is concerned with a random or chance experiment. An experiment is any well-defined action that generates data. The supposition is that the outcomes of the experiment depend on chance and therefore cannot be predicted with certainty. The outcomes or results of the experiment are then considered to be random, meaning that each outcome of the experiment has an equal opportunity of occurrence.

The set of all possible outcomes of an experiment is called the "sample space," S, of the experiment. Any single outcome in a sample space is called an "element member," or "sample point," of the sample space. The sample space may be denoted as

$$S = \{\text{all possible outcomes}\}$$

The outcomes are actually listed or identified inside the braces by letters, numbers, or whatever is most convenient for or applicable to the problem. To illustrate, consider the examples that follow.

In tossing a coin, the sample space would consist of the outcomes, head and tail. This is written as $S = \{H, T\}$, where the letters H and T identify head and tail, respectively. Numbers could have been used by letting 0 = head and 1 = tail, giving $S = \{0, 1\}$. This is an example of a discrete sample space, since the number of outcomes is finite (can all be listed). Another type of a discrete sample space is one that contains a very large number of outcomes. For example, if the experimental outcomes were the number of acceptable units produced by a high-volume manufacturing process in a given production year, then the number of elements in the associated sample space could be a very large number, $S = \{0, 1, \ldots, \text{total annual volume}\}$. It is appropriate in a situation like this to consider the space to be the whole set of natural numbers (nonnegative integers). This sample space is said to be "countably infinite." A sample space is discrete if there is either a finite or a countably infinite number of outcomes.

In rolling a die and observing the number of dots on the upturned face, the sample space is $X = \{1, 2, 3, 4, 5, 6\}$. This also is a discrete sample space. In measuring the length of a steel bar, the sample space is $S = \{X | X > 0\}$. This is the set of all possible length measures that could be made on a single steel bar. The X's are positive real numbers. Conceptually, the sample space consists of all points in an appropriate interval on a continuous measurement scale, resulting in an infinite number of possible outcomes. Such a sample space is said to be uncountable and is commonly called a "continuous sample space." The continuous sample space is sometimes approximated by a discrete sample space since the measurement devices have limited capability to make the sample space truly continuous. The measurements would be recorded in a distinct number of measurement units, that is, 1.502 cm or 15.02 mm. The precision is determined by the capability of the measuring device used. The value 1.502 cm is representing the continuous interval of values from 1.5015 to 1.5025 cm.

In the example involving the measurement of the length of the steel bars, the measures constitute a continuous sample space; however, in observing the number of steel bars outside of specification in a production lot of 10,000, the sample space would be $S = \{0, 1, 2, \ldots, 10{,}000\}$, which is a discrete sample space.

For any experiment performed, simple or complex, there is a sample space associated with it. Being able to describe this sample space is the first step in determining probabilities associated with it.

Definition of An Event

A collection of one or more sample points is called an "event." It is denoted by a capital letter, such as $E = \{\text{collection of outcomes}\}$. Some examples: In tossing a coin, consider the event a head that appears. The sample space $S = \{H, T\}$. The event $E = \{\text{head}\} = \{H\}$. Consider rolling a die and that the event is the upturned face of an even number. Here $S = \{1, 2, 3, 4, 5, 6\}$, and E = upturned face is an even number = $\{2, 4, 6\}$. In measuring the diameter of a shaft, consider that the event is that the shaft is within specification of 1.500 ± .002 cm.

$$\begin{aligned} S &= \{X | X > 0\} \\ E &= \text{shaft within specification} \\ &= \{X | 1.498 \text{ cm} \leqslant X \leqslant 1.502 \text{ cm}\} \end{aligned}$$

Often the interest is in events that are combinations of two or more events. The relationship between events can be depicted by means of Venn diagrams. The following are definitions, illustrations, and examples of these conditions:

1. **Intersection of Events.** The intersection of events A and B is denoted $A \cap B$ and is the portion of the sample space S that contains all outcomes that are common to both A and B (Exhibit 13.2.1). An example: If A is the event that a customer wishes to buy a seat on the noon flight to Denver, and if B is the event that a seat is available, $A \cap B$ is a sale.

a. The null event, sometimes called the "impossible" or "empty" event, contains no outcomes. It is denoted by the symbol ϕ. An example: If all the power output of a generating plant is being used, then the event E of additional power availability is the null event.

Exhibit 13.2.1 Venn Diagram Showing $A \cap B$

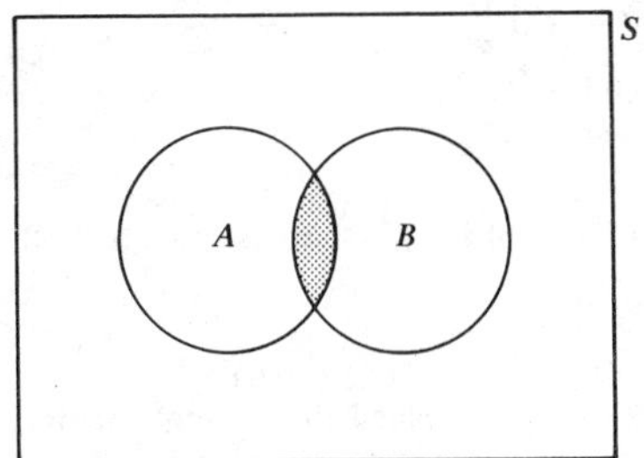

Exhibit 13.2.2 Venn Diagram Showing $A \cup B$

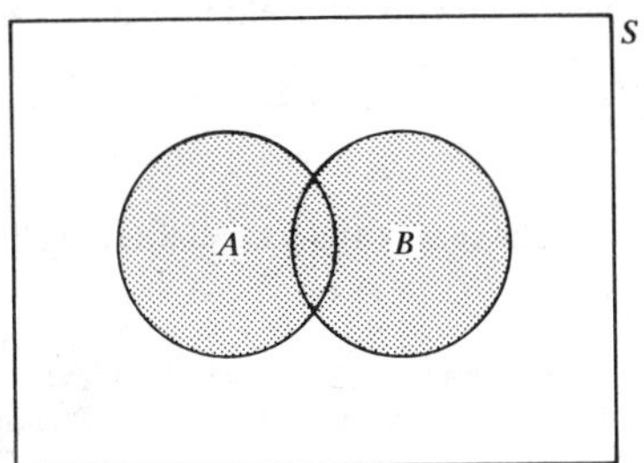

Exhibit 13.2.3 Venn Diagram Showing $\overline{A}$ (Shaded Area)

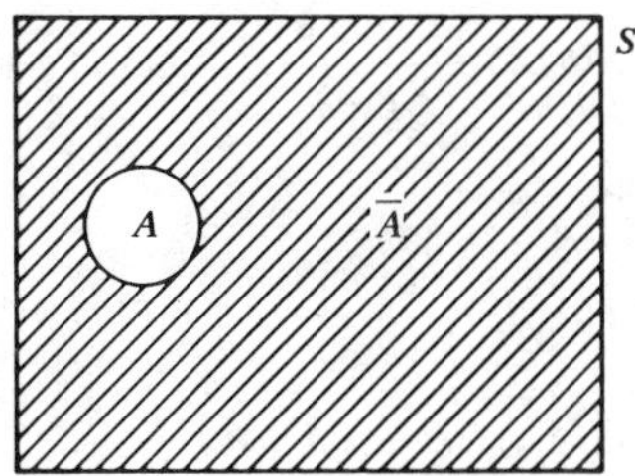

b. Mutually exclusive events are events that have no common outcomes (cannot occur simultaneously). Notationally $A \cap B = \phi$. An example: If A is the event that a passenger is on time for a flight departure, and if B is the event that a passenger is not on time for a flight departure, then A and B are mutually exclusive events.

2. Union of Events. The union of events A and B is the event containing all of the outcomes of event A or B or both. It is denoted as $A \cup B$ (Exhibit 13.2.2). An example: A system is composed of two components, X and Y. It is known that the system will not function if either or both components are defective. If A is the event that component X is defective, and if B is the event that component Y is defective, then the event $A \cup B$ is the event that X is defective or Y is defective or both are defective.

3. Complement of an Event. The complement of an event A with respect to the sample space S is the set of all outcomes of S that are not contained in A. The complement of A is denoted as $\overline{A}$ or A' (Exhibit 13.2.3). An example: All checking accounts at a local bank receive free checking if a \$200 minimum balance is maintained. If event A represents all checking customers with a balance of \$200 or more, $\overline{A}$, the complement of A, represents those customers with a balance less than \$200.

Definition of an Event Probability

An experiment is performed N times. The number of times, f, that an event E occurs is recorded, and the relative frequency, f/N, is computed. The limit of the ratio $\{f/N\}$ as N approaches infinity is the probability of the event E and is denoted as $P(E)$. In symbols

$$\lim_{N \to \infty} (f/N) = P(E)$$

Performing an experiment independently N times is often referred to as "performing N trials of the experiment." "As N approaches infinity" is the concept of performing "an infinitely long series of trials." Intuitively, this is the concept of "in the long run." This definition is simply saying that the probability of an event is the relative frequency with which the event takes place in the long run.

In addition to the relative frequency concept, there are two other avenues used in obtaining an event probability. One is a theory based on a finite set of equally likely outcomes, and the other is a subjective measure based upon the degree of belief one holds in a specified proposition.

Properties of an Event Probability

The properties of an event probability are as follows:

1. The probability of an event E is a number between 0 and 1.

$$0 \leqslant P(E) \leqslant 1$$

2. The sum of the probabilities of all possible outcomes of a given sample space is equal to one.

$$P(S) = 1$$

3. If A and B are two mutually exclusive events in S, then $P(A \cup B) = P(A) + P(B)$.
4. If E is any event in S, then P (not E) $= P(\bar{E}) = 1 - P(E)$. This is called the "complementary probability."
5. Conditional probabilities and the multiplication rule: The probability of an event B's occurring when it is known that some event A has occurred (or is certain to occur) is called a "conditional probability." It is denoted as $P(B|A)$ and is given by

$$P(B|A) = \frac{P(A \cap B)}{P(A)}, \qquad \text{if } P(A) \neq 0 \tag{1}$$

or

$$P(A|B) = \frac{P(A \cap B)}{P(B)}, \qquad \text{if } P(B) \neq 0 \tag{2}$$

If $P(A|B) = P(A)$ and $P(B|A) = P(B)$, the events A and B are said to be independent and the following applies: The events A and B are independent if and only if

$$P(A \cap B) = P(A)\,P(B) \tag{3}$$

If n events are mutually independent, then the probability of their joint occurrence is the product of their individual probabilities and is given by

$$P(A_1 \cap A_2 \cap \cdots \cap A_n) = P(A_1)\,P(A_2) \ldots P(A_n) \tag{4}$$

The multiplication rule is often referred to as the "and" rule, because in verbalization of the problem statement, the "and" connector is used.

Example 13.2.1 An assembly is made up of two independently machined components, X and Y. If A is the event that X is defective, if B is the event that Y is defective, and if $P(A) = .2$ and $P(B) = .1$, find the probability of an assembly in which both are defective. From equation 3, the probability that both components are defective in an assembly is equal to $P(A \cap B) = P(A) \times P(B) = (.2)\,(.1) = .02$. Events A and B are assumed to be independent since the products were machined on different machines and the quality of one does affect the quality of the other.

A general case for conditional probabilities is embodied in a rule called "Bayes' theorem." It has many applications in statistical analysis, especially in connection with subjective or personalistic probabilities.

The expression for Bayes' theorem is

$$P(B_k|A) = \frac{P(B_k) \times P(A|B_k)}{\sum_{i=1}^{n} P(B_i)\,P(A|B_i)} \tag{5}$$

where $B_1 \cup B_2 \cup \cdots \cup B_n = S$ and $B_i \cap B_j = \phi, i \neq j$. This rule provides a means for calculating the conditional probability $P(B_k|A)$.

$P(B_k)$ = a priori or existence probability of event B_k $(k = 1, 2, \ldots, n)$
A = sample event, a nonempty subset of the sample space—the observation.
$P(A|B_k)$ = conditional probability of event A given B_k as the source of $A (P(B_k|A)$ is the posterior probability.)

The type of problem where Bayesian statistics can be most helpful is one where an engineer is faced with many alternatives in making a decision. The technique considers the alternatives as several hypotheses, $H_1, H_2, \ldots, H_n$, and mathematically evaluates the most likely hypothesis.

6. The addition rule: If A and B are any two events, then the probability of "at least one" of them occurring is

$$P(A \text{ or } B) = P(A \cup B) = P(A) + P(B) - P(A \cap B) \tag{6}$$

If A, B, and C are any three events, then

$$\begin{aligned} P(A \text{ or } B \text{ or } C) &= P(A \cup B \cup C) \\ &= P(A) + P(B) + P(C) - P(AB) - P(AC) - P(BC) + P(A \cap B \cap C) \end{aligned} \tag{7}$$

If the two events A and B are *mutually exclusive*, that is, that it is impossible for both to happen, the addition law becomes

$$P(A \cup B) = P(A) + P(B) \tag{8}$$

For n mutually exclusive events, the rule becomes

$$P(A_1 \cup A_2 \cup \cdots \cup A_n) = P(A_1) + \cdots + P(A_n)$$

Example 13.2.2 An assembly is being made up of two components, X and Y, which are considered independent since they come from different sources. If A is the event that X is defective and $P(A) = .10$, and if B is the event that Y is defective and $P(B) = .05$, what is the probability that an assembly is defective? An assembly is defective if X or Y or both are defective.

$$\begin{aligned} P(\text{defective assembly}) &= P(A) + P(B) - P(A \cap B) \\ &= .05 + .10 - .005 = .145 \end{aligned}$$

assuming A and B are independent.

Counting Rules Used in Computing Probabilities

In calculating the probability or relative frequency f/N of an event A, it is necessary to determine f (the number of times that event A occurs) and N (the total number of outcomes of the experiment). The basic principle commonly applied to determine these values is known as "combinatorial analysis." The fundamental principle of counting states: If one experiment can result in any N possible outcomes, and another in R possible outcomes, then taken together, there are $N\,R = NR$ possible outcomes of the two experiments. In general, if there are N_1 outcomes of the first experiment, N_2 outcomes of a second, N_3 of a third, and N_k of the kth experiment, then there are $N_1\, N_2\, N_3 \ldots N_k$ possible outcomes of the k experiments. For example, if an assembly consists of three parts, A, B, and C, and if there are 4 sources for part A, 5 sources for part B, and 3 sources for part C, then there are $4 \times 5 \times 6 = 120$ possible ways of selecting parts A, B, and C.

If order is important in arrangements of a group of objects, such as in the orderings of three letters, *DGF*, then there are six possible orderings of these letters (*DGF*, *DFG*, *FGD*, *GDF*, *FDG*, *GFD*). These orderings are called "permutations."

The number of permutations of N distinct objects is $N! = (N)\,(N-1)\,(N-2) \ldots 3 \times 2 \times 1$. In our example the number of orderings of the three letters *DGF* is $3! = 3 \times 2 \times 1 = 6$.

The number of permutations of N distinct objects taken r at a time is ${}_NP_r = N!/(N-r)!$

In our example of *DGF*, if we select two letters or parts at a time from the group of three, the total number of arrangements would be

$${}_3P_2 = \frac{3!}{(3-2)!} = \frac{(3)\,(2)\,(1)}{(1)} = 6$$

or *DG*, *DF*, *GF*, *GD*, *FD*, *FG*.

If ordering is not important in arrangements, then we refer to these arrangements as "combinations." Stated another way, a combination is a permutation with the importance of order removed.

The numbered combinations of N outcomes taken r at a time are

$$\binom{N}{r} = {}_NC_r = \frac{{}_NP_r}{r!} = \frac{N!}{r!\,(N-r)!}$$

Example 13.2.3 From four nurses and six nurse's aides, determine the number of four-person teams that can be formed consisting of two nurses and two nurse's aides. Number of ways to select two nurses:

$$_4C_2 = \frac{4!}{2!\,(4-2)!} = \frac{(4)\,(3)\,(2)\,(1)}{(2)\,(1)\,(2)\,(1)} = 6$$

Number of ways to select two nurse's aides:

$$\binom{6}{2} = {}_6C_2 = \frac{6!}{2!\,(6-2)!} = \frac{(6)\,(5)\,(4)\,(3)\,(2)\,(1)}{(2)\,(1)\,(4)\,(3)\,(2)\,(1)} = 15$$

The number of teams possible is

$$6 \times 15 = 90$$

There are other general theorems and concepts relating to permutations, combinations, and partitioning of populations or sample spaces, but those just discussed have the highest utility. The reader is referred to Blank,[1] Walpole and Myers,[2] and Parzen[3] for more detailed information.

13.2.4 RANDOM VARIABLES

In most industrial applications, it is desirable to have numbers associated with each outcome of an experiment's sample space. The orderly arrangement and analysis of these numerical sample spaces is facilitated by the concept of a random variable.

A "random variable" is a function whose value is a real number determined by each element in the sample space of the statistical experiment. Capital letters, such as X, are used to denote a random variable, and the corresponding lowercase letter, that is, x, denotes one of its values. A random variable is often referred to as a rule for assigning numbers to each experimental outcome or sample point. The rule would be a counting or a computational procedure (X) that yields a number x.

A random variable may be discrete; that is, it can take on a finite or countable infinite number of distinct values. Observing the number of "undercount" bottles of aspirin in a sample of 50 bottles leads to a discrete random variable. Alternately, a random variable may taken on any value in a measurement interval. Such a variable is a continuous random variable. The maintenance repair time in a fork truck example would be described as a continuous random variable.

Random variables associated with counting are discrete; those associated with scale measurements are continuous. Correspondingly, "count" data are referred to as "discrete data," whereas "measured" data are referred to as "continuous data."

Examples of discrete and continuous random variables are:

1. The number of hours required to do routine maintenance on a fork truck in a scheduled maintenance program (continuous random variable).
2. The number of bottles of aspirin that contain less than standard count in a sample of 50 bottles selected at random from a carton (discrete random variable).
3. The number of stock runouts reported per day at a catalog warehouse (discrete random variable).
4. The amount of electric energy generated per hour at a generating plant (continuous random variable).

A function of one or more random variables is also a random variable. In the fork truck example listed first, the average of 100 such time measures is a random variable.

In practice, the choice of a random variable is one of convenience and is dictated by the questions the experimenter wants to answer. For example, in tossing a die, the outcome can be formalized as a random variable X that can take on one of the six values $X = 1, 2, 3, 4, 5, 6$. In an industrial setting, consider a work sampling study being made on a production line that contains four machines. Each time the four machines are observed, their activity state of being busy (B) or idle (I) is the experimenter's primary concern. Counting rules indicate that 2^4, or 16, different outcomes (state of the machines) are possible: *IIII*, *IIIB*, *IIBI*, *IBII*, *BIII*, *BIIB*, *BIBI*, *BBII*, *BBIB*, *BBBI*, *BIBB*, *IIBB*, *IBBB*, *IBIB*, *IBBI*, *BBBB*. These 16 outcomes constitute the sample space. The outcomes are not equally likely unless the two states B and I are equally likely. A random variable X could then be defined as the number of machines that are observed to be busy each time the experimenter visits the floor, or $X = (0, 1, 2, 3, 4)$. This type of example could be extended to n

machines, where the experiment again would be the work sampling, the sample space $S = \{2^n$ outcomes$\}$, and the random variable the number of machines observed to be busy, or $X = \{0, 1, 2, \ldots, n\}$.

If the work sampling experiment were repeated a number of times, a relative frequency for each observed value of X could be calculated. The observed relative frequencies are estimates of the probabilities associated with each and provide a means for the experimenter to study the behavior patterns of the random variable in question. The studies are done by using graphic and computational methods associated with probability distribution functions, a discussion of which follows.

Probability Distribution Functions

The mathematical relationship that associates each value of a random variable with the probability that the value will be assumed is referred to as a "probability distribution" or "probability density" function. The probability distribution function (PDF) for discrete random variables is denoted as $p(x)$ or $f(x)$. The probability density function for continuous random variables is denoted as $f(x)$.

An equivalent way to describe the relationship between random variables and their associated probabilities is by means of a cumulative distribution function (CDF), denoted as $F(x)$. A more thorough discussion of PDFs, CDFs, and their properties follows.

Definition of a Discrete PDF

The function $f(x)$ is a PDF of the discrete random variable X if, for each outcome x,

$0 \leq f(x) \leq 1$ [The $f(x)$ values are nonnegative.]
$\Sigma f(x) = 1$ (The sum of the probabilities of all the experiment's outcomes is equal to 1.)

where $P(X = x) = f(x)$.

Example 13.2.4 Discrete case: In a time study to establish production standards for machine operators, a record of the number of machine breakdowns was considered important. The number of machine breakdowns was recorded for 100 days of operation, with the results as shown in Exhibit 13.2.4.

If the random variable X indicates the number of machine breakdowns on a given day, it can assume the values 1, 2, 3, 4, where $x = 1$ is one breakdown per day, $x = 2$ is two breakdowns per day, $x = 3$ is three breakdowns per day, and $x = 4$ is four breakdowns per day. The estimated probabilities are

$$P(X = 1) = f/N = 10/100 = .10$$
$$P(X = 2) = 61/100 = .61$$
$$P(X = 3) = 20/100 = .20$$
$$P(X = 4) = 9/100 = .09$$

The PDF is written as

$$P(X = x) = \begin{cases} .10 \text{ if } x = 1 \\ .61 \text{ if } x = 2 \\ .20 \text{ if } x = 3 \\ .09 \text{ if } x = 4 \end{cases}$$

Exhibit 13.2.4 Frequency Table of Machine Breakdowns

Number of Machine Breakdowns per Day (x)	Number of Days for Each Breakdown Frequency (f)
1	10
2	61
3	20
4	9

Note. $100 = N$.

Definition of a Continuous PDF

The function $f(x)$ is a PDF of the continuous random variable X defined over the real line if

$$f(x) \geqslant 0 \qquad \text{for all } x$$

$$\int_{-\infty}^{\infty} f(x)\, dx = 1$$

$$P(a < X < b) = \int_{a}^{b} f(x)\, dx$$

Example 13.2.5 The life expectancy of an electrical component is described by the following PDF:

$$f(x) = \begin{cases} \lambda e^{-\lambda x} & \text{for } x > 0 \text{ (all positive values of } x) \\ 0 & \text{elsewhere } (x \leqslant 0) \\ 0 < \lambda = \text{failure rate} \end{cases}$$

The term "elsewhere" is used to describe the variable range that does not apply, since it is assumed that a component cannot fail prior to its use.

Find the probability of a component life between 1 and 3 hr ($1 \leqslant X \leqslant 3$).

$$P(1 \leqslant X \leqslant 3) = \int_{1}^{3} \lambda e^{-\lambda x}\, dx = e^{-\lambda} - e^{-3\lambda}$$

If $\lambda = 0.01$, then

$$\begin{aligned} P(1 \leqslant X \leqslant 3) &= e^{-0.01} - e^{-0.03} \\ &= 0.99005 - 0.97045 = .01961 \end{aligned}$$

Definition of the CDF

A CDF accumulates the probability over the range of values for the variable associated with the outcomes. It is described as the probability that the random variable X has a value less than or equal to some specific value x, that is, $F(x) = P(X \leqslant x)$.

The discrete case is given as

$$F(x) = P(X \leqslant x) = \sum_{y \leqslant x} f(y)$$

Example 13.2.6 In Example 13.2.4 about machine breakdown, what is the probability of less than four breakdowns?

$$P(X \leqslant 3) = \sum_{i=1}^{3} f(x) = f(1) + f(2) + f(3) = 0.10 + 0.61 + 0.20 = .91$$

The continuous case is given as

$$P(X \leqslant x) = F(x) = \int_{-\infty}^{x} f(y)\, dy$$

Example 13.2.7 In Example 13.2.5, what is the probability that component life is less than or equal to 3 hr?

$$\begin{aligned} P(X \leqslant 3) = F(3) &= \int_{0}^{3} \lambda e^{-\lambda x}\, dx = -e^{-\lambda x}\Big|_{0}^{3} = e^{0} - e^{-3\lambda}, \qquad \text{if } \lambda = 0.01 \\ &= 1 - e^{-0.03} = 1 - 0.97045 \\ &= .90925 \end{aligned}$$

Exhibit 13.2.5 Number of Machine Breakdowns

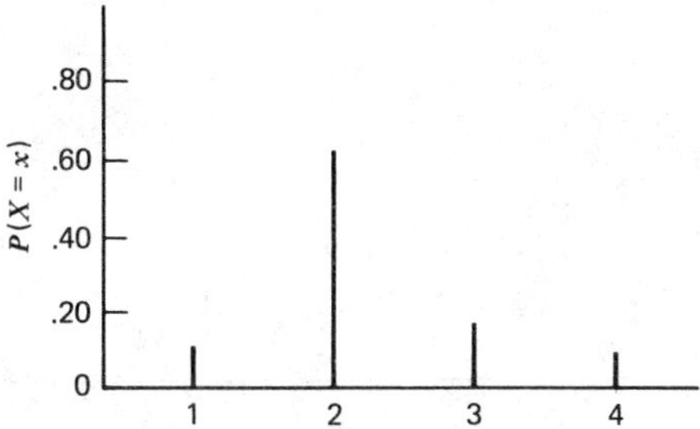

Graphics of Discrete and Continuous PDFs and CDFs

Discrete PDF. The graph of a discrete PDF, commonly called a "probability line chart," is depicted by a series of vertical lines located at the specified values of X.

Example 13.2.8 Using the data from Example 13.2.4 regarding the number of machine breakdowns, the graph would appear as shown in Exhibit 13.2.5, plotted using the points $(x, f(x))$.

Discrete CDF. The graph of a discrete CDF is a step function. It is plotted from the machine breakdown data (Exhibit 13.2.6) as illustrated in Exhibit 13.2.7, using the points $(x, F(x))$.

Probability Density Function of a Continuous Random Variable. The graph of a continuous PDF is nonnegative between $\pm\infty$. The plotting points are $(x, f(x))$.

Example 13.2.9 The PDF of the failure time of a certain electrical component is represented by

$$f(x) = \begin{cases} 0.01e^{-0.01x} & x > 0 \\ 0 & \text{elsewhere} \end{cases}$$

and is shown graphically in Exhibit 13.2.8.

Exhibit 13.2.6 Cumulative Relative Frequencies of Machine Breakdowns

x	$F(x)$
1	0.10
2	0.71
3	0.91
4	1.00

Exhibit 13.2.7 Number of Machine Breakdowns

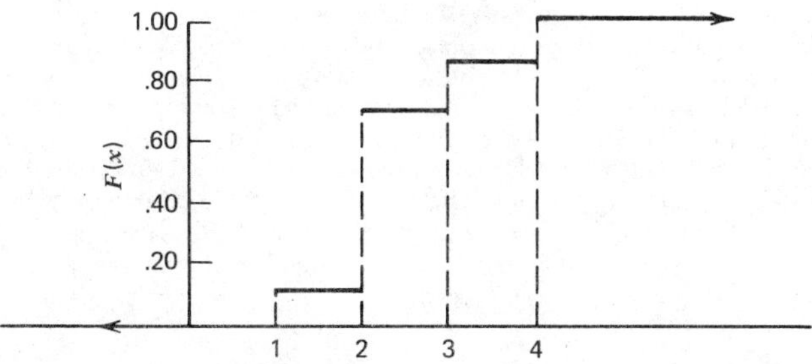

Exhibit 13.2.8 The PDF of Component Failure Times

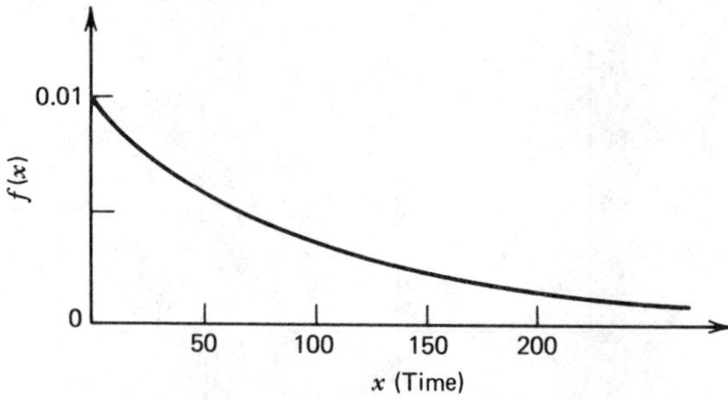

Exhibit 13.2.9 The CDF of Component Failure Times

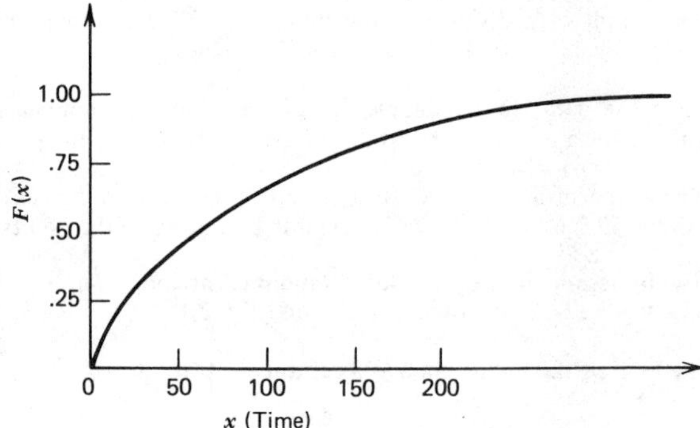

Continuous Distribution Function. The graph of a continuous CDF is represented by a continuous curve between $\pm\infty$. The plotting points are $(x, F(x))$.

Example 13.2.10 In the electrical component example (Example 13.2.5), the failure CDF is

$$F(x) = 1 - e^{-0.01x}$$

which is shown graphically in Exhibit 13.2.9.

13.2.5 EMPIRICAL DISTRIBUTIONS

Empirical distributions are the result of organizing data obtained from the outcomes of a statistical experiment. Different types of tabular and graphic formats are used to display the data in a meaningful manner for analysis.

The descriptive terms most often used in conjunction with empirical data are "relative frequency distribution" and "relative cumulative frequency distribution." The relative frequency distribution is similar to the PDF, and its graphic display is called a "histogram." A relative cumulative frequency distribution is similar to a CDF and is called an "ogive" (oh'jive). (For the necessary procedures to organize the data in these formats, see Blank[1] and Walpole and Myers.[2])

The data in the frequency table shown in Exhibit 13.2.10 represent the lengths of the repair times for machine breakdowns, measured to the nearest tenth of an hour, of a random sample of 115 repair tickets from a maintenance operation. Exhibit 13.2.11 displays the corresponding histogram, and Exhibit 13.2.12 its cumulative frequencies, the ogive.

In many applications, once the graphic representations have been determined, it is assumed that the relative frequency or probability distribution pattern as displayed can be represented by a theoretical PDF or CDF. In the application of the empirical or theoretical distributions, there are common measures of the distributions' characteristics used to describe and interpret their properties. One of these measures is that of central tendency or location, a value around which the observations tend to cluster and which characterizes their magnitude. Commonly used measures of central tendency are the arithmetic mean, the median, and the mode. A second measure of interest is that of dispersion or variation, which measures the variability among the observations. Measures

Exhibit 13.2.10 Frequency Table for Machine Repair Times (Hours)

Class Interval (Hours)	Class Midpoint	Frequency f	Cumulative Frequency	Cumulative Relative Frequency
0.5–1.4	1.2	0	0	.000
1.5–1.9	1.7	3	3	.026
2.0–2.4	2.2	8	11	.096
2.5–2.9	2.7	7	18	.157
3.0–3.4	3.2	10	28	.243
3.5–3.9	3.7	15	43	.374
4.0–4.4	4.2	27	70	.610
4.5–4.9	4.7	15	85	.740
5.0–5.4	5.2	14	99	.860
5.5–5.9	5.7	7	106	.922
6.0–6.4	6.2	6	112	.975
6.5–6.9	6.7	3	115	1.000

used to describe variation are variance, SD, and range. A discussion of each of these measures follows.

Measures of Central Tendency

There are three measures of central tendency: the mean, the median, and the mode. They are single values used to represent all the data resulting from an experiment. The "median" is the value below which and above which 50% of the frequencies lie. The "mode" is the most frequently observed value.

"Arithmetic average," "mean," and "expected value" are terms used to describe the most often used measure of central tendency. The mean of a distribution, either empirical or theoretical, is a central value about which the data balance.

For empirical or random sample data, the mean is calculated as follows:

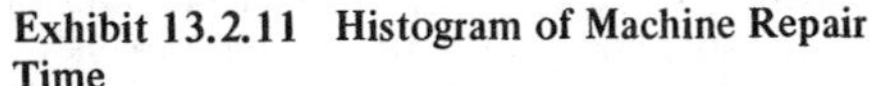

$$\overline{X} = \sum_{i=1}^{n} \frac{X_i}{n}$$

Exhibit 13.2.11 Histogram of Machine Repair Time

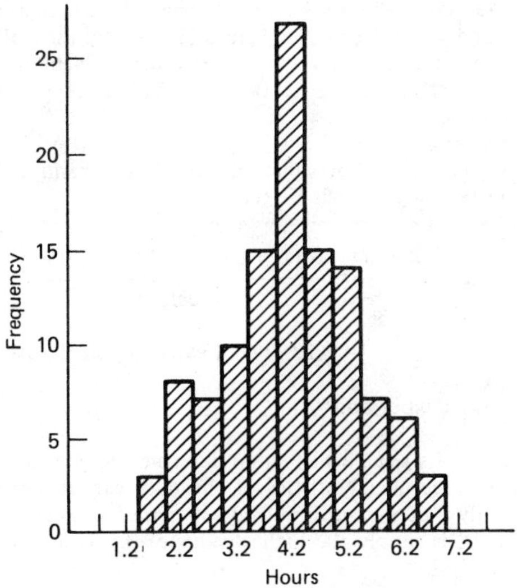

Exhibit 13.2.12 Cumulative Relative Frequency Distribution for Machine Repair Time

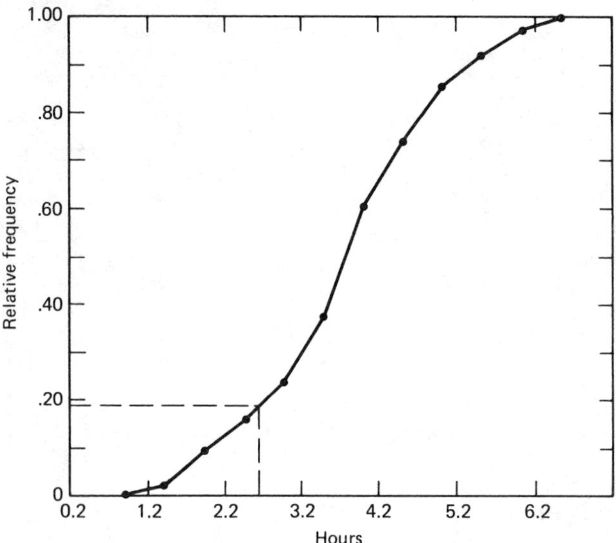

The values of X_i are summed over all $X_i (i = 1, 2, \ldots, n)$ and divided by the number of observations, n. If the data are in the form of a frequency table, as in Exhibit 13.2.10, the formula becomes

$$\bar{X} = \sum_{i=1}^{k} \frac{f_i X_i}{n}$$

where $n = \Sigma_{i=1}^{k} f_i$.

The sample data just mentioned represent a random sample of size n drawn from a population. A "population" is defined as the totality of observations with which the experimenter is concerned. Many random samples of size n are possible from the same population, and the statistics calculated from these samples may vary from sample to sample. A statistic is a function of a random variable and therefore is a random variable. In the previous formula and in those that follow, Latin capital letters are used to represent statistics, since they are random variables calculated from a sample of n independent random variables all of which have the same probability distribution. Lowercase Latin letters are used to represent observed values associated with each random variable; that is, X_i is the random variable, and x_i is the observed value. This is why the statistic $\bar{X}$ assumes the value of $\bar{x} = \Sigma_{i=1}^{n} x_i/n$ when X_1 assumes the value x_1, X_2 assumes the value x_2, and so on. Similarly, in the variation measures discussed in the next section, lowercase letters are used when observed data are substituted in their formulas.

Example 13.2.11 In the problem of machine breakdowns (Example 13.2.4), $X = 1, 2, 3, 4$. Using the frequencies corresponding to each X_i, the mean is

$$\bar{x} = \sum_{i=1}^{4} \frac{f_i x_i}{n} = \frac{10(1) + 61(2) + 20(3) + 9(4)}{100}$$

$$= 2.28 \text{ breakdowns/day on the average}$$

Measures of Variation and Dispersion

A frequency distribution and its related histogram describe the general variability pattern of the data they represent. These same characteristics of the data can also be represented by a single measure. The most commonly used measures of dispersion are the range, the SD, and the variance.

The "range," denoted as R, is simply the highest value of the data minus the lowest value of the data.

$$R = (X_H - X_L)$$

where X_H is the highest value of X and X_L is the lowest value of X. The range is used when the sample size is small and the data have not been organized into a frequency table. Considerable use is made of the range in quality control applications.

The measure most frequently used for describing dispersion is the SD, denoted S. It is defined mathematically as

$$S = \sqrt{\frac{\Sigma_{i=1}^{n} (X_i - \overline{X})^2}{n - 1}}$$

where n is the sample size. The formula can also be expressed in a computationally easier form as

$$S = \sqrt{\frac{n\Sigma_{i=1}^{n} X_i^2 - (\Sigma_{i=1}^{n} X_i)^2}{n(n - 1)}}$$

If the data are in the form of a frequency table, the formula becomes

$$S = \sqrt{\frac{n\Sigma_{i=1}^{k} f_i X_i^2 - (\Sigma_{i=1}^{k} f_i X_i)^2}{n(n - 1)}}$$

where k is the number of different values of X used in the table.

In the machine breakdown example (Example 13.2.4), s would be

$$s = \sqrt{\frac{100\ \Sigma_{i=1}^{4} f_i X_i^2 - (\Sigma_{i=1}^{4} f_i x_i)^2}{4(4 - 1)}}$$

$$= \sqrt{\frac{100[(10)(1^2) + (61)(2^2) + 20(3^2) + 9(4)^2] - (228)^2}{100(99)}}$$

$$= 0.72 \text{ breakdowns}$$

The third measure of dispersion is variance. "Variance" is the square of the SD and is denoted by S^2. In the machine breakdown problem,

$$s^2 = (\text{SD})^2 = (0.72)^2 = 0.552$$

13.2.6 THEORETICAL DISTRIBUTIONS

In the discussion of the mean, SD, and variance, the formulas and examples used were all associated with empirical distributions. Similar measures exist for theoretical probability distributions. The values $\overline{X}$, S, and S^2 are statistics calculated from sample data. The values μ, σ, and σ^2, representing similar properties of theoretical probability distributions, are constants. The comparison is shown in Exhibit 13.2.13.

The basic theoretical formulas are

$$\mu = E(X) = \Sigma\, xf(x) \qquad \text{(discrete case)}$$

$$\mu = E(X) = \int_{-\infty}^{\infty} xf(x)\, dx \qquad \text{(continuous case)}$$

$$\sigma^2 = V(X) = \Sigma\, x^2 f(x) - \mu^2 \qquad \text{(discrete case)}$$

Exhibit 13.2.13 Comparison Table

Empirical Distributions	Theoretical Distribution
$\overline{X}$ = mean	μ (mu)
S = SD	σ (sigma)
S^2 = variance	σ^2 (sigma squared)

Exhibit 13.2.14 Discrete Distributions

Theoretical Distribution	Parameters	Probability Distribution Function	Expected Value or Mean μ	Variance σ^2
Uniform	$k = 1, 2, \ldots$	$f(x; k) = \frac{1}{k} \quad x = x_1, x_2, \ldots, x_k$ $= 0 \quad$ otherwise	$\frac{\Sigma_{i=1}^{k} x_i}{k}$	$\frac{\Sigma_{i=1}^{k} (x_i - \mu)^2}{k}$
Binomial	$n = 1, 2, \ldots$ $0 \leqslant p \leqslant 1$ $q = 1 - p$	$b(x; n, p) = \binom{n}{x} p^x q^{n-x}$ $= 0 \quad$ otherwise	np	npq
Poisson	$\mu > 0$	$p(x; \mu) = \frac{\mu^x}{x!} e^{-\mu}$ $= 0$ otherwise	μ	μ
Hypergeometric	$N = 1, 2, \ldots$ $n = 1, 2, \ldots, N$ $k = 0, 1, 2, \ldots, N$	$h(x; N, n, k) = \frac{\binom{k}{x}\binom{N-k}{n-x}}{\binom{N}{n}} \quad x = 0, 1, \ldots, \min(k, n)$ $= 0$ otherwise	$\frac{nk}{N}$	$\left(\frac{N-n}{N-1}\right) n \left(\frac{k}{n}\right)\left(1 - \frac{k}{N}\right)$

Exhibit 13.2.15 Continuous Distributions

Theoretical Distribution	Parameters	Probability Density Function $f(x)$	Expected Value (Mean) μ	Variance σ^2
Normal	$-\infty < \mu < \infty$, $\sigma > 0$	$f(x) = \frac{1}{\sigma(2\pi)^{1/2}} \exp\left(-\frac{1}{2}[(x-\mu)/\sigma]^2\right)$	μ	σ^2
Exponential	$\lambda > 0$	$f(x) = \lambda e^{-\lambda x} \quad x > 0$; $= 0$ otherwise	$\frac{1}{\lambda}$	$\frac{1}{\lambda^2}$
Uniform	$-\infty < \alpha < \beta < \infty$	$f(x) = \frac{1}{\beta - \alpha}$ for $a < x < \beta$; $= 0$ otherwise	$\frac{\alpha + \beta}{2}$	$\frac{(\beta - \alpha)^2}{12}$
Gamma	$\alpha > 0, \quad \beta > 0$	$f(x) = \frac{1}{\beta^{\alpha}\Gamma(\alpha)} x^{\alpha-1} e^{-x/\beta} \quad x > 0$; $= 0$ elsewhere	$\alpha\beta$	$\sigma\beta^2$
Beta	$a > 0$, $b > 0$	$f(x) = \begin{cases} \frac{1}{B(a,b)} x^{a-1}(1-x)^{b-1} & 0 < x < 1 \\ 0 \text{ otherwise} \end{cases}$ where $B(a,b) = \int_0^1 x^{a-1}(1-x)^{b-1}\,dx$	$\frac{a}{a+b}$	$\frac{ab}{(a+b)^2\,(a+b+1)}$

$$\sigma^2 = V(X) = \int_{-\infty}^{\infty} x^2 f(x)\, dx - \mu^2$$

$$\sigma = \sqrt{\sigma^2}$$

There are hundreds of probability distributions, but only a few are used commonly. Those used most often and their corresponding distribution functions and parameters are listed in Exhibits 13.2.14 and 13.2.15.

Probability distribution functions are used in making inferences about a population. The observations from the population are considered to be numerical values of a random variable X, which has a probability function $f(x)$.

The true $f(x)$ is usually unknown, and a standard probability function, such as those in Exhibits 13.2.14 and 13.2.15 is usually used as a model. The experimenter would choose the model that gives the best approximation of the distribution associated with the population. Information in the form of sample statistics, a histogram, and/or an ogive, developed from a representative random sample of the population in question, would serve as a guide in selecting the standard model to be used for $f(x)$. It would be assumed that the mean and the variance of the chosen $f(x)$ are the mean and the variance of the correspondng population. The normal distribution is used to demonstrate some of these relationships in the next section.

The Normal Probability Distribution

It is often the tendency for a majority of sample measurements to cluster about the center or average value. In fact, for a controlled process the pattern is so repetitive that a general bell-shaped curve can be formed by smoothing over the frequency distribution or histogram. This bell-shaped curve is called the "Gaussian" or "normal distribution" (Exhibit 13.2.16). If a histogram approximates the form of the bell-shaped curve, the sample is usually considered to be from a normal distribution. The normal distribution is by far the most frequently used statistical model.

The normal distribution is symmetrical about its average value. It is precisely defined by designating its two parameters, the mean μ and the SD σ. Thus changing any one or both of these parameters defines a different normal distribution. The baseline of the curve extends to infinity in both directions from the mean. The normal distribution is the assumed population distribution for many industrial and engineering observations. Generally, the normal distribution portrays control. Most variations in manufacturing processes are considered to follow this pattern.

In Exhibit 13.2.16 the area under the theoretical normal distribution is divided into six areas, three on each side of the average. Each area is one SD in width on the baseline. If the observations are normally distributed, 99.7% of the outcomes will fall within ± 3 SD of the average.

The curve in Exhibit 13.2.16 is commonly referred to as the "standard normal distribution curve." The values along the baseline represent the distribution of the standard variate, Z. This

Exhibit 13.2.16 Approximate Probability Areas for the Standardized Normal Distribution

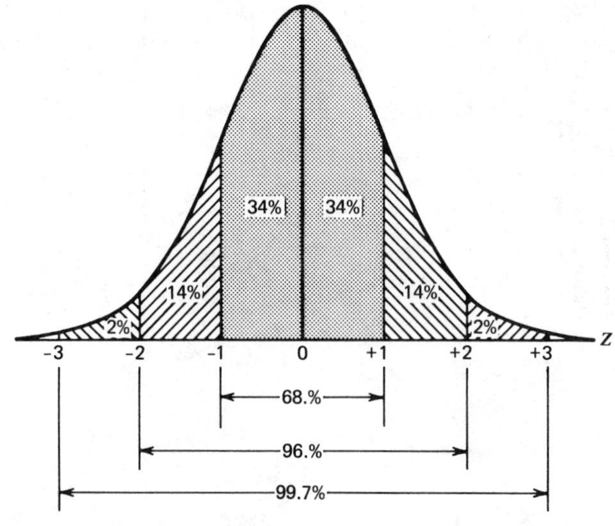

value is a result of expressing any normally distributed random variable, X, as a deviation from its mean, μ, measured in multiples of its SD, σ.

$$Z = \frac{X - \mu}{\sigma} \quad \text{(mean of } Z = 0\text{, variance of } Z = 1\text{)}$$

This simple transformed random variable has several desirable properties, namely:

1. It is a dimensionless value, since all measurement units cancel.
2. It is a small number and easy to work with 99.74% of the X values have a resulting Z value between ± 3.
3. It requires only one table of its cumulative probabilities since any normal random variable X can always be reduced to a corresponding Z value.

Exhibit 13.2.17 Unit Normal Distribution

K_α	$f(z)$	$P(z < K_\alpha)$	$P(z > K_\alpha)$	$P(\lvert z\rvert > K_\alpha)$	$P(\lvert z\rvert < K_{\alpha/2})$
0.0	.3989	.5000	.5000	1.0000	.0000
0.1	.3970	.5398	.4602	.9203	.0797
0.2	.3910	.5793	.4207	.8415	.1585
0.3	.3814	.6179	.3821	.7642	.2358
0.4	.3683	.6554	.3446	.6892	.3108
0.5	.3521	.6915	.3085	.6171	.3829
0.6	.3332	.7257	.2743	.5485	.4515
0.7	.3123	.7580	.2420	.4839	.5161
0.8	.2897	.7881	.2119	.4237	.5763
0.9	.2661	.8159	.1841	.3681	.6319
1.0	.2420	.8413	.1587	.3173	.6827
1.1	.2179	.8643	.1357	.2713	.7287
1.2	.1942	.8849	.1151	.2301	.7699
1.3	.1714	.9032	.0968	.1936	.8064
1.4	.1497	.9192	.0808	.1615	.8385
1.5	.1295	.9332	.0668	.1336	.8664
1.6	.1109	.9452	.0548	.1096	.8904
1.7	.0940	.9554	.0446	.0891	.9109
1.8	.0790	.9641	.0359	.0719	.9281
1.9	.0656	.9713	.0287	.0574	.9426
2.0	.0540	.9772	.0228	.0455	.9545
2.1	.0440	.9821	.0179	.0357	.9643
2.2	.0355	.9861	.0139	.0278	.9722
2.3	.0283	.9893	.0107	.0214	.9786
2.4	.0224	.9918	.0082	.0164	.9836
2.5	.0175	.9938	.0062	.0124	.9876
2.6	.0136	.9953	.0047	.0093	.9907
2.7	.0104	.9965	.0035	.0069	.9931
2.8	.0079	.9974	.0026	.0051	.9949
2.9	.0060	.9981	.0019	.0037	.9963
3.0	.0044	.9987	.0013	.0027	.9973
1.2816	.1755	.9000	.1000	.2000	.8000
1.6449	.1031	.9500	.0500	.1000	.9000
1.9600	.0584	.9750	.0250	.0500	.9500
2.0537	.0484	.9800	.0200	.0400	.9600
2.3263	.0267	.9900	.0100	.0200	.9800
2.5758	.0145	.9950	.0050	.0100	.9900

Exhibit 13.2.18 Shaded Area Relating to the Probability

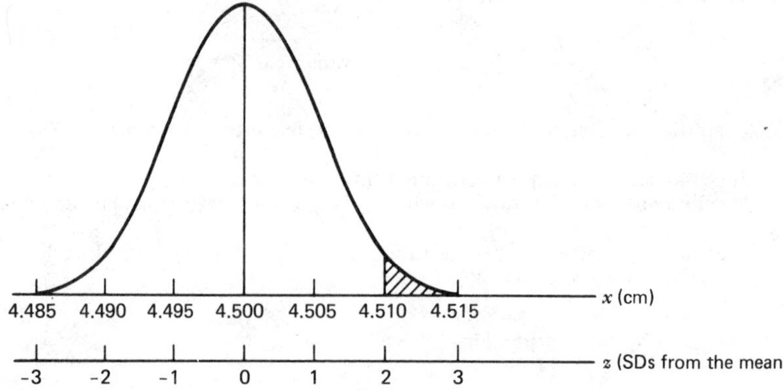

It was pointed out earlier that the area under a probability distribution indicates probability. Thus, to determine a desired probability for a normal distribution, we make use of the standardized normal variate Z and a table of areas applicable to the entire family of normal distributions (Exhibit 13.2.17).

Example 13.2.12 The finished inside diameter of a piston ring is normally distributed with a mean of 4.500 cm and an SD of 0.005 cm. What is the probability of obtaining a diameter exceeding 4.510 cm?

Given

$$\mu = 4.500 \text{ cm}$$

$$\sigma = 0.005 \text{ cm}$$

$$Z = \frac{4.510 \text{ cm} - 4.500 \text{ cm}}{0.005 \text{ cm}} = \frac{0.01}{0.005} = 2.00$$

Hence

$$\text{probability (diameter} > 4.510 \text{ cm}) = P(Z > 2.00)$$

From table of normal areas

$$P(Z > 2.00) = 0.0228$$

The solution is represented graphically in Exhibit 13.2.18. The shaded area represents the probability value sought.

Distribution of Sample Averages

Repeated samplings from a population will produce sample means, $\overline{X}$, and SDs, S, which will form distributions of their own. Distributions formed in this manner are called "sampling distributions."

The sampling distribution of $\overline{X}$ has properties that are related to those of the population from which the sample was selected. The theory of the $\overline{X}$ sampling distribution is based upon a statistical law known as the central limit theorem. This theorem states that the averages of repeated samples of size n will tend to distribute normally, regardless of the distribution from which they are selected (except in some rare cases). This is one reason why the normal distribution is frequently applicable.

Another characteristic of the sampling distribution of $\overline{X}$ is that its variation is not as great as that of the distribution of individual values. This is illustrated in Exhibit 13.2.19.

The combined effect of these results can be stated as follows:

1. The mean of the distribution of sample means is the same as the mean of the population of individual values from which the sample was taken ($\mu_X = \mu_{\overline{X}}$).

2. The SD of the distribution of sample means equals $1/\sqrt{n}$ times the SD of the universe of individual values ($\sigma_{\overline{X}} = \sigma_X/\sqrt{n}$).

Exhibit 13.2.19 Relationship Between Distributions of Averages and Distributions of Individuals

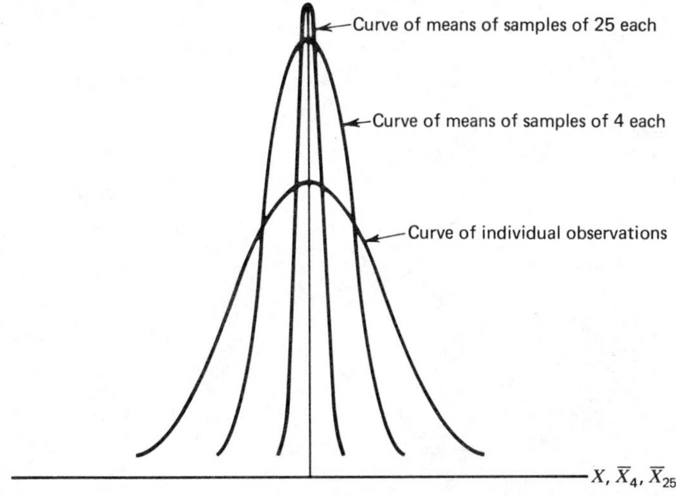

Example 13.2.13 Past data suggest that the mean diameter of bushings turned out by a process is 22.57 cm and that the SD is 0.08 cm. Estimate the probability that a sample of 16 bushings will have a mean diameter equal to or greater than 22.63 cm.

Given

$$\mu_X = 22.57 \text{ cm}$$

$$\sigma_X = 0.08 \text{ cm}$$

$$n = 16$$

Then

$$\sigma_{\overline{X}} = \frac{-\sigma_X}{\sqrt{16}} = \frac{0.08}{4} = 0.02$$

and

$$Z = \frac{22.63 - 22.57}{0.02} = \frac{0.06}{0.02} = 3.0$$

Hence $P(\overline{X} > 22.63) = P(Z > 3.00)$, and from Exhibit 13.2.17 we find that the $P(Z > 3.00) =$ 0.0013.

REFERENCES

1. L. BLANK, *Statistical Procedures for Engineering, Management, and Science*, McGraw-Hill, New York, 1980.
2. R. E. WALPOLE and R. H. MYERS, *Probability and Statistics for Engineers and Scientists*, 2nd ed., Macmillan, New York, 1978.
3. E. PARZEN, *Modern Probability Theory and Its Applications*, Wiley, New York, 1960.

BIBLIOGRAPHY

GUTTMAN, I., S. S. WILKS, and J. S. HUNTER, *Introductory Engineering Statistics*, 2nd ed., Wiley, New York, 1971.

ROSS, S., *A First Course in Probability*, Macmillan, New York, 1976.

CHAPTER 13.3

Decision Principles and Analysis

JAMES R. BUCK
The University of Iowa

13.3.1 INTRODUCTION

Decision making is at the very heart of industrial engineering and is the end product of decision formulation, analysis, and thorough consideration. The quality of the decision is a result of the thoroughness of each step along the way. Although decision results can never be guaranteed, failure to use quality procedures in decision making reduces one's chances to pure luck.

There are numerous applications of decision making in engineering. Decisions need to be made on the materials or component parts to be purchased in terms of the quantity, frequency, timing, vendors, and quality level.[1,2] Activity sequencing, operating layouts, and manufacturing, assembly, and servicing methods must be decided from among widely varying alternatives. Numerous decisions exist in the inventorying[2] and distribution of products to the ultimate buyers. These areas of application are only representative of the vast list of decision applications in industrial engineering, as shown in other chapters of this handbook. A wide variety of decision applications in practice exist in the literature. The planning, programming, budgeting (PPB) system of the U.S. Department of Defense, which started in the 1960s, is a noteworthy application.[3] Other examples of decision making on resource allocation, safety, health care delivery, new product planning, and so on, are shown in Martin[4] along with new theoretical directions.

13.3.2 DECISION FORMULATION

Formulating a decision to be addressed is perhaps the most important part of decision making and clearly the part where creativity and sound judgment enter. These features will become apparent in the theoretical and practical discussions that follow.

Theoretically, all decisions can be viewed as consisting of five component parts:

1. *Actions* or alternative courses of action under consideration.
2. *Events* that can occur and that can affect the outcomes.
3. *Beliefs* in the probable occurrence of these events for each action.
4. *Outcomes* associated with each combination of an action and an event.
5. *Evaluations* of the outcomes and the actions.

Components 1 and 2 constitute the controlled and the uncontrolled features of a decision, respectively. It is usually assumed in the theory that a list of actions can be developed in a form such that every conceivable course of action within the scope of the decision is included in a manner such that one and only one action will certainly be selected. Let A be the list of actions and $a_1, a_2, \ldots, a_i, \ldots, a_n$ be the individual actions listed in arbitrary order. Events are usually denoted as a list of discrete events such that one and only one of these events will surely occur. When events are so listed, the individual events are said to be exclusive, and the list is exhaustive. The list of events may be identified as E, with the separate events in the list as $e_1, e_2, \ldots, e_j, \ldots, e_m$. Although discrete actions and events are more typically seen in practice, either A or E or both may be described as a continuous variable if this description is appropriate for the decision at hand.

Beliefs and outcomes (components 3 and 4) are associated with every unique combination of action and event. A belief is the probability estimate (either objective or subjective—see Chapter 13.2) that the event will occur if that action is selected. Often the beliefs are subjectively estimated for each event associated with an action. Let p_{ij} denote the belief that event e_j will occur when action a_i is selected. Since beliefs are treated theoretically in the same fashion as probabilities, and the events are exclusive (only one can occur) and exhaustive (surely one will occur), then

$$\sum_{j=1}^{n} p_{ij} = 1$$

regardless of whether the beliefs are subjectively[5–8] or objectively estimated. Outcomes are also specified for each unique combination of event and action, and these outcomes are an estimate of the perceived results that will occur if that action is selected and that event occurs. An outcome may simply be a list of all the things that the event implies for the action that is relevant to the decision at hand. The list of these outcomes for an action and an event is denoted here as θ_{ij}. Perceived results include all economic, motivational, physical, and any other items of concern to the decision maker.

The final component of a decision is the evaluations, which have *two* important operations. In the first operation, the evaluation converts the perceived outcomes of each action and event into one or more numerical values that denote the value of those outcomes relative to the decision maker or the decision making body (e.g., a firm). Most typically, an economic evaluation is made where either costs or profits are used as the single attribute of the evaluated outcome. The second operation of the evaluation component is to combine some or all of the outcome evaluations of an action into an overall action evaluation. The methodologies for combining the outcome evaluation for an action differ with alternative principles of decision making, so that action comparisons can be made to reflect the decision maker's objectives.[9] These principles are discussed later. However, the outcome evaluations are symbolically represented here as $u(\theta_{ij})$, or more simply as u_{ij}, and the action evaluations as U_i.

With the preceding decomposition of the decision into components, the theory of decision making needs a means of recombining these components into a complete decision structure. Two major forms of synthesis are commonly used. The extensive or tree form of synthesis is shown in Exhibit 13.3.1*a* as a sequence of actions followed by events. However, the tree representation is not compact for easy computation, and so the matrix representation, shown in Exhibit 13.3.1*b*, is more typically employed for analysis. Example 13.3.1 illustrates a decision formulation of a typical decision situation. Since the decision maker's ultimate task is to select one of the actions listed, he or she must derive an evaluation of the entire action as U_i based on the outcome evaluations U_{ij} and/or the beliefs p_{ij}.

Example 13.3.1 Five different methods of production are under consideration as actions $a_1, \ldots, a_5$, along with the option of continuing with the existing method as a_6. Also, five different levels of sales are forecast as events $e_1, \ldots, e_5$. The following cost savings for each action are listed in

Exhibit 13.3.1 Decision Synthesis (*a*) Extensive Form and (*b*) Matrix Form

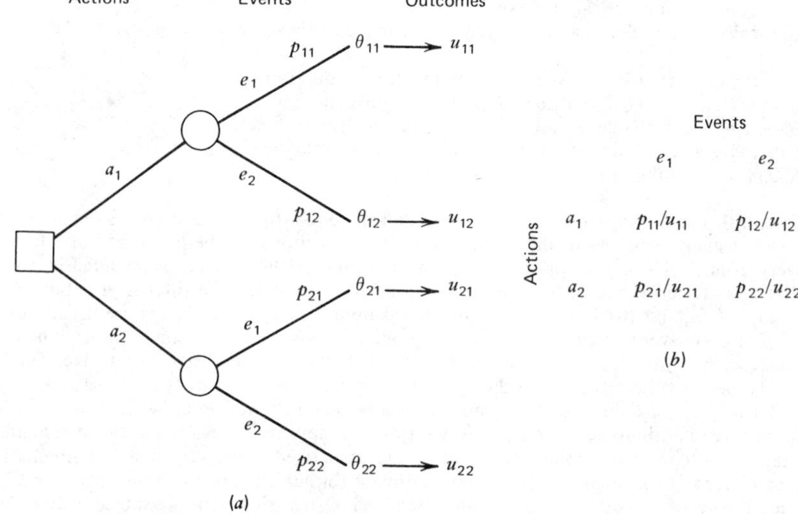

terms of thousands of dollars annually, along with estimated event beliefs:

	e_1	e_2	e_3	e_4	e_5
pi	.2	.4	.2	.1	.1
a_1	10	8	6	4	5
a_2	6	6	7	6	7
a_3	7	7	6	7	5
a_4	8	6	6	7	5
a_5	6	5	8	9	6
a_6	4	5	7	9	5

The decision at hand is to select one of the production methods.

13.3.3 TYPES OF DECISIONS AND DECISION MAKING

Decisions may be classified into various types based on particular attributes. Certainty decisions exist when only a single event appears likely, but when there are more likely events, then uncertainty or risk decisions result. Risk decisions are often distinguished from uncertainty decisions by the former's having precise belief estimates, whereas uncertainty decisions do not. This distinction is really unnecessary because uncertainty decisions are treated as if arbitrary belief estimates exist.

Another attribute for decision classification consists of repeated decisions, where decision recurrence is expected, versus one-time decisions, where there is no such explanation. Also, decisions can be classified by their type of formulation, where either or both of the actions or events are treated as values along a continuous variable or as discrete elements.

Any decision may be viewed with respect to any of these attributes for analysis. However, the viewpoint taken and the eventual formulation should be made with respect to the fidelity of the formulation relative to the decision at hand and to the importance of the decision relative to the analysis effort. Certainty and one-time decisions are almost always simpler and therefore require less analytical effort than uncertainty (or risk) and repeated decisions.

13.3.4 OUTCOME EVALUATION AND UTILITIES

A large percentage of the decisions occurring in engineering have outcomes evaluated in such economic terms as monetary returns, profits, costs, or cost savings. It should also be kept in mind that (1) these economic evaluations are either a benefit or a cost where the method of comparing action evaluations is reversed and that (2) economic benefits or costs that occur at different times should be converted into economic terms that reflect the time value of money (as shown in Section 9).

Another approach to outcome evaluation is described by "utility theory."[5,9–14] This theory provides a means for converting all the perceived outcomes into a measurement of relative worth, regardless of the number of attributes associated with the outcome, such as social, moral, economic, or psychological values. The basis of the best-known version of this theory lies with the decision maker's comparison of a series of lotteries. In each comparison the decision maker compares a specified outcome that would surely happen (a lottery certain) to a risky lottery where either the least- or most-preferred outcome could occur (the standard gamble). The decision maker is then required to specify his or her belief in the occurrence of the most-preferred outcome when he or she would be indifferent to either the lottery certain or the standard gamble. Utility values of the specified outcomes in the lottery certain are the specified beliefs at indifference. Although this utility theory approach to outcome evaluation has many appealing features and many advocates, there are both theoretical and operational difficulties.[15,16] For these reasons, utility theory is not frequently employed in decision analysis.

13.3.5 PRINCIPLES IN DECISION MAKING

Various principles have been proposed over time for creating action evaluations and comparing these evaluations for the selection of an action. Many of the principles reflect different objectives on the part of the decision maker, the degree of confidence in the belief estimates, and/or the decision maker's attitude toward risk. The discussions and examples immediately following describe and illustrate the more common decision principles.

Certainty Decisions

In certainty decisions where the outcomes have been evaluated according to a single criterion, the basis for comparing the action evaluations is straightforward, and so there is no need for a guiding principle. Simply select the action with the best evaluation.

If multiple criteria are used in a certainty decision, one may invoke the principle of "dominance" to search for an action that is best in every criterion or at least never worse than the best in any criterion. When a single action dominates all others, then clearly that action should be selected.

In those decisions where dominance does not exist, then decision analysts and decision makers usually resort to the principle of "trade-offs." In this principle there is a marginal amount of one criterion that one would be willing to sacrifice for a unit increase in another criterion. If these marginal trade-offs can be established between all the criteria, and if the trade-off rates remain constant regardless of the amount of the criterion, then the evaluation of each criterion for an action can be combined into a single linear polynomial with weighting coefficients as shown in Raiffa and Schlaifer[2] and Farrar.[17] However, extreme care should be exercised to ensure that the trade-off rates remain constant.

Uncertainty Decisions

Uncertainty decisions pose greater difficulties, and numerous principles have been proposed for use in decisions of this type. Here again the dominance principle may be applicable. If a single action has outcome evaluations such that the outcome evaluations of other actions are never better regardless of the event, then that action is said to dominate all others, and it should be selected. Although dominance of this form may not exist very often, one always ought to check the decision for dominance before considering other principles which may have lower management acceptance.

Checking for dominance in uncertainty decisions where a single criterion is used for outcome evaluations may be done by simply scanning the evaluated outcomes for each event and identifying those actions with the best evaluated outcome. When a single action is identified to be best for all events, then that action dominates. Dominance also exists when one action is never better than another, regardless of the event. An action that is never better is said to be dominated, and such actions may be ignored in order to simplify the analysis. (This is illustrated later in Example 13.3.2.) If multiple criteria are used in the evaluation, then dominance exists only when an action is at least as good as another in each criterion *and* in every event.

When dominance does not exist, then another principle must be invoked, and there are numerous ones to choose from. The remaining principles may be classified into two groups: (1) those that do not utilize the beliefs associated with the outcomes and (2) those that do. The first group is commonly known as "uncertainty principles," whereas the second is frequently called "risk principles." Among other considerations, the choice between uncertainty and risk principles should follow one's confidence in the accuracy and precision of the beliefs; relatively low confidence would cause one to favor the uncertainty principles, whereas the risk principle should be favored when there is relatively high confidence. Different uncertainty and risk principles are discussed in the remainder of this section, and numerical examples are periodically inserted to illustrate these principles.

Uncertainty Principles

Maximin. One of the best known of the uncertainty principles is maximin worth. When the evaluated outcomes are described in terms of worth where a greater evaluation is preferred, then one may evaluate each action under this principle by the *minimum* outcome evaluation over all the events. This principle then proceeds to say that one ought to select the action that has the *greatest* minimum outcome. That is, one should try to select actions that maximize the minimum evaluations. In those decisions where lower outcome evaluations are preferred, then the rule reverses to minimax cost. In either case, the worst outcome evaluation becomes the action evaluation (ignoring all other evaluated outcomes of an action). (Example 13.3.2 illustrates the operational aspects of this and other uncertainty principles where the high conservative nature of this principle is illustrated.) This principle is attributed to the eminent statistician Abraham Wald, and it is a special case of an application of game theory.

Minimax Regret. Another uncertainty principle, which was described by the famous Bayesian statistician L. J. Savage, is minimax regret. In this principle, "regret" is defined as the difference between the *best* evaluated outcomes possible under an event and the actual evaluated outcomes. With outcomes evaluated in worth values, the regret associated with action a_i and event e_j is the greatest worth in event e_j less u_{ij}. When costs are used in evaluations, then the regret is u_{ij} less the minimum cost in event e_j. Once all of the evaluated outcomes of a decision are converted into regret values by either of the methods shown, then the action evaluations consist of the maximum regret values of the action. It then follows, under this principle, that one should select that action with the least action evaluation; that is, the action in which the maximum regret is least. (This principle is demonstrated in Example 13.3.2.) It is important to note that, after converting either costs or worths to regrets, this principle operates in a consistent manner as the regrets.

Hurwicz Principle. The next uncertainty principle to be discussed here is one that the well-known economist L. Hurwicz devised. In the use of this principle, the decision maker must describe his or her degree of optimism in the decision outcome in terms of a number from 0 (where there is no optimism) to 1 (where optimism is greatest).

Let α symbolize the degree of optimism. Then, for each action, the best evaluated outcome is multiplied by α, and the worst is multiplied by $1 - \alpha$, and the *sum* of these optimism-weighted values is the Hurwicz criterion. When evaluated outcomes are in worth values, then the action with the greatest Hurwicz criterion should be selected. With cost values, select the action with the least Hurwicz criterion. (Example 13.3.2 also illustrates this principle.) It should be noted here that an α value of 0 makes this principle identical to the maximin worth principle and that an α value of 1 yields what may be called the "maximax worth principle." Therefore all intermediate values of α give a linear mixture between these principles.

Laplace Principle. A final uncertainty principle discussed here is an adaptation of the principle stated by Laplace, which is that unless there is reason to believe otherwise, assume that all the events are equally likely, and then select the action with the greatest expected worth under this assumption. If the evaluated outcomes are given as cost values, then select the action with the least expected cost, given the assumption. With m different events, the assumption is that all the events are equally likely or equivalent, and the probability of each event's occurrence is $1/m$. The expected worth associated with an action under this assumption is

$$E(U_i) = \sum_{j=1}^{m} \frac{1}{m} u_{ij} = \frac{1}{m} \sum_{j=1}^{m} u_{ij}$$

In an operational sense, the Laplace principle uses the sum of the worth (or cost) evaluated outcomes over *all* events divided by the number of events as a criterion. The action with the greatest worth criterion or the least cost criterion should be selected. (Example 13.3.2 also illustrates this principle.) It is of note here that only the Laplace principle employs *all* of the evaluated outcomes of an action in the action evaluation for uncertainty decisions.

Example 13.3.2 The principle of dominance and the four uncertainty principles are illustrated here using the decision of Example 13.3.1. A review of that example discloses that no one action has the maximum cost savings for all events; therefore no one action dominates the rest. However, action a_6 is never better than any of the other actions, regardless of the event. Therefore a_6 is dominated, and it is eliminated from further consideration.

In applying the maximin principle to actions $a_1, \ldots, a_5$, the action evaluations consist of the least cost savings as follows:

	Action				
	a_1	a_2	a_3	a_4	a_5
Action evaluations	4	6	5	5	5

Here it is observed that action a_2 has the greatest least cost savings, and it would be selected under this principle.

The minimax regret principle first requires one to compute regret values as the greatest cost savings for each event less the actual cost savings. The computed regret values are:

	e_1	e_2	e_3	e_4	e_5	Maximum Regret
a_1	0	0	2	5	2	5
a_2	4	2	1	3	0	4
a_3	3	1	2	2	2	3
a_4	2	2	2	2	2	2
a_5	4	3	0	0	1	4

Maximum regret values for each action are shown in the right-hand column of the preceding list. Action a_4 has the least maximum regret, and so it would be selected under this principle.

This decision may be analyzed under the Hurwicz principle with an optimism value of $\alpha = \frac{3}{4}$. In this case the maximum cost savings of an action are multiplied by $\frac{3}{4}$, the minimum cost savings are multiplied by $\frac{1}{4}$, and the results are summed for the Hurwicz criteria, which are:

	Action				
	a_1	a_2	a_3	a_4	a_5
Hurwicz criteria	34/4	27/4	26/4	29/4	32/4

Since action a_1 has the greatest Hurwicz criterion, that action would be selected under this principle.

The Laplace principle applied to this example would result in an expected cost savings for each action as the sum of the cost savings over all events divided by 5, or:

	Action				
	a_1	a_2	a_3	a_4	a_5
$E(U_i)$	6.6	6.4	6.4	6.4	6.8

Action a_5 would be selected under the Laplace principle because that action has the greatest expected cost savings.

Actions selected in this example under the four uncertainty principles are:

	Principle			
	Maximin	Minimax Regret	Hurwicz	Laplace
Action selected	a_2	a_4	a_1	a_5

This diversity reflects the different viewpoints and assumptions of these uncertainty principles.

Criticisms. The maximin worth principle carries the implication that the event selecting mechanism will select that event which harms the decision maker most, no matter what the action choice is. There are many situations where that implication is clearly infeasible, but this pessimistic viewpoint does fit some security decision circumstances and the theory of games.[9] A similar criticism can be leveled at the minimax regret principle because only the worst regret is used for the action evaluation. Moreover, the use of a single evaluated outcomes or of two evaluated outcomes to represent the entire action is another point of criticism for these principles and the Hurwicz principle as well. These and other logical or sufficiency traps are discussed in Raiffa and Schlaifer[2] and Morris,[18] and most of these traps are overcome by the Laplace principle.

Risk Principles

In contrast to the uncertainty principles, which do not generally employ belief values in the action evaluations, risk principles do. However, the risk principles use these beliefs in different ways, or different statistics are used to obtain action evaluations.

Maximization of Expected Worth. The most-used risk principle is the maximization of expected worth or, equivalently, the minimization of expected cost. In this principle the expected worth (or cost) is

$$E(U_i) = \sum_{j=1}^{m} p_{ij} u_{ij} \tag{1}$$

and the action with the greatest expected worth (or least expected cost) should be selected. (This is illustrated later in Example 13.3.3.) The concept of the maximum expected worth principle is that, on the average event, the best action is selected. However, this principle may select an action with widely varying evaluated outcomes.

Least Expected Regret. Another principle that has been advocated is least expected regret. In essence, the worth or cost evaluations are changed to regret, as described previously; these regret values are substituted for u_{ij} in equation 1; and the action with the least expected regret, $E(R_i)$, is selected. As it turns out, use of the least expected regret principle always leads to the identical choices of actions as the expected worth principle, and so these are dual principles. Since the calculation of the expected regret is operationally more difficult than the expected worth principle, the expected regret principle is rarely used. However, it is briefly described here because it relates to the concept of decision information, which is discussed later.

Minimizing the Worth or Cost Variance. Many people view risk as the variation in the action evaluations. This viewpoint leads to the principle of minimizing the worth or cost variance. Consequently, the action evaluations under this principle are

$$V(U_i) = \sum_{j=1}^{m} [u_{ij} - E(U_i)]^2 \, p_{ij} \tag{2}$$

After $E(U_i)$ is computed using equation 1, then this value may be substituted into equation 2 to find the variance of that action. This principle then goes on to state that the action with the least variation should be selected. (This is illustrated in Example 13.3.3.)

Farrar Principle. Whereas the principle of maximizing the expected worth ignores the outcome variability, and the principle of minimum variance ignores the expected worth other than for computation, it would seem only natural to combine these principles in some way. An intuitive way of combining these statistics was recognized by Farrar.[19] He *subtracted* the product of the square root of the worth variance (i.e., the SD) and a constant k from the expected worth. In this case the value of k is a weighting value (i.e., a trade-off rate) that denotes the decision maker's concern for outcome variation relative to the expected worth. If the outcomes are evaluated in terms of cost rather than worth, then the product of k and the SD of cost is *added* to the expected cost. The concept here is to make action evaluations in terms of a lower bound of worth or an upper bound of cost in a similar manner as confidence intervals. (Example 13.3.3 also illustrates the application of this principle.)

Semivariance Principle. Both the principle of minimum variance and the Farrar principle for combining the variance with expected worths (or expected costs) have been criticized for viewing variations as a symmetrical measure of risk while the worth or cost skewness is ignored. These critics suggest that risk is better measured as downward worth variation or upward cost variation (i.e., semivariance). In the case of worth, the semivariance about the mean worth is

$$SV_d = \sum_{j=1}^{m} [E(u_i) - u_{ij} | u_{ij} < E(u_i)]^2 p_{ij}$$

and the upward semivariance about the mean cost is

$$SV_u = \sum_{j=1}^{m} [u_{ij} - E(u_i) | u_{ij} > E(u_i)]^2 p_{ij}$$

Action evaluations can then be described, respectively, as the expected worth less k times $\sqrt{SV_d}$ or the expected cost plus k times $\sqrt{SV_u}$, where k is a weighting factor for $\sqrt{SV}$ (i.e., the semi-standard deviation of worth or cost). Often k is unity. As Exhibit 13.3.2 illustrates, the resulting action evaluation is similar to a one-sided confidence interval that is of directional (downward or

Exhibit 13.3.2 Relative Frequencies of Ordered Evaluations With Three Projects

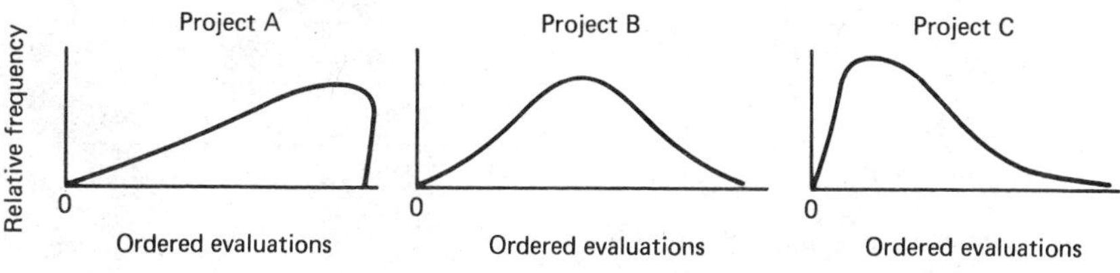

upward) interest. Although this principle ignores the opposite direction of variation, it does capture any directional worth or cost skewness effects in comparison with a symmetrical worth or cost distribution, all with the same expected value and variance. Depending upon whether one is comparing worth or cost outcome evaluations, the relative importance of project A over B and C or project C over A and B is readily seen.

Aspiration. The last risk principle discussed here is that of aspiration. This principle is highly distinct from the preceding principles because it is partially based on the concept of "satisficing."[20] Operationally, the aspiration principle requires the decision maker to denote a level of worth that he or she aspires to obtain or a cost level that one aspires to stay below. The principle then follows that one should select that action which maximizes the belief that the level of aspiration level will be achieved. Thus the level of aspiration is analogous to a satisfaction level, and so this principle serves to make satisfaction most likely. (Again, Example 13.3.3 illustrates this principle in comparison to the other risk principles stated here.)

Summary. Of the risk decision principles, the maximum expected worth or the minimum expected cost are clearly the most used. No one faults these principles as necessary, only as not being sufficient. This insufficiency naturally led to the expectation variance or the expectation semivariance principles, with the idea that more statistics are better. In some cases more statistics may create cognitive overload or may require more analytical resources than the decision deserves.

Example 13.3.3 These six risk principles are illustrated here using the decision of Example 13.3.1. The expected cost savings are computed using equation 1, with the following results:

	Action				
	a_1	a_2	a_3	a_4	a_5
$E(U_i)$	7.3	6.3	6.6	6.4	6.3

Action a_1 maximizes the expected cost savings and would therefore be selected under the maximum expected worth principle. By substituting the computed regret values shown in Example 13.3.2 for u_{ij} in equation 2, the expected regrets may be computed for each action as follows:

	Action				
	a_1	a_2	a_3	a_4	a_5
$E(R_i)$	1.1	2.1	1.8	2.0	2.0

Action a_1 has the least expected regret and thus should be selected under this principle. This illustration confirms the duality of this and the maximum expected worth principle because the order of increasing regret expectations is the same as the decreasing worth expectations.

Cost saving variances may be computed for the actions using equation 2. The expected cost savings are:

	Action				
	a_1	a_2	a_3	a_4	a_5
$V(U_i)$	3.61	0.21	0.44	0.84	2.01

Under the principle of minimizing the worth variances, action a_2 would be selected.

Farrar's expectation variation principle with $k = 1$ would have yielded the following criteria values for the actions:

	Action				
	a_1	a_2	a_3	a_4	a_5
$E(U_i) - \sqrt{V(U_i)}$	5.40	5.84	5.94	5.48	4.88

Action a_3 would be selected under this principle.

With the semivariance principle, concern would be addressed to the downward semivariances from the expected cost savings. Thus the criterion values for this example would be

	Action				
	a_1	a_2	a_3	a_4	a_5
$E(U_i) - \sqrt{SV_d}$	5.90	6.05	6.03	5.86	5.46

In this case, action a_2 would be selected, with action a_3 a close second.

The aspiration principle applied to this example would have a variety of results, depending upon the aspiration level selected:

	Action					
Level	a_1	a_2	a_3	a_4	a_5	Select
6 $p(u_{ij} \geqslant 6)$	0.8	1.0	0.0	0.0	0.6	a_2
7 $p(u_{ij} \geqslant 7)$	0.6	0.3	0.7	0.3	0.3	a_3
8 $p(u_{ij} \geqslant 8)$	0.6	0.0	0.0	0.2	0.1	a_1

The table shows the actions selected for the three aspiration levels investigated. Use of these six risk principles in this example lead to the selection of different actions, depending upon the emphasis on the variation. The first two ordered choices under the first five risk principles are:

	Principle			
	Max $E(U_i)$	Min $V(U_i)$	Max $[E(U_i)]$	Max $[E(U_i) - \sqrt{SV_d}]$
First choice	a_2	a_2	a_3	a_2
Second choice	a_3	a_3	a_2	a_3

Here actions a_2 and a_3 are the predominant choices.

13.3.6 CONTINUOUS DECISION FORMULATIONS AND PRINCIPLES

In the decision situations just formulated, both the set of actions and the events were described as discrete entities with an arbitrary order, and the decision principles were tailored to that formulation. Although the discrete formulation is particularly useful when the actions and events have multiple dimensions of difference, there are situations when the set of actions and/or the set of events each vary along a single dimension. Typical examples are decisions with continuous or nearly continuous action sets, including the amount of materials to be purchased or the safety factor to be used in a design. Event sets, too, can vary in a continuous or noncontinuous fashion, such as usage rate of an inventoried item, the degree of competition to be experienced with a new product line, the fraction lot defectives of a process, or the productivity of an operation. It may be that there really is a discrete set of actions and/or events, but that the very large number of them makes it convenient to treat the set as continuous, or it may be that the set is really continuous in nature and that analytical fidelity is sought. Regardless of the reason, continuous formulations may be made of the actions and/or events.

Clearly, a continuous formulation of either or both necessitates a similar formulation of the beliefs, outcomes, and outcome evaluations. For example, a continuous event set may be represented by the variable y, so that the event set would be $e(y)$. If the set of actions is discrete, then beliefs and evaluated outcomes may be shown, respectively, as $p_i(y)$ and $u[\theta_i(y)]$. When the action set is also represented by the continuous variable w, then the beliefs, the evaluated outcomes, and the action evaluations become $p(x, y)$, $u[\theta(x, y)]$, and $U(w)$, respectively. Four different cases of discrete and continuous formulations of the actions and events are shown in Exhibit 13.3.3.

With events represented as a continuous random variable, then extreme values of the evaluated outcomes (i.e., maximums or minimums) can be found through the use of the calculus or a one-dimensional search (see Chapter 14.3). Decision principles such as the maximin worth or minimax regret require extreme evaluation information in order to determine the action evaluations of U_i on $U(w)$. The Laplace principle and the risk principles require integration of the evaluated outcome functions, products of belief and evaluated outcome functions, or products of the belief and

Exhibit 13.3.3 Alternative Discrete/Continuous Decision Formulations

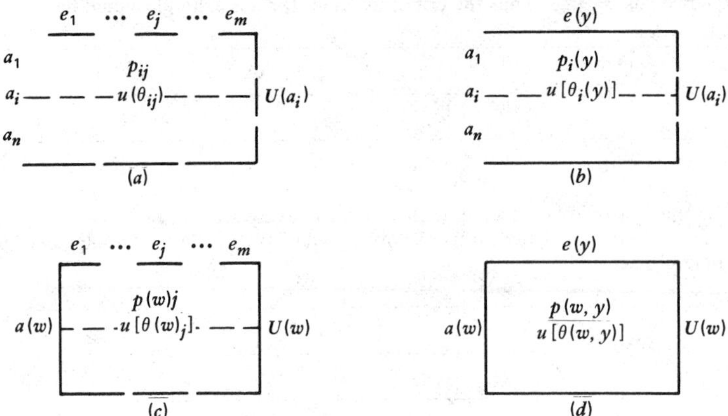

squared differences in the outcome functions. All of the principles follow naturally from the purely discrete once these continuous formulation features are considered.

When the events are given as a continuous variable y or as a series of discrete values along this variable, then the beliefs or probabilities of the events can be written as a function of y, as indicated in Exhibit 13.3.3. Sometimes it is analytically convenient to describe the belief using a standard probability function, as given in Exhibits 13.2.14 and 13.2.15 at the end of Chapter 13.2. These exhibits show the functional form of a number of these standard functions, along with formulas for computing statistics of these functions. If the statistics of the belief function $p(y)$ can be estimated from data or subjectively, then these estimated statistics can be equated to the corresponding statistical formulas in order to determine the parameter values of a chosen belief function. This procedure is known as "parameter estimation by the method of moments," and it is demonstrated in Example 13.3.4.

When events are described by a single random variable y, then the evaluated outcomes must also be so described. If the evaluated outcomes are a linear function of y (e.g., $U_i(y) = a + by$), then the expected worth (or cost) is the worth or cost at the expected value of y. Also, the variance of worth or cost is equal to the product of the square of the linear slope and the variance of y. This situation is known as a "certainty equivalence" because the statistic of y can replace y in the linear equation to have the statistic of worth or cost as

$$E[U_i] = \int (a + by)\, p(y)\, dy = a + bE(y)$$

However, certainty equivalence does not strictly hold with quadratic functions (e.g., $U_i(y) = a + by + cy^2$) where

$$E[U_i] = \int (a + by + cy^2)\, dy = a + bE(y) + c[E(y)^2 + V(y)]$$

because two statistics of y are needed to compute $E[U_i]$.

The formulation of the expectation and variance of the worth or cost function $u[\theta_i(y)]$ or $u[\theta(x, y)]$ is

$$E[U_i] = \int u[\theta_i(y)]\, p(y)\, dy$$

$$V[U_i] = \int \{u[\theta_i(y)] - E[U_i]\}^2 p(y)\, dy$$

Semivariances can be obtained with $V[U_i]$ by controlling the limits of integration. A series of numerical values at an equal spacing of y increments may then be computed, and one of the numerical integration methods shown in Spetzler and Stael von Holstein[21] can be employed to compute approximations of these worth or cost statistics, as shown in Example 13.3.4.

Example 13.3.4 Events of a decision may be defined by the fraction lot defectives in which a random variable of $0 \leqslant y \leqslant 1$ can be described by the beta probability density function. Records in the quality control department indicate that the expected percentage of defectives is 10% and that the variance of this percentage is 0.00177. The results of setting these two statistics of y equal to their corresponding formulate for the mean and variance are

$$0.10 = \frac{a}{(a+b)} \qquad 0.00177 = \frac{ab}{(a+b)^2\ (a+b+1)}$$

where a is solved to be approximately 5 and $b \cong 45$. The resulting belief function is

$$p(y) = \frac{49!}{4!\ 44!} y^4 (1-y)^{44} = 9534420 y^4 (1-y)^{44}$$

Cost savings of action a_1 are described as a quadratic function, $\theta_1(y) = 20 + 3y - 0.05y^2$, so that the expected cost savings of this action are

$$E[U_1] = \int_0^1 (20 + 3y - 0.05y^2)[9534420y^4(1-y)^{44}]\ dy$$

This integral may be numerically approximated using Simpson's rule and steps of y equal to 0.05 as follows:

y	Cost Savings	Belief Density	Product	Coefficient	Product
0.00	20.0000	0.0000	0.0000	1	0.0000
0.05	20.1499	6.2375	125.6855	4	502.7421
0.10	20.2995	9.2462	187.6938	2	375.3877
0.15	20.4489	3.7852	77.4037	4	309.6146
0.20	20.5980	0.8306	17.1080	2	34.2160
0.25	20.7469	0.1185	2.4587	4	9.8347
0.30	20.8955	0.0118	0.2467	2	0.4934
0.35	21.0325	0.0008	0.0176	4	0.0706
0.40	21.1920	0.0000	0.0000	1	0.0000
					Sum = 1232.3591

The expected cost savings are $(0.05)(1232.36)/3 \cong 20.5393$.

13.3.7 LOSS INTEGRALS

A decision principle that is applicable when the events are described by a continuous random variable y is the "minimization of the expected loss integral." If the evaluated outcomes are worth values, then for $y > y'$, the *difference* between the two actions' evaluated outcomes is a loss function.

Exhibit 13.3.4 shows a linear loss function $f(y)$ and an associated probability (density) function of y which is Gaussian (normal) or $p(y)$. The integral of the product of these two functions is

$$\int_{y'}^{+\infty} byp(y)\ dy = b \int_{y'}^{+\infty} yp(y)\ dy \tag{3}$$

which Raiffa[22] and Sheid[23] show to be a linear combination of the Gaussian ordinate and the cumulative Gaussian integral. There are tabled values in Raiffa[22] and Sheid[23] for the standardized case (i.e., the Gaussian distribution where $\mu = 0$ and $\sigma = 1$). If the actual mean and SD of the Gaussian is known, along with values of a and y', then the linear Gaussian loss integral may be directly computed. When the loss function is different from equation 3, then an approximation procedure for any reasonable loss function can be used.[24]

The meaning of the loss integral for worth values is that this integral is the expected loss if the random variable y exceeds the break-even point y'. If action a_1 has a greater worth than a_2 when $y > y'$, and if action a_2 is chosen, then this loss integral specifies the risk one assumes for that choice. In the illustration shown in Exhibit 13.3.4, action a_2 has a greater expected profit, and this action would be chosen over a_1 using the maximum expected worth principle. The loss integral portrays the expected loss with this action choice, and therefore the action with the least loss integral ought to be selected. Since loss integrals are asymmetrical by definition, this loss integral

Exhibit 13.3.4 Linear Loss Integral Concept

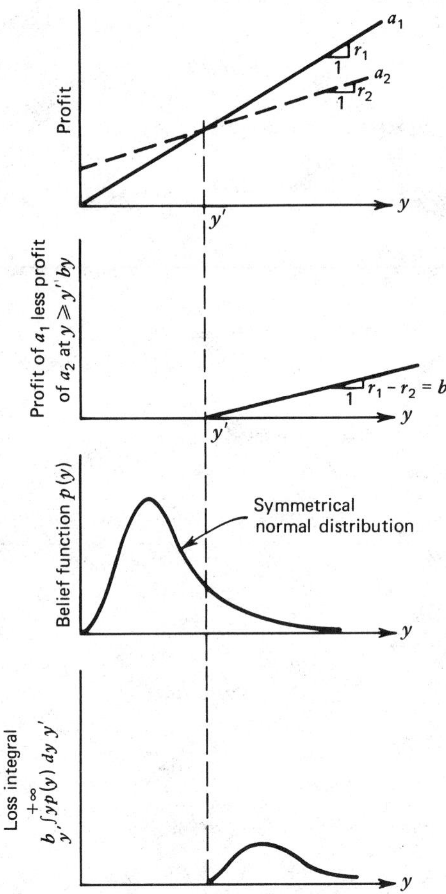

principle bears strong similarities to the expectation semivariance risk principle. However, computational complexities arise with loss integrals when there are more than two actions.

13.3.8 DECISION INFORMATION

Information in decision making is a communication to the decision maker that alters the belief associated with the event. Outside sources or management information systems (see Section 12) can provide these communications. Since beliefs may change with information, so can the choice of an action and the received worth. The expected increase in received worth is the value of this information.

If the communication maker is correct and only one event is possible, then this situation is said to be "perfect information." As it turns out, the value of perfect information is the minimum expected regret. Perfect information is rarely, if ever, available, but there often are symptoms z that are correlated with the events, but not perfectly correlated. The fraction of defectives in a sample, economic indicators, and many other things are symptoms of the events, in which case the beliefs in the values of the event variable y may be found, given the value of z observed through Bayes' equation as

$$p(y|z) = [p(y)]\,[p(z|y)]\,[\Sigma_y p(y)\,p(z|y)] \tag{4}$$

Bracketed portions on the right-hand side of equation 4 are given, respectively, the special names of

1. The a priori (or prior) belief function (i.e., the belief function without knowledge of z).
2. The evidence belief function.
3. The residue or normalizing factor.

On the left-hand side of equation 4 is the a posteriori (or posterior) belief function, which accounts for the observed symptom z.

Example 13.3.5 A coating processor may fail (e_1) or not (e_2) where the decision actions may be to continue the automated operation (a_1), put the processor on manual mode (a_2), or shut down the processor and make corrections (a_3). This decision, with evaluated outcomes, may be as follows:

	e_1 0.4	e_2 $0.6 = p_j$	$E(U_i)$
a_1	4	10	6.4
a_2	6	7	6.6
a_3	7	5	6.2

Action a_2 would be selected under the expected worth principle. With perfect information available, one would select a_3 if event e_1 were forecast; otherwise a_1 would be selected, with the expected worth of

$$E(U_i) = 0.4(7) + 0.6(10) = 8.8$$

Since this expected worth with perfect information is 2.2 worth units higher than without it, that is the value of this perfect information. However, perfect information is not available, but a symptom of a failed processor is a defective coating found in a random sample of 10 units out of the processor. When the processor was shut down, a check was made to see if the processor had failed. Samples had been taken before the shutdown, and the numbers of past occurrences of the sampling with and without a failed processor was as follows:

Sampling Condition	e_1	e_2	Sum
z_1: no defectives	10	65	75
z_2: defectives found	20	5	25
Sum	30	70	100

Evidence belief values computed from these data are

$$p(z_1|e_1) = 1/3,\ p(z_1|e_2) = 65/70,\ p(z_2|e_1) = 2/3,\ p(z_2|e_2) = 5/70$$

Since the prior beliefs of e_1 and e_2 are .4 and .6, respectively, the posterior beliefs of events, given the evidence, are computed with Bayes' equation as

$$p(e_1|z_1) = \frac{0.4[1/3]}{[0.4(1/3) + 0.6(65/70)]} = 0.193$$

$$p(e_1|z_2) = \frac{0.4[2/3]}{[0.4(2/3) + 0.6(5/70)]} = 0.862$$

$$p(e_2|z_1) = 1 - 0.193 = 0.807$$

and

$$p(e_2|z_2) = 1 - 0.862 = 0.138$$

The expected worth of each action computed with these posterior beliefs with each sampling situation is as follows:

Sampling Situation	a_1	a_2	a_3
z_1	8.842	6.807	5.386
z_2	4.828	6.138	6.724

This result indicates a decision rule that action a_1 should be selected under sampling conditions z_1 and action a_3 under z_2 otherwise. Since the frequency of the sampling condition z_1 is

$$p(z_1) = p(z_1 e_1)\, p(e_1) + p(z_1 e_2)\, p(e_2)$$

$$p(z_1) = (1/3)(0.193) + (65/70)(0.807) = 0.8137$$

Expected worths from the following plan are

$$E[\text{plan}] = 8.842(0.8137) + 6.724(1 - 0.8137) = 8.4474$$

Without fallible information, the expected worth was 6.6, so the difference in expected worth, or 1.85 worth units, represents the value of this information and the maximum one could afford to pay to obtain it.

13.3.9 CONJUGATE DISTRIBUTIONS

Whenever a prior belief function and an evidence function are of such forms that the resulting posterior function is identical in functional form to the prior function, then prior and posterior functions are said to be "natural conjugate distributions." It should be noted that only the functional forms must be identical, not the parameter values. Raiffa and Schlaifer[2] identified a number of conjugate distributions, and they and others[22] demonstrate how this concept is useful in decision making because the posterior function can be found readily from the prior by a single change in the parameter values that is supplied by the evidence. That is, one's prior belief is revised in light of the evidence to a posterior belief.

Three of the natural conjugate distributions that are particularly useful in industrial engineering are beta belief function under binomial evidence, gamma belief functions under Poisson evidence, and Gaussian belief functions under Gaussian evidence. The functional forms of these distributions (see Chapter 13.2, Exhibits 13.2.14 and 13.2.15) and products of the beta and binomial, gamma and Poisson, or two Gaussian functions can be shown to be beta, gamma, and Gaussian, respectively, as given by Holloway,[11] Raiffa,[22] and Sasieni et al.[25] In the case of the beta binomial, where the prior beta function has parameters a and b and the binomial has parameters n and x, the posterior beta has parameters $A = a + x$ and $B = b + n - x$. Accordingly, the prior belief parameters (a and b) are modified by the evidence function parameters. Since the mean and variance of the beta are specified when the parameters are known (see Exhibit 13.3.2), then the posterior beta has a mean and a variance of

$$E(y) = \frac{(a + x)}{(a + b + n)}$$

$$V(y) = \frac{(a + x)(b + n - x)}{(a + b + n)^2 (a + b + n + 1)}$$

When the binomial evidence function reflects sampling with replacement as fallible information, then a sample of size n yields x occurrences of one of the binomial results where $x = 0, 1, 2, \ldots, n$. As n is increased, the variance tends to decrease in order to make the posterior belief $p(y)$ evidence less uncertain. Also, as x changes relative to n, both the mean and the variance will change. As a result, the evidence can significantly alter the prior belief. Example 13.3.6 illustrates the effect of work sampling evidence on the belief of equipment utilization.[26]

Example 13.3.6 If the utilization of a particular machine were completely unknown, then the belief in the utilization percentage could be described as a uniform random variable over the range $0 \leqslant q \leqslant 1$. This belief function is equivalent to a beta distribution with parameters $a = 1$ and $b = 1$. A sampling plan is devised for taking 60 observations where the machine will be either observed as idle or utilized on each observation. Twenty samples were planned over each of the next 3 days for this binomial sampling, where the results and resulting posterior beta parameters were as follows:

Day	Idle Observation	Posterior Beta (Parameters/Statistics)
0	0	$a = 1,\quad b = 1,\quad E(q) = 0.50,\quad \sqrt{V(q)} = .289$
1	5	$A = 16,\quad B = 6,\quad E(q) = 0.73,\quad \sqrt{V(q)} = .093$
2	7	$A = 29,\quad B = 13,\quad E(q) = 0.69,\quad \sqrt{V(q)} = .71$
3	6	$A = 43,\quad B = 19,\quad E(q) = 0.69,\quad \sqrt{V(q)} = .058$

The results indicate that there is a progressive revision in the statistics of the beta with a constant reduction in the SD (indicating the remaining uncertainty).

The gamma Poisson case is similar to the beta binomial. A prior gamma distribution with the mean ab and the variance ab^2 is used to describe the mean of a Poisson distribution. Collecting data (evidence) on the Poisson distribution of $x_1, x_2, \ldots, x_n$ and then letting x be the sum of the sample yields a posterior gamma with parameter $A = a + x$ and $B = b/(nb + 1)$. Mean and variance statistics of the gamma posterior are then

$$E(y) = \frac{(a + x)\, b}{nb + 1}$$

$$V(y) = \frac{(ax)\, b^2}{(nb + 1)^2}$$

Here again, a larger sample size n causes a smaller posterior variance to reflect less belief uncertainty resulting from more evidence.

In the case where a process variable is known to be Gaussian (normal) distributed with a *known* variance but an uncertain average value, one's uncertainty about the average can be modified in light of evidence about that average. The uncertainty about the average may be described by a Gaussian distribution with a mean and variance of μ_1 and σ_1^2, where the subscript 1 refers to one's prior belief in the process average. If the process average is μ and the *known* process variance is σ^2, then the coefficient of prior conviction may be defined as $C = \sigma^2/\sigma_1^2$, and this coefficient denotes the relative degree of certainty about the average before evidence is collected. Evidence may be collected by sampling the process n times and finding the value of the process variable being examined where the sample mean is $\bar{x}$. It follows from the development shown by Raiffa[22] that the posterior mean and the variance of belief in the average process variable are

$$E(\mu_2) = \frac{c\mu_1 + n\bar{x}}{c + n}$$

$$V(\mu_2) = \frac{\sigma^2}{c + n}$$

where the subscript 2 denotes the posterior belief. The posterior mean of the process average is simply the weighted average between the prior mean with a conviction degree weighting and the sample mean with a sample size weighting. It is also of note that the posterior variance gets smaller with a larger n, as Example 13.3.7 illustrates.

Example 13.3.7 A refrigeration system is known to have a temperature variance of 3°F (1.8°C) over the different locations in the room. The mean temperature is believed to be 10°F (−12°C), but the variability in this prior belief of the mean temperature is described by a Gaussian (normal) function with a variance of 4°F (2.2°C). This situation results in the coefficient of prior conviction being equal to 3°/4° or 0.75. Five temperature samples gave a sample mean of 8°F (−13°C). The posterior belief expectation and variance are then

$$E(\mu_2) = \frac{[0.75(10) + 5(8)]}{5.75} = 8.3°\text{F } (-13.2°\text{C})$$

$$V(\mu_2) = \frac{3}{5.75} = 0.52°\text{F } (0.29°\text{C})$$

Because of the rather high uncertainty of the prior mean relative to the low known process variation, the posterior mean reflected the sample mean almost entirely. Even a sample of one that gave 8°F would result in a posterior mean of 8.9°F (−12.8°C) and a variance of 1.7°F (1°C).

13.3.10 DECISION STRATEGIES

Sometimes decisions repeat themselves over time and this sequence of decisions may be handled by a decision rule or strategy. However, the nature of this strategy differs if the successive decisions are independent of the previous decision (e.g., ordering perishable items) from the case where successive decisions depend on earlier choices (e.g., ordering nonperishable items). In the former case a single action may be repeated over and over again as long as the decision remains unchanged. This is a "pure strategy." Alternatively, one may choose among two or more actions on a random selection basis as a "mixed strategy." It remains for the analyst to determine the proper probabilities for selecting the actions in the mixed strategy. However, a better minimax or maximin regret strategy often can be obtained from alternative mixed strategies than from pure strategies.[11, 27] Mixed strategies also make sense in many practical security situations (e.g., patrol routing). However, mixed strategies can never create a greater expected worth than a pure strategy with independent repeated decisions.

With dependent decision sequences one should take the nature of the dependencies into account over the sequential decision points that constitute the sequence stages. The principle of Bellman[28] provides the basis for approaching dependent sequential decisions. This principle states that regardless of the decision made in prior decision stages of the sequence, the remaining decisions should constitute an optimum decision sequence. In essence, this principle directs decision makers to start at the far future end of the sequence, identify the best actions for the final decision, given every prior choice in the immediately preceding decision, and then repeat this analysis sequentially toward the initial decision. This is a "rollback" form of analysis.[11,27] This procedure eliminates nonoptimum strings of decision choices sequentially until the current decision is arrived at to disclose an optimum decision sequence. It is also of note that this procedure is appropriate for sequences of different decisions as described by decision trees such as an expanded version of Exhibit 13.3.1*a*.

A special case of decision strategies occurs with probabilistic processes. Typical examples include replacement decisions where failures are sudden or where performance deteriorates probabilistically with time,[29] decisions on backup procedures, reliability/maintainability decisions in design, organization retraining decisions, and decisions on the adaptive control of manufacturing processes with fallible information seeking. The key to this special case lies in capturing the nature of the probabilistic process in either analytic form or through a computer simulation before applying the strategy concepts shown above. Various cases of stochastic processes are shown in the literature.[4, 17, 27, 28, 30]

13.3.11 SUMMARY

Risky decision making is the very heart of industrial engineering because of the nature of this engineering discipline. Design, operation, or management activities in manufacturing or service organizations all require decisions where the future is unsure. The foregoing materials provide analytical guidance in these decision making tasks. This guidance includes such principles as the concept of decision information values, Bayesian techniques of using the information, and decision strategies. Beyond this analytical foundation lies the artistic and creative part of decision making, which includes the recognition, formulation, and implementation of decisions and the assigning of decision making resources to the analytical process. The analytical part provides a theoretical closure to these extremely important artistic and creative activities. Although one may not like to face the risk of a decision, there is an economic limit to reducing the uncertainty before a decision is made. As we see through the window of the future but dimly, we must be prepared to make the most of our observations.

REFERENCES

1. H. BIERMAN, JR., C. P. BONINI, and W. H. HANSMAN, *Quantitative Analysis for Business Decisions*, Irwin, Homewood, IL, 1969.
2. H. RAIFFA and R. SCHLAIFER, *Applied Statistical Decision Theory*, Division of Research, Graduate School of Business Administration, Harvard University, Boston, 1961.
3. J. R. BUCK and J. M. A. TANCHOCO, "Sequential Bayesian Work Sampling," *AIIE Transactions*, Vol. 6, No. 4 (1974), pp. 318–326.
4. J. J. MARTIN, *Bayesian Decision Problems and Marker Chains*, Wiley, New York, 1967.
5. G. HADLEY, *Introduction to Probability and Statistical Decision Theory*, Holden-Day, San Francisco, 1967.
6. A. KAUFMANN, *The Science of Decision Making*, McGraw-Hill, New York, 1968.
7. S. A. SCHMITT, *Measuring Uncertainty: An Elementary Introduction to Bayesian Statistics*, Addison-Wesley, Reading, MA, 1969.
8. R. W. THRALL, C. H. COOMBS, and R. L. DAVIS, Eds., *Decision Processes*, Wiley, New York, 1954.
9. M. MILNOR, *Game Theory and Related Approaches to Social Behavior*, D. Shubik, Ed., Wiley, New York, 1964.
10. E. W. ADAMS, "Survey of Bernoullian Utility Theory," in H. Solomon, Ed., *Mathematical Thinking in the Measurement of Behavior*, Part 2, The Free Press of Glencoe, IL, 1960.
11. C. A. HOLLOWAY, *Decision Making Under Uncertainty: Models and Choices*, Prentice-Hall, Englewood Cliffs, NJ, 1979.

12. R. KEENEY and H. RAIFFA, *Decisions With Multiple Objectives: Preferences and Value Trade-offs*, Wiley, New York, 1976.
13. R. D. LUCE and H. RAIFFA, *Games and Decisions*, Wiley, New York, 1957.
14. M. TRIBUS, *Rational Descriptions, Decisions, and Design*, Pergamon, New York, 1969.
15. J. HIRSCHLEIFER, "Risk, the Discount Rate, and Investment Decisions," *American Economic Review*, Vol. 51, No. 2 (1961), pp. 112–120.
16. L. J. SAVAGE, M. S. BARTLETT, J. A. BARNARD, D. R. COX, E. S. PEARSON, and C. A. B. SMITH, *The Foundations of Statistical Inference*, Wiley, New York, 1962.
17. D. FARRAR, *The Investment Decision Under Uncertainty*, Prentice-Hall, Englewood Cliffs, NJ, 1962.
18. W. T. MORRIS, *Management Science: A Bayesian Introduction*, Prentice-Hall, Englewood Cliffs, NJ, 1967.
19. I. J. GOOD, *The Estimation of Probabilities: An Essay on Modern Bayesian Methods*, Research Monograph No. 30, The MIT Press, Cambridge, MA, 1965.
20. R. O. SWALM, "Utility Theory–Insights Into Risk Taking," *Harvard Business Review*, November–December 1966, pp. 123–136.
21. C. S. SPETZLER and C. A. S. STAEL VON HOLSTEIN, "Probability Encoding in Decision Analysis," *Management Science*, Vol. 22, No. 3 (1975), pp. 340–358.
22. H. RAIFFA, *Decision Analysis, Introductory Lectures on Choices and Uncertainty*, Addison-Wesley, Reading, MA, 1968.
23. F. SHEID, *Theory and Problems of Numerical Analysis*, Schaum's Outline Series, McGraw-Hill, New York, 1968.
24. C. W. CHURCHMAN, R. L. ARNOFF, and E. L. ACKOFF, *Introduction to Operations Research*, Wiley, New York, 1957.
25. M. SASIENI, A. YASPAN, and L. FRIEDMAN, *Operations Research–Methods and Problems*, Wiley, New York, 1959.
26. J. R. CANADA, *Intermediate Economic Analysis for Management and Engineering*, Prentice-Hall, Englewood Cliffs, NJ, 1971.
27. G. A. FLEISCHER, *Risk and Uncertainty: Nondeterministic Decision Making in Engineering Economy*, Monograph Series No. 2, Engineering Economy Division, AIIE, Norcross, GA, 1975.
28. R. BELLMAN, *Dynamic Programming*, Princeton University Press, Princeton, NJ, 1957.
29. R. SCHLAIFER, *Introduction to Statistics for Business Decision*, McGraw-Hill, New York, 1961.
30. W. T. MORRIS, *Decision Analysis*, Grid Publishing, Columbus, OH, 1977.

BIBLIOGRAPHY

DEGROOT, M. H., *Optimal Statistical Decisions*, McGraw-Hill, New York, 1970.

GOETZ, B. E., "Perplexing Problems in Decision Theory," *The Engineering Economist*, Vol. 14, No. 3 (1969), pp. 129–140.

KAHNEMAN, D., and A. TVERSKY, "On the Psychology of Prediction," *Psychology Review*, Vol. 80, 1973, pp. 237–251.

SIMON, H. A., *Models of Man*, Wiley, New York, 1957.

WILDE, D. J., *Optimum Seeking Methods*, Prentice-Hall, Englewood Cliffs, NJ, 1964.

CHAPTER 13.4

Design of Experiments for Industrial Engineers

VIRGIL L. ANDERSON
Purdue University

ROBERT A. McLEAN
University of Tennessee

13.4.1 OVERVIEW

Why should industrial engineers be interested in designing experiments? They have looked at data from so-called experiments for years. In some cases the results have been confusing, or the data looked wrong when compared to theory or preconceived ideas. In such cases people have been known to rerun the experiment, forget the experiment completely, or even change some data in order to make the results look better.

A comment that statisticians hear so often is that designing experiments takes so much time and that the engineer cannot afford to take more data. In many instances data, often taken haphazardly, are already available, and it is not understood why the statistician cannot just analyze those data and interpret the results rather than take more data from a well-designed experiment.

The main reason for taking data from a designed experiment is that the investigator can place a given probability statement on the results if the experiment has been *carefully* designed. Also, engineers who use data to help them draw conclusions usually want to know how widely the results will apply (inference space).

In almost all cases dealing with data, the experimenter wants to keep the number of observations small. Carefully designed experiments will allow minimum sample size for a specified problem if the variation is known. If the variation is not known (as in almost all cases), a small sample or a pre-experiment may be used to estimate the variation before the overall experiment is run.

Is all of this magic? No! It requires thinking, cooperation, work, and a willingness to learn basic concepts such as "confounding" and "biasedness."

Let us look at the past a bit and then turn our attention to a couple of simple design examples before delving into a few good industrial engineering designs of experiments to complete this chapter. The reader must understand that, to run experiments efficiently, he or she must read and study books, such as Anderson and McLean,[1] thoroughly.

For centuries humans have run experiments to answer questions. The idea of taking a sample to draw conclusions about a much larger group (population) is not new. Cochran and Cox,[2] however, point out that randomization is a relatively new concept, and Anderson and McLean[1] indicate that recently too many investigators have not been careful enough in defining how wide the results of their experiments apply (inference space).

Design Examples

A Pattern Maker

To give direction to the thinking of whether investigators should actually take care to design experiments or not, let us consider the following example: In a small shop a pattern maker wanted to buy a new lathe. He had narrowed the decision down to two brands, and these two manufacturers offered to let him try the lathes before deciding which to buy. Company representatives brought the two lathes to his shop, and he set up an experiment to help him decide between the two.

He thought 16 different patterns (requiring approximately the same time to cut) would be enough for him to make a decision if he used the *time required to cut a pattern to specifications* as the basis for his decision (criterion). Of course this meant that he would buy the lathe that required less time per pattern since the cost of the lathes was equal.

He was a "careful" experimenter and required that each pattern be used on both machines. This would allow him to take the difference in time required to cut the pattern on each machine, making his decision easy.

To begin the experiment, he flipped a coin to decide which machine should be used first throughout the experiment. The layout for the experiment was as follows:

	Pattern				
Lathe	1	2	3	....	16
1	1	3	5	....	31
2	2	4	6	....	32

The numbers inside the table indicate the order in which the patterns were to be cut on the lathes.

It has been the authors' experience that many experiments are run this way, or they are run without the first randomization because it is easy to keep the records straight as the investigator goes through the experiment. This is not a very thoroughly designed experiment, because if the pattern maker learns how to make that given pattern on lathe 1, he will probably retain some of that knowledge when he gets to lathe 2. Hence, if it should turn out that he can, in general, cut patterns faster on lathe 2, he will not know for sure whether it was due to the lathe's being "better" or to his learning from the first cut on lathe 1. This is an example of confounding; that is, the effect of *lathe* and *learning* cannot be separated. Hence there is a biased estimate of the effect of the lathe.

To improve the design of this simple experiment, many people would completely randomize the order of cutting the patterns. One possible layout of the completely randomized designed experiment is as follows:

	Pattern															
Lathe	1	2	3	4	5	6	7	8	9	10	11	12	13	14	15	16
1	27	9	7	15	24	19	10	14	13	8	29	12	21	11	28	17
2	3	25	2	16	22	1	20	31	23	5	26	4	30	18	32	6

The numbers inside the table indicate the order in which the patterns are to be cut on the particular lathe. For example, the first cutting would be pattern number 6 on lathe 2, and the last one (32) would be pattern number 15 on lathe 2.

Although complete randomization provides a more thoroughly designed experiment, one can easily see that peculiar sequences can be obtained. Notice in this so-called completely randomized design that lathe 2 is used to cut first for the first six times and that lathe 1 is used first for the next nine times. It is not known whether this sequencing interferes with the correct decision to buy the better lathe or not, but it is known (mathematically) that complete randomization does provide unbiased estimates of the effects.

Another way to run this experiment (the best way, we think) is to use the following layout:

	Sequence	
	1 Patterns	2 Patterns
Cut	4 7 15 6 3 14 1 10	16 2 9 5 12 13 8 11
First	Lathe 2	Lathe 1
Second	Lathe 1	Lathe 2

The operational procedure for this approach is more complicated, and the need for the additional detail is hard to explain to some experimenters. It is necessary to run the same number of patterns first on lathe 1 as on lathe 2 in order to obtain an unbiased estimate of the difference in time required for the two lathes. We insist on randomly assigning the various patterns to the two different sequences so that one lathe will not be favored over the other as the result of unsuspected differences among the patterns. One possible selection is the use of patterns 4, 7, 15, 6, 3, 14, 1, and 10 for sequence 1, as is shown in the preceding layout. One additional precaution must be taken in order to guard against such effects as fatigue. This can be accomplished by randomly selecting the order in which the patterns are actually cut. This can be done by randomly drawing the numbers 1 through 16. One such sequence would be 14, 11, 6, . . . , 10. Thus pattern 14 would be cut on lathe 2 and then on lathe 1; this would be followed by pattern 11 first on lathe 1 and then on lathe 2; and so on (not necessarily alternating).

This last design is discussed in detail later and is called a "crossover" design.

A Large Manufacturing Company

Another example of a designed experiment (this one illdefined) occurred a number of years ago in a large manufacturing company. A man working in the production area set up an experiment to test a new alloy, possibly one to replace an old one in production. He ran only *one* heat (batch) of metal with the new alloy and another heat with the old one. Taking one ingot from each heat and 30 pieces of metal from each ingot, he proceeded to test each of the 60 pieces for the property in which he was interested. With the data he made a one-way analysis of variance on the alloys, using the pieces within ingots, with 58 degrees of freedom as the error. The results showed that the new alloy was "better" than the old one, and the experimenter convinced the vice president in charge of production to change the production procedures so that the new alloy would be used in the future. Since the experimenter had used a "designed experiment" and had tested the data "statistically," the vice president concluded that there could be no doubt that the new one was better.

The change cost the company $200,000, and after 2 years in the field, there was as much trouble with the product made from the new alloy as there had been with the old product. The vice president was disgusted and called one of the authors of this chapter to say he would never allow his company to use designed experiments again. After some discussion, the vice president allowed the author to talk with the experimenter to find out how the experiment had been conducted.

In wanting to keep the cost of the experiment low, the experimenter had not considered the possibility that the property in which he was interested varied considerably both from heat to heat and from ingot to ingot within a heat. From an analysis-of-variance point of view, his expected mean squares should have been as follows:

Analysis of Variance of Alloy Problem

Source	*df*	EMS
Alloys (A)	1	$\sigma_p^2 + 30\sigma_I^2 + 30\sigma_H^2 + 30\phi(A)$
Pieces (p) in ingots	58	σ_p^2
Total	59	

Hence, rather than testing that the alloy effect was zero [$\phi(A) = 0$], he was really testing that the total effect for ingots (I), heats (H), and alloy was zero [$30\sigma_I^2 + 30\sigma_H^2 + 30\phi(A) = 0$]. Since the long-run production of the new alloy did not produce the improvement seen in the experiment, $\phi(A)$ must equal zero. Thus it must have been that σ_I^2 or σ_H^2 or both were not zero. The notation utilized here is that ϕ represents a fixed effect, that is, the effect of two specific alloys, and σ^2 represents a random effect, for example, the ingots are a random sample from all possible ingots.

When the results were explained, the vice president was willing to use designed experiments again on this type of problem, but insisted upon having more than one heat for each alloy.

A Manufacturer of Synthetic Hats

Still another example of a designed experiment (an excellent one) occurred in a company fabricating men's synthetic felt hats. The manufacturer had experienced extreme difficulty in producing these hats so that the flocking appeared on the molded rubber base in a uniform fashion in order to simulate the real felt hats. In approaching this problem, a committee was formed, consisting of a development engineer, a manufacturing foreman, a chief operator, a sales representative, and a statistician. The statistician's job was to obtain from the others all possible causes of imperfect hats. Factors that were thrown out for discussion were as follows: thickness of foam rubber base, pressure of molding, time of molding, viscosity of the latex used to glue the flocking to the molded rubber base, age of the latex, nozzle sizes of several different spray guns, direction of spraying, condition of the flocking, speed of drying, and effect of location within the drying furnace.

After considerable discussion, the committee decided that the most serious problems were probably connected with the nozzle size and the pressure under which the latex was sprayed. In arriving at these factors, the committee essentially forced a review of the entire production process, which led to a better understanding of the process and to an eventual solution to the problem. In this review the chief operator brought out the standard operating levels of the nozzle size as well as the pressure under which the latex was sprayed. Talk with the chief operator revealed that the latex pressure varied considerably because of the viscosity of latex, and from this information the pressure levels were eventually obtained. Additional inquiry ascertained that the manufacturing area had two different nozzle sizes that had been used interchangeably; consequently, the two sizes became the basis for these factor levels.

The authors believe that one is almost always able to find realistic levels for all major factors through committee action of this type. Occasionally, considerable effort is needed to find out just how shoddy one's manufacturing operation really is, and this example shows that actual production operations will allow a process to operate at various levels as long as it works. The determination of the optimal levels is, of course, the desired end of the experimental investigation.

As one can imagine in the manufacturing of synthetic hats, the measuring of product quality is a very difficult task. Consequently, the method used in the experiment just described was to grade the finished hat visually on the following items: the hungry appearance of the flocking, the starchy appearance of the flocking, and the appearance of the brim. During the course of the investigation, these responses were found to be essentially independent of each other and consequently could be treated as three separate dependent variables which could be investigated one at a time. The standards for grading each of these variables were arrived at again through committee action, which eventually resulted in a visual display board and gave the inspectors a realistic means of grading each variable.

One of the most difficult problems in certain types of industrial experiments is the specification of a dependent variable. It is usually quite obvious what the variable should be; however, the methods of measurement are sometimes quite difficult. In ideal cases the value would simply be measured by some simple inspection tool, whereas in other cases the value might be almost impossible to measure and would have to be graded by one or more inspectors.

Having agreed on the variables to analyze, the committee decided to include six nozzle locations, each with a high and low latex viscosity and each with a high and low air pressure, plus two types of base material as factors and levels in the experiment. This required a 2^{13} factorial experiment where all possible combinations of the 13 factors were to be considered. After some work, a fractional replicated design was run, requiring only 256 of the (2^{13}) 8192 combinations. This design allowed complete information on all main effects and on two factor interactions. Fractional factorials are described in detail in Anderson and McLean.[1]

The experiment required certain blocking procedures (described later in this chapter), and the results were excellent. The company was able to pinpoint the difficulty in the spraying mechanism, make the necessary changes, and produce a profitable product in 6 months. An interesting sidelight to this experiment was that the company had been losing about $8/hat when the experiment began, and the committee had been given a deadline of 1 year to solve the problem or the company would discontinue the product. With the results from this one experiment, the company was able to make a profit, and a competing company bought the whole process within the year.

The next section describes the authors' basic philosophy regarding the design of experiments, which has been used successfully since 1974.

13.4.2 BACKGROUND

One approach to designing experiments in the last decade considers three essential ingredients of a well-designed experiment, which are expressed by Anderson and McLean.[1] These three ingredients, in order of importance, are (1) inference space, (2) randomization, and (3) replication.

Inference Space

"Inference space" is a term that replaces the term usually used by statisticians, "population." It has been the authors' experience that "inference space" demands more attention from the researcher. Inference space means the limits to which the investigator may use the results of an experiment. Common practice at present is for researchers to indicate, before the experiment is set up, how extensively they wish the results to apply. This requires that they define the experimental units that they are to use in the research and that will be the basis for the inferences. They must also define the time interval and the geographical area to which they wish the results to apply and must then decide which levels of all factors they want controlled in the experiment. Ordinarily, more time should be spent on this phase of designing the experiment than on either of the other two phases (randomization and replication), because without the inference space clearly defined, the so-called best-designed experiment may be worthless to the investigator. We define this ingredient (inference space) as a part of the designed experiment. Hence there can be no "best-designed experiment" without a carefully defined inference space.

Randomization

Randomization is the next most important ingredient in designing experiments. It must be present in the experiment for probability statements to be made. Fisher[3] expressed the idea that it is the physical basis of the validity of the test. It is also the basis of the validity of confidence intervals.

Included in the ingredient of randomization is another concept, "restriction" on randomization. To help the reader understand this concept, let us first explain "completely randomized," which means having no restriction on randomization with respect to the source of the experimental unit. This concept can be seen by considering a factor, t, with five levels and three experimental units treated with each of the five levels of factor t completely at random.

Assume there are 15 randomly drawn experimental units from the inference space to be used for the entire experiment. One way to obtain a completely randomized design is to select a random number between 1 and 5, say 2. Then the first experimental unit must receive treatment 2, or the second level of factor t. Select another number between 1 and 5, say 5; then the second experimental unit must receive treatment 5. Continue sampling or drawing random numbers in this manner until the 15 experimental units have all been "treated." This sampling procedure requires that each level of factor t be represented three times in the experiment; the design of this experiment is completely randomized.

The mathematical model for analyzing the data from such an experiment is

$$y_{ij} = \mu + T_i + \epsilon_{(i)j} \qquad i = 1, 2, \ldots, 5; \qquad j = 1, 2, 3 \tag{1}$$

where y_{ij} = the response from experimental unit j treated with level i of factor t
μ = overall mean
T_i = effect of the ith level of factor t
$\epsilon_{(i)j}$ = the experimental error caused by the jth experimental unit nested in the ith level of factor t

The assumptions for the analysis of the data in a model such as this are:

1. y_{ij} is a random variable.
2. The variances of the responses within levels of factor t are equal.
3. The model is additive.
4. The experimental error is normal and independently distributed, with mean zero and variance σ^2, or NID(0, σ^2).

This complete randomization assures the experimenter that the experimental units from the inference space have three opportunities for selection for each treatment. These three units for each treatment then allow a measure of the variation within the treatments, which is the representation of the variation across the entire inference space. It follows, then, that the test of significance on treatment effects (T_i in equation 1) must be based on the excessive amount of variation among the means of the treatments over the amount of variation obtained from within the treatments, where the variation within the treatments is accounted for by the variance due to $\epsilon_{(i)j}$ in equation 1. Hence in equation 1 the $\epsilon_{(i)j}$ is the correct error for evaluating T_i.

If, however, the sampling procedure were such that the first random draw (level 2) were used on the first three experimental units, the second draw (level 5) on the next three experimental units, and so on, we would have a restriction on randomization because the randomization procedures were allowed only 5 times, not 15 as is required for complete randomization. If, as is frequently the case, there tends to be similarity between adjacent units in space or time, the variation within the group of three is smaller than the variation between the groups of three. In this design, then, the variation due to treatments is not separable from the variation between groups of three units (treatments confounded with groups). If, then, there is a source of error variation between groups, it will not be possible to test for treatments, as is depicted in the model

$$y_{ij} = \mu + T_i + \delta_{(i)} + \epsilon'_{(i)j} \tag{2}$$

where $\delta_{(i)}$ is the "restriction" error, or that random component due to the ith group of units (note how the subscript is identical to the subscript to T_i, indicating complete confounding) and $\epsilon'_{(i)j}$ is NID(0, $\sigma^2_{\epsilon'}$), the error due the jth unit in the ith group. It is the variation due to $\delta_{(i)}$ that is representative of the variation across the inference space, whereas the variation due to the $\epsilon'_{(i)j}$ in equation 2 represents only a small portion across the inference space. Hence one needs an estimate of σ^2_δ to test for the effect of treatments. Of course $\delta_{(i)}$ has no degree of freedom, which indicates that this is a poor design and should not be used.

Using the algorithm for deriving the EMS described in Chapter 2 of Anderson and McLean,[1] we can show that the correct error term for testing treatments is $\delta_{(i)}$. However, there is no estimate of this error in equation 2 unless the whole experiment is repeated. Hence the restriction on the randomization has caused $\delta_{(i)}$, which in turn tells the experimenter that this sampling procedure is not a good one even before the first observation is made. This, then, allows the experimenter to change the design early, before taking data that will not give a good analysis.

Before explaining the third ingredient of designed experiments, namely, replication, a few words should be said about special cases where randomization may not be required. If an investigator can show by actual experimentation that the results are the same whether randomization takes place or not, it may not be necessary to randomize. This will happen on extremely well controlled experiments only. We are familiar with an example of a laboratory experiment on one cylinder engine gasoline consumption in which it did not matter whether speeds were randomized or taken in order. The reason for this was that the controls on speed were so precise and the setups so repeatable that the errors were identical within the capabilities of the recording equipment. In terms of the model for analyzing the data from this experiment (similar to equation 2), $\delta_{(i)}$ is approximately zero.

With this nonrandom possibility in mind, and knowing an experiment cannot be designed well without knowing the inference space, we rank inference space above randomization in importance when considering ingredients of a well-designed experiment.

Replication

The third ingredient, replication, is quite often required for an estimate of an error term or to provide the basis for making decisions on the importance of factors contributing to the response variables. In addition, as the number of observations increases on a given treatment, the more precise the estimate of the effect of the treatment becomes, or the smaller its variance becomes.

If, however, previous experimentation has shown that certain information is available, for example, that the variance is known or that higher-order terms in the model are zero, it may not be necessary to replicate the entire experiment. In fact, there are many good experiments run with fewer than the total number of levels of the factors in a "factorial" experiment. These experiments are called "fractional replicated factorials."

Because of the various well-designed experiments without complete replications, the ingredient replication is placed in third position behind inference space and randomization.

Readers of the material on designs and analyses in the next section should understand (1) basic statistical concepts, (2) distributions such as mean $\bar{y}$, t, χ^2, and F, and (3) analysis of variance and regression models.

13.4.3 DESIGNS

A definition of a designed experiment is an arrangement of the experimental material, including randomization of experimental units to the treatments, so that statistical tests of significance (and confidence intervals) on the effects and interactions of the factors being studied can be made. To accomplish this, care must be taken to set up the arrangement efficiently (keep the cost reasonably low) while covering the inference space. In the following coverage of designs, major headings are used to indicate designs encountered by the authors in engineering studies.

Block Designs

Importance of Blocking

Many authors of books on the design of experiments express the importance of "blocking," or placing all treatments or all combinations of the levels of all factors of interest in a homogeneous group (thereby removing some of the effect of an extraneous variable from the experimental error) and repeating this group or block in time and/or space with different experimental units. To show this concept, we use mathematical equations that are to be used as the basis for analyses of the data from the designed experiment.

Returning to the concepts involved in setting up equations 1 and 2, it follows that another design of the experiment is to arrange three blocks of five treatments each, where the five experimental units are randomized onto the five treatments per block. It follows that the equation to be used as the basis for the analysis is

$$y_{ij} = \mu + B_i + \delta_{(i)} + T_j + \epsilon_{ij} \qquad (3)$$

where y_{ij} and μ have the same meaning as they did in equation 2, $\delta_{(i)}$ is similar to $\delta_{(i)}$ in equation 2, and

B_i = effect of block i
T_j = effect of the jth treatment
ϵ_{ij} = error due to the jth treatment in block i (assuming there is no interaction)

Since the experimenter is interested in testing for treatment effects only, this is an excellent design because ϵ_{ij} is the basis for the test of T_i since the variation across treatments is compared to the variation due to ϵ_{ij}'s. If it should turn out that the effects of B_i and $\delta_{(i)}$ are zero, then B_i, $\delta_{(i)}$, and ϵ_{ij} may be pooled, and equation 3 becomes equation 1. This completes the demonstration that, in general, blocking is always worthwhile in experiments and should be used whenever possible.

Latin square designs, used correctly, merely extend restriction on randomization in one more dimension. Too many engineers *incorrectly* use Latin square designs as special fractional factorials (see Chapter 8 in Anderson and McLean[1]). Since it is difficult to enable the reader to understand these points in the space allowed here, we recommend that Latin squares not be used by novices.

Incorrect Use of Blocks

In the previous example, the effect of $\delta_{(i)}$, the restriction error caused by blocks, did not decrease the importance of blocking because the experimenter was interested in the effect of blocking plus the restriction error in reducing the estimated experimental error, $\hat{\epsilon}_{ij}$. In that case $\hat{\epsilon}_{ij}$ was the basis for testing treatment effects only.

One must consider the case in which the blocking concept is used incorrectly. Consider an example in evaluating microforms, where the interest was in reducing user dissatisfaction. The experimenter wanted to use two types of projection (front, F, and back, B) as blocks and to randomize four screen angles (0°, 45°, 90°, 105°) twice within each block. The measured variable was seconds required to read the material presented each time. Pictorially the design was as follows:

Types of Projection	
Front	Back
90°	45°
0°	105°
45°	0°
90°	105°
105°	90°
0°	45°
45°	0°
105°	90°

The correct degrees of freedom and the model for the analysis of the data from this design are

$$16 = 1 + 1 + 0 + 3 + 3 + 8$$

$$y_{ijk} = \mu + T_i + \delta_{(i)} + A_j + TA_{ij} + \epsilon_{(ij)k} \qquad i = 1, 2; \quad j = 1, 2, 3, 4; \quad k = 1, 2 \tag{4}$$

where y_{ijk} = seconds required to read the microform from the kth observation using the jth angle and the ith type of projection
μ = overall mean
T_i = effect of the ith type of projection (fixed)
$\delta_{(i)}$ = restriction error due to all of the angles used with the ith type of projection before the other type of projection is used
A_j = effect of the jth angle (fixed)
TA_{ij} = effect of the interaction of the ith type with the jth angle
$\epsilon_{(ij)k}$ = error due to the kth observation within the ith type jth angle, assumed NID(0, σ^2).

The analysis of variance appropriate for analyzing data using equation 4 is as follows:

Source	*df*	EMS
Types of projection (T_i)	1	$\sigma_\epsilon^2 + 8\sigma_\delta^2 + 8\phi(T)$
Screen angles (A_j)	3	$\sigma_\epsilon^2 + 4\phi(A)$
TA_{ij}	3	$\sigma_\epsilon^2 + 2\phi(TA)$
Repeats within ($T - A$)	8	σ_ϵ^2
Combination $\epsilon_{(ij)k}$		

It is apparent from the EMS column that there is no test for types of projection because there is no source with an EMS of $\sigma_\epsilon^2 + 8\sigma_\delta^2$. Hence this is an incorrect use of blocks. To obtain the source to estimate $\sigma_\epsilon^2 + 8\sigma_\delta^2$, there must be a replicate of the experiment. This concept is demonstrated in the following section.

Correct Use of Blocks

To make the experiment described in the preceding section a good one, the investigator should set up the eight treatment combinations as follows (one possible randomization):

Blocks			
1		2	
F	45°	B	105°
B	0°	F	90°
B	105°	B	45°
F	0°	F	105°
B	90°	F	45°
F	105°	B	0°
B	45°	B	90°
F	90°	F	0°

The degrees of freedom and the model for the analysis of the data from this design are as follows:

$$30 = 1 + 1 + 0 + 1 + 1 + 3 + 3 + 3 + 3 + 0$$

$$y_{ijk} = \mu + B_i + \delta_{(i)} + T_j + BT_{ij} + A_k + BA_{ik} + TA_{jk} + BTA_{ijk} + \epsilon_{(ijk)}$$

$$i = 1, 2; \quad j = 1, 2; \quad k = 1, 2, 3, 4$$

where

B_i = effect of ith block (random)
$\delta_{(i)}$ = restriction error using blocks correctly
T_j, A_k, TA_{jk} = same as T_i, A_j, TA_{ij} of equation 4
$BT_{ij}, BA_{ik}, BTA_{ijk}$ = appropriate errors for testing (note arrows), which would probably pool for a common error
$\epsilon_{(ijk)}$ = residual error, not estimable with one observation per treatment combination per block

This is explained in detail in Anderson and McLean.[1]

Repeated Measures and Crossover Designs

A researcher in an industrial plant dealing with soldering parts on electronic equipment was interested in studying paced and unpaced production. He set up the experiment in a plant that used women only and recorded the ages of the women. He grouped the women into ages as follows: young (18–23), middle (30–35), and old (52–57). He was able to obtain six women in each of these age groups, which allowed him to have a sequence (selected at random) of (paced, unpaced) and (unpaced, paced) for 2 weeks for three women in each group. The layout of the design of the experiment can be portrayed as follows:

	Ages					
	Young		Middle		Old	
	Sequence		Sequence		Sequence	
Weeks	1 Women 1 2 3	2 Women 4 5 6	1 Women 7 8 9	2 Women 10 11 12	1 Women 13 14 15	2 Women 16 17 18
1	P	U	U	P	U	P
2	U	P	P	U	P	U

If the assumptions of the analysis of variance have been met, the analysis of variance of the data (different response variables) from this experiment, which uses a crossover design, is as follows:

Source	*df*
Ages (A)	2
Sequence (S)	1
AS	2
Females in ($A - S$) cells (F)	12
Weeks (W)	1
Paced versus unpaced (P)	1
AP	2
SP	1
ASP	2
Residual	11

Refer to Grizzle,[4] Myers,[5] and Anderson and McLean[1] for details of this type of design and analysis. For designs and analyses utilizing extensions of these designs to more sequences, refer to Albert et al.[6] and Westlake.[7]

Other Designs

There are many other designs used in industrial engineering that deal with factorial experiments and fractional factorials plus response surface designs. For these designs and analyses, refer to Myers,[8] Anderson and McLean,[1] and Box et al.[9]

REFERENCES

1. V. L. ANDERSON and R. A. MCLEAN, *Design of Experiments: A Realistic Approach*, Dekker, New York, 1974.
2. W. G. COCHRAN and G. M. COX, *Experimental Designs*, 2nd ed., Wiley, New York, 1957.
3. R. A. FISHER, *The Design of Experiments*, 7th ed., Hafner, New York, 1960.
4. J. E. GRIZZLE, "The Two-Period Change-Over Design and Its Use in Clinical Trials," *Biometrics*, Vol. 21, 1965, pp. 467–480.
5. J. L. MYERS, *Fundamentals of Experimental Design*, 3rd ed., Allyn & Bacon, Boston, 1979.
6. K. S. ALBERT, S. W. BROWN, JR., K. A. DESANTO, A. R. DISANTO, R. D. STEWART, and T. T. CHEN, "Double Latin Square Study to Determine Variability and Relative Bioavailability of Methylprednisolon," *Journal of Pharmaceutical Sciences*, Vol. 68, No. 10 (October 1979), pp. 1312–1316.
7. W. J. WESTLAKE, "Statistical Aspects of Comparative Bioavailability Trials," *Biometrics*, Vol. 35, 1979, pp. 273–280.
8. R. H. MYERS, *Response Surface Methodology*, Allyn & Bacon, Boston, 1971.
9. G. E. P. BOX, W. G. HUNTER, and J. S. HUNTER, *Statistics for Experimenters*, Wiley, New York, 1978.

BIBLIOGRAPHY

WINER, B. J., *Statistical Principles in Experimental Design*, 2nd ed., McGraw-Hill, New York, 1971.

CHAPTER 13.5

Hypothesis Testing and Statistical Inference

DON T. PHILLIPS
Texas A & M University

13.5.1 INTRODUCTION

One might define the function of applied statistics to be the science and the art of collecting, tabulating, interpreting, and executing statistical inference in such a way that the conclusions reached regarding the underlying behavior and characteristics of random phenomena are consistent and unbiased. This definition implies that statistical procedures, when properly applied and executed, are independent of the decision maker and depend entirely upon the statistical methodologies, definitions, and parameters required to execute the statistical test employed.

The science of statistics is purely mathematical, with probability theory as the cornerstone. Because all statistical methods are based on probability concepts, it is necessary for one to understand fully the concept of a probability measure before undertaking statistical analysis. This does not imply that one must know all there is to know about probability theory in order to use statistics, but one must understand at least some basic concepts in probability. This basic understanding is partially contained in Chapter 13.3 of this handbook. To proceed with the current topic, a fundamental knowledge of statistical measures, random variables, probability density functions, and statistical sampling procedures is assumed.

13.5.2 STATISTICAL INFERENCE

Each (and every) random variable has one (and only one) probability distribution. That distribution is also unique. For the most part, statistics deals with these probability distributions. Industrial engineers are interested in finding factual knowledge about certain random phenomena and do this by way of the probability distributions of the variables themselves or by way of probability density functions of other variables that may be related to the variable of interest.

In basic statistics one learns that probability density functions can be defined by certain constants called "distribution parameters." These parameters, in turn, can be used to characterize random variables through measures of location, shape, and variability of random phenomena. The most important descriptions are the mean, μ, and the variance, σ^2. The mean is a measure of the center of the distribution (an analogy is the center of gravity of a mass, as in physics), and the variance is a measure of the spread (range) of a distribution (an analogy being the moment of inertia of a mass). Hence, when we speak of the mean and the variance of a random variable, we refer to the use of statistical parameters (constants) that characterize the probabilistic behavior of the random variable of interest. Recall that

$$\mu = \sum_x xp(x)$$

or

$$\mu = \int_x xf(x)$$

where $p(x)$ represents probabilities of a *discrete* random variable and $f(x)$ represents the density function of probability of a *continuous* random variable. The parameters of interest are embodied in the form of the probability density functions and are examined shortly. Also

$$\sigma^2 = \sum_x (x - \mu)^2 p(x)$$

or

$$\sigma^2 = \int_x (x - \mu)^2 f(x)$$

Mathematical statistics show that many random variables that occur in nature follow the same general form of distribution, with differences only in the parameters and the statistical quantities μ and σ^2. Some of these recurring distributions have been given special names, such as:

Binomial	Beta	Uniform
Hypergeometric	Normal	Cauchy
Poisson	Chi-square	Rayleigh
Geometric	Student's t	Maxwell
Negative binomial	F distribution	Weibull
Gamma	Exponential	Erlang

Thus it is not difficult to see why the field of probability and statistics is a discipline within itself, nor is it difficult to see why almost every discipline in existence needs a working knowledge of statistics. Random phenomena (variables) exist in all phases of activity.

When one is interested in certain random phenomena, the first requirement seems to be to develop some means of measuring them. Upon doing so, one often collects a number of observations of the random phenomena. Statistics deals with developing tools and techniques for choosing those n observations (a sample) and manipulating them in such a way that useful information is gained about the underlying random variable(s). This information is generally derived from studying probability distributions or functions of the random variables. The average (or mean) and/or the variance (or spread) of the probability distribution of the random variable obviously yield useful information.

The two most widely used "statistics" are the mean of the sample of n observations

$$\bar{x} = \sum_{i=1}^{n} \frac{x_i}{n}$$

and the variance of the sample

$$s^2 = \sum_{i=1}^{n} \frac{(x_i - \bar{x})^2}{n - 1}$$

Note that these descriptors are not theoretical in nature, but are calculated from a set of n data points.

Mathematical developments have proven that $\bar{x}$ is usually the best single (point) estimate of μ and that s^2 is usually the best single (point) estimate of σ^2, where μ and σ^2 are the mean and the variance, respectively, of the underlying random variable. Normally, the nature of μ and σ^2 are totally unknown, and statistics is used to draw inferences about their true (unknown) values.

Since knowledge of the mean and variance are of utmost importance, statistics deals extensively with developing tools and techniques for studying their behavior. Two basic objectives are dealt with in the remainder of this chapter.

1. Ways of testing to see whether or not some assumed value of μ or σ^2 is "reasonable" under normal operating or assumed conditions.
2. Ways of using μ (σ^2) and $\bar{x}$ (s^2) such that one can state with a given measure of confidence that the population parameters of these estimates fall within a given interval. For example, we may say we are 95% confident that the interval from I_1 to I_2 includes the true value of μ.

The first objective is dealt with using *statistical hypothesis testing*, whereas the second gives rise to *confidence interval estimation*.

Statistical tools also exist for estimating the difference between, or for testing an assumption(s) about (a hypothesis), the means or variances of two or more probability distributions. These tools are natural extensions of the tools developed for estimating and testing hypotheses about single populations as described previously.

All statistical methods have, as a basis, a sample of n observations on the random phenomena of interest. Such methods require that the random sample (of size n) be "representative" of the outcomes that could occur. Because of this, much is said about making sure the sample is a random sample. Many statistical calculations being made today are invalid because the data are not representative.

13.5.3 STATISTICAL HYPOTHESIS TESTING

Working with a representative (random) sample of n observations, statisticians have shown that the following function for sufficiently large values of n

$$z = \frac{\bar{x} - \mu}{\sigma/\sqrt{n}}$$

is a random variable that follows (or has) a normal probability distribution, with $\mu = 0$ and $\sigma^2 = 1$. This fact is the result of the well-known central limit theorem, which states: If $\bar{x}$ is the mean of a random sample of size n taken from a population having the mean μ and the finite variance σ^2, then

$$z = \frac{\bar{x} - \mu}{\sigma/\sqrt{n}}$$

is the value of a random variable whose probability distribution approaches that of the standard normal distribution ($\mu = 0$, $\sigma^2 = 1$) as n approaches infinity. This is undoubtedly the most amazing theorem in statistics, for it does not require that one know anything about the shape of the probability distribution of the individual observations. It requires only that the distribution of those x's have a finite mean, μ, and a variance, σ^2.

Recall that the standard normal density function is as shown in Exhibit 13.5.1. If we use the notation $z_{\alpha/2}$ ($-z_{\alpha/2}$) to indicate the value of z where $\alpha/2$ of the area under the curve falls to the right (left) of $z_{\alpha/2}$ ($-z_{\alpha/2}$), then we establish the cross-hatched area shown in Exhibit 13.5.2. The cross-hatched area represents the region in which $(1 - \alpha)$ of all random variables, characterized by the standard normal density function $f(z)$ with mean $\mu_z = 0$ and variance $\sigma_z^2 = 1$, can be expected to lie. Within the context of a hypothesis test, this area is called the "acceptance region," and the tail areas are called the "rejection regions." The points $-z_{\alpha/2}$ and $z_{\alpha/2}$ are called the "rejection points," for reasons that are shortly evidenced. Although the underlying probability density func-

Exhibit 13.5.1 Standard Normal Density Function

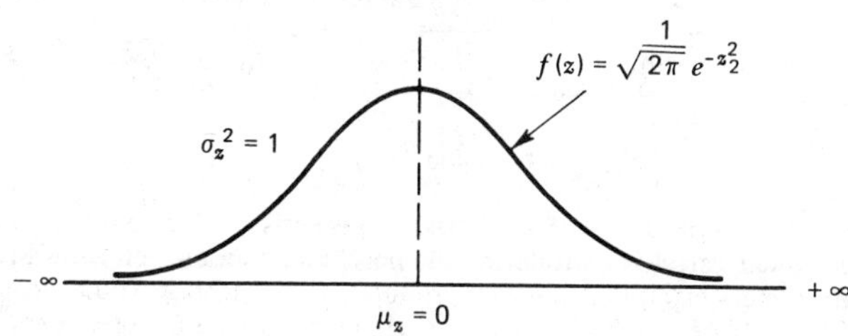

Exhibit 13.5.2 Acceptance and Rejection Regions

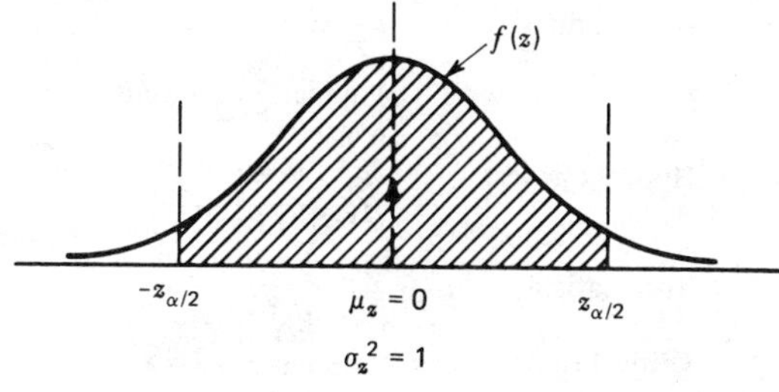

tion $f(z)$ might change, these concepts remain the same. The procedure of statistical hypothesis testing is now developed from these definitions.

13.5.4 TESTING A MEAN VALUE WITH VARIANCE KNOWN

Given a value of $\bar{x}$ computed using a sample of size n from an infinite population with known mean (μ) and variance (σ^2), the probability that the random variable $z = (\bar{x} - \mu)/(\sigma/\sqrt{n})$ falls between the points $-z_{\alpha/2}$ and $z_{\alpha/2}$ is $1 - \alpha$. Note that α is a value between zero and one and represents the probability that a random variable $\bar{x}$ that approximates the mean μ will naturally fall outside the points $-z_{\alpha/2}$ and $z_{\alpha/2}$. This interpretation of the natural behavior of the random variable $\bar{x}$, along with the distribution of the related (transformed) variable z, allows one to structure a hypothesis test concerning the true mean, μ. Assume that the value of $\alpha = 0.05$ and that the following statement is to be tested:

$$H_0: \ \mu = \mu_0$$
$$H_1: \ \mu \neq \mu_0$$

The value of μ_0 is the numerical value of μ that is assumed known or is hypothesized. H_0 is the "null" or "primary" hypothesis, and H_1 the "alternate" or "secondary" hypothesis. If a random sample of size n is extracted from the population of random variables under the study,

$$\text{then} \quad \bar{x} = \frac{\sum_{i=1}^{n} x_i}{n}$$

could be calculated. Since $\bar{x}$ is the best point estimator of μ, and μ is assumed to be equal to μ_0, one would expect the random variable $z = (\bar{x} - \mu_0)/(\sigma/\sqrt{n})$ to fall between the points $-z_{\alpha/2} = -z_{0.025}$ and $z_{\alpha/2} = z_{0.025}$ 95% of the time. The values for $-z_{0.025}$ and $z_{0.025}$ can be determined by using a standard normal table (see Exhibit 13.5A1) and are equal to ± 1.96. These values of $z_{\alpha/2}$ are called "critical values" and obviously depend upon α. Hence calculation of z yields a statistic that will cause H_0 to be believed 95% of the time and H_1 to be believed only 5% of the time if H_0 is true. Therefore α can be interpreted as the magnitude of the error of *rejecting* the *null hypothesis* when in fact it is true. This error is often referred to as an "error of type 1." In comparison, if the null hypothesis is *false*, there is still a chance that the calculated value of z will lie between ± 1.96 ($\alpha = 0.05$). This result will cause the decision analyst to *accept* the null hypothesis when in fact it is false. The magnitude (likelihood) of this error is commonly denoted by β, and this error is called an "error of type 2." The following table characterizes the decision process:

True State of Nature	Decision Based Upon Sampling Evidence	
	H_0 True	H_0 False
H_0 True	No error	Type 1 (α)
H_0 False	Type 2 (β)	No error

To illustrate the basic procedures of statistical hypothesis testing, consider the following example: An oil investment cartel is considering the purchase of an oil well from Blow Hard, Inc., in Texas. The current owners claim that the well produces on the average 100 barrels of oil/day, with a variance of 100 barrels of oil/day. To test this claim, the cartel chooses $\alpha = 0.05$ and observes daily production for 16 days. Total production over this period of time is 1690 barrels of oil. Can the owners' claim be disputed?

Assumptions
$\mu = 100$
$\sigma^2 = 100$
Population is infinite
$\alpha = 0.05$

Hypothesis test
$H_0: \ \mu = 100$
$H_1: \ \mu \neq 100$

Test statistic
$$Z = \frac{\bar{x} - \mu}{\sigma/\sqrt{n}}$$

Critical values
$\pm Z_{0.025} = \pm 1.96$

Hence

$$Z = \frac{\bar{x} - \mu}{\sigma/\sqrt{n}} = \frac{105.63 - 100}{10/4} = 2.252$$

Since the value of $Z = 2.252$ is greater than $Z_{\alpha/2} = 1.96$, one would choose to reject H_0 in favor of H_1. In other words, if H_0 is true, it is highly unlikely that a sample of size $n = 16$ would yield a value of $\bar{x}$ equal to 105.63, resulting in a value of Z as large as 2.252. What value of $\bar{x}$ would cause the decision maker to accept H_0?

Since $-z_{\alpha/2} = -1.96$, one can define the following relationship:

$$1.96 = \frac{\bar{x}_c - 100}{2.5}$$

Hence $\bar{x}_c = 104.9$ is the minimum daily production rate that will result in acceptance of H_0.

Next, let us assume that the true population mean (daily production rate) is equal to $\mu_0 = 110$ and not 100. What is the probability that the null hypothesis will be erroneously accepted? Assuming that the variance remains constant, consider Exhibit 13.5.3. Note that the probability that the null hypothesis will be accepted (erroneously!) is given by the area marked β. This area can be calculated as follows: Since

$$Z = \frac{\bar{x} - \mu_0}{\sigma/\sqrt{n}}$$

Then

$$-Z = \frac{104.9 - 110}{2.5}$$

yields the area to the left of the critical point $\bar{x} = 104.9$ with respect to the actual mean value $\mu = 110$. Hence $Z = -2.04$.

Using the table in Exhibit 13.5.A1, the probability corresponding to the β area is given by $\beta = 0.0217$. Note, however, that the β error also corresponds to that area to the *left* of $-Z_{0.025} = -1.96$ and that a value of Z falling into that area would cause a correct rejection of H_0. The $\bar{x}_c$ value corresponding to this point is $\bar{x}_c = 95.1$, and the probability of a random variable falling into this region when $\mu_0 = 110$ is determined by

$$Z - \frac{95.1 - 110}{2.5} = \frac{-14.9}{2.5} = -5.96$$

From Appendix 13.5.A1, the probability of a value 95.1 or smaller is essentially zero. Therefore $\beta = 0.0217$ is the correct value to four significant digits. In other words, the probability of accepting the null hypothesis when μ is actually $\mu_0 = 110$ is only 0.0217. Note that this probability is calculated based upon the values of $\bar{x}_c = 95.1$ and $\bar{x}_c = 104.9$, which were, in turn, uniquely determined by the chosen value of α. Clearly, one would never know what the true population mean (μ) is in any meaningful application. Hence the value of β cannot be calculated, except in reference to values different from the null hypothesis. For this example, consider the following values of β calculated

Exhibit 13.5.3 Probability of Acceptance of Null Hypothesis

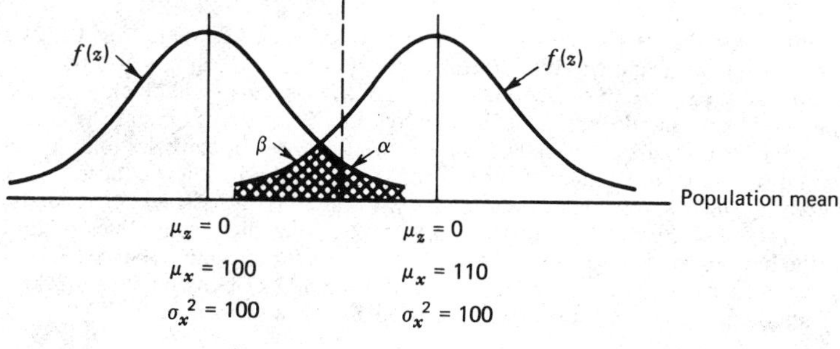

for various alternate values of μ_0:

μ_0	91	93	95	96	97	98	99
β	0.025	0.20	0.484	0.641	0.7764	0.8741	0.9315

μ_0	100	101	102	103	104	105	107	109
β	0	0.9315	0.8741	0.7764	0.641	0.484	0.20	0.025

For clarity, one should note that *in the limit* as μ_0 approaches 100, the probability of a type 2 error approaches $\beta = 1 - \alpha$. At the exact point $\mu_0 = 100$, the null hypothesis is true and a type 2 error does not exist; hence $\beta = 0$ at that single point. Exhibit 13.5.4 graphically depicts the behavior of the β error as a fraction of the true (unknown) population mean, under the original rejection criteria specified by the null hypothesis and the chosen α error. This curve is an OC curve. Observe that the preceding calculations were all performed relative to an alternative hypothesis that included two rejection points $(-z_{\alpha/2}, z_{\alpha/2})$. Such a hypothesis test is called a "two-tailed hypothesis test."

Consider once again our numerical example. Note that the null hypothesis states that well production is *exactly* 100 barrels/day. In reality, purchase of the well would be desirable if daily production met or exceeded 100 barrels/day. In that case the null and alternate hypothesis would appear as one of the following two alternatives:

1	2
H_0: $\mu \geqslant 100$ H_1: $\mu < 100$	H_0: $\mu \leqslant 100$ H_1: $\mu > 100$

Both hypothesis statements reflect the same objective, but there are significant differences in the decision criteria utilized in each hypothesis. The null hypothesis of alternative 1 assumes that the well is producing 100 or more barrels/day unless statistical evidence proves otherwise, resulting in rejection of H_0. The null hypothesis of alternative 2 assumes that the well is inferior unless production records indicate that daily output is more than 100 barrels/day, which will result in rejection of H_0. Both tests are valid and are called "one-tailed hypothesis tests" under a given type 1 error. Consider again the original data with $\alpha = 0.05$. The following calculations illustrate test applications:

	Test 1	Test 2
Assumptions	$\mu \geqslant 100$ $\sigma^2 = 25$ Population is infinite $\alpha = 0.05$	$\mu \leqslant 100$ $\sigma^2 = 25$ Population is infinite $\alpha = 0.05$
Hypothesis test	H_0: $\mu \geqslant 100$ H_1: $\mu < 100$	H_0: $\mu \leqslant 100$ H_1: $\mu > 100$
Test statistic	$Z = \dfrac{\bar{x} - \mu}{\sigma/\sqrt{n}}$	$Z = \dfrac{\bar{x} - \mu}{\sigma/\sqrt{n}}$
Critical value	$-Z_\alpha = -Z_{0.05} = -1.645$	$Z_\alpha = Z_{0.05} = 1.645$
Hence	$Z = \dfrac{\bar{x} - \mu}{\sigma/\sqrt{n}} = \dfrac{105.63 - 100}{10/4} = 2.252$	$Z = 2.252$
Conclusion	Accept H_0 and buy the well	Reject H_0 and buy the well

Note that both test statistics use the value of 100 in calculating a Z value, for it is at this single point that the type 1 error and the type 2 errors are the greatest. It should also be noted that the type 2 error now exists in only one direction, and hence the OC curve will be only one sided.

For completeness, one should note that the null hypothesis is an a priori state of nature that one chooses to believe unless statistical evidence indicates otherwise. In statistics one should not "accept" the null hypothesis, but rather "fail to reject" the hypothesis. These concepts are consistent with the underlying uncertain (stochastic) nature under which hypothesis testing is conducted. One can never be absolutely certain of statistical inference. Along these same lines of thought, it is critical that the hypothesis test be chosen *before* statistical sampling and not after. Selection of the test (one or two tailed) and the associated α error should never be chosen a posteriori to "confirm statistically" any belief. Such statistical inference is obviously improper.

Exhibit 13.5.4 Operating Characteristic Curve

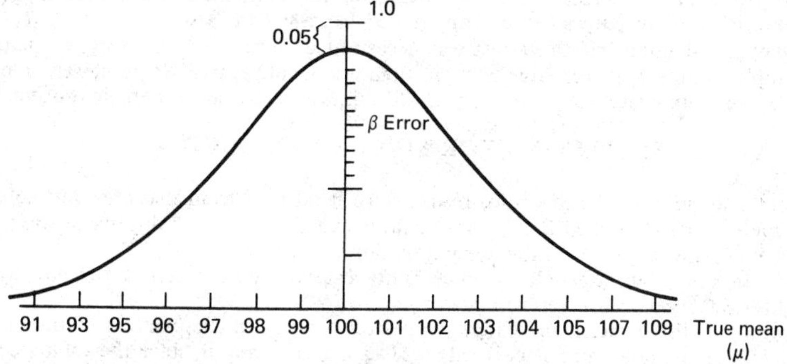

In summary, the decision maker can apply either a one-tailed or a two-tailed hypothesis test at a chosen type 1 error of α. The true type 2 (β) error is always unknown since the actual population mean is unknown. However, the β risk can be characterized by the construction of an OC curve.

13.5.5 TESTING A MEAN VALUE WITH VARIANCE UNKNOWN

Consider again the oil well example. After detailed examination, it was discovered that the theoretical variance (σ^2) was really not known, but had been "guessed at" by the seller. In order to estimate σ^2, the sample of 16 days was used to calculate S^2 in the following manner:

$$S^2 = \frac{\Sigma_{i=1}^{n} (x_i - \bar{x})^2}{n - 1}$$

Using $n = 16$, a value of $S^2 = 225$ was calculated. Under the assumption that σ^2 is unknown, the Z statistic is no longer a valid test statistic. Statisticians have shown that the following test statistic should be used:

$$t = \frac{\bar{x} - \mu}{s/\sqrt{n}}$$

This is student's t test statistic. The rejection points $t_{\alpha/2}$ and $-t_{\alpha/2}$ for a two-tailed test can be determined from an appropriate t table (see Exhibit 13.5.A2), but they now depend upon a parameter called the "degrees of freedom," which is defined by $dF = n - 1$. Consider the original two-tailed hypothesis test under the new assumption (σ^2 unknown).

Assumptions	σ^2 unknown $\mu = 100$ Population is infinite Population is normal $\alpha = 0.05$
Hypothesis test	H_0: $\mu = 100$ H_1: $\mu \neq 100$
Test statistic	$t = \frac{\bar{x} - \mu}{s/\sqrt{n}}$
Critical values	$t_{\alpha/2, dF} = t_{0.025,15} = 2.131$ $-t_{\alpha/2,dF} = -t_{0.025,15} = -2.131$

Hence

$$t = \frac{105.63 - 100}{15/4} = 1.501$$

Therefore the null hypothesis cannot be rejected, and one is lead to believe that the true well production is actually 100 barrels/day.

This example illustrates the use of the t statistic when testing hypotheses related to a population mean value (μ) when the variance (σ^2) is *unknown*. As before, both two-tailed and one-tailed tests are possible, whichever the problem situation demands. Rejection limits are set based upon a chosen type 1 (α) error, and OC curves can be constructed to reflect the associated type 2 (β) error risk. Finally, one should note that the t test was necessitated because σ^2 was being estimated by S^2 from a sample of size n. If n is large enough, then one would expect S^2 to closely approximate σ^2, and the Z test can be used anyway. For $n \geqslant 30$ this is generally an acceptable procedure.

13.5.6 HYPOTHESIS TESTING FOR SINGLE VARIANCE

Management was quite perplexed at this result, since it differed from that previously obtained by means of the original two-tailed Z test. An engineer explained that the difference was probably a result of lack of knowledge concerning the population variance, σ^2.

Since σ^2 was unknown, the induced uncertainty caused failure to reject the null hypothesis. (Note that the rejection points were also wider.)

Reflecting upon this logic, management requested a statistical examination of the sample variance to determine if the original (specified) σ^2 had indeed changed, since the estimated value of σ^2 (S^2) was greater than the original specified value. The proper statistical test is a chi-square test.

Assumptions	$\sigma^2 = 100$ Population is infinite Population is normal $\alpha = 0.05$
Hypothesis test	$H_0\colon\ \sigma^2 = 100$ $H_1\colon\ \sigma^2 \neq 100$
Test statistic	$x^2 = \dfrac{(n-1)\,S^2}{\sigma^2}$
Critical values	$x^2_{dF,\,1-\alpha/2};\quad x^2_{dF,\,\alpha/2},\quad dF = n - 1$

As in previous tests, the χ^2 rejection values are both a function of the chosen type 1 (α) error and the sample size ($dF = n - 1$). Critical values are easily obtained from χ^2 tables (see Exhibit 13.5.A3). The procedure is as follows:

Hypothesis test	$H_0\colon\ \sigma^2 = 100$ $H_1\colon\ \sigma^2 \neq 100$
Test statistic	$\chi^2 = \dfrac{(n-1)\,S^2}{\sigma^2} = \dfrac{(15)(225)}{100} = 33.75$
Critical values	$\chi^2_{15,\,0.975} = 6.262,\quad \chi^2_{15,\,0.025} = 27.488$

Hence, since $\chi^2 = 33.75$ is greater than the critical value $\chi^2_{15,0.025} = 27.488$, one is lead to believe that the underlying statistical variance of well production has changed from 100 barrels/day to something else. Of course, based upon the value of S^2, one might conclude that it has increased to somewhere around 225 barrels/day. Although further investigation would be up to the decision maker (buyer), it appears that a good course of action would be to obtain more sample data and reexamine the entire situation.

13.5.7 HYPOTHESIS TESTING FOR TWO POPULATION MEANS WITH VARIANCES KNOWN

After a rather lengthy discussion, company management reached the decision that the well should be purchased and that, because of optimistic market projections, a second well should also be purchased if one could be found that produced 100 more barrels/day on the average than the single established well. Further testing led management to believe that for comparison purposes the first well did indeed produce at a rate of 100 barrels/day and would be purchased. At this point, Blow Hard, Inc., presented a new well, which it claimed produced at a rate of 100 barrels/day more than the first well. Management again insisted upon statistical investigation, and in order to provide new, current data, a separate, independent evaluation was undertaken. Samples from both wells were obtained to compare daily production rates. Blow Hard, Inc., assured the purchasing cartel that the variance in daily production for the first well was actually 240 barrels/day and for the second well 276 barrels/day. A sample of production records for $n_1 = 12$ days from well one and $n_2 = 18$ days from well two provided average daily production of $\bar{x}_1 = 102$ and $\bar{x}_2 = 212$ barrels/day, respectively.

Since the variances of production, $\sigma_1^2 = 240$ and $\sigma_2^2 = 276$, are both assumed known, the proper test statistic is a Z test for the difference in the means of populations. The procedure is as follows:

Assumptions	σ_1^2 and σ_2^2 known Population is infinite $\alpha = 0.05$
Hypothesis test	H_0: $\mu_2 - \mu_1 = \delta$ H_1: $\mu_2 - \mu_1 \neq \delta$, $\delta = 100$ barrels/day
Test statistic	$Z = \dfrac{(\bar{x}_2 - \bar{x}_1) - (\mu_2 - \mu_1)}{\sqrt{(\sigma_1^2/n_1) + (\sigma_2^2/n_2)}}$
Critical values	$-Z_{\alpha/2}, Z_{\alpha/2}$

The numerical calculations are as follows:

$$H_0: \mu_2 - \mu_1 = 100$$
$$H_1: \mu_2 - \mu_1 \neq 100$$

$$Z = \frac{(\bar{x}_2 - \bar{x}_1) - (\mu_2 - \mu_1)}{\sqrt{(\sigma_1^2/n_1) + (\sigma_2^2/n_2)}} = \frac{(212 - 102) - (100)}{\sqrt{(240/12) + (276/18)}}$$

$$Z = \frac{10}{5.94} = 1.684$$

The critical value for this test is given by $\pm Z_{\alpha/2} = \pm Z_{0.025} = \pm 1.96$. Hence there is no statistical evidence to support rejection of the null hypothesis, and the proper decision would be to purchase both wells under the management interpretation of these results. However, the same clever engineer who questioned the first test results again questioned the validity of using σ_1^2 and σ_2^2 in the calculations. The question was then posed, "Can we perform a similar test without knowing the population variances?" A clever statistician responded "yes," since the same example used to calculate $\bar{x}_1$ and $\bar{x}_2$ could be used to estimate σ_1^2 and σ_2^2 with S_1^2 and S_2^2, respectively.

13.5.8 HYPOTHESIS TESTING FOR TWO MEANS WITH POPULATION VARIANCES UNKNOWN BUT ASSUMED EQUAL

The test statistic, assumptions, and hypothesis test are as follows:

Assumptions	$\sigma_1^2 = \sigma_2^2$ (both are unknown but assumed equal) Population is infinite $\alpha = 0.05$
Hypothesis test	H_0: $\mu_2 - \mu_1 = \delta$ H_1: $\mu_2 - \mu_1 \neq \delta$
Test statistic	$t = \dfrac{(\bar{x}_2 - \bar{x}_1) - (\mu_2 - \mu_1)}{\sqrt{(n_1 - 1) S_1^2 + (n_2 - 1) S_2^2}} \sqrt{\dfrac{n_1 n_2 (n_1 + n_2 - 2)}{n_1 + n_2}}$
Critical value	$-t_{\alpha/2, dF}$; $t_{\alpha/2, dF}$, $dF = n_1 + n_2 - 2$

Using the same set of data, it was found that $S_1^2 = 165$ and $S_2^2 = 453$. The calculations related to this example are as follows:

$$H_0: \mu_2 - \mu_1 = \delta$$
$$H_1: \mu_2 - \mu_1 \neq \delta$$

$$t = \frac{(212 - 102) - (100)}{\sqrt{11(165) + 17(453)}} \sqrt{\frac{(12)(18)(28)}{30}}$$
$$t = 1.46$$

The critical values (rejection points) for this test will be $t_{0.025, dF=28} = 2.048$ and $-t_{0.025, dF=28} = -2.048$. Since the calculated value of $t = 1.454$ is less than the upper rejection value, there is insufficient evidence to reject the null hypothesis. Finally, one should note that, under the assumption that $\sigma_1^2 = \sigma_2^2$, a pooled estimate of the variance is used from a total sample of $N = n_1 + n_2$. As in the single-parameter t test, if n_1 and n_2 are both greater than 30, one can simply use the two-parameter Z test directly.

Our same clever manager now observes that these results can be reached only if it can be assumed that the two population variances are equal. At this point, the manager asks our statistician, "Can we test this assumption?" The answer is yes, using an F test for equality of variances.

13.5.9 HYPOTHESIS TESTING FOR EQUALITY OF TWO POPULATION VARIANCES

The following assumptions, test statistic, and hypothesis test should be observed when conducting an F test:

Assumptions	Populations are normal Populations are infinite Sample is random
Hypothesis test	H_0: $\sigma_1^2 = \sigma_2^2$ H_1: $\sigma_1^2 \neq \sigma_2^2$
Test statistic	$F = \dfrac{S_1^2}{S_2^2}$
Critical value	$F_{1-\alpha/2,\nu_1,\nu_2}$, $F_{\alpha/2,\nu_1,\nu_2}$ ν_1 = degrees of freedom in numerator, $\nu_1 = n_1 - 1$ ν_2 = degrees of freedom in denominator, $\nu_2 = n_2 - 1$

The critical values for commonly used values of $F_{\alpha/2,\nu_1,\nu_2}$ are easily found in statistical tables such as those in Exhibit 13.5.A4. The values of $F_{1-\alpha/2,\nu_1,\nu_2}$ are usually calculated for the left-hand rejection point according to the following formula:

$$F_{1-\alpha,\nu_1,\nu_2} = [F_{\alpha/2,\nu_2,\nu_1}]^{-1}$$

For example, if S_1^2 were calculated using a sample of size $n_1 = 13$, and S_2^2 were calculated using a sample of size $n_2 = 21$, the rejection points for the quantity $F = S_1^2/S_2^2$ for $\alpha = 0.10$ would be given by

$$F_{\alpha/2,\nu_1,\nu_2} = F_{0.05,12,20} = 2.28$$

and

$$F_{1-\alpha/2,\nu_1,\nu_2} = [F_{0.05,20,12}]^{-1} = [2.54]^{-1} = 0.394$$

To illustrate the F test, consider the previous example. Recall that $S_1^2 = 155$, $n_1 = 11$, $S_2^2 = 453$, and $n_2 = 17$.

Assumptions	Populations are normal Populations are infinite $\alpha = 0.10$
Hypothesis test	H_0: $\sigma_1^2 = \sigma_2^2$ H_1: $\sigma_1^2 \neq \sigma_2^2$
Test statistic	$F = \dfrac{S_1^2}{S_2^2}$
Critical values	Using $\alpha = 0.10$, $n_1 = 11$, $n_2 = 17$ $F_{0.05,10,16} = 2.49$ $F_{0.95,10,16} = [F_{0.05,16,10}]^{-1} = [2.84]^{-1} = 0.352$

Using the preceding data

$$F = \frac{S_1^2}{S_2^2} = \frac{155}{453} = 0.342$$

Hence, the null hypothesis would be rejected, and one is led to assume that $\sigma_1^2 \neq \sigma_2^2$. Our astute manager, still seeking statistical evidence, now inquires as to the availability of a statistical test when $\sigma_1^2 \neq \sigma_2^2$. Fortunately, such a test is available—the t' test.

13.5.10 HYPOTHESIS TESTING FOR EQUALITY OF TWO MEANS WITH VARIANCES UNKNOWN AND NOT EQUAL

The proper statistical test for this procedure is the t' test, which is based upon the following:

Assumptions	$\sigma_1^2 \neq \sigma_2^2$ Populations are normal Populations are infinite
Hypothesis test	$H_0\colon\ \mu_1 - \mu_2 = \delta$ $H_1\colon\ \mu_1 - \mu_2 \neq \delta$
Test statistic	$t' = \dfrac{(\bar{x}_1 - \bar{x}_2) - \delta}{\sqrt{(S_1^2/n_1) + (S_2^2/n_2)}}$
Critical values	$t_{\alpha/2, dF};\quad t_{1-\alpha/2, dF}$

$$dF = \left\{ \frac{[(S_1^2/n_1) + (S_2^2/n_2)]^2}{[(S_1^2/n_1)^2/(n_1 - 1)] + [(S_2^2/n_2)^2/(n_2 - 1)]} \right\} - 2$$

For the oil well example, the following calculations illustrate the procedure:

Assumptions	$\sigma_1^2 \neq \sigma_2^2$ Populations are normal Populations are infinite $\alpha = 0.05$
Hypothesis test	$H_0\colon\ \mu_1 - \mu_2 = 100$ $H_1\colon\ \mu_1 - \mu_2 \neq 100$
Critical values	$t_{\alpha/2, dF}, -t_{\alpha/2, dF}$

$$dF = \left\{ \frac{[(145/11) + (453/17)]^2}{[(145/11)^2/10] + [(453/17)^2/16]} \right\} - 2$$

Hence $\quad dF = [25.71] - 2 = 23$

Therefore $\quad t_{0.025, 23} = 2.069$

$$-t_{0.025, 23} = -2.069$$

Test statistic

$$t' = \frac{(212 - 102) - 100}{\sqrt{(145/11) + (453/17)}} = \frac{10}{\sqrt{39.83}}$$

$$t' = 1.585$$

Since $t' = 1.585$ is less than the critical value of $t = 2.069$, statistical evidence does not support rejection of the null hypothesis.

Exhibit 13.5.5 summarizes the results to this point of the tests concerning means and variances.

13.5.11 CONFIDENCE INTERVAL ESTIMATION

A subject closely related to hypothesis testing is that of confidence interval estimation. The basic concepts are best understood by again considering the test statistic

$$Z = \frac{\bar{x} - \mu}{\sigma/\sqrt{n}}$$

Recall that the random variable Z follows a *standard normal* density function as shown in Exhibit 13.5.1. If, as before, we use the notation $Z_{\alpha/2}$ and $-Z_{\alpha/2}$ to indicate the values of the random variable Z where $\alpha/2$ of the area lies to the right of $Z_{\alpha/2}$ and $\alpha/2$ to the left of $-Z_{\alpha/2}$, then the following probability statement must be true:

$$\text{Probability}\left[-Z_{\alpha/2} \leqslant \frac{\bar{x} - \mu}{\sigma/\sqrt{n}} \leqslant Z_{\alpha/2}\right] = 1 - \alpha$$

If we rearrange the inequality in brackets and solve for the true population parameter μ, then we obtain

Exhibit 13.5.5 A Summary of Test Statistics

Hypothesis Test	Assumptions	Test Statistic	Critical Values
$H_0: \mu_1 = \mu_0$ $H_1: \mu_1 \neq \mu_0$	σ^2 is known Sample is random Population is infinite	$Z = \dfrac{\bar{x} - \mu_0}{\sigma/\sqrt{n}}$	$Z_{\alpha/2}, Z_{1-\alpha/2}$
$H_0: \mu_1 = \mu_0$ $H_1: \mu_1 \neq \mu_0$	σ^2 is unknown Population is normal Sample is random Population is infinite	$t = \dfrac{\bar{x} - \mu_0}{S/\sqrt{n}}$	$t_{\alpha/2,dF}$; $t_{1-\alpha/2,dF}$
$H_0: \mu_1 - \mu_2 = \delta$ $H_1: \mu_1 - \mu_2 \neq \delta$	σ_1^2, σ_2^2 are known Sample is random Population is infinite	$Z = \dfrac{(\bar{x}_1 - \bar{x}_2) - \delta}{\sqrt{(\sigma_1^2/n_1) + (\sigma_2^2/n_2)}}$	$Z_{\alpha/2}, Z_{1-\alpha/2}$
$H_0: \mu_1 - \mu_2 = \delta$ $H_1: \mu_1 - \mu_2 \neq \delta$	σ_1^2, σ_2^2 are unknown $\sigma_1^2 - \sigma_2^2$ Population is normal Sample is random Population is infinite	$t = \dfrac{(\bar{x}_1 - \bar{x}_2) - \delta}{\sqrt{(n_1 - 1)S_1^2 + (n_2 - 1)S_2^2}} \sqrt{\dfrac{n_1 n_2 (n_1 + n_2 - 2)}{n_1 + n_2}}$	$t_{\alpha/2,dF}$; $t_{1-\alpha/2,dF}$, $dF = n_1 + n_2 - 2$
$H_0: \mu_1 - \mu_2 = \delta$ $H_1: \mu_1 - \mu_2 \neq \delta$	σ_1^2, σ_2^2 are unknown $\sigma_1^2 \neq \sigma_2^2$ Population is normal Sample is random Population is infinite	$t^1 = \dfrac{(\bar{x}_1 - \bar{x}_2) - \delta}{\sqrt{(S_1^2/n_1) + (S_2^2/n_2)}}$	$t_{\alpha/2,dF}$; $t_{1-\alpha/2,dF}$, $dF = \left\{\dfrac{[(S_1^2/n_1) + (S_2^2/n_2)]^2}{[(S_1^2/n_1)^2/(n_1 - 1)] + [(S_2^2/n_2)^2/(n_2 - 1)]}\right\} - 2$
$\sigma_1^2 = \sigma_0$	Population is normal Sample is random Population	$x^2 = \dfrac{(n - 1)S_1^2}{\sigma_0^2}$	$x^2_{\alpha/2,dF}$; $x^2_{1-\alpha/2,dF}$, $dF = n - 1$
$\sigma_1^2 = \sigma_2^2$	Populations are normal Samples are random Populations are infinite	$F = \dfrac{S_1^2}{S_2^2}$	$F_{\alpha/2,\nu_1,\nu_2}$ $F_{1-\alpha/2,\nu_2,\nu_1}$

$$P\left[\left(\bar{x} - Z_{\alpha/2}\frac{\sigma}{\sqrt{n}}\right) \leqslant \mu \leqslant \left(\bar{x} + Z_{\alpha/2}\frac{\sigma}{\sqrt{n}}\right)\right] = 1 - \alpha$$

This last expression says that, if we take a random sample of n observations on the random phenomena of interest and calculate the interval,

$$\bar{X} - z_{\alpha/2}\frac{\sigma}{\sqrt{n}} \text{ to } \bar{X} + z_{\alpha/2}\frac{\sigma}{\sqrt{n}}$$

we are $1 - \alpha$ confident that the interval will include the true mean, μ. In other words, if we repeatedly took samples of size n and calculated the preceding interval from each sample, then in the long run $100(1 - \alpha)\%$ of the intervals will include μ. The interval is obviously calculated only once, and the resulting values of the end points constitute a $1 - \alpha$ confidence interval estimate of μ.

For a numerical illustration, consider once again the first example given in this chapter—the single oil well production problem.

Assumptions

$$\alpha = 0.05$$
$$\sigma^2 = 25$$
$$\pm Z_{\alpha/2} = \pm Z_{0.025} = \pm 1.96$$
$$\bar{x} = 105.63$$
$$n = 16$$

The 95% confidence interval for the true (unknown) population mean (μ) is given by

$$[105.63 - (1.96)(\tfrac{5}{4}) \leqslant \mu \leqslant 105.63 + (1.96)(\tfrac{5}{4})]$$

or

$$[108.08 \leqslant \mu \leqslant 103.18]$$

In other words, if we repeatedly estimated μ with $\bar{x}$ from a sample of size $n = 16$ from the same population, 95% of the time the preceding procedure will yield an interval that will contain the true population mean, μ.

Note that this 95% interval assumes that one knows σ, the SD of the random variable. Usually this is not the case, and we have to estimate σ with s. If we change the expression for z accordingly, we get a different distribution. Thus

$$t = \frac{\bar{X} - \mu}{s/\sqrt{n}}$$

is a random variable that follows a t distribution.

Now, utilizing the knowledge that we have provided about the t distribution, we write a probability statement similar to the preceding one:

$$P\left[-t_{\alpha/2} \leqslant \frac{\bar{x} - \mu}{s/\sqrt{n}} \leqslant t_{\alpha/2}\right] = 1 - \alpha$$

Upon rearranging the inequality in the brackets, we obtain

$$P\left[\left(\bar{X} - t_{\alpha/2}\frac{s}{\sqrt{n}}\right) \leqslant \mu \leqslant \left(\bar{X} + t_{\alpha/2}\frac{s}{\sqrt{n}}\right)\right] = 1 - \alpha$$

which again is a $1 - \alpha$ confidence interval estimate of μ when one does not know σ.

It should be pointed out that, when the sample size from which s is calculated exceeds 30, it makes little difference whether one uses the normal distribution with $\sigma = s$ or the more precise distribution. The probability distributions of t and z become the same at $n = \infty$.

If one is interested in calculating a $1 - \alpha$ confidence interval for the quantity $\mu_1 - \mu_2$, the difference between the means of two different distributions, one would take a sample from each distribution—n_1 from the first and n_2 from the second. From these one would calculate $\bar{X}_1$ and s_1^2 as well as $\bar{X}_2$ and s_2^2. In a manner identical to that just used, the following $1 - \alpha$ confidence interval could be constructed:

$$P\{[(\bar{X}_1 - \bar{X}_2) - z_{\alpha/2}\sqrt{(\sigma_1^2/n_1) + (\sigma_2^2/n_2)}] \leq (\mu_1 - \mu_2) \leq [(\bar{X}_1 - \bar{X}_2) + z_{\alpha/2}\sqrt{(\sigma_1^2/n_1) + (\sigma_2^2/n_2)}]\} = 1 - \alpha$$

Clearly this procedure could be repeated for *any* test statistic previously discussed in this chapter. The reader is referred to any of a number of engineering statistics texts for developments of χ^2, F, and t[1] confidence intervals.

13.5.12 TESTING FOR EQUALITY OF MEANS AND VARIANCES FOR MORE THAN TWO POPULATIONS

Testing means $\mu_1 = \mu_2 = \cdots = \mu_N$

In the case where there are N populations under consideration and the test of hypothesis is equality of population means, a different type of procedure is necessitated. This area of statistics is called "experimental design" or "analysis of variance" and is discussed in Chapter 13.6 of this handbook. Since it is covered elsewhere in some detail, its explanation is not pursued here.

Testing variance $\sigma_1^2 = \sigma_2^2 \cdots \sigma_N^2$

In the case where there are N populations and the test of equality of population variances is required, the most commonly applied test is Cochran's test for homogeneity of variances.

Assumptions	Samples are independent Populations are normal There are N populations
Hypothesis test	H_0: $\sigma_1^2 = \sigma_2^2 \cdots \sigma_N^2$ (At least one variance is different) H_1: $\sigma_1^2 \neq \sigma_2^2 \cdots \sigma_N^2$
Test statistic	$R = \dfrac{\max [S_i^2]\ i = 1, 2, \ldots, N}{\sum_{i=1}^{N} S_i^2}$, S_i^2 = unbiased point estimator of σ_i^2 for $i = 1, 2, \ldots, N$ (Each S_i^2 is calculated from a fixed sample of size n.)
Critical value	Accept the null hypothesis that $\sigma_1^2 = \sigma_N^2 \cdots \sigma_N^2$ if $R \leq RC_{\alpha,n,N}$, $RC_{\alpha,n,N}$ = a tabled critical value for chosen values of the type 1 error (α), the sample size (n), and the number of populations (N)

A table of critical values for $\alpha = 0.05$ and 0.01, for values of N up to 120 and of n up to 145, are given in Bowker and Lieberman.[1]

13.5.13 FURTHER CONSIDERATIONS

One should note that, for most of the hypothesis tests that have been discussed in this chapter, the assumption of normally distributed random variables is required. In actual practice, this may not be justified. One has two choices by which this assumption can be ignored. First, one can obtain enough samples to use a Z test (normal tables) rather than a t test (t tables). If the sample is large enough, then the assumption of normality is not required. However, in the case where larger samples cannot be obtained or cost prohibits larger samples, the use of *nonparametric statistical tests* are necessitated.

A nonparametric test is one that requires no assumptions regarding the form or shape of the underlying random variables. Usually all that is required is knowledge of the scale of measurement used in the experiment and whether the random variable is discrete or continuous. The treatment of nonparametric statistics is well beyond the scope of this introductory chapter. However, the reader should be aware of its role in hypothesis testing. Recent texts on nonparametric statistics[2-4] provide complete coverage of this topic.

Finally, it should be noted that the concepts of type 1 (α) error, type 2 (β) error, critical values, OC curves, and one- and two-tailed hypothesis tests are common to all hypothesis testing. This chapter has illustrated only tests concerning means and variances since they are the most common to industrial engineering. One might also find occasion to test hypotheses in quality control applications, proportions, percentages, and in goodness-of-fit testing. These applications and others are well documented in a host of applied engineering textbooks.

Exhibit 13.5.A1 Cumulative Normal Distribution $\Phi(t) = \int_{-\infty}^{t} \frac{1}{\sqrt{2\pi}} e^{-u^2/2}\, du$

t	0.00	0.01	0.02	0.03	0.04
0.0	0.500 00	0.503 99	0.507 98	0.511 97	0.515 95
0.1	0.539 83	0.543 79	0.547 76	0.551 72	0.555 67
0.2	0.579 26	0.583 17	0.587 06	0.590 95	0.594 83
0.3	0.617 91	0.621 72	0.625 51	0.629 30	0.633 07
0.4	0.655 42	0.659 10	0.662 76	0.666 40	0.670 03
0.5	0.691 46	0.694 97	0.698 47	0.701 94	0.705 40
0.6	0.725 75	0.729 07	0.732 37	0.735 65	0.738 91
0.7	0.758 03	0.761 15	0.764 24	0.767 30	0.770 35
0.8	0.788 14	0.791 03	0.793 89	0.796 73	0.799 54
0.9	0.815 94	0.818 59	0.821 21	0.823 81	0.826 39
1.0	0.841 34	0.843 75	0.846 13	0.848 49	0.850 83
1.1	0.864 33	0.866 50	0.868 64	0.870 76	0.872 85
1.2	0.884 93	0.886 86	0.888 77	0.990 65	0.892 51
1.3	0.903 20	0.904 90	0.906 58	0.908 24	0.909 88
1.4	0.919 24	0.920 73	0.922 19	0.923 64	0.925 06
1.5	0.933 19	0.934 48	0.935 74	0.936 99	0.938 22
1.6	0.945 20	0.946 30	0.947 38	0.948 45	0.949 50
1.7	0.955 43	0.956 37	0.957 28	0.958 18	0.959 07
1.8	0.965 07	0.964 85	0.965 62	0.966 37	0.967 11
1.9	0.971 28	0.971 93	0.972 57	0.973 20	0.973 81
2.0	0.977 25	0.977 78	0.978 31	0.978 82	0.979 32
2.1	0.982 14	0.982 57	0.983 00	0.983 41	0.983 82
2.2	0.986 10	0.986 45	0.986 79	0.987 13	0.987 45
2.3	0.989 28	0.989 56	0.989 83	0.990 10	0.990 36
2.4	0.991 80	0.992 02	0.992 24	0.992 45	0.992 66
2.5	0.993 79	0.993 96	0.994 13	0.994 30	0.994 46
2.6	0.995 34	0.995 47	0.995 60	0.995 73	0.995 85
2.7	0.996 53	0.996 64	0.996 74	0.996 83	0.996 93
2.8	0.997 44	0.997 52	0.997 60	0.997 67	0.997 74
2.9	0.998 13	0.998 19	0.998 25	0.998 31	0.998 36
3.0	0.998 65	0.998 69	0.998 74	0.998 78	0.998 82
3.1	0.999 03	0.999 06	0.999 10	0.999 13	0.999 16
3.2	0.999 31	0.999 34	0.999 36	0.999 38	0.999 40
3.3	0.999 52	0.999 53	0.999 55	0.999 57	0.999 58
3.4	0.999 66	0.999 68	0.999 69	0.999 70	0.999 71
3.5	0.999 77	0.999 78	0.999 78	0.999 79	0.999 80
3.6	0.999 84	0.999 85	0.999 85	0.999 86	0.999 86
3.7	0.999 89	0.999 90	0.999 90	0.999 90	0.999 91
3.8	0.999 93	0.999 93	0.999 93	0.999 94	0.999 94
3.9	0.999 95	0.999 95	0.999 96	0.999 96	0.999 96

Source. From W. M. HINES and D. C. MONTGOMERY, *Probability and Statistics in Engineering Science*, 2nd ed., Wiley, New York, 1980, pp. 474–475. Reprinted by permission.

Exhibit 13.5.A1 (*Continued*)

t	0.05	0.06	0.07	0.08	0.09
0.0	0.519 94	0.523 92	0.527 90	0.531 88	0.535 86
0.1	0.559 62	0.563 56	0.567 49	0.571 42	0.575 34
0.2	0.598 71	0.602 57	0.606 42	0.610 26	0.614 09
0.3	0.636 83	0.640 58	0.644 31	0.648 03	0.651 73
0.4	0.673 64	0.677 24	0.680 82	0.684 38	0.687 93
0.5	0.708 84	0.712 26	0.715 66	0.719 04	0.722 40
0.6	0.742 15	0.745 37	0.748 57	0.751 75	0.754 90
0.7	0.773 37	0.776 37	0.779 35	0.782 30	0.785 23
0.8	0.802 34	0.805 10	0.807 85	0.810 57	0.813 27
0.9	0.828 94	0.831 47	0.833 97	0.836 46	0.838 91
1.0	0.853 14	0.855 43	0.857 69	0.859 93	0.862 14
1.1	0.874 93	0.876 97	0.879 00	0.881 00	0.882 97
1.2	0.894 35	0.896 16	0.897 96	0.899 73	0.901 47
1.3	0.911 49	0.913 08	0.914 65	0.916 21	0.917 73
1.4	0.926 47	0.927 85	0.929 22	0.930 56	0.931 89
1.5	0.939 43	0.940 62	0.941 79	0.942 95	0.944 08
1.6	0.950 53	0.951 54	0.952 54	0.953 52	0.954 48
1.7	0.959 94	0.960 80	0.961 64	0.962 46	0.963 27
1.8	0.967 84	0.968 56	0.969 26	0.969 95	0.970 62
1.9	0.974 41	0.975 00	0.975 58	0.976 15	0.976 70
2.0	0.979 82	0.980 30	0.980 77	0.981 24	0.981 69
2.1	0.984 22	0.984 61	0.985 00	0.985 37	0.985 74
2.2	0.987 78	0.988 09	0.988 40	0.988 70	0.988 99
2.3	0.990 61	0.990 86	0.991 11	0.991 34	0.991 58
2.4	0.992 86	0.993 05	0.993 24	0.993 43	0.993 61
2.5	0.994 61	0.994 77	0.994 92	0.995 06	0.995 20
2.6	0.995 98	0.996 09	0.996 21	0.996 32	0.996 43
2.7	0.997 02	0.997 11	0.997 20	0.997 28	0.997 36
2.8	0.997 81	0.997 88	0.997 95	0.998 01	0.998 07
2.9	0.998 41	0.998 46	0.998 51	0.998 56	0.998 61
3.0	0.998 86	0.998 89	0.998 93	0.998 97	0.999 00
3.1	0.999 18	0.999 21	0.999 24	0.999 26	0.999 29
3.2	0.999 42	0.999 44	0.999 46	0.999 48	0.999 50
3.3	0.999 60	0.999 61	0.999 62	0.999 64	0.999 65
3.4	0.999 72	0.999 73	0.999 74	0.999 75	0.999 76
3.5	0.999 81	0.999 81	0.999 82	0.999 83	0.999 83
3.6	0.999 87	0.999 87	0.999 88	0.999 83	0.999 89
3.7	0.999 91	0.999 92	0.999 92	0.999 92	0.999 92
3.8	0.999 94	0.999 94	0.999 95	0.999 95	0.999 95
3.9	0.999 96	0.999 96	0.999 96	0.999 97	0.999 97

Exhibit 13.5.A2 Percentage Points of the t Distribution

					α						
ν^a	0.45	0.40	0.35	0.30	0.25	0.125	0.05	0.025	0.0125	0.005	0.0025
1	0.158	0.325	0.510	0.727	1.000	2.414	6.314	12.71	25.45	63.66	127.3
2	0.142	0.289	0.445	0.617	0.817	1.604	2.920	4.303	6.205	9.925	14.09
3	0.137	0.277	0.424	0.584	0.765	1.423	2.353	3.183	4.177	5.841	7.453
4	0.134	0.271	0.414	0.569	0.741	1.344	2.132	2.776	3.495	4.604	5.598
5	0.132	0.267	0.408	0.559	0.727	1.301	2.015	2.571	3.163	4.032	4.773
6	0.131	0.265	0.404	0.553	0.718	1.273	1.943	2.447	2.969	3.707	4.317
7	0.130	0.263	0.402	0.549	0.711	1.254	1.895	2.365	2.841	3.500	4.029
8	0.130	0.262	0.399	0.546	0.706	1.240	1.860	2.306	2.752	3.355	3.833
9	0.129	0.261	0.398	0.543	0.703	1.230	1.833	2.262	2.685	3.250	3.690
10	0.129	0.260	0.397	0.542	0.700	1.221	1.813	2.228	2.634	3.169	3.581

Exhibit 13.5.A2 (*Continued*)

ν^a	0.45	0.40	0.35	0.30	0.25	0.125	0.05	0.025	0.0125	0.005	0.0025
						α					
11	0.129	0.260	0.396	0.540	0.697	1.215	1.796	2.201	2.593	3.106	3.500
12	0.128	0.259	0.395	0.539	0.695	1.209	1.782	2.179	2.560	3.055	3.428
13	0.128	0.259	0.394	0.538	0.694	1.204	1.771	2.160	2.533	3.012	3.373
14	0.128	0.258	0.393	0.537	0.692	1.200	1.761	2.145	2.510	2.977	3.326
15	0.128	0.258	0.393	0.536	0.691	1.197	1.753	2.132	2.490	2.947	3.286
20	0.127	0.257	0.391	0.533	0.687	1.185	1.725	2.086	2.423	2.845	3.153
25	0.127	0.256	0.390	0.531	0.684	1.178	1.708	2.060	2.385	2.787	3.078
30	0.127	0.256	0.389	0.530	0.683	1.173	1.697	2.042	2.360	2.750	3.030
40	0.126	0.255	0.388	0.529	0.681	1.167	1.684	2.021	2.329	2.705	2.971
60	0.126	0.254	0.387	0.527	0.679	1.162	1.671	2.000	2.299	2.660	2.915
120	0.126	0.254	0.386	0.526	0.677	1.156	1.658	1.980	2.270	2.617	2.860
∞	0.126	0.253	0.385	0.524	0.674	1.150	1.645	1.960	2.241	2.576	2.807

[a] ν = degrees of freedom.
Source. From W. M. HINES and D. C. MONTGOMERY, *Probability and Statistics in Engineering Science*, 2nd ed., Wiley, New York, 1980, p. 477. Reprinted by permission.

Exhibit 13.5.A3 Percentage Points of the χ^2 Distribution

ν^a \ α	0.995	0.990	0.975	0.950	0.500	0.050	0.025	0.010	0.005
1	0.00+	0.00+	0.00+	0.00+	0.45	3.84	5.02	6.63	7.88
2	0.01	0.02	0.05	0.10	1.39	5.99	7.38	9.21	10.60
3	0.07	0.11	0.22	0.35	2.37	7.81	9.35	11.34	12.84
4	0.21	0.30	0.48	0.71	3.36	9.49	11.14	13.28	14.86
5	0.41	0.55	0.83	1.15	4.35	11.07	12.83	15.09	16.75
6	0.68	0.87	1.24	1.64	5.35	12.59	14.45	16.81	18.55
7	0.99	1.24	1.69	2.17	6.35	14.07	16.01	18.48	20.28
8	1.34	1.65	2.18	2.73	7.34	15.51	17.53	20.09	21.96
9	1.73	2.09	2.70	3.33	8.34	16.92	19.02	21.67	23.59
10	2.16	2.56	3.25	3.94	9.34	18.31	20.48	23.21	25.19
11	2.60	3.05	3.82	4.57	10.34	19.68	21.92	24.72	26.76
12	3.07	3.57	4.40	5.23	11.34	21.03	23.34	26.22	28.30
13	3.57	4.11	5.01	5.89	12.34	22.36	24.74	27.69	29.82
14	4.07	4.66	5.63	6.57	13.34	23.68	26.12	29.14	31.32
15	4.60	5.23	6.27	7.26	14.34	25.00	27.49	30.58	32.80
16	5.14	5.81	6.91	7.96	15.34	26.30	28.85	32.00	34.27
17	5.70	6.41	7.56	8.67	16.34	27.59	30.19	33.41	35.72
18	6.26	7.01	8.23	9.39	17.34	28.87	31.53	34.81	37.16
19	6.84	7.63	8.91	10.12	18.34	30.14	32.85	36.19	38.58
20	7.43	8.26	9.59	10.85	19.34	31.41	34.17	37.57	40.00
25	10.52	11.52	13.12	14.61	24.34	37.65	40.65	44.31	46.93
30	13.79	14.95	16.79	18.49	29.34	43.77	46.98	50.89	53.67
40	20.71	22.16	24.43	26.51	39.34	55.76	59.34	63.69	66.77
50	27.99	29.71	32.36	34.76	49.33	67.50	71.42	76.15	79.49
60	35.53	37.48	40.48	43.19	59.33	79.08	83.30	88.38	91.95
70	43.28	45.44	48.76	51.74	69.33	90.53	95.02	100.42	104.22
80	51.17	53.54	57.15	60.39	79.33	101.88	106.63	112.33	116.32
90	59.20	61.75	65.65	69.13	89.33	113.14	118.14	124.12	128.30
100	67.33	70.06	74.22	77.93	99.33	124.34	129.56	135.81	140.17

[a] ν = degrees of freedom.
Source. From W. M. HINES and D. C. MONTGOMERY, *Probability and Statistics in Engineering Science*, 2nd ed., Wiley, New York, 1980, p. 476. Reprinted by permission.

Exhibit 13.5.A4 Percentage Points of the F Distribution ($\alpha = 0.10$)

	ν_1								
ν_2	1	2	3	4	5	6	7	8	9
1	39.86	49.50	53.59	55.83	57.24	58.20	58.91	59.44	59.86
2	8.53	9.00	9.16	9.24	9.29	9.33	9.35	9.37	9.38
3	5.54	5.46	5.39	5.34	5.31	5.28	5.27	5.25	5.24
4	4.54	4.32	4.19	4.11	4.05	4.01	3.98	3.95	3.94
5	4.06	3.78	3.62	3.52	3.45	3.40	3.37	3.34	3.32
6	3.78	3.46	3.29	3.18	3.11	3.05	3.01	2.98	2.96
7	3.59	3.26	3.07	2.96	2.88	2.83	2.78	2.75	2.72
8	3.46	3.11	2.92	2.81	2.73	2.67	2.62	2.59	2.56
9	3.36	3.01	2.81	2.69	2.61	2.55	2.51	2.47	2.44
10	3.28	2.92	2.73	2.61	2.52	2.46	2.41	2.38	2.35
11	3.23	2.86	2.66	2.54	2.45	2.39	2.34	2.30	2.27
12	3.13	2.81	2.61	2.48	2.39	2.33	2.28	2.24	2.21
13	3.14	2.76	2.56	2.43	2.35	2.28	2.23	2.20	2.16
14	3.10	2.73	2.52	2.39	2.31	2.24	2.19	2.15	2.12
15	3.07	2.70	2.49	2.36	2.27	2.21	2.16	2.12	2.09
16	3.05	2.67	2.46	2.33	2.24	2.18	2.13	2.09	2.06
17	3.03	2.64	2.44	2.31	2.22	2.15	2.10	2.06	2.03
18	3.01	2.62	2.42	2.29	2.20	2.13	2.08	2.04	2.00
19	2.99	2.61	2.40	2.27	2.18	2.11	2.06	2.02	1.98
20	2.97	2.59	2.38	2.25	2.16	2.09	2.04	2.00	1.96
21	2.96	2.57	2.36	2.23	2.14	2.08	2.02	1.98	1.95
22	2.95	2.56	2.35	2.22	2.13	2.06	2.01	1.97	1.93
23	2.94	2.55	2.34	2.21	2.11	2.05	1.99	1.95	1.92
24	2.93	2.54	2.33	2.19	2.10	2.04	1.98	1.94	1.91
25	2.92	2.53	2.32	2.18	2.09	2.02	1.97	1.93	1.89
26	2.91	2.52	2.31	2.17	2.08	2.01	1.96	1.92	1.88
27	2.90	2.51	2.30	2.17	2.07	2.00	1.95	1.91	1.87
28	2.89	2.50	2.29	2.16	2.06	2.00	1.94	1.90	1.87
29	2.89	2.50	2.28	2.15	2.06	1.99	1.93	1.89	1.86
30	2.88	2.49	2.28	2.14	2.05	1.98	1.93	1.88	1.85
40	2.84	2.44	2.23	2.09	2.00	1.93	1.87	1.83	1.79
60	2.79	2.39	2.18	2.04	1.95	1.87	1.82	1.77	1.74
120	2.75	2.35	2.13	1.99	1.90	1.82	1.77	1.72	1.68
∞	2.71	2.30	2.08	1.94	1.85	1.77	1.72	1.67	1.63

Source. From W. M. HINES and D. C. MONTGOMERY, *Probability and Statistics in Engineering Science*, 2nd ed., Wiley, New York, 1980, pp. 482–483. Reprinted by permission.

10	12	15	20	24	30	40	60	120	∞	ν_2
60.20	60.71	61.22	61.74	62.00	62.26	62.53	62.79	63.06	63.83	1
9.39	9.41	9.42	9.44	9.45	9.46	9.47	9.47	9.48	9.49	2
5.23	5.22	5.20	5.18	5.18	5.17	5.16	5.15	5.14	5.13	3
3.92	3.90	3.87	3.84	3.83	3.82	3.80	3.79	3.78	3.76	4
3.30	3.27	3.24	3.21	3.19	3.17	3.16	3.14	3.12	3.10	5
2.94	2.90	2.87	2.84	2.82	2.80	2.78	2.76	2.74	2.72	6
2.70	2.67	2.63	2.59	2.58	2.56	2.54	2.51	2.49	2.47	7
2.54	2.50	2.46	2.42	2.40	2.38	2.36	2.34	2.32	2.29	8
2.42	2.38	2.34	2.30	2.28	2.25	2.23	2.21	2.18	2.16	9
2.32	2.28	2.24	2.20	2.18	2.16	2.13	2.11	2.08	2.06	10
2.25	2.21	2.17	2.12	2.10	2.08	2.05	2.03	2.00	1.97	11
2.19	2.15	2.10	2.06	2.04	2.01	1.99	1.96	1.93	1.90	12
2.14	2.10	2.05	2.01	1.98	1.96	1.93	1.90	1.88	1.85	13
2.10	2.05	2.01	1.96	1.94	1.91	1.89	1.86	1.83	1.80	14
2.06	2.02	1.97	1.92	1.90	1.87	1.85	1.82	1.79	1.76	15
2.03	1.99	1.94	1.89	1.87	1.84	1.81	1.78	1.75	1.72	16
2.00	1.96	1.91	1.86	1.84	1.81	1.78	1.75	1.72	1.69	17
1.98	1.93	1.89	1.84	1.81	1.78	1.75	1.72	1.69	1.66	18
1.96	1.91	1.86	1.81	1.79	1.76	1.73	1.70	1.67	1.63	19
1.94	1.89	1.84	1.79	1.77	1.74	1.71	1.68	1.64	1.61	20
1.92	1.88	1.83	1.78	1.75	1.72	1.69	1.66	1.62	1.59	21
1.90	1.86	1.81	1.76	1.73	1.70	1.67	1.64	1.60	1.57	22
1.89	1.84	1.80	1.74	1.72	1.69	1.66	1.62	1.59	1.55	23
1.88	1.83	1.78	1.73	1.70	1.67	1.64	1.61	1.57	1.53	24
1.87	1.82	1.77	1.72	1.69	1.66	1.63	1.59	1.56	1.52	25
1.86	1.81	1.76	1.71	1.68	1.65	1.61	1.58	1.54	1.50	26
1.85	1.80	1.75	1.70	1.67	1.64	1.60	1.57	1.53	1.49	27
1.84	1.79	1.74	1.69	1.66	1.63	1.59	1.56	1.52	1.48	28
1.83	1.78	1.73	1.68	1.65	1.62	1.58	1.55	1.51	1.47	29
1.82	1.77	1.72	1.67	1.64	1.61	1.57	1.54	1.50	1.46	30
1.76	1.71	1.66	1.61	1.57	1.54	1.51	1.47	1.42	1.38	40
1.71	1.66	1.60	1.54	1.51	1.48	1.44	1.40	1.35	1.29	60
1.65	1.60	1.54	1.48	1.45	1.41	1.37	1.32	1.26	1.19	120
1.60	1.55	1.49	1.42	1.33	1.34	1.30	1.24	1.17	1.00	∞

REFERENCES

1. A. H. BOWKER and G. J. LIEBERMAN, *Engineering Statistics*, 2nd ed., Prentice-Hall, New York, 1972.
2. S. SIEGEL, *Nonparametric Statistics for the Behavioral Sciences*, McGraw-Hill, New York, 1956.
3. I. MILLER and J. FREUND, *Probability and Statistics for Engineers*, 2nd ed., Prentice-Hall, New York, 1977.
4. G. E. NOETHER, *Elements of Nonparametric Statistics*, Wiley, New York, 1967.

BIBLIOGRAPHY

GIBRA, I. N., *Probability and Statistical Inference for Scientists and Engineer*, Prentice-Hall, Englewood Cliffs, NJ, 1973.

HINES, W. M., and D. C. MONTGOMERY, *Probability and Statistics in Engineering and Management Science*, Ronald, New York, 1972.

HOGG, R. V., and A. T. CRAIG, *Introduction to Mathematical Statistics*, 2nd ed., Macmillan, New York, 1965.

KIRKPATRICK, E. G., *Introductory Statistics and Probability for Engineering, Science and Technology*, Prentice-Hall, Englewood Cliffs, NJ, 1974.

MCBRIDE, V. E., class notes, University of Arkansas, Department of Industrial Engineering, Fayetteville, AR, 1968.

SNEDECOR, G. W., and W. G. COCHRAN, *Statistical Methods*, 6th ed., The Iowa State University Press, Ames, 1967.

VOLK, W., *Applied Statistics for Engineers*, 2nd ed., McGraw-Hill, New York, 1969.

WALPOLE, R. E., and R. H. MYERS, *Probability and Statistics for Engineers and Scientists*, 2nd ed., Macmillan, New York, 1978.

CHAPTER 13.6

Regression and Correlation

DOUGLAS C. CROCKER

Eastman Kodak Company

13.6.1 INTRODUCTION TO REGRESSION ANALYSIS

What is Regression Analysis?

Regression analysis is:

A technique for measuring and explaining (reducing unexplained) variability in a system.
An aid to understanding interrelationships in complex systems.
A process for building a useful model of a system.
A method for improving forecasting or prediction.
A mechanism for focusing on important phenomena.
A system for evaluating theories or beliefs.
An aid in formulating new theory.
A method for obtaining better control of variation.
A technique for estimating equation parameters.

Regression modeling involves practical problems, problems of judgment, and a good deal of art. This chapter is not intended to be a recipe book or a catalog of rules of thumb. It is intended to introduce the reader to some basic principles involved in statistical modeling while at once exposing the dangers. In this spirit, this chapter discusses many of the difficulties that may be encountered in attempting to model systems displaying statistical variation. It is intended to serve as a good blend of theoretical structure, philosophical outlook, and practical guidance.

The Profundity of the General Linear Model

An equation of the form

$$Y_i = \sum_{j=0}^{P} b_j X_{ij} \tag{1}$$

is sometimes referred to as the "general linear model." In this equation, Y is a variable whose behavior is of interest. It was once common to refer to Y as the "dependent" variable, taken from the *mathematical* concept of a function. In *statistical* modeling most authors have come to call Y the "response" variable. This is the convention adopted here.

In equation 1, Y is a linear additive function of the X variables which are P in number, $P \geqslant 1$. These X's were formerly often referred to as "independent" variables, again using the mathematical sense. They are now sometimes called "regressors" or "explanatory variables," but are more commonly called "predictors" (although "prediction" may not be the goal). The subscript j denotes *which* predictor. In this general form of equation 1, there is a dummy variable, $X_0 = 1$ ("dummy" because it does not vary), which is *not* counted as a predictor, but *is* included in the summation. Its coefficient, b_0, is the "constant" term or "intercept." It is in units of Y. The other regression coefficients, $b_j (j = 1 \text{ to } P)$, are the slopes (multipliers of their respective predictors) and are expressed in units of Y/X_j. These b_j's are unknowns that are to be determined from the analysis. The values obtained are estimates of the "true" unknown coefficients, β_j. Geometrically, equation 1 represents a line, a plane, or a hyperplane in $P + 1$ dimensional space. This process is known as "multiple linear regression (MLR) analysis." The subscript, i, denotes the series of observations in the sample going from 1 to n. Each observation provides a value for each of the variables–the predictors and the response–for each of the units or individuals in the sample. The unit of observation may be, for example, a day, a person, an automobile, a task, an event, a batch, or even a chapter in this book.

The X's can, of course, represent quite complicated transformations of originally observed bits of information about each unit. Reciprocals, powers, and logs are examples, and so are ratios or products of two (or more) predictors. It is astounding to witness how often this linear additive equation form gives a very good representation of the underlying physics that relate the response to the predictors.* That the variability–the behavior–of so many things in nature can be so well described (predicted) by this simple summation process is truly profound.

The Great Utility and the Attendant Dangers

Variation is the essence of statistical modeling. Variation is the problem. Information is contained in variation. In fact, without variation, there is no information. The activities of industrial engineers virtually always involve dealing with variation in *multiple-variable* systems. The goal may be to evaluate or explain previous events or to predict or control future events. Modeling a response variable in such systems is usually complex and difficult. Part of the difficulty arises in most cases because the data come from the existing system as it normally operates rather than being generated during a designed experiment. (See Chapter 13.4 regarding the nature of experimental design and some advantages attendant to controlled experimentation.) Such data might be called "nonexperimental" or "clinical."

Some major difficulties found in dealing with nonexperimental data result from the interrelationships naturally present among the predictors. The unwanted intercorrelations are avoided in controlled experiments by keeping the predictors uncorrelated with–orthogonal to–each other. This difficulty in dealing with clinical data is shared with many other disciplines. In fact, the exception is the analyst who *is* able to operate with "scientific" laboratory technique.

The great power of MLR lies in its ability to relate simultaneously the many intercorrelated predictors to the response–to deal with nonexperimental data. Herein also lies the main source of danger. Successful modeling of nonexperimental data is a tricky business. But not all the dangers are associated with the natural intercorrelations of nonexperiments. The variety of ways in which the analyst can encounter trouble is nearly as great as the variety of problem situations. Perhaps no other technique suffers more misuse and abuse than regression analysis. Because of this, much criticism of the general technique is offered by those who apparently do not understand its power or proper use and who misrepresent it. The dangers can be avoided or treated if they are recognized and understood. Much of the balance of this chapter deals directly or indirectly with establishing appropriate safeguards.

What Is the Question? The Importance of Goals

Multiple linear regression *should not* be a process that follows a fixed, predetermined path or that employs an established ritual for achieving a goal. That is because different goals require different analytic behavior. As illustrated by this chapter's opening list, regression goals are various. Before attempting to model a system, it is important to know what the model is supposed to do. What is the question the analysis is supposed to answer?

Because we are dealing here with practice in industrial engineering, it is important first to make the distinction between *science* and *decision making* (see Healy[1]). The statistical requirements for establishing scientific truth are much more stringent than for decision making. The manager cannot wait for the discovery of ultimate truth; he or she must decide today. Ordinarily, the industrial engineer operates in support of that process and will serve the manager best if the decision process is supported in a timely manner. This is *not* to suggest carelessness or disregard for theory. It *is* to suggest recognition of the basic fact that the manager will make the decision with or without the potential help. A responsibly derived, yet imperfect, model can be very much better than no model at all (see Chapter 13.2).

Very broadly, the various goals can be put into five categories. These represent a natural evolutionary sequence of four steps, any one of which may be the intended end use.

1. Exploration (Ex)–fishing, hypothesis finding (see Finch[2]).

2. Specification (S)–hypothesis testing (see Chapter 13.5), confirmation of the model form (rarely an end use).

3. Estimation (Es)–estimating model parameters with sufficient precision. (Estimating *future events* is referred to in this chapter as "prediction.")

*A known underlying causal relationship is *not* a *requirement* for useful statistical modeling.

4. Prediction (P)–use of the model for anticipation.
5. Control (C)–use of the model to prescribe change or to direct or guide policy or the behavior of a system.

Kinds of Models

Kinds of models seem to lie more along a continuum and are therefore less easily classified. The main continuum is closeness to causality. The scale slides from loose empiricism to exact causal representation (mechanism). How far along the scale the analyst moves may depend upon either the maturity of the corresponding physical discipline or the needs imposed by the goals.

There is another subset where causality is not an issue. These models might be called "associative." Here the response and the predictors may both be "caused" by some outside force. They behave concomitantly. An example is the use of leading indicators in economic models. Another example is the precursory use of animal characteristics or behavior to predict the severity of the winter. Presumably no one would claim that the extra hair on the woolly bear caterpillar *causes* snow to fall. (Causation might be suspected in the other direction in such cases if it were not for our belief that cause must precede in time its effect.)

This simple classification scheme thus takes the form shown in Exhibit 13.6.1.

Appropriate Use of Statistics

Statistical measures and diagnostics can and do serve an essential role in regression modeling, but they must be used appropriately. Their use must be related to goals. In general, any adequate MLR computer program system (a must–see Section 13.6.5) will list many statistics that may not be relevant in any given situation. For example, the multiple correlation coefficient, R, is universally printed. It may be of no interest. Further, even if it is of interest, its value must be judged in the context of the problem (see Section 13.6.3). It depends on the question. The analyst must know what question he or she wants answered and must use relevant statistical measures accordingly.

The Role of Assumptions

Assumptions are, in most regression articles and texts, listed as a sort of litany to precede the analysis as if they universally apply. Moreover, they are treated as if they describe the *problem setting.* They are really descriptions of the *mathematical model* whose behavioral properties are known and which is to be used as an analog to the system under study. Assumptions (model characteristics) relate to goals just as statistics do. Those that are relevant are rarely ever exactly met by the problem system. The severity of trouble the analyst may expect because of the remaining differences ("violations of assumptions") is a matter for judgment and experience and cannot be removed from the problem context.

Exhibit 13.6.2 offers a skeleton relationship of assumptions to goals in a hierarchical order. (For a more complete discussion, see Eisenhart.[3]) In practice, all variables are (almost always)

Exhibit 13.6.1 Simple Classification of Regression Models

Class	Kind	Basis
Associative	Concomitant, precursory, premonitory	Observation
Physical	Empirical	Observation
	Causal, mechanistic	Theory

Exhibit 13.6.2 Cumulative Relationships of Assumptions to Goals

Goals	Desirable Data Characteristics	Model, Process Characteristics
Exploration (Ex)	Random Y for given X	Least squares fitting
Specification (S)	"Complete" X set	"Correct" model form
Estimation (Es)	Spread, balanced Xs	b's normal by central limit theorem
Prediction (P) and control (C)	Typical X space	Specified error distribution

Exhibit 13.6.3 Relating Models to Evolutionary Goal Sequences[a]

	Kinds of Models		
	Associative	Physical	
Goal Sequences	Concomitant	Empirical	Mechanistic
Analysis			Existing theory
1	Ex	Ex → End	
2	S	S → End	
3	Es	Es	Es → End
Synthesis			
4a	P → End	P → End	P → End
4b			C → End

[a]Circled abbreviations are goals given in Exhibit 13.6.2.

known without error. Random residual variation in Y is associated with a host of small, unimportant (in context) contributions. Notice that the usual assumptions of homoscedasticity and normality are not imposed for specification and estimation. The least squares estimates of the regression coefficients provided by MLR are the most efficient, unbiased linear estimates among all *linear* estimates for uniform error variance and are still unbiased for nonuniform error variance. The central limit theorem will give very good protection—just as with ordinary averaging—allowing the normal model to be used with nonnormal data for establishing confidence intervals. Stated characteristics are *cumulative* descending the table.

Evolutionary Modeling and the Team Concept

Exhibit 13.6.3 shows the relationships of models to goals and depicts the evolutionary paths followed. There are three main paths (with circled steps numbered in the left-hand column) leading to a variety of end uses. Prediction is the only end use of associative (concomitant) models. Steps 2 and 3 (S and Es) are cyclical in all paths. Exploration may be the end use of an empirical model that is used, for instance, to direct further research and experimentation. Such experimentation may help to confirm hypotheses (specification). This may be the end use. The resulting specification may establish new theory and move into the mechanistic region where estimating model parameters (the β's) may be the end use. Estimating these constants for *existing* theoretical models is also a possibility. Either physical model path may lead to prediction as an end use. Finally, notice that only mechanistic models can serve the end use of control.* Notice also that analysis ends with the first three steps; the last two (prediction and control) represent synthesis.

In general, a thorough understanding of the system modeled will be required for success. Such knowledge *and* adequate modeling skill may not reside in the same person. This suggests a team effort, where the system expert(s) can suggest initial goals, which variables are important, and in which form they should appear. The modeler can then construct the model and interpret the diagnostic output. This will invariably generate questions leading to goal clarification, model and data changes, and/or further investigation. This evolutionary process continues until the (evolving) goals are met. Teamwork is important. The analyst cannot model in a vacuum. Problem definition (goals) and model specification evolve together.

Some Example Applications

It may be helpful to the reader to see some of the variety encountered by the author. Exhibit 13.6.4 lists categories of problem settings where regression modeling was employed, along with the number of projects undertaken in each category over approximately a 17 year span.

Individual projects varied greatly in complexity and size. A few examples may serve to illustrate the extreme variety. (The goal class for each example is abbreviated parenthetically.)

1. Hospital patient service as a function of age, patient category, hospital type, and so on. (Es)
2. Concentration of vitamin A in rat livers as a function of dosage, method, solution, and isomer form. (C)

*Keep in mind that the empirical-mechanistic distinction is a continuum. It can be argued that the purely correct mechanistic extreme is never achieved.

Exhibit 13.6.4 Variety of Regression Projects

Category	Number of Projects
Labor prediction, cost	87
Quality	63
Machine performance	49
Personnel performance	45
Energy consumption	43
Physiological	39
Cost (other)	28
Equipment utilization	24
Equipment calibration	23
Service	12
Space needs	7
Personnel planning	6
Miscellaneous other	27
Total	453

3. Blood bank blood use (16 classes) as a function of season and day of week. (P)
4. Evaluation of improvement in service with methods change masked by (covariate) production level changes. (Es)
5. Punch press setup time as a function of die characteristics. (P)
6. Mailing and shipping times as functions of distance and region. (Es, C)
7. Compression of plastic bottles related to weight, minimum wall thickness, and mold cavity. (C)
8. Test scores as an aid in job placement. (P)
9. Effectiveness of inspection related to speed, quality, and product class (see Chapter 8.4). (C)
10. Product performance characteristics related to chemical formulation. (C)
11. Human factors treadmill and force platform experiments (see Chapter 6.4). (Ex, S)
12. Time prediction model for programming numerically controlled machining (see Chapter 6.7). (P)
13. Effectiveness of alternative methods of conducting training programs (see Chapter 5.2). (C)
14. Physiological response to various shift work schedules (see Chapter 8.3). (C)
15. Energy use related to weather, production, policies (e.g., federal temperature guidelines). (S, Es, P, C)
16. Dental health of children related to trace element composition of tooth enamel (see Curzon and Crocker[4]). (S)
17. Development of standard data (see Chapter 3.6). (Es, P)

13.6.2 RELATING TWO VARIABLES

Starting Simple

The actual use of the *simple* (single-predictor) model is rare. (Real systems are rarely that simple.) However, for examining the principles involved in regression modeling, the simple model serves well. For illustration, hypothetical data representing steam consumption (Y) for a particular building are modeled here. This response variable was chosen because (1) energy use has universal relevance and global importance, (2) such a wide variety of goals can be authentically represented (see example 15 in the preceding section) in an energy system, and (3) this same problem setting can be expanded in following sections to represent more complex modeling ventures. The structure under study might be an office complex, a factory, a warehouse, a hospital, a hotel, or even a home. For demonstration, only 20 observations are contained in the sample. Each observation represents a 4 week period. Weekly data would be preferred in most cases, but to cover extremes of weather in only 20 data points, 4 week periods were chosen. The goal is to establish control of steam consumption for this building. "Excessive" use is now dismissed as being weather related.

At first it is assumed that comfort heating in this building is the major use of steam. Its use (measured in Giga-British thermal units, gbtu), should be reasonably well related to degree-days (X). This is measured relative to 65°F (degree-days F/1.8 = degree-days C) and is also reported here on a per period basis. The 20 observations are shown in Exhibit 13.6.5, with the periods numbered within year from 1 to 13 and years numbered 1, 2, and 3. The method of least squares will be employed to relate steam use to degree-days.

Exhibit 13.6.5 Hypothetical Data for Steam Consumption

i	Period Number	k degree-days $(X)^a$	gbtu $(Y)^b$
1	10-1	0.156	7.991
2	11-1	0.419	8.589
3	12-1	0.658	9.145
4	13-1	1.090	11.212
5	1-2	1.380	11.754
6	2-2	1.103	11.469
7	3-2	1.000	10.584
8	4-2	0.703	9.509
9	5-2	0.207	7.457
10	6-2	0.086	6.989
11	7-2	0.024	6.537
12	8-2	0.005	4.938
13	9-2	0.026	5.275
14	10-2	0.161	7.452
15	11-2	0.307	7.962
16	12-2	0.664	8.915
17	13-2	1.039	9.758
18	1-3	1.275	11.183
19	2-3	1.193	11.523
20	3-3	0.953	10.426

$n = 20$; $\Sigma X_i = 12.449$; $\Sigma Y_i = 178.67$;
$\Sigma X_i^2 = 12.085$; $\Sigma Y_i^2 = 1678.7$; $\Sigma X_i Y_i = 129.33$

[a] k degree-days = 10^3 degree days.
[b] gbtu = Giga-British thermal units = 10^9 btu.

Why Least Squares?

In MLR, understanding variation is the basis for problem solving. Variation in the response is made up (theoretically) of two parts:

1. The systematic variation (signal), which is associated with or is in response to changes in the predictors.
2. Leftover variation (noise), which is called "residual error" or "experimental error."

The distinction is really not so sharp. The leftover error is actually associated with a great many things that, in practice, *might* be measured (and included in the model) if analysts had sufficient time, wisdom, patience, and money. They simply choose not to try to identify all sources of variation. They will discontinue the search when there seems to be no regular pattern of errors left over *and* when either all the reasonable predictors have been adequately tested *or* the *residual error variance* is small enough—again depending on goals. In terms of the true coefficients and residual error of the theoretical model, the observed response variable may be expressed as

$$Y_i = \sum_{j=0}^{P} \beta_j X_{ij} + \epsilon_i \tag{2}$$

where ϵ_i is the "residual error" associated with Y and (theoretically) has variance σ_ϵ^2. The fitted model containing the estimates of the β_j's, then, is

$$\hat{Y}_i = \sum_j b_j X_{ij} \tag{3}$$

where the circumflex or "hat" on Y denotes the predicted or estimated value of the response. It is like an average (where a "bar" is used). In fact, it is the *conditional* average, given the location in the space defined by the X_{ij}'s. It is an estimate of the expected or true value of the response for that location or set of conditions.*

The differences between the observed and fitted values of Y are the residual errors, or, simply, "residuals,"

$$e_i = Y_i - \hat{Y}_i = \hat{\epsilon}_i \tag{4}$$

where e_i is an estimate of the "true error" ϵ_i. In practice, e_i may contain anything the analyst chooses to omit from the model. It has sample variance

$$s^2_{Y \cdot X} = s^2_e = \frac{\Sigma_{i=1}^{n} (Y_i - \hat{Y}_i)^2}{n - P - 1} = \frac{\Sigma_{i=1}^{n} e_i^2}{n - P - 1} = \hat{\sigma}^2_\epsilon \tag{5}$$

which, for the theoretical case, is an estimate of the "experimental" error variance. The subscript $Y \circ X$ ("Y dot X") means "for Y, given the model containing a particular set of X's." Thus $s^2_{Y \cdot X}$ is the sample estimate of the residual variance in Y, given the model.

The least squares method chooses values for the b_j's of equation 3, which are unbiased estimates of the β_j's of equation 2. The least squares estimates are *universally* minimum variance unbiased estimates for *normally* distributed residual errors and are minimum variance among all *linear* estimates (linear combinations of the observed Y's), regardless of the residual error distribution shape (see Eisenhart[5]). The b_j's (as well as the $\hat{Y}_i$'s) are linear combinations of the observed Y_i's. The least squares method determines the weight given to each Y value. The derivations of the least squares solution and/or associated equations used later in this chapter are shown in other sources.[6–10] In essence, the b_j's are chosen to minimize the numerator of equation 5—the sum of squares of e_i's of equation 4—hence "least squares."

Finding the Least Squares Line

In returning to the example problem, a geometric interpretation is presented first. Exhibit 13.6.6 is a plot of steam consumption versus degree-days from Exhibit 13.6.5. The regression coefficient, b_1, is represented by the slope of the least squares line. It is the tangent of the angle θ. The e_i's whose squares are to be summed to a minimum are distances measured in the Y direction from the points to the line. They are illustrated by typical distances, e_4 and e_{17}.

The least squares solutions for the simple model are

$$b_1 = \frac{SPXY}{SSX} \qquad b_0 = \overline{Y} - b_1\overline{X} \tag{6}$$

where

$SPXY$ = the (corrected) sum of products of the XY pairs
SSX = the (corrected) sum of squares of X's
$\overline{Y}$ and $\overline{X}$ = the arithmetic averages (which are also least squares estimators) of the two variables

These averages and sums (with all sums taken for i = 1 to n) are

$$\begin{aligned} \overline{X} &= \frac{\Sigma X_i}{n} & \overline{Y} &= \frac{\Sigma Y_i}{n} \\ SPXY^{\dagger} &= \Sigma(X_i - \overline{X})(Y_i - \overline{Y}) & &= \Sigma X_i Y_i - n\overline{X}\overline{Y} \\ SSX &= \Sigma(X_i - \overline{X})^2 & &= \Sigma X_i^2 - n\overline{X}^2 \end{aligned} \tag{7}$$

These equations (6 and 7) yield the following values for the example:

$$\begin{aligned} \overline{X} &= 12.449/20 = 0.623 & \overline{Y} &= 178.67/20 = 8.93 \\ SPXY &= 129.33 - 20\,(0.623)\,(8.93) & &= 18.06 \\ SSX &= 12.085 - 20\,(0.623)^2 & &= 4.32 \\ b_1 &= 18.06/4.32 = 4.18 & b_0 &= 8.93 - 4.18\,(0.632) = 6.29 \end{aligned}$$

Equation 3 then takes the form $\hat{Y}_i = 6.29 + 4.18X_i$.

*It is important to realize that, although the model may be used for predicting future Y values, $\hat{Y}$ does not predict their *individual* behavior, but estimates the conditional average about which those individuals are expected to vary.

†These are not shown in the computationally easiest form. This form demonstrates meaning. Computation for problems of substance will be done by computer.

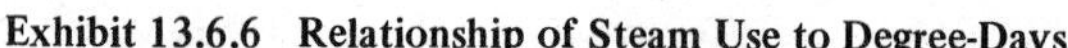

Exhibit 13.6.6 Relationship of Steam Use to Degree-Days

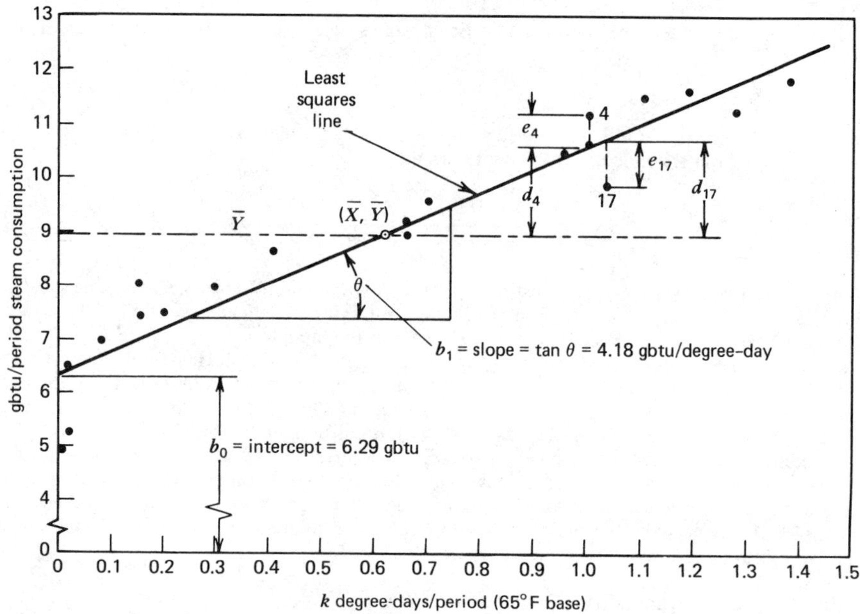

Residual Variance

For $P = 1$, equation 5 reduces to $s^2_{Y \cdot X} = SSRes/(n - 2)$, where $SSRes$ is the residual sum of squares given by

$$SSRes = SSY - SSReg \tag{8}$$

and where SSY is the (corrected) sum of squares of Y's, and $SSReg$ is the regression sum of squares. In equation 8, SSY is exactly parallel in form to SSX in equation 7, and when divided by its $n - 1$ degrees of freedom, it yields the Y mean square, which might be used to estimate the variance of Y. Regardless of the appropriateness of such an interpretation, the expression $SSY/(n - 1)$ *is* a measure of the raw variability in the response whose explanation is the goal. The contribution to SSY that is associated with X is $SSReg$. This is given by

$$SSReg = b_1 SPXY \tag{9}$$

and is the sum of squared distances from $\overline{Y}$ to the regression line as shown by typical distances d_4 and d_{17} in Exhibit 13.6.6. For this example,

$$SSY = 82.55 \qquad s_Y = \left(\frac{SSY}{19}\right)^{1/2} = 2.08$$

$$SSReg = (4.18)\,(18.06) - 75.49$$

$$SSRes = 82.55 - 75.49 = 7.06$$

$$s_{Y \cdot X} = \left(\frac{SSRes}{18}\right)^{1/2} = 0.626$$

It can be seen that $s_{Y \cdot X}$ is only 30% of s_Y. That is, the regression equation provided a 70% reduction in variation in Y. Another way to evaluate this *residual standard deviation* or *residual standard error* is to compare it to the mean of Y. In this case, it is 100 (0.626)/(8.93) = 7.0% of the mean and, depending on the goal, might represent a satisfactory reduction in variability. Still another way to measure the association between Y and X—and hence the residual lack of association—is to use the correlation coefficient that is developed next.

Correlation

The theoretical concept of correlation arises in conjunction with the bivariate normal distribution function. That function has five parameters. If the two variables are X and Y, the parameters are the means (μ_X, μ_Y) and the variances (σ_X^2, σ_Y^2) of each variate and a measure of covariation, the correlation coefficient, ρ (rho). This chapter does not deal with the theoretical bivariate (or multivariate) normal distribution. However, in practice, the sample correlation coefficient, r, is a useful measure of linear association. It is a dimensionless ratio ranging from −1.0 (perfect inverse linear agreement), through zero (orthogonal or linearly unrelated), to +1.0 (perfect direct linear agreement). The value can be obtained from equation 10 and *used as an index without any assertion whatever being made about distribution form.*

$$r_{XY} = \frac{SPXY}{[(SSX)\,(SSY)]^{1/2}} = \frac{s_{XY}}{s_X s_Y} \qquad (10)$$

The first form of the expression for r_{XY} has the same numerator as b_1 in equation 6, which shows that it is just a rescaling of the same basic information. It is easily shown that $r_{XY} = b_1 s_X/s_Y$.* In the second form in equation 10, s_{XY} is the sample *covariance* (*not* an SD). It has the same sign as $SPXY$ and r and is $SPXY/(n-1)$.

The square of r is called the "coefficient of determination." It ranges from zero to one and can be interpreted as the fraction of the variation in Y (with variation represented by SSY) that is accounted for or "explained" by variation in X. Thus (and from equation 8)

$$r_{XY}^2 = \frac{SSReg}{SSY} = 1 - \frac{SSRes}{SSY} \qquad (11)$$

Using equation 10 for the example data, $r_{XY} = 18.06/[(4.32)\,(82.55)]^{1/2} = .956$; $r_{XY}^2 = .915$. This is seen to be equal to the result of equation 11, where $r_{XY}^2 = 75.49/82.55 = .914$ (slight rounding error). So variation in X has accounted for 91% of SSY. This is *approximately* the same as claiming a 91% reduction in the variance of Y (from s_Y^2 to $s_{Y \cdot X}^2$).

Model Specification

In other circumstances, where the physics and chemistry are not so well understood (e.g., in studying the "cause" of a disease), the question may focus on the statistical significance (see Chapter 13.5) of the relationship. The analyst is attempting to decide whether the relationship seen in the sample is something real or just the result of chance association. This decision is appropriate along all goal sequences (see Exhibit 13.6.3), except where existing theory permits prior specification† of the model.

Model specification is the process of choosing an *adequate* representation of reality. To decide this question of reality, the analyst would want a test model for the behavior of the estimator, b_j, when the association is just chance. He or she could use the t test model with the null hypothesis that $\beta_j = 0$ (or some other appropriate value). The alternative hypothesis might be $\beta_j > 0$. The t distribution is appropriate by the central limit theorem. Then

$$t_j = \frac{b_j}{s_{b_j}} \qquad (12)$$

with the critical value for t of $t_{n-2,\alpha}$, where α represents the specified degree of risk of rejecting a true null hypothesis (claiming a nonexistent association). The standard errors for b_0 and b_1 are given by

$$s_{b_0} = s_{Y \cdot X}\left(\frac{\Sigma X_i^2}{nSSX}\right)^{1/2} \qquad (13)$$

*The subscript order "XY" on r is arbitrary; $r_{XY} = r_{YX}$. But the ratio s_X/s_Y implies that b_1 is for "Y on X." With s_Y/s_X, b_1 would be "X on Y," with minimization of squared errors taken in the X direction, a different line except where $r = 1.0$.

†The distinction between "specification" and "estimation" is rarely made. See Hunter and Box[11] for further discussion. Also, see Healy[1] regarding significance testing.

$$s_{b_1} = \frac{s_{Y \cdot X}}{(SSX)^{1/2}} \tag{14}$$

For the example data, $s_{b_0} = 230$ and $s_{b_1} = 0.296$. The corresponding t ratios are $t_0 = 27.5$ and $t_1 = 14.1$, indicating, as was "known" in advance, that both constants are statistically well removed from zero (highly "significant" compared to a critical value of $t_{18,0.05} = 2.10$). This information is put to more appropriate use later in this section. An intermediate precaution should concern the analyst, that of model validation.

Model Validation

The least squares method has permitted each of the data points to play a role in determining the constants b_0 and b_1. It is entirely possible (and nearly always true) that some observations in the data set contain errors (mistakes) in one or more of the variables or arise from unusual conditions that the model is not intended to represent. The "back substitution" (obtaining $Y_i - \hat{Y}_i$ values for the *development* data set) may reveal suspicious points. Generally, residual errors in excess of $\pm 2s_{Y \cdot X}$ should be viewed with suspicion. There is an extensive literature associated with this problem of detecting "outliers." One discussion that might serve as a starting point is Barnett.[12]

One possibility for testing the model's stability is to *validate* it on fresh data, that is, by back substituting data points that were *not* used in developing the model. These validation residual errors can be examined and their variance compared to the development residual variance. This is a practical matter to be judged in context (more than relying on tests for equality of variance, for example). Little faith can be placed in a model that fares poorly in the light of this comparison or that gives wild predictions for some points.

Coefficient Estimation

Suppose the model has been specified from existing theory, or by exploration, testing, and validation, and is judged adequate. Back substitution residual errors are well behaved. Now, whether the values of the b_j's themselves are of interest, or whether they are simply to be used in the equation for predicting future values of Y, the precision with which they estimate the β_j's is of concern.* *Point estimates* were obtained from equation 6, but coefficient estimation is not complete without obtaining *interval estimates*. It is not sufficient, where estimation is the goal (or a step on the path to the goal), just to have "significant" t ratios.

Use of the confidence interval (CI) concept (see Chapter 13.3) helps to contrast these two steps, specification and estimation. With $\alpha = 0.05$ risk that the CI will not contain β as is claimed, the interval is

$$100(1 - \alpha)\ CI = 95\%\ CI = b_j \pm t_{n-2,0.05} s_{b_j} \tag{15}$$

Notice that (with prescribed t) ts_b represents the maximum probable error associated with the estimate of β. This can be expressed as a percentage error (where engineers very often seek estimates that are within 5 or 10%). Using b as the base (in the absence of knowing β), let E represent the potential percentage error associated with a t value of 2, a value commonly used in stating a CI. Then

$$E = \frac{100ts_b}{b} = \frac{200s_b}{b} \tag{16}$$

But notice that s_b/b is just the inverse of the t ratio *calculated* from the sample using equation 12. Thus

$$E = \frac{200}{t} \tag{17}$$

which implies the need for calculated t values of 20 or even 40 to meet our common expectation of a 10 or 5%, respectively, error of estimation!

From equations 14 and 16 it can be seen that the error in estimating the slope is directly proportional to $s_{Y \cdot X}$ and inversely proportional to $(SSX)^{1/2}$. Thus, to achieve a prescribed value of E, either of two things must be done: (1) an improved (less noisy) model must be found to reduce

*In general, the requirements for precision will be greatest for control, least for prediction (see subsection on intercorrelated estimates in Section 13.6.3).

residual error, or (2) a larger sample must be obtained to increase SSX. (See Salem[13] for a more complete discussion of this issue.) In general, precision will improve approximately as the square root of n.

Interval Estimation for a Point on the Line

The regression equation can be used to estimate the "true" value of the response for some specified value of the predictor. This is estimating a conditional population mean of Y and is analogous to estimating (unconditionally) the population mean in a univariate setting. The CI for this case is

$$100(1-\alpha)\ CI = \hat{Y}_c \pm t_{n-2,\alpha} s_{\hat{Y}_c}$$

$$s_{\hat{Y}_c} = s_{Y\cdot X}\left[\frac{1}{n} + \frac{(X_c - \bar{X})^2}{SSX}\right]^{1/2} \tag{18}$$

where the subscript c denotes the condition—the location in X—at which the estimate is to be made. Notice that the square root of n (again) determines the interval width at the mean of X and that the interval grows wider the greater distance X_c is from $\bar{X}$, the sample mean.

In most texts this CI is presented as a pair of curved lines, implying a confidence band for the entire line. Equation 18 is meant to be used for *one* specified location. To sustain α as the risk of not containing the true value, *the entire procedure* of selecting n observations, computing the coefficients, and so on, would need to be followed for *each* X_c. Wider limits would be needed if the analyst desired limits for the entire true line. Acton[14] gives a good discussion of this and many related concepts.

Predicting a Future Value

For predicting a future value at X_c, $\hat{Y}_c$ is obtained from the regression equation just as in the CI. Here it is the estimate of the mean *about which individual values are expected to vary*. The expression for prediction limits for a single future value of Y must recognize this extra source of variation associated with individuals. The interval for prediction is here abbreviated PI and called for example, a "95% PI" for $\alpha = 0.05$.

$$100(1-\alpha)\ PI = \hat{Y}_c \pm t_{n-2,\alpha} s_{Y\cdot X}\left[1 + \frac{1}{n} + \frac{(X_c - \bar{X})^2}{SSX}\right]^{1/2} \tag{19}$$

The use of t in this expression implies the *additional* requirement that the *individuals* be normally distributed around the line. If this is not the case, some other constant (possibly with asymmetry) representing the actual distribution would be substituted for t.

Again, this process applies for a *single prediction*. If some fraction of all future values is to be included within the limits, the limits would be called "tolerance limits." (The reader is referred again to Acton[14] for more detailed discussion.) Exhibit 13.6.7 offers selected values of K_c in equation 20 for obtaining tolerance intervals (TI) around the line at $\bar{X}(K_1)$ and at $\bar{X} \pm 2s_X(K_2)$. Linear interpolation may be employed to obtain straight-line approximations of the curved tolerance limits. The values of K were obtained by inverse interpolation of the normal distribution for 0.95 confidence of including at least 95% of all future values.

$$0.95/95\%\ TI = \hat{Y}_c \pm K_c s_{Y\cdot X} \tag{20}$$

Exhibit 13.6.7 Coefficients for 0.95/95% Tolerance Limits

n	K_1	K_2
6	5.00	6.06
8	3.97	4.66
10	3.52	4.04
20	2.78	3.01
30	2.55	2.72
40	2.47	2.58
50	2.40	2.49
70	2.31	2.37
100	2.22	2.27

13.6.3 MULTIPLE LINEAR REGRESSION

Regression in the Computer Age

With the extensive availability of the high-speed, large memory, modern digital computer, two quite revolutionary changes have taken place over the last 20 years. First, large data sets in machine-readable form have become commonly available. Second, MLR analysis with abundant diagnostics and with very large models has become an economic reality. The marriage of these two conditions presents opportunities for statistical modeling that are seemingly limitless. However, hand in hand with this enormous analytical power go the associated dangers of misuse and misinterpretation.

Many of these dangers are associated with the intercorrelations found among the predictor variables in nonexperimental data sets. The predictor matrix is said to be "ill conditioned" or is carelessly referred to as "multicollinear." (Multicollinear really means the polar condition where some of the X's enter into linear combinations resulting in an indeterminant system. "Intercorrelation" is used here to describe the general case of nonorthogonality among the predictors.) Historically, such data sets have been avoided because of the difficulties they represented in hand computation; orthogonality was achieved by experimental design. Consequently, attempts to develop proper methodology for dealing with nonexperimental data were delayed until recent years. With the ready availability of the electronic computer, the arithmetic horrors of intercorrelated data sets have been pretty well forgotten, and appropriate methods of modeling are evolving.

The basic relationships and computational forms, represented in matrix notation, are shown here paralleling the equations of the simple case given earlier ($\nu = P - n - 1$).

("true" Y) $$E(\mathbf{Y}) = \mathbf{X}\boldsymbol{\beta} \tag{21}$$

(observed Y) $$\mathbf{Y} = \mathbf{X}\boldsymbol{\beta} + \boldsymbol{\epsilon} \tag{22}$$

(b_j's) $$\mathbf{b} = (\mathbf{X}'\mathbf{X})^{-1}\mathbf{X}'\mathbf{Y} \tag{23}$$

(predicted Y) $$\hat{\mathbf{Y}} = \mathbf{X}\mathbf{b} \tag{24}$$

$$SSRes = \mathbf{Y}'\mathbf{Y} - \mathbf{b}'\mathbf{X}'\mathbf{Y} \tag{25}$$

$$\sum_{j=1}^{P} SSReg_j = \mathbf{b}'\mathbf{X}'\mathbf{Y} = \sum_{j=1}^{P} b_j SPX_jY \tag{26}$$

(variance-covariance) $$\widehat{V(\mathbf{b})} = [\mathbf{X}'\mathbf{X}]^{-1}\hat{\sigma}_\epsilon^2 \tag{27}$$

$$s_{Y \cdot X}^2 = \hat{\sigma}_\epsilon^2 = \frac{SSRes}{(n - P - 1)} \tag{28}$$

(joint CI for b's) $$(\boldsymbol{\beta} - \mathbf{b})'\mathbf{X}'\mathbf{X}(\boldsymbol{\beta} - \mathbf{b}) \leqslant P s_{Y \cdot X}^2 F_{P,\nu,\alpha} \tag{29}$$

(CI for $\hat{Y}$) $$100(1 - \alpha)\ CI = \mathbf{X}_c'\mathbf{b} \pm t_{\nu,\alpha} s_{Y \cdot X}[\mathbf{X}_c'(\mathbf{X}'\mathbf{X})^{-1}\mathbf{X}_c]^{1/2} \tag{30}$$

(PI) $$100(1 - \alpha)\ PI = \mathbf{X}_c'\mathbf{b} \pm t_{\nu,\alpha} s_{Y \cdot X}[\mathbf{X}_c'(\mathbf{X}'\mathbf{X})^{-1}\mathbf{X}_c + 1]^{1/2} \tag{31}$$

Intercorrelation Effects

In regression modeling, intercorrelation affects the process in three basic ways (in addition to computational difficulties, which are ignored here, assuming the use of a computer). These three have many secondary and corollary consequences, which will be easily perceived if the basic three are understood. These three are:

1. Potentially enlarged variances of the b_j's.
2. Intercorrelated estimates of the b_j's.
3. Ambiguity in assessing the individual contributions to the regression sums of squares.

Potentially Enlarged Variances

In the theoretical case—with "correct" model and fixed residual variance—the variances of the b_j's will grow larger as intercorrelated predictors are added to the model (see Snee[15]) as a consequence of the inverse matrix in equation 27. (Notice that *estimates* of the variances of the b_j's are obtained because an *estimate* of residual error variance is used. The *true* theoretical variances result from using σ_ϵ^2.) In many cases *in practice* also, the s_b^2's will grow larger with the addition of intercorrelated predictors. This is because the increase due to the inverse matrix will more than offset the decrease due to a smaller $s^2_{Y \cdot X}$, which results from additional regression sums of squares. However, in practice, $s^2_{Y \cdot X}$ is often reduced enough by the extra predictor(s) to offset the intercorrelation effect in the inverse matrix. These considerations are at the heart of the burgeoning variety of (predictor) variables selection schemes currently appearing in the literature. A full discussion of this topic is beyond the scope of this chapter. For an introduction to the topic, see Hocking.[16]

Intercorrelated Estimates

In the left side of equation 27, in addition to the diagonal variances, there are off-diagonal covariances of pairs of b_j's. Just as with correlation between variables in equation 10, *covariance* of b_j's implies *correlation* of b_j's. Exhibit 13.6.8 depicts the joint sampling distribution for a pair of positively correlated b_j's. The distribution results from repeated samplings of n values of Y for a given X matrix. For the case of $P = 2$, the correlation of the b_j's is *equal in magnitude but opposite in sign to the intercorrelation of the* X_j's. (They *tend* to be equal and opposite also for $P > 2$; see Exhibit 13.6.13.) Notice that the unconditional sampling range of, for example, b_2 (shown by distance A) is very large compared to the *conditional* range of b_2 (shown by distance B), given the particular estimate of β_1. The important consequence of these two considerations is that *errors in estimating the* β_j's *tend to be compensating among intercorrelated predictors.* So intercorrelations may adversely affect the precision of estimate of the β_j's, *but may have little adverse effect on the use of the model for prediction.* This last conclusion depends, of course, on the intercorrelations among the predictors staying about the same in prediction as they were in the sample.

Ambiguity in Assessing Contributions

The underlying nature of the problem is easy to comprehend. (For an introductory geometric interpretation of these phenomena, see Crocker.[17,18]) Interpreting the specific consequences in a particular problem can be extremely complicated. This is true because the ambiguity can be of up to Pth order. The problem is further complicated by the existence of two basic classes of intercorrelated ambiguity, which, for $P \geqslant 3$, can simultaneously be present in all sorts of hierarchical combinations. Here, the surface will only be scratched* with an illustration contrasting the two classes for $P = 2$, the least complex intercorrelation situation. (See also the subsection on application to the example at the end of this section.)

In most references, "intercorrelated" and "confounded" are regarded as synonymous. Actually, confounding is only one of the two classes just mentioned. The other has not been given a name by others, but is here titled "resolving." The name "resolving" was chosen because the separate effects of the two or more (resolving) predictors are not "resolved" (clearly seen) until they appear in the model *together.* The contrast between confounding and resolving is shown in Exhibit 13.6.9. The circles at the bottom left represent the two predictors. Area is proportional to regression sums

Exhibit 13.6.8 Correlation of Estimates of Slopes

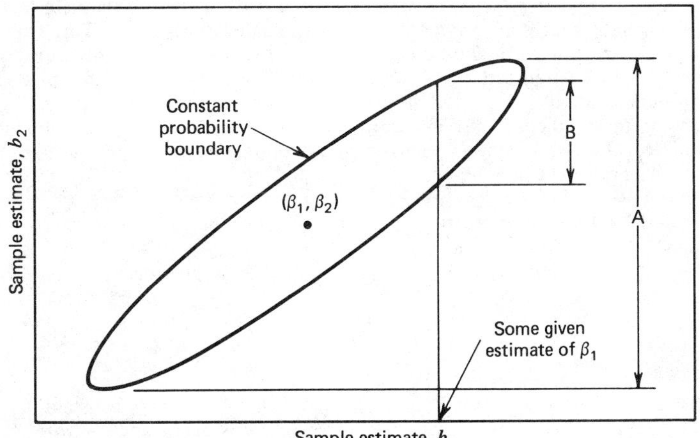

*A thorough presentation of the geometry of regression, which clarifies these intercorrelation phenomena, is in preparation for publication in *Technometrics.* A copy can be obtained from the author.

Exhibit 13.6.9 Demonstration of Intercorrelation Effects[a]

	Class			
	Confounding		Resolving	
Definition	$R^2 < r_{1Y}^2 + r_{2Y}^2$		$R^2 > r_{1Y}^2 + r_{2Y}^2$	
Model order	X_1 first	X_2 first	X_1 first	X_2 first
SSReg$_1$	10	(4)	4	(10)
SSReg$_2$	(7)	13	(13)	7

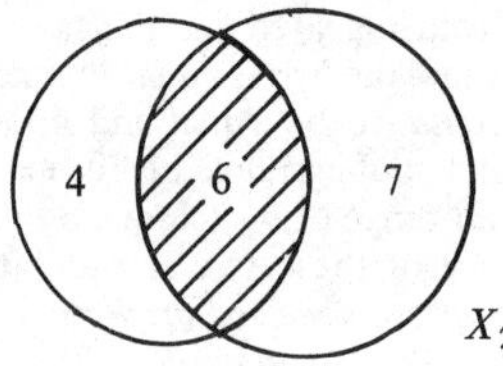

4 6 7

X_1 X_2

[a]For assumed $n = 103, SSY = 27, F_j = 10\ SSReg_j$.
For last predictor (circled entries), $F_j = t_j^2$.

of squares with values as shown. The shaded area of overlap represents intercorrelation. The exhibit shows the allocation of the 17 *SSReg* units to the two predictors for the two classes and for the two possible orderings of predictors in the model (see also Exhibit 13.6.13).

In the confounding case, the ambiguous six units are allocated to the first predictor; the second predictor accounts for the balance. In resolving, the six units are available to the *second* predictor only after the first has clarified the picture. Notice that the total information–17 units–is always the same. (Other features of this exhibit are discussed in subsequent sections as other diagnostic measures are presented.) Relevant to this allocation process, care must be taken in interpreting equation 26. This equation says that the *total* regression sum of squares can be obtained from the sums of products of the b_j's with their corresponding SPX_jY's. It does *not* assert that the *individual* $SSReg_j$'s can be found this way. As can be seen from the foregoing discussion, the *individual* $SSReg_j$'s will depend on the *order* of appearance in the model. The total is order independent.

Extreme confounding is frequently encountered in nonexperimental data sets. It is important to recognize two quite different situations that may arise. Essentially, there is a duplication of information–a redundancy in the system. In one situation it may be that the *same* information is presented twice in slightly different forms (such as two different price indexes). This represents "model redundancy" and is dealt with by removing the redundant predictor. By contrast, it might be that two really *different effects* are present, but, because in nature they are highly intercorrelated, their separate contributions cannot be discriminated statistically. This represents "data redundancy" and can clearly present a danger if one or the other predictor is arbitrarily excluded from the model, if estimation is the goal. An example is the use of R & D and capital expenditures to assess the number of technical staff needed in a business. Both effects are real, yet it would not be surprising to see them highly intercorrelated and thus inseparably confounded. This true dilemma motivates the current development of biased estimation techniques such as "ridge regression" (e.g., see Wichern and Churchill[19]).

The Meaning of Partial Correlation

For the two-predictor case, the (first-order) partial correlation is given by

$$r_{2Y\cdot 1} = \frac{(r_{2Y} - r_{12}r_{1Y})}{[(1 - r_{12}^2)(1 - r_{1Y}^2)]^{1/2}} \tag{32}$$

This gives the correlation of X_2 with Y, given X_1 (or "while holding X_1 constant" or "while first removing the effect of X_1"). For a given pair of correlations of X_1 and X_2 with Y, r_{12} can influence this expression to be larger *or* smaller in absolute value than it would be in the orthogonal case ($r_{12} = 0$). When the partial is diminished compared to the orthogonal case, confounding exists. When the partial is increased, it is resolving. Partial correlations relating to the example in Exhibit 13.6.9 would, in each case, be based upon the circled (last-position) values. Ordinary correlations would be based on the uncircled (first-position) values. The coefficient of determina-

Exhibit 13.6.10 Coefficients of Determination for Exhibit 13.6.9 Values

	r_{1Y}^2	$r_{1Y\cdot 2}^2$	r_{2Y}^2	$r_{2Y\cdot 1}^2$	$R_{Y\cdot 12}^2$	$r_{1Y}^2 + r_{2Y}^2$
Confounding	.370	.286	.481	.412	.630 <	.851
Resolving	.148	.500	.259	.565	.630 >	.407

tion (equation 11) can be used to represent these two views. The ordinary r^2 would use equation 11 as is. The partial coefficient of determination would place the circled value in the numerator and the net amount of SSY remaining–after removing the effect of the first predictor–in the denominator. Exhibit 13.6.10 shows these ratios based on $SSY = 27$.

Multiple Correlation; Practical Interpretation

The multiple correlation is represented by R. It is in fact the correlation of $\hat{Y}$ with Y, where $\hat{Y}$ is a linear combination of the X's. Of course the X's may individually have correlations with Y of either sign. Hence R is arbitrarily defined as being positive. Direct practical interpretation of R is difficult. Two transformations help to improve interpretation. One is R^2. As with the simple model, R^2 is the "coefficient of determination" and represents the fraction of SSY accounted for by the model ($R^2 = SSReg/SSY$). For orthogonal predictors, $R^2 = \Sigma_{j=1}^{P} r_{jY}^2$. For $P = 2$, if $R^2 > r_{1Y}^2 + r_{2Y}^2$, X_1 and X_2 represent a resolving pair. Where $R^2 < r_{1Y}^2 + r_{2Y}^2$, X_1 and X_2 are confounded. These relationships were shown in Exhibit 13.6.9 and evaluated in Exhibit 13.6.10. A second transformation is "% s_Y removed." The percentage reduction in s_Y is related to R as follows:

$$\% \, s_Y \text{ removed} = 100 \left\{ 1 - \left[\frac{(1 - R^2)(n - 1)}{(n - P - 1)} \right]^{1/2} \right\} \tag{33}$$

For a more extensive discussion of R and a graph of equation 33, see Crocker.[20] For small samples, a correction can be made to R^2 to recognize lost degrees of freedom. Letting $\bar{R}$ = "corrected R," the correction is

$$\bar{R}^2 = \frac{1 - (1 - R^2)(n - 1)}{(n - P - 1)} = \frac{s_Y^2 - s_{Y \cdot X}^2}{s_Y^2} \tag{34}$$

This "corrected" coefficient of determination is exactly proportional to the sampling estimate of the reduction in error variance.

The Meaning of the *t* Ratio

As shown in Section 13.6.2, the t ratio (equation 12) gives the number of standard errors the estimated value of the coefficient is away from zero. That is still a correct interpretation in the multiple case. It is still useful in assessing the precision of the estimate as per equation 17. The t ratio does *not*, however, measure the contribution, the importance, the practical significance, or even the statistical significance of the associated term in the model! To use this statistic for assessing the contribution of a predictor, it must be carefully qualified. It answers the question "What is the impact of the *unique* contribution of this predictor?" "Unique" is taken here to mean "impact after resolving." Hence it is the same as asking what the impact is for this predictor put *last* in the model.

For answering scientific questions about truth, this gives the t ratio a conservative interpretation. In terms of its influence in reducing $s_{Y \cdot X}^2$, $|t| = 1.0$ is the break-even value for any one predictor. With $|t| > 1.0$, $s_{Y \cdot X}$ is reduced by including this predictor. To have (unique) statistical significance, $|t|$ should exceed some appropriate critical value. For excellent precision in estimating β, $|t|$ should be near, say, 20 or 40 (see subsection on coefficient estimating). The analyst must be wary not to exclude an important term with small t resulting from confounding. What action is appropriate depends heavily upon goals (see Section 13.6.1) and upon intimate system knowledge.

Relating *t* and *F* in Modeling

The ordered F ratio for a single predictor is defined by the ratio of mean squares (MS), regression/residual.

$$F_j = \frac{MSReg_j}{MSRes} = \frac{SSReg_j/1}{SSRes/(n - P - 1)} \tag{35}$$

It is called "ordered" because it contains the *SSReg* of the associated predictor, and this quantity is order dependent, as illustrated in Exhibit 13.6.9. When $j = P$, $F_j = t_j^2$. Thus the t ratios are all proportional to the square roots of the respective *SSReg* obtained for each predictor *as if it were in last* position.

For the example of Exhibit 13.6.9, the denominator of equation 35 is (27–17)/100 = .1. Hence the ordered F values are the Exhibit 13.6.9 entries multiplied by 10, and for the circled values these are t^2. So, it is seen that t ratios are really "partial" t ratios and are best interpreted in terms of their relationship to last-position *SSReg* contributions.

Dealing With Interactions

Sometimes intercorrelation is carelessly referred to as "interaction." Care should be taken to distinguish these two very different concepts. Intercorrelation is a *data* phenomenon and is *not* determined by the form of the regression equation, but rather by *the particular set of observed values of the predictor variables.* Interaction is a *model* characteristic. It is represented in the model by the *product* of two or more predictors. It is put there in an attempt to measure interactive behavior in the system represented by the model. Equation 36 shows an interactive model where $X_3 = X_1 X_2$ represents a third predictor created from the first two. (Subscript i is omitted for simplicity.)

$$\hat{Y} = b_0 + b_1 X_1 + b_2 X_2 + b_3 X_3 \tag{36}$$

The meaning of "interaction" is this: The *effect* of one predictor depends on the *value* of another predictor. This is easily seen to be the case for equation 36 by factoring either X_1 or X_2. For illustration, X_1 is used.

$$\hat{Y} = b_0 + (b_1 + b_3 X_2) X_1 + b_2 X_2 \tag{37}$$

Here the coefficient of X_1 is $(b_1 + b_3 X_2)$. Therefore the *effect* of X_1 (its coefficient, $b_1 + b_3 X_2$) depends on the *value* of X_2. By symmetry, the reverse is also true.

No special steps need to be taken to evaluate an interaction. Its t ratio will assess its *additional* contribution to *SSReg* as was previously discussed. However, care is needed in interpreting the associated "main effects." In general, where the X's are in their raw original forms, the interaction term will be highly confounded with the associated main effects—the predictors from which it is formed. This will tend to depress the t ratios of these main effects even where the interaction contributes a sizable *SSReg* (thereby reducing $s^2_{Y \cdot X}$). This should be of no concern. It is purely an arbitrary scaling problem. If desired, the interaction can be made approximately orthogonal to the main effects by subtracting their respective means before forming the product. This has no effect on the statistical assessment of the interaction.

Basics for Attribute Modeling

Regression modeling is not limited to using quantitative predictors. Any categorization, classification, or logical distinction can be represented. If there is a single class, no distinction is needed. If there are two classes, (e.g., male, female), an additional X is provided to give an attribute code to distinguish the two classes: $X = 0$ if the individual is in the first class (male), $X = 1$ for the second (female). The value chosen is arbitrary, but 0, 1 coding is easiest to interpret. This is "differential coding," which means that the intercept, b_0, will represent the level in Y of the $X = 0$ group, and the coefficient for this code will estimate the *difference* in Y between the two classes.

To measure, for example, differences of each working day compared to Monday (arbitrarily chosen as the base of comparison), four extra predictors will be needed. Each will be given the value 1 only if the observation represents the associated day; otherwise it will be given the value zero. In general, the number of predictors added will be one less than the number of classes ($c - 1$). In statistically evaluating the contribution of such a categorical coding scheme, a single test statistic should be used for the $c - 1$ degrees of freedom. This is because individual (single degree of freedom) *SSReg* contributions depend on the arbitrary choice of the base of comparison and the order. The total, however, is independent of the choice of base and order. The total can be evaluated using the F ratio as shown in equation 38, assuming that these terms are last in the model.

$$F_{c-1, n-P-1, \alpha} = \frac{\sum_{j=P-c+2}^{P} SSReg_j/(c-1)}{MSRes} \tag{38}$$

Exhibit 13.6.11 Illustration of an Attribute Code Shift and Slope Change (Interaction)

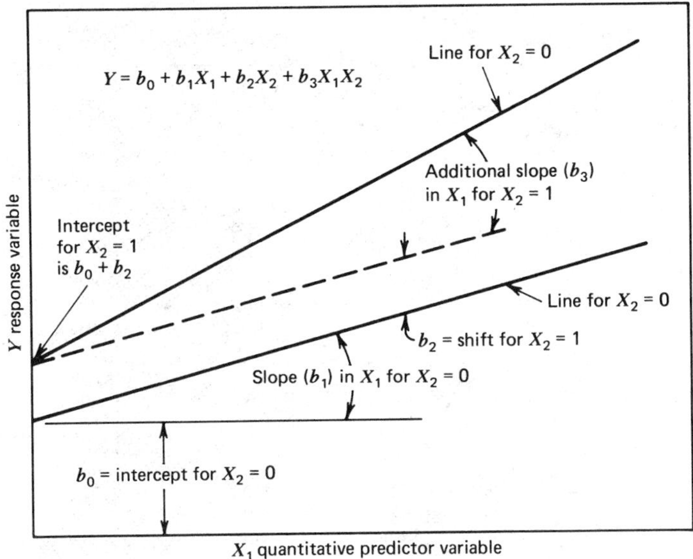

Variables selection programs that operate on individual degree-of-freedom effects are clearly inappropriate for dealing with categorical structures.

Exhibit 13.6.11 illustrates a model with a single quantitative predictor, a two-class attribute shift, and an interaction of these two. Equation 36 applies here and implies that the slope in X_1 is different for the two classes.

Dealing With Covariates

A "covariate" is a source of variation contributing to SSY that may not be of particular interest, but whose effect must be removed (1) in order to get unbiased estimates of other predictors of interest and (2) in order to reduce the noise level of the system so that predictors of interest can be more clearly seen. It may be that a covariate is confounded with a predictor of interest. The use of the t ratio in evaluating the reality of that predictor's contribution will then quite properly be conservative–discounting the information held in common with the covariate.

Where a categorical structure of three or more classes is involved in a covariate situation, special care must be taken. If the categorical group is the focus of interest, then it must be *placed at the end of the model* so that its apparent contribution, evaluated by equation 38, will have been reduced according to confounding with a covariate. If the categorical group *is the covariate*, then the term with which it is confounded will have a properly deflated t ratio independent of model position. Hence attribute code groups can always be safely placed at the end of the model.

Application to the Example

Exhibit 13.6.12 extends the original set of data in Exhibit 13.6.5 by adding two more predictors. Production (X_2) is in units per period associated with a process that uses steam for heat in the manufacturing process. A change in heating policy is represented by the attribute code (variable 3), which starts at zero where the heating level was 72°F (22°C) and goes through an adjustment (estimated by the engineer) over several periods to the new level of 65°F (18°C). The policy change should (only) affect the heating coefficient, β_1, and so is introduced as an interaction, X_3 = (degree-days) · (policy). This was suggested when the residuals from the two-predictor model displayed a slight downward trend over time.

Analysis using equation 36 provided residual errors, which were examined for pattern and excessive deviance. The fourth point was found to be at $+2.47s_{Y \cdot X}$. Further investigation revealed the malfunctioning of a steam trap in the production system, which would account for an indeterminant excess consumption of steam during the fourth period. Therefore the point was excluded and the analysis repeated for the remaining 19 observations. The residuals then looked well behaved, and the model appeared to be adequate for monitoring future consumption. A control chart (see Chapter 8.3) was then established, with $\hat{Y}$ as the center line and with limits at $\pm 2s_{Y \cdot X}$. In addi-

Exhibit 13.6.12 Example Steam Consumption–Extended Data Set

i	Period Number	Heat (X_1) k degree-days	Production (X_2) Units	Policy (V_3) Attributes	$X_3 = X_1 * V_3$ k degree-days	Steam (Y) gbtu
1	1-10	0.156	413	0.00	0.000	7.991
2	1-11	0.419	396	0.00	0.000	8.589
3	1-12	0.658	385	0.00	0.000	9.145
4	1-13	1.090	243	0.00	0.000	11.212
5	2-01	1.380	391	0.00	0.000	11.754
6	2-02	1.103	407	0.00	0.000	11.469
7	2-03	1.000	411	0.00	0.000	10.584
8	2-04	0.703	379	0.00	0.000	9.509
9	2-05	0.207	402	0.00	0.000	7.457
10	2-06	0.086	406	0.00	0.000	6.989
11	2-07	0.024	383	0.00	0.000	6.537
12	2-08	0.005	227	0.10	0.001	4.938
13	2-09	0.026	265	0.25	0.007	5.275
14	2-10	0.161	384	0.40	0.064	7.452
15	2-11	0.307	400	0.55	0.169	7.962
16	2-12	0.664	379	0.70	0.465	8.915
17	2-13	1.039	354	0.85	0.883	9.758
18	3-01	1.275	392	1.00	1.275	11.183
19	3-02	1.193	412	1.00	1.193	11.523
20	3-03	0.953	408	1.00	0.953	10.426

tion, an imprecise estimate of the effect of the policy change ($t = -1.57$) was obtained, along with good assessments of the two major rates of consumption, corrected for the policy change. The intercept represents the average consumption of all the other steam uses that are uncorrelated with the predictors used in the model. Exhibit 13.6.13 displays some of the relevant statistics from the final run. Notice that $s_{Y \cdot X}$ is only about 2.7% of the mean consumption.

13.6.4 SOME PRACTICAL CONCERNS

More on Dangers

The analyst faces a variety of dangers in practice in addition to those discussed earlier.

Avoiding Complexity

There is often pressure to "keep it simple." The danger is that, by avoiding complexity, the analyst may be seriously misled or fail to develop an adequate model. Simplicity is not necessarily a virtue.

Excessive Faith

It is easy to acquire more faith in the regression process than it deserves. Even a good model is at best a crude approximation of reality. Yet, by being computer born, it takes on a special aura, which may encourage any or all of the following:

1. Insufficient attention to validation.
2. Acceptance of results with insufficient diagnostic evaluation.
3. Overreliance on model selection algorithms.
4. Excessive optimism. (Success is not guaranteed.)
5. Inadequate *or* unrealistic use of transformations.
6. Uncritical attachment to "discovered" relationships.
7. Undue trust in the correctness of the data.
8. Improper omission of an important variable because it was prevented from varying (type 2 error).

Untested Regions

It may seem that staying within the observed ranges of the predictors for predicting future values would be safe. Not necessarily. It is easily possible to be within all the ranges and still move into a part of the *joint* space never experienced before.

Exhibit 13.6.13 Results for Final Analysis of Steam Data[a]

j	b_j	s_{b_j}	t_j	$SSReg_j$	F_j	$SSRes$	df	$MSRes$	$s_{Y \cdot X}$
0	1978	447.6	4.42	.1476E10	–	.7708E8	18	.4282E7	2069
1	3.768	0.1555	24.23	.7035E8	1241	.6730E7	17	.3959E6	629
2	12.27	1.241	9.89	.5739E7	101	.9910E6	16	.6194E5	249
3	−0.240	0.1533	−1.57	.1395E6	2.46	.8515E6	15	.5677E5	238

Correlations

jj	X vs X	b vs b	j	X vs Y
12	.396	−.374	1	.955
13	.580	−.568	2	.629
23	.165	−.086	3	.501

Statistics for Xs

$\bar{X}_j$	s_{X_j}	$R_{X \cdot X}$
597.8	477.6	.655
378.6	49.43	.404
263.6	451.2	.584

SSReg (see Section 13.6.3)

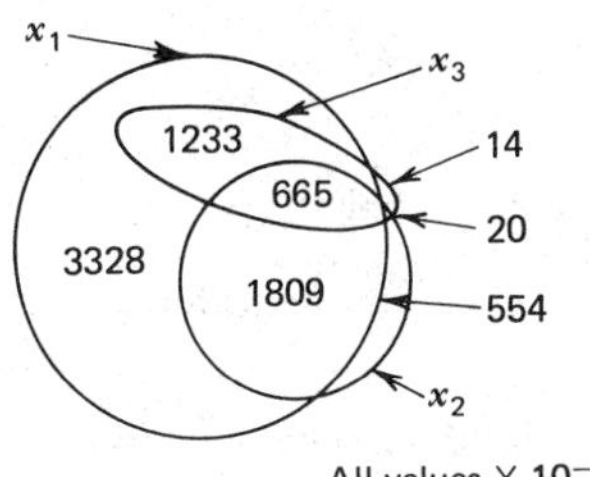

[a] $n = 19$; $\Sigma Y^2 = .1553E10$; $\bar{Y} = 8813$; $R = .994$; $R^2 = .989$; $\%s_Y Rem. = 88.5$

Concerns About Data

Right from the question of which variables to collect to the final checking of a suspicious residual, data management is a major element of any modeling venture. The collection and "scrubbing" process very often–and rightly–consumes 70% or more of the project funds and elapsed time. It bears repeating that a data set will rarely ever be free of mistakes. Some process for auditing the data must be devised at the outset or much effort will be wasted on false starts.

One effective auditing technique is the execution of a univariate analysis of each variable. Usually there is a chronology, so that a "time series" examination can be revealing. At least this process will help the analyst to become familiar with the system. Another useful auditing device is the examination of a correlation matrix of all the variables to see if the correlations make sense.

Using the Model for Prediction and Control

The model is a waste if it is not used. The person who will have responsibility for its use must be involved early enough to have understanding and to develop faith. Its use must serve an ongoing function that is desired and expected by the user's superiors or it will not survive.

Provision must be made for the timely reporting of the predictors. It is of no use to develop a prediction or control model if the necessary data cannot be obtained in a timely fashion. Results must then be reported to those who can take action. Good predictions kept in a desk serve no one.

Model Maintenance

Invariably in practice, the β's that are estimated are not in fact constant, but are creeping and shifting over time. Additionally, there will inevitably be other systems changes, which, for example, may require the inclusion of additional predictors. So, if the model is to continue in use, provision must be made for updating it. Failing this, it will begin to miss until it loses credibility and its use is discontinued.

Helpful Hints in Practice

Experience has shaped the following summary list of prerequisites for successful use of regression modeling techniques. The analyst should have and/or use:

1. Reasonably specific goals.
2. An understanding of statistical procedures.
3. Reasonable familiarity with the system modeled.
4. Restraint in transforming variables.
5. Facility for adequate diagnostic analysis.
6. A cyclical approach with documentation of decisions and choices made.
7. A willingness to validate the model and/or anticipate instability.
8. Recognition of the need for maintenance.

13.6.5 ESSENTIALS OF A GOOD COMPUTER PROGRAM

Motivation

Many different regression programs are available. Some analysts will have in-house packages. New diagnostics are constantly being invented. So, rather than to recommend any specific program, this section outlines features that a good program should have. An asterisk is placed before those features considered absolutely essential.

Input Considerations

1. Tape as well as card data input and, ideally, disk.
*2. Flexibility to specify format at execution time.
*3. End-of-data detection to avoid precounting.
4. Provision for about 100 input variables and up to 50 model variables.
*5. Essentially unlimited sample size capability.
6. Provision to operate simultaneously on two or more response variables with the same model.
7. Option to force through the origin (exclude b_0).

Data Transformation. Preferably, transformations should be handled by a subroutine that is compiled at execution time. A language such as FORTRAN is ideal, providing great flexibility. Minimum provision would include:

*1. All the ordinary arithmetic, logarithmic, trigonometric, and exponential operators.
*2. Logical examination with conditional determination of values, say, for attribute codes.
3. Blank data elimination.
4. Automatic generation of (or language capability to obtain) first differences.
5. Automatic conversion to standard form.

Output Related to Source Data

*1. Provision for appropriate titling.
*2. Complete data listing option (and first row option).
*3. Complete transformation listing option (and first row option).
*4. Means, SDs, and ranges of all transformed variables, as well as sample size.
*5. Correlation matrices (XX, XY, YY) of transformed variables.
*6. Multiple correlation among the predictors.
7. Eigenvalues of $X'X$ matrix option.
8. Inverse matrix listing option.

Output Related to Coefficient Estimation

*1. Coefficients.
*2. Standard errors of coefficients.
*3. Partial t ratios.
4. Correlation matrix of coefficients.
5. Variance inflation factor.

Output Related to Model Evaluation

*1. Individual regression sums of squares.
*2. Ordered F ratios.
*3. Successive values of $s_{Y \cdot X}$ at each step.
*4. R and/or R^2 and/or % s_Y removed.
*5. Validation data back substitution option.
6. Option to pool degrees of freedom.

Graphics

*1. Sequential plot of back substitution errors, with Y, $\hat{Y}$, $Y - \hat{Y}$ listed alongside.
2. Histogram of residuals.
3. Optional coordinate printer plots.
a. X versus X, X versus Y, Y versus Y.
b. $Y - \hat{Y}$ versus X.
c. $Y - \hat{Y}$ versus $\hat{Y}$.
d. $Y - \hat{Y}$ versus $Y - \hat{Y}$.
e. X, Y versus order (time series).

Additional Displays and Options

*1. Detection of singular matrix and identity of trigger variable.
2. Evaluation of the predictor ranges in terms of influence on Y.
3. Means and variances for any variables within cells of categorical structures.
4. Observation count for cells of two-dimensional categorical structure.
5. General evaluation of confounding and resolving.
6. Various diagnostic messages relating to assemblv errors, keypunch errors, near-singular matrices, poor scaling of data, and linear dependencies within the predictor set.
7. Interpretation of all control choices exercised.

REFERENCES

1. M. J. R. HEALY, "Is Statistics a Science?," *Journal of the Royal Statistical Society*, Vol. 141, A(1978), Part 3, pp. 385-393.
2. P. D. FINCH, "Description and Analogy in the Practice of Statistics," *Biometrika*, Vol. 66, No. 2 (1979), pp. 195–208.
3. C. EISENHART, "The Assumptions Underlying the Analysis of Variance," *Biometrics*, Vol. 3, No. 1 (1947), pp. 1-21.
4. M. E. J. CURZON and D. C. CROCKER, "Relationships of Trace Elements in Human Tooth Enamel to Dental Caries," *Archives of Oral Biology*, Vol. 23, 1978, pp. 647–653.
5. C. EISENHART, "The Meaning of 'Least' in Least Squares," *Journal of the Washington Academy of Sciences*, Vol. 54, February 1964, pp. 24–32.
6. N. R. DRAPER and H. SMITH, *Applied Regression Analysis*, Wiley, New York, 1978.
7. R. J. FREUND and P. D. MINTON, *Regression Methods*, Dekker, New York, 1979.
8. D. G. KLEINBAUM and L. L. KUPPER, *Applied Regression Analysis and Other Multivariate Methods*, Duxbury Press, North Scituate, MA, 1978.
9. G. O. WESOLOWSKY, *Multiple Regression and Analysis of Variance*, Wiley, New York, 1976.
10. M. S. YOUNGER, *A Handbook for Linear Regression*, Duxbury Press, North Scituate, MA, 1979.
11. W. G. HUNTER and G. E. P. BOX, "Experimental Studies of Physical Systems," *Technometrics*, Vol. 7, No. 1 (1965), pp. 23–42.
12. V. BARNETT, "The Study of Outliers: Purpose and Model," *Applied Statistics*, Vol. 27, No. 3 (1978), pp. 242–250.
13. M. D. SALEM, Jr., "Multiple Linear Regression Analysis for Work Measurement of Indirect Labor," *Journal of Industrial Engineering*, Vol. 18, No. 5 (1967), pp. 314–319.
14. F. S. ACTON, *Analysis of Straight Line Data*, Wiley, New York, 1959.
15. R. R. SNEE, "Some Aspects of Nonorthogonal Data Analysis," *Journal of Quality Technology*, Vol. 5, No. 2 (1973), pp. 67–79.
16. R. R. HOCKING, "The Analysis and Selection of Variables in Linear Regression," *Biometrics*, Vol. 32, No. 1 (1976), pp. 1-50.

17. D. C. CROCKER, "Intercorrelation and the Utility of Multiple Regression in Industrial Engineering," *Journal of Industrial Engineering*, Vol. 18, No. 1 (1967), pp. 79–85.

18. D. C. CROCKER, "Linear Programming Techniques in Regression Analysis: The Hidden Danger," *AIIE Transactions*, Vol. 1, No. 2 (1969), pp. 112–126.

19. D. W. WICHERN and G. A. CHURCHILL, "A Comparison of Ridge Estimators," *Technometrics*, Vol. 20, No. 3 (1978), pp. 301–311.

20. D. C. CROCKER, "Some Interpretations of the Multiple Correlation Coefficient," *The American Statistician*, Vol. 26, No. 2 (1972), pp. 31–33.

CHAPTER 13.7

Queuing Theory

M. RAGHAVACHARI

Indian Institute of Management

13.7.1 INTRODUCTION

A queue or waiting line is formed when customers arrive at a service mechanism that is busy. Queuing theory is concerned with the study of attendant delays and with the design of appropriate service mechanisms to minimize these delays and to reduce the length of the queue. Queues are commonly encountered at commercial service systems, transportation units, industrial organizations, and social service systems. The earliest applications of queuing theory were in the area of telephone engineering and were pioneered by Karl Erlang. Some of the areas of industrial engineering in which queuing theory is of great value are:

1. Operations in which articles pass along a conveyor belt and are to be packed into cartons.
2. The breakdown of one or more machines and the consequent idle time on these machines when the repair crew is busy.
3. The collection of tools by workers at a tool crib.
4. Inventory control.

Some typical areas of application of queuing theory are summarized in Exhibit 13.7.1.

A queuing system is usually specified by the following three factors:

1. The arrival pattern of customers.
2. The service mechanism.
3. The queue discipline.

13.7.2 ARRIVAL PATTERN (INPUT PROCESS)

The arrival pattern is in most cases probabilistic. It is *deterministic* or *regular* if the customers arrive at equally spaced instants. It is *probabilistic* or *stochastic* if the customers arrive according to a probability distribution. This distribution may either be assumed to be known or be completely unknown. The probability distribution sometimes specifies the distribution of the number of arrivals in a given interval of time and sometimes the distribution of time between successive arrivals. When more than one customer can enter the system at an arrival event, the situation is termed "bulk arrivals." Usually the source population from which customers or units arrive for service is infinite. A finite population source is also possible, for example, as in the case of a finite number of machines that break down or a finite number of repair persons.

Customer behavior can vary too. One case may be that of "balking," which arises when the customer does not join the queue because of its existing length. "Renege" is the customer's attitude toward leaving the system after waiting for some time. When there are multiple facilities, a customer will have an option to join any of the queues, which is called "jockeying." Several customers may be in "collusion," whereby only one person waits in the queue and all others attend to other work.

13.7.3 SERVICE MECHANISM

The service mechanism specifies when service is available, how many customers can be served at a time, and the length of service, which is usually called "service time." The service time can be known exactly, or its statistical distribution can be specified. In most queuing theory applications,

The author wishes to thank T. K. Soundar Rajan for his assistance in the preparation of this chapter.

Exhibit 13.7.1 Waiting Line Model Elements for Some Commonly Known Situations

Situation	Unit Arriving	Service or Processing Facility	Service or Process Being Performed
Ships entering a port	Ships	Docks	Unloading and loading
Maintenance and repair of machines	Broken down machine	Repair crew	Repair of machine
Assembly line, not mechanically paced	Parts to be assembled	Individual assembly operations or entire line	Assembly
Doctor's office	Patients	Doctor and doctor's staff and facilities	Medical care
Purchase of groceries at a supermarket	Customers with loaded grocery carts	Checkout counter	Tabulation of bill, receipt of payment, and bagging of groceries
Auto traffic at an intersection or bridge	Automobiles	Intersection or bridge with control points such as traffic lights or toll booths	Passage through intersection or over bridge
Inventory of items in a warehouse	Order for withdrawal	Warehouse	Replenishment of inventory
Job shop	Job order	Work center	Processing

Source. Reproduced from reference 1.

we assume that the service times of different customers are statistically independent and identically distributed random variables. The common distribution is called the "service time distribution." The "capacity" of the system is the maximum number of units that can be served at any particular time.

The service mechanism also specifies the number and configuration of servers or channels. There may be a single server or several servers. The servers can be either in parallel or arranged in series (tandem). It is also possible for a service mechanism to have both these configurations.

The distribution of service time is an important component of the service mechanism. The most common types of such a distribution are as follows:

1. **Constant Service Time.** The service time is assumed to be the same for each customer. This appears as an idealized model; however, it does give a good approximation in certain queuing processes.
2. **Exponential Service Time.** The distribution of the service time is assumed to be exponential. This means that the probability density, $f(x)$, of the service time is given by

$$f(x) = \mu e^{-\mu x}, \qquad x > 0$$

This distribution has some nice properties that are pertinent to the queuing process, and mathematical results become simpler. In this case the average service time is $1/\mu$, and μ is called the "service rate."

3. **Erlangian Service Time.** Here the service time has the distribution with density

$$\frac{1}{(k-1)!}\,\frac{k}{b_1}\left(\frac{kx}{b_1}\right)^{k-1} e^{-kx/b_1}, \qquad x \geq 0$$

where k is a positive integer and $b_1 > 0$. This is the distribution of the sum of k independent random variables, each with the exponential distribution with $\mu = b_1/k$.

There are a few more complex and general service distributions, for example, general Erlangian service time, nonstationary service time, and service time correlated with other aspects of the system.

Exhibit 13.7.2 Single-Server System

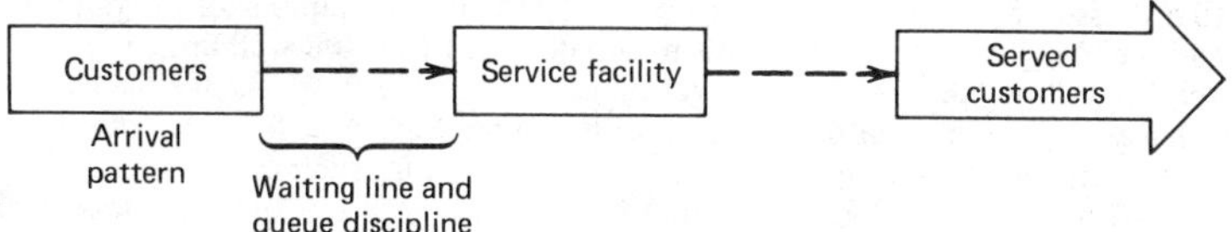

Exhibit 13.7.3 Multiserver System

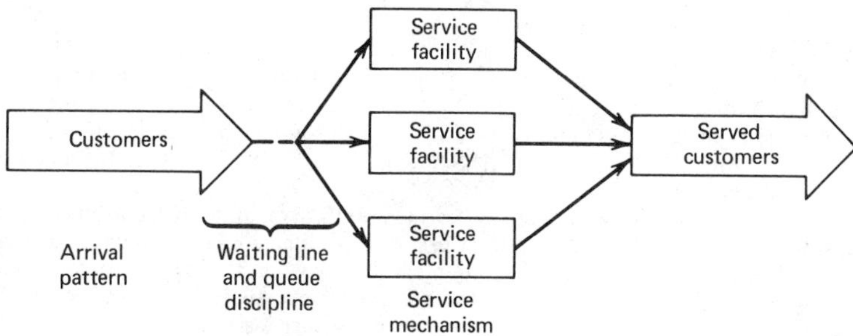

13.7.4 QUEUE DISCIPLINE

The third aspect of a queuing system is the queue discipline, which specifies how customers are to be selected for service from the customers or units awaiting service. The most common procedure is to serve customers in the order of their arrival. This is FIFO discipline. Another possibility is to select customers for service at random. At other times one uses the LIFO discipline. Finally, the queue discipline may be governed by a priority system, which leads to the study of priority queues. Other queue disciplines are considered in relation to different customer behavior as mentioned earlier.

Exhibits 13.7.2 and 13.7.3 illustrate the general description of the queuing system.

13.7.5 ANALYSIS OF QUEUING SYSTEMS

In practical applications involving waiting line phenomena, the properties are analyzed mathematically. The main properties usually include:

1. The distribution of the length of time for which a customer has to queue for service and the average length of time.
2. The distribution and the average number of customers or units at any instant.
3. The distribution and the average length of the server's busy period.
4. The economics of the system, for example, the loss due to the number of "lost" customers and the loss due to the idle time of the server and the waiting time of the customer.

The mathematical analysis generally takes two forms. One is the analytical approach, and the other is the simulation approach. In the analytical approach we describe the components of the system–for example, the arrival pattern and the service time distribution–analytically and obtain by mathematical analysis the main properties of practical interest. Often an analytical queuing problem becomes highly complicated, and simple, usable expressions are difficult to derive. The alternative method is to simulate the system. Sampling experiments are conducted on the system's various components in order to understand the physical and statistical properties, and the main properties are then investigated. This approach has become popular and feasible because of the advent of high-speed computers that enable experimentation and the simulation of complex systems.

In both the approaches a common practice is to study the properties of the system by observing the queuing process for a sufficiently long time. In the analytical approach it is usually very difficult to derive expressions at a given time. Because of this property, the "steady state," or equilibrium state, over time is considered, and the properties usually refer to this steady state.

13.7.6 NOTATION

A queuing system is usually denoted by three quantities indicating the nature of input (arrival pattern), the service mechanism, and the capacity of the service mechanism. The letter "D" stands for the deterministic case, and "M" denotes the Markovian input case, in which the customers arrive according to a Poisson process at a rate λ ($0 < \lambda < \infty$). Thus the number of arrivals during a time interval t has the Poisson distribution with parameter λt. From the well-known property of the Poisson process it follows that the interarrival times, that is, the intervals of time between successive arrivals, have the exponential distribution with density $\lambda e^{-\lambda x}$, $x > 0$. For the service mechanism, the Markovian assumption means that the service time of a customer or unit is distributed in the exponential form with rate μ. Erlangian input is denoted by "E_k," and the interarrival distribution is specified by the probability density

$$\frac{(\lambda k)^k}{(k-1)!} e^{-\lambda k u} u^{k-1}, \qquad u > 0$$

where k is a positive integer. An input of the general independent type is denoted by "GI," and a service time distribution of the general type, by "G." The capacity of the service mechanism measured by the number of servers is denoted by "S." With this notation we can think of several queuing systems, for example, M|M|1, M|M|S, M|D|S, M|G|S, M|G|1, G|G|S, GI|G|S. In these systems the queue discipline is generally assumed to be FIFO.

It is commonly assumed in theoretical studies that interarrival time distributions are stationary. In real-life situations this may not hold. Most practical problems relating to queuing processes are busy-period problems. In other words, they are most serious when the rate of arrival is the greatest. The form of interarrival time distributions that has been most studied is that corresponding to so-called random arrivals; that is, the intervals between consecutive arrivals are mutually independent random variables, all having the same probability distribution. The completely random series is likely to be a particularly good approximation when the customers are drawn from a very large pool of customers, all of whom behave independently of one another. If the calls arriving at a telephone exchange over a fairly short time are considered, they would exhibit a completely random phenomenon. If longer periods are considered, there will be changes in the behavior. An approximation to this can be best illustrated by the PDF of this series by expressing it as an exponential distribution with PDF = $\lambda e^{-\lambda x}$, $x > 0$. The exponential distribution has the characteristic "lack of memory property"; that is, the conditional probability $P(X \leq t | X \geq t_0)$ for $t > t_0$ depends only on the value $(t - t_0)$. For example, the probability that the interarrival time is ≤ 5 min given that it is ≥ 3 min is the same as the probability that the interarrival time is ≤ 800 min given that it is ≥ 798. Both the probabilities are equal to the probability that the interarrival time is ≤ 2 min. The exponential distribution assumption is a consequence of the Poisson process assumption. It appears that exponential distribution is a better approximation to the arrival distributions than to the service time distributions. In many situations the lack of memory property may not hold well for service time distributions.

The exponential distribution is likely to be a reasonable one to consider when there are a large number of customers requiring fairly short service or a small number of customers requiring long service. On the other hand, when we consider service time, it is evident that the time remaining until the completion of a service is independent of the time during which the service has already been in progress. This means that, if, for example, a customer arrives to find only one person ahead of him or her and that person is being served, the time the customer will have to queue is distributed exactly as the service time. Frequently, a Pearson type 3 distribution or a logarithmic normal distribution can be fitted to the data; however, because of the mathematical complexity, they become unwieldy to manipulate.

13.7.7 ANALYTICAL APPROACH FOR M|M|1

The simplest of the queuing models is the M|M|1 model. Recall that the input or the arrival pattern is Markovian and that the service time distribution is exponential with rate μ. The interarrival time is exponential with rate λ. The main properties can be studied mathematically, and the steady state solutions obtained. Let $(\lambda/\mu) < 1$. Define $p_n(t)$ as the probability of having n customers in the system at time t, including the customer being served. The steady state probability $p_n = \lim_{t \to \infty} p_n(t)$ is given by

$$p_n = (1 - \rho)\rho^n, \qquad n = 0, 1, 2, \ldots$$

where $\rho = \lambda/\mu$. The symbol ρ is called the "traffic intensity" or, "utilization factor" and is defined as

$$\rho = \frac{\text{mean service time of a single customer}}{\text{mean interval between arrival of successive individual customers}}$$

Exhibit 13.7.4 Relationship of Queue Length to the Utilization Factor[a]

λ	2	5	10	12	13	14	15	16
μ	16	16	16	16	16	16	16	16
ρ	0.125	0.313	0.625	0.75	0.812	0.875	0.938	1.0
L_q	0.017	0.142	1.04	2.25	3.52	6.13	14.0	∞

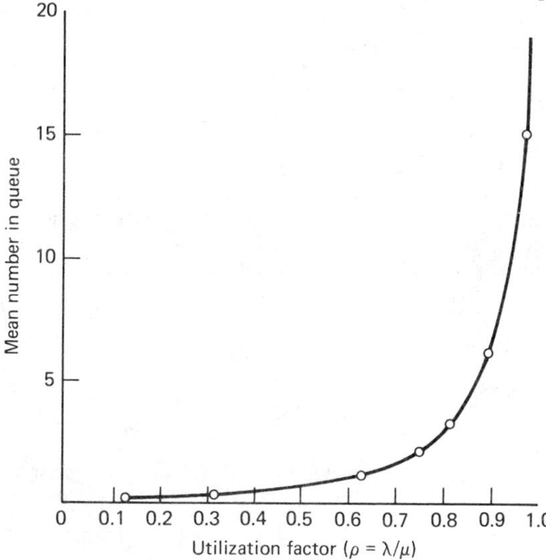

[a]Reference 4.

Thus p_n is a geometric distribution. Note that $\rho < 1$. If $\lambda = \mu$, then $\rho = 1$, and theoretically the service facility is used 100% of the time. Exhibit 13.7.4 summarizes the result of Poisson input and exponential service times. As ρ approaches unity, the number waiting in line increases rapidly and approaches infinity. This can also be verified analytically. In all cases the denominator goes to zero as ρ approaches unity, and the value of the number waiting in the queue becomes infinitely large, or, in other words, the queue explodes. It can be seen that one of the requirements of any practical system is that $\lambda < \mu$, which means $\rho < 1$, failing which an unstable system results.

If the arrivals are faster than the time in which they can be processed, the waiting line and the waiting time will increase continuously, and no steady state can be achieved. This results in the fact that there is a value to be placed on the idle time of the service facility. Whether to go in for a rapid service, and if so, at what cost, would dictate the trade-off we can have between cost and service.

Using this steady state distribution, we can verify that for the steady state

1. The average number, L, of customers in the system is $\rho/(1 - \rho)$, and the variance is $\rho/(1 - \rho)^2$.
2. The probability of finding more than N customers in the system is $\rho^{(N+1)}$. Thus the lower the value of ρ, the less unlikely that long queues will ultimately develop.
3. The average length of queue L_q is $\lambda^2/\mu(\mu - \lambda)$.
4. The average length of busy period is $(\mu - \lambda)^{-1}$.
5. The average time W (waiting time + service time) a customer spends in the system is $(\mu - \lambda)^{-1}$.
6. Average time W_q (excluding service time) a customer spends in line is $(\mu - \lambda)^{-1} - \mu^{-1} = \lambda/\mu(\mu - \lambda)$. Note that, from point 3, $L_q = \lambda W_q$. Further, $L = \lambda W$. The formula $L = \lambda W$ needs special mention. Known as "Little's formula," it states that

$$\text{average queue length} = \begin{pmatrix}\text{average number of} \\ \text{arrivals per unit time}\end{pmatrix} \times \begin{pmatrix}\text{average queue time} \\ \text{per customer}\end{pmatrix}$$

This equation is one of the basic equations and also one of the few relationships that is applicable for a large class of queuing systems. For a proof, see Little[2] and Jewell.[3]

Limited Queue Length

So far we have assumed that the length of the queue was unrestricted. Often the queue space is limited, and hence the queue length is bounded. An example is the reserved parking zone for taxicabs. Another situation in which queue length is limited is the case of a finite calling population. An example is that of the loom shed of a textile mill where the number of looms waiting for service is a finite number. The formulas will have to be modified for this case. For details of this phenomenon and the consequences for the single-server queue, the reader is referred to Ruiz-Palá.[5]

13.7.8 M|G|1 QUEUE

For the M|G|1 model the service time distribution is general. Assume that its mean is $1/\mu$ and its variance $V < \infty$. Then

1. The average number of customers in the system is

$$\rho + \frac{\lambda^2 V + \rho^2}{2(1 - \rho)}$$

This is the "Pollaczek-Khintchine formula." It shows that, even if the mean service time cannot be decreased, one can reduce the average queue size by reducing the variability of the service times. It is important to note that this average refers to instants at which a customer departs after service.

2. The average line length is

$$\frac{\lambda^2 V + \rho^2}{2(1 - \rho)}$$

3. The probability that the server is idle is $1 - \rho = 1 - \lambda/\mu$, so that ρ can still be interpreted as the utilization factor.

4. Still assuming that the queue discipline is FIFO, the average time spent by a customer in the system is

$$\frac{1}{\lambda}\left[\rho + \frac{\lambda^2 V + \rho^2}{2(1 - \rho)}\right]$$

This is also frequently called the Pollaczek-Khintchine formula. These results can be specialized to M|M|1 and M|D|1 queues. In the latter case, put $V = 0$ in the formulas.

Example 13.7.1 Consider an M|M|1 situation with $\lambda = 7$ persons/hr and $\mu = 10$ persons/hr. That is, the number of arrivals per hour has a Poisson distribution with the parameter $\lambda = 7$, and the service time has an exponential distribution with a mean of $\frac{1}{10}$ hr, or 6 min. In the steady state distribution, we have the following results: The utilization factor or traffic intensity is $\rho = \lambda/\mu = 0.70$. The expected number of persons in the system is $\rho/(1 - \rho) = 2.33$. The probability that there are more than two persons in the line is $\rho^3 = .343$. Average time spent in the queue, including service time, is $(\mu - \lambda)^{-1} = 0.33$ hr, or 20 min. Average time spent in the line before being served is $\lambda/\mu(\mu - \lambda) =$ 14 min. The waiting time distribution is exponential, with a mean of $\frac{1}{3}$ hr.

13.7.9 M|M|S QUEUE

The steady state distribution in this case is given by

$$p_n = \begin{cases} \dfrac{\rho^n}{n!}\, p_o & \text{for } n = 0, 1, \ldots, (S - 1) \\[2ex] \dfrac{\rho^n}{S!\, S^{n-S}}\, p_o & \text{for } n = S, (S + 1), \ldots \end{cases}$$

where $\rho = \lambda/\mu$ and p_o is obtained such that $\Sigma_{n=0}^{\infty} p_n = 1$. We must assume here that $\rho < S$. The probability of a busy period is

$$\frac{\rho^S}{S!\,(1 - \rho/S)}\, p_o$$

The average length of queue is

$$\frac{\rho^S \lambda \mu S}{S!\,(\mu S - \lambda)^2}\, p_o$$

The average number of customers in service is ρ.

Erlang's Loss Formula (M|M|S Queues)

Suppose that we have S servers and waiting places for only S customers. A situation of this type occurs, for example, in a telephone exchange with S trunk lines and no provision to hold subscribers who require a line, but who cannot be supplied with one. Then

$$p_n = \frac{\rho^n/n!}{\sum_{n=0}^{S} \rho^n/n!}, \qquad n \leqslant S$$

The quantity p_S gives the proportion of time for which the system is fully occupied. In the telephone example, this is the proportion of lost calls. The formula for p_S is called "Erlang's loss formula." For $\rho = 0.1$ and $S = 6$, the values of p_n are as follows:

n	p_n
0	0.4845
1	0.2907
2	0.1454
3	0.0582
4	0.0175
5	0.0035
6	0.0003

For this example, p_6 is very small. In 3 out of 10,000 calls, a subscriber would find all of the six trunk lines busy.

Example 13.7.2 Consider an M|M|S queue system with $\rho = 2$ and $S = 3$. The reader can apply the preceding formulas and obtain the following results:

Item	Results							
n	0	1	2	3	4	5	6	7
Lines busy	0	1	2	3	3	3	3	3
p_n	0.1111	0.2222	0.2222	0.1481	0.0988	0.0658	0.0439	0.0293

To illustrate the calculations for $n = 4$, we first find p_o such that $\sum_{n=0}^{\infty} p_n = 1$. Using the formula developed previously for p_n, we have

$$\sum_{n=0}^{S-1} \frac{\rho^n}{n!} p_o + \sum_{n=S}^{\infty} \frac{\rho^n}{S!\, S^{n-S}} p_o = 1$$

or

$$p_o = \frac{1}{\sum_{n=0}^{S-1} \rho^n/n! + \sum_{n=S}^{\infty} \rho^n/S!\, S^{n-S}}$$

$$= \frac{1}{\sum_{n=0}^{2} 2^n/n! + \sum_{n=3}^{\infty} (2/3)^n\, 3^3/3!}$$

$$\sum_{n=0}^{2} \frac{2^n}{n!} = 1 + 2 + 2 = 5$$

and

$$\sum_{n=3}^{\infty} \left(\frac{2}{3}\right)^n = \frac{(2/3)^3}{1 - (2/3)} = \frac{24}{27}$$

Simplifying, we have $p_o = 0.1111$.

$$p_4 = \frac{\rho^4}{S!\, S^{n-S}} p_o = \frac{2^4}{3!3} (0.1111) = \frac{16}{18} (0.1111) = 0.0988$$

Limited Queue Length

The case of limited queue length was discussed with respect to the single-server queue. Now, too, one can impose a restriction that there is a maximum capacity of the system (servers plus queue). The capacity is imposed by the physical limitation of the system, for example, customers in a department store at service channels. Another case of limited queue length arises when there is a finite calling population of customers. This case is treated later in Section 13.7.13.

Another case of the multichannel, limited queue problem is where no queue is permitted to form, for example, automobile parking.

The system characteristics and measures of effectiveness can be studied for these cases by suitable modification of the analysis. For details, the reader is referred to Ruiz-Palá.[6]

13.7.10 NETWORKS OF WAITING LINES

This is an extension of M|M|S queues where there are D departments, each being an M|M|S system, with number of servers, mean arrival rate, and mean holding time varying from department to department. A situational example is that of a machine shop that has several departments, each containing a fixed number of similar machines. Each department is a multiserver system of the M|M|S type. When a department finishes a job, the job goes either to some specified department or out of the system. If the mean arrival rates at the various departments are properly defined, a steady state distribution is obtained.

A formal statement of the problem and the main result are as follows:

1. There are D departments.
2. Department d contains $S(d)$ servers, $d = 1, 2, \ldots, D$. Within each department d, we have an M|M|S(d) system, with parameters $\lambda(d)$ and $\mu(d)$ in the usual notation.
3. Once served in department d, a customer goes (instantaneously) to department k ($k = 1, \ldots, D$) with probability p_{dk}; his or her total service is completed with probability $1 - \Sigma_k p_{dk}$.

Let $\{c(1), \ldots, c(D)\}$ denote the state of the system, with $c(k)$ denoting the number of customers waiting and in service at department k, $k = 1, \ldots, D$. Then a steady state distribution of the state of the system is given by the products

$$P\{c(1), \ldots, c(D)\} = P^{(1)}_{c(1)} P^{(2)}_{c(2)}, \ldots, P^{(D)}_{c(D)}$$

$$P^{(d)}_{c(k)} = \begin{cases} \dfrac{P^{(d)}_o [\Gamma(d)/\mu(d)]^{c(k)}}{[c(k)]}, & \text{for } c(k) = 0, 1, \ldots, S(d) \\[2ex] \dfrac{P^{(d)}_o [\Gamma(d)/\mu(d)]^{c(k)}}{[S(d)\, S(d)]^{c(k)-S(d)}}, & \text{for } c(k) > S(d) \end{cases}$$

where $\Gamma(d) = \lambda(d) + \Sigma_k p_{kd}\Gamma(k)$, or the average arrival rate of customers at department d from any source inside or outside the system. For the existence of the steady state distribution, we have to assume that

$$\Gamma(d) < \lambda(d)\, S(d) \qquad \text{for } d = 1, 2, \ldots, D.$$

13.7.11 M|E$_k$|1 QUEUES

The service time distribution is Erlang distribution with parameters μ and k. In this case, with the usual notation,

$$L_q = \frac{1+k}{2k} \frac{\lambda^2}{\mu(\mu - \lambda)}$$

$$W_q = \frac{1+k}{2k} \frac{\lambda}{\mu(\mu - \lambda)}$$

$$W = W_q + \frac{1}{\mu}$$

$$L = \lambda W$$

13.7.12 GI|G|S QUEUES

This is a general model in which the input process and the service mechanism are not Markovian. The input distribution is a general one that is identical for all interarrival intervals, and these are independently distributed. The service time distribution is arbitrary. The analytical results in this situation are quite complicated. For $S = 1$, the analysis leads to the Wiener-Hopf type of integral equations, which in turn yield the waiting time distribution. These results also have been generalized to the case when $S > 1$.

There are other variants of queuing systems, and the literature on queuing theory is quite vast. T. L. Saaty[7] has given an extensive bibliography.

13.7.13 FINITE CUSTOMER POPULATION AND MACHINE INTERFERENCE MODELS

Suppose that the source population is finite. (The Erlang's loss formula was given in this context.) One of the important applications of finite customer population is in the area of maintenance of machines by a crew of repair persons. (See Chapter 3.5 for more details of machine interference.) The formulas that follow are described in this context. Also used here are the notations used by Peck and Hazelwood,[8] who have provided extensive tables for practical application of the machine interference model. The definitions and notations are summarized as follows:

N	Population (number of machines, number of customers)
M	Number of servers (repair persons, trunk lines)
T	Average service time (Service time for each server is exponentially distributed.)
W	Average waiting time
U	Average idle time (Idle time is exponentially distributed.)
H	Average number of units being serviced = $NT/(T + W + U)$
L	Average number of units waiting for service = $NW/(T + W + U)$
J	Average number of units idle (running) = $NU/(T + W + U)$
F	Efficiency factor $(H + J)/(H + L + J)$
X	Service factor = $T/(T + U)$[Earlier this was denoted by $\rho/(1 + \rho)$.]
D	Probability that a unit will have to wait if service is required

Then, in the steady state, the probability p_n that there are n customers waiting for service is given by

$$p_n = \begin{cases} \binom{N}{n}\left(\dfrac{X}{1-X}\right)^n p_o, & n = 0, 1, \dots, M \\ \binom{N}{n}\left(\dfrac{X}{1-X}\right)^n \dfrac{n!}{M!\, M^{n-M}} p_o, & n = M+1, \dots, N \end{cases}$$

The value of p_o is obtained such that $\Sigma_{n=0}^{N} p_n = 1$.

The tables prepared by Peck and Hazelwood give the values of D and F for given values of X, M, and N. The reader can find a few numerical examples and various types of applications of the tables prepared by Peck and Hazelwood.

One of the earliest works on the machine interference model was by Ashcroft.[9] He treated a model similar to the one just described here, with the difference that the service times were assumed to be constant. A chart for determining the best machine assignment in this case, developed by Sandberg, can be found in Eilon.[10] This reference also gives a few more tables and examples for the general case of exponentially distributed service times.

Example 13.7.3 In a machine shop there are 10 identical machines. From past data it is found that the time interval between two successive breakdowns is exponentially distributed, with a mean of 3 hr. The breakdown of the machines can be assumed to follow a Poisson process. The service time required by a repair person is well approximated by an exponential distribution with a mean of 2 hr. Currently, two repair persons are available. It was found, however, that a number of these machines often were inoperative because the repair persons were attending to other machines that required service.

The company pays \$5/hour to a repair person, and it is estimated that the loss resulting from an idle machine is \$13/hour on the average. What is the optimal number of repair persons required that would minimize the total costs?

Exhibit 13.7.5 Machine Interference Example

M	F	L	H	Cost of Idle Time on Machine $13L = I$	Cost of Idle Time of Repair Persons $I' = 5(M - H)$	Total Cost $I + I'$
2	0.499	5.01	1.996	65.13	0.02	65.15
3	0.728	2.72	2.912	35.36	1.40	36.76
4	0.887	1.13	3.548	14.69	2.26	16.95
5	0.963	0.37	3.852	4.81	5.74	10.55
6	0.991	0.09	3.964	1.17	10.18	11.35

For the solution, $N = 10$, $T = 2$, $U = 3$, $X = T(T + U) = 2/5 = 0.400$. We calculate total cost for $M = 2, 3, 4, 5, \ldots$, and choose the M that gives the least value for the total cost. Note that

$H = FNX$ = average number of repair persons being utilized
L = average number of machines waiting for service = $(1 - F)N$

From Peck and Hazelwood tables corresponding to $N = 10$, we calculate the results for different values of M, as shown in Exhibit 13.7.5.

As Exhibit 13.7.5 shows, $M = 5$ gives the least total cost. The optimal number of repair persons is therefore 5. Exhibit 13.7.6 shows graphically the various costs and the optimal number of repair persons.

13.7.14 PRIORITY QUEUES

In certain queue disciplines, some classes of customers receive priority for service. This may be because the cost per unit of time of keeping certain customers in the queue may be high. The priority system amenable to the simplest mathematical analysis is the so-called nonpreemptive priority. In this system a customer, once at the service point, remains there until his or her service is complete. Then the next customer with the highest priority among those queuing is given service. In the pre-

Exhibit 13.7.6 Determination of Optimal Number of Repair Persons to Minimize Total Cost

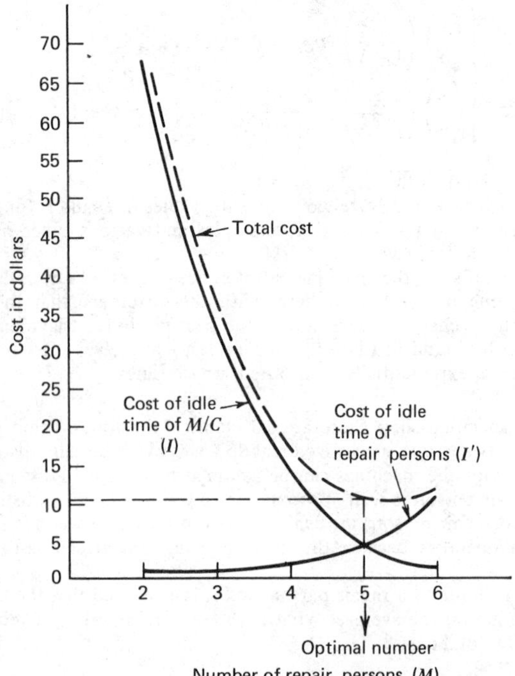

emptive priority system, a customer of high priority takes, on arrival, immediate precedence over customers of lower priority (including the one in service).

Following is a brief account of the analysis of priority queues for M|G|1 and M|M|S queue systems.

Consider the nonpreemptive priority queue discipline. Suppose that the input population is classified into r distinct types and that priorities are assigned to the types in decreasing order of importance. For example, a customer of type $(k + 1)$ has higher priority over the one of type k. Assume that each customer of type k arrives according to a Poisson process with rate λ_k, $k = 1, \ldots, r$. Assume also that the service distribution for each type k is some general distribution with mean $1/\mu_k$ and variance V_k. Define

$$\sigma_k = \sum_{j=1}^{k} \frac{\lambda_j}{\mu_j}$$

Suppose that $\sigma_r < 1$ to ensure the convergence of the system to a steady state. Then the average time in line for a type k customer right after the departure of a customer is

$$\frac{\Sigma_{j=1}^{r} \lambda_j [V_j + (1/\mu_j)^2]}{2(1 - \sigma_{k-1})(1 - \sigma_k)}$$

The average time in line for an arbitrary customer is

$$\frac{\Sigma_{k=1}^{r} \lambda_k E}{\Sigma \lambda_k}$$

where E is the time in line for type k.

For the case expected where there are S identical servers, assume further that the service time distribution is exponential, with rate μ for every type of customer. Assume $\sigma_r < S$. Then the average time in the line for the type k customer is

$$\frac{S/\mu}{(S - \sigma_{k-1})(S - \sigma_k)} P$$

where P is the probability of busy period for M|M|S queue.

13.7.15 QUEUE ANALYSIS BY SIMULATION

So far, the analytical approach to analyzing waiting line phenomena has been described. The interarrival times and the service times were postulated to follow predetermined distributions, for example, exponential or Erlang. In some situations general distributions were also considered. The analytical approach becomes intractable and unusable for many complex queuing models. As an alternative approach, the simulation method was suggested. (The reader is referred to Chapter 13.11 for a detailed description of simulation techniques.) A simple example is used here to illustrate the analysis of a waiting line problem using the simulation technique. The example involves a single-server queue at a tool crib in a factory.

Example 13.7.4 At a tool crib in a machine shop, workers arrive to collect tools from an attendant. The interarrival time distribution and the service time distribution are not known.

Suppose that the tool crib opens at 8:00 a.m. and is in operation until 9:45 a.m. If the service facility (attendant) is free, the arrival (worker) moves into the facility. If the service facility is busy, the arrival waits in the queue. The queue discipline is FIFO. The arrivals and service times are observed over 100 workers. The data are shown in Exhibits 13.7.7 and 13.7.8.

The operation of the tool crib can be simulated according to the interarrival and service time distributions given in Exhibits 13.7.7 and 13.7.8, respectively. The random numbers for arrival and service times are taken from the standard random number tables. The first two digits only are considered. If the first random number considered is 15, then looking at the cumulative probability distribution for interarrival times, 0.15 lies between 3 and 4 min. It can be interpreted that the fifteenth arrival has an interarrival time of 4 min. In the same fashion the successive interarrival and service times are simulated.

Exhibit 13.7.9 gives the complete simulation analysis for one shift of operation. From the exhibit, it can be seen that, over 20 arrivals, the average waiting time for the customer and the

Exhibit 13.7.7 Interarrival Times

Interarrival Time in Minutes	Frequency[a]	Cumulative Frequency	Probability	Cumulative Probability
3	11	11	0.11	0.11
4	15	26	0.15	0.26
5	39	65	0.39	0.65
6	20	85	0.20	0.85
7	10	95	0.10	0.95
8	5	100	0.05	1.00

[a]Total = 100.

Exhibit 13.7.8 Service Times

Service Time in Minutes	Frequency[a]	Cumulative Frequency	Probability	Cumulative Probability
3	10	10	0.10	0.10
4	20	30	0.20	0.30
5	40	70	0.40	0.70
6	20	90	0.20	0.90
7	10	100	0.10	1.00

[a]Total = 100.

attendant are given by the following:

$$\begin{aligned}\text{average waiting time in the line for customer} &= \frac{\text{total waiting time for all arrivals}}{\text{number of arrivals over the shift period}}\\ &= \frac{20}{20} = 1.00 \text{ min}\end{aligned}$$

$$\begin{aligned}\text{percentage utilization of the attendant} &= \frac{\text{shift duration} - \text{idle time of attendant}}{\text{shift duration}}(100)\\ &= \frac{(105-12)}{105}(100) = 89\%\end{aligned}$$

$$\begin{aligned}\text{average time spent by the customer in the system} &= \text{average service time of a customer} + \text{average waiting time in the line for a customer}\\ &= \frac{\Sigma_1^{20}\text{ service times}}{\text{number of arrivals}} + \text{average waiting time in the line}\\ &= \frac{93}{20} + 1 = 5.65 \text{ min}\end{aligned}$$

In the preceding calculation, it can be observed that the average service time is 4.65 min. However, the average service time calculated from the empirical distribution given in Exhibit 13.7.8 was found to be 5 min. This variation is due to the sample's small size. Over a large number of simulations, the mean service time will approximate better the actual average of 5 min. By allocating cost figures to the waiting time, the feasibility of introducing more servers can be studied.

For complex real-world queuing phenomena, the simulation technique is a powerful tool for decision making with the objective of relieving the congestion. In most cases it is the only technique available since the analytical approach is not feasible.

13.7.16 AN INVENTORY APPLICATION

The applications of queuing theory are far too many and diverse to list completely. A simple application to an inventory control situation is described here (also see Chapter 11.4).

Exhibit 13.7.9 Simulation of Tool Crib Operation

Arrival Number	Arrival Simulation: Random Number	Arrival Simulation: Interarrival Time in Minutes	Arrival Simulation: Time of Arrival	Service Simulation: Random Number	Service Simulation: Service Time in Minutes	Service Begins	Service Ends	Customer Waiting Time in Minutes	Server Idle Time in Minutes
1	15	4	8:04	46	5	8:04	8:09	–	4
2	09	3	8:07	64	5	8:09	8:14	2	–
3	41	5	8:12	09	3	8:14	8:17	2	–
4	74	6	8:18	48	5	8:18	8:23	–	1
5	00	3	8:21	97	7	8:23	8:30	2	–
6	72	6	8:27	22	4	8:30	8:34	3	–
7	67	6	8:33	29	4	8:34	8:38	1	–
8	55	5	8:38	01	3	8:38	8:41	–	–
9	71	6	8:44	40	5	8:44	8:49	–	3
10	35	5	8:49	75	6	8:49	8:55	–	–
11	41	5	8:54	10	4	8:55	8:59	1	–
12	96	8	9:02	09	3	9:02	9:05	–	3
13	20	4	9:06	70	6	9:06	9:12	–	1
14	45	5	9:11	41	5	9:12	9:17	1	–
15	38	5	9:16	40	5	9:17	9:22	1	–
16	01	3	9:19	37	5	9:22	9:27	3	–
17	67	6	9:25	21	4	9:27	9:31	2	–
18	63	5	9:30	38	5	9:31	9:36	1	–
19	39	5	9:35	14	4	9:36	9:40	1	–
20	55	5	9:40	32	5	9:40	9:45	–	–
Total					93			20	12

The "arrivals" correspond to the orders for withdrawal from inventory. The service mechanism is the process of replenishing this shortfall in the inventory. The service time is the time required to replace the items. This is the interval of time elapsing between the withdrawal of the items from inventory and the arrival of the new items. The queue is generated by the unfilled orders sent by the store possessing the inventory to the factory or wholesaler, for items to replace those sold. For simplicity in relating the inventory situation to a queuing model, note the following relationships in the terminology:

Queuing Terminology	Inventory Terminology
Service	Replacement
Arrival	Sale

The following assumptions are made with regard to the operations:

1. The store orders another unit from the factory as soon as one is sold. This means that the sum of units on hand and the unfilled replacement orders is a constant S. This is called the "maximum inventory."
2. As long as the item is in stock, sales have a Poisson distribution with rate λ per week.
3. No sales are made when the item is out of stock.
4. The replenishment time is distributed in the exponential form with mean $1/\mu$.

We have, then, an M|M|S model. An "idle" server corresponds to an item in the inventory, and a "busy" server corresponds to an out-of-stock situation. Service starts when a sale is made and ends when the replacement arrives back in stock. We can apply the queuing theory formulas to compute, for example, mean number of replacement orders outstanding, mean inventory, probability of being out of stock, and mean number of sales in replenishment time.

In the preceding model, S was fixed. One can also find the optimum value of M to satisfy some criterion.[11]

13.7.17 APPROXIMATIONS TO QUEUES

It is possible to conceptualize a queuing situation and derive a mathematical model for it. The problem may be so complex that it might become very difficult to manipulate it and obtain practicable solutions. It becomes necessary to attempt to bridge the gap between high-fidelity mathematical models and practical solutions.

The first type of approximation is to obtain operational solutions by applying models that are, a priori, the wrong ones, but that are simple to deal with. When this becomes unacceptable, simulation may be another resort. It is possible, with a great deal of work, to construct a simulation model that can be an exact replica of the real-world situation. With the advent of computers, this sort of approximation leading to simulation has become a widely acceptable tool for resolving queuing problems.

The third type of approximation is the classical one—the two techniques used are:

1. Expansion of unmanageable functions in a formula in terms of power series, ignoring the higher terms over a range of values of the variables.
2. Substitution of initial values in the formulas, followed by iteration.

It is up to the analyst to think in terms of cost and the associated trade-off in choosing a particular type of approximation.

Yet another approximation method used with advantage in queuing models is the "diffusion approximation." This method is used to analyze the equilibrium distributions and time-dependent queues. This technique has been well exploited by physicists and mathematicians for more than a century. Although queues are generally discrete, the idea is to treat them as continuous random variables and to derive the diffusion equation. For details of this technique, see Newell.[12]

13.7.18 NONEXPONENTIAL QUEUES AND PHASE-TYPE DISTRIBUTIONS

Queuing systems are generally classified as Markovian or non-Markovian, depending on whether all the underlying distributions are exponential or not. Tractable analytical or numerical results are obtained for the Markovian queues using the memoryless property of the exponential distributions representing the arrival and service times. However, there are some serious disadvantages involved in modeling real-life situations by means of exponential distributions. For example, the proper unimodality or multimodality of many real situations cannot be represented. Hence in practice it is common to use Erlang or hyperexponential distributions (finite mixtures of negative exponentials) to model random time durations that are very much different from the exponential case.

Thus in these situations we are essentially dealing with nonexponential queues. In these queues, it is not possible to get analytically tractable results for the quantities of interest since their distributions are usually given in the form of Laplace-Stieltjes, Fourier, or generating function transforms. These transforms involve several unknown parameters that are to be calculated by a situation of an auxiliary system of linear equations with complex coefficients. The coefficients of that auxiliary system depend on the roots of a transcendental equation inside some region of the complex plane. Thus a theoretical analysis in these cases is provided by using Rouche's theorem, but in practice this method is quite difficult. Hence it is desirable to have methods to study non-Markovian queues which are analytically or numerically tractable. Recently Neuts[13–15] has introduced a class of distributions that includes the Erlang and hyperexponentials as very special cases and has termed them "probability distributions of phase type" (PH-type). Further, by using the method of phases, he has also demonstrated how non-Markovian queues with PH-type arrival or service or both can be solved in a tractable form.

REFERENCES

1. E. S. BUFFA and J. S. DYER, *Management Science/Operations Research, Model Formulation and Solution Methods*, Wiley, New York, 1977, p. 287.
2. J. D. C. LITTLE, "A Proof for the Queuing Formula $L = \lambda W$," *Operations Research*, Vol. 9, 1961, pp. 383–387.
3. W. S. JEWELL, "A Simple Proof of $L = \lambda W$," *Operations Research*, Vol. 15, 1967, pp. 1109–1116.
4. BUFFA and DYER, *Management Science/Operations Research*, p. 297.
5. E. RUIZ-PALÁ, C. AVILA-BELOSO, and W. W. HINES, *Waiting Line Models*, Van Nostrand Reinhold, New York, 1967, pp.
6. RUIZ-PALÁ et al., *Waiting Line Models*, pp.
7. T. L. SAATY, *Elements of Queuing Theory*, McGraw-Hill, New York, 1961.
8. L. G. PECK and R. N. HAZELWOOD, *Finite Queuing Tables*, Wiley, New York, 1958.
9. H. ASHCROFT, "The Productivity of Several Machines Under the Care of One Operator," *Journal of Royal Statistical Society*, Series B, Vol. 12, 1950, pp. 145–151.
10. S. EILON, *Industrial Engineering Tables*, Van Nostrand Reinhold, London, 1962, pp. 151–154.
11. P. M. MORSE, *Queues, Inventories and Maintenance*, Wiley, New York, 1967.
12. G. F. NEWELL, *Applications of Queuing Theory*, Chapman and Hall, London, 1971.

13. M. F. NEUTS, "Probability Distribution of Phase Type," in *Liber Amicorum Professor Emeritus H. Florin*, Department of Mathematics, University of Louvain, Belgium, 1975.

14. M. F. NEUTS, "Computational Uses of the Method of Phases in the Theory of Queues," *Computers and Mathematics With Applications*, Vol. 1, 1975, pp. 151–166.

15. M. F. NEUTS, "Queues Solvable Without Rouche's Theorem," *Operations Research*, Vol. 27, No. 4 (July–August 1979), pp. 767–781.

BIBLIOGRAPHY

BROCKMEYER, E., H. L. HALSTROM, and A. JENSEN, *Life and Works of A. K. Erlang*, Copenhagen Telephone Company, Copenhagen, 1949.

BUFFA, E. S., *Readings in Production and Operations Management*, Wiley, New York, 1966.

CHURCHMANN, C. W., R. L. ACKOFF, and E. L. ARNOFF, *Introduction to Operations Research*, Wiley, New York, 1957.

COX, D. R., and W. L. SMITH, *Queues*, Methuen, London, 1961.

DOIG, A., "A Bibliography on the Theory of Queues," *Biometrika*, Vol. 44, 1957, pp. 490–514.

FELLER, W., *An Introduction to Probability Theory and Its Applications*, Vol. 1, 3rd ed., Wiley, New York, 1968.

HILLIER, F. S., and G. J. LIEBERMAN, *Introduction to Operations Research*, Holden-Day, San Francisco, 1974.

JACKSON, J. R., "Network of Waiting Lines," *Operations Research*, Vol. 5, August 1957, pp. 518–521.

JACKSON, J. R., "Jobshop-like Queuing Systems," *Management Science*, Vol. 10, October 1963, pp. 131–142.

JAISWAL, N. K., *Priority Queues*, Academic Press, New York, 1968.

KENDALL, D. G., "Some Problems in Theory of Queues," *Journal of Royal Statistical Society*, Vol. 13, 1951, p. 151.

KHINCHINE, A., *Mathematical Methods in the Theory of Queuing* (translated by D. M. Andrews and M. H. Quenouille), Griffin, London, 1960.

KLEINROCK, L., *Queuing Systems*, Vols. 1 and 2, Wiley, New York, 1976.

LEE, A. M., *Applied Queuing Theory*, Macmillan, London, 1966.

PRABHU, N. U., *Queues and Inventories*, Wiley, New York, 1965.

PRABHU, N. U., *Stochastic Processes*, Macmillan, New York, 1965.

SAATY, T. L., "Resume of Useful Formulas in Queuing Theory," *Operations Research*, Vol. 5, April 1957, pp. 162–200.

WAGNER, H. M., *Principles of Operations Research*, 2nd ed., Prentice-Hall, Englewood Cliffs, NJ, 1975.

CHAPTER 13.8

Markov Chains

M. RAGHAVACHARI

Indian Institute of Management, Ahmedabad

13.8.1 INTRODUCTION

The attempts to describe and study mathematically some physical and biological phenomena led to the development of Markov analysis. The idea was conceived by A. A. Markov (1856–1922), who laid the foundations of the theory of finite Markov chains. The general theory of Markov processes was developed later by a number of mathematicians.

In the fields of industrial engineering and management, the mathematical representation and analysis of how a random variable changes over time is often an important part of many decision models for inventory, queuing, replacement, maintenance, marketing, finance, and other problems. The analysis of such a problem can be classified under stochastic processes. A stochastic process is a chance process with a time dimension. Instead of just one random variable X to describe an outcome, there is a whole series of random variables, $X_1, X_2, \ldots, X_t$, and so on, associated with behavior through time t. The study of the behavior of the random variables is generally done over a state space, which is a list of all possible outcomes at any particular time.

In many real-world situations, the variable of interest, as described in the previous paragraph, is time sequenced. The variable moves from one state to another as time changes. If we can think of demand for a product being represented by a sequence of random variables $X_1, X_2, \ldots, X_t, \ldots$, where X_t represents the demand level at time period t, and X_{t-j} the demand level at the $(t-j)$th period, it is likely that the random variables so defined are dependent.

Markov processes belong to the category of the preceding description. The time variable t can be treated either as continuous or as discrete units of time. This chapter focuses mostly on the latter. This leads to the study of a "Markov chain," which is a stochastic process $\{X_n\}$, $n = 0, 1, 2, \ldots$ with a special structure. The dependence structure existing between the variables $X_1, X_2, \ldots, X_n$ is of a particular form, called "Markov dependence." The X_n's are random variables, and the dependence structure is such that *the conditional probability that the system is in a given state at any period, given past history, depends only on the state of the system in the previous period.*

13.8.2 FORMAL DEFINITION OF A MARKOV CHAIN

Consider a system that can be in one of the states E_k, $k = 1, 2, \ldots$ at a given period n, $n = 0, 1, \ldots$. Associated with every pair (E_j, E_k) is a conditional probability p_{jk}, which is the probability that the system is in state E_k in the next period, given that the system is in E_j at the present period. If X_n indicates the state of the system in the nth period, the sequence of random variables $\{X_n\}$ is a Markov chain if

$$P\{X_n = E_k | X_{n-1} = E_j, X_{n-2} = E_{j_1}, X_{n-3} = E_{j_2} \ldots\} = P_{Jk}$$

This property is equivalent to saying that the conditional probability of any event in the next period, given past events and the present state, is independent of the past events and depends only upon the present state of the process. This property is referred to as the "Markovian property." The E_k's are usually referred to as "states" of the system. They may be finite in number or infinite. The notation p_{jk} is called the "transition probability from state E_j to E_k." Note that p_{jk} does not depend on the number of the periods $(n - 1)$ and n in which the transition takes place. For example,

$$P\{X_n \text{ in } E_k | X_{n-1} \text{ is in } E_j\} = P\{X_m \text{ is in } E_k | X_{m-1} \text{ is in } E_j\}$$

The author wishes to thank T. K. Soundar Rajan for assistance in the preparation of this chapter.

The Markov chain is then called a "stationary Markov chain." The term "stationary" is henceforth omitted since this chapter covers only stationary Markov chains. The transition probabilities p_{jk} can be arranged in the form of a matrix.

From state \ To state	E_1	E_2	E_3	....
E_1	p_{11}	p_{12}	p_{13}	....
E_2	p_{21}	p_{22}	p_{23}	....
E_3	p_{31}	p_{32}	p_{33}	....
.	.	.	.	....
.	.	.	.	....
.	.	.	.	....
.	.	.	.	....

The square matrix

$$P = \begin{bmatrix} p_{11} & p_{12} & p_{13} & \cdots\cdots \\ p_{21} & p_{22} & p_{23} & \cdots\cdots \\ p_{31} & p_{32} & p_{33} & \cdots\cdots \\ \cdot & \cdot & \cdot & \cdots\cdots \\ \cdot & \cdot & \cdot & \cdots\cdots \\ \cdot & \cdot & \cdot & \cdots\cdots \end{bmatrix}$$

is called the "transition probability matrix" of the Markov chain. The matrix will be finite if the number of states is finite. The row sums of P are all equal to 1, and each $p_{jk} \geqslant 0$. Such a matrix is called a "stochastic matrix."

The number of states in a Markov chain may be finite or infinite. In the former case, we have a finite Markov chain, and in the latter case an infinite Markov chain. The three main aspects that describe a Markovian model are (1) state, (2) transition, and (3) initial conditions. The present situation in any system can usually be specified by giving values of a number of variables that describe the system. Considering an inventory example, the raw material received at different times can be of either acceptable or poor quality. The status of the raw material can be represented by the variable "quality." The possible states are (1) raw material–good and (2) raw material–poor. When we know the values of the variable, then we can infer the details of the system; in other words, the system will be specified.

Let us analyze the chance process involved in obtaining either good- or poor-quality material. We can, from past history, arrive at the behavioral pattern of the good- or poor-quality materials. As stated in the definition of a Markov process, the present state of a system is influenced only by its preceding state. In the course of time, the system changes from state to state, exhibiting a dynamic behavior. The state of the system changes from good to poor and vice versa or remains in the same state. This process of change is called "transition" or "state transition." The probabilities associated with the change from one state to another of the variable are called "transition probabilities." These may also be called "conditional probabilities," because the probability assignments to each of the states depend upon the immediately preceding state. A transition diagram can be of help in representing the transitions and associated probabilities (see Exhibit 13.8.1). The same can be represented by a transition matrix.

In industrial engineering and management areas, Markov chains serve as appropriate models in certain situations involving inventory control, waiting lines, and so on. A few examples are given here to illustrate Markov chains.

Exhibit 13.8.1 State Transition Diagram

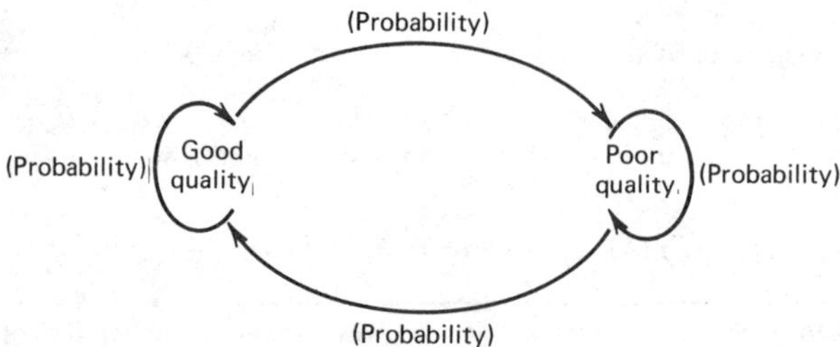

Example 13.8.1 Random Walk. A particle moves along the X axis either one unit to the right or one unit to the left. Initially it is at the origin. The probability that the particle moves to the left, given that it moved to the right in the immediately preceding trial, is $\frac{3}{4}$, and the probability that it moves to the right, given that it moved to the left in the immediately preceding trial, is $\frac{1}{2}$. This describes a Markov chain with two possible states, E_1 and E_2, defined as:

E_1 Moving to the right
E_2 Moving to the left

The transition probability matrix P is a 2 by 2 matrix.

$$P = \begin{bmatrix} \frac{1}{4} & \frac{3}{4} \\ \frac{1}{2} & \frac{1}{2} \end{bmatrix}$$

The two row sums are equal to 1.

Example 13.8.2 Random Placement of Balls. There are r cells, and at each trial a ball is placed at random in one of these r cells. Assume that the trials are independent. The states E_k of the system are defined as follows:

$$E_k\text{: Exactly } k \text{ cells are occupied}, \quad k = 0, 1, \ldots, r$$

There are $(r + 1)$ states, and we have a Markov chain with transition probabilities

$$p_{jj} = \frac{j}{r}, \quad p_{j,j+1} = \frac{(r - j)}{r}$$

and $p_{jk} = 0$ for all other combinations of j and k. The transition probability matrix when $r = 5$ is as follows:

From \ To	0	1	2	3	4	5
0	0	1	0	0	0	0
1	0	$\frac{1}{5}$	$\frac{4}{5}$	0	0	0
2	0	0	$\frac{2}{5}$	$\frac{3}{5}$	0	0
3	0	0	0	$\frac{3}{5}$	$\frac{2}{5}$	0
4	0	0	0	0	$\frac{4}{5}$	$\frac{1}{5}$
5	0	0	0	0	0	1

Example 13.8.3 Arrivals of customers follow Poisson distribution with parameter λ. Suppose that service times of successive customers are independent and identically distributed random variables. For $n \geqslant 1$, let X_n denote the number of persons waiting in line for service at the moment when the nth person to be served (on a given day) has finished being served. The $\{X_n\}$ is a Markov chain. The one-step transition probabilities are easily obtained: For $j = 0$

$$p_{0,k} = a_k$$

whereas for $j > 0$

$$p_{jk} = a_{k-j+1}$$

where a_k is the probability that k customers arrive during the service time of a customer. The transition probability matrix is

$$P = \begin{bmatrix} a_0 & a_1 & a_2 & \cdots\cdots \\ a_0 & a_1 & a_2 & \cdots\cdots \\ 0 & a_0 & a_1 & \cdots\cdots \\ 0 & 0 & a_0 & \cdots\cdots \end{bmatrix}$$

$$a_k = \int_0^\infty e^{-\lambda t} \frac{(\lambda t)^k}{k!} \, d\, F_s(t)$$

where $F_s(.)$ is the distribution function of service times. This is an imbedded Markov chain, since it corresponds to observing the stochastic process $\{N(t), t \geq 0\}$, where $N(t)$ represents the number of customers in the queue at time t at a sequence of times $\{t_n\}$ corresponding to the moments when successive customers depart service. Note that $X_n = N(t_n)$. This example is the imbedded Markov chain of an M|G|1 queue. (See Chapter 13.7 for details of M|G|1 queues.)

Example 13.8.4 Consider a queuing system, GI|M|1, discussed in Chapter 13.7. For $n \geq 1$, let X_n denote the number of persons waiting in line for service at the moment of the arrival of the nth person to arrive. Then $\{X_n\}$ is a Markov chain with state space $\{0, 1, \ldots\}$.

Example 13.8.5 Let X denote a discrete-valued random variable whose possible values are the nonnegative integers.

$$P(X = i) = p_i, \qquad p_i \geq 0, \qquad \Sigma_{i=1}^{\infty} p_i = 1$$

Let $X_1, X_2, X_3, \ldots, X_n, \ldots$ be independent observations of X. An important class of Markov chains arises from consideration of the successive partial sums S_n of X_i.

$$S_n = X_1 + \cdots + X_n$$

The $\{S_n\}$ is a Markov chain. The transition probability p_{jk} is

$$p_{jk} = P\{S_{n+1} = k \mid S_n = j\} = P\{X_{n+1} = k - j\}$$
$$= \begin{cases} p_{k-j} & \text{for } k \geq j \\ 0 & \text{for } k < j \end{cases}$$

Thus the transition probability matrix is

$$P = \begin{bmatrix} p_0 & p_1 & p_2 & p_3 & \cdots\cdots \\ 0 & p_0 & p_1 & p_2 & \cdots\cdots \\ 0 & 0 & p_0 & p_1 & \cdots\cdots \\ \cdot & \cdot & \cdot & & \cdots\cdots \\ \cdot & \cdot & \cdot & & \cdots\cdots \\ \cdot & \cdot & \cdot & & \cdots\cdots \end{bmatrix}$$

Example 13.8.6 Inventory Model: (*S*, *s*) Policy. A commodity is stocked in order to satisfy a continuing demand. Assume that replenishing of stock takes place every week t denoted by $t_1, t_2, \ldots$. The cumulative demand for the commodity during the nth week is a random variable, D_n, with distribution

$$P(D_n = k) = p_k, \qquad k = 0, 1, 2, \ldots$$

where $p_k \geq 0$ and $\Sigma_{k=0}^{\infty} p_k = 1$. Assume D_n's are independent.

The inventory policy is described as follows: The stock level is examined at the start of each period. If the available stock is less than s, immediate procurement is done so as to bring the quantity of stock on hand to level $S (S > s)$. If, however, the available stock $\geq s$, then no procurement is done. Let X_n denote the stock on hand just prior to restocking at the nth period. It is assumed that sales are lost when demand exceeds the inventory on hand. Then $\{X_n\}$ describes a Markov chain with states $0, 1, 2, \ldots, S$.

To illustrate the model, let us consider the inventory system where the commodity stocked is an electronic calculator. The demand follows a Poisson distribution with parameter $\lambda = 2$. We reorder $S = 4$ units when the stock level is less than $s = 1$ (i.e., when there are no items in inventory). The states are 0, 1, 2, 3, 4. The transition probability matrix is as follows:

$$P = \begin{bmatrix} 0.143 & 0.180 & 0.271 & 0.271 & 0.135 \\ 0.865 & 0.135 & 0 & 0 & 0 \\ 0.594 & 0.271 & 0.135 & 0 & 0 \\ 0.323 & 0.271 & 0.271 & 0.135 & 0 \\ 0.143 & 0.180 & 0.271 & 0.271 & 0.135 \end{bmatrix}$$

13.8.3 HIGHER TRANSITION PROBABILITIES

The transition probability matrix defined in the preceding examples is usually called the "one-step transition probability matrix." Usually we would like to study higher transitions. For example, we might like to find out the probability of a transition from E_j to E_k in exactly n steps. We denote by $P_{JK}^{(n)}$ the conditional probability of the system being in state E_k at the nth step, given that it is in state E_j. Clearly

$$P_{JK}^{(1)} = p_{jk}$$

$$P_{jk}^{(2)} = \sum_r p_{jr}p_{rk}$$

By induction, we can obtain $p_{jk}^{(n)}$ from $p_{jk}^{(n-1)}$ from the formula

$$P_{jk}^{(n)} = \sum_r p_{jr}p_{rk}^{n-1}$$

A more general formula is

$$p_{jk}^{(m+n)} = \sum_r p_{jr}^{(m)}p_{rk}^{(n)}$$

This identity is a special case of the Chapman-Kolmogorov identity. Just as p_{jk}'s form the transition probability matrix P, $p_{jk}^{(n)}$ for $n = 1, 2, \ldots$ also forms higher transition probability matrices: $P^{(n)}$. The (j, k) element of $P^{(n)}$ is $p_{jk}^{(n)}$. From the recursion formulas just given, it follows that

$$P^{(n)} = P^n$$

which is the nth power of the one-step transition probability matrix P.

The preceding concepts regarding higher transitions are illustrated here through the example of the random walk (Example 13.8.1).

The transition probability matrix is

$$P = \begin{bmatrix} \frac{1}{4} & \frac{3}{4} \\ \frac{1}{2} & \frac{1}{2} \end{bmatrix}$$

To derive the two-step transition probability matrix $P^{(2)}$, we have the recursive relationship

$$p_{jk}^{(2)} = \Sigma_{r=1}^{2} p_{jr}p_{rk}$$

$$P_{11}^{(2)} = p_{11} \quad p_{11} + p_{12} \quad p_{21}$$
$$= (\tfrac{1}{4}) \quad (\tfrac{1}{4}) + (\tfrac{3}{4}) \quad (\tfrac{1}{2}) = \tfrac{7}{16}$$

$$P_{12}^{(2)} = p_{11} \quad p_{12} + p_{12} \quad p_{22}$$
$$= (\tfrac{1}{4}) \quad (\tfrac{3}{4}) + (\tfrac{3}{4}) \quad (\tfrac{1}{2}) = \tfrac{9}{16}$$

$$P_{21}^{(2)} = p_{21} \quad p_{11} + p_{22} \quad p_{21}$$
$$= (\tfrac{1}{2}) \quad (\tfrac{1}{4}) + (\tfrac{1}{2}) \quad (\tfrac{1}{2}) = \tfrac{3}{8}$$

$$P_{22}^{(2)} = p_{21} \quad p_{12} + p_{22} \quad p_{22}$$
$$= (\tfrac{1}{2}) \quad (\tfrac{3}{4}) + (\tfrac{1}{2}) \quad (\tfrac{1}{2}) = \tfrac{5}{8}$$

The second-step transition probability matrix is therefore

$$P^{(2)} = \begin{bmatrix} \frac{7}{16} & \frac{9}{16} \\ \frac{3}{8} & \frac{5}{8} \end{bmatrix}$$

We can verify that $P^{(2)}$ is given by P^2.

$$P^{(2)} = PP = \begin{bmatrix} \frac{1}{4} & \frac{1}{4} \\ \frac{1}{2} & \frac{1}{2} \end{bmatrix} \begin{bmatrix} \frac{1}{4} & \frac{3}{4} \\ \frac{1}{2} & \frac{1}{2} \end{bmatrix} = \begin{bmatrix} \frac{7}{16} & \frac{9}{16} \\ \frac{3}{8} & \frac{5}{8} \end{bmatrix}$$

It can be shown analytically[1] that in general

$$P^n = \begin{bmatrix} \frac{2}{5} & \frac{3}{5} \\ \frac{2}{5} & \frac{3}{5} \end{bmatrix} + (-1)^n \left(\tfrac{1}{4}\right)^n \begin{bmatrix} \frac{3}{5} & -\frac{3}{5} \\ -\frac{2}{5} & \frac{2}{5} \end{bmatrix}$$

Conditional Probabilities

If a_j denotes the probability that the system is in E_j initially, the unconditional probability $a_k^{(n)}$ of the system being in E_k at the nth step is given by

$$a_k^{(n)} = \sum_j a_j p_{jk}^{(n)}$$

Irreducible Chains

If, for some $n > 0$, $P_{jk}^{(n)} > 0$, then we say that the state E_k can be reached from the state E_j. In other words, there is a positive probability of reaching E_k from E_j at some stage. A Markov chain is "irreducible" if and only if every state can be reached from every other state. If, for a state E_j, $p_{jj} = 1$, then we say that the state E_j is "absorbing."

Periodic and Aperiodic States

The state E_k has period $t > 1$ if $P_{KK}^{(N)} = 0$, unless $N = mt$ is a multiple of t, and t is the largest integer with this property. The state E_k is aperiodic if no such $t > 1$ exists.

Example 13.8.7 Periodic Chain. Consider a Markov chain with three states, E_1, E_2, and E_3, and transition probability matrix

$$P = \begin{bmatrix} 0 & 1 & 0 \\ 0 & 0 & 1 \\ 1 & 0 & 0 \end{bmatrix}$$

The only possible transitions are $E_1 \to E_2 \to E_3 \to E_1$. The period equals 3.

The distinction between periodic and aperiodic states is merely a technical difference; this chapter is confined only to aperiodic chains. The modification of the results to the periodic case can be obtained from the sources cited in the bibliography as well as from Feller.[1]

13.8.4 CLASSIFICATION OF STATES

Let

$f_{jk}^{(n)}$ = probability that in a chain starting from E_j the first entry to E_k occurs at the nth step

$$f_{jk}^{(0)} = 0$$

$$f_{jk} = \sum_{n=1}^{\infty} f_{jk}^{(n)}$$

$$\mu_j = \sum_{n=1}^{\infty} n f_{jj}^{(n)}$$

The f_{jk} is the probability that starting from E_j, the system will ever pass through E_k. To obtain $f_{jk}^{(n)}$, we get, recursively, $f_{jk}^{(1)}, f_{jk}^{(2)}, \ldots, f_{jk}^{(n)}$ from the formula

$$p_{jk}^{(n)} = \sum_{r=1}^{n} f_{jk}^{(r)} p_{kk}^{(n-r)}$$

Recurrent State. The state E_j is recurrent if $f_{jj} = 1$. In this case $\{f_{jj}^{(n)}\}$ is a probability distribution, and μ_j is the mean recurrence time for E_j. If $\mu_j < \infty$, then the recurrent state is called "positive recurrent state." If $\mu_j = \infty$, then the recurrent state E_j is called a "null recurrent state." Some authors refer to recurrent state as "persistent state."

Transient State. The state E_j is transient if $f_{jj} < 1$.

Ergodic State. An aperiodic recurrent state E_j with $\mu_j < \infty$ is called "ergodic."

Criteria to Determine Recurrent and Transient States

The criteria can be given in terms of the n-step transition probabilities, as follows:

1. E_j is transient if and only if

$$\sum_{n=0}^{\infty} p_{jj}^{(n)} < \infty$$

In this case

$$\sum_{n=1}^{\infty} p_{ij}^{(n)} < \infty, \qquad \text{for all } i.$$

2. E_j is a null recurrent state if and only if

$$\sum_{n=0}^{\infty} p_{jj}^{(n)} = \infty \qquad \text{but } p_{jj}^{(n)} \to 0 \qquad \text{as } n \to \infty$$

In this case $p_{ij}^{(n)} \to 0$ for all i, as $n \to \infty$.

3. An aperiodic recurrent state E_j is ergodic if and only if $\mu_j < \infty$. In this case as $n \to \infty$

$$p_{ij}^{(n)} \to f_{ij}\mu_j^{-1}$$

In summary, a few important results on the classification of states are:

1. All states of an irreducible chain are of the same type.
2. In a finite chain there are no null states, and it is impossible for all states to be transient.
3. A transition probability matrix is called "doubly stochastic" if the column sums are also equal to 1. If a Markov chain with such a transition probability matrix has a finite number M of states and is further ergodic, then $\pi_k = 1/M$, $k = 1, 2, \ldots, M$. Thus in the steady state all the states are equally probable. π_k's are defined in Section 13.8.5 below.

These concepts regarding the classification of states are illustrated here through the example of the random walk (Example 13.8.1). It can be verified that $f_{11} = f_{12} = f_{21} = f_{22} = 1$, $\mu_1 = \frac{5}{2}$, and $\mu_2 = \frac{5}{3}$. Since $f_{11} = 1$ and $f_{22} = 1$, the two states E_1 and E_2 are recurrent. Further, $\mu_1 < \infty$ and $\mu_2 < \infty$, so that E_1 and E_2 are positive recurrent. Further, as $n \to \infty$, $p_{11}^{(n)} \to \frac{2}{5}$, $p_{12}^{(n)} \to \frac{3}{5}$, $p_{21}^{(n)} \to \frac{2}{5}$, and $p_{22}^{(n)} \to \frac{3}{5}$. This can also be seen to be true from the general analytical expression for P^n.

13.8.5 LONG-RUN PROPERTIES OF MARKOV CHAINS

The n-step transition probabilities were earlier defined as $p_{jk}^{(n)}$. As n gets large, is there a limiting behavior of $p_{jk}^{(n)}$? Do they tend toward some number independent of the initial state? For irreducible, ergodic Markov chains, we can show the following:

For an irreducible, ergodic Markov chain, $\lim_{n\to\infty} p_{jk}^{(n)}$ exists and is independent of j

$$\lim_{n\to\infty} p_{jk}^{(n)} = \pi_k$$

where the π_k's uniquely satisfy the following steady state equations:

$$\pi_k > 0$$

$$\pi_k = \sum_r \pi_r p_{rk}$$

$$\sum_k \pi_k = 1$$

The π_k's are called the "steady state probabilities" of the Markov chain and are equal to the reciprocal of the mean recurrent times. We have

$$\pi_k = \mu_k^{-1}, \qquad k = 0, 1, \ldots$$

The probability distribution $\{\pi_k\}$ is called the "invariant" or "stationary" distribution of the given Markov chain. The invariance refers to the fact that the π_k's do not depend on the initial distribution. The name "stationary distribution" derives from the fact that the π_k's satisfy

$$\pi_k = \sum_r \pi_r p_{rk}^{(n)} \qquad \text{for every } n = 1, 2, \ldots$$

In the literature on Markov chains, the words "steady state," "invariance," "stationary," and "equilibrium" are synonymously used.

An irreducible, aperiodic chain possesses an invariant probability distribution $\{\pi_k\}$ if and only if it is ergodic. In this case $\pi_k > 0$, and all absolute probabilities $a_k^{(n)}$ tend to π_k as $n \to \infty$, irrespective of the initial distribution. The distribution $\{\pi_k\}$ is also called the "equilibrium distribution."

Again taking the example of the random walk, we see that the chain is ergodic. The chain therefore has an invariant distribution

$$\pi_1 = \mu_1^{-1} = \tfrac{2}{5}$$

$$\pi_2 = \mu_2^{-1} = \tfrac{3}{5}$$

and this satisfies the properties of the stationary distribution.

A summary of results on the classification of states and the existence of invariant distributions is shown in Exhibit 13.8.2.

Exhibit 13.8.2 Classification of an Aperiodic, Irreducible Markov Chain

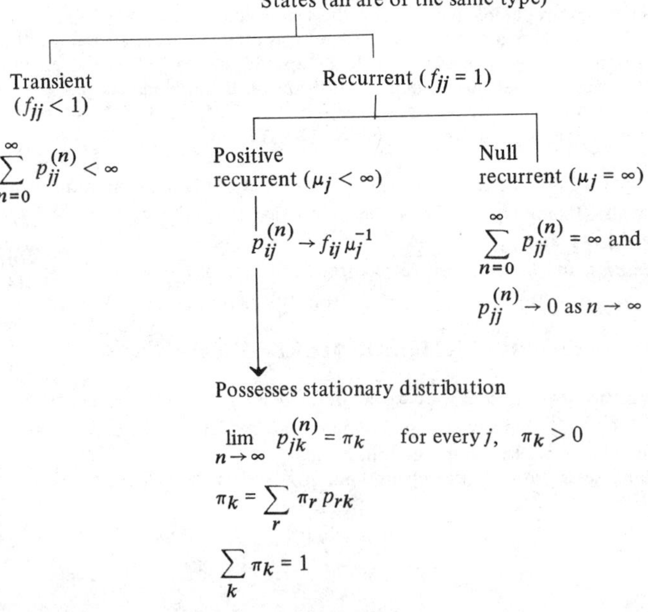

13.8.6 CONTINUOUS-TIME MARKOV CHAINS

A stochastic process is an indexed collection of random variables $\{X_t\}$, where t runs through a given set T. Usually t denotes time. In Markov chains the time parameter t is discrete, and t takes values $t = 1, 2, 3, \ldots$. If the parameter t varies continuously, we can define continuous-parameter Markov chains. An example follows.

Example 13.8.8 (see Chapter 13.7) There are M machines that are serviced by a single mechanic. A machine that breaks down is attended to by the mechanic unless he is servicing another machine. Thus a waiting line of machines to be serviced may be formed. The system is said to be in state m if m machines are not working, $m = 0, 1, \ldots, M$. The number of machines not working at time t is denoted by X_t. Under certain conditions, X_t forms a continuous-parameter Markov chain.

Denote by $X(t)$ for $t \geqslant 0$ the continuous-parameter Markov chain with $(M + 1)$ discrete states, $0, 1, \ldots, M$, and stationary transition probability function

$$p_{ij}(t) = P\{X(t + s) = j \mid X(s) = i\} \qquad \text{for } i, j = 0, 1, \ldots, M$$

The one-step transition probabilities played a major role in describing the discrete-time Markov chains. In the continuous-parameter case, a similar role is played by the transition intensities. The "intensity of transition" u_{ij} to state j from the state i is defined by

$$u_{ij} = \lim_{t \to 0} \frac{p_{ij}(t)}{t}$$

if the limit exists.

One can also consider Markov processes in continuous time with continuous state space.

13.8.7 ESTIMATION OF TRANSITION PROBABILITY MATRIX

In practical applications, the transition probability matrix of a Markov chain is not known and has to be estimated by observing the Markov chain. Suppose we observe N transitions of a Markov chain and let n_{ij} denote the number of times the transition took place from E_i to E_j. If there are M states, the sample transition count matrix is

$$\begin{bmatrix} n_{11} & n_{12} & \cdots\cdots & n_{1M} \\ n_{21} & n_{22} & \cdots\cdots & n_{2M} \\ \vdots & \vdots & & \vdots \\ n_{M1} & n_{M2} & & n_{MM} \end{bmatrix}$$

If $P = (p_{jk})$ is the transition probability matrix of the Markov chain under consideration, an estimate, $\hat{p}_{jk}$, of p_{jk} is given by

$$\hat{p}_{jk} = \frac{n_{jk}}{\sum_{r=1}^{M} n_{jr}}, \qquad \begin{matrix} j = 1, \ldots, M \\ k = 1, \ldots, M \end{matrix}$$

which is the sample proportion of transition counts. This estimate is the maximum likelihood estimate of p_{jk}.

13.8.8 FURTHER TOPICS

More Than One-Step Dependence

For a Markov chain we typically have one-step dependence. However, there are stochastic processes in which the dependence goes back more than one time unit, and in such cases we can reduce such a process to a Markov chain by approximately redefining the state space. See Cox and Miller[2] for more details.

Multivariate Markov Processes

Multivariate, discrete Markov processes can also be treated in a similar way by reducing the processes to a Markov chain by approximately redefining the state space.

Multidimensional Processes

Essentially, the theory developed for unidimensional Markov chains is applicable for Markov processes in which the state space is multidimensional. Following is a situation in which this model is applicable.

Example 13.8.9 Multidimensional Chain. Johnson Fastners is an industrial concern that produces bolts and nuts for different purposes. The company was operating on the hot forging process to produce 1 in. bolts. Later, it came to know that cold forging may also be used to produce 1 in. bolts. Hence the company has ordered and installed the cold forging equipment in its industry. Because of the convenience of the process and the method of handling, the management preferred the cold forging process.

Of course, the management did not want to dispense with the hot forging process and wanted to keep it alive so as to meet emergency situations in case the cold forging unit became inoperative (due to mechanical and environmental failures). Whenever any particular unit is inoperative, it is made operative by means of maintenance operation. It is reasonable to assume that the duration of operation of either unit and the time spent in bringing them back to the original condition are random variables with exponential distributions.

Let $X(t) = [x(t), y(t)]$ denote the state of the system at any time t, where $x(t)$ denotes the status of the cold forging process and $y(t)$ that of the hot forging process. Note that $x(t)$ can be operative (o) under repair (r) and that $y(t)$ can be operative (o) in standby (s) under repair (r). Thus the possible states of the system are (o, s), (o, r), (r, o), and (r, r). It can be shown that, with the assumptions made earlier, $\{X(t), t \geqslant 0\}$ is a Markov chain on $S = \{(o, s), (o, r), (r, o), (r, r)\}$.

Nonstationary Markov Chains

In our earlier discussion, it was assumed that the transition probability matrix of the Markov chain remains the same at every step of the transition. Thus the n-step transition probability matrix $P^{(n)}$ was shown to be P^n. When the transition probability matrix does not remain the same at every step of the transition, we have a nonstationary Markov chain. If P_i denotes the transition probability matrix at the ith transition, $i = 1, \ldots, n$, then the n-step transition probability matrix $P^{(n)}$ is given by

$$P^{(n)} = P_1, P_2, \ldots, P_n$$

When all P_is are equal to P, we have the usual formula $P^{(n)} = P^n$.

REFERENCES

1. W. FELLER, *An Introduction to Probability Theory and Its Applications*, Vol. 1, Wiley, New York, 1968.
2. D. R. COX and H. D. MILLER, *The Theory of Stochastic Processes*, Wiley, New York, 1965.

BIBLIOGRAPHY

BHAT, U. N., *Elements of Applied Stochastic Processes*, Wiley, 1972.

BUSH, R. R., and F. MOSTELLER, *Stochastic Models for Learning*, Wiley, New York, 1955.

CINLAR, E., *Introduction to Stochastic Processes*, Prentice-Hall, Englewood Cliffs, NJ, 1975.

HILLIER, F. S., and G. J. LIEBERMAN, *Introduction to Operations Research*, Holden-Day, San Francisco, 1974.

HIROMI, N., K. H. AKIRAKOSHI, and I. MAMORU, "An Application of the Markov Process to a Marine Transportation and Storage System," in K. B. Haley, Ed., *Operations Research 75*, North-Holland, Amsterdam, 1976.

HOWARD, R. A., *Dynamic Programming and Markov Processes*, The MIT Press, Cambridge, MA, 1969.

HOWARD, R. A., *Dynamic Probabilistic Systems, Volume 1, Markov Models*, Wiley, New York, 1974.

KANNAN, D., *An Introduction to Stochastic Processes*, North-Holland, Amsterdam, 1979.

KARLIN, S., and H. M. TAYLOR, *A First Course in Stochastic Processes*, Academic Press, New York, 1975.

KEMENY, J. G., and L. J. SNELL, *Finite Markov Chains*, Van Nostrand Reinhold, New York, 1960.

PARZEN, E., *Stochastic Processes*, Holden-Day, San Francisco, 1962.

CHAPTER 13.9

Time Series Forecasting

ARNOLD L. SWEET

Purdue University

13.9.1 DESIRABLE PROPERTIES OF FORECASTS

A forecast is a statement about a future event. This event should be of interest to personnel who have the responsibility for taking action based upon its occurrence. In this chapter attention is focused only on events that are numerically valued, and mathematical models are presented that will produce forecasts based on historical data. Thus the forecast will be a number, and, when using statistical models, a confidence interval can also be computed.

To choose among the models to be presented, the following characteristics of the forecast should be considered: (1) the length of time (horizon) into the future for which the forecast is to be made and the length of time between observations of the phenomena of interest, (2) the accuracy desired, (3) the amount of data available and their stability, and (4) the cost of developing and implementing the model.

As the horizon increases, the accuracy attainable with any model will decrease. If the data represent phenomena that are based on historical events that are stable, the collection of more data may produce more accurate forecasts. However, rapidly changing conditions may cause historical data to be of little value. It should also be noted that it is easier to produce accurate forecasts from aggregated data than from sparse data.

The models presented make use only of previously observed values of the variable whose value is to be forecast. Such forecasts have the advantage that the data are usually available at low cost. In addition, one feels that the data may contain considerable information about the future behavior of the phenomenon under study and should provide fairly accurate forecasts for short horizons. Models in which the variable to be forecast is assumed to be related to other variables are generally more expensive to develop, but are important for longer forecast horizons. One such class of models is that of regression models, discussed in Chapter 13.6. However, the use of such models requires that forecasts be made for each of the related variables. Thus univariate methods of forecasting must be developed before multivariate models are to be utilized. Forecasting models used in situations where numerical data are not available in sufficient quantity, or when forecasts must be made for very long time horizons, are discussed in Chapter 11.1.

13.9.2 TIME SERIES

Mathematical Framework for Time Series

If one plots the data as a function of time (a procedure that should always be followed before any computations are made), the forecaster may be tempted to try to fit the data with a deterministic function of time. Invariably, it will be seen that the curve will not pass through all of the data points, and thus a statistical approach is desirable so that an analytic statement concerning the differences between the data and the curve may be made. Thus it is necessary to invoke the use of probability theory. In this section a mathematical framework is presented that can be used to represent the variable to be forecast.

Let z_t, $t = 1, 2, \ldots, T$ be a collection of random variables. The independent variable t denotes time, and observations of z_t constitute the data to be used for the variable to be forecast. Thus the sequence of random variables z_t, $t = 1, 2, \ldots$ is sometimes called a stochastic process. Let the collection of random variables have a joint PDF for every value of t, $t = 1, 2, \ldots$. Let $\hat{z}_T(\ell)$ denote the forecast of z_t when $t = T + \ell$, when the forecast is made at $t = T$. Thus the forecast is being made for a horizon $\ell (\ell > 0)$. Since $z_{T+\ell}$ is a continuous random variable, any forecast made will be incorrect, because the probability that $z_{T+\ell}$ takes exactly any particular value equals zero. Thus it is necessary to introduce a criterion of error and to choose a procedure that will minimize the expected criterion of error. Let the error, $\epsilon_T(\ell)$, be the difference between the forecast made at time T and the actual (future) observation

$$\epsilon_T(\ell) = \hat{z}_T(\ell) - z_{T+\ell} \tag{1}$$

Two error criteria that are commonly used are the expected absolute difference

$$E|\epsilon_T(\ell)| = E|\hat{z}_T(\ell) - z_{T+\ell}| \tag{2}$$

and the expected square difference (the mean square error)

$$E\,\epsilon_T^2(\ell) = E\,(\hat{z}_T(\ell) - z_{T+\ell})^2 \tag{3}$$

If the joint PDF for the time series is known, then the value of $\hat{z}_T(\ell)$ that minimizes equation 2 or 3 can be computed. For example, it has been proved[1] that the forecast that minimizes the mean square error is the conditional expectation of $z_{T+\ell}$, given $(z_1, z_2, \ldots, z_T)$. In practice, the joint PDF is not known in advance, and often the forecaster does not have enough data to do a hypothesis test on a candidate joint PDF. Thus it is common practice to utilize forecasting models based on the first two moments of the time series, since these can usually be estimated from a reasonable amount of data. The first moment of the time series is the mean, defined as $\mu_t = E\,z_t$, $t = 1, 2, \ldots$; the second moments are the covariances, defined as $\gamma_{t,s} = E(z_t - \mu_t)(z_s - \mu_s)$, $s, t = 1, 2, \ldots$. Note that the covariance is symmetric, that is, $\gamma_{s,t} = \gamma_{t,s}$, and that, when $s = t$, we have the variance denoted by $\sigma_t^2 = E\,(z_t - \mu_t)^2$, $t = 1, 2, \ldots$. For a time series of length T, there are T means and $T(T+1)/2$ covariances (including the T variances). In practice, this is still too large a number of parameters to estimate from the data.

Stationary Time Series

Consider a time series that has the property that its behavior as a function of time is invariant with respect to the time an observer begins to observe the series. Such a time series is called (covariance) "stationary" if the mean is a constant μ and the covariance is a function only of the difference in observation times, $j = t - s$,

$$\gamma_j = \gamma_{t,t+j}, \qquad j = 0, 1, 2, \ldots$$

Becuase of the symmetry of the covariance function, $\gamma_{-j} = \gamma_{t-j,\,t} = \gamma_j$. The difference in observation times is called the "lag," and the covariance with lag zero, γ_0, is the constant variance σ^2, which is thus, like the mean, independent of time. A time series that is not stationary will be called "nonstationary." A stationary time series with T observations will have one mean and T distinct covariances which must be estimated. This is still a large number of parameters, but if the covariance itself can be modeled by a small number of parameters, the estimation problem may then become a reasonable one.

When examining data, it is easier to use a nondimensional statistic in place of the covariance, called the "autocorrelation function" (ACF) for lag j, defined for stationary series by $\rho_j = \gamma_j/\gamma_0$, $j = 0, 1, \ldots$. It can be shown that $|\rho_j| \leq 1$.

If the forecaster believes the time series to be stationary, an estimator for the mean is the sample mean

$$\bar{z} = \sum_{t=1}^{T} \frac{z_t}{T} \tag{4}$$

where T is the number of observations of the time series. An estimator for the covariance with lag j is

$$\hat{\gamma}_j = \sum_{t=1}^{T-j} \frac{(z_t - \bar{z})(z_{t+j} - \bar{z})}{T}, \qquad j = 0, 1, 2, \ldots \tag{5}$$

and an estimator for the ACF with lag j is given by

$$\hat{\rho}_j = \frac{\hat{\gamma}_j}{\hat{\gamma}_0} \tag{6}$$

From Chapter 13.6 it is known that the correlation coefficient between two random variables is a measure of the ability of a linear regression model to forecast one of the random variables, given the observation of a value for the other, when using a mean square error criterion. Hence a plot of ρ_j versus j, called an "autocorrelogram," will give the forecaster a measure of the suitability of using a linear forecasting model for the two observations of the time series when they are separated by j time units. In general, as the lag increases, the magnitude of the ACF should go to zero.

The autocorrelograms for some models of stochastic processes follow.

White Noise

Let z_t, $t = 1, 2, \ldots$ be a sequence of independent, identically distributed random variables with expectation zero and variance equal to σ_w^2. Then $\gamma_0 = \sigma_w^2$, $\rho_0 = 1$ and $\rho_j = 0$ for j not zero. It will also be assumed that z_t is a normal random variable. Hereafter, white noise will be denoted by u_t. It can be seen that white noise is a stationary time series and that only one parameter must be estimated.

An Autoregressive Model

Let z_t be generated by the difference equation

$$z_{t+1} = \phi z_t + u_{t+1}, \qquad t = 0, 1, 2, \ldots \tag{7}$$

where z_0 is known and ϕ is a constant. Equation 7 can be solved by successive substitution, with the result that

$$z_t = z_0 \phi^t + \sum_{i=1}^{t} \phi^{t-i} u_i, \qquad t = 1, 2, \ldots \tag{8}$$

If $|\phi| < 1$, taking the expected value of both sides of equation 8 and letting t go to infinity yields $\mu = 0$. Similarly, it can be shown that, as t goes to infinity,

$$\rho_j = \phi^j, \qquad j = 0, 1, 2, \ldots \tag{9}$$

Thus, if $|\phi| < 1$, z_t is a stationary process, since the effect of the initial condition, z_0, becomes negligible. On the other hand, one can show that, if $|\phi| \geqslant 1$, z_t is nonstationary, since the variance of z_t grows without bound as t increases. This model has two parameters to be estimated, ϕ and σ_w^2. If $\phi = 1$, the model is a random walk.

A plot of the observations versus time will often reveal an obvious upward or downward drift in the data and hence a nonstationary series. However, sometimes even the plot may not make the situation clear. If a plot of $\hat{\rho}_j$ versus j (called a "sample autocorrelogram") is made using equations 4, 5, and 6, the rate at which $\hat{\rho}_j$ decreases with j gives a good indication of stationarity. If $\hat{\rho}_j$ does not decrease as fast as $\hat{\rho}_j$ does in equation 9, a nonstationary model will almost surely have to be used. A method of removing nonstationarity from a set of observations is to form a new series by taking differences. Thus the new series to be formed is $w_t = z_t - z_{t-1}$. If w_t is nonstationary, it can be differenced. Differencing d times will remove a polynomial of order d in time. If the nonstationarity is due to other causes, such as $|\phi| > 1$ in equation 7, differencing will not make the series stationary. However, note that, if $\phi = 1$, differencing once will result in a stationary series.

Further Comments

Two further remarks about covariance stationary time series are appropriate. First, since a multivariate, normal joint PDF is defined by its mean and covariance matrices, one can always try to fit a normal model to a set of data for which one has chosen a model for the first and second moments. Second, the Fourier transform of the ACF for stationary series is called the "spectral density" and can be useful when choosing a model. The use of the spectral density yields a nonparametric model, and it can be estimated directly from the data. Model interpretation that takes place in the frequency, rather than the time, domain is too complex to be presented here, and the interested reader is referred to other sources listed in the bibliography.

13.9.3 EMPIRICAL METHODS

Given a data set, the ideal procedure for forecasting would be to choose a model; carry out a statistical procedure that accepted the model at some reasonable acceptance level; and then, using the model for forecasting, produce not only the forecast, but also a confidence interval for its value. However, in many situations the data set is too small to carry out such a procedure with any hope of obtaining significant values for the statistical tests. In addition, historical conditions may be changing so quickly that the past is not believed to be a good indicator of the future. In either of these two situations, a method that produces a forecast that tracks the data is desirable. Such methods are of necessity empirical, since no model for the time series is known. Some of the commonly used methods are discussed here.

Forecasting Using Moving Averages

A simple moving average (MA) model uses the arithmetic average of the N most recent observations as the forecast. The forecast is assumed valid for any horizon. Thus, for a forecast made at time t,

$$\hat{z}_t(\ell) = \sum_{i=1}^{N} \frac{z_{t-i+1}}{N}, \qquad N \geqslant 1, \qquad t \geqslant N, \qquad \ell \geqslant 1$$

Only the last N observations need to be retained, and at least N data points are needed to make the first forecast. As each new observation becomes available, the new forecast can be expressed in terms of the new observation and the most recent forecast. This expression is known as the "equation that updates the forecast." For the MA model, this equation, which can be used recursively, is

$$\hat{z}_t = \hat{z}_{t-1} + \frac{z_t - z_{t-N}}{N}, \qquad t > N \tag{10}$$

and the procedure begins with

$$\hat{z}_N = \sum_{i=1}^{N} \frac{z_i}{N}$$

Equation 10 has been applied to the data shown in Exhibit 13.9.1, using a value of $N = 3$. Thus $\hat{z}_3$ is the average of the first three data points, and $\hat{z}_4$ is computed using equation 10 with $t = 4$.

$$\hat{z}_4 = 66.7 + \frac{(51 - 40)}{3} = 70.3$$

This procedure is repeated until $t = T = 9$. If, at $t = 9$, it is desired to forecast z_{12}, the result is $\hat{z}_9(3) = 86.0$. When a new data point for $t = 10$ becomes available, this forecast could be updated to $\hat{z}_{10}(2) = 81.0$. In this example and in the ones to follow, it will always be assumed that the desired forecast is to be made at the end of the data set ($t = T = 9$), for a horizon of $\ell = 3$. However, as in Exhibit 13.9.1, the one-step-ahead forecasts, $\hat{z}_t(1)$, will also be computed, since each one-step-ahead forecast can be compared to the next data point, z_{t+1}, to evaluate the ability of the model to fit the data.

The simple MA model can be shown to be the minimum mean square error forecast if the time series model were the sum of a constant level plus white noise, and the forecast were limited to the use of only the last N observations.[2] Thus its use can be considered reasonable when tracking data with negligible trend. If the level of the data changes, the forecast will lag behind the change for a number of time intervals, and the larger N is, the longer the forecast will take to reach the new level. Thus large values of N tend to smooth out fluctuations, but also cause the forecast to respond slowly to a change in level.

If there appears to be a linear trend in the data, then a forecast based on double MAs can be constructed. Let M_t be the simple MA (computed using the right-hand side of equation 10) and let $M_t^{(2)}$ be the simple MA of M_t, given by

$$M_t^{(2)} = M_{t-1}^{(2)} + \frac{M_t - M_{t-N}}{N}, \qquad t \geqslant 2N \tag{11}$$

Assuming that the time series is a linear function of time, it can be shown (by substitution) that

$$\hat{z}_t(\ell) = 2\,M_t - M_t^{(2)} + \frac{2\ell(M_t - M_t^{(2)})}{N - 1}, \qquad t \geqslant 2N - 1 \tag{12}$$

will track a straight line. The procedure is illustrated using the data of Exhibit 13.9.1. The column labeled "$\hat{z}_t = M_t$" is the result of applying equation 10. The column labeled "$M_t^{(2)}$" is computed using values for M_t in equation 11. Then, both are used in equation 12 to compute a one-step-

Exhibit 13.9.1 Sample Computations for MA, Double Moving Average (DMA), ES, and Double ES (DES) Forecasting Models

		MA, $N = 3$	DMA, $N = 3$		ES, $\alpha = 0.10$	DES, $\alpha = 0.10$		
t	Data	$\hat{z}_t = M_t$	$M_t^{(2)}$	$\hat{z}_t(1)$	$\hat{z}_t$	S_t	$S_t^{(2)}$	$\hat{z}_t(1)$
1	40	–	–	–	40.0	26.8	–4.7	–
2	65	–	–	–	42.5	30.7	–1.1	66.0
3	95	66.7	–	–	47.8	37.1	2.7	75.3
4	51	70.3	–	–	48.1	38.5	6.3	74.3
5	55	67.0	68.0	65.0	48.8	40.1	9.7	74.0
6	87	64.3	67.2	58.5	52.6	44.8	13.2	80.0
7	124	88.7	73.3	119.3	59.7	52.7	17.1	92.3
8	65	92.0	81.7	112.7	60.3	54.0	20.8	90.8
$T = 9$	69	86.0	88.9	80.2	61.1	55.5	24.3	90.1
10	109	81.0	86.3	–	65.9	60.8	27.9	–

ahead forecast ($\ell = 1$). Thus, for example, $M_5^{(2)}$ is the average of M_3, M_4, and M_5, and

$$\hat{z}_5(1) = 2(67.0) - 68.0 + 2(1)(67 - 68)/2 = 65.0$$

As a new observation becomes available, the recursive equation for forecasting to the same time, $t + \ell$, in the future is

$$\hat{z}_t(\ell) = \hat{z}_{t-1}(\ell + 1) - \frac{2}{N-1}(M_{t-1} - M_{t-1}^{(2)}) + \frac{1}{N}\nabla^N(2z_t - M_t) + \frac{2\ell}{(N-1)N}\nabla^N(z_t - M_t), \qquad t \geqslant 2N \tag{13}$$

where $\nabla^N(\cdot)$ is a backward difference operator of order N, that is, $\nabla^N(z_t) = z_t - z_{t-N}$. Thus, to apply equation 13, it is necessary to store the previous N values of both the data and the simple MA series as well as the last forecast. For example, using equation 12 gives $\hat{z}_9(3) = 74.4$. When the next data point becomes available, M_{10} is computed, and equation 13 yields $\hat{z}_{10}(2) = 65.0$.

Forecasting Using Exponential Smoothing

The storage requirements of MA models may be prohibitive if a great many forecasts are to be made on a frequent basis. A forecasting model that requires storage of only the last forecast is the exponential smoothing (ES) model, where the forecast made at t is computed from the recursive equation

$$\hat{z}_t = \alpha z_t + (1 - \alpha)\hat{z}_{t-1}, \qquad t \geqslant 2, \qquad 0 < \alpha < 1 \tag{14}$$

The forecast is seen to depend on the latest observation and the last forecast and applied to all horizons $\ell \geqslant 1$, so ℓ is suppressed in the notation. To start the forecasting procedure, one can set $\hat{z}_1$ equal to the first observation z_1. If T observations are available, one can apply equation 14 sequentially until $t \equiv T$, and thus the forecast for all values of $t = T + \ell$ is available. A better procedure is to let $\hat{z}_1$ equal the sample mean of the T observations and then apply equation 14 sequentially until $t = T$. The solution of equation 14 is

$$\hat{z}_T = (1 - \alpha)^{T-1}\hat{z}_1 + \alpha \sum_{i=0}^{T-2}(1 - \alpha)^i z_{T-i}, \qquad T \geqslant 2 \tag{15}$$

and shows that the forecast is a weighted sum of the past observations. The weighting factor is a power of $1 - \alpha$ and causes the contribution of $\hat{z}_1$ to the forecast to become negligible as T increases. The variable α is called the "smoothing constant." Small values of α cause the forecast to respond slowly to large changes in the data. Generally, values of α between 0.01 and 0.30 are used. A method of choosing a value for α that has been suggested[3] is as follows:

Let the residual at t, defined by $\epsilon_t = z_t - \hat{z}_t$ be the error between the actual observation and the fit to the time series as computed by a forecasting model. Let the sum of squared errors, S, be defined as

$$S = \sum_{t=1}^{T} \epsilon_t^2$$

Then, S can be computed for various choices of α, and the value of α that minimizes S can be used.

It can be shown[4] that, if the time series model were the sum of a constant plus white noise, then the ES model minimizes the sum of discounted squared errors

$$DS = \sum_{t=1}^{T} (1-\alpha)^{T-t} (z_t - \hat{z}_T)^2 \tag{16}$$

It is also true[5] that the ES model minimizes the sum of squared errors when the nonstationary time series model is given by $z_t = z_{t-1} + u_t - \theta u_{t-1}$ for $|\theta| < 1$.

The result of applying equation 14 to the data of Exhibit 13.9.1, using $\alpha = 0.10$ and $\hat{z}_1 = z_1 = 40$, is shown in the column labeled "ES." Thus equation 14 yields

$$\hat{z}_2 = 0.1(65) + 0.9(40) = 42.5$$

Continued use of equation 14 yields $\hat{z}_9(3) \equiv \hat{z}_9 = 61.1$, and when a new data point at $t = 10$ becomes available, the updated forecast becomes $\hat{z}_{10}(2) \equiv \hat{z}_{10} = 65.9$.

If $\hat{z}_1$ were chosen to be the average of the nine data points, the forecasts would be quite different, since the upward trend of the data would yield a larger value for $\hat{z}_1$ than z_1 did. This illustrates that initial values can greatly affect the forecasts when data sets are small.

If the time series model is believed to have a linear trend, then a double ES model can be used. Let S_t obey the equation for ES,

$$S_t = \alpha z_t + (1-\alpha)S_{t-1}, \qquad t \geqslant 2 \tag{17}$$

where

$$S_1 = \frac{z_1 - (1-\alpha)z_2}{\alpha}$$

and let $S_t^{(2)}$ be a smoothing of S_t, namely,

$$S_t^{(2)} = \alpha\, S_t + (1-\alpha)\, S_{t-1}^{(2)}, \qquad t \geqslant 2 \tag{18}$$

with

$$S_1^{(2)} = \frac{(2-\alpha)z_1 - 2(1-\alpha)z_2}{\alpha}$$

Then the forecast model is given by

$$\hat{z}_t(\ell) = \left(2 + \frac{\alpha\ell}{\beta}\right) S_t - \left(1 + \frac{\alpha\ell}{\beta}\right) S_t^{(2)}, \qquad t \geqslant 2 \tag{19}$$

where $\beta = 1 - \alpha$. As in the case of the simple ES forecast, one can sequentially apply equations 17 and 18 to the data until $t = T$ and then make the forecast for $t = T + \ell$.

A better procedure to find initial values for use in equations 17 and 18 is to fit a straight line to the data using a linear regression model. Then, using the fitted line, solve for z_1 and z_2 and use them in equations 17 and 18.

To update a forecast made at $t = T$ for $t = T + \ell$ as a result of a new data point z_{T+1}, all that is necessary is to compute S_{T+1} using equation 17, $S_{T+1}^{(2)}$ using equation 18, and then apply equation 19 with $t = T + 1$ and ℓ replaced by $\ell - 1$. To illustrate the procedure, the first nine data points shown in Exhibit 13.9.1 were fit by a linear regression model. The result was $\hat{z}_t = 54.83 + 3.50t$. Using $\hat{z}_1 = 58.33$ and $\hat{z}_2 = 61.83$ in equations 17 and 18, and letting $\alpha = 0.10$, S_1 and $S_1^{(2)}$ were found, and their values are shown in Exhibit 13.9.1. Equations 17 and 18 were then used sequentially until $t = T = 9$. The one-step-ahead forecasts were computed using equation 19. The forecast $\hat{z}_9(3) = 97.0$ was found using equation 19, and the updated forecast $\hat{z}_{10}(2) = 101.0$ was also made using equation 19 after S_{10} and $S_{10}^{(2)}$ were found.

Forecasting Using Generalized Exponential Smoothing

It may be the case that the forecaster is not satisfied with a constant or linear trend model and wishes to use other functions of time. This may be so when the data are seasonal or when there is a desire to forecast a turning point in the time series. A class of models due to Brown[6] is presented now.

It is assumed that the time series can be represented as

$$z_t = \sum_{i=1}^{n} b_i z_i(t) + u_t, \qquad t = 1, 2, \ldots \tag{20}$$

where the $z_i(t)$ are a class of functions that are described shortly and the b_i are constants. One of the features of Brown's development is that the time origin for the functions of time used in the model will move so as always to be placed at the most recent observation, $t = T$. Thus the model to be used in the computations is written as

$$z_t = \sum_{i=1}^{n} a_i(T)\, z_i(t - T), \qquad t = 1, 2, \ldots \tag{21}$$

where the $a_i(T)$ are coefficients to be estimated from the data. Note that, when $t = T$, which is the most recent observation, $z_i(t - T) = z_i(0)$, so that these functions are at their time origin of zero. Because of the moving origin, the coefficients $a_i(T)$ are functions of time. A value for the parameter n is chosen by the forecaster, depending on the number of functions believed necessary to fit the data. The $a_i(T)$ are estimated by choosing values that minimize the sum of discounted square errors (see equation 16). It is convenient to proceed using matrix notation. In what follows, it is assumed that at least n observations of the time series are available. The forecasting model resulting from solving the minimization problem can be written as

$$\hat{z}_t(\ell) = \mathbf{z}'(\ell)\, \hat{\mathbf{a}}(t), \qquad t \geqslant n \tag{22}$$

where prime denotes matrix transposition, $\mathbf{z}(\ell)$ denotes the $n \times 1$ column vector whose ith element is $z_i(\ell)$, and $\hat{\mathbf{a}}(t)$ denotes the $n \times 1$ column vector whose ith element is $\hat{a}_i(t)$. The $\hat{\mathbf{a}}(t)$ satisfies the matrix equation

$$\hat{\mathbf{a}}(t) = \mathbf{L}'\hat{\mathbf{a}}(t-1) + \mathbf{G}^{-1}\mathbf{z}(0)\,[z_t - \hat{z}_{t-1}(1)], \qquad t > n \tag{23}$$

where -1 denotes matrix inversion. The functions $z_i(t)$ must satisfy the equation

$$\mathbf{z}(t+1) = \mathbf{L}\,\mathbf{z}(t) \tag{24}$$

where $\mathbf{L}$ is an n by n matrix and $\mathbf{G}$ is defined by

$$\mathbf{G} = \sum_{j=0}^{\infty} \beta^j \mathbf{z}(-j)\mathbf{z}'(-j), \qquad 0 < \beta < 1$$

where $\beta = 1 - \alpha$. By proper choice of $\mathbf{L}$ and $\mathbf{z}(0)$, it is possible to generate functions of time that are polynomial, trigonometric (for seasonal series), and exponential as well as products of such functions. Once having chosen $\mathbf{z}(t)$, $\mathbf{G}$ can be computed, and, for any choice of the smoothing constant α, $\mathbf{G}$ can be numerically inverted. Fortunately, the matrices $\mathbf{L}$ and $\mathbf{G}$ are available for many different functions of time.[4,7] The equation for updating the forecast is

$$\hat{z}_t(\ell) = \hat{z}_{t-1}(\ell + 1) + \mathbf{z}'(\ell)\mathbf{G}^{-1}\mathbf{z}(0)\,[z_t - \hat{z}_{t-1}(1)] \tag{25}$$

To start the forecasting procedure, one approach is to use the first n data points and solve the equation

$$\hat{\mathbf{a}}(n) = \mathbf{Z}^{-1}\mathbf{z} \tag{26}$$

where $\mathbf{z}$ is an $n \times 1$ column vector whose ith component is the observation z_i and $\mathbf{Z}$ is an n by n matrix whose (i, j)th element is $Z_{ij} = z_j(i - n)$. Having computed $\hat{\mathbf{a}}(n)$, $\hat{z}_n(1)$ can be computed using equation 22, $\mathbf{a}(n + 1)$ using equation 23, and $\hat{z}_{n+1}(1)$ using equation 22. Then equations

22 and 23 are used recursively with $\ell = 1$ until $t = T$. Another approach is to use a least squares fit of equation 21 to the T data points. Then, using the fitted equation, compute the vector **z** in equation 26 and also compute $\hat{z}(n+1)$, which is equal to $\hat{z}_n(1)$. Using equation 26 yields $\hat{\mathbf{a}}(n)$, and so the recursive procedure can begin. If n data points are not available, and instead one has chosen a function z_t to begin the recursive procedure, then this function can be used to compute the vector **z** in equation 26.

To illustrate the procedure, consider the quadratic model

$$z_t = a_1 + a_2 t + \frac{a_3 t(t-1)}{2} + u_t \tag{27}$$

Letting $\mathbf{z}(t) = [1, t, t(t-1)/2]'$ and

$$\mathbf{L} = \begin{bmatrix} 1 & 0 & 0 \\ 1 & 1 & 0 \\ 0 & 1 & 1 \end{bmatrix}$$

it can be seen by successive substitution that equation 24 generates the quadratic function of time given in equation 27. (Note that the form chosen in equation 27 and the accompanying **L** are not the only choices available for generating a quadratic function of t.) The symmetric matrix **G** is

$$\mathbf{G} = \begin{bmatrix} \alpha^{-1} & -\beta\alpha^{-2} & \beta\alpha^{-3} \\ & \beta(1+\beta)\alpha^{-3} & -\beta(1+2\beta)\alpha^{-4} \\ & & \beta(1+4\beta+\beta^2)\alpha^{-5} \end{bmatrix} \tag{28}$$

and

$$\mathbf{Z} = \begin{bmatrix} 1 & -2 & 3 \\ 1 & -1 & 1 \\ 1 & 0 & 0 \end{bmatrix} \tag{29}$$

As an example, consider the fitting of the model given in equation 27 to the data set given in Exhibit 13.9.1. A least squares fit of equation 27 to the first nine data points, using an origin at $t = 9$, yielded $\hat{z}_t = 73.815 - 8.545\ (t-9) - 1.361\ (t-9)(t-10)$. Thus $\mathbf{z}' = (\hat{z}_1, \hat{z}_2, \hat{z}_3) = (44.18, 57.41, 67.92)$. Substitution of $\mathbf{z}'$ into equation 26, and using equation 29, yielded $\hat{\mathbf{a}}(3) = (67.92, 7.79, -2.72)'$. Equation 22 (or $\hat{z}_4$ above) yielded the value of $\hat{z}_3(1)$ shown in Exhibit 13.9.2. Letting $\alpha = 0.10$ in equation 28, it was found that $\mathbf{G}^{-1}\ \mathbf{z}(0) = (0.2710, 0.0290, 0.0010)'$, and thus equation 23 was used to compute $\hat{\mathbf{a}}(4)$, whose components are shown in Exhibit 13.9.2. The use of equations 22 and 23 was continued until $t = T = 9$. Equation 22 then yielded $\hat{z}_9(3) = 38.4$, and using the observation at $t = 10$ and equation 25 yielded $\hat{z}_{10}(2) = 53.3$.

If the forecaster who uses generalized ES is willing to assume that the model of equation 20 is the correct one, then the properties of the white noise component, u_t, may be used to compute confidence intervals for the forecast. Let F be the matrix resulting from the replacement of β by β^2 (and hence α by $1-\beta^2$) in G. Then the variance of the forecast error for horizon ℓ (see equation 1) is given by

$$\sigma_\epsilon^2(\ell) = \sigma_w^2\ [1 + \mathbf{z}'(\ell)\ \mathbf{G}^{-1}\ \mathbf{F}\ \mathbf{G}^{-1}\ \mathbf{z}(\ell)] \tag{30}$$

Exhibit 13.9.2 Sample Computations for Fitting a Quadratic Function of Time Using Generalized ES, $\alpha = 0.10$

t	Data	$\hat{a}_1(t)$	$\hat{a}_2(t)$	$\hat{a}_3(t)$	$\hat{z}_t(1)$
1	40	–	–	–	–
2	65	–	–	–	–
3	95	67.92	7.79	-2.72	75.7
4	51	69.01	4.35	-2.75	73.4
5	55	68.39	1.07	-2.76	69.5
6	87	74.21	-1.19	-2.75	73.0
7	124	86.84	-2.47	-2.71	84.4
8	65	79.12	-5.71	-2.73	73.4
$T = 9$	69	72.22	-8.56	-2.72	63.6
10	109	–	–	–	–

As an example, for the ES model of equation 14, which corresponds to $n = 1$ and $z_1(t) = 1$ in equation 20, equation 30 yields

$$\sigma_\epsilon^2(\ell) = \frac{2\sigma_w^2}{2 - \alpha}$$

It can be shown (under the preceding conditions) that the forecast is an unbiased estimator of the time series. Thus the probability that an observation lies between two limits, a and b, is given by

$$P(a < z_{T+\ell} \leqslant b) = P\left(\frac{a - \hat{z}_T(\ell)}{\sigma_\epsilon(\ell)} < \frac{\epsilon(\ell)}{\sigma_\epsilon(\ell)} \leqslant \frac{b - \hat{z}_T(\ell)}{\sigma_\epsilon(\ell)}\right) \tag{31}$$

Now, since u_t is a normal random variable, $\epsilon_T(\ell)/\sigma_\epsilon(\ell)$ is a normal (0, 1) random variable, and the probability given in equation 31 can be computed if a numerical value of $\sigma_\epsilon^2(\ell)$ were available. Unfortunately, u_t is not directly observable, so σ_w^2 cannot be directly estimated and used in equation 30. However, the difference between the one-step-ahead forecast and the actual observation, denoted by $\epsilon_t(1)$, is an estimator of the residual ϵ_{t+1}. Thus an estimator of $\sigma_\epsilon^2(1)$ is given by

$$\hat{\sigma}_\epsilon^2(1) = \sum_{t=n}^{T-1} \frac{(\hat{z}_t(1) - z_{t+1})^2}{T - n} \tag{32}$$

Then equation 30, with $\ell = 1$, can be solved for σ_w^2, and this value can be used in equation 30 with $\ell > 1$.

An alternative procedure is to use ES to estimate the absolute value for the forecast error, $\hat{d}_t$, by using $\hat{d}_t = \alpha|\epsilon_t(1)| + (1 - \alpha)\hat{d}_{t-1}$. The forecast $\hat{d}_t$ converges to $(2/\pi)^{1/2}\sigma_\epsilon(1)$ and can thus be used to estimate $\sigma_\epsilon(1)$. Using the model and the data, the residuals can be used to estimate an initial value, $\hat{d}_1$, by taking the average of the absolute value of the residuals.

The forecaster should be warned that, if an incorrect model is chosen, then the residuals are not uncorrelated, and the preceding methods may result in considerable error in estimating $\sigma_\epsilon^2(\ell)$.

Returning to the results in Exhibit 13.9.2, the variance of the forecast errors can be found using equation 30 (assuming that the quadratic model was the best fit). Using equation 32 and the results shown in Exhibit 13.9.2 yields $\hat{\sigma}_\epsilon^2(1) = 708.6$. Letting $\ell = 1$ in equation 30, it was found that $\hat{\sigma}_w^2 = 573.0$. Then, using equation 30 again, it was found that $\hat{\sigma}_\epsilon^2(3) = 761.1$ for $\hat{z}_9(3)$ and $\hat{\sigma}_\epsilon^2(2) = 733.2$ for $\hat{z}_{10}(2)$. Thus, for example, equation 31 yields

$$P(10 < z_{12} < 70) = P(-1.03 < \frac{\epsilon(\ell)}{\sigma_\epsilon(\ell)} \leqslant 1.15) = .80$$

Forecasting Series With Seasonal Components

The demand for many products is known to have seasonal peaks and valleys, and such seasonal phenomena are often made apparent when inspecting a plot of the data. If desired, generalized ES models can be used to generate sums of trigonometric terms.[7] However, other models are available.

Forecasting Using Winters' Model

A model due to Winters[8] is given by $z_t = (b_1 + b_2 t)\,c(t) + u_t$, where $c(t)$ is a multiplicative seasonal factor. If there is no upward trend in the data, then b_2 may be taken to be zero. Letting the length of the season be L time units, a plot of z_t versus t should exhibit a pattern in which the peaks and valleys recur at intervals of L units of time. The seasonal factors are constrained to obey the conditions

$$\sum_{t=r}^{r+L-1} c(t) = L \tag{33}$$

and $c(t) = c(t + sL)$. The forecast is given by

$$\hat{z}_t(\ell) = [\hat{a}_1(t) + \hat{a}_2(t)\,\ell]\,\hat{c}(t + \ell - L), \qquad t \geqslant L \tag{34}$$

and the recursive relations for the parameters are

$$\hat{a}_1(t) = \frac{\alpha z_t}{\hat{c}(t-L)} + (1-\alpha)\,[\hat{a}_1(t-1) + \hat{a}_2(t-1)], \qquad t \geq 2 \tag{35}$$

$$\hat{a}_2(t) = \delta\,[\hat{a}_1(t) - \hat{a}_1(t-1)] + (1-\delta)\,\hat{a}_2(t-1), \qquad t \geq 2 \tag{36}$$

and

$$\hat{c}(t) = \frac{\gamma z_t}{\hat{a}_1(t)} + (1-\gamma)\,\hat{c}\,(t-L), \qquad t > L \tag{37}$$

In equations 35 through 37, α, δ, γ are smoothing constants which may all be different. In equation 34 it may be necessary to use the last available estimate of $c(\bullet)$ if $\ell > L$. To find initial values of the $L + 2$ parameters, the following procedure requires the use of at least the first $2L$ observations: Let $t = sL + u$, $0 < u \leq L, s \geq 0$. If mL observations are to be used ($m > 1$), the equations to be solved are

$$\bar{a}_s = \sum_{t=sL+1}^{(s+1)L} \frac{z_t}{L}, \qquad 0 \leq s < m \tag{38}$$

$$\hat{a}_2(1) = \frac{\bar{a}_{m-1} - \bar{a}_0}{(m-1)L} \tag{39}$$

$$\hat{a}_1(1) = \bar{a}_0 - 0.5L\hat{a}_2(1) \tag{40}$$

$$\hat{c}_t = \frac{z_t}{\bar{a}_s - [0.5(L+1) - u]\hat{a}_2(1)}, \qquad 1 \leq t \leq mL \tag{41}$$

$$\bar{\hat{c}}_t = \sum_{s=0}^{m-1} \frac{\hat{c}_{t+sL}}{m}, \qquad 1 \leq t \leq L \tag{42}$$

and

$$\hat{c}(t) = \bar{\hat{c}}_t L \left(\sum_{t=1}^{L} \bar{\hat{c}}_t \right)^{-1}, \qquad 1 \leq t \leq L \tag{43}$$

Equations 35 through 37 can now be solved recursively until the last data point is reached, and forecasts can then be made. After every new cycle of $c(t)$'s is estimated, it should be normalized to satisfy equation 33 by use of equation 43.

Consider the data set of Exhibit 13.9.1 again, and let $L = 4$. Using only the first observations, it follows that $m = 2$. Equation 38 yields

$$\bar{a}_0 = \sum_{t=1}^{4} \frac{z_t}{4} = 62.75$$

and

$$\bar{a}_1 = \sum_{t=5}^{8} \frac{z_t}{4} = 82.75$$

and then equations 39 and 40 yield $\hat{a}_2(1) = 5.00$ and $\hat{a}_1(1) = 52.75$. The use of equations 41 through 43 yields the values of $\hat{c}(t)$ shown in Exhibit 13.9.3. Applying equations 34 through 37 sequentially with $\alpha = \delta = \gamma = 0.10$ results in the rest of the values shown in the exhibit. The forecast $\hat{z}_9(3) = 79.5$, and the updated forecast, using the observation at $t = 10$ and equation 34, is $\hat{z}_{10}(2) = 79.6$.

Forecasting Using an Additive Model

Winters' model is a suitable candidate when the amplitude of the peaks and valleys appears to be increasing with time. If the amplitude appears not to be changing with time, a seasonal model that

Exhibit 13.9.3 Sample Computations for Fitting Winters' Model, $\alpha = \delta = \gamma = 0.10$

t	Data	$\hat{a}_1(t)$	$\hat{a}_2(t)$	$\hat{c}(t)$	$\hat{z}_t(1)$
1	40	52.75	5.00	0.73	–
2	65	57.97	5.02	1.08	–
3	95	63.20	5.04	1.46	–
4	51	68.45	5.06	0.73	53.6
5	55	73.69	5.08	0.73	85.5
6	87	78.92	5.10	1.09	122.6
7	124	84.11	5.11	1.46	64.7
8	65	89.25	5.11	0.72	69.0
$T = 9$	69	94.37	5.11	0.73	108.0
10	109	99.57	5.12	–	–

can be used is the additive seasonal model

$$z_t = b_1 + b_2 t + c(t) + u_t \tag{44}$$

where the seasonal factors $c(t)$ obey $c(t) = c(t + sL)$ and $\Sigma_{t=r}^{r+L-1} c(t) = 0$. The forecasting equation is

$$\hat{z}_t(\ell) = \hat{a}_1(t) + \hat{a}_2(t)\,\ell + \hat{c}(t + \ell - L), \qquad t \geqslant L \tag{45}$$

[where, if $\ell > L$, the last available estimate of $\hat{c}(\cdot)$ is used] and the recursive relations for the parameters are

$$\hat{a}_1(t) = \alpha[z_t - \hat{c}(t - L)] + (1 - \alpha)\,[\hat{a}_1(t - 1) + \hat{a}_2(t - 1)], \qquad t > L \tag{46}$$

$$\hat{a}_2(t) = \delta\,[\hat{a}_1(t) - \hat{a}_1(t - 1)] + (1 - \delta)\,\hat{a}_2(t - 1), \qquad t > L \tag{47}$$

and

$$\hat{c}(t) = \gamma[z_t - \hat{a}_1(t)] + (1 - \alpha)\,\hat{c}(t - 1), \qquad t > L \tag{48}$$

Using mL observations $(m > 1)$, the following equations, derived from a least square error fit of equation 44 to the data, will yield initial values for the parameters:

$$\hat{b}_1 = \frac{(4m + 1)L + 3}{m(m + 1)L^2} \sum_{t=1}^{mL} z_t - \frac{6(mL + 1)}{m(m^2 - 1)L^2} \sum_{s=1}^{m-1} s \sum_{t=1}^{L} z_{t+sL} \tag{49}$$

$$\hat{b}_2 = \frac{2[\Sigma_{t=1}^{mL} z_t - mL\hat{b}_1]}{mL(mL + 1)} \tag{50}$$

$$\hat{c}(t) = \frac{1}{m} \sum_{s=0}^{m-1} z_{t+sL} - \hat{b}_1 - \hat{b}_2 \left[t + \frac{L(m - 1)}{2}\right], \qquad 1 \leqslant t \leqslant L \tag{51}$$

$$\hat{a}_2(L) = \hat{b}_2 \tag{52}$$

and

$$\hat{a}_1(L) = \hat{b}_1 + L\hat{b}_2 \tag{53}$$

Applying equations 49 through 53 to the set of observations given in Exhibit 13.9.4, using $m = 2$ and $L = 4$, yields

$$\hat{b}_1 = \frac{39}{96} \sum_{t=1}^{8} z_t - \frac{54}{96} \sum_{t=1}^{4} z_{t+4} = 50.25$$

$$\hat{b}_2 = \frac{2[\Sigma_{t=1}^{8} z_t - 8\hat{b}_1]}{72} = 5.00$$

Exhibit 13.9.4 Sample Computations for Fitting an Additive Seasonal Model, $\alpha = \delta = \gamma = 0.10$

t	Data	$\hat{a}_1(t)$	$\hat{a}_2(t)$	$\hat{c}(t)$	$\hat{z}_t(1)$
1	40	–	–	-17.75	–
2	65	–	–	5.75	–
3	95	–	–	34.25	–
4	51	70.25	5.00	-22.25	57.5
5	55	75.00	4.98	-17.98	85.7
6	87	80.10	4.99	5.86	119.3
7	124	85.56	5.03	34.67	68.3
8	65	90.26	5.00	-22.25	77.3
9	69	94.43	4.92	-18.72	105.2
10	109	99.73	4.96	6.21	–

and the values shown in Exhibit 13.9.4. Sequential application of equations 45 through 48 were used to get the remaining values in the exhibit. The forecast $\hat{z}_9(3) = 86.6$ and the updated forecast $\hat{z}_{10}(2) = 87.1$ were both computed using equation 45.

No simple methods are known for estimating the variance of the forecast error for the two seasonal models just discussed. However, the method of generalized ES has been extended to include these two models,[9] allowing equation 30 to be used for computing the variance.

Model Evaluation

Large differences between the one-step-ahead forecast and the actual outcome can be used as a signal to the forecaster that something may be amiss. A confidence interval may be established for the forecast error (when the variance of the forecast error is known), and when one error or two consecutive errors are outside the interval, action may be required. This may involve changing the smoothing constants and/or the model. In situations where many forecasts are required, it may be preferable to have this done automatically.[10]

To choose one model from a set of likely candidates, one technique commonly used is to rank the models on the basis of how well each fits the data, using a sum of squared error or sum of absolute error criteria. However, it should be remembered that, in general (and in common with regression analysis), models with a greater number of parameters will give a better fit, but not necessarily a more statistically significant forecast.

Another common technique employed in choosing a model is to use part of the data to fit the model and then to use the model to forecast the rest of the data. A graphical comparison of the forecast and data, as well as a sum of error computation, may help the forecaster in making a choice.

Other Empirical Models

The decomposition method of forecasting is another empirical method and is used for data that are seasonal and cyclic in nature. The reader is referred to the work of Shiskin et al.[11] and McLaughlin and Boyle.[12]

The use of a leading indicator is often valuable in making forecasts. If the demand for an item depends on the production or demand of another product and lags by an interval of time that is approximately constant, observations of the demand for the leading product can be used for forecasting. Information on variables that are leading indicators can be found in the *Survey of Current Business*, published monthly by the U.S. Department of Commerce.

Listings of FORTRAN programs for the empirical methods discussed in this section are available.[2,13]

13.9.4 A STATISTICAL TECHNIQUE

A class of models called "autoregressive moving average" (ARMA) models is discussed. Statistical tests have been developed such that, after a particular model has been fitted to a set of observations, unlike the empirical models, it can be accepted or rejected at a specified significance level. Only an outline of the technique is presented here.

The ARMA Models

The class of stationary models to be discussed can be written as

$$\tilde{z}_t = \sum_{i=1}^{p} \phi_i \tilde{z}_{t-i} + u_t - \sum_{i=1}^{q} \theta_i u_{t-i} \tag{54}$$

where $\tilde{z}_t = z_t - \mu$, p and q are positive integers, ϕ_i (the AR coefficients) and θ_i (the MA coefficients) are constants, and u_t is white noise. This model is denoted as ARMA (p, q), and there are $p + q + 2$ constants to be evaluated (including μ and σ_w^2). If p or $q = 0$, all AR or MA coefficients are taken to be zero. There are restrictions on the values that the AR and MA coefficients can take. For example, it has been shown earlier that, if $p = 1$ and $q = 0$, $|\phi_1|$ must be less than 1 for z_t to be stationary. For a specified set of parameters, the ACF of z_t can be computed. Plots of the ACF for $p + q \leqslant 2$ and various ranges of the parameters can be found,[14] and a computer program is available for computing the ACF otherwise.[15] Thus, if a sample autocorrelogram is plotted and compared with various ACFs that can be generated by the ARMA models, some reasonable values for p and q can be chosen such that an ARMA (p, q) model will yield an ACF similar to it. The estimation of values for the AR and MA coefficients are discussed shortly. One difficulty encountered in this comparison process is that sampling errors cause the sample autocorrelogram to be "noisy," and hence the comparisons may be difficult. If the observations appear to be from a nonstationary process, then the comparisons cannot be carried out. Thus it may be necessary to take differences of the observations until the sample autocorrelogram appears to be from a stationary process. This leads to consideration of the next class of models.

Autoregressive Integrated Moving Average Models

Let a time series, w_t, be formed by taking the dth difference of the series z_t

$$w_t = \nabla^d z_t, \qquad d > 0, \qquad t = d + 1, d + 2, \ldots \tag{55}$$

where d is the minimum number of differences necessary to make w_t stationary. Then w_t will be modeled by an ARMA process. After the model for w_t is chosen, the original series is recovered by "integrating" equation 55, and thus z_t is called an "autoregressive integrated moving average process," denoted by ARIMA (p, d, q). For example, if $d = 1$, the integration of equation 55 yields

$$z_t = z_1 + \sum_{i=2}^{t} w_i, \qquad t = 2, 3, \ldots \tag{56}$$

Note that the integration of equation 55 will require d constants, which are evaluated by using the first d observations of the data series. It can be seen from equation 56 that, although w_t is stationary, z_t is nonstationary, since its variance must increase with t.

It has been shown that there is an equivalence between generalized ES models using polynomial trends and certain ARIMA models.[15]

Forecasting Using ARIMA Models

It can be shown[16] that the minimum mean square error forecast $\hat{z}_t(\ell)$ for the ARIMA model is the conditional expectation of $z_{t+\ell}$, conditioned on the past values $z_1, z_2, \ldots, z_t$. Thus, once a model has been chosen, the forecast can be generated directly from the equation that generates the model.

Expressions for the variance of the forecast error are available and can be used to compute confidence intervals for the forecasts.

Identification and Estimation of ARIMA Models

The successful application of ARIMA models to forecasting problems has been greatly facilitated by the availability of computer software,[17] which helps the user to choose initial models for investigation and also to perform the required computations. An outline of the procedure follows.

Plots are generated of the data and of sample autocorrelograms for the differenced and undifferenced data. To help the forecaster decide on values of p and q, plots are also made of the sample partial ACFs. Use is made of expressions for the SD of estimators of the ACF and partial ACF in order to decide if an estimate at a given lag is significant or due to sampling error. Once a model has been specified, maximum likelihood estimates of the parameters are computed, along with confidence intervals for the estimates. A time series of residuals is computed using the fitted model, and a sample ACF and partial ACF are plotted. If the hypothesis that the residual series is a white noise process can be accepted at some level of significance, the chosen model is considered the correct one for the data. The level of significance is computed, and if the forecaster feels that a better model can be chosen, another iteration takes place. To help choose the next model, large values that may appear in the sample ACF and partial ACF of the residual series are of great help in determining new values of p and q. If the forecaster is satisfied with the model, then forecasts can be made beginning at any time in the set of observations and can also be plotted. Thus the model can also be fit to the data for comparison. Confidence intervals for the forecast are also made.

If a satisfactory model cannot be found, it is sometimes helpful to form a new series by transforming the data. Taking the logarithm of the data is often found to be a useful transformation.

To use the software successfully requires practice on the part of the forecaster, and sample series from known models are available. In applications, at least 50 to 100 observations should be available, since many of the numerical procedures incorporated into the software are accurate only for "large" samples.

Seasonal Models

The ARIMA models have been generalized by Box and Jenkins[18] to account for seasonal phenomena, and these models are called "multiplicative models." Let B denote a backward shift operator, such that $B^d z_t = z_{t-d}$. If $d = 1$, the notation used is Bz_t. Then a general form for the multiplicative model can be written as

$$\sum_{i=0}^{p} \phi_i B^i \sum_{j=0}^{P} \Gamma_j B^{sj} (1 - B^s)^D (1 - B)^d \tilde{z}_t = \sum_{i=0}^{q} \theta_i B^i \sum_{j=0}^{Q} \Delta_j B^{sj} u_t \tag{57}$$

where ϕ_0, Γ_0, θ_0, and Δ_0 are equal to -1 and s is the period of the seasonal phenomena. Thus equation 57 allows differencing the data with an interval of s instead of 1 in order to remove seasonal nonstationarities from the data. The multiplicative models can be analyzed with the same computer software that is used for the ARIMA models. However, a greater variety of ACFs and partial ACFs is possible, which makes model building a more difficult task.

Other Statistical Forecasting Models

The theory for multiple series modeling has been developed, and computer software is readily available.[17] An alternate approach to the multivariate modeling and forecasting problem is the state space approach.[19] Least squares techniques have been used in econometric modeling.[20] In general, the use of multiple time series for forecasting should be considered only for forecasting variables that are important enough to warrant the additional expense involved in data collection, model building, and model maintenance.

REFERENCES

1. R. DEUTSCH, *Estimation Theory*, Prentice-Hall, Englewood Cliffs, NJ, 1965.
2. D. C. MONTGOMERY and L. A. JOHNSON, *Forecasting and Time Series Analysis*, McGraw-Hill, New York, 1976, p. 30.
3. W. G. SULLIVAN and W. W. CLAYCOMBE, *Fundamentals of Forecasting*, Reston Publishing Company, Reston, VA, 1977, p. 104.
4. MONTGOMERY and JOHNSON, *Forecasting and Time Series Analysis*, p. 49.
5. G. E. P. BOX and G. M. JENKINS, *Time Series Analysis: Forecasting and Control*, 2nd ed.. Holden-Day, San Francisco, 1976, pp. 126–170.
6. R. G. BROWN, *Smoothing, Forecasting and Prediction of Discrete Time Series*, Prentice-Hall, Englewood Cliffs, NJ, 1963.
7. BROWN, *Discrete Time Series*, pp. 174–198.
8. P. R. WINTERS, "Forecasting Sales by Exponentially Weighted Moving Averages," *Management Science*, Vol. 6, No. 3 (1960), pp. 324–342.
9. A. L. SWEET, "Adaptive Smoothing for Forecasting Seasonal Series," *AIIE Transactions*, Vol. 13, No. 3 (1981), pp. 243–248.
10. D. W. TRIGG and A. G. LEACH, "Exponential Smoothing With an Adaptive Response Rate," *Operational Research Quarterly*, Vol. 18, No. 1 (1967), pp. 53–59.
11. J. SHISKIN, A. H. YOUNG, and J. C. MUSGRAVE, *The X-11 Variant of Census Method II Seasonal Adjustment Program*, Technical Paper No. 15, Bureau of the Census, U.S. Department of Commerce, Washington, DC, 1967.
12. R. L. MCLAUGHLIN and J. J. BOYLE, *Short-Term Forecasting*, Marketing Research Technique Series No. 13, American Marketing Association, Chicago, 1962.
13. MONTGOMERY and JOHNSON, *Forecasting and Time Series Analysis*, pp. 276–296.
14. BOX and JENKINS, *Time Series Analysis*, pp. 46–84.
15. A. L. SWEET and F. MAZAHERI, "Computation of the Autocovariances of Stationary ARMA Processes," *Computers and Industrial Engineering*, Vol. 3, No. 4 (1979), pp. 313–320.
16. BOX and JENKINS, *Time Series Analysis*, pp. 126–170.

17. D. J. PACK, *A Computer Program for the Analysis of Time Series Models Using the Box-Jenkins Philosophy*, Automatic Forecasting Systems, Hatboro, PA, May 1979.
18. BOX and JENKINS, *Time Series Analysis*, pp. 300-333.
19. R. K. MEHRA and D. G. LAINIOTIS, Eds., *System Identification, Advances, and Case Studies*, Academic Press, New York, 1976.
20. M. K. EVANS, *Macroeconomic Activity: Theory, Forecasting, and Control*, Harper & Row, New York, 1969.

BIBLIOGRAPHY

Introductory texts on time series modeling and spectral analysis include:

BENDAT, J. S., and A. G. PIERSOL, *Random Data: Analysis and Measurement Procedures*, Wiley, New York, 1971.

CHATFIELD, C., *The Analysis of Time Series: Theory and Practice*, Wiley, New York, 1975.

JENKINS, G. M., and D. G. WATTS, *Spectral Analysis and Its Applications*, Holden-Day, San Francisco, 1968.

More advanced texts are:

FULLER, W. A., *Introduction to Statistical Time Series*, Wiley, New York, 1976.

KASHYAP, R. L., and A. R. RAO, *Dynamic Stochastic Models from Empirical Data*, Academic Press, New York, 1976.

Introductory texts on forecasting methods include:

SULLIVAN, W. G., and W. W. CLAYCOMBE, *Fundamentals of Forecasting*, Reston Publishing Company, Reston, VA, 1977.

WHEELWRIGHT, S. C., and S. MAKRIDAKIS, *Forecasting Methods for Management*, 2nd ed., Wiley, New York, 1977.

Introductory texts on the use of time series in forecasting include:

GILCHRIST, W., *Statistical Forecasting*, Wiley, New York, 1976.

MAKRIDAKIS, S., and S. C. WHEELWRIGHT, *Interactive Forecasting: Univariate and Multivariate Methods*, 2nd ed., Holden-Day, San Francisco, 1978.

The use of ARIMA models in forecasting is presented in the following publications:

ANDERSON, O. D., *Time Series Analysis and Forecasting, The Box-Jenkins Approach*, Butterworth, London, 1977.

GRAY, H. L., G. D. KELLEY, and D. D. MCINTIRE, "A New Approach to ARMA Modeling," *Communications in Statistics*, Vol. B7, No. 1 (1978), pp. 1–77.

JENKINS, G. M., *Practical Experiences with Modeling and Forecasting Time Series*, G. Jenkins and Partners, Jersey, Channel Islands, 1979.

MABERT, V. A., *An Introduction to Short Term Forecasting Using the Box-Jenkins Methodology*, Production Planning and Control Monograph Series No. 2, AIIE, Norcross, GA, 1975.

NELSON, C. R., *Applied Time Series Analysis for Managerial Forecasting*, Holden-Day, San Francisco, 1973.

Some recently published papers of interest are:

ADAM, E. C., W. L. BERRY, and D. C. WHYBARK, "An Experimental Comparison of Exponential and Adaptive Smoothing Forecasting Models Using Actual Operating Data," *Journal of Computers and Industrial Engineering*, Vol. 2, No. 1 (1978), pp. 91–98.

ANDERSON, O. D., "Time Series Analysis and Forecasting: Another Look at the Box-Jenkins Approach," *The Statistician*, Vol. 26, 1977, pp. 285–303.

CHATFIELD, C., "Some Recent Developments in Time Series Analysis," *Journal of the Royal Statistical Society*, Series A, Vol. 140, Part 4, 1977, pp. 492–510.

CHATFIELD, C., "Adaptive Filtering: A Critical Assessment," *Journal of the Operational Research Society*, Vol. 29, No. 9 (1978), pp. 891–896.

CHATFIELD, C., and D. L. PROTHERO, "Box-Jenkins Seasonal Forecasting: Problems in a Case Study," *Journal of the Royal Statistical Society*, Series A, Vol. 136, Part 3, 1973, pp. 295–336.

CLEVELAND, W. L., D. M. DUNN, and I. J. TARPENNING, "A Resistant Seasonal Adjustment Procedure With Graphical Methods for Interpretation and Diagnosis," in A. Zellner, Ed., *Seasonal Analysis of Economic Time Series*, Bureau of the Census, U.S. Department of Commerce, Washington, DC, 1978.

DICKINSON, J. P., "Some Statistical Results in the Combination of Forecasts," *Operational Research Quarterly*, Vol. 24, No. 2 (1973), pp. 253–260.

GOLDER, E. R., and J. G. SETTLE, "Monitoring Schemes in Short-Term Forecasting," *Operational Research Quarterly*, Vol. 27, No. 2 (1976), pp. 489–501.

GREEN, M., and P. J. HARRISON, "Fashion Forecasting for a Mail Order Company Using a Bayesian Approach," *Operational Research Quarterly*, Vol. 24, No. 2 (1973), pp. 193–205.

MCKENZIE, E., "The Monitoring of Exponentially Weighted Forecasts," *Journal of the Operational Research Society*, Vol. 29, No. 5 (1978), pp. 449–458.

ROBERTS, S. D., and R. REED, "The Development of a Self-Adaptive Forecasting Technique," *AIEE Transactions*, Vol. 1, No. 4 (1969), pp. 314–322.

TIAO, G. C., and G. E. P. BOX, *An Introduction to Applied Multiple Time Series*, TR No. 582, Department of Statistics, University of Wisconsin, Madison, October 1979.

In addition to the journals appearing in the preceding references, others containing papers on the subject of forecasting are *Biometrika*, *Decision Sciences*, *Harvard Business Review*, *IEEE Transactions on Systems, Man and Cybernetics*, *Journal of the American Statistical Association*, and *Management Science*.

CHAPTER 13.10

Control Models

SHIMON Y. NOF
THEODORE J. WILLIAMS
Purdue University

13.10.1 INTRODUCTION

Control is the fundamental engineering and managerial function whose major purpose is to measure, evaluate, and adjust the operation of a process, a machine, or a system under dynamic conditions so that it achieves desired objectives within its planned specifications under cost and safety considerations. A well-planned system can perform well without any control only as long as no variations are encountered in its own operation and in its environment. In reality, however, many changes occur over time. Machine breakdown, human error, variable material properties, and faulty information are a few examples of why systems must be controlled.

When a system is more complex and there are more potential sources of dynamic variations, a more complicated control is required. Particularly in automatic systems where human operators are replaced by machines and computers, a thorough design of control responsibilities and procedures is necessary. Control activities include automatic control of individual NC machines, materials handling equipment, and manufacturing processes. They also include control of production, inventory, quality, labor performance, and cost. Careful design of correct and adequate controls that continually identify and trace variations and disturbances, evaluate alternative responses, and result in timely and appropriate actions is therefore vital to the successful operation of a system.

The purpose of this chapter is to describe control models and explain their engineering use in the study and design of control systems. These models represent temporal relationships between the input and the resulting output or state of a system. Whether they specify the control functions in a small, self-contained mechanism or organization or in a multiechelon, distributed organization, they are based on common principles of control. Therefore explanations and examples of control principles throughout the chapter, which, for clarity, may refer specifically to a mechanical system, should be considered true for organizational systems as well. As their name implies, control models describe the control activity, and their purpose is to predict the behavior of the system being modeled over a wide range of conditions.

Why Use Control Models?

There are three main motivations for using control models, as follows:

1. In the design of a system and its controls, prediction of performance is necessary in order to evaluate alternative proposals and select the best among them. A control model provides designers with a tool *to analyze control methods and requirements and to estimate the impact and economics of the control system.* Use of control models will guide designers in structuring the designed system and in specifying its control system. For instance, analysis of a model of the control system for a machine will yield specification of characteristics of the controller for that machine. It may also have an impact on the design of the machine itself, for example, by pointing to circumstances under which the machine can or cannot operate safely. Similarly, a manual control model that represents the man-machine performance of an aircraft can be applied to design the aircraft such that its performance criteria are met while under pilot control.
2. During the operation of a working system, control models can *serve as tools in the actual control activity.* The ability of the models to predict the output state of a system can now be applied to particular actual conditions as they occur. For instance, in controlling an automated production facility, different control policies may have to be compared periodically, often just one step before operations are being performed. Each time one or several control models may have to be analyzed with the actual current state of the facility in order to decide which policy is optimal.

3. The third motivation for using control models is the need *to investigate the control ability and attributes of systems and their components.* Although such an investigation is highly related to the previous two items, control design and control implementation, it is viewed as a more research-oriented task. Of interest here are fundamental attributes such as decision logic, sensing, response time, and control quality, which characterize various types of systems. A good example are manual control models, whose purpose is to investigate and establish the capabilities of humans as controllers.

13.10.2 DEFINITIONS, TERMINOLOGY, AND TOOLS

Fundamentals of Control

"Automatic control," as the term is commonly used, is "self-correcting" or feedback control; that is, some control instrument is continuously monitoring certain output variables of a controlled process and is comparing this output with some preestablished desired value. The instrument then compares the actual and desired values of the output variable. Any resulting error obtained from this comparison is used to compute the required correction to the control setting of the equipment being controlled. As a result, the value of the output variable will be adjusted to its desired level and maintained there. This type of control is known as a "servomechanism."

At this point the reader should distinguish the servomechanism type of control from the common on-off type of control. An on-off control uses the fact of the presence or absence of an error, with no concern regarding the magnitude of that error, to so govern equipment. For example, a thermostatic control of a home refrigerator or furnace is an example of the on-off type of operation. However, it is the servomechanism type of control that is of concern here, since it is the heart of any truly automatic operation of complicated, modern production facilities.

The design and use of a servomechanism control system requires a knowledge of every element of the control loop. For example, in Exhibit 13.10.1 the engineer must know the "dynamic response," or complete operating characteristics, of each pictured device: (1) the "indicator" or "sampler," which senses and measures the actual output; (2) the controller, including both the "error detector" and the "correction computer," which contain the decision making logic; (3) the "control valve" and the transmission characteristics of the "connecting lines," which communicate and activate the necessary adjustment; and (4) the operating characteristics of the "plant," which is the process or system being controlled. "Dynamic response" or "operating characteristics" refer to a mathematical expression, for example, differential equations for the transient behavior of the process or its actions during periods of change in operating conditions. From it one can develop the "transfer function" of the process (another, often simpler representation of the differential equations) or prepare an experimental or empirical representation of the same effects.

A control engineer usually lumps all of the devices into a so-called block diagram for convenience. Such a diagram is shown in Exhibit 13.10.1.

Because of time lags due to the long communication line (typically pneumatic or hydraulic) from sensor to controller and other delays in the process, some time will elapse before knowledge of changes in an output process variable reaches the controller. When the controller notes a change, it must compare it with the variable value it desires, compute how much and in what direction the control valve must be repositioned, and then activate this correction in the valve opening. Some time is required, of course, to make these decisions and to correct the valve position.

Some time will also elapse before the effect of the valve correction on the output variable value

Exhibit 13.10.1 Block Diagram of a Typical Simple, Single Control Loop of a Process Control System

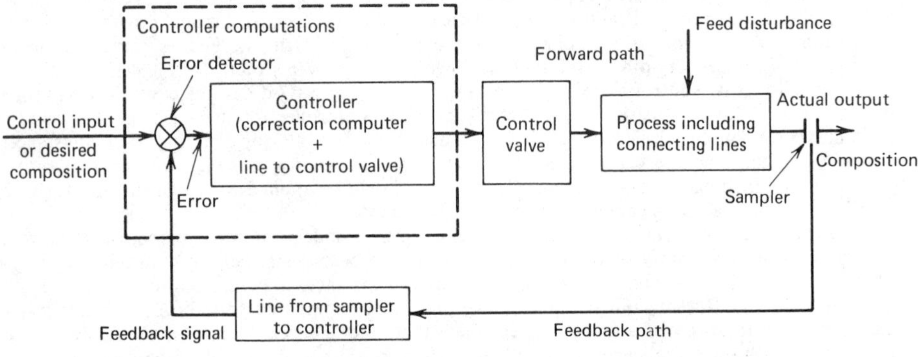

can reach the output itself and thus be sensed. It is only then that the controller will be able to know whether its first correction was too small or too large. At that time it makes a further correction, which will, after a time, cause another output change. The results of this second correction will be observed, a third correction will be made, and so on.

This series of measuring, comparing, computing, and correcting actions will go round and round through the controller and through the process in a closed chain of actions until the actual process valve is finally balanced again at the value desired by the operator. Since from time to time there are disturbances and modifications in the desired level (or type) of the output, the series of control actions never ceases. This type of control is aptly termed "feedback control." Exhibit 13.10.1 shows the direction and path of this closed series of control actions. The closed-loop concept is fundamental to a full understanding of automatic control.

Although the preceding example illustrates the basic principles involved, the actual attainment of automatic control of almost any industrial process or other complicated device will usually be much more difficult because of the speed of response, multivariable interaction, nonlinearities, response limitations, or other difficulties that may be present, as well as the much higher accuracy or degree of control that is usually desired beyond that required for the simple process just mentioned.

As defined here, automatic process control always implies the use of a feedback. This means that the control instrument is continuously monitoring certain output variables of the controlled process, such as a temperature, a pressure, or a composition, and is also comparing this output with some preestablished desired value, which is considered a "reference" or "set point" of the controlled variable. An error that is indicated by the comparison is used by the instrument to compute a correction to the setting of the process control valve or other final control element in order to adjust the value of the output variable to its desired level and to maintain it there.

If the set point is altered, the response of the control system to bring the process to the new operating level is termed that of a "servomechanism" or "self-correcting device." The action of holding the process at a previously established level of operation in the face of external disturbances operating on the process is termed that of a "regulator."

Instrumentation of an Automatic Control System

The large number of variables of a typical industrial plant constitute a wide variety of flows, levels, temperatures, compositions, positions, and other parameters to be measured by the "sensor" elements of the control system. Such devices sense some physical, electrical, or other informational property of the variable under consideration and use it to develop an electrical, mechanical, or pneumatic signal representative of the magnitude of the variable in question. The signal is then acted upon by a "transducer" to convert it to one of the standard signal levels used in industrial plants (3 to 15 psig for pneumatic systems and 1 to 4, 4 to 20, or 10 to 50 mA or 0 to 5 V for electrical systems). Signals may also be digitized at this point if the control system is to be digital. The important sensing and conversion techniques are well covered in the literature.[1]

The signals that are developed by many types of sensors are continuous representations of the sensed variables and as such are called "analog signals." When analog signals have been operated upon by an analog to digital converter, they become a series of bits, or on-off signals, and are then called "digital signals." Several bits must always be considered together in order to represent properly the converted analog signal (typically, 10 to 12 bits).

As stated previously, the resulting sensed variable signal is compared at the controller to a desired level, or set point, for that variable. The set point is established by the plant operator or by an upper-level control system. Any error (difference) between these values is used by the controller to compute the correction to the controller output, which is transmitted to the valve or other actuator of the system's parameters.

A typical algorithm by which the controller (either analog or digital) computes its correction is as follows[2]: Suppose a system comprises components that convert inputs to outputs according to relationships, called "gains," of three types, that is, proportional, derivative, and integral gains. Then the controller output is

$$\text{output} = K_p e_n + \sum_{i=1}^{n} K_R e_i + K_D(e_n - e_{n-1}) + K_1$$

where

K_p, K_D, and K_R = the proportional, derivative, and integral gains, respectively, of the controller
K_1 = a midrange constant to allow proportional control alone (i.e., $K_D = 0, K_R = 0$)
n = the last sample obtained by the controller
$(n - 1)$ = the next to last sample

The summation for the integral gain term is over all the sampling iterations until the last one, n. The error, e_n, is calculated as

$$e_n = \pm \text{ (set point } - \text{ controlled variable)}$$

The option of using either a plus or a minus sign in the error term depends on the particular control loop.

Conventional Control Methods

Analog Control

Both theoretical and practical studies of automatic control were confined principally to single-loop studies, similar to that in Exhibit 13.10.1, until the middle 1950s. Work involved continued development of the methods first proposed by researchers in the 1930s and used with important results in the succeeding two decades. Particularly important here were studies of stability (Exhibit 13.10.2) and design methods of feedback control systems, which are discussed extensively in the classical control literature: Root-Locus method, Bode-Plot representations, Nyquist diagrams, and Nichols charts.[3]

All of the preceding work involved the use of analog control systems, that is, mechanical or electrical devices that simulated the action desired. They were *analogs* of the action that a human operator would take in the same circumstance. Because of the limitations in complexity of the mechanical and electrical devices that it was possible to build in large quantities during this period, and because of the limitations in the control systems theory existing at that time, nearly all of these systems were built to control single loops, such as that in Exhibit 13.10.1.

Computer Control

Almost as soon as digital computers were developed, researchers began to consider them as candidates for automatic control systems because of their potential for breaking the limitations on complexity imposed by the earlier analog devices, as just mentioned.

Thus, following the development of capable, inexpensive, and reliable digital computers in the late 1960s, these devices became very popular elements of control systems. A digital computer can easily store the values of set points for a large number of variables; computer programs can define and quickly prepare highly complex computations that are necessary for decision making; a large number of sensors can transmit signals to the computer at the same time. Advances in computer technology have also enabled the design of hierarchical and distributed control systems. Here reliable communication between computers provides for the control of a network of machines, stations, departments, or even remote plants. In parallel, control models have been developed and advanced to represent, analyze, and design the more complex control system.

13.10.3 BASIC CONTROL MODELS

Control Modeling

Four types of modeling methodologies have been employed to represent physical components and relationships in the study of control systems:

1. Mathematical equations, in particular, differential and difference equations that are the basis of classical control theory (transfer functions are a common form of these equations).
2. Mathematical equations that are based on state variables of multivariable systems and associated with modern control theory.
3. Block diagrams.
4. Signal flow graphs.

Mathematical models are employed when detailed relationships are necessary. To simplify the analysis of mathematical equations, we usually approximate them by linear, ordinary differential equations. For instance, a characteristic differential equation of a control loop model may have the form

$$\frac{d^2x}{dt^2} + 2\alpha \frac{dx}{dt} + \beta^2 x = f(t)$$

Exhibit 13.10.2 Various Degrees of Stability in Transient Response as Achieved by Automatic Control Systems: (*a*) Stable and Overdamped, (*b*) Stable and Critically Damped, (*c*) Stable and Underdamped, (*d*) Oscillatory or Threshold of Stability, and (*e*) Unstable

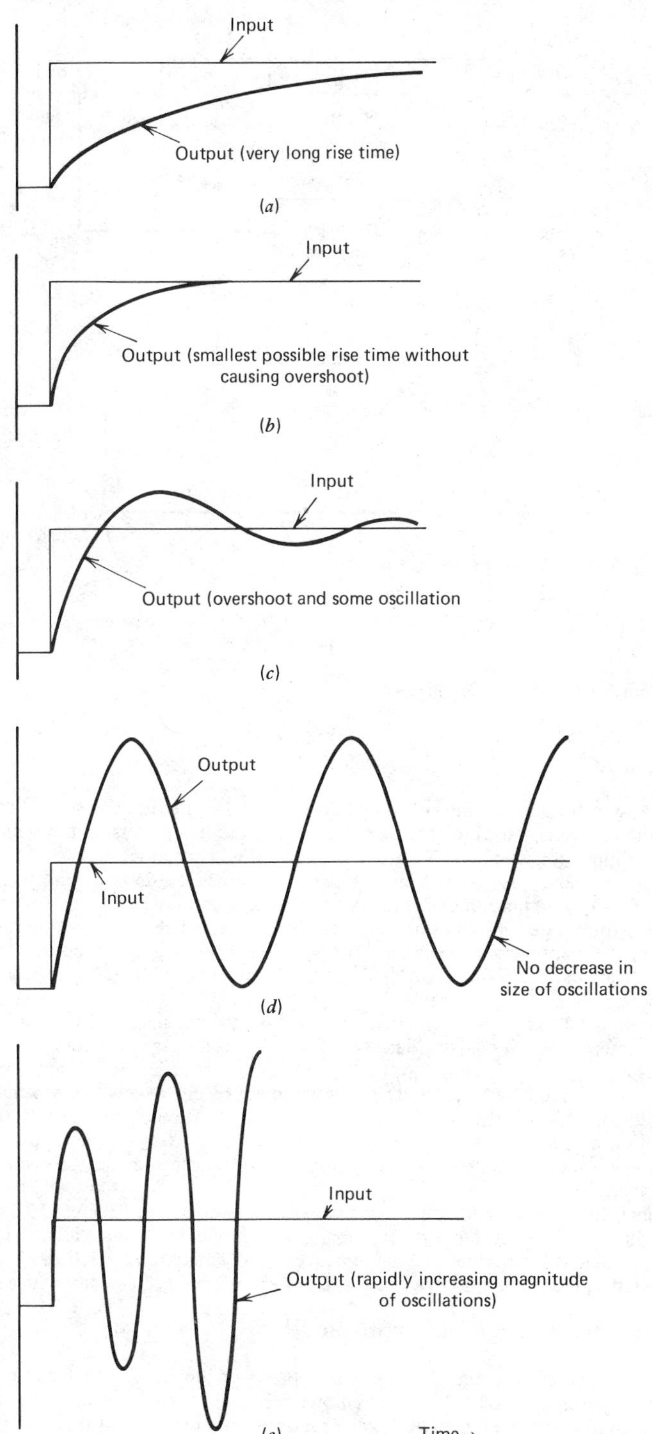

Exhibit 13.10.3 Graphic Techniques for Control Models: (*a*) Block Diagram Model of a Feedback Loop (Similar to Exhibit 13.10.1) and (*b*) Signal Flow Graph of the Same Feedback Loop

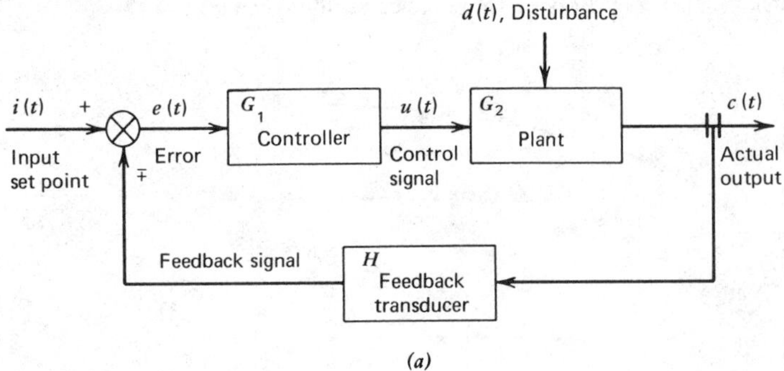

(*a*)

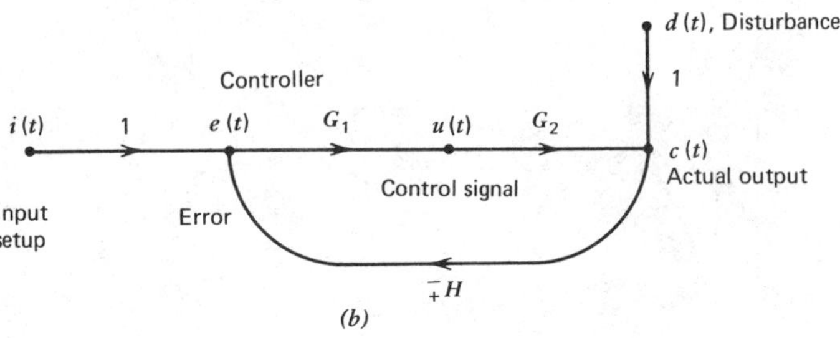

(*b*)

with initial conditions of the system given as

$$x(o) = Xo$$
$$x'(o) = Vo$$

where $x(t)$ is a time function of the controlled output variable, its first and second derivatives over time specify the temporal nature of the system, α and β are parameters of the system properties, $f(t)$ specifies the input function, and X_o and V_o are specified constants.

Mathematical equations such as this example are developed to describe the performance of a given system. Usually an equation or a transfer function is determined for each system component. Then a model is formulated by appropriately combining the individual components. This process is often simplified by applying Laplace and Fourier transforms. A graphic representation by block or signal flow diagrams (see Exhibit 13.10.3) is usually applied to define the connections between components.

Once a mathematical model is formulated, the control system characteristics can be analytically or empirically determined. The basic characteristics that are the object of the control system design are (1) response time, (2) relative stability, and (3) control accuracy. They can be expressed either as functions of frequency, called "frequency domain specifications," or as functions of time, called "time domain specifications." To develop the specifications, the mathematical equations have to be solved, a procedure that is often complicated. A relatively simpler approach can be employed by using graphic methods such as those mentioned in the previous subsection on conventional control methods.

Numerical analysis methods and computer simulation methods (see Chapter 13.11) may also be used for the solution. When systems are too complex to model mathematically, the graphic methodologies are applied to describe the numerous control details and relationships. In this case computer simulation is usually employed to evaluate the performance characteristics of the system.

Open-Loop, Feedback, and Feedforward Models

Control models can be classified as being either of the open loop or the feedback type, as defined previously. Since automatic control always requires feedback, systems in practice as well as their models involve combinations of the two types. It is quite common to find that the introduction of more automation and computer control into a system is accompanied by a process of "closing the loops," that is, providing more feedback whenever the performance of an open-loop operation is deemed unsatisfactory.

When certain knowledge about expected future input or operating conditions is available, it can be utilized by a control system to improve performance. This type of advanced preparation is called "feedforward compensation." For instance, an automatic processing station can be alerted to expected changes in material composition in order to actuate proper process variations. The feedback portion of the control system in this example will continue to be used for fine adjustment, but the overall control activity will be significantly improved.

A number of basic control models follow the three classes described previously (see Exhibit 13.10.4). With regard to computer process control, we can define three classes of application: (1) *supervisory* or *optimizing control*, as exemplified by Exhibit 13.10.5; (2) *direct digital control* (Exhibit 13.10.6); and (3) *hierarchy control* (Exhibit 13.10.7), which is a combination of the others to effect all levels of decision making simultaneously in a plant.

As can be seen from Exhibit 13.10.5, supervisory or optimizing control puts the computer in an external or secondary control loop to the primary plant control system, which remains as the conventional plant instruments and individual electronic or pneumatic analog controllers discussed previously. The computer merely changes the set point, or level of control governed by the controllers, either directly or through manual intervention. Its task then is to "trim" the plant operation to improve the economic return from its operation and not to affect its dynamic control. At the same time, a malfunction of the computer cannot adversely affect the plant control. When the computer performs the set point computation just mentioned, but does not itself adjust the plant controller and instead depends upon a plant operator for this final action, we have what is called "open-loop control." Where the computer merely samples the process variables and determines their correct operating levels, but does not perform the optimization calculations, we have what is called "process monitoring."

Direct digital control (see Exhibit 13.10.6) uses a computer to replace a group of single-loop analog controllers of the type shown in Exhibit 13.10.1 with a digital computer. This is done with the hope that the single computer will be less expensive than the many controllers it can

Exhibit 13.10.4 Summary of Control Models, Performance, and Complexity

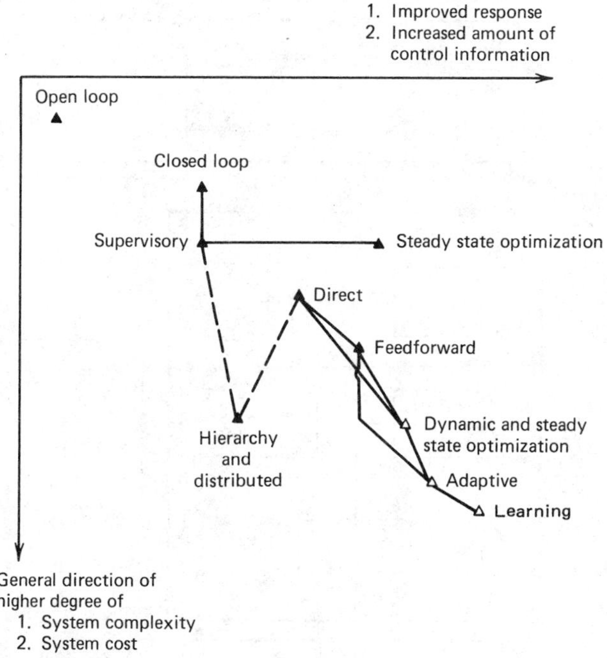

Exhibit 13.10.5 A Model of Computer Control as a Plant Optimizer and Supervisory Controller

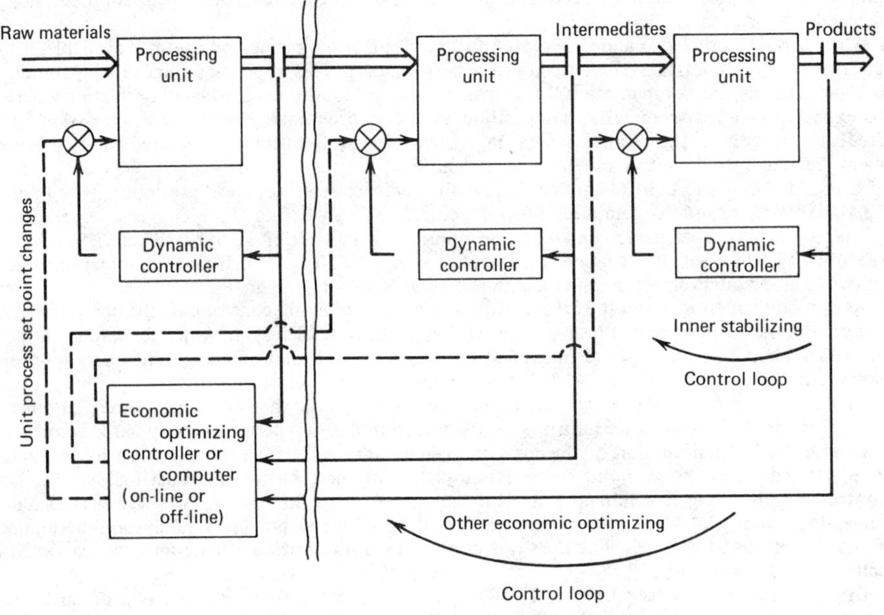

Exhibit 13.10.6 Block Diagram of a Direct Digital Control System

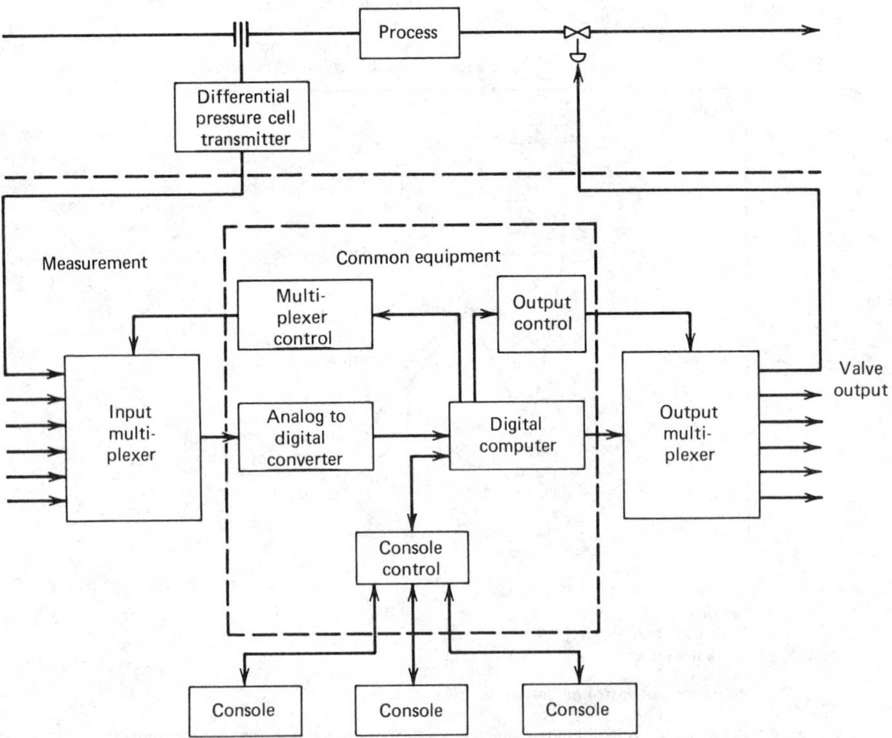

Exhibit 13.10.7 The Hierarchy Structure of the Computer Control System for a Fully Automated Industrial Plant

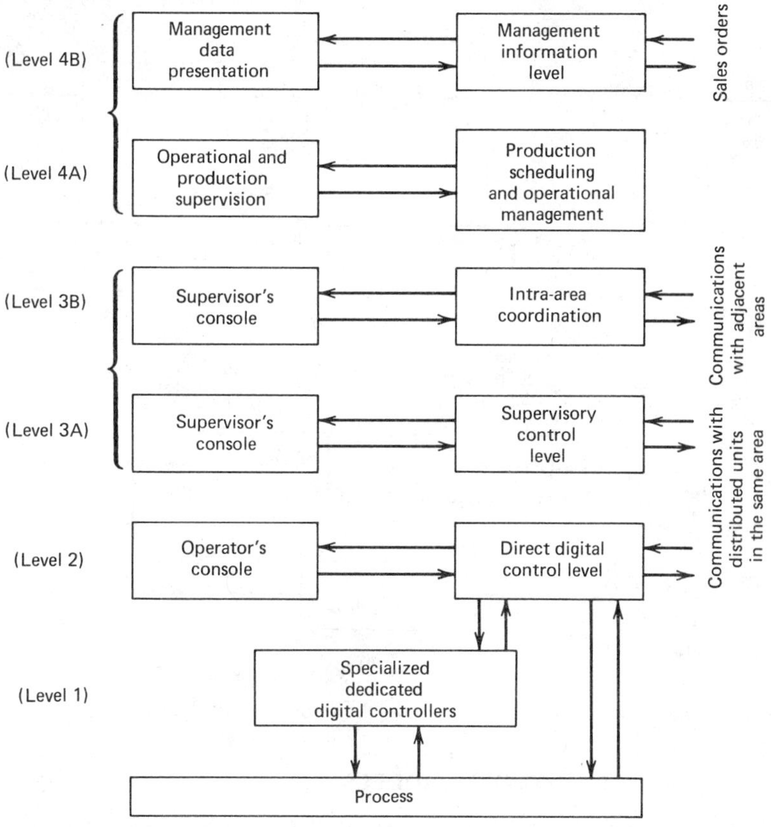

supplant.[4] The digital computer's computational ability also makes possible the application of more complex advanced control logic.

Hierarchy control, the most ambitious of models, as shown in Exhibit 13.10.7, attempts to apply computers to all plant control situations at the same time. Levels 1 and 2 machines are very similar in function, differing only in the extent of the control functions assigned to each. The dedicated digital controller of level 1 handles complex devices, such as chemical analyzers (e.g., chromatographs), and specialized feedforward and noninteracting multivariable control setups. The direct digital controller of level 2 handles a much larger number of three-mode and related types of control loops, as in Exhibit 13.10.1. It also communicates with plant operators through a console, probably composed of CRTs and a keyboard. Communication between all digital computers and with all consoles is by digital signals. Connections to analog signals are required only of the level 1 and 2 machines. Level 3 machines serve as supervisory computer control systems as previously outlined. Level 4 machines carry out production scheduling and management information functions.

When the control system in the hierarchy is implemented through a number of computer controllers in each level, a *distributed control* system results. The rapid development of microprocessor technology has made the concept of distributed control quite attractive. The use of reliable communication between a larger number of individual controllers, each responsible for its own tasks rather than for the complete operation, improves the response of the total system.

To close the list of basic models we must include certain model types that are particularly useful in manual control modeling.[5] These are the *preview model* and the *precognition model* (Exhibit 13.10.8). Both include feedforward compensation that depends on advanced knowledge of the human controller about the system. In the preview model a true display of future input set points provides the human controller with information well in advance so that he or she can effectively control the operation. In the precognition model on the other hand, the human controller can plan operations well ahead of time based on his or her precognition of the system characteristics and idiosyncrasies. Since such cognitive knowledge is usually gained by experience and is subject to

Exhibit 13.10.8 Control Models With Feedforward Compensation: (*a*) Preview Control Model and (*b*) Cognitive Control Model

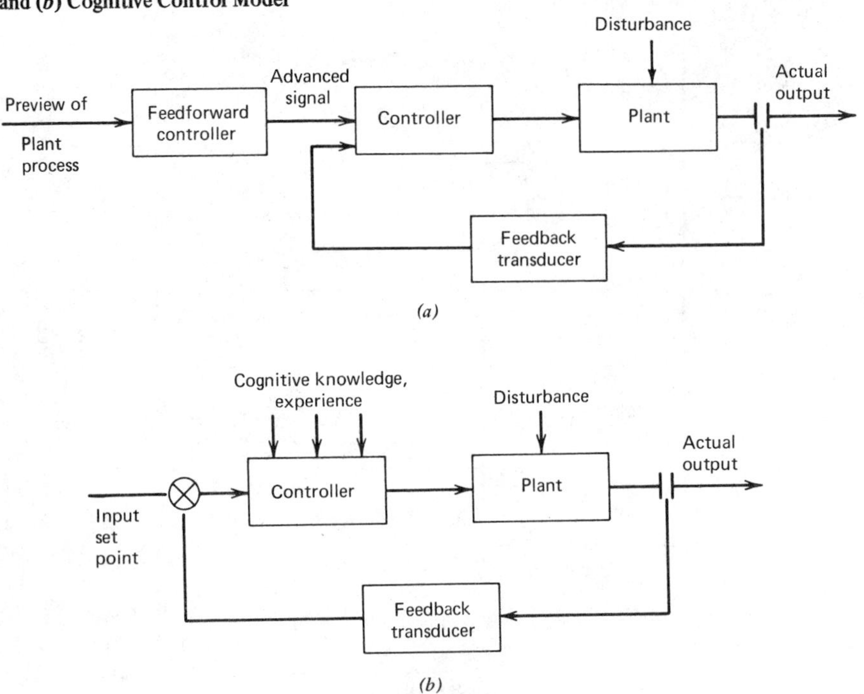

random variations, precognition may improve the response on the average, but may fail in the extremes.

13.10.4 ADVANCED CONTROL MODELS

Exhibit 13.10.1 had defined the standard feedback control techniques where the correction computation part of the control loop readjusts the position of the actuator of the process, generally a control valve. However, it became readily evident in the late 1950s that there were severe limitations on the ability of the classical control methods discussed previously and exemplified by the single-loop controllers of Exhibit 13.10.1, whether implemented by analog or digital methods. The appearance of the digital computer with its vastly increased computational ability provided the incentive for the beginning of a whole new era of development in automatic control theory. This class of developments has subsequently been labeled "modern control" in contrast to the classical control methods previously alluded to.

Important here was the application of the variational calculus to dynamic optimization problems along with the concept of dynamic programming as formulated by Bellman[6] and associates in 1956. The "maximum principle" of Pontryagin,[7] developed in 1959, and the identification techniques of the filter theory of Kalman[8] were other important discoveries during this period. This work has continued to the present day along with studies of the other important advanced control topics that are mentioned here. These developments have vastly extended the capabilities of industrial control systems in recent years. As indicated, a computer is almost always necessary to make the desired implementation.

Adaptive Control

"Adaptive control" is the capability of the control system to modify its own operation to achieve the best possible mode of operation. A general definition of adaptive control implies that an adaptive system must be capable of performing the following functions: (1) provide continuous information about the present state of the system or identify the process, (2) compare present system performance to the desired or optimum performance and make a decision to change the system so as to achieve some previously defined optimum performance, and (3) initiate a proper modification so as to drive the control system to the optimum. These three principles–identification, decision, and modification–are inherent in any adaptive system.[9] Exhibit 13.10.9 illustrates the adaptive

Exhibit 13.10.9 Adaptive Control Model: Adaptive Control Elements Form a Secondary Control Loop Around the Conventional Process Control Loop and Change the Characteristics of the Primary Controller to Provide Desired System Response

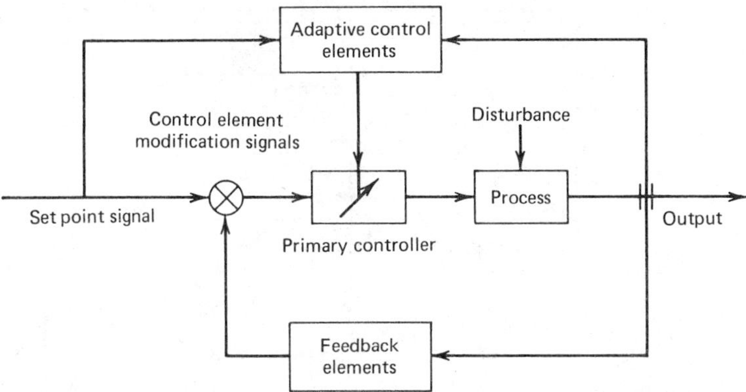

control model by picturing the adaptive controller as being in a secondary loop and modifying the operation of the regular systems controller, which is located in the primary process control loop.

Optimal Control

Compared to adaptive control, further progress toward optimal response can be achieved by methods and models of optimal control.

The "steady state optimization method" allows a computer to determine for the process a new best operating level if external conditions require such changes in order to maintain the process operation at some optimum, usually relative to economic criteria. Optimizations are computed under the assumption that the process is at some steady state and can be instantaneously transferred from one steady state to another, hence the name of this control method. Such an assumption is necessary to transforming all process operating equations to algebraic form for ready solution by a computer. In most supervisory control systems (Exhibit 13.10.5), this method can be implemented.

In "dynamic optimizing control" the control system operates in such a way that a specific performance criterion is dynamically satisfied. The criterion usually requires that the controlled system move from the original to a new state in the minimum possible time, or at the minimum total cost.[10]

Dynamic optimization and control adds one more level of sophistication to that of the steady state optimization. Thus not only is the process maintained at its optimum performance level while in the steady state regime, but the change from one operating level to another is also made to best satisfy the established overall control criteria. Models of optimal control are now mainly of academic interest because of the extremely large and powerful computing capacity that is required to implement them. However, practical attainment in the foreseeable future is certainly feasible. When it does occur, it will find particular application in systems that are highly sensitive to the process path being followed, for example, cyclical catalytic processes and batch processing of plastics. A very large computing capacity, both in speed and memory, is necessary since the equations to be solved in the mathematical model are now sets of simultaneous differential equations, as opposed to the algebraic equations of the steady state optimization.

Learning Control

"Learning control" implies that the control system contains sufficient computational ability so that over time it can develop specific representations of the control model of the system being controlled and can modify its own operaton to compensate for this newly developed knowledge. The learning control system is a further development of the adaptive controller and can apply artificial intelligence techniques.

Multivariable Noninteracting Control

"Multivariable noninteracting control" concerns large systems whose internal variables depend upon the values of other, related variables of the process. The single-loop techniques of classical con-

Exhibit 13.10.10 Model of a Feedforward and Multivariable Noninteracting Controller System

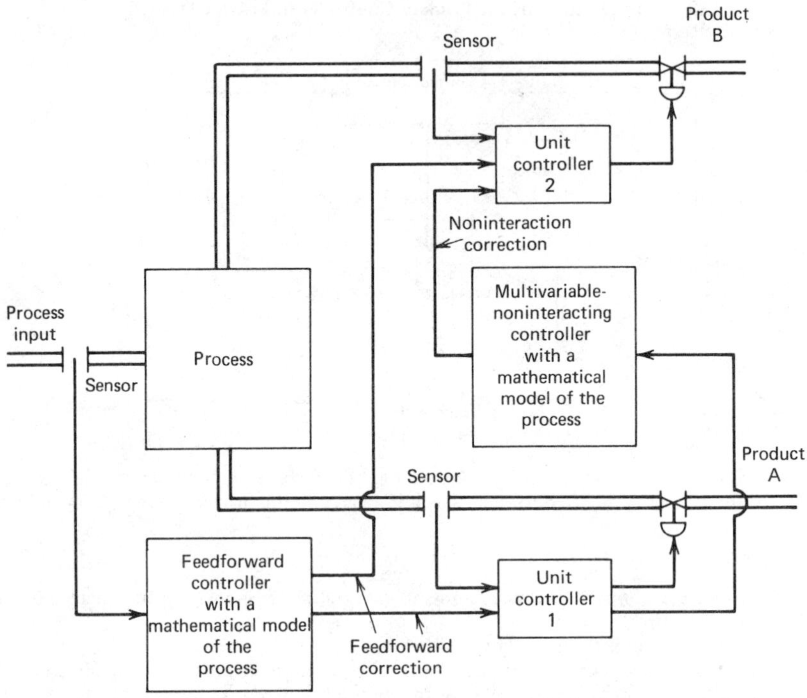

trol theory will therefore not suffice. More sophisticated techniques must be used to develop appropriate control systems for such processes.[3,11,12] Exhibit 13.10.10 shows a model of such a system. The feedforward controller compensates for the dynamics and process delays of the process itself and helps the unit process controllers compensate for input upsets. The multivariable noninteracting controller compensates for the effect of variations in flow of product A due to action of unit controller 1 on product B and thus prevents interaction of the output variables of the process.

13.10.5 APPLICATION AND SELECTION OF CONTROL MODELS IN INDUSTRIAL ENGINEERING

The purpose of this section is to describe several typical applications of control models in industrial engineering functions. The reader can by now realize that control always implies that there is a system—large or small, simple or complex—to be controlled. Three important system areas in which industrial engineers have traditionally been working with control models are manufacturing systems, man-machine systems, and information systems.

Control Models of Manufacturing Systems

Manufacturing systems probably provide the richest area for applications of control models. They include the self-contained machine or process, where a precise mathematical model can be formulated. They also require the many control activities that are less amenable to precise modeling, such as production and inventory control or quality control, where operations research and simulation techniques are applied to evaluate and plan the control system.

Models of NC Machines

A typical example is the model of an NC machine that is illustrated in Exhibit 13.10.11.[13] In this model for one machine axis, G is the transfer function (gain) of the position transducer; the velocity transducer is designed to have a transfer function of $G \cdot \alpha \cdot s$, where α specifies control parameters (damping and natural frequency) of the transducer and s is the Laplace transform variable. An input digital signal, θ_i, is passed as e_i to an amplifier with gain m, which controls a servomotor with gain G_2. This motor produces a torque, which turns a drive mechanism to a specified position dis-

Exhibit 13.10.11 A Block Diagram for Position and Velocity Control in NC Machines[a]

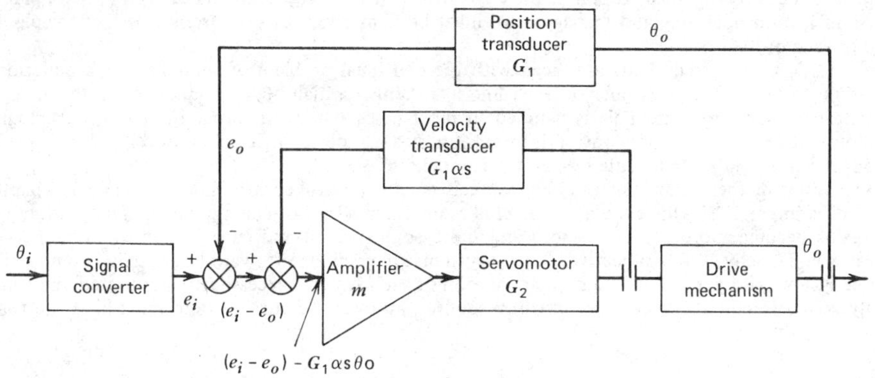

[a]Reference 13.

placement, θ_o, at a specified velocity. With certain simplifying assumptions (e.g., that the input in this example is a step function), a control model for the torque output can be formulated, using Laplace transform, as

$$\theta_o = m\left[\frac{G_2(e_i - e_o)}{s^2J + G_1\alpha s}\right]$$

In the equation, J is the system inertia. Analysis of the model can identify the specific values of the control parameters that yield the best response. For instance, the value of α determines the damping of the system, namely, how fast after every displacement command the machine can stabilize at the specified position.

Whereas the model in Exhibit 13.10.11 is based strictly on the closed-loop, feedback method, Exhibit 13.10.12 shows an adaptive control model for NC. As explained in Section 13.10.4, the adaptive control model applies a secondary loop that obtains feedback information beyond that obtained by the primary loop. In this example the primary loop involves position and velocity feedback, as in Exhibit 13.10.11, plus feedback on cutting forces, which provides for further corrections in the machine operation.

Other applications of control in manufacturing machinery are included in Section 7 of this handbook.

Models of Overall Plant and Company Control Systems

Automatic control of a large, modern industrial plant, whether achieved by a computer-based system or by conventional means, involves an extensive system for the automatic monitoring of a large number of different variables operating under a very wide range of process dynamics. It requires the development of a large number of quite complex, usually nonlinear relationships for the translation of the plant variable values into the required control correction commands. Finally, these control corrections must be transmitted to another very large set of widely scattered actua-

Exhibit 13.10.12 An Adaptive Control Model for NC Machines

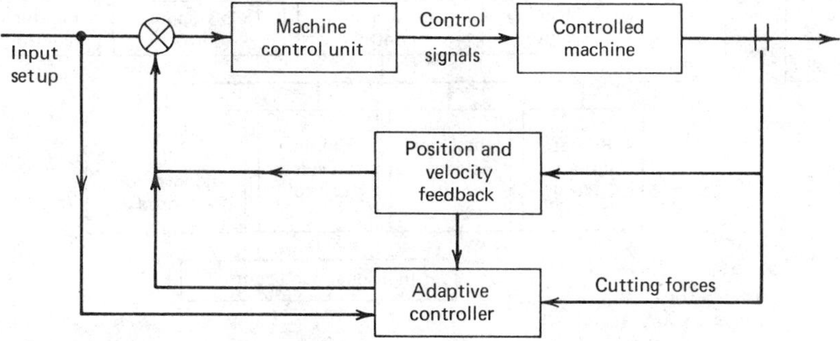

tion mechanisms of various types, which, because of the nature of manufacturing processes, may, and often do, involve the direction of the expenditure of very large amounts of energy. Also, plant personnel, both operating and management, must be kept aware of the current status of the plant and of each of its processes.

In addition, industrial plants are faced with the continual problem of adjusting the production schedule to match new customer orders while maintaining a high plant productivity at the lowest practical production costs. This is handled in most cases at present through a manual, although computer-aided, production control information system along with an in-process and finished goods inventory judged adequate by plant personnel.[14,15]

As a result of the circumstances cited, a precise, mathematical control model of an overall plant control is impractical. However, as the overall requirements for both energy savings and productivity gains become more complex, more and more sophisticated and capable control systems are necessary. To achieve this objective, control systems are gravitating toward large, digital computer-based systems. Proper design and implementation of such systems necessitate careful modeling and analysis. To obtain the necessary control responses, an overall control system must have the following capabilities:

1. A tight control of each operating unit of the plant to ensure that it is operating at its maximum efficiency of energy utilization and/or production capability. The operation is based upon the production level set by the coordination and scheduling functions listed as items 2 and 3. This control reacts directly to any emergencies that may occur in its own unit.
2. A coordination system that determines and sets the production level of all units working together between inventory locations. This system ensures that no unit is exceeding the general area level and thus using excess energy or raw materials. It responds to emergencies or upsets in any of the units under its control by shutting down or systematically reducing the output in these and related units.
3. A system capable of carrying out the scheduling function for the plant from customer orders or management decision in order to produce the required products at the optimum combination of time, energy, and raw materials suitably expressed as cost functions.

Because of the ever-widening scope of authority of each of these requirements in turn, they effectively become the distinct and separate levels of a superimposed control structure, one on top of the other. Also, in view of the amount of information that must be passed back and forth among the preceding three "levels" of control, it appears obvious that some sort of a distributed computational capability organized in a hierarchical fashion is necessary, as shown previously in Exhibit 13.10.6. A specific implementation of this model is shown in Exhibits 13.10.13 and 13.10.14 for the overall control of a steel mill.[16]

Exhibit 13.10.13 Hierarchy Arrangement of the Steel Plant Control (an Application of the Model of Exhibit 13.10.7)

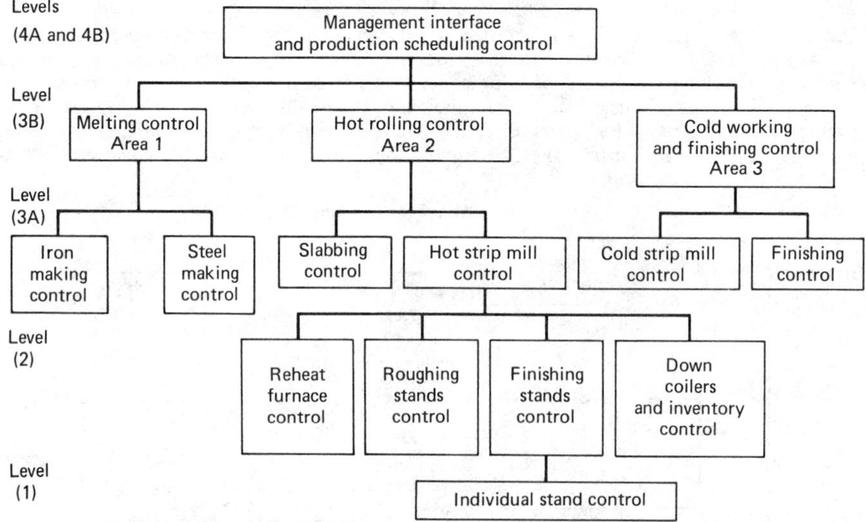

Exhibit 13.10.14 An Example Model of the Steel Mill Scheduling Procedure at the Overall Production Scheduling Level (Level 4A in Exhibit 13.10.13)

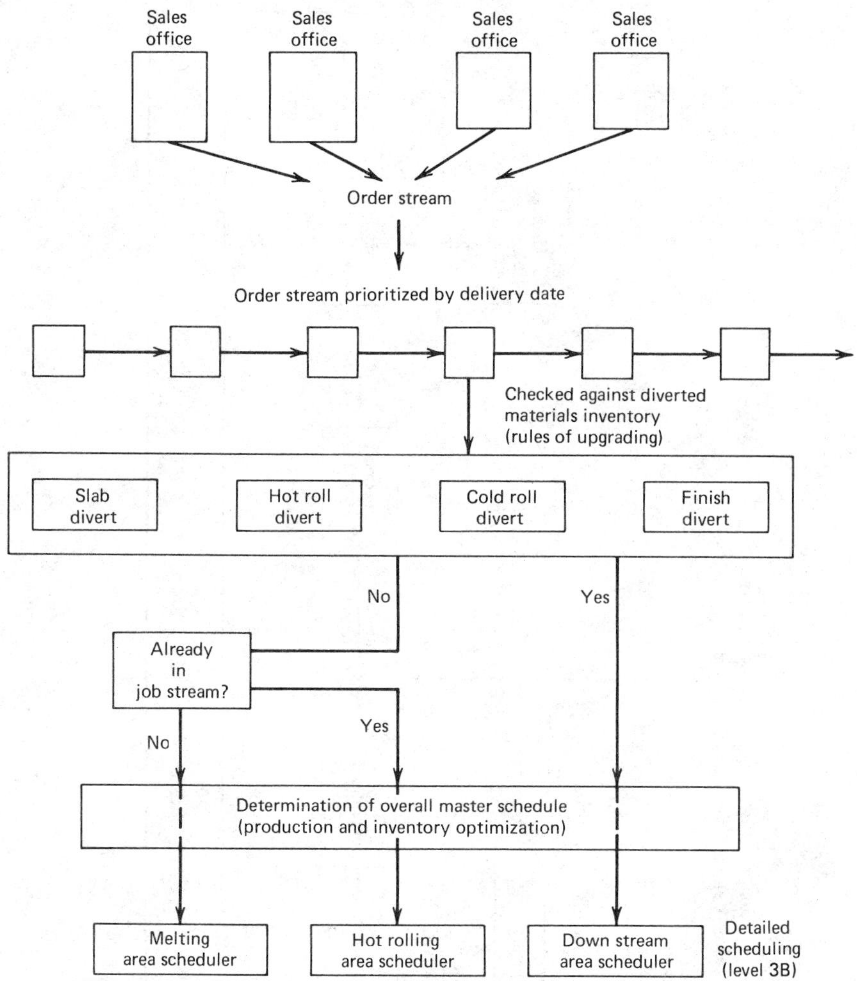

Other Models of Manufacturing Control

An example of the study of control models in computer-controlled manufacturing by operations research techniques is given in Nof et al.[17] A queuing network analysis and a simulation package are applied to evaluate alternatives of operational control, including issues of how parts are loaded into the facility, which manufacturing process to select, and how to assign parts to particular machines in the facility.

Learning control models, as previously defined, have also been suggested for overall plant control. In Nof et al.[18] the concept of the manufacturing operating system is described for the control of automatic manufacturing with certain intelligence to improve control decisions progressively.

Control Models in Man-Machine Systems

In the area of manual control, control models have been used to evaluate total man-machine performance, particularly in vehicular applications, for example, airplane piloting and automobile driving. Additionally, these models have been tested empirically in order to evaluate and measure the characteristics of the human as a controller. A large number of such models are reviewed by Sheridan and Ferrell.[5] A general review of control modeling of human behavior is provided in Rouse and Gopher.[19]

Exhibit 13.10.15 Overall Diagram of Driver-Vehicle Control System[a]

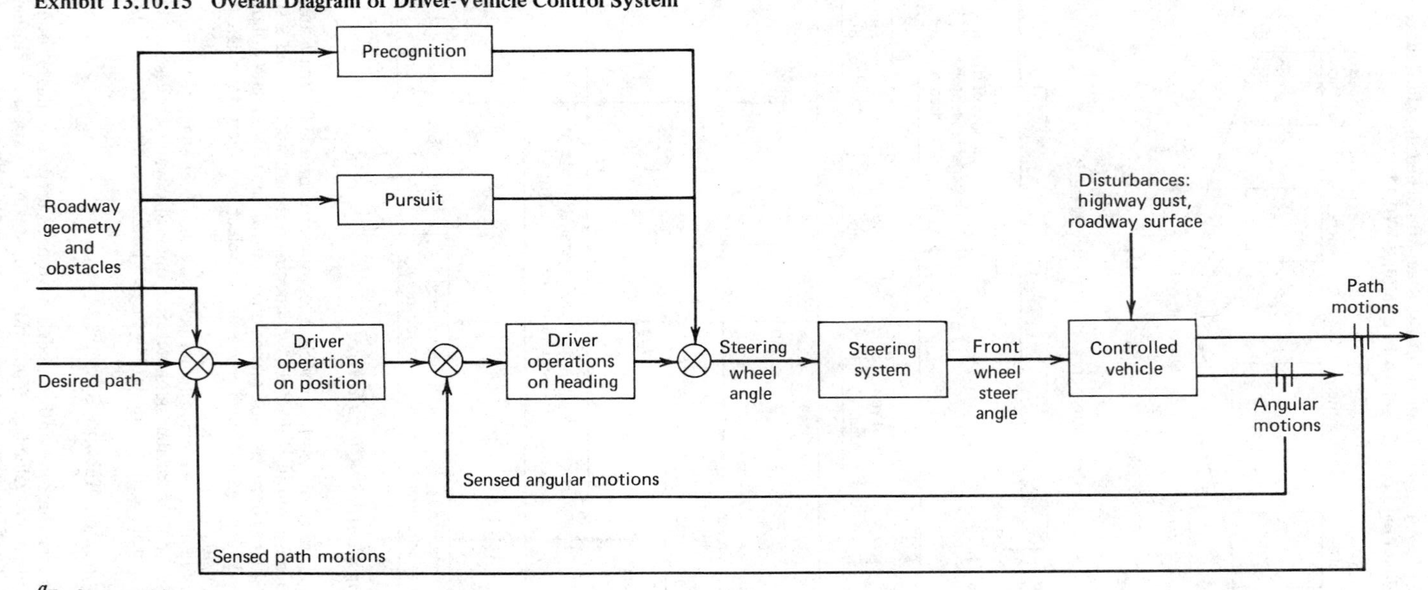

[a] Reference 20.

Driver Steering Control Models

Many control models describing the performance of a human driving a vehicle have been developed and applied to the design of bicycles, cars, and airplanes. The model illustrated in Exhibit 13.10.15 is based on a recent version.[20] This model combines feedback, open-loop, and feedforward elements.

Extensive experimentation has been carried out in order to estimate the parameters involved in these control models, so as to make them less hypothetical and more practical for analysis and design.

A basic form of the model, which describes the relationship between a displayed output, z, for example, feedback about the road, and the human control response, u, in a single-input, single-output system, is[19]

$$u(t) = \beta_1 u(t - \Delta t) + \beta_2 u(t - 2\Delta t) + \beta_3 z(t - \tau_d) + \beta_4 z(t - \tau_d - \Delta t)$$

where τ_d is the human reaction time delay and $\beta_1, \beta_2, \beta_3$, and β_4 are constants related to the backlash of the human's neuromotor systems and to the human decision making process. One very important use of control models in this area has been the study of car design for drivers' safety under a wide range of driver and road conditions.

Control Models of Display Design

Any system that involves people in control functions, such as operators, supervisors, or in any other control activities, requires some type of a display facility in order to present information to the decision maker. Examples of typical displays are instrument panels in airplanes and other vehicles and control lights and indicators in a control room of a chemical plant. A correctly designed display will significantly aid the overall performance of the controlled system and reduce the work load of the human operators. The design must specify a number of display factors, including the display format, information content and rate, and interaction patterns.

Display system design, which has been receiving increased attention, especially with the development of highly complex flight vehicles, has traditionally employed control models.[5] Some work in this area was based on frequency domain descriptions of the human operator and his or her work load, particularly with regard to visual scanning of the display. More recent works have been based on an *optimal control of the human operator*,[21] which is shown in Exhibit 13.10.16. In the previous section, optimal control models for steady state and dynamic optimization were defined. They represent automatic systems in which the controller utilizes an optimization algorithm to calculate and actuate an optimal response for dynamically changing conditions. In the case of the optimal control model for a human operator, it is hypothesized that the operator is well trained and motivated and will make an effort to respond as well as possible within human capabilities. However, the model does not imply that an optimization algorithm is used.

As shown in Exhibit 13.10.16, a human operator is assumed by the model to monitor a display, or an instrument panel. The operator is subject to observation noise, for example, errors in interpreting the display, and to motor noise, which affects neuromuscular behavior. Additionally, the model assumes a certain time delay in the human response. A typical mathematical form of the model is

$$x'(t) = Ax(t) + Bu(t) + v(t)$$

where

A = a matrix of parameters of the controlled process
B = a matrix of parameters of the controlled output
$x(t)$ = a time vector of state variables
$u(t)$ = a time vector of control variables
$v(t)$ = a vector of the noise variables

Different variants of this model have been applied for the design and evaluation of a variety of display issues.[5,22,23] As one example, such a model has been used in the design of vertical takeoff and landing aircraft. The trade-off between the level of automation in the flight control system and the display system sophistication was analyzed. The model incorporated simultaneous monitoring and control by the pilot for various design combinations, ranging from complete manual control with no monitoring to fully automatic control with monitoring only.

Control Models of Information Systems

The numerous industrial engineering applications of control models in computer information systems can be classified as being of one of two types: (1) development of information systems

Exhibit 13.10.16 Optimal Control Model of the Human Operator[a]

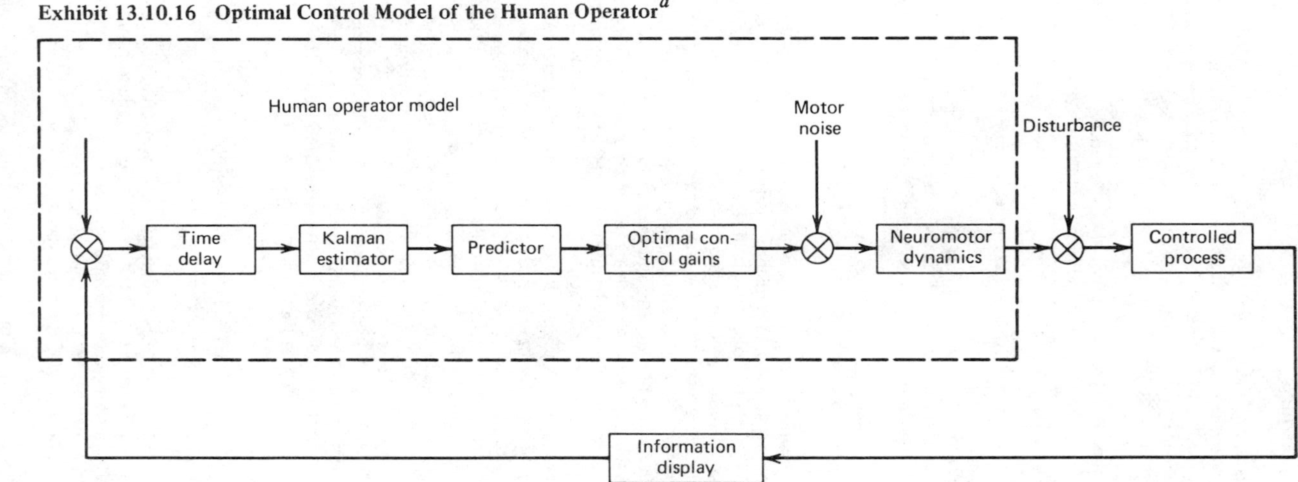

[a] Reference 21.

Exhibit 13.10.17 An Adaptive Control Model of an Information System (Application of the Model in Exhibit 13.10.8)

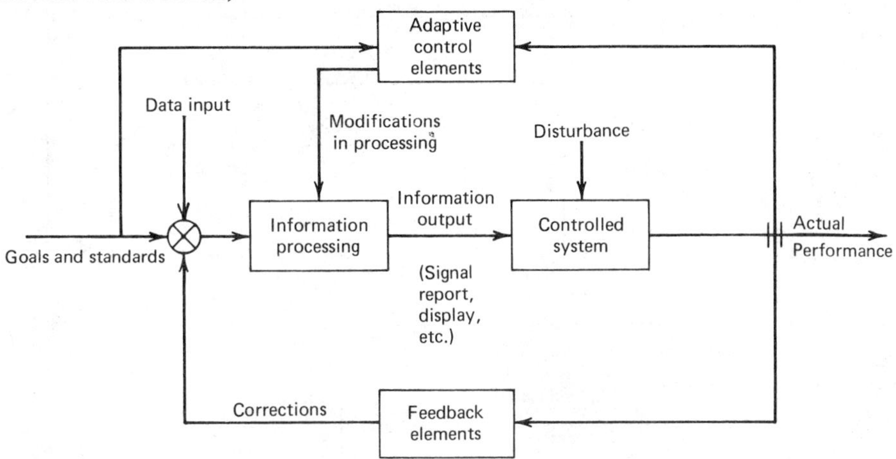

that provide the information to control operations or (2) maintenance of internal control over the quality and security of the information itself. Since information systems are usually complex, graphic models are typically being used.

Any of the control models surveyed in this chapter can essentially incorporate an information system as indicated in some of the examples given. The purpose of an information system is to provide useful, high-quality information; therefore it can be used for sound planning of operations and for preparation of realistic standards of performance. Gathering, classifying, sorting, and analyzing large amounts of data can provide timely and accurate measurement of actual performance. This can be compared to reference information and standards that are also stored in the information system, in order to immediately establish discrepancies and initiate corrective actions. Thus an information system can improve the control operation in all its major functions by measuring and collecting actual performance measures, by analyzing and comparing the actual to the desired set points, and by directing or actuating corrective adjustments.

As in other application areas, the purpose of control models of information systems is to aid in the design and analysis of control capabilities. For instance, consider the adaptive control model shown in Exhibit 13.10.17. The actual performance of a certain operation is measured by feedback elements, either manually or mechanically. This feedback, together with other data about the operation or needed by it, and with goals and standards established for it, is processed and analyzed by the information system. Information produced can be used directly as signals, or indirectly, to control the operation. In a secondary control loop, adaptive control elements measure the actual performance as well as the external conditions of the whole system, as discussed in the general adaptive control model (Section 13.10.4). When external conditions are significantly modified, as defined for the adaptive controller, modifications in the information processing system are indicated. Some of the changes can be implemented directly, whereas others may require the involvement of information system analysts.

The degree of sophistication of a given information system will determine the amount of human operator involvement. For instance, a limited data processing system will produce reports, possibly concentrating on highlighting exceptions, but a human operator will have to analyze, compare, and evaluate the information and decide how to respond. On the other hand, a decision support system will analyze the data, evaluate alternative actions, and recommend proper response to the operator. The reader is referred to Section 12 of the handbook for a specific discussion of information systems.

Control models have also been applied for internal control of information systems. Internal control includes the physical control of information assets against theft and tampering, quality control of data accuracy and integrity, and control of operation efficiency.[24,25] Exhibit 13.10.18 shows a model of internal control in an information system. Such a model is used mainly to explain the different types of internal control. More detailed models are typically employed to specify completely the necessary control functions in each particular area.

13.10.6 SOME FUTURE EXPECTATIONS

There are many possible paths that the development of control systems, and hence the development and use of control models, can take in the future. However, in many cases the paths now appear clear, as follows:

Exhibit 13.10.18 Internal Control of an Information System

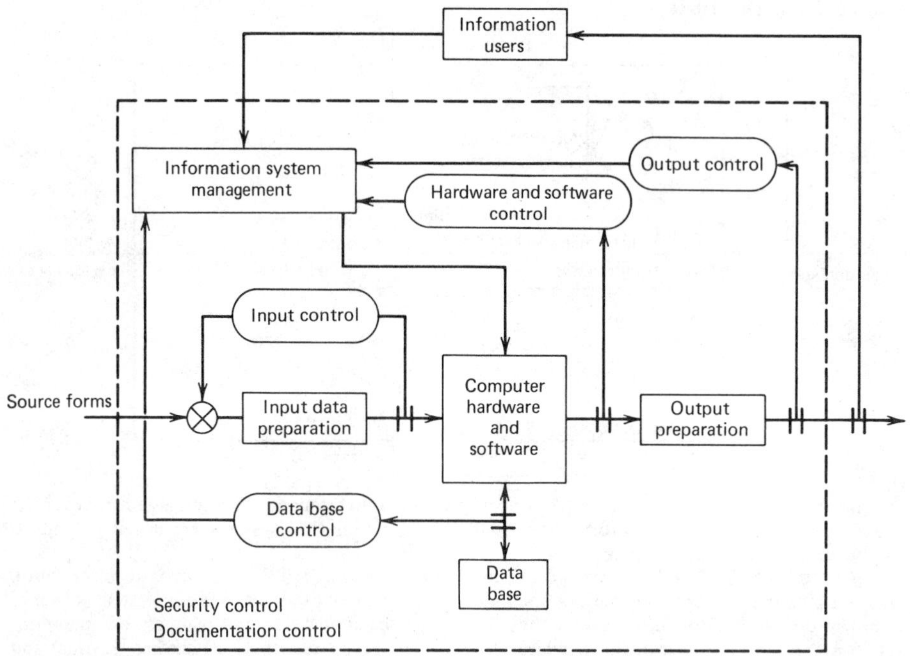

1. A major concern of all installations of computer control systems has been that of the reliability of the overall system. Not that computers are themselves any less reliable than the alternative analog systems, but their involvement in overall plant control systems in contrast to the multiple single loops of the analog system makes them more vulnerable. This will be countered by the use of multiple-redundant computer systems in place of single computers and of fault-tolerant and fault remedying techniques whenever possible.
2. Computer programming complexity is the major bottleneck in computer control system development. Company proprietary considerations of both users and vendors will hold up the development of remedies to this problem. However, programming aids and software sharing must materialize because of the extreme economic incentives involved.
3. Electronic techniques have the potential, so far unexploited, of drastically reducing the cost of industrial control systems. The implementation of these techniques, however, will be slow because of developmental capital needed by vendor companies. On the part of user companies there is also a lack of push because of the relatively small percentage that instrumentation and control constitutes of the total plant cost. Therefore, if vendors choose to compensate for lowered electronic component cost with a greater system complexity to increase reliability or to decrease programming costs, this will be highly desirable.
4. Hierarchical and distributed computer systems are the wave of the future. Their acceptance is driven by the low cost and relatively low capability of the microprocessor. At the same time they are promoting a modularity and simplification of systems programming that should be retained. Likewise, they tend to promote reliability by subdividing system tasks among several computers, thus lessening the impact of the failure of any one computer.
5. Use of artificial intelligence techniques is expected to provide new approaches to control logic specification, particularly in adaptive and learning control, once these methods become practical.

REFERENCES

1. D. M. CONSIDINE, Ed., *Process Instruments and Controls Handbook*, 2nd ed., McGraw-Hill, New York, 1974.
2. J. B. COX, L. J. HELLUMS, T. J. WILLIAMS, R. S. BANKS, and G. J. KIRK, JR., "A Practical Spectrum of DDC Chemical-Process Control Algorithms," *Instrument Society of America Journal*, Vol. 13, No. 10 (October 1966), pp. 65–72.
3. J. G. TRUXAL, *Automatic Feedback Control Synthesis*, McGraw-Hill, New York, 1955.

4. T. J. WILLIAMS and F. M. RYAN, *Progress in Direct Digital Control*, Instrument Society of America, Pittsburgh, 1969.

5. T. B. SHERIDAN and W. R. FERRELL, *Man-Machine Systems: Information, Control, and Decision Models of Human Performance*, The MIT Press, Cambridge, MA, 1974.

6. R. E. BELLMAN, *Dynamic Programming*, Princeton University Press, Princeton, NJ, 1957.

7. L. S. PONTRYAGIN, V. G. BOLTYANSKII, R. V. GAMRELIDZE, and E. F. MISCHENKO, *The Mathematical Theory of Optimal Processes*, Wiley-Interscience, New York, 1962.

8. R. E. KALMAN, *Transactions of the American Society of Mechanical Engineers*, Vol. 82D, 1960, pp. 35–45.

9. G. N. SARIDIS, *Self-Organizing Control of Stochastic Systems*, Dekker, New York, 1977.

10. M. ATHANS and P. L. FALB, *Optimal Control*, McGraw-Hill, New York, 1966.

11. P. S. BUCKLEY, *Techniques of Process Control*, Wiley, New York, 1964.

12. F. A. SHINSKEY, *Process Control Systems*, McGraw-Hill, New York, 1967.

13. R. S. PRESSMAN and J. E. WILLIAMS, *Numerical Control and Computer-Aided Manufacturing*, Wiley, New York, 1977.

14. J. J. VERZIJL, *Production Planning and Information Systems*, Halsted, New York, 1976.

15. A. K. KOCHHAR, *Development of Computer-Based Production Systems*, Halsted, New York, 1979.

16. T. J. WILLIAMS, "Hierarchy Computer Control Systems for Overall Industrial Plant Control," *Proceedings of the 13th International Conference on Systems Sciences*, University of Hawaii at Honolulu, Honolulu, 1980.

17. S. Y. NOF, M. M. BARASH, and J. J. SOLBERG, "Operational Control of Item Flow in Versatile Manufacturing Systems," *International Journal of Production Research*, Vol. 17, No. 5 (1979), pp. 479–489.

18. S. Y. NOF, W. I. BULLERS, and A. B. WHINSTON, "Decision and Control in Automatic Manufacturing," *AIIE Transactions*, Vol. 12, No. 2 (June 1980), pp. 156–169.

19. W. B. ROUSE and D. GOPHER, "Estimation and Control Theory: Application to Modeling Human Behaviour," *Human Factors*, Vol. 19, No. 4 (1977), pp. 315–330.

20. D. T. MCRUER, R. W. ALLEN, D. H. WEIR, and R. H. KLEIN, "New Results in Driver Steering Control Models," *Human Factors*, Vol. 19, No. 4 (1977), pp. 381–398.

21. D. L. KLEINMAN, S. BARON, and W. H. LEVISON, "A Control Theoretic Approach to Manned-Vehicle Systems Analysis," *IEEE Transactions on Automatic Control*, Vol. AC-16, 1971, pp. 824–832.

22. S. BARON and W. H. LEVISON, "Display Analysis With the Optimal Control Model of the Human Operator," *Human Factors*, Vol. 19, No. 5 (1977), pp. 437–458.

23. R. E. CURRY, D. L. KLEINMAN, and W. C. HOFFMAN, "A Design Procedure for Control/Display Systems," *Human Factors*, Vol. 19, No. 5 (1977), pp. 421–436.

24. J. G. BURCH, JR., and F. R. STRATER, JR., *Information Systems: Theory and Practice*, Hamilton, 1974.

25. G. B. DAVIS, *Computers and Information Processing*, McGraw-Hill, New York, 1978.

BIBLIOGRAPHY

BAILEY, S. J. and K. PLUHAR, "Flexible Manufacturing Systems, Digital Control and the Automatic Factory," *Control Engineering*, September 1979, pp. 59–64.

BIBBERO, R. J., *Microprocessors in Instruments and Control*, Wiley-Interscience, New York, 1977.

DISTEFANO, J. J., A. R. STUBBERUD, and I. J. WILLIAMS, *Feedback and Control Systems*, Schaum's Outline Series, McGraw-Hill, New York, 1967.

DORF, R. C., *Modern Control Systems*, 3rd ed., Addison-Wesley, Reading, MA, 1980.

NILSSON, N. J., *Principles of Artificial Intelligence*, Tioga, 1980.

SCHOEFFLER, J. D. and R. H. TEMPLE, *Minicomputers: Hardware, Software, and Applications*, IEEE Press, New York, 1972.

CHAPTER 13.11

Computer Simulation

JAMES R. WILSON

University of Texas

A. ALAN B. PRITSKER

Pritsker & Associates

13.11.1 INTRODUCTION TO SIMULATION MODELING

As the problems facing industrial engineers in both the public and private sectors continue to grow in size and complexity, the tools of industrial engineering and systems analysis are playing an increasingly important role in the resolution of these problems. Revolutionary advances in electronic computers over the past 30 years have given great impetus to this development and, in particular, have contributed to the emergence of simulation as one of the most powerful and widely used techniques for analyzing complex systems. Computer simulation models can be employed at four levels:

As explanatory devices to define a system or problem.

As analysis vehicles to determine critical elements, components, and issues.

As design assessors to synthesize and evaluate proposed solutions.

As predictors to forecast and aid in planning future developments.

This chapter will enable the reader to solve problems through effective simulation modeling of real-world systems. The following topics are discussed: (1) choosing an appropriate modeling approach and simulation language for a project, (2) selecting suitable input probability distributions, (3) validating the proposed model, and (4) analyzing the experimental results to obtain the desired information. To provide a basis for this discussion, we present first an overview of the simulation process.

Definition of Simulation

In its broadest sense, computer simulation is the process of designing a mathematical-logical model of a real system and experimenting with this model on a computer.[1,2] Thus simulation encompasses a model building process as well as the design and implementation of an appropriate experiment involving that model. These experiments permit inferences to be drawn about systems

Without building them, if they are only proposed systems.

Without disturbing them, if they are operating systems that are costly or unsafe to experiment with.

Without destroying them, if the object of an experiment is to determine their limits of stress.

In this way simulation models can be used for design, procedural analysis, and performance assessment.[1]

Simulation modeling assumes that we can describe a system in terms acceptable to a computer. In this regard a key concept is that of a "system state description." If a system can be characterized by a set of variables, with each combination of variable values representing a unique state or condition of the system, then manipulation of the variable values simulates movement of the system from state to state. Thus we see that simulated experimentation involves observing the dynamic behavior of a model over time; depending on the nature of the inputs, the observed output processes will be either deterministic or stochastic. Changes in the state of a system can occur continuously over time or at discrete instants in time. Although the procedures for describing the dynamic behavior of discrete and continuous change models differ, the basic concept of simulating a system by portraying the changes in the state of the system over time remains the same. We now consider the model building process in more detail.

Model Building

Models are *descriptions* of systems. The usefulness of models has been demonstrated in describing, designing, and analyzing systems. Many students are educated in their discipline by learning how to build and use models. Model building is a complex process and in most fields an art. The modeling of a system is made easier if (1) physical laws are available that pertain to the system; (2) pictorial or graphic representation can be made of the system; and (3) the variability of system inputs, elements, and outputs is manageable.[3]

The modeling of complex, large-scale systems is often more difficult than the modeling of physical systems for the following reasons: (1) few fundamental laws are available; (2) many procedural elements are involved which are difficult to describe and represent; (3) policy inputs are required which are hard to quantify; (4) random components are significant elements; and (5) human decision making is an integral part of such systems. A simulation approach circumvents many of these difficulties.

Since a model is a description of a system, it is also an abstraction of a system. Model builders must decide on the elements of the system to include in their models. To make such decisions, a purpose for model building should be established. Reference to this purpose should be made when deciding if an element of a system is significant and hence should be modeled. The success of a modeler depends on how well he or she can define significant elements and the relationship between elements.

Simulation models are ideally suited for problem solving. Simulation provides the flexibility to build either aggregate or detailed models. It also supports the concepts of iterative model building by allowing models to be embellished through simple and direct additions. The next section describes the steps involved in performing a simulation study.

Steps in a Simulation Study

As was alluded to previously, the process for the successful development of a simulation model begins with a simple model that is embellished in an evolutionary fashion to meet problem solving requirements. Within this process, the following stages of development can be identified[2,4]:

1. **Problem Formulation.** The definition of the problem to be studied, including a statement of the problem solving objective.
2. **Model Building.** The abstraction of the system into mathematical-logical relationships in accordance with the problem formulation.
3. **Data Acquisition.** The identification, specification, and collection of data.
4. **Model Translation.** The preparing of the model for computer processing.
5. **Verification.** The process of establishing that the computer program executes as intended.
6. **Validation.** The process of establishing that a desired accuracy or correspondence exists between the simulation model and the real system.
7. **Strategic and Tactical Planning.** The process of establishing the experimental conditions for using the model.
8. **Experimentation.** The execution of the simulation model to obtain output values.
9. **Analysis of Results.** The process of analyzing the simulation outputs to draw inferences and make recommendations for problem resolution.
10. **Implementation and Documentation.** The process of implementing decisions resulting from the simulation and of documenting the model and its use.

These stages of simulation development are rarely performed in a structured sequence beginning with problem definition and ending with documentation. A simulation project may involve false starts; erroneous assumptions, which must later be abandoned; reformulation of the problem objectives; and repeated evaluation and redesign of the model. If properly done, however, this iterative process should result in a simulation model that properly assesses alternatives and enhances the decision making process.[4]

13.11.2 MODELING APPROACHES

In developing a simulation model, an analyst needs to select a conceptual framework for describing the system to be modeled. The framework or perspective contains a "world view" within which the system functional relationships are perceived and described. This section summarizes the alternative world views for simulation modeling.

Discrete Simulation

Discrete simulation occurs when the dependent system variables change discretely at specified points in simulated time, referred to as "event times." The time variable may be either continuous or discrete in such a model, depending on whether the discrete changes in the dependent variable can occur at any time or only at specified times.

As an example of discrete simulation, we will examine the processing of customers by a teller at a bank. Customers arrive at the bank, wait for service by the teller if the teller is busy, are served, and then depart the system. Customers arriving at the system when the teller is busy wait in a single queue in front of the teller. To build a discrete simulation of this system, we must define the states of the system and identify the events that can change system status. Note that the state of the system is completely specified by the status of the teller and by the number of customers in the bank. The state of the system is changed by (1) the arrival of a customer at the bank and (2) the completion of service by the teller and subsequent departure of the customer. To simulate this system, we need to generate a stream of customer arrivals and their corresponding service times—perhaps by sampling from appropriate input probability distributions. Exhibit 13.11.1 summarizes the results of one sample of 10 simulated customers.

An event-oriented description of the bank teller status and the number of customers at the bank is given in Exhibit 13.11.2. In this exhibit the events are listed in chronological order. A graphic portrayal of the status variables over time is shown in Exhibit 13.11.3. These results indicate that the average number of customers at the bank in the first 40 min is 1.4525 and that the teller is idle for 20% of the time. This simple example provides a basis for the general discussion of discrete simulation that follows.

The objects within the boundaries of a discrete system, such as people, equipment, orders, and raw materials, are called "entities." There are many types of entities, and each has various characteristics or attributes. Although they engage in different types of activities, entities may have a common attribute requiring that they be grouped together. Groupings of entities are called "files." Inserting an entity into a file implies that it has some relationship to other entities in the file.

The aim of a discrete simulation model is to reproduce the activities that the entities engage in and thereby learn something about the behavior and performance potential of the system. This is done by defining the states of the system and constructing activities that move it from state to state. The state of a system is defined in terms of the numerical values assigned to the attributes of the entities.

In discrete simulation the state of the system can change only at event times. Since it remains constant between event times, a complete dynamic portrayal of the state of the system can be obtained by advancing simulated time from one event to the next. This timing mechanism is referred to as the "next event approach" and is used in most discrete simulation languages.

A discrete simulation model can be formulated by (1) defining the changes in state that occur at each event time, (2) describing the activities in which the entities in the system engage, or (3) describing the process through which the entities in the system flow. The relationship among the concepts of an event, an activity, and a process is depicted in Exhibit 13.11.4. An event takes place at a point in time at which decisions are made to start or end activities. A process is a time-ordered sequence of events and may encompass several activities. These concepts lead naturally to three

Exhibit 13.11.1 Discrete Simulation of Bank Teller

Customer Number (1)	Arrival Time (2)	Service Time (3)	Start Service Time (4)	Departure Time (5)	Time In Queue (6) = (4) – (2)	Teller Idle Time (7)
1	3.2	3.8	3.2	7.0	0.0	3.2
2	10.9	3.5	10.9	14.4	0.0	3.9
3	13.2	4.2	14.4	18.6	1.2	
4	14.8	3.1	18.6	21.7	3.8	
5	17.7	2.4	21.7	24.1	4.0	
6	19.8	4.3	24.1	28.4	4.3	
7	21.5	2.7	28.4	31.1	6.9	
8	26.3	2.1	31.1	33.2	4.8	
9	32.1	2.5	33.2	35.7	1.1	
10	36.6	3.4	36.6	40.0	0.0	0.9

Exhibit 13.11.2 Event-Oriented Description of Bank Teller Simulation

Event Time	Customer Number	Event Type	Number in Queue	Number in Bank	Teller Status	Teller Idle Time
0.0	–	Start	0	0	Idle	–
3.2	1	Arrival	0	1	Busy	3.2
7.0	1	Departure	0	0	Idle	
10.9	2	Arrival	0	1	Busy	3.9
13.2	3	Arrival	1	2	Busy	
14.4	2	Departure	0	1	Busy	
14.8	4	Arrival	1	2	Busy	
17.7	5	Arrival	2	3	Busy	
18.6	3	Departure	1	2	Busy	
19.8	6	Arrival	2	3	Busy	
21.5	7	Arrival	3	4	Busy	
21.7	4	Departure	2	3	Busy	
24.1	5	Departure	1	2	Busy	
26.3	8	Arrival	2	3	Busy	
28.4	6	Departure	1	2	Busy	
31.1	7	Departure	0	1	Busy	
32.1	9	Arrival	1	2	Busy	
33.2	8	Departure	0	1	Busy	
35.7	9	Departure	0	0	Idle	
36.6	10	Arrival	0	1	Busy	0.9
40.0	10	Departure	0	0	Idle	

alternative world views for discrete simulation modeling. These world views are commonly referred to as the "event," "activity scanning," and "process" orientations and are described in the following sections.

Event-Oriented Approach

In the event-oriented world view, a system is modeled by defining the changes that occur at event times. The task of the modeler is to determine the events that can change the state of the system and then to develop the logic associated with each event type. A simulation of the system is produced by executing the logic associated with each event in a time-ordered sequence.

To create a simulation of the bank teller problem using the event orientation, we would maintain a calendar of events and cause their execution to occur at the proper points in simulated time. The event calendar would initially contain an event notice corresponding to the first arrival event. As the simulation proceeds, additional arrival events and end-of-service events would be scheduled onto the calendar as prescribed by the logic associated with the events. Each event would be executed in a time-ordered sequence, with simulated time being advanced from one event to the next.

If the modeler employs a general-purpose language such as FORTRAN to code a discrete event model, then a considerable amount of programming effort will be directed at developing the event

Exhibit 13.11.3 Graphic Portrayal of Bank Teller Simulation

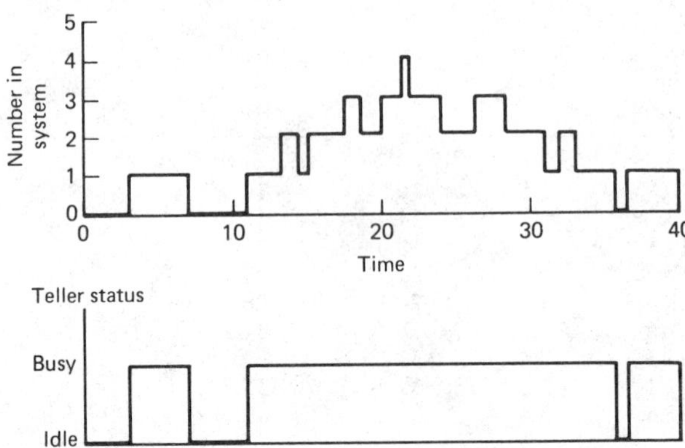

Exhibit 13.11.4 Relationship of Events, Activities, and Processes

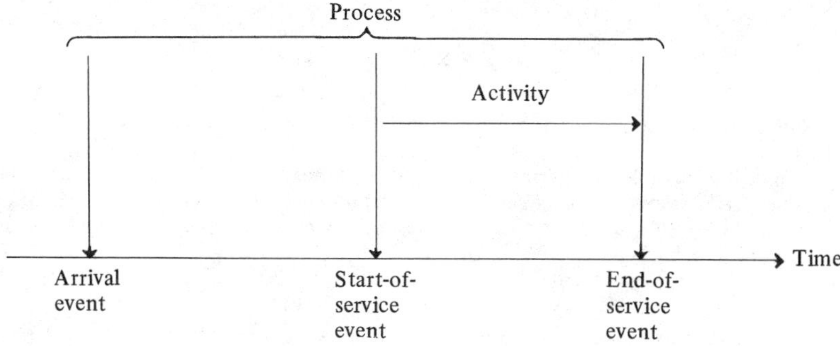

calendar and a timing mechanism for processing the events in their proper chronological order. Since this function is common to all discrete event models, a number of simulation languages have been developed that provide special features for event scheduling as well as other functions that are commonly encountered in discrete event models. The most commonly used event-oriented simulation languages are discussed in Section 13.11.3.

Process-Oriented Approach

Many simulation models include sequences of elements that occur in defined patterns, for example, a queue where entities wait for processing by a server. The logic associated with such a sequence of events can be generalized and defined by a single statement. A simulation language could then translate the statement into the appropriate sequence of events. A process-oriented language employs such statements to model the flow of entities through a system. These statements define a sequence of events that is automatically executed by the simulation language as the entities move through the process.[5]

The process orientation provides a description of the flow of entities through a process consisting of resources. Its simplicity is derived from the fact that the event logic associated with the statements is contained within the model description. The most commonly used process-oriented simulation languages are discussed in Section 13.11.3.

Activity Scanning Approach

In the activity scanning orientation, the modeler describes the activities in which the entities in the system engage and prescribes the conditions that cause an activity to start or end. The events that start or end the activity are not scheduled by the modeler, but are initiated from the conditions specified for the activity. As simulated time is advanced, the conditions for either starting or ending an activity are scanned. If the prescribed conditions are satisfied, then the appropriate action for the activity is taken. To ensure that each activity is accounted for, it is necessary to scan the entire set of activities at each time advance.

For certain types of problems, the activity scanning approach can provide a concise modeling framework. The approach is particularly well suited for situations where an activity duration is indefinite and is determined by the state of the system satisfying a prescribed condition. However, because of the need to scan each activity at each time advance, the approach is relatively inefficient when compared to the discrete event orientation. As a result, the activity scanning orientation has not been widely adopted as a modeling framework for discrete simulations.

Continuous Simulation

In a continuous simulation model, the state of the system is represented by dependent variables that change continuously over time. To distinguish continuous change variables from discrete change variables, the former are referred to as "state variables." A continuous simulation model is constructed by defining equations for a set of state variables whose dynamic behavior simulates the real system.

Models of continuous systems are frequently written in terms of differential equations. The reason for this is that it is often easier to construct a relationship for the rate of change of the state variable than to devise a relationship for the state variable directly. For example, our modeling effort might produce the following differential equation describing the behavior of the state variable s over time t together with an initial condition at time 0:

$$\frac{ds(t)}{dt} = s^2(t) + t^2$$

$$s(0) = k$$

The simulation analyst's objective is to determine the response of the variables over a specified time period.

In some cases it is possible to determine an analytical expression for the state variable, s, given an equation for ds/dt. However, in many cases of practical importance, an analytical solution for s will not be known. As a result, we must obtain the response s by integrating ds/dt over time using an equation of the following type:

$$s(t_2) = s(t_1) + \int_{t_1}^{t_2} \left(\frac{ds}{dt}\right) dt$$

How this integration is performed depends upon whether the modeler employs an analog or digital computer.

An analog computer represents the state variables in the model by electrical charges. The dynamic structure of the system is modeled using circuit components such as resistors, capacitors, and amplifiers. The principal shortcoming of an analog computer is that the quality of these components limits the accuracy of the results. In addition, the analog computer lacks the logical control functions and data storage capability of the digital computer.

A number of continuous simulation languages have been developed for use on digital computers. A digital computer performs the common mathematical operations, such as addition, multiplication, and logical testing, with great speed and accuracy, and it uses numerical methods to perform the integration operation required in continuous simulation. These methods divide the independent variable (normally time) into small slices, referred to as "steps." The values for the state variables requiring integration are obtained by employing an approximation to the derivative of the state variable over time. In this situation there is a trade-off between accuracy of state variable calculations and computer run time. A description of the various numerical integration algorithms can be found in any of the introductory texts on numerical analysis.[6]

Sometimes a continuous system is modeled using difference equations. In these models the time axis is decomposed into time periods of length Δt. The dynamics of the state variables are described by specifying an equation that calculates the value of the state variable at period $k + 1$ from the value of the state variable at period k. For example, the following difference equation could be employed to describe the dynamics of the state variable s:

$$s_{k+1} = s_k + r^* \Delta t$$

When using difference equations, the essential structure of a continuous simulation model is often reflected in the relationship between the rate r projected to period $k + 1$ and the value s_k of the state variable at period k. Continuous system simulation languages are discussed in Section 13.11.3.

Combined Discrete-Continuous Simulation

In combined discrete-continuous models, the independent variables may change both discretely and continuously. The world view of a combined model specifies that the system can be described in terms of entities, their associated attributes, and state variables. The behavior of the system model is simulated by computing the values of the state variables at small time steps and the values of attributes of entities at event times.

There are two types of events that can occur in combined simulations. "Time events" are those events scheduled to occur at specified times. They are commonly thought of in terms of discrete simulation models. In contrast, "state events" are not scheduled, but occur when the system reaches a particular state. For example, as illustrated in Exhibit 13.11.5, a state event could be specified to occur whenever state variable X crosses state variable Y in the positive direction. Note that the notion of a state event is similar to that of activity scanning in that the event is not scheduled, but is initiated by the state of the system. The possible occurrence of a state event must be tested at each time advance in the simulation.

The next section examines some special-purpose simulation languages that have been developed to implement the modeling approaches just discussed.

Exhibit 13.11.5 Example of State Event Occurrence

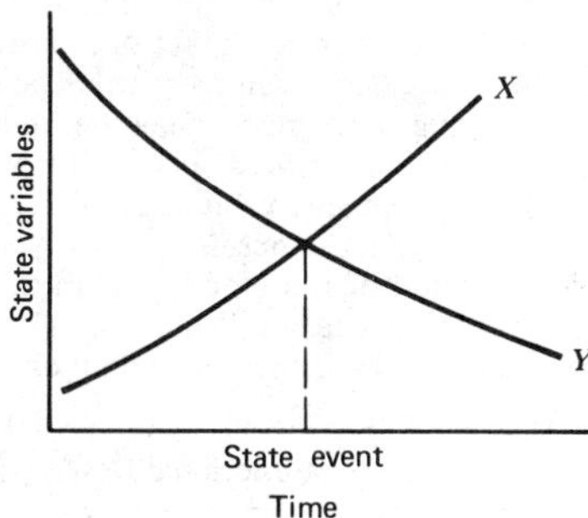

13.11.3 SIMULATION LANGUAGES

The widespread use of simulation as an analysis tool has led to the development of a number of languages specifically designed for simulation. Shannon[2] has identified the following advantages of using such a special-purpose language when performing a simulation study:

1. Reduction of the programming task.
2. Guidance in concept articulation and model formulation.
3. Aid in communication and documentation of the study.
4. Flexibility in embellishment or revision of the model.
5. Provision of the common support functions required in any simulation.

Emshoff and Sisson[7] list the following support functions as requisites for any simulation language:

1. Generation of random variates.
2. Management of the simulated clock.
3. Collection and recording of output data.
4. Summarization and statistical analysis of output data.
5. Detection and reporting of error conditions.
6. Generation of standard output reports.

This section compares the major characteristics of the most widely used simulation languages based on each modeling approach. The intent is not to provide detailed descriptions of each language, but to highlight the similarities and differences that exist among these languages. Completing the discussion is an examination of the factors that are relevant to the selection of an appropriate language.

Discrete Languages

The major process-oriented languages discussed here are GPSS and Q-GERT. The language SIMSCRIPT is primarily event-oriented. Although GASP IV and SLAM also have event-oriented capabilities, they are discussed separately as combined discrete-continuous languages.

GPSS

General Purpose Simulation System (GPSS) is a process-oriented simulation language for modeling discrete systems. It exists in a number of dialects, with GPSS/360 the most widely circulated and used version. The discussion that follows is based on this version. GPSS V is a superset of GPSS/360, and therefore programs written for GPSS/360 are compatible with GPSS V.[8] Schriber's book, *Simulation Using GPSS*,[9] is an excellent text for learning GPSS.

The principal appeal of GPSS is its modeling simplicity. A GPSS model is constructed by combining a set of standard blocks into a block diagram that defines the logical structure of the system. Dynamic entities are represented in GPSS as transactions that move sequentially from block to block as the simulation proceeds. Learning to write a GPSS program consists of learning the functional operation of GPSS blocks and how to combine the blocks logically to represent a system of interest.

The GPSS processor interprets and executes the block diagram description of a system. The language is limited in computing power and lacks a capability for floating point or real arithmetic.

As a consequence, the GPSS simulation clock is integer valued. This means that changes in the state of the system can occur only at integer points in time. For example, if we model a single-channel queuing system and select minutes as our unit of time, then the time between arrivals and the service time for the customers must be an integer number of minutes. Another consequence of an integer-valued clock is that simultaneous events can frequently occur if the selected time unit is too large, and therefore the tie breaking mechanism takes on added significance. In GPSS this problem is addressed by assigning each transaction a special attribute, called a "priority," which can be any integer between 0 and 127, with higher values assuming greater priority.

There are 48 different blocks in GPSS. Each block is pictorially represented by a stylized figure intended to be suggestive of the operation of the block. When a transaction succeeds in entering a block, the underlying event subroutine is executed; this may alter system status or modify the flow of transactions through the model. Some of the most commonly used blocks are as follows:

Block Type	Functional Description
GENERATE	Create transactions.
TERMINATE	Destroy transactions; stop a run.
ADVANCE	Freeze the movement of a transaction for a specified period of simulated time.
SEIZE	Cause a transaction to capture a facility (a single server).
RELEASE	Cause a transaction to release a facility.
ENTER	Cause a transaction to capture a specified number of units of a storage (a set of parallel servers).
LEAVE	Cause a transaction to release a specified number of units of a storage.
QUEUE	Increase the number in a queue.
DEPART	Decrease the number in a queue.
TABULATE	Record the observation of a system variable in a specified frequency distribution.

The GPSS language provides almost all of the basic simulation functions listed in the beginning of this section. In particular, it has extensive data collection and summarization capabilities. Statistics are automatically reported at the end of each run on the utilization of facilities and storages, on queue lengths, and on system sojourn times for transactions. Because GPSS is also easy to learn, the time to code and debug a GPSS model is frequently much less than would be required with a general-purpose programming language. On the other hand, GPSS models usually execute more slowly and hence are more expensive to run. It should be noted that GPSS has a limited capability for generating random variates: Apart from the built-in uniform distribution, the user must supply the cumulative distribution function for each random variable to be sampled. However, in the more recently developed versions of the GPSS processor, execution speeds have been substantially improved, and a wider variety of random variate generators has also been provided.

Q-GERT

Developed by Pritsker,[3] Q-GERT is a network-oriented simulation language. The letters "GERT" represent an acronym for Graphical Evaluation and Review Technique; the "Q" is appended to indicate that queuing systems can be modeled in graphic form.

The language employs an activity-on-branch network philosophy in which a branch represents an activity that models a processing time or delay. Nodes are used to separate branches and to model milestones, decision points, and queues. A Q-GERT network consists of nodes and branches. Flowing through the network are entities, referred to as "transactions." Different types of nodes are included in Q-GERT to allow for the modeling of complex queuing situations and project management systems.

There are 10 node types in Q-GERT. This small number is possible because the Q-GERT philosophy is to add or combine functions at nodes when they are required. In addition, Q-GERT is designed to facilitate the use of FORTRAN programming inserts at both nodes and in activities. The node types are as follows:

Node Type	Description
SOURCE	Create transactions.
REGULAR	Accumulate transactions.

Node Type	Description
STATISTICS	Accumulate transactions and collect statistics.
SINK	Accumulate transactions, collect statistics, terminate a run.
QUEUE	Determine disposition of transaction: hold or route to server.
SELECT	Determine disposition of transaction and/or server.
MATCH	Match transactions in QUEUE nodes with same attribute values and route transactions when a match occurs.
ALLOCATE	Allocate resources to transactions waiting for resources in QUEUE nodes.
FREE	Make resources available to be allocated.
ALTER	Change capacity of resources.

As an example of combining functions, note that QUEUE nodes are used in conjunction with SELECT, MATCH, and ALLOCATE nodes to model specific operations.

Attributes are used to distinguish transactions (entities) flowing through a Q-GERT model. Attribute values can be assigned at any node. Activity durations are prescribed by a distribution type and a parameter set number.

The procedures for constructing a model in Q-GERT are similar to those used in GPSS. The modeler combines the Q-GERT network elements into a network model that pictorially represents the system of interest. This network model is then transcribed onto input records for interpretation and processing by the Q-GERT analysis program.

Although Q-GERT and GPSS are similar in some respects, their differences should be noted. Both languages provide for automatic collection of statistics on many standard system entities over a single simulation run; however, Q-GERT also collects such statistics over a set of independent runs. As discussed later in Section 13.11.4, this capability facilitates the analysis of model outputs. Because of its small number of node types, Q-GERT is also quite easy to learn. Unlike GPSS, Q-GERT has a real-valued clock and provides functions to generate all of the commonly used random variates.

SIMSCRIPT

The simulation language SIMSCRIPT was originally developed at the RAND Corporation.[10] The form SIMSCRIPT II is divided into five levels:

Level 1. A simple teaching language designed to introduce programming concepts to nonprogrammers.

Level 2. Statement types that are comparable in power to FORTRAN.

Level 3. Statement types that are comparable in power to ALGOL or PL/I.

Level 4. Statement types that provide a structure for modeling using entity, attribute, and set concepts.

Level 5. Statement types for time advance, event processing, generation of samples, and accumulation and analysis of simulation-generated data.

One of the principal appeals of SIMSCRIPT as a programming and simulation language is its English-like and free-form syntax. Programs written in SIMSCRIPT are easy to read and tend to be self-documenting.

The discrete simulation modeling framework of SIMSCRIPT is primarily event oriented. In SIMSCRIPT the state of the system is defined by entities, their associated attributes, and logical groupings of entities, referred to as "sets." The dynamic structure of the system is described by defining the changes that occur at event times.

In SIMSCRIPT, two types of entities are considered. An entity that remains throughout a simulation is referred to as a "permanent entity." In contrast, "temporary entity" is used to refer to entities that are created and destroyed during execution of the simulation. In the latter case computer storage space is automatically allocated and freed during execution of the simulation as individual entities are created and destroyed. In a queuing system, each server would be modeled as a permanent entity, and the customers who arrive at and depart from the system would be modeled as temporary entities.

In SIMSCRIPT, the attributes of entities are separately named, not numbered, thereby enhancing model description. For example, we could define a temporary entity named CUSTOMER that has an attribute named MARK.TIME. Sets are also named as opposed to being numbered, further enhancing the model description. For example, a set containing customers waiting for service could be named QUEUE.

A SIMSCRIPT simulation model consists of a preamble, a main program, and event subprograms. The preamble is not part of the executable program and is used to define the elements of a model. One of the primary functions of the preamble is to define the static structure of the model by prescribing the names of permanent and temporary entities, their associated attributes, and set relationships. For each discrete event, the event name and its corresponding attributes are defined in the preamble. Declarative statements for defining all variable types and arrays are also included in the preamble. One other function of the preamble is to define variables for which statistics are to be collected.

The main program is used for initializing variables, scheduling the initial occurrence of events, and starting the simulation. The event subprograms are used for defining the logic associated with processing each event in the model. The calls to the event subprograms are scheduled by the user, but executed by the SIMSCRIPT control program.

Once the description of the static structure is completed by writing the preamble, the next step in writing a SIMSCRIPT program is to code the main program and event subprograms. These are coded by using the general-purpose programming statements of SIMSCRIPT in conjunction with special statements for creating and destroying entities, manipulating entities between sets, obtaining random samples, and scheduling events.

The language provides all of the standard support facilities previously outlined. Particularly notable are its flexible statement types for creating output reports. Functions are provided for the usual statistical computations and for random variate generation. Since the coding of the event routines is left to the user, the debugging aids in SIMSCRIPT are not as extensive as in GPSS and Q-GERT. However, the English-like structure of SIMSCRIPT facilitates communication and documentation.

Continuous Languages

Although a wide variety of special-purpose continuous system simulation languages (CSSLs) has been developed since the 1950s, the structure and functions of most CSSLs have been largely standardized in recent years. Whereas early CSSLs were block oriented, so that a continuous model was constructed using a block diagram similar to that for an analog computer, currently most CSSLs are equation oriented and have a FORTRAN-like syntax. Among the CSSLs of this second type, CSMP III, DYNAMO, and GASP IV are the most widely used.[11]

CSMP III

The Continuous System Modeling Program (CSMP) III[12] is representative of a family of CSSLs that has been developed to solve systems of first-order differential equations. The language CSMP III is available only on IBM 360 and 370 series computers. A CSMP III program is composed of three types of statements:

1. **Data statements** establish initial conditions for state variables and assign numerical values to constants and to parameters varied over multiple runs.
2. **Structural statements** specify the way in which solutions to the model equations are calculated. A number of standard functions are available for use in structural statements. For example, the integrator function computes the integral over time of a state variable subject to a specified initial condition. In addition, most standard FORTRAN mathematical function subprograms are available.
3. **Control statements** specify options for program execution and input/output formatting. For example, the TIMER statement specifies the duration of the run, the integration step width, and the data recording interval. The PRINT statement is used to obtain standard printed output.

Two recently developed CSSLs with extended features are Advanced Continuous Simulation Language (ACSL)[13] and Differential Analyses Replacement Evaluation (DARE-P.)[14] An important advantage of the DARE-P system is that the processor is written entirely in ANSI FORTRAN IV and is therefore machine independent.

DYNAMO

Systems Dynamics, as developed by Forrester, is a problem solving approach to complex problems that emphasizes the structural aspects of models of systems.[15] State variables, called "levels," are defined in difference equation form. The DYNAMO programming language[16] was developed to provide a vehicle for analyzing Systems Dynamics models. The language uses a fixed step size, Euler-type integration algorithm to evaluate the level variables over time.

A set of numbered prototype equations are defined in DYNAMO, and the user must structure his or her model to conform to these equation forms. To employ a particular form, the user codes the equation number and the combination of variables or functions required in that form. In addition to standard mathematical functions, such as the exponential and logarithmic functions, DYNAMO provides for operations involving step functions, table functions, clipping, smoothing, and delays. Printing and plotting features similar to those of CSMP III are also provided. Perhaps the most serious drawback of DYNAMO is its use of a fixed step size; if the user selects an interval that is too large, a serious loss of accuracy can occur.

Combined Discrete-Continuous Languages

Although much research has been devoted to the development of combined discrete-continuous simulation languages, General Activity Simulation Program (GASP IV) is the only such language that has achieved widespread use. The more recently developed Simulation Language for Alternative Modeling (SLAM) is based on the GASP IV design for discrete-continuous simulation, and it adds a process-oriented view together with new interface capabilities.

GASP IV

This language provides an organizational structure that allows system descriptions to be written in terms of discrete event models, continuous models, or a combination of the two.[1] This structure specifies procedures for writing differential or difference equations as well as methods for defining the logical conditions that affect system status variables. Within this framework, the GASP executive can perform the time advancement functions required by a simulation model and can call specific user-written routines to obtain system status updates. The details of operation for GASP IV are embodied within SLAM and are not given here.

A recently developed language, GASP V,[17] expands the continuous capabilities of GASP IV. The new features in GASP V involve the inclusion of different integration algorithms, which can be user selected; procedures for handling partial differential equations; and logic, memory, and generator functions. Examples of logic functions are input switches, flip-flops, and gates. Memory functions include hysteresis and delays, and generator functions for step, ramp, impulse, and dead spaces are available.

SLAM

The SLAM language incorporates the process-oriented features of Q-GERT and the combined discrete-continuous features of GASP IV.[4] In SLAM the alternate modeling world views are combined to provide a unified modeling framework. A discrete change system can be modeled within an event orientation, a process orientation, or *both*. Continuous change systems can be modeled using either differential or difference equations. Combined discrete-continuous change systems can be modeled by combining the event and/or process orientation with the continuous orientation. In addition, SLAM incorporates a number of features that correspond to the activity scanning orientation.

The process orientation of SLAM employs a network structure composed of nodes and branches. These symbols model elements in a process, such as queues, servers, and decision points. The modeling task consists of combining these symbols into a network model that pictorially represents the system of interest. The entities in the system flow through the network model. The pictorial representation of the system is transcribed by the modeler into an equivalent statement model for input to the SLAM processor.

For event-oriented simulation models, the user is required merely to code the processing logic corresponding to each event type in separate support routines. In this case the SLAM executive always advances time to the next event occurrence in order to update system status. This is accomplished by including a file or calendar of events and processing the next event whenever the processing of a current event is completed. Though standardized initialization or in subroutine INTLC, the initial events to start the simulation are prescribed and placed on the event calendar. The executive routine removes the first of these initial events and calls subroutine EVENT (IX), where IX is an event code. Subroutine EVENT transfers to the appropriate event, which is user written. In the event, the user schedules other events to occur, updates system status, and collects values for the summary report.

Models of continuous systems involve the definition of state variables by equations and of state events based on the values of state variables. The development of a SLAM continuous simulation program requires the user to write subroutine STATE for defining state equations, SEVNT statements to prescribe the conditions that define state events, and subroutine EVENT for modeling the consequences of the occurrence of a state event. In addition, the user must write a main program that calls the SLAM executive and, if necessary, subroutine INTLC for establishing the initial conditions for the simulation and RECORD statements for saving values of the system variables in summary tables and plots. When the executive routine determines that the simulation is ended, subroutine OTPUT is called. In OTPUT, the user can write a specialized output report and perform any end-of-simulation processing desired. Following the call to subroutine OTPUT, standardized summary reports are printed.

In a combined simulation, the behavior of a system model is simulated by calculating the values of the state variables at the end of steps and making changes to the values of attributes or to the number of entities in existence at event times. The step size is automatically determined by the executive function, so that events can occur only at the end of a step. The SLAM executive also considers accuracy requirements, minimum and maximum user prescribed step size values, and output reporting requirements when determining the appropriate size for the next step.

At the end of each time step, the state variables are evaluated to determine whether the conditions prescribing a state event have occurred. If a state event was passed—that is, a state variable crossed a threshold by more than an allowed amount—the step size was too large and is reduced. If a state event occurs, the model status is updated according to an event subroutine written by the user. As stated previously, the executive function automatically sets the step size so that no time event will occur within a step. This is accomplished by adjusting the size of the step so that the end of the step occurs at the time event instant.

An important aspect of SLAM is that alternate world views can be combined within the same simulation model. There are six specific interactions that can take place between the network, discrete event, and continuous world views of SLAM:

1. Entities in the network model can initiate the occurrence of discrete events.
2. Events can alter the flow of entities in the network model.
3. Entities in the network model can cause instantaneous changes to values of the state variables.
4. State variables reaching prescribed threshold values can initiate entities in the network model.
5. Events can cause instantaneous changes to the values of state variables.
6. State variables reaching prescribed threshold values can initiate events.

The ability to construct combined network-event-continuous models with interactions between each orientation greatly enhances the modeling power of the systems engineer.

Choosing a Simulation Language

The selection of a simulation language is frequently based on knowledge and availability as opposed to a formal comparison of language features. However, if the frequent use of simulation is anticipated, then a comprehensive evaluation of the available languages and anticipated modeling needs is warranted. Shannon provides a review and diagrams a procedure for making such an evaluation.[2] Exhibit 13.11.6 is a summary of important factors to consider in comparing simulation languages.[4]

13.11.4 STATISTICAL ISSUES

Running a simulation model on a computer is in essence a complex sampling experiment. Thus the procedures for designing and analyzing simulation runs are similar to the techniques used in other scientific experiments; the main difference is that the simulation analyst has greater control over the experimental conditions. An appropriate statistical analysis is a necessary part of a simulation study in order to (1) use simulation-generated data efficiently in the estimation of system performance measures and (2) reveal the scope and limitations of the conclusions based on the data. The first consideration involves the selection of a method for data collection and estimation as well as the resolution of tactical questions concerning how to start, execute, and stop the simulation. The

Exhibit 13.11.6 Features on Which to Evaluate a Simulation Language

Feature	Consideration
Training required	Ease of learning the language
	Ease of conceptualizing simulation problems
Coding considerations	Ease of coding, including random sampling and numerical integration
	Degree to which code is self-documenting
Portability	Language availability on other or new computers
Flexibility	Degree to which language supports different modeling concepts
Processing considerations	Built-in statistics gathering capabilities
	List processing capabilities
	Ability to allocate core
	Ease of producing standard reports
	Ease of producing user-tailored reports
Debugging and reliability	Ease of debugging
	Reliability of compilers, support systems, and documentation
Run-time considerations	Compilation speed
	Execution speed

second consideration reflects both the validity of the simulation model and the suitability of the overall layout specifying the runs to be performed. This section surveys the major statistical issues facing the simulation analyst in the conduct of a simulation study.

To clarify the subsequent discussion, the following classification of simulations is made with respect to output analysis:

1. **Steady State (Nonterminating) Simulations.** In this type of simulation study, we assume that, after a sufficiently long period of operation, the probability law governing the behavior of the real system will stabilize. For such a system we seek to estimate steady state or long-run average measures of system performance. Most of the statistical procedures discussed here apply to this situation.
2. **Transient (Terminating) Simulations.** When it is of interest to analyze the behavior of a system over a fixed period during which the underlying probability law changes, then the system is simulated only over that specified period. Such a terminating simulation is appropriate (1) when the corresponding real system shuts down at regular intervals or (2) when we want to study the short-term system response to certain "shocks." As is discussed subsequently, replication analysis is the appropriate technique for the investigation of terminating systems.

Choosing Input Probability Distributions

In the formulation of a simulation model, it is frequently necessary to characterize the random elements of a system by particular probability distributions. To select an appropriate distribution for an input process, the analyst must understand some of the basic properties of the common distributions and the circumstances in which those distributions arise. Pritsker[18] provides a good introduction to these topics.

In the data acquisition stage of a simulation study, empirical frequency distributions should be collected from the real system for each of the input processes to be modeled. This enables the analyst to apply goodness-of-fit tests to his or her hypothesized input distributions. A monograph by Phillips[19] describes the most popular goodness-of-fit tests and provides a FORTRAN program to evaluate an empirical frequency distribution against the most commonly used theoretical distributions.

Model Validation

Validation consists of determining that the simulation model is a reasonable representation of the real system.[20,21] Validation is normally performed in levels. The authors recommend that a validation be performed on data inputs, model elements, subsystems, and interface points. Validation of simulation models, although difficult, is a significantly easier task than validating other types of models, such as a linear programming formulation. In simulation models there is a correspondence between the model elements and system elements. Hence testing for reasonableness involves a comparison of model and system structure and comparisons of the number of times elemental decisions or subsystem tasks are performed.

Specific types of validation involve evaluating reasonableness using all constant values in the simulation model or assessing the sensitivity of outputs to parametric variation of data inputs. In making validation studies, the comparison yardstick should be both past system outputs and experiential knowledge of system performance behavior. A point to remember is that past system outputs are but one sample record of what could have happened.

Although validation of a simulation model is not solely a statistical issue, there are a number of statistical techniques that can aid in this stage of a simulation study. Shannon [2] discusses a validation procedure that is based on a comparison of input-output transformations. This involves comparing the outputs of the real system with those of the model, using inputs as nearly identical as possible. By employing an appropriate two-sample test, it is possible to evaluate the hypothesis that the two sets of responses came from the same (or nearly the same) population. A variety of tests are presented, including the nonparametric Mann-Whitney test, goodness-of-fit tests, spectral analysis tests, and the usual tests based on normality assumptions. In addition, Shannon discusses some appropriate two-sample tests when the analyst has multiple performance measures with respect to which model validation is required. It should be pointed out, however, that he emphasizes the superiority of the professional judgment of operating personnel in assessing the validity of a model of an existing system.

Estimation Methods

When a simulation model incorporates random input processes, its output performance measures are also subject to random variation, and it is necessary to take this into account when trying to make inferences about the corresponding real system. In particular, techniques are required that specify how to calculate a good estimator of a system parameter and that also provide a meaningful assessment of the reliability associated with that estimator. One measure of reliability for a simulation-based estimator is the probability that the inherent estimation error falls within some acceptable limits. In statistical language, both point and confidence interval estimators of system parameters are required.

To illustrate the estimation problems associated with simulation experiments, we refer to the bank teller system of Section 13.11.2, in which particular probability distributions have been specified for the interarrival and service times of successive customers. We seek to estimate the steady state average waiting time μ_W of a customer prior to being served by the teller. For this purpose, we might continue the simulation described in Section 13.11.2 until we have accumulated a large number of observed waiting times $\{W_j\}$ for successive customers. The application of classical statistical techniques to this data set requires that the observations be independent and that they all be sampled from the same probability distribution. In the bank system, however, there is a strong dependence between the waiting times of successive customers; moreover, the probability law governing the waiting time W_j of the jth customer depends on the customer number and approaches a limiting configuration only as j goes to infinity. The development of suitable techniques for analyzing simulation-generated data is currently an area of much active research.

This section discusses briefly the three most widely used estimation methods for simulation experiments. Throughout the discussion, the waiting time random variable is used as an example; the procedures described are appropriate for other random variables of the same type.

Replication Analysis

To estimate the steady state mean waiting time μ_W for the bank teller system, one approach is to execute, say, k independent runs of the simulation model and record the waiting times of the first b customers of each run. The independence of successive runs can be ensured by independently selecting a new set of starting seeds for all random number streams at the beginning of each run. If W_{ij} is the waiting time of the jth customer on the ith run, then the average waiting time computed over the ith run is given by

$$\overline{W}_i = \frac{1}{b} \sum_{j=1}^{b} W_{ij} \tag{1}$$

and the k observations $\{\overline{W}_i\}$ constitute a random sample of size k. If the run length b is sufficiently large, then each sample mean $\overline{W}_i$ has an expected value close to the steady state mean waiting time μ_W, and the distribution of $\overline{W}_i$ is approximately normal. These properties form the basis for replication analysis. The grand mean

$$\overline{\overline{W}} = \frac{1}{k} \sum_{i=1}^{k} \overline{W}_i \tag{2}$$

is taken as the point estimator of μ_W, and the overall sample variance

$$S^2 = \frac{1}{k-1} \sum_{i=1}^{k} (\overline{W}_i - \overline{\overline{W}})^2 \tag{3}$$

is used to construct the following $100(1-\alpha)\%$ confidence interval for μ_W:

$$\left[\overline{\overline{W}} - t_{\alpha/2,k-1} \cdot \frac{S}{\sqrt{k}}, \quad \overline{\overline{W}} + t_{\alpha/2,k-1} \cdot \frac{S}{\sqrt{k}}\right] \tag{4}$$

Note that $t_{\alpha/2,k-1}$ is the critical value cutting off a tail of size $\alpha/2$ for Student's t distribution, with $k-1$ degrees of freedom.

A disadvantage of replication analysis is that it restarts system operation anew at the beginning of each run; thus there can be a warm-up period during which atypical behavior is observed. In the bank teller system, small waiting times observed for the early customers introduce a bias into the average waiting time $\overline{W}_i$ for each run. This systematic error shifts the confidence interval given by display 4 away from the true value of the parameter μ_W; thus the net result is a drop in the reliability of this interval estimator as measured by its probability of covering the point μ_W. (See subsequent section on the design of experiments for a further discussion of the start-up problem.)

When applying replication analysis with a fixed total sample size of $n = k \cdot b$ observations, the analyst must evaluate the trade-off between the run length b and the replication count k. Increasing the run length reduces the effects of initial bias, but it also reduces the number of degrees of freedom in the overall sample variance given by equation 3, so that the confidence interval shown in display 4 becomes wider. Reducing the run length, on the other hand, will simultaneously shrink this confidence interval while shifting it away from its target point μ_W. Law[22] discusses these characteristics of replication analysis and provides guidelines for the application of this technique.

Note that in terminating simulations the run length b is fixed by the problem, and we simply seek to study the transient behavior of the system over the first b observations by averaging across a sufficiently large number of independent replications. In this case classical statistical techniques are entirely appropriate.

Subinterval Sampling

An alternative approach to the estimation of steady state waiting time is simply to execute a single run of $n = k \cdot b$ simulated customers and then to group the observations into k batches of b customers each. In effect, this experiment is equivalent to repeating a run of length b a total of k times, where the final system state for one run constitutes the initial condition for the next run. In this situation we let W_{ij} denote the waiting time of the jth customer in the ith batch, and the ith batch mean $\overline{W}_i$ is calculated according to equation 1. As in the case of replication analysis, the grand mean $\overline{\overline{W}}$ over all the batches provides a point estimator of μ_W, whereas equations 3 and 4 yield a confidence interval for μ_W.

Subinterval sampling eliminates the problem of initial bias—at least beyond the first batch. However, successive batch means $\overline{W}_i$ are no longer independent since observations inside each batch are correlated with observations inside other batches. Correlation between the batch means can seriously affect the validity of the variance estimator for $\overline{\overline{W}}$ based on equation 3 as well as the validity of the confidence interval given in display 4. The remedy for this problem is to take the batch size b large enough so that for practical purposes the batch means are independent. Fishman[23] has developed an effective procedure for determining such a batch size.

Law[22] has performed an extensive study of the relative performance of replication analysis and subinterval sampling in several types of queuing systems. He concluded that subinterval sampling generally produces superior results. This is an area of continuing investigation.

Regeneration Analysis

The bank teller system possesses the following important property: Every time an arriving customer finds the teller idle, the operation of the system "starts over" probabilistically. Such a point in time is called a "regeneration epoch," and beyond that point the future evolution of the system is independent of its past history. The portion of the process describing system behavior that occurs between two consecutive regeneration epochs is called a "cycle" or "tour." We accumulate the following quantities over each tour:

Y_i = sum of the customer waiting times during the ith tour
X_i = number of customers served during the ith tour

The regenerative property of the bank teller system ensures that the pairs (Y_i, X_i) are independent and identically distributed and that the steady state average waiting time μ_W is given by the ratio of the expected values of the variates

$$\mu_W = \frac{E[Y_i]}{E[X_i]}$$

To estimate μ_W using regenerative analysis, we simulate the operation of the system until k tours have been completed and compute the sample means

$$\overline{Y} = \frac{1}{k} \sum_{i=1}^{k} Y_i$$

$$\overline{X} = \frac{1}{k} \sum_{i=1}^{k} X_i$$

in order to obtain the following ratio estimator of μ_W:

$$\hat{r} = \frac{\overline{Y}}{\overline{X}} \tag{5}$$

In terms of the quantity

$$S^2 = \frac{1}{k-1} \sum_{i=1}^{k} (Y_i - \hat{r} X_i)^2$$

the variance of the regenerative estimator $\hat{r}$ is approximated by $S^2/(\overline{X}^2 k)$. An approximate $100(1-\alpha)\%$ confidence interval for the steady state mean waiting time μ_W is therefore given by

$$\left[\hat{r} - Z_{\alpha/2} \cdot \frac{S}{(\overline{X}\sqrt{k})}, \quad \hat{r} + Z_{\alpha/2} \cdot \frac{S}{(\overline{X}\sqrt{k})} \right] \tag{6}$$

where $Z_{\alpha/2}$ is the critical value of the standard normal distribution that cuts off a tail of size $\alpha/2$.

In general, a regenerative simulation is characterized by a tour defining state to which the system returns periodically so that successive cycles are independent replicates of steady state system behavior. This allows the application of classical statistical techniques to measurements accumulated over each cycle; in addition, the problem of initial condition bias is completely avoided. The major disadvantage of this method is revealed in systems with a large number of possible states: If the tour defining state occurs infrequently, then it may be necessary to run the simulation for a very long time in order to obtain an adequate number of cycles. It should be noted that the confidence interval given in display 6 is valid only for large samples. Crane and Lemoine[24] provide an excellent introduction to this method of simulation analysis.

Design of Experiments

A simulation run is an experiment in which an assessment of the performance of a system is estimated for a prescribed set of conditions. In the jargon of design of experiments, the conditions are referred to as "factors" and "treatments," where a treatment is a specific level of a factor. The literature in the field of design of experiments is extensive.[25-27] The purpose of this section is to present the issues relating to the design of experiments, but not the details as to how one should design a simulation experiment. The statistical techniques associated with the design of experiments are discussed in Chapter 13.4. Although the basic principles of classical experimental design also apply to simulation studies, several features of simulated experimentation distinguish it from more traditional applications. Among these distinguishing characteristics of simulation experiments are the following: (1) the simulation modeler knows in detail the structure of the process that produces the response variables; (2) additional observations of these variables are usually easy to obtain; and (3) the variance of these variables can sometimes be controlled.

The major problem involved in simulation experiments is associated with the definition of the inference space associated with the simulation model. Making a priori assessments of how widely the results obtained from the simulation model are to be applied and developing a thorough understanding of the inferences that can be made are the most neglected aspects of the design of experiments associated with simulation studies. Kleijnen[26] provides an extensive survey of these issues.

In general, the objectives of simulation experiments are:

1. To obtain knowledge of the effects of controllable factors on experimental outputs.
2. To estimate system parameters of interest.
3. To make a selection from among a set of alternatives.
4. To determine the treatment levels for all factors that produce an optimum response.

When multiple factors are involved, the approach to the first two objectives listed is to select one of the many possible experimental designs and to hypothesize a model for the analysis of variance for the experimental design selected. The experimental design specifies the combination of treatment levels along with the number of replications for each combination for which the simulation model must be exercised. Using the data obtained from the experiment, the parameters of the hypothesized model are determined along with the estimation of the error terms. The significance of each factor is then judged based on the derived model, and from this, estimates of system parameters of interest can be calculated.

In the problem of making a choice among alternatives, the statistical procedures of ranking and selection are used. Kleijnen[26] and Dudewicz[28] present state of the art reviews that summarize past research in this area and how it can be used in simulation analysis. Many procedures have been developed for specifying the sample size required in order to select the alternative whose population mean is greater than the next best population mean by a prescribed value with a given probability. The test procedures involve the computation of the sample mean based on the sample size specified and the selection of the largest sample mean observed.

A final topic relating to the design of experiments is the optimization of a simulation model. This problem differs from those previously described in that values for the controllable variables are sought that either maximize or minimize an objective function. For example, in the analysis of a periodic review inventory system, we might wish to employ simulation to determine values for the stock control level, reorder point, and time between reviews that minimize the average monthly cost of the inventory system.

Although the principles of optimization using simulation experiments are essentially the same as for optimization of mathematical expressions, there are some differences that must be considered. Since the response from a simulation typically involves random variables, the objective function or constraint equations written as a function of the simulation response will involve random variables. As a consequence, it is necessary to formulate response constraints as probability statements and to make statistical interpretations of the objective function value. Another major difference in simulation-based optimization is that neither the objective function nor its derivatives have a closed form; in fact, the objective function may not even have derivatives at some points. It should also be noted that the evaluation of a simulation-generated objective function is much more expensive than the evaluation of a mathematical expression.

There have been two basic approaches to optimization using simulation models. The first approach involves a direct evaluation of the independent variables using the simulation model. Farrell[29] divides these techniques into three categories: mathematically naive techniques, such as heuristic search, complete enumeration, and random search; methods appropriate to unimodal objective functions, such as coordinate search and pattern search; and methods for multimodal objective functions.

The second approach to optimization using simulation is response surface methodology.[30] In this method we fit a surface to experimental observations using a factorial design in the vicinity of an initial search point. We then apply an optimization algorithm, such as the gradient method, to determine the optimum values of the controllable variables relative to the fitted equation. The optimum values for the fitted surface are then used to define the next search point.

Tactical Planning

The objective of tactical planning is to make the most efficient use of the simulation model in executing the runs specified by the overall experimental design. This topic includes three major issues: (1) determining a policy for starting up the model that avoids initial bias, (2) controlling the execution of a model in order to reduce the variance of the outputs, and (3) determining a stopping point in the experimentation when a sufficient amount of data has been collected in order to ensure acceptable reliability in the final results.

Start-up Policies

Start-up policies typically prescribe a method for setting the initial conditions for a simulation model and for selecting a truncation point beyond which sample observations are to be recorded. Initial conditions are sought that will minimize the duration of the model's warm-up period, and a truncation procedure is used to try to identify the end of that warm-up period. In practice, these decisions are usually made by inspecting the results of some pilot runs of the model and by applying an intuitive test to detect the onset of steady state behavior. Several authors have attempted to formalize this procedure in order to provide automatic start-up policies that can be incorporated into the operation of a simulation model. A survey of these policies is presented in Wilson and Pritsker.[31]

It should be noted that, although the use of a data truncation procedure reduces the bias of a cumulative performance estimator, such a procedure can also inflate the variance of the estimator because of the loss of sample information. This trade-off between the bias and the variance of performance estimators when applying a start-up policy is analyzed in detail in Wilson and Pritsker.[32]

The use of start-up policies should be considered in conjunction with the estimation procedure. If estimators are to be obtained using regenerative methods, then the start-up policy decision is an easy one; that is, start the run in the regenerative state so that the first cycle starts immediately and no truncation is required. If the estimation procedure is based on a single time series, then the start-up policy is applied only once, and it is not too inefficient to truncate. However, if replication is used, the start-up policy is used repetitively, and great care is needed in establishing it.

It should be noted that theoretical studies of start-up policies have been restricted to small, well-behaved models. For such models the variability associated with sample values during start-up is not too different from the steady state variability. Thus the theoretical research tends to indicate that no truncation should be performed. Practical applications, however, indicate that this is not the case and that truncation is a reasonable policy to follow. This is especially true when dealing with job shops or conveyor systems in which many sequential operations must be performed before the system is "loaded."

Variance Reduction Techniques

To improve the precision of simulation-based performance estimators, a number of variance reduction techniques have been developed for simulation experiments. Some of these techniques are designed to exploit special information about the structure of the model, whereas other techniques actually distort the structure or operation of the model in some way. Only two of these methods are widely used in practice—common random number streams and antithetic variates.[26]

The technique of common random numbers enables the experimenter to sharpen the comparison between alternative system configurations by using exactly the same input sequence of random numbers to drive the simulation of each alternative. If the simulator can arrange for proper synchronization in the use of successive random numbers across all alternatives, then this technique ensures that all alternatives are compared under identical experimental conditions.[26] In the language of experimental design, the use of common random numbers creates a random block effect, and standard analysis procedures for block designs are appropriate when applying this variance reduction technique (see Chapter 13.4).

The basic idea behind the method of antithetic variates is to make complementary pairs of runs on the same simulation model so that the corresponding responses X, X' tend to fall on opposite sides of the mean μ_X. If this can be arranged, then the average of two antithetic responses X, X'

$$Y = \frac{X + X'}{2}$$

will have a smaller variance than the average of two independent replications. The usual method for implementing this technique is to supply complementary random number input sequences to the two runs: Where the random number r is sampled on the first run, the corresponding sampled value on the second run is $1 - r$. If k independent pairs of antithetic runs $\{(X_i, X_i')\}$ are executed, then point and confidence interval estimators of the expected value μ_X of the performance measure X can be obtained by applying replication analysis to the k observations $\{Y_i\}$.

It should be noted that the effectiveness of both common streams and antithetic sampling depends largely on the behavior of the simulation model to which they are applied; neither technique is guaranteed to improve the precision of simulation-based performance estimators. Various other variance reduction techniques have been based on concepts of regression analysis, stratified sampling, and importance sampling. Kleijnen[26] thoroughly discusses the advantages and pitfalls of each of these techniques.

Stopping Rules

If the analyst specifies beforehand that a particular level of reliability is required in the final estimator of system performance, this requirement is usually translated into a maximum half-length H for the corresponding $100(1 - \alpha)\%$ confidence interval estimator. To ensure that an adequate sample size is accumulated to satisfy this reliability requirement, a sequential stopping procedure should be used. For example, a simple stopping rule for replication analysis is to continue increasing the number of replications k and recomputing the sample standard deviation S given by equation 3 until the condition

$$t_{\alpha/2,\,k-1} \cdot \frac{S}{\sqrt{k}} \leqslant H \tag{7}$$

is finally satisfied.[2] Law and Carson[33] have developed a stopping rule based on subinterval sampling, and Fishman[34] has proposed a sequential stopping procedure for use with regenerative analysis.

In addition to determining the sample size to meet desired confidence interval specifications, there are practical issues associated with the stopping of a simulation run. Such questions involve the consideration of what to do about entities in the model at the end of a run. The answers to such questions are problem-specific. If such entities are representative of the other entities on which statistics were collected, then the further processing of them should not matter. However, if they are atypical and are of direct interest in the corresponding real system, then their processing should be considered.

13.11.5 APPLICATIONS OF SIMULATION

The applications of computer simulation have grown rapidly in the past 20 years. Surveys by Shannon and Biles[35] and Turban[36] indicate that simulation and statistics are the quantitative techniques that are most widely used in government and industry. The continuing development of simulation languages has been an important factor in this growth. Another major factor is the flexibility of simulation modeling when compared, for example, to the structural restrictions imposed by a mathematical programming formulation of a problem. Even when an analytic model can be applied to a problem, simulation is frequently used to study the practical implications of the assumptions underlying the analytic model.

It is not a difficult task to design and develop simulation models for industrial use. Typically such models are larger in size, but no more complex in concept, than the models presented in most standard textbooks on simulation. When building simulation models, analysts should bear in mind the alternative approaches available to them and avoid rigid conformity to a fixed set of modeling rules. Such modeling innovation can be found in the wide variety of applications areas discussed here. For each topic listed, references to specific studies are provided.

Manufacturing Operations

Numerous studies have been performed in the following areas:

1. Plant design.[37,38]
2. Productivity improvement.[39]
3. Manpower assignment.[40]
4. Computer-aided manufacturing.[41,42]
5. Scheduling.[43–45]
6. Materials handling.[46,47]

A particularly interesting example of the use of simulation in plant design concerned the sizing of storage tanks, evaporators, and an ion exchanger for a corn wet milling plant.[38] To model both the arrival of railroad tank cars containing raw materials and the subsequent flow of materials through the various steps of the process, a combined discrete-continuous simulation model was built. In addition to determining cost-effective sizes for the required units, the model was used to evaluate proposed manual and computerized control systems for the plant. One such control system used a linear programming blending model embedded in the simulation for setting valves in order to maximize the profitability of the end products. Thus simulation was used not only to finalize the plant design, but also to identify the most economical control strategy compatible with production quality and volume requirements.

Transportation Systems

The following topics have been extensively investigated in recent years:

1. Passenger railroad system performance.[48]
2. Bus scheduling and routing.[48]
3. Air traffic control.[49]
4. Military air terminal operations.[50]

An example of the problems and objectives of simulation studies in this area concerns the evaluation of the productive capacity of cargo processing facilities in military air terminals.[50] Specifically, management needed to determine the effects of fluctuating demands for airlift cargo on a terminal's ability to meet the demand in a timely manner. Resource utilization was also an important factor.

At a terminal, cargo arrives by truck or by aircraft. The arriving cargo is unloaded and sorted by shipment type, destination, and priority. The sorted cargo is moved to various in-process storage areas where it is held until some form of consolidation is possible. Once consolidated, it is weighed, inspected, and stored. Its status can then be classified as "movement-ready." When movement-ready cargo is selected for a mission, it is transferred to a staging area where it is combined with the other cargo assigned to the mission and defined as a load. The load is then processed by cargo loading equipment and transferred to the aircraft.

This sequence of cargo processing operations was easily modeled using the process-oriented approach in order to answer the following procedural questions: (1) Should automated load/unload equipment be installed, and where should such equipment be located? (2) How many aircraft can a terminal load simultaneously? (3) What additional resources are required to support increased air traffic during emergencies?

Project Planning and Control

Many of the applications in this area are based on a network model of a project in which arcs and nodes, respectively, represent activities and milestones in the project. Frequently these studies fall into one of the following categories:

1. Product planning.[51]
2. Marketing.[52,53]
3. Research and development.[54]
4. Construction.[55]

A recent risk analysis of pipeline construction in Alaska illustrates this type of application.[55] The construction of a pipeline basically involves (1) preparing a site for laying pipe,(2) laying the pipe, and (3) welding sections of the pipe together. Supporting operations for pipeline construction involve the building, dismantling, and moving of campsites; the construction of roads and other transportation facilities; and the relandscaping of the site. When pipeline construction is performed in Alaska, the adverse weather conditions must be considered when planning the construction project. A Q-GERT network was developed, consisting of the pipeline construction activities and transportation facility development activities. The effects of weather conditions on construction activities were also included in the model. A risk analysis was performed using Q-GERT to determine the probability of completing pipeline construction by specified due dates. A cost analysis was also performed to determine potential overrun conditions. The analysis indicated that both time and cost overruns could be expected. The effects of changing the activity schedule and construction rates were also evaluated.

Financial Planning

Work in this area includes:

1. Cash flow analysis.[56]
2. Corporate models.[57]
3. Econometric models.[58,59]

A cash flow analysis typically involves a distribution sampling experiment to estimate the mean net present value of a stochastic cash flow. Many corporate models are deterministic, discrete event simulations, which are used to compare alternative long-run management strategies. Econometric models are usually continuous simulations in which the time step is fixed and the behavior of the variables of interest is modeled in terms of algebraic and difference equations.

Environmental and Ecological Studies

Typical applications in this area have been concerned with:

1. Flood control.[60]
2. Pollution control.[61,62]
3. Energy flows.[63]
4. Marine populations.[64,65]
5. Agriculture.[66]
6. Insect control.[67]

A study of cadmium flow in a heavily industrialized urban area illustrates this work.[62] Within a 60 sq mi region of extreme northwestern Indiana, several hundred sources were found to be emitting large amounts of cadmium into both the air and the water. Answers were sought to the following questions: (1) How do system flow characteristics of emissions into the air differ from those of emissions by way of water? (2) What are the characteristics of the levels of cadmium deposited on urban structures? (3) Is the cadmium in the sludge of sewage treatment plants due more to rainfall runoff from urban structures than to effluent from industrial plants? (4) What control policies might be useful in meeting pollution standards?

To describe the flow of cadmium through the region, a continuous simulation model was used to represent cadmium flow paths within urban and rural areas. A combined discrete-continuous model was used to investigate variations in the level of cadmium on urban structures due to discretely timed rainfall events; in addition, the model was used to study the effects of the resulting runoff. To reduce cadmium levels in the municipal plant sludge, alternative control policies, such as regular street sweeping, were evaluated using the combined model.

Health Care Systems

Applications of simulation in the health care field have increased sharply in recent years. Some of the areas that have been studied include:

1. Inventory management.[68]
2. Hospital planning.[69,70]
3. Manpower planning.[71,72]
4. Materials handling.[73]

A typical application concerned the management of patient flow through the family planning clinic of a large hospital.[69] Employee work loads and patient service times both varied greatly, with periods of overwork and overcrowding followed by periods of virtual inactivity. The hospital administrators sought a solution to this problem that did not require disruption of ongoing operations or major capital improvements. A complex queuing simulation was used to evaluate the effects of alternative staffing policies and patient scheduling procedures on the observed bottlenecks. As a result of this study, a new patient scheduling procedure was adopted which eliminated the bottlenecks without any changes in the level of staffing.

13.11.6 SUMMARY

In this chapter computer simulation has been defined, and the steps in a simulation study have been discussed. Modeling techniques based on the event-, process-, and activity-oriented approaches to simulation have been developed, and several major simulation languages using each approach have been discussed. Some statistical issues relating to the design and analysis of simulation experiments have also been examined. In the last section some major application areas have been surveyed.

REFERENCES

1. A. A. B. PRITSKER, *The GASP IV Simulation Language*, Wiley, New York, 1974.
2. R. E. SHANNON, *System Simulation: The Art and Science*, Prentice-Hall, Englewood Cliffs, NJ, 1975.

3. A. A. B. PRITSKER, *Modeling and Analysis Using Q-GERT Networks*, Halstead, New York, 1979.
4. A. A. B. PRITSKER and C. D. PEGDEN, *Introduction to Simulation and SLAM*, Halstead, New York, 1979.
5. W. R. FRANTA, *The Process View of Simulation*, Elsevier North-Holland, New York, 1977.
6. B. CARNAHAN, H. LUTHER, and J. O. WILKES, *Applied Numerical Methods*, Wiley, New York, 1969.
7. J. P. EMSHOFF and R. L. SISSON, *Design and Use of Computer Simulation Models*, Macmillan, London, 1970.
8. G. GORDON, *The Application of GPSS V to Discrete Systems Simulation*, Prentice-Hall, Englewood Cliffs, NJ, 1975.
9. T. J. SCHRIBER, *Simulation Using GPSS*, Wiley, New York, 1974.
10. P. J. KIVIAT, R. VILLANUEVA, and H. MARKOWITZ, *The SIMSCRIPT II Programming Language*, Prentice-Hall, Englewood Cliffs, NJ, 1969.
11. W. GRAYBEAL and U. W. POOCH, *Simulation: Principles and Methods*, Winthrop, Cambridge, MA, 1980.
12. H. SPECKHART and W. H. GREEN, *A Guide to Using CSMP*, Prentice-Hall, Englewood Cliffs, NJ, 1976.
13. E. L. MITCHELL and J. S. GAUTHIER, "Advanced Continuous Simulation Language (ACSL)," *Simulation*, Vol. 25, 1976, pp. 72–78.
14. G. A. KORN and J. V. WAIT, *Digital Continuous-System Simulation*, Prentice-Hall, Englewood Cliffs, NJ, 1978.
15. J. W. FORRESTER, *Principles of Systems*, Wright-Allen Press, Cambridge, MA, 1971.
16. A. L. PUGH, *Dynamo II User's Manual*, MIT Press, Cambridge, MA, 1973.
17. F. CELLIER and A. E. BLITZ, "GASP V: A Universal Simulation Package," *Proceedings, IFAC Conference*, 1976.
18. PRITSKER, *Modeling and Analysis Using Q-GERT Networks*, pp. 196–209.
19. D. T. PHILLIPS, *Applied Goodness of Fit Testing*, AIIE-OR-72-1, Norcross, GA, 1972.
20. G. S. FISHMAN and P. J. KIVIAT, "Analysis of Simulation Generated Time Series," *Management Science*, Vol. 13, 1967, pp. 525–557.
21. R. L. VAN HORN, "Validation of Simulation Results," *Management Science*, Vol. 17, 1971, pp. 247–258.
22. A. M. LAW, "Confidence Intervals in Discrete Event Simulation: A Comparison of Replication and Batch Means," *Naval Research Logistics Quarterly*, Vol. 24, No. 4, 1977.
23. G. S. FISHMAN, "Grouping Observations in Digital Simulation," *Management Science*, Vol. 24, No. 5 (January 1978), pp. 510–521.
24. M. A. CRANE and A. J. LEMOINE, *Introduction to the Regenerative Method for Simulation Analysis*, Springer-Verlag, New York, 1977.
25. W. G. COCHRAN and G. M. COX, *Experimental Designs*, Wiley, New York, 1957.
26. J. P. C. KLEIJNEN, *Statistical Techniques in Simulation*, Part 1, Dekker, New York, 1974.
27. T. H. NAYLOR, *The Design of Computer Simulation Experiments*, Duke University Press, Durham, NC, 1969.
28. E. J. DUDEWICZ, *Introduction to Statistics and Probability*, American Sciences Press, Columbus, OH, 1979.
29. W. FARRELL, "Literature Review and Bibliography of Simulation Optimization," *Proceedings, 1977 Winter Simulation Conference*, 1977, pp. 116–124.
30. R. H. MYERS, *Response Surface Methodology*, Allyn & Bacon, Boston, 1971.
31. J. R. WILSON and A. A. B. PRITSKER, "A Survey of Research on the Simulation Startup Problem," *Simulation*, Vol. 31, 1978, pp. 55–58.
32. J. R. WILSON and A. A. B. PRITSKER, "A Procedure for Evaluating Startup Policies in Simulation Experiments," *Simulation*, Vol. 31, 1978, pp. 79–89.
33. A. M. LAW and J. S. CARSON, "A Sequential Procedure for Determining the Length of a Steady-State Simulation," *Operations Research*, Vol. 27, No. 5 (1979), pp. 1011–1025.
34. G. S. FISHMAN, "Achieving Specific Accuracy in Simulation Output Analysis," *Communications of the ACM*, Vol. 20, 1977, pp. 310–315.
35. R. E. SHANNON and W. E. BILES, "The Utility of Certain Curriculum Topics to Operations Research Practitioners," *Operations Research*, Vol. 18, 1970, pp. 741–745.
36. E. TURBAN, "A Sample Survey of Operations Research Activities at the Corporate Level," *Operations Research*, Vol. 20, 1972, pp. 708–721.

37. N. ADAM and J. SURKIS, "A Comparison of Capacity Planning Techniques in a Job Shop Control System," *Management Science*, Vol. 23, 1977, pp. 1011–1015.

38. R. V. SCHOOLEY, "Simulation in the Design of a Corn Syrup Refinery," *Proceedings, 1975 Winter Simulation Conference*, 1975, pp. 197–204.

39. W. HANCOCK, R. DISSEN, and A. MERTEN, "An Example of Simulation to Improve Plant Productivity," *AIIE Transactions*, Vol. 9, 1977, pp. 2–10.

40. J. E. BREDENBECK, M. G. OGDON, and H. W. TYLER, "Optimum Systems Allocation: Applications of Simulation in an Industrial Environment," *Proceedings, Midwest AIDS Conference*, 1975, pp. 28–32.

41. J. E. LENZ and J. J. TALAVAGE, *The Optimal Planning of Computerized Manufacturing Systems Simulator (GCMS)*, Report No. 7, Purdue University, W. Lafayette, IN, 1977.

42. J. RUNNER, "CAMSAM: A Simulation Analysis Model for Computer-Aided Manufacturing Systems," unpublished master's thesis, Purdue University, W. Lafayette, IN, 1978.

43. D. G. DANNENBRING, "An Evaluation of Flow Shop Sequencing Heuristics," *Management Science*, Vol. 23, No. 11 (1977), pp. 1174-1182.

44. S. EILON and S. CHOWDURY, "Duedates in Jobshop Scheduling," *International Journal of Production Research*, Vol. 14, No. 2 (1976), pp. 223–237.

45. S. K. JAIN, "A Simulation-Based Scheduling and Management Information System for Machine Shops," *Interfaces*, Vol. 6, No. 1 (1975), pp. 81–96.

46. G. L. JARVIS and R. W. WAUGH, "A GASP IV Simulation of an Automated Warehouse," *Proceedings, 1976 Winter Simulation Conference*, 1976, pp. 541–547.

47. R. M. WAUGH and R. A. ANKENER, "Simulation of an Automated Stacker Storage System," *Proceedings, 1977 Winter Simulation Conference*, 1977, pp. 769–776.

48. J. REITMAN, *Computer Simulation Applications*, Wiley, New York, 1971.

49. J. YU, W. E. WILHELM, and S. A. AKHAND, "GASP Simulation of Terminal Air Traffic," *Transportation Engineering Journal*, Vol. 100, 1974, pp. 593–609.

50. V. J. AUTERIO, "Q-GERT Simulation of Air Terminal Cargo Facilities," *Proceedings, Pittsburgh Modeling and Simulation Conference*, Vol. 5, 1974, pp. 1181–1186.

51. C. J. BELLAS and A. C. SAMLI, "Improving New Product Planning With GERT Simulation," *California Management Review*, Vol. 15, 1973, pp. 14–21.

52. M. BIRD, E. R. CLAYTON, and L. J. MOORE, "Sales Negotiation Cost Planning for Corporate Level Sales," *Journal of Marketing*, Vol. 37, 1973, pp. 7–13.

53. M. BIRD, E. R. CLAYTON, and L. J. MOORE, "Industrial Buying: A Method of Planning Contract Negotiations," *Journal of Economics and Business*, 1974, pp. 1-9.

54. L. J. MOORE and B. W. TAYLOR, "Multiteam, Multiproject Research and Development Using GERT," *Management Science*, Vol. 24, 1977, pp. 401–410.

55. Federal Power Commission Exhibit EP-237, "Risk Analysis of the Arctic Gas Pipeline Project Construction Schedule," *Transactions of the Federal Power Commission*, Vol. 167, 1976.

56. L. E. BUSSEY and G. T. STEVENS, "Net Present Value From Complex Cash Flows by Simulation," *AIIE Transactions*, Vol. 3, No. 1 (1971), pp. 81–89.

57. T. H. NAYLOR, *Corporate Planning Models*, Addison-Wesley, Reading, MA, 1979.

58. T. H. NAYLOR, *Computer Simulation Experiments with Models of Economic Systems*, Wiley, New York, 1971.

59. R. S. PINDYCK and D. L. RUBINFELD, *Econometric Models and Economic Forecasts*, McGraw-Hill, New York, 1976.

60. K. OTOBA, K. SHIBATANI, and H. KUWATA, "Flood Simulator for the River Kitakami," in J. McLeod, Ed., *Simulation*, McGraw-Hill, New York, 1968.

61. R. L. FEHR, J. R. NUCKOLS, "GASP IV Simulation of Flush Water Recycling Systems," *Proceedings, 1977 Winter Simulation Conference*, 1977, pp. 513–519.

62. J. J. TALAVAGE and M. TRIPLETT, "GASP IV Urban Model of Cadmium Flow," *Simulation*, Vol. 23, 1974, pp. 101–108.

63. A. L. SWEET and S. D. DUKET, "A Simulation Study of Energy Consumption by Elevators in Tall Buildings," *Computers and Industrial Engineering*, Vol. 1, 1976, pp. 3–11.

64. P. L. KATZ, J. W. BALSIGER, R. H. SCHAPPELLE, and T. R. SPINKER, "Economic Effects of Fluctuations in Catches and Population Levels Upon a Population of Alaska King Crab," *Proceedings of the 1977 Winter Simulation Conference*, 1977, pp. 323–329.

65. H. F. PAUL and R. L. PATTERSON, "Hydrodynamic Simulation of Movement of Larval Fishes in Western Lake Erie and Their Vulnerability to Power Plant Entrainment," *Proceedings of the 1977 Winter Simulation Conference*, 1977, pp. 305–316.

66. R. M. PEART and J. B. BARRETT, "Simulation in Crop Ecosystem Management," *Proceedings, 1976 Winter Simulation Conference*, 1976, pp. 389–402.

67. G. E. MILES, T. R. HINTS, M. C. WILFON, A. A. B. PRITSKER and R. M. PEART, "SIMAWEV II: Simulation of the Alfalfa Weevil With GASP IV," *Proceedings, Pittsburgh Conference on Modeling and Simulation*, 1974, pp. 1157–1161.

68. M. B. DUMAS and M. RABINOWITZ, "Policies for Reducing Blood Wastage in Hospital Blood Banks," *Management Science*, Vol. 23, 1977, pp. 1124–1132.

69. A. J. ALESSANDRA, T. E. GRAZMAN, R. PARAMESWARAN, and U. YAVAS, "Using Simulation in Hospital Planning," *Simulation*, Vol. 30, No. 2 (1978), pp. 62–67.

70. R. B. FETTER and J. D. THOMPSON, "The Simulation of Hospital Systems," *Operations Research*, Vol. 13, 1965, pp. 689–711.

71. C. STANDRIDGE, C. MACAL, A. A. B. PRITSKER, H. DELCHER, and R. MURRAY, "A Simulation Model of the Primary Health Care System of Indiana," *Proceedings, 1977 Winter Simulation Conference*, 1977, pp. 349–358.

72. J. SURKIS, "A Dynamic Model for Policy Studies in Community Health Service Systems," in S. J. Bernstein and W. G. Mellon, Eds., *Selected Readings in Quantitative Urban Analysis*, Pergamon, Oxford, England, 1978.

73. R. W. SWAIN and J. J. MARSH, "A Simulation Analysis of an Automated Hospital Materials Handling Systems," *AIIE Transactions*, Vol. 8, 1978, pp. 10–18.

SECTION 14

OPTIMIZATION IN INDUSTRIAL ENGINEERING

CHAPTER 14.1

Optimization: An Overview

A. RAVINDRAN
University of Oklahoma

GINTARAS V. REKLAITIS
Purdue University

14.1.1 MEANING OF OPTIMIZATION

The 1960s and 1970s saw an enormous growth in the size and complexity of industrial organizations. Management decision making has become very complex, involving large quantities of workers, materials, and equipment. The role of the industrial engineer is to help management in making these decisions on a more objective and routine basis. From the industrial engineer's point of view, a decision is a recommendation for the "best" design or operation for a given engineering system or process so that costs are minimized or profits maximized. In using the term "best," it is implied that some choice or a set of alternative courses of action is available for making the decision. The term "optimum" is generally used to denote maximum or minimum, and the general process of maximizing or minimizing is referred to as "optimization." The industrial engineer is concerned not only with optimization problems in the design of industrial and service systems, but also with problems involving manufacturing and operation of these systems after they have been designed.

Key Elements of an Optimization Study

The key elements of an optimization study are the following:

1. Definition of system boundary.
2. Criteria.
3. Independent decision variables.
4. Uncontrollable variables.

System Boundary

Before undertaking any optimization study, it is important to define clearly the boundaries of the system under investigation. In many situations it may be necessary, in order to analyze a given system, to expand the system boundaries so as to include certain interconnected systems that influence the operation of the system under study. It is important to recognize that an "optimum" is specific only to the system under investigation and may not be applicable to a larger system that encompasses the system under study.

For example, consider a paint line system where finished parts are mounted on an assembly line for painting in different colors. While studying the problem of the optimal batch size for each color and the color sequence, one has to decide whether to include in the study the fabrication department, which sends finished parts for painting. Of course, increasing the system boundaries increases the problem size and its complexity, but does not necessarily improve the results of the study. While defining the system boundaries, it is important to recognize the objectives of the optimization study, the budget available, and the time limitations.

Criteria

The criteria used in an optimization study relate to the measures of effectiveness used to compare different alternatives. Most optimization studies involve several measures of effectiveness. The most important ones are identified during the system definition and objectives-of-the-study step. In many physical and economic systems, the criterion is generally cost, profit, or productivity, but a social system may have many criteria to be optimized. Whenever a set of objectives or criteria is identified in an optimization study, conflicts among objectives are likely to occur. For example, in a paint line system, the line foreman, in order to maximize line productivity, would like long, uninterrupted production runs that would minimize color and part changes on the line, resulting in a large inventory of relatively few painted parts and colors. Marketing would like a large inventory of every color and part in order to meet customer demands. Finance would strive for lower inventory levels in order to reduce capital investment.

Independent Decision Variables

The third key element in an optimization study is the determination of the independent decision variables that are under management control. It is essential to distinguish between the *inputs*, which are given *constants* to the optimization study, and the *decision variables*, which are the different alternatives under investigation. A poor or inadequate choice of decision variables could bias the results of an optimization study and produce erroneous results. Alternatives that are not considered cannot be used, even though they might prove to be desirable.

Uncontrollable Variables

Another key element is to determine the uncontrollable variables that have an impact on the system operations. These are not under management control, but they may change, affecting the behavior of the system under study. For example, in the paint line system, equipment breakdowns and worker absenteeism affect the operation and productivity of the line. Proper allowances for these uncontrollable variables must be made while arriving at an optimum solution. Failure to do so would lead to policies that could not be implemented in practice.

Purpose of an Optimization Study

The primary purpose of all optimization studies is not always trying to determine the "true optimum" operation of the system. In practice, some of the primary benefits are associated with understanding the system under study and describing it quantitatively in terms of tables, graphs, computer programs, or mathematical equations. Such a quantitative description by itself may suggest areas of improved operations and possible bottlenecks in the system. An optimization study may identify important variables in the system (both controllable and uncontrollable) and suggest ways to handle these variables effectively. It may also pinpoint areas where a better understanding of the system may be required.

The most permanent contribution of an optimization study may well be qualitative rather than quantitative. Often one may be able to identify the problem without any numerical calculations, and this may provide insight into the nature of decisions to be made.

14.1.2 CLASSIFICATION OF OPTIMIZATION PROBLEMS

Optimization problems may be classified based on the nature of the objective function and the system constraints. Exhibit 14.1.1 classifies optimization problems primarily on the basis of whether or not they involve constraints on the decision variables. Obviously the unconstrained problems are easier to solve than the constrained ones. The unconstrained problems are further classified based on whether the objective function involves one or several design variables. Basically there are two important classes of methods for solving the unconstrained problems. The direct search methods require only that the objective function be evaluated at different points, at least through experimentation. Gradient-based methods require the analytical form of the objective function and its derivatives.

An important class of constrained optimization problems is linear programming (see Chapter 14.2), which requires both the objective function and the constraints to be linear functions. Out of all optimization models, linear programming models are the most widely used and accepted in practice. Professionally written software programs are available from all major computer manufacturers for solving very large linear programming problems. Unlike the other optimization problems which require special solution methods based on the problem structure, linear programming has just one common algorithm, known as the "simplex method," for solving all types of linear programming problems. This essentially has contributed to the successful applications of linear programming models in practice. Integer programming (see Chapter 14.4) is another important class of linearly constrained problems where some of the design variables are restricted to being discrete or integer valued.

The next class of optimization problems involves nonlinear objective functions and linear constraints. Under this class we have the following:

1. Quadratic programming, whose objective is a quadratic function.
2. Convex programming, whose objective is a special nonlinear function satisfying an important mathematical property called "convexity."
3. Linear fractional programming, whose objective is the ratio of two linear functions.

Exhibit 14.1.1 Classification of Optimization Problems

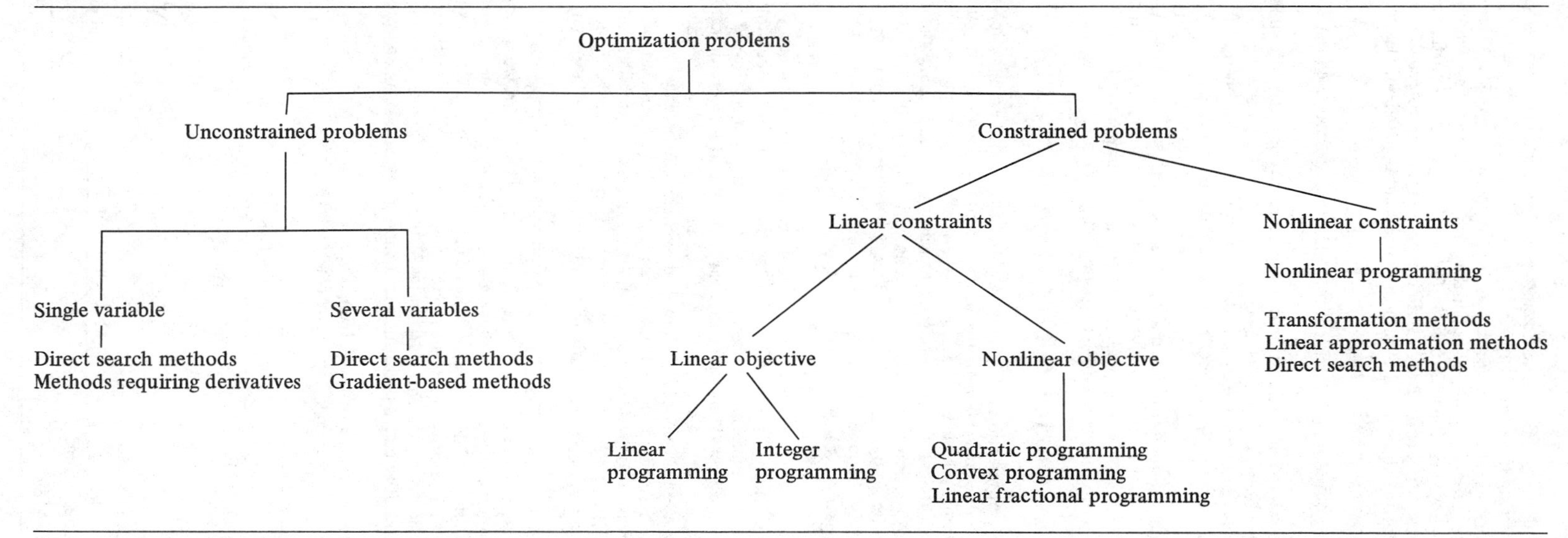

Special-purpose algorithms that take advantage of the particular form of the objective functions are available for solving these problems.

The most general optimization problems involve nonlinear objective functions and nonlinear constraints and are generally grouped under the term "nonlinear programming" (see Chapter 14.3). A number of engineering design problems fall into this class. Unfortunately, there is no single method that is best for solving every nonlinear programming problem. Hence a host of algorithms is available for solving the general nonlinear programming problem. Basically they fall into three important classes:

1. Transformation methods, which convert the problem into a sequence of unconstrained optimization problems.
2. Linear approximation methods, which use successive linear approximations of the nonlinear functions in the problem.
3. Direct search methods, which are similar to those for unconstrained problems that do not require analytical forms of the objective function and constraints.

Generally, if we have more information about the problem (such as the analytical form of the objective and constraints and their derivatives), we can devise more efficient and faster methods for its solution. Algorithms such as direct search methods, which require less information, are generally less efficient and more time consuming in solving the problem.

14.1.3 MODELS IN OPTIMIZATION[1]

Theoretically, an optimization study may be done by experimenting directly with the actual system. In other words, the system or the process is run under different policies, and by collecting data on the objective measures, decisions on the best policy are made. It may also be necessary to make some statistical analysis on the output measures in order to make the decisions.

In practice, most optimization studies are done with the help of a model or a sequence of models of the system under study. As a matter of fact, the essence of engineering optimization lies in the construction and use of models. "Model" as used here is just a simplified representation of the real system. This usage does carry with it the implication that a model in general is a representation that is less than perfect.

Why use a model? There are many conceivable reasons why one might prefer to deal with a substitute for the "real thing" rather than with the thing itself. Often the motivation is economic—to save money, time, or some valuable commodity. Sometimes it is to avoid risks associated with the tampering of a real object. Sometimes the real environment is so complicated that a representative model is needed just to understand it or to communicate with others about it. Such models are quite prevalent in the life sciences, the physical sciences, and engineering.

Given that one has something real, which we will call the "real system," and that there is some understandable reason for wanting to deal with it—that is, a "problem" related to the real system, which calls for definite "conclusions"—the modeling process can be depicted as in Exhibit 14.1.2. The broken line on the left represents what might be termed the "direct approach," for which we are seeking a substitute.

The first step is the construction of the model itself, which is indicated by the line labeled "formulation." This step requires a set of coordinated decisions as to what aspects of the real system should be incorporated in the model, what aspects can be ignored, what assumptions can and should be made, into what form the model should be cast, and so on. In some instances formulation may require no particular creative skill, but in most cases—certainly the interesting ones—it is decidedly an art. The selection of the essential attributes of the real system and the omission of the irrelevant ones require a kind of selective perception that cannot be defined by any precise algorithm.

Exhibit 14.1.2 The Modeling Process

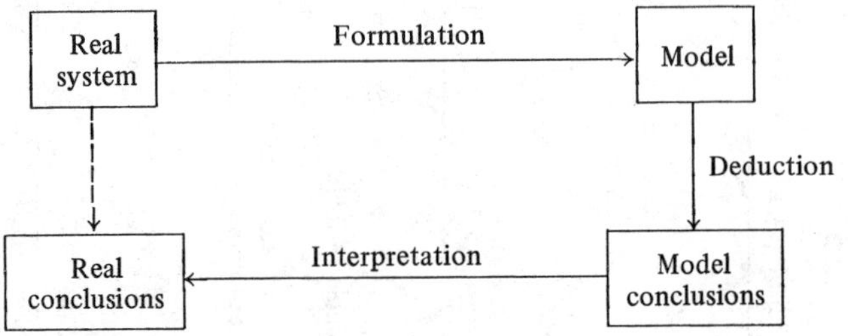

It is apparent, then, that the formulation step is characterized by a certain amount of arbitrariness, in the sense that equally competent researchers, viewing the same real system, could come up with completely different models. It is often meaningless to speak of the "right" way to formulate a model. At the same time, this does not imply that one model cannot be better than another. The "reliability" problem that follows illustrates this aspect of the formulation step.

Example 14.1.1 An electronic system consists of three components, each of which must function in order for the system to function. The reliability of the system (probability of successful performance of the system) can be improved by installing several standby units for one or more of the components. The reliability of the system is given by the product of the reliability of each component; the reliability of each component is a function of the number of standby units, as follows:

Number of Standby Units	Reliability of Component		
	1	2	3
0	0.5	0.6	0.7
1	0.6	0.75	0.9
2	0.7	0.95	1.0
3	0.8	1.0	1.0
4	0.9	1.0	1.0
5	1.0	1.0	1.0

For instance, if no standby unit is provided, the reliability of component 2 is 0.6; with two standby units, its reliability increases to 0.95. The cost and weight of each standby unit for the three components are as follows:

Component	Unit Cost ($)	Weight/Unit (kg)
1	20	2
2	30	4
3	40	6

Given a budget restriction of $150 and a weight restriction of 20 kg for all the standby units, the problem is to determine how many standby units should be installed for each component so as to maximize the total reliability of the system.

To formulate the problem, let us define the following variables:

$x_{ij} = 1$ if i standby units are installed for component j
$= 0$ otherwise

For each component, we will have a constraint of the type

$$\sum_{i=0}^{5} x_{ij} = 1 \qquad \text{for } j = 1, 2, 3$$

indicating that the number of standby units for each component is 0, 1, 2, 3, 4, or 5. The cost of component 1 is simply given by

$$20(x_{11} + 2x_{21} + 3x_{31} + 4x_{41} + 5x_{51})$$

and the constraint on the total system cost is given by

$$20 \sum_{i=0}^{5} i x_{i1} + 30 \sum_{i=0}^{5} i x_{i2} + 40 \sum_{i=0}^{5} i x_{i3} \leq 150$$

Similarly, the reliability of component 1 is given by

$$R_1 = 0.5x_{01} + 0.6x_{11} + 0.7x_{21} + 0.8x_{31} + 0.9x_{41} + 1.0x_{51} \tag{1}$$

The reliability of the system is

$$R = R_1 R_2 R_3$$

which becomes a nonlinear function to be maximized. On the other hand, if reliability of component 1 is written as

$$R_1 = (0.5^{x_{01}})(0.6^{x_{11}})(0.7^{x_{21}})(0.8^{x_{31}})(0.9^{x_{41}})(1^{x_{51}}) \tag{2}$$

then the reliability of the system $R = R_1 R_2 R_3$ can be linearized by taking logarithms and simply maximizing $\log_n R$! Even though equations 1 and 2 both give identical values for any set of feasible values of x_{i1}, using the product form (equation 2) reduces the objective function to a linear form. Thus the formative stages of the modeling process might be repeated and analyzed many times before the proper formulation becomes readily apparent. Once the problem formulation and definition is agreed upon, a more scientific step in the modeling process is begun.

Returning to Exhibit 14.1.2, the step labeled "deduction" involves techniques that depend on the nature of the model. It may involve solving equations, running a computer program, expressing a sequence of logical statements—whatever is necessary to solve the problem of interest relative to the model. Provided that the assumptions are clearly stated and well defined, this phase of modeling should not be subject to differences of opinion. The logic should be valid, and the mathematics should be rigorously accurate. All reasonable people should agree that the model conclusions follow from the assumptions, even if they do not all agree with the necessary assumptions. It is simply a matter of abiding by whatever formal rules of manipulation are prescribed by the methods in use. It is a vital part of the modeling process to rationalize, analyze, and conceptualize all components of the deductive process.

The final step, "interpretation," again involves human judgment. The model conclusions must be translated to real-world conclusions cautiously, in full cognizance of possible discrepancies between the model and its real-world referent. Aspects of the real system that were either deliberately or unintentionally overlooked when the model was formulated may turn out to be important. Since there may be no way to prove that a model has not omitted some important factor, there is room for reasonable people to disagree about the relevance of the model conclusions to the real system and to what extent the interpretation phase should be tempered by direct intuitive judgments.

The most important point revealed by Exhibit 14.1.2 is that the ties between the model and the system it represents are at best ties of plausible association and that no one, no matter how competent, can create perfection for that situation. It is part of the nature of models as simplified representations of real systems that there can be no absolute criteria by which to determine their acceptability. That is not to say, of course, that there exist no criteria by which to distinguish good models from poor ones or that model validation is not an integral part of the total modeling effort.

14.1.4 LEVELS OF MODELING

The execution of an optimization study is in most cases an expensive and time consuming activity. First, the state of reliability and robustness of nonlinear programming software is such that achievement of good results requires multiple trial runs, parameter adjustments, rescaling of variables and functions, modifications of initial variable estimates, and experimentation with constraints. Even more expensive in terms of manpower, however, is the development of the optimization model itself. In fact, the costs associated with this activity usually far outweigh the computer charges associated with preparing and performing the optimization calculations.

Given that model development is expensive, it is necessary to use considerable judgment in selecting the level of the model that is appropriate to the goals of the study and consistent with the quality of the information available about the system. It is clearly undesirable to develop a detailed dynamic model of a plant operation if an overall input-output model will suffice. Furthermore, it is useless to develop a complex model when the data available to evaluate its coefficients are sparse and unreliable. On the other hand, since it is the model that will be optimized and not the real system, it is pointless to carry out an optimization study with a simplistic model that will not adequately approximate the true system optimum.

The ideal sought in model development is sometimes called the "principle of optimum sloppiness"; the model should be only as detailed as necessary for the goals of the study for which it is constructed. Achievement of this goal, however, is always difficult. Although it is possible, through a history of successes and failures, to develop an a priori feel for the appropriate level of model complexity for specific types of systems, usually it is difficult to generalize this experience to other applications. The only foolproof way to developing a model with an optimum level of sloppiness is to proceed in steps with model development and optimization, starting with a very simple model and concluding with the version in which the refinements to the value of the optimum attained approach the accuracy of the data incorporated in the model. This, however, is a painstaking and methodical process, which goes against the preconceived "whiz kid" notions often held about the work of operations research analysts, even by analysts themselves. A common recourse to avoiding this stepwise development is to tailor the model to the optimization technique that is currently in favor or held in high esteem by colleagues or with which the analyst is most familiar at the time. This typically occurs when a new, "exotic" technique is discovered and popularized. For instance, during their heyday, the converts to geometric or dynamic programming often would attempt to recast all problems to the special mold required by the methodology.

Sometimes it may be necessary to force a problem into a particular model form. In modeling integer problems, it is essential that the problem be forced into some "tractable" form. For example, we may force the problem into a network form so that the problem can be solved successfully. The important point is to know what the impact of forcing a problem into a particular model form will be on the quality of the solution. Also, the capabilities and limitations of available optimization methods must be kept in mind when developing the model. It is clearly inappropriate to expect to be able to solve a nonlinear programming problem as large in dimensionality as can be accommodated by commercial linear programming software. However, at the modeling stage such concerns should be secondary rather than dominant.

An Example of Optimization Study

To illustrate the different levels of models that might be developed, we present an optimization study of a multiproduct polymer plant.[2–4] The plant in question consists of several serial stages of parallel processors, with in-process storage between stages. A block diagram of the process is shown in Exhibit 14.1.3. The first stage contains several nonidentical polymerization reactor trains. The second stage consists of several parallel extrusion lines of different capacity. The third stage is a blending stage, consisting of, say, five lines, each containing one or more blending units. The final processing stage is packaging, which provides three alternatives: bulk loading, packaging into boxes, or packaging into sacks.

The plant produces a number of different polymer grades and types and consequently is not operated on a continuous basis. To allow for product changeovers, each processing stage has associated with it a limited amount of in-process storage. Moreover, the blending stage has associated with it several large-capacity storage vessels, which are required to ensure adequate inventory for preparing product blends.

Suppose that a decision is made to "optimize" the process. It is suspected that, since the current product slate is substantially different from the product slate projected when the plant and its operating procedures were designed, the plant profitability could be substantially improved.

As a first step in the study, a simple linear input-output model is developed to investigate what the preferred product slate might be. The linear model is based on the average production rates and processing times required by each product in each processing stage; uses average "off-specification" production rates (those not meeting product specifications); and assumes current levels of feed availability, current market prices, and unrestricted demand. The linear model is solved using a packaged linear programming code. On the basis of several case runs, it is unequivocally concluded that some products in the current slate ought to be dropped and that the production of others ought to be increased.

Upon further study it is revealed that, as a result of increasing the production rate of some of the products and deleting others, the production of intermediate off-specification material will be increased. This material is normally recycled internally. The increased internal recycling is likely to cause bottlenecks at the extrusion stage. Consequently, the initial linear model is expanded to include the key internal material flows as shown in Exhibit 14.1.4, as well as to incorporate explicit constraints on the maximum and minimum allowable intermediate processor rates. The revised model, considerably increased in dimensionality, remains a linear model suitable for optimization by linear programming methods. The case studies conducted using the expanded model indicate that the capacities of the extrusion and blending stages do, in fact, severely limit the initially projected optimum product slate. It appears that some limited expansion of capacity would be desirable.

To determine the optimum investment in additional equipment, the model is revised to incorporate nonlinear cost terms for the investment and operating costs of the extra units as a function of their throughput. The extra production capacity is easily incorporated into the linear constraint set. However, as a result of the cost terms, the objective function becomes nonlinear. The resulting linearly constrained nonlinear programming problem is solved using available software. On the basis of such optimization, an appropriate capital expenditure is proposed for plant upgrading. At this stage it might be noted by team members that the selected additional capacity was all based on average production rates and on the assumption that the same product sequence was used as is currently being followed. It is suggested that alternate sequences and production run sizes be considered. This introduces new considerations into the plant model, such as batch sizes, run lengths, product sequencing, and changeover policies.

Exhibit 14.1.3 Conceptual Diagram of Process

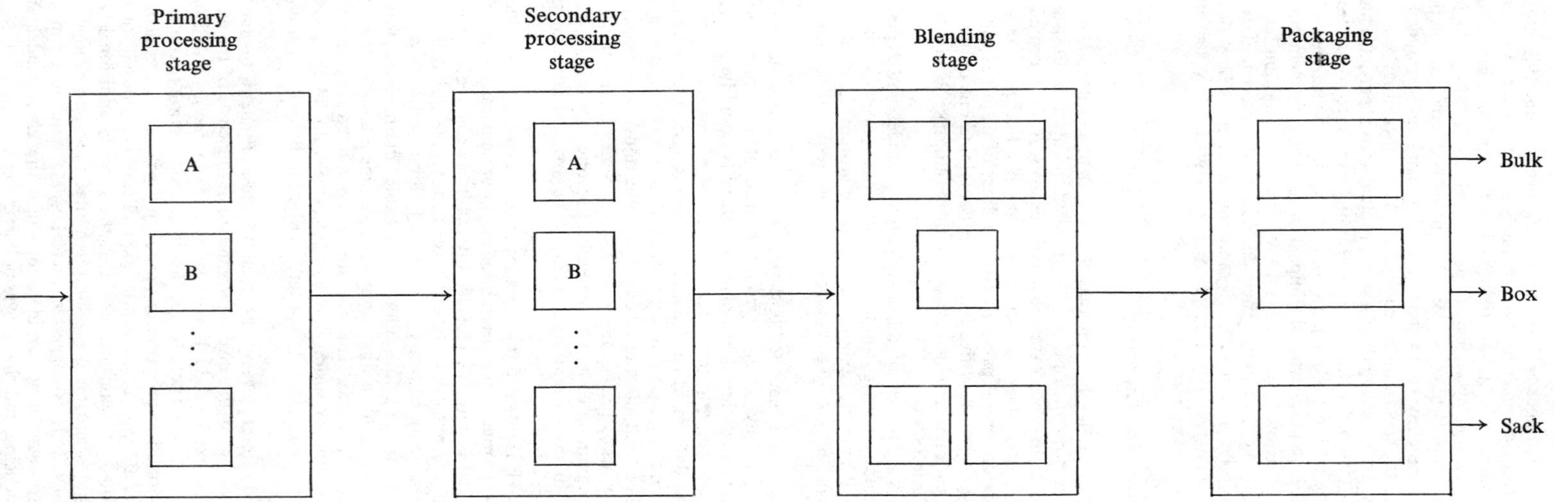

Primary processing stage

A

B

Secondary processing stage

A

B

Blending stage

Packaging stage

Bulk

Box

Sack

Exhibit 14.1.4 Process With Explicit Off-Specification Recycle

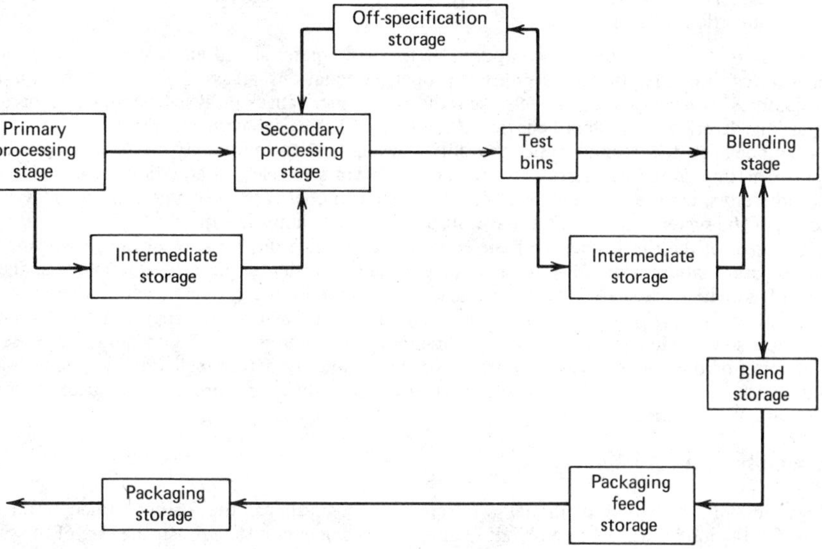

Off-specification storage
Primary processing stage
Secondary processing stage
Test bins
Blending stage
Intermediate storage
Intermediate storage
Blend storage
Packaging storage
Packaging feed storage

To incorporate these elements to even a limited degree, the model must be modified to include binary variables, which will be used to reflect the sequencing part of the problem. Computational considerations now come into play: Available mixed integer programming codes cannot accommodate nonlinearities. Consequently, each nonlinear cost term is approximated by a fixed cost and a linear cost term. The resulting mixed integer linear program is solved, and the solution might indicate that moderate changes in the production sequence, together with subdivision into shorter product runs, are attractive.

At this point it may be appropriate to include time periods in the planning model and to incorporate the regional warehouses and the distribution network in the study. Alternatively, questions could arise regarding the significance to the plant economics of unit shutdown and start-up transients as well as off-specification production during these periods. Furthermore, in-process and final product inventory levels might also need to be given closer scrutiny. If multiple time periods need to be considered, the mixed integer programming model would further be expanded. On the other hand, if transients and inventory changes do become an important concern, a completely different model would have to be formulated. A discrete event or combined discrete-continuous simulation (see Chapter 13.11) could be developed to represent the system dynamics. In this case optimization would be carried out either by case studies or by direct search over a few variables using a direct search optimization algorithm.

Finally, if it appears that fluctuations in the production rates on downtimes or changeover times of some or all of the processing stages are significant, then the simulation model may need to be refined further to include Monte Carlo sampling of these fluctuating variables. In this case it will be necessary to collect statistical data on the observed fluctuations, to identify the appropriate distribution functions, and to insert those stochastic features into the model. With this type of a model, direct optimization will normally be very costly in terms of computer time, and optimization will best be performed by case studies involving a few parameter changes.

Depending upon the depth and direction of the study, the model used in plant optimization will range from a moderately sized linear model, to a large-scale linear programming, to a large-scale mixed integer program, to a nonlinear program, to a deterministic simulation model, and finally to Monte Carlo simulation. Clearly it would be wasteful to begin the study with the most complex model if the simplest model could serve to indicate that the potential for improvement is small. Similarly, it would be meaningless to develop and use the detailed simulation model if no data were available to evaluate the stochastic model parameters.

14.1.5 DEVELOPING A MODEL

Types of Models

Having selected the desired level of detail at which to consider the system to be optimized, the engineer is next faced with the problem of actually developing the model. Basically, three types of models are commonly used in optimization studies[4]:

1. Phenomenologically based equation-oriented models.
2. Response surface models.
3. Simulation models.

Models of the first type are composed of the basic material and energy balance equations and engineering design relations and physical properties equations, all assembled as sets of equations or inequalities. Normally such models are valid over wider ranges of system conditions because the equations describe the system behavior at the level of basic engineering principles.

In models of the second type, the entire system or its component parts are represented by approximating equations derived by fitting coefficients to selected equation forms using directly or indirectly measured system response data. Such models are normally valid only over limited ranges of the system variables, but have the advantage of simplified structure.

In models of the third type, the basic equations describing the behavior of the system are grouped into separate modules or subroutines that represent a particular piece of equipment as the collection of activities associated with a change in the state of the system. Each of these modules or subroutines is usually a self-contained entity that may involve internal numerical procedures, equation solving, integration, or logical branching procedures. Simulation models are used when evaluation of the equations is complex, involving implicitly determined variables, as well as when the selection of a logical block of the calculation procedure or appropriate equations is subject to the state of the system.

Assembly of the Model

Regardless of which type of model is selected, the engineer has the further choice of the manner in which the model will be assembled: manually or using computer aids. In the case of the equation-oriented model, the simplest form of which is a linear program, the linear equations and inequalities can be written and assembled by hand, and the coefficients coded into an array suitable for processing by the linear programming codes. Alternatively, a matrix generator can be used to assemble automatically the coefficients for certain classes of constraints. For instance, constraints of the form $\Sigma x_i = 1$ could easily be generated by merely defining the indices of the variables in the sum and the numerical value of the right-hand-side constant. The matrix generator would then generate the appropriate array entries. In any application involving interconnected subsystems of various types of regular structures, it is usually efficient to use equation generators. An equation generator is written for each type of subsystem so that the assembly of the model involves merely identifying the subsystems that are present and their interconnections. For example, Bethlehem Steel Corporation has recently developed an equation generator for its linear programming model that optimizes plant operations.[5]

In the case of the response surface models, the system or its components can be employed directly to generate data from which approximating equations relating the dependent and independent variables can be derived. Fitting the coefficients of a quadratic equation to approximate some complex relationship is a typical case. Alternatively, a more complex system model (such as a simulation) can be used in conjunction with an automated approximation generator to generate the response surface.

Finally, simulation models can be assembled by coding them *ab initio* in their entirety or by making use of simulation languages or packages. Using such systems a model can be assembled using the building block concept: Individual modules provided by the simulation package are linked together by the user either through data statements or coded statements.

A number of advantages, disadvantages, and pitfalls are associated with each type of model and the manner in which it is assembled. Equation-oriented models are by far the most common in engineering applications. Typically, automated generation is employed only with linear or mixed integer programming models, although its use is also receiving attention in nonlinear process modeling applications,[2] since these involve reoccurring system activities—pumps, compressors, and other devices—whose representation in terms of equations is independent of the processing rates or conditions. Equation-oriented models are typically large in dimensionality. Consequently, during manual assembly it is tempting to eliminate variables by solving some of the equations for selected variables and substituting them out; to order the equation into sequentially solved blocks, some or all of which require iterative solution methods; or to reduce the number of inequalities by using maximum or minimum operators such as $x_4 = \min(x_1, x_2, x_3)$. In general, such devices are fraught with potential difficulties, as follows:

1. If equations were used to eliminate variables, then all bounds imposed (implicitly or explicitly) on these variables must be explicitly imposed on the expression replacing these variables. Otherwise, impossible solutions may well be generated.

2. Any iterative calculations nested within the model equations must be carried out with sufficiently tight tolerances so as not to cause errors in any numerical evaluations of model derivatives and not to cause errors in the search method logic.

3. The presence of maximum or minimum operators cannot be allowed if derivative-based optimization algorithms are used, since these operators will give rise to derivative discontinuities.

In addition, care must be taken to include adequate bounds in the equation-oriented model in order to ensure that the arguments of common functions, such as power functions with noninteger exponents, logarithms, or other transcendentals, remain within their defined ranges. In general, it is best not to insert calculation loops within the model equations. Instead, if direct solution is not possible, it is best to write the equations as explicit equality constraints. The nonlinear programming algorithm will then direct their solution toward the feasible region in the course of searching for the optimum.

14.1.6 MODEL VALIDATION

The process of acquiring the conviction that a model actually "works" is commonly called "validation." When the people involved (the users) are persuaded that a model is useful within some basic context, they will speak of it as a valid model. Its validity is of course restricted to the understood context. Even within that context, some people may refuse to accept the model's validity because they have not yet been persuaded. Thus "validation" is a considerably weaker term than "proof."

It is important that models be validated prior to implementation. The purpose of the validation procedure is to ensure that the models used in the optimization study are indeed a fair representation of the real-world system and that inferences drawn and results obtained from solving the models are compatible with the system studied. The process of validation consists of (1) verification, to ensure that a model behaves as an experimenter intends, and (2) validation, to test the agreement between model behavior and the real system behavior.

Verification is more concerned with the internal consistency of the model. For example, we may verify a computer model by counting the number of calls made to different subprograms as well as to individual statement numbers within the subprograms. These counts can then be checked with the actual estimate for such visits.

As mentioned previously, it would be futile to attempt to establish with certainty that a model is valid, but this fact does not absolve one from the responsibility of checking it against reasonable standards of appropriateness. There are a number of commonly employed techniques for doing this, depending on the nature of the model.

One method for validating predictive models is "retrospective testing," in which the model is compared against some historical standard to see if it would have accurately predicted what has since been observed to occur. For example, if a model is constructed to forecast the monthly sales of a commodity, it could be tested using historical data to compare the forecasts it would have produced to the actual sales. A similar idea, useful in cases where the model is intended to represent a class of real things, is to test it against members of the class that were not used in formulating the model. For example, if a regression model is used to fit some data, another portion of the data might be held back for later testing. Another technique, which is sometimes useful in validating certain kinds of descriptive models, is to systematically vary parameters of the real system and observe whether the model successfully "tracks" the changes. Or, the model might be subjected to artificially constructed test situations that are deliberately designed to expose weaknesses. If it performs adequately in extreme situations, one can have some confidence that it will work well under more normal circumstances.

If the model cannot be validated prior to implementation, then perhaps it can be implemented in phases for validation. For example, a new model for inventory control may be implemented for a certain selected group of items while the older system is retained for the majority of items. As the model proves itself, more items can be placed within its jurisdiction.

As a word of caution, it should be pointed out that validation can be carried too far. One might reach the point at which an enormous amount of effort is expended to increase model confidence only a small amount. Depending upon the importance of a model, it may be preferable to tolerate a lower confidence level. In some cases it may be sufficient to know that other people did something similar and that it worked for them.

Finally, it is worth remembering that real things change in time. A highly satisfactory model may very well degrade with age. Depending on how such factors affect model performance and validity, an implemented model may require anything from constant surveillance to periodic reevaluation.

14.1.7 SENSITIVITY ANALYSIS

Sensitivity analysis is concerned with studying the changes in the model results when one or more input parameters or coefficients are changed. The importance of sensitivity analysis cannot be overemphasized. Frequently it is used as part of model validation in building confidence in the results of the optimization study. An optimization study is never complete without a detailed sensitivity analysis of all model parameters and coefficients. In many cases such an analysis will be more valuable and may give more insight into the system studied than the optimal solution itself.

There are a number of reasons for doing a detailed sensitivity analysis:

1. If we can vary some of the data (i.e., that the data in some sense are controllable), then we can study the effect of these changes on the optimal solution. If it turns out that the optimal value changes (in our favor) by a considerable amount for a small change in the given parameters, then it may be worthwhile making some of the changes. For example, it may turn out that labor availability is a bottleneck in the system and that employing overtime may result in increased revenues that far exceed the added cost of overtime labor.
2. Sensitivity analysis provides valuable information regarding additions and modifications to the system under study so as to improve the overall operation of the system. For example, one can analyze the wisdom of investing in new machinery to increase the production capacity.
3. Sensitivity analysis can also clarify the effect of variations in uncontrollable parameters on the system. The uncontrollable parameter may include some constants that were estimated from past information by statistical methods. Some examples of uncontrollable parameters are cost of raw material, availability of raw material, price of energy, product demands, and equipment breakdowns. In many cases the estimates of these parameter values may not be very accurate. If it turns out through sensitivity analysis that the model results are very sensitive to these parameters, then it may be worth investing more funds to get better estimates of their values. On the other hand, if the model results are relatively insensitive to reasonable variations in certain parameter values, then there is no need to concern oneself too much about the accuracy of those parameters.

Also, a detailed sensitivity analysis of the model parameters provides credibility and better acceptance of the model results by management and thus should be a part of all optimization studies.

14.1.8 GUIDE TO SUCCESSFUL OPTIMIZATION

Some important considerations in modeling, problem formulation, and solution are as follows (see Phillips et al.[1] for a more complete discussion):

1. *Do not build a complicated model when a simple one will suffice.* One should always ask whether the cost of modeling and optimization is worth the solution. Complicated models do not necessarily give better solutions! Even with the best of motives, it is easy to get carried away by the sheer challenge of a difficult problem and thereby spend far more time and money on refining a model than the problem is worth. The power and generality of a model has little to do with its usefulness in dealing with a particular problem. In the actual practice of building models for specific purposes, the best advice is to keep it simple.
2. *Do not mold the problem formulation to fit a technique.* The assumptions in problem formulation should be consistent with the realities of the system. The choice of technique should be made only after the problem has been completely formulated. Before embarking on a sophisticated technique to determine the "true optimal solution," it may be worthwhile to look at the feasibility of using some simple heuristic methods that can give a good approximate optimal solution to the optimization problem. Interested readers may refer to Woolsey and Swanson,[6] who offer a number of "quick and dirty" techniques for solving optimization problems in production scheduling, machine sequencing, inventory control, economic analysis, and capital budgeting.
3. *A model cannot do any better than the information that goes into it.* A well-known maxim of computer programming called "GIGO" (garbage in, garbage out) also applies to modeling. A model can only manipulate the data provided to it; it cannot recognize and correct deficiencies in input. Also, models cannot *create* data, but they can *condense* data or convert it to a more useful form. Data collection is an important activity occurring at various stages of an optimization study—problem definition, model formulation, and validation. However, it is important to realize that the cost of data collection must not outweigh the expected contribution from the optimization study.
4. *Models cannot replace decision makers.* It is important that models not be oversold as replacements for decision makers. Models are also subject to human subjectivity and error. Hence the final decision at the implementation stage would invariably involve certain subjective aspects of the problem. These aspects are very apparent in problems involving conflicting objectives. Models provide a quantitative analysis of the various alternatives from which management can make an objective decision. Hence models should be viewed as aids to decision makers.

One of the best sources of descriptions of successful optimization studies carried out in practice is *Interfaces*, a quarterly journal published jointly by the Operations Research Society of America and the Institute of Management Sciences. The journal is devoted solely to the applications, practice, and implementation of operations research and management science in business, industry, and government operations. It is directed largely at those interested primarily in applications of optimization models to practical problems. In particular, readers are directed to one of the *Inter-*

faces issues that contains a good bibliography of different optimization applications by Gray and Cullinan-James.[7] Another reference, by Agin,[8] contains a good discussion of the conduct of optimization studies in practice.

REFERENCES

1. D. T. PHILLIPS, A. RAVINDRAN, and J. J. SOLBERG, *Operations Research: Principles and Practice*, Wiley, New York, 1976.
2. T. J. BERNA and A. W. WESTERBERG, "A New Approach to Optimization of Chemical Processes," *American Institute of Chemical Engineers Journal*, Vol. 26, No. 37 (1980), pp. 37–43.
3. M. C. EMBURY, G. V. REKLAITIS, and J. M. WOODS, "Scheduling and Simulation of a Staged Semi-Continuous Multiproduct Process," *Proceedings of Pacific Chemical Engineering Congress*, Vol. 2, AIChE, New York, August 1977.
4. P. V. L. N. SARMA and G. V. REKLAITIS, "Optimization of a Complex Chemical Process Using an Equation Oriented Model," in J. L. Goffin and J. M. Rousseao, Eds., *Mathematical Programming Study on Applications*, North Holland Publishers, Amsterdam, 1981.
5. B. F. SCRIBNER and D. H. BERLAU, "Optical Planning for Product/Production Facility Studies (OPPLAN)," *Proceedings of American Institute of Industrial Engineers Conference*, AIIE, Norcross, GA, 1979.
6. R. E. D. WOOLSEY and H. S. SWANSON, *Operations Research for Immediate Applications–A Quick and Dirty Manual*, Harper & Row, New York, 1975.
7. P. GRAY and C. CULLINAN-JAMES, "Applied Optimization–A Survey," *Interfaces*, Vol. 6, No. 3 (May 1976), pp. 24–41.
8. N. I. AGIN, "The Conduct of Operations Research Studies," in J. J. Moder and S. E. Elmaghraby, Eds., *Handbook of Operations Research*, Vol. 1, Van Nostrand Reinhold, New York, 1978.

CHAPTER 14.2
Linear Programming

A. RAVINDRAN
University of Oklahoma

14.2.1 INTRODUCTION

Linear programming, commonly called **LP**, merely defines a particular class of optimization problems that meet the following two conditions:

1. The objective function to be optimized can be described by a linear function of the decision variables.
2. The operating rules or constraints governing the process (e.g., limited resources) can be expressed as a set of linear equations or inequalities.

LP techniques are widely used to solve a number of military, economic, industrial, and social problems. In a 1975 survey of American companies,[1] **LP** emerged as the most often used technique (74%) among all the optimization methods. It has been reported in *Science News*[2] that about one fourth of computer time spent on scientific computation in recent years has been devoted to solving **LP** problems and their many variations! Three primary reasons for its wide use are:

1. A large variety of problems in diverse fields can be represented (within reasonable accuracy) as **LP** models.
2. Efficient techniques for solving **LP** problems are available.
3. Data variation (sensitivity analysis) can be handled with ease through **LP** models.

14.2.2 FORMULATION OF LINEAR MODELS

The three basic steps in constructing an **LP** model are as follows:

Step 1 Identify the unknown variables to be determined (design or decision variables) and represent them in terms of algebraic symbols.
Step 2 Identify all the restrictions *or* constraints in the problem and express them as linear equations or inequalities, which are linear functions of the unknown variables.
Step 3 Identify the objective or criterion and represent it as a linear function, which is to be maximized *or* minimized.

The following example illustrates these basic steps.

Example 14.2.1 (Product mix problem[3]) A company manufactures three products which require three resources–labor, material, and administration. The company's production engineering department has furnished the following data:

	Products		
	1	2	3
Labor (hr/unit)	1	1	1
Material (lb/unit)	10	4	5
Administration (hr/unit)	2	2	6
Profit ($/unit)	10	6	4

The supply of raw material is restricted to 600 lb/day. The daily availability of manpower is 100 hr. There are 300 hr of administration. Formulate a linear programming model to determine the daily production levels of the various products in order to maximize the total profit.

The formulation is as follows:

STEP 1. *Identify the decision variables.* The unknown activities to be determined are the daily rates of production on the three products. Represented by algebraic symbols, they are

$$x_1 = \text{daily production of product 1}$$
$$x_2 = \text{daily production of product 2}$$
$$x_3 = \text{daily production of product 3}$$

STEP 2. *Identify the constraints.* In this problem the constraints are the limited availability of the three resources–labor, material, and administration. Product 1 requires 1 hr of labor for each unit, and its production quantity is x_1. Hence the requirement of labor for product 1 alone will be x_1 hr (assuming a linear relationship). Similarly, products 2 and 3 will require x_2 and x_3 hr, respectively. Thus the total requirement of labor will be $x_1 + x_2 + x_3$, which should not exceed the available 100 hr. So the labor constraint becomes

$$x_1 + x_2 + x_3 \leqslant 100$$

The raw material requirements will be $10x_1$ lb for product 1, $4x_2$ lb for product 2, and $5x_3$ lb for product 3. Thus the raw material constraint is given by

$$10x_1 + 4x_2 + 5x_3 \leqslant 600$$

Similarly, the constraint for administration becomes

$$2x_1 + 2x_2 + 6x_3 \leqslant 300$$

In addition, we restrict the variables x_1, x_2, and x_3 to having only nonnegative values. This is called the "nonnegativity constraint," which the variables must satisfy. Most practical **LP** problems will have this nonnegative restriction on the decision variables.

STEP 3. *Identify the objective.* The objective is to maximize the total profit from sales. Assuming that a perfect market exists for the products such that all that is produced can be sold, the total profit from sales becomes

$$Z = 10x_1 + 6x_2 + 4x_3$$

Thus the **LP** model for our product mix problem is to find numbers x_1, x_2, x_3 that will maximize

$$Z = 10x_1 + 6x_2 + 4x_3$$

subject to the constraints

$$x_1 + x_2 + x_3 \leqslant 100$$
$$10x_1 + 4x_2 + 5x_3 \leqslant 600$$
$$2x_1 + 2x_2 + 6x_3 \leqslant 300$$
$$x_1 \geqslant 0, \quad x_2 \geqslant 0, \quad x_3 \geqslant 0$$

14.2.3 BASIC ASSUMPTIONS OF LINEAR MODELS

The **LP** approach to modeling a system under study is to decompose the system into its elementary functions, or "activities," In Example 14.2.1 there were three activities–manufacture of one unit of product 1, manufacture of one unit of product 2, and manufacture of one unit of product 3. The decision variables merely define the levels at which these activities are to be carried out. Of course the aim of the **LP** model is to determine the optimal activity levels. To change the activity level, the input and output flows into each activity have to be changed. These flows are called "items." In Example 14.2.1 the input flows were labor, material, and administration, and the output was profit in dollars.

There are two basic assumptions in the formulation of all **LP** models:

1. **Proportionality.** This guarantees that the flow of items into and out of an activity is directly proportional to its activity level. If one unit of product 1 requires 1 hr of labor, 10 lb of material,

and 2 hr of administration, then to make x units of product 1 will require x hr of labor, $10x$ lb of material, and $2x$ hr of administration for any x. Similarly, the unit profit from selling product 1 is always \$10, irrespective of how many units are sold.

2. Additivity. This assumption implies that the total usage of an item is equal to the sum of the item usages of each individual activity at its specified level. In Example 14.2.1 the total material consumption was equal to the sum of the material consumed by the individual products.

For a more detailed discussion on the input-output approach to modeling and **LP** assumptions, the reader is referred to Dantzig.[4] The proportionality and additivity assumptions imply that all the constraints of the **LP** problem can be expressed as linear equations or inequalities and that the objective is a linear function of the decision variables. It is common to find some practical problems violating one or more **LP** assumptions. In such cases, by using clever formulations or good approximations, one could still use **LP**. Some of these are discussed in the next section.

14.2.4 HANDLING NONLINEARITIES BY LINEAR PROGRAMMING[3,5]

Nonlinearities can arise in a number of ways in optimization problems in either the objective function or the constraints. Some of the nonlinearities can be handled by **LP** methods, whereas the rest have to be solved by specialized nonlinear programming methods (see Chapter 14.3).

Piecewise Linear Functions

A piecewise linear function arises when the per unit contribution (cost) depends on the level of sales (production). For example, consider a product whose profit contribution is \$10/unit for the first 40 units, \$8/unit for the next 60 units, and \$5/unit for the rest. The nonlinearity of the profit function is apparent if a graph is plotted between total profit and quantity sold. This is illustrated by Exhibit 14.2.1, which is called a piecewise linear function since it is linear in the region (0, 40), (40, 100) and (100, ∞). By partitioning the quantity sold into three activities, the profit function could be expressed as a linear function as follows:

x_1 = quantity sold at \$10/unit profit
x_2 = quantity sold at \$ 8/unit profit
x_3 = quantity sold at \$ 5/unit profit

The amount of product sold is $x_1 + x_2 + x_3$, and the objective function is to maximize $\mathbf{Z} = 10x_1 + 8x_2 + 5x_3$. Since there is a limit on how many units can be sold for a certain profit, we need the following constraints:

$$0 \leqslant x_1 \leqslant 40, \qquad 0 \leqslant x_2 \leqslant 60, \qquad 0 \leqslant x_3$$

When the objective function is maximized, it is easy to see that x_2 will not become positive until x_1 reaches its limit of 40. Similarly, x_3 cannot be positive until $x_1 = 40$ and $x_2 = 60$.

Exhibit 14.2.1 Example of a Piecewise Linear Function

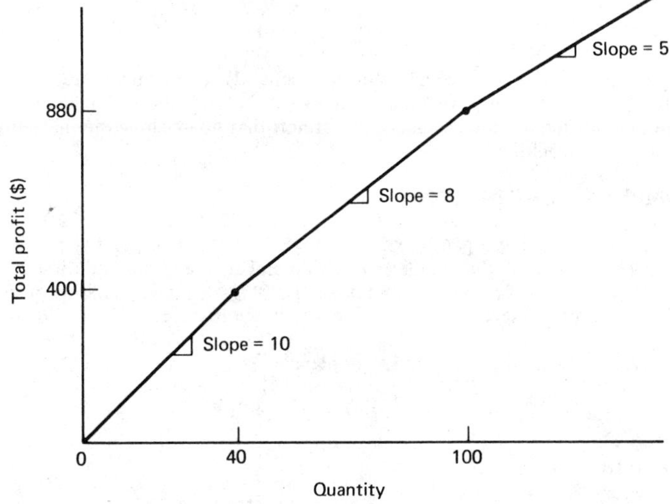

One could extend this idea to handle smooth nonlinear profit functions by approximating them into piecewise linear functions. It is important to realize that **LP** methods can be used successfully to handle piecewise linear functions as long as the following hold:

1. In a maximization problem the piecewise linear function must have decreasing slope or be a concave function (i.e., the per unit contribution must be decreasing or at least nonincreasing).
2. In a minimization problem the function must have increasing slope or be a convex function (i.e., the per unit cost must be increasing or at least nondecreasing).

If these properties do not hold, then one has to use the more complicated integer programming formulations.

Max-Min Problems

In some optimization problems one may encounter a nonlinear objective that is to maximize the minimum of several variables or functions. Consider an assembly made up of three different parts. Let x_1, x_2, and x_3 be the decision variables denoting the number of parts 1, 2, and 3 produced, respectively. If management wishes to maximize the number of assemblies, then the objective function becomes

$$\text{maximize } [\text{minimum of } (x_1, x_2, x_3)]$$

Even though this is a nonlinear function, it can be linearized as follows: Let y denote the number of assemblies made. Then the linear objective function is

$$\text{maximize } y \tag{1}$$

Since y is the minimum of x_1, x_2, x_3, we get the three additional constraints

$$y \leqslant x_1 \tag{2}$$
$$y \leqslant x_2 \tag{3}$$
$$y \leqslant x_3 \tag{4}$$

The inequalities 2, 3, and 4 in conjunction with 1 are equivalent to maximizing the minimum of x_1, x_2, and x_3.

Handling Absolute Value Functions

Absolute value sign in the constraint can be handled by replacing it by two constraints. For example, a nonlinear constraint of the type

$$|x_1 - x_2| \leqslant 30 \tag{5}$$

is equivalent to the following two linear constraints:

$$\begin{aligned} x_1 - x_2 &\leqslant 30 \\ -x_1 + x_2 &\leqslant 30 \end{aligned}$$

Nonlinear constraints of the type given in 5 occur frequently in machine balancing. If x_1 represents the daily utilization of machine 1 in minutes, and x_2 is the utilization for machine 2, then inequality 5 is equivalent to the machine balancing constraint that no machine run more than 30 min/day longer than the other machine.

14.2.5 SIMPLEX ALGORITHM

The simplex algorithm as developed by G. B. Dantzig in 1947 is an iterative procedure for solving **LP** problems. The theory of the simplex algorithm and its many computational refinements are fully presented in several outstanding textbooks (Dantzig,[4] Murty,[5] Phillips et al.,[3] Gass,[6] and Hadley[7]). The intent of this section is to describe briefly the basic principles of the simplex method.

Example 14.2.2 Consider the following **LP** problem[3]:

$$\text{minimize } \mathbf{Z} = 40x_1 + 36x_2$$

subject to
$$\begin{aligned} x_1 &\leqslant 8 \\ x_2 &\leqslant 10 \\ 5x_1 + 3x_2 &\geqslant 45 \\ x_1 \geqslant 0, \quad x_2 &\geqslant 0 \end{aligned}$$

In this problem we are interested in determining the values of the variables x_1 and x_2 that will satisfy all the restrictions and give the least value for the objective function. As a first step in solving this problem, we want to identify all possible values of x_1 and x_2 that are nonnegative and that satisfy the constraints. For example, a solution $x_1 = 8, x_2 = 10$ is positive and satisfies all the constraints. Such a solution is called a "feasible solution." The set of all feasible solutions is called the "feasible region." Solution of a linear program is nothing but finding the best feasible solution in the feasible region. The best feasible solution is called an "optimal solution" to the linear programming problem. In our example an optimal solution is a feasible solution that minimizes the objective function $40x_1 + 36x_2$. The value of the objective function corresponding to an optimal solution is called the "optimal value" of the linear program.

To represent the feasible region in a graph, every constraint may be plotted, and all values of x_1, x_2 that will satisfy these constraints can be identified. The nonnegativity constraints imply that all feasible values of the two variables will lie in the first quadrant. The constraint $5x_1 + 3x_2 \geqslant 45$ requires that any feasible solution (x_1, x_2) to the problem should be above the straight line $5x_1 + 3x_2 = 45$ (see Exhibit 14.2.2). Similarly, the constraints $x_1 \leqslant 8$ and $x_2 \leqslant 10$ are plotted. The feasible region is given by the shaded region **ABC** as shown in Exhibit 14.2.2. Obviously there are an infinite number of feasible points in this region. Our objective is to identify the feasible point with the lowest value of **Z**. The feasible points **A**, **B**, and **C** (Exhibit 14.2.2) are called the "corner points" of the feasible region.

Observe that the objective function given by $\mathbf{Z} = 40x_1 + 36x_2$ represents a straight line if the value of **Z** is fixed a priori. Changing the value of **Z** essentially translates the entire line to another straight line parallel to itself. To determine an optimal solution, the objective function line is drawn for a convenient value of **Z** such that it passes through one or more points in the feasible region. Initially **Z** is chosen as 600. By moving this line closer to the origin, the value of **Z** is further decreased (Exhibit 14.2.2). The only limitation on this decrease is that the straight line $40x_1 + 36x_2 = \mathbf{Z}$ contains at least one point in the feasible region **ABC**. This clearly occurs at the corner point **A** given by $x_1 = 8$, $x_2 = 1.6$. This is the best feasible point giving the lowest value of **Z** as 377.60. Hence $x_1 = 8$, $x_2 = 1.6$ is an *optimal solution*, and $\mathbf{Z} = 377.60$ is the *optimal value* for the linear program.

In our example one of the corner points of the feasible region (namely, **A**) was an optimal solution. As a matter of fact, the following property is true for any **LP** problem: If there exists an optimal solution to an **LP** problem, then at least one of the corner points of the feasible region will always qualify to be an optimal solution.

Exhibit 14.2.2 Graphic Solution of Example 14.2.2

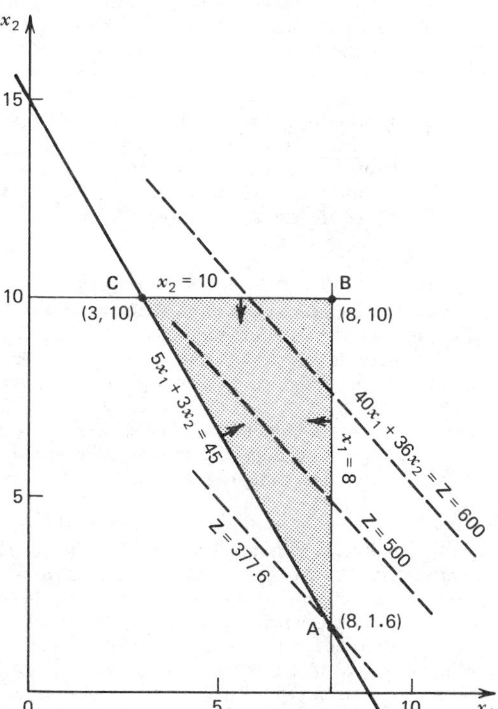

This is the fundamental property on which the simplex method for solving **LP** problems is based. Even though the feasible region of an **LP** problem contains an infinite number of points, an optimal solution can be determined by merely examining the finite number of corner points in the feasible region. In **LP** terminology the corner point feasible solutions are known as "basic feasible solutions." Hence the simplex method for solving general **LP** problems is simply an orderly procedure for generating and examining different basic feasible solutions. In problems involving just two variables, one can easily draw the feasible region in a graph and identify the corner points that are basic feasible solutions. In practice, the **LP** problems involve hundreds of constraints and several thousand variables, and we need an algebraic procedure for generating the basic feasible solutions. The simplex method uses the classical Gauss-Jordan elimination scheme for generating the basic feasible solution. The Gauss-Jordan elimination can be represented by a sequence of vector-matrix operations and hence easily implemented on a digital computer.

The general steps of the matrix-based simplex method are as follows:

1. Start with an initial basic feasible solution.
2. Improve the initial solution, if possible, by finding another basic feasible solution with a better objective function value. At this step the simplex method implicitly eliminates from consideration all those basic feasible solutions whose objective function values are worse than the present one. This makes the procedure more efficient than the naive approach, which would examine all the basic feasible solutions.
3. Continue to find better basic feasible solutions, improving the objective function values. When a particular basic feasible solution cannot be improved further, it becomes an optimal solution, and the simplex method terminates.

Computational Efficiency of the Simplex Method

The computational efficiency of the simplex method depends on (1) the number of iterations (basic feasible solutions) to go through before reaching the optimal solution and on (2) the total computer time to solve the problem. Much effort has been spent in studying the computational efficiency with regard to the number of constraints and the decision variables in the problem.

Empirical experience with thousands of practical problems shows that the number of iterations of a standard linear program with m constraints and n variables varies between m and $3m$, the average being $2m$. A practical upper bound for the number of iterations is $2(m + n)$. (Occasionally some problems have violated this bound.)

The computational time is found to vary approximately in relation to the cube of the number of constraints in the problem (m^3). For example, if problem **A** has twice as many constraints as problem **B**, then the computer time for problem **A** will be eight times that of problem **B**.

It is to be noted that the computational efficiency of the simplex method is more sensitive to the number of constraints than to the number of variables. Hence the general recommendation is to keep the number of constraints as small as possible by avoiding unnecessary or redundant constraints in the formulation of the **LP** problem.

Computational characteristics of the simplex method are summarized in a study by Cutler and Wolfe,[8] which contains the results of a set of computations involving the solution of nine **LP** problems using 30 variations of the simplex algorithm. The statistics collected show the relative efficiency of proposed variations and the important features in the efficiency of **LP** computer codes.

For a survey of the many variants of the simplex method, the reader is referred to Barnes and Crisp.[9]

14.2.6 COMPUTER SOLUTION OF LINEAR PROGRAMS

Many practical problems formulated as **LP** run into hundreds of constraints and thousands of decision variables. These invariably have to be solved using a digital computer. Since the simplex method is an iterative procedure, it can be applied mechanically to any problem. Hence it is ideally suited for computer implementation.

Even though the simplex method was developed in 1947, it was not formally published until 1951. Hence the solution of **LP** problems on a digital computer did not take place until 1952. The first successful attempt on a first-generation computer was made at the National Bureau of Standards in early 1952.

The first-generation computers used vacuum tubes with no memory to speak of. They were capable of solving only small **LP** problems with no more than 50 constraints and 100 variables and were also very slow computationally. For example, it took 15 min to solve a 10 × 20 linear program. (Modern-day computers solve such problems in 1 or 2 sec!) Thus the major industrial applications of **LP** had to wait for the advancement in computer technology. To a large extent, the advancement of **LP** and its applications has gone hand in hand with advances in computers.

The late 1950s saw the introduction of the second-generation computers. Using transistors instead of vacuum tubes, these computers were able to operate at higher speeds. Storage of data was handled by magnetic discs. The computers were able to solve moderate-size **LP** problems of the order of 300 to 400 constraints and 2000 to 3000 variables. During this phase, **LP** saw many industrial applications, but the solution of large industrial problems still had to wait for the third-generation computers.

The third-generation computers were introduced in the 1960s. They employed integrated circuits, used larger magnetic core memories, and had multiprogramming or time-sharing capability. These features increased their speeds of operation immensely, and the solution of large **LP** problems with thousands of constraints and unlimited numbers of decision variables became possible. This has resulted in the enormous success of linear programming in such diverse areas as defense, industry, commercial-retail, agriculture, education, and the environment.

Computer Codes

Commercial **LP** computer codes are available from many computer manufacturers and private companies who specialize in marketing **LP** softwares for major computer systems. Depending on their capabilities, these codes vary in their complexity, ease of use, and cost. The need to solve large **LP** problems has led to the development of very complex and sophisticated **LP** computer codes, called "mathematical programming systems" (e.g., IBM's MPSX-370, Control Data Corporation's APEX III, Management Science Systems' MPS-III). These have sophisticated data handling and analytical tools and can solve problems of the order of 8000 to 16,000 constraints and unlimited numbers of variables. An **LP** problem with 50,000 constraints has been successfully solved by MPS-III.

Linear programming problems with 5000 or more constraints are definitely large by any standard, and the success of solving them depends on the problem structure and any special properties the problem may have. Typical **LP** problems in practice would have 500 to 1500 constraints and several thousand variables. These can be solved without much difficulty by any of the advanced **LP** computing systems. These advanced **LP** systems are available on a fee basis, in addition to the normal computer rental charges. If the user is planning to solve the **LP** problem frequently, then it may be cost-effective to acquire the added **LP** software. For infrequent use, it is better to employ a local qualified computer service bureau or rent a remote job terminal connected to a major computer center that has the **LP** softwares available for its clients. An excellent survey of modern **LP** code characteristics, including data management, is given by Orchard-Hays.[10] White[11] presents a status report on computing algorithms for mathematical programming systems. All the commercial computer codes use the simplex algorithm and many of its variants as the basic method for solving the **LP** problem. Recently much interest has been generated in a new **LP** algorithm by a Russian mathematician, Khachian.[12] Even though Khachian's algorithm is theoretically an important breakthrough, it does not look promising for computer implementation (see Dantzig[13]).

Important Features of LP Softwares

It would be impractical to discuss the features of all the **LP** softwares provided by different computer manufacturers and computer service organizations. Instead, some of the important features that are common to all the advanced **LP** computing systems are briefly described here. For a more detailed description, the reader is referred to Bradley et al.[14]

Input Data

The following input data are required from the user by almost all computer codes:

1. *Names of the decision variables, constraints, and objective for easy identification by the user.* For example, instead of defining the activities as x_1, x_2, and x_3, the user may simply call them PROD 1, PROD 2, and PROD 3.
2. *Values of the constraint coefficients, objective function coefficients, and right-hand-side (RHS) constants.* It is sufficient to specify the nonzero coefficients only since, in practice, most constraint matrices have a lot of zero entries.
3. *Nature of the constraint relationship.* The user specifies the constraints as equations or inequalities ($\leqslant$ or $\geqslant$).
4. *Variable bounds.* The variables are always restricted to being nonnegative with no finite upper bounds, unless otherwise specified by the user.
5. *Multiple runs.* The user may provide several objective functions and **RHS** vectors. He or she can then instruct the program to make several **LP** runs by specifying the appropriate objective function and **RHS** vector to use. This is an effective way to study the sensitivity of the **LP** solution with respect to the input data.

Matrix Generators

Development of the input data for large **LP** problems is best done with the help of matrix generators. These are special-purpose programs, generally developed by individual users, which would transform the raw data into the LP data format required by the computer codes. They may also perform certain arithmetic calculations on tabular data to generate the values of the objective function coefficients, RHS constants, and constraint coefficients. Matrix generators can also be used to detect possible errors in the input data by performing certain consistency checks. Generally, the matrix generators allow the user to specify minimum amounts of information, usually in the form in which it is collected. Because of the large volume of data to be generated in special form, matrix generators have become indispensable while solving large LP problems. In many cases the cost of data handling, updating, and matrix generation may consume 90% of the total cost of solution by the LP models.

Advanced LP softwares in the market have their own general-purpose matrix generators as part of their LP packages. Of course individual users may also develop special matrix generators, particularly when the LP models are going to be run regularly with frequent changes in the input parameters.

User Options

Most commercial codes provide the following user options to improve the accuracy of the LP solution:

1. *Scaling*. An automatic scaling of the input data is provided as a user option in order to reduce the round-off error buildup and numerical instability in the LP calculations. Whenever the LP formulation leads to coefficients whose magnitudes vary very widely, the automatic scaling option should be exercised.
2. *Tolerances*. These dictate the degree of accuracy required in the LP solution by the user. For example, a tolerance may specify how small a number can be before it can be treated as zero.
3. *Errors*. Since the simplex method solves the LP problem in standard form (where all the constraints are expressed as equations), it is necessary to know how much error can be tolerated in satisfying the constraint equations by the LP solutions. For example, the user may tolerate an error magnitude of 1×10^{-7} between the left-hand side (LHS) and RHS of a constraint equation.

Output Features

The standard minimum output data for all LP softwares are usually the following:

1. *Variable name and its value in the optimal solution*.
2. *Optimal value of the objective function*.
3. *Shadow prices on the constraint resources*. The shadow price of a constraint gives the net change in the optimal value of the objective function per unit increase in the RHS value of that constraint. The shadow prices could be positive or negative, depending on whether an increase in the constraint resource has a positive or negative impact on the objective value. The shadow prices can be interpreted as marginal costs or marginal profits.
4. *Opportunity costs of the decision variables*. The opportunity costs have an interpretation similar to that of shadow prices, but they measure the impact with respect to the columns or variables.

In addition to the standard output, all major LP softwares are capable of providing the following optional outputs:

1. *Echo check of the input data*. The user may request either a summary or complete information on the input data. This could be used for debugging purposes.
2. *Sensitivity analysis*. This provides a postoptimality analysis of the LP solution with respect to the input parameters. Basically this provides an analysis of how sensitive the optimal value is with respect to variations in the objective function coefficients and RHS constants. Sensitivity analysis is discussed in more detail in the next section.

Report Writers

Report writers are special-purpose programs that will transform the standard LP output into user-oriented form. These are similar to the matrix generators, but are used in the output stage. These user-oriented outputs will be in a form that can be understood by shop personnel who are not familiar with the LP model. The most sophisticated LP codes also provide general-purpose report writers for users.

14.2.7 SENSITIVITY ANALYSIS IN LINEAR PROGRAMMING

In all linear programming models, the coefficients of the objective function and the constraints are supplied as input data or as parameters to the model. The optimal solution obtained by the simplex method is based on the values of these coefficients. In practice, the values of these coefficients are seldom known with absolute certainty because many of the coefficients are functions of some uncontrollable parameters. For instance, the future demands, the cost of raw materials, or the cost of energy resources cannot be predicted with complete accuracy before the problem is solved. Hence the solution of a practical problem is not complete with the mere determination of the optimal solution.

Each variation in the values of the data coefficients changes the linear programming problem, which may in turn affect the optimal solution found earlier. To develop an overall strategy to meet the various contingencies, one has to study how the optimal solution will change as a result of changes in the input (data) coefficients. This is sensitivity analysis or postoptimality analysis. Other reasons for performing a sensitivity analysis are as follows:

1. There may be some data coefficients or parameters of the linear program that are controllable, for example, availability of capital, raw material, or machine capacities. Sensitivity analysis enables one to study the effects of changing these parameters on the optimal solution. If it turns out that the optimal value (profit/cost) changes (in our favor) by a considerable amount for a small change in the given parameters, then it may be worthwhile to implement some of these changes. For example, if increasing the availability of labor by allowing overtime contributes to a greater increase in the maximum return as compared to the increased cost of overtime labor, then one might want to allow overtime production.

2. In many cases the values of the data coefficients are obtained by statistical estimation procedures on past figures, as in the case of sales forecasts, price estimates, and cost data. These estimates, in general, may not be very accurate. If we can identify which of the parameters affect the objective value most, then we can obtain better estimates of these parameters. This will increase the reliability of our model and the solution.

Practical Uses

Example 14.2.1, discussed at the beginning of this chapter, can be used to help illustrate the practical uses of sensitivity analysis. A computer output of the solution of this problem is given in Exhibit 14.2.3. Note from the optimal solution that the optimal product mix is to produce products 1 and 2 only at levels 33.33 and 66.67 units, respectively.

The shadow prices give the net impact on the maximum profit if additional units of certain resources can be obtained. Labor has the maximum impact, providing a \$3.33 increase in profit per every hour of increase in labor. Of course the shadow prices on the resources apply as long as their variations stay within the prescribed ranges on **RHS** constants given in Exhibit 14.2.3. In other words, a \$3.33/hr increase in profit is achievable as long as the labor hours are not increased beyond 150 hr. Suppose it is possible to increase the labor hours by 25% by scheduling overtime that incurs an additional labor cost of \$50. To see whether it is profitable to schedule overtime, we first determine the net increase in maximum profit due to 25 hr of overtime as (25) (3.33) = \$83.25. Since it is more than the total cost of overtime, it is economical to schedule overtime. It is important to note that, when any of the **RHS** constants is changed, the optimal solution will change. However, the optimal product mix will be unaffected as long as the **RHS** constant varies

Exhibit 14.2.3 Computer Solution of Example 14.2.1

Optimal solution	$x_1 = 33.33, x_2 = 66.67, x_3 = 0$
Optimal value	Maximum profit = \$733.33
Shadow prices	For row 1 = \$3.33, for row 2 = \$0.67, for row 3 = 0
Opportunity costs	For $x_1 = 0$, for $x_2 = 0$, for $x_3 = 2.67$

Ranges on Objective Function Coefficients

Variable	Lower Limit	Present Value	Upper Limit
x_1	6	10	15
x_2	4	6	10
x_3	$-\infty$	4	6.67

Ranges on RHS Constants

Row	Lower Limit	Present Value	Upper Limit
1	60	100	150
2	400	600	1000
3	200	300	∞

within the specified range. In other words, we will still be making products 1 and 2 only, but their quantities may change.

The ranges on the objective function coefficients given in Exhibit 14.2.3 exhibit the sensitivity of the *optimal solution* with respect to changes in the unit profits of the three products. It shows that the optimal solution will not be affected as long as the unit profit of product 1 stays between \$6 and \$15. Of course the maximum profit will be affected by the change. For example, if the unit profit on product 1 increases from \$10 to \$12, the optimal solution will be the same, but the maximum profit will increase to \$733.33 + (12 – 10) (33.33) = 799.99.

Note that product 3 is not economical to include in the optimal product mix. Hence a further decrease in its profit contribution will not have any impact on the optimal solution or maximum profit. Also, the unit profit on product 3 must increase to \$6.67 (present value + opportunity cost) before it becomes economical to produce.

Simultaneous Variations in Parameters[14]

The sensitivity analysis output on profit and **RHS** ranges is obtained by varying only one of the parameters and holding all other parameters fixed at their current values. However, it is possible to use the sensitivity analysis output when several parameters are changed simultaneously. This is done with the help of the "100% rule".

100% Rule for Objective Function Coefficients

The 100% rule for the objective function coefficients is given by

$$\sum_j \frac{\delta c_j}{\Delta c_j} \leqslant 1 \tag{6}$$

where δc_j is the actual increase (decrease) in the objective function coefficient of variable x_j and Δc_j is the maximum increase (decrease) allowed by sensitivity analysis. As long as inequality 6 is satisfied, the optimal solution to the **LP** problem will not change. For example, suppose the unit profit on product 1 decreases by \$1, but increases by \$1 for both products 2 and 3. This simultaneous variation satisfies the 100% rule, since $\delta c_1 = -1$, $\Delta c_1 = -4$, $\delta c_2 = 1$, $\Delta c_2 = 4$, $\delta c_3 = 1$, $\Delta c_3 = 2.67$, and

$$\frac{-1}{-4} + \frac{1}{4} + \frac{1}{2.67} = 0.875 < 1$$

Hence the optimal solution will not change, but the maximum profit will change by (–1) (33.33) + 1(66.67) + 1(0) = 33.34.

100% Rule for RHS Constants

The 100% rule for the **RHS** constants is given by

$$\sum_i \frac{\delta b_i}{\Delta b_i} \leqslant 1 \tag{7}$$

where δb_i is the actual increase (decrease) in the **RHS** constant of the ith constraint and Δb_i is the maximum increase (decrease) allowed by sensitivity analysis. If inequality 7 is satisfied, then the optimal product mix remains the same and the shadow prices apply, but the optimal solution and maximum profit will change. Of course the net change in the maximum profit can be obtained using the shadow prices.

Warning: The failure of the 100% rule does not automatically imply that the **LP** solution will be affected.

14.2.8 APPLICATIONS

Linear programming models are widely used to solve a number of military, economic, industrial, and social problems. The oil companies have been and still are one of the foremost users of very large **LP** models in petroleum refining, distribution, and transportation. The number of **LP** applications has grown so much in the last 20 years that it will be impossible to survey all the different applications. Instead, the reader is referred to two excellent textbooks, Gass[15] and Salkin and Saha,[16] which are devoted solely to **LP** applications in such diverse areas as defense, industry, commercial-retail, agriculture, education, and the environment. Many of the applications also contain a discussion of the experiences in using the **LP** model in practice.

Discussion of various computer solutions to large **LP** problems in industry and computational features of **LP** softwares are discussed in Driebeek,[17] Daellenbach and Bell,[18] and Bradley et al.[14] An excellent bibliography on **LP** applications is available in Gass.[6] It contains a list of references arranged by area (e.g., agriculture, industry, military, production, transportation). In the area of industrial application, the references have been further categorized by industry (e.g., chemical, coal, airline, iron and steel, paper, petroleum, railroad). For additional bibliographies on **LP** applications, readers may refer to a survey by Gray and Cullinan-James.[19] For more recent applications of **LP** in practice, readers should check the recent issues of *Interfaces*, *AIIE Transactions*, *Decision Sciences*, *European Journal of Operational Research*, *Management Science*, *Operations Research*, *Operational Research* (United Kingdom), *Naval Research Logistics Quarterly*, and *OpSearch* (India).

REFERENCES

1. F. J. FABOZZI and J. VALENTE, "Mathematical Programming in American Companies: A Sample Survey," *Interfaces*, Vol. 7, No. 1 (November 1976), pp. 93–98.
2. L. A. STEEN, "Linear Programming: Solid New Algorithm," *Science News*, Vol. 116, October 6, 1979, pp. 234–236.
3. D. T. PHILLIPS, A. RAVINDRAN, and J. J. SOLBERG, *Operations Research: Principles and Practice*, Wiley, New York, 1976.
4. G. B. DANTZIG, *Linear Programming and Extensions*, Princeton University Press, Princeton, NJ, 1963.
5. K. G. MURTY, *Linear and Combinatorial Programming*, Wiley, New York, 1976.
6. S. GASS, *Linear Programming*, 4th ed., McGraw-Hill, New York, 1975.
7. G. HADLEY, *Linear Programming*, Addison-Wesley, Reading, MA, 1962.
8. L. CUTLER and P. WOLFE, "Experiments in Linear Programming," in R. L. Graves and P. Wolfe, Eds., *Recent Advances in Mathematical Programming*, McGraw-Hill, New York, 1963.
9. J. W. BARNES and R. M. CRISP, JR., "Linear Programming: A Survey of General Purpose Algorithms," *AIIE Transactions*, Vol. 7, No. 3 (1975), pp. 212–221.
10. W. ORCHARD-HAYS, "On the Proper Use of a Powerful MPS," in R. W. Cottle and J. Krarup, Eds., *Optimizations Methods for Resource Allocation*, English University Press, London, 1974.
11. W. W. WHITE, "A Status Report on Computing Algorithms for Mathematical Programming," *Computing Surveys*, Vol. 5, No. 3 (1973), pp. 135–166.
12. L. G. KHACHIAN, "A Polynomial Algorithm in Linear Programming," *Soviet Mathematics Doklady*, Vol. 20, No. 1 (1979), pp. 191–194.
13. G. B. DANTZIG, *Comments on Khachian's Algorithm for Linear Programming*, Technical Report SOL 79-22, Department of Operations Research, Stanford University, Stanford, CA, November 1979.
14. S. P. BRADLEY, A. C. HAX, and T. L. MAGNANTI, *Applied Mathematical Programming*, Addison-Wesley, Reading, MA, 1977, pp. 220–229.
15. S. I. GASS, *An Illustrated Guide to Linear Programming*, McGraw-Hill, New York, 1970.
16. H. M. SALKIN and J. SAHA, *Studies in Linear Programming*, Elsevier, North-Holland, New York, 1975.
17. N. J. DRIEBEEK, *Applied Linear Programming*, Addison-Wesley, Reading, MA, 1969.
18. H. G. DAELLENBACH and E. J. BELL, *User's Guide to Linear Programming*, Prentice-Hall, Englewood Cliffs, NJ, 1970.
19. P. GRAY and C. CULLINAN-JAMES, "Applied Optimization—A Survey," *Interfaces*, Vol. 6, No. 3 (May 1976), pp. 24–41.

CHAPTER 14.3

Nonlinear Optimization

KATTA G. MURTY

The University of Michigan

14.3.1 SCOPE OF THIS CHAPTER

Nonlinear optimization deals with problems in which a single objective function is required to be optimized (i.e., either maximized or minimized), subject to well-defined constraints on the decision variables. It is assumed that the objective and the constraint functions are all smooth, real-valued functions of the decision variables; that at least one of them is nonlinear (if all of them are linear, the problem is a linear program, discussed in Chapter 14.2); and that all the decision variables are continuous variables (i.e., that they can assume any value in their range of variation in the feasible region). If some of the decision variables violate this condition (e.g., a decision variable such as the number of tellers employed by a bank, which can take only integer values), the problem becomes a discrete optimization problem, which is discussed in Chapter 14.4.

In some practical models, several objective functions may be required to be optimized simultaneously. In such models an optimum solution for one objective function may turn out to be a very poor solution for another objective function, and in this case it is hard to characterize what an optimum solution is for the overall problem. Such models can be tackled through the use of practical approaches such as goal programming, Pareto or vector optimality criteria, or constructing a single utility function by taking a reasonable weighted combination of the various objective functions. These multiobjective problems are not discussed here.

Sometimes one may be required to solve a dynamic problem in which the values of the decision variables have to be evaluated over an interval of time as functions of the time parameter, subject to constraints that may involve differential equations. Such dynamic optimization problems fall in the area of control theory or calculus of variations. By discretizing over time, it may be possible to approximate such dynamic problems by suitably derived static problems. This chapter considers only static problems.

In this chapter the vector $x = (x_1, \ldots, x_n)^T$ usually denotes the column vector of decision variables whose optimal values are to be found; x can be viewed as a decision point in the n-dimensional space; and x_j with subscript j denotes the jth coordinate of the decision point x. Sometimes it may be necessary to refer to several specific decision points; they are denoted by symbols with such superscripts as $x^1, x^2, \ldots, x^r, \ldots$, where $x^r = (x_1^r, \ldots, x_n^r)^T$ is the rth point in this sequence.

Exponents are also used in this chapter. Exponents are always in bold-style letters, which distinguishes them from regular superscripts. For example, if α is a real number, $\alpha^{\mathbf{2}}$ represents the square of α, and $\alpha^{\mathbf{r}}$ represents the $\mathbf{r}$th power of α.

14.3.2 SOME EXAMPLES OF NONLINEAR MODELS

The aims of this section are to give the reader an idea of what nonlinear models look like and to illustrate the range of applicability of nonlinear optimization. Most real-life problems lead to large models, and it is difficult to discuss even one of them in this brief chapter. Instead, three simple nonlinear models are discussed. In each example the decision variables are denoted by the symbols x_j, and other symbols denote data in the problem.

Optimization of a District Heating System

A region needs hot water for its heating needs. It can get this either from a nearby nuclear combined plant or by using local oil-fired plants. Heat is produced at the nuclear combined plant at

The writing of this chapter was facilitated by support from the Air Force Office of Scientific Research, Air Force Systems Command, USAF, under grant number AFOSR 78-3646. I am grateful to the many researchers who supplied me with information concerning algorithms and software.

the cost of a minor loss of electricity and is transported to the region in the form of hot water through a system of pipes. This hot water enters a heat exchanger in the region, in which the heat is transferred to the local heating system. After passing through the heat exchanger, the water is returned to the nuclear combined plant for reuse. Following are the definitions of variables and data (functional forms have been arrived at either through the heat and electrical laws or by analysis of past data):

x_1 = heat (in heat units) supplied by oil-fired plant
x_2 = heat supplied by nuclear combined plant at the plant exit
t_0 = ambient water temperature (temperature of water returning to heat exchanger through the local heating system, as well as that of water returning to nuclear combined plant from the heat exchanger)
x_3 = temperature of water leaving nuclear combined plant to the heat exchanger
x_4 = temperature of water from nuclear combined plant as it enters the heat exchanger (determined from data such as the length of pipe, its diameter and insulation, energy supplied for pumping, etc., and from using cooling laws, such as formula that follows)
x_5 = heat transferred to the water for the local heating system at the heat exchanger
x_6 = flow rate of water from the nuclear combined plant into the system
x_7 = flow rate of water through the local heating system at the heat exchanger
q = specified minimum heat units required at the region

The costs are the heat production cost at the nuclear combined plant (cost of electricity loss), the operation cost for pumps to keep water flowing between the nuclear combined plant and the heat exchanger, and the cost of heat production by the oil-fired plant; these are the three terms in the objective function that follows, in that order. The problem is to

$$\text{minimize} \quad [c_1 x_2(\alpha_0 + \alpha_1 t_0 + \alpha_2 x_3)] + (c_2 x_6^3) + (c_0 x_1)$$

subject to

$$x_1 + x_5 \geqslant q$$

$$x_5 - \frac{x_4 - t_0}{c_5[1/x_6 + 1/x_7 - 1/(x_6 + x_7)]} = 0$$

$$x_2 - c_6 x_6 (x_3 - t_0) = 0$$

$$x_3 - x_4 - \frac{c_3(x_3 - t_0) - c_4 x_6^3}{x_6} = 0$$

$$x_3 - t_0 \geqslant 0$$

$$x_4 - t_0 \geqslant 0$$

$$x_1, x_2, x_6, x_7 \geqslant 0$$

This example is taken from Fahlander et al.[1]

Helical Torsion Spring Design

This example is taken from Agrawal.[2] Helical torsion springs are used in a wide variety of applications, such as door hinge springs, springs for starters in automobiles, and springs for brush holders in electric motors. The problem is to determine optimally

x_1 = wire diameter
x_2 = mean coil diameter

to minimize weight, given torsional load, angular deflection, and material density. It is

$$\text{minimize weight} = c_1 x_1^6 + c_2 x_2 x_1^2$$

Subject to

$$\text{stress} = \frac{c_3}{x_1^{2.885} x_2^{0.115}} \leqslant 1$$

$$x_1, x_2 \geqslant 0$$

where c_1, c_2, and c_3 are given constants derived from the density of spring wire material, its modulus of elasticity, and so on, and where the formulas come from spring laws.

Loading Nonlinear Boilers

This example is from C. H. White.[3] There are five boilers in a shop operating in parallel. The problem is to decide how to share the steam load across them so as to meet a specified total steam output at the lowest fuel cost while operating all five boilers within their allowed operating range. It may be possible to get a lower overall total cost by actually shutting down one or more of the boilers and meeting the demand using only the remaining boilers. This is a discrete (or combinatorial) optimization problem which is not discussed here. For further information on discrete optimization see Chapter 14.4. For $i = 1$ to 5, let

ℓ_i, k_i = minimum and maximum loads, respectively, at which boiler i can be operated
x_i = load at which the ith boiler is operated
L = specified total load (steam output)

The data satisfy $\Sigma_{i=1}^{5} \ell_i < L < \Sigma_{i=1}^{5} k_i$, so the problem is feasible. The efficiency of the ith boiler, when it is operating, is defined to be the percentage 100 (heat content in the steam delivered)/(heat content in the input fuel), and it tends to be load dependent, so we denote it by $f_i(x_i)$. From past experience, it has been estimated that $f_i(x_i)$ can be approximated by the cubic polynomial $a_{0i} + a_{1i}x_i + a_{2i}x_i^2 + a_{3i}x^3$, where the values of the parameters a_{0i}, a_{1i}, a_{2i}, and a_{3i}, estimated using past data, are given for each $i = 1$ to 5.

Clearly $x_i/f_i(x_i)$ is a measure of the cost of obtaining a load of x_i from boiler i. So the problem is to

$$\text{minimize} \sum_{i=1}^{5} \frac{x_i}{f_i(x_i)}$$

subject to

$$\sum_{i=1}^{5} x_i = L$$

$$\ell_i \leqslant x_i \leqslant k_i, \qquad i = 1 \text{ to } 5$$

See Avriel and Dembo[4] and Balinski and Lemarechal[5] for many more examples of practical applications of nonlinear programming.

14.3.3 THE PARAMETER ESTIMATION PROBLEM IN CONSTRUCTING NONLINEAR MODELS

In constructing nonlinear models, the functional forms used are normally determined either by theoretical analysis or by practical intuition (as in the boiler problem, where boiler efficiency is assumed to be a cubic polynomial of the load). These functional forms sometimes contain unknown parameters (such as the coefficients of the cubic polynomial in the functional form for boiler efficiency) whose values are usually determined by curve fitting so that the function approximates the observed data as closely as possible. This problem of finding the best values for the parameters is called the "parameter estimation problem," or the "curve fitting problem." How this is carried out is discussed here using the boiler efficiency example presented in the previous section.

Let ξ represent the load at which the boiler is operating, and y the efficiency. The aim is to approximate $y(\xi)$ by a function of the form $a_0 + a_1\xi + a_2\xi^2 + a_3\xi^3$, where $a = (a_0, a_1, a_2, a_3)$ is the vector of parameters to be estimated. For this, collect data on the efficiency of the boiler at various load levels. Suppose we have r observations, with the tth observation at load ξ_t yielding an observed efficiency of y_t, for $t = 1$ to r. To derive the closest fit, we need to construct a measure of the deviation of the functional value from the observed, depending on the parameter vector a. Three different measures are in common use. They are

$$L_2(a) = \sum_{t=1}^{r} \left(y_t - a_0 - \sum_{r=1}^{3} a_r \xi_t^r\right)^2$$

$$L_1(a) = \sum_{t=1}^{r} \left| y_t - a_0 - \sum_{r=1}^{3} a_r \xi_t^r \right|$$

$$L_\infty(a) = \text{maximum} \left\{ \left| y_t - a_0 - \sum_{r=1}^{3} a_r \xi_t^r \right| : t = 1 \text{ to } r \right\}$$

The L_2 measure is a sum of squares, and to minimize it, we have to choose the values of $a = (a_0, a_1, a_2, a_3)$ so as to minimize $L_2(a)$, subject to any constraints on a determined by theoretical or practical considerations. This minimization problem is called a "least squares problem," and this approach for obtaining parameter values is called the "least squares approach" or the "method of least squares." If $\hat{a} = (\hat{a}_0, \hat{a}_1, \hat{a}_2, \hat{a}_3)$ is the best vector of parameter values obtained under this method, the function $\hat{a}_0 + \hat{a}_1\xi + \hat{a}_2\xi^2 + \hat{a}_3\xi^3$ is called the "least squares approximation" for the efficiency of the boiler as a function of load.

If the values of parameters are determined so as to minimize the measure $L_3(a)$, the resulting function is called the "Tchebycheff approximation."

If all the parameters appear linearly in the functional form (as in the boiler efficiency example), the problems of determining parameter values to minimize the L_1 or the L_∞ measures of deviation can both be posed as linear programs and solved by the efficient simplex method, a great advantage in this case. However, if the parameters appear nonlinearly in the functional form, the least squares method is preferred for parameter estimation.

If the measure of deviation is too large even at the best parameter values, it is necessary to review the choice of the functional form and modify it. Besides, it is possible that no functional form may provide a good approximation for all possible values of the decision variables. It is only necessary to find a good functional representation in the neighborhood of the optimum values for the decision variables if some reliable practical knowledge is available on the likely location of this optimum.

Thus even the process of constructing a mathematical model for a practical problem might itself need the application of optimization algorithms for parameter estimation.

14.3.4 VARIOUS TYPES OF NONLINEAR PROGRAMMING PROBLEMS

If it is required to maximize a function $z'(x)$ subject to some constraints, it is equivalent to the problem of minimizing $z(x) = -z'(x)$ subject to the same constraints. Thus, because every maximization problem can be transformed into an equivalent minimization problem, only algorithms for minimization are discussed here.

Unconstrained Minimization Problems. These are problems of the form

$$\text{minimize} \quad \theta(x) \tag{1}$$

Linear Equality Constrained Minimization Problems. These are problems of the form

$$\text{minimize} \quad \theta(x)$$

subject to

$$Ax = b \tag{2}$$

where A is a given matrix of order $m \times n$ and rank m. It is possible to find $n - m$ specific feasible solutions of equation 2, $\{x^1, \ldots, x^{n-m}\}$, such that every feasible solution, x, for equation 2 can be expressed as $\alpha_1 x^1 + \cdots + \alpha_{n-m} x^{n-m}$ for some real numbers $\alpha_1, \ldots, \alpha_{n-m}$. Substituting this expression for x in $\theta(x)$ transforms equation 2 into the problem of minimizing $\theta(\alpha_1 x^1 + \cdots + \alpha_{n-m} x^{n-m})$, which is now a function of $\alpha_1, \ldots, \alpha_{n-m}$ (since $x^1, \ldots, x^{n-m}$ are known points). Using this, all problems like equation 2 can be solved by algorithms for unconstrained minimization.

General Linearly Constrained Nonlinear Programs. These are problems of the form

$$\text{minimize} \quad \theta(x)$$

subject to

$$\begin{aligned} Ax &= b \\ Dx &\geqslant d \end{aligned} \tag{3}$$

where A, b, D, and d are given matrices of orders $m \times n$, $m \times 1$, $p \times n$, $p \times 1$, respectively. The inequality constraints may include bounds on the variables.

Nonlinear Equality Constrained Optimization Problems. These are problems of the form

$$\text{minimize} \quad \theta(x)$$

subject to

$$h_t(x) = 0, \qquad t = 1 \text{ to } p \tag{4}$$

where at least one $h_t(x)$ is nonlinear. In addition, there may be some bounds on the variables.

General Nonlinearly Constrained Optimization Problems. These are problems of the form

$$\text{minimize} \quad \theta(x)$$

subject to

$$\begin{aligned} g_i(x) &\geqq 0, \qquad i = 1 \text{ to } m \\ h_t(x) &= 0, \qquad t = 1 \text{ to } p \end{aligned} \tag{5}$$

Nonlinear Equation Solving Problems. Here we are asked to find a point x satisfying a system of equations

$$h_t(x) = 0, \qquad t = 1 \text{ to } p \tag{6}$$

where at least one of the $h_t(x)$ is nonlinear. Usually $p = n$, so we have as many equations as there are variables. However, sometimes $p < n$ (such a system is called an "underdetermined system") or $p > n$ (such a system is called an "overdetermined system"). Problem 6 is mathematically equivalent to the unconstrained minimization problem

$$\text{minimize} \quad f(x) = \sum_{t=1}^{p} [h_t(x)]^2 \tag{7}$$

If problem 6 has a feasible solution, that solution is optimal to equation 7, and the minimum value of $f(x)$ in equation 7 is zero. If the minimum value of $f(x)$ in equation 7 is greater than zero, problem 6 has no feasible solution, and vice versa. Even if the minimum value of $f(x)$ in equation 7 is strictly positive, if its value is reasonably small, an optimum solution of equation 7 may be used as an approximate solution of problem 6. Solving problem 6 through equation 7 is the least squares approach. In practice, this approach may not be satisfactory, as discussed later in Section 14.3.9.

14.3.5 BASIC TERMINOLOGY OF NONLINEAR PROGRAMMING

We will assume that all the functions are differentiable everywhere and that the partial derivatives are continuous. If some of these partial derivatives are not continuous, the problem is said to be a "nonsmooth optimization problem," and special algorithms have been developed for solving it.[6] If some of the functions are actually not differentiable, the problem is a "nondifferentiable optimization problem." Algorithms for such problems cannot use derivatives.[7]

The column vector of partial derivatives of a differentiable function $f(x)$ will be denoted by $\nabla f(x) = (\partial f(x)/\partial x_1, \ldots, \partial f(x)/\partial x_n)^T$, which is also known as the "gradient vector" of $f(x)$ at x.

To solve a nonlinear program on a computer, a subroutine has to be provided by the user for evaluating the objective and each of the constraint functions. If the algorithm needs the derivatives, a subroutine for evaluating each of them has to be provided too. If the functions are complicated, deriving analytical expressions for the partial derivatives may be hard, and writing subroutines for evaluating them even harder, and error prone. In this case an approximation to the partial derivative $\partial f(x)/\partial x_j$ can be obtained from

$$\frac{\partial f(x)}{\partial x_j} \simeq \frac{f(x_1, \ldots, x_{j-1}, x_j + \epsilon, x_{j+1}, \ldots, x_n) - f(x)}{\epsilon} \tag{8}$$

where ϵ is a sufficiently small positive number. This is known as a "finite difference approximation" to the partial derivative, or the "numerically computed derivative." The symbol ϵ is known as the "difference approximation step size." A careful choice of ϵ usually leads to a good gradient approximation. Nowadays most practitioners are using these numerically computed derivatives.

The matrix of second-order partial derivatives of a function $f(x)$ is known as the "Hessian" of that function, denoted by $H[f(x)] = [\partial^2 f(x)/\partial x_i \partial x_j]$. To evaluate the Hessian, n^2 subroutines have to be provided for computing all the second-order partial derivatives, and hence algorithms that require frequent evaluation of the Hessian matrix are not very popular.

Convex and Concave Functions

The function $f(x)$ is said to be a "convex function" if, for every pair of points x^1, x^2 and for all $0 \leqq \alpha \leqq 1$, we have $f[\alpha x^1 + (1 - \alpha) x^2] \leqq \alpha f(x^1) + (1 - \alpha) f(x^2)$. A convex function in one variable, y, is illustrated in Exhibit 14.3.1. The important property of convex functions is that, when you join two points on the surface of the function by a chord, the function itself lies underneath the chord on the interval joining these points. A function of a single variable is convex if its slope is monotone increasing.

The function $f(x)$ is said to be a "concave function" if $f[\alpha x^1 + (1 - \alpha) x^2] \geqq \alpha f(x^1) + (1 - \alpha) f(x^2)$ holds for all x^1, x^2, and $0 \leqq \alpha \leqq 1$. Thus the function $f(x)$ is a concave function if $-f(x)$ is a convex function.

Convex and convave functions figure prominently in nonlinear programming. Some of the algorithms for solving nonlinear programming problems can be mathematically proved to converge to a solution of the problem only under suitable convexity assumptions on the objective and constraint functions. Before using such an algorithm, the user may want to check whether the functions in the model satisfy these convexity assumptions, to be sure that the algorithm is guaranteed to work on the problem. Hence it is important to have methods for checking whether a given function is convex or not.

It is possible to classify simple functions as convex, concave, or neither. However, for complicated functions in many variables, testing whether they are convex or concave or not may be very hard. If a function is twice continuously differentiable, it is convex if its Hessian matrix is positive semidefinite for all x. For any given point x, it is possible to check whether or not the Hessian matrix is positive semidefinite very efficiently by pivotal methods,[8] but it is clearly impractical to repeat this check for every decision point x. If the Hessian matrix is positive definite at a point x, we can at least conclude that the function is locally convex at x. Similarly, if the Hessian is negative definite at a point x, the function is locally concave at x. Quite often this is all that can be checked efficiently in practice.

Various Types of Optimal Solutions

Let **K** denote the set of feasible solutions of a nonlinear program in which $\theta(x)$ is the objective function. The point $\bar{x}$ in **K** is said to be a "global (or absolute) minimum" for $\theta(x)$ over **K** if $\theta(\bar{x}) \leqq \theta(x)$ for all x in **K** or a "local (or relative) minimum" if $\theta(\bar{x}) \leqq \theta(x)$ for all x in **K** close to $\bar{x}$ (see Exhibit 14.3.2). The global maximum and local maximum are defined similarly.

Exhibit 14.3.1 Illustration of a Convex Function, $f(y)$, of one Variable, y

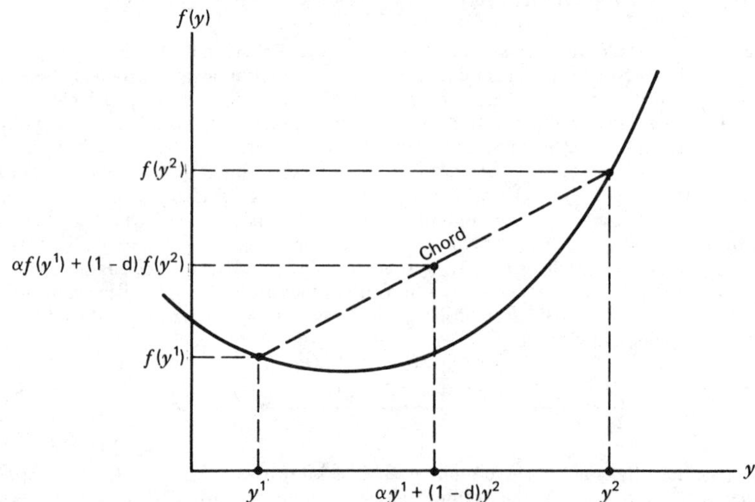

Exhibit 14.3.2 Illustration of $n = 1$ and the Feasible Region of the Interval $a \leqq x \leqq b; x^2$ Is the Global Minimum; a, x^1, x^3 Are the Other Local Minima

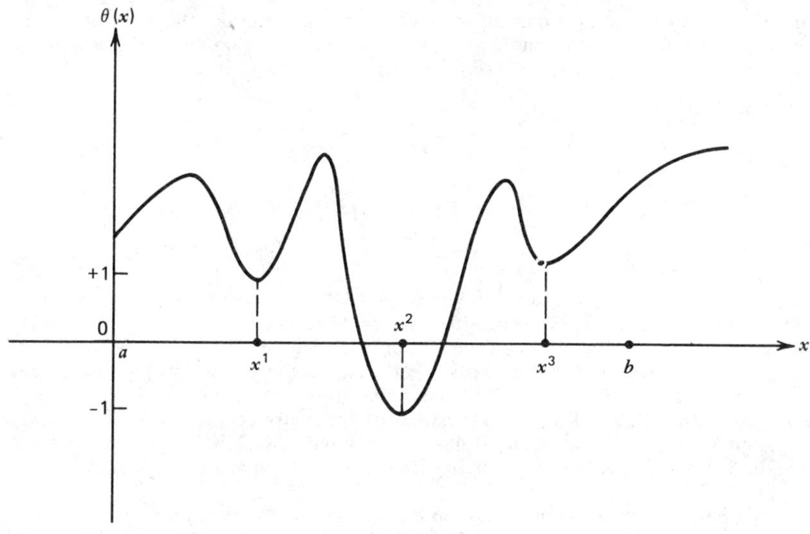

Necessary Optimality Conditions for a Local Minimum

These are mathematical conditions that every local minimum for a nonlinear program must satisfy. During the early development of nonlinear programming theory, a lot of research effort was devoted to deriving these necessary optimality conditions for various types of nonlinear programs, discussed in Section 14.3.4. These conditions are very useful to check whether a (guessed or) suggested decision point is a local minimum. (If it does not satisfy these conditions, we can conclude that it is not even a local optimum.) Also, there are algorithms that are based on these conditions. For each type of nonlinear program presented here, the simplest and most useful necessary optimality condition for that problem is provided.

Condition for the Unconstrained Minimization Problem. If $\bar{x}$ is a local minimum for expression 1, then we must have

$$\nabla\theta(\bar{x}) = 0 \tag{9}$$

Equation 9 is known as the "first-order necessary optimality condition" for expression 1. If $\theta(x)$ is twice continuously differentiable, another necessary condition for $\bar{x}$ to be a local minimum for expression 1 is that $H[\theta(\bar{x})]$ be positive semidefinite, that is, that $\theta(x)$ be locally convex at $\bar{x}$.

Notice that equation 9 is a system of n equations in n unknowns. So, if we have an algorithm for solving nonlinear equations, we can use that algorithm on equation 9 and solve the unconstrained minimization problem in expression 1.

Condition for Equality Constrained Nonlinear Programs. Consider equation 4. The derivation of necessary conditions for a local minimum in this problem requires some simple mathematical conditions on the constraint functions, called "constraint qualifications" in the literature.[9,10] If $\bar{x}$ is feasible to equation 4 at which these conditions hold, and if $\bar{x}$ is a local minimum for equation 4, then there exists a Lagrange multiplier vector $\bar{\pi} = (\bar{\pi}_1, \ldots, \bar{\pi}_p)$ that together with $\bar{x}$ satisfies

$$\frac{\partial\theta(\bar{x})}{\partial x_j} - \sum_{t=1}^{p} \bar{\pi}_t \frac{\partial h_t(\bar{x})}{\partial x_j} = 0 \quad \text{for } j = 1 \text{ to } n \tag{10}$$

These conditions are known as the "first-order necessary optimality conditions" for equation 4.

The constraints on the decision variables in equation 4, $h_t(x) = 0$ for $t = 1$ to p, together with the necessary optimality conditions in expression 10, provide a set of $n + p$ equations in $n + p$ unknowns (the n decision variables x_j, $j = 1$ to n, and the p Lagrange multipliers π_t, $t = 1$ to p, one for each constraint in equation 4); by solving this system of equations using an algorithm for solving nonlinear equations, we can solve equation 4. This is known as the "Lagrange multiplier approach" for solving the nonlinear program in equation 4.

Condition for the General Constrained Nonlinear Program. Consider the nonlinear program in equation 5. If $\bar{x}$ is a feasible solution that is a local minimum for equation 5 and is a regular point, then there exist Lagrange multipliers $\bar{\pi} = (\bar{\pi}_i)$, $\bar{\mu} = (\bar{\mu}_t)$ that together with $\bar{x}$ satisfy

$$\frac{\partial \theta(\bar{x})}{\partial x_j} - \sum_{i=1}^{m} \bar{\pi}_i \frac{\partial g_i(\bar{x})}{\partial x_j} - \sum_{t=1}^{p} \bar{\mu}_t \frac{\partial h_t(\bar{x})}{\partial x_j} = 0, \quad j = 1 \text{ to } n \tag{11}$$

$$\bar{\pi}_i \geqq 0 \quad \text{for all } i = 1 \text{ to } m$$

$$\bar{\pi}_i g_i(\bar{x}) = 0 \quad \text{for all } i = 1 \text{ to } m$$

The last set of constraints in equation 11 is known as the "complementary slackness conditions for optimality." These conditions are known as the "first-order necessary conditions for optimality" for equation 5. They were commonly known as "Kuhn-Tucker necessary optimality conditions" because they were discussed by Kuhn and Tucker in their 1951 paper. But it was recently found that they were first discussed in 1939 by W. Karush in his Master of Science thesis. Hence nowadays they are being called "Karush-Kuhn-Tucker necessary optimality conditions." A feasible solution $\bar{x}$ for equation 5 is said to be a "Karush-Kuhn-Tucker point" (KKT point, or a stationary point) for equation 5 if there exist Lagrange multiplier vectors $\bar{\pi}$, $\bar{\mu}$ such that $\bar{x}$, $\bar{\pi}$, $\bar{\mu}$ together satisfy equation 11.

The general nonlinear program in expression 8 is said to be a "convex programming problem" if $\theta(x)$ is convex, $g_i(x)$ is concave for each i, and $h_t(x)$ is affine (i.e., a function of the form $a_0 + a_1x_1 + \cdots + a_nx_n$ with $a_0, \ldots, a_n$ given constants) for all t. In a convex programming problem, every local minimum is a global minimum, and in fact every KKT point is a global minimum. In a nonconvex programming problem, this property may not hold. So there may be lots of local minima (which are not global minima) for a nonconvex program, and on such problems it is difficult to find a global minimum unless an actual enumeration is carried out over all the local minima, which is very hard. Thus, in general, when solving a nonconvex programming problem, realistically we can at best hope to find a local minimum or a KKT point obtained by descent methods.

As mentioned previously, mathematical proofs that algorithms for nonlinear programming do work are usually based on the assumption that the nonlinear program being solved is a convex programming problem. In real-life applications, these convexity assumptions may not hold; also, checking whether they hold is hard. Many practical models tend to be nonconvex, and practitioners do not normally want to waste computer time on checking for convexity. Instead, they tend to use the algorithm on the problem anyway, and the better algorithms usually lead to either a local minimum or a KKT point on a majority of practical problems.

Convergence Rate of Nonlinear Programming Algorithms

In general, algorithms for solving linear programs or combinatorial optimization problems tend to be finite algorithms. In contrast, nonlinear programming algorithms generally tend not to be finite, and for such algorithms the best that we can hope for is that they generate a sequence of points that in the limit converges to a local minimum for the problem. So if nonlinear programming algorithms are to be useful in practice, they should converge rapidly. This rate of convergence can in fact be measured mathematically. Nowadays there are several algorithms for nonlinear programs that can be expected to converge rapidly. With these algorithms, we reach very close to a local minimum after only a small number of steps, particularly if the algorithm is initiated with a point known to be in the vicinity of the optimum.

The following sections focus on the main ideas behind some of the most promising algorithms in use today. For detailed discussion of algorithms, see other sources.[6,7,9–14]

14.3.6 LINE MINIMIZATION ALGORITHMS

The line minimization problem is the problem of minimizing a function, $f(\lambda)$, of one parameter, λ, either over the whole line or over a specified interval, $\ell \leqq \lambda \leqq u$. Most algorithms for nonlinear programs in many variables use a line minimization algorithm as a subroutine repeatedly.

If $f(\lambda)$ is concave and ℓ and u are finite, the minimum is either ℓ or u, whichever gives the smallest value to $f(\lambda)$. If $f(\lambda)$ is not concave, the approach based on function approximation works with the idea that, in the neighborhood of a local minimum, the function $f(\lambda)$ can be reasonably approximated by a parabola. Once a good parabola, $P(\lambda)$, is fitted by curve fitting, the minimum of $P(\lambda)$ can be used as an approximation to the minimum of the original function $f(\lambda)$.

The approach begins with an initial point, λ_0, and first finds a finite interval containing the minimum, known as a "bracket for the minimum." Choose a positive step length Δ. Compute $f(\lambda_0)$ and $f(\lambda_1)$, where $\lambda_1 = \lambda_0 + \Delta$. If $f(\lambda_1) < f(\lambda_0)$, the direction of increasing λ is the right direction to pursue; otherwise, replace Δ by $-\Delta$ to reverse the direction and go through the procedure discussed next. Define $\lambda_r = \lambda_{r-1} + 2^{r-1}\ \Delta$ for $r = 2, 3, \ldots$ and keep on computing $f(\lambda_r)$, $r = 2, 3, \ldots$ as long as they keep on decreasing, until either the upper bound on λ is reached or a value k for r is found such that $f(\lambda_{k+1}) > f(\lambda_k)$. In this case we have $\lambda_{k-1}, \lambda_k, \lambda_{k+1}$ satisfying $f(\lambda_k) < f(\lambda_{k-1})$, $f(\lambda_{k+1}) > f(\lambda_k)$. Among the four points λ_{k-1}, λ_k, $(\lambda_k + \lambda_{k+1})/2$, and λ_{k+1}, drop either λ_{k-1} or λ_{k+1}, whichever is farther from the point in the pair $\{\lambda_k, (\lambda_k + \lambda_{k+1})/2\}$ that yields the smallest value to $f(\lambda)$. Let the remaining points be called $\lambda_a, \lambda_b, \lambda_c$, where $\lambda_a < \lambda_b < \lambda_c$. These points are equidistant, and $f(\lambda_b) \leqq f(\lambda_c)$, $f(\lambda_b) \leqq f(\lambda_a)$. So this interval λ_a to λ_c brackets the minimum.

Once the points λ_a, λ_b, λ_c bracketing the minimum are obtained, we fit a quadratic function $P(\lambda) = \alpha_0 + \alpha_1\lambda + \alpha_2\lambda^2$ with $\alpha_0, \alpha_1, \alpha_2$ determined so that $f(\lambda) = P(\lambda)$ at $\lambda = \lambda_a, \lambda_b, \lambda_c$. Since $P(\lambda)$ is a quadratic function, its minimum can easily be computed analytically. It can be verified that this minimum is $\lambda_* = \lambda_b + \delta$, where

$$\lambda_* = \frac{(\lambda_a^2 - \lambda_b^2)\, f(\lambda_c) + (\lambda_b^2 - \lambda_c^2)\, f(\lambda_a) + (\lambda_c^2 - \lambda_a^2)\, f(\lambda_b)}{2[(\lambda_a - \lambda_b)\, f(\lambda_c) + (\lambda_b - \lambda_c)\, f(\lambda_a) + (\lambda_c - \lambda_a)\, f(\lambda_b)]}$$

If $|\delta|$ is less than some specified tolerance, the algorithm is terminated by accepting λ_* as the minimum point. Otherwise, the whole procedure is repeated with a reduced step length and with the point in the pair $\{\lambda_b, \lambda_*\}$ that yields the smallest value to $f(\lambda)$ as the new initial point. In practice, only a few iterations of this method are needed to reach close to a local minimum for $f(\lambda)$.

14.3.7 UNCONSTRAINED MINIMIZATION

Consider the problem in expression 1. Two descent methods that work very well are discussed here.

Method of Conjugate Gradients

The method goes through several cycles. Each cycle consists of $n + 1$ line search steps. Cycle 1 begins with a selected initial point x^0. A general cycle is as follows:

STEP 1. Let x^0 be the starting point of this cycle. Let $y^1 = -\nabla\theta(x^0)$. Find λ_1, the minimum of $\theta(x^0 + \lambda y^1)$, over $\lambda \geqq 0$, by doing a line search, and let $x^1 = x^0 + \lambda_1 y^1$. Go to the next step.

GENERAL STEP k. Let x^{k-1} be the point obtained at the end of step $k - 1$ in this cycle. Let

$$y^k = -\nabla\theta(x^{k-1}) + \frac{[\nabla\theta(x^{k-1})]^T\,[\nabla\theta(x^{k-1})]}{[\nabla\theta(x^{k-2})]^T[\nabla\theta(x^{k-2})]}\, y^{k-1}$$

Compute λ_k, the minimum of $\theta(x^{k-1} + \lambda y^k)$, over $\lambda \geqq 0$, by doing a line search. Let $x^k = x^{k-1} + \lambda y^k$. If $\nabla\theta(x^k)$ is close to zero, terminate with x^k as a local minimum for expression 1. (In this case x^k satisfies the necessary optimality conditions in equation 9 approximately.) If $k = n + 1$, terminate this cycle, and if the change in the objective value during this cycle is small, terminate the algorithm, treating x^k as an approximation to a local minimum for expression 1; otherwise, go to the next cycle, beginning it with x^k as the initial point. On the other hand, if $k < n + 1$, go to the next step in this cycle.

The Variable Metric Method

This method belongs to the class of quasi-Newton methods. It is also commonly known as the "DFP method," after Davidon, Fletcher, and Powell who proposed it. It also goes through several cycles, each cycle consisting of $n + 1$ steps. A general cycle is as follows:

STEP 1. Let x^0 be an initial point for starting the algorithm if this is cycle 1; let it be the last point obtained in the previous cycle otherwise. Let $D_1 = I$, the identity matrix of order n. Let $y^1 = -D_1\nabla\theta(x^0)$. Compute λ_1, the minimum of $\theta(x^0 + \lambda y^1)$, over $\lambda \geqq 0$. Let $x^1 = x^0 + \lambda_1 y^1$. Go to the next step.

GENERAL STEP $k + 1$. Let x^k and D_k be the point and the matrix, respectively, at the end of the previous step. Compute $P_k = x^k - x^{k-1}$, $Q_k = \nabla\theta(x^k) - \nabla\theta(x^{k-1})$, and $y^{k+1} = -D_{k+1}\nabla\theta(x^k)$, where

$$D_{k+1} = D_k + \frac{P_k P_k^T}{P_k^T Q_k} - \frac{(D_k Q_k)(D_k Q_k)^T}{Q_k^T D_k Q_k}$$

Find λ_{k+1}, the minimum of $\theta(x^k + \lambda y^{k+1})$, over $\lambda \geqq 0$. Let $x^{k+1} = x^k + \lambda_{k+1} y^{k+1}$. If $k + 1 < n + 1$, go to the next step in this cycle. If $k + 1 = n + 1$, go to the next cycle with x^{k+1} as the initial point.

Terminate whenever the gradient vector is close to zero, or when the change in the objective value in a cycle, or the change in the solution over a cycle, are small. The final point at termination is taken as an approximation to the local minimum of expression 1.

14.3.8 LINEARLY CONSTRAINED NONLINEAR PROGRAMMING PROBLEMS

Two algorithms that seem to work very well are discussed here.

The Reduced Gradient Method

The problems in this class can be expressed in the following form by simple transformations:

$$\text{minimize } \theta(x)$$

subject to

$$Ax = b \tag{12}$$

$$\ell \leqq x \leqq u$$

where A is a matrix of order $m \times n$ and rank m, and ℓ and u are lower- and upper-bound vectors, respectively, for the variables. Using the m equality constraints in equation 12, it is possible to eliminate m of the variables and express the problem purely in terms of the remaining $n - m$ variables. Even though this is not done explicitly, the algorithm is based on this fact. For this, choose m dependent (or basic) variables corresponding to a basis, B, for equation 12. Let x_B and x_D denote the column vectors of basic and nonbasic variables, respectively. Let D denote the matrix [of order $m \times (n - m)$] of coefficients of the nonbasic variables in equation 12.

Let $\bar{x} = (\bar{x}_B, \bar{x}_D)$ be a feasible solution of the problem. Let $c = [\nabla\theta(\bar{x})]^T$ be the gradient vector of $\theta(x)$ at $\bar{x}$. Denote by c_B, c_D the vectors of partial derivatives of $\theta(x)$ at $\bar{x}$ with respect to x_B, x_D respectively. Let $\bar{c}_D = c_D - c_B B^{-1} A$. The vector $\bar{c}_D$ is known as the "reduced gradient vector" at $\bar{x}$, with x_D as the vector of nonbasic variables. For j such that x_j is a nonbasic variable, define

$$y_j = -\bar{c}_j, \text{ if } \bar{c}_j < 0 \text{ and } \bar{x}_j < u_j \text{ or if } \bar{c}_j > 0 \text{ and } \bar{x}_j > \ell_j$$
$$= 0 \text{ otherwise}$$

and let y_D be the vector of these y_j in proper order. Then y_D is a descent direction in the space of the nonbasic variables x_D. Define $y_B = B^{-1} D y_D$ and let $y = (y_B, y_D)$. If $y = 0$, the current feasible solution, $\bar{x}$, is a local minimum, and we terminate. Otherwise, determine $\hat{\lambda}$, the maximum value of λ for which $\bar{x} + \lambda y$ continues to lie within the bounds ℓ and u, and then solve the line minimization problem of minimizing $\theta(\bar{x} + \lambda y)$, over $0 \leqq \lambda \leqq \hat{\lambda}$. If $\bar{\lambda}$ is the optimum solution of this line minimization problem, $\bar{x} + \bar{\lambda} y$ is the next point, and the algorithm goes through another step with it. There are rules for changing the basic vector to make the algorithm efficient.

Algorithms Based on Active Set Strategies

Consider equation 3. If we can guess which of the inequality constraints here will hold as equations (i.e., are active) at the optimum solution, then we can solve the problem using the algorithm for handling linear equality constraints only. Suppose we are able to partition D into

$$\begin{pmatrix} D_I \\ \cdots \\ D_{\bar{I}} \end{pmatrix} \quad \text{and, correspondingly, } d \text{ into} \quad \begin{pmatrix} d_I \\ \cdots \\ d_{\bar{I}} \end{pmatrix}$$

such that an optimum solution of equation 5 satisfies $D_I x = d_I$ and $D_{\bar{I}} x > d_{\bar{I}}$. Having determined a tentative I, the algorithm makes an effort to solve the following:

$$\text{minimize } \theta(x)$$

For each $j = 1$ to n, select a grid of points in the range of possible variation of x_j and, by introducing the new variables as just described, approximate each of $\theta_j(x_j)$ and $g_{ij}(x_j)$ for $i = 1$ to m by linear functions in terms of these new variables. Using these and equation 20, expression 19 itself can be approximated into a linear program in terms of the new variables. When the functions in expression 19 are both separable and convex, this approach leads to a very reasonable approximation to the optimum solution of expression 19.

Linear Approximation Method

Consider the general nonlinear program in equation 5. This approach for solving the program was suggested by Griffith and Stewart in 1961. Given the point x^k in iteration k, the approach consists of replacing each of the functions $\theta(x)$, $g_i(x)$, and $h_t(x)$ by their linear approximations at x^k to yield the linear program

$$\text{minimize} \quad \theta(x^k) + [\nabla\theta(x^k)]^T (x - x^k)$$

subject to

$$\begin{aligned} g_i(x^k) + [\nabla g_i(x^k)]^T (x - x^k) &\leqq 0, & i &= 1 \text{ to } m \\ h_t(x^k) + [\nabla h_t(x^k)]^T (x - x^k) &= 0, & t &= 1 \text{ to } p \\ -u_j \leqq x_j - x_j^k &\leqq u_j, & j &= 1 \text{ to } n \end{aligned} \tag{21}$$

where u_j is some positive bound for each j. The linear approximations are usually reasonable only locally, that is, in a small neighborhood around the approximating point x^k. Therefore the bounds are introduced on the variables to restrict the movement of the decision point to a small neighborhood around x^k. Hence in solving the linear program in equation 21, only small steps are made, away from the current point x^k. Thus this method is called a "small-step gradient method."

Solve the linear program in equation 21 by the simplex method, and let x^{k+1} be the optimum solution for it. Repeat the process over again with x^{k+1}. Terminate when the point is nearly feasible, and the change in points between successive steps is small.

Even though this method is not generally guaranteed to converge, it is reported to be reasonably effective for solving practical problems. It is most useful in solving problems involving many variables and only a few nonlinear constraints in addition to possibly several linear ones. It is very simple, since one needs only a good linear programming code to use it. The bounds introduced on the variables greatly influence the success of the algorithm. If u_j's are too small, progress tends to be slow. If u_j's are large, the linear approximation turns out to be poor as you move away from the current point and may result in points with large infeasibilities.

Generalized Reduced Gradient Method

For applying this method, the problem is transformed into the following form:

$$\text{minimize} \quad \theta(x)$$

subject to

$$\begin{aligned} h_t(x) = 0, &\quad t = 1 \text{ to } p \\ \ell \leqq x &\leqq u \end{aligned} \tag{22}$$

where ℓ and u are the lower- and upper-bound vectors for the feasible points x, respectively. We can always assume that an initial feasible solution is available. If a feasible solution is not available, let $x^0 = (x_1^0, \ldots, x_n^0)^T$ be an initial point. Change equation 22 into

$$\text{minimize} \quad \theta(x) + \alpha x_{n+1}$$

subject to

$$h(x) - x_{n+1}h(x^0) = 0$$

$$\ell \leqq x \leqq u$$

$$0 \leqq x_{n+1} \leqq 1$$

where x_{n+1} is an additional variable and α is a large positive constant. For this modified problem, $(x_1^0, \ldots, x_n^0, 1)$ is a feasible solution, and the modified problem is in the same form as equation

subject to

$$\begin{aligned} Ax &= b \\ D_{\bar{I}}x &= d_I \end{aligned} \tag{13}$$

When an optimum solution $\bar{x}$ for equation 13 is obtained, we can also obtain the Lagrange multiplier vectors $\bar{\pi}$, $\bar{\mu}_I$ such that these vectors together satisfy the first-order necessary conditions for optimality for equation 13. If $\bar{x}$ satisfies $D_{\bar{I}}x \geqq d_{\bar{I}}$ and $\bar{\mu}_I \geqq 0$, $\bar{x}$ is a local minimum for equation 3 and we terminate. On the other hand, if there exists a $\bar{\mu}_i < 0$ in $\bar{\mu}_I$, then by dropping the ith inequality constraint from the active set, we can get a better solution. This type of strategy is known as an "active set strategy." In it, rules are developed to determine when to add an inequality constraint to the active set and when to delete an inequality constraint from the active set. This strategy is combined together with the algorithm for minimizing $\theta(x)$ subject to the current active constraints treated as equality constraints. Algorithms of this type have turned out to be very successful in solving very large nonlinear programs of the form in equation 3.

14.3.9 ALGORITHMS TO SOLVE NONLINEAR EQUATIONS

Consider the system

$$h_t(x) = 0, \qquad t = 1 \text{ to } n \tag{14}$$

As discussed in Section 14.3.4, equation 14 is equivalent in a theoretical sense to

$$\text{mimimize} \quad \theta(x) = \sum_{t=1}^{n} [h_t(x)]^2 \tag{15}$$

However, when we solve equation 15 by an unconstrained minimization algorithm, we may obtain only a local minimum, which is not a global minimum for equation 15, and this does not lead to any information about equation 14. Actually, $\theta(x)$, which is a sum of squares, tends to be hard to minimize. So, if the least squares method is to be used to solve equation 14, it is desirable to use special algorithms designed for least squares minimization.[15,16]

Newton's Method to Solve Equation 14

Assuming that each $h_t(x)$ is differentiable, the Jacobian, the matrix consisting of the gradient vectors, will be denoted as

$$J[h(x)] = \begin{bmatrix} [\nabla h_1(x)]^T \\ \vdots \\ [\nabla h_n(x)]^T \end{bmatrix}$$

Since equation 14 is a system of n equations in n unknowns, $J[h(x)]$ is a square matrix of order n. Newton's method begins with an initial estimate of the solution, x^0, and obtains a sequence of points, x^r, $r = 1, 2, \ldots$. Let x^r be the current solution. If x^r is close to a solution of equation 14, we can express that solution as $x^r + y$, where y is a point close to the origin, 0. In this case, by the mean value theorem of differential calculus, we can approximate

$$h(x^r + y) \simeq h(x^r) + J[h(x^r)]\, y \tag{16}$$

To choose the correction vector y so that $x^r + y$ is a solution of equation 14, we set the right hand side of expression 16 to zero and solve for y, which yields $y = -\{J[h(x^r)]\}^{-1}h(x^r)$, assuming that the Jacobian at x^r is nonsingular. This leads to the new point

$$x^{r+1} = x^r - \{J[h(x^r)]\}^{-1}\, h(x^r) \tag{17}$$

The sequence of points $x^1, x^2, \ldots$ is obtained recursively using equation 17 until a point in the sequence has $h(x)$ close to zero, at which time the method terminates by accepting that point as an approximate solution of equation 14.

Continuation Methods

These methods have recently been developed for solving systems of the form in equation 14. Let a be a fixed, preselected point. Define a function of x and a parameter λ varying between 0 to 1,

Exhibit 14.3.3 The Curve $x(\lambda)$ of Solutions of $g(\lambda, x) = 0$, as a Function of the Parameter λ in the Interval 0 to 1. Curve Starts at the Known Solution $x(0) = a$ for $\lambda = 0$. When You Trace This Curve and Reach $\lambda = 1$, You Get a Solution of the System $h(x) = 0$

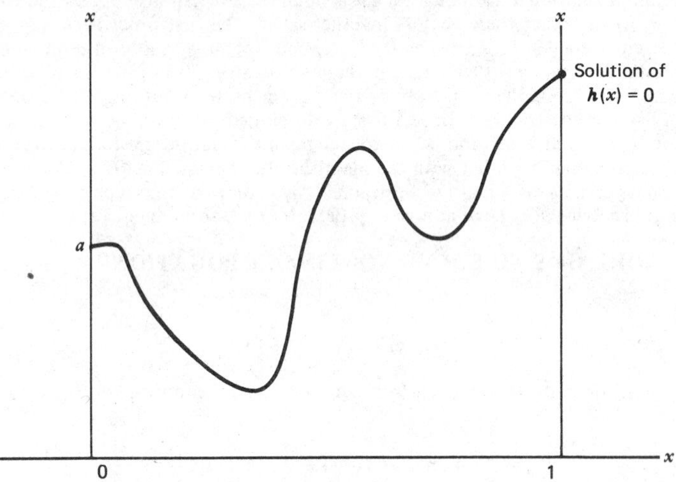

$g(\lambda, x) = \lambda h(x) + (1 - \lambda)(x - a)$. Examine the solutions of

$$g(\lambda, x) = 0 \tag{18}$$

as λ varies parametrically from 0 to 1. The point a is the unique solution when $\lambda = 0$. Any solution of equation 18 for $\lambda = 1$ is also a solution of equation 14. Denoting the solution of equation 18 as a function of the parameter λ by $x(\lambda)$, the algorithm traces the curve $x(\lambda)$, beginning with $x(0) = a$, until $\lambda = 1$ (see Exhibit 14.3.3).

Several methods have been proposed to trace the curve $x(\lambda)$. Depending on the method used, this approach is called either a "continuation method" or a "homotopy method." One method for tracing this curve uses the length of the curve as the independent parameters s, expressing λ and $x(\lambda)$ as functions of s denoted by $\lambda(s)$ and $x(s)$. We have $g[\lambda(s), x(s)] = 0$. Differentiating this identity with respect to s leads to

$$d(ds)\ \{g[\lambda(s), x(s)]\} = 0$$

$$\lambda(0) = 0$$

$$x(0) = a$$

This system can now be solved numerically using numerical methods for solving ordinary differential equations.[17] Mangasarian[18] discusses methods for transforming the general constrained nonlinear program as the problem of solving a system of nonlinear equations of the form of equation 14. Using these transformations, it is possible to solve any nonlinear program with the algorithms discussed here.

14.3.10 GENERAL CONSTRAINED NONLINEAR PROGRAMS

This section focuses on the methods used most often by practitioners.

Separable Programming

A function $f(x)$ of n decision variables $x_1, \ldots, x_n$ is said to be separable if it is a sum of n functions, each of which involves only one variable; that is, if $f(x) = f_1(x_1) + \cdots + f_n(x_n)$. For example, $f(x_1, x_2) = (x_1^3 - 2x_1^2 + x_1 + 6) + (x_2^4 - x_2)$ is a separable function, but $g(x_1, x_2) = 3x_1^2 - 2x_1x_2 + x_2^2$ is not. The nonlinear program

$$\text{minimize } \theta(x)$$

subject to

$$g_i(x) \leqq 0, \quad i = 1 \text{ to } m \tag{19}$$

is said to be a separable program if each of the functions $\theta(x), g_1(x), \ldots, g_m(x)$ are separable. Separable programs can be solved by approximating the nonlinear separable functions with piecewise linear functions and then solving the transformed problem using the simplex algorithm. A discussion of how to derive this approximation follows.

Consider a single decision variable, y, and let $f(y)$ be a function. Shift the origin so that 0 is a lower bound for y in the problem. Divide the nonnegative range of y into nonoverlapping intervals by selecting some grid points $y^1, \ldots, y^k$, which need not be equally spaced (see Exhibit 14.3.4).

Join the points $[y^t, f(y^t)]$, $[y^{t+1}, f(y^{t+1})]$ by a straight line for each $t = 0$ to $k - 1$ to get the piecewise linear approximation $P(y)$. If $f(y)$ is convex, we will always have $P(y) \geqq f(y)$, so the error in the approximation is always nonnegative. Also, by selecting the grid points closer to each other, the approximation $P(y)$ can be made as close to $f(y)$ as desired. Let c^t be the slope of $P(y)$ in the interval $[y^t, y^{t+1}]$, that is, $c^t = [f(y^{t+1}) - f(y^t)]/(y^{t+1} - y^t)$ for $t = 0$ to $k - 1$. Let U_t be the length of the interval $[y^t, y^{t+1}]$ for $t = 0$ to $k - 1$. For any value of the variable y, denote by ξ_t the length of the overlap of the interval $[0, y]$ with the interval $[y^t, y^{t+1}]$ for each $t = 0$ to $k - 1$. Then we have

$$y = \xi_0 + \xi_1 + \cdots + \xi_{k-1}$$

$$P(y) = c^0\xi_0 + c^1\xi_1 + \cdots + c^{k-1}\xi_{k-1}$$

So, when expressed in terms of these new variables, $\xi_0, \xi_1, \ldots, \xi_{k-1}$, the piecewise linear approximation $P(y)$ becomes a linear function. Of course the new variables $\xi_0, \xi_1, \ldots, \xi_{k-1}$ have to satisfy the constraints that, for $t = 0$ to $k - 1$, $0 \leqq \xi_t \leqq U_t$ and that $\xi_t = 0$ unless each of the previous variables $\xi_0, \ldots, \xi_{t-1}$ is equal to its upper bound.

Now, in the nonlinear program in expression 19, if all the functions are separable, let

$$\theta(x) = \theta_1(x_1) + \cdots + \theta_n(x_n) \tag{20}$$

and

$$g_i(x) = g_{i1}(x_1) + \cdots + g_{in}(x_n) \quad \text{for } i = 1 \text{ to } m$$

Exhibit 14.3.4 The Continuously Drawn Curve is $f(y)$. The Dashed Piecewise Linear Curve $P(y)$ Is the Approximation to $f(y)$ Obtained From the Choice of These Grid Points. The Symbols c^0, c^1 Are the Slopes of $P(y)$ in the Various Intervals Between Grid Points

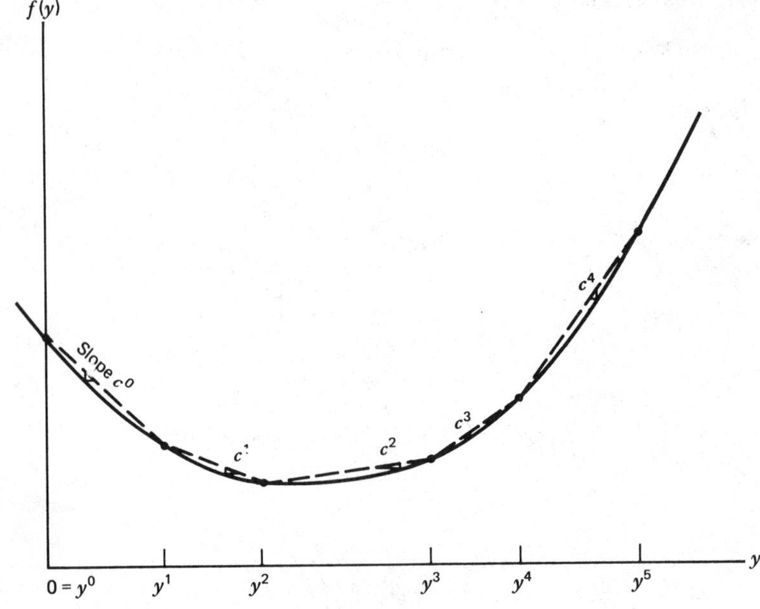

22. Also, since α is a large positive number, x_{n+1} is forced to have a value equal to zero at an optimum solution of the modified problem if the original problem in equation 22 is feasible.

Thus we will consider the original problem in equation 22 itself, but will assume that an initial feasible solution is available. The computations in a general step of this algorithm follow.

Let $\bar{x}$ be the feasible solution at the beginning of this step. As in the reduced gradient method of Section 14.3.8, the variables are partitioned into two vectors: a vector x_B consisting of p basic variables and a vector x_D of $n - p$ nonbasic variables. The basic variables are implicitly determined by the nonbasic variables through the equality constraints in equation 22. Let $\partial h(\bar{x})/\partial x_B$ denote the $p \times p$ matrix of the partial derivatives of the functions $h_t(x)$ with respect to the basic variables in x_B, evaluated at the feasible solution $\bar{x}$. The expressions $\partial h(\bar{x})/\partial x_D$, $\partial\theta(\bar{x})/\partial x_B$, and $\partial\theta(\bar{x})/\partial x_D$ have corresponding meanings. The reduced gradient at $\bar{x}$ with respect to x_D as the vector of nonbasic variables is

$$\bar{c}_D = \frac{\partial\theta(\bar{x})}{\partial x_D} - \frac{\partial\theta(\bar{x})}{\partial x_B}\left(\frac{\partial h(\bar{x})}{\partial x_B}\right)^{-1}\frac{\partial h(\bar{x})}{\partial x_D}$$

For j such that x_j is a nonbasic variable, define

$$\begin{aligned} y_j &= -\bar{c}_j \text{ if } \bar{c}_j < 0 \text{ and } \bar{x}_j < u_j \text{ or if } \bar{c}_j > 0 \text{ and } x_j > \ell_j \\ &= 0 \text{ otherwise} \end{aligned}$$

and let y_D be the vector of these y_j in proper order. If $y_D = 0$, the current feasible solution $\bar{x}$ is a local minimum for equation 22, and we terminate. Otherwise, y_D is a descent direction in the space of the nonbasic variables x_D. The x_D has to be changed since $x_D = \bar{x}_D + \lambda y_D$, where λ is a positive parameter. The corresponding values of the basic variables, $x_B(\lambda)$, are determined uniquely from

$$\begin{aligned} h[x_B(\lambda), \bar{x}_D + \lambda y_D] &= 0 \\ x_B(0) &= \bar{x}_B \end{aligned} \tag{23}$$

This generates a curve, Γ, in the feasible region. By moving along this curve from the current point $\bar{x}$, we can get a better decision point. As we move along the curve Γ, even though the nonbasic variables change linearly, the basic variables must change nonlinearly with the parameter λ, to continuously satisfy equation 23. This is hard to do. So computationally this is accomplished by finding the tangent line, L, to the curve Γ at the current point $\bar{x}$, moving linearly along this tangent line L to a point $\tilde{x}^1$ which will in general be infeasible to equation 22. Then the algorithm returns to a point $\tilde{x}$ on the curve Γ in the feasible region by a correction procedure based on Newton's method of Section 14.3.9 initiated with $\tilde{x}^1$ (see Exhibit 14.3.5).

If the new feasible solution $\tilde{x}$ is better than $\bar{x}$, go to the next step with it. If Newton's method fails to converge, or if it leads to a point $\tilde{x}$ satisfying $\theta(\tilde{x}) > \theta(\bar{x})$, then the point $\tilde{x}^1$ is retracted along L halfway towards $\bar{x}$, and Newton's method is repeated with the new $\tilde{x}^1$.

Rules have been developed for changing the basic and nonbasic vectors during the algorithm to make it more efficient.

Exhibit 14.3.5 In a Step of the Generalized Reduced Gradient Method, the Algorithm Moves From $\bar{x}$, Along the Tangent Line L, to the Infeasible Point $\tilde{x}^1$ and Then Returns to the Feasible Region at $\tilde{x}$ by Using Newton's Method

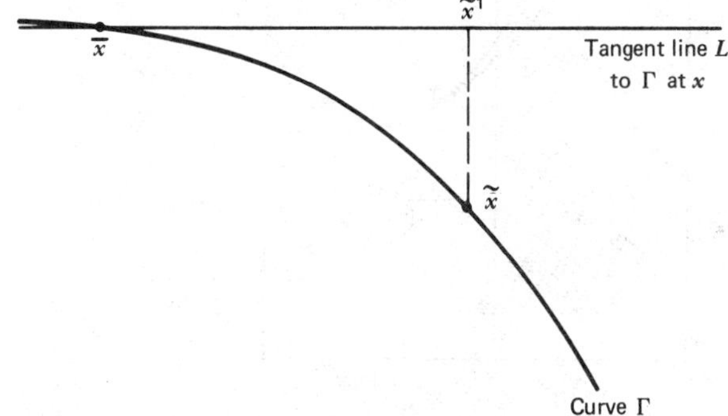

Penalty Methods

There are various methods in this class, all of which attempt to solve the constrained optimization problem by iterative approximation of it by related, but simpler, unconstrained optimization problems. The approximation is achieved by adding to the objective function penalty terms representing a high cost for violating the constraints. In these methods the severity of the penalty terms and, consequently, the closeness of the unconstrained minimization problem to the original problem are controlled by a penalty parameter, r. As r approaches zero, the approximation becomes closer (see Exhibit 14.3.6).

Consider the general nonlinear program in equation 5. The "exterior point penalty method" transforms this into the unconstrained minimization problem

$$\text{minimize } P(x, r) = \theta(x) + \frac{1}{r}\left[\sum_{t=1}^{p} [h_t(x)]^2 + \sum_{i=1}^{m} \{\text{minimum } [0, g_i(x)]\}\right] \tag{24}$$

where r is a positive penalty parameter. If equation 5 has an optimum solution, as r approaches zero an optimum solution of equation 24 tends also to be optimal to equation 5. When r becomes very small, even small changes in the values of $h_t(x)$ from zero or the value of a $g_i(x)$ in the negative range produce abrupt increases in the value of $P(x, r)$, and hence the unconstrained minimization problem in equation 24 becomes ill conditioned and computationally hard to solve. One way of handling this is to solve equation 24 parametrically with a sequence of values for r decreasing to zero. Denoting an optimum solution of equation 24 by $x(r)$, we can start with values of r sufficiently large initially, dividing it by a positive number, such as 2, after each step. The optimum solution obtained in a step can be used as the initial solution for beginning the unconstrained minimization in the next step with a reduced value for r. In this way the sequence of solutions obtained, $x(r)$, could converge to an optimum solution for equation 5.

Then there are the "interior point penalty methods" (which are also called "barrier function methods"), of which the well-known Sequential Unconstrained Minimization Technique (SUMT) is an example. These methods begin with a point that satisfies all the inequality constraints as strict inequalities and moves among points, all of which have this property. In these methods

Exhibit 14.3.6 For Any r, $P(x, r) = \theta(x)$ for All x in the Feasible Region (the Continuously Drawn Curve). Away From the Feasible Region, $P(x, r)$ Increases Rapidly, Its Steepness of Increase Going up as r Decreases to Zero. In the Infeasible Region, the Dotted Portion Might Represent $P(x, r)$ When $r = \frac{1}{2}$, and the Dashed Portion Might Be $P(x, r)$ When $r = \frac{1}{4}$

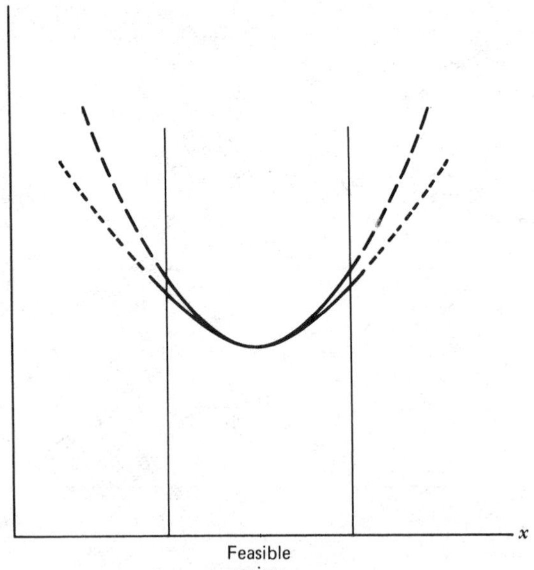

equation 5 is transformed into the unconstrained minimization problem

$$\text{minimize } B(x,r) = \theta(x) + \frac{1}{r}\left\{\sum_{t=1}^{p} [h_t(x)]^2 + \sum_{i=1}^{m} \frac{1}{g_i(x)}\right\} \tag{25}$$

Again, a sequence of problems is solved with decreasing values of r. The first problem uses an initial point that satisfies $g_i(x) > 0$ for all $i = 1$ to m. Since $1/g_i(x)$ approaches ∞ as $g_i(x)$ approaches zero, in the unconstrained minimization of equation 25, all the points obtained will satisfy $g_i(x) > 0$, for all i, even though in the limit some of these inequality constraints may hold as equations. Under mild conditions it can be shown that the sequence of optimum solutions of equation 25, $x(r)$, over values of r decreasing to zero converges to a local minimum of equation 5. The efficiency of this method can be improved by combining it with active set strategies of the type discussed in Section 14.3.8.

Recently, penalty function methods called "augmented Lagrangian methods" have been developed. These methods combine penalty terms with Lagrange multiplier terms. They are reported to avoid the ill-conditioning difficulties encountered when the penalty parameter term approaches zero.[14]

14.3.11 HOW TO USE THE ALGORITHMS

One important distinction between linear and nonlinear models should be emphasized. Practical models of the linear programming, network flow type can be very large scale, involving several thousand variables and a few thousand constraints. The only data that are needed in constructing a linear model are the coefficients of each variable in the objective and constraint functions and the right-hand side constants, which tend to be reasonably easy to obtain (at least approximately) in practical problems. On the other hand, in nonlinear models, a variable may appear in several nonlinear terms in each of the objective or the constraint functions. The functional form of each term has to be determined, and the parameters (like the exponents, coefficients, etc., as discussed in Section 14.3.3), if any, have to be estimated. Thus constructing nonlinear models tends to be hard, and for this reason many nonlinear programs constructed in practical applications usually have only a few variables. The same phenomenon seems to hold for algorithmic efficiency. A large-scale linear programming problem for the simplex algorithm might mean one involving several thousand variables and constraints; a nonlinear program with only a few hundred variables would be considered a large-scale problem for nonlinear programming algorithms.

In practical applications, unless the functions are very simple, checking for convexity tends to be very hard. Also, many practical applications lead to nonconvex models. As discussed previously, on such problems the best that we can hope for is to obtain an approximation to a local minimum. The final solution obtained by a nonlinear programming algorithm usually depends on the initial solution with which the algorithm is started. On most algorithms in use today, good results can be expected if the initial solution is chosen fairly close to the optimum, and in many applications this can be done by using the practical insights that the modeler has about the problem. Also, if it is possible to guess some bounds on the variables at the optimum solution, introducing these bounds as constraints can improve the rapid convergence of the algorithm. It is always worthwhile to carry out the algorithm several times, each time starting with a different initial point in the likely region of optimality, and to take the best of all the solutions obtained as a reasonable approximation to the minimum.

14.3.12 NONLINEAR PROGRAMMING SOFTWARE

The performance of an algorithm in solving practical problems using a digital computer depends heavily on (1) the quality of the algorithm itself and (2) the skill with which the software for the algorithm is prepared. Professional software writers use proper scaling, matrix factorizations, and many other numerical analysis tools to improve computational efficiency and to make sure that computations are carried out with adequate precision for numerical stability. Preparing software is an art in itself, and a practitioner with a problem to solve should always explore the possibility of obtaining a professionally written code to solve his or her problem. Professionally written codes normally employ a combination of ideas from various algorithms, so that they work reasonably well on a large class of problems. Nowadays several excellent codes are available in the nonlinear programming area, which have been tested and documented and are readily available to the public. A recent article[19] provides a survey of available software, with addresses of people from whom the software can be obtained. Most of these codes are available for charges ranging from \$50 to \$250, and hence they are not very expensive.

The Systems Optimization Laboratory, Operations Research Department, Stanford University, Stanford, California 94305 offers two excellent codes for large-scale models. One is MINOS, based on the reduced gradient method, for solving large-scale linearly constrained optimization problems. The other is MINOS/Augmented, for large scale nonlinear programs involving nonlinear constraints. The laboratory also hopes to make available soon a code based on projected Lagrangian methods for solving nonlinear programs.

For nonlinear programs that are not large-scale problems, the Numerical Algorithms Group (NAG) and National Physical Laboratory (NPL) libraries provide the most extensive collection of quality software. The subroutines in these libraries are all carefully written and tested and uniformly documented. The optimization subroutines in the NPL library can be obtained from the Distribution Center, Control Analysis Corporation, 285 Hamilton Avenue, Palo Alto, California 94301. The optimization subroutines in the NAG library can be obtained from NAG-U.S.A., 1250 Grace Court, Downer's Grove, Illinois 60515.

Another source of quality software for nonlinear programming (unconstrained minimization, nonlinear least squares, nonlinear equation solving problems) is the MINPACK routines available from the Applied Mathematics Division, Argonne National Laboratory, Building 221, 9700 South Cass Avenue, Argonne, Illinois 60439. Also, the Argonne Code Center, National Energy Software Center, at this address makes available the Harwell library of codes, with excellent subroutines for sparse matrix handling, unconstrained minimization, quadratic programming, and linearly constrained nonlinear programming.

For solving nonlinear least square problems, an excellent code named NL2 SOL is available through International Mathematical and Statistical Libraries, Inc. (IMSL), NBC Building, 6th Floor, 7500 Bell Air Boulevard, Houston, Texas 77036. A description of this code is contained in Dennis et al.[15]

For general constrained optimization, the general reduced gradient method and the augmented Lagrangian methods seem to work well. There are several people who offer codes based on general reduced gradient (two of them being Professor L. S. Lasdon, School of Business Administration, University of Texas, Austin, Texas 78712 and Professor K. M. Ragsdell, School of Mechanical Engineering, Purdue University, West Lafayette, Indiana, 47907).

Watson[20] and Watson and Fenner[21] contain published codes for fixed-point computing and nonlinear equation solving problems based on continuation methods.

REFERENCES

1. K. FAHLANDER, K. SVANBERY, and J. E. MARKLUND, "Optimization of a District Heating System," in K. B. Haley, Ed., *Operational Research '78*, North-Holland, Amsterdam, 1978.
2. G. K. AGRAWAL, "Helical Torsion Springs for Minimum Weight by Geometric Programming," *Journal of Optimization Theory and Applications*, Vol. 25, No. 2 (June 1978), pp. 307–310.
3. C. H. WHITE, personal communication, 1978.
4. M. AVRIEL and R. S. DEMBO, Eds., *Engineering Optimization*, Mathematical Programming Study 11, North-Holland, Amsterdam, 1979.
5. M. L. BALINSKI and C. LEMARECHAL, Eds., *Mathematical Programming in Use*, Mathematical Programming Study 9, North-Holland, Amsterdam, 1978.
6. C. LEMARECHAL and R. MIFFLIN, Eds., *Nonsmooth Optimization–Proceedings of IIASA Workshop*, Pergamon, New York, 1978.
7. M. L. BALINSKI and P. WOLFE, Eds., *Nondifferential Optimizations*, Mathematical Programming Study 3, North-Holland, Amsterdam, 1975.
8. K. G. MURTY, *Linear and Combinatorial Programming*, Wiley, New York, 1976, pp. 481–520.
9. M. AVRIEL, *Nonlinear Programming Analysis and Methods*, Prentice-Hall, Englewood Cliffs, NJ, 1976.
10. A. V. FIACCO and G. P. MCCORMICK, *Nonlinear Programming: Sequential Unconstrained Minimization Technique,"* Wiley, New York, 1968.
11. M. S. BAZARAA and C. M. SHETTY, *Nonlinear Programming Theory and Algorithms*, Wiley, New York, 1979.
12. P. E. GILL and W. MURRAY, *Numerical Methods for Constrained Optimization*, Academic Press, New York, 1974.
13. W. MURRAY, Ed., *Numerical Methods for Unconstrained Optimization*, Academic Press, New York, 1972.
14. D. A. PIERRE and M. J. LOWE, *Mathematical Programming Via Augmented Lagrangians, An Introduction With Computer Programs*, Addison-Wesley, Reading, MA, 1975.
15. J. E. DENNIS, JR., D. M. GAY, and R. W. WELSCH, *An Adaptive Nonlinear Least-Squares*

Algorithm, MRC-Technical Summary Report 2010, Mathematics Research Center, University of Wisconsin, Madison, October 1979.

16. L. NAZARETH, "Some Recent Approaches to Solving Large Residual Nonlinear Least Squares Problems," *SIAM Review*, Vol. 22, No. 1 (January 1980), pp. 1–11.

17. L. T. WATSON, "A Globally Convergent Algorithm for Computing Fixed Points of C^2 Maps," *Applied Mathematics and Computation*, Vol. 5, 1979, pp. 297–311.

18. O. L. MANGASARIAN, "Equivalence of the Complementarity Problem to a System of Nonlinear Equations," *SIAM Journal of Applied Mathematics*, Vol. 31, 1976, pp. 89–92.

19. A. D. WAREN and L. S. LASDON, "The Status of Nonlinear Programming Software," *Operations Research*, Vol. 27, No. 3 (May–June 1979), pp. 431–456.

20. L. T. WATSON, "Fixed Points of C^2 Maps," *Journal of Computational and Applied Mathematics*, Vol. 5, 1979, pp. 131–140.

21. L. T. WATSON and D. FENNER, "Chow-Yorke Algorithm for Fixed Points or Zero of C^2 Maps," to appear in *ACM Transactions on Mathematical Software.*

BIBLIOGRAPHY

ALLGOWER, E., and K. GEORG, "Simplicial and Continuation Methods for Approximating Fixed Points and Solutions to Systems of Equations," *SIAM Review*, Vol. 22, 1980, pp. 28–86.

SAIGAL, R., "Fixed Point Computing Methods," in J. Belzer, A. G. Holzman, and A. Kent (Eds.), *Encyclopedia of Computer Science and Technology*, Dekker, New York, 1976.

TODD, M. J., *The Computation of Fixed Points and Applications*, Springer-Verlag, New York, 1976.

WRIGHT, M. H., *A Survey of Available Software for Nonlinearly Constrained Optimization*, Technical Report SOL 78-4, Systems Optimization Laboratory, Stanford University, Stanford, CA, January 1978.

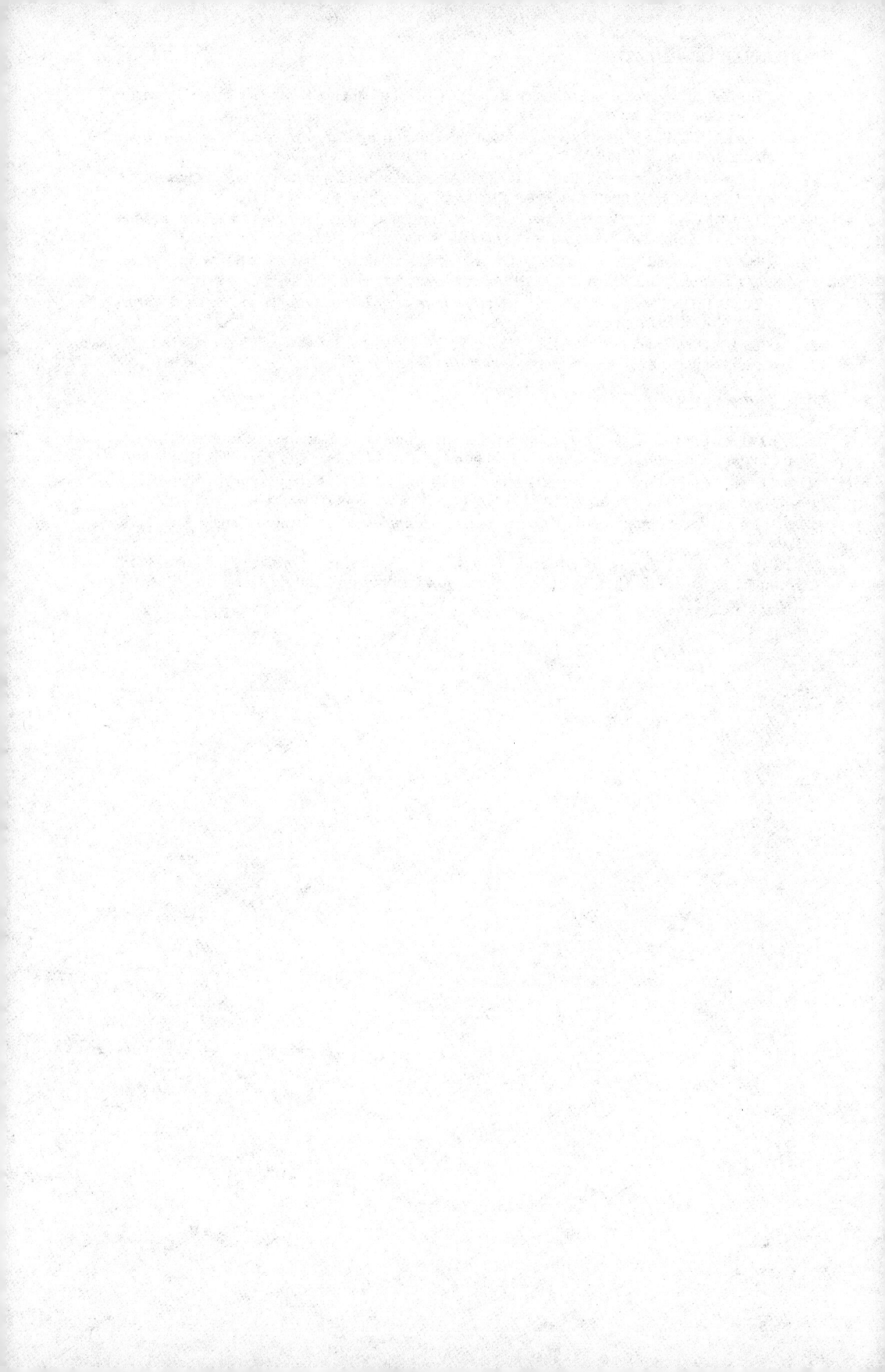

CHAPTER 14.4

Discrete Optimization

JEFFREY L. ARTHUR

Oregon State University

JAMES P. CAIE, JR., JOEL D. COHEN, ROBERT H. HARDER

General Motors Corporation

14.4.1 INTRODUCTION

Discrete optimization problems, also known as (linear) integer programs, are linear programming (LP) problems (see Chapter 14.2) in which some or all of the decision variables are restricted to being integer valued. A "pure integer program" (PIP) is one in which all of the variables are restricted to being integers, whereas a "mixed integer program" (MIP) allows some of the variables to be continuous while restricting the others to integer values. A special class of discrete optimization problems known as "binary programs" or "0-1 programs" are those in which all of the variables are restricted to taking on a value of either zero or one.

Discrete models must be employed instead of LP models when the assumption of divisibility[1] inherent in LP models is violated. Although it might seem that discrete models would be easier to solve than LP models, since the former deal with a finite (or countably infinite) number of possible solutions while the latter deal with a continuum, this is not the case (see Wagner[2] for an example).

14.4.2 FORMULATION TECHNIQUES FOR DISCRETE MODELS

Various guidelines have evolved for formulating integer programs that have generally led to more useful models, usually in the sense that solution of the model on a computer can be accomplished in a relatively short period. Some of these guidelines are:

1. If it is known a priori that an integer variable will assume a large value (usually 20 or more), then treat that variable as a continuous one and round the continuous solution.

2. The coefficients in the constraints contribute to the difficulty in solving a problem, so care should be taken to write the constraints in a form that is as uncomplicated as possible. For example, if x_1 and x_2 are required to be (nonnegative) integers, then the constraint

$$121x_1 + 164x_2 \leqslant 189$$

can be replaced by the equivalent expression

$$x_1 + x_2 \leqslant 1$$

Other similar techniques are described in Shapiro.[3]

3. Any reduction in the size of the feasible region reduces the computational effort; thus good lower and upper bounds on the variables are desirable.

Discrete models also arise in practice in order to handle certain types of formulation difficulties. Included are the following situations:

1. Consider the situation where it is necessary that at least one of two constraints be maintained. For instance, we may have a choice between two resources for the manufacture of a certain product. Such a situation involves "either-or" constraints; for example, we might encounter the following constraints: either

$$4x_1 + 5x_2 \leq 16$$

or

$$2x_1 + x_2 \leq 12$$

If we let M be a very large number and define y to be a binary (0 or 1) variable, an equivalent representation of the either-or constraints is

$$4x_1 + 5x_2 \leq 16 + My$$

$$2x_1 + x_2 \leq 12 + M(1 - y)$$

Notice that if $y = 0$ in the final solution, then the first constraint must hold, whereas the second constraint becomes irrelevant (because of the large value assigned to M). The opposite holds true if $y = 1$ in the solution.

2. For a generalization of the preceding technique, consider the situation where at least K out of L constraints must hold ($K < L$). If we write the L constraints as

$$\sum_{j=1}^{n} a_{\ell j} x_j \leq b_\ell \qquad \ell = 1, 2, \ldots, L$$

then this restriction can be written as

$$\sum_{j=1}^{n} a_{\ell j} x_j \leq b_\ell + M y_\ell \qquad \ell = 1, 2, \ldots, L$$

$$\sum_{\ell=1}^{L} y_\ell \leq L - K$$

where M is again a very large number and $y_\ell = 0$ or $1, \ell = 1, 2, \ldots, L$.

3. Occasionally one encounters a situation where a function is required to take on one of R specified values. If we write this restriction as

$$\sum_{j=1}^{n} a_{ij} x_j = b_1 \text{ or } b_2 \text{ or} \ldots \text{or } b_R$$

then an equivalent representation is

$$\sum_{j=1}^{n} a_{ij} x_j = \sum_{i=1}^{R} b_i y_i$$

$$\sum_{i=1}^{R} y_i = 1$$

where $y_i = 0$ or 1 for $i = 1, 2, \ldots, R$.

It is also worthwhile to note that PIPs in which the variables are bounded (either implicitly or explicitly) can be converted to 0-1 programs through a technique called "binary decomposition" and that nonlinear polynomial 0-1 programs can be converted to linear 0-1 programs (e.g., see Phillips et al.[4] and Glover and Woolsey,[5] respectively, for details).

14.4.3 EXAMPLES OF DISCRETE MODELS

Although there are many published examples of integer programming models, two of the most thoroughly studied are the knapsack problem and the facility location problem.

Example 14.4.1 The Knapsack Problem. Consider a knapsack with a finite capacity, V. For each item j that can be packed in the knapsack ($j = 1, \ldots, J$), let $V_j > 0$ be the volume per unit and $C_j > 0$ be the value per unit. The objective is to pack the knapsack so that its total value is maximized, subject to the limitation of its capacity. If we define X_j as the number of units of item j to pack in the knapsack, then the problem can be expressed as

$$\text{maximize} \sum_{j=1}^{J} C_j X_j$$

subject to

$$\sum_{j=1}^{J} V_j X_j \leqslant V$$

$$X_j \geqslant 0 \text{ and integer} \quad \text{for } j = 1, \ldots, J$$

This model is a generalized knapsack problem allowing any nonnegative integer value for each X_j, thus permitting more than one unit of an item to be included in the knapsack. A special case of the knapsack problem is one in which each X_j is restricted to the value 0 or 1, so that no more than one item of each type may be packed in the knapsack.

Knapsack problems are PIPs with one constraint. The knapsack problem is important for two reasons. First, a large number of integer programming problems, including capital budgeting, machine loading, and project selection problems, can be shown to be equivalent to the knapsack model. Second, efficient methods have been developed for the solution of this class of problems. Efficient solution procedures for knapsack problems have been the basis for the development of new procedures for solving general integer programming problems. For further information on knapsack problems, their properties, and their solution, see Garfinkel and Nemhauser.[6]

Example 14.4.2 The Facility Location Problem. Suppose there are several customers, each with a known demand, D_j, for a certain commodity. It is desired to manufacture this commodity at no more than three of the potential sites. We let F_i be the fixed cost for building the facility at site i, B_i the maximum capacity for a facility at site i, and C_{ij} the unit cost of transporting the commodity from site i to customer j. The objective is to determine how many facilities to build and at which sites to build them in order to meet customer demand at the lowest possible cost. Define

$$Y_i = 1 \text{ if a facility is built at site } i$$
$$= 0 \text{ otherwise}$$

$$X_{ij} = \text{amount of the commodity transported from site } i \text{ to customer } j$$

Note: We are assuming that at most one facility can be built at any site. The problem is then to

$$\text{minimize} \sum_i \sum_j C_{ij} X_{ij} + \sum_i F_i Y_i$$

subject to

$$\sum_j X_{ij} - B_i Y_i \leqslant 0 \qquad \text{for all } i$$

$$\sum_i X_{ij} = D_j \qquad \text{for all } j$$

$$\sum_i Y_i \leqslant 3$$

$$Y_i = 0 \text{ or } 1 \qquad \text{for all } i$$

$$X_{ij} \geqslant 0 \qquad \text{for all } i, j$$

This facility location problem is modeled as an MIP problem. The model can be transformed into a much more complex distribution model incorporating such features as multiple products, conditions to prevent splitting the manufacture of a customer's order, service considerations to include time limitations on routes, and multiechelon warehousing considerations.

14.4.4 ALGORITHMS FOR DISCRETE OPTIMIZATION

The algorithms that have been developed for solving discrete optimization problems are typically classified into the following categories:

1. Enumerative algorithms.
2. Cutting plane algorithms.
3. Group theoretic algorithms.

Each category is discussed here in varying degrees of detail.

Enumerative Algorithms

The central idea behind all enumerative algorithms is a logical search of the set of all possible integer solutions without having to check each possible solution individually. Such algorithms attempt to eliminate large numbers of possible solutions from consideration by employing a strategy commonly known as "divide and conquer."

Perhaps the most widely known enumerative algorithm is the modification by Dakin[7] of the basic algorithm by Land and Doig[8] for solving the general MIP problem, known as the "branch and bound algorithm."

The basic principles of the branch and bound method can best be illustrated by means of an example.

Example 14.4.3

$$\text{maximize} \quad z = x_1 + 4x_2$$

subject to

$$x_1 + 3x_2 \leqslant 7.5$$
$$x_1 \leqslant 3$$
$$x_2 \leqslant 2$$
$$x_1, x_2 \geqslant 0 \text{ and integer}$$

The algorithm begins by solving the MIP as a linear program by ignoring the integer restrictions on x_1 and x_2. The graphic solution of this linear program, LP-1, is given in Exhibit 14.4.1. At the optimal solution of $x_1 = 1.5$, $x_2 = 2$, we see that the integer restriction on x_2 is met, but that for

Exhibit 14.4.1 The LP-1 Feasible Region of Example 14.4.3

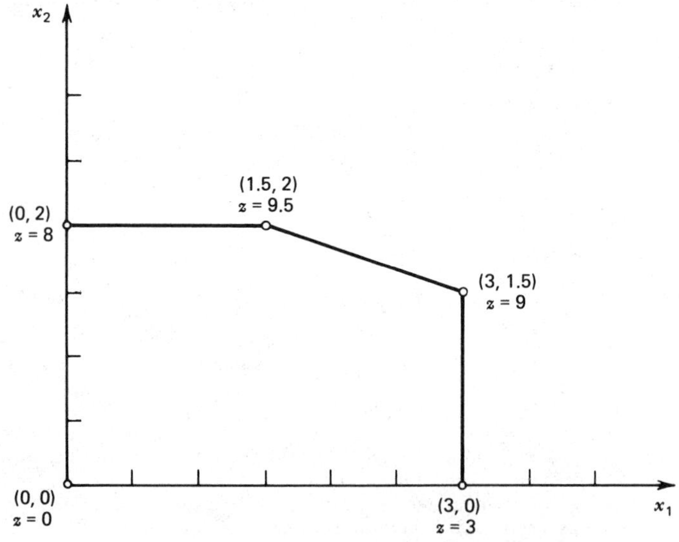

Exhibit 14.4.2 The LP-2 Feasible Region

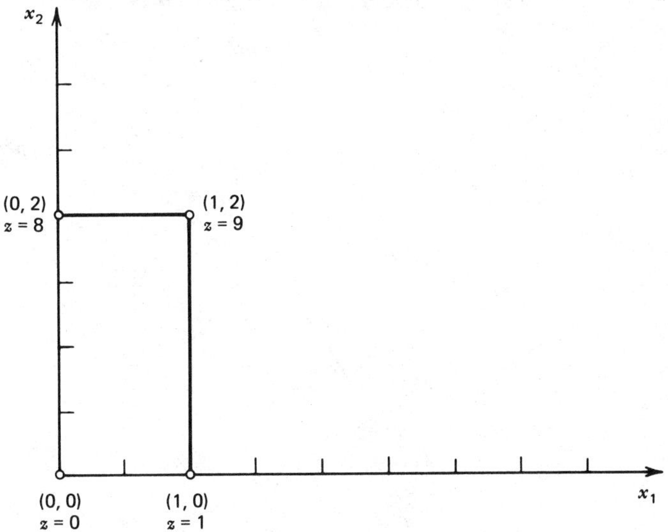

x_1 it is violated. Notice that we now have an upper bound of 9.5 on the optimal objective function value z^* for the MIP.

To try to force x_1 to integer values, we form two new LPs, LP-2 and LP-3, by adding a new constraint to LP-1. The new constraints must be such that the optimal solution to LP-1 is no longer feasible to LP-2 or LP-3, yet all integer feasible solutions must still be acounted for. The constraint $x_1 \leqslant 1$ will be added to LP-1 to form LP-2, while the constraint $x_1 \geqslant 2$ will form LP-3 from LP-1. The solutions to LP-2 and LP-3 are given in Exhibits 14.4.2 and 14.4.3, respectively.

The optimal solution to LP-2, $x_1 = 1$, $x_2 = 2$, is feasible to the MIP. Hence we have a lower bound on the optimal MIP objective function value, namely, $z^* \geqslant 9$. In addition, no further branching from LP-2 need be performed.

The optimal solution to LP-3, $x_1 = 2$, $x_2 = 1.83$, now violates the integer restriction on x_2. By adding the constraints $x_2 \leqslant 1$ and $x_2 \geqslant 2$ to LP-3, we form LP-4 and LP-5 and solve these LPs. The reader can verify that LP-4 has an optimal solution of $x_1 = 3, x_2 = 1$, which is feasible to the MIP with $z = 7$, and that LP-5 is infeasible.

Exhibit 14.4.3 The LP-3 Feasible Region

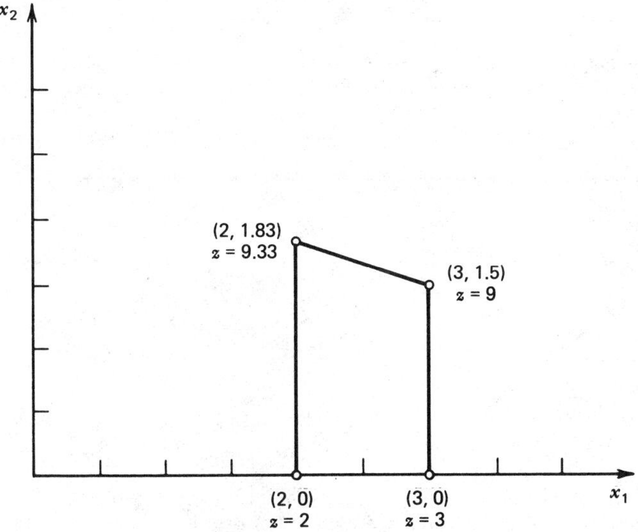

Exhibit 14.4.4 A Tree Diagram Representation of Branch and Bound Algorithm of Example 14.4.3

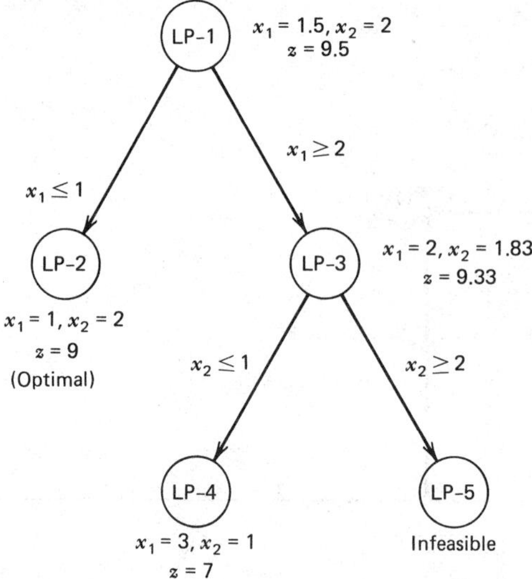

The solution procedure used to solve this MIP can be illustrated through the use of a tree diagram as shown in Exhibit 14.4.4, which reflects the basic strategy of "divide and conquer."

A special enumerative algorithm for 0-1 integer programs was developed by Balas,[9] which is commonly known as the "implicit enumeration procedure." His algorithm requires the problem to be formulated according to the following guidelines:

1. The objective must be a minimization.
2. All of the objective function coefficients must be nonnegative.
3. All constraints must be of the greater-than-or-equal to type.

Procedures for transforming a problem into this form are given in Plane and McMillan.[10]

The idea behind Balas' algorithm is to start with all x_j equal to zero and sequentially change one variable at a time to the value one until a feasible solution is found. The algorithm then "backtracks" to explore other possible solutions, continuing until all possible combinations have either explicitly or implicitly been explored. Various improvements in the basic algorithm, such as the use of surrogate constraints, are discussed in some detail in Geoffrion and Marsten.[11]

Cutting Plane Algorithms

The first approach to solving integer programming problems was offered by Gomory.[12] The motivation behind his algorithm was that one would reach the optimal integer solution by solving a sequence of linear programs in which each new program added a constraint that cut off the optimal solution to its predecessor. This additional constraint was called a "cutting plane," or simply a "cut."

The example that follows illustrates this approach.

Example 14.4.4

$$\text{maximize} \quad z = 3x_1 + x_2$$

subject to

$$x_1 + 2x_2 \leqslant 8$$

$$3x_1 - 4x_2 \leqslant 12$$

$$x_1, x_2 \geqslant 0 \quad \text{and integer}$$

After solving the initial linear program, again called LP-1, by ignoring the integer restrictions (which is done in Exhibit 14.4.5), the algorithm tests to see if all integer variables have assumed integer values. When the integrality restrictions are not met, the algorithm uses information in the

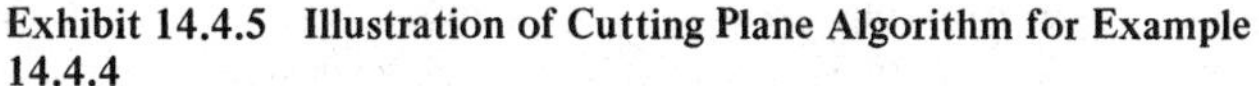

Exhibit 14.4.5 Illustration of Cutting Plane Algorithm for Example 14.4.4

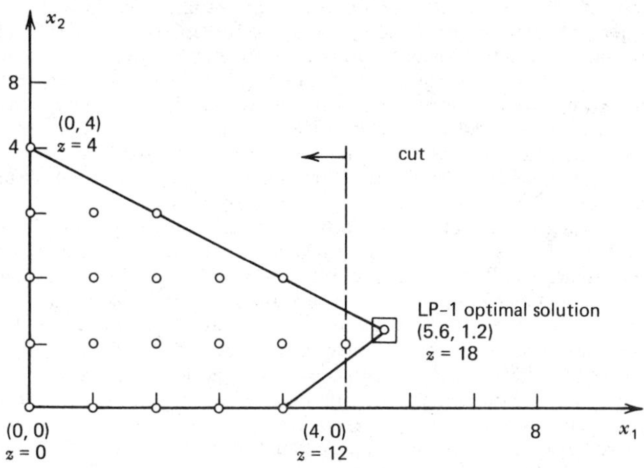

final simplex tableau from LP-1 to generate the new constraint. The computational details for generating the cut are quite involved (e.g., see Garfinkel and Nemhauser[13]), but for our example the algorithm would generate the following constraint at the end of LP-1:

$$-\tfrac{2}{5}s_1 - \tfrac{1}{5}s_2 + s_3 = -\tfrac{3}{5}$$

where s_1 and s_2 are the slack variables corresponding to the original constraints and s_3 is the new slack variable for this cut. The cut is shown in Exhibit 14.4.5 by the dashed line. Two important characteristics of all cuts are illustrated in this example:

1. The old LP optimal solution violates the cut.
2. All integer feasible solutions satisfy the new cut.

The algorithm continues by solving the augmented linear program and adding new cuts until an optimal solution to one of the programs satisfies the integer restrictions. This becomes the optimal solution to the original integer program. For our example, this will occur after the addition of one more cut of the form

$$-\tfrac{1}{2}s_1 - \tfrac{1}{2}s_3 + s_4 = -\tfrac{1}{2}$$

Further details on a variety of cutting plane algorithms, including modifications that are necessary when the problem is an MIP instead of a PIP, can be found in Garfinkel and Nemhauser.[6]

Group Theoretic Algorithms

Another category of algorithms for solving integer programming problems–and one that has received widespread attention in recent years–is the group theoretic approach first identified by Gomory[14] and expanded largely by Shapiro.[15] The group theoretic methods are quite complex in their mathematical content and hence are not discussed in great detail here; suffice it to say that they have been used mainly for PIP problems due to computational difficulties encountered when attempts have been made to extend the techniques to MIPs.

Finally, it would be incomplete to discuss integer programming solution techniques without including some mention of heuristics. Much recent work has been concerned with the development of heuristic techniques for special classes of integer programs, including the knapsack and facility location problems (see Chapter 11.11). Heuristics are also playing larger roles in the development of algorithms for solving larger and more realistic integer programming models. For an excellent discussion on the state of the art in integer programming algorithms, the reader is referred to Geoffrion and Marsten.[11]

14.4.5 COMPUTER SOLUTION OF DISCRETE OPTIMIZATION PROBLEMS

After the discrete optimization problem has been formulated, it is necessary to find the appropriate computer software for solving the model. The modeler can use a commercial integer programming code (such as IBM's MPSX-MIP), a specially developed code, or a heuristic code. The type of software to use in a given situation is not always readily apparent.

The most common method for solving a discrete optimization problem is through the use of a commercial integer programming code. The primary reason for the popularity of such codes is their availability–most computer manufacturers develop integer programming software for their customers, and a large number of consulting firms market their own software. Other reasons for the use of commercial integer programming software include the reliability of the codes (since codes presumably have been used on a wide variety of problems, we can have confidence in the solution obtained) and the fact that they are very general in nature. Many commercial codes use a version of the branch and bound algorithm, whereas others are based on modifications of Benders' decomposition.[16]

The difficulty in using commercial codes is that they are often quite slow in solving a certain problem because of their general nature. This becomes more of a problem as the size and complexity of the model increase. When the model is too large to be solved in a reasonable time using a commercial code, the modeler must resort to a specialized code or a heuristic. A specialized code that exploits certain properties of the structure of the model may already exist or may have to be developed. For instance, a number of specialized codes exist for efficiently solving various types of knapsack and facility location problems. If the specialized code has to be developed, a number of points must be considered. First, development often takes a long time and a lot of resources. Second, a specialized integer programming code can be used only on models that have the same basic structure. Therefore the modeler must be certain that the model for which the special code is to be developed is realistic and important enough to warrant a large expenditure for limited application.

If a commercial or specialized integer programming code is unsuitable for the problem at hand (possibly because of the large size and lack of exploitable structure of the model), the modeler may be forced to develop a heuristic code. The code may contain some optimization modules designed to solve specific segments of the problem. Other segments of the problem may be attacked using "quick and dirty" greedy algorithms or Monte Carlo simulation techniques. The advantage of a heuristic is that it can generate solutions to very complex problems in a short period. However, heuristic codes suffer from the same drawbacks that are disadvantages of specialized codes, plus the fact that the solution obtained by a heuristic may not even be optimal.

14.4.6 APPLICATIONS OF DISCRETE MODELS IN GENERAL MOTORS

Machine Load Planning Model

The Machine Load Planning (MLP) model is used for capacity planning in a manufacturing plant. The objective of the model is to determine a machine assignment and production cycle for each part produced in a plant during a planning period. The model develops a production plan that minimizes total setup, storage, and overtime costs consistent with plant storage and machine capacity constraints. The assumptions of the model are as follows:

1. The assignment of parts to tools is the responsibility of the user. The model considers only tool-to-machine allocations rather than part-to-tool and tool-to-machine allocations.
2. Over the planning horizon, either a tool is not used or it must be mounted, perhaps several times, on one and only one machine. Each user-defined machine has a specific capacity associated with it.
3. When the production of a part is completed, the part is stored in an area with a specific storage capacity.
4. The planning horizon is logically divisible into a number of equivalent time periods. For example, the planning horizon could cover 16 weeks.
5. For each tool and each machine applicable to the tool, the cycle is limited to some subset of 1, 2, 4, or 8 time periods. In practice, this usually amounts to saying that a tool is mounted on a machine for production once every 1, 2, 4, or 8 weeks.
6. The model trades off setup and holding costs. There are two types of setup costs–major and minor. Major setup costs include the expenses associated with removing the previous tool from a machine and mounting a new tool. Minor setup costs include the expenses associated with changing a tool insert or changing the material type to produce a new part without performing a major setup. The model assumes that all minor setups will take place in a predesignated order after the associated major setup is done.
7. Demand is constant over the planning horizon.
8. There is a one level bill of material. Once the parts are produced, they are ready for packaging and shipping.
9. Sequencing considerations for each tool on a machine are ignored and addressed by the user based upon output of the model.

Model Formulation and Solution Technique

Let t be the index for tools, m the index for machines, and k the index for possible cycles. The decision variables are

$$X_{tmk} = 1 \text{ if tool } t \text{ is set up on machine } m \text{ using the } k\text{th cycle}$$
$$= 0 \text{ otherwise.}$$

Based upon the demand for a tool, the production rate of a tool on a machine, the setup time and costs, the inventory holding costs, and the part sizes and containers used for storage, one can determine the following constants:

R_{tmk} = hours required on machine m to satisfy the planning horizon demand of tool t using the kth cycle
F_{tmk} = the total floor space required to hold the output of tool t when run on machine m using the kth cycle
C_{tmk} = the total setup and inventory holding costs over the planning horizon when tool t is run on machine m using the kth cycle

Let B_m be the hours available on machine m during the planning horizon, and let S be the floor space available for storing the production output.

The problem is now stated as an integer programming problem:

$$\text{minimize} \sum_t \sum_m \sum_k C_{tmk} X_{tmk}$$

subject to

$$\sum_m \sum_k X_{tmk} = 1 \qquad \text{for each } t \tag{1}$$

$$\sum_t \sum_k R_{tmk} X_{tmk} \leq B_m \qquad \text{for each } m \tag{2}$$

$$\sum_t \sum_m \sum_k F_{tmk} X_{tmk} \leq S \tag{3}$$

$$X_{tmk} = 0 \text{ or } 1 \qquad \text{for each } t, m, k \tag{4}$$

Constraints 1 and 2 ensure that a tool is run on only one machine using only one cycle and that the machine hours available are not exceeded. Constraint 3 ensures that the floor space available is not overused. Constraint 4 is the integer programming requirement.

The typical MLP plant application has approximately 1500 tools, 300 presses, and 4 alternative cycles, resulting in up to 10,000 integer variables. This problem, in its present form, is much too large to be solved on a timely basis. The first action to simplify the model is to eliminate dominated integer variables. An integer variable is dominated when a related integer variable associated with the same tool and machine, but with a different cycle, has a smaller cost and time requirement than the dominated variable possesses. Dominated integer variables are eliminated from the problem because they will never appear in the optimal solution. Approximately 25% of the integer variables have been eliminated because of the dominance principle, leaving about 8000 integer variables still to deal with in the typical MLP problem.

The next step is to simplify the model by subdividing the problem into separate, independent subproblems. This is done by finding groups of tools and machines that are completely independent. The basic characteristic of a tool-machine group is that the tools in the group can run only on machines in the group and that machines in the group are compatible only with tools in the group. The typical MLP problem has been subdivided into approximately six independent tool-machine subgroups, with the largest subgroup containing about 2000 integer variables.

An MLP problem with 2000 integer variables is still too large to solve on a timely basis with a commercial integer programming code. The next step is to simplify the problem further by relaxing the floor space constraint. A procedure called "subgradient optimization"[17] has been used to eliminate the floor space constraint and consider its effect in the objective function. The subgradient approach involves relaxing the original model by bringing constraints 2 and 3 into the objective function and estimating an imputed cost of floor space. This imputed cost of floor space is then used to adjust the holding cost component of the original problem's objective function.

The revised MLP model now is

$$\text{minimize} \sum_t \sum_m \sum_k C'_{tmk} X_{tmk}$$

subject to

$$\sum_t \sum_m \sum_k X_{tmk} = 1 \tag{5}$$

$$\sum_t \sum_k R_{tmk} X_{tmk} \leqslant B_m \qquad \text{for all } m \tag{6}$$

$$X_{tmk} = 0 \text{ or } 1 \quad \text{for all } t, m, k \tag{7}$$

where C'_{tmk} is C_{tmk} plus the cost of occupying floor space.

We use a specially developed integer programming code that can solve very large problems having the same structure as the revised model. This code transforms the revised model into a knapsack problem by bringing constraints 5 and 7 into the objective function using Lagrangian relaxation. The resulting knapsack problem is solved using a branch and bound method based on the work of Ross and Soland.[18]

In summary, a very large complex integer model has been transformed into a number of smaller, simpler integer models that can be solved very efficiently with a specialized integer programming code.[19] The optimization phase of a typical MLP computer run requires 1 CPU min on an IBM 370/168.

Implementation of the Model

The model has been incorporated into the solution phase of a capacity planning system called the MLP system. This section briefly describes the data base generation and report phases of the system and discusses how the system is being used to improve production planning in plastic and metal stamping environments at GM.

The data base generation phase obtains MLP data from the plant's existing production control, industrial engineering, and accounting files and edits them. The data needed to make MLP decisions can be categorized by part, tool, and machine. Part data include information associated with the production and subsequent storage of a part at a specific operation, such as blanking. Required part data include part material and labor-added values to calculate unit holding costs; part standard containers, standard pack quantities, and storage locations to determine part storage capacity requirements; average part demand data to determine total holding cost and machine time capacity requirements of various production cycles over the planning horizon.

Tool data include information associated with the setup and production characteristics of a tool at a particular operation. Required tool data include major setup man-hours, to calculate a tool's setup cost on a machine; the elapsed time for setup, to determine how long a machine is down for setup and unavailable for production; the available production cycles representing the candidate set of cycles that the model chooses from for each tool; the production rate for each tool, taking into consideration all run time and downtime allowances; and the production manpower requirements stated as the number of people needed to produce parts on a specific tool.

Required machine data include location, configuration, and capacity information. A machine may be specified as a single facility, part of a machine line, or an entire machine line. The capacity of a machine is stated in units of time over the planning horizon. A calendar is used to specify the exact capacity available over the planning horizon, taking into consideration holidays, weekends, regular time shifts, and overtime shifts.

The system also needs part-tool assignment data to specify exactly which tools make each part. The number of cavities in a tool that can produce a particular part is required information for calculating the part's production rate in pieces per hour. The system has the ability to consider minor setup data for various parts on a tool in order to determine the cost and time of changing tool inserts or the various material types being processed.

The data that drive the model are the tool-machine compatibility data, which specify all the potential tool-machine assignments along with the priority of each potential assignment. For each tool, the user must indicate all the potential machines a tool can run on and the relative preference of each potential assignment. The user must consider such things as welding configurations, toggle press locations, material isolation problems, automation, and production rate efficiencies when generating the tool-machine compatibility.

The MLP system generates a variety of reports. The machine load report displays the tools the system has allocated to each machine. For each tool on a machine, the report shows the tool's recommended cycle and the setup and production time requirements over the planning horizon. The machine load summary report summarizes machine utilization by machine, tonnage, department, and plant. The warehouse reports display projected warehouse loads in square feet and the utilization of available storage space resulting from an MLP production plan. The container reports display container requirements by container type, and the cost reports display storage, setup, and overtime cost information associated with a given production plan.

The MLP system can be used for both short- and long-range planning. In short-range planning the system can be used to support scheduling. Schedulers at one GM plant are basing their schedules on the system's recommended machine assignment and cycle for each tool. They manually consider current inventory status, recent part hot lists, and resource balance constraints and manually sequence each of the tools on their designated machines. The MLP system has greatly simplified scheduling in this application. The schedulers are unable to follow the MLP load only about 5% of the time, because of instances where MLP loaded too many hot jobs on a single machine.

The MLP system can also be used for planning beyond the current week. In long-range planning the system can be used to answer management "what if" questions dealing with how fluctuating demands affect inventory capacity requirements for machines, containers, and storage. Several plants have used the system to determine the amount and type of additional machinery necessary to meet forecast demands. Another plant dealing with the production of bulky parts has used the system to design its warehouse storage facilities. The MLP system does not take an excessive amount of computer time when it runs. In a typical application with 1500 tools and 300 presses, the system requires about 2 CPU min on an IBM 370/168. Because of these short run times, the users of the system feel free to vary parameters and make several runs when answering management "what if" type of questions.

The MLP system has recently been expanded for capacity planning in a multilevel bill of material environment. The system is currently being extended into a weekly capacity planning system, which will determine the week production starts for a tool in addition to its machine assignment and production cycle.

Material and Technology Evaluation System

An automobile manufacturer's fleet fuel economy is calculated on the basis of the sales mix of various categories, or families, of cars. These categories are determined by the EPA certification procedures. Within families, vehicles are grouped into weight classes of predetermined increments. The fuel economy ratings determined for each weight class are used to calculate the sales-weighted fleet fuel economy.

For a given sales mix and volume, a planner can improve the fleet fuel economy in two ways: (1) by changing the weight of a vehicle sufficiently to move the vehicle into another weight class or (2) by changing the technology. The first method is accomplished by adding a premium material (PM) component to the vehicle. A PM component replaces a part of the vehicle with a lighter part made from a premium material, thereby incurring a higher cost. For example, a steel hood could be replaced with an aluminum one. The second method is to add a technology (TECH) component. A TECH component employs improved technology (e.g., a new tire) to make a percentage improvement in fuel economy. It may increase or decrease the weight and has an increased cost over the standard component.

The number of ways to apply components to vehicles for a large manufacturer exceeds practical human evaluation. The Material and Technology Evaluation System (MATES) was developed to optimize the use of premium cost lightweight and fuel saving technology components to achieve a desired fleet fuel economy target.

The general statement of the problem is: Choose, in a least-cost manner, the PM and TECH components to apply to each vehicle so as to meet the fleet fuel economy target, subject to vehicle-component compatibility. In addition, tooling decisions and component capacities should also be considered.

A general model was formulated. For a representative problem with 200 vehicles, 15 components, and 6 weight classes, this formulation required more than 4200 zero-one variables and 6000 constraints. Solving a simplified general model would require hours of computer time with commercially available optimization codes and development of special-purpose codes. To implement a system quickly, the tooling and component capacity restrictions were eliminated. This would not impair use of the system for making tactical decisions, and multiple runs of the system could be used to evaluate strategic decisions.

Even after separating the problem on some logical component usage constraints, the number of 0-1 variables is large. To have a system with reasonable response time, a heuristic method was developed. The method relies on three techniques: (1) solving a series of knapsack problems, (2) considering a subset of the possible vehicle-component combinations, and (3) indicating the usage status of each component on each vehicle by means of a labeling technique.

Since the fuel economy rating is the same for any weight in a weight class, it is optimal to use as few scarce PM components as possible and still remain within the desired weight class. For a vehicle at any given weight, the weight increase to the top of its current weight class and the weight reduction to the top of the next lower weight class are known. The problem is to choose which PM and/or TECH components to remove (add) to meet the weight increase (reduction) target as closely as possible without being greater (less) than the target. Since components have discrete-valued weights, a knapsack problem is solved to find the best combination of components to remove or add as the case may be.

The knapsack technique[20] is used to create combinations of components (component packets) that most efficiently meet the weight reduction targets for each vehicle. Rather than considering all possible combinations of components to apply to a vehicle, a cost-effectiveness decision rule is used to limit the number of component packets under consideration. One example of a restriction for the heuristic is that a component packet may have at most one TECH component in it.

The heuristic method uses a ranking approach. That is, at each iteration, the best policy from a cost-effectiveness and fuel economy viewpoint is applied. The method begins by considering vehicles one at a time: Components are applied or removed as needed to get the vehicle into its desired weight range. Component packets are created using the knapsack technique and applied in cost-effectiveness order. At each stage, the cost and fuel economy effect of leaving a vehicle's weight unchanged, of increasing weight (removing components), or of applying a component packet to drop it into the next lower weight class is evaluated and acted upon. To increase flexibility, a labeling scheme is used to indicate component usage and consistency on each vehicle.

This technique can find good solutions to large problems very quickly. The heuristic technique is being used in conjunction with a CRT-oriented management information system. This allows the user to easily evaluate alternative fuel economy actions under different assumptions. The Material and Technology Evaluation System is being used extensively for current and future model year planning, to make fleet PM and TECH component usage decisions, and to determine capacity levels required for these components.

The Carline Assignment Model

The Carline Assignment Model (CLAM) is a mathematical tool designed to aid corporate management in the process of assigning carlines (products) to assembly plants. The mathematical technique used in this decision process is mixed integer programming.

The objective of the model is to determine which carlines (products) to build at each assembly plant and how many of each of the assigned carlines to build so as to minimize relevant assembly and distribution costs while satisfying demand and production constraints. Currently the model has the ability to consider:

1. Forecast demand by carline and sales area.
2. Plant production rules and restrictions (e.g., only a certain number of carlines may be assigned to a plant, or certain combinations of carlines are not allowed).
3. Current plant locations and capacities.
4. Outbound (finished vehicle to dealer) freight costs.
5. Inbound (manufacturing plant to assembly plant) freight costs.
6. Local content restrictions (Canadian/U.S. trade agreement).
7. Production costs.
   a. Tooling and rearrangement.
   b. Capital expenditure.
   c. Overtime.
   d. Assembly.

The problem is formulated as a mixed integer model where the integer variables represent the yes/no decisions of whether or not a certain carline is to be built at a plant. Exhibit 14.4.6 shows the structure of the model.

The transportation portion of the matrix is formulated as a classical network-type distribution problem. The network consists of arcs representing the plants, carline-to-plant assignments, transportation, and demand by carline by sales area. The integer variables are tied to the plant-carline production to ensure that the upper and lower production bounds of a particular plant-carline arc are active only when its corresponding integer variable is set to 1 (on).

The combinatorial constraints are equations that define the combinations of carlines that can be built at an assembly plant. Only certain combinations are allowed because the assembly of all possible combinations would be too complex for the particular plant. Shown in Exhibit 14.4.7 is an example where a particular plant has the ability to build any of four A-type vehicles and any of four X-type vehicles for the Chevrolet (CHV), Pontiac (PON), Oldsmobile (OLD), and Buick (BUK) divisions. The restrictions are that the A- and X-type cars cannot be built together and that

Exhibit 14.4.6 Model Formulation

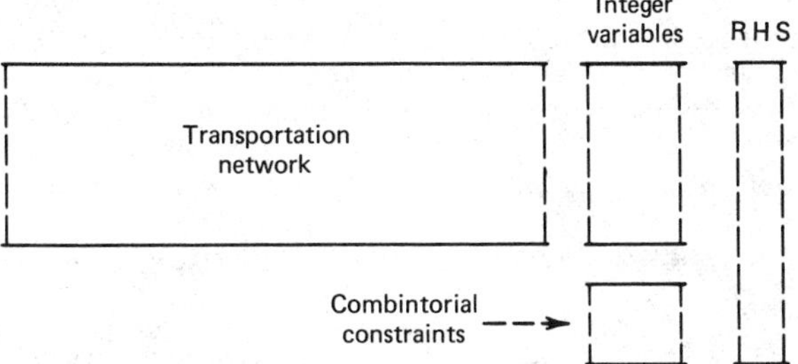

Exhibit 14.4.7 Combinatorial Constraints

CHV A	PON A	OLD A	BUK A	CHV X	PON X	OLD X	BUK X	Right-Hand Side
4				1	1	1	1	⩽4
	4			1	1	1	1	⩽4
		4		1	1	1	1	⩽4
			4	1	1	1	1	⩽4
1	1	1	1	4				⩽4
1	1	1	1		4			⩽4
1	1	1	1			4		⩽4
1	1	1	1				4	⩽4
1	1	1	1	1	1	1	1	⩽3

no more than three different carlines can be assigned to the plant. The first four combinatorial constraints ensure that, if any of the A-type cars are assigned (the A variable set to 1), all variables representing the X-type cars cannot be assigned and will be set to zero. The first four constraints also ensure that no A-type variable can be set to 1 when an X-type variable is turned on. Since these first four constraints allow an A-type variable to assume a fractional value, it is quite possible that the branch and bound type of mixed integer code may branch on an A variable even though zero is its only feasible value. The second four constraints form a stronger formulation when an X variable is set to 1 because it immediately forces all A variables to zero and eliminates unnecessary branching. The last constraint ensures that no more than three carlines are assigned to the plant.

The typical CLAM problem contains 150 integer variables, 8000 continuous variables, and 3000 constraints. This mixed integer programming problem is too large for a commerical integer programming package to obtain optimal solutions on a timely basis. International Business Machines' MPSX-MIP currently runs about 10 CPU hr on an IBM 370-158 to generate nonoptimal integer solutions.

The most important aspect of getting integer solutions for the carline assignment problem using MPSX-MIP is to ensure that a good branching strategy has been chosen. This strategy must branch first on integer variables that have the greatest amount of influence. The first integer variables to branch on are those that automatically set the values of other integer variables. Setting these variables means that they will not have to be branched on. Another important characteristic of a good branching strategy is that integer variables associated with a particular carline are not all branched on in succession. The integer variables associated with a specific carline should be spread out in the branching process to avoid assigning the same carline to too many plants because this tends to produce infeasibilities toward the end of the branching process.

The nonoptimal solutions of MPSX-MIP have offered good cost savings potential for the carline assignment project. However, the 10 hr of computer time required to generate good integer solutions is not acceptable. In conjunction with consultants, a specialized integer programming code is being developed. It will take advantage of the special network structure to obtain optimal solutions quickly.

The system requires a large data base to contain all of the outbound freight rates for all potential carline/assembly plant assignments (approximately 10,000 records) and the forecast demand by carline by sales area (approximately 15,000 records). Parameters for a particular analysis are entered through CRTs by users in a corporate production planning function. Types of analysis that the system is used for are plant location studies and assignment of carlines to plants for future years' production. Future development includes expanding the model to include other factors and costs that affect the scheduling process.

REFERENCES

1. F. S. HILLIER and G. J. LIEBERMAN, *Introduction to Operations Research*, 3rd ed., Holden-Day, San Francisco, 1980, pp. 25–26.
2. H. M. WAGNER, *Principles of Operations Research*, 2nd ed., Prentice-Hall, Englewood Cliffs, NJ, 1975, p. 472.
3. J. F. SHAPIRO, *Mathematical Programming: Structures and Algorithms*, Wiley, New York, 1979, pp. 344–346.
4. D. T. PHILLIPS, A. RAVINDRAN, and J. J. SOLBERG, *Operations Research: Principles and Practice*, Wiley, New York, 1976, p. 198.
5. F. GLOVER and R. E. D. WOOLSEY, "Converting the 0-1 Polynomial Programming Problem to a 0-1 Linear Program," *Operations Research*, Vol. 22 (1974), pp. 180–182.
6. R. S. GARFINKEL and G. L. NEMHAUSER, *Integer Programming*, Wiley, New York, 1972, pp. 214–249.
7. R. J. DAKIN, "A Tree Search Algorithm for Mixed Integer Programming Problems," *Computer Journal*, Vol. 8, No. 3 (1965), pp. 250–255.
8. A. H. LAND and A. G. DOIG, "An Automatic Method of Solving Discrete Programming Problems," *Econometrica*, Vol. 28 (1960), pp. 497–520.
9. E. BALAS, "An Additive Algorithm for Solving Linear Programs With Zero-One Variables," *Operations Research*, Vol. 13, No. 4 (1965), pp. 517–546.
10. D. R. PLANE and C. MCMILLAN, JR., *Discrete Optimization*, Prentice-Hall, Englewood Cliffs, NJ, 1971, pp. 47–49.
11. A. M. GEOFFRION and R. E. MARSTEN, "Integer Programming: A Framework and State-of-the-Art Survey," *Management Science*, Vol. 18, No. 9 (1972), pp. 465–491.
12. R. E. GOMORY, "Outline of an Algorithm for Integer Solutions to Linear Programs," *Bulletin of the American Mathematical Society*, Vol. 64 (1958), pp. 275–278.
13. GARFINKEL and NEMHAUSER, *Integer Programming*, pp. 154–213.
14. R. E. GOMORY, "On the Relation Between Integer and Noninteger Solutions to Linear Programs," *Proceedings of the National Academy of Science*, Vol. 53 (1965), pp. 260–265.
15. SHAPIRO, *Mathematical Programming*, pp. 284–354.
16. J. F. BENDERS, "Partitioning Procedures for Solving Mixed-Variables Programming Problems," *Numerische Mathematik*, Vol. 4 (1962), pp. 238–252.
17. K. G. MURTY, *Linear and Combinatorial Programming*, Wiley, New York, 1976.
18. G. T. ROSS and R. M. SOLAND, "A Branch and Bound Algorithm for the Generalized Assignment Problem," *Mathematical Programming*, Vol. 8, No. 1 (1975), pp. 91–103.
19. J. CAIE, J. LINDEN, and W. L. MAXWELL, *Solution of a Machine Load Planning Problem*, Technical Report No. 396, School of Operations Research and Industrial Engineering, Cornell University, Ithaca, NY, 1978.
20. H. GREENBERG and R. HEGERICH, "A Branch Search Algorithm for the Knapsack Problem," *Management Science*, Vol. 16, No. 5 (1970), pp. 327–332.

CHAPTER 14.5

Network Optimization

GARY E. WHITEHOUSE

University of Central Florida

14.5.1 NETWORK MODELS: WHEN AND WHERE TO USE THEM

There are a number of linear programming systems that are referred to as "network models." These systems can be exploited to allow the construction of efficient algorithms. The motivation for considering the network models is usually the need to solve larger problems than it would otherwise be possible to solve using existing technology. The first of the network models was the transportation model, which lead to widespread use of linear programming to solve logistic problems. More recently, emphasis has been placed on the development of efficient algorithms to solve particular large-scale systems.

Network models are probably the most important special structure in linear programming. This chapter (1) considers the characteristics of these models, (2) gives one approach to the solution of a network, and (3) formulates applications within the industrial engineering environment.

Some potential network applications suggested by Phillips et al.[1] are:

1. A nationwide chain store is interested in supplying a special product to its retail outlets from its various warehouses. What shipping plan would minimize the total cost of transportation?

2. In a machine shop a batch of jobs is to be assigned to a group of machines. What assignment of jobs to machines will maximize the total efficiency of the shop?

3. A pipe network distributes water from several pumping stations to various customers. If the capacities of the pipes are known, what is the maximum flow that is possible from the stations to the customers?

4. If we were to assign one-way traffic signs to a road network, what assignment would maximize the flow of traffic or the number of highway users?

5. A trucking firm has a table of distances between cities. It wants to find the shortest route and the shortest distance between all pairs of cities so as to design efficient service routes for its trucks.

6. A project consists of a large number of activities that must be done in a specified sequence. The project manager wants to determine how these activities should be scheduled and coordinated so as to minimize the project duration.

14.5.2 THE GENERAL NETWORK FLOW PROBLEM

A typical network flow problem arises in logistics and involves the distribution of products from factories (sources) to markets (destinations). The units produced at factories and demanded at the markets are known. The product does not have to be routed directly from source to destination, but may be transshipped through intermediary points. There may also be capacity restrictions on some links. The goal is to minimize the variable cost of producing and shipping products to meet consumer demand. The sources, destinations, and transshipment points are referred to as "nodes" of the network, and the transportation links connecting nodes are called "arcs."

An example of network nodes is shown in Exhibit 14.5.1. The nodes are represented by numbered circles and the arcs by arrows. The arcs are assumed to be directed. The network in Exhibit 14.5.1 shows a number of additional characteristics of the network problem. Each arc has flow capacity, and a per unit cost is specified for shipping along each arc (e.g., the flow along arc 3-5 is between 0 and 10 units at a cost of $4/unit flowing). The numbers in parentheses represent the units supplied or demanded at a node. Node 1 is a source node supplying 25 units, whereas nodes 4 and 5 are destination (sink) nodes requiring 5 and 20 units, respectively, as indicated by the negative signs. Nodes 3 and 4 are transshipment nodes. This type of problem is referred to as a "minimum-cost flow problem" or the "capacitated transshipment problem."

Exhibit 14.5.1 Minimum-Cost Flow Problem

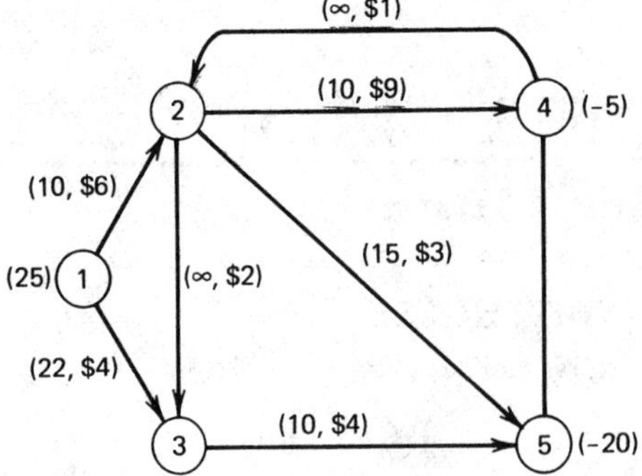

The goal is to find the minimum-cost flow pattern and fulfill demands from source nodes. The linear programming model for this network is developed as follows: Let x_{ij} be the number of units shipped from node i to j using arc $i - j$. The objective is

$$\text{minimize} \quad 6x_{12} + 4x_{13} + 2x_{23} + 9x_{24} + 3x_{25} + 4x_{35} + 1x_{42}$$

The capacity constraints are

$$\begin{aligned} 0 &\leqslant x_{12} \leqslant 10 \\ 0 &\leqslant x_{13} \leqslant 22 \\ 0 &\leqslant x_{23} \leqslant \infty \\ 0 &\leqslant x_{24} \leqslant 10 \\ 0 &\leqslant x_{42} \leqslant \infty \\ 0 &\leqslant x_{25} \leqslant 15 \\ 0 &\leqslant x_{35} \leqslant 10 \end{aligned}$$

The flow-balance equations at all nodes are

(flow out of a node) – (flow into a node) = (net supply at the node)

$$\begin{aligned} x_{12} + x_{13} &= 25 \\ x_{23} + x_{24} + x_{25} - x_{12} - x_{42} &= 0 \\ x_{35} - x_{13} - x_{23} &= 0 \\ x_{42} - x_{24} &= -5 \\ -x_{25} - x_{35} &= -20 \end{aligned}$$

It is important to recognize the special structure of these balance equations. There is one balance equation for each node in the network, and the flow variables x_{ij} have only 0, +1, and –1 coefficients in these equations. Each variable appears in exactly two balance equations, once with a +1 coefficient, corresponding to the node from which the arc emanates, and once with a –1 coefficient, corresponding to the node upon which the arc is incident. It is this particular structure that is exploited in developing specialized, efficient algorithms.

The following example illustrates the general minimum-cost flow problem with n nodes:

$$\text{minimize} \quad z = \sum_i \sum_j c_{ij} x_{ij}$$

subject to

$$\sum_j x_{ij} - \sum_k x_{ki} = b_i, \qquad i = 1, 2, \ldots, n \qquad \text{[flow balance]}$$

$$\ell_{ij} \leqq x_{ij} \leqq u_{ij} \qquad \text{[flow capacities]}$$

The summations are taken only over the arcs in the network. That is, the first summation in the ith flow-balance equation is over all nodes j such that $i-j$ is an arc of the network, and the second summation is over all nodes k such that $k-i$ is an arc of the network. The objective function summation is over arcs $i-j$ that are contained in the network and represents the total cost of sending flow over the network. The ith balance equation is interpreted as before: It states that the flow out of node i minus the flow into i must equal the net supply (demand if b_i is negative) at the node. The upper bound on arc flow is u_{ij}, which may be $+\infty$ if the capacity on arc $i-j$ is unlimited. The lower bound on arc flow is ℓ_{ij}, often taken to be zero, as in the previous example. The following section focuses on variations of this general problem in some detail.

14.5.3 SPECIAL NETWORK MODELS

A number of special cases of the minimum-cost flow models are discussed here.

The Transportation Problem

The transportation problem is a network flow model without intermediate locations.

Define

a_i = number of units available at source i
b_j = number of units required at destination j
c_{ij} = unit transportation cost from source i to destination j

Initially we will assume that the total product availability is equal to the total product requirements; that is,

$$\sum_{i=1}^{m} a_i = \sum_{j=1}^{n} b_j$$

The following example indicates what to do when this supply-demand balance is not satisfied. If we define the decision variables as x_{ij} being the number of units to be distributed from source i to destination j, we can then formulate the transportation problem as follows:

$$\text{minimize } z = \sum_{i=1}^{m} \sum_{j=1}^{n} c_{ij} x_{ij} \qquad (1)$$

subject to

$$\sum_{j=1}^{n} x_{ij} = a_i, \qquad i = 1, 2, \ldots, m \qquad (2)$$

$$\sum_{i=1}^{m} (-x_{ij}) = -b_j, \qquad j = 1, 2, \ldots, n \qquad (3)$$

$$x_{ij} \geqq 0, \qquad i = 1, 2, \ldots, m; \quad j = 1, 2, \ldots, n \qquad (4)$$

Expression 1 represents the minimization of the total distribution cost, assuming a linear cost structure for shipping. Equation 2 states that the amount being shipped from source i to all possible destinations should be equal to the total availability, a_i, at that source. Equation 3 indicates that the amounts being shipped to destination j from all possible sources should be equal to the requirements, b_j, at that destination. Usually equation 3 is written with positive coefficients and right-hand sides by multiplying through by minus one.

The network model for the transportation problem is shown in Exhibit 14.5.2. The information for the transportation model is usually expressed in a special tableau, as shown in Exhibit 14.5.3.

The Assignment Problem

The transportation model has so far been described in terms of material flow from source to destinations. The model has many additional applications. Consider that n jobs are to be assigned to n machines and that c_{ij} measures the cost of job i on machine j. If we let

x_{ij} = 1 if job i is assigned to machine j
= 0 otherwise

Exhibit 14.5.2 Network Model for the Transportation Problem

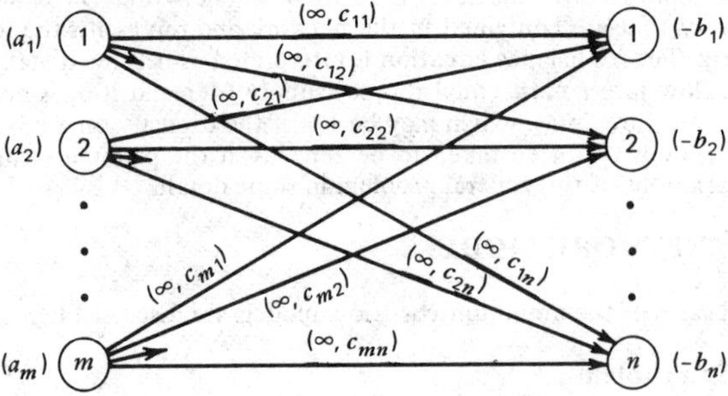

we can find the best assignment by solving the optimization problem

$$\text{minimize} \quad z = \sum_{i=1}^{n} \sum_{j=1}^{n} c_{ij} x_{ij}$$

subject to

$$\sum_{j=1}^{n} x_{ij} = 1, \qquad i = 1, 2, \ldots, n$$

$$\sum_{i=1}^{n} x_{ij} = 1, \qquad j = 1, 2, \ldots, n$$

$$x_{ij} = 0 \text{ or } 1, \qquad i = 1, 2, \ldots, n; \qquad j = 1, 2, \ldots, n$$

The first set of constraints shows that each job is to be assigned to exactly one machine, and the second set indicates that each machine is to perform one job.

The assignment problem is an integer program because the decision variables x_{ij} are restricted to being zero or one. However, if these constraints are replaced by $x_{ij} \geqq 0$, the model becomes a special case of the transportation problem, with one unit available at each source (job) and one unit required by each destination (machine). Network flow problems have integer solutions, and therefore formal specification of integrality constraints is unnecessary. Therefore application of the network flow algorithms will solve such integer problems directly.

Exhibit 14.5.3 Special Transportation Tableau

	Destinations				
Sources	1	2	. . .	n	Supply
1	c_{11} x_{11}	c_{12} x_{12}	. . .	c_{1n} x_{1n}	a_1
2	c_{21} x_{21}	c_{22} x_{22}	. . .	c_{2n} x_{2n}	a_2
⋮	⋮	⋮		⋮	⋮
m	c_{m1} x_{m1}	c_{m2} x_{m2}	. . .	c_{mn} x_{mn}	a_m
Demand	b_1	b_2	. . .	b_n	Total

The Maximal Flow Problem

For the maximal flow problem, the goal is to send as much material as possible from a source node s to a sink node t. No costs are associated with flow. If v denotes the amount of material sent from node s to node t, and x_{ij} denotes the flow from node i to node j over arc $i-j$, the formulation becomes

$$\text{maximize } v$$

subject to

$$\sum_j x_{ij} - \sum_k x_{ki} = \begin{cases} v \text{ if } i = s \text{ (source)} \\ -v \text{ if } i = t \text{ (sink)} \\ 0 \text{ otherwise} \end{cases}$$

$$0 \leqq x_{ij} \leqq u_{ij}, \qquad i = 1, 2, \ldots, n; \qquad j = 1, 2, \ldots, n$$

The summations are taken only over the arcs in the network. The upper bound u_{ij} for the flow on arc $i-j$ is taken to be $+\infty$ if arc $i-j$ has unlimited capacity. The interpretation is that v units are supplied at s and consumed at t.

The Shortest-Path Problem

The shortest-path problem is a particular network model that has received a great deal of attention for both practical and theoretical reasons. The essence of the problem can be stated as follows: Given a network with distance c_{ij} associated with each arc, find a path through the network from a particular origin (source) to a particular destination (sink) that has the shortest total distance. There are a number of important applications that can be formulated as shortest-path problems where this formulation is not obvious at the outset. These include problems of equipment replacement, capital investment, project scheduling, and inventory planning. The theoretical interest in the problem is due to its having a special structure, in addition to being a network, that results in very efficient solution procedures. Dynamic programming takes advantage of this efficiency of solution.

The formulation of the shortest-path problem is as follows:

$$\text{minimize } z = \sum_i \sum_j c_{ij} x_{ij}$$

subject to

$$\sum_j x_{ij} - \sum_k x_{ki} = \begin{cases} 1 \text{ if } i = s \text{ (source)} \\ 0 \text{ otherwise} \\ -1 \text{ if } i = t \text{ (sink)} \end{cases}$$

$$x_{ij} \geqq 0 \qquad \text{for all arcs } i-j \text{ in the network}$$

The shortest-path problem can be interpreted as a network flow problem with one unit of flow sent from the source to the sink at minimum cost.

Longest-Path Problems

If we modify the formulation of the shortest-path problem to maximize the objective function, we will find the longest path through the network. This longest-path problem is essentially the activity network analysis presented in Chapter 11.2. The algorithms presented in that chapter are good examples of efficient methods available to solve network models.

14.5.4 SPECIAL NETWORK ALGORITHMS

As we have seen, there are a number of problems that can be modeled as network systems. The networks could be optimized using the simplex method. However, there has been much effort in specializing the rules of the simplex method to take advantage of the structure of network models. The resulting algorithms are extremely efficient and permit the solution of network models so large that they would be impossible to solve by ordinary linear programming procedures. Their efficiency stems from the fact that a pivot operation for the simplex method can be carried out by simple

Exhibit 14.5.4 Machine/Job Costs

	Job 1	Job 2	Job 3
Machine 1	7	9	14
2	15	11	13
3	15	13	13

addition and subtraction without the need for maintaining and updating the usual tableau at each iteration. An added benefit of these algorithms is that the optimal solutions generated turn out to be integer if the relevant constraint data are integer.

There is insufficient space here to describe efficient algorithms for all network models described in Section 14.5.3, but an algorithm to solve the assignment problem is considered in order to illustrate the ease in solving network models using specialized algorithms. Industrial engineers interested in the solution of the other models are referred to general mathematical programming or operations research tests.[1–5]

Consider the assignment example described by Exhibit 14.5.4. The basic approach to finding the optimal assignment is based on the observation that the final answer is not affected by adding or subtracting a constant from any row or column of the cost matrix. For example, if the cost of doing any job on machine 2 increases by $3, so that the second row becomes 18, 14, and 16, it can be shown that the optimal assignment will not be affected by this change.

The solution procedure is to subtract a sufficiently large cost from the various rows or columns in such a way that an optimal assignment is found by inspection. First, examine each row (column) of the cost matrix to identify the smallest element. This quantity is subtracted from all elements of the row (column), producing a cost matrix with at least one zero in each row and column. Now try to find a feasible solution using the cells with zero costs. If this is possible, we have an optimal assignment.

For our example, develop a reduced cost matrix by subtracting 7 from row 1, 11 from row 2, and 13 from row 3. This yields the costs shown in Exhibit 14.5.5.
This exhibit gives a feasible assignment with

$$M_1 \rightarrow J_1, \quad M_2 \rightarrow J_2, \quad M_3 \rightarrow J_3$$

This is the optimal assignment.

In general, it may not be possible to find a feasible solution by row and column manipulations. Suppose we specify our problem as shown in Exhibit 14.5.6. The resulting reduced matrix is shown in Exhibit 14.5.7. There is no feasible solution for this problem. But there is a way to continue

Exhibit 14.5.5 Reduced Cost Assignment Matrix

	Job 1	Job 2	Job 3
Machine 1	0	2	4
2	4	0	2
3	2	0	0

Exhibit 14.5.6 Revised Machine/Job Costs

	Job 1	Job 2	Job 3
Machine 1	7	9	8
2	15	11	13
2	15	13	16

Exhibit 14.5.7 Reduced Matrix for the Revised Model

		Job 1	Job 2	Job 3
Machine	1	0	2	0
	2	4	0	1
	3	2	0	2

Exhibit 14.5.8 Manipulating the Reduced Matrix

		Job 1	Job 2	Job 3
Machine	1	0	2	0
	2	4	0	1
	3	2	0	2

Exhibit 14.5.9 The Reduced Matrix

		Job 1	Job 2	Job 3
Machine	1	0	3	0
	2	3	0	0
	3	1	0	0

operating on the matrix by eliminating the zero matrix positions from further consideration. Let us see how this works. We want to delete from consideration the minimum number of rows and columns such that their elimination will remove the matrix positions containing a zero. Examination of the reduced matrix clearly shows that elimination of the first row and the second column will accomplish this, and thus we draw lines through the selected row(s) and column(s). Our reduced matrix now should look like the one shown in Exhibit 14.5.8.

We now examine the remaining matrix positions to determine the minimum cell. We subtract the value of the minimum cell from each of the remaining cells and add that value to each matrix position at an intersection of a row and column elimination. Thus in our case the minimum remaining matrix position is $x_{23} = 1$; we subtract that value from x_{21}, x_{23}, x_{32}, and x_{33}; finally, we add that value to x_{12}. This operation produces the matrix shown in Exhibit 14.5.9. It is now possible to select a zero path that proves to be x_{32}, x_{23}, x_{11}, for which the total cost is

$$c_{32} + c_{23} + c_{11} = 13 + 13 + 7 = 33$$

The procedure of eliminating rows and columns is, in general, reiterative. Should its first application fail to provide a feasible zero path, it can be used as many times as may be necessary on successive matrices until a feasible zero path is found. This example has illustrated the comparative ease of solving network models using specialized algorithms.

14.5.5 APPLICATIONS OF NETWORK MODELS TO SOLVE INDUSTRIAL ENGINEERING PROBLEMS

Production Scheduling—A Transportation Example

The production scheduling problem is one that can be modeled as a transportation problem. Consider the problem of a weekly production schedule for the next 3 months. The production cost of an item is $12. The monthly demands are 400, 700, and 600, which must be met. Plant production capacity is 500 units/month. It is also possible to employ overtime help in each month. The overtime increases the monthly production by 200 units at an additional cost of $3/unit. Inventory can be stored for $2/unit/month. The question we are trying to answer is, "How should we schedule production to minimize the total costs?"

To convert this to a transportation problem, we consider the production periods as sources and the markets as weekly demands. Since overtime production is possible, there are six supply points. The decision variables are

x_{1j} = normal production in month 1 for use in month j
x_{2j} = overtime production in month 1 for use in month j
x_{3j} = normal production in month 2 for use in month j
x_{4j} = overtime production in month 2 for use in month j
x_{5j} = normal production in month 3 for use in month j
x_{6j} = overtime production in month 3 for use in month j

Since the total normal and overtime production exceeds the total demand, we must create a dummy market to absorb the extra supply. Exhibit 14.5.10 displays the transportation model for this problem. Some of the cost elements in the exhibit are set at M (an infinitely large value) to denote that shipments are impossible. That is, you cannot use production in month 2 or 3 to supply the first month's demand. The inventory cost of $2/unit/month is added to the production cost whenever an item is stored to meet future demands. Now the production scheduling can be solved using the transportation algorithm.

Production Line Scheduling—An Assignment Example

Consider a multiproduct line that is capable of producing a family of several different product models. Each product part is a discrete item with physical shape, size, weight, and composition. A finished product may consist of only one part or may be an assembly of two or more parts. The operation routing for the several products is similar, but there are some differences. The operations themselves may vary considerably from one product to another. It will be assumed that the operation routings and the content of each operation on each product have been previously established by a process design department as illustrated by Young.[6]

Consider further that demand schedules for the several products are established on a daily, weekly, or monthly basis by a production control department and that the production line has been physically arranged with a certain number of production stages and a set number of machines or work stations at each stage. The general arrangement of the stages has also been established as part of the plant layout function.

Exhibit 14.5.10 Transportation Tableau for Production Scheduling

	Month 1 Demand	Month 2 Demand	Month 3 Demand	Dummy	Supplies
Month 1 (Regular Time)	12	14	16	0	500
Month 1 (Overtime)	15	17	19	0	200
Month 2 (Regular Time)	M	13	14	0	500
Month 2 (Overtime)	M	15	17	0	200
Month 3 (Regular Time)	M	M	12	0	500
Month 3 (Overtime)	M	M	15	0	200
Demands	400	700	600	400	Total Cost

For a given demand period, a selected group of products is to be scheduled into the production line. This group may be only a part of the family of products that can be scheduled through the line. Products are to be scheduled through the line in economic lots, limited by the total demands for the period. That is, one or more batch runs of each product may be made during the demand period, but no batch run may exceed the demand for the given period.

First, it is desired to establish an optimum order for scheduling the products into the production line, considering the costs of converting the production line setup from one product to another. We will assume that these conversion costs can be estimated or computed from historical data.

In computing the conversion costs from one product to another, considering the time for tool changeover and machine resetting, it also is desirable to include machine and operator delay time costs.

Second, it is desired to find an optimum ordering of the product batch runs that will minimize the total setup costs for the entire schedule of products within a given demand period. To examine the use of the assignment model for optimizing this sequence, consider a schedule of five products with the conversion (or setup) costs shown in Exhibit 14.5.11. The numbers represent the line conversion costs from (source) product to (designation) product. (The 1000s on the diagonal are artificial and preclude a product from following itself.)

The solution matrix is shown in Exhibit 14.5.12 and reveals that the least-cost solution, arbitrarily starting with product 1, would be 1 → 3 → 4 → 2 → 5 at a cost of $300.

Occasionally this technique will give an unsatisfactory solution because the least-cost sequence of jobs will include multiple cycles of setups (e.g., 3-5-3, 1-4-2-1). If this occurs, the more complicated traveling salesman algorithm will have to be used.

Maximizing the Rate of Return on Investments—An Example of Longest-Path Analysis

Consider the case where an investor has alternate investment strategies available. For example, he or she can buy 1 year bonds yielding 10%/annum, 3 year bonds yielding 12%/annum, or 5 year bonds yielding 9%/annum. Given that the investor has a fixed horizon, his or her problem is to maximize ROI.

For our example, assume that we wish to obtain the maximum return over a 5 year horizon. The

Exhibit 14.5.11 A Setup Cost Matrix for Solution by the Assignment Model of Linear Programming

		Product to Follow (Destination)					
		1	2	3	4	5	Supply
Product to	1	1000	60	100	70	50	1
Precede	2	90	1000	110	80	30	1
(Source)	3	100	65	1000	80	40	1
	4	80	70	120	1000	50	1
	5	20	75	90	90	1000	1
Demand		1	1	1	1	1	

Exhibit 14.5.12 Solution Matrix

	Follower				
	1	2	3	4	5
Predecessor 1	950	0	[0]	0	0
2	60	950	30	10	[0]
3	60	5	910	[0]	0
4	30	[0]	20	910	0
5	[0]	35	30	30	980

Exhibit 14.5.13 Network Model for Investment Flows

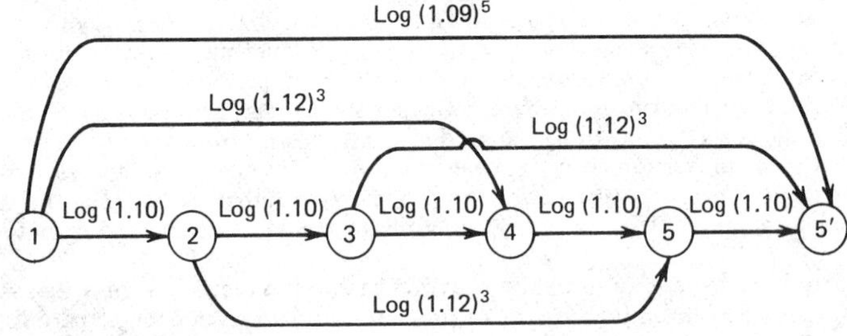

flow of investment can be modeled in network form as shown in Exhibit 14.5.13. The nodes represent the beginning of the investment year in question, and the arcs the investment opportunities. Thus, at the beginning of year 1, we can buy a 1 year, 3 year, or 5 year bond, but at the beginning of year 2, we can only buy a 1 year or 3 year bond because a 5 year bond will not mature within our 5 year planning horizon. We should note that node 5′ represents the end of year 5 and thus the end of our investment plan.

The problem is to find the path from node 1 to 5′ that maximizes the rate of return on our money. Since we are dealing with compound interest, money invested along path 1-2-5-5′ will yield a return $(1.10)\ (1.12)^3\ (1.0)$; thus the coefficients on our activities are multiplicative instead of being additive. However, we could take the log of each rate of return, and the value of the path in question becomes $\log (1.10) + 3 \log (1.12) + \log (1.0)$. Since the log increases with the size of a number, this problem can be solved by finding the longest path through the network after the log has been taken of the coefficient of each activity. The best investment strategy would be to buy a series of two 1 year bonds and a 3 year bond yielding a rate of return of 69.9%.

Distribution Systems With Capacity Limitations—A Maximal Flow Example

Most of the important applications of network flow models involve the conversion of linear programming models, but this is beyond the scope of this handbook. Elmaghraby,[3] however, has suggested a number of interesting situations in which the algorithm as presented in this chapter is directly applicable.

Suppose we wish to transport goods from a number of sources, $S_1, S_2, \ldots, S_m$, to a number of destinations, $D_1, D_2, \ldots, D_n$. There are a number of units, a_i, available at each source and a number of units, d_j, needed at each destination. Now distributions can take place only over certain routes, and these routes have capacity c_{ij}. The problem is to determine a feasible distribution pattern if one exists. Note that we are not trying to minimize cost as is often the case in problems of this sort.

Exhibit 14.5.14 represents a network approach for modeling this system. The sources are represented by nodes S_1 through S_m, and the destinations by nodes D_1 through D_n. For each permissible route (i, j), an arc with capacity c_{ij} is shown joining nodes S_i and D_j. Next, a master source, S, and a master terminal, D, are added to the flow network. The arcs $S - S_i$, which have capacity

Exhibit 14.5.14 Network Model of a Transportation System

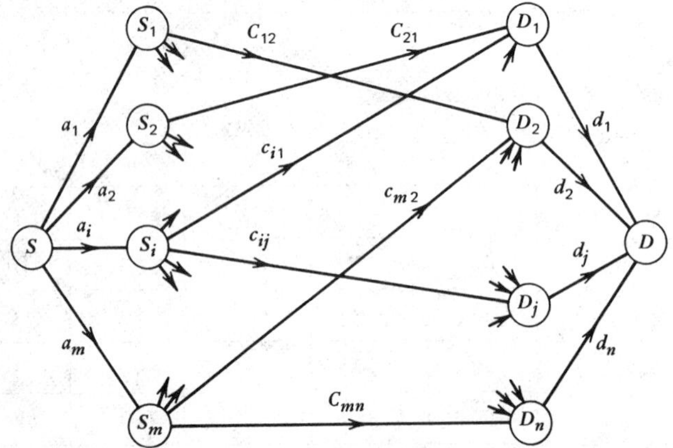

a_i, limit the amount of product sent from S_i. The needs of each destination, D_j, are limited in a similar manner by the arcs $D_j - D$, whose capacities are d_j. We next apply network flow methods to the model. If the flow from D equals the sum of the d_i, then the solution satisfies the conditions of the problem. If not, it is impossible to ship the desired commodities from sources S_i to the destinations D_j using the permissible routes.

Maximizing the Minimum Level of Efficiency in an Assignment Problem— An Example of Combined Network Concepts

The typical goal in the optimization of job assignment problems is to maximize the total efficiency in a system. Now suppose we wish to assign jobs in such a way that the lowest efficiency is as large as possible. We handle this problem as a maximal flow problem. First, an arbitrary assignment is made. The lowest efficiency in the arbitrary assignment is observed, and all possible assignments with that efficiency or lower are considered to be illegal. A flow network of the legal assignments is made, and another feasible solution is found by network flow methods. Once again, the lowest efficiency is observed, and assignments with that efficiency or lower are eliminated from consideration. The process continues until the flow method can no longer find a feasible assignment. The lowest efficiency in the last feasible assignment is the answer to the problem.

Let us assume the assignment matrix as shown in Exhibit 14.5.15. Also assume that an arbitrary assignment of

$$J_1 \to P_1 \quad (10)$$
$$J_2 \to P_5 \quad (5)$$
$$J_3 \to P_2 \quad (6)$$
$$J_4 \to P_3 \quad (8)$$
$$J_5 \to P_4 \quad (4)$$

has been made. We note that the minimum level of efficiency is $4(J_5 \to P_4)$ for this assignment. We will now remove all assignments with an efficiency of 4 or less and develop a network flow model (Exhibit 14.5.16) to test for feasibility.

In Exhibit 14.5.16 node S is the source and node D the sink. The flow from S to D represents the number of jobs assigned. Thus, if this flow is equal to the number of jobs, then a feasible assignment has been found. Nodes J_i and P_j represent the jobs and people, respectively. The paths joining S to J_i and P_j to D all have capacity of 1 and represent the fact that each job can be assigned only once and that each person can do only one job. All feasible assignments are represented by arcs joining appropriate J_i's and P_j's. For example, J_5 can be done only by P_5 if our goal is to have a minimum efficiency of 5 or better.

Applying the maximal flow analysis (8), we would find a maximal flow of 5 associated with

$$J_1 \to P_4 \quad (5)$$
$$J_2 \to P_5 \quad (5)$$
$$J_3 \to P_1 \quad (7)$$
$$J_4 \to P_3 \quad (8)$$
$$J_5 \to P_2 \quad (5)$$

Thus we have a feasible solution, and the minimum performance level is 5, which occurs on paths $J_1 \to P_4$, $J_2 \to P_5$, and $J_5 \to P_2$. If all jobs with efficiency 5 are removed, the maximal flow will be less than 5 because J_5 cannot be assigned.

Exhibit 14.5.15 Job Assignment Matrix

	Person 1	2	3	4	5
Job 1	10	9	6	5	1
2	2	3	8	4	5
3	7	6	2	1	5
4	4	8	8	4	3
5	4	5	2	4	2

Exhibit 14.5.16 Network Model of the Job Assignment Problem

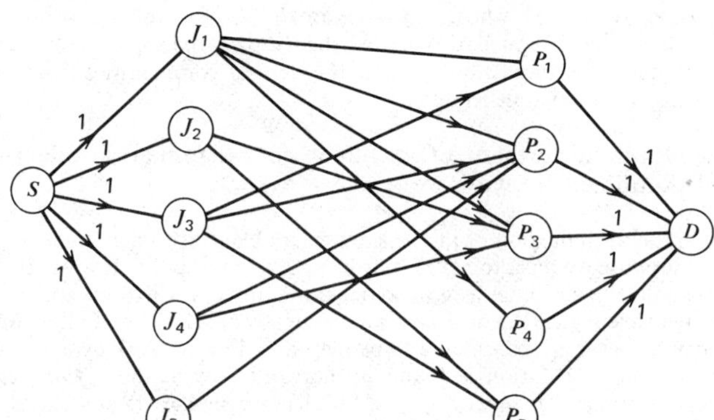

14.5.6 SUMMARY

The material discussed in this chapter is a small portion of the material available within the network of optimization area. A complete discussion of network analysis is impossible within the limitations of this chapter. However, it is hoped that the industrial engineer will recognize the potential of network modeling and solution. More complete discussions of network analysis are available in the texts listed in the references and the bibliography.

REFERENCES

1. D. T. PHILLIPS, A. RAVINDRAN, and J. J. SOLBERG, *Operations Research*, Wiley, New York, 1976.
2. S. P. BRADLEY, A. C. HAX, and T. L. MAGNANTI, *Applied Mathematical Programming*, Addison-Wesley, Reading, MA, 1977.
3. S. E. ELMAGHRABY, *Activity Networks*, Wiley-Interscience, New York, 1977.
4. G. E. WHITEHOUSE, *Systems Analysis and Design Using Network Techniques*, Prentice-Hall, Englewood Cliffs, NJ, 1973.
5. G. E. WHITEHOUSE and B. L. WECHSLER, *Applied Operations Research*, Wiley, New York, 1976.
6. H. H. YOUNG, "Optimization Problems for Production Lines," *Journal of Industrial Engineering*, Vol. 18, No. 1 (January 1967).

BIBLIOGRAPHY

ELMAGHRABY, S. E., *Network Models in Management Science*, Springer-Verlag Lecture Series on Operations Research, New York, 1970.

FORD, L. R., and D. R. FULKERSON, *Flows in Networks*, Princeton University Press, Princeton, NJ, 1962.

HU, T. C., *Integer Programming and Network Flows*, Addison-Wesley, Reading, MA, 1969.

INDEX